SLEEP: A COMPREHENSIVE HANDBOOK

SLEEP: A COMPREHENSIVE HANDBOOK

TEOFILO LEE-CHIONG, M.D.
National Jewish Medical and Research Center
University of Colorado Health Sciences Center
Denver, Colorado

**WILEY-LISS**

A John Wiley & Sons, Inc., Publication

Published by John Wiley & Sons, Inc., Hoboken, New Jersey
Published simultaneously in Canada

For general information on our other products and services or for technical support, please contact our Customer Care Department within the United States at (800) 762-2974, outside the United States at (317) 572-3993 or fax (317) 572-4002.

Wiley also publishes its books in a variety of electronic formats. Some content that appears in print may not be available in electronic formats. For more information about Wiley products, visits our web site at www.wiley.com.

Library of Congress Cataloging-in-Publication Data is available.

Lee-Chiong, Teofilo
   Sleep: A Comprehensive Handbook

ISBN-13 978-0-471-68371-1
ISBN-10 0-471-68371-X

Printed in the United States of America

10 9 8 7 6 5 4 3 2 1

CONTENTS

v

PREFACE

A textbook has to meet the needs of various readers, from the specialist who yearns to know "more and more of less and less" to the generalist who is limited by knowing "less and less of more and more." This need is nowhere more apparent than in the multidisciplinary science of sleep medicine.

Assembled in this preface, as well as in the pages of the textbook, is the collective expertise of the major authorities on contemporary sleep medicine worldwide. The authors have attempted to write a current and comprehensive text that covers the entire spectrum of adult and pediatric sleep medicine, encompassing major disease entities affecting sleep and that are, in turn, affected by sleep itself. Separate sections on sleep among women, in the elderly, and in special patient groups emphasize the unique character of their sleep.

The Science of Sleep Medicine

Normal Human Sleep. Normal human sleep is comprised of two distinct states known as non-rapid eye movement (NREM) and rapid eye movement (REM) sleep. NREM sleep is subdivided into four stages: stage 1, stage 2, stage 3, and stage 4. REM sleep may be further subdivided into two stages: phasic and tonic. Several models have been proposed to explain the regulation of sleep and wakefulness. One such model proposes that the regulation of the sleep–wake cycle is governed by two processes: a sleep-dependent homeostatic process and a sleep-independent circadian process. (Rama AN et al)

The Neurobiology of Sleep. The basic tenet of the neurobiology of sleep is that sleep is a product of the central nervous system. The neurobiology of sleep and wake-fulness can be described as a system distributed along the neuraxis from the medulla oblongata to the neocortex. While mechanisms within the brain produce sleep and wakefulness, brain mechanisms also are the targets of their influence. (Marks GA)

Physiologic Processes During Sleep. Many of the physiologic changes occurring during sleep are associated with changes in the level of activity of the autonomic nervous system. NREM sleep is characterized by a period of relative autonomic stability with sympathetic activity remaining at about the same level as during relaxed wakefulness, and parasympathetic activity increasing through vagus nerve dominance and heightened baroreceptor gain. During tonic REM sleep, a relative increase in parasympathetic activation is noted (mostly as a result of sympathetic input decline). Changes in autonomic function and inherent changes in the control exerted by the central nervous system (CNS) affect most organ systems in the body during sleep. (Rosenthal L)

Biological Rhythms and Sleep. The fundamental behavioral circadian rhythm is the rest–activity cycle. Circadian timing is an inherited adaptation and is genetically determined. The timing of the sleep–wake cycle depends on the interaction of a number of brain systems, in particular sleep–wake systems and the circadian timing system (CTS). The CTS not only coordinates the timing of the sleep–wake cycle by opposing the sleep homeostat but also serves to coordinate the timing of pacemakers and oscillators in other tissues and organs to facilitate adaptation. (Moore RY)

Biology of Dreaming. Correlation of dreaming with a specific, identifiable EEG pattern became the focus for efforts to describe the physiological processes that are the biological basis for the dreaming process. However, biological dream theories cannot provide us with the content of dreams, the meaning of dreams, the construction of dreams, or the function of dreaming. (Kramer M)

Psychology of Dreaming. The major reason for studying dreaming in the modern context is to understand the functioning of the mind, to understand consciousness. A secondary but important reason for studying dreaming is to see if such a study will unlock the mysteries of psychosis. The dream experience can be influenced by a number of factors and can be usefully quantified. The dream is orderly and organized, signal not noise, as it reflects meaningful psychological differences and responds to and reflects emotionally laden influences. (Kramer M)

The Function of Sleep. Why we sleep remains one of nature's greatest mysteries. Some tentative conclusions regarding this question can be made. While sleep may have beneficial effects on general health, its primary function concerns the brain and not the body. Sleep, in some general way, facilitates normal neuronal function. It is possible that sleep is a time when overall neuronal function is facilitated either by sleep-dependent increases in gene expression and protein synthesis, or alterations in neuronal activity. (Frank MG)

The Evolution of Sleep: A Phylogenetic Approach. There is extensive variation in both the amount and phasing of sleep across taxonomic groups. In contrast to the wealth of knowledge on mammalian species, there is a relative lack of information on sleep in reptiles, amphibians, fishes, and invertebrates. A phylogenetic evaluation of sleep demonstrates that all mammals, birds, and reptiles engage in sleep, and evidence for sleep in amphibians, fishes, and invertebrates is strong if not certain. (Lesku JA et al)

Neuropharmacology of Sleep and Wakefulness. The two main regions that modulate our sleep–wake cycles are the mesopontine reticular activating system (RAS) and the hypothalamus. In addition, the intralaminar thalamus and the basal forebrain are modulated by the RAS and hypothalamus and participate in the process of arousal and alertness, as well as in the modulation of sleep states. Neuroactive agents that modulate these regions will also modulate the level of arousal. (Garcia-Rill E et al)

Epidemiology of Sleep Disorders. The difficulty in distinguishing between normal and abnormal sleep is reflected in the evolution of the classifications and definitions of symptomatology. This evolution in classifications is also reflected in the epidemiological studies of sleep disorders. This has rendered comparisons between earlier and more recent surveys problematic. (Ohayon MM, Guilleminault C)

Classification of Sleep Disorders. The American Academy of Sleep Medicine has recently completed the *International Classification of Sleep Disorders, Version 2* (ICSD-2). Significant changes have occurred to keep up with the changing field of sleep medicine. As the science of sleep medicine is progressively developing from its early scientific underpinnings, and greater clarity is slowly evolving in many areas of patient diagnosis and treatment, that new data has been applied to the ICSD. (Chesson AL)

Insomnia

Insomnia: Prevalence and Daytime Consequences. Insomnia is a common problem. Most studies assessing the prevalence of insomnia in the general population find that between 30% and 35% of individuals have experienced some difficulty sleeping in the previous year. The disorder is associated with negative consequences including increased use of medical services, absenteeism from work, automobile and industrial accidents, poorer work performance, greater risk for depression, and negative impact on family life. (Brown WD)

Causes of Insomnia. There are many purported causes of insomnia, covering a broad range of medical, psychiatric, and behavioral factors. A model of chronic insomnia has been proposed that assumes a multifactorial etiology and categorizes causal factors according to their role in the formation of insomnia: predisposing, precipitating, or perpetuating. According to this model, all individuals have a certain level of predisposition to insomnia, and insomnia occurs when this predisposition interacts with exposure to a precipitating factor. Common precipitating factors include medical disorders, psychiatric disorders, environmental factors, medication effects, primary sleep disorders, or circadian rhythm changes that negatively affect sleep. Perpetuating factors are behavioral and cognitive changes that occur once an individual has been sleeping poorly for a period of time. (Stepanski EJ)

Medications that Can Cause Insomnia. Medications can cause insomnia, and conversely withdrawal of medications can lead to sleep symptoms, including insomnia. Certain medications are used to treat sleepiness and thereby are designed to have insomnia as a therapeutic effect. Other classes of drugs are associated with insomnia as an unwanted side effect. (Welsh CH, Fugit RV)

Fatal Familial Insomnia. Fatal familial insomnia is a rare but uniformly fatal disease characterized by sleep disturbances, autonomic dysregulation, and dementia. It is inherited in an autosomal dominant fashion, resulting from a point mutation at codon 178 of the prion gene. It is rapidly progressive once symptoms appear. There is no known specific treatment. (Polnitsky CA)

Evaluation of Insomnia. The evaluation process in the management of an insomnia complaint involves consideration of many potential clinical syndromes. Sleep-focused physical and mental status examinations are important for accurate diagnosis. Sleep diary data recorded before the initial evaluation is useful to assess sleep scheduling across nights more objectively. Data from self-report questionnaires that focus on sleepiness, anxiety, depression, general psychopathology, sleep quality, and insomnia help the clinician to appreciate the clinical issues, identify diagnoses, and select treatments. (Moul DE, Buysse DJ)

Pharmacologic Therapy of Insomnia. Hypnotic agents are primarily indicated for the treatment of transient sleep disruption such as those caused by jet lag, shift work, or acute stress, but are also used in selected persons with chronic insomnia (ideally for primary insomnia that failed to respond to behavioral therapy, or secondary insomnia that did not improve with treatment of the underlying condition). The selection of a particular hypnotic medication should be based on the characteristics of the patient, duration and timing of insomnia, and the pharmacological profile of the agent. (Lee-Chiong T, Sateia M)

Nonpharmacologic Therapy of Insomnia. Chronic insomnia often is perpetuated by dysfunctional beliefs about sleep, heightened anxiety, and sleep-disruptive compensatory practices. Nonpharmacologic insomnia therapies such as relaxation therapy, stimulus control, sleep restriction therapy, and cognitive–behavioral therapy target behavioral and psychological factors that maintain and exacerbate sleep difficulties. (Means MK, Edinger JD)

Excessive Sleepiness

Sleep Deprivation and Its Effects on Cognitive Performance. Although sleep deprivation will ultimately lead to the involuntary onset of sleep in an individual, the cognitive performance effects of sleep deprivation can be evident even before sleep occurs uncontrollably in the form of sudden microsleeps or sleep attacks. Sleep deprivation adversely affects basic cognitive processes involving speed and accuracy of attention and memory, as well as higher order cognitive processes involving executive functions. Microsleeps and behavioral lapses increase with sleep loss as a function of wake state instability. (Dorrian J, Dinges DF)

Narcolepsy. Narcolepsy is a chronic neurological disorder of excessive daytime sleepiness that characteristically has a childhood onset and is associated with a hypocretin deficiency. The cardinal features of narcolepsy are daytime somnolence, cataplexy, sleep paralysis, and hypnagogic hallucinations. Successful treatment for narcolepsy includes both behavioral and pharmacological treatments. (Pelayo R, Lopes MC)

Idiopathic Hypersomnia. The term idiopathic hypersomnia has been used to categorize individuals with prominent daytime sleepiness, but who lack the classic features of narcolepsy or evidence of another disorder known to cause daytime sleepiness, such as sleep apnea. Nocturnal sleep is prolonged and uninterrupted. Naps are usually more than an hour in duration and are nonrefreshing. No amount of sleep ameliorates the daytime sleepiness. Currently, no specific marker is available to confirm the diagnosis. (Brooks SN)

Post-Traumatic and Recurrent Hypersomnia. Commonly, patients who have suffered even minor head trauma complain of sleep disturbances, including hypersomnia. Other etiologies of hypersomnia include postinfectious hypersomnia, Kleine–Levin syndrome, idiopathic recurring stupor/ endozepine stupor, and menstrual-related hypersomnia. (D'Ambrosio CM, Baron J)

Sleeping Sickness, Human African Trypanosomiasis. Sleeping sickness is an endemic parasitic disease that is exclusively located in intertropical Africa. After a bite by the tsetse fly, the illness evolves in two stages, the hemolymphatic stage I followed by the meningoencephalitic stage II, ending with demyelinization, altered consciousness, cachexia, and death if untreated. Excessive daytime sleepiness is one of the most reported signs during stage II of the illness. (Buguet A et al)

Medications that Induce Sleepiness. Drug-induced sedation is one of the most common effects and side effects of central nervous system (CNS) active drugs. Drugs known to induce daytime sleepiness are associated with declines in daytime performance and increased rates of automobile accidents. Benzodiazepines, antidepressants, and other agents may be utilized for their sedative side effects in anxious and insomniac patients. (Pagel JF)

Evaluation of Excessive Sleepiness. The identification of pathological sleepiness begins the process of establishing a proper diagnosis and allows initiation of treatment and follow-up of the individual's response. The diagnostic process begins with and is based primarily on a thorough sleep and general medical history. Nocturnal polysomnography is indicated in the evaluation of patients with suspected respiratory disturbances during sleep, narcolepsy, and idiopathic hypersomnia, and those in whom an adequate explanation of excessive sleepiness is not reached following a thorough sleep history. The Multiple Sleep Latency Test is a validated, objective measure of the ability or tendency to fall asleep under standardized conditions. The Maintenance of Wakefulness Test is an objective, laboratory-based measure of the subject's ability to remain awake under standardized conditions for a defined period of time. (Wise MS)

Therapy for Excessive Sleepiness. Sleepiness can be associated with a wide variety of conditions. When improved alertness can be attained by improved sleep hygiene and life-style interventions, these approaches are recommended. If sleepiness is secondary to a medical, neurological, or psychiatric condition, the condition should be addressed first. Caffeine is widely used as a countermeasure for sleepiness arising from both behavioral and nonmedical origins. When sleepiness results from a primary sleep disorder affecting a presumed underlying sleep–wake mechanism, such as narcolepsy, psychostimulants and/or awake-promoting substances are generally used palliatively to manage the condition. (Hirshkowitz M)

Napping. Short periods of sleep, or naps, during the daytime or at night can be used to actively cope with the physiological need of sleepiness. A nap may also be capable of maintaining waking performance and alertness under prior sleep restrictions. (Takahashi M, Kaida K)

Sleep Loss, Sleepiness, Performance, and Safety. Sleep loss and sleepiness degrade performance efficiency and increase the likelihood of operational errors that may contribute to traumatic or catastrophic incidents. A substantial number of people regularly confront sleep loss and sleepiness because of work, other situational demands, or medical conditions. Still, we are far from a precise estimate of the magnitude of risk from excessive sleepiness or its contribution to societal loss. (Rosa RR)

Sleep Disordered Breathing Syndromes

Physiology of Sleep Disordered Breathing. Upper airway competence involves complex interactions between anatomy and physiology. Airway size is determined by both dilating and collapsing forces. Dilating forces include upper airway muscle tone, mechanical force of the airway wall structure, and positive intraluminal airway pressure. Collapsing forces include tissue mass, surface adhesive forces, and negative intraluminal pressures. The resulting difference in these forces is the distending force, which acts on the wall of the upper airway. When the distending force increases, the airway size increases; when it decreases, the airway size decreases. (Woodson BT)

Snoring. Snoring is a repetitive sound caused by vibration of upper airway structures during sleep. Snoring results from pharyngeal vibration and is triggered by conditions that increase upper airway resistance and/or compliance, such as obesity, male gender, and nasal congestion. As the ramifications of obstructive sleep apnea are becoming clearer, the perception of snoring has changed from a sometimes noxious but otherwise benign marker of slumber, to an indicator of a potentially serious breathing disorder. (Olson EJ, Park JG)

Overview of Obstructive Sleep Apnea in Adults. Obstructive apnea is defined, by convention, as cessation of nasal/oral airflow for at least 10 seconds, despite persistent ventilatory efforts. One definition of obstructive hypopnea consists of reduction of airflow by at least 30% from baseline, of at least 10 seconds in duration and accompanied by oxyhemoglobin desaturation of 4% or more. Obstructive sleep apnea–hypopnea is prevalent in the community. Acquiring objective data supporting its diagnosis, traditionally employing nocturnal polysomnography, is a prerequisite to developing a treatment strategy. (Sanders MH, Givelber RJ)

Upper Airway Resistance Syndrome. Critical upper airway narrowing can also occur during sleep, leading to recurrent arousals and sleep fragmentation, even in the absence of discrete apneas, hypopneas, hypoxemia, or clear airflow reduction. This has been widely termed the upper airway resistance syndrome. This syndrome appears to cause daytime fatigue and sleepiness and may trigger both nocturnal and diurnal hypertension in a fashion similar to discrete obstructive sleep apnea. (Ballard RD)

Central Sleep Apnea. Central apnea is due to temporary failure in breathing rhythm generation resulting in the loss of ventilatory effort, lasting at least 10 seconds. Central apneas occur in many pathophysiological conditions. Depending on the cause or mechanism, central apneas may not be clinically significant. In contrast, in some disorders, central apneas result in pathophysiological consequences. (Javaheri S)

Obesity Hypoventilation Syndrome. Obesity hypoventilation syndrome (OHS) is broadly defined as hypercapnia during wakefulness in obese persons. The exact pathophysiologic mechanisms of OHS have yet to be fully elucidated, but at least four different factors may influence its development, including morbid obesity, mechanical limitation to increase minute ventilation, blunted central chemoreceptor response to hypercapnia and hypoxemia, and the coexistence of obstructive sleep apnea. Effective therapy, may therefore depend on correction of multiple contributing factors. (Park JG)

Cardiovascular Complications of Obstructive Sleep Apnea. Patients with obstructive sleep apnea have an increased risk of hypertension, ischemic heart disease, stroke, and heart failure. Obstructive sleep apnea causes acute changes in cardiovascular regulation during sleep, disrupting the normal state of cardiovascular relaxation. Recent studies intimate an important role for sympathetic activation, inflammation and endothelial dysfunction, disordered coagulation, metabolic dysregulation, and possibly oxidative stress in the development of cardiovascular disease. (Hahn PY et al)

Pulmonary Hypertension and Sleep Disordered Breathing. Sleep disordered breathing (SDB) is associated with a variety of chronic cardiovascular sequelae, including effects on the right heart and pulmonary vasculature. The coexistence of underlying lung disease with SDB raises the risk of pulmonary hypertension, due apparently to lower baseline oxygen tensions. SBD, in the absence of concurrent lung disease, might also be associated with the development of pulmonary hypertension (PH), although the degree of PH identified in this group of patients has been in the mild range and its clinical significance is unclear. (Judd BG)

Neurocognitive and Functional Impairment in Obstructive Sleep Apnea. Persons with obstructive sleep apnea experience greater difficulty in performing everyday activities such as bathing and grocery shopping, role limitations at home and at work, more bodily pain, reduced energy levels, and perception of poorer overall health. Daytime fatigue can result in increased accidents and diminished work performance resulting in fewer promotions or loss of work. The stress of living with a chronic illness can result in increased anxiety and diminished quality of life. (Brown WD)

Sleep Apnea and Cerebrovascular Disease. Sleep apnea has been found at alarmingly high rates in patients with acute stroke and after full neurologic recovery. There are several hematologic and hemodynamic changes in sleep apnea that can play significant roles in the pathogenesis of stroke. Sleep apnea represents a modifiable risk factor, but whether treatment of sleep apnea in the acute stroke setting, the rehabilitation setting, or as primary/secondary prevention is of benefit awaits further treatment studies. (Mohsenin V, Yaggi H)

Radiographic and Endoscopic Evaluation of the Upper Airway. Imaging modalities (magnetic resonance imaging, computed tomography, nasopharyngoscopy, cephalometry, and fluoroscopy) have objectively quantified upper airway structures and identified specific craniofacial and oropharyngeal soft tissue structural risk factors for obstructive sleep apnea. Imaging studies should be considered in sleep apneic patients being evaluated for upper airway surgery or oral appliances. (Schwab RJ, Kline NS)

Evaluation of Sleep Disordered Breathing: Polysomnography. Attended full-channel polysomnography (PSG) is considered the standard assessment and evaluation for sleep disordered breathing (SDB). PSG is also recommended for positive pressure titration in order to determine the optimal therapeutic pressure. A preoperative clinical evaluation including PSG is routinely indicated to evaluate for the presence of SDB prior to upper airway surgery. Follow-up PSG is recommended after good response to oral appliance treatment in patients with moderate to severe SDB to ensure therapeutic benefit, after surgical treatment in patients with moderate to severe SDB to ensure satisfactory response, after surgical treatment of patients with SDB whose symptoms return despite initial success of treatment, after substantial weight loss or gain, and when clinical response is insufficient. (Mehra R, Strohl KP)

Evaluation of Sleep Disordered Breathing: Portable Sleep Monitoring. The number of potential patients usually exceeds the number of sleep laboratory facilities capable of performing the test. To increase access to diagnosis and potentially reduce cost, there has been an effort to produce systems that incorporate part or all of polysomnography but make it portable and ideally usable without an attendant technician. (Littner M)

Indications for Treatment of Obstructive Sleep Apnea in Adults. The primary treatment modality for obstructive sleep apnea is positive pressure therapy. In patients who do not accept positive pressure therapy despite careful attempts to optimize the treatment, second line therapy should be explored. While palatal surgery can effectively treat snoring, the effect on the apnea–hypopnea indices and daytime sleepiness is less robust. Oral appliances may help some patients. (Davé NB, Strollo PJ)

Medical Treatment of Obstructive Sleep Apnea: Life-Style Changes, Weight Reduction, and Postural Therapy. Several alternative interventions have the potential for success in those patients who fail or refuse treatment trials using continuous positive airway pressure (CPAP) therapy, dental devices, or upper airway surgery. Obesity is a risk factor and probably is a cause or a precipitant of obstructive sleep apnea. Physicians should advise patients to eliminate evening alcoholic beverages and to reduce overall amounts of alcohol. Many patients could benefit from a trial of supine avoidance treatment including obese and nonobese subjects with or without simultaneous use of CPAP or dental devices. (Kapen S)

Pharmacological Treatment of Sleep Disordered Breathing. Aminophylline, theophylline, acetazolamide, thyroid supplement, tricyclic or serotonin reuptake inhibitor antidepressants, and sedative-hypnotics all have minimal use in the treatment of obstructive sleep apnea. Estrogen replacement appears helpful in postmenopausal females. Oxygen and carbon dioxide are helpful in some patients with obstructive or central sleep apnea by stabilizing ventilatory control. (Hudgel DW)

Positive Airway Pressure Therapy for Obstructive Sleep Apnea. Positive airway pressure is generally the preferred treatment for individuals with moderate or severe obstructive sleep apnea (OSA). The most common type provides a constant pressure and is called continuous positive airway pressure (CPAP). A second type, called bilevel positive airway pressure (BPAP), provides two pressure levels, one during inhalation and a lower one during exhalation. The third type of device is called autotitrating positive airway pressure (APAP) and uses variable flow controlled by computer algorithms in an attempt to determine optimal pressure. Finally, a fourth type of positive pressure device called noninvasive positive pressure ventilation (NIPPV) places two different pressures at a set rate to entrain breathing and provide ventilatory assistance. CPAP improves airway patency during sleep, which in turn improves sleep quality, sleep continuity, daytime alertness, and overall quality of life in symptomatic patients with moderate or severe OSA. (Hirshkowitz M, Lee-Chiong T)

Upper Airway Surgery for Obstructive Sleep Apnea. Upper airway surgery for obstructive sleep apnea (OSA) modifies dysfunctional pharyngeal anatomy (e.g., ablating pharyngeal soft tissue or altering the facial skeleton from which the soft tissues are suspended) or bypasses the pharynx. (Sher AE)

Oral Devices Therapy for Obstructive Sleep Apnea. Oral devices improve airway and tongue space by repositioning the mandible both downward and forward. Indications for the use of these devices include patients who have mild to moderate sleep apnea, patients who only snore and have been diagnosed with apnea or where apnea has been ruled out, and patients who are intolerant to continuous positive airway pressure (CPAP) or might have had surgery that was deemed unsuccessful. (Bailey DR)

Circadian Rhythm Sleep Disorders

Advanced, Delayed, Irregular, and Free-Running Sleep–Wake Disorders. The delicate interplay of endogenous and exogenous factors, required to maintain normal sleep–wake rhythm, can become recurrently or chronically impaired in some individuals, leading to a group of disorders called circadian rhythm sleep disorders. They are characterized by an alteration of the circadian timing system or a misalignment between the timing of the individual's sleep–wake rhythm and the 24 hour social and physical environment. (Dagan Y et al)

Jet Lag. Jet lag refers to the lag between the time frame of the biological clock and that of the destination time zone. Because the biological clock is slow to adjust, there will be several days after arrival in the new time zone before the biological clock "catches up" with the new routine. (Monk TH)

Shift Work Sleep Disorder. Shift work is generally defined as any schedule that requires work outside a broadly defined "day shift," usually 6 a.m. to 6 p.m. Shift work is associated with both increased difficulty sleeping and with increased sleepiness during waking hours. (Richardson GS)

Neurological and Medical Disorders Associated with Circadian Rhythm Disturbances. Circadian rhythmicity may play an important role in the expression of some common neurological and medical disorders. Diurnal changes in physiology, behavior, and endocrine function may also influence the already disrupted sleep that is commonly seen in patients with neurological and medical illnesses, further exacerbating the disease process. (Gourineni R, Zee PC)

Psychiatric Disorders Associated with Circadian Rhythm Disturbances. Although there is currently no direct evidence that circadian abnormalities are causally related to any given psychiatric condition, alterations in the amplitude or phase of several circadian output variables, as well as mounting evidence that certain circadian rhythm manipulations prove therapeutically effective, suggest a potential pathophysiological role for the circadian pacemaker in major depressive disorder, bipolar disorder, and seasonal affective disorder or winter depression. (Jones S, Benca RM)

Therapy of Circadian Sleep Disorders. There are two general treatment strategies for circadian rhythm sleep disorders: (1) resetting the clock or (2) overriding the clock. With clock resetting, the phase (timing) of the circadian pacemaker is shifted (reset) so that the output signals for sleep and wake are more congruous with a person's desired sleep/wake schedule. Light exposure and/or melatonin, administered at the optimal circadian phase, are currently the most practical phase-resetting agents available for clinical use. In cases where circadian resetting is impractical or undesirable, overriding the circadian signal with a hypnotic or alerting medication may be the preferred strategy. (Sack R, Johnson K)

Parasomnias

Disorders of Arousal and Sleep-Related Movement Disorders. An arousal disorder is characterized by an incomplete awakening from sleep where the individual has voluntary movements but no awareness of their actions. There are three types of arousal disorders including confusional arousals, sleep terrors, and sleepwalking. Nocturnal leg cramps and rhythmic movement disorders are classified as sleep-related movement disorders and consist of relatively simple movements that disturb sleep. (Cavanaugh K, Friedman NR)

Sleepwalking. Sleepwalkers may engage in complex motor behaviors but lack responsiveness to others. Once they are fully awake, they typically have no memory of having been out of bed, or of what they were doing, or why. (Cartwright RD)

REM Sleep Behavior Disorder and REM-Related Parasomnias. Rapid eye movement (REM) sleep behavior disorder is characterized by the loss of physiologic skeletal muscle atonia during REM sleep in association with excessive motor activity while dreaming. Other REM-related parasomnias include nightmares, catathrenia, painful nocturnal erections, sleep paralysis, and REM sleep-related sinus arrest. (Tippmann-Peikert M et al)

Nocturnal Enuresis in Children. Nocturnal enuresis is characterized by the frequent occurrence of normal complete uncontrolled micturition during sleep in children older than 5 years of age. Pathophysiology is still uncertain and management depends cultural factors. (Challamel MJ, Cochat P)

Sleep Bruxism. Sleep bruxism is a stereotyped, sometimes intense, orofacial movement with bilateral activation of the masseter and temporal "jaw closure" muscles. Bruxism is thought to be a major cause of temporomandibular disorder with patients complaining of pain and/or dyskinesia in the temporomandibular region. (Bader G)

Sleep-Related Eating Disorders. Sleeping and eating comprise two of the basic elements that drive human behavior. When the two are simultaneously affected, a fascinating spectrum of disease states may result. The differential diagnosis of sleep-related eating disorders include nocturnal eating syndrome and sleep-related eating disorder. (Auger RR, Morgenthaler TI)

Other Parasomnias. Sleep paralysis is one of the cardinal symptoms of narcolepsy; it can also occur on its own, in which case it exists in two forms, either familial or isolated. Sudden unexplained nocturnal death syndrome (SUNDS) is defined as sudden death in healthy young adults during sleep, in particular among young Asian males. An impaired sleep related-penile erection is the inability to sustain a penile erection during sleep that would be sufficiently large or rigid enough to engage in sexual intercourse. The cause is usually organic if the reduction occurs in the presence of normal sleep architecture. Sleep-related painful erections are defined as painful erections typically during REM sleep but not during the awake state. Sleep-related abnormal swallowing syndrome is characterized by inadequate swallowing of saliva, which results in aspiration with coughing, choking, and brief arousals or awakenings from sleep. (Qureshi A)

Movement Disorders

Restless Legs Syndrome. Restless legs syndrome (RLS) is characterized by four core symptoms: an urge or sensation to move the limbs, the urge worsens when at rest, movement improves the urge at least temporarily, and symptoms worsen in the evening or at night. RLS impacts up to 10% of adults in North America and Western Europe and is the most common neurologic disorder that causes chronic insomnia. (Becker PM)

Periodic Limb Movement Disorder. Periodic limb movement disorder (PLMD) is characterized by periodic episodes of highly stereotyped and repetitive involuntary rhythmic movements that usually affect the lower extremities during sleep. PLMD is strongly associated with restless legs syndrome. (Khassawneh B)

Sleep in Infants and Children

Ontogeny of EEG Sleep from Neonatal Through Infancy Periods. Maturation of infant behavior requires careful evaluation of both waking and sleep behaviors. Serial neonatal and infant electroencephalographic (EEG)/polysomnographic studies document the ontogeny of cerebral and noncerebral physiologic behaviors. EEG patterns and other physiologic relationships serve as templates for normal brain maturation and help distinguish intrauterine from extrauterine development. (Scher MS)

Sleep in Infants and Children. Sleep occupies a major portion of the lives of newborns, infants, and children. Three distinct sleep states can be identified in the term newborn: active sleep (REM), quiet sleep (NREM), and indeterminate sleep. Sleep becomes consolidated into a long nocturnal period of approximately 10 hours in children between 2 and 5 years of age. Generally, sleep patterns of children during middle childhood resemble those of older individuals, but there is considerable individual variability. (Sheldon SH)

Sleep and Breathing During Early Postnatal Life. Instability of the breathing pattern is an inherent characteristic of the normal healthy infant during sleep. Dramatic and profound changes in respiration during sleep occur with maturation of the neurophysiological, metabolic, and mechanical components of the respiratory system. The effect of REM sleep on the various mechanisms involved in respiratory control are of particular importance during infancy. (Goldbart AD et al)

Congenital Syndromes Affecting Respiratory Control During Sleep. The spectrum of disorders of respiratory control can include abnormal integration of responses to respiratory stimuli such as in central congenital hypoventilation syndrome, or abnormal regulation of autonomic function such as familial dysautonomia of which respiratory control is one aspect; or can involve abnormal development of brainstem structures in respiratory control such as myelomeningocele. Therapeutic management of central hypoventilation syndromes usually requires mechanical assisted ventilation or close monitoring for evidence of sleep disordered breathing causing cardiopulmonary compromise. (Witmans MB et al)

Sudden Infant Deaths. Sudden infant death syndrome (SIDS) is defined as sudden death of an infant under 1 year of age that remains unexplained after a complete postmortem examination, death scene investigation, and case conference. In most industrialized countries, SIDS is the leading cause of death in infants between the ages of 1 and 12 months. Little is known on the mechanisms responsible for the deaths of these infants. (Kahn A et al)

Obstructive Sleep Apnea in Children. Obstructive sleep apnea syndrome is common in children. It is commonest among preschoolers due to adenotonsillar hypertrophy. The vast majority of children with obstructive sleep apnea syndrome have both symptomatic and polysomnographic resolution following a tonsillectomy and adenoidectomy. (Bandla P, Marcus CL)

The Sleepless Child. Common causes of sleeplessness in children include sleep-onset association disorder, night wakings, early awakening, nighttime eating/drinking disorder, separation anxiety, parental limit setting, and night fears. Although sleeplessness may be primarily caused by a medical factor, there may also be a comorbid behavioral cause, or a behavioral cause may develop from a medical one and perpetuate the sleeplessness. (Moorcroft WH)

The Sleepy Child. Childhood is a time of growth, learning, and development. In order for these processes to occur the child must be awake, alert, and able to interact with and learn from the environment. Daytime sleepiness impairs the child's ability to do this. A sleepy child falls asleep at inappropriate times. The problems that lead to excessive sleepiness in children can generally be elucidated by a careful history and eliminated by appropriate treatment. (Rosen G)

Craniofacial Syndromes and Sleep Disorders. Children and adults with craniofacial syndromes represent a population at increased risk for sleep disordered breathing, due to anatomic and, in some cases, neuromuscular differences in the upper airway. All patients with craniofacial anomalies or upper airway compromise should be screened for sleep-related symptoms in order to initiate proper evaluation and treatment. In addition, these individuals may report behavioral sleep complaints, such as sleep-onset or sleep maintenance insomnia. (Wills LM et al)

Medical Disorders. Medical disorders that impact sleep in pediatric patients include asthma, gastroesophageal reflux, and otitis media. Adequate treatment of these conditions may improve the quality of sleep for the child. (Palmer JM, Brooks LJ)

Sleep in Children with Neurological Disorders. Fragmented sleep is common in children with neurological disorders. Alterations in sleep–wake function vary, depending on the anatomic location of the neurological lesion. The severity of alterations in sleep–wake function also depends on the extent of the lesion and whether it is static or progressive. Medications used to treat neurological disease may also impact sleep architecture. (Kotagal S)

Sleep in Children with Neuromuscular Disease. Much of the morbidity and mortality in children with neuromuscular disease occurs because of respiratory muscle weakness, which impairs both ventilation and secretion clearance. Respiratory abnormalities are often first noted during sleep. Once respiratory compromise has progressed to the point of causing sleep-related daytime symptoms, respiratory failure may be imminent. (Givan DC)

Sleep in Children with Behavioral and Psychiatric Disorders. Primary behavioral and psychiatric disorders in children are frequently associated with and/or complicated by sleep disturbances, which may be related to such factors as the psychopathology of the underlying disorder, comorbid conditions, or pharmacologic treatment. Even modest improvements in sleep quality and duration may have a significant impact on neurobehavioral functioning. (Owens JA, Davis KF)

Circadian Rhythm Disorders in Infants, Children, and Adolescents. Children are in flux with regard to sleep and its timing. The major feature of circadian rhythm sleep disorders is a misalignment between the child's sleep pattern and the sleep pattern that is required by parents, day care, and school. (Herman JH)

Sleep in the Elderly

Normal Sleep in Aging. With aging, sleep, as with other physiological processes, undergoes increasingly noticeable changes. Many of the changes accompanying aging are part of a gradual process rather than an abrupt change and reflect changes in both homeostatic and circadian processes that occur throughout the life span. Chronological age by itself seems to explain very little of the observed prevalence of sleep complaints. Medical diseases and chronic illness may account for most of the changes in sleep observed in old age. (Ayalon L, Ancoli-Israel S)

Sleep Disordered Breathing in Older Adults. Sleep disordered breathing (SDB) is highly prevalent in older adults, with sleep apnea alone affecting nearly 20% of the elderly. However, sleep apnea syndrome, defined as the presence of an elevated apnea-hypopnea index (AHI) and clinical symptoms, is far less common. It is possible that SDB in the elderly is a different pathophysiologic process than SDB in younger subjects. Those with cardiovascular co-morbidities or other clinical symptoms such as excessive daytime sleepiness can suffer significant morbidity and mortality from SDB. (Gooneratne N)

Insomnia and Aging. Insomnia is a prevalent and persistent problem associated with aging. A large number of physical health conditions such as pain syndromes, primary pulmonary problems, neurologic disturbances, and dementias such as Alzheimer's disease contribute to the increase in sleep problems associated with aging. Medications used to manage chronic conditions may also contribute to insomnia in the elderly. (Friedman L)

Sleep in Institutionalized Older Adults. Nighttime sleep disruption is characteristic of nursing home residents, and is typically accompanied by daytime sleepiness. Many factors contribute to these sleep problems, including medical and psychiatric illness, medications, circadian rhythm abnormalities, sleep disordered breathing, environmental factors and lifestyle habits. There is some suggestion that these factors are amenable to treatment, particularly improving daytime activity patterns, increasing light exposure and reducing sleep-disruptive care giving practices at night. (Martin JL, Alessi C)

Sleep Among Women

Patterns of Sleep in Women: An Overview. A woman's changing hormone profile influences her sleep. In general, more disruption can be anticipated with abrupt changes and withdrawal of female hormones. Some sleep disorders, such as sleep disordered breathing and restless legs syndrome, may also be influenced by the reproductive stage. (Driver HS)

Sleep During Pregnancy and Postpartum. Hormonal changes during early pregnancy, continuous enlargement of the fetus throughout pregnancy, and a newborn with random cycles of sleeping and feeding contribute to a woman's sleep loss. Problems with sleep are experienced as early as the tenth week of pregnancy. The major concern for postpartum women is sleep loss and resulting physical fatigue, negative mood states, and cognitive impairment. (Lee KA)

Menstrual-Related Sleep Disorders. Despite considerable hormonal fluctuations over the course of the menstrual cycle, sleep architecture and circadian rhythm appear to remain relatively stable in normal women. Women with dysmenorrhea may experience sleep disturbance, especially during the menses. Complaints of poor sleep and fatigue are common among women with both premenstrual syndrome and premenstrual dysphoric disorder. Menstrual-associated insomnia and hypersomnia are rare disorders of sleep with manifestations that occur primarily during the late luteal phase. (Pien GW, Beothy EA)

Sleep Disordered Breathing in Women. Sleep disordered breathing is a relatively common disorder in women. Many women with sleep disordered breathing may remain undiagnosed. Although men are more likely to develop sleep disordered breathing, the difference should not lead to missing the diagnosis in women. (Badr MS)

Sleep During the Perimenopausal Period. Perimenopause is a transitional period occurring prior to menopause, or cessation of menses. Changing hormone levels and other physiological alterations may underlie some of the increased sleep disturbances. Other factors such as significant life events, increased weight gain, and incidence of sleep apnea, primary insomnia, and depression may also contribute to the increased incidence of sleep disturbance and dissatisfaction with sleep observed in perimenopausal women. (Rogers NL, Grunstein RR)

Sleep During Postmenopause. Though health-care practitioners and older women have a tendency to attribute the appearance of sleep disturbance to the onset of menopause or "hormone problems," the sleep complaints and changes experienced by older women are likely to be attributable to chronic physical or mental illness and related factors such as stress and caregiving. The sleep of older women may be affected by some unique factors such as chronic hot flashes. (Moe KE)

Sleep in the Respiratory Disorders

Respiratory Control During Sleep. Sleep is a regulated state in which the regulation of different elements of the respiratory system is heterogeneous. Wakefulness is associated with an ill-defined, but important excitatory stimulus, which has been called the 'wakefulness stimulus.' Removal of the wakefulness stimulus reduces the drive to breathe, but sleep modifies the processing of the chemical and mechanical signals that contribute to stable values of carbon dioxide, oxygen and pH. Those areas of the brainstem involved in the control of sleep state also seem to play an important role in responses to a variety of respiratory stimuli, such as hypercapnia, hypoxia and a variety of upper airway reflexes. (Krimsky WR, Leiter JC)

Asthma. Circadian rhythms clearly play an important role in asthma. The proposed mechanisms of nocturnal worsening of asthma include sleep-related changes in lung volume, bronchial hyperresponsiveness, cortisol and beta-adrenergic receptor responsiveness, parasympathetic tone, and airway inflammation. Perhaps the most intriguing hypothesis is that nocturnal asthma is primarily an inflammatory disorder, with a mechanism that is distinct from nonnocturnal asthma and a pathophysiology that centers on a worsening of both central and peripheral lung inflammation at night. (Beuther DA et al)

Chronic Obstructive Pulmonary Disease and Sleep. - Chronic obstructive pulmonary disease (COPD) is a progressive lung condition characterized by chronic airflow obstruction that is incompletely reversible. Increased sleep disruption is common in COPD as evidenced by increased frequency of arousals, increased frequency of sleep stage changes, and decreased total sleep time. Sleep-associated desaturation is often an expression of worsening respiratory failure during sleep. Patients with COPD who have concomitant obstructive sleep apnea may be particularly vulnerable to developing respiratory failure and secondary hemodynamic complications. (Iber C)

Sleep and Breathing in Cystic Fibrosis. In patients with cystic fibrosis, marked gas exchange abnormalities can first appear during sleep, preceding the appearance of daytime respiratory failure. Patients with cystic fibrosis and moderate to severe lung disease may exhibit marked nocturnal desaturation, especially in REM sleep. The main mechanisms responsible for this appear to be reduced respiratory drive and loss of postural muscle tone. (Piper AJ et al)

Restrictive Thoracic and Neuromuscular Disorders. Sleep disordered breathing is common in restrictive thoracic and neuromuscular diseases. Hypoxemia and hypoventilation are common during sleep, related to reductions in functional residual capacity (FRC) and blunting of central drive (either primary or secondary to progressive bicarbonate retention). These gas exchange abnormalities lead to arousals, fragmenting sleep and producing symptoms such as morning headache and daytime hypersomnolence. (Perrin C et al)

Noninvasive Ventilation and Sleep. Noninvasive ventilation has a beneficial effect on ventilation during sleep. Noninvasive ventilation appears capable of improving sleep quality and quantity as well, in patients with restrictive or obstructive ventilatory defects. Although sleep appears improved, it is rarely normalized. (Liistro G, Rodenstein D)

Sleep in the Cardiac Disorders

Hypertension and Cardiovascular Disease. Obstructive sleep apnea has adverse effects on blood pressure, cardiovascular status, and probably cardiovascular mortality. There is also evidence that effective therapy with CPAP can improve blood pressure and cardiac function in adult obstructive sleep apnea patients. (Ballard RD)

Congestive Heart Failure. Both obstructive sleep apnea and central sleep apnea with Cheyne–Stokes respiration can exist separately or together and interact in the same patient with congestive heart failure. Obstructive sleep apnea may impair cardiac function and contribute to increased morbidity and mortality in patients with concomitant heart failure. Central sleep apnea with Cheyne–Stokes respiration may be a marker for poor cardiac function and may indicate a worse prognosis in these patients. (Mazza E, Gurubhagavatula I)

Cardiac Arrhythmias and Sudden Death During Sleep. Autonomic nervous system activity and disturbed respiration during sleep are capable of provoking both atrial and ventricular arrhythmias in patients with cardiovascular disease. A significant number of atrial arrhythmias, in particular atrial fibrillation, in patients under 60 years of age and lethal ventricular arrhythmias have their onset at nighttime. (Verrier RL, Josephson ME)

Sleep in the Other Medical Disorders

Sleep and the Gastrointestinal Tract. The interaction of gastrointestinal functioning and sleep may lead to sleep complaints as well as the pathogenesis of some gastrointestinal disorders. Sleep-related gastroesophageal reflux (GER) is an important factor in the development of esophagitis and respiratory complications of GER. Patients with functional bowel disorders have an increase in sleep complaints. (Orr WC)

Renal Disease. Sleep complaints and primary sleep disorders are very prevalent in end-stage renal disease (ESRD) patients and appear to have important adverse effects on their overall health and well-being. Primary sleep disorders such as sleep apnea (SA), restless legs syndrome (RLS), and periodic limb movement disorder (PLMD) are very common. (Parker KP)

Endocrine and Metabolic Disorders and Sleep. Obesity is frequently associated with sleep disorders, including obstructive sleep apnea (OSA), sleep disruption, and daytime sleepiness and fatigue. The proinflammatory cytokines, TNFα and IL-6, may play a role in mediating sleepiness in patients with sleep apnea and obesity. Although knowledge on the association between diabetes mellitus and sleep disturbances is limited, it is plausible that there is an association between insulin resistance and sleep apnea. Administration of testosterone worsens SA, whereas female sex hormones appear to be protective of OSA. Disturbances of sleep are not uncommon in patients with acromegaly and disorders of the adrenal and thyroid glands. (Vgontzas AN et al)

Sleep in Fibromyalgia and Chronic Pain. Patients with fibromyalgia often report sleep fragmentation, early morning awakenings, unrefreshing sleep, fatigue, and insomnia. Primary sleep disorders, especially sleep apnea and restless legs syndrome with periodic limb movements of sleep, may also be found in fibromyalgia patients. There is a complex relationship between chronic pain and sleep. Pain can disrupt sleep and poor sleep can increase pain intensity. (Harding SM, Lee-Chiong T)

Sleep and the Immune Response. Sleepiness is frequently experienced during acute infections and other inflammatory diseases. These changes in sleep are part of the microbe-induced acute phase response and are mediated by cytokines. In addition, changes in the immune response are associated with sleepiness and sleep loss. (Krueger JM, Majde JA)

Sleep in the Neurologic Disorders

Alzheimer's Dementia. Alzheimer's disease is a progressive neurodegenerative disorder that accounts for approximately two-thirds of all dementias worldwide. Sleep disturbance adds an additional burden to the compromised function and quality of life directly attributable to dementia. Accurate assessment of sleep disturbances in demented patients can only be done in the context of appreciating the potential contributing associated medical disorders, current drug treatments, psychopathologies, primary sleep disorders, and behavioral and environmental conditions that may exist. (Vitiello MV)

Neurodegenerative Disorders. There are several progressive neurodegenerative dementias and all of them are characterized by a progressive cognitive and functional decline with increasing neuropathology over the course of the illness. Although the sleep disturbances associated with them may appear superficially similar, there are likely to be vast differences in their expression. The impact of homeostatic versus circadian disruption, environmental versus neurodegenerative etiologies, and the role played by aging across the continuum of age-associated sleep and circadian changes all need to be accounted for to ultimately understand the etiology of sleep disturbance associated with a particular neurodegenerative illness. (Harper DG)

Parkinson's Disease. Problems initiating or maintaining sleep in Parkinson's disease (PD) may be due to uncontrolled motor symptoms, the effects of medications, restless legs syndrome, depression, or circadian sleep–wake reversal in patients with superimposed dementia. Multiple factors interact to cause sleepiness in PD patients, including somnolence intrinsic to the disorder itself, the effects of medication, and sleep disordered breathing. REM sleep behavior disorder (RBD) is present in 15–33% of PD patients assessed in PD clinics. Hallucinations and behavioral problems at night can also occur. (Silber MH)

Seizures. Sleep states have a potent effect on the expression or suppression of epileptic seizure manifestations. NREM sleep, awakening from sleep, and other transitional arousal states are conducive to electrographic and clinically evident seizures, whereas REM sleep is not. Sleep disturbances often parallel the severity of seizure disorders. Antiepileptic drugs can ameliorate seizure-related sleep disturbances, but improvement in sleep architecture is not a critical factor in seizure control. (Shouse MN)

Headaches and Sleep. Sleep may play a role in headache genesis. Certain headache types may be associated with or evolve from certain sleep stages. Various sleep pathologies or disorders may also lead to the development of headaches arising from sleep. (Greenough GP)

Cerebrovascular Disorders. The usual sleep hours between midnight and 6 a.m. have the lowest stroke risk, as well as having the lowest blood pressure and lowest catecholamine and corticosteroid levels. The stroke risk rises with elevations in these three parameters after awakening. Specific sleep–wake cycle anomalies resulting from stroke injury to specific brain regions have been well documented. Obstructive sleep apnea (OSA) and snoring are associated with an increased risk for stroke. (Labib B, Nazarian SM)

Brain and Spinal Cord Injury. There is a complex relationship between disorders of sleep and traumatic brain injury, with evidence supporting the development of a number of sleep disorders as a result of traumatic brain injury, including insomnia, circadian rhythm disorders, periodic limb movement disorder, obstructive sleep apnea, narcolepsy, and post-traumatic hypersomnia. (Castriotta RJ)

The Blind Patient. Blind persons may have defective retinal processing or an impaired retinohypothalamic tract (that conducts photic information from the retina to the biological clock located in the suprachiasmatic nucleus of the hypothalamus) and therefore may be unable to exhibit a 24 hour pattern. There are a higher percentage of insomnia and free-running circadian patterns in blind persons compared to sighted adults. (Leger D, Metlaine A)

Sleep in the Psychiatric Disorders

Schizophrenia. Sleep disruption is a common and very debilitating comorbid symptom of schizophrenia. Sleep disruption can aggravate psychosis; conversely, increased susceptibility to external stimuli imposed by schizophrenia means that the psychopathology would increase sleep disruption. For the most part, medications used to treat schizophrenia improve problematic sleep. However, their discontinuation can, if only temporarily, worsen sleep symptoms. (Norwood RJ, Lee-Chiong T)

Mood Disorders. Sleep complaints are pervasive in those diagnosed with major depressive disorder, including difficulty falling asleep, intermittent awakenings, and early morning awakenings. Most antidepressant medications suppress REM sleep; these agents may also produce alterations in sleep consolidation and sleep architecture. (Armitage R)

Anxiety Disorders and Sleep. The cognitive and physiologic changes associated with excessive anxiety alter sleep and are made worse by the lack of sleep. The sleeplessness that anxiety can cause turns back on itself and magnifies anxiety, thus setting up a vicious cycle. (Weissberg M)

Trauma and Post-traumatic Stress Disorder. Sleep disturbances are common in post-traumatic stress disorder (PTSD). Anxiety dreams, increased REM phasic activity, increased arousals from REM sleep, increased startle response, low dream recall, and possibly elevated awakening thresholds from sleep may characterize PTSD. (Pillar G et al)

Alcohol, Alcoholism, and Sleep. Ethanol has been found to have far-reaching effects on sleep and sleep disorders. The sleep of healthy normals appears to be disturbed with acute high ethanol doses. Individuals with alcoholism commonly have sleep problems, which may occur during active drinking, acute ethanol discontinuation, and prolonged abstinence. (Hyde M et al)

Drugs of Abuse and Sleep. Nearly all drugs of abuse have considerable effects on sleep and wakefulness and on particular stages of sleep. The sleep stage most typically affected by these drugs is REM sleep. It has been hypothesized that sleep and wake changes, although not the primary reinforcing mechanisms, function as contributing factors in maintaining the compulsive and excessive drug use, and in increasing the risk for relapse. (Hyde M et al)

Sleep in Special Patient Groups

Sleep and the Caregiver. Many internal and situational factors can influence the caregiver's ability to obtain quality sleep. In fact, not getting enough sleep is a major cause of illness and stress in caregivers. Caregivers are "on call" 24 hours per day, 7 days per week. Caregiver anxiety, worry, grief, or bereavement can lead to sleep disturbances. (Carter PA)

Sleep in Patients with HIV Disease. Sleep disturbance (primarily insomnia) and fatigue are very common and often disabling symptoms for this population of individuals. Although it may begin at any point along the spectrum of human immunodeficiency virus (HIV) disease, sleep disturbance often appears very early in the course of infection and seems to contribute to a decrease in quality of life during the course of illness. (Jaffe SE)

The Patient with Cancer. Sleep disturbances are a common complaint in cancer patients. The etiologic factors contributing to the sleep problems are multiplicative and include side effects of treatment, pain, maladaptive sleep behaviors, medications, the diagnosis itself, and the specific treatment of surgery, chemotherapy, radiation, or hormonal therapies. Fatigue, mood disturbance, and a compromised immune system are possible consequences of sleep disturbance. (Engstrom CA)

Sleep in the Intensive Care Unit. Patients in the intensive care unit (ICU) are more susceptible to significant sleep deprivation. The cause appears to be multifactorial and includes the type and severity of the patient's underlying illness, medications received, use of hemodynamic and respiratory monitoring devices, use of mechanical ventilation, and the ICU environment itself. Sleep deprivation can affect cognitive behavior, as well as cellular immune function and tissue repair. (Krachman SL, Chatila W)

Sleep and the Cardiac Surgery Patient. Sleep deprivation, including decreased quantity, increased fragmentation, and decreased quality of sleep, is prevalent in adults who have undergone cardiac surgery, and it has been shown to be associated with decrements in postoperative physical function and emotional well-being. During the early postoperative period, management of environmental stimuli, including reductions in noise, lighting, and the frequency of intrusive patient care interactions, may facilitate sleep. Adequate medication for pain is also an important consideration. (Redeker NS, Hedges C)

Sleep Disturbances After Noncardiac Surgery. There are profound sleep disturbances in the postoperative period with initial slow-wave sleep and REM sleep suppression and subsequent REM rebound. It seems that the surgical trauma-induced inflammatory stress response, metabolic–endocrine stress response, circadian disturbances, and postoperative opioid use are the most important factors influencing sleep after surgery. Sleep disturbances might play a significant role in the development of cardiopulmonary instability, postoperative cognitive disturbances, and fatigue. (Gögenur I, Rosenberg J)

Relevance of Anesthesiology for Sleep Medicine. Sleep and anesthesia are altered arousal states actively generated by the central nervous system. There are no contemporary data that systematically characterize the effect of volatile anesthetics on the sleep of patients without the confounding factors of surgical insult, polypharmacy, trauma, or coexisting disease. (Lydic R, Baghdoyan HA)

Sleep at High Altitudes. Poor sleep is a prominent manifestation of rapid ascent to high altitude and is in part related to acute changes in ventilatory stimuli. Sleep at high altitude is characterized by frequent arousals and poor quality. It is worsened by the development of high altitude illnesses such as acute mountain sickness and high altitude pulmonary edema. Sleep improves with acclimatization. (Chatila W, Krachman S)

Sleep and Aviation. Modern pilots and aircrews must cope with a variety of nonstandard work schedules in order to effectively meet customer/mission demands.

Aviation personnel are likely to be faced with unpredictable work hours, long duty periods, and circadian disruptions, which all lead to difficulties obtaining adequate sleep. Progress toward the widespread development and implementation of scientifically valid sleep/rest- and performance-optimization strategies is already contributing to operational safety throughout the system. (Caldwell JA)

Sleep, Exercise, and Sports. A clear relationship between exercise and sleep remains undetermined; nonetheless, some studies suggest the intriguing possibility of utilizing exercise to improve sleep. The existence of a circadian advantage or jet lag effect on sports performance also remains inconsistently supported. (Enderlin CA, Richards KC)

Sleep, Sleep, Loss and Circadian Influences on Performance and Professionalism of Health Care Workers. Acute care nurses, physicians, interns, residents, medical students, and staff are faced with challenges of prolonged wakefulness time, chronic insufficient sleep, and the need to work at times of low circadian wakefulness stimuli and thus will likely experience times of impaired cognitive performance. The significance of the impaired performance spans lack of sensitivity and reduced compassion to medical error. (Veasey SC)

The Student with Sleep Complaints. In the first two decades of life, when one is most likely to be a student, there are profound changes taking place in physiologic sleep patterns. In addition to these biologic processes, there are profound psychosocial changes that occur especially in early student life. Sleep problems are common and underrecognized in the student group. Poor sleep quality has been linked to increased tension, irritability, depression, more frequent use of alcohol and illicit drugs, accidents, and lower academic performance. (Bijwadia J, Dexter D)

Sleep Assessment Methods

The Sleep Interview and Sleep Questionnaires. A comprehensive sleep interview with the use of appropriate questionnaires should allow clinicians to have a good idea of what sleep disorder or disorders a patient may have. There are some commonly used validated questionnaires that can help with the assessment of daytime sleepiness and fatigue: the Epworth Sleepiness Scale, the Stanford Sleepiness Scale, and the Fatigue Severity Scale. (Bae CJ, Golish JA)

Polysomnography. Polysomnography is the monitoring of physiologic signals from various organs and transduction of those signals to a recording device. Polysomnography is the standard for the diagnoses of many sleep disorders including, but not limited to, obstructive sleep apnea. Although most sleep laboratories utilize similar montages, equipment and scoring standards may vary. (Collop NA)

Pediatric Polysomnography. The polysomnographer faces multiple challenges posed by the evolution of pediatric electroencephalography and sleep staging, changes in respiratory rates and patterns with age, differing presentations and etiologies of disorders such as obstructive sleep disordered breathing in children versus adults, and even the issues of dealing with small children and their families. The complexity of pediatric polysomnography is increased by limited normative data for interpretation of its various parameters. (Griebel ML, Moyer LK)

Introduction to Sleep Encephalography. The main limitation with routine electroencephalography (EEG) with a brief 20–30 minute recording is its poor sensitivity for epilepsy. EEG-video monitoring is the highest level of epilepsy monitoring and is the gold standard. (Benbadis SR)

Monitoring Respiration During Sleep. Monitoring of respiration during sleep allows the assessment of physiological variables that are required to characterize sleep-related breathing disorders. During an obstructive respiratory event, pharyngeal collapse occurs with an increasing inspiratory effort; measuring changes in intrathoracic pressure and activity of inspiratory muscles is important for defining these events. Although the only direct method for measuring airflow is pneumotachography, indirect measurements of flow are widely used because of better patient tolerance. Measurement of thoracic volume in sleep studies is mainly performed to infer flow. Noninvasive assessments of blood gases are generally employed during sleep studies. (Magalang UJ et al)

Recording and Monitoring Limb Movements During Sleep. Proper electrode placement is essential for recording limb movements. Limb movements can be seen and recorded in the legs as well as the arms. The preferred placement to record leg movements is on the belly of the anterior tibialis muscle group. When monitoring the arms, the extensor digitorium is the best site for recording. (Scott CE)

Actigraphy. Actigraphy refers to a methodology for recording and analyzing movement from small, computerized devices worn on the body. Actigraphy can be very useful in investigations of group differences, sleep pattern variations over time, and the effects of behavioral or treatment interventions. (Acebo C)

pH Monitoring and Other Esophageal Tests. When nocturnal gastroesophageal reflux occurs, physiologic changes associated with sleep may influence the clearance of esophageal acid, upper airway protection, and other defense mechanisms sufficiently to promote an increase in the severity of both esophageal and extraesophageal complications of gastroesophageal reflux disease (GERD). Nocturnal heartburn associated with GERD can result in sleeping difficulties and impaired daytime function. Finally, GERD may result in extraesophageal manifestations that can be disruptive of sleep. Esophageal pH monitoring has become the most common technique used to investigate the potential role of GERD in producing symptoms. (Cott G)

Psychological Assessment of the Sleep Patient. The goals of a general psychological assessment of the sleep patient include evaluating psychological factors contributing to sleep problems, diagnosing or ruling out psychiatric disorders, and providing treatment recommendations. An enormous number of instruments have been developed in order to objectively describe, measure, and quantify different aspects of psychological functioning. A more specialized neuropsychological assessment may be performed when neurocognitive abilities such as attention, concentration, memory, and organization are in question. (Ikelheimer ABR, Hoyt B)

Operating and Managing a Sleep Disorders Center. - Sleep centers began in academic environments. Recognition of the huge numbers of patients with sleep disorders forced expansion beyond the academic sleep center. Standards for the professional staff, facilities, equipment, personnel, and operation of the sleep center were developed, and the American Academy of Sleep Medicine (AASM) now offers accreditation to centers that meet these standards and Standards of Practice documents to guide the practice of sleep medicine. (Rosenberg RS)

Accrediting a Sleep Program. American Academy of Sleep Medicine accreditation is considered the "gold standard" for sleep programs and it provides many advantages. It helps assure quality in delivering sleep medicine services for physicians and patients and it can also be useful in distinguishing a program from others in a competitive marketplace. (Arand D)

Painstaking research had been undertaken to assure the accuracy and timeliness of the data in this book. Nonetheless, the disciplines of sleep and dreaming are constantly evolving. Each day, new discoveries prompt us to redefine old concepts and formulate new ones—these will be incorporated in future editions of this book. Readers are encouraged to share their opinions and recommendations regarding this textbook.

Sleep is as ancient as life itself. It is a universal phenomenon that occupies nearly a third of human existence. Yet until as recently as a century ago, sleep has remained almost exclusively in the realm of fables, poetry, and theology. Advances in our understanding of the complex biology and physiology of sleep and of the various sleep disorders have radically transformed the role of sleep in our lives. Nonetheless, many fundamental questions on the nature of sleep remain unanswered. Why do we sleep? What is the function of sleep across the span of ages in one's lifetime, and in the diversity of species? How does sleep alter biologic processes?

This textbook is not meant to be the culmination of our knowledge of the science of sleep. Rather, consider it but a pause as we reflect on our place in the rapidly altering landscape of sleep medicine.

ACKNOWLEDGMENTS

I wish to express my sincere gratitude to the outstanding authors for the excellent texts they have generously provided, to the editorial board of John Wiley & Sons for their expert counsel and unwavering commitment throughout the course of this undertaking, and to my editorial associate, Dusty Christian.

Finally, I must thank my wife, Grace Zamudio, for her boundless humor and infinite patience, and my daughter, Zoe Lee-Chiong, in whose future lies mine. It is to them that I gratefully dedicate this work.

TEOFILO L. LEE-CHIONG Jr., MD

CONTRIBUTORS

Christine Acebo, E.P. Bradley Hospital/Brown Medical School, East Providence, Rhode Island

Cathy A. Alessi, Veterans Administration Greater Los Angeles Healthcare System, Los Angeles, California

Naeem Ali, The Ohio State University Sleep Disorders Center, Columbus, Ohio

Charles J. Amlaner, Jr., Indiana State University, Terre Haute, Indiana

Donna Arand, Kettering Medical Center Sleep Disorders Center, Dayton, Ohio

Roseanne Armitage, University of Michigan, Ann Arbor, Michigan

R. Robert Auger, Mayo Clinic College of Medicine, Rochester, Minnesota

Liat Ayalon, University of California San Diego and Veterans Affairs San Diego Healthcare System, San Diego, California, USA

Liat Ayalon, University of California, San Diego, California

Gaby Bader, University of Gothenburg, Gothenburg, Sweden

Charles J. Bae, The Cleveland Clinic Foundation, Cleveland, Ohio

Helen A. Baghdoyan, University of Michigan, Ann Arbor, Michigan

Dennis R. Bailey, Englewood, Colorado

Robert D. Ballard, National Jewish Medical and Research Center, Denver, Colorado

Preetam Bandla, The Children's Hospital of Philadelphia, Philadelphia, Pennsylvania

Joshua Baron, Tufts–New England Medical Center, Boston, Mssachusetts

Philip M. Becker, Sleep Medicine Associate of Texas, Dallas, Texas

Selim R. Benbadis, University of South Florida College of Medicine, Tampa, Florida

Ruth M. Benca, University of Wisconsin–Madison Medical School, Madison, Wisconsin

Elizabeth A. Beothy, M. Safwan Badr, Wayne State University, Detroit, Michigan

David A. Beuther, National Jewish Medical and Research Center, Denver, Colorado

Jagdeep Bijwadia, Mayo Health System, Eau Claire, Wisconsin

Sylvie Bisser, Centre international de recherche médicale, Franceville, Gabon

Bradley F. Boeve, Mayo Clinic College of Medicine, Rochester, Minnesota

Katy Borodkin, Sheba Medical Center, Tel Hashomer, Israel

Bernard Bouteille, Institut de neurologie tropicale, Limoges, France

Lee J. Brooks, Children's Hospital of Philadelphia, Philadelphia, Pennsylvania

Stephen N. Brooks, Stanford Sleep Disorders Clinic, Stanford, California

W. David Brown, Sleep Diagnostic, The Woodlands, Texas

Alain Buguet, Neurobiologie des états de vigilance, Lyon, France

Daniel J. Buysse, University of Pittsburgh, Pittsburgh, Pennsylvania

Peter T. P. Bye, Royal Prince Alfred Hospital, Camperdown, New South Wales, Australia.

John A. Caldwell, Air Force Research Laboratory, Brooks City–Base, Texas

Patricia A. Carter, The University of Texas at Austin School of Nursing, Austin, Texas

Rosalind D. Cartwright, Rush University Medical Center, Chicago, Illinois

Richard J. Castriotta, University of Texas Health Science Center at Houston, Houston, Texas

Keith Cavanaugh, The Children's Hospital, Denver, Colorado

Raymond Cespuglio, Neurobiologie des Etsts de vigilance, Lyon, France

Marie-Josèphe Challamel, Hôpital Debrousse, Lyon, France

Florian Chapotot, Centre de recherches du service de santé des armées, La Tronche, France

Wissam Chatila, Temple University School of Medicine, Philadelphia, Pennsylvania

Maida L. Chen, Childrens Hospital Los Angeles, Los Angeles, California

Andrew L. Chesson, Jr., Louisiana State University Medical Center, Shreveport, Louisiana

Teofilo Lee-Chiong, National Jewish Medical and Research Center, Denver, Colorado

S. Charles Cho, Stanford Sleep Disorders Clinic, Stanford, California

George P. Chrousos, National Institutes of Health, Bethesda, Maryland

Pierre Cochat, Hôpital Edouard-Herriot and Université Claude-Bernard, Lyon France

Nancy A. Collop, Johns Hopkins University, Baltimore, Maryland

Gary R. Cott, National Jewish Medical and Research Center, Denver, Colorado

Nilesh B. Davé, University of Pittsburgh, Pittsburgh, Pennsylvania

Yaron Dagan, Sheba Medical Center, Tel Hashomer, Israel

Carolyn D'Ambrosio, Tufts University School of Medicine, Boston, Massachusetts

Carolyn M. D'Ambrosio, Tufts–New England Medical Center, Boston, Mssachusetts

B. Dan, University Children's Hospital of Brussels, Brussels, Belgium I. Kato Nagoya City University Medical School, Nagoya, Japan, T. Sawaguchi Tokyo Women's Medical University School of Medicine, Tokyo, Japan, I. Kelmanson St. Petersburg State Pediatric Medical Academy, St. Petersburg, Russia.

Katherine Finn Davis, School of Medicine, Emory University, Atlanta, Georgia

Donn Dexter, Mayo Health System, Eau Claire, Wisconsin

David F. Dinges, University of Pennsylvania School of Medicine, Philadelphia, Pennsylvania

Jillian Dorrian, The University of South Australia, Adelaide, South Australia

Helen S. Driver, Kingston General Hospital, Kingston, Ontario, Canada

Jack D. Edinger, Duke University Medical Center, Durham, North Carolina

Carol A. Enderlin, University of Arkansas for Medical Sciences Little Rock, Arkansas

Christine A. Engstrom, Department of Veterans Affairs Medical Center, Baltimore, Maryland

P. Franco, University Children's Hospital of Brussels,-Brussels, Belgium I. Kato Nagoya City University Medical School, Nagoya, Japan, T. Sawaguchi Tokyo Women's Medical University School of Medicine, Tokyo, Japan, I. Kelmanson St. Petersburg State Pediatric Medical Academy, St. Petersburg, Russia.

Marcos G. Frank, University of Pennsylvania, School of Medicine, Philadelphia, Pennsylvania

Leah Friedman, Stanford University, Stanford, California

Norman R. Friedman, The Children's Hospital, Denver, Colorado

Randolph V. Fugit, Denver Veterans Affairs Medical Center, Denver, Colorado

Erik Garpestad, Tufts University School of Medicine, Boston, Massachusetts

Deborah C. Givan, James Whitcomb Riley Hospital for Children, Indianapolis, Indiana

Rachel J. Givelber, University of Pittsburgh School of Medicine, Pittsburgh, Pennsylvania

Aviv D. Goldbart, University of Louisville, Louisville, Kentucky

Ismail Gögenur, Gentofte University Hospital, Hellerup, Denmark

Joseph A. Golish, The Cleveland Clinic Foundation, Cleveland, Ohio

Cameron Good, University of Arkansas for Medical Sciences, Little Rock, Akansas

Nalaka Gooneratne, University of Pennsylvania, Philadelphia, Pennsylvania

Ramadevi Gourineni, Northwestern University Feinberg School of Medicine, Chicago, Illinois

David Gozal, University of Louisville, Louisville, Kentucky

Glen P. Greenough, Dartmouth–Hitchcock Medical Center, Lebanon, New Hampshire

May L. Griebel, Arkansas Children's Hospital, Little Rock, Arkansas

J. Groswasser, University Children's Hospital of Brussels,Brussels, Belgium I. Kato Nagoya City University Medical School, Nagoya, Japan, T. Sawaguchi Tokyo Women's Medical University School of Medicine, Tokyo, Japan, I. Kelmanson St. Petersburg State Pediatric Medical Academy, St. Petersburg, Russia.

Ronald R. Grunstein, University of Sydney, Camperdown, New South Wales, Australia

Christian Guilleminault, Stanford University, Stanford, California

Indira Gurubhagavatula, Hospital of the University of Pennsylvania, Philadelphia, Pennsylvania

Peter Y. Hahn, Mayo Clinic College of Medicine, Rochester, Minnesota

Louise Harder, Spokane, Washington

Susan M. Harding, University of Alabama at Birmingham Sleep–Wake Disorders Center, Birmingham, Alabama

David G. Harper, Harvard Medical School, Belmont, Massachusetts

Christine Hedges, Jersey Shore University Medical Center, Neptune, New Jersey

John H. Herman, UT Southwestern Medical Center, Dallas, Texas

Nicholas S. Hill, Tufts University School of Medicine, Boston, Massachusetts

Max Hirshkowitz, Baylor College of Medicine, Houston, Texas

Brian Hoyt, National Jewish Medical and Research Center, Denver, Colorado

David W. Hudgel, Henry Ford Health System, Detroit, Michigan

Maren Hyde, Henry Ford Hospital, Detroit, Michigan

Conrad Iber, Hennepin County Medical Center, Minnepolis, Minnesota

Amy B. Robinson Ikelheimer, National Jewish Medical and Research Center, Denver, Colorado

Sonia Ancoli-Israel, University of California, San Diego, California

Suzan E. Jaffe, Aventura, Florida

S. Javaheri, Sleepcare Diagnostics, Mason, Ohio

Kyle Johnson, Oregon Health & Science University, Portland, Oregon

Stephany Jones, University of Wisconsin–Madison Medical School, Madison, Wisconsin

Mark E. Josephson, Harvard Medical School, Boston Massachusetts

Brooke G. Judd, Dartmouth–Hitchcock Medical Center, Lebanon, New Hampshire

A. Kahn, University Children's Hospital of Brussels,Brussels, Belgium I. Kato Nagoya City University Medical School, Nagoya, Japan, T. Sawaguchi Tokyo Women's Medical University School of Medicine, Tokyo, Japan, I. Kelmanson St. Petersburg State Pediatric Medical Academy, St. Petersburg, Russia.

Kosuke Kaida, National Institute of Industrial Health, Kawasaki, Japan

Sheldon Kapen, Detroit VA Medical Center Detroit, Michigan

Thomas G. Keens, Childrens Hospital Los Angeles, Los Angeles, California

Basheer Y. Khassawneh, Jordan University of Science and Technology, Irbid, Jordan

Rami Khayat, The Ohio State University Sleep Disorders Center, Columbus, Ohio

Neil S. Kline, University of Pennsylvania Medical Center, Philadelphia, Pennsylvania

Suresh Kotagal, Mayo Clinic, Rochester, Minnesota

Samuel L. Krachman, Temple University School of Medicine, Philadelphia, Pennsylvania

Samuel Krachman, Temple University School of Medicine, Philadelphia, Pennsylvania

Milton Kramer, Maimonides Medical Center, Brooklyn, New York

William R. Krimsky, Dartmouth Medical School, Lebanon, Hampshire

James M. Krueger, Washington State University, Pullman, Washington

Clete A. Kushida, Stanford Sleep Disorders Clinic, Stanford, California

Bishoy Labib, Central Arkansas Veterans Healthcare System, Little Rock, Arkansas

Kathryn A. Lee, University of California, San Francisco, California

Damien Leger, Université Paris V, René Descartes, Paris, France

James C. Leiter, Dartmouth Medical School, Lebanon, Hampshire

John A. Lesku, Indiana State University, Terre Haute, Indiana

G. Liistro, Cliniques Universitaires St-Luc. Bruxelles, Belgium

Michael R. Littner, David Geffen School of Medicine at UCLA Sepulveda, California

Maria Cecilia Lopes, Stanford Sleep Disorders Clinic, Stanford University, Stanford, California

Ralph Lydic, University of Michigan, Ann Arbor, Michigan

Ulysses J. Magalang, The Ohio State University Sleep Disorders Center, Columbus, Ohio

Jeannine A. Majde, Washington State University, Pullman, Washington

Atul Malhotra, Brigham and Women's Hospital, Boston, Massachusetts

Carole L. Marcus, The Children's Hospital of Philadelphia, Philadelphia, Pennsylvania

Gerald A. Marks, University of Texas Southwestern Medical Center, Dallas, Texas

Jennifer L. Martin, Veterans Administration Greater Los Angeles Healthcare System, Los Angeles, California

Richard J. Martin, National Jewish Medical and Research Center, Denver, Colorado

Emilio Mazza, Hospital of the University of Pennsylvania, Philadelphia, Pennsylvania

Melanie K. Means, Duke University Medical Center, Durham, North Carolina

Reena Mehra, Case Western Reserve University, Cleveland, Ohio

Arnaud Metlaine, Université Paris V, René Descartes, Paris, France

Maree M. Milross, Royal Prince Alfred Hospital, Camperdown, New South Wales, Australia.

Karen E. Moe, University of Washington, Seattle, Washington

Vahid Mohsenin, Yale University School of Medicine, New Haven, Connecticut

Karlind T. Moller, Hennepin County Medical Center Minneapolis, Minnesota

Timothy H. Monk, University of Pittsburgh Medical Center, Pittsburgh, Pennsylvania

William H. Moorcroft, Colorado State University, Fort Collins, Colorado

Robert Y. Moore, University of Pittsburgh, Pittsburgh, Pennsylvania

Timothy I. Morgenthaler, Mayo Clinic College of Medicine, Rochester, Minnesota

Douglas E. Moul, University of Pittsburgh, Pittsburgh, Pennsylvania

Linda K. Moyer, Arkansas Children's Hospital, Little Rock, Arkansas

Sarkis M. Nazarian, Central Arkansas Veterans Healthcare System, Little Rock, Arkansas

Rachel J. Norwood, National Jewish Medical and Research Center, Denver, Colorado

Maurice M. Ohayon, Stanford University, Stanford, California

Eric J. Olson, Mayo Clinic, Rochester, Minnesota

Lyle J. Olson, Mayo Clinic College of Medicine, Rochester, Minnesota

William C. Orr, Lynn Institute for Healthcare Research, Oklahoma City, Oklahoma

Judith A. Owens, Brown University Medical School, Providence, Rhode Island

J. F. Pagel, University of Colorado Medical School, Boulder, Colorado

John M. Palmer, Children's Hospital of Philadelphia, Philadelphia, Pennsylvania

John G. Park, Mayo Clinic, Rochester, Minnesota

Kathy P. Parker, Nell Hodgson Woodruff School of Nursing Atlanta, Georgia

Maja Tippmann-Peikert, Mayo Clinic College of Medicine, Rochester, Minnesota

Slobodankad Pejovic, National Institutes of Health, Bethesda, Maryland

Rafael Pelayo, Stanford Sleep Disorders Clinic, Stanford University, Stanford, California

Christophe Perrin, Tufts University School of Medicine, Boston, Massachusetts

Grace W. Pien, M. Safwan Badr, Wayne State University, Detroit, Michigan

Giora Pillar, Technion–Israel Institute of Technology, Haifa, Israel

Amanda J. Piper, Royal Prince Alfred Hospital, Camperdown, New South Wales, Australia.

Charles A. Polnitsky, Regional Sleep Laboratory, Waterbury, Connecticut

Asher Qureshi, St. Francis Hospital and Medical Center, Hartford, Connecticut

Anil Natesan Rama, Stanford Sleep Disorders Clinic, Stanford, California

Niels C. Rattenborg, University of Wisconsin Medical School, Madison, Wisconsin

Nancy S. Redeker, University of Medicine and Dentistry of New Jersey, School of Nursing, Newark, New Jersey

Kathy C. Richards, University of Arkansas for Medical Sciences Little Rock, Arkansas

Gary S. Richardson, Henry Ford Hospital, Detroit, Michigan

Edgar Garcia-Rill, University of Arkansas for Medical Sciences, Little Rock, Akansas

D. Rodenstein, Cliniques Universitaires St-Luc. Bruxelles, Belgium

Timothy Roehrs, Henry Ford Hospital, Detroit, Michigan

Naomi L. Rogers, University of Sydney, Camperdown, New South Wales, Australia

Roger R. Rosa, National Institute for Occupational Safety and Health, Washington, DC

Gerald Rosen, Hannepin County Medical Center, Minneapolis, Minnesota

Jacob Rosenberg, Gentofte University Hospital, Hellerup, Denmark

Richard S. Rosenberg, American Academy of Sleep Medicine, Westchester, Illinois

Leon Rosenthal, Sleep Medicine Associates of Texas, Dallas, Texas

Thomas Roth, Henry Ford Hospital, Detroit, Michigan

Robert Sack, Oregon Health & Science University, Portland, Oregon

Mark H. Sanders, University of Pittsburgh School of Medicine, Pittsburgh, Pennsylvania

Michael Sateia, Dartmouth-Hitchcock Medical Center, Lebanon, New Hampshire

S. Scaillet, University Children's Hospital of Brussels,-Brussels, Belgium I. Kato Nagoya City University Medical School, Nagoya, Japan, T. Sawaguchi Tokyo Women's Medical University School of Medicine, Tokyo, Japan, I. Kelmanson St. Petersburg State Pediatric Medical Academy, St. Petersburg, Russia.

Mark S. Scher, Case-Western Reserve University, Cleveland, Ohio

Richard J. Schwab, University of Pennsylvania Medical Center, Philadelphia, Pennsylvania

Crintz E. Scott, St. Anthony Sleep Disorders Center, Denver, Colorado

Stephen H. Sheldon, Children's Memorial Hospital, Chicago, Illinois

Aaron E. Sher, Medical Director, Capital Region Sleep/Wake Disorders Center of St. Peter's Hospital, Capital Region Otolaryngology – Head & Neck Surgery Group, Clinical Associate Professor of Surgery and Pediatrics, Albany Medical College, Albany, New York

Margaret N. Shouse, UCLA School of Medicine, Las Angeles, California

Michael H. Silber, Mayo Clinic College of Medicine, Rochester, Minnesota

Virend K. Somers, Mayo Clinic College of Medicine, Rochester, Minnesota

Edward J. Stepanski, Rush University Medical Center, Chicago, Illinois

Kingman P. Strohl, Case Western Reserve University, Cleveland, Ohio

Patrick J. Strollo, University of Pittsburgh, Pittsburgh, Pennsylvania

James Q. Swift, Hennepin County Medical Center Minneapolis, Minnesota

Masaya Takahashi, National Institute of Industrial Health, Kawasaki, Japan

Riva Tauman, University of Louisville, Louisville, Kentucky

Sigrid Carlen Veasey, University of Pennsylvania School of Medicine, Philadelphia, Pennsylvania

Richard L. Verrier, Harvard Medical School, Boston Massachusetts

Alexandros N. Vgontzas, Penn State College of Medicine, Hershey, Pennsylvania

Michael V. Vitiello, University of Washington, Seattle, Washington

Tiffany Wallace, University of Arkansas for Medical Sciences, Little Rock, Akansas

Sally L. Davidson Ward, Childrens Hospital Los Angeles, Los Angeles, California

Michael Weissberg, University of Colorado School of Medicine, Denver, Colorado

Carolyn H. Welsh, Denver Veterans Affairs Medical Center, Denver, Colorado

Alexander White, Tufts University School of Medicine, Boston, Massachusetts

Laurel M. Wills, Hennepin County Medical Center Minneapolis, Minnesota

Merrill S. Wise, Baylor College of Medicine, Houston, Texas

Manisha B. Witmans, Childrens Hospital Los Angeles, Los Angeles, California

B. Tucker Woodson, Medical College of Wisconsin, Milwaukee, Wisconsin

Henry Yaggi, Yale University School of Medicine, New Haven, Connecticut

Phyllis C. Zee, Northwestern University Feinberg School of Medicine, Chicago, Illinois

SLEEP: A COMPREHENSIVE HANDBOOK

PART I

THE SCIENCE OF SLEEP MEDICINE

1

NORMAL HUMAN SLEEP

ANIL NATESAN RAMA, S. CHARLES CHO, AND CLETE A. KUSHIDA
Stanford Sleep Disorders Clinic, Stanford, California

Normal human sleep is comprised of two distinct states known as non-rapid eye movement (NREM) and rapid eye movement (REM) sleep. NREM sleep is subdivided into four stages (stage 1, stage 2, stage 3, and stage 4). Stages 3 and 4 are collectively referred to as slow-wave sleep. REM sleep may be further subdivided into two stages: phasic and tonic. The purpose of this chapter is to provide the reader with an overview of normal human sleep.

SLEEP ARCHITECTURE

NREM Sleep

NREM sleep accounts for 75–80% of sleep time. Stage 1 NREM sleep comprises 3–8% of sleep time. Stage 1 (Figure 1.1 and Figure 1.2) sleep occurs most frequently in the transition from wakefulness to the other sleep stages or following arousals during sleep. In stage 1 NREM sleep, alpha activity (8–13 Hz), which is characteristic of wakefulness, diminishes and a low-voltage, mixed-frequency pattern emerges. The highest amplitude electroencephalography (EEG) activity is generally in the theta range (4–8 Hz). Electromyography (EMG) activity decreases and electrooculography (EOG) demonstrates slow rolling eye movements. Vertex sharp waves (50–200 ms) are noted toward the end of stage 1 NREM sleep.

Stage 2 NREM (Figure 1.3) sleep begins after approximately 10–12 min of stage 1 NREM sleep and comprises 45–55% of total sleep time. The characteristic EEG findings of stage 2 NREM sleep include sleep spindles and K-complexes. A sleep spindle is described as a 12–14 Hz waveform lasting at least 0.5 s and having a "spindle"-shaped appearance. A K-complex is a waveform with two components: a negative wave followed by a positive wave, both lasting more than 0.5 s. Delta waves (0.5–4 Hz) in the EEG may first appear in stage 2 NREM sleep but are present in small amounts. The EMG activity is diminished compared to wakefulness.

Stage 3 and stage 4 (Figure 1.4 and Figure 1.5) NREM sleep occupy 15–20% of total sleep time and constitute slow-wave sleep. Stage 3 sleep is characterized by moderate amounts of high-amplitude, slow-wave activity; whereas stage 4 sleep is characterized by large amounts (e.g., >50% of a 30 s period) of high-amplitude, slow-wave activity. EOG does not register eye movements in stages 2–4 of NREM sleep. Muscle tone is decreased compared to wakefulness or stage 1 sleep [1]

REM Sleep

REM sleep (Figure 1.6) accounts for 20–25% of sleep time. The first REM sleep episode occurs 60–90 min after the onset of NREM sleep. EEG tracings during REM sleep are characterized by a low-voltage, mixed-frequency activity with slow alpha (defined as 1–2 Hz slower than wake alpha) and theta waves.

Based on EEG, EMG, and EOG characteristics, REM sleep can be divided into two stages—tonic and phasic. Characteristics of the tonic stage include a desynchronized EEG, atonia of skeletal muscle groups, and suppression of monosynaptic and polysynaptic reflexes. Phasic REM sleep is characterized by rapid eye movements in all directions as

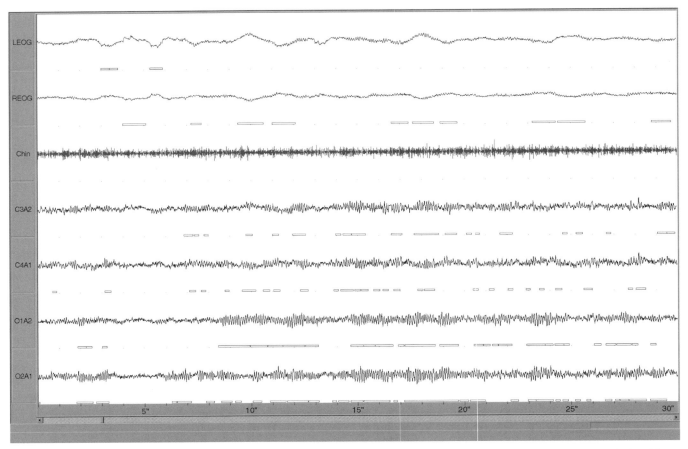

Figure 1.1 Wakefulness   Sleep is polygraphically defined by Non-Rapid Eye Movement (NREM) Stages 1, 2, 3, 4, and Rapid Eye Movement (REM) based on electroencephalographic (EEG), electrooculographic (EOG) and electromyographic (EMG) changes. The polygraph record is divided into segments of equal size (e.g., epoch length of 300 mm and a duration of 30 sec [10 mm/sec]). A single stage score is assigned to each epoch. In Stage Wakefulness (W), the EEG typically demonstrates alpha activity and/or a low voltage, mixed frequency activity. Rapid eye movements and eye blinks can be evident in the EOG. There is usually a relatively high tonic EMG.

well as by transient swings in blood pressure, heart rate changes, irregular respiration, tongue movements, and myoclonic twitching of chin and limb muscles [2–5]. Sawtooth waves, which have a frequency in the theta range and have the appearance of the teeth on the cutting edge of a sawblade, often occur in conjunction with rapid eye movements. A few periods of apnea or hypopnea may occur during REM sleep.

NREM–REM Cycle

The NREM–REM sleep cycle occurs about every 90 min and approximately four to six cycles occur per major sleep episode. The ratio of NREM sleep to REM sleep in each cycle varies during the course of the night. The early cycles are dominated by slow-wave sleep and the later cycles are dominated by REM sleep. The first episode of REM sleep may last only a few minutes and subsequent REM episodes progressively lengthen in duration during

the course of the major sleep period. In summary, slow-wave sleep is prominent in the first third of the night and REM sleep is prominent in the last third of the night. The temporal arrangement of sleep type is described graphically by a hypnogram (Figure 1.7).

SLEEP AND AGING

Sleep patterns change throughout life. Newborns may spend more than 16 h of the day asleep but intermittently sleep and awaken throughout the 24 h period. At the age of three months, infants should be able to sleep through the course of the night and take two or more daytime naps. As the child first enters school, he or she should be able to solidify a major nocturnal sleep period with perhaps a single daytime nap. As the child ages and during adulthood, the major nocturnal sleep is typically not accompanied by a daytime nap. Age-associated deterioration of

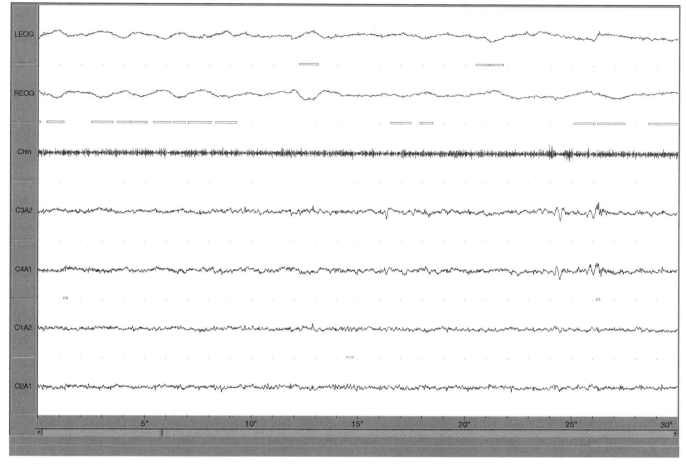

Figure 1.2 Stage 1 Sleep   In Stage 1 sleep, the EEG is relatively low voltage with mixed frequency. There is prominence of activity in the 2-7 cps range as well as a decrease in the amount, amplitude, and frequency of alpha activity. K complexes and sleep spindles are absent. Vertex sharp waves may appear. Slow eye movements may be appreciated. Tonic EMG levels are typically below that of relaxed wakefulness.

the sleep pattern results in fragmented sleep in the elderly, where more time is spent in bed but less time asleep.

Slow-wave sleep and REM sleep patterns also change throughout life. Slow-wave sleep declines after adolescence and continues to decline as a function of age. REM sleep decreases from more than 50% at birth to 20–25% during adolescence and middle age.

SLEEP NEUROPHYSIOLOGY

NREM Sleep

The transition from wakefulness to NREM sleep is associated with altered neurotransmission at the level of the thalamus, whereby incoming messages are inhibited and the cerebral cortex is deprived of signals from the outside world. NREM sleep is characterized by three major oscillations (Figure 1.8). Spindles (7–14 Hz) are generated within

thalamic reticular neurons that impose rhythmic inhibitory sequences onto thalamocortical neurons. However, the widespread synchronization of this rhythm is governed by corticothalamic projections. There are two types of delta activity [6, 7]. The first type is clock-like waves (1–4 Hz) generated in thalamocortical neurons and the second type is cortical waves (1–4 Hz) that persist despite extensive thalamectomy. However, the hallmark of NREM sleep is the slow oscillations (<1 Hz), which are generated intracortically and have the ability to group the thalamically generated spindles as well as thalamically and cortically generated delta oscillations, leading to a coalescence of the different rhythms [8, 9].

REM Sleep

Transection studies demonstrate that the pontomesencephalic region is critical for REM sleep generation [10]. When the mesopontine region is connected to rostral structures,

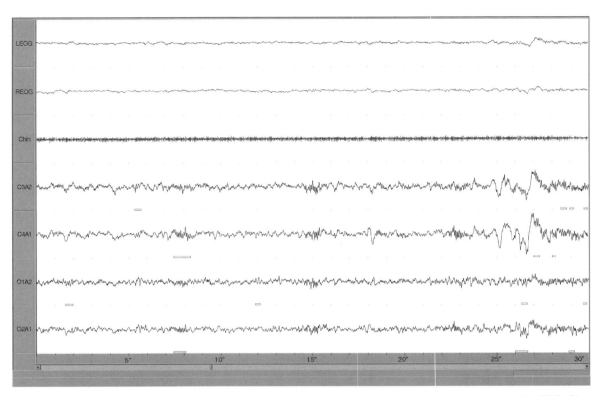

Figure 1.3 Stage 2 Sleep In Stage 2 sleep, a relatively low voltage, mixed frequency pattern, also characterizes the EEG. Sleep spindles and/or K complexes occur intermittently. A sleep spindle is defined by activity between 12 and 14 cycles per second (cps) of at least 0.5 sec duration, whereas K complexes are negative sharp wave immediately followed by a positive component with a total duration of over 0.5 sec. There is no sufficient high amplitude, slow activity that defines Stages 3 and 4 of sleep.

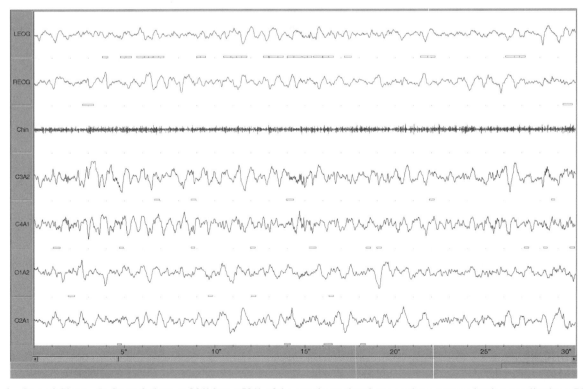

Figure 1.4 Stage 3 Sleep In Stage 3 sleep, ≥20% but <50% of the epoch consists 2 cps or slower waves having amplitudes >75 µV from peak to peak are present in the EEG.

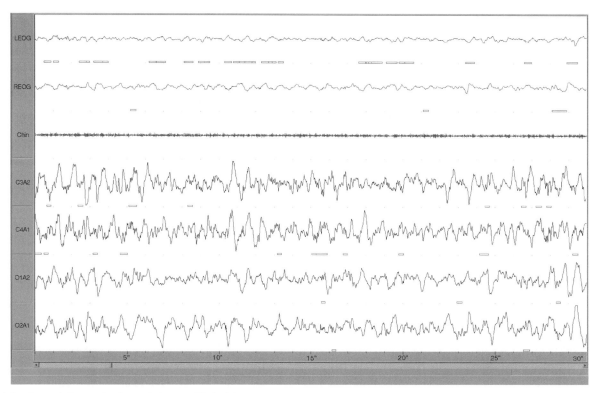

Figure 1.5 Stage 4 Sleep In Stage 4 sleep, >50% of the epoch consists 2 cps or slower waves having amplitudes >75 µV peak to peak are present in the EEG.

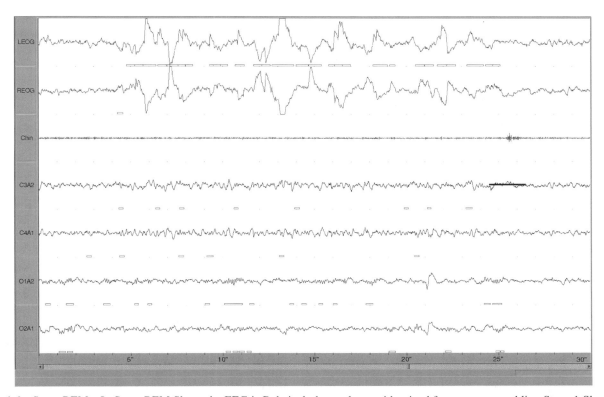

Figure 1.6 Stage REM In Stage REM Sleep, the EEG is Relatively low voltage with mixed frequency, resembling Stage 1 Sleep EEG. Episodic REMs can be appreciated in the EOG. The EMG is low in amplitude; tonic mental-submental EMG tracing almost always reaches its lowest levels.

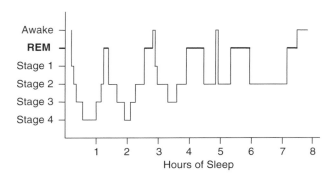

Figure 1.7 Young adult hypnogram.

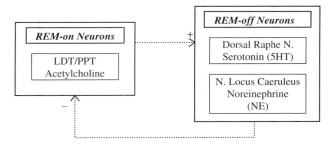

Figure 1.9 NREM–REM reciprocal interaction model.

REM sleep phenomena such as a desynchronized EEG and pontogeniculo-occipital (PGO) spikes are seen in the forebrain. When the mesopontine region is continuous with the medulla and spinal cord, REM sleep phenomena such as skeletal muscle atonia can be seen.

The pontomesencephalic area contains the so-called cholinergic "REM-on" nuclei, specifically the laterodorsal tegmental (LDT) and pedunculopontine tegmental (PPT) nuclei. The LDT and PPT nuclei project through the thalamus to the cortex, which produces the desynchronization of REM sleep. PGO spikes are a precursor to the rapid eye movements seen in REM sleep and are formed in the cholinergic mesopontine nuclei and propagate rostrally through the lateral geniculate and other thalamic nuclei to the occipital cortex [11]. The LDT and PPT nuclei project caudally via the ventral medulla to alpha motor neurons in the spinal cord, where skeletal muscle tone is inhibited during REM sleep by the release of glycine [12]. In addition, as NREM sleep transitions to REM sleep, tonic inhibition of REM-generating cholinergic pontomesencephalic nuclei by brainstem serotoninergic and adrenergic nuclei decreases, thereby allowing the development of PGO spikes and muscle atonia [13]. Thus the cholinergic REM-on nuclei of the PPT and LDT slowly activate the monoaminergic "REM-off" nuclei of the dorsal raphe and locus caeruleus that in turn inhibit REM-on nuclei (Figure 1.9).

Hypocretin has an important role in the modulation of wakefulness and REM sleep. Hypocretin neurons are located in the lateral hypothalamus and widely project to brainstem and forebrain areas, densely innervating monoaminergic and cholinergic cells. Hypocretin neurons promote wakefulness and inhibit REM sleep [14]. Elevated levels of hypocretin during active waking and in REM sleep compared to quiet waking and slow-wave sleep suggest a role for hypocretin in the central programming of motor activity [15]. Hypocretin projections to the nucleus pontis oralis may play a role in the generation of active (REM) sleep and muscle atonia [16].

AUTONOMIC NERVOUS SYSTEM

The autonomic nervous system (ANS) regulates the vital functions of internal homeostasis. The ANS is comprised of the sympathetic nervous system and parasympathetic nervous system. The essential autonomic feature of NREM sleep is increased parasympathetic activity and decreased sympathetic activity. The essential autonomic feature of REM sleep is an additional increase in parasympathetic activity and an additional decrease in sympathetic activity, with intermittent increases in sympathetic activity occurring during phasic REM (Table 1.1). For example, pupilloconstriction is seen during NREM sleep and is maintained during REM sleep with phasic dilatations noted during phasic REM sleep.

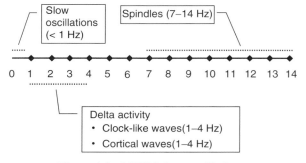

Figure 1.8 NREM sleep oscillations.

TABLE 1.1 Autonomic Nervous System Fluctuations During Normal Human Sleep

Sleep Period	Parasympathetic Nervous System	Sympathetic Nervous System
NREM sleep	Increase	Decrease
REM sleep		
Tonic	Increases further	Decreases further
Phasic		Intermittent increases

MODEL OF SLEEP REGULATION

Several models have been proposed to explain the regulation of sleep and wakefulness. One such model proposes that the regulation of the sleep–wake cycle is governed by two processes: a sleep-dependent homeostatic process (Process S) and a sleep-independent circadian process (Process C) [17].

Process S is a homeostatic process that is dependent on the duration of prior sleep and waking. This process shows an exponential rise during waking and a decline during sleep. In other words, the longer a person stays awake, the sleepier he or she becomes; conversely, the longer a person sleeps, the lower the pressure to remain asleep.

Process C is a circadian process that is independent of duration of prior sleep and waking. This process is under the control of an independent circadian oscillator, which determines the rhythmic propensity to sleep and awaken. In other words, each person has an endogenous drive to fall asleep and awaken at a certain time regardless of the duration of prior sleep or wake.

The two-process model posits that the timing of sleep and waking is determined by the interaction between Process S and Process C. Sleep onset is thought to occur when both the homeostatic and circadian drive to sleep intersect.

Other models have also been proposed, such as the opponent-process model and the three-process model of alertness regulation; however, further work is necessary to determine the biological substrates of the elements of these models and the pathways by which they interact.

CONCLUSION

Considerable research has been directed toward elucidating the basic mechanisms of normal human sleep. A firm understanding of the principles of normal human sleep is critical to understanding abnormal sleep and the disorders associated with sleep.

REFERENCES

1. Rechtschaffen A, Kales A (Eds). *A Manual of Standardized Terminology, Techniques and Scoring System for Sleep Stages of Human Subjects.* BIS/BRI, UCLA, Los Angeles, 1968.
2. Baust W, Holzbach E, Zechlin O. Phasic changes in heart rate and respiration correlated with PGO-spike activity during REM sleep. *Pflugers Arch* **331**:113–123(1972).
3. Orem J. Neuronal mechanisms of respiration in REM sleep. *Sleep* **3**(3-4):251–267(1980).
4. Oksenberg A, Gordon C, Arons E, Sazbon L. Phasic activities of rapid eye movement sleep in vegetative state patients. *Sleep* **24**(6):703–706(Sept 15, 2001).
5. Chokroverty S. Phasic tongue movements in human rapid eye-movement sleep. *Neurology* **30**(6):665–668(1980).
6. Steriade M, Nunez A, Amzica F. A novel slow (<1 Hz) oscillation of neocortical neurons in vivo: depolarizing and hyperpolarizing components. *J Neurosci* **13**:3252–3265(1993).
7. Steriade M, Contreras D, Curro Dossi R, Nunez A. The slow (<1 Hz) oscillation in reticular thalamic and thalamocortical neurons: scenario of sleep rhythm generation in interacting thalamic and neocortical networks. *J Neurosci* **13**(8):3284–3299(1993).
8. Steriade M, Amzica F. Coalescence of sleep rhythms and their chronology in corticothalamic networks. *Sleep Res Online* **1**(1):1–10(1998).
9. Amzica F, Steriade M. Electrophysiological correlates of sleep delta waves. *Electroencephalogr Clin Neurophysiol* **107**(2):69–83(1998).
10. Siegel JM. Brainstem mechanisms generating REM sleep. In: Kryger MH, Roth T, Dement WC (Eds), *Principles and Practice of Sleep Medicine.* Saunders, Philadelphia, 2000, pp. 112–133.
11. Steriade M, Pare D, Datta S, Oakson G, Curro Dossi R. Different cellular types in mesopontine cholinergic nuclei related to ponto-geniculo-occipital waves. *J Neurosci* **10**(8):2560–2579(1990).
12. Holmes CJ, Jones BE. Importance of cholinergic, GABAergic, serotonergic and other neurons in the medial medullary reticular formation for sleep–wake states studied by cytotoxic lesions in the cat. *Neuroscience* **62**:1179–1200(1994).
13. Aston-Jones G, Bloom FE. Activity of norepinephrine- containing locus coeruleus neurons in behaving rats anticipates fluctuations in the sleep–waking cycle. *J Neurosci* **1**:876–886(1981).
14. Estabrooke IV, McCarthy MT, Ko E, Chou TC, Chemelli RM, Yanagisawa M, Saper CB, Scammell TE. Fos expression in orexin neurons varies with behavioral state. *J Neurosci* **21**(5):1656–1662(2001).
15. Kiyashchenko LI, Mileykovskiy BY, Maidment N, Lam HA, Wu MF, John J, Peever J, Siegel JM. Release of hypocretin (orexin) during waking and sleep states. *J Neurosci* **22**(13):5282–5286(2002).
16. Xi MC, Fung SJ, Yamuy J, Morales FR, Chase MH. Induction of active (REM) sleep and motor inhibition by hypocretin in the nucleus pontis oralis of the cat. *J Neurophysiol* **87**(6):2880–2888(2002).
17. Borbely AA. A two process model of sleep regulation. *Hum Neurobiol* **1**(3):195–204(1982).

ADDITIONAL READING

Chokroverty S, Daroff, RB (Eds). *Sleep Disorders Medicine: Basic Science, Technical Considerations, and Clinical Aspects*, 2nd ed. Butterworth-Heinemann, Philadelphia, 1999.

Kryger MH, Roth T, Dement WC (Eds). *Principles and Practice of Sleep Medicine*, 3rd ed. Saunders, Philadelphia, 2000.

Lee-Chiong T (Ed). Sleep disorders. *Med Clin North Am* **88**(3):(May 2004).

2

THE NEUROBIOLOGY OF SLEEP

University of Texas Southwestern Medical Center, Dallas, Texas

INTRODUCTION

Progress in the understanding of the neurobiology of sleep, as with basic science in general, is greatly dependent on two circumstances. One is advance in technology and the other is discovery. Rational questions on the nature of sleep behavior have a long published history and, with little doubt, extend to prehistory. It is not surprising that much attention would be paid to a prominent human endeavor such as sleep. Yet the current condition of our knowledge of how, and for what purpose, we sleep is far from complete. The paucity of technology through most of the past to observe the operation of the central nervous system has resulted in a relatively short history of efforts to identify neural mechanisms underlying the generation and maintenance of sleep and wakefulness. A limitation of time as it impacts the capriciousness of the process of discovery is one factor. Another critical factor, which has become apparent with the accumulation of knowledge, is the complexity of the problem.

A large body of knowledge has accumulated on sleep/wake behavior including natural observation, pathological alterations, and experimental manipulation. The conclusion that constitutes the basic tenet of the neurobiology of sleep is that sleep is a product of the central nervous system. To understand how the brain produces sleep is to identify those neural mechanisms necessary and sufficient to produce it. Early investigations employing the emerging technology of modern neuroscience conceived of a brain made up of centers of localized function. Thus destruction of the "sleep center" would result in the elimination of sleep and serve to identify the structure and its function. In as much as the use of this approach has yet to yield a consensus as to what structure constitutes the "sleep center," a concept of interacting, distributed systems has emerged as a more plausible mechanism of this process.

While mechanisms within the brain produce sleep and wakefulness, brain mechanisms also are the targets of their influence, and these various mechanisms need not be mutually exclusive. Alterations in brain activity accompanying changes in state of arousal are so widespread and global in nature that they can be viewed as constituting reorganization in whole brain. In every region of brain, neurons alter their rate or pattern of firing with changes in state [1]. The specific roles played by such mechanisms are being elucidated as technological tools for study become available.

Despite the level of complexity, a great deal is known of the neural mechanisms subserving sleep. This knowledge is sufficient to impact the development of rational therapies for many sleep disorders. General concepts on the neuroscience of sleep are emerging and serious candidates for critical neural mechanisms have been identified. Through further advances in technology and discovery a full understanding will be achieved. The following is an overview of the current perspectives in the field of sleep.

PROBLEM OF DEFINITION

Defining sleep/wake behavior is not a trivial matter nor is a single definition appropriate for all cases. Definitions based on overt, gross behavior suffer from an inability to distinguish several conditions generally not recognized as sleep. The problem of definition is most acute when applied

Copyright © 2006 John Wiley & Sons, Inc.

11

across species. Circadian patterns of rest and activity are observed across phyla including single-celled organisms and plants [2]. Conservation through evolution may provide a clue to the adaptive value of temporally regulating activity and may also indicate that the mechanisms subserving this behavior probably vary with the complexity of the organism expressing them. From the viewpoint of neurobiology, definitions of sleep are couched in subservience to a nervous system. This requires that the subject not only posses a nervous system but also that the behavior be dependent on its operation.

Neurobiological investigations of sleep/wake mechanisms have been conducted almost exclusively with mammalian species. This has generated a definition of sleep based on a confluence of several electrophysiological correlates of brain activity. It should be pointed out that these indicators of state are defining properties and not chosen because of any relationship to fundamental saliency or functional significance; nor is the definition invariant. For example, a specific pattern in the electroencephalogram (EEG) is used to define sleep, yet other measures of neural activity clearly indicate the presence of sleep in a cat with its neocortex removed [3]. This flexibility in defining sleep can result in a high degree of ambiguity in interpreting effects of experimental manipulations and pathological conditions. There is no absolute agreement on how many indicators are required to identify sleep. Problems also arise when neurobiological definitions of sleep are applied to species other than mammalian and avian. Current investigations are revealing many similarities between the inactive states of the fruit fly and mammalian sleep [4]. Electrophysiological correlates of neural activity may be uncovered, but brain mechanisms implicated in the control of sleep/wake behavior in the mammalian brain do not exist in the fly. If the fly sleeps, then it will have to be defined differently than mammalian sleep. Inasmuch as the functions of sleep are currently unknown, it remains to be determined whether the sleep states of mammals and flies are even analogous and, if so, on what basis.

Among mammalian species, there appears to be a high degree of similarity in both expression and mechanism [5]. Differences exist in daily amounts and temporal distribution of sleep. Certain species' specializations exist, such as the unihemispheric sleep of some cetaceans and the single, consolidated, sleep period of some primates, including humans. Consensus among workers in the field is that the basic neural mechanisms identified in work on cats and rats likely apply to humans and other mammals.

DEFINING CHARACTERISTICS OF SLEEP AND WAKEFULNESS

In the third decade of the twentieth century, the application of the newly discovered EEG [6] began to yield insights into the altered brain activity associated with the sleep/wake cycle. Initial findings recognized clear distinctions in the EEG during wake and sleep. The low-amplitude, high-frequency activity characteristic of wakefulness increases in amplitude and decreases in frequency during sleep. At first, degrees of slowing and increased amplitude within sleep were viewed along a single dimension, depth, with the lightest sleep being most similar to wake [7]. With hindsight, many subsequent observations on sleeping subjects report data indicating multidimensional aspects to sleep. It was not, however, until the mid-twentieth century that Aserinsky and Kleitman reported the cyclic appearance of a distinct stage of sleep associated with dreaming and characterized by a wake-like EEG in the presence of rapid eye movements [8].

The pioneering work of Jouvet and colleagues utilizing cats [9] identified the presence of rapid eye movement (REM) sleep, which they called paradoxical sleep for the wake-like brain activity present. Several defining characteristics differentiated this state from that of the rest of sleep or what is now referred to as non-REM (NREM) sleep. In addition to the low-voltage, fast EEG, and rapid eye movements absent from NREM sleep, there appears an inhibition of muscle activity between paroxysmal muscle movements, wake-like activity in the hippocampus, and a unique spindling activity in the pons, now recognized as PGO waves. Based on threshold to arousal by the presentation of sensory stimuli, REM sleep is a deep sleep: thus the association of EEG amplitude and frequency with sleep depth had to be revised. Work on animals permitted neurophysiological investigations that initially took the form of gross brain lesions. Results of these studies confirmed the individual identities of the two stages of sleep by indicating reliance on different neural mechanisms for their generation [3].

In adult therian mammals studied, the general organization of sleep and wakefulness assumes a similar form [5]. In addition to a cyclic alternation of sleep and wake states, there is a more rapid alternation within sleep between NREM and REM. The occurrence of REM sleep is always preceded by NREM. The distribution of sleep/wake episodes repeats daily, thus expressing a circadian rhythm shared by many physiological functions under a common temporal influence [10]. The faster ultradian rhythm of the sleep cycle also may be served by mechanisms independent of sleep. Many observations support a basic, rest/activity cycle, with relatively fixed period, underlying several physiological functions [2]. Sleep stage amounts, temporal, daily distributions, and the period of the ultradian sleep cycle are species-specific traits. The period of the sleep cycle is highly correlated to the size of the species and, inversely, to its basal metabolic rate [5].

In addition to temporal factors controlling sleep/wake behavior, total sleep time and time in the individual stages also express homoeostatic types of regulation [11]. That is, when sleep, or a specific stage, is not permitted to be expressed, the amount lost tends to be recovered, as if a quota were being maintained. Time lost, however, is usually greater than time recovered, giving rise to the concept that sleep intensity increases, permitting recovered sleep to be more efficient. The amplitude of slow-wave activity in the EEG is an indicator of intensity of NREM sleep [12] and density of phasic activity, such as eye movements or PGO waves, has been used to reflect REM sleep intensity [13]. The rates of incurring a sleep debt and of recovery appear not only to be species specific but also characteristic of strains within a species, indicating a high degree of heritability [14]. The inverse dependence of sleep, or stage amounts, on prior expression creates another factor contributing to the oscillation among states of arousal making up the cyclic nature of sleep/wake behavior.

NATURE OF SLEEP/WAKE MECHANISMS

Two of the major characteristics of sleep are its reversibility and sensitivity to modulation by a variety of influences. In addition to inducing arousal from sleep by stimuli in any sensory modality of sufficient magnitude, amounts of sleep are affected by many factors such as ambient temperature, lighting conditions, and level of oxygen in the air, as well as by a host of wake experiences, including stress and learning. These would indicate that neural mechanisms whose primary function are not the generation of sleep and wake can control sleep and wake behavior. This raises a question as to how a sleep mechanism can be identified. Does the observation that loud sounds inhibit sleep make the auditory system a sleep mechanism? On some levels the answer is yes. Yet we know that the auditory system is not necessary for the production of sleep and wakefulness. Is necessity then the criteria for judging primacy? In a system of distributed, interactive mechanisms, it may be that no one mechanism is necessary.

Historically, it was thought that the withdrawal of sensory input produced sleep by removing excitation to the neural systems of the brain that give rise to wakefulness [15]. Studies utilizing brain transections and lesions were not successful at proving this hypothesis. They did, however, provide the antecedents to the discovery by Moruzzi and Magoun that the reticular core of the brain, when stimulated electrically, was sufficient to induce arousal. This led to the concept of the ascending reticular activating system as a primary mechanism of conscious wakefulness [16]. Additional work utilizing lesion techniques found that destruction of certain regions resulted in decreased sleep (reviewed in [3]). The conclusion was that there existed mechanisms within the brain opposed to the arousal induced by the activating system. The sleep process, then

and now, was no longer viewed as a passive result of disfacilitation, but rather as an active process subserved by active mechanisms. The discovery of the neurally active REM sleep stage firmly entrenched this view as doctrine in sleep research.

Differences in neural activity as well as in behavior among wake, NREM sleep, and REM sleep are so great that they appear to constitute discrete states of arousal. If each state is actively produced, then there may exist mechanisms exclusively subserving each state. Evidence supports such a division and most of the putatively identified brain mechanisms are categorized as such. The individual states, however, are not completely independent of each other. As mentioned previously, there is a dependence of REM sleep on prior NREM sleep. And with only three states, an increase or decrease in one will by necessity tend to have a reciprocal effect on time spent in the other states. A decrease in the efficacy of a wake-inducing mechanism, for example, may reduce wakefulness, but also will result in more sleep. There are circumstances under experimental and pathological conditions in which other than the three normal states can occur, such as coma or dissociated states, that do not conform to the definitions of any one state. However, the common and most often repeated finding with experimental destruction or pharmacological intervention of brain function is the tenacity with which only the three states appear, though possibly at altered levels, as well as the trend toward complete recovery of preintervention amounts.

Although the action potential of a single neuron can be considered an all-or-none event, neural interactions within networks are graded phenomena. The fact that neural networks produce the discrete states of arousal with rare instances of dissociation is an important clue to their organization. Saper and colleagues [17] have likened this to a switch that is only stable within one of the configurations of the confluence of processes attendant to one of the three states of arousal. Historically, this function was performed by "executive mechanisms" centralizing decision making by integrating input from multiple sources. A more egalitarian alternative consists of relatively equipotent mechanisms interacting through reciprocal connectivity. The process suggested for the "switch" is mutual inhibition. This type of interaction favors stable configurations in which only one mutually inhibitory influence dominates at one time. Inasmuch as the executive mechanisms of sleep and wakefulness have not been found and evidence is accumulating for the reciprocal connectivity of sleep and arousal centers, a view of interacting, distributed mechanisms is currently in favor. Such a system is also consistent with the difficulty with which selective destruction of individual components of the system fail to chronically eliminate any state of arousal. Putative sleep/wake mechanisms are segmentally distributed through the brain. Determination of the specific roles played by each mechanism will be needed to understand the whole.

MECHANISMS OF WAKEFULNESS

Since the original proposal of the ascending reticular activating system [16] to account for wakefulness, several systems have been implicated in contributing to this function. With the introduction of sophisticated immunological and histochemical techniques, certain aminergic systems in the brainstem were differentiated from the diffuse reticular core of the brain [18, 19]. These systems share several properties that include widespread projections and utilization of neurotransmitters associated with neuromodulation, making these systems appealing candidates for control over the global alterations accompanying state changes. The noradrenergic system of the locus coeruleus and the serotonergic midline, raphe system have been speculated to play various roles [20], but the current consensus is that these wake-active neurons are involved in setting a general preparedness for wake activity associated with alertness and sensory–motor function [21, 22]. These monoaminergic systems are virtually silent in REM sleep. The brainstem also contains a population of cholinergic neurons in the lateral dorsal tegmental nucleus and the pedunculopontine tegmental nucleus in which the majority are most active during wake and REM sleep [23, 24]. This system is thought to contribute to the activation associated with both these states. While sharing many targets with the adrenergic and serotonergic systems, cholinergic brainstem neurons differ in that they do not project directly to the neocortex [23]. Their influence on cortical activation is relayed through the thalamus and extrathalamic pathways of the hypothalamus and basal forebrain. The brainstem cholinergic and monoaminergic systems also innervate the reticular formation [19, 23].

Although much has been discovered, it is ironic that the least progress has been made in specifically identifying mechanisms of the reticular formation itself [25, 26]. Extending from the medulla oblongata to the midbrain, the complex structure has been resistant to revealing its secrets. Early stimulation and lesion experiments implicated the more rostral aspects of the reticular formation [3] and it was shown later that neurons residing in this region of the midbrain, projecting to the midline thalamus, discharge at their highest rates during the states of cortical activation, wake and REM sleep [1]. In that the majority of reticular neurons utilize the excitatory transmitter glutamate, as do the thalamic neurons that relay to the neocortex, this mechanism provides another path for cortical activation. Reticular influences also can be relayed through the extrathalamic pathways. Most sensory and motor systems collaterally innervate the reticular formation. Excitation

of the reticular formation by sensory, or electrical, stimulation probably is responsible for the rapid arousal from sleep following their presentation. The reticular formation is not a homogeneous mass with respect to its innervation, projections, or local circuitry; however, one property characteristic of its structure is the high degree of intraconnectivity. As one moves more caudal, fewer and fewer long, ascending projections of reticular neurons reach the thalamus, but rather end in more rostral regions of the reticular formation. There also is a high degree of local interconnectivity. The structure of the reticular formation is well suited for the propagation of ascending as well as descending influences. This is consistent with the findings of focal electrical stimulation and local microinjection of drugs into the reticular formation inducing global changes in arousal. The specific role played by the reticular formation in behavior during wakefulness is not clear at this time. Its intraconnectivity may aid in the integration of multiple systems [25].

Characteristic of the distributed nature of structures controlling states of arousal, wake mechanisms are located rostral to the brainstem—in the diencephalon, thalamus and hypothalamus, and the telencephalon, basal forebrain and neocortex [27, 28]. A population of neurons in the posterior lateral hypothalamus, tuberomammallary nucleus, utilizes histamine as a neurotransmitter [29]. Shared with the aminergic cell groups of the brainstem, these neurons have widespread projections and activity patterns selective to wakefulness. Antagonism of this arousal system produces the hypnotic effects of antihistamines. Also found in the posterior hypothalamus are neurons that synthesize a newly discovered peptide transmitter, orexin, also known as hypocretin [30]. Deficiency in this system is associated with the sleep disorder narcolepsy [31]. Current evidence links this system to maintenance of wakefulness. This also is supported by the finding of excitatory inputs to other known arousal mechanisms [30].

The medial nuclei of the thalamus link, though not exclusively, brainstem activation to widespread areas of the neocortex [28]. This region has been considered a rostral extension of the reticular formation. It is at least a major target of it. The entire thalamus, as well as the neocortex, undergo profound alterations in activity with changes in state. The specific alterations are dependent on mutual interactions between these structures and provide many of the defining characteristics of each state [28]. Excitation of the thalamus is critical to the accurate relay of sensory information to the cortex during wakefulness.

Cortically projecting cholinergic neurons are distributed within several nuclei of the basal forebrain and include the nucleus of the diagonal band of Broca, the substantia innominata, and the magnocellular preoptic nucleus. This appears to be a major activation system of the neocortex achieved through the release of acetylcholine [32]. More caudal arousal systems project to this region; the cholinergic neurons discharge at their highest rates during states of cortical activation, and antagonism of cholinergic transmission in the cortex is sufficient to block spontaneous activation. The role of the basal forebrain is not solely to relay excitation to the cortex. Stimulation, lesion, and drug manipulation can have great effects on the time spent in individual states. This is probably accomplished through the reciprocal connections that basal forebrain neurons make with many other arousal-related systems [27].

MECHANISMS OF NREM SLEEP

Despite the original premise that inhibition of reticular activation is the basis for the presence of active sleep mechanisms, identification of specific neural circuitry in the inhibition of the reticular formation has not been forthcoming. By some estimates, 20–25% of reticular neurons utilize the inhibitory transmitter gamma-aminobutyric acid (GABA) [26]. One possibility is that excitatory inputs to inhibitory neurons are at work; however, injection of GABA receptor agonists into the pontine reticular formation induces wakefulness [33]. If direct inhibition of the reticular formation is a mechanism of NREM sleep, identification will require an increased understanding of the reticular formation itself. Evidence in support of other active NREM sleep mechanisms is compelling.

Several sources of evidence implicate the presence of a sleep-generating mechanism in the anterior hypothalamus–basal forebrain region [34, 35]. The finding of neurons selectively active during sleep has identified several mechanisms. One of these mechanisms is comprised of a collection of neurons in the ventrolateral preoptic (VLPO) nucleus, in which the vast majority contain GABA and the inhibitory peptide transmitter galanin (reviewed in [17] and [34]). Small excitotoxic lesions of these neurons cause a reduction in sleep correlated to the amount of cell loss. Reciprocal connections have been observed between the VLPO nucleus and several wake-related structures, including the histaminergic and orexinergic neurons, locus coeruleus, dorsal raphe, and cholinergic regions of the brainstem and basal forebrain. It has been hypothesized that reciprocal inhibitory connections between wake-active centers and the sleep-active VLPO nucleus constitutes the sleep switch preventing the expression of mixed or disassociated states of arousal [17]. Additional sleep-active neurons are found throughout the hypothalamic preoptic area with a more dense aggregation in the median preoptic nucleus. These neurons share many of the properties of the VLPO nucleus in connectivity and utilization of GABA [34]. An additional property is that they are warm-sensitive and are posited to mediate the relationships between sleep and temperature.

Just anterior to the preoptic area lies the basal forebrain, which was discussed as a wake mechanism but also serves

NREM sleep [32, 35]. Distributed among the cholinergic neurons of these nuclei are a large population of GABA-containing cells. NREM sleep-active neurons are found in this region and evidence indicates that at least some are GABAergic. Some of these GABAergic neurons are projection neurons with one target being the neocortex. Thus sleep-active GABAergic neurons of the basal forebrain may serve to inhibit the wake-active cholinergic neurons and directly inhibit cortical activity in the production of NREM sleep. The GABAergic nature of sleep-promoting neurons is probably responsible for the hypnotic effects of systemically administered agents that potentiate GABA transmission such as the benzodiazepines. A role for the basal forebrain in sleep production has been supported further by the action of adenosine in this region to increase sleep [36]. Adenosine is a product of cellular energy utilization. Levels of adenosine increase with the sustained increase in activity accompanying prolonged wakefulness. The basal forebrain may be one site of this action. Sleep-active neurons of the preoptic area also are excited by adenosine. Both these regions may mediate the wake-promoting effects of caffeine, an adenosine receptor antagonist.

MECHANISMS OF REM SLEEP

The results of brain transections that isolate the medulla oblongata and pons from the rest of the brain clearly indicate that structures sufficient to produce REM sleep lie within these regions of the brainstem [37]. Additional evidence indicates that communication between these two regions is necessary for the appearance of REM sleep [37]. Many mechanisms have been identified in the pons, but medullary mechanisms remain relatively obscure.

The many physiological phenomena occurring during REM sleep are separated into two categories. They are the phasic events occurring discontinuously and sporadically and the tonic events occurring rather continuously throughout a REM sleep episode (discussed in [38]). The phasic events include autonomic irregularities, muscle twitches, rapid eye movements, and field potentials recorded at various places along the neuraxis called PGO waves. The widespread distribution of REM sleep phasic activity, in a variety of systems, depends on propagation. With the use of discrete lesions and histochemical tracing methods, some of these pathways have been identified [39]. Phasic events tend to occur at the same time within REM periods, which has raised speculation of a phasic event system with a single or a few central generators. An area in the caudal pontine reticular formation, in the subcoeruleus (below the locus coeruleus), has been putatively identified as a generator of PGO wave activity [39].

The major tonic events of REM sleep are the muscle atonia and the widespread neural activation, which includes a wake-like EEG. During NREM sleep there is a diminution of muscle activity; however, during REM sleep, there is an increase in activity in the motor centers of the brain while an active inhibition is exerted on motor neurons. The result is paralysis and atonia in the majority of the skeletal musculature. This phenomenon appears to be dependent on the activation of a population of neurons in the caudal pontine reticular formation projecting to and facilitating activity in the medial medullary reticular formation that provides the inhibition to the motor neurons [40]. Bilateral lesions in the subcoeruleus area of the pons can result in REM sleep without muscle atonia, whereby animals express a variety of integrated behaviors during this sleep state [41].

The wake-like activation of REM sleep recruits many of the mechanisms involved in wakefulness. Neurons of the reticular formation, brainstem and forebrain cholinergic neurons, thalamus, and neocortex all exhibit firing rates and levels of excitability equal to or greater in REM sleep as compared to wake [1, 24, 27, 28]. One notable exception are the aminergic systems for their almost complete silence. It is tempting to conclude that the absence of the widespread neuromodulatory influences of norepinephrine, serotonin, and histamine are the basis for all the differences between REM sleep and wake, but few demonstrations of this have been produced at the cellular level. It is hard to conceive that the turning-off of these major systems does not make a major contribution to the nature of the REM sleep state.

With the early indication from transection data of where the critical REM sleep mechanisms reside, numerous brainstem systems have been investigated but no complete picture has yet to emerge. The most enduring concept stems from the initial discovery of the brainstem cholinergic system by Shute and Lewis in 1963 [42]. Based on neuronal projections, they suggested that this was the substrate of the ascending reticular activating system. It was subsequently found in the cat that when agents potentiating cholinergic transmission were microinjected into the pontine reticular formation, they induced a dramatic, rapid onset of long-lasting, REM sleep episodes [43]. The state-related activity of cholinergic neurons is still open to question, since while some neurons in the area fire selectively in REM sleep, most discharge at their highest rates in REM sleep and wake [24, 44]. It has been suggested that a reciprocal inhibition between cholinergic REM-on cells and aminergic REM-off cells provides the mechanism for reciprocal activities and state oscillations [45]. This model differs from the switch discussed earlier in that what would have been an unstable condition now becomes the NREM sleep that intervenes between REM and wake. It has been found that acetylcholine levels are highest in the reticular formation during REM sleep [46]. This may be due to reticular formation projections from cholinergic REM-on cells or a mechanism that inhibits cholinergic release during

wake. Evidence supports such a role for the REM-off (or wake-on) noradrenergic neurons through projections to presynaptic, cholinergic terminals in the reticular formation [47]. Wake-on/REM-on cholinergic neurons provide ascending activation in REM sleep as in wakefulness, and levels of acetylcholine are high in the thalamus during both states.

It would appear that the release of acetylcholine in the pontine reticular formation is a condition sufficient to induce REM sleep. Directly or indirectly, brainstem cholinergic neurons may excite the reticular formation initiating ascending activation, excite the pontine neurons responsible for muscle inhibition, inhibit serotonin release responsible for the appearance of PGO waves [38], and provide additional ascending activation via thalamic and extrathalamic relays to the cortex. Much of this description of the role of brainstem cholinergic neurons may be shown to be true. But acetylcholine in the pontine reticular formation is not sufficient to induce REM sleep. Pontine microinjections fail to do so after ponto medullary transections [48]. There is still some undisclosed mechanism in the medulla required for REM sleep. It is not clear that the integrity of brainstem cholinergic neurons is necessary for REM sleep [49]. Excitotoxic lesions of the region produce a long-lasting decrease in REM sleep amounts correlated to the number of cholinergic cells lost. However, this effect also is correlated to the size of the lesion: the latter being consistent with a distributed system of multiple mechanisms in the region including the rostral pontine reticular formation.

While evidence supports the brainstem as sufficient in the generation of REM sleep, additional structures are implicated in its control. In the preoptic area of the hypothalamus, known as the extended VLPO nucleus, there is a population of GABAergic neurons projecting to brainstem aminergic nuclei that appear to selectively fire in REM sleep, possibly contributing to the inhibition of aminergic neurons [17, 50]. Pharmacological manipulations of the basal forebrain can effect all states. One telling finding is that microinjection of cholinergic agonists in the basal forebrain blocks the REM sleep induction by injections in the pontine reticular formation [51]. Similar to the mechanisms of NREM sleep and wake, mechanisms of REM sleep also appear to be distributed and interactive.

CONCLUSION

The view expressed in this chapter describes the neurobiology of sleep and wakefulness as a system distributed along the neuraxis from the medulla oblongata to the neocortex. The high degree of interaction among components of the system gives rise to the unique and interdependent expression of the states of arousal. Some of the concepts may not survive the final analysis, as has been the fate of several compelling notions of the past, and novel mechanisms await to be found. Sleep/wake mechanisms appear so highly integrated in the brain that a complete understanding of them will require advances in technology and the chance of discovery that only time can offer.

REFERENCES

1. Steriade M, Hobson JA. Neuronal activity during the sleep–waking cycle. *Prog Neurobiol* **6**:155–376(1976).

2. Kleitman N. Phylogenetic, ontogenetic and environmental determinants in the evoluton of sleep–wakefulness cycles. In: Kety SS, Evarts EV, Williams HL(Eds), *Sleep and Altered States of Consciousness*. Williams & Wilkins, Baltimore, 1967, Vol 45, p 30.

3. Moruzzi G. The sleep–waking cycle. *Ergebn Physiol* **64**:1–163(1972).

4. Shaw PJ, Cirelli C, Greenspan RJ, Tononi G. Correlates of sleep and waking in *Drosophila melanogaster*. *Science* **287**:1834–1837(2000).

5. Zepelin H. Rechtschaffen A. Mammalian sleep, longevity, and energy metabolism. *Brain Behav Evol* **10**:425–470(1974).

6. Berger H. Über das elektrenkephalogramm. *Arch Psychiatr Nervenkrank* **87**:527–570(1929).

7. Loomis AL, Harvey EN, Hobart G. Potential rhythms of the cerebral cortex during sleep. *Science* **81**:597–598(1935).

8. Aserinsky E, Kleitman N. Regularly occurring periods of eye motility, and concomitant phenomena, during sleep. *Science* **118**:273–274(1953).

9. Jouvet M. Recherches sur les structures nerveuses et les mécanismes responsables des différentes phases du sommeil physiologique. *Arch Ital Biol* **100**:125–206(1962).

10. Turek FW. Circadian rhythms. *Horm Res.* **49**:109–113(1998).

11. Borbély AA, Acherman P. Concepts and models of sleep regulation: an overview. *J Sleep Res* **1**:63–79(1992).

12. Tobler I, Borbély AA. Sleep EEG in the rat as a function of prior waking. *Electroencephalogr Clin Neurophysiol* **64**:74–76(1986).

13. Takahashi K. Intensity of REM sleep. In: Mallick BN, Inoué S (Eds), *Rapid Eye Movement Sleep*. Narosa Publishing House, New Delhi, 1999, pp 384–392.

14. Franken P, Chollet D, Tafti M. The homeostatic regulation of sleep need is under genetic control. *J Neurosci* **21**:2610–2621(2001).

15. Bremer F. Cerveau "isolé" et physiologie du sommeil. *C R Soc Biol* **118**:1235–1241(1935).

16. Moruzzi G, Magoun HW. Brain stem reticular formation and activation of the EEG. *Electroencephalogr Clin Neurophysiol* **1**:455–473(1949).

17. Saper CB, Chou TC, Scammell TE. The sleep switch: hypothalamic control of sleep and wakefulness. *Trends Neurosci* **24**:726–731(2001).

18. Dahlström A, Fuxe K. Evidence for the existence of monoamine neurons in the central nervous system. I. Demonstration of monoamines in the cell bodies of brain stem neurons. *Acta Physiol Scand Suppl* **62**:232–xxx(1964).

19. Fuxe K. Evidence for the existence of monoamine neurons in the central nervous system. IV. Distribution of monoamine terminals in the central nervous system. *Acta Physiol Scand Suppl* **247**:37–85(1965).

20. Jouvet M. The role of monoamines and acetylcholine-containing neurons in the regulation of the sleep–wake cycle. *Ergebn Physiol* **64**:166–307(1972).

21. Aston-Jones G, Rajkowski J, Cohen J. Role of locus caeruleus in attention and behavioral flexibility. *Biol Psychiatry* **46**:1309–1320(1999).

22. Jacobs BL, Fornal CA. Serotonin and motor activity. *Curr Opin Neurobiol* **7**:820–825(1997).

23. Woolf NJ. Cholinergic systems in mammalian brain and spinal cord. *Prog Neurobiol* **37**:475–524(1991).

24. Datta S, Siwek DF. Single cell activity patterns of pedunculopontine tegmentum neurons across the sleep–wake cycle in the freely moving rats. *J Neurosci Res* **70**:611–621(2002).

25. Hobson JA, Brazier MAB (Eds). *The Reticular Formation Revisited*. Raven Press, New York, 1980.

26. Jones BE. Reticular formation. Cytoarchitecture, transmitters and projections. In: Paxinos G (Ed), *The Rat Nervous System*. Academic Press, New South Wales, Australia, 1995, pp 155–171.

27. Jones BE. Arousal systems. *Front Biosci* **8**:s438–451(2003).

28. McCormick DA, Bal T. Sleep and arousal: thalamocortical mechanisms. *Annu Rev Neurosci* **20**:185–215(1997).

29. Lin JS, Sakai K, Jouvet M. Evidence for histaminergic arousal mechanisms in the hypothalamus of cat. *Neuropharmacology* **27**:111–122(1988).

30. Peyron C, Tighe DK, van Den Pol AN, de Lecea L, Heller HC, Sutcliffe JG, Kilduff TS. Neurons containing hypocretin(orexin) project to multiple neuronal systems. *J Neurosci* **18**:9996–10015(1998).

31. Mignot E, Lammers GJ, Ripley B, Okun M, Nevsimalova S, Overeem S, Vankova J, Black J, Harsh J, Bassetti C, Schrader H, Nishino S. The role of cerebrospinal fluid hypocretin measurement in the diagnosis of narcolepsy and other hypersomnias. *Arch Neurol* **59**:1553–1562(2002).

32. Jones BE. Activity, modulation and the role of basal forebrain cholinergic neurons innervating the cerebral cortex. *Prog Brain Res* **145**:157–169(2004).

33. Xi M-C, Morales FR, Chase MH. Evidence that wakefulness and REM sleep are controlled by a GABAergic pontine mechanism. *J Neurophysiol* **82**:2015–2019(1999).

34. McGinty D, Szymusiak R. Brain structures and mechanisms involved in the generation of NREM sleep: focus on the preoptic hypothalamus. *Sleep Med Rev* **5**:323–342(2001).

35. Szymusiak R. Magnocellular nuclei of the basal forebrain: substrates of sleep and arousal regulation. *Sleep* **18**:478–500(1995).

36. Strecker R E, Moriarty S, Thakkar MM, Porkka-Heiskanen T, Basheer R, Dauphin LJ, Rainnie DG, Portas CM, Greene RW, McCarley RW. Adenosinergic modulation of basal

forebrain and preoptic/anterior hypothalamic neuronal activity in the control of behavioral state. *Behav Brain Res* **115**:183–204(2000).

37. Siegel JM. REM sleep control mechanisms: evidence from lesion and unit recording studies. In: Mayes A (Ed), *Sleep Mechanisms and Functions*. Van Nostrand Reinhold, Workingham, UK, 1983, pp 217–231.

38. Dement WC. The biological role of REM sleep (circa 1968). In: Kales A (Ed), *Sleep Physiology and Pathology*. Lippincott, Philadelphia, 1969, pp 245–265.

39. Datta S. Cellular basis of pontine ponto-geniculo-occipital wave generation and modulation. *Cell Mol Neurobiol* **17**:341–365(1997).

40. Chase MH, Morales FR. The atonia and myoclonia of active (REM) sleep. *Annu Rev Psychol* **41**:557–584(1990).

41. Morrison AR. Paradoxical sleep without atonia. *Arch Ital Biol* **126**:275–289(1988).

42. Shute, CC, Lewis PR. Cholinesterase-containing systems of the brain of the rat. *Nature* **199**:1160–1164(1963).

43. George R, Haslett W, Jenden D. A cholinergic mechanism in the pontine reticular formation: induction of paradoxical sleep. *Int J Neuropharmacol* **3**:541–552(1964).

44. El Mansari M, Sakai K, Jouvet M. Unitary characteristics of presumptive cholinergic tegmental neurons during the sleep–waking cycle in freely moving cats. *Exp Brain Res* **76**:519–529(1989).

45. Hobson JA, McCarley RW, Wyzinski PW. Sleep cycle oscillation: reciprocal discharge by two brainstem neuronal groups. *Science* **189**:55–58(1975).

46. Kodama T, Takahashi Y, Honda Y. Enhancement of acetylcholine release during paradoxical sleep in the dorsal tegmental field of the cat brain stem. *Neurosci Lett* **114**:277–282(1990).

47. Semba K, Greene RW, Rasmusson DD, McCarley RW, Weider J. Noradrenergic presynaptic inhibition of acetylcholine release in the rat pontine reticular formation: an *in vitro* electrophysiological and *in vivo* microdialysis study. *Soc Neurosci Abstract* **23**:1065(1997).

48. Vanni-Mercier G, Sakai K, Lin JS, Jouvet M. Carbachol microinjections in the mediodorsal pontine tegmentum are unable to induce paradoxical sleep after caudal pontine and prebulbar transections in the cat. *Neurosci Lett* **130**:41–45(1991).

49. Webster HH, Jones BE. Neurotoxic lesions of the dorsolateral pontomesencephalic tegmentum cholinergic cell area in the cat. II. Effects upon sleep–waking states. *Brain Res* **458**:285–302(1988).

50. Lu J, Bjorkum AA, Xu M, Gaus SE, Shiromani PJ, Saper CB. Selective activation of the extended ventrolateral preoptic nucleus during rapid eye movement sleep. *J Neurosci* **22**:4568–4576(2002).

51. Baghdoyan HA, Spotts JL, Snyder SG. Simultaneous pontine and basal forebrain microinjections of carbachol suppress REM sleep. *J Neurosci* **13**:229–242(1993).

3

PHYSIOLOGIC PROCESSES DURING SLEEP

LEON ROSENTHAL

Sleep Medicine Associates of Texas, Dallas, Texas

Sleep is a highly organized, complex behavior characterized by a relative disengagement from the outer world and variable but specific brain activity. Under normal conditions, sleep is associated with little muscular activity, a stereotypic posture, and reduced response to environmental stimuli. Sleep may be delayed but is indispensable for the survival of the species. As such, it is endogenously generated, homeostatically regulated, and reversible.

Sleep is typically evaluated using polysomnographic techniques, which enable the simultaneous characterization of the electroencephalogram (EEG), electro-oculogram (EOG), and electromyogram (EMG). Since 1968, standardized criteria have been followed to record and score human sleep.

THE NREM–REM CYCLE

Sleep is organized into non-rapid eye movement (NREM) and rapid eye movement (REM) sleep, which are easily characterized using polysomnographic techniques. In humans, NREM sleep appears as wakefulness-maintaining mechanisms wane. NREM sleep is divided into four stages based on the pattern of the brainwaves. Stage 1 NREM sleep is a transitional phase between full wakefulness and sleep. It can also emerge briefly during transitions from sleep to wake or after brief body movements. The EEG is characterized by relatively low-voltage slow activity in the theta range (4–7 Hz). Slow eye movements may be present and the EMG usually reveals a decline in the tonic activity relative to the waking state. Stage 2 NREM sleep is marked by the appearance of EEG spindles (fast activity in the

7–14 Hz range lasting at least 0.5 s) and K-complexes, which consist of high-voltage waves with a negative sharp component followed by a positive component. Stage 2 NREM sleep is the first bona fide sleep stage; adults spend 50–60% of sleep time in this particular stage of sleep. Stage 3 and stage 4 NREM sleep are frequently combined and called delta sleep, deep sleep, or slow-wave sleep. The EEG during this period is characterized by high-amplitude waves of 0.5–2 Hz.

REM sleep represents an active form of sleep and is characterized by low-voltage, intermixed cerebral activity associated with striated muscle atonia and rapid eye movements. Most dreams are thought to occur during this phase of sleep. REM sleep can be separated into tonic and phasic components. Tonic REM sleep is associated with near paralysis of most muscular groups. Only the diaphragm, the cardiac muscle, and some sphincters at the top and the bottom of the gastrointestinal tract remain active during REM sleep. In fact, the reduction of muscle tone is actively induced by the release of glycine onto the motoneurons. Phasic REM sleep is characterized by occasional bursts of EMG activity (myoclonias), rapid eye movements, and activity of the middle ear ossicles.

Human adults typically begin sleep by progressing from stage 1 NREM sleep through stage 4 NREM sleep. The progression of sleep stages might be intermittently interrupted by changes in body posture or partial arousals. After 70–80 min of NREM sleep, the first REM phase might be initiated, which typically lasts 5–10 min. The length of the NREM–REM cycle (from the start of NREM sleep to the end of the first REM period) is about 90–110 min. The NREM–REM cycle is usually repeated four to six times during a typical

TABLE 3.1   Physiologic Characteristics of Adult Human Sleep

- NREM–REM cycle (90 min long)
- NREM sleep
  - Stage 1 NREM: diminution of alpha waves by theta waves, rolling eye movements
  - Stage 2 NREM: spindles, K-complexes
  - Stage 3/4 NREM: delta waves
- REM sleep
  - Tonic: desynchronized EEG (low-voltage fast pattern mixed with small amount of theta rhythm and, often, with "sawtooth" waves), muscular atonia, depression of monosynaptic and polysynaptic reflexes
  - Phasic: bursts of rapid eye movements, myoclonic twitchings, irregular heart beat and respiration (with variable blood pressure), spontaneous activity of the middle ear muscles
- Endogenously generated
- Regulated by homeostatic and circadian factors
- Modulated by environmental factors
- Sleep rebound follows sleep loss
- Functional impairment produced by sleep loss/deprivation

night of sleep. While stage 3 and stage 4 NREM sleep are more prevalent at the beginning of the night (usually during the first NREM–REM cycles), REM sleep is usually of short duration during the initial NREM–REM cycles and lengthens in subsequent cycles of the night (Table 3.1).

In young adults, stage 1 NREM sleep constitutes about 5–10% of the night; stage 3 and stage 4 NREM sleep about 10–20%; REM sleep 20–25%; and the largest amount of sleep time, 50–60% is spent in stage 2 NREM sleep. The most significant factor affecting total sleep time and sleep stages is age. Newborn infants, during the first months of life, sleep 17–18 h a day and spend 50% of sleep time in REM sleep. The cyclical alternation of NREM–REM sleep is also shorter in the newborn, at about 50–60 min. Also, slow-wave sleep is maximal in young children and decreases markedly with age. In the elderly, sleep requirements decrease and nighttime awakenings increase.

CIRCADIAN AND HOMEOSTATIC DETERMINANTS OF SLEEP

Sleep, as other physiological variables, is regulated by the circadian timing system. The suprachiasmatic nucleus in the hypothalamus serves as the central neural pacemaker of the circadian timing system. The dominant synchronizing input to the human circadian pacemaker is environmental light. The retinohypothalamic tract links the retina to the suprachiasmatic nucleus, conveying photic information that enables synchronization to the light–dark cycle. Humans are usually synchronized to the 24 h day, with most adult humans

sleeping at night. It is in fact the temporal interplay of the circadian pacemaker and the sleep homeostatic drive that determines alertness, neurobehavioral performance, and sleep.

The propensity to fall asleep follows a biphasic pattern during the 24 h day. Two peaks of sleepiness have been characterized—one during nocturnal hours (2–6 a.m.) and another during daytime hours (2–4 p.m.). The sleepiness rhythm parallels the circadian variation in body temperature, with shortened latencies occurring in conjunction with temperature reductions. Likewise, more difficulty falling and staying asleep is associated with the rising phase of the temperature curve.

Sleep per se is considered a basic physiologic need state. It has been likened to hunger, which is critical to the survival of the organism. The homeostatic drive for sleep increases during wakefulness and decreases during sleep. Acute sleep deprivation is followed by an increase in the propensity to fall asleep and a parallel response in the propensity to stay asleep. The homeostatic drive to sleep is impacted by the oscillations of the circadian rhythm, which, for example, enhance alertness in the early evening, even after a sleepless night.

REGULATION OF NREM SLEEP

NREM sleep expresses the unified activity of many neuronal networks. Neurons found in the solitary tract nucleus of the medulla, raphe nuclei of the brainstem, reticular thalamic nuclei, anterior hypothalamus, preoptic area, basal forebrain, and orbital cortex are involved in the generation of NREM sleep.

The anterior hypothalamus and the adjacent basal forebrain have the most significant sleep-promoting effects in the brain. Gamma-aminobutyric acid-ergic neurons are involved in the inhibition of activating systems. The reticular nucleus of the thalamus is the synchronizing pacemaker of EEG spindle oscillations. It is in fact the thalamus, the first relay station, where afferent information is blocked at sleep onset, thus enabling the preservation of sleep. Serotonin-containing neurons of the raphe nuclei provide diffuse innervation to the brain and might be important in dampening certain sensory input and attenuating cortical activation in the initiation of slow-wave sleep. Adenosine, CSF-borne factors, and opiates may also have an effect on sensory modulation and are likely to play a role in the initiation and maintenance of sleep. However, no single neurotransmitter or neuromodulator has been found to be critical or sufficient for the initiation and/or maintenance of sleep.

REGULATION OF REM SLEEP

The nucleus reticularis pontis oralis in the caudal midbrain and rostral pons is critical to the generation of REM sleep.

Many of the neurons in this area are responsive to acetylcholine and in fact microinjection of carbachol, an acetylcholine agonist, in this area results in extended periods of REM sleep. The discharge patterns of many neurons during REM sleep resemble those during active wakefulness and bursts of spike potentials arising from the pons, lateral geniculate nucleus, and occipital cortex (PGO spikes) are known to be associated with REM sleep. PGO spikes represent one of the phasic events of REM sleep that include bursts of eye movements. These spikes are generated in the pons by cholinergic neurons. However, PGO spikes are not indispensable for REM sleep, as destruction of the cells that generate PGO activity does not result in the elimination of REM sleep. In addition, other electrophysiological characteristics are associated with REM sleep, such as desynchronized cortical EEG rhythms, hippocampal theta rhythms, striated muscle atonia, rapid eye movements, brief clonic contractions of facial and extremity muscles, and dream generation.

NORMAL AUTONOMIC CHANGES IN SLEEP

Many of the physiologic changes occurring during sleep are associated with changes in the level of activity of the autonomic nervous system. NREM sleep is characterized by a period of relative autonomic stability with sympathetic activity remaining at about the same level as during relaxed wakefulness, and parasympathetic activity increasing through vagus nerve dominance and heightened baroreceptor gain. During tonic REM sleep, a relative increase in parasympathetic activation is noted (mostly as a result of sympathetic input decline). Phasic REM sleep is characterized by an increase of both sympathetic and parasympathetic activity. The status of autonomic activity during sleep can be summarized as reflecting prevalent parasympathetic influence during NREM sleep (associated with quiescence of sympathetic activity), and great variability in sympathetic activity (associated with phasic changes in tonic parasympathetic discharge) during REM sleep. Changes in autonomic function and inherent changes in the control exerted by the central nervous system (CNS) affect most organ systems in the body during sleep. A brief review of some of the relevant changes affecting the cardiovascular system, respiration, cerebral blood flow, thermal control, and endocrine and genital function follows.

Cardiac Physiology

NREM sleep is usually characterized by brief heart rate acceleration during normal inspiration to accommodate venous return. During expiration, there is a progressive decrease in heart rate. This variability in cardiac rhythm is considered a marker of cardiac health. During REM sleep, heart rate becomes variable with episodes of tachycardia and bradycardia. Phasic REM sleep might be associated with significant increases in heart rate as a result of bursts of sympathetic activity and might lead to significant arrhythmias, in particular, when associated with ventilatory instability. Likewise, striking changes in coronary blood flow occur during REM sleep and sleep-state transitions. Individuals with heart disease may experience life-threatening arrhythmias and myocardial ischemia (and/or infarction) during REM sleep as a result of sympathetic nerve activity, which is concentrated in short, irregular bursts. These bursts trigger momentary and intermittent increases in heart rate and arterial blood pressure to levels similar to wakefulness.

Respiratory Physiology

Sleep not only modifies the neural control of ventilation but also impacts its mechanical and chemical control. NREM sleep is characterized by regularity of both respiratory frequency and amplitude. There is a decrease in alveolar ventilation with a concomitant decrease in arterial P_{O_2} and increase in P_{CO_2}. During REM sleep, there is a further decline in tidal volume and minute ventilation drops to its lowest level. Central apneas and periodic breathing are more frequent during REM sleep and these are mostly associated with phasic REM sleep. Hypoxic ventilatory response is lower during NREM sleep when compared to wakefulness, although some studies have described gender differences in this response. Both men and women experience a similar decline in the hypoxic ventilatory response during REM sleep. Increases in end-tidal P_{CO_2} during sleep results in an increase in ventilation. However, this response is variable. Likewise, hypocapnia has an important inhibitory effect on respiration during sleep. It should also be noted that sleep results in a general decrease in muscle tone. This is particularly relevant to the muscles of the upper airway, which in turn have an impact on ventilation. The genioglossal muscle activity pulls the tongue down and forward enabling the airway to remain open. NREM sleep results in decreased discharge activity with further reductions noted during REM sleep. The potential result of this physiological change is the obstruction of the upper airway, which might result in a pathological condition (obstructive sleep apnea).

Cerebral Blood Flow

Cerebral blood flow (CBF) mostly decreases during NREM sleep when compared to wakefulness. During REM sleep there are significant regional changes in CBF. In general, a significant increase in CBF is associated with REM sleep. Certain areas of the brain have been described as experiencing significant increases in CBF—among these, the pontine tegmentum, the dorsal mesencephalon, thalamic nuclei, the amygdala, and the anterior cingulated and the entorhinal cortex. Interestingly, a decrease of CBF in cortical and limbic structures during post-sleep wakefulness has been described (when compared to pre-sleep wakefulness). It has been speculated that such a change might be a reflection of a resetting process by which the circulatory and metabolic activity of the brain is set at a lower level of activity. Some researchers have interpreted these findings as evidence of the "restorative" function of sleep.

Temperature Regulation

Core body temperature (T_b) shows a circadian variation. The T_b cycle is a sinusoidal-like function with a maximum in the early evening and a minimum in the early morning. The amplitude of temperature variation is about 1 °C. It should be noted that the circadian T_b variation is independent of muscular activity. Under normal conditions, the drop in T_b during nocturnal sleep is accomplished by two separate mechanisms. One is the sleep-related reduction in the body's thermal set point (the result of increased heat dissipation and decreased heat generation) and the second is the intrinsic circadian temperature variation (which is independent of sleep). The preoptic-anterior hypothalamic area is critical to the regulation of T_b. During NREM sleep T_b is regulated at a lower set point when compared to the wake state. In contrast, T_b is not regulated during REM sleep, which represents a poikilothermic state. As a result, the body temperature during REM sleep drifts toward the environmental temperature.

Endocrine Function

Sleep in humans is associated with prominent changes in the function of virtually every endocrine system in the body. It is through the various hormones secreted in the body that tissue growth is promoted, sexual development and activity are regulated, the absorption of sodium is synchronized, and, perhaps most importantly, the response to stress is modulated to preserve homeostatic balance. The plasma concentrations of many hormones display sleep-related variations. However, such correlations do not necessarily indicate a causal relationship between them. In fact, circadian regulation in many instances synchronizes these events. The endocrine systems that undergo sleep-related changes include the adrenocorticotropic hormone, thyrotropin, growth hormone, gonadotropic hormones, prolactin, and melatonin. New research also suggests that sleep duration per se might have an effect on endocrine function. Specifically, short sleep duration has been found to be associated with decreased leptin levels (which suppresses food intake) and a concurrent increase in ghrelin levels (which stimulates appetite). These findings suggest for the first time that a potential association might exist between poor sleep schedule practices (specifically insufficient sleep) and obesity.

Genital Function

Penile erections are a naturally occurring phenomenon associated with REM sleep. This phenomenon has been demonstrated to be present in all healthy males from infancy to old age. Similar clitoral erections and vaginal engorgement have been documented in women during REM sleep. These physiologic events are the result of increased parasympathetic activity, which results in local vasodilation, decreased venous outflow, and increased bulbocavernosus muscular activity. In contrast, few changes in genital function are present during NREM sleep.

THE FUNCTION OF SLEEP

The available research provides strong evidence that sleep is critical to the survival of the species. In fact, chronic sleep deprivation in rats has shown that these animals die after 2–3 weeks. REM-deprived rats survive for longer periods but end up dying as well. Unfortunately, a clear cause for the death of these animals has not been identified. While no widespread agreement exists on why sleep is important, it is clear that sleep deprivation (or chronic insufficient sleep) results in increased sleepiness and decreased functioning. In addition, several hypotheses about the function of sleep have been advanced. Perhaps a critical function of sleep is the one related to thermoregulation. The studies in sleep-deprived rats showed them to experience an inability to retain heat as their body temperature dropped during the experiment despite experiencing an increase in metabolic rate. Other theories have suggested that sleep might have a role in the conservation of metabolic energy and cognition. The ontogenetic changes documented in REM sleep across the process of maturation and a series of elegant experiments using REM deprivation during critical phases of development have enabled researchers to study the effects of these manipulations on neural maturation. In addition, REM sleep seems to have an important role in memory consolidation.

CONCLUSION

The description of REM sleep in the 1950s was instrumental in enabling the systematic study of sleep. Since that time, research in this area has progressed from a nascent research discipline to a clinical medical specialty. Critical to both research and the clinical aspects of the discipline are the characterization of the physiological manifestations of sleep and the understanding of the implications of these processes. The available evidence has established that sleep serves an important function, as evidenced by the rebound of sleep following sleep loss and the developmental, functional, and metabolic impairments produced by sleep deprivation. While no unitary theory of sleep function has explained the wealth of data on available sleep phenomena, the evidence suggests that the function of sleep is likely multidimensional and differential depending on the organism's stage of development.

ADDITIONAL READING

Kryger MH, Roth T, Dement WC (Eds). *Principles and Practice of Sleep Medicine*, 3rd ed. Saunders, Philadelphia, 2000.

Lee-Chiong TL, Sateia MJ, Carskadon MA (Eds). *Sleep Medicine*. Hanley & Belfus, Philadelphia, 2002.

Rechtschaffen A, Kales A. *A Manual of Standardized Terminology, Techniques and Scoring Systems for Sleep Stages of Human Subjects*. UCLA Brain Information Service/Brain Research Institute, Los Angeles, 1968.

Rosenthal L, Roehrs TA, Rosen A, et al. Level of sleepiness and total sleep time following various time in bed conditions. *Sleep* **16**:226–232(1993).

Schmidt MH, Schmidt HS. Sleep-related erections: several mechanisms and clinical significance. *Curr Neurol Neurosci Rep* **4**:170–178(2004).

Shaffery JP, Roffwarg HP, Speciale SG, Marks GA. Ponto-geniculo-occipital wave suppression amplifies lateral geniculate nucleus cell-size changes in monocularly deprived kittens. *Dev Brain Res* **114**:109–119(1999).

Tung A, Mendelson WB. Anesthesia and sleep. *Sleep Med* **8**(3):213–226(2004).

Spiegel K, Tasali E, Penev P, Van Cauter E. Sleep curtailment in healthy young men is associated with decreased leptin levels, elevated ghrelin levels, and increased hunger and appetite. *Ann Intern Med* **141**(11):846–850(2004).

Taheri S, Lin L, Austin D, Young T, Mignot E. Short sleep duration is associated with reduced leptin, elevated ghrelin, and increased body mass index. *PLoS Med* **1**(3):210–217(2004).

4

BIOLOGICAL RHYTHMS AND SLEEP

ROBERT Y. MOORE

University of Pittsburgh, Pittsburgh, Pennsylvania

INTRODUCTION

There are many biological rhythms. A rhythm is a repetitive biological event with three features—period, phase, and amplitude. Period is the length of the rhythm. Phase is the timing of the rhythm in relation to a stimulus, and amplitude is a measure of the amount of the rhythmic event. In this review, I will focus on circadian rhythms— rhythms with a period of approximately 24 hours. The solar cycle of light and dark is the most important recurring stimulus in our environment. Not surprisingly, living organisms have evolved adaptive mechanisms that utilize the solar cycle to promote survival and reproduction. These mechanisms, termed "circadian rhythms" (from *circa*, about, and *diem*, a day) are now known to depend on genetically controlled rhythmic molecular events that control a wide array of rhythms in cellular, system, and behavioral functions. Circadian rhythms are expressed in essentially all living organisms from prokaryotic through eukaryotic species to the human and are critical to survival. An example of a typical circadian rhythm is shown in Figure 4.1.

The fundamental behavioral circadian rhythm is the rest–activity cycle. This rhythm first appeared in animals with the evolution of the nervous system. In mammals it is expressed as a sleep–wake rhythm presenting in one of two patterns. Some mammals, particularly rodents and carnivores, are nocturnal—awake at night and asleep during the day. Nocturnal animals depend on olfaction and audition as the primary senses to perceive their environment and have evolved very elaborate peripheral sensory receptors and brain mechanisms to support these sensory modalities.

Diurnal mammals, primates are the best example, sleep at night and are awake during the day, and use vision as their primary sensory modality.

The circadian rhythm in sleep–wake behavior in humans is a fundamental component of behavioral adaptation. Placing wake behavior during the solar day ensures maximal use of vision to guide behavior. Sleeping during the night provides an optimal time for this critical restorative behavior. The functional effect of circadian regulation is to provide a temporal organization of wake and sleep to permit maximally effective adaptive waking behavior.

The following are briefly reviewed: (1) molecular mechanisms of circadian timing, (2) circadian pacemakers, (3) the circadian timing system, (4) the waking system, and (5) the interaction of circadian and homeostatic control of sleep and waking.

MOLECULAR MECHANISMS OF CIRCADIAN TIMING

Circadian timing is an inherited adaptation and, as such, is genetically determined. Over the last 15 years extensive research on the molecular basis of circadian timing has been performed, and we now have a detailed body of information. Two important generalizations have emerged: (1) the fundamental basis of circadian timing is interacting positive and negative transcriptional feedback loops [1], and (2) in mammals and other complex organisms, many tissues contain circadian clocks and pacemakers [2]. There are two important sets of genes involved in these feedback loops in the mammalian circadian pacemaker:

Sleep: A Comprehensive Handbook, Edited by T. Lee-Chiong.
Copyright © 2006 John Wiley & Sons, Inc.

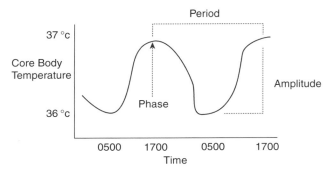

Figure 4.1 Diagram of the core body temperature rhythm. The rhythm has a peak in the late afternoon and a trough in the early morning with an amplitude of about 1 °C. The period of the rhythm is 24 h and the phase is designated as the time of the peak temperature. See text for further description.

period (*per*) genes and *cryptochrome* (*cry*) genes are part of the negative loop, while *clock* and *Bmal1* are components of the positive loop. The general organization of the molecular pathways is shown in Figure 4.2. Rhythmic transcriptional enhancement by the transcription factors, Clock and Bmal1, is essential for clock function and provides a basic drive to the molecular clock. These proteins form heterodimers in the cytoplasm that enter the nucleus to activate *per* and *cry* transcription by binding to E box enhancers.

The resultant Per and Cry proteins also form heterodimers and translocate to the nucleus to act as negative regulators through inhibition of the Clock–Bmal1 activation

of transcription. The positive component of the process involves rhythmic regulation of *Bmal1*. In order to generate a positive feedback loop, Clock–Bmal1 heterodimers both activate *per* and *cry* transcription and transcription of the orphan nuclear receptor gene, *Rev-Erbα*. The Rev-Erbα protein acts on specific response elements in the *Bmal1* promoter to repress *Bmal1* transcription. The Cry protein also inhibits transcription of the *Rev-Erbα* gene, allowing increased transcription of *Bmal1*. This indicates that the positive and negative loops are both regulated by Clock–Bmal1 heterodimers [1]. This appears to be the basic organization of the molecular clock but much remains to be learned. The significance of this is emphasized by recent findings that a familial form of advanced sleep phase syndrome is the result of a mutation in one of the *per* genes, *per2* [3].

CIRCADIAN PACEMAKERS

Circadian pacemakers are clocks that regulate the temporal organization of function in other circadian clocks or in other brain systems or tissues. The first mammalian circadian pacemaker to be identified was the suprachiasmatic nucleus (SCN) of the hypothalamus [4–6]. The SCN is comprised of neuronal oscillators that are coupled though connections to form a pacemaker [7]. Each oscillator contains the molecular machinery noted above and clock function is accomplished by coupling the molecular feedback loops to the control of neuronal membrane potential [7]. The SCN is comprised of multiple component oscillators that can dissociate and function independently [8]. In part, these are identifiable anatomically [5] (Figure 4.3). The SCN has two principal subdivisions. The core is the central part of the nucleus and contains neurons that produce GABA and a peptide, either vasoactive intestinal polypetide (VIP) or gastrin releasing peptide (GRP). The major afferent input to the core is from visual centers, primarily the retina. Retinal afferents use the excitatory transmitter

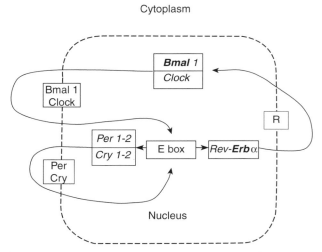

Figure 4.2 Diagram showing the positive and negative transcriptional feedback loops in the mammalian molecular clock. The positive loop includes the genes *clock* and *Bmal1* and their proteins. The regulation of this loop is partially through *Rev-Erbα* and its product [R]. The negative loop includes the *period* (*per*) and *cryptochrome* (*cry*) genes and their products. See text for description.

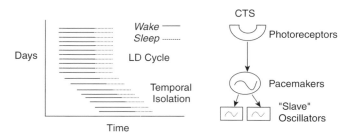

Figure 4.3 Diagrammatic representation of the sleep–wake rhythm and the circadian timing system (CTS). See text for description.

glutamate (GLU) and one of two peptides, melanopsin (melan) or pituitary adenylate cyclase activating peptide (PACAP). The other SCN division, the shell, surrounds the core and contains neurons that produce GABA and arginine vasopressin (AVP). The shell receives input from the limbic cortex, thalamus, hypothalamus, and brainstem. There are extensive intrinsic connections within the SCN. Each subdivision projects densely upon itself and the homologous contralateral SCN. The core projects on the shell but the shell has only minimal projections on the core. Thus the flow of information is from visual receptive areas and raphe to the core and from multiple forebrain and brainstem areas, and the core, to the shell. As will be described later, both subdivisions project to effector systems under circadian control in an overlapping but not identical pattern. The organization of the SCN is similar in all mammals including primates and virtually identical with respect to core and shell organization.

THE CIRCADIAN TIMING SYSTEM

The SCN is part of a set of brain structures whose function is circadian regulation and, hence, are designated the circadian timing system (CTS). The principal features of circadian rhythms determine the basic organization of the system. This is clearly presented by the sleep–wake rhythm (Figure 4.4). Under normal conditions of a light–dark cycle, the period of the rhythm is 24 h and the onset of sleep and onset of waking occur at about the same time each day. This is an entrained rhythm and the entrainment process requires input through the eyes. With removal of time cues, as in temporal isolation where subjects do not have access to time information or a regular light–dark cycle, sleep–wake cycles remain quite regular but with a period that

exceeds 24 h. This is referred to as a free-running rhythm. Entrainment requires visual photoreceptors and a visual pathway from the photoreceptors to the circadian pacemaker. The free-running rhythm is established by the intrinsic period of the pacemaker—the SCN—and is slightly longer than 24 h in humans. These two features of circadian rhythm establish the necessary components of the CTS (Figure 4.4).

The entrainment pathway is initiated with specific photoreceptors. Over the last few years, it became apparent that these were not conventional photoreceptors. For example, in transgenic mice, a knockout of the genes for both rods and cones did not affect entrainment and the process was mediated by a set of ganglion cells that contain photopigments, particularly melanopsin [9]. These ganglion cells project to the SCN core through the retinohypothalamic tract. There is an important secondary visual pathway that also projects to the SCN core from the intergeniculate leaflet, a ventral thalamic component of the lateral geniculate complex. This pathway, the geniculohypothalamic tract, contains GABA colocalized with the peptide neuropeptide Y [5] and appears to modulate the effect of retinal input on the SCN. As noted earlier, there are numerous other inputs to the SCN, particularly to the shell, that probably modulate output from that subdivision. The output of the SCN is predominantly through the hypothalamus with differing sets of projections mediating control over differing functional systems [10, 11]. These include rostral projections to autonomic centers and the neuroendocrine systems and caudal projections to the posterior hypothalamic arousal systems (Figure 4.5). With the recent rapid application of molecular methods to circadian research, it has become evident that there is a hierarchy of pacemakers and oscillators that extend from the SCN pacemaker to other brain pacemakers and oscillators to a large series of clock elements in many tissues and organs [2].

WAKE PATHWAYS

As described earlier, behavior occurs in one of two distinctive states, wake and sleep. Wake is a state in which the cerebral cortex is activated with awareness of the sensory and internal environment and a continual elaboration of adaptive behavior. The maintenance of the waking state is an active process requiring the integrity of a well-defined set of brain pathways. The first evidence for wake pathways came from a clinical pathological analysis of an unusual encephalitis that occurred in association with the influenza pandemic of 1918–1926. The encephalitis, studied by von Economo, was characterized by impaired responsiveness, extending into prolonged coma in many sufferers, and had a restricted pathology involving the upper brainstem and posterior hypothalamus which von Economo viewed

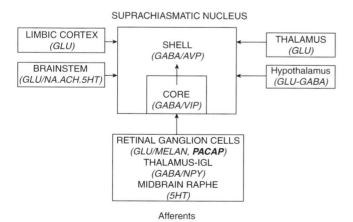

Figure 4.4 Suprachiasmatic nucleus pacemaker organization in mammals. See text for description.

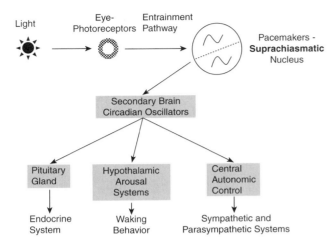

Figure 4.5 Organization of the circadian timing system. See text for description.

as critical components of waking. Although this was confirmed by animal studies with discrete lesions, research over the next 20 years on waking function was focused on the brainstem and resulted in identification of the pontine and mesencephalic reticular formation as an "ascending reticular activating system." The hypothalamus became neglected, and it was assumed that the principal waking pathways involved either direct projections from the brainstem to cerebral cortex, as with the monoamine neuron systems, or a relay through the nonspecific nuclei of the midline-intralaminar thalamus (Figure 4.6).

Recent data have emphasized the importance of hypothalamic projections in maintenance of waking. The hypothalamus has extensive neocortical projections and many of these arise from the posterior hypothalamus. The function of these projections, particularly those producing hypocretin [12] or histamine [13], is clearly to promote

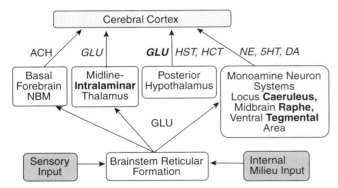

Figure 4.6 Waking pathways. Waking is maintained by pathways originating in the brainstem, hypothalamus, thalamus, and basal forebrain. See text for description. Abbreviations: ACH, acetylcholine; DA, dopamine; GLU, glutamate; HCT, hypocretin; HST, histamine; NE, norepinephrine; 5HT, serotonin.

wake. Input through the cholinergic systems is also an important component of the waking systems [14]. Hypothalamic input to the cortex is critical to maintaining the waking state but inputs from a number of other areas are important for both arousal and modulation of specific thalamic input and cortical processing. The circuitry involved in sleep regulation has been reviewed recently [15].

HOMEOSTATIC AND CIRCADIAN CONTROL OF THE SLEEP–WAKE CYCLE

It is evident to nearly everyone that the propensity to sleep increases with the time since the last sleep episode and this fits with the general concept that sleep has a restorative function. The obvious questions are why sleep–wake cycles are so regular and why the timing of waking behavior, and sleep, is so precise. The propensity to sleep as a function of time awake is referred to as homeostatic sleep drive, and we now recognize that the precise regulation of sleep–wake behavior occurs through an interaction of the CTS with homeostatic drive for sleep. The basis for homeostatic sleep drive is not fully established and is probably quite complex. There are two hypotheses that have substantial support. The first is that sleep permits restoration of energy resources in the brain [16]. The second is that homeostatic drive is a function of accumulation of a sleep-promoting substance. This is an old idea but a large body of recent evidence supports the view that increasing adenosine content, perhaps in local areas, is a critical component of the homeostat [17]. It is likely that other factors contribute.

How do circadian function and homeostatic drive interact in sleep–wake regulation? An early formulation of this was proposed by Daan and Borbely [18] and is shown in Figure 4.7. Homeostatic sleep drive is maximal at the time of sleep onset and is dissipated gradually through the sleeping period.

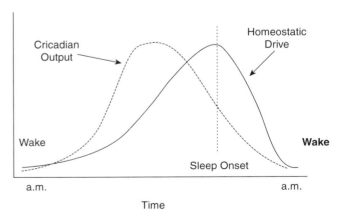

Figure 4.7 Interaction of circadian and homeostatic factors in sleep–wake regulation. See text for description.

At the time of waking, little or no homeostatic drive remains but it begins to accumulate with time awake and gradually increases over the waking period. During the waking day, increasing homeostatic drive is opposed by a circadian drive for arousal until shortly before sleep, when circadian influences gradually decrease. Although there was relatively little evidence for the Daan–Borbely model at the time it was first proposed, more recent data have supported it. This includes behavioral and electrophysiological data and studies of the circadian timing system. The current evidence indicates that the output of the SCN pacemaker is an important component of the circadian promotion of arousal. Ablation of the SCN in primates abolishes the circadian rhythm in sleep–wake behavior and increases total sleep time [19]. SCN control of the rest–activity rhythm can be mediated by a diffusible factor. The nature of this signal and the SCN signal mediating arousal effects are unknown but a recent study implicates a peptide—prokineticin [20].

SUMMARY

The timing of the sleep–wake cycle depends on the interaction of a number of brain systems, in particular, sleep–wake systems and the CTS. The CTS not only coordinates the timing of the sleep–wake cycle by opposing the sleep homeostat but also serves to coordinate the timing of pacemakers and oscillators in other tissues and organs to facilitate adaptation.

REFERENCES

1. Reppert SM, Weaver DR. Coordination of circadian timing in mammals. *Nature* **418**:935–941(2002).

2. Gachon F, Nagoshi E, Brown SA, Ripperger J, Schibler U. The mammalian circadian timing system: from gene expression to physiology. *Chromosoma* **113**:103–112(2004).

3. Toh KL, Jones CR, He Y, Eide EJ, Hinz WA, Virshup DM, Ptacek LJ, Fu YH. An hPer2 phosphorylation site mutation in familial advanced sleep phase syndrome. *Science* **291**:1040–1043(2001).

4. Hastings MH, Reddy AB, Maywood ES. A clockwork web: circadian timing in brain and periphery, in health and disease. *Nat Rev Neurosci* **4**:649–661(2003).

5. Moore RY, Speh JC, Leak RK. Suprachiasmatic nucleus organization. *Cell Tissue Res* **309**:89–98(2002).

6. Pace-Schott EF, Hobson JA. The neurobiology of sleep: genetics, cellular physiology and subcortical networks. *Nat Rev Neurosci* **3**:591–605(2002).

7. Hastings MH, Herzog ED. Clock genes, oscillators, and cellular networks in the suprachiasmatic nuclei. *J Biol Rhythms* **19**:400–413(2004).

8. Yamaguchi S, Isejima H, Matsuo T, Okura R, Yagita K, Kobayashi M, Okamura H. Synchronization of cellular clocks in the suprachiasmatic nucleus. *Science* **302**:1408–1412 (2003).

9. Berson D. Strange vision: ganglion cells as circadian photoreceptors. *Trends Neurosci* **26**:314–320(2003).

10. Chou TC, Scammell TE, Gooley JJ, Gaus SE, Saper CB, Lu J. Critical role of dorsomedial hypothalamic nucleus in a wide range of behavioral circadian rhythms. *J Neurosci* **23**:10691–10702(2003).

11. Moore R, Danchenko RL. Paraventricular–subparaventricular hypothalamic lesions affect circadian functions. *Chronobiol Int* **19**:345–360(2002).

12. Siegel JM. Hypocretin (orexin): role in normal behavior and neuropathology. *Annu Rev Psychol* **55**:125–148(2004).

13. Haas H, Panula P. The role of histamine and the tuberomammillary nucleus in the nervous system. *Nat Rev Neurosci* **4**:12–130(2003).

14. Jones BE. Activity, modulation and role of basal forebrain cholinergic neurons innervating the cerebral cortex. *Prog Brain Res* **145**:157–169(2004).

15. Hobson JA, Pace-Schott EF. The cognitive neuroscience of sleep: neuronal systems, consciousness and learning. *Nat Rev Neurosci* **3**:679–693(2002).

16. Benington JH, Heller HC. Restoration of brain energy metabolism as the function of sleep. *Prog Neurobiol* **45**:347–360(1995).

17. Basheer SR, Thakkar MM, McCarley RW. Adenosine and sleep–wake regulation. *Prog Neurobiol* **73**:379–396(2004).

18. Daan S, Beersma DG, Borbely AA. Timing of human sleep: recovery process gated by a circadian pacemaker. *Am J Physiol* **246**:R161–183(1984).

19. Edgar DM, Dement WC, Fuller CA. Effect of SCN lesions on sleep in squirrel monkeys: evidence for opponent processes in sleep–wake regulation. *J Neurosci* **13**:1065–1079(1993).

20. Cheng M, Bullock CM, Li C, Lee AG, Bermak JC, Belluzzi J, Weaver DR, Leslie FM, Zhou QY. Prokineticin 2 transmits the behavioural circadian rhythm of the suprachiasmatic nucleus. *Nature* **417**:405–410(2002).

5

BIOLOGY OF DREAMING

MILTON KRAMER

Maimonides Medical Center, Brooklyn, New York

BIOLOGICAL THEORIES OF DREAMING

Observations by Aserinsky and Kleitman [1, 2] identified that there were episodic bursts of rapid eye movements during sleep associated with dreaming, and Dement and Kleitman [3] showed that there was a particular electroencephalographic (EEG) pattern associated with these periods of rapid eye movements, which recurred regularly during the night. This combination of bursts of conjugate rapid eye movements and a relatively low-voltage mixed-frequency EEG with muscle atonia [4] has been designated rapid eye movement (REM) sleep. It is this correlation of dreaming with a specific, identifiable EEG pattern that became the focus for the efforts to describe the physiological processes that are the biological basis for the dreaming process. Furthermore, interest in the biological basis of mental function in psychoanalysis reflects Freud's conviction [5] that psychology will have a physiological basis. This leads to the idea that the current neuroscience approach may provide a way to bridge the mind/brain divide.

There are currently only two competing theories, from a neuroscience point of view, of the functional anatomy of dreaming. The activation-synthesis theory of Hobson and collaborators [6] has been extended into their activation information modulation (AIM) space state model, which is still a bottom-up, subcortical brainstem view of the origin and content of dreaming. The source for the initiation of dreaming is in the pons, which sends random signals to the cortex. The cortex responds passively to the pontine signals in that its response is determined by these random signals. The cortex elaborates the dream experience, making

the best of a bad job. The content of the dream experience is shaped by affect in the dream [7]. The model for dreaming is the verbal output seen in the productions of demented patients.

The competing view of a brain-based theory of dreaming has been that articulated by Solms [8]. This theory is based on a clinicoanatomical approach, brain lesions, using neuropsychological techniques. It is allegedly a top-down, cortically based theory. Any stimulus that activates the brain can initiate the dreaming process. It requires engaging the ventral tegmental area of Tsai, whose fibers pass through the ventromedial prefrontal area adjacent to the anterior horns of the ventricles.

In order to appreciate the value and, more importantly, the limitations of the neuroscience approach to dreaming, it is worthwhile to review the two theories [6–8]. They have major points of difference.

The orientation in the activation-synthesis (A-S) AIM (activation information modulation) space state model [6] toward dreaming is:

1. that dreaming is coextensive with REM sleep;
2. that the model for the experience of dreaming is dementia;
3. that the stimulus for cortical activation is random discharges from the pons in what has been described as pontogeniculo-occipital (PGO) waves;
4. that the cortical response to the PGO waves is a relatively passive one; and
5. the content of the dream is shaped by the affect in the dream.

Dreaming is defined in the A-S theory [6] as "mental activity occurring in sleep which is characterized by vivid sensorimotor imagery, that is experienced as waking reality, despite such distinctly cognitive features as impossibility or improbability of time, place, person, and actions; emotions, especially fear, elation, and anger predominate over sadness, shame, and guilt and sometimes reach sufficient strength to cause awakening, memory for even vivid dreams is evanescent and tends to fade quickly unless special steps are taken to retain it." This definition captures, Hobson believes, what people mean by dreaming and serves both psychological and cognitive neuroscience.

In the A-S theory [6], differences among REM, non-REM, and wake mentation will be explained by the distinctive physiology of REM sleep. REM sleep is related to dreaming because:

"Dream reports are more likely to be reported from a REM awakening than a non-REM awakening (80% versus 40%);

Dream recall decreases quickly after the REM period ends;

The word count of the dream report correlates positively with REM time and external stimuli are appropriately incorporated into the time sequence of the dreaming narrative;

Judges can tell REM reports from non-REM reports; and

Qualitatively there are REM/non-REM report differences. REM reports are generally longer, more vivid, with more movement and emotion and are less related to waking life. Non-REM is more thought like."

The aspects of REM dreaming that are rare in non-REM according to A-S theory [6] include:

"The hallucinatory nature of the experience;

That the images change rapidly and are often bizarre;

The delusional nature of the experience;

The decreased self-reflection;

The creation of a confabulatory narrative;

The effect of instinctual programs such as fight/flight that organize the cognition; and

The attenuated volitional control."

These features of dreaming, A-S theory [6] holds, will be explained by the distinctive biology of REM sleep. The A-S theory postulates an isomorphism between the biology and psychology of dreaming, which reflects either a similarity (biological meaning of isomorphism) or an identity (mathematical meaning) between the two. It is a highly reductionistic theory that explains the experience of dreaming by its biology.

The control of REM dreaming [6] is described anatomically, physiologically, cellularly, and chemically. The anatomic control is in the pons; therefore it is subcortical, that is, in the brainstem. Physiologically, it is represented by PGO waves from the pons to the lateral geniculate body to the occipital cortex. At the cellular level in the pons, the REM-on cells are in the mesopontine tegmentum. The REM-off cells are in the nucleus locus caeruleus and dorsal raphe nucleus. The chemical control of dreaming is a consequence of REM-on cells secreting acetylcholine, while the REM-off cells secrete norepinephrine and serotonin.

The AIM space state model of the A-S hypothesis [6] shows in a three-dimensional model the change in *activation* during dreaming sleep from low to high; while the *information* source shifts from external to internal; and the *modulation* shifts from high norepinephrine and serotonin to high acetylcholine.

The activation of various brain areas in A-S theory [6] is somewhat in this order:

1. "There is stimulus from the pontine and the midbrain reticular activating circuits and nuclei. This leads to an ascending arousal of multiple forebrain structures. The contribution to the dream experience is consciousness, eye movements, and motor pattern information via the PGO system."

2. "Diencephalic structures (e.g., hypothalamus and basal forebrain) are activated involving autonomic and instinctual (fight/flight) functions, and cortical arousal. The contribution to dreaming is to further support consciousness and provide instinctual elements."

3. "Anterior limbic structures are active including the amygdala, anterior cingulate, parahippocampal cortex, hippocampus, and medial frontal areas. This contributes emotional labeling of stimuli, goal directed behavior, and movements. This activation contributes to the dream's emotionality, affective salience, and movement."

4. "The dorsolateral prefrontal cortex is inactive during dreaming. This area is involved with executive functions, logic, and planning. This would explain in the dream the absence of volition, logic, orientation, and working memory."

5. "The basal ganglia become active. They are involved in the initiation of motor actions. In the dream, they may account for the initiation of fictive movement."

6. "The thalamic nuclei become active, for example, the lateral geniculate nucleus that would be involved in the relay of sensory and pseudosensory information to the cortex. For the dream experience, it transmits PGO information to the cortex."

7. "Primary motor and sensory cortices are blocked. Therefore, sensory percepts and motor commands do not occur."

8. "The inferior parietal cortex is stimulated. This area is involved in the spatial integration of processed heteromodal input. It provides the spatial organization for the dream."

9. "The visual associational cortex becomes active and involves the higher order integration of visual percepts and images."

10. "The cerebellum is activated and is involved in fine tuning of movement and in vestibular function and contributes fictive movement to the dream."

In the A-S theory, dreaming is generated by the random output of the brainstem and passively synthesized by the forebrain. The cortex, it is said, makes the best of a bad job. Dementia is the model for the dream.

The forebrain or cortical theory of Solms [8] is based on a clinicoanatomical approach in patients with localized brain lesions, utilizing neuropsychological techniques to specify the functional deficiency and CT scanning to confirm the lesion site. Dreaming is explained as a top-down process despite the nuclei of Tsai being subcortical. Solms observed that "there was a reported loss of the experience of dreaming in patients who had either (1) a bilateral mediobasal frontal cortex lesion involving fibers from the ventral tegmental area of Tsai, an appetitive center which is the source of seeking, wishing/desiring behavior; or (2) a lesion of the inferior parietal area of either side of the brain which on the right involves spatial orientation and on the left involves symbolic activity. Those individuals who reported they had lost the experience of dreaming were those who also reported poorer sleep. In those patients with a lesion in the parieto-temporo-occipital association region, the visual elements in dreaming are lost as well as the ability to create visual images from memory while awake."

Dreaming is not assumed to be REM sleep dependent in the cortical theory [6, 8]. "Dreaming is thought to be initiated by an arousing stimulus; such as REM sleep, seizures, or noise. This activation stimulates circuits that arise from cell groups in the ventral tegmental area of Tsai. It is a dopaminergic system. It is connected to frontal and limbic structures. The ventral tegmental area of Tsai circuits instigate goal-seeking behavior. It is the wanting, wishing command system. Anterior limbic structures block transmission, which interrupts goal directed behavior (e.g., voluntary motor activity) and facilitates 'back projection' processes. The dorso-lateral cortex (the voluntary executive center) and primary visual cortices (the site for perception) are inhibited. The inferior parietal cortices become active and provide the spatial (right side) and sym-

bolic (left side) aspects of dreaming. Lastly, the occipital association areas provide memories of perceptions from which the imagery of dreaming is constructed."

For Solms [8] dreaming does not isomorphically reflect simple activation of perceptual and motor areas, as these are not activated during dreaming. The imagery of dreaming is not just reproduced, it is constructed each time from memory. Dreaming is not REM bound.

What particularly are the limitations in the A-S hypothesis [6]? Is the A-S theorists' particular definition of the dream representative of the dreaming process? Dreaming is defined in the A-S theory as "vivid, sensorimotor imagery, experienced as waking reality, despite it being improbable or impossible at times, and emotions are seen as prominent." This is a highly selective and arbitrary definition of dreaming. Nielsen [9] points out that there is a continuum of cognitive experiences during sleep from so-called apex dreaming, to regular dreaming, to cognitive activity (often called sleep mentation), to cognitive processes. There is no generally accepted or standardized definition. The position put forth by the A-S theorists, that their definition captures what people mean by dreaming and that it serves both psychology and cognitive neuroscience, is questionable. Taub and colleagues [10] have shown that what people say they think a nightmare is like and what they report as a personal nightmare experience are clearly differentiable. The concept of nightmare (e.g., what people think nightmares are like) is more intense, better constructed, and reported in fewer words than their own nightmare experience. A definition of dreaming needs to be made on an empirical basis rather than out of opinions about what it is or is not. Utilizing idealized versions of experience may be suitable for literary undertakings but not for scientific ones either in psychology or in the neurosciences.

The assumptions by the A-S hypothesis theorists [6] that dreaming is coextensive with REM sleep has been widely challenged. Solms [11] has pointed out that dreaming and REM sleep are doubly dissociated; not all REM sleep yields a dream report, and dream reports can be recovered from non-REM sleep. Furthermore dreaming may not even be sleep bound as both Foulkes and Fleisher [12] and Kripke and Sonnenschein [13] have collected dream reports from subjects who were awake.

The position that dreaming is the result of the activity of the pontine generator for REM sleep is challenged by the clinicoanatomical studies of Solms [8]. He reports that patients with core brainstem lesions, which would have prevented the PGO wave from reaching the cortex and who are hypoaroused while awake, continued to report dreaming.

The idea that the model for the dream experience is the output of the demented brain is related to the description of the physiologic process that A-S theorists assume underlies dreaming. In this view, random discharges from the pons, so-called PGO waves, stimulate cortical structures that do the best they can to organize these chaotic stimuli and the results are hallucinations and narratives that are poorly organized, confabulated, and easily forgotten, allegedly like the experiences of the demented awake. The cortex makes the best of a bad job. However, Snyder [14] has pointed out, based on his large series of laboratory-collected dreams, that it is the dreams mundane nature that best characterizes them, rather than their being impossible or improbable. Heynick [15], in a systematic analysis of speech reported as part of the content of the dream experience, observes how well constructed the speech in dreams is from a grammatical and syntactical point of view. Apparently our linguistic capacity during dreaming operates with surprising efficiency and is capable of generating well-formed, often syntactically complex sentences. Kramer [16] has shown that dream content is highly ordered and where we know there are psychological differences there are dream content differences. This is true at the group level as demographic variables such as gender, age, race, marital status, and social class show dream content differences as do psychiatric illnesses. The dreams of schizophrenics are different from those of the depressed.

Dream content is highly ordered at the individual level as well, [16] and dream content varies across the REM period, and from REM period to REM period throughout the night. The dreams of one individual are different from those of another. Within an individual, dreams are different night to night, but there is a content correlation from night to night such that across 20 nights of laboratory-collected dreams, night 19 correlates 0.8 with night 20. Dreams are more predictive night to night than the physiology of sleep [17]. Dreaming is ordered, not chaotic, and certainly not random.

The assumption that the cortical response to PGO stimulation is a relatively passive one underpins the conviction that the form of the dream will be determined by the physiological stimulus and the dream will be isomorphic with the determining physiology—PGO waves. Pivik [18] has concluded from a review of the psychophysiological studies of dreaming that there is a "general absence of robust psychophysiological relationships between tonic levels of physiological activity and sleep mentation"; and that "studies. . . were unable to demonstrate a consistent correlation of phasic activity with the qualitative aspects of sleep mentation." Pivik's conclusions contradict any suggestion of isomorphism, a central tenant of the A-S hypothesis. Furthermore, as the sensorimotor cortices are not activated during dreaming and the association areas are activated, the images experienced during dreaming are constructed each time and are not the result of perception–thus decreasing the likelihood of an isomorphic relationship between the physiological and psychological aspects of dreaming.

Reiser [19] along with Hobson [7] are of the opinion "that emotion is a prominent part of the dream experience and that it plays a role in generating and shaping both the process and the content of dreaming." Dream content studies [20–22] have not found emotions to be an inevitable part of the dream experience. Hall and Van de Castle [20] found emotions in at most 56% of spontaneously reported dreams of men and 84% of the dreams of women. Strauch and Meier [21] noted that emotion was present in about half of laboratory-reported dream experiences. Kramer and Brik [22] found emotions reported in at most 37% of the laboratory-collected dreams of men. Kramer, in his selective affective theory of dream function [23], suggests that it is the change in emotion from pre-sleep to post-sleep that is related to dream content.

In summary, the specific critiques of the A-S hypothesis of dreaming [6] include: the recognition of the limited and arbitrary nature of the definition of dreaming; doubt cast on the idea that dementia is an appropriate model for dreaming; the evidence that REM sleep and the dreaming experience are not coextensive; the work showing that dreaming is not isomorphic with REM sleep; the observations that emotion in the dream may not be the shaper of dream experience; and that the narrative nature of the dream is not explained.

The anatomical cortical theory of Solms [8] is offered as a top-down theory. It, too, is not able to predict the content of the dream from the pattern of neuronal activation. It may be helpful to review aspects of the theory to see how it compares to the A-S hypothesis and in what way it may lend anatomic support to Freudian dream theory [5]. Reiser [19], however, has advised that we not look to the neurosciences to confirm or refute Freud. In his clinicoanatomical approach, Solms [8]:

1. Places the initiation of the dreaming process in a subcortical area related to goal-seeking behavior of an appetitive sort and not in the pons as in A-S theory (this suggests the Freudian wish fulfilling motive force).
2. Calls attention to the blocking of access to the sensorimotor cortex by anterior limbic structures leading to a backward (regressive) movement in dream formation (suggesting both the Freudian censor in the block and topographical regression).
3. Points out that it is the visual association areas that are activated during dreaming, not the primary visual cortex, and therefore images are constructed from memory and are not perceptions, making the isomorphism postulated by A-S theory less likely.
4. Recognizes that both spatial and symbolic activities are involved in dreaming, while A-S theory incorporates but does not address this issue.

5. Notes the loss of dreaming occurring in those who complain of sleep problems (suggesting the possible sleep protective function of dreaming).

6. Observes that dreaming is actively constructed and not passively elaborated (suggesting the Freudian dream work).

Hobson, Pace-Schott, and Stickgold [6] provide a telling and detailed critique of the anatomical cortical theory [8]. "They point out the limitations of lesion studies, for example, the recovery across time of the lost function raises questions about the role of the damaged brain area as essential to the transiently lost function. They wonder if REM awakening studies are the basis for the claim of lost dreaming (which they are not), as spontaneous memory of dreaming would be an inadequate test. They question the support Solms seeks from the leucotomy literature for the role of the medio-basal prefrontal area in dream initiation. Not all leucotomy patients, for example, reported the loss of dreaming. The surgery could well have interfered with the recovery of intrapsychic experiences. The surgery could well have destroyed connections to subcortical limbic structures as well as those from [the] ventral tegmental area. They see the role of dopamine in dreaming as problematic. They point out that dopamine has been reported as both inhibiting dreaming and enhancing it." The cortical theory is a reductive theory that also poses biology as explaining the psychological experiences of dreaming, but less so than A-S theory. It does not address, anymore than the A-S theory did, the narrative aspects of dreaming.

What we want to know about dreaming begins with our desire to know the content of the dream experience. Dream content is the base for our search for what the dream experience means, what it is made of, how it is constructed, and what it accomplishes. The functional anatomical explanations or the descriptions of the secretions from various cells of neurotransmitters or neuromodulators contribute little, if anything, to answer our questions about the content, meaning, construction, or function of the dream experience.

As McGuinn [24] has so elegantly pointed out, there are no transduction rules to go from the discharge of neurons in the central nervous system or from the secretions of neurons to the concomitant mental states. He is so pessimistic as to doubt that we have the intellectual tools to develop such a system. The biological dream theories cannot provide us with the content of dreams, the meaning of dreams, the construction of dreams, or the function of dreaming. They cannot address the differences in the content of the dream experience among individuals or within an individual night to night. These theories do not address the semantics or pragmatics of dreaming; they limit themselves, at best, to the syntax of dreaming.

CONCLUDING REMARKS

The mind/brain problem has not been resolved by the biological approaches to dreaming. It is not that biology is uninteresting, it simply does not answer the questions that are asked. Biology does not address the meaning of behavior or the goal-directed nature of behavior, nor does it explain the nature of experience (Qualia). The problem of consciousness is not illuminated. The explanatory gap between biology (body) and consciousness (mind) remains.

REFERENCES

1. Aserinsky E, Kleitman N. Regularly occurring periods of eye motility and concomitant phenomena, during sleep. *Science* **118**:273–274(1953).

2. Aserinsky E, Kleitman N. Two types of ocular motility occurring in sleep. *J Appl Physiol* **8**:11–18(1955).

3. Dement W, Kleitman N. Cyclic variations in EEG during sleep and their relationship to eye movements, bodily motility and dreaming. *Electroencephalogr Clin Neurophysiol* **9**:673–690(1957).

4. Hodes R, Dement W. Depression of electrically induced reflexes ("H-reflexes") in man during low voltage EEG "sleep." *Electroencephalogr Clin Neurophysiol* **17**:617–629(1964).

5. Freud S. *The Interpretation of Dreams*. Basic Books, New York, 1900/1955, p xviii.

6. Hobson J, Pace-Schott E, Stickgold R. Dreaming and the brain: toward a cognitive neuroscience of conscious states. *Behav Brain Sci* **23**:793–842(2000).

7. Hobson J. The new neuropsychology of sleep: implications for psychoanalysis with commentaries by M. Solms, A. Braun, M. Reiser and reply by J. Hobson and E. Pace-Schott. *Neuro-Psychoanalysis* **1**:157–225(1999).

8. Solms M. *The Neuropsychology of Dreams: A Clinico-anatomical Study*. Lawrence Erlbaum Associates, Mahway, NJ, 1997.

9. Nielsen T. Mentation in REM and non-REM sleep: a review and possible reconciliation of two models. *Behav Brain Sci* **23**:851–856(2000).

10. Taub J, Kramer M, Arand D, Jacobs G. Nightmare dreams and nightmare confabulations. *Comprehensive Psychiatry* **19**:285–295(1978).

11. Solms M. Dreaming and REM sleep are controlled by different mechanisms. *Behav Brain Sci* **23**:843–850(2000).

12. Foulkes D, Fleisher S. Mental activity in relaxed wakefulness. *J Abnorm Psychol* **84**:66–75(1975).

13. Kripke D, Sonnenschein D. A biologic rhythm in waking fantasy. In: Pope K, Singer J (Eds), *The Stream of Consciousness*. Plenum Press, New York, 1978, pp 321–332.

14. Snyder F. The phenomenology of dreaming. In: Madow L, Stone L (Eds), *The Psychodynamic Implications of the*

Physiological Studies on Dreams. Charles C Thomas, Springfield, IL, 1967.

15. Heynick F. *Language and Its Disturbances in Dreams.* Wiley, Hoboken, NJ, 1993.

16. Kramer M. The psychology of the dream: art or science? *Psychiatr J Univ Ottawa* **6**:87–100(1982).

17. Kramer M, Roth T. The stability and variability of dreaming. *Sleep* **1**:319–325(1979).

18. Pivik R. Psychophysiology of dreams. In: Kryger M, Roth T, Dement W (Eds), *Principles and Practice of Sleep Medicine.* Saunders, Philadelphia, 2000.

19. Reiser M. The dream in contemporary psychiatry. *Am J Psychiatry* **158**:351–359(2001)..

20. Hall C, Van de Castle R. *The Content Analysis of Dreams.* Appleton-Century-Crofts, New York, 1966.

21. Strauch I, Meier B. *In Search of Dreams*: *Results of Experimental Dream Research.* State University of New York Press, Albany, 1996.

22. Kramer M, Brik I. Affective processing across the night by dreams. *Sleep Suppl* **25**: A180–181(2002).

23. Kramer M. The selective mood regulatory theory of dreaming: an update and revision. In: Moffitt A, Kramer M, Hoffmann R (Eds), *The Functions of Dreaming,* State University of New York Press, Albany, 1993, pp 139–196.

24. McGuinn C. *The Mysterious Flame.* Basic Books, New York, 1999.

ADDITIONAL READING

Chalmers D. Facing up to the problem of consciousness. *J Consciousness Studies* **2**:200–219(1995).

Edelman G. *Wider than the Sky.* Yale University Press, New Haven, CT, 2004.

Foulkes D. *Children's Dreaming and the Development of Consciousness.* Harvard University Press, Cambridge, MA, 1999.

Hobson A. *Dreaming.* Oxford University Press, New York, 2002.

Kendler K. A psychiatric dialogue on the mind–body problem. *Am J Psychiatry* **158**:989–1000(2001).

Kramer M. Sigmund Freud's "The Interpretation of Dreams": the initial response. *Dreaming* **4**:47–52(1994).

Levine L. Materialism and qualia: the explanatory gap. *Pacific Philos Quarterly* **64**:354–361(1983).

Metzinger T. (Ed). *Conscious Experience.* Allen Press, Lawrence, KS, 1995.

Searle J. *Consciousness and Language.* Cambridge University Press, New York, 2002.

Panksepp J. *Affective Neuroscience.* Oxford University Press, New York, 1998.

6

PSYCHOLOGY OF DREAMING

Maimonides Medical Center, Brooklyn, New York

INTRODUCTION

The traditional interest in the dream experience had been to foretell the future. It was believed that the will of God was revealed to prophets in dreams. Some 5% of people in a scientific population survey reported that they have the conviction that dreams predict the future. The most popular dream books for many years were those that linked images that appeared in dreams to various numerical values and these values were used by people to select what number to pick in "playing the numbers," the illegal forerunner to the modern-day lottery. Psychotherapists, influenced by Freud, have used the reports of the dream experience to help them understand the dreamer, their patient. Unfortunately, there had been very little scientific basis to support the conviction that therapists had about the revelatory potential in dream analysis.

The major reason for studying dreaming in the modern context is to understand the functioning of the mind, to understand consciousness. The dream may be seen as the mind in pure culture. Dreaming may present us with a consciousness that is the least influenced by external events [1, 2]. A secondary but important reason for studying dreaming is to see if such a study will unlock the mysteries of psychosis as had been suggested by Jung, Freud [3], and Hughlings Jackson. Mental illness has been and remains a major public health problem.

There has been great reluctance to studying the dream experience in a scientific manner. Neuroscientists have been interested in the study of the functions of the brain and have seen the dream experience as epiphenomenal. The dream has been and continues to be seen by some as

a degradation or waste product. It is for them the foam on the beer of sleep. The dream experience is a first-person experience and until quite recently only third-person experiences were considered as amenable to scientific study. First-person or subjective experiences depend on the introspective ability of the subject as well as the willingness of the subject to be an honest reporter. It was thought that the difficulty in collecting reasonable size dream samples coupled with the great variability assumed possible in dreaming made scientific studies of dreaming impossible. The view that dreams could not be studied scientifically was enhanced by the belief that the reliability of dream measurement had been unexplored and that the validity of dream measurement was open to serious question. The discovery in 1953 of a physiological correlate of dreaming, rapid eye movement (REM) sleep, which occurred several times a night, made possible the collection of large dream samples and experimental paradigms became feasible. Dream reports, it was shown, could be reliably scored and dreams were found to be more predictive night to night than sleep stage scoring.

The dream as a psychological experience, reflected in the dream report, should not be confused with REM sleep, a physiological experience. The reductive algorithm to move from dreaming consciousness to brain physiology does not currently exist and may never exist. With apologies to Emerson, a foolish biological "reductionism" may be the hobgoblin of little minds. Dreaming to be the object of scientific scrutiny must be regular, that is, orderly and not random, signal not noise. The remainder of this discussion will explore studies directed at establishing the regularity of dreaming.

Sleep: A Comprehensive Handbook, Edited by T. Lee-Chiong.
Copyright © 2006 John Wiley & Sons, Inc.

DO DREAMS EXIST?

The question has been raised whether the dream experience exists as an event in time or as a confabulation either of the awakening process or of waking consciousness [1, 4]. Freud mentioned the work of Goblot who argued that the dream was elaborated during the waking process. Malcolm, an English philosopher of mind, argued, in his seminal book *Dreaming*, that on logical grounds dreaming could not be a concomitant of the sleeping state and that it is probably an event related to the waking process. In an unpublished essay entitled "The Goblot Phenomenon," the great empiricist of dream studies, the late Calvin Hall, also suggested that dreaming was an experience concomitant with the waking process.

There are experimental studies that directly and indirectly address the issue of whether the dream is an experience extended in time during sleep. Incorporation of tactile and aural stimuli presented early or late during a REM period appears appropriately in the subsequent dream reports from these awakenings. Familiar and unfamiliar names presented during a REM period are differentially incorporated into the subsequent dream report. The familiar names are more likely to be incorporated. The content of the dream report (e.g., intensity) shows a developmental course across a REM period. The direction of eye movements during a REM period has been shown in some but not all studies to be appropriate to the action described in the dream report from that REM period. Subjects awakened from a REM period are more likely to tell a "dream-like" story in response to a stimulus than when awakened from non-rapid eye movement (NREM) sleep. Subjects asked to make up a nightmare, to report a nightmare they have experienced, and to tell a dream they have had report three different experiences, with the made up nightmare being the most intense, shortest, and best organized of the three. Dreams are experiences that are concomitant with sleep and are not constructed during the awakening process.

COLLECTION AND MEASUREMENT OF DREAM REPORTS

Dream recall is a highly variable process, and to be able to study dreaming, an awareness of those factors that may influence the ability to recall the dream experience is necessary [1, 5, 6]. The *place* where the dream is experienced and reported (e.g., at home or in a laboratory) may well influence the nature of the dream experience and whether the experience is recalled to be reported. The *method of awakening* will influence the recall of the dream. The faster the awakening, the more likely a dream will be recalled. The *interpersonal situation* between the dream reporter and the dream collector will influence how much and

what will be reported. The gender similarity or difference of the pair and the status relationship of the collector and dreamer will influence recall and content. The *style of collection* will influence recall and content. For example, asking whether emotion occurred in the dream will increase the number of reports of emotion obtained. The *stage of sleep* from which the subject awakens will affect dream recall, with greater recall if the awakening is from REM sleep. The *method of recording* the dream experience will affect the recall of dreams. The dreamer writing out the dream gives shorter and better organized dream reports than telling them into a recorder. The ***type of subject*** who is reporting the dream experience will affect the frequency and nature of what is recalled. The subject who recalls dreams, who is more attuned to inner processes, and who is more verbal will provide more dream reports with greater detail than a less introspective, less verbal subject.

There are characteristics of the dream experience itself which are determining of which of the several dreams of the night are most likely to be remembered the next day. There is a *recency* effect so that it is the last dream that is most likely to be remembered. There is a ***primacy*** effect as well so that the first dream experience is more likely to be recalled than the second dream of the night. The longer the dream experience, a ***length effect***, the more likely the recall. And, the more ***dramatic*** the dream experience, the more likely it will be recalled. These characteristics have been described as the saliency theory of dream recall.

There are other factors such as gender and age that are known to influence dream recall. Women recall more dreams than men and the older one gets the less dream recall is reported. Brain damage independent of age decreases dream recall. Psychological variables such as ego strength, anxiety, repression, and field independence in most people do not co-vary with dream recall. The meaning of the dream in the context of the interpersonal collection procedure can influence the dreamer's ability to recall and his/her willingness to report the dream experience.

Serious question has been raised by those interested in the spiritual and creative aspects of dreaming that measuring the dream destroys its very essence. Examining the contribution that the scientific study of dreaming has made to our knowledge of dreaming belies such a negative assessment. There are a number of problems associated with efforts to measure the dream experience as reflected in the dream report. First, we are trying to capture an event experienced in one ***state***, sleep, reported in another state, wakefulness. Our earlier discussion offering evidence that dreams exist and are not products of the awakening process supports the idea that the dream report is an adequate representation of the dream experience. Second, the dream experience is presented as a verbal report. The *verbal abilities* of the subject may account for dream differences between subjects and a waking verbal control should be considered. Third, what in the dream report is to be counted as the *scoreable report*? A set of rules to deal with asides and redundancies is necessary, as would be the case in dealing with any effort to quantify verbal behavior. Fourth, the question of *word length* and how to deal with dream reports of differing lengths arises. A word length correction may be appropriate, but it has been suggested that some things may require more words to describe them than others. Fifth, there are *scoring systems* available for the quantification of dream reports. The reliability and validity of these systems have been established. Clinical concepts of interest can be found or created by combining aspects of an extant system, analogous to using Nissl staining for cell bodies and Golgi staining for dendrites in studying the nervous system. A series of dream reports were scored for hostility with a number of different systems. There was considerable overlap among the various systems, but some 75% of the variance was unexplained. The systems apparently had different conceptualizations of hostility and perhaps some of the prized ineffability of dreaming was in the unexplained variance.

The content of dream reports is reliable as shown by comparing home dream reports to laboratory-collected reports both scored with the same content system by two different investigators. Dream content reports are remarkably stable across time comparing home-reported dreams of college students some three decades apart and from two different institutions. Yet, dream reports of the same person are variable enough from night to night such that the dreams of one night can be distinguished from those of another night.

NORMATIVE, UNIVERSAL, REPETITIVE DREAMS AND NIGHTMARES

The quantitative work with dreams [1, 6–10] has provided a picture of the normative dream. The *average dream* has 2.6 characters, 4.8 activities, 1.4 social interactions, and 1.3 settings per dream. Other features of dreams which occur in some but not all dreams include negatives 0.77, emotions 0.70, misfortunes 0.41, failure 0.13, success 0.12, and good fortune 0.06 per dream. Aspects of dreams which do not occur in all dreams are less likely to be central to the dream experience.

The major demographic organizer of dreams is the gender of the dreamer, with age, race, marital status, and social class relatively less important. Males have fewer characters in their dreams than women, but they are mostly men and are less likely to be known to the dreamer. Women have more people in their dreams who are equally likely to be men or women and are more likely to be known to the dreamer. Men have more physical activity and aggressive social interactions than women. Women have more friendly social interactions than men but fewer sexual interactions. Men have more outdoor settings in their dream reports and women more indoor settings.

There are **typical or universal dreams** that occur to most people and in which the content is essentially the same across dreamers. Freud listed some 23 different typical dreams such as feeling embarrassed being naked, the death of a loved one, swimming, or being in a fire. He attributed an identical meaning to some (e.g., being naked and loss of a loved one) and a dreamer-specific meaning to others (e.g., swimming and being in a fire). Harris [9] has pointed out that universal dreams such as falling reflect feelings of insecurity, while dreams of being pursued reflect a feeling of being attacked and may have a differential gender frequency. Ward and co-workers [7] related universal dreams to different aspects of development. It has been observed that the frequency of the various universal dreams is quite similar across cultures.

The **repetitive dream** has been the object of some interest. It has generally been seen as the response to a similar set of troubling emotional circumstances, usually starting in childhood, involving only the dreamer, and generally having a negative tone. Domhoff [8] has suggested that dreams are a metaphorical attempt at problem resolving and that repetitive dreams may be an attempt to deal with unresolved emotional concerns of the dreamer.

There has long been a fascination with the nightmare experience. Nightmares are a universal experience as almost everyone has been awakened from a disturbing dream feeling anxious with some sense of difficulty in breathing. Bad dreams, which have a negative emotional tone but do not awaken the dreamer, are even more common. The actual nightmare experience is not nearly as intense as what people imagine a nightmare to be but is clearly distinguishable from an ordinary dream. Nightmares cannot be defined based on their content or explained by chronic nightmare sufferers being better dream recallers. There is the suggestion that the discomfort experienced by those experiencing nightmares is a reaction to the experience rather than a concomitant of the nightmare. Not all people who experience nightmares are troubled by them. Those individuals who are troubled by their nightmares and who describe their nightmares as reactive to daytime distress are more likely to seek help for their nightmares. Techniques that encourage rescripting of the nightmare experience have been shown to be helpful.

DREAMS AND PSYCHOLOGICAL DIFFERENCES

Dreams, to be ordered (i.e., nonrandom and meaningful), should reflect psychological differences in circumstances where we know such differences exist, and indeed that is what has been found. It has been shown at the *group level* that there are demonstrable dream content differences associated with demographic variables (e.g., gender, age, race, marital status, and social class) [1, 11, 12]. The major psychiatric illnesses show clear psychological differences from normals as the dreams of people with schizophrenia and depression have different dream content from each other and from normals. The schizophrenic has strangers as its common character type while the depressed have family members and normals have friends.

Psychological differences among people at the group level is also found at the *individual level*, and these differences are also found in their dreams. The dreams of individuals can be distinguished one from another. People are to some degree different from day to day and the dreams of one night can be distinguished from those of another night from the same person. The regularity in dreaming is suggested by dreams becoming increasingly predictive from night to night such that in a series of 20 consecutive nights of dream collection, the 19th night's dreams predict the 20th night's dreams at a level of 0.8.

Two *patterns of dream relationships* across the multiple dreams of a night have been described. One pattern is of a progressive-sequential (P-S) nature, in which metaphorically a problem is posed, worked on, and to some degree resolved. An illustration of a P-S dream series of a young single woman is one in which she first dreams she is a child clinging to stay in the hospital, then of being rejected by a colleague's wife; in the third dream she has a partner and they are victorious in a game; in the next dream she decides she doesn't need the doctor despite evidence to the contrary, and in the last dream she turns on the doctor. The other pattern is of a traumatic-repetitive (T-R) nature, in which a concern or problem is expressed metaphorically with different imagery in each dream of the night. An illustration of a T-R dream series from the same young woman is one in which in the first dream someone is lost and trying to call home; in the second dream she is at an orphanage and because there was no room in the car she had to go home with someone else; and in the last dream she is being left by her mother in a laboratory where there might not be enough room. The dream reports across a night's dreaming support the possibility of content change in as simple a form as the number of words in each dream report. In a series of dreams it was found that the first dream was shorter than the second and the third was shorter than the fourth, but there was no length difference between the second and the third. There were content differences as well across the night with the word content held constant. There were three content differences between the first and second dream of the night and five between the second and the third dream but none between the third and the fourth.

There is a pattern of development or *organization even within a dream*. The content of the dream shows an increase in intensity for the first 10 min and then shows a plateau for 10–20 min and resumes again. A timing pattern similar to the ebb and flow of the eye movements across a REM period is shown by the content change across the dream.

Content differences in dreams are found where we expect to find psychological differences between people. Demographic variables and psychiatric illnesses show dream content differences. Dreams of individuals are different one from the other as are dreams of different nights of the same individual. There are different patterns of dreams across the night and there is a developmental pattern of dream content within the dream itself. The dream appears to be a highly ordered experience in reflecting psychological differences.

INFLUENCES ON DREAM CONTENT

The dream experience has been described as responsive to a number of manipulations [1, 13, 14]. Tart has summarized some of these manipulations as including direct suggestion, stimuli introduced during the dream experience, hypnosis, conditioning, and REM deprivation and rebound.

The more intensely emotional experiences of the previous day are more likely to be represented in the dreams of the following night. The beginning and ending of an experience, such as sleeping for 20 consecutive nights in a sleep laboratory for dream collection, can be correctly identified in dreams while the middle of the experience cannot. References to sleeping in the unusual situation of a sleep laboratory continue to be represented at the same frequency (i.e., without adaptation) across a 20 night series. The vagaries of the interpersonal situation between the dreamer and the dream collector are reflected in the reported dreams. Dreams critical of the morning dream collector, a psychiatrist, are reported at night to the technician but not in the morning. And a very macho man does not report scary dreams with homosexual implications to the psychiatrist in the morning but does report dreams that capture his sexual relationship to women and his fighting with men. Male volunteers in the sleep laboratory reported dreams suggestive of their fears of being exploited by the male experimenters while female subjects had dreams that suggested a fear of being raped. A female subject, a nurse, dreamed about problems with an intravenous (IV) infusion that had continuing trouble when there was a female dream collector, but on nights when a male physician was the dream collector the IV ran without a problem. Patients had dreams about their therapists while sleeping in the laboratory, while their therapists who slept in another laboratory on the same night dreamed about the conference the next day where their work with the patient was to be discussed. The conference leaders dreamed about the research group and the research group probably about the research funding agency. We dream about what concerns us. Depressed patients who were given medications that alter their depressed mood had concomitant changes in their dreams showing increased motility, intimacy, and sexuality as the depression lifted. Playing the names of familiar people and unfamiliar people during the dream experience leads to a greater incorporation of the familiar person into the dream. Emotionally significant experiences from the waking state are what become represented in dreams.

DREAMS AND WAKING THOUGHT

Dreams are organized not random events as they reflect psychological differences among people and are influenced by significant emotional events [1, 11, 15]. The next question to ask is about the relationship between dreaming and waking thought.

The themes of pre- and post-sleep verbalizations have been shown to be more closely related to each other than the multiple REM dream reports of a night are to each other. Waking thought is more thematically constrained than the more fluid dreaming thought.

The themes of pre-sleep thought are more closely related to the themes of subsequent dream reports than the dream reports are to the themes of post-sleep verbalizations. Dreams appear to be more reactive to prior wakeful thought than proactive to subsequent wakeful thought.

Dreamers show both a trait and state relationship between dream reports and waking fantasy. Subjects had their REM dream reports and Thematic Apperception Test (TAT) stories scored with ten of the need press variables described by Murray. The rank order intensity scores of the ten variables were significantly related. In another study, spontaneous verbal samples were collected immediately before and after REM dream collection in the sleep laboratory. The contents of the verbal samples and dream reports were found to be significantly related. The TAT study demonstrates a trait and the verbal sample study a state relationship between dreams and waking thought.

The changes in feeling states (mood) across the night are related to the content of the intervening dreams. Mood adjective checklists done before and after REM dream collection show a relationship between who and what was dreamed about and the change in aspects of mood from night to morning. The strongest relationship is between the characters in dream reports and the change in the unhappy aspect of mood. The change in sleepiness across the night is related to the amount of NREM sleep one obtains.

Some aspects of pre-sleep mood correlate with some dream report contents and some dream report contents correlate with some aspects of post-sleep mood. However, the same aspect of pre- and post-sleep mood does not correlate with the intervening dream content. The changes in mood across the night are not a simple pass through the night's dreams.

Inferences drawn about patients from waking observations and dream reports are similar. Patients had their histories written down by their therapists and had a psychological test report prepared based on their TAT and Rorschach Tests. The patients also had five REM dreams collected. Independent judges read the history, the dream reports, and the test results and based on each data source did a 100 item Q sort describing the patient. The therapist did a Q sort as well. These Q sorts were significantly correlated among the judges. The dream led to similar inferences about the patient as one would obtain from the history and psychological testing. Yet as the amount of overlap—explained variance—was low, there are areas where different inferences could be drawn from the dream than from the other data sources.

Successful from unsuccessful treatment in long-term psychodynamic psychotherapy or psychoanalysis has been distinguished by comparing the change from the first to the last dream report in therapy. The therapist's ranking of the degree of improvement among a series of similar long-term psychotherapy treatments has been significantly correlated to a judge's ranking based on comparing the degree of change between the first and the last dream report in therapy. The degree to which patients change in the dynamic psychotherapies is reflected in changes in their dream reports.

Waking and sleeping fantasy are related. However, waking thought is more thematically constrained than the more fluid dreaming thought. Dream themes appear to be more related to pre-sleep thought than to post-sleep thought. Dreamers show both a trait and state relationship between dream reports and waking fantasy. The changes in feeling states (mood) across the night are related to the content of the intervening dreams. But the changes in mood across the night are not a simple pass through the night's dreams. Inferences drawn about patients from waking observations and dream reports are similar but not identical. Changes in dream content parallel the differential outcomes reported in the dynamic psychotherapies.

DREAM MEANING

Does dreaming have the necessary structure to make a search for the meaning of a dream reasonable? If the content of dreaming is random, more like noise than a signal, then a search for the meaning of a dream would be futile.

However, the evidence does indicate that the dream has the necessary order to support meaning [1, 3, 16–18]. Dreaming has to be related to the ongoing and changing waking concerns of the dreamer if it is to have meaning.

Dreams have order, as they are organized within a REM period, showing an increase in the intensity of content and across the REM periods of the night, and as there are differences in content between REM periods. The dreams of one night can be distinguished from those of another night of the same dreamer, but the dream reports of a dreamer are also related across nights. The dream reports of one dreamer can be distinguished from those of another dreamer. There are dream content differences related to demographic variables such as gender, age, race, marital status, education, and social class. Gender is the major demographic organizer of dream content with age and race a distant second. The dreams of schizophrenics are different from those of the depressed and both are different from normals.

Dreams have been shown to have a connection to the waking emotional concerns of the dreamer. Similar clinical inferences can be drawn from dreams as from the clinical history and psychological test reports of patients. Dream contents have both a trait and state connection to the dreamer as shown by TAT stories and verbal sample analysis. Mood change across the night is related to who and what is dreamed about and mood change in depressives treated with antidepressants has concomitant dream content change. There is a thematic connection from pre-sleep mentation to dream content to post-sleep mentation, with the dream content appearing to be more reactive than proactive. Dream content changes reflect the changes in long-term psychotherapy.

The dream to be understood needs to be approached as a figurative rather than literal statement. A psychological meaning system needs to be applied to a dream to establish its meaning. There are a number of such systems of psychological meaning such as provided by Freud, Jung, Adler, Gendlin, Delaney, Hall, Kramer, the Existentialists, and the Gestaltists.

Kramer [16] has suggested that the dream be parsed into phrases and that one begins at the beginning to use one's associations to each phrase, keeping in mind that the dream is to be understood figuratively, that meaning is contextual, and that the dream is its own context. The associations to each succeeding phrase serve to select from and narrow down the possible meanings. In the end, the understanding that requires the fewest assumptions and accounts for most of the content is accepted as the most likely meaning, recognizing that other meanings are possible.

Dreams have the essential structure, orderliness and connection to waking fantasy so that a search for meaning appears justified. Approaches to establish meaning have been suggested by many both with and without the co-operation of the dreamer.

THE FUNCTIONS OF DREAMING

An interest in the psychological function of dreaming remains to the present day [1, 3, 19, 20]. Freud saw dreaming as the disguised attempted fulfillment of an infantile wish in the service of maintaining the continuity of sleep. He recognized that a theory of dreaming did not require that it attribute a function to dreaming, but given his teleological predilections he preferred functional theories. The function of dreaming can refer either to how the dream is constructed or to what dreaming achieves, to the consequences of the dream experience. The consequence of dreaming is often related to how it is constructed. Theories of dreaming generally are either assimilative or accommodative. The assimilative theories are more likely to account for the totality of dreaming and function outside conscious awareness (i.e., without recall and secondary reworking). These theories generally have the dream achieve some corrective or reductive goal. An accommodative theory has the dreamer recalling the dream experience either alone or in therapy and, following some exploration, an understanding develops that serves to alter the dreamer. Dream theories may encompass both an assimilative and an accommodative function.

Many functions have been suggested for dreaming. Freud saw the dream as the protector of sleep; Jung saw it as a compensation for conscious exaggeration; Adler as an affective generator to support the life style; French as solving emotional problems; Giora saw the dream as a form of emotional thinking; Breger saw it as an information processing process; and Kramer as a selective affective modulator.

SUMMARY

The dream [1] is a legitimate object for scientific study. It exists as an experience in time, which is adequately captured as a verbal report. The dream experience can be influenced by a number of factors and can be usefully quantified. The dream is orderly and organized, signal not noise, as it reflects meaningful psychological differences and responds to and reflects emotionally laden influences.

REFERENCES

1. Kramer M. The psychology of the dream: art or science? *Psychiatri J Univ Ottawa* **7**:87–100(1982).
2. Foulkes D. *The Grammar of Dreams*. Basic Books, New York, 1978.
3. Freud S. *The Interpretation of Dreams*. Basic Books, New York, 1900/1955.
4. Malcolm N. *Dreaming*. Routledge and Kegan Paul, London, 1975.

5. Winget C, Kramer M. *Dimensions of the Dream.* University of Florida Press, Gainesville, 1979.

6. Hall C, van de Castle R. *The Content Analysis of Dreams.* Appleton, New York, 1966.

7. Ward C, Beck A, Rascoe E. Typical dreams: incidence among psychiatric patients. *Arch Gen Psychiatry* **5**:606–615(1961).

8. Domhoff G. The repetition of dreams and dream elements: a possible clue to a function of dreams. In: Moffitt A, Kramer M, Hoffman R (Eds), *The Functions of Dreaming.* State University of New York Press, Albany, 1993.

9. Harris I. The dream of the object endangered. *Psychiatry* **20**:15–62(1957).

10. Levin R, Fireman G. Phenomenal qualities of nightmare experience in a prospective study of college students. *Dreaming* **12**:109–120(2002).

11. Glucksman M, Kramer M. Using dreams to assess clinical change during treatment. *J Am Acad Psychoanal Dynamic Psychiatry* (to be published).

12. Winget C, Kramer M, Whitman R. Dreams and demography. *Can Psychiatr Assoc J* **17**:203–208(1972).

13. Tart C. From spontaneous event to lucidity: a review of attempts to consciously control nocturnal dreaming. In: Wolman B (Ed), *Handbook of Dreams.* Van Nostrand Reinhold, New York, 1979.

14. Fox R, Kramer M, Baldridge B, Whitman R, Ornstein P. The experimenter variable in dream research. *Dis Nerv Syst* **29**:698–701(1968).

15. Kramer M. The selective mood regulatory function of dreaming: an update and revision. In: Moffitt A, Kramer M, Hoffmann R (Eds), *The Functions of Dreaming.* State University of New York Press, Albany, 1993.

16. Kramer M. Dream translation: an approach to understanding dreams. In: Delaney G (Ed), *New Directions in Dream Interpretation.* State University of New York Press, Albany, 1993.

17. Gendlin E. *Let Your Body Interpret Your Dreams.* Chiron Publication, Basel, 1986.

18. Shulman B. An adlerian view. In: Kramer M (Ed), *Dream Psychology and the New Biology of Dreaming.* Charles C. Thomas, Springfield, IL, 1969.

19. Moffatt A, Kramer M, Hoffmann R. (Eds). *The Functions of Dreaming.* State University of New York Press, Albany, 1993.

20. Piaget J. *Play, Dreams and Imitation in Childhood.* Norton, New York, 1962.

ADDITIONAL READING

Blechner M. *The Dream Frontier.* The Analytic Press, Hillsdale, NJ, 2001

Cartwright R, Lamberg L. *Crisis Dreaming.* HarperCollins, New York, 1992.

Domhoff G. *Finding Meaning in Dreams.* Plenum Press, New York, 1996.

Foulkes D. *Dreaming: A Cognitive-Psychological Analysis.* Lawrence Erlbaum Associates, Hillsdale, NJ, 1985.

Foulkes D. *Children's Dreaming and the Development of Consciousness.* Harvard University Press, Cambridge, MA, 1999.

Hartmann E. *Dreams and Nightmares.* Plenum Press, New York, 1998.

Lippman P. *Nocturnes.* The Analytic Press, Hillsdale, NJ, 2000.

Moffatt A. Kramer M, Hoffman R (Eds). *The Functions of Dreaming.* State University of New York Press, Albany, 1993.

Ullman M, Limmer C. *The Variety of Dream Experience*, 2nd ed. State University of New York Press, Albany, 1999.

7

THE FUNCTION OF SLEEP

Marcos G. Frank

University of Pennsylvania, School of Medicine, Philadelphia, Pennsylvania

INTRODUCTION

In the last fifty years scientists have made extraordinary progress characterizing the neurobiology, regulation, and genetic underpinnings of mammalian sleep. Mechanisms generating rapid eye movement (REM) and non-rapid eye movement (NREM) sleep have been identified, genes important in the timing and intensity of sleep have been isolated, and much is known about the detrimental effects of abnormal sleep on human performance [1]. It is thus a bit ironic and even distressing that the most fundamental question about sleep is still unanswered. *Why we sleep* remains one of nature's greatest mysteries. Several representative theories about sleep function are discussed in this chapter. As will become evident, no one theory holds preeminence; rather, each theory has its merits and faults. An emphasis is therefore placed on theoretical considerations that may improve experimental approaches to understanding sleep function.

FRAMING THE PROBLEM OF SLEEP FUNCTION: LITTLE THEORIES VERSUS BIG THEORIES

Even the casual reader will be impressed by the abundance of competing theories regarding sleep function. Sleep has been proposed to be important for somatic anabolic processes, brain cooling, restoration of brain molecules, removal of neurotoxins, and higher-order functions such as brain maturation, species-specific "programming," and memory [2, 3]. These theories are "little" because they explain only a portion of sleep behavior. But are these little theories of sleep function merely incomplete descriptions of

a deeper, more basic process? Could there still be a "big," unifying theory of sleep function that has greater explanatory power than a collection of little theories? Before reviewing some representative ideas about sleep function, it is useful to consider what would constitute a unifying theory of sleep function. We can start by listing several points that should be addressed by such a theory.

1. *Phylogeny. A unifying theory of sleep function should explain the presence (or absence) of sleep across the animal kingdom.* Animals as diverse as flies to field mice all exhibit sleep-like states that are regulated by homeostatic and circadian mechanisms [1]. In particular, molecular studies have revealed several genes common to drosophila and mammals whose expression is tied to the sleep–wake cycle [4]. This raises the intriguing possibility that the function of sleep is quite ancient. Therefore a unifying theory of sleep function should explain the presence of sleep in all species in which it has been identified. In addition, such a theory should explain how and why sleep evolved in the first place by identifying those aspects of sleep that proved adaptive during the course of evolution.

2. *Ontogeny. A unifying theory of sleep function should account for the dramatic changes in sleep during development.* REM sleep amounts are greatly elevated in the first weeks of life in most mammals and rapidly decline concurrent with postnatal development. NREM sleep amounts are initially low, then rapidly rise to adult values during the same time [5]. In rodents, REM sleep does not "rebound" following total or selective REM sleep deprivation until the third to fourth postnatal week. NREM sleep, however, shows rebounds in amounts and later in EEG slow-wave

activity quite early in development [5]. A unifying theory of sleep function should account for these dramatic changes in sleep expression and should address whether the presumed function is different in developing animals, or preserved in some fashion across the life span.

3. *Regulation. A unifying theory of sleep function should relate the homeostatic regulation of sleep to the proposed function.* A sleep homeostat determines the amount and intensity of sleep based on prior sleep–wake history [1]. As discussed by Benington, the homeostatic regulation of sleep should be closely associated with its function [6]. According to this argument, for any sleep-dependent process there should be a feedback mechanism that communicates the state of progress of that process to the homeostat *if that process is central to sleep expression.* A unifying theory of sleep function must therefore indicate how that function communicates with the sleep homeostat.

4. *The Primacy of Sleep. A unifying theory of sleep function should explain why this function only (or preferentially) occurs during sleep.* The first requirement of any theory of sleep function is that it explain why sleep, and no other state (e.g., quiet wakefulness) or change in other systems (e.g., increased enzymatic activity), is required for that function. In particular, such a theory should explain why the specific somatic or nervous system changes unique to sleep, such as loss of consciousness, are required or conducive for that function.

In addition, when considering a theory of sleep function it is useful to ask:

1. *Is the theory mutually exclusive?* That is, does the theory exclude other theories if true, or does it subsume or is it subsumed by competing theories?
2. *Is the theory supported by mutually reinforcing lines of evidence?* That is, is there a convergence of evidence in support of the theory, and are these lines internally consistent with each other?
3. *Is the proposed function central to sleep expression?* Sleep has numerous effects on physical processes, but not all of these processes are necessarily linked to sleep function.
4. *Does the theory deal with REM and NREM sleep?* A unifying theory should account for both sleep states, and in cases where REM sleep does not exist (e.g., in invertebrates and possibly reptiles) it should explain why.

EVALUATING THEORIES OF SLEEP FUNCTION

No one theory of sleep function has adequately addressed all the points listed in the preceding section. However, we can organize "little" theories in a manner that may reveal larger, unifying principles. We can begin by first addressing whether sleep is primarily for the brain or for the body (somatic versus neural theories). Somatic theories of sleep function propose that sleep facilitates anabolic processes or restores some bodily function worn down by wakefulness. Neural theories, on the other hand, propose that sleep is primarily for the brain and are further subdivided into metabolic and cognitive categories. Metabolic or "housekeeping" theories propose that sleep detoxifies substances that accumulate during wake (e.g., from increased oxidative metabolism or glutamate release) [7], or restores and repairs neural substrates degraded by wakefulness. Cognitive theories propose that sleep serves higher-order functions such as neural development or memory, presumably by promoting synaptic plasticity.

Somatic Theories of Sleep Function

Sleep appears to have beneficial effects on the body and general health. Sleep amounts may influence mortality and morbidity and there are important interactions between sleep and the endocrine and immune systems [8]. Prolonged sleep deprivation (SD) in rodents, drosophila, and possibly humans is fatal—which would support a general "life-sustaining" function for sleep [3, 4, 9]. Such a function would satisfy phylogenetic considerations since the adaptive value is obvious, and evolution might favor the periodic release of anabolic substances during periods of quiescence when energy output is low. Developing infants might also have a greater need for heightened activity in endocrine and immune systems and thus a greater need for sleep. Nevertheless, it is unclear if the primary function of sleep is to facilitate bodily functions. The prolonged SD studies are intriguing, yet death may be caused by abnormalities in hypothalamic regulatory mechanisms that are *secondary* to a general loss of neural function. Interestingly, shorter periods of SD (24–72 h) have negligible effects on autonomic output and only modestly affect organ function, athletic performance, and recovery from exercise [10, 11]. A serious obstacle to a pure somatic theory of sleep function is that it does not explain the extensive neural changes accompanying sleep. It is difficult to imagine why facilitation of anabolic processes or endocrine or immune function would require the loss of consciousness and other peculiarities of NREM and REM sleep. These events could just as easily occur during periods of quiet wake when the animal is less subject to predation.

Neural Metabolic Theories: Detoxification and Regeneration

Metabolic theories are appealing because they relate sleep function to brain processes and the function they propose is intuitively sensible. Who hasn't felt more mentally alert after a good night of sleep? Metabolic theories, however, have little experimental support. There is no convincing evidence that sleep removes a toxic by-product of wakefulness. Nor does there appear to be any gross brain damage in sleep-deprived animals, even in cases where the animals are sleep-deprived to death [3, 12, 13]. Detoxification theories also fail to address the respective roles of NREM and REM sleep. Indeed, it is difficult to imagine how REM sleep could counteract the ill effects of heightened brain metabolism since it is a state characterized by intense cortical activation. Phylogenetic considerations pose an additional problem because contrary to predictions of the neurotoxin theories, metabolic rate (when corrected for body mass) is negatively correlated with sleep time [3]. Considering that the brain is an organ of unusually high metabolism, one would have instead predicted that animals with faster overall metabolism would sleep *more* not less.

There is slightly stronger evidence in support of neurorestorative theories, but it is still not clear what (if anything) sleep restores in the brain. There is no convincing evidence that REM sleep promotes the synthesis of molecules important for neuronal function or structure. NREM sleep is associated with cerebral protein synthesis and sleep in general may upregulate several genes important for neuronal membranes and other structural components [2, 14]. However, the functional consequences of these sleep-related changes in protein and genes are unknown. A recent hypothesis that sleep restores cerebral energy stores is not convincingly supported by the evidence [15–17]. A critical problem with neurorestorative theories is that they do not explain the abundance of sleep during infancy. Even though waking amounts are very low during infancy, neonates sleep much more than adults. This is especially true in precocial species such as the lamb that have fully developed REM and NREM sleep in utero with little or no wake [5]. Nor is neonatal sleep a passive response to the environment because it is regulated at very early ages [5]. Simply stated, if the primary function of sleep is to restore something depleted in wake, then infants should sleep *less* not more than adults.

Cognitive Theories of Sleep Function: Learning and Brain Development

The prevailing view among the general public is that sleep is good for learning and memory but among sleep scientists this idea has been bitterly contested for decades [2]. In contrast to metabolic theories that have foundered for lack of evidence, cognitive theories suffer from an abundance of evidence of very mixed quality. The situation has improved over the last decade and there appears to be a convergence of findings supporting a role for sleep in adult learning and plasticity and brain development [2]. Several issues need to be resolved; for example, there is little agreement over which sleep state is important for learning (REM versus NREM), nor is it clear what kind of learning is most affected by sleep (episodic versus procedural) [2]. There are three additional obstacles for a purely cognitive theory of sleep function. To begin with, it is not clear why sleep should be needed for learning in the first place. Animals certainly learn while awake, and many forms of synaptic plasticity are induced during wake (and perhaps only in wake) [2]. A common reply is that sleep is important for the consolidation of synaptic changes, which occurs after the initial induction of plasticity [18]. This may be true, but the underlying mechanisms responsible for sleep-dependent consolidation have not been identified. Until they are, the role of sleep in this process will continue to be debated. A second unresolved issue is that there is no obvious communication between learning or plasticity mechanisms and sleep homeostasis. A final point is that the main supports for a cognitive theory of sleep function, namely, the effects of sleep on brain development and the effects of sleep on adult mnemonic processes, have not been integrated in a manner that explains how sleep influences brain morphology and plasticity across the life span. It is possible that sleep function is different in developing and adult brains, but a parsimonious theory should propose mechanisms common to both.

CONCLUSIONS

For all their achievements, sleep scientists remain in the ticklish position of not knowing why we sleep. An analogous situation in the sciences is hard to find. For example, eating, like sleep, is a complex, regulated behavior governed by specific brain regions and hormones. Eating, like sleep, is ubiquitous in the animal kingdom and undergoes important transformations during ontogeny. The function of eating, however, is not disputed. In this chapter representative theories of sleep function have been reviewed in the hopes of revealing a deeper, more unifying understanding of sleep function. The evidence is equivocal, but some tentative conclusions can be made. First, it appears that while sleep may have beneficial effects on general health, its primary function concerns the brain and not the body. Second, sleep has profound effects on mental performance, which suggests that sleep in some general way facilitates normal neuronal function. Strong evidence for this are findings showing that sleep affects learning and synaptic plasticity— which are experimentally accessible manifestations of normal neuronal activity. It is therefore possible that sleep is a time when overall neuronal function is facilitated either by sleep-dependent increases in gene expression and protein synthesis or alterations in neuronal activity. Why this process primarily occurs in sleep and not in wake is but one of many unanswered questions that must await future investigation.

REFERENCES

1. Pace-Schott EF, Hobson JA. The neurobiology of sleep: genetics, cellular physiology and subcortical networks. *Nat Rev Neurosci* **3**:591–605(2002).

2. Benington JH, Frank MG. Cellular and molecular connections between sleep and synaptic plasticity. *Prog Neurobiol* **69**:77–101(2003). This article reviews arguments for and against the proposition that sleep facilitates learning and synaptic plasticity. It also presents theoretical connections between events that occur during sleep and synaptic remodeling.

3. Rechtschaffen A. Current perspectives on the function of sleep. *Perspect Biol Med* **41**:359–390(1998). This is a review by one of the leading thinkers in the field of sleep research. It contains thoughtful commentary on current theories of sleep function, as well as material on the original extended sleep deprivation studies carried out in rats.

4. Shaw PJ, Tononi G, Greenspan RJ, Robinson DF. Stress response genes protect against lethal effects of sleep deprivation in *Drosophila*. *Nature* **417**:287–291(2002).

5. Davis FC, Frank MG, Heller HC. Ontogeny of sleep and circadian rhythms. In: Zee PC, Turek FW (Eds), *Regulation of Sleep and Circadian Rhythms*., Vol. 133. Marcel Dekker, New York, 1999, pp 19–80.

6. Benington J. Sleep homeostasis and the function of sleep. *Sleep* **23**:959–966(2001).

7. Inoue S, Honda K, Komoda Y. Sleep as neuronal detoxification and restitution. *Behav Brain Res* **69**:91–96(1995).

8. Akerstedt T, Nilsson PM. Sleep as restitution: an introduction. *J Intern Med* **254**:6–12(2003).

9. Cortelli P, Gambetti P, Montagna P, Lugaresi E. Fatal familial insomnia: clinical features and molecular genetics. *J Sleep Res* **8**:23–29(1999).

10. Horne JA. Sleep function, with particular reference to sleep deprivation. *Ann Clin Res* **17**:199–208(1985).

11. Bonnet MH. Sleep deprivation. In: Kryger M, Roth T, Dement WC. (Eds). *Principles and Practice of Sleep Medicine*. Saunders, Philadelphia, 2000, pp 53–71.

12. Ramanathan L, Gulyani S, Nienhuis R, Siegel JM. Sleep deprivation decreases superoxide dismutase activity in rat hippocampus and brainstem. *Neuroreport* **13**:1387–1390(2002).

13. Gopalakrishnan A, Ji LL, Cirelli C. Sleep deprivation and cellular responses to oxidative stress. *Sleep* **27**:27–35(2004).

14. Cirelli C, Gutierrez CM, Tononi G. Extensive and divergent effects of sleep and wakefulness on brain gene expression. *Neuron* **41**:35–43(2004). This is an extremely important study that provides a comprehensive analysis of gene expression in sleep and wake.

15. Benington J, Heller HC. Restoration of brain energy metabolism as the function of sleep. *Prog Neurobiol* **45**:347–360(1995).

16. Kong J, Shepel PN, Holden CP, MacKiewicz M, Pack AI, Geiger JD. Brain glycogen decreases with increased periods of wakefulness: implications for homeostatic drive to sleep. *J Neurosci* **22**:5581–5587(2002).

17. Franken P, Gip P, Hagiwara G, Ruby NF, Heller HC. Changes in brain glycogen after sleep deprivation vary with genotype. *Am J Physiol Regul Integr Comp Physiol* **285**:R413–419(2003).

18. Stickgold R, Hobson JA, Fosse R, Fosse M. Sleep, learning and dreams: off-line memory reprocessing. *Science* **294**:1052–1057(2001).

8

THE EVOLUTION OF SLEEP: A PHYLOGENETIC APPROACH

JOHN A. LESKU
Indiana State University, Terre Haute, Indiana

NIELS C. RATTENBORG
Max Planck Institute for Ornithology—Seewiesen Starnberg, Germany

CHARLES J. AMLANER, JR.
Indiana State University, Terre Haute, Indiana

INTRODUCTION

Although scientists and physicians have been pursuing the meaning of sleep for decades, the functions of sleep remain elusive [1]. There is strong consensus that the comparative method is a powerful, yet underutilized, approach for illuminating sleep function [2–5]. Under the comparative paradigm, sleep is compared across the animal kingdom with the aim of revealing the evolutionary history of sleep. Studies utilizing this approach have demonstrated extensive variation in both the amount and phasing (e.g., monophasic—diurnal or nocturnal; or polyphasic—crepuscular or arrhythmic) of sleep across taxonomic groups [2]. Although most, if not all, vertebrates sleep, the study of sleep in these organisms is not likely to yield the fundamental function(s) for which sleep evolved. Additional adaptations likely became associated with sleep after its initial evolution. Sleep in mammals may therefore perform a tapestry of functions, making the dissection of sleep's original evolutionary function that much more difficult to understand or recognize.

The usefulness of the comparative method is clear, it is impeded by a lack of information on sleep in nonmammalian vertebrates and sleep-like behavior in invertebrates. Consequently, the electrophysiological hallmarks of mammalian sleep are often used as the "gold standard" for sleep in nonmammals, while sleep in other vertebrate classes has largely been neglected. Although the relative lack of information on sleep in reptiles, amphibians, and fishes has been the main obstacle to wide-ranging comparative sleep work, the relative wealth of knowledge on mammalian species has allowed for detailed comparisons among mammalian taxa [6].

Sleep is foremost a behavioral state. At the organismal level, sleep is readily identifiable by (1) behavioral quiescence, (2) increased arousal threshold, (3) rapid reversibility to wakefulness [7], (4) a species-specific sleep site and posture [8], (5) circadian organization, and (6) homeostatic regulation [9]. Both circadian rhythm and homeostatic regulation dictate propensity to sleep [5]. The circadian rhythm aligns sleep with a period(s) of the 24-hour day, whereas homeostatic regulation is a function of prior time awake. Sleep deprivation results in a "debt that is repaid" during successive sleep periods. In essence, sleep loss is compensated for by an increase in sleep time and intensity during recovery sleep. Intensity of sleep is measured via slow-wave activity (i.e., mean electroencephalogram power density between 0.75 and 4.0 Hz) [5] and is correlated with an increased arousal threshold following sleep deprivation.

This review aims to provide insight into variation in sleep architecture among the vertebrate classes as well as an overview of the limited, yet increasing information on sleep-like behavior in invertebrates. We focus on the evolution of sleep as measured by electrophysiological attributes using extant organisms as models for ancestral forms.

MAMMALS

Marsupials and Terrestrial Placentals

All mammals studied to date show some form of sleep [6]. Although total sleep time and the relative proportions of the sleep stages vary greatly among taxa, all exhibit an alternating cycle of slow-wave sleep (SWS; also called quiet sleep or non-rapid eye movement (NREM) sleep), punctuated by episodes of rapid eye movement. (REM) sleep also called active sleep or paradoxical sleep). SWS is characterized by an electroencephalogram (EEG) of high-amplitude, low-frequency waves, which are the result of synchronous neuronal firing between adjacent neurons in the neocortex and thalamocortical interactions. In humans, SWS refers only to stages 3 and 4 of NREM; whereas in the animal literature, SWS usually refers to all stages of sleep other than REM sleep. During SWS, heart and respiratory rate remain steady, and thermoregulation remains functional. Eye movements (measured by the electro-oculogram (EOG)) are absent, while muscle tone (measured by the electromyogram (EMG)) persists.

REM sleep is distinguished from SWS by a low-amplitude, mixed-frequency or activated EEG. Heart and respiratory rate are irregular and thermoregulation is suspended; therefore an endotherm is essentially poikilothermic during REM sleep. Rapid eye movements are present and, unlike in wakefulness, there is EMG atonia. The mammalian sleep cycle consists of an alternation between SWS and REM sleep. Normally, the sleep cycle is entered through SWS and terminates after an episode of REM sleep with a brief awakening.

Insectivores, carnivores, and ungulates engage in drowsiness, a stage intermediate between wakefulness and sleep. Although it is probable that all terrestrial mammals exhibit drowsiness to varying degrees, it is most notable in the aforementioned groups. The EEG of a drowsy animal usually shows sleep spindles or slow waves superimposed on a background of waking EEG activity. Arousal thresholds remain low and eye states are intermediate between open and closed. The function of drowsiness is unknown, drowsiness may permit vigilance during sleep in perilous environments [6].

Monotremes

Monotremes are the earliest offshoot of the mammalian evolutionary line [10]. There are only three living

representatives of monotremes: the short-beaked echidna (*Tachyglossus aculeatus*), the long-beaked echidna (*Zaglossus bruijni*), and the duck-billed platypus (*Ornithorhynchus anatinus*). EEG sleep studies have been conducted on the first and third. Understanding sleep in these egg-laying mammals may clarify the earliest sleep state from which all existing sleep states evolved.

The first sleep study on a monotreme (the short-beaked echidna) concluded that although the mammal engaged in SWS, it did not exhibit the EEG correlates and common features of REM sleep [11]. However, a reanalysis of sleep in *T. aculeatus* revealed concurrent cortical features of SWS and subcortical signs of REM sleep [12]. Specifically, the forebrain generated a high-amplitude, low- frequency EEG typical of SWS, while brainstem neurons fired with an irregular pattern similar to that observed in placental mammals during REM sleep. Thus the echidna exhibited a state composed of markers common to both SWS and REM sleep, suggesting that the two temporally distinct states arose in the placental and marsupial mammalian clade from a single, heterogeneous sleep state. Some controversy persists, however, over the characterization of sleep states in the echidna. In a subsequent study, Nicol et al. [13] reported SWS and REM sleep characterized by an activated EEG, reduced tonic EMG, intermittent EOG activity, and decreased heart rate, typical of REM sleep in placental and marsupial mammals.

The only study on sleep in the platypus identified both SWS and REM sleep [14]. As in the echidna, sleep was characterized by a high-amplitude, low-frequency EEG, typical of SWS. Although brainstem neuronal units were not recorded, the platypus showed bursts of rapid eye movements and twitching of the bill and head, similar to phasic skeletomuscle activity in other mammals during REM sleep. Arousal thresholds were higher in REM sleep than in SWS, again consistent with placental and marsupial mammals. Calculations of REM sleep time, based on the temporal distribution of eye movements and twitches, suggest that the platypus spends more time in REM sleep than any other animal studied.

Aquatic Mammals

There are three extant aquatic mammalian orders: Cetacea, Pinnipedia, and Sirenia. Aquatic mammals are conflicted by the need to simultaneously sleep, surface to breathe, and maintain vigilance. Several species of aquatic mammals appear to have overcome this conflict by engaging in unihemispheric slow-wave sleep (USWS), a unique state during which one cerebral hemisphere shows EEG activity indicative of SWS, while the other hemisphere shows activity indicative of wakefulness (Figure 8.1; reviewed in [15]). Interhemispheric asymmetries in the

A B C

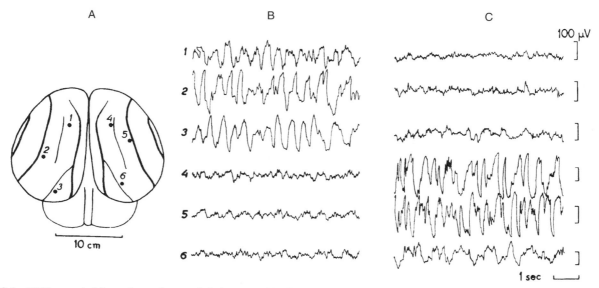

Figure 8.1 EEG recorded from the parieto-occipital cortex (A) of a bottlenose dolphin (*Tursiops truncatus*) during unihemispheric slow-wave sleep (USWS). (B) The left hemisphere (1–3) shows high-amplitude, low-frequency EEG activity denoting SWS, whereas the right hemisphere (4–6) is awake with low-amplitude, high-frequency EEG activity. (C) The interhemispheric asymmetry is reversed and the left hemisphere is awake while the right hemisphere sleeps. (Reprinted from Mukhametov et al., Interhemispheric asymmetry of the electroencephalographic sleep patterns in dolphins, *Brain Research* **134**:581–584. Copyright © 1977, with permission from Elsevier Science.)

EEG are associated with interhemispheric asymmetries in temperature, with the sleeping hemisphere having a lower parietal cortex temperature relative to the awake hemisphere [16]. During USWS, the eye contralateral to the sleeping hemisphere is usually closed, while the eye contralateral to the awake hemisphere is open.

USWS has been identified in five cetacean species: pilot whale (*Globicephala scammoni*), bottlenose dolphin (*Tursiops truncates*), common porpoise (*Phocoena phocoena*), Amazonian dolphin (*Inia geoffrensis*), and the beluga whale (*Delphinapterus leucas*). In all cetaceans examined, USWS is the predominant form of SWS; only rarely do cetaceans engage in unambiguous bihemispheric slow-wave sleep (BSWS). Interestingly, REM sleep either is absent, is greatly reduced, or occurs in a modified form in cetaceans; therefore the majority of sleep time is USWS. Bottlenose and Amazonian dolphins may swim slowly or hover at the surface of the water during USWS, while periodic fin movements maintain a stable posture. Furthermore, they breathe periodically without arousing to bilateral wakefulness. Interestingly, BSWS induced via pentobarbital administration, inhibit respiration [17], suggesting that USWS in cetaceans is, in part, an adaptation to maintain motor activity to allow surfacing to breathe. However, dolphins may also engage in USWS to monitor their environment. During USWS in a captive bottlenose dolphin, a visual stimulus presented to the open eye elicited a behavioral and electrophysiological response, such that the animal aroused to bilateral wakefulness. Indeed, Goley [18] found that Pacific white-sided dolphins (*Lagenorhynchus obliquidens*) kept their open eye on adjacent dolphins, perhaps to maintain relative position within the pod. Thus USWS may be used during long-distance migrations as it allows both vigilance and motor control concurrent with sleep.

Selective sleep deprivation studies in dolphins have been instrumental in identifying the biological targets benefiting from sleep. USWS allows for one hemisphere to be deprived of sleep while permitting sleep in the other. As deprivation continues, only the deprived hemisphere increases its attempt to fall asleep and when allowed to recover from sleep deprivation, only the deprived hemisphere exhibits a rebound in SWS, indicating that sleep is homeostatically regulated independently within each hemisphere [19] and that sleep benefits primarily the brain and not the body, a finding consistent with the fact that cetaceans may continue to swim while sleeping.

Within the order Pinnipedia there are three families: Odobenidae (walruses), Otariidae (eared seals: fur seals and sea lions), and Phocidae (true seals). Electrophysiological sleep studies have been conducted in eared and true seals, yet no such study has been attempted in walruses. Pinnipeds, unlike cetaceans, partition their time between terrestrial and aquatic environments and may therefore sleep both in and out of the water. In contrast to cetaceans, much of SWS in eared seals is BSWS, rather than USWS. Eared seals engage in USWS both in and out of the water, with the proportion of SWS composed of USWS being greater during sleep in the water. Also unlike cetaceans, eared seals engage in unequivocal REM sleep.

Interestingly, REM sleep as a percent of total sleep time decreases in the water, when compared to sleep on land. Like cetaceans, however, eared seals engage in USWS to allow respiration concurrent with sleep. For example, fur seals assume a stereotypic sleep posture in the water with three flippers in the air while one flipper paddles in order to keep the nares above the water's surface. Conversely, true seals only display BSWS and REM sleep and must hold their breath while asleep at sea. Periodic, brief awakenings permit motor control and the seal surfaces to breathe. On the other hand, elephant seals floating near the surface simply raise their heads above the water and breathe without arousing to wakefulness.

Manatees (Order: Sirenia) also engage in USWS [20]. A captive Amazonian manatee (*Trichechus inunguis*) exhibited REM sleep (1% of recording time) and SWS (27% of recording time) with 25% of SWS being unihemispheric. However, during bouts of USWS the manatee remained motionless underwater [20]. USWS in the manatee was not used to allow respiration concurrent with sleep; rather the manatees aroused to bilateral wakefulness for each respiratory act. In this instance, USWS may serve another function, such as predator detection.

Correlates of Sleep Architecture in Mammals

Correlational studies have led to various hypotheses on sleep function and factors influencing sleep architecture. For example, there are two clear energy conservation hypotheses about mammalian sleep [6]. Energy conservationists suggest that sleep limits energy expenditure by reducing metabolic rate below that accomplished by rest alone. Alternatively, sleep may enforce rest, which, in turn, limits energy expenditure. If so, one would predict that mammals with higher mass-special metabolic rates would sleep more, and although this prediction was initially supported (Table 8.1) [21], the relationship was negative after controlling for body weight via partial correlation (as mammals with higher mass-specific metabolic rates actually sleep less) [22]. Elgar et al. [22] suggest that animals with higher mass- specific metabolic rates need to

allocate more time to foraging than those with lower mass-specific metabolic rates. Foraging time may thus limit sleep time.

Potential predation may restrict REM sleep time as REM sleep is negatively correlated with an index of "overall danger" [23]. Under these conditions, prey species should not engage in lengthy bouts of REM sleep and it's associated high arousal thresholds due to increased predation threat. Elgar et al. [22] did not address the potential role of predation in influencing sleep architecture; however, they did demonstrate that geographic latitude correlated positively with cycle length after controlling for body weight. Presumably, the inhibition of thermoregulatory mechanisms (which occurs during REM sleep) limits the amount of time an animal in a temperate climate can endure long periods of uninterrupted REM sleep. Theoretically, extended bouts of REM sleep would be detrimental to an endothermic animal if the ambient temperature were significantly below that of thermoneutrality. Precocial species, those that are relatively self-sufficient at birth, exhibit less REM sleep as adults than altricial species, those that are relatively helpless upon birth or hatching [22, 23]. However, since precocial animals are generally larger than their altricial counterparts, Elgar et al. [24] reanalyzed a revised version of their 1988 dataset and revealed that although altricial families have significantly more REM sleep than precocial families, this correlation was no longer significant after controlling for body weight.

Mammals with greater mass-specific metabolic rates may sleep longer [21], leading Siegel and co-workers to hypothesize that increased sleep requirements are necessary as a consequence of increased metabolic rate and the associated production of free radicals [25, 26], which are detrimental to protein structure. Finally, since sleep is likely of principal importance to the brain rather than to the body [1], one would expect larger brains to require more sleep; however, brain weight negatively correlated with total sleep time, suggesting that larger brained organisms may be able to handle sustained periods of wakefulness better than smaller brained organisms.

TABLE 8.1 Summary of Correlational Studies[a]

Variables	Cycle Length (min)	Metabolic Rate (cm^3 O_2/g/h)	Body Weight (kg)	Brain Weight (g)	Life Span	Overall Danger Index	Geographic Latitude
Total sleep time (h/day)	−[21]	+[21]/−[22]	−[22]	−[21, 22]	−[21]	?	?
SWS (h/day)	−[21]	+[21]/−[22]	−[22, 23]	−[21–23]	−[21, 23]	−[23]	?
REM sleep (h/day)	ns [21]	+[21]/−[22]	−[22, 23]	−[21–23]	−[21, 23]	−[23]	?
Cycle length (min)	n/a	−[21]/+[22]	+[22]	+[21, 22]	+[21]	?	−[22]

[a]A plus (+) denotes a significant ($p < 0.05$) positive correlation between the two variables, a minus (−) denotes a significant ($p < 0.05$) negative correlation, and ns means that the correlation was not significant. A question mark (?) indicates that the correlation was not calculated. For the correlations of metabolic rate with various sleep characteristics, Elgar et al. [22] calculated a partial correlation between total and SWS times with metabolic rate, after controlling for adult body weight, whereas Zepelin and Rechtschaffen [21] correlated sleep characteristics with mass-specific metabolic rate.

All of these correlational studies, however, are fraught by a common concern. Zepelin and Rechtschaffen [21] and Allison and Cicchetti [23] used each species as a statistically independent unit, whereas Elgar et al. [22] pooled data at the family level. However, the simple phylogenetic framework used by Elgar et al. [22] is no longer considered valid. Thus all studies experienced pseudoreplication since phylogenetic relationships between taxa were neglected. Data from mouse to elephant, bat to human, and horse to kangaroo were weighted equally against one another. Future work should readdress these datasets under an explicit, phylogenetic context using independent contrasts to control for pseudoreplication where possible.

BIRDS

The EEG of a sleeping bird is similar to that of a sleeping mammal; and yet, birds are more closely related to extant reptiles (i.e., crocodilians) than they are to mammals. A sleeping bird meets all of the behavioral criteria listed for mammals (Figure 8.2) and exhibits both SWS and REM sleep. Birds, like mammals, may be classified as nocturnal, diurnal, crepuscular, or arrhythmic with respect to their timing of sleep. A waking bird exhibits bilateral eye opening, complex body movements, an activated, low-amplitude, mixed-frequency EEG, a highly active EOG, highly tonic

Figure 8.2 Emperor penguins (*Aptenodytes forsteri*) displaying the typical avian head postures associated with wakefulness (right) and sleep (left and middle). (Courtesy of Grass-Telefactor, An Astro-Med, Inc. Product Group.)

EMG, variable heart rate, and low arousal threshold. As in mammals, avian SWS is characterized by high-amplitude, low-frequency EEG activity (reviewed in [27]). Interestingly, sleep spindles and K-complexes during SWS are absent. Neck EMG is typically tonic without phasic events. Eye movements are infrequent with the exception of brief, high-frequency oscillations of the eye. Heart, respiratory, and metabolic rate are all stable and reduced relative to waking.

REM sleep is characterized by the highest arousal threshold and a relatively high-frequency, low-amplitude EEG. Birds typically show much less REM sleep than mammals and periods of REM sleep are usually shorter than 10 s. Eye movements are present in clusters. Hippocampal theta waves and PGO spikes have not been recorded during avian REM sleep. Heart rate is either variable or may increase or decrease. Finally, as in mammals, thermoregulation is inhibited during REM sleep in birds.

In mammals, sleep intensity is gauged by slow-wave activity (SWA). Unlike in mammals, sleep deprivation does not appear to result in an increase in SWA in subsequent bouts of sleep in birds. However, sleep duration and amount of REM sleep all increase during recovery after sleep deprivation. Interestingly, migratory songbirds, such as the white-crowned sparrow (*Zonotrichia leucophrys gambelii*), appear to reduce their amount of time sleeping by over 60% during migration [28].

The ontogenesis of avian sleep is similar to that of mammalian sleep patterns. Domestic fowl exhibit EEG components of an adult bird (i.e., SWS and REM sleep) while in the egg, but unlike most natal mammals, birds experience little REM sleep prehatch. SWS is identifiable by day 17 and is of typical form by day 18. REM sleep appears during day 18 or 19, just before hatching on day 20. REM sleep declines rapidly in chickens from 16.5% to 6.4% by the end of posthatch day 3.

Like aquatic mammals, birds engage in USWS [15, 27, 29]. Avian USWS is defined by unilateral eye closure (UEC) and associated interhemispheric asymmetries in SWS-related EEG activity. However, the degree of interhemispheric asymmetry in birds is small compared to aquatic mammals. USWS has been reported in 8 species across 6 avian orders and UEC has been reported in 29 species from 13 orders. Spooner [30] was the first to demonstrate an association between asynchronous eye closure and interhemispheric asymmetries associated with USWS. However, the function of avian USWS remained unclear until Rattenborg et al. [31, 32] showed that USWS allows birds to maintain vigilance for predators while sleeping. Using the "group edge effect" paradigm (birds on the periphery of a group perceive greater danger than those in the center), birds on the outside of the group increased their use of USWS by 150% relative to the birds in the middle and showed a strong preference for directing their open eye away from the other birds and toward potential threats. Not only does this show the adaptive significance of USWS, it also reveals the plasticity of the trait under changing predation regimes. Unihemispheric REM sleep has never been reported in birds (or mammals).

Correlates of Sleep Architecture in Birds

Only two comparative studies exist that examine the correlates of sleep architecture in birds [33, 34]. Amlaner and Ball [33] calculated correlations between total sleep time (TST) and environmental (latitude and hours of daylight) and ecological (sleep exposure, social sleep index, and an index of vulnerability or exposure to potential predation) factors thought to affect sleep architecture. However, they did not differentiate between SWS and REM sleep as their dataset was based totally on a behavioral definition of sleep. (At the time, there existed very little information on sleep architecture in wild birds—an ongoing limitation in comparative bird studies even today; see [34].) Like the correlational studies in mammals, there was great interspecies variability of TST. The average TST was 7 h. There was a negative correlation between TST and latitude and for species experiencing longer relative day length. Birds at arctic latitudes (>68.00°) during the summer, when day length is virtually 24 h, averaged a TST of 3.7 ± 1.3 h (mean $\pm$ s.e. $n = 12$). This may reflect (1) a bird's need to increase vigilance during the long daylight hours, (2) the need to accomplish more important behaviors at the expense of sleeping (a natural form of sleep deprivation), and/or (3) the direct alerting effects of light. A stepwise multiple regression based on these variables revealed that day length accounted for 49.1% ($r = -0.7$) of the variance contributing to TST; however, latitude was also a good predictor of TST ($r = -0.62$).

Schmidt [34] used an electrophysiological dataset that did differentiate between SWS and REM sleep. Like mammals, SWS time correlated negatively with resting mass-specific metabolic rate. However, the correlation between SWS time and body weight was not significant. SWS time was highly conserved between taxonomic orders, whereas REM sleep time varied significantly between orders. Since altricial mammals have more REM sleep than precocial mammals, one would expect birds to exhibit a large proportion of REM sleep, as birds are extremely altricial at birth. However, although birds have comparable amounts of SWS as mammals, they exhibit half the amount of REM sleep [34]. Interestingly, precocial birds (Order: Sphenisciformes, Anseriformes, and Galliformes) exhibited more REM sleep (but not SWS) than altricial orders (Order: Columbiformes, Psittaciformes, Strigiformes, Passeriformes). Arboreal birds require less REM sleep than either terrestrial or aquatic birds, or mammals. Most passerines appear to have extremely small amounts of REM sleep

(3–10 min per 24 h), although 16% of sleep is REM sleep in white crowned sparrows [28]. Amlaner and Ball [27] speculated that passerines engage in less REM sleep because perching, which requires muscle tone, is inhibited during REM sleep. Alternatively, Schmidt [34] suggests that since passerines have larger optic lobes (relative to other avian orders) they require less REM sleep, although details and empirical support for this hypothesis are lacking. Diurnal birds did not differ in amounts of either SWS or REM sleep relative to nocturnal or polyphasic birds.

REPTILES

The class Reptilia is composed of four orders: Crocodilia (alligators, caimans, crocodiles, and gharials), Chelonia (tortoises and turtles), Squamata (lizards and snakes), and Rhynchocephalia (tuataras). Sleep has been investigated in all orders with the exception of Rhynchocephalia and all representatives studied exhibited sleep, according to behavioral criteria. However, the electrophysiological correlates of behavioral sleep in reptiles are often inconsistent and contradictory, sometimes within the same species [2, 3, 5, 35]—thus inferring the evolutionary pathway of sleep in birds and mammals from reptilian studies has been impeded and the need for future reptilian work is apparent.

Order: Crocodilia

Animals of this order are the closest extant relatives to modern-day birds. One might therefore expect crocodilian sleep architecture to be similar to that of birds (i.e., SWS and REM sleep). Studies in the caiman (*Caiman sclerops*) have been most telling. High-voltage sharp spikes in the EEG were prominent during periods of behavioral quiescence and were reduced upon arousal [36]. After sleep deprivation, there was an increase in spike activity similar to the rebound in slow-wave activity (SWA) in mammals. Interestingly, some studies have reported SWA (i.e., high-amplitude, low-frequency activity) reminiscent of mammalian SWS [37, 38]. Warner and Huggins [37] attribute the difference between their findings and that of Flanigan et al. [36] to the presence of other caimans in their study, which, presumably, relaxed their animals. REM sleep was not observed in either study. Unilateral eye closure (UEC) has been observed in caimans, but the possibility of unihemispheric sleep was not investigated [37].

Order: Testudines

Although turtles and tortoises sleep, their sleep apparently does not resemble mammalian and avian SWS or REM sleep. Box turtles (*Terrapene carolina*) [39] and the red-footed tortoise (*Geochelone carbonaria*) [40] both meet the behavioral criteria for sleep: a species-specific sleep posture, behavioral quiescence, increased arousal threshold, rapid reversibility, and homeostatic regulation. Additionally, both had an EEG of high-amplitude spiking activity superimposed over a low-voltage background. EEG spikes disappeared with spontaneous or induced arousal. Spiking activity increased after sleep deprivation and was associated with increased arousal thresholds. Neither classical high-amplitude slow waves nor REM sleep was observed in either species. However, administration of atropine sulfate—a cholinergic blocking agent that increases mammalian slow waves—increased spiking activity in red-footed tortoises [41]. Furthermore, administration of parachlorophenylalanine—a serotonin synthesis inhibitor that suppresses mammalian slow waves—reduced spiking activity in the three tortoises studied, suggesting that these spikes are homologous to mammalian SWS [42]. UEC was reported for both box turtles and the red-footed tortoise [39, 40]. In the marginated tortoise (*Testudo marginata*), quiescence was associated with bilateral eye closure, body relaxation, increased arousal threshold, reduced EMG activity, and high-voltage slow waves. Likewise, the EEG of the yellow-footed tortoise (*Testudo denticulata*, now *Geochelone denticulata*) exhibited high-voltage spiking activity during sleep. In contrast, the loggerhead sea turtle (*Caretta caretta*) did not show spiking activity or slow waves during behavioral sleep. One study on the European pond turtle (*Emys orbicularis*) reported both SWS and REM sleep, a result that remains unreplicated.

Order: Squamata

Sleep in the desert iguana (*Dipsosaurus dorsalis*) [43] and spiny-tailed iguana (*Ctenosaura pectinata*) [44] consisted of a reduction of EEG amplitude and frequency relative to waking. Upon arousal, EEG amplitude increased in brainstem and forebrain electrodes [44]. Eye movements occurred at 4–25 min intervals, reminiscent of mammalian REM sleep [44]; however, this may be recording artifacts due to brief awakenings or nictitating membrane activity and not reflective of true REM sleep. Following sleep deprivation, the green iguana (*Iguana iguana*) and spiny-tailed iguana exhibited a rebound in sleep as indicated by total sleep time and by the increased frequency of high-voltage spikes. Chameleons (*Chamaeleo jacksoni* and *Chamaeleo melleri*) also exhibited bursts of high-voltage spikes during sleep.

REM sleep in reptiles appears to be ambiguous based on historic reports. Although REM sleep has been reported in the chameleon (*Chamaeleo* sp.) [45], desert iguana [43], and spiny-tailed iguana [44, 46], in the desert iguana, REM sleep was characterized by an increase in EEG amplitude to that of wakefulness with a concurrent atonic EMG [43]. REM sleep was temperature sensitive and appeared

with greater incidence at higher temperatures. Conversely, other studies have not observed REM sleep in either the green or spiny-tailed iguana [47] and SWS has never been reported in squamates.

Despite the pervasiveness of UEC in squamates, only one behavioral study has explicitly investigated UEC in reptiles. Mathews et al. [48] demonstrated that the frequency of UEC increased in the western fence lizard (*Sceloporus occidentalis*) following exposure to a predator. Furthermore, lizards preferentially directed their open eye toward the last known position of the predator, suggesting at least a predator-detection function for UEC in reptiles similar to that of birds [31, 32].

Reptiles exhibit unambiguous behavioral sleep that is associated with concurrent changes in brain activity. However, the inconsistencies among and within species and between studies make interpretation difficult from many perspectives. High-voltage spikes are present in most studies, yet absent in others, which may reflect underlying methodological differences between studies, such as laboratory adaptation, electrode placement, and ambient temperature. Interestingly, it has been proposed that high-voltage spikes may occur in a state that is the precursor to mammalian and avian SWS. Sleep in reptiles appears to be homeostatically regulated as deprivation results in a reduction of sleep latency and an increase in sleep time when allowed to proceed. Furthermore, following sleep deprivation there is often an increase in spiking activity, leading some to suggest that spiking may reflect sleep intensity. Along these lines, it has been suggested that the spiking activity of reptiles resembles subcortical spiking during SWS in mammals. The fact that reptiles have a comparatively small cortex has also led some to suggest that reptiles lack the neuroanatomy necessary to generate SWS-related SWA. However, the minimal amount of cortex required to generate SWA remains unclear. Finally, the prevalence of UEC in crocodilians [37], chelonians [39, 40], and squamates (Family: Chamaeleonidae [45], Iguanidae [44, 47], Phrynosomatidae [48]) suggests that reptiles are able to engage in unihemispheric sleep; however, no study to date has examined the EEG correlates of UEC in reptiles.

AMPHIBIANS AND FISHES

Evidence for the existence of sleep in both amphibians and fishes is ambiguous [3, 5]. Although some species may show signs of behavioral sleep, others do not. Three species of tree frog (*Hyla cinerea*, *H. septentrionalis*, and *H. squirrella*) were monitored in the lab with EEG. Behavioral sleep was associated with decreased amplitude and increased frequency relative to waking (opposite of the typical mammalian/avian pattern). Arousal thresholds were not measured. The EEG of the common frog (*Rana temporaria*) showed increased low-frequency activity during periods of behavioral quiescence and reduced muscle tone and bradycardia. Alternatively, captive bullfrogs (*Rana catesbeiana*) may not sleep; a tentative conclusion based on a lack of reduction in responsiveness to electrical stimuli and continuous bilateral eye opening. Furthermore, no EEG changes were noted during quiescence, although bullfrogs exhibited an EEG similar to that of sleeping tree frogs (i.e., decreased EEG activity). Sleep-like behavior in the western toad (*Bufo boreas*) was accompanied by a slowing of the EEG. However, high-voltage, fast spiking activity has also been reported during sleep-like behavior in the toad.

Behavioral sleep in amphibians may occur in the absence of changes in brain activity. Conversely, electrophysiological changes associated with sleep may be observed in the absence of accompanying behavioral measures of sleep. The few amphibian studies that report EEG correlates of behavioral state also contain significant interstudy differences that impede interpretation. It is also possible that recorded sleep patterns may not be representative of natural patterns. Experimental animals may have been stressed due to novel laboratory conditions and they may require a more prolonged acclimation period. Of course, these criticisms could be said of many laboratory studies that involve wild animals belonging to all taxa.

Electrophysiological data on fishes is scarce. Indeed only two studies have recorded neuronal activity during periods of behavioral quiescence. Sleep-like periods in the catfish (*Ictalurus nebulosus*) were defined by decreased behavioral motility and, unlike wakefulness, were associated with EEG slow-wave activity and spiking. Spiking disappeared upon arousal. Conversely, a study on the tench (*Tinca tinca*) did not find an association between behavioral state and variation in EEG activity, but observed decreased muscle tone and respiratory rate during sleep-like behavior. One laboratory reported decreased sleep latency in carp following 96 h of sleep deprivation. Similarly, sleep-deprived perch exhibited rebound in sleep during recovery. Thus, as in amphibians and reptiles, although the electrophysiological correlates of sleep in fishes are ambiguous, fish show behavioral signs of a homeostatically regulated sleep-like state.

INVERTEBRATES

It is now evident from behavioral, pharmacological, and electrophysiological data, mostly on terrestrial arthropods, that invertebrates engage in sleep or a state homologous to sleep [3, 5]. The fruit fly [49] (reviewed in [50]), cockroach, scorpion, honey bee, butterfly, locust, and mosquito all exhibit sleep-like behavior, characterized by increased arousal thresholds and associated postures. The deprivation of sleep-like behavior in cockroaches resulted in a subsequent rebound effect. Scorpions are the oldest extant arthropod group and exhibit three distinct vigilant states: (1) activity, (2) alert immobility, and (3) relaxed immobility. Arousal thresholds, measured by response to mechanical stimulation, were lowest for alert scorpions and highest for relaxed scorpions [51]. Like cockroaches, sleep-like behavior deprivation enhanced sleep-like behavior when allowed to proceed unimpeded. Honey bee (*Apis mellifera*) sleep-like behavior follows a circadian rhythm. Furthermore, it is monophasic and persists in constant darkness, indicating that it is not a direct response to photoperiod. Optomotor interneurons in the optic lobes of honey bees showed circadian rhythm in response to moving visual stimuli [52]. Sensitivity was higher during the day than during the night and corresponded with an increase and decrease in locomotor activity, respectively. Head position was the lowest and antennal immobility was the greatest during sleep-like behavior [53]. Like mammals and birds, sleep-like behavior decreased with increasing age. EEG recordings have been achieved from the mushroom bodies of sleeping honey bees [54]. Spiking activity increased with behavioral rest, characterized by antennal immobility. Following 12 h sleep deprivation via forced activity, bees showed a decreased latency and increased total time of antennal immobility, suggesting sleep in honey bees is homeostatically regulated [55].

Sleep in *Drosophila* has been characterized by behavioral quiescence, increased arousal threshold, and homeostatic regulation [49, 56]. Sleep deprivation studies using per^{01} mutant fruit flies (i.e., flies that lack a circadian rhythm) have revealed *Drosophila* sleep to be homeostatically regulated (Figure 8.3). Furthermore, stimulants and hypnotics affect *Drosophila* sleep in a manner similar to mammalian sleep. Young fruit flies exhibit significantly higher total sleep time than older fruit flies, a pattern typical of mammals and birds. Interestingly, chronic sleep deprivation results in fly death after approximately 70 h [57]. *Drosophila* also exhibit stereotypic correlates of neuronal activity depending on behavioral state. Nitz et al. [58] recorded local field potentials (LFPs) from the medial brain between the mushroom bodies of *Drosophila* in various behavioral states. Awake, interactive flies exhibited spike-like potentials in their LFPs, which disappeared with behavioral quiescence. Taken in concert, this suggests that behavioral quiescence in *Drosophila* is homologous to vertebrate sleep.

Many genes are differentially expressed in *Drosophila* due to changes in behavioral state independent of the circadian clock [59]. Sleep architecture and timing appear to have a strong genetic component [60]. Similar genes are up- and down regulated within sleep/wake states in *Drosophila* [61] and in the rat [62], further suggesting that sleep in the fruit fly is homologous to mammalian sleep. Specifically, certain "waking" genes in *Drosophila* are activated

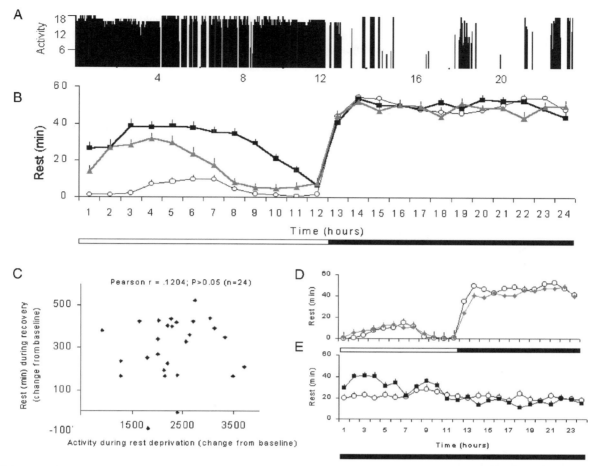

Figure 8.3 (A) The activity record for *Drosophila* maintained on a 12L:12D (open horizontal bar:dark horizontal bar) light cycle. (B) Sleep–activity cycle of undisturbed flies (circles) and flies sleep deprived via manual stimulation (squares) or by an automated system (triangles). Sleep-deprived flies showed an increase in sleep during the subsequent light period. (C) The amount of sleep during the 12 h recovery period was not correlated with the amount of activity during sleep deprivation. (D) Stimulation of flies during the light period did not result in a compensatory increase in sleep during recovery (diamonds) relative to baseline (circles). (E) Under constant darkness, per[01] flies had the same amount of sleep as under a light–dark photoperiod but sleep was evenly distributed across the 24 h (circles). Twelve hours of automated sleep deprivation resulted in an increase in sleep during the first 6 h of recovery (squares) compared to baseline (circles). (Reprinted with permission from Shaw et al. Correlates of sleep and waking in *Drosophila melanogaster. Science* **287**:1834–1837. Copyright © 2000 AAAS.)

in the first few hours after waking. These genes are functionally homologous to "waking" genes in the rat. One such shared "waking" gene is *BiP* (*Hsc*70-3) that codes for an endoplasmic reticulum chaperone protein. Chaperone proteins are thought to promote proper folding and shaping of other proteins. Their expression at the onset of wakefulness and during sleep deprivation may very well hold an important clue to sleep function [57]. Some genes are upregulated after wakefulness and more so after sleep deprivation. Three hours of sleep deprivation results in an increase in the transcription of genes coding for transcription factors and/or genes involved in energy metabolism. After 8 h of deprivation there is an upregulation of growth factors, molecular chaperones, higher mRNA levels of heat shock proteins, neurotransmitters, transporters, and

enzymes [62] and the synthesis of cholesterol, myelin structural proteins, and myelin-related receptors [61]. The increase in molecular chaperone protein expression suggests an increase in the mobilization of either newly synthesized proteins or those destined for catabolism. Furthermore, the upregulation of synaptic plasticity genes (e.g., brain-deprived neurotrophic factor (BDNF)) suggests the remodeling of neuronal configurations [62] or memory acquisition [61]. Wakefulness in *Drosophila* is also characterized by an increase in the levels of mRNA for arylalkylamine *N*-acetyltransferase (aaNAT1, also called dopamine acetyltransferase (DAT), an enzyme responsible for the catabolism of monoamines. Although rats lack this enzyme, arylsulfotransferase serves a similar function. Interestingly, an increase or decrease in gene expression can occur in brain regions where EEG correlates of sleep are reduced or nonexistent [61].

A recent study reported behavioral sleep accompanied by slow-wave activity in crayfish [63]. Behavioral sleep was associated with specific postures (i.e., floating, lying on one-side) and this "sleep-posture" was associated with the highest arousal threshold to a vibratory stimulus. Sleep deprivation resulted in a compensatory increase in total sleep time during subsequent sleep bouts. Interestingly, the authors presented a waking EEG of high-voltage spikes and a sleeping EEG of continuous slow-wave activity. Slow-wave activity disappeared upon arousal. REM sleep was not observed.

The cephalopod, *Octopus vulgaris*, displays color patterns and associated postures that correlate with rest during the nocturnal phase of the photoperiod. While resting, arms are upturned, skin texture is smooth, and chromatophores are relaxed, resulting in a gray-green color to be expressed only on the dorsal body surface and purple on the ventral arms [64]. Furthermore, specific postures are associated with elevations in arousal threshold [2]. Another study revealed a rest/activity cycle in cuttlefish. Captive cuttlefish would lie still for 10–15 min periods interrupted by flashes of bold color from their chromatophores and twitches of the tentacles resembling mammalian and avian REM sleep.

CONCLUSION

A phylogenetic evaluation of sleep demonstrates that all mammals, birds, and reptiles engage in sleep, and evidence for sleep in amphibians, fishes, and invertebrates is strong if not certain. In critically important sleep studies of rats and fruit flies, it has been shown that chronic sleep deprivation is fatal [57, 65], attesting to the necessity for sleep and to its ancient evolutionary age.

Sleep in mammals and birds can be divided into two states: SWS and REM sleep. Although it is necessarily intuitive that sleep evolved from wakefulness, it is unclear which of the two sleep states (if either) evolved first. An understanding of this sequence can have significant consequences for our understanding of the functions of sleep. For example, the SWS–REM sleep cycle may be unique to mammals and birds, suggesting an association with a shared character of the two groups, such as endothermy or an enlarged forebrain. Below we summarize three theories currently circulating on the evolution of the sleep cycle exhibited in mammals and birds (Figure 8.4) [4, 35, 66].

The first hypothesis, henceforth called *SWS-first* (Figure 8.4a) is advanced in direct opposition to the *REM sleep-first* hypothesis [35]. The first sleep study in monotremes concluded that the short-beaked echidna did not exhibit REM sleep [11], supporting the notion that REM sleep evolved after SWS in the evolutionary lineage of mammals and birds. SWS was therefore assumed to be the ancestral sleep state and REM sleep subsequently evolved twice: once in birds (or their dinosaur ancestors) and once in the placental and marsupial mammalian clade, a finding consistent with reptilian studies at the time [36, 47] (see Reptiles section). However, slow waves, which are the hallmark of SWS, originate in the mammalian neocortex. Reptiles, perhaps void of necessary telencephalic structures, may not be able to generate slow-wave activity. Nevertheless, sleep deprivation experiments and pharmacological evidence suggest that the high-voltage spikes often reported in reptilian sleep studies may occur in a state that is the precursor to mammalian and avian SWS.

The second hypothesis, REM sleep-first, states that REM sleep is the ancestral sleep state [66] and that SWS is derived in the mammalian and avian lineages through convergent evolution (Figure 8.4b). This hypothesis is supported by a great deal of correlational data leading to the

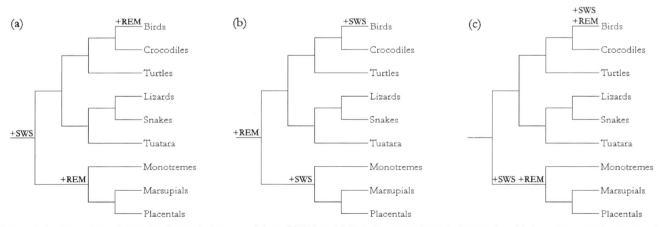

Figure 8.4 Three hypotheses for the evolutionary origins of SWS and REM sleep seen in endotherms (i.e., birds and mammals; see text for description). Phylogeny adapted from [10].

following four conclusions. (1) Monotremes, the oldest extant group of mammals, exhibit REM sleep (see Monotremes section) and, interestingly, the platypus, the oldest of the monotremes, may engage in more REM sleep than any other mammal. The identification of REM sleep in both the echidna and platypus suggests that REM sleep originated earlier in mammalian evolution than had previously been thought. Furthermore, it suggests that REM sleep, or a REM sleep-like state, was present in the reptilian ancestor to both the mammalian and avian lineage. (2) During REM sleep, an endotherm's thermoregulatory mechanisms are either suspended or impaired. Therefore an endothermic animal is essentially poikilothermic during REM sleep. Assuming that REM sleep is indeed the antecedent sleep state exhibited by our reptilian ancestors, such heterothermic animals would not be adversely affected by the inhibition of, albeit, nonexistent metabolic heat production mechanisms. However, with the evolution of endothermy, REM sleep became detrimental to survival. A second sleep state (i.e., SWS) evolved to permit thermoregulation concurrent with sleep. (3) The ontogenesis of the sleep cycle may also be used as evidence for the REM sleep-first hypothesis. REM sleep is the dominant sleep state of the mammalian fetus. After birth, REM sleep time decreases rapidly until it stabilizes at adult levels. The decrease in REM sleep is marked by a concurrent increase in SWS and wakefulness throughout early maturation. (4) Lastly, whereas the slow waves of mammalian SWS are propagated from the neocortex, REM sleep originates in the rhombencephalon (pons) of the brainstem—the most ancient caudal brain structure. Thus evolution of the SWS/REM sleep cycle may have begun with REM sleep and not SWS as once thought. The obvious drawback of this hypothesis is the absence of unequivocal REM sleep in extant reptiles. However, REM sleep or a REM sleep-like state may be highly temperature sensitive as it is in mammals, and future studies of sleep in reptiles must examine this factor further.

The existence of alternating SWS and REM sleep in the mammalian and avian lineages may be due to convergent evolution and not to inheritance from a common reptilian ancestor (Figure 8.4c), since evidence for unequivocal SWS and REM sleep in reptiles is controversial. Although REM sleep in placental and marsupial mammals is characterized by low-amplitude EEG activity, REM sleep in the platypus was associated with SWS-like cortical activity [14]. REM sleep with an activated EEG may have thus evolved in marsupial and placental mammals after divergence of the monotreme line. Moreover, due to the absence of unambiguous REM sleep and SWS in extant reptiles, SWS and REM sleep with cortical activation may have evolved twice: once in the mammalian and once in the avian clades.

Sleep differs from wakefulness on all levels of organization: behavioral, electrophysiological, cellular, molecular, and genetic. Indeed, sleep and wakefulness favor different cellular processes [61]. Present information suggests that a proximate sleep function is likely focused at the level of the neuron or synapse rather than the organ or tissue. Current criteria for sleep are based largely on changes in neural activity. In instances where such indicators are cryptic or absent (see Reptiles section and Amphibians and Fishes section), new criteria are needed, perhaps at the level of gene expression. This does not discount a functional significance for the EEG correlates of sleep in mammals and birds; it merely acknowledges that since invertebrates also sleep, the physiological criteria for sleep should not be limited to EEG correlates, which are clearly not shared by invertebrates, fishes, amphibians, and reptiles. Such investigations can shed much light on the nature and functions of sleep.

ACKNOWLEDGMENTS

We would like to thank Steven Lima and Peter Scott for their thoughtful and constructive comments on this chapter.

REFERENCES

1. Rechtschaffen A. Current perspectives on the function of sleep. *Perspect Biol Med* **41**:359–390(1998).

2. Campbell SS, Tobler I. Animal sleep: a review of sleep duration across phylogeny. *Neurosci Biobehav Rev* **8**:269–300(1984).

3. Hartse KM. Sleep in insects and nonmammalian vertebrates. In: Kryger MH, Roth T, Dement WC (Eds), *Principles and Practice of Sleep Medicine*, 2nd ed. Saunders, Philadelphia, 1994, pp 95–104.

4. Nicolau MC, Akaârir M, Gamundí A, González J, Rial RV. Why we sleep: the evolutionary pathway to the mammalian sleep. *Prog Neurobiol* **62**:379–406(2000).

5. Tobler I. Phylogeny of sleep regulation. In: Kryger MH, Roth T, Dement WC (Eds), *Principles and Practice of Sleep Medicine*, 3rd ed. Saunders, Philadelphia, 2000, pp 72–81.

6. Zepelin H. Mammalian sleep. In: Kryger MH, Roth T, Dement WC (Eds), *Principles and Practice of Sleep Medicine*, 3rd ed. Saunders, Philadelphia, 2000, pp 82–92.

7. Piéron H. *Le Problème Physiologique du Sommeil*. Masson, Paris, 1913 (in French).

8. Flanigan WF. Behavioral states and electroencephalograms of reptiles. In: Chase MH (Ed), *The Sleeping Brain. Perspectives in the Brain Sciences.* Brain Information Service/Brain Research Institute, UCLA, Los Angeles, pp 14–18.

9. Tobler I. Deprivation of sleep and rest in vertebrates and invertebrates. In: Inoué S, Borbély AA (Eds), *Endogenous Sleep*

Substances and Sleep Regulation. Japan Scientific Societies Press, Tokyo, 1985, pp 57–66.

10. Meyer A, Zardoya R. Recent advances in the (molecular) phylogeny of vertebrates. *Annu Rev Ecol Evol Syst* **34**:311–338(2003).

11. Allison T, van Twyver H, Goff WR. Electrophysiological studies of the echidna, *Tachyglossus aculeatus.* I. Waking and sleep. *Arch Ital Biol* **110**:145–184(1972).

12. Siegel JM, Manger PR, Nienhuis R, Fahringer HM, Pettigrew JD. The echidna *Tachyglossus aculeatus* combines REM and non-REM aspects in a single sleep state: implications for the evolution of sleep. *J Neurosci* **16**:3500–3506(1996).

13. Nicol SC, Andersen NA, Phillips NH, Berger RJ. The echidna manifests typical characteristics of rapid eye movement sleep. *Neurosci Lett* **283**:49–52(2000).

14. Siegel JM, Manger PR, Nienhuis R, Fahringer HM, Shalita T, Pettigrew JD. Sleep in the platypus. *Neuroscience* **91**:391–400(1999).

15. Rattenborg NC, Amlaner CJ, Lima SL. Behavioral, neurophysiological and evolutionary perspectives on unihemispheric sleep. *Neurosci Biobehav Rev* **24**:817–842(2000).

16. Kovalzon VM, Mukhametov LM. Temperature variations in the brain corresponding to unihemispheric slow wave sleep in dolphins. *J Evol Biochem Physiol* **18**:307–309(1982) (in Russian).

17. Mukhametov LM, Polyakova IG. EEG investigation of sleep in porpoises (*Phocoena phocoena*). *J High Nerve Activ* **31**:333–339(1981) (in Russian).

18. Goley PD. Behavioral aspects of sleep in Pacific white-sided dolphins (*Lagenorhynchus obliquidens*, Gill 1865). *Mar Mam Sci* **15**:1054–1064(1999).

19. Oleksenko AI, Mukhametov LM, Polyakova IG, Supin AY, Kovalzon VM. Unihemispheric sleep deprivation in bottlenose dolphins. *J Sleep Res* **1**:40–44(1992).

20. Mukhametov LM, Lyamin OI, Chetyrbok IS, Vassilyev AA, Diaz R. Sleep in an Amazonian manatee, *Trichechus inunguis. Experientia* **48**:417–419(1992).

21. Zepelin H, Rechtschaffen A. Mammalian sleep, longevity and energy metabolism. *Brain Behav Evol* **10**:425–470 (1974).

22. Elgar MA, Pagel MD, Harvey PH. Sleep in mammals. *Anim Behav* **36**:1407–1419(1988).

23. Allison T, Cicchetti DV. Sleep in mammals: ecological and constitutional correlates. *Science* **194**:732–734(1976).

24. Elgar MA, Pagel MD, Harvey PH. Sources of variation in mammalian sleep. *Anim Behav* **40**:991–995(1990).

25. Ramanathan L, Gulyani S, Nienhuis R, Siegel JM. Sleep deprivation decreases superoxide dismutase activity in rat hippocampus and brainstem. *Neuroreport* **13**:1387–1390 (2002).

26. Siegel JM. Why we sleep. *Sci Am* **Nov**:72–77(2003).

27. Amlaner CJ, Ball NJ. Avian sleep. In: Kryger MH, Roth T, Dement WC (Eds), *Principles and Practice of Sleep Medicine,* 2nd ed. Saunders, Philadelphia, 1994, pp 81–94.

28. Rattenborg NC, Mandt BH, Obermeyer WH, Winsauer PJ, Huber R, Wikelski M, Benca RM. Migratory sleeplessness in the white-crowned sparrow (*Zonotrichia leucophrys gambelii*). *PLoS Biol* **2**:924–936(2004).

29. Ball NJ, Amlaner CJ, Shaffery JP, Opp MR. Asynchronous eye-closure and unihemispheric quiet sleep of birds. In: Koella WP, Obál F, Schulz H, Visser P (Eds), *Sleep '86.* Gustav Fischer Verlag, New York, 1988, pp 151–153.

30. Spooner CE. *Observations on the Use of the Chick in the Pharmacological Investigation of the Central Nervous System.* Ph.D. dissertation, University of California, Los Angeles.

31. Rattenborg NC, Lima SL, Amlaner CJ. Half-awake to the risk of predation. *Nature* **397**:397–398(1999).

32. Rattenborg NC, Amlaner CJ, Lima SL. Unilateral eye closure and interhemispheric EEG asymmetry during sleep in the pigeon (*Columba livia*). *Brain Behav Evol* **58**:323–332(2001).

33. Amlaner CJ, Ball NJ. A synthesis of sleep in wild birds. *Behaviour* **87**:85–119(1983).

34. Schmidt DF. *A Comparative Analysis of Avian Sleep.* Ph.D. dissertation, University of Arkansas, Fayetteville, 1994.

35. Taylor L, Vana A, Givon L. The evolution of sleep: a reconsideration of the development of the quite sleep/active sleep cycle. *Med Hypotheses* **54**:761–766(2000).

36. Flanigan WF, Wilcox RH, Rechtschaffen A. The EEG and behavioral continuum of the crocodilian, *Caiman sclerops. Electroencephalogr Clin Neurophys* **34**:521–538(1973).

37. Warner BF, Huggins SE. An electroencephalographic study of sleep in young caimans in a colony. *Comp Biochem Physiol* **59**:139–144(1978).

38. Meglasson MD, Huggins SE. Sleep in a crocodilian, *Caiman sclerops. Comp Biochem Physiol* **63A**:561–567(1979).

39. Flanigan WF, Knight CP, Hartse KM, Rechtschaffen A. Sleep and wakefulness in Chelonian reptiles. I. The box turtle, *Terrapene carolina. Arch Ital Biol* **112**:227–252(1974).

40. Flanigan WF. Sleep and wakefulness in Chelonian reptiles. II. The red-footed tortoise, *Geochelone carbonaria. Arch Ital Biol* **112**:253–277(1974).

41. Hartse KM, Rechtschaffen A. Effect of atropine sulfate on the sleep-related EEG spike activity of the tortoise, *Geochelone carbonaria. Brain Behav Evol* **9**:81–94(1974).

42. Hartse KM, Rechtschaffen A. The effect of amphetamine, nembutal, alpha-methyl-tyrosine, and parachlorophenylalanine on the sleep-related spike activity of the tortoise, *Geochelone carbonaria*, and on the cat ventral hippocampus spike. *Brain Behav Evol* **21**:199–222(1982).

43. Huntley AC. Electrophysiological and behavioral correlates of sleep in the desert iguana, *Dipsosaurus dorsalis hallowell. Comp Biochem Physiol* **86A**:325–330(1987).

44. Tauber ES, Rojas-Ramírez J, Hernández-Peon R. Electrophysiological and behavioral correlates of wakefulness and sleep in the lizard, *Ctenosaura pectinata. Electroencephalogr Clin Neurophysiol* **24**:424–433(1968).

45. Tauber ES, Roffwarg HP, Weitzman ED. Eye movements and electroencephalogram activity during sleep in diurnal lizards. *Nature* **212**:1612–1613(1966).

46. Ayala-Guerrero F, Huitrón-Reséndiz S. Sleep patterns in the lizard *Ctenosaura pectinata*. *Physiol Behav* **49**:1305–1307(1991).

47. Flanigan WF. Sleep and wakefulness in iguanid lizards, *Ctenosaura pectinata* and *Iguana iguana*. *Brain Behav Evol* **8**:401–436(1973).

48. Mathews CG, Amlaner CJ. Eye states and postures of the western fence lizard (*Sceloporus occidentalis*), with special reference to asynchronous eye closure and behavioral sleep. *J Herp* **34**:472–475(2000).

49. Shaw PJ, Cirelli C, Greenspan RJ, Tononi G. Correlates of sleep and waking in *Drosophila melanogaster*. *Science* **287**:1834–1837(2000).

50. Cirelli C. Searching for sleep mutants of *Drosophila melanogaster*. *BioEssays* **25**:940–949(2003).

51. Tobler I, Stalder J. Rest in the scorpion—a sleep-like state? *J Comp Physiol A* **163**:227–235(1988).

52. Kaiser W, Steiner-Kaiser J. Neuronal correlates of sleep, wakefulness and arousal in a diurnal insect. *Nature* **301**:707–709 (1983).

53. Sauer S, Kinkelin M, Herrmann E, Kaiser W. The dynamics of sleep-like behavior in honey bees. *J Comp Physiol A* **189**:599–607(2003).

54. Schuppe H. Rhythmic brain activity in sleeping bees. *Wien Med Wochenschr* **145**:463–464(1995) (in German).

55. Sauer S, Herrmann E, Kaiser W. Sleep deprivation in honey bees. *J Sleep Res* **13**:145–152(2004).

56. Hendricks JC, Finn SM, Panckeri KA, Chavkin J, Williams JA, Sehgal A, Pack AI. Rest in drosophila is a sleep-like state. *Neuron* **25**:129–138(2000).

57. Shaw PJ, Tononi G, Greenspan RJ, Robinson DF. Stress response genes protect against lethal effects of sleep deprivation in *Drosophila*. *Nature* **417**:287–291(2002).

58. Nitz DA, van Swinderen B, Tononi B, Greenspan RJ. Electrophysiological correlates of rest and activity in *Drosophila melanogaster*. *Curr Biol* **12**:1934–1940(2002).

59. Hendricks JC, Sehgal A. Why a fly? Using drosophila to understand the genetics of circadian rhythms and sleep. *Sleep* **27**:334–342(2004).

60. Tafti M, Franken P. Functional genomics of sleep and circadian rhythm: invited review: genetic dissection of sleep. *J Appl Physiol* **92**:1339–1347(2002).

61. Cirelli C, Gutierrez CM, Tononi G. Extensive and divergent effects of sleep and wakefulness on brain gene expression. *Neuron* **41**:35–43(2004).

62. Mackiewicz M, Pack AI. Functional genomics of sleep. *Respir Physiol Neurobiol* **135**:207–220(2003).

63. Ramón F, Hernández-Falcón J, Nguyen B, Bullock TH. Slow wave sleep in crayfish. *Proc Natl Acad Sci* **101**:11857–11861(2004).

64. Mather JA, Mather DL. Skin colours and patterns of juvenile *Octopus vulgaris* (Mollusca, Cephalopoda) in Bermuda. *Vie Milieu* **44**:267–272(1994).

65. Rechtschaffen A, Gilliland MA, Bergmann BM, Winter JB. Physiological correlates of prolonged sleep deprivation in rats. *Science* **221**:182–184(1983).

66. Siegel JM, Manger PR, Nienhuis R, Fahringer HM, and Pettigrew JD. Monotremes and the evolution of rapid eye movement sleep. *Philos Trans R Soc London B* **353**:1147–1157(1998).

9

NEUROPHARMACOLOGY OF SLEEP AND WAKEFULNESS

EDGAR GARCIA-RILL, TIFFANY WALLACE, AND CAMERON GOOD
University of Arkansas for Medical Sciences, Little Rock, Arkansas

INTRODUCTION

There are two main regions that modulate our sleep–wake cycles—the mesopontine reticular activating system (RAS) and the hypothalamus. In addition, the intralaminar thalamus and the basal forebrain are modulated by the RAS and hypothalamus and participate in the process of arousal and alertness, as well as in the modulation of sleep states. Neuroactive agents that modulate these regions will also modulate the level of arousal. By the same token, disorders that manifest changes in sleep–wake states and/or affect arousal, alertness, and sleep can be expected, of necessity, to include dysregulation in the above-named regions. Many neurological and psychiatric disorders involve, and may even be presaged by, disruption of sleep–wake control regions. For example, many patients with Parkinson's disease exhibit sleep dysregulation years before the clinical signs of the disease, and almost half of REM sleep behavior disorder (RBD) patients will develop Parkinson's disease as many as 13 years after developing the symptoms of RBD [1]. Sleep dysregulation is a hallmark of psychiatric disorders such as schizophrenia, anxiety disorders, and depression [2]. The hallucinations in schizophrenia have been equated with REM sleep intrusion into waking, that is, dreaming while awake [3]. Moreover, because the process of arousal is essential to attention, and attention to learning and memory, disruption of arousal-related systems has profound effects on higher cognitive functions. That is, sleep–wake disorders, or the effects of psychoactive agents on these systems, may be at the root of decrements in cognitive performance.

CONNECTIVITY

The RAS is composed of three main nuclei—the cholinergic pedunculopontine nucleus (PPN) (and its medial partner, the laterodorsal tegmental nucleus), the noradrenergic locus coeruleus (LC), and the serotonergic raphe nucleus (RN). The RN sends inhibitory projections to the PPN and LC, and the LC inhibits the PPN while the PPN activates the LC. The RAS sends the majority of its ascending cholinergic and monoaminergic projections to the thalamus and hypothalamus, while also synapsing on other regions [4]. During waking, all three cell groups are active while, in slow-wave sleep, the cholinergic cells decrease firing while monoaminergic cells remain active. However, in REM sleep, the cholinergic cells are highly active while monoaminergic cells decrease their firing rates markedly [4]. Therefore cholinergic RAS neurons are active during waking and REM sleep, that is, during synchronization of fast cortical rhythms, but slow their firing during synchronization of slow cortical rhythms [5].

The RAS receives input from all afferent sensory systems in parallel to primary afferent sensory projections. That is, the "nonspecific" projection system to the RAS relays "arousal" information through the intralaminar thalamus (ILT) to the cortex. This system functions in parallel to the shuttling of "specific" sensory information through

Correspondence to: Edgar Garcia-Rill, Center for Translational Neuroscience, Department of Neurobiology & Developmental Sciences, University of Arkansas for Medical Sciences, 4301 W. Markham St., Little Rock, AR 72205.

the primary sensory thalamic nuclei to the cortex. It is the temporal summation of intralaminar "nonspecific" inputs (the context) with primary "specific" inputs (the content) at the level of the cortex that is thought to participate in "binding," that is, the conscious perception of a sensory event [6]. Disturbances in RAS driving of the ILT and/or of thalamocortical reverberating activity thus can be expected to lead to disturbances in perception.

The RAS also sends descending projections to postural and locomotion control systems. Such connectivity allows the RAS to act as the "fight-or-flight" control system, simultaneously activating higher centers while priming motor systems to respond appropriately to sudden stimuli. These descending projections modulate the (1) pontine inhibitory area (PIA) that is thought to control the atonia of REM sleep, (2) pontine neurons that generate the startle response, a flexor response that primes the motor system, and (3) reticulospinal systems that drive locomotion [7]. RAS projections to the dorsal subcoeruleus region also modulate the generation of pontogeniculo-occipital (PGO) waves during REM sleep. This region generates high-frequency bursts of activity (like those required for long-term potentiation) that have been proposed to promote consolidation of certain memories during REM sleep via its projections to the hippocampus [8].

The hypothalamic sleep–wake modulatory system is composed mainly of the (1) tuberomammillary nucleus (TMN) with excitatory histaminergic projections, (2) lateral hypothalamus (LH) with excitatory orexinergic projections, and (3) ventrolateral preoptic (VLPO) region with mostly inhibitory GABAergic projections. These hypothalamic sleep–wake modulatory systems are thought to help stabilize sleep–wake states [9]. On the one hand, the excitatory LH orexin projections can be thought of as an "on" switch promoting waking, especially through their activation of excitatory TMN histaminergic neurons, which are tonically active during waking (and significantly decrease firing during sleep), and of excitatory basal forebrain and RAS, especially cholinergic, neurons. The basal forebrain cholinergic projection system is especially active during waking and serves to raise the excitability of the cortex. Interestingly, acetylcholine release from the basal forebrain is greater during REM sleep than during waking [10]. On the other hand, the inhibitory VLPO GABAergic projections can be thought of as an "off" switch promoting slow-wave sleep through its inhibition of the RAS, LH, basal forebrain, and cortex.

Figure 9.1 outlines the main ascending and descending projections described. The dorsal ascending cholinergic (labeled PP for PPN) and monoaminergic (LC, RN) projections from the RAS to the ILT serve to activate the cortex via thalamocortical projections. There is also a massive set of ventral projection systems that bypass the thalamus to terminate diffusely throughout the cortex. These originate

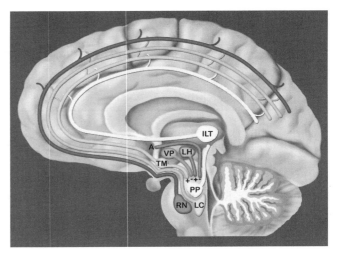

Figure 9.1

in the RAS (noradrenergic LC and serotonergic RN) and are joined by ascending projections from the TM (histaminergic TMN), LH (orexinergic), and VP (GABAergic VLPO), as well as from the basal forebrain (acetylcholine, not shown). In turn, the TMN, LH, and VLPO send descending projections to the RAS that may act reciprocally to stabilize sleep–wake states.

In addition to these transmitters, adenosine (A, diffusely localized) is a ubiquitous homeostatic factor thought to be involved in sleep–wake regulation [11]. Conditions of high metabolic activity or prolonged wakefulness lead to a buildup of adenosine, which decreases subsequent to sleep. Therefore adenosine also may modulate sleep–wake states via its inhibitory actions on most cells, but particularly on excitatory cholinergic RAS and basal forebrain neurons. Adenosine injections into the RAS are known to decrease waking, while adenosine levels in the basal forebrain (but not in the thalamus) progressively increase during prolonged wakefulness and decrease during subsequent recovery of sleep [12].

The close relationship between sleep–wake regulation and other homeostatic control functions should be remembered. These systems, especially hypothalamic sleep–wake modulatory regions, interact with the regulation of food intake, metabolism, hormone release, and temperature [13]. This means that disorders of hypothalamic sleep–wake modulatory regions, or psychoactive agents that modulate them, can be expected to also affect homeostatic control systems. For example, Kleine–Levin syndrome is a postpubertal onset disorder characterized by episodes of hypersomnia, mood disturbances (especially depression), compulsive hyperphagia (especially carbohydrates), hypersexuality, and signs of dysautonomia. This disorder points to a pathological locus bridging sleep–wake control, mood control, and homeostatic control systems.

BLOOD FLOW DURING SLEEP AND WAKING

The state of slow-wave sleep is marked by decreases in regional cerebral blood flow throughout the brain, but more significantly in the RAS, thalamus, hypothalamus, and basal forebrain [14]. This state is characterized by decreased activity of cholinergic RAS and basal forebrain neurons, of LH orexinergic cells, and of TMN histaminergic cells. That is, the major excitatory projection systems decrease their outputs during slow-wave sleep. On the other hand, REM sleep is characterized by increased blood flow in the RAS, the thalamus, and the anterior cingulate cortex, among others, with decreases in blood flow in the dorsolateral prefrontal cortex [15]. It has been proposed that the unregulated activity of RAS cholinergic neurons is responsible for REM sleep and, via unknown mechanisms, for decreased frontal lobe blood flow, or "hypofrontality" [2]. The hypofrontality of REM sleep is thought to account for the lack of critical judgment during dreaming (and during hallucinations). This state is also characterized by the generation of PGO waves (triggered by descending cholinergic PPN projections to the pons), now thought to be involved in some aspects of sleep-dependent plasticity. Unlike sleep, the process of awakening entails two stages, a rapid (5 min) reestablishment of consciousness that is marked by increases in cerebral blood flow in the RAS and thalamus, followed by a slower (15 min) increase in cerebral blood flow, primarily in anterior cortical regions [16]. Therefore psychoactive agents that affect blood flow can be expected to alter sleep–wake states. It should also be noted that cholinergic RAS neurons have some of the highest concentrations of nitric oxide in the brain. Therefore wherever the PPN projects, a corollary effect may include vasodilation.

NEUROPHARMACOLOGY

The foregoing suggests that psychoactive agents that affect the function of sleep–wake regulating systems also can have profound effects on a host of processes, from higher cognitive performance to attention, learning, and memory, to homeostatic regulation and more. The following is a brief description of the effects of only the most common agents.

Alcohol

Alcohol appears to preferentially affect small neurons, particularly granule cells throughout the cerebral and cerebellar cortices and the hippocampus, perhaps by enhancing GABAergic transmission. Direct effects on sleep–wake control regions appear to also involve potentiation of GABAergic transmission. However, at high doses, its effects are neurotoxic, mainly on basal forebrain neurons,

and thus may impair diffuse cholinergic input to the cortex. It should be noted that significant impairment in motor performance, such as driving, occurs at very low blood alcohol concentrations, an effect potentiated by sleep deprivation or sleep loss. Most alcoholic patients suffer from insomnia, which is clinically important since alcoholism can exacerbate the adverse consequences of insomnia, such as mood changes and anxiety, and because insomnia has been associated with alcohol relapse. In general, sleep loss has greater sedative effects than low doses of alcohol, but similar effects on psychomotor performance. Alcohol produces greater memory impairment than sleep loss, probably because of its marked effects on the hippocampus.

Alcohol, aside from its recreational uses, is the prototypical anxiolytic, having a calming effect on the stressed, or overstressed, organism. It is evident that alcohol intake is a form of self-medication used as an anxiolytic by patients suffering from psychiatric disorders that involve hypervigilance, for example, schizophrenia, anxiety disorders, and depression. It is likely that alcohol treatment programs could increase their success rates if patients who ingest excessive amounts of alcohol, but are not true "alcoholics," would be properly diagnosed and treated. Such patients, in the absence of appropriate treatment for their disorder, are certain to relapse into the alternative of self-medication. Because alcohol ingestion can lead to decreased frontal lobe blood flow, the ultimate effect will be to exacerbate the hypofrontality already evident in these psychiatric patients, further impairing decision-making capacity and lowering the threshold for uncritical action.

Anesthetics and Sedatives

The proposed mechanisms of action of anesthetics have typically involved a myriad of cellular effects at different sites by disparate drugs. Recent evidence suggests that the primary site of action of most anesthetics may be the sleep–wake control system [17]. The parallel manifestations between sleep and anesthesia suggest that anesthetics basically "hijack" the sleep–wake control system to induce anesthesia. This concept allows for a more rational characterization of traits that should be used to determine anesthetic level. That is, considering that arousal and alertness represent a continuum of levels from mania to coma, with physiological and behavioral concomitants, the monitoring of EEG, along with behavioral and autonomic signs, should be used routinely to assess level of anesthesia.

Most anesthetics, including barbiturates, etomide, propofol, neuroactive steroids, and volatile anesthetics, act on GABAa receptors among other receptors [18]. Sedation and natural sleep occur greatly as a result of enhanced GABAergic transmission, which in turn affects the release of a number of excitatory transmitters such as acetylcholine, excitatory amino acids, and histamine. That is, these actions may take place specifically in such regions as the RAS, TMN, and basal forebrain (all of which have local circuit GABAergic neurons and receive GABAergic input from VLPO, as described earlier), thereby regulating the level of arousal.

Most barbiturates are dangerous drugs with a narrow therapeutic index between the dose required for sedation and the dose that will cause coma and death. These agents typically decrease cerebral blood flow, although regional differences between agents are evident. Volatile and steroid anesthetics also are known to reduce cerebral blood flow along with cerebral oxygen metabolism, an effect that maintains the coupling between metabolism and flow.

The benzodiazepines act by binding to a site that modulates GABA receptors, especially GABAa receptors. These agents produce sedative, hypnotic, anxiolytic, and anticonvulsant activities. They act generally by amplifying GABAergic transmission, such that short-acting agents have been used to promote sleep in insomnia patients, although more recently, effective nonbenzodiazepine hypnotics have been developed. These agents also act to facilitate GABAa receptor function (e.g., zolpidem and zaleplon). Insomnia is a very common symptom, especially in the elderly, and has a number of causes, including physical, social, and psychiatric. Treatment of such causes, rather than symptomatic alleviation of insomnia, obviously is more desirable. Similarly, use of benzodiazepines as anxiolytic agents represents symptomatic treatment, requiring the assessment of the underlying causes in order to prevent prolonged treatment with potential for leading to dependence. Some psychiatric conditions are characterized by insomnia, mainly due to the nighttime manifestation of hypervigilance, increased REM sleep drive (decreased REM sleep latency, increased REM sleep duration, sleep fragmentation due to frequent awakenings, etc.). Identification of the underlying mechanism allows for the design of a more rational therapeutic strategy.

A naturally occurring metabolite of GABA, gammahydroxybutyrate (GHB), is a potent CNS depressant, and acute intoxication with GHB or its analogs can lead to respiratory depression and even death. Like most hypnotics, GHB can induce tolerance and produce dependence. In pharmacological doses, it is used as a sedative/anesthetic, in alcohol/opiate detoxification, and for the treatment of cataplexy in narcolepsy. Narcolepsy is a disorder marked by significant daytime sleepiness, hypnagogic hallucinations, and episodes of cataplexy, loss of consciousness, and postural collapse, which sometimes occur with affective incitement. These episodes are reminiscent of the loss of consciousness and of postural muscle tone that accompanies REM sleep, almost as if the patient transitions directly from waking to REM sleep without passing through the requisite slow-wave sleep state. Narcolepsy is thought to arise from degeneration of excitatory orexinergic neurons in the LH. How does GHB act to decrease cataplexy? The cellular mechanisms are unknown, but one possible mechanism may be through direct or indirect activation of GABAb receptors, which can inhibit PPN neurons (which induce REM sleep) and elicit slow-wave activity, including spike and wave activity via the thalamus, a form of nonconvulsive epilepsy. This agent, when administered before bedtime, appears to induce the symptoms of narcolepsy and contain them at night [19]. High doses of GHB can decrease glucose metabolism but, surprisingly, do not significantly alter global blood flow.

Antihistamines

Pathology and lesions of the TMN cause hypersomnia (recall that these neurons are highly active during waking). Histaminergic inputs to the RAS suppress slow-wave sleep and promote waking, although they do not affect REM sleep significantly [20]. Administration of antihistamines (histamine receptor blockers) results in sedation. Such an effect may result from blockade of histaminergic inputs to the RAS, basal forebrain, and/or LH. It should be noted that the cortex has the highest concentration of histamine receptors, so that widespread changes in cortical excitability are also possible through that mechanism. Antihistamines reduce blood flow in frontal cortex and midbrain, which could also account for the cognitive impairments and decrement in psychomotor function observed. Some tolerance can develop over time that can decrease such impairments. In children, first- and second-generation antihistamine intoxication can induce coma, although the newer (third-generation) pediatric formulations (e.g., fexofenadine, loratadine, cetirizine) appear to be safer.

Caffeine

The popularity of caffeine is related to its stimulant properties, which are mediated by its ability to reduce adenosine release in the brain. Caffeine appears to block adenosine A1 and A2a receptors, producing a psychomotor stimulant effect. Because of the high levels of A2a receptors in the striatum, the potential use of caffeine for the treatment of Parkinson's disease has been advanced. Since adenosine A2a receptor blockade appears to protect dopaminergic neurons from toxic agents, a neuroprotective role has been proposed for caffeine in the treatment of Parkinson's disease. Caffeine intake has also been associated with a decreased risk of Alzheimer's disease, again presumably acting as a neuroprotective agent.

Caffeine is known to lower cerebral blood flow while simultaneously inducing an increase in metabolism through inhibition of adenosine receptors, leading to a state of relative hypoperfusion for prolonged periods of time. However, its alerting effects are obviously mediated by inhibition of adenosinergic inputs to RAS and basal forebrain cholinergic neurons (see earlier discussion).

Nicotine

Inhaled nicotine in cigarette smoke is known to permeate the lungs where more than 80% of the available nicotine is absorbed into the bloodstream. The short delivery time and elimination half-lives (8 min and 2 h, respectively) assure that, within a short time, the effect can be reproduced by smoking another cigarette [21]. After absorption into the blood, nicotine readily crosses the blood–brain barrier and appears to be rapidly partitioned into brain tissue. Concentrations of nicotine in the brain have been reported to be 5–7 times higher than blood concentrations. Smokers assert that, in addition to its positive effects on concentration and attention, the primary positive effect of smoking is that it calms and relaxes. Recent findings suggest that one of the sites of action of nicotine may be in the RAS, specifically, on PPN neurons. One study found that systemic administration of nicotine, or localized injection of a nicotinic receptor agonist into the PPN, led to a dose-dependent decrease in the amplitude of the P13 potential in the rat, the rodent equivalent of the sleep state-dependent P50 potential in the human [22]. These results suggest that nicotinic agonists, at least initially, may reduce the level of arousal, as manifested by the amplitude of this waveform.

Figure 9.2 is from a study that used intracellular recordings from PPN neurons in brainstem slices. Application of a nicotinic agonist (1,1-dimethyl-4-phenyl-piperazinium—DMPP) directly (in the presence of tetrodotoxin—TTX) hyperpolarized 20% of PPN cholinergic neurons, an effect blocked by pretreatment with the nicotinic receptor blocker mecamylamine (MEC). DMPP also directly depolarized 10%, and indirectly depolarized 70%, of PPN neurons. These results suggest that nicotine, at least initially, has an inhibitory effect on cholinergic RAS neurons, which could produce the calming effect reported upon inhalation of cigarette smoke. Additional effects on other populations of cells also imply that there is a complex interaction between nicotine and the RAS. Some of the indirect effects appear to be mediated by nicotine's ability to presynaptically modulate the release of a number of transmitter systems.

The proposed inhibitory effect of nicotine on the RAS is in keeping with clinical evidence. The majority of cigarettes are consumed by the mentally ill, especially those with disorders involving hypervigilance or hyperarousal, such as schizophrenia, anxiety disorders, and depression [23]. That is, smoking may be a form of self-medication, presumably because of its calming effects. This effect (inhibition of cholinergic RAS neurons) appears to differ from the role of smoking in reducing the incidence of Parkinson's disease, which appears to be manifested as a neuroprotective action on dopaminergic neurons by nicotine.

Cerebral vasodilation is seen immediately after smoking, but chronic smokers show global reductions in cerebral blood flow. Considering that hypofrontality is present in

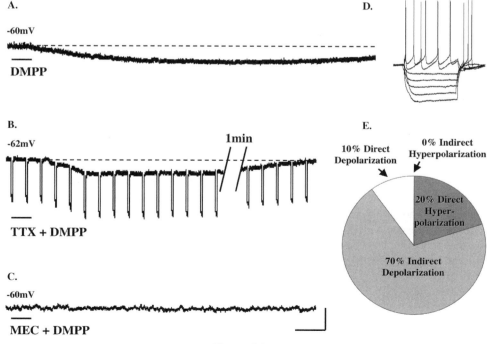

Figure 9.2

schizophrenia, anxiety disorders, and depression, the initial beneficial, calming effects of nicotine may be followed by deleterious consequences on cortical blood flow. Such an effect may drive craving for the next cigarette, creating a vicious cycle of continuous self-administration.

Stimulants

The most common stimulant, amphetamine, induces release of monoamines, especially dopamine, but also blocks reuptake and may have neurotoxic effects on nigral neurons [24] and, more recently, is suspected of inducing degeneration of certain striatal neurons [25]. Unfortunately, this agent is abused for recreational purposes and continues to be prescribed for the treatment of attention deficit disorder (ADD). Fortunately, methylphenidate does not appear to have such neurotoxic effects, although its use has decreased. The difference between these agents appears to be that methylphenidate is mainly a dopamine uptake inhibitor without major influence on release. While any psychotropic agent can have deleterious effects on brain cells, particularly if abused, great care is required when using amphetamine, especially in the young. Amphetamine psychosis occurs in two forms—acute intoxication after a single large dose (characterized by confusion and disorientation), and chronic abuse after repeated use that produces a schizophrenia-like syndrome [26]. The increased release of dopamine by amphetamine is considered a model for schizophrenia and contributed to the "dopamine theory" of schizophrenia (see later discussion).

Methamphetamine, a popular street drug similar to amphetamine, has become widely abused and probably has severe neurotoxic effects. One potential mechanism via which these agents promote hypervigilance is through activation of the striatum and disinhibition of cholinergic RAS neurons. A more direct effect would be induced release of dopamine and noradrenaline at the level of the cortex. These effects are accompanied by transient increases in cerebral blood flow (midbrain, thalamus, and frontal cortex), but abstinent abusers are known to have decreased blood flow in basal ganglia and certain cortical areas; that is, long-term effects on blood flow may be deleterious. MDMA (3,4-methylenedioxymethamphetamine) or "ecstasy" is another recreational abused amphetamine that is even more neurotoxic, has hallucinogenic properties at high doses, and has been linked to a number of deaths.

Modafinil is a newer stimulant that does not appear to act through dopaminergic mechanisms, like amphetamine. Modafinil does seems to affect structures involved in the regulation of sleep–wake states [27] and to affect a number of transmitter systems, including noradrenergic, histaminergic, and orexinergic, as well as excitatory amino acid and serotonin release. In addition, it may block GABAa receptors [28]. Modafinil has been found to be effective in the treatment of daytime sleepiness in patients with narcolepsy [29]. This agent was recently reported to block "spatial neglect" resulting from cortical stroke [30]. This revolutionary finding suggests that sensory neglect arising from stroke (right hemisphere strokes typically lead to persistent neglect of the left spatial field, whereas left hemisphere strokes lead to only transient neglect of the contralateral spatial field) can be treated successfully by "waking up" the involved cortex, implying that the consequences of stroke may be to decrease activity, blood flow, or metabolism, one or more of which modafinil may counteract. It should be noted that the potential blockade of GABAa receptors may make it undesirable for use in patients whose condition includes epilepsy or heightened susceptibility to seizures.

However, the wake-promoting property of modafinil may make it beneficial for the treatment of other disorders involving decreased activity, blood flow, or metabolism. For example, the hypofrontality in various psychiatric disorders may be amenable to such therapy, as long as the hypervigilance present in these conditions is not exacerbated by using low doses. In addition, cocaine abusers show reduced frontal cortex blood flow, a mechanism thought to promote risky or erroneous decision-making [31]. Therefore the hypofrontality of drug abuse also may be amenable to correction using modafinil, perhaps also at low doses. The RAS is thought to be damaged in about 85% of cases of coma, with the rest accounted for by hypothalamic damage. While the use of amphetamine or methylphenidate has been advocated in patients with coma, a better alternative that does not induce significant cardiovascular effects could be modafinil.

SCHIZOPHRENIA, ANXIETY DISORDER, AND DEPRESSION

The connectivity of the RAS described earlier is reviewed in Figure 9.3, including the potential sites of action of therapeutic agents aimed at alleviating some of the symptoms of these disorders, especially those related to hypervigilance and sleep dysregulation, along with hypofrontality. The serotonergic RN is known to inhibit the PPN and LC, with the cholinergic PPN exciting the LC and the noradrenergic LC inhibiting, via alpha-2 adrenergic receptors, the PPN. The PPN sends excitatory cholinergic projections to the substantia nigra (SN), which, in turn, sends dopaminergic projections to the striatum.

The treatment of depression previously included tricyclic antidepressants such as amitryptiline, imipramine, and clomipramine, agents that mainly blocked reuptake of noradrenaline and serotonin, and blocked histamine and acetylcholine release, thus accounting for increased sleepiness. The selective serotonin reuptake inhibitors (SSRIs) more selectively affect the RAS by increasing the inhibition

Reticular Activating System

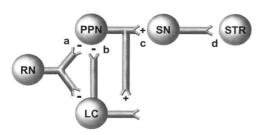

Figure 9.3

at site "a" in Figure 9.3, thus downregulating arousal levels, especially through promoting inhibition of the PPN and LC. It is not clear if the etiology of depression is related to disinhibition of the PPN and LC by a decrement in serotonergic tone, although this would seem a likely origin for the sleep–wake symptomatology of depression.

The treatment of anxiety disorder, as mentioned earlier, has involved the use of benzodiazepine amplification of GABAergic inhibition. In addition, the use of the alpha-2 noradrenergic receptor agonist clonidine produces anxiolytic effects, probably by inhibiting autoreceptors in the LC and postsynaptic receptors in the PPN (site "b" in Figure 9.3), thus downregulating vigilance. Because of the peripheral cardiovascular effects of clonidine, alpha-2 adrenergic receptor agonists without such actions would be more desirable. One study provided strong evidence for the use of the alpha-2 adrenergic receptor agonist dexmedetomidine as an anxiolytic for the treatment of anxiety disorders like post-traumatic stress disorder, panic attacks, and general anxiety disorder [32]. The etiology of anxiety disorder has been proposed to include downregulation or degeneration of LC outputs (possibly induced by stress hormones), which would act to release, or disinhibit, PPN neurons at site "b" in Figure 9.3.

The treatment of schizophrenia previously involved the use of the dopaminergic receptor blocker haloperidol, which induced tardive dyskinesia, among other serious side effects. Newer antipsychotics such as risperidone and quetiapine appear to block dopaminergic, noradrenergic, and serotonergic receptors. More striking antipsychotic effects were provided by the use of clozapine, which was designed as a muscarinic cholinergic blocker for the treatment of Parkinson's disease. The serious side effect of agranulocytosis made this a dangerous agent. However, one later-generation agent that has maintained anticholinergic activity without this side effect is olanzapine. These agents appear to have partial penetrance at serotonergic, cholinergic, and dopaminergic receptors, basically reducing muscarinic cholinergic activation of the SN (at site "c" in Figure 9.3), as well as partially blocking dopaminergic actions in the striatum (at site "d" in Figure 9.3). The etio-

logy of schizophrenia has been suggested to include increased PPN output, accounting for marked hypervigilance and hallucinations. Excessive PPN output would over-activate the SN and, in turn, increase striatal release of dopamine [33], that is, complying with the "dopamine theory" of schziophrenia. A recent review describes some of these, and additional, medications that affect sleep [34].

ADDITIONAL CONSIDERATIONS

Particular attention to hormonal conditions is warranted. After all, the first sign of puberty is pulsatile hormone (LH) release during sleep. For example, narcolepsy is tightly linked with certain human leukocyte antigen (HLA) haplotypes, suggesting that it is an autoimmune disorder. Kleine–Levin syndrome, discussed earlier, is linked to similar haplotypes, which suggests an autoimmune etiology [35]. Interestingly, in most cases of narcolepsy, Kleine–Levin syndrome, as well as schizophrenia, panic attacks, obsessive–compulsive disorder, and other disorders, the age of onset is soon after puberty. Along other lines, in about 20% of schizophrenic patients, the mother had an influenza attack during the second trimester, while narcoleptics are born predominantly during the late winter–early spring, that is, after influenza season. It has been suggested that developmental dysregulation, either pre- or perinatally (initial insult), becomes pathologically manifest after exposure to puberty and its hormonal onslaught [2, 36]. These considerations point to complex interactions between development, environment, and hormonal status, all of which seem to affect sleep–wake regulation in as yet unknown ways. These findings suggest that the effects of hormones, either prescribed or taken as dietary supplements, or abused, need to be more closely studied and considered in the design of therapeutic interventions.

ACKNOWLEDGMENT

Supported by USPHS award NS20246 and the Arkansas Biosciences Institute.

REFERENCES

1. Schenck CH, Bundlie SR, Mahowald MW. Delayed emergence of a parkinsonian disorder in 38% of 29 older men initially diagnosed with idiopathic rapid eye movement sleep behavior disorder. *Neurology* **46**:388–393(1996).

2. Garcia-Rill E. Disorders of the reticular activating system. *Med Hypoth* **49**:379–387(1997).

3. Dement WC. Studies on the effects of REM deprivation in humans and animals. *Res Publ Assoc Res Nerv Ment Dis* **43**:456–467(1967).

4. Garcia-Rill E. Mechanisms of sleep and wakefulness. In: Lee-Chiong T, Sateia MJ, Carskadon MA (Eds), *Sleep Medicine*. Hanley & Belfus, Philadelphia, 2002, pp 31–39.

5. Steriade M, Amzica F, Contreras D. Synchronization of fast (30–40 Hz) spontaneous cortical rhythms during brain activation. *J Neurosci* **16**:392–417(1996).

6. Llinas R, Ribary U, Joliot M, Wang, XJ. Content and context in temporal thalamocortical binding. In: Buzsaki G, et al. (Eds), *Temporal Coding in the Brain*. Springer-Verlag, Berlin, 1994, pp 251–272.

7. Reese NB, Garcia-Rill E, Skinner RD. The pedunculopontine nucleus—auditory input, arousal and pathophysiology. *Prog Neurobiol* **47**:105–133(1995).

8. Datta S, Patterson EH. Activation of phasic pontine wave (p-wave): a mechanism of learning and memory processing. In: Maquet J, Stickgold R, Smith C (Eds), *Sleep and Brain Plasticity*. Oxford University Press, Oxford, UK, 2003, pp 135–156.

9. Saper SB, Chou TC, Scammell TE. The sleep switch: hypothalamic control of sleep and wakefulness. *Trends Neurosci* **24**:726–731(2001).

10. Vazquez J, Baghdoyan HA. Basal forebrain acetylcholine release during REM sleep is significantly greater than during waking. *Am J Reg Integ Comp Physiol* **280**:R598–R601 (2001).

11. Greene R, Siegel J. Sleep. A functional enigma. *NeuroMolec Med* **5**:59–68(2004).

12. Porkka-Heiskanen T, Alanko L, Kalinchuk A, Stenberg D. Adenosine and sleep. *Sleep Med Rev* **6**:321–332(2002).

13. Mognot E, Taheri S, Nishino S. Sleeping with the hypothalamus: emerging therapeutic targets for sleep disoders. *Nat Neurosci Supp* **5**:1071–1075(2002).

14. Maquet P, Paeters J-M, Aerts J, Delfiore G, Delguedre C, Luxen A, Franck G. Functional neuroanatomy of human rapid-eye-movement sleep and dreaming. *Nature* **383**:163–166(1996).

15. Maquet P, Delguedre C, Delfiore G, Aerts J, Paters J-M, Luxen A, Franck G. Functional neuroanatomy of human slow wave sleep. *J Neurosci* **17**:2807–2812(1997).

16. Balkin TJ, Braun AR, Wesensten NJ, Jeffries K, Varga M, Baldwin P, Belenky G, Herschovitz P. The process of awakening: a PET study of regional brain activity patterns mediating the re-establishment of alertness and consciousness. *Brain* **125**:2308–2319(2002).

17. Lydic R, Baghdoyan HA, McGinley J. Opioids, sedation and sleep. Different states, similar traits, and the search for common mechanisms. In: Malviya S, Naughton N, Tremper KK (Eds), *Contemporary Clinical Neuroscience: Sedation and Analgesia for Diagnostic and Therapeutic Procedures*. Humana Press, Totowa, NJ, 2003, pp 1–31.

18. Nelson LE, Guo TZ, Saper CB, Franks NP, Maze M. The sedative component of anesthesia is mediated by GABAa receptors in an endogenous sleep pathway. *Nat Neurosci* **5**:979–984(2002).

19. Mamelak M, Scharf MB, Woods M. Treatment of narcolepsy with gamma-hydroxybutyrate. A review of clinical and sleep laboratory findings. *Sleep* **9**:285–289(1986).

20. Lin JS, Hou Y, Sakai K, Jouvet M. Histaminergic descending inputs to the mesopontine tegmentum and their role in the control of cortical activation and wakefulness in the cat. *J Neurosci* **16**:1523–1537(1996).

21. Benowitz NL. Pharmacology of nicotine: addiction and therapeutics. *Annu Rev Pharmacol Toxicol* **36**:597–613(1996).

22. Mamiya N, Buchanan R, Wallace T, Skinner RD, Garcia-Rill E. Nicotine suppresses the P13 auditory evoked potential by acting on the pedunculopontine nucleus in the rat. *Exp Brain Res* **164**:109–119(2005).

23. Hughes JR, Hatsukami DK, Mitchell JE, Dahlgren LA. Prevalence of smoking among psychiatric outpatients. *Am J Psychol* **143**:993–997(1996).

24. Seiden LS, Lew R, Malberg JE. Neurotoxicity of methamphetamine and methylenedioxymethamphetamine. *Neurotox Res* **3**:101–116(2001).

25. Fornai F, Lazzeri G, Lenzi P, Gesi M, Ferrucci M, Soldani P, Pellegrini A, Capobianco L, De Blasi A, Ruggieri S, Paparelli A. Amphetamines induce ubiquitin-positive inclusions within striatal cells. *Neurol Sci* **24**:182–183(2003).

26. Seiden LS, Sabol KE. Amphetamine: effects on catecholamine systems and behavior. *Annu Rev Pharmacol Toxicol* **32**:639–677(1993).

27. Engber TM, Dennis SA, Jones BE, Miller MS, Contreras PC. Brain regional substrates for the actions of the novel wake-promoting agent modafinil in the rat: comparison with amphetamine. *Neuroscience* **87**:905–911(1998).

28. Ferraro L, Antonelli T, Taganelli S, O'Connor WT, Perez de la Mora M, Mendez-Franco J, Rambert FA, Fuxe K. The vigilance promoting drug modafinil increases extracellular glutamate levels in the medial preoptic area and the posterior hypothalamus of the conscious rat: prevention by local GABAa receptor blockade. *Neuropsychopharmacology* **20**:346–356(1999).

29. US Modafinil in Narcolepsy Multicenter Study Group. Randomized trial of modafinil as a treatment for the excessive daytime sleepiness of narcolepsy. *Neurology* **54**:1166–1175(2000).

30. Woods AJ, Garcia-Rill E, Meythaler J, Mark VW, Jewell GR, Mennemeier M. Altered magnitude estimation in neglect following left-hemisphere damage improves with pharmacologic treatment for arousal. *J Int Neuropsychol Soc* **10**: 218(2004).

31. Kosten TR, Cheeves C, Palumbo J, Seibyl JP, Price LH, Woods SW. Regional cerebral blood flow during acute and chronic abstinence from combined cocaine–alcohol abuse. *Drug Alcohol Depend* **50**:187–195 (1998).

32. Miyazato H, Skinner RD, Garcia-Rill E. Locus coeruleus involvement in the effects of immobilization stress on the P13 midlatency auditory evoked potential in the rat. *Prog Neuropsychopharmacol Biol Psychiatry* **24**:1177–1201 (2000).

33. Garcia-Rill E, Biedermann JA, Chambers T, Skinner RD, Mrak RE, Husain M, Karson CN. Mesopontine neurons in schizophrenia. *Neuroscience* **66**:321–335(1995).

34. Qureshi A, Lee-Chiong T. Medications and their effects on sleep. *Med Clin North Am* **88**:751–766(2004).

35. Dauvilliers Y, Mayer G, Lecendreux M, Neidhart E, Peraita-Adrados R, Sonka K, Biliard M, Tafti M. Kleine–Levin syndrome. An autoimmune hypothesis based on clinical and genetic analyses. *Neurology* **59**:1739–1745(2002).

36. Garcia-Rill E, Skinner RD. The sleep-state-dependent P50 midlatency auditory evoked potential. In: Lee-Chiong TL, Sateia MJ, Carskadon MA (Eds), *Sleep Medicine*. Hanley & Belfus, Philadelphia, 2002, pp 697–704.

10

EPIDEMIOLOGY OF SLEEP DISORDERS

Maurice M. Ohayon and Christian Guilleminault
Stanford University, Stanford, California

INTRODUCTION

Each of us will spend about 27 years of his or her lifetime sleeping. This fact alone explains why neuroanatomists and neurophysiologists have been studying sleep for over a century. The epidemiology of sleep, however, is a relatively young field of study, although physicians have always been interested in knowing how widespread abnormal sleep phenomena are in the population. How does the population at large sleep? Well or poorly? Do we sleep too much or too little? Where is the cutoff between normal and abnormal sleep? Does sleep remain the same over the life span? Is it different between men and women and across cultures? These are some of the fundamental questions that the epidemiology of sleep disorders has been trying to answer.

THE HAZARDS OF SLEEP CLASSIFICATIONS

Classifications represent the advancement of our knowledge and understanding of sleep disorders. They are attempts to provide operationalized criteria to delineate abnormal sleep in all its forms. Abnormality, however, exists relative to a norm. This would imply that we know what constitutes normal sleep, but not abnormal sleep. Yet, the contrary is true. At this point in time, we can only say what does not constitute normal sleep. *The International Classification of Sleep Disorders* [1] was the first exhaustive attempt to classify abnormal sleep.

The difficulty in distinguishing between normal and abnormal sleep is reflected in the evolution of the classifications and definitions of symptomatology. For example, insomnia was defined by the American Institute of Medicine in 1979 as unsatisfactory sleep [2]. In the same year, the Association of Sleep Disorders Centers published its first classification [3] in which insomnia was referred to as a "heterogeneous group of conditions . . . considered to be responsible for inducing disturbed sleep or diminished sleep".

In 1987, the American Psychiatric Association for the first time devoted a section in the *Diagnostic and Statistical Manual of Mental Disorders* (DSM-III-R) to sleep disorders [4]. The section was essentially divided into dyssomnias (insomnia, hypersomnia and sleep–wake disorders) and parasomnias (sleepwalking, sleep terrors, and nightmares). Insomnia was defined as difficulty initiating sleep (DIS), difficulty maintaining sleep (DMS)—be it in the form of disrupted sleep (DS) or early morning awakening (EMA)—or nonrestorative sleep (NRS) lasting at least one month, occurring at least three times a week, and causing either distress or daytime repercussions.

In 1990, efforts by an international group of sleep researchers and sleep specialists produced the *International Classification of Sleep Disorders* (ICSD-90) [1], which listed nearly 80 sleep disorder diagnoses. In this classification, insomnia was more stringently defined by taking into account severity, frequency of symptoms, and impact on social and occupational functioning. In the DSM-IV, the latest edition of its classification published in 1994 [4], the American Psychiatric Association decided to harmonize its sleep disorder criteria and diagnoses with those of the ICSD-90. Insomnia was defined as a complaint of DIS,

DMS, or NRS lasting at least one month and causing either distress or daytime consequences.

This evolution in classifications is also reflected in the epidemiological studies of sleep disorders. This has rendered comparisons between earlier and more recent surveys problematic for three principal reasons: (1) symptomatology is defined differently across studies; (2) time frames are also different; and (3) different methodologies have been used to collect data.

If we follow a traditionally used model, sleep disorders will be divided into two large categories: dyssomnias and parasomnias. Dyssomnias are sleep disorders characterized by abnormalities in the quantity, quality, or timing of sleep. As such, they are associated with difficulty initiating or maintaining the sleep or daytime sleepiness. Parasomnias cover abnormal behavioral or physiological events occurring during sleep but do not involve the sleep mechanisms per se.

DYSSOMNIAS

Insomnia and Related Disorders

To date, there is no consensus on how to define and to measure insomnia in epidemiology. As a consequence, epidemiological findings largely varied depending on the definition used. We shall take the problem of insomnia as an example of the way epidemiologic results may be collected and divided.

Since the end of the 1970s, more than 50 epidemiological studies have assessed the prevalence of insomnia symptomatology in the general population. Methodologies have included face-to-face interviews, postal questionnaires, telephone interviews, or a combination of two of the above.

The definition of insomnia also varied considerably from one survey to another. Earlier studies evaluated insomnia based on the presence of DIS or DMS regardless of the frequency or severity of the symptom or daytime consequences. It was done simply by asking about the presence of these symptoms. Subsequently, DIS and DMS were assessed using the frequency of the symptom, an occurrence of 3 nights or more per week being necessary for the symptom to be present. Other studies asked about the severity of the symptoms, for example, being bothered "a lot" or "not at all" by the symptom.

Other studies, in addition to assessing the presence of insomnia symptoms, inquired about daytime repercussions of these symptoms such as daytime sleepiness, irritability, depressive or anxious mood, or needing to seek help. Finally, other studies inquired about dissatisfaction with sleep quantity or quality.

Table 10.1 gives the definitions used in epidemiological studies and the prevalence of insomnia.

Prevalence of Insomnia

In epidemiological studies, the binary query about the presence of insomnia symptoms gave high prevalence rates with an average around 33%. One of the earliest epidemiological surveys on insomnia symptoms was carried out by Bixler et al. [25] in the metropolitan area of Los Angeles with 1006 respondents aged 18 years or over. The overall prevalence of insomnia symptoms was 32.2% (DIS, 14.4%; DS, 22.9%; and EMA, 13.8%). Subsequent studies [13, 26–29] found a similar prevalence in the general population when inquiries were made about the presence of insomnia symptoms (Table 10.1).

Epidemiological studies using frequency to determine the prevalence of insomnia symptoms are the most common [8–11, 15, 23, 30–33]. In some studies, the subjects had to make a subjective assessment of the frequency of the symptom on a four- or five-point scale [8, 31–33]: for example, never, sometimes, often, or always; often or always being the cutoff point to determine the presence of insomnia. Mostly, however, frequency of the symptom is assessed on a weekly basis [9–11, 14, 23, 30]: for example, never, one or two nights, three or four nights, five nights or more per week; a frequency of three nights or more per week being the cutoff used to conclude the presence of insomnia. The prevalence of insomnia symptoms drops to around 16–21% when frequency is used to determine the presence of insomnia and has similar rates among countries (Table 10.1).

Epidemiological studies using severity of the symptoms (e.g., being bothered a lot; having great or very great DIS or DMS or a major complaint) gave prevalence of insomnia between 10% and 28% of the general population [34–37].

In most of the studies that assessed the prevalence of insomnia symptoms accompanied with daytime consequences, the prevalence was much lower, being about 10% [9, 10, 15, 18, 38–40]. One study provided a higher prevalence than the other studies mainly because the rate was based on lifetime estimation [38].

Dissatisfaction with the quantity of sleep can be expressed as a complaint of sleeping not enough or sleeping too much. Sleeping not enough has been reported with prevalence ranging from 20% to 41.7% in the general population [10, 41–43]. Sleeping too much is far less frequent with prevalence ranging between 2.8% and 9.5% [25, 39].

Dissatisfaction with quality of sleep had various definitions. In some studies, participants were asked to assess their level of satisfaction with their sleep. The prevalence of individuals reporting being dissatisfied with their sleep ranged from 8% to 18.5% [17, 18, 20–22, 44]. Other studies have inquired about perception of sleep as being poor or subjects considering themselves as being insomniac. Between 10% and 18.1% of the population reported being poor sleepers or being insomniacs [6, 11].

TABLE 10.1 Definition and Prevalence of Insomnia in the General Population

Authors	Year	Place	Number	Age	Definition	Prevalence (%) Male/Female
Mallon and Hetta [5]	1997	Sweden	876	65–79	Moderate or major complaints of DIS, DS, EMA	DIS: 14/30 DS: 31.4 EMA: 33.4
Asplund [6]	1998	Jamtland county, Sweden	3669 wom.	40–64	Bad night's sleep	18.1%
Maggi et al. [7]	1998	Veneto region, Italy	2398	≥65	Often or always having DIS or EMA	35.6/54.0
Ancoli-Israel and Roth [8]	1999	USA	1000	≥18	Difficulty sleeping on a frequent basis	9.0
Hoffmann [9]	1999	Belgium	1618	≥18	• Having DIS, DMS, or EMA at least 3 times per week	22.0
					• DIS, DMS, EMA + daytime consequences	9.0
Hetta et al. [10]	1999	Sweden	1996	≥18	• Having DIS, DMS, or EMA at least 3 times per week	22.0
					• DIS, DMS, EMA + daytime consequences	13.0
Vela-Bueno et al. [11]	1999	Madrid, Spain	1131	≥18	• Having DIS, DMS, or EMA at least 4 times per week	1.7/27.4
					• Considered themselves insomniacs	7.8/14.4
Chiu et al. [12]	1999	Hong Kong, China	1034	≥70	Consider themselves as having insomnia	8.6/17.5
Yamaguchi et al.	1999	Kanazawa, Japan	236	>60	Insomnia ≥ 3 nights/week	14.0/19.7
Mallon et al. [13]	2000	Sweden	1870	45–65	Presence of DIS, DMS, or EMA	25.4/36.0
Doi et al. [14]	2000	Japan	3030	≥20	Often or always DIS, DMS, or EMA	17.3
Léger et al. [15]	2000	France	12778	≥16	• Having DIS, DMS, or EMA at least 3 times per week	25.0/34.0
					• DIS, DMS, or EMA + daytime consequences	14.0/23.0
Barbar et al. [16]	2000	Hawaii, USA	3845 males	71–93	DIS, DS, EMA	32.6
Ohayon et al. [17]	2000	Paris, France	1026	≥60	Dissatisfied with sleep quality or quantity	11.5/16.0
Ohayon and Zulley [18]	2001	Germany	4115	≥15	• DIS, DMS, EMA, or NRS + daytime consequences	8.5
					• Dissatisfaction with sleep	5.6/8.2
					• DSM-IV insomnia diagnoses	6.0
Hajak [19]	2001	Germany	1913	≥18	Severe insomnia	4.0
Pallesen et al. [20]	2001	Norway	2001	≥18	DIS, DMS, EMA + daytime consequences	11.7
Ohayon et al. [21]	2001	UK, Germany, Italy	2429	≥65	DIS, DS, EMA, NRS ≥ 3 nights/week	DIS: 16.0 DS: 33.0 EMA: 16.0 NRS: 11.0
Ohayon and Smirne [22]	2002	Italy	3970	≥15	Dissatisfaction with sleep	10.1
					DSM-IV insomnia diagnoses	6.0
Ohayon and Partinen [23]	2002	Finland	982	≥18	• DIS, DMS, EMA, or NRS ≥ 3 nights/week	37.6
					• Dissatisfaction with sleep	11.9
					• DSM-IV insomnia diagnoses	11.7
Ohayon and Hong [24]	2002	South Korea	3719	≥15	• DIS, DMS, EMA, or NRS ≥ 3 nights/week	14.8/19.1
					• DSM-IV insomnia diagnoses	4.7/5.1
Kiejna et al.	2003	Poland	47924	≥15	Suffering from insomnia	18.1/28.1

DIS, difficulty initiating sleep; DS, disrupted sleep; EMA, early morning awakening; NRS, nonrestorative sleep.

Unfortunately, most of these studies did not provide any information about the chronicity of these symptoms. Studies that measured it, showed that insomnia is mostly chronic [18, 22, 44, 45]. Only 4% of subjects with insomnia symptoms reported a duration of 1 month or less. About 6% of these subjects evaluated the duration being between 1 and 6 months; 5% said the duration is between 6 and 12 months, and 85% mentioned a duration of 1 year or more (68% said it lasted 5 years or more) [45].

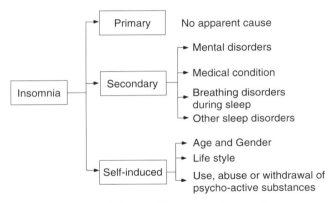

Figure 10.1 Possible causes of insomnia.

Factors Associated with Insomnia

There are many causes of insomnia. It can be divided in three main categories (Figure 10.1): (1) secondary to another physical and/or mental illness, (2) induced by use of psychoactive substances or because of life style, or (3) without apparent cause.

Figure 10.2 displays the proportions for most common causes of insomnia. Sleep-related breathing disorders such as obstructive sleep apnea syndrome or hypoventilation account for 5–9% of insomnia complaints [40, 46, 47]. Periodic limb movement disorders and/or restless legs syndrome are found in about 15% of individuals with insomnia complaints [46–49]. Medical or neurological conditions are observed in 4–11% of insomnia complaints [23,

40, 46, 47, 50]. Poor sleep hygiene or environmental factors account for about 10% of insomnia complaints and substance-induced for 3–7% [22, 23, 40, 46, 47].

Sociodemographic Factors

Gender Women are more likely than men to report insomnia symptoms [11, 15, 31–33, 37, 50–52], daytime consequences [9,10,15,39], and dissatisfaction with sleep [15, 44] and to have insomnia diagnoses [18, 23, 24]. Women/men ratios for insomnia symptoms are about 1.4. The difference between women and men increases with age, the ratio of women/men being about 1.7 after 45 years of age. Women are twice more likely than men to have an insomnia diagnosis. Some studies have found that the prevalence of insomnia increases in menopausal women as compared to their younger counterparts [53–55].

Age Almost all epidemiological studies reported an increased prevalence of insomnia symptoms with age, reaching close to 50% in elderly individuals (≥65 years old) [8–11, 18, 25, 26, 28, 44, 50, 52]. However, the prevalence of insomnia with daytime consequences and the prevalence of sleep dissatisfaction have mixed results. Other studies found lower rates in middle-aged individuals [56], while still other studies reported an increasing prevalence with age [10, 11, 15, 39, 44].

Insomnia in the elderly noninstitutionalized population has been the subject of several epidemiological studies [5, 12, 16, 17, 21, 38, 57–62]. Most of these studies were limited to insomnia symptoms; only two studies assessed sleep

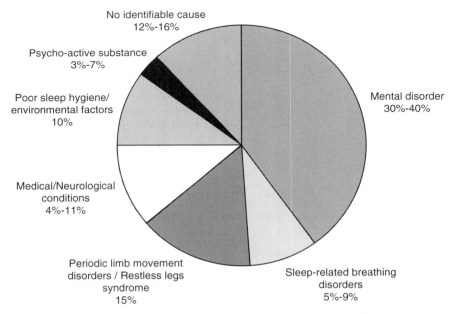

Figure 10.2 Distribution of insomnia complaints by etiological causes.

dissatisfaction [12, 17]. Prevalence based on presence/ absence of insomnia symptoms gave a very high rate (up to 65%). In elderly community-based samples, the prevalence of insomnia symptoms and sleep dissatisfaction do not significantly increase with age [7, 12, 17, 59, 60] but is higher in women than in men [7, 12, 58–60, 62]. Some studies found that insomnia symptoms without sleep dissatisfaction have a weak association with physical diseases and mental disorders [15, 41, 42].

Income and Education Prevalence of insomnia is higher in individuals with lower incomes [25, 52, 62] and in those with lower education [8, 25, 61]. However, these associations can be the result of other factors such as age. Use of the poverty index will provide a better indication of the association between insomnia and poverty.

Physical Illnesses

In the general population, subjects with insomnia symptoms were consistently found to perceive their health as being poorer than the rest of the population [7, 12, 16, 18, 22, 57, 60, 61]. Up to half of subjects with insomnia symptoms have recurrent, persistent, or multiple health problems [25, 37]. Most frequently reported associations were with upper airway diseases [18, 26, 34], rheumatic diseases [18, 34, 63, 64], chronic pain [64, 65], and cardiovascular diseases [7, 18, 66].

Mental Disorders

Epidemiological studies have consistently reported that a mental disorder is associated with 30–40% of insomnia complaints (Table 10.2). Up to 60% of subjects with insomnia symptoms were reported to have symptoms of mental disorders [37, 68]. In individuals with a current major depressive episode, the presence of insomnia symptoms was found in nearly 80% of the subjects [33, 54, 69]. Four longitudinal studies examined the relationship between the persistence of insomnia symptoms and the appearance of mental disorders (Table 10.2).

Ford and Kamerow [39] found a high co-occurrence of insomnia complaints and mental disorders (40%). Insomnia complaints were associated with a higher risk (odds ratio of 39.8) for developing a new major depressive illness if they persisted over two interviews within a 12 month interval, but were not a significant factor if they ceased by the second interview. Another study in young adults between 21 and 30 years of age [38] found that subjects with a history of insomnia were four times more likely to develop a new major depressive disorder in the 3.5 years following the initial interview. Another survey followed up 2164 individuals age 50 years and over in Alameda County (California) during a one year period [67]. The presence of major depression at the last assessment was eight times more likely to occur in individuals with insomnia on both assessments and ten times more likely to occur in those who reported insomnia only on the last assessment. However, insomnia was a less important predictor of future depression than other depressive symptoms (anhedonia, feelings of worthlessness, psychomotor agitation/retardation, mood disturbance, thoughts of death) [67].

Should the insomnia warrant a specific diagnosis? Should the insomnia be considered as part of the mental disorder manifestation? The first epidemiological study using a classification to categorize insomnia subjects was published

TABLE 10.2 Exploration of Association Between Mental Disorders and Insomnia in Epidemiological Studies

Descriptive Studies	Longitudinal Studies
• Mellinger et al., 1985, USA [37]	• Ford and Kamerow, 1989, USA (Baltimore, Durham, Los Angeles) [39]
• Henderson et al., 1995, Australia (Canberra and Queanbeyan) [60]	• Breslau et al., 1996, USA (southeast Michigan) [38]
• Foley et al., 1995, USA (E. Boston, New Haven, Iowa, Washington counties) [66]	• Roberts et al., 2000, USA (Alameda County, CA) [67]
• Newman et al., 1997, USA (Forsyth (NC), Sacramento (CA), Washington counties, Maryland and Pittsburg) [62]	• Roberts et al., 2002, USA (Houston, TX)
• Ohayon, 1997, France [40]	
• Maggi et al., 1998, Italy (Veneto region) [7]	
• Hoffmann, 1999, Belgium [9]	
• Hetta et al., 1999, Sweden [10]	
• Ohayon et al., 2000, Canada [50]	
• Ohayon et al., 2003, United-Kingdom, Germany, Italy, Portugal [45]	
30% to 60% of insomniacs have mental disorder symptoms, 36% psychiatric diagnosis	Persistence of insomnia = 4 to 8 times more likely of developing a mental disorder

in 1997 [40]. The differential diagnosis of insomnia was based on the DSM-IV classification. Overall, 5.6% of the sample had one of the DSM-IV insomnia diagnoses. "Insomnia Related to Another Mental Disorder" and "Primary Insomnia" were the most frequent diagnoses. Similar studies were conducted in the United Kingdom, Germany, Canada, Italy, and Finland [18, 22, 23, 40, 43, 45]. The European studies included 20,536 subjects age 15 years and over. The overall prevalence of insomnia symptoms accompanied by sleep dissatisfaction was 12.4%. Insomnia diagnoses were observed in 35% of the subjects with insomnia; noninsomnia sleep disorders accounted for an additional 5%. The final diagnosis resulted in a mental disorder diagnosis in 36% of the insomnia symptoms accompanied by sleep dissatisfaction.

Life Style

Epidemiological studies have reported higher risks of insomnia in individuals with high stress [18, 22–24] and among individuals without work [9, 52, 70, 71], with shift/night work [18, 22], or sleeping in a bedroom with inappropriate temperature [18].

Psychoactive Substances

Epidemiological studies have reported higher risks (odds of 1.2–2) of insomnia in individuals using tobacco [30, 72–74], antihypertensive drugs [34], and alcohol [8, 18, 75]. Alcohol was used as a sleeping aid in 4 out of 10 individuals with sleep disturbances [8, 76].

The epidemiological surveys described here clearly show that insomnia symptoms are very common in the general population. For a sizable portion of the population, these complaints represent serious sleep disorders that require medical attention.

We shall now present epidemiologic surveys on other sleep disorders.

Excessive Daytime Sleepiness and Related Disorders

Hypersomnia has reported rates varying from 0.3% to 16.3% in the United States [26, 38, 39, 77]. The Cardiovascular Health Study found a 20% prevalence of participants being "usually sleepy in the daytime" in a sample of 4578 adults aged 65 and older.

In Europe, various studies found a prevalence varying from 5% to 16% [7, 34, 78–80].

Excessive daytime sleepiness can be caused by various factors such as poor sleep hygiene, work conditions, and psychotropic medication use [77, 81]. Excessive daytime sleepiness has been found to be associated also with sleep-disordered breathing [78, 79, 81], psychiatric disorders, especially depression [19, 38, 39, 81], and physical illnesses [78, 79].

Narcolepsy   Based on representative community samples, prevalence varies from 20 to 67 per 100,000 inhabitants in Europe and North America [82, 83]. In Japan the rate is between 160 and 590 per 100,000 inhabitants [84, 85]. In Hong Kong, prevalence is estimated to be between 1 and 40 narcoleptics per 100,000 inhabitants [86]; in Saudi Arabia it is 40 per 100,000 inhabitants [87].

Sleep Breathing Disorders

In the Finnish twin cohort study, 1.4% had an AHI greater than or equal to 10, with an oxygenation desaturation index (ODI) of at least 4% [7, 64]. In Spain, the prevalence of AHI ≥ 10 was at 19% among men and 14.9% among women [88]. In Italy, 4.8% of the population between 30 and 69 years of age had an AHI greater than 5, and 3.2% had an AHI greater than 10 [89]. In the Wisconsin Sleep Cohort study [90], the prevalence of sleep apnea syndrome (daytime sleepiness and/or nonrefreshing sleep and an AHI of 5 or greater) was estimated at 4% among men and 2% among women. In another U.S. study, the prevalence of sleep apnea, defined as AHI ≥ 10 accompanied by daytime symptoms, was estimated at 3.3% among men [57, 77]. In Australia [91], the rate of obstructive sleep apnea syndrome, based on an AHI of 15 or greater, was estimated at 3.6% overall, and at 5.7% for men and 1.2% for women, and a prevalence of AHI ≥ 5 at 3.7%. In Hong Kong [92], the rate was 2.1% with AHI ≥ 5 accompanied by daytime sleepiness.

Restless Legs Syndrome (RLS)

The prevalence of RLS in the general population in European studies oscillates between 5.8% and 9.8% [49, 71, 93–95].

PARASOMNIAS

Parasomnias are sleep disorders characterized by abnormal behavioral or physiological events occurring at different sleep stages or during sleep–wake transitions. These disorders have seldom been investigated in the adult general population.

Arousal Parasomnias

In the adult general population, prevalence of sleepwalking varied between 1.9% and 3.2% [77, 96–98]; a 2.2% prevalence of night terrors has been reported [97], and a prevalence of 2.9% for confusional arousal [99].

REM Sleep Disorder Parasomnias

Nightmares have been reported to occur at least once a week in 5% of the adult population [100]; 6.2% of the population had at least one episode of sleep paralysis in their lifetime; 0.8% experienced severe sleep paralysis (at least one episode per week), and 1.4% moderate sleep paralysis (at least one episode per month) [83, 98, 101]. REM sleep behavior disorder is estimated to occur in 0.5% of the general population [102].

REFERENCES

1. Diagnostic Classification Steering Committee, Thorpy MJ, Chairman. *International Classification of Sleep Disorders: Diagnostic and Coding Manual* (ICSD). American Sleep Disorders Association, Rochester, MN, 1990.

2. American Institute of Medicine. *Sleeping Pills, Insomnia and Medical Practice.* National Academy of Sciences, Washington, DC, 1979.

3. Association of Sleep Disorders Centers. Diagnostic classification of sleep and arousal disorders. *Sleep* **2**:5–122(1979).

4. APA (American Psychiatric Association). *Diagnostic and Statistical Manual of Mental Disorders*, 4th ed (DSM-IV). The American Psychiatric Association, Washington, DC, 1994.

5. Mallon L, Hetta J. A survey of sleep habits and sleeping difficulties in an elderly Swedish population. *Ups J Med Sci* **102**:185–197(1997).

6. Asplund R. Daytime sleepiness and napping amongst the elderly in relation to somatic health and medical treatment. *J Intern Med* **239**:261–267(1996).

7. Maggi S, Langlois JA, Minicuci N, Grigoletto F, Pavan M, Foley DJ, Enzi G. Sleep complaints in community-dwelling older persons: prevalence, associated factors, and reported causes. *J Am Geriatr Soc* **46**:161–168(1998).

8. Ancoli-Israel S, Roth T. Characteristics of insomnia in the United States: results of the 1991 National Sleep Foundation Survey. I. *Sleep* **22** (Suppl 2):S347–353(1999).

9. Hoffmann G. Evaluation of severe insomnia in the general population—implications for the management of insomnia: focus on results from Belgium. *J Psychopharmacol* **13**(4 Suppl 1):S31–S32(1999).

10. Hetta J, Broman JE, Mallon L. Evaluation of severe insomnia in the general population—implications for the management of insomnia: insomnia, quality of life and healthcare consumption in Sweden. *J Psychopharmacol* **13**(4 Suppl 1):S35–S36(1999).

11. Vela-Bueno A, De Iceta M, Fernandez C. Prevalencia de los trastornos del sueno en la ciudad de Madrid. *Gac Sanit* **13**:441–448(1999).

12. Chiu HF, Leung T, Lam LC, Wing YK, Chung DW, Li SW, Chi I, Law WT, Boey KW. Sleep problems in Chinese elderly in Hong Kong. *Sleep* **22**:717–726(1999).

13. Mallon L, Broman JE, Hetta J. Relationship between insomnia, depression, and mortality: a 12-year follow-up of older adults in the community. *Int Psychogeriatr* **12**:295–306(2000).

14. Doi Y, Minowa M, Okawa M, Uchiyama M. Prevalence of sleep disturbance and hypnotic medication use in relation to sociodemographic factors in the general Japanese adult population. *J Epidemiol* **10**:79–86(2000).

15. Léger D, Guilleminault C, Dreyfus JP, Delahaye C, Paillard M. Prevalence of insomnia in a survey of 12,778 adults in France. *J Sleep Res* **9**:35–42(2000).

16. Barbar SI, Enright PL, Boyle P, Foley D, Sharp DS, Petrovitch H, Quan SF. Sleep disturbances and their correlates in elderly Japanese American men residing in Hawaii. *J Gerontol A Biol Sci Med Sci* **55**:M406–411(2000).

17. Ohayon MM, Vechierrini MF. Daytime sleepiness is an independent predictive factor for cognitive impairment in the elderly population. *Arch Intern Med* **162**:201–208(2002).

18. Ohayon MM, Zulley J. Correlates of global sleep dissatisfaction in the German population. *Sleep* **24**:780–787(2001).

19. Hajak G, SINE Study Group. Study of insomnia in Europe. Epidemiology of severe insomnia and its consequences in Germany. *Eur Arch Psychiatry Clin Neurosci* **251**:49–56(2001).

20. Pallesen S, Nordhus, IH, Nielsen GH, Havik OE, Kvale G, Johnsen BH, Skjotskift S. Prevalence of insomnia in the adult Norwegian population. *Sleep* **24**:771–779(2001).

21. Ohayon MM, Zulley J, Guilleminault C, Smirne S, Priest RG. How age and daytime activities are related to insomnia in the general population. Consequences for elderly people. *J Am Geriatr Soc* **49**:360–366(2001).

22. Ohayon MM, Smirne S. Prevalence and consequences of insomnia disorders in the general population of Italy. *Sleep Med* **3**:115–120(2002).

23. Ohayon MM, Partinen M. Insomnia and global sleep dissatisfaction in Finland. *J Sleep Res* **11**(4):339–346(2002).

24. Ohayon MM, Hong SC. Prevalence of insomnia and associated factors in South Korea. *J Psychosom Res* **53**:593–600(2002).

25. Bixler EO, Kales A, Soldatos CR, Kales JD, Healey S. Prevalence of sleep disorders in the Los Angeles metropolitan area. *Am J Psychiatry* **136**:1257–1262(1979).

26. Klink M, Quan SF. Prevalence of reported sleep disturbances in a general adult population and their relationship to obstructive airways diseases. *Chest* **91**:540–546(1987).

27. Klink ME, Quan SF, Kaltenborn WT, Lebowitz MD. Risk factors associated with complaints of insomnia in a general adult population. Influence of previous complaints of insomnia. *Arch Intern Med* **152**:1634–1637(1992).

28. Quera-Salva MA, Orluc A, Goldenberg F, Guilleminault C. Insomnia and use of hypnotics: study of a French population. *Sleep* **14**:386–391(1991).

29. Welstein L, Dement WC, Redington D, Guilleminault C. Insomnia in the San Francisco Bay area: a telephone survey.

In: Guilleminault C, Lugaresi E (Eds), *Sleep/Wake Disorders: Natural History, Epidemiology, and Long-Term Evolution.* Raven Press, New-York, 1983, pp 29–35.

30. Janson C, Gislason T, De Backer W, et al. Prevalence of sleep disturbances among young adults in three European countries. *Sleep* **18**:589–597(1995).

31. Karacan I, Thornby JI, Anch M, et al. Prevalence of sleep disturbance in a primarily urban Florida county. *Soc Sci Med* **10**:239–244(1976).

32. Karacan I, Thornby JI, William R. Sleep disturbance: a community survey. In: Guilleminault C, Lugaresi E (Eds), *Sleep/Wake Disorders: Natural History, Epidemiology, and Long-Term Evolution.* Raven Press, New-York, 1983, pp 37–60.

33. Olson LG, King MT, Hensley MJ, Saunders NA. A community study of snoring and sleep-disordered breathing—prevalence. *Am J Respir Crit Care Med* **152**(2):711–716(1995).

34. Gislason T, Almqvist M. Somatic diseases and sleep complaints: an epidemiological study of 3201 Swedish men. *Acta Med Scand* **221**:475–481(1987).

35. Husby R, Lingjaerde O. Prevalence of reported sleeplessness in northern Norway in relation to sex, age and season. *Acta Psychiatr Scand* **81**(6):542–547(1990).

36. Liljenberg B, Almqvist M, Hetta J, Roos BE, Agren H. The prevalence of insomnia: the importance of operationally defined criteria. *Ann Clin Res* **20**:393–398(1988).

37. Mellinger GD, Balter MB, Uhlenhuth EH. Insomnia and its treatment: prevalence and correlates. *Arch Gen Psychiatry* **42**:225–232(1985).

38. Breslau N, Roth T, Rosenthal L, Andreski, P. Sleep disturbance and psychiatric disorders: a longitudinal epidemiological study of young adults. *Biol Psychiatry* **39**:411–418(1996).

39. Ford DE, Kamerow DB. Epidemiologic study of sleep disturbances and psychiatric disorders. An opportunity for prevention? *JAMA* **262**:1479–1484(1989).

40. Ohayon MM. Prevalence of DSM-IV diagnostic criteria of insomnia: distinguishing between insomnia related to mental disorders from sleep disorders. *J Psychiatr Res* **31**:333–346(1997).

41. Ohayon M, Caulet M, Lemoine P. Sujets âgés, habitudes de sommeil et consommation de psychotropes dans la population française. *Encephale* **22**:337–350(1996).

42. Ohayon MM, Caulet M, Arbus L, Billard M, Coquerel A, Guieu JL, Kullmann B, Laffont F, Lemoine P, Paty J, Pecharde JC, Vecchierini F, Vespignani H. Are prescribed medications effective in the treatment of insomnia complaints? *J Psychosom Res* **47**:359–368(1999).

43. Ohayon MM, Shapiro CM, Kennedy SH. Differentiating DSM-IV anxiety and depressive disorders in the general population: comorbidity and treatment consequences. *Can J Psychiatry* **45**:166–172(2000).

44. Ohayon M. Epidemiological study on insomnia in the general population. *Sleep* **19**(3):S7–S15(1996).

45. Ohayon MM, Roth T. Place of chronic insomnia in the course of depressive and anxiety disorders. *J Psychiatr Res* **37**(1):9–15(2003).

46. Buysse DJ, Reynold CF, Hauri P, Roth T, Stepanski E, Thorpy MJ, Bixler E, Kales A, Manfredi R, Vgontzas A, Stapf BS, Houck PR, Kupfer DJ. Diagnostic concordance for DSM-IV disorders: a report from the APA/NIMH DSM-IV field trial. *Am J Psychiatry* **151**:1351–1360(1994).

47. Jacobs EA, Reynolds CF 3rd, Kupfer DJ, Lovin PA, Ehrenpreis AB. The role of polysomnography in the differential diagnosis of chronic insomnia. *Am J Psychiatry* **145**:346–349(1988).

48. Edinger JD, Fins AI, Goeke JM, McMillan DK, Gersh TL, Krystal AD, McCall WV. The empirical identification of insomnia subtypes: a cluster analytic approach. *Sleep* **19**:398–411(1996).

49. Ohayon MM, Roth T. Prevalence of restless legs syndrome and periodic limb movement disorder in the general population. *J Psychosom Res* **53**:547–554(2002).

50. Ohayon MM, Caulet M, Priest RG, Guilleminault C. DSM-IV and ICSD-90 insomnia symptoms and sleep dissatisfaction. *Br J Psychiatry* **171**:382–388(1997).

51. Lindberg E, Janson C, Gislason T, Bjornsson E, Hetta J, Boman G. Sleep disturbances in a young adult population: can gender differences be explained by differences in psychological status? *Sleep* **20**(6):381–387(1997).

52. Ohayon MM, Caulet M, Guilleminault C. Complaints about nocturnal sleep: how a general population perceives its sleep, and how this relates to the complaint of insomnia. *Sleep* **20**:715–723(1997).

53. Mitchell ES, Woods NF. Symptom experiences of midlife women: observations from the Seattle Midlife Women's Health Study. *Maturitas* **25**:1–10(1996).

54. Owens JF, Matthews KA. Sleep disturbance in healthy middle-aged women. *Maturitas* **30**:41–50(1998).

55. Punyahotra S, Dennerstein L, Lehert P. Menopausal experiences of Thai women. Part 1: Symptoms and their correlates. *Maturitas* **26**(1):1–7(1997).

56. Kageyama T, Kabuto M, Nitta H, Kurokawa Y, Taira K, Suzuki S, Takemoto T. A population study on risk factors for insomnia among adult Japanese women: a possible effect of road traffic volume. *Sleep* **20**:963–971(1997).

57. Blazer DG, Hays JC, Foley DJ. Sleep complaints in older adults: a racial comparison. *J Gerontol A Biol Sci Med Sci* **50A**:M280–M284(1995).

58. Foley DJ, Monjan AA, Brown SL, Simonsick EM, Wallace RB, Blazer DG. Sleep complaints among elderly persons: an epidemiologic study of three communites. *Sleep* **18**:425–432(1995).

59. Ganguli M, Reynolds CF, Gilby JE. Prevalence and persistence of sleep complaints in a rural older community sample: the MoVIES project. *J Am Geriatr Soc* **44**:778–784(1996).

60. Henderson S, Jorm AF, Scott LR, Mackinnon AJ, Christensen H, Korten AE. Insomnia in the elderly: its prevalence and correlates in the general population. *Med J Aust* **162**:22–24(1995).

61. Kim K, Uchiyama M, Okawa M, Liu X, Ogihara R. An epidemiological study of insomnia among the Japanese general population. *Sleep* **23**:41–47(2000).

62. Newman AB, Enright PL, Manolio TA, Haponik EF, Wahl PW. Sleep disturbance, psychosocial correlates, and cardiovascular disease in 5201 older adults: the Cardiovascular Health Study. *J Am Geriatr Soc* **45**:1–7(1997).

63. Hagen KB, Kvien TK, Bjorndal A. Musculoskeletal pain and quality of life in patients with noninflammatory joint pain compared to rheumatoid arthritis: a population survey. *J Rheumatol* **24**:1703–1709(1997).

64. Telakivi T, Partinen M, Koskenvuo M, Salmi T, Kaprio J. Periodic breathing and hypoxia in snorers and controls validation of snoring history and association with blood pressure and obesity. *Acta Neurol Scand* **76**:69–75(1987).

65. Andersson HI, Ejlertsson G, Leden I, Schersten B. Impact of chronic pain on health care seeking, self care, and medication. Results from a population-based Swedish study. *J Epidemiol Community Health* **53**:503–509(1999).

66. Foley DJ, Monjan A, Simonsick EM, Wallace RB, Blazer DG. Incidence and remission of insomnia among elderly adults: an epidemiologic study of 6,800 persons over three years. *Sleep* **22** (Suppl 2):S366–S372(1999).

67. Roberts RE, Shema SJ, Kaplan GA, Strawbridge WJ. Sleep complaints and depression in an aging cohort: a prospective perspective. *Am J Psychiatry* **157**:81–88(2000).

68. Ohayon MM, Caulet M, Lemoine P. Comorbidity of mental and insomnia disorders in the general population. *Compr Psychiatry* **39**:185–197(1998).

69. Weissman MM, Bland RC, Canino GJ, Faravelli C, Greenwald S, Hwu HG, Joyce PR, Karam EG, Lee CK, Lellouch J, Lepine JP, Newman SC, Rubio-Stipec M, Wells JE, Wickramaratne PJ, Wittchen H, Yeh EK. Cross-national epidemiology of major depression and bipolar disorder. *JAMA* **276**:293–299(1996).

70. Chevalier H, Los F, Boichut D, Bianchi M, Nutt DJ, Hajak G, Hetta J, Hoffmann G, Crowe C. Evaluation of severe insomnia in the general population: results of a European multinational survey. *J Psychopharmacol* **13**(4 Suppl 1):S21–S24(1999).

71. Ulfberg J, Nyström B, Carter N, Edling C. Prevalence of restless legs syndrome among men aged 18 to 64 years: an association with somatic disease and neuropsychiatric symptoms. *Mov Disord* **16**:1159–1163(2001).

72. Phillips BA, Danner FJ. Cigarette smoking and sleep disturbance. *Arch Intern Med* **155**:734–737(1995).

73. Revicki D, Sobal J, DeForge B. Smoking status and the practice of other unhealthy behaviors. *Fam Med* **23**:361–364(1991).

74. Wetter DW, Young TB. The relation between cigarette smoking and sleep disturbance. *Prev Med* **23**:328–334(1994).

75. Pillitteri JL, Kozlowski LT, Person DC, Spear ME. Over-the-counter sleep aids: widely used but rarely studied. *J Subst Abuse* **6**:315–323(1994).

76. Roehrs T, Hollebeek E, Drake C, Roth T. Substance use for insomnia in Metropolitan Detroit. *J Psychosom Res* **53**:571–576(2002).

77. Bixler EO, Vgontzas AN, Ten Have T, Tyson K, Kales A. Effects of age on sleep apnea in men: I. Prevalence and severity. *Am J Respir Crit Care Med* **157**:144–148(1998).

78. Janson C, Gislason T, De Backer W, et al. Daytime sleepiness, snoring and gastro-oesophageal reflux amongst young adults in three European countries. *J Intern Med* **237**:277–285(1995).

79. Ohayon MM, Caulet M, Philip P, Guilleminault C, Priest R. How sleep and mental disorders are related to complaints of daytime sleepiness. *Arch Intern Med* **157**:2645–2652(1997).

80. Nugent AM, Gleadhill I, McCrum E, Patterson CC, Evans A, MacMahon J. Sleep complaints and risk factors for excessive daytime sleepiness in adult males in Northern Ireland. *J Sleep Res* **10**:69–74(2001).

81. Hublin C, Kaprio J, Partinen M, Heikkila K, Koskenvuo M. Daytime sleepiness in an adult, Finnish population. *J Intern Med* **239**:417–423(1996).

82. Hublin C, Kaprio J, Partinen M, Koskenvuo M, Heikkila K, Koskimies S, Guilleminault C. The prevalence of narcolepsy: an epidemiological study of the Finnish Twin Cohort. *Ann Neurol* **35**:709–716(1994).

83. Ohayon MM, Zulley J, Guilleminault C, Smirne S. Prevalence and pathological associations of sleep paralysis in the general population. *Neurology* **52**:1194–1200(1999).

84. Honda Y, Asaka A, Tanimura M, Furusho T. A genetic study of narcolepsy and excessive daytime sleepiness in 308 families with a narcolepsy or hypersomnia proband. In Guilleminault C, Lugaresi E (Eds), *Sleep/Wake Disorders: Natural History, Epidemiology and Long-Term Evolution*. Raven Press, New York, 1983, pp 187–199.

85. Honda Y. Census of narcolepsy, cataplexy and sleep life among teenagers in Fujisawa City. *Sleep Res* **8**:191(1979).

86. Wing YK, Chiu HF, Ho CK, Chen CN. Narcolepsy in Hong Kong Chinese—a preliminary experience. *Aust N Z J Med* **24**:304–306(1994).

87. al Rajeh S, Bademosi O, Ismail H, Awada A, Dawodu A, al-Freihi H, Assuhaimi S, Borollosi M, al-Shammasi S. A community survey of neurological disorders in Saudi Arabia: the Thugbah study. *Neuroepidemiology* **12**(3):164–178(1993).

88. Duran J, Esnaola S, Rubio R, Iztueta A. Obstructive sleep apnea–hypopnea and related clinical features in a population-based sample of subjects aged 30 to 70 yr. *Am J Respir Crit Care Med* **163**(3 Pt 1):685–689(2001).

89. Cirignotta F, D'Alessandro R, Partinen M, et al. Prevalence of every night snoring and obstructive sleep apneas among 30–69 year old men in Bologna Italy. *Acta Neurol Scand* **79**:366–372(1989).

90. Young T, Palta M, Dempsey J, Skatrud J, Weber S, Badr S. The occurrence of sleep-disordered breathing among middle-aged adults. *N Engl J Med* **328**:1230–1235(1993).

91. Bearpark H, Elliott L, Grunstein R, et al. Snoring and sleep apnea—a population study in Australian men. *Am J Respir Crit Care Med* **151**:1459–1465(1995).

92. Ip MS, Lam B, Tang LC, Lauder IJ, Ip TY, Lam WK. A community study of sleep-disordered breathing in middle-aged Chinese women in Hong Kong: prevalence and gender differences. *Chest* **125**:127–134(2004).

93. Rothdach AJ, Trenkwalder C, Haberstock J, Keil U, Berger K. Prevalence and risk factors of RLS in an elderly population: the MEMO study. Memory and Morbidity in Augsburg Elderly. *Neurology* **54**(5):1064–1068(2000).

94. Lavigne GJ, Montplaisir JY. Restless legs syndrome and sleep bruxism: prevalence and association among Canadians. *Sleep* **17**:739–743(1994).

95. Phillips B, Young T, Finn L, Asher K, Hening WA, Purvis C. Epidemiology of restless legs symptoms in adults. *Arch Intern Med* **160**:2137–2141(2000).

96. Hublin C, Kaprio J, Partinen M, Heikkila K, Koskenvuo M. Prevalence and genetics of sleepwalking: a population-based twin study. *Neurology* **48**:177–181(1997).

97. Ohayon MM, Guilleminault C, Priest RG. Night terrors, sleepwalking, and confusional arousals in the general population: their frequency and relationship to other sleep and mental disorders. *J Clin Psychiatry* **60**:268–276(1999).

98. Téllez-Lòpez A, Sánchez EG, Torres FG, Ramirez PN, Olivares VS. Hábitos y trastornos del dormir en residentes del área metropolitana de Monterrey. *Salud Mental* **18**:14–22(1995).

99. Ohayon MM, Priest RG, Zulley J, Smirne S. The place of confusional arousals in sleep and mental disorders: general population findings (13057 subjects). *J Nerv Ment Dis* **188**:340–348(2000).

100. Ohayon MM, Morselli PL, Guilleminault C. Prevalence of nightmares and its relationship to psychopathology and daytime functioning in insomnia subjects. *Sleep* **20**:340–348(1997).

101. Goode GB. Sleep paralysis. *Arch Neurol* **2**:228–234(1962).

102. Ohayon MM, Caulet M, Priest RG. Violent behaviour during sleep. *J Clin Psychiatry* **58**:369–378(1997).

103. Brabbins CJ, Dewey ME, Copeland JRM, et al. Insomnia in the elderly: prevalence, gender differences and relationships with morbidity and mortality. *Int J Geriatr Psychiatry* **8**:473–480(1993).

11

CLASSIFICATION OF SLEEP DISORDERS

ANDREW L. CHESSON, JR.
Louisiana State University Medical Center, Shreveport, Louisiana

As this chapter is being finalized, the *International Classification of Sleep Disorders, Version 2* (ICSD-2), is being completed for publication by the American Academy of Sleep Medicine [1]. Significant changes have occurred to keep up with the changing field of sleep medicine. As the science of sleep medicine is progressively developing from its early scientific underpinnings, and greater clarity is slowly evolving in many areas of patient diagnosis and treatment, that new data has been applied to the ICSD.

The goals of ICSD-2, as indicated by the task force charged with writing the document, are:

1. To describe all currently recognized sleep and arousal disorders, and to base the descriptions on scientific and clinical evidence.
2. To present the sleep and arousal disorders in an overall structure that is rational and scientifically valid.
3. To render the sleep and arousal disorders as compatible with ICD-9 and ICD-10 as possible.
4. To provide for each sleep disorder code a number that is compatible with ICD-9-CM and ICD-10-CM.

To address the first two goals, ICSD-2 sorted sleep disorders into eight categories based on a variety of considerations, including common complaint (e.g., insomnia or hypersomnia), a presumed basic etiology (e.g., biological clock disturbances for circadian rhythm sleep disorders), or the organ system from which the problems arise, such as sleep-related breathing disorders. There is also the issue of sleep disorders involving both adult and pediatric patients. In some cases the pediatric presentation or diagnostic criteria are unique enough to warrant a specific pediatric designation. All of these factors, as well as a decade of practical use with the ground-breaking ICSD-1, have led to the remarkable efforts resulting in ICSD-2. The American Academy of Sleep Medicine Board of Directors and staff felt that the time was right to make changes. The history and process of the growth and development of the ICSD-2 are outlined for interested readers in the introduction to the ICSD-2.

The eight categories used by the ICSD-2 include:

I. Insomnias
II. Sleep-Related Breathing Disorders
III. Hypersomnias Not Due to a Sleep-Related Breathing Disorder
IV. Circadian Rhythm Sleep Disorders
V. Parasomnias
VI. Sleep-Related Movement Disorders
VII. Isolated Symptoms, Apparently Normal Variants, and Unresolved Issues
VIII. Other Sleep Disorders

To address the latter two goals, one of the practical difficulties for the ICSD is how to relate clinical classifications to functional criteria for billing. Also at issue is how to relate these classifications to other classifications in use, such as the *International Classification of Diseases* (ICD) and the *American Psychiatric Association's Diagnostic and Statistical Manual* (DSM), which contains a number of the same sleep diagnoses looked at from a different perspective.

ICD-CM (*International Classification of Diseases*) is often focused and constrained by its use for billing. When ICSD-1 was developed, ICD-9 had already been published. Some sleep disorders such as narcolepsy or restless legs syndrome had already been listed in various sections of ICD-9. ICD-9-CM is still used in the United States. ICD-10-CM is on the horizon. At the time of this writing, there were ongoing discussions between ICSD-2 and the National Center for Healthcare Statistics that assigns ICD-10-CM codes, but not all the ICD-10-CM codes proposed have been officially adopted at this point. Therefore ICSD-2 lists both the ICD-9-CM codes and the proposed ICD-10-CM codes in handy charts for interested readers.

In addition to changes concerning individual sleep disorders, which will be highlighted next, there are structural changes between ICSD-1 and 2. As indicated by the authors, the major categories of those structural changes are as follows.

1. It is only concerned with the diagnosis of sleep disorders. The "axial" system, as used in ICSD-1 containing the primary diagnosis, the procedures performed, associated and related medical disorders, and levels of severity, is no longer included.

2. A listing of procedures used to diagnose sleep disorders has been excluded.

3. The terms "intrinsic" and "extrinsic" dyssomnias were exchanged for the eight separate categories noted earlier.

4. Secondary sleep disorders (those due to mental, neurologic, or other medical disorders) are not included.

5. The text describing individual disorders does not have as complicated a list of minimal to full diagnostic criteria, substituting a single set of criteria, as well as eliminating severity criteria.

The remainder of this chapter will address the eight categories covered by the ICSD-2. For each category, there follows a list of the disorders in that category, a brief summary of key features within the category, highlights of changes from the ICSD-1, discussion points of controversy that may have occurred during the development of that category, and/or hallmark features or key clinical notes warranting special focus.

The ICSD-2 systematically classifies each listed disorder by providing an individual section for each, which includes the following: alternate names that may be found in the literature; essential features; associated features with the condition, demographics; predisposing and precipitating factors; familial patterns; onset, course, and complications; pathology and pathophysiology; polysomnographic and other objective findings; specific diagnostic criteria that define the diagnosis clinically; clinical and pathophysiological

subtypes; unresolved issues and further directions likely to occur in the future for better defining the condition; differential diagnosis; and a bibliography of key references.

I. Insomnias

Adjustment Sleep Disorder (Acute Insomnia)

Psychophysiological Insomnia

Paradoxical Insomnia (formerly Sleep State Misperception)

Idiopathic Insomnia

Insomnia Due to Mental Disorder

Inadequate Sleep Hygiene

Behavioral Insomnia of Childhood

    Sleep-Onset Association Type

    Limit-Setting Sleep Type

    Combined Type

    Unspecified Type

Insomnia Due to Medical Condition

Insomnia Due to Drug or Substance

Insomnia Not Due to Substance or Known Physiological Condition, Unspecified (Nonorganic Insomnia, NOS)

Physiological (Organic) Insomnia, Unspecified

This category includes a group of sleep disorders that are noted by a repeated difficulty with sleep initiation, duration, consolidation, or quality that occurs despite adequate time and opportunity for sleep and results in some form of daytime impairment, presumably caused by the nighttime sleep difficulty.

As the cause of insomnia is often multifactorial, diagnosis is often a complex task. In the past, classification of insomnia was often narrowly defined as a difficulty initiating, maintaining, or obtaining sufficient amounts of sleep. The ICSD-2 has pointed out that there are individuals with such sleep-related symptoms who do not experience daytime symptoms, yet there are other individuals who do evidence daytime impairment from a pattern of self-imposed sleep restriction. If the sleep restriction is removed, they manifest no sleep–wake deficits. To take into account such issues, the ICSD-2 defines insomnia using the following criteria.

General Criteria for Insomnia

A. A complaint of difficulty initiating sleep, difficulty maintaining sleep, or waking up too early, or sleep that is chronically nonrestorative or poor in quality. In children, the sleep difficulty is often reported by the caretaker and may consist of observed bedtime resistance or inability to sleep independently.

B. The sleep difficulty occurs despite adequate opportunity and circumstances for sleep.

C. At least one of the following forms of daytime impairment related to the nighttime sleep difficulty is reported by the patient:

 I. Fatigue or malaise;

 II. Attention, concentration, or memory impairment;

 III. Social or vocational dysfunction or poor school performance;

 IV. Mood disturbance or irritability;

 V. Daytime sleepiness;

 VI. Motivation, energy, or initiative reduction;

 VII. Proneness for errors or accidents at work or while driving;

 VIII. Tension, headaches, or GI symptoms in response to sleep loss;

 IX. Concerns or worries about sleep.

In the insomnias, the reorganization has combined a number of items found in the intrinsic and extrinsic sleep disorders under "dyssomnia" in ICSD-1 into a more organized category related to the symptom of insomnia. Some specific sections with more significant revisions or changes are as follows.

The new category of *paradoxical insomnia* was formally identified as sleep state misperception. This was previously problematic in that a complaint of severe insomnia occurred without evidence of objective sleep disturbance and without levels of daytime impairment of a degree that would be expected for the sleep deficits reported. While science is yet to clarify the nature and physiologic characteristics underlying this problem, the new classification better characterizes the details and gets away from the concept that the problem relates only to a misplaced perception on the part of the patient. Hopefully, these better definitions can help eventually elucidate some of the underlying physiology.

The *behavioral insomnia of childhood* helps consolidate some conditions that were more widely separated in ICSD-1. The new division focuses on difficulty falling and/or staying asleep in a child related to an identified behavioral etiology. *Sleep-onset association type* focuses on the child's dependency on specific activities or circumstances at bedtime and the *limit-setting sleep type* focuses on inadequate limit setting by a caregiver. The combinations and situations in which details are not clarified are also specified. Insomnias due to mental and medical disorders, or drug or substance abuse are characterized under their chief symptom rather than under the primary associated disorder.

II. Sleep-Related Breathing Disorders

Central Sleep Apnea Syndromes
 Primary Central Sleep Apnea

Other Central Sleep Apnea Due to Medical Condition
 Cheyne–Stokes Breathing Pattern
 High-Altitude Periodic Breathing
 Central Sleep Apnea Due to Medical Condition Not Cheyne–Stokes or High-Altitude
 Central Sleep Apnea Due to Drug or Substance
 Other Sleep-Related Breathing Disorder Due to Drug or Substance
 Primary Sleep Apnea of Infancy (formerly Primary Sleep Apnea of Newborn)

Obstructive Sleep Apnea Syndromes
 Obstructive Sleep Apnea, Adult
 Obstructive Sleep Apnea, Pediatric

Sleep-Related Hypoventilation/Hypoxemic Syndromes
 Sleep-Related Nonobstructive Alveolar Hypoventilation, Idiopathic
 Congenital Central Alveolar Hypoventilation Syndrome
 Sleep-Related Hypoventilation/Hypoxemia Due to Medical Condition
  Sleep-Related Hypoventilation/Hypoxemia Due to Pulmonary Parenchymal or Vascular Pathology
  Sleep-Related Hypoventilation/Hypoxemia Due to Lower Airway Obstruction
  Sleep-Related Hypoventilation/Hypoxemia Due to Neuromuscular or Chest Wall Disorders

Other Sleep-Related Breathing Disorders
 Sleep Apnea/Sleep-Related Breathing Disorder, Unspecified

Again, in this category, "intrinsic" and "extrinsic" sleep disorders from the ICSD-1 have been characterized with a focus on central control of breathing, obstructive elements of breathing, and sleep-related hypoventilation/hypoxemic conditions. Conditions that were rather difficult to manage under the previous ICSD-1, such as Cheyne–Stokes breathing and various causes of central apnea, such as those related to altitude, medical conditions, or drug or substance abuse, help focus on the underlying cause rather than lumping these disorders into a hard-to-characterize large basket, or a focus outside of sleep onto an associated medical condition. The category of *primary sleep apnea of infancy* expands the time frame of "newborn" and consolidates the issues of prematurity, apparent life-threatening events, and gets away from a link to SIDS, which suggests a causal relationship between apnea and that disorder which is not currently supported by the research.

There are still areas yet unresolved based on lack of science or similarities of underlying definitions. For instance, the lack of upper airway resistance syndrome will undoubtedly be unsatisfying to some readers, but the committee pointed out that the same pathophysiology and consequences appear to make it unnecessary to separate out as a separate syndrome.

III. Hypersomnias Not Due to a Sleep-Related Breathing Disorder

Narcolepsy
    Narcolepsy with Cataplexy
    Narcolepsy Without Cataplexy
    Narcolepsy Due to Medical Condition
    Narcolepsy, Unspecified
Other Hypersomnias
    Recurrent Hypersomnia
        Kleine–Levin Syndrome
        Menstrual Related Hypersomnia
    Idiopathic Hypersomnia with Long Sleep Time
    Idiopathic Hypersomnia Without Long Sleep Time
    Behaviorally Induced Insufficient Sleep Syndrome
    Hypersomnia Due to Medical Condition
    Hypersomnia Due to Drug or Substance
    Hypersomnia Not Due to Substance or Known Physiological Condition (Nonorganic Hypersomnia, NOS)
    Physiological (Organic) Hypersomnia, Unspecified (Organic Hypersomnia, NOS)

This category focuses on sleepiness defined as the inability to stay alert and awake during the major waking episode. It defines some assessments utilizing the Epworth Sleepiness Scale, the Multiple Sleep Latency Test (MSLT), and the Maintenance Wakefulness Test (MWT). Narcolepsy, long the hallmark of hypersomnia, is classified with or without features of cataplexy due to the consideration that future research may differentiate these disorders. Narcolepsy is also divided into categories due to a medical condition or as otherwise unspecified. This is clearly a "splitter" rather than a "lumper" approach, but it was felt to be beneficial with upcoming potential use of recent exploration of hypocretin and new animal models for the disease. The remainder of the hypersomnias are subdivided by elements, or by other related conditions, or by having differentiating measurable features, which again may be of consequence based on future research. A disorder previously listed elsewhere in ICSD-1, *insufficient sleep*, is now characterized under a consequence of that problem, which is often the patient's presenting complaint, that of insufficient sleep leading to daytime sleepiness.

IV. Circadian Rhythm Sleep Disorders

Primary Circadian Rhythm Sleep Disorders
    Circadian Rhythm Sleep Disorder, Delayed Sleep Phase Type
    Circadian Rhythm Sleep Disorder, Advanced Sleep Phase Type
    Circadian Rhythm Sleep Disorder, Irregular Sleep–Wake Type
    Circadian Rhythm Sleep Disorder, Free-Running (Nonentrained) Type
    Circadian Rhythm Sleep Disorders Due to Medical Condition
    Primary (Organic) Circadian Rhythm Sleep Disorders, Unspecified
    Other Physiological (Organic) Circadian Rhythm, Unspecified
Behaviorally Induced Circadian Rhythm Sleep Disorders
    Circadian Rhythm Sleep Disorder Not Due to Substance or Known Physiological Condition, Jet Lag Type
    Circadian Rhythm Sleep Disorder Not Due to Substance or Known Physiological Condition, Shift Work Type
    Circadian Rhythm Sleep Disorder Not Due to Substance or Known Physiological Condition, Delayed Sleep Phase Type
    Circadian Rhythm Sleep Disorder Not Due to Substance or Known Physiological Condition, Unspecified (Nonorganic Circadian Rhythm Sleep Disorder, NOS)
    Other Circadian Rhythm Sleep Disorder Not Due to Substance or Known Physiological Condition
    Other Circadian Rhythm Sleep Disorder due to Drug or Substance

This grouping identifies the various types of circadian sleep disorders and provides classification of each. A key feature is a persistent or recurrent pattern of sleep disturbance, which is primarily due to alteration of the circadian time-keeping system or a misalignment between the individual's circadian rhythm and external influences that affect the timing of sleep. This can lead to complaints of insomnia or excessive daytime sleepiness. This cluster of disorders was generally previously linked together under "Dyssomnias" in the ICSD-1 classification. The larger subdivision of this group relates to *primary* versus *behaviorally induced* grouping to help provide a focus on *physiology.* Perhaps the main difficulty with this category is that the names have become cumbersome and lengthy, particularly in the behaviorally induced section. More than likely, shorthand clinical notations will be utilized by most sleep specialists. *Circadian Rhythm Sleep Disorder Not Due to a Substance or Known Physiological Condition, Jet Lag Type* will almost surely become *Jet Lag.*

V. Parasomnias

Disorders of Arousal (from NREM Sleep)
　Confusional Arousals
　Sleepwalking
　Sleep Terrors
Parasomnias Usually Associated with REM Sleep
　REM Sleep Behavior Disorder
　　Parasomnia Overlap Disorder
　　Status Dissociatus
　Recurrent Isolated Sleep Paralysis
　Nightmare Disorder
Other Parasomnias
　Sleep-Related Dissociative Disorder
　Sleep-Related Enuresis
　Sleep-Related Groaning (Catathrenia)
　Exploding Head Syndrome
　Sleep-Related Hallucinations
　Sleep-Related Eating Disorder
　Parasomnia, Unspecified
　Parasomnia Due to Drug or Substance
　Parasomnia Due to Medical Condition

This category continues to be composed of undesirable physical or experiential events that accompany sleep, occurring during entry into sleep, or during arousals from sleep. Major divisions in this group include *Disorders of Arousal from NREM Sleep, Parasomnias Usually Associated with REM sleep*, and *Other Parasomnias*. Some of the disorders previously characterized as parasomnias have been moved to the *Sleep-Related Movement Disorders*, including such things as *bruxism* and *rhythmic movement disorder*. These disorders are generally ones that encompass a broad range of predominantly simple movements such as myoclonic, repetitive, rocking, rhythmic, grinding, cramping fragmentary distonic, or disconnective movements or tremors. They are usually not associated with goal-oriented or experiential concomitants as per the authors' description of some of these characterizations. The activities remaining in this group often involve complex, somewhat purposeful, and goal-directed behaviors even though the goal directive may have personal meaning to the individuals at the time, rather than logical and sound behavior characteristics.

Many of these do not require polysomnographic definition (primarily *REM sleep behavior disorder* does); however, polysomnographic monitoring can provide strong collaborative documentation. The authors did address the issue of some disorders that relate to the four basic drive states (sleep, sex, feeding, and aggression) all having been documented to merge in pathological forms from the disorders of arousal category. Although these were not all given specific individual classifications at present, many appear to be somnambulistic behavioral variations.

VI. Sleep-Related Movement Disorders

Restless Legs Syndrome (including Sleep-Related Growing Pains)
Periodic Limb Movement Sleep Disorder
Sleep-Related Leg Cramps
Sleep-Related Bruxism
Sleep-Related Rhythmic Movement Disorder
Sleep-Related Movement Disorder, Unspecified
Sleep-Related Movement Disorder Due to Drug or Substance
Sleep-Related Movement Disorder Due to Medical Condition

This section is new as a category. It is composed of sleep disorders that are characterized mainly by relatively simple, usually stereotyped movements, that disturb sleep, or by other muscle involvement. *Restless legs syndrome* (RLS) is perhaps an exception to this characterization. It's been moved out of the intrinsic sleep disorders in ICSD-1 into this category. The link to the overall characterization of RLS to this group is the close association between RLS and periodic limb movements of sleep. Also, inherent in the disorders classified in this section, there needs to be a significant sleep complaint or evidence that sleep is disturbed by the movement. The differentiation between body movements that are classified elsewhere (i.e., parasomnias) is that those movements are more complex and goal-directed. The definition of goal-directed, however, may be outside conscious awareness. The authors point out that there are some movement disorders that can occur between sleep and wakefulness. A movement disorder is classified by ICSD-2 here if the presentation during sleep is significantly different than during wakefulness. *Sleep-related bruxism* is such an example. If the disorder is generally similar during wakefulness and sleep, it is not classified in ICSD-2 as a sleep disorder. An example would be sleep-related epilepsy, which is listed under *Sleep Disorders Associated with Conditions Classifiable Elsewhere*.

Disorders that involve movements that are typical of normal sleep, but differ from the usual presentation in severity, are typically listed in the category of *Isolated Symptoms, Apparently Normal Variants, and Unresolved Issues*. *Sleep Starts, Hypnic Jerks* is such an example.

In the descriptions of these individual disorders, the reader can see that polysomography may be necessary to make a firm diagnosis and an all-night video may also be essential for clarification.

VII. Isolated Symptoms, Apparently Normal Variants, And Unresolved Issues

Long Sleeper

Short Sleeper

Snoring

Sleep Talking

Sleep Starts, Hypnic Jerks

Benign Sleep Myoclonus of Infancy

Hypnagogic Foot Tremor and Alternating Leg Muscle Activation During Sleep

Propriospinal Myoclonus at Sleep Onset

Excessive Fragmentary Myoclonus

The largest groups of these disorders were previously classified in the ICSD-1 as "proposed sleep disorders." Others generally tended to come from various classifications within the parasomnias. Some of this group, in the opinion of the ICSD-2 Committee, did not have enough scientific basis for decisions regarding whether they were pathological disorders and so it was hard to classify them elsewhere.

VIII. Other Sleep Disorders

Other Physiological (Organic) Sleep Disorder

Physiological (Organic) Sleep Disorder, Unspecified

Other Sleep Disorder Not Due to Substance or Physiological Condition

Environmental Sleep Disorder

Sleep Disorder Not Due to Substance or Physiological Condition, Unspecified

In this category are those sleep disorders that could not be classified elsewhere by the committee formulating the ICSD-2. Reasons for this include: a sleep disorder that overlaps many of the other categories; domains that have new sleep disorders that might be discovered during the lifetime of ICSD-2; and situations in which a category might fulfill the need for sleep disorder diagnoses while insufficient data are available. Examples of these include *Environmental Sleep Disorder*, so classified because sleep can be affected in many ways by the environment, including insomnia, hypersomnia, or parasomnia. *Other Sleep Disorders Not Due to Substance or Physiologic Condition* would be in an area where the code could be used temporarily when the diagnostician is dealing with a case of excessive daytime sleepiness and has ordered studies, including a polysomnogram; at that point, the differential diagnosis might include *obstructive sleep apnea, narcolepsy, or RLS*; but while pending the polysomnogram confirmation and/or a MSLT, the patient may need a temporary diagnosis.

Appendix A: Sleep Disorders Associated with Conditions Classifiable Elsewhere

Fatal Familial Insomnia

Fibromyalgia

Sleep-Related Epilepsy

Sleep-Related Headaches

Sleep-Related Gastroesophageal Reflux Disease

Sleep-Related Coronary Artery Ischemia

Sleep-Related Cardiac Arrhythmias

Sleep-Related Abnormal Swallowing, Choking, and Laryngospasm

Appendix B: Other Psychiatric/Behavioral Disorders Frequently Encountered in the Differential Diagnosis of Sleep Disorders

Mood Disorders

Anxiety Disorders

Selected Somatoform Disorders (Somatization Disorder, Hypochondriasis)

Schizophrenia and Other Psychotic Disorders

Selected Disorders Usually First Diagnosed in Infancy, Childhood, or Adolescence (Mental Retardation, Pervasive Developmental Disorders, Attention Deficit/ Hyperactivity Disorder)

Personality Disorder

The remaining portions of the classification include Appendix A and Appendix B. Included here are generally medical disorders that affect sleep or those that are affected by sleep. These include a small number of medical disorders that may be of particular importance to sleep diagnosticians and/or frequently encountered in the differential diagnosis of certain ICSD-2 sleep diagnoses. Appendix B deals with psychiatric and behavioral disorders that fit similar criteria.

Hopefully, this overview of the ICSD-2 will give the reader some familiarity with this impressive book. The American Academy of Sleep Medicine deserves considerable credit for its activities in setting up and fostering this important new document, which will undoubtedly contribute to further growth of the sleep field. The numerous contributors to development of the classification and the elucidation of the various disorders deserve recognition and appreciation for their efforts. The list of contributors is notable for many internationally recognized experts in the field. However, there are also many not named in the contributor list who contributed in direct and indirect ways to this document, whether they did so because of their knowledge, interest, and expertise in particular disorders, or because of their responsibility to the American Academy of Sleep Medicine. The AASM staff and Board of Directors spent countless hours moving the project along and making final decisions and, in some cases, providing final conflict resolution. All of these individuals should take pride in this final product.

The American Academy of Sleep Medicine has also provided permission for the utilization of this material in an unselfish effort to help, in one more way, in the education of sleep physicians.

REFERENCE

1. *International Classification of Sleep Disorders, Version 2: Diagnostic and Coding Manual.* American Academy of Sleep Medicine, Rochester, MN, 2005.

PART II

INSOMNIA

12

INSOMNIA: PREVALENCE AND DAYTIME CONSEQUENCES

W. David Brown

Sleep Diagnostics of Texas, The Woodlands, Texas

PREVALENCE

Insomnia is a remarkably common, though varied complaint. The term insomnia includes difficulty falling asleep, maintaining sleep for sufficient duration, or a sense of nonrestorative sleep. In addition, the term may be used to describe a symptom of another medical or psychiatric disorder, or the term may also be considered a distinct diagnostic entity. As a result of the nonspecific use of the term and a lack of standard diagnostic methods, it is not always clear that surveys are measuring the same or similar phenomena. However, most epidemiological studies are fairly consistent in the occurrence of the complaint.

General Population

Most studies assessing the prevalence of insomnia in the general population find that between 30% and 35% of individuals responding have experienced some difficulty sleeping in the previous year [1–5]. In addition, roughly 10–20% of the population considers their sleep complaint to be severe [1, 5–9]. Although the definition of severe can vary, it is usually equivalent to disrupted sleep every night or almost every night for at least two weeks and generally includes daytime symptoms. For example, the 1991 National Sleep Foundation Survey of 1000 Americans found that 36% of respondents reported a current sleep problem. Of this group, 27% described themselves as having occasional insomnia and 9% reported their sleep

disturbance was chronic (difficulty sleeping on a frequent basis) [1]. This is consistent with a 1989 survey that found 10.2% of a U.S. population had chronic insomnia [6]. These rates are also consistent with other populations. A survey of over 12,000 individuals in France found occasional insomnia in 29%, 19% had occasional problems with daytime consequences, and 9% had severe insomnia [4]. In Italy, insomnia symptoms were found in 27.6% of the population and insomnia disorder was reported in 7% [7]. In Germany, the insomnia rate was found to be 28.5% [10] with severe insomnia affecting between 4% and 13.5% of the population [10, 11]. In South Korea, 17% of the survey reported symptoms at least three days/week and 5% met DSM-IV criteria for insomnia disorder [8]. (See Figure 12.1.)

Primary Care

Surveys find that insomnia is more prevalent in primary care populations. A study by Shochat et al. [12] found the prevalence of insomnia was 69% with 50% reporting occasional insomnia and 19% reporting chronic insomnia. The higher rate in a medical population would be expected due to concomitant psychiatric and medical illness. In a large survey of five managed care organizations, insomnia with daytime dysfunction was reported by 32.5% of the respondents [13]. Of some note, only 0.9% of the clinic patients were seeing a physician specifically for sleep and only 11.6% of the insomnia with daytime dysfunction group was taking a prescription medication for sleep.

Sleep: A Comprehensive Handbook, Edited by T. Lee-Chiong.
Copyright © 2006 John Wiley & Sons, Inc.

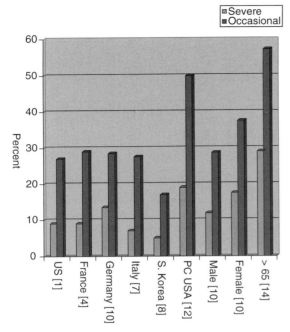

Figure 12.1 Prevalence of insomnia in various populations: selected surveys.

However, about twice that number (21.4%) was taking an over-the-counter medication for sleep. This study pointed out that while many primary care patients have significant sleep problems, few are actually being treated [13].

Gender

Most studies show that insomnia prevalence is higher in women than in men [5, 10, 13–16]. The prevalence of insomnia in women is about 1.3 times that of men [17]. For example, a survey of 1536 individuals in three rural towns in Upper Bavaria found that, for mild insomnia, men had a prevalence of 12.9% compared to women with 16.8%. For severe insomnia the rates were 8.5% and 17.5%, respectively. The gender differences were strongly influenced by age. Before the age of 30, the prevalence rates were very similar [10]. The reason for the female predominance is unknown. Women have three specific areas that separate them from men: the menstrual cycle, childbirth, and menopause. It is estimated that about 15% of women experience sleep disturbance premenstrually [18]. During pregnancy, sleep worsens progressively with between 13% and 60% reporting sleep disturbance during the first trimester and 66–97% by the third trimester [19]. Estimates of insomnia in peri- and postmenopausal women range from 36% to 50% [20] and are attributed to the arousals due to hot flashes, mood disorders, and

sleep-disordered breathing [19]. While, subjectively, women clearly report more sleep problems, apart from pregnancy, these differences have been objectively difficult to demonstrate [18, 21].

Age

The prevalence of insomnia also increases with age for a variety of reasons including retirement, loss of a spouse or close friend, illness, or side effects from medication [14]. A survey of 9000 elderly persons (>65 years old) was examined at baseline and again three years later. The study found that 57% of elderly reported at least one chronic sleep complaint and 29% had chronic difficulty falling asleep or waking too early. Only 12% reported no sleep complaints. These data are very similar to other surveys of an older population that find prevalence rates of 19–38% [6, 10, 14]. About 5% of elderly individuals without insomnia developed symptoms each year and 15% of those with insomnia resolved each year [22]. This study failed to show a difference in prevalence based on gender. However, the men had a significantly higher remission rate (52% versus 46%) and this could explain the lack of difference [22]. One study looking at race in an older population found that the prevalence of insomnia in African Americans was significantly lower than whites (19.8% versus 23.6%). For the whites the gender differences were 20.4% men and 25.5% women. In African Americans the gender differences were 16.4% and 21.8%, respectively, both lower than their white counterparts [23].

Psychiatric Patients

In patient samples, over 60% of those with major depressive disorder report sleep disturbances and between 30% and 50% of patients with chronic insomnia have a psychiatric disorder, typically a mood disorder [6, 24]. Alcoholics also have high rates of insomnia. One study documented that 61% of alcoholics had insomnia during the six months prior to treatment [25] and the insomnia may persist for weeks to months after initiating abstinence [26]. Compared to patients without insomnia, patients with insomnia were more likely to report frequent alcohol use for sleep (55% versus 28%), had worse PSG measures of sleep continuity, and had more severe alcohol dependence and depression. Insomnia remained a robust predictor of relapse after controlling for other variables. Both severity of alcohol dependence and level of depression were significantly associated with insomnia. However, patients with current depression were excluded from the study. This study did not find increased insomnia in women or with increased age. Age and gender differences may disappear in patients seeking treatment for alcoholism.

Baseline insomnia predicted relapse even after controlling for severity of dependence and depressive symptoms [25].

CONSEQUENCES

Quality of Life

Quality of life (QoL) is a concept used to describe an individual's ability to function and to derive satisfaction from doing so [27]. Quality of life is impaired in insomnia patients [9, 11, 27–30]. The 1991 Gallup survey found that 72% of insomnia subjects woke up feeling drowsy or tired [1]. Thirty percent of chronic insomnia subjects reported that their quality of life was fair or poor compared to 19% of occasional insomniacs and only 4% of no insomnia subjects [30]. Fewer of the insomnia group reported enjoying relationships and they had more problems with their spouses as well as friends. They reported feeling physically worse than a no insomnia group, their mood was worse, and they reported difficulty handling minor irritations.

In a survey of a German population, quality of life was rated as "bad" in 22% of severe insomniacs compared to only 3% of subjects without a sleep complaint. QoL was rated as "good" in only 28% of severe insomniacs versus 68% of subjects without sleep problems. Despite this finding, only 55% of the severe group had ever discussed their sleep with a doctor and only 27% were taking medication for sleep [11].

QoL scores were measured in three matched groups of severe insomnia, mild insomnia, and good sleepers. QoL scores were progressively worse from good sleepers to severe insomnia [31]. Insomnia patients report poorer cognitive ability including decreased attention, concentration, mental acuity, reasoning, and problem solving ability. Even when individuals with psychiatric illness are excluded, insomnia patients produce higher scores on depression, anxiety, and health status scales [27, 29, 31]. In general, insomnia can have a negative impact on home life and recreational functioning. Insomnia patients report decreased enjoyment of interpersonal relationships, are more likely to feel irritated by their children and are less likely to help them with their homework, and tend to watch TV more than read or exercise [27, 30, 31].

Use of Medical Services

Insomnia is associated with poorer health though the cause or effect relationship has not been established. However, it is clear that insomnia patients use more medical services—both directly for treatment of the sleep complaint and secondarily for increased medical complaints [11, 29, 32]. Moderate to severe insomnia sufferers have up to two times as many doctor visits and hospital stays as do good sleepers [10, 32]. This increased usage is not accounted for by direct treatment for insomnia. Most studies show that only a small percentage of insomnia patients seek treatment specifically for their sleep complaint [8, 11–13, 30] and only a minority of insomnia patients are taking prescription medications for sleep [9, 13, 15]. However, insomnia patients are taking more medications for other medical problems including cardiovascular, genitourinary, and gastrointestinal drugs [32]. After controlling for demographic variables and comorbid conditions, the negative association of insomnia remained significant on emergency rooms visits, calls to physicians, and OTC drug use [29]. In a study of health maintenance organization patients, mean total health care costs (all inpatient and outpatient care provided by the health plan) were approximately 60% higher in the insomnia group (Group Health Cooperative of Puget Sound). The insomnia group had greater medical comorbidity and a higher prevalence of depression [9].

In one study, 1855 elderly residents of an urban community responded to a comprehensive interview and a follow-up interview after 3.5 years. In males, insomnia was the strongest predictor of both mortality and placement in a nursing home. For females, insomnia was a borderline predictor of mortality but did not predict nursing home placement [33]. Other studies have failed to confirm these results [14, 34]. However, nursing home care for the elderly is the largest source of direct costs attributable to insomnia [17]. When a caregiver's sleep is disturbed, there is less tolerance for remaining the caregiver. Between 40% and 51% of caregivers who institutionalize an elderly person cite sleep disruption as a major factor [17].

Work Performance

Absenteeism from work is far more prevalent in severe insomniacs than in good sleepers [32]. A severe insomnia complaint may be one of the best predictors of absenteeism from work [35]. In addition, insomnia patients experience fewer promotions and pay raises and they are less optimistic about their future job opportunities. In French workers matched for professional activities and work schedule, the severe insomnia group missed twice as many workdays the previous year as did good sleepers [32]. The insomnia group also had more difficulty with concentration and finishing their work and had more work-related accidents. In a survey of a general U.S. population, the insomnia group had ten times as many absent days as did the good sleepers [27]. The insomnia group was also less enthusiastic about their future job and career opportunities. Between 1976 and 1981, data were collected on navy servicemen who entered the service at the same level. This study found that self-defined "poor sleepers" received significantly fewer promotions, remained in lower pay grades, received fewer positive recommendations, and had higher attrition rates, compared with self-defined good sleepers [36]. In a primary care setting, insomnia patients showed 3.5 days of additional disability/month [9]. Insomnia was associated with significantly greater impairment according to both self-rated (Social Desirability Schedule) and interviewer-rated (Brief Disability Questionnaire) disability measures [9]. Twenty-four percent of the insomnia patients had moderate to severe occupational role disability compared to 14% without insomnia.

Accidents

In the 1991 Gallup survey, 5% of insomnia subjects reported having an automobile accident due to sleepiness compared to only 2% of the occasional or no insomnia groups [30]. Later studies have shown mixed results. Middle-aged drivers dissatisfied with their sleep were three times as likely to have had an auto accident in the previous year compared to other drivers [7]. However, in a French population, severe insomnia patients had seven times as many industrial accidents as did good sleepers, but they did not experience more traffic accidents [32]. It was speculated that the poor sleepers may have self-limited their driving.

Cognitive Functioning

In the 1991 Gallup survey, chronic and occasional insomnia subjects reported more difficulty concentrating and remembering things than a no insomnia group [30]. Zammit noted that insomnia patients reported poorer cognitive ability including poorer attention, concentration, mental acuity, reasoning, problem solving ability, and mental reactivity. However, only the mean total score differentiated the groups, suggesting that there is no single cognitive area influenced by insomnia [27]. Consistent with this finding, a review of 18 studies examining the neuropsychological functions in insomnia patients concluded that there was no consistent evidence of cognitive dysfunction for any neuropsychological functions in insomnia patients. However, the authors noted there appeared to be a pattern of cognitive dysfunctions that is not consistent across studies. The areas that were most likely to show reduced performance in insomnia patients were attention span and vigilance performance [37].

Risk for Depression/Alcoholism

Insomnia patients have higher rates of depression [9]. Although often perceived as a symptom of depression, insomnia can also be a precursor of depression. A study of 14,915 individuals from four European countries also verified the order of appearance of current symptomatology in individuals who had both insomnia and a psychiatric disorder. The insomnia complaint tended to precede or occur with a mood disorder (first episode or relapse) but tended to appear at the same time or to follow an anxiety disorder episode [38].

Both insomnia and hypersomnia are associated with suicidal behavior in patients with major depression [39]. Sleep disturbance is associated with suicidality in patients with major depression and is a significant independent predictor of completed suicide in psychiatric patients. Insomnia may be a better predictor of suicidal behavior than is a specific plan or suicide note [40].

Cost

Chronic insomnia can be associated with increased use of medical services, decrements in work performance, family problems, emotional problems, automobile and industrial accidents, and increased use of alcohol. In the United States, based on a prevalence of 10%, there would be 25 million people with insomnia. Direct costs have been estimated at $13.9 billion with a large majority of the costs attributed to nursing home care [41]. In addition to the direct cost of medical treatment and drugs, indirect costs may include reduced productivity, increased absenteeism, accidents, and hospitalization, as well as increased medical costs due to increased morbidity and mortality, depression due to insomnia, and increased alcohol consumption [42]. Estimates of the total direct, indirect, and related costs of insomnia have been estimated to range from $30 to $35 billion [43] to a high of $92.5–107.5 billion [42].

CONCLUSION

Insomnia is a remarkably common problem and the prevalence rates are similar in most surveys. The disorder is more common in a general medical practice, in women, and in the elderly. The disorder is associated with negative consequences including increased use of medical services, absenteeism from work, automobile and industrial accidents, poorer work performance, greater risk for depression, and negative impact on family life. The direct and indirect costs associated with insomnia are estimated to be in the billions of dollars each year.

REFERENCES

1. Ancoli-Israel S, Roth T. Characteristics of insomnia in the United States: results of the 1991 National Sleep Foundation Survey. I. *Sleep* **22**(Suppl 2):S347–S353(1999).

2. Bixler EO, Kales A, Soldatos CR, Kalse JD, Healy S. Prevalence of sleep disorders: a survey of the

Los Angeles metropolitan area. *Am J Psychiatry* **136**:1257–1262(1979).

3. Mellinger GD, Balter MB, Uhlenhuth EH. Insomnia and its treatment. Prevalence and correlates. *Arch Gen Psychiatry* **42**(3):225–232(1985).

4. Leger D, Guilleminault C, Dreyfus JP, Delahaye C, Paillard M. Prevalence of insomnia in a survey of 12,778 adults in France. *J Sleep Res* **9**(1):35–42 (2000).

5. Ohayon MM. Epidemiology of insomnia: what we know and what we still need to learn. *Sleep Med Rev* **6**(2):97–111 (2002).

6. Ford DE, Kamerow DB. Epidemiologic study of sleep disturbances and psychiatric disorders. An opportunity for prevention. *JAMA* **262**(11):1479–1484(1989).

7. Ohayon MM, Smirne S. Prevalence and consequences of insomnia disorders in the general population of Italy. *Sleep Med* **3**(2):115–120 (2002).

8. Ohayon MM, Hong SC. Prevalence of insomnia and associated factors in South Korea. *J Psychosom Res* **53**(1):593–600 (2002).

9. Simon G, Vonkorff M. Prevalence, burden, and treatment of insomnia in primary care. *Am J Psychiatry* **154**(10):1417–1423(1997).

10. Weyerer S, Dilling H. Prevalence and treatment of insomnia in the community: results from the Upper Bavarian Field Study. *Sleep* **14**(5):392–398(1991).

11. Hajak G. Epidemiology of severe insomnia and its consequences in Germany. *Eur Arch Psychiatry Clin Neurosci* **251**(2):49–56 (2001).

12. Shochat T, Umphress J, Israel AG, Ancoli-Israel S. Insomnia in primary care patients. *Sleep* **22** (Suppl 2):S359–S365 (1999).

13. Hatoum HT, Kania CM, Kong SX, Wong JM, Mendelson WB. Prevalence of insomnia: a survey of the enrollees at five managed care organizations. *Am J Manag Care* **4**(1):79–86(1998).

14. Foley D, Monjan A, Brown S, Simonsick E, Wallace R, Blazer D. Sleep complaints among elderly persons: an epidemiologic study of three communities. *Sleep* **18**(6):425–432(1995).

15. Ishigooka J, Suzuki M, Isawa S, Muraoka H, Murasaki M, Okawa M. Epidemiological study on sleep habits and insomnia of new outpatients visiting general hospitals in Japan. *Psychiatry Clin Neurosci* **53**(4):515–522(1999).

16. Ohayon MM, Lemoine P. Daytime consequences of insomnia in the French general population. *Encephale* **30**(3):222–227(2004).

17. Walsh J, Ustun TB. Prevalence and health consequences of insomnia. *Sleep* **22**(Suppl 3):S427–S436(1999).

18. Manber R, Armitage R. Sex, steroids, and sleep: a review. *Sleep* **22**(5): 540–555(1999).

19. Moline ML, Broch L, Zak R. Sleep in women from adulthood through menopause. In: Lee-Chiong T, Sateia M, Carskadon M (Eds), *Sleep Medicine*. Hanley & Belfus, Philadelphia, 2002, p 108.

20. Brugge K, Kripke D, Ancoli-Israel S, Garfinkle L. The association of menopausal state and age with sleep disorders. *Sleep Res* **18**:208(1989).

21. Young T, Rabago D, Zgierska A, Austin D, Finn L. Objective and subjective sleep quality in premenopausal, perimenopausal, and postmenopausal women in the Wisconsin sleep cohort study. *Sleep* **26**(6):667–672(2003).

22. Foley DJ, Monjan A, Simonsick EM, Wallace RB, Blazer DG. Incidence and remission of insomnia among elderly adults: an epidemiologic study of 6800 persons over three years. *Sleep* **22**(Suppl 2):S366–S372(1999).

23. Foley DJ, Monjan AA, Izmirlian G, Hays JC, Blazer DG. Incidence and remission of insomnia among elderly adults in a biracial cohort. *Sleep* **22**(Suppl 2):S373–S378(1999).

24. Nowell P. Sleep in patients with mood disorders. In: Lee-Chiong TL, Sateia M, Carskadon MA (Eds), *Sleep Medicine*. Hanley & Belfus, Philadelphia, 2002, pp 541–548.

25. Brower KJ, Aldrich MS, Robinson EA, Zucker RA, Greden JF. Insomnia, self-medication, and relapse to alcoholism. *Am J Psychiatry* **158**(3):399–404 (2001).

26. Brower KJ. Insomnia, alcoholism and relapse. *Sleep Med Rev* **7**(6):523–539 (2003).

27. Zammit G, Weiner J, Damato N, Sillup G, McMillan C. Quality of life in people with insomnia. *Sleep* **22**(Suppl 2): S379–S385(1999).

28. Drake CL, Roehrs T, Roth T. Insomnia causes, consequences, and therapeutics: an overview. *Depress Anxiety* **18**(4):163–176(2003).

29. Hatoum HT, Kong SX, Kania CM, Wong JM, Mendelson WB. Insomnia, health-related quality of life and healthcare resource consumption. A study of managed-care organization enrollees. *Pharmacoeconomics* **14**(6):629–637(1998).

30. Roth T, Ancoli-Israel S. Daytime consequences and correlates of insomnia in the United States: results of the 1991 National Sleep Foundation survey. II. *Sleep* **22**(Suppl 2):S354–S358(1999).

31. Leger D, Scheuermaier K, Philip P, Paillard M, Guilleminault C. SF-36: evaluation of quality of life in severe and mild insomniacs compared with good sleepers. *Psychosom Med* **63**(1):49–55(2001).

32. Leger D, Guilleminault C, Bader G, Levy E, Paillard M. Medical and socio-professional impact of insomnia. *Sleep* **25**(6):625–629 (2002).

33. Pollak CP, Perlick D, Linsner JP, Wenston J, Hsieh F. Sleep problems in the community elderly as predictors of death and nursing home placement. *J Community Health* **15**(2):123–135(1990).

34. Althius MD, Fredman L, Langenberg PW, Magaziner J. The relationship between insomnia and mortality among community-dwelling older women. *J Am Geriatr Soc* **46**(10):1270–1273(1998).

35. Leigh JP. Employee and job attributes as predictors of absenteeism in a national sample of workers: the importance of health and dangerous working conditions. *Soc Sci Med* **33**(2):127–137(1991).

36. Johnson LC, Spinweber CL. In: Guilleminault C, Lugaresi E (Eds), *Sleep/Wake Disorders: Natural History, Epidemiology, and Long-Term Evaluation*. Raven Press, New York, 1983, pp 13–28.

37. Fulda S, Shulz H. Cognitive dysfunction in sleep disorders. *Sleep Med Rev* **5**(6):423–445 (2001).

38. Ohayon MM, Roth T. Place of chronic insomnia in the course of depressive and anxiety disorders. *J Psychiatr Res* **37**:9–15(2003).

39. Agargun MY, Kara H, Solmaz M. Sleep disturbances and suicidal behavior in patients with major depression. *J Clin Psychiatry* **58**(6):249–251(1997).

40. Hall RC, Platt DE, Hall RC. Suicide risk assessment: a review of risk factors for suicide in 100 patients who made severe suicide attempts. Evaluation of suicide risk in a time of managed care. *Psychosomatics* **40**(1):18–27(1999).

41. Walsh JK, Engelhardt CL. The direct economic costs of insomnia in the U.S. for 1995. *Sleep* **22**(Suppl 2):S386–S393(1999).

42. Stoller MK. Economic effects of insomnia. *Clin Ther* **16**(5):873–897(1994).

43. Chilcott LA, Shapiro CM. The socioeconomic impact of insomnia. An overview. *Pharmacoeconomics* **10**(Suppl 1):1–14(1996).

13

CAUSES OF INSOMNIA

EDWARD J. STEPANSKI

Rush University Medical Center, Chicago, Illinois

INTRODUCTION

Insomnia is defined as self-reported difficulty falling asleep, difficulty staying asleep, or having nonrestorative sleep, usually in association with daytime impairment. There are many purported causes of insomnia, covering a broad range of medical, psychiatric, and behavioral factors. In many instances, insomnia is assumed to be secondary to another primary medical, psychiatric, or sleep disorder. However, specific cause and effect pathways have not yet been demonstrated, and a complete understanding of the causes of insomnia remains elusive. Several models are commonly used to explain causes of insomnia.

PRIMARY VERSUS SECONDARY INSOMNIA

One typology often used to understand causes of insomnia is to differentiate whether the insomnia is secondary to another primary disorder, or is an independent disorder [1]. This distinction has been deemed important in that a diagnosis of secondary insomnia will lead to focusing treatment efforts on the primary disorder, rather than on the symptom. Psychiatric disorders, such as depression and anxiety, are an example of primary disorders commonly associated with a secondary insomnia [2, 3]. The comorbidity between insomnia and depression is particularly strong [4, 5]. Patients with severe insomnia are eight times more likely to have depression than patients without insomnia [4]. Studies of patients with medical disorders also find high rates of insomnia in these patients [4].

Research has found that secondary insomnia occurs much more frequently than primary insomnia. In one study that evaluated patients presenting to sleep centers with reports of insomnia, 20% of cases were diagnosed with primary insomnia, 44% of cases with insomnia secondary to a mental disorder, and another 8% of cases with insomnia secondary to a medical disorder or substance abuse disorder [2].

Primary insomnia is diagnosed using criteria from the *Diagnostic and Statistical Manual of Mental Disorders–IV* when the insomnia appears to be an independent disorder [1]. Various behavioral and cognitive factors are theorized as contributing to primary insomnia. This is reflected by the fact that the *International Classification of Sleep Disorders–2* provides subtypes for primary insomnia that specifically implicate behavioral and cognitive factors as causes of the insomnia [6]. Subtypes of primary insomnia categorized within the ICSD nosological system include psychophysiological insomnia, inadequate sleep hygiene, idiopathic insomnia, and paradoxical insomnia. Behavioral factors, such as spending excessive time in bed, napping during the day, drinking caffeine near bedtime, and using the bed for work, are hypothesized causes of insomnia due to inadequate sleep hygiene. Psychophysiological insomnia is thought to result from learned associations between the bed and increased tension and arousal. This category can also be thought of as "conditioned insomnia." This category is the most commonly diagnosed of the primary insomnia subtypes [2]. Idiopathic insomnia begins in childhood and is thought to result from a neurophysiological abnormality in the central nervous system. Finally, paradoxical insomnia is diagnosed when the patient reports being awake most or all of the night, although there is evidence that the patient has been asleep. Polysomnography in these patients will show normal sleep time concurrent with the patient's report of having been awake much of the night.

Limitations of the Secondary Insomnia Model

Research shows a strong association between insomnia and psychiatric and medical disorders, but it is not clear that the insomnia is always a consequence of these disorders. There are a number of studies showing that the relation between insomnia and other presumed primary conditions does not follow a straightforward cause and effect model. For example, research has shown that an episode of insomnia predicts subsequent development of new depression [3, 7]. Also, episodes of insomnia have been shown to predict a recurrence of depression in previously recovered individuals [8]. These data suggest the possibility that, at least in some cases, insomnia can be a cause of depression rather than a consequence of depression. Alternatively, insomnia and depression may both result from a common vulnerability or other factor. Epidemiological research has also shown that insomnia can precede serious medical disorders [9]. A review of ten studies with measures of sleep and cardiac events found that insomnia was a statistically significant predictor of an incident coronary event [9]. Consideration of these alternative views of the relation between insomnia and comorbid disorders is important since it may change the treatment plan. More aggressive treatment of insomnia may be indicated if this condition can lead to a subsequent episode of depression or other morbidity beyond the expected daytime fatigue.

Another difficulty with secondary insomnia is that a definitive determination that insomnia is secondary to another primary disorder is often impossible in clinical practice. Lichstein has written extensively on this issue and recommends that there must be contiguity between the origin and course of the insomnia, relative to the suspected primary disorder [10]. According to his system, both of these requirements are needed to diagnose insomnia as being definitively secondary to the associated primary condition. If either the origin or the course of the insomnia is associated with that of the primary disorder, the insomnia is only viewed as partially related to the primary condition.

A final shortcoming of viewing insomnia as a consequence of another primary disorder is that not all patients with the primary disorders associated with insomnia actually develop insomnia. For instance, one study found that half of the patients with medical disorders also reported mild or severe insomnia [4]. If the insomnia is secondary to these medical conditions, wouldn't all patients with that condition develop insomnia? Clearly other factors are at work in causing insomnia beyond the presence or absence of another "primary" disorder. Another theoretical model is needed that can account for these anomalies and provide a more thorough understanding of how various factors interact to cause insomnia.

THEORETICAL MODEL FOR CHRONIC INSOMNIA: PREDISPOSING, PRECIPITATING, AND PERPETUATING FACTORS

Determining a specific cause of chronic insomnia is often challenging because of multiple contributing factors. A model of chronic insomnia has been proposed that assumes a multifactorial etiology and categorizes causal factors according to their role in the formation of the insomnia: predisposing, precipitating, or perpetuating [11]. This model helps to improve the understanding of how various factors interact together to produce insomnia and also illustrates how an episode of acute insomnia evolves into chronic insomnia.

According to this model, all individuals have a certain level of predisposition to insomnia, and insomnia occurs when this predisposition interacts with exposure to a precipitating factor. Individuals who are highly predisposed will experience insomnia in response to minor precipitants, while others are robust sleepers with a low predisposition to insomnia and experience insomnia only in response to very significant precipitating events. Over time, the precipitating factor recedes or disappears entirely but is replaced by perpetuating factors that maintain the insomnia. The predisposing factors explain why only certain patients develop insomnia (highly predisposed) when experiencing pain or another condition that commonly precipitates insomnia, while others (with low predisposition) continue to sleep well in the presence of the same disorder.

Predisposing Factors

The presence of predisposing factors is inferred by the different thresholds observed across individuals for the development of insomnia. For example, some people experience sleep disturbance after drinking a cup of coffee at dinnertime, while others can have several cups with no apparent change in their ability to fall asleep or stay asleep. However, specific mechanisms underlying a predisposition to insomnia have not been elucidated. One theory is that physiological hyperarousal is a predisposing factor for insomnia (see Table 13.1). Research has shown that individuals with chronic insomnia, as compared to normal sleepers, have a faster increase in heart rate in response to stress [12], an increased metabolic rate [13], increased heart rate variability [14], increased beta activity in the sleep EEG [15], increased secretion of ACTH, and increased daytime alertness despite short sleep [16]. Evidence to support

TABLE 13.1   Causal Factors in the Formation of Acute or Chronic Insomnia

Predisposing factors
  Somatic hyperarousal
  Cognitive hyperarousal
  Decreased homeostatic sleep drive

Precipitating factors
  Medical disorders
  Psychiatric disorders
  Medication effects
  Substance abuse
  Circadian rhythm disorders
  Primary sleep disorders

Perpetuating factors
  Behavioral factors
  Cognitive factors

this view is increasing, but controversy continues in that it has not yet shown whether hyperarousal itself is a cause or consequence of insomnia. In addition to these measures of physiological hyperarousal, there is evidence that patients with chronic insomnia are prone to increased cognitive arousal [17] or emotional arousal [18]. These constructs are not mutually exclusive, and more research is needed to understand the role of hyperarousal in all forms related to insomnia.

Another possible predisposing factor is the presence of a decreased homeostatic drive for sleep. Sleep initiation, as well as overall regulation of a coherent sleep–wake schedule, has been shown to be related to homeostatic and circadian mechanisms. Recent studies have shown that patients with chronic insomnia do not show the same increase in slow-wave sleep following sleep deprivation that is observed in normal sleepers, consistent with decreased homeostatic drive [19, 20]. A reduced sleep drive would be expected to make it more difficult to initiate and maintain sleep under both baseline and sleep-deprived conditions. Decreased sleep drive could interact with precipitating factors to produce chronic insomnia.

Precipitating Factors

Precipitating factors are those disorders or conditions typically hypothesized as causes of secondary insomnia in that model of insomnia. Common precipitating factors include medical disorders, psychiatric disorders, environmental factors, medication effects, primary sleep disorders, or circadian rhythm changes that negatively affect sleep (see Table 12.1). Precipitating factors can be acute stressful events, such as final exams, conflicts in the workplace, or marital distress. Acute changes might also include environmental factors such as excessive noise or light in the bedroom. Sleeping in a new environment can precipitate

poor sleep, and many individuals report insomnia their first night in a hotel room. Changes in sleep–wake schedule caused by jet lag or a change in work shift can also precipitate insomnia.

Precipitants to insomnia also include chronic medical and psychiatric conditions, as discussed earlier in the context of secondary insomnia. Examples of medical factors that cause insomnia include any condition associated with pain, respiratory disorders associated with dyspnea when supine, neurodegenerative disorders, renal failure, and hyperthyroidism. Many different medications are also associated with insomnia, including stimulating antidepressants, steroids, beta blockers, bronchodilators, and decongestants. Primary sleep disorders causing insomnia include restless legs syndrome, periodic limb movement disorder, and sleep-disordered breathing. Although sleep-disordered breathing more typically presents with a complaint of excessive daytime sleepiness, insomnia can also be a presenting complaint. All of these primary sleep disorders become more prevalent with aging.

Perpetuating Factors

Perpetuating factors are behavioral and cognitive changes that occur once an individual has been sleeping poorly for a period of time. Examples of common behavioral changes include keeping an irregular sleep–wake schedule, spending excessive amounts of time in bed attempting to gain more sleep time, taking daytime naps, and engaging in stimulating activities during the night. These are changes that people often make in response to insomnia, in an effort in obtain additional sleep and rest. However, while such changes might help alleviate symptoms in the short-term (e.g., feeling more rested after a nap), they promote continued insomnia in the long-term (e.g., inability to initiate sleep at a normal bedtime due to an afternoon nap). Having a snack during the night may intuitively seem like a good idea to help a patient reinitiate sleep, but it can soon become a habit leading to increased arousal during the night and a continued tendency to spend time awake.

Cognitive changes that occur with insomnia include an increased preoccupation with sleep during daytime hours, as well as a fear of not sleeping and fear of daytime impairment. These cognitive changes lead to increased tension at bedtime, and upon awakening during the night. Patients who have experienced difficulty sleeping for weeks or months often respond to nocturnal awakenings with concern, if not outright frustration and anger, that was not present when they were sleeping well. Additionally, patients may develop irrational fears regarding the effects of insomnia. For instance, older patients may believe that their inability to sleep will lead to marked degradation of their medical status, and even death. Another type of cognitive feature occasionally present in patients with insomnia is unreasonable expectations regarding sleep. An example of this problem would be a patient who believes that he or she requires nine hours of total sleep time, even though a careful sleep history shows that they have been a lifelong seven hour sleeper.

The role of these behavioral and cognitive changes, or perpetuating factors, in maintaining chronic insomnia helps to explain another finding that conflicts with the presumed cause and effect relation between insomnia and primary disorders: the primary disorder can remit with treatment, but the insomnia continues on [21,22]. If the insomnia is truly secondary to the primary condition, it should remit in conjunction with the primary disorder. A multifactorial model of chronic insomnia with predisposing, precipitating, and perpetuating factors shows that the primary disorder initially triggers the insomnia but is not needed to cause the insomnia to continue.

SUMMARY

A traditional model of insomnia distinguishes between secondary insomnia that occurs as a consequence of another primary disorder, and primary insomnia that is an independent disorder. This model explains the high comorbidity between insomnia and psychiatric and medical disorders. However, chronic insomnia has also been shown to predict the subsequent development of psychiatric and medical disorders, and many patients with primary disorders do not exhibit insomnia. Therefore understanding the exact relation between insomnia and comorbid disorders requires additional study. At this time, chronic insomnia is best understood as resulting from a combination of predisposing, precipitating, and perpetuating factors.

REFERENCES

1. American Psychiatric Association. *Diagnostic and Statistical Manual of Mental Disorders*, 4th ed. American Psychiatric Association, Washington, DC, 1994.

2. Buysse D, Reynolds C, Kupfer D, Thorpy M, Bixler E, Manfredi R, Kales A, Vgontzas A, Stepanski E, Roth T, Hauri P, Mesiano D. Clinical diagnoses in 216 insomnia patients using the International Classification of Sleep Disorders (ICSD), DSM-IV and ICD-10 categories: a report from the APA/NIMH DSM-IV field trial. *Sleep* **17**:630–637(1994).

3. Ford DE, Kamerow DB. Epidemiologic study of sleep disturbances and psychiatric disorders: an opportunity for prevention. *JAMA* **262**:1479–1484(1989).

4. Katz DA, McHorney CA. Clinical correlates of insomnia in patients with chronic illness. *Arch Intern Med* **158**:1099–1107(1998).

5. Foley D, Ancoli-Israel S, Britz P, Walsh J. Sleep disturbances and chronic disease in older adults. Results of the 2003 National Sleep Foundation Sleep in America survey. *J Psychosom Res* **56**:497–502(2004).

6. American Academy of Sleep Medicine. *The International Classification of Sleep Disorders, 2nd ed.: Diagnostic and Coding Manual.* American Academy of Sleep Medicine, Westchester, IL, 2005.

7. Chang PP, Ford DE, Mead LA, Cooper-Patrick L, Klag MJ. Insomnia in young men and subsequent depression. The Johns Hopkins precursors study. *Am J Epidemiol* **146**:105–114(1997).

8. Perlis ML, Giles DE, Buysse DJ, Tu X, Kupfer DJ. Self-reported sleep disturbance as a prodromal symptom in recurrent depression. *J Affect Disord* **42**:209–212(1997).

9. Schwartz S, McDowell Anderson W, Cole SR, Cornoni-Huntley J, Hays JC, Blazer D. Insomnia and heart disease: a review of epidemiologic studies. *J Psychosom Res* **47**:313–333(1999).

10. Lichstein KL. Secondary insomnia. In: Lichstein KL, Morin CM (Eds), *Treatment of Late-Life Insomnia.* Sage Publications Inc., Thousand Oaks, CA, 2000.

11. Spielman A. Assessment of insomnia. *Clin Psychol Rev*, **6**:11–26(1986).

12. Stepanski E, Glinn M, Zorick F, Roehrs T, Roth T. Heart rate changes in chronic insomnia. *Stress Med* **10**:261–266(1994).

13. Bonnet MH, Arand D. Hyperarousal and insomnia. *Sleep Med Rev* **1**:97–108(1997).

14. Bonnet MH, Arand D. Heart rate variability in insomniacs and matched normal sleepers. *Psychosom Med* **60**:610–615(1998).

15. Perlis ML, Merica H, Smith MT, Giles DE. Beta EEG in insomnia. *Sleep Med Rev* **5**:363–374(2001).

16. Stepanski E, Zorick F, Roehrs T, Young D, Roth T. Daytime alertness in patients with chronic insomnia compared with asymptomatic control subjects. *Sleep* **11**:54–60(1988).

17. Harvey AG, Greenall E. Catastrophic worry in primary insomnia. *J Behav Ther Exp Psychiatry* **34**:11–23(2003).

18. Kales A, Kales JD. *Evaluation and Treatment of Insomnia.* Oxford University Press, New York, 1984.

19. Besset A, Villemin E, Tafti M, Billiard M. Homeostatic process and sleep spindles in patients with sleep-maintenance insomnia: effect of partial(21 h) sleep deprivation. *Electroencephal Clin Neurophys* **107**:122–132(1998).

20. Stepanski E, Zorick F, Roehrs T, Roth T. Effects of sleep deprivation on daytime sleepiness in primary insomnia. *Sleep* **23**:215–219(2000).

21. Zayfert C, DeViva J. Residual insomnia following cognitive behavioral therapy for PTSD. *J Trauma Stress* **17**: 69–73(2004).

22. Jinda RD, Thase ME, Fasiczka AL, Friedman ES, Buysse DS, Frank E, Kupfer D. Electroencephalographic sleep profiles in single-episode and recurrent unipolar forms of major depression: II. Comparison during remission. *Biol Psychiatry* **51**:230–236(2002).

ADDITIONAL READING

Vgontzas AN, Bixler EO, Lin HM, Prolo P, Mastorakos G, Vela-Bueno A, Kales A, Chrousos GP. Chronic insomnia is associated with nyctohemeral activation of the hypothalamic–pituitary–adrenal axis: clinical implications. *J Clin Endocrinol Metab* **86**:3787–3794(2001).

14

MEDICATIONS THAT CAN CAUSE INSOMNIA

Carolyn H. Welsh and Randolph V. Fugit
Denver Veterans Affairs Medical Center, Denver, Colorado

INTRODUCTION

Insomnia is defined as inadequate or poor-quality sleep characterized by one or more of the following: difficulty falling asleep, difficulty maintaining sleep, waking up too early in the morning, or unrefreshing sleep [1]. Insomnia is therefore a symptom rather than a specific disease, and the perception that medications induce insomnia is subjective. Often, insomnia can be a symptom of a medical or psychiatric disorder such as anxiety or depression. For example, up to 90% of patients with depression complain of insomnia, and it may be difficult to distinguish the impact of disease from that of the drug.

Symptoms of insomnia are tracked poorly in the medical literature, perhaps because sleep difficulties are not perceived to be life threatening and because good sleep histories are rarely obtained from patients. Since insomnia isn't always life threatening or of clear clinical significance, there are no absolute indications to remove medications. Insomnia can, however, be a major life-altering problem and should be addressed whenever needed. A targeted history not only of prescription medications but also herbal preparations should be sought for possible association with insomnia and to confirm need for on-going use. Therapeutically, it may be reasonable to transiently remove a medication to assess for symptomatic improvement off the medication and to confirm its association with insomnia.

Medications can cause insomnia, and conversely withdrawal of medications can lead to sleep symptoms, including insomnia. Certain medications are used to treat sleepiness and thereby designed to have insomnia as a therapeutic effect. In general, medications affecting neurotransmitters (norepinephrine, serotonin, acetylcholine, and dopamine) can alter sleep [2]. Other classes of drugs are associated with insomnia as an unwanted side effect and will also be discussed.

CENTRAL NERVOUS SYSTEM STIMULANTS

Central nervous system (CNS) stimulants include those agents classified as psychomotor stimulants (e.g., amphetamine and methylphenidate) or as products that produce stimulatory effects that occur from differing neurochemical mechanisms other than psychomotor stimulation (e.g., nicotine and caffeine).

Medications described as psychomotor stimulants have been shown to significantly enhance the concentrations of monoamines (norepinephrine, dopamine, and serotonin) in the synaptic cleft, as well as block their reuptake and thus generate increased behavioral activation, arousal, and alertness [2, 3]. The stimulatory effects ascribed to these medications are due to their effects on the dopaminergic system and the sympathomimetic effects from their noradrenergic actions. Amphetamines and methylphenidate have been used for the therapeutic indications of management or treatment of obesity, narcolepsy, and attention deficit hyperactivity disorder (ADHD). However, in spite of clinical benefit in these disease states, they have also been determined to produce a substantial decline in sleepiness, increased sleep latency, increased time to the onset of rapid eye movement (REM) sleep, as well as a decreased

Sleep: A Comprehensive Handbook, Edited by T. Lee-Chiong.
Copyright © 2006 John Wiley & Sons, Inc.

TABLE 14.1   Drugs Associated with Insomnia

Medication	Incidence of Insomnia (If Available)
Central Nervous System Stimulants	
Amphetamine	
Benzphetamine	
Dextroamphetamine	
Methamphetamine	
Methyphenidate	
Dexmethyphenidate	
Modafinil	
Pemoline	
Phentermine	
Caffeine	
Nicotine	
Psychiatric Medications Associated with Insomnia	
SELECTIVE SEROTONIN REUPTAKE INHIBITORS	
Fluoxetine	5–9%
Paroxetine	8–14%
Sertraline	7–16%
Citalopram	10%
Escitalopram	9%
Fluvoxamine	15–19%
OTHER	
Bupropion	5–19%
Venlafaxine	8%

proportion of REM sleep (Table 14.1). This has been substantiated in research by Mitler and colleagues, during which patients suffering from narcolepsy were treated with methamphetamine in doses of 20–60 mg daily for 4 days [4]. Disrupted sleep patterns were identified by polysomnography in all (*n* = 8) of the patients, with seven of these patients also describing insomnia. Additional studies with dextroamphetamine and methylphenidate in doses greater than 50–60 mg per day have also shown similar results regarding disturbed nocturnal sleep patterns [5].

Modafinil is a newer generation CNS stimulant that currently has the indication in the United States for the treatment of narcolepsy, and somewhat controversially for treatment of shift work sleep disorder and residual sleepiness in continuous positive airway pressure (CPAP) treated obstructive sleep apnea. This medication appears to induce less CNS stimulatory effects than amphetamines and methylphenidate. Anorectic drugs are used for treatment or avoidance of obesity. All except fenfluramine appear to stimulate the CNS and result in insomnia. Tiratricol used as an appetite suppressant is a weak thyromimetic and may induce insomnia by this mechanism.

TOBACCO, NICOTINE, AND CAFFEINE

As previously mentioned, nicotine and caffeine's stimulant effects on the CNS differ significantly from the psychomotor stimulant drugs, and thus their sympathomimetic and psychomotor stimulatory effects are significantly weaker. Cigarette smoking has been associated with sleep initiation insomnia in a large cross-sectional study comparing smokers to nonsmokers [6]. Active smokers have been noted to have associated sleep disturbances characterized as difficulty initiating sleep and nonrestorative sleep. A potential reason for this is a direct effect of elevated nicotine levels to disrupt sleep. Alternatively, nightly tobacco withdrawal may lead to nighttime symptoms including problems starting and staying asleep. Nicotine itself is used therapeutically to facilitate smoking cessation and has been associated with high rates of insomnia in up to 70% of normal volunteers.

Caffeine is a methylxanthine compound structurally related to theophylline, theobromine, and uric acid. Caffeine is an adenosine antagonist and stimulates the central nervous system, heart, muscles, and potentially the pressor centers responsible for the control of blood pressure. The primary literature is full of descriptions identifying that caffeine in usual doses (150–400 mg) administered prior to bedtime can significantly delay sleep onset and substantially reduce total sleep times [3].

PSYCHIATRIC MEDICATIONS

Patients with clinical psychiatric disorders already have a marked increase in the incidence of disrupted sleep patterns resulting in insomnia. This issue can be compounded by the fact that many of the medications used to treat these disease states have been associated with causing the undesirable side effect of insomnia (Table 14.1). Clinical evidence has shown the incidence of insomnia with the selective serotonin reuptake inhibitors (SSRIs) in 5–19% of patients [7]. This finding appears to be dose-related and is associated with a discontinuation rate ranging from 2% to 5%, with fluoxetine having the highest rate of discontinuation secondary to this adverse effect. Many studies evaluating the SSRIs utilizing polysomnographic data have shown decreased total sleep time (TST) and REM sleep, and increased awakenings associated with these agents. However, patients treated with citalopram for 5 weeks demonstrated a similar reduction in REM sleep, but no significant change in TST. This SSRI-associated insomnia significantly contrasts to the substantial sedative effects of the tricyclic antidepressants. In contrast, other antidepressant agents such as bupropion, which blocks the uptake of both dopamine and norepinephrine, have also shown a significant propensity to cause insomnia.

RESPIRATORY MEDICATIONS

Patients with underlying lung diseases, including asthma and chronic obstructive lung disease (COPD), sleep poorly. Much of the insomnia is from the respiratory illness itself; it is estimated that 61–74% of asthmatic persons have nocturnal awakenings. Treatment with asthma medications generally improves the quality of sleep but, less frequently, insomnia is attributable to the respiratory medications (Table 14.2) [3]. Most of these agents induce insomnia by

TABLE 14.2   Medications for Medical Conditions with Insomnia as a Reported Side Effect

Medication	Incidence of Insomnia (If Available)
Respiratory Medications	
Theophylline	1–2%
Aminophylline	
BETA AGONISTS (INHALED, ORAL, OR SUBCUTANEOUS)	
Albuterol	
Metaproterenol	
Terbutaline	
Epinephrine	
Salmeterol	1–2%
Formoterol	
CORTICOSTEROIDS	
Prednisone	
Methylprednisolone	
Dexamethasone	
Inhaled corticosteroids	Rare
ANTIHISTAMINES	Variable, usually cause somnolence
Brompheniramine	Rare
Diphenhydramine	Rare
Cardiac Medications	
Reserpine	
Diltiazem	
DIURETICS	
Furosemide	
Hydrochlorothiazide	
Bumetanide	
STATINS	
Simvastatin	<1%
ANTIARRHYTHMIC AGENTS	
Amiodarone	1–3%
BETA-BLOCKERS	Variable
Propranolol	
Metoprolol	

increasing catecholamine output either acting directly (beta agonists) or by adenosine receptor blockade (theophylline). Oral formulations of beta agonists are the worst culprits with insomnia rates as high as 4%; inhaled preparations only rarely lead to insomnia. For the long-acting beta agonist salmeterol, several studies show improved sleep with low-dose treatment. Theophylline may also increase gastroesophageal reflux, potentially leading to arousals or awakenings. Corticosteroids can either help or hinder nocturnal sleep. Improvement in airflow can lead to improved sleep efficiency, however, oral corticosteroids such as prednisone may shift circadian rhythms and thereby cause insomnia. Even inhaled corticosteroids may rarely cause insomnia. In a large asthma program, the medications leading to insomnia were most likely to be theophylline and corticosteroids [8].

Antihistamines are often used as over-the-counter sedative hypnotic drugs because they are often associated with excessive sleepiness. Rarely, these cause agitation and insomnia. The alkylamine class of antihistamines such as brompheniramine may be more likely to induce stimulation of the nervous system and insomnia. Diphenhydramine is frequently used to treat insomnia especially in elderly patients on the basis of several small, randomized trials [9]. Use of antihistamines in combination preparations with pseudoephedrine or other decongestant medications may lead to insomnia [10] but is more attributable to the pseudoephedrine than the antihistamines.

CARDIAC MEDICATIONS AND INSOMNIA

Although numerous cardiac medications have insomnia listed as a side effect, the incidence of insomnia is uncommon for these classes of drugs (Table 14.2). Furosemide and other diuretics commonly interrupt sleep due to the need for urination rather than having a direct effect on central nervous system processing of sleep. Statins and antiarrhythmic drugs have discernable but low rates of insomnia. Of the antiarrhythmic drugs, amiodarone, a class III antiarrhythmic agent, has the highest reported rate of insomnia, which can be clinically important as it is currently the most commonly prescribed drug for both atrial and ventricular arrhythmias.

Beta-blockers are noted to induce insomnia with sleep side effects in as many as 4.3% of the population [11, 12]. More lipophilic beta-blockers such as propranolol have higher rates of sleep disturbances, whereas hydrophilic drugs such as atenolol show the lowest rates of insomnia and metoprolol has an intermediate position. Reserpine increases REM sleep and is associated with nightmares and insomnia. Diltiazem has been described to cause both insomnia and somnolence although other calcium channel blockers have not been shown to have effects on sleep.

INFECTIOUS DISEASE MEDICATIONS

Insomnia is reported sporadically for antimicrobial medications. Antibiotics causing insomnia on a case report basis are listed in Table 14.3. Clinical trials indicate that insomnia may be a problem for up to 13% of patients receiving one of the fluroquinolones. Of these agents, ofloxacin has the highest reported incidence. Reports, however, are only anecdotal and mechanisms remain unknown.

TABLE 14.3 Antimicrobial and Antiviral Medications Associated with Insomnia

Medication	Incidence of Insomnia (If Available)
Antimicrobial Medications	
Atovaquone	
Isoniazid	
Meropenem	
Pentamidine	
Spectinomycin	
Sulfamethizole	
Sulfasalazine	
CEPHALOSPORINS	
Cefaclor	
Cefpodoxime	
Cefprozil	
FLUOROQUINOLONES	
Ciprofloxacin	1%
Cinoxacin	<1%
Gatifloxacin	2%
Grepafloxacin	>1%
Levofloxacin	0.3%
Ofloxacin	up to 13%
Sparfloxacin	5.7%
ANTIMALARIALS	
Mefloquine	
Antiviral Medications	
Abacavir	12%
Amantadine	14%
Didanosine	22%
Efavirenz	16%
Ganciclovir	5%
Lamivudine	11%
Lopinavir	2%
Ribavirin	
Rimantadine	1–3%
Ritonavir	<2%
Zidovudine	7%

The most common antiviral agents with a propensity to cause insomnia are amantadine and rimantadine [13]. Clinical studies assessing the incidence of insomnia with amantadine and rimantadine in doses ranging from 100 to 300 mg/day have shown insomnia occurs in approximately 5–35% and 2–15% of patients on these medications, respectively. Other sleep disturbances reported with these agents include abnormal dreams and abnormal muscle contractions. Overall, rimantadine has proved to have a lower incidence of insomnia and other sleep disturbances when compared with amantadine over all dosages studied.

The antiretroviral agents are synthetic antiviral agents used to treat patients infected with the human immunodeficiency virus (HIV). These agents have been shown in many pre- and postmarketing clinical studies to have a significantly increased incidence of insomnia (Table 14.3) [3, 14]. The exact mechanism by which this occurs is unknown. Certain medications within this class have a profound effect on the central nervous system. In the case of efavirenz, a reverse transcriptase inhibitor, insomnia and abnormal dreams occur in 16% and 6% of patients, respectively, leading to discontinuation in about 2% of patients. Symptoms appear rapidly, generally within the first two days of therapy, and may resolve spontaneously over a two to four week period if treatment is continued. The literature suggests that the symptoms may be more tolerable if efavirenz is taken just before bedtime, although this timing appears contradictory.

NEUROLOGICAL MEDICATIONS AND INSOMNIA

Drugs used to treat seizures and dementia impact sleep by causing either insomnia or hypersomnolence (Table 14.4). Not unexpectedly, hypersomnia is a more common symptom from seizure medications; however, two epilepsy medications, lamotrigine and felbamate, have a high incidence of associated insomnia [15]. These are thought to act predominantly by attenuation of glutamate excitatory neurotransmission. In one case series, lamotrigine was associated with a 6.4% rate of insomnia severe enough to require a change in therapy. Donepezil, a reversible inhibitor of the enzyme acetylcholinesterase used in dementia treatment, has a 6–14% rate of reported insomnia, with the higher rates seen when the dose is titrated over a shorter time interval (1 versus 6 weeks) [3]. Baclofen, used to treat muscle spasm, has a 2–7% rate of insomnia. Its reported mechanism of action is as a gamma-aminobutyric acid (GABA) receptor agonist and withdrawal from this medication may be the mechanism of insomnia.

Parkinson's disease is characterized by sleep complaints as well as bradykinesia and tremor. It is often difficult to determine whether Parkinson's disease or the medication used to treat it contribute to sleep fragmentation, insomnia,

TABLE 14.4 Neurological Medications Associated with Insomnia

Medication	Incidence of Insomnia (If Available)
Lamotrigine	6.4%
Felbamate	
Clobazam	
Zonisamide	
Donepezil	6–14%
Baclofen	2–7%
DOPAMINERGIC	
Levodopa	20%
Entacapone	30%
Amantadine	up to 14%
ENDOCRINE DRUGS	
Thyroxine	
Corticosteroids	
Adrenocorticotropic hormone (ACTH)	
Goserelin (in women only)	11%
Tamoxifen	up to 55%
ANTINEOPLASTIC DRUGS	
Vincristine	
Trastuzumab (herceptin, anti-human epidermal growth factor receptor-2)	24–29%
Beta-interferon	>1%
Pamidronate	<1%
Zoledronic acid	>10%

or somnolence. With short-acting L-dopa/carbidopa, the onset of periodic limb movements is thought to coincide with the lowest drug levels and may contribute to sleep interruption and sleep maintenance insomnia. The dopaminergic agonists ropinerole and pramipexole have high rates of sleepiness rather than insomnia. The catechol-*O*-methyltransferase (COMT) inhibitor entacapone, when added to L-dopa, is associated with high rates of insomnia initially (30%), which resolve with continued use. Amantadine is frequently used as an adjunct treatment in Parkinson's disease; it causes insomnia and is discussed above with the antiviral medications.

ENDOCRINE THERAPY AND INSOMNIA

Insomnia has been associated with exogenous thyroxine, ACTH, and cortisol. Thyroxine is a rare cause of insomnia mentioned in several case reports. Insomnia is described with ACTH and corticosteroid medications in hospitalized and nonhospitalized patients [16]. The mechanism of insomnia is thought to be a condition of central nervous system hyperarousal leading to difficulties initiating and maintaining sleep [17]. The antiestrogenic drug tamoxifen

is associated with rates of insomnia above 50% in breast cancer studies, although the related raloxifene does not have a high incidence of insomnia [3]. Gonadal releasing hormone (GNRH) analog use in women but not men has been associated with frequent reports of insomnia. For example, in women with endometriosis, use of such an agent, goserelin, is associated with 11% insomnia rates, higher than for the danazol control group with rates at about 4% [3]. It is speculated that the mechanism of insomnia for GNRH analogs is a manifestation of low estrogen levels.

ANTINEOPLASTIC DRUGS

The effects of cancer medications on insomnia are hard to evaluate as these are most frequently used in combination regimens and often include a corticosteroid as well to blunt nausea. Vincristine has been associated with marked insomnia in cases with single agent overdose (Table 14.4). Trastuzumab, a monoclonal antibody therapy for breast cancer that binds to the HER2 receptor, is reported to have a high rate of insomnia [3]. Agents such as zoledronic acid used to treat hypercalcemia of malignancy have insomnia as an unanticipated side effect. Beta-interferon has a 6% incidence of reported somnolence but has also been described to cause depression, anxiety, insomnia, lethargy, confusion, and psychosis [3].

ALTERNATIVE MEDICATIONS

There is only sparse data in the primary literature regarding sleep disturbances with the natural or alternative medications; however, evidence exists that some have been associated with an increased incidence of insomnia (Table 14.5) [18]. Providers should be aware of a patient's use of alternative medications through a complete medication history, as many of these agents have the potential of causing sleep disturbances. The most likely agents to be associated with insomnia include ephedra, caffeine, and ginseng. The primary alkaloid constituents of ephedra include ephedrine, pseudoephedrine, and, to a lesser degree, phenylpropanolamine, all of which can directly and indirectly stimulate the sympathetic nervous system similar to other psychomotor stimulants. Ginseng (many varieties) has been shown to cause CNS stimulation through many proposed mechanisms. Most data are derived from animal studies but have shown that ginseng increases central choline uptake and the release of acetylcholine from hippocampal tissues, as well as inhibiting GABA, glutamine, dopamine, norepinephrine, and serotonin uptake in animal tissues. These actions show a dose-dependent relationship. Guarana is a popular additive to soft drinks and herbal

TABLE 14.5   Natural or Alternative Medications Associated with Insomnia

Caffeine or caffeinated beverages
   Coffee
   Black Tea
   Green tea
Chromium
Copaiba balsam
Country mallow
Cowhage
Deanol
DHEA
Ephedra
Eyebright
Feverfew
Ginseng
   American
   Panax
   Panex pseudoginseng
   Siberian
Guarana
Khat
Khella
Marsh blazing star
Mate
Niacin
Phosphatidylserine
Policosanol
SAMe
St. John's wort
Sweet vernal grass
Tiratricol
Tonka bean
Valerian
Vitamin C (ascorbic acid)
   Acerola
   Cherokee rosehip
   Rose hip
Wormwood (aboveground parts)
Yohimbe

TABLE 14.6   Medications Where Withdrawal Is Associated with Rebound Insomnia

Opiates, narcotics
Barbiturates
Benzodiazepines
Alcohol
Gamma-hydroxybutyrate
Androgenic anabolic steroids

sedative withdrawal (Table 14.6). For the benzodiazepines, insomnia with withdrawal is agent specific. The insomnia is more common and immediately apparent with short to intermediate acting agents such as triazolam, midazolam, and lormetazepam. History of use should be sought, as the presenting symptom may be insomnia but the antecedent use of medications may not be provided. Alcohol frequently has accompanying early morning awakening rather than sleep initiation insomnia. This is discussed further in Chapter 112. Gamma-hydroxybutyrate is a drug of abuse and a drug useful in treatment of narcolepsy. Insomnia is an early and persistent symptom of its withdrawal, starting at the first day and extended over two weeks. Withdrawal of anabolic androgenic steroids can be accompanied by insomnia.

MISCELLANEOUS DRUGS THAT CAUSE INSOMNIA

Drugs that cause urination as a therapeutic effect, including all diuretics, should not be overlooked for a patient with difficulty sleeping. In a similar manner, medications inducing defecation such as lactulose can interrupt sustained sleep.

REFERENCES

1. National Center on Sleep Disorders Research, National Heart, Lung and Blood Institute. *Insomnia : Assessment and Management in Primary Care*. NIH publication No. 98-4088, 1998.
2. Spier SA. Toxicity and abuse of prescribed stimulants. *Int J Psychiatry Med* **25**:69–79(1995).
3. Thomson Healthcare. *Physicians Desk Reference*, 57th ed. Medical Economics Company, Montvale, NJ, 2003.
4. Mitler MM, Hajdukovic R, Erman MK. Treatment of narcolepsy with methamphetamine. *Sleep* **16**:306–317(1993).
5. Rechtschaffen A, Maron L. The effect of amphetamine on the sleep cycle. *Electroencephalogr Clin Neurophysiol* **16**:438–445(1964).
6. Wetter DW, Young TB. The relation between cigarette smoking and sleep disturbance. *Prev Med* **23**:328–334(1994).
7. Preskorn SH. Comparison of the tolerability of bupropion, fluoxetine, imipramine, nefazodone, paroxetine, sertraline, and venlafaxine. *J Clin Psychiatry* **56**(Suppl 6):12–21(1995).

weight loss products. The primary constituent of guarana is caffeine (discussed previously). Other alkaloids such as theophylline and theobromine have been identified in the product, both of which also have stimulatory effects.

MEDICATION WITHDRAWAL

Insomnia may be seen with withdrawal from sedatives such as benzodiazepines or barbiturates, opiates, or alcohol; this is termed rebound insomnia [19]. Withdrawal is covered in Chapter 16 but merits mention here as the task of determining reasons for insomnia involves consideration of drugs added and removed. Insomnia rates are very high after

8. Bailey WC, Richards JM, Manzella BA, Brooks CM, Windsor RA, Soong SJ. Characteristics and correlates of asthma in a university clinic population. *Chest* **98**:821–828(1990).

9. Rickels K, Morris RJ, Newman H, Rosenfeld H, Schiller H, Weinstock R. Diphenhydramine in insomniac family practice patients: a double-blind study. *J Clin Pharmacol* **23**:234–242(1983).

10. Kosoglou T, Radwanski E, Batra VK, Lim JM, Christopher D, Affrime MB. Pharmacokinetics of loratadine and pseudoephedrine following single and multiple doses of once- versus trice-daily combination tablet formulations in healthy adult males. *Clin Ther* **19**:1002–1012(1997).

11. Danjou P, Puech A, Warot D, Benoit JF. Lack of sleep-inducing properties of propranolol(80 mg) in chronic insomniacs previously treated by common hypnotic medications. *Int Clin Psychopharmacol* **2**:135–140(1987).

12. Yamada Y, Shibuya F, Hamada J, Sawada Y, Iga T. Prediction of sleep disorders induced by beta-adrenergic receptor blocking agents based on receptor occupancy. *J Pharmacokinet Biopharm* **23**:131–145(1995).

13. Hayden FG, Gwaltney JM, Van de Castle RL, Adams KF, Giordani B. Comparative toxicity of amantadine hydrochloride and rimantadine hydrochloride in healthy adults. *Antimicrob Agents Chemother* **19**:226–233(1981).

14. Ammassari A, Murri R, Pezzotti P, Trotta MP, Ravasio L, De Longis P, Lo Caputo S, Narciso P, Pauluzzi S, Carosi G, Nappa S, Piano P, Izzo CM, Lichtner M, Rezza G, Monforte A, Ippolito G, d'Arminio Moroni M, Wu AW, Antinori A, AdICONA Study Group. Self-reported symptoms and medication side effects influence adherence to highly active antiretroviral therapy in persons with HIV infection. *J Acquir Immune Defic Syndr* **28**:445–449(2004).

15. Ketter TA, Post RM, Theodore WH. Positive and negative psychiatric effects of antiepileptic drugs in patients with seizure disorders. *Neurology* **53**(Suppl):53–67(1999).

16. Radakovic-Fijan S, Furnsinn-Friedl AM, Honigsmann H, Tanew A. Oral dexamethasone pulse treatment for vitiligo. *J Am Acad Dermatol* **44**(5):814–817(2001).

17. Vgontzas AN, Bixler EO, Line H-M, Prolo P, Mastorakos G, Vela-Bueno A, Kales A, Chrousos GP. Chronic insomnia is associated with nyctohemeral activation of the hypothalamic–pituitary–adrenal axis: clinical implications. *J Clin Endocrinol Metab* **86**:3787–3794(2001).

18. Beutler JA, DerMarderosian A (Eds). *The Review of Natural Products*. Wolters Kluwer Health Inc, St. Louis, MO, 2004.

19. Roehrs T, Vogel G, Roth T. Rebound insomnia: its determinants and significance. *Am J Med* **88**:39S–42S(1990).

ADDITIONAL READING

Novak M, Shapiro CM. Drug-induced sleep disturbances: focus on nonpsychotropic medications. *Drug Safety* **16**:133–149(1997). A detailed review of the state of knowledge of medications with insomnia as a significant side effect.

Pagel JF. Medication effects on sleep. *Dental Clin North Am* **45**:855–864(2001). An overview of medications causing insomnia and somnolence.

15

FATAL FAMILIAL INSOMNIA

CHARLES A. POLNITSKY

Regional Sleep Laboratory, Waterbury, Connecticut

INTRODUCTION

Fatal familial insomnia (FFI) is a rare and recently described condition, the investigation of which is leading to new information about genetics, transmission of disease, and control of the circadian cycle. First described by Lugaresi and co-workers in 1986 [1], FFI is one of a group of human prion diseases that have received wide public notice since bovine spongiform encephalopathy (BSE, mad cow disease) became a matter of worldwide concern [2].

The human prion diseases include Creutzfeldt–Jakob disease (CJD) and its variant (vCJD), which is the human manifestation of BSE; kuru; Gerstmann–Sträussler–Scheinker (GSS) disease; and FFI. CJD and FFI are the most closely related genetically, sharing a common point mutation at a single codon. They also have sufficiently similar clinical manifestations that specific diagnosis may be problematic at presentation. In retrospect, many of the CJD diagnoses made since its identification in 1923 (and before the current body of knowledge about FFI had been acquired) may have been cases of FFI. Because of significant clinical overlap between the two conditions (and even the absence of insomnia in a few individuals with documented FFI), some workers have continued to suggest that the clinical distinction may be more one of taxonomy than strictly reflecting genetic variability [3].

PRION MOLECULAR BIOLOGY

Nonpathogenic prions (pronounced "pree-ons") are normal cellular components found mainly in neurons but that also exist in a number of other tissues. They have no known essential function but are thought to play a variety of roles in cellular signal transmission, metabolism, and response to injury [4]. Essentially, all mammalian genomes code for nonpathogenic prions. They are formed in the endoplasmic reticulum, modified in the Golgi apparatus, then transported to the cell surface [5]. Prusiner synthesizer described first the concept of pathogenic prions in 1982 [2]. His theory that they were composed entirely of protein, could transmit disease, contained no nucleic acid, and could propagate or reproduce was initially greeted with scepticism. Subsequent work in his laboratory and many others has confirmed these characteristics that violate most conventional models of human infectious disease.

Working on scrapie, a prion disease of sheep, Prusiner determined that the causative agent was a single protein, which he termed prion (PrP). Nonpathogenic PrPs can be degraded by proteinases and are termed PrP^C. The scrapie agent that was proteinase-resistant was termed PrP^{Sc} (or PrP^{res}). Other work, on the GSS disease pathogen, revealed that a single mutation at codon 102 of the prion gene (PrPNP), substituting leucine for proline, results in a significant PrP conformational change. Whereas PrP^C is configured with 4 α helices, PrP^{Sc} forms an alternative structure of β strands, which arrange themselves in β sheets.

Exactly how PrP^{Sc} is propagated is still being debated. One theory (template-directed refolding hypothesis) is that when β sheets come into contact with the α helices of PrP^C, they are able to induce a conformational change yielding additional PrP^{Sc} molecules [2]. Initially, the misfolded prions seem to be of no biologic consequence and they undergo degradation. At a critical point, the ability

Sleep: A Comprehensive Handbook, Edited by T. Lee-Chiong.

of the cell to dispose of the PrP^{Sc} is overwhelmed. The liberated PrP^{Sc} molecules induce refolding of additional PrP^C, and the sequence results in geometric accumulation of PrP^{Sc}. The alternative (seeded nucleation hypothesis) proposal describes a situation in which both PrP^C and PrP^{Sc} are universally present, but the latter in very small quantity. In this disease state, PrP^{Sc} aggregates recruit additional molecules and become the infections agent [3, 4]. The actual mechanism of the resultant extensive neuronal loss is not yet known. This model applies to all of the prion diseases.

MODE OF DISSEMINATION

The prion diseases are unique in that they can be inherited and transmitted (by ingestion, blood transfusion, tissue grafting, and administration of human biologicals, or via fomites such as surgical instruments) and can also be sporadic, resulting from spontaneous mutation. The degree of susceptibility to developing one of the diseases may also be linked to natural genotypic variations in the PrPNP [6]. Proof that they are not caused by an infectious agent in the usual sense has come from a variety of experiments. For example, when PrP^C knockout mice are inoculated with the scrapie agent, they do not develop the disease since they lack the nonpathogenic prion to serve as substrate for the process of PrP^{Sc} replication.

GENETICS OF FFI

The hereditary form of FFI is transmitted as an autosomal dominant trait with high penetrance. Both genders are equally affected. A point mutation at codon 178 of PrNP on chromosome 20, in which the base sequence GAC is replaced with AAC, results in substitution of aspartic acid by asparagine in the PrP molecule (D178N transformation). This single change establishes PrP susceptibility to β sheet conformation. The FFI phenotype will emerge in individuals in whom a natural polymorphism for either methionine or valine at codon 129 of the PrPNP results in methionine at that location. If valine occurs at that point in the amino acid sequence, the CJD phenotype will be expressed instead [7]. If methionine is also present at codon 129 of the *nonpathogenic* allele (129 met/met homozygosity), a rapidly progressive form of FFI will result; a course slightly longer and having its own clinical and pathological characteristics occurs in the case of valine being placed at the same location (129 met/val heterozygosity) [8]. This dichotomy is further elaborated on later. The approximate natural distribution at PrP codon 129 is met/met, 39%; met/val 50%; val/val 11% [9].

HISTORY AND EPIDEMIOLOGY OF FFI

The index case was described in 1986 [1]. Features included refractory insomnia, dysautonomia, and motor signs. Brain pathology revealed atrophy and gliosis in the anterior ventral (AV) and mediodorsal (MD) thalamic nuclei. The D178N transformation as the genetic basis for the condition was reported in 1992 [10]. The first report of transmission to laboratory mice appeared in 1995 [11]. Additional kindreds were subsequently identified around the world, with more than 27 pedigrees containing about 60 individuals identified to date [12, 13]. By 1998, it was reported widely that there was considerable phenotypic variability in presentation [14]. In 1999 a sporadic form of FFI, in which the D178N mutation was not present, was reported [15, 16]. Transmission to laboratory animals was also demonstrated [16]. Significant variation in degree of susceptibility in various geographic regions was recently reported [17]. Age of onset ranges from 20 to 60 years, with a mean age of about 51 [8].

CLINICAL PRESENTATION

Signs and symptoms may vary considerably. The index patient in the original report developed disturbed and abbreviated sleep of 2–3 hours a night, loss of libido and impotence, followed by urinary retention, constipation, salivation, rhinorrhea, lacrimation, diaphoresis, and fever. Inability to sleep, vivid dreaming, and increasingly frequent lapses into a dreamlike (oneiric) state with motor activity followed over the subsequent 3 months. Autonomic dysfunction worsened; dysarthria, stupor, tremors, myoclonus, diplopia, dysphagia, tachycardia, hypertension, and irregular breathing with episodic apnea became persistent. Death from pneumonia occurred at 9 months [1].

As other cases appeared in the literature, additional features were documented. Most individuals initially complained of failing attention and vigilance, memory lability, and difficulties with planning and forecasting [12]. Although sleep disruption and daytime drowsiness were consistent markers in the earlier reports, with increasing awareness of the disease and availability of confirmatory genetic testing, it became apparent that sleep disruption (especially at clinical onset) was not a necessary criterion for diagnosis [14, 18].

There are two distinct time courses spanning the interval from diagnosis to death, related to the polymorphism at codon 128 of the unaffected allele [8, 19]. Met/met homozygosity is associated with a shorter disease duration (8–11 months; mean 9 months) than met/val heterozygosity (11–72 months; mean 31 months). Phenotypic characteristics also vary. The 129 met/met homozygotes display characteristic early onset of oneiric episodes, confusion, hallucination, myoclonus, and significant autonomic dysfunction. The 128 met/val heterozygotes more characteristically present with ataxia and dysarthria and early sphincter incontinence. They are also much more likely to develop tonic–clonic seizures [8]. The conditions are uniformly fatal. While some individuals die suddenly in full consciousness, others eventually progress to stupor and coma, with death occurring from respiratory failure or infection [12].

CLINICAL TEST RESULTS

Twenty-four hour body temperature, heart rate, and mean arterial pressure (MAP) are elevated and lack normal circadian variability. Plasma norepinephrine (NE) is elevated and cortisol is at the upper normal limit, with ACTH markedly reduced. Circadian oscillations of NE, growth hormone, prolactin, FSH, and melatonin are reduced or absent [8, 20].

Reports of EEG characteristics vary. Routine EEG can be normal in early stages. Eventually the pattern changes to monomorphic flat activity occasionally punctuated by 1–2 Hz sharp waves coinciding with episodes of myoclonus. Twenty-four hour polysomnography shows progressive brain wave homogeneity, with degeneration to a state of subwakefulness characterized by a mixture of alpha and theta activity. During times in which they are awake and in contact, patients may attempt to achieve actual sleep but are unable to succeed. Sleep spindles, K-complexes, and delta activity are absent. REM sleep is reduced and fragmented and may be dissociated from normal atonia, resulting in motor activity similar to that seen in REM behavior disorder. REM episodes may occur randomly in short bursts throughout the 24 hour cycle. With progression, there is increasing prominence of the abnormal REM pattern, and increasing lack of responsiveness to stimulation. The 129 met/val heterozygotes may have preservation of a fairly normal 24 hour sleep–wake cycle initially, followed by degeneration later in the course [8].

(^{18}F)-FDG PET scans show severe reduction in glucose utilization in the thalamus and cingular cortex, and at times in other areas as well [8, 18]. Postmortem histology confirms the imaging findings, showing extensive neuronal loss, with reactive gliosis in the mediodorsal and anteroventral thalamic nuclei and the inferior olivary nuclei [1, 3, 10]. Other variable pathologic features, found throughout the brain, have also been described [18].

There are no specific findings on analysis of blood or cerebrospinal fluid, although one report on three patients indicated significantly increased 5-HIAA levels in the CSF [21]. Hypocretin-1 levels in the CSF are in the normal range [22]. Immunodetectable pathologic prion deposits are characteristically seen in involved brain areas on autopsy [3]. Quantitative immunoblot analysis done on brain tissue shows patterns that differentiate between FFI and CJD [4, 5].

CLINICOPATHOLOGIC IMPLICATIONS ABOUT SLEEP REGULATION

The thalamic nuclei are thought to integrate and relay functions between the brainstem and cerebral cortex. Transmission failure of sleep spindles from the reticular nuclei via the mediodorsal nucleus of the thalamus is considered to underlie abnormalities of sleep initiation and stage transition [23]. Additionally, there is an extensive list of brain areas that regulate sleep and are the origins of fibers that pass through the thalamic AV and MD nuclei. A direct functional role in sleep–wake regulation has also been proposed for the PrP^C itself.

Polysomnography with spectral EEG analysis has been done on individuals with and without the D178N mutation. Codon 129 polymorphism was linked to sleep regulation in both groups [24]. Increased central nervous system serotonergic system activity may also underlie sleep disturbance [12]. Disruption in the MD nucleus of signal transmission between the limbic cortex and hypothalamus may be responsible for autonomic and endocrine dysregulation seen throughout the 24 hour sleep–wake cycle [3, 25].

PREVENTION AND TREATMENT

Avoidance of exposure to sources of PrP^{Sc} remains the only available preventive measure. Pathogenic prions have significant stability and resistance to thermal, ionizing, and chemical degradation; sterilization procedures are not effective alternatives. Experience with iatrogenic and food-borne CJD indicates that identification of potential donor sources and vectors is essential, although logistics of surveillance and choice of test method are still under debate. In context of the long latent period and the relatively rapid course to debility and death once diagnosis is made, detection in the subclinical stage will likely be important to maximize therapeutic effectiveness.

Specific treatment of established prion diseases will need to utilize one or more of the following approaches: blocking transformation of PrP^C to PrP^{Sc}; facilitating cellular clearance or sequestration of PrP^{Sc}; inhibition of PrP^C synthesis or transport; rendering PrP^{Sc} into a protease-sensitive form [26]. Investigations on use of thorazine, quinacrine, and pentosan polysulfate, as well as other agents, are under way [27, 28, 29].

No directly effective medical therapy is currently available. General supportive measures, in context of realistic outcome expectations and individual end-of-life preferences, are all that can be offered to FFI victims at present. The possibility of some slow-wave sleep may be offered with administration of sodium oxybate (gamma hydroxybutyrate). REM abnormalities do not appear to be influenced, however [29]. Several PrP^{Sc}-specific antibodies have been developed in animals. Possible future applications include screening, vaccination, and immunotherapy tools [28, 30].

CONCLUSION

Fatal familial insomnia is a rare but uniformly fatal disease characterized by sleep disturbances, autonomic dysregulation, and dementia. It is one of a group of unique diseases in which the causative agent is the misfolded variation of a prion, a naturally occurring intracellular structure composed entirely of protein. FFI is inherited in an autosomal dominant fashion, resulting from a point mutation at codon 178 of the prion gene. Transmission is possible by ingestion or parenteral inoculation of tissue containing pathogenic prions. A spontaneously appearing form has also been identified. Phenotypic characteristics vary considerably. FFI is rapidly progressive once symptoms appear. Avoidance is the only current mode of prevention. There is no known specific treatment.

REFERENCES

1. Lugaresi E, Medori R, Montagna P, Baruzzi A, Cortelli P, Lugaresi A, et al. Fatal familial insomnia and dysautonomia with selective degeneration of thalamic nuclei. *N Engl J Med* **315**(16):997–1003(1986).

2. Prusiner SB. The prion diseases. *Sci Am* **Jan**:86–93(1995).

3. King CY, Diaz-Avalos R. Protein-only transmission of three yeast prion strains. *Nature* **428**:319–323(2004).

4. Tanaka M, Chien P, Naber N, Cooke R, Weissman JS. Conformational variations in an infectious protein determine prion strain difference. *Nature* **428**:323–328(2004).

5. Collins S, McLean CA, Masters CL. Gerstmann–Sträussler–Scheinker syndrome, fatal familial insomnia, and kuru: a review of these less common human transmissible spongiform encephalopathies. *J Clin Neurosci* **8**:387–397(2001).

6. Kovacs GG, Voigtlander T, Gelpi E, Budka H. Rationale for diagnosing human prion disease. *World J Biol Psychiatry* **5**(2):83–91(2004).

7. Parchi P, Petersen RD, Chen SG, Autillo-Gambetti L, Capellari S, Monari L, et al. Molecular pathology of fatal familial insomnia. *Brain Pathol* **8**:539–548(1998).

8. Collinge J, Palmer MS, Dryden AJ. Genetic predisposition to iatrogenic Creutzfeldt–Jacob disease. *Lancet* **8**:1441–1442(1991).

9. Monari L, Chen SG, Brown P, Parchi P, Petersen RB, Mikol J, et al. Fatal familial insomnia and familial Creutzfeldt–Jacob disease: different prion proteins determined by a DNA polymorphism. *Proc Nat Acad Sci* **91**:2839–2842(1994).

10. Montagna P, Cortelli P, Avoni P, Tinuper P, Plazzi P, Gallassi R, et al. Clinical features of fatal familial insomnia: phenotypic variability in relation to a poymorphism at codon 129 of the prion protein gene. *Brain Pathol* **8**:515–520(1998).

11. Knight RSG, Will RG. Prion diseases. *J Neurol Neurosurg Psychiatry* **75**(Suppl 1):i36–i42(2004).

12. Medori R, Tritschler H-J, LeBlanc A, Villare F, Manetto V, Hsiao YC, et al. Fatal familial insomnia, a prion disease with a mutation at codon 178 of the prion protein gene. *N Engl J Med* **326**:444–449(1992).

13. Tateishi J, Brown P, Kitamoto T, Hoque ZM, Roos R, Wollman R, et al. First experimental transmission of fatal familial insomnia. *Nature* **376**:434–435(1995).

14. Montagna P, Gambetti P, Cortelli P, Lugaresi E. Familial and sporadic fatal insomnia. *Lancet Neurol* **2**:167–176(2003).

15. Spacey SD, Pastore M, McGillivray B, Fleming J, Gambetti P, Feldman H. Fatal familial insomnia. First account in a family of Chinese descent. *Arch Neurol* **61**:122–125(2004).

16. Zerr I, Giese A, Windl O, Kropp S, Schulz-Schaeffer W, et al. Phenotypic variability in fatal familial insomnia (D178N-129M) genotype. *Neurology* **51**:1398–1405(1998).

17. Parchi P, Capellari S, Chin S, Schwarz HB, Schecter NP, Butts JD, et al. A subtype of sporadic prion disease mimicking fatal familial insomnia. *Neurology* **52**:1757–1763(1999).

18. Mastrianni JA, Nixon R, Layzer R, Telling GC, Han D, et al. Prion protein conformation in a patient with sporadic fatal insomnia. *N Engl J Med* **320**:1630–1638(1999).

19. Soldevila M, Calafell F, Andres AM, Yague J, Helgason A, et al. Prion susceptibility and protective alleles exhibit marked geographic differences. *Hum Mutat* **22**:104–105(2003) (full text published online; DOI: 10.1002/humu.9157).

20. Bar K-J, Hager F, Nenadic I, Opfermann T, Brodhun M, Tauber RF, et al. Serial positron emission tomographic findings in an atypical presentation of fatal familial insomnia. *Arch Neurol* **59**:1815–1818(2002).

21. Goldfarb L, Peterson RB, Tabaton M, Brown P, LeBlanc AC, Montagna P, et al. Fatal familial insomnia and familial Creutzfeldt–Jacob disease: disease phenotype determined by a DNA polymorphism. *Science* **258**:806–808(1992).

22. Portaluppi F, Cortelli P, Avoni P, Vergnani L, Maltoni P, Pavani A, et al. Dissociated 24-hour patterns of somatotrophin and prolactin in fatal familial insomnia. *Neuroendocrinology* **61**:731–737(1995).

23. Cortelli P, Polinsky R, Montagna P, Lugaresi E. Alteration of the serotonergic system in fatal familial insomnia. *Ann Neurol* **50**:421–422(2001).

24. Martinez-Rodriguez JE, Sanchez-Valle R, Saiz A, Lin L, Iranzo A, Mignot E, et al. Normal hypocretin-1 levels in the cerebrospinal fluid of patients with fatal familial insomnia. *Sleep* **26**:1068(2003).

25. Sforza E, Montagna P, Tinuper T, Cortelli P, Avoni P, Ferrillo F, et al. Sleep–wake cycle abnormalities in fatal familial insomnia. Evidence of the role of the thalamus in sleep regulation. *Electroencephalogr Clin Neurophysiol* **94**:398–405(1995).

26. Plazzi G, Montagna P, Beelke M, Nobili L, De Carli F, Cortelli P, et al. Does the prion protein gene 129 codon polymorphism influence sleep? Evidence from a fatal familial insomnia kindred. *Clin Neurophysiol* **113**:1948–1953(2002).

27. Benarroch EE, Stotz-Potter E. Dysautonomia in fatal familial insomnia as an indicator of the potential role of the thalamus in autonomic control. *Brain Pathol* **8**:527–530(1998).

28. Rossi G, Salmona M, Forloni G, Bugiani O, Tagliavini F. Therapeutic approaches to prion diseases. *Clin Lab Med* **23**:187–208(2003).

29. Prusiner SB. Detecting mad cow disease. *Sci Am* **July**:86–93(2004).

30. Nunzianta M, Gilch S, Schatzl HM. Prion diseases: from molecular biology to intervention strategies. *ChemBioChem* **4**:1268–1284(2003). (This extensively documented review covers prion metabolism, biological action, mechanisms of disease, and correlations with treatment strategies.)

31. Benito-Leon J. Compassionate use of quinacrine in Creutzfeldt-Jacob disease fails to show significant effects. *Neurology* **64**:1824(2005).

32. Reder AT, Mednick AS, Brown PP, Spire JP, Van Cauter E, Wollmann RL, et al. Clinical and genetic studies of fatal familial insomnia. *Neurology* **45**:1068–1075(1995).

33. Bradbury J. PrP^{Sc}-specific antibodies offer new hopes in prion diseases. *Lancet* **361**:1964(2003).

ADDITIONAL READING

Cortelli P, Gambetti P, Montagna P, Lugaresi E. Fatal familial insomnia: clinical features and molecular genetics. *J Sleep Res* **8**(Suppl 1):23–29(1999). This review by the group who published the index case covers most of the subsequently reported patients and synthesizes the FFI clinical presentation.

Cortelli P, Perani D, Parchi P, Grassi F, Montagna P, De Martin M, et al. Cerebral metabolism in fatal familial insomnia: relation to duration, neuropathology, and distribution of protease-resistant prion protein. *Neurology* **49**:126–133(1997). This paper compares PET scan findings in met/met homozygotes with met/val heterozygotes and describes histopathologic correlations.

Gambetti P, Parchi P, Chen SG. Hereditary Creutzfeldt–Jacob disease and fatal familial insomnia. *Clin Lab Med* **23**:43–64(2003). This is one of a number of extensive reviews contained in this issue, which is devoted entirely to prion diseases.

Glatzel M, Stoeck K, Seeger H, Luhrs T, Aguzzi A. Human prion diseases. Molecular and clinical aspects. *Arch Neurol* **62**:545–552(2005). A recent and concise summary that is notable for its clarity.

Prusiner SB. Prion diseases and the BSE crisis. *Science* **278**:245–251(1997). This is a well-illustrated technical review of prion molecular biology.

16

EVALUATION OF INSOMNIA

Douglas E. Moul and Daniel J. Buysse

University of Pittsburgh, Pittsburgh, Pennsylvania

INTRODUCTION

The evaluation of insomnia is a process that may extend throughout the course of a patient's medical, psychiatric, or psychological treatment. This process requires two basic perspectives. The first focuses on medical classification of the patient's syndrome, providing overall structure to how the complaint of insomnia may be addressed medically. The second focuses on the individualized particulars in a patient's presentation and choices, and provides the evaluation with individually unique data that aids in selecting particular treatment strategies.

MEDICAL CLASSIFICATION

Insomnia is a complaint of inability to obtain enough sleep or restorative sleep despite enough time given for sleep. Insomnia may be present even if one has the daytime complaint of subjective sleepiness, provided that one is not actually able to sleep. Classification into categories of insomnia from the *Diagnostic and Statistical Manual of Mental Disorders*, 4th edition (DSM-IV) [1], or *International Classification of Sleep Disorders*, 2nd edition (ICSD-2) [2] establishes the insomnia complaint in relation to key clinical features, which are cross-coded to the International Classification of Diseases. The range of interventions available to a patient will be determined by how the insomnia is classified.

The DSM-IV defines the diagnosis of primary insomnia as that in which insomnia is the central and only feature. This diagnosis of primary insomnia requires the absence of another psychiatric, medical, or sleep-disorder cause of poor sleeping. A presenting complaint of difficulty initiating or maintaining sleep or of poor sleep quality that has lasted for at least one month is required, along with the presence of a daytime impairment (e.g., difficulty concentrating, moodiness, irritability, or fatigue) attributed to the sleep difficulty. Daytime consequences separate patients with primary insomnia from those with simple dissatisfaction with their sleep.

By contrast, the diagnoses of psychophysiological insomnia, idiopathic insomnia, and paradoxical insomnia are classified separately within the ICSD-2, but would be subsumed under primary insomnia in the DSM-IV (see Table 16.1). Psychophysiological insomnia occurs when psychological conditioning factors have made the stimulus conditions of the bed or bedroom discriminant stimuli for wakefulness, to evoke arousal responses that impair sleep. Idiopathic insomnia is a childhood-onset insomnia that has continued into adulthood. Paradoxical insomnia is an insomnia complaint in the presence of normal conventionally-scored polysomnography (PSG) it is not considered a delusional disorder because usual PSG is not a perfect measure of sleep. Both the DSM-IV and ICSD classifications of insomnias assume that the patient has given himself or herself adequate opportunity for sleep. While it may seem logical that insomnias might be classified by whether they concern separable difficulties with initiating or maintaining sleep, or with early morning awakening, studies have not established that these distinctions define stable subtypes [3].

Difficulties with sleeping may be present in many sleep, medical, or psychiatric disorders, discussed elsewhere in this volume. An outline of the differential diagnosis of an

Sleep: A Comprehensive Handbook, Edited by T. Lee-Chiong.
Copyright © 2006 John Wiley & Sons, Inc.

TABLE 16.1 Differential Diagnosis of a Presenting Complaint of Insomnia

Pure Insomnias	Manic Episode (Bipolar Disorder)
Primary Insomnia (DSM–IV)	Psychotic Episode
Psychophysiological (ICSD-2)	Alcohol Dependence/Abuse
Paradoxical (ICSD-2)	Caffeine-Induced Sleep Disorder
Idiopathic (ICSD-2)	Nicotine Dependence
Behavioral (ICSD-2 types)	Stimulant Dependence/Abuse
Adjustment Sleep Disorder	Sedative/Hynotic Withdrawal
Inadequate Sleep Hygiene	Opiate Withdrawal
Limit-Setting Sleep Disorder	Post-Traumatic Stress Disorder
Sleep-Onset Association Disorder	Panic Disorder
Primary Sleep Disorders (ICSD-2)	Somatoform Disorder
Restless Legs Syndrome	Pain Disorder
Periodic Limb Movement Disorder	Tourette's Disorder
Delayed Sleep Phase Syndrome	Separation Anxiety Disorder
Advanced Sleep Phase Syndrome	Adjustment Disorder
Shift-Work Sleep Disorder	General Medical Conditions
Jet-Lag Syndrome	Hyperthyroidism
Nightmare Disorder	Estrogen/Progesterone Deficiency
Sleep Apnea Syndromes	Hypercortisolemia
Primary Neurological Disorders	Renal Failure
Head Trauma	Pregnancy
Stroke	Fibromyalgia
Headache Disorders	Chronic Fatigue Syndrome
Seizure Disorder	Iron Deficiency
Parkinson's Disease	Nocturnal Dyspnea
Dementias	Cardiac Failure
Fatal Familial Insomnia	Gastroesophageal Reflux Disease
Primary Psychiatric Disorders	Pain
Major Depressive Episode	Chronic Toxin Exposures
Generalized Anxiety Disorder	Medication-Induced Insomnia

insomnia complaint is presented in Table 15.1. Many diagnoses can be identified with history taking along with physical and mental status examinations. When obstructive sleep apnea, periodic limb movement disorder, narcolepsy, or a parasomnia are not suspected, polysomnography is unlikely to provide helpful initial diagnostic information [4]. However, after several intervention approaches have proven unsuccessful or if there is significant diagnostic uncertainty, then PSG may be more useful. For example, the presence of a short REM latency may suggest the diagnosis of major depression in someone with a severe insomnia complaint. Some clinicians find actigraphy useful in clarifying patterns of sleep and wakefulness, particularly in less verbal patients.

Epidemiological studies estimate one-third of the population has insomnia during a one year period, with 10% having chronic insomnia difficulties [5]. In primary care settings, chronic insomnias from multiple causes may be present in more than 30%. The prevalence of various insomnia-related disorders will make various diagnoses more or less likely in any particular clinic, depending on how patients are selected into that clinic. Thus the positive predictive value of a presenting complaint of insomnia in relation to specific diagnoses will depend on the referral pattern. For example, an insomnia complaint in a pulmonary sleep clinic is more likely to be due to sleep apnea, whereas in a general psychiatry clinic it is more likely to be related to an affective, anxiety, or substance abuse diagnosis. Nonetheless, it is important to be aware that the presenting complaint of insomnia does not necessarily classify a patient one way or another, even if knowledge of the clinic's referral prevalence helps to structure and prioritize evaluation processes.

Table 16.2 presents an outline for an initial evaluation for sleep disorders. It is best to obtain a history both from the patient and his or her bedpartner. Sleep-focused physical and mental status examinations are important for accurate diagnosis. Sleep diary data recorded before the initial evaluation is useful to assess sleep scheduling across nights more objectively and serves as a counterweight against

TABLE 16.2   Outline for a Sleep Disorders Assessment in Relation to a Chief Complaint of Chronic Insomnia

Identifying Information and Chief Complaint
  Demographic risk factors
  Initial stated complaint
  Clinician making referral

History of Present Illness
  Predisposing and initiating factors
  Diagnostic studies
  Therapy trials (e.g., drugs, behaviors)
  Course of illness
  Complicating disorders
  Obvious life stressors

Sleep Habits
  When sleep attempted in 24 hours
  When sleep occurs in 24 hours (include nighttime
    awakenings)
  Review of sleep logging over weeks
  Sleep scheduling on weekends
  Bedtime routines and cognitions
  Influences over sleep scheduling (time-zone travel,
    shift work, etc.)

Sleep Disturbances
  Adverse bed conditions
  Problems with bedpartner
  Symptoms of another sleep disorder

Daytime Functioning
  Daytime inability to nap
  Symptoms of lowered concentration, fatigue, irritability,
    moodiness

Health Habits
  Review of prescribed and over-the-counter medications
  Use of caffeine, alcohol, nicotine, and street drugs
  Diet patterns, preferences, or restrictions
  Exercise regimen

Past Medical History (also see Table 15.1)
  Sources of pain or discomfort

Neurological syndromes
Relevant medical syndromes

Past Psychiatric History (see Table 15.1)
  Hospitalizations and rehabilitations
  Suicidal/self-harm attempts
  Diagnoses and treatments
  Past and current impairments

Family History
  Sleep disorders
  Substance abuse disorders
  Psychiatric disorders
  Relevant medical disorders

Social and Occupational History
  Level of schooling
  Past and present quality of primary relationships
  Occupational exposures and stresses
  Significant losses
  Life aspirations

Assessment of Self-Report Questionnaire Data
  Sleepiness
  Depression
  Anxiety
  General psychopathology
  Sleep quality
  Insomnia specific

Brief Physical Examination
  Risk factors for sleep apnea
  Signs of neurological illness
  Lab tests as needed

Mental Status Examination
  Quality of alertness and concentration
  Symptoms of depression, anxiety, psychotic, or
    obsessive–compulsive disorders
  Report of any blank spells or other loss of disorders
  Impaired insight into psychological aspects of problem

semantic difficulties associated with summary self-reports. Data from self-report questionnaires [6] (see Chapter 132) that focus on sleepiness, anxiety, depression, general psychopathology, sleep quality, and insomnia help the clinician to appreciate the clinical issues, identify diagnoses, and select treatments. There is a selection of sleep diaries and other questionnaires available to clinics. Each clinic should select a set of standard self-report instruments for all sleep disorder patients to complete, in accordance with the clinical population served. The selection and interpretation of these self-report measures is beyond the scope of this chapter. However, the use of standard self-report measures greatly enhances the efficiency of both medical classification and individualized evaluation.

INDIVIDUALIZED COMPONENTS OF EVALUATION

Since psychiatric conditions are often associated with an insomnia complaint, adopting a psychiatric approach to evaluating patients who complain of insomnia is prudent, especially in avoiding the assumption that a patient's statements must be taken as literally true or false. For example, dramatic statements like "Doctor, I haven't sleep at all" may not be intended to be taken literally, but be a variant of "My poor sleep *really* bothers me." While sometimes time-consuming, obtaining data about the concrete details of the patient's sleep helps to clarify the context of the insomnia complaint. Sometimes a complaint about insomnia may be partly intended as a hint for the clinician to ask about more stigmatized or feared diagnoses. Keeping an open mind assists in accurate diagnosis and treatment. Using an initial loosely structured approach to history taking and examination will afford opportunities for branching to key rule-in primary diagnoses. This is especially true when thought distortions (e.g., negative thoughts in depressed patients and fearful thoughts in anxiety patients) or gross deficits in reality testing (e.g., delusions in psychotic patients) are present. Yet even in non complaining adults, self-reports about sleep can be inaccurate.

Individualizing diagnosis and treatment implies a review of collateral problems that may contribute to the maintenance or severity of insomnia. The high prevalence of collateral clinical issues in insomnia patients may make the strict classification of insomnias by duration or specific symptoms unhelpful. For example, in the pain patient, it will be insufficient to focus solely on insomnia complaints when collateral issues of pain control contribute to the overall clinical picture. Similar cautions are needed in relation to psychiatric syndromes, neurological syndromes, or other medical syndromes.

The patient's individualized history of insomnia contributes strategically to case formulation. Spielman's classification [7, 8] of *predisposing*, *initiating*, and *maintaining* factors or causes (see also Chapter 13) of insomnia in the clinical history often helps with clinical problem solving. If insomnia had a childhood onset, then the diagnosis of idiopathic insomnia may be more actively considered as a form of strong *predisposition* to insomnia. A number of patients will report that they are temperamentally sensitive to noise, light, or the movements of the bedpartner. Since factors such as age, gender, and menopausal status can influence the nature of a patient's capacity for sleep and of his or her style of reporting, case formulation should take into account these demographic factors.

Initiating factors often help to describe the patient's difficulties. Insomnia after head injury, stroke, multiple sclerosis, or other neurological events can lead one to investigate which brain sleep nuclei may have been damaged. In other cases, patients may convincingly describe an initiating event (e.g., parent's death, other psychological trauma) or period (e.g., pregnancy, particular job) that predated the development of chronic insomnia. Chronic illnesses such as mononucleosis, chronic fatigue, migraine, fibromyalgia, progressive back pain, and other conditions may also to some degree serve as initiating events. The personal historical effects of these events and periods cannot be undone. Yet previously unappreciated neurological conditions may be treated. And when emotional coping strategies in relation to initiating psychological events remain impaired and contribute to insomnia, their treatment may improve the insomnia. Sometimes the simple identification of the initiating event as significantly contributing to insomnia

will be therapeutic in itself because it reduces the perplexity and worry the patient may experience about his or her insomnia.

Factors that *maintain* insomnia may be quite modifiable. While this discussion focuses on treatment of the primary insomnia patient, the principles remain the same for depression, anxiety, pain, and many other disorders where insomnia is a complaint. The psychology of the insomnia patient, with whatever comorbidity, can serve to aggravate and amplify the difficulties of having a frail sleep system. The insomnia patient may have such *preoccupation with suffering* that he or she fails to address basic *environmental issues* concerning nocturnal noise or light exposure or *difficulty with a bedpartner* who snores loudly or thrashes in bed. Occasionally, improved bedding, ear plugs, or eye shades will help. In some cases, patients will adopt *adverse practices* in relation to sleep, such as exercising immediately before going to bed or in the middle of the night.

The patient's *medication and substance-use regimen* are also by definition potentially modifiable, both in regard to type of substance and timing. Medications and substances potentially causing insomnia are discussed in Chapter 13. In the case of caffeine, it may be important to get the patient to try minimizing or eliminating its use for a few weeks to test if it is affecting sleep adversely. In the case of alcohol, it is important to educate the patient that it often promotes initial deep sleep at sleep onset, but then leads to fragmented sleep later in the night. Counseling against alcohol use for sleep induction is especially important for the abstinent alcoholic with insomnia. If pain is affecting sleep, pain management is a priority. In some cases, patients may benefit from rescheduling normally taken sedating medications toward bedtime.

Even when physical and medication factors may be helpfully modified, they may not be as powerful or as relevant in some patients as the *psychological factors* that motivate poor sleep practices. These factors may be grouped into those pertaining to erroneous or distorted ideas about *sleep time management, sleep habit training, consequences from insomnia*, and *arousal management*.

The *sleep time management* and *habit training practices* of some insomnia patients can often work in opposition to predictable sleeping. The well-supported two-process model [9] of normal sleep regulation states that continuous time spent awake and circadian timing govern a person's propensity to sleep. No doubt many people with robust sleep systems can violate these principles without subjective distress. However, in insomnia patients with frail sleep systems, the practices of daytime napping or of "nighttime napping" work against obtaining single, deep and continuous nocturnal sleep periods. Time spent lying down for sleep across 24 hours should be consolidated and generally limited to the amount of time that the patient's brain can actually sustain sleep. Irregularly timed awake periods tend to produce irregularly timed and unpredictable sleep. Second, under normal conditions, the circadian system has a forbidden zone for sleep [10] in the early evening, and a high sleep propensity period from 3 a.m. until dawn. Thus attempting to sleep during the early evening will be frustrating, and the circadian advantages of pre-dawn sleep diminished, if the patient attempts to sleep too early. The plan of going to bed early to permit more sleep may seem rational but is empirically counterproductive. An additional reason for obtaining a history and prospective sleep logging is to observe if the patient may have important variability across 2 or 3 weeks, to see if the equivalent of nightly jet lagging is occurring by how the patient is scheduling sleep. Variability in sleep timing not only undercuts stable biological rhythm timing but also undermines the capacity of the patient's behavioral habit system to regulate the normal, unconscious, biologically based scheduling of sleep and wakefulness. By attempting to schedule sleep adaptively in a self-conscious day-to-day manner, many patients unknowingly over manage the behavioral sleep habit system, when this habit system often needs 3–4 weeks of behavioral consistency in order to work properly.

The day-to-day manner of managing sleep scheduling is often motivated by *distorted ideas about the consequences* of insomnia [11]. The presence of cognitive distortions means that certain thoughts are exaggerations of perceptions or distortions of inferences that a person has in relation to the exact situational truth. While insomnia researchers believe there really are daytime consequences from chronic insomnia, these consequences are not like what insomniacs sometimes imagine them to be. For example, some insomniacs believe that if they don't get a good night's sleep, they will be a virtual total failures at work; whereas the exact truth is more likely that they will not be at their best, but still put in good performances. The burden of chronically saying to oneself that one will be a failure at work if sleep is not near-perfect can only produce an intense sense of sleep worry coupled with an impulsive, impatient style of managing sleep. As part of the diagnostic evaluation process, it is often helpful to ask patients what they may be thinking about their sleep, in order to catch some of these unhelpful cognitions and attempt to modify them to reflect actual circumstances. This cognitive–behavioral style of insomnia evaluation is the major basis for cognitive–behavioral therapy. In a related way, the evaluation may also concern how the patient generally handles ideas about performances. Self-compulsively perfectionistic persons may use an "all-or-none" cognitive management style that generally works poorly for managing sleep [12].

Hyperarousal in insomnia patients is partly an oversimplification as the cause of insomnia in many cases, but is nonetheless a useful heuristic for clinical problem solving. Arousal is not a single entity, and there is no agreed-upon physiological definition of arousal, even though a variety of metrics suggest biological hyperarousal [13]. Yet the usefulness of the notion of hyperarousal in many insomnia patients is that they often have arousing configurations of their lives, their minds, and their bedrooms that seem to prevent predictable sleep. A key focus for evaluation can be to appraise various arousal issues, so as to assist the insomnia patient with behavioral problem solving (but not in an "all-or-none" way). Over-arousal can be a central problem for those patients who lead very time-pressured or conflict-laden lives. Often life-style change can be currently difficult or impossible, or be inconsistent with one's self-concept. For example, caregiving for a spouse with Alzheimer's disease can be over arousing, yet reflect deep commitments to the spouse while posing problems for good sleeping. The stresses and strains that patients may face can increase arousal. Nonetheless, some evaluation whether some aspects of the life style might be favorably modified to improve sleep predictability and quality can present therapeutic opportunities. For example, finding opportunities for daytime "time-outs" from daily stresses may be possible.

Patients' hyperaroused style of thinking about their stresses may also be modifiable. If a person spends most of the day frightened by various possible adverse and time-pressured consequences, his or her emotional system may be chronically overaroused due to an unmodulated and anxious pattern of thinking even in the absence of a depressive or anxiety disorder. Since evidence points to nighttime arousal mirroring daytime arousal [14], modulation of daytime mental life can be important for some insomnia patients even when the daytime arousal does not concern sleep at all. But *when* one thinks about stressors is important too. Many busy insomniacs find that bedtime is the only time when time-pressured demands are not imminent, and find themselves thinking or worrying in bed, about past or future events or demands, or about their sleep. Since cognitive activity is an arousal stimulus that likely encourages the sleep regulation system [15] to remain in "wake mode," bedtime patterns of cognitive and physical arousal can make sleep difficult to achieve or maintain. Even if the patient gets to sleep easily, the sleep system can be too habit-focused on wakefulness for the person to remain asleep easily. Hence, it is important to ask patients what is going on in their minds while in bed. Not all patients ruminate in bed, but many do. For those who do, the intervention may be partly focused on providing education about thought-stream and arousal monitoring, particularly in bed.

Not all arousal may be consciously accessible, but may nonetheless be linked to features of the bed, bedroom, or bedtime. To evaluate this possibility, it is useful to ask if the bed is used for activities other than sleeping or sex, including just being awake. This is the perspective of stimulus control therapy (see Chapter 18). The stimulus control theory postulates that the insomnia patient has developed a pattern of habitual responding to bedroom stimuli with counterproductive arousal responses. For example, if the bed is the location where heated family arguments occur, or where engrossing mental activities occur, then the person's emotional habit system learns to interpret the bed, bedroom, or bedtime as a stimulus for becoming more aroused. Not surprisingly, some insomnia patients may sleep better away from home bedroom stimuli, where the key arousing stimuli are not present. Since stimulus control therapy is one of the best-documented behavioral interventions for insomnia, evaluation of the stimulus conditions surrounding the bed and bedtime is often useful.

EVALUATION IN RELATION TO TREATMENT PLANNING

Treatment actually begins at the start of evaluation, and evaluation continues with treatment. New information from or about the patient may arise after several clinic visits. As part of the evaluation process, the patient's life style and treatment preferences are important to appreciate. The patient's life style will provide parameters for discussions about possible interventions. For example, social activities late on the weekends may interfere with the regularity of sleep scheduling, but be an important constraint that needs to be factored into treatment planning. Another aspect is the patient's treatment preferences. Many patients will prefer to utilize behavioral therapies for their insomnia [16], hoping to avoid needing to depend on a pill to get to sleep. However, others will disfavor behavioral methods for enhancing sleep and instead be interested in using a sedative-hypnotic medication for sleep. Having some assessment of patient preferences will assist with guiding the interventions toward those the patient may accept. An assessment of the patient's capacity to understand the treatment rationale and for the need for prospective sleep logging will guide the selection of how complicated a rationale to provide and of how to monitor treatment effectiveness. For instance, some patients will not do sleep logging.

Since the effect of a particular treatment strategy for a particular patient cannot be known in advance, there is a need to evaluate treatment effectiveness prospectively. This is best done when the patient can do sleep logging and can participate in clinical problem solving with the careful review of the sleep log data along the lines of the treatment rationale(s). This style of evaluation encourages the patient to adopt a more objective perspective about his or her own sleep across several weeks, and teaches a better form of self-evaluation. Intermittent reviews also provide opportunities for the clinician to learn of new information the patient may come to provide to the evaluation process.

SUMMARY

The evaluation process in the management of an insomnia complaint involves consideration of many potential clinical syndromes other than DSM-IV primary insomnia or the specific insomnias in the ICSD-2. Insomnia is a highly prevalent complaint in primary care and will be part of the clinical picture of up to 30% of primary care patients. In cases where insomnia is a presenting complaint, adoption of a circumspect approach to evaluation will optimize accurate formal diagnosis as well as individualized treatment planning.

REFERENCES

1. American Psychiatric Association. Task Force on DSM-IV. *Diagnostic and Statistical Manual of Mental Disorders: DSM-IV*, 4th ed. American Psychiatric Association, Washington, DC, 1994.

2. American Sleep Disorders Association. *The International Classification of Sleep Disorders: Diagnostic and Coding Manual*, 2nd ed. American Sleep Disorders Association, Rochester, MN, 2005.

3. Hohagen F, Kappler C, Schramm E, Riemann D, Weyerer S, Berger M. Sleep onset insomnia, sleep maintaining insomnia and insomnia with early morning awakening—temporal stability of subtypes in a longitudinal study on general practice attenders. *Sleep* **17**(6):551–554(1994).

4. Chesson A Jr, Hartse K, Anderson WM, Davila D, Johnson S, Littner M, et al. Practice parameters for the evaluation of chronic insomnia. An American Academy of Sleep Medicine report. Standards of Practice Committee of the American Academy of Sleep Medicine. *Sleep* **23**(2):237–241(2000).

5. Ohayon MM. Epidemiology of insomnia: what we know and what we still need to learn. *Sleep Med Rev* **6**(2):97–111(2002).

6. Moul DE, Hall M, Pilkonis PA, Buysse DJ. Self-report measures of insomnia in adults: rationales, choices, and needs. *Sleep Med Rev* **8**(3):177–198(2004).

7. Spielman AJ, Saskin P, Thorpy MJ. Treatment of chronic insomnia by restriction of time in bed. *Sleep* **10**(1):45–56(1987).

8. Spielman AJ, Yang C-M, Glovinsky PB. Assessment techniques for insomnia. In: Kryger MH, Roth T, Dement WC (Eds), *Principles and Practice of Sleep Medicine*. Saunders, New York, 2000, pp 1239–1250. (This chapter in a standard text for sleep medicine provides a slightly different perspective and additional resources regarding the evaluation of insomnia.)

9. Borbély AA, Achermann P, Trachsel L, Tobler I. Sleep initiation and initial sleep intensity: interactions of homeostatic and circadian mechanisms. *J Biol Rhythms* **4**(2):149–160(1989).

10. Lavie P. Ultrashort sleep–waking schedule. III. "Gates" and "forbidden zones" for sleep. *Electroencephalogr Clin Neurophysiol* **63**(5):414–425(1986).

11. Morin CM. *Insomnia: Psychological Assessment and Management.* Guilford Press, New York, 1993. (Written from the perspective of cognitive–behavioral therapy, this book provides not only specific ideas about assessment, but describes how the assessment can be integrated with treatment over many weeks.)

12. Vincent NK, Walker JR. Perfectionism and chronic insomnia. *J Psychosom Res* **49**(5):349–354(2000).

13. Bonnet MH, Arand DL. 24-Hour metabolic rate in insomniacs and matched normal sleepers. *Sleep* **18**(7):581–588(1995).

14. Ribeiro S, Gervasoni D, Soares ES, Zhou Y, Lin SC, Pantoja J, et al. Long-lasting novelty-induced neuronal reverberation during slow-wave sleep in multiple forebrain areas. *PLoS Biol* **2**(1):126–137(2004).

15. Steriade M, Jones EG, Llinás RR. *Thalamic Oscillations and Signaling.* Wiley, Hoboken, NJ, 1990.

16. Morin CM, Gaulier B, Barry T, Kowatch RA. Patients' acceptance of psychological and pharmacological therapies for insomnia. *Sleep* **15**(4):302–305(1992).

ADDITIONAL READING

Hauri P, Linde SM. *No More Sleepless Nights.* Wiley, Hoboken, NJ, 1990. A useful resource book for patients.

Lacks P. *Behavioral Treatment for Persistent Insomnia.* Pergamon Press, New York, 1987. Another book-length discussion, adopts a more behavioral approach to assessment and treatment.

Moul DE, Hall M, Pilkonis PA, Buysse DJ. Self-report measures of insomnia in adults: rationales, choices, and needs. *Sleep Med Rev* **8**(3):177–198(2004). This review article of self-report measures summarizes a wide scope of available questionnaires for use in clinical and research activities regarding insomnia.

17

PHARMACOLOGIC THERAPY OF INSOMNIA

TEOFILO LEE-CHIONG
National Jewish Medical and Research Center, Denver, Colorado

MICHAEL SATEIA
Dartmouth-Hitchcock Medical Center, Lebanon, New Hampshire

NEUROCHEMISTRY OF SLEEP

Sleep is an active process that is generated and maintained by a number of specific nuclei and governed by various sleep stage-specific neurotransmitters, hormones, and peptides. The solitary tract nucleus, anterior hypothalamus–preoptic area, nonspecific thalamic nuclei, and basal forebrain are important in the initiation of slow-wave sleep. Serotonin and gamma-aminobutyric acid (GABA) are the main neurotransmitters associated with non-rapid eye movement (NREM) sleep. Adenosine is believed to play a key role in modulation of homeostatic (slow-wave) sleep drive. On the other hand, the pedunculopontine nuclei and the laterodorsal tegmental nuclei located dorsolaterally at the pontine midbrain junction have major roles in the generation of rapid eye movement (REM) sleep. The main neurotransmitter related to the generation of REM sleep is acetylcholine. This neurochemistry underlies much of the pharmacologic activity of the various hypnotic agents [1].

CURRENT USE OF HYPNOTIC AGENTS

The goals of pharmacologic therapy of insomnia consist of alleviation of nighttime sleep disturbance and relief of its daytime consequences. Clinicians should correctly identify any underlying secondary causes of insomnia, such as sleep disordered breathing, restless legs syndrome, or mood disorders, and manage them accordingly. Incorporation of nonpharmacologic interventions, including proper sleep hygiene practices and cognitive–behavioral therapy, is recommended.

Hypnotic agents are primarily indicated for the treatment of transient sleep disruption such as those caused by jet lag, shift work, or acute stress but are also used in selected persons with chronic insomnia (ideally for primary insomnia that failed to respond to behavioral therapy, or secondary insomnia that did not improve with treatment of the underlying condition). Until quite recently, no hypnotic medications have been approved by the U.S. Food and Drug Administration (FDA) for such long-term usage. As of this writing, eszopiclone, a nonbenzodiazepine, will soon be released without a restriction to short-term usage. Medications used in the treatment of insomnia include the benzodiazepines, nonbenzodiazepine benzodiazepine receptor agonists (NBBRAs), antidepressants, and the nonprescription hypnotic agents, such as histamine antagonists, melatonin, and herbal compounds.

HYPNOTIC MEDICATIONS

The selection of a particular hypnotic agent from among the numerous available compounds should be done cautiously with consideration of its hypnotic efficacy, absorption and elimination profile, onset and duration of action, effect on sleep stages, risks of tolerance, dependency, and withdrawal, abuse potential, possible drug interactions, and

Sleep: A Comprehensive Handbook, Edited by T. Lee-Chiong.
Copyright © 2006 John Wiley & Sons, Inc.

TABLE 17.1   Currently Available Agents for the Treatment of Insomnia: Benzodiazepines

Agent	Adult Dosages	Duration of Action	Primary Metabolism/ Excretion	Important Adverse Effects	Important Drug Interactions	Not Recommended	Comments
Estazolam	1–2 mg at bedtime; 0.5 mg for (1) elderly or (2) debilitated patients	Intermediate-acting	Hepatic/ renal	Dizziness Drowsiness Headache	Alcohol Azole antifungals Barbiturates Muscle relaxants Opioid analgesics Sodium oxybate	Benzodiazepine hypersensitivity Chronic pulmonary insufficiency Elderly or debilitated patients Pregnancy or lactation Sleep apnea	Can cause withdrawal symptoms after abrupt discontinuation. Caution if history of substance abuse or severe depression.
Flurazepam	15–30 mg at bedtime; 15 mg for (1) elderly and (2) debilitated patients	Long-acting	Hepatic/ renal	Dizziness Drowsiness Headache	Alcohol Azole antifungals Barbiturates Muscle relaxants Opioid analgesics Sodium oxybate	Benzodiazepine hypersensitivity Chronic pulmonary insufficiency Pregnancy or lactation Severe depression Severe liver dysfunction Sleep apnea	Hangover sedation on the morning after nighttime administration. Potential for dependency.
Quazepam	7.5–15 mg at bedtime; 7.5 mg for elderly patients	Long-acting	Hepatic/ renal	Dizziness Drowsiness Fatigue Headache	Alcohol Azole antifungals Barbiturates Muscle relaxants Opioid analgesics	Benzodiazepine hypersensitivity Chronic pulmonary insufficiency Elderly or debilitated patients Pregnancy Sleep apnea	Hangover sedation on the morning after nighttime administration. Reduction in dosage in hepatic disease.
Temazepam	15 mg at bedtime; 7.5 mg for (1) elderly or (2) debilitated patients Maximum recommemded dose: 30 mg	Intermediate-acting	Hepatic/ renal	Dizziness Drowsiness Fatigue Headache	Alcohol Azole antifungals Barbiturates Muscle relaxants Opioid analgesics Sodium oxybate	Benzodiazepine hypersensitivity Chronic pulmonary insufficiency Pregnancy or lactation	Can cause withdrawal symptoms after abrupt discontinuation.
Triazolam	0.25 mg at bedtime; 0.125 mg for (1) elderly, (2) debilitated patients, or (3) liver dysfunction Maximum recommended dose: 0.5 mg	Short-acting	Hepatic/ renal	Confusion Drowsiness Fatigue Headache	Alcohol Azole antifungals Barbiturates Muscle relaxants Opioid analgesics Sodium oxybate	Benzodiazepine hypersensitivity Chronic pulmonary insufficiency Pregnancy Severe depression Severe liver dysfunction	Can cause rebound insomnia or withdrawal symptoms after drug discontinuation. Potential for dependency.

cost (Tables 17.1 and 17.2). It is advisable to select an agent with minimal risk of daytime residual effects. Specific characteristics of the patient, including the presence of medical or psychiatric illnesses, age, pregnancy or lactation, and occupation, should also be taken into account.

Duration of action differs among the various hypnotic preparations and this distinction has been used to select the appropriate agent for the specific insomnia complaint. Short-acting agents are most appropriate for persons who have difficulty falling asleep. Patients with both sleep-

TABLE 17.2   Currently Available Agents for the Treatment of Insomnia: Nonbenzodiazepines

Agent	Adult Dosages	Primary Metabolism/ Excretion	Important Adverse Effects	Drug Interactions	Not Recommended	Comments
Eszopiclone	Initial dose: 2 mg before bedtime; 1 mg before bedtime for (1) elderly or (2) use with strong CYP3A4 inhibitor Maximum recommended dose: 3 mg (2 mg for elderly)	Hepatic/renal	Headache Unpleasant taste Xerostomia	Azole antifungals Clarithromycin Ciprofloxacin Diclofenac Doxycycline Erythromycin Ethanol Isoniazid Ketoconazole Nefazodone Olanzapine Quinidine Verapamil	Hypersensitivity to eszopiclone Pregnancy or lactation (use caution) Depression Severe hepatic impairment	Should be administered immediately prior to bedtime or after bedtime if the patient has difficulty falling asleep. Tablet should not be crushed or broken. Avoid taking after a heavy meal.
Zaleplon	Initial dose: 10 mg at bedtime; 5 mg for (1) elderly, (2) debilitated, (3) low body weight, or (4) mild to moderate hepatic impairment Maximum recommended dose: 20 mg	Hepatic/renal, gastrointestinal	Dizziness Headaches Somnolence	Cimetidine Ethanol Rifampin	Hypersensitivity to zaleplon Pregnancy or lactation Severe hepatic impairment	Has little effect on sleep stages. Reduces sleep latency.
Zolpidem	10 mg at bedtime; 5 mg for (1) elderly, (2) debilitated or (3) hepatic impairment Maximum recommended dose: 10 mg Dosage adjustment needed during dialysis	Hepatic/renal	Dizziness Drowsiness Headache Nausea	Antidepressants Ethanol Ketoconazole Rifampin Ritonavir	Hypersensitivity to zolpidem	Reduce sleep latency. Decreases number of awakenings. Increases total sleep time. Delays onset of REM. Increases stages 3 and 4 sleep time.
Trazodone	Initial dose: 50 mg at bedtime	Hepatic/renal	Arrhythmias Blurred vision Delirium Dizziness Drowsiness Hypotension Priapism	Antihypertensives Chlorpromazine Droperidol Monoamine Oxidase inhibitors Trifluoperazine Warfarin	Hypersensitivity to trazodone Pregnancy or lactation Cardiac disease or arrhythmias	Not FDA approved for therapy of insomnia.

onset and maintenance insomnia may be given agents with intermediate action. Finally, long-acting compounds, including flurazepam and quazepam, may be useful for some patients with both early morning awakenings and daytime anxiety, although their long duration of action is associated with accumulation and daytime performance impairments. Hence caution is advised with such agents, particularly in the elderly.

It is generally recommended that sedative-hypnotic therapy be limited to short-term use. Whenever these agents are used for longer periods, consideration should be given to intermittent usage, and indications should be reassessed on a regular basis. Monitoring for effectiveness of the drug, adverse reactions, self-escalation of the medication dose, and alterations in medical or psychiatric status must occur.

Older hypnotic agents, such as barbiturates and chloral hydrate, have been superceded by newer, more effective, and safer agents. In addition to the danger of psychological and physical dependency, barbiturates can interact with numerous other medications via their induction of liver enzymes. Overdose is an ever-present danger with barbiturates. The popularity of chloral hydrate has waned primarily due to the rapid development of tolerance as well as the potential for development of rashes, gastric discomfort, and hepatic toxicity following its use.

Benzodiazepines and the nonbenzodiazepine agonist agents act at the GABA–benzodiazepine receptor complex. Benzodiazepines act nonselectively at two central receptor sites—omega (1) and omega (2). The sedative action of benzodiazepines is related primarily to omega (1) receptors, whereas omega (2) receptors are believed to be responsible for their effects on memory and cognitive functioning. The hypnosedative action of the nonbenzodiazepine agents, zopiclone, zolpidem, and zaleplon, is comparable with that of benzodiazepines. These agents interact preferentially with omega (1) receptors. These newer agents are less likely than benzodiazepines to impair daytime performance and memory due to their relatively short duration of action and their low potential for residual effect [2]. Unfortunately, most studies on benzodiazepines and nonbenzodiazepine receptor agonists are of relatively short duration and lack data regarding long-term follow-up.

Benzodiazepines

In general, benzodiazepines are effective and safe in the treatment of various types of insomnia. They typically increase total sleep time, reduce sleep latency, and decrease the frequency of awakenings in patients with insomnia. In addition to their hypnotic properties, benzodiazepines are potent anxiolytics, muscle relaxants, and anticonvulsants. Benzodiazepines can be classified into three groups with respect to duration of action: short half-life (<3 hours), medium half-life (8–24 hours), and long half-life (>24 hours) (Table 17.1).

Benzodiazepines, although typically well-tolerated, may be associated with several adverse consequences. Short-acting agents may cause rebound daytime anxiety and greater withdrawal symptoms following cessation of their use. The effect of agents with long elimination half-lives may persist into the next day, producing sleepiness, incoordination, and cognitive impairment. Patients should be cautioned against operating motor vehicles or performing tasks that require vigilance and alertness when using these drugs. The residual effects of long-acting hypnotics are particularly of concern among the elderly. The presence of preexisting memory impairment, reduced clearance of the agent, and possibly increased central nervous system sensitivity may increase the risk for confusion and falls among elderly patients. When prescribing hypnotic agents to elderly patients, one should start at a low dose and monitor its effects closely.

Benzodiazepines should generally be avoided in pregnant women and in breast-feeding mothers. Due to the theoretical dangers of respiratory depression, benzodiazepines should be given cautiously, if at all, to patients with untreated sleep apnea and profound obstructive and restrictive ventilatory impairment, including emphysema and obesity–hypoventilation syndrome. Lethality with overdose of benzodiazepines, when ingested alone, is low, but rises with coingestion of other compounds such as alcohol and other central nervous system depressants.

Other potential disadvantages of benzodiazepines include the development of tolerance and risk for dependence and abuse, as well as withdrawal symptoms and rebound insomnia following cessation of chronic use, particularly with short-acting agents. However, the limited data on long-term usage precludes firm conclusions regarding the true prevalence of such complications.

Abrupt discontinuation following long-term use of benzodiazepine use may result in withdrawal symptoms (agitation, anxiety, confusion, irritability, restlessness, tremulousness), relapse (recurrence of insomnia), or rebound (worsening of disturbed sleep compared to pretreatment baseline) [3]. The severity of benzodiazepine withdrawal syndrome is related to the medication dosage, duration of treatment, and the rapidity of tapering of drug use. In cases of severe withdrawal, significant morbidity or death can ensue.

Relapse after benzodiazepine discontinuation is common, regardless of the type of intervention used to aid withdrawal. In one study, 46% of 76 patients with persistent insomnia and prolonged benzodiazepine use had relapsed at the end of a 24 month follow-up [4]. The long-term relapse rate following hypnotic discontinuation is substantially higher than that seen after a course of cognitive–behavioral therapy.

Rebound insomnia with deterioration of sleep quality and daytime well-being can occur among patients with insomnia following withdrawal of hypnotic agents, primarily short-acting medication. The duration of sleep deterioration can be protracted with marked internight variability. Although rebound insomnia can develop following short-term therapy with benzodiazepines, it is particularly prominent after chronic treatment with rapidly metabolized agents. Rebound insomnia can be minimized by intermittent use of hypnotic agents and by gradual reduction of the dose administered.

Nonbenzodiazepine Benzodiazepine Receptor Agonists

Zolpidem is a nonbenzodiazepine imidazopyridine that binds preferentially to the omega (1) subtype of benzodiazepine receptors. In contrast to the benzodiazepines, it possesses no anticonvulsant or muscle relaxant activity. It has a quick onset of action and has a short half-life of approximately 2.4 hours. The appeal of zolpidem as a hypnotic agent rests in its relative lack of any appreciable withdrawal symptoms, tolerance, rebound insomnia, or residual daytime cognitive and motor impairment. Compared to the benzodiazepines, it is less likely to disrupt normal sleep architecture. The dose of zolpidem is commonly 5–10 mg at bedtime. The lower dose is recommended for elderly patients and individuals that have underlying medical disorders.

Zaleplon, another nonbenzodiazepine benzodiazepine receptor agonist, has minimal effects on sleep architecture, a rapid onset of action, and a short duration of action with half-life of only about 1 hour. Walsh and co-workers examined the hypnotic efficacy of zaleplon (10 mg) over a period of 35 nights in 113 primary insomniacs. Zaleplon significantly shortened sleep latency with no evidence of tolerance to its sleep-promoting effects and no rebound insomnia upon its discontinuation [5].

Zopiclone, not available in the United States, is a cyclopyrrolone hypnosedative that potentiates GABA-mediated neuronal inhibition. It is generally at least as effective as the benzodiazepines in the treatment of insomnia. Zopiclone is well tolerated, possesses minimal tolerance potential, and is associated with a low risk of residual clinical effects. Clinical trials found no evidence for significant rebound insomnia and indicated that the risk of withdrawal reactions with therapeutic doses of zopiclone is very low [6].

Eszopiclone is a cyclopyrrolone, nonbenzodiazepine agent that has recently received FDA approval. Long-term pharmacologic treatment of chronic primary insomnia has been studied using eszopiclone. Throughout a 6 month period, eszopiclone (3 mg) produced significant and sustained improvements in sleep latency, wake time after sleep onset, number of awakenings, number of nights per week with awakenings, total sleep time, and quality of sleep as well as next-day function, alertness, and sense of physical well-being compared with placebo. There was no evidence of tolerance [7].

Compared to conventional benzodiazepines, zolpidem, zaleplon, and zopiclone are less likely to cause significant rebound insomnia or tolerance. Although zolpidem and zopiclone are relatively safe drugs, they are not devoid of abuse and dependence potential. Risk of abuse remains a concern, particularly among patients with a history of abuse or dependence on alcohol or other drugs, and those with psychiatric disease [8].

Antidepressants

Antidepressants have been increasingly utilized over the past decade in the treatment of insomnia. However, despite their widespread use to aid sleep, there is limited data on their appropriate use among persons with insomnia, particularly in patients without mood disorders. Serotonin-specific antidepressants such as trazodone have fewer adverse effects than the older tricyclic agents, including doxepin and amitriptyline, and have surpassed the latter's popularity as hypnotic agents. The use of tricyclic antidepressants (e.g., amitriptyline, trimipramine, and doxepin) is associated with greater anticholinergic reactions, cardiotoxicity, and orthostatic hypotension.

Trazodone is a widely prescribed medication for insomnia. A 5-HT (2) and alpha (1) receptor antagonist, it possesses both anxiolytic and sedative properties. In one placebo-controlled crossover study, the acute effects of 100 mg of trazodone were investigated in 11 patients with insomnia related to a depressive episode or recurrent depressive disorder. Compared to placebo, trazodone increased sleep efficiency, total sleep time, total sleep period, and slow-wave sleep and decreased wakefulness, early morning awakening, and stage 2 NREM sleep. Improvements in subjective sleep quality were noted [9]. It can also lengthen REM latency and decrease REM sleep. Trazodone does not appear to possess any significant tolerance or dependence potential. Although widely prescribed for individuals with chronic primary insomnia, its effectiveness in this application has not been established. Side effects associated with the use of trazodone include atrial and ventricular arrhythmias and priapism. Serotonin syndrome can result with coadministration with similar serotonin-specific agents.

Mirtazapine, a noradrenergic and specific serotonergic antidepressant, acts by antagonizing central alpha 2-adrenergic and 5-HT (2) and 5-HT (3) receptors. Significant improvements in sleep latency, sleep efficiency, and wake after sleep onset were noted during mirtazapine administration in patients with major depression and insomnia [10].

Other Prescription Agents

Quetiapine is an atypical antipsychotic with antihistaminergic, antidopaminergic, and antiadrenergic properties. It has been noted to increase total sleep time, sleep efficiency, stage 2 NREM sleep, and subjective sleep quality when given to healthy subjects [11]. It is commonly employed as a sleep aid for patients with psychiatric illness, although its effectiveness for this indication has not been established.

Nonprescription Hypnotic Agents

Patients with insomnia commonly self-administer nonprescription sleep agents, including alcohol, to manage their sleep disturbances. In a survey of 176 ambulatory elderly subjects, aged 60 years or more, 48% reported that they had used, within the past year, one or more therapies for sleep, including nonprescription products (50% of therapies), prescription products (17%), and nonpharmacologic activities such as walking or drinking milk (34%). Frequently used nonprescription products included acetaminophen, diphenhydramine, alcohol, and herbal products. Seventy-nine percent took them at least 1 day per week and 32% took them daily. Respondents stated that these products subjectively improved sleep latency, number of nighttime awakenings, and total hours of sleep [12].

Histamine Antagonists A majority of over-the-counter hypnotic agents are composed of antihistamines. Aside from their actions on histamine H1 receptors, these agents may also act on serotonergic, cholinergic, and central alpha-adrenergic receptors. The first-generation histamine

H1-antagonists, such as diphenhydramine and chlorpheniramine, can induce sedation by virtue of their ability to cross the blood–brain barrier. In contrast, use of second-generation agents, including loratadine and fexofenadine, is less likely to result in sedation. There is little published data on the efficacy of nighttime administration of antihistamines as sleep aids for insomnia. In general, the effectiveness of antihistamines for treatment of chronic insomnia is not well demonstrated. Tolerance to the hypnotic effects of diphenhydramine may develop rapidly. In a short-term double-blind, placebo-controlled, crossover study, 111 patients with insomnia were given diphenhydramine (50 mg) or placebo at bedtime. Diphenhydramine improved sleep latency to a significantly greater degree than did placebo. Patients who took diphenhydramine also reported feeling more rested the following morning [13].

Melatonin   The secretion of endogenous melatonin by the pineal gland is synchronized to the circadian rhythm with increased production at nighttime. Melatonin has been evaluated primarily for the therapy of insomnia secondary to circadian rhythm sleep disturbances. Studies on its use for primary insomnia are more limited. In a study in which 10 patients with primary insomnia received melatonin (0.3 or 1.0 mg) or placebo 60 minutes before bedtime, no significant group differences were noted in sleep electroencephalographic (EEG) records, or the amount and subjective quality of sleep based on sleep logs and analogue-visual scales [14]. In a separate study, subjects with actigraphically confirmed decreases in sleep efficiency received three melatonin doses (0.1, 0.3, and 3.0 mg) or placebo orally 30 min before bedtime. Sleep was improved by all three melatonin doses. Melatonin (0.3 mg) restored sleep efficiency, acting primarily in the middle third of the night, and normalized plasma melatonin levels. However, melatonin (3.0 mg) induced hypothermia and caused plasma melatonin levels to remain elevated into the daylight hours [15]. Other studies have demonstrated improvement in sleep following evening administration of melatonin in older individuals with reduced endogenous melatonin levels.

Botanical Compounds   Self-medication with herbal preparations is common. Additional studies are required to better define the roles of these compounds in the management of both transient and chronic insomnia.

Kava   There has been little published data on the efficacy of kava (*Piper methysticum*), a psychoactive agent belonging to the pepper family, in the therapy of insomnia. In one randomized clinical study, 61 patients with sleep disturbances associated with anxiety, tension, and restlessness received daily doses of 200 mg of a kava extract (WS 1490) or placebo over a period of 4 weeks. Statistically significant differences in favor of kava compared to baseline

were noted for sleep questionnaire scores for "Quality of sleep" and "Recuperative effect after sleep" ($P = 0.007$ and $P = 0.018$, respectively) [16]. Kavapyrones, the active constituents of kava, possess centrally acting skeletal muscle relaxant and anticonvulsant properties and might possibly have an effect on GABA receptors as well. Adverse effects include dizziness, mild gastrointestinal disturbances, and allergic skin reactions. Kava has been removed from the market in a number of countries as a result of concern about numerous reported cases of hepatotoxicity. A scaly dermatitis known as kava dermopathy can follow chronic use of kava in supratherapeutic dosages [17].

Valerian   Valerian is a popular botanical sleep remedy and is often found as one of the ingredients for herbal compounds marketed for the therapy of insomnia. The sedative properties of valerian (*Valeriana officinalis*) have been ascribed to the possible interaction of its valepotriates and sesquiterpene constituents with GABA, adenosine, or barbiturate receptors [17]. Polysomnography conducted after administration of valerian has demonstrated an increase in slow-wave sleep, a decrease in stage 1 NREM sleep, and variable effects on sleep efficiency, sleep onset, time awake after sleep onset, and REM sleep [18]. Stevinson and Ernst, reviewing the evidence for the effects of valerian on insomnia based on nine randomized, placebo-controlled, double-blind trials, concluded that there was inconclusive evidence for the efficacy of valerian as a treatment for insomnia [19]. Abdominal pain, chest tightness, tremor, lightheadedness, mydriasis, and fine hand tremors have been reported following overdoses with valerian [17].

Other Botanical Agents   Other natural products that have been used as mild sedatives include passionflower (*Passiflora incarnata*) and skullcap (*Scutellaria laterifolia*) [17]. These agents may be combined with valerian or kava in commercial botanical products. No clinical trials of the use of skullcap or passionflower for insomnia have been published in the medical literature.

Newer Agents

A number of agents with novel and promising mechanisms of action are under development.

Indiplon   Indiplon is a nonbenzodiazepine $GABA_A$ receptor modulator that possesses sedative-hypnotic properties. Studies have demonstrated indiplon-induced improvements in latency to persistent sleep and sleep efficiency. Walsh and co-workers studied 79 patients with primary insomnia and difficulty with sleep maintenance in a randomized trial that compared four indiplon modified-release doses (10, 20, 30, and 35 mg) and placebo. Indiplon (20, 30, and 35 mg) produced an increase in mean sleep efficiency and reduced wakefulness after sleep onset. Latency to persistent sleep was significantly reduced for all indiplon doses [20].

TAK-375 TAK-375 is a melatonin receptor agonist that is highly selective for ML1 receptors located mainly in cells of the suprachiasmatic nucleus. They are postulated to mediate the effects of melatonin on circadian rhythms. It has little affinity for other ML receptor subtypes, including ML2. Studies have confirmed the sleep-promoting effects of TAK-375. In a multicenter, randomized trial, 375 healthy adults received either TAK-375 (16 or 64 mg) or placebo 30 minutes before bedtime. In this first-night-effect model of transient insomnia, both TAK-375 groups had a significantly shorter mean latency to persistent sleep compared to placebo. Mean total sleep time was also longer in subjects who were given TAK-375. No significant group differences for time spent in any sleep stage were noted [21]. In a separate double-blind, placebo-controlled study, 107 subjects with chronic insomnia were each randomized to TAK-375 (4, 8, 16, and 32 mg) or placebo given 30 minutes before bedtime on two consecutive nights. There were five treatment periods, with a 5 or 12 day washout period between treatments. Compared to placebo, significant increases in mean total sleep time and sleep efficiency, and reductions in latency to persistent sleep were seen in all TAK-375 groups. No residual drowsiness was evident [22].

DISCONTINUING LONG-TERM USE OF HYPNOTIC AGENTS

The long-term use of hypnotic agents for chronic insomnia, although not recommended, is a clinical reality.

Long-term use of hypnotics is common in both primary insomnia, when conservative and nonpharmacological treatments have been unsuccessful, unavailable, or simply ignored, and in secondary insomnia, as an adjunct to treatment of the primary condition, or when such treatment has failed to correct the insomnia. Patients receiving hypnotic agents chronically should be informed that this is an "off-label" use of the medications.

As previously noted, there is little literature that establishes the effectiveness of chronic benzodiazepine administration for insomnia. One study involving 192 long-term users of benzodiazepine hypnotics, aged ≥65 years, compared the sleep characteristics and withdrawal symptoms of patients who wished to withdraw and those who desired to continue their use of benzodiazepines. Benzodiazepines had been used continuously for >10 years (60% of patients) and for >20 years (27%). Eighty percent had successfully withdrawn 6 months later. Improvements in performance on several cognitive–psychomotor tasks were noted among the withdrawers at 24 or 52 weeks. Withdrawers did not differ in sleep or benzodiazepine withdrawal symptoms

compared to those who continued taking benzodiazepine agents [23]. As described, recent literature has suggested that some nonbenzodiazepines (eszopiclone) may maintain long-term effectiveness without significant safety problems.

The usual clinical management for withdrawal from chronic benzodiazepine use is gradual tapering. A combination of cognitive–behavioral therapy and benzodiazepine tapering may be superior to tapering alone in the management of patients with insomnia and chronic benzodiazepine use [24].

SUMMARY

The management of insomnia should address not only the perceived difficulty with nighttime sleep but also its daytime consequences. The sleep disturbance related to insomnia may be due to many causes. One should therefore attempt to identify any factors that may precipitate or perpetuate these complaints and initiate appropriate corrective measures. It is of paramount importance to prevent the progression of transient complaints into chronic, unrelenting insomnia. Most patients benefit from a combination of sleep hygiene counseling, behavior modification, and the judicious administration of hypnotic agents. Pharmacotherapeutic management is generally effective for transient insomnia due to jet lag or acute stressors. It may also be used intermittently in patients with more chronic complaints. The selection of a particular hypnotic medication should be based on the characteristics of the patient, duration and timing of insomnia, and the pharmacological profile of the agent. It is advisable to use the lowest effective dose and to monitor carefully both the therapeutic response as well as its side effects.

REFERENCES

1. Qureshi A, Lee-Chiong T. Medications and their effects on sleep. *Med Clin N Am* **88**:751–766(2004).
2. Terzano MG, Rossi M, Palomba V, Smerieri A, Parrino L. New drugs for insomnia: comparative tolerability of zopiclone, zolpidem and zaleplon. *Drug Saf* **26**(4):261–282(2003).
3. Schweizer E, Rickels K. Benzodiazepine dependence and withdrawal: a review of the syndrome and its clinical management. *Acta Psychiatr Scand Suppl* **393**:95–101(1998).
4. Bélanger L, Morin CM, Bastien CH, Guay B, Leblanc J, Vallières A, Radouco-Thomas M. Benzodiazepine discontinuation in chronic insomnia: a survival analysis over a 24-month follow-up [abstract]. *Sleep* **26**(Suppl):A308–A309(2003).
5. Walsh JK, Vogel GW, Scharf M, Erman M, William Erwin C, Schweitzer PK, Mangano RM, Roth T. A five week, polysomnographic assessment of zaleplon 10 mg for the treatment of primary insomnia. *Sleep Med* **1**(1):41–49(2000).

6. Noble S, Langtry HD, Lamb HM. Zopiclone. An update of its pharmacology, clinical efficacy and tolerability in the treatment of insomnia. *Drugs* **55**(2):277–302(1998).

7. Krystal AD, Walsh JK, Laska E, Caron J, Amato DA, Wessel TC, Roth T. Sustained efficacy of eszopiclone over 6 months of nightly treatment: results of a randomized, double-blind, placebo-controlled study in adults with chronic insomnia. *Sleep* **26**(7):793–799(2003).

8. Hajak G, Muller WE, Wittchen HU, Pittrow D, Kirch W. Abuse and dependence potential for the non-benzodiazepine hypnotics zolpidem and zopiclone: a review of case reports and epidemiological data. *Addiction* **98**(10):1371–1378(2003).

9. Saletu-Zyhlarz GM, Abu-Bakr MH, Anderer P, Gruber G, Mandl M, Strobl R, Gollner D, Prause W, Saletu B. Insomnia in depression: differences in objective and subjective sleep and awakening quality to normal controls and acute effects of trazodone. *Prog Neuropsychopharmacol Biol Psychiatry* **26**(2):249–260(2002).

10. Winokur A, DeMartinis NA 3rd, McNally DP, Gary EM, Cormier JL, Gary KA. Comparative effects of mirtazapine and fluoxetine on sleep physiology measures in patients with major depression and insomnia. *J Clin Psychiatry* **64**(10):1224–1229(2003).

11. Cohrs S, Rodenbeck A, Guan Z, Pohlmann K, Jordan W, Meier A, Ruther E. Sleep-promoting properties of quetiapine in healthy subjects. *Psychopharmacology* (*Berl*) **174**(3):421–429(2004).

12. Sproule BA, Busto UE, Buckle C, Herrmann N, Bowles S. The use of non-prescription sleep products in the elderly. *Int J Geriatr Psychiatry* **14**(10):851–857(1999).

13. Rickels K, Morris RJ, Newman H, Rosenfeld H, Schiller H, Weinstock R. Diphenhydramine in insomniac family practice patients: a double-blind study. *J Clin Pharmacol* **23**(5–6):234–242(1983).

14. Almeida Montes LG, Ontiveros Uribe MP, Cortes Sotres J, Heinze Martin G. Treatment of primary insomnia with melatonin: a double-blind, placebo-controlled, crossover study. *J Psychiatry Neurosci* **28**(3):191–196(2003).

15. Zhdanova IV, Wurtman RJ, Regan MM, Taylor JA, Shi JP, Leclair OU. Melatonin treatment for age-related insomnia. *J Clin Endocrinol Metab* **86**(10):4727–4730(2001).

16. Lehrl S. Clinical efficacy of kava extract WS 1490 in sleep disturbances associated with anxiety disorders. Results of a multicenter, randomized, placebo-controlled, double-blind clinical trial. *J Affect Disord* **78**(2):101–110(2004).

17. Fugh-Berman A, Jerry M, Cott J. Dietary supplements and natural products as psychotherapeutic agents. *Psychosom Med* **61**:712–728(1999).

18. Donath F, Quispe S, Diefenbach K, Maurer A, Fietze I, Roots I. Critical evaluation of the effect of valerian extract on sleep structure and sleep quality. *Pharmacopsychiatry* **33**(2):47–53(2000).

19. Stevinson C, Ernst E. Valerian for insomnia: a systematic review of randomized clinical trials. *Sleep Med* **1**(2):91–99(2000).

20. Walsh JK, Lankford DD, Krystal A, Roth T, Jochelson P, Garber M, Alexander T, Burke J. Efficacy and tolerability of four doses of indiplon (NBI-34060) modified-release in elderly patients with sleep maintenance insomnia [abstract]. *Sleep* **26**(Suppl):A78(2003).

21. Roth T, Walsh J. Phase II study of the selective ML-1 receptor agonist TAK-375 in a first night effect model of transient insomnia [abstract]. *Sleep* **26**(Suppl):A294(2003).

22. Erman M, Seiden D, Zammit G. Phase II study of the selective ML-1 receptor agonist TAK-375 in subjects with primary chronic insomnia [abstract]. *Sleep* **26**(Suppl):A298(2003).

23. Curran HV, Collins R, Fletcher S, Kee SC, Woods B, Iliffe S. Older adults and withdrawal from benzodiazepine hypnotics in general practice: effects on cognitive function, sleep, mood and quality of life. *Psychol Med* **33**(7):1223–1237(2003).

24. Baillargeon L, Landreville P, Verreault R, Beauchemin JP, Gregoire JP, Morin CM. Discontinuation of benzodiazepines among older insomniac adults treated with cognitive-behavioural therapy combined with gradual tapering: a randomized trial. *CMAJ* **169**(10):1015–1020(2003).

ADDITIONAL READING

Jindal RD, Buysse DJ, Thase ME. Maintenance treatment of insomnia: What can we learn from the depression literature? *Am J Psychiatry* **161**:19–24(2004).

Mendelson WB, Roth T, Cassella J, Roehrs T, Walsh JK, Woods JH, Buysse DJ, Meyer RE. The treatment of chronic insomnia: drug indications, chronic use and abuse liability. Summary of a 2001 New Clinical Drug Evaluation Unit Meeting Symposium. *Sleep Med Rev* **8**(1):7–17(2004).

Montplaisir J, Hawa R, Moller H, Morin C, Fortin M, Matte J, Reinish L, Shapiro CM. Zopiclone and zaleplon vs benzodiazepines in the treatment of insomnia: Canadian consensus statement. *Hum Psychopharmacol* **18**(1):29–38(2003).

Neubauer DN. Pharmacologic approaches for the treatment of chronic insomnia. *Clin Cornerstone* **5**(3):16–27(2003).

Roehrs T, Roth T. Hypnotics: an update. *Curr Neurol Neurosci Rep* **3**(2):181–184(2003).

18

NONPHARMACOLOGIC THERAPY OF INSOMNIA

MELANIE K. MEANS AND JACK D. EDINGER
VA & Duke University Medical Centers, Durham, North Carolina

INTRODUCTION

Insomnia is a prevalent disorder characterized by difficulty initiating or maintaining sleep or by chronically poor sleep quality. Accompanying nocturnal sleep disruption are daytime complaints (e.g., fatigue, poor concentration, lowered social functioning) that can significantly compromise daily functioning, health status, and quality of life. Sleep difficulties may arise from a variety of events, conditions, or circumstances, such as stress, environmental factors, changes to the sleep–wake cycle, medical or psychiatric illnesses, or ingestion of sleep-disrupting substances. Regardless of the precipitating factors, insomnia may assume a chronic course perpetuated by psychological and behavioral factors such as dysfunctional beliefs about sleep, heightened anxiety, and sleep-disruptive compensatory practices [1]. Although sedative hypnotic medications are often prescribed for insomnia, such treatment is symptom-focused and fails to address underlying behavioral and cognitive factors sustaining the sleep problems. In contrast, nonpharmacologic or behavioral insomnia therapies specifically target behavioral and conditioning factors with the goals of eradicating these perpetuating mechanisms and restoring normal sleep–wake functioning. This chapter reviews behavioral insomnia therapies, the most common of which are summarized in Table 18.1.

STIMULUS CONTROL

Stimulus control is a well-established insomnia therapy consisting of a structured behavioral regimen designed to disassociate a problem behavior or conditioned autonomic response from a specific environmental setting [2]. This approach is based on the assumption that both the timing (bedtime) and setting (bed/bedroom) are associated with repeated unsuccessful sleep attempts and over time become conditioned cues that perpetuate insomnia. As a result, the goal of this treatment is that of reassociating the bed and bedroom with successful sleep attempts. Stimulus control achieves this endpoint by curtailing sleep-incompatible activities in the bed and bedroom and by enforcing a consistent sleep–wake schedule. In practice, this therapy requires instructing the patient to (1) go to bed only when sleepy; (2) establish a standard wake-up time; (3) get out of bed whenever awake for more than 15–20 min; (4) avoid reading, watching TV, eating, worrying, and other sleep-incompatible behaviors in the bed and bedroom; and (5) refrain from daytime napping. From a theoretical perspective, it is probable that strict adherence to this regimen not only corrects aberrant, sleep-disruptive conditioning but also likely reestablishes a normal sleep drive and sleep–wake rhythm. Because the treatment recommendations may appear counterintuitive, it is important that the insomnia patient understands the therapeutic rationale. Stimulus control instructions usually can be administered in one visit; however, follow-up visits are beneficial to facilitate compliance and achieve optimal success.

SLEEP RESTRICTION

Sleep restriction therapy reduces nocturnal sleep disturbance primarily by restricting the time allotted for sleep

TABLE 18.1   Common Behavioral Therapies for Insomnia

Stimulus Control   A structured behavioral regimen designed to establish the bedroom as a stimulus for sleep by eliminating behaviors that are incompatible with sleep.

Sleep Restriction   A behavioral strategy in which time in bed at night is restricted in order to encourage a more consolidated sleep pattern.

Cognitive–Behavioral Insomnia Therapy   A multifaceted treatment approach designed to reduce dysfunctional beliefs about sleep and correct sleep-disruptive habits.

Progressive Muscle Relaxation   A relaxation strategy where the patient alternately tenses and relaxes muscle groups in order to reduce muscle tension. Typically, muscles are tensed and relaxed in a systematic order during a 15–20 min session.

each night so that the time spent in bed closely matches the individual's actual sleep requirement [3, 4]. Some insomnia sufferers attempt to alleviate their sleep difficulties by spending excessive time in bed. However, allotting too much time for sleep fragments the sleep pattern and creates excessive time awake each night. By restricting time in bed, mild sleep deprivation may be induced. As a result, sleep drive is increased, wakefulness is reduced, and the sleep pattern is consolidated. This treatment typically begins by calculating the individual's average total sleep time (ATST) from a sleep log that is kept for at least 2 weeks. An initial time-in-bed (TIB) prescription may either be set at the ATST or at a value equal to the ATST plus an amount of time that is deemed to represent normal nocturnal wakefulness (e.g., ATST + 30 min). However, unless persuasive evidence suggests the individual has an unusually low sleep requirement, the initial TIB prescription is seldom set below 6 hours per night. On subsequent visits, the TIB prescription is increased by 15–20 min increments following weeks wherein the patient, on average, is sleeping >85% or 90% of the TIB and continues to report daytime sleepiness. Conversely, TIB is usually reduced by similar increments following weeks wherein the patient, on average, sleeps less than 80% of the time spent in bed. Since TIB adjustments often are necessary, sleep restriction typically entails an initial visit to introduce treatment instructions and follow-up visits to alter TIB prescriptions.

Sleep compression therapy [5] shares the same therapeutic goal and rationale as sleep restriction therapy but achieves this through an alternate methodology. Instead of immediately restricting TIB to an amount close to ATST, the therapist gradually reduces TIB over a number of weeks until this value approximates ATST.

SLEEP HYGIENE

Sleep hygiene connotes a loosely defined set of recommendations targeting life style and environmental factors.

Patients are educated about healthy sleep behaviors and sleep-conducive environmental conditions [6]. For example, insomnia patients may be encouraged to exercise daily, eliminate the use of caffeine, alcohol, and nicotine, eat a light snack at bedtime, and ensure that the sleeping environment is quiet, dark, and comfortable. Sleep hygiene is seldom used as a primary intervention but is often included with other interventions such as stimulus control or sleep restriction.

RELAXATION THERAPIES

A variety of relaxation strategies, including progressive muscle relaxation, passive relaxation, autogenic training, biofeedback, imagery training, meditation, and hypnosis, have been used to treat insomnia [5, 7–9]. Common to these approaches is their focus on factors such as performance anxiety and bedtime arousal, which often perpetuate sleep difficulties. Accordingly, the goal of these therapies is to reduce or eliminate the physiological (e.g., muscle tension) and/or cognitive (e.g., racing thoughts) arousal that disrupts sleep. Regardless of the specific relaxation strategy employed, treatment typically entails conducting specific treatment exercises and teaching relaxation skills over multiple treatment sessions. The patient is encouraged to practice at home in order to gain mastery and facility with self-relaxation. Once the patient achieves sufficient relaxation skills, these can be applied to facilitate sleep initiation by reducing sleep-related anxiety and bedtime arousal.

PARADOXICAL INTENTION

Designed mainly to address the excessive performance anxiety that contributes to sleep onset difficulties, paradoxical intention requires the insomnia patient to remain awake as long as possible after retiring to bed [10]. The patient is instructed to purposefully engage in the feared activity (staying awake) in order to reduce performance anxiety that confounds associated goal-directed behavior (falling asleep). This method alleviates both the patient's excessive focus on sleep and anxiety over not sleeping; as a result, sleep becomes less difficult to initiate. Paradoxical intention is seldom used albeit seemingly effective. Like the other treatments, an initial visit to provide treatment instructions and follow-up sessions to support the patient and assure compliance are usually recommended when administering this intervention.

COGNITIVE THERAPY

Cognitive therapy is a psychotherapeutic approach that alters dysfunctional cognitions contributing to emotional arousal and maladaptive behaviors. During the therapeutic process, dysfunctional beliefs are first identified and monitored. The validity of these beliefs is challenged, and then the beliefs are replaced with more adaptive and realistic substitutes. When applied specifically to insomnia patients, cognitive therapy targets unrealistic expectations about sleep as well as misconceptions or misattributions regarding the causes of insomnia, the consequences of insomnia, the ability to control and predict sleep, and sleep behaviors [11]. The goal of this therapy is to reduce the cognitive arousal and anxiety contributing to insomnia by helping the patient adopt a more adaptive "mental set." Whether provided through formalized patient education modules or via the cognitive restructuring method similar to that commonly used in cognitive therapy with clinically depressed individuals [12], cognitive therapy involves multiple treatment sessions with a skilled therapist.

COGNITIVE–BEHAVIORAL THERAPY

Cognitive–behavioral therapy (CBT) is a multicomponent treatment approach consisting of cognitive therapy strategies used in combination with behavioral therapies such as stimulus control, sleep restriction, and sleep hygiene [11]. One presumed advantage of this treatment is that it includes treatment components addressing the range of cognitive and behavioral mechanisms that perpetuate insomnia. Although CBT is a seemingly more complex treatment than those previously described, in practice, this intervention usually requires no more therapist or patient treatment time than do the other treatments. Morin [11] has developed an eight-session treatment protocol, but some CBT models utilize as few as two sessions [13] in their clinical applications. As with the other therapeutic approaches, multiple treatment sessions to provide patients with sufficient support and follow-up are recommended.

OTHER APPROACHES

In efforts to maximize treatment response, the effects of combining pharmacotherapy with behavioral treatment have been investigated. These preliminary studies indicate that combination treatment equals or slightly outperforms medication-only and behavioral treatment-only conditions [14, 15]. At long-term assessments, however, sleep improvements are sustained in behavioral treatment but not combination treatment. Clearly, more research is needed to ascertain the potential benefits of combining behavioral interventions with hypnotic medications.

When insomnia is caused by a desynchronization of the patient's sleep schedule with his/her biological circadian rhythm, therapeutic approaches such as chronotherapy or bright light therapy may be indicated.

Acupuncture is presumed to promote sleep by stimulating the release of neurotransmitters involved in the sleep–wake system [16]. Although sufficient evidence does not exist to recommend acupuncture as a primary treatment for insomnia, some recent studies have shown this technique improves sleep patterns [16, 17].

EFFECTIVENESS OF NONPHARMACOLOGIC THERAPIES

Nonpharmacologic insomnia treatments produce significant improvements in sleep patterns that are durable over time [18]. Practice parameters issued by the American Academy of Sleep Medicine [19] recognize stimulus control, sleep restriction, progressive muscle relaxation, biofeedback, paradoxical intention, and CBT as effective therapies for treating insomnia, whereas sleep hygiene, imagery training, and cognitive therapy lack sufficient evidence for recommending these approaches as a stand-alone therapy. Compared to other nonpharmacologic approaches, stimulus control and sleep restriction therapies demonstrate the most robust and sustained effects on sleep outcomes [18, 20]. Because most multicomponent CBT treatments include stimulus control and sleep restriction interventions, CBT is highly effective as well. Furthermore, these behavioral treatments are generally preferred by patients over pharmacologic approaches [18].

CONCLUSION

Chronic insomnia often is perpetuated by dysfunctional beliefs about sleep, heightened anxiety, and sleep-disruptive compensatory practices. Medications frequently prescribed to treat insomnia are symptom-focused and fail to address these psychological and behavioral problems. In contrast, nonpharmacologic insomnia therapies such as relaxation therapy, stimulus control, sleep restriction therapy, and cognitive–behavioral therapy target behavioral and psychological factors that maintain and exacerbate sleep difficulties. These therapies are efficacious in improving sleep patterns and represent the optimal treatment for chronic insomnia.

REFERENCES

1. Spielman AJ, Caruso LS, Glovinsky PB. A behavioral perspective on insomnia treatment. *Psychiatr Clin North Am* **10**:541–553(1987).

2. Bootzin R. Stimulus control treatment for insomnia. *Proceedings of the 80th Annual Convention of the American Psychological Association* **7**:395–396(1972).

3. Spielman AJ, Saskin P, Thorpy MJ. Treatment of chronic insomnia by restriction of time in bed. *Sleep* **10**:45–56(1987).

4. Wohlgemuth WK, Edinger JD. Sleep restriction therapy. In: Lichstein KL, Morin CM (Eds), *Treatment of Late-Life Insomnia*. Sage Publications, Thousand Oaks, CA, 2000; pp 147–166.

5. McCrae CS, Lichstein KL. Secondary insomnia: a heuristic model and behavioral approaches to assessment, treatment, and prevention. *Appl Prev Psychol* **10**:107–123(2001).

6. Hauri PJ. Sleep hygiene, relaxation therapy, and cognitive interventions. In: Hauri PJ (Ed), *Case Studies in Insomnia*. Plenum Publishing, New York, 1991, pp 65–84.

7. Borkovec TD, Fowles DC. Controlled investigation of the effects of progressive and hypnotic relaxation on insomnia. *J Abnorm Psychol* **82**:153–158(1973).

8. Stepanski EJ. Behavioral therapy for insomnia. In: Kryger MH, Roth T, Dement WC (Eds), *Principles and Practice of Sleep Medicine*, 3rd ed. Saunders, Philadelphia, 2000, pp 647–656.

9. Nicassio P, Bootzin R. A comparison of progressive relaxation and autogenic training as treatments for insomnia. *J Abnorm Psychol* **83**:253–260(1974).

10. Turner RM, Ascher LM. Controlled comparison of progressive relaxation, stimulus control, and paradoxical intention therapies for insomnia. *J Consult Clin Psychol* **47**:500–508(1979).

11. Morin CM. *Insomnia: Psychological Assessment and Management*. Guilford Press, New York, 1993. (This comprehensive and clinically oriented treatment manual for insomnia provides detailed guidance for evaluating and treating insomnia complaints with a multifaceted cognitive–behavioral approach.)

12. Beck AT, Rush AJ, Shaw BF, Emery G. *Cognitive Therapy of Depression*. Guilford Press, New York, 1979.

13. Edinger JD, Sampson WS. A primary care "friendly" cognitive behavioral insomnia therapy. *Sleep* **26**:177–182(2003).

14. Morin CM, Wooten V. Psychological and pharmacological approaches to treating insomnia: critical issues in assessing their separate and combined effects. *Clini Psychol Rev* **16**:521–542(1996).

15. Morin CM, Hauri PJ, Espie CA, Spielman AJ, Buysse DJ, Bootzin RR. Nonpharmacologic treatment of chronic insomnia. An American Academy of Sleep Medicine review. *Sleep* **22**:1134–1156(1999). (This extensive systematic literature review of efficacy data for behavioral insomnia interventions also provides a comprehensive summary table of research studies evaluated.)

16. Lin Y. Acupuncture treatment for insomnia and acupuncture analgesia. *Psychiatry Clin Neurosci* **49**:119–120(1995).

17. Larzelere MM, Wiseman P. Anxiety, depression, and insomnia. *Primar Care* **29**:339–360(2002).

18. Means MK, Edinger JD. Behavioral treatment of insomnia. *Expert Rev Neurotherapeutics* **2**:127–137(2002).

19. Chesson AL Jr, Anderson WM, Littner M, Davila D, Hartse K, Johnson S, Wise M, Rafecas J. Practice parameters for the nonpharmacologic treatment of chronic insomnia. An American Academy of Sleep Medicine report. Standards of Practice Committee of the American Academy of Sleep Medicine. *Sleep* **22**:1128–1133(1999).

20. Harvey L, Inglis SJ, Espie CA. Insomniacs' reported use of CBT components and relationship to long-term clinical outcome. *Behav Res Ther* **40**:75–83(2002).

ADDITIONAL READING

Espie CA. Insomnia: conceptual issues in the development, persistence, and treatment of sleep disorder in adults. *Annu Rev Psychol* **53**:215–243(2002). This article reviews conceptual models of insomnia and presents an integrated psychobiological model of sleep.

Lichstein KL, Morin CM (Eds). *Treatment of Late-Life Insomnia*. Sage Publications, Thousand Oaks, CA, 2000. This book discusses the assessment, diagnosis, and treatment of insomnia as it applies to older adults. Both pharmacological and behavioral treatments are reviewed, and special emphasis is given to sleep medication withdrawal, secondary insomnia, and institutional settings.

Perlis ML, Lichstein KL. *Treating Sleep Disorders: Principles and Practice of Behavioral Sleep Medicine*. Wiley, Hoboken, NJ, 2003. This recently published book covers diagnostic and treatment issues related to nonpharmacologic interventions for insomnia.

PART III

EXCESSIVE SLEEPINESS

19

SLEEP DEPRIVATION AND ITS EFFECTS ON COGNITIVE PERFORMANCE

JILLIAN DORRIAN
The University of South Australia, Adelaide, South Australia

DAVID F. DINGES
University of Pennsylvania School of Medicine, Philadelphia, Pennsylvania

INTRODUCTION

The daily homeostatic drive for sleep is among the most pervasive behavioral control in complex animals, including modern humans. Without adequate sleep, waking brain functions and their expression in behavior change systematically. Conspicuous among the changes is the increased initiation of sleep, manifested by increasingly frequent intrusions of sleep into wakefulness and reduced latencies to sleep onset and to sleep consolidation. Sleep, it appears, cannot be postponed indefinitely. Neurobiological mechanisms deep in the brain initiate sleep as sleep deprivation progresses [1]. These initiations can occur during waking performance and can have serious consequences for performance, even under the most motivating of circumstances.

Sleep deprivation can occur in a number of ways. It can result from not sleeping for a long enough period to ensure normal (internally driven) wakefulness occurs without sleepiness. It can result from fragmented or disturbed sleep due to an untreated sleep disorder or medical condition or other circumstances. It can be acute (lasting a night or two) or chronic (many days, weeks, months, or years). It can be

voluntary or involuntary. However it occurs, it has profound consequences for behavior.

Although sleep deprivation will ultimately lead to the involuntary onset of sleep in an individual, the cognitive performance effects of sleep deprivation can be evident even before sleep occurs uncontrollably in the form of sudden microsleeps or sleep attacks. More than a century of research investigating the effects of sleep deprivation on human behavior has established several sources of variation in cognitive performance capability. These can be summarized as follows:

- Dynamic shifts in performance occur across time such that performance is influenced (1) by the number of hours that an individual has gone without sleep, (2) by the duration of sleep, (3) by the consolidation of the sleep obtained, and (4) by the circadian phase (time-of-day) at which cognitive performance is assessed [2, 3].
- Performance in a sleep-deprived individual is punctuated by periods of nonresponding, or performance lapses [4].
- Between lapses, a general slowing in behavioral responses is observed [4].
- Performance impairment becomes increasingly marked with time-on-task [4].

Correspondence to: David F. Dinges, Division of Sleep and Chronobiology, Department of Psychiatry, University of Pennsylvania School of Medicine, 1013 Blockley Hall, 423 Guardian Drive, Philadelphia, PA 19104-6021.

- Not only do sleep-deprived individuals exhibit performance lapses or errors of omission (i.e., failure to respond to a stimulus), but they also exhibit errors of commission (i.e., responses in the absence of a stimulus) [5].
- Thus cognitive performance variability is the hallmark of sleep deprivation, and such variability is believed to reflect wake state instability [5].
- Sleep deprivation also produces impairment in executive or higher order functioning [6].
- There are substantial interindividual differences in cognitive timing and severity of cognitive effects from sleep deprivation [7].

While these observations have been recorded primarily during laboratory investigations, it is clear from simulator and field research that they are also apparent in "real world" performance [8–12]. The implications of this are clear. The sleep-deprived individual in the home, in the workplace, or on the road is likely to experience impaired performance and a subsequent increase in risk of errors or accidents. This chapter will discuss sources of performance variation during sleep deprivation, highlighting the ways in which they have influenced both theoretical and practical understanding of the effects of sleep deprivation on human cognitive performance functions.

TEMPORALLY DYNAMIC PERFORMANCE SHIFTS

Since the first study of sleep deprivation and human performance, investigators have noted performance changes, not only as a result of time spent awake, but also as a result of time-of-day and the interaction of these two factors. In biological terms, cognitive performance capability at any point in time is largely dependent on the interaction between two systems: the homeostatic sleep drive and the circadian system [2]. The homeostatic drive for sleep escalates with increasing time awake. The circadian system produces oscillations in functioning, which have an approximate frequency of one cycle every 24 hours. These cycles, commonly referred to as circadian rhythms, can be measured in almost all areas of biology and behavior including many hormone levels, body temperature, immune responses, appetite and digestion, kidney function, sleep–wake behavior, and a wide range of neurocognitive functions [13]. In the human brain, circadian timing is internally (endogenously) controlled by the suprachiasmatic nuclei (SCN) in the anterior hypothalamus. External (exogenous) factors can also influence circadian timing. The most powerful exogenous influence is the light–dark cycle, which serves to synchronize human functioning with environmental night and day [14].

Under conditions of sleep deprivation, the combined effects of the homeostatic sleep drive and the endogenous circadian pacemaker on performance are particularly evident. The two systems interact to produce dynamic changes in neurobehavioral functions across the day. After awakening in the morning, performance capability increases across the day, despite increasing time awake. If the morning awakening is earlier than usual, a dip in alertness can occur in midafternoon. This siesta dip is commonly referred to as the postprandial or postlunch dip, but it is largely unrelated to food intake, reflecting the interaction of the homeostatic and circadian drives. Following the siesta dip, alertness and cognitive performance continue to increase to a circadian peak in the evening. A decrease in alertness is then observed across the remainder of the night, until a nadir is reached in the early hours of the morning (see Figure 19.1 for an illustration of performance variability with time-of-day). With increasing hours of wakefulness, the escalating drive for sleep amplifies the circadian oscillation in performance, such that, over successive days, the level of cognitive impairment at and after the circadian nadir (i.e., between the hours of 4 a.m. and 8 a.m.) becomes greater [2, 14].

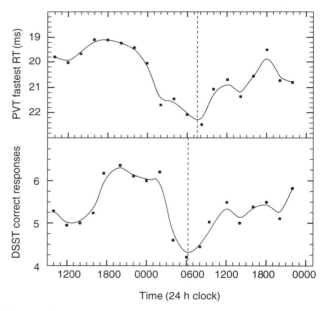

Figure 19.1 Illustration of performance variation over time. Average performance is shown for five subjects on the psychomotor vigilance task (PVT) fastest 10% of response times and digit symbol substitution task (DSST). Data were collected during a 36 h constant routine protocol (i.e., subjects remained in bed, in dim light for a 36 h period). Fit-lines indicate a distance-weighted least squares function. The circadian nadir is indicated by the vertical dotted lines. (Adapted from [2].)

Temporally dynamic performance shifts produced by the interaction of sleep homeostatic and circadian drives have consistently been recorded throughout the 100 year history of studies of sleep deprivation and human performance. These two processes have formed the basis of the most common theoretical and mathematical models of the changes in cognitive performance over time [15]. In recent years, the subcortical neurobiological mechanisms underlying these two processes have begun to be elucidated [1, 16], although the basic physiological mechanisms relevant to their effects on cortically mediated cognitive processes during sleep loss have yet to be identified [13].

BEHAVIORAL LAPSES

As noted earlier, cognitive performance in a sleep-deprived individual is punctuated by periods of nonresponding called performance lapses [4]. However, early approaches to understanding how performance changed during sleep loss did not recognize lapses and instead focused on finding functional lesions. That is, the first scientific studies of the effects of severe sleep deprivation on human cognitive performance sought to identify areas of cognitive capability that were completely eliminated by sleep loss [3, 17].

Despite excellent experimental methods, no loss of cognitive or motor functions was found in these experiments. Sleep-deprived subjects could still perform all tasks, but not as well as when they were not sleep deprived [3–5]. What was clear from all of these early experiments was that, even after days without sleep, there were transient periods when some subjects could transiently perform at non-sleep-deprived levels, but that overall performance was highly variable and worse than baseline.

Based on these observations, some investigators concluded that sleep deprivation affected performance via a reduction in motivation—subjects could perform the tasks but did not wish to do so. However, ensuing research did not lend support to either the functional lesion or motivational approaches to explaining the effects of sleep deprivation on performance. Rather, results suggested that sleep deprivation was associated with increased performance variability from moment to moment. One of the most influential concepts to describe the increased variability in performance during sleep deprivation was the lapse hypothesis [4, 5]. Research conducted in the 1930s and 1940s identified blocks in performance, or periods of nonresponsiveness, which were associated with changes in brain activity and eye movement, typically considered to be general indicators of sleep onset. Consistent with these ideas, in 1959, the lapse hypothesis was proposed to describe performance variability during sleep deprivation [4]. It was proposed that intermittent lapses in cognitive performance in sleep-deprived subjects were the result of very brief,

involuntary sleep periods, or microsleeps [4, 5]. The notion that microsleeps underlie performance lapses in sleep-deprived subjects was a seminal advance, but the lapse hypothesis could not explain all cognitive performance effects from sleep deprivation. Other things occurred in sleep-deprived adults trying hard to perform that were not accounted for by lapses per se.

RESPONSE SLOWING

According to the lapse hypothesis, responses between lapses should be near optimal. However, research has demonstrated a slowing of the fastest portion of reaction times, independent of performance lapses. Specifically, studies using visual and auditory sustained attention tasks have found that the fastest 10–25% of reaction times are also significantly affected by sleep deprivation [4]. This phenomenon was recognized even by those who posited the lapse hypothesis. It is described as response slowing [4] and it is a fundamental neurobehavioral consequence of sleep deprivation.

TIME-ON-TASK DECREMENTS

The effects of sleep deprivation on cognitive performance typically worsen as the duration of the task increases [4, 5]. As with response slowing, this well-established time-on-task decrement is not accounted for by the lapse hypothesis. One of the earliest demonstrations of a time-on-task effect worsening during sleep deprivation was conducted by Kleitman [3, 18], who referred to the phenomenon as loss of endurance. Sleep-deprived subjects performing a color-naming task became slower and less accurate as the task duration was extended from 1 to 12 min [18]. Since then, a large number of cognitive tasks have shown sensitivity to sleep deprivation when the duration of the tasks is extended and measurements are taken of changes in performance as a function of time on task. This is especially the case for attention-demanding tasks, such as vigilance [4].

ERRORS OF COMMISSION

While the lapse hypothesis accurately predicts that sleep deprivation will increase behavioral lapses, which are errors of omission (i.e., failure to respond to a salient signal or failure to respond in a timely manner), it does not predict that sleep loss will increase the likelihood of errors of commission (i.e., responses in the absence of salient stimuli). Yet the latter do occur during sleep deprivation. In fact, errors of omission and errors of commission co-vary during

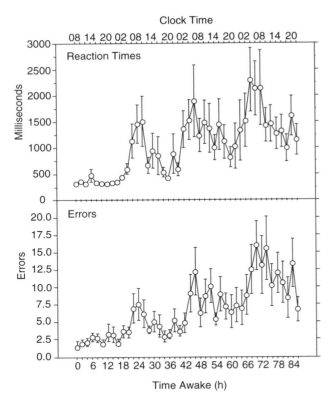

Figure 19.2 Relationship between long reaction times (errors of omission) and false alarm errors (errors of commission) on the psychomotor vigilance task (PVT) in 13 subjects during 88 h of sleep deprivation. (Adapted from [5].)

sleep deprivation [4, 5] (see Figure 19.2 for an illustration of this covariation across 88 h of sleep deprivation).

It is likely that increases in errors of commission (e.g., false alarms on a vigilance task) occur as a result of the sleep-deprived subject attempting to prevent lapses [5]. Viewed in this way, increasing errors of commission during sleep deprivation reflect increased compensatory effort (i.e., increased motivation to perform) to counteract the effects of sleep loss [5]. The strategy fails, however, as it merely trades one kind of cognitive fault (errors of omission) with another type of fault (errors of commission). Therefore, regardless of the performance strategy adopted by the sleep-deprived person, overall performance efficiency will suffer.

WAKE STATE INSTABILITY

The increased propensity for sleep to occur quickly in the sleep-deprived individual—even when sleep is being resisted during cognitive performance—as well as the tendency for performance to show behavioral lapses, response slowing, time-on-task decrements, and errors of commission are signs that sleep-initiating mechanisms deep in the brain [1, 16] are activating during wakefulness [13, 19]. Thus the cognitive performance variability that is the hallmark of sleep deprivation [4, 5] appears to reflect wake state instability [1, 5]. Sleep-initiating mechanisms repeatedly interfere with wakefulness, making cognitive performance increasingly variable and dependent on compensatory mechanisms (e.g., increased motivation to perform, walking and talking). However, the compensatory effort to resist sleep ultimately cannot prevent intrusions of sleep initiation into wakefulness. Such intrusions are evident in reports of sleep-deprived subjects "semidreaming" while engaged in verbal cognitive tasks [3], and reports of healthy sleep-deprived people falling asleep while engaged in dangerous activities [19, 20]. Wake state instability means that at any given moment in time the cognitive performance of the sleep-deprived individual is unpredictable and a product of interactive, reciprocally inhibiting neurobiological systems mediating sleep initiation and wake maintenance [1, 16].

EXECUTIVE FUNCTIONS

Sleep deprivation adversely affects basic cognitive processes involving speed and accuracy of attention and memory, but its effects also extend to higher order cognitive processes involving executive functions [6], such as creative thinking, word fluency, nonverbal planning [20], verb generation, and response inhibition [21]. These measures are considered to measure cognitive flexibility. It is believed that these adverse effects of sleep deprivation on cognitive flexibility are primarily due to its effects on dysfunction of the prefrontal cortex [6]. The neurobiological mechanisms that may subserve these putative effects have not yet been identified.

INTERINDIVIDUAL DIFFERENCES

Theoretical constructs discussed earlier, such as performance lapses, state instability, and frontal lobe dysfunction, have been used to explain overall trends in aggregate data across groups of sleep-deprived individuals. However, it is evident that individuals may differ substantially in their responses to sleep deprivation. While some individuals may be highly susceptible to the negative cognitive effects of sleep loss, others may remain much less affected [5]. These differences are especially apparent when subjects undergo mild to moderate sleep deprivation—all people are impaired by severe levels of sleep loss. It appears that a person's vulnerability to cognitive impairment from sleep loss may be more predictable than previously thought.

Recent studies of individual differences in cognitive performance responses to sleep deprivation have found that the stability of individual responses across repeated exposures to sleep deprivation was remarkably stable. Those who were more impaired cognitively on one exposure to a night of sleep loss were more impaired on a second exposure to sleep loss, and those who were less impaired cognitively on one exposure to sleep loss were less impaired on a second exposure [7].

PERFORMANCE DEFICITS IN THE REAL WORLD

Cognitive performance impairments found in laboratory experiments on sleep deprivation make it clear that performance impairments can be expected to occur in sleep-deprived people engaged in work (e.g., flying an airplane) and everyday performance (e.g., driving a car) in the real world. This has been the case whenever sleep deprivation has been studied in real world situations, or catastrophes due to sleep loss have been investigated [8–12, 19, 20]. Sleep deprivation is a major risk factor for errors and accidents, especially in safety-sensitive occupations. Moreover, due to its dramatic effects on attention and reaction time, sleep deprivation is particularly dangerous when operating a motor vehicle [9–11, 23].

CONCLUSION

Sleep deprivation adversely affects basic cognitive processes involving speed and accuracy of attention and memory, as well as higher order cognitive processes involving executive functions. Microsleeps and behavioral lapses increase with sleep loss as a function of wake state instability. While these functional manifestations of brain impairment from inadequate sleep have consistently been demonstrated by over a century of research, the neurobiological processes subserving them are only now being elucidated. The critical nature of the effect of sleep deprivation on neurobehavioral functions and cognitive performance is also manifest in the major role sleep deprivation has in lowered productivity and increased risks of errors, accidents, and catastrophic outcomes in society at large.

REFERENCES

1. Saper CB, Chou TC, Scammell TE. The sleep switch: hypothalamic control of sleep and wakefulness. *Trends Neurosci* **24**(12):726–731(2001).

2. Van Dongen HPA, Dinges DF. Circadian rhythms in fatigue, alertness and performance. In: Kryger MH, Roth T, Dement WC (Eds), *Principles and Practice of Sleep Medicine.* Saunders, Philadelphia, 2000, pp 391–399.

3. Kleitman N. *Sleep and Wakefulness.* University of Chicago Press, Chicago, 1963.

4. Dinges DF, Kribbs NB. Performing while sleepy: effects of experimentally-induced sleepiness. In: Monk TH (Ed), *Sleep, Sleepiness and Performance.* Wiley, Chichester, UK, 1991, pp 97–128.

5. Doran SM, Van Dongen HPA, Dinges DF. Sustained attention performance during sleep deprivation: evidence of state instability. *Arch Ital Biol* **139**:253–267(2001).

6. Horne JA. Human sleep, sleep loss and behavior. Implications for the prefrontal cortex and psychiatric disorder. *Br J Psychiatry* **162**:413–419(1993).

7. Van Dongen HPA, Baynard MD, Maislin G, Dinges DF. Systematic inter-individual variability differences in neurobehavioral impairment from sleep loss: evidence of trait-like differential vulnerability. *Sleep* **27**(3):423–433(2004).

8. Monk TH (Ed). *Sleep, Sleepiness and Performance.* Wiley, Chichester, UK, 1991.

9. Mitler MM, Carskadon MA, Czeisler CA, Dement WC, Dinges DF, Graeber RC. Catastrophes, sleep, and public policy: consensus report. *Sleep* **11**(1):100–109(1988).

10. Dinges DF. An overview of sleepiness and accidents. *J Sleep Res* **4**(2):4–11(1995).

11. Stutts JC, Wilkins JW, Osberg JS, Vaughn BV. Driver risk factors for sleep-related crashes. *Accident Anal Prev* **35**:321–331(2003).

12. Baldwin DC, Daugherty SR. Sleep deprivation and fatigue in residency training: results of a national survey of first- and second-year residents. *Sleep,* **27**(2):217–223(2004).

13. Durmer JS, Dinges DF. Neurocognitive consequences of sleep deprivation. *Semin Neurol* **25**(1):117–129(2005).

14. Czeisler CA, Khalsa SBS. The human circadian timing system and sleep–wake regulation. In: Kryger MH, Roth T, Dement WC (Eds), *Principles and Practice of Sleep Medicine* Saunders, Philadelphia, 2000, pp 353–376.

15. Mallis MM, Mejdal S, Nguyen LT, Dinges DF. Summary of the key features of seven biomathematical models of human fatigue and performance. *Aviation Space Environ Med* **75**(3):A4–A14(2004).

16. Mignot E, Taheri S, Nishino S. Sleeping with the hypothalamus: emerging therapeutic targets for sleep disorders. *Nat Neurosci* **5**(Suppl):S1071–S1075(2002).

17. Patrick GT, Gilbert JA. On the effects of loss of sleep. *Psychol Rev* **3**:469–483(1896).

18. Lee MAM, Kleitman N. Studies on the physiology of sleep: II. Attempts to demonstrate functional changes in the nervous system during experimental insomnia. *Am J Physiol* **67**:141–152(1923).

19. Torsvall L, Akerstedt T. Sleepiness on the job: continuously measured EEG changes in train drivers. *Electroencephalogr Clin Neurophysiol* **66**:502–511(1987).

20. Nansen F. *Farthest North: The Incredible Three-Year Voyage to the Frozen Latitudes.* Random House, New York, 1999.

21. Horne JA. Sleep loss and "divergent" thinking ability. *Sleep* **11**(6):528–536(1988).

22. Harrison Y, Horne JA. Sleep loss impairs short and novel language tasks having a prefrontal focus. *J Sleep Res* **7**(2):95–100(1998).

23. Pack AI, Pack AM, Rodgman E, Cucchiara A, Dinges DF, Schwab CW. Characteristics of crashes attributed to the driver having fallen asleep. *Accident Analy Prev* **27**(6):769–775(1995).

ADDITIONAL READINGS

Dinges DF, Baynard M, Rogers NL. Chronic sleep deprivation. In: Kryger MH, Roth T, Dement WC (Eds), *Principles and Practices of Sleep Medicine*, 4th ed. Saunders, Philadelphia, 2005, pp 67–76.

Van Dongen HPA, Dinges DF. Circadian rhythms in sleepiness, alertness, and performance. In: Kryger MH, Roth T, Dement WC (Eds), *Principles and Practices of Sleep Medicine*, 4th ed. Saunders, Philadelphia, 2005, pp 435–443.

20

NARCOLEPSY

RAFAEL PELAYO AND MARIA CECILIA LOPES

Stanford Sleep Disorders Clinic, Stanford University, Stanford, California

INTRODUCTION

Narcolepsy is a chronic neurological disorder of excessive daytime sleepiness, which characteristically has a childhood onset and is associated with a hypocretin deficiency [1–3]. Historically, the word "narcolepsy" was first coined by Gélineau in 1880 to designate a pathological condition characterized by irresistible episodes of sleep of short duration recurring at close intervals. The symptoms of narcolepsy can be conceptualized as a blurring of the boundaries between the awake, sleeping, and dreaming brain. The awake narcoleptic may feel sleepy. The sleeping narcoleptic may have disturbed sleep due to arousals. Dreaming phenomenon may occur while the patient is awake. The onset of narcolepsy is typically in the second decade of life but may begin at a younger age or as an adult. Narcolepsy may have an abrupt or insidious onset. In the latter situation, the full syndrome may take 12 years to develop. During this time, patients may be misdiagnosed, simply labeled as lazy or depressed. It may not be until after the child is an adult that the correct diagnosis is made [4]. It is therefore important for healthcare providers to consider narcolepsy as a possibility in any person with unexplained excessive daytime sleepiness.

CLINICAL SYMPTOMS

The cardinal features of narcolepsy are daytime somnolence, cataplexy, sleep paralysis, and hypnagogic hallucinations. These symptoms were called the "tetrad" of narcolepsy by Yoss and Daly [5]. Rechtschaffen et al. [6] reported the presence of abnormal sleep-onset REM sleep periods in narcolepsy, which allowed for objective physiological measurements. The disturbed nocturnal sleep has been described as an important complaint and may compose a fifth symptom associated with narcoleptic syndrome. The mnemonic aide CHESS may help the reader remember the five symptoms of narcoplesy: Cataplexy, Hypnagogic hallucinations, Excessive daytime sleepiness, Sleep paralysis, and Sleep disruption.

Cataplexy is characterized by the sudden loss of muscle tone while awake, typically triggered by a strong positive emotion such as laughter or surprise. It can also be triggered less commonly by anger or fear. Cataplexy is virtually a pathognomonic symptom of narcolepsy. When an experienced clinician witnesses a cataleptic attack, confirmatory sleep laboratory testing for narcolepsy might not be necessary [1]. Narcoleptic patients remain conscious during the attack and are able to remember the details of the event afterward. The episodes are typically brief and may only last a few seconds. Some patients can have other narcoleptic symptoms manifest during an episode of cataplexy, such as hypnagogic hallucinations or sleep paralysis, or may simply fall asleep. Cataplexy may involve only certain muscles or the entire voluntary musculature. Most typically, the jaw sags, the head falls forward, the arms drop to the side, and the knees buckle. The severity and extent of cataplexy can range from a state of absolute powerlessness, which seems to involve the entire body, to no more than a fleeting sensation of weakness. Although the extraocular muscles are supposedly not involved, the patient may complain of blurred vision. Respiration may become irregular during an attack, which may be related to weakness of the abdominal or intracostal muscles. Complete loss of muscle tone, which results in a fall with risk of serious injuries, might occur during a cataplectic attack. The attacks may also be subtle and not noticed by nearby individuals. An attack may consist only of a slight buckling of the knees. Patients may perceive this abrupt and short-lasting weakness and may simply sit or stand against a wall. Speech may be slurred. If it involves the upper limbs, the individual may be described as clumsy due to dropping cups or spilling liquids when surprised or laughing. A patient may present with repetitive falls that cannot be easily explained. This may result in a misdiagnosis of atonic seizures [4]. The duration of each cataplectic attack, partial or total, is highly variable. They usually range from a few seconds to 2 minutes and rarely up to 30 minutes. The term *status cataplecticus* can be applied to prolonged attacks. Attacks can be elicited by emotion, stress, fatigue, or heavy meals. Laughter and anger seem to be the most common triggers, but a feeling of elation while listening to music, reading a book, or watching a movie can also induce the attacks. Merely remembering a funny situation may induce cataplexy, and it may also occur without obvious precipitating acts or emotions. In children it often occurs while playing with others. The potential emotional impact of cataplexy cannot be overstated [7]. Individuals may be misdiagnosed with a psychiatric disorder before recognition of narcolepsy.

Hypnagogic hallucinations are fragments of auditory or dream-like imagery that occur at sleep onset. Similar episodes upon awakening are called hypnopompic hallucinations. This is a nonspecific symptom of narcolepsy. It probably occurs clinically more frequently in any sleep-deprived individual. Hypnagogic hallucinations should not be confused with psychosis. Auditory hallucinations are common. The patient may be frightened by the imagery.

Sleep paralysis is a self-descriptive term that can be a terrifying experience. It is a transient inability to move usually upon awakening. Patients find themselves suddenly unable to move the limbs, to speak, or even to breathe deeply. This state is frequently accompanied by hallucinations. During episodes of sleep paralysis, particularly the first occurrence, the patient may have extreme anxiety associated with fear of dying. This anxiety is often greatly intensified by the hallucinations, sometimes terrifying, that may accompany the sleep paralysis. Patients may be reluctant to talk about these events and these experiences can be so frightful that the patient may resist going to bed to sleep. Sleep paralysis, like hypnagogic hallucinations, may occur in any sleep-deprived individual in the absence of narcolepsy. They do not occur in all narcoleptics and may be transitory. Patients with narcolepsy may, with time, learn to recognize hypnagogic hallucinations and sleep paralysis as frightening but otherwise essentially benign phenomena and be better able to tolerate them.

The most common complaint in narcoleptic patients is daytime sleepiness or an inability to maintain wakefulness in daytime. In young children this may not be quickly recognized as abnormal. The parents may not complain if their children sleep more than usual or may view the sleepiness as a normal phase of development. Certainly most adolescents falling asleep in the classroom do not have narcolepsy. Delays in diagnosis are usually related to an absence of clinical suspicion for narcolepsy. In the 24 hour cycle, established narcolepsy patients do not sleep more than average. The increase in daytime sleep is countered by impaired nighttime sleep.

School impairment is not unusual in children with narcoleptic syndrome. Children and adolescents with narcolepsy may report embarrassment, academic decline, and loss of self-esteem. Patients may be misdiagnosed with attention deficit disorder [4,8] Psychosocial problems are not uncommon. The differences in clinical manifestations between children and adults are summarized in Table 20.1. It can be difficult to recognize daytime sleepiness in very young children, but cataplexy is a more obvious symptom in the young. A potential need for supportive psychotherapy or counseling should be anticipated when developing a therapeutic plan [7].

EPIDEMIOLOGY

Narcolepsy is not a rare disorder. A survey in the United States found an incidence rate per 100,000 persons per year of 1.37. The incidence rate was highest in the second decade, followed in descending order by the third, fourth, and first decades. The prevalence was estimated at 56.3 per 100,000 persons. Approximately 36% of prevalence cases did not have cataplexy [9]. These results are consistent with previous narcolepsy prevalence estimates of 1 out of 2000 of the general population. Onset as young as age 2 has been reported [4].

PATHOPHYSIOLOGY

An animal model for narcolepsy was developed using dogs. Using positional cloning, an autosomal recessive mutation responsible for narcolepsy was discovered in this canine model [10]. Independent work using a gene knock-out mice model confirmed the role of hypocretin in narcolepsy [11]. The hypocretins, which are also called orexins, are neuropeptides. Two novel neuropeptides have been described (Hcrt-1 and Hcrt-2), derived from the same precursor genes that are synthesized by neurons located exclusively in the lateral, posterior, and perifornical hypothalamus. Hypocretin-containing neurons have widespread projections throughout the central nervous system

TABLE 20.1

Symptom	Children	Adults
Cataplexy	Seen in most narcoleptic children, because it is easy to recognize. Often occurs while playing with other children; frequently the first symptom recognized. Differential diagnosis with other causes of drop attacks in children.	Around 70% of narcoleptic patients experience cataplexy. Usually occurs in association with laugher; it is the first symptom in 5–8%. Can be described similar to drop attacks. It may appear on average 6 years after the onset of EDS.
Hypnagogic hallucinations	Symptoms such as nightmares and hypnagogic hallucinations were considered as part of normal childhood. These facts may be difficult to diagnose.	Present in 30% of adults. Visual hallucinations usually consist of simple forms. Auditory hallucinations are also common.
EDS	Sometimes may be difficult to recognize. More usual symptom is falling asleep in class. Sleepiness can be hidden behind other abnormal behavior such as hyperactivity.	Easier to recognize than in children. Difficult to concentrate in slow activities. Problems in work that lead to sleep episodes in monotonous situations.
Sleep paralysis	Frequently accompanied by hallucinations. Children dislike to talk about these events.	Sometimes can be a terrifying experience. May occur in healthy subjects. Symptom decreases with age.
Disturbed nocturnal sleep	This symptom may not affect children; sometimes a transitory complaint.	Worse during adulthood. May have influence in the treatment response.

EDS, excessive daytime sleepiness.

(CNS) with particularly dense excitatory projections to monoaminergic centers such as the noradrenergic locus caeruleus, histaminergic tuberomammillary nucleus, serotoninergic raphe nucleus, and dopaminergic ventral tegmental area. The hypocretins were originally believed to be primarily important in the regulation of appetite; however, a major function emerging from research on these neuropeptides is the regulation of sleep and wakefulness. Deficiency in hypocretin neurotransmission results in the sleep disorder narcolepsy in mice, dogs, and humans.

Canine narcolepsy is an autosomal recessive disease caused by disruption of the hypocretin receptor 2 gene (*Hcrtr2*). In most humans with narcolepsy the Hcrtr receptor is intact but the actual peptide may be deficient or absent. Hypocretin levels can be measured in cerebrospinal fluid (CSF) and has been found to be either very low or absent in most narcoleptics [3, 12]. This difference may account for genetic transmission differences between the canine model and humans. In the canine model, narcolepsy can be predicted through breeding. In humans this is not the case. Although familial studies indicate a 20–40 times increased risk of narcolepsy in first-degree relatives, only 25–31% of monozygotic twins are concordant for narcolepsy [2, 13, 14].

Since most identical human twins are discordant for narcolepsy, an additional pathophysiological mechanism has been sought to explain the development of narcolepsy. The possibility that an autoimmune mechanism is damaging the hypocretin-producing cells in the hypothalamus has been raised [11, 15]. This is supported by the tight linkage of a subtype of human leukocyte antigen (HLA) DQ1 with humans with narcolepsy with cataplexy. Similar linkages have been found in other autoimmune disorders

such as multiple sclerosis [16, 17]. However, extensive searches for physical evidence of an autoimmune process have been inconclusive or negative [18].

The mechanisms through which hypocretin deficiency results in narcolepsy are unknown. A cholinergic/monoaminergic imbalance underlying the symptomatology of narcolepsy has been established and is most likely caused by an absence of hypocretin signaling (see Figure 20.1). The widespread projections of hypocretin neurons make it difficult to elucidate its exact functional importance [19, 20].

The hypocretin system is consistently involved in the vast majority of patients with primary narcolepsy with cataplexy. There have been reports of secondary narcolepsy, particularly associated with head injuries and neoplasms [21, 22].

DIAGNOSIS

At this time, the diagnostic criteria for narcolepsy in children is the same as in adults [1]. Within the context of the appropriate clinical history, an overnight polysomnogram with a short REM latency followed by a multiple sleep latency test (MSLT) with a maximum average sleep latency of 8 minutes and two or more sleep-onset REM periods (SOREMPs) has been considered diagnostic for narcolepsy. A urine toxicology screen is typically performed on the morning of the MSLT. If an experienced clinician observes unequivocal cataplexy, the overnight polysomnogram and MSLT are not required to establish the diagnosis. During an unequivocal bout of cataplexy the deep tendon reflexes would be expected to be absent and not brisk when elicited with a reflex hammer.

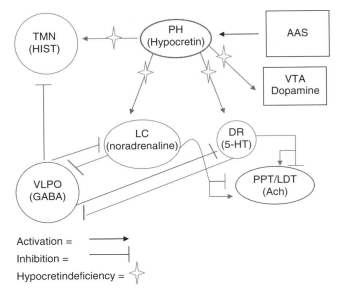

Figure 20.1 Activity in the hypothalamic branch of the ascending arousal system and sleep regulation. The ascending arousal system (AAS) sends projections from the brainstem and posterior hypothalamus (PH) throughout the forebrain. Neurons of the laterodorsal tegmental nucleus and pedunculopontine tegmental nucleus (LDT and PPT) send cholinergic fibers (Ach) to many forebrain targets, including the thalamus, which then regulate cortical activity. Aminergic nuclei diffusely project throughout much of the forebrain, regulating the activity of cortical and hypothalamic targets directly. Neurons of the tuberomammillary nucleus (TMN) contain histamine (HIST), neurons of the raphé nuclei contain 5-HT (serotoninergic neurons), neurons of the locus caeruleus (LC) contain noradrenaline (NA), and neurons of the ventral tegmental area (VTA) contain dopamine. Sleep-promoting neurons of the ventrolateral preoptic nucleus (VLPO) contain GABA. This figure represents the hypocretin deficiency, and increased GABAergic system and cholinergic connections in PPT and LDT may explain several symptoms in narcolepsy syndrome. Activation, ⟶; inhibition, ⊣; hypocretin deficiency, ✧.

The established diagnostic criteria may not always be applicable to children since the MSLT was not validated for children younger than 8 years. If a person is taking medication that suppresses REM sleep, such as most antidepressants, the MSLT results may not be reliable. If a patient abruptly stops such antidepressants a phenomenon of REM rebound may occur giving a false positive study for narcolepsy. A similar situation can occur if the MSLT is started at a time in the morning that a person usually sleeps; this scenario may occur in the context of an adolescent with a delayed sleep phase pattern. A false positive MSLT for narcolepsy can occur if the subject is jet lagged, sleep deprived, or has subtle sleep disordered breathing.

As our understanding of the pathophysiology of narcolepsy improves, modifications to the diagnostic criteria

would be expected. Direct assays of hypocretin levels are more specific for the diagnosis of narcolepsy. Hypocretin can be measured in cerebrospinal fluid. The absence of hypocretin in cerebrospinal fluid is diagnostic of narcolepsy [3, 23]. CSF hypocretin level measurements may be particularly useful in complex or ambiguous clinical situations. However, the presence of normal CSF levels of hypocretin does not exclude the possibility of narcolepsy. A reliable blood serum equivalent test is not yet available. Its future development would be expected to change dramatically the diagnostic approach for narcolepsy. This would potentially decrease the delays between the onset of symptoms and conclusive diagnosis, which currently characterize the diagnosis.

TREATMENT

Successful treatment for narcolepsy includes both behavioral and pharmacological treatments [24]. The situation is analogous to juvenile diabetes mellitus, where a combination of diet with medication can control the condition. Behavioral treatment begins with the patient having as thorough an understanding of the condition as possible. Having the patient and family members meet other people with narcolepsy may be helpful. Volunteer support organizations such as the Narcolepsy Network have been established. Patients need to understand that developing healthy sleep habits are important for the rest of their lives. Two brief naps a day of about 20 minutes each should be strongly encouraged. For school-aged children, the healthcare provider should serve as an advocate on behalf of the patient to encourage the school to allow the child to nap in a safe and comfortable environment. Failure to consider or properly apply nondrug treatments as part of the comprehensive management may lead to unsatisfactory results for the patient and the family. These factors can result in patients with narcolepsy who are not properly managed due to either underdosing or overdosing of medication or incorrect medication selection.

Drug therapy must take into account possible side effects, with the fact kept in mind that narcolepsy is a life-long illness and patients will have to receive medication for years. Tolerance or addiction may be seen with some compounds. Treatment of narcolepsy thus balances avoidance of side effects, including tolerance, with maintenance of an active life. There are no double-blind placebo-controlled trials of medication specifically for children with narcolepsy. The medications that are commonly prescribed are not specifically FDA approved for narcolepsy in children.

The drugs most widely used to treat excessive daytime sleepiness are the central nervous system stimulants. However, a number of side effects including irritability, anxiety, nervousness, headache, psychosis, tachycardia, hypertension, nocturnal sleep disturbances, tolerance, and drug dependence may arise.

There are two drugs with different modes of action that have been our first-line treatment for narcolepsy. The first one, modafinil, is considered more as a "somnolytic" than a nonspecific stimulant [25, 26]. The initial dose should be relatively low to avoid headaches. Headaches are the most common side effect of modafinil and may be avoided or minimized by gradually increasing the dose. Modafinil should be considered the initial pharmacological agent used to treat the excessive daytime sleepiness of narcolepsy in children.

The second drug is gamma-hydroxybutyrate, GHB, which is the first substance ever approved specifically for cataplexy [27]. The drug generic name is sodium oxybate. In the popular media it has the infamous name of the "date rape drug." Illegal use of this substance for recreational purposes has been of great concern. Important CNS adverse events associated with abuse of GHB include seizure, respiratory depression, and profound decreases in level of consciousness, with instances of coma and death. Sodium oxybate has powerful CNS depressant effects. GHB can increase slow-wave sleep. This medication when given at bedtime maybe of value to reduce cataplexy. Unlike most other medications discussed, sodium oxybate is available as a liquid. Patients may prefer this medication over other medications used for cataplexy, particularly if insomnia is also present. Dosing guidelines for patients younger than 16 years old are not established.

Cataplexy seems to respond best to medications with noradrenergic reuptake blocking properties. There are no systematic trials of anticataplexy drugs on children. Postpubertal teenagers are usually treated as young adults. In this group, two medications have been more commonly used—clomipramine and fluoxetine. Both of these drugs have active noradrenergic reuptake blocking metabolites (desmethylclomipramine and norfluoxetine). It is through these metabolites that the therapeutic effect may be mediated. New medications with noradrenergic reuptake blocking are promising.

Given the putative autoimmune pathophysiology for narcolepsy, the possibility of using immunosuppressive therapy has been considered [15, 28]. Immunosuppressive treatment could, in theory, minimize or reverse the development of narcolepsy if initiated very early in the disease process. This potential therapy underscores the importance that healthcare providers have a high index of suspicion for narcolepsy when a patient presents with excessive daytime sleepiness.

With the recent discovery of a neuropeptide responsible for narcolepsy, novel potential therapeutic approaches may be discovered. Hypocretin analogs may be potentially useful to treat narcolepsy [29]. A further theoretic treatment possibility would be transplantation of hypocretin-producing cells.

REFERENCES

1. Thorpy M. *International Classification of Sleep Disorders: Diagnostic and Coding Manual.* American Sleep Disorders Association, Rochester, MN, 1990.

2. Melberg A, Ripley B, Lin L, Hetta J, Mignot E, Nishino S. Hypocretin deficiency in familial symptomatic narcolepsy. *Ann Neurol* **49**(1):136–137(2001).

3. Mignot E, Chen W, Black J. On the value of measuring CSF hypocretin-1 in diagnosing narcolepsy. *Sleep* **26**(6):646–649(2003).

4. Guilleminault C, Pelayo R. Narcolepsy in prepubertal children. *Ann Neurol* **43**(1):135–142(1998).

5. Yoss RE, Daly DD. On the treatment of narcolepsy. *Med Clin North Am* **52**(4):781–787(1968).

6. Rechtschaffen A, Dement W. Studies on the relation of narcolepsy, cataplexy, and sleep with low voltage random EEG activity. *Res Publ Assoc Res Nerv Ment Dis* **45**:488–505(1967).

7. Dahl RE, Holttum J, Trubnick L. A clinical picture of child and adolescent narcolepsy. *J Am Acad Child Adolesc Psychiatry* **33**(6):834–841(1994).

8. Guilleminault C, Pelayo R. Narcolepsy in children: a practical guide to its diagnosis, treatment and follow-up. *Paediatr Drugs* **2**(1):1–9(2000).

9. Silber MH, Krahn LE, Olson EJ, Pankratz VS. The epidemiology of narcolepsy in Olmsted County, Minnesota: a population-based study. *Sleep* **25**(2):197–202(2002).

10. Lin L, Faraco J, Li R, et al. The sleep disorder canine narcolepsy is caused by a mutation in the hypocretin (orexin) receptor 2 gene. *Cell* **98**(3):365–376(1999).

11. Taheri S, Zeitzer JM, Mignot E. The role of hypocretins (orexins) in sleep regulation and narcolepsy. *Annu Rev Neurosci* **25**:283–313(2002).

12. Ripley B, Overeem S, Fujiki N, et al. CSF hypocretin/orexin levels in narcolepsy and other neurological conditions. *Neurology* **57**(12):2253–2258(2001).

13. Khatami R, Maret S, Werth E, et al. Monozygotic twins concordant for narcolepsy–cataplexy without any detectable abnormality in the hypocretin (orexin) pathway. *Lancet* **363**(9416):1199–1200(2004).

14. Dauvilliers Y, Maret S, Bassetti C, et al. A monozygotic twin pair discordant for narcolepsy and CSF hypocretin-1. *Neurology* **62**(11):2137–2138(2004).

15. Hecht M, Lin L, Kushida CA, et al. Report of a case of immunosuppression with prednisone in an 8-year-old boy with an acute onset of hypocretin-deficiency narcolepsy. *Sleep* **26**(7):809–810(2003).

16. Krahn LE, Pankratz VS, Oliver L, Boeve BF, Silber MH. Hypocretin (orexin) levels in cerebrospinal fluid of patients with narcolepsy: relationship to cataplexy and HLA DQB1*0602 status. *Sleep* **25**(7):733–736(2002).

17. Ebrahim IO, Sharief MK, De Lacy S, et al. Hypocretin (orexin) deficiency in narcolepsy and primary hypersomnia. *J Neurol Neurosurg Psychiatry* **74**(1):127–130(2003).

18. Overeem S, Steens SC, Good CD, et al. Voxel-based morphometry in hypocretin-deficient narcolepsy. *Sleep* **26**(1):44–46(2003).

19. Fujiki N, Morris L, Mignot E, Nishino S. Analysis of onset location, laterality and propagation of cataplexy in canine narcolepsy. *Psychiatry Clin Neurosci* **56**(3):275–276(2002).

20. Kilduff TS, Peyron C. The hypocretin/orexin ligand–receptor system: implications for sleep and sleep disorders. *Trends Neurosci* **23**(8):359–365(2000).

21. Rosen GM, Bendel AE, Neglia JP, Moertel CL, Mahowald M. Sleep in children with neoplasms of the central nervous system: case review of 14 children. *Pediatrics* **112**(1 Pt 1):46–54(2003).

22. Marcus CL, Trescher WH, Halbower AC, Lutz J. Secondary narcolepsy in children with brain tumors. *Sleep* **25**(4):435–439(2002).

23. Mignot E, Lammers GJ, Ripley B, et al. The role of cerebrospinal fluid hypocretin measurement in the diagnosis of narcolepsy and other hypersomnias. *Arch Neurol* **59**(10):1553–1562(2002).

24. Pelayo R, Chen W, Monzon S, Guilleminault C. Pediatric sleep pharmacology: you want to give my kid sleeping pills? *Pediatr Clin North Am* **51**(1):117–134(2004).

25. Littner M, Johnson SF, McCall WV, et al. Practice parameters for the treatment of narcolepsy: an update for 2000. *Sleep* **24**(4):451–466(2001).

26. Tafti M, Dauvilliers Y. Pharmacogenomics in the treatment of narcolepsy. *Pharmacogenomics* **4**(1):23–33(2003).

27. Borgen LA, Cook HN, Hornfeldt CS, *Fuller* DE. Sodium oxybate (GHB) for treatment of cataplexy. *Pharmacotherapy* **22**(6):798–799(2002).

28. Boehmer LN, Wu MF, John J, Siegel JM. Treatment with immunosuppressive and anti-inflammatory agents delays onset of canine genetic narcolepsy and reduces symptom severity. *Exp Neurol* **188**(2):292–299(2004).

29. Schatzberg SJ, Cutter-Schatzberg K, Nydam D, et al. The effect of hypocretin replacement therapy in a 3-year-old Weimaraner with narcolepsy. *J Vet Intern Med* **18**(4):586–588(2004).

ADDITIONAL READING

Mignot E. An update on the pharmacotherapy of excessive daytime sleepiness and cataplexy. *Sleep Med Rev* **8**(5):333–338(2004). This article reviews recent management of narcolepsy including the new agent atomoxetine for cataplexy.

21

IDIOPATHIC HYPERSOMNIA

STEPHEN N. BROOKS

Stanford Sleep Disorders Clinic, Stanford, California

INTRODUCTION

The characterization of primary disorders of excessive daytime somnolence (EDS) is a work in progress, dating back to the late nineteenth century. As understanding of sleep and wakefulness has evolved, definitions of recognized disorders have been refined and new disorders have been described. Narcolepsy with cataplexy is currently the best characterized disorder of the group. The developing story of the hypocretins has led to exciting new insights into the pathogenesis of narcolepsy with cataplexy. Yet even here the picture is incomplete. Idiopathic hypersomnia (which has also been known as idiopathic CNS hypersomnia, primary hypersomnia, functional hypersomnia, and non-rapid eye movement narcolepsy) is a relatively new addition to the nosology of disorders of EDS. Generally, the term has been used to categorize individuals with prominent EDS, but who lack the classic features of narcolepsy or evidence of another disorder known to cause EDS (such as sleep apnea). The imprecise definition of idiopathic hypersomnia (IH) has led to diagnostic difficulties. Without doubt, many patients have been diagnosed with idiopathic hypersomnia, when, in fact, they suffered from other disorders, such as narcolepsy without cataplexy (or prior to the onset of cataplexy), delayed sleep phase syndrome, or upper airway resistance syndrome [1]. Increased understanding of sleep disorders along with improved investigational methods have reduced but not eliminated diagnostic imprecision. Absence of specific biologic markers or animal models of IH adds to the challenge. Nonetheless, the classification of sleep disorders will continue to evolve, and the understanding of IH as an entity (or a group of entities) will deepen in kind.

HISTORY OF IDIOPATHIC HYPERSOMNIA

In the 1960s a number of researchers attempted to distinguish patients with narcolepsy from others who presented with significant EDS but without other features of the narcoleptic syndrome, such as cataplexy or sleep onset REM periods (SOREMPs). They proposed terms such as "essential narcolepsy," "non-REM narcolepsy," and "hypersomnia." In 1972, Roth, Nevsimalova and Rechtschaffen [2] described the condition of "hypersomnia with sleep drunkenness." The term "sleep drunkenness" was used to describe a condition of marked difficulty with coming to full alertness upon awakening from sleep, including symptoms of confusion, disorientation, and poor motor coordination. A few years later, Roth [3] described monosymptomatic (EDS) and polysymptomatic (EDS, prolonged nocturnal sleep time, sleep drunkenness) forms of idiopathic hypersomnia. The Diagnostic Classification of Sleep and Arousal Disorders [4] referred to IH as "idiopathic central nervous system (CNS) hypersomnolence"; this term was dropped in the subsequent edition [5].

EPIDEMIOLOGY OF IDIOPATHIC HYPERSOMNIA

Idiopathic hypersomnia is believed to be less common than narcolepsy, but estimation of prevalence is obviously elusive, because strict diagnostic criteria are lacking, and no specific biological marker has been identified. Most estimates derive from calculating the ratio of IH to narcolepsy in sleep clinic populations. Bassetti and Aldrich [6] estimate that, in Caucasians, IH occurs in 2–5/100,000 and is even less common in African Americans. IH occurs equally in males and females.

As denoted, the etiology of the disorder is unknown, but viral illnesses, including Guillain–Barré syndrome, hepatitis, mononucleosis, and atypical viral pneumonia, may herald the onset of sleepiness in a subset of patients. EDS may occur as part of the acute illness, but it persists after the other symptoms subside. Mignot et al. [7] found a significant increase in HLA-DQß1*0602 in subjects with IH (52% versus 17% in controls). Familial cases are known to occur, with increased frequency of HLA-Cw2 and HLA-DR11 [8]. Most cases of IH are associated with neither a family history nor an obvious viral illness.

DEFINITION AND CLINICAL FEATURES OF IDIOPATHIC HYPERSOMNIA

The *International Classification of Sleep Disorders* (*ICSD*), *Revised: Diagnostic and Coding Manual* Second Edition [9] lists the following diagnostic criteria for IH:

Idiopathic Hypersomnia with Long Sleep Time

A. The patient has a complaint of excessive daytime sleepiness occuring almost daily for at least three months.

B. The patient has prolonged nocturnal sleep time (more than 10 hours) documented by interview, actigraphy or sleep logs. Waking up in the morning or at the end of naps is always laborious.

C. Nocturnal polysomnography has excluded other causes of daytime sleepiness.

D. The polysomnogram demonstrates a short sleep latency and a major sleep period that is prolonged to more than 10 hours in duration.

E. If an MSLT is performed following overnight polysomnography, a mean sleep latency of less than eight minutes is found and fewer than two SOPEMPs are recorded. Mean sleep latency in idiopathic hypersomnia with long sleep time has been shown to be 6.2 ± 3.0 minutes.

F. The hypersomnia is not better explained by another sleep disorder, medical or neurological disorder, mental disorder, medication use, or substance use disorder.

Idiopathic Hypersomnia without Long Sleep Time

A. The patient has a complaint of excessive daytime sleepiness occuring almost daily for at least three months.

B. The patient has normal nocturnal sleep time (greater than six hours but less than 10 hours) documented by interview, actigraphy or sleep logs.

C. Nocturnal polysomnography has excluded other causes of daytime sleepiness.

D. Polysomnography demonstrates a major sleep period that is normal in duration (greater than six hours but less than 10 hours).

E. An MSLT following overnight polysomnography demonstrates a mean sleep latency of less than eight minutes and fewer than two SOPEMPs. Mean sleep latency in idiopathic hypersomnia has been shown to be 6.2 ± 3.0 minutes.

F. The hypersomnia is not better explained by another sleep disorder, medical or neurological disorder, mental disorder, medication use, or substance use disorder.

In what might be termed "classic" IH (Roth's "polysymptomatic" form), patients experience the insidious onset of EDS over a period of weeks or months, beginning in adolescence or early adulthood. In some cases, the onset of hypersomnia follows a period of insomnia. Daytime sleepiness is prominent and continuous but not irresistible (in contrast to the "sleep attacks" that may occur in narcolepsy). Naps are usually of more than an hour in duration and are nonrefreshing (again in contrast to narcolepsy). "Microsleeps," with or without automatic behavior, may occur throughout the day. Nocturnal sleep is prolonged and uninterrupted. No amount of sleep ameliorates the daytime sleepiness.

There is difficulty with awakening from sleep, which may assume the proportions of "sleep drunkenness." Psychiatric symptoms, especially depression, are prevalent in patients suffering from IH [2, 3]. It remains to be determined whether these symptoms constitute essential features of IH or represent responses to the chronic, debilitating disorder.

Bassetti and Aldrich [6] found that hypnagogic hallucinations and sleep paralysis occurred in approximately 40% of their series of 42 patients with IH, a proportion similar to what has been found in narcolepsy without cataplexy. Some patients with IH experience associated symptoms suggesting autonomic nervous system dysfunction, including orthostatic hypotension, syncope, vascular type headaches, and Raynaud-like phenomena.

Patients with IH may differ in one or more ways from the "classic" form. There is a subgroup of patients with IH who only complain of EDS (monosymptomatic IH). Bassetti and Aldrich [6] found that restless sleep with frequent awakenings occurred in nearly half their patients. They also reported that, in a few of their patients, daytime sleepiness assumed a variable form with irresistible sleep episodes similar to the narcoleptic pattern. Lesperance et al. [10] have reported patients who presented with a complaint of EDS, high scores on the Epworth sleepiness scale [11], and no complaints of nocturnal sleep disturbance, but who exhibited reduced sleep efficiency on nocturnal polysomnography and reduced mean sleep latency on MSLT. Interestingly, administration of melatonin at night improved EDS without improving nocturnal sleep.

The symptoms of IH are severe enough to cause affected individuals to encounter significant problems with work, school, and social functioning. Almost all patients encounter difficulties with driving because of sleepiness. Some patients note increased hypersomnia with excessive physical activity, heavy meals, alcohol, psychological stress, or menses. Many, however, report no obvious aggravating factors.

Although IH has been regarded as a lifelong disorder with stable symptoms after the progressive onset, spontaneous improvement or even complete resolution of EDS has occurred in some patients [6, 12, 13].

Such variations in clinical course, along with the diversity of symptoms and test results among individual patients, support the notion that IH represents a heterogeneous group of disorders with overlapping features rather than a single entity [6, 14].

POLYSOMNOGRAPHIC AND NEUROPHYSIOLOGIC STUDIES

Nocturnal polysomnographic studies of patients with IH usually reveal shortened initial sleep latency, normal or increased total sleep time, and normal sleep architecture (in contrast to narcoleptic patients, who exhibit significant sleep fragmentation). Ideally, prolonged polysomnographic monitoring across 24 hour period(s) would be more useful in characterizing the sleep–wake patterns in these patients. This is not practical in clinical practice, and even in research settings, methodological and standardization issues need to be addressed. Mean sleep latency on MSLT is usually reduced, often in the 8–10 min range, but SOREMPs are not typically seen. Using spectral analysis, Sforza et al. [15] found reduced sleep pressure, as evidenced by decreased slow-wave activity during the first two NREM episodes of nocturnal sleep in patients with IH. The investigators suggest that, in some patients with IH, EDS may be related to nonrestorative nocturnal sleep. Vankova et al. [16] demonstrated increased REM density in patients with the polysymptomatic form of idiopathic hypersomnia (as well as in patients with narcolepsy–cataplexy) compared to controls. Bove et al. [17] studied sleep spindle density in patients with narcolepsy, IH, and controls. They found that average sleep spindle density (number of spindles per minute of stage 2 sleep) was higher in patients with narcolepsy and IH and highest in IH. The highest sleep spindle density occurred during the first part of the night in patients with narcolepsy and during the second half of the night in patients with IH. This suggests that

excessive thalamic blockade may occur at the end of the night in patients with IH, perhaps relating to the difficulty with awakening that these patients experience. A study measuring evoked potentials found that subjects with idiopathic hypersomnia or severe obstructive sleep apnea had prolonged visual P300 latency compared to normals or subjects with narcolepsy; subjects with idiopathic hypersomnia or obstructive sleep apnea had longer auditory P300 latency than normals; subjects with idiopathic hypersomnia had reduced auditory P300 amplitude compared to subjects with narcolepsy [18]. Such studies may lead to identification of a specific neurophysiological marker for IH.

PATHOPHYSIOLOGY

Little is known about the pathophysiology of idiopathic hypersomnia. No animal model is available for study, and no specific biologic markers have been identified. Neurochemical studies using CSF have suggested that individuals with idiopathic hypersomnia may have a derangement in the noradrenergic system [19–21].

Measurements of CSF hypocretin (Hcrt) levels in patients with IH have yielded variable results. Dauvilliers et al. [22] reported CSF levels of Hcrt-1 in seven subjects with IH: six were normal and one was elevated. Bassetti et al. [23] reported normal CSF levels of Hcrt-1 in five subjects with IH. Ebrahim et al. [24] found detectable but decreased CSF levels of Hcrt-1 and Hcrt-2 in six subjects with IH. In Kanbayashi's series [25], three of twelve subjects with IH had intermediate CSF levels of Hcrt-1 (levels were normal in the other nine). Mignot et al. [7] measured CSF levels of Hcrt-1 in 29 subjects with IH and found normal values in 28 and an intermediate value in one. As yet, no meaningful conclusions can be drawn from these findings.

Nevsimalova et al. [26] studied ten subjects with polysymptomatic IH and found a phase delay in the circadian rhythm of melatonin compared to controls. There was a shift in elevated nocturnal melatonin levels to morning hours as well as a (nonsignificant) prolongation of the melatonin signal in the IH subjects, perhaps contributing to difficulty with awakening.

DIFFERENTIAL DIAGNOSIS

Other disorders producing EDS (such as sleep disordered breathing, periodic limb movement disorder, psychiatric illness, insufficient sleep syndrome, or circadian rhythm disorders) must be ruled out before the diagnosis of IH is made. Of course, EDS may have multiple causes in an individual patient, and IH may still be a consideration in cases where EDS persists despite what appears to be adequate treatment of other identified disorders.

Patients with upper airway resistance syndrome [1] may present with EDS. Snoring may or may not occur with this disorder. Anatomic features of upper airway crowding and multiple cortical arousals during nocturnal polysomnography are important clues to the diagnosis. Monitoring of esophageal pressure during polysomnography, demonstrating increased respiratory effort leading to cortical arousals, provides definitive diagnosis. If this technique is not available, the use of nasal cannulae may be helpful to demonstrate the presence of flow limitation associated with cortical arousals.

Narcolepsy without cataplexy (or prior to the onset of cataplexy) may present a difficult diagnostic challenge. Several clinical features are useful in distinguishing narcolepsy without cataplexy from IH. Patients suffering from narcolepsy without cataplexy tend to have disrupted nocturnal sleep with multiple awakenings. Daytime sleepiness tends to be more compelling, and naps tend to be briefer and more refreshing than in IH. The presence of SOREMPs during the MSLT in patients with narcolepsy without cataplexy is very helpful in making the distinction. Total sleep time across the 24 h period is generally normal in narcolepsy, but this may also be the case in some patients with IH. Since symptoms of sleep paralysis and/or hypnagogic hallucinations may occur in either disorder, their presence is not a distinguishing feature. HLA typing is not helpful. The association with HLA-DQß1*0602 is less robust in narcoleptic patients without cataplexy than in those with cataplexy. This allele may also be found in a significant number of patients with IH [7] and in up to 25% of the general population. The measurement of CSF Hcrt-1 is useful if the peptide is decreased or undetectable, but this occurs in only a minority (10–20%) of patients with narcolepsy without cataplexy.

Patients with psychiatric disorders (particularly depression) often complain of hypersomnia. In such cases, the severity of daytime sleepiness tends to vary more from one day to the next than in patients with IH. Nocturnal sleep is often poor as well in patients with psychiatric disorders. In a study comparing subjects with mood disorders to those with IH, Billiard et al. [27] found that mean sleep latency on MSLT was longer and total nocturnal sleep time was shorter in subjects with mood disorders compared to those with IH. Vgontzas et al. [28] examined nocturnal and daytime sleep in IH, psychiatric hypersomnia, and controls using polysomnography. During the nocturnal studies, subjects with psychiatric hypersomnia showed longer sleep latency, longer wake time after sleep onset, and shorter total sleep time than the other groups. During daytime sleep, there were similar findings in subjects with psychiatric hypersomnia, as well as a higher percentage of stage 1 sleep and a lower percentage of stage 2 sleep. The investigators concluded that psychiatric hypersomnia represents a disorder of hyperarousal and that IH represents a disorder of hypoarousal.

Delayed sleep phase syndrome (DSPS) may be mistaken for IH because of similar age of onset and some shared features. Patients with DSPS may be profoundly sleepy during the day and difficult to arouse from sleep. In pure form, however, there should also be times during the 24 h period when alertness is normal. Total sleep time, sleep latency, sleep architecture, and ease of awakening should be normal in DSPS, if the affected individual is allowed to follow the dictates of his/her circadian clock. Things become less clear, however, when (as most often occurs) school or work schedules mandate variance from the individual's ideal circadian rhythm. Further confounding the picture is the very common comorbidity of insufficient sleep syndrome during adolescence and early adulthood, leading to marked variations in sleep time (especially during weekends), the need for naps, and the possibility that the individual will experience some degree of sleepiness throughout the day. Nocturnal polysomnography and MSLT may actually be misleading if performed at inappropriate (i.e., phase advanced) times. In such situations, prolonged sleep latency may occur on the nocturnal polysomnogram, and MSLT may demonstrate decreased mean sleep latency and even SOREMPs. Sleep diaries (with actigraphy in difficult cases) may be quite helpful in demonstrating changes in sleep patterns and daytime symptoms during weekends or vacation periods. Measurement of dim light melatonin onset is useful in defining the individual's underlying circadian rhythm.

So-called "long sleepers" are individuals who require more sleep (10 or more hours) per 24 h period than what is typically considered normal, but whose sleep is otherwise normal in architecture and physiology [9]. They may experience EDS if not able to obtain their biologic quota of sleep time. Basically, this amounts to insufficient sleep syndrome at one end of the spectrum of sleep need. In fact, Billiard et al. [29] have suggested that the symptoms of IH are a consequence of chronic sleep deprivation in very long sleepers. Questioning the patient about sleep schedules and daytime symptoms during periods during which they were able to sleep as much as desired may yield useful information. Adjustment of schedule to allow increased total sleep time should be adequate to reduce EDS and clarify the diagnosis.

TREATMENT

Treatment of IH is often less than satisfactory. Life-style and behavioral modifications, including good sleep hygiene, are appropriate, but treatment with stimulant medication or modafinil [30] is usually necessary.

As mentioned earlier [10], the administration of melatonin at night has been useful in treating some patients with features of IH. Montplaisir and Fantini [31] gave melatonin at bedtime to ten patients with the polysymptomatic form of IH, and in five of these patients the symptoms improved. These findings suggest an intriguing possible addition to the limited treatment options for IH.

SUMMARY

The definition of IH has evolved over time and will continue to do so as understanding of the neurological underpinnings of sleep and wakefulness deepens. It is believed to be less common than narcolepsy (perhaps by a factor of 10). The differential diagnosis of IH can be challenging even to experienced clinicians. Currently, no specific marker is available to confirm the diagnosis. Clinically, the category of idiopathic hypersomnia is heterogeneous, including individuals with EDS but with or without one or more of the other features of Roth's polysymptomatic form [14]. Whether IH represents a collection of distinct disorders with overlapping features but different neurological mechanisms or a single pathophysiological entity with variable manifestations among affected individuals remains to be determined. Even so, more refined characterization of subgroups of patients will likely broaden treatment options.

REFERENCES

1. Guilleminault C, Stoohs R, Clerk A, Cetel M, Maistros P. A cause of excessive daytime sleepiness. The upper airway resistance syndrome. *Chest* **104**:781–787(1993).

2. Roth B, Nevsimalova S, Rechtschaffen A. Hypersomnia with "sleep drunkenness." *Arch Gen Psychiatry* **26**:456–462(1972).

3. Roth B. Narcolepsy and hypersomnia. *Arch Neurol Psychiatry* **119**:31–41(1976).

4. Diagnostic Classification of Sleep and Arousal Disorders. *Sleep* **2**:1–137(1979).

5. *International Classification of Sleep Disorders: Diagnostic and Coding Manual.* American Sleep Disorders Association, Rochester, Mn, 1990.

6. Bassetti C, Aldrich MS. Idiopathic hypersomnia. A series of 42 patients. *Brain* **120**:1423–1435(1997).

7. Mignot E, Lammers GJ, Ripley B, Okun M, Nevsimalova S, Overeem S, Vankova J, Black J, Harsh J, Bassetti C, Schrader H, Nishino S. The role of cerebrospinal fluid hypocretin measurements in the diagnosis of narcolepsy and other hypersomnias. *Arch Neurol* **59**:1553–1562(2002).

8. Montplaisir J, Poirier G. HLA in disorders of excessive sleepiness without cataplexy in Canada. In: Honda Y, Juji T (Eds), *HLA in Narcolepsy.* Springer-Verlag, Berlin, 1988, p 186.

9. *International Classification of Sleep Disorders, Revised: Diagnostic and Coding Manual.* Second Edition American Academy of Sleep Medicine. Westchester, IL, 2005.

10. Lesperance P, Lapierre O, Gosselin A, Montplaisir J. Effect of exogenous melatonin on excessive daytime somnolence associated with disturbed sleep: an open clinical trial. *Sleep Res* **26**:111(1997).

11. Johns M. A new method for measuring daytime sleepiness: the Epworth sleepiness scale. *Sleep* **14**(6):540–545(1991).

12. Bruck D, Parkes JD. A comparison of idiopathic hypersomnia and narcolepsy–cataplexy using self-report measures and sleep diary data. *J Neurol Neurosurg Psychiatry* **60**:576–578(1996).

13. Billiard M, Dauvillies Y. Idiopathic hypersomnia. *Sleep Med Rev* **5**(5):351–360(2001).

14. Aldrich MS. The clinical spectrum of narcolepsy and idiopathic hypersomnia. *Neurology* **46**:393–401(1996).

15. Sforza E, Gaudreau H, Petit D, Montplaisir J. Homeostatic sleep regulation in patients with idiopathic hypersomnia. *Clin Neurophysiol* **111**(2):277–282(2000).

16. Vankova J, Nevsimalova S, Sonka K, Spackova N, Svejdova-Blaejova K. Increased REM density in narcolepsy-cataplexy and the polysymptomatic form of idiopathic hypersomnia. *Sleep* **24**(6):707–711(2001).

17. Bove A, Culebras A, Moore JT, Westlake RE. Relationship between sleep spindles and hypersomnia. *Sleep* **17**:449–455(1994).

18. Sangal RB, Sangal JM. P300 latency: abnormal in sleep apnea with somnolence and idiopathic hypersomnia, but normal in narcolepsy. *Clin Electroencephalogr* **26**:146–153(1995).

19. Montplaisir J, De Champlain J, Young SN, Missala K, Sourkes TL, Walsh J, Remilliard G. Narcolepsy and idiopathic hypersomnia: biogenic amines and related compounds in CSF. *Neurology* **32**:1299–1302(1982).

20. Faull KF, Guilleminault C, Berger PA, Barchas JD. Cerebrospinal fluid monoamine metabolites in narcolepsy and hypersomnia. *Ann Neurol* **13**:258–263(1983).

21. Faull KF, Thiemann S, King RJ, Guilleminault C. Monoamine interactions in narcolepsy and hypersomnia: a preliminary report. *Sleep* **9**:246–249(1986).

22. Dauvilliers Y, Baumann CR, Carlander B, Bischof M, Blatter T, Lecendreux M, Maly F, Besset A, Touchon J, Billiard M, Tafti M, Bassetti CL. CSF hypocretin-1 levels in narcolepsy, Kleine–Levin syndrome, and other hypersomnias and neurological conditions. *J Neurol Neurosurg Psychiatry* **74**:1667–1673(2003).

23. Bassetti C, Gugger M, Bischof M, Mathis J, Sturzenegger C, Werth E, Radanov B, Ripley B, Nishino S, Mignot E. The narcoleptic borderland: a multimodal diagnostic approach including cerebrospinal fluid levels of hypocretin-1 (orexin A). *Sleep Med* **4**(1):3–4(2003).

24. Ebrahim IO, Sharief MK, de Lacy S, Semra YK, Howard RS, Kopelman MD, Williams AJ. Hypocretin(orexin) deficiency in narcolepsy and primary hypersomnia. *J Neurol Neurosurg Psychiatry* **74**:127–130(2003).

25. Kanbayashi T, Inoue Y, Chiba S, Aizawa R, Saito Y, Tsukamoto H, Fujii Y, Nishino S, Shimizu T. CSF hypocretin-1(orexin-A) concentrations in narcolepsy with and without cataplexy and idiopathic hypersomnia. *J Sleep Res* **11**:91–93(2002).

26. Nevsimalova S, Blazejova K, Illnerova H, Hajek I, Vancova J, Pretl M, Sonka K. A contribution to pathophysiology of idiopathic hypersomnia. *Suppl Clin Neurophysiol* **53**:366–370(2000).

27. Billiard M, Dolenc L, Aldaz C, Ondze B, Besset A. Hypersomnia associated with mood disorders: a new perspective. *J Psychosom Res* **38**(Suppl 1):41–47(1994).

28. Vgontzas AN, Bixler EO, Kales A, Criley C, Vela-Bueno A. Differences in nocturnal and daytime sleep between primary and psychiatric hypersomnia: diagnostic and treatment implications. *Psychosom Med* **62**:220–226(2000).

29. Billiard M, Rondouin G, Espa F, Dauvilliers Y, Besset A. Physiopathology of idiopathic hypersomnia. Current studies and new orientations. *Rev Neurol* **157**(11 Pt 2):S101–S106(2001).

30. Bastuji H, Jouvet M. Successful treatment of idiopathic hypersomnia and narcolepsy with modafinil. *Prog Neuropsychopharmacol Biol Psychiatry* **12**(5):695–700(1988).

31. Montplaisir J, Fantini L. Idiopathic hypersomnia: a diagnostic dilemma. A commentary on "Idiopathic hypersomnia"(M. Billiard and Y. Dauvilliers). *Sleep Med Rev* **5**(5):361–362(2001).

ADDITIONAL READING

Billiard M. Idiopathic hypersomnia. *Neurol Clin* **14**(3):573–582(1996).

Billiard M, Merle C, Carlander B, Ondze B, Alvarez D, Besset A. Idiopathic hypersomnia. *Psychiatry Clin Neurosci* **52**(2):125–129(1998).

Guilleminault C, Mondini S. Mononucleosis and chronic daytime sleepiness: a long term follow-up study. *Arch Intern Med* **146**:1333–1335(1986).

22

POST-TRAUMATIC AND RECURRENT HYPERSOMNIA

CAROLYN M. D'AMBROSIO AND JOSHUA BARON
Tufts–New England Medical Center, Boston, Massachusetts

POST-TRAUMATIC HYPERSOMNIA AND RELATED DISTURBANCES OF SLEEP

Head trauma is a major cause of morbidity and mortality, associated with a yearly incidence of two million per year in the United States. The sequelae of head trauma span the neuropsychological spectrum, ranging from headaches and poor concentration to significant disability and coma. Commonly, patients who have suffered even minor head trauma complain of sleep disturbances, including hypersomnia and insomnia. While sleep complaints are nonspecific, several studies have shown an association between traumatic brain injury (TBI) and obstructive sleep apnea, the upper airway resistance syndrome, and narcolepsy. Post-traumatic hypersomnia has also been described. This is defined as subjective sleepiness, objective proof of sleepiness on an MSLT (latency of ≤10 min with fewer than 2 REM-onset naps), and an unremarkable polysomnogram, all occurring after traumatic brain injury.

In a study of 184 adults with a history of head and neck trauma, all participants complained of daytime sleepiness, with 181 patients noting an inability to work because of their sleepiness. Objective data from MSLT studies in these patients revealed significant daytime sleepiness in the majority of these patients, with 28% of patients having an MSLT ≤5 min and 55% of patients having an MSLT over 5 min but less than 10 min. Factors that predicted an MSLT of 5 or fewer minutes included the presence of coma of at least 24 h and skull fracture. Fifty-nine patients (32%), including all patients with whiplash injuries, had sleep-disordered breathing (either OSA or UARS). Five patients (3%) had three or more sleep-onset REM periods

on MSLT in the absence of cataplexy. Four of these patients had HLA typing, and two of the four were positive for the DR15 DQ6 haplotype, implying a propensity to sleep-onset REM. The majority of patients with an MSLT score of ≤10 min had no explanation, however, for their sleepiness and qualified for the diagnosis of post-traumatic hypersomnia [1].

Castriotta and Lai [2] examined ten patients with a history of TBI and excessive daytime sleepiness with polysomnography and MSLT. These patients scored themselves as high on the Epworth Sleepiness Scale, with a mean score of 15. Additionally, the mean daytime sleep latency on MSLT was 6 min. All of these patients had abnormal sleep studies, with OSA found in 60%, upper airway resistance syndrome in 10%, narcolepsy in 20%, and post-traumatic hypersomnia in 10% [2].

Cohen et al. [3] examined two groups of patients who suffered traumatic brain injury with subsequent coma. One group was comprised of 22 patients admitted to a rehabilitation facility, the other a group of 77 discharged patients who were seen in follow-up two to three years after their initial injury. These individuals were administered questionnaires with the goal of eliciting complaints of initiating or maintaining sleep, excessive somnolence, changes in sleep–wake patterns, and parasomnias. Of the hospitalized patients, 73% complained of disturbances of sleep. Of those with sleep disturbances, 82% of patients complained of disorders of initiating and maintaining sleep. Fourteen percent of patients with sleep complaints complained of excessive somnolence. Of patients who had been discharged, 37% complained of excessive somnolence and 3% complained of difficulty initiating and

Sleep: A Comprehensive Handbook, Edited by T. Lee-Chiong.
Copyright © 2006 John Wiley & Sons, Inc.

maintaining sleep, with 52% of patients identifying a sleep disturbance. Neither age nor duration of coma seemed to predict the incidence of sleep complaints [3].

Several reports of delayed sleep phase syndrome after trauma exist in the literature [4–6]. These include adolescents and adults who, after sustaining traumatic brain injury, were unable to fall asleep until the early morning hours. One such study demonstrated a 12 h delay in melatonin peak concentration, which could be phase-advanced to its normal time with the use of 5 mg of endogenous melatonin [6]. Chronotherapy, which is useful for idiopathic delayed sleep phase syndrome, has also been attempted, but the compliance rate for this treatment is poor [4, 5]. While the pathophysiology of post-traumatic delayed sleep phase syndrome has yet to be elucidated, the postulated mechanism is damage to the suprachiasmatic nucleus and its connections [4].

HYPERSOMNIA ASSOCIATED WITH MOOD DISORDERS

Although depression is frequently associated with the development of insomnia, hypersomnia is considered one of the "atypical" symptoms. As such, it is present in only the minority of patients with unipolar depression. It is thought to be significantly more common in patients with bipolar disorder and childhood depression [7, 8].

When a series of 661 patients with unipolar and bipolar depression were questioned, hypersomnia was found to be present in 14% of men, with a significantly greater percentage, 21%, present in women. Additionally, hypersomnia was significantly more common in patients under the age of 30 than in older patients. Despite suggestions that hypersomnia may present more frequently in patients with mild disease, this relationship was not confirmed by this study [9].

Those patients who complain of hypersomnia may not necessarily be hypersomnolent on objective measures. When 36 patients with mood-related complaints of hypersomnia were compared to those patients with idiopathic hypersomnia, depressed patients were found to have a significantly longer MSLT than those with idiopathic disease. Additionally, when patients with mood disorders and complaints of hypersomnia were given the opportunity to sleep freely, they slept for a significantly shorter period of time than counterparts with idiopathic hypersomnia (7.7 h versus 9.9 h, respectively) [10]. A similar study reviewed the MSLT in 25 patients with bipolar disorder who complained of hypersomnia and compared these results to those of 23 patients with narcolepsy. The patients with bipolar disorder had significantly longer sleep latencies than narcoleptic patients. Likewise, the patients with bipolar disorder had mean sleep latencies that exceeded

10 min for each nap [11]. These findings suggest that the degree of hypersomnia may be overreported in patients with unipolar and bipolar depression. This may be due to reporting extended time in bed as evidence of increased sleep.

The incidence of hypersomnia in seasonal affective disorder has also been investigated. While patients complained of increased sleep during the winter months (9.9 h) and a normal duration of sleep during the summer (7.4 h), patients' sleep logs revealed that only 6% of patients truly slept more than 9 h during the winter [12]. Likewise, light box treatment was reported to reduce hypersomnia on subjective measures, yet there was no effect of treatment on duration of sleep when sleep diaries were examined [13].

POSTINFECTIOUS HYPERSOMNIA

In 1931 von Economo described the syndrome of encephalitis lethargica, which was thought to have an infectious or postinfectious etiology. The key symptoms consisted of drowsiness progressing to coma, along with ophthalmoplegia, oculogyric crises, and basal ganglia or cerebellar signs. Patients also presented with disorders of central respiration, psychiatric disturbances, and akinetic mutism. Though there are several descriptions of similar symptom complexes, the syndrome is now quite rare and few of these syndromes fit the original description of encephalitis lethargica [14].

Although there is not a definitive treatment for encephalitis lethargica, one study showed improvement with methylprednisolone treatment [15].

KLEINE–LEVIN SYNDROME

The history of this fascinating disorder begins in 1925, when Kleine first described two adolescent boys with periodic hypersomnia, accompanied by confusion and excessive food consumption [16]. Levin added a third such case in 1929 in his article "Periodic somnolence and morbid hunger: a new syndrome." This child was also an adolescent male with pathological hunger and long periods of sleep [17]. Critchley and Hoffman [18], in 1942, described 9 additional cases, reviewed 15 further cases from the literature, and assigned the eponym of Kleine–Levin syndrome.

The Kleine–Levin syndrome is a rare disorder that occurs most commonly in adolescent boys with a male to female ratio of 3:1 [19]. It is characterized by the triad of hypersomnia, hyperphagia, and hypersexuality, all co-occurring in periodic episodes. More variably, patients may develop delusions or hallucinations, behavioral disinhibition, confusion, and amnesia for the spells. They may also experience autonomic symptoms such as facial flushing and bradycardia. Each episode occurs from 3 days to 2 weeks at a time, and the interval between episodes lasts from a matter of weeks to months. Cognitive function and sleep are always normal between attacks [19, 20].

While patients are often unable to identify factors that provoked their initial Kleine–Levin attack, approximately half of patients identify factors that trigger subsequent attacks. The majority of these patients identify a febrile illness preceding some of their attacks. Other factors that have been described include extremes of ambient temperature, alcohol use, head trauma, and emotional stress [19, 20].

The polysomnographic features of patients with Kleine–Levin syndrome are variable. While there is agreement about an increase in total sleep time (701 ± 270 min in one study) and decreased sleep efficiency during an attack [21], the architecture of sleep both during and between attacks is discrepant among studies. While one study notes an increase in REM sleep and a decrease in stages 1, 3, and 4 during an attack [19], a more recent study notes a decrease in REM and slow-wave sleep with a proportionate increase in stage 1 sleep [21]. Polysomnograms obtained between attacks typically show a similar decrease in sleep efficiency and mild decreases in REM and slow-wave sleep. Electroencephalography is usually normal between attacks and may show generalized slowing during a Klein–Levin attack. Imaging, including head CT and MRI, and lumbar puncture are normal in patients [19, 21].

The etiology of Klein–Levin syndrome (KLS) is unknown. Suggestions of a genetic basis arise from a proven association with the HLA-DQb1 haplotype [21] and a pair of siblings with KLS, each with the HLA-DR2, DQ1, and DR5 haplotypes [22]. Despite this, familial groupings of KLS are few and far between. An anatomic anomaly is suggested by a SPECT scan on one patient showing hypoperfusion of the frontal and temporal lobes during and between attacks [23]. Neuropathology on four individuals who suffered from KLS showed inflammatory microglia in the thalamus and a small locus caeruleus with decreased substantia nigra pigmentation. These findings were thought to be secondary to a viral infection [24–26]. The fact that symptoms of KLS are similar to those seen in mood disorders would imply a primary psychiatric disorder; however, hypersexuality and hyperphagia often occur during manic attacks and do not coexist with the hypersomnia that may exist in atypical depression.

A more convincing etiology would be a primary neuroendocrinologic disturbance, as patients with KLS have symptoms similar to individuals with hypothalamic–pituitary masses or damage. Likewise, six patients showed diminished cortisol responses to insulin-induced hypoglycemia during attacks [27].

Treatment for Kleine–Levin syndrome has included light therapy, stimulants (methylphenidate, ephedrine), and anticonvulsants (valproic acid, carbamazepine) with variable success [28–30]. Lithium, in doses of 300 mg/day, is the current mainstay of treatment [31].

IDIOPATHIC RECURRING STUPOR/ENDOZEPINE STUPOR

Several cases of a periodic stupor or periodic coma have been described. These episodes are similar to KLS attacks in their periodicity, hypersomnia, and increased frequency in males. They differ, however, on several characteristics. While KLS tends to occur in adolescent males, the mean age of onset for idiopathic recurring stupor (IRS) attacks is the middle to upper 40s and only two instances of adolescent onset have been described [32–35].

In almost all of the reported instances of IRS, the onset of an episode is heralded by dysarthria, ataxia (patients are described as "staggering about as if drunk"), and feelings of malaise or fatigue, which precede the episode by hours to a few days. The patient then descends into a state of obtundation, lasting 2 hours to 5 days, from which he can be aroused only with noxious stimuli. Once aroused, a patient may be able to perform simple cognitive tasks before resuming a stuporous state. When the episode ends, patients do not recollect the event and may even confabulate about the events of the prior days. They may also remain "stunned" for hours after the episode ends. The frequency of these episodes ranges from every 3–365 days and, in many cases, there are no precipitating events [32–35].

During episodes, patients show diffuse hypotonia with diminished reflexes. The characteristic EEG of patients during an IRS attack is slow beta activity (14–16 Hz) most prevalent anteriorly. The EEG is normal between episodes. Polysomnography between episodes is also normal; however, some patients have shown sleep fragmentation and reduced efficiency of sleep, with decreased slow-wave and REM sleep [32–35].

In early reports of IRS, Tinuper et al. [32] attempted reversal of the ictal state with 0.4–1 mg of flumazenil. Despite negative toxicology for benzodiazepines, all patients who were given a similar dose awakened spontaneously and answered simple questions. Similarly, the EEG reverted to a normal alpha rhythm [32]. This prompted analysis of plasma cerebrospinal fluid collected from patients during an IRS attack. This yielded significantly increased concentrations of a benzodiazepine-like substance, endozepine-4, compared to patients' interictal concentration and to normal controls (40–300 times greater than controls) [36]. The etiology of elevated endozepine-4 levels remains unclear, however.

TABLE 22.1

Disorder	Age of Onset/Sex	Interictal EEG	Ictal EEG	PSG	Treatment
Kleine–Levin syndrome	Adolescence, M > F	Normal	Generalized slowing	Increased TST and decreased sleep efficiency during attack	Lithium, anticonvulsants, stimulants
Idiopathic recurring stupor	Fifth decade, M > F	Normal	14–16 Hz background	14–16 Hz background during attack	Flumazenil
Menstrual-related hypersomnia	Variable: early adolescence to fifth decade	Poorly organized posterior rhythm	Low-amplitude theta	Increase in TST during attack	Conjugated estrogens

MENSTRUAL-RELATED HYPERSOMNIA

Menstruation-linked hypersomnia is another type of recurrent hypersomnia, initially thought to be related to KLS. This syndrome was first described in 1975 by Billiard, Guilleminault, and Dement [37]. In the initial description, a 13-year-old girl developed periodic hypersomnia, which preceded her menses by 2–3 days. During each episode, the child would awake only to void, and she had significantly decreased intake of food and fluids. Hypersomnia would continue until 2 days after her period stopped, at which time she would recover spontaneously. Oftentimes, the onset of hypersomnia was preceded by the patient becoming hostile and withdrawn [37]. Other reports note similar symptoms, with onset from early adolescence to the early fifth decade [38, 39]. (See Table 22.1.)

EEG during periods of hypersomnia reveals low-amplitude theta activity with interictal EEGs showing a poorly organized posterior rhythm or less prominent theta. Polysomnography between episodes is normal, and during an event, there is an increase in total sleep time with no significant alteration in sleep architecture in most cases (a 42-year-old woman showed decreased REM sleep and alpha-delta sleep) [37–39]. Interestingly, individuals who develop menstrual-related hypersomnia have shown abnormalities in monoamine metabolites. One patient, after stimulation with probenicid, had an abnormal increase in 5-HIAA and homovanillic acid [37]. Another patient showed abnormally high levels of HVA and 5-HIAA on day 25 of her menstrual cycle and an abnormally high level of 5-HIAA on day 14 of her cycle, though she was asymptomatic [38].

Estrogen, which blocks progesterone release and inhibits ovulation, has been used successfully for the prevention of menstrual-related hypersomnia, with return of symptoms upon return of ovulatory cycles [37, 38]. This has led to the hypothesis that progesterone precipitates hypersomnia in patients prone to menstrual-related hypersomnia. One patient, for whom conjugated estrogens had no effect, responded to symptomatic treatment with methylphenidate [39].

Two reports of premenstrual sleep terrors and sleepwalking have been reported and may be related to menstrual-related hypersomnia. One such patient, with borderline mental retardation, developed episodes at the age of 17; another patient, who had a history of unprovoked parasomnias, developed episodes at the age of 46. In each case, the onset of sleepwalking–sleeptalking and sleep terrors would precede menstruation by 4–6 nights. Both patients responded to treatment with benzodiazepines [40].

REFERENCES

1. Guilleminault CG, Yuen KM, Gulevich MG, Karadeniz D, Leger D, Philip P. Hypersomnia after head–neck trauma: a medicolegal dilemma. *Neurology* **54**:653–659(2000).

2. Castriotta RJ, Lai JM. Sleep disorders associated with traumatic brain injury. *Arch Phys Med Rehabil* **82**:1403–1406(2001).

3. Cohen M, Oksenberg A, Snir D, Stern MJ, Groswasser Z. Temporally related changes of sleep complaints in traumatic brain injured patients. *J Neurol Neurosurg Psychiatry* **55**: 313–315(1992).

4. Quinto C, Gellido C, Chokroverty S, Masdeu J. Posttraumatic delayed sleep phase syndrome. *Neurology* **54**:250–252(2000).

5. Patten SB, Lauderdale WM. Delayed sleep phase disorder after traumatic brain injury. *J Am Acad Child Adolesc Psychiatry* **31**:100–102(1992).

6. Nagtegaak JE, Kerkhof GA, Smits MG, Swart AC, van de Meer YG. Traumatic brain injury-associated delayed sleep phase syndrome. *Funct Neurol* **12**:345–348(1997).

7. Kupfer DJ, Himmelhoch JM, Swartzburg M, Anderson C, Byck R, Detre TP. Hypersomnia in manic–depressive diseases (a preliminary report). *Dis Nerv Syst* **33**:720–724(1972).

8. Detre T, Himmelhoch J, Swartzburg M, Anderson CM, Byck R, Kupfer DJ. Hypersomnia in manic–depressive disease. *Am J Psychiatry* **128**:1303–1305(1972).

9. Posternak MA, Zimmerman M. Symptoms of atypical depression. *Psychiatry Res* **104**:175–181(2001).

10. Billiard M, Dolenc L, Aldaz C, Ondze B, Besset A. Hypersomnia associated with mood disorders: a new perspective. *J Psychosomat Res* **38**:41–47(1994).

11. Nofzinger EA, Thase ME, Reynolds CF, Himmelhoch JM, Mallinger A, Houck P, Kupfer DJ. Hypersomnia in bipolar depression: a comparison with narcolepsy using the multiple sleep latency test. *Am J Psychiatry* **148**:1177–1181(1991).

12. Shapiro CM, Devins GM, Feldman B, Levitt AJ. Is hypersomnolence a feature of seasonal affective disorder? *J Psychosom Res* **38**:49–54(1994).

13. Lam RW, Buchanan A, Mador JA, Corral MR. Hypersomnia and morning light therapy for winter depression. *Biol Psychiatry* **31**:1062–1064(1992).

14. Howard RS, Lees AJ. Encephalitis lethargica: a report of four recent cases. *Brain* **110**:19–33(1987).

15. Blunt SB, Lane RJ, Turjanski N, Perkin GD. Clinical features and management of two cases of encephalitis lethargica. *Mov Disord* **12**:354–359(1997).

16. Kleine W. Periodisch Schlafsucht. *Mschr Psychiatr Neurol* **57**:285–298(1925).

17. Levin M. Periodic somnolence and morbid hunger. *Brain* **49**:494–594(1936).

18. Critchley M, Hoffman HL. The syndrome of periodic somnolence and morbid hunger, Kleine–Levin syndrome. *Br Med J* **1**:137–139(1942).

19. Gadoth N, Kesler A, Vainstein G, Peled P, Lavie P. Clinical and polysomnographic characteristics of 34 patients with Kleine–Levin syndrome. *J Sleep Res* **10**:337–341(2001).

20. Papacostas SS, Hadjivasilis V. The Kleine–Levin syndrome. Report of a case and review of the literature. *Eur Psychiatry* **15**:231–235(2000).

21. Dauvilliers Y, Mayer G, Lecendreux M, Neidhart E, Peraita-Adrados R, Sonka K, Billiard M, Tafti M. Kleine–Levin syndrome: an autoimmune hypothesis based on clinical and genetic analyses. *Neurology* **59**:1738–1745(2002).

22. Katz JD, Ropper AH. Familial Kleine–Levin syndrome: two siblings with unusually long hypersomnic spells. *Arch Neurol* **59**:1959–1961(2002).

23. Landtblom A-M, Dige N, Schwerdt K, Safstrom P, Granerus G. A case of Kleine–Levin syndrome examined with SPECT and neuropsychological testing. *Acta Neurol Scand* **105**:318–321(2002).

24. Carpenter S, Yassa R, Ochs R. A pathologic basis for Kleine–Levin syndrome. *Arch Neurol* **39**:25–28(1982).

25. Koerber RK, Torkelson R, Haven G, Donaldson J, Cohen SM. Increased cerebrospinal fluid 5-hydroxytryptamine and 5-hydroxyindoleacetic acid in Kleine–Levin syndrome. *Neurology* **34**:1597–1600(1984).

26. Fenzi F, Simonati A, Crosato F, Ghersini L, Rizzuto N. Clinical features of Kleine–Levin syndrome with localized encephalitis. *Neuropediatrics* **24**:292–295(1993).

27. Malhotra S, Das MK, Gupta N, Murualidharan R. A clinical study of Kleine–Levin syndrome with evidence for hypothalamic–pituitary axis dysfunction. *Biol Psychiatry* **42**:299–301(1997).

28. Crumley FE. Valproic acid for Kleine–Levin syndrome. *J Am Acad Child Adolesc Psychiatry* **36**:868–869(1997).

29. Crumley FE. Light therapy for Kleine–Levin syndrome. *J Am Acad Child Adolesc Psychiatry* **37**:1245(1998).

30. Mukkades NM, Meltem Erocal K, Bilge S. Carbamazepine for Kleine–Levin syndrome. *J Am Acad Child Adolesc Psychiatry* **38**:791–792(1999).

31. Goldberg M. The treatment of Kleine–Levin syndrome with lithium. *Can J Psychiatry* **28**:491–493(1983).

32. Tinuper P, Montagna P, Cortelli P, Avoni P, Lugaresi A, Schoch P, Bonetti EP, Gallassi R, Sforza E, Lugaresi E. Idiopathic recurring stupor: a case with possible involvement of the gamma-aminobutyric acid(GABA)ergic system. *Ann Neurol* **31**:503–506(1992).

33. Tinuper P, Montagna P, Plazzi G, Avoni P, Cerullo A, Cortelli P, Sforza E, Bonetti EP, Schoch P, Rothstein JD, Guidotti A, Lugaresi E. Idiopathic recurring stupor. *Neurology* **44**:621–625(1994).

34. Soriani S, Carozzi M, DeCarlo L, Plazzi G, Provini F, Rothstein JD, Tinuper P, Bouquet F, Lugaresi E, Montagna P. Endozepine stupor in children. *Cephalalgia* **17**:658–661(1997).

35. Lugaresi E, Montagna P, Tinuper P, Plazzi G, Gallassi R, Wang TL, Markey SP, Rothstein JD. Endozepine stupor: recurring stupor linked to endozepine-4-accumulation. *Brain* **121**:127–133(1998).

36. Rothstein JD, Guidotti A, Tinuper P, Coretelli P, Avoni P, Plazzi G, Lugaresi E, Schoch P, Montagna P. Endogenous benzodiazepine receptor ligands in idiopathic recurring stupor. *Lancet* **340**:1002–1004(1992).

37. Billiard M, Guilleminault C, Dement WC. A menstruation-linked periodic hypersomnia. *Neurology* **25**:436–443(1975).

38. Sachs C, Persson HE, Hagenfeldt K. Menstrual-related periodic hypersomnia: a case study with successful treatment. *Neurology* **32**:1376–1379(1982).

39. Bamford CR. Menstrual-associated sleep disorder: an unusual hypersomniac variant associated with both menstruation and amenorrhea with a possible link to prolactin and metoclopramide. *Sleep* **16**(5):484–486(1993).

40. Schenk CH, Mahowald MW. Two cases of premenstrual sleep terrors and injurious sleep walking. *J Psychosom Obstet Gynecol* **16**:79–84(1995).

23

SLEEPING SICKNESS

ALAIN BUGUET
Neurobiologie des états de vigilance, Lyon, France

FLORIAN CHAPOTOT
Centre de recherches du service de santé des armées, La Tronche, France

RAYMOND CESPUGLIO
Neurobiologie des états de vigilance, Lyon, France

SYLVIE BISSER
Centre international de recherches médicales, Franceville, Gabon

BERNARD BOUTEILLE
Institut de neurologie tropicale, Limoges, France

INTRODUCTION

Sleeping sickness, or human African trypanosomiasis (HAT), is an endemic parasitic disease that is exclusively located in intertropical Africa, being rooted between the latitudes of 15° N and 15° S by the geographical distribution of the tsetse fly (genus *Glossina*). Although HAT has been known for centuries, as shown by the description of the disease by the Arab historian Ibn Khaldoun in 1406 [1], it became epidemic at the end of the nineteenth and beginning of the twentieth centuries. Colonization ended tribal skirmishes, installed a rational administration, built roads, and favored population exchanges. A devastating epidemic developed, reaching a peak between 1915 and 1930, killing an estimated one million people. The colonial administration issued biomedical missions and, within a few years, the medical services had found the cause of the disease, its epidemiological and clinical characteristics, and diagnostic criteria and had proposed specific therapeutic approaches. The creation of mobile teams allowed the progressive control of the disease. By the 1960s, its prevalence had dropped to approximately 0.01%. However, the large federal colonial entities were replaced by several independent states, which soon lacked financial support, and the national health services were not afforded the highest priority. Political turmoil, wars, and civil unrest contributed to disorganization. Today, the disease is considered as a resurgent epidemic in the Sudan, Uganda, Democratic Republic of Congo, and Angola (Figure 23.1) and a public health concern in several other countries of Western and Central Africa. The World Health Organization (WHO) [2] estimates that among the 60 million people exposed to tsetse flies, no more than five million are benefiting from clinical and epidemiological surveillance. Therefore only 40,000 new cases are reported each year. The WHO estimates the actual number of patients between 300,000 and 500,000. Today, sleeping sickness is considered to rank among the most neglected diseases [3]. Another characteristic is that resurgence always seems to originate from historical well-identified foci, explaining some WHO strategies.

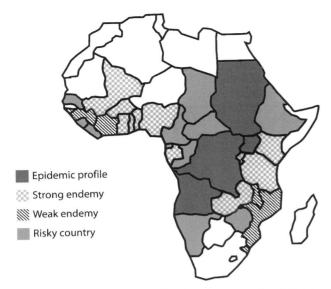

Figure 23.1 Map of Africa showing the geographical distribution of human African trypanosomiasis. (Adapted from WHO African trypanosomiasis web pages, http://www.who.int/health-topics/afrtryps.htm.)

Legend:
- Epidemic profile
- Strong endemy
- Weak endemy
- Risky country

On October 7, 1994, the WHO appealed for international solidarity [4]. The international organization obtained support from the French and Belgian governments, OCEAC (Organisation pour la lutte contre les endémies en Afrique Centrale, Yaoundé, Cameroon), and nongovernmental organizations such as Doctors without Borders. In 2000, the WHO used HAT as a model for the concept of "orphan medication." Since then, private corporations have collaborated with the WHO: funds are being raised and medications are being made freely available to health services. The WHO is able to organize, consolidate, and coordinate medical actions. Concomitantly, in 2001, African governments decided to develop and coordinate their actions for the control of the disease by creating the Pan African Tsetse and Trypanosomiasis Eradication Campaign (PATTEC), under the auspices of the African Unity Organization. Indeed, trypanosomiasis is not only a human disease, but also an animal pathology, which induces terrible economical losses, depriving African developing countries of meat.

TSETSE FLIES AND TRYPANOSOMES

The bloodthirsty tsetse flies infect their hosts with trypanosomes of the *Trypanosoma brucei* species, which is itself divided into several subspecies of which only two are infective to humans. Infected by *T. brucei brucei*, wild and domesticated animals develop nagana. Humans are resistant to *T. brucei brucei*. Exposure to normal human serum triggers trypanolysis within a few minutes [5]. The trypanolytic factor was identified as a haptoglobin-related protein [6, 7].

The human disease is schematically divided into two separate geographical entities. In Western and Central Africa, *T. brucei gambiense* is transmitted mainly by *G. palpalis*, a forest tsetse fly. The protozoa reservoir is most exclusively human. In Eastern Africa, *T. brucei rhodesiense* has an almost solely animal reservoir, including wild big game and domestic animals. Humans are accidentally infected. *Trypanosoma brucei rhodesiense* is most likely transmitted by *G. morsitans* in the savannah, by *G. pallidipes* at the forest edge, or by *G. fuscipes* in riverine and swamplands. The Rhodesian parasites are mainly differentiated from *T. brucei brucei* by the human serum resistance test.

African trypanosomes are extracellular parasites characterized by their ability to alter the composition and organization of their plasma membrane to escape immune defenses of the host, the latter being elicited within a few days after infection [8]. Antibodies from the host bind to the parasite surface, activate complement, and destroy the trypanosome. The parasites are mantled with variant surface glycoproteins (VSGs). Although only one variable antigen type (VAT) is expressed at a given time, it can be replaced by another type of VSG. The sequential expression of VSG genes realizes the so-called antigenic variation. Therefore the trypanosome glycoprotein shell is subject to constant variation of exposed epitopes. Antibodies against one VAT provoke trypanolysis. However, some trypanosomes expressing another VSG escape the host reaction. They are able to proliferate and initiate a new wave of parasitemia, until the host produces an adapted antibody response. Once again, antigenic variation will allow some other trypanosomes to express a new VSG, and so on. Trypanosomes possess about 1000 VSG genes, representing about 10% of their genome.

IMMUNOPATHOLOGY

Major immunological changes are observed in HAT, being marked by lymphadenopathy, splenomegaly, and hypergammaglobulinemia, as well as autoimmune reactions and immunodepression. At the hemolymphatic stage of the disease (stage I), trypanosomes (via the VSGs) express VATs, leading to a specific antibody response in the lymph and plasma [9]. With each new wave of trypanosomes, a corresponding wave of immunoglobulins induces a massive hypergammaglobulinemia, mainly IgMs, which are not all VSG-specific. A marked polyclonal B-cell activation generates specific and nonspecific antibodies, autoantibodies, and immune complexes [10]. Autoantibodies are associated with the stage of meningoencephalitis (stage II). Their detection in the serum and cerebrospinal fluid (CSF) could be a marker of central nervous system (CNS) involvement.

Macrophages play a key role in specific immunity, as antigen presenting cells in synergy with antibodies and cytokines are involved in immunosuppressive and immunopathological phenomena. Macrophages are activated by interferon-γ (IFN-γ), itself produced by CD8$^+$ T cells stimulated by the trypanosome lymphocyte triggering factor [11], and also by a lipopolysaccharide-like product released by trypanosomes. Activated macrophages produce tumor necrosis factor-α (TNF-α), which, in conjunction with the latter two substances, should lead to the production and release of nitric oxide (NO), a trypanostatic and trypanocidal molecule. However, in animals and humans, our group has shown that blood NO decreases due to the inhibitory effect of interleukin-10 (IL-10) on TNF-α and the activation of arginase, which deprives the macrophagic NO synthase of its substrate, L-arginine (see [12]). A profound deregulation of the cytokine network is therefore involved in the pathogenic mechanisms of HAT. On the contrary, in the brain, NO is increased due to the infiltration of macrophages [12]. TNF-α is also produced by astrocytes and microglial cells in the CNS [13]. TNF-α and other cytokines contribute to the generation of somnogenic molecules such as IL-1 [14].

Precisely how the trypanosomes enter the CNS is still unknown. The parasites occupy the meninges soon after the infection, inducing meningitis. The crossing of the blood–brain barrier by trypanosomes is not yet elucidated, whether it is directly due to the parasite or to substances released locally at the choroid plexus level. Then, the trypanosomes reach the Virchow–Robin spaces and penetrate the cortex. They may also exfiltrate from damaged or permeable meningeal and parenchymal vessels, especially at the level of mesodiencephalic regions. The mechanism by which trypanosomes damage the CNS remains unanswered. Toxin production by trypanosomes, leading to inflammation and vascular permeability, may contribute to it. It may be related to elevated cytokine concentrations found in both patients and experimental models. However, the most important mechanism may be related to autoantibody production against CNS components, a consistent feature of HAT [15].

CLINICAL ASPECTS

Just after the bite by the tsetse fly, a chancre appears at the point of inoculation. The illness then evolves in two stages, the hemolymphatic stage I followed by the meningoencephalitic stage II, ending with demyelinization, altered consciousness, cachexia, and death if untreated [16]. The acute Rhodesian form of HAT evolves toward death in weeks, while the chronic Gambian form will do so in several months or years [17].

The trypanosomes reach the lymphatic and blood circulations. Clinical symptoms are diverse and nonspecific and the diagnosis is difficult. One of the characteristic signs is the classical cervical adenopathies (firm, mobile, painless). Fever is irregular, recurrent, oscillating in 3–5 day cycles. Cardiac symptoms with arrhythmia or even pancarditis may be observed, especially in the Rhodesian form, persistent tachycardia being seen in the Gambian form. Other symptoms are identified, such as cutaneous eruptions, pruritus with skin lesions being common, facial edema, and rare hepatosplenic disorders. Stage I is considered to end when neurological symptoms begin and/or when trypanosomes and/or mononuclear inflammatory cells appear in the CSF, marking the beginning of CNS invasion.

Among several neurological and psychiatric symptoms developing insidiously, excessive daytime sleepiness is one of the most reported signs at stage II. However, apart from the signs already described for stage I, any kind of neurological or psychiatric symptoms can be observed, especially in the Gambian disease: headaches, sensory disturbances with uncomfortable diffuse superficial or deep painful sensations (hyperpathia), sensory deficit, presence of primitive reflexes (palmomental reflex, sucking reflex), exaggerated deep tendon reflexes, tremor (fine and diffuse without any myoclonic jerk at rest or during movement), abnormal movements (upper members, choreic gesticulations), pyramidal alterations with Babinski sign, alterations in muscle tone, numbness, cerebellar signs (ataxia), and psychiatric disorders (indifference, absent gaze, mutism, confusion, mood swings, euphoria, agitation, aggressive behavior).

DIAGNOSIS

An accurate identification of the evolutionary stage of HAT is crucial, as treatment for stage II remains toxic. As yet, there are no specific clinical signs or blood tests for stage determination [18–20]. The finding of trypanosomes in the blood, lymph, or CSF is the only direct means to positively diagnose HAT. Several ways of improving the direct microscopic examination have been proposed: for example, blood sample centrifugation, centrifugation of blood in a capillary tube, use of a mini anion exchange centrifugation technique, and double CSF centrifugation.

Following the WHO recommendations [2], the diagnosis of stage II is based on the CSF examination to search for (1) the presence of trypanosomes; (2) elevated white blood cell counts—the cut-off proposed by the WHO is $<5/\mu L$ for stage I; and (3) determination of total protein concentration; however, according to the technique used, different cut-offs have been proposed and vary from 250 to 450 mg/L.

Immunologically based techniques are not yet totally reliable, although they have greatly improved field diagnosis of HAT. This is the case of the card agglutination trypanosomiasis test (CATT) for *T. brucei gambiense* and the procyclic agglutination trypanosomiasis test (PATT) for *T. brucei rhodesiense*. Indirect immunofluorescence tests, enzyme linked immunosorbent assay (ELISA), polymerase chain reaction (PCR), and the most recent surface-enhanced laser desorption-ionization time-of-flight mass spectrometry [21] are sensitive but cannot be used as field diagnosis techniques. There is hope that a new latex technique using given VSGs or recombinant antigens or synthetic peptides will be available for use in the field [22, 23].

TREATMENT

Although there are few available active medications to treat HAT [24], the WHO [25] has obtained commitment from pharmaceutical companies to build up stocks of all trypanocides for at least five years. Trypanocides can therefore be obtained via the WHO in Geneva. However, what makes the therapeutics of HAT difficult is that the operational treatment of stage II is based on the use of melarsoprol, an arsenical derivative issued in 1949 [26]. Melarsoprol may cause dramatic side effects, which are potentially lethal, emphasizing the need for accurate stage diagnosis. On the contrary, treatment of stage I uses relatively well-tolerated medications such as pentamidine or suramin.

Pentamidine (pentamidine isethionate BP, Pentacarinat®), introduced in 1936, is the treatment of choice for Gambian infections at stage I [27]. The commonly recommended dosage is 4 mg/kg per day by intramuscular injection either every day or every other day, for a series of 7–10 injections. Pentamidine is generally well tolerated, although various adverse effects can be observed, such as reversible renal toxicity [28].

Suramin (Bayer 205, Germanin®), derived from trypan red and commercially available since 1916, is preferred for treating Rhodesian infections at stage I. The usual dosage is 20 mg/kg by a series of five intravenous injections at 5–7 day intervals. Suramin has few side effects.

Melarsoprol (Mel B, Arsobal®) remains the principal medication for the treatment of stage II HAT, although concentrations in the CSF were found to be 50–100 times lower than that in plasma [29]. The commonly accepted protocol consists of daily slow intravenous injections, realized during three to four consecutive days weekly, throughout a three-week period, with a maximum of 3.6 mg/kg injected daily. The number of series of injections is often based on the CSF white blood cell count [30]. A continuous treatment schedule at 2.2 mg/kg per day for 10 days has recently been shown to be effective [31]. Melarsoprol is primarily neurotoxic, with a risk of fatal encephalopathy [27, 32], but other undesirable side effects may occur. Overall fatality ranges from 2% [33] to 9.8% [34] for *T. brucei*

gambiense-infected patients and from 3.4% [35] to 12% [36] in Rhodesian infections. Furthermore, 10–30% of patients may be resistant to melarsoprol, especially in Uganda and Angola [37, 38]. In such a case, a second melarsoprol administration may be effective, but an alternative treatment (eflornithine or nifurtimox) is recommended.

Eflornithine (difluoromethyl-ornithine, DFMO, Ornidyl®), used against stage II HAT since 1992, is an irreversible specific inhibitor of the trypanosome ornithine decarboxylase [39], efficient in both stages of the *T. brucei gambiense* infection [40], although results are disappointing in Rhodesian forms [41]. The standard intravenous treatment schedule is hard to set up in field conditions and consists of 100 mg/kg of DFMO given in slow infusion every 6 h during 14 consecutive days. This costly and complicated treatment should be reserved for patients refractory to melarsoprol. Furthermore, DFMO is not devoid of side effects, which are similar to those of anticancer drugs.

Nifurtimox (Bayer 2505, Lampit®), used for the treatment of American trypanosomiasis, Chagas' disease, is active *per os* (three daily doses of 5 mg/kg each, during 2–3 weeks) in both stages of *T. brucei gambiense* infection. Its efficacy against *T. brucei rhodesiense* is unknown. Since it has not been approved for use in HAT, initial trials have involved patients refractory to melarsoprol with no other treatment alternatives. The incidence and severity of neurological side effects increase with the duration of treatment.

No specific remedies for HAT have been developed since melarsoprol first appeared in 1949. Trypanocides are not considered by pharmaceutical companies as being a priority, particularly because of the small and unpredictable market. The only trypanocidal drug currently under clinical trial, DB-289, an orally administered pentamidine substitute, is reserved for stage I [42].

SLEEP IN SLEEPING SICKNESS

Clinical Descriptions of Sleep–Wake Disturbances

Sleep alterations in sleeping sickness have been described by physicians for more than a century. Most reports describe somnolence, but insomnia is not rare. Mackensie [43] had ruled out hypersomnia having observed that the patients slept often, in short sleep bouts, during the day as well as at night. However, the most widely accepted description was that of patients being sleepy by day and restless by night [44]. Although Mackensie's description evoked a major disturbance in the 24 h alternation of the sleep–wake cycle, this aspect was not specifically studied. In 1910, Lhermitte [45], and concomitantly Van Campenhout [46], described the suddenness with which the patients fall asleep and established a parallel with the sleep crises of narcolepsy.

Polysomnographic Approach of Sleep–Wake Alterations: Nighttime or Daytime Recordings

Polysomnographic (PSG) (electroencephalogram, EEG; electro-oculogram, EOG; electromyogram, EMG) recordings were conducted early after the development of this new technique. Polysomnography represents the only objective means to distinguish between wakefulness, rapid eye movement (REM) sleep, and non-rapid eye movement (NREM) sleep and its four stages, with sleep stages 3 and 4 representing slow-wave sleep [47]. However, PSG recordings were only taken at night [48] or during diurnal naps [49].

The largest study of nocturnal sleep conducted by Bert et al. [48] in Dakar (Senegal) revealed abnormalities in 16 out of 17 stage II patients, the only patient at stage I, and one patient clinically ranked as being at early stage II showing a normal sleep structure. In their meningoencephalitic patients, the authors had difficulties in scoring intermediary stages. However, stage 4 and REM sleep were normal in most patients. They noted the scarcity of vertex sharp waves, spindles, and K-complexes. The authors stressed the parallel between anomalies of the sleep traces and severity of the clinical state. They also discussed the similarity of sleep structure with that of narcoleptic patients, showing one hypnogram revealing sleep onset REM periods (SOREMPs), but they did not attribute a special importance to the phenomenon.

In Paris, Schwartz and Escande [49] examined a Lebanese patient from Dakar, who had contracted HAT in Senegal and was at an advanced stage of meningoencephalitis. The patient was treated with melarsoprol. During the treatment procedure, he underwent three standard wake recordings in the morning and one PSG in the afternoon (between 14:00 h and 16:00 h). During the four following months, PSG was performed at 3 week intervals: one nocturnal PSG, performed 2 months after treatment, was preceded and followed by two afternoon PSG recordings. Overall, sleep episodes occurred by night and by day. The authors noted three SOREMPs in the afternoon tests and a short REM sleep latency in the night recording. They stressed the resemblance with narcoleptic sleep organization. They also experienced sleep stage scoring difficulties and noted the paucity of vertex spikes, K-complexes, and spindles. Sleep patterns improved with time.

To our knowledge, besides our published recordings, the literature only reports on three other patients recorded by PSG. Sanner et al. [50] reported on a German patient, who presented nocturnal insomnia and daytime hypersomnia, starting 12 days after returning from a 20 day trip to Zambia, Zimbabwe and Tanzania. The patient rapidly developed multiorgan failure attributed to *T. brucei rhodesiense* infection. She recovered in a few days after suramin treatment. Nocturnal PSG recordings were performed on days 7, 9, and 15, and 6 months after the onset of the disease. Normal amounts of REM sleep were observed, but slow-wave sleep almost disappeared in the early days to be restored to normal 6 months later. In a review paper on waking disorders, Billiard and Ondzé [51] presented the 24 h hypnogram of a 26-year-old Cameroon patient showing one SOREMP and altered sleep–wake cycle. The same hypnogram was presented in a book by Billiard and Callander [52]. The third patient, a 39-year-old Zaïre man (Democratic Republic of the Congo), was examined in Paris [53]. He had a complicated 6 year long history with several psychiatric hospitalizations and was even sent to jail. The diagnosis of trypanosomiasis was finally made and the patient was treated with eflornithine. All clinical symptoms (meningoencephalitis, confusion, and epileptic seizures) disappeared rapidly. He underwent two 48 h PSG recordings—15 days after completion of the treatment procedure and 2 months later. The patient presented disrupted distribution of sleep and wake episodes and a shortened REM latency indicating the presence of SOREMPs.

The 24 h Polysomnographic Methods Used in Our Investigations in Africa

The first 24 h sleep recording in sleeping sickness was performed in 1988 in Niamey (Niger) on a migrant worker who had contracted the disease in Côte d'Ivoire [54]. Although the EEG traces were loaded with slow waves, sleeping, waking, and REM sleep were identifiable. The 24 h recording revealed the disappearance of the normal distribution of sleeping and waking, which occurred indifferently throughout the nychthemeron.

Our team then used PSG to analyze sleep and wake modifications, first in stage II patients and more recently in stage I patients. Altogether, in 8 field trials, 106 PSG recordings were taken on 43 stage II patients, 52 recordings on 35 stage I patients, and 44 recordings on 44 healthy villagers who had volunteered as controls. All patients had been infected by *T. brucei gambiense*. In addition, two Caucasian stage II patients, who had contracted Rhodesian HAT in Rwanda, were also followed-up after melarsoprol treatment in France over a period of 11 months. Informed consent to participate in the study was always obtained from either the patients or their families, as was the agreement from national health authorities (Angola, Congo, Côte d'Ivoire, Niger). In all investigations, the diagnosis of sleeping sickness was confirmed by finding trypanosomes in the blood, lymph gland fluid, and/or CSF.

The techniques used have evolved over this 16 year period. A total of fifty-three 24 h PSG paper recordings were taken on 8- or 10-channel polygraphs following a previously published procedure [55]. Basically, two channels were devoted to the EEG, while the EMG and EOG were taken on separate channels. Other channels served to record the electrocardiogram and the respiratory cycle (nasal and buccal airflow with Cu–Ct thermocouples; chest movements with a strain gauge). Paper speed was 15 mm/s, thus determining 20 s scoring periods. In most patients, classical scoring of sleep states and wakefulness could be performed [47]. In a few patients at very advanced stage II, the scoring technique had to be adapted following Schwartz and Escande [49] and Buguet et al. [54], because of the invasion by ubiquitous EEG slow waves. In these patients, NREM sleep was thus divided into light sleep and SWS. In all patients, wakefulness and REM sleep were easily identified.

The remaining PSG recordings used portable Holter-type equipment (Figures 23.2, 23.3). As none of the patients recorded on paper had shown any respiratory disturbance, most of the recordings included only EEG, EOG, and EMG traces. Recently, analog portable recorders were replaced by ambulatory digital recording systems, which allow for quantitative EEG analyses.

Eighteen of our meningoencephalitic patients participated in a study on hormonal secretory rhythms. They were catheterized in the median basilic vein, and 10 mL

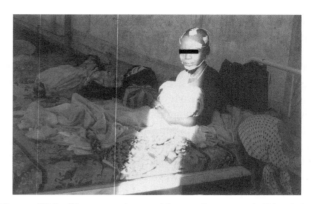

Figure 23.3 Young patient with meningoencephalitis being recorded with an ambulatory digital acquisition system under the hospital tent set up in an Angolan village.

blood samples were removed each hour over a 24 h period in eight of them. Blood was continuously withdrawn through peristaltic pumps in the other ten patients. The overall 24 h blood withdrawal was always kept below 300 mL. A similar investigation with continuous blood withdrawal was conducted in Abidjan (Yopougon University Hospital) on six healthy subjects. In these three investigations, designated to study hormonal secretion, blood sampling was terminated either after 24 h or upon request from the patients or if the patient's hemoglobin dropped below 10 g/dL.

SLEEP PATTERNS IN HEALTHY SUBJECTS

In villages from Côte d'Ivoire and Angola, 44 healthy subjects volunteered to serve as controls for the HAT villagers and were recorded by 24 h ambulatory PSG. All healthy volunteers had a major sleep episode at night and sometimes a small nap in the afternoon or late morning [56]. They did not show any disturbance in their sleep–wake cycle nor any SOREMPs (Figure 23.4).

In the six healthy subjects, who slept with an indwelling catheter as controls for endocrine studies, sleep efficiency was impaired due to frequent awakenings, resulting in a reduction in total sleep time. REM sleep was also reduced. These data obtained in Africa were concordant with the observations of Adam [57] in seven healthy Scots also sleeping with a catheter.

THE POLYSOMNOGRAPHIC SYNDROME IN STAGE II PATIENTS

The meningoencephalitic patients showed a major disturbance in the 24 h distribution of sleep and wakefulness episodes, the extension of which increased with the severity

Figure 23.2 Voluntary subject from an Angolan village being recorded with an ambulatory digital acquisition system.

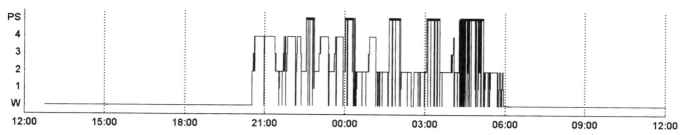

Figure 23.4 Hypnogram showing a 24 h distribution of sleep states (PS, REM sleep; 1–4, stages 1–4 of NREM sleep) and wakefulness (W) in a healthy subject in Angola. As in the rest of the world, the subject sleeps at night, with a well-structured sleep.

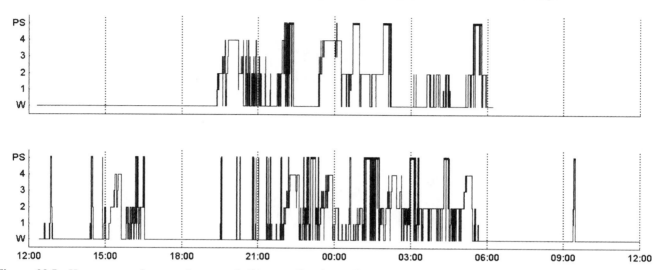

Figure 23.5 Hypnograms of two meningoencephalitic stage II patients. The upper graph represents 24 h sleep and wakefulness distribution with nocturnal episodes of destructured sleep, the alterations being essentially shown by restless sleep and the presence of sleep onset REM periods (SOREMPs), especially in the second sleep episode. The lower graph is that of an advanced meningoencephalitic patient, demonstrating the complete polysomnographic syndrome: disturbed distribution of sleep and wakefulness throughout the nycthemeron; presence of SOREMPs in several sleep episodes.

of the disease (Figure 23.5). Severely ill patients experienced sleep episodes throughout the nychthemeron. As there was no increase in the 24 h total sleep time (i.e., no hypersomnia), the average duration of wakefulness and sleep episodes was inversely related to the severity of the disease. Therefore the sleep–wake alternation occurred in shorter cycles. The occurrence of REM sleep episodes throughout the nychthemeron is the best example of such a disturbance in 24 h rhythms.

The deregulation of sleep and wakefulness 24 h distribution was completed by an alteration of the structure of sleep episodes. The latter started often with REM sleep phases (SOREMPs). The number of SOREMPs increased in relation to the severity of clinical symptomatology.

The sleep disturbances were not aggravated in the patients examined in the Daloa clinic (Côte d'Ivoire) whether they had a catheter or not [58]. It seems that these hyperpathic patients had a lowered sensitivity to environmental stimuli.

Contrary to our expectations, the numerous alterations of the EEG were not specific for HAT [59]. During sleep,

normal features were seen. However, transient activation phases were decreased in the patients. Four patients presented monophasic frontal delta bursts predominant during slow-wave sleep along with paroxysmal hypnopompic hypersynchronic events during slow-wave sleep. It was therefore concluded that in sleeping sickness patients, although dampened, the waking process remains responsive and slows down only at a late stage of meningoencephalitis.

THE POLYSOMNOGRAPHIC SYNDROME AS A DIAGNOSTIC MEAN

The PSG alterations may be determinant in diagnosing the passage from stage I to stage II. We explored this hypothesis in two investigations conducted in Angola, during which we examined 11 patients at stage II and 24 at stage I, as well as 15 healthy volunteers. All stage II patients exhibited the complete sleep–wake syndrome. Among stage I patients, 13 showed abnormal PSG signs, especially sleep episodes with SOREMPs (Figure 23.6). Healthy

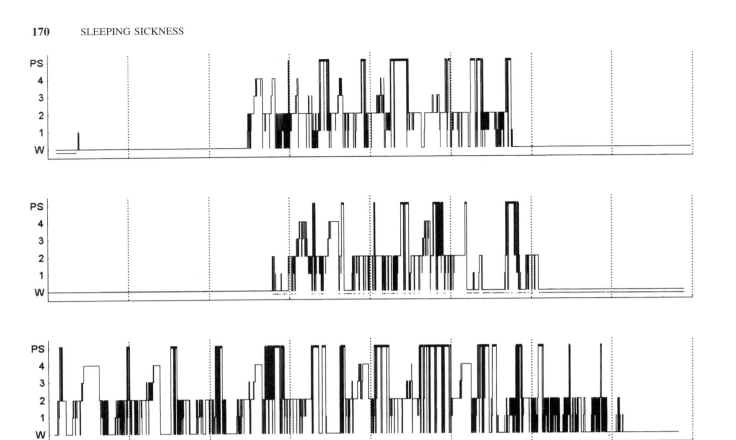

Figure 23.6 Hypnograms of three patients diagnosed clinically and biologically as being at stage I of human African trypanosomiasis. The upper graph shows a hypnogram demonstrating a normal sleep pattern. The middle graph represents a slightly disturbed nocturnal sleep pattern, with the occurrence of a phase of REM sleep after a long wakefulness episode. The lower graph demonstrates a completely disturbed sleep pattern showing an obvious alteration of sleep–wakefulness function.

volunteers had absolutely normal sleep patterns and distribution. The presence of SOREMPs in patients with HAT may therefore indicate the passage of trypanosomes into the CNS or the induction of CNS reactions through the action of trypanosome-induced chemokines across the blood–brain barrier.

TREATMENT EFFECT ON THE POLYSOMNOGRAPHIC SYNDROME

In the investigation conducted in Brazzaville [60] on ten meningoencephalitic patients, 24 h PSG was recorded before treatment with melarsoprol, and then weekly during the 3 week treatment procedure. In all patients, the disturbed sleep–wake alternation improved after treatment. The number of SOREMPs also normalized, especially in the less severely sick patients.

The treatment-related improvement of the sleep–wake 24 h alternation was also observed in two Caucasian patients with severe meningoencephalitis. These had been infected by *T. brucei rhodesiense* in Rwanda. They were treated in Marseille (France) with melarsoprol and had PSG recorded 4, 6, and 11 months later [61]. Although their sleep disturbances were improved by treatment, normalization was only obtained 11 months later in one of the two patients. Reversibility of the PSG syndrome may therefore take several months to be completed.

SUMMARY

The major findings of PSG in stage II patients with African trypanosomiasis can be summarized as follows. The 24 h alternation of sleep and wakefulness is altered proportionally to the severity of the disease. It is reversed by melarsoprol treatment. Sleep structure alterations appear early in stage II, with frequent SOREMPs. SOREMPs recede or disappear after melarsoprol treatment.

The occurrence of SOREMP-like episodes has also been shown in a rat model of African trypanosomiasis [62]. The animals were infected by *T. brucei brucei* and followed with PSG until death, which occurred on an average 3 weeks after the inoculation of the parasites. SOREMPs appeared immediately after the decline of food intake and body weight (12–14 days after infection), which also corresponded to the presence of trypanosomes in the CNS [63].

GENERAL CONCLUSION

As stated recently by Peter De Raadt [1], who worked for the WHO for over 30 years, history repeats itself. Today, sleeping sickness has come back. It is considered to be a reemergent disease, menacing 60 million people living in the tsetse belt of Africa, one of the poorest areas of the world [64]. It is also an orphan disease with a limited therapeutic arsenal. However, despite a past disinterest by government and international institutions and little funding, research on HAT has developed considerably [17]. Furthermore, African trypanosomiasis is not only deadly for humans; it represents a major plague infecting and killing domestic cattle in an economically sensitive continent [65]. The recent Congress on the tsetse fly and trypanosomiasis held in Brazzaville in March 2004 [66] insisted on the fact that both animal and human African trypanosomiases represent major obstacles for the economic development of the African continent. There is, however, hope due to the recent promotion of cooperation between pharmaceutical groups and governments from Africa and the rest of the world, under the auspices of the WHO [25]. Sleeping sickness has found its place within the concept of neglected diseases and patients can now have access to free medications provided by Aventis (pentamidine, melarsoprol, eflornithine) and Bayer (suramin). At Lomé (Togo) in 2001, the African governments committed themselves to eliminate the disease. The objective is not only to develop new medications to treat the patients, but also to obtain new means to improve the stage diagnosis of HAT in order to optimize treatment strategies.

Research on sleep in sleeping sickness belongs to this logic. It has provided new knowledge, which may be used for the determination of the evolutionary stages of HAT, and also for the follow-up of treatment efficacy.

ACKNOWLEDGMENTS

This work was supported by the World Health Organization (grant TDR/ID 910048), the French Ministry of Cooperation, the Société Elis, companies UTA and Air Afrique and Région Rhône-Alpes grant under the Priority Thematic Program "Emerging and transmissible diseases, parasitic diseases" (grants 00816028 and 00816053). Institutions that coparticipated in this work were as follows: Defence Research and Development Toronto, Canada; EA 3734 Neurobiologie des états de vigilance, Université Claude-Bernard, Lyon, France; EA 3174 Institut de neurologie tropicale, Université de Limoges, France; Laboratoire de physiologie, Faculté de médecine, Abidjan, Côte d'Ivoire; Laboratoire de physiologie et de psychologie environnementales, CNRS, Strasbourg, France; Projet de recherches cliniques sur la trypanosomiase, Daloa, Côte d'Ivoire; Service des grandes endémies, Brazzaville, Congo; Service de neurologie, CHU de Brazzaville, Congo. The software used for ambulatory studies was provided by PhiTools Company (Strasbourg, France).

REFERENCES

1. De Raadt P. Historique de la maladie du sommeil. *Med Trop* **64**:116–117(2004).

2. WHO Expert Committee on control and surveillance of African trypanosomiasis. *WHO Technical Report Series, No. 881.* World Health Organization, Geneva, 1998.

3. Stich A, Barrett MP, Krishna S. Waking up to sleeping sickness. *Trends Parasitol* **19**:195–197(2003).

4. Louis F, Bouteille B, Pere P, Buguet A. La résurgence de la maladie du sommeil. In: Haut comité pour la défense civile (Ed), *Le risque biologique.* Masson, Paris, 2004, pp 85–102.

5. Lorenz P, Betschart B, Owen JS. *Trypanosoma brucei brucei* and high-density lipoproteins: old and new thoughts on the identity and mechanism of the trypanocidal factor in human serum. *Parasitol Today* **11**:348–352(1995).

6. Raper J, Nussenzweig V, Tomlinson S. The main lytic factor of *Trypanosoma brucei brucei* in normal human serum is not high density lipoprotein. *J Exp Med* **183**:1023–1029(1996).

7. Vanhamme L, Pays E. The trypanosome lytic factor of human serum and the molecular basis of sleeping sickness. *Int J Parasitol* **34**:887–898(2004).

8. Pays E. Antigenic variation in African trypanosomes. In: Dumas M, Bouteille B, Buguet A (Eds), *Progress in Human African Trypanosomiasis, Sleeping Sickness.* Springer Verlag, Paris, 1999, pp 31–52.

9. Barry JD, Emergy DL. Parasite development and host responses during the establishment of *Trypanosoma brucei* infection transmitted by tsetse fly. *Parasitology* **88**:67–84(1984).

10. Greenwood BM, Whittle HC. The pathogenesis of sleeping sickness. *Trans R Soc Trop Med Hyg* **74**:716–725(1980).

11. Bakhiet M, Olsson T, Ljungdahl Å, Höjeberg B, van der Meide P, Kristensson K. Induction of interferon-γ, transforming growth factor-β, and interleukin-4 in mouse strains with different susceptibilities to *Trypanosoma brucei brucei. J Interferon Cytokine Res* **16**:427–433(1996).

12. Buguet A, Banzet S, Bouteille B, Vincendeau P, Tapie P, Doua F, Bogui P, Cespuglio R. NO a cornerstone in sleeping sickness: voltammetric assessment in mouse and man. In: Moravec J, Takeda N, Singal PK (Eds), *Adaptation Biology and Medicine, Volume 3. New Frontiers.* Narosa Publishing House, New Delhi, India, 2002, pp 222–230.

13. Chao C, Hu S, Petersen PK. Glia, cytokines, and neurotoxicity. *Crit Rev Neurobiol* **9**:189–205(1995).

14. Pentreath VW. Neurobiology of sleeping sickness. *Parasitol Today* **7**:215–218(1989).

15. Okomo-Assoumou MC, Geffard M, Daulouède S, Chaugier C, Lemesre JL, Vincendeau P. Circulating antibodies directed against tryptophan-like epitopes in sera of patients with human African trypanosomiasis. *Am J Trop Med Hyg* **52**:461–467(1995).

16. Dumas M, Bisser S. Clinical aspects of human African trypanosomiasis. In: Dumas M, Bouteille B, Buguet A (Eds), *Progress in Human African Trypanosomiasis, Sleeping Sickness.* Springer Verlag, Paris, 1999, pp 215–233.

17. Dumas M, Bouteille B, Buguet A. *Progress in Human African Trypanosomiasis, Sleeping Sickness.* Springer Verlag, Paris, 1999.

18. Bisser S, Lejon V, Preux PM, Bouteille B, Stanghellini A, Jauberteau MO, Büscher P, Dumas M. Blood–cerebral fluid barrier and intrathecal immunoglobulins compared to field diagnosis of central nervous system involvement in sleeping sickness. *J Neurol Sci* **193**:127–135(2002).

19. Lejon V, Boelaert M, Jannin J, Moore A, Büscher P. The challenge of *Trypanosoma brucei gambiense* sleeping sickness diagnosis outside Africa. *Lancet Infect Dis* **3**:804–808(2003).

20. Lejon V, Claes F, Tran T, Büscher P. Des progrès dans le domaine du diagnostic de la trypanosomiase humaine africaine. *Med Trop* **64**:124–125(2004).

21. Papadopoulos MC, Abel PM, Agranoff D, Stich A, Tarelli E, Bell BA, Planche T, Loosemore A, Saadoun S, Wilkins P, Krishna S. A novel and accurate diagnostic test for human African trypanosomiasis. *Lancet* **363**:1358–1363(2004).

22. Büscher P, Draelants F, Magnus E, Vervoort T, Van Meirvenne N. An experimental latex agglutination test for antibody detection in human African trypanosomiasis. *Ann Soc Belge Med Trop* **71**:267–273(1991).

23. Lejon V, Büscher P. Stage determination and follow-up in sleeping sickness. *Med Trop* **61**:355–360(2001).

24. Bouteille B, Oukem O, Bisser S, Dumas M. Treatment perspectives for human African trypanosomiasis. *Fund Clin Pharmacol* **17**:171–181(2003).

25. Bauquerez R, Jannin J. La vitalité de nouveaux engagements. *Med Trop* **64**:125–126(2004).

26. Friedheim EAH. Mel B in the treatment of human trypanosomiasis. *Am J Trop Med Hyg* **29**:173–180(1949).

27. Apted FIC. Present status of chemotherapy and chemoprophylaxis of human trypanosomiasis in the eastern hemisphere. *Pharmacol Ther* **11**:391–413(1980).

28. Coulaud J, Caquet R, Froli G, Saimot G, Pasticier A, Payet M. Atteintes rénale et pancréatique sévères au cours d'un traitement par la pentamidine d'une trypanosomiase africaine. *Ann Med Interne* (*Paris*) **10**:665–669(1975).

29. Burri C, Baltz T, Giroud C, Doua F, Welker HA, Brun R. Pharmacokinetic properties of the trypanocidal drug melarsoprol. *Chemotherapy* **39**:225–234(1993).

30. Neujean G. Contribution à l'étude des liquides rachidiens et céphaliques dans la maladie du sommeil à *Trypanosoma gambiense. Ann Soc Belge Med Trop* **30**:1125–1387 (1950).

31. Burri C, Nkunku S, Merolle A, Smith T, Blum J, Brun R. Efficacy of a new, concise schedule for melarsoprol in treatment of sleeping sickness caused by *Trypanosoma brucei gambiense*: a randomised trial. *Lancet* **355**:1419–1425(2000).

32. Pépin J, Milord F, Khonde A, Niyonsenga T, Loko L, Mpia B. Risk factors for encephalopathy and mortality during melarsoprol treatment of *Trypanosoma brucei gambiense* sleeping sickness. *Trans R Soc Trop Med Hyg* **89**:92–97(1995).

33. Dutertre J, Labusquière R. La thérapeutique de la trypanosomiase. *Med Trop* **26**:342–356(1966).

34. Bertrand E, Rive J, Serié F, Kone I. Encéphalopathie arsenicale et traitement de la trypanosomiase. *Med Trop* **33**:385–390(1973).

35. Buyst H. The treatment of *T. rhodesiense* sleeping sickness with special reference to its physio-pathological and epidemiological basis. *Ann Soc Belge Med Trop* **55**:95–104(1975).

36. Apted FIC. Four years' experience of melarsen oxide/BAL in the treatment of late-stage Rhodesian sleeping sickness. *Trans R Soc Trop Med Hyg* **5**:75–86(1957).

37. Legros D, Evans S, Maiso F, Enyaru JCK, Mbulamberi D. Risk factors for treatment failure after melarsoprol for *T. b. gambiense* trypanosomiasis in Uganda. *Trans R Soc Trop Med Hyg* **93**:439–442(1999).

38. Stanghellini A, Josenando T. The situation of sleeping sickness in Angola: a calamity. *Trop Med Int Health* **6**:330–334(2001).

39. Bitonti AJ, Mc Cann PP, Sjoerdsma A. Necessity of antibody response in the treatment of African trypanosomiasis with alpha-difluoromethylornithine. *Biochem Pharmacol* **35**:331–334(1986).

40. Milord F, Pépin J, Loko L, Ethier L, Mpia B. Efficacy and toxicity of eflornithine for treatment of *Trypanosoma brucei gambiense* sleeping sickness. *Lancet* **340**:652–655(1992).

41. Iten M, Matovu E, Brun R, Kaminsky R. Innate lack of susceptibility of Ugandan *Trypanosoma brucei rhodesiense* to DL-α-difluoromethylornithine(DFMO). *Trop Med Parasitol* **46**:190–195(1995).

42. Donkor IO, Assefa H, Rattendi D, Lane S, Vargas M, Goldberg B, Bacchi C. Trypanocidal activity of dicationic compounds related to pentamidine. *Eur J Med Chem* **36**:531–538(2001).

43. Mackensie S. La maladie du sommeil en Afrique. *Mercredi Med* **47**:597–598(1890).

44. Manson-Bahr P. *Manson's Tropical Diseases. A Manual of Diseases of Warm Climates.* Cassel, London, 1942.

45. Lhermitte J. *La maladie du sommeil et les narcolepsies*, Severeyns L (Ed). Etablissements d'Imprimerie, Paris, 1910.

46. Van Campenhout. La maladie du sommeil et les narcolepsies: notes concernant la maladie du sommeil (Trypanosomiase). *Rev Neurol* 200–203(1910).

47. Rechtschaffen A, Kales A. *A Manual of Standardized Terminology, Techniques and Scoring System for Sleep Stages of Human Subjects.* US Government Printing Office,

Washington, DC, 1968, National Institute of Health Publ 204, Public Health Service.

48. Bert J, Collomb H, Fressy J, Gastaud H. Etude électroencéphalographique du sommeil nocturne au cours de la trypanosomiase humaine africaine. In: Fischgold H (Ed), *Le sommeil de nuit normal et pathologique. Etudes électroencéphalographiques.* Masson, Paris, 1965, pp 334–352.

49. Schwartz BA, Escande C. Sleeping sickness: sleep study of a case. *Electroencephalogr Clin Neurophysiol* **29**:83–87(1970).

50. Sanner BM, Büchner N, Kotterba S, Zidek W. Polysomnography in acute African trypanosomiasis. *J Neurol* **415**:878–879(2000).

51. Billiard M, Ondzé B. Troubles de l'éveil. Deuxième partie: troubles secondaires de l'éveil. *Rev Neurol* **157**:480–496(2001).

52. Billiard M, Callander B. Other hypersomnias. In: Billiard M (Ed), *Sleep, Physiology, Investigations and Medicine.* Kluwer Academic/Plenum Publishers, New York, 2003, pp 447–456.

53. Monge-Strauss MF, Millet AL, Bicakova-Rocher A, Mellerio F, De Recondo A, Florencio S. Désorganisation du rythme circadien veille/sommeil dans la trypanosomiase africaine humaine: recueil de deux enregistrements continus pendant 48 h à 2 mois d'intervalle chez un sujet traité par alpha-difluorométhylornithine(DFMO). *Neurophysiol Clin* **27**:157–158(1997).

54. Buguet A, Gati R, Sèvre JP, Develoux M, Bogui P, Lonsdorfer J. 24 hour polysomnographic evaluation in a patient with sleeping sickness. *Electroencephalogr Clin Neurophysiol* **72**:471–478(1989).

55. Buguet A, Rivolier J, Jouvet M. Human sleep patterns in Antarctica. *Sleep* **10**:374–382(1987).

56. Buguet A, Tapie P, Bisser S, Chapotot F, Banzet S, Bogui P, Cespuglio R. Sleep in tropical Africa: at the laboratory and in villages without electrical power. *J Sleep Res* **11** (Suppl 1): 30(2002).

57. Adam K. Sleep is changed by blood sampling through an indwelling venous catheter. *Sleep* **5**:154–158(1982).

58. Buguet A, Bert J, Tapie P, Tabaraud F, Doua F, Lonsdorfer J, Bogui P, Dumas M. Sleep–wake cycle in human African trypanosomiasis. *J Clin Neurophysiol* **10**:190–196(1993).

59. Tapie P, Buguet A, Tabaraud F, Bogui P, Doua F, Bert J. EEG and polygraphic features in 24-hour recordings in sleeping sickness and healthy Africans. *J Clin Neurophysiol* **13**:339–344(1996).

60. Buguet A, Tapie P, Bert J. Reversal of the sleep–wake cycle disorder of sleeping sickness after trypanosomicide treatment. *J Sleep Res* **8**:225–235(1999).

61. Montmayeur A, Brosset C, Imbert P, Buguet A. Cycle veille-sommeil au décours d'une trypanosomose humaine africaine à *Trypanosoma brucei rhodesienne* chez deux parachutistes français. *Bull Soc Pathol Exot* **87**:368–371(1994).

62. Darsaud A, Bourdon L, Mercier S, Chapotot F, Bouteille B, Cespuglio R, Buguet A. Twenty-four hour disruption of the sleep–wake cycle and sleep-onset REM-like episodes in a rat model of African trypanosomiasis. *Sleep* **27**:42–46(2004).

63. Darsaud A, Bourdon L, Chevrier C, Keita M, Bouteille B, Queyroy A, Canini F, Cespuglio R, Dumas M, Buguet A. Clinical follow-up in the rat experimental model of African trypanosomiasis. *Exp Med Biol* (*Maywood*) **228**:1355–1362(2003).

64. Cox FEG. History of sleeping sickness (African trypanosomiasis). *Infect Dis Clin North Am* **18**:231–245(2004).

65. Kristjanson PM, Swallow BM, Rowlands GJ, Kruska RL, De Leeuw PN. Measuring the costs of African animal trypanosomiasis, the potential benefits of control and returns to research. *Agr Syst* **59**:79–98(1999).

66. Milleliri JM, Louis FC, Buguet A. Premier congrès international de Brazzaville sur la mouche tsé-tsé et les trypanosomoses, 23–25 mars 2004. *Med Trop* **64**:115–126(2004).

24

MEDICATIONS THAT INDUCE SLEEPINESS

J. F. PAGEL

University of Colorado Medical School, Denver, Colorado

INTRODUCTION

Drug-induced "sleepiness" is one of the most common effects and side effects of central nervous system (CNS) active drugs. Antidepressants, antipsychotics, antihypertensives, antiepilectics, and other CNS active agents may have sedative effects utilized clinically in selected patients. However, an increase in motor vehicular accidents is associated with drugs that induce daytime sleepiness. This side effect of waking sedation can be the limiting factor in the use of many CNS active agents. Validated psychological and performance tests have not yet been utilized to assess alterations in daytime performance associated with many of the agents known to induce daytime sedation as a side effect. The newer nonbenzodiazepine sedative hypnotic agents have side effect profiles that are more benign than those associated with many of the OTC medications commonly used to treat insomnia. These agent have Low addictive potential, low toxicity, and minimal residual day time sleepiness.

MEDICATIONS THAT INDUCE SLEEPINESS

Sleepiness is perhaps the most commonly reported side effect of CNS active pharmacological agents (the 1990 Drug Interactions and Side Effect Index of the *Physicians' Desk Reference* lists drowsiness as a side effect of 584 prescription or over-the-counter (OTC) preparations). Unfortunately, the terminology describing daytime sleepiness, generally considered to be "the subjective state of sleep need," is poorly defined, interchangeably including such contextual terminology as sleepiness, drowsiness, languor, inertness, fatigue, and sluggishness [1]. Subjective behavioral complaints may not accurately reflect the results of tests for physiological sleepiness [2].

Psychological tests utilized in assessing daytime sleepiness include validated questionnaires, spontaneously occurring psychological variations, and performance measures. The most widely used questionnaires for sleepiness include the Stanford Sleepiness Scale (SSS) and the Epworth Sleepiness Scale, assessing patient reports of sleepiness induced limitations on waking behavior. The most prominent tests developed to utilize spontaneously occurring psychological variations are the Multiple Sleep Latency Test (MSLT) and the Maintenance of Wakefulness Test (MWT). Both of these tests use modified polysomnography to assess sleep onset latency during a series of waking nap periods. The effects of sleepiness on daytime performance can be assessed by tests of complex reaction and coordination, or by tests that assess complex behavioral tasks likely to be affected by sleepiness (i.e., tests of driving performance) [3]. Many of these tests have been shown in general and controlled (i.e., clinical trial) populations to have sensitivity to low-level sedation, highly reproducible results, and correlations with effects on the actual real life analog of driving performance and accident data [4]. This data has been shown to correlate with epidemiological studies suggesting that several groups of drugs known to induce daytime sleepiness are associated with increased rates of automobile accidents (e.g., sedating antihistamines, long-acting benzodiazepines, and sedating antidepressants) [4, 5]. Few of these tests have, however, been generally applied to the vast majority of drugs noted to induce sleepiness

Sleep: A Comprehensive Handbook, Edited by T. Lee-Chiong.
Copyright © 2006 John Wiley & Sons, Inc.

as a side effect. Since performance measures are susceptible to nontask-related influences of motivation, distraction, and comprehension of instructions, the results of performance and questionnaire rating tests do not always correlate with the results obtained from the MSLT and MWT [1].

POPULATION GROUPS WITH ABNORMALITIES OF ALERTNESS/SLEEPINESS

Diagnoses that lead to alternations in sleep and alertness are quite common. Obstructive sleep apnea (OSA) with its well-described effect of daytime somnolence affects approximately 5% of the population. Some of these patients with OSA also have chronic insomnia, yet treatment of these patients with sedative/hypnotic medications can cause respiratory depression, increased apnea, and worsened sleep in these patients. Patients with narcolepsy often report improved sleep with daytime amphetamine use. Periodic limb movement disorder (PLMD) may positively respond to benzodiazepine treatment yet increase in intensity with the use of some antidepressants. Increased daytime alertness typifies a spectrum of common diagnoses including chronic insomnia, anxiety disorder, and post-traumatic stress disorder. Such patients may demonstrate altered responses to medications inducing alertness and/or sleepiness with unexpected results. Stimulants may induce sleepiness in some patients while hypnotics may induce agitation and insomnia even when used in anesthetic settings and dosages [3].

NEUROTRANSMITTER SYSTEMS MODULATING SLEEPINESS

The effects of most sedative–hypnotic drugs on sedation and arousal are secondary to selective neurotransmitter effects rather than through nonspecific CNS depression. The neuronal systems modulating waking and sleep are contained within the isodendritic core of the brain extending from the medulla though the brainstem and hypothalamus up to the basal forebrain. Multiple factors and systems are involved, with no single chemical neurotransmitter identified as necessary or sufficient for modulating sleep and wakefulness. Neurons are likely to utilize more than one neurotransmitter and secondary signaling systems [6]. The ubiquitous and fast-acting neurotransmitters of the CNS have a multiplicity of highly site-specific roles that may result in net changes in behavioral states and cortical arousal that directly contrasts with their effects at the microscopic level. Most drugs with clinical sedative/hypnotic effects or sedative side effects can be shown to affect one or more of the widely dispersed central neurotransmitters

important in the neuromodulation of sleep and wakefulness including dopamine, epinephrine and norepinephrine, acetylcholine, serotonin, histamine, glutamate, gamma-aminobutyrate (GABA), and adenosine. Other neuropeptides known to affect sleep and wakefulness include substance P, corticotrophin releasing factor, thyrotrophin releasing factor, vasoactive intestinal peptide, and neurotensin [7].

MEDICATIONS FOR THE TREATMENT OF INSOMNIA—SEDATIVE/HYPNOTICS

Historically, sedative/hypnotics have been some of the most commonly prescribed drugs. Chloral hydrate was the original "Mickey Finn" slipped into the drinks of unsuspecting marks for the purposes of criminal activity. Unfortunately, the LD_{50} (potentially fatal dose) for chloral hydrate is quite close to the therapeutic dose, and murders rather than robberies were often the result. In the years leading up to the 1960s, barbiturates were commonly utilized for their sedative effects. Unfortunately, these medications can be drugs of abuse and have a significant danger of overdose. Marilyn Monroe, Elvis Presley, and Jim Morrison, among others, were celebrities who died during this era from overdoses of sleeping pills. These medications and similar barbiturate-like medications (methaqualone (Quaalude, Sopor), glutethimide (Doriden), ethchlorvynol (Placidyl), methyprylon (Nodudar)) are still available but should be used sparingly because of their potential for abuse and overdose [8].

In the 1970s benzodiazepines became available for the treatment of insomnia. These drugs act at GABA neuroreceptors and have far less overdose danger and abuse potential than previous medications used for sleep. The many drugs in this class are best viewed therapeutically based on their pharmacodynamics (Table 24.1). Rapid onset of action is characteristic of flurazepam (Dalmane) and triazolam (Halcion), indicating that both these agents have excellent sleep-inducing effects. Flurazepam, like diazepam (Valium) and clorazepate (Tranzene), has the characteristic of having active breakdown products. This results in an extraordinarily long active half-life that can approach 11 days. This prolonged effect in the elderly has been associated with increased auto accidents and falls with hip fractures [4, 9]. Withdrawal from these long-acting agents can be difficult, with an initial syndrome of insomnia followed by persistent anxiety that may extend beyond the half-life of the agent.

Benzodiazepines are rapid eye movement (REM) sleep suppressant medications, and withdrawal often results in episodes of increased REM sleep (REM sleep rebound). REM sleep is known to have a role in learning and memory consolidation. For short-acting agents such as triazolam (Halcion), this rebound occurs during the same night in

TABLE 24.1 Sedative–Hypnotics

Class	Drug	Sleep Stage Effects	Significant Side Effects	Indications
		Benzodiazepines		
Short onset short half-life	Triazolem (Halcion)	Decreased amplitude for stages 3 and 4; increased stage 2 (all) Shortened sleep latency, in night REM sleep rebound	Loss of effect with chronic use, dependence Antegrade amnesia	Transient insomnia
Short onset; medium half-life	Estazolam (ProSom)	Shortened sleep latency, decreased REM sleep	Daytime sleepiness	Transient Insomnia
Short onset; long half-life	Flurazepam (Dalmane)	Shortened sleep latency, decreased REM sleep, withdrawal REM sleep rebound	Daytime sleepiness, chronic buildup (car accidents, hip fractures)	Transient insomnia, anxiety
Medium onset; Medium half-life	Temezepam (Restoril), Clonazepam (Klonopin)	Decreased REM sleep	Daytime sleepiness, poor sleep induction	Transient insomnia, anxiety
		GABA Receptor Agents		
Short onset, medium half-life	Zolpidem (Ambien), Eszopiclone (Estorra)	Shortened sleep latency, increased total sleep time, benzo effects with dose above that normally prescribed	Idiosyncratic daytime sleepiness	Transient, shift work chronic insomniacs, depression with insomnia
Short onset, short half-life	Zaleplon (Sonata), Indiplon	Shortened sleep latency, may redose in night	Idiosyncratic poor sleep induction	Transient, shift work and chronic insomniacs
		Other agents		
Chloral hydrate	Chloral hydrate	Short sleep latency, decreased REM sleep, withdrawal REM sleep rebound	Low lethal dose, loss of effect with chronic use	Transient insomnia in controlled settings
Barbiturates and barbiturate-like agents	Phenobarbitol, methaqualone, glutethimide ethchlorvynol, methylprylon	REM sleep suppression, short sleep latency, decreased REM sleep, withdrawal REM sleep rebound	Addiction, low lethal dose, loss of effect with chronic use	No sleep indications
Sedating antihistamine H_1 blockers	Diphenhydramine	Decreased sleep latency in some patients	Daytime sedation, confusion, not indicated for use in the elderly	OTC, insomnia

which the medication is taken and has been associated with daytime memory impairment, particularly at higher dosages [10]. Temezepam (Restoril) and estazolam (ProSom) have half-lives compatible with an 8 h night of sleep. Temeze-pam, because of its slower onset of action, is less effica-cious as a sleep-inducing agent than other drugs used as hypnotics in this class. All benzodiazepines can result in respiratory depression in patients with pulmonary disease and may lose sleep-inducing efficacy with prolonged use [11, 12].

The newer hypnotics zolpidem (Ambien), zaleplon (Sonata), eszopiclone (LUNESTA), and indiplon are benzodiazepine-like agents, exerting effects at the same GABA receptors. Benzodiazepine withdrawal is not blocked by these agents. These agents have excellent efficacy with minimal side effects. Although any agent used to induce sleep can result in a dependence on that agent to induce sleep, abuse potential for these agents is minimal. Idiosyncratic reactions of persistent daytime somnolence and/or memory loss have been reported in some patients. Tachyphylaxis is unusual, and these agents can be used on a long-term basis. Sleep is altered minimally, and REM rebound is not associated with these agents. (Table 24.1) [8, 10]. Zolpidem and eszopiclone have a 6–8 h half-life, while zaleplon and indiplon are shorter acting (3–4 h). Clinical comparison of these agents suggests that zolpidem and eszopiclone may have greater sleep-inducing efficacy and zaleplon fewer side effects. After extensive long-term clinical trials eszapaclone has been approved for extended use in patients who have chronic insomnia.

In the last thirty years, although the drugs for treatment of insomnia have become safer, the number of sedative/hypnotics prescribed in the United States has declined. This decrease most likely reflects the public and medical community's concern as to the side effects, limitations, and costs of the available hypnotic drugs. Most hypnotic medications, in general, can be safely utilized on a short-term basis for the treatment of transient insomnia. For persistent chronic insomnia due to anxiety disorders, idiopathic insomnia (persistent life-long insomnia without other sleep-associated diagnoses), and persistent agitated depression, chronic hypnotic use with the newer nonbenzodiazepine hypnotics can be justified and is indicated if medication use leads to improvement in waking performance [10, 13]. These newer hypnotic agents are less likely to have deleterious side effects than most OTC treatments for insomnia [14, 15].

OTHER SEDATIVE AGENTS

Ethanol is probably the most widely used hypnotic medication. In patients with chronic insomnia, 22% of patients report using ethanol as a hypnotic [13, 16]. Unfortunately, chronic use to induce sleep can result in tolerance, dependence, and diminished sleep efficiency and quality. When used in excess with other sedative/hypnotic agents, overdose can be fatal.

Over-the-counter sleeping pills contain sedating H_1 antihistamines, usually diphenhydramine, hydroxyzine, or triprolidine. These agents are varyingly effective, generally inducing sedation with acute use, but often inducing increased daytime sleepiness, anticholinergic effects, and cognitive impairment persisting into the day following nighttime use [4]. These agents are not recommended for use in the elderly [11]. Seizure thresholds can be lowered

by their use in epileptic patients. Both ethanol and sedating antihistamines are associated with decreased performance on daytime driving tests and an increase in automobile accidents. Sedation is infrequent with H_2 antagonists (e.g., cimetidine, ranitidine, famotidine, and nizatidine), but somnolence as a side effect is evidently reproducible in susceptible individuals.

ANTIDEPRESSANTS

Sedating antidepressants are often used to treat insomnia. A significant percentage of individuals with chronic insomnia and/or daytime sleepiness also have depressive symptoms. Chronic insomnia itself can lead to depression [16]. Depression associated with insomnia is likely a different diagnostic entity than depression without insomnia, and treatment of the former with nonsedating antidepressants may produce no improvement in sleep even when the underlying depression resolves. Use of antidepressants is limited by side effects (anticholinergic effects, daytime hangover, etc.) and danger with overdose (particularly the tricyclics) [17]. Sedating antidepressants include the tricyclics (amitryptiline, imipramine, nortryptiline, etc.), trazadone (Deseryl), and the newer agents mirtazapine (Remeron) and nefazodone (Serzone). The SSRIs fluoxetine (Prozac) and sertraline (Zoloft) have a tendency to induce insomnia; however, in some patients, paroxetine (Paxil) may induce mild sedation. Most other antidepressants are noted to induce somnolence in some patients (Table 24.2). Depression-related insomnia responds to sedating antidepressants more rapidly and with lower doses compared to other symptoms of depression [15]. In patients with insomnia and concomitant depression, antidepressants are often used in combination with sedative–hypnotic medications. Use of sedating antidepressants has been associated with declines in daytime performance and driving test performance, and an increased potential for involvement in motor vehicular accidents [5].

ANTIEPILECTIC AGENTS

Sedation is one of the most common side effects of older anti-epileptic drugs reported at levels of 70% with phenobarbitol, 42% with carbamazepine and valproate, and in 33% of patients using phenytoin and primidone [18]. Both phenobarbitol and carbamazepine have been shown to induce sleepiness on the Multiple Sleep Latency Test (MSLT). In newer antiepileptic agents, sedation was reported at levels of 15–27% for topiramate and at levels of 5–10% in the clinical trials for gabapentin, lamotrigine, vigabatrin, and zonisamide. The neurochemical basis for

TABLE 24.2   Antidepressants

Class	Drug	Sleep Stage Effects	Indications
Tricyclic	**Trimipramine, nortriptyline, doxepin, amoxapine, amitryptiline, imipramine, protriptyline**[a]	Increased REM sleep latency; Decreased REM sleep (++), SWS latency, deep sleep, sleep latency	Depression with insomnia, REM sleep and SWS suppression, chronic pain, fibromyalgia, enuresis, etc.
Nontricyclic sedating	**Desimprinine, maprotiline, mirtazapine**	Increased REM sleep latency; Decreased SWS latency, REM sleep (++), sleep latency	Depression, depression with insomnia, REM sleep suppression
MAOI	Phenelzine, tranylcypromine	Increased stage 4; Decreased REM sleep latency, REM sleep (+++)	Depression, REM sleep suppression
SSRI	Fluoxetine[a] **paroxetine,** sertraline, fluvoxamine, citalopram, HBR	Increased REM sleep latency, sleep latency, stage 1; Decreased REM sleep	Depression, PTSD, obsessive compulsive disorder, phobias, cataplexy, etc.
SSRI + tricyclic	Venlafaxine	Increased REM sleep latency; Decreased sleep latency, REM sleep	Depression
DA-NA-SSRI	Bupropion	Increased REM sleep latency, sleep latency	Depression, nicotine withdrawal
NONTRICYCLIC non-SSRI	**Nefazodone**	Increased REM sleep; Decreased sleep latency	Depression, depression with insomnia and anxiety
5HT 1a agonist	Buspirone	Increased REM sleep latency; Decreased REM sleep	Anxiety

Key: (++), higher levels of effect; sedating agents in bold type; insomnia-inducing agents in italics.
[a] Documented as respiratory stimulant.

the sedation induced by many of these agents remains undefined except for those agents known to have GABA agonist effects (e.g., gabapentin, phenobarbitol).

ANTIHYPERTENSIVE AGENTS

The complaints of tiredness, fatigue, and daytime sleepiness (2–4.3%) associated with beta-blocker use may occur secondary to disturbed sleep or direct action of the drug. Sleep disturbance appears to be more common with the lipophilic beta blockers (e.g., propranolol). The newer beta-blocking drugs with vasodilating properties (e.g., carvedilol, labetalol) are also associated with reported fatigue and somnolence (3–11%). Sedation is the most common side effect reported for the alpha-2 agonists clonidine and methyldopa (30–75%). Alpha-1 antagonists (e.g., terazosin, prazosin) are sometimes associated with transient sedation. Tests of cognitive and psychomotor performance in patients using these drugs have shown inconsistent results, with improved function in some patients and worsened function in others. Both the sedation and insomnia associated with use of antihypertensive agents are presumed to be secondary to effects on adrenergic receptors involved in neuroregulation of sleep and wakefulness.

ANTIPSYCHOTIC AGENTS

Sedation is a common side effect of the traditional antipsychotics, with chlorpromazine and thioridazine somewhat more sedating than haloperidol. Clinical studies have shown a high incidence of persistent sedation with clozepine (46%) with less frequent reports of sedation with risperidone, olanzapine, sertindole, and quetiapine. Despite the significant levels of sedation associated with clozapine use, improvement in daytime cognitive performance has been documented in selected patients. The sedation associated with these agents is postulated to be primarily associated with effects on histaminic receptors.

NONPRESCRIPTION AGENTS INDUCING DAYTIME SEDATION

In addition to the sedating antihistamines, a variety of nonprescription and herbal agents are marketed as hypnotics. Melatonin is a neural hormone effective in resetting circadian rhythms of sleep and body core temperature through its actions on the suprachiasmatic nucleus. For individuals with insomnia secondary to disruptions in circadian rhythms, melatonin can act as a hypnotic and is a useful adjunct to treatment that often includes cognitive therapies, light exposure, and other hypnotics. The best data supporting the sedative effect of an herbal agent is for valerian [19]. Evidence supporting the hypnotic efficacy of other herbal agents, including kava, chamomile, passionflower, and skullcap, is limited.

Among drugs of abuse, marijuana has significant hypnotic effects. The central sedative effects of barbiturates, barbiturate-like agents, benzodiazepines, and opioids can induce fatal respiratory suppression at higher doses particularly when abuse is coupled with ethanol. Adolescents abusing amphetamines and cocaine may paradoxically present with the complaint from parents of persistent daytime sedation on returning home after long episodes of drug-induced wakefulness.

OTHER AGENTS INDUCING DAYTIME SEDATION

Most agents with CNS activity are reported to induce sleepiness as a side effect in some patients (Table 24.3) [20]. The sedative side effects of some of these agents are used

TABLE 24.3 Medications Reported in Clinical Trials and Case Reports to Have Sleepiness as a Side Effect

Medication Class	Neurochemical Basis for Sleepiness	Specific Medications	Specific Medications
Antihistamines	Histaminine receptor blockade	Azatadine (Optamine), chlorpheniramine (Chlor-Trimeton) Dexbrompheniramine (Polaramine), Clemastine (Tavist), Cyproheptadine (Periactin)	Diphenhydramine (Benadryl), doxylamine (Unisom), promethazine (Phenergan), triprolidine (Actifed)
Antiparkinsonian agents	Dopamine receptor agonists	Benztropine (Cogentin), biperiden (Akineton)	Procyclidine (Kemadrin), trihexiphendyl (Artane)
Antimuscarinic/ antispasmodic	Varied effects	Atropine, belladona, dicyclomine (Bentyl), glycopyrrolate (Robinul), Hyoscyamine	Ipratropium bromide (Atrovent), Mepenzolate bromide (Cantil), methscopolamide bromide (Pamine), scopolamide
Skeletal muscle relaxants	Varied effects	Baclofen (Lioresal), Carisoprodol (Soma), chlorzoxazone (Parafon Forte), cyclobenzaprine (Flexeril)	Dantroline (Dantrium), metaxalone (Skelaxin), methocarbamol (Robaxin), orphenadrine (Norflex)
Alpha-adrenaergic blocking agents	Alpha-1 adrenergic antagonists	Doxazosin (Cardura), prazosin (Minipress)	Terazosin (Hytrin)
Beta-adrenergic blocking agents	Beta-adrenergic antagonists	Acebutolol (Sectral), atenolol (Tenormin), betaxolol (Kerlone), bisoprolol (Zobeta), carvedilol (Coreg), esmolol (Brevibloc)	Labetalol (Normodyne), metoprolol (Lopressor), nadolol (Corgard), pindolol (Visken), propranolol (Inderal), sotalol (Betapace), timolol (Blocadren)
Opiate agonists	Opioid recptor agonists (general CNS depression)	Codeine, fentanyl (Sublimaze), hydrocodone hydromorphine Dilaudid), levomethadyl (Orlamm), levorphanol (Levodromoran), meperidine (Demerol)	Methadone, morphine, opium, oxycodone, oxymorphone, (Numorphan), propoxyphene (Darvon), sufentanil (Sufenta), tramadol (Ultram)
Opiate partial agonists	Opioid recptor agonists (general CNS depression)	Buprenorphine (Buprenex), butorphanol (Stadol)	Nalbuphine (Stadol), pentazocine (Talwin)
		Anticonvulsants	
Barbiturates	GABA receptor agonists	Mephobarbital (Mebaral)	Phenobarbital (Luminal), primidone (Mysoline)
Benzodiazepines	GABA receptor agonists	Elonazepam (Klonopin)	
Hydantoins	General effects?	Ethotoin (Peganone), phenytoin (Dilantin)	Fosphenytoin (Cerebyx)
Succinimides	General effects?	Ethosuximide (Zarontin)	Methsuximide (Celontin)
Other	Varied effects including GABA potentiation	Carbamazine (Tegretol), felbamate (Felbatol), gabapentin (Neurontin), lamotrigine (Lamictal), levetiracetam (Keppra)	Oxacarbazepine (Trleptal), tigabine (Gabitril), topiramate (Topamax), valproic acid (Depakene), zonisamide (Zonegran)

TABLE 24.3 (*Continued*)

Medication Class Specific Medications	Neurochemical Basis for Sleepiness	Specific Medications		Specific Medications

Antidepressants

Medication Class	Neurochemical Basis	Specific Medications	Specific Medications
MAOI	Norepinephrine, 5HT and dopamine effects	Phenelzine (Nardil)	Tranylcypromine (Parnate)
Tricyclic	Acetylcholine blockade, norepinephrine and 5HT uptake inhibition	Amitryptyline (Elavil), clomipramine (Anafranil), desimpramine (Norpramin), doxepin (Sinequan), imipramine (Tofranil)	Maprotiline (Ludiomil), nortriptyline (Pamelor), protriptyline (Vivactil), tripramine (Surmontil)
SSRI	5HT uptake inhibition	Citalopram (Celexa), escitalopram (Lexapro), fluoxetine (Prozac)	Fluvoxamine (Luvox), paroxetine (Paxil), sertraline (Zoloft)
Others Antidepressants	5HT, dopanine, and norepinephrine effects	Bupropion (Wellbutrin), mirtazapine (Remeron), nafazodone (Serazone)	Trazadone (Deseryl), venlafaxine (Effexor)
Antipsychotics	Dopamine receptor blockade, varied effects on histaminic, cholinergic, and alpha-adrenergic receptors	Fluphenazine (Prolixin), mesoridazine (Serentil), perphenazine (Trilafon), prochlorperazine (Compazine), thioridazine (Mellaril), trifluperazine (Stelazine), aripiprazole (Abilify), clozapine (Clozaril)	Haloperidol (Haldol), loxapine (Loxitane), molidone (Moban), olanzapine (Zyprexia), pimozide (Orap), quetiapine (Seroquel), risperidone (Risperdal), thiothixene (Navane), ziprasidone (Geodone)
Barbiturates	GABA agonists	Amobarbital (Amytal), butabarbital (Butisol), mephobarbital (Mebaral), pentobarbitol (Nembutal)	Phenobarbitol (Luminal), secobarbitol (Seconal), secobarbitol/amobarbitol (Tuinal)
Benzodiazepines	GABA agonists	Alprazolam (Xanax), chlordiazepoxide (Librium), clorazepate (Tranxene), diazepam (Valium), estazolam (ProSom), flurazepam (Dalmane)	Lorazepam (Ativan), midazolam (Versed), oxazepam (Serax), quazepam (Doral), emezepam (Restoril), triazolam (Halcion)
Anxiolytics, miscellaneou sedatives and hypnotics	GABA agonists, varied effects	Buspirone (Buspar), chloral hydrate, dexmedetomidine (Precedex), droperidol (Inapsine), hydroxyzine (Vistaril, Atarax)	Meprobamate (Equanil,Miltown), promethazine (Phenergan), zaleplon (Sonata), zolpidem (Ambien), eszopiclone (Estorra)
Antitussives	General?	Benzonatate (Tessalon)	Dextromethorphan (Robitussin)
Antidiarrhea agents	Opioid, general?	Diphenoxylate (Lomotile)	Loperamide (Imodium)
Antiemetics	Antihistamine and varied effects	Dimenhydrinate (Dramamine), diphenidol (Vontrol), meclizine (Antivert), prochlorperazine (Compazine)	Triethylperazine (Torecan), trimethobenzamide (Tigan), metoclopramide (Reglan)
Genitourinary smooth muscle relaxants	General?	Flavoxate (Urispas), oxybutynin (Ditropan)	Tolterodine (Detrol)

clinically in specific situations (e.g., muscle relaxants in patients at bed rest, beta blockers in patients with performance anxiety, and opiates in patients unable to sleep secondary to chronic pain). However, sleepiness is a common and often unwanted side effect for many types of prescription medications including commonly used antitussives, antiemetics, antidiarrhea agents, and genitourinary smooth muscle relaxants (Table 23.3). These drug-induced side effects can limit the use of these agents in patients in which the level of persistent daytime sleepiness affects waking. Drug-induced sedation has particularly deleterious effects on the performance of complex tasks such as the operation of motor vehicles. Patients should be advised of the negative effects of potentially sedating drugs on waking performance.

CONCLUSION

Drug-induced sedation is one of the most common effects and side effects of CNS active drugs. The newer nonbenzodiazepine sedative–hypnotic agents have lower addictive potential and toxicity than older agents and can be utilized on a long-term basis in patients with chronic insomnia. Benzodiazepines, antidepressants, and other agents may be utilized for their sedative side effects in anxious and insomniac patients. Drugs known to induce daytime sleepiness are associated with declines in daytime performance and increased rates of automobile accidents. This side effect of waking sedation can be the limiting factor in the use of many CNS active agents.

REFERENCES

1. Buysse DJ. Drugs affecting sleep, sleepiness and performance. In Monk TM (Ed), *Sleep, Sleepiness, and Performance* Wiley, West Sussex, England, 1991, pp 4–31.

2. Roehrs T, Carskadon MA, Dement WC, Roth T. Daytime sleepiness and alertness. In: Kryger M, Roth T, Dement WC(Eds), *Principles and Practice of Sleep Medicine*, 3rd ed. Saunders, Philadelphia, 2000, pp 43–53.

3. Pivik RT. The several qualities of sleepiness: psychophysiological considerations. In: Monk TM(Ed), *Sleep, Sleepiness and Performance* Wiley, West Sussex, England, 1991, pp 3–38.

4. O'Hanlon JF, Vermeeren A, Uiterwijk MMC, Van Veggel LMA, Swijgman HF. Anxiolytics' effects on the actual driving performance of patients and healthy volunteers in a standardized test. *Neuropsychobiology* 31:81–88(1995).

5. Volz HP, Sturm Y. Antidepressant drugs and psychomotor performance. *Neuropsychobiology* 31:146–155 (1995).

6. Schwartz JH. Neurotransmitters. In: Kandel ER, Schwartz JH, Jessell TM(Eds), *Principles of Neural Science*, 4th ed. McGraw Hill, New York, 2000, pp 280–297.

7. Jones BE. Basic mechanisms of the sleep–wake states. In: Kryger M, Roth T, Dement WC(Eds), *Principles and Practice of Sleep Medicine*, 3rd ed. Saunders, Philadelphia, 2000, pp 134–154.

8. Mitler NM. Nonselective and selective benzodiazepine receptor agonists: Where are we today? *Sleep* 23(Suppl 1): S39–S47(2000).

9. Ray WA, Griffen MR, Downey W. Benzodiazepines of long and short elimination half life and the risk of hip fracture. *JAMA* 262:3303–3307(1989).

10. Doghramji K. The need for flexibility in dosing hypnotic agents. *Sleep* 23(Suppl 1):S16–S20(2000).

11. Ashton H. The effects of drugs on sleep. In: Cooper R(Ed), *Sleep*. Chapman & Hall Medical, London, 1994, pp 174–207.

12. George CFP. Perspectives in the management of patients with chronic respiratory disorders. *Sleep* 23(Suppl 1):S31–S35(2000).

13. Sateia MJ, Doghramji K, Hauri PJ, et al. Evaluation of chronic insomnia. *Sleep* 23:243–314(2000).

14. O'Hanlon JF, Ramaekers JG. Antihistamine effects on actual driving performance in a standard driving test: a summary of Dutch experience, 1989–94. *Allergy* 50:234–242(1995).

15. Pagel JF, Parnes BL. Medications for the treatment of sleep disorders: an overview. Primary Care companion. *J Clin Psychiatry* 3(3):118–125(2001).

16. Breslau N, Roth T, Rosenthal L, et al. Sleep disturbance and psychiatric disorders: a longitudinal epidemiological study of young adults. *Biol Psychiatry* 39:411–418(1996).

17. Pagel JF. The treatment of insomnia. *Am Fam Physician* 49:1417–1422(1994).

18. Schweitzer PK. Drugs that induce sleep and wakefulness. In: Kryger M, Roth T, Dement WC(Eds), *Principles and Practice of Sleep Medicine*, 3rd ed. Saunders, Philadelphia, 2000, pp 441–461.

19. Stevinson C, Ernst E. Valerian for insomnia: a systematic review of randomized clinical trials. *Sleep Med* 1(2):91–99(2000).

20. *AHFS Drug Information*. American Society of Health-System Pharmacists, Bethesda, MD, 2003.

25

EVALUATION OF EXCESSIVE SLEEPINESS

MERRILL S. WISE

Baylor College of Medicine, Houston, Texas

INTRODUCTION

Characterization of excessive sleepiness is one of the most important challenges faced by the sleep medicine specialist. Pathological sleepiness occurs in association with a variety of sleep disorders, medical and psychiatric diseases, and inadequate sleep and due to medication side effects. Excessive sleepiness is defined as sleepiness that occurs in a situation when an individual would usually be expected to be awake and alert [1]. It is a chronic problem for approximately 5% of the general population, and it is associated with significant morbidity and increased mortality risk to the individual and others [2, 3]. The identification of pathological sleepiness begins the process of establishing a proper diagnosis and allows initiation of treatment and follow-up of the individual's response.

The diagnostic process begins with and is based primarily on a thorough sleep and general medical history. Since some patients are not fully aware of the extent of their excessive sleepiness, the sleep specialist may need to query the patient's spouse, co-workers, or others who are in daily contact with the patient. Questionnaires such as the Epworth Sleepiness Scale provide information about the individual's subjective perception of sleepiness on a daily basis [4] (see Tables 25.1 and 25.2). Sleep diaries provide documentation of the individual's typical sleep–wake schedule. Selective use of nocturnal polysomnography provides useful objective data regarding sleep architecture and continuity, respiratory disturbance, limb movements or abnormal behaviors in sleep, and other intrinsic sleep pathology. The Multiple Sleep Latency Test is performed during the day to quantify the degree of sleepiness and to identify sleep-onset REM periods as part of the evaluation of individuals with possible narcolepsy [5]. Several other techniques have been used to characterize daytime sleepiness such as pupillography and continuous EEG/video monitoring, but these tools are not widely used in routine clinical practice.

CLINICAL HISTORY AND EXAMINATION

During the patient encounter, the examiner may observe clinical features of sleepiness such as ptosis, yawning, blank facial expression, loss of postural tone, and overt sleep. Patients with pathological sleepiness are notorious for overestimating or underestimating their degree of sleepiness. A helpful approach involves having the patient review the frequency and severity of sleepiness in a variety of circumstances. Individuals with mild excessive sleepiness tend to fall asleep primarily in sedentary situations such as riding in an automobile or during long lectures. Individuals with moderate to severe sleepiness fall asleep in more active situations such as while sitting and talking with another person, or while driving. Those with severe pathological sleepiness may fall asleep suddenly in the midst of an active conversation, while standing up, while taking a shower, or while eating.

The sleep history should include a detailed review of the patient's usual sleep–wake schedule in order to identify problems with inadequate nocturnal sleep or circadian problems. Individuals who perform shift work, college students, and those with medical or military professions are especially prone to irregular sleep patterns and inadequate

Sleep: A Comprehensive Handbook, Edited by T. Lee-Chiong.
Copyright © 2006 John Wiley & Sons, Inc.

TABLE 25.1   The Epworth Sleepiness Scale

How likely are you to doze off or fall asleep in the following situations, in contrast to feeling just tired? This refers to your usual way of life in recent times. Even if you have not done some of these things recently, try to work out how they would have affected you. Use the following scale to choose the *most appropriate number* for each situation:

0 = would never doze
1 = slight chance of dozing
2 = moderate chance of dozing
3 = high chance of dozing

Situation[a]	Chance of Dozing
Sitting and reading	_____
Watching TV	_____
Sitting, inactive in a public place (e.g., a theater or a meeting)	_____
As a passenger in a car for an hour without a break	_____
Lying down to rest in the afternoon when circumstances permit	_____
Sitting and talking to someone	_____
Sitting quietly after a lunch without alcohol	_____
In a car, while stopped for a few minutes in traffic	_____

[a] The numbers in the eight situations are added together to give a global score between 0 and 24. Table 24.2 summarizes scores for various conditions.
Source: Used with permission from [4].

TABLE 25.2   Groups of Experimental Subjects, Ages, and Epworth Sleepiness Scale Scores

Subjects/ Diagnoses	Number (M/F)	Age in Years (Mean ± SD)	ESS Scores (Mean ± SD)
Normal controls	30 (14/16)	36.4 ± 9.9	5.9 ± 2.2
Primary snoring	32 (29/3)	45.7 ± 10.7	6.5 ± 3.0
Obstructive sleep apnea syndrome	55 (53/2)	48.4 ± 10.7	11.7 ± 4.6
Narcolepsy	13 (8/5)	46.6 ± 12.0	17.5 ± 3.5
Idiopathic hypersomnia	14 (8/6)	41.4 ± 14.0	17.9 ± 3.1
Insomnia	16 (6/12)	40.3 ± 14.6	2.2 ± 2.0
Periodic limb movement disorder	18 (16/2)	52.5 ± 10.3	9.2 ± 4.0

Source: Used with permission from [4].

sleep. For the patient with inadequate nocturnal sleep, having the patient increase the duration of sleep on a consistent basis can be diagnostically helpful. The patient who continues to experience pathological sleepiness despite improved sleep hygiene and adequate sleep duration is likely to have a primary sleep disorder or medical disorder.

TABLE 25.3   Conditions Associated with Excessive Sleepiness

Extrinsic causes	Insufficient sleep syndrome
	Inadequate sleep hygiene
	Environmental sleep disorder
	Circadian rhythm problems
	Jet lag syndrome
	Shift work syndrome
	Medication-related hypersomnia
Intrinsic causes	Obstructive sleep apnea syndrome
	Narcolepsy
	Idiopathic hypersomnia
	Periodic limb movement disorder
	Restless legs syndrome
	Periodic hypersomnia
	Circadian rhythm sleep disorders
	Delayed sleep phase syndrome
	Irregular sleep–wake pattern
	Non-24 h sleep–wake disorder
	Structural CNS lesions
	Obstructive hydrocephalus
	Hypothalamic and diencephalic lesions

Table 25.3 summarizes conditions that are commonly associated with excessive sleepiness.

A thorough review of the patient's medication history will sometimes uncover a sedating medication, or evidence of medication toxicity. Medications that may be associated with excessive sleepiness include barbiturates, benzodiazepines, trazodone, tricyclic antidepressants, antipsychotic medications, lithium, clonidine, propranolol, and traditional antihistamines such as diphenhydramine. Exposure to toxins such as heavy metals may produce an encephalopathy that includes somnolence.

The terms used by patients to describe their symptoms may require clarification. The patient who complains of fatigue or tiredness is generally describing a feeling of listlessness, loss of energy, or weakness, but these individuals tend not to report overt sleepiness. Fatigue should be differentiated from excessive sleepiness, which is characterized by lapses in awareness, impaired cognitive and motor function, and a propensity for transition into overt sleep. This latter history is suggestive of a sleep disorder, whereas chronic fatigue or "feeling tired" generally indicates a medical or psychiatric disorder. Table 25.4 summarizes conditions associated with chronic fatigue. Excessive sleepiness should be differentiated from lethargy or stupor, problems that generally indicate a neurological disturbance due to a medical disorder or a structural central nervous system lesion. This distinction may be difficult in patients with severe respiratory disturbance during sleep or the patient in an ICU setting with chronic severe sleep deprivation.

TABLE 25.4 Conditions Associated with Chronic Fatigue

Psychiatric conditions	Depression
	Anxiety with or without insomnia
	Personality disorders
Medical conditions	Chronic infection
	Malignancy
	Autoimmune disorders
	Hypothyroidism
Neurological conditions	Myasthenia gravis
	Multiple sclerosis
	Parkinson's disease
	Myopathies and neuropathies
Chronic fatigue syndrome	
Pregnancy	

Individuals with excessive sleepiness should be questioned carefully about respiratory problems during sleep. Sleepiness with persistent loud snoring, especially snoring in all sleeping positions, is strongly suggestive of obstructive sleep apnea. Bed partner reports of gasping or choking sounds or observed apnea are also features that strongly suggest obstructive sleep apnea. These symptoms indicate the need for overnight polysomnography to assess respiratory function during sleep.

The patient with narcolepsy reports chronic daily sleepiness regardless of the quantity of nocturnal sleep obtained. Narcolepsy is generally associated with moderate to severe daily sleepiness that consists of periodic overwhelming sleep attacks superimposed on a background of chronic sleepiness across the day. If the narcolepsy patient is allowed to sleep, he/she generally reports feeling refreshed for one to several hours, followed by the return of overwhelming sleepiness. The history should incorporate questions regarding auxiliary symptoms of narcolepsy, including cataplexy, sleep paralysis, and hypnagogic hallucinations [6].

Cataplexy is characterized by sudden bilateral loss of tone, often precipitated by sudden emotion such as laughter or fright [6]. The loss of tone may range from mild loss of facial expression or head nodding, to severe loss of posture and falling to the ground. Cataplexy may require differentiation from atonic seizures, syncope, transient ischemic attacks, and pseudoseizures [6]. Sleep paralysis is characterized by inability to move or speak around the time of transition into or out of sleep, with intact consciousness. This phenomenon is often frightening to patients, and it may end suddenly with tactile stimulation or a sudden sound. Hypnagogic hallucinations are usually visual or auditory experiences that occur during the transition into or out of sleep. The auxiliary symptoms of narcolepsy are thought to represent the abnormal intrusion of REM sleep phenomena into the waking state [6]. The individual with narcolepsy often experiences fragmented nocturnal sleep,

and narcolepsy patients are not spared from the possibility of coexisting obstructive sleep apnea, periodic limb movement disorder, or other sleep disorders.

A special challenge exists with regard to individuals with structural brain lesions leading to somnolence. Patients with obstructive hydrocephalus generally experience progressive deterioration in level of consciousness, headache, nausea, and possibly vomiting. The patient with intracranial hemorrhage such as subdural hematoma or expanding epidural hematoma may complain of headache and may manifest focal neurological deficits. Occasionally, sleepiness may be confused with the lapse of awareness that can occur with seizures, including complex partial or absence seizures, or the fugue state associated with nonconvulsive status epilepticus. An electroencephalogram (EEG) or prolonged EEG/video monitoring study may be helpful in identifying and characterizing seizures.

The sleep history should include questions about the patient's work and social habits, including shifts in circadian rhythm. The individual with late work hours is predisposed to the delayed sleep phase syndrome or inadequate nocturnal sleep. Pilots and flight attendants who travel frequently across multiple time zones are at risk for jet lag and other circadian problems [6]. Shift workers often struggle to obtain adequate sleep, and the health effects of frequently changing shifts is of practical importance. A history of excessive alcohol use or drug abuse is important, and in certain cases performance of urine drug screening may be indicated to identify illicit drug abuse as the cause of sleepiness.

Regardless of the eventual diagnosis, patients with excessive sleepiness should be counseled regarding the potential risks associated with sleepiness, including risk of motor vehicle accidents, work-related accidents, and loss of productivity. Individuals with excessive sleepiness may struggle with emotional and cognitive problems, including problems with attention and concentration, distractability, memory difficulties, depressed mood, and interpersonal problems due to irritability and poor coping skills.

Excessive sleepiness may present in unique ways in children and adolescents. Because children are often teased or disciplined for sleepiness in the classroom, they may deny having sleepiness, or they may attempt to conceal sleepiness through a variety of strategies [7]. Young children often manifest a paradoxical overactivity during drowsiness, possibly in an effort to self-stimulate. Excessive sleepiness may be associated with a variety of behavior problems including irritability and emotional lability, acting out behaviors, attention deficit/hyperactivity disorder symptoms, and learning problems. Children with excessive sleepiness may miss key social or athletic opportunities, and they often show a tendency to become socially isolated or withdrawn [7].

ROLE OF POLYSOMNOGRAPHY AND THE MULTIPLE SLEEP LATENCY TEST

Nocturnal polysomnography is indicated in the evaluation of patients with suspected respiratory disturbances during sleep, narcolepsy, and idiopathic hypersomnia, and those in whom an adequate explanation of excessive sleepiness is not reached following a thorough sleep history [8]. Polysomnography should be initiated within 30 min of the patient's usual sleep time. The Multiple Sleep Latency Test (MSLT) is a validated, objective measure of the ability or tendency to fall asleep under standardized conditions [5]. For proper interpretation, the MSLT must be preceded by overnight polysomnography to document the patient's total sleep time and sleep stage distribution. The MSLT is indicated for evaluation of suspected narcolepsy and possibly for suspected idiopathic hypersomnia [5]. The MSLT is not routinely indicated for patients with excessive sleepiness due to other sleep disorders or medical or psychiatric disorders. For example, many patients with mood disorders report excessive sleepiness or fatigue, and the MSLT is not indicated in this population unless there is concern regarding narcolepsy. The MSLT is not routinely indicated for evaluation of the sleepiness associated with obstructive sleep apnea or to document response to nasal CPAP. If there is clinical suspicion of narcolepsy in the patient with documented obstructive sleep apnea, the patient should undergo appropriate treatment of sleep apnea followed by repeat polysomnography (with CPAP if being used to treat sleep apnea), followed by the MSLT.

The rationale for the MSLT is based on the premise that the speed at which an individual falls asleep is an indication of the severity of sleepiness [1, 5]. The MSLT consists of a series of five nap opportunities at 2 h intervals during the patient's usual period of wakefulness. Nap opportunities begin 1.5–3 h after the patient's usual rise time, and the patient is observed and not allowed to sleep between nap opportunities. The patient is placed in a dimly lit, quiet room in the reclining position and given the instruction to "Please lie quietly, assume a comfortable position, keep your eyes closed and try to fall asleep." The patient is given a 20 min opportunity to sleep. Sleep latency is measured from the time of lights out until the first epoch of any stage of sleep. If the patient falls asleep, he/she is allowed to sleep for 15 min to provide an opportunity for entry into REM sleep (sleep-onset REM period). If the patient does not sleep during a 20 min nap opportunity, sleep latency for that opportunity is scored as 20 min. Because sleep latency is influenced by the quantity and quality of sleep the previous night, the patient must be studied with overnight polysomnography for proper interpretation. For a more detailed discussion, the reader is referred to the recently published practice parameters for clinical use of

the MSLT, which were developed by the American Academy of Sleep Medicine (AASM) [5].

The clinical usefulness of the MSLT in diagnosing narcolepsy is based on two physiological parameters measured by the test [1, 5]. The mean sleep latency value (the arithmetic mean of sleep latencies from all nap opportunities) provides objective documentation of pathological sleepiness under standardized conditions. Analysis of data from four papers that were reasonably free of inclusion bias (39 subjects with narcolepsy) indicates that the weighted mean sleep latency among narcolepsy subjects was 3.0 ± 3.1 min [1]. Control subjects without evidence of sleep disorders demonstrated a weighted mean sleep latency of 10.5 ± 4.6 min [1]. This difference is statistically significant ($p < 0.001$). These findings indicate that most patients with narcolepsy have objective evidence of hypersomnia as determined by a mean sleep latency of less than 5 min on the MSLT.

The second useful parameter for diagnosis of narcolepsy using the MSLT involves identification of sleep-onset REM periods (SOREMPs) [1, 5]. In a review of nine studies that are reasonably free of inclusion bias, SOREMPs were identified frequently in patients with narcolepsy [1]. The presence of two or more SOREMPs was associated with a sensitivity of 0.78 and a specificity of 0.93 [1]. It is important to note that SOREMPs do not occur exclusively in patients with narcolepsy. SOREMPs are observed in patients with obstructive sleep apnea, or any condition associated with reduced nocturnal REM sleep leading to "REM rebound" during the day. Recent withdrawal of REM suppressing medications such as selective serotonin reuptake inhibitors (SSRIs) may lead to REM rebound during the day.

Diagnostic sensitivity and specificity of the MSLT are optimal when standardized procedures are followed and when interpretation is meticulous. Occasionally, for various reasons, patients with suspected narcolepsy do not meet polygraphic diagnostic criteria. Repeat MSLT testing may be indicated when the initial test is affected by extraneous circumstances or when appropriate study conditions were not present during initial testing, when ambiguous or uninterpretable findings are present, or when the patient is suspected to have narcolepsy on clinical grounds but earlier MSLT evaluation did not provide polygraphic confirmation [5]. It is well documented that patients with evolving narcolepsy may not meet MSLT diagnostic criteria initially, but upon repeat testing in 6–12 months criteria are present. The International Classification of Sleep Disorders allows individuals to be diagnosed with narcolepsy on clinical grounds when unequivocal cataplexy is present in addition to chronic pathological sleepiness [6]. However, in North America and Europe most centers utilize routine polygraphic evaluation to provide objective confirmation of SOREMPs. Since narcolepsy is a serious and lifelong disorder with a major impact on the patient and others, the routine use of the MSLT for confirmation seems justified. In addition, MSLT diagnostic criteria for narcolepsy must be met when the patient has no history of cataplexy [6].

Establishing normative mean sleep latency values for the MSLT is complicated due to several factors. Mean sleep latency values are influenced by physiological, psychological, and test protocol variables [1]. The quantity and continuity of nocturnal sleep prior to the MSLT clearly influence mean sleep latency values. Achievement of at least 6 h of sleep the preceding night is required for a valid MSLT. Although the MSLT is a validated measure of sleepiness, there is no large systematically collected repository of normative data for the MSLT [1]. Methodological variations may exist from center to center, and deviation from the standard protocol has the potential to alter sleep latencies significantly. Normative studies vary with regard to how rigorously control subjects are screened, how carefully subjects maintained a consistent sleep–wake pattern prior to the MSLT, whether subjects were allowed to use caffeine or medications, and whether urine drug screens were performed [1]. Delineation of normative ranges is limited by the large standard deviation in mean sleep latency values, as well as ceiling and floor effects that indicate that mean sleep latency values are not normally distributed. These latter factors result in overlap between values among healthy controls and individuals with excessive sleepiness [1].

Pooled data from normal subjects across all ages on the MSLT give a mean sleep latency of 10.4 ± 4.3 min when using the four nap protocol, and 11.6 ± 5.2 min when using the five nap protocol [1, 5]. Based on the traditional 2 standard deviations from the mean approach, 95% of the values from control populations on the four nap test would fall between 1.8 and 19 min, and for the five nap test, the mean sleep latency value would fall between 1.2 and 20 min [1]. Using this approach, the MSLT does not discriminate well between clinical and control populations. However, within the clinical population the MSLT is useful for diagnostic purposes. Table 25.5 summarizes MSLT mean sleep latency values from normal control subjects using the four nap and five nap protocols. For comparison, sleep latency values are also provided for a group of patients with narcolepsy.

TABLE 25.5 Summary of Control and Narcolepsy Mean Sleep Latency Values on MSLT (from Pooled Data) [1, 5]

Test Protocol	Mean ± SD (minutes)
MSLT (4 nap protocol)	10.4 ± 4.3
MSLT (5 nap protocol)	11.6 ± 5.2
MSLT in patients with narcolepsy	3.1 ± 2.9

Based on these data, the mean sleep latency should not be the sole criterion for determining the presence or severity of excessive sleepiness, certifying a diagnosis, or determining response to treatment. Global assessment should integrate the clinical history, objective test results, and other medical information [1, 5].

The Maintenance of Wakefulness Test (MWT) is an objective, laboratory-based measure of the subject's ability to remain awake under standardized conditions for a defined period of time [5]. The MWT is potentially useful in assessing response to treatment of disorders associated with excessive sleepiness. The test is not used to diagnose a condition. Table 25.6 compares and summarizes the MSLT and MWT protocols. The MSLT and MWT measure different aspects of sleepiness and wakefulness during the day, and as a consequence mean sleep latency values do not correlate well between the two tests.

The MWT is performed during the day after a typical night of sleep, and when the individual is treated for a condition causing excessive sleepiness. The MWT consists of four tests at 2 h intervals. With each test the subject is placed in a dimly lit, quiet room with a 7.5 watt lamp positioned three feet lateral to the patient's head and one foot off the floor. The subject is given instructions to "Please sit still and remain awake for as long as possible. Look directly ahead of you, and do not look directly at the light." Sleep onset is defined as the first epoch of greater than 15 s of cumulative sleep in a 30 s epoch. There are two protocols for the MWT based on the duration of each test. Twenty minute and 40 min protocols can be used, but the 40 min protocol has been recommended by the AASM in recently published practice parameters [5].

As with the MSLT, establishing normative mean sleep latency values for the MWT has been difficult. Whereas the MSLT shows a "floor effect" in subjects with severe sleepiness, the MWT shows a "ceiling effect" in subjects with normal levels of wakefulness, with many subjects remaining awake during each trial [1, 5]. Table 25.7 summarizes findings regarding the MWT 40 protocol based on the systematic control study performed by Doghramji and colleagues [9]. The recent review by the MSLT and MWT Task Force commissioned by the American Academy of Sleep Medicine contains a variety of survival plots for the MWT 20 min and MWT 40 min protocols, as well as percentiles for mean sleep latencies on the 40 min protocol [1].

CONCLUSIONS

Characterization of excessive sleepiness begins with and is based primarily on a carefully taken clinical history. Questionnaires such as the Epworth Sleepiness Scale and

TABLE 25.6 Comparison of the Multiple Sleep Latency Test and the Maintenance of Wakefulness Test [5]

	MSLT	MWT
Objective	To measure individual's ability or tendency to fall asleep and propensity for entry into REM sleep	To measure individual's ability to remain awake (usually after treatment of underlying sleep disorder)
Time of test	During usual period of wakefulness	During usual period of wakefulness
Protocol	Four or Five nap opportunities at 2 h intervals beginning 1.5–3 h after waking	Four tests at 2 h intervals
	20 min opportunity to nap	20 or 40 min test
Instructions to patient	"Please lie quietly, assume a comfortable position, keep your eyes closed and try to fall asleep"	"Please sit still and remain awake for as long as possible"
Rule for termination	End nap opportunity after 20 min if no sleep; end test after 15 min of sleep	End test at first epoch of any stage of sleep, or after 20 or 40 min of wakefulness
Rule for determination of sleep latency	Arithmetic mean of all nap opportunities	Arithmetic mean of all four tests
Reporting results	Mean sleep latency (minutes) Number of SOREMPs	Mean sleep latency (minutes)
Recommendations regarding CNS active medications	*Optimal*: Patient tapered off all CNS active medications at least 14 days before test	Patient on his/her usual dosage of medication or CPAP at the time of the test
	Suboptimal: Patient continues on usual medications without change	
	Unacceptable: Patient tapered abruptly off CNS active medication just before test (may lead to REM rebound)	

sleep–wake diaries may provide additional data. Overnight polysomnography is indicated as part of the evaluation of respiratory disturbances in sleep, of narcolepsy, and for characterization of certain abnormal movements or behaviors in sleep. The MSLT is indicated for evaluation of suspected narcolepsy in order to quantify the degree of excessive sleepiness and to document sleep-onset REM periods. The MSLT is not routinely indicated for evaluation of all patients with sleepiness. The MWT is used selectively when objective data are necessary to document response to treatment for conditions that cause excessive sleepiness.

TABLE 25.7 Maintenance of Wakefulness Test (40 min Protocol) Control Findings [5, 9]

Mean sleep latency, using latency to first epoch of sleep (any stage)	30.4 ± 11.2 min
Upper limit of 95% confidence interval	40.0 min

The determination of a patient's response to treatment is especially important when the person's work involves public safety. Examples include pilots, air traffic controllers, bus, train, or subway drivers, and military and medical professionals.

REFERENCES

1. Arand D, Bonnet M, Hurwitz T, Mitler M, Rosa R, Sangal RB. Clinical use of the MSLT and MWT. *Sleep* **28**:123–144 (2005).

2. Bixler EO, Kales A, Soldatos CR, Kales JD, Healey S. Prevalence of sleep disorders in the Los Angeles metropolitan area. *Am J Psychiatry* **136**:1257–1262(1979).

3. Lavie P. Sleep habits and sleep disturbances in industrial workers in Israel: main findings and some characteristics of workers complaining of excessive daytime sleepiness. *Sleep* **4**:147–158 (1981).

4. Johns MW. A new method for measuring daytime sleepiness: the Epworth sleepiness scale. *Sleep* **14**:540–545(1991).

5. Littner MR, Kushida C, Wise M, Davila DG, Morgenthaler T, Lee-Chiong T, Hirshkowitz M, Loube DL, Bailey D, Berry RB, Kapen S, Kramer M. Practice parameters for clinical use of the Multiple Sleep Latency Test and the Maintenance of Wakefulness Test. An American Academy of Sleep Medicine Report: Standards of Practice Committee of the American Academy of Sleep Medicine. *Sleep* **28**:113–121(2005).

6. American Sleep Disorders Association. *International Classification of Sleep Disorders, Revised*: *Diagnostic and Coding Manual*. American Sleep Disorders Association, Rochester, MN, 1997.

7. Wise MS. Childhood narcolepsy. *Neurology* **50**(Suppl 1):S37–S42(1998).

8. Chesson AL, Ferber RA, Fry JM, Grigg-Damberger M, Hartse KM, Hurwitz TC, Johnson S, Littner M, Kader GA, Rosen G, Sangal B, Schmidt-Nowara W, Sher A. The indications for polysomnography and related procedures. *Sleep* **20**:423–487(1997).

9. Doghramji K, Mitler M, Sangal RB, Shapiro C, Taylor S, Walsleben J, Belisle C, Erman MK, Hayduk R, Hosn R, O'Malley EB, Sangal JM, Schutte SL, Youakim JM. A normative study of the maintenance of wakefulness test (MWT). *Electroencephalogr Clin Neurophysiol* **103**:554–562(1997).

26

THERAPY FOR EXCESSIVE SLEEPINESS

MAX HIRSHKOWITZ

Baylor College of Medicine, Houston, Texas

Sleepiness can occur naturally or it may be induced. It can be defined in terms of self-report, physiological response, or observation. Clinically significant sleepiness interferes with activities of daily living and produces either a struggle to remain awake, inappropriate lapses into sleep, or both. In general, sleepiness may be viewed as a serious, noncontiguous, potentially life-threatening condition that affects not only the afflicted individual but also their family, friends, co-workers, and society at large. Innumerable traffic and industrial accidents are either directly caused by or are contributed to by sleepiness. In this chapter, the causes of sleepiness will be discussed and the approaches currently available for treating sleepiness and improving an individual's alertness will be reviewed.

SLEEPINESS IN THE GENERAL POPULATION AND ADENOSINE ANTAGONISM

Treating sleepiness can be easy or difficult, depending on the cause, the cooperation of the patient, or both. Sleep deprivation represents the leading cause of sleepiness. The internal sleep homeostatic mechanism generally accrues sleep debt in response to sleep loss. When schedule permits, the individual naps or extends their sleep time (e.g., "sleeping-in" on weekends). If daytime alertness improves and sleepiness is reduced with these simple behavioral interventions, then the problem is likely nonmedical. Further life-style advice, sleep hygiene improvements, and avoidance of soporific substances or intoxicants are recommended. The judicious use of caffeinated substances may provide acute relief because methylxanthine stimulants

counteract sleepiness. Caffeine and theobromine, usually consumed as coffee or chocolate, are widely traded commodities used the world over. In this country, major sources of caffeine also include tea, cola, and other sodas.

Adenosine is an inhibitory neurotransmitter in the central nervous system (CNS). Caffeine is a CNS stimulant that acts by inhibiting this inhibitory transmitter, thereby producing activation. Caffeine is a plant alkaloid with the formula $C_8H_{10}N_4O_2$ with a mean half-life of 5 h (ranging from 1.5 to 9.5 h) [1]. Caffeine is rapidly absorbed (99% within 1 h). A typical 10 ounce mug of drip-brewed coffee will contain an average of 240 mg of caffeine (which is equivalent to a 6 ounce cup of espresso). A typical can of cola soda (12 ounces) will have 40 mg caffeine while 4 ounces of chocolate ranges from an average of 24 (milk), to 80 (dark), to 140 (Baker's). Caffeine can produce acute blood pressure elevations (5–15 mmHg systolic, 5–10 mmHg diastolic) with even more exaggerated increases in people experiencing stress. Caffeine is thought to account for as much as 14% of coronary heart disease and 20% of stroke-related deaths; however, the pharmacologically lethal oral dose in humans is estimated at 10–15 g [2].

SLEEPINESS IN THE PATIENT POPULATION

While caffeine is the stimulant of choice for individuals who self-medicate insufficient sleep syndrome, it is also self-administered by individuals suffering from a wide array of disorders of excessive sleepiness. These disorders include obstructive forms of sleep-disordered breathing, narcolepsy, idiopathic hypersomnia, circadian rhythm

TABLE 26.1 Sleep Disorders Associated with Excessive Sleepiness

Type of Condition	Sleep Disorder	Comments
Primary sleep disorder	Obstructive sleep-disordered breathing	Sleepiness is thought to result from frequent awakenings and arousals; that is, sleep fragmentation and sleep loss.
	Narcolepsy and idiopathic hypersomnia	Sleepiness is thought to relate to dysfunction in basic CNS sleep-control mechanisms.
	Circadian rhythm disorders	Sleepiness is thought to be caused by disentrainment of circadian sleep–wake mechanisms and the behavioral schedule.
Secondary sleep disorder	Dyssomnias associated with medical conditions	Sleepiness can arise because of sleep disturbances associated with pain, GE reflux, fever, metabolic conditions, cardiopulmonary diseases, endocrine disorders.
	Dyssomnias associated with neurological conditions	Focal lesions, myotonic dystrophy, encephalitis, cerebral palsy, Parkinson's Disease, multiple, sclerosis, head injuries, Prader–Willi syndrome, Kleine–Levin syndrome.
	Dyssomnias associated with psychiatric conditions	Atypical depression, seasonal affective disorder, schizophrenia, bipolar disorder.

disorders, and sleepiness secondary to medical, neurological, and psychiatric conditions (see Table 26.1). Sleepiness can also be iatrogenic.

Sleepiness associated with particular sleep disorders is treated with specifically targeted treatments. In such sleep disorders, the sleepiness is theorized as a link in a chain of causality resultant from sleep disturbance that in turn results from a sleep-related pathophysiology. In such cases, it may be possible to correct the sleep-disrupting pathophysiology. By contrast, in sleep disorders where the sleepiness is presumably related to an underlying core neurological deficit, focused treatments are unavailable and therapy tends to be palliative.

TREATING SLEEPINESS IN OBSTRUCTIVE FORMS OF SLEEP-DISORDERED BREATHING

Obstructive forms of sleep-disordered breathing include disorders ranging from upper airway resistance syndrome to severe obstructive sleep apnea. These sleep-related breathing impairments are caused by airway obstruction. By contrast, nonobstructive cessations and/or respiratory reduction may result from central (brainstem) changes in ventilatory control, metabolic factors, or heart failure. During obstructive events, respiratory effort continues but airflow stops due to reduced or loss of airway patency [3].

Many treatments are available for sleep-disordered breathing, including weight loss, positive airway pressure therapy, oral appliances, and surgery. Weight loss is difficult to achieve and maintain; therefore it is recommended but seldom relied on. Currently, positive airway pressure is the preferred and most widely used therapy; it comes in three varieties—continuous, bilevel, and sleep adjusting. Nasal continuous positive airway pressure (nCPAP) is the most common. nCPAP works by delivering a fan or turbine generated flow at a set pressure to the nares. This positive pressure creates a "pneumatic splint" and thereby maintains airway patency. It is highly effective in most patients; however, it requires nightly utilization. Patients with more severe sleep-disordered breathing or who are sleepier at baseline are more adherent to therapy. Sleep normalization can be immediate and impressive. Sleep, in a patient with significant sleep-disordered breathing, is marked by frequent brief arousals. These arousals are needed to return ventilatory control to the voluntary system so that breathing will resume after airway closure. Once an appropriate pressure is reached that maintains airway patency, these constant sleep disruptions disappear, permitting the patient to sleep uninterrupted, possibly for the first time in decades [4].

TREATING SLEEPINESS IN NARCOLEPSY AND IDIOPATHIC HYPERSOMNIA

Ever since Doyle and Daniels [5] reported ephedrine's efficacy for treating sleepiness in narcolepsy, stimulant medications have provided the therapeutic mainstay. Ephedrine's adverse effects, cost, and tachyphylaxis prompted therapeutic evolution to the use of amphetamines [6], where it remained for two decades. Methylphenidate's introduction provided another option that came to dominate first line treatment of narcolepsy [7] until modafinil's introduction

TABLE 26.2   Drugs Used to Counteract Sleepiness

Name	Brand	Therapeutic Dose	Comment
Modafinil	Provigil, Alertic	1–600 mg	Novel wake-promoting substance with a 10–12 h half-life that is chemically different from amphetamines. Originally promoted for once-daily dosing but often used bid in clinical practice. Mechanism of action not known but thought to work via histamine and GABA pathways.
Methylphenidate	Ritalin	10–80 mg	Structurally similar to amphetamine and is rapidly absorbed. As with amphetamines, it works through DA release and reuptake blockade. Has a half-life of 2–4 h and is usually taken three or four times daily in divided doses.
Amphetamine, methamphetamines	Desoxyn	5–60 mg	This is the classic form of amphetamine—it is a racemic mixture of levo- and dextroamphetamine isomers. It has less central and more peripheral action than dextroamphetamine and is usually taken three or four times daily in divided doses.
Dextroamphetamine	Adderal Biphetamine	20–60 mg	D-isomer of amphetamine. Moderately long half-life at 8–12 h. Usually taken in divided doses, twice or three times daily.
Levoamphetamine	n/a	20–60 mg	L-isomer of amphetamine. Not available in the U.S. and not considered advantageous over the D-isomer form.
Selegiline	Eldepryl	10 mg	Monoamine oxidase type B inhibitor that increases DA levels. Catabolizes to levoamphetamine and metamphetamine. Can provoke hallucinations and contraindicated for concurrent administration with 5HT reuptake inhibitors.
Mazindol	Mazanor, Sanorex	2–12 mg	Chemically different from amphetamines. Mixed reports on efficacy but known tolerance and abuse potential.
Pemoline	Cylert	25–100 mg	Chemically different from amphetamines, this once popular stimulant is seldom used now because of possible acute liver necrosis and several reported fatalities.

Notation: Milligrams (mg); twice daily (bid); dopamine (DA); hour (h).

in the 1990s. Other stimulants used to treat sleepiness in narcolepsy and idiopathic hypersomnia in the past (but seldom now) include pemoline and mazindol [8]. Table 26.2 shows the wide assortment of drugs used to counteract sleepiness.

For narcolepsy, the current American Academy of Sleep Medicine standards of practice advises that modafinil is effective for treating daytime sleepiness [9]. This recommendation is expressed as a "standard," which means that it is a generally accepted patient care strategy reflecting a high degree of clinical certainty. This highest form of practice parameter is based on the drug's favorable benefit-to-risk ratio reported in three level I, grade A published studies [10–13] with confirmation from additional lower grade studies. Amphetamine, methamphetamine, dextroamphetamine, and methylphenidate are also recommended for use

in treating sleepiness in narcolepsy as a "guideline" based on three level II, grade B and four level V, grade C studies. The guideline is also based on many years of clinical practice experience; however, it is warned that the benefit-to-risk ratio is not well documented. Using selegiline is also endorsed as a "guideline" based on two level II, grade B and one level IV, grade C studies. Guidelines are patient care strategies reflecting a moderate degree of clinical certainty. No practice guidelines are available for treating sleepiness in idiopathic insomnia; however, most clinicians use a similar approach to that used for narcolepsy [14].

Amphetamine and related stimulants have a large array of side effects, many as a direct result of sympathetic nervous system activation [15, 16]. These drugs as a class increase activity, including talking. They can produce tremor, anorexia, insomnia, GI complaints, irritability, and headaches. The sympathetic activation puts stress on and may adversely affect the cardiopulmonary system. Amphetamine psychosis, paranoia, and hallucinations can also occur. Amphetamines tend to produce euphoria and therefore have a high abuse potential; thus they are schedule II substances. Tolerance may develop with chronic use, more so at higher doses. Tolerance and other side effects are reportedly less common with methylphenidate than other amphetamines. Abrupt discontinuation is discouraged, although little or no data are available concerning rapid withdrawal. Amphetamine discontinuation is usually characterized by the individual being profoundly sleepy and having dramatic sleep rebound.

In contrast to amphetamines and its congeners, modafinil has relatively few side effects (most notably headache and nausea) when administered in the therapeutic dose range. High doses (greater than 800 mg) may cause tachycardia and hypertension. Modafinil does not appear to induce euphoria until doses exceeding therapeutic levels are reached (800–1000 mg). Little or no tolerance was seen in extended trials.

TREATING SLEEPINESS IN CIRCADIAN RHYTHM DISORDERS

One of the principal mechanisms regulating the sleep–wake cycle involves an internal 24 hour biological clock. This clock is responsible for our circadian rhythm and in an optimal schedule, hours in bed must coincide with the sleepy phase of its cycle. In jet lag, an individual rapidly relocates to a different time zone. If the new time zone has desired awake time coincident with the sleepy phase of the circadian cycle, excessive sleepiness may result. In another circadian disorder, advanced sleep phase, the circadian rhythm is shifted earlier. Therefore the sleepiness cycle is advanced with respect to clock time making the individual drowsy in the evening and wanting to retire to bed early. By contrast, when the biological clock is shifted later than the desired schedule, delayed sleep phase results. Individuals with delayed sleep phase are more alert in the evening and early nighttime, stay up later, and are sleepy in the morning [17, 18]. In the past, chronotherapy was used to reentrain the circadian rhythm; however, it has largely been replaced by bright light therapy [19]. Bright light appears to be the critical factor in controlling the biological clock. With precise timing of bright light exposure, the biological clock phase advanced, phase delayed, or even stopped and reset. In general, bright light in the evening will delay the sleep phase and bright light in the morning will advance the sleep phase. It also appears that melatonin or melatonin receptor agonists may be effective chronobiotics. Melatonin serves as a signal of darkness to the brain;

therefore, when administered in the evening, it will advance the sleep phase but when given in the morning it will delay the sleep phase.

Perhaps the worst of the circadian disorders is shift work sleep disorder (SWSD). The individual with SWSD continually struggles to remain awake during his/her night shift and to sleep during the day. If his/her internal biological clock does not shift with the work schedule, he/she remains in a perpetual state of circadian disentrainment. Night shift work can be dangerous and excessive sleepiness on the job further contributes to this hazard. Shift-workers face unique challenges to their sleep integrity. The sleep environment must be kept dark. Blackout curtains or eyeshades are helpful in this regard. Additionally, the room must be kept cool and quiet. Earplugs can be useful. Sleep time must be given priority and not violated by intrusions by family and friends. If the bedroom has a telephone, it should be unplugged and the clock should be turned toward the wall or put under the bed to avoid clock watching. Regularity is critical, the patient should strive to arise at the same time daily and get an adequate amount of sleep (8 h for most individuals). Bright light should be avoided for several hours before scheduled bedtime. A low dose of pharmaceutical grade melatonin or a melatonin agonist may promote sleep. Finally, if the patient drinks caffeinated beverages, then particular attention should be paid to the timing of ingestion with respect to sleep schedule. Capital intensiveness and our ever-growing 24 hour society intensify our need to find a treatment for this problem; however, drug therapies remain controversial. Critics argue that stimulant treatment amounts to pharmacologically abetting life-style choices. Advocates note that hospital emergency rooms and nuclear reactors remain open all night and workers need to be awake and alert. Nonetheless, randomized clinical trials with modafinil found improved alertness at night among patients with SWSD and an indication for treatment was granted by the regulatory authorities.

TREATING SLEEPINESS IN MEDICAL, NEUROLOGICAL, AND PSYCHIATRIC CONDITIONS

In cases where the sleepiness is attributed to an underlying medical, neurological, or psychiatric condition, the standard approach is to treat the cause. Thus, if pain is disturbing sleep and that sleep disruption produces daytime sleepiness, then better pain management is recommended. In some cases, even with optimized treatment, excessive sleepiness will persist. Some research indicates efficacy of augmentation therapies with modafinil or stimulants. Such an approach should be pursued with the utmost care and only when the additional medication is not contraindicated by the patient's current drug regimen and/or comorbidity factors. Augmentation studies have evaluated patients with depression [20], seasonal affective disorder [21], Parkinson's disease [22–24], pain-related fatigue [25], myotonic dystrophy [26], multiple sclerosis [27], and obstructive sleep apnea [28].

TREATING IATROGENIC SLEEPINESS

The three major sources of iatrogenic sleepiness involve (1) soporific side effects of drug therapy, (2) direct sleepiness-promoting effects of a nondrug therapeutic intervention (e.g., radiation therapy), and (3) sleepiness produced by insomnia resultant from the primary intervention (e.g., postsurgical pain). Drugs known to promote sleepiness as their primary effect include sedative–hypnotics and minor tranquilizers, while others may induce sleepiness as an unwanted side effect. Patients using sedative–hypnotics may experience sleepiness if a medication's duration of action extends beyond the scheduled sleep period; that is, there may be a "hangover" effect. Switching to a shorter acting substance and/or lowering the dose may alleviate this problem. When the sleepiness is an unwanted side effect, the problem may be unavoidable; however, medication review often provides an opportunity to switch to less soporific substances with similar therapeutic efficacy. The following medications are associated with sleepiness [29]: anxiolytics (diazepam (Valium), alprazolam (Xanax)), central-acting antihistamines (diphenhydramine (Benadryl)), antidepressants (trazodone (Desyrel), amitriptyline (Elavil), doxepin (Adapin, Sinequan), mirtazepine (Remeron)), antipsychotics (chlorpromazine (Thorazine), haloperidol (Haldol), thioridazine (Mellaril), olazipine (Zyprexa), risperidone (Risperdal), quetiapine (Seroquel)), antihypertensives (clonidine (Catapres)), anticonvulsants (carbamazepine (Tegretol), phenytoin (Dilantin), gabapentin (Neurontin)), narcotics (meperidine (Demerol), codeine, oxycodo (Percodan), hypdrocodone (Vicodin)), and some steroids.

CONCLUSION

Sleepiness can be associated with a wide variety of conditions. When improved alertness can be attained by improved sleep hygiene and life-style interventions, these approaches are recommended. Caffeine is widely used as a countermeasure for sleepiness arising from both behavioral and nonmedical origins. When sleepiness results from a primary sleep disorder affecting a presumed underlying sleep–wake mechanism (e.g., narcolepsy), psychostimulants and/or awake-promoting substances are generally used palliatively to manage the condition. These "scheduled" medications may have serious side effects and carry a potential for abuse. Therefore they should be used properly and only under physician supervision. If sleepiness is secondary to a medical, neurological, or psychiatric condition, the condition should be addressed first and if residual sleepiness persists, augmentation therapy may be appropriate. Finally, medication review should be considered in patients being treated concurrently for other conditions in order to rule out iatrogenic factors.

REFERENCES

1. Vanderveen JE, Armstrong LE, Butterfield GE, et al. *Caffeine for the Sustainment of Mental Task Performance*. National Academy Press, Washington, DC, 2001.

2. James JE. Critical review of dietary caffeine and blood pressure: a relationship that should be taken more seriously. *Psychosom Med.* **66**:63–71(2004).

3. Guilleminault C, Partinen M, Querasalva MA, Hayes B, Dement WC, Ninomurcia G. Determinants of daytime sleepiness in obstructive sleep-apnea. *Chest* **94**:32–37(1988).

4. Hirshkowitz M, Littner M, Kuna ST, et al. *Sleep-Related Breathing Disorders: Sourcebook*, 2nd ed. HAIG, Milwaukee, WI, 2003.

5. Doyle JB, Daniels LE. Symptomatic treatment for narcolepsy. *JAMA* **96**:1370–1372(1931).

6. Prinzmetal M, Bloomberg W. Use of benzedrine for the treatment of narcolepsy. *JAMA* **105**:2051–2054(1935).

7. Yoss RE, Daly D. Treatment of narcolepsy with Ritalin. *Neurology* **9**:171–173(1959).

8. Mitler MM, Nelson S, Hajdukovic R. Narcolepsy: diagnosis, treatment, and management. *Psychiatr Clin North Am* **10**:593–606(1987).

9. Littner M, Johnson SF, McCall WV, et al. Practice parameters for the treatment of narcolepsy: an update for 2000. *Sleep* **24**:451–466(2001).

10. US Modafinil in Narcolepsy Multicenter Study Group. Randomized trial of modafinil as a treatment for the excessive daytime somnolence of narcolepsy. *Neurology* **54**:1166–1175(2000).

11. US Modafinil in Narcolepsy Multicenter Study Group. Randomized trial of modafinil for the treatment of pathological somnolence in narcolepsy. *Ann Neurol* **43**:88–97(1998).

12. Broughton RJ, Fleming JAE, George CFP, et al. Randomized, double-blind, placebo-controlled crossover trial of modafinil in the treatment of excessive daytime sleepiness in narcolepsy. *Neurology* **49**:444–451(1997).

13. Billiard M, Besset A, Montplaisir J, et al. Modafinil: a double-blind multicentric study. *Sleep* **17**:S107–S112(1994).

14. Billiard M, Dauvilliers Y. Idiopathic hypersomnia. *Sleep Med Rev* **5**:349–358(2001).

15. Mitler MM, Aldrich MS. Stimulants: efficacy and adverse effects. In: Kryger MH, Roth T, Dement WC (Eds), *Principles and Practice of Sleep Medicine*, 3rd ed. Saunders, Philadelphia, 2000, pp 429–440.

16. Ballas CA, Evans DL, Dinges DF. Psychostimulants in psychiatry: amphetamine, methylphenidate, and modafinil. In: Schatzberg AF, Nemeroff CB (Eds), *Textbook of Psychopharmacology*, 3rd ed. American Psychiatric Publishing, Washington, DC, 2004, pp 671–684.

17. Moore-Ede MC, Sulzman FM, Fuller CA. *The Clocks That Time Us*. Harvard University Press, Cambridge, MA, 1982.

18. Zammit GK. Delayed sleep phase syndrome and related conditions. In: Pressman MR, Orr WC (Eds), *Understanding Sleep: the Evaluation and Treatment of Sleep Disorders*. American Psychological Association, Washington, DC, 1997, pp 229–248.

19. Czeisler CA, Kronauer RE, Allan JS, et al. Bright light induction of strong (type 0) resetting of the human circadian pacemaker. *Science* 244:1328–1333(1989).

20. DeBattista C, Lembke A, Solvason HB, et al. A prospective trial of modafinil as an adjunctive treatment of major depression. *J Clin Psychopharmacol* 24:87–90(2004).

21. Lundt L. Modafinil treatment in patients with seasonal affective disorder/winter depression: an open-label pilot study. *J Affect Disord* 81:173–178(2004).

22. Happe S. Excessive daytime sleepiness and sleep disturbances in patients with neurological diseases: epidemiology and management. *Drugs* 63:2725–2737(2003).

23. Rye DB. Sleepiness and unintended sleep in Parkinson's disease. *Curr Treat Options Neurol* 5:231–239(2003).

24. Adler CH, Caviness JN, Hentz JG, et al. Randomized trial of modafinil for treating subjective daytime sleepiness in patients with Parkinson's disease. *Mov Disord* 18:287–293(2003).

25. Fishbain DA, Cutler RB, Lewis J, et al. Modafinil for the treatment of pain-associated fatigue: review and case report. *Pain Palliat Care Pharmacother* 18:39–47(2004).

26. MacDonald JR, Hill JD, Tarnopolsky MA. Modafinil reduces excessive somnolence and enhances mood in patients with myotonic dystrophy. *Neurology* 59:1876–1880(2002).

27. Rammohan KW, Rosenberg JH, Lynn DJ, et al. Efficacy and safety of modafinil (Provigil) for the treatment of fatigue in multiple sclerosis: a two centre phase 2 study. *J Neurol Neurosurg Psychiatry* 72:179–183(2002).

28. Schwartz JR, Hirshkowitz M, Erman MK, Schmidt-Nowara W. Modafinil as adjunct therapy for daytime sleepiness in obstructive sleep apnea: a 12-week, open-label study. *Chest* 124:2192–2199(2003).

29. *Drugdex*. Micromedex Healthcare Series Vol 122. Online resource from Thomson Healthcare, Inc.

ADDITIONAL READING

Abad VC, Guilleminault C. Emerging drugs for narcolepsy. *Expert Opin Emerg Drugs* 9:281–291(2004). If you want to know what may be coming around the bend, this is the place to find out what's new and what's in the pipeline. While many of these approaches may never "pan out," someone is betting on them.

Bassetti C, Aldrich MS. Idiopathic hypersomnia. A series of 42 patients. *Brain* 120:1423–1435(1997). This large series of cases of a fairly rare sleep disorder reviews features and treatment approaches.

Berry RB. Medical therapy. In: Johnson JT, Gluckman JL, Sanders MH (Eds), *Management of Obstructive Sleep Apnea*. Martin Dunitz, xx, UK, 2002, pp 89–118. This comprehensive and clinically oriented book chapter clearly delineates the established medical therapies for treating sleepiness in patients with obstructive forms of sleep-disordered breathing.

Mitler MM, Hajdukovic R. Relative efficacy of drugs for the treatment of sleepiness in narcolepsy. *Sleep* 14:218–220(1991). This article compares assorted stimulants and wake-promoting substances from a variety of trials and synthesizes the information to characterize each drug's relative efficacy for treating sleepiness in patients with narcolepsy.

Wagner DR. Disorders of the circadian sleep–wake cycle. *Neurol Clin* 14:651–670(1996). This easy-to-read chapter describes sleepiness disorders resulting from circadian factors. Practical therapeutic approaches are provided for these circadian dysrhythmias.

27

NAPPING

MASAYA TAKAHASHI[1] AND KOSUKE KAIDA[1,2]
1. National Institute of Industrial Health, Kawasaki, Japan
2. Japan Society for the Promotion of Sciences, Tokyo, Japan

INTRODUCTION

A monophasic pattern of sleep is one of the main features of human behavior. However, short periods of sleep, or naps, during the daytime or at night can be used to actively cope with the physiological need for sleepiness. A nap may also be capable of maintaining waking performance and alertness under prior sleep restrictions. Evidence regarding the role of napping has been accumulating rapidly, and a summary of the research findings may contribute to the further understanding and investigation of effective napping strategies.

In this chapter, the function of napping is addressed in terms of the timing of a nap and possible target populations (the healthy young, the healthy elderly, and insomniacs). In addition, sleep inertia is discussed as one of the negative side effects of napping.

DAYTIME NAPS

Healthy People

Reduced alertness in the early afternoon disturbs our daily activities. Indeed, working populations are at risk for errors and accidents in the workplace. Traffic accidents in the community occur frequently during this particular time zone (2:00–4:00 p.m.). Taking a brief nap at the peak of daytime sleepiness may be a reasonable solution.

A 20 min nap taken at 2:00 p.m. has been shown to produce improved performance and alertness, compared to no nap [1]. Young subjects (three men, four women; mean age 21 years) were asked to nap for a mean of 19.9 min of total sleep time (TST) or to stay awake, sitting on a chair. On subsequent performance tests that involved logical reasoning, addition, and auditory vigilance, the number of correct responses under the nap condition was significantly higher than that under the no-nap condition. The accuracy of the vigilance task was improved for up to 2 h after napping. Subjective sleepiness was significantly lower after napping than after no nap, but fatigue and motivation scores were similar between the two conditions. Electroencephalogram (EEG) alpha activity, used as an objective measure of alertness, paralleled the performance results, with a significantly higher degree of alertness under the nap condition.

Cognitive neuroscientists have increasingly recognized that sleep plays an essential role in the learning process. In this context, one may presume that a nap could also improve learning. The hypothesis that a daytime nap scheduled at 2:00 p.m. reverses the decline in performance throughout the day was tested in undergraduates ($n = 129$) using a visual texture discrimination task (TDT) [2]. TDT performance deteriorated from the morning through the evening in a no-nap group. In contrast, TDT performance was maintained after a 30 min nap and was improved after a 60 min nap. Differences in perceptual processing between the no-nap group and the 30 min nap or the 60 min nap group were significant at 4:00 p.m. and 7:00 p.m. A subsequent study, which compared the effects of a 60 min nap, and a 90 min nap, and no nap, demonstrated that texture discrimination can be facilitated if a nap contains both deep non-rapid eye movement (NREM) sleep and REM sleep [3]. Perceptual deterioration, however, appears to be reversed by a nap with deep NREM sleep but not REM sleep. When measured at 9:00 a.m. the next morning (24 h after the first test), TDT performance in the 90 min nap group was equivalent to the level in a group who were tested after two nights of sleep without daytime naps.

Taking a nap in the early afternoon is difficult in industrial settings, even if napping has been shown to be effective for improving daytime functioning. Instead, a postlunch rest may be another suitable time period for napping. This possibility was examined for three groups of young subjects ($n = 10$ in each group, 20–30 years old) [4]. A 15 min nap opportunity started at 12:30 p.m. (mean TST = 7.3 min) was found to produce significantly increased alertness as judged by P300 latency, compared to no nap or a 45 min nap opportunity (mean TST = 30.1 min). Both the 15 and 45 min nap groups showed significantly higher levels of subjective alertness than the no-nap group. Performance of a transcription task in the evening significantly improved after the 15 min nap opportunity, compared to no nap.

The above-mentioned laboratory findings have also been verified in the workplace [5]. Factory workers (six men, two women; mean age 49 years) took a 15 min nap opportunity at 12:30 p.m. on a reclining chair during the nap week, but remained awake during the no-nap week. The participants then determined whether or not she/he would nap during a follow-up week. Polysomnography (PSG) was not performed during the nap opportunity. Perceived alertness increased significantly in the afternoon after napping at the end of the nap week, compared to the no-nap week. Similar effects were seen during the follow-up week, during which almost half of the subjects chose to nap. Choice reaction time (RT) performance measured at 3:00 p.m. was not affected by napping. No significant differences in actigraphic measures of nocturnal sleep were observed among the participants during the three weeks of the study. The participants' reports revealed that the nap opportunity eased their body, made them feel less sleepy while engaged in a monotonous job, and reduced the amount of yawning in the afternoon. Some participants wished to have a longer naptime.

Sleep-Deprived Healthy People

Sleep deprivation elevates daytime sleepiness and modifies the effects of napping. A 15 min nap opportunity (mean TST = 10.2 min) during a postlunch rest was reported to maintain performance and alertness during the midafternoon among young subjects (seven men, five women; mean age 22 years) who had only slept for 3.5 h the night before [6]. However, a 15 min nap opportunity at around 3:00 p.m. (mean TST = 7.0 min) plus a pre-nap intake of caffeine (150 mg) was shown to be superior at reducing afternoon sleepiness and driving impairments for 2 h than was napping, caffeine intake, or no nap, among young students who had only slept for 5 h during the previous night (six men, six women; mean age 23 years) [7]. As to the minimum duration of a nap, 10 min of TST has been proposed to be needed to ensure recuperative effects under sleep-restricted conditions [8]. Taken together, the currently available findings support the use of a brief nap in counteracting the negative impacts of sleep deficit. Furthermore, the effectiveness of napping seems to be continued for 2–3 h following napping.

Elderly People

Aging considerably affects sleep and wakefulness. A large percentage of elderly people experience poor sleep at night and degraded alertness during the daytime. The latter problem might be solved by taking a brief nap, according to the findings of the previous research involving younger subjects. This seems to be the case, based on experimental results that a 34 min nap opportunity scheduled at 1:00 p.m. (mean TST = 24 min), compared to no nap, significantly increased the number of correct responses in a visual detection task and significantly enhanced alertness at both objective and subjective levels in elderly subjects (five men, five women; mean age 73 years) [9]. A 30 min nap taken by 3:00 p.m., combined with light exercise in the evening, has been explored to see if such an intervention would improve the sleep of elderly people. The data obtained so far seems to support the validity and reliability of this intervention [10].

A highly controlled study of elderly subjects (four men, five women; mean age 79 years) demonstrated a significant improvement in evening sleepiness as measured by longer sleep latency, reduced TST, earlier wake-time, and no significant changes in body temperature rhythm after a 90 min nap opportunity (1:30–3:00 p.m., mean TST = 57 min), compared to no nap [11]. Considering the differences in the nap duration among the studies mentioned, it may be reasonable to assume that favorable consequences can be expected if a nap opportunity of less than 30 min is scheduled in the early afternoon.

Patient Groups

Very little data is available for the clinical utility of napping to treat or ameliorate sleep disorders. Although patients suffering from insomnia, sleep-disordered breathing, narcolepsy, and other sleep disorders could manage their conditions by taking an appropriately controlled nap [12], the focus here is on chronic insomnia.

A recent study of self-reported primary sleep onset insomniacs revealed significant improvements in performance and alertness after a 10 min nap, compared to no nap [13]. Participants (three men, seven women; mean age 40 years) reported that their mean sleep latency was longer than 30 min at least four times a week, that their symptoms had lasted for at least 6 months, and that they experienced negative daytime consequences from their sleep problem. The subjects slept for approximately 5 h at night before the nap and no-nap trials. A 10 min nap was scheduled at 3:00 p.m. The results of performance tests showed significantly better scores in the symbol-digit substitution task, the letter cancellation task, and reaction time (RT) speed 2.6 h after the nap, compared to the results of these tasks performed without a nap. Subjective ratings of sleepiness and fatigue were not significantly different between the two conditions. Objective alertness as assessed by sleep onset latency increased significantly, and this effect was maintained until 2 h after napping compared to no nap. Interestingly, all but one participant did not acknowledge having been asleep, despite a mean total sleep time of 9.7 min.

These findings suggest that people with insomnia may be able to receive the benefits of a brief nap, in terms of performance and objective alertness, if they could initiate and maintain sleep during a daytime nap opportunity. Sleep clinicians sometimes advise insomniacs to refrain from a daytime nap to avoid further difficulty in falling asleep during the main sleep period. Whether taking a 10 min nap during the daytime would disturb subsequent sleep at night is uncertain, but this seems unlikely, based on data obtained in healthy subjects [5, 9]. If so, impaired waking function resulting from chronic insomnia could be counteracted by a brief nap.

NIGHTTIME NAPS

Laboratory Research

The appropriate timing and duration of a nap are very hard to select, particularly in subjects working night shifts, in whom dramatic reductions in performance and alertness are anticipated because of the dual influences of circadian trough and extended wakefulness. In a simulated night shift study of oil refinery workers (14 men, aged 31–52 years), either a 30 min nap opportunity at 1:20 a.m. (mean TST = 24.5 min) or a 50 min nap opportunity at 1:00 a.m. (mean TST = 38.1 min) was reported to significantly reduce an objective measure of sleepiness (the Repeated Test of Sustained Wakefulness) at 2:40 a.m., compared to no nap [14]. However, these alerting effects were not detected at 5:30 a.m. following a 30 min nap opportunity at 4:10 a.m. (mean TST = 27.5 min) or a 50 min nap opportunity at 3:50 a.m. (mean TST = 46.6 min). Lapses in a choice RT

task after any of the nap opportunities, compared to no nap, were reduced by half at the end of the shift.

While a number of countermeasures against problems associated with night shifts have been proposed, napping and bright light (BL) exposure are the two major candidates of interest. BL techniques may need at least three days to induce a phase shift in the circadian system, allowing a complete adaptation to night shift schedule. Hence BL techniques cannot be applied to rapidly rotating shift schedules, like the one to two successive night shift schedules of air traffic control specialists (ATCSs) in the United States. A simulated night shift study of ATCSs (28 men, 31 women; mean age 33 years) revealed that, overall, a 2 h nap opportunity taken at 1:45 a.m. produced significantly better results in a vigilance task performed during the late period of the night shift, compared to a 45 min nap opportunity at 3:00 a.m. or no nap [15]. However, the 45 min nap condition significantly shortened the response times in a simulated air traffic control task, compared to no nap.

Field Research

Most previous research on nighttime naps has only addressed napping on the first night shift. Little is thus known about the effects of napping during consecutive night shifts. According to a worksite study of aircraft maintenance engineers (12 men, mean age 35 yeas) working two 12 h night shifts in a row, a 20 min nap opportunity between 1:00 and 3:00 a.m. seemed to yield significantly shorter response times in a vigilance task at the end of the fist night shift than no nap [16]. However, this improvement after napping was not observed during the second night shift. The nap opportunity did not affect subjective fatigue when measured at the end of each shift. The reported duration of the nap was about 20 min for each nap opportunity, although the PSG was not recorded. These findings suggest that a brief nap may be useful in reducing performance deterioration on the first night shift where prior wakefulness tends to be prolonged. However, whether workplace napping has any benefits during the second night shift remains uncertain.

When the extent to which shift workers (12 men, mean age 37 years) napped during night shifts was examined over the course of one year, six bimonthly evaluations showed that a mean of 67% (58–72%) of the workers took a nap during a 1 h period of rest between 11:30 p.m. and 3:30 a.m. [17]. The estimated duration of the nap was reported to be 31 min. Participants gave favorable responses to napping: 88% felt that napping "reduced fatigue and sleepiness" and "increased energy."

Although many problems are encountered while conducting field research, laboratory findings regarding napping must be examined in the actual workplace. This process should enable possible strategies of naps to be refined, making the implementation of nap strategies more practical.

SLEEP INERTIA FOLLOWING NAPPING

A number of barriers have been recognized when implementing a nap in operational settings, despite the fact that napping produces favorable influences on waking function. The most notable factor is sleep inertia—or transient decreases in performance and alertness immediately after awakening from a nap [18]. Sleep inertia might disturb the quality of post-nap activities for a certain period. Since naps of any length seem to cause sleep inertia, maximizing the gains of napping and minimizing the losses of sleep inertia are essential.

We can control the length of a nap, but not its architecture. Since sleep inertia has a more detrimental effect after awakening from deep NREM sleep, the length of a nap should be shorter than 20 min or 90–100 min [12]. Subjects are not expected to enter deep NREM sleep during the former nap time, and the latter nap time should enable the subject to awake from REM sleep or shallow NREM sleep in theory. Generally, sleep inertia is shown to continue for approximately 30 min after awakening [18]. Thus designing a nap regime that includes this period of inertia seems appropriate. Special attention must also be paid to tasks requiring a critically important decision immediately upon awakening.

Recent studies have explored alternative techniques to coping with sleep inertia in a more active, practical way. The alertness-enhancing effects of a 20 min nap have been shown to be strengthened by a pre-nap intake of caffeine (200 mg) or by post-nap exposure to bright light (2000 lux, 1 min) or a post-nap face washing at a temperature of 25 °C, with the strongest enhancement seen for the caffeine intake condition [19]. Spontaneous awakening at a time decided before napping—or self-awakening—can be another method, since self-awakening has been proved to be more effective in maintaining post-nap levels of alertness than forced awakening using external means [20].

CONCLUSION

The research findings described here clearly indicate the potential of napping for facilitating the performance and alertness of healthy people and for alleviating the negative effects associated with night shifts or chronic insomnia. One important assignment is to determine how the advantages and disadvantages of napping can be balanced. Currently, limited data is available regarding the effects of napping in middle-aged populations. Repeated, multiple nap schedules may also require more attention.

REFERENCES

1. Hayashi M, Watanabe M, Hori T. The effects of a 20 min nap in the mid-afternoon on mood, performance and EEG activity. *Clin Neurophysiol* **110**:272–279(1999).

2. Mednick SC, Nakayama K, Cantero JL, Atienza M, Levin AA, Pathak N, Stickgold R. The restorative effect of naps on perceptual deterioration. *Nat Neurosci* **5**:677–681(2002).

3. Mednick S, Nakayama K, Stickgold R. Sleep-dependent learning: a nap is as good as a night. *Nat Neurosci* **6**:697–698(2003).

4. Takahashi M, Fukuda H, Arito H. Brief naps during post-lunch rest: effects on alertness, performance, and autonomic balance. *Eur J Appl Physiol Occup Physiol* **78**:93–98(1998).

5. Takahashi M, Nakata A, Haratani T, Ogawa Y, Arito H. Post-lunch nap as a worksite intervention to promote alertness on the job. *Ergonomics* **47**:1003–1013(2004).

6. Takahashi M, Arito H. Maintenance of alertness and performance by a brief nap after lunch under prior sleep deficit. *Sleep* **23**:813–819(2000).

7. Reyner LA, Horne JA. Suppression of sleepiness in drivers: combination of caffeine with a short nap. *Psychophysiology* **34**:721–725(1997).

8. Tietzel AJ, Lack LC. The recuperative value of brief and ultra-brief naps on alertness and cognitive performance. *J Sleep Res* **11**:213–218(2002).

9. Tamaki M, Shirota A, Hayashi M, Hori T. Restorative effects of a short afternoon nap (<30 min) in the elderly on subjective mood, performance and EEG activity. *Sleep Res Online* **3**:131–139(2000).

10. Tanaka H, Shirakawa S. Sleep health, lifestyle and mental health in the Japanese elderly; ensuring sleep to promote a healthy brain and mind. *J Psychosom Res* **56**:465–477(2004).

11. Monk TH, Buysse DJ, Carrier J, Billy BD, Rose LR. Effects of afternoon "siesta" naps on sleep, alertness, performance, and circadian rhythms in the elderly. *Sleep* **24**:680–687(2001).

12. Takahashi M. The role of prescribed napping in sleep medicine. *Sleep Med Rev* **7**:227–235(2003).

13. Brooks-Tietzel AJ. *An Investigation into Brief Afternoon Naps as a Countermeasure to Daytime Sleepiness.* Flinders University, Adelaide, 2004.

14. Sallinen M, Harma M, Akerstedt T, Rosa R, Lillqvist O. Promoting alertness with a short nap during a night shift. *J Sleep Res* **7**:240–247(1998).

15. Della Rocco PS, Comperatore C, Caldwell L, Cruz C. *The Effects of Napping on Night Shift Performance.* Federal Aviation Administration Civil Aeromedical Institute, Oklahoma City, 2000, DOT/FAA/AM-00/10.

16. Purnell MT, Feyer AM, Herbison GP. The impact of a nap opportunity during the night shift on the performance and alertness of 12-h shift workers. *J Sleep Res* **11**:219–227 (2002).

17. Bonnefond A, Muzet A, Winter-Dill AS, Bailloeuil C, Bitouze F, Bonneau A. Innovative working schedule: introducing one short nap during the night shift. *Ergonomics* **44**:937–945 (2001).

18. Tassi P, Muzet A. Sleep inertia. *Sleep Med Rev* **4**:341–353(2000).

19. Hayashi M, Masuda A, Hori T. The alerting effects of caffeine, bright light and face washing after a short daytime nap. *Clin Neurophysiol* **114**:2268–2278(2003).

20. Kaida K, Nakano E, Nittono H, Hayashi M, Hori T. The effects of self-awakening on heart rate activity in a short afternoon nap. *Clin Neurophysiol* **114**:1896–1901(2003).

ADDITIONAL READING

Anthony WA (Ed). *The Art of Napping*. Biblio Distribution, Cheshire, CT, 1997. This book, written in plain language, addresses the daily use of napping based on both a line of scientific evidence and funny stories about great persons who loved napping.

Anthony WA, Anthony CW (Eds). *The Art of Napping at Work*: *The No-Cost, Natural Way to Increase Productivity and Satisfaction*. Biblio Distribution, Cheshire, CT, 1999. This is a companion book to *The Art of Napping* and focuses specifically on workplace napping to improve the quality of work.

Dinges DF, Broughton RJ (Eds). *Sleep and Alertness: Chronobiological, Behavioral, and Medical Aspects of Napping*. Raven Press, New York, 1989. This comprehensive book opens the door to the science of napping and covers broad ranges of topics relevant to napping.

Stampi C (Ed). *Why We Nap*: *Evolution, Chronobiology, and Functions of Polyphasic and Ultrashort Sleep*. Birkhäuser, Boston, 1992. This innovative book focuses on polyphasic patterns of sleep/naps and highlights the importance of sleep management.

28

SLEEP LOSS, SLEEPINESS, PERFORMANCE, AND SAFETY

ROGER R. ROSA

National Institute for Occupational Safety and Health, Washington, DC

Almost all modern adults lose sleep or occasionally feel sleepy, even extremely sleepy, during their waking hours. These events are irregular or circumstantial for many of us. A significant minority of the population, however, must confront sleep loss or sleepiness on a daily or weekly basis because of work demands, child or elder care, or medical conditions that compromise the sleep–wake system. Work schedules and other work demands probably are the largest contributors to sleep loss among healthy adults, and even among some adolescents attending school and working part-time jobs. It is well established that work schedules requiring night or rotating shift work, early morning awakening, or extended, irregular, or on-call hours reduce the opportunity to obtain adequate, good quality sleep (see [1] for review). Reducing either the quantity or quality of sleep increases sleepiness during waking hours and may increase the likelihood of performance errors or adverse incidents.

PREVALENCE OF SLEEP LOSS AND SLEEPINESS

There are no population-based estimates of the prevalence of sleep loss or excessive sleepiness recorded routinely in national health databases. In random-sample surveys conducted by the National Sleep Foundation in 2001 and 2002, 37–40% of adult respondents reported that sleepiness interfered with their daily activities at least a few days per month. Sleepiness reports in those surveys were more frequent among females, shift workers, caregivers to children or other dependents, and people reporting depression or marital dissatisfaction.

The numbers of people potentially experiencing sleep loss or sleepiness from demanding work schedules are significant. In the United States, 8.4% of all full-time workers, or 8.4 million people, were estimated to be working night, rotating, or irregular shifts in 2001 (data from the U.S. Bureau of Labor Statistics). The number of people working extended hours or overtime schedules is less well documented, but those schedules are receiving considerable research and public policy attention given their use among industry sectors critical to public safety, such as health care, police and fire protection, or transportation. The prevalence of persistent sleep disorders associated with extreme sleepiness during time awake (e.g., narcolepsy, sleep-disordered breathing) is estimated at 12–13 million people in the United States (data from the National Heart, Lung, and Blood Institute).

LABORATORY ASSESSMENTS

Dose-related increases in sleepiness after sleep loss have been demonstrated repeatedly in the laboratory with physiological measures, subjective reports, and standardized behavioral tests that reflect operationally significant aspects of human function. Physiological sleep tendency, as measured by the time taken to fall asleep on sleep latency tests, decreases after total or partial sleep loss, sleep fragmentation, or a shift of waking activities to the overnight hours [2]. On average, these latencies are 50–60% shorter after one night of total sleep loss, after five to seven consecutive days of reduced sleep (2–3 h), or after severely fragmented sleep induced by forced arousals every minute [3].

Sleep: A Comprehensive Handbook, Edited by T. Lee-Chiong.

Standardized performance tests measuring perceptual–motor responsiveness (e.g., reaction time tasks), eye–hand coordination (e.g., visual tracking tasks), or ability to maintain attention or concentration (e.g., vigilance or monitoring tasks) also degrade with sleep deprivation. Belenky et al. [4] demonstrated a dose-related pattern of these effects characterized by progressively slower reaction times and more frequent missed signals with increasing amounts of sleep restriction (3, 5, 7, or 9 h per night during seven nights in the laboratory). Performance losses also have been demonstrated with more complex measures of cognitive function, such as memory, decision making, creative problem solving, and planning skills, but these effects vary across studies and may or may not indicate deficits in cerebral executive functions [5].

Regardless of the underlying mechanisms, the overall pattern of performance decrements associated with sleep deprivation in the laboratory, and parallel increases in subjective (see [3]) or physiological sleepiness, suggest increased potential for operational errors or adverse incidents in real-world settings. Studies comparing performance during sleep loss to performance during alcohol use support this suggestion. In two representative studies [6, 7], one night of sleep loss produced performance impairments similar to 0.08–0.10% blood alcohol concentration, the legal limit in most industrialized countries. These impairments were especially acute in the late-night/early morning hours when the endogenous downturn in the circadian rhythm of alertness compounds decrements associated with lost sleep [8].

Performance during laboratory simulations of more realistic tasks often declines with sleep loss, but not always. Decrements in simulated driving performance with sleep loss have been demonstrated repeatedly. Roge et al. [9], for example, reported greater variability in driving control and narrowing of peripheral vision (measured with a secondary task) after sleep loss. Horne et al. [10] and Fairclough and Graham [11] demonstrated increased safety-critical lane deviations after sleep loss that were similar to alcohol impairment. Horne et al. [10] further demonstrated that moderate sleep restriction (5 h of nighttime sleep), combined with moderate alcohol intake (legal blood alcohol concentrations of less than 0.06%), significantly worsened critical lane drifting when compared with either sleep loss or alcohol alone.

Performance during simulations of more complex professional tasks after sleep loss has been less consistent. Taffinder et al. [12], for example, observed a 14% slowing of performance and 20% more errors during simulated laproscopic surgery after one night of sleep loss. Howard et al. [13], on the other hand, observed no major decrements after one night of sleep loss in the clinical performance of anesthesiologists participating in simulated surgery. Taken together, these studies suggest that some tasks are more sensitive to sleep loss than others, and that performance can be modulated by other contributing factors such as degree of training or professional engagement (observed in the anesthesia task but not in driving), the degree of monotony (more apparent in driving), task duration, or other impairing influences (such as alcohol).

IN SITU STUDIES

Recognition of the contribution of sleep loss and sleepiness to actual traumatic or catastrophic incidents has increased dramatically in the past two decades, fueled by an expanding set of compelling laboratory demonstrations and perhaps stimulated by a 1988 public policy consensus statement from the Association of Professional Sleep Societies [14]. Since that consensus statement, a growing number of published field studies of safety risks associated with sleep loss and sleepiness has been paralleled by increasing awareness of sleepiness, fatigue, and adequacy of rest issues in public policy discussions. The National Transportation Safety Board has included rest, sleep, and fatigue as critical issues on its "most wanted" list for several years. Relatedly, the Federal Motor Carrier Safety Administration proposed revision to the hours of service regulations for commercial drivers extends the mandatory uninterrupted rest period from 8 to 10 h to increase the opportunity for a full night of sleep. Several recent studies of the real-world risks of drowsy driving support these policy concerns despite considerable variability in the prevalence estimates of sleepy drivers in either the commercial or general driving population. The contribution of sleepiness to motor vehicle collisions varies from 1% to 20% depending on the estimate method. The lower limit of this range relies on an explicit attribution of sleepiness in an official crash report while the upper limit is a mathematical extrapolation based on the argument that sleepiness often is not recognized in official reports [15]. Other estimates within those limits are based on questionnaire surveys of sleepiness and its contribution to collisions.

Systematic surveys suggest that driving while drowsy is widespread in the general population and not captured sufficiently by focusing on crashes. In a comprehensive, random-sample survey, McCartt et al. [16] observed that 55% of drivers reported drowsy driving in the past year, 23% reported falling asleep without crashing, 3% had crashed after falling asleep, and 2% had crashed while driving drowsy. Factors contributing to drowsy driving in that study included demographics (being younger, male, and more educated), driving more miles annually, getting fewer hours of sleep, having trouble staying awake during the day, getting sleepy more quickly while driving, working shifts, and driving as part of the job. Other frequently cited risk factors include driving during the late-night or early morning hours, use of alcohol or sedating medications, or the presence of sleep disorders or other medical disorders [15, 17].

Studies focusing on specific risk groups, such as sleep disorder patients [18], shift workers [19], or on-call medical residents [20], reinforce the general surveys by reporting a higher incidence of drowsiness-related crashes or near crashes. Young et al. [18] observed at least a three fold higher odds of a motor vehicle crash among men with mild sleep apnea (apnea–hypopnea index or AHI > 5) compared to subjects without sleep-disordered breathing. Among men and women with more severe apnea (AHI > 15), the odds of experiencing multiple car crashes over a 5 year period were seven times higher. These results, from a population-based sample of over 900 people, are consistent with earlier reports of greater crash risk based on small clinical samples of sleep apnea patients. In a survey of emergency medicine residents, Steele et al. [20] reported more frequent reports of motor vehicle crashes and near crashes while driving after night shift, especially among residents reporting poor tolerance of night work and difficulty overcoming drowsiness. Similar results were reported by Gold et al. [19] in a sample of nurses working rotating shifts.

Sleep loss and sleepiness also are significant contributing factors to unintentional injuries not related to driving. Most of the studies demonstrating these associations examined work-related injuries but there is at least one study examining a broader range of injuries. Goldberg et al. [21] surveyed the sleep habits and complaints of seventy patients treated in an emergency department for a variety of injuries, such as lacerations, bruises, sprains, and fractures, from a variety of events, including falls and motor vehicle collisions. The average reported sleep duration of these patients immediately prior to injury was significantly less than their usual sleep on weekdays or weekends. In addition, 77% of these patients reported symptoms and disturbances (e.g., snoring, breathing pauses, restless legs) that suggested moderate to high risk for a sleep disorder.

Studies of work-related injuries offer the opportunity to examine sleep and sleepiness in combination with a multitude of risk factors potentially associated with traumatic injuries or catastrophic incidents. In a cohort of over 7000 Dutch workers in a variety of occupations, self-reported fatigue and need for recovery were associated with increased risk of occupational injuries after controlling for age, smoking, alcohol consumption, educational level, shift work, and work environment [22]. Studies concentrating on high-risk occupations, such as construction workers, loggers, or firefighters, report similar results. Chau et al. [23], for example, examined a variety of job and personal factors (including sociodemographic data, job, safety training, smoking habit, alcohol consumption, sporting activities, physical disabilities, hearing, vision, and sleep

disorders) for their contribution to injuries in a sample of 880 construction workers. Among other factors, reports of having a sleep disorder (defined by sleep duration, "sleeping badly," or regular use of sleeping pills) doubled the odds of an injury associated with a moving object.

Working conditions, especially work schedules, can influence sleep duration and sleepiness independently of a worker's personal attributes or the presence of sleep disorders. Night shift is the work schedule factor most associated with sleep loss and sleepiness [24], and also the factor most associated with risk of injuries when other day–night differences are held constant in the analysis. Shifts that begin very early in the morning (4:00–7:00 a.m.) also are a concern as sleep taken before the shift often is truncated, resulting in increased sleepiness during subsequent waking hours [1, 24]. To date, however, there are no published studies directly examining early morning starts for their effect on injuries or adverse incidents.

A meta-analysis combining injury data from several studies demonstrates the risk associated with night shift [25]. Compared to morning/day shift, injury risk increased by 18% during the afternoon/evening shift and 34% during the night shift. Relative risk also increased across consecutive shifts with a more precipitous increase observed across night shift compared to morning/day shift. These results are consistent with our own worksite observations of increased subjective sleepiness and decreased reaction time during night shifts, and progressive decreases in total sleep time from early to late in the workweek [1].

COUNTERMEASURES

The best countermeasure for acute sleep loss or sleepiness is to go to sleep when it is safe to do so. The best countermeasure for chronic sleep loss or sleepiness is to maintain a routine of optimal sleep (usually 7–8 h) taken at regular times. Individuals with sleep disorders should consult with a physician to obtain proper treatment to improve their sleep. Maintaining sleep, alertness, and safety while working shifts or other demanding schedules requires careful design of the schedule and the work environment by the organization and conscientious planning by the worker [26]. When adequate sleep is not feasible because of work or personal demands, then a short nap may help return sleep to optimal daily levels and improve alertness [27]. If napping is not practical, moderate amounts of caffeine can temporarily increase alertness with minimal side effects [28]. Stronger stimulants are best avoided unless medically indicated. Popular behavioral countermeasures, such as physical activity, cold air or water, loud noise, or listening to music, usually have very transient effects that disappear after a few minutes.

CONCLUSION

Sleep loss and sleepiness degrade performance efficiency and increase the likelihood of operational errors that may contribute to traumatic or catastrophic incidents. A substantial number of people regularly confront sleep loss and sleepiness because of work, other situational demands, or medical conditions. Still, we are far from a precise estimate of the magnitude of risk from excessive sleepiness or its contribution to societal loss. Nonetheless, the public health burden from sleep loss and sleepiness is sufficient to motivate prevention efforts to reduce risk through improved environmental designs (e.g., alerting rumble strips on highways), administrative controls and policies (e.g., redesigned work schedules), and training and awareness campaigns to alert individuals to the benefits of good sleep.

REFERENCES

1. Rosa RR. Examining work schedules for fatigue: it's not just hours of work. In: Hancock PA, Desmond PA (Eds), *Stress, Workload, and Fatigue*. Lawrence Erlbaum Associates, Mahwah, NJ, 2001, pp 513–528.

2. Arand D, Bonnet M, Hurwitz T, Mitler M, Rosa RR, Sangal B. The clinical use of the MSLT and MWT. *Sleep* **28**(1):123–144(2005).

3. Bonnet MH, Arand DL. Clinical effects of sleep fragmentation versus sleep deprivation. *Sleep Med Rev* **7**:297–310(2003).

4. Belenky G, Wesensten NJ, Thorne DR, Thomas ML, Sing HC, Redmond DP, Russo MB, Balkin TJ. Patterns of performance degradation and restoration during sleep restriction and subsequent recovery: a sleep dose–response study. *J Sleep Res* **12**:1–12 (2003).

5. Jones K, Harrison Y. Frontal lobe function, sleep loss and fragmented sleep. *Sleep Med Rev* **5**:463–475(2001).

6. Dawson D, Reid K. Fatigue, alcohol and performance impairment. *Nature* **388**:235(1997).

7. Williamson AM, Feyer AM. Moderate sleep deprivation produces impairments in cognitive and motor performance equivalent to legally prescribed levels of alcohol intoxication. *Occup Environ Med* **57**(10):649–655(2000).

8. Åkerstedt T, Folkard S. Validation of the S and C components of the three-process model of alertness regulation. *Sleep* **18**:1–6 (1995).

9. Roge J, Pebayle T, El Hannachi S, Muzet A. Effect of sleep deprivation and driving duration on the useful visual field in younger and older subjects during simulator driving. *Vision Res* **43**:1465–1472(2003).

10. Horne JA, Reyner LA, Barrett PR. Driving impairment due to sleepiness is exacerbated by low alcohol intake. *Occup Environ Med* **60**:689–692(2003).

11. Fairclough SH, Graham R. Impairment of driving performance caused by sleep deprivation or alcohol: a comparative study. *Hum Factors* **41**:118–128(1999).

12. Taffinder NJ, McManus IC, Gul Y, Russell RC, Darzi A. Effect of sleep deprivation on surgeons dexterity on laparoscopy simulator. *Lancet* **352**(9135):1191(1998).

13. Howard SK, Gaba DM, Smith BE, Weinger MB, Herndon C, Keshavacharya S, Rosekind MR. Simulation study of rested versus sleep-deprived anesthesiologists. *Anesthesiology* **98**:1345–1355(2003).

14. Mitler MM, Carskadon MA, Czeisler CA, Dement WC, Dinges DF, Graeber RC. Catastrophes, sleep, and public policy: consensus report. *Sleep* **11**:100–109(1988).

15. Horne J, Reyner L. Vehicle accidents related to sleep: a review. *Occup Environ Med* **56**:289–294(1999).

16. McCartt AT, Ribner SA, Pack AI, Hammer MC. The scope and nature of the drowsy driving problem in New York State. *Accid Anal Prev* **28**:511–517(1996).

17. Lyznicki JM, Doege TC, Davis RM, Williams MA. Sleepiness, driving, and motor vehicle crashes. Council on Scientific Affairs, American Medical Association. *JAMA* **279**:1908–1913(1998).

18. Young T, Blustein J, Finn L, Palta M. Sleep-disordered breathing and motor vehicle accidents in a population-based sample of employed adults. *Sleep* **20**:608–613(1997).

19. Gold DR, Rogacz S, Bock N, Tosteson TD, Baum TM, Speizer FE, Czeisler CA. Rotating shift work, sleep, and accidents related to sleepiness in hospital nurses. *Am J Public Health* **82**:1011–1014(1992).

20. Steele MT, Ma OJ, Watson WA, Thomas HA Jr, Muelleman RL. The occupational risk of motor vehicle collisions for emergency medicine residents. *Acad Emerg Med* **6**:1050–1053(1999).

21. Goldberg R, Shah SJ, Halstead J, McNamara RM. Sleep problems in emergency department patients with injuries. *Acad Emerg Med* **6**:1134–1140(1999).

22. Swaen GM, Van Amelsvoort LG, Bultmann U, Kant IJ. Fatigue as a risk factor for being injured in an occupational accident: results from the Maastricht Cohort Study. *Occup Environ Med* **60**(Suppl 1):88–92(2003).

23. Chau N, Gauchard GC, Siegfried C, Benamghar L, Dangelzer JL, Francais M, Jacquin R, Sourdot A, Perrin PP, Mur JM. Relationships of job, age, and life conditions with the causes and severity of occupational injuries in construction workers. *Int Arch Occup Environ Health* **77**:60–66(2004).

24. Åkerstedt T. Shift work and disturbed sleep/wakefulness. *Occup Med (Lond)* **53**:89–94(2003).

25. Folkard S, Tucker P. Shift work, safety and productivity. *Occup Med (Lond)* **53**:95–101(2003).

26. Knauth P, Hornberger S. Preventive and compensatory measures for shift workers. *Occup Med (Lond)* **53**:109–116 (2003).

27. Takahashi M. The role of prescribed napping in sleep medicine. *Sleep Med Rev* **7**:227–235(2003).

28. Wesensten NJ, Belenky G, Kautz MA, Thorne DR, Reichardt RM, Balkin TJ. Maintaining alertness and performance during sleep deprivation: modafinil versus caffeine. *Psychopharmacology (Berl)* **159**:238–247(2002).

ADDITIONAL READING

Åkerstedt T (Ed). Work hours, sleepiness and accidents. *J Sleep Res* **4**(Suppl 2):(1995). A series of articles on work-related sleepiness and safety risks.

Carskadon MA (Ed). *Adolescent Sleep Patterns: Biological, Social, and Psychological Influences.* Cambridge University Press, Cambridge, UK, 2002. Includes information on sleep loss and sleepiness in adolescents.

Caruso CC, Hitchcock EM, Dick RB, Russo JM, Schmit JM. *Overtime and Extended Work Shifts: Recent Findings on Illnesses, Injuries, and Health Behaviors.* DHHS (NIOSH) Publication No. 2004-143. National Institute for Occupational Safety and Health, Cincinnati, OH, 2004. Annotated bibliography of recent studies on the health and safety effects of demanding work schedules.

Cluydts R, De Valck E, Verstraeten E, Theys P. Daytime sleepiness and its evaluation. *Sleep Med Rev* **6**:83–96(2002). Review of measurements of sleepiness.

Landrigan CP, Rothschild JM, Cronin JW, Kaushal R, Burdick E, Katz JT, Lilly CM, Stone PH, Lockley SW, Bates DW, Czeisler CA. Effect of Reducing Interns' Work Hours on Serious Medical Errors in Intensive Care Units, *N Engl J Med* **351**(18):1883–1848(2004). Reducing hours improved sleep and reduced attentional failures and medical errors.

McConnell CF, Bretz KM, Dwyer WO. Falling asleep at the wheel: a close look at 1269 fatal and serious injury-producing crashes. *Behav Sleep Med* **1**:171–183(2003).

NCSDR/NHTSA Expert Panel on Driver Fatigue and Sleepiness. *Drowsy Driving and Automobile Crashes.* Joint publication of the Department of Transportation, National Highway Traffic Safety Administration and the Department of Health and Human Services, National Heart, Lung, and Blood Institute, National Center on Sleep Disorders Research, Washington, DC, 1998.

Pratt SG. *Work-Related Roadway Crashes Challenges and Opportunities for Prevention.* DHHS (NIOSH) Publication No. 2003-119. National Institute for Occupational Safety and Health, Cincinnati, OH, 2003. Reviews occupational crash statistics, contributing factors, and approaches to prevention.

Rosa RR, Colligan MJ. *Plain Language About Shiftwork.* DHHS (NIOSH) Publication No. 97-145. National Institute for Occupational Safety and Health, Cincinnati, OH, 1997. Information booklet for a nontechnical audience on shift-work health and safety issues.

PART IV

SLEEP DISORDERED BREATHING SYNDROMES

29

PHYSIOLOGY OF SLEEP DISORDERED BREATHING

B. TUCKER WOODSON

Medical College of Wisconsin, Milwaukee, Wisconsin

INTRODUCTION

Obstructive sleep apnea (OSA) is a common disorder resulting from collapse of the pharyngeal airway during sleep. The cause and mechanisms of this collapse are multifactorial but ultimately result from the combination of a structurally vulnerable upper airway combined with a sleep related loss of muscle tone. Our understanding of this collapse and obstruction has evolved. Historically, two basic schools of thought existed to describe the genesis of airway collapse—"active" versus "passive" mechanisms. The active theory proposed by Weitzman and co-workers [1] in 1978 resulted from observations of spasmodic closure of the lateral pharyngeal walls and closure of the velopharynx timed at the end of expiration. This sphincteric closure "apparently by active muscle contraction" was maintained for the duration of inspiration and was followed by airway openings occurring following arousals. Since, in humans, electromyographic studies of pharyngeal constrictors fail to demonstrate collapse combined with expiratory muscle activity, the concept of active muscular contraction causing airway closure in OSA has been replaced in favor of other mechanisms [2].

As an alternative to the active theory, a theory based on passive mechanisms does not require active neuromuscular contraction of pharyngeal muscle to close the airway. Obstruction instead results from the interaction of loss of dilating activity of pharyngeal muscles, the mass of the tongue and other tissues, and negative inspiratory intraluminal pressures [3]. As early as 1976, Sauerland and Harper [4] observed a loss of both phasic and tonic electromyographic

tongue activity during NREM sleep in OSA subjects but not snoring subjects. Both OSA and snoring subjects contrasted with normal individuals who did not snore who did not have significant electromyographic activity in the genioglossus muscle during wakefulness or NREM sleep. Their findings suggested that muscle activity was important in maintaining stability in OSA subjects but that loss of muscle activity alone was inadequate to create airway closure. Additional, unknown structural features "such as a large tongue" contributed to collapse, which when combined with negative inspiratory airway pressures obstructed the airway. This basic description still holds true; however, why the human airway is vulnerable and what the physiologic and structural perturbations are that create collapse continue to be investigated.

The human upper airway has a complex task. It has the challenge of maintaining ventilation while simultaneously allowing for alimentation, phonation, and speech. In other mammals, the larynx is positioned near the skull base, creating a physiologically separate respiratory and alimentary pathway [5]. Anatomically, the tongue is an oral structure. The intimate relationship to the skull base results in a highly stable airway that is independent of muscle tone. In adult humans, the larynx resides in the neck separated from the bony enclosure of the craniofacial skeleton (Figure 29.1). The tongue is oral and pharyngeal and is more critical in supporting ventilation. A longer pharyngeal airway further increases with age, males, and OSA. The laryngeal descent process (klinoraphy), while facilitating speech development, predisposes the soft tissue supraglottic pharyngeal airway to obstruction and requires compensatory

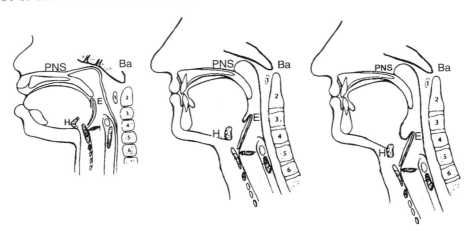

Figure 29.1 Differences in upper airway length between infants (*left*), nonapneic adults (*middle*), and OSA adults (*right*) are shown. In infants, the tongue compromises a shorter segment of the pharyngeal airway and the larynx may reside at the level of the second cervical vertebra. In nonapneic adults, the larynx may reside at the fourth cervical vertebra and both airway length and tongue area are greater in obstructive sleep apnea.

mechanisms to maintain stability [6, 7]. In such a setting, the addition of otherwise nonpathologic structural or physiologic changes may lead to upper airway collapse during sleep.

Multiple mechanisms help maintain upper airway stability. Neuromuscular tone, ventilatory control, sleep and arousal effects, upper airway reflexes and peripheral nervous system changes, craniofacial and soft tissue structure, body position, vascular tone, surface tension forces, lung volume effects, and expiratory collapse may all contribute. Integrating these requires a model of the balance of forces as well as understanding of basic mechanics of the Starling resistor, and dynamic and passive changes altering the upper airway.

BALANCE OF FORCES AND STARLING RESISTOR

A potentially large number of anatomic and physiologic processes must be integrated into a model of upper airway obstruction during sleep. One method that allows this integration is the concept of "balance of forces." The balance of forces model allows an accurate description of how multiple variables alter upper airway size (Figure 29.2). Airway size is determined by both dilating and collapsing forces. Dilating forces include upper airway muscle tone, mechanical force of the airway wall structure, and positive intraluminal airway pressure. Collapsing forces include tissue mass, surface adhesive forces, and negative intraluminal pressures. The resulting difference in these forces is the distending force, which acts on the wall of the upper airway. When the distending force increases, the airway size increases; when it decreases, the airway size decreases.

The distending force of the upper airway is the transmural pressure (P_{tm}) of the airway. The equation $P_{tm} = P_{out} - P_{in}$ defines transmural pressure where P_{out} represents the sum of the upper airway dilating forces and P_{in} represents the sum of the collapsing forces. Another more clinically relevant means to conceive of the forces

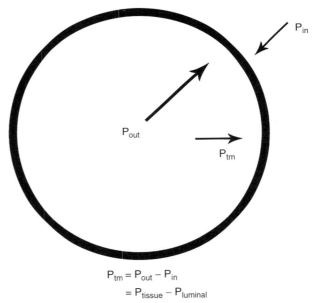

$$P_{tm} = P_{out} - P_{in}$$
$$= P_{tissue} - P_{luminal}$$

Figure 29.2 The principle of the balances of forces is diagrammed. Transmural forces (P_{tm}) on the upper airway are depicted. As P_{tm} increases, the airway enlarges; as P_{tm} decreases, the airway collapses. P_{tm} is described by the basic equation $P_{tm} = P_{out} - P_{in}$. P_{tm} may also be described as the difference between tissue forces (P_{tissue}) and luminal ($P_{luminal}$) or airway forces (i.e., $P_{tm} = P_{tissue} - P_{luminal}$) (see text for details).

that act on the upper airway is by considering the skeletal airway structure as a constant and describing the dynamic forces as being either tissue pressures or luminal pressures ($P_{tm} = P_{tissue} - P_{luminal}$). Tissue pressure includes the forces from tissue mass, tissue elastance, surface tension, and neuromuscular dilating and collapsing forces. Luminal pressures include the segmental airway pressure (P_{airway}) and pressures relating to airflow (P_{flow}). Airway pressures may be dilating (i.e., such as positive pressures during expiration or with the application of external positive pressure—CPAP) or collapsing (negative inspiratory pressure). Although seemingly esoteric, such a model ($P_{tm} = P_{luminal} - P_{tissue}$) provides a means of quantifying and describing upper airway collapse. Studies have been able to replicate a syndrome identical to OSAS in non-OSA subjects by applying negative intraluminal airway pressures to the upper airway during sleep.

The compliance (dA/dP) of the upper airway represents the tendency of the upper airway to collapse and can be calculated, allowing measurement of the intrinsic collapsibility of the upper airway. These measures, however, require controlling for both airflow velocity and muscular tone. Airflow velocity effects on luminal pressures are described by Bernoulli's equation and if velocity is zero flow effects on the luminal pressure are eliminated ($P_{luminal} = P_{airway} + 0$). When neuromuscular tone is held constant, then tissue forces are constant ($P_{tissue} = k$). In this situation, measured airway pressure represents the distending or transmural pressure of the upper airway ($P_{tm} = P_{airway} - k$) and, combined with measures of upper airway size, allows for calculation of airway compliance (dA/dP) independent of physiologic influences. Furthermore, airway pressure can be measured and manipulated (such as with nasal continuous positive airway pressure—CPAP) to assess changes in airway size and compliance.

STARLING RESISTOR

In contrast to a structurally patent airway, where the pressure difference across the tube wall is of minimal importance and flow is determined by driving pressure, in sleep disordered breathing the tube wall characteristic becomes the major determinate of airway cross-sectional area and airflow. The upper airway is a collapsible conduit with its flow described by the Starling resistor concept (Figure 29.3). The Starling resistor concept builds upon Poiseuille's law, which describes flow in noncollapsible tubes. Poiseuille's law states $V = P_1 - P_2 / R$ (where $V =$ flow, $P_1 =$ pressure upstream, $P_2 =$ pressure downstream, $P_1 - P_2$ driving pressure). The resistance component R is determined by length of the tube (L), fluid viscosity (η), and the radius (r) of the tube ($R =$ resistance $= 8\eta L/\pi r^4$). Changes in resistance, however, primarily relate to changes

in the area since viscosity and length are usually constant. For a rigid tube, flow is determined by the driving pressure across the tube ($P_1 - P_2$), Ohm's law ($V = I/R$). In contrast, in a collapsible tube, flow may be independent of driving pressure. Resistance changes and is determined by airway size, which, in turn, fluctuates with airway luminal pressures, airflow, and all the other surrounding forces acting on the airway (such as described by the balance of forces). In the most basic form of collapsible tube, "a simple collapsible tube" (defined as a tube having a wall without intrinsic structural forces), variations in airway size are determined primarily by the upstream and downstream airway pressures, flow, and the transmural pressures of the upper airway.

A simple collapsible tube is often modeled with a soft thin walled rubber drain placed between two rigid tubes. Three possible conditions of flow across this tube exist and may include unimpeded flow, flutter, and obstruction. These roughly correlate with the three basic clinical patterns of normal breathing, snoring, and obstruction.

In the human upper airway, the supraglottic pharynx is the collapsible tube with a transmural pressure (P_{tm}); the downstream pressure (P_{ds}) is negative inspiratory pressure (trachea), and upstream pressure (P_{us}) is ambient pressure (nose). During wakefulness, low negative inspiratory intraluminal pressures (i.e., 5 cm H_2O) combined with a large upper airway (a positive transmural pressure) result

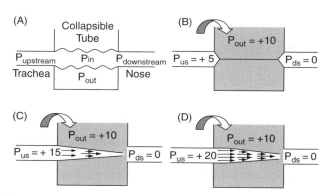

Figure 29.3 Characteristics of a Starling resistor are shown as a model basin with two attached rigid tubes spanned by a collapsible segment (A). Behavior of an ideal Starling resistor is depicted for differing conditions of upstream pressure (P_{us}) in (B)–(D). The pressure difference (i.e., transmural pressure = $P_{in} - P_{out} = P_{tm}$) across the airway determines airway size. In (B), fluid fills the basin and the pressure outside the tube (P_{out}) is greater than pressure inside ($P_{out} > P_{in}$); the tube collapses and no flow occurs. In (C), upstream pressure is increased. When dilating pressures are greater than collapsing pressures ($P_{in} > P_{out}$), the tube is patent and flow occurs. In (D), flow increases with increased positive upstream pressures and unchanged downstream pressures. The driving pressure (downstream − upstream pressure) does not determine flow. Flow is determined by upstream pressure.

in unimpeded flow. During sleep, although the balance of forces changes, in normals without snoring or apnea, a structurally larger and stable upper airway remains patent. Since transmural pressure is greater than both downstream and upstream pressures ($P_{tm} > P_{us} > P_{ds}$), the airway behaves more or less as a rigid tube; collapse does not occur and airway size and resistance are not altered. In contrast, in apneic subjects with a structurally smaller and more unstable upper airway, transmural pressures become less than both downstream and upstream pressures ($P_{us} > P_{ds} > P_{tm}$). With a negative transmural pressure more negative than the closing pressure of the airway during inspiration, airway size is zero. No flow occurs regardless of the driving pressure across the tube.

An intermediate condition occurs in snorers. In snorers, upstream (ambient) pressures are higher than transmural pressures but downstream pressures ($P_{us} > P_{tm} > P_{ds}$) are less. This results in flutter due to a choke point of the airway exposed to alternating negative transmural pressure (i.e., P_{ds}) and positive transmural pressure (i.e., P_{us}). When the airway is open, it is exposed to downstream pressure, which acts to collapse the choke point. When the choke point is closed, it suddenly is no longer exposed to negative downstream pressures; instead the segment is exposed to more positive upstream pressures. These pressures dilate and open the airway. An open airway is now exposed to negative downstream pressures and the airway collapses. Repeating of this cycle creates snoring at the choke point.

STATIC AND DYNAMIC COLLAPSE

When conceptualizing the upper airway, it is often very easy to oversimplify the many complex interactions. Dynamic collapse occurs during inspiration when negative inspiratory pressure and airflow factors predominate. Major dynamic forces include phasic neuromuscular tone and airflow [8, 9]. Other dynamic components may include surface tension forces, which increase collapse and impede opening, vascular compliance, which may alter upper airway size, and segmental interactions of the upper airway during inspiration and expiration [10–12]. Aberrations in inspiratory flow result in hypoxemia, arousal, and the morbidity of OSA [13].

Collapse and increased resistance occur during both inspiration and expiration in OSA [14, 15]. Expiratory collapse is a more static process. Dividing various forces into static forces include structure (craniofacial and soft tissue), tonic muscle tone, and passive changes in airway luminal pressures. Expiratory collapse's contribution is underemphasized yet critical with maximum airway collapse occurring at end expiration and preceding dynamic obstruction (Figure 29.4).

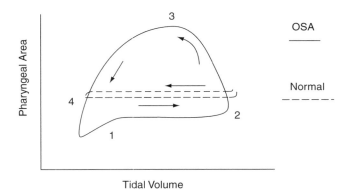

Figure 29.4 Pattern of collapse as a function of the ventilatory cycle in both obstructive sleep apnea syndrome (solid line) and normals (dashed line) is shown. (1) Phasic activation at the onset of inspiration slightly increases upper airway size. (2) Early expiration is associated with an increase in airway size in OSA. (3) Airway collapse begins in later expiration. (4) The smallest airway size is at end expiration. Note normals have little airway size variability during the ventilatory cycle.

NEUROMUSCULAR TONE

Neuromuscular tone dilates and stiffens the airway walls. The genioglossus muscle is considered prototypical but others also contribute (such as the geniohyoid, ala nasi, tensor and levator palatini, stylopharyngeus, and styloglossus). Muscle tone is determined by multiple factors, including voluntary activity, postural tone, the drive from central respiratory neurons, sleep state (wake, NREM and REM sleep), ventilation (hypercarbia and hypoxia), and upper airway mechanoreceptors.

Ventilatory related muscle tone is determined by central respiratory neuron drive. This drive is affected by sleep state (wake, NREM and REM sleep), chemical control (hypercarbia and hypoxia), and upper airway mechanoreceptors. Central respiratory innervation of the diaphragm and upper airway muscles is via the phrenic, vagus, glossopharyngeal, and hypoglossal nerves. Activation is nonuniform and activity is hierarchical. Ventilatory activity is highest in the diaphragm and is reduced or absent in most upper airway muscles (inversely proportional to postural activation) unless ventilatory drive is increased. For this reason, the effects of the sleep/wake state on upper airway muscles are not uniform [16]. Postural upper airway tonic muscle activity decreases progressively with depth of sleep. A portion of this activity is maintained during NREM (but not REM) sleep (Figure 29.5). In OSA patients, muscle tone is increased during wake compared to nonapneic "normals." This likely compensates for a structurally smaller and more collapsible upper airway. This augmented muscle tone is reduced during sleep (Figure 29.6).

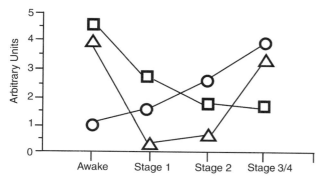

Figure 29.5 A sleep related decrease in tonic muscle tone (squares) and its associated increase in upper airway resistance (circles) with differing levels of NREM sleep are shown. Conceptually, loss of phasic upper airway reflexes (triangle) occurs acutely at sleep onset and is reduced in stage 2; increased reflex activity in stage 3/4 stabilizes ventilation despite high airway resistance.

Upper airway phasic muscle activity is linked to the ventilatory cycle and is activated during inspiration. It results from both inherent central respiratory neuron activity and reflexes mediated via peripheral nervous system mechanoreceptors. Central mechanisms preactivate upper airway muscles during inspiration. The resulting upper airway muscle activity stabilizes the upper airway prior to negative inspiratory forces generated by the diaphragm [17]. This activity is independent of upper airway mechanoreceptors.

Upper airway preactivation but not diaphragm activity is suppressed by various sedative medications including alcohol and benzodaizepines. With the loss of preactivation, the upper airway may be unable to compensate for negative airway pressure and other collapsing forces and upper airway obstruction may worsen. Upper airway mechanoreceptors primarily located at the level of the epiglottis react to negative airway pressure and drive phasic upper airway muscle activity [8, 18]. Both OSA and non-OSA individuals have the potential to demonstrate this reflex during wakefulness; however, in the normal airway, the reflex is not active at rest and requires augmented breathing (such as exercise). In OSA patients, this reflex primarily mediates increased upper airway muscle activity that is present during wakefulness. This reflex is also state dependent and in OSA patients is acutely lost following the transition from wake to NREM sleep [19]. Decreased phasic upper airway muscle tone worsens airway obstruction.

Wake, rapid eye movement (REM), and NREM sleep differ in arousal mechanisms and ventilatory sensitivity to hypoxia and hypercapnia [20]. Differences in these may be present between those with OSA and normals. Central ventilatory instability may result from normal sleep mechanisms, from the cyclic alternating potential (CAP), and from abnormal posthypoxic drive. Hypoxic ventilatory drive decreases from wake to NREM and is lowest in REM sleep. Hypercapneic drive decreases from wake to NREM and REM sleep and may result in hypoventilation during sleep if uncompensated. Ventilation during NREM sleep is primarily mediated by carbon dioxide and chemical control. Both classic arousals and oscillations in the cyclic alternating potential (which alter arousal threshold sensitivity) affect ventilation, ventilatory drive, and airway stability (Figure 29.7).

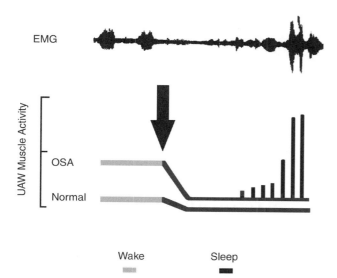

Figure 29.6 Changes in muscle tone in patients with OSA (upper) compared to normal subjects (lower). Sleep onset (arrow) is associated with decrease in tonic muscle tone in both groups. Loss of wake results in loss of phasic muscle tone in OSA (not depicted). Increased negative inspiratory pressure and arousals associated with apneas result in augmented phasic inspiratory EMG activity in OSA (seen in upper graph).

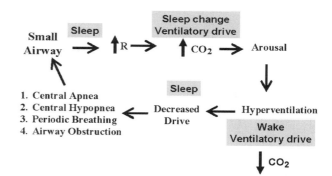

Figure 29.7 The interaction of small upper airways, increased resistance (with associated increase in carbon dioxide), and the effects of sleep on muscle tone are diagrammed. Changes in sleep state and the associated changes in central ventilatory control augment normal physiologic reflexes, which result in decreased central respiratory drive and worsening of upper airway obstruction (see text for details).

Arousal acutely activates and stabilizes the upper airway muscles in sleep. Cyclic arousals and ventilatory changes may worsen airway collapse. Brief awakenings rapidly shift central CO_2 sensitivity to the lower levels of wakefulness. This increases ventilation and quickly reduces CO_2. Rapid resumption of sleep (or in the case of CAP to more stable NREM sleep) results in CO_2 levels below "stable NREM's" critical hypocapneic ventilatory threshold. A ventilatory "overshoot" occurs and ventilatory drive decreases. Ventilatory undershoots may contribute to central apneas, central hypopneas, periodic breathing, and oscillating decreases and increases in upper airway muscle activity. In structurally vulnerable upper airways, periodic breathing destabilizes the upper airway due to the nonuniformity of neuromuscular drive to lower and upper airway muscles. Since the genioglossus and other upper airway muscles are less tightly linked to central motor neurons than the diaphragm, a decrease in drive that has little effect on the diaphragm function may eliminate motor activity to upper airway muscles and cause decreased upper airway size, increased airway resistance, and overt obstruction.

PERIPHERAL NERVOUS SYSTEM EFFECTS

Interactions of the peripheral nervous system contribute to OSA via altering arousal, compensatory reflexes, and muscle size and function. Ultimately, it is the increased work of breathing and mechanoreceptor stimulation and not asphyxia or hypoxia that results in brainstem, spinal cord, or cortical arousal. These in turn cause autonomic activation (blood pressure, heart rate), limb movements, and sleep fragmentation [13].

Negative inspiratory pressure reflex stimulation of upper airway muscles is reduced or eliminated by application of nasal CPAP or topical pharyngeal anesthesia. In OSA patients during wakefulness, CPAP or topical upper airway anesthesia markedly decreases genioglossus EMG activity, whereas normal subjects have no change [21, 22].

Mechanoreceptor mediated reflexes appear to be critical in compensation for OSA but not normal subjects. Damage to these afferent mechanisms may worsen OSA. In fact, abnormalities such as pharyngeal nerve damage and decreases in pharyngeal tactile sensitivity are observed in OSA.

Immunohistopathologic muscle fiber type changes observed in OSA may be due to muscle denervation and reinnervation [23–25]. Muscle fiber hypertrophy and degenerative changes resulting from denervation and reinnervation and eccentric contraction during stretch may worsen OSA by affecting muscle elastance, strength of contraction, and force of dilation. Changes to a greater percentage of type II muscle fiber have been observed as well as muscle hypertrophy, which, if progressive, would contribute to

impingement on airway size [26]. Vestibular afferents may also affect sleep disordered breathing in both children and adults. Vestibular system abnormalities affecting the autonomic nervous system have been demonstrated. The vestibular system has also been shown to mediate changes in upper airway muscle tone with positional changes.

STRUCTURE

Structure significantly predisposes to development and is a fundamental abnormality of OSA [27, 28]. Structurel affects airway size, compliance, and shape, which are critical to flow. Although the pharynx is the core, no single morphologic abnormality exists (Figure 29.8). The three major structural determinates are obesity, soft tissue morphology, and skeletal morphology [29]. Within individuals, structures associated with sleep disordered breathing vary, with differences also affected by ethnicity, gender, obesity, and age [30, 31].

In children, adenotonsillar hypertrophy predominates as a cause of OSA. It is not unique and other craniofacial and soft tissue abnormalities also are present in childhood [32, 33]. In fact, these abnormalities may predispose to obstruction, and adenotonsillar hypertrophy may only complete it. In adults, no single structural abnormality has been identified. Multiple anatomic features are commonly associated with OSA. The anatomy is disproportionate and the site of obstruction variable. Obesity increases the risk and severity of OSA and the severity of hypoxemia during sleep. Obesity and OSA are each independently associated with one another, with each contributing to the other's severity. Fat distribution around the neck and airway has long been postulated to compromise the airway; however, obesity effects on metabolism, ventilation, and lung volume may be more important than volume and mass effects [34–36]. Leptin and various inflammatory cytokines, which are either produced or influenced by adiposity, increase CO_2 response and may contribute to central ventilatory sensitivity.

CRANIOFACIAL CHARACTERISTICS

Most studies assessing structure lack appropriate control groups with groups differing not only in OSA but often also age, weight, ethnicity, and gender. "Normal" may be a subjective and not objective definition. However, few would argue that the skeletal and cartilage frameworks, which support the soft tissues that ultimately determine airway characteristics, are not critical. Framework abnormalities vary and a constellation of abnormalities is consistently observed. Compared to nonsnoring controls, craniofacial variables associated with OSA include

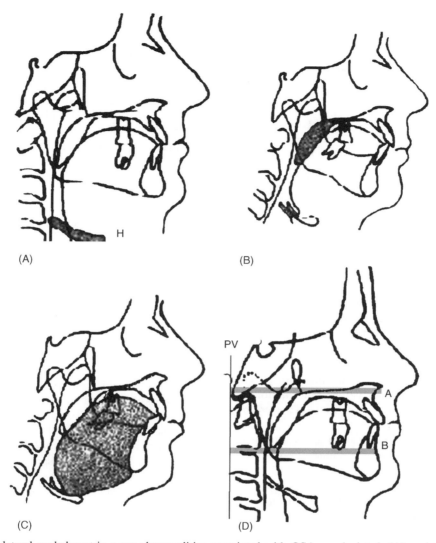

Figure 29.8 Common lateral cephalometric x-ray abnormalities associated with OSA are depicted: (A) an inferior placed hyoid, (B) increased length and width of the soft palate, (C) increased size of the tongue and increased apposition of the tongue and palate with compromised oropalatal airway, and (D) decreased projection of the maxilla (PV-A) and mandible (PV-B). (PV = porion vertical, A = subspinale, B = supramentale, H = hyoid).

increased distance of the hyoid bone from the mandibular plane, a decreased mandibular and maxillary projection (in contrast to anterior growth), a downward and posterior rotation "dolichocephalic" of the mandible and maxilla, increased vertical length of the face and upper airway, and increased cervical angulation. In the Wisconsin Cohort of Sleep Disordered Breathing study, two-thirds or more of the variability of the apnea hypopnea index was explained by facial structure and obesity [29]. In nonobese subjects, the major contributor was facial structure. Abnormalities in maxillary position and width explained much of the OSA in all patient groups. It may be postulated that OSA is primarily a disorder of maxillary not mandibular

development. An inferiorly based hyoid position has been consistently associated with sleep disordered breathing. This may represent abnormalities in airway length or increased tongue size.

NOSE

Nasal obstruction contributes to the presence and severity of OSA [37]. Nasal blockage may (1) reduce nasal afferent reflexes, which help to maintain muscular tone, (2) augment the tendency for mouth opening, which destabilizes the lower pharyngeal airway (by posterior rotation, vertical opening, and inferior displacement of the hyoid), (3) reduce humidification, increase mucus viscosity, and increase surface tension forces, and (4) increase upstream airway resistance, thus increasing downstream airway collapse [14, 38]. A multitude of pathologies cause nasal obstruction and warrant appropriate evaluation.

PHARYNGEAL SOFT TISSUES

In adults no single soft tissue structure contributes to OSA. The relative contribution of soft tissue size differs among individuals and particularly between ethnic groups [39]. Current genetic research associates abnormalities in soft tissues to African American populations with skeletal abnormalities more common in Caucasians and Asian populations [40, 41].

The size as well as position of the tongue are important considerations in OSA. In the supine position, the tongue projects posteriorly and is counteracted by the tone of the genioglossal muscle. MRI volumetric studies have identified tongue size as a major predictor of OSA [42].

The soft palate is both longer and wider in OSA than in normals. The tongue is larger; the oral to palatal airspace is smaller; and the posterior airspace behind both the tongue and palate is narrower. In some the epiglottis is posteriorly placed.

Cross-sectional shape of the airway in apneics tends to be more elliptical rather than more circular. The elliptical shape increases the surface area of the airway and frictional resistance compared to a more circular conduit.

Additionally, when the airway is oriented with its long axis in an anterior–posterior direction (in the midsagittal plane), contraction of major airway dilators (such as the genioglossus muscle) is less effective.

BODY/JAW POSITION/GRAVITY

Body position alters airway size and collapsibility. Airway size decreases following movements from sitting to supine as well as lateral decubitus to supine [43]. Changes are greater in OSA. Tissue mass, change in lung volume, tracheal tug, and vascular volume may contribute. Gravity affects the lower pharynx and retroepiglottic airway more than other segments during parabolic flight [43]. Gravity has minimal effect on the position and airway of the upper pharynx and upper tongue base in nonapneic individuals. In apneic subjects, the isolated effects of gravity have not been evaluated and whether gravity contributes to obstruction in the upper pharynx in addition to the lower pharynx is unknown. It may be speculated that, since gravity alters the airway by affecting mass, as mass increases gravity effects will increase.

VASCULAR EFFECTS

Blood volume changes in the head and neck may affect upper airway size. In human subjects, pharyngeal upper airway size may be altered by changes in leg elevation [11]. This effect is likely mediated through changes in central venous pressure. The location and exact cause of this change in size and its relationship to OSA are unknown. However, indirect evidence suggests that abnormalities in blood vessels or blood volume contribute. The soft tissue of the lateral pharyngeal wall is a major abnormality in OSA [44]. Anatomically, they are both enlarged and abnormally compliant in OSA. Why is not known, yet potential abnormalities may involve muscle, blood vessels, and fat. Given that CPAP's structural effectiveness is mediated primarily by dilation of the lateral wall at low CPAP pressures (5–15 cm H_2O), venous blood volume likely contributes significantly to these observed changes. Whether vascular abnormalities contribute to the development of OSA awaits further study [45].

SURFACE TENSION

Other mechanisms contributing to airway narrowing may include tissue surface adhesive forces [46]. Increased surface tissue adhesion worsens airway collapse, and lubricants decrease airway collapse. Upper airway pressure flow characteristics during NREM sleep demonstrate hysteresis. For a given pressure flow parameter, flow is decreased in late inspiration compared to early inspiration. Similar airflow findings have been observed in the lower airway and have been explained using collapse from surface adhesive forces.

LUNG VOLUME

Changes in lung volume significantly alter pharyngeal upper airway size. This "lung volume dependence" of pharyngeal airway size occurs during wakefulness and sleep [45]. Increased lung volumes increase pharyngeal size and decreased lung volumes contribute to pharyngeal collapse. Although when initially observed, reflex activation of upper airway dilator muscles was speculated, subsequent studies have demonstrated that changes are a mechanical effect of tracheal and thoracic traction. Thoracic traction, commonly referred to as "tracheal tug," markedly influences pharyngeal size and patency and is mediated through the mediastinum, intrathoracic pressures, and the trachea. Changes are independent of neuromuscular activity or upper airway muscle support, with a likely mechanism being passive tracheal traction altering pharyngeal collapsibility by increasing longitudinal tension and stability on the pharyngeal wall (Figure 29.9) [47].

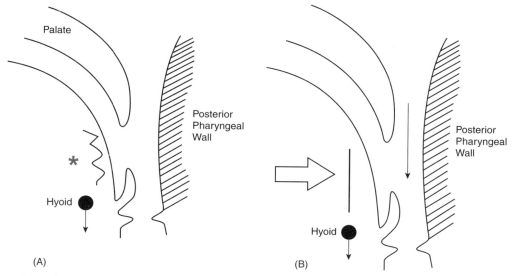

Figure 29.9 The effects of tracheal tug and longitudinal tension on the airway are demonstrated. Inferior displacement of the trachea and/or hyoid bone increases longitudinal tension of the pharyngeal airway and decreases collapsibility.

EXPIRATION

Expiratory obstruction and flow limitation are common in adults who snore and OSA patients but not normals [48, 49]. Expiratory obstruction increases the work of breathing due to the need to overcome the effects of positive airway pressure (auto-PEEP) [50]. How this alters airway collapse during sleep is unknown. More directly, both nonapneic and apneic individuals demonstrate that positive expiratory intraluminal pressure

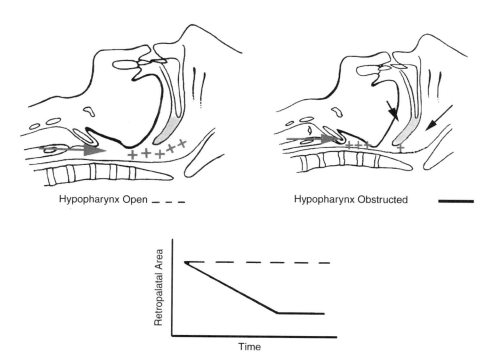

Figure 29.10 Behavior of multielement model of the pharynx is depicted for expiration. Without lower pharyngeal obstruction (*left*), positive pressure contributes to stability of both lower and upper pharyngeal segments. With lower pharyngeal obstruction (*right*), positive expiratory pressure (+) is not transmitted to upper pharyngeal segments. Cross-sectional retropalatal airway size collapses more on obstructed (solid line) than nonobstructed (dotted line) breaths (lower portion of graph).

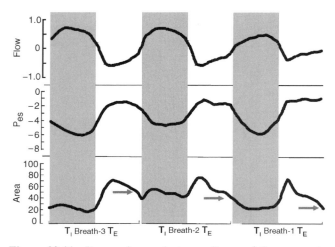

Figure 29.11 Progressive expiratory collapse of the upper airway preceding apnea is shown (T_I = inspiratory time as grey, T_E = expiratory time as white). Note flow and pressure (upper two panels) remain constant until apnea at end of tracing. Cross-sectional airway size at end expiration (arrows) progressively decreases in the three breaths (B-3 to B-1) preceding apnea. Airway size during each inspiration is stable until the actual apnea.

dilates the airway during both wakefulness and sleep [51, 52].

Patients with OSA tongue base levels of obstruction lose the dilating effects of positive expiratory pressures, resulting in greater retropalatal airway collapse (Figure 29.10) [12]. Decreased retropalatal size during expiration further predisposes the upper airway to obstruction on subsequent inspiratory breaths.

Airway size is ventilatory cycle dependent. Changes are greatest in OSA and minimal in normals. In OSA, cross-sectional area increases during early expiration and narrows during the last half of expiration. The magnitude of the expiratory collapse increases and is progressively worse in the several breaths before apnea (Figure 29.11) [53]. Ultimately, this progressive narrowing either completely obstructs the airway or results in a critical size where the combination of negative inspiratory pressure, Bernoulli forces, and surface adhesive forces combine to create airway closure during the following inspiration. In OSA, only the inspiratory breath immediately preceding apnea demonstrates abnormal resistance and collapse.

AIRWAY SHAPE

Airway shape contributes to OSA. Early observations of airway shape focused on the role of muscle action on the upper airway. Patients with severe OSA, obesity, and marked lateral wall collapse demonstrated sagittally oriented airways that were speculated to result in unfavor-

able muscular mechanics to reopen the airway [54]. Others have assessed how airway shape alters the distribution of frictional and streamlined flow in the airway. Circular tubes are more efficient than those that are flat [55]. How shape determines collapse is unknown, with either or both mechanisms contributing.

CONCLUSION

Upper airway competence involves complex interactions between anatomy and physiology. For most, OSA is an abnormality of a structurally small and abnormally collapsible upper airway interacting with normal physiologic mechanisms. The mechanisms individually or in combination may or may not be pathologic, yet together airway instability and ventilation are abnormal during sleep. Understanding the pathophysiology provides insight into why and how to treat the disorder but also expanding on simplistic models of airway collapse may lead to potential novel new medical and surgical treatments for OSA.

REFERENCES

1. Weitzman ED, Pollak CP, Borowiecki BB, Shprintzen R, Rakoff S. The hypersomnia–sleep apnea syndrome: site and mechanism of upper airway obstruction. In: Guilleminault C, Dements WC (Eds), *Sleep Apnea Syndromes*, Kroc Foundation Series Vol 11. Alan R Liss, New York, 1978, Chap 15.

2. Guilleminault C, Hill M, Simmons FB, et al. Passive constriction of the upper airway during central apneas: fiberoptic and EMG investigations. *Respir Physiol* **108**:11–22(1997).

3. Remmers JE, deGroot WJ, Sauerland EK, Anch AM. Pathogenesis of upper airway occlusion during sleep. *J Appl Physiol* **44**:931–938(1978).

4. Sauerland EK, Harper RM. The human tongue during sleep: electromyographic activity of the genioglossus muscle. *Exp Neurol* **51**:160–170(1976).

5. Lieberman DE, McCarthy RC. The ontogeny of cranial base angulation in humans and chimpanzees and its implications for reconstructing pharyngeal dimensions. *J Hum Evol* **36**:487–517(1999).

6. Malhotra A, Pillar G, Fogel RB, et al. Pharyngeal pressure and flow effects on genioglossus activation in normal subjects. *Am J Respir Crit Care Med* **165**:71–77(2002).

7. Pae EK, Lowe AA, Fleetham, JA. A role of pharyngeal length in obstructive sleep apnea patients. *Am J Orthod Dentofacial Orthop* **111**:12–17(1997).

8. Malhotra A, Huang Y, Gogel R, et al. The male predisposition to pharyngeal collapse: importance of airway length. *Am J Respir Crit Care Med* **166**:1388–1395(2002).

9. Fogel R, Malhotra A, Pillar G, et al. Genioglossal activation in patients with obstructive sleep apnea versus control subjects— mechanisms of muscle control. *Am J Respir Crit Care Med* **164**:2025–2030(2001).

10. Jokic R, Klimaszewski A, Mink J, Fitzpatrick M. Surface tension forces in sleep apnea: the role of a soft tissue lubricant—a randomized double-blind, placebo-controlled trial. *Am J Respir Crit Care Med* **157**:(1522)–1525(1998).

11. Shepard J, Pevernagie D, Stanson A, et al. Effects of changes in central venous pressure on upper airway size in patients with obstructive sleep apnea. *Am J Respir Crit Care Med* **153**:250–254(1996).

12. Woodson BT. Expiratory pharyngeal airway obstruction during sleep: a multiple element model. *Laryngoscope*, **113**: 1450–1459(2003).

13. Kimoff RJ, Cheong TH, Olha AE, et al. Mechanisms of apnea termination in obstructive sleep apnea: role of chemoreceptor and mechanoreceptor stimuli. *Am J Respir Crit Care Med* **149**: 707–714(1994).

14. Schwab RJ, Gefter WB, Hoffman EA, Gupta KB, Pack AI. Dynamic upper airway imaging during respiration in normal subjects and patients with sleep disordered breathing. *Am Rev Respir Dis* **148**:1385–1400(1993).

15. Tuck S, Remmers J. Mechanical properties of the passive pharynx in Vietnamese pot-bellied pigs. II. Dynamics. *J Appl Physiol* **92**:2236–2244(2002).

16. Sériès F. Upper airway muscles awake and asleep. *Sleep Med Rev* **6**:195–212(2002).

17. Strohl KP, Hensley MJ, Hallet M, Saunders NA, Ingram RH. Activation of upper airway muscles before onset of inspiration in normal humans. *J Appl Physiol* **53**:87–98(1983).

18. Mezzanote WS, Tangle DJ, White DP. Influence of sleep onset on upper-airway muscle activity in apnea patients versus normal controls. *Am Rev Respir Crit Care Med* **153**:1880–1887(1996).

19. Mezzanote WS, Tangle DJ, White DP. Waking genioglossal electromyogram in sleep apnea patients versus normal controls (a neuromuscular compensation mechanism). *J Clin Invest* **89**:1571–1579(1992).

20. Dempsey J, Smith C, Harms C, Chow C, Saupe K. Sleep and breathing state of the art review: sleep-induced breathing instability. *Sleep* **19**(3):236–247(1996).

21. Deegan PC, Nolan P, Carey M, McNicholas WT. Effects of positive airway pressure on upper airway dilator muscle activity and ventilatory timing. *J Appl Physiol* **81**(1):470–479(1996).

22. Liistro G, Stanescu D, Veriter C, Rodenstein D, D'Odemont J. Upper airway anesthesia induces airflow limitation in awake humans. *Am Rev Respir Dis* **146**:581–585(1992).

23. Woodson BT, Garancis JC, Toohill RJ. Histopathologic changes in snoring and obstructive sleep apnea syndrome. *Laryngoscope* **101**:1318–1322(1991).

24. Petrof BJ, Pack AI, Kelly x, Eby J, Hendricks JC. Pharyngeal myopathy of loaded upper airway in dogs with sleep apnea. *J Appl Physiol* **76**(4):1746–1752(1994).

25. Friberg D. Heavy snorer's disease: a progressive local neuropathy. *Acta Otolaryngol* **119**(8):925–933(1999).

26. Sériès F, Cote C, St Pierre S. Dysfunctional mechanical coupling of upper airway tissues in sleep apnea syndrome. *Am J Respir Crit Care Med* **159**:1551–1555(1999).

27. Isono S, Remmers J, Tanaka A, et al. Anatomy of the pharynx in patients with obstructive sleep apnea and in normal subjects. *J Appl Physiol* **82**:1319–1326(1997).

28. Galvin JR, Rooholamini SA, Standford W. Obstructive sleep apnea: diagnosis with ultrafast CT. *Radiology* **171**:775–778(1989).

29. Dempsey JA, Skatrud JB, Jacques AJ, Ewanowski SJ, Woodson T, Hanson PR, Goodman B, Young T. Anatomical determinates of sleep disordered breathing across the spectrum of clinical and non-clinical subjects. *Chest* **122**: 840–851(2002).

30. Lyberg T, Krogstad O, Djupesland G. Cephalometric analysis in patients with obstructive sleep apnea syndrome: skeletal morphology. *J Laryngol Otol* **103**:287–292(1989).

31. Will MJ, Ester MS, Ramirez SG, Tiner BD, McAnear JT, Epstein L. Comparison of cephalometric analysis with ethnicity in obstructive sleep apnea syndrome. *Sleep* **18**:873–875(1995).

32. Aren R, McDonough JM, Costarino AT, Mahboubi S, et al. Magnetic resonance imaging of the upper airway structure of children with obstructive sleep apnea syndrome. *Am J Respir Crit Care Med* **164**:698–703(2001).

33. Zucconi M, Caprioglio A, Calori G, Ferini-Strambi, Oldani A, et al. Craniofacial modifications in children with habitual snoring and obstructive sleep apnoea: a case control study. *Eur Respir J* **13**:411–417(1999).

34. Mortimore I, Marshall I, Wraith P, et al. Neck and total body fat deposition in non-obese and obese patients with obstructive sleep apnea compared with that in control subjects. *Am J Respir Crit Care Med* **157**:280–283(1998).

35. Bradley T, Brown I, Grossman R, et al. Pharyngeal size in snorers, nonsnorers, and patients with obstructive sleep apnea. *N Engl J Med* **315**:1327–1331(1986).

36. Kryger M, Felipe L, Holder D, et al. The sleep deprivation syndrome of the obese patient—a problem of periodic nocturnal upper airway obstruction. *Am J Med* **56**:531–539(1974).

37. Young T, Finn L, Kim H. Nasal obstruction as a risk factor for sleep-disordered breathing. *Allergy Clin Immunol* **99**:757–762(1997).

38. Meurice J, Marc I, Carrier G, Sériès F. Effects of mouth opening on upper airway collapsibility in normal sleeping subjects. *Am J Respir Crit Care Med* **153**:255–259 (1996).

39. Redline S, Tishler PV, Hans MG, et al. Differences in sleep disordered breathing in African Americans and Caucasians. *Am J Respir Crit Care Med* **155**:186–192(1997).

40. Schwab RJ, Parirstein M, Pierson R, Mackley A, et al. Identification of upper airway anatomic risk factors for obstructive sleep apnea with volumetric MRI. *Am J Respir Crit Care Med* **x**:x–x(2003).

41. Martin S, Marshall I, Douglas N. The effect of posture on airway caliber with the sleep-apnea/hypopnea syndrome. *Am J Respir Crit Care Med* **152**:721–724(1995).

42. Do K, Ferreyra H, Healy J, et al. Does tongue size differ between patients with and without sleep-disordered breathing? *Laryngoscope* **110**:1552–1555(2000).

43. Beaumont M, Fodil R, Isabey D, Lofaso F, Touchard D, Harf A, Louis B. Gravity effects on upper airway area and lung volumes during parabolic flight. *J Appl Physiol* **84**(5):1639–1645(1998).

44. Schwab RJ, Gupta KB, Gefter WB, Metzger LJ, Hoffman EA, Pack AI. Upper airway and soft tissue anatomy in normal subjects and patients with sleep-disordered breathing. Significance of the lateral pharyngeal walls. *Am J Respir Crit Care Med* **152**:1673–1689(1995).

45. Kuna ST, Deepak GB, Ryckman C. Effect of nasal airway positive pressure on upper airway size and configuration. *Am Rev Respir Dis* **138**:969–975(1988).

46. Van der Touw T, Crawford ABH, Wheatley JR. Effects of a synthetic lung surfactant on pharyngeal patency in awake human subjects. *J Appl Physiol* **79**:78–85(1997).

47. Van de Graaff W. Thoracic traction on the trachea: mechanisms and magnitude. *J Appl Physiol* **70**(3):1328–1336(1991).

48. Sériès F, Marc I. Influence of lung volume dependence of upper airway resistance during continuous negative airway pressure. *J Appl Physiol* **77**:840–844(1994).

49. Stanescu D, Kostinavev S, Sonna A, Liistro G, Veriter CI. Expiratory flow limitation during sleep in heavy snorers and obstructive sleep apnea patients. *Eur Respir J* **9**:2116 (1996).

50. Lofasa F, Lorino AM, Fodil R, et al. Heavy snoring with upper airway resistance syndrome may induce positive end-expiratory pressure. *J Appl Physiol* **85**:860(1998).

51. Badr SM, Dawak A, Skatrud JB, Morrell MJ, Zahn BR, Babcock MA. Effect of induced hypocapnic hypopnea on upper airway patency in humans during NREM sleep. *Respir Physiol* **110**:33–45(1997).

52. Rowley JA, Sannders CS, Zahn BR, Badr SM. Effect of REM sleep on retroglossal cross-sectional area and compliance in normal subjects. *J Appl Physiol* **x**:x–x(2001).

53. Morrell MJ, Arabi Y, Zahn B, Badr MS. Progressive retropalatal narrowing preceding obstructive apnea. *Am J Respir Crit Care Med* **158**:1974–1981(1998).

54. Rodenstein DO, Dooms G, Thomas Y, Liistro G, Stanesco DC, Culle C, Aubert-Tulkens G. Pharyngeal shape and dimensions in healthy subjects, snorers, and patients with obstructive sleep apnoea. *Thorax* **45**:723–727(1990).

55. Leiter JC. Upper airway shape. Is it important in the pathogenesis of obstructive sleep apnea? *Am J Respir Crit Care Med* **153**:894–898(1996).

30

SNORING

Eric J. Olson and John G. Park
Mayo Clinic, Rochester, Minnesota

INTRODUCTION

Snoring is a repetitive sound caused by vibration of upper airway structures during sleep. It is a very common phenomenon, occurring habitually in 44% of men and 28% of women in the Wisconsin Sleep Cohort, a longitudinal study of middle-aged state employees [1]. Evaluation and management typically occur when snoring is perceived to be a cause of bedpartner sleep disruption and social embarrassment, or a marker of obstructive sleep apnea–hypopnea syndrome (OSAHS). As interest in OSAHS has grown, so has attention to snoring since it is the most common clinical feature of OSAHS. The shared attribute of upper airway instability explains the close clinical and pathophysiologic connections between snoring and OSAHS. The reader is referred to Chapters 30–38 for additional information.

PATHOPHYSIOLOGY

The upper airway, which stretches from the nares to the larynx, is a complex structure that is an integral component of respiration, phonation, and deglutition. As such, it should not be considered as a simple rigid tube, but more as a dynamic conduit that is extremely challenging to evaluate. Complicated models have been developed to try to describe upper airway behavior and they predict that this flexible tube will begin to vibrate during sleep given certain gas flow, wall compliance, and airway dimension parameters [2]. In practical terms, sleep-associated decreases in upper airway muscle tone increase upper airway collapsibility and

airflow turbulence, which results in audible soft tissue vibration in individuals with anatomically and/or functionally susceptible upper airways. Various upper airway imaging techniques demonstrate that snorers have reduced airway area at the level of the soft palate, tongue base, and hyoid bone [3]. Yet it is not clear whether these changes are a cause of snoring or a result of vibration-induced upper airway mucosal edema and inflammation. Snorers' upper airways tend to be more compliant (collapsible) than nonsnorers, but not as compliant as apneics [4]. Gender differences in upper airway collapsibility, likely attributable to hormonal differences, probably explain the male predominance in snoring.

Acquired anatomic or functional factors that increase upper airway resistance and lead to greater subatmospheric inspiratory pressures within the airway, or that jeopardize pharyngeal wall stability during sleep, will increase snoring risk. Weight gain exacerbates anatomic upper airway narrowing, as do supine sleep posture and mouth breathing, which reposition the tongue and soft palate. Snoring becomes louder and more frequent during pregnancy, peaking in the third trimester and subsiding after delivery. The odds ratio for habitual snoring with chronic nasal congestion at night was 3 in the Wisconsin Sleep Cohort [5]. Alcohol, sedatives/hypnotics, and opioid analgesics amplify upper airway compliance. Sleep stage modulates airway mechanics. Regular, loud snoring is often seen in nonrapid eye movement (NREM) stages 3–4 (slow-wave sleep) with its stable elevation of airway resistance.

Direct visualization confirms that snoring can occur with vibration of any nonrigid component of the upper airway,

Sleep: A Comprehensive Handbook, Edited by T. Lee-Chiong.
Copyright © 2006 John Wiley & Sons, Inc.

including the soft palate, uvula, posterior tongue base, and pharyngeal walls. Multiple sites may vibrate simultaneously, which accounts for the wide snoring sound spectrum. Snoring typically occurs during inspiration but may occur solely during exhalation, or both. Nasal, mouth, or oronasal breathing may be accompanied by snoring. Night-to-night snoring differences can be due to variations in sleep position, nasal congestion, alcohol or drug use, or sleep indebtedness.

The intermediate collapsibility of the nonapneic's upper airway relative to normals and OSAHS patients has led to the concept that snoring and OSAHS are part of a spectrum of sleep-related breathing disorders. Lying within the continuum are respiratory effort-related arousals (RERAs). A RERA is a series of breaths occurring over at least 10 seconds associated with ever-increasing respiratory effort against a narrowed upper airway that terminates with an arousal before criteria for an apnea or hypopnea are met [6]. With esophageal pressure monitoring, RERAs are marked by progressively negative esophageal pressure deflections (reflecting increasing work of breathing) during the breaths immediately preceding an arousal (Figure 30.1). Loud snoring often, but not invariably, accompanies RERAs. Upper airway resistance syndrome (UARS) was coined by Guilleminault et al. [7] to describe the condition of excessive sleepiness associated with 10 or more RERAs per hour. Use of this term is not recommended in the latest *International Classification of Sleep Disorders* [8]. Instead, RERAs are to be considered part of OSAHS.

ASSESSMENT

The evaluation of a snorer should characterize the symptom, elucidate snoring risk factors, pursue OSAHS to the extent of clinical suspicion, and establish treatment options.

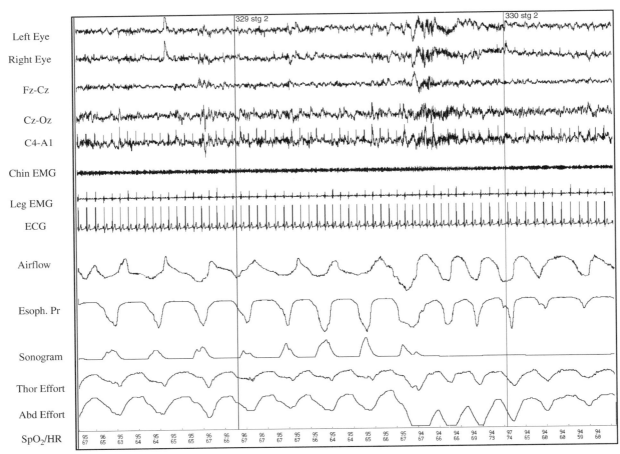

Figure 30.1 Respiratory effort-related arousal. Depicted on this 60 s polysomnogram section is an arousal preceded by crescendo snoring and progressively negative esophageal pressure. Airflow and oxyhemoglobin saturation are preserved so the event cannot be scored as a hypopnea. Esoph Pr = esophageal pressure; Thor Effort = thoracic wall motion by inductive plethysmography; Abd Effort = abdominal wall motion by inductive plethysmography; SpO$_2$/HR = oxyhemoglobin saturation/heart rate.

TABLE 30.1   The History and Physical Exam of the Snoring Patient

History	Determine conditions of increased upper airway resistance
Establish extent of snoring	General inspection
Frequency	Craniofacial trauma
Intensity	Craniofacial anomalies
Position-dependence	Cushing's syndrome (exogenous or endogenous)
Impact on bedpartner	Nose
Search for OSAHS clues	Septal deviation
Gasping, choking	Polyps
Snort arousals	Turbinate hypertrophy
Witnesses apneas	Mucosal inflammation/edema
Nonrestorative sleep	Neoplasm
Excessive daytime sleepiness	Mouth
Identify risk factors	Macroglossia
Gender	Down syndrome
Alcohol	Acromegaly
Sedatives/hypnotics	Hypothyroidism
Opioid analgesics	Amyloidosis
Psychotropic medications	Low-hanging soft palate
Sleep deprivation	Elongated/erythematous/edematous uvula
Chronic rhinitis	Tonsillar hypertrophy
Allergic	Overbite/overjet
Infectious	Jaws
Perennial nonallergic	Retrognathia
Miscellaneous	Micrognathia
Smoking	Midface hypoplasia
Family history	Neck
Physical Exam	Increased circumference
Note vitals	
Obesity	
Hypertension	

History

The history focuses on the description of the snoring, OSAHS clues, and identification of snoring risk factors (Table 30.1). Questioning should ideally be done in the presence of the bedpartner since he/she often prompt the evaluation, and the snorer frequently lacks insight or downplays the condition. Collateral reports of snoring intensity from others in the snorer's sleep environment (e.g., children in the home, travel partners) make it less likely that the bedpartner is overly noise sensitive or that the snoring dissatisfaction is a surrogate concern for more profound relationship issues. The mere presence of snoring lacks specificity for OSAHS. In the Wisconsin Sleep Cohort, only one-tenth of the habitual snorers had OSAHS [1]. Habitual, socially disruptive snoring with witnessed apneas terminated by snorts, gasps, or choking increases diagnostic accuracy. Similarly, excessive daytime sleepiness is not specific for OSAHS. The more common explanation of insufficient sleep time must be considered. Smokers have a relative risk of snoring of 2.29 compared to nonsmokers [9], presumably because of irritant-induced upper airway inflammation. Snoring frequently runs in families but the genetics have not been well characterized.

Physical Exam

The physical exam concentrates on head and neck conditions that compromise upper airway patency (Table 29.1). Typical but nonspecific findings in snorers are a crowded oropharynx, redundant neck tissue, and excess body weight. The uvula may appear erythematous and edematous as a result of repetitive vibratory trauma. Healthy nasal mucosa should appear pink and glistening. Boggy and erythematous nasal mucosa may be seen with perennial nonallergic rhinitis. While the exam cannot definitively separate snorers with and without OSAHS, it can hone OSAHS risk determination. Neck circumference over 17 inches has been highly correlated with OSAHS [10]. When controlling for body mass index and neck circumference, tonsillar enlargement (defined as lateral impingement of greater than 50% of the posterior pharyngeal airspace) and narrowing of the airway by the lateral pharyngeal walls (defined as encroachment of greater than 25% of the pharyngeal space by peritonsillar tissues, excluding the tonsils) are also predictive [10].

Differential Diagnosis

Snoring is commonly recognized, yet other sleep-related sounds may occasionally be misconstrued as snoring. Stridor is a harsh, high-pitched inspiratory sound due to laryngeal obstruction that, in the chronic state, may be due to neoplasm, relapsing polychondritis, laryngomalacia, cricoarytenoid arthritis, or vocal cord dysfunction from recurrent laryngeal nerve disease or neurologic disease (namely, multiple system atrophy). Sleep groaning (catathrenia) describes recurrent episodes of expiratory moaning or grunting lasting 2–20 s that may occur during NREM stage 2 and rapid eye movement sleep [11]. Wheezing is a more musical sound during exhalation and/or inhalation that signals narrowing of central or peripheral airways and can be due to a wide range of conditions, most commonly asthma and chronic obstructive pulmonary disease.

Adjunctive Evaluations

The history and exam parameters may be applied to prediction models that have been proposed to help clinicians determine the probability that a snorer has OSAHS. These models are sensitive but not specific because the cardinal features of OSAHS, namely, snoring, excessive sleepiness, and obesity, are highly prevalent in the general population. In a survey of multiple adult primary care practices, the prevalence of snoring, excessive sleepiness, and increased risk for OSAHS was nearly 40% [12]. Encouraging figures have been reported with the Berlin Questionnaire, a survey that assigns high or low risk for OSAHS on the basis of snoring characteristics, sleepiness, body mass index, and hypertension. When applied in a multiclinic, primary care setting with portable polysomnography to assess validity, classification as high risk by the instrument predicted a respiratory disturbance index >5 with a sensitivity of 0.86, sensitivity of 0.77, and positive predictive value of 0.89 [12]. There is no consensus on the optimal prediction formula and their role remains to be fully defined.

For snorers who did not begin their evaluation with an otorhinolaryngologist, referral for a thorough upper airway exam with fiberoptic pharyngoscopy by such a specialist is indicated in patients seeking surgical treatment options for snoring, when input is needed regarding the medical management of potentially contributory upper airway disorders (e.g., chronic rhinitis), or when the initial head and neck exam reveals an obvious (e.g., tonsillar hypertrophy) or unexplained abnormality. Neither pharyngoscopy during inspiratory efforts against an occluded airway (Müller's maneuver) nor upper airway imaging with computed tomography or magnetic resonance imaging have been consistently shown to predict snoring responses to surgery and are therefore not routinely indicated. Laboratory testing for associated conditions such as hypothyroidism should be performed when clinical features are suggestive.

TABLE 30.2   Indications for Polysomnography in Snorers

Clinical features suggestive of OSAHS
Breathing disturbances during sleep
Daytime dysfunction
Difficulties maintaining sleep
odesity with increased neck girth (>17 inches in men; >16 inches in women)
Comorbidities whose management would be optimized by treating OSAHS
Surgery for snoring considered
Concurrent sleep disorders suspected

Sleep Laboratory Testing

Indications for polysomnography for snoring are listed in Table 30.2. The prospect that finding and treating OSAHS in a snorer may improve control of his/her hypertension, dysrhythmias, headaches, nocturnal angina, pulmonary hypertension, seizure disorder, or nocturnal gastroesophageal reflux may provide additional leverage for polysomnography. Most sleep specialists recommend polysomnography for asymptomatic snorers before they undergo upper airway surgery to confirm the severity of the symptom, exclude occult OSAHS for which surgery may not be appropriate, and establish a baseline for postoperative comparisons.

The standard polysomnogram is a laboratory-based, technologist-conducted, multimodality recording of sleep and breathing. Snoring definitions and measurement techniques are not standardized. Not all sleep facilities objectively assess snoring during polysomnography, but the attending technologist usually makes a subjective rating of snoring frequency and intensity. Microphone options include contact systems that measure snoring via detection of neck vibration, or sound sensors attached to the patient at, for instance, the collar. Sound detecting microphones may also be suspended above the patient, but sensitivity may vary depending on the patient's sleep position.

Patients may find polysomnography inconvenient and difficult to access, so portable pulse oximetry monitoring during sleep may be used to screen snorers for OSAHS. Obstructive apneas and hypopneas result in repetitive oscillations in the oxyhemoglobin saturation. In practice, this may be used to exclude OSAHS in a snorer when index of suspicion is low. However, a recent exhaustive review of portable monitoring for OSAHS conducted jointly by the American Academy of Sleep Medicine (AASM), American Thoracic Society, and American College of Chest Physicians concluded there was insufficient evidence to support oximetry as a singular technique for ruling-in or ruling-out OSAHS [13]. RERAs are not detectable by pulse oximetry since the events terminate before oxyhemoglobin desaturation occurs.

TABLE 30.3 Treatment Options for Snoring

Life-style modifications
    Weight loss
    Smoking cessation
    Sleep position restriction
    Eliminate alcohol consumption before bed
    Decrease use of snoring risk factor medications
Nasal-directed therapies
    Antirhinitis therapies
        Allergen avoidance
        Nasal corticosteroids
        Antihistamines
        Sympathomimetics
        Allergy immunotherapy
        Nasal anticholinergics
    Nonprescription products
        External nasal dilator strips
        Internal nasal dilators
        Oronasal lubricants
    Surgery
        Septoplasty
        Turbinate reduction
        Polypectomy
Oral appliances
Surgery
    Uvulopalatopharyngoplasty
    Laser-assisted uvulopalatoplasty
    Radiofrequency ablation
Bedpartner-directed therapies
    Background noise
    Earplugs
    Separate bedrooms

TREATMENT

The treatment goals are to eliminate bedpartner sleep disruption and social awkwardness for the patient. A wide variety of treatment options are available (Table 30.3). Choices are influenced by whether snoring occurs in the context of OSAHS. Treatment studies often have significant shortfalls, including selection biases, the admixture of OSAHS patients, and, perhaps most importantly, the lack of objective outcome measures. The limitation of reliance on subjective endpoints was demonstrated in a study of 69 couples in which the snorer underwent uvulopalatopharyngoplasty [14]. Postoperative snoring questionnaire responses were compared to pre-and postoperative objective measures of snoring. Nearly 80% of patients reported reduced snoring and improved sleep quality, and 26% of bedpartners no longer reported interference with their sleep, yet no objective differences were found in the snoring.

Life-Style Modifications

Although evidence supporting these maneuvers is limited, it stands to reason that weight loss, smoking cessation, avoiding sleep on the back, and minimizing use of agents that relax the pharyngeal musculature will decrease upper airway resistance by increasing pharyngeal cross-sectional area or decreasing compliance. It is difficult to predict the impact of these life-style changes on snoring and they are admittedly challenging to attain and maintain, but they should be recommended given the beneficial effects on general health. The "tee-shirt with tennis balls," a relatively tight-fitting nightshirt with a pouch containing tennis balls along the length of the spine, can promote sleep off the back.

Nasal-Directed Therapies

The nose produces two-thirds of total airway resistance [15]. Further reductions in cross-sectional area by allergic, infectious, or chronic nonallergic (idiopathic) rhinitis predispose to upper airway collapse by amplifying the pressure differential between the atmosphere and the thoracic cavity, resulting in greater flow turbulence in the pharynx, or by promoting oral breathing, which may decrease nasal receptor-derived stimulation of ventilation and alter phasic activity of upper airway muscles [15]. Data are limited, but experience would indicate that snoring may be mitigated by intranasal or systemic rhinitis therapies tailored to the etiology. As-needed use of topical decongestants may suffice for snoring exacerbated by acute infectious rhinitis. Allergen avoidance plus nasal steroids or antihistamines are first-line therapies for allergic nasal congestion. Chronic, nonallergic rhinitis may respond to oral sympathomimetics or intranasal ipratropium.

A variety of nonprescription products to be used on or in the nose are available to treat snoring, such as external nasal dilator strips (ENDS), internal nasal dilators (INDs), and lubricants. The Clinical Practice Committee of the AASM recently reviewed the evidence for these treatments [16] and drew the following conclusions on the basis of very limited data: ENDS are safe and may be effective in mild, nonapneic snoring, but predictors of response are not known; INDs may decrease snoring intensity, but long-term therapy adherence is poor; and there are insufficient data regarding oronasal lubricants.

Nasal surgery, including septoplasty, turbinate reduction, and polypectomy, has been used in snorers. Reductions in snoring have been reported in up to 75% of patients, but typically in small, uncontrolled series. Predicting postoperative response is difficult.

Oral Appliances

Oral appliances (OAs) have been developed for the purpose of relieving snoring by mechanically enlarging or stabilizing the upper airway by advancing the mandible or tongue. The mandible may be set forward to 50–75% of maximal protrusion. A 1995 review of OAs [17] concluded that subjective improvements in snoring were reported in nearly almost all case series. Subsequent studies, mostly comparing OAs to continuous positive airway pressure (CPAP) or to other OAs, have consistently revealed subjective and, in some cases, objective snoring improvements [18]. OAs are indicated for snorers with or without mild OSAHS who "do not respond to or are not appropriate candidates for treatment with behavioral measures such as weight loss or sleep-position change" [19]. Side effects, including excessive salivation, xerostomia, teeth and gum discomfort, temporomandibular joint (TMJ) pain, unintended displacement during sleep, and an altered sense of morning occlusion, are common but usually mild. Contraindications include insufficient teeth to anchor the device, active dental disease, TMJ disorder, minimal protrusive range, and childhood age. Close collaboration with the dentist is crucial for proper patient assessment, device fabrication and titration, and monitoring for possible long-term dental and skeletal changes.

Surgery

Surgical interventions offer the promise of freedom from a nightly therapy such as an OA, but their impact may be limited because of the multisite origin of snoring. Subjective and objective snoring outcome data are often discrepant. Reliable factors predicting a favorable outcome remain frustratingly elusive.

Uvulopalatopharyngoplasty (UPPP). Introduced the same year as CPAP [20], UPPP modifies the retropalatal airway by excision of the uvula, a portion of the soft palate, and tonsils (if present) under general anesthesia. Subjective success may be 80% or greater, but corroborative, long-term, objective follow-up data are minimal. Predicting response is difficult; young, nonobese patients with tonsillar hypertrophy are probably the best UPPP candidates. Acute postoperative pain is pronounced. Potential long-term side effects include velopharyngeal insufficiency (nasal reflux), voice change, and globus sensation.

Laser-Assisted Uvulopalatoplasty (LAUP). This procedure is a sequential, office-based trimming of the uvula and adjacent soft palate typically by carbon dioxide laser under local anesthesia first described in 1990 [21]. Postprocedure pain and complication rates are probably similar to UPPP. An AASM review concluded that "the long-term effectiveness of LAUP on treatment of snoring has not been convincingly established" and that "LAUP appears comparable to UPPP in relieving snoring" [22]. Reported snoring improvement rates are 40–90% at varying follow-up lengths.

Radiofrequency Ablation (RFA). This also is a sequential, office-based procedure performed under local anesthesia in which a handpiece is used to focally apply a temperature-controlled radiofrequency signal, most commonly to the soft palate, and thereby reduce tissue volume and stiffen via subsequent fibrosis. A systematic review of the 22 studies of RFA of the soft palate since the original description in 1998 [23] revealed significant subjective improvements in snoring [24]. Pain intensity and duration, as well as complications, are less than with UPPP or LAUP. Side effects are rare and primarily limited to mucosal erosion.

Interventions to decrease soft palatal fluttering via injection of sclerosing agents and stents have been proposed but are insufficiently studied.

Continuous Positive Airway Pressure (CPAP)

CPAP will eliminate snoring, but it is of minimal practical value for snorers without OSAHS. Nonapneic snorers generally will not accept CPAP. Furthermore, most medical insurers will not cover the cost of CPAP for simple snoring.

Bedpartner-Directed Therapies

Under certain circumstances, snoring management may include modifications to the home sleep environment to lessen the impact of snoring on the bedpartner, including background "white noise" (e.g., fan) to cancel the snoring, use of ear plugs, or sleeping in separate rooms. However, these interventions have often been tried and judged inadequate before the snorer presents for evaluation.

CONCLUSION

As the ramifications of OSAHS are becoming clearer, the perception of snoring has changed from a sometimes noxious, but otherwise benign marker of slumber, to an indicator of a potentially serious breathing disorder. Snoring results from pharyngeal vibration and is triggered by conditions that increase upper airway resistance and/or compliance, such as obesity, male gender, and nasal congestion. Snoring is common in adults and about one-tenth of habitual snorers will have OSAHS. It may be challenging to distinguish nonapneic from apneic snorers, but this distinction is important since some snoring treatments are inappropriate for OSAHS. A polysomnogram is generally performed when there is clinical suspicion for OSAHS and/or surgical treatment for snoring is being considered. Treatment options for snoring are broad and tailored to identify snoring risk factors. Life-style modifications should be advised in all snorers. More aggressive options, such as OAs or upper airway surgery, may not eliminate snoring, but commonly produce subjective snoring improvements.

REFERENCES

1. Young T, Palta M, Dempsey J, Skatrud J, Weber S, Badr S. The occurrence of sleep-disordered breathing among middle-aged adults. *N Engl J Med* **328**:1230–1235(1993).

2. Huang L, Williams JE. Neuromechanical interaction in human snoring and upper airway obstruction. *J Appl Physiol* **86**:1759–1763(1999).

3. Ayappa I, Rapoport DM. The upper airway in sleep: physiology of the pharynx. *Sleep Med Rev* **7**:9–33(2002).

4. Gleadhill IC, Schwartz AR, Schubert N, Wise RA, Permutt S, Smith PL. Upper airway collapsibility in snorers and in patients with obstructive hypopnea and apnea. *Am Rev Respir Dis* **143**:1300–1303(1991).

5. Young T, Finn L, Kim HC. Nasal obstruction as a risk factor for sleep-disordered breathing. *J Allergy Clin Immunol* **99**:S757–S762(1997).

6. American Academy of Sleep Medicine. Sleep-related breathing disorders in adults: recommendations for syndrome definition and measurement techniques in clinical research. *Sleep* **22**:667–689(1999).

7. Guilleminault C, Stoohs R, Clerk A, Cetel M, Maistros P. A cause of excessive daytime sleepiness. The upper airway resistance syndrome. *Chest* **104**:781–787(1993).

8. American Academy of Sleep Medicine. *International Classification of Sleep Disorders*, 2005.

9. Wetter DW, Young TB, Bidwell TR, Badr MS, Palta M. Smoking as a risk factor for sleep-disordered breathing. *Arch Intern Med* **154**:2219–2224(1994).

10. Schellenberg JB, Maislin G, Schwab RJ. Physical findings and the risk for obstructive sleep apnea: the importance of oropharyngeal structures. *Am J Respir Crit Care Med* **162**:740–748(2000).

11. Vetrugno R, Provini F, Plazzi G, Vignatelli L, Lugaresi E, Montagna P. Catathrenia (nocturnal groaning): a new type of parasomnia. *Neurology* **56**:681–683(2001).

12. Netzer NC, Stoohs RA, Netzer CM, Clark K, Strohl KP. Using the Berlin Questionnaire to identify patients at risk for the sleep apnea syndrome. *Ann Intern Med* **131**:485–491(1999).

13. Chesson AL, Berry RB, Pack A. Practice parameter for the use of portable monitoring devices in the investigation of obstructive sleep apnea in adults. *Sleep* **26**:907–913(2003).

14. Miljeteig H, Mateika S, Haight JS, Cole P, Hoffstein V. Subjective and objective assessment of uvulopalatopharyngoplasty for treatment of snoring and sleep apnea. *Am J Respir Crit Care Med* **150**:1286–1290(1994).

15. Rappai M, Collop N, Kemp S, deShazo R. The nose and sleep-disordered breathing: what we know and what we do not know. *Chest* **124**:2309–2323(2003).

16. Meoli AL, Rosen CL, Kristo D, Kohrman M, Gooneratne N, Aguillard RN, Fayle R, Troell R. Nonprescription treatments of snoring or obstructive sleep apnea: an evaluation of products with limited scientific evidence. *Sleep* **26**:619–624(2003).

17. Schmidt-Nowara W, Lowe A, Wiegand L, Cartwright R, Perez-Guerra F, Menn S. Oral appliances for the treatment of snoring and obstructive sleep apnea: a review. *Sleep* **18**:501–510(1995).

18. Ferguson KA. The role of oral appliance therapy in the treatment of obstructive sleep apnea. *Clin Chest Med* **24**:355–364(2003).

19. American Sleep Disorders Association. Practice parameters for the treatment of snoring and obstructive sleep apnea with oral appliances. *Sleep* **18**:511–513(1995).

20. Fujita S, Conway W, Zorick F, Roth T. Surgical correction of anatomic abnormalities in obstructive sleep apnea syndrome: uvulopalatopharyngoplasty. *Otolaryngol Head Neck Surg* **89**:923–934(1981).

21. Kamami Y. Laser CO_2 for snoring: preliminary results. *Acta Otorhinolarygol Belg* **44**:451–456(1990).

22. Littner M, Kushida CA, Hartse K, Anderson WM, Davila D, Johnson SF, Wise MS, Hirshkowitz M, Woodson BT. Practice parameter for the use of laser-assisted uvulopalatoplasty: an update for 2000. *Sleep* **24**:603–619(2001).

23. Powell NB, Riley RW, Troell RJ, Li K, Blumen MB, Guilleminault C. Radiofrequency volumetric tissue reduction of the palate in subjects with sleep-disordered breathing. *Chest* **113**:1163–1174(1998).

24. Stuck BA, Maurer JT, Hein G, Hormann K, Verse T. Radiofrequency surgery of the soft palate in the treatment of snoring: a review of the literature. *Sleep* **27**:551–555(2004).

31

OVERVIEW OF OBSTRUCTIVE SLEEP APNEA IN ADULTS

MARK H. SANDERS AND RACHEL J. GIVELBER
University of Pittsburgh School of Medicine, Pittsburgh, Pennsylvania

Increasing awareness of the nature and prevalence of obstructive sleep apnea–hypopnea (OSAH) among the public as well as growing understanding of its health consequences mandate that the medical community maintain a high level of suspicion for this disorder in daily practice. In addition, it is essential for providers to be knowledgeable with regard to the differential diagnosis, diagnostic strategies, and therapeutic modalities for OSAH patients. This chapter will present an overview of the current fundamentals related to the clinical identification and management of these patients. Readers are referred to the specific topic chapters for more detailed discussions and literature citations.

DEFINITION OF TERMS AND DESCRIPTORS

Sleepiness and fatigue are pervasive complaints in modern society. It is clear that OSAH is but one potential etiology for these complaints and it is essential that the clinician understand the epidemiology of OSAH in order to efficiently direct diagnostic strategies. Obstructive apnea is defined, by convention, as cessation of nasal/oral airflow for at least 10 seconds, despite persistent ventilatory efforts (Figure 31.1). Although there is ongoing discussion regarding the most clinically and physiologically relevant definition of an obstructive hypopnea, a Committee of the American Academy of Sleep Medicine concluded that these events reflect reduction of airflow by at least 30% from baseline, of at least 10 seconds in duration and accompanied by oxyhemoglobin desaturation of 4% or more [1] (Figure 31.2). An "index" reflects the hourly frequency of a particular sleep-related event (e.g., Apnea Index, AI:

average number of apneas per hour of sleep; Hypopnea Index, HI: average number of hypopneas per hour of sleep; Apnea + Hypopnea Index, AHI: average number of apneas plus hypopneas per hour of sleep). The American Academy of Sleep Medicine has defined mild, moderate, and severe sleep apnea hypopnea as AHI 5–15, 16–30, >30, respectively [2]. Some of the literature also utilizes the term Respiratory Disturbance Index (RDI). Although this term may be incorrectly used synonymously with the AHI, it more appropriately reflects the average hourly frequency of sleep apneas + hypopneas + events in which there is a central nervous system arousal associated with an obstructive upper airway event that doesn't meet the criteria for apnea or hypopnea.

PATHOPHYSIOLOGY OF OSAH

Although at first thought, the upper airway may be considered to be a simple tube, further reflection reveals that it must be sufficiently stiff to provide a conduit for ventilation, but sufficiently supple to allow for deglutition and speech. Thus the upper airway is a complex anatomic structure with an intricate control system that remains incompletely understood. Normally, the forces that promote upper airway closure during sleep are balanced by forces that promote patency. The forces promoting upper airway occlusion (apnea) or non-occlusive obstruction (hypopnea) include negative intraluminal pressure and compliant (floppy) pharyngeal walls, while activation of upper airway dilator muscles, which may also stiffen the airway walls, resist mural collapse. An anatomically narrow upper airway (e.g., due to

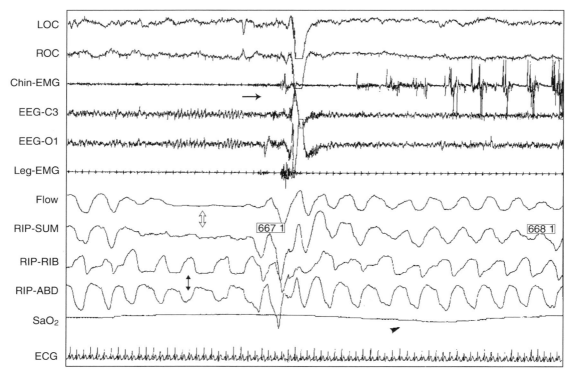

Figure 31.1 Obstructive apnea. LOC = left electro-oculogram; ROC = right electro-oculogram; Chin EMG = submental electromyogram; EEG = electroencephalogram; Leg EMG = anterior tibialis electromyogram; Flow = airflow; RIP-SUM = respiratory inductance plethysmography recording of the sum of the rib cage plus abdominal excursion (qualitatively reflecting ventilatory effort); RIP-ABD = respiratory inductance plethysmography recording of the rib cage excursion; RIP-ABD = respiratory inductance plethysmography recording of the sum of the abdominal excursion; Sao_2 = oxyhemoglobin saturation by pulse oximetry; ECG = electrocardiogram (usually a modified chest lead). (From: Rowley JA, Badr MS. Obstructive sleep apnea: new insights aid in diagnosis. *J Respir Dis* **21**:505–515(2000).)

redundant soft tissue, macroglossia, retro- or micrognathia, or reduced upper airway dilator muscle activity) (Figure 31.3) predisposes to an elevated resistance to airflow, which permits generation of abnormally negative intraluminal pressure during inspiration, thereby favoring collapse unless offset by augmented upper airway dilator muscle activity. Alternatively, inadequately activated upper airway dilator muscles in the presence of a normal or minimally increased upper airway resistance may predispose to instability. Figure 31.4 is a schematic representation of this balance of forces and how it may pathophysiologically relate to OSAH.

EPIDEMIOLOGY AND RISK FACTORS

By any standard, OSAH is prevalent in the community with 9% of women and 27% of men between 30 and 60 years of age having an AHI > 5; and 2% of women and 4% of men having AHI > 5 *and* daytime sleepiness. The prevalence of OSAH increases with age until approximately the seventh to eighth decade, when it plateaus [3]. The risk of OSAH also increases with increasing body weight and neck circumference [4, 5]. A 10% increase in weight is associated with a sixfold increase in risk for developing an AHI ≥ 15 [4]. Male gender is an independent risk factor for OSAH compared with premenopausal women, but the prevalence in woman increases with menopause. Interestingly, some data indicate that hormone replacement therapy reduces OSAH risk in menopausal women [6]. The data are not sufficiently compelling, however to warrant use of these agents specifically for this purpose, especially in view of recent information regarding adverse effects. The relation between ethnicity and risk for OSAH is unclear. Two studies of African Americans who were less than 25 years and greater than 65 years of age, respectively, were at increased risk for OSAH [7, 8], but the Sleep Heart Health Study, which evaluated a population of over 5000 community-dwelling adults, did not find this to be the case [3]. There is increasing evidence for a hereditable component to OSAH with substantially increased risk when a primary family member is affected [9, 10].

It is essential to recognize, however, that epidemiological features in isolation of the clinical history, physical examination, and objective assessment of breathing during sleep are insufficiently sensitive or specific to establish a

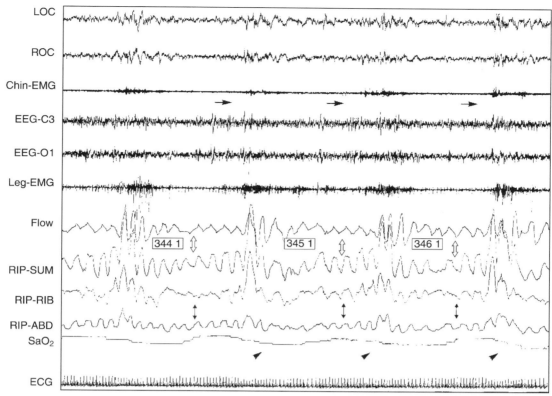

Figure 31.2   Obstructive hypopnea. LOC = left electro-oculogram; ROC = right electro-oculogram; Chin EMG = submental electromyogram; EEG = electroencephalogram; Leg EMG = anterior tibialis electromyogram; Flow = airflow; RIP-SUM = respiratory inductance plethysmography recording of the sum of the rib cage plus abdominal excursion (qualitatively reflecting ventilatory effort); RIP-ABD = respiratory inductance plethysmography recording of the rib cage excursion; RIP-ABD = respiratory inductance plethysmography recording of the sum of the abdominal excursion; Sao_2 = oxyhemoglobin saturation by pulse oximetry; ECG = electrocardiogram (usually a modified chest lead). (From: Rowley JA, Badr MS. Obstructive sleep apnea: new insights aid in diagnosis. *J Respir Dis* 21:505–515(2000).)

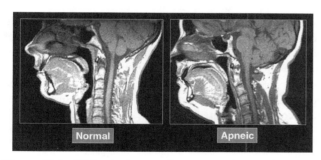

Figure 31.3   Midsagittal magnetic resonance image of the upper airway of a normal individual (*left panel*) and a patient with obstructive sleep apnea–hypopnea (*right panel*). Note the patients narrower airway lumen. (From: Schwab R, Kuna S, Remmers JE. Anatomy and physiology of upper airway obstruction. In: Kryger M, Roth T, Dement WC (Eds), *Principles and Practice of Sleep Medicine*, 4th ed. Saunders, Philadelphia, 2005, Chap 87.)

diagnosis of OSAH. The epidemiologic features should rather be considered as clinical clues, which should prompt consideration of further exploration of the possibility of OSAH.

SYMPTOMS AND PHYSICAL SIGNS ASSOCIATED WITH OSAH

Certain *symptoms and physical signs* provide a basis for an increased suspicion for OSAHS. Although common and nonspecific in this regard, daytime sleepiness or fatigue may provide a clue to the presence of OSAH and this should be included in the differential diagnosis. Suspicion of OSAH should be heightened by bedpartner reports of apnea, periods of snoring interrupted by relative silence terminated by snoring ("resuscitative snoring"), and patient complaints of awakening with a sensation of choking, smothering, or gasping. In our opinion, patients complaining of what would otherwise be thought to represent paroxysmal nocturnal dyspnea (PND) should also be suspected of having sleep apnea–hypopnea (obstructive or central). Over and above obesity and a large neck circumference, physical findings such as a narrow oropharynx (e.g., a high, arched palate or short intermolar distance, Figure 31.5), a crowded oropharynx due to redundant soft tissue and/or relative macroglossia, retro/

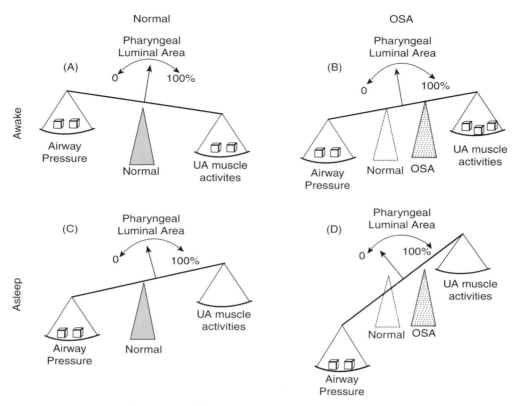

Figure 31.4 Schematic representation of the balance of forces that may influence upper airway patency. UA = upper airway; airway pressure is negative or positive intraluminal pressure. (From: Schwab R, Kuna S, Remmers JE. Anatomy and physiology of upper airway obstruction. In: Kryger M, Roth T, Dement WC (Eds), *Principles and Practice of Sleep Medicine*, 4th ed. Saunders, Philadelphia, 2005, Chap 87.)

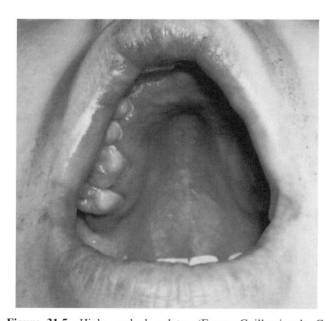

Figure 31.5 High, arched palate. (From: Guilleminault C, Bassiri A. Clinical features and evaluation of obstructive sleep apnea–hypopnea syndrome and upper airway resistance syndrome. In: Kryger M, Roth T, Dement WC (Eds), *Principles and Practice of Sleep Medicine*, 4th ed. Saunders, Philadelphia, 2005, Chap 87.)

micrognathia, and enlarged tonsils represent risk factors for OSAH.

Increasingly considered a likely consequence of OSAH, the presence of systemic arterial hypertension should heighten clinical suspicion, especially in the presence of other risk factors. Similarly, difficult to control hypertension should prompt questions exploring the possibility of OSAH. Other comorbid conditions that appear to predispose to OSAH include untreated hypothyroidism, acromegaly, congestive heart failure, and nasal obstruction.

Table 31.1 provides a list of potential causes of daytime sleepiness and Table 31.2 provides a more comprehensive list of risk factors for OSAH as well as clinical clues, including symptoms and signs. It is important to recognize that many of the symptoms and signs are neither sensitive nor specific for OSAH and their diagnostic value lies in the clinical context in which they exist.

ESTABLISHING A DIAGNOSIS OF OSAH

Since clinical signs and symptoms are insufficiently sensitive and specific for OSAH, acquiring objective data that support a diagnosis of OSAH is a prerequisite to developing

TABLE 31.1   Causes of Daytime Sleepiness

Insufficient sleep
Narcolepsy
Sleep apnea–hypopnea
Insomnia (there are multiple etiologies of insomnia)
Medical/neurologic diagnoses
Sleep disruption due to pain or environmental factors
Depression and other psychiatric diagnoses
Medication/drugs (including alcohol and over-the-counter agents)
Parasomnias and movement disorders during sleep including
    periodic leg movement disorder, sleep-related seizures
Shift work
Circadian rhythm disorders
Post-traumatic hypersomnolence
Kleine–Levin syndrome
Idiopathic hypersomnolence

TABLE 31.2   Risk Factors For, and Clues to, the Presence of Obstructive Sleep Apnea–Hypopnea (OSAH)

Obesity (especially "central obesity"); large neck circumference
Male gender, postmenopausal females
Increasing age
Abnormal upper airway and/or craniofacial anatomy
    Crowded oropharynx
    Redundant soft tissue
    Relative macroglossia
    Retrognathia/micrognathia
    Narrow oropharynx (e.g., high, arched palate)
    Nasal obstruction
Family history of OSAH
Medical diagnoses
    Hypothyroidism
    Acromegaly
    Systemic arterial hypertension
Drugs (including alcohol, benzodiazepines, and narcotics)

Symptoms and signs of OSAH
    Daytime sleepiness; nonrefreshing sleep
    Awakening with a sensation of choking, gasping, or smothering
    Snoring interrupted by periods of silence that are terminated by
        loud snorting ("resuscitative snorting")
    Apnea observed during sleep
    Nocturia
    Altered cognitive performance (e.g., reduced ability to concentrate)
    Personality alterations (e.g., irritability)
    Morning headaches that tend to remit with time after awakening
    Abnormal body movements during sleep (e.g., twitches or "jerks")

a treatment strategy. The traditionally employed diagnostic procedure is nocturnal polysomnography (PSG), which records the following:

- parameters of sleep quality, architecture, and continuity (the electroencephalogram or EEG; left and right electro-oculograms or LEOG/REOG; submental electromyogram or submental EMG),

- nasal/oral airflow,

- qualitative or quantitative metrics of breathing effort (usually respiratory inductance plethysmography or RIP or recording esophageal pressure fluctuations as reflections of intrathoracic pressure swings, respectively),

- oxyhemoglobin saturation by pulse oximetry (Sp_{O_2}),

- the electrocardiogram, usually employing a modified chest lead,

- body position using a specific position sensor, video monitor,

- the EMG of the anterior tibialis muscle is often recorded as well.

The channels that record sleep quality, architecture, and continuity permit the clinician to determine if a negative study might be due to absence of sleep. In addition, recording sleep architecture may identify indicate absence of recorded rapid eye movement (REM) sleep, REM is considered to be the sleep stage most susceptible to abnormal breathing events and oxyhemoglobin desaturation. Therefore, if the patient is not recorded during REM sleep, it is possible that the severity of OSAH will be underestimated or may even result in a false negative study. Assessment of breath-by-breath airflow permits identification of apneas and hypopneas (identification of the latter events also requires meeting a desaturation criterion provided by the Sp_{O_2} channel). Parameters of breathing effort permit classification of apneas and hypopneas as obstructive as opposed to "central". In addition to providing a criterion in the definition of hypopneas, the Sp_{O_2} also gives the clinician a metric by which to judge the physiologic severity of abnormal breathing events. The electrocardiogram may reveal significant dysrhythmias and also provides important information regarding the physiologic as well as clinical severity of apneas and hypopneas. Body position is also usually recorded because OSAH may be more prevalent and severe during sleep in the supine position compared with the lateral recumbent position, making this information relevant to clinical interpretation of the PSG. Most sleep laboratories also record the anterior tibialis EMG to assess for abnormal motor activity such as periodic limb movements during sleep. Examples of a PSG "montage" are shown in Figures 30.1 and 30.2. Performing nocturnal PSG requires a specialized sleep laboratory with trained technicians. It also obviously requires the patient to spend the night outside his/her home environment. These factors mandate that overnight PSG is a time and resource intensive diagnostic procedure, which is associated with not inconsequential cost. Furthermore, the specialized nature of PSG and finite number of facilities limit accessibility to diagnostic services. The latter is particularly important in light of the high prevalence of OSAH in the community.

In an effort to reduce cost and improve accessibility to diagnostic services for patients suspected of having OSAH, a number of strategies have been proposed. These include recording a limited number of channels and/or recording in the patients home rather than a full, in-laboratory PSG. A complete, in-laboratory PSG that is attended by a sleep technician is designated a "type 1" study. A complete PSG that is performed in the patient's home is designated a "type 2" study. There are three "types" of limited channel studies: "type 3" studies record a minimum of four channels, including at least two channels of respiratory movement, or respiratory movement and airflow, heart rate or EKG, and SpO_2 and "type 4" studies record one or two parameters. Type 3 and type 4 studies are usually performed in the patients home, unattended by a technician, to reduce inconvenience to the patient and avoid delays associated with sleep laboratory bed availability. Some clinicians believe that there is an advantage conferred by monitoring the patient in his/her usual sleep environment with the data being more representative of the patients nightly experience compared with monitoring in a sleep laboratory. The validity of this argument remains to be demonstrated, however. Currently, there is considerable controversy regarding the role of these "portable cardiopulmonary sleep studies" including the most appropriate parameters to record and whether patient outcome suffers from absence of full PSG data. Although these home monitoring strategies have strong conceptual appeal, there is insufficient evidence on which to base universal adoption. In addition, adequate validation studies have not been performed in patient populations with comorbidities that conceivably could impact on sensitivity and specificity. Consequently, a joint task force of the American Thoracic Society, American College of Chest Physicians, and American Academy of Sleep Medicine indicated that unattended type 3 monitors are not recommended for use to rule in or rule out OSAN or provide information regarding the probability that the ANI is greater or less than 15. The task force suggested that type 3 studies performed in an attended in-laboratory environment may rule in or exclude OSAH in patients without significant comorbid conditions if the records are scored by a knowledgeable technician or clinician, and that a type 1 study (in-laboratory PSG) should be performed in symptomatic patients with a negative type 3 study. The use of type 4 monitors in either an attended or unattended environment was not recommended [11–13] to provide information regarding the probability of an ANI greater or less than 15.

HEALTH OUTCOMES OF OSAH

Cardiovascular and Cerebrovascular Disease

Several population studies have confirmed an association between OSAH and cardiovascular disorders including systemic arterial hypertension, coronary heart disease (angina pectoris, myocardial infarction, history of coronary artery bypass, percutaneous transluminal angioplasty), and congestive heart failure [14–16]. Moreover, it appears that even relatively low RDI confers increased risk for cardiovascular disease (Figure 31.6). These data demonstrate an

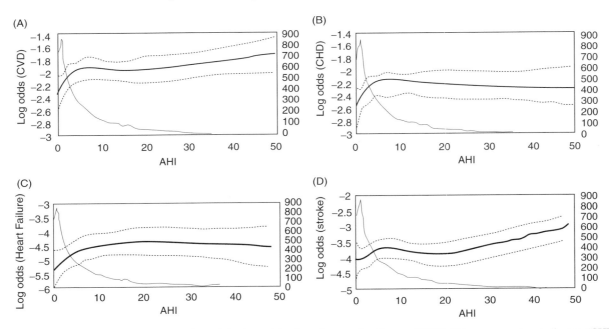

Figure 31.6 Restricted cubic spline regression of the log odds of cardiovascular disease (CVD) (A), coronary heart disease (CHD) (B), heart failure (C), and stroke (D) with increasing Apnea + Hypopnea Index (AHI). (From [15].) The heavy solid line indicates the predicted log odds which reflects risk, the dotted lines represent the 95% confidence intervals, the light solid line represents the number of participants at each level of AHI and is referenced to the values on the right-sided "Y-axis".

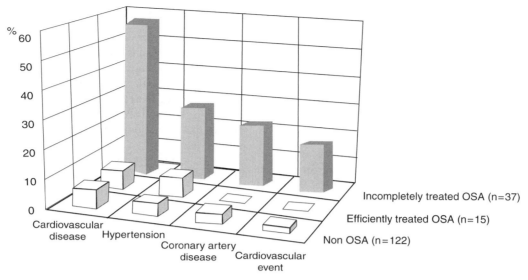

Figure 31.7 Percentage of middle-age men with ineffectively or untreated OSAH, effectively treated OSAH, and without OSAH and who were otherwise healthy at baseline, who had an incident cardiovascular event over 7 years' follow-up. (From [18].)

association, but not necessarily cause-and-effect, between OSAH and cardiovascular and cerebrovascular disease. On the other hand, a longitudinal study demonstrated that the higher the RDI at baseline, the greater the odds for having systemic arterial hypertension 4–8 years later [17], and another demonstrated that OSAH is an independent risk factor for developing cardiovascular disease over 7 years of follow-up in middle-aged males [18] (Figure 31.7). It is noteworthy that the latter study also found that effective therapy of OSAH reduced the risk for developing cardiovascular disease over the follow-up interval.

Neurocognition and Performance

Issues related to neurocognition and performance include those that deal with executive function (e.g., decision making, planning, organization), working memory, sustained attention or vigilance. In addition, the response characteristics to stimuli that are presented repeatedly over time (including reaction time, failure to respond to a stimulus, and responses made without a prompting stimulus) are also considered to be a performance metrics. Compared with non-OSAH control subjects, altered cognitive processing was observed in patients with AHI > 15. Although decrements in executive function may not be great in community-recruited populations including subjects with OSAH, clinic populations of OSAH patients have impaired executive function including planning and constructional power. This difference may represent selection bias but nonetheless supports an association between OSAH and altered executive function in at least some patients. Importantly, and perhaps not unexpectedly, OSAH patients have deficits in the ability to sustain attention to tasks over time.

These patients exhibit increased reaction time as well as false responses and lapses. This and other neurocognition abnormalities may be related to OSAH-related sleep fragmentation with consequent sleepiness with reduced vigilance and/or perturbations due to OSAH-related hypoxia. Regardless of the mechanism, there are important implications with regard to vehicular and occupational safety (discussed next).

Vehicular Safety

OSAH patients are at increased risk for motor vehicle crashes. Studies employing driving simulators have demonstrated substantially greater accident-prone features in OSAH patients than in control subjects [19, 20]. It is noteworthy that successful treatment of OSAH is not only associated with improved performance on a driving simulator but also with reduced traffic crashes [21].

Metabolic Consequences

Recent studies have demonstrated that independent of obesity, OSAH is associated with insulin resistance and diabetes mellitus, [22–24]. Insulin resistance is a key component of the metabolic syndrome, which is also characterized by central obesity, systemic arterial hypertension, and abnormal lipid metabolism. Insulin resistance and metabolic syncronic are recognized risk factors for cardiovascular disease in adults. In light of the previously described association between OSAH and obesity as well as hypertension, it has been proposed that the increased cardiovascular risk conferred by OSAH is mediated through the metabolic syndrome. Support for this linkage is provided by a recent investigation indicating that insulin sensitivity is improved after a short period of treatment of OSAH with continuous positive airway pressure (CPAP) [25]. On-going research is exploring this potentially important pathophysiologic pathway.

TREATMENT OF OSAH

A fundamental element of therapeutic intervention is avoidance of behaviors and practices that increase risk for and promote upper airway instability during sleep. Obese OSAH patients should be encouraged in the strongest possible terms to lose weight. Furthermore, patients should be counseled to *avoid alcohol, anxiolytics* (e.g., *benzodiazepines*), *hypnotics*, and *narcotics*, which can enhance upper airway collapsibility during sleep. Patients should be counseled to notify all healthcare providers of their OSAH in order to avoid inadvertent prescription of medications that can adversely influence upper airway stability. In some patients, *restriction of body position during sleep to lateral recumbency or head elevation* may be advantageous, although this must be objectively confirmed with overnight monitoring [26, 27]. It has also been recommended that efforts be made to *reduce nasal airway resistance* (e.g., topical corticosteroids), but in general, if there is a favorable effect of OSAH, it is therapeutically insufficient. Patients should be explicitly counseled not to operate a motor vehicle or other potentially dangerous equipment/machinery if they are not alert and local laws should be followed with regard to notification of administrative authorities.

Weight Reduction

Although there is a remarkable paucity of randomized clinical trials [28], it is evident both from the literature and clinical experience that weight reduction usually confers improved upper airway stability [4, 29]. The benefits of weight reduction on OSAH appear to be present regardless of whether it is achieved by dietary modification or bariatric surgery.

Positive Airway Pressure (PAP) Therapy

Delivery of airflow under low levels of pressure via a nasal mask or an oral–nasal mask is the mainstay of medical therapy for OSAH [30]. PAP is almost always successful in maintaining adequate upper airway patency and oxyhemoglobin saturation as well as improving sleep architecture and continuity. The degree of patient benefit with regard to improved daytime alertness is directly proportional to adherence with therapy. In this regard, maximizing patient adherence is the greatest challenge for the clinician.

Oral Appliance Therapy

There are two types of oral appliances, both worn during sleep: (1) mandibular advancement appliances (MAAs)—devices that (there are many versions) are fitted to the maxillary and mandibular dentition and protrude the mandible with the goal of increasing oropharyngeal patency and stability [31, 32]; and (2) tongue-retaining devices (TRDs)—devices that do not require intact dentition for retention and hold the tongue in an anterior position and prevent relapse into the posterior oropharynx during sleep. In general, MAAs have been widely studied and TRDs less extensively evaluated. MAAs are usually less successful in alleviating upper airway obstruction during sleep than PAP therapy, but patients may be more adherent to treatment with them. Side effects of MAAs are usually not severe and include excessive salivation, temporomandibular joint pain, or discomfort. Possibly the most notable side effect is the potential for altered changing of dental occlusion. Dental sleep experts should be involved in the assessment and management of patients with regard to oral appliance therapy.

Upper Airway and Craniofacial Surgery

Surgical procedures for OSAH may be useful in patients with identifiable anatomic upper airway or craniofacial abnormalities. The procedures include uvulopalatopharyngoplasty and laser palatoplasty, which are intended to stabilize the upper airway in the retropalatal region; genioglossal advancement and hyoid myotomy, which are intended to stabilize the retrolingual airway; and radiofrequency tissue volume reduction of the base of the tongue [33]. It is not possible to accurately predict which patients will have a successful surgical outcome. In general, these procedures have measurable risks and side effects and have not been shown to be as uniformly effective in treating OSAH patients as PAP [34–38]. Mandibular–maxillary advancement, a substantially invasive and costly surgical procedure, which addresses both the retropalatal and retrolingual regions, has been shown to be highly effective in some but not all studies. Further research to optimize patient selection for each type of surgical intervention is needed and will probably improve outcomes.

CONCLUSION

OSAH is a prevalent disorder in the community with important individual and societal health consequences. The diagnosis of OSAH requires that clinicians maintain a high index of suspicion and explore pertinent historical and anatomic clues. Highlighting the importance of establishing the diagnosis is that OSAH is a treatable disorder with consequent improvement in daytime function, quality of life, and reduced cardiovascular risk.

REFERENCES

1. Meoli AL, Casey KR, Clark RW, Coleman JA, Fayle RW, Troell RJ, Iber C. Hypopnea in sleep-disordered breathing in adults. American Academy of Sleep Medicine Position Paper. *Sleep* **24**:469–470(2001).

2. American Academy of Sleep Medicine Task Force. Sleep-related breathing disorders in adults: recommendations for syndrome definition and measurement techniques in clinical research. *Sleep* **22**:667–689(1999).

3. Young T, Shahar E, Nieto FJ, Redline S, Newman AB, Gottlieb DJ, Walslaben JA, Enright PL, Samet JM. Predictors of sleep-disordered breathing in community-dwelling adults: the Sleep Heart Health Study. *Arch Intern Med* **162**:893–900(2002).

4. Peppard PE, Young T, Palta M, Dempsey J, Skatrud J. Longitudinal study of moderate weight change and sleep-disordered breathing. *J Am Med Assoc* **284**:3015–3021(2000).

5. Flemons WW. Obstructive sleep apnea. *N Engl J Med* **347**:498–504(2002).

6. Shahar E, Redline S, Young T, Boland LL, Baldwin CM, Nieto FJ, O'Connor GT, Rapoport DM, Robbins JA. Hormone replacement therapy and sleep-disordered breathing. *Am J Respir Crit Care Med.* **167**:1186–1192(2003).

7. Redline S. Epidemiology of sleep-disordered breathing. *Semin Respir Crit Care Med* **19**:113–122(1998).

8. Ancoli-Israel S, Klauber M, Stepnowsky C, Estline E, Chinn A, Fell R. Sleep-disordered breathing in African-American elderly. *Am J Respir Crit Care Med* **152**:1946–1949(1995).

9. Redline S, Tishler PV, Tosteson TD, Williamson J, Kump K, Browner I, Ferrette V, Krejci P. The familial aggregation of obstructive sleep apnea. *Am J Respir Crit Care Med* **151**: 682–687(1995).

10. Carmelli D, Colrain IM, Swan GE, Bliwise DL. Genetic and environmental influences in sleep-disordered breathing in older male twins. *Sleep* **27**:917–922(2004).

11. Collop NA, Shepard JW Jr, Strollo PJ Jr, representing the ATS/ACCP/AASM Taskforce Steering Committee. Executive summary on the systematic review and practice parameters for portable monitoring in the investigation of suspected sleep apnea in adults. *Am Rev Respir Dis* **169**:1160–1163(2004).

12. Flemons WW, Littner MR, Gay P, Anderson WM, Hudgel DW, McEvoy RD, Loube DI. Home diagnosis of sleep apnea: a systematic review of the literature: an evidence review cosponsored by the American Academy of Sleep Medicine, the American College of Chest Physicians, and the American Thoracic Society. *Chest* **124**:1543–1579(2003).

13. Chesson AL Jr, Berry RB, Pack A. Practice parameters for the use of portable monitoring devices in the investigation of suspected obstructive sleep apnea in adults. *Sleep* **26**:907–913 (2003).

14. Nieto FJ, Young TB, Lind BK, et al. Association of sleep-disordered breathing, sleep apnea, and hypertension in a large community-based study. Sleep Heart Health Study. *J Am Med Assoc* **283**:1829–1836(2000).

15. Shahar E, Whitney CW, Redline S, Lee ET, Newman AB, Javier Nieto F, O'Connor GT, Boland LL, Schwartz JE, Samet JM. Sleep-disordered breathing and cardiovascular disease: cross-sectional results of the Sleep Heart Health Study. *Am J Respir Crit Care Med* **163**(1):19–25(2001).

16. Peker Y, Kraiczi H, Hedner J, Loth S, Johansson A, Bende M. An independent association between obstructive sleep apnoea and coronary artery disease. *Eur Respir J* **14**(1):179–184 (1999).

17. Peppard PE, Young T, Palta M, Skatrud J. Prospective study of the association between sleep-disordered breathing and hypertension. *N Engl J Med* **342**:1378–1384(2000).

18. Peker Y, Hedner J, Norum J, Kraiczi H, Carlson J. Increased incidence of cardiovascular disease in middle-age men with obstructive sleep apnea. *Am J Respir Crit Care Med* **166**: 159–165(2002).

19. Barbé F, Pericás J, Muñoz A, Findley L, Maria Anto J, Agusti AGN, de Lluc JM. Automobile accidents in patients with sleep apnea syndrome: an epidemiological and mechanistic study. *Am J Respir Crit Care Med* **158**:18–22(1998).

20. George CF, Smiley A. Sleep apnea & automobile crashes. *Sleep* **22**:790–795(1999).

21. George CF. Reduction in motor vehicle collisions following treatment of sleep apnoea with nasal CPAP. *Thorax* **56**:508–512(2001).

22. Elmasry A, Lindberg E, Berne C, Janson C, Gislason T, Awad Tageldin MA, Boman G. Sleep-disordered breathing and glucose metabolism in hypertensive men: a population-based study. *J Intern Med* **249**:153–161(2001).

23. Ip MSM, Bing LAM, Ng MMT, Lam WK, Tsang KWT, Lam KSL. Obstructive sleep apnea is independently associated with insulin resistance. *Am J Respir Crit Care Med* **165**:670–676 (2002).

24. Punjabi NM, Shahar E, Redline S, Gottlieb DJ, Givelber R, Resnick HE, for the Sleep Heart Health Study Investigators. Sleep-disordered breathing, glucose intolerance, and insulin resistance. The Sleep Heart Health Study. *Am J Epidemiol* **160**:521–530(2004).

25. Harsch IA, Schahin SP, Radespiel-Tröger M, Weintz O, Jahreiβ H, Fuchs FS, Wiest GH, Hahn EG, Lohmann T, Konturek PC, Ficker JH. CPAP treatment rapidly improves insulin sensitivity in patients with OSAS. *Am J Respir Crit Care Med* **169**:156–162(2004).

26. Cartwright RD. Effect of sleep position on sleep apnea severity. *Sleep* **7**:110–114(1984).

27. Neill AM, Angus SM, Sajkov D, McEvoy RD. Effects of sleep posture on upper airway stability in patients with obstructive sleep apnea. *Am J Respir Crit Care Med* **155**:199–204(1997).

28. Smith PL, Gold AR, Meyers DA, Haponik EF, Bleecker ER. Weight loss in mildly to moderately obese patients with obstructive sleep apnea. *Ann Intern Med* **103**:850–855(1985).

29. Schwartz AR, Gold AR, Schubert N, Stryzak A, Wise RA, Permutt S, Smith PL. Effect of weight loss on upper airway

collapsibility in obstructive sleep apnea. *Am Rev Respir Dis* **144**:494–498(1991).

30. Grunstein R, Sullivan C. Continuous positive airway pressure for obstructive sleep apnea–hypopnea breathing disorders. In: Kryger M, Roth T, Dement WC (Eds), *Principles and Practice of Sleep Medicine*, 4th ed. Saunders, Philadelphia, 2005.

31. Hoekema A, Stegenga B, De Bont LG. Efficacy and co-morbidity of oral appliances in the treatment of obstructive sleep apnea-hypopnea: a systematic review. *Crit Rev Oral Biol Med* **15**:137(2004).

32. Ferguson KA, Lowe AA. Dental appliances for the treatment of snoring and obstructive sleep apnea. In: Kryger M, Roth T, Dement WC (Eds), *Principles and Practice of Sleep Medicine*, 4th ed. Saunders, Philadelphia, 2005.

33. Powell NB, Riley RW, Troell RJ, Blumen MB, Guilleminault C. Radiofrequency volumetric tissue reduction of the palate in subjects with sleep-disordered breathing. *Chest* **113**:1163–1174(1998).

34. Pazos G, Mair EA. Complications of radiofrequency ablation in the treatment of sleep-disordered breathing. *Otolaryngol Head Neck Surg* **125**:462(2001).

35. Terris DJ, Chen V. Occult mucosal injuries with radiofrequency ablation of the palate. *Otolaryngol Head Neck Surg* **125**:468(2001).

36. Powell NB, Riley RW, Guilleminault C. Surgical management for sleep-disordered breathing. In: Kryger M, Roth T, Dement WC (Eds), *Principles and Practice of Sleep Medicine*, 4th ed. Saunders, Philadelphia, 2005.

37. American Sleep Disorders Association. Practice parameters for the treatment of obstructive sleep apnea in adults: the efficacy of surgical modifications of the upper airway. *Sleep* **19**:152–155(1996).

38. Sher AE, Schechtman KB, Piccirillo JF. The efficacy of surgical modifications of the upper airway in adults with obstructive sleep apnea syndrome. An American Sleep Disorders Association review. *Sleep* **19**:156–177(1996).

32

UPPER AIRWAY RESISTANCE SYNDROME

ROBERT D. BALLARD

National Jewish Medical and Research Center, Denver, Colorado

BACKGROUND INFORMATION

Obstructive sleep apnea is routinely defined by total cessations in airflow that last 10 seconds (apneas), and/or discrete reductions in airflow that last at least 10 seconds and trigger 4% reductions in oxygen saturation (hypopneas). The two major sequelae of apneas and hypopneas are event-associated hypoxemia and subsequent arousal or awakening from sleep. However, critical upper airway narrowing can also occur during sleep, leading to recurrent arousals and sleep fragmentation (Figure 32.1), even in the absence of discrete apneas, hypopneas, hypoxemia, or clear airflow reduction. This has been widely termed the upper airway resistance syndrome, although to date no widely accepted definition exists.

HISTORY OF THE UPPER AIRWAY RESISTANCE SYNDROME

The first reference to upper airway resistance syndrome (UARS) can be traced to 1982, when Guilleminault and colleagues described 25 pediatric patients who presented with snoring, daytime sleepiness, and behavioral problems [1]. These investigators compared these patients to 25 normal controls, evaluating all patients with polysomnography, incorporating the insertion of an esophageal balloon for monitoring intrathoracic pressure. Neither the affected patients nor the healthy controls demonstrated sleep associated hypoxemia or obstructive sleep apnea. However, during sleep symptomatic patients had substantially greater

swings in esophageal pressure during inspiration (range of -30 to -53 cm H_2O) in comparison to healthy controls (range of -11 to -20 cm H_2O). This suggested that the symptomatic patients had greater sleep-associated narrowing of the pharynx, necessitating increased inspiratory effort against the increased respiratory resistive load. Subsequent therapy with resection of the tonsils and adenoids was observed to improve sleep and daytime alertness in all 25 affected patients. This is likely the first description of a population of patients who did not have clear-cut obstructive sleep apnea, but rather a sleep-associated increase in upper airway resistance that contributed to sleep disruption and daytime sleepiness.

In 1991 Guilleminault and colleagues described another group of 15 adult patients with heavy snoring, who had mild daytime sleepiness, but no clear-cut obstructive sleep apnea by standard polysomnography [2]. When these patients were reevaluated after the insertion of an esophageal balloon for intrathoracic pressure monitoring, they demonstrated frequent 2–10 second arousals that occurred subsequent to increasing inspiratory esophageal pressure swings. In these patients CPAP therapy reduced the frequency of arousals during sleep, while subsequently improving daytime alertness as assessed by multiple sleep latency testing. These same investigators later reported another population of 15 patients thought to have UARS based on the observation of brief arousals following intervals of increased inspiratory esophageal pressure swings in association with mildly diminished airflow [3]. Of interest, 2 of these 15 patients gave no history of snoring, while 3 of the 15 reported snoring only occasionally, and then lightly.

Sleep: A Comprehensive Handbook, Edited by T. Lee-Chiong.

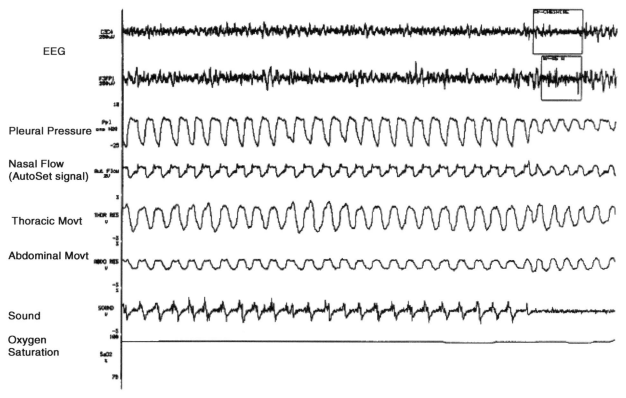

EEG

Pleural Pressure

Nasal Flow
(AutoSet signal)

Thoracic Movt

Abdominal Movt

Sound

Oxygen
Saturation

Figure 32.1 Example of an upper airway narrowing event occurring during sleep. Note the presence of heavy snoring in association with increased respiratory effort, denoted by increased pleural pressure swings. Flow appears to be well maintained (although the inspiratory flow profile has a flattened contour) with stable oxygen saturation. The event is terminated by an arousal (boxes on EEG channels), which leads to a reduction in pleural pressure swings to presleep baseline and a rounding of the inspiratory flow profile. (From [9], with permission.)

The investigators reported that all 15 of these patients had narrowing at the base of the tongue detectable by cephalometric radiographs.

These investigators have therefore described a condition with sequelae similar to those of obstructive sleep apnea, although differing substantively by the absence of discrete apneas and hypopneas. The presumed pathophysiology is similar to that of obstructive sleep apnea in that sleep onset is associated with reduction in upper airway dilator muscle activity (Figure 32.2), which subsequently allows upper airway narrowing. This narrowing yields airflow limitation, with the patient maintaining a fixed or even somewhat diminished airflow despite increasing inspiratory effort in response to afferent traffic from receptors localized in the chest wall, inspiratory muscles, and airways. Once inspiratory effort reaches a critical threshold level, this triggers an arousal from sleep, which not only restores upper airway dilator muscle activity with subsequent widening of the airway, but also triggers an adrenergic surge in association with a transient increase in blood pressure. Widening of the airway, the adrenergic surge, and the restoration of wakefulness to facilitate breathing, until sleep onset recurs, perpetuate this cycle.

UARS: PATHOPHYSIOLOGY

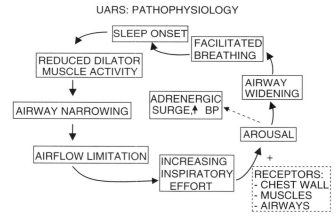

Figure 32.2 Proposed pathophysiology for recurrent airway narrowing events in UARS.

SEQUELAE

The hallmark clinical indicator of UARS has traditionally been daytime somnolence. In the report by Guilleminault and colleagues in 1982, 25 pediatric patients with daytime sleepiness, despite the absence of discrete obstructive sleep apnea, were felt to have upper airway resistance syndrome [1]. Tonsillectomy and adenoid resection improved daytime alertness, as demonstrated by improvement in mean sleep latency measured my multiple sleep latency testing from pre-to postintervention. The same laboratory used similar techniques to demonstrate that mean sleep latency improved from 11.3 to 14.6 minutes after administering CPAP to 15 adult patients with suspected UARS [1]. The somnolence observed in these patients is likely attributable to the frequent arousals and sleep disruption resulting from critical, sleep associated narrowing of the upper airway.

Recent evidence has also accumulated suggesting that, in a fashion similar to obstructive sleep apnea, UARS can also trigger both diurnal and nocturnal hypertension. Lofaso and associates reported 105 patients with snoring, but no discrete obstructive sleep apnea [4]. Fifty-five of these snorers were observed to have more than ten arousals per hour, and this indicator of sleep disruption was associated with an increase in diurnal diastolic blood pressure when compared to the 50 snorers with less than 10 arousals per hour.

Guilleminault and colleagues reported seven suspected UARS patients with normal blood pressures during wakefulness [5]. Continuous monitoring of blood pressure during sleep revealed the presence of increases in both systolic and diastolic blood pressure that occurred during arousals and within segments of "labored breathing." In this same report, the investigators described six suspected UARS patients with daytime hypertension. These patients had a significant reduction in average diurnal systolic and diastolic blood pressures after one month of therapy with nasal CPAP. The observations from these two studies therefore suggest that UARS can contribute to both nocturnal and diurnal increases in blood pressure in a fashion similar to that observed with obstructive sleep apnea. These reports also suggest that effective therapy of UARS, in this case with nasal CPAP, can subsequently improve diurnal blood pressure, similar to benefits observed from the effective therapy of obstructive sleep apnea.

SIGNS AND SYMPTOMS

It is now widely accepted that, for the diagnosis of UARS, the affected patient must complain of daytime fatigue or sleepiness, with fatigue possibly being more common. Whereas we once thought that snoring was an important part of this syndrome, it now appears that snoring is common, but may or may not be present in affected patients. Affected patients are often leaner than classic obstructive sleep apnea patients, often demonstrating a body mass index that is less than 25 kg/m [2]. It has also been widely observed that affected patients are frequently younger and might include a larger proportion of pediatric patients. It appears that hypertension can result from upper airway resistance syndrome, and this potential diagnosis should be considered when evaluating patients for new onset hypertension. Conversely, it has been reported that hypotension can also be commonly seen in UARS [6].

Gold and associates have reported the prevalence of a variety of constitutional symptoms in patients with UARS or traditional obstructive sleep apnea of varying severity [7]. These researchers observed that patients with UARS were more likely than traditional obstructive sleep apnea patients to present with other sleep related complaints or manifestations of aberrant sleep (Figure 32.3). UARS patients were more likely to complain of sleep onset insomnia, recurrent headaches, and irritable bowel syndrome. During sleep studies they were also more likely to demonstrate "alpha–delta sleep," as well as bruxism, both potential

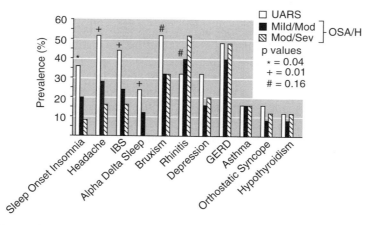

Figure 32.3 Comparative incidence of symptoms and signs in patient groups with UARS (AHI < 10), mild/moderate OSA (AHI 10–39), and moderate/severe OSA (AHI ≥ 40). (From [7], with permission.)

indicators of aberrant sleep. At present we lack information as to whether or not effective intervention for UARS subsequently improves these symptoms and aberrant findings during sleep.

DIAGNOSIS

The clinical diagnosis of UARS has been limited by a lack of standardization of nomenclature, criteria, and measurement techniques. A variety of different definitions of UARS have been utilized in published reports, varying substantially from report to report, even when reports originate from the same laboratory. For example, Guilleminault and colleagues in 1991 conducted polysomnographic studies of suspected patients using both esophageal pressure balloons for monitoring inspiratory pressure and pneumotachographs for monitoring airflow [2]. Arousals from sleep were identified using standard criteria. Arousals were felt to result from upper airway resistance if inspiratory esophageal pressure swings during the one to two breaths preceding arousal were the lowest since the last previous arousal, there was a simultaneous decrement in airflow, and inspiratory esophageal pressure swings became less negative in association with an increase in airflow during the two breaths after arousal.

One year later, the same investigators published a report in which criteria for diagnosis of UARS in somnolent patients included snoring for more than 10% of total sleep time, and frequent 3–14 second arousals that were preceded by inspiratory esophageal pressure excursions that were greater than those observed during baseline wakefulness, with subsequent reductions in inspiratory esophageal pressure swings on the breaths immediately following arousals [8]. However, in 1993, Guilleminault and colleagues described a third set of criteria for UARS, in which subjects demonstrated at least ten 3–14 second arousals per hour, with the esophageal pressure nadir in the breath before arousal being at least one standard deviation below inspiratory esophageal pressure nadirs measured during quiet, supine wakefulness [3]. In addition, the inspiratory esophageal pressure nadir had to be most negative during a snoring period immediately preceding the arousal, becoming less negative in the breath immediately following an arousal. It is apparent from these reports that even the investigators from a single laboratory had difficulty establishing consistent criteria by which to define this syndrome.

Other investigators have proposed differing criteria. Rees and associates defined upper airway resistive events as a plateauing of inspiratory nasal flow with increasingly negative pleural pressure swings for two or more breaths compared with the preceding baseline in the absence of

hypopnea [9]. Hypopneas were defined as at least 10 seconds of at least a 50% reduction in thoracoabdominal movements. Lofaso and associates provided a somewhat more comprehensive definition, in that they required patients to have a discrete apnea hypopnea index that was less than five events per hour, in association with at least ten arousals per hour of sleep, and evidence of daytime sleepiness as manifested by an Epworth Sleepiness Score of >10 [10]. Upper airway resistance events were defined by the detection of flow limitation by pneumotachography, in association with a concomitant decrease in inspiratory esophageal pressure nadir. Affected patients had to demonstrate more than ten of these upper airway resistance events per hour of sleep.

More recent studies have suggested that the use of nasal pressure monitoring provides a more accurate tool for the monitoring of airflow than previously used thermistors. Montserrat and colleagues demonstrated in 1997 that the continuous recording of nasal pressure yielded findings that were comparable to those obtained from continuous monitoring via a pneumotachograph [11]. At approximately the same time, Norman and associates demonstrated that the detection of flow reduction and flattening per nasal pressure tracing is much more sensitive than by thermistor (Figure 32.4) for the detection of both obstructive sleep apnea and upper airway resistance events [12]. These findings suggested that monitoring of nasal pressure might provide an acceptable and more palatable alternative technique in esophageal pressure monitoring.

Hosselet and co-investigators subsequently used nasal pressure to define flow-limited events as at least two consecutive, flow-limited breaths [13]. When comparing ten symptomatic (one snorer, five obstructive sleep apnea, and four UARS patients) verses four asymptomatic control patients, all UARS patients had at least 30 flow-limited events per hour. These investigators proposed this as acceptable defining criteria for UARS.

Epstein and colleagues subsequently compared the sensitivity of nasal pressure versus thermistor for the monitoring of airflow in patients across a spectrum of sleep disorder breathing, ranging from primary snoring to frank obstructive sleep apnea. These investigators found that the use of nasal pressure monitoring substantially increased the detection of UARS, although it was not clear that this technique yielded clinically significant advantages in patients with obstructive sleep apnea [21]. At present, it is therefore widely accepted that nasal pressure monitoring is an acceptable replacement for pneumotachography in the assessment of patients with UARS, and clearly superior to nasal/oral flow thermistors [14]. However, many suggest that, for a definitive diagnosis, an esophageal pressure balloon should be placed for the monitoring of inspiratory pressure swings [15]. Obviously, this latter technique has not been widely accepted by all patients.

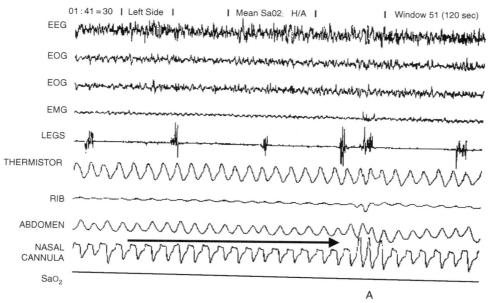

Figure 32.4 Two minute tracing from a patient with UARS. Note the persistent flattening of the nasal pressure (cannula) tracing (arrow) in the presence of an unchanging thermistor signal. With a brief arousal from sleep (**A**), there is an obvious increase in the nasal pressure signal, but minimal increase in the thermistor signal. (From [12], with permission.)

THERAPY

Surgical Therapy

As previously described, in 1982, Guilleminault and colleagues reported 25 pediatric cases of UARS that improved symptomatically in response to subsequent tonsillectomy and adenoid resection [1]. This response was demonstrated by an increase in mean sleep latency from testing conducted before and after intervention.

More recently, Newman and colleagues reported nine patients with suspected UARS that were treated with a variety of upper airway surgeries intended to increase airway patency during sleep [16]. These nine patients demonstrated a mean reduction in the Epworth sleepiness score from 12 points presurgery to 3.4 points after surgery ($p = 0.001$). In a similar study, Utley and associates treated 11 UARS patients with laser-assisted uvulopalatoplasty. These authors observed that 82% of the treated patients reported improvement in daytime sleepiness, and that the Epworth sleepiness score for the group decreased from 13.5 preintervention to 8 after surgical procedure [22].

Powell and colleagues described 22 patients with "mild obstructive sleep apnea," although 14 actually appeared to be UARS patients (RDI < 5) [17]. All patients underwent radiofrequency reduction of the palate. Patients subsequently reported improvement in snoring, perceived sleep efficiency, and reduced Epworth sleepiness scores.

Although none of these studies were controlled or blinded, the results suggest that standard surgical interventions may improve symptoms of daytime fatigue and sleepiness in patients with UARS.

Oral Appliance Therapy

Yoshida and colleagues recently investigated the efficacy of oral appliance therapy for UARS [18]. These investigators identified 32 UARS patients, manifested by an apnea–hypopnea index of less than five events per hour, an arousal index of greater than ten per hour, and a score on the Epworth Sleepiness Scale in excess of ten. These researchers observed that therapy with an oral appliance yielded significant improvements in Epworth sleepiness score, sleep latency measured by mean sleep latency testing, arousal index during sleep, sleep efficiency, and minimum oxygen saturation observed during sleep. Although this is an unblinded, uncontrolled case series, the results suggest that oral appliance therapy may be a useful therapeutic alternative in patients with UARS.

CPAP Therapy

Guillemenault and colleagues reported in 1991 from 15 adult patients with UARS that CPAP reduced the frequency of arousals during polysomnography and improved sleep latency as measured by subsequent multiple sleep latency testing [2]. However, these investigators also observed that none of the 15 patients wanted to continue long-term CPAP therapy, despite the apparent clinical benefit.

In a subsequent study, the same investigators again observed that effective CPAP therapy improved mean sleep latency from 5.1 to 13.5 minutes in patients with UARS [3]. In the patients who were most somnolent during the daytime, these investigators reported that CPAP reduced arousal frequency from 31 to 8 per hour.

It is therefore apparent that CPAP can improve sleep quality and daytime alertness in patients with UARS. However, CPAP compliance appears to remain a problem in this population, perhaps greater than observed with conventional obstructive sleep apnea patients. Krieger and colleagues reported 98 "nonapneic snorers," defined by having an apnea–hypopnea index that was less than 15 per hour [19]. These investigators reported that only 33 (34%) of these patients initially accepted CPAP therapy, and that only 60% of these patients remained compliant with CPAP after three years of prescribed therapy. However, the average nightly use of CPAP in the persistently compliant patients was 5.6 hours.

In an even less encouraging report, Rauscher and associates assessed the utility of CPAP in "nonapneic snorers," defined by having daytime sleepiness, at least 20 arousals per hour during sleep, and an apnea–hypopnea index of less than 5 per hour [20]. Only 11 (19%) of these patients initially accepted CPAP, and the average nightly use of CPAP in these patients was only 2.8 hours after six months of therapy. These studies concur that, although CPAP effectively addresses the anatomic narrowing of the upper airway that occurs during sleep and appears to improve sleep quality, relatively few patients can be expected to comply with this therapy on a long-term basis.

CONCLUSIONS

Substantial evidence has accumulated that UARS is a distinct entity in the spectrum of sleep disordered breathing. This syndrome appears to cause daytime fatigue and sleepiness and may trigger both nocturnal and diurnal hypertension in a fashion similar to discrete obstructive sleep apnea. There is additional evidence that UARS may be linked to a variety of somatic complaints, including headaches, irritable bowel syndrome, and sleep-onset insomnia.

Although it appears that the use of an esophageal pressure monitoring of inspiratory pressure and the measurement of nasal pressure are both sensitive to upper airway resistance events, at present there is no standard for the definition or diagnosis of upper airway resistance syndrome. Limited studies suggest that CPAP is effective in treating the sleep-associated narrowing of the upper airway and

associated sleep disruption, but long-term CPAP compliance remains uncertain. A few reports suggest that upper airway surgery and oral appliance therapy may be more effective in this population than in conventional obstructive sleep apnea patients, but these interventions at present remain inadequately studied.

REFERENCES

1. Guilleminault C, Winkle R, Korobkin R, Simmons B. Children and nocturnal snoring: evaluation of the effects of sleep related respiratory resistive load and daytime functioning. *Eur J Pediatr* **139**(3):165–171(1982).
2. Guilleminault C, Stoohs R, Duncan S. Snoring (I). Daytime sleepiness in regular heavy snorers. *Chest* **99**(1):40–48(1991).
3. Guilleminault C, Stoohs R, Clerk A, Cetel M, Maistros P. A cause of excessive daytime sleepiness. The upper airway resistance syndrome. *Chest* **104**(3):781–787(1993).
4. Lofaso F, Coste A, Gilain L, Harf A, Guilleminault C, Goldenberg F. Sleep fragmentation as a risk factor for hypertension in middle-aged nonapneic snorers. *Chest* **109**(4):896–900(1996).
5. Guilleminault C, Stoohs R, Shiomi T, Kushida C, Schnittger I. Upper airway resistance syndrome, nocturnal blood pressure monitoring, and borderline hypertension. *Chest* **109**(4):901–908(1996).
6. Bao G, Guilleminault C. Upper airway resistance syndrome—one decade later. *Curr Opin Pulm Med* **10**(6):461–467(2004).
7. Gold AR, Dipalo F, Gold MS, O'Hearn D. The symptoms and signs of upper airway resistance syndrome: a link to the functional somatic syndromes. *Chest* **123**(1):87–95(2003).
8. Guilleminault C, Stoohs R, Clerk A, Simmons J, Labanowski M. From obstructive sleep apnea syndrome to upper airway resistance syndrome: consistency of daytime sleepiness. *Sleep* **15**(6 Suppl):S13–S16(1992).
9. Rees K, Kingshott RN, Wraith PK, Douglas NJ. Frequency and significance of increased upper airway resistance during sleep. *Am J Respir Crit Care Med* **162**(4 Pt 1):1210–1214(2000).
10. Lofaso F, Goldenberg F, d'Ortho MP, Coste A, Harf A. Arterial blood pressure response to transient arousals from NREM sleep in nonapneic snorers with sleep fragmentation. *Chest* **113**(4):985–991(1998).
11. Montserrat JM, Farre R, Ballester E, Felez MA, Pasto M, Navajas D. Evaluation of nasal prongs for estimating nasal flow. *Am J Respir Crit Care Med* **155**(1):211–215(1997).
12. Norman RG, Ahmed MM, Walsleben JA, Rapoport DM. Detection of respiratory events during NPSG: nasal cannula/pressure sensor versus thermistor. *Sleep* **20**(12):1175–1184(1997).
13. Hosselet JJ, Norman RG, Ayappa I, Rapoport DM. Detection of flow limitation with a nasal cannula/pressure transducer system. *Am J Respir Crit Care Med* **157**(5 Pt 1):1461–1467(1998).
14. Ayappa I, Norman RG, Krieger AC, Rosen A, O'Malley RL, Rapoport DM. Non-invasive detection of respiratory effort-

related arousals (REras) by a nasal cannula/pressure transducer system. *Sleep* **23**(6):763–771(2000).

15. Guilleminault C, Poyares D, Palombini L, Koester U, Pelin Z, Black J. *Variability of Respiratory Effort in Relation to Sleep Stages in Normal Controls and Upper Airway Resistance Syndrome Patients.* Elsevier, Stanford, CA, 2001.

16. Newman JP, Clerk AA, Moore M, Utley DS, Terris DJ. Recognition and surgical management of the upper airway resistance syndrome. *Laryngoscope* **106**(9 Pt 1):1089–1093(1996).

17. Powell NB, Riley RW, Troell RJ, Li K, Blumen MB, Guilleminault C. Radiofrequency volumetric tissue reduction of the palate in subjects with sleep-disordered breathing. *Chest* **113**(5):1163–1174(1998).

18. Yoshida K. Oral device therapy for the upper airway resistance syndrome patient. *J Prosthet Dent* **87**(4):427–430(2002).

19. Krieger J, Kurtz D, Petiau C, Sforza E, Trautmann D. Long-term compliance with CPAP therapy in obstructive sleep apnea patients and in snorers. *Sleep* **19**(9 Suppl):S136–S143(1996).

20. Rauscher H, Formanek D, Zwick H. Nasal continuous positive airway pressure for nonapneic snoring? *Chest* **107**(1):58–61(1995).

21. Epstein MD, Chicoine SA, Hanumara RC. Detection of upper airway resistance syndrome using a nasal cannula/pressure transducer. *Chest.* **2000** Apr; 117(4):1073–1077.

22. Utley DS, Shin EJ, Clerk AA, Terris DJ. A cost-effective and rational surgical approach to patients with snoring, upper airway resistance syndrome, or obstructive sleep apnea syndrome. *Laryngoscope.* **1997** Jun; 107(6):726–734.

33

CENTRAL SLEEP APNEA

S. Javaheri

University of Cincinnati, Cincinnati, Ohio

INTRODUCTION

Central apnea is due to temporary failure in breathing rhythm generation resulting in the loss of ventilatory effort, lasting at least 10 seconds. During central apnea, there is no medullary inspiratory neural output to the diaphragm and other inspiratory thoracic pump muscles. Therefore, polygraphically, central apnea is characterized by the absence of naso-oral airflow and thoracoabdominal excursions.

An obstructive apnea is due to relaxation of the oropharyngeal muscles and occlusion of the upper airway in the presence of continual rhythmic contractions of inspiratory thoracic pump muscles. Polygraphically, obstructive apnea is characterized by the absence of naso-oral airflow in spite of thoracoabdominal excursions.

Central apnea, as a polysomnographic finding, could be due to a number of physiological conditions and pathological causes (Table 33.1). In part because central apneas may occur in normal individuals (physiological central apnea), the minimum number of events required during sleep to represent a distinct disorder is not clear. More importantly, this issue is further compounded (1) by difficulty in accurately distinguishing central hypopneas from obstructive hypopneas, events that should be added to the number of central apneas and reported as central apnea–hypopnea index, and (2) that both central and obstructive hypopneas, like central and obstructive apneas, commonly present together in the same subject. For those reasons, in patients with systolic heart failure, we have used arbitrary polysomnographic criteria [1–3] to classify the disordered breathing into either predominant central or obstructive sleep apnea.

But what the minimum central apnea–hypopnea index should do to define the presence of a "clinically significant" sleep-related breathing disorder is not clear. In our studies of patients with heart failure, arbitrarily, an apnea–hypopnea index of 15 per hour or greater has been used, but this does not mean that it is the appropriate threshold. Further studies using outcome are necessary to answer this question.

In the present chapter, I briefly review some of the conditions associated with or causing central sleep apnea (Table 32.1). My hope is that the proposed classification, which is based on physiological mechanisms and pathological categories of disorders, makes it easier to understand central sleep apnea. In this regard, in spite of varied causes of central apnea, there is a unified concept that explains the mechanism of genesis of central apnea during sleep in most (but not all) conditions noted in Table 32.1. This has to do with the profound effect of sleep on control of breathing, the concept of apneic threshold, and the proximity of the prevailing (eucapnic) P_{CO_2} to the apneic threshold P_{CO_2}.

CONTROL OF BREATHING

Three distinct control systems influence breathing. These include (1) automatic/metabolic control of breathing, (2) behavioral control of breathing, and (3) a wakefulness drive to breathe.

Conceptually, the automatic/metabolic control system reflects the underlying automaticity of breathing rhythm and that this system is coupled to the metabolic rate. The

Sleep: A Comprehensive Handbook, Edited by T. Lee-Chiong.
Copyright © 2006 John Wiley & Sons, Inc.

TABLE 33.1 Central Sleep Apnea

I. Physiological CSA
 A. Sleep-onset
 B. Postarousal/postsigh
 C. Phasic REM sleep
II. Eucapnic–hypopcapnic (nonhypercapnic) CSA
 A. Systolic heart failure
 B. Idiopathic
 C. Idiopathic pulmonary arterial hypertension
 D. High altitude
 E. Cerebrovascular disorders
III. Hypercapnic CSA
 A. Alveolar hypoventilation with normal pulmonary function
  1. Congenital and primary
  2. Brainstem and spinal cord disorders (encephalitis, tumor, infarcts, cervical cordotomy, anterior cervical spinal artery syndrome)
 B. Neuromuscular disorders
  1. Muscular disorders (myotonic and Duchenne dystrophies, acid maltase deficiency)
  2. Neuromuscular junction disorders (myasthenia gravis)
  3. Spinal cord/peripheral nerve disorders (amyotrophic lateral sclerosis, multiple sclerosis, polio)
 C. Opioids
IV. CSA with endocrine disorders
 A. Acromegaly
 B. Hypothyrodism
V. CSA with obstructive sleep apnea
 A. A minor component of OSA
 B. With CPAP therapy of OSA/post-tracheostomy
 C. Postuvulopalatopharyngoplasty
VI. CSA with upper airway disorders

REM = rapid eye movement; CSA = central sleep apnea; OSA = obstructive sleep apnea.

various components of this system include the potomedullary respiratory centers (the dorsal and ventral respiratory groups and a central pacemaker, presumably located in the pre-Botzinger complex) and a number of receptors (e.g., the peripheral arterial chemoreceptors—the carotid bodies located between internal and external carotid arteries—and the medullary central chemoreceptors) whose activities are regulated by a variety of physiological variables, specifically P_{O_2} and P_{CO_2}/H^+ [4]. This complex system is responsible for the act of breathing at all times and maintaining arterial P_{O_2} and $[H^+]$ relatively constant by coupling alveolar ventilation to metabolic rate (metabolic control). Any disturbance in breathing that changes P_{O_2}, P_{CO_2}, or $[H^+]$ will alter the activity of the system in such a way that the initial disturbance is minimized (a negative feedback system). Under normal circumstances, P_{CO_2} is the most important variable controlling breathing, such that any small change in P_{CO_2} affects ventilation.

The behavioral control of breathing is best exemplified by the circumstances when the respiratory system is used for functions such as talking, swallowing, or laughing.

Wakefulness also influences breathing (wakefulness drive) and this is mediated through the ascending reticular activating system, located in the brainstem, and part of the brain arousal system. However, the wakefulness drive to breathe along with the behavioral control are absent during sleep, and therefore breathing comes under the metabolic control system. This switch is critical, because the metabolic control system is sensitive to small changes in P_{CO_2}, and sleep unmasks a very P_{CO_2}-sensitive apneic threshold.

APNEIC THRESHOLD AND THE MECHANISMS OF GENESIS OF CENTRAL APNEA DURING SLEEP

The mechanisms involved in the genesis of central apnea relate specifically to the state of sleep and removal of the wakefulness drive on breathing. This unmasks the apneic threshold P_{CO_2} [5–8], a P_{CO_2} level below which rhythmic breathing ceases resulting in central apnea.

Normally with onset of sleep, ventilation decreases and P_{CO_2} rises by few mmHg. If the spontaneous prevailing P_{CO_2}, referred to as the eucapnic P_{CO_2}, decreases below the apneic threshold P_{CO_2}, breathing ceases and central apnea occurs. As a result of central apnea, P_{CO_2} rises and after it exceeds the apneic threshold P_{CO_2}, breathing resumes. Therefore the difference between two P_{CO_2} set points the prevailing P_{CO_2}, the P_{CO_2} at the apneic threshold minus, is a critical factor for development of central apnea. The less this difference, the greater the likelihood of occurrence of central apnea. This is because small increases in ventilation could lower the prevailing P_{CO_2} below the apneic threshold. On the contrary, when apneic threshold P_{CO_2} is far away from the eucapnic P_{CO_2}, large ventilatory changes are necessary to lower the P_{CO_2} below the apneic threshold P_{CO_2}. Therefore the likelihood of developing central apnea decreases. It must be emphasized that what is important in the genesis of central apnea is the difference between the two P_{CO_2} set points, rather than the actual value of the prevailing P_{CO_2}. As will be discussed later, this is the unifying mechanism mediating development of central apnea under a variety (but not all) of conditions.

I should reiterate that central apnea is precipitated by sleep, because of unmasking of the apneic threshold. It is rare to observe central apnea during wakefulness, independent of P_{CO_2} level. When central apnea occurs during "wakefulness," it is usually at a time when the patient is very relaxed and almost dosing. If such patients were being monitored electroencephalographically when central apnea occurs, slowing of brain waves are commonly observed. Central apnea is then followed by hyperpnea resulting in a shift in frequency of brain waves.

With this background, I will discuss the mechanisms and causes of central sleep apnea. Emphasis, however, is placed on congestive heart failure, which is the most common cause of central sleep apnea in the general population.

CENTRAL SLEEP APNEA (CSA)

Physiologic Central Apnea

The conditions (Table 32.1) causing or associated with central sleep apnea in this category are considered normal sleep phenomena and, not surprisingly, the frequency of occurrence of such central apneas is normally minimal. We observe such central apneas with onset of sleep, after an arousal or a sigh, and occasionally during phasic rapid eye movement (REM) sleep.

Sleep-Onset Central Apnea   As noted earlier, with removal of wakefulness drive and onset of sleep, the CO_2-sensitive apneic threshold is exposed and therefore a higher than awake Pco_2 level is necessary to maintain rhythmic breathing. This is because, normally, the apneic threshold Pco_2 is close to the level of awake Pco_2. Consequently, with sleep onset, few central apneas (sleep-onset central apneas) occur, and along with a decrease in ventilation, Pco_2 rises. When Pco_2 is above apneic threshold, rhythmic breathing, at ventilation less than that during wakefulness, is maintained.

Postarousal/Postsigh Central Apnea   During an arousal, the prevailing sleeping Pco_2 (which is higher than awake Pco_2) is considered hypercapnic for the wake brain; therefore ventilation rises and Pco_2 decreases (negative feedback system). When the arousal is over and sleep resumes, the prevailing Pco_2 is considered hypocapnic for the sleeping brain. Therefore central apneas occur, as they do with sleep onset. Similarly, after a sigh, even without an arousal, if Pco_2 decreases below apneic threshold, central apnea occurs.

Phasic REM Sleep   One of the unique features of phasic rapid eye movement (REM) sleep is the variability in breathing pattern. The source of the variability does not depend on variations in chemoreceptor activity. These variations are intermittent and in animal studies are associated with pontogeniculo-occipital (PGO) waves influencing medullary respiratory centers [9]. In cat, intermittent inhibition of the diaphragm, referred to as fractionation, has been observed [9].

In humans during phasic REM sleep, commonly irregular breathing and sometimes central (Figure 33.1) and obstructive apneas are observed. These events most probably have to do with pontomedullary electrophysiological events that occur during phasic REM sleep. With total inhibition of rhythmic breathing, central apneas occur. However, when there is predominant inhibition of inspiratory

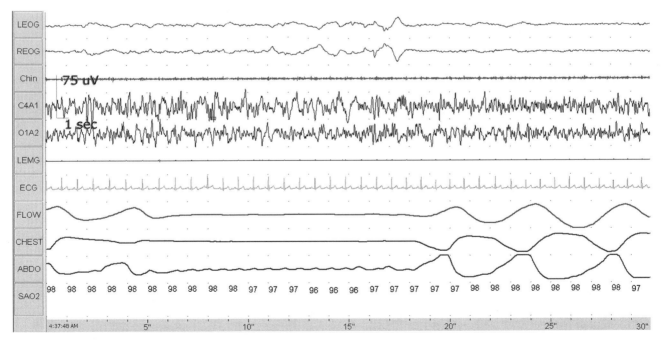

Figure 33.1   An epoch of a polysomnogram showing an episode of central sleep apnea during phasic rapid eye movement sleep. Cardiogenic oscillations are observed on the abdominal tracing. As noted in the text, an irregular breathing pattern, central apneas, or obstructive apneas may occur during phasic REM sleep.

drive to the muscles of the upper airway, with relative preservation of diaphragmatic inspiratory activity, obstructive apneas occur. Following these events, commonly an arousal, characterized by appearance of alpha waves, and an increase in chin EMG occur, but soon sleep continuity with tonic REM sleep follows.

Eucapnic–Hypocapnic (Nonhypercapnic) CSA

These disorders are generally characterized by (1) an awake steady-state $Paco_2$ that is either within or less than ($<36\,mmHg$) normal value and (2) increased ventilatory response to changes in Pco_2 and perhaps also to Po_2.

During sleep, the prevailing Pco_2 may fall below the apneic threshold (resulting in a central apnea), either because of inability to increase the Pco_2 or because the apneic threshold Pco_2 rises (e.g., due to hypoxemia), or a combination of the two. In addition, in these disorders (Table 32.1), the hypercapnic ventilatory response is invariably increased. The increase in ventilatory response will increase the likelihood of developing central apnea. This is because anytime an arousal occurs, the immediate prearousal Pco_2 becomes hypercapnic for the aroused brain, and therefore intense hyperventilation occurs, which could drive the prevailing Pco_2 below apneic threshold, causing a central apnea as sleep is resumed. Following central apnea, Pco_2 increases until an arousal recurs. In this way, the cycle of apnea–hyperpnea is perpetuated. The disorders in this category are briefly discussed below.

Systolic Heart Failure   Heart failure is a highly prevalent syndrome [10]. It is estimated that about 5 million Americans, about 2% of the population, have heart failure with an approximate annual incidence of half a million. It is projected that the prevalence of heart failure will continue to rise into the twenty-first century. Because heart failure is highly prevalent and central sleep apnea is common in the setting of the failing heart, heart failure is the most common cause of central sleep apnea in the population.

As will be discussed later, the pattern of periodic breathing in this disorder is unique in that it has long crescendo–decrescendo arms of changes in thoracoabdominal excursions and airflow. This is due to a long arterial circulation time, a pathophysiological feature of systolic heart failure.

Periodic breathing and central apnea have been known to occur in patients with heart failure since two centuries ago [11–14]. Recognition of this disorder preceded that of obstructive sleep apnea. John Cheyne [13] and William Stokes [14] have been credited with description of periodic breathing in heart failure, hence the eponym Cheyne–Stokes breathing. However, 37 years before John Cheyne's description, John Hunter, a British surgeon, had observed this unique pattern of breathing [11, 12]. Describing the case history of Mr. Boyde who apparently suffered from

atrial fibrillation and congestive heart failure, Hunter wrote: "The pulse was irregular, as usual, and quick; but his breathing was very particular: he would cease breathing for twenty or thirty seconds, and then begin to breathe softly, which increased until he breathed extremely strong, or rather with violent strength, which gradually died away till we could not observe that he breathed at all. He could not lie down without running the risk of being suffocated, therefore he was obliged to sit up in his chair."

Cheyne–Stokes breathing (CSB) is a form of periodic breathing, commonly with an intervening central apnea (or hypopnea) situated between the waxing and waning arms. As noted earlier, a distinctive feature of CSB is a long "recovery phase," which reflects a prolonged arterial circulation time. This prolonged waxing phase of CSB distinguishes it from other forms of periodic breathing with central sleep apnea such as the idiopathic form, in which the recovery arm is somewhat abrupt and short rather than smooth and prolonged.

Prevalence of Central Apnea in Systolic Heart Failure
Studies [1–3, 15–26] of patients with stable heart failure and left ventricular systolic dysfunction show that 40–80% have an apnea–hypopnea index ≥ 15 per hour. These indices include both central sleep apnea and obstructive sleep apnea, events that commonly occur together during sleep in a patient with heart failure.

The largest prospective study [18] involved 100 ambulatory male patients with stable, treated heart failure. In this study, 114 consecutive eligible subjects who were followed in a cardiology and a primary care clinic were asked to participate (88% recruitment). Importantly, no information was sought about symptoms or risk factors for sleep apnea. Using an apnea–hypopnea index of 15 per hour or greater as the threshold, 49 subjects (49% of all patients) had moderate to severe sleep apnea–hypopnea with an average index of 44 ± 20 (1 SD) per hour. This index included both central and obstructive events.

Because both obstructive and central sleep apneas and hypopneas usually occur together in the same patient, and at times polysomnographic distinction is difficult, investigators have used various arbitrary polysomnographic criteria to define the predominant form of sleep apnea in a given patient. With this in mind, 5–40% of all patients with systolic heart failure have predominantly obstructive sleep apnea, and 30–60% have central sleep apnea. In our prospective study of 100 subjects with systolic heart failure [18], about 12% had predominantly obstructive sleep apnea–hypopnea, and 37% had predominantly central sleep apnea–hypopnea.

Mechanisms of Central Sleep Apnea and Periodic Breathing in Heart Failure   For simplicity, I make a distinction between mechanisms of central sleep apnea in heart failure (which have to do with apneic threshold) and the mechanisms mediating periodic breathing (which have to do with pathological processes of heart failure). These are discussed separately later, but I emphasize that this distinction is for simplicity, and some of the mechanisms mediating both central apnea and periodic breathing overlap.

Normally, with onset of sleep, ventilation decreases and P_{CO_2} increases. This maintains the prevailing P_{CO_2} above the apneic threshold P_{CO_2}, and rhythmic breathing occurs. However, in some patients with heart failure, the prevailing awake P_{CO_2} does not rise during sleep [27, 28]. Because of the proximity of the prevailing P_{CO_2} to the exposed apneic threshold [8, 27], central sleep apnea occurs. The reason for the lack of normally observed sleep-induced rise in P_{CO_2} in some patients with heart failure is not clear. I believe, however, that it could be due to the lack of normally observed sleep-induced decrease in ventilation. Presumably, in heart failure patients with severe left ventricular diastolic dysfunction (and stiff left ventricle), which invariably accompanies systolic dysfunction, when venous return increases, pulmonary capillary pressure could rise. This results in a small increase in respiratory rate and ventilation, preventing the normally observed rise in P_{CO_2}. Since changes in respiratory rate are small, a large number of patients need to be studied for such changes to become statistically significant; alternatively, measurements of central hemodynamic and Pa_{CO_2} are needed to prove this hypothesis.

Several studies [29–31] have shown that subjects with heart failure and low Pa_{CO_2} have a high probability of developing central sleep apnea. The reason for this is not clear, but it may also have to do with the severity of the left ventricular diastolic dysfunction that invariably accompanies systolic dysfunction. Heart failure patients with hypocapnia have a more severe left ventricular diastolic dysfunction and a higher wedge pressure than eucapnic heart failure patients [19]. Diastolic dysfuction will be further unmasked during supine position as venous return rises. Consequently, such patients may not be able to normally decrease their ventilation with sleep onset and their P_{CO_2} remains close to apneic threshold.

It is emphasized that although a low awake Pa_{CO_2} is highly predictive of central sleep apnea, it is not a prerequisite. Many patients with heart failure and central sleep apnea have normal awake Pa_{CO_2} [15, 32]. What is important is the proximity of the apneic threshold to the Pa_{CO_2} [8, 27].

Another implication of the difference between apneic threshold and the prevailing P_{CO_2} may relate to the gender difference in prevalence of central apnea in congestive heart failure. Combining the results of several studies (reviewed in [3]), 40% of the male subjects have central sleep apnea, which is significantly higher than the 18% prevalence of central sleep apnea in female subjects. However, in women with congestive heart failure and systolic dysfunction, risk of central sleep apnea is six times higher in those

60 years or older when compared to those less than 60 years old [21]. Premenopausal women have a lower apneic threshold than men [33], and this may be in part the reason for a lower prevalence of central apnea in premenopausal females than in males with heart failure. However, the lower apneic threshold in women than in men could be mediated by both female [33] and male hormones [34].

Another mechanism that increases the likelihood of developing central apnea during sleep, and also periodic breathing, is enhanced ventilatory response to CO_2 [35], which will be discussed later.

So far, I have discussed the mechanisms of central sleep apnea in heart failure. However, as noted earlier, in heart failure, central apnea occurs in the background of periodic breathing. This unique pattern of periodic breathing is characterized by long crescendo–decrescendo ventilation arms. The mechanisms underlying this unique pattern of periodic breathing are to some extent distinct from those mediating development of central apnea, though some mechanisms overlap.

Mathematical and experimental models [35–45] of the negative feedback system controlling breathing homeostasis predict that increased arterial circulation time (which delays the transfer of information regarding changes in P_{O_2} and P_{CO_2} from pulmonary capillary blood to the chemoreceptors), enhanced gain of the chemoreceptors (enhanced CO_2/O_2 chemosensitivity), and enhanced plant gain (a large change in Pa_{CO_2} for a small change in ventilation) collectively increase the likelihood of periodic breathing. All three factors may be present in systolic heart failure. It is therefore not surprising that heart failure is so conducive to development of period breathing. First, in systolic heart failure, effective arterial circulation time could be increased for a variety of reasons, such as pulmonary congestion, left atrial and ventricular enlargement, and diminished stroke volume. Second, plant gain is increased because of a low functional residual capacity. Functional residual capacity could be decreased because of pleural effusion, cardiomegaly, pulmonary congestion, or edema. Third, in some patients with heart failure, hypercapnic ventilatory response is increased [35]. The latter has been shown to be one of the distinguishing features between those heart failure patients with or without significant periodic breathing and central apnea during sleep. In individuals with increased sensitivity to carbon dioxide, the chemoreceptors elicit a large ventilatory response whenever the partial pressure of carbon dioxide rises. The consequent intense hyperventilation, by driving the P_{CO_2} below the apneic threshold, results in central apnea. Due to central apnea, P_{CO_2} rises. Therefore the cycles of hyperventilation and hypoventilation are maintained.

It is emphasized that these alterations in the negative feedback system controlling breathing are not necessarily state (sleep or wake) specific. This is why periodic breathing

TABLE 33.2 Treatment of CSA in Systolic Heart Failure

Optimization of cardiopulmonary function
Heart transplantation
Nocturnal administration of nasal supplemental oxygen
Theophylline
Acetazolamide
Positive airway pressure devices
Atrioventricular pacing devices

The above modalities including acetazolamide (unpublished observations) have been used in the treatment of central sleep apnea (CSA) in systolic heart failure. However, controlled long-term studies are necessary to determine if therapy affects the natural history (morbidity and mortality) of systolic heart failure.

may occur during both wakefulness and sleep, though most frequently during sleep. Sleep promotes periodic breathing because of both the assumption of supine position and its own specific effects. For example, in supine position and during sleep, cardiac output decreases (further prolonging arterial circulation time), and also both functional residual capacity and metabolic rate decrease enhancing the plant gain. All these alterations increase the likelihood of developing periodic breathing during sleep.

Treatment of central sleep apnea in heart failure may be achieved in several ways (Table 32.2) [1–3, 46–55]. Since heart failure is the fundamental reason for development of periodic breathing and central sleep apnea, we first maximize the therapy of heart failure. If periodic breathing persists, several therapeutic options are available (Table 32.2). Details on various treatments of central sleep apnea are discussed elsewhere [1–3].

Idiopathic Central Sleep Apnea   This is an uncommon disorder [56, 57]. Polysomnographically, it is characterized by repetitive episodes of central apnea. Patients are commonly older males and may present with complaints of restless sleep, insomnia, and/or daytime symptoms such as sleepiness and fatigue related to sleep fragmentation caused by arousals. The latter symptoms are similar to those with obstructive sleep apnea–hypopnea syndrome.

By definition, idiopathic central apnea is a diagnosis of exclusion and other causes of central apnea noted in Table 32.1 need to be ruled out. Particularly in older patients, silent congestive heart failure and central nervous system disorders such as stroke and multiple small infarctions should be excluded.

Patients with idiopathic central apnea commonly have a low arterial Pco_2 and increased hypercapnic ventilatory response [56, 57] during wakefulness. As noted earlier, a heightened ventilatory response to CO_2 facilitates development of central apnea during arousals from sleep. Though I suspect that patients with idiopathic central sleep apnea

may have a narrow difference between their eucapnic and apneic threshold Pco_2 levels, no systematic study on the relationship between prevailing Pco_2 and apneic threshold Pco_2 has been reported.

There are two studies [58, 59] on the use of acetazolamide in treatment of idiopathic central sleep apnea. In both studies, acetazolamide was effective in decreasing central apneas. Acetazolamide causes metabolic acidosis and decreases $Paco_2$. In spite of a lower (than normal) $Paco_2$, central apneas decrease. This is because the apneic threshold Pco_2 decreases more than that of $Paco_2$. This results in a widening of the two Pco_2 set points [8], which will decrease the likelihood of developing central apnea as discussed earlier. Continuous positive airway pressure devices, supplemental nasal oxygen, and theophylline are other alternatives, but no systematic studies are available. In a case report, use of a bilevel positive airway pressure device resulted in worsening of central apnea, presumably by lowering the prevailing Pco_2 below apneic threshold [60].

Idiopathic Pulmonary Arterial Hypertension   This disorder is characterized by pulmonary arterial hypertension with normal pulmonary capillary pressure. It is a disorder of unknown etiology, although most recently certain genetic factors have been found to be associated with the familial form of the idiopathic pulmonary hypertension [61].

In the more advanced hemodynamic state of this disorder, there is severe pulmonary arterial hypertension, right ventricular failure, and diminished cardiac output. In one study [62] of 20 patients with idiopathic pulmonary artery hypertension (3 men and 17 women), 6 had moderately severe sleep apnea with an apnea–hypopnea index of 37/hour resulting in desaturation. The mean $Paco_2$ did not significantly differ between the two groups, but patients with periodic breathing had significantly more hemodynamic abnormalities than those without. Presumably, diminished stroke volume and increased arterial circulation time were the underlying mechanisms in mediating periodic breathing in this disorder.

It is emphasized that in the study of Schultz et al. [62], 3 out of the 3 men, but only 3 of 17 women had periodic breathing. This gender distribution is similar to that in systolic heart failure and may have to do with gender differences in apneic threshold Pco_2 [33, 34].

For treatment of central sleep apnea in idiopathic pulmonary arterial hypertension, we recommend nocturnal supplemental nasal oxygen [62]. Interestingly, however, in one patient with idiopathic pulmonary hypertension, hypocapnia, and low cardiac output, central sleep apnea was eliminated after lung transplantation [63]. This case report emphasizes the importance of altered hemodynamics mediating periodic breathing. In contrast, in another case report [64], use of bilevel positive airway pressure therapy was associated with death; although cause and effect cannot be proved, several possibilities including worsening of central sleep apnea and reduction in cardiac output with use of the bilevel device may be speculated.

It is emphasized, however, that obstructive sleep apnea is a known cause of secondary pulmonary hypertension (for review see [65]). Therefore, in patients with pulmonary arterial hypertension, obstructive sleep apnea should be ruled out by polysomnography. If obstructive sleep apnea is present, use of a continuous positive airway pressure (CPAP) device is the treatment of choice. Pulmonary arterial hypertension may improve with effective treatment of obstructive sleep apnea with CPAP [65].

High Altitude   Recognition of periodic breathing at high altitude followed the similar observation in patients with heart failure [66]. This was not surprising, because the pattern of periodic breathing at high altitude resembles Cheyne–Stokes breathing in heart failure, with one exception. The cycle of periodic breathing is short. As noted earlier, periodic breathing in heart failure has long crescendo–decrescendo arms because of increased circulation time, a pathological feature of systolic heart failure.

The mechanism of periodic breathing at high altitude has been studied in humans. The underlying mechanism is hypoxemia that occurs at high altitude. It has been shown that hypoxemia narrows the difference between spontaneous P_{CO_2} minus apneic threshold P_{CO_2} and increases hypocapnic chemosensitivity below the apneic threshold [8]. Furthermore, the quantity of periodic breathing during sleep at high altitude tends to be more in individuals with enhanced ventilatory response to both hypercapnia and hypoxia [67]. The relation with hypercapnic ventilatory response resembles that reported in patients with heart failure [35].

In any case, at high altitude during sleep when apneic threshold is unmasked, central apnea occurs as spontaneous P_{CO_2} falls below apneic threshold. With apnea, P_{CO_2} rises and breathing resumes. As breathing increases, the consequent reduction in P_{CO_2} perpetuates the cycle.

Inhalation of supplemental oxygen and a small amount of CO_2 decreases periodic breathing. Furthermore, it has also been shown that administration of acetazolamide improves desaturation and ameliorates the symptoms of acute mountain sickness in humans at high altitude [68, 69]. As discussed earlier, acetazolamide [8] widens the difference between the two P_{CO_2} set points (in contrast to hypoxemia), resulting in improvement of periodic breathing at high altitude.

Cerebrovascular Disorders   A number of studies [70–78] have shown that patients with stroke (acute, chronic, ischemic, nonischemic) have obstructive and central sleep apneas. In regard to obstructive sleep apnea, this disorder could either precede (cause or contribute to development of) or be caused by stroke. In either case, obstructive sleep apnea may influence the outcome of patients with stroke.

In contrast to obstructive sleep apnea, central apneas are most probably caused by the stroke and, in one large study [75], central apneas decreased when patients were restudied later on.

A pattern of breathing similar to Cheyne–Stokes breathing has also been reported in patients with stroke [76–79]. This pattern of breathing appears to have no relation to site of pathology. The mechanisms of central sleep apnea and periodic breathing in stroke are not well understood, though it has been shown that both supplemental nasal oxygen and theophylline are effective therapeutic modalities. These findings are similar to those in systolic heart failure [48, 49].

Hypercapnic CSA

These heterogeneous disorders are characterized by daytime steady-state hypercapnia. According to the alveolar ventilation equation, P_{aCO_2} is proportional directly to CO_2 production and inversely to alveolar ventilation. In the disorders in this category, the rise in P_{aCO_2} is due to decreased global ventilation, and therefore the term hypoventilation is appropriate [4].

Effects of sleep on these disorders are multiple. In fact, the full impact of these disorders as manifested by pronounced hypercapnia becomes unmasked during sleep when the wakefulness drive to breathe is absent, and ventilation decreases.

Normally, with sleep onset and removal of the wakefulness drive, ventilation decreases and P_{CO_2} rises 2–4 mmHg. However, such a small physiological decrease in ventilation results in a large increase in P_{CO_2} when steady-state awake hypercapnia is already present. This is dictated by the alveolar ventilation equation and has to do with the hyperbolic relation of P_{CO_2} with alveolar ventilation [4]. Similarly, in the presence of hypercapnia, if an arousal occurs, because of the return of the wakefulness drive to breathe, ventilation increases somewhat but P_{CO_2} decreases considerably. This increases the probability of developing central apnea, if apneic threshold P_{CO_2} is also elevated. An increase in the apneic threshold and its proximity to a chronically elevated P_{CO_2} is critical and not yet proved. However, it is a likely assumption, because it has been shown that steady-state acute hypercapnia results in an elevated apneic threshold P_{CO_2} [80].

The second effect of sleep on breathing in disorders in this category has to do with the nature of the pathophysiological process. In some of the disorders in which either the pathological process involves brainstem medullary respiratory centers responsible for automatic breathing, or metabolic control is defective (e.g., due to use of opioids), central apneas may also occur during sleep, when the wakefulness drive to breath is absent.

The third effect of sleep on these disorders has to do with neurophysiological changes of REM sleep. During REM sleep, there is costal muscle atonia. Therefore, in neuromuscular disorders involving the diaphragm, which normally is the only inspiratory thoracic pump muscle active during REM sleep, airflow ceases, and thoracoabdominal tracings look like a central apnea.

The last but not the least impact of sleep in these disorders is the development of severe desaturation. This also relates to chronic steady-state hypercapnia and the hyperbolic nature of Pa_{CO_2} in relation to alveolar ventilation noted earlier. With sleep onset, ventilation decreases and P_{CO_2} rises considerably. This in turn results in severe hypoxemia, given the reciprocal relation between alveolar P_{CO_2} and P_{O_2}.

Alveolar Hypoventilation Syndromes with Normal Pulmonary Function

These central nervous system disorders are characterized by daytime hypercapnia, diminished or absent CO_2 chemosensitivity, and normal pulmonary function.

These disorders causing chronic alveolar hypoventilation could be either genetic (congenital central hypoventilation syndrome and perhaps primary hypoventilation syndrome) or acquired (a variety of brainstem and spinal cord disorders). The unique feature of all of these disorders, however, is the failure of automatic/metabolic control of breathing, which becomes manifest during sleep (Onidine's curse), when the wakefulness drive to breathe is absent. As noted earlier, during sleep breathing is controlled by the automatic/metabolic pathway. If this pathway is defective, ventilation will decrease dramatically and central apneas could occur.

Congenital central hypoventilation syndrome (CCHS), first described by Mellins et al. [81], is a rare genetic disorder [82–88] commonly associated with other neurocristopathies such as Hirschsprung's disease, which is due to segmental colonic aganglionosis. The disorder manifests itself after birth.

The genetic nature of CCHS was suspected by its familial occurrence and more importantly by the concordance in monozygotic twins. More recently, however, several groups of investigators have reported a high prevalence of heterozygous de novo mutations in the homebox gene *PHOX2B* in most (but not all) patients with CCHS [85–88]. Most of the mutations consist of an expansion of a polyalanine stretch (five to nine alanine expansions within a 20 residue polyalanine tract), with the remaining being frameshift or nonsense mutations after the *PHOX2B* homedomain. The counterparts of human CCHS in mice include the knockout mice ($PHOX2B^{-/-}$), which die during the embryonic period, and the heterozygous ($PHOX2B^{+/-}$) mice, which show decreased hypercapnic ventilatory response [89]. This phenotype could be the result of maldevelopment of chemoafferent neurons.

In addition, targeted gene inactivation of several other transcription factors in mice have been found to affect the development of specific groups of brainstem neurons involved in central breathing control [85]. These mutants may eventually help in understanding other perhaps genetic disorders of respiratory control such as central sleep apnea of prematurity, sudden infant death syndrome, and idiopathic alveolar hypoventilation syndrome, an adult disorder. One specific example involves a mutation in the transcription factor of *MafB*, which affects the development of respiratory neurons in pre-Botzinger complex [90]. As noted earlier, this group of neurons is involved in respiratory rhythmogenesis. *MafB*-deficient mice have fatal central apnea at birth.

Treatment of CCHS may be achieved by diaphragmatic pacing or mechanical ventilation by mask, via tracheostomy, or by negative pressure ventilation [82, 91–93]. If diaphragmatic pacing or negative pressure ventilation unmask or result in upper airway occlusion during sleep, tracheostomy may be necessary, although a positive airway pressure device should be the first choice. Respiratory stimulants are generally ineffective [94].

Primary (idiopathic) alveolar hypoventilation syndrome is usually a disorder of adult males diagnosed in the third or fourth decade. This disorder is characterized by chronic hypercapnia in an adult without any demonstrable neuromuscular, thoracic, pulmonary, or central nervous system pathology [95]. The mechanisms leading to chronic hypercapnia in primary alveolar hypoventilation syndrome are not understood and it could have a genetic basis.

Treatment of idiopathic alveolar hypoventilation syndrome should be individualized and sometimes two modalities are complementary. Treatment options include nocturnal oxygen, negative pressure ventilation, and noninvasive nocturnal mechanical ventilation or diaphragmatic pacing [95–101]. As noted in the treatment of CCHS, if with negative pressure ventilation or diaphragmatic pacing, upper airway occlusion occurs, positive airway pressure devices should be used; otherwise tracheostomy with mechanical ventilation may be applied [102].

Brainstem disorders result in severe hypoventilation and central apnea during sleep. This is not surprising, because central chemoreceptors and respiratory centers are located in this region. Furthermore, afferent inputs from peripheral chemoreceptors are also processed in this location. Various pathological processes such as compression, ischemia, infarct, tumor, and encephalitis (viral, bacterial, Leigh disease) involving brainstem have been associated with central sleep apnea [103–111].

It is important to emphasize that respiratory centers are bilateral and widespread. For this reason, infarcts causing apnea are usually bilateral. Unilateral infarcts rarely cause central apneas unless they are extensive, involving medullary nerves above and below medullary neuronal decussation [106].

Cervical cordotomy [112] and anterior cervical spinal artery syndrome [113] result in automatic failure of breathing during sleep (Ondine's curse). In these two conditions, the process involves the descending pathways subservient to the automatic control of breathing. In the past, spinothalamic cervical cordotomy was performed for intractable pain. These individuals later developed Ondine's curse. Anterior spinal artery syndrome results in motor palsy and dissociated sensory loss below the level of the lesion, along with bladder dysfunction. Ondine's curse occurs when the anterior cervical spine is involved bilaterally, for example, because of thrombosis of the anterior spinal artery [113].

For treatment of central apneas and hypoventilation in this category, therapy should be individualized. A number of modalities including bilevel ventilation in time mode, tracheostomy with mechanical ventilation, or diaphragm pacing could be used.

Neuromuscular Disorders This category (Table 32.1) includes a large number of neuromuscular disorders that may affect respiratory muscles (muscular dystrophies such as myotonic dystrophy, idiopathic diaphragmatic paralysis), the neuromuscular junction (myasthenia gravis), phrenic and intercostal nerves (amyotrophic lateral sclerosis), and the brainstem [114].

Depending on the site of pathology, a specific pathophysiological breathing disorder could occur during sleep. For example, with diaphragmatic involvement, REM sleep hypoventilation or polygraphically what appears to be a central apnea (discussed earlier) may occur. With involvement of pharyngeal muscles, obstructive apneas and hypopneas may predominate during sleep.

As respiratory muscle weakness progresses and daytime hypercapnia develops, sleep-related breathing disorders manifest with more severe hypercapnia, more severe desaturation, and perhaps more central apneas. Both of these two phenomena have to do with where P_{CO_2} resides on the hyperbolic curve of the alveolar ventilation equation discussed earlier.

Treatment of sleep-related breathing disorders in neuromuscular diseases should be individualized [114–118]. Two important factors determining the modality of therapy are (1) if there is impairment in rhythmogenesis due to a pathological process involving respiratory centers and (2) the presence of upper airway obstruction during sleep. Use of bilevel ventilation could be extremely helpful for patients with hypercapnia if the brainstem is not involved. If rhythmogenesis is impaired and upper airway obstruction is present during sleep, tracheostomy with assisted ventilation may become necessary.

Opioids Ventilatory depression during wakefulness is a well-known effect of opioid drugs. This ventilatory depression is characterized by decreased tidal volume and minute ventilation, increased P_{aCO_2}, and diminished hypercapnic and hypoxic ventilatory responses [119, 120].

The effects of these drugs during sleep have not been well studied. In one study [121], overnight administration of hydromorphone to normal individuals did not result in a significant change in apnea–hypopnea index. However, in a study [122] of ten subjects from a methadone maintenance program, six had a central apnea index >5/hour, and in four of these patients, the central apnea index was >10/hour. In another study [123] of three patients (BMI = 26, 36 and 51 kg/m²) who were receiving long-term opioid medications, polysomnography showed long episodes of obstructive disordered breathing and also central apneas.

The mechanisms of opioid-induced central apneas have not been studied. However, opioid-induced respiratory depression during wakefulness is characterized by reduction in hypoxic and hypercapnic ventilatory responses. It is therefore not surprising that with removal of the wakefulness drive to breathing, during sleep when rhythmic breathing is metabolically controlled, respiratory depression is manifested by more severe hypercapnia and cessation of breathing with central apneas. It is also conceivable that opioids may suppress arousal reflex that could prolong episodes of apneas, hypopneas, and periods of hypoventilation. Presence of hypercapnia could also promote central apnea by way of an increased plant gain discussed earlier.

Central Sleep Apnea in Endocrine Disorders

Central sleep apnea has been observed in two endocrine disorders, acromegaly and hypothyroidism. In both disorders, however, obstructive sleep apnea is the predominant form.

Acromegaly Several studies have reported a relatively high prevalence of sleep apnea in patients with acromegaly [124–126]. Both obstructive and central sleep apneas occur. The mechanism of obstructive sleep apnea is primarily related to excess pharyngeal soft tissue and macroglossia. In acromegaly, the viseromegaly is caused by cellular hyperplasia, excess connective tissue, and extracellular water. Obstructive sleep apnea may contribute to the high prevalence or progression of cardiovascular disease in patients with acromegaly [124]. Treatment of acromegaly with octreotide (a somatostatin analog) could result in tissue regression and improvement in sleep apnea [126–128]. However, this may take several months, and sleep apnea may persist in spite of therapy of acromegaly. A trial of CPAP should be the first approach.

Studies by Grunstein and colleagues [124, 129] show that patients with acromegaly also suffer from central sleep apnea, and this correlated with levels of hGH, IGF-1 hypercapnic ventilatory response. As discussed earlier, an enhanced ventilatory response could increase the likelihood of developing central sleep apnea.

In the study of Grunstein et al. [127], treatment of acromegaly with octreotide decreased both obstructive and central sleep apnea.

Hypothyroidism   Patients with hypothyroidism, particularly those with myxedema, may develop obstructive sleep apnea due to excess pharyngeal soft tissue, similar to patients with acromegaly [126, 130]. However, central apnea may also be observed [131] though it is much less frequent than obstructive apneas. The mechanism remains unclear.

Meanwhile, the treatment of choice is a trial of CPAP. Obstructive sleep apnea may persist after adequate therapy of hypothyroidism, and therefore therapy with CPAP should continue, though the pressure may need to be adjusted, particularly if weight loss occurs.

CSA with Obstructive Sleep Apnea

Central apneas are observed in polysomnograms of patients with obstructive sleep apnea, and also after treatment of obstructive sleep apnea with either CPAP or tracheostomy. However, there are no systematic studies. Therefore I briefly discuss primarily my personal observations.

Central apneas are frequently observed in polysomnograms of patients with obstructive sleep apnea–hypopnea syndrome. Because they are few in number, such central apneas have no clinical significance. However, I have observed them more frequently during the night of CPAP titration, particularly at sleep onset, after arousals, and in those patients who have mixed apneas. In regard to the latter, during CPAP titration, the obstructive component of the mixed apnea is eliminated by CPAP and central apneas remain. At times as CPAP level is increased, central apneas occur more frequently or lengthen. I have also observed central apneas during CPAP titration in patients with obstructive sleep apnea who have undergone uvulopalatopharyngoplasty (UPPP). However, what appears to be a central apnea could be prolonged expiration, as nasal positive pressure is readily transmitted to the oral cavity resulting in prolonged oral leak, simulating a central apnea.

I have also seen a few patients who underwent tracheostomy for treatment of obstructive sleep apnea. Post-tracheostomy central apneas may be observed [132]. Because tracheostomy is rarely used for treatment of obstructive sleep apnea, no systematic studies are available to determine the frequency and clinical significance of central apneas after tracheostomy.

CSA with Upper Airway Disorders

There are many receptors in the nose, larynx, and pharynx [133, 134] and animal studies have demonstrated that stimulation of upper airway receptors [134–136] may cause central apnea. Apnea may be produced by water, chemical, or mechanical stimulation of the receptors. Central apnea has also been produced by stimulation of the superior laryngeal nerve [137].

Human studies show that preterm infants develop central apnea in response to oropharyngeal stimulation by water or saline [138]. Systematic studies in adults, however, are not available. Suratt et al. [139] showed that, in normal adults, induced nasal obstruction resulted in central apneas, although the predominant effect was obstructive sleep apnea. Similarly, Zwillich and associates [140] showed a significant increase in the number of central apneas in normal man during nasal obstruction. In contrast, however, in seven subjects with allergic rhinitis who were studied both when asymptomatic and during exacerbation, the central apnea index did not change significantly [141], although the obstructive apnea index increased significantly, though mildly.

Further evidence of upper airway receptors mediating central apnea in humans stems from effects of high frequency (30 Hz) low-pressure ventilation aborting central apneas [142]. Similarly, application of CPAP improves central apneas from both idiopathic causes [143, 144] and due to systolic heart failure [1–3].

Based on the studies of induced nasal obstruction causing central apnea and high frequency and positive airway pressure devices eliminating this breathing disorder, it has been postulated that oropharyngeal closure may induce central apnea via upper airway receptor stimulation.

SUMMARY

Central apneas occur in many pathophysiological conditions. Depending on the cause or mechanism, central apneas may not be clinically significant. In contrast, in some disorders central apneas result in pathophysiological consequences. Under such circumstances, diagnosis and treatment of central sleep apnea may improve quality of life, morbidity, and perhaps mortality. Overall, however, we know much less about the clinical significance of central sleep apnea than obstructive apnea. I hope this writing stimulates further research in the area.

REFERENCES

1. Javaheri S. Heart failure. In: Kryger MH, Roth T, Dement WC (Eds), *Principles and Practices of Sleep Medicine*, 4th ed. Saunders, Philadelphia, 2005, pp 1208–1217.

2. Javaheri S. Central sleep apnea in congestive heart failure: prevalence, mechanisms, impacts and therapeutic options. *Respir Crit Care Med* **26**(1):44–45(2005).

3. Javaheri S. Sleep-related breathing disorders in heart failure. In: Mann DL (Ed), *Heart Failure, A Companion to Braunwald's Heart Disease*. Saunders, Philadelphia, 2004, pp 471–487.

4. Javaheri S. Determinants of carbon dioxide tension. In: Gennari X, et al. (Eds), *Acid–Base Disorders*. Marcel Dekker/CRC Press, Boca Raton, FL, 2005.

5. Dempsey JA, Skatrud JB. Fundamental effects of sleep state on breathing. *Curr Pulmonol* **9**:267–304(1988).

6. Skatrud JB, Dempsey JA. Interaction of sleep state and chemical stimuli in sustaining rhythmic ventilation. *J Appl Physiol* **55**:813–822(1983).

7. Dempsey JA, Skatrud JB. A sleep-induced apneic threshold and its consequences. *Am Rev Respir Dis* **133**:1163–1170 (1986).

8. Nakayama H, Smith CA, Rodman JR, et al. Effect of ventilatory drive on CO_2 sensitivity below eupnea during sleep. *Am J Crit Care Med* **165**:1251–1258(2002).

9. Orem J, Kubin L. Respiratory physiology: central neural control. In: Kryger MH, Roth T, Dement WC (Eds), *Principles and Practices of Sleep Medicine*, 3rd ed. Saunders, Philadelphia, 2000, pp 205–228.

10. American Heart Association. *Heart and Stroke Statistical Update—2004*. American Heart Association, Dallas, 2003.

11. Ward M. Periodic respiration. A short historical note. *Ann R Coll Surg Engl* **52**:330–334(1973).

12. Allen E, Turk JL, Murley R. *The Case Books of John Hunter FRS*, Parthenon Publishing Group, New York, 1993, pp 29–30.

13. Cheyne J. A case of apoplexy, in which the fleshy part of the heart was converted into fat. *Dublin Hosp Rep Commun* **2**:216–223(1818).

14. Stokes W. Observations on some cases of permanently slow pulse. *Dublin Q J Med Sci* **2**:73–85(1846).

15. Javaheri S, Parker TJ, Wexler L, et al. Occult sleep-disordered breathing in stable congestive heart failure. *Ann Intern Med* **122**:487–492(1995). [Erratum, *Ann Intern Med* **123**:77 (1995).]

16. Hanly PJ, Millar TW, Steljes DG, et al. Respiration and abnormal sleep in patients with congestive heart failure. *Chest* **96**:480–488(1989).

17. Javaheri S, Parker TJ, Liming JD, et al. Sleep apnea in 81 ambulatory male patients with stable heart failure: types and their prevalences, consequences, and presentations. *Circulation* **97**:2154–2159(1998).

18. Javaheri S. Sleep disorders in 100 male patients with systolic heart failure. A prospective study. *Int J Cardiol* (in press).

19. Tremel F, Pépin J-L, Veale D, et al. High prevalence and persistence of sleep apnea in patients referred for acute left ventricular failure and medically treated over 2 months. *Eur Heart J* **20**:1201–1209(1999).

20. Solin P, Bergin P, Richardson M, et al. Influence of pulmonary capillary wedge pressure on central apnea in heart failure. *Circulation* **99**:1574–1579(1999).

21. Sin DD, Fitzgerald F, Parker JD, et al. Risk factors for central and obstructive sleep apnea in 450 men and women with congestive heart failure. *Am J Respir Crit Care Med* **160**:1101–1106(1999).

22. Yasuma F, Nomura H, Hayashi H, et al. Breathing abnormalities during sleep in patients with chronic heart failure. *Jpn Circ J* **53**:1506–1510(1989).

23. Lofaso F, Verschueren P, Rande JLD, et al. Prevalence of sleep-disordered breathing in patients on a heart transplant waiting list. *Chest* **106**:1689–1694(1994).

24. Fries R, Bauer D, Heisel A, et al. Clinical significance of sleep-related breathing disorders in patients with implantable cardioverter defibrillators. *Pace* **22**:223–227(1999).

25. Staniforth AD, Kinnear WJM, Starling R, et al. Nocturnal desaturation in patients with stable heart failure. *Heart* **79**:394–399(1998).

26. Lanfranchi PA, Braghiroli A, Bosimini E, et al. Prognostic value of nocturnal Cheyne–Stokes respiration in chronic heart failure. *Circulation* **99**:1435–1440(1999).

27. Xie A, Skatrud JB, Puleo DS, et al. Apnea–hypopnea threshold for CO_2 in patients with congestive heart failure. *Am J Respir Crit Care Med* **165**:1245–1250(2002).

28. Tkacova R, Hall ML, Luie PP, et al. Left ventricular volume in patients with heart failure and Cheyne–Stokes respiration during sleep. *Am J Respir Crit Care Med* **156**:1549–1555(1997).

29. Hanly P, Zuberi N, Gray R. Pathogenesis of Cheyne–Stokes respiration in patients with congestive heart failure. Relationship to arterial P_{CO_2}. *Chest* **104**:1079–1084(1993).

30. Naughton M, Bernard D, Tam A, et al. Role of hyperventilation in the pathogenesis of central sleep apneas in patients with congestive heart failure. *Am Rev Respir Dis* **148**:330–338(1993).

31. Javaheri S, Corbett WS. Association of low Pa_{CO_2} with central sleep apnea and ventricular arrhythmias in ambulatory patients with stable heart failure. *Ann Intern Med* **128**:204–207(1998).

32. Javaheri S. Central sleep apnea and heart failure. Letter to the Editor. *Circulation* **342**:293–294(2000).

33. Zhou XS, Shahabuddin S, Zahn BR, et al. Effect of gender on the development of hypocapnic apnea/hypopnea during NREM sleep. *J Appl Physiol* **89**:192–199(2000).

34. Zhou XS, Rowley JA, Demirovic F, Diamond MP, Badr MS. Effect of testosterone on the apnea threshold in women during NREM sleep. *J Appl Physiol* **94**:101–107(2003).

35. Javaheri S. A mechanism of central sleep apnea in patients with heart failure. *N Engl J Med* **341**:949–954(1999).

36. Cherniack NS. Respiratory dysrhythmias during sleep. *N Engl J Med* **305**:325–330(1981).

37. Khoo MCK, Kronauer RE, Strohl KP, et al. Factors inducing periodic breathing in humans: a general model. *J Appl Physiol* **53**:644–659(1982).

38. Cherniack NS, Longobardo GS. Cheyne–Stokes breathing: an instability in physiologic control. *N Engl J Med* **288**:952–957(1973).

39. Carley DW, Shannon DC. A minimal mathematical model of human periodic breathing. *J Appl Physiol* **65**:1400–1409(1988).

40. Cherniack NS. Apnea and periodic breathing during sleep. *N Engl J Med* Sep **23**;341(13):985–987(1999).

41. Khoo MCK. Theoretical models of periodic breathing in sleep apnea. In: Bradley TD, Floras JS (Eds), *Implications in Cardiovascular and Cerebrovascular Disease.* Marcel Dekker, New York, 2000, pp 146, 335–384.

42. Guyton AC, Crowell JW, Moore JW. Basic oscillating mechanisms of Cheyne–Stokes breathing. *Am J Physiol* **187**:395–398(1956).

43. Younes M. The physiologic basis of central apnea and periodic breathing. *Curr Pulmonol* **10**:265–326(1989).

44. Millar TW, Hanly PJ, Hunt B, et al. The entrainment of low frequency breathing periodicity. *Chest* **98**:1143–1148(1990).

45. Hall MJ, Xie A, Rutherford R, et al. Cycle length of periodic breathing in patients with and without heart failure. *Am J Respir Crit Care Med* **154**:376–381(1996).

46. Javaheri S. Treatment of central sleep apnea in heart failure. *Sleep* **23**:S224–S227 (2000).

47. Javaheri S, Abraham WT, Brown C, Nishiyama H, Giesting R, Wagoner LE. Prevalence of obstructive sleep apnea and periodic limb movement in 45 subjects with heart transplantation. *Eur Heart J* **25**:260–266(2004).

48. Javaheri S. Pembrey's dream: the time has come for a long-term trial of nocturnal supplemental nasal oxygen to treat central sleep apnea in congestive heart failure. *Chest* **123**:322–325(2003).

49. Javaheri S, Parker TJ, Wexler L, et al. Effect of theophylline on sleep-disordered breathing in heart failure. *N Engl J Med* **335**:562–567(1996).

50. Javaheri S. Effects of continuous positive airway pressure on sleep apnea and ventricular irritability in patients with heart failure. *Circulation* **101**:392–397(2000).

51. Naughton MT, Liu PP, Benard DC, et al. Treatment of congestive heart failure and Cheyne–Stokes respiration during sleep by continuous positive airway pressure. *Am J Respir Crit Care Med* **151**:92–97(1995).

52. Sin DD, Logan AG, Fitzgerald FS, et al. Effects of continuous positive airway pressure on cardiovascular outcomes in heart failure patients with and without Cheyne–Stokes respiration. *Circulation* **102**:61–66(2000).

53. Teschler H, Döhring J, Wang YM, et al. Adaptive pressure support servo-ventilation. *Am J Respir Crit Care Med* **164**:614–619(2001).

54. Javaheri S. Heart failure and sleep apnea: emphasis on practical therapeutic options. Chest Clinics of North America. *Clin Chest Med* **24**:207–222(2003).

55. Garrigue S, Bordier P, Jaïs P, et al. Benefit of atrial pacing in sleep apnea syndrome. *N Engl J Med* **346**:404–412(2002).

56. Bradley TD, McNicholas WT, Rutherfored R, Popkin J, Zamel N, Phillipson EA. Clinical and physiologic heterogeneity of the central sleep apnea syndrome. *Am Rev Respir Dis* **134**:217–221(1986).

57. Bradley TD, Phillipson EA. Central sleep apnea. *Clin Chest Med* **13**:493–505(1992).

58. White DP, Zwillich CW, Pickett CK, Douglas NJ, Findley LJ, Weil JV. Central sleep apnea; improvement with acetazolamide therapy. *Arch Intern Med* **142**:1816–1819(1982).

59. DeBacker WA, Verbraecken J, Willemen M, Wittesaele W, DeCock W, deHeyning PV. Central apnea index decreases after prolonged treatment with acetazolamide. *Am J Respir Crit Care Med* **151**:87–91(1995).

60. Hommura F, Nishimura M, Oguri M, Makita H, Hosokaw K, Saito H, Miyamoto K, Kawakami Y. Continuous versus bilevel positive airway pressure in patient with idiopathic central sleep apnea. *Am J Respir Crit Care Med* **155**:1482–1485(1997).

61. Farber HW, Loscalzo J. Pulmonary arterial hypertension. *N Engl J Med* **351**:1655–1665(2004).

62. Schulz R, Baseler G, Ghofrani HA, Grimminger F, Olschewski H, Seeger W. Nocturnal periodic breathing in primary pulmonary hypertension. *Eur Respir J* **19**:658–663(2002).

63. Schulz R, Fegbeutel C, Olschewski H, Rose F, Schafers HJ, Seager W. Reversal of nocturnal periodic breathing in primary pulmonary hypertension after lung transplantation. *Chest* **125**:344–347(2004).

64. Shiomi T, Guilleminault C, Sasanabe R, Oki Y, Hasegawa R, Otake K, Banno K, Usui K, Meakawa M, Kanayama H, Takahashi R, Kobayashi T. Primary pulmonary hypertension with central sleep apnea—sudden death after bilevel positive airway pressure therapy. *Jpn Circ J* **64**:723–726(2000).

65. Young T, Javaheri S. Systemic and pulmonary hypertension in obstructive sleep apnea. In: Kryger MH, Roth T, Dement WC (Eds), *Principles and Practices of Sleep Medicine*, 4th ed. Saunders, Philadelphia, 2005.

66. Weil JV. Sleep at high altitude. In: Kryger MH, Roth T, Dement WC (Eds), *Principles and Practices of Sleep Medicine*, 3rd ed. Saunders, Philadelphia, 2000, pp 204–253.

67. White DP, Gleeson K, Pickett CK, Rannels AM, Cymerman A, Weil JV. Altitude acclimatization: influence on periodic breathing and chemoresponsiveness during sleep. *J Appl Physiol* **63**:401–412(1987).

68. Sutton JR, Houstion CS, Mansell AL, McFadden MD, Rigg JRA, Powles ACP. Effect of acetazolamide on hypoxemia during sleep at high altitude. *N Engl J Med* **301**:1329–1331(1979).

69. Greene MK, Kerr AM, McIntosh IB, Prescott RJ. Acetazolamide in prevention of acute mountain sickness: a double-blind controlled cross-over study. *Br Med J* **283**:811(1981).

70. Yaggi H, Mohsenin V. Obstructive sleep apnoea and stroke. *Lancet Neurol* **3**:333–342(2004).

71. Hermann DM, Bassetti CL. Sleep-disordered breathing and stroke. *Curr Opin Neurol* **16**:87–90 (2003).

72. Yaggi H, Mohsenin V. Sleep-disordered breathing and stroke. *Clin Chest Med* **24**:223–237(2003).

73. Dyken ME, Somers VD, Yamada T, Ren Z-Y, Zimmerman MB. Investigating the relationship between stroke and obstructive sleep apnea. *Stroke* **27**:401–407(1996).

74. Shahar E, Whitney CW, Redline S, Lee ET, Newman AB, Nieto FJ, et al. Sleep-disordered breathing and cardiovascular disease: cross-sectional results of the Sleep Heart Health Study. *Am J Respir Crit Care Med* **163**:19–25(2001).

75. Parra O, Arboix A, Bechich S, Garcia-Eroles L, Montserrat JM, López JA, et al. Time course of sleep-related breathing

disorders in first-ever stroke or transient ischemic attack. *Am J Respir Crit Care Med* **161**:375–380(2000).

76. Brown HW, Plum F. The neurologic basis of Cheyne–Stokes respiration. *Am J Med* **30**:849–860(1961).

77. Nachtmann A, Siebler M, Rose G, Sitzer M, Steinmetz H. Cheyne–Stokes respiration in ischemic stroke. *Neurology* **45**:820–821(1995).

78. North JB, Jennett S. Abnormal breathing patterns associated with acute brain damage. *Arch Neurol* **31**:338–344(1974).

79. Heyman A, Birchfield RI, Sieker HO. Effects of bilateral cerebral infarction on respiratory center sensitivity. *Neurology* **8**:694–700(1958).

80. Boden AG, Harris MC, Parkes MJ. Apneic threshold for CO_2 in the anesthetized rat: fundamental properties under steady-state conditions. *J Appl Physiol* **85**:898–907 (1998).

81. Mellins RB, Balfour HH, Turino GM, Winters RW. Failure of automatic control of ventilation (Ondine's curse). *Medicine* **49**:487–504(1970).

82. Vanderlaan M, Holbrook CR, Wang M, Tuell A, Gozal D. Epidemiologic survey of 196 patients with congenital central hypoventilation syndrome. *Pediatr Pulmonol* **37**:217–229 (2004).

83. Trochet D, O'Brien LM, Gozal D, Trang H, Nordenskjöld A, Laudier B, et al. *PHOX2B* genotype allows for prediction of tumor risk in congenital central hypoventilation syndrome. *Am J Hum Genet* **76**:421–426(2005).

84. Trang H, Dehan M, Beaufils F, Zaccaria I, Amiel J, Gaultier C. The French congenital central hypoventilation syndrome registry: general date, phenotype, and genotype. *Chest* **127**:72–79(2005).

85. Blanchi B, Sieweke MH. Mutations of brainstem transcription factors and central respiratory disorders. *Trends Mol Med* **11**:23–30(2005).

86. Amiel J, et al. Polyalanine expansion and frameshift mutations of the paired-like homeobox gene *PHOX2B* in congenital central hypoventilation syndrome. *Nat Genet* **33**:459–461 (2003).

87. Sasaki A, et al. Molecular analysis of congenital central hypoventilation syndrome. *Hum Genet* **114**:22–26(2003).

88. Weese-Mayer DE, et al. Idiopathic congenital central hypoventilation syndrome: analysis of genes pertinent to early autonomic nervous system embryologic development and identification of mutations in *PHOX2B*. *Am J Med Genet* **123**:267–278(2003).

89. Dauger S, et al. *PHOX2B* controls the development of peripheral chemoreceptors and afferent visceral pathways. *Development* **130**:6635–6642(2003).

90. Blanchi B, et al. *MafB* deficiency causes defective respiratory rhythinogenesis and fatal central sleep apnea at birth. *Nat Neurosci* **6**:1091–1100(2003).

91. Cirignotta F, Schiavina M, Mondini S, Lugaresi A, Fabiani A, Cortelli P, Lugaresi E. Central alveolar hypoventilation (Ondine's curse) treated with negative pressure ventilation. *Monaldi Arch Chest Dis* **51**:22–26(1996).

92. Flageole H, Adolph VR, Davis GM, Laberge JM, Nguyen LT, Guttman FM. Diaphragmatic pacing in children with conge-

nital central alveolar hypoventilation syndrome. *Surgery* **118**:25–28(1995).

93. Garrido-Garcia H, Alvarez JM, Escribono PM, La Banda F, Cambarrutta C, Gracia ME, Labarta C, Arroyo O, De la Cruz FS, Gutierrrez R, Moreno JG. Treatment of chronic ventilatory failure using a diaphragmatic pacemaker. *Spinal Cord* **36**:310–314(1998).

94. Oren J, Newth CJL, Hung CE, Brouillette RT, Bachand RT, Shannon DC. Ventilatory effects of almitrine bismesylate in congenital central hypoventilation syndrome. *Am Rev Respir Dis* **134**:917–919(1986).

95. Reichel J. Primary alveolar hypoventilation. *Clin Chest Med* **1**:119–124(1980).

96. Bubis MJ, Anthonisen NR. Primary alveolar hypoventilation treated by nocturnal administration of O_2. *Am Rev Respir Dis* **118**:947–953(1978).

97. McNicholas WT, Carter JL, Rutherford R, Zamel N, Phillipson EA. Beneficial effect of oxygen on primary alveolar hypoventilation with central sleep apnea. *Am Rev Respir Dis* **125**:773–775(1982).

98. Guilleminault C, Stoohs R, Schneider H, Podszus T, Peter JH, von Wichert P. Central alveolar hypoventilation and sleep: treatment by intermittent positive-pressure ventilation through nasal mask in an adult. *Chest* **96**:1210–1212 (1989).

99. Man GCW, Jones RL, MacDonald GF, King EG. Primary alveolar hypoventilation managed by negative-pressure ventilators. *Chest* **76**:219–221(1979).

100. Garay SM, Turino GM, Goldring RM. Sustained reversal of chronic hypercapnia in patients with alveolar hypoventilation syndromes. *Am J Med* **70**:269–274(1981).

101. Farmer WC, Glenn WW, Gee JBL. Alveolar hypoventilation syndrome: studies of ventilatory control in patients selected for diaphragm pacing. *Am J Med* **64**:39–49(1978).

102. Glenn WWL, Gee JBL, Cole DR, Farmer WC, Shaw RK, Beckman CB. Combined central alveolar hypoventilation and upper airway obstruction. *Am J Med* **64**:50–60(1978).

103. Cohn JE, Kuida H. Primary alveolar hypoventilation associated with western equine encephalitis. *Ann Intern Med* **56**:633–644(1962).

104. White DP, Miller F, Erickson RW. Sleep apnea and nocturnal hypoventilation after western equine encephalitis. *Am Rev Respir Dis* **127**:132–133(1983).

105. Devereaux MW, Keane JR, Davis RL. Automatic respiratory failure associated with infarction of the medulla: report of two cases with pathologic study of one. *Arch Neurol* **29**:46–52(1973).

106. Levin BE, Margolis G. Acute failure of automatic respirations secondary to a unilateral brainstem infarct. *Ann Neurol* **1**:583–586(1977).

107. Kraus J, Heckmann JG, Druschky A, Erbguth F, Neundörfer B. Ondine's curse in association with diabetes insipidus following transient vertebrobasilar ischemia. *Clin Neurol Neurosurg* **101**:196–198(1999).

108. Manning HL, Leiter JC. Respiratory control and respiratory sensation in a patient with a ganglioglioma within the dorsocaudal brain stem. *Am J Respir Crit Care Med* **161**:2100–2106(2000).

109. Cummiskey J, Guilleminault C, Davis R, Duncan K, Golden J. Automatic respiratory failure: sleep studies and Leigh's disease. *Neurology* **37**:1876–1878(1987).

110. Schulz R, Fegbeutel C, Althoff A, Traupe H, Grimminger F, Seeger W. Central sleep apnoea and unilateral diaphragmatic paralysis associated with vertebral artery compression of the medulla oblongata. *J Neurol* **250**:503–505(2003).

111. Yglesias A, Narbona J, Vanaclocha V, Artieda J. Chiari type 1 malformation, glossopharyngeal neuralgia and central sleep apnea in a child. *Child Neurol* **38**:1126–1130(1996).

112. Severinghaus JW, Mitchell RA. Onidine's curse: failure of respiratory center automacity while awake. *Clin Res* **10**:122(1962).

113. Manconi M, Mondini S, Fabiani A, Rossi P, Ambrosetto P, Cirignotta F. Anterior spinal artery syndrome complicated by the Ondine curse. *Arch Neurol* **60**:1787–1790(2003).

114. George CFP. Neuromuscular disorders. In: Kryger MH, Roth T, Dement WC (Eds), *Principles and Practices of Sleep Medicine*, 3rd ed. Saunders, Philadelphia, 2000, pp 1087–1092.

115. Kerby GR, Mayer LS, Pingleton SK. Nocturnal positive pressure ventilation via nasal mask. *Am Rev Respir Dis* **135**:738–740(1987).

116. Ellis ER, Bye PTP, Bruderer JW, Sullivan CE. Treatment of respiratory failure during sleep in patients with neuromuscular disease: positive-pressure ventilation through a nose mask. *Am Rev Respir Dis* **135**:148–152(1987).

117. Newson-Davis IC, Lyall RA, Leigh PN, Moxham J, Goldstein LH. The effect of non-invasive positive pressure ventilation (NIPPV) on cognitive function in amyotrophic lateral sclerosis: a prospective study. *J Neurol Psychiatry* **71**:482–487(2001).

118. Goldstein RS, Molotiu N, Skrastins R, Long S, De Rosie J, Contreras M, Popkin J, Rutherford R, Phillipson EA. Reversal of sleep-induced hypoventilation and chronic respiratory failure by nocturnal negative pressure ventilation in patients with restrictive ventilatory impairment. *Am Rev Respir Dis* **135**:1049–1055(1987).

119. Santiago TV, Edelman NH. Opioids and breathing. *J Appl Physiol* **59**:1675–1685(1985).

120. Weil JV, McCullough RE, Kline JS, Sodal IE. Diminished ventilatory response to hypoxia and hypercapnia after morphine in normal man. *N Engl J Med* **292**:1103–1106(1975).

121. Robinson RW, Zwillich CW, Bixler EO, Cadieux RJ, Kales A, White DP. Effects of oral narcotics on sleep-disordered breathing in healthy adults. *Chest* **91**:197–203(1987).

122. Teichtahl H, Prodromidis A, Miller B, Cherry G, Kronborg I. Sleep-disordered breathing in stable methadone programme patients: a pilot study. *Addiction* **96**:395–403(2001).

123. Fareny RJ, Walker JM, Cloward TV, Rhondeau S. Sleep-disordered breathing associated with long-term opioid therapy. *Chest* **123**:632–639(2003).

124. Grunstein RR, Ho KKY, Sullivan CE. Sleep apnea in acromegaly. *Ann Intern Med* **115**:527–532(1991).

125. Colas A, Fergone D, Marzullo P, Lombardi G. Systemic complications of acromegaly: epidemiology, pathogenesis, and management. *Endocr Rev* **25**:102–152(2004).

126. Rosenow F, McCarthy V, Caruso AC. Sleep apnea in endocrine diseases. *J Sleep Res* **7**:3–11(1998).

127. Grunstein RR, Ho KKY, Sullivan CE. Effect of octreotide, a somatostatin analog, on sleep apnea in acromegaly. *Ann Intern Med* **121**:478–487(1994).

128. Herrmann BL, Wessendorf TE, Ajaj W, Kahlke S, Teschler H, Mann K. Effects of octreotide on sleep apnea and tongue volume (magnetic resonance imaging) in patients with acromegaly. *Eur J Endocrinol* **151**:309–315(2004).

129. Grunstein RR, Ho KKY, Berthon-Jones M, Stewart D, Sullivan CE. Central sleep apnea is associated with increased ventilatory response to carbon dioxide and hypersecretion of growth hormone in patients with acromegaly. *Am J Respir Crit Care Med* **150**:496–502(1994).

130. Rajagopal KR, Abbrecht PH, Derderian SS, Pickett C, Hofeldt F, Tellis CJ, Zwillich CW. Obstructive sleep apnea in hypothyroidism. *Ann Intern Med.* 101:491–494(1984).

131. Meyrier A. Central sleep apnea in hypothyroidism. *Am Rev Respir Dis* **127**:504–507(1983).

132. Fletcher EC. Recurrence of sleep apnea syndrome following tracheostomy. A shift from obstructive to central sleep apnea. *Chest* **96**:205–209(1969).

133. Widdicombe J. Airway receptors. *Respir Physiol* **125**:3–15(2001).

134. James JEA, Daly MB. Nasal reflexes. *Proc R Soc Med* **62**:1287–1293(1969).

135. Storey AT, Johnson P. Laryngeal receptors initiating apnea in lamb. *Exp Neurol* **47**:42–55(1975).

136. Harms CA, Zeng Y-J, Smith CA, Vidruk EH, Dempsey JA. Negative pressure-induced deformation of the upper airway causes central apnea in awake and sleeping dogs. *J Appl Physiol* **80**:1528–1539(1996).

137. Lawson EE. Prolonged central respiratory inhibition following reflex-induced apnea. *J Appl Physiol: Respirat Environ Exercise Physiol* **50**:874–879(1981).

138. Davis AM, Koenig JC, Thack BT. Upper airway chemoreflex responses to saline and water in preterm infants. *J Appl Physiol* **64**:1412–1420(1988).

139. Suratt PM, Turner BL, Wilhoit SC. Effect of intranasal obstruction on breathing during sleep. *Chest* **90**:324–329(1986).

140. Zwillich CW, Pickett C, Hanson FN, Weil JV Jr. Disturbed sleep and prolonged apnea during nasal obstruction in normal man. *Am Rev Respir Dis* **124**:158–160(1981).

141. McNicholas WT, Tavlo S, Cole P, Zamel N, Rutherford R, Griffin D, Phillipson EA. Obstructive apneas during sleep in patients with seasonal allergic rhinitis. *Am Rev Respir Dis* **126**:625–628(1982).

142. Henke KG, Sullivan CE. Effects of high-frequency pressure waves applied to upper airway on respiration in central apnea. *J Appl Physiol* **73**:1141–1145(1992).

143. Issa FG, Sullivan CE. Reversal of central sleep apnea using nasal CPAP. *Chest* **90**:165–171(1986).

144. Hoffstein V, Slutsky AS. Central sleep apnea reversed by continuous positive airway pressure. *Am Rev Respir Dis* **135**:1210–1212(1987).

34

OBESITY HYPOVENTILATION SYNDROME

JOHN G. PARK

Mayo Clinic, Rochester, Minnesota

INTRODUCTION

Obesity hypoventilation syndrome (OHS) is broadly defined as hypercapnia during wakefulness in obese persons. The American Academy of Sleep Medicine has incorporated the definition of OHS into a broader term of sleep hypoventilation syndrome [1].

Previous nomenclature for OHS, also commonly referred to as pickwickian syndrome, includes obesity cardiopulmonary syndrome and postural syndrome. The term "pickwickian" refers to a character in Charles Dickens' novel *The Posthumous Papers of the Pickwick Club*; the character, Joe, is described as a "wonderfully fat boy" who is hypersomnolent, edematous, and "red-faced" (suggestive of being polycythemic) [2]. Although the term "pickwickian" was first alluded to in 1889 by Heath in reference to an obese patient who probably had obstructive sleep apnea [2], the first descriptions of OHS as it is currently defined were reported by Auchincloss and co-workers and Sieker and colleagues in 1955 [3]. Subsequently, Burwell et al. [3] coined the term "pickwickian syndrome" in 1956, referring to those with clinical features of marked obesity, hypersomnolence, periodic respiration, polycythemia, cyanosis, and right ventricular hypertrophy and failure.

The prevalence of OHS is from 5% to 31% of obese adults, and it is uncommon but possible in obese children [4–8]. Common associated clinical findings include pulmonary and systemic hypertension, hypersomnolence, obstructive sleep apnea, hypoxemia, polycythemia, peripheral edema, and right and left heart failure ("obesity cardiomyopathy") [3, 4, 8–10]. Not surprisingly, patients with OHS require increased health care resources and have increased mortality compared with body mass index (BMI) matched controls without hypoventilation [8, 11].

PATHOPHYSIOLOGY

The exact pathophysiologic mechanisms of OHS have yet to be fully elucidated, but at least four different factors may influence the development of OHS. The first factor is morbid obesity, defined as a BMI $\geq 35 \, \text{kg/m}^2$. Numerous studies, however, suggest that weight alone does not correlate with OHS, and, furthermore, OHS is somewhat uncommon among obese persons [12, 13]. Nowbar et al. [8], however, have shown that the incidence of OHS does increase with further increase in weight. Regardless, obesity increases metabolic demand because of the increased muscle work required to move the excess weight. This increased metabolic demand results in increased oxygen consumption and carbon dioxide production. Thus, in obese persons, minute ventilation is increased to maintain normocapnia.

The required increase in minute ventilation may be hampered by mechanical limitation, the second factor contributing to overall hypoventilation. To adapt to the increased work of breathing due to increased elastic and nonelastic resistance, respiratory rate increases and tidal volume decreases, both in patients with OHS and in otherwise normal subjects [14]. This adaptive mechanism is limited in

Sleep: A Comprehensive Handbook, Edited by T. Lee-Chiong.
Copyright © 2006 John Wiley & Sons, Inc.

obese persons because of their decreased expiratory reserve volume, which is then exaggerated in the supine position. The decrease in lung volume is partly a result of an elevated diaphragm due to truncal obesity [15]. Thus, because of increased resistive load and limited lung volume, the total work of breathing in patients with OHS has been shown to be 3 times that in normal-weight subjects and 1.3 times that in subjects with simple obesity (obesity without OHS) [15]. These results have led to the proposal that rather than continue to increase energy expenditure in breathing, patients with OHS tend to breathe with much lower tidal volume [13]. A decrease in tidal volume results in increased dead-space ventilation (airflow in regions of the airway not involved in gas exchange), which then worsens ventilation–perfusion mismatch [13]. These adaptive measures ultimately result in poor ventilation and further increase in Pa_{CO_2}.

The standard equation for Pa_{CO_2} is

$$Pa_{CO_2} = K \frac{\dot{V}_{CO_2}}{\dot{V}_A}$$

where K is a constant, $\dot{V}_{CO_2}$ represents carbon dioxide production, and $\dot{V}_A$ represents alveolar ventilation. This equation essentially states that arterial carbon dioxide is dependent on production of carbon dioxide and its elimination by effective ventilation. Thus obese persons are at risk for hypercapnia as a result of both increased carbon dioxide production and decreased ventilation.

The third factor contributing to OHS is blunted central chemoreceptor response to hypercapnia and hypoxemia. Leech et al. [16] showed that subjects with OHS can voluntarily hyperventilate to normalize their Pa_{CO_2}. Furthermore, progesterone, a respiratory stimulant, was shown to be efficacious in normalizing or at least decreasing Pa_{CO_2} [13]. These findings suggest that mechanical limitation is not the sole cause of hypercapnia. Others have shown that persons with OHS have decreased ventilatory drive in response to hypoxia and hypercapnia (one-sixth and one-third of normal persons, respectively) [13]. To show that hypercapnia was due to decreased neural output, rather than mechanically limited inadequate response to adequate neural output, diaphragm activity in response to hypercapnia was assessed. When subjects were exposed to hypercapnic conditions, those with OHS had decreased diaphragmatic activity compared with normocapnic obese controls [13].

The fourth factor contributing to OHS may be the coexistence of obstructive sleep apnea (OSA). Some have suggested that OHS is a severe form of OSA and that, due to repeated hypoventilation during apneic phases, OSA is a causative factor in the development of OHS. Other studies, however, have failed to show that OHS is related to the frequency, number, duration, or type of episodes of apnea or hypopnea [13]. Furthermore, up to 25% of patients with OHS do not have OSA [7]. However, because OSA can be treated effectively with nasal continuous positive airway pressure (nCPAP) devices, which improves hypercapnia in some patients, OSA may indeed contribute to the development of daytime hypercapnia in certain persons [17, 18]. Nevertheless, hypercapnia does not respond to nCPAP or tracheostomy in all patients, so OSA most likely is not the sole cause of OHS [7, 18].

In summary, the true pathophysiologic mechanisms of OHS may vary from person to person. Depending on the person's susceptibility, one or all of the abovementioned factors may contribute to or ultimately result in chronic hypoventilation in the setting of obesity. Effective therapy, therefore, may depend on correction of multiple contributing factors.

EVALUATION

In evaluating a patient with hypercapnia, multiple causes must be considered before attributing the cause to OHS [6]. Among the many possible causes of hypercapnia are problems with ventilatory control, anatomical defects, and electrolyte abnormalities (Table 34.1). Thus a complete medical history, a physical examination, and an arterial blood gas determination should establish a foundation from which to guide the physician in further testing of the patient. For example, if the history indicates the patient is a lifelong smoker with considerable dyspnea on exertion and chronic cough, pulmonary function testing and chest radiography may be most appropriate to evaluate for possible chronic obstructive pulmonary disease. Likewise, if the examination suggests paradoxical respiration, further neurologic evaluation may be necessary to investigate the reason for diaphragmatic paralysis. Guided by the initial evaluation, pulmonary function testing may help assess the functional capacity of the lungs, radiologic assessment may identify structural abnormalities, blood chemistries may identify electrolyte abnormalities, polysomnography will identify sleep-disordered breathing, and ventilatory response to hypercapnia and hypoxemia will help assess the integrity of the respiratory control system. In addition, further neurologic or rheumatologic evaluation may be necessary, depending on the clinical suspicion.

THERAPY

Several different courses of therapy have been used to control OHS. Progesterone has been tried with some short-term success in the past [13], but, because of lack of proven long-term efficacy and potential adverse effects, no pharmacologic agents are currently recommended for the treatment of OHS.

TABLE 34.1 Causes of Chronic Hypercapnia

Mechanical limitations
  Kyphoscoliosis
  Chronic obstructive lung disease
  Advanced interstitial lung disease
  Postpneumonectomy
  Post-thoracoplasty
  Myopathies
    Muscular dystrophies
    Myotonic dystrophies
    Acid-maltase deficiency
  Obesity hypoventilation syndrome
Neuropathic limitations
  Myasthenia gravis
  Eaton–Lambert syndrome
  Bilateral diaphragm paralysis
  Spinal cord injury
  Motor neuron disease (e.g., amyotrophic lateral sclerosis)
  Peripheral neuropathies (e.g., Guillain–Barré syndrome)
Central control limitations
  Primary alveolar hypoventilation
  Cerebrovascular accident
  Central nervous system neoplasm
  Arnold–Chiari malformation
  Bulbar poliomyelitis
  Neurosarcoidosis
  Carotid body dysfunction or trauma
  Prolonged hypoxia
  Obesity hypoventilation syndrome
  Drugs (e.g., narcotics, sedatives)
Metabolic causes
  Hypothyroidism/myxedema
  Electrolyte abnormalities (e.g., hypokalemia, hypophosphatemia)
  Chronic metabolic alkalosis

Initially, nCPAP was shown to be effective in decreasing daytime P_{CO_2} in some patients with OHS [17]. Other noninvasive positive pressure ventilators have become the preferred mode of respiratory support since their development. Use of such noninvasive ventilators, such as biphasic positive airway pressure, has been shown to be more effective than nCPAP in improving hypercapnia, edema, hypersomnolence, and dyspnea [18].

Some patients, however, may not respond completely to noninvasive ventilation. The most consistent benefit appears to result from surgical weight loss. According to the National Institutes of Health consensus statement in 1991 regarding morbid obesity, patients with a $BMI \geq 35\,kg/m^2$ and an obesity-related comorbid condition or those with a $BMI \geq 40\,kg/m^2$ without a comorbid condition are recommended for surgical treatment for obesity [19]. Surgery for weight loss has been shown to normalize lung volumes, normalize P_{CO_2} and Pa_{O_2}, significantly improve OSA, improve pulmonary hypertension, and improve

polycythemia [7, 12]. The benefits of such drastic measures, however, must be considered against the increased morbidity associated with surgery in obese persons.

Studies are just emerging regarding the role of leptin resistance or deficiency in both obesity and OHS, and its potential role in the treatment of OHS. Leptin is a 16 kD protein produced by white adipose tissue that may have a role in appetite suppression and respiratory stimulation [20]. Animal studies suggest it may have a role in the treatment of OHS [20], but more studies are needed before these findings can be translated into effective medical therapy.

REFERENCES

1. Anonymous. Sleep-related breathing disorders in adults: recommendations for syndrome definition and measurement techniques in clinical research. The Report of an American Academy of Sleep Medicine Task Force. *Sleep* **22**(5):667–689(1999).

2. Lavie P. Nothing new under the moon. Historical accounts of sleep apnea syndrome. *Arch Intern Med* **144**(10):2025–2028(1984).

3. Burwell C, Robin E, et al. Extreme obesity associated with alveolar hypoventilation—a pickwickian syndrome. *Am J Med* **211**:811–818(1956).

4. Alexander J, Amad K, et al. Observations on some clinical features of extreme obesity, with particular reference to cardiorespiratory effects. *Am J Med* **32**:512–524(1962).

5. Riley D, Santiago T, et al. Complications of obesity-hypoventilation syndrome in childhood. *Am J Dis Child* **130**:671–674(1976).

6. Glauser FL, Fairman RP, et al. The causes and evaluation of chronic hypercapnea. *Chest* **91**(5):755–759(1987).

7. Sugerman HJ, Fairman RP, et al. Long-term effects of gastric surgery for treating respiratory insufficiency of obesity. *Am J Clin Nutr* **55**(2 Suppl):597S–601S (1992).

8. Nowbar S, Burkart KM, et al. Obesity-associated hypoventilation in hospitalized patients: prevalence, effects, and outcome. *Am J Med* **116**(1):1–7(2004).

9. Alpert MA. Obesity cardiomyopathy: pathophysiology and evolution of the clinical syndrome. *Am J Med Sci* **321**(4):225–236(2001).

10. Kessler R, Chaouat A, et al. The obesity-hypoventilation syndrome revisited: a prospective study of 34 consecutive cases. *Chest* **120**(2):369–376(2001).

11. Berg G, Delaive K, et al. The use of health-care resources in obesity-hypoventilation syndrome. *Chest* **120**(2):377–383(2001).

12. Emirgil C, Sobol B. The effects of weight reduction on pulmonary function and the sensitivity of the respiratory center in obesity. *Am Rev Respir Dis* **108**:831–842(1973).

13. Luce JM. Respiratory complications of obesity. *Chest* **78**(4):626–631(1980).

14. McIlroy M, Eldridge F, et al. The effect of added elastic and non-elastic resistances on the pattern of breathing in normal subjects. *Clin Sci* **15**:336–344(1956).

15. Kopelman P. Sleep apnoea and hypoventilation in obesity. *Int J Obes* **16**(Suppl 2):S37–S42(1992).

16. Leech J, Onal E. et al. Voluntary hyperventilation in obesity hypoventilation. *Chest* **100**(5):1334–1338(1991).

17. Sullivan CE, Berthon-Jones M, et al. Remission of severe obesity-hypoventilation syndrome after short-term treatment during sleep with nasal continuous positive airway pressure. *Am Rev Respir Dis* **128**(1):177–181(1983).

18. Masa JF, Celli BR, et al. The obesity hypoventilation syndrome can be treated with noninvasive mechanical ventilation. *Chest* **119**(4):1102–1107(2001).

19. Anonymous. NIH conference. Gastrointestinal surgery for severe obesity. Consensus Development Conference Panel. *Ann Intern Med* **115**(12):956–961(1991).

20. Fitzpatrick M. Leptin and the obesity hypoventilation syndrome: a leap of faith? *Thorax* **57**(1):1–2(2002).

ADDITIONAL READING

Barrera F, Hillyer P, et al. The distribution of ventilation, diffusion, and blood flow in obese patients with normal and abnormal blood gases. *Am Rev Respir Dis* **108**:819–830(1973). This article reviews the mechanism of ventilation/perfusion mismatch resulting in hypoxemia and hypercapnia in those with OHS.

Martin TJ, Sanders MH. Chronic alveolar hypoventilation: a review for the clinician. *Sleep* **18**(8):617–634(1995). This article presents a comprehensive review of causes of chronic hypoventilation and separates the causes into "can't breathe" and "won't breathe" categories. It also reviews evaluation and treatment of chronic hypoventilation.

Zwillich CW, Sutton FD, et al. Decreased hypoxic ventilatory drive in the obesity-hypoventilation syndrome. *Am J Med* **59**:343–348(1975). This article discusses the details of decreased hypoxic and hypercapnic respiratory drive in those with OHS.

35

CARDIOVASCULAR COMPLICATIONS OF OBSTRUCTIVE SLEEP APNEA

PETER Y. HAHN, LYLE J. OLSON, AND VIREND K. SOMERS

Mayo Clinic College of Medicine, Rochester, Minnesota

INTRODUCTION

Cardiovascular disease is highly prevalent in the United States, affecting 23% of the population. It is a significant source of decreased productivity, health care expenditure, and impaired quality of life. Most importantly, it results in considerable morbidity and mortality with approximately 38.5% of all deaths in the United States attributable to cardiovascular disease. There is growing awareness that obstructive sleep apnea (OSA) may be an important and treatable risk factor for cardiovascular disease (Table 35.1). Cross-sectional data show that patients with OSA have increased rates of hypertension, ischemic heart disease, heart failure, and stroke. Recent studies suggest that OSA may act through multiple mechanisms, including sympathetic activation, inflammation, endothelial dysfunction, disordered coagulation, metabolic dysregulation, and oxidative stress (Table 35.2), to elicit cardiovascular dysfunction and disease.

PHYSIOLOGIC EFFECTS OF OSA

In general, sleep is characterized by a state of cardiovascular "relaxation." A decline in blood pressure (BP), heart rate (HR), systemic vascular resistance (SVR), and cardiac output (CO) is commonly observed with sleep onset. With deeper levels of NREM sleep, these measures decline further and likely reflect a further decrease in sympathetic activity. During REM sleep, however, sympathetic activity may increase resulting in lability of heart rate and blood pressure, both of which can approach waking levels. In OSA, this state of cardiovascular relaxation is disrupted. Hypoxemia, hypercapnia, increased sympathetic activity, and dramatic swings in intrathoracic pressure during apneic events can lead to acute and chronic adverse cardiovascular effects [1].

During recurrent apneas, hypoxemia and retained CO_2 can stimulate chemoreceptors, leading to an increase in sympathetic nervous activity (SNA). SNA progressively rises during obstructive apneas and peaks with arousal. This rise in SNA increases systemic vascular resistance (SVR) by peripheral vasoconstriction. Systemic blood pressure subsequently also increases gradually during the apnea. Although cardiac output (CO) and heart rate (HR) generally decrease during the apnea, both increase dramatically with arousal and apnea termination. The resumption of breathing at arousal contributes to heightened blood pressure, likely reflecting the increase in cardiac output in the face of severe peripheral vasoconstriction. Oxyhemoglobin saturation also decreases during apnea and recovers only slowly after apnea termination. The sudden increase in cardiac output at arousal means that myocardial oxygen demand increases at a time when oxyhemoglobin saturation is still low.

In addition to increased sympathetic activity and its effects on HR, BP, SVR, and CO, the large swings in intrathoracic pressure during apneas also have acute cardiovascular effects. Inspiratory effort against a collapsed upper airway can result in impressive swings in intrathoracic

Sleep: A Comprehensive Handbook, Edited by T. Lee-Chiong.

TABLE 35.1   Cardiovascular Diseases Potentially Associated with OSA

Hypertension
Ischemic heart disease
Heart Failure
    LV systolic dysfunction
    LV diastolic dysfunction
    Congestive heart failure
Bradyarrhythmias
    Sinus bradycardia
    Sinus pause
    Sinoatrial arrest
Atrial fibrillation
Stroke
Pulmonary hypertension

TABLE 35.2   Potential Mechanisms Linking OSA and Cardiovascular Disease

Increased sympathetic activation
Oxidative stress
Inflammation
Endothelial dysfunction
Disordered coagulation
Metabolic dysregulation

pressure, which can be as low as $-80 \, cm \, H_2O$. Such dramatic negative intrathoracic pressure changes can result in distortion of intrathoracic structures and can impair cardiac filling and cardiac function. The negative intrathoracic pressure directly results in an increased left ventricular (LV) transmural pressure gradient, which effectively acts to increase LV afterload. LV relaxation is also affected, further impeding LV filling and preload. Increased afterload and reduced preload together lead to a reduction in stroke volume and cardiac output during the apnea. With arousal and resumption of breathing, venous return increases, potentially distending the right ventricle and causing the interventricular septum to shift to the left, resulting in impaired LV compliance and LV diastolic filling.

MECHANISMS LINKING OSA AND CARDIOVASCULAR DISEASE

Sympathetic Activation

Decreased heart rate variability and increased blood pressure variability have been shown to elevate cardiovascular risk. Normotensive individuals with decreased heart rate variability have a higher risk for future hypertension. Increased blood pressure variability is associated with target organ damage [2]. As well as increased blood pressure variability. This is likely due to the heightened sympathetic drive that is present in patients with OSA even during awake normoxic conditions [3]. This high level of tonic sympathetic excitation may in part be due to increased chemoreflex activation. Deactivating chemoreflex drive by having OSA patients breathe 100% O_2 results in a significant decrease in sympathetic activity, heart rate, and blood pressure [4].

Oxidative Stress, Inflammation, and Endothelial Dysfunction

Reactive oxygen species (ROS) and oxidative stress have been implicated in the development and progression of cardiovascular disease. ROS are highly reactive molecules that react with proteins, nucleic acids, and lipids resulting in cell injury. Obstructive sleep apnea is characterized by repetitive episodes of hypoxia followed by reoxygenation. The repetitive hypoxia/reoxygenation in OSA may be analogous to tissue reperfusion injury, which is associated with the formation of reactive oxygen species and endothelial damage. The role of reactive oxygen species and oxidative stress, however, in the development of cardiovascular disease in patients with OSA remains controversial. Some studies have found increased levels of ROS breakdown products in patients with OSA, whereas others have not. Reactive nitrogen species (RNS) and nitrosative stress have also been implicated as possible mechanisms for the development of cardiovascular disease. However, it has recently been shown that OSA may not be accompanied by increased free nitrotyrosine, a marker for nitrosative stress.

There is increasing evidence that inflammation plays an important role in the development of atherosclerosis and cardiovascular disease. C-reactive protein (CRP) and serum amyloid-A (SAA), both nonspecific markers of systemic inflammation, may be risk factors for cardiovascular disease. Patients with OSA have increased levels of both CRP and SAA. It has been suggested that CRP may decrease after CPAP therapy. The increased levels of these inflammatory markers suggest that OSA represents a state of heightened systemic inflammation. Further evidence of this is the observation that various cytokines including TNF-α, IL-8, and IL-6 have all been found to be increased in OSA patients when compared to controls. IL-6 induces production of C-reactive protein from the liver. The primary stimulus for this state of heightened inflammation is unclear. ROS may play a role by directly stimulating endothelial cells to produce proinflammatory cytokines and upregulate adhesion molecules. Low oxygen tension itself has also been shown to directly stimulate circulating leukocytes. Adherence of leukocytes to the endothelium may be an important event in the development of endothelial dysfunction.

Endothelial dysfunction is considered to be a nascent form of atherosclerosis and is thought to play a vital role in the development of a wide array of cardiovascular diseases. Endothelial dysfunction can be seen in patients with hypertension, hyperlipidemia, diabetes, and smoking. Patients with OSA who are free of any other overt cardiac or vascular disease have been shown to have impaired endothelial function. This suggests that OSA itself may be an independent risk factor for the development of impaired endothelial function [5]. Adhesion of leukocytes to vascular endothelium may interfere with the ability of endothelial cell derived nitric oxide (NO) to relax vascular smooth muscle. This results in impairment of endothelium-dependent vasodilation. Animals pretreated with blocking antibodies directed against adhesion molecules have shown attenuation of this impairment. Adhesion receptor expression (ICAM, VCAM, E-selectin) is significantly increased in patients with OSA compared to obese controls [6, 7]. Although some studies have measured circulating adhesion receptors, this increase suggests that the endothelium is activated and "primed" for binding with leukocytes. Monocytes isolated from OSA patients have increased expression of adhesion molecules (CD15, CD11c) and demonstrate increased adherence to human endothelial cells in culture. Levels of adhesion receptors, proinflammatory cytokines, and monocyte adherence have been shown to decrease after therapeutic CPAP [6, 7]. Interestingly, leukocytes isolated from OSA patients show higher levels of ROS production, which may play a role in directly inactivating NO [8]. Studies by several groups have also shown that NO breakdown products are reduced in OSA, suggesting an impairment of NO synthesis.

Other abnormalities of the vascular endothelium associated with OSA have been described. Endothelin-1 is a potent long-acting vasoconstrictor important for regulating vascular tone and is elevated in OSA, and may contribute to sustained vasoconstriction. Furthermore, endothelin-1 concentration decreases with CPAP therapy. CRP may also contribute to vascular disease by its direct effects on the endothelium. Elevated homocysteine levels are also associated with endothelial dysfunction and the development of cardiovascular disease. However, homocysteine levels are not elevated in healthy OSA patients compared with controls.

Vascular endothelial growth factor (VEGF) is a glycoprotein that plays an important role in angiogenesis by stimulating normal and abnormal vessel growth. A recent study found that, in patients with OSA, VEGF levels are increased and related to the severity of OSA. VEGF levels have been shown to decrease with CPAP therapy [9]. A protective role for VEGF against cardiovascular disease has been postulated.

Disordered Coagulation

Recent evidence suggests that OSA results in abnormalities of coagulation that may be important in the development of cardiovascular disease. Interpretation of many of these studies, however, presents difficulties due to the inclusion of patients with concomitant disease processes such as hypertension, hypercholesterolemia, and smoking, which may also have effects on coagulation. Several studies have shown that platelet activation and aggregation are increased in patients with OSA. Increased platelet activation appeared to be related to increased arousal index [10]. This relationship may reflect sympathetic activation and increased nocturnal levels of catecholamines associated with arousals. Total serum fibrinogen and whole blood viscosity levels are also elevated in OSA, which together with platelet activation may contribute to a predisposition to clot formation. Some studies suggest that several coagulation factors including factors XIIa and VIIa are higher in OSA patients. Patients with OSA may also have increased levels of thrombin breakdown products, D-dimer and thrombin–antithrombin (TAT) complex, again implicating an underlying state of hypercoagulability. Fibrinolytic activity may also be reduced in OSA. Many of these abnormalities in coagulation including platelet activation appear to be alleviated after CPAP therapy. However, a recent study showed that elevated levels of factors XIIa and VIIa, TAT complex, and soluble P-selectin did not fall with 1 month of therapeutic CPAP [11].

Metabolic Dysregulation

Diabetes and obesity are important risk factors for cardiovascular disease. Recent studies have investigated the possible association between OSA and glucose intolerance and found that OSA patients had higher levels of fasting blood glucose, insulin, and glycosylated hemoglobin compared with matched controls. These changes appeared to be independent of body weight, suggesting that OSA impairs glucose tolerance. The severity of OSA appears to correlate with the degree of insulin resistance and severe OSA has been shown to greatly increase the risk of overt diabetes mellitus [12, 13]. The mechanism by which OSA is linked to impaired glucose tolerance is not fully understood but may be related to sympathetic activation and sleep deprivation. Leptin resistance may also play a role. Leptin, an adipocyte-derived hormone, suppresses appetite and promotes satiety. Obesity is associated with elevated leptin levels suggesting resistance to the metabolic effects of leptin. Interestingly, men with OSA have higher leptin levels than matched obese individuals without OSA, suggesting an even greater resistance to leptin [14]. Resistance to leptin in OSA patients may predispose to weight gain. Leptin may also predispose to platelet aggregation and has been implicated as an independent marker of increased cardiovascular risk [15]. Recent evidence has shown an independent association between leptin and CRP, supporting a possible inflammatory role for leptin.

CARDIOVASCULAR DISEASES ASSOCIATED WITH OSA

Hypertension

Several studies have demonstrated compelling evidence that implicates OSA in the development of hypertension. Experimentally induced OSA in dogs using intermittent and repetitive airway occlusion during sleep resulted in a 15% increase in both nocturnal and daytime blood pressure within 5 weeks [16]. In humans, four large cross-sectional population based studies and one prospective population based study have found associations between AHI and daytime hypertension. The Sleep Heart Health Study examined the association between OSA and blood pressure in over 6000 subjects. Results suggested that OSA was independently associated with hypertension. The odds ratio for the most severe group was 1.37, showing an overall small to moderate effect. More specifically, the prevalence of hypertension was found to progressively increase with severity of OSA (increasing AHI) [17]. The Sleep Heart Health Study and other cross-sectional studies have also found stronger associations between OSA and hypertension in less obese and younger subjects.

Evidence that implicates OSA in the development of hypertension has come from the prospective findings of the Wisconsin Sleep Cohort. In this study, an AHI of 15 or greater was associated with a threefold increased risk of developing new hypertension when subjects were evaluated 4 years after the initial sleep study [18]. Even subjects with minimally elevated baseline AHI (0 < AHI < 5) had a 42% increased odds of developing hypertension over the 4 year follow-up. The most recent set of guidelines from the Joint National Committee on the Detection, Prevention, Evaluation and Treatment of High Blood Pressure (JNC VII) lists OSA as the first of the identifiable causes of hypertension.

The mechanisms linking OSA with hypertension are complex and not fully elucidated. As discussed earlier, however, increased sympathetic activity, inflammation-mediated endothelial dysfunction, and predisposition to

weight gain due to leptin resistance likely play important roles in the pathogenesis of hypertension in OSA.

Although there is general consensus that CPAP treatment reduces nocturnal blood pressure in patients with OSA, the effect on daytime blood pressure is less clear. The effectiveness of CPAP in reducing daytime blood pressure in hypertensive patients has recently been evaluated in several studies. Two randomized placebo-controlled trials using subtherapeutic or "sham" CPAP as a control demonstrated that several months of CPAP resulted in a reduction in daytime blood pressure of between 1.3 and 5.3 mmHg [19, 20]. The effect seemed to be greatest in patients with more severe OSA compared with those with mild OSA. A consistent decrease in diastolic blood pressure after 24 hours of CPAP has also been observed.

Ischemic Heart Disease

There is a high prevalence of OSA in patients with coronary artery disease (CAD). This association is supported by several case-controlled and cross-sectional epidemiologic studies. In the Sleep Heart Health Study cohort, OSA was found to be an independent risk factor for coronary artery disease. The presence of obstructive sleep apnea in patients with CAD may also be a prognostic indicator [21]. There is also evidence supporting an association between OSA and myocardial infarction. OSA has been shown to be very common in patients with prior MI. In these patients, OSA appears to be as strong a risk factor as obesity, smoking, and hypertension.

OSA is associated with intermittent hypoxemia, CO_2 retention, sympathetic activation, surges in blood pressure, and increases in left ventricular afterload. Not suprisingly, all of these may predispose to the development of nocturnal myocardial ischemia. Studies have noted nocturnal ST-segment changes consistent with myocardial ischemia in patients with OSA who do not have clinically significant coronary artery disease. Patients with severe OSA demonstrate more frequent ST-segment depression and CPAP has been shown to reduce the total duration of ST-segment depression in these patients [22]. The ST-segment changes may be related to increased myocardial oxygen demand during the postapneic surge in blood pressure and heart rate at the time when the oxyhemoglobin saturation remains low. Although some patients with OSA may experience nocturnal ischemia, a recent study found that cardiac troponin T was not elevated in patients with severe OSA and coexistent CAD, suggesting that myocardial injury is not taking place [23].

Heart Failure

Sleep-disordered breathing in patients with heart failure can be primarily obstructive, primarily central (Cheyne–Stokes respirations, central sleep apnea), or a combination of both. Cheyne–Stokes respirations and central sleep apnea are reviewed in another chapter. Epidemiologic studies suggest an association between OSA and heart failure (HF). The Sleep Heart Health Study demonstrated that patients with OSA (defined as an AHI $\geq$ 11) were more than twice as likely to have heart failure. In this group of subjects, the odds ratio of having heart failure was higher than that for all other cardiovascular diseases. In other case-control studies, 11%–37% of heart failure patients undergoing polysomnography (PSG) were found to have OSA. Although OSA has been associated with both systolic and diastolic dysfunction, patients with diastolic dysfunction may have an especially high likelihood of OSA. Diastolic dysfunction may be related to longstanding hypertension and the effects of increased afterload and transmural wall stress associated with the direct effects of recurrent apneas over a lengthy period. Acute exacerbations in HF can also occur due to the acute effects of OSA on left ventricular function.

OSA may contribute directly to the development of HF by its effects on sympathetic drive, endothelial function, hypertension, and ischemic heart disease—all of which are known to be important risk factors for HF. Activation of inflammation in OSA may also be an important factor as systolic dysfunction can be induced directly by inflammatory cytokines via their effects on myocardial contractility [24].

Heart failure may itself predispose to the development of OSA. During sleep, the supine position may lead to redistribution of edema from the lower extremities to the upper airway. Upper airway soft tissue edema can result in increased upper airway resistance, increased inspiratory force, and potential collapse. Pulmonary congestion and stretch of pulmonary vagal receptors also may lead to hyperventilation and daytime hypocapnia predisposing to periodic breathing during sleep. During periodic breathing, central respiratory drive and drive to the pharyngeal dilator muscles decline, which may result in collapse of the vulnerable upper airway. Interestingly, during PSG many of these patients will demonstrate predominantly periodic breathing in the first half of the night followed by predominantly obstructive apneas during the latter half. Therefore the coexistence of OSA and HF may result in a vicious cycle [25].

Preliminary data suggest that treatment of OSA in patients with heart failure may have important beneficial effects. The physiologic effects of CPAP in OSA include reducing LV afterload, increasing stroke volume, reducing myocardial oxygen demand, and reducing cardiac and peripheral sympathetic tone. In some respects, the effects of CPAP are analogous to the effects of beta-blockers. Several small short-term studies of patients with heart failure and OSA have shown a modest improvement of both ejection fraction and functional class after treatment with CPAP for as little as 1 month [26, 27]. Although these results are encouraging, larger randomized trials need to be done to determine whether specific treatment of OSA in heart failure patients will result in a reduction of long-term morbidity and mortality.

Cardiac Arrhythmias

OSA patients undergoing polysomnography are frequently noted to have disturbances of cardiac rhythm. Sinus bradycardia, sinus pauses, and sinoatrial block are the most frequently observed nocturnal disturbances in rhythm. These bradyarrhythmias are associated with the apneic event and likely represent a form of the "diving response" whereby apnea and hypoxemia trigger a reflex increase in vagal tone. Effective treatment of OSA with CPAP results in resolution of these bradyarrhythmias. Supraventricular tachycardias have also been shown in patients with OSA. These appear to resolve after CPAP therapy. Evidence implicating OSA in the development of other serious arrhythmias including ventricular tachycardia remains unclear. Several studies have shown conflicting results regarding the association between OSA and ventricular tachycardia. Although it is generally thought to be more prevalent with desaturations of <65%, methodological limitations, small sample sizes, comorbid conditions, and lack of control groups have allowed only limited interpretation.

More recently, an association between OSA and atrial fibrillation (AF) has been noted. OSA appears to play a role in the recurrence of AF after cardioversion. One study showed that untreated OSA in patients cardioverted for AF doubled the likelihood of recurrence of AF within 12 months compared to patients receiving effective CPAP treatment [28]. Another recent study found that close to half of patients with AF are likely to have OSA [29]. This study also showed that the association of OSA with AF was even greater than the association of OSA with hypertension, body mass index, or neck circumference. Adrenergic activation, acute blood pressure changes, and cardiac distortion induced by obstructive apneas may play a role in the association between OSA and atrial fibrillation.

CONCLUSION

Cardiovascular disease is a major cause of morbidity and mortality. Large epidemiologic studies show that patients with OSA have an increased risk of hypertension, ischemic heart disease, stroke, and heart failure. OSA causes acute changes in cardiovascular regulation during sleep, disrupting the normal state of cardiovascular relaxation. Recent studies intimate an important role for sympathetic activation, inflammation and endothelial dysfunction, disordered coagulation, metabolic dysregulation, and possibly oxidative stress in the development of cardiovascular disease. However, whether treatment of OSA attenuates cardiovascular risk remains to be determined.

REFERENCES

1. Shamsuzzaman AS, Gersh BJ, Somers VK. Obstructive sleep apnea: implications for cardiac and vascular disease. *JAMA* **290**:1906–1914(2003).

2. Palatini P, Penzo M, Racioppa A, et al. Clinical relevance of nighttime blood pressure and of daytime blood pressure variability. *Arch Intern Med* **152**:1855–1860(1992).

3. Somers VK, Dyken ME, Clary MP, Abboud FM. Sympathetic neural mechanisms in obstructive sleep apnea. *J Clin Invest* **96**:1897–1904(1995).

4. Narkiewicz K, van de Borne PJ, Pesek CA, Dyken ME, Montano N, Somers VK. Selective potentiation of peripheral chemoreflex sensitivity in obstructive sleep apnea. *Circulation* **99**:1183–1189(1999).

5. Kato M, Roberts-Thomson P, Phillips BG, et al. Impairment of endothelium-dependent vasodilation of resistance vessels in patients with obstructive sleep apnea. *Circulation* **102**:2607–2610(2000).

6. Ohga E, Nagase T, Tomita T, et al. Increased levels of circulating ICAM-1, VCAM-1, and L-selectin in obstructive sleep apnea syndrome. *J Appl Physiol* **87**:10–14(1999).

7. Ohga E, Tomita T, Wada H, Yamamoto H, Nagase T, Ouchi Y. Effects of obstructive sleep apnea on circulating ICAM-1, IL-8, and MCP-1. *J Appl Physiol* **94**:179–184(2003).

8. Schulz R, Mahmoudi S, Hattar K, et al. Enhanced release of superoxide from polymorphonuclear neutrophils in obstructive sleep apnea. Impact of continuous positive airway pressure therapy. *Am J Respir Crit Care Med* **162**:566–570(2000).

9. Lavie L, Kraiczi H, Hefetz A, et al. Plasma vascular endothelial growth factor in sleep apnea syndrome: effects of nasal continuous positive air pressure treatment. *Am J Respir Crit Care Med* **165**:1624–1628(2002).

10. Hui DS, Ko FW, Fok JP, et al. The effects of nasal continuous positive airway pressure on platelet activation in obstructive sleep apnea syndrome. *Chest* **125**:1768–1775(2004).

11. Robinson GV, Pepperell JC, Segal HC, Davies RJ, Stradling JR. Circulating cardiovascular risk factors in obstructive sleep apnoea: data from randomised controlled trials. *Thorax* **59**:777–782(2004).

12. Punjabi NM, Sorkin JD, Katzel LI, Goldberg AP, Schwartz AR, Smith PL. Sleep-disordered breathing and insulin resistance in middle-aged and overweight men. *Am J Respir Crit Care Med* **165**:677–682(2002).

13. Ip MS, Lam B, Ng MM, Lam WK, Tsang KW, Lam KS. Obstructive sleep apnea is independently associated with insulin resistance. *Am J Respir Crit Care Med* **165**:670–676 (2002).

14. Phillips BG, Kato M, Narkiewicz K, Choe I, Somers VK. Increases in leptin levels, sympathetic drive, and weight gain in obstructive sleep apnea. *Am J Physiol Heart Circ Physiol* **279**:H234–H237(2000).

15. Wallace AM, McMahon AD, Packard CJ, et al. Plasma leptin and the risk of cardiovascular disease in the west of Scotland coronary prevention study (WOSCOPS). *Circulation* **104**:3052–3056(2001).

16. Brooks D, Horner RL, Kozar LF, Render-Teixeira CL, Phillipson EA. Obstructive sleep apnea as a cause of systemic hypertension. Evidence from a canine model. *J Clin Invest* **99**:106–109(1997).

17. Nieto FJ, Young TB, Lind BK, et al. Association of sleep-disordered breathing, sleep apnea, and hypertension in a large community-based study. Sleep Heart Health Study. *JAMA* **283**:1829–1836(2000).

18. Peppard PE, Young T, Palta M, Skatrud J. Prospective study of the association between sleep-disordered breathing and hypertension. *N Engl J Med* **342**:1378–1384(2000).

19. Pepperell JC, Ramdassingh-Dow S, Crosthwaite N, et al. Ambulatory blood pressure after therapeutic and subtherapeutic nasal continuous positive airway pressure for obstructive sleep apnoea: a randomised parallel trial. *Lancet* **359**:204–210(2002).

20. Becker HF, Jerrentrup A, Ploch T, et al. Effect of nasal continuous positive airway pressure treatment on blood pressure in patients with obstructive sleep apnea. *Circulation* **107**:68–73(2003).

21. Milleron O, Pilliere R, Foucher A, et al. Benefits of obstructive sleep apnoea treatment in coronary artery disease: a long-term follow-up study. *Eur Heart J* **25**:728–734(2004).

22. Peled N, Abinader EG, Pillar G, Sharif D, Lavie P. Nocturnal ischemic events in patients with obstructive sleep apnea syndrome and ischemic heart disease: effects of continuous positive air pressure treatment. *J Am Coll Cardiol* **34**:1744–1749(1999).

23. Gami AS, Svatikova A, Wolk R, et al. Cardiac troponin T in obstructive sleep apnea. *Chest* **125**:2097–2100(2004).

24. Finkel MS, Oddis CV, Jacob TD, Watkins SC, Hattler BG, Simmons RL. Negative inotropic effects of cytokines on the heart mediated by nitric oxide. *Science* **257**:387–389(1992).

25. Bradley TD, Floras JS. Sleep apnea and heart failure. Part I: Obstructive sleep apnea. *Circulation* **107**:1671–1678(2003).

26. Sin DD, Logan AG, Fitzgerald FS, Liu PP, Bradley TD. Effects of continuous positive airway pressure on cardiovascular outcomes in heart failure patients with and without Cheyne–Stokes respiration. *Circulation* **102**:61–66(2000).

27. Kaneko Y, Floras JS, Usui K, et al. Cardiovascular effects of continous positive airway pressure in patients with heart failure and obstructive sleep apnea. *N Engl J Med* **348**:1233–1241(2003).

28. Kanagala R, Murali NS, Friedman PA, et al. Obstructive sleep apnea and the recurrence of atrial fibrillation. *Circulation* **107**:2589–2594(2003).

29. Gami AS, Pressman G, Caples SM, et al. Association of atrial fibrillation and obstructive sleep apnea. *Circulation* **110**:364–367(2004).

ADDITIONAL READING

Bradley TD, Floras JS. Sleep apnea and heart failure. *Circulation* **107**:1671–1678(2003). A comprehensive review of the literature addressing OSA and heart failure.

Parish JM, Somers VK. Obstructive sleep apnea and cardiovascular disease. *Mayo Clin Proc* **79**:1036–1046(2004). A detailed recent review addressing cardiovascular disease and OSA.

Von Kanel R, Dimsdale JE. Hemostatic alterations in patients with obstructive sleep apnea and the implications for cardiovascular disease. *Chest* **124**:1956–1967(2003). A literature review summarizing important findings regarding OSA and coagulation.

36

PULMONARY HYPERTENSION AND SLEEP DISORDERED BREATHING

BROOKE G. JUDD

Dartmouth–Hitchcock Medical Center, Lebanon, New Hampshire

INTRODUCTION

The cardiovascular consequences of sleep disordered breathing (SDB) have attracted significant attention in recent years, although the primary focus has been the effects on the left heart and its associated vasculature. In comparison, relatively little research has looked at the effects of SDB on the pulmonary vasculature and the right heart. The goal of this chapter is to review the role that SDB plays in the development of pulmonary hypertension (PH), the mechanisms that may lead to PH, and the response of the PH to treatment of SDB.

SDB encompasses a number of overlapping disorders, including obstructive sleep apnea syndrome (OSAS), hypoventilation syndromes, and central sleep apnea (CSA). OSAS is the most prevalent of these disorders, thus the majority of the work assessing cardiovascular changes, including pulmonary hemodynamics, has been performed in patients with OSAS. Therefore the predominant theme in this chapter will be the relationship between OSAS and PH.

INCIDENCE OF PULMONARY HYPERTENSION AND RIGHT HEART DYSFUNCTION IN SDB

While OSAS is often cited as a risk factor for developing PH [1, 2], the relative contribution has been unclear. A review of this topic by Kessler et al. [3] in 1996 outlines the earlier work performed in this area. The prevalence of

persistent PH in studies included in their review ranged from 17% to 73% [3–10]. A number of studies are of limited general applicability due to multiple factors— small sample size, biased sample selection, and lack of uniform use of right heart catheterization to perform pulmonary arterial pressure measurements. For example, the study demonstrating a PH prevalence of 73% included only patients with known concurrent underlying lung disease [9].

In an attempt to limit selection bias, several studies have been performed with consecutive, unselected patients diagnosed with OSAS by polysomnography [8, 10–18]. A summary of much of the pertinent data from these studies can be found in Table 36.1. In these studies, the incidence of PH in association with SDB ranged from 17% to 41%. Table 36.1 also includes a summary of the pertinent data derived from right heart studies. These are discussed in more detail later. A number of these studies, however, also included patients with evidence of underlying lung disease. Furthermore, evidence of obstructive lung disease on pulmonary function testing and/or arterial blood gas abnormalities were found to correlate with the development of PH [8, 12, 17] or clinical cor pulmonale [19], making the relationship between SDB and PH more difficult to interpret.

As the presence of underlying lung disease appears to significantly contribute to the development of PH, other investigators more actively excluded patients with evidence of obstructive lung disease from their studies. These investigators included patients without known underlying cardiac or pulmonary disease and normal spirometry, [10,

Sleep: A Comprehensive Handbook, Edited by T. Lee-Chiong.
Copyright © 2006 John Wiley & Sons, Inc.

TABLE 36.1   Prevalence of Pulmonary Hypertension or Right Ventricular Dysfunction in Obstructive Sleep Apnea Syndrome

Reference	Patients (n)	Patient Selection	RV Dysfunction (%)	PH (%)
Alachantis et al. [11]	29	No cardiac/lung disease		21
Apprill et al. [12]	46	Unselected, consecutive		20
Bady et al. [13]	44	No COPD		27
Chaouat et al. [8]	220	Unselected, consecutive		17
Hawrylkiewicz et al. [14]	67	Normal PFT, ABG		0
Sajkov et al. [15]	32	Normal PFT, ABG		34
Sajkov et al. [10]	27	Normal PFT, ABG		41
Sanner et al. [16]	92	Normal PFT, ABG		20
Shinozaki et al. [17]	25	Unselected, consecutive		32
Yamakawa et al. [18]	37	No cardiac/lung disease		22
Bradley et al. [19]	50	Unselected, consecutive	12[a]	
Hanly et al. [20]	51	Normal ABG	NS[b]	
Noda et al. [21]	51	Not specified	11.8/21.4[c]	
Sanner et al. [22]	107	Normal PFT	18	

[a]Percent of patients with clinical signs/symptoms. No direct RV measurement.

[b]NS, no significant difference from OSAS group.

[c]All subjects in study with AHI $\geq$ 20 events per hour. AHI, apnea–hypopnea index.

Abbreviations: ABG, arterial blood gas; COPD, chronic obstructive pulmonary disease; PFT, pulmonary function testing; RV, right ventricle.

11, 13–16, 18], using either right heart catheterization or Doppler echocardiography to assess pulmonary arterial pressures. The summary data of these studies are depicted in Table 36.2. When underlying cardiopulmonary disease is ruled out, the incidence of PH ranged from 0% to 41%. Importantly, the degree of pulmonary hypertension observed remained in the mild range across all of the studies. No clearly reproducible factors were associated with an increased risk for pulmonary hypertension, including severity of OSAS or BMI. Greater sleep time with oxyhemoglobin saturation (Sao_2) below 90%, however, was found to correlate in two studies [13, 14].

The evidence to date thus does support the concept that patients with isolated OSAS are at risk of developing PH, although the degree of PH identified in all studies has been in the mild range despite severe OSAS.

RIGHT HEART STUDIES

While the main objective of this chapter is to evaluate the role of SDB in the development of PH, it is also important to consider the effects on the right heart, as the clinical manifestations of PH may include right ventricular (RV) dysfunction and cor pulmonale. There have been a small number of studies that have attempted to directly assess the relationship of SDB to alterations in right heart structure and function [19–22]. The pertinent data are summarized in Table 36.1. It should be noted that there is very little information on the presence of clinically apparent RV dysfunction (i.e., cor pulmonale) in patients with SDB but without underlying lung disease.

The largest study to date examining the right heart in SDB is derived from the Framingham Heart Study [23],

TABLE 36.2   Prevalence of Pulmonary Hypertension in OSAS when Cardiopulmonary Disease Excluded

Reference	Patients (n)	Method	AHI (events per hour) +PH	AHI (events per hour) −PH	PH (%)	mPAP (in PH+ patients) (mmHg)
Alchanatis et al. [11]	29	Doppler echo			20%	25.6
Bady et al. [13]	44	RH cath	53.4	43.3	27%	28.5
Hawrylkiewicz et al. [14]	67	RH cath		62	0%	NA
Sajkov et al. [10]	27	Doppler echo	56.8	51.9	41%	22.8
Sajkov et al. [15]	32	Doppler echo	44.7	47.1	34%	23.6
Sanner et al. [16]	92	RH cath	43.5	38.6	20%	21.9
Yamakawa et al. [18]	37	Doppler echo	61.2	44.3	22%	Not reported

Abbreviations: AHI, apnea–hypopnea index; Doppler echo, Doppler echocardiography; mPAP, mean pulmonary arterial pressure; PH, pulmonary hypertension; RH cath, right heart catheterization.

in which subjects at the Framingham Heart Study site for the Sleep Heart Health Study were evaluated with echocardiography. A major difference between this study and other studies described is the community-based nature of the population, thus eliminating any sleep laboratory referral bias. The study found an independent positive association between RV wall thickness and severity of OSAS. The clinical implications of this finding, however, remain unclear. A significant limitation in this, as well as other studies utilizing echocardiography to assess the right ventricle, is the measurement variability inherent in the quantitation of right heart structure and function.

ETIOLOGY OF PULMONARY HYPERTENSION IN SDB

The effects of SDB on the pulmonary vasculature are presumed primarily related to hypoxic vasoconstriction of the pulmonary arterial bed. Hypoxia is a well-documented potent pulmonary arterial vasoconstrictor and it is well established that chronic daytime hypoxia can lead to PH and cor pulmonale [24]. The effects of the episodic hypoxia often seen in OSAS, however, are less clear. While animal models have demonstrated a sustained increase in the right ventricular pressure [25] and pulmonary arterial pressure [26] after transient exposures to hypoxic–hypercarbic mixtures (mimicking obstructive respiratory events), the duration of hypoxia required for the development of persistent pulmonary arterial hypertension and/or right ventricular changes in humans remains elusive. It is hypothesized that the transient nocturnal oxygen desaturations observed on OSAS can lead to more permanent persistent PH by inducing vascular remodeling [10, 25]. This hypothesis is supported by evidence demonstrating that the development of PH is correlated with lower baseline waking arterial oxygen tension (Pao_2) and longer periods of nocturnal Sao_2 <90% [8, 10, 13, 16]. The first manifestation of pulmonary vascular change may be the exercise-induced PH observed in the absence of resting PH demonstrated in OSA patients [14].

Left ventricular dysfunction is another common cause of PH in the general population and may contribute to the development of PH in patients with SDB/OSAS. Left ventricular hypertrophy and/or elevated pulmonary wedge pressures have been reported in patients with OSAS and no other evidence of underlying cardiac disease [21, 27, 28], although this has not been a consistent finding [8, 20].

TREATMENT OF PULMONARY HYPERTENSION IN SDB

Very few studies have examined the effect of nasal continuous positive airway pressure (nCPAP) therapy on pulmonary arterial pressures in patients with OSAS, and the results have conflicted [29–31]. When patients with underlying lung disease were excluded, one study [31] did demonstrate significant declines in the daytime pulmonary arterial pressures after 4 months of treatment with nCPAP. Another finding of this study was a decrease in the pulmonary vascular response to hypoxia after therapy, suggesting possible improvements in pulmonary endothelial function. Clearly, however, more work needs to be done in assessing the pulmonary vascular response to treatment.

SUMMARY

It is becoming increasingly clear that SDB is associated with a variety of chronic cardiovascular sequelae, including effects on the right heart and pulmonary vasculature. While it is well established that daytime hypoventilation can lead to PH, the duration of isolated nocturnal hypoxemia necessary to cause PH remains unclear. It is evident that the coexistence of underlying lung disease with SDB raises the risk of PH, due apparently to lower baseline oxygen tensions. There is also significant evidence that suggests that OSAS in the absence of concurrent lung disease is associated with the development of PH, although the degree of PH identified in this group of patients has been in the mild range and the clinical significance is unclear. Additionally, although the development of PH does not appear to correlate with the severity of OSAS as defined by the AHI, there is some correlation with the severity and length of nocturnal oxygen desaturations. Thus, although there is no distinct cutoff value for any of these factors that will predict which patients with SDB may develop PH, it is reasonable to consider a continuum of disease in which increases in the pulmonary arterial pressures will become more profound as the severity or number of risk factors increase.

REFERENCES

1. Murray J. Disorders of the pulmonary circulation: general principles and diagnostic approach. In: Murray J, Nadel J (Eds), *Textbook of Respiratory Medicine*. Saunders, Philadelphia, 2000, pp 1485–1502.

2. Phillipson EA. Sleep apnea. In: Braunwald E, Fauci A, Kasper DL, Hauser SL, Longo DL, Jameson JL (Eds), *Harrison's Principles of Internal Medicine*. McGraw-Hill, New York, 2001, pp 1520–1523.

3. Kessler R, Chauoat A, Weitzenblum E, et al. Pulmonary hypertension in the obstructive sleep apnoea syndrome: prevelance, causes and therapeutic consequences. *Eur Respir J* **9**:787–794(1996).

4. Podszus T, Bauer W, Mayer J, Penzel T, Peter JH, Wichert P. Sleep apnea and pulmonary hypertension. *Klin Wschr* **64**:131–134(1986).

5. Weitzenblum E, Kreiger J, Apprill M, et al. Daytime pulmonary hypertension in patients with obstructive sleep apnea syndrome. *Am Rev Respir Dis* **138**:345–349(1988).

6. Krieger J, Sforza E, Apprill M, et al. Pulmonary hypertension, hypoxemia and hypercapnia in obstructive sleep apnea patients. *Chest* **96**:729–737(1989).

7. Laks L, Lehrharft B, Grunstein RR, et al. Pulmonary hypertension in obstructive sleep apnea. *Eur Respir J* **8**:537–541(1995).

8. Chaouat A, Weitzenblum E, Krieger J, Oswald M, Kessler R. Pulmonary hemodynamics in the obstructive sleep apnea syndrome: results in 220 consecutive patients. *Chest* **109**:380–386(1996).

9. Fletcher EC, Schaaf JM, Miller J, Fletcher JG. Long-term cardiopulmonary sequelae in patients with sleep apnea and chronic lung disease. *Am Rev Respir Dis* **135**:525–533(1987).

10. Sajkov D, Cowie RJ, Thornton AT, Espinoza H, McEvoy RD. Pulmonary hypertension and hypoxemia in obstructive sleep apnea syndrome. *Am J Respir Crit Care Med* **149**:416–422(1994).

11. Alchanatis M, Tourkohoriti G, Kakouros S, et al. Daytime pulmonary hypertension in patients with obstructive sleep apnea: the effect of continuous positive airway pressure on pulmonary hemodynamics. *Respiration* **68**:566–572(2001).

12. Apprill M, Weitzenblum E, Krieger J, et al. Frequency and mechanism of daytime pulmonary hypertension in patients with obstructive sleep apnoea syndrome. *Cor et Vasa* **33**:42–49(1991).

13. Bady E, Achkar A, Pascal S, Orvoen-Frija E, Laaban JP. Pulmonary arterial hypertension in patients with obstructive sleep apnea syndrome. *Thorax* **55**:934–939(2000).

14. Hawrylkiewicz I, Sliwinski P, Palasiewicz G, Plywaczewski R, Zielinski J. Effect of nocturnal hypoxia on pulmonary hemodynamics in patients with obstructive sleep apnea syndrome. *Pneumonol Alergol Pol* **68**:28–36(2000).

15. Sajkov D, Wang T, Saunders NA, Bune AJ, Neill AM, McEvoy RD. Daytime pulmonary hemodynamics in patients with obstructive sleep apnea without lung disease. *Am J Respir Crit Care Med* **159**:1518–1526(1999).

16. Sanner BM, Doberauer C, Konermann M, Sturm A, Zidek W. Pulmonary hypertension in patients with obstructive sleep apnea syndrome. *Arch Intern Med* **157**:2483–2487(1997).

17. Shinozaki T, Tatsumi K, Sakuma T, et al. Daytime pulmonary hypertension in the obstructive sleep apnea syndrome. *Jpn J Thoracic Dis* **33**:1073–1079(1995).

18. Yamakawa H, Shiomi T, Sasanabe R, et al. Pulmonary hypertension in patients with severe obstructive sleep apnea. *Psychiatr Clin Neurosci* **56**:311–312(2002).

19. Bradley TD, Rutherford R, Grossmann RF, et al. Role of daytime hypoxemia in the pathogenesis of right heart failure in the obstructive sleep apnea syndrome. *Am Rev Respir Dis* **131**:835–839(1985).

20. Hanly P, Sasson Z, Zuberi N, Alderson M. Ventricular function in snorers and patients with obstructive sleep apnea. *Chest* **102**:100–105(1992).

21. Noda A, Okada T, Yasuma F, Nakashima N, Yokota M. Cardiac hypertrophy in obstructive sleep apnea syndrome. *Chest* **107**:1538–1544(1995).

22. Sanner BM, Konermann M, Sturm A, Müller X, Zidek W. Right ventricular function in patients with obstructive sleep apnoea syndrome. *Eur Respir J* **10**:2079–2083(1997).

23. Guidry UC, Mendes LA, Evans JC, et al. Echocardiographic features of the right heart in sleep-disordered breathing. *Am J Respir Crit Care Med* **164**:933–938(2001).

24. Malik A, Vogel S, Minshall R, Tiruppathi C. Pulmonary circulation and regulation of fluid balance. In: Murray J, Nadel J (Eds), *Textbook of Respiratory Medicine*. Saunders, Philadelphia, 2000, pp 129–131.

25. Nattie EE, Bartlett D Jr, Johnson K. Pulmonary hypertension and right ventricular hypertrophy caused by intermittent hypoxia and hypercapnia in the rat. *Am Rev Respir Dis* **118**:653–658(1978).

26. McGuire M, Bradford A. Chronic intermittent hypercapnic hypoxia increases pulmonary arterial pressure and haematocrit in rats. *Eur Respir J* **18**:279–285(2001).

27. Alchanatis M, Paradellis G, Pini H, et al. Left ventricular function in patients with obstructive sleep apnoea before and after treatment with nasal continuous positive airway pressure. *Respiration* **67**:367–371(2000).

28. Alchanatis M, Tourkohoriti G, Kosmas EN, et al. Evidence for left ventricular dysfunction in patients with obstructive sleep apnoea syndrome. *Eur Respir J* **20**:1239–1245(2002).

29. Sforza E, Krieger J, Weitzenblum E, Apprill M, Lampert E, Ratomaharo J. Long-term effects of treatment with nasal continuous positive airway pressure on daytime lung function and pulmonary hemodynamics in patients with obstructive sleep apnea. *Am Rev Respir Dis* **141**:866–870(1990).

30. Chaouat A, Weitzenblum E, Kessler R, et al. Five-year effects of nasal continuous positive airway pressure in obstructive sleep apnoea syndrome. *Eur Respir J* **10**:2578–2582(1997).

31. Sajkov D, Wang T, Saunders NA, Bune AJ, McEvoy RD. Continuous positive airway pressure treatment improves pulmonary hemodynamics in patients with obstructive sleep apnea. *Am J Respir Crit Care Med* **165**:152–158(2002).

NEUROCOGNITIVE AND FUNCTIONAL IMPAIRMENT IN OBSTRUCTIVE SLEEP APNEA

W. DAVID BROWN

Sleep Diagnostics of Texas, The Woodlands, Texas

The primary events of obstructive sleep apnea (OSA) include repetitive episodes of hypoxemia, hypercapnia, and subsequent sleep fragmentation. Apnea can also be associated with daytime sleepiness, transient blood pressure elevations, changes in chemistry, and possibly changes in brain morphology [1]. Both the primary events and secondary consequences of obstructed breathing during sleep could theoretically lead to cognitive, emotional, and behavioral changes. Studies have documented many negative effects that apnea has on functioning. There remains a question as to the cause and permanence of the impairment.

NEUROCOGNITIVE FUNCTIONING

In an early description of apnea patients, Guilleminault et al. [2] reported the clinical symptomatology of 50 patients with OSA. They noted that major symptoms of the disorder were intellectual deterioration and difficulty in focusing attention and concentrating. These problems were most troublesome in the morning and early afternoon hours.

Since that early observation, many efforts have been made to document cognitive change in individuals with sleep apnea. Although impairment is often found, there is a great deal of discrepancy between studies and the clinical significance of the changes is rarely discussed. This may reflect the difficulty in measuring a broad range of "cognitive" abilities, cognitive test limitations, difficulty translating measured changes to practical importance, study design limitations, differing severity of the study groups,

concomitant medical conditions, and age effects. Attempting to attribute the measured changes to a specific primary cause such as sleep fragmentation or hypoxemia has also proved difficult, perhaps because of the interrelatedness of these events.

Table 37.1 summarizes the findings of studies that examined the effects of OSA on cognitive functioning. General intellectual functioning is the broadest measure that attempts to describe cognitive changes. It generally refers to an IQ score or a general intelligence and is typically assessed with all or part of the Wechsler Adult Intelligence Scale (WAIS-R) [3]. Most studies that have used this inclusive measure find that OSA has a negative impact on intellectual functioning [4–8] and hypoxemia is the proposed mechanism causing the impairment [9]. However, a recent meta-analysis [10] found that general intelligence and verbal intelligence are unaffected by OSA. The authors noted that, on both norm-referenced and control-referenced data, the OSA subjects performed better than published norms. They suggested this might be due to a selection bias or the tendency for the average mean scores to drift upward over time. Visual intelligence produced a much greater effect size but had considerable variability.

Frequent arousals from sleep, sleep stage deprivation, and transient hypoxemia have all been associated with memory impairment [1, 11, 12]. As such, there is a great deal of interest in the role of apnea on memory. Memory is, however, a complex process, that is divided into different components such as short- and long-term memory, verbal, and visual. Short-term memory can be further divided into

TABLE 37.1 Summary of Neuropsychological Test Results, Mechanism of Action, and Treatment Effects

Function	Impairment	Mechanism	Treatment
(1) Memory	Impaired [4, 5, 7, 13–17] Inconsistent [10]	EDS [4, 5, 13, 14, 21] Hypox [7, 15, 16]	R [14, 21] P [19]
(2) Executive Functioning	Impaired [4, 5, 8, 10, 14, 21, 59] Not Impaired [15, 18]	Hypox [58, 20, 21]	P [14, 21, 59]
(3) General Intelligence Unimpaired [8]	Impaired [6, 11, 17,22,26]	Hypox [6,11,17,22]	
(4) Verbal Functioning Unimpaired [13, 36,8]	Impairment [11]	Hypox [11]	
(5) Attention/Vigilance	Impaired [1, 6,8,22,26,35,54]	Hypox [22, 26,54] EDS [6]	R [7, 37,50]
(6) Motor Skills	Impaired [12, 26,54]	Hypox [12, 26]	

a. Speed
Unimpaired [8, 13]
b. Fine Skills/Drawing
Impaired [8, 35]
   Hypox [7]
    P [7, 35]
EDS = Excessive Sleepiness
Hypox = Hypoxemia
R = Reversible
  P = Permanent

working memory, and this memory can have further subdivisions [3]. Most studies have looked at long-term memory and find decrements in OSA patients [3–5, 13, 14]. The majority of studies attribute the memory changes to hypersomnolence [4, 5, 13, 14], but other studies have attributed the changes to hypoxemia [7, 15, 16]. In an effort to help determine the relative contribution of sleepiness, sleep fragmentation, or hypoxemia on various neuropsychological changes, Adams et al. [16] performed neuropsychological tests on a large group of mild to moderately severe apnea patients who were healthy except for complaints of snoring. The subjects were screened for psychological impairment or medical conditions that could alter their psychological functioning. Each subject also completed a MSLT to assess the role of sleepiness on test results. Analysis also included breathing variables including the AHI and lowest oxygen saturation and time the oxygen saturation was <90%, as well as an arousal index. The results indicated there was a small, but significant association between polysomnographic measures of sleep disordered breathing (SDB) and a range of neuropsychological functions including working memory, declarative memory, and signal discrimination. The apnea severity indices of time <90% or RDI predicted the neuropsychological outcomes in a dose–response manner. Sleepiness and arousals did not predict the variables. Vigilance was negatively associated with sleepiness and was not explained by polysomnographic variables [16].

Even mild apnea patients (AHI of 10–30 episodes/hour without significant hypoxemia) demonstrated vigilance and working memory deficits. The memory deficits were mild and the authors suggested that the more profound changes were attributed to the more severe hypoxemia, sleep fragmentation, or both [17]. However, a more recent study failed to detect any changes in STM or any cognitive functions measured in a group of mild to moderately severe OSA subjects. There was no correlation found between the severity of the OSA and cognitive measures [18].

Naegelle et al. [15] showed that short-term memory was altered regardless of the nature of the information used, verbal or visual, or the lag between presentation and retrieval. The researchers speculated that the STM changes observed were due to chronic intermittent oxyhemoglobin desaturations rather than sleepiness [15]. In a later paper [19], the authors found that STM deficits remained even after successful treatment with nasal CPAP. The study did not perform MSLTs but the subjects subjectively noted improved alertness. Since oxygen saturations were improved with CPAP, hypoxemia was not considered the primary cause of the persisting problem. It was speculated that the STM deficits might reflect the cumulative detrimental action of both hypoxemia and sleep fragmentation [19].

In contrast to the previous study, a meta-analysis [10] found the effects of OSA on short-term verbal memory was small and not significant. The analysis indicated that the control-referenced data suggested moderate impairments in both short- and long-term visual memory but these impairments were not seen in norm-referenced studies. The analysis also showed that long-term verbal memory showed moderate impairments in norm-referenced samples but not in the control referenced samples. These results are difficult to interpret and the authors suggested that poor psychometric characteristics of some of the memory tests used might be contributing to the confusion.

Unlike memory, most studies are consistent in showing that OSA has pronounced effects on attention and vigilance. The meta-analysis found that control-referenced studies show OSA had markedly impaired sustained attention [10]. The cause of the impairment remains controversial. Greenberg et al. [8] compared OSA patients with hypoxemia to other groups of sleepy patients and found that the changes in hypoxemic OSA patients exceeded that of either sleepiness or aging. Therefore the changes in attention were caused by hypoxemia [8]. Montplaisir et al. [20] attributed the changes in attention to EDS. However, they also found that the best predictor of daytime alertness and EDS was minimum Sao$_2$ [20]. Attention returns to normal following treatment.

Motor skills have also been found to be impaired in OSA patients. In control-referenced studies there were moderate to large effect sizes but these effects were much smaller and not significant in norm-referenced studies [10]. The major effect was seen in fine motor skills as opposed to gross motor speed. Manual dexterity did not return to normal following treatment, suggesting irreversible anoxic central nervous system (CNS) damage [20].

Executive functions include skills such as problem solving, goal-oriented behavior, and mental flexibility [3]. Most studies find some impairment of executive functioning [5, 8, 10, 14, 20]. A meta-analysis suggested that the effect was moderate to large in both norm-referenced and control-referenced samples [10]. Other studies have failed to find significant changes [18] or found the changes to be smaller than in individuals with frontal lobe lesions [15]. Most authors consider the executive changes to be caused by hypoxemia and the functioning does not completely return following treatment [14, 21]. This suggests that the impairment may be a permanent result of anoxic CNS damage.

In summary, obstructive sleep apnea appears to cause measurable changes in neurocognitive functioning. The most robust negative effects are seen in attention, vigilance, and executive functioning. Fine motor skills and manual dexterity are more impaired than gross motor speed. The disorder has less clear effects on general intelligence and verbal fluency. The effects on memory are more difficult to interpret. Even mild OSA appears to negatively effect memory. Short-term visual memory appears to be most sensitive to the disorder but long-term memory may also be impaired. The finding that some functioning returns to normal following treatment of OSA with nasal CPAP while others remain suggests that sleep disruption, daytime sleepiness, and hypoxemia have differential effects on cognitive functioning and may result in permanent anoxic CNS damage [20].

MOOD

Many studies have examined the psychological correlates of obstructive sleep apnea. The tests typically used to assess emotional functioning are the MMPI, the Profile of Mood States (POMS), the Beck Depression Inventory (BDI), and the Symptom Checklist (SCL- 90). Studies that used the MMPI typically find that apnea patients have elevated scores on hypochondriasis (Scale 1), depression (Scale 2), hysteria (Scale 3), or social introversion (Scale 0) [13, 22–24]. The pattern suggests mild chronic dysphoria, tendency to focus on somatic problems, lowered activity level, difficulty expressing emotions, emotional apathy, and subjective helplessness. SCL-90 scores often are consistent with the MMPI, showing somatic concerns and dysphoric moods suggesting the apnea patients present with a "somatic–depressive" personality pattern [13].

Watson et al. [25] found that of 102 apneics, 51% had BDI scores that reflected minimal or no depression, 26% had scores reflecting a mild depression, 20% had scores reflecting moderate depression, and 4% fell into the category for severe depression. These ratings correlated with the total number of obstructive apneas and with the AHI [25].

Reynolds et al. [26] studied 25 consecutive male patients meeting criteria for OSA to determine the prevalence of diagnosable psychiatric disorders in this population. Of the 25 patients, 60% did not meet Research Diagnostic Criteria (RDC) for any present or past psychiatric diagnosis. However, 20% met criteria for past major depressive episodes or chronic intermittent depression, 16% for alcohol abuse, and 4% for cyclothymia [26]. In addition, the prevalence of sleep apnea in individuals with major depression was relatively uncommon [27, 28]. These studies suggest that apnea may not be a primary cause of psychiatric disorders but that apnea worsens mood. In fact, higher depression scores are seen in sleep disorders other than apnea [29].

While theory and most studies suggest that a negative mood should be associated with obstructive sleep apnea, not all studies are consistent with the idea. Pillar and Lavie [30] used the SCL-90 to assess the association between sleep apnea and psychiatric symptoms. They studied 2271 patients and, in the male population, they found no association with the existence or severity of sleep apnea with depression or anxiety [30].

There is some concern over the methods of determining the presence and severity of psychiatric illness in apnea patients. Sleep disruption is a common symptom of both depression and apnea. Some of the differences found may simply reflect the number of items on the instrument that reflect sleep quality and daytime sleepiness [31, 32].

Mood generally improves with treatment of obstructed breathing with nasal CPAP. However, this is not always the case. A recent study suggested that there might be a placebo effect working. The investigators used a placebo CPAP group and found that both the CPAP treatment and placebo groups showed significant improvement in mood states (POMS) after 7 days of CPAP use [32]. One study failed to find improvement in emotional status after 3 months and 1 year of CPAP treatment [33].

Most studies confirm that sleep apnea is associated with mild chronic dysphoria and somatic concern. It is not clear if the obstructed breathing is the cause of the mood changes or may perhaps worsen an existing tendency. The instruments used to assess mood changes may be misleading in that they may be heavily influenced by questions regarding sleep. Finally, it is still not clear if CPAP treatment actually improves mood.

QUALITY OF LIFE

As previously noted, obstructive sleep apnea can have a negative effect on cognitive functioning and emotion. The immediate symptoms of loud snoring, disturbed sleep, and daytime sleepiness, as well the stress of living with a chronic illness, can all conspire to diminish quality of life. It is increasingly more important to assess the patient's view of an illness and the benefits from treatment, as these views will ultimately affect the utility of treatment [34]. Among the most common concerns raised by patients with severe sleep apnea are abnormal fatigue and somnolence, obesity, snoring, depression and use of alcohol and antidepressants, frequent nocturnal awakenings, problems with CPAP (nasal mask and noise), relationships and sexual problems, loss of memory, and fear of death [35].

MARITAL PROBLEMS

Several common symptoms of obstructive sleep apnea can lead to adverse consequences in the family. Loud snoring, the most common symptom of OSA, is often the primary reason for patients seeking help. It has a disruptive effect on a spouse's sleep and frequently results in couples sleeping in separate bedrooms and puts an added burden on a marriage. In contrast to a divorced group, married sleep apnea patients are more depressed, exhausted, and socially isolated [36]. In a unique demonstration of the effect snoring has on the bedpartner, Beninati et al. [37] measured the sleep of snorers and their bedpartners simultaneously.

Using a split night protocol, they found that the patient's AHI fell from a median score of 26 episodes/hour to 7 episodes/hour. The spouse's arousal index fell from 21 arousals/hour to 12 arousals/hour and their sleep efficiency improved from 74% to 87%.

OSA has a major impact on marriage for obese individuals. OSA and frequent sleepiness were important predictors of divorce and the rate of divorce was two to three times higher in subjects with OSA and sleepiness. In men, the OSA had to be severe enough to cause daytime sleepiness before the divorce rate was affected. OSA had a more pronounced effect on divorce rate in women. In women with OSA and frequent sleepiness there was a seven times higher likelihood of being divorced at least twice. Women with OSA were more likely to report anxiety-like symptoms and the authors speculated that these symptoms may have a negative effect on relationships or perhaps men are less tolerant of snoring and sleepy partners [38].

Erectile dysfunction (ED) and other sexual problems are also common in men with OSA [39]. ED has been reported to range from 30% to 68% of OSA patients [40–42]. The ED may be caused by secondary problems from OSA such as hypertension or the use of hypertensive medication [43] or perhaps to low testosterone levels [44]. However, a recent study suggests that ED is related, at least in part, to nerve alterations related to the severity of the apnea, and in particular to the degree of hypoxia [40]. In a large-scale survey study of men with ED, there was no indication that men with ED were more likely to present with risk factors for OSA than those without ED. The authors speculated that any relation between ED and OSA would be subtle [45].

WORK AND HEALTH

Daytime sleepiness associated with apnea often impairs the patient's ability to work, to solve problems at work or at home, or even perform simple tasks. Many have either lost jobs or are on the verge of losing jobs due to their inability to adequately perform at the workplace. OSA patients perceive themselves as having poorer health [38] and as a result have more sick leave than other workers. These patients may also limit their social activities due to embarrassment about falling asleep.

A number of studies [46–50] have systematically looked at the quality of life in social, emotional, and physical domains using the Sickness Impact Profile (SIP) [51] and the Medical Outcomes Study SF-36 (SF-36) [52]. These studies clearly indicate that OSA can be associated with severely impaired quality of life. Even mild apnea may lead to diminished quality of life in a number of areas [47]. Mild to moderate apnea is associated with lower vitality, whereas severe apnea is more broadly associated with poorer quality of life [53].

Summarizing these studies, individuals with moderate to severe apnea described themselves as excessively sleepy. They reported difficulty with attention, concentration, and planning. They were working shorter hours or were working less efficiently and acknowledged declines in recreation time and hobbies as well as increases in interpersonal difficulties. They experience greater difficulty in performing everyday activities such as bathing and grocery shopping, they have role limitations at home and at work, more bodily pain, and reduced energy levels, and they perceive their overall health to be poorer than the general U.S. population. These results are similar to published normative data for other chronic conditions such as hypertension and type II diabetes [53]. In an obese population with OSA, subjects reported more impaired work performance and increased sick leave. Frequent sleepiness was the best predictor of sick leave. The OSA patients with sleepiness averaged 5 weeks more sick leave than subjects without OSA. This finding was independent of other medical conditions such as hypertension or diabetes [38].

Studies also show that quality of life measures improve with treatment. The largest effects are in the areas of sleeping better and feeling more rested and in recreation, pastimes, and social functioning. There were moderately large improvements in psychological functioning, performing household tasks, and work performance [34].

Sleep apnea is a chronic disorder with associated physical conditions including obesity, hypertension, congestive heart failure, cardiac arrhythmias, diabetes, and obstructive airways disease [54]. Untreated sleep apnea has increased mortality. The experience of having apnea is one of having a serious and potentially life-threatening disorder. In addition, the treatment of apnea may require the use of nasal CPAP, which could be viewed not only as an encumbrance but also as a reminder that the person has a potentially serious medical condition possibly resulting in death [35].

OSA is also associated with increased accidents. Compared to people without SDB, men who were habitual snorers or had an AHI > 5 were at least three times as likely to have had at least one accident and men and women combined with an AHI > 15 were seven times as likely to have had multiple accidents in a 5 year period. These results are independent of age, average miles drive per year, alcohol use, BMI, and education [55].

Quality of life issues may ultimately be the most interesting consequence of obstructive sleep apnea. Loud snoring, erectile dysfunction, and daytime fatigue can place a significant strain on marriages resulting in marital problems or divorce. Daytime fatigue can result in increased accidents and diminished work performance resulting in fewer promotions or loss of work. The stress of living with a chronic illness can result in increased anxiety and diminished quality of life. These factors may be the initial concerns that lead individuals to seek treatment and improvement in life quality may ultimately determine compliance with treatment.

REFERENCES

1. Macey PM, Henderson LA, Macey KE, Alger JR, Frysinger RC, Woo MA, Harper RK, Yan-Go FL, Harper RM. Brain morphology associated with obstructive sleep apnea. *Am J Respir Crit Care Med* **166**(10):1382–1387(2002).

2. Guilleminault C, van der Hoed J, Mitler M. Clinical overview of the sleep apnea syndromes. In: Guilleminault C, Dement WC (Eds), *Sleep Apnea Syndromes*. Alan R. Liss, New York, 1978, pp 1–12.

3. Decary A, Rouleau I, Montplaisir J. Cognitive deficits associated with sleep apnea syndrome: a proposed neuropsychological test battery. *Sleep* **23**(3):369–381(2000).

4. Bedard MA, Montplaisir J, Richer F, Rouleau I, Malo J. Obstructive sleep apnea syndrome: pathogenesis of neuropsychological deficits. *J Clin Exp Neuropsychol* **13**:950–964(1991).

5. Berry DT, Webb WB, Block AJ, Bauer RM, Switzer DA. Nocturnal hypoxia and neuropsychological variables. *J Clin Exp Neuropsychol* **8**(3):229–238(1986).

6. Cheshire K, Engelman H, Deary I, Shapiro C, Douglas NJ. Factors impairing daytime performance in patients with sleep apnea/hypopnea syndrome. *Arch Intern Med* **152**:538–541(1992).

7. Findly LJ, Barth JT, Powers DC, Wilhoit SC, Boyd DG, Suratt PM. Cognitive impairment in patients with obstructive sleep apnea and associated hypoxemia. *Chest* **90**:686–690(1986).

8. Greenberg GD, Watson RK, Deptula D. Neuropsychological dysfunction in sleep apnea. *Sleep* **10**(3):254–262(1987).

9. Sateia MJ. Neuropsychological impairment and quality of life in obstructive sleep apnea. *Clin Chest Med* **24**(2):249–259(2003).

10. Beebe DW, Groesz L, Wells C, Nichols A, McGee K. The neuropsychological effects of obstructive sleep apnea: a meta-analysis of norm-referenced and case controlled data. *Sleep* **26**(3):298–307(2003.

11. Bonnet MH. Cognitive effects of sleep and sleep fragmentation. *Sleep* **16**:S65–S67(1993).

12. Roth T, Costa e Silva JA, Chase MH. Sleep and cognitive (memory) function: research and clinical perspectives. *Sleep Med* **2**(5):379–387(2001).

13. Kales, A, Caldwell A, Cadieux R, Vela-Bueno A, Ruch L, Mayes S. Severe obstructive sleep apnea-II: Associated psychopathology and psychosocial consequences. *J Chronic Dis* **38**(5):427–434(1985).

14. Valencia-Flores M, Bliwise DL, Guilleminault C, Cilveti R, Clerk A. Cognitive function in patients with sleep apnea after acute nocturnal nasal continuous positive airway pressure (CPAP) treatment: sleepiness and hypoxemia effects. *J Clin Exp Neuropsychol* **18**:197–210(1996).

15. Naegele B, Thouvard V, Pepin J, Levy P, Bonnet C, Perret J, Pellat J, Feuerstein C. Deficits of cognitive executive functions in patients with sleep apnea syndrome. *Sleep* **18**(1):43–52(1995).

16. Adams N, Strauss M, Schluchter M, Redline S. Relation of measures of sleep-disordered breathing to neuropsychological functioning. *Am J Respir Crit Care Med* **163**:1626–1631(2001).

17. Redline S, Strauss M, Adams N, Winters M, Roebuck T, Spry K, Rosenberg C, Adams K. Neuropsychological function in mild sleep-disordered breathing. *Sleep* **20**(2):160–167(1997).

18. Boland LL, Shahar E, Iber C, Knopman DS, Kuo TF, Nieto FJ. Measures of cognitive function in persons with varying degrees of sleep-disordered breathing: the Sleep Heart Health Study. *J Sleep Res* **11**(3):265–272(2002).

19. Naegele B, Pepin J, Levy P, Bonnet C, Pellat J, Feuerstein C. Cognitive dysfunction in patients with obstructive sleep apnea syndrome(OSAS) after CPAP treatment. *Sleep* **21**(4):392–397(1998).

20. Montplaisir J, Bedard MA, Richer F, Rouleau I. Neurobehavioral manifestations in obstructive sleep apnea syndrome before and after treatment with continuous positive airway pressure. *Sleep* **15**(6):S17–S19(1992).

21. Bedard MA, Montplaisir J, Richer F, Malo J, Rouleau I. Persistent neuropsychological deficits and vigilance impairments in sleep apnea syndrome after treatment with continuous positive airway pressure. *J Clin Exp Neuropsychol* **15**:330–341(1993).

22. Klonoff H, Fleetham J, Taylor R, Clark C. Treatment outcome of obstructive sleep apnea: physiological and neuropsychological concomitants. *J Nerv Ment Dis* **175**(4):208–212(1987).

23. Aikens JE, Mendelson WB. A matched comparison of MMPI responses in patients with primary snoring or obstructive sleep apnea *Sleep* **22**(3):355–359(1999).

24. Aikens JE, Caruana-Montaldo B, Vanable PA, Tadimeti L, Mendelson WB. MMPI correlates of sleep and respiratory disturbance in obstructive sleep apnea. *Sleep* **22**(3):362–369(1999).

25. Watson R, Greenberg G, Bakos L. Sleep apnea and depression. *Sleep Res* **16**:293(1987).

26. Reynolds CF, Kupfer DJ, McEachran AB, Taska LS, Sewitch DE, Coble PA. Depressive psychopathology in male sleep apneics. *Clin Psychiatry* **45**(7):287–290(1984).

27. Gierz M, DuPont R, Morehouse R, Irwin M, Gillin JC. Oxygen saturation during sleep in a psychiatric population. *Sleep Res* **17**:180(1988).

28. Reynolds CF, Coble PA, Spiker PG, Neil JF, Holzer BC, Kupfer DJ. Prevalence of sleep apnea and nocturnal myoclonus in major affective disorders: clinical and polysomnographic findings. *J Nerv Ment Dis* **170**:565–567(1982).

29. Vandeputte M, de Weerd A. Sleep disorders and depressive feelings: a global survey with the Beck depression scale. *Sleep Med* **4**(4):343–345(2003).

30. Pillar G, Lavie P. Psychiatric symptoms in sleep apnea syndrome: effects of gender and respiratory disturbance index. *Chest* **114**(3):697–703(1998).

31. Weaver TE. Outcome measurement in sleep medicine practice and research. Part 2: assessment of neurobehavioral performance and mood. *Sleep Med Rev* **5**(3):223–236(2001).

32. Yu BH, Ancoli-Israel S, Dimsdale JE. Effect of CPAP treatment on mood states in patients with sleep apnea. *J Psychiatr Res* **33**(5):427–432(1999).

33. Borak J, Cieslicki JK, Koziej M, Matuszewski A, Zielinski J. Effects of CPAP treatment on psychological status in patients with severe obstructive sleep apnea. *J Sleep Res* **5**(2):123–127(1996).

34. Weaver, TE. Outcome measurement in sleep medicine practice and research. Part 1: assessment of symptoms, subjective and objective daytime sleepiness, health-related quality of life and functional status. *Sleep Med Rev* **5**(2):103–128(2001).

35. Veale D, Poussin G, Benes F, Pepin JL, Levy P. Identification of quality of life concerns with obstructive sleep apnoea at the time of initiation of continuous positive airway pressure: a discourse analysis. *Qual Life Res* **11**(4):389–399(2002).

36. Cartwright RD, Knight S. Silent partners: the wives of sleep apnea patients. *Sleep* **10**(3):244–248(1987).

37. Beninati W, Harris CD, Herold DL, Shepard JW Jr. The effect of snoring and obstructive sleep apnea on the sleep quality of bed partners. *Mayo Clin Proc* **74**(10):955–958(1999).

38. Grunstein RR, Stenlof K, Hedner JA, Sjostrom L. Impact of self-reported sleep-breathing disturbances on psychosocial performance in the Swedish obese subjects (SOS) study. *Sleep* **18**(8):635–643(1995).

39. Guilleminault C. Clinical features and evaluation of obstructive sleep apnea. In: Kryger M, Roth T, Dement WC (Eds), *Principles and Practice of Sleep Medicine*, 2nd ed. Saunders, Philadelphia, 1994, pp 667–677.

40. Fanfulla F, Malaguti S, Montagna T, Salvini S, Bruschi C, Crotti P, Casale R, Rampulla C. Erectile dysfunction in men with obstructive sleep apnea: an early sign of nerve involvement. *Sleep* **23**(6):775–781(2000).

41. Pressman MR, Diphilippo MA, Kendrick JI, Conroy K, Fry JM. Problems in the interpretation of nocturnal penile tumescence studies: disruption of sleep by occult sleep disorders. *J Urol* **136**:595–598(1985).

42. Schmidt HS, Wise HA. Significance of impaired penile tumescence and associated polysomnographic abnormalities in the impotent patient. *J Urol* **126**:348–351(1981).

43. Hirshkowitz M, Karacan I, Gurakar A, Williams RL. Hypertension, erectile dysfunction and occult sleep apnea. *Sleep* **12**(3):223–232(1989).

44. Santamaria JD, Prior JC, Fleetham JA. Reversible reproductive dysfunction in men with obstructive sleep apnea. *Clin Endocrin* **28**:461–470(1988).

45. Seftel AD, Strohl KP, Loye TL, Bayard D, Kress J, Netzer, NC. Erectile dysfunction and symptoms of sleep disorders. *Sleep* **25**(6):643–647(2002).

46. Akashiba T, Kawahara S, Akahoshi T, Omori C, Saito O, Majima T, Horie T. Relationship between quality of life and mood or depression in patients with severe sleep apnea syndrome. *Chest* **122**(3):861–865(2002).

47. D'Ambrosio C, Bowman T, Mohsenin V. Quality of life in patients with obstructive sleep apnea. Effect of nasal

continuous positive airway pressure—a prospective study. *Chest* **115**(1):123–129(1999).

48. Fornas C, Ballester E, Arteta E, Ricou C, Diaz A, Fernandez A, Alonso J, Montserrat JM. Measurement of general health status in obstructive sleep apnea hypopnea patients. *Sleep* **18**(10):876–879(1995).

49. Gall R, Isaac L, Kryger M. Quality of life in mild obstructive sleep apnea. *Sleep* **16**(8):S59–S61(1993).

50. Jenkinson C, Stradling J, Peterson S. Comparison of three measures of quality of life outcome in the evaluation of continuous positive airway pressure for sleep apnoea. *J Sleep Res* **6**:199–204(1997).

51. Bergner M, Bobbit R, Carter WB, Gilson BS. The sickness impact profile: development and final revision of a health status measure. *Med Care* **19**(8):787–805(1981).

52. Ware J, Snow K, Kosinski M. *SF-36 Health Survey. Manual and Interpretation Guide.* The Health Institute, New England Medical Center, Boston,(1993).

53. Baldwin CN, Griffith KA, Nieto J, O'Connor GT, Walsleben JA, Redline S. The association of sleep-disordered breathing and sleep symptoms with quality of life in the Sleep Heart Health Study. *Sleep* **24**(1):96–105(2001).

54. Smith R, Ronald J, Delaive K, Walld R, Manfreda J, Kryger M. What are obstructive sleep apnea patients being treated for prior to this diagnosis? *Chest* **121**(1):164–172(2002).

55. Young T, Blustein J, Finn L, Palta M. Sleep-disordered breathing and motor vehicle accidents in a population-based sample of employed adults. *Sleep* **20**(8):608–613(1997).

56. Block AJ, Berry D, Webb W. Nocturnal hypoxemia and neuropsychological deficits in men who snore. *Eur J Respir Dis* **69**:405–408(1986).

57. Ramos Platon MJ, Espinar Sierra J. Changes in psychopathological symptoms in sleep apnea patients after treatment with nasal continuous positive airway pressure. *Int J Neurosci* **62**:173–195(1992).

58. Watson R, Greenberg G, Deptula D. Neuropsychological performance following treatment of sleep apnea. *Sleep Res* **14**:136(1985).

59. Ferini-Strambi L, Baietto C, Di Gioia MR, Castaldi P, Castronovo C, Zucconi M, Cappa SF. Cognitive dysfunction in patients with obstructive sleep apnea(OSA): partial reversibility after continuous positive airway pressure(CPAP). *Brain Res Bull* **61**(1):87–92(2003).

38

SLEEP APNEA AND CEREBROVASCULAR DISEASE

Vahid Mohsenin and Henry Yaggi
Yale University School of Medicine, New Haven, Connecticut

INTRODUCTION

Stroke is the third leading cause of death and a major cause of serious long-term disability in the United States. There are several well established and modifiable risk factors for the development of stroke. These include arterial hypertension, cardiac disease, dyslipidemia, diabetes mellitus, and smoking among others. Sleep apnea has been found at alarmingly high rates (>50%) in patients with acute stroke and after full neurologic recovery. Sleep apnea is highly prevalent in the general population with a frequency of 2–4% and is associated with high incidence of obesity, coronary artery disease, and hypertension. There are several hematologic and hemodynamic changes in sleep apnea that can play significant roles in the pathogenesis of stroke. Several recent large epidemiological studies have shown a strong association between these two disorders independent of known risk factors for stroke. There is convincing evidence to believe that sleep apnea is a modifiable risk factor for stroke; however, prospective studies are needed to establish the cause-and-effect relationship. Understanding the link between obstructive sleep apnea and stroke may provide a novel preventative and therapeutic approach in the management of stroke.

OVERVIEW OF THE PUBLIC HEALTH IMPORTANCE OF STROKE

Stroke is a heterogeneous disorder that encompasses various subtypes (Table 38.1) [1]. Ischemic stroke is the most common form of stroke. Within this type is a self-limited form of stroke known as transient ischemic attack (TIA), for which the neurologic symptoms last less than 24 hours. There are several modifiable risk factors for stroke (Table 38.2) [2]. Treatment and prevention of these risk factors reduce the incidence of stroke. However, stroke continues to exact a heavy toll in death and disability [3]. In the United States alone, 750,000 strokes occur annually, and of these 150,000 are fatal. Despite effective therapies for stroke, it remains the third leading cause of death and the leading cause of disability [3]. Many survivors are left with mental and physical impairment, often resulting in periods of prolonged hospitalization and further lengthy convalescence in a stroke rehabilitation center.

OVERVIEW OF THE PUBLIC HEALTH IMPORTANCE OF SLEEP APNEA

Sleep apnea occurs when the upper airway repeatedly closes during sleep and is associated with a constellation of symptoms and objective findings. The obstructive sleep apnea syndrome (OSAS) is present when there is more than five apneas or hypopneas per hour of sleep (apnea–hypopnea index, AHI) in association with excessive daytime sleepiness and symptoms of cognitive dysfunction. OSAS occurs in 4% of men and 2% of women who are between 30 and 60 years of age [4]. The public health impact of OSAS is great, and an increased mortality has been demonstrated in patients with untreated OSAS [5, 6]. There are strong associations with some of the leading causes of mortality

Correspondence to: Vahid Mohsenin, 290 Congress Avenue, New Haven, Connecticut 06519.

TABLE 38.1 Classification of Cerebrovascular Disease

Ischemic stroke		86%
Transient ischemic attack	(20%)	
Thrombotic	(48%)	
Embolic	(18%)	
Hemorrhagic stroke		14%
Intracerebral	(8%)	
Subarachnoid	(6%)	

TABLE 38.2 Modifiable Risk Factors and Their Prevalence in Stroke

Risk Factor	Estimated Relative Risk	Prevalence (%)
Hypertension	4.0–5.0	25–40
Cardiac disease	2.0–4.0	10–20
Atrial fibrillation	5.6–17.6	0.5–1.0
Diabetes mellitus	1.5–3.0	4.0–8.0
Smoking	1.5–2.9	20–40
Alcohol abuse	1.0–4.0	5.0–30
Dyslipidemia	1.0–2.0	6.0–50

including coronary artery disease, congestive heart failure, hypertension, and stroke.

SLEEP DISORDERED BREATHING AND STROKE

There are several epidemiologic studies showing an association between snoring and stroke. Despite some methodologic problems such as self-reporting of snoring and potential for recall bias and misclassification, they demonstrated a strong association, with relative risk ranging from 1.33 to 10.3, similar in degree to the traditional risk factors for stroke, independent of age, gender, smoking, hypertension, and obesity [7–14]. Several large prospective studies corroborate these case-control and cross-sectional studies showing relative risk of 1.33–2.0 [9].

An early study of ten patients recovering from hemispheric stroke revealed a high prevalence (80%) of obstructive sleep apnea compared to age and BMI-matched controls with similar frequency of hypertension and smoking without stroke [15]. Another case-control study of consecutively admitted inpatients with stroke [16] showed high prevalence of obstructive sleep apnea. This study compared the polysomnograms of 27 healthy age- and gender-matched controls recruited from the local population to 24 inpatients with recent stroke confirmed by neurologic examination and imaging studies of the brain. Overall, obstructive sleep apnea was diagnosed in 19% of the controls and 71% of the stroke patients. The mean lowest arterial oxygen saturation level was 91% in the control group

and 85% in the stroke group, and the mean AHI was 4 events per hour for controls and 26 events per hour for stroke patients. Perhaps the strongest epidemiologic evidence demonstrating the association between sleep disordered breathing and cerebrovascular disease comes from the initial results of the Sleep Heart Health Study conducted in the United States [17]. This study explored the cross-sectional association between sleep disordered breathing and prevalent self-reported cardiovascular disease (myocardial infarction, angina, coronary revascularization procedures, heart failure, or stroke) in a large cohort of 6424 individuals who underwent unattended overnight polysomnography at home. This study showed a dose–response relationship between the severity of sleep apnea expressed as AHI quartiles and increasing risk for the development of cardiovascular events. Furthermore, this study demonstrated that even mild to moderate obstructive sleep apnea was significantly associated with the development of coronary artery disease, congestive heart failure, and stroke independent of known cardiovascular risk factors. Interestingly, hypoxemia appeared to explain 10–40% of the AHI effect, and sleep fragmention per se, as measured by the arousal index, was not associated with cardiovascular disease in this study. If the associations observed in this study are indeed causal, it appears that even a modestly elevated risk of stroke coupled with the high prevalence of mild to moderate sleep disordered breathing will have considerable public health implications.

Another approach used to gain some insight into the temporal relationship between sleep disordered breathing and stroke prospectively followed 161 consecutive patients with first-ever stroke or TIA admitted to a stroke unit [18]. Here, TIA was strictly defined according to the National Institute of Neurologic Diseases and Stroke classification [19]. In this study, previously validated portable respiratory recordings were performed within 48–72 hours after admission (acute phase) and after 3 months (stable phase). The important findings of this study were that there were no significant differences in obstructive sleep apnea severity according to stroke subtype (TIA, ischemic stroke, or hemorrhagic stroke) or location of the stroke. The study also found that the frequency of obstructive apneas did not significantly decline from the period immediately after stroke to 3 months later. Patients with stroke and evidence of obstructive sleep apnea had significantly lower functional status, as assessed by the Barthel Index (a multifaceted scale measuring mobility and activities of daily living), compared to patients with stroke and no evidence of sleep disordered breathing at discharge, at 3 months, and at 12 months. Similarly, a recent study [20] of 61 patients admitted to a stroke rehabilitation unit revealed that patients with sleep apnea had significantly worse functional capacity at discharge and a significantly longer stay in the rehabilitation unit even after adjustment for stroke severity.

PATHOPHYSIOLOGIC MECHANISMS IN SLEEP APNEA LEADING TO STROKE

During obstructive sleep apnea, repetitive episodes of airway occlusion result in hypoxemia, hypercapnia, significant changes in intrathoracic pressure, and arousals from sleep. Consequently, this elicits autonomic, hemodynamic, coagulopathic, and vascular injury processes that serve as plausible mechanisms whereby obstructive sleep apnea may cause stroke.

Individual episodes of sleep apnea cause surges in blood pressure at apnea termination with as much as a 40 mmHg rise in mean arterial pressure (Figure 38.1) [21]. In addition to blood pressure swings at night, sustained diurnal hypertension can arise from chronic exposure to recurrent obstructive apneas. The Sleep Heart Health Study [23] demonstrated that sleep disordered breathing was associated with prevalent hypertension even after controlling for potential confounders such as age, gender, body mass index (and other measures of adiposity), alcohol, and smoking. The odds of hypertension increase with increasing severity of OSAS in a dose–response fashion. The presence of sleep disordered breathing at baseline was accompanied by a substantially increased risk for future hypertension at 4 year follow-up with an odds ratio = 2.89 (95% confidence interval 1.6–5.64) for subjects with an AHI > 15 compared to those without any nocturnal apnea [24].

Marked changes in cerebral blood flow and intracranial pressure occur during individual episodes of obstructive sleep apnea [25]. The mechanical effects of increased intracranial pressure may impede cerebral blood flow, thus predisposing to cerebral ischemia.

Cardiac arrhythmia is a known cause of stroke. OSAS is associated with cardiac arrhythmias, conduction abnormalities such as second-degree atrioventricular block [26], and potentially life-threatening arrhythmias such as ventricular tachycardia [27]. In a population of 81 patients with stable heart failure, atrial fibrillation was present approximately four times more frequently in heart failure patients with sleep apnea than in those without [28], thus increasing the risk of cardioembolic stroke with this arrhythmia.

Patent foramen ovale (PFO) is present in 20–54% of subjects with ischemic stroke [29] and 69% of patients with OSAS [30]. PFO can potentially give rise to ischemic stroke by means of paradoxical embolization. Two studies have demonstrated that obstructive sleep apnea may predispose to right-to-left shunting through a PFO [30, 31]. Right-to-left shunting was present in 9 out of 10 patients with

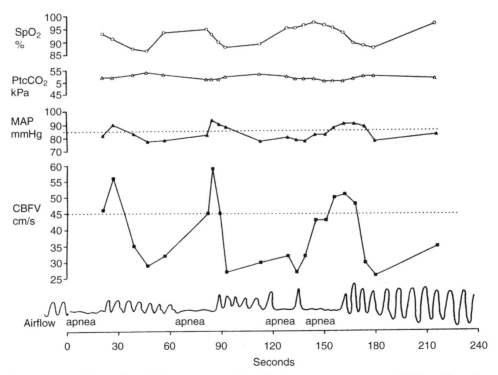

Figure 38.1 Simultaneous recordings of arterial oxygen saturation (Sao_2), transcutaneous arterial Pco_2 ($P_{tc}co_2$), mean arterial blood pressure (MAP), cerebral blood flow velocity (CBFV), and respiratory airflow during sleep in a patient with obstructive sleep apnea. The recording shows prolonged periods of low CBFV compared to baseline (dashed line) during the obstructive apnea with a steep rise at the end of the apnea paralleling rises in MAP. (Adapted from [22].)

PFO during obstructive apneas of greater than 17 seconds in duration [31]. No right-to-left shunting was detected during normal breathing during wakefulness.

Elevated plasma fibrinogen levels are believed to be associated with increased risk of stroke and other cardiovascular events [32–34]. Patients with OSAS have been shown to have elevated morning levels of fibrinogen and platelet aggregability [35–37]. Fibrinogen level was positively correlated to AHI and length of respiratory events and negatively correlated to minimal and average minimal oxygen saturation measured [38]. There was a reduction of platelet reactivity following the application of continuous positive airway pressure (CPAP).

Atherosclerosis is a leading cause of stroke. As noted previously, the strong association between sleep apnea and cardiovascular disease is at least, in part, related to overlapping risk factors for atherosclerosis that track with OSAS, namely, obesity, hypertension, diabetes mellitus, and hyperlipidemia [39]. However, epidemiologic studies derived from the general population suggest that sleep apnea constitutes a significant risk for cardiovascular disease independent of these known risk factors [17, 24, 40]. Furthermore, in these studies, there is a dose–response relationship between the severity of OSAS and the cardiovascular morbidities. This implies that the link between cardiovascular disease and obstructive sleep apnea is specific to the syndrome of sleep apnea.

Experimental evidence has linked hypoxia with atherosclerosis through a variety of pathways: direct injury due to reactive oxygen species, decreased nitric oxide bioavailability, which plays a key role in the regulation of vascular tone, increased endothelin-1, a potent vasoconstrictor, increased adhesiveness of leukocytes, increased homocysteine concentrations, and an increase in oxidized LDL and foam cell formation, which eventually all lead to endothelial dysfunction [41]. C-reactive protein (a nonspecific marker of inflammation and risk factor for atherosclerosis) and IL-6 (a proinflammaory cytokine that is also implicated in the pathogenesis of atherosclerosis) were significantly higher in patients with OSAS compared with weight-matched controls [42].

THERAPY FOR SLEEP APNEA IN STROKE

The availability of CPAP, the main medical therapy for patients with OSAS since the early 1980s, has been demonstrated to be effective in decreasing daytime sleepiness and improving quality of life [43, 44], improving left ventricular function in patients with congestive heart failure and sleep disordered breathing [45], decreasing sleeping and daytime blood pressure in patients with hypertension [46], decreasing motor vehicle accidents [47], and even possibly improving mortality [6]. Two short-term CPAP treatment

trials of patients with OSAS after acute stroke have been published that have provided insight into the acceptance of CPAP in this setting [48, 49]. The results are encouraging as they demonstrate beneficial effects and comparable compliance rates to OSAS patients without stroke. In a non randomized short-term CPAP treatment trial in patients with stroke and OSAS there was a significant reduction in nocturnal blood pressure (8 mmHg) after 10 days of treatment in comparing CPAP compliant and noncompliant patients [49]. Interestingly, in a logistic regression model, aphasia and the severity of motor disability as quantified by the Barthel Index were significant negative predictors of acceptance of CPAP. The second CPAP study was a randomized treatment trial [48], and it demonstrated that depressive symptoms are reduced in patients treated with nasal CPAP at 7 and 28 days compared to untreated controls. There was no significant improvement in delirium, activities of daily living, or cognitive functioning. Compliance was lower in this study (∼50%) perhaps, in part, related to the fact that the cohort was from an older population.

Overall, the primary acceptance of CPAP appears comparable to patients with OSAS without stroke, and CPAP appears to exert a beneficial influence in terms of subjective perception of well-being, hypertension, and depression. However, long-term compliance is not certain especially in a population of patients with more functional and cognitive disability.

PRACTICAL GUIDELINES FOR THE MANAGEMENT OF PATIENTS WITH STROKE: SLEEP EVALUATION

The patients with TIA and stroke should undergo a thorough sleep history interview and physical examination. The most common presentation of OSAS is excessive daytime sleepiness and unrefreshing sleep. Intermittent snoring with breath holding terminated by loud snorts and body movements are typical features that patients may not be able to report. Other related complaints include restless sleep, choking or coughing during sleep, nocturia, and headaches. Physical examination of the upper airways may disclose a deviated nasal septum or swollen turbinates, retrognathia, an enlarged tongue, a hypertrophic uvula, a redundant soft palate, or paralyzed vocal cords. Limb weakness also appeared to be an independent predictor of OSAS in acute stroke, but other stroke characteristics such as severity and subtype do not appear to be predictive of upper airway obstruction [50]. In view of the high prevalence of OSAS and nonspecificity of symptoms in the setting of stroke, every patient should be screened. Likewise, patients with OSAS should also be evaluated for sleep-related breathing disorders.

CONCLUSION

OSAS is strongly associated with stroke independent of known cardiovascular risk factors. The mechanisms underlying this risk of stroke are multifactorial and include hypertension, changes in cerebral hemodynamics, paradoxical embolism through PFO, increased hypercoagulability, and increased risk of atherosclerosis. In view of the high prevalence of OSAS in stroke patients and the effectiveness of CPAP, patients with stroke should undergo screening for sleep disordered breathing. Sleep apnea represents a modifiable risk factor, but whether treatment of sleep apnea in the acute stroke setting, the rehabilitation setting, or as primary/secondary prevention is of benefit awaits further treatment studies.

REFERENCES

1. National Institute of Neurological Disorders and Stroke. Special Report: Classification of Cerebrovascular Disease. *Stroke* **21**:637–676(1990).

2. Sacco R. Risk factors and outcomes for ischemic stroke. *Neurology* **45**:S10–S14(1995).

3. American Heart Association. *2002 Heart and Stroke Statistical Update*. American Heart Association, Dallas, TX.

4. Young T, Palta M, Dempsey J, Skatrud J, Weber S, Badr S. The occurrence of sleep-disordered breathing among middle-aged adults. *N Engl J Med* **328**:1230–1235(1993).

5. He J, Kryger M, Zorick F, Conway W, Roth T. Mortality and apnea index in obstructive sleep apnea. Experience in 385 male patients. *Chest.* **94**:9–14(1988).

6. Marti S, Sampol G, Munoz X, et al. Mortality in severe sleep apnoea/hypopnoea syndrome patients: impact of treatment. *Eur Respir J* **20**:1511–1518(2002).

7. Hu F, Willet W, Manson J, et al. Snoring and the risk of cardiovascular disease in women. *J Am Coll Cardiol* **35**:308–313(2000).

8. Jennum P, Schultz-Larsen K, Davidsen M, Christensen N. Snoring and risk of stroke and ischaemic heart disease in a 70 year old population. A 6-year follow-up study. *Int J Epidemiol* **23**:1159–1164(1994).

9. Koskenvuo M, Kaprio J, Telakivi T, Partinen M, Heikkila K, Sarna S. Snoring as a risk factor for ischaemic heart disease and stroke in men. *Br Med J (Clin Res Ed)* **294**:16–19(1987).

10. Neau J, Meurice J, Paquereau J, Chavagnat J, Ingrand P, Gil R. Habitual snoring as a risk factor for brain infarction. *Acta Neurol Scand* **92**:63–68(1995).

11. Palomaki H. Snoring and the risk of ischemic brain infarction. *Stroke.* **22**:1021–1025(1991).

12. Partinen M, Palomaki H. Snoring and cerebral infarction. *Lancet* **2**:1325–1326(1985).

13. Smirne S, Palazzi S, Zucconi M, Chierchia S, Ferini-Strambi L. Habitual snoring as a risk factor for acute vascular disease. *Eur Respir J* **6**:1357–1361(1993).

14. Spriggs D, French J, Murdy J, Curless R, Bates D, James O. Snoring increases the risk of stroke and adversely affects prognosis. *Q J Med* **83**:555–562(1992).

15. Mohsenin V, Valor R. Sleep apnea in patients with hemispheric stroke. *Arch Phys Med Rehabil* **76**:71–76(1995).

16. Dyken M, Somers V, Yamada T, Ren Z, Zimmerman M. Investigating the relationship between stroke and obstructive sleep apnea. *Stroke.* **27**:401–407(1996).

17. Shahar E, Whitney C, Redline S, et al. Sleep-disordered breathing and cardiovascular disease: cross-sectional results of the Sleep Heart Health Study. *Am J Respir Crit Care Med* **163**:19–25(2001).

18. Parra O, Arboix A, Bechich S, et al. Time course of sleep-related breathing disorders in first-ever stroke or transient ischemic attack. *Am J Respir Crit Care Med* **161**:375–380(2000).

19. Whisnant J, Busford J, Bernstein E, Cooper E, Dyken M, Eastone D. National Institute of Neurologic Disease and Stroke special report: classification of cerebrovascular disease. *Stroke* **21**:637–676(1990).

20. Kaneko Y, Hajek V, Zivanovic V, Raboud J, Bradley T. Relationship of sleep apnea to functional capacity and length of hospitalization following stroke. *Sleep* **26**:293–297(2003).

21. Leung R, Bradley T. Sleep apnea and cardiovascular disease. *Am J Respir Crit Care Med* **164**:2147–2165(2001).

22. Balfors E, Franklin K. Impairment in cerebral perfusion during obstructive sleep apneas. *Am J Respir Crit Care Med* **150**:1587–1591(1994).

23. Neito F, Young T, Lind B, et al. Association of sleep-disordered breathing, sleep apnea, and hypertension in a large community based study. *JAMA* **283**:1829–1836(2000).

24. Peppard P, Young T, Palta M, Skatrud J. Prospective study of the association between sleep-disordered breathing and hypertension. *N Engl J Med* **342**:1378–1384(2000).

25. Shiomi T, Guilleminault C, Stoohs R, Schnittger I. Leftward shift of the intraventricular septum and pulsus paradoxus in obstructive sleep apnea syndrome. *Chest* **100**:894–902(1991).

26. Zwillich C, Devlin T, White D, Douglas N, Weil J, Martin R. Bradycardia during sleep apnea. Characteristics and mechanism. *J Clin Invest* **69**:1286–1292(1982).

27. Fichter J, Bauer D, Arampatzis S, Fries R, Heisel A, Sybrecht G. Sleep-related breathing disorders are associated with ventricular arrhythmias in patients with an implantable cardioverter defibrillator. *Chest* **122**:558–561(2002).

28. Javaheri S, Parker T, Liming J, et al. Sleep apnea in 81 ambulatory male patients with stable heart failure: types and their prevalences, consequences, and presentations. *Circulation* **97**:2154–2159(1998).

29. Lechat P, Mas J, Lascault G, et al. Prevalence of patent foramen ovale in patients with stroke. *N Engl J Med* **318**:1148–1152(1988).

30. Shanoudy H, Soliman A, Raggi P, Liu J, Russell D, Jarmukli N. Prevalence of patent foramen ovale and its contribution to hypoxemia in patients with obstructive sleep apnea. *Chest* **113**:91–96(1998).

31. Beelke M, Angeli S, Del Sette M, et al. Obstructive sleep apnea can be provocative for right-to-left shunting through a patent foramen ovale. *Sleep* **25**:856–862(2002).

32. Resch K, Ernst E, Matrai A, Paulsen H. Fibrinogen and viscosity as risk factors for subsequent cardiovascular events in stroke survivors. *Ann Intern Med* **117**:371–375(1992).

33. Toss H, Lindhaul B, Siegbahn A, Wallentin L. Prognostic influence of increased fibrinogen and C-reactive protein levels in unstable coronary artery disease. Frisc Study Group. Fragmin during instability in coronary artery disease. *Circulation* **96**:4204–4210(1997).

34. Wilhelmsen L, Svardsudd K, Kristoffer K, Larsson B, Wellin L, Tiblin G. Fibrinogen as a risk factor for stroke and myocardial infarction. *N Engl J Med* **311**:501–505(1984).

35. Bokinsky G, Miller M, Ault K, Husband P, Mitchell M. Spontaneous platelet activation and aggregation during obstructive sleep apnea and its response to therapy with nasal continuous positive airway pressure: a preliminary investigation. *Chest* **108**:625–630(1995).

36. Chin K, Ohi M, Kita H, et al. Effects of NCPAP therapy on fibrinogen levels in obstructive sleep apnea syndrome. *Am J Respir Crit Care Med* **153**:1972–1976(1996).

37. Toffler G, Brezinski D, Shafer A, et al. Concurrent morning increase in platelet aggregability and the risk of myocardial infarction and sudden cardiac death. *N Engl J Med* **316**:1514–1518(1987).

38. Wessendorf T, Thilmann A, Wang Y, Schreiber A, Konietzko N, Teschler H. Fibrinogen levels and obstructive sleep apnea in ischemic stroke. *Am J Respir Crit Care Med* **162**:2039–2042(2000).

39. Newman A, Nieto J, Guidry U, et al. Relation of sleep-disordered breathing to cardiovascular disease risk factors. *Am J Epidemiol* **154**:50–59(2001).

40. Young, Peppard P, Palta M, et al. Population-based study of sleep-disordered breathing as a risk factor for hypertension. *Arch Intern Med* **157**:1746–1752(1997).

41. Lavie L. Obstructive sleep apnoea syndrome: an oxidative stress disorder. *Sleep Med Rev* **7**:35–51(2003).

42. Yokoe T, Minoguchi K, Matsuo H, et al. Elevated levels of C-reactive protein and interleukin-6 in patients with obstructive sleep apnea syndrome are decreased by nasal continuous positive airway pressure. *Circulation* **107**:1129–1134(2003).

43. D'Ambrosio C, Bowman T, Mohsenin V. Quality of life in patients with obstructive sleep apnea. Effect of nasal continuous positive airway pressure: a prospective study. *Chest* **115**:123–129(1999).

44. Young T, Peppard P, Gottlieb D. Epidemiology of obstructive sleep apnea: a population health perspective. *Am J Respir Crit Care Med* **165**:1217–1239(2002).

45. Kaneko Y, Floras J, Usui K, et al. Cardiovascular effects of continuous positive airway pressure in patients with heart failure and obstructive sleep apnea. *N Engl J Med* **348**:1233–1241(2003).

46. Pepperell J, Ramdassingh-Dow S, Crosthwaite N, et al. Ambulatory blood pressure after therapeutic and subtherapeutic nasal continuous positive airway pressure for obstructive sleep apnoea: a randomised parallel trial. *Lancet* **359**:204–210(2002).

47. Findley L, Smith C, Hooper J, Dineen M, Suratt P. Treatment with nasal CPAP decreases automobile accidents in patients with sleep apnea. *Am J Respir Crit Care Med* **161**:857–859(2000).

48. Sandberg O, Franklin KA, Bucht G, Eriksson S, Gustafson Y. Nasal continuous positive airway pressure in stroke patients with sleep apnoea: a randomized treatment study. *Eur Respir J* **18**:630–634(2001).

49. Wessendorf T, Wang Y, Thilmann A, Sorgenfrei U, Konietzko N, Teschler H. Treatment of obstructive sleep apnoea with nasal continuous positive airway pressure in stroke. *Eur Respir J* **18**:623–629(2001).

50. Turkington P, Bamford J, Wanklyn P, Elliott M. Prevalence and predictors of upper airway obstruction in the first 24 hours after acute stroke. *Stroke* **33**:2037–2042(2002).

39

RADIOGRAPHIC AND ENDOSCOPIC EVALUATION OF THE UPPER AIRWAY

RICHARD J. SCHWAB AND NEIL S. KLINE

University of Pennsylvania Medical Center, Philadelphia, Pennsylvania

Modern upper airway imaging techniques have significantly advanced our understanding of the pathophysiology and biomechanics of obstructive sleep apnea. These imaging modalities (magnetic resonance imaging (MRI), computed tomography (CT), nasopharyngoscopy, cephalometry, and fluoroscopy) have objectively quantified upper airway structures and identified specific craniofacial and oropharyngeal soft tissue structural risk factors for obstructive sleep apnea.

This chapter will begin with a brief overview of the anatomy of the upper airway. This will be followed by a discussion of the imaging modalities that have been utilized to study the upper airway in subjects with obstructive sleep apnea (OSA). The advantages and disadvantages of each modality will be reviewed. New imaging techniques, such as digital morphometrics, will be introduced. This will be followed by a review of important observations from static, dynamic, and state-dependent upper airway imaging that have advanced our understanding of the pathogenesis of OSA. Finally, the physiologic effects of the primary treatment options (continuous positive airway pressure (CPAP), weight loss, oral mandibular repositioning devices, and upper airway surgery) for sleep apnea will be assessed using these imaging modalities.

The pathogenesis of OSA involves both an anatomic and a neurologic component. However, this chapter will only discuss the anatomical issues as they relate to this disorder. There is strong evidence that upper airway anatomy plays

This work was supported by the National Institutes of Health Grants HL-57843, HL-60287, HL-67948, HL-62408.

an integral role in the pathophysiology and biomechanics of OSA [1–3].

UPPER AIRWAY ANATOMY

The pharynx performs three crucial physiologic functions (breathing, swallowing, vocalization) that impact on each other. Each of these functions is influenced by upper airway anatomy. The upper airway is subdivided into three sections: (1) the nasopharynx (the nasal turbinates to the hard palate), (2) the oropharynx, subdivided into retropalatal (the hard palate to the caudal margin of the soft palate) and retroglossal (the caudal margin of the soft palate to the base of the epiglottis) regions; and (3) the hypopharynx (the base of the tongue to the larynx) (see Figures 39.1–39.3). Below the hypopharynx lies the larynx, which is attached to the relatively nonflexible trachea [1, 2].

The boundaries of the oropharynx include the soft palate superiorly and anteriorly, the dorsum of the tongue inferiorly, the palatoglossal and palatopharyngeal arches on the lateral aspects, and the posterior wall [2, 4]. The posterior walls are formed by the palatopharyngeus, superior constrictor, and middle constrictor muscles. The lateral pharyngeal walls consist of several muscles including the hyoglossus, palatoglossus, palatopharyngeus, stylohyoid, styloglossus, and the stylopharyngeus. Additionally, palatine tonsils are located in the lateral walls of the oropharynx. The tonsils often become enlarged in children, leading to obstruction of the airway and sleep apnea. The

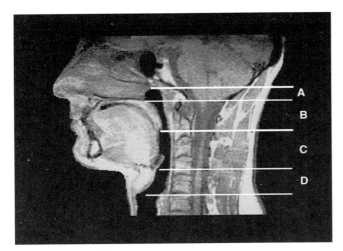

Figure 39.1 Midsagittal MRI of a normal subject demonstrating the upper airway regions: (A) nasopharynx, (B) retropalatal region, (C) retroglossal region, and (D) hypopharynx.

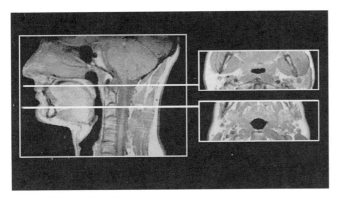

Figure 39.2 Midsagittal MRI with corresponding axial images of the retropalatal and retroglossal region in a normal subject.

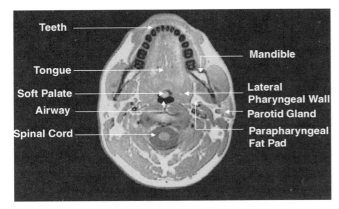

Figure 39.3 Axial MRI in the retropalatal region demonstrating upper airway anatomy of a normal subject.

mandible bounds all of the soft tissue structures that are lateral to the airway in the oropharynx [1, 2] (see Figure 39.3). The base of the tongue, the epiglottis, and the larynx form the hypopharynx, which is inferior to the oropharynx.

UPPER AIRWAY IMAGING MODALITIES

There are several imaging modalities that have been utilized to study the upper airway. Below is a discussion of the advantages and disadvantages of nasopharyngoscopy, fluoroscopy, acoustic reflection, cephalometry, computed tomography (CT), magnetic resonance imaging (MRI), and digital morphometrics (see Table 39.1); all of these imaging techniques have provided important information that has advanced our understanding of upper airway anatomy and physiology.

Nasopharyngoscopy

Nasopharyngoscopy utilizes fiberoptics to examine the upper airway. This technique is frequently used by otorhinolaryngologists to evaluate the nasal and oropharyngeal passages. It allows for visualization of the lumen of the upper airway without emitting ionizing radiation. Dynamic and state-dependent changes in the upper airway can also be obtained with nasopharyngoscopy. However, nasopharyngoscopy is invasive, sometimes difficult to tolerate, and does not allow for visualization of the soft tissue and skeletal structures beyond the lumen of the airway.

The Müller maneuver, inspiration against an occluded mouth and nares, can be utilized during nasopharyngoscopy to potentially demonstrate collapsibility of the oropharynx [5]. It has been proposed that apneics who demonstrate collapsibility of the retropalatal region of the oropharynx may have a better outcome after uvulopalatopharyngoplasty (UPPP) [4]. Conversely, patients who show collapsibility of the retroglossal region may benefit from a sliding genioplasty or maxillomanibular advancement. However, most of the studies examining the Müller maneuver have not objectively quantified upper airway changes.

Fluoroscopy

Fluoroscopy utilizes standard radiographic techniques to examine the upper airway. It allows for continuous, dynamic, uniplanar visualization in the lateral and anterior–posterior planes of the upper airway. However, because of the continuous radiation that is emitted, it is not routinely used to examine the upper airway in patients with sleep apnea. Additionally, it does not allow for quantitative or three-dimensional analysis of the upper airway and surrounding soft tissue structures.

TABLE 39.1   Advantages and Disadvantages of Different Upper Airway Imaging Modalities

Advantages	Disadvantages
Acoustic Reflection	
• Inexpensive	• Does not display information about upper airway structures beyond the lumen
• Noninvasive	
• No radiation emission	• Does not display anatomical structures
• Measures airway areas and distances	• Opening mouth alters anatomy
• Dynamic imaging	• Not useful for studying state-dependent changes
• No weight limitations	
Fluoroscopy	
• Dynamic imaging	• Emits ionizing radiation
	• Only uniplanar visualization in lateral or anteroposterior view
	• Unable to quantify sizes of soft tissue structures
Nasopharyngoscopy	
• Direct visualization using fiberoptics	• Invasive
• Easily performed	• Does not display information about upper airway structures beyond the lumen
• Minimal complications/discomfort	
• Dynamic imaging capability	
• State-dependent imaging	
• Using Müller maneuver may predict best surgical procedure (UPPP vs. sliding genioplasty)	
• No weight limitations	
Cephalometry	
• Inexpensive	• Does not provide cross-sectional, three-dimensional, or volumetric data
• No weight limitations	
• Allows for quantification of skeletal structures and their angles	• Subject must be in standardized position, sitting upright in rigid head stabilizer
• Allows for evaluation of effect of oral appliances on airway caliber	• Unable to perform dynamic or state-dependent imaging
Computed Tomography (CT)	
• Good resolution of soft tissue and skeletal structures	• Expensive (compared to cephalometry)
	• Emits ionizing radiation
• Allows for cross-sectional and three-dimensional reconstructions	• Poor fat tissue resolution
	• 300 pound weight limit
• Electron beam CT allows fast, dynamic imaging	
• Less expensive than MRI	
Magnetic Resonance Imaging (MRI)	
• Excellent soft tissue and adipose tissue resolution	• Most expensive imaging modality
	• Difficult for subjects to initiate and maintain sleep in loud MRI scanner
• Three-dimension and volumetric reconstruction	• 300 pound weight limit
• Cross-sectional representation with quantification of structure sizes	• Exclusion of some subjects with metallic implants (i.e., pacemakers)
• No ionizing radiation	
• Dynamic imaging capability with ultrafast MRI	
Digital Morphometrics	
• Inexpensive	• Does not provide cross-sectional, three-dimensional, or volumetric data
• Easy to perform	
• Imaging in all body positions	• Only allows for visualization of the lumen and not the surrounding structures
• Noninvasive and safe	• Allows for quantification with parallel laser instrument
	• Subject must be in standardized position
	• Unable to perform dynamic or state-dependent imaging
	• Has not been validated with independent investigations
	• Opening mouth alters anatomy

Acoustic Reflection

Acoustic reflection relies on the reflection of sound waves through the mouth and pharynx to calculate upper airway area [6, 7]. The technique involves inserting a mouthpiece between the incisors and sound waves are emitted from the mouthpiece and reflect back. This technique is inexpensive, easy to perform, and noninvasive. In addition, acoustic reflection is a rapid technique and allows for dynamic airway evaluation. However, it has primarily been used in research settings.

Because acoustic reflection uses sound waves, it only allows for the calculation of upper airway areas and distances. It does not provide an actual image of the upper airway nor does it provide data about structures that are beyond the lumen of the upper airway [6, 7].

Cephalometry

Cephalometry is a two-dimensional radiograph that examines soft tissue and craniofacial structures that bound the upper airway. The size of upper airway craniofacial structures can be accurately measured with a cephalometric radiograph. The angles between these craniofacial structures can also be identified with cephalometry. However, standardized imaging and analysis parameters must be utilized in making a cephalometric radiograph and the head needs to be placed in a rigid structure for stabilization [8–12]. Therefore a cephalometric radiograph is usually obtained with the subject in a sitting position.

Cephalometry is inexpensive, readily available, and easy to perform. It allows for objective measurements of the mandible, maxilla, and hyoid and the angles that relate to these structures. However, three-dimensional representations of the head and neck are not possible with standard cephalometric approaches. It is also not suited for the evaluation of dynamic or state-dependent changes of the upper airway. In addition, cephalometry only provides uniplanar information primarily from a lateral view.

Studies utilizing cephalometry have demonstrated that subjects with small mandibles, retrognathia, and/or micrognathia are at increased risk for obstructive sleep apnea [8–12]. An inferiorly placed hyoid bone has also been shown to be a risk factor for obstructive sleep apnea with cephalometrics [12]. Cephalometry has also been used to evaluate the effect of oral appliances on the mandible and the airway.

Computed Tomography

Computed tomography (CT) provides high-resolution imaging of the upper airway and surrounding soft tissue and craniofacial structures [13–18]. By compiling axial CT images, three-dimensional reconstructions can be created of the upper airway and surrounding soft tissue and craniofacial structures [16, 18]. Newer technology using helical CT has allowed for direct volumetric measurements of the upper airway structures. To investigate the dynamic changes of the upper airway, electron beam CT has been used, providing 50 ms interval pictures [14].

Because CT incorporates the emission of radiation, it is not ideally suited for studying the upper airway changes during sleep (or any study in which repeated imaging is required). Additionally, CT is relatively expensive compared to the aforementioned modalities. Moreover, at least compared to MRI, CT has relatively poor resolution for adipose tissue. Despite these shortcomings, studies utilizing CT imaging have provided important information about the biomechanics and anatomy of the upper airway.

Magnetic Resonance Imaging

Important insights into the pathogenesis of OSA have also been made with MRI [3,19–23]. MRI provides superior upper airway soft tissue resolution (particularly adipose tissue) compared to CT [4, 14]. In addition to cross-sectional imaging, three-dimensional volumetric reconstructions of the head and neck structures can be compiled with MRI illuminating the size and shape of specific upper airway structures (e.g., airway, parapharyngeal fat pads, tongue, soft palate, lateral pharyngeal walls and mandible) [3, 22] (see Figure 39.4). Since MRI does not expose subjects to radiation, multiple sets of images can be obtained from the same subject without risk.

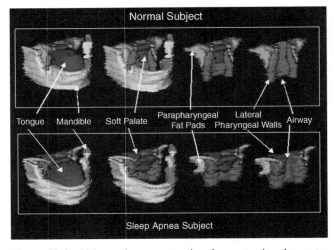

Figure 39.4 Volumetric reconstruction demonstrating the anatomical differences between a normal subject and a patient with sleep apnea. The patient with sleep apnea has a smaller airway, larger soft palate, and larger lateral pharyngeal walls than the normal subject.

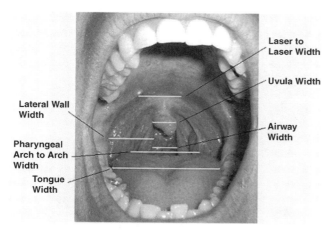

Figure 39.5 Digital morphometric photograph of a normal subject demonstrating pharyngeal anatomy that can be objectively measured with this technique.

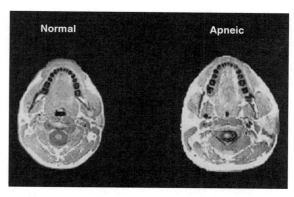

Figure 39.6 Axial MR image of a normal subject and a patient with sleep apnea demonstrating a significantly smaller airway area in the apneic.

Unfortunately, MRI is not available at all institutions and is expensive. Because MRI uses powerful magnets, subjects with some metallic implants (i.e., pacemakers) are excluded from investigation. Additionally, analogous to CT imaging, MRI also has weight limitations.

Digital Morphometrics

Digital morphometrics involves obtaining photographs of the oropharynx using a digital camera. Structures that can be visualized using this technique include the uvula, lateral pharyngeal walls, tongue, airway, and soft palate. Additionally, this technique may be used to image extra-oral structures including submental tissue and the mandible [24].

This technology is inexpensive, noninvasive, and easy to perform. The Mallampati pharyngeal airway classification can be objectively measured with this imaging technique [24]. There are no risks to the subject with digital morphometrics. Measurements can be obtained with the subject in all body positions but these measurements can only be obtained during wakefulness with the mouth opened. The morphometric images provide surface information only and do not allow for evaluation of the soft tissue and bony structures beyond the lumen of the airway (Figure 39.5). Although this is an interesting new imaging modality, investigations must be performed to determine the value of digital morphometrics as an imaging tool for examining upper airway anatomy in patients with sleep apnea.

INSIGHTS INTO THE PATHOPHYSIOLOGY OF OBSTRUCTIVE SLEEP APNEA

Static Imaging During Wakefulness

Upper airway imaging during wakefulness has provided significant insights into the pathophysiology and pathogenesis of obstructive sleep apnea. The majority of these studies have utilized MRI and CT imaging.

These studies have demonstrated that upper airway caliber during wakefulness is smaller (primarily in the retropalatal region) in subjects with obstructive sleep apnea compared to normals [2–4]. The smaller upper airway in apneics is secondary to enlargement of the surrounding soft tissues and/or changes to the craniofacial structures (see Figures 39.4 and 39.6). Recently, it has been demonstrated that the volume of the upper airway soft tissue structures (tongue, lateral pharyngeal walls, soft palate, parapharyngeal fat pads; see Figure 39.4) was significantly greater in apneics than normals and that this enlargement was a significant risk factor for sleep apnea [3]. The enlargement of the soft tissue structures surrounding the upper airway in apneics is thought to be secondary to the effects of obesity, gender, and genetics.

Obesity has been shown to increase the size of the parapharyngeal fat pads and may increase the size of the muscular upper airway soft tissue structures [1, 3]. Recently, heritability of upper airway soft tissue structures has been demonstrated, suggesting that genetic factors are important in mediating the size of these structures. Finally, imaging studies investigating gender differences have determined that soft tissue structures of the oropharynx are larger in men. Such studies have also shown that the airway diameter is smaller in women compared to men [25].

In addition to the enlargement of upper airway soft tissue structures in apneics, changes in craniofacial structure have also been shown to be important in the pathogenesis of obstructive sleep apnea [12]. In general, these cephalometric studies have demonstrated that the mandible is smaller and the hyoid bone is inferiorly positioned in patients with obstructive sleep apnea [8–12].

Dynamic Upper Airway Imaging

Dynamic upper airway imaging provides a better framework than static imaging for understanding of the sequence

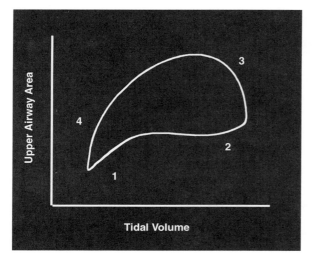

Figure 39.7 Diagram demonstrating the decrease in upper airway volume at the end of expiration (phase 4). Phases 1 and 2 represent inspiration. Phase 3 represents early expiration.

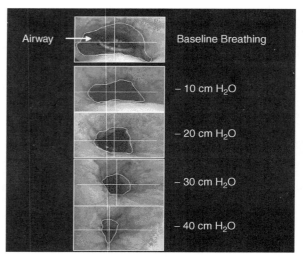

Figure 39.8 Nasopharyngoscopy in a normal subject during a Müller maneuver demonstrating the collapsibility of the pharynx with greater negative pressure generated during subsequent Müller maneuvers.

of events during pharyngeal obstruction and apnea [5, 26, 27]. Dynamic imaging studies during respiration, utilizing MRI and CT, have demonstrated important changes in upper airway caliber during respiration with airway size being relatively constant in inspiration, the largest in early expiration, and the smallest at the end of expiration [26, 27]. The end of expiration may be a particularly vulnerable time for airway narrowing in patients with obstructive sleep apnea [Figure 39.7]. Airway closure in obstructive sleep apnea, however, has been shown to occur during inspiration or expiration [13]. Dynamic changes in upper airway mechanics can also be examined with nasopharyngoscopy during a Müller maneuver [5, 28]. Such studies have shown that the upper airway is smaller and more compliant in apneics compared to normal subjects [5, 28] (see Figure 39.8). These studies have also demonstrated that there is a difference in the collapsibility of the upper airway in different airway regions (i.e., between the retropalatal and retroglossal regions) [5]. The retropalatal region is more likely to collapse under negative pressure generated with the Müller maneuver than the retroglossal region [5, 28].

State-Dependent Imaging

The majority of upper airway imaging studies in patients with sleep apnea have been obtained during wakefulness [17–20]. Although such studies have provided important insights into the pathophysiology of obstructive sleep apnea, sleep apnea is a disorder that occurs during sleep. State-dependent studies have demonstrated decreases in cross-sectional area in the retropalatal region during sleep

[21] (see Figure 39.9). Such studies have also shown that the retropalatal region is more likely to occlude during sleep than the retroglossal region in patients with obstructive sleep apnea [21]. The soft palate, tongue, and lateral pharyngeal walls all have been shown to contribute to the closure of the upper airway during sleep [21]. The airway closes in both anterior–posterior and lateral dimensions during sleep. State-dependent reductions in the lateral airway dimension in the retropalatal region may be secondary to thickening of the lateral pharyngeal walls [21] (Figure 39.9).

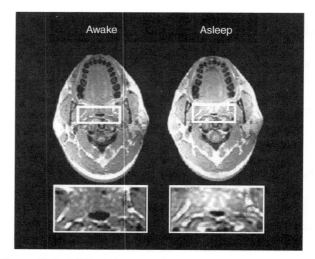

Figure 39.9 MRI of the retropalatal region of a normal subject during wakefulness (*left*) and during sleep (*right*). Airway caliber is smaller during sleep.

UPPER AIRWAY IMAGING AND THE TREATMENT OF OBSTRUCTIVE SLEEP APNEA

Not only has upper airway imaging provided seminal insights into the pathogenesis of obstructive sleep apnea but it has also provided a foundation that has led to a better understanding of the physiologic effects of the primary treatments (CPAP, weight loss, oral appliances, upper airway surgery) for OSA.

Obesity is a major risk factor for sleep apnea, and weight loss has been shown to clinically improve sleep disordered breathing [29, 30]. The specific mechanism by which obesity increases the risk for sleep apnea is not known. However, upper airway imaging studies have shown that the volume of the parapharyngeal fat pads is enlarged in patients with sleep apnea [3]. Thus it has been hypothesized that weight loss would decrease the size of these fat pads, which surround the upper airway. In fact, imaging studies have shown that an 18% weight loss increases upper airway caliber and decreases lateral pharyngeal wall and parapharyngeal fat pad volumes (Figures 39.10 and 39.11) [23]. Such data suggest that weight loss reduces the size of adipose and nonadipose tissue surrounding the upper airway [23].

Although weight loss should be advocated in all patients with sleep apnea, it is often difficult to achieve in this patient population. Therefore the primary treatment for patients with sleep apnea is CPAP since it has been shown to be highly effective and is noninvasive [31]. CPAP acts as a pneumatic splint to enlarge the upper airway in both normals and apneics. The volume of the upper airway increases as CPAP increases [19, 32] (Figure 39.12). Upper airway imaging studies with CPAP have demonstrated that CPAP increases airway diameter more in the lateral dimension than in the anterior–posterior dimension

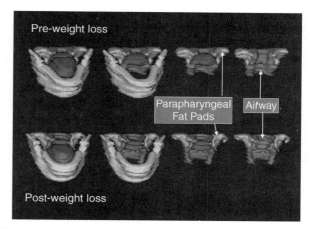

Figure 39.11 Volumetric reconstruction in a normal subject demonstrating the upper airway anatomical differences between pre-weight loss and post-weight loss. The parapharyngeal fat pads and lateral pharyngeal walls are larger pre-weight loss.

(Figure 39.12). The increase in the lateral airway dimension with CPAP is secondary to a reduction in lateral pharyngeal wall thickness [19, 20, 32].

CPAP is the primary treatment option for patients with OSA; however, compliance with CPAP is not always ideal [31]. For patients who are unable to tolerate CPAP, an alternative option is an oral appliance. Mandibular repositioning devices have been the most utilized oral appliances [12, 33]. It has been shown that mandibular repositioning devices are an effective, noninvasive alternative to CPAP in patients with mild to moderate sleep disordered breathing [33]. Although mandibular repositioning devices have been shown to increase the posterior airway space, the specific biomechanical alterations that explain the increase in airway caliber with these devices have not been well characterized. Furthermore, each oral appliance (there are over 50 appliances currently available) may

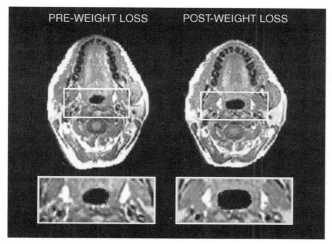

Figure 39.10 MRI in a normal subject demonstrating the increase in upper airway area in the retropalatal region before and after weight loss.

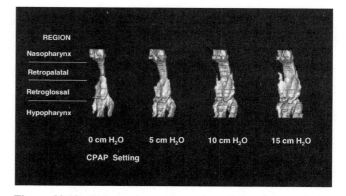

Figure 39.12 Volumetric reconstruction of the pharynx during different CPAP settings in a normal subject. The retropalatal and retroglossal regions increase in the lateral dimension with increasing CPAP.

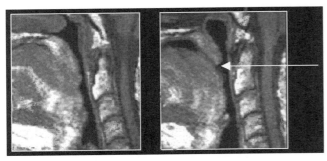

Figure 39.13 Midsagittal MR images demonstrating a pre-UPPP apneic on the left and the same patient, post-UPPP, on the right. The airway caliber is increased in the region where the soft palate is resected (see arrow).

have a different mechanism of action. Nonetheless, upper airway imaging studies have shown that mandibular advancing devices increase airway caliber more in the retropalatal region (posterior and lateral increases in airway dimensions) than in the retroglossal region [33].

In patients with obstructive sleep apnea who are unable to tolerate CPAP or an oral appliance, upper airway surgery is a viable treatment option. Unfortunately, the success rate in patients undergoing the most common surgery for sleep apnea—uvulopalatopharyngoplasty (UPPP) surgery—is only 50% [34]. A UPPP involves the removal of the tonsils (if present), uvula, distal margin of the soft palate, and any excessive pharyngeal tissue. Sleep apnea patients who demonstrate retropalatal closure of the airway during a Müller maneuver have a better outcome with uvulopalatopharyngoplasty than those patients who demonstrate retroglossal collapse [4, 35]. However, the biomechanical changes that underlie the efficacy or lack of efficacy of

UPPP have not been determined. Preliminary upper airway imaging studies in patients undergoing UPPP have shown that airway caliber increases in the region in which the soft palate is resected (Figures 39.13 and 39.14). However, these preliminary studies have shown that airway caliber remains small in the nonresected portion of the soft palate.

CONCLUSIONS

Upper airway imaging has provided important information about the biomechanics and pathogenesis of obstructive sleep apnea. These imaging techniques can objectively quantify oropharyngeal soft tissue and craniofacial structures. This has led to a better understanding of the anatomic risk factors that contribute to obstructive sleep apnea. Currently, there are no approved clinical indications for imaging subjects with suspected sleep apnea. However, imaging studies should be considered in sleep apneic patients being evaluated for upper airway surgery or oral appliances.

REFERENCES

1. Hudgel DW. The role of upper airway anatomy and physiology in obstructive sleep apnea. *Clin Chest Med* **13**:383–398(1992).

2. Schwab RJ, Gupta KB, Gefter WB, Metzger LJ, Hoffman EA, Pack AI. Upper airway and soft tissue anatomy in normal subjects and patients with sleep-disordered breathing. Significance of the lateral pharyngeal walls. *Am J Respir Crit Care Med* **152**:1673–1689(1995).

3. Schwab RJ, Pasirstein M, Pierson R, Mackley A, Arens R, Maislin G, Pack AI. Identification of upper airway anatomic risk factors for obstructive sleep apnea with volumetric MRI. *Am J Respir Crit Care Med* **168**:522–530(2003).

4. Schwab RJ, Goldberg AN. Upper airway assessment: radiographic and other imaging techniques. *Otolaryngol Clin North Am* **31**:931–968(1998).

5. Ritter CT, Trudo FJ, Goldberg AN, Welch KC, Maislin G, Schwab RJ. Quantitative evaluation of the upper airway during nasopharyngoscopy with the Müller maneuver. *Laryngoscope* **109**:954–963(1999).

6. Fredberg JJ, Wohl ME, Glass GM, Dorkin HL. Airway area by acoustic reflections measured at the mouth. *J Appl Physiol* **48**:749–758(1980).

7. Bradley TD, Brown IG, Grossman RF, et al. Pharyngeal size in snorers, nonsnorers, and patients with obstructive sleep apnea. *N Engl J Med* **315**:1327–1331(1986).

8. Bacon WH, Turlot JC, Krieger J, Stierle JL. Cephalometric evaluation of pharyngeal obstructive factors in patients with sleep apneas syndrome. *Angle Orthod* **60**:115–122(1990).

9. DeBerry-Borowiecki B, Kukwa A, Blanks RH. Cephalometric analysis for diagnosis and treatment of obstructive sleep apnea. *Laryngoscope* **98**:226–234(1988).

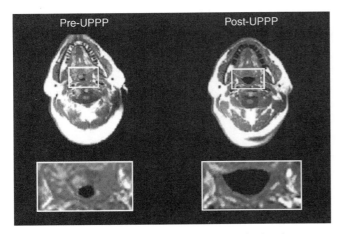

Figure 39.14 MRI demonstrating the increase in the airway area (in the region where the soft palate was resected) in a patient with sleep apnea who underwent uvulopalatopharyngoplasty.

10. Guilleminault C, Riley R, Powell N. Obstructive sleep apnea and abnormal cephalometric measurements. Implications for treatment. *Chest* **86**:793–794(1984).

11. Lowe AA, Fleetham JA, Adachi S, Ryan CF. Cephalometric and computed tomographic predictors of obstructive sleep apnea severity. *Am J Orthod Dentofacial Orthop* **107**:589–595(1995).

12. Miles PG, Vig PS, Weyant RJ, Forrest TD, Rockette HE Jr. Craniofacial structure and obstructive sleep apnea syndrome—a qualitative analysis and meta-analysis of the literature. *Am J Orthod Dentofacial Orthop* **109**:163–172(1996).

13. Burger CD, Stanson AW, Daniels BK, Sheedy PF II, Shepard JW Jr. Fast-computed tomographic evaluation of the effect of route of breathing on upper airway size and function in normal men. *Chest* **103**:1032–1037(1993).

14. Galvin JR, Rooholamini SA, Stanford W. Obstructive sleep apnea: diagnosis with ultrafast CT. *Radiology* **171**:775–778(1989).

15. Haponik EF, Smith PL, Bohlman ME, Allen RP, Goldman SM, Bleecker ER. Computerized tomography in obstructive sleep apnea. Correlation of airway size with physiology during sleep and wakefulness. *Am Rev Respir Dis* **127**:221–226(1983).

16. Ryan CF, Lowe AA, Li D, Fleetham JA. Three-dimensional upper airway computed tomography in obstructive sleep apnea. A prospective study in patients treated by uvulopalatopharyngoplasty. *Am Rev Respir Dis* **144**:428–432(1991).

17. Caballero P, Alvarez-Sala R, Garcia-Rio F, et al. CT in the evaluation of the upper airway in healthy subjects and in patients with obstructive sleep apnea syndrome. *Chest* **113**:111–116(1998).

18. Lowe AA, Gionhaku N, Takeuchi K, Fleetham JA. Three-dimensional CT reconstructions of tongue and airway in adult subjects with obstructive sleep apnea. *Am J Orthod Dentofacial Orthop* **90**:364–374(1986).

19. Abbey NC, Block AJ, Green D, Mancuso A, Hellard DW. Measurement of pharyngeal volume by digitized magnetic resonance imaging. Effect of nasal continuous positive airway pressure. *Am Rev Respir Dis* **140**:717–723(1989).

20. Ryan CF, Lowe AA, Li D, Fleetham JA. Magnetic resonance imaging of the upper airway in obstructive sleep apnea before and after chronic nasal continuous positive airway pressure therapy. *Am Rev Respir Dis* **144**:939–944(1991).

21. Trudo FJ, Gefter WB, Welch KC, Gupta KB, Maislin G, Schwab RJ. State-related changes in upper airway caliber and surrounding soft-tissue structures in normal subjects. *Am J Respir Crit Care Med* **158**:1259–1270(1998).

22. Schotland HM, Insko EK, Schwab RJ. Quantitative magnetic resonance imaging demonstrates alterations of the lingual musculature in obstructive sleep apnea. *Sleep* **22**:605–613(1999).

23. Welch KC, Foster GD, Ritter CT, Schellenberg JB, Wadden TA, Arens R, Maislin G, Schwab RJ. A novel volumetric magnetic resonance imaging paradigm to study upper airway anatomy. *Sleep* **25**:532–542(2002).

24. Kline N, Nkwuo E, Pack AI, Kuna S, Schwab R. Quantitative upper airway digital morphometrics and the risk for obstructive sleep apnea. *Sleep Abstracts Ed* **27**:A230(2004).

25. Brooks LJ, Strohl KP. Size and mechanical properties of the pharynx in healthy men and women. *Am Rev Respir Dis* **146**:1394–1397(1992).

26. Welch KC, Ritter CT, Gefter WB, Schwab RJ. Dynamic respiratory related upper airway imaging during wakefulness in normal subjects and patients with sleep disordered breathing using MRI. *Am J Respir Crit Care Med* **157**:A54(1998).

27. Schwab RJ, Gefter WB, Hoffman EA, Gupta KB, Pack AI. Dynamic upper airway imaging during respiration in normal subjects and patients with sleep disordered breathing. *Am Rev Respir Dis* **148**:1385–1400(1993).

28. Ritter CT, Trudo FJ, Goldberg AN, Welch KC, Maislin G, Schwab RJ. Quantitative evaluation of the upper airway changes in normals and apneics during Müller maneuver. *Am J Respir Crit Care Med* **157**:A54(1998).

29. Strobel RJ, Rosen RC. Obesity and weight loss in obstructive sleep apnea: a critical review. *Sleep* **19**:104–115(1996).

30. Wittels EH, Thompson S. Obstructive sleep apnea and obesity. *Otolaryngol Clin North Am* **23**:751–760(1990).

31. Loube D, Gay P, et al. Indications for positive airway pressure treatment of adult obstructive sleep apnea patients: a consensus statement. *Chest* **115**:863–866(1999).

32. Schwab RJ, Pack AI, Gupta KB, Metzger LJ, Oh E, Getsy JE, Hoffman EA, Gefter WB. Upper airway and soft tissue structural changes induced by CPAP in normal subjects. *Am J Respir Crit Care Med* **154**:1106–1116(1996).

33. Liu Y, Zeng X, Fu M, Huang X, Lowe AA. Effects of a mandibular repositioner on obstructive sleep apnea. *Am J Orthod Dentofacial Orthop* **118**:248–256(2000).

34. Sher AE, Schechtman KB, Piccirillo JF. The efficacy of surgical modifications of the upper airway in adults with obstructive sleep apnea syndrome. An American Sleep Disorders Association review. *Sleep* **19**:156–177(1996).

35. Lanois SH, Feroah TR, Campbell WN, et al. Site of pharyngeal narrowing predicts outcome of surgery for obstructive sleep apnea. *Am Rev Respir Dis* **147**:182–189(1993).

40

EVALUATION OF SLEEP DISORDERED BREATHING: POLYSOMNOGRAPHY

REENA MEHRA AND KINGMAN P. STROHL
Case Western Reserve University, Cleveland, Ohio

INTRODUCTION

Sleep Disordered Breathing

Sleep disordered breathing (SDB) is a process characterized by repetitive partial or complete upper airway occlusion often associated with oxygen desaturation and arousals. It may be accompanied by several signs and symptoms adversely affecting quality of life and health status. These include snoring, excessive daytime somnolence, neurocognitive deficits, irritability, functional impairment, poor work performance, depression, automobile accidents due to drowsy driving, morning headaches, nocturnal enuresis, bedpartner sleep disruption, and cardiovascular morbidity. SDB encompasses a spectrum of disorders including obstructive sleep apnea, central sleep apnea, upper airway resistance syndrome (UARS), and sleep hypoventilation syndrome. The prevalence of sleep apnea syndrome (defined as an apnea–hypopnea index (AHI) of at least 5 events/hour and excessive daytime somnolence) has been identified in 2% of women and 4% of men in a group of middle-aged adults [1], but there are pockets of higher prevalence in patients who are obese and have chronic cardiovascular disease. Failure to recognize SDB is costly both to the individual and to society: underdiagnosis is thought to cost the United States $3.4 billion in additional medical costs per year [2]. In addition, the cost of losses in productivity and auto accidents must be considered [3, 4].

Polysomnography

Attended full-channel polysomnography (PSG) is considered the standard assessment and evaluation for SDB. In PSG, measurements are made of the electroencephalogram (EEG), electro-oculogram (EOG), and electromyogram (EMG) for muscle tone (these first three variables required to allow stage of sleep to be determined), respiratory airflow, respiratory effort, arterial oxygen saturation, snoring intensity, electrocardiogram, and EMG of the anterior tibialis muscles. The respiratory parameters not only provide useful information in the setting of uncomplicated SDB but also track sleep-related problems in patients with underlying lung disease such as obstructive airway diseases (asthma, chronic obstructive airway disease), restrictive airway disease, pulmonary vascular disease, and congestive heart failure. Such respiratory measurements include monitoring of airflow, snoring, respiratory muscle activity, oxygenation, and at times carbon dioxide levels.

INDICATIONS FOR POLYSOMNOGRAPHY IN THE SETTING OF SUSPECTED SDB

Stratification of patients with suspected sleep apnea may be based on the following four symptoms: habitual snoring, excessive daytime somnolence, a body mass index (BMI) $> 35 \, kg/m^2$, and observed/witnessed apneas. Each of these factors by itself has low sensitivity and specificity.

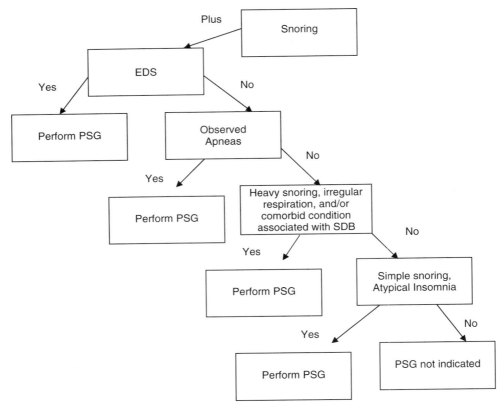

Figure 40.1 Indications for PSG for SDB based on symptom presentation. (Adapted from [5].)

On the other hand, patients with all four of the above symptoms may be placed in a high-risk group and have an approximately 70% likelihood of having an apnea–hypopnea index >10 events/hour [5]. Other risk factors include male gender, postmenopausal women, age, race, obesity (BMI 28–35), and craniofacial abnormalities. Figure 40.1 provides a reasonable symptom-based algorithm of PSG-related decision making in the setting of suspected SDB.

Snoring

Snoring appears to be strongly associated with a subsequent diagnosis of SDB with this relationship more pronounced in men than women. Results from 1409 patients in a sleep clinic indicated that nearly all levels of estimated snoring frequency were associated with a greater likelihood of sleep apnea [6]. In the sample from this clinic, sensitivities approximating 90% were obtained in men, and specificities approximating 90% were obtained in women, but high diagnostic accuracy (high specificity in men; high sensitivity in women) could not be achieved with the three snoring questions used. Generally, associations between snoring and sleep apnea were independent of age and sex [6]. A study has also demonstrated that, in preschool children

and those 5–7 years of age, the parental report of frequent snoring in both age groups was highly sensitive and specific, indicating scores derived from parental-report questionnaires of children's snoring and other sleep and wake behaviors can be used as surrogate predictors of snoring or sleep-disordered breathing in children [7]. Snoring is also the most frequent symptom of sleep apnea, occurring in 70–95% of patients [8], but because it is so common in the general population it is a poor predictor of sleep apnea [9]. A study has shown that the absence of snoring makes SDB unlikely; only 6% of patients with SDB do not report snoring [10]. Interestingly, the majority of patients who deny snoring actually do snore when this is measured objectively [11].

Witnessed Apneas

Witnessed or observed apneas are a common reason for referral to the sleep clinic often due to concern of the patient's bedpartner. Bedpartners, however, often do not provide a reliable account about apneas during sleep, and even trained medical staffs are suboptimal when diagnosing respiratory events in patients with SDB through clinical observation [12]. Gender-related reporting bias may occur, as female patients with SDB are less likely to report nocturnal apneas [13, 14]. The patient may report waking up with acute panic and choking. Episodes usually only last for a few seconds, but can cause considerable distress, both to the patient and the partner. These events can be differentiated from other causes of nocturnal breathlessness such as paroxysmal nocturnal dyspnea in the patient with congestive heart failure, panic disorder, sleep-related asthma, sleep-related laryngospasm, and sleep-related choking syndrome.

Excessive Daytime Somnolence

Excessive daytime sleepiness is caused by fragmented sleep related to frequent arousals. Like snoring, sleepiness is common and by itself a poor discriminator. Thirty percent to 50% of the general population without sleep apnea report moderate to severe sleepiness [1, 15]. It is important to attempt to differentiate sleepiness (the urge or pressure to fall sleep) from reports of fatigue, lethargy, malaise, or exhaustion. Patients themselves may underreport their sleepiness [16], either because they are not aware of it or because there are social pressures to deny that it is a problem. Other common causes of EDS such as insufficient sleep, poor sleep hygiene, drugs, and shift work should always be asked about in the history.

Clinical Prediction Rules

Several utility tools and symptom score clinical prediction rules have been developed in order to better identify those individuals with a high pretest probability of SDB, thereby prompting further evaluation with diagnostic PSG [9, 10, 17, 18]. A recent study has also examined four clinical prediction rules; however, the sensitivities of these rules ranged from 33% to 39% and specificities ranged from 87% to 93% and therefore were not sufficiently accurate to discriminate between patients with and without SDB, but may be useful in prioritizing patients for split-night PSG [19]. The Berlin questionnaire is a survey tool that has demonstrated efficacy in the setting of SDB. Survey items address the presence and frequency of snoring behavior, waketime sleepiness or fatigue, and history of obesity or hypertension. Patients with persistent and frequent symptoms in any two of these three domains are considered to be at high risk for sleep apnea [20]. Several tools are available for measuring sleepiness both subjectively and objectively. There is no gold standard, but the easiest and most practical is the Epworth Sleepiness Scale (ESS) [21]. Drawbacks include poor correlation with the severity of sleep apnea and the disadvantages that accompany any self-evaluated test such as misperception of sleep episodes. The input of the partner can be useful [22]. The major advantages of the ESS are that it is simple, quick, and inexpensive and has high test–retest reliability [23].

EVIDENCE-BASED DATA REGARDING USE OF PSG IN THE SETTING OF SUSPECTED SDB

Standard Full PSG for the Diagnosis of SDB

According to the American Parameters for the Indications for Polysomnography and Related Procedures, full-night attended PSG is indicated for the diagnosis of sleep-related breathing disorders [24].

Positive Pressure Therapy Titration PSG for SDB Therapy

PSG is also recommended for positive pressure titration (i.e., continuous positive airway pressure (CPAP), or bilevel airway pressure (BiPAP)) in order to determine the optimal therapeutic pressure. Full-night PSG with titration is indicated in patients with documented diagnosis of SDB for which therapy is indicated. PSG with titration is recommended if (1) an apnea index (AI) is at least 20/hour or an AHI > 30/hour irrespective of the patient's symptoms; (2) an AHI or respiratory arousal index is at least 10/hour in a patient with excessive daytime somnolence [24]. More recent guidelines have also suggested that comorbidity such as hypertension, diabetes mellitus, stroke, and cardiovascular disease should also warrant treatment with positive pressure therapy, and therefore titration study in the setting of diagnosed SDB. An optimal titration study should be defined as sufficient sleep time to determine the optimal pressure, reduction of the AHI, adequate REM–supine time during the optimal pressure setting, reduction of snoring, oxygen saturation greater than 90%, and, in the case of respiratory arousals suggesting UARS, reduction of respiratory arousals and crescendo snore arousals. An ideal titration study should include optimal titration in the supine position in REM sleep; additional titration may be necessary to address EEG arousals that occur in the context of residual SDB [24].

Split-Night PSG for the Diagnosis and Therapy of SDB

A split-night study involves an initial diagnostic PSG followed by positive pressure titration during PSG during the same study. This is a viable alternative to one full night of PSG followed by a second full night of titration. The American Academy of Sleep Medicine Practice Parameters for Polysomnography and Related Procedures recommend specifically that performing a split-night study is ideal when the following four criteria are met: (1) An AHI of at least 40/hour is documented during a minimum of 2 hours of diagnostic PSG. Split-night studies may also be considered if the AHI is 20–40/hour in the context of clinical judgment such as long events or severe oxygen desaturations. (2) Positive pressure therapy titration is carried out for more than 3 hours. (3) PSG documents that positive pressure titration eliminates or nearly eliminates the respiratory events during REM and NREM sleep, including REM sleep in the supine position. (4) A second study with positive pressure therapy titration is performed if the titration during the split-night study is less than 3 hours or if there was not resolution or near-resolution of events during REM and REM–supine sleep [24].

Several studies have been performed to assess the utility of split-night studies. In one study, the scoring of 90 minutes of the recording had a low sensitivity (42%) and high specificity (100%) for detecting an AHI > 10 [25]. Another study suggested that the diagnosis of SDB and CPAP titration could be carried out in one night in the majority of patients with an AI > 20 and suggested potential cost savings compared to one night of diagnosis and one night of titration [26]. Also, conclusions from another study were that a positive PSG with an AHI ≥ 5 in approximately 3 hours of recording time yielded high probability of the same or higher AHI for the full night; however, if the study was negative for SDB (defined as AHI < 5) then there was potential for false negatives [27]. Another study evaluated 107 patients in a nonrandomized controlled trial and concluded that if the AHI > 40 and the titration period was at least 3 hours, then a split-night study was similar to a full-night titration [28].

Number of Nights of PSG Required for the Diagnosis of SDB

Multiple studies to assess the diagnostic usefulness of one night versus two nights of PSG monitoring in the setting of SDB have been performed [29–35]. There are no studies that have looked directly at the variability of SDB in a diagnostic study over more than two nights. An epidemiologic study involving patients 30–60 years old demonstrated no difference in diagnostic outcome between PSG recordings obtained on the first and second nights [1]. A study of elderly men demonstrated no systematic difference between the first and second nights, but five of the fourteen individuals would have been reclassified. There was, however, no predictable direction of change [36]. In patients with moderate to severe SDB, the reproducibility of the respiratory parameters from night to night is good [37]. For milder SDB, a single negative study may not exclude SDB, and a second study should be considered [38, 39]. Sleep position, acclimatization to a foreign sleep environment, concurrent respiratory tract infections, and variable alcohol and drug use are thought to be responsible for night to night variability in both respiratory and sleep parameters.

PSG in SDB Prior to Upper Airway Surgery

A preoperative clinical evaluation including PSG is routinely indicated to evaluate for the presence of SDB prior to upper airway surgery. This is also a way to provide a preoperative baseline for the patient so that recrudescence of symptoms prompting a postoperative PSG may be compared to the baseline [24].

PSG as a Follow-Up Tool in SDB

Follow-up PSG is recommended after good response to oral appliance treatment in patients with moderate to severe SDB to ensure therapeutic benefit, after surgical treatment in patients with moderate to severe SDB to ensure satisfactory response, after surgical treatment of patients with SDB whose symptoms return despite initial success of treatment, after substantial weight loss or gain, and when clinical response is insufficient. Follow-up PSG is not recommended or routinely indicated in patients treated with positive pressure therapy whose symptoms continue to be resolved with positive pressure treatment [24].

COMPONENTS OF PSG USED FOR THE EVALUATION OF SDB

PSG involves the recording of multiple physiologic parameters during sleep. PSG and clinical sleep medicine originated during the late 1950s with the discoveries of sleep initially described by Kleitman, Aserinsky, and Dement in Chicago and subsequently by Jouvet, Michel, and Mounier in France. Observations regarding the relationships between upper airway collapsibility and disrupted sleep during PSG were discovered thereafter. The following is a brief overview of PSG channels used to assess respiratory variables.

Airflow Monitoring

Most airflow sensors detect apneas reliably, but the detection and quantification of decreased flow needed to diagnose hypopneas depend on the type of sensor used. Hypopneas make up the majority of obstructive respiratory events [40] and therefore measurement needs to be reliable.

Pneumotachometer This method of airflow monitoring provides a direct quantitative measurement of airflow or tidal volume, however, it requires connection to a sealed mask placed over the nose or mouth. It is considered the reference standard for obstructive apnea and hypopnea detection. Therefore, although this provides a beneficial and accurate research tool, in the clinical setting this technique is somewhat cumbersome with the exception of application or titration of positive pressure therapy. This measures the flow rate of gases during breathing. The breath is passed through a short tube in which there is a fine mesh, which presents a small resistance to the flow. Flow (V') is derived from the pressure difference over a small, fixed resistance, offered by a fine metal mesh. The pressure drop across the resistance relates linearly to flow at relatively low flows, when the flow pattern is laminar. Higher flows give rise to a turbulent flow pattern, when the pressure drop across the resistance changes more than proportionally with flow. Accurate measurements are best performed when the flow pattern is laminar and flow is linearly related to pressure drop.

Intranasal Pressure/Tranducer This technique provides an indirect measurement of airflow by detecting pressure changes with an excellent response to airflow profile and is capable of detecting airflow limitation. Changing pressures require a transducer, which can respond to rapid changes. Nasal pressure transducers provide a significantly more sensitive measure of airflow than temperature-based transducers; and many believe that the pressure transducers may provide a measure of upper airway resistance as inspiration and expiration provide transducer signal fluctuations similar to airflow [41, 42]. In one study, the nasal cannula/pressure transducer has been found to provide a noninvasive reproducible detector of all events in SDB. In particular, it detects the same events as esophageal manometry (respiratory effort-related arousals, RERAs) with an intraclass correlation coefficient of 0.96 [43]. The new transducers provide additional information for scoring hypopneas. If used for evaluation of sleep-related breathing disorders, the new level of sensitivity may lead to scoring of many more events than are typically scored with other methods of airflow detection. These events may be as significant as conventionally scored apneas, but at present virtually all of the clinical literature is based on temperature-based airflow transduction. Nasal pressure monitoring is not recommended for patients who are predominantly mouth breathers or have nasal obstruction, in which case airflow may be underestimated [44]. Nasal pressure sensors connected to the nose via nasal prongs are more accurate than thermoelements in detecting hypopneas [44]. However, nasal pressure is falsely increased in the presence of nasal obstruction, and there is a nonlinear relation between nasal pressure and nasal flow. Square root linearization of nasal pressure greatly increases the accuracy for quantifying hypopneas and detecting flow limitation [45, 46]. Mouth breathing can affect the measurement, but pure mouth breathing is uncommon [47].

Nasal Thermocouple/Thermistor Nasal and/or oral thermocouple is an inexpensive way to assess airflow via indirect semiquantitative assessment detecting increased temperature of expired air, with only directional changes providing reliable results. Thermocouples are commonly used for temperature measurement as they are highly accurate and operate over a broad range of temperatures. They consist of two different metal wires that are welded together at one end (A). These wires generate a thermoelectric voltage between their open ends that changes according to the temperature difference between the two ends, that is, between the junction (A) and the reference (R). Thermistors consist of an electronic component (semiconductor material) that exhibits a large change in resistance in proportion to a small change in temperature. In comparison to thermocouples, thermistors have a limited (smaller) temperature range; however, they are highly sensitive within this range. The resistance of these devices often changes in a nonlinear fashion with temperature and additional instruments are required to linearize the reading. In laboratory models that have compared thermistors and thermocouples to a pneumotachograph, the thermal sensors have been shown to be nonlinearly related to airflow, generally providing an overestimation of ventilation [48]. Therefore they cannot be used to determine hypopneas reliably. Furthermore, their accuracy varies greatly depending on the position of the sensors, the sleep position of the patient, the presence of nasal obstruction, and the type of thermoelement used [49]. For these reasons, the ASDA Task Force does not recommend thermoelements for the detection of obstructive respiratory events [23]. Despite this, they continue to be used for flow detection in many commercially available sleep diagnostic systems.

Respiratory Inductance Plethysmography The literature supports that respiratory inductance plethysmography (RIP) is acceptable for the semiquantitative measurement of ventilation assessed by thoracic and abdominal pressure changes [49]. With this technique, transducers are placed at the level of the nipples and at the umbilicus to monitor cross-sectional changes reflected by changes in inductance or resistance to change in flow of the transducers [50]. The sum of the signals may provide an estimate of tidal volume and respiratory pattern during sleep. RIP is based on the two-compartment model of thoracoabdominal movement during respiration [51]. Accurate initial calibration and constancy during body movements and changes during respiration are imperative to obtaining valid measurements [52]. Measurement inaccuracies may occur due to slippage (displacement of transducer bands) and position changes [50]. RIP detects changes in the volume of the chest and abdomen during inspiration and expiration and, when properly calibrated, the sum of the two signals can provide an estimate of tidal volume [53]. However, calibration may be difficult to maintain throughout the night [54]. RIP allows an acceptable semiquantitative measurement of ventilation and therefore hypopneas. The ASDA Task Force recommends the use of RIP or measurement of nasal pressure using nasal cannulae to detect airflow and ventilation [23].

Snore Monitoring

Microphone The microphone is used to detect snoring during PSG, providing an output signal with easily identifiable waveforms. This tracing in conjunction with sleep technician comments may assist in the diagnosis of primary snoring when there is lacking evidence for SDB. In addition, snore monitoring is of use during positive pressure titration to determine the optimal pressure setting [55].

Respiratory Muscle Monitoring

Surface Diaphragmatic EMG This modality provides an indirect measurement of respiratory effort via electrodes placed on the chest wall; however, it is not an ideal measurement for detecting RERAs or central respiratory events [23]. Reliable recordings are difficult to obtain, and it is prone to electrocardiography artifact and nonrespiratory muscle artifact is difficult to eliminate. There are no data on accuracy, reliability, or correlation with long-term outcomes with respect to this technique [23].

Esophageal Balloon Manometry The measurement of esophageal pressure with continuous overnight monitoring is the reference standard for measuring respiratory effort during PSG. Respiratory efforts are associated with changes in pleural pressure, which can be accurately measured using esophageal manometry. This method is useful when distinguishing central versus obstructive apneas and is useful for detecting RERAs in the setting of UARS during which there is increasingly more negative esophageal pressures immediately preceding an arousal, subsequent to which the esophageal pressure fairly rapidly returns to normal levels [56–59].

Piezosensors, Strain Gauges, Magnetometers Other methods that have been used to assess airflow include piezosensors, strain gauges, and magnetometers. Piezosensors may measure qualitative changes in airflow; however, they are not reliable in distinguishing central from obstructive respiratory events. In one study, 37% of the apneas scored as central based on strain gauge measurements were reclassified as obstructive or mixed based on esophageal balloon measurements, indicating a high misclassification rate [60].

Pulse Oximetry Monitoring

The fundamental physical property that allows the measurement of arterial oxygen saturation is that blood changes color with saturation. Hemoglobin, in its reduced form or oxygenated state, absorbs light at wavelengths below approximately 630 nm, which includes the entire part of the visible spectrum aside from the red region. The opposite situation occurs in the near-infrared region (810–1000 nm), where hemoglobin absorbs more light when it is desaturated. Pulse oximeters usually are designed with two emitters (usually light-emitting diodes): one designed to emit light in the red region (~660 mn) and the other in the near-infrared region (~925 or 940 nm). In order to measure absorption of arterial blood only, without interference from venous blood, skin, bone, and so on, and to minimize scatter effect, a differential absorption is calculated by dividing the small change in intensity by the total intensity of the output light. Chromophores other than oxyhemoglobin and reduced hemoglobin, such as carboxyhemoglobin and methemoglobin, may cause falsely elevated readings for the arterial oxygen saturation. Of note, oximeters may be calibrated using functional or fractional oxygen saturation with the former reading slightly higher (1–3%). Oximeters may be prone to artifact, such as during states of poor perfusion, intense ambient light excessive patient motion (particularly at the probe site), and electrical noise, and also may be affected by changes in heart rate and circulation time [61]. Overall, pulse oximetry is easy to use, inexpensive, readily available, and noninvasive and permits continuous monitoring of oxygen saturation. Pulse oximetry sensitivity is improved with shorter sampling intervals and minimal filtering in order to achieve the most rapid response [62, 63].

Carbon Dioxide Monitoring

Expired End-Tidal CO_2 Monitoring This modality works by drawing a stream of air from the nose or mouth to a chamber in which a light is shone through the air. The degree of absorption at a certain frequency of infrared light is proportional to the concentration of CO_2. The light may be split, with half passing through a reference cell. The light may also be "chopped" so that it is not continuously heating the gas in the reference cell.

Continuous measurement of CO_2 reflects the excretion pattern of carbon dioxide from the lung. Values for CO_2 are near or at zero on inspiration and show an abrupt rise until the end of expiration when there is a plateau in the CO_2 level. The end-expiratory value is correlated with arterial P_{CO_2} provided that there is complete gas emptying to functional residual capacity and little effect of ventilation–perfusion mismatch. End-tidal CO_2 monitoring may help in identifying hypoventilation in obesity hypoventilation syndrome, as well as chronic obstructive pulmonary disease, congestive heart failure, and neurologic diseases that produce neuromuscular weakness. In addition, end-tidal

CO_2 values can be helpful in assessing disorders of chronic hyperventilation, distinguishing pathophysiologic from psychogenic causes by the persistence or resolution of hypocapnia during sleep.

A limitation of end-tidal CO_2 monitoring includes the inability to measure levels in the setting of CPAP or bilevel pressure therapy in order to assess response to treatment. The value of capnography (breath-by-breath CO_2 measurements) is twofold, but limited in scope compared to direct measures of airflow. First, capnography may detect absence of expiratory airflow and signal an apnea. There are no established criteria, but an apnea event is considered to occur when there is a failure for end-expiratory CO_2 to fall after expiration with subsequent decline in CO_2 over the next 10 seconds to values at or near zero. The end of the apnea is heralded by the occurrence of a rapid rise in CO_2 with expiration; the length of an event is from the peak of the last CO_2 rise to the peak of the next CO_2 rise if greater than 10 seconds. A cardiac oscillation in the capnography signal indicates a patent airway during a central apnea (but the frequency of this is only recorded at an anecdotal level of evidence). There can be expiratory puffs after an obstructed inspiratory effort, but again the frequency of this is only recorded at an anecdotal level of evidence. Second capnography can be used to identify obstructive hypopneas. In this instance there are persistent efforts characterized by asynchronous chest wall movements of the ribcage and abdomen with a rising end-tidal CO_2 over time. Again the frequency of this kind of pattern is only recorded at an anecdotal level of evidence.

Scoring of events from the CO_2 monitor should be directed at apneas, and to the extent possible correlated with ribcage/abdominal movements and CO_2 fluctuation in synchrony with EKG (central event) or in synchrony with respiratory efforts (expiratory puffs signaling an obstructive event). The length of an event is from the peak of the last CO_2 rise to the peak of the next CO_2 rise, if greater than 10 seconds. Obstructive hypopneas would be detected by a series of breaths with asynchronous ribcage/abdominal motion associated with an increasing end-expiratory CO_2 level.

Transcutaneous CO_2 There are two methods that may be employed in transcutaneous CO_2 monitoring: the first using a silver electrode that measures CO_2 that has diffused from the skin through a gas-permeable membrane into solution (response time less than 1 minute), and the other uses an infrared capnometer that analyzes CO_2 in the gas phase (response time more than 2 minutes). Studies have assessed end-tidal and transcutaneous monitoring of CO_2 and concluded that neither of these measurements were an accurate reflection of CO_2 levels and therefore should not be used during routine PSG [63, 64].

CLASSIFICATION AND SCORING OF RESPIRATORY EVENTS

Obstructive Apneas. Obstructive apneas are defined as a clear decrease (defined as >50%) from baseline in the amplitude of a valid measure of breathing during sleep lasting at least 10 seconds associated with continued thoracoabdominal respiratory effort. Baseline is described as the mean amplitude of stable breathing and oxygenation in the preceding 2 minutes prior to the event in individuals who have a stable breathing pattern during sleep or the mean amplitude of the three largest breaths in the 2 minutes preceding the onset of the event in individuals without a stable breathing pattern [23].

Central Apneas. Central apneas are defined as a clear decrease (defined as >50%) from baseline in the amplitude of a valid measure of breathing during sleep lasting at least 10 seconds associated with lack of thoracoabdominal respiratory effort [23].

Mixed Apneas. Mixed apneas are defined as a clear decrease (defined as >50%) from baseline in the amplitude of a valid measure of breathing during sleep lasting at least 10 seconds associated with initial lack of thoracoabdominal respiratory effort, but presence of effort during the latter part of the respiratory event.

Hypopneas. A distinct amplitude reduction (30–50%) of a validated measure during sleep lasting at minimum 10 seconds that does not meet the definition criteria for apnea, but is associated with an oxygen desaturation of greater than 3% or an arousal [23].

Cheyne–Stokes Respirations. Respiratory monitoring indicates at least three consecutive cycles of cyclical crescendo and decrescendo change in breathing pattern with cycle length variable, but more commonly 60 seconds with one or more of the following: (1) five or more central apneas or hypopneas per hour of sleep, and/or (2) the cyclic crescendo–decrescendo change in breathing amplitude with a duration of at least 10 minutes [23].

Sleep-Related Hypoventilation. This is defined as one or both of the following: (1) an increase of P_{CO_2} levels by 10 Torr compared to awake supine values, and/or (2) oxygen desaturation during sleep not explained by apneas or hypopneas. This is designated severe if the oxygen saturation is less than 85% for greater than 50% of the sleep period or in the setting of cor pulmonale/biventricular failure [23].

See Figures 40.2–40.5 for typical PSG recordings of different SDB events.

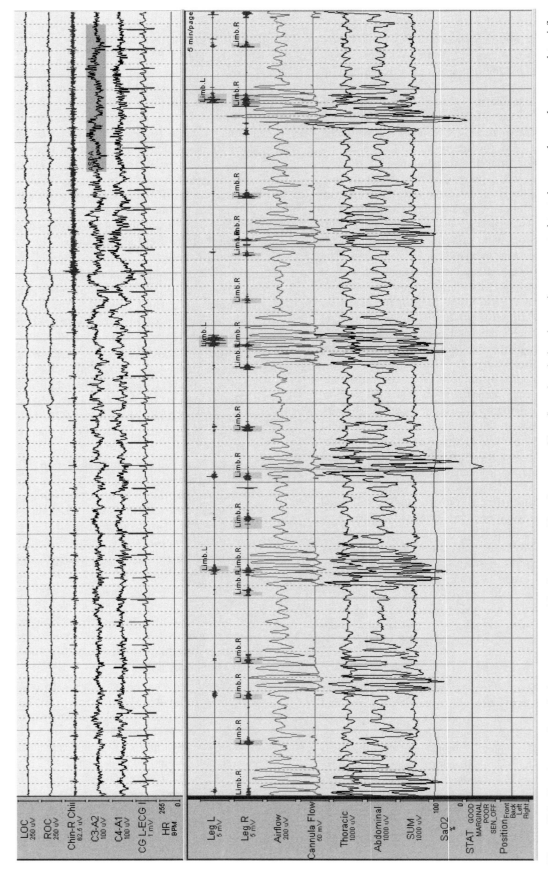

Figure 40.2 Respiratory channels demonstrate repetitive hypopneas. Note the more evident flattening during the respiratory events detected per nasal cannula compared to airflow detected by thermistor.

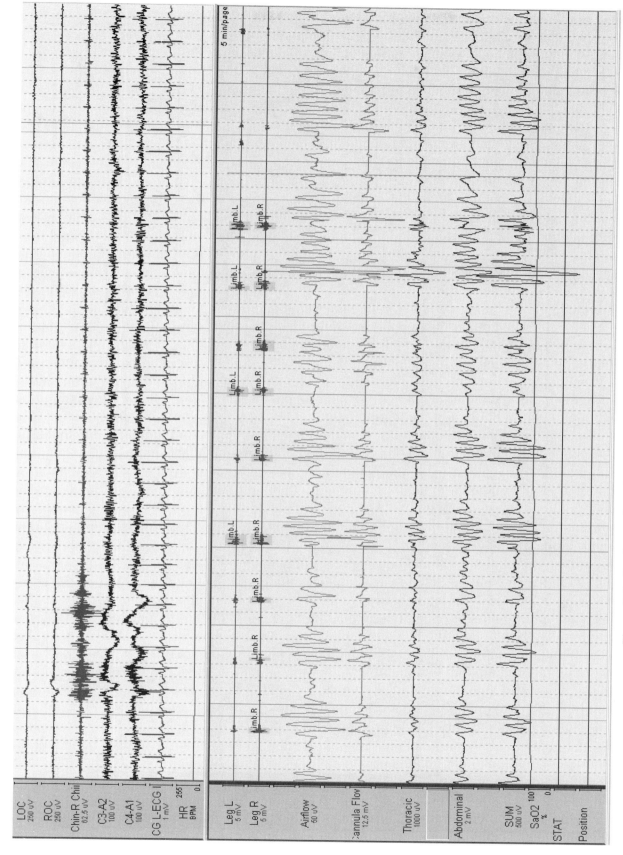

Figure 40.3 Respiratory channels demonstrate repetitive obstructive hypopneas and apnea.

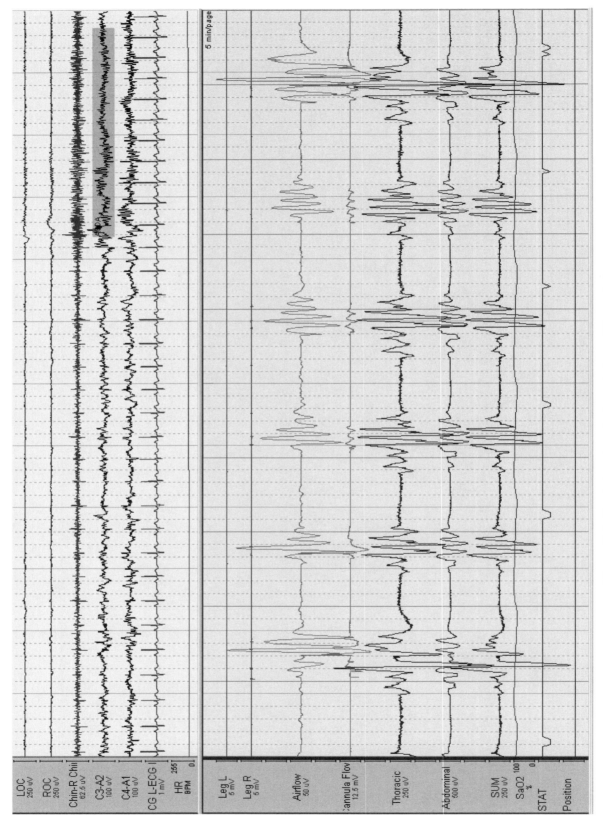

Figure 40.4 Respiratory channels demonstrate repetitive central apneas.

312

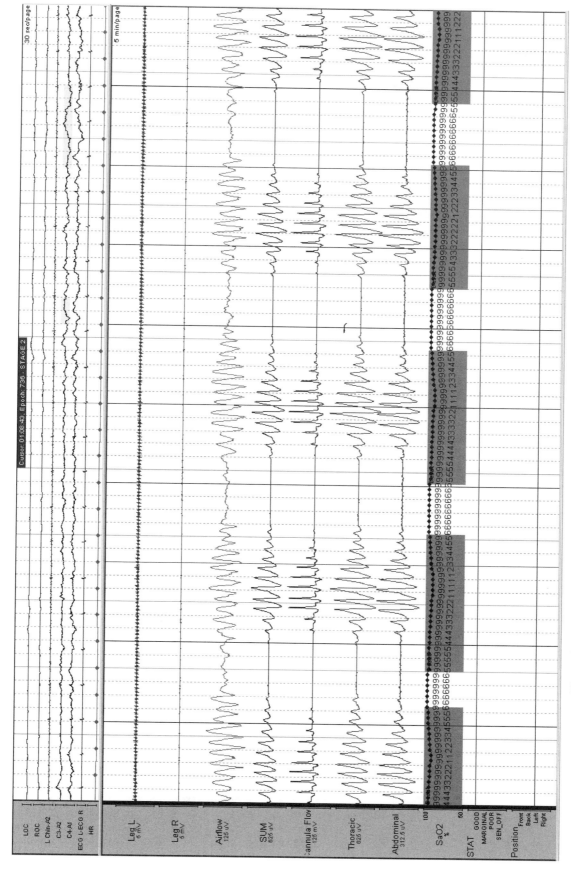

Figure 40.5 Respiratory channels demonstrate Cheynes–Stokes respirations and central sleep apnea pattern.

REFERENCES

1. Young T, et al. The occurrence of sleep-disordered breathing among middle-aged adults. *N Engl J Med* **328**(17):1230–1235(1993).

2. Kapur V, et al. The medical cost of undiagnosed sleep apnea. *Sleep* **22**(6):749–755(1999).

3. Findley L, et al. Vigilance and automobile accidents in patients with sleep apnea or narcolepsy. *Chest* **108**(3):619–624(1995).

4. Rodenstein DO. Sleep apnoea syndrome: the health economics point of view. *Monaldi Arch Chest Dis* **55**(5):404–410(2000).

5. Chesson AL Jr, et al. The indications for polysomnography and related procedures. *Sleep* **20**(6):423–487(1997).

6. Bliwise DL, Nekich JC, Dement WC. Relative validity of self-reported snoring as a symptom of sleep apnea in a sleep clinic population. *Chest* **99**(3):600–608(1991).

7. Montgomery-Downs HE, et al. Snoring and sleep-disordered breathing in young children: subjective and objective correlates. *Sleep* **27**(1):87–94(2004).

8. Whyte KF, et al. Clinical features of the sleep apnoea/hypopnoea syndrome. *Q J Med* **72**(267):659–666(1989).

9. Flemons WW, et al. Likelihood ratios for a sleep apnea clinical prediction rule. *Am J Respir Crit Care Med* **150**(5 Pt 1):1279–1285(1994).

10. Viner S, Szalai JP, Hoffstein V. Are history and physical examination a good screening test for sleep apnea? *Ann Intern Med* **115**(5):356–359(1991).

11. Hoffstein V, Mateika S, Anderson D. Snoring: is it in the ear of the beholder? *Sleep* **17**(6):522–526(1994).

12. Haponik EF, et al. Evaluation of sleep-disordered breathing. Is polysomnography necessary? *Am J Med* **77**(4):671–677(1984).

13. Young T, et al. The gender bias in sleep apnea diagnosis. Are women missed because they have different symptoms? *Arch Intern Med* **156**(21):2445–2451(1996).

14. Redline S, et al. Gender differences in sleep disordered breathing in a community-based sample. *Am J Respir Crit Care Med* **149**(3 Pt 1):722–726(1994).

15. Duran J, et al. Obstructive sleep apnea–hypopnea and related clinical features in a population-based sample of subjects aged 30 to 70 yr. *Am J Respir Crit Care Med* **163**(3 Pt 1):685–689(2001).

16. Engleman HM, Hirst WS, Douglas NJ. Under reporting of sleepiness and driving impairment in patients with sleep apnoea/hypopnoea syndrome. *J Sleep Res* **6**(4):272–275(1997).

17. Crocker BD, et al. Estimation of the probability of disturbed breathing during sleep before a sleep study. *Am Rev Respir Dis* **142**(1):14–18(1990).

18. Maislin G, et al. A survey screen for prediction of apnea. *Sleep* **18**(3):158–166(1995).

19. Rowley JA, Aboussouan LS, Badr MS. The use of clinical prediction formulas in the evaluation of obstructive sleep apnea. *Sleep* **23**(7):929–938(2000).

20. Netzer NC, et al. Using the Berlin Questionnaire to identify patients at risk for the sleep apnea syndrome. *Ann Intern Med* **131**(7):485–491(1999).

21. Johns MW. A new method for measuring daytime sleepiness: the Epworth sleepiness scale. *Sleep* **14**(6):540–545(1991).

22. Kingshott RN, et al. Self assessment of daytime sleepiness: patient versus partner. *Thorax* **50**(9):994–995(1995).

23. Sleep-related breathing disorders in adults: recommendations for syndrome definition and measurement techniques in clinical research. The Report of an American Academy of Sleep Medicine Task Force. *Sleep* **22**(5):667–689(1999).

24. Practice parameters for the indications for polysomnography and related procedures. Polysomnography Task Force, American Sleep Disorders Association Standards of Practice Committee. *Sleep* **20**(6):406–422(1997).

25. Scharf SM, et al. Screening for subclinical sleep-disordered breathing. *Sleep* **13**(4):344–353(1990).

26. Iber C, et al. Single night studies in obstructive sleep apnea. *Sleep* **14**(5):383–385(1991).

27. Sanders MH, et al. Diagnosis of sleep-disordered breathing by half-night polysomnography. *Am Rev Respir Dis* **144**(6):1256–1261(1991).

28. Yamashiro Y, Kryger MH. CPAP titration for sleep apnea using a split-night protocol. *Chest* **107**(1):62–66(1995).

29. Mendelson WB. Use of the sleep laboratory in suspected sleep apnea syndrome: is one night enough? *Cleve Clin J Med* **61**(4):299–303(1994).

30. Meyer TJ, et al. One negative polysomnogram does not exclude obstructive sleep apnea. *Chest* **103**(3):756–760(1993).

31. Kader GA, Griffin PT. Reevaluation of the phenomena of the first night effect. *Sleep* **6**(1):67–71(1983).

32. Aber WR, et al. Consistency of respiratory measurements from night to night during the sleep of elderly men. *Chest* **96**(4):747–751(1989).

33. Altose M, et al. The validity of polysomnographic data obtained during 24-hour ambulatory blood pressure monitoring. *Sleep* **18**(4):272–275(1995).

34. Mosko SS, Dickel MJ, Ashurst J. Night-to-night variability in sleep apnea and sleep-related periodic leg movements in the elderly. *Sleep* **11**(4):340–348(1988).

35. Wittig RM, et al. Night-to-night consistency of apneas during sleep. *Am Rev Respir Dis* **129**(2):244–246(1984).

36. Dealberto MJ, et al. Factors related to sleep apnea syndrome in sleep clinic patients. *Chest* **105**(6):1753–1758(1994).

37. Chediak AD, et al. Nightly variability in the indices of sleep-disordered breathing in men being evaluated for impotence with consecutive night polysomnograms. *Sleep* **19**(7):589–592(1996).

38. Littner M. Polysomnography in the diagnosis of the obstructive sleep apnea–hypopnea syndrome: where do we draw the line? *Chest* **118**(2):286–288(2000).

39. Le Bon O, et al. Mild to moderate sleep respiratory events: one negative night may not be enough. *Chest* **118**(2):353–359(2000).

40. Redline S, et al. Effects of varying approaches for identifying respiratory disturbances on sleep apnea assessment. *Am J Respir Crit Care Med* **161**(2 Pt 1):369–374(2000).

41. Hosselet JJ, et al. Detection of flow limitation with a nasal cannula/pressure transducer system. *Am J Respir Crit Care Med* **157**(5 Pt 1):1461–1467(1998).

42. Norman RG, et al. Detection of respiratory events during NPSG: nasal cannula/pressure sensor versus thermistor. *Sleep* **20**(12):1175–1184(1997).

43. Ayappa I, et al. Noninvasive detection of respiratory effort-related arousals (RERAs) by a nasal cannula/pressure transducer system. *Sleep* **23**(6):763–771(2000).

44. Series F, Marc I. Nasal pressure recording in the diagnosis of sleep apnoea hypopnoea syndrome. *Thorax* **54**(6):506–510 (1999).

45. Montserrat JM, et al. Evaluation of nasal prongs for estimating nasal flow. *Am J Respir Crit Care Med* **155**(1):211–215(1997).

46. Thurnheer R, Xie X, Bloch KE. Accuracy of nasal cannula pressure recordings for assessment of ventilation during sleep. *Am J Respir Crit Care Med* **164**(10 Pt 1):1914–1919(2001).

47. Ballester E, et al. Nasal prongs in the detection of sleep-related disordered breathing in the sleep apnoea/hypopnoea syndrome. *Eur Respir J* **11**(4):880–883(1998).

48. Farre R, et al. Accuracy of thermistors and thermocouples as flow-measuring devices for detecting hypopneas. *Eur Respir J* **11**(1):179–182(1998).

49. Berg S, et al. Comparison of direct and indirect measurements of respiratory airflow: implications for hypopneas. *Sleep* **20**(1):60–64(1997).

50. Kryger MH, Roth T, Dement WC (Eds). *Principles and Practice of Sleep Medicine.* Saunders, Philadelphia, 2000, pp 1217–1230.

51. Konno K, Mead J. Measurement of the separate volume changes of rib cage and abdomen during breathing. *J Appl Physiol* **22**(3):407–422(1967).

52. Chadha TS, et al. Validation of respiratory inductive plethysmography using different calibration procedures. *Am Rev Respir Dis* **125**(6):644–649(1982).

53. Cantineau JP, et al. Accuracy of respiratory inductive plethysmography during wakefulness and sleep in patients with obstructive sleep apnea. *Chest* **102**(4):1145–1151 (1992).

54. Whyte KF, et al. Accuracy of respiratory inductive plethysmograph in measuring tidal volume during sleep. *J Appl Physiol* **71**(5):1866–1871(1991).

55. Ayappa I, et al. Relative occurrence of flow limitation and snoring during continuous positive airway pressure titration. *Chest* **114**(3):685–690(1998).

56. Guilleminault C, et al. A cause of excessive daytime sleepiness. The upper airway resistance syndrome. *Chest* **104**(3):781–787(1993).

57. Guilleminault C, Stoohs R, Duncan S. Snoring (I). Daytime sleepiness in regular heavy snorers. *Chest* **99**(1):40–48(1991).

58. Guilleminault C, et al. Upper airway sleep-disordered breathing in women. *Ann Intern Med* **122**(7):493–501 (1995).

59. Guilleminault C, et al. Upper airway resistance syndrome, nocturnal blood pressure monitoring, and borderline hypertension. *Chest* **109**(4):901–908(1996).

60. Boudewyns A, et al. Assessment of respiratory effort by means of strain gauges and esophageal pressure swings: a comparative study. *Sleep* **20**(2):168–170(1997).

61. West P, George CF, Kryger MH. Dynamic in vivo response characteristics of three oximeters: Hewlett-Packard 47201A, Biox III, and Nellcor N-100. *Sleep* **10**(3):263–271(1987).

62. Phillips BA, Anstead MI, Gottlieb DJ. Monitoring sleep and breathing: methodology. Part I: Monitoring breathing. *Clin Chest Med* **19**(1):203–212(1998).

63. Clark JS, et al. Noninvasive assessment of blood gases. *Am Rev Respir Dis* **145**(1):220–232(1992).

64. Sanders MH, et al. Accuracy of end-tidal and transcutaneous P_{CO_2} monitoring during sleep. *Chest* **106**(2):472–483(1994).

41

EVALUATION OF SLEEP DISORDERED BREATHING 2: PORTABLE SLEEP MONITORING

MICHAEL R. LITTNER

David Geffen School of Medicine at UCLA Sepulveda, California

INTRODUCTION

The obstructive sleep apnea/hypopnea syndrome (OSA) is marked predominantly by daytime somnolence and nighttime snoring often in obese individuals [1, 2]. The diagnosis is confirmed by demonstrating a significant number of obstructive apneas (absence of airflow with continued respiratory effort) and/or obstructive hypopneas (reduction in airflow despite sufficient respiratory effort to produce normal airflow) [1]. The daytime somnolence appears to be, in large part, the result of short, amnestic arousals that fragment and reduce the efficiency of sleep. OSA appears to affect about 4% of men and 2% of women between 30 and 60 years of age [3]. OSA is associated with systemic hypertension, myocardial infarction, motor vehicle accidents, and cerebrovascular accidents [4–7].

Daytime somnolence is a nonspecific symptom and may be due to narcolepsy, insufficient sleep, and idiopathic hypersomnia among other conditions [2]. In addition, snoring is a nonspecific finding; for example, 67% of obese patients (body mass index (BMI) $\geq$ 30) who snored loudly (patient report) had OSA [8]. The general nonspecificity of daytime sleepiness and snoring requires objective measurement of apneas and hypopneas during sleep for confirmation of OSA.

Typically, the approach to confirmation involves an overnight sleep study while monitoring a number of respiratory channels, sleep staging by electroencephalogram (EEG), electro-oculogram (EOG), and chin electromyogram (EMG), as well as leg movements that may also produce frequent arousals [9]. The study is attended by a technician to perform and observe the study, ensure quality and safety, and make needed interventions including application of the most frequently used therapy, continuous positive airway pressure (CPAP). This approach is called polysomnography (PSG).

The number of potential patients usually exceeds the number of sleep laboratory (lab) facilities capable of performing the test. The labor intensity of the attendant, scoring and interpretation of the study, and cost of the space and equipment make PSG relatively expensive, costing typically $1000 or more a study [10].

To increase access to diagnosis and potentially reduce cost, there has been an effort to produce systems that incorporate part or all of the PSG but make it portable and ideally usable without an attendant technician. The ideal system would measure the minimum number of channels necessary, be self-contained and self-administered by the patient, be amenable to rapid and accurate scoring, and provide information that would confirm OSA with identical specificity and sensitivity to the PSG. This chapter will evaluate the ability of various methods to achieve this goal in adults.

TABLE 41.1 American Academy of Sleep Medicine Classification of Levels of Studies of Sleep Apnea Evaluation (Modified) (6 hours Overnight Recording Minimum)

	Level I Attended PSG	Level II Unattended PSG	Level III Modified Portable Sleep Apnea Testing	Level IV Continuous Single or Dual Bioparameter Recording
Measures	Minimum of seven, including EEG, EOG, chin EMG, ECG, airflow, respiratory effort, oxygen saturation	Minimum of seven, including EEG, EOG, chin EMG, ECG or heart rate, airflow, respiratory effort, oxygen saturation	Minimum of four, including ventilation (at least two channels of respiratory movement, or respiratory movement and airflow), heart rate or ECG, oxygen saturation	Minimum of one: oxygen saturation, flow, or chest movement
Body position	Documented or objectively measured	Possible	Possible	No
Leg movement	EMG or motion sensor desirable but optional	Optional	Optional	No
Personnel	Yes	No	No	No
Interventions	Possible	No	No	No

Abbreviations: EEG, electroencephalography; EOG, electro-oculography; EMG, electromyography; ECG, electrocardiography; patterned after Reference [11].

CLASSIFICATION OF METHODS FOR DIAGNOSIS OF SLEEP DISORDERED BREATHING

The American Academy of Sleep Medicine (AASM), formerly known as the American Sleep Disorders Association, in 1994 [11, 12] classified diagnostic sleep equipment into four levels (Table 41.1). Attended PSG has already been described and is Level I. Unattended PSG is Level II. Measurement of a minimum of four channels, which must include oximetry, one or more channels of respiratory effort or movement, heart rate, and usually a measure of airflow, is Level III. A single- or two-channel system typically including oximetry is Level IV. For purposes of this review, systems that do not meet minimum criteria for a Level III will be classified as Level IV. The classification is essentially one of lesser and lesser channels that are typically part of the PSG.

Portable monitoring systems are generally designed to be used unattended, usually in the patient's home. However, the systems can also be used attended in the sleep laboratory and this will also be reviewed.

For purposes of this chapter, attended PSG will be the reference for comparison of portable monitoring systems.

WHAT IS THE PROPER STUDY DESIGN TO VALIDATE A PORTABLE MONITOR?

As discussed in a recent review [13], validation of a particular device involves comparison to attended PSG with determination of the sensitivity and specificity of the portable monitor. This comparison should be made in a patient population that is representative of the population in which the method is to be used. Patient selection should be consecutive without undue referral biases or at least with the referral bias clearly defined and uninfluenced by the investigator or a small group of providers. In addition, the prevalence of OSA in the study population should be typical of the population for which the device is ultimately to be used. For example, if a method tests only high-probability patients for validation, the results cannot be confidently extrapolated to populations of moderate or low probability.

There are two approaches that should be used in every validation study. First, the sensitivity and specificity under ideal conditions should be determined in a simultaneous comparison with attended PSG. This must be done blinded and without intervention by a technician to repair or correct possible data loss from the portable monitor. This provides the sensitivity and specificity for the diagnosis in direct comparison during the same real-time period as the PSG. The report should include the apneas and hypopneas during various patient positions for the PSG and for the portable system. Ideally, the portable system should have a position monitor. If the system does not perform well in this setting, the system is of questionable use. This comparison is of benefit in validation for attended in-lab use only.

The second step in the validation process is to compare the in-lab PSG to the portable monitor used in the intended environment, usually in the patient's home. The study should be blinded and randomized and the PSG and portable monitor should be applied for every patient. The interval between studies should be short, preferably a week or less. Variables that may affect the results are body position, total sleep time, rapid eye movement (REM) sleep time, and environmental conditions such as room temperature and extraneous noise. These contribute to normal night-to-night variability [14], which may differ between the lab and portable monitoring environment.

A strategy to deal with variability that is not an intrinsic characteristic of the portable monitoring device is to also conduct the PSG simultaneous with the portable monitor on a second and third night in the lab. Ideally, a fourth night should also be performed outside the lab in order to determine the night-to-night variability of the unattended portable monitor. This information would help separate the effects of night-to-night variability on the results from those due to intrinsic differences between PSG and portable monitor. To date, one study of a Level III monitor has adopted much of this approach [15].

The methods should include full disclosure of the PSG and portable monitor sensors and channels, definition of apneas and hypopneas, oximeter sampling and recording rates, and funding for the study.

WHAT CAN BE EXPECTED FROM A COMPARISON OF A PORTABLE MONITOR AND PSG?

The hypothesis that portable monitoring can be as diagnostically effective as PSG rests on the assumption that not all of the PSG monitored channels are necessary to make a diagnosis of OSA. That is, some of the channels are either redundant or measure variables that are not essential to the diagnosis. For this to be valid, the definition of what constitutes a confirmatory study for OSA is critical. The definition of an apnea is the cessation of airflow for 10 seconds or more that cannot be attributed to another cause or artifact. The definition of a hypopnea is more difficult. To date, there is no universally accepted definition. A report of a task force of the American Academy of Sleep Medicine on research methods [16] provided several alternative definitions that include a reduction in airflow and may or may not include an arterial oxygen desaturation or EEG arousal. To further complicate matters, a respiratory effort related arousal (RERA) was included as a respiratory event consistent with OSA. A RERA is an increase in respiratory effort from partial upper airway obstruction that is less than a hypopnea. A RERA does not require any obvious reduction in airflow or arterial oxygen desaturation and is associated with an arousal. An added wrinkle to the definition of a hypopnea is that the Centers of Medicare & Medicaid Services (CMS) (i.e., Medicare) require a 4% desaturation during sleep in addition to airflow reduction [17]. The

Medicare criteria require that sleep be measured using traditional sensors in a sleep lab, making most if not all portable systems currently unacceptable as diagnostic devices for Medicare purposes.

Without judging what is a clinically relevant hypopnea (an unresolved question at this time), the design of a portable system is potentially limited by the goals of measurement. For example, if the goal is to define OSA by a combination of hypopneas associated with 4% desaturations and clear-cut apneas, a two-channel system may be sufficient if the issues of sleep, central apneas (apneas without continued respiratory effort), and body position are not clinically relevant. On the other hand, if arousals are part of the definition of hypopneas, such a system would fall short and if RERAs are required, the two-channel system would be totally inadequate. These types of considerations have not been well evaluated in most previous studies. Some studies are weighted to favor the portable system by defining respiratory events identically between PSG and portable monitor with the exception of use of sleep time in the PSG and recording time (often minus artifact) in the portable monitor. In summary, the more types of events that are acceptable to make a diagnosis of OSA, the less likely that the portable monitor will detect most of the events.

With these considerations in mind, the following evaluates the evidence to support or not support the use of portable monitors to diagnose OSA.

WHAT IS THE EVIDENCE TO DATE?

The reader should be aware that there are a large number of studies that have used portable monitors without direct comparison to PSG for a variety of epidemiologic and diagnostic purposes. However, these will not be reviewed since they provide little or no insight into the sensitivity and specificity of portable monitoring compared to PSG in an individual patient. Based on the evidence to be discussed, Level II and IV portable monitors are not sufficiently accurate or validated to recommend for use at this time, particularly unattended in the home. Level III monitors are useful attended in the lab and of possible usefulness unattended in either the lab or the home.

In October 2003, a joint task force of the AASM, the American College of Chest Physicians (ACCP), and the American Thoracic Society (ATS) published an evidence-based review (Joint Review) of portable monitors [13]. Fifty-one publications with 54 studies were reviewed. Sensitivities and specificities were calculated in 49 of these studies. Since then there have been at least 16 publications (one Level II, six Level III, six Level IV, and three of a hybrid Level IV system) with 19 studies (three had both simultaneous lab as well as home to lab studies). In what follows, the apnea/hypopnea index per hour of sleep is designated as AHI for PSG and the respiratory disturbance index per hour of recording or equivalent is designated as RDI for portable monitors unless otherwise indicated.

Many studies, particularly Level IV, required different thresholds for AHI and/or RDI to achieve the highest possible sensitivity and specificity pairs (best sensitivity and specificity). This left many patients with a nondiagnostic RDI, which would require a subsequent evaluation including potentially an attended PSG. Despite the use of best values, many studies failed to achieve an acceptable pair for diagnostic purposes. This was defined in the Joint Review as a likelihood ratio (LR) pair of ≥ 5 to increase post-test probability (i.e., increasing the positive predictive value) of OSA with a positive test and ≤ 0.2 to decrease post-test probability (i.e., increasing the negative predictive value) with a negative test. These LR values indicate a modest improvement in diagnostic accuracy [18] over no test at all. The reader is referred to Reference [18] for a more detailed discussion of LRs.

The Joint Review classified evidence based on the following Grades:

I—blinded comparison, consecutive patients, reference standard (i.e., PSG) performed on all patients;

II—blinded comparison, nonconsecutive patients, reference standard performed on all patients;

III—blinded comparison, consecutive patients, reference standard not performed on all patients;

IV—reference standard was not applied blindly or independently.

The definition of hypopnea and the threshold AHI to define OSA varied from study to study but was consistent within each study. That is, the evidence can be used to determine the performance of portable monitors compared to PSG but cannot easily be used to define what is an acceptable AHI or RDI to identify OSA across all studies.

There were a total of three papers on Level II monitors of evidence Grades II, IV, and IV [19–21]. In addition, there is one study published since the Joint Review of Grade II evidence [22]. The study suggests that similar data can be obtained from home compared to a telemetry monitored and partially attended in-hospital study but the failure rate of home monitoring was unacceptably high at 23.4%. In addition, the telemetry monitored studies had an 11.2% failure rate. Of 99 subjects, evaluable data was available in 65 for both nights. Using the telemetry monitored studies as the reference standard, the sensitivity and specificity were 94.9% and 80.8% with LRs of 4.95 and 0.063 for the 65 subjects (calculated from data presented in the publication). The paucity of data does not allow one to reach any conclusion regarding the utility of these systems in

the diagnosis of OSA. In concept, Level II should be the most accurate. In practice, as indicated by one of the publications [22], the complexity of these systems makes patient set up and subsequent data loss a potential problem.

Nine of nine studies of a Level III monitor done simultaneously and attended in the sleep lab had an acceptable LR pair from the Joint Review [13]. Only one of the studies had a group of nondiagnostic RDIs (36%). Of four studies comparing home to lab, two had an acceptable LR pair with 22% and 37% nondiagnostic RDI values. Data loss, when reported, was under 10% for those with an acceptable LR pair.

Table 41.2 summarizes the data in the simultaneous studies from the Joint Review [13]. In addition, Table 40.2 includes three studies not yet published at the time the Joint Review was closed [15, 23, 31]. All had acceptable LRs and were Grade I, II, and IV evidence. There were 18% nondiagnostic studies in one study, 41% in another, and 12% data loss in one. Of note, from Reference [23], LR pairs (manual scoring and calculated from the data presented in the paper) at five AHI values (5, 10, 15, 20, 30) were acceptable at AHI values of 5, 10, 20, and 30. The LR pair at AHI of 15 was 4.7 and 0.116.

Table 41.3 summarizes data from four home to lab studies from the Joint Review [13]. Table 40.3 includes two studies [15, 31] not yet published at the time that the Joint Review was closed. These two studies had acceptable LRs but data loss was 14% and 18% and one had 36% nondiagnostic studies. In addition, there is an unpublished Level 3 study in manuscript form available on the Internet [37]. The LRs were acceptable at AHI thresholds of 5 and 15.

Of 25 studies of a Level IV monitor done simultaneously in the sleep lab, 14 had an acceptable LR pair [13]. Nine of the 14 studies had nondiagnostic RDI values ranging from 11% to 67%. Of eight studies comparing home to lab, one had an acceptable LR pair with 49% nondiagnostic RDIs. Data loss, when reported, was under 10% for those with an acceptable LR pair.

Since the Joint Review, at least six Level IV monitor studies have been published [38–42], two simultaneous, one on different nights for oximetry and PSG in the lab, and three home to lab. The results of these six studies had a spectrum of sensitivities and specificities with PSG. One simultaneous lab study [39] using a fast Fourier analysis of the spectrum of the heart rate and saturation from the pulse oximeter had acceptable LRs and, if reproducible in a home to lab study, may show promise. On the other hand, in one study [42], 40% of patients with a normal home oximetry had significant OSA (AHI > 15/h) on PSG. However, this study used a 12 second oximeter recording setting, which has been documented to substantially underestimate the number of arterial oxygen desaturations [43–45]. In another Level IV study of 31 subjects using a system that records

TABLE 41.2 Level III Monitors In-Lab

Study	AHI	RDI	Prevalence (%)	Sensitivity (%)	Specificity (%)	LR (H)	LR (L)	Non (%)	PPV	NPV	Evidence Grade	Comment
24	10	6	24	89	92	11.1	0.12	0	78	96	II	Reduction in airflow plus 3% reduction in arterial oxygen desaturation or an arousal used to determine AHI. Reduction in airflow plus 3% reduction in arterial oxygen desaturation used to determine RDI.
25	15	15	50	86	95	17	0.15	0	94	87	IV	Reduction in airflow plus 4% reduction in arterial oxygen desaturation or an arousal used to determine AHI.
26	5	5	62	95	96	24	0.05	0	98	92	I	Reduction in airflow used to determine AHI and RDI. Compressed time frame for scoring RDI but not AHI.
27	10	10	57	97	100	**	0.03	0	100	96	II	Reduction in thoracoabdominal movement of 50% plus 4% reduction in arterial oxygen desaturation for AHI. Discernable reduction in airflow plus 4% reduction in arterial oxygen desaturation for RDI.
28	15	15	27	86	95	17	0.15	0	86	95	II	Reduction in airflow used to determine AHI and RDI.
29	10	10	84	95	100	**	0.05	0	100	80	II	Reduction in airflow plus 4% reduction in arterial oxygen desaturation or 2% reduction in arterial oxygen saturation plus arousal for AHI. Reduction in airflow plus 2% reduction in arterial oxygen desaturation for RDI.
30	10	10	47	92	96	25	0.08	0	93	95	I	Reduction in airflow used to determine AHI and RDI.

TABLE 41.2 *(Continued)*

Study	AHI	RDI	Prevalence (%)	Sensitivity (%)	Specificity (%)	LR (H)	LR (L)	Non (%)	PPV	NPV	Evidence Grade	Comment
31	10 for sensitivity 20 for specificity	10 for sensitivity 20 for specificity	63.3 for sensitivity 43.3 for specificity	100 (64 corr. specificity)	88 (77 corr. sensitivity)	6.5	0	36	83	100	II	Reduction in airflow plus 4% reduction in arterial oxygen desaturation (Denver) or 2% reduction in arterial oxygen desaturation (Los Angeles) or arousal for AHI and RDI. Arousals were measured indirectly with PM. Compressed time frame for scoring RDI but not AHI.
32	10	10	66	100	100	**	0	0	100	100	I	Reduction in airflow used to determine AHI and RDI. Sleep stages not used for either PSG or PM.
15 (new study)	15	15	48 (est.)	95	91	10.6	0.06	0	91	96	I	Reduction in airflow and 2% desaturations with automatic scoring for AHI and RDI. 12% data loss for PM.
33 (new study)	15	10 for sensitivity 20 for specificity	62	100 (corr. specificity 67)	93 (corr. sensitivity 88)	12.6	0	18	95	100	II	RDI reduction in thoracoabdominal movement and reduction in nasal pressure. AHI reduction in thoracoabdominal movement for hypopneas. 3% data loss
23 (new study)	5 levels (5, 10, 15, 20, 30)	6.7 for sensitivity 27.6 for specificity	86 for AHI 5 (RDI 6.7) 44 for AHI 30 (RDI 27.6)	97.1 (corr. specificity 90.9)	90.9 (corr. sensitivity 88.6)	10.7	0.10	41	98.5	90.9	IV	Automatic scoring had unacceptable results. PSG scoring used arousals. Data loss under 10%. Evidence Grade IV since blinding of scoring not reported.

Abbreviations: AHI, apnea/hypopnea index per hour of sleep with PSG; RDI, apnea/hypopnea index per hour of recording unless otherwise indicated for portable monitor; LR, likelihood ratio; L, low; H, high; Non(%), percent of nondiagnostic tests; PPV, positive predictive value = true positives/true positives plus false positives (%); NPV, true negatives/true negatives plus false negatives (%); **, cannot be calculated due to division by 0; PSG, attended polysomnography; PM, portable monitor; corr., corresponding sensitivity or specificity when best sensitivity and specificity at different RDI or AHI thresholds. When this occurs, there are a number of nondiagnostic tests.

TABLE 41.3 Level III Home–Lab Studies

Study	AHI	RDI	Prevalence (%)	Sensitivity (%)	Specificity (%)	LR (H)	LR (L)	Non (%)	PPV	NPV	Evidence Grade	Comment
31	10 for sensitivity 20 for specificity	10 for sensitivity 20 for specificity	61.4 for sensitivity 41.4 for specificity	91 (corr. specificity 70.4)	82.9 (corr. sensitivity 86)	5.1	0.13	22	78	83	II	Reduction in airflow plus 4% reduction in arterial oxygen desaturation (Denver) or 2% reduction in arterial oxygen desaturation (Los Angeles) or arousal for AHI and RDI. Arousals were measured indirectly with PM. Compressed time frame for scoring RDI but not AHI.
34	10	10	74	100	66	2.9	0	0	89	100	IV	Reduction in airflow for AHI and RDI. Two minute epochs used for PM
35	15	10 for sensitivity 20 for specificity	55	94 (corr. specificity 35)	89 (corr. sensitivity 38)	3.26	0.179	55	64	54	IV	Reduction in thoracoabdominal movement of AHI. Reduction in chest movement for RDI. Only patients with RDI < 30 included in analysis.
36	10	8 for sensitivity 23 for specificity	84	95 (corr. specificity 33)	93 (corr. sensitivity 63)	9	0.15	37	98	55	II	Reduction in airflow or thoracoabdominal paradox with an arousal or cyclical reduction in arterial oxygen desaturation for AHI. Same with cyclical 2% reduction in arterial oxygen desaturation and no arousals criteria. One month between studies.
15 (new study)	15	15	48 (est.)	91	83	5.35	0.11	0	83	91	I	Reduction in airflow and 2% desaturations with automatic scoring for AHI and RDI. 14% data loss for PM. 91% split-night studies in sleep laboratory, up to 3 nights averaged for home.
33 (new study)	15	10 for sensitivity 20 for specificity	76	100 (corr. specificity 75)	100 (corr. sensitivity 61)	**	0	36	100	100	I	RDI and AHI reduction in thoracoabdominal movement. Manual scoring better than automatic. 18% data loss for PM.
37 (new study)	5 15	5 15	80 70	100 86	100 100	** **	0 0.14	0 0	100 100	100 100 75	IV	Not obviously blinded. Automatic scoring with review for the portable device. Patient selection not well described.

Abbreviations: AHI, apnea/hypopnea index per hour of sleep with PSG; RDI, apnea/hypopnea index per hour of recording unless otherwise indicated for portable monitor; LR, likelihood ratio; L, low; H, high; Non(%), percent of nondiagnostic tests; PPV, positive predictive value; NPV, negative predictive value = true positives/true positives plus false positives; NPV, true negatives/true negatives plus false negatives; **, cannot be calculated due to division by 0; PSG, attended polysomnography; PM, portable monitor; corr., corresponding sensitivity or specificity when best sensitivity and specificity at different RDI or AHI thresholds. When this occurs, there are a number of nondiagnostic tests.

oronasal sound and airflow, eight normal PSG studies were classified as positive and one classified as moderate and one as severe on PSG were normal on the portable study [46].

There is at least one system that uses an alternative technology. This monitor is a hybrid with an oximeter, an actigraph, and a measurement of radial artery pulse volume. The studies to date on this monitor show promise [47–49] and one validation study comparing both in-lab and home monitoring with sensitivity and specificity at specific thresholds is available [49]. The LRs in this study are acceptable at several AHI thresholds.

WHAT ARE LIMITATIONS OF PSG AS A REFERENCE STANDARD?

There are limitations to PSG implementation and interpretation. Sleep staging is reasonably well-standardized according to published rules [50] but these were developed before OSA was well recognized. For example, arousals were not well defined [50] and while there are subsequent published recommendations [51], there are no universally accepted or easily reproducible definitions, making interscorer reliability poor between clinical centers.

Scoring of hypopneas is in evolution. Although research definitions have been proposed [16], the correlation between these definitions and clinical outcomes is essentially unknown at this time. This leads to difficulty in determining a threshold AHI to confirm OSA.

Night-to-night variability of the AHI or RDI can be substantial and is due to a number of factors, including body position and the amount of REM sleep (supine and REM AHI values are almost always higher than NREM and lateral position AHI value). While the mean AHI in a group of OSA patients does not change substantially, individual patients may have large increases or decreases [14]. For this reason, more than one night of PSG may be necessary to clarify whether a patient has OSA. This variability also makes it difficult to know how much of the difference between a portable monitor and PSG result is normal variability and how much is from the limited set of monitored variables attended or unattended during sleep.

The use of a single AHI to characterize the entire night's study is simplistic. For example, the classification of OSA does not take into account a number of variables that may well have clinical relevance such as supine and REM AHI values and the degree of arterial oxygen desaturation.

SLEEP STAGING

Portable monitors do not generally provide a measure of REM sleep and many do not provide body position. This makes it difficult to fully characterize the RDI result. For example, a patient who snores and has severe daytime sleepiness may sleep mostly in stages 2 to 4 NREM sleep and have a RDI of 4 on one night but have normal REM on a second night with a RDI of 15. Most portable monitors do not have sleep staging and the interpretation of these two RDI values would be difficult. On the other hand, a PSG with sleep stages would provide important information in the interpretation of the study. In particular, an AHI of 4 in the first case would potentially prompt a second baseline study but in the case of the portable monitor might be interpreted as nonsignificant and not be followed up.

WHAT IS THE APPROPRIATE AHI DEFINITION OF OSA BY PORTABLE MONITORING?

Historically, hypopneas (decreased airflow) have been used to characterize OSA and studies have suggested that hypopneas may have the same clinical significance as apneas in many patients [52]. However, the standard method of measuring airflow with a thermistor has come under scrutiny and it is clear that many hypopneas have been unrecognized by this technique [16]. In addition, partial upper airway obstruction that leads to increased amplitude of intrathoracic pressure can trigger an arousal (i.e., a RERA) and such arousals may produce daytime sleepiness [16, 53].

Methods to capture more subtle hypopneas and measure airflow more quantitatively have become available. These currently focus around nasal pressure measurement, which is an indirect measure of airflow and more sensitive than thermistors [16]. The nasal pressure has been favorably compared against pneumotachograph airflow in OSA and appears more accurate than thermistor airflow [54, 55]. In addition, the use of an esophageal balloon or tube to measure intrathoracic pressure swings is recommended to determine the presence of RERAs [16].

Based on this technology, definitions of hypopnea and respiratory events for research purposes have been proposed including syndrome definition using a composite AHI ≥ 5 for confirmation of OSA [16]. However, almost all previous OSA studies used thermistors and none of the new definitions have been adequately validated against thermistors in patients with OSA or against non-OSA controls. Given the newer, more sensitive technology to detect respiratory events, it is possible, even likely, that a diagnostic AHI will be much higher than previously observed and many individuals who were considered on a combination of clinical evaluation and PSG results to *not* have OSA will now have an AHI in the OSA range of at least 5 and possibly much higher.

PSG is potentially capable of capturing all of the currently recommended respiratory events, whereas portable monitors, in general, capture only disturbances in airflow and saturation leading to a RDI that frequently underestimates the number of potential respiratory disturbances during sleep (i.e., apneas, hypopneas, desaturations, arousals, and RERAs). In addition, depending on the technology and definitions used, the RDI may vary considerably on the same night in the same patient.

To confuse the matter further, Medicare, as mentioned, has published criteria for scoring hypopneas on PSG for purposes of qualifying for CPAP [17]. These require a ≥30% decrease in airflow associated with a 4% desaturation from baseline during recorded sleep of ≥2 hours duration. The PSG must be performed in a facility-based sleep study lab and not in the home or in a mobile facility. Without the sleep requirement, it is likely that a portable monitor could more readily replicate this definition. Of note, several Local Medical Review Policies (LMRP) have substituted recording time for sleep time (e.g., http://www.tricenturion.com/content/Doc_View.cfm?type=bull&File=ACF-38E8.doc). Medicare criteria require an AHI of at least 5 with symptoms of OSA such as daytime sleepiness or an AHI of 15, irrespective of symptoms.

The user of a portable monitor should be aware of the operating characteristics of the monitor and not rely on computer-generated scoring. In addition, since the portable monitor does not measure a number of events that may be recorded on the PSG, does not usually measure sleep, and may not measure position, a negative study should not be accepted to exclude a diagnosis of OSA. On the other hand, since the portable monitor is generally less sensitive than the PSG, a positive study with a properly validated monitor, if technically adequate, should generally be accepted as confirmatory in the appropriate clinical setting.

WHAT ARE DIFFERENTIAL DIAGNOSTIC CONSIDERATIONS?

Patients with Cheyne–Stokes respiration may mimic OSA but with a combination of airflow, respiratory movement, and saturation measurements, this should be apparent on a portable monitor. Patients with COPD may have periods of desaturation that occur typically during REM sleep [56]. Since the portable monitor does not measure REM sleep, studies in patients with severe COPD should be avoided if attempting to diagnose OSA.

As mentioned, daytime sleepiness can occur in sleep disorders other than OSA [2]. The typical Level III portable monitor is of little use in these cases and patients with daytime sleepiness and a negative portable monitor study should have the cause of the daytime sleepiness characterized. This will often require a PSG and possibly a multiple sleep latency test (MSLT), which requires measurement of sleep staging [2, 57].

TECHNICAL CONSIDERATIONS

The type of sensors may impact the results. For example, use of a thermistor is excellent for detection of apneas but relatively insensitive for detection of modest reductions in airflow [16]. Thoracoabdominal movement by inductance plethysmography appears more sensitive for detection of hypopneas but the belts may lose calibration or shift during the study. Nasal pressure appears to be very sensitive to reductions in airflow but data loss may be a problem due to loss of signal or mouth breathing [16].

Several recent studies have documented that the method of sampling the saturation signal with an oximeter is important in accurately measuring reductions in arterial oxygen saturation [43–45]. For example, an oximeter set at a 3 second recording rate produced almost twice as many 3% desaturations as a 12 second recording rate [43]. Furthermore, desaturations stored in oximeter memory substantially underestimated desaturation displayed in real time online at any recording rate [44].

The method of scoring—manual versus computer—is also a consideration. Without the ability to manually review data, results will always be suspect since artifact may often mimic respiratory events. In general, computer scoring has been less accurate than manual scoring but the time involved is considerably greater with manual scoring [58]. In addition, the ability to independently calibrate and test the equipment is desirable to ensure that equipment failure is not producing erroneous results.

WHAT CAN BE SUPPORTED BY THE EVIDENCE?

As discussed previously, based on the current evidence, a Level III system with a minimum of airflow, oximetry, respiratory movement, and heart rate measurements is recommended. Strongly recommended is an additional sensor to measure body position. Also recommended is a sensor to measure snoring.

The use of an attended Level III portable monitor to diagnose OSA would appear from both evidence and strategic analyses to be more appropriate rather than to exclude patients with OSA since:

1. A positive portable study, if properly performed in a patient with clinical features of OSA, has a high degree of specificity and positive predictive value.
2. A negative or nondiagnostic portable study should be followed, usually with an attended PSG, since the portable monitor study (a) is less likely to detect other evidence of OSA including RERAs and subtle hypopneas and will not allow the determination of REM AHI; and (b) will not diagnose other disorders

contributing to the patient's clinical presentation such as periodic limb movement disorder.

Based on considerations similar to the above, the AASM/ATS/ACCP task force guidelines [58] recommend that attended Level III studies are acceptable for diagnosis with careful follow-up of negative studies including, in most cases, a PSG for confirmation.

Up to this point, this chapter has concentrated on the diagnosis of OSA without considering that PSG is used to monitor CPAP titration during sleep. To date, there appears to be only one study that examined a Level III portable monitoring montage to titrate CPAP during an attended study [59]. In addition, the use of an attended portable monitor to make a diagnosis during the first half of the night followed by a CPAP titration during the second half of the night (split-night study) has not been examined. For these reasons, use of a portable monitor to both diagnose and titrate CPAP cannot be well supported by evidence.

WHAT OTHER OPTIONS MAY BE CONSIDERED?

The evidence is lacking to support unattended use of a portable monitor in the patient's home as a standalone approach to diagnosis of OSA. However, in the proper setting, with appropriate patient selection, and careful follow-up including ready access to attended PSG, home portable studies are feasible. Based on an integration of the evidence available, the following conditions would appear to be necessary:

1. A high pretest probability (i.e., a high prevalence of OSA in the patient population), ideally to exceed 70%. There are a number of equations that use readily available data such as BMI, sex, history of snoring, and neck circumference or more complicated data such as x-rays of the upper airway with cephalometric measurements [57, 60–65].
2. The availability of attended PSG for patients with a strong clinical history and a negative or nondiagnostic portable monitoring study.
3. The availability of treatment including PSG titration for CPAP.
4. An experienced sleep practitioner who is capable of evaluating both the clinical and portable monitoring information.

The approach to CPAP titration is beyond this chapter but there has been a trend to use autotitrating CPAP (APAP) machines unattended in the patient's home. The reader is referred to an evidence-based review of the topic and guidelines published by the AASM [66, 67], which indicated that unattended use for CPAP titration is not established for CPAP naïve patients. Subsequent to publication of the guidelines, at least one study has provided evidence that APAP can lead to favorable outcomes in CPAP naïve patients [68]. In general, such an approach should only be carried out with the knowledge that the evidence for the efficacy of unattended home CPAP titration in CPAP naïve patients is in evolution.

COST EFFECTIVENESS

This is a complicated topic since the costs must be weighed against the accessibility of patients to diagnostic studies. If there are sufficient resources to study all patients who are identified as candidates, then the cost of the attended PSG, often a split-night study, must be balanced against the cost of the portable study and the potential need for a second study for CPAP treatment. The lower sensitivity of the portable study for OSA, particularly if the research criteria [16] are used for comparison, and the night-to-night variability of any test for OSA require careful evaluation of negative and nondiagnostic studies with strong consideration given to proceeding to a subsequent attended PSG. In addition, local reimbursements are also an issue and the Medicare rules essentially exclude portable monitoring attended or unattended as an option for confirming a diagnosis of OSA.

If there are not sufficient resources, then portable monitoring with the possibility of APAP becomes a potential option, recognizing all the limitations of portable monitoring and the use of APAP machines to titrate and determine treatment for patients.

Of note, the use of a Level III portable monitor attended in the lab is another potential addition to the overall diagnostic strategy and at least one analysis suggests that this may be more cost-effective than performing attended PSG on all patients [69].

Although a comprehensive answer cannot be given, the following, at a minimum, should assessed:

1. What is the cost of the PSG equipment, supplies, space, utilities, technician time, physician time, and the like?
2. What is the cost of portable monitoring equipment, supplies, time spent with patient, interpretation time, and the like?
3. What are the number of studies that are nondiagnostic and require follow-up PSG?
4. What is the strategy for CPAP titration? Does it include an attended PSG, a portable monitoring titration, an APAP, or some combination?
5. What are the adherence and compliance with CPAP with any of the strategies?

6. What is an acceptable wait time for a test and, if too long, how does this impact the quality of life of the patients?

7. Based on the acceptable wait time, what are the resources necessary for each of the possible strategies?

8. What is the patient population to be studied? What is the prevalence of OSA? What is the likelihood that other diagnoses are present such as periodic limb movement disorder or narcolepsy?

REFERENCES

1. Bassiri AG, Guilleminault C. Clinical features and evaluation of obstructive sleep apnea–hypopnea syndrome. In: kryger MH, Roth T, Dement WC (Eds), *Principles and Practice of Sleep Medicine*, 3rd ed. Saunders, Philadelphia, 2000, pp 869–878.

2. American Sleep Disorders Association. *International Classification of Sleep Disorders, Revised: Diagnostic and Coding Manual*. American Sleep Disorders Association, Rochester, MN, 1997.

3. Young T, Palta M, Dempsey J, Skatrud J, Weber S, Badr S. The occurrence of sleep-disordered breathing among middle-aged adults. *N Engl J Med* **328**:1230–1235(1993).

4. Findley LJ, Unverzagt ME, Suratt PM. Automobile accidents involving patients with obstructive sleep apnea. *Am Rev Respir Dis* **138**:337–340(1988).

5. Peppard PE, Young T, Palta M, Skatrud J. Prospective study of the association between sleep-disordered breathing and hypertension. *N Engl J Med* **342**:1378–1384(2000).

6. Hung J, Whitford EG, Parsons RW, Hillman DR. Association of sleep apnoea with myocardial infarction in men. *Lancet* **336**:261–264(1990).

7. Dyken ME, Somers VK, Yamada T, Ren ZY, Zimmerman MB. Investigating the relationship between stroke and obstructive sleep apnea. *Stroke* **27**:401–407(1996).

8. Resta O, Foschino-Barbaro MP, Legari G, Talamo S, Bonfitto P, Palumbo A, Minenna A, Giorgino R, De Pergola G. Sleep-related breathing disorders, loud snoring and excessive daytime sleepiness in obese subjects. *Int J Obes Relat Metab Disord* **25**:669–675(2001).

9. Hening W, Allen R, Earley C, Kushida C, Picchietti D, Silber M. The treatment of restless legs syndrome and periodic limb movement disorder. An American Academy of Sleep Medicine Review. *Sleep* **22**:970–999(1999).

10. Chervin RD, Murman DL, Malow BA, Totten V. Cost–utility of three approaches to the diagnosis of sleep apnea: polysomnography, home testing, and empirical therapy. *Ann Intern Med* **130**:496–505(1999).

11. Ferber R, Millman R, Coppola M, Fleetham J, Murray CF, Iber C, McCall V, Nino-Murcia G, Pressman M, Sanders M, Strohl K, Votteri B, Williams A. Portable recording in the assessment of obstructive sleep apnea. *Sleep* **17**:378–392(1994).

12. Standards of Practice Committee of the American Sleep Disorders Association. Practice parameters for the use of portable recording in the assessment of obstructive sleep apnea. *Sleep* **17**:372–377(1994).

13. Flemons WW, Littner MR, Rowley JA, Gay P, Anderson WM, Hudgel DW, McEvoy RD, Loube DI. Home diagnosis of sleep apnea: a systematic review of the literature: an evidence review cosponsored by the American Academy of Sleep Medicine, the American College of Chest Physicians, and the American Thoracic Society. *Chest* **124**:1543–1579(2003).

14. Le Bon O, Hoffmann G, Tecco J, Staner L, Noseda A, Pelc I, Linkowski P. Mild to moderate sleep respiratory events: one negative night may not be enough. *Chest* **118**:353–359(2000).

15. Reichert JA, Bloch DA, Cundiff E, Votteri BA. Comparison of the NovaSom QSG, a new sleep apnea home-diagnostic system, and polysomnography. *Sleep Med* **x**:213–218(2003).

16. Anonymous. Sleep-related breathing disorders in adults: recommendations for syndrome definition and measurement techniques in clinical research. The Report of an American Academy of Sleep Medicine Task Force. *Sleep* **22**:667–689(1999).

17. http://www.aptweb.org/pdf/cmscpap.pdf (last accessed Dec 13, 2003).

18. Flemons W, Littner MR. Measuring agreement between diagnostic devices. *Chest* **124**:1535–1542(2003).

19. Orr WC, Eiken T, Pegram V, et al. A laboratory validation study of a portable system for remote recording of sleep related respiratory disorders. *Chest* **105**(1):160–162(1994).

20. Mykytyn IJ, Sajkov D, Neill AM, et al. Portable computerized polysomnography in attended and unattended settings. *Chest* **115**:114–122(1999).

21. Portier F, Portmann A, Czernichow P, et al. Evaluation of home versus laboratory polysomnography in the diagnosis of sleep apnea syndrome. *Am J Respir Crit Care Med* **162**:814–818(2000).

22. Gagnadoux F, Pelletier-Fleury N, Philippe C, Rakotonanahary D, Fleury B. Home unattended vs hospital telemonitored polysomnography in suspected obstructive sleep apnea syndrome: a randomized crossover trial. *Chest* **121**:753–758(2002).

23. Calleja JM, Esnaola S, Rubio R, Duran J. Comparison of a cardiorespiratory device versus polysomnography for diagnosis of sleep apnoea. *Eur Respir J* **20**:1505–1510(2002).

24. Ballester E, Solans M, Vila X, et al. Evaluation of a portable respirator recording device for detecting apneas and hypopnoeas in subjects from a general population. *Eur Respir J* **16**:123–127(2000).

25. Claman D, Murr A, Trotter K. Clinical validation of the Bedbugg in detection of obstructive sleep apnea. *Otolaryngol Head Neck Surg* **125**:227–230(2001).

26. Emsellem H, Corson W, Rappaport B, et al. Verification of sleep apnea using a portable sleep apnea screening device. *South Med J* **83**:748–752(1990).

27. Ficker JH, Wiest GH, Wilpert J, Fuchs FS, Hahn EG. Evaluation of a portable recording device (Somnocheck) for use in patients with suspected obstructive sleep apnoea. *Respiration* **68**:307–312(2001).

28. Man G, Kang B. Validation of a portable sleep apnea monitoring device. *Chest* **108**:388–393(1995).

29. Redline S, Tosteson T, Boucher M, et al. Measurement of sleep-related breathing disturbances in epidemiologic studies. Assessment of the validity and reproducibility of a portable monitoring device. *Chest* **100**:1281–1286(1991).

30. Verse T, Pirsig W, Junge-Hulsing B, et al. Validation of the POLY-MESAM seven-channel ambulatory recording unit. *Chest* **117**:1613–1618(2000).

31. White D, Gibb T, Wall J, et al. Assessment of accuracy and analysis time of a novel device to monitor sleep and breathing in the home. *Sleep* **18**:115–126(1995).

32. Zucconi M, Ferini-Strambi L, Castronovo V, et al. An unattended device for sleep-related breathing disorders: validation study in suspected obstructive sleep apnoea syndrome. *Eur Respir J* **9**:1251–1256(1996).

33. Dingli K, Coleman EL, Vennelle M, Finch SP, Wraith PK, Mackay TW, Douglas NJ. Evaluation of a portable device for diagnosing the sleep apnoea/hypopnoea syndrome. *Eur Respir J* **21**:253–259(2003).

34. Ancoli-Israel S, Mason W, Coy T, et al. Evaluation of sleep disordered breathing with unattended recording: the Nightwatch System. *J Med Eng Tech* **21**:10–14(1997).

35. Whittle AT, Finch SP, Mortimore IL, et al. Use of home sleep studies for diagnosis of the sleep apnoea/hypopnoea syndrome. *Thorax* **52**:1068–1073(1997).

36. Parra O, Garcia-Esclasans N, Montserrat J, et al. Should patients with sleep apnoea/hypopnoea syndrome be diagnosed and managed on the basis of home sleep studies? *Eur Respir J* **10**:1720–1724(1997).

37. Carter GS, Coyle MA, Mendelson WB. Validity of a portable cardio-respiratory system to collect data in the home environment in patients with obstructive sleep apnea. http://www. cms.hhs.gov/coverage/download/id110a.pdf (last accessed Sept 4, 2004).

38. Golpe R, Jimenez A, Carpizo R. Home sleep studies in the assessment of sleep apnea/hypopnea syndrome. *Chest* **122**:1156–1161(2002).

39. Zamarron C, Gude F, Barcala J, Rodriguez JR, Romero PV. Utility of oxygen saturation and heart rate spectral analysis obtained from pulse oximetric recordings in the diagnosis of sleep apnea syndrome. *Chest* **123**:1567–1576(2003).

40. Raymond B, Cayton RM, Chappell MJ. Combined index of heart rate variability and oximetry in screening for the sleep apnoea/hypopnoea syndrome. *J Sleep Res* **1**:53–61(2003).

41. Roche N, Herer B, Roig C, Huchon G. Prospective testing of two models based on clinical and oximetric variables for prediction of obstructive sleep apnea. *Chest* **12**:747–752(2002).

42. Hussain SF, Fleetham JA. Overnight home oximetry: can it identify patients with obstructive sleep apnea–hypopnea who have minimal daytime sleepiness? *Respir Med* **97**:537–540(2003).

43. Davila DG, Richards KC, Marshall BL, O'Sullivan PS, Osbahr LA, Huddleston RB, Jordan JC. Oximeter's acquisition parameter influences the profile of respiratory disturbances. *Sleep* **26**:91–95(2003).

44. Davila DG, Richards KC, Marshall BL, O'Sullivan PS, Gregory TG, Hernandez VJ, Rice SI. Oximeter performance: the influence of acquisition parameters. *Chest* **122**:1654–1660 (2002).

45. Wiltshire N, Kendrick A, Catterall J. Home oximetry studies for diagnosis of sleep apnea/hypopnea syndrome. Limitation of memory storage capabilities. *Chest* **120**:384–389(2001).

46. Liesching TN, Carlisle C, Marte A, Bonitati A, Millman RP. Evaluation of the accuracy of SNAP technology sleep sonography in detecting obstructive sleep apnea in adults compared to standard polysomnography. *Chest* **125**:886–891(2004).

47. Pillar G, Bar A, Betito M, Schnall RP, Dvir I, Sheffy J, Lavie P. An automatic ambulatory device for detection of AASM defined arousals from sleep: the WP100. *Sleep Med* **4**:207–212(2003).

48. Bar A, Pillar G, Dvir I, Sheffy J, Schnall RP, Lavie P. Evaluation of a portable device based on peripheral arterial tone for unattended home sleep studies. *Chest* **123**:695–703(2003).

49. Pittman SD, Ayas NT, MacDonald MM, Malhotra A, Fogel RB, White DP. Using a wrist-worn device based on peripheral arterial tonometry to diagnose obstructive sleep apnea: in-laboratory and ambulatory validation. *Sleep* **27**:923–933 (2004).

50. Rechtschaffen A, Kales A. *A Manual of Standardized Terminology, Techniques, and Scoring System for Sleep Stages of Human Subjects*. Brain Information Service, Los Angeles, 1968.

51. Anonymous. EEG arousals: scoring rules and examples: a preliminary report from the Sleep Disorders Atlas Task Force of the American Sleep Disorders Association. *Sleep* **15**:173–184(1992).

52. Gould GA, Whyte KF, Rhind GB, Douglas D, et al. The sleep hypopnea syndrome. *Am Rev Respir Dis* **137**:895–898(1988).

53. Guilleminault C, Stoohs R, Clerk A, Cetel M, Maistros P. A cause of excessive daytime sleepiness. The upper airway resistance syndrome. *Chest* **104**:781–787(1993).

54. Heitman SJ, Atkar RS, Hajduk EA, Wanner RA, Flemons WW. Validation of nasal pressure for the identification of apneas/hypopneas during sleep. *Am J Respir Crit Care Med* **166**:386–391(2002).

55. Norman RG, Ahmed MM, Walsleben JA, Rapoport DM. Detection of respiratory events during NPSG: nasal cannula/pressure sensor versus thermistor. *Sleep* **20**:1175–1184(1997).

56. Littner MR, McGinty DJ, Arand DL. Determinants of oxygen desaturation in the course of ventilation during sleep in chronic obstructive pulmonary disease. *Am Rev Respir Dis* **122**:849–857(1980).

57. Chesson AL, Ferber RA, Fry JM, Grigg-Damberger M, Hartse KM, Hurwitz TD, Johnson S, Kader GA, Littner M, Rosen G, Sangal RB, Schmidt-Nowara W, Sher A. Standards of Practice Committee Task Force. Practice parameters for the indications for polysomnography and related procedures. *Sleep* **20**:406–422(1997).

58. Chesson AL Jr, Berry RB, Pack A. Practice parameters for the use of portable monitoring devices in the investigation of suspected obstructive sleep apnea in adults. *Sleep* **26**:907–913(2003).

59. Montserrat JM, Alarcón A, Lloberes P, Ballester E, Fornas C, Rodriguez-Roisin R. Adequacy of prescribing nasal continuous positive airway pressure therapy for the sleep apnoea/hypopnoea syndrome on the basis of night time respiratory recording variables. *Thorax* **50**:969–971(1995).

60. Gurubhagavatula I, Maislin G, Pack AI. An algorithm to stratify sleep apnea risk in a sleep disorders clinic population. *Am J Respir Crit Care Med* **164**:1904–1909(2001).

61. Flemons WW, Whitelaw WA, Brant R, Remmers JE. Likelihood ratios for a sleep apnea clinical prediction rule. *Am J Respir Crit Care Med* **150**:1279–1285(1994).

62. Viner S, Szalai JP, Hoffstein V. Are history and physical examination a good screening test for sleep apnea? *Ann Intern Med* **115**:356–359(1991).

63. Netzer NC, Stoohs RA, Netzer CM, et al. Using the Berlin Questionnaire to identify patients at risk for the sleep apnea syndrome. *Ann Intern Med* **131**:485–536(1999).

64. Kushida CA, Efron B, Guilleminault C. A predictive morphometric model for the obstructive sleep apnea syndrome. *Ann Intern Med* **127**:581–587(1997).

65. Tsai WH, Remmers JE, Brant R, Flemons WW, Davies J, MacArthur C. A decision rule for diagnostic testing in obstructive sleep apnea. *Am J Respir Crit Care Med* **167**:1427–1432(2003).

66. Berry RB, Parish JM, Hartse KM. The use of auto-titrating continuous positive airway pressure for treatment of adult obstructive sleep apnea. An American Academy of Sleep Medicine review. *Sleep* **25**:148–173(2002).

67. Littner M, Hirshkowitz M, Davila D, Anderson WM, Kushida CA, Woodson BT, Johnson SF, Merrill SW. Standards of Practice Committee of the American Academy of Sleep Medicine. Practice parameters for the use of auto-titrating continuous positive airway pressure devices for titrating pressures and treating adult patients with obstructive sleep apnea syndrome. An American Academy of Sleep Medicine report. *Sleep* **25**:143–147(2002).

68. Senn O, Brack T, Matthews F, Russi EW, Bloch KE. Randomized short-term trial of two autoCPAP devices versus fixed continuous positive airway pressure for the treatment of sleep apnea. *Am J Respir Crit Care Med* **168**:1506–1511(2003).

69. Reuven H, Schweitzer E, Tarasiuk A. A cost-effectiveness analysis of alternative at-home or in-laboratory technologies for the diagnosis of obstructive sleep apnea syndrome. *Med Decis Making* **21**:451–458(2001).

42

INDICATIONS FOR TREATMENT OF OBSTRUCTIVE SLEEP APNEA IN ADULTS

Nilesh B. Davé and Patrick J. Strollo
University of Pittsburgh, Pittsburgh, Pennsylvania

INTRODUCTION

Obstructive sleep apnea (OSA) is associated with significant daytime sleepiness, reduced quality of life, insulin resistance, motor vehicle crashes, as well as vascular morbidity and mortality [1–3]. The medical therapy of choice for OSA, positive pressure via a mask, is unique [4]. The treatment is administered in one of the most intimate settings, the bedroom. Positive pressure therapy substantially decreases snoring, snoring related arousals, and hypoxia. Treatment favorably affects daytime sleepiness, driving risk, vascular function, vascular risk (e.g., risk for strokes, myocardial infarctions), and quality of life [5–9].

The severity of OSA is graded by the apnea–hypopnea index (AHI). The American Academy of Sleep Medicine recommends classifying sleep apnea as mild (AHI 5–15), moderate (AHI 15–30), and severe (AHI > 30) [10]. This index statistically correlates the presence of sleepiness, neurocognitive impairment, and vascular risk [11–13]. It is relatively easy to treat patients with severe, symptomatic OSA. The difficulty with regard to treatment frequently occurs when patients with severe OSA are asymptomatic or when patients are profoundly symptomatic with a low AHI.

PATIENT ASSESSMENT

Successful treatment cannot be accomplished without proper patient assessment. It is very helpful to understand what the patient hopes to gain from the evaluation. This is best handled by seeing the patient prior to polysomnography. The clinician can then understand what is driving the evaluation: the complaint of snoring, the complaint of fatigue and/or daytime sleepiness, or the concern of vascular risk. It is also helpful to understand if it is the patient, spouse, or referring physician who is most concerned about OSA.

If it is the patient who is most concerned with the possibility of OSA and he or she is subjectively sleepy, there is a good chance that medical therapy with positive pressure will be accepted. These patients are good candidates for split-night polysomnography [14–16]. If the patient does not complain of daytime fatigue or sleepiness, and/or does not regard snoring as a significant problem, acceptance and adherence to positive pressure therapy may be difficult to establish, and split-night polysomnography may not be the best approach [17, 18]. In this circumstance, it is generally best to obtain a full night of diagnostic polysomnography data and review the findings prior to a trial of positive pressure.

The clinician needs to know if insufficient sleep and/or depression are contributing to the complaint of daytime sleepiness or fatigue [19]. Does shift work or a possible sleep phase shift contribute to daytime impairment? Could concomitant narcolepsy without cataplexy or idiopathic hypersomnolence be present? Does the patient have difficulty with sleep maintenance unrelated to OSA? If so, adequate therapy may involve treatment of insomnia or restless leg syndrome. Can non-sleep-related pathology, such as chronic pain, be contributing to alterations in sleep architecture and continuity?

Before positive pressure therapy is attempted, a number of issues that are likely to impact on acceptance and adherence of positive pressure should be considered. Is the patient familiar with positive pressure therapy? If not, an educational intervention is needed prior to the introduction of therapy [20, 21]. Is nasal obstruction present? If so, medical and possibly mechanical treatment of the nose may be necessary for effective treatment [22–24]. Is the patient claustrophobic? An attempt at desensitization may be beneficial prior to instituting therapy if claustrophobia is present [25].

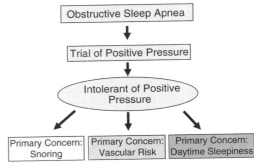

Figure 42.1 Focusing the treatment to the primary patient complaint.

TAILORING THE TREATMENT TO A GIVEN PATIENT

Not all patients with the same severity of OSA are affected similarly. Treatment of the minimally symptomatic patient with severe OSA can be challenging. Treatment may be difficult to accept or adhere to and some treatment options are not uniformly effective. The long-term impact of treatment is uncertain.

In the absence of definitive long-term outcome data, there is uncertainty regarding how hard to push therapy in the patients with mild to moderate OSA with minimal symptoms [26]. Patients who are profoundly symptomatic with relatively mild OSA may not accept positive pressure therapy. The long-term effect of alternative treatments to positive pressure is unknown but they may be of value in select circumstances.

Once the decision has been made that the patient could benefit from a trial of therapy, the first intervention in conjunction with life-style recommendations (i.e., avoiding alcohol and sedatives, proper sleep hygiene, smoking cessation, and fitness) should be a trial of positive pressure via a mask [4, 27]. This trial is best accomplished in the laboratory with an attendant technician. Attended positive pressure titrations allow for further patient education and reassurance by the technical staff as well as proper mask fit, optimal modality (i.e., continuous positive airway pressure or bilevel pressure), and an accurate pressure prescription [20]. The decision to do a split- or full-night study will depend on the considerations discussed earlier.

In-line heated humidification may be particularly useful in elderly patients and in patients with nasal congestion and mouth leaks [28, 29]. It should be prescribed for patients being treated with systemic anticoagulation to decrease the risk of epistaxis [30]. Chin straps and oral–nasal masks may be tried for mouth leaks but are poorly tolerated compared to nasal interfaces with heated humidification [31, 32].

SECOND LINE THERAPY: ALTERNATIVES TO POSITIVE PRESSURE

Despite adequate preparation and an effective attended titration, there are a number of patients with an elevated AHI and/or daytime symptoms who will not accept or adhere to positive pressure therapy. This problem highlights the need for follow-up with objective measurements of adherence to positive pressure therapy. In these patients, it is important to revisit the primary complaint driving the evaluation in the context of the severity of OSA and underlying vascular risk (see Figure 42.1).

SNORING

In patients with mild OSA (AHI 5–15) with minimal symptoms of fatigue/daytime sleepiness and the primary complaint of snoring, a trial of an oral appliance or a palatal procedure is a reasonable option [27].

Patients may prefer an oral appliance to positive pressure [33]. The response to treatment is not complete, mandating follow-up [8]. Despite expert adjustment, treatment with oral appliance therapy may be limited by tooth movement and bite discomfort [34, 35]. The long-term outcomes with oral appliance therapy are not well characterized.

Palatal procedures include conventional scalpel technique uvulopalatopharyngoplasty (UPPP), laser-assisted uvulopalatoplasty (LAUP), and radiofrequency treatment of the palate (somnoplasty) [36–38]. The pros and cons of the palatal procedures are discussed in detail elsewhere in this volume. Overall, palatal procedures alone can be an effective treatment for snoring. If a tonsillectomy is included, mild OSA can be favorably impacted; however, as in the case of oral appliances, the response to treatment may not be complete and follow-up is mandatory [39].

Optimal treatment of nasal pathology can favorably modify snoring and may be an important contribution to the treatment plan. This may require medical interventions (i.e., antihistamines, nasal steroids, or leukotriene antagonists) [22, 24]. Mechanical treatment of nasal obstruction may provide added value. Radiofrequency treatment of the nasal turbinates can be effective and may avoid an operation [23].

VASCULAR RISK

Patients with untreated OSA are at risk for vascular morbidity and/or mortality [40–43]. Emerging evidence supports the concept that intermittent hypoxia may be the primary determinant of vascular risk related to OSA [44]. The ischemia–reperfusion injury caused by repeated episodes of hypoxia result in elevated oxidative stress that leads to elevated circulating levels of radical oxygen species (ROS) that contribute further to endothelial dysfunction [45, 46]. This may be mediated, in part, by reactive oxygen species that are precipitated by an ischemia–reperfusion insult related to the intermittent cell hypoxia [47]. In animal experiments, intermittent hypoxia has been shown to upregulate sympathetic tone, resulting in catecholamine release and elevated blood pressure [48]. Intermittent hypoxia (IH), an innate feature of OSA, promotes an inflammatory cascade [44, 49, 50]. During IH, these endothelial cells experience recurrent states of low and normal oxygen tension similar to ischemia–reperfusion injury. There is a growing body of literature supporting the initiation of the production of inflammatory mediators during hypoxia.

Nocturnal oxygen may be accepted in patients who do not tolerate positive pressure therapy [51]. Although definitive evidence is lacking, it is biologically plausible that nocturnal oxygen would favorably affect vascular risk. One current limitation to this treatment option is the inconvenience of transporting oxygen concentrators that are noisy (up to 50 decibels), bulky, and on average weigh between 20 and 50 pounds [52]. If vascular comorbidity is present in the absence of significant daytime impairment, treatment with positive pressure may not be accepted [18]. Similar difficulty may be encountered with oral appliance therapy. There is no definitive data to support surgery other than tracheostomy, as an effective treatment option to impact vascular comorbidities related to OSA [53, 54].

DAYTIME SYMPTOMS

It is always helpful to determine the response of impaired daytime function (i.e., fatigue and sleepiness) to positive pressure therapy. Unfortunately, this effect is difficult to asses when patients are unwilling to accept treatment with positive pressure. Confounders such as depression

and chronic sleep deprivation (the most common cause of daytime impairment) should be excluded [19]. An objective assessment of daytime sleepiness, such as the multiple sleep latency test (MSLT), is helpful in determining the degree of daytime impairment and providing insight into the possibility of a concomitant diagnosis of narcolepsy without cataplexy or idiopathic hypersomnolence [55].

A judicious trial of a daytime stimulant may be in order to improve quality of life. This trial is best accomplished in conjunction with positive pressure therapy. Certain patients may have continued daytime sleepiness despite treatment with continuous positive airway pressure (CPAP) or bilevel pressure. Recent animal data indicates that long-term intermittent hypoxia may damage wake-promoting regions of the brain [56]. Pack et al. [57] have reported success with modafinil as adjunctive therapy for daytime sleepiness in OSA in patients compliant with CPAP therapy; however, the addition of modafinil did not normalize MSLT scores.

Nonamphetamine daytime stimulants appear to be reasonably safe as an adjunct to treatment with positive pressure for daytime sleepiness [58]. Stimulant therapy alone cannot be recommended for patients with sleep apnea (AHI > 5) [57, 59]. If the patient does not accept positive pressure therapy, second line therapy for OSA should be pursued—whether medical, surgical, or dental—prior to contemplating adjunctive stimulant treatment. Follow-up monitoring of blood pressure is necessary to monitor the vascular risk due to continued episodes of intermittent hypoxia.

SPECIAL CIRCUMSTANCES

Upper Airway Resistance Syndrome

There is uncertainty regarding the use of stimulant therapy alone in patients with upper airway resistance syndrome (UARS) [60–62]. Ideally, a trial of treatment with positive pressure would be advisable. Unfortunately, a significant percentage of these patients may not accept treatment with positive pressure. This approach is frequently hampered by the fact that third-party payers will not reimburse homecare companies for a positive pressure treatment trial of UARS, and the patient may be unwilling to bear the cost on his/her own.

Down Syndrome

Patients with Down syndrome have upper airway abnormalities that place them at risk for sleep disordered breathing [63]. In the adult Down syndrome patient, the challenge is therapeutic not diagnostic [64]. Many of these patients have difficulty accepting positive pressure therapy. Oxygen may be easier to tolerate and worth trying if CPAP or bilevel pressure is not an option [51]. It is essential that the patient's caregiver be trained to help the patient adjust to the prescribed therapy.

Hospitalized Patients

OSA may be identified in medical patients hospitalized with another primary diagnosis. Clinical experience dictates that the prevalence of OSA is increased compared to healthy outpatients. This is undoubtedly reflective of the high incidence of obesity, cardiovascular disease, cerebrovascular disease, and diabetes in this patient population. These patients are challenging to diagnose and treat while acutely hospitalized. The need for monitoring and the use of intravenous medications pose problems for the sleep laboratory that may not have nursing personnel available to provide this level of care. The patient may be reluctant to pursue treatment with positive pressure during the hospitalization. In addition, sleep deprivation, the use of sedatives and narcotics, and suboptimal volume status (usually volume overload) may worsen the severity of the underlying OSA. It may be helpful to screen for OSA acutely via an overnight oximetry study, but definitive treatment with CPAP or bilevel pressure may be best reserved when the patient is stabilized as an outpatient. Head of bed elevation and supplemental oxygen are better tolerated acutely [51, 65–67].

Elderly

The elderly (age > 80 years), much like hospitalized patients, are challenging to treat. Major abnormalities of the sleep schedule are frequently present. Concomitant insomnia with or without advanced phase disorders make it challenging to assess a response to positive pressure if OSA is present [68]. Many of these patients have significant vascular risk and treatment makes good clinical sense. Second line therapy with oxygen and head of bed elevation is frequently the best fit in these patients and may provide significant benefit [51, 65–67]. In debilitated patients, a hospital bed may be a worthwhile consideration to facilitate head of bed elevation.

Hypoventilation Syndromes

Hypercapnia is common in OSA but frequently overlooked. One recent series found that 17% of patients referred for polysomnography had evidence of daytime hypercapnia [69]. There is uncertainty whether CPAP is contraindicated. If the patient complains of frequent morning headache or has evidence of persistent right heart failure and/or hypercapnia at the time of follow-up, a bilevel pressure titration should be considered [70].

SUMMARY

The cumulative epidemiology data provide a convincing argument that patients with OSA are at risk for impaired daytime performance (sleepiness and/or fatigue), insulin resistance, automobile crashes, and vascular complications. It has also become evident that a "dose–response" relationship exists with regard to the AHI and vascular risk for the group as a whole; however, patients may have variable clinical effects [11–13].

The primary treatment modality for OSA is positive pressure therapy. Differential susceptibility to daytime sleepiness and vascular risk exists. In patients who do not accept positive pressure therapy despite careful attempts to optimize the treatment, second line therapy should be explored.

A careful assessment of the primary treatment concern should guide further interventions. While palatal surgery can effectively treat snoring, the effect on the AHI and daytime sleepiness is less robust. Oral appliances may help some patients [33]. Recent longitudinal data suggests that the durability of oral appliance therapy is uncertain and subject to frequent dental complications [34, 35].

Treatment with oxygen should be considered in patients who will not accept positive pressure therapy and are felt to be at increased risk for vascular complications [51] and in patients acutely hospitalized until they are stable and able to be evaluated with attended or unattended polysomnography [71]. The current generation of oxygen concentrators are noisy and difficult to transport. This limits the utility of this treatment option in highly mobile patients [52]. Patients with Down syndrome, hospitalized patients, and the elderly may be more accepting of treatment with oxygen via nasal cannula alone. While this approach makes clinical and biologic sense, definitive outcome evidence is lacking.

REFERENCES

1. Ip MS, et al. Obstructive sleep apnea is independently associated with insulin resistance.[Comment]. *Am J Respir Crit Care Med* **165**:670–676(2002).

2. Punjabi NM, et al. Sleep-disordered breathing and insulin resistance in middle-aged and overweight men.[Comment]. *Am J Respir Crit Care Med* **165**:677–682(2002).

3. Strollo PJ Jr, Rogers RM. Obstructive sleep apnea. *N Engl J Med* **334**:99–104(1996).

4. Strollo PJ Jr, Sanders MH, Atwood CW. Positive pressure therapy. *Clin Chest Med* **19**:55–68(1998).

5. George CF. Reduction in motor vehicle collisions following treatment of sleep apnoea with nasal CPAP.[Comment]. *Thorax* **56**:508–512(2001).

6. Pepperell JC, et al. Ambulatory blood pressure after therapeutic and subtherapeutic nasal continuous positive airway

pressure for obstructive sleep apnoea: a randomised parallel trial. [Comment]. *Lancet* **359**:204–210(2002).

7. Jenkinson C, Davies RJ, Mullins R, Stradling JR. Comparison of therapeutic and subtherapeutic nasal continuous positive airway pressure for obstructive sleep apnoea: a randomised prospective parallel trial.[Comment]. *Lancet* **353**:2100–2105(1999).

8. White J, Cates C, Wright J. Continuous positive airway pressure for obstructive sleep apnoea.[update of Cochrane Database Syst Rev 2000;(2):CD001106; PMID: 10796595]. *Cochrane Database Syst Rev* CD001106(2002).

9. Sin DD, Mayers I, Man GC, Pawluk L. Long-term compliance rates to continuous positive airway pressure in obstructive sleep apnea: a population-based study. *Chest* **121**:430–435(2002).

10. Anonymous. Sleep-related breathing disorders in adults: recommendations for syndrome definition and measurement techniques in clinical research. The Report of an American Academy of Sleep Medicine Task Force. [Comment]. *Sleep* **22**:667–689(1999).

11. Gottlieb DJ, et al. Relation of sleepiness to respiratory disturbance index: the Sleep Heart Health Study. *Am J Respir Crit Care Med* **159**:502–507(1999).

12. Shahar E, et al. Sleep-disordered breathing and cardiovascular disease: cross-sectional results of the Sleep Heart Health Study. [Comment]. *Am J Respir Crit Care Med* **163**:19–25(2001).

13. Peppard PE, Young T, Palta M, Skatrud J. Prospective study of the association between sleep-disordered breathing and hypertension. [Comment]. *N Engl J Med* **342**:1378–1384(2000).

14. Strollo PJ Jr, et al. Split-night studies for the diagnosis and treatment of sleep-disordered breathing. *Sleep* **19**:S255–S259(1996).

15. McArdle N, et al. Split-night versus full-night studies for sleep apnoea/hypopnoea syndrome. *Eur Respir J* **15**:670–675(2000).

16. Sanders MH, Constantino JP, Strollo PJ, Studnicki K, Atwood CW. The impact of split-night polysomnography for diagnosis and positive pressure therapy titration on treatment acceptance and adherence in sleep apnea/hypopnea. *Sleep* **23**:17–24(2000).

17. Meslier N, et al. A French survey of 3,225 patients treated with CPAP for obstructive sleep apnoea: benefits, tolerance, compliance and quality of life. *Eur Respir J* **12**:185–192(1998).

18. McArdle N, et al. Long-term use of CPAP therapy for sleep apnea/hypopnea syndrome. *Am J Respir Crit Care Med* **159**:1108–1114(1999).

19. Douglas NJ. "Why am I sleepy?": sorting the somnolent. *Am J Respir Crit Care Med* **163**:1310–1313(2001).

20. Hoy CJ, Vennelle M, Kingshott RN, Engleman HM, Douglas NJ. Can intensive support improve continuous positive airway pressure use in patients with the sleep apnea/hypopnea syndrome? *Am J Respir Crit Care Med* **159**:1096–1100(1999).

21. Chervin RD, Theut S, Bassetti C, Aldrich MS. Compliance with nasal CPAP can be improved by simple interventions. *Sleep* **20**:284–289(1997).

22. Berger WE. Treatment update: allergic rhinitis. *Allergy Asthma Proc* **22**:191–198(2001).

23. Powell NB, et al. Radiofrequency treatment of turbinate hypertrophy in subjects using continuous positive airway pressure: a randomized, double-blind, placebo-controlled clinical pilot trial. *Laryngoscope* **111**:1783–1790(2001).

24. Zozula R, Rosen R. Compliance with continuous positive airway pressure therapy: assessing and improving treatment outcomes. *Curr Opin Pulm Med* **7**:391–398(2001).

25. Edinger JD, Radtke RA. Use of in vivo desensitization to treat a patient's claustrophobic response to nasal CPAP. *Sleep* **16**: 678–680(1993).

26. Barbe F, et al. Treatment with continuous positive airway pressure is not effective in patients with sleep apnea but no daytime sleepiness: a randomized, controlled trial.[Comment]. *Ann Intern Med* **134**:1015–1023(2001).

27. Henderson JH II, Strollo PJ Jr. Medical management of obstructive sleep apnea. *Prog Cardiovasc Dis* **41**:377–386(1999).

28. Martins De Araujo MT, Vieira SB, Vasquez EC, Fleury B. Heated humidification or face mask to prevent upper airway dryness during continuous positive airway pressure therapy. [Comment]. *Chest* **117**:142–147(2000).

29. Massie CA, Hart RW, Peralez K, Richards GN. Effects of humidification on nasal symptoms and compliance in sleep apnea patients using continuous positive airway pressure. [Comment]. *Chest* **116**:403–408(1999).

30. Strumpf DA, Harrop P, Dobbin J, Millman RP. Massive epistaxis from nasal CPAP therapy. *Chest* **95**:1141(1989).

31. Bachour A, Hurmerinta K, Maasilta P. Mouth closing device (chinstrap) reduces mouth leak during nasal CPAP. *Sleep Med* **5**:261–267(2004).

32. Sanders MH, et al. CPAP therapy via oronasal mask for obstructive sleep apnea. *Chest* **106**:774–779(1994).

33. Ferguson KA, Ono T, Lowe AA, Keenan SP, Fleetham JA. A randomized crossover study of an oral appliance vs nasal-continuous positive airway pressure in the treatment of mild–moderate obstructive sleep apnea.[Comment]. *Chest* **109**:1269–1275(1996).

34. Petit FX, et al. Mandibular advancement devices: rate of contraindications in 100 consecutive obstructive sleep apnea patients. *Am J Respir Crit Care Med* **166**:274–278(2002).

35. Rose EC, Staats R, Virchow C Jr, Jonas IE. Occlusal and skeletal effects of an oral appliance in the treatment of obstructive sleep apnea. *Chest* **122**:871–877(2002).

36. Li KK, Powell NB, Riley RW, Troell RJ, Guilleminault C. Radiofrequency volumetric reduction of the palate: an extended follow-up study. *Otolaryngol Head Neck Surg* **122**:410–414(2000).

37. Troell RJ, Powell NB, Riley RW, Li KK, Guilleminault C. Comparison of postoperative pain between laser-assisted uvulopalatoplasty, uvulopalatopharyngoplasty, and radiofrequency volumetric tissue reduction of the palate. *Otolaryngol Head Neck Surg* **122**:402–409(2000).

38. Ho WK, Wei WI, Chung KF. Managing disturbing snoring with palatal implants: a pilot study. *Arch Otolaryng Head Neck Surg* **130**:753–758(2004).

39. McGuirt WF Jr, Johnson JT, Sanders MH. Previous tonsillectomy as prognostic indicator for success of uvulopalatopharyngoplasty. *Laryngoscope* **105**:1253–1255(1995).

40. Partinen M, Guilleminault C. Daytime sleepiness and vascular morbidity at seven-year follow-up in obstructive sleep apnea patients. *Chest* **97**:27–32(1990).

41. Parra O, et al. Sleep-related breathing disorders: impact on mortality of cerebrovascular disease. *Eur Respir J* **24**:267–272(2004).

42. Marti S, et al. Mortality in severe sleep apnoea/hypopnoea syndrome patients: impact of treatment. *Eur Respir J* **20**:1511–1518(2002).

43. Milleron O, et al. Benefits of obstructive sleep apnoea treatment in coronary artery disease: a long-term follow-up study. [see Comment]. *Eur Heart J* **25**:728–734(2004).

44. Neubauer JA. Invited review: physiological and pathophysiological responses to intermittent hypoxia. *J Appl Physiol* **90**:1593–1599(2001).

45. Lavie L. Obstructive sleep apnoea syndrome—an oxidative stress disorder. *Sleep Med Rev* **7**:35–51(2003).

46. Griendling KK, FitzGerald GA. Oxidative stress and cardiovascular injury: Part I: basic mechanisms and in vivo monitoring of ROS. *Circulation* **108**:1912–1916(2003).

47. Prabhakar NR. Sleep apneas: an oxidative stress?[Comment]. *Am J Respir Crit Care Med* **165**:859–860(2002).

48. Sica AL, Greenberg HE, Ruggiero DA, Scharf SM. Chronic-intermittent hypoxia: a model of sympathetic activation in the rat. *Respir Physiol* **121**:173–184(2000).

49. Naldini A, Carraro F, Silvestri S, Bocci V. Hypoxia affects cytokine production and proliferative responses by human peripheral mononuclear cells. *J Cell Physiol* **173**:335–342(1997).

50. Gozal D, Lipton AJ, Jones KL. Circulating vascular endothelial growth factor levels in patients with obstructive sleep apnea. *Sleep* **25**:59–65(2002).

51. Landsberg R, Friedman M, Ascher-Landsberg J. Treatment of hypoxemia in obstructive sleep apnea. [Erratum appears in *Am J Rhinol* **16**(1):67(2002)]. *Am J Rhinol* **15**:311–313(2001).

52. Kacmarek RM. Delivery systems for long-term oxygen therapy. *Respir Care* **45**:84–92; discussion 92–94(2000).

53. He J, Kryger MH, Zorick FJ, Conway W, Roth T. Mortality and apnea index in obstructive sleep apnea. Experience in 385 male patients. *Chest* **94**:9–14(1988).

54. Bridgman SA, Dunn KM. Surgery for obstructive sleep apnoea. *Cochrane Database Syst Rev* CD001004(2000).

55. Carskadon MA, et al. Guidelines for the multiple sleep latency test(MSLT): a standard measure of sleepiness. *Sleep* **9**:519–524(1986).

56. Veasey SC, et al. Long-term intermittent hypoxia in mice: protracted hypersomnolence with oxidative injury to sleep–wake brain regions. [see Comment]. *Sleep* **27**:194–201(2004).

57. Pack AI, Black JE, Schwartz JR, Matheson JK. Modafinil as adjunct therapy for daytime sleepiness in obstructive sleep apnea. *Am J Respir Crit Care Med* **164**:1675–1681(2001).

58. Arnulf I, Homeyer P, Garma L, Whitelaw WA, Derenne JP. Modafinil in obstructive sleep apnea–hypopnea syndrome: a pilot study in 6 patients. *Respiration* **64**:159–161(1997).

59. Kingshott RN, et al. Randomized, double-blind, placebo-controlled crossover trial of modafinil in the treatment of residual excessive daytime sleepiness in the sleep apnea/hypopnea syndrome. *Am J Respir Crit Care Med* **163**:918–923(2001).

60. Douglas NJ. Upper airway resistance syndrome is not a distinct syndrome. [Comment]. *Am J Respir Crit Care Med* **161**:1413–1416(2000).

61. Guilleminault C, Chowdhuri S. Upper airway resistance syndrome is a distinct syndrome.[Comment]. *Am J Respir Crit Care Med* **161**:1412–1413(2000).

62. Rees K, Kingshott RN, Wraith PK, Douglas NJ. Frequency and significance of increased upper airway resistance during sleep. *Am J Respir Crit Care Med* **162**:1210–1214(2000).

63. Marcus CL, Keens TG, Bautista DB, von Pechmann WS, Ward SL. Obstructive sleep apnea in children with Down syndrome. *Pediatrics* **88**:132–139(1991).

64. Smith DS. Health care management of adults with Down syndrome. *Am Fam Physician* **64**:1031–1038(2001).

65. McEvoy RD, Sharp DJ, Thornton AT. The effects of posture on obstructive sleep apnea. *Am Rev Respir Dis* **133**:662–666(1986).

66. Neill AM, Angus SM, Sajkov D, McEvoy RD. Effects of sleep posture on upper airway stability in patients with obstructive sleep apnea. *Am J Respir Crit Care Med* **155**:199–204(1997).

67. Hakala K, Maasilta P, Sovijarvi AR. Upright body position and weight loss improve respiratory mechanics and daytime oxygenation in obese patients with obstructive sleep apnoea. *Clin Physiol* **20**:50–55(2000).

68. Avidan AY. Sleep changes and disorders in the elderly patient. *Curr Neurol Neurosci Rep* **2**:178–185(2002).

69. Resta O, et al. Hypercapnia in obstructive sleep apnoea syndrome. *Netherlands J Med* **56**:215–222(2000).

70. Resta O, et al. Prescription of nCPAP and nBIPAP in obstructive sleep apnoea syndrome: Italian experience in 105 subjects. A prospective two centre study. *Respir Med* **92**:820–827(1998).

71. Chesson AL Jr, et al. Practice parameters for the use of portable monitoring devices in the investigation of suspected obstructive sleep apnea in adults. *Sleep* **26**:907–913(2003).

43

MEDICAL TREATMENT OF OBSTRUCTIVE SLEEP APNEA: LIFE-STYLE CHANGES, WEIGHT REDUCTION, AND POSTURAL THERAPY

SHELDON KAPEN

Detroit VA Medical Center, Detroit, Michigan/Department of Neurology, Wayne State University Medical School, Detroit, Michigan

Obstructive sleep apnea (OSA) is a major public health hazard. The Wisconsin Sleep Cohort Study found the incidence in middle-aged individuals (aged 30–60) to be 24% in males and 9% in females when OSA was defined as having an apnea–hypopnea index (AHI) of 5 or more [1]. In those with an AHI of 5 or more plus symptomatology (excessive daytime sleepiness (EDS)), the incidence was 4% in males and 2% in females. This would be considered obstructive sleep apnea syndrome (OSAS). The presence of sleepiness is important because EDS is associated with a high rate of occupational and traffic accidents and OSA also leads to cardiovascular and cerebrovascular morbidity including hypertension, coronary artery disease, and stroke [2]. OSA probably causes excess mortality also although the evidence is not as strong as for morbidity [3, 4].

The etiology and pathogenesis of OSA are multifactorial and complicated and consequently so is the management. Although attention has been focused on defects in the upper airway as the immediate cause of sleep-related airflow obstruction, many risk factors influence both the incidence and severity of the disease. Generally, they ultimately act on the upper airway to influence its size, shape, and collapsibility. The knowledge of the presence of significant risk factors provides important clues to decisions regarding treatment. Tracheostomy was the first and only therapy available for OSA when the latter condition was first recognized in the 1960s and 1970s [5]. Tracheostomy bypasses the site of obstruction in the upper airway and should provide a complete cure. However, its disadvantages are obvious and can be summarized as follows: (1) There are major technical problems involved in performing tracheostomy in obese patients, particularly in patients with upper body obesity (male pattern) whose necks are crowded with adipose tissue. (2) Postoperative complications are common including infection and dehiscence. (3) The care of a tracheostomy demands time, effort, and skill on the part of the patient. (4) Patients find it difficult to adjust to the cosmetic disfigurement associated with tracheostomy.

In 1981, Fujita at Henry Ford Hospital in Detroit introduced the procedure called uvulopalatopharyngoplasty (UPPP), which avoided the necessity of a neck opening by amputating the uvula, excising the tonsils, and tightening the pharyngeal mucosa, thereby widening the pharyngeal cross section [6]. This procedure was enthusiastically received by the sleep medicine community, especially by ENT surgeons, and a number of case series appeared in the literature. However, the procedure's popularity soon dampened when it was realized that at best it resulted in only a bare 50% reduction in the AHI and even this figure was not only unpredictable but there was quite a bit of recidivism in one year or longer follow-up studies [7]. For this reason, and in the face of the finding that a large percent of subjects had their site of narrowing or obstruction at the level of the base of the tongue at a distance from the site of the UPPP, this surgery is now performed much less frequently.

Continuous positive airway pressure (CPAP) was introduced in 1981 [8]. Continuous positive airway pressure is now the treatment of choice for sleep disordered breathing, and consequently thousands of patients are using these machines and a whole industry has grown up centered around their manufacture, maintenance, and supply of replacement parts. The air pressure generated by CPAP acts as a mechanical splint, ensuring patency of the upper airway during sleep. However, this therapy is hardly a perfect solution to the treatment conundrum in OSAS. First, many patients refuse CPAP even prior to a home trial. Second, in those who initially tolerate and accept a home trial of CPAP, 46% do not meet minimal standards for compliance, defined as usage for 4 hours or more per night for at least 70% of nights [9]. Some of the reasons for noncompliance include side effects such as nasal congestion, dryness, mask discomfort, noise, and inconvenience. Consequently, other treatment options are required, particularly for milder cases (AHI of 15 or less) and those without EDS.

This chapter will describe the results for several alternative interventions, which have the potential for success in those patients who fail or refuse treatment trials using CPAP, dental devices, or upper airway surgery. The discussion will be limited to adults and will not enter into pediatric applications. Pharmacological options are reviewed in another chapter.

LIFE-STYLE CHANGES

Alcohol

Alcohol has a deleterious effect on sleep and alertness. Acutely, alcohol acts as a hypnotic and decreases sleep latency but disrupts sleep continuity in the latter half of the night [10]. Because of its effect on sleep latency, it is often used inappropriately as a sleep aid [11]. Its effect on sleep architecture has been known for a long time, namely, suppression of REM sleep and of wakefulness. These acute changes are followed by REM rebound in the latter part of the sleep period [10]. When a chronic alcoholic abruptly withdraws from alcohol, withdrawal symptoms appear, which may include tremulousness, anxiety, and insomnia. Delirium tremens results when the withdrawal syndrome is severe and is associated with confusion, disorientation, hallucinations, and a major REM sleep rebound [12]. Motor manifestations are prominent in place of the usual inhibition of somatic musculature, so-called REM without atonia. Such a syndrome is also seen in cats with lesions in the dorsolateral pons [13], in human narcolepsy [14], and in REM behavior disorder [15]. REM sleep is a state in which sleep disordered breathing events are longer and HbO_2 desaturation is more severe so REM rebound would be especially dangerous for patients with OSAS [16]. However, this effect of alcohol withdrawal on OSAS patients has not been reported in the literature either clinically or experimentally.

Sleep disruption is common in chronic alcoholic patients, both in those who continue to drink and after long periods of abstinence. The incidence of insomnia is very high in alcoholics and is reflected in poor sleep continuity including measures such as long sleep latency, increased number of arousals and awakenings, increased wake after sleep onset, and sleep efficiency [17]. Slow-wave sleep is reduced while REM sleep is relatively well preserved. These changes in sleep architecture are deleterious for sleep disordered breathing since sleep deprivation and sleep disruption lead to elevated thresholds for arousal and longer durations of apneas and hypopneas [18]. Indeed, comparison of chronic alcoholics and nonalcoholic control subjects reveals that the alcoholics have a greater incidence of OSA and their disordered breathing events tend to be more severe in terms of both duration and desaturation [19, 20].

Alcohol also has a direct effect on breathing during sleep. The first controlled experiment to study the response to acute administration of alcohol prior to bedtime used 3 ounces of 80% proof beverage on one night and placebo on the second in six OSA patients [21]. The number of hypoxic breathing events in five moderately severe patients increased from 538 on the placebo night (total for all five subjects) to 863 on the alcohol night. Hemoglobin O_2 desaturation was also worse with alcohol than with placebo. The sixth patient had more severe OSA with his lowest saturation being 65% at baseline. When alcohol was compared with placebo in this patient, it was found that the number of events associated with desaturation below 76% increased from 3 with placebo to 25 with alcohol. Alcohol had no effect on sleep breathing in the control group. Similar results were recorded in other more recent studies, including the conversion of snoring to frank apnea during the first hour of sleep [22–25].

In addition to the acute effects of alcohol on respiratory control during sleep, alcohol has a tonic effect in chronic drinkers, including during periods of abstinence [19]. Among 188 individuals undergoing detoxification, sleep disordered breathing was common and the incidence of AHI values of 10 or more was higher than in a group of 87 control subjects. Age and BMI were significant risk factors in a multiple regression analysis in both of these two groups. In those alcoholics older than 60 years of age, 50% had OSA as defined above. Sleep disruption, sleepiness, and hypoxia were more prominent in the alcoholics with sleep disordered breathing than in the alcoholics with an AHI less than 10.

Some of the mechanisms by which alcohol could disrupt breathing during sleep have been studied. Administration of alcohol prior to bedtime increased pharyngeal resistance from 1.9 ± 0.5 cm $H_2O/L/S$ on placebo nights to

3.3 ± 0.8 [26]. Nasal resistance was also elevated. Alcohol diminishes electromyographic activity in the genioglossus muscle in men, and in women, it has a similar effect during the follicular phase of the menstrual cycle but not during the luteal phase when progesterone secretion is high [27]. Another study confirmed that vodka before sleep increased the rise in inspiratory resistance, which occurs with sleep onset. It had the same effect in snoring and nonsnoring subjects and augmented the response to an inspiratory load in the P0.1 test, which correlated with the inspiratory resistance [28]. Finally, alcohol administration increases heart rate, the low frequency component of heart rate variability (a marker for relative sympathetic activation), and muscle nerve sympathetic activity [29]. These findings help to explain the adverse effects of alcohol in OSA, in particular, the increase in AHI and the exacerbation of HbO_2 desaturation.

There have been no studies specifically addressing the treatment of OSAS by reduction of alcohol intake. However, it should be the practice of every physician who manages OSA patients to advise them to eliminate evening alcoholic beverages and to reduce overall amounts of alcohol.

Smoking

A limited number of studies have reported more disturbed sleep in cigarette smokers using questionnaires and sleep diaries. Phillips and Danner found a greater likelihood of difficulties with sleep onset, sleep maintenance, and daytime sleepiness in both high school students and adults [30]. A similar trend was seen with sleep diaries in a small group of smokers matched with nonsmokers [31]. The smokers had poorer sleep than the nonsmokers but they also had higher alcohol and caffeine intake, which may have confounded the conclusions. Abstinence in chronic smokers leads to a deterioration of sleep [32]. Polysomnography reveals an increase in the number of arousals and awakenings and a rise in daytime sleepiness as represented by the multiple sleep latency test (MSLT) compared with the usual smoking period. These effects are accompanied by adverse psychological effects such as irritability, anxiety, and tension. A case report also describes a syndrome called "nocturnal sleep-disturbing nicotine craving" characterized by awakening with a necessity for cigarette smoking before the patient is able to return to sleep [33].

Cigarette smoking is an independent risk factor for snoring as has been demonstrated in the Tucson Epidemiologic Study of Obstructive Airways Disease [34] and for OSA as demonstrated in the Wisconsin Sleep Cohort Study [35]. In the latter survey, logistic regression analysis revealed an odds ratio of 4.44 for moderate or worse sleep disordered breathing and the odds ratio increased to 40.47 in those individuals with a 2 pack per day smoking habit. In both the Tucson and Wisconsin studies, former smokers no longer were independently at risk for snoring or OSA when compared with nonsmokers. A French study confirmed the association of smoking and snoring in 15–20 year old students [36] and a more recent report confirmed that smoking is a risk factor for snoring in adults [37]. Kashyap and co-workers found that 35% of OSA patients (>10 AHI) as opposed to 18% of nonapneics (<5 AHI) were smokers, while former smokers were again found to be not at risk [38]. In a large number of clinic patients (as opposed to the general population), the more severe OSA patients were heavier smokers but multiple regression analysis eliminated smoking as an independent risk factor [39]. Although Casasola and co-workers reported a similar AHI between otherwise healthy nonobese smokers and nonsmokers, the number of subjects was small (18 smokers, 20 nonsmokers); but regardless, HbO_2 saturation was lower in the smoking group [40].

Nicotine has actually been used in clinical trials in OSA with the rationale that it stimulates breathing and upper airway muscle activity [41–43]. No clinically significant effects on measures of OSA were described. A negative correlation between duration of sleep disordered breathing events and serum nicotine levels was described in one study along with a positive correlation of serum nicotine with the lowest HbO_2 saturation [41]. Nausea and vomiting were frequent in this study. In short, nicotine administration does not have an important role in the therapy of OSA.

Exercise

Numerous papers have been published on the effects of exercise on sleep and sleep architecture. When moderate exercise is carried out at a time not too close to sleep onset, there is a modest positive influence on slow-wave sleep, particularly in already fit individuals [44]. Such a result is generally interpreted as evidence in favor of a restorative function for slow-wave sleep. Of course, exercise can also assume an important role as part of a weight loss program, thus incorporating it into a life-style approach to the treatment of OSA. However, is there any evidence that exercise has a direct effect on any measures of sleep disordered breathing separate from its impact on body weight, respiratory function, overall fitness, and sleep? Several studies have addressed this question and have agreed that a chronic exercise program has a modest positive effect on sleep disordered breathing. Giebelhaus and co-workers described the results of a 6 month exercise protocol on 11 subjects with moderate to severe sleep disordered breathing in which the AHI fell from 32.8 to 23.6 [45]. Weight loss was not significant and did not correlate with the improvement in OSA. Similarly, Norman and co-workers conducted a 6 month study in 9 mild to moderate OSA patients whose AHI decreased even more (21.7 to 11.8) [46]. There was a reduction in AHI in every patient in this study and several dropped below 5. Although there was a fall in BMI of $1.6 \, kg/m^2$, the change in BMI showed no correlation with the change in AHI. Norman and co-workers also found improvements in measures of quality of life, mood, and daytime sleepiness (Epworth Sleepiness Scale). Finally, data from the Wisconsin Sleep Cohort Study were consistent with an independent protective effect of exercise on the development of OSA in middle-aged individuals [47].

An organized exercise program has an integral role in health maintenance and helps to prevent many diseases such as diabetes, heart disease, and stroke. To these conditions should be added obstructive sleep apnea syndrome, which in turn has many interrelationships to the above chronic illnesses. It is clear that exercise alone cannot be an effective therapy for OSAS but it should be part of an overall treatment plan, and in the milder cases or in those patients who are resistant or intolerant to the major treatment modalities, such as CPAP, exercise may assume a much more important role.

WEIGHT REDUCTION

At an early stage of recognition of OSAS as an entity, it was obvious that obesity was a common accompaniment of the disease. A series reported by Guilleminault and colleagues had 70% of patients overweight [48]. Obesity is a risk factor and probably is a cause or a precipitant of OSAS; a history of recent weight gain is a common thread in the anamnesis of recently diagnosed patients. In any large series, there is usually a not inconsiderable number of massively obese individuals with body mass index (BMI) > 40. Therefore weight reduction would seem to be a viable therapeutic option.

Nonsurgical Weight Reduction

Browman and co-workers described an obese patient who had polysomnographic studies during a period of 3 years, during which time his weight fluctuated over a range of 26 kg [49]. His AHI was 59.6 at a weight of 111 kg and 3.1 at 85 kg. This case showed a remarkable sensitivity for a reduction in AHI secondary to relatively small degrees of weight loss.

An early series of diet-induced weight loss included 15 hypersomnolent patients who had an average AHI of 55 ± 7.5 and a starting weight of $106.2 \pm 7.3 \, kg$ [50]. After a mean loss of 9.6 kg, the apnea frequency fell to 29.2 ± 7.1 during NREM sleep and the mean HbO_2 saturation during apneas rose by 4%. In addition, sleep architecture improved (reduction of stage 1) and EDS was partially relieved.

Suratt and co-workers described the results of diet-induced weight loss in 8 OSA patients [51]. The mean weight loss in these patients was $20.6 \pm 12.8 \, kg$; this was accompanied by improvements in blood gases measured during wakefulness; desaturation episodes per hour of sleep; and movement arousals. The AHI showed a significant drop in 6 of 8 patients and this was correlated with the BMI. These investigators also reported positive effects of weight loss on collapsibility in the upper airway using pulse flow resistance.

A Canadian group from Hoffstein's laboratory reported on 12 obese OSA patients who lost $26 \pm 18 \, kg$ by means of diet; the AHI changed from 57 ± 29 before weight reduction to 14 ± 10 after and the lowest HbO_2 saturation rose from $54 \pm 20\%$ to $80 \pm 8\%$ [52]. Certain aspects of pharyngeal function also improved. This series of patients was one of the most successful examples of the effects of weight loss on the severity of OSA.

Pasquali and co-workers reported on a series of 23 patients with BMI up to 61.0 but also included a few patients who were close to normal [53]. A mean weight loss of $18.5 \pm 14.7 \, kg$ was seen but those with greater obesity had a larger drop in weight. The AHI fell from 66.5 ± 23.0 to 33.0 ± 26.2 and the HbO_2 saturation also improved. The degree of weight loss correlated with the change in AHI ($r = -0.55$).

Comparable results to the previous earlier studies were obtained by other authors including Schwartz et al. [54], Suratt et al.[55], Herrendorf et al. [56], Kajaste et al. [57], and Pedro-Botet Pons et al. [58]. These patients were very obese (over 100 kg or BMI over 40) and weight loss ranged from 9.2 to 32.7 kg. The range of the pretreatment AHI (in 2 cases HbO_2 desaturation 4% index [ODI4]) was 31–90. Post-treatment AHI or ODI4 was 14–62 for a reduction of 17–51.8 (range). Some of the studies reported a correlation between the magnitude of weight loss and improved apnea ratings (generally AHI). Specific methods to induce weight loss were the following: a diet of 1000 kcal per day; very-low-calorie diets (VLCD); so-called Optifast program, using behavior therapy based weight reduction; and a diet of 600–800 kcal/day.

Surgical Weight Reduction

Hoffmeister and co-workers published the first case report of bariatric surgery in an OSA patient in 1978 [59]. The patient was a 34 year old male complaining of shortness of breath who was a heavy snorer who frequently slept in a chair because of disturbed sleep and had a history of EDS. His presurgical weight was 382 pounds and his height was 5'10". Polysomnography was not performed. His weight dropped to 324 pounds at 4½ weeks following surgery and to 234 at 6 months. At one month after surgery and at the later follow-up, the patient reported undisturbed sleep, disappearance of snoring, and resolution of sleepiness.

Most ensuing studies have utilized polysomnography in order to provide objective evidence of improvement if any [60–70]. Unlike noninvasive induction of weight loss, bariatric surgery has been limited to those OSA patients who were morbidly obese, defined roughly as having a body mass index (BMI) 40 or above. When BMI has been reported or can be extrapolated from the data, the group average is 50.0 and comparable measures of obesity have been reported when expressed as "excess weight". In these groups of morbidly obese patients, the results have been generally positive. Weight reduction has ranged between 32% and 75% for both of the main surgical procedures, gastroplasty and gastric bypass. This degree of weight loss is associated with significant improvements in parameters of sleep-disordered breathing, sleep architecture and quality of sleep, and severity of obesity-related diseases such as diabetes and hypertension. Marked reductions in AHI have been reported such as from 88.8 (apnea index) prior to surgery to 8.0 afterwards with 12/15 patients classified as "cured" (under 35 apneic episodes per night) [60]. Similarly, measures of HbO_2 saturation are elevated including mean and lowest saturations.

Some of the series report only subjective results and lack objective verification by polysomnography. This includes the series with the greatest number of patients, that of Dhabuwala and co-workers in 2000 [71]. The largest series with polysomnography is that of Sugerman with 110 participants [70]. The mean BMI for Sugerman's patients prior to surgery was 56 while the BMI following surgery was 38 after 4 ½ years. The surgical treatment was responsible for a reduction of AHI from 64 to 26.

The literature on bariatric surgery for OSA has major limitations in that none of the publications include control groups and none of the patient series were randomized. Although generally unstated, the patients were presumably selected according to criteria that included better prognosis and higher motivation. Thus conclusions as to the efficacy of the weight reduction surgery cannot be applied automatically to the entire population of morbidly obese OSA patients. Shneerson and co-workers said it clearly in the *Cochrane Database Systematic Review*: "There is a need for randomized controlled trials of these commonly used treatments in obstructive sleep apnoeas. These should identify which subgroups of patients with sleep apnoeas benefit most from each type of treatment and they should have clear and standardized outcome measures" [72].

Mechanisms of Improvement of OSA with Weight Loss

Following weight loss, there is a reduction of upper airway collapsibility and a fall in resistance to airflow in awake patients [54, 73]. This is accompanied by a reduction in P_{crit} in obese subjects after weight loss [54]; P_{crit} is the level of negative pressure at which a Starling resistor collapses—the lower the P_{crit}, the less the collapsibility. Also, pharyngeal function improves as measured by the reduction of pharyngeal cross-sectional area in relation to lung volume [52]. The reader is referred to a recent review for discussion of the pulmonary findings in obesity and the effects of weight loss [74].

POSTURAL CHANGES

Snoring often appears for the first time or is louder during sleep when a subject is supine as opposed to a lateral or prone position. In the supine posture, the tongue tends to fall back due to gravitational forces unless these forces are counteracted by enhanced activity of the genioglossus muscle, which pulls the tongue forward [75]. However, in OSA patients, the genioglossus is already overactivated during wakefulness [76, 77], putting this muscle at a disadvantage at the onset of sleep. Consequently, one would expect an increased AHI and a decreased HbO_2 saturation nadir in OSA patients on their backs during sleep compared with a side or prone position with perhaps a further improvement when elevated in a chair and, indeed, this has been described by many investigators. A criterion for posture-dependent sleep apnea was set by Cartwright and co-workers as a ratio of 0.5 or below for AHI in the lateral compared with AHI supine [78]. Using this criterion, Oksenberg and co-workers found that 55.9% of a large number of patients had posture-dependent OSA [79].

Patients with posture dependency, as defined by Cartwright [78], tend to have a less severe form of sleep apnea. Oksenberg's patients had at least 30 sleep disordered events in both supine and lateral positions. In this group, posture-dependent patients had lower AHI values (27.8 vs 44.0); were younger (52.9 vs 54.9 years of age); were less obese (BMI 29.4 vs 31.9); and had a longer sleep latency on the MSLT (9.6 vs 8.4 minutes) [79]. Sleep quality was also better in the positional group. Thus less severely affected patients would be more likely to respond to positional therapy, allowing for a useful treatment option in the milder group of OSA patients who frequently are less tolerant of CPAP than the more severely affected. However, there are no definite predictive factors for the success of positional therapy because of the large overlap in the above parameters between positional and nonpositional patients, thus mandating consideration of positional therapy in all patients with OSA when discussing treatment.

Even in those individuals whose changes in AHI do not meet criteria for positional dependency, measures of severity are worse on the back than on the side. AHI is higher on the back; the duration of individual apneas is longer;

minimum HbO$_2$ saturation is lower; arousals at the termination of apneas are longer; snoring is louder; and the change in heart rate at the time of the arousals is greater [80]. Therefore this is another reason why posture is important for all OSA patients, not just those with defined positional dependency.

Generally, the severity of OSA is worse during REM sleep than during NREM sleep, in particular, apnea duration and the degree of HbO$_2$ desaturation. The explanation for this probably relates to the higher threshold for arousals and a weaker response to blood gases in REM sleep [16]. Would there then be a difference in the response to postural change between REM and NREM sleep? The evidence is mixed on this point. George et al. [81] and Pevernagie et al. [82] found a lesser tendency for sleep disordered breathing to be alleviated during REM sleep on the side while Cartwright et al. [83] reported no difference between sleep stages in this respect. In light of these data, it would be imperative to test for the effects of postural changes on OSA parameters in all sleep stages before recommending this mode of therapy and to be cautious in patients with little or no REM sleep during baseline since there might be a REM rebound with the institution of therapy.

In many patients, combination therapy might be appropriate such as CPAP or a dental device together with prevention of the supine posture. Sleep laboratories generally titrate CPAP pressure separately in both the back and side positions, if feasible. The optimum pressure tends to be lower on the side [84]. As for dental devices such as those advancing the mandible, they improve OSA in both positions although, surprisingly, Yoshida reported a lack of effect in the lateral position versus prone and supine [85].

As mentioned earlier, when an individual is supine, gravitational forces cause the tongue to fall back and to reduce the oropharyngeal lumen. This is counteracted by contraction of the genioglossus muscle acting to resist the gravitational forces and to drive the tongue forward [75], but this compensatory activity of the genioglossus is less effective in apneic subjects [76, 77]. Indeed, the cross-sectional area and the anterior–posterior dimension of the oropharynx decreases in the supine position when compared with a more upright or seated posture in both nonapneic snorers and apneics [86]. Furthermore, P$_{crit}$ (critical closing pressure), the negative luminal pressure in response to which the upper airway collapses and becomes occluded, increases on the back [87]. Other findings in the supine position include an inverse correlation of the volume of the nasal passage 2–4 cm from the nares and the AHI in snorers versus nonsnoring control subjects [88] and a higher nasal resistance using posterior rhinomanometry in snorers and OSA patients [89]. Although most of these data are found in both apneics and snoring and nonsnoring controls,

there is also narrowing of the retropalatal space only in apneics [90]. This finding is critical because, in OSA patients, the retropalatal space is the most constricted area of the upper airway even in the upright position [91].

Most of the physiological findings mentioned earlier have compared the supine with *upright* positions. Little information is available regarding supine versus *lateral*. In one of the few studies on this issue, Isono and coworkers produced paralysis of upper airway musculature with a muscle relaxant in order to determine the degree of collapsibility of the upper airway in response to negative pressures in isolation from neurogenic factors. When subjects were in the lateral position, there was a decreased closing pressure and an increased maximum cross-sectional area compared with supine [92].

In summary, from both review of the literature and from experience of sleep medicine practitioners, avoidance of the supine posture during sleep is an effective, inexpensive, and simple treatment option for patients with obstructive sleep apnea. Posture therapy is also supported by investigations of the anatomy and pathophysiology of the upper airway. The evidence suggests that many patients could benefit from a trial of supine avoidance treatment including obese and nonobese subjects with or without simultaneous use of CPAP or dental devices. However, only minor attention has been directed to postural therapy as a serious treatment option and it is usually suggested to the patient by an incidental comment such as "try to stay off your back." A more serious approach would be to offer one of the methods available to prevent sleeping on the back, for example, the tennis ball technique, as a serious option for the less severe OSA patient such as those with AHI under 30 with lesser degrees of sleepiness or cardiovascular events. A randomized, controlled trial using the tennis ball method along with a standardized placement of the tennis balls is needed. In the meantime, education of sleep medicine providers and their patients along with attention to compliance would be helpful.

REFERENCES

1. Young T, Palta M, Dempsey J, Skatrud J, et al. The occurrence of sleep disordered breathing among middle-aged adults. *N Engl J Med* **328**(17):1230–1235(1993).

2. Shahar E, Whitney CW, Redline S, Lee ET, et al. sleep disordered breathing and cardiovascular disease: cross-sectional results of the Sleep Heart Health Study. *Am J Respir Crit Care Med* **163**(1):19–25(2001).

3. He JMK, Zorick FJ, Conway W, Roth T. Mortality and apnea index in obstructive sleep apnea: experience in 385 male patients. *Chest*, **94**:9–14(1988).

4. Partinen M AJ, Jamieson A, Guilleminault C. Long-term outcome for obstructive sleep apnea syndrome patients: mortality. *Chest* **94**:1200–1204(1988).

5. Coccagna G, Mantovani M, Brignani F, Parchi C, et al. Tracheostomy in hypersomnia with periodic breathing. *Bull Physiopathol Respir* (*Nancy*), **8**(5):1217–1227(1972).

6. Fujita S. Surgical correction of anatomic abnormalities in obstructive sleep apnea syndrome: uvulopalatopharyngoplasty. *Otolaryngol Head Neck Surg* **89**(6):923–934(1981).

7. Walker-Engstrom ML, Tegelberg A, Wilhelmsson B, Ringqvist I. 4-year follow-up of treatment with dental appliance or uvulopalatopharyngoplasty in patients with obstructive sleep apnea: a randomized study. *Chest* **121**(3):739–746(2002).

8. Sullivan CE. Reversal of obstructive sleep apnoea by continuous positive airway pressure applied through the nares. *Lancet* **18**(8225):862–865(1981).

9. Kribbs N. Objective measurement of patterns of nasal CPAP use by patients with obstructive sleep apnea. *Am J Respir Crit Care Med* **147**(4):887–895(1993).

10. Roehrs T, Roth T. Sleep, sleepiness, sleep disorders and alcohol use and abuse. *Sleep Med Rev* **5**(4):287–297(2001).

11. Johnson EO, Roehrs T, Roth T, Breslau N. Epidemiology of alcohol and medication as aids to sleep in early adulthood. *Sleep* **21**(2):178–186(1998).

12. Kotorii T, Nakazawa Y, Yokoyama T, Ohkawa T, et al. Terminal sleep following delirium tremens in chronic alcoholics–polysomnographic and behavioral study. *Drug Alcohol Depend* **10**(2-3):125–134(1982).

13. Henly K. A re-evaluation of the effects of lesions of the pontine tegmentum and locus coeruleus on phenomena of paradoxical sleep in the cat. *Acta Neurobiol Exp* **34**:215–232(1974).

14. Schenck CH. Motor dyscontrol in narcolepsy: rapid-eye-movement (REM) sleep without atonia and REM sleep behavior disorder. *Ann Neurol* **32**:3–10(1992).

15. Schenck CH, Patterson AL, Mahowald MW. Rapid eye movement sleep behavior disorder. A treatable parasomnia affecting older adults. *JAMA* **25**(13):1786–1789(1987).

16. Weiss JW, Anand A. Cardiorespiratory changes in sleep disordered breathing. In: Kryger MH, Roth T, Dement WC (Eds), Principles and Practice of Sleep Medicine. Saunders, Philadelphia, 2000, pp 859–860.

17. Cohn TJ, Foster JH, Peters TJ. Sequential studies of sleep disturbance and quality of life in abstaining alcoholics. *Addict Biol* **8**(4):455–462(2003).

18. White DP, Douglas NJ, Pickett CK, Zwillich CW, et al. Sleep deprivation and the control of ventilation. *Am Rev Respir Dis* **128**(6):984–986(1983).

19. Aldrich MS, Brower KJ, Hall JM. Sleep disordered breathing in alcoholics. *Alcohol Clin Exp Res* **23**(1):134–140(1999).

20. Vitiello MV, Personius JP, Nuccio MA, Koerker RM, et al. Nighttime hypoxemia is increased in abstaining chronic alcoholic men. *Alcohol Clin Exp Res* **14**(1):38–41(1990).

21. Scrima L, Broudy M, Nay KN, Cohn MA. Increased severity of obstructive sleep apnea after bedtime alcohol ingestion: diagnostic potential and proposed mechanism of action. *Sleep* **5**(4):318–328(1982).

22. Scrima L, Hartman PG, Hiller FC. Effect of three alcohol doses on breathing during sleep in 30–49 year old nonobese snorers and nonsnorers. *Alcohol Clin Exp Res* **13**(3):420–427(1989).

23. Tsutsumi W, Miyazaki S, Itasaka Y, Togawa K. Influence of alcohol on respiratory disturbance during sleep. Psychiatry Clin Neurosci, **54**(3):332–333(2000).

24. Scanlan MF, Roebuck T, Little PJ, Redman JR, et al. Effect of moderate alcohol upon obstructive sleep apnoea. *Eur Respir J* **16**(5):909–913(2000).

25. Issa FG, Sullivan CE. Alcohol, snoring and sleep apnea. *J Neurol Neurosurg Psychiatry* **45**(4):353–359(1982).

26. Robinson RW, White DP, Zwillich CW. Moderate alcohol ingestion increases upper airway resistance in normal subjects. *Am Rev Respir Dis* **132**(6):1238–1241(1985).

27. Leiter JC, Doble EA, Knuth SL, Bartlett D Jr. Respiratory activity of genioglossus. Interaction between alcohol and the menstrual cycle. *Am Rev Respir Dis* **135**(2):383–386(1987).

28. Dawson A, Bigby BG, Poceta JS, Mittler MM. Effect of bedtime alcohol on inspiratory resistance and respiratory drive in snoring and nonsnoring men. *Alcohol Clin Exp Res* **21**(2):183–190(1997).

29. van de Borne P, Mark AL, Montano N, Mion D, et al. Effects of alcohol on sympathetic activity, hemodynamics, and chemoreflex sensitivity. *Hypertension* **29**(6):1278–1283(1997).

30. Phillips BA, Danner FJ. Cigarette smoking and sleep disturbance. *Arch Intern Med* **155**(7):734–737(1995).

31. Lexcen FJ, Hicks RA. Does cigarette smoking increase sleep problems. *Percept Mot Skills* **77**(1):16–18(1993).

32. Prosise GL, Bonnet MH, Berry RB, Dickel MJ. Effects of abstinence from smoking on sleep and daytime sleepiness. *Chest* **105**(4):1136–1141(1994).

33. Rieder A, Kunze U, Groman E, Kiefer I, et al. Nocturnal sleep-disturbing nicotine craving: a newly described symptom of extreme nicotine dependence. *Acta Med Austriaca* **28**(1):21–22(2001).

34. Bloom JW, Kaltenborn WT, Quan SF. Risk factors in a general population for snoring. Importance of cigarette smoking and obesity. *Chest* **93**(4):678–683(1988).

35. Wetter DW, Young TB, Bidwell TR, Badr MS, et al. Smoking as a risk factor for sleep disordered breathing. *Arch Intern Med* **154**(19):2219–2224(1994).

36. Delasnerie-Laupretre N, Patois E, Valatx JL, Kauffmann F, et al. Sleep, snoring and smoking in high school students. *J Sleep Res* **2**(3):138–142(1993).

37. Kauffmann F, Annesi I, Neukirch F, Oryszczyn MP, et al. The relation between snoring and smoking, body mass index, age, alcohol consumption and respiratory symptoms. *Eur Respir J* **2**(7):599–603(1989).

38. Kashyap R, Hock LM, Bowman TJ. Higher prevalence of smoking in patients diagnosed as having obstructive sleep apnea. *Sleep Breath* **5**(4):167–172(2001).

39. Hoflstein V. Relationship between smoking and sleep apnea in clinic population. *Sleep* **25**(5):519–524(2002).

40. Casasola GG, Alvarez-Sala JL, Marques JA, Sanchez-Alarcos JM, et al. Cigarette smoking behavior and respiratory alterations during sleep in a healthy population. *Sleep Breath* **6**(1):19–24(2002).

41. Davila DG, Hurt RD, Offord KP, Harris CD, et al. [German] Acute effects of transdermal nicotine on sleep architecture,

snoring, and sleep disordered breathing in nonsmokers. *Am J Respir Crit Care Med* **150**(2):469–474(1994).

42. Zevin S, Swed E, Cahan C. Clinical effects of locally delivered nicotine in obstructive sleep apnea syndrome. *Am J Ther* **10**(3):170–175(2003).

43. Hein H, Kirsten D, Jugert C, Magnussen H. Nicotine as therapy of obstructive sleep apnea? *Pneumologie* **49** (Suppl 1):185–186(1995).

44. Youngstedt SD, O'Connor PJ, Dishman RK. The effects of acute exercise on sleep: a quantitative synthesis. *Sleep* **20**(3):203–214(1997).

45. Giebelhaus V, Lormes W, Lehmann M, Netzer N. Physical exercise as an adjunct therapy in sleep apnea—an open trial. *Sleep Breath* **4**(4):173–176(2000).

46. Norman JF, von Essen SG, Fuchs RH, McElligott M. Exercise training effect on obstructive sleep apnea syndrome. *Sleep Res Online* **3**(3):121–129(2000).

47. Peppard PE, Young T. Exercise and sleep-disordered breathing: an association independent of body habitus. *Sleep* **27**(3):480–484(2004).

48. Guilleminault C, Tilkian A, Dement WC. The sleep apnea syndromes. *Annu Rev Med* **27**:465–484(1976).

49. Browman CP, Sampson MG, Yolles SF, Gujavarty KS, et al. Obstructive sleep apnea and body weight. *Chest* **85**(3):435–438(1984).

50. Smith PL, Gold AR, Meyers DA, Haponik EF, et al. Weight loss in mildly to moderately obese patients with obstructive sleep apnea. *Ann Intern Med* **103**(6 Pt 1):850–855(1985).

51. Suratt PM, McTier RF, Findley LJ, Pohl SL, et al. Changes in breathing and the pharynx after weight loss in obstructive sleep apnea. *Chest* **92**(4):631–637(1987).

52. Rubinstein I, Colapinto N, Rotstein LE, Brown IG, et al. Improvement in upper airway function after weight loss in patients with obstructive sleep apnea. *Am Rev Respir Dis* **138**(5):1192–1195(1988).

53. Pasquali R, Colella P, Cirignotta F, Mondini S, et al. Treatment of obese patients with obstructive sleep apnea syndrome (OSAS): effect of weight loss and interference of otorhinolaryngoiatric pathology. *Int J Obes* **14**(3):207–217(1990).

54. Schwartz AR, Gold AR, Schubert N, Stryzak A, et al. Effect of weight loss on upper airway collapsibility in obstructive sleep apnea. *Am Rev Respir Dis* **144**(3 Pt 1):494–498(1991).

55. Suratt PM, McTier RF, Findley LJ, Pohl SL, et al. Effect of very-low-calorie diets with weight loss on obstructive sleep apnea. *Am J Clin Nutr* **56**(1 Suppl): 182S–184S(1992).

56. Herrendorf G, Hajak G, Rodenbeck A, Simen S, et al. [German] Ambulatory, comprehensive, behavior therapy-oriented weight reduction program (Optifast Program). An alternative therapy in obstructive sleep apnea. *Nervenarzt* **67**(8):695–700(1996).

57. Kajaste S, Telakivi T, Mustajoki P, Pihl S, et al. Effects of a cognitive-behavioural weight loss programme on overweight obstructive sleep apnoea patients. *J Sleep Res* **3**(4):245–249(1994).

58. Pedro-Botet Pons J, Roca Montanari A. [Spanish]. Efficacy of weight loss in the treatment of obstructive sleep apnea syn-

drome. Experience in 135 patients. *Med Clin (Barc)* **98**(2):45–48(1992).

59. Hoffmeister JA, Cabatingan O, McKee A. Sleep apnea treated by intestinal bypass. *J Maine Med Assoc* **69**(3):72–74(1978).

60. Charuzi I, Ovnat A, Peiser J, Saltz H, et al. The effect of surgical weight reduction on sleep quality in obesity-related sleep apnea syndrome. *Surgery* **97**(5):535–538(1985).

61. Charuzi I, Fraser D, Peiser J, Ovnat A, et al. Sleep apnea syndrome in the morbidly obese undergoing bariatric surgery. *Gastroenterol Clin North Am* **16**(3):517–519(1987).

62. Charuzi I, Lavie P, Peiser J, Peled R. Bariatric surgery in morbidly obese sleep-apnea patients: short- and long-term follow-up. *Am J Clin Nutr* **55**(2 Suppl):594S–596S(1992).

63. Guardiano SA, Scott JA, Ware JC, Schechner SA. The long-term results of gastric bypass on indexes of sleep apnea. *Chest* **124**(4):1615–1619(2003).

64. Kurono T, Sakuma T, Shinozaki T, Kouchiyama S, et al. [Japanese] A case of obstructive sleep apnea syndrome remarkably improved by gastric restriction surgery. *Nihon Kyobu Shikkan Gakkai Zasshi* **28**(5):767–772(1990).

65. Peiser J, Lavie P, Ovnat A, Charuzi I. Sleep apnea syndrome in the morbidly obese as an indication for weight reduction surgery. *Ann Surg* **199**(1):112–115(1984).

66. Pillar G, Peled R, Lavie P. Recurrence of sleep apnea without concomitant weight increase 7.5 years after weight reduction surgery. *Chest* **106**(6):1702–1704(1994).

67. Rasheid S, Banasiak M, Gallagher SF, Lipska A, et al. Gastric bypass is an effective treatment for obstructive sleep apnea in patients with clinically significant obesity. *Obes Surg* **13**(1):58–61(2003).

68. Reimao R, Lemmi H, Cowan G, Golden EB. [Portuguese] The role of gastroplasty in the treatment of obstructive sleep apnea. *Arq Neuropsiquiatr* **44**(1):38–43(1986).

69. Victor DW Jr, Sarmiento CF, Yanta M, Halverson JD. Obstructive sleep apnea in the morbidly obese. An indication for gastric bypass. *Arch Surg* **119**(8):970–972(1984).

70. Sugerman HJ, Fairman RP, Sood RK, Engle K, et al. Long-term effects of gastric surgery for treating respiratory insufficiency of obesity. *Am J Clin Nutr* **55**(2 Suppl):597S–601S(1992).

71. Dhabuwala A, Cannan RJ, Stubbs RS. Improvement in co-morbidities following weight loss from gastric bypass surgery. *Obes Surg* **10**(5):428–435(2000).

72. Shneerson J, Wright J. Lifestyle modification for obstructive sleep apnoea. *Cochrane Database Syst Rev* **2001**(1):CD002875.

73. Hakala K, Maasilta P, Sovijarvi AR. Upright body position and weight loss improve respiratory mechanics and daytime oxygenation in obese patients with obstructive sleep apnoea. *Clin Physiol* **20**(1):50–55(2000).

74. Koenig S. Pulmonary complications of obesity. *Am J Med Sci* **321**(4):249–279(2001).

75. Malhotra A, Fogel R, Stanchina M, Patel SR, et al. Postural effects on pharyngeal protective reflex mechanisms. *Sleep* **27**(6):1105–1112(2004).

76. Suratt P. Upper airway muscle activation is augmented in patients with obstructive sleep apnea compared with that in normal subjects. *Am Rev Respir Dis* **137**:889–894(1983).

77. Mezzanotte W. Waking genioglossal EMG in sleep apnea patients versus normal controls (a neuromuscular compensatory mechanism). *J Clin Invest* **89**:1571–1579(1992).

78. Cartwright RD. Effect of sleep position on sleep apnea severity. *Sleep* **7**(2):110–114(1984).

79. Oksenberg A, Silverberg DS, Arons E, Radwan H. Positional vs nonpositional obstructive sleep apnea patients: anthropomorphic, nocturnal polysomnographic, and multiple sleep latency test data. *Chest* **112**(3):629–639(1997).

80. Oksenberg A, Khamaysi I, Silverberg DS, Tarasiuk A. Association of body position with severity of apneic events in patients with severe nonpositional obstructive sleep apnea. *Chest* **118**(4):1018–1024(2000).

81. George CF, Millar TW, Kryger MH. Sleep apnea and body position during sleep. *Sleep* **11**(1):90–99(1988).

82. Pevernagie DA, Shepard JW Jr. Relations between sleep stage, posture and effective nasal CPAP levels in OSA. *Sleep* **15**(2):162–167(1992).

83. Cartwright RD, Diaz F, Lloyd S. The effects of sleep posture and sleep stage on apnea frequency. *Sleep* **14**(4):351–353(1991).

84. Oksenberg A, Silverberg DS, Arons E, Radwan H. The sleep supine position has a major effect on optimal nasal continuous positive airway pressure: relationship with rapid eye movements and non-rapid eye movements sleep, body mass index, respiratory disturbance index, and age. *Chest* **116**(4):1000–1006(1999).

85. Yoshida K. Influence of sleep posture on response to oral appliance therapy in sleep apnea syndrome. *Sleep* **24**(5):538–544(2001).

86. Battagel JM, Johal A, Smith AM, Kotecha B. Postural variation in oropharyngeal dimensions in subjects with sleep disordered breathing: a cephalometric study. *Eur J Orthod* **24**(3):263–276(2002).

87. Penzel T, Moller M, Becker HF, Knaack L, et al. Effect of sleep position and sleep stage on the collapsibility of the upper airways in patients with sleep apnea. *Sleep* **24**(1):90–95(2001).

88. Virkkula P, Maasilta P, Hytonen M, Salni T, et al. Nasal obstruction and sleep-disordered breathing: the effect of supine body position on nasal measurements in snorers. *Acta Otolaryngol* **123**(5):648–654(2003).

89. Desfonds P, Planes C, Fuhrman C, Foucher A, et al. Nasal resistance in snorers with or without sleep apnea: effect of posture and nasal ventilation with continuous positive airway pressure. *Sleep* **21**(6):625–632(1998).

90. Yildirim N, Fitzpatrick MF, Whyte KF, Jalleh R, et al. The effect of posture on upper airway dimensions in normal subjects and in patients with the sleep apnea/hypopnea syndrome. *Am Rev Respir Dis* **144**(4):845–847(1991).

91. Tsuiki S, Almeida FR, Bhalla PS, Lowe AA, et al. Supine-dependent changes in upper airway size in awake obstructive sleep apnea patients. *Sleep Breath* **7**(1):43–50(2003).

92. Isono S, Tanaka A, Nishino T. Lateral position decreases collapsibility of the passive pharynx in patients with obstructive sleep apnea. *Anesthesiology* **97**(4):780–785(2002).

44

PHARMACOLOGICAL TREATMENT OF SLEEP DISORDERED BREATHING

DAVID W. HUDGEL

Henry Ford Health System, Detroit, Michigan

INTRODUCTION

There are three types of sleep disordered breathing (SDB) for which pharmacological therapy has been evaluated: obstructive sleep apnea (OSA), central sleep apnea (CSA), and obesity–hypoventilation syndrome (OHS). OSA is characterized by sleep-associated, repetitive inspiratory collapse of the upper airway in the face of persistent and increasing inspiratory muscle effort. CSA is cyclic cessation of inspiratory effort leading to lack of gas exchange. Both OSA and CSA are cyclic oscillatory events, interrupted by arousal, and a decrease in restful sleep and often excessive daytime sleepiness. In OHS decreased inspiratory effort during sleep is inadequate to maintain normal gas exchange in the face of the large mechanical load placed on the ventilatory system by obesity. These three conditions can exist separately or in combination in a given patient. Even within a given patient during sleep, the mixture of these three entities can vary, depending on body position and different sleep stages. For instance, in non-rapid eye movement (NREM) sleep, lying on his/her side, an individual may have CSA, only to experience OSA when he/she turns supine or enters REM sleep. Oftentimes, arterial oxygen desaturation occurs only in REM sleep without apneic events, representing a component of OHS at that particular time.

Based on the above, review of diagnostic polysomnograms (PSGs) often demonstrates that two or all three of these SDB varieties exist in a single patient. Therefore stepwise treatment of each of these components must be considered. Symptomatic improvement is often obtained with treatment of the most common entity, such as OSA. In this situation, use of nasal continuous positive airway pressure (CPAP) may be adequate to relieve symptoms, such as snoring and excessive daytime sleepiness. Whether persistence of other components, such as CSA and/or OHS, in such a patient is detrimental is unknown. Surely combination therapies can be considered, such as CPAP plus supplemental oxygen (OSA + OHS), or CPAP plus acetazolamide (OSA + CSA). Whether these combination therapies are advantageous is unknown at the present time since no published investigations evaluating combination therapy exist. Basic to the literature on the pharmacological treatment of SDB is the assumption that patients are only suffering from one the components, which is not totally realistic.

In this chapter, the use of non-CPAP therapies for primarily OSA and CSA will be reviewed. Inhaled and oral agents will be considered.

SUPPLEMENTAL OXYGEN

Supplemental low-flow nasal oxygen administration has been shown to improve oxygenation and decrease the number of OSA events, although often lengthening individual apneas. [1, 2]. Because of the lengthening of apnea events, it was hypothesized that COPD patients with sleep disordered breathing might have a worsening of hypercapnic acidosis with nocturnal oxygen administration. However, different from this reasoning, an objective evaluation showed that oxygen resulted in improved sleep quality and sleep oxygenation without an increase in the number of, or lengthening of, sleep disordered breathing events [3, 4]. No worsening of hypercapnia was produced in these patients with the oxygen, even in those with severe hypercapnia. Therefore it was concluded that supplemental oxygen might be beneficial, not detrimental, for patients with SDB, regardless of whether they had chronic hypercapnic respiratory failure or not. However, Fleetham et al. [5] did not find improvement in sleep quality in COPD patients given an oxygen supplement. Although oxygenation during sleep improved in the patients they studied, total sleep time and arousals did not improve. Interestingly, in patients with both OSA and CSA, **Smith et al**. [2] found that oxygen improved central and mixed apneas (central followed by an obstructive component), but did not resolve pure obstructive apneas, nor did it improve arterial oxygen saturation in REM sleep. After little attention for some time, this topic was revisited recently by Pokorski and Jernajczyk [6], who used 30% oxygen, balanced with nitrogen administered at 4 L/min via nasal prongs in OSA patients. The apnea/hypopnea index (AHI) decreased significantly from 53 to 39 events/hour of sleep. Somewhat surprisingly, overall apnea duration was shorter on the oxygen administration night than on the compressed air night, different from previous observations. Nocturnal blood pressure (BP), especially diastolic BP, and heart rate decreased on oxygen versus compressed air. Most likely, the beneficial mechanism of action of oxygen for OSA, CSA, and periodic breathing is similar to that in high-altitude sleep-induced periodic breathing and CSA. By decreasing post-apnea hyperpnea and the subsequent hypocapnia, oxygen reduces the wide fluctuations in arterial carbon dioxide tension, thereby keeping the arterial CO_2 tension above the CO_2 apneic threshold.

Therefore oxygen administration will occasionally be helpful for SBD, especially periodic breathing in OSA patients, with and without COPD. Because not all studies have shown positive results, oxygen administration for OSA alone has not evolved as a commonly used therapeutic modality. Comfort can be taken in that it has been adequately demonstrated that using oxygen in those with significant sleep-related hypoxemia who also have SDB will not worsen the coexisting SDB.

SUPPLEMENTAL CARBON DIOXIDE INHALATION

By stabilizing ventilation, and therefore preventing the hypopneic portion of the oscillatory, periodic breathing pattern characteristic of OSA, an increase in the inhaled carbon dioxide tension to maintain a subarousal constant degree of hypercapnia and narrowing of the range of arterial

CO_2 tension fluctuations diminished OSA considerably [7]. Unfortunately, no controlled trials have been conducted supporting this observation. Mild hypercapnia has been shown to improve CSA in CHF and non-CHF patients [8, 9] and resolved CSA that appeared in a tracheostomized OSA patient [10]. In all these instances, the apneas and periodic breathing were completely or nearly completely eliminated. However, there were some changes in sleep architecture with CO_2 administration. Sleep efficiency (time asleep/time in bed) and REM sleep time both decreased, even though arousals in light sleep were reduced. Objective daytime sleepiness, assessed by the multiple sleep latency test (MSLT, a series of five daytime 15–20 min naps, 2 hours apart), demonstrating baseline hypersomnolence, improved minimally. In addition, daytime performance tests improved minimally. Unfortunately, the mask through which the CO_2 was administered was only worn on the CO_2 night and was reported to cause considerable discomfort, leading to premature abandonment of the trial in some patients. In this situation, it might be considered a positive finding that daytime performance testing did not worsen. Most likely, CO_2 therapy has not "caught on" and been used frequently because of potential sleep disruption, and the possibility of dangerous hypercapnia and hypoxemia occurring in the unattended home environment.

SUPPLEMENTAL FEMALE SEX HORMONE REPLACEMENT

The prevalence of sleep disordered breathing in females has recently been studied extensively [11, 12] and reviewed by White [13]. There is a progressive increase in OSA prevalence with advancing years through the menopausal years in females studied in the Wisconsin State Employee Sleep Cohort Study. An analysis of the Sleep Heart Health Study data showed an OSA prevalence of 12% in postmenopausal females with data being controlled for age and BMI. Prevalence of an AHI $\geq$ 15 events/hour of sleep was approximately 50% lower for women using hormone replacement therapy (HRT) versus those postmenopausal women not on HRT. Interestingly, this relationship held up in different weight-based strata. OSA was more prevalent in older women, but fewer of these women used HRT. Bixler et al. [14] found a fourfold increase in OSA in postmenopausal women not on HRT relative to premenopausal women, but OSA was still not as common in postmenopausal females as in men. The presence of OSA in these females was associated with the presence of central obesity, characterized by neck and abdominal fat distribution. The prevalence of OSA in women on HRT was not found to be elevated in these epidemiologic studies. Therefore these results would suggest that HRT would be beneficial for postmenopausal female OSA patients.

The mechanism of action of estrogen in postmenopausal women is unclear, although several possibilities exist. Upper airway inspiratory muscle activity has been shown to increase in postmenopausal women given HRT [15]. Estrogen has been shown to increase brain and carotid body blood flow, which may change central ventilatory drive. It may stabilize sleep, resulting in fewer apneic arousals, and therefore fewer postarousal apneas. Estrogen upregulates brain progesterone receptors, possibly contributing to the increased ventilatory drive associated with progesterone administration.

Controlled therapeutic trials with HRT in OSA, in distinction from epidemiological studies reviewed earlier, have produced mixed results. Use of 30 mg of progesterone alone only decreased apnea length, not frequency [16]. A comparison of the effects of estrogen, with and without progesterone, was conducted by Keefe et al. [17] in five women. Both estrogen and the combination improved OSA. Presumably, the effect noted was due to the estrogen. In a larger study, estrogen improved OSA in postmenopausal women [18]. Unfortunately, these investigators used an indirect method of apnea assessment—the pressure-sensitive bed. These studies indicate that the inclusion of estrogen in a postmenopausal HRT program is important in those with OSA. Whether or not HRT treatment of OSA in postmenopausal females is justified considering the potential health risks of this therapy has not been evaluated. Presumably, it would be a potential treatment in a particular woman whose other health risks from HRT are low.

In a series of predominantly uncontrolled studies, high-dose progesterone has been used to treat OHS. The rationale was to stimulate ventilatory control to increase ventilation in these obese individuals. Oftentimes, these patients also had OSA. In one study, those who responded to medroxyprogesterone were hypercapnic [19]. This finding was not substantiated in another study, which was a randomized, double-blind controlled study [20]. Normocapnic OSA patients were generally unresponsive [16, 20, 21].

RANDOMIZED CONTROLLED TRIALS (RCTs) AND NON-RCTs OF SPECIFIC AGENTS

In the past, one evidence-based review of the pharmacological treatment of OSA was published [22]. Recently, the Cochrane Library published a comprehensive document compiling and critiquing the RCTs conducted on the pharmacological treatment of OSA [23]. Two of the authors independently extracted data from 51 articles identified by electronic search. Of these 51 articles only 9 were blinded RCTs. Of the remaining 42 articles, some were review articles, and others were inadequately controlled trials, without control groups and sometimes without a placebo arm. Treatment of central sleep apnea was not included in this analysis.

Of the nine studies selected for review, most had a limited number of subjects and methodological items, such as inclusion and exclusion criteria were not commonly identified. All studies were crossover, but three had no washout interval between the placebo and the test agent. In only two studies was the use of an identical placebo documented. One study was single blind, others were double blind.

Three RCTs evaluated protriptyline in 23 patients. Trials were 2–3 weeks of intervention on each arm of the study. No improvement in apnea/hypopnea index (AHI) or oxygenation occurred [24–26]. No improvement in sleep disruption was found. Improvement in subjective daytime sleepiness was found in two studies, but not in the third study, which used a visual analog scale for symptom rating [26]. In the latter study, protriptyline was administered for only two weeks and REM sleep time was not decreased, possible evidence of insufficient dose or therapeutic duration.

Four non-RCT trials have been conducted with protriptyline, including 47 patients [27–30]. When REM sleep time was decreased, as would be anticipated, OSA was improved since more severe OSA usually occurs in REM sleep. Without the changes in REM sleep time and associated improvement in OSA, no significant changes were noted.

Anticholenergic side effects are significant with protriptyline, especially in the urinary tract, making it extremely difficult for most men to tolerate it long term. Therefore its potential use is restricted to REM-specific OSA in females. However, it has not been objectively evaluated in this circumstance. Likely, a selective serotonin reuptake inhibitor (SSRI) would be more tolerable (see below).

Acetazolamide was examined in 10 subjects in one RCT over 2 weeks [26]. A significant decrease in AHI was found, but there was no improvement in the number of arousals or daytime sleepiness. These findings differ from uncontrolled studies, which showed either no improvement in OSA or actual worsening (see below). There appears to be a dose-dependent effect, with higher doses decreasing obstructive events, and lower doses improving central events, but this has not been evaluated objectively. Dose-dependent side effects, especially digit paresthesias, limit patient acceptability.

One non-RCT was conducted with acetazolamide in nine OSA patients with AHI values ranging from 5 to 57 events/hour [31]. The drug was effective in both REM and NREM sleep with improvement in oxygenation in NREM sleep only. Another study of both CSA ($N = 6$) and OSA patients ($N = 4$) showed no improvement in CSA or OSA, and worsening of OSA in two patients [32]. Surely, it is our experience and that of others that acetazolamide is a much more effective drug for CSA and periodic breathing than OSA [33, 34]. Likely, it is not an agent that will be useful in OSA.

Single studies evaluated aminophylline and theophylline [35, 36] by RCT. One study was an intravenous single-night study and the other was a 1 month theophylline oral trial. In this latter study, 3/12 subjects dropped out. Intravenous aminophylline improved central but not obstructive apneas. Neither preparation improved AHI or oxygenation, and significantly disturbed sleep.

In 15 male congestive heart failure patients with periodic breathing and central apneas during sleep, 5 days of oral theophylline improved central apnea index from 26 to 6 events per hour [37]. However, because of the stimulating characteristics of theophylline, the sleep pattern was not improved. Thus, although the breathing pattern can be improved with these adenosine antagonists, sleep remains disturbed, diminishing the therapeutic value of these compounds.

Single RCTs examined the effect of medroxyprogesterone [20], clonidine [38], sabeluzole [39], and buspirone [40]. No differences between active product and placebo were identified in AHI, oxygenation, or sleep duration, except for sabeluzole, a glutamate antagonist. In a well-designed study with 4 week arms, and a 2 week washout period between drug and placebo, Hedner et al. [39] evaluated oxygen desaturation index in 13 OSA patients. Although there was no overall effect of the drug on the group oxygenation, those with higher blood levels of sabeluzole had improvement in their sleep oxygenation. No further trials with this drug exist.

Thus, of these trials, only acetazolamide and sabeluzole were found to be effective, but neither drug was shown to resolve all of the pathological changes that accompany OSA: apneic events, intermittent hypoxemia, and excessive daytime sleepiness. However, addressing these studies in a general, scientific design surely could be improved. In none of the studies is there enough statistical power to show that these drugs are not effective in OSA. "No evidence of an effect" does not mean "evidence of no effect". However, in the clinical sleep research community, very few investigators are continuing to examine the potential benefit of pharmacological treatment of OSA with drugs that effect ventilation and its control, or alter neuromediator activity. Although inadequate to rule out a potential benefit, the abovementioned studies have dampened enthusiasm for this type of treatment, and these results surely have diminished funding for such research.

OTHER PHARMACOLOGICAL AGENTS

Some agents have only been studied with non-RCT methodology. The serotonin precursor L-tryptophan improved OSA in 12 patients [41]. In an open label study, fluoxetine improved AHI from 57 to 34 events/hour and was moderately well tolerated [30]. These results were not confirmed with paroxetine [42]. Although serotonin may be involved in the hypoglossal motoneuron stimulation, the SSRIs tested to date in OSA do not seem to be remarkably effective.

By enhancing sleep continuity, benzodiazepines improved sleep disordered breathing, especially CSA [43, 44]. Worsening of the apnea, as might have been predicted by some via depressing ventilatory drive activity, was not produced by these agents. Neither apnea length nor arterial oxygen saturation was worsened.

Metoprolol and cilazapril, a beta-adrenergic inhibitor and ACE inhibitor, respectively, both improved AHI from 40 to approximately 27 events/hour in one study [45]. Presumably, the mechanism of action of these agents was through the autonomic nervous system. No further studies exist for these agents. Clinically, patients on these drugs for hypertension surely present with severe OSA, so their impact on OSA must be minimal, at best.

Use of thyroxine generated early excitement in the treatment of those with OSA and hypothyroidism. Early case reports and a more formal study demonstrated reversal of OSA with thyroid hormone replacement [46–48]. However, these patients were myxedematous and quite ill. As more studies were conducted, it appeared the obesity and increasing age accompanying hypothyroidism were instrumental in the OSA of hypothyroidism [49]. In a study by Grunstein and Sullivan [50], six of eight obese hypothyroid OSA patients did not improve on adequate thyroid hormone replacement, and three patients worsened. These latter data are consistent with clinical observations; thyroid hormone replacement does not usually resolve the OSA in hypothyroid patients. Likely, there is a distinct difference in the profoundly myxedematous patients originially reported and the chemically hypothyroid, nonmyxedematous patients usually seen today. None of these thyroxine trials were blinded or controlled.

INFLAMMATORY MEDIATOR INHIBITORS

A new area of investigation is emerging, examining the effect of various types of inflammatory cytokine inhibition. Interleukin-6 (IL-6) and tumor necrosis factor alpha (TNF-α) are two inflammatory mediators that have been identified as being elevated in OSA patients. These cytokines are also elevated in those with central obesity [51]. Recently, these cytokines were found to be elevated after one night of complete sleep deprivation, or during a week in which young volunteers underwent mild sleep restriction from 8 to 6 hours of sleep in 24 hours for one week. Both conditions resulted in objective daytime sleepiness, fatigue, and impaired vigilance [52, 53]. These mediators have also been associated with the development of insulin resistance and cardiovascular disease [51, 54–56]. They have been found to be elevated in OSA, independent of obesity [51]. Other mediators, such as C-reactive protein, have also been found to be elevated in those with an increased risk of cardiovascular disease [57]. Thus obesity, insulin resistance, and sleep restriction are all variables associated with OSA and with elevated circulating levels of cytokines. The significance of the elevated cytokines and the inflammation they represent as a cause of OSA, or even as a cause of the complications OSA produces, like excessive daytime sleepiness, is unsubstantiated by these observations. The elevated cytokine levels may simply be a consequence of the OSA or its sequelae.

However, if either a cytokine agonist or antagonist altered the outcome variable, then a more certain cause-and-effect relationship could be drawn. Indeed, that has been done in that Vgontzas et al. [58] demonstrated that etanercept, a TNF-α antagonist, partially inhibited the daytime sleepiness in 8 male obese (BMI = 39 kg/m^2) OSA patients. Although it was a double-blind, placebo-controlled study, it was a fixed order, nonrandomized study with placebo followed by active drug. Although there was no change in subjective sleepiness by the active drug, the MSLT sleep latency increased significantly. These results are concerning because the MSLT sleep latencies were not abnormal at baseline, in that no mean nap sleep latency was below 10 minutes. In addition, we do not know what the drug would do to the sleep latencies during an MSLT in nonapneic control subjects. Possibly, it is alerting. With etanercept there was a small, but significant, decrease in the AHI, from 53 ± 11 to 44 ± 10 events per hour of sleep. There were no changes in serum insulin, blood sugar, and adioponectin, findings inconsistent with the notion that TNF-α contributes to the presence of insulin resistance. Possibly, an insulin–hypoglycemic clamp study would be needed to demonstrate a change in the insulin resistance. Because of the above concerns, these data do not support the concept that inflammation is involved in the etiology of excessive daytime sleepiness in OSA, and, more importantly, there is minimal support for the hypothesis that inflammation is involved in the basic pathophysiology of OSA itself. Although further studies need to be conducted to better define the role of inflammation in the etiology of OSA, these results do not demonstrate a large role of inflammation in the etiology of OSA. Whether etanercept, or similar compounds, will be useful therapeutically in OSA awaits further study.

PRACTICAL CONSIDERATIONS: THE NEED FOR ACCURATE DIAGNOSIS

If sleep physicians are to consider treating more than one component of SDB, as often exists, then our diagnostic equipment must be accurate enough to reliably make these specific diagnoses. This is best done with in-laboratory PSGs, conducted with carefully calibrated diagnostic instruments. The need for minimally invasive measurements, such as esophageal pressure measurement, is undetermined. Surely, in a given, patient this measurement might be definitive. Ambulatory monitoring of SDB is not specific enough to lead to accurate diagnoses within the realm of SDB, and therefore its use, other than for a gross screening tool, is to be discouraged. In fact, if we are to consider treatment of SBD components, follow-up PSGs will be necessary, obviously impacting the wait lists for in-laboratory sleep beds. It is the contention of this author that more judicious use of in-laboratory beds in the evaluation of snoring will help provide bed availability for objective stepwise treatment of the many patients with more than one component of the SDB spectrum.

SUMMARY

Pharmacological treatment of OSA remains in its infancy. Multiple pilot-type studies, both RCTs and non-RCTs, have been conducted, but results have not been promising enough to stimulate large adequately powered, well-controlled follow-up studies. Although the majority of investigations conducted to date are not adequately powered to prove that a particular pharmacological agent is or is not effective, investigators have been discouraged by initial results. Institutional review boards and granting agencies have been reluctant to approve and fund extensive studies that may delay the institution of proven effective therapy, such as CPAP.

Studies have demonstrated that some agents may be effective in particular situations. Central sleep apnea is easier to control pharmacologically than obstructive apnea. Acetazolamide may be particularly useful in this regard but has the drawback of potentially worsening obstructive events. Therefore a polysomnographic evaluation of the therapeutic response to its use is recommended 2–4 weeks after initiation of treatment. Theophylline reduces periodic breathing and central apneas, especially in heart failure patients; however, because of the sleep disruption produced by this drug, it may not help daytime sleepiness in these patients. Therefore theophylline is not recommended for long-term use. Interestingly, by stabilizing sleep, benzodiazepine and nonbenzodiazepine sleep aids decrease the number of arousals and thereby reduce postarousal apneas, leading to both improved sleep quality and normalized sleep respiration.

There is minimal evidence that OSA can be treated pharmacologically. In specific REM-related OSA, antidepressants that decrease REM sleep time may be useful, although use of these agents in this particular situation has not been analyzed in detail. This treatment is especially useful when the whole-night AHI does not meet third-party coverage criteria for CPAP therapy. Possibly, new anti-inflammatory agents will be useful, but inadequate data are available at this time on these drugs. However, these studies on inflammation and its interaction with sleep and sleep disorders open new doors. Obviously, the goal of the endeavor to identify effective pharmacological treatments of the components of SDB is to decrease our field's dependency on CPAP, a treatment with which insufficient compliance is an all too common problem. The availability of a set of effective, safe drugs for various indications in SDB patients would be a huge step forward in patient convenience and satisfaction.

REFERENCES

1. Martin RJ, Sanders MH, Gray BA, et al. Acute and long-term ventilatory effects of hyperoxia in the adult apnea syndrome. *Am Rev Respir Dis* **125**:175–180(1982).

2. Smith PL, Haponik EF, Bleecker ER. The effects of oxygen in patients with sleep apnea. *Am Rev Respir Dis* **130**:958–963(1984).

3. Kearley R, Wynne JW, Block AJ, et al. The effect of low flow oxygen on sleep-disordered breathing and oxygen desaturation. A study of patients with chronic obstructive lung disease. *Chest* **78**:682–685(1980).

4. Wynne JW, Block AJ, Hunt LA, et al. Disordered breathing and oxygen desaturation during daytime naps. *Johns Hopkins Med J* **143**:3–7(1978).

5. Fleetham F, West P, Mezon B, et al. Sleep, arousals, and oxygen desaturation in chronic obstructive pulmonary disease—the effect of oxygen therapy. *Am Rev Respir Dis* **126**:429–433(1982).

6. Pokorski M, Jernajczyk U. Nocturnal oxygen enrichment in sleep apnoea. *J Int Med Res* **28**:1–8(2000).

7. Hudgel DW, Hendricks C, Dadley A. Alteration in obstructive apnea pattern induced by changes in O_2 and CO_2 inspired concentrations. *Am Rev Respir Dis* **138**:16–19(1988).

8. Xie A, Rankin F, Rutherford R, et al. Effects of inhaled CO_2 and added dead space on idiopathic central sleep apnea. *J Appl Physiol* **82**(3):918–926(1997).

9. Steens RD, Millar TW, Xiaoling S, et al. Effect of inhaled 3% CO_2 on Cheyne–Stokes respiration in congestive heart failure. *Sleep* **17**(1):61–68(1994).

10. Badr MS, Grossman JE, Weber SA. Treatment of refractory sleep apnea with supplemental carbon dioxide. *Am J Respir Crit Care Med* **150**(2):561–564(1994).

11. Young T, Finn L, Austin D, et al. Menopausal status and sleep-disordered breathing in the Wisconsin Sleep Cohort Study. *Am J Respir Crit Care Med* **167**:1181–1185(2003).

12. Shahar E, Redline S, Yount T, et al. Hormone replacement therapy and sleep-disordered breathing. *Am J Respir Crit Care Med* **167**:1186–1192(2003).

13. White D. The hormone replacement dilemma for the pulmonologist (Editorial). *Am J Respir Crit Care Med* **167**:1165–1166(2003).

14. Bixler EO, Vgontzas AN, Lin HM, et al. Prevalence of sleep-disordered breathing in women. *Am J Respir Crit Care Med* **163**:608–613(2001).

15. Popovic RM, White DP. Upper airway muscle activity in normal women: influence of hormonal status. *J Appl Physiol* **84**(3):1055–1062(1998).

16. Block AJ, Wynne JW, Boysen PG, et al. Menopause, medroxyprogesterone and breathing during sleep. *Am J Med* **70**:506–510(1981).

17. Keefe DL, Watson R, Naftolin F. Hormone replacement therapy may alleviate sleep apnea in menopausal women: a pilot study. *Menopause* **6**(3):196–200(1999).

18. Polo-Kantola P, Rauhala E, Helenius H, et al. Breathing during sleep in menopause: a randomized, controlled, crossover trial with estrogen therapy. *Am Coll Obst Gynecol* **102**(1):68–75(2003)

19. Strohl KP, Hensley NJ, Saunders SM, et al. Progesterone administration and progressive sleep apnea. *JAMA* **245**:1230–1232(1981).

20. Cook WR, Benich JJ, Wooten SA. Indices of severity of obstructive sleep apnea syndrome do not change during medroxyprogesterone acetate therapy. *Chest* **96**:262–266(1989).

21. Rajagopal KR, Abbrecht PH, Jabbari B. Effects of medroxyprogesterone acetate in obstructive sleep apnea. *Chest* **90**:815–821(1986).

22. Hudgel DW, Thanakitcharu S. Pharmacologic treatment of sleep-disordered breathing. *Am J Respir Crit Care Med* **158**:691–699(1998).

23. Smith I, Lasserson TJ, Wright J. Drug treatments for obstructive sleep apnoea. *The Cochrane Library* **2**:1–31(2004).

24. Brownell L, West P, Sweatmen X, et al. Protriptyline in obstructive sleep apnoea: a double blind trial. *N Engl J Med* **307**(17):1037–1042(1982).

25. Stepanski E, Conway W, Young D, et al. A double-blind trial of protriptyline in the treament of sleep apnoea syndrome. *Henry Ford Hosp Med J* **36**(1):5–8(1988).

26. Whyte K, Gould G, Shapiro AA, et al. Role of protriptyline and acetazolomide in the sleep apnea/hypopnea syndrome. *Sleep* **11**(5):463–472(1988).

27. Clark RW, Schmidt HS, Schaal SF, et al. Sleep apnea treatment with protriptyline. *Neurology* **29**:1287–1292(1979).

28. Conway WA, Zorick F, Piccione P, et al. Protriptyline in the treatment of sleep apnea. *Thorax* **37**:47–53(1982).

29. Smith PL, Haponik EF, Allen RP, et al. The effects of protriptyline in sleep-disordered breathing. *Am Rev Respir Dis* **127**:8–13(1983).

30. Hanzel DA, Proia NG, Hudgel DW. Response of obstructive sleep apnea to fluoxetine and protriptyline. *Chest* **100**:416–421(1991).

31. Tojima H, Kunitomo F, Kimura H, et al. Effects of acetazolamide in patients with sleep apnoea syndrome. *Thorax* **43**:113–119(1988).

32. Sharp JT, Druz WS, Sousa VD, et al. Effect of metabolic acidosis upon sleep apnea. *Chest* **87**:619–624(1985).

33. White DP, Zwillich CW, Pickett CK, et al. Central sleep apnea—improvement with acetazolamide therapy. *Arch Intern Med* **142**:1816–1819(1982).

34. DeBacker WA, Verbraecken J, Willemen M, et al. Central apnea index decreases after prolonged treatment with acetazolamide. *Am J Respir Crit Care Med* **151**:87–91(1995).

35. Espinoza H, Antic R, Thornton A, et al. The effects of aminophylline on sleep and sleep-disordered breathing in patients with obstructive sleep apnoea. *Am Rev Respir Dis* **136**:80–84(1987).

36. Mulloy E. Theophylline in obstructive sleep apnoea. *Chest* **101**(3):753–757(1992).

37. Javaheri ST, Parker J, Wexler L, et al. Effect of theophylline on sleep-disordered breathing in heart failure. *N Engl J Med* **335**:562–567(1996).

38. Issa FG. Effect of clonidine in obstructive sleep apnea. *Am Rev Respir Dis* **145**(2):435–439(1992).

39. Hedner J, Grunstein R, Eriksson B, et al. A double blind, randomized trial of sabeluzole—a putative glutamate antagonist—in obstructive sleep apnoea. *Sleep* **19**(4):287–289(1996).

40. Mendelson WB, Maczaj M, Holt J. Buspirone administration to sleep apnea patients. *J Clin Psychopharmacol* **11**:71–72(1991).

41. Schmidt HS. L-Tryptophan in the treatment of impaired respiration in sleep. *Bull Eur Physiolpathol Respir* **19**:625–629(1983).

42. Kraiczi H, Dahlof P, Carlson J, et al. Paraxetine reduces sleep disordered breathing in obstructive sleep apnea patients (Abstract). *Am J Respir Crit Care Med* **155**:C47(1997).

43. Guilleminault C, Crowe C, Quera-Salva MA, et al. Periodic leg movement, sleep fragmentation and central sleep apnoea in two cases: reduction with clonazepam. *Eur Respir J* **1**:762–765(1988).

44. Bonnet MH, Dexter JR, Arand DL. The effect of Triazolam on arousal and respiration in central sleep apnea patients. *Sleep* **13**(1):31–41(1990).

45. Weichler U, Herres-Mayer B, Mayer J, et al. Influence of antihypertensive drug therapy on sleep pattern and sleep apnea activity. *Clin Pharmacol* **78**:124–130(1991).

46. Massumi RA, Winnacker JL. Severe depression of the respiratory center in myxedema. *Am J Med* **36**:876–882(1964).

47. Orr WC, Males JL, Imes NK. Myxedema and obstructive sleep apnea. *Am J Med* **70**:1061–1066(1981).

48. Rajagopal KR, Abbrecht PH, Derderian SS, et al. Obstructive sleep apnea in hypothyroidism. *Ann Intern Med* **101**:491–494(1984).

49. Lin C, Tsan K, Chen P. The relationship between sleep apnea syndrome and hypothyroidism. *Chest* **102**:1663–1667(1992).

50. Grunstein RR, Sullivan CE. Sleep apnea and hypothyroidism: mechanisms and management. *Am J Med* **85**:775–779(1988).

51. Vgontzas AN, Papanicolaou DA, Bixler EO, et al. Sleep apnea and daytime sleepiness and fatigue: relation to visceral obesity, insulin resistance and hypercytokinemia. *J Clin Endocrinol Metab* **85**:1151–1158(2000).

52. Vgontzas AN, Zoumakis E, Bixler EO, et al. Adverse effects of modest sleep restriction on sleepiness, performance and inflammatory cytokines. *J Clin Endocrinol Metab* **89**(5):2119–2126(2004).

53. Shearer WT, Reuben JM, Mullington JM, et al. Soluble TNF-α receptor 1 and IL-6 plasma levels in humans subjected to the sleep deprivation model of spaceflight. *J Allergy Clin Immunol* **107**:165–170(2001).

54. Vgontzas AN, Bixler EO, Papanicolaou DA, et al. Chronic systemic inflammation in overweight and obese adults. *JAMA* **283**:2235–2236(2000).

55. Chrousos GP. The role of stress and the hypothalamic–pituitary–adrenal axis in the pathogenesis of the metabolic syndrome: neuro-endocrine and target tissue-related causes (Review). *Int J Obes* **24**:S50–S55(2000).

56. Ridker PM, Rifai N, Stampfer MJ, et al. Plasma concentration of interleukin-6 and the risk of future myocardial infarction among apparently healthy men. *Circulation* **101**:1767–1772(2000).

57. Luc G, Bard JM, Juhan-Vague I, et al. C-reactive protein, interleukin-6 and fibrinogen as predictors of coronary heart disease: the PRIME study. *Arterioscler Thromb Vasc Biol* **23**:1255–1261(2003).

58. Vgontzas AN, Zoumakis E, Lin H-M, et al. Marked decrease in sleepiness in patients with sleep apnea by etanercept, a tumor necrosis factor-α antagonist. *J Clin Endocrinol Metab* **89**(9):4409–4413(2004).

45

POSITIVE AIRWAY PRESSURE THERAPY FOR OBSTRUCTIVE SLEEP APNEA

MAX HIRSHKOWITZ

Baylor College of Medicine and the Michael E. DeBakey VAMC, Houston, Texas

TEOFILO LEE-CHIONG

National Jewish Medical and Research Center, Denver, Colorado

INTRODUCTION

Positive airway pressure therapy is the treatment of choice for most patients with obstructive sleep apnea (OSA). It involves providing a fan or turbine generated airflow through the nose (and sometimes the mouth) in order to maintain airway patency. There are four basic types of positive airway pressure devices. The most common type provides a constant pressure and is called continuous positive airway pressure (CPAP). A second type provides two pressure levels, one during inhalation and a lower one during exhalation, and is called bilevel positive airway pressure (BPAP). The pressure drop during exhalation is designed to increase comfort for patients who have trouble exhaling against an incoming pressure. The third type of device is called autotitrating positive airway pressure (APAP) and uses variable flow controlled by computer algorithms in an attempt to determine optimal pressure. Finally, a fourth type of positive pressure device called noninvasive positive pressure ventilation (NIPPV) places two different pressures at a set rate to entrain breathing and provide ventilatory assist.

CPAP for treating patients with OSA was initially described in 1981 by Sullivan and colleagues [1]. In their report, five patients with severe OSA were treated with CPAP applied through the nares. Low-pressure levels (ranging from 4.5 to 10 cm H_2O) completely prevented upper airway occlusion, allowing uninterrupted sleep [1].

With CPAP, the patient sleeps with a single constant pressure throughout the night. Several methods have been developed for pressure selection. These include the following.

1. Attended full-night laboratory polysomnography during which a technologist adjusts the pressure until all or most of the obstructive respiratory events are eliminated.
2. Attended split-night laboratory polysomnography during which the initial two, or more, hours are used to diagnose obstructive sleep apnea and the remainder of the night is used to adjust pressures until all or most of the obstructive respiratory events are eliminated.
3. Formulas to derive pressures from clinical, polysomnographic, and/or anthropometric variables. The calculated pressure can also be used as a starting CPAP pressure for laboratory titration.
4. Unattended home or laboratory recordings using autotitrating devices.

The current standard of practice in the United States is attended laboratory polysomnography with technologist-attended pressure titration using either full- or split-night protocol. During the sleep evaluation, sleep stages and respiratory variables are recorded with the goal of obtaining a fixed single pressure that eliminates apnea, hypopnea,

Sleep: A Comprehensive Handbook, Edited by T. Lee-Chiong.
Copyright © 2006 John Wiley & Sons, Inc.

snoring, and respiratory effort related arousals (RERAs) in all body positions and sleep stages. In general, higher pressures are required to prevent airway occlusion during rapid eye movement (REM) sleep than during non-rapid eye movement (NREM) sleep, except in patients with congestive heart failure. Additionally, sleep disordered breathing is usually worse when an individual sleeps supine. During attended sleep studies, trained sleep technologists adjust CPAP to meet changing pressure requirements, intervene when mask leak or patient discomfort problems arise, and intercede if potentially fatal cardiac arrhythmias or sustained hypoxemia occurs. The optimal pressure determined during this titration procedure is thereafter administered therapeutically nightly at home.

MECHANISM OF ACTION

It is generally held that the positive pressure creates a pneumatic splint for the vulnerable portions of the nasopharyngeal airway. One explanatory model posits upper airway functions as a Starling resistor with a collapsible segment in the oropharynx. Airway collapse occurs when the intraluminal pressure is less than the critical opening pressure (P_{crit}, or the pressure required to keep it open). As nasal pressure is raised above P_{crit} by positive airway pressure, inspiratory airflow increases in proportion to the level of positive pressure applied until airway occlusions are abolished [2].

Upper airway (UA) patency is presumably maintained by the activity of muscles in the head and neck. These include cervical muscles that provide caudal traction on the UA. Inspiratory increases in UA patency cannot be attributed solely to activity of UA muscles. The thorax also applies caudal traction to the UA [3]. Tonic thoracic traction on the trachea increases with inspiration and has been shown to improve patency of the upper airway. This force acting through the pull of mediastinal and pulmonary structures is transmitted through the carina; intrathoracic pressure changes acting independently either draw the trachea into or push the trachea out of the thorax [4].

Hoffstein and co-workers [5] examined the relationship between lung volume and pharyngeal cross-sectional area in nine obese patients with OSA patients and ten age-matched, obese control subjects without OSA. Measurements were made in the upright-seated posture using acoustic reflection. They observed abnormally small pharyngeal cross-sectional area and considerable variation with changes in lung volume in the OSA group. Differences in pharyngeal area were significant at all lung volumes below total lung capacity (TLC), as was the difference in the magnitude of change in pharyngeal area with change in lung volume between the patients and control subjects [5].

EFFICACY IN PATIENTS WITH OBSTRUCTIVE SLEEP APNEA

Overall, CPAP is extremely safe and very effective in patients with moderate and severe OSA. Furthermore, the beneficial effects of CPAP are sustained over time. CPAP-related subjective and clinical outcomes observed at 1 month follow-up are present 5 months later [6]. However, although efficacy is well established in patients with moderate and severe disease, the extent to which patients with mild OSA benefit from CPAP is less certain. The minimal diagnostic threshold at which CPAP provides beneficial outcomes is not well characterized. One study in patients with mild OSA (AHI = 5.0–14.9 per hour) found improved symptom scores, mental flexibility, and depression rating. However, no change in subjective and objective sleepiness was observed [7]. By contrast, Engleman and colleagues [8] randomized a placebo-controlled crossover trial in 34 patients with mild OSA (AHI = 5–15) and daytime sleepiness and found that CPAP improved symptom score, subjective sleepiness, cognitive function, depression score, and five subscales of the SF-36 health/functional status questionnaire. Nonetheless, only 14 of the 34 patients preferred CPAP [8]. Redline and co-workers [9] studied 111 middle-aged snorers (age 25–65 years) with relatively low levels of OSA (respiratory disturbance index (RDI) of 5–30) who did not report sleepiness. Subjects were randomized to either CPAP or conservative therapy and 49% in the CPAP group improved on measures of well-being, mood, and functional status compared to 26% of those randomized to conservative therapy. A beneficial effect of CPAP over CT was most evident for patients with hypertension, diabetes, and without sinus problems. No relationship to baseline sleepiness or apnea severity was found [9].

Sleep Quality

OSA is defined and indexed according to specific sleep pathophysiologies, including the number of apnea episodes, hypopnea episodes, CNS arousals, and desaturation events. There are also concomitant disturbances in sleep integrity, continuity, and architecture. Much of the original therapeutic work, not surprisingly, aimed to reduce these sleep-related abnormalities; however, most of it was nonrandomized and not controlled. Later, more rigorously designed clinical trials confirmed improvements associated with CPAP therapy. A randomized, double-blind, placebo-controlled trial comparing CPAP to placebo CPAP (CPAP at an ineffective pressure) found improved sleep quality in 48 CPAP-naive patients with OSA. CPAP lowered the respiratory disturbance index (RDI) and number of arousals and increased arterial oxygen saturation (Sao_2). However, short-term CPAP was no different than placebo in improving sleep quality as assessed by sleep architecture, sleep efficiency, total sleep time, and wake after sleep-onset time [10].

Daytime Sleepiness

Sleepiness and other OSA-related clinical symptoms decrease, and quality of life improves in patients receiving CPAP and conservative therapy (sleep hygiene and weight loss) compared to those receiving only conservative therapy [11]. Similarly, sleepiness (measured both by self-report (Epworth Sleepiness Scale) and objectively (Osler wakefulness test)) decreased, and quality of life (measured with SF-36) improved in patients treated for 1 month with CPAP versus subtherapeutic CPAP in a randomized, double-blind, controlled, parallel group trial [12]. Furthermore, because sleepiness and impaired vigilance are thought to play a significant role in motor vehicle accidents (MVAs) and because patients with OSA have higher MVA rates, driving simulations have also been used to assess CPAP-related changes. In patients with OSA, CPAP therapy improves driving simulator steering performance (road position, length of drive before "crashing," and number of off-road events) and reduces reaction time to target stimuli [13]. Attempts to find physiological correlates for CPAP-related improvements have been disappointing. Standard polysomnographic baseline variables, including AHI and microarousal frequency, are poor predictors of improvement in daytime function. Weak to moderate correlations are reported for hypoxemia, wakefulness test sleep latency, OSA symptoms, quality of life, and reaction time. However, significant improvements in self-ratings of daytime function were related to the amount of CPAP usage [14]. In another study, CPAP improved self-reported symptoms of OSA (including snoring, restless sleep, and irritability) but did not improve objective (Multiple Sleep Latency Test) or subjective (Epworth Sleepiness Scale) measures of sleepiness in patients with mild OSA (AHI = 5–30; mean, 12.9). Furthermore, no benefit over placebo was found for any test of neurobehavioral function, quality of life (SF-36 and Functional Outcomes of Sleep Questionnaire [FOSQ]), mood score (Profile of Moods States and Beck Depression Index), or 24 hour blood pressure [15].

Patient Satisfaction

A questionnaire mailed to 148 patients and phone calls to 42 patients with OSA indicated that 105 patients self-reported continued CPAP use 17 ± 11 (mean $\pm$ SD) months later after initiation of therapy. The majority (81%) perceived CPAP as effective for treating the disorder, 5% were unsure, and 14% believed that CPAP was ineffective, notwithstanding resolution of sleep apnea on polysomno-

graphy. Improvement was also noted by the family members in 83% of the patients [16].

Healthcare Utilization

Treating patients with OSA is associated with a significant reduction in physician claims and hospital stays. Changes in health care utilization (physician claims and hospitalizations) were documented 2 years after diagnosis and treatment of OSA patients. Utilization data for the period of the study were matched to that of controls from the general population by gender, age, and geographic location. The difference in physician claims between the patients and their matched controls 2 years after diagnosis and treatment was significantly less than the difference in the year before diagnosis. Changes were only significant among patients adhering to CPAP treatment [17].

Cardiovascular Effects

A cause-and-effect relationship has been suggested between OSA and systemic hypertension (HTN). In a pioneering study, Lavie and colleagues [18] observed a linear increase in blood pressure and number of patients with hypertension as a function of sleep apnea severity (indexed by the apnea–hypopnea index). In their study, multiple logistic regression analysis showed that each additional apnea event per hour of sleep increased the odds of hypertension by about 1%, whereas each 10% decrease in nocturnal oxygen saturation increased the odds by 13% [18]. CPAP also affects the circadian profiles of blood pressure (BP) and heart rate (HR) in patients with OSA. Average BP (systolic/diastolic) and HR decreased significantly in patients with hypertension in response to CPAP therapy; however, BP did not decrease in normotensive patients with OSA [19]. By contrast, other studies link CPAP therapy with a reduction in blood pressure, even in normotensive patients with OSA [20]. This is especially the case when there is significant nocturnal oxygen desaturation. A small decrease in 24 hour diastolic blood pressure was seen in 68 patients with the greatest fall occurring between 2:00 a.m. and 9:59 a.m. Reduction in diastolic blood pressure was greater in patients using CPAP ≥ 3.5 h per night and in patients with more than twenty 4% desaturations per hour at baseline. The frequency of desaturation was the best predictor of diastolic blood pressure fall with CPAP therapy [20].

In another study, arterial BP in 12 men with OSA and HTN was determined at baseline and after CPAP therapy for 6 months. Antihypertensive medication was discontinued 1 week before baseline. Weight did not change during the 6 month study period. After 6 months of therapy, the apnea indices, systolic and diastolic blood pressures, blood pressure variability, heart rate, and heart rate variability decreased. Blood pressure declined during wakefulness periods, with a further decline during NREM and REM sleep [21].

Patients with hypertension can be classified based on their circadian pattern of blood pressure as "dippers" or "nondippers." "Dippers"are defined by an average reduction of at least 10 mmHg systolic and 5 mmHg diastolic BP at night compared to daytime values. Nondipping is reportedly common among patients with OSA; furthermore, CPAP restores the normal circadian "dipper" pattern. CPAP also significantly lowers average daytime and nighttime BP. Sixty-eight percent of subjects who were "nondippers" before CPAP treatment reversed to become "dippers" [22]. In another study, CPAP therapy in patients with OSA decreased nighttime (10 p.m. to 6 a.m.) mean arterial blood pressure to a much greater extent than placebo; however, groups did not differ with respect to daytime (6 a.m. to 10 p.m.) blood pressure decrease [23].

However, not all investigators report similar findings. Rauscher and colleagues [24] argue that hypertension in OSA is more closely linked to weight loss than CPAP-related decreased sleep apnea. In their study of 60 hypertensive patients with OSA, hypertension severity decreased in 40% of patients in the combined nasal CPAP and weight loss group, in 58% of patients in the weight loss only group, in 29% of patients in the nasal CPAP only group, and in 6% of the patients in the no treatment group. Multivariate analysis of variance revealed that only the percentage change in body mass index significantly contributed to the course of hypertension [24].

In addition to Cheyne–Stokes respiration (with central sleep apnea), OSA can also accompany congestive heart failure (CHF). Two mechanisms are posited as means by which OSA contributes to CHF: one mechanical and one neural. Mechanically, left ventricular afterload develops secondary to the combined effects of increased peripheral resistance (associated with systemic blood pressure elevation) and exaggerated negative intrathoracic pressure (produced by inhaling against a closed airway). The neural mechanism involves activation of the sympathetic nervous system by hypoxemia and repeated CNS arousals disturbing sleep. Consequently, CPAP improves cardiac function (increasing left ventricular ejection fraction) in patients with CHF [25].

Mortality

In a sample of 385 men with OSA, greater mortality occurred in those with 20 or more apneas per hour of sleep compared to those with less. This difference in apnea-associated mortality was more evident in younger patients (age less than 50 years) in whom all-cause mortality is less frequent [26]. Treatment with CPAP reversed the increase in apnea-associated mortality. These data strongly suggest that OSA is a serious, potentially life-threatening condition that can be effectively managed with positive airway pressure therapy.

By contrast, some individuals contend that the health consequences of OSA are exaggerated and CPAP's efficacy for improving health outcomes has been poorly evaluated. John Wright and colleagues [27] reviewed the association between OSA and key health outcomes and the efficacy of CPAP for treating OSA. The authors dismissed most of the data linking OSA to adverse health-related outcomes, cardiac arrhythmias, ischemic heart disease, cardiac failure, systemic or pulmonary hypertension, and stroke as poorly designed, weak, and contradictory [27]. These findings have been widely refuted by experts in the field and by more recent controlled studies.

THERAPEUTIC ADHERENCE

CPAP therapy is only as beneficial as its utilization. Thus less than optimal CPAP utilization is a significant problem in clinical practice. To a large extent, the level of use has as much to do with the individual as it does with the therapy. While a noticeably beneficial therapy may lead to somewhat greater adherence in willing patients, a noxious intervention can discourage use in even the most ardent and willing participant. Therefore correcting mask problems, discomfort, nasal allergies, and other barriers to utilization, especially during the first few months, is critical to achieving therapeutic adherence. The first month of therapy will also predict later use because it provides a baseline for how much an individual is likely to accept any therapy or be driven off by its associated difficulties. Marketing theory predicts greatest sales (or utilization) in those who believe a product is beneficial and also believe that they can change their behavior to use the product. Any doubts about the product's usefulness or about one's ability to use the product act as barriers to utilization.

Use of nasal CPAP therapy was studied in 17 chronically treated ((820 ± 262 days) patients with OSA. A CPAP system was used for 94% of the monitored days and the mean effective daily rate of use was 7.1 ± 1.1 hours. Sixty percent of patients used their device nightly [28]. In another study, 35 patients were issued CPAP machines that monitored usage more precisely than hour-meters. These systems had built-in microprocessors that could sense CPAP mask pressure. This surveillance system found that patients attempted CPAP use 66 ± 37% of the days monitored with mean duration of use equal to 4.88 ± 1.97 hours on nights used. Self-reports overestimated CPAP use by 69 ± 110 minutes. The majority (60%) of patients claimed to use CPAP nightly; however, only 46% used CPAP for 4 or more hours on 70% of monitored days. Patients with more education employed professionally were more likely to be regular users. Also, as expected, the frequency and duration of CPAP use in the first month reliably predicted its later use [29]. A large European multicenter study reported better therapeutic adherence with 80% of the patients using the machine 4 or more hours nightly on more than 70% of nights. Seventy-seven percent, 82%, and 79% of patients adhered to therapy at 1, 2, and 3 months, respectively. Finally, adherent patients reported greater improvements in minor symptoms [30]. Similarly, Rosenthal and associates [31] found that patients who regularly used CPAP during the first week of treatment continued using CPAP for the entire first year. Specifically, hours of use during the first week correlated with hours of use in the first year. By contrast, patients with mild OSA (between 5 and 25 respiratory events per hour) had a high rate of CPAP discontinuation. A marked improvement in daytime sleepiness was seen in patients who used CPAP for more than 4 hours per night [31].

A study of CPAP usage patterns found a bimodal distribution. Approximately half the patients studied were consistent CPAP users (more than 90% of nights with a mean ($\pm$SD) usage duration of 6.22 ($\pm$1.21) hours per night). The remainder were intermittent users with a wide range of nightly use (mean = 3.45 ± 1.94 hours per night). The percent of days skipped correlated with decreased nightly usage duration. Patterns of nightly use were discernable by the fourth treatment day [32]. To determine the consequences of intermittent CPAP use, Kribbs and colleagues [33] evaluated 15 patients with OSA (1) before CPAP treatment (pretreatment), (2) after 30–237 days post-treatment during a night of CPAP use (on CPAP), and (3) during a night without CPAP (off CPAP). CPAP effectively eliminated apneas and hypopneas. Significant reduction in both objective and subjective sleepiness were realized following CPAP treatment nights; however, sleeping without CPAP reversed most of the sleep and daytime alertness gains realized with CPAP [33].

In a large study of 1211 consecutive patients with OSA, 52 (4.5%) refused CPAP treatment (more often female and current smokers) while 1103 patients took CPAP home. Of those patients who took CPAP home, 20% discontinued CPAP citing lack of benefit as the reason. Sixty-eight percent of patients continued treatment 5 years or more. Factors influencing long-term use include snoring history, severity of illness (indexed with AHI), and self-reported sleepiness (measured with Epworth Sleepiness Scale (ESS)). Eighty-six percent of patients with ESS > 10 and an AHI $\geq$ 30 were still using CPAP at year three. As expected, average nightly CPAP use during the first 3 months predicted later use [34]. In another study, CPAP usage in patients who underwent split-night CPAP titration was 3.8 ± 2.9 hours/night compared to 5.2 ± 2.2 hours/night in patients who had traditional full-night titrations. Subjects were matched for age, sex, body mass index, sub-

jective sleepiness, and severity of OSA. The Epworth Sleepiness Scale on the initial clinic visit, however, did not predict usage at week 4–6 [35].

INCREASING CPAP USAGE

Adherence to CPAP therapy can be improved with education, airway humidification, proper selection of the CPAP interface, and prompt and aggressive management of adverse effects related to CPAP use. By reducing discomfort and immediately attending to difficulties, the clinician can remove barriers to regular use. Improving fit and feel of the mask will help optimize use. In one study, the most common complaints reported by patients on CPAP therapy were nocturnal awakenings (46% of patients) and nasal problems, such as dryness, congestion, and sneezing (44% of patients) [16].

CPAP Education

Eighty newly diagnosed patients with OSA were randomized to receive either usual support (standard) or more intensive support. The intensive support program included nursing CPAP education at home (that involved the patient's partner), a 3-night trial of CPAP in the sleep center, and additional home visits once CPAP therapy commenced. The group receiving intensive support had greater improvements over 6 months in OSA symptoms, mood, and reaction time compared to the group receiving standard support [36]. Thus providing CPAP education and a support program can improve functional outcomes. Participation in a group clinic designed to encourage CPAP use likewise reported improved CPAP adherence. Outpatient, 2 hour groups scheduled every 6 months provided education, support, symptom treatment, and equipment monitoring for patients with OSA. Patients attending the sessions had increased CPAP use that was sustained long term. Twenty-nine percent of patients increased nightly CPAP use by 2 hours or more [37]. Simple interventions can also improve CPAP utilization. Examples of simple interventions include (1) weekly phone calls to detect problems and encourage use and (2) providing written information about OSA and the importance of regular CPAP use. The interventions are particularly effective when they are provided during the first month of CPAP treatment [38]. By contrast, other investigators did not find education and support related improvements in CPAP usage. One hundred eight patients with OSA received either standard support (educational brochures on OSA and CPAP, CPAP education by nurses, CPAP acclimatization, and review by physicians and nurses at weeks 4 and 12) or intensive support (more education, including a videotape, telephone support by nurses, and early review at weeks 1 and 2). No group differences were found for CPAP use, ESS, or cognitive function [39].

Nasal Versus Full-Face Masks

Mouth leaks represent a fairly common situation leading to discomfort and dissatisfaction with CPAP. Chinstraps can sometimes correct this problem; however, at other times a full-face mask is needed. Additionally, full-face masks may help in patients who are obligate mouth breathers or who are unable to tolerate nasal masks. Full-face and nasal masks were compared in patients with OSA who had and had not previously undergone uvulopalatopharyngoplasty (UPPP). Nightly use was marginally greater for nasal masks, notwithstanding greater mouth leakage [40]. Overall, nasal masks appear to be preferred by most patients.

Humidifiers

A large very large percentage of individuals using CPAP complain of nasal symptoms. Nasal resistance is affected by both mouth leakage and the use of humidification. A study in six subjects using nasal CPAP demonstrated increased nasal resistance (measured with posterior rhinomanometry) after a deliberate mouth leak was created for 10 minutes. No change in nasal resistance occurred when subjects breathed through their noses; however, the mouth leak produced a large increase in resistance. Furthermore, the magnitude and duration of this mouth leak induced resistance were greatly attenuated by a heated humidifier. Using a cold pass-over humidifier provided little benefit [41]. In another study, temperature and relative humidity of the inspired air in the mask were measured either during a night's sleep or during a daytime study that simulated mouth leaks. Nasal masks (with or without humidification) or full-face masks were used. Heated humidification significantly increased the relative humidity of the inspired air both when the mouth was closed and during mouth leaks. By contrast, the full-face mask prevented changes in relative humidity both when subjects had their mouths closed and open [42].

Nasal dryness, rhinorrhea, nasal congestion, sneezing, or epistaxis are common problems in patients using CPAP therapy and each potentially adversely affects optimal utilization. Humidification can remedy these problems and consequently has become an important adjunct to CPAP therapy. Humidifying inspired air reduces CPAP-related drying of nasal passages [43]. CPAP utilization may be further enhanced when heated humidification is provided. In a randomized study using 38 patients in a crossover design, CPAP use and nasal symptoms were compared for heated humidification, cold pass-over humidification, and no humidification. CPAP use was greater with heated humidity than either cold pass-over or no humidity (which did not differ). Dry mouth, throat, and/or nose were less frequently reported when CPAP was used with heated humidity compared to no humidity. Although patient satisfaction was greater with either heated or cold pass-over humidity, only heated humidity improved feeling refreshed on awakening. Finally, no significant differences were found between groups for a global adverse side effect score [44]. A study attempting to predict who would need cold or heated humidification followed 82 patients with OSA for as long as 530 days. Factors increasing the possible need for adding a heated humidifier include age >60 years, drying medications, presence of chronic mucosa disease, and previous UPPP. Upper airway symptoms appeared to be the main factor for adding a cold pass-over humidifier to the airflow circuit. As in other studies, CPAP use increased with the heated but not the cold humidifier [45].

Nasal Lubrication

A variety of salves, nose drops, and oils are available that purportedly decrease upper airway dryness. In a comparison of heated humidifiers to oily nose drops (Colda-Stop, Desitin, Inc., Germany), 24 patients with OSA complaining of CPAP-related upper airway dryness were tested. Heated humidification reduced airway dryness more than oily nose drops. Every patient (100%) treated with heated humidification reported improvement in the degree and frequency of upper airway dryness and greater comfort when using CPAP. By contrast, 42% of patients treated with oily nose drops reported improvement in the degree of upper airway dryness and greater comfort when using CPAP. Only 25% reported improvement in the frequency of upper airway dryness. Patients intending to discontinue CPAP therapy due to dryness all continued therapy when provided heated humidification but did not when provided oily nose drops [46].

Pressure Ramp

Many CPAP machines come equipped with a "pressure ramp." This feature allows pressure to be reset to a low level and then gradually increased over time. The amount of time and rate of increase are usually adjustable. The pressure ramp is proposed to improve patient comfort by theoretically allowing a patient sufficient time to fall asleep before higher pressures are reached. Some patients find the ramp feature useful; however, others do not. Having lower pressures at sleep onset may permit obstructive events during subtherapeutic CPAP levels and allow sleep-onset central apnea episodes to occur. Moreover, repeated activation of the ramp can decrease effective CPAP therapy [47].

Using Sedative-Hypnotics

Sedative-hypnotics are sometimes used by clinicians in patients who have difficulty sleeping with CPAP. Sometimes they are used to assist patients during the acclimatization phase of initial CPAP use. Other times they are used when CPAP therapy successfully unmasks insomnia once the longstanding accumulated apnea-related sleepiness resolves. However, great caution should be taken whenever sedative-hypnotics are used for this purpose. For example, administering triazolam (0.25 mg, taken before bedtime) to 12 patients with severe OSA increased arousal threshold, thereby prolonging apnea duration and increasing oxyhemoglobin desaturations [48].

OTHER MODES OF POSITIVE AIRWAY PRESSURE

Autotitrating PAP

Optimal pressure for maintaining airway patency may vary significantly over the course of a single night. The progression through different sleep stages and variations associated with changes in body position more often than not produce differing pressure requirements. Additionally, night-to-night variability can be provoked by a host of factors. Using a fixed single pressure based on a sample usually obtained when the patient was sleeping supine during REM sleep (when pressure requirement is often maximal) may amount to using more pressure than needed for large portions of the night. In one study, 49.3% of home treatment time on APAP was spent at a pressure equal to or less than the effective pressure level determined during a polysomnographic recording [49]. The disadvantage of using more pressure than needed is that higher pressures increase the propensity of mask leaks, mouth leaks, and pressure intolerance. Any of these factors can reduce CPAP acceptance and therapeutic adherence. Another application of APAP technology involves using the device to identify a fixed single pressure for subsequent treatment with a conventional CPAP device (APAP titration). Several studies have compared APAP titration to conventional CPAP titration.

APAP efficacy in autoadjusting mode (APAP treatment) matches conventional CPAP. Twenty-five patients with OSA randomized to treatment with APAP and then CPAP, or vice versa, did not differ for AHI, awakening/arousal index, slow-wave sleep duration, nocturnal oxygen saturation, self-reported sleepiness, or nightly usage. The mean pressure required was significantly lower with APAP than with CPAP [50]. Similar results were found in a study of 20 patients with AHI greater than 15 episodes per hour of sleep. Following a baseline diagnostic polysomnographic evaluation, patients underwent a manually adjusted CPAP titration on one night and APAP titration on another

night. The pressure required to abolish apneas and CNS arousals was significantly lower during APAP titration than during conventional CPAP titration. No significant differences between the two titration approaches were found for AHI and arousal index reduction. Except for SaO_2 nadir being lower during APAP titration, no significant titration differences were found in any measure of oxygenation or sleep architecture. Finally, it should be noted that, during APAP titration, failure to increase pressure and failure to maintain minimum pressure occurred in 7 of the 20 subjects. The system self-corrected in two subjects but manual resetting of the system was required in five subjects [51]. Substituting APAP titration for conventional CPAP titration was also studied in 122 patients during their titration night in a hospital sleep laboratory. Subsequent CPAP acceptance by patients was not affected by APAP titration compared to conventional CPAP titration; the percentages of patients successfully established on CPAP at week six were 73% and 64% for the APAP- and CPAP-titration groups, respectively [52].

Two important issues face the clinician using APAP devices for titration and treatment. The first is the unknown level of equivalence in performance of devices from different manufacturers and different models of machines. Research outcomes with one APAP brand and model do not necessary generalize to other devices because each manufacturer has different computer algorithms controlling pressure adjustment. Thus certain APAP systems might be less useful in patients with mild to moderate OSA (e.g., see [53]). The second issue concerns using APAP titration to determine later fixed pressure or the use of APAP to replace standard laboratory titration. While some literature suggests that APAP holds promise, the level of evidence was not sufficient to support a standard of practice guideline. Thus using such an approach clinically is either (1) possibly substandard care or (2) experimental. If it is being used experimentally, ethical guidelines require informed consent. In either case, if an adverse event occurs during therapy (e.g., the patient falls asleep at the wheel and has a motor vehicle accident) the clinician is in a vulnerable position.

The Standards of Practice Committee of the American Academy of Sleep Medicine has developed practice parameters as a guide to the appropriate use of APAP. Recommendations include: (1) the presence of OSA must be established by an acceptable diagnostic method; (2) APAP titration and treatment are not currently recommended for patients with congestive heart failure, significant lung disease such as chronic obstructive pulmonary disease, daytime hypoxemia and respiratory failure from any cause, or prominent nocturnal arterial oxygen desaturation due to conditions other than OSA (e.g., obesity hypoventilation syndrome); (3) patients who do not snore (either due to palate surgery or naturally) should not be titrated with an APAP device that relies on vibration or sound in the device's algorithm; (4) APAP devices are not currently recommended for split-night studies; (5) certain APAP devices may be used during attended titration to identify by PSG a single pressure for use with standard CPAP for treatment of OSA; (6) once an initial successful attended CPAP or APAP titration has been determined by PSG, certain APAP devices may be used in the self-adjusting mode for unattended treatment of patients with OSA; (7) use of unattended APAP to either initially determine pressures for fixed CPAP or for self-adjusting APAP treatment in CPAP-naïve patients is not currently established; (8) patients being treated with fixed CPAP on the basis of APAP titration or being treated with APAP must be followed to determine treatment effectiveness and safety; and (9) a reevaluation and, if necessary, a standard attended CPAP titration should be performed if symptoms do not resolve or the CPAP or APAP treatment otherwise appears to lack efficacy [54]. These practice parameters are based on the evidence-based medicine review compiled by its task force [55]. The task force reviewed 30 pertinent articles published in peer review journals at that time.

Bilevel Positive Airway Pressure

Bilevel positive airway pressure (BPAP) devices provide two pressure levels, one during inhalation and a lower one during exhalation. Complaints of dyspnea or discomfort during CPAP use, especially during expiration (against the continuous pressure), may represent a barrier to CPAP adherence in some patients. To study whether decreased expiratory pressure would increase hours of nightly CPAP use, 83 patients with OSA were randomized to receive either CPAP or BPAP. Masks were fitted with sophisticated surveillance microprocessors to determine actual usage times. No treatment group differences were noted in usage duration and both groups had equivalent complaints regarding mask discomfort, machine noise, and nasal stuffiness [56].

Notwithstanding CPAP therapy, oxygen desaturation due to hypoventilation persists in some patients. Patients with OSA resistant to CPAP therapy (defined as a persistent AHI ≥ 5 or a mean nocturnal $Sao_2 < 90\%$) are often morbidly obese and have abnormal awake blood gas values (significantly lower Pao_2 and higher $Paco_2$). In one study, the percentage of time with $Sao_2 < 90\%$ has been independently associated with CPAP resistance. These patients improve when treated with BPAP in control mode. BPAP in control mode was adequate for nocturnal ventilation and improved awake blood gas values [57].

Noninvasive Ventilation Positive Pressure Ventilation

Nocturnal nasal noninvasive positive pressure ventilation (NIPPV) has been tried in patients with REM sleep hypo-

ventilation and CO_2 retention notwithstanding treatment with CPAP and supplemental oxygen. Piper and associates [58] reported their experience with NIPPV in 13 obese patients with severe OSA and hypercapnia who failed to respond initially to CPAP therapy. In their study, short-term NIPPV provided lasting benefits that permitted most patients to resume CPAP therapy. All patients tolerated NIPPV, which was provided using a volume-cycled ventilator. Significant improvements in daytime arterial blood gas values were achieved after 7–18 days of NIPPV. Sixty-nine percent of the patients were successfully placed on CPAP after this period. Three patients required a longer period (up to 3 months) to maintain adequate nocturnal ventilation. One patient required NIPPV long term. The investigators postulated that effective nasal ventilation leads to an overall improvement in spontaneous ventilation and blood gas values both awake and asleep as a result of improved central ventilatory drive [58].

CONCLUSION

Positive airway pressure is generally the preferred treatment for individuals with moderate or severe OSA. Positive pressure devices come in several varieties; however, continuous positive airway pressure (CPAP) is most commonly prescribed. CPAP improves airway patency during sleep, which in turn improves sleep quality, sleep continuity, daytime alertness, and overall quality of life in symptomatic patients with moderate or severe OSA. Further studies are needed to assess the benefits of CPAP therapy for patients with less severe OSA and to better specify positive cardiovascular outcomes. The effectiveness of CPAP is compromised because a large proportion of patients cannot tolerate or are nonadherent to regular use of the mask and machine. Adherence reportedly can be improved by increased mask comfort, patient education, heated humidification, and prompt correction of mask and machine related adverse events. Automatic self-adjusting positive pressure devices (APAP) show great promise; however, at present they are not recommended as a replacement for laboratory titration. Overall, CPAP is a safe and effective treatment for patients with OSA.

REFERENCES

1. Sullivan CE, Issa FG, Berthon-Jones M, Eves L. Reversal of obstructive sleep apnoea by continuous positive airway pressure applied through the nares. *Lancet* **1**(8225):862–865(1981).
2. Smith PL, Wise RA, Gold AR, Schwartz AR, Permutt S. Upper airway pressure–flow relationships in obstructive sleep apnea. *J Appl Physiol* **64**(2):789–795(1988).

3. Van de Graaff WB. Thoracic influence on upper airway patency. *J Appl Physiol* **65**(5):2124–2131(1988).

4. Van de Graaff WB. Thoracic traction on the trachea: mechanisms and magnitude. *J Appl Physiol* **70**(3):1328–1336(1991).

5. Hoffstein V, Zamel N, Phillipson EA. Lung volume dependence of pharyngeal cross-sectional area in patients with obstructive sleep apnea. *Am Rev Respir Dis* **130**(2):175–178(1984).

6. Jenkinson C, Davies RJ, Mullins R, Stradling JR. Long-term benefits in self-reported health status of nasal continuous positive airway pressure therapy for obstructive sleep apnoea. *QJM* **94**(2):95–99(2001).

7. Engleman HM, Martin SE, Deary IJ, Douglas NJ. Effect of CPAP therapy on daytime function in patients with mild sleep apnoea/hypopnoea syndrome. *Thorax* **52**(2):114–119(1997).

8. Engleman HM, Kingshott RN, Wraith PK, Mackay TW, Deary IJ, Douglas NJ. Randomized placebo-controlled crossover trial of continuous positive airway pressure for mild sleep apnea/hypopnea syndrome. *Am J Respir Crit Care Med* **159**(2):461–467(1999).

9. Redline S, Adams N, Strauss ME, Roebuck T, Winters M, Rosenberg C. Improvement of mild sleep-disordered breathing with CPAP compared with conservative therapy. *Am J Respir Crit Care Med* **157**(3 Pt 1):858–865(1998).

10. Loredo JS, Ancoli-Israel S, Dimsdale JE. Effect of continuous positive airway pressure vs placebo continuous positive airway pressure on sleep quality in obstructive sleep apnea. *Chest* **116**(6):1545–1549(1999).

11. Ballester E, Badia JR, Hernandez L, Carrasco E, de Pablo J, Fornas C, Rodriguez-Roisin R, Montserrat JM. Evidence of the effectiveness of continuous positive airway pressure in the treatment of sleep apnea/hypopnea syndrome. *Am J Respir Crit Care Med* **159**(2):495–501(1999).

12. Jenkinson C, Davies RJ, Mullins R, Stradling JR. Comparison of therapeutic and subtherapeutic nasal continuous positive airway pressure for obstructive sleep apnoea: a randomised prospective parallel trial. *Lancet* **353**(9170):2100–2105(1999).

13. Hack M, Davies RJ, Mullins R, Choi SJ, Ramdassingh-Dow S, Jenkinson C, Stradling JR. Randomised prospective parallel trial of therapeutic versus subtherapeutic nasal continuous positive airway pressure on simulated steering performance in patients with obstructive sleep apnoea. *Thorax* **55**(3):224–231(2000).

14. Kingshott RN, Vennelle M, Hoy CJ, Engleman HM, Deary IJ, Douglas NJ. Predictors of improvements in daytime function outcomes with CPAP therapy. *Am J Respir Crit Care Med* **161**(3 Pt 1):866–871(2000).

15. Barnes M, Houston D, Worsnop CJ, Neill AM, Mykytyn IJ, Kay A, Trinder J, Saunders NA, Douglas McEvoy R, Pierce RJ. A randomized controlled trial of continuous positive airway pressure in mild obstructive sleep apnea. *Am J Respir Crit Care Med* **165**(6):773–780(2002).

16. Hoffstein V, Viner S, Mateika S, Conway J. Treatment of obstructive sleep apnea with nasal continuous positive airway pressure. Patient compliance, perception of benefits, and side effects. *Am Rev Respir Dis* **145**(4 Pt 1):841–845(1992).

17. Bahammam A, Delaive K, Ronald J, Manfreda J, Roos L, Kryger MH. Health care utilization in males with obstructive sleep apnea syndrome two years after diagnosis and treatment. *Sleep* **22**(6):740–747(1999).

18. Lavie P, Herer P, Hoffstein V. Obstructive sleep apnoea syndrome as a risk factor for hypertension: population study. *BMJ* **320**(7233):479–482(2000).

19. Suzuki M, Otsuka K, Guilleminault C. Long-term nasal continuous positive airway pressure administration can normalize hypertension in obstructive sleep apnea patients. *Sleep* **16**(6):545–549(1993).

20. Faccenda JF, Mackay TW, Boon NA, Douglas NJ. Randomized placebo-controlled trial of continuous positive airway pressure on blood pressure in the sleep apnea–hypopnea syndrome. *Am J Respir Crit Care Med* **163**(2):344–348(2001).

21. Mayer J, Becker H, Brandenburg U, Penzel T, Peter JH, von Wichert P. Blood pressure and sleep apnea: results of long-term nasal continuous positive airway pressure therapy. *Cardiology* **79**(2):84–92(1991).

22. Akashiba T, Minemura H, Yamamoto H, Kosaka N, Saito O, Horie T. Nasal continuous positive airway pressure changes blood pressure "non-dippers" to "dippers" in patients with obstructive sleep apnea. *Sleep* **22**(7):849–853(1999).

23. Dimsdale JE, Loredo JS, Profant J. Effect of continuous positive airway pressure on blood pressure: a placebo trial. *Hypertension* **35**(1 Pt 1):144–147(2000).

24. Rauscher H, Formanek D, Popp W, Zwick H. Nasal CPAP and weight loss in hypertensive patients with obstructive sleep apnoea. *Thorax* **48**(5):529–533(1993).

25. Naughton MT, Bradley TD. Sleep apnea in congestive heart failure. *Clin Chest Med* **19**(1):99–113(1998).

26. He J, Kryger MH, Zorick FJ, Conway W, Roth T. Mortality and apnea index in obstructive sleep apnea. Experience in 385 male patients. *Chest* **94**(1):9–14(1988).

27. Wright J, Johns R, Watt I, Melville A, Sheldon T. Health effects of obstructive sleep apnoea and the effectiveness of continuous positive airway pressure: a systematic review of the research evidence. *BMJ* **314**(7084):851–860(1997).

28. Fleury B, Rakotonanahary D, Hausser-Hauw C, Lebeau B, Guilleminault C. Objective patient compliance in long-term use of nCPAP. *Eur Respir J* **9**(11):2356–2359(1996).

29. Kribbs NB, Pack AI, Kline LR, Smith PL, Schwartz AR, Schubert NM, Redline S, Henry JN, Getsy JE, Dinges DF. Objective measurement of patterns of nasal CPAP use by patients with obstructive sleep apnea. *Am Rev Respir Dis* **147**(4):887–895(1993).

30. Pepin JL, Krieger J, Rodenstein D, Cornette A, Sforza E, Delguste P, Deschaux C, Grillier V, Levy P. Effective compliance during the first 3 months of continuous positive airway pressure. A European prospective study of 121 patients. *Am J Respir Crit Care Med* **160**(4):1124–1129(1999).

31. Rosenthal L, Gerhardstein R, Lumley A, Guido P, Day R, Syron ML, Roth T. CPAP therapy in patients with mild OSA: implementation and treatment outcome. *Sleep Med* **1**(3):215–220(2000).

32. Weaver TE, Kribbs NB, Pack AI, Kline LR, Chugh DK, Maislin G, Smith PL, Schwartz AR, Schubert NM, Gillen KA, Dinges DF. Night-to-night variability in CPAP use over the first three months of treatment. *Sleep* **20**(4):278–283(1997).

33. Kribbs NB, Pack AI, Kline LR, Getsy JE, Schuett JS, Henry JN, Maislin G, Dinges DF. Effects of one night without nasal CPAP treatment on sleep and sleepiness in patients with obstructive sleep apnea. *Am Rev Respir Dis* **147**(5):1162–1168(1993).

34. McArdle N, Devereux G, Heidarnejad H, Engleman HM, Mackay TW, Douglas NJ. Long-term use of CPAP therapy for sleep apnea/hypopnea syndrome. *Am J Respir Crit Care Med* **159**(4 Pt 1):1108–1114(1999).

35. Strollo PJ Jr, Sanders MH, Costantino JP, Walsh SK, Stiller RA, Atwood CW Jr. Split-night studies for the diagnosis and treatment of sleep-disordered breathing. *Sleep* **19**(10 Suppl):S255–S259(1996).

36. Hoy CJ, Vennelle M, Kingshott RN, Engleman HM, Douglas NJ. Can intensive support improve continuous positive airway pressure use in patients with the sleep apnea/hypopnea syndrome? *Am J Respir Crit Care Med* **159**(4 Pt 1):1096–1100(1999).

37. Likar LL, Panciera TM, Erickson AD, Rounds S. Group education sessions and compliance with nasal CPAP therapy. *Chest* **111**(5):1273–1277(1997).

38. Chervin RD, Theut S, Bassetti C, Aldrich MS. Compliance with nasal CPAP can be improved by simple interventions. *Sleep* **20**(4):284–289(1997).

39. Hui DS, Chan JK, Choy DK, Ko FW, Li TS, Leung RC, Lai CK. Effects of augmented continuous positive airway pressure education and support on compliance and outcome in a Chinese population. *Chest* **117**(5):1410–1416(2000).

40. Mortimore IL, Whittle AT, Douglas NJ. Comparison of nose and face mask CPAP therapy for sleep apnoea. *Thorax* **53**(4):290–292(1998).

41. Richards GN, Cistulli PA, Ungar RG, Berthon-Jones M, Sullivan CE. Mouth leak with nasal continuous positive airway pressure increases nasal airway resistance. *Am J Respir Crit Care Med* **154**(1):182–186(1996).

42. Martins De Araujo MT, Vieira SB, Vasquez EC, Fleury B. Heated humidification of face mask to prevent upper airway dryness during continuous positive airway pressure therapy. *Chest* **117**(1):142–147(2000).

43. Brown LK. Back to basics: if it's dry, wet it: the case for humidification of nasal continuous positive airway pressure air. *Chest* **117**(3):617–619(2000).

44. Massie CA, Hart RW, Peralez K, Richards GN. Effects of humidification on nasal symptoms and compliance in sleep apnea patients using continuous positive airway pressure. *Chest* **116**(2):403–408(1999).

45. Rakotonanahary D, Pelletier-Fleury N, Gagnadoux F, Fleury B. Predictive factors for the need for additional humidification during nasal continuous positive airway pressure therapy. *Chest* **119**(2):460–465(2001).

46. Wiest GH, Lehnert G, Bruck WM, Meyer M, Hahn EG, Ficker JH. A heated humidifier reduces upper airway dryness during continuous positive airway pressure therapy. *Respir Med* **93**(1):21–26(1999).

47. Pressman MR, Peterson DD, Meyer TJ, Harkins JP, Gurijala L. Ramp abuse. A novel form of patient noncompliance to administration of nasal continuous positive airway pressure for treatment of obstructive sleep apnea. *Am J Respir Crit Care Med* **151**(5):1632–1634(1995).

48. Berry RB, Kouchi K, Bower J, Prosise G, Light RW. Triazolam in patients with obstructive sleep apnea. *Am J Respir Crit Care Med* **151**(2 Pt 1):450–454(1995).

49. Meurice JC, Marc I, Series F. Efficacy of auto-CPAP in the treatment of obstructive sleep apnea/hypopnea syndrome. *Am J Respir Crit Care Med* **153**(2):794–798(1996).

50. d'Ortho MP, Grillier-Lanoir V, Levy P, Goldenberg F, Corriger E, Harf A, Lofaso F. Constant vs. automatic continuous positive airway pressure therapy: home evaluation. *Chest* **118**(4):1010–1017(2000).

51. Sharma S, Wali S, Pouliot Z, Peters M, Neufeld H, Kryger M. Treatment of obstructive sleep apnea with a self-titrating continuous positive airway pressure (CPAP) system. *Sleep* **19**(6):497–501(1996).

52. Stradling JR, Barbour C, Pitson DJ, Davies RJ. Automatic nasal continuous positive airway pressure titration in the laboratory: patient outcomes. *Thorax* **52**(1):72–75(1997).

53. Fleury B, Rakotonanahary D, Hausser-Hauw C, Lebeau B, Guilleminault C. A laboratory validation study of the diagnostic mode of the Autoset system for sleep-related respiratory disorders. *Sleep* **19**(6):502–505(1996).

54. Standards of Practice Committee of the American Academy of Sleep Medicine. Practice Parameters for the Use of Auto-Titrating Continuous Positive Airway Pressure Devices for Titrating Pressures and Treating Adult Patients with Obstructive Sleep Apnea Syndrome. An American Academy of Sleep Medicine Report. *Sleep* **25**(2):143–147(2002).

55. Berry RB, Parish JM, Hartse KM. The use of auto-titrating continuous positive airway pressure for treatment of adult obstructive sleep apnea. An American Academy of Sleep Medicine review. *Sleep* **25**(2):148–173(2002).

56. Reeves-Hoche MK, Hudgel DW, Meck R, Witteman R, Ross A, Zwillich CW. Continuous versus bilevel positive airway pressure for obstructive sleep apnea. *Am J Respir Crit Care Med* **151**(2 Pt 1):443–449(1995).

57. Schafer H, Ewig S, Hasper E, Luderitz B. Failure of CPAP therapy in obstructive sleep apnoea syndrome: predictive factors and treatment with bilevel-positive airway pressure. *Respir Med* **92**(2):208–215(1998).

58. Piper AJ, Sullivan CE. Effects of short-term NIPPV in the treatment of patients with severe obstructive sleep apnea and hypercapnia. *Chest* **105**(2):434–440(1994).

46

UPPER AIRWAY SURGERY FOR OBSTRUCTIVE SLEEP APNEA

AARON E. SHER

Medical Director, Capital Region Sleep/Wake Disorders Center of St. Peter's Hospital
Capital Region Otolaryngology – Head & Neck Surgery Group
Clinical Associate Professor of Surgery and Pediatrics, Albany Medical College, Albany, New York

Upper airway surgery for obstructive sleep apnea syndrome (OSAS) attempts to modify dysfunctional pharyngeal anatomy or bypass the pharynx (tracheotomy). Some modifications of the pharynx diminish the bulk of soft tissue structures that abut the air column, whereas others place them under tension or alter their spatial interrelationships. The former goal is achieved through ablating pharyngeal soft tissue. The latter goals are accomplished by modifying the facial skeleton from which the soft tissues are suspended.

In fewer than 2% of adult OSAS patients with specific space-occupying pathological lesions (such as benign or malignant neoplasms, inflammatory lesions, lesions resulting in nasal obstruction), surgical correction may achieve the desired results. However, the vast majority of adult patients do not demonstrate clearly identifiable lesions. It is conceptualized that, in this group, "disproportionate anatomy" of the upper airway results from unfavorable anatomic features of the surrounding soft tissue structures and underlying maxillomandibular skeleton. It is the disproportion that is believed to predispose to OSAS. The configuration of the pharyngeal lumen is determined by the size and shape of soft tissue structures that abut it (such as faucial and lingual tonsils, tongue, soft palate) and their spatial orientation in relation to each other. The spatial orientation is determined by the orientation of underlying muscle planes established through their origins and insertions in the vertebral and craniofacial skeleton. The latter will depend, ultimately, on the craniofacial skeletal characteristics of the patient [1].

The pharynx has properties associated with a collapsible biological conduit. Collapse occurs at a discrete (<1 cm) locus. Data derived from endoscopic, fluoroscopic, and computed tomographic (CT) scan procedures, and manometric evaluation awake, asleep, and narcotized suggest that the pattern of static pharyngeal narrowing and/or dynamic pharyngeal collapse is localized and patient specific, although it may shift with body position, sleep state, and following upper airway surgery [2].

Failure of surgical procedures aimed at a limited locus within the pharynx is believed to result from residual or secondary airway compromise at a remote locus not addressed. The pharynx is conceptually divided into two functional loci: retropalatal (located posterior to the soft palate) and retrolingual (located posterior to the vertical portion of the tongue). The pharynx is preoperatively classified into three categories: Type I (only the retropalatal region is compromised), Type II (both retropalatal and retrolingual regions are compromised), and Type III (only the retrolingual region is compromised). While a number of awake and asleep methods of endoscopic, radiometric, and manometric identification have been applied clinically in defining these three patterns of airway collapsibility, awake lateral cephalometry and endoscopy are currently the most commonly applied methods of pharyngeal classification (Table 46.1)[3, 4].

Classical surgical techniques to enlarge the nasal or pharyngeal airway (e.g., nasal septal reconstruction, turbinate mucosal cauterization, turbinate outfracture, and

TABLE 46.1   Surgical Procedures: Mechanism and Primary Locus of Action

Mechanism of Action	Locus of Primary Action	Procedure
Soft tissue ablation	Retropalatal	Uvulopalatopharyngoplasty (UPPP), laser-assisted uvulopalatoplasty (LAUP)
	Retrolingual	Laser midline glossectomy/lingualplasty (LMG), radiofrequency tongue base ablation (RFTBA), tongue base reduction with hyoepiglottoplasty (TBRHE)
	Retropalatal and retrolingual	Uvulopalatopharyngoglossoplasty (UPPGP)
Skeletal modification (soft tissue repositioning)	Retropalatal	Transpalatal advancement pharyngoplasty (TPAP)
	Retrolingual	Mandibular advancement (MA), genioglossal advancement (GA), hyoid myotomy and suspension of hyoid from mandible (HM-1), hyoid myotomy and attachment of hyoid to thyroid cartilage (HM-2)
	Retropalatal and retrolingual	Maxillomandibular advancement (MMA)
Bypass upper airway		Tracheotomy

tonsillectomy) frequently fail to correct OSAS in adults. On the other hand, tonsillectomy and/or adenoidectomy has a high success rate in children with OSAS who have demonstrably enlarged tonsils and adenoids (and in the unusual adult who has OSAS and markedly hypertrophic tonsils and/or adenoids). Arising from frequent inadequacy of these procedures to address adult OSAS, new surgical approaches were developed to ablate pharyngeal soft tissue or modify the position of pharyngeal soft tissue structures through mobilization and repositioning of underlying skeletal structures (Table 46.1)[4].

Following are brief descriptions of some of these surgical procedures [5–17].

Uvulopalatopharyngoplasty (UPPP). Enlarges the retropalatal airway by excision of the tonsils (if present), trimming and reorientation of the posterior and anterior tonsillar pillars, and excision of the uvula and posterior portion of the soft palate (Figure 46.1).

Laser-Assisted Uvulopalatoplasty (LAUP). Enlarges the retropalatal airway by ablation of the uvula and posterior margin of the soft palate with carbon dioxide laser. While tonsil ablation can be accomplished with the laser, LAUP as reported frequently does not include tonsil ablation. LAUP is generally performed under topical and local anesthesia in the physician's office (Figure 46.2).

Uvulopalatopharyngo-glossoplasty (UPPGP). Combines UPPP with limited resection of the tongue base, enlarging retropalatal and retrolingual portions of the airway.

Laser Midline Glossectomy (LMG) and Lingualplasty. Enlarges the retrolingual airway by laser extirpation of a 2.5 cm by 5 cm midline rectangular strip of posterior tongue and, in lingualplasty, additional lateral wedges. Laser lingual tonsillectomy, reduction of the aryepiglottic folds, and partial epiglottectomy are performed in selected patients (Figure 46.3).

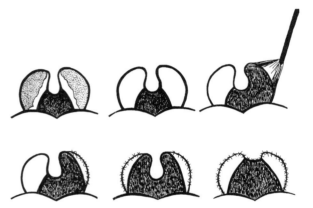

Figure 46.1   Uvulopalatopharyngoplasty (UPPP). Surgical steps are depicted from left to right, top row followed by bottom row. Tonsils are removed. Posterior tonsillar pillars are divided vertically from uvula to level of upper pole of tonsillar fossae, rotated across the fossae, and sutured to trimmed anterior tonsillar pillars. Sutures may tack the posterior pillars to the midportion of the fossa. Uvula and posterior soft palate are transected at approximately the upper pole of the tonsillar fossae.

Radiofrequency Tongue Base Ablation (RFTBA). Enlarges the retrolingual airway by applying radiofrequency energy to the tongue base with a needle electrode. Radiofrequency tongue base ablation is

Figure 46.2   Laser-assisted uvulopalatoplasty (LAUP). The posterior tonsillar pillars are surgically ablated, as are the uvula and posterior portion of the soft palate.

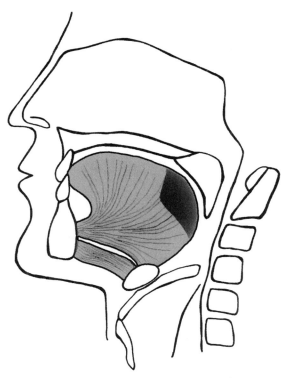

Figure 46.3 Laser midline glossectomy (LMG). A midline rectangular strip of posterior tongue is extirpated.

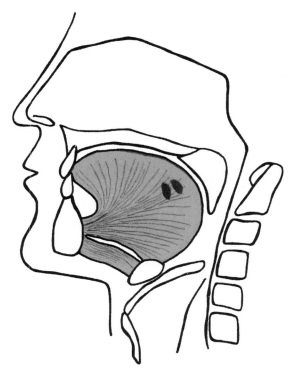

Figure 46.4 Radiofrequency tongue base ablation (RFTBA). Two lesions have been created in the muscle mass of the tongue using a radiofrequency needle electrode. Insulation of the proximal portion of the needle maintains the integrity of the mucosa.

generally performed under topical and local anesthesia in the physician's office (Figure 46.4).

Tongue Base Reduction with Hyoepiglottoplasty (TBRHE). Enlarges the retrolingual airway by excision of tongue base. The excision is performed through a transcervical lesion. The neurovascular bundle is identified and protected. The hyoid is suspended from the mandible under tension. Unlike LMG, lingualplasty, and RFTBA, which are performed transorally, TBRHE is performed through a cervical incision.

Transpalatal Advancement Pharyngoplasty (TPAP). Enlarges the retropalatal airway by resection of the posterior hard palate with advancement of the soft palate in an anterior direction into the defect.

Mandibular Advancement (MA). Enlarges the retrolingual airway utilizing sagittal mandibular osteotomies to effect anterior mobilization of the insertion of the tongue at the genial tubercle. There must be significant antecedent mandibular deficiency and dental malocclusion to permit the requisite degree of anterior movement of the mandible and mandibular teeth (Figure 46.5).

Maxillomandibular Advancement (MMA). Provides maximal enlargement of the retrolingual airway and some

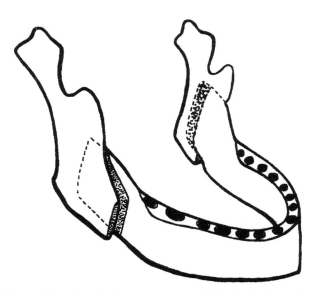

Figure 46.5 Mandibular advancement (MA). Sagittal mandibular osteotomies permit anterior mobilization of the anterior mandible in cases with antecedent mandibular deficiency and malocclusion.

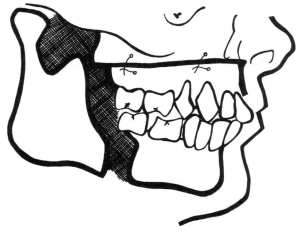

Figure 46.6 Maxillomandibular advancement (MMA). The maxilla and mandible are advanced by Le Fort I maxillary and sagittal-split mandibular osteotomies.

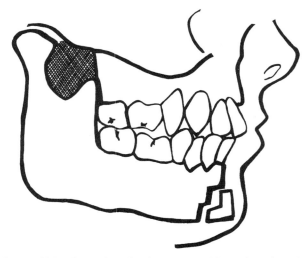

Figure 46.7 Genioglossal advancement (GA) (lateral projection). Limited parasagittal mandibular osteotomy permits anterior advancement of the genial tubercle.

enlargement of the retropalatal airway. It permits significant mandibular advancement in patients lacking maxillomandibular disproportion. The maxilla and mandible are both advanced by means of Le Fort I maxillary and sagittal-split mandibular osteotomies. The degree of mandibular advancement achieved, if performed without maxillary advancement, would result in mandibular prognathism and dental malocclusion. The exception would be the patient with severe mandibular deficiency but normal maxillary development (who would then be a candidate for MA rather than MMA). Details of MMA depend on the patient's dental occlusion (Figure 46.6).

Genioglossal Advancement (GA). Places the tongue under traction. It does so without altering dental occlusion and is achieved by performing limited parasagittal mandibular osteotomy and anterior advancement of the genial tubercle (Figures 46.7 and 46.8). This procedure has undergone various modifications over time.

Hyoid Myotomy and Suspension (Two Variations, HM-1 and HM-2). Tends to enlarge the retrolingual airway and exerts anterior traction on the tongue, hyoid, and suprahyoid musculature, with release of the infrahyoid muscles. Two techniques have been described: (1) suspension of the hyoid from the mandible by a fascial strip (HM-1) and (2) suspension of the hyoid from the superior margin of the thyroid cartilage by permanent suture (HM-2) (Figures 46.9 and 46.10).

Tracheotomy. Creates a percutaneous opening into the trachea. The tracheotomy is usually stented and maintained by inserting a rigid or semirigid hollow tube.

The patient breathes through the tube when the external end is unplugged. Since the tracheotomy opens into the airway proximal to the pharynx, it bypasses the region of collapse. When the patient is awake, the external end of the tube is plugged. The tube is of sufficiently small diameter that, when plugged, it permits air to enter the trachea from the pharynx and larynx, passing around the tube. The increased resistance to airflow created by the presence of the plugged tube occupying a portion of the tracheal lumen can be decreased by using a tube that has distal ventilating

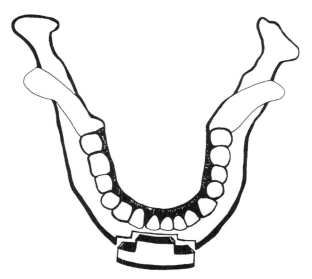

Figure 46.8 Genioglossal advancement (GA) (viewed from above). Limited parasagittal mandibular osteotomy permits anterior advancement of the genial tubercle.

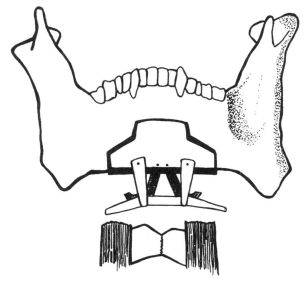

Figure 46.9 Hyoid myotomy and suspension, Type 1 (HM-1). The infrahyoid muscles are released and the hyoid is suspended from the mandible by a fascial strip.

holes through which air can flow. Alternatively, it may be possible to fashion a tracheotomy that does not require a cannula, or utilize a stomal plug instead of cannula.

Surgical outcomes can be defined in terms of improvement or lack of improvement in a variety of physiological and/or behavioral parameters. Generally, criteria for surgical success have been defined by arbitrary (investigator

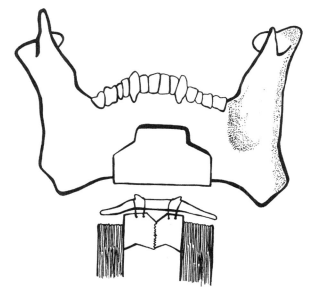

Figure 46.10 Hyoid myotomy and suspension, Type 2 (HM-2). The infrahyoid muscles are released and the hyoid is suspended from the superior margin of the thyroid cartilage by permanent sutures.

designated) degrees of improvement in polysomnography (PSG) measured variables of respiratory physiology (Table 46.2). In more recent publications, criteria of success incorporate data on improvement in PSG measures of sleep architecture. Most recent has been the inclusion of preoperative and postoperative outcome data utilizing standardized instruments, which measure quality of life. Limited agreement between degree of improvement in PSG measures and degree of improvement in subjective quality of life measures is sometimes cited (with subjective outcomes frequently more favorable than objective physiological outcomes). This discrepancy is the subjective of considerable current interest. Placebo effect is cited as potentially contributory, but possible other contributing factors remain uncertain.

In general, there is a wide range of individual response to surgical procedures for OSAS: from resolution, to amelioration, to lack of improvement, to worsening of the severity of OSAS. The percentage of patients achieving "success" depends on the specific criterion selected to define success. The more restrictive the criterion, the lower tends to be the success rate. Thus a higher success rate is likely to result if the definition of success is 50% reduction from preoperative to postoperative RDI (respiratory disturbance index, or apnea–hypopnea index) than if the definition of success is 50% reduction from preoperative to postoperative RDI *and postoperative RDI fewer than 20 apneas and hypopneas per hour*. On the other hand, the rate of success appears to depend also on the degree to which patients are preselected for specific surgical procedures by techniques that attempt to define anatomical or functional airway properties believed to impact favorably or unfavorably on surgical impact (such as Type I, II, or III classification). For example, the rate of success of UPPP (determined by the following PSG criteria: postoperative RDI decreased by at least 50% from preoperative RDI to a level below 20 apneas and hypopneas per hour *or* postoperative AI decreased by at least 50% from preoperative AI to a level below 10 apneas per hour) is 55% when UPPP is applied to patients with Type I classification of the pharynx, whereas it is 5% when applied to patients with Type II or III classification. The greater likelihood of success with UPPP in Type I patients is felt to reflect the high degree of localized anatomical impact of UPPP only on the retropalatal airway (ablating tonsils and posterior soft palate). As noted in Table 46.2, other operations have been developed to address the retrolingual airway differentially. Procedures that address the retropalatal airway and procedures that address the retrolingual airway are frequently applied in the same patient (contemporaneously or sequentially) if the patient is classified as having Type II pharyngeal anatomy. The composite outcome of the combination of procedures is shown to surpass that of the individual procedures in such patients [2, 4].

TABLE 46.2 Criteria Defining Surgical Success

Criterion	Definition		
1A	$\frac{\text{Postop RDI}}{\text{Preop RDI}} \leq 50\%$		
1B	$\frac{\text{Postop A1}}{\text{Preop A1}} \leq 50\%$		
2A	$\frac{\text{Postop RDI}}{\text{Preop RDI}} \leq 50\%$	*and*	Postop RDI $\leq$ 20 apneas and hypopneas per hour
2B	$\frac{\text{Postop RDI}}{\text{Preop RDI}} \leq 50\%$	*and*	Postop RDI $\leq$ 15 apneas and hypopneas per hour
2C	$\frac{\text{Postop RDI}}{\text{Preop RDI}} \leq 50\%$	*and*	Postop RDI $\leq$ 10 apneas and hypopneas per hour
2D	$\frac{\text{Postop A1}}{\text{Preop A1}} \leq 50\%$	*and*	Postop A1 $\leq$ 10 apneas per hour
3	$\frac{\text{Postop RDI}}{\text{Preop RDI}} \leq 40\%$	*and*	$\frac{\text{Postop A1}}{\text{Preop A1}} \leq 40\%$ *and* $<$ Postop A1 $<$ 10 apneas per hour *or* Postop RDI $<$ 15 apneas and hypopneas per hour and Postop A1 $<$ 15 apneas per hour

Measurement of airway collapsibility demonstrates the impact of upper airway surgery for OSAS and the incremental impact of multiple surgical modifications. A model of the upper airway likens it to a simple collapsible tube. The tendency of the upper airway to collapse can be expressed quantitatively in terms of a critical pressure (P_{crit}), which is the pressure surrounding the area of collapse. If atmospheric pressure is designated zero, then airway collapse will occur whenever P_{crit} is a positive number (indicating that it is higher than atmospheric pressure). P_{crit} levels are higher during sleep than during wakefulness in both normal individuals and OSAS patients. In normals, P_{crit} rises from awake values that are more negative than -41 cm H_2O to asleep values of -13 cm H_2O. The means that, in normals, atmospheric pressure is greater than P_{crit} even during sleep, and the pharynx will not collapse. In OSAS patients, the spectrum of awake values of P_{crit} is -40 to -17 cm H_2O, and P_{crit} during sleep is $+2.5$ cm H_2O. Although the pharyngeal airway of awake OSAS patients tends to be more collapsible than that of awake normals, P_{crit} does not cross the critical line of zero (i.e., atmospheric pressure) except when the individual with OSAS has sleep onset, and OSAS results. Patients who have varying degrees of partial pharyngeal collapse have intermediate, but negative levels of P_{crit} during sleep: -6.5 cm H_2O for asymptomatic snorers and -1.6 cm H_2O for patients with hypopneas but no apneas. In general, P_{crit} must be below -5 cm H_2O to eliminate sleep disordered breathing. P_{crit} for an OSAS patient can, alternatively, be defined as the lowest level of nasal CPAP at which airflow is maintained.

Examples of the decrement in P_{crit} that can be achieved by nonsurgical interventions are -6 cm H_2O through the loss of 15% of body weight; -3 to -4 cm H_2O through protriptyline treatment; and -4 to -5 cm H_2O through the avoidance of sleeping in the supine position. P_{crit} decrement in response to upper airway surgery has been documented for UPPP and TPAP. When 13 patients undergo UPPP, P_{crit} decreases from a level of 0 to a level of -3 cm H_2O ($P = 0.016$). In those patients who have greater than 50% decrease in RDI in NREM sleep, P_{crit} decreases from -1 to -7 cm H_2O ($P = 0.01$). The degree of improvement in sleep disordered breathing is correlated significantly with the change of P_{crit} ($P = 0.001$), and the decrease in RDI is determined by the magnitude of the decrease in P_{crit} rather than by the initial level of P_{crit}. No significant change in P_{crit} is detected in nonresponders. Sequential performance of TPAP after UPPP results in incremental decrease in P_{crit} to a level below that resulting from UPPP. Four patients underwent TPAP after previous UPPP. Mean post-UPPP P_{crit} of 5 is decreased after TPAP to -4 cm H_2O ($P < 0.01$). TPAP increases the retropalatal airway cross-sectional area by 321% compared to the post-UPPP cross-sectional area (29–95 cm^2, $P < 0.01$). It is likely that each surgical procedure decreasing OSAS severity also decreases P_{crit}, and that procedures performed concomitantly or in sequence result in incremental decreases in P_{crit}. It is suggested that the application of P_{crit} as an indirect measure of OSAS treatment efficacy might prove valuable in patient selection, treatment selection, and treatment evaluation [14, 18, 19].

In an attempt to achieve the maximum degree of improvement in each individual patient while minimizing the extensiveness of surgical intervention (e.g., degree of invasiveness of procedures applied and number of procedures applied), protocols have been established by which surgery is staged in planned sequential phases, the nature of the procedures applied in each phase determined by the anatomical classification (Type I, II, or III). The Stanford protocol is widely applied in this manner. If the patient does not achieve a level of improvement designated as "success" after the first phase of surgery (usually documented by PSG six months after surgery), he/she is offered a second attempt at surgical intervention (involving one or more additional surgical interventions). The procedure(s) offered in the second phase may (or may not, depending on circumstances) be surgically more aggressive than those performed in the first phase. The key factor is that procedures done in the second phase are chosen to be anatomically and functionally complementary to those done in the first phase. In general (with exceptions), phase I surgery consists of UPPP and/or GA and/or HM, depending on specific circumstances. In general (with exceptions), phase II surgery consists of MMA. The Stanford protocol results in a "successful" outcome after phase I surgery in about 50% of patients (higher or lower, depending on initial severity of OSAS). The remainder are offered a second phase of additional surgical intervention with ultimate "success" achieved in 75–100% of cases [20–22].

There is limited evidence that surgical improvement after some procedures (i.e., UPPP) may wane at sequential follow-up evaluations performed at increasing lapsed time since surgery. In some cases (though not all) increasing body weight during the lapsed time period is cited as the explanation for deteriorating outcome. Limitation of data precludes definitive generalization regarding long-term surgical effectiveness or lack thereof [23–25].

In general, subjective claims of improvement in sleep quality, degree of daytime sleepiness and its consequences, and other aspects of quality of life exceed the documented degree of improvement in PSG respiratory and sleep architecture parameters. While placebo effect of surgery is likely at least a partial explanation for this discrepancy, the degree to which other unidentified factors contribute remains unknown.

Some patients seek surgical therapy for snoring rather than for sleep apnea or its sequelae (excessive daytime sleepiness, cognitive deficiency, cardiovascular morbidity). The complaint may relate only to a social/domestic issue: bed-partner leaves the bed in response to sleep-disruptive snoring. The patient may offer no classical OSA related complaints except socially disruptive snoring. Surgical elimination of obstructive lesions is appropriate in such cases, and correction of nasal obstruction may result in amelioration of snoring. However, there is most often no clearly identifiable obstructing pathology. Surgical procedures frequently applied in this clinical scenario include LAUP and radiofrequency palate ablation (RFPA). Both procedures (which alter palate structure) have proved applicable for the treatment of snoring in the absence of OSA, as both can be performed readily in the physician's office under topical and local anesthesia. Lack of requirement for hospitalization and general anesthesia makes it easier for patients to afford the cost of these procedures without third party coverage (which generally does not recognize snoring as adequate indication for surgical intervention and, therefore, does not reimburse for snoring remedies). However, in such cases, lack of third party coverage for diagnostic preoperative PSG (snoring with no other symptoms of OSAS is generally not regarded as adequate indication for diagnostic PSG) creates a dilemma

for patient and surgeon alike. Many patients who seek surgery for snoring prove (if studied by PSG) to have significant OSAS. Satisfactory amelioration of snoring by LAUP and RFPA is reported in 65–90% of cases, although the data does not suggest that either of these procedures is adequate for significant OSAS. Therefore, if one of these procedures converts a snoring OSAS patient to a nonsnoring OSAS patient, the inadequate surgical response will remain unidentified unless the correct diagnosis of OSAS was preoperatively established by PSG [26, 27].

As opposed to LAUP and RFPA (generally applied for snoring), surgical procedures for OSAS are usually performed under general anesthesia in the operating room. The length of hospital stay and the intensity of postoperative in-hospital monitoring required once the patient is released from the postanesthesia care unit (PACU or recovery room) remain controversial. Early postoperative concerns include adequacy of analgesia, ventilation, and hydration. Considerable throat pain and edema may result from surgery for OSA. OSAS patients have intrinsically compromised upper airways. The surgery may or may not adequately correct the functional airway compromise. Edema induced by the surgery may exacerbate preexisting airway compromise not adequately relieved by the surgical procedure, thereby exacerbating postoperative OSAS. Drugs used to induce and maintain general anesthesia may continue to depress ventilation in the early postoperative period. Postoperative analgesics may depress ventilation and predispose to OSAS. In response to these concerns, some advocate in-patient observation for one or more nights, with utilization of intensive physiological monitoring in the early postoperative period. The duration and degree of intensity of monitoring would depend on preoperative OSAS severity, extent of surgical intervention, and patient comorbidity. On the other hand, fiscal constraints mandate allocation of such resources to situations requiring them. The most severe complications of upper airway surgery for OSAS (including death) may result from inadequate postoperative ventilation. Application of nasal CPAP (particularly if the patient has been preoperatively conditioned to use it) may diminish this risk. In some cases, prophylactic tracheotomy is performed to assure adequacy of airway management in the early postoperative period [28–31].

REFERENCES

1. Sher AE. Obstructive sleep apnea syndrome: a complex disorder of the upper airway. *Otolaryngol Clin North Am* **92**:593–608(1990).

2. Sher AE, Schechtman KB, Piccirillo JF. The efficacy of surgical modifications of the upper airway in adults with obstructive sleep apnea syndrome. *Sleep* **19**:156–177(1996).

3. Fujita S. Midline laser glossectomy with lingualplasty: a treatment of sleep apnea syndrome. *Op Tech Otolaryngol Head Neck Surg* **2**:127–131(1991).

4. Sher AE. Upper airway surgery for obstructive sleep apnea. *Sleep Med Rev* **6**:195–212(2002).

5. Chabolle F, Wagner I, Blumen M. Tongue base reduction with hyoepiglottoplasty: a treatment for severe obstructive apnea. *Laryngoscope* **109**:1273–1280(1999).

6. Fujita S, Conway W, Zorick F. Surgical correction of anatomical abnormalities in obstructive sleep apnea syndrome: uvulopalatopharyngoplasty. *Otolaryngol Head Neck Surg* **89**:923–924(1981).

7. Kamami Y-V. Out-patient treatment of sleep apnea syndrome with CO_2 laser: laser-assisted UPPP. *J Otolaryngol* **23**:395–399(1994).

8. Kuo PC, West RA, Bloomquist DS, McNeill RW. The effect of mandibular osteotomy in three patients with hypersomnia sleep apnea. *Oral Surg Oral Medi Oral Pathol* **48**:385–392(1979).

9. Miljeteig H, Tvinnereim M. Uvulopalatopharyngoglossoplasty (UPPGP) in the treatment of the obstructive sleep apnea syndrome. *Acta Otolaryngol (Stockh)* **492**(Suppl):86–89(1992).

10. Riley RW, Powell NB, Guilleminault C. Inferior mandibular osteotomy and hyoid myotomy suspension for obstructive sleep apnea: a review of 55 patients. *J Oral Maxillofac Surg* **47**:159–164(1989).

11. Riley RW, Powell NB, Guilleminault C. Maxillofacial surgery and nasal CPAP. A comparison of treatment for obstructive sleep apnea syndrome. *Chest* **98**:1421–1425(1990).

12. Riley RW, Powell NB, Guilleminault C. Obstructive sleep apnea and the hyoid. A revised surgical procedure. *Otolaryngol Head Neck Surg* **111**:717–721(1994).

13. Weitzman ED, Kahn E, Pollak CP. Quantitative analysis of sleep and sleep apnea before and after tracheostomy in patients with the hypersomnia–sleep apnea syndrome. *Sleep* **2**:407–423(1980).

14. Winakur SJ, Smith PL, Schwartz AR. Pathophysiology and risk factors for obstructive sleep apnea. *Semin Respir Crit Care Med* **19**:999–1012(1998).

15. Woodson BT, Fujita S. Clinical experience with lingualplasty as part of the treatment of severe obstructive sleep apnea. *Otolaryngol Head Neck Surg* **107**:40–48(1992).

16. Woodson BT, Toohill RJ. Transpalatal advancement pharyngoplasty for obstructive sleep apnea. *Laryngoscope* **103**:269–276(1993).

17. Woodson BT, Nelson L, Mickelson S. A multi-institutional study of radiofrequency volumetric tissue reduction for OSAS. *Otolaryngol Head Neck Surg* **125**:303–318(2001).

18. Schwartz AR, Schiebert N, Rothman W. Effect of uvulopalatopharyngoplasty on upper airway collapsibility in obstructive sleep apnea. *Am Rev Respir Dis* **145**:527–532(1992).

19. Woodson BT. Retropalatal airway characteristics in uvulopalatopharyngoplasty compared with transpalatal advancement pharyngoplasty. *Laryngoscope* **107**:735–740(1997).

20. Bettega G, Pepin J-L, Veale D. Obstructive sleep apnea syndrome: fifty-one consecutive patients treated by maxillofacial surgery. *Am J Respir Crit Care Med* **162**:641–649(2000).

21. Prinsell JR. Maxillomandibular advancement surgery in a site-specific treatment approach for obstructive sleep apnea in 50 consecutive patients. *Chest* **116**:1519–1529(1999).

22. Riley RW, Powell NB, Guilleminault C. Obstructive sleep apnea syndrome: a review of 306 consecutively treated surgical patients. *Otolaryngol Head Neck Surg* **108**:117–125(1993).

23. Janson C, Gislason T, Bengtsson H. Long-term follow-up of patients with obstructive sleep apnea treated with uvulopalatopharyngoplasty. *Arch Otolaryngol Head Neck Surg* **123**:257–262(1997).

24. Larsson LH, Carlsson-Nordlander B, Svanborg E. Four-year follow-up after uvulopalatopharyngoplasty in 50 unselected patients with obstructive sleep apnea syndrome. *Laryngoscope* **104**:1362–1368(1994).

25. Lu S-J, Chang S-Y, Shiao G-M. Comparison between short-term and long-term post-operative evaluation of sleep apnoea after uvulopalatopharyngoplasty. *J Laryngol Otol* **109**:308–312(1995).

26. Littner M, Kushida C, Hartse K, McDowell A, Davila D, Johnson S, Wise M, Hirshkowitz M, Woodson BT. Practice parameters for the use of laser-assisted uvulopalatoplasty: an update for 2000. *Sleep* **24**:603–619(2001).

27. Sher AE, Flexon PB, Hillman D, Emery B, Swieca J, Smith TL, Cartwright R, Dierks E, Nelson L. Temperature-controlled radiofrequency tissue volume reduction in the human soft palate. *Otolaryngol Head Neck Surg* **125**:312–318(2001).

28. Esclamado RM, Glenn MG, McCulloch TM. Perioperative complications and risk factors in the surgical treatment of obstructive sleep apnea syndrome. *Laryngoscope* **99**:1125–1129(1989).

29. Haavisto L, Suonpaa J. Complications of uvulopalatopharyngoplasty. *Clin Otolaryngol* **19**:243–247(1994).

30. Mickelson SA, Hakim I. Is postoperative intensive care monitoring necessary after uvulopalatopharyngoplasty? *Otolaryngol Head Neck Surg* **119**:352–356(1998).

31. Riley RW, Powell NB, Guilleminault C, Pelayo R, Troell RJ, Li KI. Obstructive sleep apnea surgery: risk management and complications. *Otolaryngol Head Neck Surg* **117**:648–652(1997).

47

ORAL DEVICES THERAPY FOR OBSTRUCTIVE SLEEP APNEA

DENNIS R. BAILEY

Englewood, Colorado

INTRODUCTION

Oral devices (oral appliances) (ODs) are becoming recognized as a reasonable means by which sleep-related breathing disorders (SRBDs) may be managed. The indications for the use of these devices is for those patients who have mild to moderate sleep apnea, only snore, and have been diagnosed with apnea, or where apnea has been ruled out and for patients who are intolerant to CPAP or may have had surgery and that has also been unsuccessful. These devices have an impact on the airway by virtue of their ability to improve airway space as well as tongue space through repositioning of the mandible both downward (open) and forward. In the past 5–10 years the use of these devices has steadily increased and their effectiveness has also improved. The future use of ODs will only become more prevalent as the number of patients who are diagnosed with apnea and SRBDs grows and the need for treatment options likewise increases.

EVOLUTION OF ORAL DEVICES FOR MANAGEMENT OF SLEEP APNEA

OD therapy for the management of SRBDs began in the late 1970s with the development of the tongue retaining device (TRD) by Dr. Samelson, who had a loud snoring problem and wanted to find a solution for it [1]. He developed the TRD, a device that was fabricated out of a flexible material and had a bulb-like receptacle at the front that would securely hold the tongue forward during sleep. The general principle was predicated on the belief that it was the tongue that would collapse back into the airway during sleep and would obstruct the airway. With the tongue held forward, the airway was now opened, allowing for the free flow of air during sleep and the opportunity for obstruction of the airway was alleviated as was the snoring.

In the 1980s there was the growing use of a type of appliance known as the functional appliance (FA) or activator in orthodontics. This appliance was designed to fit both the upper and lower teeth for the purpose of repositioning the mandible to enhance and promote lower jaw growth. The FA was a one-piece design often referred to as a monobloc appliance because it was one piece for the upper and lower teeth together that actually stabilized the mandible into an open and forward position with the result being lower jaw growth and improvement of the malocclusion. A side benefit of these devices was the improvement in the patient's sleep pattern, an improvement in their ability to breathe, and a reduction in their snoring when this had preceded the use of the appliance. In the 1960s and 1970s researchers such as Linder-Aronson in Sweden and Woodside in Canada had demonstrated the effect of the FA on mandibular growth as well as the improvement in the airway [2]. This FA design became well known and evolved into the first OD designed to cause mandibular repositioning for the management of SRBD. One of the early designs patterned after the FA was the nocturnal airway patency appliance or NAPA developed by Dr. Peter George [3]. This was a one-piece OD that was found to be quite

Sleep: A Comprehensive Handbook, Edited by T. Lee-Chiong.
Copyright © 2006 John Wiley & Sons, Inc.

successful which was secured to the upper and lower teeth and repositioned the mandible. It was designed such that it separated the teeth a minimal amount, usually 2–5 millimeters, and advanced the mandible forward. Other appliances in the 1980s were introduced and were similar in design and concept. All of them were a single OD or monobloc design that stabilized the mandible in a predetermined position much like the NAPA.

In the 1990s alternate appliance designs became available and the evolutionary change at this time was the creation of an OD that was in two separate pieces, one for the upper and one for the lower teeth, that were attached in some fashion and allowed for movement of the mandible during sleep to varying degrees. The significant difference in the ODs was the manner by which the upper and lower components were joined and to what degree the mandible could move. In addition, the use of an OD that is in two pieces allows for a greater degree of ability to adjust and modify the appliance with the intent of achieving the optimum effect on the airway and hence the SRBD. The treating practitioner now has the ability to create a change in the jaw position to address the needs of the patient for both comfort as well as optimum effectiveness.

Another related condition that the two-piece OD design has is the ability to address sleep bruxism as well as the SRBD. It has been shown that there is a significant correlation between sleep bruxism and SRBD [4]. Many of the ODs have support designed and built into the posterior aspect of the appliance so in those patients who also are known to have bruxism the appliance will jointly manage both conditions.

For the OD to be of optimum function the qualities of the appliance should adhere to some specific traits in its design. The criteria that follow are recommended to maximize the efficacy of the appliance, to reduce side effects, and to allow for changes when necessary as the management of the SRBD continues.

1. *Adjustability.* This allows for modification of the OD as needed in specific situations such as:

- If dental work is done and the appliance no longer fits over the teeth. The OD can be adjusted (modified) or it can be relined to adapt to the treatment that was done, such as a crown or large restoration (filling).
- If the OD becomes loose-fitting, it can be relined to improve retention.
- If a tooth is lost and that area needs to be sealed or made more secure.

2. *Titrateability.* This relates to the future need to alter the vertical opening of the OD or if more or less advancement is required. Unlike CPAP, it is not practical to titrate

for effectiveness of the OD during the sleep study. This needs to be done over time as the patient becomes accustomed to using the appliance and changes are indicated to improve the results. These changes are of two forms:

- To affect the degree of advancement it may be necessary to modify or otherwise alter the mechanism that is responsible for this movement. This can range from the need to alter an elastic type force to the adjustment of a screw.
- To affect the vertical opening the most customary manner by which this is done is by the adjustment of the posterior pads. Frantz demonstrated in a case report that when the pads were increased in thickness, this facilitated an increase in the vertical opening, which resulted in a significant improvement in the effectiveness of the OD [5].

3. *Full Tooth Coverage.* It is important that all of the existing teeth be covered by the OD in the upper as well as the lower portion. This prevents the potential for undesirable tooth movement when the appliance is in use. Tooth movement may occur despite the best attempts to prevent it but when it has been found to result from use of the OD it is often minimally significant [6, 7].

4. *Posterior Support.* Posterior support is customarily flat pads in the upper and lower components that contact when the appliance is in place. This support is intended to:

- Stabilize and support the temporomandibular joint (TMJ) with the appliance in place. It has been reported that the occurrence of TMJ discomfort/dysfunction with the use of an OD is not a major concern and does not impact treatment of the sleep apnea [8].
- The support in the posterior portion allows the appliance to function much like a bite splint does for the management of sleep for sleep bruxism in those patients who have this as a concurrent finding.

5. *Jaw Mobility.* This allows for mandibular movement during sleep. Patients, even if they do not have sleep bruxism, experience jaw movement during sleep in the form of swallowing or licking the lips. In the 1980s those patients who had a one-piece OD (monobloc) often were unable to tolerate the OD because of jaw pain associated with the inability to have mandibular movement regardless of whether they had bruxism.

6. *Patent Nasal Passages.* It is essential that the patient be able to breathe through the nose when the appliance is in place and to have minimal nasal resistance. This improves their oxygen levels during sleep, has been shown to reduce mouth breathing, may be beneficial in reducing the RDI, and thus improves the level of success with the appliance. The

nasal airway may also be improved with the use of nasal dilation strips and in some cases the use of nasal sprays such as saline or in more severe cases prescription medications [9, 10].

Statement about improved nasal airway when dorsum of tongue allows soft palate to drop nasal resistance. Trial of Jaw Ripo effect on nose.

INDICATIONS FOR ORAL DEVICES

In 1995 the Standards of Practice Committee of the American Sleep Disorders Association (currently the American Academy of Sleep Medicine or AASM) published the practice parameters as well as a review of the use of ODs for patients with sleep apnea [11, 12].

The generally accepted indications are:

1. For patients where sleep apnea has been ruled out and the major concern is the snoring, which may be more disruptive to the bedpartner than the snorer.
2. For patients diagnosed with mild sleep apnea where weight loss and/or positional therapy is not an option.
3. For patients with moderate to severe sleep apnea who are intolerant of CPAP therapy or refuse this as a means of managing their sleep apnea.
4. For patients who have failed surgery, are not candidates for surgery, or refused surgery.

CONTRAINDICATIONS FOR ORAL DEVICES

The guidelines from the AASM Standards of Practice Committee as cited also published situations in which ODs are not indicated for use. They are:

1. In patients who are diagnosed with only central sleep apnea.
2. In patients who are compromised dentally. This would include people with loose teeth that have inadequate support, with teeth that are diseased or are otherwise compromised, with broken teeth or where there is an inadequate number of teeth to support the OD. The dental status of the patient needs to be evaluated thoroughly prior to the fabrication of the appliance.
3. In patients who have TMJ dysfunction or other types of orofacial complaints. In many cases the OD may also be helpful in managing the TMJ dysfunction and needs to be considered accordingly. Additionally, the practitioner who is providing the appliance should be capable of addressing a TMJ situation should it arise during the treatment. It has been found that sleep position may be a factor in causing a TMJ disc disor-

der. One study demonstrated that sleep position affected the ipsilateral joint and had no effect on the contralateral joint [13].

An additional contraindication to the use of the OD is when the patient is unable to nose breathe. This should be evaluated and addressed if it is the case prior to the fabrication of the appliance and the patient should have the proper medical care to attempt to improve this situation.

AVAILABLE ORAL DEVICES FOR MANAGEMENT OF SLEEP APNEA

Currently it is estimated that there are over 40 different devices available and there have been estimates that there are more than 50. The most popular and frequently utilized ODs are those that have received clearance from the Federal Drug Administration (FDA) for the management of snoring and sleep apnea. In April of 2002 the FDA revised the classification for ODs from class I to class II, indicating that these devices now have special controls [14]. This change indicates that these devices are viewed as regulated medical devices and as such reasonably assure their safety as well as a reasonable level of effectiveness. Not all of the appliances that are available for the management of snoring are also cleared for sleep apnea. Table 47.1 is a listing of the some of the appliances that are currently available and is by no means comprehensive and does not represent ODs that have not been reviewed or received clearance by the FDA but are being marketed nor does it include appliances that are one of a kind and are typically made by a dentist or a dental laboratory usually for a specific patient. This table indicates the current status of the appliances from the FDA for snoring and/or sleep apnea.

Presently the FDA recognizes ODs in three distinct categories: tongue retaining devices (TRDs), mandibular repositioners, and palatal lifting devices. At this time the mandibular repositioners are the most commonly utilized devices. The TRD is utilized by some practitioners in select situations such as those with a compromised dentition and in edentulous patients. The palatal lifting devices are not used at all at this time but remain as one of the categories because of the FDA guidelines.

ORAL DEVICES FOR SPECIAL CIRCUMSTANCES

There are two distinct patient types that require the design and fabrication of more custom type ODs. These are children or adolescents who have sleep apnea and edentulous patients. These two patient groups are not good candidates for an oral appliance such as the ones listed in Table 47.1. On occasion the TRD may be an effective alternative but many patients do not tolerate this on a long-term basis.

TABLE 47.1   Oral Devices Utilized for the Management of SRBD in Sleep Medicine

Appliance (Common Name)	FDA Clearance For	
	Snoring	Sleep Apnea
Elastic mandibular advancement (EMA)	Yes	Yes
Herbst	Yes	Yes
Hilson adjustable appliance	Yes	No
Klearway	Yes	Yes
Nocturnal oral airway dilator (NORAD)	Yes	Yes
Nocturnal airway patency appliance (NAPA)	Yes	Yes
OASYS oral airway system	Yes	Yes
PM positioner (adjustable)	Yes	Yes
Silencer	Yes	Yes
Silent night	Yes	No
Snoreguard	Yes	No
Therasnore	Yes	No
Thornton anterior positioner (TAP appliance)	Yes	Yes
Tongue retaining device (TRD)	Yes	No
Tongue stabilizing device	Yes	Yes

In the adolescent population the use of the OD is usually a design that is similar to the monobloc appliance that is used to address a malocclusion and to promote jaw growth. These types of appliances are the FA or activator appliance as discussed previously. They typically will need to be remade over time as the individual grows to address the changes in the dentition and the occlusion. As the adolescent patient acquires all of his/her adult teeth, the possibility of utilizing one of the FDA recognized devices becomes possible. This is an area that needs to be explored more and will no doubt receive more attention in the years to come.

In the edentulous patient a form of dentures without teeth with support in the posterior only may be utilized. The posterior aspect has flat surfaces that contact to support the mandible in a vertical position to improve the tongue space and to better support the airway. An alternate design is to bevel the posterior surfaces of these dentures so that the mandible is supported in the vertical dimension but also has a tendency to be advanced when the posterior surfaces come in contact.

At times the patient who has a SRBD may also experience improvement in the airway by simply wearing the dentures while sleeping. Many patients have been advised to remove their dentures during sleep to allow the tissues to rest. However, this may lead to a compromise of the airway and worsening of the apnea. Two studies have demonstrated that when the dentures are worn during sleep the number of apnea events is significantly reduced and the blood oxygen levels also can improve [15, 16]. If improvement by simply wearing the present dentures during sleep is not satisfactory, then a specific OD can be fabricated in an effort to reduce the apnea and improve the blood oxygen levels during sleep.

THE FUNCTION OF ORAL DEVICES

The primary function of ODs is to open and improve the airway during sleep by repositioning the mandible and the tongue to prevent these structures from collapsing into the airway during sleep. A number of theories have been proposed regarding the exact mechanism for this but in the end a clear understanding of this action is not currently available. For the OD to be successful, the lateral dimension of the airway is the critical area where improvement needs to occur.

In 1985 White et al. [17] found that the airway of the apnea patient during sleep was narrower than the airway of those who did not have sleep apnea. Schwab [18] has utilized imaging techniques to demonstrate that the lateral walls of the airway have a significant role in the improvement of the airway in patients with sleep apnea. Another study utilized cine computerized tomography to determine the effect on the airway of the OD that mainly protruded the jaw [19]. With apnea the measurements at the retropalatal and retroglossal areas decreased the most. With the appliance in place these two areas showed improvement; the lateral aspects of the retropalatal and retroglossal levels of the pharynx were affected more than in the sagittal plane.

EFFECTIVENESS OF ORAL DEVICES

Over the last ten years many studies have looked at the use of ODs to manage SRBD and the results are steadily improving. In 1999 Schmidt-Nowara [20] examined the use of ODs and reported that these devices should no longer be viewed as "experimental" and that the role of the OD has been demonstrated adequately.

There are numerous studies that could be reviewed and cited here. One that is of interest was done as a crossover study and compared a device that advanced the mandible to an oral plate that did not advance the mandible and acted as a control [21]. The outcome was that of the 28 patients who used the advancement appliance 96% had subjective improvement. The oral plate had no impact on the RDI or the blood oxygen levels. The overall success rate for patients who had advancement was stated as 62.5% and included individuals with moderate and severe apnea.

In 2000 Mehata et al. [22] published results regarding the elastic mandibular advancement appliance (EMA), indicating that it was effective for the management of sleep apnea. Overall, patients who had a RDI of 52.6 ± 28.2 prior to the use of the EMA had a RDI of 21.9 ± 19.3 with the appliance.

In 1996 Henke et al. [23] reported that when 23 of 29 patients had a follow-up sleep study with the use of a fixed position (monobloc) mandibular advancement appliance there was improvement. The pre-appliance RDI went from 37 ± 23 to a RDI of 18 ± 20 with this device. The study indicated that 16 of 23 (69%) of the patients were considered to be responders because the RDI was reduced by 50% or more and was less than 20. In addition, these same 16 patients continued to use the appliance for 3 years or more.

In 2001 Marklund et al. [24] reported that out of 33 patients with a RDI of less than 10 a satisfactory outcome was seen in 19 of the patients. At about 5 years the RDI decreased from 22 ± 17 to 4.9 ± 5.1, demonstrating that with continued use and ongoing adjustment or replacement as needed there was continued success. The study also reported that with long-term follow-up improved outcomes and success were evident. In 2004 Marklund et al. [25] published a study that was designed to help determine treatment success. The outcome of the study determined that ODs are best recommended for women with sleep apnea, men who have supine-dependent sleep apnea, and snorers who do not have apnea.

Comparison of ODs to CPAP. A number of studies have compared ODs to CPAP [26–28]. The general consensus is that CPAP is more effective when and if it is adequately tolerated. These studies further indicated that ODs were preferred by more patients when compared to CPAP and in some instances there was improved compliance for a greater length of time.

Comparison of ODs to Surgery. It has been shown that ODs have an improved outcome over surgery and are a reasonable alternative in the management of sleep apnea when surgery has failed [29, 30]. It may be prudent to utilize the OD prior to surgery because it is less invasive and is reversible.

DETERMINING POTENTIAL EFFECTIVENESS OF ORAL DEVICES

It is important to be able to determine the potential effectiveness for the OD prior to its use. In some instances the patient can simply open the mouth so that there is 5 mm or so between the upper and lower front teeth and advance the mandible to determine if it is easier to breathe. At the same time they may have more difficulty making a snoring sound or cannot make one at all.

A study utilizing a flexible fiberoptic endoscope demonstrated that the principal area of obstruction was at the tongue base and during inspiration there was additional obstruction in the area of the epiglottis [31]. With a jaw-thrust maneuver these two areas improved and the obstruction was significantly reduced, further giving credibility to testing the patient's subjective perception of breathing with mandibular repositioning.

Another option that is growing in popularity is the use of sound waves to acoustically image the airway (pharyngometry) [32]. The sound waves are projected into the airway through a soft elastic-like mouthpiece and once they contact the structures in the airway they are reflected back where a computer is then able to interpret the anatomical location and determine the area of narrowing. Then, with the mandible repositioned, the same test is done again to determine if there is improvement in the airway and specifically if the site of narrowing is improved. This technology can be helpful in determining the optimum position for the mandible when utilizing an OD.

SIDE EFFECTS AND THEIR MANAGEMENT

Use of the OD may cause some side effects for the user, most often at the early stage of the treatment. The most common side effects and their management are:

1. *Excess Salivation.* This often is present at the initiation of the treatment and usually resolves within the first week, once the patient is accustomed to the device. An adequate lip seal with the appliance in place can help to prevent this.
2. *Dry Mouth.* This is usually related to ongoing mouth breathing. Having a lip seal with the appliance in place and being able to nose breathe generally resolve this.
3. *Tooth Movement.* Many times any tooth movement that occurs is minor in nature and usually does not pose a long-term problem [6]. Often this is associated with mandibular repositioning that is comprised of mostly advancement. If the repositioning is more in the vertical plane, open this potentially is less likely

to occur. Also, many times patients feel as if there is tooth movement when in fact they have experienced minor bite changes.

4. *Bite Changes*. These changes are most common in the first half-hour or less upon removal of the OD in the morning. The changes are usually subtle and resolve spontaneously. If they persist, there are exercises that can be done, and the use of a mild muscle relaxer can be helpful. Medication, if required, is usually only necessary for a short period of time until the patient is accustomed to the appliance. Regardless-nud to be followed on an ongoing basis.

5. *Jaw Pain and/or TMJ Discomfort*. This may occur in the early stages of the use of the OD. Many times this is due to strain on the muscles of the head and neck and will resolve over time. Often the pain from the muscles is felt in the area of the TMJ due to pain referral patterns [33]. Exercises, muscle relaxer medications, and at times physical therapy may be indicated to manage the pain. Actual joint problems are rarely of concern as reported earlier. Specially compounded transdermal medications that are applied on the skin in the area of the pain may also be helpful on a short-term or even a long-term basis.

CONCLUSION

The use of an OD to manage mild to moderate sleep apnea and SRBD is becoming more common, is a more widely accepted alternative to CPAP or surgery, and is gaining in patient acceptance as well as demand. As more research is done and as these appliances become more sophisticated, their impact on the management of sleep apnea will continue to improve. The means by which ODs can be most successful is with continued monitoring over time as the patient's sleep patterns change and/or alterations in life style occur so that the OD can be modified accordingly.

REFERENCES

1. Cartwright R, Samelson C. The effects of a nonsurgical treatment for obstructive sleep apnea. *JAMA* **248**:705–709(1982).

2. McNamarra JA (Eds). *Naso-Respiratory Function and Craniofacial Growth*. Center for Human Growth and Development, The University of Michigan, Monograph Number 9. Michigan University Press, Ann Arbor, MI, 1979, pp 27–40.

3. George PT. A modified functional appliance for treatment of obstructive sleep apnea. *J Clin Orthod* **XXI**(3):171–175(1987).

4. Ohayaon M, Li K, Guilleminault C. Risk factors for sleep bruxism in the general population. *Chest* **119**:53–61(2001).

5. Frantz D. The difference between success and failure. *Sleep Rev* **2**:20–23(2001).

6. Fransoon A, Tegelberg A, Svenson X, Lennartsson B, Isacsson G. Influence of mandibular protruding device on airway passages and dentofacial characteristics in obstructive sleep apnea and snoring. *Am J Orthod Dentofacial Orthop* **122**:371–379(2002).

7. Rose E, Staats R, Virchow C, Jonas I. Occlusal and skeletal effects of an oral appliance in the treatment of obstructive sleep apnea. *Chest* **122**:871–877(2002).

8. Ribeiro De Alameida F, Bittncourt LR, Ribeiro de Almeida CI, Tsuiki S, Lowe AA, Tufik S. Effects of mandibular posture on obstructive sleep apnea severity and the temporomandibular joint in patients fitted with an oral appliance. *Sleep* **25**(5):507–513(2002).

9. McLean HA, Urton AM, Driver HS, Tan AK, Day AG, Munt PW, Fitzpatrick MF. Effect of treating severe nasal obstruction on the severity of obstructive sleep apnea. *Eur Respir J* **25**(3):521–527(2005).

10. Li HY, Wang PC, Hsu CY, Liou CC, Chen NH. Nasal resistance in patients with obstructive sleep apnea. *Otorhinolaryngol Relat Spec* **67**(2):70–74(2005).

11. Schmidt-Nowara W, Lowe A, Wiegand L, Cartwright R, Perez-Guerra F, Menn S. Oral appliances for the treatment of snoring and obstructive sleep apnea: a review. *Sleep* **18**:501–510(1995).

12. Schmidt-Nowara W, Lowe A, Wiegand L, Cartwright R, Perez-Guerra F, Menn S. Practice parameters for the treatment of snoring and obstructive sleep apnea with oral appliances. *Sleep* **18**:511–513(1995).

13. Hibi H, Ueda M. Body posture during sleep and disc displacement in the temporomandibular joint: a pilot study. *J Oral Rehabil* **32**(2):85–89(2005).

14. Center for Devices and Radiological Health, U.S. Food and Drug Administration. Class II Special Controls Guidance Document: Intraoral Devices for Snoring and/or Obstructive Sleep Apnea: Guidance for Industry and FDA. Bulletin Nov 12, 2002.

15. Bucca C, Carossa S, Pivwtti S, Gai V, Rolla G, Preti G. Edentulism and worsening of obstructive sleep apnea. *Lancet* **353**:121(1999).

16. Endeshaw YW, Katz S, Ouslander JG, Bliwise DL. Association of denture use with sleep-disordered breathing among older adults. *J Public Health Dent* **64**(3):181–183(2004).

17. White D, et al. Pharyngeal resistance in normal humans: influence of gender, age and obesity. *J Appl Physiol* **58**:365–371(1985).

18. Schwab R. Imaging for the snoring and sleep apnea patient. In: Attanasio R, Bailey D (Eds), *Sleep Disorders: Dentistry's Role. Dental Clinics of North America*. Saunders, Philadelphia, 2001, pp 759–796.

19. Kyung SH, Park YC, Pae EK. Obstructive sleep apnea patients with the oral appliance experience pharyngeal size and shape changes in three dimensions. *Angle Orthod* **75**(1):15–22(2005).

20. Schmidt-Nowara W. Recent developments in oral appliance therapy of sleep disordered breathing. *Sleep Breath* **3**:103–106(1999).

21. Mehata M, Qian J, Petocz P, Darendeliler, Cistulli P. A randomized, controlled study of a mandibular advancement splint for obstructive sleep apnea. *Am J Respir Crit Care Med* **163**:1457–1461(2001).

22. Mehata M, Frantz D, Kuna S. An oral elastic mandibular advancement device for obstructive sleep apnea. *Am J Respir Crit Care Med* **161**:420–425(2000).

23. Henke K, Loube D, Morgan T, Mitler M, Berger J, Erman M. The mandibular repositioning device: role in the treatment of obstructive sleep apnea. *Sleep* **19**:794–800(1996).

24. Marklund M, Sahlin C, Stenlund H, Persson M, Franklin K. Mandibular advancement device in patients with obstructive sleep apnea. *Chest* **120**:162–169(2001).

25. Marklund M, Stenlund H, Franklin KA. Mandibular advancement devices in 630 men and women with obstructive sleep apnea and snoring. *Chest* **125**:1270–1278(2004).

26. Tan Y, L'estrange P, Luo Y, et al. Mandibular advancement splints and continuous positive airway pressure in patients with obstructive sleep apnea: a randomized cross-over trial. *Eur J Orthod* **24**:239–249(2002).

27. Fergusson K, Ono T, Lowe A, Keenan S, Fleetham J. A randomized crossover study of an oral appliance vs nasal-continuous positive airway pressure in the treatment of mild–moderate obstructive sleep apnea. *Chest* **109**:1269–1275(1996).

28. Clark G, Blumenfeld I, Yoffe N, Peled E, Lavie P. A crossover study of continuous positive airway pressure with anterior mandibular positioning devices on patients with obstructive sleep apnea. *Chest* **109**:1477–1483(1996).

29. Millaman R, Rosenberg C, Carlisle C, Kramer N, Kahn D, Bonitati A. The efficacy of oral appliances in the treatment of persistent sleep apnea after uvulopalatopharyngoplasty. *Chest* **113**:992–996(1998).

30. Wilhelmsson B, Teglberg A, Walker-Engstrom M, et al. A prospective randomized study of a dental appliance compared with uvulopalatopharyngoplasty in the treatment of obstructive sleep apnea. *Acta Otolaryngol* **119**:503–509(1999).

31. Uzun L, Ugur MB, Altunkaya H, Ozer Y, Ozkocak I, Demirel CB. Effectiveness of the jaw-thrust maneuver in opening the airway: a flexible fiberoptic endoscopic study. *J Otorhinolaryngol Relat Spec* **67**(1):39–44(2005).

32. Shepard JW, Gefter WB, Guilleminault C, Hoffman E, Hoofstein V, Hdgel DW, Suratt PM, White DP. Evaluation of the upper airway in patients with obstructive sleep apnea. *Sleep* **14**(4):361–371(1991).

33. Travel J, Simons D. *Myofascial Pain and Dysfunction: The Trigger Point Manual.* Williams & Wilkins, Baltimore, 1983, pp 183–281 and pp 305–320.

PART V

CIRCADIAN RHYTHM
SLEEP DISORDERS

48

ADVANCED, DELAYED, IRREGULAR, AND FREE-RUNNING SLEEP–WAKE DISORDERS

YARON DAGAN, KATY BORODKIN

Sheba Medical Center, Tel Hashomer, Israel

LIAT AYALON[*]

University of California San Diego and Veterans Affairs San Diego Healthcare System, San Diego, California, USA

INTRODUCTION

In humans, sleep and wake episodes occur at regular times that match the 24 hour day–night cycle. This temporal organization is known as circadian rhythmicity (from the Latin *circa*, "about", *dies*, "day"), which is present in many behavioral and physiological functions, such as fluctuations of body temperature and melatonin secretion. The circadian timing system is thought to play a central role in the generation, maintenance and synchronization of circadian rhythms to each other and to the environmental 24 hour period. The core component of this system is the suprachiasmatic nucleus (SCN) of the hypothalamus, which generates the endogenous circadian rhythms of the organism. The SCN is entrained by environmental factors, the most prominent of which is light.

The delicate interplay of endogenous and exogenous factors, required to maintain normal sleep–wake rhythm, can become recurrently or chronically impaired in some individuals, leading to a group of disorders called circadian rhythm sleep disorders (CRSDs). CRSDs are characterized by an alteration of the circadian timing system or a misalignment between the timing of the individuals sleep–wake rhythm and the 24 hour social and physical environment. In patients with CRSD, sleep episodes occur at inappropriate times, often causing wake periods to occur at

undesired times. Consequently, the patient complains of insomnia or excessive daytime sleepiness and impairment in various areas of functioning.

The second edition of the *International Classification of Sleep Disorders* of the American Academy of Sleep Medicine [1] refers to CRSDs according to their etiology as behaviorally induced or physiological disorders. CRSD of physiologic origin can occur as a primary condition or as a result of other medical disorders or drug use. These disorders are considered to be a malfunction of the biological clock per se and presumably can be distinguished from the behaviorally induced CRSD by inflexibility of the sleep–wake cycle [2]. Behaviorally induced CRSDs are viewed to be a consequence of maladaptive behaviors, such as an individual's voluntary choice to create a temporal mismatch between internal and external factors, or environmental conditions, as occurs in shift work and jet lag. The main focus of this chapter will be the clinical manifestations, prevalence, pathophysiology, diagnosis, and treatment of primary and secondary CRSDs of physiologic origin.

PRIMARY CIRCADIAN RHYTHM SLEEP DISORDERS

Delayed Sleep Phase Type

Delayed sleep phase type (DSPT), also known as delayed sleep phase syndrome, is characterized by habitual

[*]NIMH 5 T32 MH18399-17

Correspondence to: Yaron Dagan, Institute for Fatigue and Sleep Medicine, Sheba Medical Center Tel Hashomer 52621, Israel.

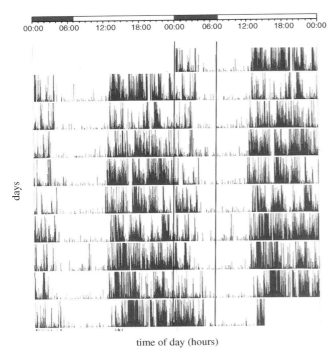

Figure 48.1 Actogram of delayed sleep phase type. Rest-activity patterns were monitored by actigraphy (see Diagnosis section). White and black areas correspond to sleep and wake episodes, respectively. The 24 hour period is double-plotted in a raster format. Night hours are indicated by two continuous vertical lines along the actogram. Note that the timing of sleep onset and offset is dlayed relative to night hours.

sleep–wake times that are delayed usually more than 2 hours relative to conventional or socially acceptable times. Sleep onset is typically delayed until 2 a.m. to 6 a.m. and wake time occurs in the late morning or afternoon (see Figure 48.1). Although delayed, the sleep–wake cycle in these patients is stable, with little day-to-day variability in sleep-onset times. Attempts to advance the sleep–wake phase to earlier hours by enforcing conventional sleep and wake times yield little permanent success. Usually, sleep architecture is reported to be normal for age. Disturbance of the sleep–wake cycle is strongly associated with severe impairment of social, occupational, and domestic functioning [1]. DSPT is frequently associated with learning disabilities and personality disorders [3].

Limited data are available at present regarding the prevalence of DSPT in the general population. This condition is more common in adolescents and young adults, with estimated prevalence of 7–16% [1]. DSPT is the most frequent diagnosis (83.5%) among patients with CRSDs approaching sleep clinics for help [2].

The underlying mechanisms involved in the pathophysiology of DSPT are largely unknown. Several studies suggest that DSPT is associated with phase changes of the circadian timing system [4]. Hereditary trends were also explored in DSPT. Familial DSPT was described, in which an autosomal dominant mode of inheritance was evident [5]. Structural polymorphisms in the human period 3 gene (*hper3*) were implicated in the pathogenesis of DSPT [6a].

Advanced Sleep Phase Type

Advanced sleep phase type (ASPT), also known as advanced sleep phase syndrome, is a stable advance of the major sleep period, characterized by habitual sleep-onset and wake-up times that are several hours earlier than desired or socially acceptable. Typical sleep-onset times are between 6 p.m. and 9 p.m. and wake times are between 2 a.m. and 5 a.m. (see Figure 48.2). Patients with ASPT complain of early morning insomnia and excessive evening sleepiness. No abnormalities in sleep architecture were reported, once the patients are allowed to maintain an advanced schedule [1].

ASPT is thought to be less common than DSPT and very rare in adolescents and young adults. A prevalence of 1% was reported in middle-aged adults [6] and of 1.2% in patients with CRSDs who approached sleep clinics [2]. However, the prevalence of this disorder might be underestimated since ASPT is better tolerated than DSPT.

It is assumed that the pathophysiology of ASPT involves altered phase relationships of endogenous circadian rhythms. Several pedigrees of familial ASPT with advanced melatonin and temperature rhythms were reported, in which the ASPT phenotype segregated as an autosomal dominant

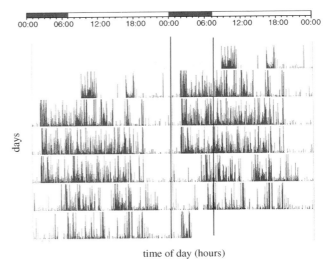

Figure 48.2 Actogram of advanced sleep phase type. Note that, in ASPT, the timing of sleep onset and offset is advanced relative to the circadian night (indicated by two continuous vertical lines along the actogram).

mode of inheritance [7]. Although a mutation of human period 2 (*hper2*) gene was identified in a large family with ASPT [6], other findings indicate genetic heterogeneity in this disorder [21].

Free-Running (non-entrained) Type

Free-running type, also known as non-entrained type or non-24-hour sleep–wake syndrome, is marked by a sleep–wake cycle usually longer than the 24 hour period. Due to this, sleep and wake episodes are delayed each day to later hours, thus alternating between synchrony and complete asynchrony with the environmental schedule (see Figure 48.3). Attempts to adopt regular sleep–wake times are associated with difficulties initiating sleep at night coupled with daytime sleepiness. The inability to adhere to a scheduled life style often leads to severe impairment of educational, occupational, social, and domestic functioning [1]. This condition is associated with personality disorders [3].

The incidence of the disorder in the general population is unknown. Non-entrained type is thought to occur in over half of totally blind individuals [1]. Among patients with CSRDs who approach sleep clinics, 12.1% receive a diagnosis of non-entrained type [2]. Onset may occur at any age in blind people.

In blind people, a lack of photic input to the circadian pacemaker may readily account for the free-running rhythms. In sighted individuals, reduced sensitivity to the entraining influences of light and endogenous circadian period longer than 24 hours were suggested to underlie the disrupted sleep–wake cycle [1]. Altered phase relationships of endogenous circadian rhythms were reported in this disorder [4]. Thus far, genetic screening analyses did not yield any persistent association with the disorder.

Irregular Sleep–Wake Type

Irregular sleep–wake type, also known as irregular or disorganized sleep–wake pattern, is characterized by lack of a clearly defined sleep–wake circadian rhythm. The timing and the length of sleep and wake episodes are variable and unpredictable throughout the 24 hour period (see Figure 48.4). These patients are likely to manifest inability to initiate and maintain sleep at night, frequent daytime napping, and excessive daytime sleepiness. Other circadian rhythms, such as endocrine and body temperature, may also show a loss of diurnal variability [1].

Irregular sleep–wake type is a rare condition, which can begin at any age. It is associated with neurological disorders such as dementia and in children with mental

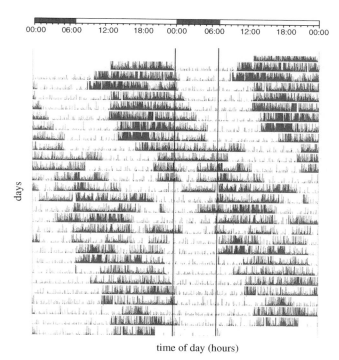

Figure 48.3 Actogram of non-entrained type. Note that in non-entrained type, sleep onset and offset are delayed each day by approximately 1 hour. As a result of this pattern, sleep episodes and night hours sometimes coincide and sometimes are completely out of phase.

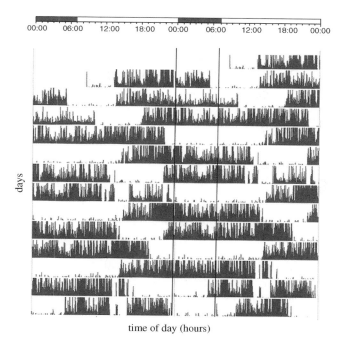

Figure 48.4 Actogram of irregular sleep–wake type. Note the variability in the timing and length of sleep and wake episodes characteristic of this disorder.

retardation. Anatomical or functional abnormalities of the endogenous pacemaker or weakened environmental entraining factors may be involved in the pathology of this disorder [1].

SECONDARY CIRCADIAN RHYTHM SLEEP DISORDERS

In addition to CRSD as a primary condition, disruption of the circadian sleep–wake cycle may arise from an underlying primary medical or neurological condition. These disorders are classified as *Other CRSD due to a Known Physiological Condition* [1]. Patients may display a variety of symptoms, including insomnia and excessive sleepiness. Disruption of the sleep–wake cycle may range from altered phase to irregular sleep–wake patterns. Impairment in major areas of functioning and quality of life is evident in these patients. Demographics are largely unknown [1]. Dementia, movement disorders, hepatic encephalopathy [1], traumatic head injury [9], and brain tumors [10] were reported to play a causative role in this type of CRSD.

Secondary CRSD can occur also as a result of psychoactive substance abuse or dependence, or as a side effect of a drug. These disorders are classified as *Other CRSD Due to a Drug* [1]. Patients may display a variety of disturbances in sleep–wake patterns. Treatment with haloperidol was reported to disrupt the sleep–wake rhythms of patients with schizophrenia [11] and Gilles de la Tourette's syndrome [12]. Additionally, administration of fluvoxamine was associated with emergence of DSPT in patients with obsessive–compulsive disorder [13]. It was suggested that these drugs alter the circadian sleep–wake cycle through their effects on serotonin and melatonin levels.

DIAGNOSIS

Diagnosis of CRSD is largely based on recognizing the characteristics and patterns described earlier. A clinical interview should refer to the patient's sleep–wake habits and difficulties in various areas of functioning (e.g., social, educational, occupational, domestic). Questioning the patient about his/her hours of alertness, hunger times, heredity, ability to adjust the sleep–wake cycle to changes in environmental demands, and history of head injury or brain tumors and use of psychoactive and neuroleptic medication can also prove helpful in formulation of the diagnosis.

In addition to clinical interview, the diagnostic procedures should include 7–4 days of monitoring sleep-wake patterns by means of sleep log and/or actigraphy. The actigraph is a watch-sized device, worn on the wrist, that samples hand motion. A computerized algorithm can provide highly reliable data on sleep and wake periods of the patient [14]. The documentation of sleep–wake cycles requires monitoring for at least several days; therefore actigraphy is the most appropriate objective tool for diagnosing CRSD, and in most cases polysomnography is not required. Importantly, actigraphic monitoring must be conducted in free conditions, since sleep–wake schedule obtained under forced conditions can mask the pattern of the schedule, thus misleading the diagnosis.

TREATMENT

Therapies of CRSD aim to synchronize the circadian clock of the individual with the environmental light–dark cycle. Several treatment options were suggested, such as chronotherapy and vitamin B_{12} administration. At present, the most accepted methods are light therapy and melatonin administration.

Light therapy (also referred to as phototherapy) has become increasingly popular with the emerging recognition of the importance of light in resetting the circadian timing system. Scheduled light exposure at specific times of the 24 hour period can result in a phase shift in the endogenous circadian rhythms of a variety of functions, such as melatonin secretion, body temperature, and sleep propensity. Morning bright light exposure induces a phase advance, whereas evening bright light exposure induces phase delay. Bright light applied by light devices at the intensities of 2000–4000 lux has been successfully used to realign the circadian phase of patients with DSPT and ASPT, and some evidence supports its effectiveness in treatment of nonentrained type, jet lag, shift work, and dementia [15]. Recommended intensities and time limits for phototherapy in the treatment of these disorders have recently been provided by the American Academy of Sleep Medicine [15].

While phototherapy may be a demanding treatment for many patients, frequently leading to compliance problems, melatonin administration is a relatively simple and easy treatment modality. Melatonin is thought to be one of the central endogenous time cues, which is able to reset and entrain the circadian pacemaker in a time-dependent mode. Phase advance is produced by melatonin administered in the evening, whereas melatonin administration in the morning induces phase delays. Beneficial effects of 0.5–5 mg/day of melatonin were demonstrated in several types of CRSDs [16–18]. Although some recent welldesigned studies indicate that even relatively large doses of melatonin (10 mg/day for a month) have no toxicological effects [19], its long-term effects remain to be fully researched and resolved.

In patients for whom all of these treatment modalities fail to help, a rehabilitative approach is recommended. The patients should be guided to accept that their condition is permanent and should be encouraged to consider changes in life style that will be congruent with their sleep–wake cycle [20].

CONCLUSION

CRSDs are sleep pathologies frequently unfamiliar to doctors. Although CRSDs are relatively easily diagnosed and several promising treatment modalities are available, many cases of patients with CRSDs are erroneously diagnosed as psychophysiological insomniacs. Consequently, these patients are unsuccessfully treated, usually with sedative-hypnotic drugs, which can lead to potentially harmful psychological and adjustment consequences. It is of great importance to raise the awareness of these disorders on the part of pediatricians, family doctors, psychiatrists, neurologists, and psychologists.

REFERENCES

1. American Academy of Sleep Medicine. *International Classification of Sleep Disorders*, 2nd ed. AASM, Rochester, MN, 2005.

2. Dagan Y, Eisenstein M. Circadian rhythm sleep disorders: toward a more precise definition and diagnosis. *Chronobiol Int* **16**:213–222(1999).

3. Dagan Y, Sela H, Omer H, Hallis D, Dar R. High prevalence of personality disorders among circadian rhythm sleep disorders (CRSD) patients. *J Psychosom Res* **41**:357–363(1996).

4. Uchiyama M, Okawa M, Shibui K, Kim K, Tagaya H, Kudo Y, Kamei Y, Hayakawa T, Urata J, Takahashi K. Altered phase relation between sleep timing and core body temperature rhythm in delayed sleep phase syndrome and non-24-hour sleep–wake syndrome in humans. *Neurosci Lett* **294**:101–104(2000).

5. Ancoli-Israel S, Schnierow B, Kelsoe J, Fink R. A pedigree of one family with delayed sleep phase syndrome. *Chronobiol Int* **18**:831–840(2001).

6. Ando K, Kripke DF, Ancoli-Israel S. Estimated prevalence of delayed and advanced sleep phase syndromes. *Sleep Res* **24**:509(1995).

6a. Ebisawa T, Uchiyama M, Kajimura N, Mishima K, Kamei Y, Katoh M, Watanabe T, Sekimoto M, Shibui K, Kim K, Kudo Y, Ozeki Y, Sugishita M, Toyoshima R, Inoue Y, Yamada N, Nagase T, Ozaki N, Ohara O, Ishida N, Okawa M, Takahashi K, Yamauchi T. Association of structural polymorphisms in the *human period3* gene with delayed sleep phase syndrome. *EMBO Rep* **2**:342–346(2001).

7. Jones CR, Campbell SS, Zone SE, Cooper F, DeSano A, Murphy PJ, Jones B, Czajkowski L, Ptacek LJ. Familial advanced sleep-phase syndrome: a short-period circadian rhythm variant in humans. *Nat Med* **5**:1062–1065 (1999).

8. Satoh K, Mishima K, Inoue Y, Ebisawa T, Shimizu T. Two pedigrees of familial advanced sleep phase syndrome in Japan. *Sleep* **26**:416–417(2003).

9. Boivin DB, James FO, Santo JB, Caliyurt O, Chalk C. Non-24-hour sleep–wake syndrome following a car accident. *Neurology* **60**:1841–1843(2003).

10. Borodkin K, Kanety H, Ayalon L, Dagan Y. Dysregulation of circadian rhythms following prolactin-secreting pituitary microadenoma. *Chronobiol Int* **22**(1):145–156(2005).

11. Wirz-Justice A, Haug HJ, Cajochen C. Disturbed circadian rest–activity cycles in schizophrenia patients: an effect of drugs? *Schizophr Bull* **27**:497–502(2001).

12. Ayalon L, Hermesh H, Dagan Y. Case study of circadian rhythm sleep disorder following haloperidol treatment: reversal by risperidone and melatonin. *Chronobiol Int* **19**:947–959(2002).

13. Hermesh H, Lemberg H, Abadi J, Dagan Y. Circadian rhythm sleep disorders as a possible side effect of fluvoxamine. *CNS Spectrums* **6**:511–513(2001).

14. Ancoli-Israel S, Cole R, Alessi C, Chambers M, Moorcroft W, Pollak CP. The role of actigraphy in the study of sleep and circadian rhythms. *Sleep* **26**:342–392(2003).

15. Chesson AL Jr, Anderson WM, Littner M, Davila D, Hartse K, Johnson S, Wise M, Rafecas J. Practice parameters for the non-pharmacologic treatment of chronic insomnia. An American Academy of Sleep Medicine report. Standards of Practice Committee of the American Academy of Sleep Medicine. *Sleep* **22**:1128–1133(1999).

16. Dagan Y, Yovel I, Hallis D, Eisenstein M, Raichik I. Evaluating the role of melatonin in the long-term treatment of delayed sleep phase syndrome (DSPS). *Chronobiol Int* **15**:181–190(1998).

17. Nagtegaal JE, Kerkhof GA, Smits MG, Swart AC, Van Der Meer YG. Delayed sleep phase syndrome: a placebo-controlled cross-over study on the effects of melatonin administered five hours before the individual dim light melatonin onset. *J Sleep Res* **7**:135–143(1998).

18. Sack RL, Brandes RW, Kendall AR, Lewy AJ. Entrainment of free-running circadian rhythms by melatonin in blind people. *N Engl J Med* **343**:1070–1077(2000).

19. Seabra ML, Bignotto M, Pinto LR Jr, Tufik S. Randomized, double-blind clinical trial, controlled with placebo, of the toxicology of chronic melatonin treatment. *J Pineal Res* **29**:193–200(2000).

20. Dagan Y, Abadi J. Sleep–wake schedule disorder disability: a lifelong untreatable pathology of the circadian time structure. *Chronobiol Int* **18**:1019–1027(2001).

21. Toh KL, Jones CR, He Y, Eide EJ, Hinz, WA, Virshup DM, Ptacek LJ, Fu YH. An hPer2 phosphorylation site mutation in familial advanced sleep phase syndrome. *Science* **291**(5506):1040–1043(2001).

ADDITIONAL READING

Cajochen C, Kraüchi K, Wirz-Justice A. Role of melatonin in the regulation of human circadian rhythms and sleep. *Neuroendocrinol* **15**:432–437(2003). This article describes the role of melatonin in the regulation of sleep, and the use of melatonin to treat circadian rhythm sleep disorders.

Cermakian N, Boivin DB. A molecular perspective of human circadian rhythm disorders, *Brain Res Rev* **42**:204–220(2003). This thorough article reviews the molecular mechanisms underlying circadian rhythmicity and the genetic mutations involved in human circadian rhythm disorders.

Reiter RJ. Melatonin: clinical relevance. *Best Practice Res Clin Endocrinol Metab* **17**:273–285(2003). This article reviews the cellular mechanisms of melatonin production and the clinical uses of melatonin.

Van Gelder RN. Recent insights into mammalian circadian rhythms, *Sleep* **27**:166–171(2004). This short review describes recent progress in understanding the mammalian circadian clock.

49

JET LAG

TIMOTHY H. MONK

University of Pittsburgh Medical Center, Pittsburgh, Pennsylvania

INTRODUCTION TO THE HUMAN BIOLOGICAL CLOCK

The human biological clock, or "circadian system," is located in the suprachiasmatic nucleus (SCN) of the hypothalamus [1]. The function of the biological clock is to prepare the individual for restful sleep at night and active wakefulness during the day. This is done through daily or "circadian" rhythms, which are generated by the SCN and which can be observed in many different physiological measures (e.g., body temperature, plasma cortisol level, plasma melatonin level) as well as in psychological measures (e.g., mood, alertness, performance ability). One measure used extensively by scientists is the body temperature rhythm, which under normal conditions shows a trough at around 4–6 a.m., a peak at around 8–10 p.m., and a difference of about 1.5 E'F between the two [2]. Another key rhythm is that of melatonin (as measured in the blood or saliva), which is usually at zero level for most of the day; production of melatonin by the pineal gland then gets switched on at about 8–11 p.m., peaks during the night, and then gets switched off at around 4–5 a.m. [3, 4].

There are three properties of the biological clock that are important when considering the effects of travel across several time zones ("jet lag"). First, the clock is endogenous and self-sustaining, with a momentum of its own. This means that the SCN will continue to generate circadian rhythms, even if the individual is continuously awake and is unaware of the time of day. Thus, circadian rhythms are *not* simply the result of changes in waking state, ambient light, posture, or activity, although these factors do have some influence on the observed rhythms. For some rhythms

(e.g., core body temperature) the influence of external factors can be equivalent to the influence of the SCN; for other rhythms (e.g., cortisol) they are minimal. Second, the SCN is *resistant* to changes in timing, taking several days to become realigned to a new routine (as is required when crossing time zones). While this process of realignment is taking place, the biological clock will not be functioning appropriately, leading to impairments in both nocturnal sleep and daytime wakefulness. There may also be malaise and irritability resulting from a loss in the temporal harmony of the different component rhythms (see later discussion).

WHAT IS JET LAG?

The label "jet lag" refers to the lag between the time frame of the biological clock and that of the destination time zone. Because the biological clock is slow to adjust, there will be several days after arrival in the new time zone before the biological clock "catches up" with the new routine [5]. During this adjustment process sleep is impaired, with nocturnal awakenings related to hunger, the need to void, or a simple lack of sleepiness. During the day, there may be a lack of alertness and performance ability, not only because of the partial sleep loss, but also because the biological clock is (inappropriately) preparing the individual for sleep. There is also a malaise and loss of concentration, which may result from a loss of the normal phase interrelationships between the various rhythmic processes. This has been likened to a symphony orchestra after the arrival of a second conductor, beating a different time [6]. Until all of the instruments switch to the new conductor, there is a cacophony of noise. In biological rhythm terms this is described as "internal desynchrony."

IS JET LAG ALL IN THE MIND?

Some people advance the view that jet lag is simply a function of being in a strange bed and having partaken of unfamiliar foods and alcoholic beverages. This is not true. While there are indeed effects due to unfamiliarity with sleeping quarters, diet, and alcohol, there are also very powerful effects stemming from the change in time zone itself.

This was nicely illustrated in a laboratory jet lag simulation [7] in which eight middle-aged volunteers experienced a 6 h phase advance without knowing it. Throughout the 15 day time isolation study, they had no clocks or time cues but were told when to go to bed, get up, and take meals. Body temperature was continuously measured, all sleeps were recorded polygraphically, and performance and mood were regularly assessed. The 6 h phase advance (made, surreptitiously, during the sixth night of sleep) was equivalent to a flight from New York to Paris. Most subjects were unaware that they had been phase shifted and were surprised that they felt tired. However, the effects on circadian rhythms and sleep were profound, lasting for most of the week after the change. The circadian temperature rhythm amplitude was halved, and sleep efficiency was disrupted by 30% immediately following the phase shift. There was then a gradual recovery, linear for rhythm amplitude and in a zigzag pattern for sleep. Thus, even when there is no travel and the bed and daytime activities remain the same, jet lag can become a very real agent of disruption for sleep and daytime functioning.

A similar pattern of results is found when sleep is measured in field studies of jet lag [8, 9]. It is noteworthy that after eastward flight a zigzag recovery pattern, similar to that found in the laboratory, is often observed, with more disruption on the *second* night after arrival, for example, than after the first. Disruptions of sleep are *not* usually present after equivalent north–south flights [10].

IS EASTBOUND WORSE THAN WESTBOUND?

In the early 1970s a group of German researchers led by Karl Klein and Hans-Martin Wegmann performed an elegant series of studies involving volunteers who were flown across the Atlantic [11]. These studies showed quite conclusively that the westward direction involved fewer overall jet lag effects than the eastward direction. A review of many different phase shift studies [12] concluded that each postflight day allowed 90 min of recovery toward the new zone in the westbound direction, but only 60 min in

the eastbound direction. Thus, for example, a 6 h phase delay (Paris to New York) would require 4 days to complete recovery, while a 6 h phase advance (New York to Paris) would require 6 days. This adjustment rate has been broadly confirmed in a study of night workers aboard an isolated offshore oil rig. When their melatonin rhythms were studied, an adjustment rate of 90 min of phase delay per day was observed [13].

IS OUTBOUND WORSE THAN HOMEBOUND?

Klein, Wegmann, and Hunt [11] showed that as compared to the effects due to direction (east versus west), the effects due to traveling away versus traveling home were minimal. This assumes, of course, that the individual spends enough time (more than a week) at the destination for complete adjustment to that time zone to occur.

ARE THERE AGE EFFECTS?

Anecdotally many middle-aged and elderly travelers report experiencing more jet lag complaints than they did when young adults. This ties in with the coping problems experienced by many late-middle-aged night workers [14]. Unfortunately, there is very little hard evidence from field studies regarding age effects in jet lag, as most studies have been concerned with young adults. There is, however, some evidence of age effects in airline pilots [5, 15]. Also, laboratory simulations [16] have shown that middle-aged men suffer more jet lag symptoms than young men to the same 6 h phase advance in routine. Interestingly, circadian manipulations (bright lights) designed to assist the adjustment to night work were found to be more efficacious in young adults than in middle-aged adults [17]. At even further advanced age (over 70 years) sleep appears to be even more disrupted, even though the pacemaker can phase adjust at more or less the same rate [18]. Also, the same directional asymmetry (advance more difficult than delay) observed in young and middle-aged subjects has been confirmed in the elderly [19].

CHRONIC JET LAG

There is a chronic jet lag syndrome experienced by many business executives, whose jobs require them to travel to many different time zones per month, and by airline personnel, the essence of whose job is jet travel [20]. As with rotating shift workers, it is likely that such individuals are in a semipermanent state of circadian desynchrony. The health consequences of rotating shift work include gastrointestinal disorders, cardiovascular disorders, cancer, and depression [21, 22]. It is likely that patients suffering from chronic jet lag syndrome are equally at risk for these disorders. Certainly, recent work by Cho [23] has demonstrated cognitive deficits in such individuals and has even shown brain changes (temporal lobe atrophy) associated with such deficits. Sufferers from chronic jet lag are, moreover, often in late middle-age when they are particularly vulnerable to mental and physical illness.

WHICH JET LAG COUNTERMEASURES WORK?

A number of different countermeasures for jet lag have been proposed including the jet lag diet, aroma therapy, and various devices of sometimes dubious scientific validity. This chapter will focus on three promising areas: behavioral techniques, use of bright light, and use of medications including sleeping pills and pills containing the pineal hormone melatonin. Undoubtedly, there may be other valid countermeasures, but it is often difficult to separate placebo effects from genuine remedies.

BEHAVIORAL COUNTERMEASURES

Most experts agree on a package of behavioral countermeasures that will undoubtedly lessen the jet lag experienced [24]. First, the traveler should, if possible, anticipate the phase change required by the trip by retiring to bed and getting up earlier for several days before an eastward trip and later before a westward trip. This should be done without truncating the sleep episode. Upon boarding the plane, the traveler should change his/her wristwatch to the destination time zone and attempt to follow a schedule indicated by that time (i.e., be awake if it is "day," asleep if it is "night"). Caffeine and alcohol should be avoided, and plenty of fluids should be taken.

It is important for the traveler to recognize the potential fragility of his/her sleep after arrival, and to adopt good sleep hygiene principles accordingly [25]. These should include avoiding caffeine within 5 hours of bedtime, eating and drinking only in moderation, and avoiding heavy physical exercise close to bedtime. The traveler should have earplugs and a snack (not chocolate) available to counter the effects of noisy air-conditioners and midnight hunger. It is also helpful to pack a night-light or dim flashlight so that bathroom use can be accomplished without switching on bright room lights.

The traveler should select a schedule, if possible, that allows an early evening arrival. If this is impossible, he/she should remain awake until 10 p.m. local time. If sleep is irresistible, a *short* nap is permitted in the early afternoon. An alarm should be set to prevent this nap lasting more than 2 hours. Light exercise in the daylight is encouraged (see later discussion) and the traveler should avoid remaining indoors wherever possible. Klein and Wegmann [26] showed that requiring people to remain indoors after arrival significantly worsened the jet lag they experienced.

A strategy that may work if the trip is a short one (<3 days) is to remain on home-based time, scheduling sleep periods, meals, and meetings to coincide with the "daytime" of home time zone. This has been shown empirically to be of benefit in a study of airline personnel [27]. Daylight should be avoided and the home time zone routine strictly adhered to.

BRIGHT LIGHT (AND GOGGLES)

Daylight illumination levels are powerful sychronizers of the human biological clock, suggesting the simple strategy of being out and about in the daylight as much as possible at one's destination (see earlier discussion). It is possible to be more sophisticated in one's use of light to lessen jet lag effects, as Daan and Lewy [28] have suggested. This approach makes use of the fact that bright light at the end of the biological day tends to delay circadian rhythms, and at the beginning of the biological day tends to advance them. Thus appropriately timed exposure to daylight (or a bank of very bright artificial lights) can be used to move the timing of the biological clock to the desired phase position. This can be even more powerful if used in conjunction with dark goggles to restrict daylight exposure at the other end of the day. A book on countering the effects of jet lag [29] has utilized this approach to develop various light exposure strategies to minimize the amount of jet lag experienced.

MELATONIN PILLS

Melatonin is a pineal hormone that is a very important component of the circadian system [1]. As noted earlier, on a normal routine, production of the hormone is "switched on" at around 8–11 p.m. and off again at about 4–5 a.m. When circulating in the bloodstream, melatonin can cause drowsiness, and its use to counter jet lag effects is thus appealing, particularly for administration around bedtime. Additionally, melatonin pills are one of the few pharmacological agents shown definitively to be a true chronobiotic, that is, a compound known to directly affect the timing of the biological clock [30, 31]. Although not as powerful as daylight, melatonin pills can, when administered at the correct circadian time, effect a phase change in the SCN [32].

A number of field studies [33–36] have demonstrated the efficacy of melatonin pills (typically in a dose around 5 mg) in alleviating jet lag. However, melatonin pills have not always been successful as a countermeasure [37, 38], and some experts [39] discourage their regular use.

SLEEPING PILLS

Sleeping pills (hypnotics) are best used infrequently to counter transient insomnia, and a number of studies have shown them to be effective in consolidating phase-shifted sleep [40, 41]. At one time there was a hope derived from hamster studies [42] that one particular hypnotic, triazolam, might also have a direct clock phase-shifting effect as well. It now appears, however, that the effect was mediated by changes in the activity pattern of the animals, rather than any more direct effect on the SCN. Thus triazolam, like other hypnotics, may help consolidate sleep but appears *not* to be influencing the *rate* at which the biological clock adjusts to the new time zone.

In their review of sleep-promoting agents in transmeridian travel, Stone and Turner [43] note that hypnotics with a duration of action around 3–5 h are likely to be helpful in alleviating jet lag symptoms. Thus the new short half-life hypnotics such as zolpidem, zopiclone, and zaleplon may be useful [44]. However, having reviewed the effects of various hypnotics, Stone and Turner [43] suggest that hypnotics only be used sparingly and in conjunction with behavioral countermeasures.

CONCLUSIONS

Jet lag involves sleep disruptions and impairments in mood and daytime functioning. A number of possible treatment strategies are available, including behavioral plans, sleep hygiene, timed daylight exposure, sleeping pills, and melatonin.

ACKNOWLEDGMENTS

Sincere thanks are due to Melissa L. Clark for research assistance. Primary support for this work was provided by NASA Grants NAG9-1234 and NNJ04HF76G; National institute on Aging Grants AG-13396 and AG-020677; and by GCRC Grant RR-00056.

REFERENCES

1. Moore RY. Circadian rhythms: basic neurobiology and clinical applications, *Annu Rev Med* **48**:253–266(1997).

2. Colquhoun WP. *Biological Rhythms and Human Performance*. Academic Press, London, 1971.

3. Reiter RJ. The melatonin rhythm: both a clock and a calendar. *Experientia* **49**:654–664(1993).

4. Shanahan TL, Kronauer RE, Duffy JF, Williams GH, Czeisler CA. Melatonin rhythm observed throughout a three-cycle bright-light stimulus designed to reset the human circadian pacemaker. *J Biol Rhythms* **14**:237–253(1999).

5. Graeber RC. Jet lag and sleep disruption. In: Kryger MH, Roth T, Dement WC (Eds), *Principles and Practice of Sleep Medicine*. Saunders, Philadelphia, 1989.

6. Monk TH. Coping with the stress of jet lag. *Work & Stress* **1**:163–166(1987).

7. Monk TH, Moline ML, Graeber RC. Inducing jet lag in the laboratory: patterns of adjustment to an acute shift in routine. *Aviat Space Environ Med* **59**:703–710(1988).

8. Sasaki M, Endo S, Nakagawa S, Kitahara T, Mori A. A chronobiological study on the relation between time zone changes and sleep. *Jikeikai Med J* **32**:83–100(1985).

9. Seidel WF, Cohen SA, Bliwise NG, Dement WC. Jetlag after eastward and westward flights. *Sleep Res* **16**:639(1987).

10. Hauty GT, Adams T. Phase shifts of the human circadian system and performance deficit during the periods of transition: I. East–west flight. *Aerospace Med* **July**:668–674(1966).

11. Klein KE, Wegmann HM, Hunt BI. Desynchronization of body temperature and performance circadian rhythms as a result of out-going and homegoing transmeridian flights. *Aerospace Med* **43**(2):119–132(1972).

12. Aschoff J, Hoffman K, Pohl H, Wever RA. Re-entrainment of circadian rhythms after phase-shifts of the zeitgeber. *Chronobiologia* **2**:23–78(1975).

13. Barnes RG, Deacon SJ, Forbes MJ, Arendt J. Adaptation of the 6-sulphatoxymelatonin rhythm in shiftworkers on offshore oil installations during a 2-week 12-h night shift. *Neurosci Lett* **241**:9–12(1998).

14. Foret J, Bensimon G, Benoit O, Vieux N. Quality of sleep as a function of age and shift work. In: Reinberg A, Vieux N, Andlauer P (Eds), *Night and Shift Work: Biological and Social Aspects*. Pergamon Press, Oxford, 1981.

15. Preston FS, Bateman SC, Meichen FW, Wilkinson R, Short RV. Effects of time zone changes on performance and physiology of airline personnel. *Aviation Space Environ Med* **47**(7):763–769(1976).

16. Moline ML, Pollak CP, Monk TH, Lester LS, Wagner DR, Zendell SM, Graeber RC, Salter CA, Hirsch E. Age-related differences in recovery from simulated jet lag. *Sleep* **14**(5):x–x(1991).

17. Campbell SS. Effects of timed bright-light exposure on shift-work adaptation in middle-aged subjects. *Sleep* **18**:408–416(1995).

18. Monk TH, Buysse DJ, Reynolds CF, Kupfer DJ. Inducing jet lag in older people: adjusting to a 6-hour phase advance in routine. *Exp Gerontol* **28**:119–133(1993).

19. Monk TH, Buysse DJ, Carrier J, Kupfer DJ. Inducing jetlag in older people: directional asymmetry. *J Sleep Res* **9**:101–116(2000).

20. Monk TH. Disorders relating to shift work and jet lag. In: Oldham JM (Ed), *Sleep Disorders Section: Volume 13 Annual*

Review of Psychiatry. American Psychiatric Press, Washington, DC, 1994.

21. Scott AJ, LaDou J. Shiftwork: effects on sleep and health with recommendations for medical surveillance and screening. *Occup Med* **5**:273–299(1990).

22. Costa G. Shift work and occupational medicine: an overview. *Occup Med (London)* **53**:83–88(2003).

23. Cho K. Chronic "jet lag" produces temporal lobe atrophy and spatial cognitive deficits. *Nat Neurosci* **4**:567–568(2001).

24. Redfern P, Minors D, Waterhouse J. Circadian rhythms, jet lag, and chronobiotics: an overview. *Chronobiol Int* **11**:253–265(1994).

25. Zarcone VP. Sleep hygiene. In: Kryger MH, Roth T, Dement WC (Eds), *Principles and Practice of Sleep Medicine* (Vol 2). Saunders, Philadelphia, 1994.

26. Klein KE, Wegmann HM. The resynchronization of human circadian rhythms after transmeridian flights as a result of flight direction and mode of activity. In: Scheving LE, Halberg F, Pauly JE (Eds), *Chronobiology.* Igaku Shoin Ltd, Tokyo, 1974.

27. Lowden A, Akerstedt T. Retaining home-base sleep hours to prevent jet lag in connection with a westward flight across nine time zones. *Chronobiol Int* **15**:365–376(1998).

28. Daan S, Lewy AJ. Scheduled exposure to daylight: a potential strategy to reduce "jet lag" following transmeridian flight. *Psychopharmacol Bull* **20**:566–568(1984).

29. Oren DA, Reich W, Rosenthal NE, Wehr TA. *How to Beat Jet Lag—A Practical Guide for Air Travelers.* Henry Holt, New York, 1993.

30. Lewy AJ, Ahmed S, Latham Jackson JM, Sack RL. Melatonin shifts human circadian rhythms according to a phase–response curve. *Chronobiol Int* **9**(5):380–392(1992).

31. Sack RL, Lewy AJ. Melatonin as a chronobiotic: treatment of circadian desynchrony in night workers and the blind. *J Biol Rhythms* **12**:595–603(1997).

32. Sharkey KM, Fogg LF, Eastman CI. Effects of melatonin administration on daytime sleep after simulated night shift work. *J Sleep Res* **10**:181–192(2001).

33. Arendt J, Aldhous M, Marks V. Alleviation of jet lag by melatonin: preliminary results of a controlled double blind trial. *Br Med J* **292**:1170(1986).

34. Arendt JH, Aldhous M, English J, Marks V, Marks M, Folkard S. Some effects of jet lag and their alleviation by melatonin. *Ergonomics* **30**:1379–1393(1987).

35. Arendt J. Jet lag/night shift, blindness and melatonin. *Trans Med Soc London* **114**:7–9(1997).

36. Petrie K, Dawson AG, Thompson L, Brook R. A double-blind trial of melatonin as a treatment for jet lag in international cabin crew. *Biol Psychiatry* **33**(7):526–530(1993).

37. Spitzer RL, Terman M, Williams JB, Terman JS, Malt UF, Singer F, Lewy AJ. Jet lag: clinical features, validation of a new syndrome-specific scale, and lack of response to melatonin in a randomized, double-blind trial. *Am J Psychiatry* **156**:1392–1396(1999).

38. Edwards BJ, Atkinson G, Waterhouse J, Reilly T, Godfrey R, Budgett R. Use of melatonin in recovery from jet lag following an eastward flight across 10 time-zones. *Ergonomics* **43**:1501–1513(2000).

39. Samuel A, Wegmann H, Vejvoda M, Maass H, Gundel A, Schutz M. Influence of melatonin treatment on human circadian rhythmicity before and after a simulated 9-hr time shift. *J Biol Rhythms* **6**(3):235–248(1991).

40. Walsh JK, Muehlbach MJ, Scweitzer PK. Acute administration of triazolam for the daytime sleep of rotating shift workers. *Sleep* **7**:223–229(1984).

41. Seidel WF, Roth T, Roehrs T, Zorick F, Dement WC. Treatment of a 12-hour shift of sleep schedule with benzodiazepines. *Science* **224**:1262–1264(1984).

42. Turek FW, Losee-Olson S. A benzodiazepine used in the treatment of insomnia phase-shifts the mammalian circadian clock. *Nature* **321**:167–168(1986).

43. Stone BM, Turner C. Promoting sleep in shiftworkers and intercontinental travelers. *Chronobiol Int* **14**:133–143(1997).

44. Terzano MG, Rossi M, Palomba V, Smerieri A, Parrino L. New drugs for insomnia: comparative tolerability of zopiclone, zolpidem and zaleplon. *Drug Saf* **26**:261–282(2003).

50

SHIFT WORK SLEEP DISORDER

GARY S. RICHARDSON

Henry Ford Hospital, Detroit, Michigan

INTRODUCTION

A discussion of shift work sleep disorder (SWSD) necessarily begins with a discussion of shift work itself. Shift work is generally defined as any schedule that requires work outside a broadly defined "day shift," usually 6 a.m. to 6 p.m. As of 1997, 16.8% of full-time U.S. workers regularly worked shifts other than the day shift [1]. The biological implications of specific shifts within this broad category vary significantly, however. Evening shifts, typically those between 2 p.m. and midnight, are generally well tolerated, allowing later times of arising and longer typical sleep durations than even typical day shifts [2]. Night shifts, those between 9 p.m. and 8 a.m. are less well tolerated. Schedules that regularly include night shifts (straight night and rotating shift schedules) together comprise 6.4%, or 5.8 million full-time U.S. workers [1].

The biological difficulties posed by the night shift stem from the modulation of physiological processes, including those governing sleep and wakefulness, by the endogenous circadian clock. When the circadian orientation of night workers is examined, it is out of phase with both day and night schedules and is much more variable than that of day workers [3]. This is because light provides an intensity-dependent orienting, or phase-setting, signal for the internal clock, and even incidental exposure to bright sunlight, such as might occur during the drive home after the night shift, or in the fulfillment of family or household responsibilities throughout the day, will undermine efforts to entrain the circadian clock to a nocturnal orientation. Only extraordinary control of the light–dark environment both at home and at work allows stable and complete circadian adaptation to the night shift [4].

CONSEQUENCES OF SHIFT WORK

Given the impediments to physiological adaptation to night work, it is not surprising that epidemiological studies have established that there is significant morbidity associated with shift work exposure. Most prominently, shift work is associated with both increased difficulty sleeping and with increased sleepiness during waking hours [5]. Sleepiness on the job, reflecting both the effects of sleep deficits accumulated during day sleep and the clock-mediated impairment of alertness during the night, results in a significantly greater risk of accidents [6]. Shift workers are also at increased risk for a variety of adverse health outcomes including cardiovascular disease [7], ulcer disease [8], and breast cancer in women [9]. Unlike the link between shift work and accidents where sleepiness clearly plays an important role, it is unclear how shift work may predispose to these other medical problems. Both chronic sleep deprivation, with attendant changes in immunological activity [10], for example, and circadian misalignment with alterations in melatonin secretion [11] have been proposed as possible pathological mechanisms. Resolving the specific mechanisms involved in the increased risk for medical disease has important implications for the future management of shift work related sleep complaints.

SHIFT WORK SLEEP DISORDER

Although epidemiological studies suggest the difficulty with sleep and alertness at work is common among shift workers, a subset experience difficulty severe enough to warrant the clinical diagnosis of shift work sleep disorder (SWSD). Diagnostic criteria for SWSD established by the American Academy of Sleep Medicine [12] call for the presence of either (1) insomnia during day sleep or (2) a complaint of excessive sleepiness during night work hours. The complaint (or complaints) should not be adequately explained by another medical, psychiatric, or primary sleep disorder and must be temporally associated with the shift exposure.

Using this definition, recent work using a representative U.S. sample, aged 18–65, has established that the prevalence of SWSD is 14.1% among night workers, and 8.1% among workers on rotating shift schedule, an overall prevalence of 10% [13]. This study also shows that those with SWSD have significantly higher rates of sleepiness-related accidents, ulcer disease, and depression than did workers on the same schedule without SWSD. These findings provide an important link to the epidemiologic studies of shift work, suggesting that the SWSD subpopulation may contribute differentially to at least some of the associated morbidity. Furthermore, these data suggest that treatments that successfully improve the sleep–wake complaints of SWSD may mitigate the long-term consequences of shift work exposure.

RISK FACTORS FOR SHIFT WORK SLEEP DISORDER

The limited epidemiology available for SWSD does not support a formal evaluation of risk factors. The work referenced earlier suggests that SWSD is more common in night workers than in those on rotating shift schedules, though the difference was not statistically significant. Other potential risk factors for SWSD, such as age and gender, were not examined.

Some studies have used "shift work intolerance" or dissatisfaction with the shift schedule, rather than SWSD, as the endpoint in risk factor assessments. Problems with subjective sleep quantity and quality are important contributors to shift work intolerance [14], suggesting that these measures may provide reasonable proxies for SWSD. These studies have implicated a number of factors including female gender, increasing age, and duration of shift work exposure as possible risk factors [15, 16].

These findings are generally consistent with clinical impressions, which generally support the assertion that any factor that might limit the duration or quality of daytime sleep, and/or impair alertness during working hours,

will increase the risk of SWSD. For example, studies have shown that female shift workers obtain significantly less sleep than their male counterparts, apparently because additional home and family responsibilities limit the time available for day sleep [17].

The specific shift schedule may also be a risk factor for SWSD. For example, it has generally been held that forward rotating shifts (day to evening to night) are more readily tolerated than backward rotation (day to night to evening) [18], and that slow rotation (changing every week) is less adverse than a rapidly changing schedule. However, more recent data have challenged the significance of these features of the shift schedule [19, 20]. Instead, it appears that more specific features of the schedule (start times, number of consecutive days off) are much more important in predicting worker satisfaction and sleep parameters than is the direction or speed of rotation [21].

MAKING THE DIAGNOSIS OF SHIFT WORK SLEEP DISORDER

The diagnosis of SWSD is made on the basis of a clinical history of significant insomnia and/or significant sleepiness during work hours occurring in the setting of shift work. The significance of the complaint is generally assessed in terms of interference with function on the job, or family and social activities. As outlined previously, symptoms should be temporally associated with the shift, meaning that sleep and alertness should improve during periods of day work or time off. Sleep diaries are particularly helpful in establishing the temporal pattern of the complaint.

Polysomnography of day sleep and multiple sleep latency test (MSLT) assessments at normal work times are included in the formal definition of SWSD [12], but the absence of relevant normal values for either measure at these times substantially limits their diagnostic utility. In clinical practice, these measures are typically used only to evaluate the possible contribution of other sleep disorders. Actigraphy can be helpful if available and may also be useful in assessing the impact of interventions (see later discussion).

MANAGEMENT OF SHIFT WORK SLEEP DISORDER

Management of SWSD should follow an ordered approach that includes three distinct steps (Table 50.1). The first step is to verify the diagnosis and identify contributory medical, psychiatric, or sleep disorders. While the existence of other primary sleep disorders would formally exclude the diagnosis of SWSD, in clinical practice this distinction is often difficult to make. Symptoms related to mild sleep apnea, for example, may be exacerbated by shift work, resulting

TABLE 50.1   Approach to the Patient with Shift Work Sleep Disorder

I. Verify diagnosis and rule out other primary sleep disorders.
   A. If present, treat primary disorder and reevaluate.
II. If no primary sleep disorder, or if disorder adequately treated:
   A. Optimize day sleep.
      (1) Evaluate sleep conditions and general sleep hygiene.
      (2) Consider alternate sleep timing and/or addition of second sleep episode.
III. Consider pharmacological treatment.
   A. Improve duration and quality of day sleep (e.g., use short-acting BzRA such as zaleplon or zolpidem)
   B. Improve alertness during work at night (e.g., use modafinil before night shift).

in apparent temporal coincidence with the shift schedule. Similarly, it is often difficult to determine whether difficulty sleeping during the day reflects a contribution of underlying primary insomnia. In practical application, this means retaining a high index of suspicion for coexisting sleep disorders and adding disease-specific treatment (e.g. weight loss or CPAP for obstructive sleep apnea) or behavioral therapy for primary insomnia to the SWSD regimen.

The second step in managing SWSD is to optimize the duration and quality of day sleep. This step recognizes that among the many factors that can limit day sleep are several that are readily remediable. This process should begin with a "sleep audit" to establish the degree of sleep limitation imposed by the work schedule. A modified sleep diary, the sleep audit focuses on timing and duration of attempted sleep over a representative sample of the work schedule. In rotating or variable schedules, the audit should be continued until at least one full rotation of the schedule is completed. The average number of hours per day for each shift serves as a rough measure of the severity of sleep limitation. For comparison, large surveys spanning multiple shift work settings suggest an average daily sleep duration of approximately 6 hours during the night shift [2, 22]. The audit should be reviewed with the SWSD patient to understand the rationale for specific sleep scheduling decisions.

Once the practical limits on sleep times are understood, an alternate timing strategy can be devised to increase time available for sleep. Most commonly, this involves the addition of a second sleep episode, or nap, in the evening before work, but it should be emphasized that there is no "correct" solution to day sleep. Morning sleep, afternoon sleep, and split sleep schedules may all be used to help the shift worker obtain a reasonable amount of sleep in the setting of competing practical limitations. The goal is to maximize sleep duration, and empirical trials may be necessary to find the optimal solution for each patient. Repeat sleep audits are used to assess results.

The quality of sleep obtained during the day is also a problem for shift workers. Treatment of poor sleep quality in the shift worker utilizes many of the same sleep hygiene strategies that apply to primary insomnia. The notable exception is that there should be no proscription against napping. Attention to caffeine and nicotine use, as well as identification and elimination of counterproductive sleep treatments, is essential. Some additional practical advice is more specific for the day sleeper. Phone ringers and cell phones should be readily switched off, and answering machine/voicemail messages explaining the schedule should be prepared. Family, neighbors, and friends need to be educated about the preeminent importance of adequate, uninterrupted day sleep.

A persistent area of controversy in the practical management of SWSD is the extent to which light exposure should be regulated. While maintenance of a dark sleeping room has clear benefit, if only for its direct impact on sleep quality, it is less clear whether shift workers should be counseled to limit light exposure for the rest of the day. While limitation of light exposure during the day and enhancement of light exposure at night can produce circadian phase positions that are significantly more conducive to day sleep [23, 24], shift workers are generally resistant to such restrictions, perceiving flexibility with their time off as one of the few advantages of night work. A reasonable compromise is to recommend that light exposure during days between night shifts be as consistent as possible, fitting as much as possible into a routine that includes the scheduled sleep hours. This will increase the consistency of the circadian phase position and make designing day sleep schedules a more reliable process.

The third step in managing SWSD is the use of pharmacologic interventions. Reserved for patients in whom symptoms do not respond to reasonable efforts to optimize day sleep, pharmacologic options include both hypnotics to increase the quality and duration of day sleep, and wake-enhancing medications to increase alertness during night work hours. In the former group, hypnotic benzodiazepine receptor agonists (BzRAs) have been shown to improve day sleep in objective assessments, but the impact on alertness the following night is limited [25, 26]. Newer selective agents, particularly those with shorter half-lives such as zolpidem or zaleplon, are now the drugs of choice in this setting.

Melatonin, a hormone with sedative and chronobiotic (direct circadian phase shifting) properties [27], offers theoretical advantages in the treatment of sleep disruption associated with SWSD. However, while there is substantial support for its use [28, 29], many of the field and laboratory trials are difficult to interpret because they have also incorporated strict control of light exposure. When the contribution of the two interventions is separated, there appears to be little benefit of melatonin treatment [30]. Consistent with this interpretation, placebo-controlled trials in real shift work settings without control of light–dark exposure have generally been negative [31].

Wake-enhancing medications are the most common pharmacologic intervention if the category is expanded to include caffeine. Caffeine can clearly have a beneficial effect on alertness [32], but its use is limited by tolerance and side effects. Stimulants such as D-amphetamine, methyl-phenidate, and pemoline are also limited by side effects, as well as a significant risk of abuse [33]. More recently, modafinil has been approved for use in the treatment of SWSD [34]. Clinical experience with modafinil in SWSD remains limited, but the substantial advantages of this drug over older stimulants in the risk of abuse and side effects dictate that it is the treatment of choice for this indication.

CONCLUSION

Shift work sleep disorder (SWSD) affects approximately 10% of full-time workers on night or rotating shifts. Risk factors for SWSD include increasing age, female gender, and any predisposition to sleep disturbance or impaired alertness. Recent data demonstrate that SWSD increases the risk of morbidity associated with shift work, such as ulcer disease and sleepiness-related accidents. Treatment for SWSD should follow a stepwise approach in which (1) the contribution of coexisting medical, psychiatric, or sleep disorders is first addressed, (2) day sleep quality and quantity are optimized, and (3) pharmacologic inter-ventions to improve day sleep and enhance nocturnal alert-ness are then employed.

REFERENCES

1. Beers TM. Flexible schedules and shift work: Replacing the "9-to-5" workday? *Monthly Labor Rev* **23**(6):33–40(2000).

2. Tepas DI, Carvalhais AB. Sleep patterns of shiftworkers. *Occup Med* **5**(2):199–208(1990).

3. Sack RL, Blood ML, Lewy AJ. Melatonin rhythms in night shift workers. *Sleep* **15**(5):434–441(1992).

4. Czeisler CA, Johnson MP, Duffy JF, Brown EN, Ronda JM, Kronauer RE. Exposure to bright light and darkness to treat physiologic maladaptation to night work. [see Comments]. *N Engl J Med* **322**(18):1253–1259(1990).

5. Akerstedt T. Work hours and sleepiness. *Neurophysiol Clin* **25**(6):367–375(1995).

6. Folkard S, Tucker P. Shift work, safety and productivity. *Occup Med (London)* **53**(2):95–101(2003).

7. Knutsson A, Akerstedt T, Jonsson BG, Orth-Gomer K. Increased risk of ischaemic heart disease in shift workers. *Lancet* **2**(8498):89–92(1986).

8. Tuchsen F, Jeppesen HJ, Bach E. Employment status, non-daytime work and gastric ulcer in men. *Int J Epidemiol* **23**(2):365–370(1994).

9. Hansen J. Increased breast cancer risk among women who work predominantly at night. *Epidemiology* **12**(1):74–77(2001).

10. Vgontzas AN, Zoumakis E, Bixler EO, et al. Adverse effects of modest sleep restriction on sleepiness, performance, and inflammatory cytokines. *J Clin Endocrinol Metab* **89**(5):2119–2126(2004).

11. Hansen J. Light at night, shiftwork, and breast cancer risk. *J Natl Cancer Inst* **93**(20):1513–1515(2001).

12. American Academy of Sleep Medicine. *The International Classification of Sleep Disorders, Revised: Diagnostic and Coding Manual.* American Academy of Sleep Medicine, Rochester, MN, 2001.

13. Drake C, Roehrs T, Richardson GS, Walsh JK, Roth T. Shift-work sleep disorder: Prevalence and consequences beyond that of symptomatic day workers. *Sleep* **27**(8):1453–1464(2004).

14. Axelsson J, Akerstedt T, Kecklund G, Lowden A. Tolerance to shift work—how does it relate to sleep and wakefulness? *Int Arch Occup Environ Health* **77**(2):121–129(2004).

15. Andlauer P, Reinberg A, Fourre L, Battle W, Duverneuil G. Amplitude of the oral temperature circadian rhythm and the tolerance to shift-work. *J Physiol* **75**(5):507–512(1979).

16. Oginska H, Pokorski J, Oginski A. Gender, ageing, and shift-work intolerance. *Ergonomics* **36**(1-3):161–168(1993).

17. Dekker DK, Paley MJ, Popkin SM, Tepas DI. Locomotive engineers and their spouses: coffee consumption, mood, and sleep reports. *Ergonomics* **36**(1-3):233–238(1993).

18. Lavie P, Tzischinsky O, Epstein R, Zomer J. Sleep–wake cycle in shift workers on a "clockwise" and "counter-clockwise" rotation system. *Israel J Med Sci* **28**(8–9):636–644(1992).

19. Tucker P, Smith L, Macdonald I, Folkard S. Effects of direction of rotation in continuous and discontinuous 8 hour shift systems. *Occup Environ Med* **57**(10):678–684(2000).

20. Cruz C, Detwiler C, Nesthus T, Boquet A. Clockwise and counterclockwise rotating shifts: effects on sleep duration, tim-ing, and quality. *Aviat Space Environ Med* **74**(6 Pt 1):597–605(2003).

21. Smith PA, Wright BM, Mackey RW, Milsop HW, Yates SC. Change from slowly rotating 8-hour shifts to rapidly rotating 8-hour and 12-hour shifts using participative shift roster design. *Scand J Work Environ Health* **24** (Suppl 3):55–61(1998).

22. Colligan MJ, Tepas DI. The stress of hours of work. *Am Ind Hyg Assoc J* **47**(11):686–695(1986).

23. Boivin DB, James FO. Circadian adaptation to night-shift work by judicious light and darkness exposure. *J Biol Rhythms* **17**(6):556–567(2002).

24. Czeisler CA, Dijk DJ. Use of bright light to treat maladaptation to night shift work and circadian rhythm sleep disorders. *J Sleep Res* **4**(S2):70–73(1995).

25. Porcu S, Bellatreccia A, Ferrara M, Casagrande M. Perfor-mance, ability to stay awake, and tendency to fall asleep during

the night after a diurnal sleep with temazepam or placebo. *Sleep* **20**(7):535–541(1997).

26. Walsh JK, Schweitzer PK, Anch AM, Muehlbach MJ, Jenkins NA, Dickins QS. Sleepiness/alertness on a simulated night shift following sleep at home with triazolam. *Sleep* **14**(2):140–146(1991).

27. Richardson G, Tate B. Hormonal and pharmacological manipulation of the circadian clock: recent developments and future strategies. *Sleep* **23** (Suppl 3):S77–S85(2000).

28. Arendt J, Skene DJ, Middleton B, Lockley SW, Deacon S. Efficacy of melatonin treatment in jet lag, shift work, and blindness. *J Biol Rhythms* **12**(6):604–617(1997).

29. Sack RL, Lewy AJ. Melatonin as a chronobiotic: treatment of circadian desynchrony in night workers and the blind. *J Biol Rhythms* **12**(6):595–603(1997).

30. Burgess HJ, Sharkey KM, Eastman CI. Bright light, dark and melatonin can promote circadian adaptation in night shift workers. *Sleep Med Rev* **6**(5):407–420(2002).

31. Wright SW, Lawrence LM, Wrenn KD, Haynes ML, Welch LW, Schlack HM. Randomized clinical trial of melatonin after night-shift work: efficacy and neuropsychologic effects. (See Comments.) *Ann Emerg Med* **32**(3 Pt 1):334–340(1998).

32. Walsh JK, Muehlbach MJ, Humm TM, Dickins QS, Sugerman JL, Schweitzer PK. Effect of caffeine on physiological sleep tendency and ability to sustain wakefulness at night. *Psychopharmacology* **101**(2):271–273(1990).

33. Akerstedt T, Ficca G. Alertness-enhancing drugs as a countermeasure to fatigue in irregular work hours. *Chronobiol Int* **14**(2):145–158(1997).

34. Sleepiness versus sleeplessness: shift work and sleep disorders. *J Clin Psychiatry* **65**(7):1007–1015(2004).

ADDITIONAL READING

Akerstedt T. Shift work and disturbed sleep/wakefulness. *Sleep Med Rev* **2**(2):117–128(1998). This report reviews extensive data on the link between shift work and sleep disturbances, with emphasis on objective and field studies.

Moore-Ede MC, Richardson GS. Medical implications of shift-work. *Annu Rev Med* **36**:607–617(1985). An older reference that introduces the concept of shift work intolerance and identifies the characteristic pathologies that are increased in workers with difficulty adapting to shift work.

Richardson GS, Malin HV. Circadian rhythm sleep disorders: pathophysiology and treatment. *J Clin Neurophysiol* **13**(1):17–31(1996). A review of circadian rhythm sleep disorders including shift work sleep disorder, with focus on treatment.

Wagner DR. Disorders of the circadian sleep–wake cycle. *Neurol Clin* **14**(3):651–670(1996). A review of circadian rhythm sleep disorders with emphasis on pathophysiology.

51

NEUROLOGICAL AND MEDICAL DISORDERS ASSOCIATED WITH CIRCADIAN RHYTHM DISTURBANCES

RAMADEVI GOURINENI AND PHYLLIS C. ZEE
Northwestern University Feinberg School of Medicine, Chicago, Illinois

INTRODUCTION

Circadian rhythms in physiology and behavior are ubiquitous in all living organisms from single cells to humans. The most obvious of these circadian rhythms is the daily sleep–wake cycle. These endogenous circadian rhythms have a period of oscillation that is approximately 24 hours and are genetically regulated [1]. In mammals, the suprachiasmatic nucleus (SCN), a paired structure located in the hypothalamus, is the site of a master circadian pacemaker [2, 3]. The SCN not only generates circadian rhythms, but also maintains synchronization of internal circadian rhythms to the external physical, social, and work schedules. This chapter will focus on the relation of circadian rhythms with common neurological and medical disorders.

SEIZURE DISORDER

It is well known that there is a diurnal rhythm to seizure frequency, so that in some epilepsies, seizures occur predominantly at night, while in other types, seizures occur more commonly during the day [4]. The temporal distribution of these seizures may be due to state-dependent changes in neuronal excitability during sleep and drowsiness and/or to modulation by the circadian system. Substantial evidence exists supporting the role of the sleep–wake cycle in this diurnal variation of seizure frequency. It is not yet known,

however, whether circadian modulation, independent of sleep, also plays an important role.

The timing of seizures varies with the type of epilepsy. For example, partial seizures of temporal lobe origin occur more frequently in the mid-afternoon [4], while other epilepsy syndromes such as benign rolandic epilepsy of childhood, autosomal dominant frontal lobe epilepsy, and electrical status epilepticus of sleep (ESES) occur during the night (in association with sleep). Several of the generalized epilepsies that begin in childhood characteristically occur on awakening. The prototype of these awakening generalized seizures is juvenile myoclonic epilepsy (JME) [5] (Figure. 51.1). Therefore, in clinical practice, the timing of seizure occurrence can help the clinician to distinguish between the various epilepsy types.

As mentioned earlier, the diurnal distribution of seizures has a strong relation with the sleep and wake states. To start with, the sleep or drowsiness states may play a major role in the diurnal propensity to the awakening generalized seizures. In addition, NREM sleep has a marked effect on the timing, duration, and rate of generalization of partial seizures [6]. This may be due to the synchronization of the electroencephalogram (EEG), which together with phasic arousal events, such as K-complexes, may combine to promote electrographic seizure propagation during this state [5]. There is also evidence that the circadian system may play a role in modulating seizures. Interictal epileptiform discharges increase at night and this persists despite wakefulness, suggesting the presence of an endogenous

Sleep: A Comprehensive Handbook, Edited by T. Lee-Chiong.
Copyright © 2006 John Wiley & Sons, Inc.

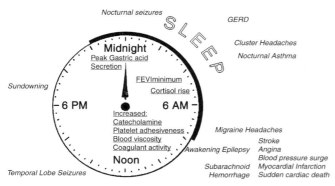

Figure 51.1 Circadian distribution of various physiological functions (inside clock in bold) and various neurological and medical disorders (in italics outside the clock) in relation to the time and the sleep–wake cycles. Nocturnal seizures are most prominent in the first third of the sleep period, with awakening epilepsy around the time of waking up and temporal lobe seizures in the midafternoon. Sundowningmost prominently starts in the evening. GERD (gastroesophageal reflux disease) is most prominent during the early evening and night and correlates with the time of peak gastric acid secretion. Cluster headaches peak during stage REM sleep and nocturnal asthma around 4 a.m., when the FEVI (forced expiratory volume in 1 second) is minimum. Stroke and cardiovascular events and vascular headaches (migraines) occur shortly after waking up and correlate with the platelet adhesiveness, blood viscosity, coagulant activity, and catecholamine levels, all of which peak in the morning.

circadian rhythm of seizure activity [7]. This evidence suggests that both the sleep–wake cycle and the circadian system contribute to the diurnal variability in seizure propensity.

In addition to the relation that seizures have with the sleep and circadian systems, seizures also disrupt sleep [8] and circadian rhythms. Epileptic patients frequently report sleep problems such as insomnia and hypersomnia. Although there are multiple potential etiologies, interictal discharges at night may alter sleep regulation and provoke sleep disruption [9], contributing to these problems. There is also evidence of circadian changes in sleep habits. For instance, Dieter Janz described the sleep habits in his patients with JME, who reported staying up late at night, difficulty falling asleep, having their deepest sleep toward morning, and being drowsy for some time after awakening [5]. Interestingly, their sleep and wake pattern resembles that of patients with a circadian rhythm sleep disorder of the delayed sleep phase type. Nocturnal melatonin production is also elevated and associated with a later phase in patients with untreated epilepsy. Therefore patients with epilepsy may have a disruption in sleep and circadian rhythms, which may further exacerbate seizure frequency.

Studies have also shown that electrically induced temporal lobe seizures in hippocampal kindled rats cause transient changes in the phase of circadian rhythms and the

direction of the phase shift depends on the timing of the seizures [10]. However, the effect of seizures on human circadian rhythms has not been well studied. There is some evidence of short-term changes in the amplitude of temperature rhythms following electroconvulsive therapy (ECT) for depression [11]. ECT also results in prolongation of REM latency and normalization in the timing of the circadian rhythms of adrenocorticotrophic hormone, cortisol, and growth hormone in depressed patients [12]. Interestingly, normalization in circadian rhythms correlates with an improvement in depression.

CEREBROVASCULAR AND CARDIOVASCULAR DISORDERS

There is a prominent circadian variation in the timing of strokes and cardiovascular events. Several studies have demonstrated a higher frequency of ischemic strokes in the morning after awakening [12–14]. It has been postulated that circadian variation in platelet aggregability, coagulant activity, blood viscosity, catecholamine levels, and blood pressure, all of which rise in the morning, may explain this preferential timing of strokes [15, 16] (Fig. 51.1). In addition to circadian factors, which contribute to changes in coagulation and cardiovascular function, assumption of the upright posture (associated with awakening) plays an important role in the morning incidence of vascular events. This postural change is associated with platelet activation [15], as well as elevated catecholamine levels and plasma rennin activity, which contribute to the morning rise in blood pressure [15]. The degree of the morning surge in blood pressure has been shown to be related to the risk of silent cerebral infarction in elderly hypertensives, with a similarly increased risk following the afternoon siesta [17]. Therefore circadian rhythms in coagulation and cardiovascular function, together with the postural change associated with waking likely explain the increased incidence of stroke during the morning hours.

Not only do circadian rhythms influence the timing of stroke, but there is also evidence that circadian rhythms may be disrupted following acute stroke. A loss of the normal nocturnal dipping of blood pressure or even reverse dipping may be seen [18], and this may be clinically important in the acute phase, when careful blood pressure management is critical [18]. Alteration in the amplitude of the circadian rhythm of melatonin has also been reported in relation to stroke, so that there is a reduction in the nocturnal levels of melatonin [19]. Given the possible neuroprotective and immunological functions of melatonin, it is possible that the alteration in melatonin secretion may contribute to the increased morbidity in patients with cerebrovascular disease [20].

In contrast to ischemic stroke, much less is known about the timing of intracranial hemorrhage. The timing of intracerebral and subarachnoid hemorrhage (SAH) has been shown to have a seasonal distribution, with higher incidence in the winter [21]; a circadian modulation, however, has not been established. Limited evidence suggests that SAH events tend to peak between 9 and 10 a.m. in hypertensive patients with cerebral aneurysms when compared to those without hypertension [22] (Figure 51.1), suggesting that blood pressure elevation in the morning hours may increase the risk of hemorrhage.

Cardiovascular events like angina, acute myocardial infarction, cardiac arrhythmias, and sudden cardiac death also occur with a higher frequency in the morning [23, 24] (Figure 51.1). Pathophysiological mechanisms similar to the ones implicated in stroke may be responsible. In addition, a loss of morning endothelium-dependent vasodilation (EDV) may be seen in individuals with established coronary heart disease [25]. EDV may normally have a cardioprotective role in counteracting potentially adverse time-dependent changes in hemodynamic parameters such as thrombosis and fibrinolysis [26].

NEURODEGENERATIVE DISORDERS

Circadian rhythm disturbances have been associated with several neurodegenerative disorders, such as dementia and Parkinson's disease. Sleep disturbances, such as frequent awakenings, daytime sleepiness, and evening agitation, are common in patients with dementia and often precipitate nursing home placement [27, 28]. In addition to comorbid medical, psychiatric, and sleep disorders, alteration in circadian rhythmicity contributes to the high prevalence of sleep disturbances in this population [29, 30]. An association also exists between the level of sleep and circadian rhythm disruption and dementia severity [31, 32].

The types of circadian rhythm dysregulation in patients with Alzheimer's disease and other dementias range from alterations in sleep phase, decreased amplitude, to lack of discernable circadian rhythms. Decrease in the amplitude of the circadian rhythms of melatonin [33] growth hormone, thyroid stimulating hormone, rennin, aldosterone, estradiol, testosterone, and cortisol [34] as well as dampening in the diurnal variability of blood pressure and heart rate [35] have been reported in AD. A delay in the timing or phase of circadian rhythms can also be seen in patients with AD. A 4–5 hour delay has been reported in severely demented patients [36], with the most substantial delays in individuals with "sundowning" [37]. Finally, irregular sleep–wake pattern, in which there is a loss of a clear circadian rhythm of the sleep–wake cycle, has been reported in patients with dementia [32].

Circadian rhythm abnormalities in dementia may be due to degeneration of the central circadian clock as evidenced by reduction in the number of vasopressin SCN neurons [38]. Furthermore, in certain populations, like the institutionalized elderly, there is a lack of exposure to external synchronizing agents such as light and social schedules, which may act as predisposing as well as precipitating factors in the development of circadian rhythm disturbances [39].

Several studies have shown that increasing timed exposure to light or social structure can improve sleep–wake consolidation in patients with dementia [40, 41]. In addition to behavioral approaches, there is some evidence that treatment with melatonin in the evening may improve sleep and reduce behavioral disturbances in individuals with AD [42, 43]. However, a recent multicenter placebo-controlled trial of melatonin for the treatment of sleep disturbance in AD failed to demonstrate its effectiveness [44].

Sleep disturbance, excessive sleepiness, and fatigue are also common and often debilitating symptoms that significantly affect the quality of life in patients with Parkinson's disease (PD). It has been estimated that approximately 60% of patients with PD have sleep complaints consisting mostly of frequent nocturnal awakenings [45] and daytime sleepiness [46]. Sleep disruption in PD is most likely due to nocturnal motor symptoms and primary sleep disorders, such as restless legs, sleep apnea, and parasomnias. However, there is some evidence that circadian rhythms may also influence the symptoms of PD [47].

Limited data also indicates that circadian rhythms are altered in Parkinson's disease. For instance, heart rate variability and blood pressure are blunted [48, 49]. Renal fluid handling is also disturbed [50], and this may be due to cell loss in the supraoptic and paraventricular nuclei [51], which are involved in the regulation of fluid balance. This may account for the orthostatic hypotension and loss of circadian renal water excretion seen in PD. Diurnal fluctuation in motor disability can also be seen in individuals receiving chronic dopaminergic therapy [52]. These motor symptoms, referred to as "wearing off" or "on–off" phenomena, are characterized by worsening in the afternoon or evening. It is unclear whether this fluctuation in motor function is due to the effects of pharmacological treatment and/or related to disease progression per se.

HEADACHES

Headaches and sleep disorders commonly occur in the same people, and this may be related to the similarity of the neurotransmitters (serotonin and histamine) involved in the regulation of sleep, which are also implicated in the pathogenesis of certain types of headaches [53, 54]. In addition to the relationship of headaches with sleep, a circadian and circannual rhythm has been described for cluster headaches. Cluster headaches are more frequent during seasons when the day lengths are changing (spring and autumn). In addition, headaches tend to recur during the same time of the day or night [55], particularly during stage REM sleep [56] (Figure 51.1). Anatomically, the centers responsible for cluster headaches and circadian rhythms are closely related. The anterior hypothalamus (which contains the SCN) plays a central role not only in maintaining circadian rhythms, but also in measuring seasonal changes in day length [57]. Recent functional imaging data suggests that the posterior hypothalamus plays a role in the pathogenesis of cluster headaches [58]. The SCN projects to this area of the hypothalamus [59], suggesting a strong relation for cluster headaches with the centers involved in maintaining circadian rhythmicity.

There is also evidence that circadian rhythms are disrupted during cluster headache, particularly the endocrine rhythms. Nocturnal melatonin and testosterone levels are reduced, and the 24 hour cortisol levels are increased with a delay in the timing of the cortisol minimum during the cluster period. In addition to these changes, the 24 hour mean prolactin levels are lower during both the cluster and remission periods [60]. The finding of such circadian alterations during the remission period suggests that these changes are not simply the result of pain, but may be part of the pathology underlying cluster headaches.

Migraine headaches also exhibit a circadian periodicity, and the peak of attacks occurs between 8 and 10 a.m. (Figure 51.1), with reduced frequency between 8 p.m. and 4 a.m. [61]. The circadian rhythm of migraine onset parallels that of myocardial infarction and stroke. These associations may be related to the alteration of vasomotor tone due to the changes in platelet aggregability, plasma cortisol, and plasma catecholamines at this time, which may also be involved in the initiation of migraine attacks [61] (Figure 51.1).

GASTROINTESTINAL DISORDERS

Gastroesophageal reflux disorder (GERD) exhibits a diurnal pattern in its symptomatology, so that approximately 74% of individuals with chronic GERD report nocturnal heartburn [62]. This nocturnal predilection may be related to the circadian variation in gastrointestinal functioning. Gastric acid secretion exhibits a circadian rhythm, with an increase in basal acid output in the late evening hours and reduction in the morning hours [63] (Figure 51.1). This finding, together with the physiologic changes associated with the sleep state such as changes in posture, reduced arousal threshold and delayed acid clearance by the esophagus, can promote nocturnal acid reflux [64] (Figure 51.1). GERD in turn may further disrupt nocturnal

sleep [65] and has been shown to exacerbate symptoms of bronchial asthma at night [66].

Sleep and circadian disturbances are also common in hepatic encephalopathy and other forms of liver disease. Cirrhotic patients report sleep with a delayed sleep phase pattern [67] and delayed melatonin rhythms are seen in these individuals [68]. Animal studies have also demonstrated similar circadian disturbances [69], which improve with low-protein diet or neomycin (common therapies for hepatic encephalopathy) [70], suggesting that these disturbances are caused by the liver dysfunction itself. It is not clear, however, whether the circadian rhythm abnormalities associated with liver disease reflect a change in the output of the SCN or from changes in melatonin metabolism resulting from liver dysfunction.

RESPIRATORY DISORDERS

Many patients with respiratory disorders note increase in the severity of symptoms during the night or in the early morning. One disorder in which this is characteristic is nocturnal asthma (NA). While it is not clear whether this is a qualitatively different disorder with unique circadian characteristics, or simply a more severe form of bronchial asthma, circadian modulation does appear to have an important role in its pathogenesis. Circadian variations in airway and lung function are likely contributing factors. In most normal individuals, the peak values of respiratory function occur at 4 p.m., and the lowest values at 4 a.m. (during sleep) [71, 72] (Figure 51.1), and NA is characterized by an exaggeration of this variation by more than 15% [73]. Although this may be an effect of sleep, the endogenous circadian rhythm of pulmonary function may also play a role, independent of sleep [72]. This is associated with an increase in the markers of inflammation (interleukin 1β, lymphocyte, circulating bronchoalveolar and alveolar eosinophil levels) [74, 75] at night in NA compared to asthmatic subjects with a similar severity of daytime symptoms. Therefore the nocturnal symptomatology in NA may be due at least in part to circadian variation in pulmonary function and inflammation. Circadian modulation of respiratory function in patients with NA may also explain the sometimes observed decreased response to treatment during the early morning. For example, inhibition of lymphocyte proliferation by steroids at 4 a.m. is reduced, requiring up to a tenfold increase in dose [76].

CONCLUSION

Circadian rhythms in physiology and behavior play an important role in the regulation of sleep and maintenance of health. Although a causal relationship between alterations in circadian rhythms and neurological or medical conditions has not been clearly established, there is increasing evidence that circadian rhythmicity may play an important role in the expression of some common neurological and medical disorders. Diurnal changes in physiology, behavior, and endocrine function may also influence the already disrupted sleep that is commonly seen in patients with neurological and medical illnesses, further exacerbating the disease process. Therefore understanding circadian variation in the manifestation of these disorders not only will improve our understanding of the pathogenesis of some common neurological and medical disorders, but could also lead to the development of better diagnostic and therapeutic strategies.

REFERENCES

1. Czeisler CA, et al. Stability, precision, and near-24-hour period of the human circadian pacemaker. *Science* **284**(5423):2177–2181(1999).

2. Moore RY, Lenn NJ. A retinohypothalamic projection in the rat. *J Comp Neurol* **146**(1):1–14(1972).

3. Stephan FK, Zucker I. Circadian rhythms in drinking behavior and locomotor activity of rats are eliminated by hypothalamic lesions. *Proc Natl Acad Sci USA* **69**(6):1583–1586(1972).

4. Quigg M, et al. Temporal distribution of partial seizures: comparison of an animal model with human partial epilepsy. *Ann Neurol* **43**(6):748–755(1998).

5. Janz D, Epilepsy with grand mal on awakening and sleep–waking cycle. *Clin Neurophysiol* **111** (Suppl 2):S103–S110(2000).

6. Bazil CW. Sleep and epilepsy. *Curr Opin Neurol* **13**(2):171–175(2000).

7. Binnie CD, et al. Temporal characteristics of seizures and epileptiform discharges. *Electroencephalogr Clin Neurophysiol* **58**(6):498–505(1984).

8. Mendez M, Radtke RA. Interactions between sleep and epilepsy. *J Clin Neurophysiol* **18**(2):106–127(2001).

9. Vaughn BV, D'Cruz OF. Sleep and epilepsy. *Semin Neurol* **24**(3):301–313(2004).

10. Quigg M, et al. Seizures induce phase shifts of rat circadian rhythms. *Brain Res* **913**(2):165–169(2001).

11. Szuba MP, Guze BH, Baxter LR Jr. Electroconvulsive therapy increases circadian amplitude and lowers core body temperature in depressed subjects. *Biol Psychiatry* **42**(12):1130–1137(1997).

12. Linkowski P, et al. 24-hour profiles of adrenocorticotropin, cortisol, and growth hormone in major depressive illness: effect of antidepressant treatment. *J Clin Endocrinol Metab* **65**(1):141–152(1987).

13. Marsh EE III, et al. Circadian variation in onset of acute ischemic stroke. *Arch Neurol* **47**(11):1178–1180(1990).

14. Wroe SJ, et al. Diurnal variation in incidence of stroke: Oxfordshire community stroke project. *BMJ* **304**(6820):155–157(1992).

15. Brezinski DA, et al. Morning increase in platelet aggregability. Association with assumption of the upright posture. *Circulation* **78**(1):35–40(1988).

16. Krantz DS, et al. Circadian variation of ambulatory myocardial ischemia. Triggering by daily activities and evidence for an endogenous circadian component. *Circulation* **93**(7):1364–1371(1996).

17. Kario K, et al. Morning surge in blood pressure as a predictor of silent and clinical cerebrovascular disease in elderly hypertensives: a prospective study. *Circulation* **107**(10):1401–1406(2003).

18. Jain S, et al. Loss of circadian rhythm of blood pressure following acute stroke. *BMC Neurol* **4**(1):1(2004).

19. Fiorina P, et al. Impaired nocturnal melatonin excretion and changes of immunological status in ischaemic stroke patients. *Lancet* **347**(9002):692–693(1996).

20. Pierpaoli W, Yi C. The involvement of pineal gland and melatonin in immunity and aging. I. Thymus-mediated, immunoreconstituting and antiviral activity of thyrotropin-releasing hormone. *J Neuroimmunol* **27**(2-3):99–109(1990).

21. Nyquist PA, et al. Circadian and seasonal occurrence of subarachnoid and intracerebral hemorrhage. *Neurology* **56**(2):190–193(2001).

22. Kleinpeter G, Schatzer R, Bock F. Is blood pressure really a trigger for the circadian rhythm of subarachnoid hemorrhage? *Stroke* **26**(10):1805–1810(1995).

23. Muller JE, et al. Circadian variation in the frequency of onset of acute myocardial infarction. *N Engl J Med* **313**(21):1315–1322(1985).

24. Muller JE, et al. Circadian variation in the frequency of sudden cardiac death. *Circulation* **75**(1):131–138(1987).

25. Shaw JA, et al. Diurnal variation in endothelium-dependent vasodilatation is not apparent in coronary artery disease. *Circulation* **103**(6):806–812(2001).

26. Walters J, et al. Biological rhythms, endothelial health and cardiovascular disease. *Med Sci Monit* **9**(1):RA1–RA8(2003).

27. Bliwise DL. Sleep disorders in Alzheimer's disease and other dementias. *Clin Cornerstone* **6** (Suppl 1A):S16–S28(2004).

28. Pollak CP, et al. Sleep problems in the community elderly as predictors of death and nursing home placement. *J Commun Health* **15**(2):123–135(1990).

29. Gehrman P, et al. The timing of activity rhythms in patients with dementia is related to survival. *J Gerontol A Biol Sci Med Sci* **59**(10):M1050–M1055(2004).

30. Shochat T, Loredo J, Ancoli-Israel S. Sleep disorders in the elderly. *Curr Treat Options Neurol* **3**(1):19–36(2001).

31. Vitiello MV, Prinz PN. Alzheimer's disease. Sleep and sleep/wake patterns. *Clin Geriatr Med* **5**(2):289–299(1989).

32. Ancoli-Israel S, et al. Variations in circadian rhythms of activity, sleep, and light exposure related to dementia in nursing-home patients. *Sleep* **20**(1):18–23(1997).

33. Mishima K, et al. Melatonin secretion rhythm disorders in patients with senile dementia of Alzheimer's type with disturbed sleep-waking. *Biol Psychiatry* **45**(4):417–421(1999).

34. Van Gool, WA, Mirmiran M. Aging and circadian rhythms. *Prog Brain Res* **70**:255–277(1986).

35. Otsuka A, et al. Absence of nocturnal fall in blood pressure in elderly persons with Alzheimer-type dementia. *J Am Geriatr Soc* **38**(9):973–978(1990).

36. Satlin A, et al. Circadian locomotor activity and core-body temperature rhythms in Alzheimer's disease. *Neurobiol Aging* **16**(5):765–771(1995).

37. Volicer L, et al. Sundowning and circadian rhythms in Alzheimer's disease. *Am J Psychiatry* **158**(5):704–711(2001).

38. Swaab DF, et al. Biological rhythms in the human life cycle and their relationship to functional changes in the suprachiasmatic nucleus. *Prog Brain Res* **111**:349–368(1996).

39. Wever RA, Polasek J, Wildgruber CM. Bright light affects human circadian rhythms. *Pflugers Arch* **396**(1):85–87(1983).

40. Ancoli-Israel S, et al. Effect of light treatment on sleep and circadian rhythms in demented nursing home patients. *J Am Geriatr Soc* **50**(2):282–289(2002).

41. Mishima K, et al. Morning bright light therapy for sleep and behavior disorders in elderly patients with dementia. *Acta Psychiatr Scand* **89**(1):1–7(1994).

42. Asayama K, et al. Double blind study of melatonin effects on the sleep–wake rhythm, cognitive and non-cognitive functions in Alzheimer type dementia. *J Nippon Med Sch* **70**(4):334–341(2003).

43. Brusco LI, et al. Effect of melatonin in selected populations of sleep-disturbed patients. *Biol Signals Recept* **8**(1-2):126–131(1999).

44. Singer C, et al. A multicenter, placebo-controlled trial of melatonin for sleep disturbance in Alzheimer's disease. *Sleep* **26**(7):893–901(2003).

45. Tandberg E, Larsen JP, Karlsen K. A community-based study of sleep disorders in patients with Parkinson's disease. *Mov Disord* **13**(6):895–899(1998).

46. Partinen M. Sleep disorder related to Parkinson's disease. *J Neurol* **244**(4 Suppl 1):S3–S6(1997).

47. Bliwise DL, et al. Disruptive nocturnal behavior in Parkinson's disease and Alzheimer's disease. *J Geriatr Psychiatry Neurol* **8**(2):107–110(1995).

48. Kallio M, et al. Nocturnal cardiac autonomic regulation in Parkinson's disease. *Clin Auton Res* **14**(2):119–124(2004).

49. Sakata M, et al. Mesolimbic dopaminergic system is involved in diurnal blood pressure regulation. *Brain Res* **928**(1-2):194–201(2002).

50. Hineno T, et al. Disappearance of circadian rhythms in Parkinson's disease model induced by 1-methyl-4-phenyl-1,2,3,6-tetrahydropyridine in dogs. *Brain Res* **580**(1-2):92–99(1992).

51. Ansorge O, Daniel SE, Pearce RK. Neuronal loss and plasticity in the supraoptic nucleus in Parkinson's disease. *Neurology* **49**(2):610–613(1997).

52. Fertl E, et al. Circadian secretion pattern of melatonin in de novo parkinsonian patients: evidence for phase-shifting properties of L-dopa. *J Neural Transm Park Dis Dement Sect* **5**(3):227–234(1993).

53. Blau JN. Resolution of migraine attacks: sleep and the recovery phase. *J Neurol Neurosurg Psychiatry* **45**(3):223–226(1982).

54. Lance JW, et al. Brainstem influences on the cephalic circulation: experimental data from cat and monkey of relevance to the mechanism of migraine. *Headache* **23**(6):258–265(1983).

55. Ekbom K. Nitrolglycerin as a provocative agent in cluster headache. *Arch Neurol* **19**(5):487–493(1968).

56. Cohen AS, Kaube H. Rare nocturnal headaches. *Curr Opin Neurol* **17**(3):295–299(2004).

57. Sumova A, et al. Seasonal molecular timekeeping within the rat circadian clock. *Physiol Res* **53** (Suppl 1):S167–S176(2004).

58. May A, et al. Hypothalamic activation in cluster headache attacks. *Lancet* **352**(9124):275–278(1998).

59. Chou TC, et al. Afferents to the ventrolateral preoptic nucleus. *J Neurosci* **22**(3):977–990(2002).

60. Waldenlind E, Gustafsson SA. Prolactin in cluster headache: diurnal secretion, response to thyrotropin-releasing hormone, and relation to sex steroids and gonadotropins. *Cephalalgia* **7**(1):43–54(1987).

61. Solomon GD. Circadian rhythms and migraine. *Cleve Clin J Med* **59**(3):326–329(1992).

62. Farup C, et al. The impact of nocturnal symptoms associated with gastroesophageal reflux disease on health-related quality of life. *Arch Intern Med* **161**(1):45–52(2001).

63. Moore JG. Circadian dynamics of gastric acid secretion and pharmacodynamics of H2 receptor blockade. *Ann NY Acad Sci* **618**:150–158(1991).

64. Orr WC. Sleep and gastroesophageal reflux: what are the risks? *Am J Med* **115** Suppl 3A:109S–113S(2003).

65. Shaker R, et al. Nighttime heartburn is an under-appreciated clinical problem that impacts sleep and daytime function: the results of a Gallup survey conducted on behalf of the American Gastroenterological Association. *Am J Gastroenterol* **98**(7):1487–1493(2003).

66. Sontag SJ, et al. Asthmatics have more nocturnal gasping and reflux symptoms than nonasthmatics, and they are related to bedtime eating. *Am J Gastroenterol* **99**(5):789–796(2004).

67. Cordoba J, et al. High prevalence of sleep disturbance in cirrhosis. *Hepatology* **27**(2):339–345(1998).

68. Steindl PE, et al. Changes in the 24-hour rhythm of plasma melatonin in patients with liver cirrhosis—relation to sleep architecture. *Wien Klin Wochenschr* **109**(18):741–746(1997).

69. Zee PC, et al. Portacaval anastomosis disrupts circadian locomotor activity and pineal melatonin rhythms in rats. *Brain Res* **560**(1-2):17–22(1991).

70. Steindl PE, et al. A low-protein diet ameliorates disrupted diurnal locomotor activity in rats after portacaval anastomosis. *Am J Physiol* **271**(4 Pt 1):G555–G560(1996).

71. Hetzel MR, Clark TJ. Comparison of normal and asthmatic circadian rhythms in peak expiratory flow rate. *Thorax* **35**(10):732–738(1980).

72. Spengler CM, Shea SA. Endogenous circadian rhythm of pulmonary function in healthy humans. *Am J Respir Crit Care Med* **162**(3 Pt 1):1038–1046(2000).

73. Calhoun WJ. Nocturnal asthma. *Chest* **123**(3 Suppl):399S–405S(2003).

74. Jarjour NN, Busse WW. Cytokines in broncho-alveolarlavage fluid of patients with nocturnal asthma. *Am J Respir Crit Care Med* **152**(5 Pt 1):1474–1477(1995).

75. Mackay TW, et al. Role of inflammation in nocturnal asthma. *Thorax* **49**(3):257–262(1994).

76. Kraft M, et al. Nocturnal asthma is associated with reduced glucocorticoid receptor binding affinity and decreased steroid responsiveness at night. *J Allergy Clin Immunol* **103**(1 Pt 1):66–71(1999).

52

PSYCHIATRIC DISORDERS ASSOCIATED WITH CIRCADIAN RHYTHM DISTURBANCES

STEPHANY JONES AND RUTH M. BENCA

University of Wisconsin–Madison Medical School, Madison, Wisconsin

INTRODUCTION

A variety of physiological and behavioral variables display circadian rhythmicity in humans such as sleep–wake cycle organization, subjective alertness, mood, body temperature, and the secretion of several hormones. Mammalian circadian rhythms are governed by a master oscillator in the suprachiasmatic nucleus (SCN) of the hypothalamus that is coupled to peripheral oscillators throughout the body. The central circadian oscillator can function autonomously, but synchronization to a precise 24 hour period requires entrainment to the solar day by the light–dark cycle. This synchronization of the circadian pacemaker ensures the proper organization of physiological and behavioral functions and allows for the proper adaptation of the organism to day/night and seasonal cycles.

Although there is currently no direct evidence that circadian abnormalities are causally related to any given psychiatric condition, alterations in the amplitude or phase of several circadian output variables, as well as mounting evidence that certain circadian rhythm manipulations prove therapeutically effective, suggest a potential pathophysiological role for the circadian pacemaker in major depressive disorder, bipolar disorder, and seasonal affective disorder (SAD) or winter depression.

Circadian abnormalities have long been hypothesized to play a role in the pathophysiology of many forms of psychiatric illness. Disturbances in the sleep–wake cycle, hormonal rhythms, and temperature regulation have been widely described in the symptomatology of mood disorders.

It is not surprising that circadian rhythm abnormalities have been associated with various mood disorders, since mood regulation in humans shows circadian variability. In normal subjects, mood is generally best at the peak of the core body temperature in the late afternoon/early evening, whereas mood is at its lowest point coinciding with the trough in core body temperature in the early morning hours [1]. Sleep loss in normal subjects generally results in decrements in subjective mood [2], and a recent analysis of subjective mood in healthy volunteers reported mood state to be influenced interactively by both circadian phase and duration of prior wakefulness. Specifically, the timing of self-reports of lowest subjective happiness moved toward later circadian phases as duration of wakefulness increased [3]. Whether the circadian pacemaker directly regulates mood, or whether the circadian effect on mood is secondary to other circadian outputs (e.g., neuroendocrine factors, sleep patterns) that in turn modulate mood is not specifically known, however.

MAJOR DEPRESSIVE DISORDER

Biological markers of circadian rhythms have been studied more extensively in major depression than in any other psychiatric disorder. Major depressive disorder is a mood disorder with a lifetime prevalence of 15% in women and 8% in men. The *Diagnostic and Statistical Manual of Mental Disorders*, 4th edition (DSM-IV) [4] outlines two diagnostic subtypes of major depression, melancholic and atypical, with each subtype differing substantially in its clinical profile. Although both subtypes exhibit marked mood disturbance, atypical depression is characterized by hypersomnia, lethargy, and fatigue while symptoms of melancholic depression include insomnia, loss of appetite, and, often, hypothalamic–pituitary–adrenal (HPA) axis and sympathetic nervous system activation. Several clinical features of major depression, including diurnal mood variation and alterations in hormonal secretion, suggest an involvement of the circadian system in this disorder, and abnormalities in sleep and sleep stage distribution are perhaps the most consistently observed circadian abnormality in major depression.

Dysregulation of the HPA axis is one of the most replicated findings in at least a subset of patients with major depressive disorder. Frequently reported findings include elevated cortisol and corticotropin-releasing hormone (CRH), nonsuppression of cortisol on the dexamethasone suppression test, and/or a blunted adrenocorticotropic hormone (ACTH) response to CRH [5].

Given that cortisol secretion is regulated in large measure by the circadian system, much research has focused on potential abnormalities in the circadian regulation of the HPA axis in depression. Evidence for disruptions in the circadian period of the cortisol secretory period in major depression has been mixed. After administration of metyrapone, a cortisol synthesis inhibitor, Young and colleagues [6, 7] reported increased pituitary output of adrenocorticotropic hormone (ACTH) in the evening compared to the morning in clinically depressed patients, and interpreted this result as support for the notion of circadian alterations in HPA axis function. In a more recent study, however, they examined the 24 hour secretory patterns of both ACTH and cortisol and found no evidence of alterations in the circadian or ultradian parameters of the HPA axis in currently depressed women compared to controls [8]. In general, evidence for circadian modulation of cortisol secretion in depression has failed to support depression-associated phase shifts in this rhythm, but the question of whether or not these abnormalities exist is still unclear. Disruptions in the sleep–wake cycle, which are integral to the depressive state, could at least partially account for cortisol abnormalities reported during the depressive phase of the disorder, since sleep deprivation has been shown to increase activity of the HPA axis [9].

Several studies have also reported abnormalities, including phase advances and/or decreases in the nocturnal amplitude of melatonin rhythm, in the circadian profile of melatonin secretion in patients with major depressive disorder, although these findings have not been observed in all studies. A recent study, which controlled for medication status, age, season, and gender of participants, reported a phase shift in the pineal secretion of melatonin in depressed patients relative to controls; depressed patients showed a delay of 77 minutes in the onset of melatonin relative to the control group matched for age and season [10]. The study failed to assess hypercortisolemia—an abnormality common in depression and known to influence melatonin secretion—and it did not clearly report on the affective state of depressed participants across samplings.

Some of the best-documented clinical findings in major depression are alterations in sleep architecture. Disturbed sleep (e.g., prolonged sleep latency, early morning awakenings, reduced sleep efficiency), loss of slow-wave sleep, and several abnormalities in rapid eye movement (REM) sleep have been widely described in groups of depressed patients (reviewed in [11]). Alterations in REM sleep have particular relevance as a marker of circadian abnormalities in depression since the amount and timing of REM sleep are governed in large part by a circadian process. In healthy subjects, REM sleep propensity normally reaches its maximum in the latter portion of the sleep period. In contrast, depressed patients often exhibit abnormally early REM sleep onset relative to sleep onset (i.e., reduced REM sleep latency), a larger proportion of REM sleep during the first third of the night, and increased eye movements during the REM sleep period. Although these alterations of REM sleep are more profound during periods of clinical illness, there is evidence that these abnormalities persist during clinical remission and have been described in unaffected first-degree relatives of depressed patients [12].

Alterations in REM sleep timing and distribution in depression have been hypothesized to reflect abnormally advanced circadian rhythms relative to the sleep–wake cycle. Wehr and Wirz-Justice [13] proposed the "internal coincidence model," which postulates a critical period in the early morning hours when sleep is suggested to exert a depressogenic effect in vulnerable individuals. It has been hypothesized that if depressed patients indeed sleep at the wrong circadian time, shifting the sleep–wake period to earlier in the evening would synchronize out-of-phase rhythms, resulting in mood improvement. Several studies employing sleep phase advances have proved therapeutically effective for depressed patients [14]. Sleep phase advances have also been used successfully to preserve the antidepressant effects of total sleep deprivation [15, 16]. Although sleep phase advances are unquestionably effective in symptom alleviation in some depressed patients, these treatments do not invariably produce a clinical response. Furthermore, there is currently no evidence that these sleep–wake manipulations alter REM sleep latency [17]. The lack of evidence for remediation of the purported circadian abnormality calls into question whether or not these circadian abnormalities are pathophysiologically significant or are merely epiphenomena of the depressed state, and further research is needed to clarify this issue. Nevertheless, the efficacy of sleep phase advance treatment in depressed patients and recent evidence for an interactive effect between sleep and the circadian system on mood regulation in normal subjects [18] highlight the importance of proper alignment of endogenous circadian phase and sleep–wakefulness timing for mood regulation.

Although there is not consistent support for uniform disruption of circadian variables in depression, there is a substantial body of support for circadian abnormalities in at least some depressive patients. Given the increasing recognition of etiological heterogeneity in depressive disorders, it is likely that only a subset of depressive patients will present with circadian abnormalities. Identifying particular groups of patients with circadian abnormalities and determining whether such abnormalities are causally related to depressive symptomatology is a critical next step.

BIPOLAR DISORDER

Bipolar disorder is a mood disorder characterized by recurrent episodes of mania typically accompanied by major depressive episodes. The disorder is reported to have a lifetime prevalence of approximately 1% in the general population with an estimated 5–10% of patients presenting symptoms with a seasonal pattern of recurrence. The high degree of seasonal presentation of symptoms in bipolar disorder and the profound impact of sleep–wake cycle timing in the regulation of mood in both the manic and depressive states suggest that a subset of bipolar patients may have baseline circadian abnormalities and/or altered photoperiodic sensitivities that disrupt the normal entrainment of the clock to the natural environment. The changing clinical profile of patients with bipolar disorder represents a special challenge to empirical analyses of circadian variables in this disorder. Behavioral alterations during the clinical state can influence the output of the circadian pacemaker and, to further complicate issues, studies of depression have often mixed patients with major depressive disorder and those with bipolar disorder.

Many of the circadian output variables that are disrupted in patients suffering from major depression, however, have also been reported to be abnormal in patients with bipolar disorder during the depressive phase. Sleep abnormalities commonly reported in major depression, including decreased sleep continuity, reduced slow-wave sleep (SWS), and reduced REM sleep latencies have been observed in both bipolar depressive and manic episodes.

The relationship between circadian variables and sleep disturbance in bipolar disorder is complex and incompletely understood. As is the case in major depression, sleep phase advances have been effectively used to treat the depression associated with bipolar illness [19], but sleep manipulations are also known to elicit mania in susceptible patients. Shifts in the timing of sleep have been shown to induce mood changes in bipolar patients, suggesting that an abnormal phase relationship between sleep and the endogenous circadian rhythm or abnormal circadian control of sleep might be of pathophysiological significance in this disorder [12]. Extending the period of sleep and darkness has proved effective in preventing mania and rapid cycling in susceptible individuals [20], although it is not entirely clear if the critical therapeutic feature in this case is the extension of the dark period or the stabilization of the timing of the sleep–wake cycle.

Substantial attention has been given to potential melatonin abnormalities in bipolar disorder, but few studies have focused on patients in all phases of the disorder. One analysis of the melatonin secretory profiles in bipolar patients in the manic, depressed, and euthymic states reported no phase differences between patients and controls but did report lower melatonin levels in patients relative to controls in each phase of the disorder [21]. More recent analyses of melatonin abnormalities have focused on melatonin sensitivity to light in bipolar patients. There is evidence of a supersensitivity to light in some bipolar patients, and it has been hypothesized that this sensitivity might result in circadian phase instability. Nurnberger and colleagues [22] compared melatonin suppression levels in response to light in euthymic bipolar and unipolar patients; they reported no overall difference in suppression levels between the two groups, but when bipolar patients alone were compared to controls, they exhibited increased dark-adjusted melatonin suppression as well as a significantly later melatonin peak.

SEASONAL AFFECTIVE DISORDER

Winter depression, or seasonal affective disorder (SAD), is a clinical subtype of major depressive disorder or bipolar disorder in which patients experience recurrent depressive episodes during the fall/winter months with spontaneous remission of symptoms in spring/summer. Bipolar patients with winter depression may show evidence of hypomania in the spring and summer months. Sadness, irritability, and anxiety are characteristic features of SAD, as are atypical symptoms including increased sleep and appetite as well as carbohydrate craving. The prevalence of SAD in the United States is estimated to be approximately 6% with a greater prevalence among females (reviewed in [18, 23]. Although the etiology of SAD is unclear, the seasonal presentation of depressive symptoms and the efficacy of light therapy in symptom alleviation implicate circadian rhythm and potential photoperiodic abnormalities in SAD pathogenesis.

Bright light exposure induces marked improvement of depressive symptomatology in SAD patients [2]. The clinical efficacy of light exposure in these patients led to the early hypothesis that the abbreviated light period in winter was responsible for the onset of depression. Early studies using bright light exposure in the morning and evening, effectively extending the photoperiod, proved effective in symptom reduction and seemed to support the shortened photoperiod hypothesis of SAD. Subsequent evidence indicating that daily pulses of light were equally as effective as photoperiodic extension in symptom reduction proved inconsistent with the hypothesis [24], but the efficacy of light therapy has fostered increasingly refined research on the potential role of circadian and circannual abnormalities in SAD.

One theory of circadian dysfunction in SAD, initially proposed by Lewy, posits that the endogenous circadian system in SAD patients is phase-delayed relative to the sleep–wake cycle [25]. Several studies have sought to test this hypothesis by examining the circadian phase of melatonin onset under dim-light conditions in SAD patients and they have generally produced conflicting results. Thompson et al. [26] reported no difference between the melatonin onset of SAD patients compared to controls, whereas Lewy and colleagues [27] found significant differences between the two groups. In a recent study using a forced desynchrony protocol, a procedure that disassociates the sleep–wake cycle from the endogenous circadian rhythm and effectively unmasks the function of the clock, Koorengevel et al. [28, 29] found no differences in the phase of the melatonin secretion between SAD patients and controls, nor did they find evidence for differences in other circadian parameters such as body temperature rhythm or diurnal mood variation.

Although the mood-enhancing effect of phototherapy in SAD is essentially uncontested, the exclusive therapeutic effect of morning phototherapy has been a matter of debate. If the phase-delay hypothesis of SAD were correct, morning phototherapy would be expected to result in a phase advance of the circadian system and greater clinical improvement. Although many studies have found that a phase advance appears to be involved in the antidepressant response to morning light [30, 31], there is also evidence to suggest that evening phototherapy, which should result in phase delay and consequent mood deterioration, also has a therapeutic effect in SAD patients [32]. The efficacy of light in reducing or preventing depression in SAD patients adds credence to the supposition that interactions between photoperiodic mechanisms and the circadian system are etiologically significant in SAD, but evidence for phase delay of circadian rhythms is not unequivocally supported from the data on light therapy.

Wehr et al. [33] have recently helped to characterize the nature of circadian and circannual abnormalities in SAD, reporting that SAD patients generated a seasonal change in the duration of their dim-light melatonin secretion profile. Specifically, SAD patients produced a longer duration

of nocturnal melatonin secretion in winter relative to summer while normal subjects exhibited no such seasonal alteration in the melatonin profile. These data suggest that SAD patients, contrary to normal subjects, exhibit a change-of-season signal similar to that used by other mammals that show significant seasonal changes in behavior. The authors suggest that this signal potentially mediates the pathogenesis of winter depression, although the mechanism through which the melatonin signal leads to affective deterioration is still unclear. SAD patients may indeed have circadian abnormalities, but alterations in melatonin production and sensitivity to light may be seasonally dependent.

The conflicting evidence for involvement of circadian abnormalities in SAD could reflect a number of factors. Although circadian output variables such as melatonin secretion and body temperature are easily measured, interpretation of these measures as reliable indices of clock function is often empirically confounded. Analyses of circadian output variables carried out under conditions of normal lighting and sleep–wakefulness timing ultimately provide limited information about the functioning of the circadian pacemaker. Drug status, quality of light stimulus used, age of participants, and season are not consistently specified or controlled for in all studies, and each of these variables can potentially influence the output of the circadian pacemaker. Lastly, clinical SAD research is also hampered by small sample sizes, diagnostic inconsistencies, as well as potential etiological heterogeneity. Further use of protocols that specifically manipulate light exposure and sleep–wake cycles as well as greater attention to patient selection criteria will continue to enhance our understanding of circadian abnormalities in SAD.

CONCLUSION

Given the prevalence of circadian abnormalities in mood disorders, a clearer understanding of the role of the circadian pacemaker in normal and abnormal mood regulation is critical for understanding the biological underpinnings of the disease process. Studies of circadian physiology in mental illness have been plagued by small sample size, inadequate protocols, and, perhaps most importantly, patient heterogeneity. The notion of a unitary cause for any complex disease is increasingly recognized as insupportable; thus it is likely that only a subset of patients will present with circadian abnormalities. Identifying these groups of patients and determining whether circadian abnormalities are causally related to symptomatology is a critical next step. Recent progress in characterizing the molecular mechanisms of the human clock function will likely provide new insights on the etiology and pathophysiology of psychiatric illness.

REFERENCES

1. Monk TH, Buysse DJ, Reynolds CFD, Jarrett DB, Kupfer DJ. Rhythmic vs homeostatic influences on mood, activation, and performance in young and old men. *J Gerontol* **47**(4):221–227(1992).

2. Brendel DH, Reynolds CF III, Jennings JR, Hoch CC, Monk TH, Berman SR, et al. Sleep stage physiology, mood and vigilance responses to total sleep deprivation in healthy 80-year-olds and 20-year-olds. *Psychophysiology*, **27**:677–685 (1990).

3. Boivin DB, Czeisler CA, Dijk DJ, Duffy JF, Folkard S, Minors DS, et al. Complex interaction of the sleep–wake cycle and circadian phase modulates mood in healthy subjects. *Arch Gen Psychiatry* **54**(2):145–152(1997).

4. *Diagnostic and Statistical Manual of Mental Disorders*, 4th ed. American Psychiatric Association, Washington, DC, 1994.

5. Gold PW, Chrousos GP. Organization of the stress system and its dysregulation in melancholic and atypical depression: high vs low CRH/NE states. *Mol Psychiatry* **7**(3):254–275(2002).

6. Young EA, Haskett RF, Grunhaus L, Pande A, Weinberg VM, Watson SJ, et al. Increased evening activation of the hypothalamic–pituitary–adrenal axis in depressed patients. *Arch Gen Psychiatry* **51**(9):701–707(1994).

7. Young EA, Lopez JF, Murphy-Weinberg V, Watson SJ, Akil H. Normal pituitary response to metyrapone in the morning in depressed patients: implications for circadian regulation of corticotropin-releasing hormone secretion. *Biol Psychiatry* **41**(12):1149–1155(1997).

8. Young EA, Carlson NE, Brown MB. Twenty-four-hour ACTH and cortisol pulsatility in depressed women. *Neuropsychopharmacology* **25**(2):267–276(2001).

9. Spiegel K, Leproult R, Van Cauter E. Impact of sleep debt on metabolic and endocrine function. *Lancet* **354**(9188):1435–1439(1999).

10. Crasson M, Kjiri S, Colin A, Kjiri K, L'Hermite-Baleriaux M, Ansseau M, et al. Serum melatonin and urinary 6-sulfatoxymelatonin in major depression. *Psychoneuroendocrinology* **29**(1):1–12(2004).

11. Benca RM, Obermeyer WH, Thisted RA, Gillin JC. Sleep and psychiatric disorders: a meta-analysis. *Arch Gen Psychiatry* **49**:651–668(1992).

12. Riemann D, Voderholzer U, Berger M. Sleep and sleep–wake manipulations in bipolar depression. *Neuropsychobiology* **45**(Suppl 1):7–12(2002).

13. Wehr TA, Wirz-Justice A. Internal coincidence model for sleep deprivation and depression. In: Koella WP (Ed), *Sleep* 1980. Karger, Basel, 1981 pp 26–33.

14. Wirz-Justice A, Van den Hoofdakker RH. Sleep deprivation in depression: what do we know, where do we go? *Biol Psychiatry* **46**(4):445–453(1999).

15. Riemann D, Konig A, Hohagen F, Kiemen A, Voderholzer U, Backhaus J, et al. How to preserve the antidepressive effect of sleep deprivation: a comparison of sleep phase advance and sleep phase delay. *Eur Arch Psychiatry Clin Neurosci* **249**(5):231–237(1999).

16. Voderholzer U, Valerius G, Schaerer L, Riemann D, Giedke H, Schwarzler F, et al. Is the antidepressive effect of sleep deprivation stabilized by a three day phase advance of the sleep period? A pilot study. *Eur Arch Psychiatry Clin Neurosci* **253**(2):68–72(2003).

17. Wehr TA, Wirz-Justice A. Circadian rhythm mechanisms in affective illness and in antidepressant drug action. *Pharmacopsychiatria* **15**(1):31–39(1982).

18. Lam RW, Levitan RD. Pathophysiology of seasonal affective disorder: a review. *J Psychiatry Neurosci* **25**(5):469–480(2000).

19. Benedetti F, Barbini B, Campori E, Fulgosi MC, Pontiggia A, Colombo C. Sleep phase advance and lithium to sustain the antidepressant effect of total sleep deprivation in bipolar depression: new findings supporting the internal coincidence model? *J Psychiatr Res* **35**(6):323–329(2001).

20. Wehr TA, Turner EH, Shimada JM, Lowe CH, Barker C, Leibenluft E. Treatment of rapidly cycling bipolar patient by using extended bed rest and darkness to stabilize the timing and duration of sleep. *Biol Psychiatry* **43**(11):822–828(1998).

21. Kennedy SH, Kutcher SP, Ralevski E, Brown GM. Nocturnal melatonin and 24-hour 6-sulphatoxymelatonin levels in various phases of bipolar affective disorder. *Psychiatry Res* **63**(2-3):219–222(1996).

22. Nurnberger JI Jr, Adkins S, Lahiri DK, Mayeda A, Hu K, Lewy A, et al. Melatonin suppression by light in euthymic bipolar and unipolar patients. *Arch Gen Psychiatry* **57**(6):572–579(2000).

23. Magnusson A, Boivin D. Seasonal affective disorder: an overview. *Chronobiol Int* **20**(2):189–207(2003).

24. Terman M, Terman JS, Quitkin FM, McGrath PJ, Stewart JW, Rafferty B. Light therapy for seasonal affective disorder. A review of efficacy. *Neuropsychopharmacology* **2**(1):1–22(1989).

25. Lewy AJ, Sack RA, Singer CL. Assessment and treatment of chronobiologic disorders using plasma melatonin levels and bright light exposure: the clock-gate model and the phase response curve. *Psychopharmacol Bull* **20**(3):561–565(1984).

26. Thompson C, Childs PA, Martin NJ, Rodin I, Smythe PJ. Effects of morning phototherapy on circadian markers in seasonal affective disorder. *Br J Psychiatry* **170**:431–435(1997).

27. Lewy AJ, Bauer VK, Cutler NL, Sack RL, Ahmed S, Thomas KH, et al. Morning vs evening light treatment of patients with winter depression. *Arch Gen Psychiatry* **55**(10):890–896(1998).

28. Koorengevel KM, Beersma DG, den Boer JA, van den Hoofdakker RH. A forced desynchrony study of circadian pacemaker characteristics in seasonal affective disorder. *J Biol Rhythms* **17**(5):463–475(2002).

29. Koorengevel KM, Beersma DG, den Boer JA, van den Hoofdakker RH. Mood regulation in seasonal affective disorder patients and healthy controls studied in forced desynchrony. *Psychiatry Res* **117**(1):57–74(2003).

30. Sack RL, Lewy AJ, White DM, Singer CM, Fireman MJ, Vandiver R. Morning vs evening light treatment for winter

depression. Evidence that the therapeutic effects of light are mediated by circadian phase shifts. *Arch Gen Psychiatry* **47**(4):343–351(1990).

31. Terman JS, Terman M, Lo ES, Cooper TB. Circadian time of morning light administration and therapeutic response in winter depression. *Arch Gen Psychiatry* **58**(1):69–75(2001).

32. Terman M, Terman JS, Quitkin FM, McGrath PJ, Stewart JW, Rafferty B. Light therapy for seasonal affective disorder: a review of efficacy. *Neuropsychopharmacology* **2**:1–22(1989).

33. Wehr TA, Duncan WC Jr, Sher L, Aeschbach D, Schwartz PJ, Turner EH, et al. A circadian signal of change of season in patients with seasonal affective disorder. *Arch Gen Psychiatry* **58**(12):1108–1114(2001).

53

THERAPY OF CIRCADIAN SLEEP DISORDERS

ROBERT SACK AND KYLE JOHNSON
Oregon Health & Science University, Portland, Oregon

INTRODUCTION

Circadian rhythm sleep disorders (CRSDs), defined and described in the preceding chapters, have a common underlying mechanism—a misalignment between the timing of internal circadian rhythms and the normal, desired or required time for sleep and wake. In some CRSDs, the misalignment stems from abnormalities of endogenous circadian mechanisms. For example, advanced sleep phase syndrome (ASPS) appears to be caused by a circadian pacemaker (internal body clock) that has an excessively short period (less than 24 hours). In another example of endogenous pathology, a failure of photic input to the circadian pacemaker causes non-24-hour free-running rhythms in totally blind people. Alternatively, in some CRSDs, the circadian system is normal, but external circumstances exceed the limits of adaptation. For example, in shift work sleep disorder (SWSD), the circadian system cannot adapt to a sudden inversion of the sleep–wake schedule.

This chapter will describe some of the currently available treatments for CRSDs. There are two general treatment strategies: (1) resetting the clock or (2) overriding the clock. With clock resetting, the phase (timing) of the circadian pacemaker is shifted (reset) so that the output signals for sleep and wake are more congruous with a person's desired sleep–wake schedule. Clock resetting strategies draw on an understanding of the principles of entrainment derived from circadian science. Light exposure and/or melatonin, administered at the optimal circadian phase,

are currently the most practical phase-resetting agents available for clinical use.

In cases where circadian resetting is impractical or undesirable, overriding the circadian signal with a hypnotic or alerting medication may be the preferred strategy (in circadian terminology, this is termed *masking*). As will be explained, clock resetting and clock override can be used concurrently.

BRIEF REVIEW OF CIRCADIAN SCIENCE RELEVANT TO THE TREATMENT OF CRSDs

As described in previous chapters, mammalian circadian rhythms are generated by intracellular protein transcriptional feedback mechanisms in the neurons of the suprachiasmatic nucleus (SCN) of the hypothalamus (Figure 53.1). Output signals from the SCN modulate not only daily rhythms in alertness but also core body temperature and the secretion of certain hormones such as melatonin and cortisol (see Figure 53.2). In humans [1], the intrinsic rhythm of the clock is slightly longer than 24 hours so that precise synchronization to a 24 hour day (entrainment) depends on exposure to environmental circadian time signals (zeitgebers)—most importantly, the solar light–dark cycle. In the absence of timing signals or light (e.g., in totally blind people), circadian rhythms typically "free-run" on a non-24-hour cycle (average = 24.5 h), expressing the intrinsic rhythm of the clock. In a sense, phase resetting is a normal, daily, function of the circadian system, and the therapeutic use of light exposure is an extension of this process. Nonphotic time cues (e.g.,

Correspondence to: Robert Sack, 3181 S.W. Sam Jackson Park Road, Portland, Oregon 97239-3098.

Sleep: A Comprehensive Handbook, Edited by T. Lee-Chiong.
Copyright © 2006 John Wiley & Sons, Inc.

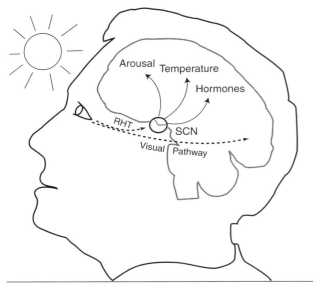

Figure 53.1 Light and the circadian system. The suprachiasmatic nucleus (SCN) of the hypothalamus contains the circadian clock. The intrinsic period of the clock is slightly longer than 24 hours. Entrainment to an exact 24 hour period involves photic input to the clock via a special pathway, the retinohypothalamic tract (RHT) that is distinct from the visual pathway. Efferent pathways from the clock are widespread and regulate the circadian rhythms of sleep–wake (level of arousal), core body temperature, and the secretion of certain hormones such as melatonin and cortisol.

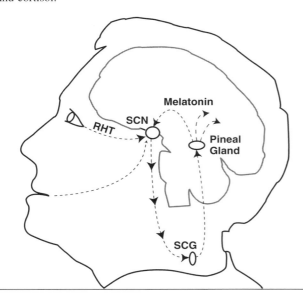

Figure 53.2 Melatonin and the circadian system. The timing of melatonin secretion by the pineal gland is controlled by the circadian pacemaker, located in the suprachiasmatic nucleus (SCN) of the hypothalamus. Efferents pass from the SCN to the superior cervical ganglion (SCG) in the spinal cord, and postsynaptic sympathetic neurons enervate the pineal gland. Melatonin receptors on SCN neurons are activated by circulating melatonin from the pineal gland or from oral administration.

scheduled sleep and activity) may have some influence on the clock, but their potency, compared to light, appears to be relatively weak.

It has recently been discovered that the circadian visual system utilizes specialized (nonrod, noncone) receptors in the ganglion cells of the retina that are sensitive to environmental brightness. Photic information is conveyed to the SCN by a specific pathway (the retinohypothalamic tract) that is separate from the pathway mediating vision (see Figure 53.1). The SCN also has a dense population of melatonin receptors that presumably play a role in normal circadian regulation and mediate the phase-resetting effects of melatonin administration.

Phase-Resetting Effects of Light: Dependence on Intensity, Duration, Timing, and Wavelength

Intensity Compared to some other species (e.g., rodents), humans are relatively insensitive to light [2]. Consequently, effective therapeutic phase shifting requires light exposure that approaches solar intensity. In addition to absolute intensity, the contrast between therapeutic light exposure and background ambient light may be important. For example, even relatively dim light can promote phase shifting if the exposure occurs in the context of almost complete darkness [3]. If solar light is available, phase shifting can be promoted by simply going outside, even on cloudy days.

If therapeutic light exposure is needed during the hours of darkness, an artificial light source can be used. A list of commercially available fixtures can be obtained from the Society for Light Therapy and Biological Rhythms (http://www.sltbr.org). The light sources vary in size from a large light box to a compact visor. In addition, a device that mimics the gradually increasing intensity of dawn sunlight (called a *dawn simulator*) has been shown to cause phase shifting [4].

Appropriately timed avoidance of light can also promote phase resetting. For example, Eastman and colleagues [5] have shown that wearing goggles on the commute home from a night shift can promote an adaptive phase delay. The goggles presumably block the expected phase advance produced by morning sunlight, thereby permitting a greater phase delay.

Duration The duration of light exposure remains an issue for further research. The initial exposure to a bright light stimulus may be the most potent component with diminishing returns as the duration of the stimulus is prolonged. In clinical studies, durations have varied from 15 minutes to 2 hours. It has been shown that intermittent light exposure can be about as effective as continuous light exposure [6]. In clinical settings, it is important that the prescribed duration of treatment be realistic and consistent with the life style of the subject. Bright light fixtures can be placed in settings where the patient can carry out other activities; for example, eating breakfast, reading, watching television, or doing homework.

Timing The phase adjusting (shifting) effects of light and melatonin on the circadian clock are critically dependent on the time of day (the phase of the 24 hour cycle). Light exposure around dusk (in normally entrained individuals) shifts the clock later, producing a phase delay. Light exposure around dawn shifts the clock earlier, producing a phase advance. Light exposure in the middle of the day has little phase shifting effect. These time-dependent effects of light can be plotted as a phase–response curve (PRC), as shown in Figure 53.3. In the natural environment, these differential phase shifting effects are thought to maintain entrainment by correcting for any drift away from a 24 hour cycle.

If an individual's clock has been reset, the light PRC will be shifted as well; consequently, light exposure can have unexpected effects. For example, in the case of a night worker who has made a significant circadian adaptation (i.e., resetting the clock to be congruent with his night-work/day-sleep schedule) light exposure at dawn may *hit* the phase delay, rather than a phase advance, portion of the light PRC. If the phase of the pacemaker is unknown, it may be necessary to measure a circadian marker (e.g., dim light melatonin onset) in order to predict the effects of a light stimulus.

People ordinarily sleep in darkened rooms with their eyes closed, thereby limiting their exposure to light. Consequently, sleep times can reset the clock indirectly, by gating a person's exposure to light. In some cases of CRSD, it may be difficult to know if the disorder is primary (caused by an intrinsic abnormality of the circadian system), or secondary (related to behavior patterns and associated sleep schedule that resets the clock by gating the time of light exposure).

Wavelength The recent discovery that retinal ganglion cells have an essential role in mediating the circadian effects of light [7] has been followed by intense research on melanopsin and cryptochrome, two photopigments that may be involved. The action spectrum of these candidate photopigments indicates a peak sensitivity of about 500 nanometers (blue-green light). The clinical implications of these findings have not yet been fully explored [8].

There is no scientific basis for using *full spectrum* light sources (as has been suggested in the past); in fact, these sources emit significant amounts of UV light, which may be toxic to the retina, although plexiglass diffusers, typically used in fixtures, filter out most of the UV spectrum.

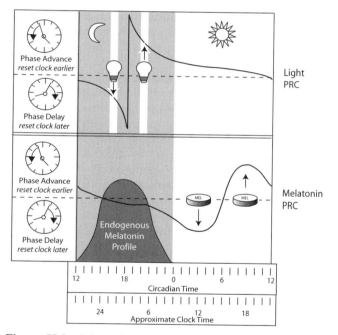

Figure 53.3 Schematic phase response curves (PRCs) for light exposure and melatonin administration. The effects of light and melatonin on the circadian system are dependent on the time of administration. This relationship can be shown as a phase–response curve. In summary, light exposure late in the day or melatonin administration in the morning will shift the circadian clock later (cause a phase delay), while light exposure in the morning or melatonin late in the day will cause the circadian clock to shift earlier (cause a phase advance). Both PRCs have an inflexion point, the time that advance responses become delay responses. According to convention, circadian time 0 is the beginning of the light phase (daytime) of the 24 hour cycle, and circadian time 12 is the beginning of the dark phase (night).

Clock Resetting with Melatonin

Melatonin is a hormone normally produced by the pineal gland at night in the dark. Plasma levels rise sharply at about 9 p.m. (± about 2 hours) and remain elevated for about 12 hours. The measurement of the melatonin profile in either plasma or saliva (under dim light conditions) has become the preferred research method for assessing circadian phase.

Redman et al. [9] were the first to show that a daily injection of melatonin could entrain rats that were free-running in constant dim light conditions. Subsequently, Sack et al. [10] showed that a daily oral dose of melatonin could entrain free-running rhythms in totally blind people. Lewy et al. [11] described a melatonin PRC in sighted people that was essentially 180° out of phase with the light PRC; in other words, exogenous melatonin appears to be a darkness signal, causing a phase delay at the same times that light causes a phase advance (and vice versa).

Initial human trials with melatonin employed doses (3–10 mg) that produced plasma levels many times higher than physiologic concentrations. However, Lewy et al. [12] showed that entrainment of blind free-runners was possible with much lower doses (less than 0.5 mg) that produced plasma levels in the physiologic range. If there is a dose–response curve for melatonin, it appears to be relatively flat. It is probably more useful to think of melatonin as a timing signal with change in circulating levels being more important than absolute concentrations.

Although melatonin appears to have significant phase shifting (*chronobiotic*) properties, it remains unclear whether it can overcome the effects of a competing bright light stimulus. In sighted people, melatonin administration probably works best in situations where the competing light signals are minimized.

In the United States, melatonin is not classified as a drug; it is sold as a nutritional supplement. The FDA does not license it as a treatment for any disorder. Melatonin is thought to be safe with only one report of potential complications when melatonin administration was associated with increased seizure frequency in children with severe neurological deficits and epilepsy [13]. Although melatonin does not appear to be implicated in pubertal onset in humans, this remains a theoretical concern given the importance of melatonin in seasonally breeding mammals. Therefore melatonin should be used with caution in children approaching puberty.

Clock Resetting with Scheduled Rest and/or Activity

A few studies have suggested that vigorous exercise can shift circadian rhythms, and some field trials have shown that physical training improved sleep quality and performance on the job. However, a recent study showed no added benefit of exercise when combined with appropriately timed, bright light exposure [14].

OVERRIDING THE CIRCADIAN SYSTEM

Hypnotic Agents

In certain CRSDs, when sleep is attempted at an unfavorable circadian phase (e.g., jet lag, short runs of night work), circadian desynchrony is self-limited and hypnotics may be used short term to promote sleep. Rapidly metabolized hypnotics with a short half-life are preferred in order to minimize residual drowsiness during the subsequent wake period. In some instances, long-term use of hypnotics for CRSDs, ideally on an intermittent schedule, may be justified for patients who are unable to reset their clocks to fit their required sleep schedule.

Alerting Agents

Although sleep promotion with hypnotics can be useful, awake-time sleepiness may remain a problem because of the persistence of a misaligned circadian system. The addition of alerting agents may be required to promote wakefulness.

Caffeine is the most widely used alerting drug in the world and has been shown to improve alertness in simulated night work. Night workers (especially transportation workers) sometimes use amphetamine drugs (e.g., methylphenidate, dextroamphetamine) illicitly to maintain alertness. Tolerance can develop, as well as mood disorders. When these drugs are discontinued, rebound sleepiness is to be expected.

Modafinil, a nonamphetamine alerting drug initially licensed for the treatment of narcolepsy, has recently been approved by the FDA for the treatment of shift work sleep disorder. The risk potential appears to be significantly lower than with amphetamines.

The Homeostat–Circadian Interaction

In addition to the circadian clock, the sleep–wake cycle is driven by the build-up of sleep drive that is dependent on the duration of prior wakefulness (the homeostatic process of sleep regulation) [15]. In a sense, the homeostat is like a clock because the normal build-up of sleep drive sufficient to exceed the threshold for sleep takes about 16 hours and is discharged during the sleep period (like an hourglass). Normally, the homeostatic and circadian processes operate in tandem. The circadian alerting process drops away at bedtime, the same time that a substantial drive for sleep has accumulated. However, in CRSDs, the circadian and homeostatic mechanisms are, by definition, uncoordinated, resulting in shortened, nonrefreshing sleep, which leads to the accumulation of homeostatic sleep drive.

COUNSELING AND EDUCATION

Most people are intuitively aware of the homeostatic regulation of sleep (not getting enough sleep will make you sleepy), but the circadian regulation of sleep can be more difficult for patients to grasp. It may come as a surprise that, after staying awake for the entire night, it is not necessarily easy to sleep the following day (because of the circadian alerting process). Therefore it is worthwhile to spend some time educating patients about the circadian system prior to initiating treatment.

TREATMENT SUGGESTIONS FOR SPECIFIC DISORDERS

Shift Work Sleep Disorder

Although unconventional work and sleep schedules are extremely common in modern society, most workers do not seek medical assistance. That may change as an awareness of treatment availability becomes more widespread. Treatments for shift work sleep disorder (SWSD) often need to be customized because work schedules vary so widely, ranging from an occasional or intermittent night duty to a steady five (or more) nights per week. In many occupations (including medicine), night work can be unpredictable and erratic. Furthermore, night workers typically adopt day-active schedules on their days off so that even regular night-workers alternate between a diurnal and nocturnal orientation; as a result, circadian adaptation during a run of night work can quickly reverse during the off days. Evening (swing) shift workers often adopt a delayed sleep phase, citing a need to "wind down" after getting home from work.

Bright light exposure has been repeatedly shown to promote phase resetting in simulated night shift work and to benefit adaptation in some field studies. Because the human circadian clock has a period longer than 24 hours, the light exposure should be timed to promote a phase delay. As mentioned earlier, avoidance of bright light in the morning is useful. Sometimes, bright light exposure can be arranged in actual work situations.

Melatonin administration is convenient although it may not be able to override a strong light stimulus. Some studies have shown accelerated adaptation in both simulated night work and actual night workers [16, 17]. Melatonin may benefit daytime sleep in night workers by an additional direct sleep-promoting action. A number of studies have shown that melatonin shortens sleep latency and increases sleep duration in young subjects who are mildly sleep deprived and who are sleeping during the day (a description that fits many night workers). We have hypothesized that melatonin may promote sleep by counteracting the circadian alerting process [18].

Delayed Sleep Phase Syndrome

Delayed sleep phase syndrome (DSPS) may be primary (perhaps related to a circadian period that is very long) or secondary (e.g., consistently staying up late and sleeping late in order to avoid school). In either case, the treatment challenge is to shift the sleep and circadian cycle to a time frame congruent with the patient's wishes and role in society. Treatment success will depend heavily on the patient's motivation so that time spent explaining the mechanism involved in DSPS, and the rationale for treatment, is well worthwhile. It is quite helpful for teenage patients to be seen with parents and to offer a biological explanation that reframes the problem and minimizes blame and guilt [19].

The first described therapy for DSPS, termed *chronotherapy* [20], was based on the formulation that patients with DSPS had an exceptionally long circadian period that made it much easier for them to delay than to advance. In this treatment, patients are prescribed a sleep schedule that regularly shifts later by about 3 hours per day, around the clock, until the sleep is occurring at the desired time. This can be an effective treatment for adolescents during the school year when absences need to be minimized; however, it requires the full involvement of the parents.

Subsequently, reported treatments for DSPS have used bright light in the morning [21] and melatonin [22] in the evening to promote phase advances to a desired sleep time. At our clinic we use the protocol outlined in Table 53.1. It is based on the assumption that, because

---

TABLE 53.1   Patient Instructions for Delayed Sleep Phase Syndrome

1. Before attempting to change your sleep times, obtain a baseline assessment of your sleep–wake schedule.
    (a) Keep a sleep diary.
    (b) Wear an actigraph (if available).
2. Next, concentrate on keeping a consistent sleep schedule; for example, if your preferred time in bed is 3 a.m. to 11 a.m., be consistent about this sleep time (this may not be possible for people who have school or employment obligations).
3. Then, reset your internal body clock.
    (a) Get at least 30 minutes of bright light exposure promptly upon awakening.
        (i) Go for a walk in the sunlight.
        (ii) Sit by a bright light fixture.
    (b) Take melatonin (0.5–3.0 mg) two hours before your bedtime.
    (c) Shift your sleep schedule earlier using a predetermined plan; for example, move bedtime and wake time 15 minutes earlier every other day.
4. If you are still awake 30 minutes after lights out, take a sleeping medication (prescribed by your doctor).
5. After you are sleeping regularly at the desired time, keep your schedule, even on weekends and vacation. Use the sleeping medications as little as possible.
6. Continue to keep a sleep diary to track changes.
7. If you relapse, start at step #2 again.

phase advances in DSPS are difficult, they must be gradual, and that the use of a hypnotic drug may be indicated during the transition from a delayed to a desired phase. The rate of change is negotiated with the patient (from 15 minutes every 3 days, to 30 minutes every other day). Patients are given a spreadsheet that provides a schedule for medications and bed times. Timed light exposure and melatonin administration are used for clock resetting, while a hypnotic is used as a "back-up" to ensure sleep at the desired time. In adolescents, this treatment strategy is best initiated over winter or summer school breaks.

Advanced Sleep Phase Syndrome

Advanced sleep phase syndrome (ASPS) is uncommon, although a tendency for ASPS often occurs in older people.

Because early morning awakening is not likely to interfere with employment or academic obligations, patients are less likely to seek attention. The recommended treatment is essentially the mirror image of the treatment for DSPS, involving bright light exposure in the evening and, if indicated, melatonin in the morning. The dose of morning melatonin should be very low (0.5 mg or lower) to minimize daytime sedation.

Non-24-hour Sleep–Wake Syndrome

In this disorder, circadian rhythms fail to entrain and *free-run* on a non-24-hour cycle, expressing the intrinsic period of the body clock. Some degree of relative coordination to a 24 hour rhythm may occur. The pattern is readily demonstrated by serial assessments of the endogenous melatonin

TABLE 53.2 Patient Recommendations for Jet Lag

1. *Before your trip:* If possible, start the adaptation process 2–3 days before leaving on a long journey. For example, if traveling eastward, start shifting your lights out and wakeup time *earlier* by about 30 minutes per day. Obtain bright light exposure upon awakening, either by going outside into the sunlight or, if it is still dark outside, with a bright artificial light. If traveling westward, shift your sleep time 30 minutes *later* each day, and get as much light exposure in the evening as possible. If the trip is shorter (three time zones or less), and you won't be gone for long, it may be better to try to keep your sleep synchronized to your home time zone while away.

2. *While in flight:* Because sleep deprivation is one of the causes of jet lag, consider using a sleeping medication to help you sleep during the flight. Zaleplon is probably a good choice because it is short-acting (out of your system in about 3–4 hours). If you use a sleeping medication, you should not drink alcohol. The compression of your legs from sitting (and sleeping) in an airplane seat for a long time can increase the risk of blood clot formation (deep vein thrombosis). Taking an aspirin and occasionally walking about the cabin will reduce the risk.

3. *Upon arrival at your destination:*
   (a) Eastward flight:
      (i) *Light exposure*: In general,[a] morning light exposure will facilitate resetting your internal body clock to an earlier time. Go for a walk outside in the morning when you first get up.
      (ii) *Melatonin:* Take melatonin (0.5–3.0 mg) about 2 hours prior to the local bedtime to help reset your body clock to local time.
      (iii) *Sleeping medication:* You may want to take a sleeping pill at bedtime until you have adapted to the local time.
      (iv) *Alerting medication:* Modafinil (100 mg), an alerting medication, can be taken in morning. Do not take it later in the day; it might keep you awake at night.
   (b) Westward flight:
      (i) *Light exposure:* In general,[a] evening light exposure will facilitate resetting your internal body clock to a later time. If possible, stay outdoors until dusk and/or be in a brightly lit indoor space until bedtime.
      (ii) *Melatonin:* Take melatonin (0.5 mg) if you wake up too early (the usual problem with westward travel). This will help reset your clock later. Do not take melatonin at bedtime; it can inhibit the adaptation process. Taking melatonin in the morning hours may cause some daytime sleepiness, but if the dose is very low (0.5 mg or less), the hypnotic effect is minimal.
      (iii) *Sleeping medication*: If you wake up early and cannot get back to sleep (and it is several hours before you expect to be up), take a short-acting sleeping medication (e.g., zaleplon).
      (iv) *Alerting medication:* Modafinil (100 mg) can be taken in the morning to promote daytime alertness. Do not take it later in the day—it might keep you awake at night.

4. *After a few days in your destination:* It takes about a day per time zone of travel to reset your body clock, so, depending on the distance and direction of travel, it may be a week (or more) before you are completely adapted to local time. Jet lag symptoms will recede as you become adapted to local time. However, sleeping in an unfamiliar environment may be difficult and justify the use of a sleeping medication for the duration of your trip. When a sleeping medication is discontinued, you may expect a few nights of lighter sleep before returning to baseline.

[a]If travel involves crossing more than eight time zones, this rule may not be valid because the timing of sunlight at either dawn or dusk may be unfavorable (resetting your clock in the wrong direction). To compensate for this problem, if you are traveling eastward more than eight time zones, avoid light at dawn for the first few days after arrival. If you are traveling westward more than eight time zones, avoid light at dusk for the first few days. On subsequent days, apply the recommendations described above.

profile. The vast majority of patients are totally blind [23] and the failure to entrain is related to the lack of photic input to the circadian clock. Treatment, using low doses of melatonin at the same time of day, has been found to be very effective in the majority of these cases.

There are also a handful of case reports of sighted individuals with this disorder. The etiology is unclear but it has been suggested that these patients have an endogenous circadian period that is beyond the range of entrainment. In sighted patients, bright light exposure as well as melatonin administration timed to produce phase advances have been shown to be helpful [24].

Irregular Sleep–Wake Pattern

Most patients with irregular sleep–wake pattern have no discernable pathology of the circadian system although the pattern of sleep and wake can resemble that seen in animals after a lesion in the circadian clock. Treatment involves emphasizing good sleep hygiene combined with melatonin administration in the evening and bright light in the morning to reinforce their circadian rhythms.

Time Zone Change (Jet Lag) Syndrome

Jet lag is self-limited; however, travel time is precious and treatments based on circadian principles are rational and have been shown to have benefit when tested in clinical trials [25]. In some people (e.g., airline crews), jet lag may be recurrent or even chronic. Although appropriately timed light exposure and melatonin treatment can accelerate circadian adaptation, there is an inevitable lag period until clock resetting is complete. Consequently, hypnotic medications, and possibly alerting agents, can help to control symptoms until circadian synchronization has been restored. Table 53.2 summarizes some recommendations provided in our clinic to counteract jet lag.

SUMMARY

The treatment of CRSDs can be based on a solid foundation of circadian rhythm science. As an understanding of the circadian system grows, treatments will become more precise and effective.

REFERENCES

1. Dijk DJ, Lockley SW. Integration of human sleep–wake regulation and circadian rhythmicity. *J Appl Physiol* **92**:852–862(2002).
2. Lewy AJ, Wehr TA, Goodwin FK, Newsome DA, Markey SP. Light suppresses melatonin secretion in humans. *Science* **210**:1267–1269(1980).
3. Wright KP Jr, Kronauer RE, Jewett ME, Ronda JM, Klerman EB, Czeisler CA. Human circadian entrainment to the 24 hr day in a dim light–dark cycle. *Sleep Res Online* **2** (Suppl 1):739(1999).
4. Danilenko KV, Wirz-Justice A, Krauchi K, Weber JM, Terman M. The human circadian pacemaker can see by the dawn's early light. *J Biol Rhythms* **15**:437–446(2000).
5. Eastman CI, Martin SK. How to use light and dark to produce circadian adaptation to night shift work. *Ann Med* **31**:87–98(1999).
6. Gronfier C, Wright KP Jr, Kronauer RE, Jewett ME, Czeisler CA. Efficacy of a single sequence of intermittent bright light pulses for delaying circadian phase in humans. *Am J Physiol* **287**:E174–E181(2004).
7. Berson DM. Strange vision: ganglion cells as circadian photoreceptors. *Trends Neurosci* **26**:314–320(2003).
8. Lockley SW, Brainard GC, Czeisler CA. High sensitivity of the human circadian melatonin rhythm to resetting by short wavelength light. *J Clin Endocrinol Metab* **88**:4502–4505(2003).
9. Redman J, Armstrong S, Ng KT. Free-running activity rhythms in the rat: entrainment by melatonin. *Science* **219**:1089–1091(1983).
10. Sack RL, Brandes RW, Kendall AR, Lewy AJ. Entrainment of free-running circadian rhythms by melatonin in blind people. *N Engl J Med* **343**:1070–1077(2000).
11. Lewy AJ, Bauer VK, Ahmed S, Sack RL. The human phase response curve (PRC) to melatonin is about 12 hours out of phase with the PRC to light. *J Biol Rhythms* **14**:227–236(1999).
12. Lewy AJ, Bauer VK, Hasler BP, Kendall AR, Pires LN, Sack RL. Capturing the circadian rhythms of free-running blind people with 0.5 mg melatonin. *Brain Res* **918**:96–100(2001).
13. Sheldon SH. Pro-convulsant effects of oral melatonin in neurologically disabled children. *Lancet* **351**:1254(1998).
14. Burgess HJ, Crowley SJ, Gazda CJ, Fogg LF, Eastman CI. Pre-flight adjustment to eastward travel: 3 days of advancing sleep with and without morning bright light. *J Biol Rhythms* **18**:318–328(2003).
15. Edgar D, Dement W, Fuller C. Effect of SCN lesions on sleep in squirrel monkeys: evidence for opponent processes in sleep–wake regulation. *J Neurosci* **13**:1065–1079(1993).
16. Sack RL, Lewy AJ. Melatonin as a chronobiotic; treatment of circadian desynchrony in night workers and the blind. *J Biol Rhythms* **12**:595–603(1997).
17. Sharkey KM, Eastman CI. Melatonin phase shifts human circadian rhythms in a placebo-controlled simulated night-work study. *Am J Physiol* **282**:R454–R463(2002).
18. Sack RL, Hughes RJ, Edgar DM, Lewy AJ. Sleep promoting effects of melatonin: at what dose, in whom, under what conditions, and by what mechanisms. *Sleep* **20**:908–915(1997).
19. Garcia J, Rosen G, Mahowald M. Circadian rhythms and circadian rhythm disorders in children and adolescents. *Semin Pediatr Neurol* **8**:229–240(2001).
20. Czeisler C, Richardson G, Coleman R, Zimmerman J, Moore-Ede M, Dement W, Weitzman E. Chronotherapy: resetting the

circadian clocks of patients with delayed sleep phase insomnia. *Sleep* **4**:1–21(1981).

21. Rosenthal NE, Joseph-Vanderpool JR, Levendosky AA, Johnston SH, Allen R, Kelly KA, Souetre E, Schultz PM, Starz KE. Phase-shifting effects of bright morning light as treatment for delayed sleep phase syndrome. *Sleep* **13**:354–361(1990).

22. Nagtegaal JE, Kerkhof GA, Smits MG, Swart AC, Van Der Meer YG. Delayed sleep phase syndrome: a placebo-controlled cross-over study on the effects of melatonin administered five hours before the individual dim light melatonin onset. *J Sleep Res* **7**:135–143(1998).

23. Sack RL, Lewy AJ, Blood ML, Keith LD, Nakagawa H. Circadian rhythm abnormalities in totally blind people: incidence and clinical significance. *J Clin Endocrinol Metab* **75**:127–134(1992).

24. McArthur AJ, Lewy AJ, Sack RL. Non-24-hour sleep–wake syndrome in a sighted man: circadian rhythm studies and efficacy of melatonin treatment. *Sleep* **19**:544–553(1996).

25. Sack RL. Shift work and jet lag. In: Lee-Chiong TL, Sateia MJ, Carskadon MA (Eds), *Sleep Medicine*. Hanley and Belfus, Philadelphia, 2002.

PART VI

PARASOMNIAS

54

DISORDERS OF AROUSAL AND SLEEP-RELATED MOVEMENT DISORDERS

KEITH CAVANAUGH AND NORMAN R. FRIEDMAN
The Children's Hospital, Denver, Colorado

An arousal disorder is characterized by an incomplete awakening from sleep where the individual has voluntary movements but no awareness of their actions. There are three types of arousal disorders including confusional arousals, sleep terrors, and sleepwalking (somnambulism) (discussed in Chapter 55). Nocturnal leg cramps and rhythmic movement disorders are classified as sleep-related movement disorders and consist of relatively simple movements that disturb sleep.

AROUSAL DISORDERS

Confusional Arousals

Confusional arousals are characterized by episodes of mental confusion on arousal or awakening [1]. They consist of confusion during and following arousal from sleep, usually from deep sleep in the first part of the night. They have also been labeled sleep drunkenness and excessive sleep inertia.

Typically individuals will display disorientation to time and space and disruption of speech. Memory impairment is present with both retrograde and anterograde types seen. Other clinical manifestations include perceptual impairment, inappropriate behavior, aggressive behavior, and errors of logic. These episodes are typically not associated with fear, walking behavior, or intense hallucinations.

Although prevalence rates for other parasomnias like night terrors and somnambulism have been estimated between 1% and 30% during childhood, the prevalence of confusional arousals is unknown [2]. Less common in older children and rarely reported in adults, they are almost universally seen in young children before 5 years of age. Gender has not been shown to play a role.

Overall, they are generally benign and tend to cease by adolescence. However, personal injury has been reported. Confused individuals may resist and become aggressive when restrained.

Predisposing factors can be any factor that deepens sleep and impairs ease of awakening. This includes but is not limited to young age, recovery from sleep deprivation, circadian rhythm sleep, medications, and encephalopathic conditions. Abnormalities in areas impairing arousal (periventricular gray, the midbrain reticular area, and the posterior hypothalamus) have been identified in rare organic cases. Spontaneous confusional episodes can be induced by force arousal.

Differential diagnosis would include sleep terrors, somnambulism, REM sleep behavior disorder, and Sleep-related epileptic seizures (i.e., complex partial associated with ictal discharges).

Polysomnographic studies show confusional events occur during slow-wave sleep. They typically occur during the first third of the night but have been seen in awakening for lighter stages of sleep and very rarely in REM awakenings. EEG monitoring during the confusion period may show brief episodes of delta activity, stage 1 theta patterns, repeated microsleeps, or a diffuse and poorly reactive alpha rhythm [3].

Treatment primarily involves reassurance, avoiding facilitating causes (sleep deprivation, CNS depressants, stress), and medications. Mild stimulants are utilized to lighten deep sleepers.

Sleep Terrors

A child who wakes up within the first few hours of the night with a piercing scream or cry is most likely experiencing a sleep terror. The child appears to be in fear of something. Tachycardia, tachypnea, skin flushing, diaphoresis, dilated pupils, and increased muscle tone may be present. The child is unresponsive to his/her parents and is inconsolable. Inconsolability is a key feature. Except for some brief dream images, the child is amnesic to the episode. The memories include the need to act against monsters or other threats. He/she may have some incoherent vocalizations or call out to the parents but they are not awake. The attacks may last 30 seconds to 5 minutes and usually occur within the first third of the night.

The typical age of onset is 4–12 years with spontaneous resolution with onset of puberty. The prevalence is 3% for children and 1% for adults and is more common in males. Ninety percent of the patients have a family history of either sleepwalking or sleep talking.

It is not associated with a higher incidence of psychopathology. However, the study by Laberge et al. [4] has recently demonstrated higher anxiety scores in children experiencing night terrors. Sleep disordered breathing, restless legs syndrome, sleep deprivation, fever, distended bladder, or CNS depressant medications are potential precipitants.

On polysomnography, one has an abrupt awakening during slow-wave sleep with sympathetic hyperactivity.

The differential diagnosis includes nightmares, confusional arousals, and sleep-related epilepsy. The distinction between sleep terrors and nightmares is straightforward. A child who experiences a nightmare will remember the dream content, is consolable, and does not have a large sympathetic discharge. Nightmares are preferential to REM sleep so they are more common during the last one-third of the night. A confusional arousal does not have the associated terror.

To manage sleep terrors one should treat the predisposing factors. Identify and treat sleep disordered breathing. Avoid excessive tiredness. Prompted night awakenings have been effective. During an attack, parents should protect their child from injury but should not try to awaken the child. Make soothing comments and speak calmly and repetitively. Typically, a child does not want to be touched, although some seem to feel better when being held. Another treatment option is prompted nighttime awakenings.

SLEEP-RELATED MOVEMENT DISORDERS

Nocturnal Leg Cramps

Nocturnal leg cramps involve intensely painful sensation accompanied by muscle tightness or tension that occurs during sleep in the calf with a majority being unilateral and occasionally involving the foot [5]. Duration of symptoms vary from a few seconds up to 30 minutes and spontaneously remit. The sensation leads to arousal or awakening from sleep. Frequency is usually one to two episodes nightly occurring several times a week. Some individuals experience them primarily in the daytime and do not report sleep disturbances. They are also referred to as a charley horse, nocturnal leg pain, and muscle hardness.

Symptoms of nocturnal leg cramps have been identified between 16% and 37% [6] of healthy individuals [7], particularly following significant exercise. Although incidence is increased in the elderly, the exact prevalence of leg cramps is not known. Symptoms can wax and wane over many years' duration. Gender influence seems to suggest female predominance, as they are often reported during pregnancy.

Predisposing factors include pregnancy, diabetes mellitus, metabolic disorders, prior vigorous activity, fluid and electrolyte abnormalities, endocrine disorders, neuromuscular disorders, peripheral vascular disease, and disorders of decreased mobility (i.e., Parkinson's disease and arthritis). Medications that have been reported to cause leg cramps include oral contraceptives, diuretics, nifedepine, ß-agonists, steroids, morphine, cimetidine, penicillamine, statins, and lithium [8]. Some familial cases have been reported with an autosomal dominant pattern [9]. Muscle membrane overexcitability and abnormalities in calcium metabolism have been suggested but not confirmed [10].

Differential diagnosis includes chronic myelopathy, peripheral neuropathy, akathisia, restless legs syndrome, muscular pain-fasciculation syndromes, and disorders of calcium metabolism.

Treatment involves massage, application of heat, or slow stretching. Patients should be advised about general measures to improve sleep, such as not going to bed until sleepy, ensuring a comfortable environment for sleep, and avoidance of alcohol and caffeine-containing beverages before bed. Cramps can be aborted by making use of reciprocal inhibition reflexes, in which contracting a group of muscles forces relaxation of the antagonistic group [11].

Quinine, an alkaloid originally produced from the bark of the cinchona tree, has been used to treat leg cramps since 1940. However, studies examining its benefit have shown mixed results [11–13]. The Federal Drug Administration ordered the discontinuation of its use for leg cramps in 1995 after concluding the risk of side effects outweighed the benefits [14]. Side effects included potentially fatal hypersensitivity reaction, particularly quinine-induced thrombocytopenia, pancytopenia, hemolytic uremia syndrome, hepatitis, and cinchonism, a condition manifested by tinnitus, visual disturbances, vertigo, nausea, vomiting, abdominal pain, and deafness [15]. Other medications that have been studied with promise include natridrofuryl, orphenadrine, and calcium channel antagonists.

Polysomnographic studies show nonperiodic bursts of gastrocnemius EMG activity [16].

Rhythmic Movement Disorder

A rhythmic movement disorder (RMD) is a common condition that rarely receives much attention. A RMD is present when the infant has repetitive stereotypic movements of large muscle groups around sleep onset. Two-thirds of infants at 9 months of age have some form of a rhythmic activity. "Head banging" (movement of the head in an anterior–posterior direction) is the most commonly recognized variant; however, many body areas may be involved, including:

- Head rolling (lateral movement of the head and neck)
- Body rocking (rocking of the entire body in an antero-posterior direction as the child rises onto his/her hands and knees)
- Body rolling (rotation of the entire torso)
- Leg banging or rolling

Typically, the onset is prior to 1 year of age and resolves by 3 years. For some, the body movement may persist into childhood. The relationship with a RMD and developmental delay is rare. A recent study did find an association of higher anxiety scores in children who exhibited body rocking [4]. The condition is more common in males.

Postulated predisposing factors include environmental stress or lack of environmental stimulation. Some hypothesize that RMD represents a vestibular form of self-stimulation.

The diagnosis is by history. One may confirm the diagnosis by having the parents make a videotape of the behavior. Diagnostic testing is rarely necessary. If the movements are atypical, prolonged, or violent, one should consider an overnight polysomnogram with an expanded EEG montage and video recording to exclude seizures.

The management of RMD is supportive. One may consider padding the headboard or a protective helmet for violent rhythmic activity.

MISCELLANEOUS PHYSICAL ACTIVITIES

Sleep Starts

Sleep starts are sudden brief contractions of the legs or arms that occur at sleep onset. They usually consist of a single asymmetric muscle contraction during the transition from wakefulness to asleep. They are also known as hypnic jerks. The jerks are often associated with at least one of the following: subjective feeling of falling, sensory flash (i.e., a flash of light or a loud bang), or a dream.

The prevalence is upwards of 70%. Predisposing factors include nicotine, intense evening exercise, stress, or intake of stimulants especially excessive caffeine. They are benign unless they produce repeated awakenings. Rarely, a person may develop sleep-onset insomnia. One must distinguish sleep starts from the following conditions: epileptic seizures (myoclonus occurs during wakefulness and sleep and is associated with EEG epileptiform discharges); periodic limb movement (movements occur after sleep onset and have a periodicity); fragmentary myoclonus (brief jerks or twitches that are symmetrical and bilateral and occur during all sleep stages); and neonatal sleep myoclonus (marked twitching of the fingers, toes, and face during infant's sleep).

Polysomnographic monitoring during an episode might demonstrate a brief, high-amplitude muscle potential during transition from wakefulness to sleep, possibly associated with arousals and tachycardia following an intense episode.

Sleep Talking

Sleep talking or somniloquy is the utterance of sounds or speech during sleep without simultaneous subjective detailed awareness of the event [1]. While it is typically brief, infrequent, and devoid of signs of emotional stress, occasionally it is a nightly occurrence. More frequent episodes can involve longer episodes and be infused with anger and hostility. It can be spontaneous or induced by conversation with the sleeper.

The exact prevalence is unknown but is presumed to be very common. Overall, its course is benign and self-limited. It can last from a few days to many years. When present in individuals over 25 years of age, it can be associated with comorbid psychopathologic or medical illness. Concordance of speech content with recalled mental activity after awakening supports the assertion that utterances reflect ongoing mental activity during sleep [3]. It may be more common in males than in females but overall is not considered to have a gender preference.

Precipitating factors include emotional stress, febrile illness, or other sleep disorders (i.e., sleep terrors, confusional arousals, obstructive sleep apnea syndrome, and REM sleep behavior disorder).

Polysomnographic studies have shown sleep talking present in all stages of sleep, but it is more common in stages I, II, and REM [17]. The timing of occurrence can vary if influenced by other sleep disorders. In obstructive sleep apnea syndrome, it can occur during arousal from sleep.

In REM behavior sleep disorder, it tends to occur out of REM sleep. In somnambulism it occurs during arousals out of slow-wave sleep.

No specific treatment exists. However, recognizing and treating an underlying associated condition that precipitates its occurrence should be targeted. Avoidance of stress may be helpful.

REFERENCES

1. American Sleep Disorders Association. *International Classification of Sleep Disorders, Revised: Diagnostic and Coding Manual*. American Sleep Disorders Association, Rochester, MN, 1997, pp 177–180.

2. Ohayen MM, et al. Night terrors, sleep walking, and confusional arousals in the general population: their frequency and relationship to other sleep and mental disorders. *J Clin Psychiatry* **60**(4):286–276(1999).

3. Gastaut H, et al. A clinical and polygraphic study of episodic phenomena during sleep. *Rec Adv Biol Psychiatry* **7**:197(1965).

4. Laberge L, Tremblay RE, Vitaro F, Montplaisir J. Development of parasomnias from childhood to early adolescence. *Pediatrics* **106**(1 Pt 1):67–74(2000).

5. Leung AK, et al. Leg cramps in children: incidence and clinical characteristics. *J Natl Med Assoc* **91**:329 (1999).

6. Naylor JR, et al. A general population survey of rest cramps. *Age Ageing* **23**: 418–420(1994).

7. Norris FH. An electromyographic study of induced and spontaneous muscle cramps. *Electroencephalogr Clin Neurophysiol* **9**:139(1957).

8. Eaton JM. Is this really a muscle cramp? *Postgrad Med J* **86**:227–232(1989).

9. Ricker K, et al. Autosomal dominant cramping disease. *Arch Neurol* **47**:810(1990).

10. Cutler P. Cramps in the legs and feet. *JAMA* **252**:2332 (1980).

11. Fowler AW. Relief of cramps. *Lancet* **i**:99(1973).

12. Man-Son-Hing M, Wells G, Lau A. Quinine for nocturnal leg cramps: a meta-analysis including unpublished data. *J Gen Intern Med* **13**:600–606(1998).

13. Daniel HW. Simple cure for nocturnal leg cramps. *N Engl J Med* **301**:216(1979).

14. US Department of Health and Human Services. Stop to marketing of quinine for night leg cramps. *FDA Consumer* **29** (July-Aug):1–2 (1995).

15. Butler JV, et al. Nocturnal leg cramps in older people. *Postgrad Med J* **78**:596–598 (2002).

16. Saskin P, et al. Sleep and nocturnal leg cramps. *Sleep* **11**:307(1988).

17. Arkin AM. Sleep talking: a review. *J Nerv Ment Dis* **143**:101(1966).

ADDITIONAL READING

Guilleminault C, Palombini L, Pelayo R, Chervin RD. Sleepwalking and sleep terrors in prepubertal children: what triggers them? *Pediatrics* **111**(1):17–25(2003). PMID: 12509590.

Frank NC, Spirito A, Stark L, Owens-Stively J. The use of scheduled awakenings to eliminate childhood sleepwalking. *J Pediatr Psychol.* **22**(3):345–353(1997). PMID: 9212552

55

SLEEPWALKING

ROSALIND D. CARTWRIGHT

Rush University Medical Center, Chicago, Illinois

INTRODUCTION

Many young children sleepwalk. The best estimate is 15% walk often but if those who only walk rarely are included this rate may be as high as 30%. The preteen years are when this is most prevalent. These episodes become less frequent, or stop altogether, in late adolescence. A few will continue to walk into the adult years when this may be a more serious problem as they may hurt themselves or become aggressive against others. There is an inherited vulnerability to this sleep disorder, which is increased if both parents were childhood sleepwalkers. Good sleep hygiene rules, especially avoiding sleep deprivation, is the most important guideline for avoiding episodes in those at risk.

SLEEPWALKING IN CHILDREN

Sleepwalking is one of a group of sleep disorders that are called the parasomnias. This classification groups together sleepwalking, sleep terrors, and confusional arousals. All three share a common feature that is of an incomplete awakening from deep sleep, usually within the first 3 hours of the night. The arousals tend to be recurrent and to consist of a mixed brain state of partial sleep and partial waking. Although while walking, the sleepwalkers have their eyes open, to an observer they look "different." They are described as having a blank stare. Also, they may engage in complex motor behaviors, such as preparing food, repairing a motorbike, or directing traffic, but lack responsiveness to others. Once they are fully awake they typically have no

memory of having been out of bed, or of what they were doing, or why. Both boys and girls sleepwalk, especially between the ages of 4 and 6 when, it is estimated, that 15–30% of all children will have episodes of this kind. As children, both sexes are equally affected and these episodes can begin anytime after the child has learned to walk. The frequency varies a good deal, with some who are severely affected having more than one episode per night every night. The frequency gradually reduces through the teenage years. Most often, this resolution occurs without any formal treatment.

A pediatrician, if consulted, will usually reassure the parent and explain this disorder as an immaturity of the central nervous system, which they will "grow out of" on their own. Clearly there is some neurological difficulty in switching from deep sleep (also called stage 4, slow-wave sleep, or delta sleep) to the active rapid eye movement (REM) sleep. This shift from deep sleep to REM sleep normally happens approximately every 90 minutes, with the change being most dramatic in the first two cycles of the night, when sleep is at its deepest. (See Figure 55.1.) If the child is monitored in a sleep laboratory, he/she may show an arousal just before entering the first or second REM period. At that time the brain waves are a mixture of high-amplitude delta waves, alpha rhythm, and waking activity.

There appears to be a genetic component to this disorder, as those with first degree relatives, who were themselves sleepwalkers, are ten times more likely to sleepwalk than those with no such relatives [1]. The transmission, however, is not by way of a single dominant gene. Both twin studies [2] and family studies [1] suggest that what is inherited is a

Sleep: A Comprehensive Handbook, Edited by T. Lee-Chiong.
Copyright © 2006 John Wiley & Sons, Inc.

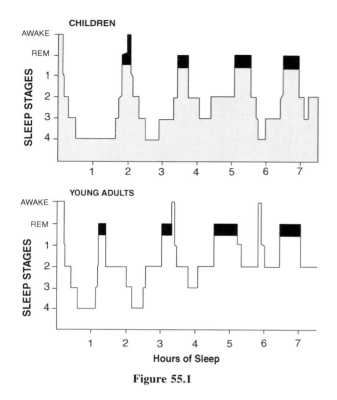

Figure 55.1

flaw in the deep or delta sleep system, making it more vulnerable to partial arousals. The genetic susceptibility to sleepwalking has been linked to the *DQB1* genes already implicated in two other sleep disorders—narcolepsy and REM behavior disorder [3]. HLA-DQB1 typing was done on 60 sleepwalkers and 60 controls. These investigators suggest that the common link is to motor control during sleep. In another approach to understanding the pathophysiology of sleepwalking, power density analysis has been applied to the sleep recordings of those with a history of sleepwalking and normal controls. Previously, the standard sleep stage scoring [4] did not identify any distinctive indicator to support a sleepwalking diagnosis unless there was an observed behavioral arousal in the first or second cycle. There are now three studies reporting the delta waves have a lower than normal power density at the beginning of the night in the sleepwalkers than in those of age-matched controls [5–7]. This reduced delta power has been shown in both children and adults who continue to sleepwalk. There is one brain imaging study [8], which captured a sleepwalking episode during a sleep study. This occurred 59 minutes after sleep onset in a 16 year old male. He had a strong history of sleepwalking several times each week. His grandmother, mother, and younger brother also were known to sleepwalk. The brain imaging study showed a decrease in the normal blood flow patterns in the prefrontal areas and an increased flow in the thalamus and cingulate cortex. These differences from normal blood flow

patterns roughly relate to the observations of the sleepwalker's lack of awareness at the time and later failure to recall these episodes, as well as the mental confusion and emotional behaviors that occur during an event.

The slow-wave sleep at the beginning of the night, which is the point of vulnerability to this disorder, is strongly associated with a peak in the output of growth hormone. As adolescents reach full stature, there is a natural reduction in this stage of sleep in the first and second sleep cycles. This may account for the reduction in sleepwalking episodes around this age. However, the genetic vulnerability will persist and with additional triggers these events can continue to appear. These may then take on a more serious character.

Sleepwalking is closely related to another parasomnia, sleep terrors. In these the child typically sits up abruptly and screams as if in panic. These too appear to spring out of the first or second slow-wave sleep cycle of the night. The child may then fall back to sleep or get up into a sleepwalking or running behavior. They are usually resistant to comforting in this state and may strike out at a parent. If this happens frequently and disrupts the household or if the child is able to get out of the house unobserved while sleepwalking, the child and the parents may need more than reassurance. Consultation with an accredited sleep disorders service is then in order.

TREATMENT FOR SLEEPWALKING IN CHILDREN

The first step in dealing with this disorder is to make sure the child has a regular routine for sleep. An irregular schedule with several late nights in a row can be a trigger for a sleepwalking event. Sleep deprivation is followed by a deepening of the first few hours of sleep at the next opportunity and this makes the transition to REM sleep more problematic. The last hour before bedtime should not be physically strenuous: no rough and tumble play. Quiet relaxing activity—story time, warm bath, reassuring closeness—eases the child into sleep. Safety precautions in the home, such as a baby gate across the stairs or an electric eye across the bedroom door that will sound an alarm, are good adjuncts. If the child is heard snoring and is tired and cranky during the day, a physical examination looking into the amount of room in the upper airway is worthwhile. If the tonsils are enlarged, it is possible that they can interfere with free breathing and this in turn can precipitate a parasomnia event (a partial arousal). A study by Guilleminault et al. [9] found sleep disordered breathing (SDB) in 49 of 84 children with parasomnia diagnoses. These were mostly of sleepwalking and sleep terrors. Twenty-nine of those with both these parasomnias and SDB had a positive family history of parasomnias.

Forty-three of the 49 with SDB were treated surgically with tonsillectomy, adenoidectomy, and/or turbinate revision. Surgical treatment of the sleep disordered breathing eliminated the parasomnias by report and follow-up sleep studies performed 3–4 months later showed no SDB. The six children who went untreated for the breathing problem continued to have sleepwalking events. The same study implicated periodic leg movements of sleep (PLMD) as a second, although less common, sleep disorder that can contribute to sleepwalking in children. This too when treated resolved the episodes of parasomnias. Parents need not "wait out" what may be many years of sleepwalking in a young child if there are treatable triggering factors that can be addressed. On occasion a particular food such as sodas containing caffeine is the precipitating factor, which can be eliminated with good effect. Children whose sleep is fragmented by short arousals will have less energy and more difficulty learning. If they snore or kick at night it is well worthwhile to have a sleep study performed. Children have less obvious signs of oxygen desaturation and lengthy pauses in their breathing while asleep than do adults who suffer from obstructive sleep apnea (OSA). However, even a very mild degree of respiratory disturbance in sleep can act as a trigger for sleepwalking.

The landmark study of the parasomnias in children reported by Broughton [10] included nocturnal enuresis along with sleepwalking and sleep terrors as also occurring in the transition from slow-wave sleep to REM sleep. Although bedwetting can occur at any point in the sleep cycle, there is a type that does take place in the first third of the night and which aggregates in families along with sleepwalking and sleep terrors. In fact, Broughton used loading the child with fluids prior to a sleep test as a probe to elicit sleepwalking behavior in the laboratory.

Because the parasomnias often overlap in the same child and sometimes on the same night, and because they may be triggered by a variety of internal and external factors, they call for a variety of treatment approaches. Before choosing a treatment modality it is important to rule out the possibility that the nocturnal behavior is due to a seizure rather than a parasomnia event. Complex partial seizures may lead to walking about in a dissociated state. If these occur at any time of night, or if the behaviors are repetitive and stereotyped, the child should be worked up by an epileptologist. We have noted the importance of low stress and calming activities before sleep as well as regular sleep times. We have also mentioned surgery for those who have enlarged tonsils or adenoids, which prevent free breathing in sleep and which may trigger sleepwalking in a vulnerable child. If there is no respiratory problem or leg kicking, but the sleepwalker is over 5 years of age and wets the bed within the first few hours of sleep that should be addressed directly. There are both behavioral and pharmacological treatments for nocturnal enuresis. Other treatments that

have helped control sleepwalking are hypnosis [11, 12] and the use of benzodiazepines [13]. Clonazapam in very small doses has two effects that are helpful: it lightens deep sleep making the transition to REM sleep easier and it has a muscle relaxing effect making arising to walk less likely. Many patients have been controlled on this medication over many years without the development of tolerance.

ADULT SLEEPWALKING

The prevalence of sleepwalking in adults is estimated at 3% of the general population. One of the many unknown facts about adult sleepwalking is whether this sleep disorder can occur in adulthood without there being a prior childhood history of this or one of the other parasomnias. What is apparent from those who apply for treatment is that the sex ratio in the adult cases is no longer equal but heavily favors males and that they may then exhibit aggressive behaviors toward others, to property, or to themselves. They may also engage in inappropriate sexual behavior and/or sleep eating. An epidemiological study set the prevalence rate of sleepwalking that involved aggression at 2.1% in a large survey study of adults [14]. Because this disorder may result in serious injury, it is important that a differential diagnosis to rule out a seizure disorder be considered. A recent study of 22 sleepwalkers found abnormal spiking activity in the temporal region during the sleep study in 12 (55%) and, of those, 11 responded to anticonvulsant medication with a complete cessation of sleepwalking [15]. The same precipitating factors that were reported for childhood cases are also implicated in adults: sleep deprivation, psychological distress, and a change in sleep schedule. In addition, it is now reported that drug abuse including alcohol, caffeine, and other drugs as well as psychotropic medications are related to more serious violence while sleepwalking [16]. Although there has been a repeated claim that sleepwalking in adults is more likely to be accompanied by psychopathology, there has been no one psychological profile identified. Many adult patients have no indications of abnormality on psychological tests. The exception has been some with a tendency to be rather passive or overcontrolled, and underexpressive of negative emotions when provoked. Serious violent acts during sleepwalking are more likely to occur in males, who were sleepwalkers or had sleep terrors as children or a strong family history for these disorders, when they are stressed and use excessive caffeine. Their sleep recordings when scored using the traditional sleep stages show very little stage 4 (delta) sleep [16]. The same factors were reported in a recent paper reviewing a case of a sleepwalking murder [17]. When the sleep recordings are subject to power density analysis, they show reduced delta power in the first two sleep cycles of the night. This suggests that these people are more easily aroused from their deep sleep, but not into a fully conscious state. They clearly have dysfunctional face-recognition and then misperceive loved ones as dangerous persons to be attacked. The profound amnesia for these events makes it difficult to gain some understanding of these phenomena. Not all adult sleepwalkers are aggressive. Some engage in attempts to rescue loved ones from what they believe to be a dangerous situation such as a fire. One common thread appears to be a sense of emergency requiring a fight or flight response. Adult sleepwalking behaviors vary from mild nonaggressive to seriously violent acts. The behavior of females is more often of the mild, nonaggressive type such as cooking, eating, or telephoning while males are more often sexually active or aggressive to others. Both sexes display exploratory behaviors, as if searching for something. These behaviors suggest these acts are driven by basic needs, which are not controlled by the executive functions of the prefrontal cortex, which appears in the study by Bassetti et al. [8] to be less active during sleepwalking than in normal sleep.

TREATMENT OF ADULT SLEEPWALKING

The same rules apply to reduce the likelihood of these events in adults as in children. Most important is a medical examination to rule out any disorder other than a parasomnia. This should be followed by a referral to an accredited sleep disorders service, which will instruct the patient on the importance of a regular sleep schedule, avoiding excessive caffeine and alcohol, learning stress reduction techniques, and removing dangerous objects and hazards. A sleep study may then be ordered to determine if there is a sleep disorder such as sleep apnea or periodic limb movements, which might act as a precipitating factor. The records should be scored by power density analysis to reveal whether there is abnormally low delta power. If apnea or PLMD is present, it should be treated. If events continue and are at all aggressive, the patient should be prescribed a benzodiazepine and followed by home logs. Physicians should routinely include questions about a history of parasomnias in their work-up of all new patients.

REFERENCES

1. Kales A, Soldatos C, Bixler E. Hereditary factors in sleepwalking and night terrors. *Br J Psychiatry* **137**:111–118(1980).
2. Hublin C, Kaprio J, Partinen M, Heikkila K, Koskenvuo M. Prevelence and genetics of sleepwalking. A population-based twin study. *Neurology* **48**:177–181(1997).
3. Lecendreux M, Bassetti C, Dauvillers Y, Mayer G, Neidhart E, Tafti M. HLA and genetic susceptibility to sleepwalking. *Mol Psychiatry* **8**:114–117(2003).
4. Rechtschaffen A, Kales A. *A Manual of Standardized Terminology, Techniques and Scoring System for Sleep Stages of Human Subjects.* Brain Information Service/Brain Research Institute, University of California, Los Angeles, 1968.
5. Espa F, Ondze B, Deglise P, Billiard M, Besset A. Sleep architecture, slow wave activity, and sleep spindles in adult patients with sleepwalking and sleep terrors. *Clin Neurophysiol* **111**:929–939(2000).
6. Guilleminault C, Poyares D, Abat F, Palombini L. Sleep and wakefulness in somnambulism a spectral analysis study. *J Psychosom Res* **51**:411–416(2001).
7. Gaudreau H, Joncas S, Zadra A. Montplaisir J. Dynamics of slow-wave activity during the NREM sleep of sleepwalkers and control subjects. *Sleep* **23**:755–760(2000).
8. Bassetti C, Vella S, Donati S, Weilepp P, Weder B. SPECT during sleepwalking. *Lancet* **356**:484–485(2000).
9. Guilleminault C, Palombini L, Delayo R, Chervin R. Sleepwalking and sleep terrors in prepubertal children: What triggers them? *Pediatrics* **111**:17–25(2003).
10. Broughton R. Sleep disorders: disorders of arousal? *Science* **159**:1070–1077(1968).
11. Reid W, Ahmed I, Levie C. Treatment of sleepwalking: a controlled study. *Am J Psychother* **35**:27–37(1981).
12. Hurwitz TD, Mahowald MW, Schenck CH, Schluter JL, Bundlie SR. A retrospective outcome study and review of hypnosis as treatment of adults with sleep walking and sleep terror. *J Nerv Ment Dis* **179**:228–233(1991).
13. Schenck C, Mahowald M. Long term, nightly benzodiazepine treatment of injurious parasomnias and other disorders of disrupted nocturnal sleep in 170 adults. *Am J Med* **100**:333–337(1996).
14. Ohayon M, Caulet M, Priest R. Violent behavior during sleep. *J Clin Psychiatry* **58**: 369–376(1997).
15. Atay T, Karacan I. A retrospective study of sleep walking in 22 patients: clinical and polysomnographic findings. *Sleep Hypnosis* **2**: 1121–1129(2000).
16. Moldofsky H, Gilbert R, Lue F, MacLean A. Sleep related violence. *Sleep* **18**:731–739(1995).
17. Cartwright R. Sleepwalking violence: a sleep disorder, a legal dilemma and a psychological challenge. *Am J Psychiatry* **161**:1149–1158(2004).

ADDITIONAL READING

Broughton R. NREM arousal parasomnias. In: Kryger M, Roth T, Dement W (Eds), *Principles and Practice of Sleep Medicine*, 3rd ed. Saunders, Philadelphia, 2000, pp 693–706. This is a very good review of the three parasomnias that arise out of the first 3 hours of sleep: confusional Arousals, Sleep Terrors and Sleepwalking. The illustrations are excellent, showing sleepwalking recordings in two young children.

Frank N, Spirito A, Stark L, Owens-Stively J. The use of scheduled awakenings to eliminate childhood sleepwalking. *J Pediatr Psychol* **22**:345–353(1997). This paper describes a behavioral intervention which was successful in stopping sleepwalking in three children. This control was maintained for 6 months. It required the parents to wake the child before their usual sleepwalking time. It suggests that scheduled awakenings may abort the abnormal arousal and is a simple procedure to try at home.

Robinson A, Guilleminault C. Disorders of arousal. In: Chokroverty S, Hening W, Walters A (Eds), *Sleep and Movement Disorders*. Butterworth-Heinemann, Philadelphia, 2003, pp 265–272. This is a summary of recent studies on the genetics, differential diagnosis, and treatment of these disorders and has an excellent set of references.

56

REM SLEEP BEHAVIOR DISORDER AND REM-RELATED PARASOMNIAS

Maja Tippmann-Peikert, Timothy I. Morgenthaler, Bradley F. Boeve, and Michael H. Silber
Mayo Clinic College of Medicine, Rochester, Minnesota

INTRODUCTION

Rapid eye movement (REM) sleep behavior disorder (RBD) is characterized by the loss of physiologic skeletal muscle atonia during REM sleep in association with excessive motor activity while dreaming [1, 2]. The disorder most often presents when activity is disruptive or has resulted in harm to the patient or bedpartner [3]. Even though the earliest descriptions of RBD date back to around 1600 (Miguel de Cervantes in *Don Quixote de La Mancha*) [4], the first formal definition of the disorder was provided by Schenck et al. in 1986 [2].

The prevalence of RBD in the general population is estimated to be between 0.38 and 0.5% [5, 6]. It most commonly affects men (87–90%) with a disease onset in the sixth or seventh decade [1, 3, 5, 6]. RBD can present in an acute form, most often related to drug withdrawal, or more commonly in a chronic form. Although up to half of patients with RBD present without other neurologic diagnoses, there is a strong relationship to concurrent or eventual development of synucleinopathies such as dementia with Lewy bodies (DLB), multiple system atrophy (MSA), and Parkinson's disease (PD). Since RBD can precede the diagnosis of these diseases by several years, close follow-up is merited. Most often RBD can be treated satisfactorily with modifications to the sleep environment and pharmacologic agents, making accurate diagnosis very important. Other REM-related parasomnias include nightmares, catathrenia, painful nocturnal erections, sleep paralysis, and REM sleep-related sinus arrest. Some of these disorders represent a blurring of the boundaries between REM sleep and wakefulness, while others appear to represent exaggerated manifestations of REM-related physiology.

REM SLEEP BEHAVIOR DISORDER

Clinical Features

Three large case series describe the clinical and laboratory findings in RBD [1, 7, 8]. A prodrome consisting of sleep-talking, yelling, and nocturnal limb twitching may precede the syndrome by months to years in up to one-quarter of all patients. The lack of daytime symptoms frequently results in a delay of the diagnosis by several years or even decades [1, 3].

Dream enactment behavior during REM sleep represents the clinical hallmark of the disorder and may consist of vocalizations including talking, swearing, yelling, laughing, and simple or complex motor behaviors ranging from limb jerks, gesturing, reaching, grabbing, and flailing to punching, kicking, leaping out of bed, and running. Patients appear to be acting out distinctly altered, unpleasant dreams with recurrent themes involving being chased by or fighting off a human or animal attacker. Animals frequently involved include snakes, spiders, and dogs. Patients occasionally report sports or adventure dreams. Very rarely does dream content involve aggression by the sleeper [1]. The observed behaviors are usually driven by dream

Sleep: A Comprehensive Handbook, Edited by T. Lee-Chiong.

content, which the patient will often recall if aroused during or immediately after an episode. The frequency of abnormal nocturnal behaviors varies from several times nightly to once in several months or years. The behaviors typically occur during the latter half of the night when most REM sleep occurs.

Self-injury from dream enactment behavior has been reported in 32–77% of patients in different series and may result in minor ecchymoses, lacerations, fractures, and even subdural hematomas [1–3]. Assaults on the bed-partner occur in about two-thirds of cases and in 16% result in injuries from punching, kicking, pulling hair, and attempted strangulation. The contrast between violent nocturnal behavior and placid or pleasant daytime character has been highlighted by some authors [3]. Excessive daytime sleepiness is reported by a significant number of patients in the setting of other coexisting sleep disorders, such as obstructive sleep apnea syndrome [1].

Polysomnographic Findings

The hallmark polysomnographic feature in RBD is the loss of normal REM sleep atonia, in the form of persistent elevation of tonic EMG activity or more commonly excessive amounts of phasic muscle activity (Figure 56.1). Different muscle groups and limbs are variably involved. Periodic limb movements of sleep (PLMs) and prominent aperiodic limb movements in non-rapid eye movement (NREM) sleep are frequently encountered, implicating a generalized dysregulation of the motor system during all sleep stages [1, 3]. Although some reports suggest that patients may have over-representation of slow-wave sleep for age, most often sleep

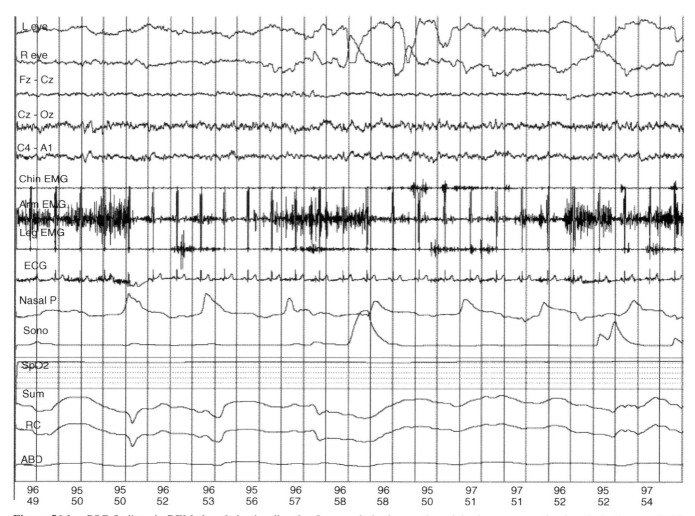

Figure 56.1 PSG findings in REM sleep behavior disorder. Increased phasic muscle activity is present predominantly in the arm–EMG lead, less pronounced in the chin–EMG and leg–EMG leads (30 second epoch).

architecture is normal, unless disturbed by an accompanying medical or sleep disorder, such as in narcolepsy or PD.

Patients with neurodegenerative disorders and RBD may occasionally exhibit ambiguous sleep (a pattern that shares features of wakefulness, NREM and REM sleep) and in advanced disease the EEG of all three states may be indistinguishable, a phenomenon known as status dissociatus. Some patients with neurodegenerative disorders may have the incidental polysomnographic finding of REM sleep without atonia (RSWA) but without a history of abnormal dream enactment behavior [9, 10]. Since it is uncertain whether this phenomenon is a subclinical form or prodrome of RBD, it is best denoted as RSWA rather than RBD.

Differential Diagnosis

When evaluating patients with nocturnal parasomnias, careful attention should be paid to eliciting a history of disruptive, unusual behaviors and to any relationship with dreaming. Because acute forms of RBD have been associated with use of medications, a careful drug history is always necessary (Table 56.1). Any symptoms or signs suggestive of a central nervous system disease, especially a neurodegenerative disorder, should be noted. A full neurologic examination is indicated and a neuropsychological evaluation may be of value if cognitive impairment is suspected. Polysomnography with additional upper extremity EMG derivations and time synchronized video recording links episodic nocturnal motor behaviors to REM sleep and documents increased REM sleep-related tonic and/or phasic skeletal muscle activity.

The differential diagnosis of RBD includes a variety of nocturnal motor behaviors. Nocturnal seizures, particularly frontal lobe seizures, may mimic RBD behaviors and if suspected should be evaluated with an additional 16–20 EEG derivations. Untreated obstructive sleep apnea (OSA) can present with agitated REM sleep-related arousals resulting in motor behaviors indistinguishable from RBD. Periodic

TABLE 56.1   Substances Associated with Acute RBD

Withdrawal:	Alcohol
	Amphetamines
	Cocaine
	Barbiturates
	Meprobamate
	Pentazocine
	Nitrazepam
Medication/intoxication:	Biperiden
	Tricyclic antidepressants
	MAO inhibitors
	Serotonin reuptake inhibitors
	Venlafaxine
	Caffeine

limb movements of sleep can be confused clinically with motor behaviors of RBD. NREM sleep parasomnias such as sleepwalking, sleeptalking, night terrors, or confusional arousals also need to be differentiated. Nightmares typically lack the motor activity seen in RBD. Rhythmic movement disorders may mimic RBD but do not involve dream content and uncommonly occur during REM sleep. Nocturnal panic attacks present with anxiety, palpations, and a sensation of impending doom after arousing from sleep while awareness is fully preserved. Nocturnal psychogenic dissociative states can be clearly distinguished from RBD because the behaviors arise during wakefulness after arousal from sleep.

Diagnostic Criteria

The diagnostic criteria for RBD in the *International Classification of Sleep Disorders-2* are (1) the presence of REM sleep without atonia on a PSG reflected by the EMG finding of excessive phasic or tonic muscle activity on submental or limb derivations; (2) the presence of sleep-related injurious, potentially injurious, or disruptive behaviors either by history or observed during polysomnography; and (3) the absence of EEG epileptiform activity during REM sleep unless RBD can be clearly distinguished from any concurrent REM sleep-related seizure disorder. Thus a PSG is needed for the definitive diagnosis of RBD, but probable RBD can be diagnosed on clinical grounds if a PSG cannot be performed or if no REM sleep was recorded on a PSG. However, the ability of the clinical history alone to distinguish RBD from other sleep disorders, especially OSA, may be low [11] and thus a PSG should be performed whenever possible.

Comorbidities

About half of the cases of RBD reported in the literature initially present with an associated chronic neurologic illness, most commonly a neurodegenerative disorder [1–3, 12]. A less frequently identified comorbidity is narcolepsy [13]. The frequency of RBD in narcolepsy has not been established, but RBD and narcolepsy share disturbances of sleep state boundary control between REM sleep and wakefulness [13]. It is also possible that RBD may manifest in patients with narcolepsy who are treated with tricyclic antidepressants (TCAs) or selective serotonin reuptake inhibitors (SSRIs), both of which may cause REM sleep without atonia [14] and probably clinical RBD (Table 56.1) [13].

Although RBD was initially thought to be associated with a wide variety of dementing disorders, more recent studies reveal that the strongest association exists with MSA, PD, and DLB, collectively known as the synucleinopathies. In one study of 84 patients with neuro-degenerative disease and RBD, 77 had a synucleinopathy as did an additional 45 of 46 patients with REM sleep without atonia [15]. Comparison of cognitive abilities in patients with dementia and RBD versus a group of controls with Alzheimer's disease but no RBD showed that the RBD/dementia group met diagnostic criteria for DLB [16]. Overall, RBD is estimated to be present in 15–33% of patients with PD, and 69–90% of patients with MSA [9, 17–19]. The positive predictive values for RBD indicating a synucleinopathy in patients with cognitive or parkinsonian disorders has been estimated at greater than 90% [15]. In addition, 58% of patients with PD and 90–95% of patients with MSA exhibit RSWA [9, 17–19]. In one series of patients initially diagnosed with idiopathic RBD, 65.4% developed parkinsonism or dementia at a mean of 13.3 years after onset of RBD symptoms [20]. Another study described neuropsychological evaluation abnormalities similar to those found in patients with DLB in 17 patients with idiopathic RBD [21]. Autopsy studies of patients with RBD have all revealed concurrent synucleinopathies. Although Alzheimer changes have been found in addition to Lewy bodies in some patients with RBD, there have been no reported autopsy cases of tauopathies alone, including Alzheimer's disease, frontotemporal dementia, progressive supranuclear palsy, or corticobasal ganglionic degeneration [18]. A single case is reported of an 84 year old patient with longstanding idiopathic RBD and a normal neurologic examination with subsequent autopsy findings of brainstem-predominant Lewy body disease [22]. These observations suggest that apparently idiopathic RBD may be an early disease manifestation of an underlying synucleinopathy and that patients with RBD should be followed for the development of neurodegenerative disorders [15, 23].

Pathopysiology of RBD

The pathogenesis of RBD is not entirely clear. In normal REM sleep, muscle atonia is promoted by the pontine peri-locus caeruleus (peri-LC) alpha neurons, which excite the gigantocellularis nucleus of the reticular formation (GCRF) neurons in the medulla. The GCRF inhibits the spinal alpha motor neurons via the ventrolateral reticulospinal tract. In addition, locomotor generators located in the pons are inhibited in REM sleep. The combined effect of these two processes normally prevents dream enactment behaviors. One can speculate that lesions in the pathways responsible for REM sleep atonia may result in REM sleep without atonia but without any motor behaviors. However, if both the pathways promoting REM sleep atonia as well as the center suppressing locomotion are affected, then REM sleep atonia may be accompanied by excessive movements resulting in RBD.

Several brainstem structures mainly located in the pons, including the peri-LC region, the pedunculopontine nucleus (PPN), and the laterodorsal tegmental nucleus, appear to be involved in the pathophysiology of RBD [24, 25]. At least in cats, bilateral peri-LC lesions lead to a loss of REM sleep muscle atonia associated with simple or complex motor behaviors depending on the location of the lesion [26, 27]. There is currently debate whether PPN lesions alone are sufficient to cause RBD [28, 29]. Recently, lesions in the ventral mesopontine junction were found to increase phasic REM sleep motor activity [30].

Imaging studies have provided conflicting results regarding the pathogenesis of RBD. A small uncontrolled series reported six RBD patients with abnormal brainstem signal suggestive of small vessel ischemic changes on magnetic resonance imaging (MRI), but most authors have reported normal brainstem imaging studies [1, 8, 31, 32]. Positron emission tomography (PET) and single photon emission computed tomography (SPECT) scan studies of patients with idiopathic RBD have shown decreased dopamine activity in the striatum compared to normal control subjects, but less reduction than in patients with PD [31, 33, 34]. It is uncertain whether these findings have direct relevance to the pathogenesis of RBD or indicate that the patients were in a preclinical stage of PD.

Therapy

Patients with RBD are predisposed to self-injury and violent behavior. Therefore basic measures to create a safe bedroom environment are of utmost importance and should be undertaken by all patients suffering from RBD. These include removal of all furniture from the bedside and placing of padding or mattresses on the floor next to the bed. Some patients have used creative self-confinement devices to tether themselves to the bed in order to avoid injury. Pillow barricades or other barrier devices in the bed may prevent injuries to the bedpartner.

Because the etiology of RBD remains unknown, the therapy to date has been symptomatic. The mainstay of pharmacologic therapy is clonazepam. It has been proved to be safe and effective in controlling the motor behaviors as well as the unpleasant dreams in almost 90% of patients when used as a long-term therapeutic agent [1, 8, 35]. The abuse potential appears negligible and tolerance does not generally develop. There are, however, no randomized controlled double-blind trials to prove the efficacy of the drug. Caution must be exercised if clonazepam is given to elderly, demented, or parkinsonian patients as side effects include confusion, dizziness, gait unsteadiness, impotence, and daytime sedation. Clonazepam may also worsen OSA, if left untreated. The mechanism of action of clonazepam in RBD is unknown. Clonazepam decreases the amount of phasic muscle activity during REM sleep but tonic EMG activity persists [8, 36]. RBD symptoms typically recur immediately after discontinuation of therapy. Recently, there have been a few open label trials suggesting a beneficial effect of melatonin on symptoms of RBD. Melatonin (3–12 mg at bedtime) resulted in symptomatic improvement in 26 of 29 patients [37–39]. Follow-up PSG of patients treated with melatonin showed a decreased percentage of REM sleep without atonia [38]. Although randomized, controlled treatment trials are clearly needed, melatonin may be a valuable drug for treatment of patients who cannot tolerate clonazepam or who experience an incomplete therapeutic response. A recent study has suggested that pramipexole may also be effective [40].

Future Directions

RBD is a disorder wide open to further investigation. Further autopsy and imaging studies are needed to investigate the pathology of both "idiopathic" RBD and RBD associated with neurologic diseases. This may provide insight into the pathophysiology of not only RBD but also the frequently associated neurodegenerative disorders. RBD may also present an opportunity to test potential early therapies for synucleinopathies.

OTHER REM SLEEP PARASOMNIAS

Nightmares

Nightmares occur when frightening dream content leads to awakening from REM sleep. The lack of associated motor activity, dream enactment behavior, vocalization, confusion, and prominent autonomic activation differentiates nightmares from RBD, sleep terrors, and other disorders characterized by arousals from NREM sleep. In adults the prevalence of recurrent nightmares is about 4% [41]. Prevalence is higher in children and in adults with psychiatric comorbidities, such as post-traumatic stress disorder or substance abuse. Medications that are known to induce nightmares include levodopa, beta-blockers, SSRIs, and TCAs. Treatment involves the correction of underlying medical or psychiatric conditions and, if necessary, psychotherapy, cognitive–behavioral therapy, and hypnosis.

Sleep Paralysis

Sleep paralysis represents intrusion of REM sleep atonia into wakefulness leading to complete paralysis of skeletal muscles (except respiratory and extraocular muscles) while the patient is fully conscious. The phenomenon most often occurs in the morning upon awakening from REM sleep, especially when supine, but is rarely seen in the evening during sleep onset [42]. Patients describe the

experience as extremely frightening and sometimes experience a sense of dyspnea. The duration of paralysis ranges from seconds to several minutes. External stimuli such as being called or touched by someone may abort an attack. The lifetime prevalence in the general population is estimated to be 6–60% [43, 44]. A higher prevalence has been reported in patients with narcolepsy with cataplexy in whom sleep paralysis occurs at a frequency of 60%, and prevalence may be increased in patients with obstructive sleep apnea syndrome [42]. Treatment of sleep paralysis is not usually indicated, but tricyclic antidepressants have been reported to be beneficial.

Nocturnal Groaning (Catathrenia)

Catathrenia, also known as nocturnal groaning, is characterized by a high-pitched expiratory groaning noise of 5–20 seconds duration that occurs in clusters for several minutes to one hour [45]. Catathrenia typically occurs in the second half of the night and is most prominent during REM sleep, but may persist in NREM stage 2 sleep. The noises tend to disappear with position changes associated with EEG arousals. Respiratory distress, motor activity, and dreaming are not associated. This condition often presents during the first or second decade and may persist nightly for years. Patients are usually unaware of the disorder, but significant bedpartner or family disturbance is frequent. The etiology remains unclear but partial glottic closure has been implicated. General physical, neurologic, pulmonary, otorhinolaryngologic (including wake laryngoscopy), and psychologic examinations are typically normal [45]. The long-term prognosis is unknown. No successful therapies have been identified [46].

REM Sleep-Related Sinus Arrest

The condition is characterized by sinus pauses up to 9 seconds in duration during REM sleep in otherwise healthy, young individuals [47]. Vocalization may accompany prolonged sinus arrest. Careful cardiac evaluations, including Holter monitoring and cardiac catheterization, have failed to identify pathology, thus implicating sleep-related autonomic dysfunction as a possible etiology. A potential risk is sudden nocturnal death from prolonged sinus arrest. Although treatment is not usually indicated, prevention of sudden death includes consideration of permanent pacemaker insertion.

REM Sleep-Related Painful Erections

The condition affects men with a mean age of about 40 years old and is characterized by recurrent nocturnal awakenings from REM sleep due to painful erections [48]. Significant sleep disruption and daytime hypersomnolence may result. Penile and psychiatric pathology is typically absent, but erectile dysfunction complaints are not uncommon. No effective therapy has been identified, but consideration may be given to REM sleep suppressing agents such as SSRIs and TCAs, or beta blockers, such as propranolol.

REFERENCES

1. Olson EJ, Boeve BF, Silber MH. Rapid eye movement sleep behaviour disorder: demographic, clinical and laboratory findings in 93 cases. *Brain* **123**(Pt 2):331–339(2000).

2. Schenck CH, et al. Chronic behavioral disorders of human REM sleep: a new category of parasomnia. *Sleep* **9**(2):293–308(1986).

3. Schenck CH, Mahowald MW. REM sleep behavior disorder: clinical, developmental, and neuroscience perspectives 16 years after its formal identification in SLEEP. *Sleep* **25**(2):120–138(2002).

4. Ormsby J, *Don Quixote (1605, 1615). Traducido por John Ormsby.* 1885, http://www.csdl.tamu.edu/cervantes/V2/CPI/index.html: London.

5. Ohayon MM, Caulet M, Priest RG. Violent behavior during sleep. *J Clin Psychiatry* **58**(8):369–376(1997), quiz 377.

6. Chiu HF, et al. Sleep-related injury in the elderly—an epidemiological study in Hong Kong. *Sleep* **23**(4):513–517(2000).

7. Sforza E, Krieger J, Petiau C. REM sleep behavior disorder: clinical and physiopathological findings. *Sleep Med Rev* **1**:57(1997).

8. Schenck CH, Mahowald MW. A polysomnographic, neurologic, psychiatric and clinical outcome report on 70 consecutive cases with REM sleep behavior disorder (RBD). *Clev Clin J Med* **57**(Suppl):S9–23(1990).

9. Gagnon JF, et al. REM sleep behavior disorder and REM sleep without atonia in Parkinson's disease. *Neurology* **59**(4):585–589(2002).

10. Bliwise DL, et al. Inter-rater reliability for identification of REM sleep in Parkinson's disease. *Sleep* **23**(5):671–676(2000).

11. Schenck CH, et al. A polysomnographic and clinical report on sleep-related injury in 100 adult patients. *Am J Psychiatry* **146**(9):1166–1173(1989).

12. Turner RS. Idiopathic rapid eye movement sleep behavior disorder is a harbinger of dementia with Lewy bodies. *J Geriatr Psychiatry Neurol* **15**(4):195–199(2002).

13. Schenck CH, Mahowald MW. Motor dyscontrol in narcolepsy: rapid-eye-movement (REM) sleep without atonia and REM sleep behavior disorder (comment). *Ann Neurol* **32**(1):3–10(1992).

14. Winkelman J, James L. Serotonergic antidepressants are associated with REM sleep without atonia. *Sleep* **27**(2):317–321(2004).

15. Boeve BF, et al. Association of REM sleep behavior disorder and neurodegenerative disease may reflect an underlying synucleinopathy. *Mov Disord* **16**(4):622–630(2001).

16. Ferman TJ, et al. REM sleep behavior disorder and dementia: cognitive differences when compared with AD. *Neurology* **52**(5):951(1999).

17. Plazzi G, et al. REM sleep behavior disorders in multiple system atrophy. *Neurology* **48**(4):1094–1097(1997).

18. Comella CL, et al. Sleep-related violence, injury, and REM sleep behavior disorder in Parkinson's disease. (See Comment.) *Neurology* **51**(2):526–529(1998).

19. Plazzi G. REM sleep behavior disorders in Parkinson's disease and other Parkinsonian disorders. *Sleep Med* **5**(2):195–199(2004).

20. Schenck CH, Bundlie SR. MW, REM behavior disorder (RBD): delayed emergence of parkinsonism and/or dementia in 65% of older men initially diagnosed with idiopathic RBD, and an analysis of the minimum & maximum tonic and/or phasic electromyographic abnormalities found during REM sleep. *Sleep* **26**(Suppl):A316(2003).

21. Ferini-Strambi L, et al. Neuropsychological assessment in idiopathic REM sleep behavior disorder (RBD): does the idiopathic form of RBD really exist? *Neurology* **62**(1):41–45(2004).

22. Uchiyama M, et al. Incidental Lewy body disease in a patient with REM sleep behavior disorder. (See Comment.) *Neurology* **45**(4):709–712(1995).

23. Boeve BF, et al. Synucleinopathy pathology and REM sleep behavior disorder plus dementia or parkinsonism. *Neurology* **61**(1):40–45(2003).

24. Benarroch EE. Brainstem in multiple system atrophy: clinicopathological correlations. *Cell Mol Neurobiol* **23**(4–5):519–526(2003).

25. Schenck CH, et al. A case of REM sleep behavior disorder with autopsy-confirmed Alzheimer's disease: postmortem brain stem histochemical analyses. *Biol Psychiatry* **40**(5):422–425(1996).

26. Hendricks JC, Morrison AR, Mann GL. Different behaviors during paradoxical sleep without atonia depend on pontine lesion site. *Brain Res* **239**(1):81–105(1982).

27. Jouvet M. Neurophysiology of the states of sleep. *Physiol Rev* **47**(2):117–177(1967).

28. Rye DB. Contributions of the pedunculopontine region to normal and altered REM sleep. (See Comment.) *Sleep* **20**(9):757–788(1997).

29. Morrison AR. The pathophysiology of REM-sleep behavior disorder (comment). *Sleep* **21**(5):446–449(1998).

30. Boeve BF, et al. REM Sleep behavior disorder in Parkinson's disease, dementia with Lewy bodies, and multiple system atrophy. In: *Mental and Behavior Dysfunction in Movement Disorders*. Humana Press, Totawa, NJ, 2003, pp 383–397.

31. Culebras A, Moore JT. Magnetic resonance findings in REM sleep behavior disorder. *Neurology* **39**(11):1519–1523(1989).

32. Boeve BF, et al. REM sleep behavior disorder and degenerative dementia: an association likely reflecting Lewy body disease. *Neurology* **51**(2):363–370(1998).

33. Eisensehr I, et al. Reduced striatal dopamine transporters in idiopathic rapid eye movement sleep behaviour disorder. Comparison with Parkinson's disease and controls. *Brain* **123**(Pt 6):1155–1160(2000).

34. Albin RL, et al. Decreased striatal dopaminergic innervation in REM sleep behavior disorder. *Neurology* **55**(9):1410–1412(2000).

35. Schenck CH, Hurwitz TD, Mahowald MW. Symposium: normal and abnormal REM sleep regulation: REM sleep behaviour disorder: an update on a series of 96 patients and a review of the world literature. *J Sleep Res* **2**(4):224–231(1993).

36. Lapierre O, Montplaisir J. Polysomnographic features of REM sleep behavior disorder: development of a scoring method. (See Comment.) *Neurology* **42**(7):1371–1374(1992).

37. Takeuchi N, et al. Melatonin therapy for REM sleep behavior disorder. *Psychiatry Clin Neurosci* **55**(3):267–269(2001).

38. Kunz D, Bes F. Melatonin as a therapy in REM sleep behavior disorder patients: an open-labeled pilot study on the possible influence of melatonin on REM-sleep regulation. *Mov Disord* **14**(3):507–511(1999).

39. Boeve BF, Silber MH, Ferman TJ. Melatonin for treatment of REM sleep behavior disorder in neurologic disorders: results in 14 patients. *Sleep Med* **4**(4):281–284(2003).

40. Fantini ML, et al. The effects of pramipexole in REM sleep behavior disorder. *Neurology* **61**(10):1418–1420(2003).

41. Janson C, et al. Prevalence of sleep disturbances among young adults in three European countries. *Sleep* **18**(7):589–597(1995).

42. Cheyne JA. Situational factors affecting sleep paralysis and associated hallucinations: position and timing effects. *J Sleep Res* **11**(2):169–177(2002).

43. Dahlitz M, Parkes JD. Sleep paralysis. *Lancet* **341**(8842):406–407(1993).

44. Ohayon MM, et al. Prevalence and pathologic associations of sleep paralysis in the general population. *Neurology* **52**(6):1194–1200(1999).

45. Vetrugno R, et al. Catathrenia (nocturnal groaning): a new type of parasomnia. *Neurology* **56**(5):681–683(2001).

46. Pevernagie DA, et al. Vocalization during episodes of prolonged expiration: a parasomnia related to REM sleep. *Sleep Med* **2**(1):19–30(2001).

47. Guilleminault C, et al. Sinus arrest during REM sleep in young adults. *N Engl J Med* **311**(16):1006–1010(1984).

48. Ferini-Strambi L, et al. Sleep-related painful erections: clinical and polysomnographic features. *J Sleep Res* **5**(3):195–197(1996).

57

NOCTURNAL ENURESIS IN CHILDREN

MARIE-JOSÈPHE CHALLAMEL
Hôpital Debrousse, Lyon, France

PIERRE COCHAT
Hôpital Edouard-Herriot and Université Claude-Bernard, Lyon France

Bedwetting is a very common condition in school-age children with a prevalence rate of 15–20% at age 5, 8–10% at age 6, 3–4% at age 10, and up to 1–2% in young adults [1–3]. Two-thirds of them experience monosymptomatic nocturnal enuresis, mainly boys, and one-third suffer from voiding disorder, mainly girls, sometimes leading to persistent troubles toward adulthood.

DEFINITIONS

Nocturnal enuresis is characterized by the frequent occurrence of normal complete uncontrolled micturition during sleep in children older than 5 years of age. Voiding disorders include a wide number of functional problems, mainly bladder and urethral instability. Bladder instability is due to uninhibited contractions of the detrusor during bladder filling. Urethral instability is characterized by inadequate urethral sphincter relaxation during bladder filling in the absence of detrusor contraction.

Both nocturnal enuresis and voiding disorders may be of primary (absence of wetting-free period) or secondary (onset of wetting after a significant period of dryness) onset. Primary troubles are far more frequent than secondary ones, in an approximate ratio of 10:1.

Correspondence to: Marie-Josèphe Challamel, Exploration sommeil de l'enfant, Bâtiment A, Niveau 2, Hôpital Debrousse, 29 rue Sœur Bouvier, 69005 Lyon, France.

CLINICAL PRESENTATION

Differentiating Monosymptomatic Nocturnal Enuresis from Voiding Disorders or Other Causes of Nocturnal Enuresis

Bladder disorders, of which bladder instability is the most frequent, are out of the scope of this chapter; however, they may mimic nocturnal enuresis.

In most cases, the diagnosis procedure of primary nocturnal enuresis is limited to the patient's history. Specific questions will help in differentiating monosymptomatic nocturnal enuresis from voiding disorders (Table 57.1 and Figure 57.1) [1]

Clinical Investigations

If physical examination is normal, which is usually the case, no further investigation is required in the presence of a typical picture of primary monosymptomatic nocturnal enuresis. However, routine clinical parameters must be assessed, that is, height and growth velocity, blood pressure, and urinalysis using dipstick (glucose, leukocytes and nitrite, density, pH, protein), in order to exclude other specific diagnoses (e.g., diabetes mellitus, congenital tubular disorders, chronic renal insufficiency, central or nephrogenic diabetes insipidus, chronic tubulointerstitial nephritis, urinary tract malformation, obstructive sleep apnea syndrome, NREM sleep parasomnias, sickle cell disease,

Sleep: A Comprehensive Handbook, Edited by T. Lee-Chiong.
Copyright © 2006 John Wiley & Sons, Inc.

TABLE 57.1   Differentiating Monosymptomatic Nocturnal Enuresis from Voiding Disorders

Clinical Symptoms and History	Monosymptomatic Enuresis	Voiding Disorders
Family history of nocturnal enuresis?	Present	Usually absent
Daytime wetting?	Absent	Present
Nighttime wetting?	Daily	Less frequent
Environmental influence?	Yes	Usually absent
Use of diapers?	Frequent	Less frequent
Sleep arousal because of voiding call?	Absent	Frequent
Failure of anticholinergic agents	Yes	No
Urinary tract infections?	Absent	Frequent
Constipation?	Usually absent	Frequent
Pollakiuria?	Absent	Frequent

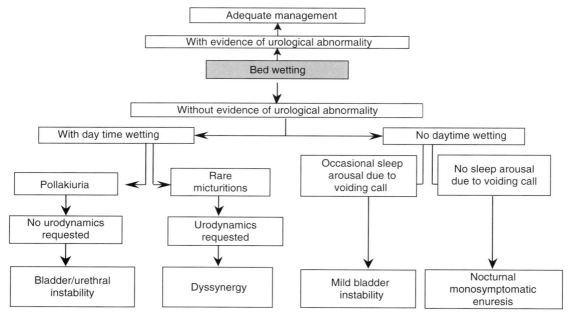

Figure 57.1   Clinical assistance for decision making in the presence of bedwettings. (From [1].)

neurological disorders, medications). In selected cases, anatomical/physiological assessments, such as frequency–volume charts, evaluation of kidney and bladder function, measurement of bladder capacity, measurement of plasma vasopressin concentrations, and notation of arousal difficulties, have been suggested.

NOCTURNAL ENURESIS AND PSYCHOLOGICAL OR PSYCHIATRIC DISORDERS

For a long time, primary nocturnal enuresis has been regarded as a symptom of an underlying psychopathological problem. Recent surveys show that it is a common behavioral problem without a psychiatric component. In primary nocturnal enuresis, the likelihood of significant psycho-

pathology is less than 10% [4]. Enuretic children do tend to exhibit anxiety and problematic conduct but the symptoms are more likely to be the result than the cause of enuresis; psychotherapy is indicated only if a psychopathological condition is present, or if enuresis itself has major psychological consequences for the child or the family.

In secondary monosymptomatic enuresis, enuresis may occur after personal or familial disturbance (e.g., moving, school problems, parental divorce, grandparent death, sudden illness in a family member, or new arrival of a sibling).

Nocturnal enuresis, sometimes associated with diurnal enuresis, may be a part of psychiatric disorders, including attention-deficit hyperactivity disorder [5], autism, and other severe disturbances. In such conditions, enuresis is a minor medical symptom but a major problem with respect to the patient's quality of life, with an associated risk of punishment and abuse.

PATHOPHYSIOLOGY

Enuresis is a multifactorial disorder and probably comprises several types. Pathophysiological findings suggest that four interactive factors are involved [1, 6, 7]:

- Genetic factors.
- Sleep factors and arousal factors, which explain the patient's inability to waken in response to signals from a full bladder.
- Relative nocturnal polyuria due to a maturational blunting of the normal vasopressin peak concentration during nocturnal sleep.
- Reduced nocturnal functional bladder capacity, especially in some cases who do not respond to desmopressin treatment.

Genetic Factors

Heredity is one of the most important factors contributing to enuresis. The probability of having enuresis increases as a function of the closeness and number of blood relative with a history of enuresis. The percentage of bedwetters increases from about 15% of children without any family history of enuresis to about 45% when one of the parents and about 75% when both parents used to wet his/her bed [2]. Recent evidence points to the involvement, in some cases, of a single autosomal dominant gene [6].

Sleep in Enuretic Children

Enuretic children often appear to their parents as deep sleepers. Sleep enuresis had been thought to be a disorder of sleep and formerly classified as an arousal disorder. However, more recent studies have challenged this concept and showed that enuretic children were normal sleepers and that the macrostructure of sleep was normal [8]. However, Goodwin et al. [9] have shown that children with enuresis had significantly greater sleep time, and greater time in stage 2 than children without enuresis. On the other hand, Hunsballe [10] has shown increased delta waves during NREM sleep, which could be indicative of a greater depth of sleep.

Bedwetting Episodes and Sleep Stages

Bedwetting episodes may occur at any sleep stage as well as during nocturnal awakenings [11]. They may occur during all NREM sleep stages according to the duration of each stage. Enuretic episodes are more frequent during light NREM sleep than during deep sleep [12, 13]. Interestingly, Rathke and Jovanovic [14] have shown that among 64 enuretic episodes, 39% occurred during stage 2, only

20% during stage 3–4, and 9% during REM sleep. Actually, enuretic episodes seem to be more related to the period of the nighttime than to sleep stages, as enuretic children wet their beds more frequently in the first two-thirds of the night [8, 13]. In addition, Kalo and Bella [15] have shown that out of 83 Saudi children with enuretic episodes during sleep, 30% wet their beds during daytime sleep as well.

Arousal Mechanisms and Enuresis

A high arousal threshold may be one of the pathogenic factors underlying nocturnal enuresis, but few laboratory sleep studies have quantitatively analyzed this phenomenon. A defect in arousal mechanisms to full bladder capacity probably plays a role in monosymptomatic nocturnal enuresis in children. However, more specific studies are needed to define this role. A maturational delay in the development of the arousal response to afferent visceral activity caused by bladder distension has been suggested. In nonenuretic children as well as in adults, sleep arousal occurs when maximal bladder capacity has been reached ("voiding call "). During the first decade, children normally experience a high arousal threshold. Alon et al. [16] have shown in children with diabetes mellitus that waking up response to full bladder is age-dependent: that is, the younger the child the higher the risk of nocturnal enuresis. Wille [17] showed, using a waking test, that enuretic children were more difficult to arouse at night compared to controls. Wolfish et al. [18] indicated an arousal defect in enuretics: 15 male enuretic children aged 7–12 years were compared to 18 controls; arousal was successful in only 9.3% of attempts in enuretic children compared to 39.7% in controls. In contrast, Boyd [19], comparing the arousability of 100 school-age enuretic children with age- and sex-matched controls, showed that there was no significant difference in the depth of sleep between the two groups. These contradictory results might be explained by the fact that these studies were done in heterogeneous populations of enuretic children and different environmental conditions.

Antidiuretic Hormone Secretion During Sleep

Normal subjects show a marked circadian variation in urine output and osmolarity with a reduced volume of concentrated urine during nighttime. This circadian rhythm is normally regulated by vasopressin. Urine production during nighttime is about half that of daytime; it has therefore been suggested that a substantial number of enuretic patients may lack physiological circadian variation in urine excretion rate and urine osmolarity. In a study of 11 subjects with monosymptomatic nocturnal enuresis, Nørgaard et al. [20] demonstrated that all produced larger volumes of diluted urine during nighttime than their own bladder could contain. Such patients have subsequently been shown to lack the normal nocturnal increase in vasopressin, with a significant lower morning concentration of vasopressin than nonenuretic children [21].

Further studies are needed to establish whether the lack of vasopressin production during the night is an important pathogenic factor in nocturnal enuresis, the balance between bladder capacity and urine production being a major determinant of enuresis. In enuretic children without nocturnal polyuria who had normal circadian variation in urine production, it has been suggested that enuresis may be due to a low bladder functional capacity [6].

Nocturnal Enuresis and Sleep Breathing Disorders (SBDs)

Nocturia and bedwetting are symptoms of obstructive sleep apnea syndrome in adults. In adults it is likely to be related to an increased atrial natriuretic peptide concentration and a decrease in plasma renin activity and aldosterone level. There are no data to support this in children. The relationship between nocturnal enuresis and SBDs in children has been inconsistent, with some studies finding an association and others not finding one. In a recent study from Goodwin et al. [9], which analyzed the relationship between parasomnias and SBDs in a population of preadolescent children 6–11 years of age, children with a Respiratory Disturbance Index (RDI) of more than one were more likely to have enuresis than children with a RDI less than one: 11.3% versus 6.3% ($P < 0.08$). Moreover, in this study, enuresis was strongly associated with frequent loud snoring ($P < 0.007$), a known correlate of SBDs in children. In another study from Brooks and Topol [22], children with a RDI ≤ 1 had a significantly lower prevalence of enuresis (17%), than did children with a RDI > 1 (47%) ($P < 0.05$). In this study there was no significant difference in the prevalence of enuresis in children with a RDI $= 1$–5, 5–15, or >15 ($P < 0.92$).

In children the relationship between nocturnal enuresis and upper airway obstruction needs to be elucidated. Nocturnal enuresis may be secondary to increased intra-abdominal pressure from paradoxical breathing associated with increased respiratory efforts. The improvement in enuresis after surgery may also be due to the disappearance of sleep fragmentation with subsequent normalization of arousal mechanisms.

MANAGEMENT

Several considerations have to be taken into account: conventional treatment should take place along with psychotherapy, natural medicine, physiotherapy, hypnotherapy, and possibly surgery. Access to the treatment of enuresis

could be perceived as a medical privilege, because no treatment is available in developing countries.

Any kind of urinary dysfunction may have a psychosocial component (anxiety, guilt feelings, opposition) and its treatment has been shown to improve self-esteem.

Results of adequate treatment in compliant patients are sometimes difficult to assess, since some studies report a spontaneous cure rate of approximately 15% per year (although others report an apparent resistance to all available treatments). Placebo is beneficial in around one-third of the patients. For these reasons, an overall 70–80% success rate is usually expected.

The main therapeutic options in children with nocturnal enuresis are summarized in Figure 57.2 [1]. In any case, regardless of the age of the child, an enuresis diary will allow determination of the severity of bedwetting and further interpretation of outcome under treatment.

Behavioral and Educational Interventions

Nocturnal continence involves several skills that are generally not attained before the age of 4–5 years: awareness of urgency, ability to initiate urination, ability to inhibit urination while awake, and ability to inhibit urination while asleep.

Simple Behavioral Methods Treatment should be adapted to the child's age. Children under the age of 6 years may be managed by reassurance: explaining to the child and his/her family that the child has different patterns of sleep and does not sense the need to urinate can help to relieve the psychological consequences of enuresis. Preventing irregular sleep–wake patterns and sleep deprivation may control bedwetting in young children. For older children, understanding and support are the most important attitude toward helping the patient; but the child should not be left in diapers at night and should be assigned household responsibilities associated with the problem.

Independent of the age of the patient, reward systems (star charts given for dry nights), lifting or waking the child at night to urinate, retention control training to enlarge bladder capacity, and fluid restriction may be effective for some children and could therefore be tried as first-line therapy.

If no change in the nocturnal wetting pattern occurs after a delay of about 3 months, when the child is motivated, the addition of an enuresis alarm system can be considered. Long-term pharmacological treatment should be limited to children older than 6 years who did not respond to behavioral approaches.

Alarm Systems Alarm interventions are highly effective for the treatment of nocturnal bedwetting and have more lasting effect than pharmacological treatment. Desmopressin

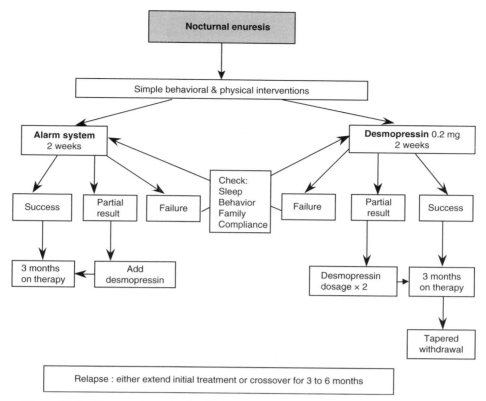

Figure 57.2 Therapeutic option in children with monosymptomatic enuresis. (From [1].)

offers better short-term results but an enuresis alarm is significantly more efficient in the long-term. The combination of alarm with behavioral and educational interventions offers better results than an alarm on its own.

Pharmacological Treatment

Tricyclics The mechanism of action of tricyclics (e.g., imipramine) on nocturnal enuresis is unknown; it probably acts at several levels. The drug may work by its effect on sleep (arousal effects). It may also reduce premature contractions of the detrusor muscle following partial filling of the bladder by its anticholinergic effect. Tricyclic agents have a positive effect on monosymptomatic nocturnal enuresis compared to placebo but it is generally accepted that imipramine, a potentially dangerous drug, should not be recommended in a benign condition like monosymptomatic nocturnal enuresis.

Use of tricyclics should be limited to older children and adolescents and only for short periods of time.

The daily dose of imipramine ranges between 1 and 2.5 mg/kg BW, or 25–75 mg depending on age and BW. Some patients do not respond, whereas others develop tolerance. In addition, the relapse rate after withdrawal approximates 40–60%. Side effects include restlessness,

sleep disturbance, attention disorders, headache, abdominal pain, constipation, weight loss, and acrocyanosis. In addition, one should keep in mind that life-threatening adverse events have been reported such as sudden death, coma, convulsions, and heart or liver failure.

Desmopressin Desmopressin is a synthetic agonist of the neuropeptide vasopressin with enhanced antidiuretic properties. Thanks to its safety, it has replaced imipramine in the pharmacological approach to enuresis. Desmopressin mainly acts by reducing nocturnal urine production via direct vasopressin effects [12]. It may also act by improving the patient's ability to awaken and by improving bladder instability. Given orally at a dose of 0.2–0.4 mg (tablets 0.1 and 0.2 mg) or intranasally at a dose of 20–40 μg (2–4 sprays) for a period of about 3 months, desmopressin is effective in 50–80% of patients with monosymptomatic nocturnal enuresis. Unfortunately, such a treatment is associated with a high relapse rate. It is recommended to limit fluid intake (<250 mL), while desmopressin is effective. Desmopressin has minor side effects including headache, mild abdominal cramping, nausea, nasal congestion, epistaxis, and reversible body weight gain. Water intoxication with hyponatremia and convulsions have been reported in a few cases under special conditions.

Children with a family history of nocturnal enuresis and a high nocturnal urine output seem to be good responders to desmopressin. Nørgaard et al. [20] have shown that 91% of enuretic patients with a family history of enuresis compared to only 71% of those without are responders to desmopressin.

CONCLUSION

Monosymptomatic nocturnal enuresis is a very common finding in school-age children. However, the medical approach to enuresis lacks sufficient data: pathophysiology is still uncertain and management depends on both cultural and financial abilities. Precise studies on arousal mechanisms and urodynamic changes during sleep may be important issues for future research. Research on sleep enuresis is difficult to promote since there is no available experimental animal model, and conflicting ethical issues limit clinical research in humans because nocturnal enuresis only affects "healthy" children.

REFERENCES

1. Challamel MJ, Cochat P. Enuresis: pathophysiology and treatment. Review article. *Sleep Med Rev* **3**:313–324(1999).

2. Jarvelin MR, Vikeväinen-Tervonen L, Moilanen I, Huttenen NP. Enuresis in seven year old children. *Acta Paediatr Scand* **77**:148–153(1988).

3. Laberge L, Tremblay RE, Vitaro F, Montplaisir J. Development of parasomnias from childhood to early adolescence. *Pediatrics* **106**:67–74(2000).

4. Friman PC, Handwerk ML, Swearer SM, et al. Do children with primary nocturnal enuresis have clinically significant behavior problems? *Arch Adolesc Med* **152**:537–539(1998).

5. Robson WL, Jackson HP, Blackhurst D, Leung AK. Enuresis in children with attention-deficit hyperactivity disorder. *South Med J* **90**:503–505(1997).

6. Nørgaard JP, Djurhuus JC. The pathophysiology of enuresis in children and young adults. *Clin Pediatr Spec Issue* 5–9(1993).

7. Nevus T, Bader G, Sillen U. Enuresis, sleep and desmopressin treatment. *Acta Paediatr* **91**:1121–1125(2002).

8. Kales A, Soldatos CR, Kales JD. Sleep disorders: insomnia, sleepwalking, night terrors, nightmares and enuresis. *Ann Intern Med* **106**:582–592(1987).

9. Goodwin JL, Kaemingk KL, Fregosi RF, Rosen GM, Morgan WJ, Smith T, Quan F. Parasomnias and sleep disordered breathing in caucasian and hispanic children. The Tucson children's assessment of sleep apnea study. *BMC Med* **2**:14(2004).

10. Hunsballe JM. Increased delta component in computerized sleep electroencephalographic analysis suggests abnormally deep sleep in primary monosymptomatic nocturnal enuresis. *Scand J Urol Nephrol* **34**:294–302(2000).

11. Mikkelsen EJ, Rapoport JL, Nee L, et al. Childhood enuresis; I. Sleep patterns and psychopatholgy. *Arch Gen Psychiatry* **37**:1139–1144(1980).

12. Nevus T, Stenberg A, Lackgren G, Tuvemo T, Hetta J. Sleep of children with enuresis: a polysomnographic study. *Pediatrics* **103**:1193–1197(1999).

13. Bader G, Nevus T, Kruse S, Sillen U. Sleep of primary enuretic children and controls. *Sleep* **25**:579–583(2002).

14. Rathke W, Jovanovic UJ. Sleep profile and ultradian sleep periodicity in child bedwetters. *Waking Sleeping* **1**:313–317(1977).

15. Kalo BB, Bela H. Enuresis: prevalence and associated factors among primary school age children in Saudi Arabia. *Acta Paediatr* **85**:1217–1222(1996).

16. Alon U, Woodward V, Howard CP. Urine volume, age and nocturnal enuresis: a prospective study on newly diagnosed children with diabetes mellitus. *Child Hosp* **Q4**:157–160(1992).

17. Wille S. Nocturnal enuresis sleep disturbance and behavioural patterns. *Acta Paediatr* **4**:772–774(1994).

18. Wolfish NM, Pivik RT, Busby KA. Elevated sleep arousal thresholds in enuretic boys: clinical implications. *Acta Paediatr* **86**:381–384(1997).

19. Boyd MMM. The depth of sleep in enuretic school-children and in nonenuretic controls. *J Psychosom Res* **4**:274–281(1960).

20. Noorgaard JP, Djurhuus JC, Watanabe H, et al. Experience and current status of research into the pathophysiology of nocturnal enuresis. *Br J Urol* **79**:825–835(1997).

21. Rittig S, Knudsen UB, Norgaard JP, et al. Abnormal diurnal rhythm of plasma vasopressin and urinary output in patients with enuresis. *Am J Physiol* **25**:664–671(1989).

22. Brooks LJ, Topol H. Enuresis in children with sleep apnea. *J Pediatr* **142**:515–518(2003).

58

SLEEP BRUXISM

GABY BADER

University of Gothenburg, Gothenburg, Sweden

The etiology of bruxism, which is defined as a parafunction by dentists, is controversial and includes both occlusal and psychological factors. Bruxism can occur during both wakefulness and sleep.

Sleep bruxism (SB), previously referred to as "nocturnal bruxism," is a stereotyped, sometimes intense, orofacial movement with bilateral activation of the masseter and temporal "jaw closure" muscles. The sleeping subject is unaware of the repeated grinding or clenching of the teeth although he/she may experience temporomandibular and jaw pain as a result. The sound of grinding teeth often impairs the sleep of bedpartners. Under current sleep medicine diagnostic criteria, SB is classified as a parasomnia related movement disorder [1]. Sleep bruxism can occur in otherwise normal subjects or can be associated with dental, mandibular, or maxillary malfunction, such as malocclusion.

Bruxism is thought to be a major cause of temporomandibular disorder (TMD), with patients complaining of pain and/or dyskinesia in the temporomandibular region. TMD, if frequent, may lead to excessive tooth destruction or even periodontal tissue damage as well as limitation of mandibular movement.

ETIOLOGY AND PATHOPHYSIOLOGY

The etiology of SB is probably different from the parafunctional activity of the jaw muscles while awake. Clenching and grinding activity can be separate disorders with different etiologies [2]. Bruxism can be primary or idiopathic, or can be secondary to other medical conditions, often neurological (e.g., Parkinson's disease, Tourette syndrome, Rett syndrome, dyskinesia) or psychiatric (e.g., mental retardation, dementia, depression). It can also be iatrogenic, associated with smoking, alcohol [3], medication (SSRI [4], antipsychotic [5]), and drugs (cocaine [6], amphetamine [7]).

It had long been believed that sleep bruxism was dependent on peripheral mechanisms including discrepancies of dental occlusion and other morphological orofacial changes. However, this view has been abandoned since no relationships have been found between tooth wear and ongoing SB level [8]. Grinding sounds do not occur during all episodes of bruxism and oromotor activity associated with SB does not always occur with tooth contact. Peripheral sensory inputs from the oral mucosa and muscles or the periodontium, which are known to affect the activity of the jaw muscles during wakefulness, can have an impact on the sleep–wake mechanisms and on SB. In a recent study of 31 SB patients, both with and without complaints of chronic facial pain, episodes of bruxism were more frequent and of longer duration in patients without chronic pain. No differences in sleep quality were reported. It was suggested that chronic facial pain exerted an inhibitory effect on the muscle hyperactivity [9].

Although peripheral and sensory mechanisms play a certain role in the etiology of SB, today SB is thought to be mainly centrally controlled with contributions from biological and psychosocial factors. Among the arguments supporting a central control of the disorder [10] are the many reports of altered psychological functioning, such as elevated stress, that are encountered in SB patients. Sjöholm et al. [11] have reported that bruxers have an increased total duration time of body movements during sleep and we have shown that patients with bruxism present a different pattern of body movements during sleep compared with nonbruxers [12], with an increase of body movements of short duration (twitches, jerks, or any sudden brusque movements of the extremities) lacking the periodicity encountered in, for example, periodic limb movements. These findings reinforce the hypothesis of a central and common etiology for bruxism and movement disorders during sleep. Furthermore, SB, like other sleep movement disorders, seems to be modulated by neurotransmitters in the central nervous system. Notably, disorders of dopaminergic functioning are often linked to bruxism [27], and catecholamine-related medication has been shown to reduce teeth grinding [13].

During sleep, motor suppression usually causes the jaw to remain open and most likely tooth contact is associated with arousals. Is SB related to arousals? Recent studies have shown that SB develops following a sequence of events including sympathovagal changes. Increased sympathetic mediated cardiac output can be observed with cortical arousals [14, 15] many seconds before grinding occurs, and tongue displacement occurs approximately 1 second before the orofacial activity. Most bruxism episodes end with swallowing. This suggests that SB is involved in the sleep arousal responses.

Orofacial and rhythmic jaw movements (RJMs) occur normally during sleep and are usually related to breathing, swallowing, and chewing (characterized by alternating activity of the jaw-opening and jaw-closing muscles). The same brainstem structures are implicated in both the control of muscle tone during sleep and RJMs, and the same neurotransmitters (catecholamines, gamma aminobutyric acid) modulate these activities.

Rhythmic masticatory muscle activity (RMMA), characterized by coactivation of both jaw-opening and jaw-closing muscles, has been observed during sleep in approximately 60% of normal subjects. RMMA during sleep has often been associated with swallowing and can correspond to chewing automatism observed in normal subjects with NREM sleep (stages 1 and 2). When awake, chewing movements stimulate salivary flow [16]. During sleep, salivary flow is normally reduced [17], and this reduction is associated with a low frequency of swallowing ($<$10 per hour of sleep, approximately one-third of waking frequency) [18]. Chewing can increase the production of saliva and hence the frequency of swallowing. It has been hypothesized that RMMA increases salivary flow during sleep and that the resulting lubrication of the mouth and the esophageal tract [19] may be associated with a change in airway patency. RMMA during sleep occurs more often and with greater amplitude in SB patients than in normal subjects. It has therefore been suggested that SB is an amplified RMMA associated with transient microarousals [20].

EPIDEMIOLOGY

Most people at some time during their lives experience tooth clenching or grinding. However, estimating the prevalence of sleep bruxism is difficult since the subject is not aware of his/her condition and therefore reports can only be initiated by a bedpartner, or by a dentist or other health professional if unusual wear to the teeth is noted or complaints of TMD pain or discomfort are investigated. Rarely, the patient consults for headache upon awakening in the morning.

SB is reported by approximately 5–8% of the adult population [21]. Since the majority of the prevalence reports are based on questionnaires, and many orofacial motor activities (swallowing, tics, chewing, and other automatisms) can be misinterpreted, these results have to be considered cautiously.

Risk Factors

Sleep bruxism has long been associated with psychological and stress factors. However, strong relationships with daily stress, mood, or physical activity have not yet been demonstrated [22]. Although sleep bruxism appears to be common in children with mental deficiency, it does not seem to be a major risk factor for development of SB [23]. Subjects who smoke have been reported to be at higher risk of developing sleep bruxism [24].

Gender and Familial Predisposition

Although gender differences have been reported in daytime clenching, there are no reports in the SB literature of significant gender differences in SB. A nationwide Finnish study of sleep bruxism in twin children using self-reports showed that women presented slightly more bruxism than men [25]. Familial cases have rarely been reported.

Development

SB can develop at an early age in childhood, soon after the eruption of the teeth, at about 1 year of age [1]. Children have a higher prevalence of SB than adults, up to 14–20% [26, 27]. Bruxism appears to develop into a chronic disorder, but the course of the disorder is not well known. In a recent 20 year longitudinal study of children with oral parafunctions—including bruxism—the disorder was shown to persist in adulthood in many of the subjects [28]. SB recognized in childhood persisted through adulthood in 35% of the subjects [29], while a study on twins shows the persistence of the trait in 86% of the adult subjects [25]. The prevalence seems to decline over age and in the elderly population only a few are aware of having SB (3% in people over 60 years of age [21]). However, many elderly wear dentures, which may be removed during the night, thus eliminating the mechanism of grinding.

CLINICAL FEATURES

The frequency of occurrence of SB varies over short periods of time [30], subjects can sleep many nights or weeks without bruxism. The main reason for subjects to seek medical consultation is the occurrence of tooth-grinding sounds reported by the room/bedpartner. Other reasons include complaints of pain, tenderness, or discomfort at the temporomandibular region, temporal headaches, or tooth hypersensitivity, particularly on awakening. However jaw pain and headache symptoms are not strongly related to bruxism [22]. Reduction in salivary flow can be observed in SB.

SB can also be diagnosed based on the patient's history and orofacial examination. Tooth wear can easily be recognized. Hypertrophy of the jaw muscles, particularly the masseter muscles, can be seen, especially visible when the subject voluntarily clenches his/her teeth. Tongue indentation can rarely be observed. A comprehensive dental examination is compulsory. However, in most cases, no evident jaw abnormality is found, and tooth wear is not a reliable sign of SB.

Patients with SB, though they have higher occurrence of RMMA during sleep than normal subjects, usually remain good sleepers. Subjectively reported sleep quality is not necessarily impaired in this parasomnia, and bruxism has not been found to be significantly related to sleep quality [22]. Sleep macrostructure is often not affected, although, in severe cases, deterioration of sleep can occur.

SB patients can exhibit decreased attention and memory and other disorders such as insomnia; anxiety can be observed concomitantly. There have been many reports describing abnormal indexes of personality in patients with SB and TMD. Even though statistical links between bruxism and psychosocial factors, including stress, subject's physical activity, or mood [22], are still unclear, these remain important issues to assess during clinical evaluations.

Polysomnography (PSG) Recording

Traditional polysomnography (EEG, chin and intercostal EMG, EOG, recording of the respiratory function) along with audio–video recordings of the subject's sleep is the optimal technique to diagnose SB and to identify associated disturbances of sleep (Figure 58.1). Extra electromyography electrodes should be placed on the masseter and temporalis muscles to record movements of the jaw muscles. This is often done bilaterally. However, recent studies showed that bilateral differences in SB are small and hence unilateral masseter muscle recording and scoring is satisfactory, provided it is done in conjunction with video recordings [31, 32]. EMG from the anterior tibialis is recommended to monitor occurrence of periodic limb movements. Preferably, two nights should be investigated to allow for habituation and particularly since there can be a very high night-to-night variability in teeth grinding. Calibration should be obtained at the beginning of the recording by asking the subjects to achieve a maximal voluntary clench. They should also swallow, cough, and move the jaw. These movements will help differentiate the various EMG patterns.

PSG Features

The sleep macrostructure in SB patients (sleep latency, sleep efficiency, duration of sleep stages, number of awakenings after sleep onset, total sleep time) does not generally differ markedly from controls, although it has been reported that patients with SB can have a tendency toward shallower sleep, with increased light sleep stages and decreased deep sleep, increased stage shifts, and nocturnal awakenings.

Bruxism is expressed by repeated phasic increases of the tone of the jaw muscles (masseter, temporalis) and observed in the PSG as rhythmic discharges at the jaw EMG electrodes. The motor activity of the jaw muscles can either be sustained (tonic) or rhythmical (phasic), bursts occurring with a frequency of 0.5–1.5 per second.

When analyzing a sleep recording, three types of SB episodes should be identified: tonic activity (EMG bursts lasting more than 2 seconds), phasic activity (at least 2 EMG bursts lasting more than 0.25 second and less than 2 seconds), and mixed activity (including both tonic and phasic activities within an interval of 3 seconds). To be scored as an event, the amplitude of the signal should exceed 25% of the maximal voluntary clench when awake. SB episodes should be separated by at least 3 second intervals. Bursts lasting less than 0.25 second observed in the masseter EMG should not be scored as bruxism episodes. The occurrence of the bursts should be quantified (total number of SB episodes) as well as the mean number of SB per episode and per hour of sleep. The mean burst duration, and if possible the mean RMS burst amplitude, per episode should also be calculated. An index (number of episodes per hour of sleep) of more than 5 is considered as pathological and representative of SB.

SB episodes appear most often in light sleep stages (1 and 2) [33] than in REM sleep, although there are reports of occurrence only in REM sleep [34]. SB occurs also during transient changes in the sleep EEG, which shows a natural arousal rhythm. Like apneas, sleep bruxism episodes are more prominent when the subject sleeps in a supine position [35].

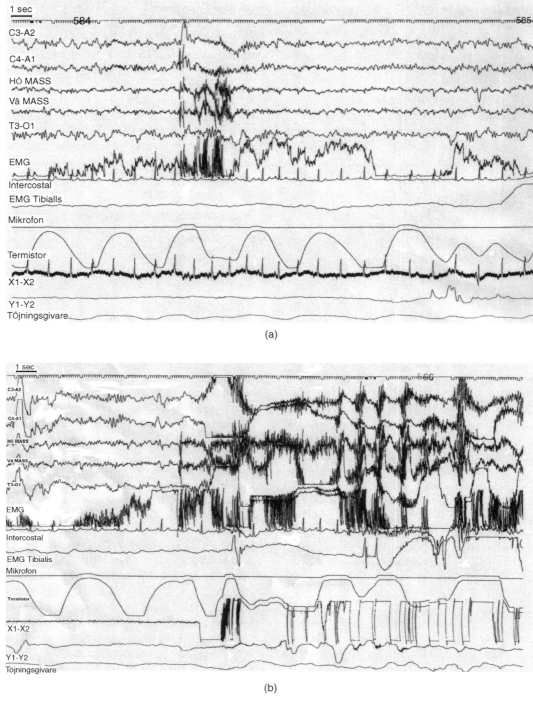

(a)

(b)

Figure 58.1   Two polysomnographic records of bruxism: a short episode (a) and a long episode (b) with phasic bursts where rhythmical patterns can be observed with movement artifacts. Termistor: air flow; Vä mass: left masseter; Hö mass: right masseter Mikrofon: microphone; Töjningsgivare: abdominal gauge; Intercostal: intercostal EMG; X1-X2: horizontal EOG; Y1-Y2: vertical EOG.

While a reduction in salivary flow can be observed in SB patients, rhythmic masticatory and swallowing movements during sleep seem to occur more frequently in SB than in normal subjects (SB patients can have up to six times more RMMA episodes per hour of sleep than normal subjects [36]).

Bruxers also seem to have more total duration time of body movements during sleep than nonbruxers [11] and

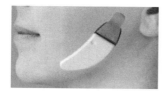

Figure 58.2 The BiteStrip™.

an increase in the number of brief movements such as twitches or jerks of the extremities [12]. Microarousals are a common trait in SB patients. Movements are often preceded by a shift of sleep stage [14] and the observed increase of brief movements reflects the increased tendency for arousal preceding episodes of bruxism. An autonomic "arousal" response is observed prior to the SB episodes with acceleration of the heart rate [14], as is observed preceding body movements during sleep. Furthermore, alpha activity often develops during the SB episodes. K-complexes as well as K-alpha events have also been reported to be much less frequent than in controls [36]. Transient arousals can periodically interrupt the background activity of the NREM sleep. These transient arousals, described as cyclic alternating pattern (CAP) [37], are EEG patterns containing clusters of phasic activity (K-complexes, delta bursts, K-alpha complexes) lasting about 10 seconds, separated by intervals with the same duration composed of tonic theta and delta activity of sleep. CAP reflects unstable sleep associated with transient variation of muscle tone and neurovegetative activities [38], hence the arousal fluctuations [39, 40]. Most SB episodes are observed in the active phase of the CAP.

Although today there are reliable ambulatory recorders which allow high quality sleep studies at home, it is highly recommended that SB patients are investigated in the sleep laboratory since audio and video monitoring is essential to help discriminate SB from other orofacial activities. Recently, a new simple disposable device for screening bruxism, the BiteStrip™, has become commercially available (Figure 58.2). It is a miniature electronic screener with self-contained EMG electrodes. The device identifies contractions of the masseter muscles exceeding a predetermined individual threshold. The overall results are read at the end of the recording on a LED scale. The obtained scores have been shown to be highly correlated with masseter EMG scored for bruxing activity during sleep [41].

ASSOCIATED CLINICAL DISORDERS

Breathing Disorders

In recent epidemiological surveys studying associated sleep symptoms and disorders, obstructive sleep apnea (OSA) was reported to be a high risk factor for tooth grinding during sleep. However, any association between breathing disorders in sleep and SB remains questionable. There are early studies reporting occurrence of SB together with apneas [42], but others have not found significant breathing disorders or oxygen desaturation in a SB population [14].

Movement Disorders

SB patients have been reported to have more body movements during sleep than controls, especially movements of short duration, which increase after a few hours of sleep [12]. Since SB patients have frequent microarousals [14] and since a shift of sleep stage often precedes a movement, it is thought that these short movements may reflect the tendency for arousal for these SB subjects.

The occurrence of SB with concomitant body movements and movement disorders during sleep seems to be frequent [52], and some authors found that up to one-third of SB episodes are associated with body movements [14, 43]. However, others could not show a correlation between body movements and increased masseter activity [12].

Restless legs syndrome (RLS) with related periodic limb movement disorders (PLMDs) are seen in approximately 10% of persons with SB, while up to 17% of subjects with RLS-related symptoms complain of tooth grinding [21, 44]. RLS-related PLM symptoms and SB may occur concomitantly but seem to be independent disorders. They might develop coincidentally as a response to an arousal reaction known to facilitate occurrence of motor activation. The clinical history should rule out REM behavior disorders.

Pain and Epilepsy

Many subjects with SB report diffuse muscle pain (20–30%) [14], which was suggested to be either secondary to intense use of the jaw muscles or to a less specific condition similar to fibromyalgia [45]. SB can occur with temporal lobe epilepsy.

DIAGNOSTIC

Although sleep is generally considered a period of relative calm, non-SB-related orofacial movements resulting from somniloquism, snoring, or abnormal swallowing pattern can occur, and it is therefore essential to discriminate SB from these physiological or unusual orofacial motor activities. In most cases sleep bruxism is easily recognized in the clinical history.

Asymptomatic rhythmic masticatory muscle activity, corresponding to chewing automatism, should be distinguished from the phasic–tonic contractions of the jaw-closing muscles observed in SB. RMMA occurs with a frequency of less than 2 episodes per hour of sleep, while SB has a frequency of >5 episodes. The EMG activity in SB is more than 30% greater than RMMA in normal subjects and RMMA is not associated with tooth grinding.

Swallowing can easily be differentiated on the sleep records from grinding, but not always from short-lasting clenching. If the salivary flow is greatly reduced, medical disorders such as Sjögren's syndrome should be ruled out.

Nocturnal groaning (NG) is a newly described and very seldom seen sleep-related respiratory phenomenon [46] that has a familial link [47]. It is characterized by a deep inspiration, followed by a short expiratory phase and a long period with reduced breathing movements. These hypopnea-like events produce a typical monotonous vocal sound, whose pitch and timbre vary among the subjects but are rather constant within each subject. The pathophysiology of NG remains unexplained. A comorbidity of NG and SB has been found in a few patients.

Epileptic seizures or discharges during sleep can be associated with tooth grinding or rhythmic jaw movements. In a study of 45 patients with mesial temporal lobe epilepsy, sleep bruxism was identified in 18% of them [48]. In a patient with temporal lobe seizure during sleep, it was described as the predominant epileptic-related motor event [49]. Epileptic discharges are usually identified in the EEG recordings.

As previously mentioned, subjects taking drugs (cocaine) or certain types of medication (SSRI) can develop sleep bruxism; hence it is essential to rule out these conditions.

MANAGEMENT TREATMENT

There is no specific treatment for bruxism other than management through orthodontic, behavioral, and pharmacological interventions to prevent damage to the teeth or other oral structures and to minimize the impairment of sleep quality.

Orthodontic

Any "causal anatomic" anomalies should be treated. Symptomatic treatment of the masticatory muscle and the temporomandibular region should be performed as well as adjustment of occlusion and restoration of tooth contours (fillings, etc.) if necessary.

Various oral devices have been used to protect the teeth and prevent TMD dysfunction or pain. A rubber mouth guard protector over the teeth or a hard plastic bite splint may help to prevent further dental damage. However, a soft mouth guard should be used only for the short term due to rapid degradation. A bite splint may improve the condition but can aggravate sleep apnea.

The efficacy of occlusal appliances (splint, palatal control device) in reducing SB is still controversial. In a recent study these devices have been shown to reduce muscle activity associated with SB [50]. Treatment compliance is relatively low, with many patients discontinuing use of these devices within a year due to esthetic concerns or discomfort. Paradoxically, an increase of oromotor activity in some patients has been noticed after using either soft or hard oral devices. Hence caution has to be observed when recommending them.

Behavioral

Adequate sleep hygiene is fundamental and the patients should be given careful instructions as well as information about bruxism. Stress, if present, should be addressed by appropriate counseling or psychotherapy. Relaxation techniques have been reported to improve SB, while the utility of biofeedback and nocturnal alarms using EMG activity of the masseter muscles is questionable. Hypnosis has been used successfully to treat parasomnias, however, there is no controlled study about its effect on SB.

Pharmacological

Various pharmacological approaches have been proposed for the treatment of SB. Botulinum toxin, used to treat cervicofacial dystonia, has also been suggested to treat SB and has been administered to some patients with masseter hypertrophy [51]. The results are controversial and one has to be aware of the risks associated with such therapy. Benzodiazepines and muscle relaxants have been reported to reduce bruxism-related motor activity. However, there are no satisfactory placebo-controlled studies. A recent single-blind, nonrandomized study of 10 patients with SB has shown that the condition improved with acute clonazepam therapy (1 mg prior to lights-off) [52].

Since sleep bruxism is a movement disorder, attempts have been made to treat patients with catecholamine precursors such as L-dopa [53] or other dopamine-related medications. The results were mild and not yet conclusive [54].

Beta-adrenergic blockers decrease sympathetic tone and have been suggested as a means to reduce tooth grinding. Indeed, propranolol, a beta-adrenergic antagonist thought to act both centrally and peripherally, was reported to reduce the number of masseter muscle contractions [5, 13]. However, in a recent controlled blind study, no differences were found between propranolol and placebo [55].

Since bruxism is associated with arousals, it was hypothesized that centrally acting alpha-agonists, which depress the central nervous system, might affect the motor activity in SB. Recently, attempts to treat bruxism with clonidine, an antihypertensive alpha-2 receptor agonist, have been reported. Orofacial and SB movements were significantly decreased in the nine treated subjects. Clonidine was also shown to suppress REM sleep and, probably as a consequence, increase stage 2 sleep. Hypotension was observed as a side effect in two patients [56].

Antidepressants such as tricyclics have been tried in SB. However, serotonin agents had no effects on SB [57] and selective serotonin reuptake inhibitors (SSRIs) have been associated with grinding and daytime clenching. Hence clinicians should be aware that prescribing these agent to patients with depression might result in an exacerbation of their SB.

Clinicians have to be cautious when prescribing drugs that can exacerbate motor activity during sleep and other sleep-related disorders such as apnea. When sleep bruxism is related to breathing disorders (apnea/hypopneas), the successful treatment of these disorders with, for example, CPAP may eliminate SB [58].

CONCLUSIONS

The etiology and pathophysiology of sleep bruxism remain unclear. Does SB represent one or many entities? Is it necessarily an abnormal motor activity, such as in PLMD, even when it does not disturb sleep, induce discomfort, or wear teeth? Movements of the mouth occur normally during sleep. The function of rhythmic masticatory movements may be to increase salivary flow in order to lubricate the oroesophageal tract and thereby maintain airway patency. SB may be such a motor transient activity, however exaggerated, and may express, like other movement-related events during sleep, an arousal reaction, as observed in the EEG and the autonomic response. The reasons for this "exaggeration" can be multifactorial and remain to be clarified.

REFERENCES

1. *ICSD—International Classification of Sleep Disorders: Diagnostic and Coding Manual.* Diagnostic Classification Steering Committee, Thorpy MJ, Chairman. American Sleep Disorders Association, Rochester, MN, 1990.

2. Rugh JD. Association between bruxism and TMD. In: McNeill C (Ed), *Current Controversies in Temporomandibular Disorders.* Quintessence, Chicago, 1992, p 29.

3. Hartman E. Alcohol and bruxism. *N Engl J Med* **301**:304(1979).

4. Ellison JA, Stanziani P. SSRI-associated nocturnal bruxism in four patients. *J Clin Psychiatry* **54**:432(1993).

5. Amir I, Hermesh H, Gavish A. Bruxism secondary to antipsychotic drug exposure: a positive response to propanolol. *Clin Neuropharmacol* **20**:86(1997).

6. Cardoso FEC, Jankovic J. Cocaine related movement disorders. *Mov Disord.* **8**:175(1993).

7. Ashcroft GW, Eccleston D, Waddell JL. Recognition of amphetamine addicts. *Br Med J* **1**:57(1965).

8. Baba K, Haketa T, Clark GT, Ohyama T. Does tooth wear status predict ongoing sleep bruxism in 30-year-old Japanese subjects? *Int J Prosthodont* **17**:39(2004).

9. Camparis CM, Siqueira JTTS, Bittencourt LRA, Tufik S. Chronic facial pain and its relation to sleep bruxism. *J Sleep Res* **13**(Suppl 1):117(2004).

10. Lobbezoo F, Naeije M. Bruxism is mainly regulated centrally, not peripherally. *J Oral Rehabil* **28**:1085(2001).

11. Sjöholm TT, Polo OJ, Alihanka JM. Sleep movements in teethgrinders. *J Craniomandib Disord Facial Oral Pain* **6**:184 (1992).

12. Bader G, Kampe T, Tagdae T. Body movement during sleep in subjects with longstanding bruxing behaviour. *Int J Prosthodontics* **13**:327(2000).

13. Sjöholm T, Lehtinen I, Piha SJ. The effect of propranolol on sleep bruxism: hypothetical considerations based on a case study. *Clin Auton Res* **6**:37(1996).

14. Bader G, Kampe T, Tagdae T, Karlsson S, Blomqvist M. Descriptive physiological data on a sleep bruxism population. *Sleep* **20**:982(1997).

15. Smirne S, Zucconi M, Oldani A, Castronovo V. Autonomic function in sleep bruxism. *J Sleep Res* **7**:254(1998).

16. Dodds MWJ, Johnson DA. Influence of mastication on saliva, plaque pH and masseter muscle activity in man. *Arch Oral Biol* **38**:623(1993).

17. Schneyer LH, Pigman W, Hanahan L, Gilmore RW. Rate of flow of human parotid, sublingual, and submaxillary secretions during sleep. *J Dent Res* **35**:109(1956).

18. Lichter I, Muir RC. The pattern of swallowing during sleep. *Electroencephalogr Clin Neurophysiol* **38**:427(1975).

19. Lavigne GJ, Manzini C. Bruxism. In: Kryger MH, Roth T, Dement WC (Eds), *Principles and Practice of Sleep Medicine,* 3d ed. Saunders, Philadelphia, 2000, p 773.

20. Kato T, Montplaisir JY, Guitard F, Sessle BJ, Lund JP, Lavigne GJ. Evidence that experimentally induced sleep bruxism is a consequence of transient arousal. *J Dent Res.* **82**:284(2003).

21. Lavigne G, Montplaisir JV. Restless legs syndrome and sleep bruxism: prevalence and association among Canadian. *Sleep* **17**:739(1994).

22. Watanabe T, Ichikawa K, Clark GT. Bruxism levels and daily behaviors: 3 weeks of measurement and correlation. *J Orofac Pain* **17**:65(2003).

23. Richmond G, Rugh JD, Dolfi R, Wasilewsky JW. Survey of bruxism in an institutionalized mentally retarded population. *Am J Ment Defic* **88**:418(1984).

24. Lavigne GJ, Lobbezoo F, Rempre PH, et al. Cigarette smoking as a risk factor or an exacerbating factor for restless legs syndrome and sleep bruxism. *Sleep* **20**:290(1997).

25. Hublin C, Kaprio J, Partinen M, Koskenvuo M. Sleep bruxism based on self-report in a nationwide twin cohort. *J Sleep Res* **7**:61(1998).

26. Widmalm SE, Christiensen RL, Gunn SM. Oral parafunctions as temporomandibular disorder risk factors in children. *J Craniomandib Pract* **13**:242(1995).

27. Satoh T, Harada Y. Electrophysiological study on tooth-grinding during sleep. *Electroencephalogr Clin Neurophysiol* **35**:267(1973).

28. Carlsson GE, Egermark I, Magnusson T. Predictors of bruxism, other oral parafunctions, and tooth wear over a 20-year follow-up period. *J Orofac Pain* **17**:50(2003).

29. Abe K, Shimakawa M. Genetic and developmental aspects of sleep talking and teethgrinding. *Acta Paedopsychiatr* **33**:339(1966).

30. Rugh JH, Harlan J. Nocturnal bruxism and temporomandibular disorders. In: Jankovic J, Tolosa E (Eds), *Advances in Neurology*. Raven Press, New York, 1988, p 329.

31. Oksenberg A, Arons E, Gavish A, Hadas N, Lavie P, Shochat T. Bilateral comparisons of online signals of a disposable electronic bruxism device and masseter EMG bruxism events. *Sleep* **27**(Abstr Suppl):A358(2004).

32. Rompre PH, Guitard F, Montplaisir JY, Lavigne GJ. Comparison between right and left masseter muscle activity in sleep bruxism patients. *Sleep* **27**(Abstr Suppl):A287(2004).

33. Reding GR, Zepelin H, Robinson JE Jr, Zimmerman SO, Smith VH. Nocturnal tooth-grinding: all night psychophysiological studies. *J Dent Res* **47**:786(1968).

34. Ware JC, Rugh J. Destructive bruxism: sleep stage relationship. *Sleep* **11**:172(1988).

35. Miyawaki S, Lavigne GJ, Pierre M, Guitard F, Montplaisir JY, Kato T. Association between sleep bruxism, swallowing-related laryngeal movement, and sleep positions. *Sleep* **26**:461(2003).

36. Lavigne GJ, Rompre PH, Guitard F, Sessle BJ, Kato T, Montplaisir JY. Lower number of K-complexes and K-alphas in sleep bruxism: a controlled quantitative study. *Clin Neurophysiol* **113**:686(2002).

37. Samson-Dollfus D. L'electro-encephalogramme du premature jusqu'a l'age de trois mois et du nouveau ne a terme. Thesis, Paris, 1955.

38. Terzano MG, Parrino L, Boselli M, Spaggiari MC, Di Giovanni G. Polysomnographic analysis of arousal responses in OSAS by means of the cyclic alternating pattern (CAP). *J Clin Neurophysiol* **13**:145(1996).

39. Terzano MG, Parrino L, Boselli M, Spaggiari MC, Ferrillo F, Di Giovanni G. The cyclic alternating pattern (CAP) as a physiological mechanism of EEG synchronisation during NREM sleep. *J Sleep Res* **3**(Suppl 1):251(1994).

40. Zucconi M, Oldani A, Ferini-Strambi L, Smirne S. Arousal fluctuations in non-rapid eye movement parasomnias: the role of cyclic alternating pattern as a measure of arousal instability. *J Clin Neurophysiol* **12**:147(1995).

41. Gavish A, Tzichinsky O, Hadas N, Lavie P, Shochat T. Identification of bruxism in treatment seeking patients and controls. *Sleep* **27**(Abstr Suppl):A362(2004).

42. Okeson JP, Phillips BA, Berry DTR, Cook YR, Cabelka JF. Nocturnal bruxing events in subjects with sleep-disordered breathing and control subjects. *J Craniomandib Disord Facial Oral Pain* **5**:258(1991).

43. Sjöholm T. Sleep bruxism: pathophysiology, diagnosis and treatment. Academic dissertation. *Ann Univ Turkuensis Ser D* **181**:105(1995).

44. Lavigne GJ, Velly-Miguel AM, Montplaisir J. Muscle pain, dyskinesia and sleep. *Can J Physiol Pharmacol* **69**:678(1991).

45. Moldofsky H. The contribution of sleep–wake physiology to fibromyalgia. *Adv Pain Res Ther* **17**:227(1990).

46. Pevernagie DA, Boon PA, Mariman AN, Verhaeghen DB, Pauwels RA. Related articles, links vocalization during episodes of prolonged expiration: a parasomnia related to REM sleep. *Sleep Med* **2**:19(2001).

47. Ferini-Strambi L, Oldani A, Manconi M, Castronovo V, Zucconi M. Familial nocturnal groaning: relationship with other parasomnias. *J Sleep Res* **13**(Suppl 1):222(2004).

48. Almeida CAV, Lins OG, Lins SG, Valenca MM. Sleep in mesial temporal lobe epilepsy. *J Sleep Res* **13**(Suppl 1):777(2004).

49. Meletti S, Cantalupo G, Volpi L, Rubboli G, Magaudda A, Tassinari CA. Rhythmic teeth grinding induced by temporal lobe seizures. *Neurology* **62**:2306(2004).

50. Dube C, Rompre PH, Manzini C, Guitard F, de Grandmont P, Lavigne GJ. Quantitative polygraphic controlled study on efficacy and safety of oral splint devices in tooth-grinding subjects. *J Dent Res* **83**:398(2004).

51. Smyth AG. Botulinum toxin treatment of bilateral masseter hypertrophy. *Br J Oral Maxillofac Surg* 29(1993).

52. Saletu B, Parapatics S, Saletu A, Anderer P, Gruber G, Saletu-Zyhlarz G. Sleep bruxism: controlled polysomnographic and psychometric studies with clonazepam. *J Sleep Res* **13** (Suppl 1):638(2004).

53. Lobbezoo F, Lavigne GJ, Gosselin A, Montplaisir J. Acute administration of L-dopa in sleep bruxism: polysomnographic study with a double-blind crossover design. Abstract (2d International Congress of the World Federation of Sleep Research Societies, Nassau 1995). *Sleep Res* **24**A:132(1995).

54. Eztel KR, Stocksill JW, Rugh JD, Fischer JG. Tryptophan supplementation for nocturnal bruxism: report of negative results. *J Craniomandib Disord Facial Oral Pain* **5**:115(1991).

55. Huynh N, Guitard F, Manzini C, Montplaisir J, de Champlain J, Lavigne GJ. Lack of effect of propranolol on sleep bruxism : a controlled double blind study. *Sleep* **27**(Abstr Suppl); A286(2004).

56. Huynh NT, Lavigne GJ, Guitard F, Rompre PH, Montplaisir J, de Champlain J. Clonidine reduces sleep bruxism in a double-blind randomized crossover trial: preliminary findings. *J Sleep Res* **13**(Suppl 1):350(2004).

57. Mohamed SE, Christensen LV, Penchas J. A randomized double-blind clinical trial of the effect of amitriptyline on nocturnal masseteric motor activity (sleep bruxism). *J Craniomandib Pract* **15**:326(1997).

58. Oksenberg A, Arons E. Sleep bruxism related to obstructive sleep apnea: the effect of continuous positive airway pressure. *Sleep Med* **3**:513(2002).

59

SLEEP-RELATED EATING DISORDERS

R. Robert Auger and Timothy I. Morgenthaler
Mayo Clinic College of Medicine, Rochester, Minnesota

INTRODUCTION

Sleeping and eating comprise two of the basic elements that drive human behavior. When the two are simultaneously affected, a fascinating spectrum of disease states may result. The differential diagnosis of sleep-related eating disorders encompasses a wide variety of clinical syndromes (Table 59.1). These include nocturnal (night) eating syndrome (NES) and sleep-related eating disorder (SRED). The latter two conditions will be the primary focus of this chapter.

Stunkard et al. [1] first reported NES as an association of nocturnal hyperphagia, insomnia, and morning anorexia in the absence of a daytime eating disorder. Although the International Classification of Sleep Disorders (ICSD-see reference 20) definition differs from Stunkard's in some respects, both describe nocturnal eating during established wakefulness, either prior to initiating sleep or after sleep onset. In the latter instance, arousals arise from non-rapid eye movement (NREM) stages 1 and 2 sleep [2, 3]. Alertness is preserved during episodes, with full recall the following morning. To date, most reports regarding NES have originated from investigators whose primary research interests have been in eating disorders rather than sleep disorders.

SRED was initially described by Schenck et al. [4] in 1991. As it shares a considerable number of overlapping features with NES, it has been suggested that the two represent a continuum of pathophysiology [5]. Historical distinguishing features include a preponderance of partial or complete amnesia for the eating episodes in SRED, as well as occasional ingestion of unusual food items, often prepared carelessly, at times resulting in injuries. As in NES, SRED events usually arise from NREM sleep and occur during alpha rhythm. In contrast to NES, however, a large number of SRED episodes arise exclusively from slow-wave sleep (SWS).

While concomitant sleep disorders do not occur in NES, SRED is strongly associated with other sleep disorders, particularly somnambulism and restless legs syndrome/periodic limb movements of sleep (RLS/PLMS). Additionally, multiple apparently drug-induced cases of SRED have been described [6–9].

Other features common to both conditions include an increased prevalence of psychopathology, a female predominance, and a temporal connection between onset and stressful events. Moreover, both result in social embarrassment, pronounced sleep disruption, morning anorexia with abdominal distention, and ultimately health problems secondary to obesity.

The pathophysiology of both processes is unclear. In the case of SRED, the high prevalence of coexisting sleep disorders may provoke arousals with varying levels of consciousness, prompting somnambulism and sleep-related eating in predisposed individuals [5]. Neuroendocrine studies performed on patients with NES suggest that dysregulation of leptin and melatonin may account for impairment of appetite suppression and sleep disruption [10]. However, a subsequent study implied that circadian factors might have confounded these results, emphasizing the need for further clarification [11].

Sleep: A Comprehensive Handbook, Edited by T. Lee-Chiong.
Copyright © 2006 John Wiley & Sons, Inc.

TABLE 59.1   Syndromes of Nocturnal Eating

Nocturnal/night-eating syndrome
Sleep-related eating disorder
Kleine–Levin syndrome
Dissociative disorder
Bulimia nervosa with nocturnal eating
Binge eating disorder

Treatment of these patients can be challenging. In SRED, associated sleep disorders should be identified and treated, and short-acting hypnotics should be discontinued and substituted with intermediate or long-acting agents. Although no controlled pharmacological trials are published, successful treatment with topiramate and dopaminergic agents has been described, either as monotherapy or in combination with opioids and sedatives [9, 12–14]. Recently, various experts have reported encouraging results with zonisamide, although no studies have been published to date. Treatment efficacy has been reported in NES patients with d-fenfluramine, and success with selective serotonin reuptake inhibitors (SSRIs) has been reported in both conditions [3, 9, 15]. Behavioral interventions used in isolation appear to be generally unhelpful [9].

NIGHT (NOCTURNAL) EATING SYNDROME

Historical Background

Stunkard and colleagues [1] originally described NES ("night eating syndrome") in 1955 in 25 patients with refractory obesity. Over 90% of the patients were female, and the median age was 35. Sixty-four percent of these individuals exhibited the triad of nocturnal hyperphagia, insomnia, and morning anorexia, with the conspicuous absence of daytime eating disorders. Although polysomnographic recording was not clinically available at this time, the behavior was reported as occurring during established wakefulness, usually resulting in a complaint of sleep-onset insomnia due to hunger. The patients described the consumption of high-caloric foods, and at least 25% of their daily caloric intake occurred after the evening meal. The frequency of episodes varied, and an association was noted with stressful life events and periods of weight gain. Psychopathology was quite prominent, particularly depression and anxiety.

Clinical Characteristics

Despite the distinctiveness of this clinical syndrome, only nine reports of nocturnal eating appeared in the literature during the 36 years following its original description, few of which referred to Stunkard's NES criteria [1, 16]. In 1990, the entity of Nocturnal Eating (Drinking) Syndrome (also referred to as NES) was described in the ICSD (see reference 21). Although the two descriptions are similar in many respects, the ICSD (see reference 20) definition primarily emphasizes the awakenings of hungry infants in the context of a sleep-onset association disorder. Importantly, both definitions stipulate that other medical, psychiatric, or sleep disorders are not contributing to the sleep disturbance. Despite its description in the ICSD (see reference 20), studies frequently refer to NES in terms more similar to Stunkard's criteria, resulting in nosologic confusion and a lack of diagnostic clarity.

Demographics and Associated Features

Based on survey studies involving both community and patient samples, the estimated prevalence of NES is 1.5% in the general adult population and 27% in the adult obese population [17]. Approximately 6% of subjects presenting to a sleep disorders center with a complaint of insomnia report symptoms typical of NES, roughly 70% of whom are female [2]. The average duration from onset to clinical presentation is 10 years [2]. Greater than 50% of patients correlate onset of the condition with stressful life events, and nearly 28% have concurrent major psychiatric diagnoses.

According to a case control study of 10 overweight NES patients (mean BMI 28.5, mean age 57.3, 80% women), more than 50% of daily energy intake is consumed between the hours of 8 p.m. and 6 a.m., compared with 15% consumed during the same time period by controls, as recorded by actigraphy and sleep diaries [10]. Approximately 70% of evening caloric intake in NES patients consists of carbohydrates, as compared to roughly 47% during the day. Nighttime awakenings are also more common in the afflicted, with over 50% of awakenings occurring in association with eating episodes.

Polysomnographic Findings

A polysomnographic study involving seven patients with NES showed no sleep abnormalities other than decreased sleep efficiency. All patients were observed to eat during periods of arousal, reportedly in association with NREM sleep, although the authors did not provide sleep stage affiliations. All patients reported full wakefulness during the episodes, with a compulsion to eat in the absence of hunger [2]. NES subjects were differentiated from a group of binge eaters and "nonspecific" nocturnal eaters by shorter periods of "eating latency" and "latency in recapturing sleep after eating episode," suggesting a close association between arousal and eating.

SLEEP-RELATED EATING DISORDER

Historical Background

Schenck and colleagues [4] first reported SRED in 19 adult patients in 1991. This was extended to a cumulative series of 38 patients, 95% of whom were evaluated with polysomnography [9]. While the condition observed shared many overlapping features with NES, it was distinguished by a preponderance of partial or complete amnesia for eating episodes, a strong association with other sleep disorders, and a high frequency of arousals from SWS, characterizing it as a variant of a NREM arousal parasomnia. Furthermore, none of the patients had eating disturbances prior to sleep onset or complained of the initial insomnia typical of Stunkard's description of NES.

Demographics and Associated Features

The prevalence of SRED is uncertain and varies among the studies that have addressed this question. One study estimated a prevalence of nearly 5% in the general population, and roughly 9–17% in those with eating disorders [18]. According to studies by Schenck et al. [9] and Winkelman [5], the average age of onset of SRED is approximately 22–27 years, with a mean of approximately 12–16 years before clinical presentation. Greater than 65% of patients are female, and more than 40% are overweight. More than 80% of patients describe a diminished level of consciousness during eating episodes, with a varying degree of amnesia for the events.

Nightly eating is reported in 71%, usually in the absence of hunger or thirst during the sleep period. More than 65% ingest unpalatable substances such as buttered cigarettes and frozen foods. Injuries from careless food preparation occur in approximately one-third of patients [9]. Forty-four percent attribute their overweight status exclusively to nocturnal eating [9]. Winkelman's series described a 35% prevalence of associated eating disorder diagnoses, whereas only 5% of Schenck's series had a lifetime diagnosis, a discrepancy perhaps affected by selection bias. Only one patient in both series combined described purging after a sleep-related eating episode.

Prior sleep diagnoses are reported in 78% of subjects, with most reporting RLS/PLMS or somnambulism, either exclusively or in combination with another parasomnia [5]. Major stressful events are associated with onset of symptoms in approximately 16% of cases [5, 9]. Associated psychiatric disease is commonly present, including affective disorders in 37%, anxiety disorders in 18%, and prior substance abuse in 24% [9]. In Winkelman's series of 37 patients, 70% of patients were taking psychotropic drugs at the time of the study. In Schenck's series, psychotropic drug use was also reported, but it is unclear at what frequency they were used [9].

Polysomnographic Findings

In two published series, polysomnography demonstrated findings consistent with somnambulism in roughly 50–70% of patients, PLMS in approximately 25%, and obstructive sleep apnea (OSA) in 10–14% [5, 9]. In five of six patients in whom nocturnal eating was observed, all arousals emanated during NREM sleep, and exclusively during SWS in three. One patient had seven episodes during one night, six of which originated from NREM sleep (two from SWS), and one from rapid eye movement (REM) sleep. All subjects exhibited electroencephalographically (EEG) defined wakefulness during the episodes [5, 9].

Medication-Induced SRED

Many drugs have been implicated in the initiation of SRED, including triazolam, amitriptyline, olanzapine, and risperidone [6, 8, 9]. One series demonstrated an association with zolpidem, particularly in those with underlying sleep disorders [7].

Genetic Factors

Consistent with other arousal disorders, a familial relationship has been described in 5–26% of subjects in two case series [5, 9]. De Ocampo and colleagues [19] described a woman with SRED whose fraternal twin sister and father were also affected.

PATHOPHYSIOLOGY

Birketvedt and colleagues [10] studied neuroendocrine parameters in a group of 12 female NES patients (mean age 53 years) and compared them to 21 controls. Patients were admitted to a Clinical Research Center for 24 hours, given scheduled meals with a specified caloric content, and underwent serial blood samples. Comparisons were made between obese and nonobese subjects from each group (mean BMI of obese NES and control subjects 36.1 and 30.4, respectively). Both nocturnal melatonin levels and the nocturnal rise in leptin were lower in NES subjects compared to controls. The authors suggested that dysregulation of these hormones may result in inadequate suppression of appetite and impaired sleep consolidation and could ultimately reflect underlying dysfunction of the hypothalamic–pituitary–adrenal axis.

A separate group analyzed 24 hour endocrine samples in seven subjects (one female, mean age 21.7 years, mean BMI 22.9) in a randomized protocol with crossover to either a diurnal or nocturnal life style, the latter of which conformed to the eating pattern described in those with NES [11]. In the nocturnal group, melatonin and leptin analyses revealed the same patterns as the Birketvedt study, suggesting that the latter group's findings may have been due to the nocturnal life style alone, with the confounding effects of light and its suppression of melatonin, in addition to the entrainment of leptin to meal timing.

Physiologic studies have not been performed in patients with SRED. It has been suggested that the high prevalence of coexisting sleep disorders may provoke arousals with varying levels of consciousness, prompting somnambulism and sleep-related eating in predisposed individuals [5].

CLASSIFICATION AND DIFFERENTIAL DIAGNOSIS OF SYNDROMES OF NOCTURNAL EATING

Nocturnal eating occurs in a variety of syndromes, and SRED and NES should be differentiated from other conditions (Table 59.1). Kleine–Levin syndrome may be associated with compulsive eating during waking periods but occurs predominantly in adolescent males, and in concert with hypersexuality, behavioral abnormalities, and periodic hypersomnolence. Nocturnal eating associated with dissociative disorders occurs with varying levels of awareness during EEG-defined wakefulness, but individuals commonly exhibit other psychopathology associated with the primary psychiatric illness.

Controversy exists as to whether NES in particular represents an extension of a daytime eating disorder. In the case of binge eating disorder and bulimia nervosa with nocturnal eating, the eating disorder continues to manifest itself throughout the day. In further support of its distinction from daytime eating disorders, one study compared NES patients with a group of four subjects with bulimia nervosa (BN). The BN group did not report nighttime awakenings, had less eating episodes per day, and consumed, on average, approximately seven times more than the night-eating subjects [10].

As alluded to previously, there is a legitimate debate as to whether NES and SRED should be classified as independent entities or whether they should be considered as a continuum of a single condition involving eating urges and sleep disorders. A review of the literature reflects this controversy, and sleep eating is often difficult (if not impossible) to classify with the current description of night eating as provided by the ICSD (see reference 20).

A study that adhered quite rigorously to this description of NES demonstrates the challenges imposed by the restrictive criteria. Spagiarri and colleagues [3] investigated ten subjects (60% female) with adult-onset nocturnal eating utilizing polysomnography. The majority of their findings are typical of prior descriptions of NES, including full recall for eating events, and a strong compulsion to consume food in the absence of hunger [3].

Similar to the aforementioned study [2], polysomnographic data revealed frequent awakenings, the majority of which were eating-related, with overall poor sleep efficiency. Eating episodes arose strictly from NREM sleep, with 80% arising from NREM stages 1 and 2, and the remaining 20% from SWS [3]. Nevertheless, 50% of subjects failed to meet formal ICSD (see reference 20), criteria due to the presence of narcolepsy in one patient and PLMS (with or without RLS) in 40%. Based on these findings, the authors suggested major revisions to the ICSD (see reference 20) criteria, including omission of the exclusionary criterion of coexisting sleep disorders.

As few studies on NES include detailed sleep histories or utilize polysomnography, the distinction between SRED and NES in many instances may be arbitrary. Moreover, many NES studies do not describe whether or not amnesia for the events was present, or the level of alertness during episodes. This last point is particularly important, as one SRED series noted significant variability in level of alertness during eating episodes, both within and between nighttime periods [5]. Perhaps in support of these arguments, the ICSD (see reference 21) recognizes the less restrictive concept SRED and eliminated NES from sleep lexicon (Table 59.2).

TABLE 59.2   (ICSD-see reference 21) Diagnostic Criteria for Sleep-Related Eating Disorder (SRED)

Sleep Related Eating Disorder

A. Recurrent episodes of involuntary eating and drinking occur during the main sleep period.
B. One or more of the following must be present with the recurrent episodes of involuntary eating and drinking:
  i. Consumption of peculiar forms or combinations of food or inedible or toxic substances
  ii. Insomnia related to sleep disruption from repeated episodes of eating, with a complaint of nonrestorative sleep, daytime fatigue, or somnolence
  iii. Sleep related injury
  iv. Dangerous behaviors performed while in pursuit of food or while cooking food
  v. Morning anorexia
  vi. Adverse health consequences from recurrent binge eating of high-caloric foods
C. The disturbance is not better explained by another sleep disorder, medical or neurological disorder, mental disorder, medication use, or substance use disorder.

EVALUATION

A thorough sleep history is essential to recognition and diagnosis of sleep-related eating disorders. Most patients will not describe the complaint spontaneously, and therefore direct questioning may be required. The timing, frequency, and description of food ingested during eating episodes should be elicited, as well as an estimate of the percentage of events for which the patient is amnestic. A history of concurrent psychiatric, medical, and sleep disorders must also be sought and evaluated. A complete medication history, including when pharmaceuticals were started, is required. Other features of the history and examination are described elsewhere.

In some patients (e.g., those with associated RLS), the history alone may be sufficient, and an empiric therapeutic trial may be appropriate without additional studies. However, if other sleep disorders are suspected (e.g., OSA), or if the episodes are stereotypic, or have resulted in injury, polysomnography should be performed. Ideally, food should be placed at the bedside, and a synchronized video should be recorded so that sleep state during eating episodes can be confirmed. When seizures are a consideration, a full EEG montage should be employed.

TREATMENT

Treatment reports of both NES and SRED are anecdotal at this juncture, as no controlled trials have been published. In a series of seven NES patients, d-fenfluramine induced complete recovery in one patient, and ≥50% reduction in the number of nocturnal eating episodes in an additional five subjects [3]. As this medication was removed from the market, it no longer represents a viable treatment option. An open-label trial of sertraline demonstrated success in treating 17 NES patients, and remission of nocturnal eating has also been described in 2 SRED patients treated with fluoxetine [9, 15].

Pharmacotherapies and nasal continuous positive airway pressure (nCPAP) (where needed for OSA) were effective treatments in 85% of patients in Schenck's SRED series, with benefit maintained at 7 years of follow-up [9]. Notably, hypnotherapy, psychotherapy, and various behavioral techniques, including environmental manipulation, were ineffective [9].

Treatment targeted to the underlying sleep disorder is essential. Five patients with RLS/PLMS and one patient with somnambulism and PLMS had remission of their SRED with combinations of carbidopa/L-dopa, codeine, and clonazepam. Two patients with OSA had resolution of their symptoms with nCPAP therapy. Clonazepam monotherapy was effective in 50% of patients with coexisting somnambulism [9].

Interestingly, dopaminergic agents as monotherapy were effective in 25% of the SRED–sleepwalking subgroup, and in combination with benzodiazepines (mainly clonazepam), opiates, or both in approximately 87% of subjects [9]. Success with combinations of dopaminergic and opioid drugs, with the occasional addition of sedatives, has also been described in a series of seven patients without associated sleep disorders [13]. In those for whom opioids and sedatives are relatively contraindicated (e.g., in those with histories of substance abuse), two case reports have described success with a combination of bupropion, levodopa, and trazodone [12].

In a case series of four patients ($n = 2$ SRED, 2 NES) administered topiramate, one patient with NES had complete remission of nocturnal eating, and the remainder had either a moderate or marked response to treatment. Notable weight loss was observed in all patients, and benefits were maintained at a mean of 8.5 months [14]. Various experts have described promising results with zonisamide, but no formal reports have been published to date.

Behavioral strategies should complement the overall treatment plan and include deliberate placement of food to avoid indiscriminate wandering, maintenance of a safe sleep environment, and education regarding proper sleep hygiene and stress management. Offending medications must also be eliminated.

ADDITIONAL CLINICAL CONSIDERATIONS

Beyond those already discussed, a variety of clinical scenarios merit special attention in the night-eating population. The presence of indiscriminate eating necessitates careful deliberation before prescribing the monoamine oxidase inhibitors. Caution is also warranted in those with food allergies. Finally, one must also be aware that nocturnal eating can complicate management of diabetes mellitus and dyslipidemia, and jeopardize procedures that mandate overnight fasting [16].

CONCLUSION

Nocturnal eating is an element of a variety of syndromes. NES and SRED are of particular interest to the sleep specialist. There is debate as to whether the two conditions represent separate entities or simply two poles of a pathophysiologic continuum.

Evaluation of these conditions consists of a detailed sleep history, in conjunction with a complete medical and psychiatric history. Workup frequently requires polysomnography with synchronized video and a full EEG montage. Food placed at the bedside is recommended.

Nonsomatic therapies in isolation appear unhelpful in treating either condition. Treatment of underlying sleep disorders is essential. Although no controlled pharmacological trials have been published, anecdotal reports suggest efficacy with topiramate and dopaminergic agents, either alone or in combination with opioids and sedatives. Success with SSRIs has been reported in both conditions. Behavioral interventions comprise a comprehensive treatment plan.

Beyond the obvious social embarrassment afforded by indiscriminate nocturnal eating, particular concern is warranted in those with food allergies and with concurrent use of monoamine oxidase inhibitors. The resultant obesity can also complicate the management of other medical conditions, including dyslipidemia and diabetes mellitus.

REFERENCES

1. Stunkard A, Grace W, Wolff H. The night-eating syndrome. A pattern of food intake among certain obese patients. *Am J Med* **19**:78–86(1955).

2. Manni R, Ratti M, Tartara A. Nocturnal eating: prevalence and features in 120 insomniac referrals. *Sleep* **20**(9):734–738(1997).

3. Spaggiari M, Granella F, Parrino L, Marchesi C, Melli I, Terzano M. Nocturnal eating syndrome in adults. *Sleep* **17**(4):339–344(1994).

4. Schenck C, Hurwitz T, Bundlie S, Mahowald M. Sleep-related eating disorders: polysomnographic correlates of a heterogeneous syndrome distinct from daytime eating disorders. *Sleep* **14**(5):419–431(1991).

5. Winkelman J. Clinical and polysomnographic features of sleep-related eating disorder. *J Clin Psychiatry* **59**:14–19(1998).

6. Lu M, Shen W. Sleep-related eating disorder induced by risperidone (letter). *J Clin Psychiatry* **65**(2):273–274(2004).

7. Morgenthaler T, Silber M. Amnestic sleep-related eating disorder associated with zolpidem. *Sleep Med* **3**:323–327(2002).

8. Paquet V, Strul J, Servais L, Pelc I, Fossion P. Sleep-related eating disorder induced by olanzapine (letter). *J Clin Psychiatry* **63**(7):597(2002).

9. Schenck C, Hurwitz T, O'Connor K, Mahowald M. Additional categories of sleep-related eating disorders and the current status of treatment. *Sleep* **16**(5):457–466(1993).

10. Birketvedt G, Florholmen J, Sundsfjord J, Osterud B, Dinges D, Bilker W, Stunkard A. Behavioral and neuroendocrine characteristics of the night-eating syndrome. *JAMA* **282**(7):657–663(1999).

11. Qin L, Li J, Wang Y, Wang J, Xu J, Kaneko T. The effects of nocturnal life on endocrine circadian patterns in healthy adults. *Life Sci* **73**(19):2467–2475(2003).

12. Schenck C, Mahowald M. Combined bupropion–levodopa–trazodone therapy of sleep-related eating and sleep disruption in two adults with chemical dependency (letter). *Sleep* **23**(5):587–588(2000).

13. Schenck C, Mahowald M. Dopaminergic and opiate therapy of nocturnal sleep-related eating disorder associated with sleepwalking or unassociated with another nocturnal disorder (abstract). *Sleep* **25**:A249–A250(2002).

14. Winkelman J. Treatment of nocturnal eating syndrome and sleep-related eating disorder with topiramate. *Sleep Med* **4**:243–246(2003).

15. O'Reardon J, Stunkard A, Allison K. Clinical trial of sertraline in the treatment of night eating syndrome. *Int J Eat Disord* **35**:16–26(2004).

16. Schenck C, Mahowald M. Review of nocturnal sleep-related eating disorders. *Int J Eat Disord* **15**(4):343–356(1994).

17. Rand C, Macgregor A, Stunkard A. The night eating syndrome in the general population and among postoperative obesity surgery patients. *Int J Eat Disord* **22**:65–69(1997).

18. Winkelman J, Herzog D, Fava M. The prevalence of sleep-related eating disorder in psychiatric and non-psychiatric populations. *Psychol Med* **29**:1461–1466(1999).

19. De Ocampo J, Foldvary N, Dinner D, Golish J. Sleep-related eating disorder in fraternal twins. *Sleep Med* **3**:525–526(2002).

20. American Academy of Sleep Medicine, International Classification of Sleep Disorders, Revised: Diagnostic and Coding Manuel, 1–170; (2000), Westchester, IL.

21. American Academy of Sleep Medicine, International Classification of Sleep Disorders, 2nd ed: Diagnostic and Coding Manual, (2005), Westchester, IL.

ADDITIONAL READING

Mahowald, MW, Schenck CH. NREM sleep parasomnias. *Neurol Clin* **14**(4):675–696(1996).

McElroy SL, Kotwal R, Hudson JI, Nelson EB, Keck PE Jr. Zonisamide in the treatment of binge-eating disorder: an open-label, prospective trial. *J Clin Psychiatry* **65**:50–56(2004).

Miyaoka T, Yasukawa R, Tsubouchi K, Miura S, Shimizu Y, Sukegawa T, Maeda T, Mizuno S, Kameda A, Uegaki J, Inagaki T, Horiguchi J. Successful treatment of nocturnal eating/drinking syndrome with selective serotonin reuptake inhibitors. *Int Clin Psychopharmacol* **18**:175–177(2003).

Schnenck CH, Mahowald MW. Parasomnias. Managing bizarre sleep-related behavior disorders. *Postgrad Med* **107**(3):145–156(2000).

Yager J. Nocturnal eating syndromes. To sleep, perchance to eat. *JAMA* **282**(7):689–690(1999).

60

OTHER PARASOMNIAS

ASHER QURESHI

St. Francis Hospital and Medical Center, Hartford, Connecticut

Parasomnias are undesirable physical phenomena that occur primarily during sleep. They can be classified according to the phase of sleep in which they predominate. Hence the *International Classification of Sleep Disorders* (ICSD) classifies the parasomnias into Arousal Disorders, Sleep–Wake Transition Disorders, and REM Sleep-Related Parasomnias. Certain parasomnias can be encountered in all stages of sleep. This chapter will focus on the parasomnias that have not been covered in the other chapters. These include REM-related parasomnias such as sleep paralysis and sleep-related penile problems, and others that do not have a sleep state predominance such as sleep-related nocturnal swallowing syndrome and sudden unexplained nocturnal death syndrome.

SLEEP PARALYSIS

Sleep paralysis is a very common condition that is encountered across the globe. It is one of the cardinal symptoms of narcolepsy. However, it is also encountered in healthy people. The feelings of ineffectuality or paralysis are a common feature of dreaming and especially nightmares [1]. Most of the times these symptoms are sporadic in healthy people and are therefore not reported. However, when the frequency of these episodes becomes more than once per day, the symptom is usually brought to the attention of a physician. Sleep paralysis can also occur in conjunction with hypnagogic hallucinations or nightmares that can be a very frightening feeling for the patient. Sleep paralysis nightmare [2] is the term used to describe this condition. The clinical syndrome of sleep paralysis can also occur

on its own, in which case it exists in two forms. It can be either familial or isolated. In either case it usually occurs at sleep onset or upon awakening. Different cultures have different names for this condition. It is called kanashibari in Japan, ghost oppression in China, and old hag attack in Newfoundland.

Definition

The *International Classification of Sleep Disorders* (ICSD) *Diagnostic and Coding Manual* defines sleep paralysis as "a period of inability to perform voluntary movements at sleep onset or upon awakening either during the night or in the morning" [3].

Prevalence

Isolated sleep paralysis can occur at least once in a lifetime in 40–50% of normal people [3, 4], While 17–40% of narcoleptics have sleep paralysis. It is much less common as a chronic complaint. Multiple sleep paralysis episodes occur often or always in 0–1% of young adults and sometimes in 7–8% of young adults [5]. Familial sleep paralysis in patients lacking sleep attacks and cataplexy is extremely rare. Other studies have reported rates of 21–34% for the prevalence of sleep paralysis [6–9]. The difference in prevalence may be explained by the wording of the questionnaire [8]. Cultural differences also exist in the perception of this symptom as a disease. Canadians consider sleep paralysis as a part of a dream, whereas Japanese have a specific name and significance for it [10].

Sleep: A Comprehensive Handbook, Edited by T. Lee-Chiong.
Copyright © 2006 John Wiley & Sons, Inc.

The onset of sleep paralysis can occur at any age; however, mostly it begins in adolescence or young adulthood [11]. Wing et al. [12] cite a peak age of onset of 16–17 years. There is no gender specificity in isolated cases. The familial cases are more frequent in women. Although there is no familial pattern in isolated cases, familial sleep paralysis is transmitted as an X-linked dominant trait. Nongenetic HLA factors and environmental factors are also implicated in sleep paralysis [12].

Pathophysiology

No specific pathology or central nervous system defect has been found to date to explain the cause of this syndrome. The defect probably lies in the mechanisms controlling the normal motor paralysis of rapid eye movement (REM) sleep. This defect could be microstructural or related to neurochemical or neuroimmunologic dysfunction. Further research is needed in this regard. Several predisposing factors have been identified which can increase the frequency of the episodes of sleep paralysis. These include factors that disturb sleep–wake schedule as insomnia, shift work, jet lag, or stress.

Clinical Features

The usual presentation is a feeling of an inability to move or perform a motor activity at the time of waking up or falling asleep. Limb, trunk, and head movements are not possible. Ocular and respiratory movements are intact. Sensorium is usually clear. In some instances the patient may experience hallucinations, which make the episode even more frightening. Occasionally patients may experience difficulty in breathing. Episodes last from one to several minutes. They are usually terminated spontaneously or upon external stimulation of the patient by touch or movement induced by a bedpartner. In some patients repeated efforts to move or vigorous eye movements can also abort the sleep paralysis. In isolated cases the episodes of sleep paralysis occur on awakening from sleep, whereas in familial cases or patients with narcolepsy they occur at sleep onset.

Acute anxiety is commonly seen with the attacks. In chronic cases, depression and chronic anxiety can be induced by these episodes. Dream-like mentation or threatening hypnagogic or hypnopompic imagery can also be experienced by the patient. No injury or complication has been reported with the episodes of sleep paralysis.

The attacks of sleep paralysis can be induced by any disturbance of the sleep–wake rhythm. Irregular sleep habits and sleep deprivation can worsen the frequency of attacks for all forms of sleep paralysis. The isolated sleep paralysis can occur during shift work and jet lag [13, 14]. Nurses, medical students, and interns have been reported to have a higher incidence of sleep paralysis due to their shift work and disturbed sleep–wake schedules [14–16]. Takeuchi et al. [17] were able to induce isolated sleep paralysis episodes by a schedule of sleep interruptions producing sleep-onset REM (SOREM) periods. Other predisposing factors include mental stress, overtiredness, and supine posture during sleep [6]. Two personality factors that are most often found in people with sleep paralysis are imaginativeness and vividness of nightmare imagery [6].

Supine posture is associated with the occurrence of sleep paralysis. Episodes occur three to four times more commonly in supine sleep [18]. According to Chyene, sleep paralysis is more frequent during the beginning and middle of sleep and may be associated with arousals. The intensity of hallucinations that accompany sleep paralysis is also more severe in the beginning and middle of sleep [18]. Insomnia is also one of the major precipitants of sleep paralysis [14]. Stress was also found to be a predisposing factor in a study of Nigerian students by Ohaeri [19]. Isolated sleep paralysis was significantly associated with GHQ-12 and life events scores in these students [19].

Polysomnographic Features

Polysomnographic recordings during episodes of sleep paralysis show suppression of muscle tone in submental and axial or peripheral electromyogram (EMG). The electroencephalogram (EEG) is that of wakefulness, which is associated with waking eye movements and blinks on the electro-oculogram (EOG). H reflex studies have shown suppression of anterior motor neuron excitability that is similar to that seen in cataplexy and REM sleep. Direct transitions to and from REM sleep to episodes of sleep paralysis have been documented. Occasionally a drowsy stage I sleep pattern with slow rolling eye movements occur with sleep paralysis [20, 21]. Sleep paralysis is therefore considered a dissociated REM state in which the motor atonia of REM sleep is present in isolation.

A multiple sleep latency test (MSLT) is usually recommended along with overnight polysomnography to exclude the presence of narcolepsy and to demonstrate the REM association of the episodes of sleep paralysis. No HLA correlation exists for sleep paralysis. However, in patients with narcolepsy who have sleep paralysis, HLA antigens as DQB1 and DR15 may be present.

Differential Diagnosis

Narcolepsy is the main differential in the diagnosis of sleep paralysis as the symptom is one of the cardinal manifestations of narcolepsy. The presence of other associated features of narcolepsy such as excessive daytime sleepiness, hypnagogic hallucinations, and cataplexy are sufficient to differentiate between the two. Although cataplexy also causes an inability to move due to motor inhibition, its onset is usually triggered by emotional stimuli and it occurs when the patient is awake.

Other differentials include atonic generalized epileptic seizures, which occur during daytime waking states, the atonic drop attacks of vertebrobasilar insufficiency, which occur in elderly and during the day, and hysteric and psychotic states with immobility, in which there is other psychopathology often evident. Another important differential is hypokalemic paralysis, in which the attacks occur at rest and on awakening. The attacks can be precipitated by ingestion of alcohol or a high carbohydrate diet. Hypokalemia is present during the attacks and the condition is reversed by correction of the potassium level. This condition is more common in adolescent males and also has a familial transmission. Localized palsies such as Saturday night palsy is due to compression of a nerve by sleeping in an abnormal posture and is easily distinguished by its localized nature.

Diagnostic and Severity Criteria

The features needed to diagnose sleep paralysis include an inability to move the body at sleep onset or upon awakening, brief episodes of complete or partial skeletal muscle paralysis, and absence of other medical or psychiatric illnesses that can explain the symptom. Association with hypnagogic hallucinations or dream-like mentation may be seen with episodes of sleep paralysis. Polysomnographic monitoring usually shows suppression of skeletal muscle tone, sleep-onset REM period, and dissociated REM sleep.

It can be mild, moderate, or severe depending on the frequency of the symptom [3]. Sleep paralysis is mild if episodes occur less than once per month; moderate if episodes occur more than once per month but less than weekly, and severe if episodes occur at least once per week.

Treatment

No specific pharmacologic treatment is required for isolated sleep paralysis, especially if the episodes are infrequent. Maintaining good sleep hygiene is very important and sufficient in mild cases, as a disturbed sleep-wake cycle predisposes to sleep paralysis. Avoidance of stress and shift work is also recommended. In cases where the episodes are frequent, chronic, and severe, several different kinds of antidepressants and sedatives have been used. Only anecdotal reports are cited in the literature for the efficacy of these drugs. No randomized clinical trials are available due to the rarity of familial sleep paralysis with frequent symptoms. Selective serotonin reuptake inhibitors (SSRIs), tricyclic antidepressants (clomipramine and desipramine), monoamine oxidase inhibitors (phenelzine), and benzodiazepines have been used for treating sleep paralysis [22, 23]. Another option for treatment that is quoted in the literature is hypnosis [24].

Course and Prognosis

The condition is usually benign and other than causing anxiety does not cause any injuries or harm. The isolated cases usually occur sporadically. However, familial forms and those associated with narcolepsy occur chronically and may predispose to anxiety and depression. Predisposing factors can increase the frequency of the episodes in either forms and good sleep hygiene is recommended to keep these factors under control.

SUDDEN UNEXPLAINED NOCTURNAL DEATH SYNDROME (SUNDS)

It is defined as sudden death in healthy young adults during sleep. A high incidence of sudden unexplained nocturnal deaths has been reported in young Asian males. These deaths are known as pokkuri in Japan, bangungut in the Philippines, and sudden unexplained nocturnal death in the United States. This disorder was first recognized in the United States in the 1970s, which was coincident with the increased immigration of Southeast Asians. In Asian countries this disorder has been described since the 1950s.

Prevalence

The prevalence of this disorder in the United States is 90 cases/100,000 [25]. The highest incidence has been in the Hmong, an ethnic subgroup from the highlands of northern Laos. Among the Southeast Asians settled in the United States the prevalence is 92 cases/100,000 in Hmong Laotians, 82 cases/100,000 in other Laotians, and 59 cases/100,000 in Kampucheans [26]. The range of this disorder in Asian countries varies from 25 cases/100,000 in northeastern Thailand to 574 cases/100,000 in Laotian and Cambodian refugees living in Thailand [27].

There is a male predominance and the median age of the victims is 33 years (range 24–44). In younger victims the etiology is elusive but in older men a cardiovascular pathology is more likely to be found. Asians are at risk and in the United States most of the deaths have occurred in the first 2 years of immigration (median length of stay 17 months). No familial pattern has been identified [26].

Clinical Features

The patients are usually healthy and are found dead in the morning, with the death itself being unwitnessed. Of those resuscitated ventricular fibrillation has been found [28]. Those resuscitated have reported sensation of airway obstruction, chest discomfort, or pressure and numb or weak limbs [29]. Postmortem analysis has demonstrated cardiac conduction defects in many of the victims. In some observed episodes moaning, thrashing, screaming, violent motor activity, perspiration, and labored breathing have been noted.

Careful review of the terminal events surrounding these deaths suggests that the victims suffered from night terrors. Night terrors are characterized by vocalization, motor activity, an unarousable state, and severe autonomic discharge. The recognition of both night terrors and cardiac anomalies in these patients offers a pathophysiologic mechanism for their sudden death [30]. Sudden unexplained nocturnal death syndrome (SUNDS) has also been recognized by the people in the northeastern part of Thailand for years. Older people in these regions have described SUNDS victims as making loud groans, showing signs of difficulty in breathing or labored respiration, becoming rigid, and dying. Relatives or those who have witnessed the episodes of death corroborate the descriptions of the elderly. These victims were also unresponsive and difficult to arouse. The instability of the physiological systems, especially respiration in REM phase, may play a role in precipitating the sudden death [31].

Pathophysiology

Sudden unexplained nocturnal death syndrome, in some victims, has been characterized by an abnormal electrocardiogram with ST-segment elevation in leads V1, V2, and V3 and sudden death due to ventricular fibrillation. Vatta et al. [32] enrolled ten families and screened them for mutations. Mutations were seen in SCN5A in three families. One mutation, R367H, was in the first P segment of the pore-lining region between the DIS5 and DIS6 transmembrane segments of SCN5A. A second mutation, A735V, was in the first transmembrane segment of domain II (DIIS1) close to the first extracellular loop between DIIS1 and DIIS2. The third mutation, R1192Q, lay in domain III. Analysis of these mutations in *Xenopus* oocytes showed that the R367H mutant channel did not express any current. The effect of this mutation was to depress peak current due to the loss of one functional allele. The A735V mutation expressed currents with steady-state activation voltage shifted to more positive potentials. The R1192Q mutation accelerated the inactivation of the sodium channel current. Both these mutations caused reduction in the sodium channel current at the end of phase 1 of the action potential [32].

Myoglobin, myosin creatine kinase MM, creatine kinase BB in cardiac muscle, and H chain of myosin in atrial and ventricular muscle were studied by Zhu et al. [33] in eight patients who died of SUNDS to investigate the role of early myocardial ischemia in the syndrome. The results showed loss of creatine kinase MM, creatine kinase BB, myoglobin, and myosin from cardiac muscle cells. This indicates that occurrence of SUNDS is closely associated with acute myocardial ischemia [33].

Kirschner et al. [34] performed autopsies on 18 SUND victims. Of the 18 hearts examined, 14 showed slight to significant cardiomegaly, which is characteristic of increased cardiac workload. The reasons for the cardiomegaly were not known. Conduction system anomalies were present in all but one heart. These abnormalities included persistent fetal dispersion of the atrioventricular node and/or bundle of His (14 hearts), accessory conduction fiber connections (13 hearts), and congenital heart block (1 heart). These abnormalities were associated with variations in the structure of the base of the heart, which suggests an aberrant developmental process. Although the functional significance of these findings has not been established, the conduction system anomalies may be the reason for cardiac arrhythmias and sudden death in sleep [34]. The mechanism of death in SUNDS is thought to be ventricular fibrillation, which may have been precipitated by a sudden sympathetic discharge.

Sudden unexplained nocturnal death has been known to occur in the same population and areas where hypokalemic periodic paralysis, endemic distal renal tubular acidosis, and renal stones are endemic. SUND has occurred in families of patients with endemic distal renal tubular acidosis and hypokalemic periodic paralysis. It can present as sudden onset of muscle paralysis with potentially lethal cardiac arrhythmias and respiratory failure from severe hypokalemia that occurs in the middle of the night. Surveys have shown deficiency of serum and urinary potassium. The main factor responsible for SUND and hypokalemic periodic paralysis is probably potassium deficiency. Low urinary citrate concentrations and the high prevalence of acidification defects are also seen in this population and indicate that potassium deficiency is also responsible for the prevalence of endemic distal renal tubular acidosis and renal stones [35].

Other studies have found thiamine as a possible factor. Nonspecific pulmonary edema with increased lung weight was reported by Pollanen et al. [28]. Between 1973 and 1989, 14 cases of SUNDS were reviewed by the coroner in the Commonwealth of the Northern Marianas and Guam. Except for one Yapese male, they were all healthy, male Filipinos aged 23–55, who were either found dead in bed or described by their colleagues as having nocturnal seizure activity. This activity consisted of gurgling, frothing, and tongue biting immediately prior to death. Autopsy findings showed no abnormal anatomic findings to account for death. Serum and urine drug analyses were also negative. Ten of the patients showed absence of significant atherosclerosis or grossly detectable structural cardiac anomaly. Four showed cardiomegaly. Migrants from

Southeast Asia have a predisposition to this syndrome, which appears to decline with longer residence in the new country. Studies suggest that at least some deaths may be associated with an abnormal cardiac conduction system. Acute pancreatitis has also been seen in some series. As to why the condition is limited to males and occurs in sleep has not been adequately explained. Stress and depression are believed to be predisposing factors as well [36].

Treatment

No specific treatments or preventions are available. Longer residence in a new country tends to decrease the incidence of this condition. The role of thiamine and potassium supplementation in certain subsets may be helpful. Further investigation and understanding of this condition is needed to provide viable treatment strategies.

IMPAIRED SLEEP-RELATED PENILE ERECTIONS

It is the inability to sustain a penile erection during sleep that would be sufficiently large or rigid enough to engage in sexual intercourse. The cause is usually organic if the reduction occurs in the presence of normal sleep architecture. All diseases that compromise vascular, neural, and endocrine function can be potential predisposing factors for erectile dysfunction. It should be kept in mind that the sleep-related tumescence decreases with age when considering this disorder. In organic causes of impotence, penile circumference may increase without an increase in penile rigidity.

Prevalence

Ten percent of adult males in the United States have chronic erectile dysfunction; 60–70% are organic in origin. Impaired sleep-related penile erections could occur at any age. The proportion of organic basis for the erectile complaints after age 45 increases dramatically. By age 70, about 70–85% of impotent men will have a diminished sleep-related erection, impaired penile rigidity, or both. No familial pattern exists other than that for the underlying cause [37].

Pathophysiology

The cause varies with the disease affecting the patient. Vascular insufficiency, autonomic nervous system dysfunction, peripheral neuropathy, venous leakage, decreased testosterone, and elevated prolactin can be the mechanisms behind impotence. The common diseases associated with impotence are diabetes mellitus, hypertension, cancer, heart disease, renal failure, spinal cord injury, epilepsy, obstructive sleep apnea, multiple sclerosis, alcoholism, and pelvic injury. Many drugs can cause impotence too. These include the antihypertensives, antidepressants, antipsychotics, digoxin, methadone, disulfiram, and amphetamines. Illicit drugs such as heroin, alcohol, and cigarette smoking can also contribute to impotence. Prostate surgeries can also cause impotence by causing nerve injury. If left untreated, mental, social, and marital problems can result. Anxiety and depression can be associated with this condition.

Diagnosis and Polysomnographic Features

The nocturnal penile tumescence test is used to differentiate organic from nonorganic causes of impotence. Two or three nights of recording are necessary. For a valid interpretation the patient should have no severe sleep fragmentation, 180 minutes of NREM sleep and 30 minutes of REM sleep, and at least one REM period of 15 minutes or more. Sleep-related penile erection is considered abnormal if the longest full tumescence episode has a duration of less than 5 minutes, largest increase at coronal sulcus does not exceed 4 mm, and the buckling force is less than 500 grams [37, 38].

Other investigations include hormonal evaluation, psychologic and psychiatric evaluation, hemodynamic monitoring, bulbocavernous reflex latency, penile sensory responses, and other tests of autonomic nervous system function.

Treatment and Prognosis

Organic impotence seldom improves without treatment. The course and response to treatment vary with the cause of the impotence. The treatment options are variable depending on the cause of the impotence. They include hormonal replacement, medications to improve penile blood flow, and surgeries. These are beyond the scope of this chapter.

SLEEP-RELATED PAINFUL ERECTIONS

Sleep-related painful erections are defined as painful erections typically during REM sleep. These patients report recurrent awakenings with pain associated with partial or full erections. Repeated awakenings can produce insomnia, excessive sleepiness, anxiety, and irritability. Peyronie's disease and phimosis may be present concurrently but do not account for the sleep-related painful penile erections. Of note, awake state erections in these patients are not accompanied by pain.

Its prevalence is rare and it occurs in only 1% of patients who present with erectile problems or sexual dysfunction. It occurs typically after the age of 40. The disorder gets more severe with age. No familial pattern exists [39].

Polysomnogram shows awakening during an episode of sleep-related penile tumescence. The major differentials are Peyronie's disease, infections, and phimosis. Treatment usually focuses on elimination of the cause.

SLEEP-RELATED ABNORMAL SWALLOWING SYNDROME

This disorder is characterized by inadequate swallowing of saliva, which results in aspiration with coughing, choking, and brief arousals or awakenings from sleep. It is rare and typically starts in middle age. The exact cause is unknown but it is presumed that these patients have an inability to swallow saliva during sleep [40].

Clinical Features

The common complaint is a sense of choking and blocked breathing at night. The sleep is usually disrupted and restless. The return to sleep is usually without any problems. The episodes subside with awakening. Respiratory infections are common, which are due to aspiration. Use of hypnotics and central nervous system depressants can predispose to abnormal swallowing. Only two cases have been reported in the literature by Guilleminault et al. [41].

Mechanism of Swallowing in Sleep

The pattern of swallowing during sleep was investigated by Lichter and Muir [42], who did a study in ten normal subjects. Sleep was staged by means of EEG, EOG, and EMG, and an external sensing device was attached to the neck to monitor swallowing. The study showed that, during sleep, swallowing was episodic, with long swallow-free periods. Swallowing almost exclusively occurred with movement arousals. These arousals were most frequent during stages REM, 1, and 2 of sleep [42].

Clinical evidence indicates that swallowing may be impaired in sleep. Studies have shown that swallowing frequency is decreased during sleep probably due to decreased salivary flow [43, 44]. To address this issue, swallowing during sleep was studied by Anderson et al. [45] in cats by injecting water through a nasopharyngeal tube. Swallowing occurred in both non-rapid eye movement (NREM) sleep and rapid eye movement (REM) sleep. In NREM sleep, the injections caused arousals followed by swallowing, but, in the majority of cases, swallowing occurred in NREM sleep before arousal. The swallowing in NREM sleep was comparable to swallowing in wakefulness. In contrast, the injections in REM sleep were less likely to cause arousal, and the swallows occurred as hypotonic events. Apneas were sometimes elicited by the injections in REM sleep. There was also repetitive swallowing on arousal [45].

Polysomnographic Features

Frequent awakenings from sleep are usually of short duration (5–10 minutes). Slow-wave sleep is absent. Swallowing studies show abnormal swallowing during sleep with accumulation of saliva in the hypopharynx and aspiration into the trachea. Coughing and gagging following periods of gurgling are seen. EEG arousals are seen with these episodes. No apneas are documented [40].

Differential Diagnosis

Differential diagnosis includes obstructive sleep apnea, gastroesophageal reflux disease, sleep terrors, sleep choking syndrome, and sleep-related laryngospasm.

Treatment

It is important to exclude other common causes of choking as obstructive sleep apnea and gastroesophageal reflux. Other mechanisms that need to be considered include unusual salivary production, muscle dysmotility, or structural abnormality of the upper airway. No specific treatments have been cited in the literature.

REFERENCES

1. Liddon SC. Sleep paralysis and hypnagogic hallucinations. Their relationship to the nightmare. *Arch Gen Psychiatry* **17**:88–96(1967).

2. Broughton RJ. Neurology and dreaming. *Psychiatr J Univ Ottawa* **7**:101–110(1982).

3. Sleep paralysis. In: *International Classification of Sleep Disorders, Revised. Diagnostic and Coding Manual*. American Academy of Sleep Medicine, Rochester, MN, 2001, pp 166–169.

4. Buzzi G, Cirignotta F. Isolated sleep paralysis: a web survey. *Sleep Res Online* **3**(2):61–66(2000).

5. Nielsen TA, Zadra A. Dreaming disorders. In: Kryger MH, Roth T, Dement WC (Eds), *Principles and Practice of Sleep Medicine*, 3rd ed. Saunders, Philadelphia, 2000, pp 753–772.

6. Spanos NP, McNulty SA, DuBreuil SC. The frequency and correlates of sleep paralysis in a university sample. *J Res Pers* **29**:285–305(1995).

7. Fukuda K, Ogilvie R, Takeuchi T. The prevalence of sleep paralysis among Canadian and Japenese college students. *Dreaming* **8**:59–66(1998).

8. Fukuda K. One explanatory basis for the discrepancy of reported prevalences of sleep paralysis among healthy respondents. *Percept Mot Skills* **77**:803–807(1993).

9. Nielsen TA, Zadra A, Germain A, Montplaisir J. The 55 typical dreams questionnaire: assessment of 200 sleep patients. *Sleep* **21**(Suppl):286(1998).

10. Fukuda K, Ogilvie RD, Takeuchi T. Recognition of sleep paralysis among normal adults in Canada and in Japan. *Psychiatry Clin Neurosci* **54**(3):292–293(2000).

11. Dahlitz M, Parkes JD. Sleep paralysis. *Lancet* **341**(8842):406–407(1993).

12. Wing YK, Lee ST, Chen CN. Sleep paralysis in Chinese: ghost oppression phenomenon in Hong Kong. *Sleep* **17**(7):609–613(1994).

13. Snyder S. Isolated sleep paralysis after rapid time zone change ("jet lag") syndrome. *Chronobiology* **10**:377–379(1982).

14. Kotorii T, Uchimura N, Hashizume Y, Shirakawa S, Satomura T, Tanaka J, Nakazawa Y, Maeda H. Questionnaire relating to sleep paralysis. *Psychiatry Clin Neurosci* **55**(3):265–266(2001).

15. Everett HC. Sleep paralysis in medical students. *J Nerv Ment Dis* **136**:283(1963).

16. Penn NE, Kripke DF, Scharff J. Sleep paralysis among medical students. *J Psychol* **107**:247–252(1981).

17. Takeuchi T, Miyasita A, Sasaki Y, et al. Isolated sleep paralysis elicited by sleep interruption. *Sleep* **15**:217–225(1992).

18. Cheyne JA. Situational factors affecting sleep paralysis and associated hallucinations: position and timing effects. *J Sleep Res* **11**(2):169–177(2002).

19. Ohaeri JU. The prevalence of isolated sleep paralysis among a sample of Nigerian civil servants and undergraduates. *Afr J Med Med Sci* **26**(1-2):43–45(1997).

20. Hishikawa Y. Sleep paralysis. In: Guilleminault C, Dement WC, Passouant P (Eds), *Narcolepsy*. Spectrum, New York, 1976, p 97.

21. Nan'no H, Hishikawa Y, Koida H, et al. A neurophysiological study of sleep paralysis in narcoleptic patients. *Electroencephalogr Clin Neurophysiol* **28**:382–390(1970).

22. Snyder S, Hams G. Serotoninergic agents in the treatment of isolated sleep paralysis. *Am J Psychiatry* **139**(9):1202–1203(1982).

23. Roth B. *Narcolepsy and Hypersomnia*. Karger, Basel, 1980.

24. Nardi TJ. Treating sleep paralysis with hypnosis. *Int J Clin Exp Hypn* **29**(4):358–365(1981).

25. Baron RC, Thacker SB, Gorelkin L, et al. Sudden death among Southeast Asian refugees. An unexplained phenomenon. *JAMA* **250**:2947–2951(1983).

26. American Sleep Disorders Association. *International Classification of Sleep Disorders, Revised: Diagnostic and Coding Manual*. American Sleep Disorders Association, Rochester, MN, 1997, pp 193–195.

27. Aldrich MS. *Sleep Medicine*. Oxford University Press, New York, 1999, p 280.

28. Pollanen MS, Chiasson DA, Cairns J, Young JG. Sudden unexplained death in Asian immigrants: recognition of a syndrome in metropolitan Toronto. *Can Med Assoc J* **155**:537–540(1996).

29. Verrier RL, Mittleman MA. Sleep-related cardiac risk. In: Kryger M, Roth T, Dement W (Eds), *Principles and Practice of Sleep Medicine*, 3rd ed. Saunders, Philadelphia, 2000, p 1007.

30. Melles RB, Katz B. Night terrors and sudden unexplained nocturnal death. *Med Hypotheses* **26**(2):149–154(1988).

31. Anchaiswad W. Is sudden unexplained nocturnal death a breathing disorder? *Psychiatry Clin Neurosci* **49**(2):111–114(1995).

32. Vatta M, Dumaine R, Varghese G, Richard TA, Shimizu W, Aihara N, Nademanee K, Brugada R, Brugada J, Veerakul G, Li H, Bowles NE, Brugada P, Antzelevitch C, Towbin JA. Genetic and biophysical basis of sudden unexplained nocturnal death syndrome (SUNDS), a disease allelic to Brugada syndrome. *Hum Mol Genet* **11**(3):337–345(2002).

33. Zhu BL, Guan DW, Li DX, Tohru O, Taizo N. Changes of myocardial myoglobin, myosin and creatine kinase in cases of sudden nocturnal death syndrome. *Chin Med J* **107**(1):36–40(1994).

34. Kirschner RH, Eckner FA, Baron RC. The cardiac pathology of sudden, unexplained nocturnal death in Southeast Asian refugees. *JAMA* **256**(19):2700–2705(1986).

35. Nimmannit S, Malasit P, Chaovakul V, Susaengrat W, Vasuvattakul S, Nilwarangkur S. Pathogenesis of sudden unexplained nocturnal death (lai tai) and endemic distal renal tubular acidosis. *Lancet* **338**(8772):930–932(1991).

36. Park HY, Weinstein SR. Sudden unexpected nocturnal death syndrome in the Mariana Islands. *Am J Forensic Med Pathol* **11**(3):205–207(1990).

37. American Sleep Disorders Association. *International Classification of Sleep Disorders, Revised: Diagnostic and Coding Manual*. American Sleep Disorders Association, Rochester, MN, 1997, pp 169–173.

38. Ware JC. Evaluation of impotence. Monitoring periodic penile erections during sleep. *Psychiatr Clin North Am* **10**:675–686(1987)

39. American Sleep Disorders Association. *International Classification of Sleep Disorders, Revised: Diagnostic and Coding Manual*. American Sleep Disorders Association, Rochester, MN, 1997, pp 173–175.

40. American Sleep Disorders Association. *International Classification of Sleep Disorders, Revised: Diagnostic and Coding Manual*. American Sleep Disorders Association, Rochester, MN, 1997, pp 188–190.

41. Guilleminault C, Eldridge FL, Phillips JR, Dement WC. Two occult causes of insomnia and their therapeutic problems. *Arch Gen Psychiatry* **33**:1241–1245(1976).

42. Lichter I, Muir RC. The pattern of swallowing during sleep. *Electroencephalogr Clin Neurophysiol* **38**(4):427–432(1975).

43. Lear CSC, Flanagan JB, Moorees CFA. The frequency of deglutition in man. *Arch Oral Biol* **10**:83–96(1965).

44. Dergachyov Oy, Gutterman LA, Mareev DV, Burikov AA. Swallowing motions in man during sleep. *Sleep Res* **26**:195(1997).

45. Anderson CA, Dick TE, Orem J. Swallowing in sleep and wakefulness in adult cats. *Sleep* **18**(5):325–329(1995).

PART VII

MOVEMENT DISORDERS

61

RESTLESS LEGS SYNDROME

PHILIP M. BECKER

Sleep Medicine Associate of Texas, Dallas, Texas

INTRODUCTION

Restless legs syndrome (RLS) is a common neurologic disorder with sensory and motor components which has both primary and secondary causation. The CNS pathophysiological mechanisms that cause RLS remain uncertain. As defined by the International Restless Legs Syndrome Study Group and then refined by a NIH Consensus Conference in 2002, RLS is characterized by four core symptoms: an urge or sensation to move the limbs, the urge worsens when at rest, movement improves the urge at least temporarily, and symptoms worsen in the evening or at night [1]. Sir Thomas Willis provided a description in 1685 that speaks to the distressing experience of the patient: "...when being a bed, they take themselves to sleep, presently in the legs and arms, leapings and contractions of the tendons, and so great a restlessness and tossing of their members ensue, that the diseased are no more able to sleep, then if they were in a place of the greatest torture" [2]. Karl Ekbom [3] was the first modern-day physician to investigate the disorder, being the first to name the disorder restless legs syndrome. Some investigators substitute Ekbom's syndrome as another name for RLS. Over the last 20 years, research has been successful in defining the characteristics and therapy of the disorder, while the etiology of RLS continues to be investigated.

CHARACTERISTICS OF RLS

Studies in North America and Western Europe that have utilized the NIH consensus definition of RLS have consistently reported that approximately 10% of adults experience symptoms of RLS on one or more nights per month [4, 5]. Clinically relevant RLS that is deserving of treatment is estimated to affect 2–5% of the adult population [3].

Approximately 60% of adult RLS patients are female [4]. Forty percent (40%) of primary RLS patients report onset of some symptoms before age 20. The frequency of RLS increases with each decade of life, being infrequent as a nightly disorder in children under age 10 and reaching its highest prevalence after age 50. Prior to menopause women who have had three or more pregnancies have more complaints of RLS than nulliparous women [5]. RLS appears in <2% in adults from Asia. A study completed in Turkey indicated that affected individuals fell between the low frequency of Asians and the high frequency of Western Europeans. Although no study has yet been completed, it is a clinical impression that African-Americans rarely report primary RLS.

The presenting complaint of RLS is often one of sleep-onset or sleep maintenance insomnia that is frequently severe. Periodic leg movements while awake (PLMW) present in the transition in or around sleep and represent the most specific and sensitive electrographic evidence in RLS patients [6]. Periodic leg movements of sleep (PLMS) is the polysomnographic finding that has been reported to define the severity of restless legs syndrome. Not all RLS patients have PLMS. Up to 20% of RLS patients will be found to have little or no periodic leg movements of sleep on a single night of polysomnography. The majority of patients with the four core symptoms of RLS generally do not require sleep testing unless other medical disorders of sleep such as apnea or narcolepsy are

Sleep: A Comprehensive Handbook, Edited by T. Lee-Chiong.
Copyright © 2006 John Wiley & Sons, Inc.

suspected. Current standards of practice do not require polysomnography to arrive at the diagnosis [7].

Primary Versus Secondary RLS

Primary RLS is thought to be a familial/genetic disorder that more commonly shows its presentation before age 30. Secondary RLS is seen more commonly after age 50 and arises from various medical conditions, particularly iron deficiency and renal disease.

Primary RLS is thought to have an autosomal dominant mode of inheritance. Through analysis of a large familial cohort of French-Canadians, Desautels and colleagues [8] found a locus on chromosome 12q, while a separate study of a three-generation Italian family identified linkage to chromosome 14q [9]. In view of these differences, as well as studies that did not demonstrate linkage to 12q or 14q, additional work must be done before the genetics of primary RLS will be well understood.

Secondary RLS arises from multiple causes. Reduction of iron stores, as measured by serum and CSF ferritin, has been extensively investigated by Earley and Allen at Johns Hopkins [10]. Renal failure predisposes individuals to the development of restless legs syndrome. It is estimated that 35% of patients undergoing renal dialysis will have symptoms of RLS. After renal transplant, the symptoms commonly resolve. Pregnancy is associated with the presentation of varying degrees of RLS in 10–30% of expectant mothers, the majority of whom report that their RLS resolves quickly after delivery. Axonal neuropathies, such as CMT 2, produce RLS in up to 37% of patients [11]. RLS symptoms may also be more common in patients with rheumatologic disorders and diabetes, although there are investigators who dispute this association.

Etiology

Although the etiology of RLS remains uncertain, there are suggestions of causation from clinical research and therapeutics. Connor et al. [12] had the opportunity to study donated brains of RLS patients and then compare them to controls. Typical markers for dopamine did not demonstrate differences between patients and controls. Analysis of the substantia nigra showed reduction in total iron with reduced H-ferritin and increased transferrin, signs of iron deficiency [12].

A majority of functional imaging studies have found mild reductions of D-2 receptor activity in the basal ganglia or associated brain structures. Dopamine has also been implicated through studies demonstrating that dopamine antagonists exacerbate RLS whether patients are pretreated with either levodopa or oxycodone. There is much interest in areas 10 and 11 of the midbrain, as these rudimentary structures send long dopaminergic projections into the

spinal cord and influence gating of peripheral nociceptive input [13]. It proves of interest that dopaminergic therapy is very successful in managing RLS symptoms, but the specific neurologic sites of action and mechanism of benefit remain to be elucidated.

Diagnostic Assessment

RLS is considered to be a clinical disorder that is principally diagnosed by history. Examination may provide opportunities to rule out masking disorders, while sleep testing is best used when other sleep disorders are suspected.

The challenge for a clinician is to recognize that patients have difficulty describing the sensation in their legs. About one-third to one-half of patients will offer the classic report of "creepy, crawly, jumpy, anxious" feelings inside the calves. A lesser percentage will describe the leg sensation as being painful, but the pain is generally not incapacitating or limiting to function. Compared to other painful disorders, movement relieves the sensation, at least temporarily.

To assist a clinician in diagnosis, the acronym U-R-G-E can prove helpful in remembering the four core symptoms of RLS.

U = urge or sensation to move the legs
R = rest or stillness of the legs worsens the urge to move
G = going is good (at least temporary relief of the urge occurs with movement)
E = evening or nighttime worsening of symptoms

As to disorders that might result in confusion with RLS, nocturnal leg cramps meet the four criteria. Patients are easily able to distinguish between the muscular spasm that usually awakens them from sleep and the symptoms of RLS. As both disorders are common, not infrequently they coexist. The descriptions of pins, needles, and numbness of neuropathy represent a separate disorder of the sensory axons. Discerning whether symptoms are present in the morning best makes the distinction between RLS and neuropathy. Neuropathy is present upon awakening, while in RLS the urge to move is rarely noted at this time of day.

Patients will commonly complain about their difficulties of sleep onset or maintenance when they consult with a physician. Reported sleep time of 3–5 hours per night is common. Even though patients are sleep deprived, RLS often makes it difficult for these patients to nap. If significant sleep-onset delay is reported, the physician should ask the patient if he/she is ever bothered with "funny feelings in your legs around bedtime." When patients report the four core symptoms of RLS (URGE acronym), the physician should determine whether the patient requires a CBC, CHEM panel, folate, B_{12}, and iron status/ferritin. Serum ferritin should always be evaluated in women who continue with menstruation, patients with known disorders leading to blood loss, or the older patient who reports recent development or significant exacerbation of RLS.

The position of the American Academy of Sleep Medicine is that RLS is a clinical disorder that can be diagnosed by history alone. Polysomnography should be considered if other sleep disorders, such as sleep disordered breathing, are suspected or when the symptoms of RLS appear in an atypical manner. Although periodic leg movements of sleep are found in at least 80% of RLS patients during polysomnography, the high rate of PLMS of five or more per sleep hour in the general population makes the findings of leg movements during polysomnography nonspecific and of limited value in arriving at a correct diagnosis [7].

A more specific and sensitive test is the suggested immobilization test (SIT) [6]. The SIT is conducted in the 1 hour before normal sleep. Patients arrive at the laboratory, receive placement of electromyographic sensors over the muscle body of each anterior tibialis, and then lie in bed making every effort not to move consciously for 1 hour. At 5 minute intervals the patient is asked to rate the level of sensation in their legs on a 10 cm visual analogue scale. EMG discharges of 2–10 seconds occurring during wakefulness (PLMW) are then counted. Patients often have to end the SIT after 30–45 minutes because the intensity of the sensation becomes so distressing that they can no longer lie in bed.

Because the degree of restlessness, sleep disturbance, and PLMS varies significantly from night to night, the use of lower extremity actigraphy has gained interest as a diagnostic tool. Actigraphic monitoring can be done with a variety of devices and algorithms, so it is important that the clinician understand the features and limitations of particular devices. Assessment of movement in all directional planes becomes important to assure correlation to body position, presumed sleep disturbance, movement frequency, and movement intensity. Traditionally, research about PLMS has utilized comprehensive diagnostic polysomnography, particularly the ruling out of movements related to breathing changes such as upper airway resistance. Actigraphic studies of PLMS are most helpful during management of treatment, since actigraphy can provide cost-effective, objective evidence of reduction of movements over three or more days.

If patients meet the criteria for RLS but also have symptoms of neuropathy, it would be appropriate to complete electromyographic and nerve conduction velocity studies on the lower extremities. Although rarely used clinically, there are biopsy studies that show that about one-third of patients with RLS have small fiber neuropathy. Future work utilizing neuroimaging techniques are expected to add to the knowledge of RLS.

TREATMENT OF RLS

RLS has been reported to respond to many different types of therapy. This broad therapeutic profile of potential agents produces confusion about treatment planning. Recent research shows that pharmacological intervention for RLS produces a sizable placebo response that is comparable to studies that have been conducted for management of pain or major depressive disorder. Interpretation of sensation requires processing by the frontal and temporal lobes with significant variability among individuals regarding their threshold for discomfort. In view of this "power of the mind to help," evidence of benefit for any therapy of RLS is best assessed through studies utilizing double-blind, placebo-controlled trials.

To assist treatment planning, the Medical Advisory Board of the Restless Legs Syndrome Foundation proposed an algorithm to assist physicians in the management of RLS patients [16]. Goals for treatment need to be defined for both patient and physician. Characterization of the severity and frequency of RLS symptoms into three categories—intermittent/mild, daily/moderate, and refractory/severe cases—will assist selection of initial treatment.

Treatment Goals

Treatment should be focused on a set of target symptoms. The goal is to reduce symptoms to the lowest possible level while minimizing side effects. Behavioral interventions need to be included along with the use of pharmacological agents to increase the efficacy of therapy. The urge or sensation to move the legs should be reduced to the lowest possible level throughout the 24 hours. Periodic leg movements when awake and asleep should be reduced to an asymptomatic level for both the patient and bedpartner. Sleep quality should be restored so the patient experiences more rapid sleep onset and improved sleep continuity, thereby enhancing daytime alertness. Side effects from medication should be minimized, particularly the augmentation that arises from dopaminergic therapy.

Treatment Classes

Historically, Willis in 1685 described the benefit of laudanum, an opiate derivative, for his patients who were "in the place of greatest torture" [2]. Ekbom in his 1945 paper identified the benefits of codeine and iron as well as barbiturates [3]. Because of the identification of PLMS in RLS patients, the use of clonazepam, a common therapy of the 1960s and 1970s for myoclonic epilepsy, became a standard therapy, although there is only one recent placebo-controlled trial of clonazepam (1 mg) that showed improvement in 10 RLS and 16 PLMD patients with acute presentation [14]. Although old literature had suggested that

TABLE 61.1 Pharmacotherapy of RLS—Nondopaminergic

Class	Agent	Starting Dose (mg)	Dose Range (mg)	Most Common Side Effect
Benzodiazepine agonists	Clonazepam	0.5	1–3	Sedation
	Temazepam	7.5–15	15–30	Sedation
	Triazolam	0.125–0.25	0.125–0.25	Sedation/withdrawal
	Zolpidem	5–10	5–10	Sedation, if awakened at hour 3
Anticonvulsants	Gabapentin	300	300–3000	Sedation
	Carbamazepine	25–50	50–300	Sedation/drying
	Valproate	500	500–1500	Sedation/weight gain
Opiates	Codeine	30–60	30–120	Side effects for any opiate:
	Propoxyphene	65	65–130	constipation, dizziness,
	Tramadol	50	50–300	nausea, sedation, vomiting
	Hydrocodone	5	5–40	
	Oxycodone	5	5–30	
	Levorphanol	2	2–8	
	Methadone	5	5–30	
Minerals	Iron	65 (elem)	65 TID	Constipation
	Magnesium	125	250 TID	Diarrhea

levodopa worsened RLS (before recognition of augmentation), Akpinar from Turkey reported in 1982 that levodopa significantly improved the symptoms of RLS [15]. Since the mid-1980s extensive research has been conducted on the benefits of dopaminergic therapy, defining its efficacy and side effects. Placebo-controlled trials on moderate to severe patients who meet NIH consensus criteria for RLS have been and are currently being conducted for ropinirole, pramipexole, pergolide, and other dopamine agonists. The role of anticonvulsants such as gabapentin and valproate is also being defined, as is the use of opiates for the severe patient with RLS.

Based on clinical experience, therapeutic classes for the treatment of RLS include benzodiazepine agonists, dopaminergic agents, mineral supplementation, anticonvulsants, and opiates. Table 61.1 lists common agents for the management of RLS, while Table 61.2 defines the pharmacology of the most commonly used dopaminergic therapies. Only dopaminergic agents have had large-scale, controlled study in RLS patients using current diagnostic criteria. In view of the lack of well-controlled trials for other therapeutic agents, the Restless Legs Syndrome Foundation Medical Advisory Board (RLS MAB) has decided to base recommendations on available research that has been filtered through their broad clinical experience. Recommendations regarding treatment are initiated based on the severity of the urge to move and its frequency of presentation [16].

Side effects of dopaminergic therapy are similar across the class. Compared to Parkinson's disease patients, side effects are much less problematic in RLS patients and only infrequently result in discontinuation of therapy. Nausea, less frequent vomiting, dizziness, muscle aches, headaches, and occasional sedation may occur. Less commonly, patients report nasal stuffiness, insomnia, constipation, or leg edema. In patients treated for Parkinson's disease, excessive daytime sleepiness has been reported from these agents and there has been concern about potential sudden sleep onset. Although less than 12% of RLS patients in clinical trials have reported any degree of sleepiness or lethargy on dopaminergic therapy, continued observation for and precaution about hypersomnia arising from dopamine therapy is recommended.

TABLE 61.2 Pharmacodynamics of Common Dopaminergic Agents

Agent	Starting Oral Dose (mg)	Minutes to Therapeutic Effect	Common Timing of Dosage Adjustment	Recommended Daily Maximum Dose (mg)
Carbidopa/levodopa				
Regular	25/100	30–60	One every 1–3 days	300[a]
CR	25/100	60–120	One every 1–3 days	300[a]
Pramipexole	0.125	60–120	One every 2–3 days	1.5–3.0
Ropinirole	0.25–0.5	45–60	One every 2–3 days	4–8
Pergolide	0.05	60–150	One every 3–4 days	1–2
Others	Review individual agents to assess dosing and administration			

[a]Doses of carbidopa/levodopa have been reported up to 1200 mg/day; a limit of 300 mg daily should trigger assessment of augmentation.

Augmentation

Although dopaminergic agents offer significant clinical benefit with only limited side effects, the primary limiting factor to long-term therapy is augmentation. Augmentation is defined as the earlier onset of RLS symptoms at least 2 hours before presentation prior to the initiation of therapy.

Augmentation might also be diagnosed if any of two other symptoms are present: increased symptoms on increased dosage of medication; decreased symptoms upon decrease of dosage; spread of symptoms into other body parts; shortening of therapeutic benefit when compared to initiation of therapy; or periodic leg movements of awake either presenting or worsening [1].

Various methods have been proposed to manage augmentation. Augmentation is common on levodopa and is best managed by gradually tapering the medication and substituting another therapeutic agent, most commonly ropinirole, pramipexole, or gabapentin. Dopamine agonists are also thought to cause augmentation, although the frequency appears lower than with levodopa and perhaps with less intensity. Other management strategies include dividing a dose and providing the lower divided dosage at earlier times of the day. Others are comfortable increasing the total dosage by giving a standard dosage of the dopamine agonist in the 1 or 2 hours before symptom onset throughout the day [16].

When these strategies prove unsuccessful, particularly when they exacerbate the presentation of augmentation, the use of high-potency, long-acting opiates such as methadone and levorphanol prove quite helpful. It is also currently thought that long-acting dopamine agonists such as carbergoline and perhaps rotigitine may have a lower propensity to cause augmentation, although some researchers would caution that observation over a number of years will be needed to assure that long-acting dopamine agonists do not produce augmentation.

Treatment of the Intermittent/Mild Patient

Patients with mild RLS may benefit from behavioral strategies of intervention. Avoidance of substances that overly stimulate the nervous system such as caffeine, chocolate, monosodium glutamate, and decongestants are best avoided or minimized past 3:00 p.m. Aerobic exercise may prove beneficial after conditioning has occurred. When first beginning exercise, some patients report a temporary worsening of their RLS. Many RLS patients may find that antidepressants, particularly selective serotonin reuptake inhibitors, exacerbate their symptoms, while bupropion and trazodone seem to have a lesser propensity of exacerbation. Many patients have discovered counterstimuli that improve their RLS symptoms, examples being kneehigh socks, riding a stationary bike or walking on a treadmill, stretching in or out of bed, hot baths or showers, ice packs, massage, vigorous rubbing, and many others that are specific to the patient.

The RLS MAB recommendations for mild RLS allow pharmacologic therapy when patients have sufficient distress from their urge to move that it makes it difficult to sleep, resulting in reduced function on the following day. Some mild patients may have particularly disruptive restlessness occurring only one or two times per month, while other mild patients will report disturbed sleep up to three times per week from the symptoms of RLS. If the behavioral interventions above do not offer sufficient relief, the first line of therapy is from the dopaminergic class.

As therapy will be on an as-needed basis, patients have to be able to identify those situations that usually result in worsening RLS (sitting or confinement in a plane, automobile, movie, play, concert). Initial options include pretreatment with ropinirole 0.25–0.5 mg one hour before confinement, with pramipexole 0.125–0.25 mg two hours before confinement, or with carbidopa/levodopa 25/100 mg 30–60 minutes before confinement. For those patients who are unable to predict when the symptoms of RLS will appear, rapidly acting dopaminergic therapy such as levodopa/carbidopa 25/100 mg at one-half to one tablet or ropinirole 0.25–0.5 mg is most appropriate.

Any of the other treatment classes are allowed by the RLS MAB recommendations, although it is more common to utilize these agents for moderate or severe RLS patients.

Treatment of Moderate/Daily RLS

Patients with moderate severity of RLS will generally experience significant symptoms in the 1–3 hours before or at bedtime on most nights of the week. Occasionally, patients report falling asleep and then waking 20–60 minutes later with their symptoms of restlessness. They show significant delays of sleep onset and intermittent sleep maintenance problems that lead to daytime fatigue, irritability, and decrease in function. More often than not such patients will benefit from pharmacotherapy for RLS.

The RLS MAB recommendation is management by dopamine agonists. The most widely utilized agents are the nonergot agonists, pramipexole and ropinirole [16]. As the initial open-label studies of pramipexole were first reported, it is commonly utilized, but ropinirole has the advantage of large double-blind, placebo-controlled clinical trials that demonstrate the efficacy and safety of ropinirole. The 1 hour onset of action of ropinirole makes it easier for patients to remember as bedtime approaches. The longer half-life of pramipexole allows treatment of patients who develop their symptoms two or more hours before bedtime, often resulting in a dosing of medication at supper and again at bedtime. Treatment is commonly initiated at the lowest tablet size of 0.25 mg of ropinirole or 0.125 mg of pramipexole. Unless nausea or other side effects intervene, the medication can be increased every 2–3 days until relief of the urge to move, sleeplessness, and movement occur.

Therapeutic dosing of ropinirole in over 900 RLS patients has consistently demonstrated that a majority respond to dosages of 1.5–2 mg. The double-blind, placebo-controlled studies had a maximum dose of 4 mg with approximately 10–15% of patients receiving this higher dosage [17–19]. There are anecdotal reports of the use of ropinirole up to 8 mg per day in the most severe RLS patients.

Studies of pramipexole occurred prior to ropinirole [20] and have been of smaller size for placebo-controlled trials, although long-term follow-up, open-label studies have also been done. A large industry-sponsored, double-blind, placebo-controlled study is ongoing to provide a comprehensive understanding of pramipexole in the treatment of RLS. Therapeutic dosages of pramipexole range from 0.375 to 1.5 mg, although clinical practice has seen dosages as high as 4 mg per day. Again, the goal is to initiate therapy before the onset of symptoms so as to minimize the sensation, sleep disturbance, and side effects.

In any of the well-studied dopaminergic therapies, side effects (nausea, achiness, headache, or lightheadedness) have been of mild to moderate significance. In the majority of patients the side effects could be managed by a temporary reduction of the dosage before again increasing the dosage in a more gradual manner.

The recommendations also discuss the potential value of other therapeutic classes in the management of RLS and sleep disturbance [16]. Gabapentin may be used as initial therapy in moderate RLS, particularly when it is associated with pain. Gabapentin may also be used to enhance sleep onset and maintenance. One crossover, double-blind trial of gabapentin showed a mean dosage of 1855 mg to offer therapeutic effect [21]. At common therapeutic dosages, a significant subset of patients will complain of excess sedation or other side effects.

Carbamazepine has demonstrated efficacy in a double-blind, placebo-controlled trial from 1984 [22]. Other anticonvulsants may be of potential benefit but they have received inadequate study. It is the consensus opinion of experts that dopamine agonists and gabapentin probably are more effective with lesser side effect [16].

Opiates can be used in the patient with moderate RLS, although it is typical to use them for intermittent exacerbation or in rotation between primary therapy that is complicated by augmentation or other side effect. The use of opiates will be described later in the section on the severe RLS patient.

It is common for patients to require more than a single agent. Up to 60% of patients with moderate or severe RLS can be managed with one medication, most commonly pramipexole or ropinirole. The other patients will typically benefit from the addition of a sedative/hypnotic agent or gabapentin to enhance sleep. As patients typically experience their problems of sleep in the first half of the night, Zolpidem 5–10 mg is commonly offered. Patients who experience sleep disturbance later in the night might also find value in gabapentin 100–900 mg p.o.h.s. As patients with RLS appear to have a moderately higher prevalence of generalized anxiety, the use of clonazepam 0.5–2 mg p.o. one hour before bedtime can assist sleep onset, maintenance, and daytime anxiety. A separate section later will cover the treatment of depression in patients with RLS.

Treatment of Severe/Refractory RLS

Patients with severe RLS experience nearly nightly symptoms with the urge to move often presenting in the afternoon or early evening. It is the severe patient who can have symptoms throughout the 24 hours. In such patients it is important to explore symptoms at the very earliest stage of RLS onset, since such patients commonly report the urge to move presented around bedtime when their symptoms first started.

In the severe patient, the daytime intensification of the sensation often appears rapidly and patients are compelled to walk, exercise, or massage/pound their limbs. The lower extremities are most commonly affected, but severe patients are more likely to have involvement of the upper extremities. RLS of greatest severity impacts the entire body, although occasional patients have it localized to the feet or only in one limb.

Profound insomnia is common. Sleep deprivation and the resulting fatigue can create a cyclical pattern of worsening of the RLS, and then worsening sleep, that worsens fatigue, and so on. Such patients are typically very distressed, reporting that they might be "better off dead" or they ask the doctor to "just cut off my legs."

Severe RLS does not necessarily require higher dosages of medication or the use of combination therapy. A significant number of severe RLS patients will respond to a single agent. Split dosing may be required. Examples include ropinirole at 0.5–2 mg at approximately 6:00 p.m. and 1 hour before bedtime or pramipexole 0.25–1 mg in the early evening and approximately 1–2 hours before bedtime.

Severe patients require regular monitoring of their serum ferritin since fluctuations of iron stores can result in exacerbation and potentially reduce response to therapeutic intervention. It also must be noted that severe patients are more likely to require higher total dosages of therapy with more common usage of combination therapy.

It is the current opinion that severe patients are also more likely to develop augmentation on dopaminergic therapy, particularly levodopa. It is thought that severe patients who develop augmentation on any dopaminergic agent will have a higher likelihood of augmentation on another dopaminergic agent. Severe patients may also require therapy earlier in the day, including the morning. They will be exposed to higher dosages of medication over the 24 hours, increasing the potential for side effects such as nausea, vomiting, headaches, myalgia, daytime sleepiness, and orthostatic hypotension [16].

The treatment of severe patients requires the initiation of therapy prior to the onset of symptoms. Ropinirole, pramipexole, or other dopamine agonists should be administered 1–2 hours prior to typical time of symptom onset. Titration to the highest effective level that results in the least amount of side effect is recommended. To offer the greatest degree of improvement in sleep and daytime function, a combination of various therapeutic classes is often needed. The most common combination therapy is a dopamine agonist with either a sedative/hypnotic (clonazepam, temazepam, zolpidem) or gabapentin to enhance sleep while minimizing carryover sedation. The usage of high-potency opiates proves particularly important in therapy of severe patients. Opiates are particularly needed when augmentation occurs in this group since they are more likely to continue augmenting even as dosages are given earlier or at higher levels.

Selection of high-potency opiates is based on onset of action and half-life of the selected agent. Table 61.3 lists a variety of opiate agents and their pharmacokinetic properties. Lower potency, short-acting agents such as propoxyphene, codeine, tramadol, hydrocodone, and perhaps oxycodone might best be considered for as-required use or for patients of lesser severity. Hydromorphone 2–4 mg allows for a fairly rapid onset of action but its half-life of approximately 4 hours may limit its utility. Longer acting, high-potency opiates that include sustained release opiate preparation (oxycodone CR or fentanyl patch), methadone 5–40 mg per day, or levorphanol tartrate 2–8 mg per day often prove highly efficacious. Although monitoring of abuse must occur with opiates, the severe RLS patient is deserving of effective therapy when other treatments fail due to reasons of reduced efficacy or side effects.

It is of interest that the opiates are fairly well tolerated in this group with only occasional problems of daytime sedation, nausea, or constipation. Severe patients treated with opiates are at risk for the development of sleep disordered breathing, particularly upper airway resistance, so that the physician must monitor any complaints of increasing daytime sleepiness and consider polysomnographic testing with pressure airflow transduction in the patient who develops new symptoms even as RLS is controlled.

RLS AND DEPRESSION

Major depressive disorder is common in patients who have RLS. Reports of depression have varied from a low of 20% up to 70% of patients with RLS and PLMD. The initiation of antidepressant therapy often proves a complication for RLS patients. Significant exacerbation of RLS occasionally occurs with tricyclic antidepressants. Patients appear to tolerate low-dosage tricyclic antidepressants, such as amitriptyline or imipramine, when offered to improve sleep quality. When standard tricyclic antidepressant dosages are offered, it is less certain how RLS will change for better or worse.

There is a strong consensus that the SSRIs (fluoxetine, paroxetine, sertraline, venlafaxine, citalopram, escitalopram) and serotonin modulators (mirtazapine, buspirone, perhaps nefazodone, perhaps trazodone) exacerbate RLS in the majority of patients. Bupropion in its various formulations does not exacerbate, and may even improve, RLS and PLMS, particularly when it is given earlier in the day.

Other patients will report significant dysphoria even though they do not meet the clinical criteria for the diagnosis of major depressive disorder. Anxiety occurs more often in RLS patients. The usage of benzodiazepines, particularly clonazepam, or gabapentin may offer significant improvements in mood for such patients. Again, carryover sedation must be followed when these medications are utilized during the daytime.

TABLE 61.3 Pharmacokinectics of Opioid Agonists

Opioid	Onset (min)	Peak (h)	Duration (h)	Adult ½ (h)	Interval (h)	P.O. Dose (mg)[a]
Codeine	15–30	0.5–1	4–6	2–3	3–6	200
Hydrocodone	ND[b]	ND	4–6	3–4.5	ND	ND
Hydromorphone	15–30	0.5–1	4–6	2–4	3–6	7.5
Levorphanol	30–90	0.5–1	4–8	12–16	8–12	4
Morphine	15–60	0.5–1	3–6	2–4	3–6	60
Methadone	30–60	0.5–1	>8 (chronic)	15–30	6–12	20
Oxycodone	15–30	0.5–1	4–6	3–4	3–6	30
Oxymorphone	5–15	0.5–1	3–6	ND	4–6	10
Propoxyphene	30–60	2–2.5	4–6	3.5–15	4–8	130–150

[a]Milligram dosing is a comparison for pain relief (not RLS efficacy).
[b]ND = no or insufficient data.

CONCLUSION

Restless legs syndrome is a disorder of unknown neurological origin that is common and arises from primary and secondary causes. Primary RLS appears to have a familial origin and presents most commonly before the age of 30. Secondary RLS is common in patients who are iron deficient or have renal disease. Neuropathy is a separate disorder, although RLS and neuropathy can coexist, particularly in patients with an axonal neuropathy.

Severity follows a continuum from the intermittent, mild patient to a patient who has refractory symptoms in all limbs throughout the day. Behavioral and life-style changes should be part of the initial intervention, particularly for the mild patient. As nocturnal sleep, daytime dysfunction, and mood are impacted by the patient's RLS, intervention with medications is appropriate. The RLS Foundation Medical Advisory Board has developed a treatment algorithm that emphasizes the primary role of dopamine agonists such as ropinirole and pramipexole in the treatment of mild, moderate, or severe RLS. Gabapentin, clonazepam, and mild opiates might also have a role. As severity of RLS increases, patients will benefit from combination therapy of dopamine agonists, sedative/hypnotic medications, gabapentin or other anticonvulsants, and opiates.

The primary complication to dopaminergic therapy, particularly with levodopa, is augmentation of the symptom. The intensification and earlier presentation of RLS symptoms can be managed in a number of ways, but the most severe cases of RLS typically will require high-potency, long-acting opiates. Side effects require close monitoring and if patients become sleepy on any of the therapies, caution must be exercised when operating machinery or a motor vehicle. Further clinical investigation in the sleep laboratory may be appropriate.

Restless legs syndrome impacts up to 10% of adults in North America and Western Europe and is the most common neurologic disorder that causes chronic insomnia. Effective relief is available in patients who greatly appreciate therapy that relieves the urge to move, the need to walk, the inability to relax in sedentary activities, the sleep disturbance, the periodic leg movements of wake and sleep, and the daytime fatigue, anxiety, and depression.

REFERENCES

1. Allen RP, Hening WA, Montplaisir J, Picchietti D, Trenkwalder C, Walters AS, et al. Restless legs syndrome: Diagnostic criteria, special considerations, and epidemiology: a report from the RLS Diagnosis and Epidemiology Workshop at the National Institutes of Health. *Sleep Med* **4**:101–119(2003).

2. Willis T. *The London Practice of Physick*. Bassett, Dring, Harper, and Crook, London, 1692.

3. Ekbom K. Restless legs: a clinical study. *Acta Med Scand Suppl* **158**:1–123(1945).

4. Phillips B, Young T, Finn L, Asher K, Hening WA, Purvis C. Epidemiology of restless legs syndrome in adults. *Arch Intern Med* **160**:2137–2141(2002).

5. Rothdach AJ, Trenkwalder C, Haberstock J, Keil U, Berger K. Prevalence and risk factors of RLS in an elderly population: the MEMO study. *Neurology* **54**:1064–1068(2000).

6. Michaud M, Lavigne G, Desautels A, Poirier G, Montplaisir J. Effects of immobility on sensory and motor symptoms of restless legs syndrome. *Mov Disord* **17**(1):112–115(2002).

7. Chesson AL, Wise M, Davila D, et al. Standards of Practice Committee of the American Academy of Sleep Medicine. Practice parameters for the treatment of restless legs syndrome and periodic limb movement disorder: an American Academy of Sleep Medicine Report. *Sleep* **22**:961–968(1999).

8. Desautels A, Turecki G, Montplaisir J, Rouleau GA. Reply to Kock et al. *Am J Hum Genet* **71**:208–209(2002).

9. Bonati MT, Ferini S, Aridon P, Oldani A, Zucconi M, Casari G. Autosomal dominant restless legs syndrome maps on chromosome 14q. *Brain* **126**(Pt 6):1485–1492(2003).

10. Earley CJ, Connor JR, Beard JL, Malecki EA, Epstein DK, Allen RP. Abnormalities in CSF concentrations of ferritin and transferrin in restless legs syndrome. *Neurology* **54**(8):1698–1700(2000).

11. Ondo W, Jankovic J. Restless legs syndrome: clinicoetiologic correlates. *Neurology* **47**(6):1435–1441(1996).

12. Connor JR, Boyer PJ, Menzies SL, Dellinger B, Allen RP, Ondo WG, et al. Neuropathological examination suggests impaired brain iron acquisition in restless legs syndrome. *Neurology* **61**(22):304–309(2003).

13. Ondo WG, He Y, Rajasekaran S, Le WD. Clinical correlates of 6-hydroxydopamine injections into A11 dopaminergic neurons in rats: a possible model for restless legs syndrome. *Mov Disord* **15**(1):154–158(2000).

14. Saletu M, Anderer P, Saletu-Zyblarz G, et al. Acute placebo-controlled sleep laboratory studies with clonazepam. *Eur Neuropsychopharmacol* **11**:153–161(2001).

15. Akpinar S. Treatment of restless legs syndrome with levodopa plus benserazide. *Arch Neurol* **39**(11):739(1982).

16. Silber MH, Ehrenberg BL, Allen RP, Buchfuhrer MJ, Earley CJ, Hening WA, Rye DB. An algorithm for the management of restless legs syndrome. *Mayo Clin Proc* **79**(7):916–922(2004).

17. Walters A, Ondo W, Sethi K, Dreykluft T, Grunstein RS. Ropinirole versus placebo in the treatment of restless legs syndrome (RLS): a 12-week multicenter double-blind placebo-controlled study conducted in six countries (abstract). *Sleep* **26**(Abstr Suppl):A344(2003).

18. Garcia-Borreguero D, Montagna P, Trenkwalder C, et al. Ropinirole is effective in the treatment of restless legs syndrome (RLS): a double-blind placebo-controlled 12-week study

conducted in 10 countries (abstract). *Neurology* **60**(Suppl 1):A11–A12(2003).

19. Allen RP, Becker PM, Bogan R, Fry JM, Poceta J. Restless legs syndrome: the efficacy of ropinirole in the treatment of RLS patients suffering from periodic leg movements of sleep (abstract). *Sleep* **26**(Abstr Suppl):A341(2003).

20. Becker PM, Ondo W, Sharon D. Encouraging initial response of restless legs syndrome to pramipexole. *Neurology* **51**(4): 1221–1223(1998).

21. Garcia-Borreguero D, Larrosa O, de la Llave Y, Verger K, Masramon X, Hernandez G. Treatment of restless legs syndrome with gabapentin: a double-blind, cross-over study. *Neurology* **59**(10):1573–1579(2002).

22. Telstad W, Sorensen O, Larsen S, Lillevold PE, Stensrud P, Nyberg H. Treatment of the restless legs syndrome with carbamazepine: a double blind study. *Br Med J (Clin Res Ed)* **288**(6415):444–446(1984).

62

PERIODIC LIMB MOVEMENT DISORDER

BASHEER Y. KHASSAWNEH

Jordan University of Science and Technology, Irbid, Jordan

INTRODUCTION

Periodic limb movement disorder (PLMD) is a movement disorder that occurs only during sleep. It was originally described as nocturnal myoclonus by Symonds in 1953 [1]. Recently, the term nocturnal myoclonus has been replaced with PLMD. The etiology of this disorder is not well established. PLMD can be a primary disorder or can be associated with other sleep disorders. The diagnosis of PLMD is established when the affected individual has insomnia or excessive daytime sleepiness (EDS) and confirmed by polysomnography (PSG). Treatment is symptomatic and includes dopaminergic agents, benzodiazepines, and opiates.

EPIDEMIOLOGY

PLMD is a prevalent sleep disorder. A European study using telephone interviews with the sleep-EVAL system showed that the prevalence of PLMD in the general population was 3.9% and that it was more prevalent in women [2]. PLMD is uncommon under the age of 50 years and more prevalent in individuals older than 65 years. In the pediatric population the prevalence of PLMD is 11.9% and highly associated with attention deficit hyperactivity disorder (ADHD) and OSA [3]. PLMD is strongly associated with restless legs syndrome (RLS). More than 80% of all patients with RLS have PLMD during sleep [4]. PLMD is associated with other sleep disorders and medical, neurological, and psychiatric illnesses. PLMD has been reported with the use of certain medications or withdrawal from sedative and hypnotics (Table 62.1).

PATHOPHYSIOLOGY

The exact etiology of PLMD is not known. The location of the generator of PLMS is not known. Various hypotheses have been postulated to explain this well-established sleep disorder. The presence of PLMD in patients with complete transection of the spinal cord suggests a spinal origin of the generator or failure of inhibition of the spinal cord reflexes by the pyramidal tract or higher centers [5]. High-resolution functional magnetic resonance imaging (fMRI) suggested that the red nucleus and the reticular system of the brainstem are involved in the generation of periodic limb movements in patients with restless legs syndrome [6]. FDOPA PET scans showed nigrostriatal presynaptic dopaminergic hypofunction in patients with PLMD and RLS [7]. Emission computer tomography (SPECT) studies of patients with PLMD and Parkinson's disease showed striatal dopaminergic nerve loss [8]. In children with PLMD, the presence of PLMS is frequently associated with low serum iron and a tendency toward low serum ferritin levels [9].

CLINICAL PRESENTATION

PLMD is characterized by periodic episodes of highly stereotyped and repetitive involuntary rhythmic movements that usually affect the lower extremities during sleep (PLMS). The upper limbs can also be affected. PLMS is

Sleep: A Comprehensive Handbook, Edited by T. Lee-Chiong.

TABLE 62.1 Conditions Associated with Periodic Limb Movement Disorder

Obstructive sleep apnea
Narcolepsy
Parkinson's disease
Spinal cord injury
Multiple sclerosis
Huntington's disease
Isaacs' syndrome
Peripheral neuropathy
Stiff person syndrome
Attention deficit hyperactivity disorder
Chronic uremia
Anemia with low serum iron
Tricyclic antidepressants
Monoamine oxidase inhibitors
Withdrawal from benzodiazepines and barbiturates

usually bilateral but it can be unilateral and alternating. The classic PLMS is extension of the big toe and dorsiflexion of the ankle and flexion of the knee. The more frequent movements are either dorsiflexion of the ankle, extension of the big toe and dorsiflexion of the ankle, or extension of the big toe and dorsiflexion of the ankle followed by any other movements [10]. PLMD is unusual during wakefulness. These repetitive movements can go unnoticed for years. Occasionally the bedpartner will notice it. The presenting complaints are usually the impact of PLMS on sleep architecture. Symptomatic patients will usually have insomnia or excessive daytime sleepiness. The presence of theses symptoms is essential to establish the diagnosis of PLMD [11].

DIAGNOSIS

Clinical data are not sufficiently predictive of the presence of PLMD. Polysomnography is required to establish the diagnosis of PLMD in patients with insomnia or hypersomnia and to rule out the presence of other sleep disorders [12]. During standard polysomnography, PLMS is recorded using surface electromyography (EMG) of both tibialis anterior muscles. According to the Atlas Task Force of The American Sleep Disorders Association and the International Classification of Sleep Disorders (ICSD), PLMS is defined as bursts of muscle activity during sleep with duration of 0.5–5 seconds and occurring in series of four movements or more with an interval of 5–90 seconds in between. Each movement has to be at least 25% of the amplitude of the EMG calibration signal. Simultaneous movements in both legs are counted as one movement. PLMS can occur during all stages of sleep and during wakefulness [4, 11]. They are more common during stages 1 and 2, less frequent during delta sleep, and rare during REM sleep.

The presence of theses arousals should be considered in differentiating PLMD from RLS during PSG.

PLMS is classified according to the presence of an arousal or a respiratory disturbance. An arousals that occur concurrently with a PLMS or within 2 seconds after the termination of a PLMS is considered a PLMS-associated arrousal. A PLMS-like leg movement that occurs at the end of an apnea or hypopnea or other obstructive event is usually not counted as a PLMS [4]. When compared with patients with primary PLMD, patients with RLS have more spontaneous arousals not PLMS-associated arousals [13].

One study showed a strong association between PLMD and upper airway resistance syndrome (UARS) [14]. A high percentage of PLMS with arousals correlated with respiratory event-related arousals of UARS [14]. This association should be taken into consideration when interpreting PSG of a patient with suspected PLMD. Periodic leg movements during sleep should be differentiated from sleep starts, fragmentary myoclonus, REM sleep behavior disorder, restless leg activity, myoclonic epilepsy, and nocturnal epileptic seizures.

During a standard PSG, PLMS is reported in a number of ways (Table 62.2). The PLM index (PLMI) is the total number of periodic limb movements divided by the total sleep time in hours. The PLM arousal index (PLMAI) is the total number of arousal-associated periodic limb movements divided by the total sleep time in hours. According to the revised ICSD, mild PLMD is defined as mild insomnia or hyersomnia with a PLMI of 5–34, moderate PLMD as moderate insomnia or hypersomnia with PLMI of 25–49, and severe PLMD as severe insomnia or hypersomnia with a PLMI of more than 50 or PLMAI of more than 5.

KickStrip is a low-cost device proposed for PLMD testing. It includes a movement sensor, a central processing unit that reports the recorded movements (as Kscore). Compared with standard PSG, the Kscore has excellent agreement with the PLM index and increased accuracy with the severity level. One study concluded that the KickStrip is a valuable tool for PLMD testing for both clinical and research purposes [15].

Patients who have PLMD, especially in association with restless legs syndrome, should have serum urea, creatinine, iron, ferritin, folate, and magnesium levels obtained [16].

TABLE 62.2 PLM Parameters Reported During Standard Polysomnography

Number of PLMS (total number of periodic limb movements during sleep)
Number of arousal-associated periodic limb movements (PLMA)
Number of periodic limb movements associated with respiratory events
PLM index (PLMI, number of periodic limb movements per hour of sleep)
PLM arousal index (PLMAI, number of arousal-associated periodic limb movements per hour of sleep)

TREATMENT

Instituting proper sleep hygiene by adhering to a fixed bedtime and waketime and ensuring adequate amount of sleep is the first step. Avoiding substances that aggravate PLMD, such as alcohol or SSRI antidepressants and neuroleptic medications, is also imperative [16]. Medical therapy does not modify the disease. The goal of therapy in patients with PLMD is symptomatic relief—a decrease in the number of periodic limb movements during sleep. Quality of life improvement is important in patients with PLMD in association with RLS. Pharmacological treatment of periodic limb movement disorder should be limited to patients who meet specific diagnostic criteria [11].

According to the current ICSD, the existence of primary and secondary forms of PLMD should be taken into consideration. PLMD secondary to other disorders can be resolved with specific treatment of the primary coexisting disorder [17]. Patients who are being treated with medications should be closely followed and monitored for side effects, augmentation, and tolerance of their symptoms [17].

Dopaminergic agents, opiates, and benzodiazepines are all effective in the treatment of periodic limb movement disorder [17]. The use of these medications has been associated with reduced symptoms, improved sleep quality, and improved quality of life [17].

A number of significant complications of treatment have been noted, especially with use of dopaminergic agents. Rebound effects, with the tendency of symptoms to worsen at the end of a dosing period, lead to late night or morning recurrence of symptoms and periodic limb movements in sleep. Augmentation, which is the apparent worsening of symptoms seen with long-term use of the medications, is most common with levodopa and is increased with higher doses. Symptoms of augmentation may present as progressively earlier daily onset of restless legs syndrome symptoms, may consist of expansion of the symptoms beyond the legs (involving trunk and upper limbs), and may be seen as an increase in symptom intensity or decrease in drug efficacy [17]. Levodopa (L-dopa) with a peripheral decarboxylase inhibitor (carbidopa) has been shown to suppress PLMS in patients with PLMD and RLS. The usual L-Dopa/carbidopa starting dose is 10/100 mg or 25/100 mg given 1–2 hours before sleep. The dose can be increased or a second tablet can be given in the middle of the night. Controlled-release formulation can be given for patients who require a middle-of-the-night dosage or who develop the rebound phenomenon. Long-term therapy with L-dopa/carbidopa especially at higher dosages has been associated with augmentation phenomena. Switching to another medication rather than increasing the dosage is the proper way to overcome augmentation. The dopamine agonist pergolide has been associated with reduction of RLS symptoms and PLMS. The usual dose is 0.05–5 mg given 2 hours before bedtime [16]. Pergolide is superior to L-dopa/carbidopa and less likely to result in augmentation. Bromocriptine at an average dosage of 7.5 mg at bedtime has been associated with improvements of PLMS [16]. Ropinirole is a nonergoline dopamine agonist that resulted in improvement of the PLMI.

The benzodiazepine clonazepam has been shown to suppress PLMS and to improve the quality of sleep in patients with PLMD [3]. The usual starting dose is 0.5 mg given at bedtime. The dosage can be increased gradually up to 4 mg daily. Long-term use of clonazepam may result in tolerance. Oxycodone is considered an alternative therapy for nonresponding patients. The average dosage of 10–15 mg at bedtime has been associated with decreased number of periodic limb movements, decreased arousals, and improved sleep efficiency. Therapy in pregnant women and children requires caution. There are no current evidence-based recommendations for these patients [3].

REFERENCES

1. Symonds CP. Nocturnal myoclonus. *J Neurosurg Psychiatry* **16**:166–171(1953).

2. Ohayon MM. Prevalence of restless legs syndrome and periodic limb movement disorder in the general population. *J Psychosom Res* **53**:547–554(2002).

3. Staisny K, Oertel WH, Trenkwalder C. Clinical symptomatology and treatment of restless legs syndrome and periodic limb movement disorder. *Sleep Med Rev* **6**:253–265(2002).

4. Atlas Task Force of the American Sleep Disorders Association. Recording and scoring leg movements. *Sleep* **16**:749–759(1993).

5. Bara-Jimenez W, Aksu M, Graham B, Sato S, Hallett M. Periodic limb movements in sleep: state-dependent excitability of the spinal flexor reflex. *Neurology* **54**:1609–1616(2000).

6. Bucher SF, Seelos KC, Oertel WH, Reiser M, Trenkwalder C. Cerebral generators involved in the pathogenesis of the restless legs syndrome. *Ann Neurol* **41**:639–645(1997).

7. Ruottinen HM, Partinen M, Hublin C, Bergman J, Haaparanta M, Solin O, Rinne JO. An FDOPA PET study in patients with periodic limb movement disorder and restless legs syndrome. *Neurology* **54**:502–504(2000).

8. Happe S, Pirker W, Klosch G, Sauter C, Zeitlhofer J. Periodic leg movements in patients with Parkinson's disease are associated with reduced striatal dopamine transporter binding. *J Neurol* **250**:83–86(2003).

9. Simakajornboon N, Gozal D, Vlasic V, Mack C, Sharon D, McGinley BM. Periodic limb movements in sleep and iron status in children. *Sleep* **26**:735–738(2003).

10. Weerd AW, Rijsman RM, Brinkley A. Activity patterns of leg muscles in periodic limb movement disorder. *J Neurosurg Psychiatry* **75**:317–319(2004).

11. American Academy of Sleep Medicine. *International Classification of Sleep Disorders, Revised: Diagnostic and Coding Manual.* American Academy of Sleep Medicine, Rochester, MN, 2001, pp 65–68.

12. Hilbert J, Mohsenin V. Can periodic limb movement disorder be diagnosed without polysomnography? A case–control study. *Sleep Med* **4**:35–41(2003).

13. Eisensehr I, Ehrenberg BL, Noachtar X. Different sleep characteristics in restless legs syndrome and periodic limb movement disorder. *Sleep Med* **4**:147–152(2003).

14. Exar EN, Collop NA. The association of upper airway resistance with periodic limb movements. *Sleep* **24**:188–192(2001).

15. Shochat T, Oksenberg A, Hadas N, Molotsky A, Lavie P. The KickStrip: a novel testing device for periodic limb movement disorder. *Sleep* **26**:480–483(2003).

16. Rama AN, Kushida CA. Restless legs syndrome and periodic limb movement disorder. *Med Clin North Am* **88**:653–667(2004).

17. American Academy of Sleep Medicine. Practice parameters for the treatment of restless legs syndrome and periodic limb movement disorder. *Sleep* **22**:961–968(1999).

PART VIII

SLEEP IN INFANTS AND CHILDREN

63

ONTOGENY OF EEG SLEEP FROM NEONATAL THROUGH INFANCY PERIODS

MARK S. SCHER*

Case-Western Reserve University, Cleveland, Ohio

Electrographic and polygraphic recordings of newborns and infants have been performed for almost a half-century. Pioneering studies by multiple researchers worldwide offer neurophysiologic information concerning the developing central nervous system [1–8]. Earlier investigations predated the creation of the modern neonatal intensive care unit; however, these seminal works described for the first time electrographic patterns and physiologic behaviors that define the rudimentary state of the preterm neonate. Given the higher rate of neonatal mortality, particularly in the premature infant, the clinical neurophysiologist had a more limited consultative role in the neurologic care of the sick neonate. With the creation of the modern day tertiary care neonatal intensive care unit, sophistication of medical care, including technological improvements in physiologic recordings, now offers the neurological consultant a more active role in neonatal neurophysiological assessments for medical care.

The decline in neonatal morbidity and mortality has concentrated renewed attention on the neurological performance both during the acute and convalescent periods in the days to weeks after birth. Given the immature clinical repertoire of the newborn and infant, as well as limited access to neonates in a busy intensive care setting, EEG-polygraphic studies extend the clinician's abilities to document functional brain maturation, as well as the presence and severity of encephalopathic states. Neonatal survivors also require close supervision during infancy, as successive stages in brain maturation occur.

Maturation of infant behavior requires careful evaluation of both waking and sleep behaviors. By combining neurophysiological monitoring with systematic behavioral assessments through history-taking, functional brain maturation during infancy can be better evaluated. The clinician can apply knowledge of sleep ontogeny to evaluate different pediatric populations who may be at risk for developmental delay or specific dysfunctional performance during sleep or wakefulness.

Computer-assisted tools can extend our abilities to examine physiologic relationships between cerebral and noncerebral measures, and explore associations with representative outcome variables [9–11].

CAVEATS CONCERNING NEUROPHYSIOLOGIC INTERPRETATION OF STATE

A number of caveats will assist the neurophysiologist in an understanding of the application of sleep interpretation from the neonatal through infancy periods. Maturational changes of EEG-polygraphic patterns emerge at successively older postconceptional ages: neurophysiologic maturity of a neonate can be estimated within 2 weeks for the preterm infant and 1 week for the full-term infant, reflecting the postconceptional age (PCA) of the infant independent of birth weight. Temporal coincidence or concordance among physiologic behaviors emerge with increasing maturity, similar to fetal behavioral states.

*Supported in part by NS01110, NS26793, and NS34508.

Significant reorganization of state occurs at 30, 36, and 48 weeks postconceptional ages. Finally, serial neurophysiologic studies rather than a single recording more accurately document normal ontogeny or the evolution of encephalopathic changes reflective of a brain disorder. Subsequent developmental stages also occur during infancy with sleep reorganization after 3, 9, and 12 months of age.

The clinician needs to develop a confident style of neurophysiologic pattern recognition and clinical correlation by repetitive experiences with a wide variety of EEG-polygraphic recordings. Before an accurate interpretation can be offered to the referring clinician, knowledge of the child's postconceptional age as well as the range of behavioral phenomena that are displayed during the recording are needed; this requires close communication between the electrodiagnostic technologist and the neurophysiologist.

GENERAL COMMENTS ON RECORDING TECHNIQUES AND INSTRUMENTATION FOR NEONATES AND INFANTS

Appropriate recording techniques will yield high-quality EEG-polygraphic studies. The neurophysiologist should apply a minimum of ten EEG electrodes in addition to a full complement of noncerebral polygraphic electrodes, given that specific regional and hemispheric electrographic patterns need to be correlated with other noncerebral physiologic behaviors. Placement of electrodes by either paste or collodion must be achieved with ease and efficiency by the technologist, who is always cognizant of the fragile state of the neonate within the busy NICU environment. While double interelectrode distances may be required for the infant <36 weeks EGA, to better visualize electrographic patterns, a complete set of EEG electrodes will be required for monitoring the full-term newborn and older infant.

Adjustments in sensitivity, paper speed, and filter settings will facilitate electrographic/polysomnographic interpretation. Sensitivity settings should begin with standard $7 \mu V/mm$ but may need to be periodically adjusted during the recording. Lower frequency filter settings are preferred between 0.25 and 0.5 for neonatal recordings, to avoid the elimination of commonly occurring slow frequency waveforms. Slower paper speeds (such as 15 mm/second) will permit easier visualization of slowly reoccurring normal features, such as EEG discontinuity and asynchrony, or abnormal features such as seizures and periodic discharges. Adjustment to a lower filter settings of 1 hertz and a paper speed of 30 mm/second may be preferred, for infants after 6–8 weeks of age. State-of-the-art digital equipment facilitates these adjustments when viewed offline, following the completion of the study.

Motility, cardiorespiratory, and eye movements are essential noncerebral physiologic behaviors to record.

Noncerebral physiologic observations are relevant for both state identification as well as corroboration of a clinical observation that may have prompted the request for the study. Sources of artifact are also more readily identified and eliminated with the consistent use of noncerebral channels, supplemented by the technologist's comments.

Frequent and accurate annotations by the technologist throughout the study are strongly advised. Eye-opening and eye-closure as well as repositioning of the patient's head are common annotations that are essential for proper interpretation. Information from the medical record should be recorded by the technologist for the physician's use regarding the child's gestational and postconceptional ages, as well as states of arousal, medications, and medical procedures. Skull defects, vital signs, and pertinent laboratory studies should all be described since certain factors may affect neurophysiologic interpretation.

MATURATION OF ELECTROGRAPHIC PATTERNS IN THE NEONATE

A number of principles should be applied by the neurophysiologist for an accurate interpretation of a neonatal EEG sleep study. The neurophysiologist's ability to interpret expected age-appropriate neurophysiologic patterns is essential before recognition of encephalopathic features [12]. Changes in EEG-polygraphic patterns occur for neonates at increasing postconceptional ages up to term and into early infancy. Postconceptional age is calculated simply as the infant's estimated gestational age at birth plus the number of weeks of postnatal life (i.e., estimated gestational age at birth plus postnatal age equals postconceptional age in weeks). The neurophysiologist should approximate the electrical maturity of the preterm infant within 2 weeks of other estimates of maturity, and 1 week for a term infant. Preterm neonates recorded at postconceptional ages up to term will express similar EEG patterns for a child born at that comparable level of maturity; subtle differences may also be present because of functional brain adaptation to prematurity, as will be discussed in a subsequent section.

Two studies exemplify how neurophysiologic estimates of gestational maturity can be achieved by pattern recognition of EEG sleep recordings for either healthy or sick preterm cohorts [13, 14]. Such neurophysiologic estimates of maturity were offered even without accurate clinical examination criteria, fetal sonographic data, or other obstetrical information regarding gestational maturity; for both the healthy and sick preterm groups, assessment of neurophysiologic gestational maturity was as accurate as clinical and/or anatomical estimates. Such neurophysiologic information may be essential in problematic situations in which gestational maturity is not accurately assessed by other methods; particularly in high-risk pregnancies, which may not include accurate information with respect to gestational maturity; in symptomatic infants who are too medically ill to assess postural tone or levels of arousal; or in infants too premature who do not exhibit postural tone, primitive reflexes, or behavioral alterations during state transition to accurately estimate brain maturity.

Regional and hemispheric electrographic patterns for the preterm and full-term neonate will be initially discussed, emphasizing major features at successively older postconceptional ages. Specific aspects of temporal, spatial, and state organization of EEG-polygraphic recordings are subsequently highlighted, but this brief review should be supplemented by more detailed discussions in standard texts [1, 4, 6–8, 15].

EEG Discontinuity

Alternating segments of EEG activity and inactivity (i.e., quiescence) commonly occur in preterm neonates and have been described as EEG discontinuity or trace discontinue [16]. For the child less than 30 weeks PCA, neonatal recordings consist of predominantly discontinuous EEG patterns. Varying durations of interburst intervals define this quiescence and have been described by various authors [17–21]. For the healthy preterm infant, an interburst interval should follow the "30–20 rule": an interburst interval should not exceed 30 seconds in duration on multiple

occasions for the child less than 30 weeks estimated gestational age. As the child matures beyond 30 weeks PCA, the interburst interval should be less than 20 seconds in duration. Longer periods of EEG continuity interrupt quiescent intervals after 28 weeks PCA.

For the preterm infant less than 30 weeks PCA, electrographic activities predominate in the vertex, central, and occipital regions; bitemporal attenuation is commonly observed and reflect underdeveloped frontal and temporal regions of the brain at that level of brain maturity (Figure 63.1a).

Synchrony/Asynchrony

The electrophysiologic description known as asynchrony [22] refers to similarly appearing waveforms of EEG activity in homologous head regions (e.g., left and right temporal regions) that are separated by at least 1.5 seconds in time. Preterm neonates express a varying degree of physiologic asynchrony. While the infant less than 30 weeks postconceptional age commonly exhibits "hypersynchrony" because of the extreme cortical immaturity, physiologic asynchrony emerges after 30 weeks PCA and persists until 36 weeks PCA. Asynchrony in the child at 30–32 weeks PCA may be as much as 50% of the discontinuous portion of the sleep cycle, but after 36 weeks PCA the occurrence of asynchrony rapidly drops to 0% by postconceptional term age.

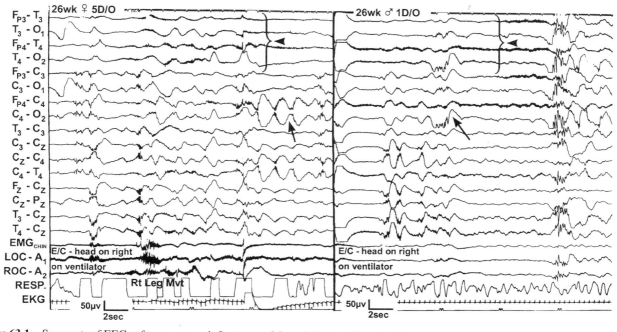

Figure 63.1 Segments of EEGs of two preterm infants—a <26-week 5-day-old and a 26-week 1-day-old, respectively. Note the prominent bitemporal attenuation (arrowheads both panels), the rhythmic delta with superimposed delta brushes in the central regions (panel one arrow), and the hypersynchronous burst in the second panel. Isolated occipital delta with superimposed occipital theta are also noted in the second panel (arrow).

Delta Brush Patterns

An admixture of fast and slow rhythms appear in the preterm EEG record as morphologically discrete waveforms that are identified with preterm neonates at varying postconceptional ages. Random or briefly rhythmic 0.3–1.5 Hz delta activity of 50–250 μV is associated with a superimposed rhythm of low to moderate amplitude faster frequencies of 10–20 Hz. Historically, different authors have described these complexes as spindle delta bursts, brushes, spindle-like fast waves, and ripples of prematurity. For infants less than 28 weeks PCA, delta brush patterns are seen in the central and midline locations with only occasional expression in the occipital regions. After 28 weeks PCA, brushes appear more abundantly in the occipital, followed by the temporal regions. Brushes can be asynchronous or asymmetric, while at other times they may be symmetrical. By term PCA, brush patterns are occasionally noted during the NREM quiet sleep or transitional sleep segments. Persistent expression or attenuation of brush rhythms in one region or hemisphere may reflect structural lesions.

Occipital Theta/Alpha Rhythms

Other patterns that can help estimate gestational maturity are the monorhythmic alpha and theta activities located in the occipital region of neonates less than 28 weeks PCA, commonly referred to as the STOP rhythm (Figure 63.2) [23]. This pattern usually persists for 6–10 seconds, can

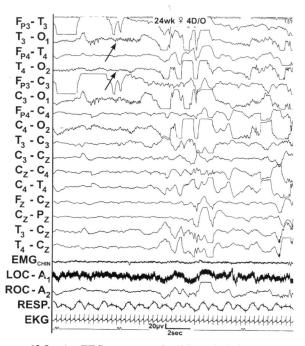

Figure 63.2 An EEG segment of a 24-week 4-day-old female with prolonged occipital theta alpha that is asymmetric in amplitude (arrows), characteristic of the STOP rhythm.

be asynchronous or asymmetric, but also may be synchronous (Figure 63.1c). Such a pattern together with midline/central brushes are electrographic features that are associated with extremely premature infants.

Temporal Theta Rhythm

A third useful developmental marker that estimates brain maturity is the burst of rhythmic 4.5–6 Hz activities noted in the midtemporal regions. Temporal theta bursts are rarely apparent in infants less than 28 weeks PCA, but become maximally expressed between 28 and 32 weeks PCA (Figure 63.3). Historically, this feature has been described as a "temporal sawtooth wave" [4], with amplitudes ranging from 20 to 200 μV. After 32 weeks PCA, its incidence rapidly diminishes [24].

Delta Wave Patterns

Rhythmic waveforms consisting of delta activity can also help estimate gestational maturity of the asymptomatic preterm neonate. Delta patterns in the central midline location are predominant for the infant less than 28 weeks gestation, together with bitemporal attenuation, as previously described. Other delta rhythms occur in the temporal and occipital locations, particularly after 28 weeks gestation (Figures 63.4 and 63.5). Between 30 and 34 weeks PCA, temporal and occipital delta rhythms become quite prominent and rhythmic, with durations that may exceed 30 seconds to 1 minute (Figures. 63.2 and 63.3).

Midline Theta/Alpha Activity

A waveform pattern [25] appears both on recordings of preterm and full-term infants, particularly during transitional or quiet sleep segments (Figure 63.6). This commonly observed pattern is sharply contoured and usually of low to moderate amplitude (Figure 63.7) in the alpha or theta ranges. While it is morphologically similar to a sleep spindle, classical spindles do not appear in the central regions until 2–4 months of age [26]. While this pattern may appear sharply contoured, this age-appropriate electrographic rhythm does not reflect a pathological or encephalopathic state.

MATURATION OF NONCEREBRAL PHYSIOLOGIC BEHAVIORS THAT DEFINE STATE IN THE PRETERM INFANT

State transitions in preterm infants less than 36 weeks PCA are not as easily identified as with the term infant. Sleep reorganization is expected to occur at, or around, 36 weeks PCA, which is similar to the in utero age when coalescence of physiologic behaviors is documented on abdominal sonography of fetal behavior of both primates and

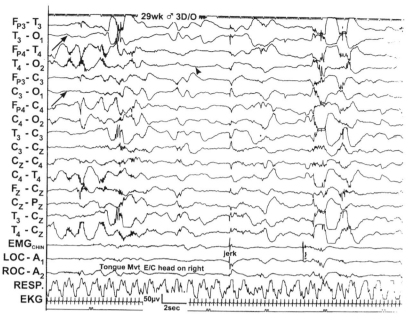

Figure 63.3 Segments of EEGs of two preterm infants approximately 29 weeks PCA, depicting abundant delta in multiple head regions as well as temporal theta activity (first panel arrow), temporo-occipital, vertex, and central brush patterns (first and second panels, arrowheads), and rhythmic occipital delta (second panel, arrow).

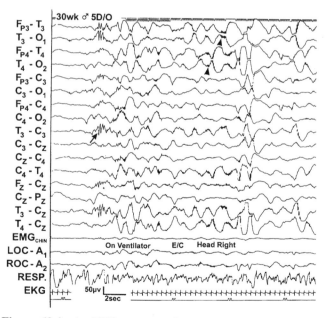

Figure 63.4 An EEG segment of a nearly 30-week 5-day-old PCA male, depicting the onset of continuous EEG segment with a left temporal theta burst (arrow), prominent delta, and superimposed delta brushes in the temporal regions (arrowhead). Note the temporal delta is more rhythmic than in Figure 63.3.

humans [27]. As a rule, state organization in the preterm infant is rudimentary and underdeveloped [28, 29]. The following summary serves as an introduction to a discussion of specific physiologic behaviors, which highlight state differentiation in the preterm infant at increasing postconceptional ages.

Rapid eye movements represent one of the main identifying features of rudimentary active sleep in the preterm infant. Eye movement phenomena become more consistently time-locked to continuous EEG activities as early as 30–31 weeks gestation [28]. Using fetal sonography [30], eye movements of the fetus during active sleep can also be documented. REM activity is not a random rhythm and does occur in a predictable interval despite brain immaturity [31]. Various classes of REM have been described during different states of sleep in the neonate, and the number and types of REMs evolve with brain maturation [32, 33]. A recent study of multiple physiologic behaviors during sleep in the preterm infant correlated the occurrence of REM with more continuous EEG tracings, while negatively correlated with discontinuous EEG segments [34] in preterm infants as early as 30 weeks PCA (Figures 63.8 and 63.9).

Motility patterns are also an integral part of state definition for the neonate but differ between preterm and full-term infants. Different motility patterns emerge at increasing postconceptional ages, both for the fetus as well as the extrauterine-reared neonate [35, 36]. Myoclonic and whole-body movements predominate for the preterm

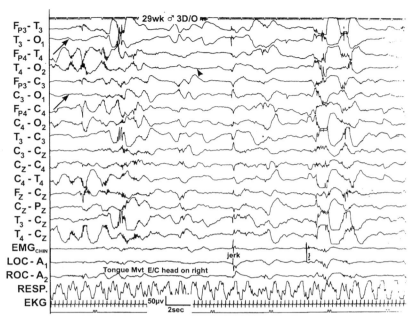

Figure 63.5   An EEG segment of a 29-week 3-day-old male with a shifting asymmetry between the left temporal central region (arrow), and the right temporal region (arrowhead), characteristic of physiologic interhemispheric asynchrony. Also note the prominent temporal theta and brushes as well as diffuse delta slowing during this discontinuous portion of the EEG.

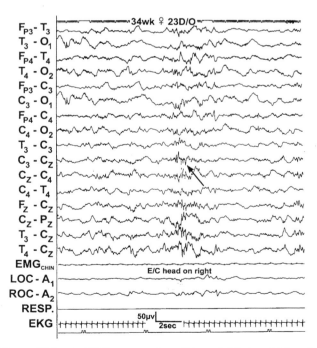

Figure 63.6   An EEG segment for a 34-week 23-day-old female with a prominent vertex and parasagittal burst of theta/alpha activity (arrows). Note the rare delta brushes and absent temporal theta at a postconceptional age of 37 weeks.

infant [37, 38] while smaller, slower segmental body movements are seen in the full-term neonate. State-specific decreases in the number of small and large body movements have been correlated with increasing EEG discontinuity in preterm infants [34], while increased head and facial movements are associated with only active sleep between 30 and 36 weeks PCA.

Maturational changes in cardiorespiratory behavior have also been studied in the preterm infant. Periodic breathing and respiratory pauses are physiological events that commonly occur in preterm infants [39–40]. Using spectral analyses, decreased variability of cardiorespiratory behavior during quiet sleep is seen at increasing postconceptional ages [41]. However, EEG measures appear to be an alternative for state prediction to noncerebral measures, such as cardiorespiratory behavior. In a recent study of multiple sleep behaviors in the preterm infant less than 36 weeks PCA, rapid eye movements rather than cardiorespiratory, motility, and temperature changes predictably varied with EEG changes, suggesting that specific brain regions physiologically coalesce with EEG activities before other neuronal systems [42].

ASSESSMENT OF STATE ORGANIZATION IN THE FULL-TERM INFANT

Extensive information has been published over the last half-century with respect to the functional significance of the

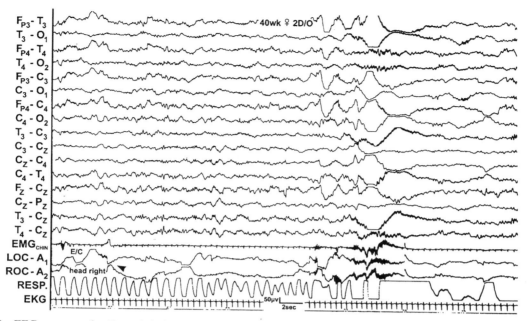

Figure 63.7 An EEG segment of a 40-week 2-day-old female, depicting mixed-frequency active sleep, characterized by continuous EEG, body movements, rapid eye movements (arrowhead), and irregular respirations and heart rate. Note the onset of a spontaneous arousal coincident with a temporary flattening of the EEG background.

relatively short ultradian sleep rhythm in the neonate [43]. For older infants, the human sleep cycle is an ultradian period with an interval of 75–90 minutes. The full-term neonate expresses an ultradian cycle that is approximating 30–70 minutes in duration [44]. Sleep segments that comprise the neonatal sleep cycle also differ from older individuals, comparing EEG and polysomnographic behaviors. Two active and two quiet sleep segments as well as transitional or indeterminate sleep segments have been described. Arousal periods, defined as reactivity, occur both

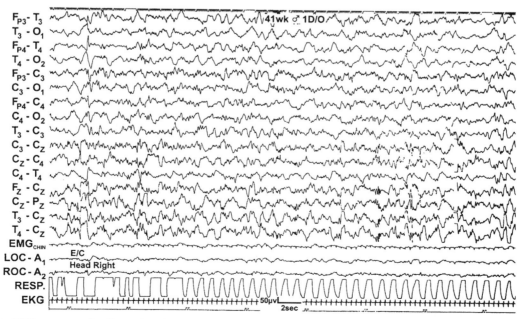

Figure 63.8 An EEG segment of a 40-week 1-day-old female that documents high-voltage slow quiet sleep. Regular respirations and the absence of rapid eye movements are noted.

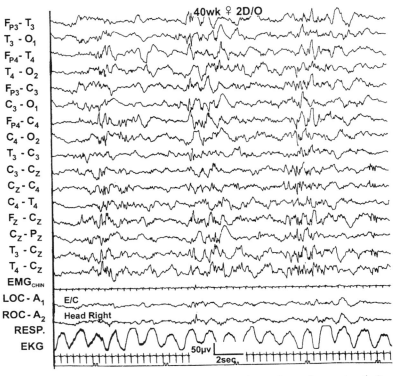

Figure 63.9   An EEG segment of a 40-week 15-day-old female, documenting a discontinuous tracé alternant, quiet sleep segment.

within and between the sleep segments. Indeterminate or transitional sleep as well as the arousal phenomena are important expressions of sleep continuity in the immature brain.

State definitions in the term infant traditionally require the temporal coalescence of specific physiologic behaviors. Based on visual analyses, comparisons between cerebral and noncerebral behaviors are temporally observed to determine state for either adults or children [45]. Visual interpretations of EEG sleep states are also easily identified for the full-term neonate [8]. Active, or REM, sleep for the full-term neonate is traditionally associated with the coalescence of rapid eye movements, increased variability of cardiorespiratory rates, low muscle tone in the context of low-voltage or mixed frequency continuous EEG patterns, and the abundance of body movements. Conversely, quiet or NREM sleep is associated with the absence of rapid eye movements, fewer body movements, higher muscle tone, and decreased variability in respiratory rates in the context of continuous high-voltage slow or discontinuous EEG patterns. The above-described patterns are not expressed until after 36 weeks postconceptional age and are no longer seen beyond 46–48 weeks postconceptional age. Typically the ultradian sleep cycle begins an active sleep after sleep onset in over 50% of newborns. This initial active sleep segment is a mixed-frequency EEG segment, which comprises 25–30% of the total sleep cycle. This active sleep segment is then followed by a brief high-voltage slow quiet sleep

segment, which is approximately 3–5% of the sleep cycle. Subsequently, a discontinuous quiet sleep period (historically described as a tracé alternant pattern) now comprises approximately 25% of the sleep cycle of the neonate. Finally, a post-quiet sleep active sleep segment known as low-voltage irregular comprises approximately 15% of the cycle. Transitional or indeterminate sleep comprises between 10% and 15% of the sleep cycle. While the child does not yet express a strong diurnal or circadian rhythmicity of sleep, wakefulness is distributed over a 24 hour period; as many as 6–8 hours of waking sleep over a 24 hour period in the neonate may occur.

Two biorhythmic processes define the temporal organization of sleep in the neonate—a weak circadian sleep wave rhythm is present, and a stronger ultradian REM and NREM rhythm is also active [46]. Both biorhythms evolve with increasing age. Internal "biologic clocks" become better organized around environmental cues, such as light/dark cycle, temperature, noise, and social interaction [47]. For the normal full-term neonate, sleep alternates with waking states in a 3–4 hour cycle, both during the night and day. This has historically been referred to as the basic rest/activity cycle (i.e., BRAC).

Within the first month or two of life after birth for the full-term infant, sleep/wake state reorganization begins, particularly with a more dominant diurnal effect to environmental cues. Circadian rhythmicity of body temperature and heart rate is noted in approximately 50% of preterm infants at 29–35 weeks PCA [48]. Yet stronger ultradian rhythms over a 3–4 hour duration correspond to social intervention such as feeding [45]. Increases in body movement activities as well as heart rate, and decreases in rectal and skin temperatures are noted during interventions, reflecting changes in the infant's microenvironment and the infant–caretaker interaction. The length of the ultradian EEG sleep cycle increases with maturing postconceptional age, demonstrating a positive correlation between cycle length and increasing PCA [34].

SLEEP ONTOGENESIS: STATE MATURATION FROM FETAL THROUGH INFANCY PERIODS

Reasons for the continuity of fetal state expression from intrauterine through neonatal ages prior to 46 weeks EGA remain obscure but may reflect the need for physiologic homeostasis of the fetus during the transition from intrauterine to extrauterine environments, requiring approximately a postnatal month of brain development before infancy sleep patterns begin to emerge.

State development involves neuronal networks that are actively maturing during fetal life. Beginning as early as 10 weeks gestational age, the human fetus displays spontaneous movements, as visualized on ultrasonography. These movements now are more clearly visualized with three-dimensional ultrasonography, which more readily can document eye opening and closure and rhythmic body movements, while fetal heart rate is electronically recorded. All these behaviors allow estimation of fetal state transitions [49]. Rhythmic cycling of motor activity has been described in fetuses as young as 20–28 weeks gestation [50], with the fetal rest–activity pattern for long quiescent periods lasting minutes to hours, during which time no respiratory movements are noted [51]. Cycle times vary but usually range between 40 and 60 minutes [52]. Behavioral estimations of quiet (i.e., NREM) sleep approximates 53% in the 30 week conceptional age study, increasing to 60% by near term ages.

State studies of fetal baboons documented similar coalescence among cerebral and noncerebral behaviors while in the intrauterine environment, for this primate species is similar to humans [53]. These same temporal relationships are expressed by preterm neonates in an extrauterine environment. These physiologic interrelationships defining state persist to 4–6 weeks of postnatal life, after which infant sleep patterns gradually emerge to resemble adult sleep rhythms between the first and second years of life. The coalescence of cerebral and noncerebral components of state follows approximately one month before term ages.

Specific features regarding sleep organization occur after 46–48 weeks EGA [54, 55]. Lengthening of the overall sleep cycle as well as reorganization of sleep architectural segments are expressed; gradual reductions in REM sleep percentage are noted, while NREM sleep becomes more abundant. Rather than a sleep-onset active or REM sleep after wakefulness, NREM sleep segments first appear. NREM sleep stages I–IV as defined [45] do not fully develop until late infancy. High-voltage delta slow NREM sleep remains the predominant electrographic expression of this segment of the sleep cycle, similar in EEG frequency distribution to the high-voltage slow quiet sleep segment of the neonate. Reductions in arousals, body movements, and REMs are noted as the child matures past 46–48 weeks PCA.

During the first 3 months of life, rapid maturation of electrical activities in the brain occur, such as the disappearance of tracé alternant, the emergence of sleep spindle activity, and the emergence of "adult-like" delta wave activity [56–59]. Using quantitative assessments of spectral EEG analyses, increases in theta power by 9 months of age [60, 61] herald the emergence of the S1 and S2 segments of the NREM sleep segment, codified for adult subjects by Rechtschaffen–Kales [45] sleep state criteria. There also is a continual decrease in total sleep time, rapid eye movement sleep, and indeterminate sleep, while concomitant increases in waking time and NREM sleep, particularly stages I–II NREM sleep.

Sleep organization for 15 normal infants as studied while at home during six 24 hour sleeping periods over 12–24 months from birth [62]. Sleep staging was scored according to adult criteria [45] with modifications for children by Guillenminault and Souquet [63]. While these authors reconfirmed earlier reported changes in percentages of total sleep time, REM sleep, NREM sleep, indeterminate sleep, and wakefulness, the authors also reported age and day/night effects on sleep ontogenesis. Modifications with age were more precocious and more pronounced for the diurnal part of the 24 hour cycle, especially as regards REM sleep. For the nocturnal part of the 24 hour cycle, there was a significant increase in sleep efficiency during the REM period after 12 months of age. The authors went on to demonstrate that the total sleep duration and the number of awakenings decreased. These authors point to the high stability of the percentage of slow-wave sleep during the first 2 years of life. Until 12 months of age, stage II / REM sleep ratio equals one, and sleep changes occur earlier during the diurnal part of the 24 hour cycle. These examples of sleep ontogeny suggest how developmental neurophysiologic changes occur within neuronal networks that are responsible for sleep expression. These data also highlight the emergence of a well-developed circadian rhythm after 3 months of age, prior to the maturation of nocturnal sleep organization. Those brain structures responsible for circadian cycling predate other regions that are responsible for generation of S-2 sleep and the decrease in REM sleep. Nine months of age appears to represent an important developmental age for sleep maturation. During the night, significant reductions in REM sleep and increases in S-2 occur after this age. After this age, rapid acceleration in brain myelination, dendritic arborization, and synaptogenesis occur, resulting in increased neuronal interactions between brainstem and thalamocortical structures [64].

Ontogeny of Autonomic Behavior During Sleep: Heart Rate Variability

As the neonate matures into infancy and early childhood ages, changes in cardiovascular functions during sleep reflect changes in autonomic nervous system dominance of neuronal networks subserving cardiac activity. In general, sympathetic nerve activity, blood pressure, and heart rates are lower during non-rapid eye movement sleep than in wakefulness. During rapid eye movement sleep, sympathetic nerve activity increases, reaching values greater than those measured during wakefulness [65], reflecting increased sympathetic control of cardiovascular function. Because short-term oscillations of heart rate (i.e., heart rate variability (HRV)) reflect autonomic nervous system activity, these values can be useful for assessing autonomic control under various physiologic and pathologic conditions. Spectral analysis of HRV can provide quantitative estimations of the balance between sympathetic and parasympathetic control [66].

Short-term HRV spectra distinguish three main power components. The higher frequency (HF) component (range 0.15–0.40 Hz), corresponding to heart rate and blood pressure oscillations induced by respiratory activity, mediated by the vagal branch of the autonomic nervous system, is considered a marker for parasympathetic activity. Lower frequency (LF) components (0.04–0.15 Hz) reflect baroreflex control of systemic blood pressure, providing a measure of sympathetic activity.

Using time-domain and frequency-domain analyses of HRV signals, researchers have reported parasympathetic predominance during NREM sleep, while increased sympathetic activity is expressed during REM sleep [67].

Sleep stage and age both significantly influence short-term HRV during sleep in both healthy infants and children [68]. Greater parasympathetic control during sleep is observed more for children than infants. This difference may reflect autonomic nervous system maturation that takes place over the first several years of life [69].

In summary, sleep ontogenesis during infancy gradually evolves into adult sleep organization over the first 2 years of life, distinguishing specific neuronal networks. Circadian rhythms appear after 3 months of age, followed by expression of an adult ultradian sleep cycle after 9 months of age. There is a lengthening of the ultradian sleep cycle after 12 months of age. Reductions in arousals, motility, rapid eye movements, and sympathetic control reflect developmental changes within multiple brain regions that are responsible for sleep initiation and maintenance.

BRAIN ADAPTATION TO STRESS AS REFLECTED IN SLEEP REORGANIZATION

Endogenous or exogenous factors can alter the ontogenesis of specific physiologic behaviors during sleep. This is exemplified by neurophysiologic studies that compare differences between preterm and full-term infants at matched postconceptional term ages concerning sleep architecture, continuity, phasic, spectral, cardiorespiratory, and temperature behaviors [44, 70–73]. Unlike the full-term infant, sleep of the preterm infant adapts to an extrauterine environment by expressing a one-third longer sleep cycle, a greater percentage of quiet sleep, fewer movements, and shorter arousals. Preterm infants also exhibit higher rectal temperatures over the ultradian cycle, with less change from NREM to REM segments. Greater cardiorespiratory irregularity is noted during quiet sleep, and lower spectral EEG energies are observed during specific sleep segments. These differences reflect conditions of prematurity on brain maturation; adaptation of brain function for the preterm infant in an extrauterine environment represents physiologic dysmaturity to biological and/or environmental stresses [74, 75]. Dysmature EEG sleep measures may also help predict neurodevelopmental performance for neonates with clinical risk factors other than prematurity, such as prenatal substance exposure [76], chronic lung disease [77], or general medical complications [78]. Documentation of the persistence or resolution of dysmature sleep behaviors during infancy for a clinical risk group needs to be better addressed.

COMPUTER-ASSISTED ANALYSES OF EEG SLEEP ORGANIZATION IN NEONATES AND INFANTS

Relationships among multiple physiologic processes are certainly less developed in the preterm infant. State transitions are more difficult to recognize, particularly over short recording intervals. Even with longer recording times, less well-developed associations among physiologic variables may not be obvious by visual analysis. Automated systems for EEG sleep analyses can complement visual inspection [10]. (Figure 63.10). Computer analyses better characterize relationships among electrographic and polysomnographic components over extended recording intervals, and better detect rudimentary sleep behaviors. Studies that compare computer and visual analyses of neonatal EEG recordings through infancy have ascertained which physiologic

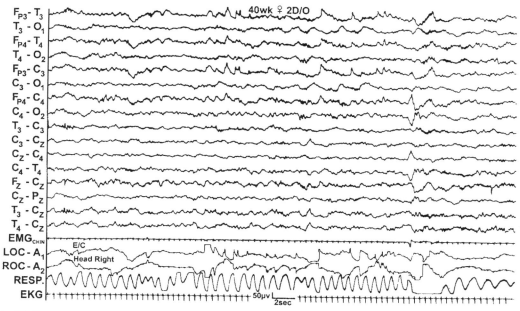

Figure 63.10   An EEG segment of a 40-week 2-day-old female, documenting a low-voltage irregular active sleep segment. Note prominent sucking and REMs.

relationships best represent state expression: spectral EEG energies and REM best define maturational trends when compared to other measures in the preterm infant [41–44] (Figure 63.11). Conversely, other noncerebral measures such as cardiorespiratory, motility, and temperature changes predict either maturational trends or state transitions less accurately. Computer algorithms may detect diurnal or nocturnal rhythmicities more accurately than visual inspection [10].

Comparatively less attention has been directed to automated analyses of neonatal EEG sleep studies compared with older persons [79]. Since an earlier review of this topic [10], further advancements in the development of both computerized devices and mathematical programming have been achieved. To succeed in the development of an automated state detector for neonates, technical innovations must recognize the unique neurophysiologic expressions of state transitions of the newborn that do not exist for the older patient. A short list of these unique electrographic/polysomnographic behaviors include a shorter sleep cycle, prominent EEG delta rhythms in different regional locations, intra- and interhemispheric electrographic asynchrony, discrete neonatal waveform patterns (i.e., delta brush and theta burst patterns), a high percentage of periodic breathing, a greater number and heterogeneity of rapid eye movements, and unique motor patterns that reflect fetal postural reflexes that precede the expression of more sophisticated movement patterns.

Previous neonatal sleep studies initially applied automated techniques to assess functional brain maturation using analyses that were based on assumptions of stationarity, without consideration of time-dependent changes [18, 21, 60, 80–85]. The preferred methodological approach has been fast Fourier transform analyses, studied initially with full-term neonates [86–90], followed by more recent reports in preterm infants [91–97]. Similar calculations, based primarily on assumptions of stationarity, were also described for specific neonatal and infant risk groups for sudden infant death syndrome [98], apnea [95], hyperbilirubinemia [99], white matter necrosis [100], and asphyxia [101] applying power analyses to one particular physiologic behavior, with little attention to the multiple neuronal networks that contemporaneously express state transitions. Single-channel monitoring devices have demonstrated that important maturational trends can be documented using standard spectral values [102] without regional or hemispheric specificity.

Few reports have combined cerebral and noncerebral measures to more comprehensively study newborn sleep states [103, 104]. One research group has applied automated analysis methods of neonatal sleep to both cerebral and noncerebral measures, combining computations to detect and quantify both stationary and nonstationary signal behaviors. Simultaneous assessment of multiple cerebral and noncerebral measures are emphasized to define neonatal state [44]. Spectral analyses of EEG [71, 105–108], cardiorespiratory behavior [109–111], arousal behavior [44, 105, 112], and REMs [44, 108, 113] establish that there are important physiologic differences during sleep between healthy preterm and full-term cohorts. Nonlinear

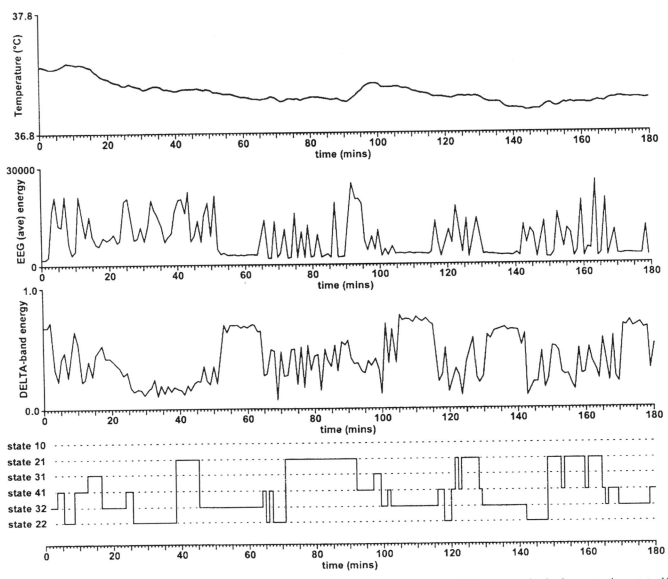

Figure 63.11   A 3 hour summary of physiological behaviors constituting neonatal sleep at full-term age, in the lower tracing: state 10 awake, state 21 mixed-frequency active sleep, state 31 high-voltage slow quiet sleep, state 41 indeterminate sleep, state 32 tracé alternant quiet sleep, state 22 low-voltage irregular active sleep. Spectral delta and total EEG energies in panels 2 and 3 illustrate changes in these values, depending on the segment of the neonatal sleep cycle. Note the minimum total EEG energy and maximum delta energy during tracé alternant quiet sleep. The top panel illustrates a slower multiple-hour temperature rhythm, which changes over multiple sleep cycles.

computations for feature extraction of EEG signals [114], arousals [112], and state/outcome prediction [115] have also been suggested as part of the overall strategy to develop an automated neonatal state detector.

Differences in the functional brain organization between neonatal cohorts have been incorporated into a statistical model that offers a mathematical paradigm to define physiologic brain dysmaturity of preterm neonates at corrected full-term ages. This dysmaturity index is based on seven selected physiologic measures [108, 112, 116–119] that best represent differences in functional brain organization and maturation between healthy preterm and full-term cohorts. This statistical model characterizes any particular physiologic behavior of the preterm infant as either delayed or accelerated in relation to full-term controls. Automated methodologies, which can capture these selected behaviors over time, offer an opportunity to characterize the process of developmental neuroplasticity within the immature brain of a neonate who has been stressed by environmental or disease conditions.

Computer analyses of EEG sleep during infancy also have helped demonstrate the physiologic ontogenesis of the neuronal macronetworks [120–125]. EEG frequencies with maturation, based on power spectral analyses, document alterations in all frequency bandwidths particularly at the higher frequency ranges for human EEG (i.e., the alpha and beta ranges). These changes are surrogate markers of cognitive and behavioral development, especially in the frontal lobe1 [124, 126]. Also, deviations in the ontogenesis of spectral signals differentiate specific at-risk populations of children [127–129]. Relatively little attention has been devoted to very high frequency spectral bandwidths (i.e., >40–1000 Hz), which have been studied primarily in adult populations.

Spectral analyses have also been performed involving sleep studies for noncerebral physiologic parameters, particularly cardiorespiratory measures, as discussed under the section for cardiorespiratory maturation. Changes in the balance between sympathetic and parasympathetic influences during sleep can be assessed by the spectral analysis of heart rate variability [68]. Few studies extend these evaluations up through infancy. Most studies dealing with maturation of cardiorespiratory behavior do not include ages beyond 6 months.

Sleep Ontogenesis and Neural Plasticity

Advances in developmental neuroscience over the last fifteen years have expanded our knowledge base regarding the sequential steps in brain maturation. The later developmental stages of this complicated process of maturation encompass remodeling or resculpting of the brain, sometimes termed plasticity or activity-dependent development, which include signaling processes within neuronal cells and networks. Use or disuse of specific neuronal populations or networks will lead to pruning and remodeling of the brain's neuronal circuitry. Apoptosis or programmed cell death also contributes to modifying brain structure or function, during both prenatal and postnatal periods [130, 131]. During the last trimester of pregnancy and into the first year of life, dendritic arborization, synaptogenesis, myelinization, and neurotransmitter development rapidly evolve in the immature brain [132], during which adverse conditions of prematurity (i.e., both prenatal and postnatal time periods), medical illnesses, and environmental stresses collectively alter the process of activity-dependent development and apoptosis on specific neuronal circuitry. Given that remodeling of neuronal connectivity is ultimately required for the expression of complex neurobehaviors including cognitive abilities at older ages [133], aberrant remodeling will alternatively be expressed as neurocognitive and neurobehavioral deficits. Automated neurophysiologic methodologies can assess brain organization and maturation in the newborn, offering a surrogate marker for activity-dependent

development of the fetal and neonatal brain. Computational algorithms applied to selected physiologic measures of neonatal sleep can provide insights into the process by which neuronal networks change and adapt over longer periods of time during extrauterine life under adverse medical and socioeconomic conditions, and in the context of genetic endowment. Applications and methods of nonlinear dynamics to experiments in neurobiology will help better characterize the biologic process of neuroplasticity [134]. Computational analyses of complex stimuli, which reflect changes in neuronal circuitry, can enhance our understanding of the encoding and transmission of information by neuronal networks that subserve complex functions that range from sleep to cognitive performance. The application of these processing techniques to both neonatal intensive care and pediatric sleep laboratory settings will permit better assessment of EEG sleep state organization and maturation, and transform neonatal and pediatric monitoring capabilities into neurointensive care and state-of-the-art monitoring facilities.

CONCLUSION

Serial neonatal and infant electroencephalographic/polysomnographic studies document the ontogeny of cerebral and noncerebral physiologic behaviors, based on visual inspection or computer analyses. EEG patterns and other physiologic relationships serve as templates for normal brain maturation and also help distinguish intrauterine from extrauterine development. Such strategies will ultimately improve our diagnostic skills for the care of the high-risk fetus, neonate, and infant.

EEG sleep studies remain the only bedside neurodiagnostic procedure that provides a continuous record of cerebral function over long periods of time. While other advanced methods of anatomical or functional inquiry, such as cranial sonography and magnetic resonance imaging, report brief snapshots of cerebral anatomy, neurophysiologic studies provide a functional perspective into brain ontogeny. Sleep ontogenesis in asymptomatic neonates and infants documents expected brain maturation, to better anticipate deviations from these biologically programmed processes under stressful and/or pathological conditions.

REFERENCES

1. Anders T, Ende R, Parmelee A. *A Manual of Standardized Terminology, Technique, and Criteria for Scoring of States of Sleep and Wakefulness in Newborn Infants.* UCLA Brain Information Service NINDS, Neurological Information Network, Los Angeles, 1971.

2. Parmelee AH, Stern R. Development of states in infants. In: Clemente DC, Purpurer DP, Mayer EE (Eds), *Sleep and the*

Maturing Nervous System. Academic Press, New York, 1972, pp 199–228.

3. Ellingson RJ. Studies of the electrical activity of the developing human brain. In: Himwich WA (Ed), *The Developing Brain—Progress in Brain Research*. Elsevier, Amsterdam, 1964, pp 26–53.

4. Dreyfus-Brisac C. Neonatal electroencephalography. In: Scarpelli EM, Cosmie EV (Eds), *Reviews in Perinatal Medicine*, Vol III. Raven Press, New York, 1979, pp 397–430.

5. Prechtl HFR. The behavioral states of the newborn infant. *Brain Res* **76**:185–212(1974).

6. Lombroso CT. Neonatal electroencephalography. In: Niedermeyer E, Lopez-Desilva F (Eds), *Electroencephalography, Basic Principles, Clinical Applications in Related Fields*. Urban and Schwarzenberg, Baltimore, 1989, pp 599–637.

7. Hrachovy RA, Mizrahi EM, Kellaway P. Electroencephalography of the newborn. In: Daly DD, Pedley TA (Eds), *Current Practice of Clinical Electroencephalography*, 2nd ed. Raven Press, New York, 1990, pp 201–242.

8. Pope JJ, Werner SJ, Bickford RG. *Atlas of Neonatal Electroencephalography*. Raven Press, New York, 1992.

9. American Electroencephalographic Society Guidelines in EEG and Evoked Potentials. Ajmone-Marsan C (Ed). *J Clin Neurophysiol* **3**(Suppl 1):1–152(1986).

10. Scher MS, Sun M, Hatzilabrou GM, Greenberg N, Cebulka G, Sclabassi R. Computer analyses of EEG sleep in the neonate: methodological considerations. *J Clin Neurophysiol* **7**:417–441(1990).

11. Scher MS, Loparo KA, Turnbull JP. Automated state analyses: proposed applications to neonatal neurointensive care. *J Clin Neurophysiol* x:x–x(2004).

12. Scher MS. Neonatal encephalopathies as classified by EEG-sleep criteria. Severity and timing based on clinical/pathologic correlations. *Pediatr Neurol* **11**:189–200(1994).

13. Scher MS, Martin J, Steppe DA, Banks DL. Comparative estimates of neonatal gestational maturity by electrographic and fetal ultrasonographic criteria. *Pediatr Neurol* **11**:214–218(1994).

14. Scher MS, Barmada A. Estimation of gestational age by electrographic, clinical and anatomical criteria. *Pediatr Neurol* **3**:256–262(1987).

15. Curzi-Dascalova L, Mirmiran M. *Manual of Methods for Recording and Analyzing Sleep–Wakefulness States in Preterm and Full-Term Infant*. Les Editions INSERM, Paris, 1996.

16. Dreyfus-Brisac C. Sleep ontogenesis in early human prematurity from 24 to 27 weeks of conceptional age. *Dev Psychobiol* **1**:162–169(1968).

17. Benda GI, Engel RCH, Zhang Y. Prolonged inactive phases during the discontinuous pattern of prematurity in the electroencephalogram of very-low-birthweight infants. *Electroencephalogr Clin Neurophysiol* **72**:189–197(1989).

18. Connell JA, Oozeer R, Dubowitz V. Continuous 4-channel EEG monitoring: a guide to interpretation with normal values in preterm infants. *Neuropediatrics* **18**:138–145(1987).

19. Hughes JR, Fino J, Gagnon L. Periods of activity and quiescence in the premature EEG. *Neuropediatrics* **14**:66–72(1983).

20. Hughes JR, Fino JJ, Hart LA. Premature temporal theta. *Electroencephalogr Clin Neurophysiol* **67**:7–15(1987).

21. Eyre JA, Nanei S, Wilkinson AR. Quantification of changes in normal neonatal EEGs with gestation from continuous five-day recordings. *Dev Med Child Neurol* **30**:599–607(1988).

22. Lombroso CT. Neonatal polygraphy in full-term and preterm infants: a review of normal and abnormal findings. *J Clin Neurophysiol* **2**:105–155(1985).

23. Hughes JR, Miller JK, Fino JJ, Hughes CA. The sharp theta rhythm on the occipital areas of prematures (STOP): a newly described waveform. *Clin Electroencephalography* **21**:77–87(1990).

24. Scher MS, Bova JM, Dokianakis SG, Steppe DA. Positive temporal sharp waves on EEG recordings of healthy neonates: a benign pattern of dysmaturity in preterm infants at postconceptional term ages. *Electroencephalogr Clin Neurophysiol* **90**:173–178(1994).

25. Hayakawa F, Watanabe K, Hakamada S, Kuno K, Aso K. FZ theta/alpha bursts: a transient EEG pattern in healthy newborns. *Electroencephalogr Clin Neurophysiol* **67**:27–31(1987).

26. Lenard, HC. Sleep studies in infancy: facts, concepts, and significances. *Acta Paedriatr Scand* **59**:572–581(1970).

27. Myers MM, Schulze KF, Fifer WP, Stark RI. Methods of quantifying state-specific patterns of EEG activity in fetal baboons and immature human infants. In: LeCanuet J, Fifer WP, Krasnesor NA, Smotherman WP (Eds), *Fetal Development: A Psychological Perspective*. Lawrence Erlbaum, Hillsdale, NJ, 1995, pp 35–49.

28. Curzi-Dascalova L, Peirano P, Morel-Kahn Inserm F. Development of sleep states in normal premature and full-term newborns. *Dev Psychobiol* **21**:431–444(1988).

29. Curzi-Dascalova L, Figueroa JM, Eiselt M, Christova E, Virassamy A, D'Allest AM, Guimaraes H, Gaultier C, Dehan M. Sleep state organization in premature infants of less than 35 weeks' gestational age. *Pediatr Res* **34**:624–628(1998).

30. Prechtl HFR, Nijhuis JG. Eye movements in the human fetus and newborn. *Behav Brain Res* **10**:119–124(1983).

31. Dittrichova J, Paul K, Pavlikova E. Rapid eye movements in paradoxical sleep in infants. *Neuropaediatrie* **3**:248–257(1972).

32. Ersyukova II. Oculomotor activity and autonomic indices of newborn infants during paradoxical sleep. *Hum Physiol* **6**:57–64(1980).

33. Lynch JA, Aserinsky E. Developmental changes of oculomotor characteristics in infants when awake and in the active state of sleep. *Behav Brain Res* **20**:175–183(1986).

34. Scher MS, Steppe DA, Dokianakis SG, Guthrie RD. Maturation of phasic and continuity measures during sleep in preterm neonates. *Pediatr Res* **36**:732–737(1994).

35. Robertson SS. Intrinsic temporal patterning in the spontaneous movement of awake neonates. *Child Dev* **53**:1016–1021(1982).

36. Robertson SS. Human cyclic motility: fetal–newborn continuities and newborn state differences. *Dev Psychobiol* **20**:425–442(1987).

37. Fukumoto M, Mochizuki N, Takeishi M, Nomura Y, Segawa M. Studies of body movements during night sleep in infancy. *Brain Dev* **3**:37–43(1981).

38. Prechtl HFR, Fargel JW, Weinmann HM, Backter HH. Postures, motility and respiration of low risk preterm infants. *Dev Med Child Neurol* **21**:3–27(1979).

39. Martin RJ, Miller MJ, Carlo WA. Pathogenesis of apnea in preterm infants. *J Pediatr* **109**:733–741(1986).

40. Glotzbach SF, Tansey PA, Baldwin RB, Ariango RL. Periodic breathing cycle duration in preterm infants. *Pediatr Res* **25**:258–261(1989).

41. Scher MS, Steppe DA, Banks DL, Guthrie RD, Sclabassi RJ. Maturational trends of EEG-sleep measures in the healthy preterm neonate. *Pediatr Neurol* **12**:314–322(1995).

42. Scher MS, Dokianakis SG, Steppe DA, Banks D, Sclabassi RJ. Computer classification of state in healthy preterm neonates. *Sleep* **20**:132–141(1997).

43. Hildebrandt G. Functional significance of ultradian rhythms and reactive periodicity. *J Interdisc Cycle Res* **17**:307–319(1986).

44. Scher MS, Steppe DA, Dahl RE, Asthana S, Guthrie RD. Comparison of EEG-sleep measures in healthy full-term and preterm infants at matched conceptional ages. *Sleep* **15**:442–448(1992).

45. Rechtschaffen A, Kales A (Eds). *A Manual of Standardized Terminology, Techniques and Scoring System for Sleep Stages of Human Subjects.* Brain Research Institute/Brain Information Services, University of California, Los Angeles, 1968.

46. Glotzbach SF, Edgar DM, Ariagno RL. Biological rhythmicity in preterm infants prior to discharge from neonatal intensive care. *Pediatrics* **95**:231–237(1995).

47. Anders TF, Sadeh A, Appareddy V. Normal sleep in neonates and children. In: Ferber R, Kryger M (Eds), *Principles and Practice of Sleep Medicine in the Child.* Saunders, Philadelphia, 1995.

48. Mirimiran M, Kok JH. Circadian rhythm in early human development. *Early Hum Dev* **262**:121–128(1991).

49. Nijhius JG, Martin CB, Prechtl HFR. Behavioral status of the human fetus. In: Prechtl HFR (Ed), *Clinics in Developmental Medicine. No. 98: Continuity of Neural Functions from Prenatal to Postnatal Life.* London Spastic International Medical Publications, London, 1984, pp 65–78.

50. Parmelee AH, Wenner WH, Akiyama Y, Schultz M, Stern E. Sleep states in premature infants. *Dev Med Child Neurol* **9**:70–77(1967).

51. Dawes GS, Fox HE, Leduc BM, Liggins GC, Richards RT. Respiratory movements and rapid eye movement sleep in the foetal lamb. *J Physiol* **220**:119–193(1972).

52. Sterman MB, Hoppenbrauwers T. The development of sleep–waking and rest–activity patterns from fetus to adult in man. In: Sterman MB, McGinty DJ, Adinolfi AM (Eds), *Brain Development and Behavior.* Academic Press, New York, 1971, pp 203–225.

53. Myers MM, Stark RI, Fifer WP, Grieve PG, Haiken J, Leung K, Schulze KF. A quantitative method for classification of EEG in the fetal baboon. *Am J Physiol* **265**:R706–714(1993).

54. Kahn A, Dan B, Groswasser J, Franco P, Sottiaux M. Normal sleep architecture in infants and children. *J Clin Neurophysiol* **13**:184–197(1996).

55. de Weerd AW, van den Bossche AS. The development of sleep during the first months of life. *Sleep Med Rev* **7**:179–191(2003).

56. Curzi-Dascalova L. EEG de veille et de sommeil du nourisson normal avant 6 mois de'age. *Rev EEG Neurophysiol* **7**:316–326(1977).

57. Ellingson RJ. The EEGs of prematures and full-term newborns. In: Klass DW, Daly DD (Eds), *Current Practice of Clinical Electroencephalography.* Raven Press, New York, 1979, pp 149–177.

58. Louis J, Zhang JX, Revol M, Debilly G, Challamel MJ. Ontogenesis of nocturnal organization of sleep spindles: a longitudinal study during the first 6 months of life. *Electroencephalogr Clin Neurophysiol* **83**:289–296(1992).

59. Schechtman VL, Harper RK, Harper RM. Distribution of slow-wave EEG activity across the night in developing infants. *Sleep* **17**:316–322(1994).

60. Sterman MP, Harper RM, Havens B, Hoppenbrouwers T, McGinty DJ, Hodgman JE. Quantitative analysis of infant EEG development during quiet sleep. *Electroencephalogr Clin Neurophysiol* **43**:371–385(1977).

61. Samson-Dollfus D, Nogues B, Menard JF, Bertoli-Lefever I, Geffroy D. Delta, theta, alpha and beta power spectrum of sleep electroencephalogram in infants aged two to eleven months. *Sleep* **6**:376–383(1983).

62. Louis J, Cannard C, Bastus H, Challamel M. Sleep ontogenesis revisited: a longitudinal 24-hour home polygraphic study on 15 normal infants during the first two years of life. *Sleep* **20**:323–333(1997).

63. Guilleminault C, Souquet M. Sleep states and related pathology. In: Korobkin R, Guilleminault C (Eds), *Advances in Perinatal Neurology.* SP Medical and Scientific Books, New York, 1979, pp 225–247.

64. Van der Knaap MS, Valk J. MR imaging of the various stages of normal myelination during the first year of life. *Neuroradiology* **31**:459–470(1990).

65. Somers VK, Dyken ME, Mark AL, Abboud FM. Sympathetic nerve activity during sleep in normal subjects. *N Engl J Med* **328**:303–307(1995).

66. Baharav A, Kotagal S, Gibbons V, Rubin BK, Pratt G, Karin J, Akselrod S. Fluctuations in autonomic nervous activity during sleep displayed by power spectrum analysis of heart rate variability. *Neurology* **45**:1183–1187(1995).

67. Gaultier C. Cardiorespiratory adaptation during sleep in infants and children. *Pediatr Pulmonol* **19**:105–117(1995).

68. Villa MP, Calcagnini G, Pagani J, Paggi B, Massa F, Ronchetti R. Effects of sleep stage and age on short-term heart rate variability during sleep in healthy infants and children. *Chest* **117**:460–466(2000).

69. Chatow U, Davidson S, Reichman BL, Akselrod S. Development and maturation of the autonomic nervous system in premature and full-term infants using spectral analysis of heart rate fluctuation. *Pediatr Res* **37**:294–302(1995).

70. Scher MS, Sun M, Steppe DA, Banks DL, Guthrie RD, Sclabassi RJ. Comparisons of EEG sleep state-specific spectral values between healthy full-term and preterm infants at comparable postconceptional ages. *Sleep* **17**:47–51(1994).

71. Scher MS, Dokianakis SG, Sun M, Steppe DA, Guthrie RD, Sclabassi RJ. Rectal temperature changes during sleep state transitions in full-term and preterm neonates at postconceptional term ages. *Pediatr Neurol* **10**:191–194(1994).

72. Scher MS, Steppe DA, Dokianakis SG, Sun M, Guthrie RD, Sclabassi RJ. Cardiorespiratory behavior during sleep in full-term and preterm neonates at comparable postconceptional term ages. *Pediatr Res* **36**:738–744(1994).

73. Scher MS, Sun M, Steppe DA, Guthrie RD, Sclabassi RJ. Comparisons of EEG spectral and correlation measures between healthy term and preterm infants. *Pediatr Neurol* **10**:104–108(1994).

74. Scher MS. Neurophysiological assessment of brain function and maturation II. A measure of brain dysmaturity in healthy preterm neonates. *Pediatr Neurol* **16**:287–295(1997).

75. Scher MS, Steppe DA, Salerno DG, Beggarly ME, Banks DL. Temperature differences during sleep between full-term and preterm neonates at matched conceptional ages. *Clin Neurophysiol* **114**:17–22(2003).

76. Scher MS, Richardson GA, Coble PA, Day NL, Stoffer DS. The effects of prenatal alcohol and marijuana exposure: disturbances in neonatal sleep cycling and arousal. *Pediatr Res* **24**:101–105(1988).

77. Hahn JS, Tharp BR. The dysmature EEG pattern in infants with bronchopulmonary dysplasia and its prognostic implications. *Electroencephalogr Clin Neurophysiol* **76**:106–113(1990).

78. Beckwith L, Parmelee AH Jr. EEG patterns of preterm infants, home environment, and later IQ. *Child Dev* **57**:777–789(1986).

79. Agarwal R, Gotman J. Digital tools and polysomnography. *J Clin Neurophysiol* **19**:136–143(2002).

80. Bes F, Baroncini P, Dugovic C, Fagioli I, Schulz H, Franc B, Salzarulo P. Time course of night sleep EEG in the first year of life: a description based on automatic analysis. *Electroencephalogr Clin Neurophysiol* **69**:501–507(1988).

81. Giaquinto S, Marciano F, Monod N, Wolfe G. Applications of statistical equivalence to newborn EEG recordings. *Electroencephalogr Clin Neurophysiol* **42**:406–413(1977).

82. Havlicek V, Chiliaeva R, Chernick V. EEG frequency spectrum characteristics of sleep states in full-term and preterm infants. *Neuropaediatrie* **6**:24–40(1975).

83. Kuks JBM, Vos JE, O'Brien MJ. EEG coherence functions for normal newborns in relation to their sleep state. *Electroencephalogr Clin Neurophysiol* **69**:295–302(1988).

84. Lombroso CT. Quantified electrographic scales on 10 preterm healthy newborns followed up to 40–43 weeks of conceptional age by serial polygraphic recordings. *Electroencephalogr Clin Neurophysiol* **46**:460–474(1979).

85. Willekens H, Oumermuth G, Duc G, Mieth D. EEG spectral powers and coherence analysis in healthy full-term neonates. *Neuropediatrics* **15**:180–190(1984).

86. Ktonas PY, Fagioli I, Salzarulo P. Delta (0.5–1.5 Hz) and sigma (11.5–15.5 Hz) EEG power dynamics throughout quiet sleep in infants. *Electroencephalogr Clin Neurophysiol* **95**:90–96(1995).

87. Witte H, Putsche P, Eiselt M, Hoffman K, Schack B, Arnold M, Jager H. Analysis of the interrelations between a low-frequency and a high-frequency signal component in human neonatal EEG during quiet sleep. *Neurosci Lett* **236**:175–179(1997).

88. Lehtonen J, Kononen M, Purhonen M, Partanen J, Saarikoski S, Launiala K. The effect of nursing on the brain activity of the newborn. *J Pediatr* **132**:646–651(1998).

89. Field T, Diego M, Hernandez-Reif M, Schanberg S, Kuhn C. Relative right versus left frontal EEG in neonates. *Dev Psychobiol* **41**:147–155(2002).

90. Eiselt M, Schindler J, Arnold M, Witte H, Swiener U, Frenzel J. Functional interactions within the newborn brain investigated by adaptive coherence analysis of EEG. *Neurophysiol Clin* **31**:104–113(2001).

91. Sawaguchi H, Ogawa T, Takano T, Sato K. Developmental changes in electroencephalogram for term and preterm infants using an autoregressive model. *Acta Paediatr Jpn* **38**:580–589(1996).

92. Myers MM, Fifer WP, Grose-Fifer J, Sahni R, Stark RI, Schulze KF. A novel quantitative measure of trace-alternant EEG activity and its association with sleep states of preterm infants. *Dev Psychobiol* **31**:167–174(1997).

93. Eiselt M, Schendel M, Witte H, Dorschel J, Curzi-Dascalova L, D'Allest AM, Zwiener U. Quantitative analysis of discontinuous EEG in premature and full-term newborns during quiet sleep. *Electroencephalogr Clin Neurophysiol*, **103**:528–534(1997).

94. Holthausen K, Breidbach O, Scheidt B, Frenzel J. Brain dysmaturity index for automatic detection of high-risk infants. *Pediatr Neurol* **22**:187–191(2000).

95. Schramm D, Scheidt B, Hubler A, Frenzel J, Holthausen K, Breidbach O. Spectral analysis of electroencephalogram during sleep-related apneas in pre-term and term born infants in the first weeks of life. *Clin Neurophysiol* **111**:1788–1791(2000).

96. Kuhle S, Klebermass K, Olischar M, Hulek M, Prusa AR, Kohlhauser C, Birnbacher R, Weninger M. Sleep–wake cycles in preterm infants below 30 weeks of gestational age. Preliminary results of a prospective amplitude-integrated EEG study. *Wien Klin Wochenschr* **113**:219–223(2001).

97. Vanhatalo S, Tallgren P, Andersson S, Sainio K, Voipio J, Kaila K. DC-EEG discloses prominent, very slow activity patterns during sleep in preterm infants. *Clin Neurophysiol* **113**: 1822–1825(2002).

98. Schechtman VL, Harper RK, Harper RM. Aberrant temporal patterning of slow-wave sleep in siblings of SIDS victims. *Electroencephalogr Clin Neurophysiol* **94**:95–102(1995).

99. Gurses D, Kilic I, Sahiner T. Effects of hyperbilirubinemia on cerebrocortical electrical activity in newborns. *Pediatr Res* **52**:125–130(2002).

100. Inder TE, Buckland L, Williams CE, Spencer C, Gunning MI, Darlow BA, Volpe JJ, Gluckman PD. Lowered electroencephalographic spectral edge frequency predicts the presence of cerebral white matter injury in premature infants. *Pediatrics* **111**:27–33(2003).

101. Hellström-Westas L. Comparison between tape recorded amplitude integrated EEG monitoring and sick newborn infants. *Acta Paediatr* **81**:812–819(1992).

102. Burdjalov VF, Baumgart S, Spitzer AR. Cerebral function monitoring: a new scoring system for the evaluation of brain maturation in neonates. *Pediatrics* **112**:855–861(2003).

103. Pan XL, Ogawa T. Microstructure of longitudinal 24 hour electroencephalograms in healthy preterm infants. *Pediatr Int* **41**:18–27(1999).

104. Regalado MG, Schechtman VL, Khoo MC, Bean XD. Spectral analysis of heart rate variability and respiration during sleep in cocaine-exposed neonates. *Clin Physiol* **21**:428–436(2001).

105. Scher MS, Sun M, Steppe DA, Guthrie RD, Sclabassi RJ. Comparisons of EEG spectral and correlation measures between healthy term and preterm infants. *Pediatr Neurol* **10**:104–108(1994).

106. Scher MS, Steppe DA, Banks DL, Guthrie RD, Sclabassi RJ. Maturational trends of EEG-sleep measures in the healthy preterm neonate. *Pediatr Neurol* **12**:314–322(1995).

107. Scher MS. Normal electrographic-polysomnographic patterns in preterm and full-term infants. *Semin Pediatr Neurol* **3**:12(1996).

108. Scher MS, Jones BL, Steppe DA, Cork DL, Seltman HJ, Banks DL. Functional brain maturation in neonates as measured by EEG-sleep analyses. *Clin Neurophysiol* **114**:875–882(2003).

109. Scher MS, Steppe DA, Dokianakis SG, Sun M, Guthrie RD, Sclabassi RJ. Cardiorespiratory behavior during sleep in full-term and preterm neonates at comparable postconceptional term ages. *Pediatr Res* **36**:738–744(1994).

110. Scher MS, Dokianakis SG, Sun M, Steppe DA, Guthrie RD, Sclabassi RJ. Rectal temperature changes during sleep state transitions in full-term and preterm neonates at postconceptional term ages. *Pediatr Neurol* **10**:191–194(1994).

111. Scher MS, Steppe DA, Salerno DG, Beggarly ME, Banks DL. Temperature differences during sleep between full-term and preterm neonates at matched conceptional ages. *Clin Neurophysiol* **114**:17–22(2003).

112. Scher MS, Kelso RS, Turnbull JP, Johnson MW, Loparo KA. Automated arousal detection in neonates. *Sleep* **26**(Suppl):A143(2003).

113. Scher MS, Dokianakis SG, Sun M, Steppe DA, Guthrie RD, Sclabassi RJ. Computer classification of sleep in preterm and full-term neonates at similar postconceptional term ages. *Sleep* **19**:18–25(1996).

114. Turnbull JP, Loparo KA, Johnson MW, Scher MS. Automated detection of tracé alternant during sleep in healthy full term neonates using discrete wavelet transform. *Clin Neurophysiol* **112**:1893–1900(2001).

115. Turnbull JP, Johnson MW, Loparo KA, Scher MS. Nonlinear dynamical system analyses of neonatal sleep state. *Sleep* **26**(Suppl):A404(2003).

116. Scher MS, Dokianakis SG, Steppe DA, Banks DL, Sclabassi RJ. Computer classification of state in healthy preterm neonates. *Sleep* **20**:132–141(1997).

117. Scher MS. Neurophysiological assessment of brain function and maturation. I. A measure of brain adaptation in high risk infants. *Pediatr Neurol* **16**:191–198(1997).

118. Scher MS. Neurophysiological assessment of brain function and maturation. II. A measure of brain dysmaturity in healthy preterm neonates. *Pediatr Neurol* **16**:287–295(1997).

119. Scher MS, Alvin J, Painter MJ. Uncoupling of clinical and EEG seizures after antiepileptic drug use in neonates. *Pediatr Neurol* **28**:277–280(2003).

120. Harmony T. Neurometric assessment of brain dysfunction in neurological patients. *Funct Neurosci* **3**:338–375(1984).

121. Bell MA, Fox NA. The relations between frontal brain electrical activity and cognitive development during infancy. *Child Dev* **63**:1142–1163(1992).

122. Harper RM, Leake B, Miyahara L, Hoppenbrouwers T, Sterman MB, Hodgman J. Development of ultradian periodicity and coalescence at 1 cycle per hour in electroencephalographic activity. *Exp Neurol* **73**:127–143(1981).

123. Woodruff DS. Brain electrical activity and behavior relationships over the life span. *Life Span Dev Behav* **1**:111–179(1979).

124. Dawson G, Panagiotides H, Grofer-Klinger LG, Hill D. The role of frontal lobe functioning in the development of infant self-regulatory behavior. *Brain Cogn* **20**:152–175(1992).

125. Marshall PJ, Bar-Haim Y, Fox NA. Development of the EEG from 5 months to 4 years of age. *Clin Neurophysiol* **113**:1199–1208(2002).

126. Thatcher RW. Maturation of the human frontal lobes: physiological evidence for staging. *Dev Neuropsychol* **7**:397–419(1991).

127. Hauser E, Seidl R, Rohrbach D, Hartl I, Marx M, Wimmer M. *Electroencephalogr Clin Neurophysiol* **87**:284–290(1993).

128. Shibagaki M, Kiyono S, Takeuchi T. Nocturnal sleep in mentally retarded infants with cerebral palsy. *Electroencephalogr Clin Neurophysiol* **61**:465–471(1985).

129. Shibagaki M, Kiyono S. Cyclic variation of integrated delta components during nocturnal sleep in mentally retarded

children. *Electroencephalogr Clin Neurophysiol* **56**:190–193(1983).

130. Bredesen DE. Neural apoptosis. *Ann Neurol* **38**:839–851(1995).

131. Hughes PE, Alexi T, Walton M, et al. Activity and injury-dependent expression of inducible transcription factors, growth factors and apoptosis-related genes within the central nervous system. *Prog Neurobiol* **57**:421–450(1999).

132. Goldman-Rakic PS. Development of cortical circuitry and cognitive function. *Child Dev* **58**:601–622(1987).

133. Caviness VS Jr. Normal development of cerebral neocortex. In: Evrard P, Minkowski A (Eds), *Developmental Neurobiology*. Raven Press, New York, 1989, pp 1–10.

134. Arabanel ADI, Rabinovich MI. Neurodynamics: nonlinear dynamics and neurobiology. *Curr Opin Neurobiol* **11**:423–430(2001).

64

SLEEP IN INFANTS AND CHILDREN

STEPHEN H. SHELDON

Children's Memorial Hospital, Chicago, Illinois

NEWBORNS, INFANTS, AND CHILDREN

Sleep occupies a major portion of the lives of newborns, infants, and children. A newborn infant typically sleeps about 70% of every 24 hours. In contrast, adults spend 25–30% of their lives sleeping. Because sleep takes up such a significant portion of a child's life, a major building block of childhood is likely sleep.

Sleep in normal infants varies significantly from normal sleep in adults. Premature infants frequently reveal a lack of concordance between electrophysiological parameters of sleep and behavioral variables. This may also be true in some term infants [1, 2]. Controversy has existed when defining sleep and waking states in infants since most accepted criteria cannot be fulfilled. Solutions have been suggested by a number of investigators. Prechtl and Beintema [3] suggested a sleep and waking state definition based on observable behaviors; Anders, Emde, and Parmelee [4] have suggested utilization of behavioral and polygraphic features; and Hoppenbrouwers [5] suggested state definition based on polygraphic features, with observational criteria used only as supplemental information. Sleep in infants and children is significantly different from sleep in adults and may serve different functions in the developing infant and child.

SLEEP IN THE PREMATURE INFANT

Periods of activity and quiescence can be identified in the human fetus by 28–32 weeks gestation [1]. However, clearly definable sleep states cannot be identified in premature neonates between 24 and 26 weeks gestation [6]. By 28–30 weeks postconception, active sleep can be identified by the presence of eye movements, body activity, and irregular respiration. Chin muscle hypotonia is very difficult to evaluate in the fetus and premature infant since there are so few periods of tonic activity before 36 weeks gestation [1]. Quiet sleep cannot be clearly identified at this time and active sleep comprises the vast majority of time the premature infant is asleep. Quiet sleep can be identified by the development of a tracé discontineau EEG pattern at about 32 weeks postconception and can clearly be identified by a tracé alternant EEG pattern at approximately 36 weeks gestation. Once quiet sleep can be identified, this state increases steadily and becomes the dominant state (equivalent to NREM sleep) at approximately 3 months of postnatal life.

Behaviorally, fetal movements can first be identified between 10 and 16 weeks gestation. Rhythmic cycling of activity can be recorded by 20 weeks. At 28–30 weeks, very brief quiet periods begin to appear. By 32 weeks postconception, body movements are absent in 53% of 20 second epochs during 2–3 hour sleep recordings [1]. The number of "no movement epochs" increases to 60% at term.

Maturational patterns can be demonstrated in EEG recordings of premature infants as early as 24 weeks gestation. Conflicting evidence exists concerning the independence of the maturation of sleep and the EEG with respect to intrauterine stage. Very young premature infants and full-term neonates have similar EEG patterns when compared at the same conceptual age. On the other hand, it has been shown that when the premature infant has

Sleep: A Comprehensive Handbook, Edited by T. Lee-Chiong.
Copyright © 2006 John Wiley & Sons, Inc.

reached 40 weeks postconception age, the infant still may not have attained a degree of EEG organization as significant as that of the full-term newborn [2]. Premature infants show spindle development that is about 4 weeks in advance of that seen in full-term infants and a statistical difference between the length of quiet sleep in the term and premature infant has been demonstrated, when measured at the same postconception age [7]. However, extrauterine development of the premature infant occurs in a somewhat artificial environment. Significant medical problems often coexist and often frequent medical interventions are required. Effects of constant light and medical treatment regimens on the development of sleep cycling and the developing central nervous system have not yet been definitely described.

TERM INFANTS: BIRTH TO TWELVE MONTHS

Three distinct sleep states can be identified in the term newborn: active sleep (REM), quiet sleep (NREM), and indeterminate sleep. Indeterminate sleep is defined as a state in which criteria for neither REM nor NREM can be identified [4]. Sucking movements that occur during wakefulness can continue during sleep and are common during active sleep.

Fine twitches, grimaces, facial movements, and occasional tremors characterize active sleep. Intermittent large limb movements, stretching, and vocalizations occur. Bursts of phasic muscle activity and respiratory irregularity are present and can occur in conjunction with phasic eye movements.

Quiet sleep is characterized by minimal movements. Muscle tone is somewhat decreased from waking levels but is increased above the level seen during active sleep.

During the first 3 months of life, substantial changes occur. Ten to 12 weeks of age appears to be a critical period of reorganization, when infantile sleep behavior and physiology matures. Sleep–wake patterns change as well. At birth, total sleep time is about 16–17 hours. Total sleep time gradually decreases, reaching 14–15 hours by 16 weeks of age, and 13–14 hours by 6–8 months.

Development of attentive behaviors during wakefulness occurs concomitantly with the development of quiet sleep and sustained sleep patterns. These changes suggest continued development of inhibitory and controlling feedback mechanisms secondary to the increasing complexity of neural networks and neurochemical maturation. By 3 months, a relatively stable distribution of sleep and wake occurs across the 24 hour day. There is a remarkably regular alternation of active and quiet sleep. Periodic respiration, common until 3 weeks of age, becomes uncommon after the first 2 months of life [8].

During the first 6 months of life, consolidation of sleep during nocturnal hours occurs. Major changes seen are in the duration of sleep periods and when they occur in the 24 hour day. At 3 weeks of age, the average length of the longest sleep period is about 3.5 hours. By 6 months of age, the longest sleep period averages about 6 consecutive hours [9].

Between 3 and 6 weeks, sleep periods lengthen considerably, and by 6 weeks of age, the longest sleep period is no longer randomly distributed throughout the day, but occurs during nocturnal hours. At 3 months, the pattern has become more consistent so that by 12–16 weeks of age the longest sleep period occurs during nighttime and the longest wake period occurs during daytime [7]. At 6 months of age, the long sleep period immediately follows the longest wake period [9].

After about 3 months of age, there is continuing development and daytime sleep becomes consolidated into discrete daytime naps.

Brief awakenings from sleep are more frequent during the first 2 months of life, than at older ages [5]. In addition, infants 1–2 months of age are more likely to awaken from active sleep than from quiet sleep. Bowe and Anders reported that this variable helped discriminate between infant sleep at 2 and 9 months of age [10].

Significant maturation can be seen in the EEG during this period of development. A *tracé alternant* pattern of quiet sleep can first be identified at 32–34 weeks gestation. This pattern is most often well developed by 36–38 weeks. The tracé alternant pattern gradually disappears over the first month of life and is replaced by continuous high-voltage slow-wave activity.

Sleep spindles appear at about 4–8 weeks of age. The shape of these spindles changes impressively early in development. At 2 months of age, sleep spindles are difficult to differentiate from background EEG activity. When first present, spindles may be quite long and last 2–4 seconds. Duration decreases continuously to about an average of 0.5–1.0 second by the end of the second year. Spindle intervals become greater with increasing age.

True slow-wave activity appears at approximately 8–12 weeks of age [9] and by 16–24 weeks of age, quiet sleep becomes differentiated into more mature and distinct NREM sleep states. By 3 months of age, quiet sleep is almost twice that of active sleep [7], and by 8 months of age, active sleep occupies about 30% of the total sleep time. Adult percentages are reached between 3 and 5 years of age [5, 11].

Sleep onset is characteristically through REM sleep in the newborn infant, meaning the first REM period typically occurs within the first 15 minutes after sleep onset. At 3 weeks of age, an infant is likely to have two-thirds of sleep periods beginning with REM sleep [9]. Infants less than 3 months of age reveal REM latencies that are predominantly less than 8 minutes in length [12]. During the first 12 weeks of life, latency from sleep onset to the first REM period gradually changes until sleep onset occurs predominantly through NREM sleep. By 6 months, the percentage of sleep episodes beginning with REM sleep is approximately 18% [9].

The ratio of active sleep to quiet sleep is sometimes considered an indicator of maturation [5]. Active sleep time exceeds quiet sleep time during the first months of life. A reversal of this relation is noted in 60% of infants at 3 months and 90% of infants at 6 months of age.

Specific changes in REM sleep percentage occurs during this period of development. During the first 6 months of life there is a marked reduction in the total REM sleep time. This represents a redistribution of sleep stages, since only a relatively mild decrease in the total sleep time occurs during the first year. This change may be an important indicator of central nervous system maturation [7].

TWO YEARS TO FIVE YEARS

In contrast to the rapid evolution that takes place during the first year of life, changes in sleep structure during this period are more gradual. Growth and development are steady. Sleep becomes consolidated into a long nocturnal period of approximately 10 hours. During the first 2–3 years, daytime sleep occurs in somewhat short daytime naps. Morning naps are typically given up first, and by the end of the fifth year, sleep is generally consolidated into a single nocturnal period.

During the latter half of the first year of life, REM sleep averages about 30–35% of the total sleep time. REM sleep and NREM sleep are evenly distributed across the nocturnal sleep period. Small and large body movements associated with REM sleep during infancy become less frequent. REM periods are of approximately uniform length. As the child continues to develop, a gradual change is seen in the uniformity and duration of these REM periods. The first REM period of the night becomes shorter, while succeeding periods become longer and associated with more intense phasic activity. There is also a slight lengthening of the overall cycle length. Two to three year old children still show a cycle length of about 60 minutes, with the first REM period occurring 1 hour after sleep onset. By 4–5 years of age, the cycle lengthens gradually to 60–90 minutes.

Between 3 and 5 years of age, REM percentage gradually decreases from 30–35% of the total sleep time to an adult level of 20–25%. There appears to be a close relationship between these changes and the augmented periods of wakefulness during the daytime.

Typically, children in this age range have approximately seven cycles during each nocturnal sleep period [13]. Sleep-onset latency averages about 15 minutes in the younger children but lengthens to between 15 and 30 minutes in the older age groups. Slow-wave sleep predominantly

occurs during the first third of the night and as much as 2 hours may be spent in slow-wave sleep. EEG voltage is also very high during this period. Stage 2 first appears from 3 to 4 minutes after the child falls asleep; stage 3 appears about 11 minutes after sleep onset, and stage 4 first appears about 4 minutes later [13].

FIVE TO TEN YEARS

Growth and development continue to be constant and gradual during middle childhood. Searching, exploration, and increasingly mature thinking behaviorally characterize this time. It is a period of trial and error.

Sleep continues to develop into a more mature pattern. Although sleep patterns of children during middle childhood resemble those of older individuals, there is considerable individual variability. There is a certain stability of the pattern for given individuals and a fairly consistent amount of time spent in each sleep stage and the number of sleep stages from night to night [13]. When compared with adult sleep patterns, total sleep time in middle childhood is approximately 2.5 hours longer with unequal distribution of the added time to each of the sleep stages. Stages in children of this age group tend to be longer in duration than in adults, but the sleep architecture seems to be as stable.

Though body movements during sleep decrease in frequency, they are generally more often seen in this age group than in adolescents and young adults. Stage 4 volume decreases from approximately 2 hours in the preschool child to 75–80 minutes in the latter portion of middle childhood. There does appear, however, to be a gender-related difference in slow-wave sleep. Males tend to exhibit a significantly greater volume of slow-wave sleep than females of comparable age [14].

Naps during this period of development are unusual. Consistent daytime napping during middle childhood may represent a pathological process. Prepubescent children are generally very alert throughout the entire day. Mean sleep-onset latencies of preadolescent (Tanner stage 1) children are greater than 15 minutes, which is an extremely alert and vigilant level.

REFERENCES

1. Parmelee AH, Stern E. Development of states in infants. In: Clemente CD, Purpura DP, Mayer FE (Eds), *Sleep and the Maturing Nervous System*. Academic Press, New York, 1972.

2. Dreyfus-Brisac C. Ontogenesis of sleep in human prematures after 32 weeks of conceptual age. *Dev Psychobiol* **3**:91(1970).

3. Prechtl HFR, Beintema D. The neurological examination of the full term newborn infant. In: *Clinics in Developmental Medicine, Vol 12*. Spastics Society and Heinemann, London, 1964.

4. Anders T, Emde R, Parmelee A (Eds). *A Manual of Standardized Terminology, Techniques and Criteria for Scoring of States of Sleep and Wakefulness in Newborn Infants*. UCLA Brain Information Service, NINDS Neurological Information Network, Los Angeles, 1971.

5. Hoppenbrouwers T. Sleep in infants. In: Guilleminault C (Ed), *Sleep and Its Disorders in Children*. Raven Press, New York, 1987.

6. Dreyfus-Brisac C. Sleep ontogenesis in early human prematurity from 24 to 27 weeks of conceptual age. *Dev Psychobiol* **1**:62(1968).

7. Stern E, et al. Sleep cycle characteristics in infants. *Pediatrics* **43**:65(1969).

8. Metcalf D. The ontogenesis of sleep–awake states from birth to 3 months. *Electroencephalogr Clin Neurophysiol* **28**:421(1979).

9. Coons S. Development of sleep and wakefulness during the first 6 months of life. In: Guilleminault C (Ed), *Sleep and Its Disorders in Children*. Raven Press, New York, 1987.

10. Bowe TR, Anders TF. The use of semi-Markof model in the study of the development of sleep–wake states in infants. *Psychophysiology* **16**:41(1979).

11. Anders TF, Keener M. Developmental course of nighttime sleep–wake patterns in full term and premature infants during the first years of life. *Sleep* **8**:173(1985).

12. Schulz H, et al. REM latency: development in the first year of life. *Electroencephalogr Clin Neurophysiol* **56**:316(1983).

13. Ross JJ, et al. Sleep patterns in pre-adolescent children: an EEG–EOG study. *Pediatrics* **42**:324(1968).

14. Coble PA, et al. EEG sleep of normal healthy children. Part I. Findings using standard measurement methods. *Sleep* **7**:289(1984).

65

SLEEP AND BREATHING DURING EARLY POSTNATAL LIFE

AVIV D. GOLDBART, RIVA TAUMAN, AND DAVID GOZAL
University of Louisville, Louisville, Kentucky

During fetal life, breathing is discontinuous and coincides with REM-like sleep (active sleep). After birth, respiratory rhythm is established as a continuous activity to maintain cellular oxygen and carbon dioxide homeostasis. Instability of the breathing pattern is an inherent characteristic of the normal healthy infant during sleep. Dramatic and profound changes in respiration during sleep occur with maturation of the neurophysiological, metabolic, and mechanical components of the respiratory system [1]. The effect of REM sleep on the various mechanisms involved in respiratory control are of particular importance during infancy, considering that REM sleep occupies more than 50% of total sleep time in newborns and this percentage is even higher in preterm newborns, thereby explaining the unique vulnerability of cardiorespiratory functions during this particular sleep state. As REM sleep percentage decreases to approximately 25% of total sleep time at 6 months of age [2, 3], the susceptibility of respiratory functions to waking–sleep state transitions progressively abates. In infants, sleep states are traditionally divided into quiet (NREM), active (REM), transitional, and indeterminate sleep states. Quiet sleep is usually characterized by the absence of REM coupled by the occurrence of tracé alternant. REM sleep is associated with rapid eye movements and continuous, irregular low voltage on the EEG, while transitional sleep reflects short (1–3 minutes) epochs interspersed between quiet and

REM sleep. Indeterminate sleep is a state that cannot be described by any other definition.

Several developmental differences between the infant and the adult respiratory systems may further account for the emergence of sleep-associated disruption of normal gas exchange in early postnatal life, and include the following:

- Infants have greater difficulty switching from nasal to oral route of breathing, and the majority of younger infants can be considered as obligatory or almost obligatory nasal breathers.

- In infants, reflexes originating in the upper airway (laryngeal chemoreflex) can induce profound apnea and bradycardia. This respiratory depressant component of the laryngeal chemoreflex decreases with maturation, and prolonged apnea is more prominent in preterm than in full-term infants [3].

- The chest wall compliance is increased in infants, thereby requiring dynamic, rather than passive, maintenance of functional residual capacity of the lungs. Furthermore, infants have "barrel shaped" rib cages and the rib cage contribution to tidal breathing is smaller compared to older children and adults.

- Paradoxical breathing especially during REM sleep is common in newborns. The duration of paradoxical breathing during sleep decreases as postnatal age increases. Paradoxical breathing is rare or absent after 3 years of age [4].

Correspondence to: David Gozal, 570 S. Preston Street, Suite #204, Louisville, KY 40202.

Sleep: A Comprehensive Handbook, Edited by T. Lee-Chiong.

- The respiratory rate is high in the neonatal period and decreases during infancy and early childhood. Respiratory rate decreases exponentially with increasing body weight. Respiratory rate is higher during REM sleep compared to NREM sleep.

- Apneas of short duration (less than 10 seconds) are common during the early period of life. Apneas are more frequent in REM sleep than NREM sleep. Apneas are mostly central and will decrease in number with advancing postnatal age. Obstructive and mixed apneas are more frequently seen in preterm than in full-term infants, possibly reflecting developmental changes in pharyngeal, laryngeal, and central airway collapsibility.

- Periodic breathing, defined as three episodes of apnea lasting longer than 3 seconds and separated by continued respiration over a period of 20 seconds or less, is a common respiratory pattern in preterm infants and may also be highly prevalent in full-term infants [5, 6]. However, periodic breathing decreases in frequency during the first year of life and is usually not considered to be of any specific pathological significance. Nevertheless, environmental factors such as sleep state transitions, arousals, hypoxia, and hyperthermia can all increase periodic breathing in infants and may lead to destabilization of cardiorespiratory homeostasis.

- Apneas during sleep in infants are associated with a fall in heart rate particularly during NREM sleep. The presence of hypoxemia will enhance this reflex bradycardia during apnea.

- Arousal from sleep is thought to be a major determinant for termination of apnea, and therefore arousal deficits have been implicated in the pathophysiology of sudden infant death syndrome. However, fewer than 10% of apneic episodes will be terminated by a full-fledged electroencephalographic arousal in infants. Autonomic arousals are nevertheless quite frequent during the period surrounding the termination of an apneic event. Moreover, while hypercapnia is a potent stimulus of arousal, hypoxia and particularly rapidly developing hypoxia are much less effective in inducing arousal. Furthermore, prone position, sleep deprivation, and prenatal/postnatal exposure to cigarette smoking are accompanied by decreased arousability in infants.

- Healthy full-term infants will usually maintain oxyhemoglobin saturation values of 92–100% during sleep in the first 4 weeks of life. Blood oxygen levels are lowest during the first week of life and increase over the next 1–3 months, such that all infants will have values of 97–100% after the age of 2 months. Basal values for arterial carbon dioxide tension during sleep are generally between 36 and 42 mmHg in newborn and infants.

While instability of respiratory patterning is a physiological characteristic of the normal healthy infant during sleep, increased awareness of factors that can suddenly convert a physiological phenomenon into a pathological condition is obviously important. Examples of predisposing factors that can precipitate apnea in infants include hyperthermia, viral infection and more specifically the one caused by respiratory syncytial virus, gastroesophageal reflux, seizures, anatomically narrow upper airway structures, and parental cigarette smoking. Identification of such risk factors is crucial for prevention, early intervention, and treatment. Furthermore, when sleep disordered breathing (SDB) is suspected in infants, polysomnography should be performed since early diagnosis and treatment will prevent the occurrence of morbid consequences associated with SDB such as poor somatic growth, neurocognitive and developmental deficits, cardiorespiratory failure, or even sudden and unexpected death.

APNEA OF PREMATURITY (AOP)

As indicated previously, the prevalence of apnea relates to the postconceptional maturity of the infant and is thus commonly seen in premature infants [7]. Although the definition of apnea may vary, it is commonly defined as cessation of airflow for 20 seconds or greater, or shorter events that are followed by bradycardia and/or desaturation. While apnea may occur in any sleep stage, it is more likely to occur during REM sleep. The premature central nervous system is characterized by reduced numbers of synaptic connections that also tend to be more inhibitory than excitatory, thereby decreasing excitatory drive to the brainstem cardiorespiratory centers and spinal cord motor neurons. The elastic recoil of the chest wall is also reduced due to lack of bone mineralization, leading to an increased recoil and higher propensity for lung volume loss. During REM sleep, the diaphragm is recruited more intensely than during waking, and it has been postulated to lead to a diaphragmatic pattern "fatigue" in preterm infants, as characterized from histological and biochemical methods as well as from electromyographic recordings. Upper airway obstruction may also be an important risk factor for apnea of prematurity, since apnea frequency and severity decrease with the use of continuous positive airway pressure (CPAP). Furthermore, upper airway resistance is increased during REM sleep due to loss of the phasic tone in the adductor muscles of the pharynx. Under these conditions, the Hering–Breuer reflex (sustained lung distention leads to a decrease in inspiratory drive, which is much stronger in infants) can lead to apnea, such that the overall breathing pattern in preterm infants will manifest as enhanced breath-to-breath variability with alternating periods of breathing and apneic intervals. Sighs (large augmented breaths) are another mechanism utilized by premature babies to recruit lung volume, and thus prevent the occurrence of hypoxemia and apnea, especially during REM sleep. Neonates exhibit a biphasic response to hypoxia that starts with an initial increase in ventilation lasting 1–2 minutes that is secondary to peripheral chemoreceptor stimulation and is followed by a decline often to below baseline ventilation, most likely resulting from a central depression of respiratory centers as well as from metabolic changes induced by hypoxia.

One of the methods commonly used to distinguish between the types of apneic events consists of the assessment of the cardiac oscillation waveform in the respiratory flow trace [8]. In central apnea, the oscillations are generally present and will be absent during obstructive events. However, in many central apneic events there is progressive loss of upper airway patency over time such that upper airway obstruction may ensue.

Of note, preterm infants have significantly more frequent apneic pauses and periodic breathing when they reach term postconceptional age compared to infants who were born at term. Furthermore, extremely preterm infants (24–28 weeks) who experience apneic and bradycardic episodes will manifest persistence of such events for longer periods of time than symptomatic infants born at term. In preterm infants, 83% of bradycardic episodes are accompanied by an apnea and 86% of bradycardic events are associated with desaturation. Thus maintaining a higher baseline Sao_2 may help prevent bradycardia.

The mainstay therapy for AOP in addition to supplemental oxygen consists of administration of methylxanthines, particularly caffeine, due to its high absorbability and prolonged half-life allowing for once daily dosage. These compounds are believed to be effective through their central effects as competitive adenosine receptor antagonists. Other therapy modalities for AOP, such as creatine and carnitine, failed to show any changes in the course of the condition. While AOP will usually resolve by 36–40 weeks postconceptional age, it can persist in a subset of infants, and the latter may benefit from home cardiorespiratory monitoring. Periodic assessments of these vulnerable infants and determination of the optimal timing for discontinuation of the home monitoring device will usually require an individualized approach aiming at reducing the duration of monitor use while maintaining the safety of the infant.

OBSTRUCTIVE SLEEP APNEA SYNDROME (OSAS) IN INFANTS

As mentioned earlier, neonates have greater difficulty switching respiration from the nasal to oral route and may therefore develop upper airway obstruction whenever a mild nasal congestion is present. Alternatively, infants born with a relatively narrow upper airway are also at increased risk of obstructive sleep apneas during sleep. Indeed, OSAS in infancy is usually associated with anatomic abnormalities that reduce the patency of the upper airways [9]. Causes of narrow upper airways in infants include craniofacial abnormalities, such as micrognathia, laryngomalacia, cleft palate, choanal atresia, Pierre Robin syndrome, mucopolysaccharide storage disease, hypothyroidism, Down syndrome, and spinal cord compression particularly in the presence of an Arnold–Chiari malformation. Infants with neuromuscular disorders are also at risk of obstructive apnea [10]. Spontaneous neck flexion was also found to cause upper airway obstruction in premature infants.

Of interest, while much less frequent in the infant compared to older toddlers or school-aged children, OSAS due to adenotonsillar hypertrophy does also occur in infants. The characteristics of OSAS in this age group include the presence of male predominance, high incidence of preterm infants, failure to gain weight, and a high recurrence rate after surgery [10].

Several reports deal with the association between OSAS and ALTE/SIDS during early infancy [11]. Studies have documented evidence of more frequent obstructive apneas in future SIDS victims, and frequent clinical symptoms of obstruction among first-degree relatives of SIDS or ALTE (apparent life-threatening event) cases. Indeed, an ALTE may be an indication of a sleep disordered breathing syndrome and infants presenting with an ALTE should have clinical evaluation of craniofacial features and upper airway patency.

The clinical symptoms of sleep disordered breathing in infants are laborious breathing, noisy breathing (not limited to snoring), night sweating, and poor growth that can lead in some cases to failure to thrive. When the diagnosis of sleep disordered breathing (SDB) is suspected in infants, overnight polysomnography should be performed in optimal environmental conditions, including natural sleep without sedation. Early diagnosis and treatment of this overall quite frequent entity will protect these young patients from its potential long-term consequences, namely, poor growth, neurocognitive and developmental deficits, cardiorespiratory failure, or even unexpected death.

GASTROESOPHAGEAL REFLUX AND SLEEP IN INFANCY

Gastroesophageal reflux (GER) is another frequent risk factor that has been associated with obstructive apnea during infancy. GER is common in both term and premature infants, but its relationship to apnea is more frequently reported in the preterm infant. GER may provoke respiratory events in infants. These events may occur solely during sleep and manifest as SDB, and thus SDB can be the sole clinical manifestation of underlying GER, presenting as apparent life-threatening events (ALTEs). However, the diagnosis of SDB secondary to GER is not always clear.

Furthermore, the causal relationship between apnea and GER is both complex and conflictive. Although studies have shown a temporal association between GER and apnea, other studies have failed to uncover this temporal association. Apnea may result from potent activation of the laryngeal chemoreflex that will in turn induce respiratory pauses, airway closure, and swallowing immediately after regurgitation to the upper airway. On the other hand, obstructive apnea per se may induce GER due to the increased negative thoracic pressures that will allow for opening of the lower esophageal sphincter and reflux of stomach contents to the pharynx. Thus diagnosis of SDB secondary to GER is not simple and can easily be missed during polygraphic recordings. On the other hand, even the presence of GER as shown by esophageal pH monitoring or other techniques in an infant with polygraphically proven SDB does not necessarily demonstrate that GER is the cause of the respiratory disturbance during sleep. Hence simultaneous recording of sleep measures and of esophageal pH is usually necessary to enable detection of tentative associations between GER and SDB and may allow for formulation of more effective management in these infants.

Of note, hypoventilation, oxygen desaturation, and frank apnea and bradycardia have also been documented during nutritive sucking and ascribed to the immaturity of the normal mechanisms that coordinate breathing, sucking, and swallowing. In the normal healthy infant, as fluid enters the pharynx or larynx, breathing ceases. This normal response protects the airway and prevents aspiration. However, this protective reflex is excessive in some infants due to the immaturity of the nervous system, resulting in prolonged apnea, and this mechanism can be enhanced by respiratory syncytial virus infection. With advancing maturation, feeding-related apneas become less frequent and eventually disappear.

SLEEP AND BPD

Bronchopulmonary dysplasia (BPD), also termed chronic lung disease of infancy, arises in infants and young children as the result of neonatal lung injury and its treatment [12]. The term BPD reflects the need for supplemental oxygen at 28 days of life in a baby with a birth weight below 1500 grams. The diagnosis is currently based on radiological, clinical, and pathological criteria, and while surfactant therapy has led to substantial reductions in BPD, the increasing survival rates of extreme premature babies has led to a resurgence of this problem.

Earlier studies have shown that infants with BPD will develop episodic hypoxemia when asleep even if their awake Sao_2 is within the normal range. These episodes are usually more frequent and severe during REM, thereby reflecting the unique mechanical disadvantage afforded by this sleep state in the context of underlying respiratory chest wall and parenchymal abnormalities. In addition, these infants are uniquely vulnerable to hypoxemia due to the presence of abnormal peripheral chemoreceptor function.

Oxygen supplementation via nasal cannula is considered to be beneficial in BPD infants, improves central respiratory drive, increases total sleep duration, and promotes somatic growth, particularly when Sao_2 is maintained above 92% [12]. In those cases in which O_2 supplementation is insufficient to ameliorate gas exchange abnormalities during sleep, nocturnal ventilatory support will be needed and should increase awareness to the potential presence of upper airway obstruction during sleep, resulting from either upper airway anatomical issues or from tracheomalacia.

SLEEP AND CONGENITAL CENTRAL HYPOVENTILATION SYNDROME

Congenital central hypoventilation syndrome (CCHS) is a relatively rare autosomal recessive condition associated with mutations in genes that regulate the embryonic development of the autonomic nervous system, particularly the *PHOX2B* gene. In its classic form, CCHS is characterized by adequate ventilation during wakefulness with onset of alveolar hypoventilation during sleep, particularly during NREM sleep. CCHS is suspected in the absence of primary neuromuscular, lung, or cardiac disease, or an identifiable brainstem lesion [13]. The diagnosis is further confirmed by demonstration of absent or near absent ventilatory response to hypercapnic challenges and the presence of a mutation in *PHOX2B* gene or more rarely in *RET*, *HASH1*, or *GDNF* genes. In infants, tracheostomy and mechanical ventilatory support is the preferred approach; however, consideration of other management modalities such as noninvasive ventilation (positive or negative pressure) and of phrenic nerve pacing may be necessary under special circumstances. Regardless of the method of ventilatory support, the major goals are to optimize oxygenation ($SaO_2 > 95\%$) and alveolar ventilation (end-tidal carbon dioxide in the 30–45 mmHg range). A comprehensive medical and surgical management plan with periodic reevaluation is usually required due to the multisystem involvement in this condition and the complexity of the care required for optimal outcomes [14].

ACKNOWLEDGMENTS

DG is supported by National Institute of Health grants; RT is the recipient of a Ohio Valley American Heart Association Fellowship award.

REFERENCES

1. Gaultier C. Cardiorespiratory adaptation during sleep in infants and children. *Pediatr Pulmonol* **19**:105–117(1995).

2. Curzi-Dascalova L, Peirano P, Morel-Kahn F. Development of sleep states in normal premature and full term newborns. *Dev Psychobiol* **2**:431–444(1988).

3. Dreyfus-Brisac C. Ontogenesis of brain bioelectrical activity and sleep organization in neonates and infants. In: Falkner F, Tanner JM (Eds), *Human Growth*, Vol 3. Plenum, London, 1979, pp 157.

4. Fisher JT, SantAmbrogio G. Airway and lung receptors and their reflex effects in the newborn. *Pediatr Pulmonol* **1**:112–126(1985).

5. Gaultier C, Praud JP, Canet E, Delaperche MF, DAllest AM. Paradoxical inward rib cage motion during rapid eye movement sleep in infants and young children. *J Dev Physiol* **9**:391–397(1987).

6. Kelly DH, Stellwagen LM, Kaitz E, Shannon DC. Apnea and periodic breathing in normal full term infants during the first twelve months. *Pediatr Pulmonol* **1**:215–219(1985).

7. Martin RJ, Abu-Shaweesh JM, Baird TM. Apnea of prematurity. *Paediatr Respir Rev* **4**:S375–S380(2004).

8. Lemke RP, Al-Saedi SA, Alvaro RE, Wiseman NE, Cates DB, Kwiatkowski K, Rigatto H. Use of a magnified cardiac airflow oscillation to classify neonatal apnea. *Am J Respir Crit Care Med* **154**:1537–1542(1996).

9. Gaultier C. Sleep apnea in infants. *Sleep Med Rev* **3**:303–312(1999).

10. Greenfeld M, Tauman R, DeRowe A, Sivan Y. Obstructive sleep apnea syndrome due to adenotonsillar hypertrophy in infants. *Int J Pediatr Otolaryngol* **67**:1055–1060(2003).

11. Kahn A, Blum D, Rebuffat E, Sottiaux M, Levitt J, Bochner A, Alexander M, Grosswasser J, Muller MF. Polysomnographic studies of infants who subsequently died of sudden infant death syndrome. *Pediatrics* **82**:721–727(1988).

12. Allen J, Gaultier C, Greenoough A, Klijanowicz A, Martinez F, Ozdemir A, Panitch HB, Phelps D, Nickerson BG, Stein MT, Tomezsko J, Van den Anker J. Statement on the care of the child with chronic lung disease of infancy and childhood. *Am J Respir Crit Care Med* **168**:356–396(2003).

13. Gozal D. Congenital central hypoventilation syndrome: an update. *Pediatr Pulmonol* **26**:273–282(1998).

14. Vanderlaan M, Holbrook CR, Wang M, Tuell A, Gozal D. Epidemiologic survey of 196 patients with congenital central hypoventilation syndrome. *Pediatr Pulmonol* **37**:217–229(2004).

66

CONGENITAL SYNDROMES AFFECTING RESPIRATORY CONTROL DURING SLEEP

Manisha B. Witmans, Maida Lynn Chen, Sally L. Davidson Ward, and Thomas G. Keens
Childrens Hospital Los Angeles, Los Angeles, California

INTRODUCTION

In 1962, Severinghaus and Mitchell [1] published the first clinical description of patients who were not able to breathe effectively during wake and/or sleep secondary to respiratory control abnormalities. Our understanding of respiratory control has increased significantly in the past four decades. Advances in molecular genetics have enabled scientists to study the genetic basis of some of the disorders of respiratory control. Syndromes that affect respiratory control during sleep, although uncommon, are now more frequently diagnosed. These disorders of primary respiratory control range in frequency, severity, and clinical presentation. Advances in technology have enabled prompt diagnosis and comprehensive treatment to enable this vulnerable group of individuals to survive and have a better quality of life. Disorders and syndromes associated with abnormal respiratory control including congenital central hypoventilation syndrome, familial dysautonomia, myelomeningocele, and Prader–Willi syndrome will be reviewed.

CONGENITAL CENTRAL HYPOVENTILATION SYNDROME

Congenital central hypoventilation syndrome (CCHS) is a potentially life-threatening disorder, characterized by the failure of autonomic control of breathing, present from birth in affected individuals [2–5]. The exact prevalence of CCHS is not known, but it is presumed to be rare. Individuals with CCHS often have autonomic dysfunction of other organ systems in addition to impaired ventilatory control. The severity of disordered ventilatory control may range from relatively mild hypoventilation during quiet sleep with adequate ventilation during wakefulness, to complete apnea during sleep and severe hypoventilation during wakefulness. Breathing during NREM sleep is almost entirely controlled by the autonomic nervous system; therefore ventilation is most severely affected during quiet sleep [2], but milder abnormalities are also present during REM sleep and wakefulness [4, 5]. Other signs of autonomic dysfunction may be present but are not essential to make the diagnosis of CCHS [2–5].

Pathophysiology

CCHS appears to represent a primary physiologic abnormality of integration of chemoreceptor input at the level of the nucleus tractus solitarii to central respiratory controllers and altered sensitivity to stimuli, rather than abnormalities in the peripheral or central chemoreceptors themselves [3, 6, 7]. The lack of appropriate responses to hypoxia and hypercarbia in CCHS has significant implications for affected individuals. In contrast to normal individuals who can augment their respiratory drive and physiological responses during stress or illness, children with CCHS have absent or insignificant chemoreceptor responses to both

Correspondence to: Thomas G. Keens, Division of Pediatric Pulmonology, Childrens Hospital Los Angeles, 4650 Sunset Boulevard, Box # 83, Los Angeles, California 90027

Sleep: A Comprehensive Handbook, Edited by T. Lee-Chiong.

hypercapnia and hypoxia both awake and asleep [7] even under normal circumstances. Although some CCHS patients are able to maintain adequate ventilation during wakefulness [6], all require assisted ventilation while sleeping; 25–35% of patients require full-time ventilatory support.

The lack of responsiveness to abnormalities in gas exchange has implications for potentially life-threatening crises for even minor fluctuations from baseline. Individuals with CCHS have compromised ability for effective gas exchange both during inactivity (i.e., watching television) as well as during any physical exertion [8, 9]. Various abnormalities in gas exchange occur with exercise [8, 9] because individuals with CCHS are not able to effectively increase their minute ventilation to meet their metabolic needs. These patients do not perceive asphyxia or dyspnea [5]. Furthermore, they do not display objective signs of respiratory distress such as tachypnea and nasal flaring or retractions, even in the presence of profound cyanosis. Without awareness and objective monitoring, the severity of the ventilatory abnormality may only be appreciated when severe cyanosis or central nervous system depression is present [5].

Etiology

Advances in the molecular genetic basis of abnormal respiratory control have contributed new insights to this disorder. Recent evidence indicates that abnormalities in CCHS and associated autonomic dysfunction result from de novo mutations, namely, an expansion mutation in a polyalanine tract within exon 3 of the *PHOX2B* gene located on 4p12 [10, 11]. *PHOX2B* encodes a homeobox transcription factor, which is involved in migration or differentiation of neural crest cells in utero. CCHS patients have one allele containing the normal 20 repeats of the polyalanine sequence and one mutated allele with an increased number of repeats, ranging from 25 to 33. CCHS may be a component of a larger neurocristopathy because of the association between CCHS and Hirschsprung's disease, tumors of neural crest origin, and other autonomic dysfunction. The CCHS recurrence risk for subsequent children in a family with a CCHS child is generally low. However, mother–child transmission of CCHS has been reported. Based on current information, the child of a CCHS parent has a 50% risk of inheriting the disorder.

Diagnosis and Clinical Course

The clinical presentation of CCHS may vary, depending on the severity of the disorder [5]. Most CCHS patients present in the newborn period with apnea or cyanosis, which is worse during sleep. These infants do not increase minute ventilation in response to hypoxia or hypercarbia and may not arouse if sleeping. There may be associated diaphoresis and tachycardia. Infants with CCHS not diagnosed during the neonatal period may have a later presentation with frank apnea, respiratory arrest, or an apparent life-threatening event (ALTE). Some infants may also present with pulmonary hypertension or cor pulmonale because of unrecognized chronic hypoxia resulting from abnormal respiratory control. Infants may also present with other features of autonomic dysfunction (feeding difficulty, breath-holding spells, or bradycardia) before the ventilatory abnormality is appreciated. The other features of autonomic dysfunction may include Hirschsprung's disease (in 15–20%), decreased heart rate variability, vasovagal syncope, pupillary abnormalities, poor heat tolerance, and esophageal dysmotility [5, 12].

The diagnosis of CCHS is based on evidence of hypoventilation during sleep without associated ventilatory muscle dysfunction, cardiac, lung, or metabolic disease, and absence of intracranial pathology [5, 13]. A variety of inborn errors of metabolism (Leigh disease, pyruvate dehydrogenase deficiency, and carnitine deficiency) may cause apnea or hypoventilation. The diagnosis depends on documentation of hypoventilation; therefore the information collected must include respiratory parameters during wakefulness, REM sleep, and NREM sleep. A polysomnogram, during sleep and wakefulness, with appropriately collected respiratory data (respitrace inductance plethysmography for chest and abdomen movement, airflow recordings, oxygen saturation with waveform, end-tidal carbon dioxide, electrocardiogram, and/or transcutaneous P_{O_2}/P_{CO_2} monitoring), can be used to make the diagnosis noninvasively. Invasive monitoring often results in arousal, which may make interpretation of the data for diagnosis more difficult. The investigations involved in making a diagnosis of CCHS may include chest x-ray, fluoroscopy of the diaphragm, MRI scans of the brain and brainstem, echocardiogram, Holter monitoring, and muscle biopsy. A rectal biopsy is not routinely performed for Hirschsprung's disease in this group. Genetic testing for mutations of the polyalanine expansion sequence of the *PHOX2B* gene appears to be highly sensitive and specific for the diagnosis of CCHS [10, 11], but other possible causes of hypoventilation should still be definitively ruled out.

Management

The goal of treatment for patients with CCHS is to ensure adequate oxygenation and ventilation because the patient is unable to achieve adequate gas exchange by spontaneous breathing [5, 13]. Trying to compensate for an individual who does not breathe adequately is particularly challenging, especially in young children. Treatment requires mechanical assisted ventilation, as no pharmacological respiratory stimulants have been shown to be effective [5, 13]. CCHS does not resolve spontaneously; therefore chronic ventilatory support at home is necessary in order for these patients to leave the hospital. Advances in technology have provided several options for ventilatory support in children with CCHS, which include positive pressure ventilators via tracheostomy, bilevel positive airway pressure via a nasal/face mask, negative pressure ventilators, or diaphragm pacing via phrenic nerve stimulation. The type of support should be tailored for the individual child based on the age and ventilatory requirements of the patient. Supplemental oxygen alone is not an appropriate treatment for CCHS because hypoventilation will persist without assisted ventilation, increasing the risk for pulmonary hypertension. Usually, infants are ventilated with positive pressure via a tracheostomy in infancy to ensure optimal oxygenation and ventilation while minimizing the risk of hypoxic injury. The transition to a home portable ventilator is made in the hospital with extensive, multidisciplinary discharge planning. Issues related to discharge include arrangements for a home ventilator, back-up ventilator, in-home nursing support (usually 16–24 hours per day), instruction for the family, medical supplies, and equipment (oxygen, pulse oximeter, end-tidal CO_2 monitoring). Objective measurements of adequate gas exchange (oxygen and carbon dioxide) are essential for home, particularly in this group of children because they do not show the signs associated with respiratory distress during hypoxia or hypercarbia. This can be a life-threatening disorder if not appropriately recognized and treated.

CCHS patients lack critical protective physiologic responses; therefore CCHS infants may be very unstable. The provision of adequate ventilation to mimic the natural physiological responses of breathing at all times is very difficult. Even minor pulmonary infections can result in profound respiratory failure without evidence of dyspnea; therefore vigilance during any illness is warranted. Children with CCHS do gradually become more stable as they mature because the respiratory system matures (lung growth, chest wall stiffness, and strength of respiratory muscles). The underlying lack of responsiveness to hypoxia and hypercarbia does not resolve and is lifelong. As a result, there are times during the child's life when ventilatory instability will occur and should be anticipated. Although older children and adolescents are more stable, these children should continue to be followed closely. Adolescence can be a particularly difficult time if the adolescent experiments with drugs or becomes nonadherent to therapy, predisposing the individual to cardiorespiratory compromise, even death.

Another potentially overlooked issue in managing these patients includes perioperative care because of primary or secondary involvement of various organ systems in CCHS.

Perioperative monitoring should be performed with vigilance since this group of patients does not respond to hypoxia and hypercapnia appropriately. Although not formally studied, it is presumed that patients with CCHS are more likely to be sensitive to respiratory depressant effects of inhalational anesthetics, narcotics, and benzodiazepines. Furthermore, the presence of dysautonomia will affect the technique for anesthesia and the medication options perioperatively. It is imperative that the anesthetist be closely involved in planning surgery in this vulnerable group.

Prognosis

Children with CCHS can have prolonged survival and good quality of life. Nevertheless, the risk of compromise or sudden death should not be underestimated. Although these patients do not regain their ventilatory responses, about two-thirds are able to come off assisted ventilation during wakefulness [14]. Most CCHS children attain adequate growth and nutrition [14] but may require gastrostomy feeding tubes for optimal growth during infancy. Intermittent hypoxemic episodes can result in permanent neurologic sequelae. Children with CCHS generally have a decreased range of mental processing abilities, compounded by significant learning disabilities. Intelligence is usually in the normal or low-normal range. Cognitive outcome appears to be related to the severity of the CCHS and control of hypoventilation.

Thus far, CCHS is a lifelong disorder without a known cure. Increasing awareness of the syndrome, earlier diagnosis and treatment, and appropriate and vigilant management of ventilation will likely continue to improve the future outlook of this group of children.

FAMILIAL DYSAUTONOMIA

Familial dysautonomia, also known as Riley–Day syndrome, is a recessively inherited disorder, involving sensory and autonomic neuropathy, resulting in autonomic dysfunction [15]. The disorder affects predominantly persons of Ashkenazi Jewish descent with an estimated incidence of 1 in 3600 in this population [15]. The autonomic dysfunction can involve any organ system, and it is generally considered to be a potentially life-threatening disorder with high mortality.

Clinical Respiratory Features

The onset is at birth and is progressive. Earliest markers include feeding problems and hypotonia. Recurrent aspiration either from swallow dysfunction or gastroesophageal reflux disease can lead to chronic lung disease. Other clinical features include decreased pain and temperature perception, absence of tears, and episodes of protracted vomiting and nausea. "Dysautonomic crises," characterized by agitation, tachycardia, and hypertension, can occur in association with abnormal response to stressors and vasomotor instability [15].

The most significant problem related to respiratory control is insensitivity to hypoxemia and hypercarbia [15, 16]. The hypoxemia resulting from traveling to high altitudes or pneumonia can lead to hypotension, bradyarrythmia, and even syncope. Recurrent apneic spells and hypoventilation have warranted chronic mechanical assisted ventilation within sleep in some children with familial dysautonomia [15]. The ventilatory abnormalities in this disorder may be unrecognized or unappreciated, resulting in premature or sudden death. In a recently reported survival analysis of patients with familial dysautonomia [17], sudden death was reported to be the most frequent cause of death and the majority (68%) occurred during sleep. Pulmonary complications are also a frequently reported cause of death, such as pneumonia. Despite the risk of sudden death, many are surviving into adulthood due to improved diagnosis and recognition of the disorder. Approximately half of the reported 551 patients have survived to at least age 20 years [17].

Diagnosis

Diagnosis of this disorder is based on a unique constellation of features, which include an absent axonal flare with injection of histamine, lack of overflow tearing, absent lingual fungiform papillae, and depressed deep tendon reflexes in an individual of Ashkenazi Jewish descent. Advances in molecular genetics have made it possible to make the diagnosis by DNA testing. The most common mutation involves a noncoding mutation in the I-κB kinase-complex associated protein (*IKBKAP*) gene.

Management

At the present time, there is no known cure for this disorder. Treatment is supportive and aimed at the specific problems. A high index of suspicion should be maintained for pneumonia and other complications of aspiration. Breath-holding spells may occur and be confused for seizures. The family should be provided with instructions regarding air travel and management during illness. The role of screening for sleep-related hypoxemia has not been clearly defined in this group. Nevertheless, sleep architecture abnormalities have been reported in addition to central hypopneas and apneas that result in profound desaturation [16]. It is possible that some of the sudden deaths during sleep may be related to autonomic crises occurring during sleep.

MYELOMENINGOCELE WITH ARNOLD–CHIARI MALFORMATION

Arnold–Chiari malformation (ACM) is a complex deformity of the central nervous system and can be associated with central hypoventilation. ACM type II, commonly associated with myelomeningocele, is characterized by displacement of the cerebellar vermis, caudal brainstem, and fourth ventricle through the foramen magnum, resulting in progressive hydrocephalus. Respiratory complications associated with ACM and myelomeningocele can be substantial, ranging from altered central respiratory control (central hypoventilation) to upper airway obstruction (bilateral vocal cord paralysis from traction on the recurrent laryngeal nerve) [13]. Theories on ACM's etiology involve abnormal migration of the neural crest cells leading to disruption or dysplasia of the brainstem nuclei involved in respiratory control.

Clinical Respiratory Features

The clinical presentation of infants with myelomeningocele can vary greatly, from sleep apnea, hypoventilation, and obstructive apnea to breath-holding spells [18, 19]. Upper airway obstruction can be a significant cause of abnormal ventilation. Bilateral vocal cord paralysis at times is due to increased intracranial pressure. Ventriculoperitoneal shunt (VPS) placement for hydrocephalus may relieve the symptoms of upper airway obstruction and/or respiratory symptoms. Posterior fossa decompression surgery may be helpful if VPS placement does not result in complete resolution of respiratory symptoms.

Several studies have reported ventilatory abnormalities including central sleep apnea, hypoventilation, and hypoxia in infants with ACM. Ventilatory responses in children with ACM are abnormal. Children with myelomeningocele have decreased ventilatory responses to hypoxia and hypercarbia [19]. The ventilatory abnormality may result from abnormal central chemoreceptor sensitivity or inappropriate integration of the response. The level of the myelomeningocele does not affect the presence of the abnormalities. Additional risk factors for sleep disordered breathing include thoracic or thoracolumbar lesions, obesity, and a previous history of posterior fossa decompression surgery. A striking feature about patients with ACM is that many of these infants or children are clinically asymptomatic, yet still demonstrate abnormal ventilatory patterns during sleep [13]. Little is known about the timing of the presentation of the abnormalities and whether screening should be routinely implemented.

Presentation

Clinical presentation of ventilatory dysfunction can be variable in timing and clinical symptoms (apneas, bradycardia, aspiration, stridor, and vocal cord paralysis). Infants and children with ACM also have a higher reported incidence of sleep disordered breathing, and sudden death has been reported to be as high as 69% [18]. Compression hydrocephalus as a cause of ventilatory abnormalities must be aggressively investigated in this population. Unrecognized sleep disordered breathing can be a significant cause of morbidity and mortality.

Diagnosis and Management

Children with myelomeningocele and stridor should undergo laryngoscopy and/or flexible bronchoscopy to evaluate vocal cord movement. Neuroimaging studies to rule out other causes of vocal cord paralysis should be obtained; that is, posterior fossa herniation may be identified and potentially surgically corrected, resolving the bilateral vocal cord paralysis. Tracheostomy has traditionally been the treatment of choice for bilateral vocal cord paralysis, but this stance has recently been challenged and conservative management is recommended in selected cases. Bilevel positive airway pressure ventilation in a timed mode is useful for those with myelomeningocele and obstructive sleep apnea syndrome and/or central sleep apnea. However, the presence of chronic hypoventilation both during wakefulness and sleep requires full-time ventilatory support. Deterioration can occur unexpectedly from progressive neurological problems despite mechanical ventilation and is associated with a poor prognosis even with mechanical ventilation [18]. The decision for mechanical ventilation should be weighed carefully against the child's overall neurological condition and prognosis and discussed with the family.

PRADER–WILLI SYNDROME

Prader–Willi syndrome (PWS) is a genetic disorder with an estimated prevalence rate of 1:10,000 to 1:25,000 live births. This disorder is associated with mutations involving chromosome 15 (15q11–q13). A mutation in the small nuclear ribonuclear protein polypeptide N (*SNRPN*) gene at 15q12 has been detected in most cases. The consequence of this sporadic mutation is failure of hypothalamic development, particularly development of the anterior hypothalamus.

Clinical Respiratory Features

Clinical features include infantile hypotonia, developmental delay, hypogonadism, and subsequent hypothalamic obesity, in addition to behavioral and sleep disorders. Patients with PWS do not have central classical hypoventilation as described previously, but they do have other abnormalities of ventilatory control [20].

Patients with PWS often exhibit significant sleep-related problems including excessive daytime sleepiness, abnormalities in organization of REM sleep, abnormalities of arousal, obstructive sleep apnea, restless movements during sleep, hypoventilation, and hypoxia [13, 20]. Some patients exhibit excessive daytime sleepiness in spite of excellent sleep quality and quantity, indicating that sleepiness may be intrinsic to PWS. Some of the characteristic features of PWS are similar to features of obstructive sleep apnea, but the degree of hypersomnolence does not correlate with the severity of sleep disordered breathing [20]. Obesity itself may further worsen the ventilatory abnormalities in PWS. Interestingly, despite the severity of obesity in this group, the incidence of obstructive sleep apnea in this group is less than expected. Although cataplexy is a common feature between narcolepsy and PWS, there is no known genetic overlap between the two syndromes.

Physiological ventilatory responses are altered in PWS. Hypercapnic ventilatory response is blunted in obese individuals with PWS. The link between obesity and responsiveness to hypercapnia has not been defined in this group. Irrespective of body habitus, hypoxic ventilatory responses are blunted or absent in PWS, most likely explained by their decreased peripheral chemoreceptor function [20]. The abnormality may be related to central hypothalamic dysfunction involving the reticular activating system, serving as one facet of the overall syndrome.

Management

In PWS patients, the main clinical problem related to ventilatory control appears to be sleep disordered breathing. Risk factors for sleep disordered breathing in this group of patients include obstructive sleep apnea (facial dysmorphism, obesity, hypotonia), alveolar hypoventilation, and restrictive lung disease (muscle weakness or scoliosis). One can reduce the severity of this problem with avoidance of obesity, which can be very challenging. Identification of sleep disordered breathing by polysomnogram will characterize the respiratory abnormality and help determine treatment. Early identification and intervention may reduce the morbidity and mortality associated with PWS. However, it is not clear when these children should be evaluated for sleep disordered breathing as the prevalence rate of sleep disordered breathing has not been defined. Obstructive sleep apnea, when present, should be treated and may require CPAP, bilevel positive airway pressure, adenotonsillectomy, or tracheostomy. Patients with PWS do not routinely require mechanical ventilation but may need to be considered on a case-by-case basis. Alveolar ventilation can be treated with positive pressure ventilation. Respiratory stimulants such as medroxyprogesterone [20], although used in select cases, are not indicated for sleep disordered breathing in PWS. These patients have poor responses to hypoxia and therefore require close monitoring during illness.

ACHONDROPLASIA AND OTHER SKELETAL DYSPLASIAS

Achondroplasia is an autosomal dominant syndrome involving inhibition of endochondral bone formation, which causes short-limbed dwarfism. It is caused by mutations in the fibroblast growth factor receptor 3 gene (*FGFR-3*) [13].

Clinical Respiratory Features

These children do not have classical hypoventilation syndromes but are at risk for respiratory compromise as a result of their syndrome. Affected infants may present with obstructive sleep apnea, waking cyanotic episodes, and chronic respiratory insufficiency or failure. Characteristics specific to this syndrome resulting in symptoms include midfacial hypoplasia, upper airway obstruction, occipital dysplasia, cervicomedullary cord compression secondary to a small foramen magnum, and thoracic cage restriction. Collectively, these can result in significant respiratory control abnormalities including hypoventilation, apnea, and sudden death.

Other skeletal dysplasias may also affect the brainstem area and/or result in thoracic restriction. The impact of thoracic restriction cannot be underestimated as the size of the thorax has significant implications for lung growth, and subsequent ability of attaining appropriate ventilation.

Assessment of these individuals involves a multisystem approach based on the risk factors. Investigations may include neuroimaging for hydrocephalus and cord compression, monitoring of gas exchange both during wakefulness and sleep, overnight polysomnography, and pulmonary function testing. A recent study suggested that there may be distinctive phenotypic presentations of respiratory problems in patients with achondroplasia [21, 22]. Nearly half of the 88 subjects with achondroplasia had abnormalities documented by overnight polysomnography, with the most common abnormality being hypoxemia [21].

Management

Treatment should be based on findings of multisystem investigations as none of the tests in isolation may identify the abnormality. Treatment options include VPS placement for hydrocephalus, cervicomedullary decompression for central respiratory control abnormalities and other neurologic dysfunction, oxygen for hypoxemia, adenotonsillectomy for obstructive sleep apnea syndrome (OSAS), or bilevel positive airway pressure for hypoventilation and/or OSAS. Those with severe obstructive sleep apnea may require a tracheostomy. Patients who present with chronic respiratory insufficiency should also be screened for obstructive lung disease, gastroesophageal lung disease, and aspiration [22]. Some of these severely affected children may require mechanical ventilation if the hypoventilation is compounded by restrictive lung disease. Children with achondroplasia have normal intelligence and will do well with appropriate therapy.

LEIGH DISEASE

Leigh disease, or subacute necrotizing encephalomyelopathy, represents a group of inherited disorders in infancy and childhood characterized by a progressive clinical course of deterioration in brainstem function [13]. Several different de novo mutations in the respiratory chain, including mitochondrial ATPase deficiency, have been reported. Patients may appear normal during infancy but develop progressive neurologic symptoms, which include poor feeding, vomiting, apnea, alveolar hypoventilation, and regression of developmental milestones. Hypoventilation may precede other neurologic symptoms. Other symptoms of brainstem dysfunction such as nystagmus, bizarre eye movements, pupillary changes, hypotonia, seizures, and sleep/wakefulness disturbances may also occur. Findings on CT or MRI of the brainstem that suggest this diagnosis include bilateral, often asymterical, brainstem lesions. Sleep architecture abnormalities may include loss of REM sleep and reduced stage deep sleep. Preferential gray matter involvement with vascular proliferation, endothelial swelling, and progressive neuronal destruction and loss of myelin may be found. Changes are predominantly seen in the midbrain, pons, periaqueductal gray matter, posterior colliculi, medulla, floor of the fourth ventricle, and posterior olive. Autopsy results have shown involvement of the dorsal respiratory group. There is no specific treatment or cure for Leigh disease. Chronic ventilatory support is an option for chronic respiratory failure, but must be weighed in the context of neurological deterioration and poor prognosis.

JOUBERT'S SYNDROME

Joubert's syndrome is a rare autosomal recessive disorder due to agenesis of the cerebellar vermis. The clinical features include hypotonia, ataxia, developmental delay, abnormal eye movements, and abnormal respiratory pattern (episodes of tachypnea, up to 100–200 breaths/minute) alternating with prolonged apneas during sleep or wakefulness during infancy [13]. The tachypnea spontaneously resolves after infancy. It is not clear whether assisted ventilation is required during infancy. The disorder itself is often progressive, and the ventilatory pattern abnormalities can cause death. However, case reports exist of patients with Joubert's syndrome in their mid-30s.

MANAGEMENT OF RESPIRATORY FAILURE IN CHILDREN WITH RESPIRATORY CONTROL DISORDERS

Currently, in children with respiratory control disorders, there is little that can be done to augment central respiratory drive and for all these disorders, there is no known cure. Pharmacological respiratory stimulants have not been shown to work. Medroxyprogesterone has been used in isolated cases but cannot be recommended for widespread use. One often-overlooked aspect is the inhibition of central respiratory drive by metabolic imbalance, such as chronic metabolic alkalosis. Iatrogenic alkalosis can suppress the respiratory drive. Therefore serum chloride concentrations should be maintained greater than 95 mEq/dL, to avoid alkalosis and thus optimize ventilation. Sedative medications should be avoided. During illnesses, these patients should be monitored cautiously. Perioperative procedures should be planned carefully in these patients who are at risk for significant compromise. With increased awareness of disorders of ventilation and development of sophisticated, portable technology, these medically complex children are able to survive longer and transition home.

Children with chronically elevated P_{CO_2} greater than 55–60 mmHg, due to abnormal respiratory drive, will likely develop progressive pulmonary hypertension. Supplemental oxygen therapy improves the Pa_{O_2} and relieves cyanosis, but this treatment alone is inadequate, as hypoventilation persists with subsequent pulmonary hypertension. Thus these children require mechanical ventilation.

Candidates for Chronic Ventilatory Support at Home

Children with chronic respiratory failure with either primary respiratory control disorders or secondary acquired respiratory control disorders who require assisted ventilation can transition home. Previously, the technology and complex medical needs prevented these individuals from leaving the hospital or a chronic care facility. A multidisciplinary team should be involved with patients who require long-term mechanical ventilation for discharge planning and follow-up. Children with respiratory control disorders are generally good candidates for chronic home mechanical ventilation as long as coexisting pulmonary disease is sufficiently stable and frequent adjustments in ventilator settings to maintain adequate gas exchange are not required. Our

general parameters for successful home mechanical ventilation are the following: (1) F_IO_2 should be <40% to maintain $Sa_{O_2} \geq 95\%$; and (2) peak inspiratory pressure (PIP) to achieve adequate ventilation should be <40 cm H_2O. In contrast to previous reports, the need for positive end-expiratory pressure (PEEP) is no longer a contraindication for home ventilation. The availability of portable ventilators has enabled even very complex children to transition home [23].

Philosophy of Chronic Ventilatory Support

The philosophy for home mechanical ventilation, is vastly different from the acute mechanical ventilation, where the goal in the latter is to wean the patient completely off ventilation. For most children going home with chronic ventilatory support secondary to respiratory failure, weaning from the ventilator is not a realistic goal. Children are much more likely to be physically active at home than in the hospital and may require additional respiratory support to enable them to engage in activities with their families. Ideally, the ventilators are adjusted to completely meet the ventilatory demands of these children, so they have energy available for other activities, in turn improving their quality of life. For children with respiratory control disorders, ventilators are adjusted to optimize gas exchange and ventilation by providing $P_{ET}CO_2$ 30–35 mmHg and $Sa_{O_2} > 95\%$. Optimal ventilation also prevents atelectasis and the development of concurrent lung disease. We purposefully aim for lower end-tidal carbon dioxide levels because data show children who are hyperventilated at night have better spontaneous ventilation while awake than those who are ventilated to higher P_{CO_2} levels [13]. It also provides these children with some reserve so that a minor respiratory illness does not result in respiratory failure. Although not formally studied, anecdotally, children with respiratory control disorders have fewer complications, and generally do better clinically, with hyperventilation during assisted ventilation. The duration of mechanical ventilation will most likely be prolonged as none of these respiratory disorders have a known cure. Regular follow-up with a multidisciplinary team is recommended for preventing both short-term and long-term complications associated with assisted ventilation.

Although most of the children do continue to need ventilatory support, sprinting off the ventilator should be evaluated and considered on an individual basis. Sprinting strategies for children requiring long-term mechanical ventilation are encouraged to help the child breathe spontaneously during the day and optimize mobility and quality of life. "Sprint weaning" is performed by first adjusting the ventilator settings to completely meet the child's ventilatory demands [13]. The child is then removed from the ventilator for short periods of time during wakefulness two-to-four times per day. Sometimes, these initial sprints may last only a few minutes. Supplemental oxygen can be added during sprinting. The child is carefully monitored noninvasively during sprints to prevent hypoxia or hypercapnia. Guidelines for terminating sprints should be well outlined and provided as orders. Indications for terminating sprints include hypoxia, rising $P_{ET}CO_2 > 45$–50 torr, or signs of distress/discomfort [13]. The length of each sprint is increased gradually as the child tolerates the duration of the sprint. Physicians should avoid the temptation to increase the sprint length too rapidly, as this often hinders the progress of weaning and may even result in requiring more ventilation. Initially, sprinting should be performed only during wakefulness. Certain patients with respiratory control disorders requiring mechanical ventilation 24 hours per day, such as severe CCHS, are not candidates for sprinting.

Modalities of Home Mechanical Ventilation

The ideal ventilators for home use are different from those used in hospitals for the treatment of acute respiratory failure. However, advances in the technology for home mechanical ventilation have enabled physicians to choose from a diverse range of options. Because children with respiratory control disorders usually do not have severe lung disease, they have the greatest number of options for different techniques to provide chronic ventilatory support at home. Different modalities are briefly reviewed below.

Portable Positive Pressure Ventilator Via Tracheostomy Portable positive pressure ventilators are the most common method of providing home mechanical ventilation for infants and children. More sophisticated ventilators permit ventilating the child at home even during intercurrent illnesses. However, this should only be done under the direction of a physician and in the presence of objective data from noninvasive monitoring equipment. Hospitalization may still be required to ensure ventilation during illnesses, especially in young infants, or patients with coexistent lung disease. Often the ventilatory requirements may change during an intercurrent illness and should be evaluated to ensure adequate oxygenation and ventilation. A smaller, uncuffed tracheostomy is preferred and the ventilator should be adjusted as necessary to compensate for the leak [13]. The rationale and preference for the smaller uncuffed tracheostomy tube are (1) to minimize the risk of tracheomalacia or tracheal mucosal damage, (2) to allow a large expiratory leak so that the child may speak, and (3) to provide a margin of safety because the child may still be able to ventilate around the tracheostomy tube. Small uncuffed tracheostomy tubes are associated with leaks that can be large and variable, especially in young children. Because of this, pressure ventilation is preferred and can be utilized on newer portable home ventilators. However, older portable ventilators are volume preset ventilators. A significant portion of the ventilator delivered breath escapes in the leak around the uncuffed tracheostomy. The tracheostomy leak can be compensated for by using the ventilator in a pressure-limited modality, also known as pressure plateau ventilation [13]. The pressure limit on the volume preset ventilator is adjusted to the desired peak inspiratory pressure, and a tidal volume setting is chosen that is sufficient to inflate the lungs, compensate for tubing compliance, and accommodate the leak. This technique is very successful in home mechanical ventilation of infants and small children. The development of newer portable ventilators allows more sophisticated adjustments of ventilation settings to minimize patient asynchrony and also allow more medically complicated patients to be ventilated at home [13].

Bilevel Positive Airway Pressure and Positive Pressure Ventilation by Mask Noninvasive intermittent positive pressure ventilation is delivered via a nasal mask or face mask using a bilevel positive airway pressure ventilator [13]. This has been generally used in older children but has been used successfully in infants as well. Bilevel ventilators are smaller, less expensive, and generally easier to use than conventional ventilators. Newer models have pressure and apnea alarms and adjustable rise times, and can be battery operated. Bilevel ventilators can provide variable continuous flow through a blower (fan), have a fixed leak preventing CO_2 retention, and can compensate for leaks around the mask. Inspiratory positive airway pressure (I-PAP) and expiratory positive airway pressure (E-PAP) can be adjusted independently. The I-PAP to E-PAP difference is proportional to tidal volume. Heated humidification and supplemental oxygen can be added to the circuit. Only the Spontaneous/Timed and Timed modes guarantee breath delivery and should be used in CCHS patients, because these patients cannot be trusted to generate their own adequate respirations. Bilevel ventilation should not be used 24 hours a day, because the mask interferes with daily activities and social interaction, and has been associated with some midface hypoplasia (although not clear if the mask was causative). Children who are unable to remove the mask should not have a full face mask because of the risk of aspiration. Most side effects related to BPAP are minimal, such as rhinitis, aerophagia, conjunctival injection, and skin breakdown.

Negative Pressure Ventilation Negative pressure ventilation (NPV) works by generating a negative inspiratory pressure outside the chest and abdomen to cause an inspiration noninvasively [13]. The ventilator rate and amount of negative pressure are individually selected. The negative pressure is proportional to the tidal volume but may be limited by leaks around the chest shell or wrap. However, airway occlusion can occur when breaths are generated by a negative pressure ventilator during sleep, especially in young children and infants. The effectiveness of negative pressure ventilation depends on the ability to move the chest wall; therefore those with marked chest wall deformities are poor candidates for NPV. Nevertheless, NPV does allow some children to avoid tracheostomy. Unfortunately, negative pressure ventilators are large, not battery operated, and not portable. Those patients who have historically used negative pressure ventilators tend to now choose bilevel ventilation for ease of care.

Diaphragm Pacing　Diaphragm pacing generates breathing using the child's own diaphragm as the respiratory pump [13]. Commercially available diaphragm pacing systems have a battery-operated external transmitter. An antenna is taped on the skin over subcutaneously implanted receivers. The transmitter generates a train of pulses for each breath, which is transmitted through the antenna to the receiver under the skin, similar to radio transmission. The receiver converts this energy to standard electrical current, which is directed to a phrenic nerve electrode by lead wires. The electrical stimulation of the phrenic nerve causes a diaphragmatic contraction, which generates the breath. The amount of electrical voltage is proportional to the diaphragmatic contraction, or tidal volume. In children, simultaneous bilateral diaphragm pacing is generally required to achieve optimal ventilation. In general, pacers are not recommended for children under age 5. Use of pacers requires that the phrenic nerves and the diaphragm function appropriately to enable effective ventilation. Ventilatory muscle weakness and phrenic neuropathies are contraindications to pacer use. Pacers can be used for daytime support of ambulatory children requiring full-time ventilatory support, in combination with positive pressure ventilation at nighttime. Obstructive apnea can be a complication because there is no synchronous upper airway skeletal muscle contraction with inspiration. However, this can often be overcome by adjusting settings on the pacers to lengthen inspiratory time and decrease the force of inspiration. In general, diaphragm pacers can only be used up to 14 hours a day and cannot be used for 24 hours continuously. Thus patients requiring ventilatory support 24 hours a day should have an alternate form of ventilation for part of the day if pacers are used.

Hospital Management in Preparation for Discharge

Coordination of discharge planning by a multidisciplinary team is vital for the successful transition home from hospital. Children should receive routine immunizations, including the influenza vaccine, prior to discharge. Ventilators for home should be equipped with a disconnect alarm so the caregiver can be alerted in the event of an accidental disconnection. Nursing care during the night is important to address any problems with ventilation that may arise, especially in children who are not able to signal for help. Use of apnea and bradycardia monitors is indicated in this setting and can be used as an adjunct. Children with central hypoventilation may not show the typical signs of respiratory distress; therefore they should have noninvasive monitoring equipment readily available, including a pulse oximeter and/or an end-tidal carbon dioxide monitor for home.

It is recommended that the patient's respiratory status be stable for a minimum of 2 weeks prior to the initial discharge date after mechanical ventilation is initiated. The patient should be evaluated on home ventilation settings and be observed to ensure effective gas exchange is achieved prior to discharge. The family must be educated in aspects of the child's care prior to discharge. The primary caregivers within the family must show competency in tracheostomy care, cardiopulmonary resuscitation, administration of medications, and equipment operation as well as chest physiotherapy. Families should also be taught how to recognize respiratory distress, including subtle signs of deterioration (lethargy, edema, pallor) and be given instructions about when to contact the physician.

Long-term Care of the Mechanically Ventilated Child

Routine evaluation of ventilator settings should be performed on a regular basis so that ventilation meets the changing requirements of the growing child. The frequency of the checks depends on the instability of the condition and the individual patient. The settings will need to be reevaluated after any infection or hospitalization. Some children may require supplemental oxygen either during spontaneous ventilation or during mechanical ventilation. Oxygen supplementation is not a substitute for adequate ventilation, nor a replacement for children with chronic respiratory failure who need assisted ventilation.

Despite careful monitoring, even the most successfully managed patients may be exposed to periods of alveolar hypoxia and hypoventilation because ventilators may not be able to meet ventilatory requirements at all times. As a result, all ventilator-assisted children with respiratory control disorders are at risk for cardiopulmonary compromise and pulmonary hypertension. Echocardiography can be useful for evaluating heart function. If pulmonary hypertension is found, it is due to inadequate ventilation until proved otherwise [13].

Any illness can significantly impact a child who is mechanically ventilated. Children with central hypoventilation syndromes cannot increase respiratory effort, tidal volume, or respiratory rate, even though a respiratory infection might be causing increased secretions and/or pneumonia, resulting in worsening P_{O_2} and P_{CO_2} levels. Despite the use of preventive and therapeutic measures directed at these problems, even a relatively trivial upper respiratory infection may compromise the ventilator-assisted child. Ventilator adjustments with an increased level of support are often needed. Young children ordinarily requiring ventilation only during sleep often need 24 hour per day support during illnesses. Although some children with central hypoventilation are safely managed at home, during minor illnesses, these patients may require hospitalization for blood gas monitoring and frequent ventilator changes. The ability to manage these children at home depends on the ability of the caregiver, proximity of the hospital, and the illness itself.

The prognosis for children on mechanical ventilation has improved significantly in the hands of specialized centers and support staff capable of caring for these children. Mechanical ventilation is a specialized treatment option and requires expertise in providing safe and effective treatment for children with respiratory control disorders. Lack of vigilance to the issues involving mechanical ventilation can lead to significant morbidity and mortality.

CONCLUSION

Central hypoventilation syndromes involve abnormal physiological responses to respiratory stimuli and can result in significant morbidity and mortality without appropriate care. A wide array of disorders has associated respiratory control abnormalities and the disorders vary in their clinical presentation and severity. There is no cure and no effective pharmacological treatments for primary central hypoventilation disorders. The evolving genetic information has shed new insights for these disorders and may one day lead to improved treatment strategies. Currently, therapeutic management usually requires mechanical assisted ventilation or close monitoring for evidence of sleep disordered breathing causing cardiopulmonary compromise. The prognosis for children with syndromes affecting respiratory control during sleep depends primarily on associated neurological problems, the extent of central nervous system injury or involvement, the nature of the underlying disorders, and the severity of hypoventilation. Some children with central hypoventilation syndromes, such as CCHS, can have prolonged survival, associated with a good quality of life with mechanical assisted ventilation. Mechanical ventilation in the presence of severe progressive neurological deterioration must be weighed carefully. Early recognition and awareness of the respiratory control issues will enable better neurological outcomes by preventing hypoxemia and cardiopulmonary compromise.

REFERENCES

1. Severinghaus J, Mitchell RA. Ondine's curse—failure of respiratory center automaticity while awake. *Clin Res* **10**:122(1962).

2. Gaultier C, Trang-Pham H, Praud JP, Gallego J. Cardiorespiratory control during sleep in the congenital central hypoventilation syndrome. *Pediatr Pulmonol* **23**:140–142(1997).

3. Gozal D. Congenital central hypoventilation syndrome: an update. *Pediatr Pulmonol* **26**:273–282(1998).

4. Weese-Mayer DE, Silvestri JM, Menzies LJ, Morrow-Kenny AS, Hunt CE, Hauptman SA. Congenital central hypoventilation syndrome: diagnosis, management, and long-term outcome in thirty-two children. *J Pediatr* **120**:381–387(1992).

5. Weese-Meyer DE, Shannon DC, Keens TG, Silvestri JM. American Thoracic Society Consensus Statement. Idiopathic congenital central hypoventilation syndrome: diagnosis and management. *Am J Respir Crit Care Med* **160**:368–373(1999).

6. Gozal D, Marcus CL, Shoseyov D, Keens TG. Peripheral chemoreceptor function in children with congenital central hypoventilation syndrome. *J Appl Physiol* **74**:379–387(1993).

7. Paton JY, Swaminathan S, Sargent CW, Keens TG. Hypoxic and hypercapneic ventilatory responses in awake children with congenital central hypoventilation syndrome. *Am Rev Respir Dis* **140**:368–372(1989).

8. Paton JY, Swaminathan S, Sargent CW, Hawksworth A, Keens TG. Ventilatory response to exercise in children with congenital central hypoventilation syndrome. *Am Rev Respir Dis* **147**:1185–1191(1993).

9. Silvestri JM, Weese-Mayer DE, Flanagan EA. Congenital central hypoventilation syndrome: cardiorespiratory responses to moderate exercise simulating daily activity. *Pediatr Pulmonol* **20**:89–93(1995).

10. Amiel J, Laudier B, Attie-Bitach T, Trang H, du Pontual L, Gener B, Trochet D, Etchevers H, Ray P, Simmoneau M, Vekemans M, Munnich A, Gaultier C, Lyonnet S. Polyalanine expansion and frameshift mutations of paired-like homeobox gene *PHOX2B* in congenital central hypoventilation syndrome. *Nature Genet* **33**:459–460(2003).

11. Weese-Mayer DE, Berry-Kravis EM, Zhou L, Maher BS, Silvestri JM, Curran ME, Marazita ML. Idiopathic congenital central hypoventilation syndrome: analysis of genes pertinent to early autonomic nervous system embryologic development and identification of mutations in *PHOX2b*. *Am J Med Genet* **123A**:267–278(2003).

12. Woo MS, Woo MA, Gozal D, Jansen MT, Keens TG, Harper RM. Heart rate variability in congenital central hypoventilation syndrome. *Pediatr Res* **31**:291–296(1992).

13. Keens TG, Davidson Ward SL. Syndromes affecting respiratory control during sleep. In: Loughlin GM, Carroll JL, Marcus CL (Eds), *Sleep and Breathing in Children. A Developmental Approach*. Marcel Dekker, New York, 2000, pp 525–553.

14. Marcus CL, Jansen MT, Poulsen MK, Keens SE, Nield TA, Lipsker LE, Keens TG. Medical and psychosocial outcome of children with congenital central hypoventilation syndrome. *J Pediatr* **119**:888–895(1991).

15. Axelrod FB, Maayan C. Familial dysautonomia. In: Chernick V, Boat TF, Kendig EL (Eds), *Kendig's Disorders of the Respiratory Tract in Children*. Lippincott-Raven, Philadelphia, 1998, pp 1103–1107.

16. Axelrod FB. Familial dysautonomia. *Muscle Nerve* **29**:352–363(2004).

17. Axelrod FB, Goldber JD, Ye XY, Maayan C. Survival in familiary dysautonomia: impact of early intervention. *J Pediatr* **141**:518–523(2002).

18. Marcus CL. Sleep-disordered breathing in children. *Am J Crit Care Med* **164**:16–30(2001).

19. Swaminathan S, Paton JY, Davidson Ward SL, Jacobs RA, Sargent CW, Keens TG. Abnormal control of ventilation in adolescents with myelodysplasia. *J Pediatr* **115**:898–903(1989).

20. Nixon GM, Brouillette RT. Sleep and breathing in Prader–Willi syndrome. *Pediatr Pulmonol* **34**:209–217(2002).

21. Mogayzel PJ, Carroll JL, Loughlin GM, Hurko O, Francomano CA, Marcus CL. Sleep disordered breathing in children with achondroplasia. *J Pediatr* **131**:667–671(1998).

22. Tasker RC, Dundas I, Laverty A, Fletcher M, Lane R, Stocks J. Distinct patterns of respiratory difficulty in young children with achondroplasia: a clinical, sleep and lung function study. *Arch Dis Child* **79**:99–108(1998).

23. Make BJ, Hill NS, Goldberg AI, Bach JR, Dunne PE, Heffner JE, Keens TG, O'Donohue WJ, Oppenheimer EA, Dominique R. Report of a Consensus Conference of the American College of Chest Physicians. Mechanical ventilation beyond the intensive care unit. *Chest* **113**:322–336S(1998).

67

SUDDEN INFANT DEATHS

A. Kahn, P. Franco, J. Groswasser, S. Scaillet, and B. Dan
University Children's Hospital of Brussels, Brussels, Belgium

I. Kato
Nagoya City University Medical School, Nagoya, Japan

T. Sawaguchi
Tokyo Women's Medical University School of Medicine, Tokyo, Japan

I. Kelmanson
St. Petersburg State Pediatric Medical Academy, St. Petersburg, Russia

INTRODUCTION AND DEFINITIONS

In most industrialized countries, sudden infant death syndrome (SIDS) is the leading cause of death in infants between the ages of 1 and 12 months. The most widely accepted definition of SIDS is that of the sudden death of an infant under 1 year of age that remains unexplained after a complete postmortem examination, death scene investigation, and case conference. New definitions have been suggested to specify the completeness of postmortem investigations [1].

The reported rates of SIDS differ greatly among various countries, ranging from 0.05 in Japan to 0.75 in the United States. The discrepancy could result from a variety of causes. Incomplete or absent postmortem examination, inexperienced pathologists, and differences in classification of causes of death can lead to potential misclassification of causes of death.

Little is known about the mechanisms responsible for the deaths of the infants. In this chapter we consider some potential physiological mechanisms for SIDS that develop in response to both infant and environmental stressors.

Correspondence to: A. Kahn, Pediatric Sleep Unit, University Children's Hospital Reine Fabiola, av. JJ Crocq 15, B-1020 Brussels, Belgium.

A POTENTIAL MODEL FOR SIDS AND THE ROLE OF ENVIRONMENTAL STRESSORS

Models have been proposed to explain the deaths, according to which a vulnerable infant exposed to an endogenous or exogenous stress at a critical developmental period is at risk of dying. Risk factors for SIDS have been identified by epidemiological and physiological studies [2–4].

Endogenous risk factors include the age of the infant, between 2 and 6 months, male gender, prematurity, or the presence of immature cardiorespiratory control mechanisms [5]. Known environmental stressors include a previous sleep deprivation [6, 7], the prone sleep position [8–10], a high room temperature [9, 11], sleeping with the face covered [11], or prenatal maternal smoking [5, 10] or drug addiction [5].

Other stressors that modify the infant's ability to cope with the environment include cosleeping. Bedsharing was postulated to protect against SIDS by favoring the infant's arousal, facilitating breastfeeding and maternal control over the sleeping child [12]. Alternatively, a greater risk of SIDS was reported in smoking cosleeping mothers [10, 13]. Cosleeping was also reported to increase the risk for an infant's death, through suffocation, asphyxia, entrapment, thermal stress, or overlaying [14], in particular if parents have consumed alcohol [13].

Some environmental factors have been associated with a decreased risk for SIDS. Such protective factors include the use of a sleeping bag, a firm bedding, breastfeeding, or the use of a pacifier [10].

The importance of environmental stress factors in the development of SIDS is highlighted by the drop in SIDS incidence measured in most countries following the prevention campaigns to inform professionals and families on the preventable risk factors for SIDS [15].

Development of Vital Control Systems

Most deaths from SIDS occur in the first 6 months of life, with a specific peak between 2 and 4 months of age [5]. This age distribution of SIDS corresponds to a period of the infant's life when significant changes occur in sleep–wake, breathing, autonomic controls, and immunological maturation. The deaths occur at a time when the infant is supposed to be sleeping, mainly during the early morning hours (between midnight and 6:00 a.m.). It could thus be hypothesized that abnormal regulation of some vital physiological control mechanisms develops during sleep. Such mechanisms are considered next, and the influence of endogenous and exogenous stressors is described.

Respiratory Controls   Central and obstructive apnea are seen during the sleep of healthy infants. The frequency of obstructive apnea decreases significantly between 8 and 12 weeks of life. Boys had more obstructive sleep apneas than girls, particularly during the peak age for SIDS [16]. Although unexplained, this finding is reminiscent of male preponderance among the victims of SIDS [5].

Future SIDS victims have been shown to have fewer short central apneas than control infants [17, 18], but more frequent mixed and obstructive sleep apneas [18, 19]. Future SIDS victims could have inappropriate breathing controls during sleep.

Postmortem findings support the development of frequent hypoxic events, possibly related to airway obstructions. These include a thickening of the basement membrane of the vocal cords [20], the presence of intrathoracic petechiae in about 80% of the victims, excessive retention of periadrenal brown fat, increased extramedullary hematopoiesis, or abnormal astroglial proliferation, especially in the brainstem [4, 21]. Additional postmortem evidence of premortem hypoxia included growth retardation, elevated blood cortisol levels, brain lactate and H^+ concentrations, elevated hypoxanthine levels in the vitreous humor, pulmonary neuroendocrine cell hyperplasia, or vascular endothelial growth factor in the cerebrospinal fluid [4, 22, 23].

The mechanisms responsible for the obstructive sleep apnea are complex. In infants, obstructive sleep apneas are most frequently due to narrowed upper airways due to nasal infection, anatomic abnormalities, neurological lesions impairing muscle contractions, or soft tissue infiltration [24, 25]. Narrowed upper airways could be inherited, as sleep apneas and smaller airways were also found in some SIDS family members [24]. Obstructive sleep apneas could also be associated with abnormal autonomic control of the upper airways [25] that can be prevented by atropine [26].

Infants sleeping prone and face down on soft bedding show episodes of airway obstruction [27]. The frequency of obstructive sleep apnea, however, is not associated with body position in infants, although the duration of the apnea increases when the infants sleep prone [28].

The development of apnea could depend on other environmental factors. Healthy infants develop a greater frequency of obstructive sleep apneas when they are exposed to some risk conditions for SIDS, such as being born from a smoking mother [29], being treated with sedative medications [30], or having been sleep deprived [31, 32].

Infants with obstructive sleep apnea develop symptoms that could reflect abnormal autonomic controls. These include, when awake, a greater frequency of breathholding spells and fatigue during feeding, while during sleep, these infants exhibited snoring, noisy breathing, and profuse sweating [33]. Similar symptoms of abundant night sweats have been found in future victims [34].

Cardiovascular and Autonomic Controls   Future SIDS victims exhibit symptoms during sleep that reflect a subtle dysautonomia. These include episodes of profuse sweating during sleep [34], tachycardia, bradycardia, higher overall heart rate, or a reduced heart rate variability [17]. Analysis of the heart rate variability in future SIDS victims showed findings compatible with a decrease in parasympathetic activity, an increase in sympathetic activity, or a combination of both conditions [17, 35]. Future victims of SIDS have a higher peak of sympathetic tonus, desynchronized from the parasympathetic peak activity during the late hours of the night, when most sudden deaths occur [35]. Such imbalance in cardiac autonomic control has been postulated to induce prolongation of the QTc interval in SIDS victims [36].

The greater sympathovagal control seen in infants at risk of SIDS could result from a delayed maturational process and from repetitive hypoxia, which modify brainstem, cerebellar, or cortical differences in autonomic controls [2].

Autonomic cardiac controls are also dependent on environmental factors. Increases in sympathovagal controls have been measured following the prenatal exposure to cigarette smoke [37], prone sleep [38], previous sleep deprivation [39], sleeping in high ambient temperatures [40], or with the face covered by a bedsheet [41]. Breastfeeding and the use of a pacifier, two factors associated with a lower risk for SIDS, were characterized by a reduction of the heart rate sympathovagal ratio [42].

Controls of Arousals   When an infant is exposed to a life-threatening challenge during sleep, arousal represents a protective mechanism [43]. Several studies have reported a developmental delay in sleep organization and a reduced frequency of awakenings from sleep in future SIDS victims [17, 18]. It has been shown that infants who became victims of SIDS not only aroused less from sleep than control infants but their arousal characteristics were different [44]. SIDS victims had significantly less complete arousals (cortical arousals), but more incomplete arousals (subcortical activation) than control infants. The data are suggestive of incomplete arousal processes in infants who eventually died.

Prematurity is also associated with greater arousal thresholds [45]. In utero environmental conditions can modify arousal responses. Infants of substance-abusing mothers aroused after longer exposure to hypoxia than control subjects [46]. More infants of cigarette smoking mothers than control infants failed to awake to environmental challenges [37, 47, 48].

Postnatal environmental factors also influence arousability from sleep. Viral infections of the airways, administration of sedative drugs, a previous sleep deprivation [32], sleeping prone [38, 49], sleeping with the face covered [50], or sleeping in high room temperatures [51] increase arousal thresholds.

CONCLUSIONS AND THE PREVENTION OF SIDS

Various infant and environmental factors thus modify the vital cardiocirculatory, respiratory, and arousal controls in healthy infants (See Figure 67.1). Similar changes in cardiorespiratory and autoresuscitative responses have been found in the analysis of sleep recordings of victims of SIDS. It is not known why some infants died, while others show similar changes but survive the first year of life. The death could be due to the degree of the initial immature controls, to the severity of the additional challenge, or to a combined effect of inadequate autoresuscitative mechanisms and the cumulative influence of infant and/or environmental stressors.

To understand why some infants are particularly vulnerable, further studies are required of the neurophysiological mechanisms associated with both normal infant development and infants at risk for sudden death. Continuous evaluation is mandatory as risk factors, and protective factors, change with modifications in child rearing habits and mortality.

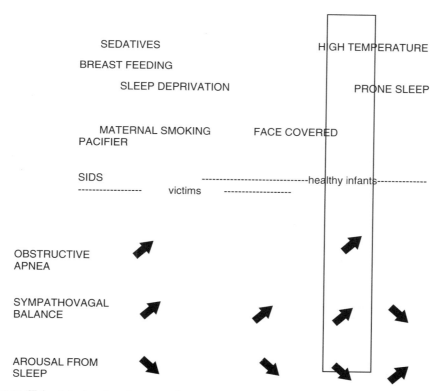

Figure 67.1   Infant and environmental stressors and their effects on vital control mechanisms.

Most environmental risk factors are modifiable risk factors. They are present in 30–80% of future SIDS victims. Their avoidance contributes to the development of safe sleep environments and reduces the risk for SIDS when health professionals and the public are kept informed. Special efforts must be directed at populations at higher risk, such as those poorly influenced by the usual prevention campaigns [52]. These include mothers of low education, immigrant families, or families of lower socioeconomic status. Attention should also be focused on most determinant risk factors, such as maternal smoking.

REFERENCES

1. Henry F. Krous, J. Bruce Beckwith, Roger W. Byard, Torleiv O. Rognum, Thomas Bajanowski, Tracey Corey, Ernest Cutz, Randy Hanzlick, Thomas G. Keens, and Edwin A. Mitchell Sudden Infant Death Syndrome and Unclassified Sudden Infant Deaths: A Definitional and Diagnostic Approach. *Pediatrics* **114**:234–238(2004).

2. Harper RM, Bandler R. Finding the failure mechanism in sudden infant death syndrome. *Nature Med* **4**:157–158(1998).

3. Kahn A, Groswasser J, Franco P, Scaillet S, Sawaguchi T, Kelmanson I, Dan B. Sudden infant death: stress, arousal and SIDS. *Pathophysiology* **10**:241–252(2004).

4. Valdes-Dapena M. The sudden infant death syndrome: pathologic findings. In: Hunt CE (Ed), *Clinics in Perinatology: Apnea and SIDS*. Saunders, Philadelphia, 1992, pp 701–717.

5. Hoffman HJ, Hillman LS. Epidemiology of the sudden infant death syndrome: maternal, neonatal, and postnatal risk factors. In: Hunt CE (Ed). *Clinics in Perinatology: Apnea and SIDS*. Saunders, Philadelphia, 1992, pp 717–738.

6. de Jonge GA, Lanting CI, Brand R, Ruys JH, Semmerkrot BA, van Wouwe JP. Sudden infant death syndrome in child care settings in The Netherlands. *Arch Dis Child* **89**:427–430(2004).

7. Emery JL. Epidemiolgy of "sudden, unexpected or rapid" deaths in children. *Br Med J* **2**:925(1959).

8. Adelson L, Kinney ER. Sudden and unexpected death in infancy and childhood. *Pediatrics* **17**:663–697(1956).

9. Fleming PJ, Blair PS, Bacon C, Bensley D, Smith I, Taylor E, et al. Environment of infants during sleep and risk of the sudden infant death syndrome: results of 1993-95 case–control study for confidential inquiry into stillbirths and deaths in infancy. *BMJ* **313**:191–195(1996).

10. Mitchell EA, Taylor BJ, Ford RPK, et al. Four modifiable and other major risk factors for cot death: the New Zealand Study. *J Paediatr Child Health* **28**:3–8(1992).

11. Ponsonby AL, Dwyer T, Couper D, Cochrane J. Association between use of a quilt and sudden infant death syndrome: a case–control study. *BMJ* **316**:195–196(1998).

12. Mosko S, Richard C, McKenna J, Drummond S. Infant sleep architecture during bedsharing and possible implications for SIDS. *Sleep* **19**:677–684(1996).

13. Carpenter RG, Irgens LM, Blair PS, England PD, Fleming P, Huber J, Jorch G, Schreuder P. Sudden unexplained infant death in 20 regions in Europe: case control study. *Lancet* **363**:185–191(2004).

14. Nakamura S, Wind M, Danello MA. Review of hazards associated with children placed in adult beds. *Arch Pediatr Adolesc Med* **153**:1019–1023(1999).

15. Mitchell EA, Brunt JM, Everard C. Reduction in mortality from sudden infant death syndrome in New Zealand: 1986–1992. *Arch Dis Child* **70**:291–294(1994).

16. Kato I, Franco P, Groswasser J, Kelmanson I, Togari H, Kahn A. Prevalence of obstructive and mixed sleep apneas in 1023 infants. *Sleep* **23**:487–492(2000).

17. Schechtman VL, Harper RM, Kluge KA, Wilson AJ, Hoffman HJ, Southall DP. Cardiac and respiratory patterns in normal infants and victims of the sudden infant death syndrome. *Sleep* **11**:413–424(1988).

18. Kahn A, Groswasser J, Rebuffat E, Sottiaux M, Blum D, Foerster M, Franco P, Bochner A, Alexander M, Bachy A, Richard P, Verghote M, Le Polain D, Wayenberg JL. Sleep and cardiorespiratory characteristics of infant victims of sudden death: a prospective case–control study. *Sleep* **15**:287–292(1992).

19. Kato I, Groswasser J, Franco P, Scaillet S, Kelmanson I, Togari H, et al. Developmental characteristics of apnea in infants who succumb to sudden infant death syndrome. *Am J Respir Crit Care Med* **164**:1464–1469(2001).

20. Shatz A, Hiss J, Arensburg B. Basement membrane thickening of the vocal cords in sudden infant death syndrome. *Laryngoscope* **101**:484–486(1991).

21. Takashima S, Armstrong D, Becker LE, et al. Cerebral hypoperfusion in the sudden infant death syndrome, brainstem gliosis and vasculature. *Ann Neurol* **4**:257–262(1978).

22. Jones KL, Krous HF, Nadeau J, Blackbourne B, Zielke HR, Gozal D. Vascular endothelial growth factor in the cerebrospinal fluid of infants who died of sudden infant death syndrome: evidence for antecedent hypoxia. *Pediatrics* **111**:358–363(2003).

23. Rognum TO, Saugstad OD. Hypoxanthine levels in vitreous humor: evidence of hypoxia in most infants who died of sudden infant death syndrome. *Pediatrics* **87**:306–310(1991).

24. Guilleminault C, Stoohs R. From apnea of infancy to obstructive sleep apnea syndrome in the young child. *Chest* **102**:1065–1071(1992).

25. Sullivan CE, Grunstein RR, Marrone O, et al. Sleep apnea-pathophysiology: upper airway and control of breathing. In: Guilleminault C, Partinnen M (Eds), *Obstructive Sleep Apnea Syndrome: Clinical Research and Treatment*. Raven Press, New York, 1990, pp 49–69.

26. Kahn A, Rebuffat E, Sottiaux M, Muller MF, Bochner A, Groswasser J. Prevention of airway obstructions during sleep in infants with breath-holding spells by means of oral belladonna: a prospective double-blind crossover evaluation. *Sleep* **14**:432–438(1991).

27. Tonkin S. Sudden infant death syndrome: hypothesis of causation. *Pediatrics* **55**:650–661(1975).

28. Groswasser J, Simon T, Scaillet S, Franco P, Kahn A. Reduced arousals following obstructive apneas in infants sleeping prone. *Pediatr Res* **49**:402–406(2001).

29. Kahn A, Groswasser J, Sottiaux M, Kelmanson I, Rebuffat E, Franco P, Dramaix M, Wayenberg JL. Prenatal exposure to cigarettes in infants with obstructive sleep apneas. *Pediatrics* **93**:778–783(1994).

30. Kahn A, Hasaerts D, Blum D. Phenothiazine-induced sleep apneas in normal infants. *Pediatrics* **75**:844–847(1985).

31. Canet E, Gaultier C, D'Allest A, et al. Effects of sleep deprivation on respiratory events during sleep in healthy infants. *J Appl Physiol* **66**:1158–1163(1989).

32. Franco P, Seret N, Van Hees JN, Scaillet S, Vermeulen F, Groswasser J, Kahn A. Decreased arousals in healthy infants following short-term sleep deprivation. *Sleep* **114**:e192–e197(2004).

33. Kahn A, Groswasser J, Sottiaux M, Rebuffat E, Sunseri M, Franco P, Dramaix M, Bochner A, Belhadi B, Foerster M. Clinical symptoms associated with brief obstructive sleep apnea in normal infants. *Sleep* **16**:409–413(1993).

34. Kahn A, Blum D, Muller F, Montauk L, Bochner A, Monod N, Plouin P, Samson-Dollfus D, Delagree EH. Sudden infant death syndrome in a twin: comparison of sibling histories. *Pediatrics* **78**:146–150(1986).

35. Franco P, Szliwowski H, Dramaix M, Kahn A. Polysomnographic study of the autonomic nervous system in potential victims of sudden infant death syndrome. *Clin Auton Res* **8**:243–249(1998).

36. Schwartz PJ, Stramba-Badiale M, Segantini P, Austoni P, Bosi G, Giorgetti R, Grancini F, Marni ED, Perticone F, Rosti D, Salice P. Prolongation of the QT interval and the sudden infant death syndrome. *N Engl J Med* **338**:1709–1714(1998).

37. Franco P, Kahn A, et al. Prenatal exposure to cigarette smoking is associated with a decrease in arousal in infants. *J Pediatr* **135**:34–48(1999).

38. Franco P, Pardou A, Hassid S, Lurquin P, Groswasser J, Kahn A. Auditory arousal thresholds are higher when infants sleep in the prone position. *J Pediatr* **132**:240–243(1998).

39. Franco P, Seret N, Van Hees JN, Lanquart JL, Groswasser J, Kahn A. Cardiac autonomic changes during sleep in sleep-deprived infants. *Sleep* **26**:845–848(2003).

40. Franco P, Szliwowski H, Dramaix M, Kahn A. Influence of ambient temperature on sleep characteristics and autonomic nervous control in healthy infants. *Sleep* **23**:401–407(2000).

41. Franco P, Lipshut W, Valente F, Adams S, Groswasser J, Kahn A. Cardiac autonomic characteristics in infants sleeping with the head covered by bedclothes. *Pediatr Res* **109**:1112–1127(2002).

42. Franco P, Chabanski S, Scaillet S, Groswasser J, Kahn A. Pacifier use modifies infant's cardiac autonomic controls during sleep. *Early Hum Dev* **77**:99–108(2004).

43. Philipson EA, Sullivan CE. Arousal: the forgotten response to respiratory stimuli. *Am Rev Respir Dis* **118**:807–809(1978).

44. Kato I, Franco P, Groswasser J, Scaillet S, Kelmanson I, Togari H, Kahn A. Incomplete arousal processes in infants who were victims of sudden death. *Am J Respir Crit Care Med* **168**:1298–1303(2003).

45. Horne RSC, Sly DJ, Cranage SM, Chau B, et al. Effects of prematurity on arousal from sleep in the newborn infant. *Pediatr Res* **47**:468–474(2000).

46. Davidson-Ward SL, Bautista DB, Woo MS, Chang M, Schuetz S, Wachsman L, Sehgal S, Bean X. Responses to hypoxia and hypercapnia in infants of substance-abusing mothers. *J Pediatr* **121**:704–709(1992).

47. Horne RSC, Ferens D, Watts AM, Vitkovic J, Lacey B, Andrew B, Cranage SM, Chau B, Greaves R, Adamson TM. Effects of maternal tobacco smoking, sleeping position, and sleep state on arousal in healthy term infants. *Arch Dis Child Fetal Neonatol Ed* **87**:F100–F105(2002).

48. Lewis KW, Bosque EM. Deficient hypoxia awakening response in infants of smoking mothers: possible relationship to sudden infant death syndrome. *J Pediatr* **127**:691–699(1995).

49. Horne RSC, Ferens D, Watts AM, Vitkovic J, Lacey B, Andrew S, Cranage SM, Chau B, Adamson TM. The prone sleep impairs arousability in healthy term infants. *J Pediatr* **138**:811–816(2001).

50. Franco P, Lipshutz W, Valente F, Adams S, Scaillet S, Groswasser J, Kahn A. Decreased arousals in infants sleeping with the face covered by bedclothes. *Pediatrics* **109**:112–117(2002).

51. Franco P, Scaillet S, Valente F, Chabanski S, Groswasser J, Kahn A. Ambient temperature is associated with changes in infants' arousability form sleep. *Sleep* **24**:325–329 (2001).

52. Kahn A, Bauche P, Groswasser J, Dramaix M, Scaillet S, et al. Maternal education and risk factors for sudden death in infants. *Eur J Pediatr* **160**:505–508(2001).

68

OBSTRUCTIVE SLEEP APNEA IN CHILDREN

PREETAM BANDLA AND CAROLE L. MARCUS
The Children's Hospital of Philadelphia, University of Pennsylvania School of Medicine, Philadelphia, Pennsylvania

INTRODUCTION

Obstructive sleep apnea syndrome (OSAS) is common in children. It is characterized by recurrent episodes of partial or complete upper airway obstruction during sleep, resulting in the disruption of normal ventilation and sleep patterns [1]. Its symptoms, polysomnographic findings, pathophysiology, and treatment are significantly different from the condition in adults (Table 68.1).

EPIDEMIOLOGY

OSAS occurs in children of all age groups. It is commonest among preschoolers due to adenotonsillar hypertrophy. The prevalence of OSAS in children is estimated to be about 2%. It occurs equally among both sexes [2]. This is in contrast to adults, where there is a male preponderance of the disease.

PATHOPHYSIOLOGY

Obstructive sleep apnea in children occurs as a result of a combination of multiple factors. These include abnormal airway structure, neuromuscular control, and other factors such as hormonal and genetic influences.

Structural Factors

Structural factors play a major role in the pathophysiology of OSAS. Most children with OSAS have some degree of upper airway narrowing as a result of either one or a combination of the following: adenotonsillar hypertrophy, craniofacial anomalies, or excess adipose tissue due to obesity. In otherwise normal children, tonsillectomy and

adenoidectomy usually lead to the resolution of symptoms, suggesting that adenotonsillar hypertrophy is a major contributing factor to childhood OSAS [3].

Achondroplasia and many congenital syndromes, including Down, Crouzon, Pierre Robin, Treacher Collins, and Cornelia de Langes syndrome, among others, have associated OSAS as a result of craniofacial anomalies such as midfacial hypoplasia, micro- or retrognathia, macroglossia, and/or associated obesity or hypotonia that result in narrowing of the upper airway.

Neuromotor Factors

A number of factors suggest that there are additional neuromotor abnormalities that play a role in the development of OSAS. (1) Children with OSAS only obstruct while asleep and not during wakefulness. (2) A small percentage of otherwise normal children continue to have persistent OSAS even after an adenotonsillectomy [3]. (3) Guilleminault et al. [4] reported a small cohort of children with OSAS who had undergone adenotonsillectomy with resolution of disease, but then went on to develop recurrence of OSAS during adolescence.

Although normal children have a smaller upper airway than adults, their airways are less collapsible and they snore less and have fewer obstructive apneas. This suggests that normal children compensate for a smaller upper airway by an increased ventilatory drive to their upper airway muscles [5]. It is thought that children with OSAS may have abnormal centrally mediated activation of their upper airway muscles that results in increased collapsibility of the upper airway [6].

Sleep: A Comprehensive Handbook, Edited by T. Lee-Chiong.
Copyright © 2006 John Wiley & Sons, Inc.

TABLE 68.1 Differences Between Childhood and Adult OSA

	Children	Adults
	Clinical Characteristics	
Peak age	Preschoolers	Elderly
Sex ratio	Equal	Male predominance, post menopausal females
Etiology	Adenotonsillar hypertrophy, obesity, craniofacial anomalies	Obesity
Body habitus	Failure to thrive, normal, obese	Obese
Excessive daytime somnolence	Uncommon	Very common
Neurobehavioral	Hyperactivity, developmental delay, cognitive impairment	Cognitive impairment, impaired vigilance
	Polysomnographic Characteristics	
Obstruction	Cyclic obstruction or prolonged obstructive hypoventilation	Cyclic obstruction
Sleep architecture	Normal	Decreased delta and REM sleep
State with OSAS	REM	REM or NREM
Cortical arousal	<50% of apneas	At termination of most apneas
	Treatment	
Surgical	Tonsillectomy and adenoidectomy (majority)	Uvulopharyngoplasty (selected cases)
Medical	CPAP (occasionally)	CPAP

CLINICAL FEATURES

The clinical features of OSAS include nocturnal symptoms such as snoring, labored breathing, paradoxical respiratory effort, observed apnea, restlessness, sweating, unusual sleep positions (e.g., sleeping sitting up, hyperextension of the neck), and secondary enuresis.

Daytime symptoms may include mouth breathing related to adenotonsillar hypertrophy, frequent upper respiratory tract infections, excessive daytime somnolence, morning headaches, fatigue, hyperactivity, aggression, and social withdrawal.

Children with OSAS are usually of normal height and weight but obesity has been increasingly recognized as a risk factor [7]. Failure to thrive and developmental delay can occur in rare cases with longstanding OSAS. Other physical examination findings may include mouth breathing, nasal voice quality, retrognathia, or micrognathia. Tonsillar hypertrophy is a common physical finding in children with OSAS, although its absence does not exclude the diagnosis.

A constellation of physical findings including a small steep mandibular plane, a high arched hard palate and an elongated soft palate, retroposition of the mandible, and a long face have been associated with OSAS [8]. The size of the adenoids and tonsils has been shown to correlate with the severity of obstructive apneas on polysomnography [9]; however, there is a large amount of clinical variability, so this cannot be used to establish a diagnosis.

Rarely, untreated OSAS resulting in pulmonary hypertension may manifest as a loud pulmonary component of the second heart sound.

COMPLICATIONS

OSAS if left untreated can result in serious morbidity from various adverse sequelae that occur as a result of chronic nocturnal hypoxemia, acidosis, and sleep disturbance.

Growth impairment can occur with OSAS and, in severe cases, may result in failure to thrive. Following adenotonsillectomy, children with OSAS frequently have a growth spurt [10]. This appears to be due to decreased caloric expenditure secondary to decreased work of breathing and an increase in the secretion of insulin-like growth factor- I following adenotonsillectomy.

Cardiovascular complications such as pulmonary hypertension, cor pulmonale, and heart failure used to be common presentations of OSAS in children, but these are now rare. Treatment of the OSAS reverses the cor pulmonale. OSAS in children can result in cardiac remodeling and hypertrophy of both the right and left ventricles, though the exact mechanism of left ventricular hypertrophy is unclear [11]. A recent study showed that children with OSAS have dysregulation of systemic blood pressure in the form of a greater mean blood pressure variability during wakefulness and sleep, a higher night-to-day systolic blood pressure, and smaller nocturnal dipping of the mean blood pressure [12]. The blood pressure dysregulation correlated with the severity of the OSAS. Increased blood pressure variability and decreased nocturnal blood pressure dipping have been shown to be associated with end-organ damage and an increased risk for cardiovascular diseases.

Untreated OSAS may result in neurocognitive deficits, learning problems, behavioral problems, and attention deficit hyperactivity disorder. Gozal [13] demonstrated a high incidence of sleep disordered breathing in poorly performing first-grade students. Children treated with tonsillectomy and adenoidectomy had significantly improved grades compared to untreated children, who demonstrated no change.

EVALUATION

The gold standard for diagnosing childhood OSAS is polysomnography. This can be performed in infants and children of any age and must be scored and interpreted using age-appropriate criteria [1]. Polysomnography can differentiate between primary snoring (i.e., snoring not associated with apnea, excessive arousals or gas exchange abnormalities) and OSAS. Children have a different pattern of upper airway obstruction compared with adults and will often desaturate with relatively short apneas. This is due to a lower functional residual capacity (FRC) and a higher respiratory rate compared to adults. Therefore obstructive apneas of any length are scored, as compared to the 10 second duration in adults [1, 14]. An apnea index of 5, while considered normal in adults, is indicative of significant OSAS in children. Many children have partial upper airway obstruction associated with hypercapnia and hypoxemia, rather than discrete obstructive apneas. This pattern has been termed "obstructive hypoventilation" [1] (Figure 68.1). Though normative data exist for childhood sleep apnea, it is yet unclear as to the degree

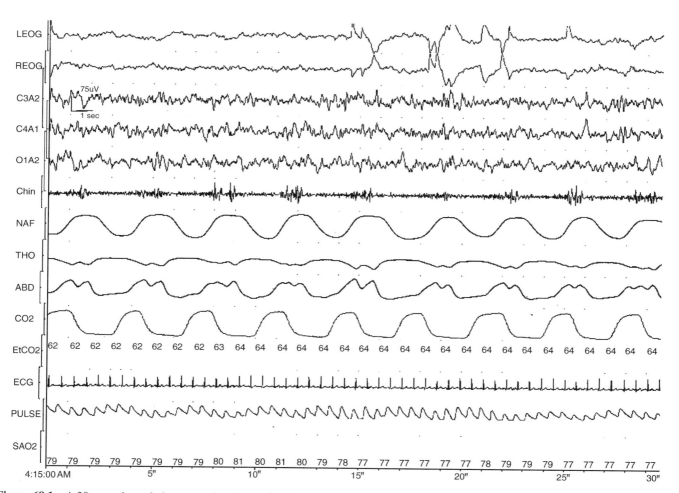

Figure 68.1 A 30 second epoch demonstrating obstructive hypoventilation in a child. Note snoring (on the chin EMG), paradoxical breathing but no apnea, associated with oxyhemoglobin desaturation and hypercapnia. LEOG, left electro-oculogram; REOG, right electro-oculogram; EEG leads (C3A2, C4A1, O1A2); Chin, chin electromyogram; NAF, oronasal thermistor; THO, thoracic wall movement; ABD, abdominal wall movement; CO2, end-tidal Pco2 waveform; EtCO2, numerical value of end-tidal Pco2; Pulse, pulse oximeter waveform; SAO2, arterial oxygen saturation.

of polysomnographic abnormalities (e.g., apnea index) that warrants intervention [15].

Screening tests such as nocturnal videotaping, pulse oximetry, and nap polysomnograms, although indicative of OSAS when positive, have limited utility because of a high false-negative rate.

TREATMENT

The vast majority of children with OSAS have both symptomatic and polysomnographic resolution following a tonsillectomy and adenoidectomy [3]. Even children with associated medical conditions such as Down syndrome [16] or obesity [17] tend to improve after adenotonsillectomy, although additional treatment may be needed. This is due to the fact that OSAS results from the relative size and structure of upper airway components, rather than the absolute degree of adenotonsillar hypertrophy.

In patients in whom adenotonsillectomy is contraindicated, or in those patients who continue to be symptomatic following adenotonsillectomy, continuous positive airway pressure (CPAP) delivered via an appropriate mask interface may be used to treat OSAS successfully in infants and children [18, 19]. This can be challenging, especially in very young or developmentally delayed children. Rarely, a tracheostomy may be necessary in very young patients, patients with craniofacial anomalies or neuromuscular syndromes, or patients who cannot tolerate CPAP or bilevel positive airway pressure following the failure of adenotonsillectomy to resolve symptoms.

Supplemental oxygen may be used in certain select patients with OSAS, either as a transitional intervention such as in neonates with mild craniofacial abnormalities who are expected to improve with growth, or in patients where all other therapeutic interventions fail and a tracheostomy is refused. Supplemental oxygen has been shown to improve oxygenation in patients with OSAS, although it does not alter the increased work of breathing or sleep fragmentation [20]. A few individuals develop a marked rise in their Pco_2 in response to supplemental oxygen [20]. Therefore, when indicated, supplemental oxygen must be started under controlled circumstances while monitoring Pco_2.

PROGNOSIS

The long-term prognosis and the natural history of childhood OSAS is unknown. It is not known whether children with OSAS will develop OSAS as adults, or whether these are two discrete entities. A study by Guilleminault et al. [4] demonstrated a 13% recurrence in adolescents who had been successfully treated for childhood OSAS. This suggests that children with OSAS, despite treatment, may be at a higher risk for the development of adult OSAS if they acquire additional risk factors such as androgen secretion at puberty, weight gain, or excessive alcohol ingestion.

Most of the complications of OSAS including cor pulmonale, behavioral problems, and growth impairment are reversible after successful treatment.

REFERENCES

1. Standards and indications for cardiopulmonary sleep studies in children. American Thoracic Society. *Am J Respir Crit Care Med* **153**(2):866–878(1996).
2. Redline S, et al. Risk factors for sleep-disordered breathing in children. Associations with obesity, race, and respiratory problems. *Am J Respir Crit Care Med* **159**(5 Pt 1):1527–1532(1999).
3. Suen JS, Arnold JE, Brooks LJ. Adenotonsillectomy for treatment of obstructive sleep apnea in children. *Arch Otolaryngol Head Neck Surg* **121**(5):525–530(1995).
4. Guilleminault C, et al. Morphometric facial changes and obstructive sleep apnea in adolescents. *J Pediatr* **114**(6):997–999(1989).
5. Marcus CL, Lutz J, Katz ES, Black CA, Galster P, Carson KA. Developmental changes in upper airway dynamics. *J Appl Physiol* **97**:98–108(2004).
6. Marcus CL, Lutz J, Black CA, Galster P, Carson KA. Upper airway dynamic responses in children with the obstructive sleep apnea syndrome. *Pediatr Res* **57**:99–107(2005).
7. Marcus CL, et al. Evaluation of pulmonary function and polysomnography in obese children and adolescents. *Pediatr Pulmonol* **21**(3):176–183(1996).
8. Guilleminault C, et al. Recognition of sleep-disordered breathing in children. *Pediatrics* **98**(5):871–882(1996).
9. Brooks LJ, Stephens BM, Bacevice AM. Adenoid size is related to severity but not the number of episodes of obstructive apnea in children. *J Pediatr* **132**(4):682–686(1998).
10. Marcus CL, et al. Determinants of growth in children with the obstructive sleep apnea syndrome. *J Pediatr* **125**(4):556–562(1994).
11. Amin RS, et al. Left ventricular hypertrophy and abnormal ventricular geometry in children and adolescents with obstructive sleep apnea. *Am J Respir Crit Care Med* **165**(10):1395–1399(2002).
12. Amin RS, et al. Twenty-four-hour ambulatory blood pressure in children with sleep-disordered breathing. *Am J Respir Crit Care Med* **169**(8):950–956(2004).
13. Gozal D. Sleep-disordered breathing and school performance in children. *Pediatrics* **102**(3 Pt 1):616–620(1998).
14. Indications and standards for cardiopulmonary sleep studies. American Thoracic Society. Medical Section of the American Lung Association. *Am Rev Respir Dis* **139**(2):559–568 (1989).

15. Cardiorespiratory sleep studies in children. Establishment of normative data and polysomnographic predictors of morbidity. American Thoracic Society. *Am J Respir Crit Care Med* **160**(4):1381–1387(1999).

16. Marcus CL, et al. Obstructive sleep apnea in children with Down syndrome. *Pediatrics* **88**(1):132–139(1991).

17. Kudoh F, Sanai A. Effect of tonsillectomy and adenoidectomy on obese children with sleep-associated breathing disorders. *Acta Otolaryngol Suppl* **523**:216–218(1996).

18. Waters KA, et al. Obstructive sleep apnea: the use of nasal CPAP in 80 children. *Am J Respir Crit Care Med* **152**(2):780–785(1995).

19. Marcus CL, et al. Use of nasal continuous positive airway pressure as treatment of childhood obstructive sleep apnea. *J Pediatr* **127**(1):88–94(1995).

20. Marcus CL, et al. Supplemental oxygen during sleep in children with sleep-disordered breathing. *Am J Respir Crit Care Med* **152**(4 Pt 1):1297–1301(1995).

69

THE SLEEPLESS CHILD

William H. Moorcroft

Northern Colorado Sleep Consultants, Fort Collins, Colorado

INTRODUCTION

Sleeplessness in children resembles insomnia in adults but has important differences [1]. For both children and adults, the problem may be one or more of the following: initially getting to sleep, staying asleep, or awakening too early. The result is a negative effect on waking behavior [2] and on medical and psychological states [3]. But in children, more than in adults, sleeplessness also affects the entire family [4] and parent–child relationships [5].

Another important distinction of child insomnia from adult insomnia is that the agents-of-change are the parents or caregivers of the poor sleeper. Likewise, the decision to make a change rests with the parents or caregivers and is somewhat subjective; what is sleeplessness for one set of parents may not be for another set [6].

It is also advisable to keep in mind that developmental milestones can affect the sleep of children [4]. Teething, crawling, walking, as well as cognitive development can suddenly change a good sleeper into a sleepless one. For example, the development of independence and autonomy in toddlers may increase bedtime resistance [4]. Usually the change is transient, but it may lead to reactions that perpetuate sleeplessness.

The most common causes of sleeplessness in children—sleep-onset association problems, night wakings, early awakening, problems of nighttime eating/drinking, separation anxiety, parental limit setting, and night fears—are discussed here. Sleeplessness in children may also be caused by problems discussed elsewhere in this volume. These include nightmares and bad dreams, delayed sleep phase syndrome, parasomnias, poor sleep hygiene, obstructive sleep apnea, restless legs syndrome, periodic limb movement disorder, and narcolepsy. Ironically, insufficient sleep may also be the culprit [6]. Indeed, parents have noted hyperarousal occurring in their overtired child. Additionally, in the contemporary Western World, television, videos, and computer games have become a major source of sleeplessness in children. Finally, even though sleeplessness may be primarily caused by a medical factor, there may also be a comorbid behavioral cause or a behavioral cause may develop from a medical one [4] and perpetuate the sleeplessness.

Two additional points: First, it is well established that most childhood sleeplessness can be reduced or even prevented by parental education [7]. This is better and more cost effective than treating the problems once they are developed. Second, it appears that the behavioral methods described later also work in children with special needs such as physical illness, psychological problems, or intellectual disabilities [8].

EARLY CHILDHOOD

Sleeplessness in early childhood is common [3, 9, 10]. For example, the National Sleep Foundation's Sleep in America Annual Poll of 2004 found that 71% of parents report that their infants woke up and needed help or attention in the past two weeks. Twenty-one percent stated that this happened more than three times per week.

Sleep: A Comprehensive Handbook, Edited by T. Lee-Chiong.
Copyright © 2006 John Wiley & Sons, Inc.

Behavioral Insomnia of Childhood, Sleep-Onset Association Type (ICSD V69.5)

Sleep-onset association disorder [11] occurs when children are unable to fall asleep by themselves; rather, they regularly need external help, such as being nursed or rocked or watching television. It can also be the result of separation anxiety [12] that peaks at 18–24 months [4], but in most cases it is created by parents who are active in helping their child get to sleep at bed and nap times [4].

The manifestation of the problem most noticed by parents is night awakenings. Typically, the real issue is not the awakenings per se, since it is known that all young children awaken during the night [4], but whether the child is a "soother" or a "signaler." Most of the time parents are not aware that soothers awaken because they lay quietly until they fall back to sleep. It is the signalers who are said to have a problem because they cry and fuss until their parents respond to them before they will fall back to sleep. When the child learns to associate parental attention with sleep onset, they have difficulty falling asleep or returning to sleep without it [2]. Colicky infants are more likely to be treated in this way and thus more likely to acquire problems with sleep-onset association [13].

Other causes of sleep-onset association problems are inadequate sleep and parents who respond to every sound the child makes at night [4]. There can be night-to-night and week-to-week variability in night wakings with signaling [4].

Treatment for sleep-onset association problems is aimed at allowing the child to form a new set of associations for falling asleep [10] without ignoring their genuine needs. The initial step is to provide circumstances for these children that are conducive to falling asleep. A good sleep environment is an important first step [6]. (Some parents like to have a sound system in the child's bedroom that plays soothing music designed for children at sleep onset. The problem with this is that it also needs to be present when the child wakes up and at naptime because the child will associate falling asleep to this music.) Next, put the child to sleep when drowsy but still awake [10]. A child should be able to "sleep through the night" by about 3 months of age so parents should be advised to start putting the child to bed when drowsy but awake at about 2–4 months [4].

It is also important to establish and consistently follow a regular schedule of bed and nap times that are consonant with a child's natural schedule [4]. A regular bedtime routine that is quiet (such as reading) and takes place in the space where the child sleeps is also beneficial [4]. Transitional objects (such as a blanket or stuffed animal) that is present at bedtime, naptime, and when the child awakens during the night can be of great help [4]. A pacifier can serve this purpose in infants but is discouraged after about 5–6 months of age since it frequently falls out during the night and the baby cannot retrieve it. For children older than 3 years of age, positive reinforcement for bedtime behaviors appropriate to falling asleep can also be helpful [4].

Once this baseline of bedtime routine is established, the next step involves actively promoting self-soothing. Here there are several options available to the parent. At one extreme is total extinction whereby the child is put to bed before falling asleep and without the rocking, nursing, holding, or whatever the child had become accustomed to [6, 14]. Typically this will result in loud and prolonged crying when first attempted. However, this response will usually subside over a few days as the child learns to self-sooth. Extinction is a well-established treatment for sleep-onset association problems [7]. It has the advantage of being quick (a few days) and simple [14]. However, many parents cannot endure their child's prolonged crying or feel guilty about it and eventually give in and comfort the child [15]. In the end, giving in is highly counterproductive, because it only serves to reinforce the signaling behavior.

There is also some controversy about this procedure or any procedure that allows the child to cry for even brief periods. Sears [16] maintains that allowing the child to cry without a caregiver responding psychologically damages the child. However, this is a minority opinion. There is no evidence to back up the claims made by Sears, while there is evidence of completely normal psychological development when children are allowed to cry as long as they receive love and attention during their waking hours [4, 14].

A method generally more tolerated by parents and generally more widely recommended [7] is what has been dubbed fading [6] or graduated extinction [7] that is championed by Ferber [17]. Some crying is allowed, but also some soothing. The soothing should be largely verbal in a voice that is slow and quiet [10]. The soothing is gradually reduced in intensity and duration as the child learns to become more self-soothing. Regular checking on the crying child is encouraged, but the interval between checks is progressively lengthened (such as by 5 minute increments). During the checking, the parents verbally encourage but do not help the child to sleep. Direct contact should be minimal if at all. This is enough to reassure the child but not to reinforce the signaling. Refer to Ferber [17] for more detailed descriptions of how this should be done.

Pantley [18] has developed another version of fading. For example, if the child needs to be nursed to fall asleep, the amount of time nursing the child before sleep is gradually reduced (or if bottle-fed, the amount of liquid in the bottle is gradually reduced). A similar approach is taken for eliminating holding or rocking the child to sleep.

Fading is thought to have several advantages [15] as well as being "probably efficacious" [7]. It can be less trying for the parents. It also is flexible and can be varied to suit individual needs. It is thought to be especially good for children experiencing separation anxiety. However, it can take longer to arrive at the goal of making the child self-soothing. As a result, the parents may abandon it too soon. For this method to work, caregivers need to be persistent and consistent. It is also more complex than extinction and requires making appropriate adjustments depending on the response and progress of the child.

It should be pointed out that following successful extinction or fading, an "extinction burst" is a common occurrence. That is, some time (days or weeks) later, crying at bedtime or following awakening reoccurs, often with great intensity. However, if not overreacted to, it will usually quickly subside. Parents need to be made aware of this probability ahead of time so that they do not overreact or sink into despair because it seems that their efforts have failed. See Stores [6] for additional information on handling relapses.

Another, but controversial, approach for reducing night awakenings in children is to implement scheduled awakenings [15]. Here the parents wake the child before the typical times of self-awakening. Following this early waking, the child usually quickly falls asleep without later awakening. The scheduled wakings are then gradually reduced until eliminated. This method has been found to be "probably efficacious" [7] but can be quite disrupting and difficult for parents.

Whatever procedure is used, it is thought best to begin by initiating it only at bedtime. Often what the child learns at bedtime quickly generalizes to night awakenings and naptime. However, if generalization does not occur, then later applying the same approach during night awakenings and naps can be implemented.

Another approach taken by some parents is cosleeping (bed sharing, family bed). Although a controversial sleeping arrangement for young children in Caucasian Western societies [19], it is a common practice with some non-Caucasians in non-Western parts of the world. Some parents start cosleeping from day one, but others resort to it later out of desperation in response to signaled awakenings by their child [4]. However, it reduces the parents' privacy, can disrupt the sleep of one or both parents, and may not suit both partners [15].

Cosleeping may only delay confronting the problem of night wakings, because at some point the child has to transfer to sleeping in his/her own bed or crib, at which time the child may still have difficulty falling asleep on his/her own. However, some children have been reported to make the transition without problem. The transition process may be done abruptly by putting the child to bed or crib in his/her own room, or by first shifting from the parents' bed to the crib next to the parents' bed, then gradually moving the crib further and further away in the parents' bedroom, then into a separate room [18].

Nighttime Eating/Drinking Problem (Behavioral Insomnia of Childhood (ICSD V69.5) 780.52-8)

Nighttime eating/drinking problem occurs when the child, after 2 months of age, has "frequent and reoccurring awakenings to eat or drink" that are followed by normal sleep [11]. A soaked diaper during the night can be an indicator of this problem. Nighttime eating/drinking problem can result from the child learning to feel hunger without actually needing nourishment [5].

Night feedings are not physiologically needed after about 6 months, but night awakenings may continue to occur in up to 50% of babies. In most babies this gradually diminishes, but a small number do not seem to "outgrow" this and may need behavioral intervention [4]. Nighttime eating/drinking problem is more likely to occur in breastfeeders [4]. Thus it is best to wean children from night breastfeedings by 6 months and not put them to bed with a bottle after this time [4].

Treatment for nighttime eating/drinking problem is to gradually lengthen the interval between feedings over a week or two and/or gradually reduce the amount of fluid in the bottle [5]. Sudden stopping is not recommended.

Sleep Problems at Naptime

Some children may only have a sleep problem at naptime. This could be a problem with the nap schedule [4]. Having children try to take a nap before they are ready is going to be unsuccessful. Likewise, having children try to nap way beyond the time when they need to nap will cause arousal from overtiredness and make falling asleep difficult. Difficulties with naps may also be due to separation anxiety [18] or a problem of poor sleep associations [4].

Separation Anxiety

If separation anxiety is the main source of the child's sleeplessness, encourage parents to increase daytime nurturing but also to do brief separations throughout the day. They can occasionally go to another room while singing or whistling so that the child knows the out-of-sight parent is still there. Also, quiet time just prior to naptime or bedtime should be encouraged. A large photo of Mom and Dad can be placed near the child's bedside. If the child is awake when put down for a nap, instruct the parent not to sneak away but to cheerfully say "good night" on their way out. However, a parent should respond quickly to their child's nighttime calls with "I'm here and everything's OK."

Early Awakening

Some children have a problem with early awakening [6]. There is no clear definition of early awakening, it will vary from one family to another. However, as a rule of thumb, the criterion of awakening before 5 a.m. is sometimes used. The source of the problem may be feeding or watching TV upon awakening, allowing a child to join parents in bed in the morning, or some other reinforcing event that leads to conditioned awakening [4, 6]. Additional possible causes include regular morning environmental disturbance, circadian phase advance, or less need for sleep than average [12]. Treatment should be directed at modifying what is causing the problem.

Colic

"Infantile colic," typically starting at 2–3 weeks of age and resolving in a few months, can cause sleeplessness [6]. The sleeplessness may continue if the child has learned to expect a lot of attention from his/her parents. The actual cause of colic has not been established [6]. Although antispasmodic drugs may relieve the colic, giving help to parents to be able to cope with the crying is at least necessary and often preferable [6].

LATER CHILDHOOD

Behavioral Insomnia of Childhood, Limit Setting Type (ICSD V69.5)

In later childhood the problem of sleeplessness is more commonly due to a problem of parental limit setting that tends to remain chronic [4]. It can occur at bedtime, naptime, or even after night awakenings. Limits are not set or enforced or are enforced inconsistently by parents [4]. Sometimes there are conflicting parenting styles, but the environment (e.g., loud TV, older siblings who go to bed later) can also be the problem. Other sources include phase delay syndrome, sensory integration problems, or being developmentally delayed accompanied by a problem of self-soothing [4]. Additional factors might include inadequate sleep hygiene, medical conditions, family issues, generalized oppositional defiant disorder (ODD), anxiety, or even restless legs syndrome [4]. The problem can manifest itself as outright refusal to go to bed to more subtle "curtain calls."

The parents are the key to treatment. They need to take on the role of being responsible for setting and enforcing a reasonable bedtime. They need to be firm, clear, and consistent, yet calm in denying or ignoring delaying tactics or protests to avoid reinforcing bedtime resistance. At the same time, they need to avoid trying to scare the child with punishments or threats. The goal is to teach the child a better way to fall asleep by increasing positive behaviors more than it is to eliminate negative ones [4].

The parent should never *ask* the child to go to bed; rather give gentle, but firm, commands. An alternative is to provide narrow choices for the child from which to choose. If the child is upset or crying when sent to bed, the parent should do brief (1 minute) checks in a reassuring but boring way [4]. Gradually lengthen these intervals to several minutes. If the child comes out of his/her bedroom, immediately return the child. Praise him/her for staying in bed.

Additional measures may be necessary. Placing a gate in the child's bedroom doorway, or closing the door when the child gets out of bed, but opening it again when he/she returns to bed can be effective. Tell the child that the gate will be removed (or door will gladly be opened) when he/she stops trying to leave the room. In any case, the parent should not go into the room, but stand on the other side of the gate/door out of sight and speak to the child in regular intervals (such as every few seconds, if necessary) in a calm voice. Gradually lengthen these intervals to several minutes. It is ok if the child falls asleep by the gate or door. Greater detail on what can be done and how to do it can be found in Associated Professional Sleep Societies pamphlets [5].

When faced with a limit-setting problem, it is critical for the parents and other caregivers to remain strictly consistent. Giving in or letting up on occasion will only cause the child to try longer and harder to avoid bed and will greatly prolong the process of changing the child's behavior. Parents should also be made aware of the likelihood of an "extinction burst" (see previous discussion) some days or even weeks after successful training.

Sleep-Onset Association Disorder (ICSD 307.42-5)

A good treatment regime for parents to follow for sleep-onset association problems in older children is desensitization. It usually takes about 1–3 weeks. They should tell their child that they will sit in a chair near the bed (not in or on the bed) until the child falls asleep. After a few successful nights, they should gradually move the chair further and further from the bed toward the door. Eventually, they should move the chair out of the door but leave the door open if the child does not get out of bed; briefly close it if the child does get out of bed. Even more effective is to couple this entire procedure with positive reinforcement such as stars or small prizes for staying quietly in bed. Regardless, praise the child the next morning for success.

Night Fears

Night fears may also be a cause of sleeplessness in children. The child typically has an aversion to being alone in his/her bedroom at night, especially without a light [6]. A variation is when the child worries that his/her parents may come to harm during the night. It is important to note that some children feign being afraid as a delaying tactic. Night fears may be seen in toddlers but peak incidence is between ages 3–6, when children's imaginations and fantasies are maturing [4], but can also occur at later ages. Night fears are common but usually benign. Parents need to respond to their child's night fears with reassuring and nurturing while being careful not to reinforce the child's fears by overreacting. If the fear is strong and resilient, the child may need help from a behavioral sleep specialist to overcome the fear [6].

Rumination

A common cause of trouble getting to sleep in older children is rumination about recent happenings [10]. Extra attention from a caregiver and conversation at bedtime are the best remedies.

INEFFECTIVE AND CONTRAINDICATED APPROACHES

Making changes in the sleepless child's diet or feeding schedule has not been found to be effective [7]. Likewise, simply telling parents that the child will outgrow the problem is not helpful.

Some things that are contraindicated for treating the sleepless child include manipulating naps and giving drugs to induce sleep. Some parents may be tempted to try reducing naptime in order to make the child sleepier in the hope that the tired child will sleep more. This does not work. The excessively sleepy child becomes overaroused (as many parents have observed), which makes sleep onset more difficult for the child and even adds to the problems of nighttime awakening. Parents should also be cautioned that napping too close to bedtime can make it more difficult for the child to fall asleep and stay asleep.

Hypnotics and over-the-counter medications containing antihistamines should not be used to induce sleepiness in children. Such treatments have not been studied and have not been given FDA approval [4]. Clinical experience suggests that the drug treatment approach has yielded mixed results with only short-term [6] and weak improvement. At best, it only temporarily covers up the underlying problem.

A serious problem with using a drug to induce sleep in the sleepless child is that the medication may be given to the child for the benefit of the parent. Also, it sends the wrong message to the child about how to treat a problem with sleep and may actually hinder the child from learning self-soothing [6]. While drugs may be appropriate in some situations such as, when behavioral treatment is not helpful, it is thought "better to find and treat the cause of the sleep problem" [4]. If drugs are used, they should be accompanied by good sleep hygiene and supplemented with behavioral treatment. The clinician should also be aware that parents may give their children OTC drugs or herbal remedies and fail to mention this to their health care provider because they do not realize that these drugs can themselves be the source of problems [4].

REFERENCES

1. Ferber R. Sleeplessness in children. In: Ferber R, Kryger MH (Eds), *Principles and Practice of Sleep Medicine in Child.* Saunders, Philadelphia, 1995, pp 79–89.

2. Ward T, Mason TBA. Sleep disorders in children. *Nursing Clin North Am* **37**:693–706(2002).

3. Stein MA, Mendelsohn J, Obermeyer WH, Amromin, J, Benca R. Sleep and behavior problems in school-aged children. *Pediatrics* **107**:e60(2001).

4. Mindell JA, Owens JA. *A Clinical Guide to Pediatric Sleep: Diagnosis and Management of Sleep Problems.* Lippincott, Williams & Wilkins, Philadelphia, 2003.

5. Associated Professional Sleep Societies. *My Child Can't Sleep.* Associated Professional Sleep Societies, Westchester, IL, 2000.

6. Stores G. *A Clinical Guide to Sleep Disorders in Children and Adolescents.* Cambridge University Press, New York, 2001.

7. Mindell JA. Emprically supported treatments in pediatric psychology: bedtime refusal and night wakings in young children. *J Pediatr Psychol* **24**:465–481(1999).

8. Wiggs L, France K. Behavioral treatments for sleep problems in children and adolescents with physical illness, psychological problems or intellectual disabilities. *Sleep Med Rev* **4**:299–314(2000).

9. National Sleep Foundation. Children and sleep. *Sleep in America Poll, 2004.* http://www.sleepfoundation.org/polls/2004SleepPollFinalReport.pdf (accessed July 30, 2004).

10. Garcia J, Wills L. Sleep disorders in children and teens; helping parents and their families get some rest. *Postgrad Med* **107**:161–164, 170–171, 175–178(2000).

11. *International Classification of Sleep Disorders* 2nd ed., American Society of Sleep Medicine, Westchester, IL, 2005.

12. Stores G, Wiggs L. *Sleep Disturbance in Children and Adolescents with Disorders of Development: Its Significance and Management.* Mac Keith Press, London, 2001.

13. Thiedke CC. Sleep disorders and sleep problems in childhood. *Am Fam Physician* **63**:277–284(2001).

14. Weissbluth M. *Healthy Sleep Habits, Happy Child.* Ballantine Books, New York, 2003.

15. Huntley R. *The Sleep Book for Tired Parents.* Parenting Press, Seattle, WA, 1991.

16. Sears W. *Nighttime Parenting: How to Get Your Baby and Child to Sleep.* Plume, New York, 1999.

17. Ferber R. *Solve Your Child's Sleep Problems.* Simon & Schuster, New York, 1985.

18. Pantley E. *The No-cry Sleep Solution.* Contemporary Books, Chicago, 2002.

19. Moorcroft WM. *Understanding Sleep and Dreams.* Kluwer Academic/Plenum Press, New York, 2003.

ADDITIONAL READING

Mindell JA. *Sleeping Through the Night: How Infants, Toddlers and Their Parents Can Get a Good Night's Sleep.* Harper-Collins, New York, 1997. Recommended by many as one of the best books for parents for helping younger children sleep better.

70

THE SLEEPY CHILD

GERALD ROSEN

Hannepin County Medical Center, Minneapolis, Minnesota

INTRODUCTION

Before discussing the sleepy child, it is important to distinguish him/her from the fatigued child. A sleepy child falls asleep at inappropriate times, times that one would reasonably expect the child to remain awake and alert; the fatigued child does not. The reason for making this distinction explicit is that fatigue and sleepiness are different, but parents and health care professionals alike often confuse the two. In this chapter, fatigue is used to describe a subjective, nonspecific feeling that has many causes and is difficult to measure; sleepiness is defined as the propensity to fall asleep and can be objectively quantified using the multiple sleep latency test (MSLT). Children who are sleepy do feel fatigued; but children who are fatigued are not necessarily sleepy—sleepiness being just one of many causes of fatigue. This chapter will address the sleepy child.

Sleepiness is a cumulative symptom, which results from one or more of the following problems:

1. Insufficient sleep quantity.
2. Poor sleep quality (sleep fragmentation).
3. Attempting to remain awake during the circadian sleep time.
4. As a primary neurologic symptom.
5. As a result of the use and/or withdrawal from some psychotropic medications.

Table 70.1 lists some of the more common causes for each of these problems that are seen in children.

NORMAL SLEEP IN A CHILD

A "normal" child is very alert during the day, rarely exhibiting daytime sleepiness except at regular nap times; conversely, when asleep he/she appears quiet and peaceful. Children normally have a high sleep efficiency, an abundance of slow-wave sleep, no obstructive apneas, and few periodic leg movements, or respiratory-related or behavioral arousals. Normally, children transition to wakefulness rapidly at the end of their sleep period and fall asleep quickly at bedtime and naptime [1].

DEVELOPMENTAL CHANGES IN THE SLEEP OF CHILDREN

Predictable developmental changes occur in children's sleep from birth through adolescence [2] (see Table 70.2). At birth, infants sleep up to 19 hours/day and have no clear circadian organization of their sleep–wake patterns. A newborn's sleep is described as active (rapid eye movement (REM) sleep precursor), quiet (NREM sleep precursor), or indeterminate. Active sleep accounts for about half of the infants total sleep time and is the state into which the child transitions to sleep out of wakefulness. By 6 months of age children's sleep can be staged using the Rechtshaffen–Kales sleep stage scoring criteria. They transition into NREM sleep from wake and are physiologically capable of consolidating 8 hours of sleep without a behavioral awakening, which allows for the long awaited developmental milestone of "sleeping through the night." This is actually

Sleep: A Comprehensive Handbook, Edited by T. Lee-Chiong.

TABLE 70.1   Sleep Problems and Their Causes that Lead to Sleepiness in Children

Sleep Problem	Causes
Insufficient sleep quantity	Acute, chronic
Poor sleep quality	Sleep apnea, periodic movements of sleep, seizures
Attempting to remain awake during the circadian sleep time	Delayed sleep phase, advanced sleep phase, irregular sleep/wake times, jet lag
As a primary neurologic symptom	Narcolepsy with /without cataplexy, idiopathic hypersomnia, myotonic dystrophy, seizures, central nervous system pathology (tumor, trauma, stroke, infection, postinfection), Prader–Willi Syndrome, Möbius' Syndrome, Smith–Magenis Syndrome, fragile X Syndrome, Niemann–Pick disease, recurrent hypersomnia
As a result of use and/or withdrawal of psychotropic medications	Stimulants, antidepressants, antipsychotics, antihistamines, alcohol, opiates, sedatives, antiepileptics, alpha agonists, beta blockers

a misnomer, insofar as everyone—children and adults alike—have numerous normal, brief, spontaneous awakenings every night that are not associated with a complete behavioral awakening. At 6 months of age most children are sleeping 11–12 hours at night and another $3\frac{1}{2}$ hours during the day divided between a morning and an afternoon nap. At $1\frac{1}{2}$ years of age, nighttime sleep remains at about 11–12 hours, but daytime naps have decreased to once a day for 1–2 hours duration, generally in the afternoon. Most children discontinue their daytime naps between 3 and 6 years of age without increasing their nighttime sleep, which remains at about 11 hours. There is a gradual decrease in nighttime sleep duration from 11 hours to 10 hours that occurs between 6 years of age to the beginning of adolescent sexual development. Preadolescent children are usually very alert during the day, with mean MSLT scores generally above 18 minutes. Adolescents become sleepier during the day at about the time of sexual maturation, at Tanner stage III, and they also have a phase delay in their preferred time of sleep onset. If they are not sleep deprived, the increase in daytime sleepiness during adolescence is modest with an decrease in mean MSLT scores from 18 minutes to 15 minutes [3].

TABLE 70.2   Average Sleep Duration: Birth–Adolescence

Age	Average Number of Hours of Sleep/Day	Range (h/day)	Naps
Infants (2–12 months)	14.5 h	10–16 h	3
Toddlers (1–3 years)	13.5 h	9–16 h	2
Preschool (3–5 years)	11 h	8–12 h	1
School age, preadolescent (5–13 years)	10 h	8–10 h	0
Adolescent (delay in sleep phase, increase in daytime sleepiness)	9 h	8–12 h	0

If children are allowed to sleep ad lib, in an environment that is appropriate, that does not have excessive light exposure at night, and is conducive to sleep, without an externally imposed schedule, and they have no primary sleep problems, they generally will establish a regular sleep–wake pattern and a consistent sleep duration that is a reflection of the synchronization of their homeostatic sleep needs and their preferred circadian sleep times. Though there is a great deal of intraindividual variation in a child's sleep duration and preferred sleep times; both of these traits tend to be stable over time in an individual child. Though there are age-related changes in these traits as were described earlier, these changes occur within the context of the child's stable homeostatic and circadian traits. This means that children who were short sleepers as toddlers, typically remain short sleepers as they get older; and the same is true for children who were long sleepers. The preferred circadian timing for sleep and wake is also a relatively stable individual trait. Toddlers who are night owl's, preferring to go to sleep after their parents do, remain night owls as they get older. For a child to have the best and longest sleep possible, the circadian and homeostatic processes must be synchronized [4]. This occurs largely through how we schedule our lives, the amount of time we allow for sleep, and the exposure to light after dusk and at dawn. When the circadian and homeostatic systems are properly synchronized, sleep onset occurs quickly, sleep efficiency is high, arousal in the morning is spontaneous, and the level of daytime alertness is high if there are no other intervening sleep problems.

EXCESSIVE SLEEPINESS IN CHILDREN

The problems and causes of excessive daytime sleepiness (EDS) are listed in Table 70.1. In addition to the defining symptom of falling asleep at inappropriate times, children who are sleepy have behavioral symptoms—yawning, eye rubbing, irritability—and cognitive symptoms [5]—slower reaction times, poorer learning. Though most of these symptoms occur concomitantly, there is not a tight correlation among them and there is a great deal of individual variability in their expression. The causes of daytime sleepiness are cumulative, and many children will have more than one cause that is contributing to their daytime symptoms of sleepiness. The symptoms of sleepiness are similar regardless of their cause. As described in Table 70.1, children become sleepy if they have had an inadequate amount of sleep; if they have a problem that leads to multiple awakenings that may fragment sleep, such as sleep disordered breathing, periodic movements of sleep, and sleep-related seizures; if they attempt to remain awake during their circadian sleep time or have an erratic sleep–wake schedule; if they have a primary sleep problem such as narcolepsy, with and without cataplexy, seizures, a brain tumor, CNS infections/strokes/trauma, idiopathic hypersomnia, myotonic dystrophy, Kleine–Levin syndrome, or Prader–Willi syndrome ; or if they are taking or withdrawing from a psychotropic drug or alcohol. Narcolepsy is discussed in Chapter 20. Recurrent, post-traumatic, postinfectious, and menstrual-related hypersomnia are discussed in Chapter 83.

DIAGNOSTIC APPROACH TO THE SLEEPY CHILD

The first step in the evaluation of a child with sleepiness is a complete medical/neurologic/sleep history, which, in most cases, will point toward one or more of the causes of EDS listed in Table 70.1. Since sleepiness is a cumulative symptom, if a child has more than one cause for the EDS, the symptoms will persist until all of the causes are treated. Consequently, it is important to gather a complete sleep history on all children who present with EDS. The most common cause of sleepiness in developed countries is chronic sleep insufficiency, which is often associated with a sleep phase delay. Sleep logs are often sufficient to define this problem, though occasionally actigraphy is necessary. The characteristic finding in these cases is of chronic sleep curtailment when the child needs to adhere to a defined sleep schedule, with sleep extension when the schedule permits, typically on the weekends. The best way of evaluating whether sleep insufficiency is the cause for daytime sleepiness is to allow the child to sleep ad lib, without any scheduling constraints for 2 weeks. A sleep log or ideally actigraphy should be used to document the timing and amount of sleep. If the daytime sleepiness resolves with sleep extension, then the cause is most likely chronic sleep insufficiency. Some children have a very long sleep requirement, up to 12–14 hours/day, which if not met results in sleepiness. If the sleepiness does not resolve with sleep extension, than a polysomnogram (PSG) is necessary for defining the presence and severity of sleep

fragmentation from periodic limb movement disorder (PLMD), obstructive sleep apnea (OSA), and seizures and/or for establishing the diagnosis of narcolepsy or idiopathic hypersomnia; the PSG must be obtained before a MSLT. A cautionary note is important in the interpretation of the MSLT. Sleep deprivation and/or withdrawal from psychotropic drugs can mimic the MSLT findings of narcolepsy. In adolescents it is advisable to always obtain at least a sleep log and preferably a 2 week actigraphic recording before the MSLT and a urine drug toxicology screen the night of the PSG.

NARCOLEPSY WITH AND WITHOUT CATAPLEXY IN CHILDREN

Narcolepsy with and without cataplexy (see Chapter 20) affects about 0.02% of the U.S. population. Onset is rare before 5 years of age and typically occurs between the ages of 10 and 25 years. Excessive daytime sleepiness, defined as falling asleep at inappropriate times, is usually the first symptom. Cataplexy, hypnagogic hallucinations, and sleep onset paralysis are less common at the time of presentation, and a history of these symptoms may be difficult to elicit from young children. Narcolepsy will most often manifest itself with the resumption of daytime napping in a child who had previously discontinued naps, when there is an adequate sleep quantity and no evidence of other sleep problems. A PSG and MSLT are important in the diagnosis of narcolepsy in children, but the studies may initially show hypersomnolence with mean MSLT scores below 8 minutes, but without the characteristic two or more REM onset naps. However, over time with repeat testing the MSLTs invariably do show the REM sleep abnormalities characteristic of narcolepsy. Classical narcolepsy is caused by the loss of hypocretin secreting cells in the lateral hypothalamus. Secondary narcolepsy has the same PSG findings but is seen in children with known neurologic disease. The most common pediatric cause of secondary narcolepsy is hypothalamic injury secondary to a brain tumor. The daytime sleepiness of narcolepsy is treated with stimulant medication and cataplexy is usually treated with tricyclic antidepressants; both of which are remarkably effective.

IDIOPATHIC HYPERSOMNIA WITH AND WITHOUT A LONG SLEEP REQUIREMENT

Idiopathic hypersomnia with and without a long sleep requirement is characterized by a constant and severe daytime sleepiness regardless of how much sleep is obtained. Similar to narcolepsy, the onset is typically adolescence to early adulthood. At the time of presentation, many children who ultimately will prove to have narcolepsy will be diagnosed with idiopathic hypersomnia because of the absence of REM sleep abnormalities on their PSG/MSLT. Treatment of idiopathic hypersomnia is the same as for narcolepsy with stimulant medication.

SLEEP DISORDERED BREATHING IN CHILDREN

Sleep disordered breathing is present in 2–5% of children. In adults with obstructive sleep apnea (OSA), excessive daytime sleepiness is a common symptom. However, sleep disordered breathing is different in children than in adults, and one of the ways it is different is in the prevalence of excessive daytime sleepiness. Children with OSA will all have snoring, and many also have observed apnea and neurocognitive deficits; but EDS among children with OSA is less clear-cut. Though many children with OSA, and their parents, will complain of excessive daytime sleepiness [6], most do not meet the stringent criteria of short mean sleep latencies on the MSLT that has been used in this chapter to define excessive daytime sleepiness [7]. Sleep disordered breathing in children is addressed more extensively in Chapter 68.

Childhood is a time of growth, learning, and development. In order for these processes to occur the child must be awake, alert, and able to interact with and learn from the environment. Excessive daytime sleepiness impairs the child's ability to do this. The problems that lead to EDS in children can generally be elucidated by a careful history and eliminated by appropriate treatment. For these reasons, the symptom of EDS in children should be taken seriously and investigated thoroughly.

REFERENCES

1. Sheldon S. Sleep in infants and children. In: Lee-Chiong T, Sateia M, Carskadon M (Eds), *Sleep Medicine*. Hanley & Belfus, Philadelphia, 2002, pp 99–104.

2. Iglowstein I, Jenni O, Molinari L, Largo R. Sleep duration from infancy to adolescence: reference values and generational trends. *Pediatrics* **111**:302–307(2003).

3. Carskadon MA, Acebo A. Regulation of sleepiness in adolescents: updates, insight's, and speculation. *Sleep* **25**:606–614(2002).

4. Dijk D, Lockley SW. Integration of human sleep–wake regulation and circadian rhythmicity. *J Appl Physiol* **92**:852(2002).

5. Fallone G, Owens J, Deane J. Sleepiness in children and adolescents: clinical implications. *Sleep Med Rev* **6**:287–306(2002).

6. Goodwin J, Kaeming K, Fregosi R, Rosen G, Morgan W, Sherrill D, Quan SF. Clinical outcomes associated with sleep-disordered breathing in Caucasian and Hispanic children—the Tucson children's assessment of sleep apnea study (TUCSA). *Sleep* **26**:587–591(2003).

7. Gozal D, Wang M, Pope D. Objective sleepiness measures in pediatric obstructive sleep apnea. *Pediatrics* **108**:693–697(2001).

71

CRANIOFACIAL SYNDROMES AND SLEEP DISORDERS

LAUREL M. WILLS, JAMES Q. SWIFT, AND KARLIND T. MOLLER
University of Minnesota, Minneapolis, Minnesota

INTRODUCTION

Children and adults with craniofacial syndromes represent a population at increased risk for sleep disordered breathing, due to anatomic and, in some cases, neuromuscular differences in the upper airway. Less well documented, but still a common clinical chief complaint, is the tendency in these individuals to report behavioral sleep complaints, such as sleep onset or sleep maintenance insomnia. This chapter will review the predisposing risk factors, the prevalence, and consequences of sleep disordered breathing for this group of patients, as well as the techniques for evaluation and intervention, both invasive and noninvasive, currently available. The reader is referred to other chapters of this text for information on behavioral sleep complaints in children with chronic health conditions and developmental concerns.

Definition and Scope of Problem

Numerous review papers discussing the topic of snoring and obstructive sleep apnea (OSA) in children include craniofacial syndromes, along with adenotonsillar hypertrophy, obesity, and neuromuscular diseases as conditions that place children at increased risk for sleep disordered breathing (SDB). Patients requiring evaluation by an interdisciplinary craniofacial clinic team may include those with congenital syndromes affecting bony and/or soft tissue structures of the cranium, face, and neck that compromise upper airway patency. They may also include patients with craniofacial trauma, vascular or lymphatic malformations (such as cystic hygroma), head and neck tumors, or new-onset of OSA symptoms following pharyngeal flap surgery or other pharyngoplasties to improve velopharyngeal closure for speech. Other patients *without* obvious facial dysmorphism may be referred for evaluation of persistent snoring or OSA, despite having undergone adenotonsillectomy.

Children with or without craniofacial differences can exhibit a broad spectrum of sleep disordered breathing, from primary snoring to "upper airway resistance syndrome" to severe OSA associated with marked oxygen desaturation. In the newborn period, severe nasal or upper airway compromise may present as a surgical emergency requiring tracheostomy in some babies. Less severely affected infants may present with intermittent apneic events or ALTEs (acute life-threatening events). The degree of obvious facial disfigurement in babies and older children does not necessarily correlate with the severity of OSA. It is necessary for clinicians to have an increased index of suspicion, however, with this population in order to routinely screen for and identify sleep-related symptoms, particularly as part of presurgical planning and perioperative risk management.

Despite being acknowledged as "at-risk," snoring and OSA continue to go unrecognized and untreated in many children with craniofacial conditions. These children may then experience various equally well-documented consequences of OSA, including compromised attention span, concentration, memory, and learning ability, hyperactivity or, alternatively, excessive daytime sleepiness, irritable mood, secondary enuresis, and, in severe cases, pulmonary hypertension (cor pulmonale), increased intracranial pressure, and failure to thrive. Since some, but not all, craniofacial syndromes are associated with neurodevelopmental disabilities or medical problems related to other organ system malformations, it is crucial to optimize sleep quality as part of overall health maintenance, as well as to avoid clouding the neuropsychological assessment of these children. In typical pediatric patients with OSA, removal of the adenoids or tonsils is often curative; however, children with craniofacial conditions may require more extensive, and often staged, surgical and/or orthodontic intervention. The craniofacial team approach to comprehensive evaluation and coordinated treatment planning for these complex patients is discussed in detail later in this chapter.

PATHOPHYSIOLOGY

The pathophysiology of OSA is thought to be related to any of a variety of factors (such as enlargement of soft tissues, abnormally formed bony structures, and/or poor neuromuscular control) that can cause narrowing of the upper airway. Further narrowing or collapse of that already-small passage occurs when the activity of pharyngeal dilator muscles, that maintain airway patency while awake, is ineffective during sleep. When these dilator muscles relax upon transition into sleep or are weakened, such as in patients with neuromuscular disorders, negative intraluminal inspiratory forces result in upper airway compromise, and thus resistance to airflow.

Individuals with differences in craniofacial morphology may demonstrate compromise at one or more sites or levels along the upper airway, from anterior nasal structures to the hypopharyngeal and laryngeal structures. Airway compromise may occur as the result of narrowing in lateral, vertical, or anterior–posterior dimensions. While many craniofacial anomalies are readily apparent, many others are quite subtle or minor, yet have a marked impact on upper airway structure and function. As the facial structures grow and develop over time, so may the respiratory mechanics change, thus requiring careful follow-along and periodic reevaluation.

CLINICAL SCENARIOS: ANATOMIC SITES AND CAUSES OF AIRWAY OBSTRUCTION

Choanal atresia or stenosis (i.e. complete or partial bony obstruction) of the nasal passages can occur as an isolated feature or as part of a cluster of features, such as in

Figure 71.1 CHARGE is an acronym for *c*oloboma, *h*eart defects, *a*tresia of the choanae, *r*enal anomalies, *g*rowth retardation, and *e*ar anomalies or hearing loss. Previously referred to as an "association" of the above features, it is now considered a recognizable "syndrome" caused by a mutation in the *CHD7* gene, discovered in 2004.

CHARGE syndrome (Figure 71.1), and may be life-threatening in a newborn, since neonates are obligate nose-breathers. Enlarged adenoids or nasal turbinates, nasal polyps, or septal deviation may also constrict the nasal airway. A high-arched palate and narrowed midface in the lateral dimension can be associated with vertically elongated and diminished nasal passages. Chronic mouth-breathing from nasal airway obstruction has been shown to result, over time, in the "long face syndrome," also referred to as "adenoid facies."

Maxillary and/or mandibular hypoplasia (or diminished growth of the bony structures of the midface and/or jaw, respectively) represent the most common contributing skeletal abnormalities to OSA in craniofacial patients. Syndromes associated with midface (maxillary) hypoplasia include the syndromic craniosynostoses (Crouzon Apert, Pfeiffer and Saethre–Chotzen syndromes) and Stickler, Antley–Bixler, and Down syndromes (Figures 71.2–71.4). These children usually have a flattened or even scaphoid facial profile, which shifts the usual position of the palate posteriorly toward the back of the nasopharynx and, in turn, can result in a wide range of severity in snoring and sleep apnea symptoms.

Mandibular (jaw) hypoplasia causes OSA symptoms as a result of the retropositioning of the base of the tongue and other soft tissues of the upper airway against the posterior

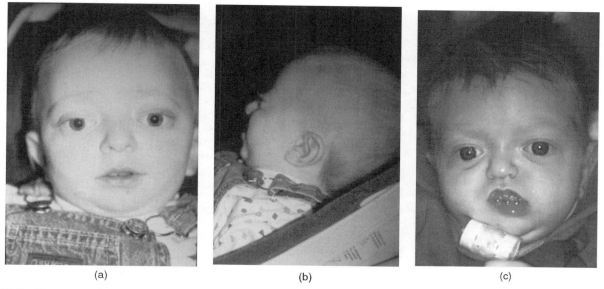

(a)   (b)   (c)

Figure 71.2   Crouzon syndrome: (a) front view and (b) side view showing relatively flat facial profile. (c) This infant required tracheostomy for severe OSA.

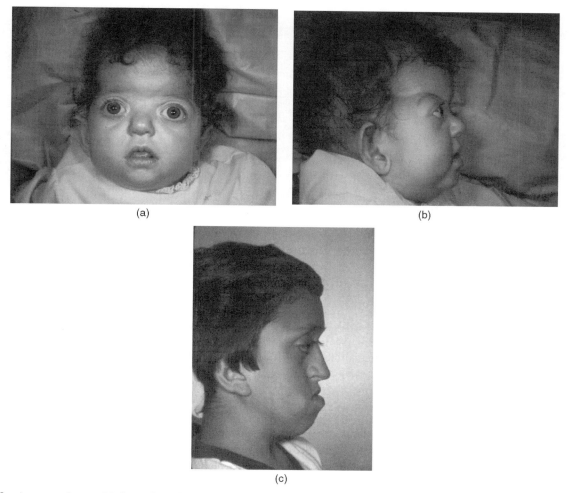

(a)   (b)

(c)

Figure 71.3   Apert syndrome: (a) front view; (b) and (c) side views showing severe "scaphoid" midfacial hypoplasia.

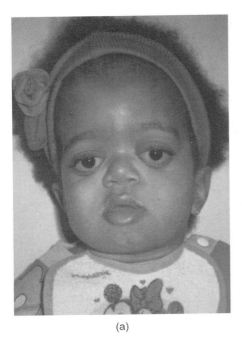

(a)

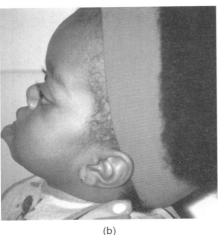

(b)

Figure 71.4   Pfeiffer sequence: (a) front view and (b) side view showing midfacial hypoplasia.

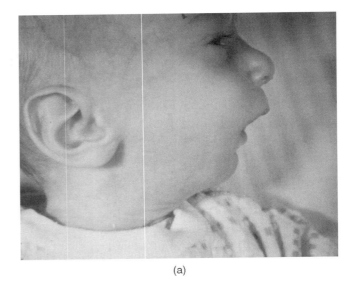

(a)

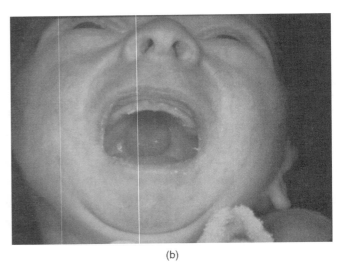

(b)

Figure 71.5   Pierre Robin sequence: side view sequence showing (a) severe mandibular hypoplasia and (b) widely cleft palate.

pharynx. Pierre Robin sequence (Figure 71.5) Treacher Collins syndrome (Figures 71.6 and 71.7), Melnick–Needles syndrome, and Nager syndrome are classic examples of mandibular hypoplasia. Asymmetrical or unilateral mandibular hypoplasia, with associated ear malformations on the affected side, is seen in oculo-auricular-vertebral (OAV) spectrum disorder, also known as Goldenhar syndrome or hemifacial/craniofacial microsomia (Figure 71.8). Though often less obviously disfiguring than some of the syndromes mentioned, OAV spectrum disorder may be deceptive in that it can be associated with quite severe OSA. Babies who were lying in utero in a transverse breech position, with the neck hyperextended and the forehead and

face compressed against the uterine wall, often develop a "positional deformation" (as opposed to a genetic or embryonic malformation) of the mandible, resulting in retrognathia (or receding point of the chin) (Figure 71.9).

Thickened or redundant soft tissues that line the airway are seen in the mucopolysaccharidoses, such as Hunter's or Hurler's syndrome. Tongue enlargement (macroglossia), for example, associated with Beckwith–Wiedemann syndrome, can cause airway obstruction. Postoperative new-onset snoring or apnea, as a complication of pharyngeal flap surgery or sphincter pharyngoplasty (done to treat velopharyngeal dysfunction in patients with cleft palate), is a rare "iatrogenic" cause of OSA. Obese patients, for example,

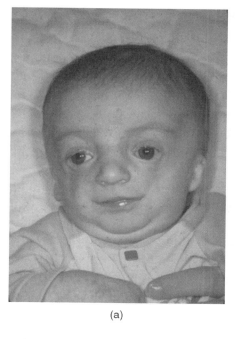

(a)

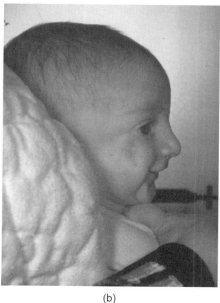

(b)

Figure 71.6 Treacher Collins syndrome: (a) front view and (b) side view showing mandibular hypoplasia.

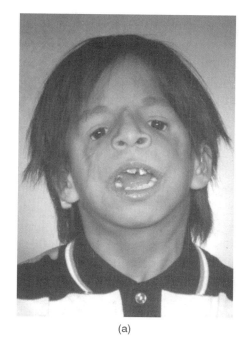

(a)

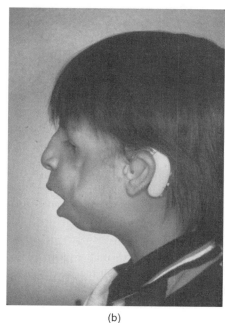

(b)

Figure 71.7 Treacher Collins syndrome: (a) front view and (b) side view showing mandibular hypoplasia.

many of those with Prader–Willi syndrome, should be evaluated for nocturnal hypoventilation (or Pickwickian syndrome") as well as OSA, with monitoring for both CO_2 retention and O_2 desaturation during a sleep study. Tonsillar or adenoidal hypertrophy in the setting of a craniofacial syndrome can present a greater hazard than in a child with a typical airway size and shape, as can a simple viral upper airway infection. Children with craniofacial syndromes and OSA may not achieve resolution of their OSA symptoms after tonsillectomy and/or adenoidectomy. Similarly, children *without* obvious facial dysmorphism who have persistent snoring and apnea despite removal of tonsils and adenoids should be evaluated further for subtle craniofacial or neuromuscular disorders.

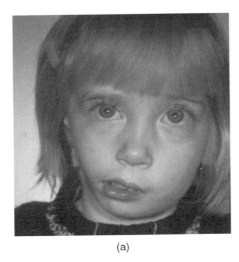

(a)

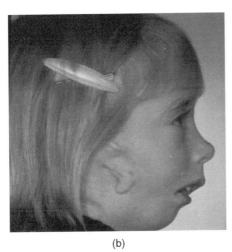

(b)

Figure 71.8 Oculo-auricular-vertebral spectrum: (a) front view (note the lateral displacement of the point of the chin) and (b) side view showing mandibular hypoplasia and ear anomaly on the right side.

Associated features of several congenital craniofacial syndromes may place patients at increased risk for complications of obstructive or central sleep apnea. For example, scoliosis or a chest wall deformity can cause restrictive lung disease, which can increase the severity of O_2 desaturations from OSA. Decreased range of motion of the cervical vertebrae or neck muscles may complicate airway management, particularly in perioperative situations, and may predispose to obstruction. A rare instance of both decreased range of motion of the neck and an endocrine abnormality (functional menopause) that predisposes to OSA is found in Turner syndrome. Some patients may also have abnormalities in the cartilaginous or connective tissues (laryngomalacia or tracheomalacia), or in the lower cranial nerve supply to the oropharyngeal musculature, for example, in those with brainstem malformations. Many patients

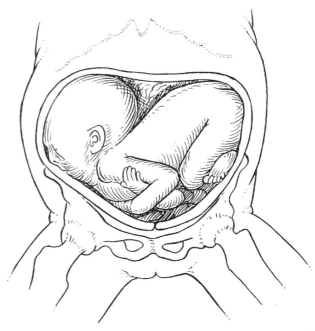

Figure 71.9 Transverse breach baby with retrognathia, as a positional "deformation" rather than an embryonic or genetic "malformation". (Reprinted from Smith's Recognizable Patterns of Human Deformation, 2/e, Figure 2.35, pg (1988) with permission from Elsevier)

with one or more of the above physical findings may not meet precise criteria for a specific syndrome diagnosis, and patients with a known, diagnosed syndrome can vary greatly in the expression of phenotypic or clinical features.

APPROACH TO EVALUATION

Comprehensive evaluation and care coordination can be provided by a craniofacial team, which ideally includes experts (or ready access to experts) in the following specialties: pediatric dentistry (to review preventive care, growth, and development), oral and maxillofacial surgery, craniofacial/plastic surgery, orthodontics, prosthodontics, otolaryngology, clinical genetics, developmental–behavioral pediatrics or child neurology, sleep medicine, audiology, speech–language pathology, clinical psychology, neuropsychology, neurosurgery, and ophthalmology. The interdisciplinary team evaluation allows for an important range of perspectives and richness in history taking and data gathering. The authors are members of the University of Minnesota Craniofacial Team, which convenes as a group once per month to evaluate and plan a course of treatment with new and follow-up patients and their parents. Additional appointments, sleep studies, or other evaluations, procedures, or operations are scheduled in the interim with individual team members based on the coordinated treatment plan. The patient's primary care provider or "medical home" is included in the information sharing and planning process. The team approach offers the family a "one-stop shopping" opportunity, with integrated follow-up outlined by an experienced team coordinator based on information gleaned from the interdisciplinary case conference at the end of the clinic day. Conservation of time and travel and a comprehensive evaluation and management fee make such a team clinic approach economically justifiable.

EVALUATION OF SLEEP-RELATED SYMPTOMS

The patient and/or parent is interviewed and given a written questionnaire regarding symptoms of sleep disordered breathing, other sleep disorders, and sleep-related daytime symptoms at home, school, and work. The physical exam includes growth parameters/body mass index, blood pressure, and thorough exam of the head and neck, with particular attention to nasal and oral structures, palate, dental occlusion and function, and tendency to mouth-breathe. Careful visual inspection of facial features as viewed from the front, sides, top (looking down from behind the patient over the forehead), and up from the chin will be necessary to detect and describe differences. Neuromotor exam of cranial nerves and an "oral mechanism" exam done with the speech–language pathologist can provide crucial information on functioning of the vocal tract. A general physical exam is indicated to assess for cardiopulmonary sequelae when OSA is suspected.

The laboratory technician-attended polysomnogram (PSG) is currently considered the "gold standard" for evaluating symptoms of sleep disordered breathing, and for most children with craniofacial conditions, a PSG would be the test of choice, given the risks and complexities of these cases. The main advantage of the multichannel PSG is the ability to simultaneously record and integrate the neurological sleep data (scalp EEG, eye movements, airway muscle tone) with the cardiorespiratory data (chest and abdomen excursion, work of breathing, oxygen saturation, transcutaneous or end-tidal CO_2 monitoring, and ECG) obtained. Esophageal pressure, another useful indicator of airway resistance, and autonomic nervous system activation are measured in some labs but are less routine. The attendant observes for increased work of breathing, snoring, breathing-related arousals, and unusal motor behavior in sleep. Unattended home cardiorespiratory sleep studies or screening pulse oximetry alone have been used, but pitfalls include the inability to determine if the child is actually asleep without the neurologic (e.g., EEG) channels, and missed cases of milder OSA that result mainly in increased work of breathing or fragmentation of sleep quality due to breathing-related arousals, without accompanying oxygen desaturation. The PSG allows for the supervised therapeutic trial of nasal CPAP (continuous positive airway pressure) equipment, when indicated, with baseline and trial portions performed during a single night in the sleep lab. An attended PSG is also the test of choice for tracheostomy predecannulation assessments of children's sleep. The child who is felt to be ready for decannulation has the PSG performed in controlled circumstances with the tracheostomy capped. Repeat PSG may be indicated a few months postoperatively to evaluate changes after craniofacial surgery or decannulation. While polysomnographic studies can help with assessing presence and severity of OSA or SDB, they cannot determine the anatomic site or source of the obstructed breathing.

IMAGING STUDIES

The lateral cephalometric radiograph (Figure 71.10) is the mainstay of conventional imaging for craniofacial conditions, demonstrating the relative positions of key structures at baseline, and following surgical interventions. Airway compromise may be readily visible due to retrognathic positioning of the jaw or recessed midfacial structures with resting soft palate in close approximation to the posterior pharyngeal wall. Enlarged adenoids and/or tonsils may be evident as well. Two-dimensional CT, MRI, and reconstructed three-dimensional (3-D) CT imaging (Figure 7.11) of the head and neck provide views that enable more detailed localization of craniofacial differences and potential airway obstruction, and hence better surgical planning. A newer technology, stereolithographic analog (SLA), produces hard plastic 3-D models of the skull (Figure 71.12). These models can help tremendously with preoperative anesthetic airway management, diagnostic visualization, and surgical treatment simulation for more complicated patients.

NASENDOSCOPY

Children with craniofacial conditions can experience airway obstruction at various anatomic levels (anterior nasal, nasopharynx, oropharynx, and laryngopharynx). Nasendoscopy can be performed in the waking and artificially induced sleeping states, under light general anesthesia in an operating room, by means of a tiny fiberoptic scope in real-time, and may enable visualization and site localization of the airway obstruction, thus assisting in guiding surgical decisions. This is a somewhat controversial procedure, in that there is some debate as to whether the findings truly correlate with the disordered breathing that occurs during natural sleep states, for example, distinguishing

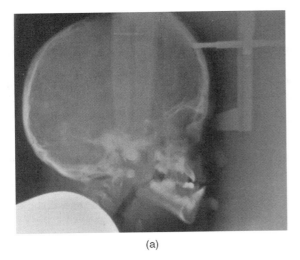

(a)

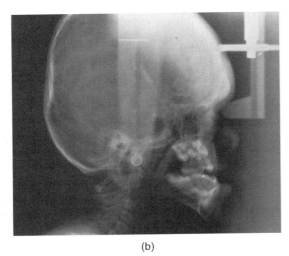

(b)

Figure 71.10   Apert sydrome: lateral cephalometric radiographs: (a) before midfacial advancement and (b) after midfacial advancement, showing expansion of the upper airway space.

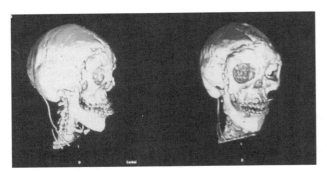

Figure 71.11   Three-dimensional CT scan of the skull.

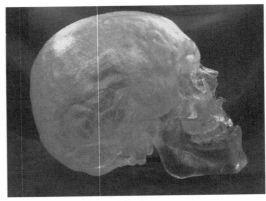

Figure 71.12   Stereolithographic analog model of the skull.

the difference between the breathing problems that might occur in REM as opposed to NREM sleep.

INTERVENTIONS

A broad variety of treatment options are available for children with sleep disorders and craniofacial conditions, many of which are similar to those used for typical pediatric patients and the reader is referred to those chapters of this text as well. Pharmacologic management of snoring from adenoidal hypertrophy may include a trial of nasal steroids or, when indicated, a course of antibiotics when there is bacterial infection of the adenoids, tonsils or sinuses. Non-invasive ventilatory support with nasal CPAP or bilevel PAP is indicated in some children with non-life-threatening, but significant, OSA and may allow reasonable postponement of craniofacial and/or orthognathic surgery until the facial bones and teeth develop more fully. Practical issues with CPAP include patient comfort and compliance; however, many children who have used CPAP from a young age become easily habituated to its use—but it is rarely considered a long-term solution. An orthodontic approach has been introduced recently that effectively treats OSA from nasal airway obstruction in some patients with a high arched palate and midface hypoplasia in the lateral dimension by use of an upper arch dental expander device. Dental repositioning devices used for mild apnea in adults are usually not applied to this population as the dental development is incomplete for children; however, there may be a role in some cases for other prosthodontic appliances.

Surgical interventions for children with craniofacial anomalies are often complex and thus require careful interdisciplinary planning and staging. Close perioperative and postoperative monitoring and risk management aim to prevent airway compromise from edema, secretions, bleeding, or inflammation or respiratory drive suppression from analgesic or sedative medications—all complications for which these patients are at increased risk. Procedures may involve soft tissues of the head and neck and/or the bony structures of the skull and face. Routine tonsillectomy and adenoidectomy may cure or significantly ameliorate OSA symptoms, just as in more typical patients; however, children with craniofacial anomalies require regular postoperative reassessment for persistence or recurrence of those symptoms. They may also be at increased risk for speech impairment due to velopharyngeal insufficiency after adenoidectomy or after a combination of adenoidectomy and subsequent midface advancement. There are no reported indications or applications for uvulopalatopharyngoplasty (UPPP) in this patient population. Nasal stenting, septoplasty, bony or soft tissue (turbinate) reduction, and tongue reduction surgeries have been used successfully in select cases. Tracheostomy bypasses upper airway obstruction for children with more severe involvement; however, it is also not a long-term solution in the majority of cases. Orthognathic surgery or craniofacial reconstructive procedures, such as midface, maxillary, and/or mandibular advancement, may prevent the need for tracheostomy by enlarging the upper airway at the site of obstruction. LeFort osteotomies at various facial levels, with either immediate repositioning and stabilization or with incremental advancement via distraction osteogenesis, have been used widely for children with congenital midface hypoplasia, such as in Apert or Crouzon syndromes and in patients with acquired midface hypoplasia, such as in posterior clefting, which has been surgically repaired. Mandibular distraction osteogenesis, which lengthens and may also change the angle of the jaw, can be performed in very young infants, in toddlers, as well as in older patients. Bony distraction is usually begun using internal or external devices about 1 week after surgery and case reports indicate postoperative changes on the order of 1–2 cm of advancement. This and other techniques, such as hyoid suspension, address the OSA caused by retropositioning of the tongue and other soft tissues in a child with congenital retrognathia or micrognathia (e.g., Pierre Robin or Treacher Collins syndrome) by pulling those structures forward and expanding the airway. One group reported improvement, by nasendoscopic measurement, in the lateral dimension of the airway, as well as the expected anterior–posterior dimension, after midface advancement. These surgical procedures are relatively newly described over the last decade and are still evolving in their application and specific techniques.

Interventions for OSA in this population, whether noninvasive or invasive, may require input and assistance from the child psychologists on the team to help the patient adjust to or cope with new equipment, the stress of hospitalization, physical discomfort or fear, and/or changes in his/her physical appearance. These professionals can assist the team in understanding developmental, behavioral, and life-style issues that impact the prognosis for an individual patient.

SUMMARY AND RECOMMENDATIONS

Obstructive sleep apnea is common in children and adults with craniofacial conditions. All patients with craniofacial anomalies or upper airway compromise should be screened for sleep-related symptoms in order to initiate proper evaluation and treatment. Careful patient monitoring at the time of procedures or operations that involve anesthesia or sedation is critical, due to the increased risk of respiratory complications for this population. Sleep medicine professionals are encouraged to collaborate with local or regional craniofacial teams. In this way, sleep clinic and laboratory personnel will gain the experience necessary to optimally evaluate these complex patients and craniofacial teams will routinely include a sleep history and evaluation as part of their comprehensive care.

BIBLIOGRAPHY

Abramson DL, Marrinan EM, Mulliken JB. Robin sequence: obstructive sleep apnea following pharyngeal flap. *Cleft Palate Craniofacial J* **34**(3):256–260(1997).

Caulfield H. Investigations in paediatric obstructive sleep apnoea: do we need them? *Int J Pediatr Otorhinolaryngol* **67**(Suppl 1):S107–S110(2003).

Contencin P, Guilleminault C, Manach Y. Long-term follow-up and mechanisms of obstructive sleep apnea (OSA) and related syndromes through infancy and childhood. *Int J Pediatr Otorhinolaryngol* **67**(Suppl 1):S119–S123(2003).

Crysdale WS, Djupesland P. Nasal obstruction in children with craniofacial malformations. *Int J Pediatr Otorhinolaryngol* **49**(Suppl 1):S63–S67(1999).

Finkelstein Y, Wexler D, Berger G, Nachmany A, Shapiro-Feinberg M, Ophir D. Anatomical basis of sleep-related breathing abnormalities in children with nasal obstruction. *Arch Otolaryngol Head Neck Surg* **126**(5):593–600(2000).

Gibson SE, Myer CM III, Strife JL, O'Connor DM. Sleep fluoroscopy for localization of upper airway obstruction in children. *Ann Otol Rhinol Laryngol* **105**(9):678–683(1996).

Guilleminault C, Abad VC. Obstructive sleep apnea. *Curr Treatment Options Neurol* **6**(4):309–317(2004).

Hoeve HL, Joosten KF, van den Berg S. Management of obstructive sleep apnea syndrome in children with craniofacial malformation. *Int J Pediatr Otorhinolaryngol* **49**(Suppl 1):S59–S61(1999).

Hoeve LJ, Pijpers M, Joosten KF. OSAS in craniofacial syndromes: an unsolved problem. *Int J Pediatr Otorhinolaryngol* **67**(Suppl 1):S111–S113(2003).

Li KK. Surgical management of obstructive sleep apnea. *Clin Chest Med* **24**(2):365–370(2003).

Li KK, Guilleminault C, Riley RW, Powell NB. Obstructive sleep apnea and maxillomandibular advancement: an assessment of airway changes using radiographic and nasopharyngoscopic examinations. *J Oral Maxillofac Surg* **60**(5):526–530 (discussion 531)(2002).

Marcus CL, Loughlin GM. Obstructive sleep apnea in children. *Semin Pediatr Neurol* **3**(1):23–28(1996).

Myatt HM, Beckenham EJ. The use of diagnostic sleep nasendoscopy in the management of children with complex upper airway obstruction. *Clin Otolaryngol.* **25**(3):200–208(2000).

Morovic CG, Monasterio L. Distraction osteogenesis for obstructive apneas in patients with congenital craniofacial malformations. *Plast Reconstr Surg* **105**(7):2324–2330(2000).

Nishikawa H, Pearman K, Dover S. Multidisciplinary management of children with craniofacial syndromes with particular reference to the airway. *Int J Pediatr Otorhinolaryngol.* **67**(Suppl 1):S91–S93(2003).

Rosen CL. Obstructive sleep apnea syndrome (OSAS) in children: diagnostic challenges. *Sleep.* **19**(10 Suppl):S274–S277(1996).

Sloan GM. Posterior pharyngeal flap and sphincter pharyngoplasty: the state of the art. *Cleft Palate Craniofac J* **37**(2):112–122(2000).

Sculerati N, Gottlieb MD, Zimbler MS, Cibbaro P, McCarthy JG. Airway management in children with major craniofacial anomalies. *Laryngoscope* **108**(12):1806–1812(1998).

Wheatley JR, Amis TC. Mechanical properties of the upper airway. *Curr Opin Pulm Med* **4**(6):363–369(1998).

72

MEDICAL DISORDERS

John M. Palmer and Lee J. Brooks

Children's Hospital of Philadelphia, Philadelphia, Pennsylvania

Sleep that knits up the ravelled sleave of care
The death of each day's life, sore labor's bath,
Balm of hurt minds, great nature's second course,
Chief nourisher in life's feast.
　　　　—William Shakespeare, *Macbeth*, Act II, Scene ii

Sleep is affected by a number of conditions. Some congenital syndromes, neurological and neuromuscular disorders, craniofacial syndromes, neuromuscular disorders, psychiatric disorders, and their relation to sleep are discussed in other chapters. Of the medical disorders that affect sleep, some are seen primarily in adults and only rarely in children. These include cardiovascular disease, rheumatologic conditions, endocrine disorders and renal disease. In this chapter, we will focus on more common medical disorders that impact sleep in pediatric patients. These include asthma, gastroesophageal reflux, otitis media, and some miscellaneous medical conditions.

ASTHMA

Asthma is the leading cause of hospitalization, chronic disease, and school absenteeism in children [1]. Asthma symptoms are the result of bronchoconstriction of the airways, typically accompanied by airway inflammation. There are many triggers for asthma, including upper respiratory infections, tobacco smoke, and seasonal allergens. Treatment revolves around anti-inflammatory preventive therapy, often with inhaled corticosteroids, and quick relief medications such as beta agonists during exacerbations.

Effects of Sleep on Asthma

Asthma symptoms worsen at night. Up to 90% of asthma patients report that cough or difficult breathing disturbs their sleep at least once a week [2], although this effect is less pronounced in patients less than 4 years of age [3].

Pulmonary function testing supports these subjective data. Patients with asthma have a 31% drop in forced expiratory volume in 1 second (FEV_1) from bedtime to rise time [4]. Patients with asthma have reduced morning peak expiratory flow (PEF) as well as a higher evening to morning drop in PEF compared to controls. This is true even in patients whose asthma is categorized as "well-controlled"[5]. Sleep itself may play a direct role in the nocturnal worsening of asthma. In two comparable populations, the peak flow dropped more in patients who slept compared to those who remained awake at night [6]. There is no indication that the observed drop in FEV_1 is related to any particular stage of sleep [4, 7–9].

There are several possible reasons for asthma worsening at night. These are related to circadian variation in hormones, airway inflammation, the recumbent position, gastroesophageal reflux, and exposure to bedroom allergens such as dust mites.

Circadian changes are noted in hormones such as melatonin as well as in inflammatory mediators. Patients with asthma have higher serum melatonin levels in the early hours of the morning compared to controls [10]. Plasma cortisol levels also rise through the evening, peaking in the early morning [11]. Cortisol levels have been evaluated as a marker for asthma, but this is controversial. Although cortisol increases in response to physiologic stressors [12],

Sleep: A Comprehensive Handbook, Edited by T. Lee-Chiong.
Copyright © 2006 John Wiley & Sons, Inc.

it may not be specific for the type or severity of asthma [11, 13]. Additionally, some studies show decreased salivary [14] or serum [15] levels of cortisol in asthma patients.

Airway inflammation is one of the hallmarks of asthma. Asthma patients have increased neutrophils and eosinophils in bronchoalveolar lavage fluid at night compared to daytime measures and normal controls [16]. Poorly controlled asthmatics have twice the plasma histamine levels that well-controlled asthma patients have at 4 a.m. [17]. These may be markers of and/or contribute to airway inflammation.

The *recumbent position* reduces lung volumes and alters respiratory patterns due to a shift of abdominal contents, and these changes are more severe in patients with asthma [18]. Intrapulmonary pooling of blood also increases in the recumbent position, which can decease lung compliance and inhibit gaseous diffusion [19].

Gastroesophageal reflux (GER) is also common in patients with asthma and is worse in the supine position [20]. Seventy-seven percent of patients with asthma complain of "heartburn," and 41% experience respiratory symptoms associated with reflux [21]. In children, there is a positive correlation between reflux score and nighttime-associated wheezing [22]. Pulmonary function in patients with asthma and GER disease is worse than in patients with either disease alone (Figure 72.1) [54]. The bronchoconstriction that occurs may be due to a vagally mediated reflex elicited by acid in the distal esophagus, or due to microaspiration.

Exposure to allergens can trigger asthma symptoms [23]. In considering the bedroom environment, these exposures may include dust mites [24], cockroaches [25], pets [26], and mice [27]. These exposures may be mitigated by modifying the environment with dust covers for mattresses, removal of carpet from the bedroom, and prohibiting pets from entering sleep areas. This is difficult to

completely implement, however, and some triggers may persist as a result of unwillingness or lack or awareness about allergens [28].

Effects of Asthma on Sleep

Questionnaire data indicate that young children with asthma are frequently awakened at night by their symptoms (e.g., cough). Nocturnal awakening occurred in one-third of the children with mild to moderate asthma even during a month of relatively stable symptoms [29]. In older patients, there is a complaint of less restorative sleep as well as decreased duration of sleep [2]. There is a correlation between frequency of sleep disturbances and worsening asthma symptomatology [29]. Patients with asthma also report a higher incidence of insomnia. In hospitalized asthma patients, there is a decrease in the oxyhemoglobin nadir during exacerbations of symptoms [30].

This sleep disturbance has daytime effects as well. There is a decrement on some tests of memory and cognition in children with poor sleep due to asthma. Additionally, these patients had more psychological problems when compared to controls [31, 32]. Improving asthma control improves sleep quality, with a subsequent improvement in the scores obtained on tests of memory and cognition as well as psychological functioning [31].

Medications used in the treatment of asthma, such as beta agonists and theophylline, may disrupt sleep. Several medications used in treatment of other conditions may promote sleepiness (Table 72.1).

GASTROESOPHAGEAL REFLUX (GER)

Independent of its relationship to asthma noted earlier, GER impacts on sleep quality and is itself changed by sleep.

Effect of Sleep on GER

Several protective physiologic mechanisms that function during wakefulness are diminished with sleep. This leads to prolonged acid contact times [33]. During sleep, patients have decreased response to the sensation of heartburn, decreased frequency of swallowing, and suppression of salivary secretion. Concurrent with this is the risk of pulmonary aspiration of gastric contents. GER without aspiration may cause bronchoconstriction in pediatric patients; aspiration of gastric contents causes marked bronchoconstriction. Frequency of GER is related to sleep position as well. The right lateral decubitus and supine positions are associated with the worst GER. The most favorable position for preventing GER is the left lateral decubitus position [34].

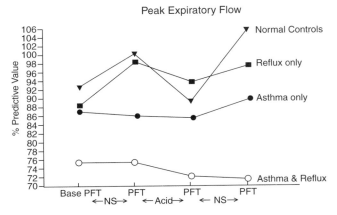

Figure 72.1 Peak expiratory flow: groupeffect, $P < 0.014$; treatment phase effect, $P < 0.003$; group by treatment effect, $P < 0.095$.

TABLE 72.1   Medications that May Affect Sleep

Class	Reported Side Effects on Sleep	Medication
Androgens	May potentiate sleep apnea in some male patients	Testosterone
Antidepressants		
Dopamine reuptake inhibitors	Increases REM latency, increases REM activity in first REM period	Bupropion
Monoamine oxidase inhibitor	REM suppression	Isocarboxazid, moclobemide, phenelzine, selegiline, tranylcypromine
SSRIs	REM suppression	Citalopram, escitalopram, fluoxetine, fluvoxamine, olanzepine,
Serotonin reuptake inhibitors	REM suppression, increases slow-wave sleep (SWS)	Nefazadone, trazadone
Serotonin/norepinephrine reuptake inhibitors	REM suppression, increases stage 1	Duloxetine, venlafaxine
Tetracyclics	REM suppression, increases stage 2, daytime sedation	Maprotiline
TCA, secondary amines	REM suppression, insomnia	Amoxapine, desipramine, nortriptyline, protriptyline, paroxetine, sertraline
TCA, tertiary amines	REM suppression	Amitriptyline, clomipramine, doxepin, imipramine, trimipramine
Anticonvulsants	Increases stage 2, decreases stage 4, increases REM sleep	Levetiracetam, oxycarbazepine, zonisamide
Antiemetics	Sedation, increases NREM sleep, incr apnea	Promethazine
Antihistamines	Sedation, increases NREM sleep	Diphenhydramine, chlorphenhydramine, tripelennamine, cetirizine
Antipsychotics	Increases SWS, decreases stage 2 sleep	Olanzapine, ouetiapine, risperidone
Anxiolytic/hypnotics	Decreases sleep latency, increases sleep duration	Zaleplon, zolpidem
Barbiturates	Slight decrease REM, increases stage 2 sleep	Secobarbital
Benzodiazepines	Decreases REM, increases NREM sleep, decreased sleep latency	Chlorazepate, estazolam, flurazepam, lorazepam
CNS depressants	Decreases REM, alpha intrusion; increases SWS	Sodium oxybate
Dopamine agonists	Decreases REM and stage 4 sleep	Apomorphine
Hormones	Worsening of OSA; decreases REM latency and increases REM in deficient patients	Human growth hormone
Hypnotics	Decreases sleep latency, increases total sleep time	Zopiclone
Narcotics	Decreases stage 2 and REM sleep	Morphine, codeine
Neutraceuticals	Phase-shifting and sleep-promoting properties	Melatonin, 5-hydroxytryptophan
Parkinson's agents	Decreases RLS, suppression of REM rebound reported	Pergolide, Pramipexole, ropinirole
Sedative hypnotics	Decreases sleep latency, insomnia associated with withdrawal	Chloral hydrate, aldesleukin, brimonidine

Effect of GER on Sleep

GER is a strong arousal stimulus. In a study of infants with reflux to the proximal esophagus, 76% of all arousals were associated with reflux events documented by pH probe [35]. This occurred far more frequently in rapid eye movement (REM) sleep than in other stages of sleep [35, 36]. As a result, these patients had more fragmented sleep than controls. Infants and young children with GER have five to six times more nighttime awakenings than controls, with an associated increase in the amount of parental intervention needed during the night. These patients also have delayed sleep onset throughout the night and a greater prevalence of daytime sleep beyond 24 months of age [37].

With adequate treatment of GER, patients have less fragmented sleep [37], decreased risk for mucosal injury [33], decrease in asthma exacerbations for those patients who have asthma, and decrease in required parental interventions.

OTITIS MEDIA AND CONDITIONS WITH ASSOCIATED PAIN OR DISCOMFORT

In the United States, approximately 75% of all children experience at least one episode of acute *otitis media*.

One-third will have three or more episodes before 3 years of age [38]. Otitis media frequently presents with pain and irritability, although not universally. In a study of 302 children younger than 4 years of age [39], 40% of the children with acute otitis media never complained of or never had symptoms of an earache. However, approximately half of the patients had sleep disturbance due to pain. This parallels findings in other disease processes, where pain causes sleep disturbance in up to 45% of patients [40].

These observations may be generalized to other childhood conditions where pain is an associated symptom, such as postoperative pain [41]; 47% of these patients had sleep disturbances. Use of effective analgesia promotes a good nights sleep, even in patients who have undergone more invasive procedures such as abdominal surgery [42].

Colic is a very common pediatric entity and is often mistaken for pain. Colic in a baby is sustained episodes of crying in excess of 3 hours per day for no apparent reason. About 20% of babies cry enough to meet the definition of colic. The timing varies, but colic usually begins at about 3 weeks of age and peaks between 4 and 6 weeks of age. It has been suggested that colic relates to temperment [43] or autonomic nervous system function, but this is not universally accepted [44, 45]. In comparing the sleep patterns of children with colic to those in a control group, there are differences based on both diaries and polysomnography [46]. Total reported sleep time was shorter in infants with colic compared with a control group at 5 weeks of age. Polysomnography data showed similar sleep architecture, movements, and breathing patterns between the study groups at 2 months of age, but the colicky infants had slightly more obstructive apneas in REM sleep than controls [46]. Despite the shorter reported daily sleep times, the polygraphic data did not suggest infantile colic to be associated with a sleep disorder.

Atopic dermatitis also affects sleep quality and duration. Children with atopic dermatitis often have disrupted sleep as well as excessive daytime sleepiness [47]. This observation was made even in those children whose atopic dermatitis was in remission; their sleep fragmentation was tied directly to scratching in just 15% of cases [48]. The sleep fragmentation may be related to decreased dopamine levels in the brain, which impacts wakefulness [49].

OTHER MEDICAL CONDITIONS

Children with neurologic conditions are at an increased risk for sleep disturbances. As many as 50% of autistic children (average age 5 years) experience severe sleep disturbances [50]. Fifty to 65% of severely mentally handicapped children may experience sleep disturbances that persist for at least 3 years [51]. Blind children are more likely to achieve nocturnal sleep duration of less than 7 hours and excessive daytime sleepiness than sighted controls [52]. Parents of children with ADHD are more likely to report problems with their children settling and going to sleep, disruptions during sleep, and disturbances of morning activities than parents of controls [53], although these disruptions are not always confirmed on polysomnography. Sleep disturbances in children in these high-risk populations are often persistent.

SUMMARY

A variety of medical conditions can affect sleep in pediatric patients, including asthma, gastroesophageal reflux, processes where pain is present, and neurological disorders. Adequate treatment of these conditions may improve the quality of the sleep for the patient.

REFERENCES

1. Mannino DM, et al. Surveillance for asthma—United States, 1980–1999. *MMWRCDC Surveill Summ* **51**(1):1–13(2002).

2. Vir R, Bhagat R, Shah A. Sleep disturbances in clinically stable young asthmatic adults. *Ann Allergy Asthma Immunol* **79**(3):251–255(1997).

3. Tirosh E, et al. Sleep characteristics of asthmatics in the first four years of life: a comparative study. *Arch Dis Child* **68**(4):481–483(1993).

4. Morgan AD, et al. Breathing patterns during sleep in patients with nocturnal asthma. *Thorax* **42**(8):600–603(1987).

5. Sadeh A, et al. Sleep and pulmonary function in children with well-controlled, stable asthma. *Sleep* **21**(4):379–384(1998).

6. Catterall JR, et al. Effect of sleep deprivation on overnight bronchoconstriction in nocturnal asthma. *Thorax* **41**(9):676–680(1986).

7. Kales A, et al. Sleep studies in asthmatic adults: relationship of attacks to sleep stage and time of night. *J Allergy* **41**(3):164–173(1968).

8. Ballard RD, et al. Effect of sleep on nocturnal bronchoconstriction and ventilatory patterns in asthmatics. *J Appl Physiol* **67**(1):243–249(1989).

9. Bellia V, et al. Relationship of nocturnal bronchoconstriction to sleep stages. *Am Rev Respir Dis* **140**(2):363–367(1989).

10. Sutherland ER, et al. Elevated serum melatonin is associated with the nocturnal worsening of asthma. *J Allergy Clin Immunol* **112**(3):513–517(2003).

11. Morrison JF, Pearson SB. The effect of the circadian rhythm of vagal activity on bronchomotor tone in asthma. *Br J Clin Pharmacol* **28**(5):545–549(1989).

12. Kapoor U, et al. Plasma cortisol levels in acute asthma. *Indian J Pediatr* **70**(12):965–968(2003).

13. Kraft M, Pak J, Martin RJ. Serum cortisol in asthma: marker of nocturnal worsening of symptoms and lung function? *Chronobiol Int* **15**(1):85–92(1998).

14. Fei GH, et al. Alterations in circadian rhythms of melatonin and cortisol in patients with bronchial asthma. *Acta Pharmacol Sin* **25**(5):651–656(2004).

15. Buske-Kirschbaum A, et al. Blunted cortisol responses to psychosocial stress in asthmatic children: a general feature of atopic disease? *Psychosom Med* **65**(5):806–810(2003).

16. Mackay TW, et al. Role of inflammation in nocturnal asthma. *Thorax* **49**(3):257–262(1994).

17. Szefler SJ, et al. Plasma histamine, epinephrine, cortisol, and leukocyte beta-adrenergic receptors in nocturnal asthma. *Clin Pharmacol Ther* **49**(1):59–68(1991).

18. Jonsson E, Mossberg B. Impairment of ventilatory function by supine posture in asthma. *Eur J Respir Dis* **65**(7):496–503(1984).

19. Shepard JW Jr. Gas exchange and hemodynamics during sleep. *Med Clin North Am* **69**(6):1243–1264(1985).

20. Bohadana AB, Hannhart B, Teculescu DB. Nocturnal worsening of asthma and sleep-disordered breathing. *J Asthma* **39**(2):85–100(2002).

21. Harding SM. Nocturnal asthma: role of nocturnal gastroesophageal reflux. *Chronobiol Int* **16**(5):641–662(1999).

22. Karaman O, et al. Results of the gastroesophageal reflux assessment in wheezy children. *Indian J Pediatr* **66**(3):351–355(1999).

23. Oddera S, et al. Airway eosinophilic inflammation and bronchial hyperresponsiveness after allergen inhalation challenge in asthma. *Lung* **176**(4):237–247(1998).

24. Sears MR, et al. A longitudinal, population-based, cohort study of childhood asthma followed to adulthood. *N Engl J Med* **349**(15):1414–1422(2003).

25. Rullo VE, et al. Daycare centers and schools as sources of exposure to mites, cockroach, and endotoxin in the city of Sao Paulo, Brazil. *J Allergy Clin Immunol* **110**(4):582–588(2002).

26. Svanes C, et al. Pet-keeping in childhood and adult asthma and hay fever: European community respiratory health survey. *J Allergy Clin Immunol* **112**(2):289–300(2003).

27. Matsui EC, et al. Mouse allergen exposure and mouse skin test sensitivity in suburban, middle-class children with asthma. *J Allergy Clin Immunol* **113**(5):910–915(2004).

28. Eggleston PA, Control of environmental allergens as a therapeutic approach. *Immunol Allergy Clin North Am* **23**(3):533–547, viii–ix(2003).

29. Strunk RC, et al. Nocturnal awakening caused by asthma in children with mild-to-moderate asthma in the childhood asthma management program. *J Allergy Clin Immunol* **110**(3):395–403(2002).

30. Smith TF, Hudgel DW. Arterial oxygen desaturation during sleep in children with asthma and its relation to airway obstruction and ventilatory drive. *Pediatrics* **66**(5):746–751(1980).

31. Stores G, et al. Sleep and psychological disturbance in nocturnal asthma. *Arch Dis Child* **78**(5):413–419(1998).

32. Fitzpatrick MF, et al. Morbidity in nocturnal asthma: sleep quality and daytime cognitive performance. *Thorax* **46**(8):569–573(1991).

33. Orr WC, Sleep and gastroesophageal reflux: what are the risks? *Am J Med* **115**(Suppl 3A):109S–113S(2003).

34. Khoury RM, et al. Influence of spontaneous sleep positions on nighttime recumbent reflux in patients with gastroesophageal reflux disease. *Am J Gastroenterol* **94**(8):2069–2073(1999).

35. Kahn A, et al. Arousals induced by proximal esophageal reflux in infants. *Sleep* **14**(1):39–42(1991).

36. Ramet J. Cardiac and respiratory reactivity to gastroesophageal reflux: experimental data in infants. *Biol Neonate* **65**(3-4):240–246(1994).

37. Ghaem M, et al. The sleep patterns of infants and young children with gastro-oesophageal reflux. *J Paediatr Child Health* **34**(2):160–163(1998).

38. Berman S. Otitis media in children. *N Engl J Med* **332**(23):1560–1565(1995).

39. Heikkinen T, Ruuskanen O. Signs and symptoms predicting acute otitis media. *Arch Pediatr Adolesc Med* **149**(1):26–29(1995).

40. Miser AW, et al. Pain as a presenting symptom in children and young adults with newly diagnosed malignancy. *Pain* **29**(1):85–90(1987).

41. Kain ZN, et al. Sleeping characteristics of children undergoing outpatient elective surgery. *Anesthesiology* **97**(5):1093–1101(2002).

42. Klamt JG, et al. Epidural infusion of clonidine or clonidine plus ropivacaine for postoperative analgesia in children undergoing major abdominal surgery. *J Clin Anesth* **15**(7):510–514(2003).

43. Canivet C, Jakobsson I, Hagander B. Infantile colic. Follow-up at four years of age: still more "emotional". *Acta Paediatr* **89**(1):13–17(2000).

44. Lehtonen L, Korhonen T, Korvenranta H. Temperament and sleeping patterns in colicky infants during the first year of life. *J Dev Behav Pediatr* **15**(6):416–420(1994).

45. Kirjavainen J, et al. The balance of the autonomic nervous system is normal in colicky infants. *Acta Paediatr* **90**(3):250–254(2001).

46. Kirjavainen J, et al. Infants with colic have a normal sleep structure at 2 and 7 months of age. *J Pediatr* **138**(2):218–223(2001).

47. Dahl RE, et al. Sleep disturbances in children with atopic dermatitis. *Arch Pediatr Adolesc Med* **149**(8):856–860(1995).

48. Reuveni H, et al. Sleep fragmentation in children with atopic dermatitis. *Arch Pediatr Adolesc Med* **153**(3):249–253(1999).

49. Friedman EH. Neurobiology of sleep disturbances in children with atopic dermatitis. *Arch Pediatr Adolesc Med* **150**(3):329–330(1996).

50. Wiggs L, Stores G. Sleep patterns and sleep disorders in children with autistic spectrum disorders: insights using parent report and actigraphy. *Dev Med Child Neurol* **46**(6):372–380(2004).

51. Quine L. Sleep problems in children with mental handicap. *J Ment Defic Res* **35**(Pt 4):269–290(1991).

52. Leger D, et al. Sleep disorders in children with blindness. *Ann Neurol* **46**(4):648–651(1999).

53. Day HD, Abmayr SB. Parent reports of sleep disturbances in stimulant-medicated children with attention-deficit hyperactivity disorder. *J Clin Psychol* **54**(5):701–716(1998).

54. Harding SM, Richter JE. The role of gastroesophageal reflux in chronic cough and asthma. *Chest* **111**(5):1389–1402(1997).

73

SLEEP IN CHILDREN WITH NEUROLOGICAL DISORDERS

SURESH KOTAGAL

Mayo Clinic, Rochester, Minnesota

GENERAL CONCEPTS

Alterations in sleep–wake function vary, depending on the anatomic location of the neurological lesion. For example, patients with Chiari malformations are likely to exhibit a sleep-related breathing disturbance due to a developmental anomaly in the medulla oblongata, whereas those with mental retardation may exhibit cortical dysfunction in the form of daytime or nighttime seizures. The severity of alterations in sleep–wake function also depends on the extent of the lesion and whether it is static or progressive. Medications used to treat neurological disorders may also impact sleep architecture, for example, suppression of rapid eye movement (REM) sleep with barbiturates and benzodiazepines that are used in the treatment of seizures. Fragmented sleep is overall quite common in children with neurological disorders. In a study of 214 intellectually impaired children, Bartlett et al. [1] documented parental reports of "sleep problems" in 86% of subjects under 6 years of age, 81% of subjects 6–11 years, and 77% of subjects who were 12–16 years old. Settling difficulties and night awakening were the most common problems. Treatment of the sleep–wake disorder may enhance the quality of life in neurologically impaired children [2].

SLEEP–WAKE FUNCTION IN SPECIFIC NEUROLOGICAL DISORDERS

Cerebral Palsy

Cerebral palsy has an incidence of approximately 1.2 per 1000 live births, in large part linked to the increased survival of premature infants weighing less than 1500 grams at birth. There is an underlying, static insult to the developing central nervous system, which may be acquired prenatally, in the perinatal period, or during early infancy. There are associated abnormalities of muscle tone, resting posture, muscle coordination, and joints. Cerebral palsy may be of the spastic, dyskinetic, or hypotonic types. Spastic cerebral palsy may be further subcategorized into the hemiplegic, diplegic, or quadriplegic types.

The various types of sleep–wake problems encountered by children with cerebral palsy are listed in Table 73.1. Sleep-related breathing problems are by far the most common complication. Children with spastic cerebral palsy may exhibit daytime irritability, fragmented sleep with frequent nighttime awakenings, and oxygen desaturation from obstructive sleep apnea due to upper airway collapse/adenotonsillar hypertrophy [3]. They are unable to compensate for the disordered breathing by changes in body position, which makes obstructive sleep apnea an especially dangerous condition in this group of patients. Obstructive sleep apnea in cerebral palsy may be linked to macroglossia, pharyngeal muscle incoordination, adenotonsillar hypertrophy, or a diminished ability to compensate for disordered breathing by rolling onto one side. Periodic change of body position at night by the caretaker might be indicated. Additionally, seizures tend to increase the fragmentation of night sleep. Phenytoin should be avoided in children with cerebral palsy owing to the potential side effect of tonsillar hypertrophy and consequent disordered breathing.

Okawa and Sasaki [4] have described how an "acerebrate" stage related to severe brain damage in the perinatal period can produce highly disorganized or undifferentiated

Sleep: A Comprehensive Handbook, Edited by T. Lee-Chiong.
Copyright © 2006 John Wiley & Sons, Inc.

TABLE 73.1 Anatomical Sites and Their Association with Sleep Disturbances in Cerebral Palsy

Anatomical Site	Sleep Disturbance
Cerebral hemispheres	(a) Arousals related to nocturnal seizures or antiepileptic drugs (b) Circadian rhythm abnormalities (delayed sleep phase syndrome, irregular sleep–wake rhythms)
Brainstem	(a) Sleep-related hypoventilation (b) Central, mixed, or obstructive sleep apnea
Upper airway	(a) Obstructive sleep apnea from a collapse of a hypotonic airway (b) Obstructive sleep apnea due to airway obstruction from secretions, macroglossia, glossoptosis, associated craniofacial abnormalities, and/or adenotonsillar hypertrophy (c) Obstructive sleep apnea due to a combination of above factors
Gastrointestinal	Gastroesophageal reflux leading to sleep fragmentation and obstructive sleep apnea
Pulmonary	Recurrent pulmonary aspiration leading to oxygen desaturation
Musculoskeletal	Increased arousals due to a decreased ability to change body position

sleep with absence of sleep spindles and lack of a clear distinction between REM and NREM sleep, or for that matter between sleep and wakefulness. This is presumably due to severe dysfunction of the brainstem, thalamic, and forebrain structures that regulate sleep stages.

Mental Retardation

Petre-Quadens and Jouvet [5] reported on prolongation in the initial REM latency and suppression in the proportion of time spent in REM sleep in patients with severe mental retardation. In general, the greater the level of mental retardation, the less is the time spent in REM sleep. Piazza et al. [6] used momentary time sampling to study sleep in 51 young people 3–21 years of age who displayed sleep and behavioral problems. As compared to controls, they found delays in sleep onset, reduced total sleep time, excessive night awakenings, and early awakenings. In a questionnaire survey of 209 children with severe intellectual impairment aged 5–16 years, Wiggs and Stores [7] queried the parents on a wide range of sleep problems (sleeplessness, excessive sleepiness, and parasomnias). They found that sleeplessness predominated, however, in the form of difficulty settling, night awakenings, or early morning awakening in 44%. Depression may coexist with mild mental retardation, in

which case it may lead to early morning awakenings. Physical problems like obstructive sleep apnea and seizures may also coexist. It is imperative to exclude these problems before attempting behavioral treatments like extinction procedures (e.g., ignoring the child).

Down Syndrome

Patients with Down syndrome have very complicated sleep problems. They develop obstructive sleep apnea as a consequence of macroglossia, midface hypoplasia, and hypotonic upper airway musculature [8–10]. Obesity may coexist. Superimposed on this is an element of chronic hypoventilation (leading to CO_2 retention) from hypotonic intercostal and diaphragmatic musculature. Sleep architecture is frequently disrupted, with decreased sleep efficiency, increased arousals, and suppression of slow-wave and REM sleep. Behavioral problems in Down syndrome may in part be linked to these abnormalities in respiratory control. Central sleep apnea has also been reported in Down syndrome [11]; however, measurement of respiratory effort in this study was by thoracic impedance, which is not a particularly sensitive technique for evaluating respiratory effort. Once nocturnal polysomnography has confirmed obstructive sleep apnea or obstructive hypoventilation, a stepwise management approach is recommended: adenotonsillectomy is usually the first step, followed by repeat polysomnographic reassessment and consideration of the use of a continuous positive airway pressure (CPAP) or bilevel positive airway pressure (BiPAP) device for any residual obstructive sleep apnea. Behavioral conditioning techniques may need to be instituted to facilitate acceptance of the mask and CPAP by the patient.

Autistic Spectrum Disorders

Autism, Asperger's syndrome, and pervasive developmental disorder not otherwise specified are called autistic spectrum disorders (ASDs) and are classified as pervasive developmental disorders in the *Diagnostic and Statistical Manual on Mental Disorders* [12,13]. The incidence of sleep–wake problems in ASDs has been estimated as between 44 and 83% [13]. In a sleep diary study, Richdale [13] found that those with higher intelligence quotient (IQ) had more severe sleep problems. Anxiety is a frequent comorbidity in children with ASDs, and this may underlie at least some of the sleep initiation and maintenance difficulties in this population. Polygraphic studies have shown prolonged initial sleep latency, decreased sleep efficiency, and early morning awakenings. The spontaneous arousal threshold may also be lower in ASDs. The issue of dysregulation of melatonin secretion contributing to sleep initiation and maintenance difficulty has also been considered, but this needs further study.

Rett Syndrome

This is a severe, X-linked dominant, neurodegenerative disorder in girls that is characterized by speech and cognitive regression in the early childhood in association with stereotypic hand wringing movement and gait apraxia following apparent normal psychomotor development during the first 6–18 months of life. It occurs exclusively in girls and appears to be lethal in males. In about 80% of Rett syndrome cases, there is a mutation in the *MECP2* gene (located on Xq28 or Xp22) which regulates transcriptional silencing. Sekul and Percy [14] have reported that over 80% of children with Rett syndrome develop sleep problems, with irregular sleep–wake rhythms being the most common. Nighttime screaming, crying, and episodes of laughter have also been reported [15]. As a group, patients with Rett syndrome may show less sleep at night, with increased sleep fragmentation, combined with carryover sleepiness into the daytime. Challamel et al. [16] feel that there is progressive disintegration of the circadian system underlying the sleep–wake problems of Rett syndrome. There have been isolated case reports about the success of melatonin in ameliorating this sleep disruption. Rett syndrome patients also display episodic hyperventilation during wakefulness, but breathing is normal during sleep. About 50% of patients manifest partial or generalized seizures, some of which may include apnea as an ictal manifestation.

Prader–Willi Syndrome (PWS)

This syndrome of congenital hypotonia, hypogonadism, and cognitive dysfunction is linked to microdeletion of the paternally contributed region of chromosome 15q11.2-q13. During the neonatal period and infancy, patients show severe anorexia and feeding difficulties to the point of requiring nasogastric tube feedings. By early childhood, however, there is a steady increase in appetite to the point that it becomes almost voracious, and associated with morbid obesity. Studies of resting energy expenditure show decreased basal metabolic rate and sleeping metabolic rate, and lower fat-free mass in PWS subjects as compared to gender and bone age-matched obese controls [17]. Though there have been no systematic studies of its prevalence, mild to moderate daytime sleepiness is very common in PWS, affecting perhaps half the subjects. It may be associated with sleep-onset REM periods on the multiple sleep latency test [18]. Results of nocturnal polysomnography are highly variable. Vgontzas et al. [19] have reported increased slow-wave sleep at night. Sleep apnea seems to be relatively infrequent, however, and does not play a role in the pathogenesis of daytime sleepiness, which is most likely secondary to hypothalamic dysfunction. In support of this is the finding of low levels of cerebrospinal fluid (CSF) hypocretin-1 in a patient with combined PWS and Kleine–Levin syndrome, who was tested at the time of increased sleepiness [20].

Lately, it has been recognized that treatment of PWS with growth hormone promotes an increase in muscle mass and motor development [21–23]. The long-term impact of growth hormone therapy on PWS has not been clearly established. Infants and young children with PWS may need to be evaluated for serious complications like sleep-related hypoventilation [24, 25].

Blindness

Blindness associated with loss of light perception due to lesions of the eye or the anterior visual pathway can disrupt circadian rhythms and neuroendocrine functions. The resultant "free running" sleep–wake cycles are longer than 24.5 hours and tend to shift to progressively later and later times around the clock [26]. The basis for the free-running cycles seems to be dysregulation of the secretion of melatonin, which is a light-sensitive hormone secreted by the pineal gland and dependent on an intact retinohypothalamic pathway. In sighted individuals, melatonin has very low secretion during the daytime, but its plasma levels rise immediately prior to bedtime and during the night. It has important sleep induction and maintenance properties. Children with severe retinopathy of prematurity, congenital bilateral glaucoma, septo-optic dysplasia, or severe bilateral optic neuritis may show "free-running" or non-24-hour sleep–wake cycles owing to lack of light perception. In a questionnaire survey of 77 blind children ranging in age from 3 to 18 years, and sighted controls, Leger et al. [27] found that 17% of blind children reported sleeping less than 7 hours at night as compared to 2.6 of controls, with blind children awakening much earlier in the morning, and also exhibiting increased daytime sleepiness. In turn, daytime sleepiness may impact attention, concentration, and behavior of blind children. The presence of multiple associated physical and neurological handicaps may further complicate management. Administration of 0.5–5 mg of melatonin 1 hour prior to bedtime seems to facilitate sleep onset, increase total sleep time, and reduce awake time at night [26, 28]. Manipulation of nonphotic "zeitgebers" like food, music, physical activity, and exercise might also be of some value in blind children with disrupted circadian rhythms.

Arnold–Chiari Malformations

Myelomeningocele is invariably associated with Arnold–Chiari malformation type II, in which a segment of the medulla and the fourth ventricle are located below the plane of the foramen magnum (generally in the cervical 1–2 vertebral level), along with hydrocephalus. In a questionnaire survey of 107 children with myelomeningocele and Chiari type II malformation, Waters et al. [29] found that 62% of the patients gave a history of breathing abnormalities, with 20% having moderate to severe abnormalities. Mechanisms underlying the sleep-related breathing disturbance in these patients include developmental malformations involving brainstem respiratory areas, mechanical compression of the brainstem from a small posterior fossa combined with hydrocephalus, unilateral or bilateral vocal cord paralysis, obstructive apnea from adenotonsillar hypertrophy or collapse of the hypotonic upper airway, as well as hypoventilation from obesity and intercostal muscle paralysis. A combination of mechanisms may also be operative. Patients with a thoracic or thoracolumbar myelomeningocele, those who have a history of a posterior fossa decompression, those with brainstem malformations, and those with pulmonary function abnormalities seem to be prone to develop moderate to severe sleep-related breathing abnormalities [29], which can consist of obstructive sleep apnea, obstructive hypoventilation, or central sleep apnea. Obstructive sleep apnea is most common. Respiratory abnormalities are generally more severe during REM sleep than during NREM sleep. The management is complicated, and at times discouraging. Though adenotonsillectomy may be the initial step in subjects with obstructive sleep apnea, most patients end up needing continuous positive airway pressure (CPAP) or bilevel positive airway pressure breathing devices. Counseling of preteens about the importance of avoiding obesity around adolescence is also recommended.

Combined obstructive and central sleep apnea may also be seen in Chiari type I malformation, characterized by descent of the cerebellar tonsils below the plane of the foramen magnum, with a normally positioned fourth ventricle and no hydrocephalus. There have been isolated case reports of obstructive sleep apnea in such patients responding successfully to posterior fossa decompression [30].

Miscellaneous Childhood Neurological Disorders

Leigh disease or subacute necrotizing encephalomyopathy is a mitochondrial disorder with symmetric and disproportionately greater involvement of the basal ganglia, brainstem, and spinal cord. Onset is generally in infancy or early childhood. It is most often secondary to deficiency of pyruvic dehydrogenase complex or cytochrome *c* oxidase. Hypotonia, lethargy, seizures, poor weight gain, and a weak cry are the usual clinical manifestations. Patients may show suppression of slow-wave sleep, absence of REM sleep, recurrent central apneas, or hypoventilation as a consequence of brainstem involvement [31]. *Joubert's syndrome* is characterized by severe motor and cognitive developmental delays along with severe cerebellar vermis hypoplasia. Patients exhibit periods of hyperpnea and apneustic breathing (prolonged inspiratory pause) [32].

Certain lysosomal storage disorders like the *mucopolysaccharidoses*, especially Hurler's syndrome, have been associated with severe obstructive sleep apnea that resolves subsequent to reduction in macroglossia and augmentation of upper airway diameter with bone marrow transplantation. Patients with *Niemann–Pick type C* disease frequently manifest fragmented night sleep, fragmentary myoclonus, and shortened latencies on the multiple sleep latency test. Two or more sleep-onset REM periods consistent with narcolepsy are, however, infrequent. Some patients have also shown CSF hypocretin deficiency, suggesting involvement of the hypothalamus from the storage disorder [33]. *Neuronal ceroid lipofuscinosis (Batten disease)* is a neurodegenerative disorder characterized by variable age of onset, blindness and retinal degeneration, seizures, and cognitive decline. Patients exhibit severe sleep initiation and maintenance difficulty, nightmares, and night terrors. The irregular and disrupted sleep does not appear to be solely on the basis of disrupted circadian rhythms [34] and may reflect a widespread neurodegenerative process. Patients with *brainstem gliomas, trauma, or encephalitis* may show impaired neural control of breathing, with presence of central apneas or central alveolar hypoventilation (Figure 73.1).

THE RELATIONSHIP BETWEEN SLEEP AND EPILEPSY

The onset of sleep may be associated with increased interictal spiking, as well as an increased propensity to clinical seizures. Frontal and temporal lobe seizures are especially prone to occur during sleep and to exhibit secondary generalization. In some children with new onset of seizures, interictal epileptiform discharges are observed only during sleep [35]. Nocturnal seizures are most likely to occur during stage 2, followed by stage 1, stage 3, and stage 4 of NREM sleep, in that order, and least likely to occur during REM sleep. *Landau–Kleffner syndrome* is characterized by regression in language function (auditory–verbal agnosia) in association with continuous epileptiform activity during more than 85% of nocturnal REM and NREM sleep [36]. Sleep deprivation is associated with activation of seizures. Ellingson et al. [37] observed that clinical seizures occurred after sleep deprivation in 19 of 788 (2.4%) of otherwise healthy subjects.

Patterns of epileptiform abnormality are considerably influenced by sleep. For example, the 3 per second spike and wave complexes of absence seizures are replaced by a single spike and waves or by polyspikes and wave complexes in sleep. The hypsarrhythmia of infantile spasms is replaced during sleep by brief periods of generalized voltage attenuation. Patients with Lennox–Gastaut syndrome may manifest long runs of generalized spike and wave discharges. Benign rolandic epilepsy of childhood is an

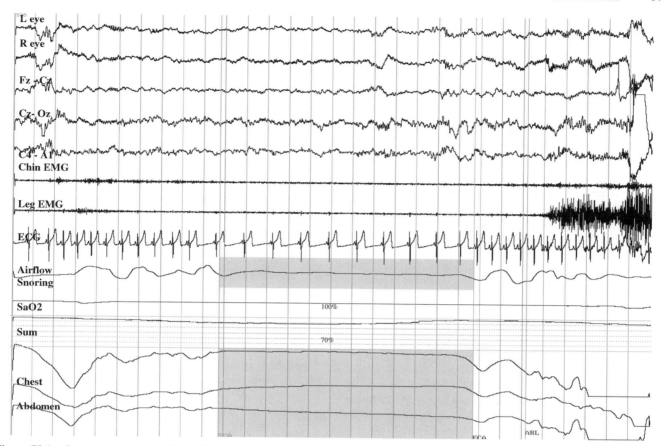

Figure 73.1 Central apnea in a child with brainstem glioma who requires noninvasive nocturnal ventilation. The patient had presented with sleep initiation and maintenance difficulties.

autosomal dominant disorder that is associated with nocturnal seizures during stages 1 or 2 of NREM sleep. The relatively infrequent seizures in this epileptic syndrome are characterized by ictal aphasia, twitching around the mouth, and presence of spikes over the rolandic or midtemporal regions.

Seizures influence sleep architecture by suppressing REM sleep, with a corresponding increase in slow-wave sleep. This effect may persist for as long as 24 hours after the seizure event. Frequent nocturnal seizures also tend to disrupt sleep, with an increased number of arousals. Patients with Lennox–Gastaut syndrome frequently exhibit tonic seizures during sleep. Conversely, sleep disorders can also adversely impact seizure control. Patients with obstructive sleep apnea may manifest poor seizure control, which is improved following correction of obstructive sleep apnea with CPAP, positional therapy, or tracheostomy [38]. In some instances, daytime somnolence may be mistaken for being a side effect of antiepileptic therapy, when in fact it may be the consequence of an underlying sleep disorder.

Antiepileptic drugs in general lead to stabilization of sleep, with a decrease in the amount of sleep fragmentation

[39]. Phenobarbital therapy is associated with suppression in the proportion of time spent in REM sleep and increased stage 3 and 4 NREM sleep (slow-wave sleep). Phenytoin and carbamazepine increase slow-wave sleep at the expense of stage 1–2 NREM sleep. Benzodiazepines increase slow-wave sleep at the expense of REM sleep. Lamotrigine therapy is associated with an increase in the proportion of time spent in REM sleep, with fewer sleep stage shifts. Felbamate is associated with insomnia in about 9% of subjects.

REFERENCES

1. Bartlett LB, Rooney V, Spedding S. Nocturnal difficulties in a population of mentally handicapped children. *Br J Ment Subnormality.* **31**:54–59(1985).

2. Stores G, Wiggs L. Sleep disturbance: a serious, widespread, yet neglected problem in disorders of development. In: Stores G, Wiggs L (Eds), *Sleep Disturbance in Children and Adolescents with Disorders of Development: Its Significance and Management.* Mac Keith Press, London, 2001, pp 3–9.

3. Kotagal S, Gibbons VP, Stith JA. Sleep abnormalities in patients with severe cerebral palsy. *Dev Med Child Neurol* **36**:304–311(1994).

4. Okawa M, Sasaki H. Sleep disorders in mentally retarded and brain-impaired children. In: Guilleminault C (Ed), *Sleep and Its Disorders in Children*. Raven Press, New York, 1987, pp 269–290.

5. Petre-Quadens O, Jouvet M. Sleep in the mentally retarded. *J Neurol Sci* **4**:354–357(1967).

6. Piazza CC, Fischer WW, Kahng SW. Sleep patterns in children and young adults with mental retardation and severe behavioral disorders. *Dev Med Child Neurol* **38**:335–344(1996).

7. Wiggs L, Stores G. Severe sleep disturbance and daytime challenging behavior in children with severe learning disabilities. *J Intellect Disabil Res* **40**:518–528(1996).

8. Zucconi M. Sleep disorders in children with neurological diseases. In: Loughlin GM, Carroll JL, Marcus CL (Eds), *Sleep and Breathing in Children. A Developmental Approach*. Marcel Dekker, New York, 2000, pp 363–383.

9. Marcus CL, Keens TG, Bautista DB, von Pechman WS, Davidson Ward SL. Obstructive sleep apnea in children with Down syndrome. *Pediatrics* **88**:132–139(1991).

10. Stebbens VA, Dennis J, Samuels MP, Croft CB, Southhall DP. Sleep-related upper airway obstruction in Down syndrome. *Arch Dis Child* **66**:1333–1338(1991).

11. Ferri R, Curzi-Dascalova L, Del Graco S, Elia M, Musumeci SA, Stefanini M. Respiratory patterns during sleep in Down syndrome: importance of central apneas. *J Sleep Res* **6**:134–141(1997).

12. American Psychiatric Association. *Diagnostic and Statistical Manual of Mental Disorders*, 4th ed. American Psychiatric Association, Washington, DC, 1994.

13. Richdale AL. Sleep problems in autism: prevalence, cause, and intervention. *Dev Med Child Neurol* **41**:60–66(1999).

14. Sekul E, Percy A. Rett syndrome: clinical features, genetic considerations, and the search for a biological marker. *Curr Neurol* **12**:173–200(1992).

15. Roane HS, Piazza CC. Sleep disorders and Rett syndrome. In: Stores G, Wiggs L (Eds), *Sleep Disturbance in Children and Adolescents with Disorders of Development: Its Significance and Management*. Mac Keith Press, London, 2001, pp 83–86.

16. Challamel MJ, Nevsimalova S, Pretl M. Sleep in Rett and Prader–Willi syndromes. Presentation to the Satellite Meeting of the Development and Sleep Committee Meeting, 15th Congress of the European Sleep Research Society, Istanbul, September 2000.

17. van Mil EA, Westerterp KR, Gerver WJ, Schrander-Stumpel CT, Saris WH. Energy expenditure at rest and during sleep in children with Prader–Willi syndrome is explained by body composition. *Am J Clin Nutr* **71**(3):752–756(2000).

18. Manni R, Politini L, Nobili L, Ferrilo F, Livieri C, Veneselli E, Biancheri R, Martinetti M, Tartara A. Hypersomnia in the Prader–Willi syndrome: clinical-electrophysiological features and underlying factors. *Clin Neurophysiol* **112**(5):800–805(2001).

19. Vgontzas AN, Bixler EO, Kales A, et al. Daytime sleepiness and REM abnormalities in Prader–Willi syndrome: evidence of generalized hypoarousal. *Int J Neurosci* **87**(3–4):127–139(1996).

20. Dauvilliers Y, Bauman CR, Carlander B, Bischof M, Blatter T, Lecendreux M, Maly F, Besset A, Touchon J, Billiard M, Tafti M, Basetti CL. CSF hypocretin-1 levels in narcolepsy, Kleine–Levin syndrome,and other hypersomnias and neurological conditions. *J Neurol Neurosurg Psychiatry* **74**(12):1667–1673(2003).

21. Bauman P, Ritzen EM, Lindgren AC. Endocrine dysfunction in Prader–Willi syndrome: a review with special reference to GH. *Endocr Rev* **22**:787–799(2001).

22. Carrel AL, Myers SE, Whitman BY, Allen DB. Benefits of long-term GH therapy in Prader–Willi syndrome: a 4-year study. *J Clin Endocrinol Metab* **87**:1581–1585(2002).

23. Whitman B, Carrel A, Bekx T, Weber C, Allen D, Myers S. Growth hormone improves body composition and motor development in infants with Prader–Willi syndrome after six months. *J Pediatr Endocrinol Metab* **17**(4):591–600(2004).

24. Eiholzer U, Nordman Y, L'Allemand D. Fatal outcome of sleep apnoea in PWS during the initial phase of growth hormone treatment. A case report. *Horm Res* **58** (Suppl 3):24–26(2002).

25. Schrander-Stumpel CT, Curfs LM, Sastrowijoto P, et al. Prader–Willi syndrome: causes of death in an international series of 27 cases. *Am J Med Genet* **124A**(4):333–338(2004).

26. Sack RL, Brandes RW, Kendall AR, Lewy AJ. Entrainment of free-running circadian rhythms by melatonin in blind people. *N Engl J Med* **343**(15):1070–1077(2000).

27. Leger D, Prevot E, Phillip P, Yence C, Labaye N, Paillard M, Guilleminault C. Sleep disorders in children with blindness. *Ann Neurol* **46**(4):648–651(1999).

28. Fischer S, Smolnik R, Herms M, Born J, Fehm HL. Melatonin acutely improves the neuroendocrine architecture of sleep in blind individuals. *J Clin Endocrinol Metab* **88**(11):5315–5320(2003).

29. Waters KA, Forbes P, Morielli A, Hum C, O'Gorman A, Vernet O, Davis MG, Tewfik TL, Ducharme F, Brouillette RT. Sleep-disordered breathing in children with myelomeningocele. *J Pediatr* **132**(4):672–681(1998).

30. Yoshimi A, Nomura K, Furune S. Sleep apnea syndrome associated with type I Chiari malformation. *Brain Dev* **24**(1):49–51(2002).

31. Yasaki E, Saito Y, Nakano K, Katsumori H, Hayashi K, Nishikawa T, Osawa M. Characteristics of breathing abnormality in Leigh and its overlap syndromes. *Neuropediatrics* **32**(6):299–306(2001).

32. Saito Y, Ito M, Ozawa Y, et al. Changes of neurotransmitters in the brainstem of patients with respiratory pattern disorders during childhood. *Neuropediatrics* **30**(3):133–140(1999).

33. Vankova J, Stepanova I, Jech R, Elleder M, Ling L, Mignot E, Nishino S, Nevsimalova S. Sleep disturbances and hypocretin deficiency in Niemann–Pick disease type C. *Sleep* **26**(4):427–430(2003).

34. Heikkila E, Hatonen TH, Telakivi T, Laasko ML, Heiskala H, Salmi T, Alila A, Santavouri P. Circadian rhythm studies in

neuronal ceroid lipofuscinosis (NCL). *Am J Med Genet* **57**(2):229–234(1995).

35. Shinnar S, Kang H, Berg AT, Goldensohn ES, Hauser WA, Moshe SL. EEG abnormalities in children with a first unprovoked seizure. *Epilepsia* **35**(3):471–476(1994).

36. Robinson RO, Baird G, Robinson G, Simonoff E. Landau–Kleffner syndrome: course and correlates with outcome. *Dev Med Child Neurol* **43**(4):243–247(2001).

37. Ellingson RJ, Wilken K, Bennett DR. Efficacy of sleep deprivation as an activation procedure in epilepsy patients. *J Clin Neurophysiol* **1**(1):83–110(1984).

38. Malow BA, Levy K, Maturen K, Bowers R. Obstructive sleep apnea is common in medically refractory epilepsy patients. *Neurology* **55**:1002–1007(2000).

39. Bourgeois B. The relationship between sleep and epilepsy in children. *Semin Pediatr Neurol* **3**:29–35(1996).

74

SLEEP IN CHILDREN WITH NEUROMUSCULAR DISEASE

DEBORAH C. GIVAN

James Whitcomb Riley Hospital for Children, Indianapolis, Indiana

INTRODUCTION

Much of the morbidity and mortality in children with neuromuscular disease (NMD) occurs because of respiratory muscle weakness, which impairs both ventilation and secretion clearance [1]. Breathing impairment is most noticeable during activities such as exercise and illness that increases the demands on the respiratory system. Less obvious, but equally demanding, are the normal physiologic demands during sleep [2]. Respiratory abnormalities are often first noted during sleep, especially in the child with limited ability to move. Once respiratory compromise has progressed to the point of causing sleep-related daytime symptoms, respiratory failure may be imminent [3, 4]. Physician knowledge of the symptoms and progression of respiratory failure as well as techniques for evaluation and treatment of these children can markedly reduce both morbidity and mortality. Appropriate intervention can allow the children to achieve an improved quality of life. Intervention prior to the onset of daytime symptoms can markedly slow the progression and even stabilize respiratory abnormalities and disability in many children with NMD [5–7].

This chapter will review mechanisms and measurement of breathing and normal changes in ventilation during sleep. How these changes contribute to worsening of respiratory function during sleep will be illustrated. The relationship of growth and development to sleep-related respiratory abnormalities will be discussed. Congenital neuromuscular diseases will be reviewed with emphasis on unique aspects of certain disease states that contribute to respiratory dysfunction. Recommendations for evaluation, including

screening tests, and treatment options that prevent or slow decline in pulmonary function will be discussed.

MECHANISMS AND MEASUREMENTS OF BREATHING IN NMD

Respiration is dependent on the central nervous system control and feedback mechanisms and the respiratory pump (chest wall and muscles of respiration). Respiration is both volitional, driven by the higher cortical centers, and automatic, responsive to feedback from arterial chemoreceptors, metabolic activity, and intrapulmonary receptors [8]. During wakefulness, both systems are operant, but respiration is automatic during NREM sleep, dependent primarily on metabolic activity. Behavioral and arousal stimuli are the dominant drives for respiration during REM sleep [9]. The majority of respiratory abnormalities in patients with NMD occur because of the respiratory muscle weakness. The respiratory muscles can be divided into four groups: (1) the upper airway muscles such as the genioglossus and other pharyngeal dilators; (2) the accessory muscles of respiration, such as the sternocleidomastoid and intercostal muscles; (3) abdominal muscles; and (4) the diaphragm, which is the primary muscle of inspiration. A few types of NMD may also have abnormalities of central control and feedback mechanisms.

Weakness of the upper airway muscles leads to airway obstruction during sleep, problems with mastication, and problems with swallowing. Impairment of the accessory muscles results in decrease in inspiration (they assist the diaphragm) and diminished expiratory force. Abdominal muscle dysfunction results in a weakened cough. The ineffective cough results in secretion retention. Ventilation mismatch and diminished oxygenation are a consequence of the resulting microatelectasis. These weakened chest wall muscles also cannot maintain adequate opposition to the normal elastic recoil of the lung, resulting in a restricted lung. Weakness of the chest wall muscles also leads to the development of scoliosis, which in turn further restricts chest wall excursion and impairs respiratory mechanics [10–12].

The diaphragm is the primary muscle of inspiration and the strength of this muscle is preserved in many neuromuscular diseases. However, if the strength of the diaphragm exceeds that of the muscles of the upper airway, obstructive sleep apnea is more likely to occur. If the diaphragm is weak or paralyzed, then hypoventilation and desaturation are likely to occur, especially during rapid eye movement (REM) sleep, when the skeletal muscles are atonic.

Measurements of pulmonary function can be used to assess muscle weakness and can predict impending respiratory failure, but are not helpful in predicting nocturnal sleep hypoventilation. Pulmonary function tests in NMD typically show a restrictive defect with a decrease in total lung capacity (TLC), decreased vital capacity (VC), decreased maximal expiratory pressure (Pe_{max}), and decreased cough pressure. Residual volume (RV)—the amount of air left in the lung after maximal expiration—is elevated. The functional residual capicity (FRC)—the amount of air left in the lung following a normal expiration—is decreased so that tidal breathing operates near RV, leaving little pulmonary reserve. Maximal inspiratory pressure (Pi_{max}) primarily reflects diaphragm strength. In diaphragm paralysis, one can also see a large drop in vital capacity when one changes from sitting or standing to the supine position. Maximal voluntary ventilation (MVV) can be measured with cooperative patients to assess muscle endurance; however, this measurement may also be reduced in conditions associated with lower airway disease [10]. The relationship of the various normal lung volumes and capacities to lung measurements in NMD is illustrated in the Figure 74.1.

Central control problems, such as congenital central hypoventilation or acquired traumatic hypoventilation, are fortunately rare, but these and other types of central respiratory drive abnormalities are often difficult to distinguish from end organ (muscle) failure. None of the routine measurements of pulmonary function that are noted above can separate central drive problems from end organ defects. The mouth occlusion pressure has been used as a way to distinguish between these defects. This is performed by measuring the negative pressure at the mouth in the first 0.1 second of inspiration against a closed airway [13]. This measurement is not commonly or easily obtained.

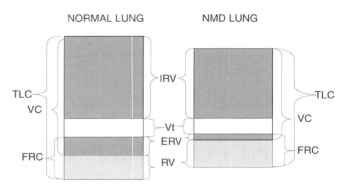

Figure 74.1 Schematic representation of lung parameters for the normal lung and the lung in chronic neuromuscular disease. The lung in neuromuscular disease shows diminished total lung capacity (TLC) with an overall restrictive defect. All lung volumes are decreased as a result. These losses occur by a decrease in the functional residual capacity (FRC)—the amount of air left in the lung after a normal exhalation—which is a combination of the residual volume (RV)—the amount of air left in the lung after maximum exhalation—and expiratory reserve volume (ERV). The tidal volume (Vt) is preserved but because of the decreased FRC, the Vt operates close to the RV with little or no cushion. VC, vital capacity, is the amount of air measured from maximal inspiration to maximal exhalation.

In general, pulmonary function tests are problematic in the young child and the individual with a developmental delay, since the accuracy of these results are predicated on active patient participation. Other measurements of lung function such as arterial blood gases reflect advanced abnormalities of respiratory function as they do not change during wakefulness until the problem has significantly progressed.

GROWTH AND DEVELOPMENT EFFECTS ON LUNG FUNCTION AND NMD

Infants normally have an increase in the RV and decrease in the FRC of the lung compared to older children and adults, because of a highly compliant chest wall. The diaphragm of the infant is horizontally placed. These two factors decrease lung reserve, make respiration inefficient, and markedly increase the work of breathing even in the healthy infant.

For at least the first 6 weeks of life, the infant is a preferential nasal breather. The upper airway structures of the infant are also small (especially the nares) and very compliant, which predisposes to upper airway collapse and obstruction. Some infants may even have initial but temporary nasal obstruction if the nose is compressed during delivery. Infants spend a great deal of time in the supine position and spend up to 20 hours per day sleeping. Infants and children also spend as much as 50% of sleep time in rapid eye movement (REM) sleep. Respiratory illness causes respiratory compromise even in the normal infant. When the stress of a respiratory illness is added to children with NMD, the work of breathing can easily exceed the child's capabilities [14].

BREATHING DURING SLEEP IN THE CHILD

With the onset of sleep and transition into non-rapid eye movement (NREM) sleep, there is a reduction of chemoreceptor sensitivity as well as a higher stimulation threshold for arousal. The arousal response originates in the respiratory nuclei located in the medulla and cortex. These changes in turn cause a reduction in respiratory drive and mute the response of the respiratory system to changes in oxygen and carbon dioxide blood levels. An increase in upper airway resistance also occurs with sleep onset from a fall in the activity of pharyngeal dilator muscles. These factors lead to a decrease in tidal volumes and an increase in respiratory rate, which results in overall decreased minute ventilation and elevation of baseline carbon dioxide levels. Carbon dioxide levels are then maintained at this higher level. Normally there is an increase in the activity of the diaphragm, intercostal muscles, and genioglossus that maintains airway patency and stabilizes ventilation [2].

REM sleep is characterized by a marked reduction in skeletal muscle tone with the important exceptions of the eye muscles and the diaphragm. During REM sleep, there is a further decrease in tidal ventilation, an increase in upper airway resistance, and marked irregularity of breathing. This marked reduction in tone of the muscles of the upper airway and accessory muscles, the dependence of breathing on diaphragmatic function, and deregulation of respiratory control characterizes REM sleep. Infants and children with neuromuscular disease will invariably experience difficulty with breathing during this phase of sleep, difficulty that usually progresses throughout the sleep stages and worsens with time.

RESPIRATORY PROBLEMS DURING SLEEP IN NMD

There are problems during sleep with breathing that are common to all children with neuromuscular disease. For example, as noted above, all children have a small upper airway with a high resistance. Secretions, tonsillar and adenoidal hypertrophy, nasal congestion, and obesity may narrow the airway, predisposing to collapse during inspiration. These children are at high risk for obstructive sleep apnea. The work of breathing is greater because of the effort needed to overcome airway resistance.

Normal infants and most children with neuromuscular disease have a highly compliant chest wall and a horizon-tally placed diaphragm, which increases the work of breathing. These mechanical deficiencies are compensated for by adopting rapid shallow breathing. This compromise results in a diminished respiratory reserve that increases the risk of hypoventilation during sleep.

As noted previously, infants and children typically sleep longer and spend a greater percentage of that sleep time in REM sleep, a factor that increases risk of respiratory failure during sleep. Conversely, REM sleep may be reduced or absent in individuals with sleep disordered breathing or respiratory abnormalities, such as diaphragm paralysis, possibly as a "protective" phenomenon [15]. Obstruction, desaturation, and hypoventilation, in addition to the direct consequences of these problems, also fragment sleep. Sleep fragmentation has also been implicated in causing some daytime symptoms such as hypersomnolence, cognitive dysfunction, and fatigue. Reduced total sleep time and diminished sleep efficiency have been reported frequently in individuals with NMD [16].

FEATURES OF SPECIFIC MUSCLE DISEASES THAT INCREASE SLEEP RISK

Characteristics of some neuromuscular diseases predispose children to problems that present primarily during sleep. Children with cerebral palsy, myelomeningocele, some congenital myopathies, and hereditary sensory and motor neuropathies are more likely to develop obstructive sleep apnea, because they have decreased upper airway tone compared to diaphragmatic muscle tone. These problems first appear during REM sleep with the onset of skeletal muscle atonia.

Individuals with Duchenne muscular dystrophy may develop diaphragm weakness during the course of the disease, while others, children with spinal muscular atrophy type 2, are relatively spared. Diaphragm weakness or paralysis can be a feature of certain diseases such as distal spinal muscular atrophy or Charcot–Marie–Tooth disease. Children with neuromuscular diseases characterized by diaphragm weakness or paralysis are at highest risk for respiratory abnormalities during REM sleep, because of the absence of support from the accessory and intercostal muscles. The paralysis of these muscles results in desaturation and hypoventilation, starting during REM sleep.

A few types of neuromuscular diseases have abnormal respiratory drive. Individuals with myelomeningocele and Arnold–Chiari malformation often exhibit diminished responsiveness to hypercarbia and hypoxia that appears to be of central origin. This problem has also been attributed to those with myotonic dystrophy and to some individuals with congenital myopathies, although more sensitive testing suggests that the likely problem is end organ (muscle) weakness. Many of these patients actually exhibit an increased respiratory drive [13].

EVALUATION

Symptoms such as orthopnea, sleep disruption, and daytime somnolence highly correlate with nocturnal desaturation [3]. By the time these symptoms occur, however, respiratory compromise is usually advanced. Infants and children have limited ability to communicate these symptoms. Physicians must rely on parental reports for this information. Some children respond to hypersomnolence by increasing activity, increasing the diagnostic difficulty presented by this problem. Symptoms are frequently masked by this type of paradoxical response.

Nocturnal desaturation and hypoventilation precede daytime respiratory failure. Identification of those individuals at risk for nocturnal hypoventilation has been attempted using pulmonary function testing. Some authors have been able to show at least a weak correlation between VC, supine VC, forced expiratory volume in the first second of exhalation (FEV_1), Pe_{max}, and FRC and nocturnal hypoventilation [3]. Studies in adults with amyotrophic lateral sclerosis suggest that a vital capacity of less than 1 liter is more predictive of outcome and survival than is the presence of nocturnal hypoventilation [17]. Infants and children may not be able to perform function testing in a reliable manner, complicating the usefulness of these measurements in children with NMD. However, pulmonary function tests are extremely useful in documenting decline of respiratory function in the cooperative patient.

Daytime blood gases have also been examined but are not a reliable predictor of impending respiratory failure. Normal daytime blood gases have been described on the morning prior to death in a patient with Duchenne muscular dystrophy [18].

The polysomnogram (PSG) is the gold standard for detection of nocturnal hypoventilation and sleep disruption. Once diagnosed, the PSG is also useful for following the progression of respiratory abnormalities and the impact of treatment. The limited availability and high cost of this method of assessment have led to substitution of other methods of evaluation. To decrease the cost, methods for selecting individuals more likely to have sleep disordered breathing have been explored. Correlation with some pulmonary function parameters do exist. For example, 50% of patient with a VC less than 1 liter require ventilator support within 2 years. A Pi_{max} of less than -30 mm H_2O has been associated with hypercapnia in 80% of patients with NMD. These screening tests are valuable and can help select patients but are difficult to obtain in children and cannot substitute for actual measurement of sleep and respiratory parameters [19].

The PSG is extremely useful for adjustment of ventilator settings with growth. At least yearly reevaluation of these parameters is suggested.

TREATMENT

Research into the molecular causes of muscle disease has progressed significantly in the past few years. Possibly in the near future, specific therapy for the neuromuscular diseases may become available. Until then, attention to regular preventative care and nutrition are crucial to maintenance of good health. Several recently published overviews of children and NMD deal with this in more detail [5, 10, 20].

Yearly assessment of cardiac stability with electrocardiogram and echocardiogram is a useful adjunct to lung function testing. Routine assessment (every 4 months) with pulmonary function tests including the MVV, Pi_{max}, and Pe_{max}, even in children with ventilatory support, allows for anticipation of pulmonary problems.

At the present time, however, the most crucial intervention is to prevent or slow the decline in pulmonary function. Devices that temporarily increase lung volumes used in conjunction with devices or techniques that assist cough are extremely helpful in preventing atelectasis and pneumonia [21].

Sufficient data now exist to show that ventilatory support can be lifesaving and is a valid choice for those with NMD. Both quality and quantity of life are increased with the use of ventilators. In many children with NMD, nocturnal support alone is sufficient to decrease illness frequency and stabilize pulmonary function. Improvement in daytime symptoms and metabolic parameters occur soon after institution of ventilation [18]. The preferred method of ventilation is noninvasive using a nasal or full face mask interface. This is effective for most children with neuromuscular disease and circumvents the need for invasive ventilation with a tracheostomy for all but the most impaired patients [7, 20].

Every effort should be made to discuss respiratory prognosis and the pros and cons of ventilation with patients and their families before a crisis occurs. Institution of ventilation is not easy. A strong commitment from patient and family is crucial to successful institution of this therapy. We highly recommend the fitting of a mask and the institution of ventilation in the sleep laboratory or intensive care unit. Here continuous monitoring of oxygen saturation and carbon dioxide is available, as are personnel trained in the assessment of respiratory function and adjustment of respiratory equipment.

Physically able individuals underestimate the quality of life for those who are disabled. Many centers that care for children with NMD still fail to discuss respiratory interventions with patient or their families. Good patient care requires the dissemination of information about potential therapies in a nonjudgmental way, to all families in order for them to make educated decisions about all interventions, especially ventilation [22, 23].

REFERENCES

1. Seddon PC, Khan Y. Respiratory problems in children with neurological impairment. *Arch Dis Child* **88**:75–78(2003).

2. McNicholas WT. Impact of sleep on respiratory muscle function. *Monaldi Arch Chest Dis* **57**:277–280(2002).

3. Hukins DA, Hillman DR. Daytime predictors of sleep hypoventilation in Duchenne muscular dystrophy. *Am J Respir Crit Care Med* **161**:166–170(2000).

4. Ragette R, Mellies U, Schwake C, Voit T, Teschler H. Patterns and predicators of sleep disordered breathing primary myopathies. *Thorax* **57**:724–728(2002).

5. Gilgoff RL, Gilgoff IS. Long-term follow-up of home mechanical ventilation in young children with spinal cord injury and neuromuscular conditions. *J Pediatr* **142**:476–480(2003).

6. Katz S, Selvadurai H, Keilty K, Mitchell M, Maclusky I. Outcome of non-invasive positive pressure ventilation in paediatric neuromuscular disease. *Arch Dis Child* **89**:121–124(2004).

7. Sritippayawan S, Kun SS, Keens TG, Davidson Ward SL. Initiation of home mechanical ventilation in children with neuromuscular disease. *J Pediatr* **142**:481–485(2003).

8. White DP. Ventilation and the control of respiration during sleep: normal mechanisms, pathologic nocturnal hypoventilation, and central sleep apnea. In: Martin RJ (Ed), *Cardiorespiratory Disorders During Sleep*. Futura, Mount Kisco, NY, 1990, pp 53–108.

9. Phillipson EA. Control of breathing during sleep. *Am Rev Respir Dis* **118**:909–939(1978).

10. Birnkrant DJ. The assessment and management of the respiratory complication of pediatric neuromuscular disease. *Clin Pediatr* **41**:301–308(2002).

11. Bourke SC, Gibson GJ. Sleep and breathing in neuromuscular disease. *Eur Respir J* **19**:1194–1201(2002).

12. Perrin C, Unterborn JN, D'Ambrosio C, Hill N. Pulmonary complications of chronic neuromuscular disease and their management. *Muscle Nerve* **29**:5–27(2004).

13. Rochester DF. Respiratory muscles and ventilatory failure: 1993 perspective. *Am J Med Sci* **305**:394–402(1993).

14. Givan DC. Sleep and breathing in children with neuromuscular disease. In: Loughlin GM, Carroll JL, Marcus CL (Eds), *Sleep and Breathing Children: A Developmental Approach*. Marcel Dekker, New York, 2000, pp 691–735.

15. Schramm CM. Current concepts of respiratory complications of neuromuscular disease in children. *Curr Opin Pediatr* **12**:203–207(2000).

16. Redding GJ, Okamoto GA, Guthrie RD, Rollevson D, Milstein JM. Sleep patterns in non-ambulatory boys with Duchenne muscular dystrophy. *Arch Phys Med Rehabil* **66**:818–821(1985).

17. Gay PC, Westbrook PR, Daube JR, Litchy WJ, Windebank AJ, Iverson R. Effects of alterations in pulmonary function and sleep variables on survival in patients with amyotrophic lateral sclerosis. *Mayo Clin Proc* **66**:686–694(1991).

18. Barbe F, Quera-Salva MA, de Lattre J, Gajdos P, Agusti AGN. Long-term effects of nasal intermittent positive-pressure ventilation on pulmonary function and sleep architecture in patients with neuromuscular disease. *Chest* **110**:1179–1183(1996).

19. Weinberg J, Klefbeck B, Borg J, Svanborg E. Polysomnography in chronic neuromuscular disease. *Respiration* **70**:349–359(2003).

20. Mellies U, Ragette R, Schwake CD, Boehm H, Voir T, Teschler H. Long-term noninvasive ventilation in children and adolescents with neuromuscular disorders. *Eur Respir J* **22**:631–636(2003).

21. Miske W, Hickey EM, Kolb SM, Weiner DJ, Panitch HB. Use of the mechanical in-exsufflator in pediatric patients with neuromuscular disease and impaired cough. *Chest* **124**:1406–1412(2004).

22. Bach JR, Chaudhry SS. Standards of care in MDA Clinics. *Am J Phys Med Rehabil* **79**:193–196(2000).

23. American Thoracic Society. Respiratory care of the patient with Duchenne muscular dystrophy. *Am J Respir Crit Care Med* **170**:456–465(2004).

75

SLEEP IN CHILDREN WITH BEHAVIORAL AND PSYCHIATRIC DISORDERS

JUDITH A. OWENS
Brown University Medical School, Providence, Rhode Island

KATHERINE FINN DAVIS
School of Medicine, Emory University, Atlanta, Georgia

INTRODUCTION

Increased recognition of the high prevalence of both sleep complaints and diagnosed sleep disorders in the pediatric population has been accompanied by heightened awareness of the protean emotional and behavioral manifestations of sleep disturbances in children and adolescents. Clinical experience, as well as empirical evidence from numerous studies and case reports, has demonstrated that childhood sleep disorders arising from both intrinsic processes (e.g., obstructive sleep apnea) and extrinsic factors (e.g., sleep onset association disorder) may present primarily with daytime neurobehavioral symptoms. These sleep-related consequences range from mood lability to attentional problems to school failure. Experimental studies, which have examined the relationship between sleep disruption and sleep loss and neurobehavioral deficits in more detail in the laboratory setting, have also contributed significantly to our understanding of the relationship between sleep and mood and behavior in children.

Alternatively, primary behavioral and psychiatric disorders in children are frequently associated with and/or complicated by sleep disturbances, which may be related to such factors as the psychopathology of the underlying disorder, comorbid conditions, or pharmacologic treatment. This chapter will describe this bidirectional relationship between sleep and neurobehavioral deficits in both experimental settings and normal pediatric and clinical populations as well as emphasize the importance of evaluating and treating sleep disturbances in all children presenting with behavioral and academic concerns.

CONCEPTUAL FRAMEWORK

As in adults, there are several basic mechanisms that form the theoretical basis for the link between disturbed sleep and neurobehavioral deficits in children. From the perspective of sleep as the primary variable, sleep deprivation resulting from inadequate sleep duration or sleep loss, fragmented or disturbed sleep related to frequent or prolonged arousals, and primary disorders of excessive daytime somnolence, all result in manifestations of excessive daytime sleepiness (EDS). Unlike adults, however, EDS in children may not be characterized by such overt behaviors as yawning, complaining about fatigue, and so on, but may rather be associated with a host of subtler or even "paradoxical" behavioral manifestations. These range from emotional lability and low frustration tolerance ("internalizing" behaviors) to neurocognitive deficits and behavioral disinhibition such as increased aggression ("externalizing" behaviors). In turn, functional deficits in mood, attention, cognition, and behavior may lead to performance deficits in the home, school, and social setting.

Alternatively, evidence suggests that neurobehavioral/emotional disruptions may lead to sleep disturbances, as is often the case with stressful life events. Some authors [1] have postulated an underlying mechanism for this complex relationship between sleep and behavior that involves linked central nervous system regulatory systems for sleep and arousal, attention, and emotion.

EXPERIMENTAL STUDIES

A number of studies that have examined the specific neuro-behavioral effects of both acute full- or partial-night sleep deprivation and chronic partial-night sleep deprivation (sleep restriction) in older children and adolescents under experimental conditions have documented relative deficits in a number of domains. An increase in behavioral signs of sleepiness as well as subjective reports of sleepiness are the most consistent and reproducible findings in these types of studies [2,3]. Other domains affected include neuropsychological measures of reaction time, memory, problem-solving ability, and creativity (so-called "executive" or higher level cognitive processes), but it should be emphasized that most of the deficits found were at a modest level. Although neuropsychological measures of attention have not been demonstrated to be significantly impaired overall, parent-reported attentional problems have been reported. Academic performance as well as positive mood has also been shown to be affected in similar studies.

Notably, however, the data on the neurobehavioral consequences of restricted sleep in children under experimental conditions is generally both smaller in quantity and less robust than similar studies in adults, at least in part due to the ethical and practical limitations posed by the pediatric population. In addition, as is frequently the case in pediatric research, developmental considerations related to chronological age and cognitive/emotional development present significant methodological challenges and often complicate the interpretation of study results. Finally, there are also likely to be individual differences in vulnerability to sleep deprivation found in children that are affected by a variety of risk and protective factors.

SLEEP AND BEHAVIOR IN NORMAL PEDIATRIC POPULATIONS

A number of studies examining the prevalence of parent- and child-reported sleep complaints in large samples of children and adolescents have attempted to further delineate the association between disrupted sleep and behavioral concerns. Most of these studies have utilized either study-specific parent questionnaires or broad-based pediatric sleep surveys to assess for a variety of sleep disturbances, ranging from bedtime resistance to prolonged night wakings to parasomnias. In general, children with sleep complaints were more likely to manifest EDS, behavioral problems, moodiness, and school and learning problems. Studies have shown that preschoolers with persistent sleep disorders are significantly more likely to have behavior problem and to be difficult for parents to manage. Additionally, a strong association has been found between poor sleep and externalizing behaviors such as hyperactivity, aggression, and general misconduct [4].

Some 20–40% of school-age children suffer from parent- or self-reported sleep disturbances and consequently experience high rates of behavioral, emotional, somatic, and social problems. In one recent survey of sleep disturbances in kindergarten through fourth grade students, teachers reported that 10% of their students manifested behavioral evidence of significant daytime sleepiness in the classroom setting [5]. Another study examining sleep duration, wake times, and daytime functioning in fifth graders found a significant correlation between early rise times and self-reported difficulties in attention and concentration [6]. Also, sleep disturbances in early childhood have been associated with social and emotional difficulties in early adolescence, suggesting a predictive quality. However, as most of these studies are cross-sectional and rely on parental (or self) report for description of both sleep and behavioral variables, the results must be interpreted cautiously in terms of sleep as a causal factor.

In similar survey studies, adolescents who reported disturbed or inadequate sleep were also more likely to report mood disturbances, subjective sleepiness, and performance deficits in both social and academic spheres [7]. Adolescents may be at increased risk for sleep disturbances and inadequate sleep for a number of reasons, including pubertal changes that are associated with both increased daytime sleepiness and a relative sleep phase delay, and social/environmental factors that contribute to both later bedtimes and earlier rise times. The resultant decrease in total sleep time in adolescents has been associated with poorer grades in school, as well as depressed mood, anxiety, and decreased energy. In about one-third of adolescents surveyed, significant sleep disturbances were reported along with increased levels of self- and parent-reported externalizing behavioral problems [8].

NEUROBEHAVIORAL DEFICITS IN PRIMARY PEDIATRIC SLEEP DISORDERS

Another important line of research, which has examined the link between sleep and behavioral disturbances, has focused on the identification of neurobehavioral deficits in clinical populations of children with diagnosed sleep disorders. Although at times anecdotal in nature and often confounded by multiple other variables, an understanding of behavioral and cognitive problems associated with clinical sleep disorders in the naturalistic setting sheds light on the long-term consequences of chronically disturbed sleep in children and clearly has relevance for the clinician diagnosing and treating these disorders. The pediatric sleep disorders, which have been studied from this perspective, include obstructive sleep apnea syndrome (OSAS)/sleep disordered breathing (SDB), restless legs syndrome/periodic limb movement disorder (RLS/PLMD), and narcolepsy. Both OSAS and RLS/PLMD share the common feature of sleep fragmentation resulting from frequent nocturnal arousals, which in turn leads to excessive daytime sleepiness (EDS) and associated neurobehavioral disturbances that are also frequently seen in primary disorders of EDS such as narcolepsy. Repeated episodes of nocturnal hypoxia most likely comprise an important additional etiologic factor in OSAS.

OBSTRUCTIVE SLEEP APNEA SYNDROME/SLEEP DISORDERED BREATHING

Although less is known about the neurobehavioral consequences of OSAS in children than in adults, most pediatric studies have supported a similar range of deficits in children in attention, memory, and executive functions, as well as an increase in subjective sleepiness and mood disturbance. A higher prevalence of parent-reported externalizing behavior problems, including impulsivity, decreased attention span, hyperactivity, and aggression, has been consistently reported in studies of children with either polysomnographically diagnosed OSAS or symptoms suggestive of SDB, such as snoring [7]. Other studies have also found higher levels of internalizing behaviors (somatic complaints, social withdrawal) as well, and children with chronic snoring and other symptoms of SDB report an increase in depressive symptoms (e.g., anhedonia, poor self-esteem, negative mood) and decreased quality of life (e.g., impaired emotional, social, and school functioning) [9]. Studies that have compared neuropsychological functions in children with OSAS have found impairments on tasks involving reaction time, vigilance, attention, executive functions, motor skills, and memory, as well as overall school performance [10]. Evidence of a correlation between disease severity and level of performance deficits has recently been reported [11].

Studies that have looked at changes in behavior and neuropsychological functioning in children following treatment (usually adenotonsillectomy) for OSAS/SDB have also documented significant improvements in daytime sleepiness, behavior, and academic performance post-treatment [12]. In addition to these more subjective reports of improvement, which may be influenced by expectations regarding treatment outcome, objective improvements in neuropsychological measurements of attention, vigilance, reaction time, and cognitive functions have also been described.

Alternatively, the prevalence of SDB symptoms in children with identified behavioral and academic problems has also been examined in several studies. One sample of first graders performing academically in the lowest 10th percentile found a prevalence of 19% of significant SDB symptoms [13]. Moreover, young children with SDB may continue to be at high risk for poor academic performance several years after the symptoms have resolved, suggesting there may be long-term effects on intellectual and cognitive abilities. A number of other studies have similarly found an increased prevalence of snoring in children with behavioral concerns. Reports document a significant increase in OSAS or SDB symptoms specifically in children being evaluated for or diagnosed with attention deficit hyperactivity disorder (ADHD) [14] and as many as 25% of ADHD diagnoses may be linked to symptoms of SDB, such as habitual snoring.

RESTLESS LEGS SYNDROME/PERIODIC LIMB MOVEMENT DISORDER

Although the prevalence of RLS/PLMD in the pediatric population is unknown, retrospective reports given by adults with these disorders suggest that symptoms (such as restless sleep and "growing pains") frequently first appear in childhood and can result in significant sleep disturbance, including difficulty falling asleep and night wakings. Additional diagnostic clues may include a positive family history of RLS/PLMD and exacerbation of symptoms by caffeine intake. As in adults, RLS symptoms in children and adolescents are typically worse in the evening and may be exacerbated by inactivity, thus contributing to significantly delayed sleep onset, although these are generally poorly articulated by the children themselves. The brief, repetitive jerks primarily of the lower extremities during stages 1 and 2 of non-rapid eye movement (NREM) sleep that characterize PLMD frequently accompany RLS. This jerking activity results in multiple nocturnal arousals and awakenings and, consequently, sleep fragmentation. This sleep fragmentation along with the RLS-associated shortened sleep duration exacerbate symptoms of EDS.

Neurobehavioral consequences related to RLS/PLMD may then occur and, as described in several studies, may present as symptoms of ADHD [15] and other neuro behavioral problems. In one study, 64% of children evaluated for ADHD in a pediatric neurology clinic were found to have clinical and polysomnographic evidence of RLS/PLMD [16]. The finding of PLMD was associated with a history of bedtime refusal and problematic night wakings. Furthermore, treatment with dopamine antagonists has been shown to result not only in improved sleep quality and quantity, but also improvement in "ADHD" behaviors previously resistant to treatment with psychostimulants.

PRIMARY DISORDERS OF EXCESSIVE DAYTIME SLEEPINESS (EDS)

Narcolepsy is another sleep disorder that is rarely diagnosed in children but that retrospective surveys nonetheless suggest frequently first presents in late childhood and early adolescence. In particular, the cardinal feature of narcolepsy, repeated episodes of profound and often irresistible sleepiness, is usually the first symptom to emerge, but is the one that is most likely to be confused with other neurobehavioral disorders, such as ADHD. Adolescents with narcolepsy often delay in seeking medical attention and are frequently labeled as having mood disorders, learning problems, and academic failure before the underlying etiology is identified, as long as several decades later [17].

The other pathognomonic symptoms of narcolepsy, such as cataplexy (the sudden complete or partial loss of muscle tone in response to an emotional stimulus) and hynagogic hallucinations may be more easily recognized as abnormal by the patient and family, and thus may be more likely to be brought to the attention of a health care professional. Nonetheless, these symptoms are also frequently misdiagnosed as part of a psychiatric or neurologic disorder, such as psychosis or a conversion reaction. In addition, with psychopharmacologic agents treatment of symptoms of depression or ADHD (SSRIs, psychostimulants) may in fact improve underlying narcoleptic symptoms, further confusing the clinical picture. Appropriate and timely treatment of both the EDS and other symptoms, however, can result in reversal or amelioration of at least some of the neurobehavioral consequences, thus making early recognition of narcolepsy an important goal for clinicians.

SLEEP AND PSYCHIATRIC DISORDERS IN CHILDREN

As mentioned earlier, sleep disturbances also have a significant impact on the clinical presentation and symptom severity, as well as management of psychiatric disorders in children and adolescents. As in adults, pediatric sleep disturbances may be a feature commonly associated with the symptom constellation of some psychiatric disorders, such as early morning awakening in depression or nightmares in post-traumatic stress disorder (PTSD), or may be frequently present as a comorbid condition, such as delayed sleep onset in anxiety disorders. The increased use, even in younger children, of psychotropic medications, which often have significant negative effects on sleep,

further complicates the issue. As is also the case in adults, it is often very difficult from a clinical standpoint to clearly distinguish between primary and secondary sleep disorders in children with behavioral and emotional problems. Therefore mental health providers should evaluate and appropriately treat both, since improvement in one is likely to have a positive impact on the related condition.

Several studies have evaluated the prevalence of sleep disturbances in samples of children and adolescents with a variety of psychiatric disorders. The results suggest an increase in a wide range of reported sleep disturbances in these mixed clinical populations, including parasomnias such as nightmares and night terrors, difficulty falling asleep, frequent and prolonged night wakings, sleep-related anxiety symptoms (fear of the dark, etc.), restless sleep, subjective poor sleep quality, and daytime fatigue [18]. Similarly, an association has been reported between sleep disturbances and psychiatric disorders, including affective disorders, ADHD, and conduct disorder in surveys of children and adolescents from the general population. The relationship between sleep disorders and psychiatric disorders is postulated to be reciprocal. That is, emotional and psychiatric difficulties can produce sleep disturbances, and similarly, sleep deprivation can alter mood and behavior, thus creating a bidirectional cycle where one disorder contributes to the presence of the other and vice versa.

MOOD DISORDERS AND SLEEP

Sleep disturbances, especially "insomnia" (usually defined as difficulty initiating and/or maintaining sleep), are very commonly reported in children and adolescents with depression and anxiety disorders [19,20]. Studies of children with major depressive disorder, for example, have reported a prevalence of insomnia of up to 75% (severe insomnia of 30%) and sleep onset delay in one third of depressed adolescents; although, it should be noted that objective data (polysomnography, etc.) does not always support these subjective complaints. In one study of psychiatrically hospitalized children and adolescents, poorer sleep quality as assessed by actigraphy, was correlated with self-reported depression and hopelessness [21]. Treatment of the associated sleep disturbances usually results in some degree of improvement in mood and fatigue, although treatment of the underlying depression may actually exacerbate the sleep disturbance because of the sleep-disrupting effects of some antidepressant medications, including SSRIs.

Sleep appears to be particularly sensitive to stressors. Reports of sleep complaints, especially bedtime resistance, refusal to sleep alone, increased nighttime fears, and nightmares are common in children who have experienced severely traumatic events (including physical and sexual abuse) [22]. These sleep complaints are also frequently associated with less dramatic but nonetheless stressful life events such as brief separation from a parent, birth of a sibling, or school transitions. Sleep disturbances, such as difficulty settling, may be a consequence of physiologic and emotional hyperarousal mechanisms set off by stressful events. Alternatively, *increased* sleep under stressful conditions may be more of a reflection of coping mechanisms, which involve shutting off distressing external stimuli. Sleep disturbances are not universally found in all children experiencing varying degrees of stress. Variables such as level of exposure and physical proximity to the traumatic situation, previous exposure, and the opportunity for habituation to the stress may play important roles in either mitigating or exacerbating associated sleep disturbances. Other developmental considerations, such as the age and temperament of the child, as well as variables such as the presence of parental psychopathology, also clearly have an important influence.

Nighttime fears, which are often a reflection of normal developmental processes, are also extremely common in young children and may lead to significant bedtime resistance and problematic night wakings. These fears may also have a learned or conditioned component, especially if they are inadvertently reinforced by parental attention. It is important to distinguish, however, between developmentally appropriate and thus usually time-limited nighttime anxiety-related behaviors and sleep-disrupting fears associated with more global daytime anxiety symptoms that could be indicative of an anxiety disorder. Research suggests that as much as 10% of children with nighttime fears also have one or more associated DSM-III-R anxiety disorders [23]. In which case, cognitive–behavioral techniques (positive reinforcement, self-control procedures, graduated extinction) often used successfully to treat isolated nighttime fears may be ineffective and more intensive intervention strategies (psychotherapy, anxiolytics) may be warranted.

SLEEP AND ADHD

Given that attention deficit hyperactivity disorder is one of the most common psychiatric disorders found in childhood, affecting some 3–5% of the population, and that sleep issues frequently are part of the clinical presentation, it is not surprising that the relationship between sleep and ADHD has been the object of increasing scientific scrutiny. As noted earlier, not only do a number of primary sleep disorders including OSAS and RLS/PLMD frequently include ADHD-like symptoms as part of their clinical presentation, but there is considerable overlap between the diagnostic

features of ADHD (inattention, hyperactivity, impulsivity) and neurobehavioral deficits associated with any significant sleep disruption or restriction in children. This relationship has led to speculation about a common underlying central nervous system pathophysiology for both ADHD and sleep deprivation that has been postulated to involve the executive functions of the prefrontal cortex, as well as dopaminergic neurotransmitter system dysfunction. Thus a better understanding of the relationship between sleep and ADHD may contribute significantly to our understanding of the relationship between specific brain functions, sleep, and behavior.

The lines of evidence supporting or refuting an increased prevalence of sleep disturbances in children with clinically diagnosed ADHD basically come from three sources, which do not always concur in their findings: anecdotal clinical experience, parent and self-report surveys of sleep disturbances in children with ADHD, and more objective measures (polysomnography, actigraphy) of sleep parameters in these children. Clinicians who evaluate and treat children with ADHD frequently report sleep disturbances, especially difficulty initiating and maintaining sleep, and, in fact, restless and disturbed sleep was at one time part of the DSM diagnostic criteria for ADHD. Surveys of parents and children with ADHD also consistently report an increased prevalence of sleep disturbances, including delayed sleep onset, poor sleep quality, restless sleep, frequent night wakings, and shortened sleep duration [24], compared with healthy controls, siblings, or children with other psychiatric, behavioral and learning problems. More objective methods of examining sleep and sleep architecture, however, have overall disclosed minimal or inconsistent differences between children with ADHD and controls on such variables as sleep duration, sleep efficiency, sleep onset latency, and percentage of REM sleep.

This inconsistency in findings both between objective and subjective measures of sleep in children with ADHD and across studies, however, is not necessarily surprising. Many of the sleep variables measured by the more objective methods, such as polysomnographic studies, may be viewed as reflective of fundamentally different sleep domains than those measured by parental ratings (sleep "architecture" vs. sleep "behavior"), especially given the situation of a child in the artificial environment of the sleep laboratory. In addition, sleep disturbances in children with ADHD are likely to be multifactorial in nature, ranging from psychostimulant-mediated sleep onset delay in some children to bedtime resistance related to a comorbid anxiety or oppositional defiant disorder in others [25]. In some children, settling difficulties at bedtime may be related to deficits in sensory integration associated with ADHD, while in others, a circadian phase delay may be the primary etiologic factor in bedtime resistance.

From a clinical standpoint, then, an important treatment goal in managing the individual child with ADHD should be the further delineation of any comorbid sleep disturbances, followed by appropriate behavioral and/or pharmacologic intervention targeted toward eventual improvement in daytime symptomatology. Behavioral treatment strategies may range from interventions such as a token positive reinforcement system for a child with oppositional defiant disorder and a limit setting sleep disorder resulting in bedtime resistance, to relaxation training for a child with anxiety-based sleep onset delay. Difficulty falling asleep related to psychostimulant use may respond to adjustments in the dosing schedule, as in some children the sleep onset delay is due to a "rebound" effect of the medication wearing off coincident with bedtime, rather than a direct stimulatory effect of the medication itself. Alpha agonists and antidepressants with sedating properties have also been used successfully in children with ADHD who appear to have primary settling problems, although no empirical evidence currently exists to support efficacy and safety of these drugs in treating ADHD-related sleep problems.

CONCLUSION

This discussion provides evidence for the interface between sleep and psychiatric disorders in children, that the relationship of sleep disturbance to behavioral problems is both a complex and important one from a scientific as well as a clinical standpoint. Furthermore, even modest improvements in sleep quality and duration may have a significant impact on neurobehavioral functioning [26]. All children presenting to health care practitioners with learning, attention, behavioral, or emotional concerns should be rigorously assessed for underlying or comorbid sleep disorders as part of the routine evaluation, particularly since parents and the children themselves may not recognize the connection and thus may fail to spontaneously volunteer such information. Furthermore, because parents of older children and adolescents, in particular, may not be aware of any existing sleep difficulties, it is also prudent to directly question the patient about sleep issues as well. The key areas of inquiry that should be included in screening for sleep disturbances in children and adolescents are (1) bedtime resistance and delayed sleep onset, (2) frequent and/or prolonged night wakings, (3) regularity, pattern, and duration of sleep, (4) snoring and other symptoms of sleep disordered breathing, (5) sleep-related anxiety behaviors (nightmares, nighttime fears), and (6) excessive daytime sleepiness (difficulty in morning awakening, naps, etc.). A number of brief parent and self-report sleep survey tools have been developed, which can facilitate the screening process and yield important information about the nature and severity of any coexisting sleep complaints.

REFERENCES

1. Dahl RE. The regulation of sleep and arousal: development and psychopathology. *Dev Psychopathol* **8**:3–27(1996).

2. Randazzo AC, Muehlback MJ, Schwtizer PK, Walch JK. Cognitive function following acute sleep restriction in children ages 10–14. *Sleep* **15**:861–868(1998).

3. Fallone G, Owens J, Deane J. Sleepiness in children and adolescents: clinical implications. *Sleep Med Rev* **6**(4):287–306(2002).

4. Chervin RD, Archbold KH, Dillon JE, et al. Inattention, hyperactivity, and symptoms of sleep-disordered breathing. *Pediatrics* **109**(3):449–456(2002).

5. Owens J, Spirito A, McGuinn M, Nobile C. Sleep habits and sleep disturbance in school-aged children. *J Dev Behav Pediatr* **21**(1):27–36(2000).

6. Epstein R, Chillag N, Lavie P. Starting times of school: effects of daytime functioning of fifth-grade children in Israel. *Sleep* **21**(3):250–256(1998).

7. Wolfson AR, Carskadon MA. Sleep schedules and daytime functioning in adolescents. *Child Dev* **69**:875–887(1998).

8. Morrison DN, McGee R, Stanton WR. Sleep problems in adolescence. *J Am Acad Child Adolesc Psychiatry* **31**(1):94–99(1992).

9. Crabtree VM, Varni JW, Gozal D. Health-related quality of life and depressive symptoms in children with suspected sleep-disordered breathing. *Sleep* **27**(6):1131–1138(2004).

10. O'Brien LM, Mervis CB, Holbrook CR, et al. Neurobehavioral implications of habitual snoring in children. *Pediatrics* **114**(1):44–49(2004).

11. O'Brien LM, Tauman R, Gozal D. Sleep pressure correlates of cognitive and behavioral morbidity in snoring children. *Sleep* **27**(2):279–282(2004).

12. Ali NJ, Pitson D, Stradlin JR. Sleep disordered breathing: effects of adenotonsillectomy on behavior and psychological function. *Eur J Pediatr* **155**:56(1996).

13. Gozal D, Pope DW. Snoring during early childhood and academic performance at ages thirteen to fourteen years. *Pediatrics* **107**(6):1394–1399(2001).

14. LeBourgeois MK, Avis K, Mixon M, Olmi J, Harsh J. Snoring, sleep quality, and sleepiness across attention-deficit/hyperactivity disorder subtypes. *Sleep* **27**(3):520–525(2004).

15. Chervin RD, Archbold KH, Dillon JE, et al. Associations between symptoms of inattention, hyperactivity, restless legs, and periodic leg movements. *Sleep* **25**(2):213–218(2002).

16. Picchietti DL, Waters AS. Restless legs syndrome and periodic limb movement disorder in children and adolescents: comorbidity with attention-deficit hyperactivity disorder. *Child Adolesc Psychiatr Clin North Am* **6**:729(1996).

17. Dahl RE, Holtum J, Trubnick L. A clinical picture of child and adolescent narcolepsy. *J Am Acad Child Adolesc Psychiatry* **6**:834(1994).

18. Simonds JF, Parraga H. Sleep behaviors and disorders in children and adolescents evaluated at psychiatric clinics. *J Dev Behav Pediatr* **5**(1):6–10(1984).

19. Roberts RE, Lewinsohn PM, Seeley JR. Symptoms of DSM-III-R major depression in adolescence: evidence from an epidemiological survey. *J Am Acad Child Adolesc Psychiatry* **34**(12):1608–1617(1995).

20. Paavonen EJ, Solantaus T, Almqvist F, Aronen ET. Four-year follow-up study of sleep and psychiatric symptoms in preadolescents: relationship of persistent and temporary sleep problems to psychiatric conditions. *Dev Behav Pediatr* **24**(5):307–314(2003).

21. Sadeh A, McGuire JPD, Sachs H, et al. Sleep and psychological characteristics of children on a psychiatry inpatient unit. *J Am Acad Child Adolesc Psychiatry* **34**:813–819(1995).

22. Sadeh A. Stress, trauma and sleep in children. *Child Adolesc Psychiatr Clin North Am* **5**:685–700(1996).

23. Muris P, Merckelbach H, Ollendick TH, King NJ, Bogie N. Children's nighttime fears: parent–child ratings of frequency, content, origins, coping behaviors and severity. *Behav Res Ther* **39**:13–28(2001).

24. Corkum P, Tannock R, Moldofsky H. Sleep disturbances in children with attention-deficit/hyperactivity disorder. *J Am Acad Child Adolesc Psychiatry* **37**:6(1998).

25. Mick E, Biederman J, Jetton J, Faraone SV. Sleep disturbances associated with attention deficit hyperactivity disorder: the impact of psychiatric comorbidity and pharmacotherapy. *J Child Adolesc Psychopharmacol* **10**(3):223–231(2000).

26. Sadeh A, Gruber R, Raviv A. The effects of sleep restriction and extension on school-age children: what a difference an hour makes. *Child Dev* **74**(2):444–455(2003).

76

CIRCADIAN RHYTHM DISORDERS IN INFANTS, CHILDREN, AND ADOLESCENTS

JOHN H. HERMAN

UT Southwestern Medical Center, Dallas, Texas

This chapter follows the format of *The International Classification of Sleep Disorders* (ICSD) (2001) [1]. It adapts the symptoms and presenting complaints seen in adults to those observed in children. Original terminology has been preserved when appropriate for the sake of clarity. Clinicians evaluating children should follow the established format for each circadian rhythm disorder in ICSD because of its universal acceptance.

It is an interesting exercise to convert disorders seen in adults to symptoms observed in children. For example, an irregular sleep–wake pattern is a disabling sleep disorder in adults interfering in all aspects of life. In young infants it is normal. But by 1 year of age it becomes a pattern that is of concern. Likewise, delayed sleep phase syndrome in an adolescent may be anything from a normal teenage attribute to a debilitating and intractable disorder. This chapter also includes some findings not available at the time of publication of ICSD in 2001, such as the identification of the gene and gene product for advanced sleep phase syndrome or the discovery of melanopsin, the retinal pigment related to circadian rhythms [2].

In this chapter more emphasis is placed on intrinsic versus extrinsic factors influencing circadian rhythms in children than ICSD placed in its description of adults. That is because children are always in a household where parents have, or should have, influence on their bedtimes and wakeup times (extrinsic factor). One of the great joys of becoming an adult is no longer having anyone tell you when to go to bed or wake up. But this also implies that children's circadian rhythm disorders may be more subject to exogenous influences since parents set these rules.

Children are in flux with regard to sleep and its timing. The long sleeping infant may become a short sleeping child. The child who awakens before the household may become the adolescent who sleeps to noon. Children may be more or less resilient than adults; some children and adolescents sleep remarkably few hours on weekdays with little or no apparent detrimental effect on daytime functioning. Others cannot tolerate the loss of 1 or 2 hours of sleep. Diagnosis of circadian rhythm disorders in children and adolescents should be based on *chronic symptoms of impaired daytime functioning*. Below is a list of circadian rhythm disorders described in this chapter and their ICD-9-CM reference numbers:

1. Irregular Sleep–Wake Pattern of Infancy and Childhood (307.45-3)
2. Delayed Sleep-Phase Syndrome (780.55-0)
   (a) Limit-setting Delayed Sleep Phase Syndrome subtype
   (b) Exogenous Delayed Sleep Phase Syndrome subtype
3. Advanced Sleep-Phase Syndrome (780.55-1)
   (a) Limit-setting Advanced Sleep Phase Syndrome subtype
   (b) Exogenous Advanced Sleep Phase Syndrome subtype
4. Non-24-Hour Sleep–Wake Disorder (780.55-2)

Sleep: A Comprehensive Handbook, Edited by T. Lee-Chiong.
Copyright © 2006 John Wiley & Sons, Inc.

(a) Limit-setting Non-24-Hour Sleep Phase Syndrome subtype

(b) Exogenous Non-24-Hour Sleep Phase Syndrome subtype

CIRCADIAN RHYTHM SLEEP DISORDERS: INTRINSIC AND EXTRINSIC FACTORS

The major feature of circadian rhythm sleep disorders is a misalignment between the child's sleep pattern and the sleep pattern that is required by parents, day care, and school. When the principal reason for the circadian rhythm disorder appears to be internal factors, such as a fast or slow circadian pacemaker, an intrinsic circadian disorder may be diagnosed.

If the principal reason for the circadian rhythm disorder appears to be external factors, such as late bedtimes or midnight exam-cramming or web-cruising, an extrinsic circadian disorder may be diagnosed. Three of the circadian rhythm sleep disorders have intrinsic and extrinsic subtypes: delayed sleep-phase syndrome (DSPS), advanced sleep-phase syndrome (ASPS), and non-24-hour sleep–wake disorder. Parents and/or social factors frequently play more of a role in the appearance of a child's circadian rhythm disorder than in other childhood sleep disorders. Since the body's internal clock has a cycle length greater than 24 hours in most individuals [3], zeitgebers (timing cues, such as sunlight) are required for a child to remain in synchrony with a desired sleep and wake schedule [4]. If parents do not provide sufficient entraining clues to facilitate the child's entrainment, then the child's circadian rhythm disorder is said to be extrinsic.

Intrinsic Circadian Rhythm Factors

Circadian rhythm disorder, including inability to awaken at desired hour, despite consistent bedtime and wakeup time.

Circadian rhythm disorder despite normal exposure to sunlight throughout day.

Inability for a child or adolescent to initiate sleep at a desired hour despite consistent bedtime and wakeup time.

Circadian rhythm disorder despite no exposure to light during normal sleeping hours.

Extrinsic Circadian Rhythm Factors

Delayed bedtime on weekends and vacations.

Delayed wakeup time on weekends and vacations.

Exposure to late (post-bedtime) light greater than 40 lux, including computer screens and video monitors.

Delayed or lack of exposure to light and activity on weekends and vacations.

Sleeping environment enables continuous darkness.

Post-bedtime social activities (telephone, social contacts).

Child or adolescent's family exhibits collective circadian rhythm disorder.

Mixed Intrinsic and Extrinsic Characteristics

In most cases a combination of exogenous and endogenous factors contributes to a child's circadian rhythm disorder. In some instances it may be apparent that exogenous factors prevail. To the contrary, some children and adolescents have great difficulty adhering to an expected schedule despite reasonable efforts on their part and their family's efforts to provide entraining cues. Such individuals would be said to exhibit an endogenous circadian rhythm disorder. In instances in which the parents appear to be capitulating to a child or adolescent's desire to sleep at socially unacceptable hours, the term limit-setting circadian rhythm disorder may be employed.

Psychiatric and Psychosocial Complications

Frequently, an Axis I disorder such as depression, bipolar disorder, an anxiety disorder, or dysthymia complicates the circadian picture [5]. For example, the child or adolescent may become unacceptably hostile when asked to awaken or go to sleep at an hour that requires accommodation. Similarly, children with other psychiatric presentations such as symptoms consistent with conduct disorder, oppositional-defiant disorder, ADHD, or a thought disorder may be virtually impossible to treat for a circadian rhythm disorder. When a psychiatric disorder exists that better explains what appears to be a circadian rhythm disorder, only the primary psychiatric disorder should be diagnosed.

Shared Characteristics of Circadian Rhythm Disorders

In most circadian rhythm sleep disorders, the underlying problem is that the child cannot sleep when sleep is desired, needed, or expected. As a result of sleep episodes occurring at inappropriate times, the corresponding wake periods may occur at undesired times. Therefore the child complains of insomnia or excessive daytime sleepiness or both. In children and adolescents, once sleep is initiated, the major sleep episode is of normal duration with normal REM and NREM cycling. However, intermittent sleep episodes may occur in some disorders, including the irregular sleep–wake pattern.

Other Disorders that Mimic Circadian Rhythm Disorders

The appropriate timing of sleep within the 24 hour day can be disturbed in other childhood sleep disorders, particularly sleep-onset association disorder, limit-setting disorder, or the childhood-onset insomnias. In such cases, only the primary sleep disorder should be diagnosed. For example, infants and children with sleep-onset association disorder can have a sleep pattern identical to delayed sleep-phase syndrome.

Important Caveat—Don't Overlook Sleep Hygiene

Some disturbance of sleep timing is a common feature of children who have a diagnosis of inadequate sleep hygiene. Many children sleep far too few hours on schooldays and attempt to recuperate weekends. Only if the timing of sleep is the predominant cause of the sleep disturbance and is outside the societal norm would a diagnosis of a circadian rhythm sleep disorder be stated. Limit-setting sleep disorder also is associated with an altered time of sleep within the 24 hour day. However, the timing of sleep in this disorder is not within the child's control nor is it intrinsically induced. Many children's bedrooms contain more display monitors than a video arcade. It is little wonder that some of these children are up at night. If the child's parents are not setting appropriate limits, the sleep disorder is more appropriately diagnosed within the extrinsic subsection of the dyssomnias (i.e., as a limit-setting sleep-disorder).

IRREGULAR SLEEP–WAKE PATTERN OF INFANCY AND CHILDHOOD (307.45-3)

An irregular sleep–wake pattern consists of temporally disorganized and variable episodes of sleeping and waking behavior.

Normal for Newborns, Pathological by 6 Months

Although infants and children with irregular sleep–wake patterns may have a total 24 hour average sleep time that is within normal limits for age, no single sleep period is of a regular length, and the likelihood of being asleep at any particular time of day is unpredictable. This is considered normal in newborn infants and is not of diagnostic significance until 6 months of age, by which age parents are exhausted. By 6 months there should be a *propensity* to sleep at night. Preschool and some school-age children may exhibit two or more major sleep episodes at nonpredictable times. Such infants and children may be awake much of the night, requiring parental supervision, and may sleep for long periods during the day.

This disorder shows no recognizable ultradian or circadian pattern in a well-kept sleep–wake log, which distinguishes it from advanced sleep-phase, delayed sleep-phase, and non-24-hour syndromes. Instead, sleep is broken up into several short blocks. Each block of sleep is almost random.

Predisposing Factors and Complications

Intrinsic factors include diffuse brain dysfunction. Extrinsic factors include the lack of a regular and unvarying daily routine, or spending excessive time in bed and napping frequently. Significant problems with parents and inability to attend school may result.

Differential Diagnosis

The inability to nap in the daytime separates many, if not most, children with insomnia associated with other causes. Occasionally, children with clear-cut narcolepsy display a sleep and nap pattern that resembles the irregular sleep–wake pattern.

Treatment

Emphasis on a rigorous 7 day/week sleep–wake schedule is central. Exposure to the light of dawn is critical in setting regular circadian rhythms and must be emphasized [6]. As much outdoor light during waking hours and dim to no light during sleeping hours are beneficial [7]. A regular schedule of morning exercise and routine times for meals and social activities are helpful [8]. In certain individuals low-dose melatonin (<1 mg) 1–2 hours before sunset might be helpful.

DELAYED SLEEP-PHASE SYNDROME (780.55-0)

Delayed sleep-phase syndrome is a disorder in which the major sleep episode of the child or adolescent is delayed in relation to the desired clock time, resulting in symptoms of sleep-onset insomnia or impossibility in awakening at the desired time (Figure 76.1).

Delayed sleep-phase syndrome (DSPS) is marked by (1) sleep-onset and wake times that are later than desired, (2) sleep-onset that occurs at approximately the same time, (3) little difficulty in maintaining sleep once it has begun, (4) extreme difficulty awakening at the desired time in the morning, and (5) an inability to advance sleep to earlier hours by enforcing specific bedtime and wake time.

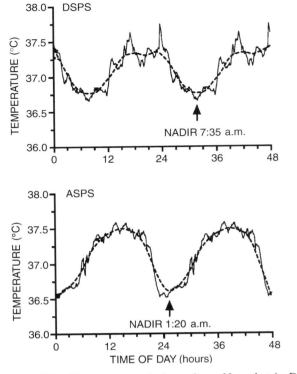

Figure 76.1 Temperature and sleep phase. Note that in DSPS temperature nadir is at 7:35 a.m., after the hour of awakening for most children, leading to morning struggles and morning sleepiness. In ASPS core body temperature begins to decline in the early evening.

Typically, the child complains of chronic difficulty in falling asleep until midnight to 4:00 a.m. or difficulty awakening when desired. Major conflicts occur with parents at bedtime and wakeup time. Daytime sleepiness, especially in the morning hours, occurs in school. On weekends and during vacations, the child sleeps to very late hours.

Associated Features

Families of children with DSPS are usually puzzled or irritated that they cannot find a way for their child to fall asleep more quickly and awaken with less conflict. The family's complicated and even severe efforts to advance the timing of their child's sleep yields little permanent success. Children with DSPS typically are in their best mood, function best, and are most alert near or after bedtime. Sleep–wake logs obtained during weekends or vacations show "late" sleep and arising times. Many children with DSPS exhibit some degree of psychopathology, but they do not share a single diagnosis. In adolescents, failure to cooperate with a plan to reschedule their sleep may be a sign of clinical depression or may be normal adolescence ("I'm going to fight... for my right... to sleeeep late!"). Whether DSPS

results directly in clinical depression, or vice versa, is unknown, but many children with DSPS exhibit symptoms of depression.

Predisposing Factors

Many children's DSPS symptoms began after a period of late-night studying or after a summer vacation, following which they cannot resume sleeping on a school-night schedule.

Prevalence and Age of Onset

The exact prevalence is unknown, but the disorder is not uncommon, probably representing 2–10% of children and adolescents. Some children with DSPS appear to function normally despite limited sleep on weekdays. It is not known if such children should be included in prevalence data.

Adolescence appears to be the most common period of life for the onset of DSPS, but childhood cases abound, as early as the start of school, at which time previously existent late sleeping first becomes of concern.

Familial Pattern

There is no known familial pattern, but typically one first-degree relative displays symptoms similar to the child's, although perhaps not as severe.

Pathology

It is presumed there exists a gene for DSPS that is yet to be identified, part of the clock–gene circuitry. Children with DSPS are thought to have a relatively weak ability to phase advance their circadian systems in response to normal environmental time cues (zeitgebers) and a strong ability to delay their circadian system even with one morning of late sleep.

Complications

School dysfunctions of varying degrees are a typical accompaniment of DSPS and are often the major complaint that brings the child to clinical attention. Absenteeism and chronic tardiness are poorly tolerated in most schools, and many children with DSPS come to be regarded as lazy or unmotivated by their families and teachers. Many are home-schooled.

Differential Diagnosis

Delayed sleep is a common feature of childhood and adolescence. It may be attributable to a personality disorder, to a major mental disturbance, to poor sleep hygiene, to inadequate zeitgebers, or to poor limit setting. If it is due to a mental disorder, the delayed sleep pattern is usually transient and covaries with the mental symptoms. A chronic pattern of sleep-phase delay is sometimes seen in children with nocturnal panic attack or in phobic, avoidant, and introverted children. Another important differentiation is from non-24-hour sleep–wake syndrome, which is characterized by incremental delays of the sleep phase. Frequently, cases of insufficient nocturnal sleep, in which sleep curtailment is due to too late a bedtime, may be confused with DSPS. In general, insufficient sleepers do not stay up as late as do children with DSPS, although they may sleep as late into the day on weekends. A number of sleep disorders that are nearly exclusive to childhood must be differentiated from childhood-onset DSPS in children with complaints of sleep-onset insomnia [9]. These sleep disorders include limit-setting sleep disorder, sleep-onset association disorder, and childhood-onset idiopathic insomnia.

Treatment

The crux of treatment is a 7 day/week regular bedtime and wakeup time. The child or adolescent may remain up later some weekend nights but should arise at the same time. Morning light and exercise should be maximized and evening light exposure (>40 lux, or dim restaurant lighting) should be minimized [10]. The child's bedroom ideally faces east and has no window covers. A bright light box is critical during winter months for some children and adolescents. Some prefer light visors or dawn simulation lamps. All these products may be found on the web from various manufacturers. One to 3 mg melatonin for a brief period administered 1 hour before the desired hour of sleep may be beneficial.

ADVANCED SLEEP-PHASE SYNDROME (780.55-1)

Advanced sleep-phase syndrome is a disorder in which the major sleep episode is advanced in relation to the desired clock time (Figure 76.1), resulting in symptoms of compelling evening sleepiness, an early sleep onset, and an awakening that is earlier than desired [11].

Advanced sleep-phase syndrome is marked by a child's inability to delay the onset of evening sleep or extend sleep later into the morning hours by enforcing more conventional social sleep and wake times. The major presenting complaint may concern either the inability to stay awake in the evening, or early morning awakening insomnia, or both. Unlike other sleep maintenance disorders, the early morning awakening occurs after a normal amount of otherwise undisturbed sleep. Unlike other causes of excessive

sleepiness, daytime schoolwork is not affected by ASPS. Typical sleep-onset times are between 6 p.m. and 8 p.m., and no later than 9 p.m., and wake times are between 1 a.m. and 3 a.m., and no later than 5 a.m.

Associated Features

Negative personal or social consequences may occur due to leaving activities early (Cub Scout meetings or Little League baseball) or not being able to complete homework due to evening sleepiness. Afflicted children and adolescents who attempt to stay up late cannot do so. If children are chronically forced to stay up later for social reasons, term papers, or exams, the early-wakening aspect of the syndrome could lead to chronic sleep deprivation and daytime sleepiness or napping.

Prevalence and Age of Onset

Prevalence is apparently rare. This condition usually first becomes of concern when children are expected to remain awake as late as other children for school or social reasons.

New Advances in Pathology

Children with advanced sleep-phase syndrome have only rarely come to clinical attention. The gene and gene product have recently been identified by association analysis of one large family with many afflicted individuals, showing the phenotype may be attributed to a missense mutation in a clock component, h*PER2*, which alters the circadian period [12].

Differential Diagnosis

Normal infants may awaken early. A mild degree of phase-advanced sleep is normal in the preadolescent child (Saturday morning cartoons start at 5:00 a.m.). The inability to sleep late differentiates children with advanced sleep-phase syndrome from those with insufficient nocturnal sleep, who sleep late.

Treatment

Treatment is the same in principle but the opposite in administration to that for delayed sleep-phase syndrome. A bright light box may be employed in winter months to extend evening bright light exposure. Products may be purchased on the web. The child should be exposed to as much evening light as possible and the bedroom should remain as dark as possible in the morning. Evening exercise should be encouraged. Melatonin at a low dose (150–300 micrograms) is administered when the child awakens early.

NON-24-HOUR SLEEP–WAKE SYNDROME (780.55-2)

Non-24-hour sleep–wake syndrome consists of a chronic pattern of incremental delays in sleep-onset and wake times. Sleep-onset and wake times occur at a period of approximately 25 hours. But cycle lengths may vary greatly from day to day. The child's sleep periodically travels in and out of phase with conventional hours for sleep. When "in phase," the child may have no sleep problem and normal daytime alertness. The child will alternate between being symptomatic and asymptomatic, as his/her hours of sleep and wakefulness drift in and out of synchrony with the family. Attempts by the family to have their child sleep and wake at conventional social times produce sleep deprivation, with secondary daytime sleepiness interfering with functioning at school. This condition is most frequently observed in blind children [13] but it is also observed in sighted individuals [14].

Associated Features

Typically, children or adolescents with this condition are unable to attend school. Most of these children are blind, either congenitally or on an acquired basis; some are mentally retarded. Less commonly, a severely schizoid or avoidant personality disorder may accompany the condition. One child who was initially described as having this disorder was later discovered to have a large pituitary adenoma that involved the optic chiasma.

Prevalence and Age of Onset

This disorder is apparently rare in the general population but appears frequently in the blind. The syndrome has been described in congenitally blind infants and children. Onset in normal-sighted individuals appears variable. The non-24-hour syndrome may be chronic and intractable or may respond well to the institution of strict and regular 24 hour time cues. Some blind individuals respond to strict 24 hour scheduling of strong social time cues.

Recent Elucidation of Pathology

The environmental light–dark cycle, acting through the retinohypothalamic tract on the suprachiasmatic nucleus, is the major source of 24 hour time information. Rods and cones initiate visual information to the cortex (Figure 76.2). Retinal cells containing the photopigment melanopsin transmit light information *only* to the hypothalamus [2]. Cortical blindness may be present in an individual with an intact retinohypothalamic tract. Such an individual would have normal circadian light information. Some sighted individuals have lesions of their retinohypothalamic tract and become arrhythmic. Most individuals with this disorder are blind, with all photoreceptors nonfunctional. In rare cases personality factors appear to be paramount.

Differential Diagnosis

A carefully kept sleep–wake log that is recorded for a lengthy period of time is essential to making this diagnosis. Non-24-hour sleep–wake syndrome should be differentiated from delayed sleep-phase syndrome and irregular sleep–wake pattern. In the delayed sleep-phase syndrome, stable entrainment to a 24 hour schedule with sleep at a delayed phase from conventional hours is present during vacations. Children with the non-24-hour sleep–wake

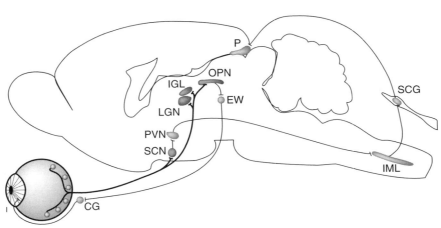

Figure 76.2   Melanopsin pathways transmitting light information from retina to suprachiasmatic nucleus (SCN) and pineal (P) gland via retinohypothalamic tract. (From Berson DM, Strange vision: ganglion cells as circadian photoreceptors. *Trends Neurosci* **26**(6):314–320(2003).)

syndrome continue in a pattern of progressive delays of sleep. The diagnosis should be suspected in any blind individual with sleep or somnolence complaints.

Treatment

Emphasis on a rigorous 7 day/week sleep–wake schedule is central. Exposure to the light of dawn is critical in setting regular circadian rhythms and must be emphasized. As much outdoor light during waking hours and dim to no light during sleeping hours are beneficial. A regular schedule of morning exercise and routine times for meals and social activities are helpful. In certain children low-dose melatonin (<1 mg) 5 p.m. might be helpful. In adolescents, 3 mg melatonin has been found to be beneficial. In blind children, melatonin at the same doses as above is administered 1 hour before desired bedtime. A well-organized and regular schedule of wakeup time, bedtime, meals, and activities is equally important to provide nonphotic zeitgebers to blind children.

REFERENCES

1. *The International Classification of Sleep Disorders, Revised Diagnostic and Coding Manual.* American Academy of Sleep Medicine, Westchester, IL, 2001.

2. Hattar S, Liao H-W, Takao M, Berson DM, Yau K-W. Melanopsin-containing retinal ganglion cells: architecture, projections, and intrinsic photosensitivity. *Science.* xx:x–x(2002).

3. Czeisler CA, Duffy JF, Shanahan TL, Brown EN, Mitchell JF, Rimmer DW, Ronda JM, Silva EJ, Allan JS, Emens JS, Dijk DJ, Kronauer RE. Stability, precision, and near-24-hour period of the human circadian pacemaker. *Science* **284**(5423):2177–2181(1999).

4. Park SJ, Tokura H. Effects of different light intensities during the daytime on circadian rhythm of core temperature in humans. *Appl Hum Sci* **17**:253–257(1998).

5. Dahl RE, Lewin DS. Pathways to adolescent health: sleep regulation and behavior. *J Adolesc Health* **31**(6 Suppl):175–184(2002).

6. Czeisler CA, Wright KP. Influence of light on circadian rhythmicity in humans. In: Turek FW, Zee PC (Eds), *Regulation of Sleep and Circadian Rhythms*, Vol 133 of Lung Biology in Health and Disease. Marcel Dekker, New York, 1999, pp 149–180.

7. Boivin DB, Czeisler CA. Resetting of circadian melatonin and cortisol rhythms in humans by ordinary room light. *Neuroreport* **9**:779–782(1998).

8. Duffy JF, Kronauer RE, Czeisler CA. Phase-shifting human circadian rhythms: influence of sleep timing, social contact and light exposure. *J Physiol* **495**:289–297(1996).

9. Weitzman ED, Czeisler CA, Coleman RM, et al. Delayed sleep-phase syndrome. A chronobiological disorder with sleep-onset insomnia. *Arch Gen Psychiatry* **38**:737–746(1981).

10. Czeisler CA, Richardson GS, Coleman RM, et al. Chronotherapy: resetting the circadian clock of children with delayed sleep phase insomnia. *Sleep* **4**:1–21(1981).

11. Kamei R, Hughes L, Miles L, Dement W. Advanced-sleep-phase syndrome studied in a time isolation facility. *Chronobiologia* **6**:115(1979).

12. Toh KL, Jones CR, He Y, Eide EJ, Hinz WA, Virshup DM, Ying-Hui F. An h*Per2* phosphorylation site mutation in familial advanced sleep phase syndrome. *Science* **291**:1040–1043(2001).

13. Miles LE, Wilson MA. High incidence of cyclic sleep–wake disorders in the blind. *Sleep Res* **6**:192(1977).

14. Weber AL, Cary MS, Conner N, Keyes P. Human non-24-hour sleep–wake cycles in an everyday environment. *Sleep* **2**:347–354(1980).

PART IX

SLEEP IN THE ELDERLY

77

NORMAL SLEEP IN AGING

LIAT AYALON AND SONIA ANCOLI-ISRAEL
University of California, San Diego, California

Aging seems to be the only available way to live a long life.
—Daniel Francois Esprit Auber

BACKGROUND

With aging, sleep, as with other physiological processes, undergoes increasingly noticeable changes. Many of the changes accompanying aging are part of a gradual process rather than an abrupt change and reflect changes in both homeostatic and circadian processes that occur throughout the life span. It is important, however, to make a distinction between changes that are part of normal aging and changes that might be considered pathological.

Sleep disturbances are common in older adults [1, 2]. However, much evidence suggests medical diseases and chronic illness may account for most of the changes in sleep observed in old age. In fact, chronological age by itself seems to explain very little of the observed prevalence of sleep complaints [3].

Our intention is not only to focus on sleep pathology, but also to focus on the changes in sleep to better understand what can be expected as part of normal aging. Distinguishing one from the other may enable an active approach in treating sleep disorders when required as well as providing an understanding of when these changes are part of the natural process of aging. We will review changes in sleep architecture, circadian rhythms, and the ability to sleep, as well as other age-related physiological changes that may affect sleep and place older adults at higher risk for sleep disturbances.

SLEEP ARCHITECTURE

The most consistent age-related change in sleep architecture is a decrease in slow-wave sleep (SWS, stages 3–4), with delta waves being reduced in both amount and amplitude in the older adult [4]. This decrease in percentage of time spent in stages 3–4 is a gradual process, beginning earlier in life, at around 20 years of age. Some investigators have suggested that these changes may be dependent on the definition of slow-wave sleep and that a revised definition that takes age-related EEG changes into account is needed. An age-related decrease in amount, frequency, and amplitude of sleep spindles has also been demonstrated [5]. Although the amount of REM sleep remains relatively constant through adulthood, it is often associated with a redistribution of REM sleep periods such that the first REM period seems to be lengthened and occurs earlier in the sleep cycle (i.e., phase advancement of REM) in the elderly [6].

Generally, older adult sleep is characterized by more sleep stage shifts and more awakenings [6] and hence increased stage 1 sleep. In addition to the changes in sleep architecture, older adults are significantly sleepier throughout the day than younger adults, as confirmed by the multiple sleep latency test (MSLT), an objective test of daytime sleepiness.

Correspondence to: S. Ancoli-Israel, University of California San Diego, Department of Psychiatry 116A, VASDHS, 3350 La Jolla Village Drive, San Diego, CA 92161.

TABLE 77.1   Subjective Reports and Objective Findings

Subjective Reports	Objective Findings
Spend too much time in bed	Decrease in deep sleep (stages 3 and 4)
Spend less time asleep	Decrease in rapid eye movement sleep
Increased number of awakenings	Significant increase in awakenings
Increased time to fall asleep	Increased frequency of sleep disorders
Less satisfied with sleep	Reduced sleep efficiency
More tired during the day	Increased daytime sleepiness
Longer and more frequent naps	Increased number of naps

TABLE 77.2   Age-Related Circadian Changes

Rhythm amplitude attenuation (sleep–wake, body temperature, melatonin)
Reduction in nighttime melatonin levels
Increase in cortisol level at its circadian minimum
Diminished sleep-related growth hormone release
Circadian rhythm phase advance (earlier bedtime and wakeup time, advanced temperature rhythm)
Reductions in retinal sensitivity to light
Reduced bright-light exposure and activity levels

THE NEED FOR SLEEP

Both subjective reports and objective measurements of sleep have suggested that when compared to younger adults, older adults take longer to fall asleep and have lower sleep efficiency (defined as the amount of sleep given the amount of time in bed), although older adults who do not suffer from sleep pathology typically show high sleep efficiencies [7]. Older adults also have more nighttime awakenings, wake up earlier than they would like in the morning, and require more daytime naps (see Table 77.1 for a list of subjective complaints and objective findings).

One of the main changes in the elderly is redistribution of sleep around the 24 hour day so that nighttime sleep is usually reported to be decreased with daytime napping being more frequent. Prolonged time spent in bed may be related to the disruption of sleep and to its redistribution. Although the architecture of sleep as well as its distribution around the 24 hour day may change, it seems that the need for sleep remains about the same. Healthy older adults rarely complain of difficulties with sleep [3].

Although the changes discussed may result in lighter sleep, more often, the decreased ability to sleep has been associated with changes in circadian rhythms, specific sleep disorders, medical and psychiatric illnesses, psychological factors, and medication use [8].

CIRCADIAN RHYTHMS AND AGING

Numerous changes in overt rhythmicity appear to be associated with aging and are considered a reflection of aging (see Table 77.2). The most consistently reported change in circadian systems with increasing age is amplitude attenuation. This attenuation has been noted in sleep–wake rhythms, body temperature rhythms, and melatonin rhythms. There is considerable evidence indicating a decrease in the amplitude of the sleep–wake rhythm with aging, suggesting that older people may be sleepier than younger people during the daytime and exhibit lower sleep efficiency with more awakenings during nighttime [9]. A decrease in the amplitude of the temperature rhythm with increasing age has been also noticed. The amplitude of core body temperature in older subjects is about 40–85% that of younger subjects. This has been found in both entrained and free-running (temporal isolation) conditions [9]. However, there is some evidence that larger amplitude differences are observed in older adults with sleep difficulties than in older adults with no sleep difficulties [10]. Some endocrine rhythms also show a decrease in amplitude with aging while others show an increase (which may be interpreted as a compensatory process) [11]. Prinz and colleagues [12] reported that the age-related increase in cortisol level at its circadian minimum is involved in impaired sleep, more beta activity during sleep, and earlier times of arising. This may support findings of a disrupted hypothalamic–pituitary–adrenal axis in subjects suffering from insomnia [13]. Sleep-related growth hormone release is also diminished in older adults [14]. Age-related differences in melatonin rhythm are probably the most documented endocrine change. Most studies report a reduction in nighttime melatonin levels with increasing age with the decrease seen in both amplitude and average levels [10].

As people age, the circadian rhythm phase advances. That is, older people tend to get sleepier earlier in the evening and wake at an earlier time of the morning than do younger adults, resulting in a higher prevalence of early morning awakening and difficulty maintaining sleep. Changes also occur in the phase relationship between core body temperature and sleep–wake rhythms. The phase advance of the temperature rhythm in aging results in the temperature nadir falling at the midpoint of the sleep period. Therefore the second half of sleep for aged persons occurs against a rising temperature rhythm with its associated drive to wakefulness.

To stay robust, circadian rhythms are entrained by time cues, called zeitgebers. The most significant zeitgeber is the daily cycle of light and dark, but other factors such as physical activity, the timing of meals, and social interactions have also been shown to be important regulators of endogenous rhythms. The rhythm is hence dependent on integrity of both the endogenous pacemaker and the strength of entrainment to zeitgebers. It seems that both are altered in aging with some of these circadian changes being attributed to deterioration of the circadian pacemaker, while other changes may result from a decline in either entrainment mechanisms or systemic processes that are controlled by the circadian clock. For example, changes in social interactions, routine, and psychosocial factors may all affect both bright-light exposure and activity levels and hence sleep–wake rhythms.

In addition to the deterioration in the function of the circadian pacemaker mechanisms, aging is associated with decreased efficacy of exogenous zeitgeber cues. Hence aging may be related to both reductions in retinal sensitivity to light and decreased bright-light exposure and activity as part of a life style. A study of independent elderly identified that they received on average only 1 hour of natural light each day [15]. This may decrease to less than 10 minutes a day for institutionalized elderly [16]. The capacity of ocular bright-light exposure to strengthen and/or phase shift the circadian timing system among older populations has been well demonstrated [15] with several studies reporting associated increases in the quality of subsequent nighttime sleep [17].

THE EFFECTS OF AGE-RELATED NORMAL PHYSIOLOGIC CHANGES ON SLEEP

Although many age-related physiological changes may affect sleep, two of the most common ones are changes in voiding during the night and menopause (see Table 77.3).

Frequent voluntary voiding of urine during the night (nocturia) as well as production of an abnormally large volume of urine during sleep (nocturnal polyuria) are common causes of awakening during the night in elderly. Although nocturia is often perceived as causing nocturnal

TABLE 77.3 Physiological Changes that May Affect Sleep

Nocturia and nocturnal polyuria
Menopause
Decreased respiratory muscle strength
Decreased expiratory flow rates
Diminished compliance of the chest wall
Physiological changes of the lung
Changes in soft palate tissue
Increased use of medications
Conditions causing pain
Neurological disorders
Psychiatric conditions (mood disorders, anxiety)

awakenings, it may also be that the fragmented/light sleep experienced by the older adult leads to more awareness of the need to void. Age-related changes in the circadian rhythm of urine excretion (e.g., fall in renal concentrating ability, sodium conservation and secretion of renin–angiotensin–aldosterone, circadian changes in arginine vasopressin (AVP) secretion, and increased output of atrial natriuretic hormone) are associated with nocturnal polyuria. In addition, with normal aging, the bladder's ability to store urine successfully decreases. Age-related lower urinary tract problems, such as detrusor overactivity, can affect frequency and urgency. Prostatic hypertrophy in men and decreased urethral resistance in women as a result of hormonal changes are also involved [18]. Interestingly, increased urinary urgency has also been associated with sleep apnea.

Menopause is associated with hormonal, physiologic, and psychological changes that affect sleep and play a pivotal role in modulating both the presence and the degree of sleep disorder. Lower sleep efficiency and multiple arousals during the night are associated with menopausal vasomotor symptoms (hot flashes) [19]. Insomnia becomes common during this time and may be partially related to psychological factors such as depression and anxiety [20]. Mild sleep apnea is more common in postmenopausal women and may be explained by the increased body mass index (BMI) often accompanying menopause.

AGE-RELATED FACTORS ASSOCIATED WITH SLEEP DISORDERS

Sleep Apnea

The prevalence of sleep apnea in the elderly has been shown to be higher than in younger adults. Ancoli-Israel and colleagues [21] reported that 25% of community dwelling elderly have an apnea index (the number of apneas per hour of sleep) of 5 or more and 62% have a respiratory disturbance index (the number of apneas plus hypopneas, or respiratory events, per hour of sleep) of 10 or more.

Several factors that are associated with aging may partially explain this high prevalence in elderly (see Table 77.3). These include decreased respiratory muscle strength, decreased expiratory flow rates, diminished compliance of the chest wall, and multiple physiological changes of the lung. Recent evidence also suggests the presence of aging-associated changes in the biomechanical properties of the tissues of the soft palate [22]. Importantly, some of the sedative hypnotic drugs commonly used for the treatment of insomnia in elderly act as central nervous system depressants and may adversely affect the control of ventilation during sleep. Prolonged use of these drugs may worsen sleep-related breathing disorders.

Periodic Limb Movements and Restless Legs Syndrome

Periodic limb movement in sleep (PLMS) is characterized by clusters of repeated leg jerks, which occur approximately every 20–40 seconds during the night, with each jerk causing a brief awakening. PLMS prevalence increases significantly with age with estimates at 45% in older adults compared to 5–6% in younger adults [23]. Another disorder, often comorbid with PLMS, is restless legs syndrome (RLS). RLS is characterized by dysesthesia in the legs, usually described by patients as "a creeping crawling sensation" or as "pins and needles," which can only be relieved with movement.

The reasons for high prevalence of these conditions in older adults are not fully understood. PLMS and RLS are often associated with a higher incidence of neuropathy, lower ferritin levels, and changes of iron metabolism. It may also manifest as secondary to various rheumatologic conditions that are more common in older age. Further causally related conditions include chronic obstructive pulmonary disease, asthma, fibromyalgia, diabetes mellitus, cancer, and neurodegenerative disorders, such as Parkinson's disease. There is some evidence for association with coffee intake, sleep apnea syndrome or snoring, stress, and the presence of mental disorders. Some medications (e.g., tricyclics, serotonin reuptake inhibitors, dopamine receptor blocking agent) can worsen PLMS and RLS.

Insomnia

Insomnia may be a primary illness or a result of a specific sleep disorder. However, it may also be a consequence of other age-related factors, such as medical and psychiatric illnesses, concomitant drug use, circadian rhythm changes, and behavioral and psychosocial factors (see Table 77.3) [8].

Complaints of difficulty sleeping may often be related to the many comorbid medical conditions common in the older adult. For example, insomnia can be caused by conditions causing pain (such as arthritis and malignancies), neurological disorders (such as restless legs, Parkinson's disease, stroke, or dementia), or organ system failure disorders (such as pulmonary disease, congestive heart failure, asthma, and gastrointestinal disorders) [8]. Insomnia is also related to many psychiatric disorders, particularly mood disorders and generalized anxiety disorder. In particular, insomnia may be a diagnostic symptom for major depression. Given the large number of potential comorbidities disrupting sleep in the elderly, treating the sleep problems should also focus on the underlying primary medical condition [8]. Medication use may also contribute to poor sleep [8]. Alerting or stimulating drugs taken late in the day may cause difficulty falling asleep at night. In particular, CNS stimulants, beta-blockers, bronchodilators, calcium channel blockers, corticosteroids, decongestants, stimulating antidepressants, stimulating antihistamines, and thyroid hormones are all known to contribute to insomnia. Sedating drugs, on the other hand, when taken early in the day may lead to excessive daytime sleepiness and daytime napping behavior, which may contribute to sleep-onset insomnia or may further exacerbate and maintain the existing insomnia [8].

It is believed that much of the insomnia resulting from early morning awakenings reflects changes in circadian rhythm. As mentioned before, the synchronization of the sleep–wake cycle by both internal and external rhythms is reduced in older adults. Older adults with advanced sleep phase syndrome are usually sleepy early in the evening and wake up too early in the morning. Although sleepy early in the evening, they do not usually go to bed early and rather take inadvertent naps that often lead to greater difficulty falling asleep later in the night. Increase in daytime napping, inactivity and bed rest, reduced outdoor light exposure, as well as greater susceptibility to external arousal may interact to predispose elderly subjects to poor sleep. Psychological factors such as bereavement, retirement, Holocaust trauma, fear of death in sleep, anxiety, and depression are also correlated with low sleep efficiency.

WHY GOOD SLEEP IS IMPORTANT IN ELDERLY

Older adults are significantly sleepier throughout the day than younger adults, as confirmed by the MSLT. In a large epidemiological study, 20% of participants reported being "usually sleepy in the daytime." Sleepiness was associated with a greater number of nocturnal awakenings, depression, sleep apnea, the presence of congestive heart failure, use of medications like digitalis and diuretics, sedentary life style, and limitations in mobility [24]. Daytime sleepiness can be a very debilitating symptom, causing social and occupational difficulties, reduced vigilance, and cognitive deficits, including decreased concentration, slowed response time, and memory and attention difficulties. These symptoms may be particularly serious in older adults who already have mild or moderate cognitive impairment.

Although very common, daytime sleepiness is not an inevitable part of aging. Many healthy older adults show intact daytime alertness on the MSLT [25]. In addition, there is an important role of subjective appraisal of the older person's symptoms. Is daytime napping and lower sleep efficiency perceived as sleep disorder or simply being accepted as a normal part of aging? Are these changes related to changes in sleep processes per se or simply secondary manifestations of aging? It is essential to distinguish between primary age-related changes in sleep and changes in sleep that are secondary to alterations in other physiologic systems. A main challenge in sleep medicine in the elderly is to recognize age-related changes and distinguish them from primary or secondary sleep disorders so we can learn to accept the former and treat the latter.

ACKNOWLEDGMENTS

Supported by NIA AG08415, NIMH 5 T32 MH18399-17, NCI CA85264, GCRC M01 RR00827, the Department of Veterans Affairs VISN-22 Mental Illness Research, Education and Clinical Center (MIRECC), and the Research Service of the Veterans Affairs San Diego Healthcare System.

REFERENCES

1. Ayalon L, Liu L, Ancoli-Israel S. Diagnosing and treating sleep disorders in the older adult. *Med Clin North Am* **88**:737–750(2004). A good review of causes and treatment of sleep problems in the elderly.

2. Bliwise DL. Normal aging. In: Kryger MH, Roth T, Dement WH (Eds), *Principles and Practice of Sleep Medicine*. Saunders, Philadelphia, 2000, pp 26–42. An excellent review of normal changes in sleep associated with aging.

3. Foley DJ, Monjan AA, Brown SL, Simonsick EM, Wallace RB, Blazer DG. Sleep complaints among elderly persons: an epidemiologic study of three communities. *Sleep* **18**:425–432(1995). A large epidemiological study of sleep in three samples of elderly.

4. Feinberg I, Koresko RL, Heller N. EEG sleep patterns as a function of normal and pathological aging in man. *J Psychiatr Res* **5**:107–144(1967).

5. Wauquier A. Aging and changes in phasic events during sleep. *Physiol Behav* **54**:803(1993).

6. Weitzman ED. Sleep and aging. In: Katzman R, Terry RD (Eds), *The Neurology of Aging*. Davis, Philadelphia, 1983, pp 167–188.

7. Hoch CC, Dew MA, Reynolds CFI, Monk TH, Buysse DJ, Houck PR, Machen MA, Kupfer DJ. A longitudinal study of laboratory- and diary-based sleep measures in healthy "old old" and "young old" volunteers. *Sleep* **17**:489–496(1994).

8. Ancoli-Israel S. Insomnia in the elderly: a review for the primary care practitioner. *Sleep* **23**:S23–S30(2000). A summary of causes and consequences of poor sleep in the elderly.

9. Weitzman ED, Moline ML, Czeisler CA, Zimmerman JC. Chronobiology of aging: temperature, sleep–wake rhythms and entrainment. *Neurobiol Aging* **3**:299–309(1982).

10. Myers BL, Badia P. Changes in circadian rhythms and sleep quality with aging—mechanisms and interventions. *Neurosci Biobehav Rev* **19**:553–571(1995).

11. Touitou Y, Haus E. Alterations with aging of the endocrine and neuroendocrine circadian system in humans. *Chronobiol Int* **17**:369–390(2000).

12. Prinz PN, Bailey SL, Woods DL. Sleep impairments in healthy seniors: roles of stress, cortisol, and interleukin-1 beta. *Chronobiol Int* **17**:391–404(2000).

13. Backhaus J, Junghanns K, Hohagen F. Sleep disturbances are correlated with decreased morning awakening salivary cortisol. *Psychoneuroendocrinology* **29**:1184–1191(2004).

14. Merriam GR, Schwartz RS, Vitiello MV. Growth hormone-releasing hormone and growth hormone secretagogues in normal aging. *Endocrine* **22**:41–48(2003).

15. Campbell SS, Kripke DF, Gillin JC, Hrubovcak JC. Exposure to light in healthy elderly subjects and Alzheimer's patients. *Physiol Behav* **42**:141–144(1988).

16. Shochat T, Martin J, Marler M, Ancoli-Israel S. Illumination levels in nursing home patients: effects on sleep and activity rhythms. *J Sleep Res* **9**:373–380(2000).

17. Ancoli-Israel S, Gehrman PR, Martin JL, Shochat T, Marler M, Corey-Bloom J, Levi L. Increased light exposure consolidates sleep and strengthens circadian rhythms in severe Alzheimer's disease patients. *Behav Sleep Med* **1**:22–36(2003).

18. Ali A, Snape J. Nocturia in older people: a review of causes, consequences, assessment and management. *Int J Clin Practice* **58**:366–373(2004).

19. Woodward S, Freedman RR. The thermoregulatory effects of menopausal hot flashes on sleep. *Sleep* **17**:497–501(1994).

20. Young T, Rabago D, Zgierska A, Austin D, Laurel F. Objective and subjective sleep quality in premenopausal, perimenopausal, and postmenopausal women in the Wisconsin Sleep Cohort Study. *Sleep* **26**:667–672(2003).

21. Ancoli-Israel S, Kripke DF, Klauber MR, Mason WJ, Fell R, Kaplan O. Sleep disordered breathing in community-dwelling elderly. *Sleep* **14**(6):486–495(1991).

22. Veldi M, Vasar V, Hion T, Kull M, Vain A. Ageing, soft-palate tone and sleep-related breathing disorders. *Clin Physiol* **21**:358–364(2001).

23. Ancoli-Israel S, Kripke DF, Klauber MR, Mason WJ, Fell R, Kaplan O. Periodic limb movements in sleep in community-dwelling elderly. *Sleep* **14**(6):496–500(1991).

24. Whitney CW, Enright PL, Newman GC, Bonekat W, Foley D, Quan SF. Correlates of daytime sleepiness in 4578 elderly persons: The Cardiovascular Health Study. *Sleep* **21**:27–36(1998).

25. Reynolds CFI, Jennings RJ, Hoch CC, Monk TH, Berman SR, Hall FT, Matzzie JV, Buysse DJ, Kupfer DJ. Daytime sleepiness in the healthy "old old": a comparison with young adults. *Am Geriatr Soc* **39**:957–962(1991).

78

SLEEP DISORDERED BREATHING IN OLDER ADULTS

Nalaka Gooneratne

University of Pennsylvania, Philadelphia, Pennsylvania

INTRODUCTION

While most of the research done to date has examined sleep disordered breathing (SDB) in young or middle-aged adults, there is a growing body of work exploring SDB in older adults. Several studies, such as the Sleep Heart Health Study, consist of large cohorts of elders and thereby allow for consideration of a broad range of factors; this is of particular importance given the multi-factorial nature of pathophysiology in elders. This interest in geriatric SDB is in part fueled by demographic changes characterized by an increased percentage of older adults, but it is also motivated by an increased awareness of the complexity of SDB in the elderly in terms of pathophysiology, management, and consequences. Many of the topics relevant to SDB in older adults, such as cardiovascular disease and SDB, have or will be discussed in other chapters. Thus this chapter will not seek to provide a general overview of SDB in the elderly, but will instead focus on key considerations that are unique to elders.

EPIDEMIOLOGY

One of the most striking aspects of SDB in the elderly is the markedly increased prevalence rate when compared to younger populations. The Sleep Heart Health Study of 5615 subjects found that the prevalence rate of SDB defined as an AHI $\geq$ 5 events/hour was 29% in subjects aged 39–49 but increased to 53% for subjects aged 60 or above [1]. Using a more strict criterion of an AHI $\geq$ 15 events/hour,

10% of subjects aged 39–49 had SDB compared to 20% of subjects aged 60 or above [1]. Another population-based study of the prevalence of sleep apnea, this time in 741 men, found a similar increase in the prevalence of sleep apnea with age: an AHI $\geq$ 5 events/hour was present in 30.5% of men over the age of 65 as compared to 7.9% of subjects aged 20–44 [2].

Interestingly, after the age of 65, this age-related increase in SDB prevalence appears to plateau and remain stable with an approximately 20% prevalence for an AHI $\geq$ 15 events/hour [1]. Thus, from a public-health perspective, most of the age-related rise in the prevalence of SDB occurs before age 65. In parallel to this is the observation that, on average, there is little increase in the AHI when measured longitudinally in the same group of elderly subjects: in one study, the mean AHI was unchanged at baseline, 3.6 years and 8.5 years of average follow-up [3]. One potential explanation is that the lack of progression in SDB noted in large epidemiology studies is largely due to a survivor effect. Subjects with severe SDB would pass away early, thus tending to reduce the mean level of SDB in the study cohort [1].

When comparing SDB across gender, there is a male predominance of the disorder even in older adults: 51% of older males have an AHI $\geq$ 20 events/hour as compared to 39% of older females in one population-based study of 385 elders [4]. This is largely due to the fact that, prior to menopause, women are less likely to develop sleep apnea. After menopause, this gender difference resolves and the incidence of sleep apnea in women is approximately identical to that in men [5]. However, the lower number of women with SDB prior to menopause means that the prevalence rate in older women remains lower than in older men. An additional important demographic difference is race. African-Americans are 2.55 times more likely than Caucasians to have SDB, even after controlling for sex, body mass index (BMI), and age [6].

The majority of SDB in these and other studies has been obstructive sleep apnea (OSA). However, some have found that the age-related increase in SDB is due largely to an increase in central sleep apnea (CSA), with relatively stable levels of OSA [2]. In this study, CSA was not present in young subjects, occurred in 1.7% of middle-aged subjects, and rose to a prevalence of 12.1% in older subjects. This finding has not been universally noted, though, with the Sleep Heart Health Study cohort having relatively stable levels of CSA as a function of age [1].

SLEEP DISORDERED BREATHING VERSUS SLEEP APNEA SYNDROME

Despite the high prevalence of SDB in older adults based on polysomnography (PSG) analysis alone, some researchers have observed that the presence of sleep apnea syndrome, which requires both PSG and clinical complaints (such as daytime sleepiness), is actually lower in older adults than in middle-aged persons [2]. Using PSG combined with a clinical history in a study of males, the prevalence of sleep apnea syndrome was identified as 4.7% in middle-aged males yet decreased to 1.7% in older male subjects [2].

The observation that SDB prevalence increases with age while the occurrence of sleep apnea syndrome decreases has led some to propose that sleep apnea in older adults is a fundamentally different process than in younger subjects. Bliwise has suggested that sleep apnea can be considered as either an age-related disorder (i.e., it occurs primarily during a specific age period) or an age-dependent disorder (i.e., the pathophysiology underlying the condition becomes more common with advancing age) [7]. In young/middle-aged adults, it may be an *age-related* disorder that peaks at about age 55. This age-related SDB results in significant daytime sleepiness and medical sequelae and is typically treated in sleep disorder clinics [7]. Sleep apnea in older subjects is different in that it is an *age-dependent* disorder. This later variant is less likely to have significant clinical consequences. It is important to emphasize that the observation that mild to moderate SDB may be relatively benign applies primarily to otherwise healthy elderly. Older adults with comorbidities such as cardiovascular disease may be at higher risk of worsening of these comorbid conditions as a result of sleep apnea and thus warrant treatment even for mild sleep apnea [2, 7].

SNORING

Another form of SDB is snoring, and this too has been found to be highly prevalent in older adults: 33% of men and 19% of women over the age of 65 reported a history of snoring in the Cardiovascular Health Study Cohort of 5201 older adults [8]. With increasing age over 75, the prevalence of snoring actually decreased, reaching 17% in those over 80 [8]. This may have been due in part to lack of awareness of snoring as unmarried subjects were more likely to respond "Do not know" to this question. In addition, this complaint of snoring was not associated with cardiovascular disease. For these reasons, snoring alone is not felt to be a major pathophysiologic concern in the elderly.

EVALUATION

The evaluation of SDB has already been addressed in prior chapters. However, there are several important issues related to older subjects that merit attention. A clinical history of snoring, witnessed apneas, or an elevated BMI are often used to identify sleep apnea patients. However, in older adults, these are often less reliable indicators of SDB and the reliance on these history elements would increase the risk of failing to diagnose SDB in elders [1]. This may be due to the fact that young/middle-aged subjects are more likely to have bedpartners who could provide this history. In addition, the observation that BMI is less closely linked to SDB in elders supports the theory that SDB in older adults is a different process than in younger subjects, as mentioned earlier [1].

PSG is generally considered the gold standard for the diagnosis of SDB in elders. However, there are some important caveats. While an AHI ≥ 5 events/hour is often used as a diagnostic criterion for the presence of SDB, the identification of an appropriate cutoff for SBD in older subjects is less well-established. This is in part due to the extremely high prevalence of SDB in the elderly when using this low cutoff, leading some to propose the establishment of an alternate AHI criterion [2]. Furthermore, there is no significant difference in subjective measures of mood and sleep perception/daytime sleepiness in older adults using this criterion [9]. Indeed, no absolute AHI criterion was found to strongly correlate with these measures [9] and even 5-year follow-up failed to reveal significant differences using an AHI criterion of 5 events/hour [10]. For these reasons, a higher AHI threshold may be a more appropriate PSG criterion for sleep apnea, especially in healthy elderly. While an AHI of 15 or 20 events/hour can be used, there is insufficient evidence to support a particular threshold at this time. In elders with comorbid conditions, such as cardiovascular disease, an AHI criterion of 5 events/hour may be warranted based on data showing an increased rate of morbidity/mortality in this population (discussed later).

Another important consideration is the night-to-night variability of the AHI as derived from PSG. One study that conducted two serial polysomnograms on consecutive nights in elderly subjects noted that, using an AHI criteria of 5 events/hour to define SDB, 37% of subjects would be classified differently based on the first night compared to the second night [11]. For example, on the first night, they would have an AHI > 5 events/hour and be considered to have sleep apnea, but on the subsequent night, they would have an AHI < 5 events/hour and thus be considered as nonapneics. Similar findings have been noted with studies using longer intervals between sleep studies. In one study with a mean 8.5 years of follow-up, while the mean AHI changed little, there were significant interindividual variations between the different time points [3]. For example, the predictive value was only 50% for an AHI $\geq$ 15 events/hour at baseline predicting an AHI $\geq$ 15 events/hour at 8.5 years follow-up.

Other diagnostic modalities have been used to identify sleep apnea in younger subjects, such as oximetry or nasal flow measurements. In general, these systems have rarely been tested in older subjects and cannot be considered as adequate surrogates at this time for PSG. Portable, at-home PSG is another option; however, its utility may be limited in older adults because of the high prevalence rates of comorbidities such as arthritis that would impair the patient's ability to apply or adjust the equipment.

Despite the high prevalence of SDB in older adults, relatively few physicians order sleep studies on their elderly patients. One study noted that while 73% of geriatricians felt that sleep disorders were important, 80% of them rarely or never ordered a sleep study to evaluate the condition [12]. Factors cited as barriers to performing sleep studies included the financial costs (24.4%), limited access to facilities (20%), perceived infrequency of diagnosable conditions (17.8%), unclear therapeutic benefits of the testing/treatment (15.6%), and feasibility in elderly (6.7%).

ETIOLOGY

The etiology of SDB in general has been discussed in previous chapters. Many similar factors are at work in the elderly population as well. When considering predictors of SDB such as age, gender, race, and body habitus, one of the most notable findings in older adults is that body habitus risk factors become less prominent with advancing age. Analysis of the Sleep Heart Health Study data found that the odds ratio for BMI and an AHI $\geq$ 15 events/hour was 2.0 for a 40 year old subject, but dropped to only 1.3 for an 80 year old subject [1]. The risk associated with a large neck circumference was also attenuated by age, becoming insignificant by age 70 [1].

The reduced significance of body habitus implies that other factors play a greater role in the pathophysiology of SDB in elders. One potential candidate is upper airway muscle changes. Research in aged Wistar rats, for example, has found age-related changes in fiber-type distribution of the genioglossus (tongue) that may decrease muscle endurance and thereby increase the risk of SDB [13]. Additionally, there are changes in the respiratory drive of older adults, which result in reduced respiratory effort during an upper airway occlusion [14]. When considering central sleep apnea, there is a large body of evidence suggesting links between central sleep apnea and congestive heart failure, a disorder that is most common in older adults; this is discussed further in other chapters. As noted earlier, SDB is more common in African-Americans than in Caucasians; however, the specific etiology underlying this difference has yet to be elucidated.

CONSEQUENCES

Sleep apnea may contribute to or exacerbate a broad range of different medical conditions. Several of these disorders are addressed in more detail in separate chapters, such as the relationship of SDB and cardiopulmonary disease. With respect to geriatric patients, a general trend is that SDB does increase the risk of these diseases, albeit to a lesser extent than in younger subjects. For example, data from the Sleep Heart Health Study shows that in fully adjusted models of SDB and hypertension (HTN) risk, the presence of an AHI $\geq$ 30 events/hour will increase the likelihood of having HTN by 64% in younger subjects, but by only 23% in older subjects [15]. Similar findings have been noted for other cardiovascular diseases as well, such as stroke or coronary heart disease: controlling for advancing age weakens the association between SDB and cardiovascular disease more than other parameters such as gender, BMI, or HDL [16]. This may in part be due to the fact that many diseases have an increased prevalence with advancing chronological age alone and also to the fact that the high rates of SDB in the elderly may tend to weaken associations between SDB and associated conditions. Furthermore, these findings may be due to the high numbers of healthy elderly in many of the study cohorts examined to date.

In addition to cardiovascular consequences, other effects of SDB are important to consider in the elderly, such as effects on neurocognitive function. There is clear evidence to suggest that increased daytime sleepiness in older adults is associated with impairments in a broad range of functional domains, such as vigilance, social activities, and general productivity [17]. However, the data linking daytime sleepiness (as diagnosed by the multiple sleep latency tests) to SDB is conflicting, especially when considering mild–moderate SDB. This may be due to coexisting diseases: studies of healthy elderly show little daytime sleepiness from SDB [18], whereas studies of elders with comorbid conditions show a more profound effect [19]. The effects of SDB on specific cognitive performance tasks, in turn, have been found to be mediated strongly by the presence or absence of daytime sleepiness in one study [20]. The relationship between SDB and advanced cognitive impairment, such as dementia, is addressed in a later chapter.

Sleep quality itself is often impaired due to SDB. In geriatric patients, however, there are numerous age-related changes in sleep quality that occur independent of SDB; most evidence suggests that these age-related changes have a more significant detrimental effect on sleep efficiency than SDB itself [21]. Depression, which can also influence sleep quality and daytime function, has been found to be increased in older adults with sleep apnea [4].

Other conditions that are of concern in geriatric patients that may be influenced by SDB include nocturia and incontinence. SDB can increase nocturnal urine production by 200–300 cc, leading to an increased frequency of nocturia [22]. In addition, the abnormal respiratory dynamics associated with SDB could lead to increased abdominal pressure and episodes of nocturnal incontinence [23].

Relatively few studies have examined the effects of SDB on mortality in large part because of the challenges of conducting such longitudinal studies in elders. Ancoli-Israel et al. [24] conducted the largest study to date when they reported mortality data on a cohort of 426 seniors. They observed that an AHI ≥ 30 events/hour was associated with a shorter survival, but when controlling for other factors such as age and cardiovascular disease, it was no longer a statistically significant predictor of mortality. This led them to conclude that SDB is most likely not a proximate cause of death, but may act through associated diseases, such as cardiovascular disease, to increase mortality. It is important to note that other studies in the elderly have found no increased mortality related to sleep apnea; however, these studies used smaller cohorts that were predominantly healthy and had AHI criteria of 5 [10] and 15 events/hour [25].

Additional evidence that comorbidities may be a significant mediator of SDB-related mortality comes from studies that have included elderly with a history of stroke or myocardial infarction. These studies have demonstrated an increased mortality rate from SDB, in several patient groups, such as stroke patients with SDB [26], and patients with coronary artery disease and SDB [27], even when controlling for other risk factors. Overall, these findings suggest that, in healthy elderly, SDB is a relatively benign condition and higher thresholds for treatment may be appropriate, although the specific age-adjusted thresholds have yet to be established. However, in subjects with comorbid conditions, such as cardiovascular disease, SDB should be more aggressively treated based on data showing an increased rate of mortality in these populations.

TREATMENT

The treatment of geriatric SDB is largely similar to that in younger subjects in that it relies on noninvasive positive pressure ventilation. One area of possible concern in the elderly is adherence to treatment. However, one study found that approximately 64% of older males adhered to treatment, a rate similar to that seen in younger cohorts [28]. The major risk factors for nonadherence in the elderly were nocturia, cigarette smoking, age at time of initial diagnosis (advanced age at time of diagnosis tended to lower adherence), and inadequate symptom resolution with CPAP [28].

There is little data on the effects of other SDB therapies in an elderly population. Other treatment options include oral appliances that alter upper airway morphology to minimize SDB; they reduced AHI from 25.1 to 14.7 events/hour in elderly heart failure patients [29]. Unfortunately, for older adults who are edentulous, this is not a viable option. Nasal oxygen has also been used for patients who cannot tolerate CPAP and have heart failure, and it has been found to have some limited benefit (AHI decreased from 49 to 29 events/hour) [30]. The risks/benefits of the different surgical therapies for SDB have not been studied in the elderly. Indeed, advanced age may be considered a relative contraindication for surgical management in some cases.

CONCLUSION

Sleep disordered breathing is highly prevalent in older adults, with sleep apnea alone affecting nearly 20% of the elderly. However, sleep apnea syndrome, defined as the presence of an elevated AHI and clinical symptoms, is far less common. For this reason, it is possible that SDB in the elderly is a different pathophysiologic process than SDB in younger subjects. Additionally, the consequences of SDB in healthy elderly appear to be relatively mild; however, older subjects with cardiovascular comorbidities or other clinical symptoms such as excessive daytime sleepiness can suffer significant morbidity and mortality from SDB. The threshold for treatment is thus tailored to the patient's overall condition. Indeed, it is this multifactorial nature of sleep in older adults that makes determining an effective treatment regimen a challenging, yet potentially rewarding process. In many cases, SDB is not fully evaluated or treated in the elderly, thus missing an excellent opportunity to substantially improve an elder's quality of life and reduce morbidity/mortality, especially in those with comorbid diseases.

REFERENCES

1. Young T, Shahar E, Nieto FJ, Redline S, Newman AB, Gottlieb DJ, Walsleben JA, et al. Predictors of sleep-disordered breathing in community-dwelling adults: the Sleep Heart Health Study. *Arch Intern Med* **162**:893–900(2002).

2. Bixler E, Vgontzas A, Ten H, T, Tyson K, Kales A. Effects of age on sleep apnea in men: I. Prevalence and severity. *Am J Respir Crit Care Med* **157**:144–148(1998).

3. Ancoli-Israel S, Kripke DF, Klauber MR, Parker L, Stepnowsky C, Kullen A, Fell R. Natural history of sleep disordered breathing in community dwelling elderly. *Sleep* **16**:S25–S29 (1993).

4. Ancoli-Israel S, Kripke DF, Klauber MR, Mason WJ, Fell R, Kaplan O. Sleep-disordered breathing in community-dwelling elderly. *Sleep* **14**:486–495(1991).

5. Tishler PV, Larkin EK, Schluchter MD, Redline S. Incidence of sleep-disordered breathing in an urban adult population: the relative importance of risk factors in the development of sleep-disordered breathing. *JAMA* **289**:2230–2237(2003).

6. Ancoli-Israel S, Klauber MR, Stepnowsky C, Estline E, Chinn A, Fell R. Sleep-disordered breathing in African-American elderly. *Am J Respir Crit Care Med* **152**:1946–1949(1995).

7. Bliwise DL. Normal aging. In: Kryger MH, Roth T, Dement W (Eds), *Principles and Practice of Sleep Medicine*. Saunders, Philadelphia, 2000, pp 26–42.

8. Enright PL, Newman AB, Wahl PW, Manolio TA, Haponik EF, Boyle PJ. Prevalence and correlates of snoring and observed apneas in 5,201 older adults. *Sleep* **19**:531–538(1996).

9. Dickel MJ, Mosko SS. Morbidity cut-offs for sleep apnea and periodic leg movements in predicting subjective complaints in seniors. *Sleep* **13**:155–166(1990).

10. Phillips BA, Berry DT, Lipke-Molby TC. Sleep-disordered breathing in healthy, aged persons. Fifth and final year follow-up. *Chest* **110**:654–658(1996).

11. Mosko SM, Dickel MJ, Ashurst J. Night to night variability in sleep apnea and sleep-related periodic leg movements in the elderly. *Sleep* **11**:340–348(1988).

12. Haponik E. Sleep disturbances of older persons: physicians' attitudes. *Sleep* **15**:168–172(1992).

13. Oliven A, Carmi N, Coleman R, Odeh M, Silbermann M. Age-related changes in upper airway muscles morphological and oxidative properties. *Exp Gerontol* **36**:1673–1686(2001).

14. Krieger J, Sforza E, Boudewijns A, Zamagni M, Petiau C. Respiratory effort during obstructive sleep apnea: role of age and sleep state. *Chest* **112**:875–884(1997).

15. Nieto FJ, Young TB, Lind BK, Shahar E, Samet JM, Redline S, D'Agostino RB, et al. Association of sleep-disordered breathing, sleep apnea, and hypertension in a large community-based study. Sleep Heart Health Study. *JAMA* **283**:1829–1836(2000).

16. Shahar E, Whitney CW, Redline S, Lee ET, Newman AB, Javier Nieto F, O'Connor GT, et al. Sleep-disordered breathing and cardiovascular disease: cross-sectional results of the Sleep Heart Health Study. *Am J Respir Crit Care Med* **163**:19–25(2001).

17. Gooneratne NS, Weaver TE, Cater JR, Pack FM, Arner HM, Greenberg AS, Pack AI. Functional outcomes of excessive daytime sleepiness in older adults. *J Am Geriatr Soc* **51**:642–649(2003).

18. Phillips BA, Berry DT, Schmitt FA, Magan LK, Gerhardstein DC, Cook YR. Sleep-disordered breathing in the healthy elderly. Clinically significant? *Chest* **101**:345–349(1992).

19. Valencia-Flores M, Campos RM, Mendez J, Haro R, Schenkel E, Bliwise D, Guilleminault C. Multiple sleep latency test (MSLT) and sleep apnea in aged women. *Sleep* **16**:114–117 (1993).

20. Dealberto M, Pajot N, Courbon D, Alperovitch A. Breathing disorders during sleep and cognitive performance in an older community sample: the EVA Study. *J Am Geriatr Soc* **44**:1287–1294(1996).

21. Redline S, Kirchner HL, Quan SF, Gottlieb DJ, Kapur V, Newman A. The effects of age, sex, ethnicity, and sleep-disordered breathing on sleep architecture. *Arch Intern Med* **164**:406–418(2004).

22. Umlauf MG, Chasens ER, Greevy RA, Arnold J, Burgio KL, Pillion DJ. Obstructive sleep apnea, nocturia and polyuria in older adults. *Sleep* **27**:139–144(2004).

23. Bliwise DL, Adelman CL, Ouslander JG. Polysomnographic correlates of spontaneous nocturnal wetness episodes in incontinent geriatric patients. *Sleep* **27**:153–157(2004).

24. Ancoli-Israel S, Kripke DF, Klauber MR, Fell R, Stepnowsky C, Estline E, Khazeni N, et al. Morbidity, mortality and sleep-disordered breathing in community dwelling elderly (see comments). *Sleep* **19**:277–282(1996).

25. Mant A, King M, Saunders NA, Pond CD, Goode E, Hewitt H. Four-year follow-up of mortality and sleep-related respiratory disturbance in non-demented seniors. *Sleep* **18**:433–438 (1995).

26. Parra O, Arboix A, Montserrat JM, Quintó LZ, Bechich S, García-Eroles L. Sleep-related breathing disorders: impact on mortality of cerebrovascular disease. *Eur Respir J* **24**:267–272(2004).

27. Peker Y, Hedner J, Kraiczi H, Loth S. Respiratory disturbance index: an independent predictor of mortality in coronary artery disease. *Am J Respir Crit Care Med* **162**:81–86(2000).

28. Russo-Magno P, O'Brien A, Panciera T, Rounds S. Compliance with CPAP therapy in older men with obstructive sleep apnea. *J Am Geriatr Soc* **49**:1205–1211(2001).

29. Eskafi M, Cline C, Petersson A, Israelsson B, Nilner M. The effect of mandibular advancement device on pharyngeal airway dimension in patients with congestive heart failure treated for sleep apnoea. *Swed Dent J* **28**:1–9(2004).

30. Javaheri S, Ahmed M, Parker TJ, Brown CR. Effects of nasal O_2 on sleep-related disordered breathing in ambulatory patients with stable heart failure. *Sleep* **22**:1101–1106(1999).

ADDITIONAL READING

Aloia MS, Di Dio L, Ilniczky N, Perlis ML, Greenblatt DW, Giles DE. Improving compliance with nasal CPAP and vigilance in older adults with OAHS. *Sleep Breath* **5**:13–21(2001). This study demonstrates the effectiveness of a simple cognitive behavioral intervention to improve compliance with CPAP treatment for SDB in older adults.

Ancoli-Israel S, Coy T. Are breathing disturbances in elderly equivalent to sleep apnea syndrome? *Sleep* **17**:77–83(1994). A more detailed discussion that nicely lays out different points of view regarding the impact of geriatric SDB.

Baldwin CM, Griffith KA, Nieto FJ, O'Connor GT, Walsleben JA, Redline S. The association of sleep-disordered breathing and sleep symptoms with quality of life in the Sleep Heart Health Study. *Sleep* **24**:96–105(2001). A comprehensive assessment of the relationship between quality of life, SDB, and comorbidities.

Bliwise DL, Benkert RE, Ingham RH. Factors associated with nightly variability in sleep-disordered breathing in the elderly. *Chest* **100**:973–976(1991). This analysis includes evaluation of several anatomic factors.

Young T. Sleep-disordered breathing in older adults: is it a condition distinct from that in middle-aged adults? *Sleep* **19**:529–530(1996). This editorial provides a concise review of the relationship between sleep apnea syndrome and SDB.

79

INSOMNIA AND AGING

LEAH FRIEDMAN
Stanford University, Stanford, California

SLEEP COMPLAINTS IN OLDER AGE

Insomnia is a subjective complaint or a symptom complex, not a specific disease entity, and definitions of insomnia vary from very broad descriptions such as "complaint of poor sleep" [1] to somewhat more specifically described sleep complaints such as problems with sleep initiation or sleep maintenance insomnia. Unlike younger people who complain primarily of problems with sleep initiation, older people complain more frequently of difficulty in maintaining sleep. As people grow older sleep continuity is often compromised as they tend to wake more frequently during the night and/or wake too early in the morning without achieving sufficient sleep to feel rested, or they suffer from a combination of both sleep initiation and maintenance insomnia.

The subjective complaint of older people of loss of sleep continuity has been validated by objective measurement showing a decline in the deeper stages of sleep, increase in the proportion of the night spent in lighter stages of sleep, and an increase in the number of nocturnal awakenings associated with a decline in sleep efficiency and an increase in daytime napping [2]. Recently, polysomnographic data has demonstrated that it may be the number of transitions from sleep to wake rather than from wake to sleep that is most problematic for older people. Low sleep efficiency resulted from the increased number not the increased length of wake bouts. In fact, older subjects were found to fall back to sleep at the same rate as younger subjects [3].

Prevalence and Persistence

Insomnia is the most common sleep complaint of older adulthood and it is also highly prevalent. Reported prevalence rates vary considerably depending in part on differences in sample composition, questions asked, and definitions of terms. According to several epidemiologic studies, the prevalence of serious insomnia in the elderly ranges from 20% to 40% [4]. It has been suggested that prevalence rates of poor sleep may be even higher than those generally reported because of the tendency of healthy older men and women to view even objectively disturbed sleep as acceptable for their age and to thus underreport sleep disturbance [5].

Increasing age is not only associated with greater prevalence of insomnia but also with a decline in remission rates and greater persistence of the problem [6]. Thus, with aging, there is also an increase in chronic insomnia. Many older people complain of having had insomnia for years with some claiming life-long insomnia. In the National Institute on Aging's epidemiologic study, 52% of the 6899 older men and women followed 3 years later retained symptoms of insomnia [7].

Gender Differences

Even though on objective measures older women's sleep has been found to be better than that of older men, prevalence rates of insomnia generally are higher in older women. A recent study found a high correspondence between older men's subjective evaluation of their sleep

Sleep: A Comprehensive Handbook, Edited by T. Lee-Chiong.

and objective measures of their sleep but there was a much weaker correspondence between women's objective and subjective sleep. This raises the question of whether the standard objective sleep measures are appropriate to measure women's experience of poor sleep quality [8].

IMPACT OF AGING ON SLEEP

The role which aging itself plays in the insomnia of old age has been questioned. It has been argued that much of the insomnia found in the elderly is actually secondary to various disease conditions and not to aging *per se*. Evidence for this perspective is supplied by the findings of the National Institute on Aging's follow-up study, that most incident cases of insomnia were associated with medical, psychological, social, or other risk factors [7]. Indeed, when coexisting psychiatric and medical disease are not present, prevalence rates of insomnia in the elderly have been found to be low [9]. It is also likely that many elderly insomniacs do not begin to have sleep problems later in life but arrive in old age with a background of poor sleep [10].

The distinction between primary insomnia and insomnia secondary to other factors with its underlying implication of causality is particularly salient to understanding and treating insomnia in older adults. Primary insomnia is used to describe the presence of sleep problems in otherwise physically and mentally healthy people. In secondary insomnia, disrupted or inadequate sleep is seen as a consequence of or secondary to a number of possible factors. The risk factors for insomnia are numerous and range from medical, psychiatric, and psycho/social factors to specific sleep disorders.

The incidence of physical conditions and mental disorders as well as the use of pharmacological agents increase greatly with aging. A large number of physical health conditions such as pain syndromes (fibromyalgia, osteoarthritis, rheumatoid arthritis), primary pulmonary problems (chronic obstructive disease, bronchitis, asthma), neurologic disturbances (headaches, cerebrovascular accidents, Parkinson's disease), and dementias such as Alzheimer's disease contribute to the increase in sleep problems associated with aging [11]. In turn, the medications used to manage chronic conditions may also contribute to insomnia in the elderly [12].

SLEEP DISORDERS

The incidence of specific physiologically based sleep disorders such as periodic limb movement disorder (continuous and repetitive leg jerks lasting between 0.5 and 5 seconds sometimes resulting in arousals) and obstructive sleep apnea (breathing cessation caused by complete or partial airway collapse) increases greatly with age [11]. While there is some dispute about the sleep impact of periodic limb movements, there is general agreement that sleep apnea is often very disruptive of sleep.

MOOD DISORDERS

Depression and anxiety disorders are prevalent in the elderly. The losses concomitant with bereavement, change in family roles, retirement, and illness are among the psychosocial hallmarks of aging and are frequently associated with insomnia as well as depression and heightened anxiety. Depressed mood has been found to be as important a factor as physical illness in the insomnia of the elderly and is frequently associated with severe sleep disturbance. There is evidence that the relationship between depression and insomnia is bidirectional. The chain of causality is difficult to disentangle but in samples of medical patients there was a stronger association between insomnia and depression than with other medical conditions. A complicating factor increasing insomnia risk in depressed patients is that medications prescribed for treatment of depression often have sleep disruptive effects [13].

NOCTURIA

Sleep interruption caused by the need to void is believed to be a common factor contributing to insomnia in the elderly. In fact, in a large sample study of older adults it was the most common complaint associated with sleep maintenance problems [14]. Nocturia can be caused by various age-associated conditions such as prostate enlargement and diabetes. Recent work suggests a relation between sleep disordered breathing and nocturia such that increased sleep disordered breathing has been found to be associated with an increased number of nocturia episodes [15].

CIRCADIAN RHYTHMS AND AGING

Specific age-related changes in the circadian regulation of sleep and wake have been proposed as important factors contributing to elderly insomnia. According to this thesis, with aging there is a weakening in the strength of the endogenous circadian drive with a concomitant dampening of circadian amplitude and changed temperature cycle timing so that the second half of the sleep cycle takes place during the rising phase of the temperature curve associated with greater wakefulness. These endogenous changes are accompanied by a decline in the effectiveness of exogenous circadian factors because of decreased retinal sensitivity to light and reductions in levels and length of daily light exposure along with similar curtailment in daily activity and exercise levels [16]. However, whether these circadian rhythm changes necessarily accompany normal aging has been disputed [17].

INSOMNIA IMPACT ON AGING

A major concern about insomnia in aging is its impact on health, well-being, and ultimate mortality. Insomnia does not necessarily indicate shortened sleep. There was little or no association between insomnia as measured by questionnaire response and excess mortality in a sample of over a million men and women ranging from 30 to 102 years of age. However, there are indicators that at the extremes sleep duration matters since both those who slept more than 8 hours and those who slept 6 hours or less had increased mortality risk [18].

It has been argued that sleep need does not decrease with age and the increase in daytime sleepiness that so frequently accompanies poor nighttime sleep is evidence for its continued need [12]. On the other hand, others argue that the evidence is not definitive and that alternative conclusions can be drawn from the evidence collected to date, ranging from a decrease in sleep need to a change in the type of sleep need with aging [17]. But poor quality and inadequate sleep in the elderly has been associated with a number of serious negative consequences.

Quality of Life

These consequences of poor sleep have been found to be associated with decrements in quality of life. Sleepiness and lack of energy to perform daily tasks or social and pleasurable activities are natural consequences of insomnia. In a large sample study of a Wisconsin community, as the number of insomnia symptoms increased there was a parallel decrease in respondents' quality of life [19].

Effects of Hypnotic Use

Older people make disproportionately high use of both prescribed and nonprescribed hypnotic medications to ameliorate their poor sleep and they continue to use these medications over extended periods of time. It is well-known that there are many risks associated with use of these medications in this age group, given altered pharmacokinetics, increased sensitivity to CNS effects of these medications, and interactions with other drugs taken for comorbid conditions. Use of these medications in the elderly has been associated with a high incidence of falls and accidents [12]. Frequent use of sleeping medications also has been found to be associated with a significantly increased mortality risk [18].

TREATMENTS FOR INSOMNIA

Because of the complex etiology of insomnia in the elderly, no treatment decisions should be made prior to a thorough sleep evaluation. When coexisting mental or physical health conditions are present, treatment of these conditions should be the first line of action [12]. Although shorter-acting hypnotics are now available, there is also growing support for the use of behavioral treatments for the elderly either alone or in conjunction with pharmacologic treatment. This is not only because of the potential negative side-effect profiles of many hypnotic medications (such as the exacerbation of sleep apnea with benzodiazepines) [11] but also because insomnia tends to be chronic and recurring and thus requires treatments with long-lasting effects [20]. Change in underlying sleep-related behaviors is one approach to long-term sleep improvement. Behavioral treatments are particularly well-suited to the needs of older insomniacs who often develop poor sleep-related behaviors such as the loss of structure for sleep–wake timing and reduction in bright-light exposure and exercise in response to changes in health and to reduced demands of work and family with aging. These issues can be addressed by specific behavioral treatments. A bonus of a number of components of these treatments, for example, bright light and exercise, is the promotion of general good health and mood. (See Chapter 18, Nonpharmacologic Therapy of Insomnia, for a description of the most commonly used behavioral treatments for insomnia.)

REFERENCES

1. Morgan K, Dallosso H, Ebrahim S, Arie T, Fentem PH. Characteristics of subjective insomnia in the elderly living at home. *Age Ageing* **17**:1–7(1988).

2. Prinz PN, Vitiello MV, Raskind MA, Thorpy MJ. Geriatrics: sleep disorders and aging. *N Engl J Med* **323**:520–526(1990).

3. Klerman EB, Davis JB, Duffy JF, Dijk DJ, Kronauer RE. Older people awaken more frequently but fall back asleep at the same rate as younger people. *Sleep* **27**:793–798(2004).

4. Sateia MJ, Doghramji K, Hauri PJ, Morin CM. Evaluation of chronic insomnia. An American Academy of Sleep Medicine review. *Sleep* **23**:243–308(2000).

5. Buysse DJ, Reynolds CF III, Monk TH, Hoch CC, Yeager AL, Kupfer DJ. Quantification of subjective sleep quality in healthy elderly men and women using the Pittsburgh Sleep Quality Index (PSQI). *Sleep* **14**:331–338(1991).

6. Buysse DJ. Insomnia, depression and aging. Assessing sleep and mood interactions in older adults. *Geriatrics* **59**:47–51(quiz 52)(2004).

7. Foley DJ, Monjan AA, Simonsick EM, Wallace RB, Blazer DG. Incidence and remission of insomnia among elderly

adults: an epidemiologic study of 6,800 persons over three years. *Sleep* **22**:S366–S372(1999).

8. Vitiello MV, Larsen LH, Moe KE. Age-related sleep change. Gender and estrogen effects on the subjective–objective sleep quality relationships of healthy, noncomplaining older men and women. *J Psychosom Res* **56**:503–510(2004).

9. Vitiello MV, Moe KE, Prinz PN. Sleep complaints cosegregate with illness in older adults: clinical research informed by and informing epidemiological studies of sleep. *J Psychosom Res* **53**:555–559(2002).

10. Morgan K, Healey DW, Healey PJ. Factors influencing persistent subjective insomnia in old age: a follow-up study of good and poor sleepers aged 65 to 74. *Age Ageing* **18**:117–122 (1989).

11. Avidan AY. Sleep changes and disorders in the elderly patient. *Curr Neurol Neurosci Rep* **2**:178–185(2002).

12. Ancoli-Israel S. Insomnia in the elderly: a review for the primary care practitioner. *Sleep* **23** (Suppl 1):S23–S30(discussion S36–S38)(2000).

13. Benca RM, Ancoli-Israel S, Moldofsky H. Special considerations in insomnia diagnosis and management: depressed, elderly, and chronic pain populations. *J Clin Psychiatry* **65** (Suppl 8):26–35(2004).

14. Middelkoop HA, Smilde-van den Doel DA, Neven AK, Kamphuisen HA, Springer CP. Subjective sleep characteristics of 1,485 males and females aged 50–93: effects of sex and age, and factors related to self-evaluated quality of sleep. *J Gerontol A Biol Sci Med Sci* **51**:M108–M115(1996).

15. Endeshaw YW, Johnson TM, Kutner MH, Ouslander JG, Bliwise DL. Sleep-disordered breathing and nocturia in older adults. *J Am Geriatr Soc* **52**:957–960(2004).

16. Hood B, Bruck D, Kennedy G. Determinants of sleep quality in the healthy aged: the role of physical, psychological, circadian and naturalistic light variables. *Age Ageing* **33**:159–165(2004).

17. Bliwise DL. Normal aging. In: Kryger MH, Roth T, Dement WC (Eds), *Principles and Practice of Sleep Medicine*, 3rd ed. Saunders, Philadelphia, 2000, pp 26–42.

18. Kripke DF, Garfinkel L, Wingard DL, Klauber MR, Marler MR. Mortality associated with sleep duration and insomnia. *Arch Gen Psychiatry* **59**:131–136(2002).

19. Schubert CR, Cruickshanks KJ, Dalton DS, Klein BEK, Klein R, Nondahl DM. Prevalence of sleep problems and quality of life in an older population. *Sleep* **25**:48–52(2002).

20. Reynolds CF III, Buysse DJ, Kupfer DJ. Treating insomnia in older adults: taking a long-term view. *JAMA* **281**:1034–1035 (1999).

80

SLEEP IN INSTITUTIONALIZED OLDER ADULTS

JENNIFER L. MARTIN AND CATHY A. ALESSI
Veterans Administration Greater Los Angeles Healthcare System, Los Angeles, California
University of California, Los Angeles

CHARACTERISTICS OF INSTITUTIONALIZED OLDER ADULTS

As the number of older adults increases, the long-term and rehabilitative care needs of this growing segment of the population are often met in institutional settings such as nursing homes (NHs). In 1999, approximately 1.5 million older U.S. adults, 5% of those over age 65, resided in NHs on any given day. The typical NH resident is white (88%), widowed (63%), female (75%), and over age 75 (86%). Nearly all residents (97%) require assistance in one or more basic activities of daily living (e.g., toileting, bathing, dressing). Residents most commonly are admitted from acute care hospitals (44%) or directly from home (32%). The average length of stay in a NH is 2.4 years, and the most common reasons for discharge are death (27%) and acute hospital admission (28%), generally due to deteriorating health or acute medical emergency. Only 29% of residents are discharged because they recover or are sufficiently stabilized to return home [1]. Tables 80.1 and 80.2 summarize the most common primary medical conditions and the services most commonly received by NH residents.

SLEEP–WAKE PATTERNS IN NURSING HOMES

Nearly any NH visitor will see evidence of sleep–wake pattern disturbance among residents. Residents are often in bed and asleep intermittently at all hours of the day, even during mealtime periods. Although older adults in the community are more likely to take naps compared to younger adults, the typical NH resident shows a different pattern in which wakefulness is frequently interrupted by brief periods of sleep. This daytime "wake fragmentation" is accompanied by nighttime "sleep fragmentation," and although older adults in the community report difficulties with nighttime sleep, the sleep disruption is much more severe in NH residents.

Research using either 24 hour behavioral observations of sleep or wrist actigraphy has shown that the sleep of NH residents is distributed across the 24 hour day rather than being consolidated to the nighttime hours, and residents seldom have a continuous hour of sleep at any time [2, 3]. Although the specific causes of sleep–wake pattern disruption vary from person to person, there are several factors that are common. These include primary sleep disorders, medical conditions, medications/polypharmacy, psychiatric disorders, and circadian rhythm disruption. Environmental factors (e.g., noise and light during the night, low daytime indoor illumination, little time spent outdoors) and behavioral factors (e.g., physical inactivity, extended time spent in bed) also appear to disrupt NH residents' sleep–wake patterns.

PRIMARY SLEEP DISORDERS IN THE NURSING HOME

No large-scale epidemiological studies have been conducted to examine the prevalence of primary sleep disorders in the NH; however, a few studies have reported the rates of sleep disordered breathing (SDB) in samples of NH residents. Depending on the precise criteria used,

TABLE 80.1 Percent of Residents with the Six Most Common Primary Medical Diagnoses at Admission to U.S. Nursing Homes [1]

Primary Diagnosis (ICD-9 codes)	Total	Men	Women
Diseases of the circulatory system (390–459)	25.8%	25.8%	25.8%
Mental disorders (290–319)	15.6%	16.3%	15.4%
Diseases of the nervous system and sense organs (320–389)	13.1%	13.4%	13.0%
Injury and poisonings (800–999)	8.0%	4.4%	9.2%
Endocrine, nutritional, and metabolic diseases and immunity disorder (240–279)	6.3%	6.7%	6.2%
Diseases of the respiratory system (460–519)	6.3%	8.7%	5.5%

one-half to three-quarters of NH residents have at least mild SDB, which has been associated with cognitive impairment, agitated behaviors, and increased mortality risk in NH residents [4, 5].

The prevalence of other primary sleep disorders among NH residents has not been reported. One could assume, however, that sleep disorders that increase in prevalence with advancing age would be at least as common in NHs as in community settings (e.g., periodic limb movement disorder (PLMD), REM sleep behavior disorder). The absence of such prevalence information is, in part, due to the difficulty in conducting polysomnographic recordings in NH residents, particularly in individuals with dementia or extreme frailty.

To date, no studies have systematically examined treatment of SDB or PLMD in NH residents. In addition, the ability of NH residents to use known effective treatments (e.g., continuous positive airway pressure) is unknown. In general, treatment of primary sleep disorders in NH residents should closely parallel the treatment of frail older adults in the community, with the caveat that associated improvements in functional status, cognition, and quality of life are paramount in the NH setting. Clearly, further research on the treatment of primary sleep disorders in the NH setting is needed.

The term "insomnia" is sometimes used to describe the sleep problems seen in NH residents; however, use of this term is somewhat misleading. This terminology can be problematic when sleep problems in NH residents are treated with sedative–hypnotic medications as if the resident suffers from primary insomnia, without taking into account the multiple contributing factors. Formal diagnosis of primary insomnia requires that potential underlying causes (e.g., medication side effects) are ruled out and, as mentioned earlier and discussed later, most NH residents have multiple factors contributing to abnormal sleep and sedative–hypnotics are associated with adverse outcomes in NH residents (see later discussion).

MEDICAL ILLNESSES, PSYCHIATRIC DISORDERS, AND MEDICATIONS

NH residents are frequently in poor physical health; many suffer from dementia, depression or both, and nearly all take multiple medications to manage medical and psychiatric conditions. Studies have reported that nursing home residents take, on average, five to eight different medications each day, although many residents take ten or more medications per day [6, 7]. Certain medications may be particularly problematic when taken near bedtime, such as diuretics or stimulating agents (e.g., sympathomimetics, bronchodilators). Use of sedating medications during the daytime (e.g., antihistamines, anticholinergics, sedating antidepressants) may contribute to daytime drowsiness and further disrupt sleep–wake cycles. Certain medications for the treatment of depression, Parkinson's disease, and hypertension can induce nightmares or impair sleep as well. Examples of common medical conditions among NH residents that may contribute to sleep difficulty include pain (e.g., from arthritis), paresthesias, nighttime cough, dyspnea from cardiac or pulmonary illness, gastroesophageal reflux, and nighttime urination.

There is increasing evidence of sleep abnormalities with neurological illnesses (e.g., Alzheimer's disease, Parkinson's disease), many of which are common among NH residents. In the NH setting, residents are often in the late stages of these neurological illnesses. Research suggests demented patients generally have more sleep disruption,

TABLE 80.2 Percent of NH Residents Receiving the Eight Most Common Services in U.S. Nursing Homes [1]

Service Received	Total	Men	Women
Nursing services	97.3%	97.5%	97.3%
Prescribed or nonprescribed medicines	93.6%	93.6%	93.6%
Medical services	91.1%	91.6%	90.6%
Personal care	91.0%	91.1%	91.0%
Nutritional services	74.0%	75.0%	73.7%
Social services	70.3%	69.6%	70.5%
Equipment or devices	50.7%	50.7%	50.7%
Physical therapy	27.2%	29.7%	26.4%

lower sleep efficiency, more stage 1, less stage 3/4, and perhaps less REM sleep compared to nondemented older people [8]. "Sundowning," the term used to describe a worsening of confusion and behavior problems in the evening or night in people with dementia, may have an underlying neurological basis and is associated with circadian rhythm disruption [6]. The sleep abnormalities, excessive daytime sleepiness, and parasomnias (e.g., REM sleep behavior disorder) associated with Parkinson's disease may be related to the pathology of the disorder and/or to its medication treatment. Problems may be even more common among NH residents with advanced disease [9].

CIRCADIAN RHYTHM DISRUPTION

Circadian rhythm disruption also contributes to sleep problems in NH residents. In older people, circadian rhythms may be blunted in amplitude and can be shifted to abnormal times. In one study, NH residents had less stable circadian rhythms of activity compared to older people living at home, regardless of cognitive status [10]. Circadian activity rhythm abnormalities have also been associated with shorter survival in NH residents [11].

In addition to the often-cited advance of circadian rhythms commonly found in older adults, environmental factors also come into play in the NH. In particular, daytime light levels in NHs are quite low, and residents are seldom taken outdoors. Typically, NH residents are exposed to only a few minutes of bright light each day—even less than older adults living in the community [12]. Since light exposure is the strongest known *zeitgeber* (time cue) in humans, this lack of daytime light may contribute to circadian disregulation. NH residents also spend extended periods in bed and are physically inactive during the daytime, which also contributes to their circadian rhythm abnormalities.

THE NURSING HOME AT NIGHT

NHs are more similar to in-patient hospital settings than to home environments. Typically, residents share rooms, and the nighttime environment is not conducive to sleep due to frequent noise and light interruptions [13]. NH residents must endure these interruptions on an extended, nightly basis. Research has shown that much of the noise produced in the NH is caused by staff, often while they provide incontinence and other personal care to residents [13, 14]. In addition to noise, nighttime exposure to room-level light of even a few hundred lux has the potential to suppress melatonin, disrupt sleep, and shift circadian rhythms.

PHARMACOLOGICAL TREATMENT OF SLEEP PROBLEMS

Although sedative–hypnotic medications can be appropriate and useful for the treatment of insomnia, particularly insomnia of short duration (e.g., after a major life stressor), the sleep problems in NH residents are generally chronic and may not be appropriate for, nor responsive to, pharmacological management with sleeping pills. In spite of this, sedative–hypnotic medications are sometimes given on an ongoing basis to NH residents with chronic, longstanding sleep problems. To our knowledge, there are no published studies examining the effectiveness of sleep medications used in NH residents in this way. There are, however, several studies documenting the increased risk of falls associated with use of sedating medications.

Extensive NH reforms enacted with the Omnibus Budget Reconciliation Act (OBRA) of 1987 (which became effective in 1991) included limits on the use of psychoactive medications in the NH setting. Although OBRA regulations specifically target antipsychotic medications, the interpretive guidelines that accompany these regulations also limit the use of anxiolytic agents and sedative–hypnotics, but not antidepressants. Use of regulated psychoactive medications must be documented in the medical record as necessary to treat an acceptable specific condition, with daily dose limits, requirements for monitoring treatment and adverse reactions, and attempts at dose reductions and discontinuation of the medication, if possible. The guidelines also provide the clinician options for using these psychoactive medications outside the stated limits when such use is clearly clinically indicated. Since implementation of OBRA guidelines, research has shown substantial decreases in use of antipsychotics among nursing home residents, with no increase in sedative–hypnotics or anxiolytics, but increased use of antidepressants [15].

NONPHARMACOLOGICAL TREATMENT OF SLEEP PROBLEMS

Given the potential risks of sedative–hypnotic medications and the common underlying causes of sleep disruption, several groups of investigators have undertaken studies to test the effectiveness of nonpharmacological interventions to improve sleep in the NH setting. Several studies have tested timed exposure to bright artificial light as a means to improve sleep–wake patterns. In randomized controlled trials, NH residents exposed to bright light showed improved sleep relative to participants who received placebo interventions [7, 16–18]. Researchers have also examined the effectiveness of supplemental melatonin, but results are mixed. Several aspects of the use of melatonin to improve sleep in NH residents, specifically, administration timing, dose, and preparation (acute versus sustained release), are not clear. A few studies have attempted to increase daytime activity levels, and results are, again, mixed. Some studies show sleep improves, while others show minimal or no changes in sleep [19]. Studies have also attempted to reduce nighttime noise and light in resident rooms. These studies have shown that it is extremely difficult to change the NH environment, and despite considerable efforts by researchers, the environment remains quite noisy at night [7, 19, 20].

An alternative approach is to use multicomponent interventions to address both internal physiological causes of sleep disturbance and external environmental factors. One such study tested a short-term (5 day) intervention combining daytime light exposure and physical activity, a structured and regularly timed bedtime routine, reduced time in bed during the day, plus provision of nighttime nursing care in a manner that minimizes disruption to sleep [7]. This intervention was successful in reducing daytime sleepiness, and residents who received the intervention were more engaged and active during the day than residents receiving usual care; however, nighttime noise and light were not significantly reduced, and there was minimal effect on nighttime sleep. Significant improvements in sleep in the NH setting likely require multiple factors be addressed simultaneously, perhaps for long periods of time (i.e., weeks or months).

One final area for intervention, which has largely been overlooked, is working at the facility level to change staff training, policies, and caregiving practices that impact resident sleep. Qualitative research is currently underway to identify ways to work with NH staff to improve the environmental factors that disrupt nighttime sleep and increase daytime sleepiness. The addition of sleep-promoting practices and the removal of unnecessary sleep-disruptive behaviors by staff may lead to meaningful changes for all residents in a facility. Real change will require administrators and other staff to buy into the notion that sleep is important and that encouraging better quality sleep–wake patterns is beneficial to both residents and staff over the long-term.

SUMMARY OF CURRENT KNOWLEDGE AND UNANSWERED QUESTIONS

In summary, nighttime sleep disruption is characteristic of NH residents and is typically accompanied by daytime sleepiness. Many factors contribute to these sleep problems, including medical and psychiatric illness, medications, circadian rhythm abnormalities, SDB, environmental factors, and life-style habits. There is some suggestion that these factors are amenable to treatment, particularly improving daytime activity patterns, increasing light exposure, and reducing sleep-disruptive caregiving practices at night. Further research is needed to determine whether treating SDB and other primary sleep disorders is feasible and results in functional or quality of life improvements among NH residents. Additional work is also needed to understand the administrative and policy factors that might lead to systemic changes in how sleep is viewed and sleep problems are addressed in NH settings.

ACKNOWLEDGMENTS

Supported by National Institute on Aging (R29 AG13885, PI: Alessi); VA Health Services Research and Development (IIR 01-053-1, PI: Alessi; and Associate Investigator Award, AIA 03- PI: Martin); and the VA Greater Los Angeles Healthcare System Geriatric Research, Education and Clinical Center (GRECC).

REFERENCES

1. National Center for Health Statistics, Gabrel CS. Characteristics of elderly nursing home current residents and discharges: data from the 1997 national nursing home survey. *Vital Health Stat* 312(2000).

2. Bliwise DL, Bevier WC, Bliwise NG, Edgar DM, Dement WC. Systemic 24-hour behavior observations of sleep and wakefulness in a skilled-care nursing facility. *Psychol Aging* **15**:16–24(1990).

3. Pat-Horenczyk R, Klauber MR, Shochat T, Ancoli-Israel S. Hourly profiles of sleep and wakefulness in severely versus mild-moderately demented nursing home patients. *Aging Clin Exp Res* **10**:308–315(1998).

4. Ancoli-Israel S, Klauber MR, Kripke DF, Parker L, Cobarrubias M. Sleep apnea in female patients in a nursing home: increased risk of mortality. *Chest* **96**(5):1054–1058(1989).

5. Gehrman PR, Martin JL, Shochat T, Nolan S, Corey-Bloom J, Ancoli-Israel S. Sleep disordered breathing and agitation in institutionalized adults with Alzheimer's disease. *Am J Geriatr Psychiatry* **11**:426–433(2003).

6. Martin J, Marler MR, Shochat T, Ancoli-Israel S. Circadian rhythms of agitation in institutionalized patients with Alzheimer's disease. *Chronobiol Int* **17**:405–418(2000).

7. Alessi CA, Martin JL, Webber AP, Kim EC, Harker JO, Josephson KR. Randomized controlled trial of a nonpharmacological intervention to improve abnormal sleep/wake patterns in nursing home residents. *J Am Geriatr Soc* **53**:803–810(2005).

8. Bliwise DL. Review: sleep in normal aging and dementia. *Sleep* **16**:40–81(1993).

9. Friedman JH, Chou KL. Sleep and fatigue in Parkinson's disease. *Parkinsonism Relat Disorders* **10**:S27–S35(2004).

10. Van Someren EJW, Hagebeuk EEO, Lijzenga C, et al. Circadian rest activity rhythm disturbances in Alzheimer's disease. *Biol Psychiatry* **40**:259–270(1996).

11. Gehrman PR, Marler M, Martin JL, Shochat T, Corey-Bloom J, Ancoli-Israel S. The timing of activity rhythms in patients with dementia is related to survival. *J Gerontol Med Sci* (in press).

12. Shochat T, Martin J, Marler M, Ancoli-Israel S. Illumination levels in nursing home patients: effects on sleep and activity rhythms. *J Sleep Res* **9**:373–380(2000).

13. Schnelle JF, Ouslander JG, Simmons SF, Alessi CA, Gravel MD. The nighttime environment, incontinence care, and sleep disruption in nursing homes. *J Am Geriatr Soc* **41**:910–914(1993).

14. Schnelle JF, Cruise PA, Alessi CA, Al-Samarrai N, Ouslander JG. Sleep hygiene in physically dependent nursing home residents. *Sleep* **21**:515–523(1998).

15. Lantz MS, Giambanco V, Buchalter EN. A ten-year review of the effect of OBRA-87 on psychotropic prescribing practices in an academic nursing home. *Psychiatr Serv* **47**:951–955(1996).

16. Ancoli-Israel S, Martin JL, Kripke DF, Marler M, Klauber MR. Effect of light treatment on sleep and circadian rhythms in demented nursing home patients. *J Am Geriatr Soc* **50**:282–289(2002).

17. Ancoli-Israel S, Gehrman PR, Martin JL, et al. Increased light exposure consolidates sleep and strengthens circadian rhythms in severe Alzheimer's disease patients. *Behav Sleep Med* **1**:22–36(2003).

18. Van Someren EJW, Kessler A, Mirmiran M, Swaab DF. Indirect bright light improves circadian rest–activity rhythm disturbances in demented patients. *Biol Psychiatry* **41**:955–963 (1997).

19. Alessi CA, Yoon EJ, Schnelle JF, Al-Samarrai NR, Cruise PA. A randomized trial of a combined physical activity and environmental intervention in nursing home residents: Do sleep and agitation improve? *J Am Geriatr Soc* **47**:784–791 (1999).

20. Schnelle JF, Alessi CA, Al-Samarrai NR, Fricker RD, Ouslander JG. The nursing home at night: effects of an intervention on noise, light and sleep. *J Am Geriatr Soc* **47**:430–438(1999).

ADDITIONAL READING

Alessi CA, Schnelle JF. Approach to sleep disorders in the nursing home setting. *Sleep Med Rev* **4**:45–56(2000). A summary of factors impacting sleep and a discussion of approaches to treatment of sleep problems in nursing home residents.

McCurry SM, Reynolds CF, Ancoli-Israel S, Teri L, Vitiello MV. Treatment of sleep disturbances in Alzheimer's disease. *Sleep Med Rev* **4**:603–628(2000). A systematic review of published treatment–outcomes studies focused on nonpharmacological management of sleep–wake disturbances in Alzheimer's disease patients.

Yesavage JA, Friedman L, Ancoli-Israel S, et al. Development of diagnostic criteria for defining sleep disturbance in Alzheimer's disease. *J Geriatr Psychiatry Neurol* **16**:131–139(2003). Description of a method to assess and diagnose sleep problems in Alzheimer's disease patients.

PART X

SLEEP AMONG WOMEN

81

PATTERNS OF SLEEP IN WOMEN: AN OVERVIEW

HELEN S. DRIVER

Kingston General Hospital and Queen's University, Kingston, Ontario, Canada

INTRODUCTION

Women's reproductive cycles vary across their lifespan giving additional biological rhythms, over the daily circadian (24 hour) rhythm, that can affect sleep. The complex reproductive changes are associated with varying levels of two steroid hormones in particular—estrogen and progesterone. Through the reproductive years, with menstrual cycles there is a constant flux of estrogen and progesterone levels with a pattern that repeats about every 28 days. During pregnancy, the levels of both hormones increase, then fall rapidly after delivery. With menopause, the levels decline and it is the loss of the effects of estrogen and progesterone that may underlie many of the physical and psychological symptoms women experience. The interaction of the changing hormones on the body and the brain should ideally cause minimal disruption to sleep or even facilitate it, for example, with the physical changes during pregnancy.

Good quality sleep is important for optimal daytime performance and mood. A recent Canadian survey found that 24% of the population aged 15 years and older responded that they regularly had trouble going to sleep or staying asleep [1]. Subjective complaints of insufficient or nonrestorative sleep affect between 10% and 35% of the general population and are more common in women than in men [1, 2], yet young women also report having a greater sleep need (see [3]). Reproductive status should be considered as a contributor to complaints of poor sleep in women. Possible factors are described in this chapter, with more detail provided in the five subsequent, complementary chapters.

MENSTRUAL CYCLES

During menstrual cycles prominent changes in reproductive hormones and body temperature occur [3] (see Chapter 83). The follicular phase is when estrogen is the predominant hormone; after ovulation, the luteal phase lasts 14–16 days and is when concentrations of estrogen and progesterone are high and body temperature is elevated by about 0.4 °C compared to before ovulation [3, 4]. The withdrawal of both estrogen and progesterone precedes menstruation. It is during the late-luteal (premenstrual) phase and the first few days of menstruation that most negative menstrual symptoms are experienced [5].

Based on a few controlled laboratory studies in young women with no menstrual-associated complaints and for ovulatory cycles, sleep across the menstrual cycle is remarkably stable [3, 5–7]. There is a small variation in rapid eye movement (REM) sleep, which tends to decrease in the luteal phase compared to the follicular phase [4, 6]. Although there is no clear-cut difference in sleep architecture, effects on sleep spindles have been observed, with increased EEG power density in the frequency range of sleep spindles (around 14 Hz) during NREM sleep in the luteal compared with the follicular phase [6]. This effect on sleep spindles has been proposed to be an influence of progesterone via the γ-aminobutyric acid (GABA$_A$) receptor [3, 6].

Using self-report data, about 70% of women report that their sleep is affected by menstrual symptoms such as bloating, tender breasts, headaches, and cramps, on average 2.5 days every month [8]. Even young women without significant menstrual-associated complaints report poorer sleep quality 3–6 days premenstrually and during 4 days of menstruation compared to other times of the menstrual cycle [5]. Mood, discomfort, and pain can affect sleep during this period.

Premenstrual Symptoms and Premenstrual Syndrome (PMS)

Many women experience premenstrual disturbances that vary in severity and type of symptom. Approximately 60% of women experience mild symptoms of PMS, and an estimated 20% have moderate PMS that they feel requires treatment; but for 3–8% of women the cyclical symptoms are severe and acknowledged as a clinical mood disorder—premenstrual dysphoric disorder (PMDD). Common symptoms that occur in the last week of the luteal phase and lessen after the onset of menstruation include irritability/anger, anxiety/tension, depression and mood swings, change in appetite, bloating and weight gain, and fatigue [7]. Sleep disturbances include insomnia, hypersomnia, unpleasant dreams, awakenings during the night, failure to wake at the expected time, and tiredness in the morning [9]. However, no significant, reproducible effects on sleep have been found in the few studies with small sample sizes on women with PMS/PMDD [7].

Painful Menstrual Conditions: Dysmenorrhea and Endometriosis

With both these conditions, pain and discomfort disturb sleep. Women who suffer from dysmenorrhea experience painful uterine cramps that debilitate them during menstruation every month [3, 10]. Women with endometriosis have misplaced tissue, of the same type that lines the inside of the uterus, which grows elsewhere in the abdominal and pelvic area and follows the menstrual cycle. Women with dysmenorrhea complain of poorer sleep quality and higher anxiety during menstruation compared to symptom-free women. There is only one published study in the sleep literature on painful menstrual conditions and sleep [10], which found reduced subjective sleep quality, sleep efficiency, and REM sleep compared with pain-free phases of the menstrual cycle and compared with controls. In turn, the disturbed sleep may worsen mood and alter the pain threshold.

Polycystic Ovarian Syndrome (PCOS)

As described in the chapter on sleep disordered breathing in women (Chapter 84), women with PCOS are more likely to develop obstructive sleep apnea (OSA)—a condition associated with snoring, repeated cessation of breathing, and daytime sleepiness. In POCS menstrual cycles are irregular or absent and the ovaries produce too much of the male sexual hormones (androgens), which causes infertility, facial hair, and weight gain. Increased sleep disordered breathing has been correlated with waist–hip ratio and testosterone in women with PCOS.

Sleep Disordered Breathing (SDB) and the Menstrual Cycle

Being female appears to reduce the risk of developing sleep disordered breathing (SDB), at least premenopausally. The prevalence of sleep apnea in premenopausal women is low (0.6%) [11]. However, compared with earlier clinical reports, the prevalence of OSA in middle-aged women in the general population at ~2% is higher than initially suggested. Polysomnographically, women with OSA tend to have a clustering of events during REM sleep, the frequency of which is related to body mass index (BMI) [12]. There may also be an association of SDB with menstrual phase. In premenopausal women with OSA, the severity of SDB was worse during REM sleep in the follicular compared with the luteal phase [13]. We have recently demonstrated a reduction in upper airway resistance during sleep in the luteal compared with the follicular phase in healthy women without any sleep complaints [14]. These findings support an increased propensity to upper airway obstruction during the follicular phase of the menstrual cycle, or more protection from SDB in the luteal phase. Although the clinical significance of these findings is not yet certain, it is conceivable that in some women OSA may only be manifest in the follicular phase of ovulatory cycles and that variability of OSA severity depending on menstrual phase might alter disease management. A more detailed overview of SDB in women is provided in Chapter 84.

Oral Contraceptive (OC) or Birth Control Pills

OC pills contain synthetic estrogen and progestin with 21 days of active hormone and the last 7 days inactive. In monophasic pills the same dosage of hormones is provided all through the entire active cycle; triphasic pills give different dosage levels during each week of the month and are designed to more closely duplicate a woman's natural hormonal pattern. These are called combined pills (containing estrogen and progesterone) whereas "minipills" contain progestin only.

Women taking monophasic oral contraceptives have persistently raised body temperatures, similar to those of naturally cycling women in the luteal phase (likely due to progesterone) [4]. While taking the active synthetic progestin and estrogen, women were found to have more stage 2 sleep compared to the inactive placebo phase [4]. When compared to naturally cycling women in the luteal phase, women taking OC pills had less deep sleep. In contrast to this small effect on sleep in women who had no sleep- or menstrual-associated complaints, for some women with premenstrual and menstrual symptoms, regularization of the menstrual cycle with OC pills may reduce their symptoms and thereby improve sleep.

PREGNANCY AND THE EARLY POSTPARTUM PERIOD

Getting enough sleep is especially important during pregnancy. This topic has recently been reviewed [15, 16] and is discussed in Chapter 82. During the first trimester, sleepiness increases due to the rise in progesterone, but it also brings on sleep disruption due to morning sickness—waking with nausea, increased urinary frequency, and breast tenderness. The second trimester has been described as more of a settling in period when sleep can improve. However, at this time snoring may start, some women experience heartburn, and leg cramps or restless legs syndrome (RLS) may begin. The third and final trimester is when sleep is most disrupted. Problems include difficulty getting comfortable (many women will sleep on their side with a pillow between their knees), heartburn, leg cramps, snoring, increased need to urinate, more time awake, and morning fatigue.

Polysomnographic studies confirm that women have more frequent awakenings and wake time starting from the first trimester and most evident in the third trimester (see [7, 15, 16]). However, it is in the first month following delivery that the greatest degree of maternal sleep disruption is found. There is a gradual increase in maternal sleep time over the next 2–4 months with maturing of the infant's circadian rhythm, but studies indicate that sleep efficiency continues to be lower than preprgenancy. The decline in sleep efficiency is greater in first-time mothers compared to mulitparous women. Breast-feeding compared with bottle-feeding has been found to influence sleep, with increased slow-wave sleep, possibly due to high prolactin levels [17].

Snoring and Obstructive Sleep Apnea (OSA)

While anatomical changes during pregnancy such as weight gain, decreased respiratory functional reserve capacity, and rhinitis (due to estrogen) predispose women to developing SDB, physiological changes—importantly increased respiratory drive (due to progesterone) and a preference for sleeping on their side—may offer protection [16]. Some women begin to snore during pregnancy [18]. Snoring, with complaints of sleep disruption and/or excessive daytime sleepiness, should be treated very seriously due to a higher risk for developing preeclampsia (high blood pressure, swelling especially in the ankles, protein in the urine, headaches) and sleep apnea. OSA may start or worsen during pregnancy, which is of concern as the periods when breathing stops lead to disrupted sleep and decreased blood oxygen levels that can also adversely affect the fetus [16, 18].

Restless Legs Syndrome (RLS) and Periodic Limb Movement in Sleep (PLMS)

Leg cramps can develop during pregnancy with about 15% of women reporting symptoms of RLS in the first trimester to 23% in the third trimester [16]. RLS is an irresistible urge to move the legs, especially in the evening and with rest. The feeling of leg discomfort is reduced by movement (e.g., getting up and walking), so that people experience difficulty getting to sleep. If the movements and twitches continue through the night with PLMS, sleep becomes fragmented and this in turn leads to daytime sleepiness. These symptoms generally go away with childbirth. Iron and folate deficiencies are known causes of RLS. Women who develop RLS during pregnancy should have their iron status checked and should probably be prescribed a multivitamin preparation containing folic acid [16].

MENOPAUSE

Between the ages of 45 and 55 years (average 51 years), a woman's production of estrogen and progesterone starts to decrease and the menstrual cycles become irregular (Chapter 85). This transitional or perimenopausal period occurs over a few years (about 4–8 years) and is when up to 80% of women experience hot flashes—suddenly feeling hot then flushed enough to sweat [7, 19]. Hot flashes can be extremely uncomfortable: they can occur during sleep—night sweats that can soak bedclothes followed by chills as the body cools down—and lead to sleep disruption. Only when menstrual periods have stopped for a year is menopause confirmed.

Hot flashes, insomnia, and other symptoms can start in the perimenopause but may continue into postmenopause (Chapter 86). Polysomnographic studies have not consistently found an association of decreased sleep efficiency with hot flashes, and the efficiency of estrogen therapy in relieving hot flashes at night has been variable [7]. Other symptoms that can be disruptive to sleep, either directly or indirectly, include mood changes (mood swings, anxiety, irritability, depression), vaginal dryness and irritation, urinary problems (more bathroom trips at night), and weight gain [19].

Insomnia

Complaints of insomnia are higher in menopausal than in premenopausal women (see [7]). Insomnia complaints include difficulty falling asleep, repeated awakenings, and waking too early in the morning and have been associated with hot flashes, palpitations, and mood swings particularly during perimenopause. Aside from hot flashes, depression and anxiety as well as SDB may be significant sleep-disrupting factors. Subjective improvements in sleep with hormone replacement therapy (HRT) have not been consistently reported in laboratory studies.

Sleep Disordered Breathing

Menopause increases the risk of breathing disorders during sleep [11]. The hormone-related changes, weight gain with a change in fat distribution, and increased age are all contributing factors. Older, overweight women with high blood pressure, sleep disturbances, or "fatigue" should be considered high risk for having OSA. Women who use HRT have OSA less frequently than postmenopausal women not on HRT [11]. Many other factors need to be considered before recommending HRT to treat apnea in women. The focus should be on using standard therapy such as continuous positive airway pressure (CPAP) or an oral appliance (for milder OSA) and weight loss.

CONCLUSION

A woman's changing hormone profile influences her sleep. In general, more disruption can be anticipated with abrupt changes and the withdrawal of female hormones. Five sleep disorders specifically associated with female reproductive stages have been proposed. Two are associated with the menstrual cycle (viz., premenstrual insomnia and premenstrual hypersomnia), two forms are pregnancy-associated and characterized by either excessive sleepiness or insomnia that develops over the course of pregnancy, and the fifth is menopausal insomnia [20]. Some sleep disorders, such as sleep disordered breathing and restless legs syndrome, may also be influenced by the reproductive stage.

REFERENCES

1. Sutton DA, Moldofsky H, Badley EM. Insomnia and health problems in Canadians. *Sleep* **24**:665–670(2001).

2. National Sleep Foundation (2002). Sleep in America. http://www.sleepfoundation.org/img/2002SleepInAmericaPoll.pdf.

3. Driver HS, Baker FC. Menstrual factors in sleep. *Sleep Med Rev* **2**:213–229(1998).

4. Baker FC, Waner JI, Vieira EF, Taylor SR, Driver HS, Mitchell D. Sleep and 24-hour body temperatures: a comparison in young men, naturally-cycling women, and in women taking hormonal contraceptives. *J Physiol* **530**:565–574(2001).

5. Baker FC, Driver HS. Self-reported sleep across the menstrual cycle in young, healthy women. *J Psychosom Res* **56**:39–243(2004).

6. Driver HS, Dijk D-J, Werth E, Biedermann K, Borbély A. Menstrual cycle effects on sleep EEG in young healthy women. *J Clin Endocrinol Metab* **81**:728–735(1996).

7. Moline ML, Broch L, Zak R, Gross V. Sleep in women across the life cycle from adulthood through menopause. *Sleep Med Rev* **7**:155–178(2003).

8. National Sleep Foundation (1998). Women and sleep poll. http://www.sleepfoundation.org/publications/1998women poll.html

9. Parry BL, Mendelson WB, Duncan WC, Sack DA, Wehr TA. Longitudinal sleep EEG, temperature and activity measurements across the menstrual cycle in patients with premenstrual depression and in age-matched controls. *Psychiatry Res* **30**:285–303(1989).

10. Baker FC, Driver HS, Rogers G, Paiker J, Mitchell D. High nocturnal body temperatures and disturbed sleep in women with primary dysmenorrhea. *Am J Physiol* **277**:E1013–E1021(1999).

11. Bixler EO, Vgontzas AN, Lin H-O, Ten Have T, Rein J, Vela-Bueno A, Kales A. Prevalence of sleep-disordered breathing in women. Effects of gender. *Am J Respir Crit Care Med* **163**:608–613(2001).

12. Ware JC, McBrayer RH, Scott JA. Influence of sex and age on duration and frequency of sleep apnea events. *Sleep* **23**:165–169(2000).

13. Edwards N, Wilcox I, Sullivan C. Haemodynamic responses to obstructive sleep apnoeas in premenopausal women. *J Hypertens* **17**:603–610(1999).

14. Driver HS, McLean H, Kumar DV, Farr N, Day A, Fitzpatrick MF. Influence of the menstrual cycle on upper airway resistance and breathing during sleep. *Sleep* **28**:449–456(2005).

15. Lee KA. Alterations in sleep during pregnancy and postpartum: a review of 30 years of research. *Sleep Med Rev* **2**:231–242(1998).

16. Pien GW, Schwab RJ. Sleep disorders during pregnancy. *Sleep* **27**:1405–1417(2004).

17. Blyton DM, Sullivan CE, Edwards N. Lactation is associated with an increase in slow-wave sleep in women. *J Sleep Res* **11**:297–303(2002).

18. Franklin KA, Holmgren PA, Jonsson F, et al. Snoring, pregnancy-induced hypertension, and growth retardation of the fetus. *Chest* **117**:137–141(2000).

19. Krystal AD, Edinger J, Wohlgemuth W, Marsh GR. Sleep in peri-menopausal and post-menopausal women. *Sleep Med Rev* **2**:243–254(1998).

20. *International Classification of Sleep Disorders—Revised, Diagnostic and Coding Manual.* American Sleep Disorders Association, Rochester, MN, 1997.

ADDITIONAL READING

Three "self-help" books for the general public describe changes through the female reproductive cycle and discuss sleep problems as well as treatment strategies; they are listed here in alphabetical order by author.

Kryger M. *Can't Sleep Can't Stay Awake. A Woman's Guide to Sleep Disorders.* HarperCollins Publishers, New York, 2004.

Walsleben JA, Baron-Faust R. *A Woman's Guide to Sleep.* Three Rivers Press, New York, 2000.

Wolfson AR. *The Woman's Book of Sleep. A Complete Resource Guide.* New Harbinger Publications, Oakland, CA, 2001.

82

SLEEP DURING PREGNANCY AND POSTPARTUM

KATHRYN A. LEE

University of California, San Francisco, California

SLEEP COMPLAINTS DURING PREGNANCY

Problems with sleep are experienced as early as the tenth week of pregnancy. Although most women expect to feel nauseated, fatigue is often their first symptom of pregnancy and 10–15% report disturbed sleep in the first trimester due to nausea, vomiting, backaches, or urinary frequency. In the second trimester, fetal movements and heartburn begin to disrupt sleep, and the majority (66%) suffer from urinary frequency, backaches, shortness of breath, leg cramps, itchy skin, and frightening dreams or nightmares that worsen in the third trimester [1–4]. During the third trimester, women report an average of 2.6 awakenings during the night and about 7.5 hours of sleep, but some report sleeping as little as 3–4 hours [5]. With more objective measures, such as wrist actigraphy or polysomnography, women sleep about 7 hours and are awake about 12% of the night during the third trimester [6, 7]. Common reasons for awakenings were compiled from a review of the literature and are presented in Table 82.1 by frequency of occurrence in the first and third trimesters. The anatomical and physiological reasons for these complaints are discussed in the following section.

ANATOMICAL AND PHYSIOLOGICAL CHANGES DURING PREGNANCY

Beginning early in pregnancy, there is a dramatic rise in the level of estrogen, progesterone, and prolactin hormones secreted from the placenta. Progesterone, the hormone essential for maintaining the pregnancy, also has inhibitory effects on the nervous system, enhances the effects of anesthetics and analgesics [8], and exerts a calming effect in which women feel drowsy, sedated, or "mellow."

Progesterone's inhibitory effects on smooth muscle cause women to experience urinary frequency early in the first trimester, long before the growing fetus begins to press on the bladder and reduce bladder capacity. By the third trimester, the discomfort and resultant sleep disturbance are most frequently attributed to the large gravid uterus, which also makes it difficult to turn or change positions without an increase in the length of arousals during sleep. To make room for the enlarging uterus, the diaphragm moves up and breathing becomes more shallow, minute ventilation increases, and women feel short of breath. The intestines and esophageal sphincter are displaced, causing esophageal reflux and complaints of heartburn, particularly in a flat, supine position [9].

Fluid volume increases by as much as 7 liters during pregnancy. While increased estrogen is likely responsible for vascular changes, the net result is nasal congestion and swelling in the extremities [9]. Changes in connective tissue and joints can be uncomfortable, and sudden leg cramps are common. Of a more serious nature, however, is an increased risk for obstructed airways and sleep disordered breathing because of the congested upper airway or additional upper body weight. The following section reviews sleep architecture as well as the more common sleep disorders associated with pregnancy.

Sleep: A Comprehensive Handbook, Edited by T. Lee-Chiong.

TABLE 82.1 Reasons for Nighttime Awakenings During Pregnancy

Reason	First Trimester (2–14 wk gestation)	Third Trimester (28–42 wk gestation)
Urination	51%	47%
Leg cramps	13%	73%
Joint pain	4%	23%
Bedpartner	9%	12%
Dreams/nightmares	3%	5%
Anxiety	3%	1%
Fetal movement	0%	5%

Changes in Sleep Stages During Pregnancy

Table 82.2 contains sleep-related factors that change during pregnancy and key issues to consider in clinical evaluation of potential sleep disorders. During the last month of pregnancy, women experience considerably more awake time and less slow-wave sleep (SWS) compared to controls and compared to their own prepregnancy baseline [6, 10–12]. Changes in REM sleep are less clear. When studied in their own home environment, sleep efficiency remains fairly stable at about 90% throughout pregnancy, REM sleep does not change, and there is a slight decrease in SWS from 17% in the first trimester to 14% in the third trimester [13]. The reduction in SWS during pregnancy is greater when examined from the baseline sleep pattern before conception [6].

Sleep Disordered Breathing

During pregnancy, shortness of breath is common, particularly when functional residual capacity is reduced by about 20% during the last trimester, before the fetus descends into the pelvis in preparation for birth. Gas exchange is even more compromised while supine, and women often find it more comfortable to sleep in a sitting position. Despite potential compromised pulmonary function, oxygen saturation remains stable during sleep in nonobese women [13, 14]. This is likely due to the increased ventilatory drive effects of progesterone and overall increase in minute ventilation [9]. Since these changes in lung function during normal pregnancy result in a resting state of respiratory alkalosis, there is a greater risk of hypoxemia and compromised oxygen delivery to the fetus during even very brief

TABLE 82.2 Typical Sleep Changes During Pregnancy and Postpartum

	First Trimester	Second Trimester	Third Trimester	Postpartum
Sleep changes	Fragmented sleep due to frequent urination	Sleep less fragmented. First onset of snoring	Sleep fragmented by frequent urination, leg cramps, heartburn, nasal congestion, irregular contractions, shortness of breath, breast tenderness, carpal tunnel/joint pain	Sleep fragmented by infant feeding. More fragmented sleep for first-time mothers than experienced mothers. Longer wake episodes for breast-feeding mothers, but they may have more deep sleep than nonlactating mothers. More opportunity to nap if not employed outside the home
	Less deep sleep (Stages 3–4 SWS) compared to preconception	Less deep sleep	Less deep sleep	More deep sleep
Daytime symptoms	Fatigue/sleepiness. Morning or evening nausea	More energy. Nasal congestion	Increased fatigue/sleepiness	Fatigue/sleepiness
Clinical evaluation	Check serum iron and folate levels for risk of restless legs syndrome	Assess for sleep disordered breathing	Assess for restless legs syndrome and sleep disordered breathing. Ask about plans for infant sleeping arrangements	Assess for excessive daytime sleepiness, cognitive dysfunction, and postpartum depression. Ask about infant's sleep and parenting activities during the night

apneic events if uterine blood flow is restricted by placental insufficiency or by positioning on the back and allowing the gravid uterus to compress the vena cava (supine hypotensive syndrome) [9].

In addition to increased minute ventilation and feeling short of breath, 10–30% report a new onset of snoring during pregnancy, compared to less than 5% who snored before conception [15, 16]. Snoring is at one end of the continuum of sleep disordered breathing, while frequent apneic events are at the severe end of the continuum. Sleep disordered breathing in pregnancy may be associated with a higher rate of fetal and newborn complications such as intrauterine growth retardation and low birth weight [17, 18].

Pregnancy-Induced Hypertension

Pregnancy-induced hypertension (PIH), also known as preeclampsia, is a disorder associated with endothelial cell dysfunction [19]. PIH occurs in about 5–10% of pregnancies and is characterized by high blood pressure with a flattened circadian rhythm [20]. PIH is associated with increased incidence of periodic leg movements as well as significantly narrowed upper airways and limited airflow during sleep, most likely due to pharyngeal edema. Women with PIH have a significantly larger neck circumference compared to healthy pregnant women, and 75% of those

with PIH reported snoring, compared with less than 30% of healthy pregnant women [21]. Since PIH is associated with restricted airflow and hypertension, frequent arousals from sleep are noted [22] and, as a result, nasal continuous positive airway pressure (CPAP) may improve sleep, reduce daytime sleepiness, and lower blood pressure. Although CPAP may not change the time spent in SWS or REM sleep, it may reduce nocturnal hypoxic episodes, improve placental perfusion, and extend gestation closer to full term. Unlike antihypertensive medications, CPAP does not appear to adversely affect the mother or fetus [20, 23].

Periodic Leg Movements and Restless Legs

The most frequent type of leg complaint from pregnant women during the night is sudden awakenings due to severe muscle cramps. Awakenings due to sudden leg cramps were reported prospectively by about 10% of women before and after pregnancy, but increased to 13% in their first trimester, 49% in their second trimester, and 73% in their third trimester [24]. Whether leg cramps are associated with periodic leg movements during sleep is not known, but women with twins or triplets are more likely to have periodic leg movements during sleep than women with a singleton pregnancy [3, 16]. The sensation of jerking arms or legs at sleep onset

TABLE 82.3 Medication Risk During Pregnancy and Lactation[a]

Medication	Low Risk (Class B)	Some Risk (Class C)	High Risk (Class D)	Contraindicated (Class X)
Antidepressants	Sertraline	Fluoxetine Paroxetine Trazadone Venlafaxine	Amitriptyline Imipramine	
Dopaminergics		Carbamazepine Carbidopa Levodopa		
Hypnotics	Diphenhydramine Zolpidem	Clonazepam Zaleplon	Alcohol (ethanol) Alprazolam Diazepam Lorazepam Midazolam Secobarbital	Flurazepam Temazepam Triazolam
Opioids	Meperidine Oxymorphone	Codeine Morphine		
Stimulants	Caffeine Pemoline	Dextroamphetamine Methamphetamine Mazindol Modafinil		

[a]Dosages should begin with half of the lowest recommended dosage due to increased sensitivity during pregnancy; lactating women should consider alternatives to breast-feeding since long-term effects on the newborn's growth and neurodevelopment remain unknown.

Class A (fetal harm is remote).

Class B (no risk in animal studies but no human studies, or risk in animals documented but no risk to fetus in controlled human studies).

Class C (animal studies document teratogenic effects but no studies in women or animals).

Class D (risk to fetus is present but use in pregnancy may outweigh risk of serious danger or disease to mother).

Class X (studies demonstrate high risk to fetus; drug is contraindicated regardless of risk to mother).

is consistently reported to be between 27% and 36% across preconception, pregnancy, and postpartum [24].

Restless legs syndrome (RLS) is associated with iron deficiency anemia, a common condition of pregnancy. Low serum ferritin and folate levels at preconception, even though within what are considered normal limits, are thought to contribute to this disorder [24]. RLS is often treated with opioids or dopamine agonists that would be undesirable to use during pregnancy or lactation (see Table 82.3). The incidence of RLS reaches its peak of 19–23% by the third trimester and typically resolves with the birth of the infant. In addition to experiencing restless legs at bedtime, sleep onset is significantly delayed and mood is more depressed [24]. Women who experience RLS in one pregnancy are more likely to experience it with subsequent pregnancies, but it can be totally absent between pregnancies [16, 25].

LABOR AND DELIVERY

It is well documented that sleep quality diminishes as pregnancy moves closer to term (40–42 weeks gestation). Women who sleep less than 6 hours at night during the few weeks prior to delivery have, on average, about 12 hours longer labor and are 4.5 times more likely to have a cesarean delivery than women who sleep more than 7 hours [26]. In the early stages of labor, when contractions do not seem to be regular and progressing, morphine sulfate has been commonly administered to induce sleep and reduce uterine contractions. The laboring woman then typically awakens in more active labor [27]. Sleep loss as a result of being in labor during the night can result in a higher risk of emotional distress during the early postpartum period [28].

During active labor, prolactin levels, which were high throughout pregnancy, begin to fall and reach a nadir about 2 hours before delivery. During a normal vaginal delivery, prolactin levels spike for 4–6 hours and then return to a normal circadian pattern [29]. Women who have emergency cesarean deliveries do not have the prolactin increase that normally occurs about 30 minutes after the onset of breast-feeding during the first few days postpartum [29]. How these labor and delivery factors affect mother and infant sleep remains unknown.

POSTPARTUM SLEEP PATTERNS

Postpartum is usually defined as the first 6 months after delivery. It begins with the birth of the infant and placenta and typically continues throughout lactation and until the infant is sleeping through the night. When new mothers have their infants in the same room on their first postpartum

night after vaginal delivery, they have lower sleep efficiency (74 ± 16.6%) and shorter REM latency (70.8 ± 23.2 minutes) compared to healthy controls, but no differences in amount of SWS or REM sleep [30]. Sleep disruption continues from the day of delivery through the first 3 months postpartum and is particularly problematic for primiparas (74.4%) and for women after cesarean delivery (73%) compared to vaginal delivery (57%) [31].

During the first few months of postpartum recovery, novice mothers (primiparas) have substantially more interrupted sleep than experienced (multipara) mothers [6]. Whereas sleep efficiency averages about 90% during the third trimester, it drops to about 77% during the first postpartum month in novice mothers but only drops to about 84% in experienced mothers. Sleep efficiency remains lower for primiparas compared to multiparas, even at 3 months postpartum [6]. However, in Japanese primiparas, who often share a futon with their infant, sleep efficiency is more similar to Caucasian multiparas [32]. Regardless of parity, SWS is increased and light sleep (stages 1 and 2) is decreased [6]. SWS returns to preconception levels by 1–3 months postpartum [6, 10, 12, 32].

Lactation and Breast-Feeding

There is little research on differences in sleep between women who breast-feed and formula feed during the postpartum period. For lactating women, basal levels of prolactin are high and there is a burst of prolactin secretion at the onset of each breast-feeding event, regardless of when sleep occurs, but the bursts are of a higher magnitude in the evening compared to morning. Within 24 hours of weaning, prolactin levels return to low basal levels and to the circadian sleep-associated patterns found in healthy adults. Both basal levels as well as bursts diminish to preconception levels by about 3 months postpartum [33, 34].

Lactating women have more SWS, less light sleep (stages 1 and 2), and fewer arousals compared to nonlactating postpartum women, but no difference, however, in the total sleep time or amount of REM sleep [35]. Following birth, Petre-Quadens and DeLee [36] did note a gradual decrease in REM sleep for formula-feeding compared to breast-feeding mothers in their small sample. Type of infant feeding has not been associated with mothers' perceptions of their sleep quality [37, 38].

Cosleeping During the Postpartum Period

In western cultures such as the United States, adults are socialized to value independence and, beginning at birth, find it more desirable to have a baby sleep alone in a separate room [39]. In most cultures, however, various forms of cosleeping, or sleeping together, exist and range from mother and baby sharing the same bed (bed-sharing) to mother and baby sleeping in the same room (room-sharing). Proponents of bed-sharing claim it facilitates breast-feeding and maternal–infant bonding, while opponents raise concerns about infant safety and risk of smothering from heavy blankets or obese parents [40]. McKenna and colleagues [41] studied three women between 2 and 4 months postpartum and found more arousals when mothers were bed-sharing with their infant, but no reduction in sleep efficiency when compared to sleeping alone on previous nights.

Population-based studies in the United States indicate that bed-sharing has increased from 6% to 13% during the past 10 years [42]. It does not appear to be associated with breast-feeding or crowded households [43]. While bed-sharing is thought to be more common for young single mothers with lower incomes and less education, recent research with affluent new parents found that over 40% were bed-sharing during the first month after delivery regardless of ethnicity or education, and the practice of bed-sharing decreased to about 20% by 3 months postpartum [44]. Only 7% of these couples indicated that they planned to bed-share after the baby was born and, therefore, it is unlikely that many new parents are counseled about health risks of bed-sharing.

Postpartum Depression

Women have a 10–20% incidence of a major depressive episode at some point in life. It is about 10–20% during pregnancy and postpartum as well, but pharmacological treatment during the childbearing years is of concern due to potential growth and development risks to the fetus and newborn. In an Italian sample of primiparous women, over 30% were still complaining of sleep loss and fatigue at 15 months postpartum, and more than 50% were reporting depressive symptoms [45]. It may be difficult for new mothers, family members, or clinicians to distinguish signs and symptoms of sleep deprivation from signs and symptoms of depression.

One hallmark of depression in the general population is an earlier onset of the first REM period after falling asleep and more REM sleep during the night. Compared to healthy new mothers, postpartum women with depressive symptoms have significantly shorter REM latency as well, but they also average 60 minutes less total sleep and about 12% lower sleep efficiency [46]. Due to the substantial sleep loss associated with caring for a new infant, most mothers easily fall asleep once they turn out the light. Particularly important hallmarks of postpartum depression include a new mother's complaint of not being able to fall asleep easily at bedtime (initiation insomnia) [47] as well as a low serum prolactin level [48].

Nonpharmacological treatments for women with postpartum depression can include late partial sleep deprivation, in which women sleep only until about 2:00 a.m. [49] to limit their amount of REM sleep, or bright light therapy, which may take up to 4 weeks to be effective [50]. Pharmacological treatment options for women with insomnia or postpartum depression are presented in Table 82.3 along with their associated risk to the fetus or newborn [51].

SUMMARY

Sleep patterns are disturbed as early as the tenth week of pregnancy, when complaints of urinary frequency and fatigue are first noticed. Compared to prepregnancy and pregnancy data, or compared to healthy controls, sleep efficiency is lowest during the first postpartum month, particularly for first-time mothers compared to experienced mothers, but significant sleep loss is evident for all new parents. The major concern for postpartum women is sleep loss and resulting physical fatigue, negative mood state, and cognitive impairment. How this sleep loss affects women's health, relationships with family, or the health of new fathers has not been a focus of research or clinical practice.

REFERENCES

1. Schweiger MS. Sleep disturbance in pregnancy. *Am J Obstet Gynecol* **114**:879–882(1972).
2. Wolfson AR, Crowley SJ, Anwer U, Bassett JL. Changes in sleep patterns and depressive symptoms in first-time mothers: Last trimester to one-year postpartum. *Behav Sleep Med* **1**:54–67(2003).
3. Pien GW, Schwab RJ. Sleep disorders during pregnancy. *Sleep* **27**:1405–1417(2004).
4. Baratte-Beebe KR, Lee K. Sources of mid-sleep awakenings in childbearing women. *Clin Nursing Res* **8**:386–397(1999).
5. Greenwood KM, Hazendonk KM. Self-reported sleep during the third trimester of pregnancy. *Behav Sleep Med* **2**:191–204(2004).
6. Lee KA, Zaffke ME, McEnany G. Parity and sleep patterns during and after pregnancy. *Obstet Gynecol* **95**:14–18(2000).
7. Gay CL, Lee KA, Lee S. Sleep patterns and fatigue in new mothers and fathers. *Biol Res Nursing* **5**:311–318(2004).
8. Herrmann WM, Beach RC. Experimental and clinical data indicating the psychotropic properties of progestogens. *Postgrad Med J* **54**:82–87(1978).
9. Sahota PK, Jain SS, Dhand R. Sleep disorders in pregnancy. *Curr Opin Pulm Med* **9**:477–483(2003).
10. Karacan I, Heine W, Agnew HW, Williams RL, Webb WB, Ross JJ. Characteristics of sleep patterns during late pregnancy and postpartum periods. *Am J Obstet Gynecol* **101**:579–586(1968).

11. Driver HS, Shapiro CM. A longitudinal study of sleep stages in young women during pregnancy and postpartum. *Sleep* **15**:449–453(1992).

12. Coble PA, Reynolds CF, Kupfer DJ, Houck PR, Day NL, Giles DE. *Comp Psychiatry* **35**:215–224(1994).

13. Edwards N, Middleton PG, Blyton DM, Sullivan CE. Sleep disordered breathing and pregnancy. *Thorax* **57**:555–558(2002).

14. Maasilta P, Bachour A, Teramo K, Polo O, Laitinen LA. Sleep-related disordered breathing during pregnancy in obese women. *Chest* **120**:1448–1454(2000).

15. Loube DI, Poceta JS, Morales MC, Peacock MD, Mitler MM. Self-reported snoring in pregnancy: association with fetal outcome. *Chest* **109**:885–889(1996).

16. Hedman C, Pohjasvaara T, Tolonen U, Suhonen-Malm AS, Myllyla VV. Effects of pregnancy on mothers' sleep. *Sleep Med* **3**:37–42(2002).

17. Schutte S, Del Conte A, Doghramji K, Gallagher K, Gallagher E, Oliver R, Rose C, Greuninger W, De Los Santos L, Youakim J. Snoring during pregnancy and its impact on fetal outcome. *Sleep Res* **23**:325(1994).

18. Franklin KA, Holmgren PA, Jönsson F, Poromaa N, Stenlund H, Svanborg E. Snoring, pregnancy-induced hypertension, and growth retardation of the fetus. *Chest* **117**:137–141(2000).

19. Taylor RN, Roberts JM. Endothelial cell dysfunction. In: Linheimer MD, Roberts JM (Eds), *Chesley's Hypertension Disorders in Pregnancy*. Appleton & Lange, Stamford, CT, 1999, pp 395–429.

20. Edwards N, Blyton DM, Kirjavainen T, Kesby GJ, Sullivan CE. Nasal continuous positive airway pressure reduces sleep-induced blood pressure increments in preeclampsia. *Am J Respir Crit Care Med* **162**:252–257(2000).

21. Izci B, Riha RL, Martin SE, Vennelle M, Liston WA, Dundas KC, Calder AA, Douglas NJ. The upper airway in pregnancy and pre-eclampsia. *Am J Respir Crit Care Med* **167**:137–140(2002).

22. Ekholm EM, Polo O, Rauhala ER, Ehblad UU. Sleep quality in preeclampsia. *Am J Obstet Gynecol* **167**:1262–1266(1992).

23. Guilleminault C, Kreutzer M, Chang JL. Pregnancy, sleep disordered breathing and treatment with nasal continuous positive airway pressure. *Sleep Med* **5**:43–51(2004).

24. Lee, KA, Zaffke ME, Barette-Beebe K. Restless legs syndrome and sleep disturbance during pregnancy: the role of folate and iron. *J Women's Health Gender-Based Med* **10**:335–341(2001).

25. Suzuki K, Ohida T, Sone T, Takemura S, Yokoyama E, Miyake T, Hara S, Motojima S, Suga M, Ibuka E. The prevalence of restless legs syndrome among pregnant women in Japan and the relationship between restless legs syndrome and sleep problems. *Sleep* **26**:673–677(2003).

26. Lee KA, Gay CL. Sleep in late pregnancy predicts length of labor and type of delivery. *Am J Obstet Gynecol* **191**:2041–2046(2004).

27. Conklin KA. Obstetric analgesia and anesthesia. In: Hacker NF, Moore JG (Eds), *Essentials of Obstetrics and Gynecology*. Saunders, Philadelphia, 1998, pp 169–170.

28. Wilkie G, Shapiro CM. Sleep deprivation and the postnatal blues. *J Psychosom Res* **36**:309–316(1992).

29. Heasman L, Spencer JAD, Symonds ME. Plasma prolactin concentrations after caesarean section or vaginal delivery. *Arch Dis Child Fetal Neonatal Ed* **77**:F237–F238(1997).

30. Zaffke ME, Lee KA. Sleep architecture in a postpartum sample: a comparative analysis. *Sleep Res* **21**:327(1992).

31. Tribotti S, Lyons N, Blackburn S, Stein M, Withers J. Nursing diagnoses for the postpartum woman. *J Obstet Gynecol Neonatal Nurs* **17**:410–417(1988).

32. Nishihara K, Horiuchi S, Eto, H, Uchida S. Comparisons of sleep patterns between mothers in post-partum from 9 to 12 weeks and non-pregnant women. *Psychiatry Clin Neurosci* **55**:227–228(2001).

33. Noel GL, Suh HK, Frantz AG. Prolactin release during nursing and breast stimulation in postpartum and nonpostpartum patients. *J Clin Endocrinol Metab* **38**:413–423(1974).

34. Nissen E, Uvnas-Moberg K, Svensson K, Stock S, Widstrom A-M, Winberg J. Different patterns of oxytocin, prolactin but not cortisol release during breastfeeding in women delivered by caesarean section or by the vaginal route. *Early Hum Dev* **45**:103–118(1996).

35. Blyton DM, Sullivan CE, Edwards N. Lactation is associated with an increase in slow-wave sleep in women. *J Sleep Res* **11**:297–303(2002).

36. Petre-Quadens I, DeLee C. Sleep-cycle alterations during pregnancy, postpartum and the menstrual cycle. In: Ferin M, Halberg F, Richart RM, Van Wiele RL (Eds), *Biorhythms and Human Reproduction*. John Wiley & Sons, Hoboken, NJ, 1974, pp 335–351.

37. Wambach KA. Maternal fatigue in breastfeeding primiparae during the first 9 weeks postpartum. *J Hum Lactation* **14**:219–229(1998).

38. Quillan SI. Infant and mother sleep patterns during 4th postpartum week. *Iss Compr Pediatr Nurs* **20**:115–123(1997).

39. Morelli GA, Rogoff B, Oppenhein D, Goldsmith D. Cultural variation in infants' sleeping arrangements: questions of independence. *Dev Psychol* **28**:604–613(1992).

40. Nakamura SW. Are cribs the safest place for infants to sleep? Yes: Bed sharing is too hazardous. *West J Med* **174**:300(2001).

41. McKenna JJ, Thoman EB, Anders TF, Sadeh A, Schechtman VL, Glotzbach SF. Infant–parent co-sleeping in an evolutionary perspective: implications for understanding infant sleep development and the sudden infant death syndrome. *Sleep* **16**:263–282(1993).

42. Willinger M, Ko C-W, Hoffman HJ, Kessler RC, Corwin MJ. Trends in infant bed sharing in the United States, 1993–2000. *Arch Pediatr Adolesc Med* **157**:43–49(2003).

43. Brenner RA, Simons-Morton BG, Bhaskar B, Revenis M, Das A, Clemens JD. Infant–parent bed sharing in an inner-city population. *Arch Pediatr Adolesc Med* **157**:33–39(2003).

44. Gay CL, Ward TM, Lee KA. Parent–newborn co-sleeping in the San Francisco Bay area. *Sleep* **27**(Abstr Suppl):A356 (2004).

45. Romito P, Saurel-Cubizolles MJ, Cuttini M. Mothers' health after the birth of the first child: the case of employed women in an Italian city. *Women Health* **21**:1–22(1994).

46. Lee KA, McEnany G, Zaffke ME. REM sleep and mood state in childbearing women: sleepy or weepy? *Sleep* **23**:877–885 (2000).

47. Kennedy HP, Beck CT, Driscoll JW. A light in the fog: caring for women with postpartum depression. *J Midwifery Women's Health* **47**:318–330(2002).

48. Abou-Saleh M, Ghubash R, Karim L, Krymski M, Bhai I. Hormonal aspects of postpartum depression. *Psychoneuroendocrinology* **23**:465–475(1998).

49. Parry BL, Curran ML, Stuenkel CA, Yokimozo M, Tam L, Powell KA, Gillin JC. Can critically timed sleep deprivation be useful in pregnancy and postpartum depressions? *J Affect Disord* **60**:201–212(2000).

50. Corral M, Kuan A, Kostaras D. Bright light therapy's effect on postpartum depression. *Am J Psychiatry* **157**:303–304(2000).

51. Briggs GG, Freeman RK, Yaffe SJ. *Drugs in Pregnancy and Lactation*. Lippincott/Williams & Wilkins, Philadelphia, 2002.

83

MENSTRUAL-RELATED SLEEP DISORDERS

GRACE W. PIEN AND ELIZABETH A. BEOTHY
Hospital of the University of Pennsylvania, Philadelphia, Pennsylvania

Over the course of the normal menstrual cycle, a complex series of interactions between the reproductive hormones of the hypothalamus, pituitary, and ovaries leads to the development and release of a mature oocyte for potential fertilization, the preparation of a receptive uterine lining, and its desquamation should fertilization fail to occur [1]. During ovulatory menstrual cycles, predictable changes in reproductive hormone levels and daily body temperature occur that may lead to alterations in sleep and circadian rhythm. Normal changes in the sleep cycle may be disrupted, however, among women with menstrual cycle disorders. Sleep disorders related to the menstrual cycle have also been described. This chapter will examine changes in sleep and the circadian rhythm during normal menstrual cycles and will review how oral contraceptives and disorders of the menstrual cycle such as dysmenorrhea and premenstrual syndrome affect sleep. Finally, we will describe the features of menstrual-related sleep disorders and review therapies for these conditions.

THE NORMAL MENSTRUAL CYCLE

By convention, the cycle begins on the first day of menses with the follicular phase, during which a small cohort of ovarian follicles begins to mature. In conjunction with rising estradiol and progesterone levels, a surge of luteinizing hormone occurs; the luteal phase begins at the peak of this surge and is followed by the release of a single mature oocyte [1]. Mean body temperature changes over the course of ovulatory menstrual cycles, with lower daily body temperatures during the follicular phase and higher tempera-

tures during the luteal phase [2]. Each phase lasts for approximately 14 days, with a mean normal cycle length of 28 days; women generally have regular menstrual cycles between menarche and the onset of menopause [1].

SLEEP AND CIRCADIAN RHYTHM DURING THE NORMAL MENSTRUAL CYCLE

A number of studies have examined self-reported sleep habits and sleep quality in normal women over the course of the menstrual cycle. Among healthy women, the duration of self-reported nocturnal sleep time has been noted to be shortest around the time of ovulation and longest prior to menses [3]. Although some investigators have found no difference in sleep quality at different phases of the menstrual cycle [4], others have described higher self-reported sleep efficiency and better sleep quality during the follicular phase than the luteal phase [5]. In animal models, progesterone has been noted to have a soporific effect [6] and in healthy women, subjective sleepiness has been observed to increase during the luteal phase, when progesterone levels are highest [7]. However, an increased self-reported latency to sleep onset during the luteal phase has also been described [5]. Despite self-reported changes in sleep over the menstrual cycle, differences between the follicular and luteal phases in electroencephalographically measured sleep propensity, total sleep time, and latency to sleep onset have not been observed [4, 7–10].

An early study of sleep architecture during the menstrual cycle suggested that the percentage of rapid eye movement (REM) sleep increases in the week prior to menses [11].

However, subsequent studies have either demonstrated more REM sleep during the follicular phase than the luteal phase [2, 4, 12], or no significant differences in the amount of REM sleep over the cycle [8, 9, 13]. Time spent in the NREM sleep stages has likewise been observed not to change significantly over the course of the cycle [2, 9] or to increase slightly during the luteal phase due to an increased percentage of stage 2 sleep [4]. Other work examining sleep architecture over the normal menstrual cycle has demonstrated that sleep spindle frequency and EEG power density in the upper frequency range of sleep spindles increase during the luteal phase [4, 10], and that during menses, the latency to the onset of slow-wave sleep increases [14].

The effect of the menstrual cycle on circadian rhythm has also been examined. In animal models, estrogen appears to advance circadian rhythm, while progesterone delays it [15–17]. However, menstrual cycle phase effects on circadian rhythm have not been identified in normal women when assessed with frequent salivary or serum melatonin sampling [18–20].

Studies of the effects of the menstrual cycle on sleep and circadian rhythm have been limited by relatively small sample sizes and have often focused on younger women. Nevertheless, they suggest that sleep architecture and circadian phase remain relatively stable over the normal menstrual cycle despite significant fluctuations in hormone levels and metabolism and the perception of changes in sleep by many women.

EFFECTS OF ORAL CONTRACEPTIVES ON SLEEP AND CIRCADIAN RHYTHM

Estrogen–progestin combination oral contraceptives and, less consistently, progestin-only contraceptives inhibit the midcycle luteinizing hormone surge, preventing ovulation [1]. Although the circadian rhythms of melatonin secretion and body temperature remain robustly entrained among contraceptive users when observed using frequent sampling techniques [18], oral contraceptive users demonstrate higher mean melatonin levels, higher mean body temperatures during sleep, and smaller amplitudes in circadian temperature rhythm than naturally cycling women [2, 18]. However, alertness and reaction times are not adversely affected [18].

Whether women taking oral contraceptives experience subjective changes in sleep quality has not been examined. However, the relationship between body temperature changes due to oral contraceptive use and sleep architecture has been described in several studies [2, 9, 22, 23]. Decreased variability in sleep architecture without changes in overall amounts of slow-wave and REM sleep [22], decreased slow-wave sleep [2, 23], and a shortened latency

to REM sleep onset [2, 9] have all been observed in oral contraceptive users compared to naturally ovulating women. Marked variation over time and between formulations in the estrogen and progestin dosages included in oral contraceptives [1] complicates these assessments. Given the small number of studies and subjects in this area, it is difficult to draw definitive conclusions about the impact of oral contraceptive use on sleep.

SLEEP AND CIRCADIAN RHYTHM IN WOMEN WITH MENSTRUAL CYCLE DISORDERS

Disorders of the menstrual cycle that may affect sleep include dysmenorrhea, premenstrual syndrome (PMS), and premenstrual dysphoric disorder (PMDD). Estimates of the prevalence of dysmenorrhea, which is characterized by painful cramping in the lower abdomen during menses, vary widely, from 3% to 90% of women [24]. Women with PMS experience physical and behavioral symptoms including fatigue, abdominal bloating, and irritability that occur only during the second half of the menstrual cycle and are severe enough to interfere with work or personal relationships [24]. Approximately 40% of women experience PMS symptoms at one time or another [24], but only 3–5% have premenstrual dysphoric disorder, the most severe form of PMS [25]. Among women with PMDD, severe mood symptoms are prominent and cause significant impairment in occupational or social function [25]. Problems with sleep, including either insomnia or hypersomnia, are among the 11 potential symptoms of PMDD described by the *Diagnostic and Statistical Manual of Mental Disorders*, 4th edition [26].

Among women with primary dysmenorrhea, subjective fatigue has been reported to be greater and sleep quality to be worse during menses than during pain-free times in the menstrual cycle and when compared to normal women [14]. Women who describe pain during menstruation have not, however, reported higher rates of insomnia than other women [27]. Using objective measures, sleep efficiency during menses is lower among women with dysmenorrhea than during other phases of the cycle and compared to normal women [14]. Dysmenorrheic women may also have less REM sleep throughout the menstrual cycle than normal women [14]. Although corroboratory work is needed, these findings suggest that women with dysmenorrhea experience significant sleep disruption.

Women with PMS describe more sleep-related complaints and the perception of worse sleep compared to normal women [11, 28, 29]. Furthermore, young women with premenstrual symptoms may be more likely than normal women to experience insomnia [27]. In addition to symptoms such as nocturnal awakenings, unpleasant dreams, and body movements, symptoms that affect daytime function during the week before menses are common and include heightened physical tension, morning tiredness, and daytime sleepiness [30].

Despite these complaints, consistent changes in sleep architecture among women with PMS and PMDD have not been observed. Sleep efficiency and sleep architecture remain stable over the menstrual cycle in women with PMS and do not differ significantly from sleep in normal women [28, 29]. Among women with PMDD, more stage 2 and less REM sleep have been demonstrated throughout the menstrual cycle compared to normal women [8], complementing similar observations in women with negative luteal phase mood symptoms [9, 31]. However, a subsequent study by the same investigators failed to replicate these findings [12]. As these studies have been limited by small sample sizes and variable diagnostic criteria for inclusion, additional work to elucidate the impact of PMS and PMDD on sleep is needed.

Among women with PMDD, decreased nocturnal melatonin concentrations have been observed compared to normal women [20, 32]. During the luteal phase, when PMDD symptoms occur, changes in melatonin secretion and metabolism include later nocturnal melatonin onset and decreased area under the curve, amplitude, and mean levels compared to the follicular phase [32]. Observations of analogous changes in melatonin among patients with affective illness have generated speculation about whether decreased melatonin levels may increase luteal phase vulnerability for depressive symptoms among women with PMDD [32].

SLEEP DISORDERS RELATED TO THE MENSTRUAL CYCLE

Menstrual-Associated Sleep Disorder

The second edition of the *International Classification of Sleep Disorders* lists menstrual related hypersomnia as a subtype of recurrent hypersomnia [1]. This condition is characterized by episodes of persistent, excessive sleepiness occurring in temporal association with the menstrual cycle [1] Menstrual related hypersomnia appears to be a rare disorder and this diagnosis should be made only in the absence of other menstrual cycle or sleep disorders, including PMS.

The episodes of hypersomnolence and prolonged sleep periods that characterize menstrual-related hypersomnia generally begin in the late luteal phase and end within a few days of the onset of menses, in the early follicular phase, with normal sleep–wake behavior between episodes [2, 3, 4]. Episodes of hypersomnolence are accompanied by behavioral or emotional changes such as irritability, apathy or anorexia [2, 3, 4]. Two case reports have involved adolescent girls with normal hormonal findings, including regular ovulatory cycles [3, 4]. Similarities between

menstrual related hypersomnia and Kleine–Levin syndrome, in which adolescent males exhibit periodic hypersomnia and increased appetite, have been noted [3]. However, menstrual related hypersomnia seems to be a distinct disorder given its cyclical variation with female menstrual phase.

As the literature on menstrual related hypersomnia is limited to case reports, it is difficult to assess the effectiveness of potential therapies. In light of the association between symptoms and menstrual phase, women with this disorder may benefit from suppression of ovulation using estrogen–progestin combination oral contraceptives. Fixed-dose formulations may offer a theoretical benefit over multiphasic oral contraceptives by minimizing fluctuations in hormone levels. Oral contraceptives have been successfully utilized for treatment of menstrual related hypersomnia [4]; one woman failed, however, to respond [2]. Menstrual related hypersomnia has also been treated with progestin-only therapy [3]. The response of menstrual related hypersomnia to stimulant medication has been variable [2, 3].

Due to the rarity of menstrual related hypersomnia and the dearth of systematic research, the etiology, risk factors, range of severity, long-term course, evolution with respect to menopause, and heritability of this condition all remain undefined. Whether a single or multiple pathophysiologic mechanisms underlie the clinical presentation is similarly unknown [2]. Descriptions of the presentation, management, and clinical course in additional cases or a larger case series are needed to shed light on this unusual disorders.

Menstrual-Related Insomnia

Recurrent sleep maintenance insomnia in the premenstrual period has been described [5]. Oral contraceptives were successfully used for treatment [5]. However, documentation of additional cases is needed for validation as a distinct entity. Among women with insomnia, significant menstrual phase effects on sleep have not been observed when measured using actigraphy [6].

Premenstrual Parasomnia

Two cases of recurrent premenstrual sleep terrors and sleepwalking have been reported, raising the possibility of a separate designation of premenstrual parasomnia [7]. As with menstrual related insomnia and hypersomnia, symptoms occur primarily during the luteal phase. Both cases responded to self-hypnosis; one woman also required treatment with clonazepam for improved symptom control [7]. Although these cases invite the question of whether menstrual phase affects the frequency of episodes of sleep terrors and somnambulism in the general female population, such an association has not been described.

CONCLUSION

Despite considerable hormonal fluctuations over the course of the menstrual cycle, sleep architecture and circadian rhythm appear to remain relatively stable in normal women. Although oral contraceptive use increases nocturnal melatonin levels, entrained circadian temperature and melatonin rhythms remain evident and consistent changes in sleep have not been observed. Women with dysmenorrhea may experience sleep disturbance, especially during menses. Complaints of poor sleep and fatigue are common among women with both PMS and PMDD; objective changes in sleep have been documented in some women with PMDD but require further corroboration. Menstrual-associated insomnia and hypersomnia are rare disorders of sleep with manifestations that occur primarily during the late luteal phase. Their etiology is unknown. A great deal remains to be learned about the relationship between the menstrual cycle and sleep, not only under normal conditions, but especially among women with menstrual-related disorders of sleep and menstrual cycle disorders. As knowledge about these conditions increases, these women may benefit from interventions targeted at reducing sleep disruption and changes in the sleep–wake pattern related to the menstrual cycle.

REFERENCES

1. Mishell J. Reproductive endocrinology. In: Stenchever MA, Droegemueller W, Herbst AL, Mishell J (Eds), *Comprehensive Gynecology*, 4th ed. Mosby, St Louis, MO, 2001, pp 71–124.
2. Baker FC, Waner JI, Vieira EF, Taylor SR, Driver HS, Mitchell D. Sleep and 24 hour body temperatures: a comparison in young men, naturally cycling women and women taking hormonal contraceptives. *J Physiol* **530**:565–574(2001).
3. Patkai P, Johannson G, Post B. Mood, alertness and sympathetic–adrenal medullary activity during the menstrual cycle. *Psychosom Med* **36**:503–512(1974).
4. Driver HS, Dijk DJ, Werth E, Biedermann K, Borbely AA. Sleep and the sleep electroencephalogram across the menstrual cycle in young healthy women. *J Clin Endocrinol Metab* **81**:728–735(1996).
5. Manber R, Bootzin RR. Sleep and the menstrual cycle. *Health Psychol* **16**:209–214(1997).
6. Manber R, Armitage R. Sex, steroids, and sleep: a review. *Sleep* **22**:540–555(1999).
7. Shibui K, Uchiyama M, Okawa M, Kudo Y, Kim K, Kamei Y, Hayakawa T, Akamatsu T, Ohta K, Ishibashi K. Diurnal fluctuation of sleep propensity across the menstrual cycle. *Psychiatry Clin Neurosci* **53**:207–209(1999).
8. Parry BL, Mendelson WB, Duncan WC, Sack DA, Wehr TA. Longitudinal sleep EEG, temperature, and activity measurements across the menstrual cycle in patients with premenstrual depression and in age-matched controls. *Psychiatry Res* **30**:285–303(1989).
9. Lee KA, Shaver JF, Giblin EC, Woods NF. Sleep patterns related to menstrual cycle phase and premenstrual affective symptoms. *Sleep* **13**:403–409(1990).
10. Ishizuka Y, Pollak C, Shirakawa S, Kakuma T, Azumi K, Usui A, Shiraishi K, Fukuzawa H, Kariya T. Sleep spindle frequency changes during the menstrual cycle. *J Sleep Res* **3**:26–29(1994).
11. Hartmann E. Dreaming sleep (the D-state) and the menstrual cycle. *J Nerv Ment Dis* **143**:406–416(1966).
12. Parry BL, Mostofi N, LeVeau B, Nahum HC, Golshan S, Laughlin GA, Gillin JC. Sleep EEG studies during early and late partial sleep deprivation in premenstrual dysphoric disorder and normal control subjects. *Psychiatry Res* **85**:127–143 (1999).
13. Cluydts R, Visser P. Mood and sleep. I. Effects of the menstrual cycle. *Waking Sleeping* **4**:193–197(1980).
14. Baker FC, Driver HS, Rogers GG, Paiker J, Mitchell D. High nocturnal body temperatures and disturbed sleep in women with primary dysmenorrhea. *Am J Physiol* **277**:E1013–E1021 (1999).
15. Albers HE, Gerall AA, Axelson JF. Effect of reproductive state on circadian periodicity in the rat. *Physiol Behav* **26**:21–25(1981).
16. Axelson JF, Gerall AA, Albers HE. Effect of progesterone on the estrous activity cycle of the rat. *Physiol Behav* **26**:631–635(1981).
17. Morin LP, Fitzgerald KM, Zucker I. Estradiol shortens the period of hamster circadian rhythms. *Science* **196**:305–307(1977).
18. Wright KP Jr, Badia P, Czeisler CA. Effects of menstrual cycle phase and oral contraceptives on alertness, cognitive performance, and circadian rhythms during sleep deprivation. *Behav Brain Res* **103**:185–194(1999).
19. Ito M, Kohsaka M, Fukuda N, Honma K, Honma S, Katsuno Y, Honma H, Kawai I, Morita N, Miyamoto T. Effects of menstrual cycle on plasma melatonin level and sleep characteristics. *Jpn J Psychiatry Neurol* **47**:478–479(1993).
20. Parry BL, Berga SL, Kripke DF, Klauber MR, Laughlin GA, Yen SS, Gillin JC. Altered waveform of plasma nocturnal melatonin secretion in premenstrual depression. *Arch Gen Psychiatry* **47**:1139–1146(1990).
21. Mishell J. Family planning. In: Stenchever MA, Droegemueller W, Herbst AL Mishell J (Eds), *Comprehensive Gynecology*, 4th ed. Mosby, St Louis, MO, 2001, pp 295–358.
22. Henderson A, Nemes G, Gordon NB, Roos L. The sleep of regularly menstruating women and of women taking an oral contraceptive. *Psychophysiology* **7**:337(1970).
23. Ho A. Sex hormones and the sleep of women. *Sleep Res* **1**:184(1972).

24. Stenchever MA. Primary and secondary dysmenorrhea and premenstrual syndrome. In: Stenchever MA, Droegemueller W, Herbst AL Mishell J (Eds), *Comprehensive Gynecology*, 4th ed. Mosby, St Louis, MO, 2001, pp 1065–1078.

25. Johnson SR. Premenstrual syndrome, premenstrual dysphoric disorder, and beyond: a clinical primer for practitioners. *Obstet Gynecol* **104**:845–859(2004).

26. APA. *Diagnostic and Statistical Manual of Mental Disorders*, 4th ed. American Psychiatric Association, Washington, DC, 1994.

27. Sheldrake P, Cormack M. Variations in menstrual cycle symptom reporting. *J Psychosom Res* **20**:169–177(1976).

28. Chuong CJ, Kim SR, Taskin O, Karacan I. Sleep pattern changes in menstrual cycles of women with premenstrual syndrome: a preliminary study. *Am J Obstet Gynecol* **177**:554–558(1997).

29. Mauri M. Sleep and the reproductive cycle: a review. *Health Care Women Int* **11**:409–421(1990).

30. Mauri M, Reid RL, MacLean AW. Sleep in the premenstrual phase: a self-report study of PMS patients and normal controls. *Acta Psychiatr Scand* **78**:82–86(1988).

31. Lee KA, McEnany G, Zaffke ME. REM sleep and mood state in childbearing women: sleepy or weepy? *Sleep* **23**:877–885(2000).

32. Parry BL, Berga SL, Mostofi N, Klauber MR, Resnick A. Plasma melatonin circadian rhythms during the menstrual cycle and after light therapy in premenstrual dysphoric disorder and normal control subjects. *J Biol Rhythms* **12**:47–64(1997).

33. AASM. *International Classification of Sleep Disorders, Revised: Diagnostic and Coding Manual*. American Academy of Sleep Medicine, Westchester, IL, 2000, pp 295–297.

34. Bootzin RR, Bamford CR. Premenstrual insomnia: a case study. *Sleep Res* **19**:196(1990).

35. Billiard M, Guilleminault C, Dement WC. A menstruation-linked periodic hypersomnia. Kleine–Levin syndrome or new clinical entity? *Neurology* **25**:436–443(1975).

36. Sachs C, Persson HE, Hagenfeldt K. Menstruation-related periodic hypersomnia: a case study with successful treatment. *Neurology* **32**:1376–1379(1982).

37. Bamford CR. Menstrual-associated sleep disorder: an unusual hypersomniac variant associated with both menstruation and amenorrhea with a possible link to prolactin and metoclopramide. *Sleep* **16**:484–486(1993).

38. Schenck CH, Mahowald MW. Two cases of premenstrual sleep terrors and injurious sleep-walking. *J Psychosom Obstet Gynecol* **16**:79–84(1995).

84

SLEEP DISORDERED BREATHING IN WOMEN

M. Safwan Badr

Wayne State University, Detroit, Michigan

Sleep apnea affects both men and women and should be recognized and treated in both genders. This chapter will address the gender differences in physiology, epidemiology, clinical presentation, and polysomnographic features of sleep disordered breathing in women versus men.

GENDER DIFFERENCES IN THE PREVALENCE OF SLEEP DISORDERED BREATHING

Obstructive sleep apnea (OSA) is more common in men than in women. However, clinic-based studies markedly overestimated the gender difference in the prevalence of sleep apnea [1, 2] relative to community-based epidemiologic studies [3–5]. The discrepancy between community prevalence and clinic prevalence indicates either a difference in clinical features or referral bias (see later discussion). Consequently, many women with sleep disordered breathing may remain undiagnosed for years following the initial evaluation [6].

Several large epidemiologic studies addressed community prevalence of sleep disordered breathing [3–5, 7–10]. The Wisconsin Sleep Cohort Study [3] investigated the occurrence of sleep disordered breathing in the general population. Participants were men and women ranging in age from 30 to 60, recruited from the community and tested using a full-night polysomnogram. The study revealed that 24% of men and 9% of women had sleep disordered breathing based on an apnea/hypopnea index of >5 events per hour of sleep. However, the prevalence of sleep apnea *syndrome* was 2% in women and 4% in men. Sleep apnea/hypopnea syndrome was defined as an apnea–

hypopnea index (AHI) $\geq$ 5 events/hour, combined with daytime sleepiness [3].

In another large epidemiologic study, Bixler et al. [4, 5] studied 1000 women and 741 men using full-night polysomnography. Prevalence of sleep apnea syndrome (defined as AHI $\geq$ 10 and daytime symptoms) in women was 1.2% versus 3.9% for men. The male–female ratio remained stable regardless of the definition and cutoff values used to identify the disorder. The prevalence of sleep apnea was quite low in premenopausal women relative to postmenopausal women. Similarly, central sleep apnea was rare in women relative to men. Thus gender and menopausal state both contributed to the prevalence of sleep apnea.

The consistent difference in prevalence between clinical and community-based studies indicates underdiagnosis in women [10–12]. Failure to pursue the diagnosis can occur if there are differences in clinical presentation. For example, Redline et al. [7] have suggested that women report snoring or other socially undesirable symptoms less than men do. Conversely, Young et al. [13] found that women with sleep apnea reported symptoms similar to those of men with the same level of sleep apnea, raising the possibility that health care providers may disregard typical symptoms in women. Furthermore, women in the working population enrolled in the Wisconsin Sleep Cohort Study reported *more* daytime sleepiness even in cases of mild or no sleep apnea. Moreover, in subjects with AHI $\geq$ 5 events/hour of sleep, 22.6% of the women and 15.5% of the men reported the frequent occurrence ($\geq$2 days per week) of excessive daytime sleepiness (Figure 84.1) [14]. Finally, some studies have shown that women report more fatigue [15] or more difficulty in initiating and maintaining sleep

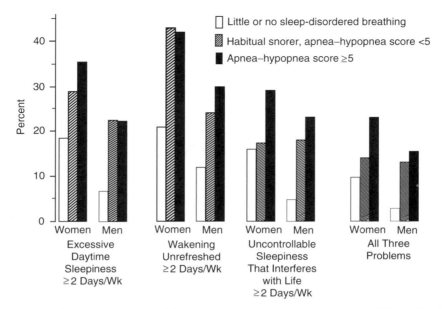

Figure 84.1 Proportion of men and women who reported daytime sleepiness, according to category of sleep disordered breathing. Note that women report more sleepiness for all levels of sleep disordered breathing.

[16]. The available studies taken together suggest that clinicians should be cognizant of the protean manifestations of sleep apnea in women and be prepared to pursue the diagnosis with sleep laboratory investigation, especially if sleepiness is present.

GENDER DIFFERENCES IN POLYSOMNOGRAPHY

Several studies have investigated whether polysomnographic features are different in men versus women [17, 18]. There is agreement that *AHI is lower in women relative to men when matched for body weight.* This is mostly due to higher AHI during NREM in men relative

to women (Figure 84.2). Interestingly, apneas and hypopneas tend to occur during REM sleep in women, which accounts for the minimal gender difference in AHI during REM. In contrast, men tend to demonstrate more supine dependence than women do [19]. There is no clear explanation for the difference in polysomnography between genders but it may reflect differences in the relative contribution of different mechanisms in both genders as well as differences between REM and NREM sleep.

Sleep disordered breathing may also manifest as cyclical occurrence of snoring, inspiratory flow limitation, and repetitive arousal without frank apneas or hypopneas. Several authors view this as a distinct entity, upper airway resistance syndrome (UARS), and a component of the

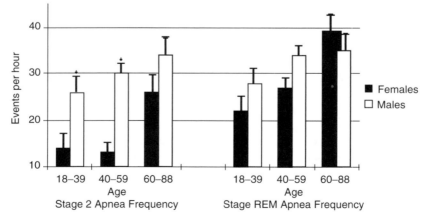

Figure 84.2 Apnea–hypopnea index (AHI) in men and women. Note higher AHI for stage 2 NREM sleep and the similar AHI during REM sleep. (From [18].)

continuum of sleep disordered breathing [20]. Interestingly, UARS is a common manifestation of sleep disordered breathing in women [6, 19, 21, 22]. It is plausible that differences in load responsiveness [23] or central breathing stability [24] mitigates the development of obstructive or central sleep apnea, respectively. Thus rhythmic breathing persists, albeit with flow limitation, high work of breathing, and recurrent arousals. Some authors suggest that sleep fragmentation associated with snoring and inspiratory flow limitation may contribute to the fatigue and generalized aches noted in fibromyalgia and may offer an etiologic link to functional somatic syndromes [21, 22].

MECHANISMS OF GENDER DIFFERENCES IN THE OCCURRENCE OF SLEEP APNEA

Upper Airway Structure and Function

The pharyngeal airway is a collapsible tube; thus patency of the pharyngeal airway depends on the length of the collapsible segment, the caliber of the pharyngeal lumen, the stiffness of the pharyngeal wall, and the pressure gradient across the pharyngeal wall. The sleep state is conducive to pharyngeal narrowing or closure by virtue of increased upper airway resistance and collapsibility.

A difference in upper airway dimensions between men and women may contribute to pharyngeal closure during sleep. The literature is inconsistent regarding the gender difference in pharyngeal cross-sectional area. Some studies [25] but not others [26, 27] have shown a larger cross-sectional area during wakefulness in men relative to women. When the volume of soft tissue surrounding the pharyngeal lumen was measured, total neck soft tissue volume is greater in men relative to women [28]; however, no difference in the volume of fat was found between the sexes [28]. Thus the lack of a consistent gender difference in cross-sectional area of the pharyngeal airway could not explain the gender difference in the prevalence of sleep disordered breathing.

Airway length is another important determinant of pharyngeal patency. Malhotra et al. [29] measured pharyngeal airway length using MRI scanning (Figure 84.3). They demonstrated that the pharyngeal airway is longer in men relative to women, even when corrected for height. Given the fact that the pharyngeal airway is the section vulnerable to collapse, a longer airway is a more vulnerable airway.

The current understanding of the determinants of upper airway patency during sleep suggests a complex interplay between intrinsic mechanical properties of the pharynx and neural regulation of pharyngeal dilator muscle activities. Therefore the effects of gender on pharyngeal structure and function during sleep defy simple

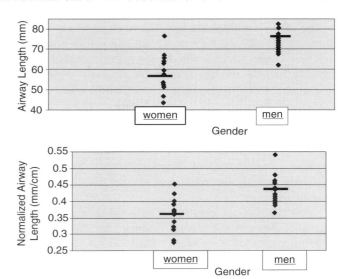

Figure 84.3 Airway length compared in men and women. (From [29].)

descriptors. For example, upper airway resistance, collapsibility, compliance, and critical closing pressure are distinct properties of the pharyngeal airway with different determinants. There is no conclusive evidence of a gender difference in upper airway resistance or collapsibility during NREM sleep [30–32]. In contrast, Pillar et al. [23] found that hypoventilation in response to an external inspiratory load is more pronounced in men relative to women. Finally, pharyngeal compliance during sleep is higher in men relative to women, suggesting a higher susceptibility to collapse [33]; this difference in compliance is due to differences in neck circumference between both genders.

In summary, investigation of static and dynamic determinants of upper airway patency has not yielded consistent results with two exceptions: *increased airway length and/or increased neck circumference. Both observations are likely to contribute to increased susceptibility to collapse in men relative to women.*

Ventilatory Control

Ventilatory motor output is an important determinant of upper airway patency during sleep [34]. Reduced ventilatory motor output is associated with upper airway obstruction in susceptible individuals [35–37]. Chemoresponsiveness is a commonly tested facet of ventilatory control. Conventional tests of chemoresponsiveness involve stimulation with hypoxia or hypercapnia and correlating changes in ventilation with changes in Pa_{CO_2} or Pa_{O_2}. There are no consistent gender differences in hypoxic and hypercapnic chemoresponsiveness during wakefulness [38] or NREM sleep [39, 40–43]. Thus the response to chemical stimuli per se is not affected by gender.

The aforementioned studies addressed elevation in the chemical stimuli and increased ventilatory motor output. However, sleep-related withdrawal of the wakefulness drive to breathe renders respiration critically dependent on chemical stimuli, especially $Paco_2$. Central apnea occurs if arterial Pco_2 is lowered below a highly sensitive "apneic threshold" [44]. Hypocapnia during sleep is an ubiquitous, powerful, and consistent inhibitory factor that can develop following transient arousal or brief hypoxia. Thus the response to withdrawal of chemical stimuli (hypocapnia) may have significant relevance to the development of sleep apnea.

Nasal mechanical ventilation is a suitable method to decrease $Paco_2$ until central apnea develops. Using nasal mechanical ventilation, Zhou et al. [24] have found that the hypocapnic apneic threshold is higher in men relative to women (Figure 84.4). Accordingly, a more pronounced hyperventilation and lower arterial Pco_2 is required in women to develop central sleep apnea. This effect does not seem to be due to progesterone [24]. Instead, it is likely due to the effect of male sex hormones. In a subsequent study, the Zhou et al. [45] evaluated the effect of male sex hormone on the apneic threshold in healthy premenopausal women for 12 days. Increased serum testosterone level elevated the apneic threshold and diminished the magnitude of hypocapnia required for induction of central apnea during NREM sleep [45]. In fact, the apneic threshold in women after testosterone administration was identical to the apneic threshold in men [45]. Thus male sex hormones seem to play a critical role in the susceptibility to develop central apnea during NREM sleep.

The existing clinical case series or epidemiologic studies support the conclusions derived from physiologic studies. There is evidence that central sleep apnea is rare in premenopausal women [5]. Likewise, male gender is a risk factor for the development of Cheyne–Stokes respiration and central sleep apnea [46], in patients with congestive heart failure (CHF). Central sleep apnea in premenopausal women seems to be distinctly infrequent. The occurrence of central sleep apnea in a premenopausal woman should alert the clinician to the possibility of comorbid conditions such as hypothyroidism, congestive heart failure, or hormonal abnormalities.

The effect of testosterone on the apneic threshold may provide a physiologic explanation for increased prevalence of sleep apnea in women with polycystic ovarian syndrome (PCOS) [47–49]. Irregular menses, obesity, elevated serum testosterone levels, and hirsutism characterize this condition. Fogel et al. [48] found that obese women with PCOS are at increased risk of OSA when compared with matched reproductively normal women. Women with PCOS had a higher AHI than controls (Figure 84.5). Interestingly, AHI correlated with

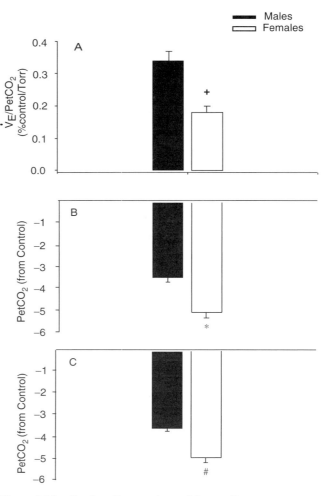

Figure 84.4 Gender effect on slope of the ventilatory response to hypocapnia (index of chemoresponsiveness) (A), and the magnitude of hypocapnia required to induce central apnea (predicted, panel B; and measured, panel C). The slope and predicted and actual values were significantly different between the two groups ($^+P = 0.001$; $^*P = 0.004$; $^\#P < 0.001$). Values are means ± SE. (From [24].)

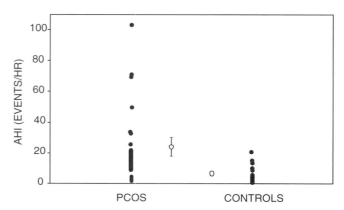

Figure 84.5 Note higher AHI in patients with PCOS compared to control subjects. (From [48].)

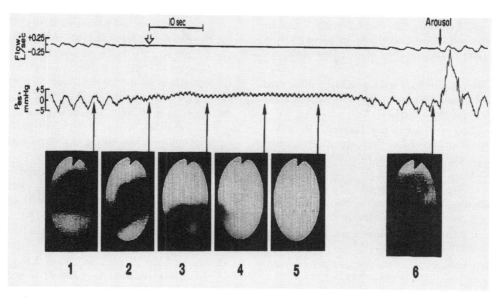

Figure 84.6 Upper airway images obtained with fiberoptic nasopharyngoscopy during sleep. Central sleep apnea is confirmed by the absence of negative pressure deflection on the espophageal pressure tracing (P$_{es}$). Note complete upper airway occlusion during central sleep apnea (images 4 and 5) with no generation of inspiratory effort. (From [50].)

waist–hip ratio, serum testosterone, and unbound testosterone in women with PCOS.

The occurrence of central apnea may contribute to recurrent central sleep apnea and periodic breathing, by triggering several processes perpetuating recurrent apnea. These include hypoxia, transient arousal, subsequent ventilatory overshoot, hypocapnia, and hence recurrent central sleep apnea. A less recognized phenomenon is that central sleep apnea may also influence the development of obstructive sleep apnea (OSA). There is evidence that central sleep apnea is associated with pharyngeal narrowing or occlusion in patients with susceptibility to upper airway collapse [50] (Figure 84.6). Thus many apneas that are central in etiology and morphology are associated with complete upper airway obstruction. Once upper airway obstruction occurs during a central apnea, mucosal adhesion forces and gravitational factors may impede pharyngeal opening and necessitate a substantial increase in drive that eventually leads to hyperventilation, hypocapnia, and recurrence of apnea. Thus the pathogenesis of central and obstructive sleep apnea seems to be inextricably linked.

SUMMARY

Sleep disordered breathing is a relatively common disorder in women. Although men are more likely to develop SDB, the difference should not lead to missing the diagnosis in women. Instead, it can illuminate potential mechanistic processes and guide research to decipher the pathophysiology of the condition.

REFERENCES

1. Guilleminault C, Quera-Salva MA, Partinen M, Jamieson A. Women and the obstructive sleep apnea syndrome. *Chest* **93**: 104–109(1988).

2. Redline S, Kump K, Tishler PV, Browner I, Ferrette V. Gender differences in sleep disordered breathing in a community-based sample. *Am J Respir Crit Care Med* **149**:722–726(1994).

3. Young T, Palta M, Dempsey J, Skatrud J, Weber S, Badr S. The occurrence of sleep-disordered breathing among middle-aged adults. *N Engl J Med* **328**:1230–1235(1993).

4. Bixler EO, Vgontzas AN, Ten Have T, Tyson K, Kales A. Effects of age on sleep apnea in men. I. Prevalence and severity. *Am J Respir Crit Care Med* **157**:144–148(1998).

5. Bixler EO, Vgontzas A N, Lin HM, Ten Have T, Rein J, Vela-Bueno A, Kales A. Prevalence of sleep-disordered breathing in women: effects of gender. *Am J Respir Crit Care Med* **163**: 608–613(2001).

6. Guilleminault C, Stoohs R, Kim YD, Chervin R, Black J, Clerk A. Upper airway sleep-disordered breathing in women. *Ann Intern Med* **122**:493–501(1995).

7. Redline S, Kump K, Tishler PV, Browner I, Ferrette V. Gender differences in sleep disordered breathing in a community-based sample. *Am J Respir Crit Care Med* **149**:722–726(1994).

8. Redline S, Tishler PV, Hans MG, Tosteson TD, Strohl KP, Spry K. Racial differences in sleep-disordered breathing in African-Americans and Caucasians. *Am J Respir Crit Care Med* **155**:186–192(1997).

9. Olson LG, King MT, Hensley MJ, Saunders NA. A community study of snoring and sleep-disordered breathing. Prevalence. *Am J Respir Crit Care Med* **152**:711–716(1995).

10. Redline S, Kump K, Tishler PV, Browner I, Ferrette V. Gender differences in sleep-disordered breathing in a community-based sample. *Am J Respir Crit Care Med* **149**:722–726(1994).

11. Young T. Analytic epidemiology studies of sleep disordered breathing—what explains the gender difference in sleep disordered breathing. *Sleep* **16**:S1–S2(1993).

12. Larsson LG, Lindberg A, Franklin KA, Lundback B. Gender differences in symptoms related to sleep apnea in a general population and in relation to referral to sleep clinic. *Chest* **124**:204–211(2003).

13. Young T, Hutton R, Finn L, Badr S, Palta M. The gender bias in sleep apnea diagnosis. Are women missed because they have different symptoms? *Arch Intern Med* **156**:2445–2451(1996).

14. Young T, Palta M, Badr MS. Sleep-disordered breathing (letter). *N Engl J Med* **329**:1429–1430(1993).

15. Chervin RD. Sleepiness, fatigue, tiredness, and lack of energy in obstructive sleep apnea. *Chest* **118**:372–379(2000).

16. Baldwin CM, Griffith KA, Nieto FJ, O'Connor GT, Walsleben JA, Redline S. The association of sleep-disordered breathing and sleep symptoms with quality of life in the Sleep Heart Health Study. *Sleep* **24**:96–105(2001).

17. O'Connor C, Thornley KS, Hanly PJ. Gender differences in the polysomnographic features of obstructive sleep apnea. *Am J Respir Crit Care Med* **161**:1465–1472(2000).

18. Ware JC, McBrayer RH, Scott JA. Influence of sex and age on duration and frequency of sleep apnea events. *Sleep* **23**:165–170(2000).

19. Mohsenin V. Gender differences in the expression of sleep-disordered breathing: role of upper airway dimensions. *Chest* **120**:1442–1447(2001).

20. Guilleminault C, Stoohs R, Clerk A, Cetel M, Maistros P. A cause of excessive daytime sleepiness. The upper airway resistance syndrome. *Chest* **104**:781–787(1993).

21. Gold AR, Dipalo F, Gold MS, O'Hearn D. The symptoms and signs of upper airway resistance syndrome: a link to the functional somatic syndromes. *Chest* **123**:87–95(2003).

22. Gold AR, Dipalo F, Gold MS, Broderick J. Inspiratory airflow dynamics during sleep in women with fibromyalgia. *Sleep* **27**:459–466(2004).

23. Pillar G, Malhotra A, Fogel R, Beauregard J, Schnall R, White DP. Airway mechanics and ventilation in response to resistive loading during sleep: influence of gender. *Am J Respir Crit Care Med* **162**:1627–1632(2000).

24. Zhou XS, Shahabuddin S, Zahn BK, Babcock MA, Badr MS. Effect of gender on the development of hypocapnic apnea/hypopnea during NREM sleep. *J Appl Physiol* **89**:192–199(2000).

25. Brown IG, Zamel N, Hoffstein V. Pharyngeal cross-sectional area in normal men and women. *J Appl Physiol* **61**:890–895(1986).

26. Martin SE, Mathur R, Marshall I, Douglas NJ. The effect of age, sex, obesity and posture on upper airway size. *Eur Respir J* **10**:2087–2090(1997).

27. Schwab RJ, Gefter WB, Hoffman EA, Gupta KP, Pack AI. Dynamic upper airway imaging during awake respiration in normal subjects and patients with sleep disordered breathing. *Am Rev Respir Dis* **148**:1385–1400(1993).

28. Whittle AT, Marshall I, Mortimore IL, Wraith PK, Sellar RJ, Douglas NJ. Neck soft tissue and fat distribution: comparison between normal men and women by magnetic resonance imaging. *Thorax* **54**:323–328(1999).

29. Malhotra A, Huang Y, Fogel RB, Pillar G, Edwards JK, Kikinis R, Loring SH, White DP. The male predisposition to pharyngeal collapse: importance of airway length. *Am J Respir Crit Care Med* **166**:1388–1395(2002).

30. Rowley JA, Zhou ZS, Vergine I, Shkoukani BS, Badr MS. The influence of gender on upper airway mechanics: upper airway resistance and Pcrit. *J Appl Physiol* **91**:2248–2254(2001).

31. Thurnheer R, Wraith PK, Douglas NJ. Influence of age and gender on upper airway resistance in NREM and REM sleep. *J Appl Physiol* **90**:981–988(2001).

32. Trinder J, Kay A, Kleiman J, Dunai J. Gender differences in airway resistance during sleep. *J Appl Physiol* **83**:1986–1997(1997).

33. Rowley JA, Sanders CS, Zahn BK, Badr MS. Gender differences in upper airway compliance during NREM sleep: role of neck circumference. *J Appl Physiol* **92**:2535–2541 (2002).

34. Badr MS. Effect of ventilatory drive on upper airway patency in humans during NREM sleep. *Respir Physiol* **103**:1–10(1996).

35. Warner G, Skatrud JB, Dempsey JA. Effect of hypoxia-induced periodic breathing on upper airway obstruction during sleep. *J Appl Physiol* **62**:2201–2211(1987).

36. Onal E, Burrows DL, Hart RH, Lopata M. Induction of periodic breathing during sleep causes upper airway obstruction in humans. *J Appl Physiol* **61**:1438–1443(1986).

37. Hudgel DW, Chapman KR, Faulks C, Hendricks C. Changes in inspiratory muscle electrical activity and upper airway resistance during periodic breathing induced by hypoxia during sleep. *Am Rev Respir Dis* **135**:899–906(1987).

38. Aitken ML, Franklin JL, Pierson DJ, Schoene RB. Influence of body size and gender on control of ventilation. *J Appl Physiol* **60**:1894–1899(1986).

39. Berthon-Jones M, Sullivan CE. Ventilatory and arousal responses to hypoxia in sleeping humans. *Am Rev Respir Dis* **125**:632–639(1982).

40. Douglas NJ, White DP, Weil JV, Pickett CK, Zwillich CW. Hypercapnic ventilatory response in sleeping adults. *Am Rev Respir Dis* **126**:758–762(1982).

41. Douglas NJ, White DP, Weil JV, Pickett CK, Martin RJ, Hudgel DW, Zwillich CW. Hypoxic ventilatory response decreases during sleep in normal men. *Am Rev Respir Dis* **125**:286–289.

42. White DP, Douglas NJ, Pickett CK, Weil JV, Zwillich CW. Hypoxic ventilatory response during sleep in normal premenopausal women. *Am Rev Respir Dis* **126**:530–533(1982).

43. Tarbichi AG, Rowley JA, Shkoukani MA, Mahadevan K, Badr MS. Lack of gender difference in ventilatory chemoresponsiveness and post-hypoxic ventilatory decline. *Respir Physiol Neurobiol.* **137**:41–50(2003).

44. Skatrud JB, Dempsey JA. Interaction of sleep state and chemical stimuli in sustaining rhythmic ventilation. *J Appl Physiol* **55**:813–822(1983).

45. Zhou XS, Rowley JA, Demirovic F, Diamond MP, Badr MS. Effect of testosterone on the apneic threshold in women during NREM sleep. *J Appl Physiol* **94**:101–107(2003).

46. Sin DD, Fitzgerald F, Parker JD, Newton G, Floras JS, Bradley TD. Risk factors for central and obstructive sleep apnea in 450 men and women with congestive heart failure. *Am J Respir Crit Care Med* **160**:1101–1106(1999).

47. Gopal M, Duntley S, Uhles M, Attarian H. The role of obesity in the increased prevalence of obstructive sleep apnea syndrome in patients with polycystic ovarian syndrome. *Sleep Med* **3**:401–404(2002).

48. Fogel RB, Malhotra A, Pillar G, Pittman SD, Dunaif A, White DP. Increased prevalence of obstructive sleep apnea syndrome in obese women with polycystic ovary syndrome. *J Clin Endocrinol Metab* **86**:1175–1180(2001).

49. Vgontzas AN, Legro RS, Bixler EO, Grayev A, Kales A, Chrousos GP. Polycystic ovary syndrome is associated with obstructive sleep apnea and daytime sleepiness: role of insulin resistance. *J Clin Endocrinol Metab* **86**:517–520(2001).

50. Badr MS, Toiber F, Skatrud JB, Dempsey J. Pharyngeal narrowing/occlusion during central sleep apnea. *J Appl Physiol* **78**:1806–1815(1995).

85

SLEEP DURING THE PERIMENOPAUSAL PERIOD

Naomi L. Rogers and Ronald R. Grunstein
Institute of Medical Research, Camperdown, New South Wales, Australia

Perimenopause is a transitional period occurring prior to menopause, or cessation of menses, and is defined as the time during which changes in the menstrual cycle occur. The duration of perimenopause may vary from 2 to 10 years, with a median age of onset in the early forties, although it may occur between the ages of 35 and 50 years.

While much research and attention have focused on the effects of menopause and postmenopause on sleep and physiological changes, there are few reports on the sleep changes that manifest during the perimenopausal period. The primary physiological changes associated with perimenopause include changes in the length and frequency of the menstrual cycle, changes in blood sugar levels, weight gain, decreased sex drive, headaches, increased premenstrual syndrome (PMS) symptoms, night sweats, hot flashes, fatigue, and sleep disturbances.

SLEEP DISTURBANCE DURING PERIMENOPAUSE

The prevalence of insomnia and sleep disturbance during perimenopause has been reported to increase as women enter the menopausal transition period [1–4]. For example, a number of population-based studies have reported a link between menopause and subjective sleep problems. In a recent study, just over 40% of perimenopausal subjects experienced problems with their sleep [5]. The sleep disturbances commonly reported during perimenopause include increased sleep latency, increased awakenings [3, 6], including an increase in movement-associated arousals, and increased sleepiness during the day [6]. Despite these and other studies, however, it is still not clear that the

increased incidence of sleep disturbance in middle-aged females is wholly attributable to the perimenopause and menopause. For example, it has been suggested that the findings from some population-based studies may be confounded by age-related problems that impact sleep, such as depression, anxiety, pain, and medical disorders that disturb sleep [5, 7, 8].

In a recently published study examining over 600 pre-, peri-, and postmenopausal women, as part of the Wisconsin Sleep Cohort Study, the authors reported an increased difficulty in initiating sleep and dissatisfaction with sleep in perimenopausal women [1]. The authors did not believe this increased incidence in sleep problems is associated with perimenopause per se. In fact, in this study the perimenopausal women had less stage 1 sleep and more slow-wave sleep (SWS) compared to the premenopausal women. The authors concluded that although menopause was associated with decreased satisfaction with sleep, it was not associated with increased reports of insomnia or sleepiness.

The underlying causes of the increased incidence of sleep disturbance and insomnia during perimenopause have not been clearly defined. The insomnia may be secondary to the physiological alterations occurring during this time, in particular, the endocrine or thermoregulatory changes, or increased severity of PMS. Another possibility is that the sleep disturbance is due to an increase in the occurrence of sleep apnea in females in this age group, possibly mediated via the increased weight gain and changes in blood sugar metabolism. Alternatively, it has been proposed that the increased incidence of insomnia in this population is not directly related to menopause onset, but is simply coincident. Possible causes for the increased insomnia

Sleep: A Comprehensive Handbook, Edited by T. Lee-Chiong.
Copyright © 2006 John Wiley & Sons, Inc.

include an increased incidence of mood disturbances, due to depression and anxiety, medical conditions, and general life circumstances.

CLIMACTERIC SYMPTOMS

During perimenopause the onset of climacteric symptoms, in particular, hot flashes and increased sweating, occurs. These symptoms have been linked with the endocrine changes that also occur during this period. Hot flashes are reported to occur in approximately 80% of females during perimenaopuse. It is well recognized that thermoregulation and sleep–wake behavior are intimately related. Decreasing core body temperature facilitates sleep onset and increases sleep propensity, and a decrease in core body temperature occurs immediately following sleep onset [9].

Although there are a number of reports of increased subjective sleep disturbance and decreased sleep quality and increased levels of objective sleep disturbance associated with high frequencies of hot flashes, not all studies find evidence to support this relationship. For example, when hot flashes occur nocturnally during sleep, it has been demonstrated that awakening from sleep and increasing core body temperature precede the hot flash [10]. In addition, one study reported that hot flashes were associated with increased amounts of SWS and total sleep time [11]. Consequently, there is not a current consensus as to the role that thermoregulatory changes play in the sleep changes observed during perimenopause.

Hot flashes have been associated with increased brain noradrenaline levels [11]. It is possible that increases in central noradrenaline levels may produce an elevation in arousal levels and alertness, and thereby potentially mediate awakenings that are temporally associated with hot flashes in some individuals.

ENDOCRINE CHANGES AND SLEEP DISTURBANCE

Changes in the secretory profiles of many endocrine variables can start between 7 and 10 years prior to the cessation of menses. During the perimenopausal period there is a gradual increase in serum levels of follicle stimulating hormone (FSH) and to a lesser degree luteinizing hormone (LH). In addition, there is a decrease in secretion of prostaglandin and estradiol. Consequently, there is a subsequent increase in the ratio between estradiol and estrone also observed. Around the time of menopause onset there is also a small decrease in testosterone, androstenedione, and sex hormone binding globulin (SHBG) levels. Prior to and following the onset of menopause an inverse relationship between SHBG and body mass index (BMI) has

been reported [12]. Thus the decrease in SHBG likely contributes to the increased weight gain that occurs during perimenopause and menopause.

A number of studies have suggested that the increased incidence of sleep disturbance during perimenopause may be due to the decreasing levels of sex steroids [10]. Furthermore, it has been suggested that hormone replacement therapy (HRT) may reverse some of the effects of changing hormone levels including the increased degree of sleep disturbance in this population. It has been reported that women on estrogen replacement therapy (ERT) experience less sleep disturbance and insomnia [13, 14], as well as decreased sleep latency, nocturnal restlessness, and wake after sleep onset (WASO) [15]. It has been hypothesized that estrogen exerts its beneficial effects on sleep by reducing climacteric symptoms, in particular, hot flashes [16]. Reductions in hot flashes and sweating have been reported to be the best predictors of improvements in sleep [15].

A decrease in climacteric symptoms may not be the only mechanism of action for estrogen, however. ERT has also been demonstrated to increase sleep in asymptomatic women during perimenopause [15]. There is a suggestion that elevated estrogen levels may improve sleep by decreasing the impact of stress or nocturnal disturbances [15, 17]. In a recent study, women taking ERT experienced less sleep disturbance during a period of nocturnal blood draws during their sleep period, via an indwelling catheter, compared with women not on ERT [16]. Subjectively, the women taking ERT experienced higher sleep quality compared with the other women. Additionally, objective measures of sleep efficiency, sleep latency, WASO, and SWS were better in the ERT group compared with the non-ERT group of subjects. Furthermore, during exposure to chronic sleep restriction, female subjects have been reported to have greater sleep efficiency [18, 19], with reduced sleep latencies, increased total sleep time, and increased percentages of SWS [18], and to be more resistant to sleep disturbance [19]. Therefore it appears that estrogen may improve or protect sleep via more than one mechanism.

MENOPAUSE AND SLEEP APNEA

It has long been recognized that obstructive sleep apnea (OSA) is more common in men than in women but that these marked differences were predominantly confined to premenopausal women. This led to the concept that hormonal changes during menopause are a risk factor for OSA [20]. Recent investigation cross sectionally and longitudinally has confirmed a gradual increase in the prevalence of OSA through the perimenopausal period with a substantial increase in postmenopausal women. In the Wisconsin Sleep Cohort ($n = 589$ women), postmenopausal women were

2.59 times more likely to have OSA (apnea–hypopnea index $\geq$15) than premenopausal women [21]. Data from the Sleep Heart Health Study ($n = 2852$) has shown that the prevalence of OSA (apnea–hypopnea index $\geq$15) in postmenopausal women was 11–12% [22]. Other cross-sectional analyses from this study found a 40–50% reduction in OSA prevalence (apnea–hypopnea index $\geq$15) in postmenopausal women receiving hormone replacement therapy (HRT; estrogen or estrogen plus progesterone) when compared with women without such replacement. This confirmed earlier data from a cross-sectional study from a large Hershey, Pennsylvania cohort that observed that postmenopausal women without HRT had a fourfold risk of OSA, but this increased risk was not present in those treated with HRT [23]. Some have even conjectured that a later development of OSA may, in part, explain the longer life span enjoyed by women [24].

There is strong evidence from physiological experiments that support an effect of menopause (or the change in female hormone concentrations accompanying menopause) in increasing the prevalence of sleep apnea. Most of the studies focus on the differences in respiratory physiology between men and premenopausal women and a few on similarities between men and postmenopausal women. Studies have shown differences between men and premenopausal women in ventilatory control (hypoxic/hypercapnic ventilatory responses) [25] and sleep apnea P_{CO_2} thresholds [26], pharyngeal fat distribution [27] and airway length [28], upper airway muscle activation [29], and airway collapsibility during sleep [30].

These epidemiological findings appear strong but there are significant confounding variables that can only be partly controlled in this type of epidemiological study [31]. Menopause is not a true binary categorical variable but may involve a 10 year transition of change in female hormones. Age effects may be confused with menopausal effects and the change in fat distribution that occurs with menopause may also cloud analysis of epidemiological studies. Moreover, despite the apparent beneficial effect of HRT in reducing the risk of OSA, therapeutic studies using HRT for women with OSA have been disappointing [32]. HRT use tends to be greater in healthier women, and this would contribute to an overestimation of the protective effect of HRT. HRT use may be a marker for women in the earlier stages of the menopausal transition [31].

From a clinical point of view, it is important to recognize that sleep-related problems in women in the peri- and postmenopausal period are more likely to involve sleep breathing disorders. Symptomatology may differ from that reported in men and is more likely to overlap with reports from groups such as patients with insomnia or chronic fatigue. This, in turn, should influence investigatory thresholds for the use of sleep or nocturnal respiratory investigation for clinical assessment in these groups.

COINCIDENT SLEEP DISTURBANCE AND INSOMNIA

In addition to the increased incidence of sleep disturbance observed during the perimenopausal period, a number of studies have reported decreased mood in perimenopausal women [33, 34]. While sleep disturbance or insomnia may produce decreases in mood, dysphoric mood may also produce increased levels of sleep disturbance. Bromberger et al. [33] reported an increase in feelings of irritability, nervousness, and mood changes in perimenopausal women, which were associated with increased sleep disturbance.

It has been suggested that the increase in depression and anxiety commonly reported in perimenopausal women may in fact be temporally and not directly related to the perimenopausal transition [33]. Increased depression may result from life circumstances, grief, or other issues common to this age group, including medical disorders, retirement, and increased pain, which coincide with perimenopause. This depression may then mediate perimenopausal insomnia or exacerbate sleep disturbance attributable to perimenopausal symptoms. In support of this theory, results from the Seattle Midlife Women's Health Study found that insomnia symptoms were stable across a 3 year period in subjects. They concluded that this finding suggested that these symptoms were associated with underlying factors, such as medical conditions that are common to this age group [35]. Other factors may also play a role in the increased sleep disturbance in this population. For example, studies have reported a link between insomnia and social factors, such as a low sense of well-being or increased caffeine consumption [34].

One interesting hypothesis that has been proposed suggests that menopause and the menopausal transition may be the trigger for primary insomnia [20]. These authors theorized that menopausal symptoms, for example, hot flashes, may produce sleep disturbance and this then acts as a precipitating factor for insomnia in these individuals. Following alleviation of the symptom, the insomnia then persists, due to behavioral conditioning.

CONCLUSION

Although there is considerable evidence for an increase in insomnia and sleep disturbance that is temporally associated with perimenopause, it is not clear that this increased incidence in sleep disturbance can be fully attributed to the onset of menopausal symptoms. While changing hormone levels and other physiological alterations may underlie some of the increased sleep disturbances in this population, other factors such as significant life events, increased weight gain and incidence of sleep apnea, primary insomnia, and depression associated with this age group may also contribute to the increased incidence of sleep disturbance and dissatisfaction with sleep observed in perimenopausal women.

REFERENCES

1. Young T, Rabago D, Zgierska A, Austin D, Laurel F. Objective and subjective sleep quality in premenopausal, perimenopausal, and postmenopausal women in the Wisconsin Sleep Cohort Study. (See Comment.) *Sleep* **26**(6):667–672(2003).

2. Kuh DL, Wadsworth M, Hardy R. Women's health in midlife: the influence of the menopause, social factors and health in earlier life. *Br J Obstet Gynaecol* **104**(8):823–833(1997).

3. Owens JF, Matthews KA. Sleep disturbance in healthy middle-aged women. *Maturitas* **30**(1):41–50(1998).

4. Kravitz HM, Ganz PA, Bromberger J, Powell LH, Sutton-Tyrrell K, Meyer PM. Sleep difficulty in women at midlife: a community survey of sleep and the menopausal transition. *Menopause* **10**(1):19–28(2003).

5. Shaver JL, Giblin E, Paulsen V. Sleep quality subtypes in mid-life women. *Sleep* **14**(1):18–23(1991).

6. Baker A, Simpson S, Dawson D. Sleep disruption and mood changes associated with menopause. *J Psychosom Res* **43**(4): 359–369(1997).

7. Dennerstein L, Dudley EC, Hopper JL, Guthrie JR, Burger HG. A prospective population-based study of menopausal symptoms. *Obstet Gynecol* **96**(3):351–358(2000).

8. Moe KE. Reproductive hormones, aging, and sleep. *Semin Reprod Endocrinol* **17**(4):339–348(1999).

9. Murphy PJ, Campbell SS. Nighttime drop in body temperature: a physiological trigger for sleep onset? *Sleep* **20**(7):505–511 (1997).

10. Moline ML, Broch L, Zak R. Sleep in women across the life cycle from adulthood through menopause. *Med Clin North Am* **88**(3):705–736(2004).

11. Woodward S, Freedman RR. The thermoregulatory effects of menopausal hot flashes on sleep. *Sleep* **17**(6):497–501(1994).

12. Rannevik G, Jeppsson S, Johnell O, Bjerre B, Laurell-Borulf Y, Svanberg L. A longitudinal study of the perimenopausal transition: altered profiles of steroid and pituitary hormones, SHBG and bone mineral density. *Maturitas* **21**(2):103–113(1995).

13. Vitiello MV, Larsen LH, Moe KE. Age-related sleep change. Gender and estrogen effects on the subjective–objective sleep quality relationships of healthy, noncomplaining older men and women. *J Psychosom Res* **56**:503–510(2004).

14. Wiklund I, Karlberg J, Mattsson LA. Quality of life of postmenopausal women on a regimen of transdermal estradiol therapy: a double-blind placebo-controlled study. *Am J Obstet Gynecol* **168**(3 Pt 1):824–830(1993).

15. Polo-Kantola P, Erkkola R, Helenius H, Irjala K, Polo O. When does estrogen replacement therapy improve sleep quality? *Am J Obstet Gynecol* **178**(5):1002–1009(1998).

16. Moe KE, Larsen LH, Vitiello MV, Prinz PN. Estrogen replacement therapy moderates the sleep disruption associated with nocturnal blood sampling. *Sleep* **24**(8):886–894(2001).

17. Burleson MH, Malarkey WB, Cacioppo JT, Poehlmann KM, Kiecolt-Glaser JK, Berntson GG, et al. Postmenopausal hormone replacement: effects on autonomic, neuroendocrine, and immune reactivity to brief psychological stressors. *Psychosom Med* **60**(1):17–25(1998).

18. Rogers NL, Coble M, Maislin G, Gordon E, Grunstein RR, Dinges DF. Gender differences in response to sleep loss. *J Sleep Res* **13**(Suppl. 1):619(2004).

19. Vgontzas AN, Zoumakis E, Bixler EO, Lin H-M, Follett H, Chrousos GP. Healthy young women compared to men are more resilient to sleep loss and sleep disturbance than age-matched men: potential protective role of estrogen. In: *The Endocrine Society Conference*, 2003.

20. Krystal AD, Edinger J, Wohlgemuth W, Marsh GR. Sleep in peri-menopausal and post-menopausal women. *Sleep Med Rev* **2**(4): 243–253(1998).

21. Young T, Finn L, Austin D, Peterson A. Menopausal status and sleep-disordered breathing in the Wisconsin Sleep Cohort Study (comment). *Am J Respir Crit Care Med* **167**(9):1181–1185(2003).

22. Shahar E, Redline S, Young T, Boland LL, Baldwin CM, Nieto FJ, et al. Hormone replacement therapy and sleep-disordered breathing. *Am J Respir Crit Care Med* **167**(9):1186–1192(2003).

23. Bixler EO, Vgontzas AN, Lin HM, Ten Have T, Rein J, Vela-Bueno A, et al. Prevalence of sleep-disordered breathing in women: effects of gender. *Am J Respir Crit Care Med* **163**(3 Pt 1):608–613(2001).

24. Lavie P. Sleep apnea syndrome: is it a contributing factor to the sex differential in mortality? *Med Hypotheses* **21**(3):273–276(1986).

25. White DP, Douglas NJ, Pickett CK, Weil JV, Zwillich CW. Sexual influence on the control of breathing. *J Appl Physiol* **54**(4):874–879(1983).

26. Zhou XS, Shahabuddin S, Zahn BR, Babcock MA, Badr MS. Effect of gender on the development of hypocapnic apnea/hypopnea during NREM sleep. *J Appl Physiol* **89**(1):192–199(2000).

27. Whittle AT, Marshall I, Mortimore IL, Wraith PK, Sellar RJ, Douglas NJ. Neck soft tissue and fat distribution: comparison between normal men and women by magnetic resonance imaging. *Thorax* **54**(4):323–328(1999).

28. Malhotra A, Huang Y, Fogel RB, Pillar G, Edwards JK, Kikinis R, et al. The male predisposition to pharyngeal collapse: importance of airway length. *Am J Respir Crit Care Med* **166**(10):1388–1395(2002).

29. Popovic RM, White DP. Influence of gender on waking genioglossal electromyogram and upper airway resistance. *Am J Respir Crit Care Med* **152**(2):725–731(1995).

30. Pillar G, Malhotra A, Fogel R, Beauregard J, Schnall R, White DP. Airway mechanics and ventilation in response to resistive loading during sleep: influence of gender. *Am J Respir Crit Care Med* **162**(5):1627–1632(2000).

31. Young T, Peppard PE, Gottlieb DJ. Epidemiology of obstructive sleep apnea: a population health perspective. *Am J Respir Crit Care Med* **165**(9):1217–1239(2002).

32. Cistulli PA, Barnes DJ, Grunstein RR, Sullivan CE. Effect of short-term hormone replacement in the treatment of obstructive sleep apnoea in postmenopausal women. *Thorax* **49**(7):699–702(1994).

33. Bromberger JT, Assmann SF, Avis NE, Schocken M, Kravitz HM, Cordal A. Persistent mood symptoms in a multiethnic community cohort of pre- and perimenopausal women. *Am J Epidemiol* **158**(4):347–356(2003).

34. Hollander LE, Freeman EW, Sammel MD, Berlin JA, Grisso JA, Battistini M. Sleep quality, estradiol levels, and behavioral factors in late reproductive age women. *Obstet Gynecol* **98**(3):391–397(2001).

35. Zapantis G, Santoro N. The menopausal transition: characteristics and management. *Best Practice Res Clin Endocrinol Metab* **17**(1):33–52(2003).

86

SLEEP DURING POSTMENOPAUSE

KAREN E. MOE

University of Washington, Seattle, Washington

Women and men experience many of the same age-related changes in sleep, though the extent of these changes appears to vary with gender. More older women than men have sleep complaints. Objective measures show that women are less likely to experience a decline in slow-wave sleep (SWS) with age and more likely to experience problems with sleep maintenance. Sleep changes in older women are often associated with chronic physical or mental illness and related factors such as caregiving. In addition, the sleep of older women may be affected by some unique factors such as chronic hot flashes.

AGE-RELATED SLEEP CHANGES IN WOMEN

Sleep complaints and disorders are prevalent among older individuals. In one study of over 9000 individuals aged 65 or older [1], over half of the participants reported having one or more of five common sleep complaints most of the time: (1) trouble falling asleep, (2) waking up during the night, (3) waking up too early, (4) needing a nap, and (5) not feeling rested. Less than 20% of the participants rarely or never had any complaints. The average sleep complaint score was significantly higher among women in this study. This gender difference was due primarily to a difference in insomnia (defined as trouble falling asleep or waking up too early): the odds for insomnia were about 50% for women compared to men.

This is consistent with other epidemiological studies. In one large group of men and women aged 50–93, women were more likely than men to report poor sleep quality, long sleep latency, nighttime awakenings, and the use of sedative–hypnotic drugs [2]. In the ongoing Sleep Heart Health Study of 6400 older individuals, women compared with men felt less rested and had more difficulties with falling asleep, staying asleep, early morning awakening, and insufficient sleep [3].

However, objective measurements of sleep (e.g., actigraphy, polysomnography, quantitative electroencephalography (EEG)) have yielded mixed results about a decline in sleep quality with age in healthy women. The Sleep Heart Health Study showed that, unlike men, women did not experience a decline in the percentage of slow-wave sleep (SWS) with age [4]. This has been a consistent finding across studies [5]. Other aspects of sleep may be more negatively affected by age in women than in men. In a study comparing the sleep of healthy "young old" (age 62–79) individuals with the sleep of the healthy "old old" (age 80–91), sleep maintenance declined with age in women more than in men [5]. In addition, the sleep of healthy older women may be more susceptible to disruption. Intravenous catheterization and frequent nocturnal blood sampling decreased the percentage of SWS in women more than in men [6].

These data collectively raise the interesting issue of the significance of disparities between subjective and objective measures of sleep quality. Such disparities seem to be most common in postmenopausal and older women. One interpretation of this provocative finding is that what we consider to be objective measures of good quality sleep may be appropriate for older men but that older women may be evaluating their sleep quality using other criteria [7].

In short, a clear picture has yet to emerge about how age affects sleep in older women. A more complete

Sleep: A Comprehensive Handbook, Edited by T. Lee-Chiong.
Copyright © 2006 John Wiley & Sons, Inc.

understanding will require a multivariate approach that incorporates several factors that interact with age to affect the sleep of older women. One of these factors is physical and mental health. It has become clear from studies in the past decade that a considerable proportion of the sleep complaints by older individuals are associated with chronic physical and mental health burden [1, 8].

Several common concomitants of aging make women more vulnerable to sleep problems, independent of menopause. These include weight gain, reduced physical activity, changes in physiological stress reactivity, and increased incidence of several diseases and chronic conditions that affect sleep (depression, thyroid dysfunction, cancer, arthritis, nocturia, etc.). One important example is the increased susceptibility of postmenopausal women to sleep disordered breathing.

SLEEP DISORDERED BREATHING

After menopause, women are more likely to develop sleep disordered breathing (SDB), including obstructive sleep apnea. SDB is abnormal breathing that occurs during sleep, ranging from increased breathing effort caused by increased airway resistance, to reduced airflow, to repeated pauses in breathing. SDB is associated with frequent nocturnal awakenings, lighter sleep, decreased SWS, and excessive daytime sleepiness and fatigue.

The well-documented gender gap in the prevalence of SDB begins to narrow with menopause and continues to narrow with increased age. In one study of 1741 men and women aged 20–100 years [9], the female/male ratio for clinically defined apnea fell from 1:3.3 for the entire study group to 1:1.44 for postmenopausal women and men matched by age and body mass index. This study provided strong support for the hypothesis that menopause increases the risk of SDB, as the prevalence of mild SDB was 3.2% for premenopausal women but 9.7% for postmenopausal women who were not using hormone therapy (HT). The prevalence of more severe SDB was 0.6% for premenopausal women versus 2.7% for postmenopausal women not using HT. SDB prevalence remained significantly higher for postmenopausal women even after adjusting for known risk factors of age and body mass index.

The longitudinal Wisconsin Sleep Cohort Study has reported similar findings [10, 11]. One strength of this study was the careful use of multivariate models to adjust for several known risk factors for SDB, such as body mass index, smoking, and alcohol use. Even with the inclusion of these risk factors, menopausal status and age remained strong and significant risk factors for SDB in women.

Menopause-related declines in estrogen and/or progesterone may be one reason that older women are at increased risk for SDB. This hypothesis is supported by studies showing positive effects of progesterone alone or of combined hormone therapy (HT; estrogen plus progesterone) on ventilatory drive and other SDB-related measures.

Age-related increases in obesity and hypertension may also contribute to older women's vulnerability to SDB. SDB is strongly associated with hypertension, obesity, and visceral adiposity. This has led to suggestions that SDB is a manifestation of the metabolic syndrome [12]. The metabolic syndrome is a constellation of closely related symptoms (visceral adiposity, hypertension, insulin resistance, elevated glucose, dyslipidemia) that together result in a significant risk of cardiovascular disease [13]. These features of the metabolic syndrome are much more common in women after menopause [13]. In particular, aging in women is associated with a decrease in lean body mass, and with a preferential increase in intra-abdominal or visceral deposition of fat relative to other areas. Visceral fat correlates strongly with SDB and is believed by some to be the principal culprit leading to SDB.

Women may therefore be doubly at risk for SDB compared with younger women, because of hormone-related changes in ventilatory drive and because of increased visceral adiposity. Hypertension, obesity, and visceral adiposity in older women should be considered high risk factors for SDB and associated sleep disturbance, as well as for other serious health problems such as cardiovascular disease and type II diabetes.

Hormone therapy (HT) may have some beneficial effects on respiration and visceral adiposity. It is associated with a lower prevalence of SDB in older women [14]. However, it is premature and inappropriate to conclude that HT is an effective or desirable treatment for SDB in older women. No large clinical trials have been conducted. Also, there are significant and serious health risks associated with HT. Standard treatments (e.g., nasal continuous positive airway pressure, surgery, and weight loss and/or exercise to reduce adiposity) remain the most viable options for older women with SDB.

HOT FLASHES IN OLDER WOMEN

Hot flashes are popularly assumed to end within a few years after the last menstrual period. However, secondary descriptive data from several large cross-sectional studies show that a significant number of postmenopausal women continue to report having hot flashes for many years, even decades, after menopause. In one multisite randomized clinical trial of hormone therapy (HT), the participants were 55–79 years old, with a mean age of 67 years. On average, menopause had occurred 18 years prior to the start of the study. About 16% of the 2763 women in the study reported that they had somewhat frequent or very frequent hot flashes during the previous week [15]. Though the younger women were more likely to report hot flashes, hot flashes were also common in the older part of the study's age range.

These studies suggest that continued, chronic, hot flashes may contribute to otherwise unexplained sleep disturbance in some older postmenopausal women. Anecdotal reports suggest that hot flashes are not always labeled as "hot flashes" by older women. They may be described in terms of waking up during the night while "feeling too warm," and feeling the need to throw off the bedcovers. This possibility should be kept in mind when exploring insomnia complaints with older women.

Up to 60% of older women have abruptly discontinued their use of estrogen or hormone therapy [16], because of the findings of the Women's Health Initiative (WHI). This large national clinical trial showed that use of a common HT regimen for 1–7 years significantly increased the risk of breast cancer, stroke, heart disease, and vascular dementia [17]. The sleep effects of withdrawal from ET or HT are unknown. However, two recent studies have shown that abrupt estrogen withdrawal can result in significant hot flashes [18, 19], which suggests that there might be sleep consequences to ET or HT discontinuation. There are no research-based clinical protocols or guidelines for women to follow when they discontinue HT. Many women are unaware that stopping HT is likely to be associated with emergence of menopausal symptoms. Anecdotal reports in the popular press have suggested that as many as 50% of women who stopped HT after the first results from the WHI study became widely disseminated started taking HT again, albeit at a lower dose. Along with hot flashes, insomnia was stated as a major, intolerable side effect of HT withdrawal for many of the women. Though ET/HT remains the most effective treatment for hot flashes, a growing number of other therapies are now available [20], including antidepressants of the selective serotonin reuptake inhibitor (SSRI) class, gabapentin, clonidine, herbal remedies, and relaxation therapies.

CONCLUSION

Though health care practitioners and older women have a tendency to attribute the appearance of sleep disturbance to the onset of menopause or "hormone problems," the sleep complaints and changes experienced by older women are likely to be attributable to chronic physical or mental illness and related factors such as stress and caregiving. Thus, as with men, sleep symptoms and their treatment should be viewed in the wider context of aging and environment. For example, fatigue or sleep complaints combined with hypertension and excessive adiposity should prompt consideration of SDB.

Finally, the well-documented mismatch of subjective sleep complaints with relative lack of objective changes in older women strongly suggests that current objective measures of sleep quality fail to capture some essential element of subjectively experienced sleep quality.

REFERENCES

1. Foley DJ, Monjan AA, Brown SL, Simonsick EM, Wallace RB, Blazer DG. Sleep complaints among elderly persons: an epidemiologic study of three communities. *Sleep* **18**:425–432(1995).

2. Middelkoop HAM, Smilde-van den Doel DA, Knuistingh Neve A, Kamphuisen HAC, Springer CP. Subjective sleep characteristics of 14,85 males and females aged 50–93: effects of sex and age, and factors related to self-evaluated quality of sleep. *J Gerontol Med Sci* **51A**:M108–M115(1996).

3. Baldwin CM, Kapur VK, Holberg CJ, Rosen C, Nieto FJ. Associations between gender and measures of daytime somnolence in the Sleep Heart Health Study. *Sleep* **27**:305–311(2004).

4. Redline S, Kirchner HL, Quan SF, Gottlieb DJ, Kapur V, Newman A. The effects of age, sex, ethnicity, and sleep-disordered breathing on sleep architecture. *Arch Intern Med* **164**:406–418 (2004).

5. Reynolds CF III, Monk TH, Hoch CC, Jennings JR, Buysse DJ, Houck PR, Jarrett DB, Kupfer DJ. Electroencephalographic sleep in the healthy "old old": a comparison with the "young old" in visually scored and automated measures. *J Gerontol Med Sci* **46**:M39–M46(1991).

6. Vitiello MV, Larsen LH, Moe KE, Borson S, Schwartz RS, Prinz PN. Objective sleep quality of healthy older men and women is differentially disrupted by nighttime periodic blood sampling via indwelling catheter. *Sleep* **19**:304–311(1996).

7. Vitiello MV, Larsen LH, Moe KE. Age-related sleep change: gender and estrogen effects on the subjective–objective sleep quality relationships of healthy, noncomplaining older men and women. *J Psychosom Res* **56**:503–510(2004).

8. Vitiello MV, Moe KE, Prinz PN. Sleep complaints cosegregate with illness in older adults. *J Psychosom Res* **53**:555–559 (2002).

9. Bixler EO, Vgontzas AN, Lin H-M, Ten Have T, Rein J, Vela-Bueno A, Kales A. Prevalence of sleep-disordered breathing in women. *Am J Respir Crit Care Med* **163**:608–613(2001).

10. Young T, Rabago D, Zgierska A, Austin D, Finn L. Objective and subjective sleep quality in premenopausal, perimenopausal, and postmenopausal women in the Wisconsin Sleep Cohort Study. *Sleep* **26**:667–672(2003).

11. Young T, Finn L, Austin D, Peterson A. Menopausal status and sleep-disordered breathing in the Wisconsin Sleep Cohort Study. *Am J Respir Crit Care Med* **167**:1181–1185(2003).

12. Vgontzas AN, Bixler EO, Chrousos GP. Metabolic disturbances in obesity versus sleep apnea: the importance of visceral obesity and insulin resistance. *J Intern Med* **254**:32–44(2003).

13. Carr MC. The emergence of the metabolic syndrome with menopause. *J Clin Endocrinol Metab* **88**:2404–2411(2003).

14. Manber R, Kuo TF, Catabldo N, Colrain IM. The effects of hormone replacement therapy on sleep-disordered breathing in postmenopausal women: a pilot study. *Sleep* **2**:163–168(2003).

15. Barnabei VAM, Grady D, Stovall DW, Cauley JA, Lin F, Stuenkel CA, Stefanick ML, Pickar JH. Menopausal symptoms in older women and the effects of treatment with hormone therapy. *Obstet Gynecol* **100**:1209–1218(2002).

16. Hersh AL, Stefanick ML, Stafford RS. National use of postmenopausal hormone therapy: annual trends and response to recent evidence. *JAMA* **291**:47–53(2004).

17. Writing Group for the Women's Health Initiative Investigators. Risks and benefits of estrogen plus progestin in healthy postmenopausal women. *JAMA* **288**:321–333(2002).

18. Simon JA, Mack CJ. Counseling patients who elect to discontinue hormone therapy. *Int J Fertil Womens Med* **48**:111–116(2003).

19. Grady D, Ettinger B, Tosteson ANA, Pressman A, Macer JL. Predictors of difficulty when discontinuing postmenopausal hormone therapy. *Obstet Gynecol* **102**:1233–1239(2003).

20. North American Menopause Society. Treatment of menopause-associated vasomotor symptoms: position statement of The North American Menopause Society. *Menopause* **11**:11–33(2004).

ADDITIONAL READING

Armitage R, Hoffmann RF. Sleep EEG, depression and gender. *Sleep Med Rev* **5**:237–246(2001). The review provides a critical evaluation of the literature on gender differences in sleep and depression. It includes a discussion of the theoretical and clinical implications of the findings.

Byles JE, Mishra GD, Harris MA, Nair K. The problems of sleep for older women: changes in health outcomes. *Age Aging* **32**:154–163(2003). This longitudinal survey of over 10,000 women explored the relationship between sleeping difficulties, health status, and the use of sleeping medications.

Kapsimalis F, Kryger MH. Gender and obstructive sleep apnea syndrome, part 1: clinical features. *Sleep* **25**:412–419(2002). Also: Part 2: mechanisms. *Sleep* **25**:499–506(2002). These two articles provide a comprehensive review of gender differences in sleep-disordered breathing in women, particularly the differences in clinical presentation, polysomnographic findings, and possible mechanisms.

Moe KE. Hot flashes and sleep in women. *Sleep Med Rev* **8**:487–497(2004). This is a comprehensive detailed review of the impact of hot flashes on the sleep of women. It includes an extensive discussion of hot flashes in older postmenopausal women.

Morin CM. *Insomnia: Psychological Assessment and Management.* Guilford Press, New York, 1993. This comprehensive and clinically oriented treatment manual for insomnia provides detailed guidance for evaluating and treating insomnia complaints with a multifaceted cognitive–behavioral approach

Polo O. Sleep in postmenopausal women: better sleep for less satisfaction. *Sleep* **26**:652–653(2003). This editorial cogently discusses the discrepancy between subjective and objective measures of sleep quality in older women.

PART XI

SLEEP IN THE RESPIRATORY DISORDERS

87

RESPIRATORY CONTROL DURING SLEEP

WILLIAM R. KRIMSKY AND JAMES C. LEITER
Dartmouth Medical School, Lebanon, New Hampshire

INTRODUCTION

Among the automatic processes that make life possible, the control of respiration is unusual. First, there is both volitional and automatic control of respiratory activity; and second, even the most basic respiratory output requires the activity of the central nervous system (CNS). The state of consciousness—wakefulness or sleep—alters CNS processes that regulate respiration in fundamental ways. Wakefulness is associated with an ill-defined but important excitatory stimulus, which has been called the "wakefulness stimulus." Removal of the wakefulness stimulus reduces the drive to breathe, but sleep also modifies the processing of the chemical and mechanical signals that contribute to stable values of carbon dioxide, oxygen, and pH. Furthermore, sleep abrogates the volitional aspects of respiratory control. Sleep is not a homogenous state. Sleep is divided into non-rapid eye movement (NREM) and rapid eye movement (REM) sleep, and respiratory control differs from wakefulness during sleep, but also differs between NREM and REM sleep.

CENTRAL CO₂ CHEMORECEPTORS

The value of carbon dioxide is probably the dominant respiratory stimulus in normal waking individuals at rest. There is a small, stepwise increase in end-tidal P_{CO_2} of ~2–3 torr as individuals progress from wakefulness to NREM sleep and a further ~2–3 torr rise in end-tidal P_{CO_2} in REM sleep. The metabolic rate declines during sleep, but the rise in P_{CO_2} has been attributed to a decrease in alveolar ventilation that is disproportionately large relative to the reduction in the metabolic rate. The increase in P_{CO_2} should theoretically increase minute ventilation in sleep, but this is not seen—ventilation and the slope of the ventilatory response to increasing inspired CO_2 both decline. The reduction in the ventilatory response to CO_2 is modest during NREM sleep and greater during REM sleep. The apparent decrease in the ventilatory sensitivity to hypercapnia has been attributed to a reduction in CO_2 chemosensitivity during sleep. However, the change in sensitivity may originate from either peripheral mechanical factors or central neural factors. As a result of decreased activation of upper airway and accessory muscles, the mechanical load on the respiratory system rises during NREM and REM sleep, and the effectiveness of inspiratory pump muscle activity declines. Thus increased work of breathing, a peripheral factor, reduces ventilation for any level of central respiratory drive whether or not central and peripheral CO_2 chemosensitivity change at all.

Central CO_2 chemosensitivity probably also declines during sleep, and the decrement in CO_2 sensitivity is greater during REM sleep than NREM sleep [1]. Each sleep state is actively controlled, and the neural systems responsible for the control of wakefulness and sleep are connected to the respiratory control system. Sleep is initiated by increased activity of GABAergic, "sleep-on" neurons in the ventrolateral preoptic nucleus of the hypothalamus, and this descending GABAergic inhibition reduces the activity of medullary and pontine respiratory-related nuclei. Furthermore, sleep is associated with a loss of excitation from the reticular activating system. Thus enhanced inhibition and reduced excitation during sleep may both account for

Sleep: A Comprehensive Handbook, Edited by T. Lee-Chiong.
Copyright © 2006 John Wiley & Sons, Inc.

the decrease in CO_2 sensitivity and associated decrease in minute ventilation [2].

There are multiple sites of central CO_2 chemosensitivity in the brainstem in animals, but a small number of studies in patients suggest that multiple sites of chemosensitivity may also exist in humans [3]. The contribution of these CO_2 sensitive regions within the brainstem varies according to state of consciousness—some sites contribute more during NREM sleep, and other sites may be more active during wakefulness. In patients with damage to dorsolateral areas within the medulla, respiration was periodic with prolonged central apneas during both REM and NREM sleep, but not during wakefulness [3, 4]. The ventilatory response to CO_2 was diminished during wakefulness in these patients, but the ventilatory responses to hypoxia or exercise were intact. Similar studies in animals indicate that specific areas of the brainstem alter the respiratory response to CO_2 only during certain states of consciousness. Stimulation of the caudal raphe in rats, for example, affected ventilation only during NREM sleep [5]. Thus some of the CO_2 chemosensory sites within the brainstem seem to contribute to the ventilatory responses to CO_2 in sleep-state specific ways; but the exact process whereby these multiple sites interact and coordinate the response of the whole animal to CO_2 remains unresolved.

The foregoing discussion emphasizes that the sleep associated reduction in the ventilatory response to hypercapnia originates from increased work of breathing and decreased central chemoreceptor sensitivity, but this assumes that the central chemosensory stimulus remains constant across sleep states. This assumption may not be warranted.

The ventilatory response is typically expressed relative to the end-tidal CO_2, but central chemoreceptors probably respond to the intracellular pH of the specific chemosensory neurons rather than the pH or CO_2 level of the blood [6]. The intracellular pH in the neurons depends on the level of CO_2, which depends on the ratio of CO_2 production (metabolism) to local perfusion (rate of CO_2 removal). Both metabolism and blood flow probably change during sleep. For example, cerebral metabolism is high during REM sleep, and blood flow to the brain increases significantly during REM sleep. The increase in blood flow is relatively greater than the increase in metabolism, and tissue CO_2 levels actually fall within some regions of the brain during REM sleep. Thus the arterial or end-tidal CO_2 may not accurately portray the changes in stimulus at the level of the chemosensory neurons, and the decrease in ventilation during sleep may reflect local changes in the chemosensory stimulus within different regions of the brainstem rather than a true alteration in chemosensitivity of the neurons.

We have emphasized the response to hypercapnia, but despite a decrease in ventilatory responsiveness to elevated CO_2 and relative hypoventilation during sleep compared to

wakefulness, CO_2 remains an important respiratory stimulus during sleep. This is apparent from the marked decline in respiratory output that accompanies even mild hypocapnia during sleep. The loss of the excitatory drive to breathe greatly depresses activity of upper airway muscles, but even diaphragmatic activity can be completely suppressed by surprisingly modest hypocapnia (particularly when compared to the waking state in which it is difficult to suppress respiratory activity even with profound hypocapnia).

In summary, the ventilatory response to CO_2 declines as one progresses from wakefulness to NREM sleep and then REM sleep. This apparent change in sensitivity may reflect an increase in the work of breathing, a change in the central chemosensory stimulus, and a change in the relative weighting of inputs from different chemosensory sites, as well as a genuine sleep-related change in central chemosensory sensitivity. Nonetheless, the central chemosensory drive to ventilation remains potent during sleep, and any loss of that drive, as a result of relative hyperventilation and hypocapnia, can be associated with a marked reduction in the respiratory output.

PERIPHERAL CHEMORECEPTORS

The carotid bodies are the main peripheral chemosensors in humans. The aortic bodies, though sensitive to hypoxia, are activated only by very severe hypoxia. The carotid bodies are affected by changes in arterial oxygen partial pressure (not oxygen saturation or content). The neural output from the carotid bodies increases slowly as the Pa_{O_2} falls to values less than ~75 torr in humans and dramatically increases as the Pa_{O_2} falls below approximately 55 torr. The carotid bodies are also sensitive to CO_2, and the interaction between hypoxia and hypercapnia is multiplicative within the carotid body. Even though the hypoxic response of the carotid bodies to the normal arterial P_{O_2} (~100 torr) is small, the carotid bodies contribute 40–55% between 10–40% of total drive to ventilation at rest because of the conjoint sensitivity to CO_2 [7].

The response to hypoxia diminishes as a function of the particular sleep state, although the effect of sleep state on the ventilatory response to hypoxia is less consistent than the reduction of the ventilatory response to hypercapnia during sleep. In some animal studies, sleep state has no effect on the hypoxic ventilatory response. In men and women, REM sleep reduces the ventilatory response to hypoxia. During NREM sleep on the other hand, the ventilatory response to hypoxia decreases in men but not in women [8, 9]. However, the waking response to hypoxia may also be larger in men than in women, so the reduced ventilatory response to hypoxia in NREM sleep in men may reflect gender differences in the hypoxic ventilatory response during wakefulness rather than a specific gender–NREM sleep interaction. The decline in ventilatory responses to hypoxia during sleep may represent a central phenomenon, a modification of peripheral sensitivity, or a combination of both factors. In contrast to studies of central chemosensitivity, there are relatively few animal studies exploring the central mechanisms whereby sleep might modify hypoxic sensitivity, and those studies in which focal areas of the brainstem were inhibited or ablated have not shown consistent sleep-state-related excitatory or inhibitory effects on ventilatory responsiveness to hypoxia. The analysis is complicated because the same brainstem areas involved in the control of respiration also modify the pattern of sleep cycling among wakefulness, NREM sleep, and REM sleep. Thus it can be difficult to interpret responses to interventions; they may reflect specific sleep state effects on respiratory control, but changes may also reflect simply changes in the control of sleep states.

During sleep, the response to hypoxia or hypercapnia has a facet missing from studies during wakefulness. As part of any ventilatory response to respiratory stimulation, in addition to increasing ventilation during sleep, the individual may arouse from sleep and restore waking control of respiration, which usually constitutes a more vigorous response to the stimulus. Restoration of waking mechanisms that support ventilatory responses to hypoxia and hypercapnia is particularly important in restoring or sustaining airway patency. The waking pattern of upper airway muscle activation is often more vigorous and maintains upper airtway patency more effectively. Thus arousal represents an important second line of defense during sleep. Both hypercapnia and hypoxia activate the sympathetic nervous system and are potent arousing stimuli, but the threshold of arousal varies as a function of sleep state. The arousal response to hypoxia tends to occur at lower oxyhemoglobin saturation levels in REM sleep compared to NREM sleep. The arousal threshold for hypoxia is surprisingly low, and a decrease in oxygen saturation to values less than 70% Sao_2 may be required to elicit an arousal from sleep [10]. Arousal thresholds for CO_2 are higher in stages 3 and 4 NREM sleep compared to REM sleep and stages 1 and 2 NREM sleep [8, 11]. The arousal threshold for CO_2 is usually between 55 and 65 torr.

In summary, the ventilatory response to hypoxia is reduced during NREM sleep in men and during REM sleep in both men and women. Arousal from sleep constitutes an important additional response to respiratory stimuli, but arousal thresholds tend to increase as individuals move from NREM sleep to REM sleep. Thus ventilatory responsiveness, whether one considers minute ventilation or arousals, seems to be diminished during sleep.

VAGAL AND UPPER AIRWAY RECEPTORS

Vagal reflexes seem to be more important in respiratory control in animals; the effect of vagal reflexes is less apparent in humans. For example, lung transplant patients, in whom the lungs are denervated, have remarkably normal respiratory responses during wakefulness and sleep [12]. There are modest changes in respiratory pattern during some conditions, but they do not have the respiratory pattern that one might predict based on studies of vagotomized animals—a slow frequency and large tidal volume. On the other hand, when investigators used mechanical ventilators to perturb the respiratory control system in awake and sleeping subjects, the effect of volume-related feedback is apparent in the breathing patterns of normal subjects during wakefulness and sleep [13, 14]. However, complicated bioengineering approaches have to be used to expose the subtle effect of the Hering–Breuer reflex, for example, on the respiratory pattern in normal waking humans. There are two related issues here. Stimuli of sufficiently large magnitude to elicit a big response are sensed by subjects and also elicit a cortically mediated behavioral response, which may mask the vagally mediated reflex responses. Therefore investigators have used very subtle stimuli during wakefulness or studied the responses during sleep or anesthesia. The studies of waking subjects indicate that vagal feedback modulates the respiratory pattern, but the modulation is small. During sleep, unfortunately, the threshold of activation of the Hering–Breuer reflex is probably elevated [15]. Thus it has been difficult to know how large a role vagally mediated reflexes play in respiratory control. Nonetheless, perturbations of the respiratory pattern by mechanical ventilation during sleep clearly reveal differences between normal subjects and lung transplant patients, who lack vagal innervation of the lungs [16]. Even if the respiratory pattern is not changed much, sensory information from the lungs and chest wall is processed by the central nervous system on a breath-by-breath basis since subjects are capable of highly accurate estimates of the relative sizes of different breaths. Thus vagal feedback probably contributes to waking control of ventilation, but the influence of vagal inputs is probably reduced during sleep.

CONTROL OF ACCESSORY MUSCLES OF RESPIRATION, INCLUDING THE UPPER AIRWAY

Although the diaphragm is the primary muscle of respiration, accessory muscles of the chest wall and abdomen and muscles in the upper airway are activated in synchrony with respiration. The muscles of the chest wall augment each inspiratory effort, and upper airway muscles stiffen the airway so that the negative transmural pressures present in the extrathoracic airway during inspiration do not collapse the upper airway. The activation of these accessory muscles is generally reduced during NREM sleep and muscle activity is actively inhibited during REM sleep.

The REM-related inhibition is greatest in axial postural muscles. Whether because of active inhibition or the withdrawal of excitation of chest wall and upper airway muscles, the chest wall becomes more compliant and upper airway resistance rises during sleep. These mechanical factors increase the work of breathing and reduce minute ventilation for any given level of central respiratory drive.

Breuer and Hering first demonstrated that volume-related sensory information arising from expansion of the lungs inhibited respiration by shortening inspiratory time. Expiratory time was shortened as well as the lung deflated; failure of the lung to deflate (persistent expiratory volume-related feedback) prolonged expiratory time. This was, so far as we are aware, the first description of a negative feedback loop in biology. There have been thousands of studies of the effect of the Hering–Breuer reflex on respiratory timing, but this emphasis on respiratory timing may be misplaced. Vagal feedback also modifies that pattern of activity of upper airway muscles. When lung volume increases during inspiration, the activity of upper airway muscles is inhibited, but when there is no volume-related feedback (e.g., the airway is obstructed during inspiration), there is little or no inhibition of upper airway muscle activity and the activation of the upper airway muscles is more vigorous. Presumably this is an effort to relieve the upper airway obstruction. Similar modulation of diaphragmatic muscle activity can be demonstrated, but it is much more modest [17, 18]. A similar effect of increasing tidal volume can be seen as a decrease of respiratory muscle activity in sleeping mechanically ventilated patients [19]. Thus the function of the Hering–Breuer reflex and lung volume-related feedback may be less concerned with respiratory timing and far more important in the control of activation of the muscles of the upper airway and respiratory pump muscles when the upper airway is mechanically loaded or inspiration or expiration is impaired.

The muscles of the upper airway, including the genioglossus, the sternohyoid, and the tensor pallitini, dilate or stiffen the extrathoracic airways and maintain airway patency when activated. The upper airway of humans is unusual in that the hypopharynx is long and relatively compliant compared to other mammals. Therefore the activity of upper airway muscles is particularly important in maintaining upper airway patency and ventilation during both wakefulness and sleep. The activity of the genioglossus, which is relatively easy to measure, has been studied most extensively. Hypoxia and hypercapnia increase genioglossal activity during wakefulness, but the response of the genioglossus to increasing levels of carbon dioxide is reduced during sleep. The genioglossal response to CO_2 has not been studied in REM sleep in humans, but given the effect of REM sleep on nondiaphragmatic skeletal muscle, it is likely that genioglossal activity during REM sleep is diminished even further. In rats, genioglossal activity increased significantly during NREM sleep only at the highest levels of inspired CO_2 (7–9%), whereas the diaphragmatic response to increasing inspired CO_2 occurred at much lower values. In REM sleep, genioglossal activity was almost completely abolished even at inspired CO_2 concentrations as high as 9%. Thus the CO_2 threshold for activation of the genioglossus was elevated compared to wakefulness and elevated compared to the activity of the diaphragm. Finally, the response of the genioglossus to hypoxia is diminished during NREM sleep [20]. Similar sleep-related changes in muscle activity probably occur in the other respiratory-related muscles of the upper airway.

In addition to the chemical inputs mentioned earlier, upper airway muscle activity is also modulated by mechanical factors: distortion of the upper airway, the pattern of inspiratory flow, air temperature, and the particular phase, inspiration versus expiration, of the respiratory cycle all modify patterns of upper airway muscle activity. Thus upper airway muscle activity is greater during inspiration than during expiration, and the transmural pressure necessary to collapse the upper airway is more negative during inspiration. In addition, the critical pressure needed to collapse the airway was reduced (the airway was less collapsible) during nasal breathing compared to breathing through a tracheostomy [21]. Airflow through the nose stimulates upper airway muscle activity more effectively than airflow through the mouth in humans. Thus a variety of stimuli (hypercapnia, nasal breathing, wakefulness, negative pressure within the airway, etc.) elicit compensatory reflex activation of upper airway muscles and enhance upper airway stability. To the extent that any of these is compromised (e.g., mouth breathing as is typical of snorers, cessation of upper airway airflow, hypocapnia, changing sleep states), the probability of respiratory instability will be enhanced. These reflex mechanisms maintaining airway patency are probably all blunted during sleep.

Although the activity of upper airway muscles is altered by a variety of stimuli, the effect of upper airway muscle activity on the mechanical characteristics of the upper airway is not always obvious. Because any given value of airway resistance reflects both the stimuli sensed within the airway and the response of muscle activity to those stimuli, airway resistance is not well correlated with upper airway muscle activity [22, 23]. Thus there is little change in the resistance of the upper airways during REM sleep versus NREM sleep despite the decrease in muscle tone associated with the former. However, one cannot know whether the decrease in upper airway muscle tone is permitted because airway resistance did not change much or other factors maintained airway patency so that upper airway resistance did not change as a consequence of the decline in upper airway muscle activity.

SUMMARY

Recent advances in studies of sleep and respiratory control indicate that sleep is a regulated state in which the regulation of different elements of the respiratory system is heterogeneous. Thus upper airway muscle activity seems more dependent on waking mechanisms to maintain activity than is the diaphragm. Those areas of the brainstem involved in the control of sleep state also seem to play an important role in responses to a variety of respiratory stimuli, such as hypercapnia, hypoxia, and a variety of upper airway reflexes.

REFERENCES

1. Rist KE, Daubenspeck JA, McGovern JF. Effects of non-REM sleep upon respiratory drive and the respiratory pump in humans. *Respir Physiol* **63**:241–256(1986).

2. Joseph V, Pequignot JM, Van Reeth O. Neurochemical perspectives on the control of breathing during sleep. *Respir Physiol Neurobiol* **130**:253–263(2002).

3. Morrell MJ, Heywood P, Moosavi SH, et al. Unilateral focal lesions in the rostro-lateral medulla influence chemosensitivity and breathing during wakefulness, sleep, and exercise. *J Neurol Neurosurg Psychiatry* **67**:637–645(1998).

4. Manning HL, Leiter JC. Respiratory control and respiratory sensation in a patient with a ganglioma within the dorsocaudal brain stem. *Am J Respir Crit Care Med* **161**:2100–2106(2000).

5. Nattie EE, Li A. CO_2 dialysis in the medullary raphe of the rat increases ventilation in sleep. *J Appl Physiol* **90**:1247–1257 (2001).

6. Putnam RW, Filosa JA, Ritucci NA. Cellular mechanisms involved in CO_2 and acid sensing in chemosensitive neurons. *Am J Physiol* **287**:C1493–C1526(2004).

7. Pan LG, Forster HV, Martino P, et al. Important role of carotid afferents in control of breathing. *J Appl Physiol* **85**:1299–1306(1998).

8. Berthon-Jones M, Sullivan CE. Ventilatory and arousal responses to hypoxia in sleeping humans. *Am Rev Respir Dis* **125**:632–639(1982).

9. Douglas NJ, White DP, Weil JV, et al. Hypoxic ventilatory response decreases during sleep in normal men. *Am Rev Respir Dis* **125**:286–289(1982).

10. Berry RB, Gleeson K. Respiratory arousal from sleep: mechanisms and significance. *Sleep* **20**:654–675(1997).

11. Berthon-Jones M, Sullivan CE. Ventilation and arousal responses to hypercapnia in normal sleeping humans. *J Appl Physiol* **57**:59–67(1984).

12. Sanders MH, Costantino JP, Owens GR, et al. Breathing during wakefulness and sleep after human heart–lung transplantation. *Am Rev Respir Dis* **140**:45–51(1989).

13. Henke KG, Arias A, Skatrud JB, et al. Inhibition of inspiratory muscle activity during sleep: chemical and nonchemical influences. *Am Rev Respir Dis* **138**:8–15(1988).

14. Simon PM, Dempsey JA, Landry DM, et al. Effect of sleep on respiratory muscle activity during mechanical ventilation. *Am Rev Respir Dis* **147**:32–37(1993).

15. Iber C, Simon P, Skatrud JB, et al. The Breuer–Hering reflex in humans. *Am J Respir Crit Care Med* **152**:217–224(1995).

16. Simon PM, Habel AM, Daubenspeck JA, et al. Vagal feedback in the entrainment of respiration to mechanical ventilation in sleeping humans. *J Appl Physiol* **89**:760–769(2000).

17. Bartlett D Jr, St John WM. Influence of lung volume on phrenic, hypoglossal and mylohyoid nerve activities. *Respir Physiol* **73**:97–110(1988).

18. Kuna ST. Inhibition of inspiratory upper airway motoneuron activity by phasic volume feedback. *J Appl Physiol* **60**:1373–1379(1986).

19. Wilson CR, Satoh M, Skatrud JB, et al. Non-chemical inhibition of respiratory motor output during mechanical ventilation in sleeping humans. *J Physiol* (*Lond*) **518**:605–618(1999).

20. Parisi RA, Santiago TV, Edelman NH. Genioglossal and diaphragm EMG responses to hypoxia during sleep. *Am Rev Respir Dis* **138**:610–616(1988).

21. Schneider H, Boudewyns A, Smith PL, et al. Modulation of upper airway collapsibility during sleep: influence of respiratory phase and flow regimen. *J Appl Physiol* **93**:1365–1376 (2002).

22. Henke KG, Badr MS, Skatrud JB, et al. Load compensation and respiratory muscle function during sleep. *J Appl Physiol* **72**:1221–1234(1992).

23. Wiegand DA, Latz B, Zwillich CW, et al. Upper airway resistance and geniohyoid muscle activity in normal men during wakefulness and sleep. *J Appl Physiol* **69**:1252–1261(1990).

ADDITIONAL READING

Bellingham MC, Ireland MF. Contribution of cholinergic systems to state-dependent modulation of respiratory control. *Respir Physiol Neurobiol* **131**:135–144(2002).

Berger AJ, Mitchell RA, Severinghaus JW. Regulation of respiration. *N Engl J Med* **297**:92–97;138–143;194–201(1977).

Haxhiu MA, Mack SO, Wilson CG, Feng P. Sleep networks and the anatomic and physiologic connections with respiratory control. *Front Biosci* **8**:d946–d962(2003).

Nattie EE. Central chemosensitivity, sleep, and wakefulness. *Respir Physiol Neurobiol* **129**:257–268(2001).

88

ASTHMA

DAVID A. BEUTHER AND RICHARD J. MARTIN

National Jewish Medical and Research Center, Denver, Colorado

INTRODUCTION

For centuries, it has been observed that the symptoms of asthma tend to worsen at night. In 1698, Dr. John Floyer stated about his own asthma, "I have observed the fit always to happen after sleep in the night, when the nerves are filled with windy spirits and the heat of the bed has rarefied the spirits and humours" [1]. In more recent times, population-based surveys have demonstrated that a majority of asthmatics have nocturnal asthma symptoms, and that a disproportionate number of deaths from asthma occur at night and in the early morning hours. In a survey of 7729 British asthmatic patients, 74% reported awakening at least one night per week with asthma symptoms [2]. Perhaps more impressive is that most deaths from asthma occur between the hours of 6 p.m. and 3 a.m. [3]. Nocturnal asthma symptoms are additionally a marker of inadequate asthma control [4]. While nocturnal asthma has a distinct and complex pathogenesis that is still incompletely understood, optimal management of both nonnocturnal and nocturnal asthmatics will depend on a thorough understanding of the mechanisms responsible for the worsening of asthma at night.

CIRCADIAN RHYTHMS IN ASTHMA: PHYSIOLOGY, HORMONAL, STEROID, AND β-2 ADRENERGIC

The circadian variation in peak expiratory flow rates, FEV_1, and bronchial hyperresponsiveness in asthmatics and

Correspondence to: Richard J. Martin, National Jewish Medical and Research Center, 1400 Jackson Street, Denver, CO 80206.

Sleep: A Comprehensive Handbook, Edited by T. Lee-Chiong.

normal subjects has been well established by several [5–7], with peak lung function usually occurring at approximately 4 p.m. and minimum lung function at approximately 4 a.m. (Figure 88.1). The cause of this circadian variation is probably multifactorial. Kraft and colleagues [8] measured serum cortisol levels in nocturnal asthmatics, nonnocturnal asthmatics, and normal controls and found that while there was a significant difference in FEV_1 among the groups, the differences in cortisol levels at different time points were not significant. However, Sutherland et al. [9] showed that nocturnal asthmatics have higher corticotropin levels than nonnocturnal asthmatics or controls, but not so for cortisol levels. This study suggests that nocturnal asthma may be marked by a blunted adrenal responsiveness to corticotropin. Furthermore, Kraft and colleagues [10] have data to suggest that reduced steroid responsiveness at night might also contribute to nocturnal asthma. They took macrophages from 4 a.m. and 4 p.m. nocturnal asthmatic bronchoalveolar lavage specimens, and measured dexamethasone-induced suppression of IL-8 and TNF-α. The nocturnal asthmatics exhibited less suppression of IL-8 and TNF-α at 4 a.m. compared to 4 p.m. There was no difference found in macrophages obtained from nonnocturnal asthmatics. Epinephrine also varies in a circadian fashion [11], with trough levels during the early morning hours. Additionally, there is a reduction in beta-receptor density and function at night in patients with nocturnal asthma [12]. Turki and colleagues [13] went on to show that there is a genetic polymorphism in the β_2-adrenergic receptor (a glycine substitution at position 16, or Gly 16) that causes accelerated downregulation of the receptor, and that nocturnal asthmatics carry this polymorphism more

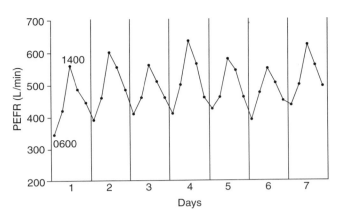

Figure 88.1 Circadian change in peak expiratory flow rate (PEFR) in an asthmatic patient. Measurements were made at 06:00, 10:00, 14:00, 18:00, and 22:00 hours over 7 days. (Adapted from [5], with permission.)

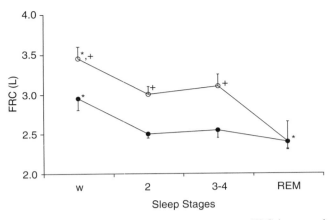

Figure 88.2 Effect of different sleep stages on FRC in normal subjects ($\bullet$) and asthmatic subjects ($\circ$). Error bars, SE. $^*p < 0.05$, wakefulness (W) and REM sleep versus all other stages in asthmatic subjects and W versus all other stages in normal subjects; $^+p < 0.05$, asthmatic subjects versus normal subjects. (Adapted from [16], with permission.)

frequently than nonnocturnal asthmatics. Thus reduced steroid and beta-receptor responsiveness may play a major role in the pathophysiology of nocturnal asthma.

Melatonin, a key hormonal regulator of circadian rhythm, has recently been shown to act as a proinflammatory hormone in nocturnal asthma [14, 15]. In the 2002 study, melatonin-stimulated peripheral blood mononuclear cells produced significantly elevated levels of IL-1, IL-6, and TNF-α in nocturnal versus nonnocturnal asthmatics. Furthermore, Sutherland et al. [15] showed that, in nocturnal asthmatics, melatonin concentrations were elevated and phase delayed compared to nonnocturnal asthmatics and controls, and that elevated melatonin levels correlated inversely to the overnight change in FEV_1, demonstrating another mechanism contributing to a nocturnal worsening of asthma.

AIRWAY RESISTANCE AND LUNG VOLUME DURING SLEEP

During sleep, both asthmatics and normal control subjects experience a significant decrease in lung volume [16] (Figure 88.2). While asthma is characterized by airflow obstruction and hyperinflation in the awake state, Ballard's study [16] shows that asthmatics have a greater decline in lung volume than normal controls, and that during REM sleep, their lung volume falls to that of the normal controls. This more dramatic fall seen in asthmatic lung volumes may be the result of additional nocturnal factors associated with the sleeping state.

Lung volume is not the only determinant of worsening lung function at night in asthma. To further investigate the role of posture and lung volume in the pathogenesis of nocturnal asthma, Ballard and colleagues [17] measured spirometry and methacholine responsiveness in eight asthmatic

subjects after daytime chest strapping to produce their previously measured nocturnal functional residual capacity (FRC). The experiment was performed in both the supine and upright postures, while awake. Bronchial hyperresponsiveness was unaffected by this treatment, but FEV_1 was significantly decreased after strapping, and to a greater degree in subjects maintained in the supine position. This suggests that something about the supine posture itself, independent of lung volume, might be an additional contributor to the decline in lung function seen in nocturnal asthma.

These studies have shown that, in nocturnal asthma, lung volume falls to a greater degree than in normal controls, and that worsening of lung function is not entirely dependent on changes in volume. One explanation for these findings is that nocturnal asthma is characterized by a worsening of inflammation at night, and that lung volume changes are a result, and not a cause, of nocturnal asthma.

In a study by Martin and colleagues [18], restoration of normal lung volume during sleep through the use of a negative pressure poncho cuirass did not prevent nocturnal asthma. There was no change in the overnight decline in FEV_1, and the increase in bronchial hyperresponsiveness was unchanged. The authors concluded that the fall in functional residual capacity at night was a result, not a cause, of a worsening of lung function at night. Therefore the pathogenesis of nocturnal asthma may not depend on a decrease in functional residual capacity at night; there may be a physiologic, sleep-related uncoupling of the airway and parenchyma, possibly due to a worsening of airway inflammation at night.

Typically, awake asthmatics experience an increase in FRC along with an increase in peripheral airway resistance during an attack, as a result of worsening airflow limitation and gas trapping. However, Irvin and colleagues [19] have demonstrated that there is a loss of dependence of airway resistance on lung volume in nocturnal asthma, and they hypothesize that the nocturnal elevation of airway resistance is due to some factor associated with sleep, and that the respiratory system, by allowing functional residual capacity to fall, fails to defend airway caliber. Thus it appears that the fall in lung volume with sleep may not be essential in its pathogenesis, and that there appear to be other factors driving nocturnal asthma that may act through an uncoupling of the airway and pulmonary parenchyma. One of these factors may be inflammation.

AIRWAY INFLAMMATION

One explanation for this uncoupling of the airway and parenchyma in nocturnal asthma is airway inflammation. Asthma is an inflammatory disorder, and a worsening of symptoms at night could be due to an increase in inflammation. Martin and colleagues [20] showed that there is an increase in cellular inflammation in nocturnal asthmatics (Figure 88.3). They found elevated numbers of total cells, leukocytes, neutrophils, and eosinophils in bronchoalveolar lavage specimens from nocturnal asthmatics at 4 a.m., compared to nonnocturnal asthmatics. Kraft et al. [21] obtained 4 a.m. and 4 p.m. endobronchial brushing specimens from proximal and distal airways in nocturnal asthmatics and nonnocturnal asthmatics. Specimens were stained for epithelial cell markers of inflammation. They found that CD51 (vitronectin and fibronectin receptor) was significantly increased at 4 a.m. in the distal brushings from nocturnal asthmatics, and that expression of CD51 correlated

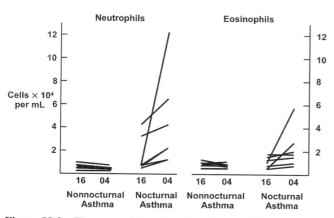

Figure 88.3 The neutrophil and eosinophil counts for each subject are shown. Between 16:00 and 04:00 hours, the nocturnal asthma group had a significant increase in neutrophils ($p < 0.05$) and eosinophils ($p < 0.05$). Similarly, between groups at 04:00 hours, the nocturnal asthmatics had a significant increase in neutrophils ($p < 0.01$) and eosinophils ($p < 0.05$). (Adapted from [20], with permission.)

to the degree of airway obstruction. These studies support the role of increased airway inflammation at night in the worsening of asthma, and this inflammation may be the source of the uncoupling of airway and parenchyma suggested in the lung volume studies previously mentioned.

Peripheral airway tissue inflammation was further documented by Kraft et al. [22], using transbronchial biopsies performed at 4 a.m. and 4 p.m. Nocturnal asthmatics were found to have a significantly greater number of asthma controller cells—CD4$^+$ T-lymphocytes, effector cells, and eosinophils—in alveolar tissue than in nonnocturnal asthmatics, suggesting that distal lung inflammation may play an important role in nocturnal asthma. Furthermore, they performed measurements of distal airway resistance using a wedged bronchoscope technique [23] and were able to show that nocturnal asthmatics had a higher peripheral airway resistance than nonnocturnal asthmatics or controls. These and other studies have suggested that nocturnal asthma may be a peripheral inflammatory process, involving more than just the central airways.

VENTILATORY DRIVE DURING SLEEP

One contributor to the increase in nocturnal symptoms and higher mortality associated with nocturnal asthma might be a blunted ventilatory drive during sleep. In normal sleeping subjects, ventilatory drive in response to hypercapnia is blunted in all stages of sleep, especially REM sleep [24]. Sleep deprivation, a common complaint among patients with nocturnal symptoms of asthma, may contribute to the decreased ventilatory response to hypercapnia [25]. However, this study was in nonasthmatic subjects. It is unclear whether nocturnal asthmatics demonstrate an excessively blunted ventilatory drive during sleep.

PARASYMPATHETIC SYSTEM

An increase in vagal parasympathetic tone can contribute to an increase in airway resistance. For example, Morrison and colleagues studied the effect of atropine given at 4 a.m. on lung function in asthmatics [26, 27]. This group found that atropine significantly reverses the fall in lung function at night. As will be discussed later, other disorders that may increase vagal tone have been implicated in contributing to nighttime bronchoconstriction, including obstructive sleep apnea, postnasal drip causing laryngeal irritation, and gastroesophageal reflux disease.

AIRWAY SECRETIONS

Many patients with asthma also suffer from allergic rhinitis. Rhinitis often becomes more bothersome at night and could worsen asthma through aspiration of secretions or from irritation of the upper airway, leading to reflex bronchoconstriction [28]. Asthmatic patients also demonstrate reduced mucociliary clearance [29], a problem that worsens even more during sleep. Additionally, the cough reflex is also blunted during REM sleep [30], which could lead to excessive accumulation of secretions, causing airway narrowing.

GASTROESOPHAGEAL REFLUX

Gastroesophageal reflux disease (GERD) has commonly been associated with asthma, but some studies strongly question its significance. The prevalence of GERD in asthma is high and has been reported to be as high as 80% [31]. In this investigation by Sontag and colleagues, asthmatics had significantly decreased lower esophageal sphincter pressures, greater esophageal acid exposure times, more frequent episodes of GERD, and longer esophageal acid clearance times. Other investigators [32] have documented that episodes of GERD identified by pH probe were associated with an increase in airway resistance in asthmatics, and that the degree of bronchoconstriction for each event was related to the duration of the reflex episode. The mechanisms proposed to explain how GERD might cause bronchoconstriction include microaspiration of stomach contents leading to airway irritation, as well as the stimulation of esophageal vagal receptors by stomach acid. In a study by Harding et al. [33], asthmatics demonstrated a decrease in peak expiratory flow rate in response to esophageal acid instillation, but there was no demonstration of aspiration by proximal esophageal pH monitoring. Therefore, if there is a pathophysiologic link between asthma and GERD, it is likely to be a result of esophageal acid stimulation of vagal receptors and resultant vagal-mediated bronchoconstriction.

Several other studies have significantly questioned the importance of GERD in nocturnal asthma. For example, direct installation of acid into the lower esophagus during sleep does not predictably cause worsening bronchoconstriction [34], contradicting the vagal stimulation hypothesis. In another study of 42 moderate to severe asthmatics with GERD [35], there was no association between reflux, as measured by pH probe at the proximal and distal esophagus, and asthma symptoms or peak expiratory flows. Furthermore, in a double-blind, placebo-controlled study of 20 asthmatics with GERD treated with omeprazole 40 mg daily or placebo [36], there was no significant difference in FEV_1, forced vital capacity (FVC), bronchial hyper-responsiveness, or diurnal variation in peak flows between the treated and placebo groups. In this study, while GERD symptoms were significantly improved with omeprazole, there was no improvement in symptoms of asthma or beta-agonist use.

AIRWAY COOLING

Core body temperature also has a circadian variability, with a decrease of approximately 1 °C between 2 a.m. and 4 a.m. [37]. Airway temperature changes are thought to be important in causing exercise-induced bronchospasm, and there is some evidence that experimentally lowering core body temperature by as little as 0.7 °C can trigger an asthma attack [38]. Interestingly, this effect is blocked by the inhalation of warm, humidified air [39]. One could presume that decreased body temperature should at least contribute to worsening airflow limitation at night, and that this effect may be more prominent in patients who bypass the heated humidification function of the nose and nasopharynx, such as those with severe rhinitis or those who demonstrate significant mouth breathing at night.

CHRONOTHERAPEUTIC TREATMENT APPROACHES

A better understanding of chronobiology and nocturnal asthma translates into unique and improved ways to treat nocturnal asthma. Most of these interventions focus on chronopharmacology, where drug dosing and schedule are altered to achieve maximum therapy during the night, while attempting to minimize drug toxicities.

Corticosteroid dosing can be adjusted to maximize time-dependent variations in efficacy while minimizing known side effects, such as adrenal suppression. If corticosteroid administration is timed to the natural peak in production of endogenous corticosteroid production, efficacy may be maximized while minimizing adrenal suppression. Reinberg et al. [40] gave 12 asthmatic boys a single variably timed 40 mg dose of methylprednisolone or placebo. Peak expiratory flows improved more when the steroid was given at 3 p.m. or 7 a.m., compared to 7 p.m. or 3 a.m. The same investigators [41] performed a similar study in eight asthmatic adults and found that dosing at 8 a.m. and 3 p.m. was more effective than dosing at 3 p.m. and 8 p.m., again suggesting that steroid dosing in the morning or early afternoon was superior to dosing in the evening. Beam and colleagues [42] showed that a single 50 mg dose of prednisone at 3 p.m. reduced the overnight fall in FEV_1 and airway inflammation in a group of seven asthmatics, while dosing at 8 a.m. or 8 p.m. was not effective in preventing the nocturnal worsening of lung function or inflammation.

Adrenal suppression was not worsened by this alternate dosing strategy; in fact, single daily dosing minimizes this adverse effect. In six normal men, Grant and colleagues [43] administered 8 mg triamcinolone either as a single 8 a.m. dose or in four 2 mg doses divided throughout the day. The divided dosing was associated with a reduction in morning plasma and 24 hour urinary 17-hydroxycorticosteroid levels, while the single dosing regimen showed no suppression.

Similar dose timing studies have looked at inhaled steroids, with similar chronopharmacologic effects. Pincus and colleagues [44] found that when inhaled triamcinolone was given as either a single 800 μg dose at 3 p.m. or four divided 200 μg doses throughout the day, there was no difference in efficacy or adverse systemic effects, suggesting that single-daily dosing in the afternoon might improve compliance without sacrificing efficacy. Further studies by the same group [45] showed that once-daily dosing of inhaled triamcinolone at 5:30 p.m. was superior to once-daily dosing at 8 a.m., but no different than four-times daily divided dosing, either in efficacy or systemic side effects. Thus optimal dosing of once-daily inhaled steroids is between 3 p.m. and 5:30 p.m.

Other therapies directed at improving nocturnal airflow limitation have been shown to be helpful. Martin and colleagues [46] have shown that a one-time dose of a long-acting theophylline at 7 p.m. results in higher drug levels and improved lung function overnight compared to traditional twice-daily dosing. These patients did not have any worsening of side effects or difficulty with sleep and had lower daytime theophylline levels. This suggests that, in nocturnal asthma, a once-daily dose timed to achieve peak levels in the early morning hours may be superior to achieving a continuous 24 hour therapeutic level of theophylline. This is another example of how dosage timing can maximize therapeutic efficacy, while minimizing toxic side effects.

In the past, inhaled bronchodilators have been limited in their efficacy in nocturnal asthma due to their short duration of action. However, salmeterol and other long-acting beta agonists have the potential to be effective for this population of asthmatics. In a study of ten patients with nocturnal asthma treated with salmeterol 100 μg twice daily or placebo [47] the salmeterol-treated subjects had significant improvement in nocturnal awakenings and beta-agonist free days. Although both the morning and evening peak flow measurements were increased, the percent overnight fall remained the same.

TREATMENT OF OTHER CONTRIBUTING CONDITIONS

In the treatment of nocturnal asthma, other nonpulmonary contributing conditions should be treated as aggressively as possible. In the patient with allergic rhinitis, aggressive nasal anti-inflammatory therapy such as topical steroids should be used along with nasal saline rinses (250 mL b.i.d.) to minimize sinus drainage and irritation at night. While GERD may only play a small role in nocturnal

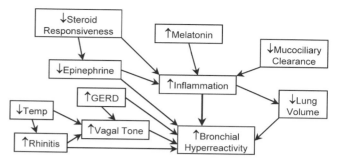

Figure 88.4   Nocturnal asthma is caused by a complex interaction of circadian and noncircadian factors.

asthma, it should be managed aggressively with behavioral modification, weight loss, elevation of the head of the bed, and medical or surgical therapy, if necessary. Since there is evidence that airway cooling may worsen nocturnal asthma, an effort should be made to warm and humidify the patient's bedroom air. While these efforts do not replace standard asthma therapy or the chronotherapeutic approaches mentioned earlier, in certain patients they may make a significant improvement in nocturnal asthma control.

CONCLUSIONS

Circadian rhythms clearly play an important role in asthma. The pathophysiology of nocturnal asthma is complex, with many factors contributing to a worsening of lung inflammation and airflow limitation at night (Figure 88.4). While nocturnal changes in lung volume and airway diameter are probably important, there appears to be something that happens during sleep that independently contributes to a worsening of lung function at night. Certainly other comorbid processes such as rhinitis and GERD can contribute to a worsening of symptoms, but they probably do not represent the primary cause of nocturnal asthma. Based on the previously mentioned studies, perhaps the most intriguing hypothesis to explain this phenomenon is that nocturnal asthma is primarily an inflammatory disorder, with a mechanism that is distinct from nonnocturnal asthma, and a pathophysiology that centers around a worsening of both central and peripheral lung inflammation at night.

REFERENCES

1. Floyer J. *A Treatise of the Asthma*. R Witkin and W Inngs, London, 1698.

2. Turner-Warwick M. Epidemiology of nocturnal asthma. *Am J Med* **85**(1B):6–8(1988).

3. Douglas NJ. Asthma at night. *Clin Chest Med* **6**(4):663–674(1985).

4. Dales RE, Schweitzer I, Kerr P, Gougeon L, Rivington R, Draper J. Risk factors for recurrent emergency department visits for asthma. *Thorax* **50**:520–524(1995).

5. Soutar CA, Costello J, Ijaduola O, Turner-Warwick M. Nocturnal and morning asthma. *Thorax* **30**:436–440(1975).

6. Hetzel MR, Clark TJH. Comparison of normal and asthmatic circadian rhythms in peak expiratory flow rate. *Thorax* **35**:732–738(1980).

7. Martin RJ, Cicutto LC, Ballard RD. Factors related to the nocturnal worsening of asthma. *Am Rev Respir Dis* **141**:33–38 (1990).

8. Kraft M, Pak J, Martin RJ. Serum cortisol in asthma: marker of nocturnal worsening of symptoms and lung function? *Chronobiol Int* **15**(1):85–92(1998).

9. Sutherland ER, Ellison MC, Kraft M, Martin RJ. Altered pituitary–adrenal interaction in nocturnal asthma. *J Allergy Clin Immunol* **112**:52–57(2003).

10. Kraft M, Hamid Q, Chrousos GP, Martin RJ, Leung DYM. Decreased steroid responsiveness at night in nocturnal asthma. *Am J Respir Crit Care Med* **163**:1219–1225(2001).

11. Barnes P, Fitzgerald G, Brown M, Dollery C. Nocturnal asthma and changes in circulating epinephrine, histamine, and cortisol. *N Engl J Med* **303**(5):263–267(1980).

12. Szefler SJ, Ando R, Cicutto LC, Surs W, Hill MR, Martin RJ. Plasma histamine, epinephrine, cortisol, and leukocyte beta-adrenergic receptors in nocturnal asthma. *Clin Pharmacol Ther* **49**(1):59–80(1991).

13. Turki J, Pak J, Green SA, Martin RJ, Liggett SB. Genetic polymorphisms of the β_2-adrenergic receptor in nocturnal and non-nocturnal asthma. *J Clin Invest* **95**:1635–1641(1995).

14. Sutherland ER, Martin RJ, Ellison MC, Kraft M. Immunomodulatory effects of melatonin in asthma. *Am J Respir Crit Care Med* **166**:1055–1061(2002).

15. Sutherland ER, Ellison MC, Kraft M, Martin RJ. Elevated serum melatonin is associated with the nocturnal worsening of asthma. *J Allergy Clin Immunol* **112**:513–517(2003).

16. Ballard RD, Irvin CG, Martin RJ, Pak J, Pandey R, White DP. Influence of lung volume in asthmatic patients and normal subjects. *J Appl Physiol* **68**(5):2034–2041(1990).

17. Ballard RD, Pak J, White DP. Influence of posture and sustained loss of lung volume on pulmonary function in awake asthmatic subjects. *Am Rev Respir Dis* **144**:499–503(1991).

18. Martin RJ, Pak J, Irvin CG. Effect of lung volume maintenance during sleep in nocturnal asthma. *J Appl Physiol* **75**(4):1467–1470(1993).

19. Irvin CG, Pak J, Martin RJ. Airway–parenchyma uncoupling in nocturnal asthma. *Am J Respir Crit Care Med* **161**:50–56(2000).

20. Martin RJ, Cicutto LC, Smith HR, Ballard RD, Szefler SJ. Airways inflammation in nocturnal asthma. *Am Rev Respir Dis* **143**:351–357(1991).

21. Kraft M, Striz I, Georges G, Umino T, Takigawa K, Rennard S, Martin RJ. Expression of epithelial markers in nocturnal asthma. *J Allergy Clin Immunol* **102**:376–381(1998).

22. Kraft M, Martin RJ, Wilson S, Djukanovic R, Holgate ST. Lymphocyte and eosinophil influx into alveolar tissue in nocturnal asthma. *Am J Respir Crit Care Med* **159**:28–234(1999).

23. Kraft M, Pak J, Martin RJ, Kaminsky D, Irvin C. Distal lung dysfunction at night in nocturnal asthma. *Am J Respir Crit Care Med* **163**:1551–1556(2001).

24. Berthon-Jones M, Sullivan CE. Ventilation and arousal responses to hypercapnia in normal sleeping humans. *J Appl Physiol* **57**(1):59–67(1984).

25. Shiffman PL, Trontell MC, Mazar MF, Edelman NH. Sleep deprivation decreases ventilatory response to CO_2 but not load compensation. *Chest* **84**(6):695–698(1983).

26. Morrison JFJ, Pearson SB, Dean HG. Parasympathetic nervous system in nocturnal asthma. *BMJ* **296**:1427–1429(1988).

27. Morrison JFJ, Pearson SB. The effect of the circadian rhythm of vagal activity on bronchomotor tone in asthma. *Br J Clin Pharmacol* **28**:545–549(1989).

28. Javaid A, Sykes AP, Ayres JG. Does aspiration of saliva trigger nocturnal asthma? *J Pak Med Assoc* **44**:60–61(1994).

29. Bateman JR, Pavia D, Clarke SW. The retention of lung secretions during the night in normal subjects. *Clin Sci Mol Med Suppl* **55**(6):523–527(1978).

30. Sullivan CE, Murphy E, Kozar LF, Phillipson EA. Waking and ventilatory responses to laryngeal stimulation in sleeping dogs. *J Appl Physiol Respir Environ Exercise Physiol* **45**(5):681–689(1978).

31. Sontag SJ, O'Connell S, Khandelwal S, Miller T, Nemchausky B, Schnell TG, Serlovsky R. Most asthmatics have gastroesophageal reflux with or without bronchodilator therapy. *Gastroenterology* **99**(3):613–620(1990).

32. Cuttitta G, Cibella F, Visconti A, Scichilone N, Bellia V, Bonsignore G. Spontaneous gastroesophageal reflux and airway patency during the night in adult asthmatics. *Am J Respir Crit Care Med* **161**:177–181(2000).

33. Harding SM, Schan CA, Guzzo MR, Alexander RW, Bradley LA, Richter JE. Gastroesophageal reflux-induced bronchoconstriction. *Chest* **108**:1220–1227(1995).

34. Tan WC, Martin RJ, Pandey R, Ballard RD. Effects of spontaneous and simulated gastroesophageal reflux on sleeping asthmatics. *Am Rev Respir Dis* **141**:1394–1399(1990).

35. Ekström T, Tibling L. Gastro-oesophageal reflux and triggering of bronchial asthma: a negative report. *Eur J Respir Dis* **71**:177–180(1987).

36. Teichtahl H, Kronborg IJ, Yeomans ND, Robinson P. Adult asthma and gastro-oesophageal reflux: the effects of omeprazole therapy on asthma. *Aust N Z J Med* **26**(5):671–676(1996).

37. Webb P. The physiology of heat regulation. *Am J Physiol* **268**: R838–R850(1995).

38. Chen WY, Horton DJ. Airways obstruction in asthmatics induced by body cooling. *Scand J Respir Dis* **59**(1):13–20(1978).

39. Chen WY, Chai H. Airway cooling and nocturnal asthma. *Chest* **81**(6):675–680(1982).

40. Reinberg A, Halberg F, Falliers C. Circadian timing of methylprednisolone effects in asthmatic boys. *Chronobiologia* **1**:333–347(1974).

41. Reinberg A, Gervais P, Chaussade M, Fraboulet G, Duburque B. Circadian changes in effectiveness of corticosteroids in

eight patients with allergic asthma. *J Allergy Clin Immunol* **71**:425–433(1983).

42. Beam WR, Weiner DE, Martin RJ. Timing of prednisone and alterations of airways inflammation in nocturnal asthma. *Am Rev Respir Dis* **146**:1524–1530(1992).

43. Grant S, Forsham P, DiRaimondo V. Suppression of 17-hydroxycorticosteroids in plasma and urine by single and divided doses of triamcinolone. *N Engl J Med* **273**:1115–1118(1965).

44. Pincus DJ, Szefler SJ, Ackerson LM, Martin RJ. Chronotherapy of asthma with inhaled steroids: the effect of dosage timing on drug efficacy. *J Allergy Clin Immunol* **95**:1172–1178(1995).

45. Pincus DJ, Humeston TR, Martin RJ. Further studies on the chronotherapy of asthma with inhaled steroids: the effect of dosage timing on drug efficacy. *J Allergy Clin Immunol* **100**:771–774(1997).

46. Martin RJ, Cicutto LC, Ballard RD, Goldenheim PD, Cherniack RM. Circadian variations in theophylline concentrations and the treatment of nocturnal asthma. *Am Rev Respir Dis* **139**:475–478(1989).

47. Kraft M, Wenzel SE, Betinger CM, Martin RJ. The effect of salmeterol on nocturnal symptoms, airway function, and inflammation in asthma. *Chest* **111**:1249–1254(1997).

CHRONIC OBSTRUCTIVE PULMONARY DISEASE AND SLEEP

CONRAD IBER

Hennepin County Medical Center, Minneapolis, Minnesota

DEFINITION

In 1998, the World Health Organization and the National Heart, Lung and Blood Institute jointly published the initial findings of the Global Initiative for Chronic Obstructive Lung Disease (GOLD) [1]. The GOLD consensus defined chronic obstructive pulmonary disease (COPD) as "a disease state characterized by airflow limitation that is not fully reversible. The airflow limitation is usually both progressive and associated with an abnormal inflammatory response of the lungs to noxious particles or gases." Incorporated into the definition is a staging of disease that is based on clinical and physiologic patterns and a deemphasis of the older terminologies of chronic bronchitis and emphysema.

PATHOPHYSIOLOGY

Lung injury in COPD occurs as a result of oxidant stress from exogenous agents such as cigarette smoke and an imbalance between endogenous proteinases and antiproteinases within the lung. Chronic inflammatory changes center in the smaller airways though the elastic skeleton of the lung and the vasculature are also damaged. Patients with airflow obstruction demonstrate an increased number of macrophages, activated neutrophils, and CD8$^+$ lymphocytes within the lung [2, 3]. Permanent pathologic changes occur in both the small airways and gas exchanging surfaces of the lung tissue in COPD. The pathologic changes are heterogeneous with varying degrees of airway inflammation, mucus hypersecretion, airway remodeling, and alveolar destruction with coalescence into poorly functioning emphysematous areas.

When there is a predominance of alveolar destruction and coalescence in COPD, the pathologic change is termed emphysema. When airway narrowing predominates in COPD, the condition is called chronic bronchitis. Chronic bronchitis is also an epidemiologic term used to describe the accompanying symptom complex of daily productive cough of at least three months duration in two consecutive years. Many chronic lung conditions are not associated with COPD. Chronic inflammatory conditions such as asthma or tuberculosis may produce chronic bronchitis without causing the irreversible airflow obstruction of COPD. Similarly, conditions such as kyphosis may produce dyspnea and respiratory failure without causing COPD.

The most common mechanism for the development of chronic obstructive pulmonary disease is smoking, although air pollution, occupational exposures, noxious gases, and hereditary deficiency of alpha-1-antitrypsin are also implicated. COPD may also occur as a result of chronic inflammatory airway disease such as cystic fibrosis. All of the structural changes of COPD can result in the development of progressive dyspnea and impairment of the elimination of carbon dioxide from the blood and delivery of oxygen into the blood.

The worsening of gas exchange in COPD is a function of the severity of the components of airway narrowing and alveolar destruction. Increased resistance within the airways also produces increased work of breathing, and incomplete emptying of the lung results in hyperinflation and inefficiency of respiratory muscles. Dynamic hyperinflation during rapid respiratory rates substantially contributes to subjective dyspnea [4].

EPIDEMIOLOGY AND PUBLIC HEALTH

COPD is a progressive irreversible lung condition that becomes symptomatic primarily in middle-aged and elderly adults. National statistics of physician-diagnosed disease suggest that the prevalence of COPD rises with age and that chronic bronchitis is more frequently diagnosed in females while emphysema is more commonly diagnosed in men [5]. Population surveys may more accurately reflect disease prevalence though often include all obstructive diseases in analyses. Recently, the Third National Health and Nutrition Examination Survey (NHANES III) completed a survey of 20,050 adults in the United States and identified obstructive lung disease (OLD) including asthma in 8.5% of the population and spirometric evidence of airflow obstruction in 6.8% [6]. In the year 2000, chronic lower respiratory diseases (largely COPD) were the fourth leading cause of mortality [7]. It is estimated that by the year 2020, COPD will rank fifth as a cause for disability in the United States [8].

Typically, symptoms of limiting dyspnea become evident in chronic obstructive pulmonary disease in the fifth decade, although patients may have cough earlier and may have symptomatic dyspnea decades earlier if they are smokers with alpha-1-antitrypsin deficiency or if they have destructive lung conditions such as cystic fibrosis, dysmotile cilia syndrome, panbronchiolitis, or inhalational injuries. COPD cannot be presumed on the basis of smoking alone as individual smokers may show variable or no evidence of airflow obstruction. Genetic predisposition may constitute an independent risk in smokers who develop COPD [9]. Longitudinal studies suggest that active asthmatic patients are at perhaps a 13-fold increased risk of developing COPD [10]. Progression of COPD is accelerated by continued smoking but to a lesser extent by chronic recurring infections.

DIAGNOSTIC TESTS

The diagnosis of COPD is based on clinical syndrome of cough and dyspnea associated with irreversible airflow obstruction. Diagnostic tests are used to confirm the presence of airflow obstruction or structural changes of emphysema. The differential diagnosis of chronic unremitting airflow obstruction also includes tracheal narrowing and incompletely treated asthma. Assessment of the severity of airflow obstruction in COPD may be determined

by spirometry, while diffusion [11] or high-resolution computerized tomographic (HRCT) imaging [12] may provide a better assessment of lung destruction. In population studies, airflow obstruction is often defined as a ratio of the forced expiratory volume in one second to the forced vital capacity (FeV_1/FVC) of <0.70 [6]. In the setting of COPD, the severity of spirometric obstruction correlates with the development of hypercapnia and mortality. HRCT of the chest is more specific than chest radiographs in defining the presence and severity of emphysema within the lung.

CONSEQUENCES OF RESPIRATORY FAILURE

The cascade of respiratory consequences from COPD include chronic cough, dyspnea, hypoxemia, and hypercapnia. The severity of dyspnea and alterations in gas exchange are determined by the extent of the structural changes, fatigue and efficiency of the respiratory muscles, and ventilatory drive [13, 14]. Respiratory muscle strength may be augmented by ventilatory rest [15] or by training [16]. Ventilatory drive is extremely variable and may be genetically determined and modified by narcotics, changes in state, and adaptation to chronic respiratory failure [17]. Patients with high ventilatory drive may experience severe dyspnea and are less likely to have hypercapnia. Patients with low ventilatory drive frequently have less prominent symptoms of dyspnea or signs of respiratory distress, though they may have profound hypercapnia and associated hypoxemia.

As in all forms of respiratory failure, COPD restrains the ability to ventilate and impairs the exchange of gas across the lung surface. Acute stressors such as infection may not only worsen airflow obstruction and gas exchange but may tip the balance to produce fatigue of breathing muscles. Chronic progression of the disease results in increasing frequency of episodes of respiratory failure and decreasing functional status. The hypercapnia and hypoxemia that accompany respiratory failure may ultimately produce (1) vasoconstrictive pulmonary hypertension and right ventricular failure and (2) a sodium avid state with hypervolemia and edema.

Progressive weight loss is a relatively common complication of respiratory failure in COPD. Loss of lean body mass occurs in approximately 20% of patients with COPD [18] and has been ascribed to systemic inflammatory response (TNF-α), increased energy expenditure at rest and with activity, and the calorigenic effects of beta-adrenergic agents.

PERMISSIVE EFFECTS OF SLEEP ON RESPIRATORY FAILURE

The physiology of sleep permits a more complete expression of respiratory failure in all diagnostic conditions

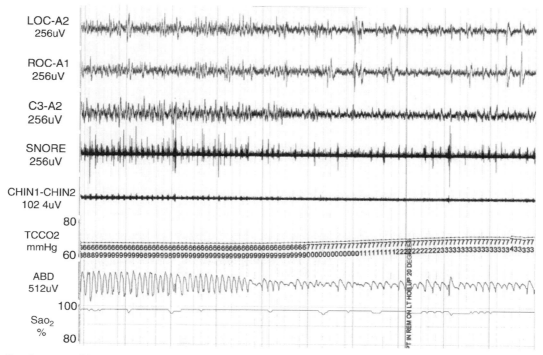

Figure 89.1 Development of irregular breathing and worsening of hypoventilation during a 5 minute transition from slow-wave to REM sleep in a hypercapnic COPD patient (FEV$_1$ 1.2 L; awake Pco$_2$ 74 mmHg). Observations were made while the patient received oxygen; light snoring occurred prior to the transition.

associated with severe lung disease. Several mechanisms act in concert to promote desaturation and hypercapnia during sleep as compared to wakefulness: (1) decreasing resting lung volume, (2) decrease in ventilatory drive, (3) increased upper airway obstruction, (4) decreased compensation for loads, and (5) irregular breathing and ribcage inhibition during REM sleep (Figure 89.1). Ventilatory responsiveness to hypoxemia and hypercapnia is decreased during sleep as compared to wakefulness and is most substantially depressed during REM sleep. Upper airway resistance increases during sleep even in the absence of obstructive sleep apnea. As a result of the profound effects of REM sleep on ventilatory responsiveness and respiratory muscles, oxygen desaturation in COPD is most severe during REM sleep.

SLEEP-ASSOCIATED DESATURATION WITHOUT OBSTRUCTIVE SLEEP APNEA SYNDROME

It is important to emphasize that the majority of patients with COPD do not have obstructive sleep apnea (OSA) [19] and that sleep-associated desaturation is more often an expression of worsening respiratory failure during sleep. Small case series suggest that sleep-associated desaturation may occur in up to 80% of patients with severe COPD [19, 20]. Sleep-associated desaturation is more

common in patients with daytime hypercapnia and typically occurs during REM sleep. Nocturnal desaturation during REM sleep is typically associated with decreasing tidal volume and is not associated with upper airway obstruction (Figure 89.2). Intensity of desaturation may vary over time as demonstrated in one study showing a 17% night-to-night variation in time of oxygen desaturation.

The mechanism of REM desaturations in COPD is multifactorial (Table 89.1). Hypercapnia, reduction in resting lung volume (FRC), and ventilation–perfusion mismatching have all been implicated. Rising CO$_2$ in the alveolus displaces oxygen, making hypoventilation an attractive explanation for sleep-associated desaturation. The fall in tidal volume and ventilation during periods of desaturation [21] would be expected to cause transient hypercapnia. Given the disadvantageous position of the diaphragm in hyperinflated COPD patients and the known effects of REM sleep on the overtaxed ribcage muscles, the observed drop in tidal volume during REM sleep would seem predictable. Adaptive mechanisms in respiratory muscle structure, however, may lessen the impact of hyperinflation on ventilatory stability [22]. In one study of 54 hypercapnic COPD patients, a sleep-associated rise of Pco$_2$ of at least 10 mmHg was seen in 43% of patients and was related to baseline Pco$_2$, percent REM sleep, and body mass index (BMI) [22a]. The predictive value of BMI and percent REM sleep was also substantiated by a second study.

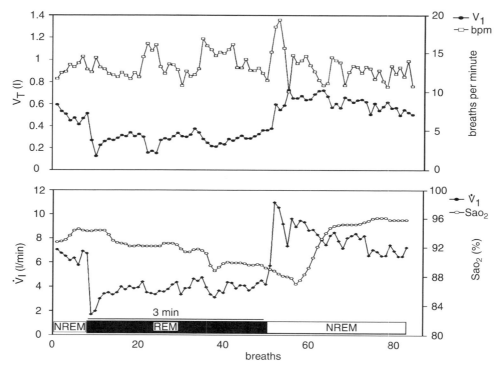

Figure 89.2   Reduction in tidal volume (V_T) but not breathing rate (bpm) during REM sleep in a patient with COPD produces a decrease in ventilation (V_I) and sustained oxygen desaturation. (Reproduced with permission from [21]).

It is not clear that all discrete episodes of sleep-associated desaturation are related to periods of worsening hypoventilation. In two small case series, the modest rise in arterial CO_2 during desaturation (average 2.3 mmHg) was insufficient to explain the severity of sleep-associated desaturation [23, 24], suggesting that ventilation–perfusion mismatching may play a greater role in sleep-associated desaturations. In a more recent study, a rise of transcutaneous CO_2 of 7.5 mmHg with sleep was noted in 12 hypercapnic patients, though the rise in CO_2 did not predict severity of oxygen desaturation [25].

A reduction in lung oxygen stores during sleep may also explain sleep-associated desaturation. The resting lung volume at end expiration (FRC) is lower in the supine than erect position and falls further during sleep. Since

most of the oxygen stored in the body is in the FRC, lower resting lung volumes may accentuate desaturation during sleep. Indeed, the most severe oxygen desaturation is seen in COPD patients with the greatest reductions in sleep-associated functional residual capacity [26]. In addition to reducing lung oxygen stores, reduction in FRC also contributes to regional airway closure, aggravating ventilation–perfusion mismatching.

It is also important to emphasize that oxygen desaturation may occur with modest decreases in arterial P_{O_2} and increases in P_{CO_2} when the sleep baseline P_{O_2} is near or below 60 mmHg, the "shoulder" of the oxyhemoglobin dissociation curve [27]. It is not surprising that COPD patients with awake oxygen desaturation are more likely to have nocturnal desaturation [28].

TABLE 89.1   Causes of Sleep-Associated Desaturation in COPD

Hypercapnia
   Sleep-associated decreased ventilatory drive
   REM-associated intercostal/accessory muscle inhibition
   Increased upper airway resistance
   Obstructive sleep apnea
Ventilation–Perfusion Mismatch
   Decreased lung volumes with airway closure
   Decreased resting lung volumes with reduced oxygen stores
Low position on oxyhemoglobin dissociation curve

COPD WITH OSAS

In a minority of patients with COPD, partial upper airway obstruction [29] or discrete OSAS episodes are noted. In one study of nonobese patients, only one of 20 had an apnea–hyponea index of >2 [19]. By contrast, in one clinical series of OSAS referred to a sleep laboratory, 28% of patients had COPD [30]. The substantial range of prevalence of COPD in case series of OSAS patients likely reflects variations in case definition and selection bias. In a large prospective study of 265 patients identified with OSAS, 11% had concomitant COPD [31], and this subset were noted to have more hypoxemia, hypercapnia, and pulmonary hypertension. Patients with COPD who have concomitant OSAS may therefore be particularly vulnerable to developing respiratory failure and secondary hemodynamic complications. Polysomnography is indicated in this setting to verify effective positive airway pressures. Ventilatory support with bilevel positive airway pressure should be considered in these patients if severe hypercapnia or respiratory distress persists despite correction of upper airway obstruction.

EFFECT OF COPD ON SLEEP STRUCTURE

COPD increases the frequency of somatic complaints such as cough, wheezing, and breathlessness. Respiratory complaints may contribute to changes in the structure of sleep and sleep complaints. A large cross-sectional study of patients with COPD revealed that when symptoms of wheezing and cough were present, 53% of patients reported insomnia and 23% percent complained of daytime sleepiness [32]. Increased sleep disruption is common in COPD as evidenced by increased frequency of arousals, increased frequency of sleep stage changes, and decreased total sleep time [33, 34].

TREATMENT

Oxygen

The mortality in COPD is related to the severity of airflow obstruction and associated morbidity. In one study, sleep-associated oxygen desaturation was a better predictor of mortality than spirometric evidence of airflow obstruction [35]. Oxygen administration is the only therapy that has consistently been shown to reduce mortality in COPD [36, 37]. In the landmark Nocturnal Oxygen Therapy Trial, administration of oxygen for at least 12 hours a day improved mortality in COPD patients with awake desaturation as defined by a Po_2 of ≤ 55 mmHg [36]. Current consensus recommendations include an arterial Po_2 (Pao_2) $\leq$ mmHg or an arterial oxygen saturation $\leq 88\%$. If there is evidence of cor pulmonale or erythrocytosis, the threshold is raised to a Pao_2 of 56 mmHg or a Sao_2 of 89%. These criteria for oxygen administration have been adopted by the Centers for Medicare and Medicaid Services for reimbursement for oxygen administration in COPD. Reimbursement for long-term oxygen is also covered for nocturnal desaturation associated with restlessness, insomnia, or cognitive changes.

Patients with COPD may have sleep-associated desaturation and yet not meet these criteria for oxygen use. In a case series, COPD patients requiring oxygen during sleep to correct nocturnal desaturation had higher daytime Pco_2 and lower FeV_1 [38]. It is not clear that correcting nocturnal desaturation benefits COPD patients who do not have awake desaturation or heart failure. In a randomized prospective trial of 76 patients with COPD and an awake Pao_2 of ≥ 56 mmHg, nocturnal oxygen conferred no benefit in terms of mortality or pulmonary hemodynamics [39]. These negative findings are supported by a recent evidence review [40].

Preventing desaturation may confer other benefits not measured in these negative studies. Oxygen administration has been shown to delay the onset of respiratory muscle fatigue and to improve subjective dyspnea. These beneficial effects may be responsible for improvements in arousal frequency and sleep architecture noted in COPD patients who are given oxygen for nocturnal desaturation [33].

Bronchodilators

Effective bronchodilator therapy is certainly indicated for daytime symptoms in COPD. Longer acting agents would seem particularly appropriate if there are nocturnal symptoms of dyspnea or wheezing. Bronchodilators have variable and somewhat unpredictable beneficial effects on nocturnal gas exchange and sleep architecture [41–43].

Respiratory Muscle Training

Unlike normal subjects whose resting ventilation is 5–7% of their maximal capacity, patients with severe COPD often exceed maximal sustainable levels of ventilation and develop fatigue of the breathing muscles. It has been recognized for nearly 30 years that respiratory muscles can be trained after weeks of regular exercise to achieve greater strength or endurance [44]. Recent studies have verified years of slowly accumulating evidence that the effect of respiratory muscle training translates into modest but significant improvements in exercise tolerance and the severity of dyspnea [16]. One case series suggested that inspiratory muscle training improved nocturnal gas exchange in COPD patients [45].

Ventilatory Stimulants

Ventilatory stimulation with progesterone resulted in an 8 mmHg drop in arterial Pco_2 and improvement in alveolar ventilation and oxygen saturation during NREM sleep in one study of hypercapnic COPD patients [46], while almitrine [47] and acetazolamide have generally produced more modest effects. Ventilatory stimulation runs the risk of increasing severity of breathlessness in severely dyspneic patients and may be most appropriate for those with low ventilatory drive.

Positive Airway Pressure

Chronic ventilatory support can substantially improve sleep-associated hypercapnia and hypoxemia that accompanies respiratory failure. Continuous positive airway pressure has been shown to decrease work of breathing in COPD and would be expected on the basis of this to prevent respiratory muscle fatigue. Improvement in respiratory muscle fatigue may be responsible for the observation that nocturnal CPAP has been shown to improve daytime respiratory muscle strength and functional status as measured by the 12 minute walk [15].

Noninvasive bilevel positive airway pressure (PAP) is a form of ventilatory support that has been shown to reduce mortality in acute respiratory failure complicating COPD [48]. In restrictive lung disease, case series show improvement in sleep quality and both nocturnal and daytime gas exchange with noninvasive nocturnal ventilatory support. Bilevel PAP has been proposed as a method of improving sleep quality and gas exchange in patients with chronic respiratory failure due to COPD. Four studies have addressed the nocturnal use of bilevel PAP in hypercapnic COPD patients. Unlike titration of positive pressure for sleep apnea, none of these studies employed titration of pressure to a targeted effect on nocturnal gas exchange or sleep parameters. In addition, rather small ventilating pressures were used in one of the studies, which may be an inadequate method in COPD patients who have high airway impedances and may require higher ventilating pressures.

Three of these studies show no effect of bilevel PAP on daytime gas exchange [49–51], and one study [52] demonstrated improvement in both nocturnal and daytime Pao_2 and $Paco_2$ that was associated with improvement in quality of life scores. Total sleep time and sleep efficiency were improved in one study that used an average ventilating pressure of 19 cm H_2O [51] and deteriorated in a second study that used relatively low ventilating pressure of 5 cm H_2O [50]. A significant drawback to the chronic use of noninvasive ventilation in COPD is the apparent low compliance rates reported in published case series [53, 54]. The utility of chronic bilevel positive airway pressure in severe COPD has yet to be determined though selected hypercapnic patients may benefit if effective pressures are demonstrated to improve sleep and/or gas exchange parameters.

REFERENCES

1. National Heart, Lung and Blood Institute and World Health Organization. Global initiative for chronic obstructive lung disease. WHO, 1998. URL: www.goldcopd.com/.

2. O'Shaughnessy T, Ansari TW, Barnes NC, Jeffery PK. Inflammation in bronchial biopsies of subjects with chronic bronchitis: inverse relationship of CD8$^+$ T lymphocytes with FEV_1. *Am J Respir Crit Care Med* **155**:857–857(1997).

3. Peleman R, Rytila PH, Kips JC, Joos GF, Pauwels RA. The cellular composition of induced sputum in chronic obstructive pulmonary disease. *Eur Respir J* **12**:839–843(1999).

4. O'Donnell DE, Webb KA. Exertional breathlessness in patients with chronic airflow limitation: the role of lung hyperinflation. *Am Rev Respir Dis* **148**:1351–1357(1993).

5. National Heart, Lung and Blood Institute of the National Institutes of Health. *Morbidity and Mortality: 2004. Chartbook on Cardiovascular, Lung and Blood Diseases*. NIH, Bethesda, MD, 2004, pp 58–79.

6. Mannino DM, Gagnon RC, Petty TL, Lydick E. Obstructive lung disease and low lung function in adults in the United States: data from the National Health and Nutrition Examination Survey. *Arch Intern Med* **160**:1683–1689(2000).

7. National Center for Health Statistics. Data on chronic obstructive pulmonary disease. NCHS, Hyattsville, MD, 2004. URL:www.cdc.gov/nchs/data/factsheets/copd.pdf.

8. Murray CJL, Lopez AD. Evidence-based health policy—lessons from the global burden of disease study. *Science* **274**(5288):740–743(1996).

9. Silverman EK, Palmer LJ, Mosley JD, Barth M, Senter JM, et al. Genomewide linkage analysis of quantitative spirometric phenotypes in severe early-onset chronic obstructive pulmonary disease. *Am J Hum Genet* **70**:1229–1239(2002).

10. Silva GE, Sherrill DL, et al. Asthma as a risk factor for COPD in a longitudinal study. *Chest* **126**:59–65(2004).

11. Sanders C, Nath PH, Bailey WC. Detection of emphysema with computed tomography. Correlation with pulmonary function tests and chest radiography. *Invest Radiol* **23**:262–266(1988).

12. Yamaguchi K, Matsubara. Computed tomographic diagnosis of chronic obstructive pulmonary disease. *Curr Opin Pulm Med* **6**:92–98(2002).

13. Killian KL, Campbell ME. In: Roussos X. (Ed), *Dyspnea. The Thorax*. Marcel Dekker, New York, 1995, pp 1709–1747.

14. Marin JM, Montes de Oca M, Rassulo J, Celli BR. Ventilatory drive at rest and perception of exertional dyspnea in severe COPD. *Chest* **115**:1293–1300(1999).

15. Mezzanotte WS, Tangel DJ, Fox AM, Ballard RD, White DP. Nocturnal nasal continuous positive airway pressure in patients with chronic obstructive pulmonary disease. Influence on waking respiratory muscle function. *Chest* **106**:1100–1108 (1994).

16. Weiner P, Magadle R, Beckerman M, Weiner M, Berar-Yanay N. Comparison of specific expiratory, inspiratory, and combined muscle training programs in COPD. *Chest* **124**:1357–1364(2003).

17. Annane D, Quera-Salva MA, Lofaso F, Vercken JB, Lesieur O, Fromageot C, Clair B, Gajdos P, Raphael JC. Mechanisms underlying effects of nocturnal ventilation on daytime blood gases in neuromuscular diseases. *Eur Respir J* **13**:157–162(1999).

18. Englen MPK, Schols A, Baken WC, Wesseling GJ, Wouters EF. Nutritional depletion in relation to respiratory and peripheral skeletal muscle function in outpatients with COPD. *Eur Respir J* **7**:1793–1797(1994).

19. Catterall JR, Douglas NJ, Calverly PM. Transient hypoxemia during sleep in chronic obstructive pulmonary disease is not a sleep apnea syndrome. *Am Rev Respir Dis* **128**:24–29 (1983).

20. Douglas NJ. Nocturnal hypoxemia in patients with chronic obstructive pulmonary disease. *Clin Chest Med* **13**(3):523–532(1992).

21. Becker HF, Piper AJ, Flynn WE, McNamara SG, Grunstein RR, Peter JH, Sullivan CE. Breathing during sleep in patients with nocturnal desaturation. *Am J Respir Crit Care Med* **159**:x–x(1999).

22. Gauthier AP, Verbank S, Estenne M, Segebarth C, Macklem PT, Pavia M. Three dimensional reconstruction of the *in vivo* human diaphragm shape at different lunch volumes. *J Appl Physiol* **76**:495(1994).

22a. O'Donoghue F, Catcheside P, Ellis E, et al. Sleep hypoventilation in hypercapnic chronic obstructive pulmonary disease: prevalence and associated factors. *European Respiratory Journal* **21**:977–9 84(2003).

23. Fletcher EC, Levin DC. Cardiopulmonary hemodynamics during sleep in subjects with chronic obstructive pulmonary disease. The effect of short and long term oxygen. *Chest* **85**(1):6–14(1984).

24. Catterall JR, Calverley PM, MacNee W, Warren PM, Shapiro CM, Douglas NJ, Flenley DC. Mechanism of transient nocturnal hypoxemia in hypoxic chronic bronchitis and emphysema. *J Appl Physiol* **59**(6):1698–1703(1985).

25. Mulloy E, McNicholas WT. Ventilation and gas exchange during sleep and exercise in severe COPD. *Chest* **109**:387–394(1996).

26. Hudgel DW, Martin RJ, Capehart M, Johnson B, Hill P. Contribution of hypoventilation to sleep oxygen desaturation in chronic obstruction pulmonary disease. *J Appl Physiol Respir Environ Exercise Physiol* **55**(3):669–677(1983).

27. Koo KW, Sax DS, Snider GL. Arterial blood gases and pH during sleep in chronic obstructive pulmonary disease. *Am J Med* **58**(5):663–670(1975).

28. DeMarco FJ, Wynne JW, Block AJ, Boysen PG, Taasan VC. Oxygen desaturation during sleep as a determinant of the "Blue and Bloated" syndrome. *Chest* **79**(6):621–625(1981).

29. Littner MR, McGinty DJ, Arand D. Determinants of oxygen desaturation in the course of ventilation during sleep in chronic obstructive pulmonary disease. *Am Rev Respir Dis* **122**(6):849–857(1980).

30. de Miguel J, Cabello J, Sanchez-Alarcos JM, Alvarez-Sala R, Espinos D, Alvarez-Sala JL. Long-term effects of treatment with nasal continuous positive airway pressure on lung function in patients with overlap syndrome. *Sleep Breathing* **6**:3–10(2002).

31. Chaouat AE, Weitzenblum E, Krieger J, Ifoundza T, Oswald M, Kessler R. Association of chronic obstructive pulmonary disease and sleep apnea syndrome. *Am J Respir Crit Care Med* **151**(1):82–86(1995).

32. Klink ME, Dodge R, Quan SE. The relation of sleep complaints to respiratory symptoms in a general population. *Chest* **105**:151–154(1994).

33. Calverley PM, Brezinova V, Douglas NJ, Catterall JR, Flenley DC. The effect of oxygenation on sleep quality in chronic bronchitis and emphysema. *Am Rev Respir Dis* **126**(2):206–210(1982).

34. Fleetham J, West P, Mezon B, Conway W, Roth T, Kryger M. Sleep arousals and oxygen desaturation in chronic obstructive pulmonary disease. The effect of oxygen therapy. *Am Rev Respir Dis* **126**(3):429–433(1982).

35. Kimura H, Suda A, Sakuma T, Tatsumi K, Kawakami Y, et al. Nocturnal oxyhemoglobin desaturation and prognosis in chronic obstructive pulmonary disease and late sequelae of pulmonary tuberculosis. Respiratory failure research group in Japan. *Intern Med* **37**(4):354–359(1998).

36. Nocturnal Oxygen Therapy Trial Group. Continuous or nocturnal oxygen therapy in hypoxemic chronic obstructive lung disease: a clinic trial. *Ann Intern Med* **93**:391–398(1980).

37. Report of the Medical Research Council working party. Long-term domiciliary oxygen therapy in chronic hypoxic cor pulmonale complicating chronic bronchitis and emphysema. *Lancet* **1**:681–685(1981).

38. Sergi M, Rizzi M, Andreoli A, Pecis M, Bruschi C, Fanfulla F. Are COPD patients with nocturnal REM sleep-related desaturations more prone to developing chronic respiratory failure requiring long-term oxygen therapy? *Respiration* **63**:117–122 (2002).

39. Chaouat A, Weitzenblum E, et al. A randomized trial of nocturnal oxygen therapy in chronic obstructive pulmonary disease patients. *Eur Respir J* **14**(5):1002–1008(1999).

40. Crockett AJ, Cranston JM, Moss JR, Alpers JH. Domiciliary oxygen for chronic obstructive pulmonary disease. *Cochrane Database Systematic Rev* **4**:CD001744(2000).

41. Martin RJ, Pak J. Overnight theophylline concentrations and effects on sleep and lung function in chronic obstructive pulmonary disease. *Am Rev Respir Dis* **145**(3):540–544 (1992).

42. Mulloy E, McNicholas WT. Theophylline improves gas exchange during rest, exercise, and sleep in severe chronic obstructive pulmonary disease. *Am Rev Respir Dis* **148**(4 Pt 1):1030–1036(1993).

43. Martin RJ, Bartelson BL, Smith P, Hudgel DW, Lewis D, Pohl G, Koker P, Souhrada JF. Effect of ipratropium bromide treatment on oxygen saturation and sleep quality in COPD. *Chest* **115**:1338–1345(1999).

44. Leith DE, Bradley M. Ventilatory muscle strength and endurance training. *J Appl Physiol* **41**:508–516(1976).

45. Heijdra YF, Dekhuijzen PN, van Herwaarden CL, Folgering HT. Nocturnal saturation and respiratory muscle function in patients with chronic obstructive pulmonary disease. *Thorax* **50**:610–612(1995).

46. Skatrud JB, Dempsey JA, Iber C, Berssenbrugge A. Correction of CO_2 retention during sleep in patients with chronic obstructive pulmonary diseases. *Am Rev Respir Dis* **124**(3):260–268(1981).

47. Gothe B, Cherniack NS, Bachand RT, Szalkowski MB, Bianco KA. Long-term effects of almitrine bismesylate on oxygenation during wakefulness and sleep in chronic

obstructive pulmonary disease. *Am J Med* **84**:436–444 (1988).

48. Lightowler JV, Wedzicha JA, Elliott MW, Ram FS. Non-invasive positive pressure ventilation to treat respiratory failure resulting from exacerbations of chronic obstructive pulmonary disease: Cochrane systematic review and meta-analysis. *BMJ* **326**:185(2003).

49. Gay PC, Hubmayr RD, Stroetz RW. Efficacy of nocturnal nasal ventilation in stable, severe chronic obstructive pulmonary disease during a 3 month controlled trial. *Mayo Clin Proc* **71**(6):533–542(1996).

50. Lin C. Comparison between nocturnal nasal positive pressure ventilation combined with oxygen therapy and oxygen mono-therapy in patients with severe COPD. *Am J Respir Crit Care Med* **154**:353–358(1996).

51. Krachman S, Quaranta AJ, Berger TJ, Criner GJ. Effects of noninvasive positive pressure ventilation on gas exchange and sleep in COPD patients. *Chest* **112**:623–628(1997).

52. Meecham Jones D, Paul EA, Jones PW, Wedzicha JA. Nasal pressure support ventilation plus oxygen compared with oxygen therapy alone in hypercapnic COPD. *Am J Respir Crit Care Med* **152**:538–544(1995).

53. Leger P, Bedicam JM, Cornette A, Reybet-Degat O, Langevin B, Polu JM, Jeannin L, Robert D. Nasal intermittent positive pressure ventilation. Long-term follow-up in patients with severe chronic respiratory insufficiency. *Chest* **105**:100–105 (1994).

54. Simonds AK, Elliott MW. Outcome of domiciliary nasal intermittent positive pressure ventilation in restrictive and obstructive disorders. *Thorax* **50**:604–609(1995).

90

SLEEP AND BREATHING IN CYSTIC FIBROSIS

Amanda J. Piper, Maree M. Milross, and Peter T. P. Bye
Royal Prince Alfred Hospital, Camperdown, New South Wales, Australia

INTRODUCTION

For most of us, sleep is a quiescent time, giving the body and mind the opportunity to rest and repair. However, for those individuals with abnormal chest wall mechanics or altered gas exchange capabilities, sleep can place additional demands on an already "overburdened" system.

Patients with end-stage cystic fibrosis (CF) are one group where the normal physiologic changes associated with sleep can exaggerate any awake alterations in gas exchange and respiratory mechanics. Cystic fibrosis is a genetic disorder in which an ion transport defect leads to lung infection and inflammation. For these patients, pulmonary complications leading to respiratory failure are the major cause of morbidity and mortality. Mucus hypersecretion in the lungs causes obstruction of the small airways, promoting the development of chronic lung infection and inflammation. In this environment, ongoing lung damage occurs resulting in more airway obstruction, bronchiectasis, and chronic bacterial colonization. As a consequence, ventilation–perfusion mismatch (V_A/Q) is worsened, while changes in the compliance and resistance of the respiratory system increase the work of breathing. Patients can also develop poor nutritional status and when this occurs on the background of lung hyperinflation and high work of breathing, tidal volume may drop, further worsening gas exchange, with the development of hypoxemia and hypercapnia (Figure 90.1). What is not widely appreciated in this group is that marked gas exchange abnormalities can first appear during sleep, preceding the appearance of daytime respiratory failure. Hypoxia has been associated with the development of pulmonary hypertension and cor pulmonale, both linked to poorer survival outcomes [1]. The appearance of hypercapnia in CF is considered to be an unfavorable prognostic feature [2]. Therefore identification and treatment of gas exchange abnormalities during sleep would appear to be an important therapeutic goal in the overall management of these patients. This is particularly so now that lung transplantation is a realistic treatment option for many patients with end-stage disease. Identifying who is at risk for nocturnal respiratory failure could assist in providing early intervention to maximize the transplant window for these individuals.

Despite the impact of hypoxemia and hypercapnia in patients with CF, the extent to which abnormal nocturnal breathing contributes to worsening awake respiratory failure and its impact on clinical outcome has not been well studied. In this chapter we will review what is known about alterations in breathing during sleep in subjects with CF, which patients are at risk of developing nocturnal hypoxemia, and what is the impact of treatment on outcome.

BREATHING DURING SLEEP IN CYSTIC FIBROSIS

Impact of Sleep on Ventilation and Gas Exchange

It is well documented that patients with CF and advanced lung disease develop hypoxia and hypercapnia during sleep, especially during episodes of rapid eye movement (REM) sleep [3–6a]. For some time, there has been interest in investigating the mechanisms underlying this abnormal gas exchange during sleep. Sleep disordered breathing

Sleep: A Comprehensive Handbook, Edited by T. Lee-Chiong.

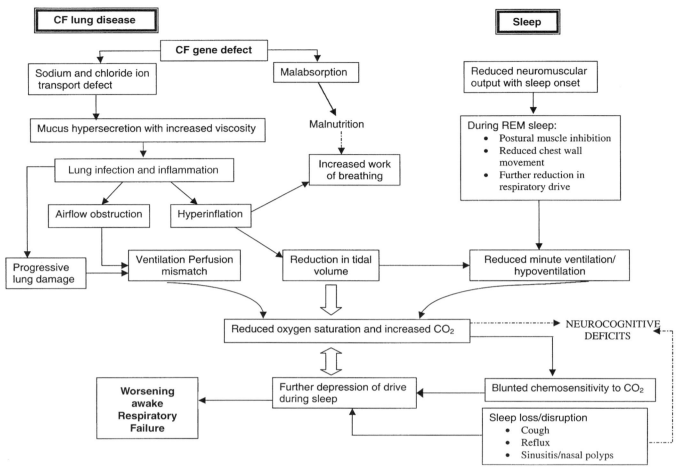

Figure 90.1   Possible interactions between sleep-related events and CF lung disease. When the normal respiratory changes of sleep are superimposed on abnormal lung mechanics and diminished gas exchange capabilities, exaggeration of hypoventilation may occur, promoting the development of respiratory failure.

and oxyhemoglobin desaturation in patients with CF are not generally characterized by snoring or frank apneic events. In fact, respiratory events such as obstructive or central apnea are uncommon [5], even in patients with moderate to severe lung disease. Nocturnal respiratory events are usually confined to REM sleep and consist of hypopneas related to periods of phasic eye movement [6].

In normal subjects, ventilation has been shown to decrease in the range of 10–15% with the onset of sleep [7]. This results in a small fall in oxyhemoglobin saturation and rise in carbon dioxide tension. The changes in ventilation occurring in REM sleep are more variable and related to the intensity of the phasic activity, but overall minute ventilation (V_E) in this sleep stage has been shown not to differ from non-rapid eye movement (NREM) sleep.

Numerous studies investigating the mechanisms of sleep-related hypoxemia and hypercapnia in CF patients have examined changes in pulmonary mechanics and pattern of ventilation occurring during sleep versus the awake state. There are two main mechanisms that have been proposed to explain the gas exchange abnormalities that occur during sleep in these patients: V_A/Q inequality and alveolar hypoventilation.

Muller and colleagues [3] studied 20 patients with CF and found REM sleep to be associated with a marked loss of intercostal and diaphragmatic tonic muscle activity, as measured with surface electrodes. There was also a decrease in the baseline position of the ribcage and abdomen, as recorded by magnetometers [3], suggesting a decrease in functional residual capacity (FRC) during REM sleep associated with the postural muscle atonia normally seen in this sleep stage. It was postulated that the decrease in FRC during REM sleep would lead to airway closure in dependant lung regions, V_A/Q mismatching and shunting, therefore reducing oxyhemoglobin saturation (Sao_2). In addition, they found the largest decreases in Sao_2 occurred during REM sleep following short episodes (<20 s) of partial phasic intercostal and diaphragmatic inhibition shown by either reduced electromyographic activity or by decreases in the movement of the thoracic and abdominal walls. They postulated that phasic episodes of hypoventilation on the background of V_A/Q mismatch would further aggravate sleep desaturation.

Using respiratory inductance plethysmography, Spier et al. [6a] and Tepper et al. [5] showed a substantial reduction in the ribcage contribution to tidal volume especially during REM sleep. These changes in chest wall movement occurred in phasic REM and were accompanied by a decrease in V_E, tidal volume, and mean inspiratory flow compared to the awake state [5]. A large breath-to-breath variability in tidal volume was also noted. These original studies of ventilation during sleep in CF subjects used magnetometers or respiratory inductance plethysmography rather than direct measures of ventilation. The accuracy of such techniques has been questioned in patients with lung disease and during sleep [8].

Two studies have been published that have used more direct measures of ventilation during sleep in CF. Ballard and associates [8] used body plethysmography and a face mask to measure alterations in ventilation during sleep in five patients with CF. They confirmed falls in tidal volume from wakefulness to NREM sleep, accounting for the concurrent decrements in minute ventilation seen. These falls in tidal volume and minute ventilation were accompanied by reductions in esophageal occlusion pressures, suggesting a reduction in respiratory drive. No change in either upper or lower airway resistance from wakefulness to NREM sleep was seen, nor was a fall in FRC detected. These findings supported the notion that a significant proportion of the fall in saturation seen in NREM sleep was associated with decrements in respiratory neuromuscular output, rather than a change in pulmonary mechanics or V_A/Q inequality. Unfortunately, no REM sleep was seen in any patient studied, so it was not possible to determine if alterations in lung volumes or changes in airway caliber would account for the hypoxemia usually occurring in this sleep stage.

Using a pneumotachograph attached to a nasal mask to allow direct measurement of ventilation, Milross et al. [9] demonstrated that V_E in REM sleep was significantly lower than during NREM sleep or wakefulness in CF patients with severe lung disease. The decrease in V_E resulted in poorer oxygenation during REM sleep than NREM sleep with a concurrent rise in transcutaneous carbon dioxide ($Tcco_2$). The reduction in V_E in REM sleep appeared to be due to a reduction in tidal volume from NREM to REM sleep.

Breathing pattern is highly variable in patients with CF compared to normal controls, especially during REM sleep [5]. This irregular, low tidal volume breathing would contribute further to falls in oxygen saturation when superimposed on V_A/Q mismatching. Differences in reports regarding the degree of sleep disordered breathing seen in patients with CF may relate to the severity of the underlying lung disease. Patients with more impaired lung function may already be on the steeper portion of the oxyhemoglobin dissociation curve, so that any alteration in sleep breathing would lead to a relatively greater degree of desaturation compared to those patients with better preserved lung function.

Although V_A/Q inequality contributes to hypoxemia, current evidence suggests that hypoventilation during sleep is the major contributing factor responsible for the falls in saturation seen in patients with CF and severe lung disease. This seems to be particularly the situation during REM sleep. The increased respiratory variability associated with REM sleep together with the reduced ribcage movement, fall in tidal volume, and reduced inspiratory flows all suggest hypoventilation-related hypoxemia.

Impact on Sleep Quality

There are a number of reasons why patients with CF may have poor sleep quality. In the presence of abnormal breathing and oxygen desaturation, arousal from sleep acts as a defense mechanism that enables recovery of ventilation and maintenance of adequate gas exchange. Interestingly, it has been shown that in patients with CF, arousal often occurred independently of falls in oxygen saturation and did not change with oxygen therapy [6a, 9]. Increased work of breathing related to hyperinflated bronchiectatic lungs may also contribute to sleep disturbance. Other factors that could cause arousal from sleep include coughing, acid reflux, upper airway obstruction from nasal polyps, or the necessity to use medications such as beta agonists during the night. Despite numerous reasons for sleep loss or disruption in this population, little information is available regarding how common these problems actually are and who is likely to be affected.

While some studies have reported sleep architecture and efficiency to be similar to that seen in normal young adults [4, 10], other studies report abnormalities. In those with severe CF lung disease, lower sleep efficiency [6a], less time in REM sleep [6a], and increased time awake after sleep onset compared with controls [11] have been reported. Episodes of cough during stages 1 and 2 sleep have been shown to delay progression to deeper stages of sleep [4]. Differences in reported findings regarding the quality of sleep in patients with CF are likely to be attributable to disease severity.

Irrespective of polysomnographically measured sleep quality, approximately 40% of adults with CF and moderate to severe lung disease perceive their sleep quality as poor [12]. A simple sleep questionnaire, the Pittsburgh Sleep Quality Index (PSQI), has been shown to correlate well with some indices of monitored sleep [12]. The PSQI is a self-rated questionnaire that assesses sleep quality and disturbances over a 4 week period. Using the PSQI, Milross and colleagues [12] found that those patients rating themselves as "good" sleepers had better sleep efficiency and more REM sleep than those with "poor" sleep quality. In addition, lower minimum sleep saturations were significantly associated with worse overall sleep quality. Other factors such as cough or reflux may make a major contribution to sleep disruption. These preliminary results with the PSQI indicate that it may provide useful clinical information.

NOCTURNAL HYPOXEMIA IN CF

Given the possible impact poor gas exchange and abnormal breathing events during sleep could have on sleep quality, neurocognitive function, and development of pulmonary hypertension and cor pulmonale, several studies have looked at daytime clinical measures that would predict nocturnal desaturation in CF patients (Table 90.1). Although daytime resting oxygenation [13] and lung function, as measured by forced expiratory volume in 1 second (FEV_1) [14], have been shown to be associated with nocturnal desaturation, these parameters appear to have only a modest ability to predict patients likely to desaturate during sleep. Using the criteria of more than 5% of the night with a $Sao_2 < 90\%$ to define significant nocturnal desaturation, Frangolias et al. [14] were able to correctly predict only 26% of cases likely to desaturate at night based on awake Sao_2 and FEV_1 values. By using evening Pao_2 and morning $Paco_2$, Milross and colleagues [6] found they could account for 74% of the variability seen in the nocturnal Sao_2 average percentage. Although other measures of lung function were significantly correlated with measures of sleep-related oxygenation, they did not add significantly to the overall predictive ability of these two variables.

Measures of exercise capacity and oxygenation during exercise have also been examined to determine whether these can improve identification of patients with nocturnal desaturation. These studies have shown that desaturation and hypercapnia occur to a greater degree during sleep than during exercise [10, 14], and that measures of exercise

TABLE 90.1   Predictors of Nocturnal/Sleep Desaturation in CF

$Sao_2 < 94\%$ most predictive of nocturnal desaturation [13, 14]
Nocturnal desaturation uncommon when $FEV_1 > 65\%$ predicted [14]
Evening and morning arterial blood gas tensions predictive of nocturnal desaturation [6]
Exercise desaturation does not predict sleep desaturation [10]
Nonhypoxemic patients desaturate more in sleep than exercise [10]

capacity and oxygenation did not confer any additional ability to predict desaturation during sleep.

Current evidence suggests that patients with an awake resting $Sao_2 > 94\%$ and a $FEV_1 > 65\%$ predicted are unlikely to desaturate at nighttime [13, 14]. Patients with FEV_1 and Sao_2 below these values appear to be at risk for nocturnal desaturation and should be targeted for further evaluation, either with arterial blood gas tension measurements, overnight oximetry, or full polysomnography, depending on clinical circumstances.

Nocturnal Sleep and Breathing During Acute Exacerbations

In the majority of studies examining sleep and breathing in CF, subjects have been studied during periods of clinical stability. Less attention has been paid to sleep and breathing during acute exacerbations of their lung disease. Not surprisingly, it has been found that sleep-related hypoxemia worsens during infective exacerbations [15, 16]. It is well known that treatment of the exacerbation results in improvements in lung function, but more recent work suggests that it also improves sleep quality and nocturnal gas exchange [15, 16]. Although individuals with poorer clinical scores on admission and lower initial awake saturation were more likely to demonstrate nocturnal desaturation, clinical awake measurements at exacerbation onset were not good predictors of nocturnal gas exchange [15]. While it is fairly clear that acute exacerbations impair sleep and nocturnal gas exchange, it has yet to be shown that identification of the problem either changes management of the patient or alters the clinical outcome.

Metabolic Alkalosis in Nocturnal Desaturation and Respiratory Failure

Recently, it was reported that 71% of CF patients with an acute exacerbation of lung disease had a mixed respiratory acidosis and metabolic alkalosis [17]. Several mechanisms for this have been postulated including increased body temperature and fluid losses causing hypochloremia on a background of chronic malnutrition. From a respiratory view point, the resulting metabolic alkalosis could be of significance, producing a depression of respiratory drive, resulting in hypoventilation and retention of carbon dioxide [18]. This same group has reported a significant association between hypochloremic metabolic alkalosis and nocturnal oxygen desaturation in clinically stable CF patients [19]. They have proposed that metabolic alkalosis may exacerbate nocturnal hypoventilation given that ventilation is under metabolic control during NREM sleep. However, these observations have yet to be confirmed and the clinical implications fully explored.

CONSEQUENCES OF NOCTURNAL HYPOXEMIA AND SLEEP LOSS

Awake Respiratory Failure and Cor Pulmonale

Gas exchange becomes compromised in CF due to severe lung damage with repeated infection—the most common cause of both morbidity and mortality in patients with CF is persistent lung infection and inflammation, with progression to chronic suppurative lung disease. Cor pulmonale is the most frequently recognized cardiovascular sequelae of the chronic lung disease of CF. Right ventricular hypertrophy is due to chronic pulmonary hypertension resulting from hypoxic pulmonary vasoconstriction and medial hypertrophy of pulmonary arterioles, as well as structural remodeling of the pulmonary microcirculation, including bronchopulmonary arterial anastomoses [20]. Eventually the sustained pulmonary hypertension will lead to right ventricular failure, a complication associated with a poor prognosis [1]. Previous studies have demonstrated that the pulmonary hypertension associated with CF can be corrected, or at least reduced, with oxygen supplementation [20]. In addition, pulmonary artery systolic pressure has recently been shown to correlate with awake, postexercise, and sleep Sao_2 in CF [1]. This group also confirmed that a higher mortality rate is seen in CF patients with pulmonary hypertension. Despite this, a long-term randomized trial of nocturnal home oxygen therapy in hypoxemic CF patients showed no significant effect on the right ventricular ejection fraction response to exercise, or on mortality [21]. This negative result may relate to the relentless progression of lung disease in CF despite oxygen therapy, with death due to severe infection. Alternatively, the negative result may relate in part to small patient numbers and/or a short duration of oxygen therapy [mean 6.2 hours/day).

Quality of Life

Both physical and psychosocial factors will impact on how an individual perceives his/her quality of life and the degree to which the disease impacts on it. Sleep loss or disturbance can severely affect daytime function and sense of well-being and is an important aspect when considering quality of life issues.

A large proportion of patients with CF and moderate to severe lung disease report poor quality sleep [12], with those individuals with more severely impaired lung function and gas exchange reporting worse quality sleep. Latency to sleep and sleep disturbance are two factors that appear to contribute most to poorer perceived sleep quality in this population. Other investigators have reported significantly poorer quality sleep measured on polysomnogram (PSG) in those patients with very severe lung disease

compared to normal controls [6a, 11]. However, there is little information regarding the impact of perceived or objectively measured sleep disturbance on quality of life or daytime function.

In a group of adults with CF and severe lung disease, PSG showed that sleep efficiency was significantly worse in the CF group compared with healthy age-matched controls, while time spent awake after sleep onset was greater [11]. Participants in this study also completed a mood scale based on the Profile of Mood states, a Stanford Sleepiness Scale, and an Epworth Sleepiness Scale (ESS) and underwent a Multiple Sleep Latency Test (MSLT). Although there was no significant difference between the two groups with respect to ESS or MSLT sleep latencies, the CF group reported significantly lower levels of activation and happiness, and greater levels of fatigue compared with controls. Furthermore, it was found that in the CF group, activation correlated with sleep efficiency, while both activation and happiness were negatively correlated with wakefulness after sleep onset. Zinman and colleagues [21] applied a psychosocial battery of standardized tests assessing mood, self-esteem, and cognitive function in a group of 28 adolescent and young adult patients with moderate to severe lung disease (FEV_1 34–39% predicted) and an awake arterial oxygen tension <65 mmHg prior to randomization to long-term home oxygen or room air. In contrast to the findings of Dancey and colleagues [11], measures of mood and self-esteem were within normal limits at baseline and demonstrated no change over the first year of follow-up [21]. However, those patients treated with oxygen therapy continued to attend school or work on a daily basis, while the air-treated group decreased their participation in these activities substantially. These authors suggested that the continued school or work attendance in the oxygen group may have been due to improved sleep quality with oxygen therapy resulting in improved sense of well-being and hence participation rates. However, no sleep measurements were made so that limited conclusions were possible. There is a need for further studies to clarify the effect of therapy on sleep quality and daytime activity in those with advanced disease.

Neurocognitive Function

Despite the recognition that significant sleep and breathing abnormalities occur in patients with CF and severe lung disease, there is scant data on how these abnormalities might impact on neurocognitive function in this population.

In the study by Dancey and colleagues [11], CF patients demonstrated deficits in neurocognitive performance tasks to a level that was only 60% of the control group, despite similar degrees of daytime sleepiness. In addition, in contrast to control subjects, performance in the CF group did not improve throughout the day. In this study there was a lack of association between neurocognitive function in CF patients and either nocturnal or daytime oxygenation. These authors speculated that neurocognitive function was less sensitive to hypoxemia in young adult CF patients than in older patients with chronic obstructive pulmonary disease. Furthermore, they suggested that this may explain the previous finding of the failure of oxygen therapy to improve neurocognitive function in this population [21]. The slowing of response rate seen in the CF subjects in the study of Dancey and co-workers [11], in order to maintain accuracy, was similar to how sleep-deprived individuals respond on comparable performance tasks. These investigators concluded that the neurocognitive deficits seen in CF patients were related to the long-term impact of chronic sleep deprivation rather than to hypoxia. These interesting results warrant further investigation to confirm the findings and to determine which, if any, interventions improve daytime performance and quality of life.

Endocrine Abnormalities

Sleep disruption and nocturnal hypoxemia may also have health consequences apart from the recognized ones on the cardiorespiratory system. In other patient groups, disturbances of sleep have been shown to affect the endocrine, immune, and metabolic systems. Insulin and glucose levels are sensitive to changes in sleep and breathing [22, 23]. Almost a third of patients with CF develop diabetes. The presence of sleep disturbance and nocturnal hypoxemia may further exaggerate this risk and contribute to difficulties controlling the problem. It has also been shown that immune function can be modulated by sleep loss [22]. If sleep does play a role in assisting host defenses during infective episodes, interruption of such sleep could, potentially, affect recovery from infection.

TREATMENT

When nocturnal hypoxemia is identified, oxygen therapy is usually added to the patient's treatment regimen in the belief that correction of nocturnal desaturation may lower pulmonary artery pressures and delay the progression to cor pulmonale. Despite the widespread use of oxygen therapy in these patients, there is a paucity of data confirming the benefits of such an approach in this population and its effect on quality of life, morbidity, or mortality.

In a single-night comparison study, 10 patients with severe, stable CF (FEV_1 25% $\pm$ 9% predicted) were randomly assigned to either humidified nocturnal low-flow oxygen or compressed air via nasal prongs at 2 L/min [6]. The progressive fall in Sao_2 from awake to REM sleep during room air breathing was almost eliminated with the use of oxygen, with those patients with the lowest baseline

Sao_2 showing the greatest improvements with oxygen. However, no differences between the oxygen and air nights with respect to sleep quality, awakenings, arousals, or sleep stage distribution were noted, apart from a reduction in the number of state changes per hour during oxygen therapy. While arterial oxygenation improved significantly with oxygen therapy, it also caused an increase in $Tcco_2$ compared with room air breathing, most notably during REM sleep.

Only one study has looked at the longer-term use of nocturnal oxygen in hypoxemic patients. Zinman and colleagues [21] undertook a randomized, double-blind trial to document the effect of nocturnal oxygen therapy on mortality and morbidity rates and on disease progression in 28 subjects with advanced CF. Subjects received either humidified oxygen or room air nocturnally and were followed for an average of 26 $\pm$ 9 months. Oxygen therapy had no significant effect on the frequency of hospitalizations or mortality rates. Group comparisons for the first year found no significant differences in measures reflecting disease progression, namely, nutritional status, exercise ability, or pulmonary function. As discussed previously, the only difference identified was that school or work attendance was maintained in the oxygen group while it deteriorated in the air group. These authors suggested the relative ineffectiveness of oxygen therapy in their CF patients might relate to the development of pulmonary hypertension from destruction of the pulmonary vascular bed due to repeated infection and inflammation rather than from vasoconstriction. Despite these rather disappointing results with respect to nocturnal oxygen as an effective therapy in slowing down disease progression and improving clinical outcome, further long-term studies have not been undertaken to confirm or refute these findings. Nevertheless, nocturnal oxygen therapy is still widely used as first line intervention in patients with CF.

Given the effectiveness of nocturnal mask ventilation (NIV) in other patient groups with nocturnal hypoxemia and hypercapnia, this technique has also been used in patients with CF. Although there are a number of uncontrolled studies, case reports, and audits in the literature, only two controlled trials have been conducted to date, both of which were short-term interventions. Gozal [24] studied six patients with severe CF lung disease and significant nocturnal gas exchange abnormalities over three separate nights: an initial baseline room air night followed by either NIV or supplemental oxygen randomly assigned over the next two nights. Like previous studies [6a], the administration of supplemental oxygen was associated with significant improvements in mean oxygen saturation throughout the night. However, these improvements were also accompanied by significant increases in $Tcco_2$, with two patients reporting morning headache. As with oxygen therapy, mean nocturnal Sao_2 was significantly improved with NIV compared to the baseline room air night.

However, in contrast with oxygen therapy, significant decreases rather than increases in Tc_{CO_2} occurred with NIV in NREM and REM sleep. While both therapies produced a significant increase in total REM duration, no other parameter of sleep architecture including sleep latency, total sleep time, or arousal index was significantly altered. Gozal [24] concluded that NIV was as effective as supplemental oxygen in improving nocturnal hypoxemia, with the additional benefit of preventing marked rises in CO_2. It was postulated that NIV might have a clinical benefit over oxygen therapy by improving alveolar ventilation during sleep.

Milross and colleagues [9] sought to further clarify these findings by directly measuring ventilation during sleep in 13 patients with moderate to severely impaired lung function (FEV_1 32% $\pm$ 11%). Patients underwent three sleep studies, where they were randomly assigned to room air, oxygen therapy, or NIV $\pm$ oxygen. During each night the patient wore a mask attached to a pneumotachograph in order to measure ventilation during sleep. Although no significant fall in minute ventilation occurred from awake to NREM sleep during either the room air or oxygen nights, there was a significant fall from NREM to REM sleep and from awake to REM sleep under both conditions. In contrast, bilevel support prevented this fall in minute ventilation from NREM to REM sleep, primarily by maintaining tidal volume in this sleep stage. Although a relative fall in minute ventilation from awake to REM sleep during bilevel support was still seen, the overall minute ventilation was similar in REM sleep on bilevel support to that occurring in NREM sleep with room air. As seen in the Gozal study [24], use of bilevel support was able to improve oxygenation to a similar extent as oxygen therapy, while attenuating the rise in Tc_{CO_2} seen with REM sleep compared to both room air and oxygen conditions. Arterial blood gas sampling showed that subjects were significantly more acidotic after a night of oxygen therapy compared with bilevel support.

Although hypoventilation on a background of V_A/Q appears to explain the sleep-related desaturation seen in patients with moderate to severe CF lung disease, therapies aimed at reversing these abnormalities, while improving saturation and ventilation, do not appear to modify sleep architecture or arousal index [9, 24]. This raises the possibility that other factors such as cough or reflux may impact on sleep quality and maintenance. However, this area has been poorly studied to date. Furthermore, controlled studies regarding the long-term impact of nocturnal oxygen therapy and NIV on quality of life, neurocognitive function, and disease progression are required.

CONCLUSION

In summary, patients with CF and moderate to severe lung disease may exhibit marked nocturnal desaturation, especially in REM sleep. The main mechanisms responsible for this appear to be reduced respiratory drive and loss of postural muscle tone. As a consequence, chest wall movement is reduced, resulting in a fall in minute ventilation, primarily due to a reduction in tidal volume. Recognizing the extent and magnitude of sleep disruption and nocturnal desaturation in these patients is important in planning intervention. However, daytime clinical measures have only a modest ability to predict patients likely to desaturate during sleep. Those patients with a $FEV_1 <$ 65% and an awake $Sa_{O_2} <$ 94% appear to be the group at risk for nocturnal desaturation. Both oxygen therapy and NIV improve oxygenation during sleep in acute intervention studies but NIV is superior in maintaining ventilation and controlling any rise in CO_2 tension. Monitoring of sleep and breathing should be an integral part of disease management in this population. Despite the benefits of these treatments in acute studies, more research is required into the optimum timing of long-term intervention, and randomized controlled trials are needed to determine the impact, if any, on morbidity and mortality.

REFERENCES

1. Fraser KL, Tullis DE, Sasson Z, Hyland RH, Thornley KS, Hanly PJ. Pulmonary hypertension and cardiac function in adult cystic fibrosis. *Chest* **115**(5):1321–1328(1999).

2. Kerem E, Reisman J, Corey M, Canny GJ, Levison H. Prediction of mortality in patients with cystic fibrosis (see comments). *N Engl J Med* **326**(18):1187–1191(1992).

3. Muller NL, Francis PW, Gurwitz D, Levison H, Bryan AC. Mechanism of hemoglobin desaturation during rapid-eye-movement sleep in normal subjects and in patients with cystic fibrosis. *Am Rev Respir Dis* **121**(3):463–469(1980).

4. Stokes DC, McBride JT, Wall MA, Erba G, Strieder DJ. Sleep hypoxemia in young adults with cystic fibrosis. *Am J Dis Child* **134**(8):741–743(1980).

5. Tepper RS, Skatrud JB, Dempsey JA. Ventilation and oxygenation changes during sleep in cystic fibrosis. *Chest* **84**(4):388–393(1983).

6. Milross MA, Piper AJ, Norman M, Grunstein RR, Sullivan CE, Bye PTP. Predicting oxygen desaturation during sleep in patients with cystic fibrosis. *Chest* **120**(4):1239–1245(2001).

6a. Spier S, Rivlin J, Hughes D, Levison H. The effect of oxygen on sleep, blood gases, and ventilation in cystic fibrosis. *Am Rev Respir Dis* **129**(5):712–718(1984).

7. Douglas NJ, White DP, Pickett CK, Weil JV, Zwillich CW. Respiration during sleep in normal man. *Thorax* **37**(11):840–844(1982).

8. Ballard RD, Sutarik JM, Clover CW, Suh BY. Effects of non-REM sleep on ventilation and respiratory mechanics in adults

with cystic fibrosis. *Am J Respir Crit Care Med* **153**(1):266–271(1996).

9. Milross MA, Piper AJ, Norman M, Becker HF, Willson GN, Grunstein RR, Sullivan CE, Bye PTP. Low flow oxygen and bilevel ventilatory support: effects on ventilation during sleep in cystic fibrosis. *Am J Respir Crit Care Med* **163**(1):129–134(2001).

10. Bradley S, Solin P, Wilson J, Johns D, Walters EH, Naughton MT. Hypoxemia and hypercapnia during exercise and sleep in patients with cystic fibrosis. *Chest* **116**(3):647–653(1999).

11. Dancey DR, Tullis ED, Heslegrave R, Thornley K, Hanly PJ. Sleep quality and daytime function in adults with cystic fibrosis and severe lung disease. *Eur Respir J* **19**(3):504–510(2002).

12. Milross M, Piper AJ, Norman M, Dobbin CJ, Grunstein RR, Sullivan CE, Bye PTP. Subjective sleep quality in cystic fibrosis. *Sleep Med* **3**(3):205–212(2002).

13. Versteegh FG, Bogaard JM, Raatgever JW, Stam H, Neijens HJ, Kerrebijn KF. Relationship between airway obstruction, desaturation during exercise and nocturnal hypoxemia in cystic fibrosis patients. *Eur Respir J* **3**(1):68–73(1990).

14. Frangolias DD, Wilcox PG. Predictability of oxygen desaturation during sleep in patients with cystic fibrosis. *Chest* **119**(2):434–441(2001).

15. Allen MB, Mellon AF, Simmonds EJ, Page RL, Littlewood JM. Changes in nocturnal oximetry after treatment of exacerbations in cystic fibrosis. *Arch Dis Child* **69**(2):197–201(1993).

16. Dobbin CJ, Milross MA, Norman M, Piper AJ, Grunstein RR, Bye PTP. The effects of a pulmonary exacerbation on sleep in adult patients with cystic fibrosis—a pilot study (abstract only). *Respirology* **7** (Suppl):A44(2002).

17. Holland AE, Wilson JW, Kotsimbos TC, Naughton MT. Metabolic alkalosis contributes to acute hypercapnic respiratory failure in adult cystic fibrosis. *Chest* **124**:490–493(2003).

18. Javaheri S, Kazemi H. Metabolic alkalosis and hypoventilation in humans. *Am Rev Respir Dis* **136**:1011–1016(1987).

19. Young AC, Wilson JW, Kotsimbos TC, Holland AE, Naughton MT. Hypochloremic metabolic alkalosis is associated with nocturnal desaturation in cystic fibrosis (abstract). *Respirology* **9**(Suppl A70)(2004).

20. Moss AJ, Harper WH, Dooley RR, Murray JF, Mack JF. Cor pulmonale in cystic fibrosis of the pancreas. *J Pediatr* **67**(5):797–807(1965).

21. Zinman R, Corey M, Coates AL, Canny GJ, Connolly J, Levison H, Beaudry PH. Nocturnal home oxygen in the treatment of hypoxemic cystic fibrosis patients. *J Pediatr* **114**(3):368–377(1989).

22. Spiegel K, Leproult R, Van Cauter E. Impact of sleep debt on metabolic and endocrine function. *Lancet* **354**(9188):1435–1439(1999).

23. Polotsky VY, Li J, Punjabi NM, Rubin AE, Smith PL, Schwartz AR, O'Donnell CP. Intermittent hypoxia increases insulin resistance in genetically obese mice. *J Physiol* **552**(Pt 1): 253–264(2003).

24. Gozal D. Nocturnal ventilatory support in patients with cystic fibrosis:comparison with supplemental oxygen. *Eur Respir J* **10**(9):1999–2003(1997).

91

RESTRICTIVE THORACIC AND NEUROMUSCULAR DISORDERS

CHRISTOPHE PERRIN, CAROLYN D'AMBROSIO, ALEXANDER WHITE, ERIK GARPESTAD, AND NICHOLAS S. HILL
Tufts University School of Medicine, Boston, Massachusetts

INTRODUCTION

Sleep is a complex physiological process that is primarily restorative but also increases vulnerability to disordered breathing that may adversely affect gas exchange or sleep quality. Many pulmonary disorders predispose to sleep disordered breathing, contributing to hypoxemia during sleep and poor sleep quality as well as morbidity and even mortality. This chapter discusses sleep-related breathing abnormalities associated with thoracic cage disorders and neuromuscular diseases (restrictive thoracic disorders), the mechanisms by which abnormal sleep leads to respiratory insufficiency, the evaluation of sleep disordered breathing in such patients, and principles of clinical management.

SLEEP AND BREATHING IN NORMAL SUBJECTS

Control of Breathing During Sleep

Normal sleep affects central respiratory control, airway resistance, and muscular tone and contractility (Figure 91.1). Sleep onset reduces respiratory center responsiveness to chemical (hypercapnia, hypoxia) and mechanical (lung inflation) stimuli [1]. Deeper stages of sleep, particularly rapid eye movement (REM), further blunt ventilatory responsiveness. Although the diaphragm is less affected than the

Correspondence to: Nicholas S. Hill, Division of Pulmonary, Critical Care, and Sleep Medicine, Tufts-New England Medical Center, Tufts University School of Medicine, 750 Washington Street, #257, Boston, MA, 02111-1526.

Sleep: A Comprehensive Handbook, Edited by T. Lee-Chiong.

accessory muscles, the responsiveness of respiratory muscles to respiratory center output is also reduced. Minute ventilation falls during non-rapid eye movement (NREM) sleep, partly related to the diminished ventilatory responsiveness and partly to a drop in metabolic rate that occurs during sleep. Minute ventilation further declines during REM sleep, particularly during phasic REM when muscle tone and inspiratory drive are further suppressed. These physiological changes are associated with only minor alterations in gas exchange among normal subjects but may contribute to profound hypoventilation and hypoxemia in patients with respiratory insufficiency.

Airway Resistance During Sleep

Upper airway resistance increases during sleep compared to wakefulness as upper airway tone decreases. In addition, circadian changes in airway caliber slightly increase resistance in lower airways during sleep [2], predisposing to upper airway occlusion and obstructive sleep apnea, particularly in individuals with anatomic or physiologic susceptibilities to sleep apnea.

Ribcage and Diaphragm Interactions During Sleep

As though pulling on a bucket handle, the diaphragm expands the ribcage during quiet breathing in upright subjects. Because of an alteration in the shape of the diaphragm during NREM sleep, intercostal muscles are recruited to maintain ribcage expansion [3]. In contrast, during REM

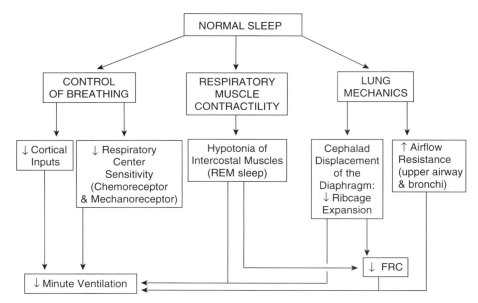

Figure 91.1   Schematic diagram of the effects of normal sleep on ventilation. The onset of sleep reduces central respiratory drive and respiratory muscle tone and increases airway resistance. These changes are accentuated during rapid eye movement (REM) sleep. The net effects are reduced functional residual capacity (FRC) and minute volume. See text for more details.

sleep, intercostal muscle activity is markedly reduced by active supraspinal inhibition of alpha motoneuron drive and selective depression of fusimotor function, leaving the diaphragm as the main ventilatory muscle.

Functional Residual Capacity During Sleep

Functional residual capacity (FRC) falls during sleep in healthy adults during both NREM and REM sleep, but not enough to substantially alter ventilation/perfusion relationships [4]. This decrease is likely related to cephalad displacement of the diaphragm in the supine position, respiratory muscle hypotonia, central pooling of blood, an increase in lung elastance, or some combination of these factors.

EFFECTS OF THORACIC RESTRICTION ON BREATHING DURING SLEEP

Restrictive thoracic disorders amplify the effects of normal sleep on gas exchange, especially when the diaphragm is weakened [5] (Figure 91.2). The mild hypoventilation that occurs at the onset of normal sleep is intensified in patients with restrictive physiology. Decreased chest wall compliance increases work of breathing and potentiates hypoventilation when inspiratory muscle tone drops during REM sleep. Abnormal chest wall mechanics also creates instability and uncoordinated chest wall motion that, along with

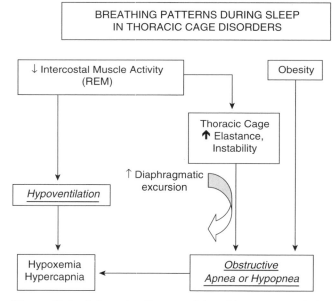

Figure 91.2   Schematic diagram of breathing patterns during sleep in patients with thoracic cage disorders. The effects of normal responses to sleep are intensified by the abnormalities in chest wall mechanics and the suppression of intercostal muscle activity that occurs during REM sleep. Chest wall instability may predispose to obstructive apneas. Comorbidities like obesity contribute to the problem. A low FRC and ventilation/perfusion ratio (V/Q) mismatch related to the abnormal chest wall mechanics predispose to even worse hypoxemia during sleep. See text for more details.

anatomic abnormalities of the upper airway, may predispose to obstructive events.

Restricted patients with daytime hypoxemia may develop life-threatening hypoxemia during sleep because a drop in PaO_2 will be associated with a large drop in arterial oxygen saturation if the oxygen tension is already on the steep part of the oxyhemoglobin dissociation curve. Furthermore, abnormalities in ribcage and abdominal mechanical properties predispose to disruption of V/Q matching and exacerbate the hypoxemia.

Hypoxemia and/or hypercapnia related to sleep disordered breathing evoke arousal responses that stimulate ventilation and limit the severity of gas exchange abnormalities. However, these arousals also promote sleep fragmentation that impairs sleep quality and blunts ventilatory responsiveness [6]. The reduced ventilatory responsiveness increases tolerance to hypercapnia and hypoxemia and may contribute to the progression of both daytime and nocturnal hypoventilation by resetting central chemoreceptors. The reduced ventilatory responsiveness has the effect of reducing sleep fragmentation, but at the cost of worsening hypoventilation and further depression of the arousal mechanism. A vicious cycle can ensue with progressive loss of the arousal mechanism, continued sleep fragmentation, and progressive hypercapnia.

ASSESSMENT OF BREATHING DURING SLEEP IN RESTRICTIVE THORACIC DISORDERS

A high index of suspicion is essential in diagnosing sleep disordered breathing in restrictive thoracic disorders because the associated symptoms (Table 91.1) are nonspecific and may be insidious in onset. Some patients with neuromuscular disease have minimal symptoms despite significant apneas and severe nocturnal oxygen desaturations. Others may report symptoms that suggest nocturnal hypoventilation, but these may be misinterpreted as part of the progressive deterioration of the underlying neuromuscular disorder. Furthermore, daytime measurements of respiratory function may be poor predictors of nocturnal breathing events. Arterial blood gases and thyroid function should be checked in any patient with pulmonary restriction

TABLE 91.1 Common Symptoms of Hypoventilation Syndromes

Morning headaches
Daytime hypersomnolence
Fatigue
Weakness
Difficulty concentrating
Depression
Enuresis
Dyspnea

and symptoms consistent with hypoventilation. Several observations may help in determining whether more extensive nocturnal monitoring should be done. First, a drop in vital capacity of 25% or more between the upright and supine positions is a more sensitive indicator of diaphragm weakness than upright measurements of vital capacity and total lung capacity alone. The percentage fall in vital capacity from the erect to the supine position correlates with the lowest SaO_2 value during REM sleep [7]. Second, maximum inspiratory and expiratory pressures are more sensitive measures to detect respiratory muscle weakness than vital capacity or total lung capacity and indicate significant respiratory muscle weakness when values are <60% predicted [8]. Third, patients who have thoracic cage disorders, a vital capacity of <1–1.5 L, and a high angle of curvature (>120°) are at high risk of developing respiratory failure. Patients with restrictive thoracic disorders who develop these findings should be considered for nocturnal polysomnography. Summary recommendations for assessment of breathing during sleep in thoracic cage disorders and neuromuscular diseases are presented in Table 91.2.

The most widely used tests of global inspiratory and expiratory muscle strength are the static maximum pressures measured at the mouth (PImax and PEmax) [8]. These tests have the advantages of being noninvasive, easy, and inexpensive to perform and normal values have been established for both adults and children. However, they are effort dependent and failure to get a good air seal around the mouthpiece may cause erroneous measurements, especially when the orofacial muscles are weak. Transdiaphragmatic pressure, measured using esophageal and gastric manometry balloons, can be used to assess diaphragmatic function, but aspiration is a risk in patients with severe neuromuscular impairment, particularly if bulbar structures are involved. For these reasons, simplified maneuvers or measures that require no subject effort have been proposed as alternative means of measuring respiratory muscle strength. The maximal sniff pressure test is

TABLE 91.2 Recommendations for a Polysomnogram in Thoracic Restrictive Disorders

- Symptoms of nocturnal hypoventilation[a]
- Risk factors for OSA including snoring, obesity, macroglossia, and retrognathia[a]
- Unexplained gas exchange abnormalities (i.e., with forced vital capacity (FVC) >50% predicted)[a]
- Abnormal nocturnal oximetry[a]
- Pressure titration when initiating NPPV[b]
- Assessment for failure to respond to NPPV therapy

[a]A polysomnogram is not required in patients meeting criteria for reimbursement of noninvasive positive pressure ventilation (see Table 91.3). These recommendations assume that criteria for NPPV have not been met, so a diagnostic study is necessary.
[b]Pressures determined during polysomnogram to eliminate respiratory events may not be adequate to augment ventilation.

easier to perform than the PImax maneuver for most subjects. Thus inspiratory muscle strength is often better reflected by maximal sniff pressure (SNIP) than by the PImax maneuver. SNIP is measured through a plug occluding one nostril during sniffs while the other nostril remains patent. Waxed plugs are hand-fashioned around the tip of a polyethylene catheter to eliminate nasal leaks. Some form of nocturnal monitoring is recommended in patients with neuromuscular disease with a vital capacity ≤55% of predicted. However, factors such as obesity increase the likelihood of detecting significant sleep disordered breathing, even at higher lung function levels.

The polysomnogram is considered the gold standard for identifying sleep disordered breathing, but oximetry alone or in combination with nocturnal respiratory monitoring may be useful in screening for sleep disordered breathing. However, the use of oximetry as a screening tool may lead to overestimation of respiratory events when compared with polysomnography. Indications for a polysomnogram in patients with restrictive thoracic diseases are listed in Table 91.2.

PATTERNS OF SLEEP DISORDERED BREATHING IN THORACIC CAGE AND SPECIFIC NEUROMUSCULAR DISORDERS

Thoracic Cage Disorders

Severe chest wall deformity predisposes to the development of respiratory failure and death (Figure 91.2). Prior chest surgery, especially thoracoplasty used in the past to treat pulmonary tuberculosis, was a common cause of chest wall deformity leading to respiratory failure and cor pulmonale. The current prevalence of respiratory failure associated with chest wall deformity is unknown, but it is undoubtedly much less than in the past because of the advent of successful chemotherapy for tuberculosis and advances in therapies to prevent the development of severe kyphoscoliosis.

Thoracic cage abnormalities not only alter chest wall mechanics and increase work of breathing but also disrupt V/Q distribution and reduce the surface area for gas diffusion. Additional factors associated with thoracoplasty are pleural thickening, variable degrees of scoliosis, and impaired inspiratory muscle function related to the distorted chest wall anatomy. In the past, parenchymal destruction resulting from prior tuberculosis predisposed to V/Q maldistribution and airflow obstruction, but this is uncommonly seen today.

By placing the respiratory muscles at a mechanical disadvantage, chest wall deformity reduces maximal strength and endurance, thereby increasing susceptibility to fatigue [9]. The supine position has an additional negative impact on respiratory muscle function in these patients and predisposes to oxygen desaturations related to hypopneas during REM sleep caused by reduced inspiratory muscle activity.

Because it is at a considerable mechanical disadvantage, diaphragm function is often impaired in severe kyphoscoliosis, and accessory muscles are recruited to maintain ventilation. During REM sleep, intercostal and accessory muscle activity is reduced by inhibition of neural output from the bulbar reticular formation [10]. This decrease in intercostal muscle tone contributes to hypoventilation during sleep as well as to a further reduction in FRC, exacerbating hypoxemia via disruption of V/Q relationships [5]. Chest wall instability can also occur and leads to paradoxical chest wall motion, further reducing the efficiency of ventilation and disrupting V/Q relationships.

Hypoventilation during sleep may have other physiologic consequences. The associated alveolar hypoxia raises pulmonary arterial pressure via hypoxic pulmonary vasoconstriction. If sustained, alveolar hypoxia causes pulmonary vascular remodeling and pulmonary hypertension leading to cor pulmonale. Hypoxemia and hypercapnia associated with hypoventilation also impair respiratory muscle function, increasing fatiguability and susceptibility to the development of respiratory failure.

Neuromuscular Diseases

Sleep disordered breathing is very common in neuromuscular diseases, occurring in an estimated 42% of individuals overall [11], and becomes more prevalent as the disease progresses. The nature and severity of sleep disordered breathing vary between the different neuromuscular syndromes, depending on the pattern and severity of respiratory muscle involvement. Weakness of the diaphragm and other respiratory muscles, deformities of the ribcage and spine, upper airway muscle involvement, obesity, and craniofacial abnormalities are all associated with neuromuscular disorders and may contribute to obstructive respiratory events, nocturnal desaturations, and sleep fragmentation. As respiratory muscle function deteriorates, worsening nocturnal hypoventilation leads to oxygen desaturation and sleep disruption, nonrestorative sleep, daytime sleepiness, impaired concentration, fatigue, lethargy, and, eventually, diurnal hypoventilation that is the hallmark of respiratory failure [12, 13].

The pattern of sleep disordered breathing in patients with neuromuscular disease reflects the distribution of respiratory muscle involvement. If the upper airway or intercostal muscles are weak but diaphragm strength is intact, then obstructive apneas or hypopneas are more likely to occur. When patients have severe diaphragm dysfunction, suppression of intercostal and accessory muscles during REM sleep leads to hypoventilation [6]. In some forms of neuromuscular disease, such as myotonic dystrophy, primary abnormalities in ventilatory control may also contribute to sleep disordered breathing, often complicated by nocturnal or even diurnal hypoventilation.

Obstructive respiratory events during sleep are common with neuromuscular disorders. Risk factors for obstructive sleep apnea associated with neuromuscular disease include snoring, a high body mass index (BMI), or anatomic abnormalities such as retrognathia or macroglossia. Such abnormal sleep events have been reported in patients with Duchenne's muscular dystrophy, amyotrophic lateral sclerosis, and myotonic dystrophy. Even in the absence of frank apneas, increased upper airway resistance during REM sleep may contribute to obstructive hypopneas. Weak pharyngeal or laryngeal muscles may not be able to sufficiently stabilize the upper airway during inspiration, predisposing to airway narrowing and increased airway resistance [14].

The large variability in the reported prevalence of obstructive events during sleep in patients with the same underlying neuromuscular condition and similar respiratory function is related, at least in part, to differences in how these events are diagnosed and recorded. For example, obstructive events are usually defined as continued or increased chest wall movements in association with reduced or absent airflow. In patients with severely weakened inspiratory muscles, however, chest wall motion during obstructive events may be too diminished to be detected, at least by standard plethysmography techniques, causing the events to be misclassified as central [15]. Esophageal manometry is a more sensitive way to detect these diminished inspiratory efforts, but it is not available in most U.S. sleep laboratories.

In addition to obstructive respiratory events, a number of other sleep-related breathing problems are very common in neuromuscular syndromes. Central apneas occur more commonly in certain neuromuscular syndromes like myotonic dystrophy. The most common form of sleep disordered breathing in patients with neuromuscular disease is nocturnal hypoventilation due to a reduction in tidal volume attributable to inspiratory muscle weakness. Usually worse during REM sleep, it occurs in many neuromuscular syndromes including bilateral diaphragm paralysis, amyotrophic lateral sclerosis, and myotonic dystrophy [12, 16, 17]. Sustained hypoventilation occurs as respiratory muscle weakness progresses. Respiratory failure, the most common cause of death in neuromuscular diseases that involve the respiratory muscles, may occur gradually over a number of years in slowly progressive neuromuscular disorders or precipitously in more rapidly progressive disorders, particularly when complicated by atelectasis or respiratory infection with secretion retention. Scoliosis and obesity also accelerate the development of respiratory failure. As with chest wall deformities, cor pulmonale is also a consequence of the respiratory failure associated with neuromuscular

syndromes. The various patterns of sleep disordered breathing in individual neuromuscular diseases are discussed in more detail later.

Of course, other factors besides sleep disordered breathing may contribute to poor sleep quality and daytime sleepiness in patients with neuromuscular disease. The inability to change position during sleep causes pain, discomfort, and frequent disruptions during sleep. Problems with secretion clearance, anxiety, and depression may also impair sleep quality in these patients.

Strokes

Cerebrovascular accidents predispose to sleep disordered breathing. The prevalence of sleep disordered breathing after a hemispheric stroke is as high as 77% for men and 64% for women, associated with disrupted sleep architecture including shorter durations of slow-wave and REM sleep. This may be related to involvement of upper airway musculature by large strokes. In addition, sleep apnea increases the risk for stroke. The profound hypoxemia occurring during apneic episodes, changes in cerebrovascular tone in association with abrupt swings in blood gases that might predispose to atherosclerosis, and an increased platelet aggregability associated with sleep apnea could predispose to strokes. These associations justify a low threshold for evaluating sleep disordered breathing in patients with strokes, and consideration of continuous positive airway pressure (CPAP) therapy among patients found to have sleep apnea, although no studies have established the effectiveness of such an approach.

Parkinson's Disease

Up to 65% of patients with Parkinson's disease complain of sleep disorders [18], related largely to abnormalities of upper airway and respiratory muscles. Muscle rigidity predisposes to upper airway obstruction and chest wall restriction, both of which may contribute to gas exchange disturbances during sleep. Oxygen desaturations and hypoventilation occur commonly during sleep, and up to two-thirds of patients report sleep-related problems in patient surveys. Upper airway muscle abnormalities cause fluttering and irregular, abrupt changes on the flow volume loop. Rhythmic oscillations that correspond to tremors of upper airway muscles are the cause of the fluttering and may increase upper airway resistance as well. Lower respiratory muscles also function abnormally, with less efficiency and inability to sustain as high resistance loads as normals. Obstructive sleep apnea per se is not thought to be associated with idiopathic Parkinson's disease, but multiple system atrophy, a similar condition that is associated with a number of extrapyramidal findings including

Shy–Drager syndrome, is associated with sleep disordered breathing such as nocturnal stridor and central apneas.

Myotonic Dystrophy

Myotonic dystrophy is associated with various patterns of sleep disordered breathing including obstructive apneas and hypopneas, central apneas, and frank central hypoventilation. Daytime hypersomnolence, nocturnal desaturations, and diurnal hypoventilation are also common and may not be explained by either sleep disordered breathing or disturbances of sleep architecture. These symptoms have been related to an irregular breathing pattern during wakefulness and light sleep that does not persist during slow-wave sleep, suggesting that there may be a brainstem abnormality affecting respiratory control [19]. A recent histopathological study showing less neuronal density in the medullary arcuate nucleus of myotonic dystrophy patients with central hypoventilation than in myotonic dystrophy patients without hypoventilation or in controls with other neurologic conditions supports the idea that this results from a primary dysfunction of the central nervous system [20].

Amyotrophic Lateral Sclerosis

Amyotrophic lateral sclerosis (ALS) has protean effects on breathing during sleep, depending on the pattern of neuromuscular involvement. The upper airway muscles are usually involved, predisposing to obstructive events, although frank obstructive sleep apnea is less common than might be predicted, perhaps because the muscles are too weak to generate inspiratory pressures low enough to collapse the upper airway. Total sleep time is less and arousals and sleep disordered breathing are more frequent than in age-matched controls, although the sleep disordered breathing is relatively mild and without frank apneas. REM-associated hypoventilation also occurs, similar to that seen in non-ALS patients with respiratory muscle weakness. The largest unselected cross-sectional study of sleep in ALS observed a mean of only 11.3 events/hour [16] and the principal cause of nocturnal oxygen desaturation was hypoventilation. Surprisingly, bulbar involvement has not been associated with the severity of sleep disordered breathing.

Diaphragm dysfunction also plays an important role in the respiratory impairment seen in ALS. Patients with diaphragm dysfunction typically recruit other inspiratory muscles to breathe, including intercostals and accessory muscles, and this recruitment is thought to be a major contributor to dyspnea in ALS. Diaphragm dysfunction increases the likelihood of major sleep-related respiratory abnormalities, particularly during REM sleep, when maintenance of ventilation is critically dependent on diaphragm function. ALS-related diaphragm dysfunction has also been associated with a dramatic reduction in REM sleep and reduced survival. The presence of orthopnea predicts responsiveness to therapy with noninvasive positive pressure ventilation in ALS, further underscoring the importance of this connection.

Spinal Cord Injury

The level of spinal cord injury determines the severity of pulmonary impairment. Patients with C5–6 injuries or lower have intact diaphragm function and, unless encumbered with secretions from a respiratory infection, are able to ventilate without difficulty. However, the importance of obstructive sleep apnea in patients with lesions at C5 or higher has been recognized for some time [21]. Factors that predispose to sleep apnea in such patients include paralyzed intercostal and abdominal muscles, impaired activation of the diaphragm, increased upper airway resistance, sleep in the supine position, antispasmodic medication, traumatized and weakened laryngeal muscles by a previous intubation, and disrupted feedback afferents from ribcage receptors that alter compensatory reflexes to ventilatory loading.

The sleep apnea syndrome in tetraplegic patients is caused mainly by obstructive events associated with an increase in neck circumference [21]. This latter finding is thought to be related to a redistribution of body fat caused by a pathological insulin sensitivity due to altered sympathetic function following high spinal cord injury [22]. In quadriplegic patients, the enlargement of neck circumference is a better predictor of sleep apnea syndrome than body mass index. Although sleep apneas and hypopneas are usually obstructive in patients with high spinal cord injury, central and mixed apneas may also occur, suggesting that additional central factors may be involved in the pathogenesis of the sleep apnea syndrome in quadriplegia.

Duchenne's Muscular Dystrophy

Sleep disordered breathing is quite common with Duchenne's muscular dystrophy (DMD) and has traditionally been thought to consist mainly of central apneas. However, some studies on nonambulatory patients have reported that up to 60% of apneas are obstructive [23]. Older patients had more nocturnal hypoxemic periods related to central apneas, or possibly "pseudocentral" apneas that were misclassified obstructive events because of the difficulty in detecting inspiratory muscle activity. Measurement of transdiaphragmatic pressure during sleep would theoretically resolve the issue but is difficult for several reasons including the potential hazard of placing manometry balloons in patients with severe cough impairment and the profound respiratory muscle weakness that complicates proper positioning of the catheters because the diaphragm may not generate a pressure difference. A study of six patients with DMD aged 16–22 years showed that hypoventilation could occur in all sleep stages, and those with diaphragmatic dysfunction were especially vulnerable to oxygen desaturation during REM sleep [24]. On the whole, however, sleep architecture was better preserved in DMD than in subjects with ALS with a similar degree of respiratory function impairment.

Vital capacity and total lung capacity are poor predictors of sleep hypoventilation in DMD [15]. On the other hand, a forced expiratory volume in 1 second (FEV_1) <40% of the predicted value correlates with sleep hypoxemia, daytime base excess, and $Paco_2$ elevation. In addition, a base excess of ≥ 4 mmol/L and a $Paco_2 \geq 45$ mmHg are sensitive indicators of sleep-related hypoventilation. Hence a daytime arterial blood gas and possibly a polysomnogram are recommended when patients with DMD have a FEV_1 $\leq 40\%$ of predicted. Awaiting symptomatic evidence of daytime hypoventilation is no longer recommended because sleep-related hypoventilation may long precede diurnal hypoventilation.

Myasthenia Gravis

Sleep disordered breathing and nocturnal oxygen desaturations in myasthenia gravis are more pronounced during REM sleep and the majority of respiratory events are central [25]. As might be anticipated, patients with more advanced disease as evidenced by a lower total lung capacity and abnormal daytime arterial blood gases, advanced age, and a higher body mass index are the most susceptible to sleep disordered breathing [25].

MANAGEMENT OF RESPIRATORY INSUFFICIENCY DURING SLEEP

Respiratory insufficiency caused by restrictive thoracic disorders usually responds well to appropriate therapy, with improvements in functional status and life expectancy. These disorders are characterized by hypoventilation and, as such, oxygen therapy alone is not only usually ineffective but may even be hazardous, often aggravating the severity of CO_2 retention. However, the simplest therapy that is likely to be effective should be sought, starting with pharmacological therapy, moving on to CPAP, using noninvasive positive pressure ventilation (NPPV) when ventilatory assistance is indicated, and reserving invasive positive pressure ventilation as a therapy of last resort.

Pharmacological Therapy

Pharmacological therapy should be considered when hypoventilation is worse than can be explained by the severity of

pulmonary dysfunction (i.e., $FEV_1 > 1.2$ L), and polysomnography shows evidence of central hypoventilation or apnea but not obstructive respiratory events. The authors have encountered such patients with myotonic dystrophy who have responded well to progestational agents (megestrol 40 mg orally three times daily) (personal observation). However, central stimulants are not recommended for patients with severe ventilatory defects and might even exacerbate dyspnea.

Continuous Positive Airway Pressure

Continuous positive airway pressure (CPAP) alone is indicated in patients with no more than milld hypoventilation who have polysomnograms showing obstructive respiratory events. Obstructive or mixed sleep apneas occurring in the presence of normal or nearly normal daytime pulmonary function in patients with mild neuromuscular disease often respond favorably to CPAP alone. A trial of CPAP is warranted in such patients, but if the condition fails to respond after a reasonable trial or hypoventilation is more than mild, noninvasive ventilation should be considered.

Noninvasive Ventilation

In the past, negative pressure ventilation predominated as the noninvasive method of choice for patients with chronic respiratory failure due to restrictive thoracic disorders and neuromuscular diseases. However, negative pressure ventilation has been supplanted by noninvasive positive pressure ventilation (NPPV) for a number of reasons. NPPV is more convenient to use, easier to apply, more portable, and, for most patients, at least as comfortable. In addition, neuromuscular disease patients treated with negative pressure ventilation have sleep fragmented by frequent arousals. Furthermore, episodes of severe oxygen desaturation persist, particularly during REM sleep, associated with obstructive events that may be induced by the negative pressure ventilation. Because of its advantages over negative pressure ventilation, NPPV is the NIV mode of first choice for patients with restrictive thoracic disorders. Rarely, negative pressure ventilation or another noninvasive approach might be considered for a patient who cannot tolerate NPPV.

Indications and Contraindications for NPPV Table 91.3 lists indications for NPPV in thoracic cage disorders and chronic neuromuscular diseases. Most clinicians agree that patients should have symptoms attributable to respiratory insufficiency, because the desire for symptom relief is a strong incentive to adhere to therapy. The most common symptoms are those associated with nocturnal sleep disruption and are listed in Table 91.1. Gas exchange criteria are also important, including evidence of daytime

TABLE 91.3 Indications for Noninvasive Positive Pressure Ventilation in Restrictive Thoracic Disorders

Symptoms (such as fatigue, dyspnea, morning headache, orthopnea) and one of the following physiologic criteria

Physiologic criteria (one of the following)

$Paco_2 > 45$ mmHg (6 kPa)

Nocturnal oximetry demonstrating oxygen saturation $\leq 88\%$ for >5 consecutive minutes

For progressive neuromuscular disease, maximal inspiratory pressure <60 cm H_2O or FVC <50% predicted

---

($Paco_2 \geq 45$ mmHg) or nocturnal hypoventilation (detected by sustained oxygen desaturation—$Sao_2 \leq 88\%$ for >5 consecutive min) [26]. Initiation of NPPV is warranted when nocturnal hypoventilation develops and before the onset of daytime hypoventilation, so nocturnal monitoring is indicated when nocturnal hypoventilation is suspected. In progressive neuromuscular diseases, NPPV is recommended if symptoms are present with a forced vital capacity <50% predicted or PImax < 60 cm H_2O [27]. With more rapidly progressive neuromuscular syndromes like ALS, orthopnea has been shown to be a strong predictor of NPPV success and should be added to the list of symptoms justifying NPPV initiation.

Absolute contraindications to noninvasive ventilation are few but include the inability to fit a mask or cooperate (Table 91.4). Relative contraindications include a lack of motivation and a compromised ability to protect the airway because noninvasive ventilation, unlike invasive ventilation, affords no direct access to the airways for suctioning. Thus invasive ventilation should be considered in patients with severe bulbar involvement, weakness of expiratory muscles, or secretion retention. Invasive mechanical ventilation should also be considered in patients with neuromuscular disease requiring nearly continuous ventilatory support. However, some clinicians advocate noninvasive ventilation even in patients who require continuous ventilatory support,

TABLE 91.4 Contraindications to Noninvasive Ventilation in Patients with Restrictive Thoracic Disorders

Relative contraindications
  Weakened cough
  Swallowing dysfunction
  Upper airway obstruction (depends on severity)
  Poorly motivated patient or family
  Inadequate financial resources
  Inadequate family/caregiver support
  Need for continuous ventilatory assistance
Absolute contraindications
  Anatomic abnormalities that preclude interface fitting
  Inability to cooperate
  Uncontrollable secretion retention

to avoid the complications and greater care burden of ventilation via a tracheostomy.

Patients with thoracic cage disorders and neuromuscular diseases and their families should be counseled about the options for ventilator support well in advance of the development of respiratory insufficiency so that they are prepared for the possible outcomes. With appropriate counseling and monitoring, respiratory crises leading to unanticipated and sometimes unwanted intubation should be a rare occurrence.

Invasive Mechanical Ventilation

Invasive positive pressure ventilation is indicated in patients who have contraindications for or who have failed noninvasive ventilation and desire aggressive support. It has long been used to treat respiratory failure and prolong survival in patients with restrictive thoracic disorders or neuromuscular diseases, but it has significant disadvantages. Invasive mechanical ventilation requires a high level of care including tracheostomy maintenance and suctioning, and its initiation may necessitate institutionalization for some patients because of inadequate caregiver resources in the home. It is also associated with a number of complications including possible interference with speech and swallowing, repeated respiratory and soft-tissue infections, tracheal stenosis, and tracheomalacia due to excessive cuff pressures.

Technical Aspects of Administering Noninvasive Positive Pressure

CPAP and NPPV are administered using a device that delivers pressurized gas to the lungs through an interface (mask or mouthpiece) affixed to the nose, mouth, or both. For long-term outpatient applications, nasal masks are the most widely used interface for administration of CPAP or NPPV. Nasal masks are available from many manufacturers in multiple sizes and shapes. Because skin irritation and ulceration are problems with nasal masks, numerous modifications to reduce pressure on the face and enhance comfort are available, including soft silicon and gel seals that may enhance comfort. Also available to optimize comfort and minimize claustrophobia are "mini-masks" that fit over the tip of the nose or nasal "prongs" that consist of small cones inserted directly into the nares. Because of the plethora of commercially available mask types and sizes, custom-molded masks are now rarely used.

Nasal interfaces are also limited by air leaking through the mouth, a problem that may be ameliorated by chin straps or switching to a full face (or oronasal) mask that covers both the nose and the mouth. Used more often for acute respiratory failure, full face masks are available that are specifically designed for long-term noninvasive ventilation. However, nasal masks are likely to remain the preferred interface because they are rated as more comfortable by patients. Also, asphyxiation may be a concern with full face masks in neuromuscular patients who are unable to remove the mask in the event of ventilator malfunction or power failure. Furthermore, interference with speech, eating, and expectoration, claustrophobic reactions, and the theoretical risk of aspiration and rebreathing are greater with full face than nasal masks. On the other hand, full face masks may be preferred in patients who are unable to close their mouths during nasal ventilation, such as those with bulbar dysfunction.

Mouthpieces held in place by lipseals can be used to provide up to 24 h/day of NPPV. Commercially available mouthpieces are simple to apply and inexpensive. Alternatively, custom-fitted mouthpieces that may enhance comfort and efficacy, and may be strapless, are also available at some centers. On average though, patients adapt more readily to nasal than to mouth interfaces. Mouthpieces can also be attached to gooseneck clamps on wheelchairs to serve as a convenient means of providing daytime noninvasive ventilatory assistance that enhances patient mobility.

Ventilators for NPPV

Portable pressure- or volume-limited ventilators or those that offer both modes are available to deliver NPPV. Volume-limited ventilators are used in the assist-control mode and are set to deliver large tidal volumes (10–15 mL/kg) to compensate for air leaks. Portable pressure-limited ventilators cycle between two levels of positive airway pressure using either flow or time triggers (bilevel ventilation). Some also offer spontaneous/timed modes that deliver backup time-cycled inspiratory and expiratory pressures with adjustable inspiratory/expiratory ratios at a preset rate. Backup settings just below spontaneous breathing rates may be desirable to ameliorate nocturnal hypoventilation; patients with advanced neuromuscular disease often allow the ventilator to control their breathing during sleep. The "bilevel" ventilators are best suited for patients requiring part-time ventilatory assistance, such as nocturnal only or intermittently during the daytime. Newer versions of these devices are suitable for patients requiring continuous ventilation, however, as long as adequate alarms and backup batteries have been incorporated.

Controlled data are lacking to guide selection of ventilator settings, but inspiratory pressures between 12 and 22 cm H_2O are often used. Higher pressures are sometimes needed for patients with high respiratory impedance, such as those with advanced scoliosis or morbid obesity. Expiratory pressure, usually 3–6 cm H_2O, is also applied during pressure-limited ventilation to minimize CO_2 rebreathing, because many of these ventilators have single-tube ventilator circuits. It is important to remember that if expiratory pressure is increased, inspiratory pressure must be raised equally to maintain the same level of inspiratory assistance (or pres-

sure support). In the authors' experience, patients are usually intolerant of higher pressures initially and must be started at subtherapeutic levels (inspiratory 8–10 cm H_2O; expiratory 2–3 cm H_2O) in order to facilitate adaptation. Therapeutic pressures can usually be attained within a matter of weeks. It is important to realize that pressures titrated in a sleep laboratory to eliminate respiratory events may not augment ventilation sufficiently and gas exchange must be monitored during adaptation (see later discussion).

Pressure-limited ventilators are most commonly used for chronic ventilatory support because they are inexpensive, well tolerated, and highly portable. Volume-limited ventilators may be preferred in patients with advanced disease because of their better monitoring capabilities, backup batteries and longer battery lives, and ability to "stack" breaths to assist with coughing. A newer generation of "hybrid" ventilators offers a variety of volume- and pressure-limited modes, sophisticated alarms, and a backup battery, but they are more expensive than the pressure-limited devices. Newer devices also offer sophisticated monitoring capabilities that can be downloaded via telephones wirelessly.

Effect of Noninvasive and Invasive Positive Pressure Ventilation on Sleep

Patients usually report better sleep soon after initiating nocturnal NPPV. This improvement may not be immediate, because some patients require days to weeks to successfully adapt to NPPV. Nonetheless, sleep is almost always symptomatically improved by nocturnal NPPV when patients are using the device for at least 4–5 hours nightly. Although no randomized controlled prospective trials have been performed, several studies have examined gas exchange or sleep quality during temporary withdrawal of noninvasive ventilation from patients with restrictive thoracic diseases who had previously responded favorably to it. In one study, nocturnal hypoventilation worsened during the period of temporary NPPV withdrawal [27], indicating that NPPV ameliorates nocturnal hypoventilation. Another showed that sleep quality deteriorated during the period of withdrawal, including reduced sleep efficiency and total sleep time, supporting the idea that NPPV improves sleep quality [28]. On the other hand, NPPV use is associated with frequent arousals in some patients in association with air leaks through the mouth [29]. Thus NPPV improves sleep in most patients with restrictive thoracic disorders but may contribute to sleep fragmentation in some. Sleep studies during invasive mechanical ventilation in patients with neuromuscular disease or chest wall deformity are lacking. In a survey of patients who had experienced both noninvasive and invasive forms of ventilation, sleep was rated as better during invasive mechanical ventilation [30].

Monitoring Noninvasive Positive Pressure Ventilation

Patients should be seen at least every few weeks during initiation of NPPV. Morning headaches and daytime hypersomnolence are usually the first symptoms to respond. Daytime arterial blood gases (or end-tidal PCO_2 if there is no parenchymal disease) should be checked and should gradually drop over several weeks as duration of usage increases. Resetting of the respiratory center sensitivity to CO_2, along with improved respiratory muscle strength or endurance, is thought to explain the improvement in daytime spontaneous ventilation associated with nocturnal ventilatory assistance [27]. If $PaCO_2$ doesn't fall into the desired range (40's to 50's) and symptoms fail to respond, inspiratory pressure and duration of NPPV should be increased. Once a plateau is reached, daytime $PaCO_2$ may remain elevated but tends to be stable for long periods of time, depending on the natural history of the underlying disorder. This improvement in gas exchange is usually sufficient to allow patients to be completely free from supplemental oxygen or ventilation during the daytime.

Improvement of nocturnal hypoventilation occurs promptly once the patient is using NPPV through the night, but some patients require several weeks or more time before they reach this point. Nocturnal oximetry should be done at that time to assess adequacy of nocturnal ventilation as evidenced by maintenance of a saturation of at least 90% overnight (in the absence of supplemental oxygen). Follow-up polysomnography is usually reserved for those patients whose ventilatory status fails to improve after an adaptation period and attempts at adjustments in ventilator settings.

SUMMARY AND CONCLUSIONS

Sleep disordered breathing is common in restrictive thoracic and neuromuscular diseases. The specific sleep-related abnormalities depend on the distribution, rate of progression, and severity of the chest wall and neuromuscular defects. Hypoxemia and hypoventilation are common during sleep, related to reductions in FRC and blunting of central drive (either primary or secondary to progressive bicarbonate retention). These gas exchange abnormalities lead to arousals, fragmenting sleep and producing symptoms such as morning headache and daytime hypersomnolence. Sleep quality is adversely affected, as the arousals reduce sleep efficiency, total sleep time, and the percentage of time spent in slow-wave or REM sleep. The therapies of sleep disordered breathing and hypoventilation in restrictive thoracic disorders include CPAP alone that may be effective in patients with obstructive sleep apnea syndrome. For most, however, NPPV is the therapy of choice. NPPV not only maintains upper airway patency but also actively assists inspiration using positive pressure, augmenting tidal volume and reversing hypoventilation. Invasive mechanical ventilation is used only if the patient is an unsuitable candidate for NPPV or NPPV fails and the patient desires aggressive support.

REFERENCES

1. Berthon-Jones M, Sullivan CE. Ventilation and arousal responses to hypercapnia in normal sleeping humans. *J Appl Physiol Respir Environ Exercise Physiol* **57**:59–67(1984).

2. Hetzel MR, Clark TJH. Comparison of normal and asthmatic circadian rhythms in peak flow rate. *Thorax* **35**:732–738 (1980).

3. Tabachnik E, Muller NL, Bryan AC, Levison H. Changes in ventilation and chest wall mechanics during sleep in normal adolescents. *J Appl Physiol* **51**:557–564(1981).

4. Hudgel DW, Martin RJ, Johnson BJ, Hill P. Mechanics of the respiratory system and breathing pattern during sleep in normal humans. *J Appl Physiol Respir Environ Exercise Physiol* **56**:133–137(1984).

5. Mezon BL, West P, Israels J, Kryger M. Sleep breathing abnormalities in kyphoscoliosis. *Am Rev Respir Dis* **122**:617–621(1980).

6. Schiffman PL, Trontell MC, Mazar MF, Edelman NH. Sleep deprivation decreases ventilatory response to CO_2 but not load compensation. *Chest* **84**:695–698(1983).

7. Bye PTP, Ellis ER, Issa FG, Donnelly PM, Sullivan CE. Respiratory failure and sleep in neuromuscular disease. *Thorax* **45**:241–247(1990).

8. Bruschi C, Cerveri I, Zoia MC, Fanfulla F, Fiorentini M, Casali L, Grassi M, Grassi C. Reference values of maximal inspiratory mouth pressures: a population based study. *Am Rev Respir Dis* **146**:790–793(1992).

9. Goldstein RS, De Rosie JA, Avendano MA, Dolmage TE. Influence of noninvasive positive pressure ventilation on inspiratory muscles. *Chest* **99**:408–415(1991).

10. Magoun HS, Rhines R. An inhibitory mechanism in the bulbar reticular formation. *J Neurophysiol* **9**:165–171(1946).

11. Labanowski M, Schmidt-Nowara W, Guilleminault C. Sleep and neuromuscular disease: frequency of sleep-disordered breathing in a neuromuscular disease clinic population. *Neurology* **47**:1173–1180(1996).

12. Finnimore AJ, Jackson RV, Morton A, Lynch E. Sleep hypoxia in myotonic dystrophy and its correlation with awake respiratory function. *Thorax* **49**:66–70(1994).

13. Barbé F, Quera-Salva MA, McCann C, Gajdos P, Raphael JC, de Lattre J, Agusti AGN. Sleep-related respiratory disturbances in patients with Duchenne muscular dystrophy. *Eur Respir J* **7**:1403–1408(1994).

14. Dematteis M, Pépin JL, Jeanmart M, Deschaux C, Labarre-Villa A, Levy P. Charcot–Marie–Tooth disease and sleep apnea syndrome: a family study. *Lancet* **357**:267–272(2001).

15. Smith PEM, Calverley PMA, Edwards RHT. Hypoxemia during sleep in Duchenne muscular dystrophy. *Am Rev Respir Dis* **137**:884–888(1988).

16. Gay PC, Westbrook PR, Daube JR, Litchy WJ, Windebank AJ, Iverson R. Effects of alterations in pulmonary function and sleep variables on survival in patients with amyotrophic lateral sclerosis. *Mayo Clin Proc* **66**:686–694(1991).

17. White JES, Drinnan MJ, Smithson AJ, Griffiths CJ, Gibson GJ. Respiratory muscle activity and oxygenation during sleep in patients with muscle weakness. *Eur Respir J* **8**:807–814(1995).

18. Garcia-Borreguero D, Larrosa O, Bravo M. Parkinson's disease and sleep. *Sleep Med Rev* **7**:115–129(2003).

19. Veale D, Cooper BG, Gilmartin JJ, Walls TJ, Griffith CJ, Gibson GJ. Breathing pattern awake and asleep in patients with myotonic dystrophy. *Eur Respir J* **8**:815–818(1995).

20. Ono S, Takahashi K, Kanda F, Jinnai K, Fukuoka Y, Mitake S, Inagaki T, Kurisaki T, Nagao K, Shimizu N. Decrease of neurons in the medullary arcuate nucleus in myotonic dystrophy. *Acta Neuropathol* **102**:89–93(2001).

21. McEvoy RD, Mykytyn I, Sajkov D, Flavell H, Marshall R, Antic R, Thornton AT. Sleep apnea in patients with quadriplegia. *Thorax* **50**:613–619(1995).

22. Karlsson AK. Insulin resistance and sympathetic function in high spinal cord injury. *Spinal Cord* **37**:494–500(1999).

23. Khan Y, Heckmatt JZ. Obstructive apneas in Duchenne muscular dystrophy. *Thorax* **49**:157–161(1994).

24. Smith PEM, Edwards RHT, Calverley PMA. Ventilation and breathing pattern during sleep in Duchenne muscular dystrophy. *Chest* **96**:1346–1351(1989).

25. Quera-Salva MA, Guilleminault C, Chevret S, Troche G, Fromageot C, Crowe McCann C, Stoos R, de Lattre J, Raphael JC, Gajdos P. Breathing disorders during sleep in myasthenia gravis. *Ann Neurol* **31**:86–92(1992).

26. Consensus Conference: Clinical indications for noninvasive positive pressure ventilation in chronic respiratory failure due to restrictive lung disease, COPD, and nocturnal hypoventilation—a consensus conference report. *Chest* **116**:521–534(1999).

27. Hill NS, Eveloff SE, Carlisle CC, Goff SG. Efficacy of nocturnal nasal ventilation in patients with restrictive thoracic disease. *Am Rev Respir Dis* **145**:365–371(1992).

28. Masa Jimenez JF, Sanchez de Cos Escuin J, Disdier Vicente C, Hernandez Valle M, Fuentes Otero F. Nasal intermittent positive pressure ventilation. Analysis of its withdrawal. *Chest* **107**:382–388(1995).

29. Meyer TJ, Pressman MR, Benditt J, McCool FD, Millman RP, Natarajan R, Hill NS. Air leaking through the mouth during nocturnal nasal ventilation: effect on sleep quality. *Sleep* **20**:561–569(1997).

30. Bach JR, Alba AS, Saporito LR. Intermittent positive pressure ventilation via the mouth as an alternative to tracheostomy for 257 ventilator users. *Chest* **103**:174–182(1993).

ADDITIONAL READING

Hill NS (Ed). Pulmonary complications of neuromuscular diseases. *Seminars in Respiratory Critical Care Medicine* Vol 23, 2002. The entire volume is devoted to respiratory complications of neuromuscular diseases. It includes an excellent chapter on sleep disordered breathing by Amanda Piper as well as more detailed descriptions of breathing abnormalities during sleep related to a number of specific neuromuscular diseases.

Mehta S, Hill NS. Noninvasive ventilation. *Am J Respir Crit Care Med* **163**:540–577(2001). This state-of-the-art review explores the trends in the use of noninvasive ventilation and then provides a current perspective on applications in patients with acute and chronic respiratory failure. The rationale for its use, currently available techniques and equipment, evidence for efficacy, selection of patients, and guidelines for application are discussed.

Phillipson EA. Control of breathing during sleep. *Am Rev Respir Dis* **118**:909–939(1978). This state-of-the-art review integrates advances into a comprehensive picture of respiratory control during sleep. Although it was written over two decades ago, the physiological concepts presented are still relevant.

92

NONINVASIVE VENTILATION AND SLEEP

G. LIISTRO AND D. RODENSTEIN
Cliniques Universitaires St-Luc, Bruxelles, Belgium

INTRODUCTION

Until 1952 assisted mechanical ventilation was almost exclusively applied noninvasively with intermittent negative pressure devices. Since that date, and until the middle of the 1980s, assisted mechanical ventilation was confined to the intensive care units and was almost exclusively applied through endotracheal or tracheotomy tubes. Noninvasive assisted mechanical ventilation using intermittent positive pressure devices had to await the experience gained with the use of nasal interfaces to treat, with continuous positive airway pressure, patients with obstructive sleep apnea [1]. In 1987 Sullivan and co-workers, who first described this treatment, popularized the use of the same type of interfaces to connect patients with severe respiratory failure to intermittent positive pressure ventilators [1].

The use of noninvasive mechanical ventilation has become more widespread over the last decade. Used in the intensive care unit (ICU), the emergency room, or the pneumology unit for acute respiratory failure or as a long-term treatment for patients with respiratory and ventilatory failure, this technique has gained success in reducing the need for invasive ventilation (tracheal intubation and tracheostomy) and in prolonging the survival in patients with chronic restrictive respiratory disorders [2]. Although some severe patients need a constant ventilatory support, nocturnal noninvasive ventilation is beneficial for the majority of patients with chronic respiratory failure who can still breathe spontaneously for the rest of the day.

In this chapter, we will focus on nocturnal noninvasive ventilation, but before reviewing its effects on sleep, we would like to clarify some definitions.

Ventilation. Movement of air (gases) in and out of the lung.

Mechanical Ventilation. Technique used to support or replace the ventilatory pump. According to this definition, continuous positive airway pressure (CPAP) used in patients with obstructive sleep apnea syndrome (OSAS) is NOT a mode of ventilation. Mechanical ventilation may be applied via a positive pressure into the airways (PPV, positive pressure ventilation) or via a negative pressure applied around the thorax (i.e., the iron lung).

Noninvasive Intermittent Positive Pressure Ventilation (NPPV). This is a technique of mechanical ventilation administered via a nasal or face mask, that may be used continuously or only during sleep: intermittent or nocturnal NPPV.

Respiratory Failure. Situation in which the lungs fail to oxygenate the arterial blood adequately.

Ventilatory Failure. One type of respiratory failure occurring when, for any reason, breathing is not strong enough to rid the body of CO_2.

SLEEP AND NORMAL VENTILATION

Ventilation decreases during sleep, but this decrease is not only attributable to the decrease in metabolic rate [3]. Sleep increases upper airway resistance but simultaneously affects the response to this increased pressure load [4]. The net result is a decrease in minute ventilation of 10–15% [5]. As ventilation decreases, arterial partial

Sleep: A Comprehensive Handbook, Edited by T. Lee-Chiong.
Copyright © 2006 John Wiley & Sons, Inc.

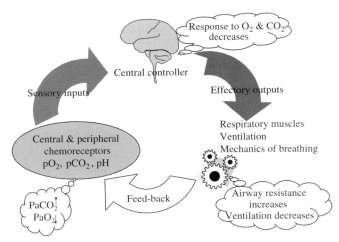

Figure 92.1 Summary of the effects of sleep on breathing. The balloons show the effects of sleep on the respiratory control system.

pressure of CO_2 (Pa_{CO_2}) rises about 3–7 mmHg, whereas a decrease of 4–9 mmHg in Pa_{O_2} and of 2% in Sa_{O_2} is observed [6]. During wakefulness, these blood gas changes would result in a stimulation of the central and peripheral chemoreceptors, feeding information to the central controller, in turn ordering the respiratory muscles to increase ventilation and reducing perturbations of the sensors (negative feedback).

Sleep modifies the control of ventilation (decreasing the ventilatory responses to hypoxia and to hypercapnia) but also changes selectively the action of the respiratory muscles. Upper airway resistance is increased essentially as a result of a reduction of upper airway muscle activity. Moreover, activity of the "accessory" muscles of ventilation is specifically affected during rapid eye movement (REM) sleep. Uncoupling of the accessory muscles and diaphragm leaves the diaphragm alone to face the ventilatory needs and further impedes the maintenance of an adequate ventilation. Figure 92.1 gives a schematic representation of the effects of sleep on breathing.

SLEEP, VENTILATION, AND DISEASE

The first sleep-related breathing disorder, in terms of prevalence and physician's awareness, is rarely treated by NPPV. Indeed, with the exception of highly selected patients, the vast majority of obstructive sleep apnea patients respond to nasal CPAP application.

Starting with the sleep apnea patient, where the ventilatory defect is specifically related to sleep, we can move toward various diseases in which failure of the ventilatory pump is more pronounced during sleep or in which the sleep-related problems represent an exaggeration of

the preexisting respiratory disorders (i.e., diseases of the pulmonary parenchyma).

As discussed earlier, we can easily understand that a patient with a ventilatory pump disorder (neuromuscular diseases, skeletal disorders) who is well-compensated during daytime can experience severe arterial oxygen desaturations during sleep or more specifically during REM sleep. Sleep position itself can impair the maintenance of an acceptable ventilation. Patients with bilateral diaphragmatic paralysis may be unable to breathe only during recumbency, because in this body position, the diaphragm encounters an upward (or cephalad) push of the abdominal contents. A functional diaphragm pushes caudally on the abdominal contents during each inspiration. However, a paralyzed muscle is able neither to resist the abdominal pressure nor to oppose the "suction" of these viscera into the thorax when the accessory respiratory muscles, replacing the diaphragm, provoke a negative intrathoracic pressure.

Finally, patients presenting with borderline Sa_{O_2} levels during wakefulness (e.g., morbidly obese subjects) or patients with chronic obstructive pulmonary disease are on the verge of the abyss: the small physiological drop in Pa_{O_2} during sleep is large enough to precipitate their Sa_{O_2} beyond acceptable values, because of the shape of the oxyhemoglobin dissociation curve.

RESPIRATORY AND VENTILATORY FAILURES AND SLEEP

Patients with chronic obstructive pulmonary disease (COPD) with moderate blood gas impairments present a normal proportion of stages 3 and 4 NREM sleep and REM sleep [7]. However, a reduced total sleep time and a decrease in both sleep efficiency and stability, with more frequent arousals and awakenings during the sleep period, are frequent findings.

In patients with restrictive ventilatory defects without respiratory or ventilatory failure, total sleep time was found to be decreased with respect to normal subjects (less than 300 min). The REM percentage was within normal limits (above 20% [8]). Complete suppression of REM sleep has been associated with severe diaphragmatic weakness [9]. Figure 92.2 presents a summary of sleep studies performed on one patient with amyotrophic lateral sclerosis (ALS), showing the poor quality of sleep during the diagnostic night and its improvement with NPPV.

In patients with respiratory and ventilatory failure, Collard et al. [10] found a somewhat reduced REM sleep (16% ± 8%). The arousal index was within normal limits [10]. Other groups reported marked sleep fragmentation with frequent arousals in patients with ALS [11].

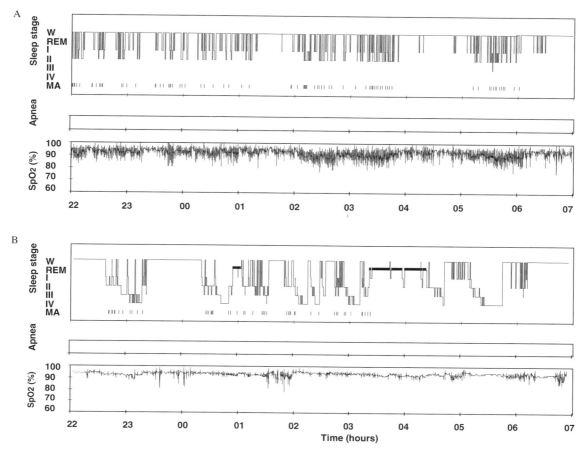

Figure 92.2 Polysomnography in a patient with chronic respiratory failure and a restrictive respiratory defect (amyotrophic lateral sclerosis) during the diagnostic night (A, upper part) and during NPPV, 2 months after initiation of the treatment (B, lower part). The X-axis shows the hours of the recordings. The upper box represents the hypogram. Wakefulness (W), rapid eye movement (REM) sleep, stages 1–4 NREM sleep (I–IV), and the microarousals (MA) are shown. The second box represents the apneas; none was recorded in this patient. The third box shows the oxygen saturation measured by pulse oximetry (SpO$_2$). During the diagnostic night, sleep architecture is disturbed with frequent awakenings. Sleep is almost exclusively represented by stages 1 and 2 NREM sleep with only rare very short periods of REM sleep. Oxygen saturation is low and extremely unstable. By contrast, both sleep architecture and oxygen saturation are improved during the treatment night.

NONINVASIVE VENTILATION: EFFECTS ON SLEEP AND BREATHING

Very few studies have been published concerning the characteristics of sleep during noninvasive mechanical ventilation specifically. A number of studies are mainly concerned with the ventilatory aspects of noninvasive ventilation during sleep, and sleep data are given as secondary information. Of course, the main reason to institute noninvasive mechanical ventilation is to correct abnormal ventilation but one should at least be sure that patients do indeed sleep and that their sleep is at least improved (if not normalized) with respect to the previous situation. For an exhaustive review of the literature, see the paper by Gonzalez et al. [12].

It is useful to stress here that nocturnal noninvasive ventilation is not in general a treatment for nocturnal respiratory failure, but a treatment for permanent respiratory failure applied during sleep.

Long-Term NPPV in COPD Patients

The first study giving detailed data on sleep architecture during noninvasive ventilatory assistance concerns a small group of seven patients with severe COPD studied by full-night polysomnography during spontaneous breathing and during treatment with noninvasive positive pressure ventilation using a nasal mask and a bilevel positive airway pressure device in the controlled mode [13]. In this randomized, crossover design study, total sleep time and sleep efficiency were low during both spontaneous breathing and NPPV. Sleep latency was increased, but not significantly, during NPPV. REM sleep proportion decreased, and REM latency increased with NPPV, but again these changes were not significant, probably because of the small number of patients.

In a meta-analysis on NPPV and stable COPD patients based on four studies, Wijkstra et al. [14] concluded that neither gas exchange nor sleep efficiency was modified by NPPV, but the overall sample size was limited, indicating a need for larger-sized studies.

The preceding studies were done on stable COPD patients, but the usefulness of NPPV has largely been demonstrated in patients with *acute* respiratory and ventilatory failure (whatever its origin) as a short-term treatment and in *chronic* ventilatory failure as a long-term therapy in patients with restrictive pulmonary disorders.

NPPV and Acute Respiratory Failure

The usefulness of noninvasive mechanical ventilation in acute respiratory failure has clearly been demonstrated [15] with detailed data on physiological and clinical outcomes. Despite the large number of studies already published, not a single one has assessed sleep. The vast majority of these studies were performed in ICU settings or in the emergency room [16]. If one considers the profound influence of leaks on the efficiency of noninvasive ventilation, and if one considers that leaks are mainly apparent during sleep, it appears that this field needs extensive investigation in the future. Sleep has seldom been studied in the ICU setting in patients requiring invasive mechanical ventilation. This is understandable on several grounds. Sleep disruption is a common experience in patients staying in ICUs, but because of the (generally) short periods of time involved (days), it carries no vital risks and has therefore less priority for ICU physicians. Perhaps the long-term consequences of chronic sleep disruption are not adequately perceived by physicians knowledgeable in rapidly evolving conditions usually lasting only from days to weeks. Moreover, sedative drugs and paralyzing agents are widely used in ICU settings, especially in intubated patients, making the assessment of sleep more difficult. However, patients sleep very badly in the ICU, with its noisy environment and permanent artificial lighting, in which nurses frequently perform procedures and inspections, irrespective of the time of day or night [17]. Parthasarathy and Tobin [18] have studied the effects of two different modes of invasive ventilation in the ICU. They observed severe sleep fragmentation under pressure support ventilation, that was reduced (but also abnormally high) during assist controlled ventilation. Central apneas were very frequent under pressure support, especially in the presence of heart failure. This study was done on only 11 patients. The introduction and rapid spread of noninvasive ventilation in ICUs, however, make it desirable to study sleep during ventilation in ICU patients.

NPPV and Chronic Ventilatory or Respiratory Failure

One problem that makes previous reports difficult to interpret in the context of this chapter is that most studies included patients with different causes of respiratory or ventilatory failure, and sometimes patients with central sleep apnea. However, we selected the most relevant of them.

Seventeen patients with respiratory failure secondary to restrictive ventilatory defects (neuromuscular diseases or severe chest wall deformities) were studied by Collard and co-workers [10] during spontaneous breathing and at the institution of NPPV with a volumetric ventilator in the controlled mode via nasal mask interfaces. At baseline, patients had a normal sleep architecture (but with an increase in stage 1 NREM sleep percentage and a reduced sleep efficiency) and normal total sleep time despite moderate to severe hypoxic and hypercapnic respiratory failure. During treatment with NPPV, these authors found no significant changes in sleep architecture or efficiency or total sleep time. The only significant change was a decrease in the movement arousal index from 17.1 on average to 10.6 per hour of sleep.

Barbé et al. [19] studied seven patients with neuromuscular diseases treated long-term at home with nocturnal noninvasive ventilation using volumetric ventilators in the controlled mode. Noninvasive ventilation significantly improved sleep efficiency and stages 3 and 4 NREM sleep, whereas total sleep time and REM sleep improved nonsignificantly.

Fourteen patients with neuromuscular diseases or chest wall deformities were studied by Annane et al. [20] at baseline, and after 1 and 3 years of treatment with NPPV using a volumetric ventilator in the assist/controlled mode. Ten patients were ventilated noninvasively and four through a tracheotomy. The authors found a significant increase in sleep efficiency and in stages 3 and 4 NREM sleep percentage after 1 and 3 years.

More recently, Gonzalez et al. [21] found that NPPV improved daytime blood gases and symptoms of 16 patients with chronic ventilatory insufficiency due to kyphoscoliosis. After 6 months, there was no significant change in sleep architecture, albeit sleep efficiency increased from 59.8% $\pm$ 22.7% to 74.9% $\pm$ 13.4% ($p > 0.05$). The REM sleep percentage was comparable to that reported by Collard et al. [10].

Of course, these studies are heterogeneous and the number of patients included is small. However, when improvements of sleep quality were observed in patients treated by NPPV, these changes were significant after long periods of time only.

DISCUSSION

The main result of this review of the literature on the effects of noninvasive ventilation on sleep is that less than 200 patients have been studied in some detail with appropriate methodological instruments (i.e., full-night polysomnography). When one considers that probably hundreds of patients have been sent home with a mechanical ventilator and instructed to use it all night every night, it appears that little attention has been paid to the question: Will these patients have a restful night's sleep while under noninvasive ventilation?

Patients with respiratory failure included in these studies have, during spontaneous breathing (baseline studies), abnormally shortened total sleep times, a decreased sleep efficiency, an excess of stages 1 and 2 NREM sleep percentage, a decrease in stages 3 and 4 NREM sleep and a decrease in REM sleep percentage. The arousal index is either normal or only slightly to moderately increased.

Noninvasive ventilation appears capable of improving sleep quality and quantity, in patients with restrictive or obstructive ventilatory defects. However, this is more uniformly found in the former, whereas in the latter results are variable from one study to the next. Although sleep appears improved, it is rarely normalized. In no study with available data did total sleep time during NPPV exceed 7 hours on average. Most studies give no data on the microstructure of sleep: only four studies report on the arousal index.

Fortunately, noninvasive ventilation appears to have a beneficial effect on ventilation during sleep, the primary reason for instituting such a treatment. Again, this is more robustly and consistently seen in patients with restrictive ventilatory defects.

The benefits of NPPV can also be obtained during the daytime, as shown by Shönhofer et al. [22]. These authors did not verify if these subjects slept during NPPV application. Nevertheless, it seems reasonable to apply NPPV during nocturnal sleep for two reasons. First, using NPPV during nocturnal sleep allows for free daytime activities, whereas the need for daytime NPPV would greatly limit free available time. Second, sleep is an especially challenging physiological period for the respiratory system. Even if no one has shown that nocturnal respiratory failure is detrimental for patients (it appears rather that permanent respiratory failure is the main problem), it seems logical to assist the respiratory system when it most needs assistance.

Leaks, a pervasive problem in NPPV, have been investigated in a couple of studies. Teschler et al. [23] are the first to have shown that leaks lead not only to a decrease in sleep quality but also to a decrease in the efficiency of NPPV and that correction of leaks not only increases the efficiency of NPPV but also improves sleep quality by increasing REM sleep percentage and decreasing arousals. Leaks seem to be very common during sleep. Clinically, they are much more common during sleep than during wakefulness (although this has not yet been rigorously studied). This is easy to understand, since most leaks take place through the mouth, which a conscious subject may keep closed, whereas the same person asleep will lose control of the oral airway. Since leaks may jeopardize the efficiency of noninvasive ventilation, it seems important to monitor the efficiency of noninvasive ventilation during sleep, in order to make sure that the treatment is as useful during sleep as during wakefulness (where most of the habituation and adjustment procedures are performed). One may discuss whether this implies the necessity of performing full-night polysomnography, or whether nocturnal monitoring is sufficient for these purposes. The answer will depend on the accessibility of polysomnography and other local conditions. However, it seems clear that the ideal solution would be to have polysomnography performed under NPPV, either at the time this treatment is instituted or at the earlier available time. Alternatively, one might perform nocturnal ventilatory monitoring and assess sleep behaviorally. If leaks, undetected during wakefulness, become evident during sleep with a decrease in NPPV efficiency, then the ventilator parameters could be adjusted to try to reduce leaks, or the mask interface could be changed. Doing this during polysomnography has the added advantage that one can verify the eventual improvements in sleep quality after correction of leaks. This is far from usual practice in most centers. However, the small amount of evidence available suggests that both sleep and ventilatory assistance quality could be enhanced by such a policy.

Clinical and experimental studies have shown that noninvasive ventilation leads to glottic narrowing and that sleep exerts a great influence on this glottic response, which in turn determines at least partly the appearance of leaks and the efficiency of noninvasive ventilation [24]. It has also been shown that adjustment of ventilator parameters (both volumetric and bilevel positive pressure ventilators) can partly circumvent this problem by changing the delivered inspiratory flow, tidal volume, or inspiratory pressures [25]. This is another reason for not relying on controls obtained only during wakefulness and to perform final adjustments of ventilator settings during sleep.

CONCLUSIONS

We can conclude that noninvasive mechanical ventilation appears useful not only to improve ventilation during sleep but also to improve sleep quantity and quality in patients with respiratory failure. Staff training is critical and, as with any new technique, initially time-consuming. NPPV setting during sleep allows the doctor/nurse/chest physician to acquire and maintain skills in both mechanical ventilation and online interpretation of sleep studies. The control of leaks may lead to further improvements in ventilation and sleep. Patients with respiratory failure secondary to restrictive ventilatory defects appear to benefit from NPPV both from the ventilatory point of view and from the sleep perspective, more uniformly than do patients with obstructive ventilatory defects. Finally, every effort should be made to monitor and record full-night polysomnography during NPPV titration to ensure both adequate sleep and adequate ventilation during sleep.

REFERENCES

1. Sullivan CE, Issa FG, Berthon Jones M, Eves L. Reversal of obstructive sleep apnea by continuous positive airway pressure applied through the nares. *Lancet* **1**:862–865(1981).

2. Simonds AK, Elliott MW. Outcome of domiciliary nasal intermittent positive pressure ventilation in restrictive and obstructive disorders. *Thorax* **50**:604–609(1995).

3. White DP, Weil JV, Zwillich CW. Metabolic rate and breathing during sleep. *J Appl Physiol* **59**:384–391(1985).

4. Lopes JM, Tabachnik E, Muller NL, Levison H, Bryan AC. Total airway resistance and respiratory muscle activity during sleep. *J Appl Physiol* **54**:773–777(1983).

5. Douglas NJ, White DP, Pickett CK, Weil JV, Zwillich CW. Respiration during sleep in normal man. *Thorax* **37**:840–844(1982).

6. Krieger J. Respiratory physiology: breathing in normal subjects. In: Kryger MH, Roth T, Dement WC (Eds), *Principles and Practice of Sleep Medicine*, 3rd ed. Saunders, Philadelphia, 2000, pp 229–253.

7. Cormick W, Olson LG, Hensley MJ, Saunders NA. Nocturnal hypoxaemia and quality of sleep in patients with chronic obstructive lung disease. *Thorax* **41**:846–854(1986).

8. Sawicka EH, Branthwaite MA. Respiration during sleep in kyphoscoliosis. *Thorax* **42**:801–808(1987).

9. White JE, Drinnan MJ, Smithson AJ, Griffiths CJ, Gibson GJ. Respiratory muscle activity and oxygenation during sleep in patients with muscle weakness. *Eur Respir J* **8**:807–814(1995).

10. Collard P, Dury M, Delguste P, Aubert G, Rodenstein DO. Movement arousals and sleep-related disordered breathing in adults. *Am J Respir Crit Care Med* **154**:454–459(1996).

11. Ferguson KA, Strong MJ, Ahmad D, George CF. Sleep-disordered breathing in amyotrophic lateral sclerosis. *Chest* **110**:664–669(1996).

12. Gonzalez MM, Parreira VF, Rodenstein DO. Non-invasive ventilation and sleep. *Sleep Med Rev* **6**:29–44(2002).

13. Strumpf DA, Millman RP, Carlisle CC, Grattan LM, Ryan SM, Erickson AD, Hill NS. Nocturnal positive-pressure ventilation via nasal mask in patients with severe chronic obstructive pulmonary disease. *Am Rev Respir Dis* **144**:1234–1239(1991).

14. Wijkstra PJ, Lacasse Y, Guyatt GH, Casanova C, Gay PC, Meecham JJ, Goldstein RS. A meta-analysis of nocturnal noninvasive positive pressure ventilation in patients with stable COPD. *Chest* **124**:337–343(2003).

15. Brochard L, Mancebo J, Wysocki M, Lofaso F, Conti G, Rauss A, Simonneau G, Benito S, Gasparetto A, Lemaire F. Noninvasive ventilation for acute exacerbations of chronic obstructive pulmonary disease. *N Engl J Med* **333**:817–822(1995).

16. Thys F, Roeseler J, Reynaert M, Liistro G, Rodenstein DO. Noninvasive ventilation for acute respiratory failure: a prospective randomised placebo-controlled trial. *Eur Respir J* **20**:545–555(2002).

17. Freedman NS, Kotzer N, Schwab RJ. Patient perception of sleep quality and etiology of sleep disruption in the intensive care unit. *Am J Respir Crit Care Med* **159**:1155–1162(1999).

18. Parthasarathy S, Tobin MJ. Effect of ventilator mode on sleep quality in critically ill patients. *Am J Respir Crit Care Med* **166**:1423–1429(2002).

19. Barbé F, Quera-Salva MA, de Lattre J, Gajdos P, Agusti A. Long-term effects of nasal intermittent positive-pressure ventilation on pulmonary function and sleep architecture in patients with neuromuscular diseases. *Chest* **110**:1179–1183(1996).

20. Annane D, Quera-Salva MA, Lofaso F, Vercken JB, Lesieur O, Fromageot C. Clair B, Gajdos P, Raphael JC. Mechanisms underlying effects of nocturnal ventilation on daytime blood gases in neuromuscular diseases. *Eur Respir J* **13**:157–162(1999).

21. Gonzalez C, Ferris G, Diaz J, Fontana I, Nunez J, Marin J. Kyphoscoliotic ventilatory insufficiency: effects of long-term intermittent positive-pressure ventilation. *Chest* **124**:857–862(2003).

22. Schönhofer B, Geibel M, Sonneborn M, Haidl P, Kohler D. Daytime mechanical ventilation in chronic respiratory insufficiency. *Eur Respir J* **10**:2840–2846(1997).

23. Teschler H, Stampa J, Ragette R, Konietzko N, Berthon-Jones M. Effect of mouth leak on effectiveness of nasal bilevel ventilatory assistance and sleep architecture. *Eur Respir J* **14**:1251–1257(1999).

24. Jounieaux V, Aubert G, Dury M, Delguste P, Rodenstein DO. Effects of nasal positive-pressure hyperventilation on the glottis in normal sleeping subjects. *J Appl Physiol* **79**:186–193(1995).

25. Parreira VF, Delguste P, Jounieaux V, Aubert G, Dury M, Rodenstein DO. Effectiveness of controlled and spontaneous modes in nasal two-level positive pressure ventilation in awake and asleep normal subjects. *Chest* **112**:1267–1277(1997).

PART XII

SLEEP IN THE CARDIAC DISORDERS

93

HYPERTENSION AND CARDIOVASCULAR DISEASE

ROBERT D. BALLARD

National Jewish Medical and Research Center Denver, Colorado

INTRODUCTION

Extensive evidence has accumulated during the last several years linking sleep disordered breathing to cardiovascular disease. The two major types of sleep disordered breathing with relevance to cardiovascular disease are obstructive sleep apnea and Cheyne–Stokes respiration. Obstructive sleep apnea is much more common than Cheyne–Stokes respiration and has been much more widely investigated.

OBSTRUCTIVE SLEEP APNEA

Obstructive sleep apnea is brought on by the collapse of the pharyngeal airway during sleep. This collapse appears to result at least partly from sleep-associated reduction in the activity of pharyngeal dilator muscles. In the patient with an anatomic predisposition toward narrowing of the pharynx (such as due to a large tongue or long palate), this can result in total or partial occlusion during sleep, despite continuing respiratory efforts (Figure 93.1).

Potential Pathophysiologic Links Between Obstructive Sleep Apnea and Cardiovascular Disease

Recent studies have provided strong evidence that untreated obstructive sleep apnea can be associated with several physiologic processes that could increase the risk for cardiovascular disease. First, it has been well established that obstructive sleep apnea is associated with increased sympathetic activation. A number of studies have determined that obstructive sleep apnea patients have increased sympathetic

activation of peripheral vasculature [1]. This is most pronounced in association with obstructive apneas and hypopneas during sleep, when affected patients demonstrate associated acute increases in systemic blood pressure. However, obstructive sleep apnea patients also have evidence of augmented sympathetic activity that persists throughout wakefulness [2]. There is also evidence that effective therapy for obstructive sleep apnea can reduce the level of sympathetic activation, both during sleep and during wakefulness [3].

Recent studies also suggest that obstructive sleep apnea patients may have vascular endothelial dysfunction. Ultrasound studies have revealed that obstructive sleep apnea patients frequently have thickening of peripheral arterial walls, as well as impaired endothelial dependent vascular dilation in response to reperfusion [4]. Such findings strongly suggest a role for obstructive sleep apnea in the promotion of vascular pathology.

In addition, evidence is accumulating that obstructive sleep apnea represents a state of oxidative stress, with increased levels of reactive oxygen species that could trigger a proinflammatory cytokine cascade [5]. In particular, it has been shown that proinflammatory cytokines such as TNFα, IL-6, and IL-8, as well as the adhesion molecules ICAM-1 and VCAM-1, are all increased in the presence of obstructive sleep apnea. There is also evidence that effective therapy of obstructive sleep apnea can reduce these proinflammatory cytokines. It has recently been noted C-reactive protein is a strong marker of cardiovascular risk, if not a direct participant in vascular endothelial damage. It has also recently been shown that C-reactive protein levels are elevated in patients with obstructive

Sleep: A Comprehensive Handbook, Edited by T. Lee-Chiong.
Copyright © 2006 John Wiley & Sons, Inc.

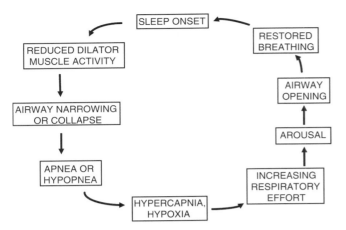

Figure 93.1 Pathophysiologic sequence by which obstructive apneas and hypopneas occur during sleep, leading to hypoxia, hypercapnia, and sleep disruption.

sleep apnea [6], signifying another potential association between obstructive sleep apnea and cardiovascular risk. Since cardiovascular disease is now widely thought to involve inflammatory processes at least at the site of localized vascular lesions, the potential role of obstructive sleep apnea as a promoter of inflammation warrants further investigation.

There is also evidence that obstructive sleep apnea may represent a "hypercoagulable state," which could predispose patients to the development of vascular disease. Studies have demonstrated that obstructive sleep apnea is associated with platelet activation and increased levels of fibrinogen, which could denote augmented clotting tendencies [7]. Finally, recent studies have demonstrated that obstructive sleep apnea can be associated with insulin resistance and glucose intolerance in a manner that is directly proportional to the severity of obstructive sleep apnea [8, 9]. Observations such as these may explain previous reports of an increased risk for the diagnosis of diabetes in obstructive sleep apnea patients, even after controlling for other risk factors. When taken together, these findings offer credible pathways by which the presence of obstructive sleep apnea could promote increased cardiovascular risk.

Obstructive Sleep Apnea and Cardiovascular Risk

As already noted, it is well established that obstructive apneas and hypopneas can trigger significant surges in systemic blood pressure in association with discrete, sleep-associated respiratory events. However, there is also compelling evidence that obstructive sleep apnea increases the risk for diurnal hypertension. Three large studies assessed the risk for hypertension in patients sampled from the general population or referred to a sleep disorders center with a suspicion of sleep apnea [10–12]. These

studies concurred that the presence of obstructive sleep apnea increases the relative risk for hypertension by a factor of 1.5 to 3, even after correcting for other risk factors such as age, gender, and body mass index. In particular, one of these studies followed patients longitudinally, with repeat assessments for sleep apnea and the development of hypertension at 4 year intervals [10]. This study revealed that the risk for developing hypertension during a 4 year follow-up period increased in a linear relationship to the apnea–hypopnea index (AHI), such that those patients with an AHI of at least 15 respiratory events per hour had a 2.89-fold greater risk of developing hypertension over 4 years of follow-up than did those patients with no respiratory events.

The apparent association between obstructive sleep apnea and hypertension might at least partially explain other studies that have reported an increased risk for cardiovascular disease in patients with obstructive sleep apnea. Partinen and Guilleminault [13] followed 198 obstructive sleep apnea patients over 7 years after treatment either with a tracheotomy or with encouragement to lose weight, although as a group they did not lose significant weight. The relative risk for developing new cardiovascular disease in the "weight-loss" group after adjustment for body mass index was 2.3 in comparison with the tracheotomy group. Of additional interest, 56% of all obstructive sleep apnea patients had an already diagnosed cardiovascular disorder at the time of entry into this study, emphasizing the need for early diagnosis and therapy of sleep apnea. These findings were further supported by results from the Sleep Heart Health Study, in which 6424 free living individuals underwent overnight polysomnography [14]. Those patients with an AHI > 11 had a relative risk of 1.42 for the prevalence of cardiovascular disease in comparison with those patients demonstrating an AHI ≤ 1.3. This increase in risk persisted despite correction for age, race, gender, smoking status, presence of self-reported diabetes and hypertension, body mass index, total cholesterol, and HDL cholesterol levels.

In this same study [14], it was also found that an AHI of at least 11 respiratory events per hour increased the risk for self-reported congestive heart failure 2.38-fold in comparison to patients with an AHI ≤ 1.3 or less. Obstructive sleep apnea was more strongly associated with self-reported congestive heart failure than either stroke or coronary heart disease. Numerous other studies have also suggested an association between obstructive sleep apnea and congestive heart failure. Combining the results of six case series in which a total of 680 congestive heart failure patients underwent nighttime polysomnography (Table 93.1), 29% of the study patients were found to have obstructive sleep apnea [15]. This link is further supported by observations from numerous case series of unselected obstructive sleep apnea patients in whom left ventricular dysfunction is common, and in whom ventricular function typically

TABLE 93.1 Data from Six Studies in Which Congestive Heart Failure Patients Were Evaluated With Formal Sleep Studies

	N	SDB	OSA	CSR
Lofaso (1994)	29	9 (45)	1 (5)	8 (40)
Chan (1997)	20	11 (55)	7 (35)	4 (20)
Javaheri (1998)	81	41 (51)	9 (11)	32 (40)
Lanfranchi (1999)	66	50 (76)	4 (6)	46 (70)
Tremel (1999)	34	28 (82)	7 (20)	21 (62)
Sin (1999)	450	316 (70)	168 (37)	148 (33)
Summary	680	455 (67)	196 (29)	259 (38)

Abbreviations: SDB, sleep disordered breathing; OSA, obstructive sleep apnea; CSR, Cheyne–Stokes respiration.

improves in response to effective therapy with continuous positive airway pressure (CPAP).

Therapy of Obstructive Sleep Apnea and Its Effect on Cardiovascular Disease

At present, the most effective therapy for obstructive sleep apnea continues to be CPAP. This incorporates the administration through a nasal or full-face mask of positive airway pressure that pneumatically splints the pharynx open during sleep, thereby preventing obstructive apneas and hypopneas. Although compliance with CPAP therapy has been an issue, it is widely accepted that if one complies with this therapy it will control obstructive sleep apnea more than 90% of the time.

Two studies have assessed the impact of therapy with effective levels of CPAP versus subtherapeutic, or "sham," CPAP upon blood pressure. Becker and colleagues [16] observed no important change in blood pressure after 9 weeks of "sham" CPAP therapy, whereas mean, diastolic, and systolic blood pressures all decreased significantly (mean reductions of approximately 10 mmHg), both at night and during the day, in patients treated with effective levels of CPAP. In a similar study, Pepperell and associates [17] observed that 1 month of effective CPAP therapy reduced 24 hour mean arterial blood pressure by 2.5 mmHg, whereas subtherapeutic CPAP allowed blood pressure to increase 0.8 mmHg during the same period. Finally, there was an intriguing report by Logan and colleagues [18] that obstructive sleep apnea was detected in 84% of patients with refractory hypertension. Two months of CPAP therapy yielded an 11 mmHg reduction in 24 hour mean systolic blood pressure in affected patients.

These observations strongly suggest that not only is obstructive sleep apnea associated with increased prevalence of hypertension and increased risk for developing new hypertension, but that effective therapy of obstructive sleep apnea can significantly reduce blood pressure, even in those patients with diagnosed refractory hypertension.

In addition, two recent studies have suggested that CPAP therapy in patients with obstructive sleep apnea and concurrent congestive heart failure can lead to improvement in both cardiac function and quality of life. Kaneko and colleagues [19] followed 12 obstructive sleep apnea patients with congestive heart failure who were treated for 1 month with effective level CPAP. The treated patients demonstrated a 10 mmHg reduction in daytime systolic blood pressure in conjunction with an improvement in left ventricular ejection fraction from 25% to 34%. No such changes were observed in a control group of 12 similar patients. In a similar fashion, Mansfield and colleagues [20] evaluated the effects of 3 months of effective CPAP therapy on quality of life and cardiac function in 19 obstructive sleep apnea patients with congestive heart failure. CPAP therapy led to both a 5% improvement in left ventricular ejection fraction and improvement in quality of life measured by SF36 and chronic heart failure questionnaires, whereas the control group of 21 patients remained unchanged.

All of these observations may explain the interesting observations of Peker and associates [21], who monitored the incidence of cardiovascular disease over 7 years in middle-aged men diagnosed with obstructive sleep apnea. Of interest, cardiovascular disease was already present in 37% of patients formally diagnosed with obstructive sleep apnea versus only 7% of those patients referred for the evaluation of sleep apnea, but with a subsequently negative polysomnogram. Over the next 7 years, only 7% of effectively treated obstructive sleep apnea patients developed new cardiovascular disease, as opposed to 57% of inadequately treated sleep apnea patients. These results suggest that effective treatment of obstructive sleep apnea reduces the excess risk of cardiovascular disease in this population.

CHEYNE–STOKES RESPIRATION

The crescendo/decrescendo periodic breathing pattern of Cheyne–Stokes respiration was first observed in association with congestive heart failure by Dr. John Cheyne in 1818. This respiratory pattern has been widely described since that time, and recent evidence suggests that it is the most common respiratory abnormality in patients with congestive heart failure (Figure 93.2). This periodic breathing pattern appears to result when a low baseline carbon dioxide level leads to a "hypocapnic apnea" with transition to sleep, during which carbon dioxide levels rise to the apneic threshold. Once this occurs, respiratory effort resumes but is typically followed by "ventilatory overshoot," which leads to recurrence of hypocapnia and often a subsequent arousal from sleep (Figure 93.3). With return to sleep, the cycle repeats itself. Potential

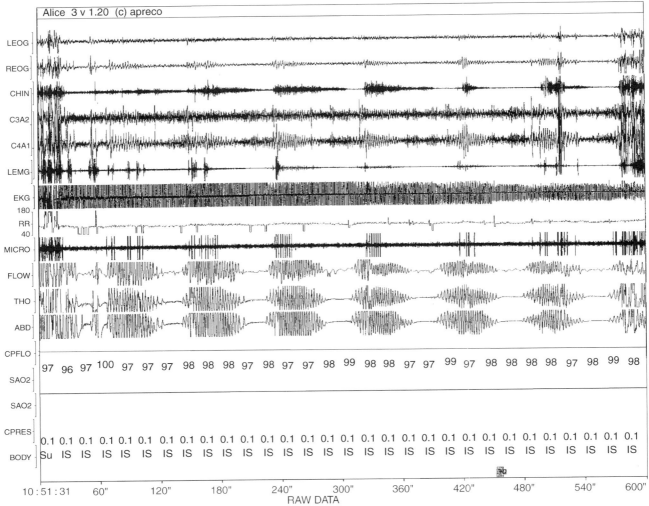

Figure 93.2   A 5 minute sample of sleep study data recording acquired from a patient with Cheyne–Stokes respiration and congestive heart failure. The flow channel again represents combined oral and nasal airflow, while THO and ABD channels represent thoracic and abdominal inspiratory effort, respectively. Notice the crescendo/decrescendo respiratory pattern terminating in central apneas, reflected in all three channels.

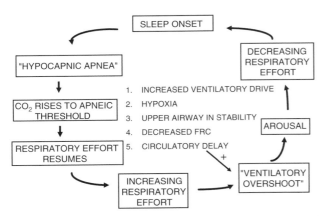

Figure 93.3   Pathophysiologic sequence by which sleep is thought to trigger Cheyne–Stokes respiration in patients with congestive heart failure.

contributors to this pathophysiologic process could include (1) an increased ventilatory drive often seen in congestive heart failure, (2) hypoxia as a result of ventilation–perfusion abnormalities and low lung volumes, (3) reduction in functional residual capacity with subsequent "underdamping," (4) upper airway instability, and (5) the circulatory time delay between the alveolar units of the lungs and the peripheral chemoreceptors that is typical of congestive heart failure. The relative contributions of these different mechanisms are likely quite variable, but it is clear that Cheyne–Stokes respiration is common in patients with congestive heart failure. If one combines the results of six different studies that performed polysomnography in unselected patients with congestive heart failure (Table 93.1), 259 out of 680 subjects, or 38%, were found to have Cheyne–Stokes respiration [15].

Morbidity of Cheyne–Stokes Respiration in Congestive Heart Failure

Several studies have demonstrated that Cheyne–Stokes respiration can impair sleep quality, and two studies by Hanley and colleagues [22, 23] have demonstrated this condition to be associated with a reduction in sleep duration, increased sleep disruption, a tendency toward lighter stages of sleep, and daytime sleepiness. Cheyne–Stokes respiration is also associated with cyclic increases in blood pressure and heart rate in a fashion quite similar to that observed with obstructive sleep apnea [24]. It has also been observed that Cheyne–Stokes respiration in congestive heart failure is associated with increases in sympathetic activity that exceed those observed with congestive heart failure alone [25].

Several studies have suggested that Cheyne–Stokes respiration can be associated with increased mortality in the congestive heart failure population. Hanley and colleagues [23] followed 16 congestive heart failure patients for up to 4.5 years. They observed that in 9 patients with Cheyne–Stokes respiration there was a 55% mortality and 2 patients required urgent heart transplants during less than 4.5 years of follow-up. In comparison, the 7 patients without Cheyne–Stokes respiration demonstrated only a single death (14% mortality). Lanfranchi and colleagues [26] followed 62 congestive heart failure patients for up to 28 months. Fifteen of these 62 patients (24%) died of cardiac causes during this follow-up. Their analysis revealed that Cheyne–Stokes respiration with an AHI of at least 30 events per hour and left atrium enlargement were the only significant independent predictors of mortality.

Therapy of Cheyne–Stokes Respiration and Its Effect on Cardiac Disease

Numerous studies have assessed the effects of CPAP on Cheyne–Stokes respiration in congestive heart failure. These studies have predictably demonstrated that 1–3 months of therapy with CPAP at 10–12 cm H_2O can yield significant improvement in sleep disordered breathing, while also improving cardiac function and quality of life indicators. Sin and colleagues [27] conducted an interesting study in which 66 patients with congestive heart failure were randomized to receive either nightly CPAP or usual therapy. For the entire group of patients, a 60% relative risk reduction in mortality and cardiac transplantation rate was observed for those patients who complied with CPAP therapy. Effective use of CPAP yielded significant improvement in left ventricular ejection fraction after 3 months of therapy and a relative risk reduction of 81% in the combined mortality and cardiac transplantation rate for those patients with Cheyne–Stokes respiration. These studies therefore suggest that CPAP therapy can improve cardiac function in congestive heart failure patients with Cheyne–Stokes respiration and can also apparently reduce the combined mortality and cardiac transplantation rates in affected patients.

Oxygen Therapy for Cheyne–Stokes Respiration

Although it has not been as widely studied, there is also evidence that nocturnal supplemental oxygen therapy can improve the respiratory pattern during sleep in congestive heart failure patients with Cheyne–Stokes respiration. Franklin and associates [28] treated 20 patients with Cheyne–Stokes respiration (16 of whom also had congestive heart failure) with supplemental oxygen via nasal canula at 1–5 L/min or via a Venturi mask at 60% inspired oxygen. Oxygen therapy in this study reduced the median AHI from 33.5 to 5 events per hour, with 17 out of 20 patients demonstrating at least 50% reduction in frequency of central apneas. However, it was not clear that sleep quality was improved by this therapy.

Krachman and associates [29] subsequently evaluated 9 congestive heart failure patients with Cheyne–Stokes respiration, in random order administering supplemental oxygen via nasal canula at 2 L/min or nasal CPAP at 9 cm H_2O. The oxygen therapy reduced the AHI from 44 events per hour to 18 events per hour, while CPAP therapy reduced the AHI to 15 events per hour, with no significant difference between the two modalities. There were slight reductions in total sleep time and sleep efficiency in those patients treated with CPAP.

These studies suggest that both CPAP and supplemental oxygen therapy can be useful in treating sleep disordered breathing in congestive heart failure patients with Cheyne–Stokes respiration, although only CPAP therapy has been observed to benefit both cardiac function and quality of life. More recent studies have also evaluated the efficacy of bilevel positive airway pressure [30] and "adaptive servo-ventilation" [31], both of which appear to be as effective as CPAP therapy. A variety of medications have also been utilized to treat Cheyne–Stokes respiration in congestive heart failure, but the only possibly useful medication is theophylline (Figure 93.4). Unfortunately, this medication has also been observed in numerous studies to disrupt sleep.

In conclusion, it is clear that obstructive sleep apnea has adverse effects on blood pressure, cardiovascular status, and probably cardiovascular mortality. There is also evidence that effective therapy with CPAP can improve blood pressure and cardiac function in adult obstructive sleep apnea patients. Obstructive sleep apnea likely occurs in 30% of all congestive heart failure patients, while Cheyne–Stokes respiration occurs in at least 40% of congestive heart failure patients. The pattern of

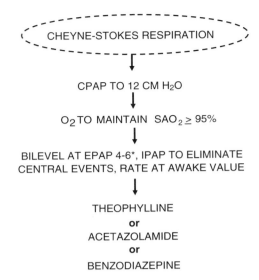

Figure 93.4 Proposed treatment algorithm for patients with Cheyne–Stokes respiration occurring in association with congestive heart failure.

Cheyne–Stokes respiration with congestive heart failure appears to be associated with increased morbidity and excess mortality. Both CPAP and supplemental oxygen appear to be effective therapies for Cheyne–Stokes respiration in congestive heart failure, but only CPAP has been shown to improve cardiac function and quality of life.

REFERENCES

1. Somers VK, et al. Sympathetic neural mechanisms in obstructive sleep apnea. *J Clin Invest* **96**(4):1897–904(1995).

2. Narkiewicz K, et al. Sympathetic activity in obese subjects with and without obstructive sleep apnea. *Circulation* **98**(8):772–776(1998).

3. Narkiewicz K, et al. Nocturnal continuous positive airway pressure decreases daytime sympathetic traffic in obstructive sleep apnea. *Circulation* **100**(23):2332–2335(1999).

4. Ip MS, et al. Endothelial function in obstructive sleep apnea and response to treatment. *Am J Respir Crit Care Med* **169**(3):348–353(2004).

5. Lavie L, Vishnevsky A, Lavie P. Evidence for lipid peroxidation in obstructive sleep apnea. *Sleep* **27**(1):123–128(2004).

6. Shamsuzzaman AS, et al. Elevated C-reactive protein in patients with obstructive sleep apnea. *Circulation* **105**(21):2462–2464(2002).

7. von Kanel R, Dimsdale JE. Hemostatic alterations in patients with obstructive sleep apnea and the implications for cardiovascular disease. *Chest* **124**(5):1956–1967(2003).

8. Ip MS, et al. Obstructive sleep apnea is independently associated with insulin resistance. *Am J Respir Crit Care Med* **165**(5):670–676(2002).

9. Punjabi NM, et al. Sleep-disordered breathing and insulin resistance in middle-aged and overweight men. *Am J Respir Crit Care Med* **165**(5):677–682(2002).

10. Peppard PE, et al. Prospective study of the association between sleep-disordered breathing and hypertension. *N Engl J Med* **342**(19):1378–1384(2000).

11. Nieto FJ, et al. Association of sleep-disordered breathing, sleep apnea, and hypertension in a large community-based study. Sleep Heart Health Study. *JAMA* **283**(14):1829–1836(2000).

12. Lavie P, Herer P, Hoffstein V. Obstructive sleep apnoea syndrome as a risk factor for hypertension: population study. *BMJ* **320**(7233):479–482(2000).

13. Partinen M, Guilleminault C. Daytime sleepiness and vascular morbidity at seven-year follow-up in obstructive sleep apnea patients. *Chest* **97**(1):27–32(1990).

14. Shahar E, et al. Sleep-disordered breathing and cardiovascular disease: cross-sectional results of the Sleep Heart Health Study. *Am J Respir Crit Care Med* **163**(1):19–25(2001).

15. Lanfranchi PA, Somers VK. Sleep-disordered breathing in heart failure: characteristics and implications. *Respir Physiol Neurobiol* **136**(2-3):153–165(2003).

16. Becker HF, et al. Effect of nasal continuous positive airway pressure treatment on blood pressure in patients with obstructive sleep apnea. *Circulation* **107**(1):68–73(2003).

17. Pepperell JC, et al. Ambulatory blood pressure after therapeutic and subtherapeutic nasal continuous positive airway pressure for obstructive sleep apnoea: a randomised parallel trial. *Lancet* **359**(9302):204–210(2002).

18. Logan AG, et al. Refractory hypertension and sleep apnoea: effect of CPAP on blood pressure and baroreflex. *Eur Respir J* **21**(2):241–247(2003).

19. Kaneko Y, et al. Cardiovascular effects of continuous positive airway pressure in patients with heart failure and obstructive sleep apnea. *N Engl J Med* **348**(13):1233–1241(2003).

20. Mansfield DR, et al. Controlled trial of continuous positive airway pressure in obstructive sleep apnea and heart failure. *Am J Respir Crit Care Med* **169**(3):361–366(2004).

21. Peker Y, et al. Increased incidence of cardiovascular disease in middle-aged men with obstructive sleep apnea: a 7-year follow-up. *Am J Respir Crit Care Med* **166**(2):159–165(2002).

22. Hanly PJ, et al. Respiration and abnormal sleep in patients with congestive heart failure. *Chest* **96**(3):480–488(1989).

23. Hanly PJ, Zuberi-Khokhar NS. Increased mortality associated with Cheyne–Stokes respiration in patients with congestive heart failure. *Am J Respir Crit Care Med* **153**(1):272–276(1996).

24. Trinder J, et al. Pathophysiological interactions of ventilation, arousals, and blood pressure oscillations during Cheyne–Stokes respiration in patients with heart failure. *Am J Respir Crit Care Med* **162**(3 Pt 1):808–813(2000).

25. Mansfield D, et al. Raised sympathetic nerve activity in heart failure and central sleep apnea is due to heart failure severity. *Circulation* **107**(10):1396–4000(2003).

26. Lanfranchi PA, et al. Prognostic value of nocturnal Cheyne–Stokes respiration in chronic heart failure. *Circulation* **99**(11):1435–1440(1999).

27. Sin DD, et al. Effects of continuous positive airway pressure on cardiovascular outcomes in heart failure patients with and without Cheyne–Stokes respiration. *Circulation* **102**(1):61–66(2000).

28. Franklin KA, et al. Reversal of central sleep apnea with oxygen. *Chest* **111**(1):163–169(1997).

29. Krachman SL, et al. Comparison of oxygen therapy with nasal continuous positive airway pressure on Cheyne–Stokes respiration during sleep in congestive heart failure. *Chest* **116**(6):1550–1557(1999).

30. Kohnlein T, et al. Assisted ventilation for heart failure patients with Cheyne–Stokes respiration. *Eur Respir J* **20**(4):934–941(2002).

31. Pepperell JC, et al. A randomized controlled trial of adaptive ventilation for Cheyne–Stokes breathing in heart failure. *Am J Respir Crit Care Med* **168**(9):1109–1114(2003).

94

CONGESTIVE HEART FAILURE

EMILIO MAZZA AND INDIRA GURUBHAGAVATULA
Hospital of the University of Pennsylvania, Philadelphia, Pennsylvania

INTRODUCTION

Congestive heart failure (CHF) is a major public health problem of growing magnitude, with approximately 5 million individuals in the United States carrying the diagnosis of CHF, and an additional 500,000 new cases being identified each year [1]. Perhaps still more concerning than its rapidly increasing incidence is the association between CHF and death, an outcome that has been linked to overactivity of the sympathetic nervous system [2, 3]. Accordingly, beta adrenergic antagonists and angiotensin converting enzyme (ACE) inhibitors have become cornerstones of pharmacologic therapy. With the goal of countering the neurohormonal alterations of CHF, use of these classes of pharmacologic agents has mitigated the rates of mortality associated with this disorder [4–6], albeit only to a limited degree. Despite the use of these agents, nearly 300,000 patients continue to die each year from CHF, and this number continues to grow steadily [1].

These limitations in conventional pharmacotherapy have led some to question whether other coexisting conditions might interact with CHF and contribute toward the excess mortality associated with CHF. One such condition is sleep disordered breathing (SDB), which encompasses two distinct clinical conditions: obstructive sleep apnea (OSA) and central sleep apnea–Cheyne–Stokes respiration (CSA–CSR). As is the case for CHF, SDB affects multiple organ systems, including the respiratory, cardiovascular, and neurohormonal systems, and activates the sympathetic nervous system. Both OSA and CSA–CSR can exist separately or together and interact in the same patient with CHF [7]. In this chapter, we will review (1) the prevalence of

OSA and CSA–CSR in CHF, (2) the cardiovascular effects of OSA and CSA–CSR in CHF, and (3) the current evidence regarding the effects of treatment of SDB on heart failure, with emphasis on the outcome of mortality.

PREVALENCE OF SLEEP DISORDERED BREATHING IN CONGESTIVE HEART FAILURE

OSA and CHF

The largest study to date to evaluate the association between OSA and CHF is the Sleep Heart Health Study, a cross-sectional, multicenter evaluation of 6424 state employees who underwent full overnight polysomnography in the home [8]. This study found that even individuals who had mild degrees of OSA (apnea–hypopnea index (AHI) > 11 events/h) suffered more than double the risk of CHF, with a relative odds of having CHF equaling 2.38. Javaheri and colleagues [9] found that among 81 consecutive ambulatory male patients with stable CHF, 11% of patients had AHI ≥ 15 events/h without symptoms of OSA. In a separate cohort of 450 ambulatory men and women with stable CHF and symptoms of OSA, 37% had an AHI ≥ 10 events/h [10]. Among CHF patients with diastolic dysfunction and preserved systolic function, OSA was found in 35% [11]. No differences have been described in frequency of occurrence between men and women (38% versus 31%) [10]. In contrast, among healthy, middle-aged individuals, OSA defined by AHI ≥ 5 events/h alone occurs among 24% of men, while OSA defined by AHI ≥ 5 events/h together with symptoms occurs in 4%

of men and 2% of women [12]. Thus the prevalence of OSA in patients with heart failure appears to be greater than that of OSA in a general, community-based population.

Several authors have attempted to elucidate the risk factors for developing OSA in CHF [10]. Male gender is considered a major risk factor. Among men, obesity defined by body mass index (BMI) ≥ 30 kg/m^2 is also a major risk factor, while among women, age ≥ 60 years is significant. Whether the increased risk of OSA in association with advancing age in women is related to changes in body fat distribution, in levels of circulating sex hormones, or due to some still undefined mechanism remains unclear.

CSA–CSR

Several studies suggest that CSA–CSR may be even more common than OSA in CHF, ranging from 33% to 40% [9, 10]. Male gender has been identified as a major risk factor, and CSA–CSR is rarely seen in women. Additionally, awake $P_{CO_2} \leq 38$ mmHg, presence of atrial fibrillation, and age ≥ 60 years are other risk factors. However, in contrast to OSA, obesity is not a risk factor for CSA–CSR. These results are not entirely unexpected, given the different pathophysiologic mechanisms for OSA compared with CSA–CSR. (Please refer to Chapters 31, 35, and 93 on the pathophysiology of OSA and CSA in this volume.)

CARDIOVASCULAR EFFECTS OF SDB

Effects of OSA

OSA may increase blood pressure, impair ventricular function, and increase pulmonary artery pressure. Each of these may, in turn, exacerbate CHF. While the mechanisms underlying these effects are likely to be complex and still remain to be elucidated, an important premise is that sympathetic activation associated with OSA is a key factor contributing to these cardiovascular effects, as well as to increased mortality in CHF [2, 3, 13].

Hypertension Two large, population-based studies [8, 14, 15] have suggested that OSA is an independent risk factor for hypertension, even after controlling for important confounders like age, gender, and obesity. The finding of this association is an important result in the context of CHF, because the Framingham Heart Study identified hypertension as the single most common risk factor for CHF [16]. Thus OSA may be a contributing factor to the development of CHF through its effects on blood pressure. While reasons for blood pressure elevation in OSA remain unclear, one plausible mechanism is via activation of the sympathetic nervous system. This activation may be related to intermittent hypoxia, arousals from sleep at apnea termination, or to

wide swings in intrathoracic pressure during efforts to inspire against an intermittently closed upper airway. Importantly, this increase in sympathetic tone during sleep persists into wakefulness [13, 17]. Sympathetic activation in OSA may indeed be responsible for increased mortality in patients with CHF as well. Some of the most powerful evidence for this comes in the form of several large, randomized trials, in which pharmacological blockade of the sympathetic nervous system led to significantly improved survival compared with the administration of placebo [5, 6].

Reduced Cardiac Output OSA exacerbates left ventricular dysfunction in patients with CHF through changes in afterload, preload, and ventricular contractility (see Figure 94.1). How OSA does this is the topic of much investigation, with a central focus being placed on cyclic, intermittent hypoxia, and additional evidence suggesting a role for the large, negative drops in intrathoracic pressure during obstructive apnea events.

Intermittent hypoxia associated with OSA impairs both myocardial contractility and relaxation [18, 19] and increases ventricular afterload by raising mean pulmonary artery pressure not only during sleep, but even during the day [20, 21]. In addition to intermittent hypoxia, OSA patients also experience increasingly negative intrathoracic pressure during obstructive events. This enhances venous return and may result in right ventricular dilation and septal bowing into the left ventricle with resultant decrease in left ventricular end-diastolic volume and preload. This effect can be simulated by the Müller maneuver, in which an awake subject attempts to inspire against a closed glottis, and the effects on septal bowing may be viewed in real-time by echocardiography. The large, negative intrathoracic pressure during obstructive apneas also increases afterload. The combination of raised afterload, lowered contractility, and diminished preload has the overall effect of reducing cardiac output [22].

Early Evidence for Increased Mortality Evidence for an association between OSA in CHF and increased mortality is still in its early stages. Preliminary data suggests an increased release of cardiac peptides in OSA patients during sleep, presumably due to right ventricular dilation. Kita et al. [23] showed that OSA patients have a significant increase in brain natriuretic peptide (BNP) during sleep compared to individuals without OSA. The importance of this is that elevated levels of BNP are not only a marker for the degree of pulmonary hypertension and both right and left ventricular failure, but may be a predictor of sudden death in patients with chronic CHF [24, 25].

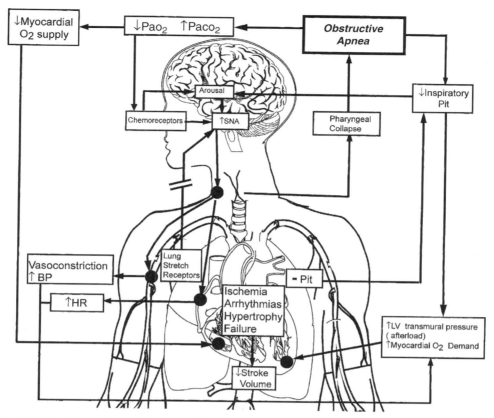

Figure 94.1 Effects of OSA on the cardiovascular system. Obstructive apneas increase left ventricular (LV) transmural pressure (i.e., afterload) through the generation of negative intrathoracic pressure (P_{it}) and elevations in systemic blood pressure (BP) secondary to hypoxia, arousals from sleep, and increased sympathetic nervous system activity (SNA). Apnea also suppresses the sympathetic inhibitory effects of lung stretch receptors, further enhancing SNA. The combination of increased LV afterload and increased heart rate (HR) secondary to increased SNA increases myocardial O_2 demand in the face of a reduced myocardial O_2 supply. These conditions predispose a patient acutely to cardiac ischemia and arrhythmias and chronically could contribute to LV hypertrophy and, ultimately, failure. The resultant fall in stroke volume can further augment SNA. (From Bradley TD, Floras JS. *Circulation* **107**:1671–1678(2003), with permission.)

Effects of CSA–CSR

Evidence suggests that CSA–CSR, when occurring in the setting of CHF, is associated with increased mortality [26–28]. One prospective case–control study of 353 inpatients at a VA hospital showed that patients with CHF and CSA–CSR had a median survival of 1.9 years, compared to 3.7 years among patients with CHF alone [28]. Another prospective cohort study examined clinically stable CHF patients with a left ventricular ejection fraction $\leq 35\%$ and found that the combination of AHI ≥ 30 events/h and left atrial diameter ≥ 25 cm^2 was associated with 2 year mortality of 86% [26]. Importantly, however, patients with CHF in both studies were not receiving beta blocker therapy, which has now become a standard part of treatment of CHF [5, 6], and which has been shown to reduce the occurrence of CSA–CSR, as well as reduce mortality in CHF. However, a more recent randomized controlled trial [27] of patients receiving beta blocker and ACE inhibitor therapy showed that patients with both CHF and CSA–CSR still experienced

higher mortality and cardiac transplantation rates (RR 2.53) compared to patients with CHF alone.

Whether CSA–CSR actually contributes to mortality risk or is simply a marker of the degree of severity of CHF remains unknown. As for the case of OSA, CSA–CSR is also associated with increased sympathetic activity. However, the degree of risk specifically attributable to CSA–CSR (via, say, intermittent hypoxia) rather than to reduced left ventricular end-diastolic volume (LVEDV) or clinical severity of CHF itself has not been quantified [29, 30].

The pathophysiologic effects of CSA–CSR on cardiovascular function are limited in comparison to OSA. Unlike OSA, negative intrathoracic pressure is not generated during central apneas so there is a less direct effect on cardiac afterload and LVEDV. Elevation of LVEDV is more likely a cause rather than an effect of CSA–CSR [31, 32]. This has been demonstrated using pulmonary capillary wedge pressure (PCWP) as a surrogate for LVEDV. When the PCWP is reduced through intensive medical therapy, there is concomitant drop in the AHI [32]. Elevations in PCWP are also associated with hypocapnia [32], which is a risk factor for occurrence of CSA–CSR [33]. Thus a proposed mechanism for CSA–CSR is that elevations in LVEDV raise pulmonary vascular volume, which stimulates pulmonary stretch receptors and increases vagal output, and subsequently increases ventilation. This in turn results in hypocapnia and induction of apnea. Supporting this is recent data showing that in a subset of CHF patients with CSA–CSR, normalization of ejection fraction through cardiac transplantation results in a reduction of the AHI [30]. In the remaining subset, however, the AHI does not normalize completely despite transplantation, suggesting that additional mechanisms may be at play; one such purported mechanism is increased peripheral chemosensitivity to carbon dioxide [34]. The hypocapnia that occurs in patients with both CHF and CSA–CSR has been associated with a 20 times increased prevalence of ventricular arrhythmia, compared to the prevalence in eucapnic counterparts [33].

TREATMENT OF SDB AND EFFECTS ON CHF

The usual treatment for SDB is continuous positive airway pressure (CPAP) therapy, which acts as a pneumatic splint to prevent airway collapse during sleep and associated episodic desaturation. Even among patients with CHF *without* SDB, CPAP has significant beneficial effects, including reduction of LV afterload [35], increase in cardiac output in patients with elevated PCWP [36], and, in the short term, reduction in cardiac sympathetic output [37] and in myocardial oxygen consumption [38].

OSA

Based on the cardiac effects of CPAP in persons with CHF who do not have SDB, we would expect patients who have both OSA and CHF to experience treatment benefit from CPAP which exceeds that experienced by patients who have OSA alone. Two recent randomized controlled trials explored the cardiovascular effects of CPAP in CHF patients who also have OSA. Kaneko et al. [39] evaluated 24 patients with systolic dysfunction (EF $\leq$ 45%) and OSA (AHI $\geq$ 30 events/h) who were randomized to receive medical therapy for CHF plus continuous positive airway pressure (CPAP) therapy for OSA versus medical therapy for CHF alone for 1 month. Patients receiving CPAP experienced a marked reduction in the AHI, which was associated with reduced daytime systolic blood pressure, reduced left ventricular end-systolic dimension, and a significant improvement in left ventricular ejection fraction. In another study [40], 55 subjects with OSA (AHI $\geq$ 5 events/h) and CHF (EF $\leq$ 55%) were randomized to CPAP versus no treatment. After 3 months, in addition to a reduction in

AHI, there was also improvement in left ventricular ejection fraction, a reduction in sympathetic activity, and improvement in quality of life scores. Thus CPAP appears to provide significant benefit to patients with CHF and OSA in the short term. Since degree of ventricular dysfunction and sympathetic hyperactivity were two markers of reduced survival in the Framingham Heart Study [16], larger, longer-term studies should assess whether CPAP-associated reduction in sympathetic nervous system activity and improvement in left ventricular function translate into a clear survival benefit.

CSA–CSR

Improvements in CSA–CSR may occur with treatment of CHF itself, with supplemental oxygen, or with CPAP. In patients with CSA–CSR and CHF, the first consideration is optimal treatment of CHF. In contrast to the case for OSA [41], intense treatment of CHF itself results in reduction of AHI in CSA. Nocturnal oxygen therapy may reduce the AHI by $\geq$50% in eucapnic patients [42], perhaps because elevated Pao_2 may blunt Pco_2 sensitivity and prevent the hyperventilation phase of CSA–CSR. In a later study, Lorenzi-Filho and colleagues [43] demonstrated that hypocapnia triggers central apneas in CSA–CSR and that inhalation of CO_2 but not supplemental O_2 reversed central apneas.

As among patients with OSA, CPAP used in patients with CSA–CSR can improve cardiac function by decreasing afterload or reducing sympathetic activity. Additional evidence also suggests that mitral regurgitant flow and level of circulating atrial natriuretic peptide decrease [44]. In a small randomized controlled trial examining the effect of CPAP in CHF patients with CSA–CSR, patients treated with CPAP had a significantly greater transplant-free survival when compared to CHF patients treated with standard medical therapy alone [27]. This finding remains to be verified in larger, controlled trials.

CONCLUSIONS

Heart failure imposes a major population health burden, and approximately half of patients with CHF experience OSA or CSA–CSR. OSA may impair cardiac function and contribute to increased morbidity and mortality in CHF. CSA–CSR may instead be a marker for poor cardiac function and indicate a worse prognosis and mortality in patients with concomitant heart failure. For both syndromes, CPAP therapy may improve left ventricular dysfunction and reduce sympathetic hyperactivity, each of which otherwise are markers of increased mortality in CHF, according to the results of the Framingham Heart Study. Whether CPAP ultimately improves survival in patients with SDB and CHF must remain a priority for future investigation by well-controlled, sufficiently powered prospective investigations with clearly defined endpoints.

REFERENCES

1. Hunt HA, Baker DW, Chin MH, et al. ACC/AHA guidelines for the evaluation and management of chronic heart failure in the adult: executive summary. *Circulation* **104**:2996–3007(2001).

2. Cohn JN, Levine TB, Olivari MT, et al. Plasma norepinephrine as a guide to prognosis in patients with chronic congestive heart failure. *N Engl J Med* **311**:819–823(1984).

3. Kaye DM, Lefkovits J. Neurochemical evidence of cardiac sympathetic activation and increased central nervous system norepinephrine turnover in severe congestive heart failure. *J Am Coll Cardiol* **23**:570–578(1994).

4. CONSENSUS Trial Study Group. Effects of enalapril on mortality in severe congestive heart failure. Results of the Cooperative North Scandinavian Enalapril Survival Study (CONSENSUS). *N Engl J Med* **316**:1429–1435(1987).

5. Merit HF Study Group. Effect of metoprolol CR/XL in chronic heart failure: metoprolol CR/XL randomized intervention trial in congestive heart failure (MERIT-HF). *Lancet* **353**:2001–2009(1999).

6. Packer M, Bristow MR, Cohn JN, et al. The effect of carvedilol on morbidity and mortality in patients with chronic heart failure. US Carvedilol Heart Failure Study Group. *N Engl J Med* **334**:1349–1355(1996).

7. Tkacova R, Niroumand M, Lorenzi-Filho G, Bradley DT. Overnight shift from obstructive to central sleep apneas in patients with heart failure. Role of P_{CO_2} and circulatory delay. *Circulation* **103**:238–243(2001).

8. Sleep Heart Health Study Research Group. Sleep-disordered breathing and cardiovascular disease. *Am J Respir Crit Care Med* **163**:19–25(2001).

9. Javaheri S, Parker TJ, Liming JD, et al. Sleep apnea in 81 ambulatory male patients with stable heart failure. Types and their prevalences, consequences, and presentations. *Circulation* **97**:2154–2159(1998).

10. Sin DD, Fitzgerald F, Parker JD, Newton G, Floras JS. Risk factors for central and obstructive sleep apnea in 450 men and women with congestive heart failure. *Am J Respir Crit Care Med* **160**:1101–1106(1999).

11. Chan J, Sanderson J, Chan W, et al. Prevalence of sleep-disordered breathing in diastolic heart failure. *Chest* **111**:1488–1493(1997).

12. Young T, Paltra M, Dempsey J, Skatrud J, Weber S, Badr S. The occurrence of sleep disordered breathing among middle-aged adults. *N Engl J Med* **328**:1230–1235(1993).

13. Carlson JT, Hedner J, Elam M, Ejnell H, Sellgren J, Wallin BG. Augmented resting sympathetic activity in awake patients with obstructive sleep apnea. *Chest* **103**:1763–1768(1993).

14. Nieto FJ, Young TB, Lind BK, et al. Association of sleep-disordered breathing, sleep apnea, and hypertension in a large community-based study. *JAMA* **283**:1829–1836(2000).

15. Peppard PE, Young T, Palta M, Skatrud J. Prospective study of the association between sleep-disordered breathing and hypertension. *N Engl J Med* **342**:1378–1384(2000).

16. Levy D, Larson MG, Vasan RS, Kannel WB, Ho KK. The progression from hypertension to congestive heart failure. *JAMA* **275**:1557–1562(1996).

17. Somers VK, Dyken ME, Clary MP, Abboud FM. Sympathetic neural mechanisms in obstructive sleep apnea. *J Clin Invest* **96**:1897–1904(1995).

18. Kusuoka H, Weisfeldt ML, Zweier JL, Jacobus WE, Marban E. Mechanism of early contractile failure during hypoxia in intact ferret heart: evidence for modulation of maximal Ca^{2+} activated force by inorganic phosphate. *Circ Res* **59**:270–282(1986).

19. Cargill RI, Kiely DG, Lipworth BJ. Adverse effects of hypoxemia on diastolic filling in humans. *Clin Sci* **89**:165–169(1995).

20. Sanner BM, Doberauer C, Konermann M, Sturm A, Zidek W. Pulmonary hypertension in patients with obstructive sleep apnea syndrome. *Arch Intern Med* **157**:2483–2487(1997).

21. Alchanitis M, Tourkohoriti G, Kakouros S, Kosmas E, Podaras S, Jordanoglou JB. Daytime pulmonary hypertension in patients with obstructive sleep apnea. *Respiration* **68**:566–572(2001).

22. Bradley TD. Hemodynamic effects of simulated obstructive apneas in humans with and without heart failure. *Chest* **119**:1827–1835(2001).

23. Kita H, Ohi M, Chin K, et al. The nocturnal secretion of cardiac natriuretic peptides during obstructive sleep apnoea and its response to therapy with nasal continuous positive airway pressure. *J Sleep Res* **7**:199–207(1998).

24. Berger R, Huelsman M, Strecke RK, et al. B-type natriuretic peptide predicts sudden death in patients with chronic heart failure. *Circulation* **105**:2369–2379(2002).

25. Leuchte HH, Neurohr C, Baumgartner R, et al. Brain natriuretic peptide and exercise capacity in lung fibrosis and pulmonary hypertension. *Am J Respir Crit Care Med* **170**:360–365(2004).

26. Lanfranchi PA, Braghiroli A, Bosimini E, et al. Prognostic value of nocturnal Cheyne–Stokes respiration in chronic heart failure. *Circulation* **99**:1435–1440(1999).

27. Sin DD, Logan AG, Fitgerald FS, Liu PP, Bradley DT. Effects of continuous positive airway pressure on cardiovascular outcomes in heart failure patients with and without Cheyne–Stokes respiration. *Circulation* **102**:61–66(2000).

28. Ancoli-Israel S, DuHamel E, Stepnowsky C, Engler R, Cohen-Zion M, Marler M. The relationship between congestive heart failure, sleep apnea, and mortality in older men. *Chest* **124**: 1400–1405(2003).

29. Solin P, Kaye DM, Little PJ, Bergin P, Richardson M, Naughton MT. Impact of sleep apnea on sympathetic nervous system activity in heart failure. *Chest* **123**:1119–1126(2003).

30. Mansfield DR, Solin P, Roebuck T, Bergin P, Kaye DM, Naughton MT. The effect of successful heart transplant treatment of heart failure on central sleep apnea. *Chest* **124**:1675–1681(2003).

31. Tkacova R, Hall MJ, Liu PP, Fitzgerald FS, Bradley TD. Left ventricular volume in patients with heart failure and Cheyne–Stokes respiration during sleep. *Am J Respir Crit Care Med* **156**:1549–1555(1997).

32. Solin P, Bergin P, Richardson M, Kaye DM, Walters EH, Naughton MT. Influence of pulmonary capillary wedge pressure on central apnea in heart failure. *Circulation* **99**:1574–1579(1999).

33. Javaheri S, Corbett WS. Association of low P_{CO_2} with central sleep apnea and ventricular arrythmias in ambulatory patients with stable heart failure. *Ann Intern Med* **128**:204–207(1998).

34. Javaheri S. A mechanism of central sleep apnea in patients with heart failure. *N Engl J Med* **341**:949–954(1999).

35. Naughton MT, Rahman MA, Hara K, Floras JS, Bradley TD. Effect of continuous positive airway pressure on intrathoracic and left ventricular transmural pressures in patients with congestive heart failure. *Circulation* **91**:1725–1731(1995).

36. Bradley TD, Holloway RM, McLaughlin R, Ross BL, Walters J, Liu PP. Cardiac output responses to continuous positive airway pressure in congestive heart failure. *Am Rev Respir Dis* **145**:377–382(1992).

37. Kaye DM, Mansfield D, Aggarwal A, Naughton MT, Esler MD. Acute effects of continuous positive airway pressure on cardiac sympathetic tone in congestive heart failure. *Circulation* **103**:2336–2338(2001).

38. Kaye DM, Mansfield D, Naughton MT. Continuous positive airway pressure decreases myocardial oxygen consumption in heart failure. *Clin Sci* **106**:599–603(2004).

39. Kaneko Y, Floras JS, Usui K, et al. Cardiovascular effects of continuous positive airway pressure in patients with heart failure and obstructive sleep apnea. *N Engl J Med* **348**:1233–1241(2003).

40. Mansfield DR, Kaye DM, Richardson M, Bergin P, Naughton MT. Controlled trial of continuous positive airway pressure in obstructive sleep apnea and heart failure. *Am J Respir Crit Care Med* **169**:361–366(2004).

41. Kraiczi H, Hedner J, Peker Y, Grote L. Comparison of atenolol, amlodipine, enalapril, hydrochlorothiazide, and losartan for antihypertensive treatment in patients with obstructive sleep apnea. *Am J Respir Crit Care Med* **161**:1423–1428(2000).

42. Franklin KA, Eriksson P, Sahlin C, Lundgren R. Reversal of central sleep apnea with oxygen. *Chest* **111**:163–169(1997).

43. Lorenzi-Filho G, Rankin F, Bies I, Bradley TD. Effects of inhaled carbon dioxide and oxygen on Cheyne–Stokes respiration in patients with heart failure. *Am J Respir Crit Care Med* **159**:1490–1498(1999).

44. Tkacova R, Liu PP, Naughton MT, Bradley TD. Effect of continuous positive airway pressure on mitral regurgitant fraction and atrial natriuretic peptide in patients with heart failure. *J Am Coll Cardiol* **30**:739–745(1997).

ADDITIONAL READING

Bradley TD, Floras JS. Sleep apnea and heart failure. Part I: obstructive sleep apnea. *Circulation* **107**:1671–1678(2003). This article reviews the pathophysiology of obstructive sleep apnea and heart failure and recent data on effects of treatment.

Bradley TD, Floras JS. Sleep apnea and heart failure. Part II: central sleep apnea. *Circulation* **107**:1822–1826(2003). An excellent review of the pathophysiology of central sleep apnea–Cheyne–Stokes respiration as well as recent data on treatment and outcomes.

Lanfranchi PA, Somers VK. Sleep-disordered breathing in heart failure: characteristics and implications. *Respir Physiol Neurobiol* **136**:153–165(2003). Well-written general review of the subject.

Leung R, Bradley TD. Sleep apnea and cardiovascular disease. *Am J Respir Crit Care Med* **164**:2147–2165(2001). A very detailed, state-of-the-art article reviewing the interaction between clinical cardiovascular disease and sleep disordered breathing.

Shamsuzzaman ASM, Gersh BJ, Somers VK. Obstructive sleep apnea. Implications for cardiac and vascular disease. *JAMA* **290**:1906–1914(2003). Review of 154 original articles examining the cardiovascular effects of obstructive sleep apnea.

Yan AT, Bradley TD, Liu PP. The role of continuous positive airway pressure in the treatment of congestive heart failure. *Chest* **120**:1675–1685(2001). This article reviews the applicability and physiologic benefits of CPAP therapy in congestive heart failure.

95

CARDIAC ARRHYTHMIAS AND SUDDEN DEATH DURING SLEEP

RICHARD L. VERRIER Ph.D. AND MARK E. JOSEPHSON M.D.
Harvard Medical School, Boston, Massachusetts

INTRODUCTION

Heart disease affects 13.5 million Americans and promotes a heterogeneous myocardial substrate with potentially severe consequences in terms of cardiac arrhythmias. A significant number of atrial arrhythmias (in particular, atrial fibrillation) in patients under 60 years of age and 15% of lethal ventricular arrhythmias have their onset at nighttime. Streamlined technology for concurrent monitoring of sleep state, electrocardiogram, respiration, and oxygen desaturation is expected to improve diagnosis and therapy of nocturnal arrhythmias.

NOCTURNAL VENTRICULAR ARRHYTHMIAS

A significant proportion of sudden cardiac deaths (15%), implantable cardioverter-defibrillator discharges (15%), and myocardial infarctions (20%) occurs during sleep in patients with ischemic heart disease (Figure 95.1) [1]. This decrement in the nocturnal incidence of ventricular arrhythmias, myocardial infarction, and myocardial ischemia is likely attributable to the decreased metabolic demands of non-rapid eye movement (NREM) sleep, which occupies ~80% of sleep time. Importantly, the nighttime distribution of these events is nonuniform [1], suggesting provocation by physiologic triggers, particularly autonomic activity and disturbed respiration, which could be monitored for improved diagnosis and therapy.

Autonomic nervous system activity fluctuates dramatically during sleep, and REM-induced bursts in cardiac sympathetic nerve activity have been implicated in nocturnal ventricular arrhythmias both with and without myocardial ischemia. The presence of myocardial ischemia or changes in cardiac substrate and mechanical function due to disease, infarction [3, 4], or aging can amplify electrical instability. Frequent or complex arrhythmias are also common in hypertensive patients whose blood pressure does not decrease at night. Q–T interval dispersion is increased at night among survivors of sudden cardiac death [5], suggesting possible increased vulnerability to cardiac arrhythmias at nighttime. Sleep apnea, which triggers surges in arterial blood pressure and sympathetic nerve activity due to oxygen desaturation, has been found to provoke nighttime ventricular tachycardia in cardiac patients following myocardial infarction [6] or heart failure. (See Chapter 35 in this volume.) REM-related nocturnal arrhythmogenesis may have a significant affective component, as REM sleep dreams may be vivid, bizarre, and emotionally intense and may generate anger and/or fear. Arrhythmias may also occur during NREM sleep, when latent automatic foci are exposed by the generalized reduction in heart rate following withdrawal of overdrive suppression, or when hypotension exacerbates impaired coronary perfusion.

Autonomic activity has significant potential to provoke ventricular arrhythmias during the first weeks following myocardial infarction (MI) [3, 4]. Cardiac sympathetic

Sleep: A Comprehensive Handbook, Edited by T. Lee-Chiong.
Copyright © 2006 John Wiley & Sons, Inc.

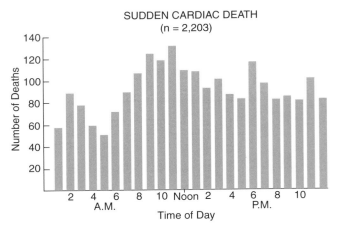

Figure 95.1 Time of day versus number of out-of-hospital sudden cardiac deaths (<1 h from onset of symptoms to death) for 2203 persons dying in Massachusetts in 1983. A significant circadian rhythm is present (*P* < 0.001) with a morning peak between 7 and 11 a.m. and a secondary peak between 5 and 6 p.m. (From [2].)

nerve activity is increased and/or parasympathetic nerve activity is decreased during NREM as well as REM sleep [4]. Sleep patterns are also significantly disturbed. Average heart rates are increased and nighttime ischemic events are more numerous and longer lasting. Patients whose left ventricular function is impaired may experience tachycardia, ventricular premature beats, and ST-segment changes during bouts of nocturnal oxygen desaturation [6]. Within 6 months after MI, the incidence of ventricular tachycardia during sleep is significantly lessened. The risk of fatal events remains high if patients' nighttime heart rate averages >90 bpm.

In some cases, nocturnal onset of ventricular arrhythmias, particularly in patients with structural heart disease, may indicate an electrically unstable substrate. Beta-adrenergic receptor blockade therapy may prove useful in blunting surges in sympathetic nerve activity. It is important to avoid medications that disrupt sleep. In treating hypertensive patients, special attention should be paid to the hemodynamic effects of antihypertensive drugs and vasodilators to avoid precipitating cardiac events by inducing profound hypotension, with risk of thrombosis and embolism in patients with stenotic lesions in the heart or brain or of myocardial infarction. It is important to rule out "white coat" hypertension, as more than 30% of individuals with elevated blood pressure readings in the physician's office or hospital prove to be normotensive during daily life as documented by ambulatory blood pressure monitoring [7]. Diagnosis of the etiology of nocturnal ventricular arrhythmias also includes monitoring for disturbed breathing, which can be treated by continuous positive airway pressure (CPAP).

NOCTURNAL ASYSTOLE AND Q–T INTERVAL PROLONGATION

Pauses in sinus rhythm of >2 s, prolonged atrioventricular (AV) conduction, Wenkebach AV block, and bradycardia are not unusual in individuals who are young or physically fit, such as athletes and heavy laborers, and are attributed to effects of increased parasympathetic activity on AV node conduction [8]. Guilleminault and colleagues [9] reported more extreme cases of asystole in young adults with apparently normal cardiac function who experienced periods of sinus arrest of up to 9 s during REM sleep. These nocturnal asystoles were attributed to exaggerated, if not abnormally elevated, vagal tone, due to the fact that muscarinic receptor blockers did not prevent the heart rate pauses although reducing their duration.

Nocturnal asystolic events can set the stage for ventricular arrhythmias in patients with cardiac disease by facilitating the development of early afterdepolarizations and the lethal arrhythmia torsades de pointes, especially if potassium channel blocking agents are employed to contain the arrhythmias. It is therefore appropriate to consider alternatives to class III antiarrhythmic drugs (potassium channel blockers) in treating patients with nocturnal heart rate pauses. In patients with coronary atherosclerosis, the acetylcholine released by nocturnal surges in vagus nerve activity could induce vasoconstriction rather than vasodilation in damaged endothelium due to impaired release of endothelium-derived relaxing factor.

ATRIAL FIBRILLATION

Vagal influences are likely to facilitate atrial fibrillation in 10–25% of cases of this arrhythmia, which afflicts 2.5 million Americans. Nocturnal peaks in onset of paroxysmal atrial fibrillation have been reported by several investigators, but none has discussed an association with either REM or NREM sleep. A significant midnight to 2:00 a.m. peak in atrial fibrillation onset and higher average nocturnal incidence of the arrhythmia were documented by Rostagno and colleagues [10] in their review of 10 years of records of mobile coronary care units staffed by cardiologists in Florence, Italy. An investigation in a Japanese population ≤60 years old found a peak in frequency of onset of atrial fibrillation at midnight, with maximum duration of the arrhythmia (77 ± 27 min/episode) greatest between midnight and 6:00 a.m. (Figure 95.2). These are periods when NREM sleep is dominant, and the arrhythmia has been termed "vagally mediated atrial fibrillation."

Other investigators characterized a 4:00 to 5:00 a.m. peak in onset of paroxysmal atrial fibrillation in patients with implantable cardioverter-defibrillators. The recorded

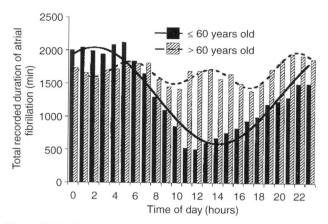

Figure 95.2 Hourly total duration of paroxysmal atrial fibrillation in younger (≤60 years old, solid bars) and older patients (hatched bars). The single harmonic fit of the data in the younger patients is shown by the unbroken line. The triple harmonic fit in the older patients is shown by the broken line. A prominent monophasic circadian rhythm is present in younger patients in contrast to a toneless triphasic rhythm in older patients. (From [11].)

episodes exhibited an atrial rate of >220 beats/min and lasted >1 min before termination by pacing or spontaneous reversion, but the arrhythmias were refractory to pharmacologic therapy.

Heart rate variability studies and the presence of bradycardia in individuals with structurally normal hearts indicate that nocturnal atrial fibrillation is provoked during periods of intense vagus nerve activity. Enhanced adrenergic activity may interact in a complex manner with changes in vagal tone to affect atrial refractoriness and dispersion of repolarization and to alter intra-atrial conduction to increase propensity for this arrhythmia. The high level of vagus nerve tone typical of slow-wave sleep may exacerbate atrial fibrillation in patients whose atria are particularly prone to the arrhythmogenic influence of acetylcholine.

Risk of atrial fibrillation is doubled in patients with disordered breathing during sleep, as apnea can provoke nocturnal hypoxemia, sympathetic nerve activity, and hemodynamic stress. Surges in blood pressure induced by obstructive apnea may distend atrial chambers and activate stretch receptors. Therapy should be directed to controlling heart rate or terminating the arrhythmia pharmacologically or with an atrial cardioverter-defibrillator. As Somers and colleagues have recently provided additional evidence that nearly one-half of the patient population with atrial fibrillation suffers from sleep apnea, monitoring for the presence of sleep disordered breathing, which can be effectively treated by CPAP, should be pursued.

IMPACT OF RESPIRATORY CHALLENGES

Obstructive sleep apnea and central sleep apnea affect 2–4% of the population, or 5–10 million Americans, particularly unfit individuals over 50 years old, a group that is also at increased risk for cardiovascular disease. Obstructive sleep apnea is known to contribute to the development of hypertension and myocardial infarction. In patients with cardiovascular disease, apneic episodes and oxygen desaturation have been found to provoke bradyarrhythmias, sinus arrest, asystole, and supraventricular and ventricular arrhythmias [6]. Central sleep apnea is also common in patients with systolic heart failure, in whom it is associated with atrial and ventricular arrhythmias. (See Chapter 94 in this volume.)

It may be useful to screen patients with nocturnal arrhythmias for the presence of apnea by home oximetry [12]. Therapeutic approaches include weight loss to a normal range, CPAP, nasal dilators, surgery or electrical stimulation of the upper airway, oral appliances, and abstinence from alcohol or tobacco before bedtime. CPAP has been found effective in minimizing or eliminating apnea-provoked atrial and ventricular arrhythmias, improving fibrinogen levels and insulin responsiveness, lessening myocardial ischemia, and reducing apnea-induced elevations in arterial and pulmonary blood pressure and sympathetic nerve activity. CPAP has also been reported to obviate the need for pacemaker implantation for nocturnal bradyarrhythmias, sinus arrest and heart block.

SUDDEN INFANT DEATH SYNDROME

Sudden infant death syndrome (SIDS) is the leading cause of mortality in infants between 1 week and 1 year of age and occurs during sleep. It may be attributable to a variety of etiologies that challenge the developing cardiorespiratory system, as the syndrome is diagnosed by exclusion to include all causes that "remain unexplained after a thorough case investigation, including performance of a complete autopsy, examination of the death scene, and review of the clinical history." SIDS took a toll of 2234 infants in 2001 in the United States or 8.1% of infant deaths. The fatal event is characterized by hypotension and bradycardia [13] and appears to result from a deficit in the normal reflex coordination of heart rate, arterial blood pressure, and respiration during sleep, so that there is a failure to respond to cardiorespiratory challenges. A potential anatomical basis is a binding deficit in the arcuate nucleus of SIDS infants, as muscarinic cholinergic activity in this structure at the ventricular medullary surface is postulated to be involved in cardiorespiratory control [14]. Some families with apnea also exhibit a significant incidence of deficits in ventilatory responses to hypoxia. Altered autonomic control is suggested by the fact that heart rates in infants who later died of SIDS are generally higher and exhibit a reduced range compared to normal infants and by evidence of autonomic instability during NREM sleep in infants with aborted SIDS events.

Repolarization abnormalities have also been reported. A 19 year prospective multicenter observational study of 34,442 infants concluded that significant prolongation (≥35 ms) of the Q–T interval in the electrocardiogram characterized the 24 (0.07% of enrolled infants) who died of SIDS [15]. The investigators suggested that SIDS results from a genetic defect that alters repolarization to increase the risk of ventricular arrhythmia due to a developmental abnormality in cardiac sympathetic innervation. Infants and children with the long Q–T syndrome genotype linked to chromosome 3 (LQT3) experience a high rate of nocturnal demise. The arrhythmias and reduced heart rates have been traced to a defect in the sodium channel gene *SCN5A*.

Monitoring for repolarization abnormalities including Q–T interval prolongation [15] and T-wave alternans [16], an electrocardiographic indicator of vulnerability to sudden cardiac death [17], can improve assessment of vulnerability to lethal ventricular arrhythmias. The latter phenomenon has been documented in infants with Q–T prolongation who became SIDS victims. Sodium channel blockade has been suggested as a potential therapy for LQT3 patients, whose death commonly occurs at rest or during sleep, but prospective studies are required.

The incidence of SIDS is increased by environmental influences including the winter season. Conflicting evidence has been provided regarding the relative increase in incidence attributable to prone (face-down) sleeping [18]. Maternal cigarette smoking during gestation and passive smoking are highly significant modifiable risk factors, and their elimination could reduce the number of SIDS deaths by 61% [19]. Preterm birth and low birth weight are established risk factors for SIDS and increase risk more than 15-fold among smokers but not at all among nonsmokers. A plausible mechanistic explanation is that nicotine blunts the infants' response to hypoxia, which is known to affect chemoreceptor activation of respiration adversely. Further support for this hypothesis is provided by the finding of nicotine and its metabolites in the pericardial fluid of SIDS infants at autopsy, although the sensitivity and specificity of this finding are unknown. Epicardial nicotine is associated with hypopnea and affects the sinoatrial node and epicardial neural fibers to induce hypotension and bradycardia, the documented symptomatology of the final event in SIDS infants [13]. Use of illegal drugs increases risk of SIDS by more than fourfold, by impairing chemoreceptor responsiveness due to infants' decreased sensitivity to carbon dioxide and may also induce the lethal arrhythmia torsades de pointes. The opportunities for intervention are straightforward and include placing infants in a supine (face-up) position for sleeping and avoiding maternal smoking during gestation and passive smoking during infancy.

BRUGADA SYNDROME AND THE SUDDEN UNEXPLAINED NOCTURNAL DEATH SYNDROME

Western men diagnosed with the Brugada syndrome and young, apparently healthy Southeast Asian men with the sudden unexplained nocturnal death syndrome (SUNDS) experience sudden cardiac death predominantly during sleep. In Southeast Asia, SUNDS is the leading cause of death, apart from accidents, of men under the age of 50. ST-segment elevation in two anterior precordial leads is considered diagnostic for the presence of the genotype [20, 21]. A sodium-channel mutation in the *SCN5A* gene has been identified by investigation of an eight-generation kindred with a high incidence of nocturnal SCD, Q–T interval prolongation, and Brugada-like electrocardiogram. Presynaptic sympathetic cardiac dysfunction has been suggested based on abnormal ^{123}I-MIBG uptake, with bradycardia-dependent Q–T prolongation, intrinsic sinus node dysfunction, conduction abnormalities, and absence of ventricular ectopy.

Autopsies of SUNDS cases have established that cardiovascular disease is absent, but, in some instances, that conduction pathways are developmentally abnormal [21]. Bradycardia is the dominant arrhythmia, but ventricular fibrillation has been documented as the final event. Companions have reported that the immediate symptoms are onset of agonal respirations during sleep along with vocalization, violent motor activity, nonarousability, rapid irregular deep breathing, perspiration, heart rate surges, and severe autonomic discharge. Several victims revived by vigorous massage reported sensations of airway obstruction, chest discomfort or pressure, and numb and weak limbs. Recurrence of these symptoms within weeks to months culminated in death.

Implantation of cardioverter-defibrillators (ICDs) is indicated as the most effective approach in preventing sudden death in both Brugada [20] and SUNDS [21] patients, although debate abounds regarding the selection of patients at high risk.

SLEEP-DISRUPTING EFFECTS OF CARDIAC MEDICATIONS

Several widely prescribed medications for cardiac patients, including antihypertensive agents and beta-receptor blocking agents, have the potential to disrupt sleep (Figure 95.3). In

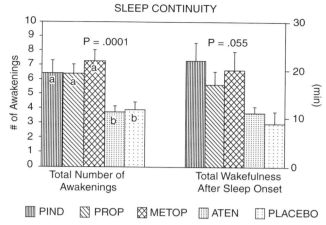

SLEEP CONTINUITY

Figure 95.3 Polysomnographic measures of sleep continuity in 30 healthy male subjects during 1 night following 1 week of treatment with beta-blockers or placebo. Lipophilic beta-blockers pindolol (PIND), propranolol (PROP), and metoprolol (METOP) significantly disturbed sleep, as indicated by the number of awakenings, which was significantly reduced with nonlipophilic atenolol (ATEN). Bars with the same letter are not significantly different. Values are reported as means ± standard error of the means [22].

particular, the lipophilic beta-blockers (pindolol, propranolol, carvedilol, and metoprolol), which have been proved to reduce the risk of sudden cardiac death, increase both the total number of awakenings and total minutes of wakefulness compared with placebo and with the nonlipophilic atenolol. These distinctions may become less apparent with prolonged therapy and penetration of the blood–brain barrier. Pindolol, which has intrinsic sympathomimetic activity, increases REM latency and, as a result, decreases REM sleep time. Beta-blocking agents deplete endogenous melatonin, a key sleep-regulating hormone that modulates sympathetic nerve activity. Sleep disturbance has also been documented in conjunction with the widely employed antiarrhythmic agent amiodarone.

Daytime fatigue and lethargy are widely reported by patients taking beta-blockers and may result from sleep disruption and prompt discontinuation of the medication or noncompliance. An additional important side effect of these beta-blockers is provocation of nightmares.

SUMMARY

Improved concurrent monitoring of cardiac patients for autonomic nervous system activity, cardiac electrical instability, and breathing disturbances can improve diagnosis of patients with nocturnal arrhythmias as well as their overall arrhythmia risk. Moreover, monitoring of these additional characteristics will allow the physician to approach sleep as an autonomic stress test for the heart to be used in conjunction with clinical history. Noninvasive

ECG monitoring of autonomic nervous system tone and reflexes can now be achieved by analysis of heart rate variability and heart rate turbulence, which is based on the pattern of heart rhythm recovery following a ventricular premature beat [23]. Cardiac electrical instability can be assessed with Q–T interval dispersion [5] or with T-wave alternans [17]. Apnea can be monitored accurately by oxygen saturation and respiratory patterns [12].

ACKNOWLEDGMENTS

Supported by grant ES 08129 from the National Institutes of Environmental Health, National Institutes of Health, Bethesda, Maryland. The authors thank Sandra Verrier for her editorial contributions.

REFERENCES

1. Lavery CE, Mittleman MA, Cohen MC, Muller JE, Verrier RL. Nonuniform nighttime distribution of acute cardiac events: a possible effect of sleep states. *Circulation* **5**:3321–3327(1997).

2. Muller JE, Ludmer PL, Willich SN, Tofler GH, Aylmer G, Klangos I, Stone PH. Circadian variation in the frequency of sudden cardiac death. *Circulation* **75**:131–138(1987).

3. Broughton R, Baron R. Sleep patterns in the intensive care unit and on the ward after acute myocardial infarction. *Electroencephalogr Clin Neurophysiol* **45**:348–360(1978).

4. Marchant B, Stevenson R, Vaishnav S, Wilkinson P, Ranjadayalan K, Timmis AD. Influence of the autonomic nervous system on circadian patterns of myocardial ischaemia: comparison of stable angina with the early postinfarction period. *Br Heart J* **71**:329–333(1994).

5. Molnar J, Rosenthal JE, Weiss JS, Somberg JC. QT interval dispersion in healthy subjects and survivors of sudden cardiac death: circadian variation in a 24-hour assessment. *Am J Cardiol* **79**:1190–1193(1997).

6. Galatius-Jensen S, Hansen J, Rasmussen V, Bildsoe J, Therboe M, Rosenberg J. Nocturnal hypoxemia after myocardial infarction: association with nocturnal myocardial ischaemia and arrhythmias. *Br Heart J* **72**:23–30(1994).

7. Myers MG, Reeves RA. White coat effect in treated hypertensive patients: sex differences. *J Hum Hypertens* **9**:729–733(1995).

8. Bjerregaard P. Mean 24-hour heart rate, minimal heart rate and pauses in healthy subjects 40–79 years of age. *Eur Heart J* **4**:44–51(1983).

9. Guilleminault C, Pool P, Motta J, Gillis AM. Sinus arrest during REM sleep in young adults. *N Engl J Med* **311**:1006–1010(1984).

10. Rostagno C, Taddei T, Paladini B, Modesti PA, Utari P, Bertini G. The onset of symptomatic atrial fibrillation and paroxysmal supraventricular tachycardia is characterized by different circadian rhythms. *Am J Cardiol* **71**:453–455(1993).

11. Yamashita T, Murakawa Y, Hayami N, Sezaki K, Inoue M, Fukui E, Omata M. Relation between aging and circadian variation of paroxysmal atrial fibrillation. *Am J Cardiol* **82**:1364–1367(1998).

12. Series F, Marc I, Cormier Y, La Forge J. Utility of nocturnal home oximetry for case finding in patients with suspected sleep apnea hypopnea syndrome. *Ann Intern Med* **119**:449–453(1993).

13. Meny RG, Carroll JL, Carbone MT, Kelly DH. Cardiorespiratory recordings from infants dying suddenly and unexpectedly at home. *Pediatrics* **93**:43–49(1994).

14. Kinney HC, Filiano JJ, Sleeper LA, Mandell F, Valdes-Dapena M, White WF. Decreased muscarinic receptor binding in the arcuate nucleus in sudden infant death syndrome. *Science* **269**:1446–1450(1995).

15. Schwartz PJ, Stramba-Badiale M, Segantini A, Austoni P, Bosi G, Giorgetti R, Grancini F, Marni ED, Perticone F, Rosti D, Salice P. Prolongation of the QT interval and the sudden infant death syndrome. *N Engl J Med* **338**:1709–1714(1998).

16. Weintraub RG, Gow RM, Wilkinson JL. The congenital long QT syndromes in childhood. *J Am Coll Cardiol* **16**:674–680(1990).

17. Verrier RL, Nearing BD, LaRovere MT, Pinna GD, Mittleman MA, Bigger JT, Schwartz PJ for the ATRAMI Investigators. Ambulatory ECG-based tracking of T-wave alternans in post-myocardial infarction patients to assess risk of cardiac arrest or arrhythmic death. *J Cardiovasc Electrophysiol* **14**:705–711(2003).

18. Klonoff-Cohen HS, Edelstein SL. A case–control study of routine and death scene sleep position and sudden infant death syndrome in Southern California. *JAMA* **273**:790–794(1995).

19. Klonoff-Cohen HS, Edelstein SL, Lefkowitz ES, Srinivasan IP, Kaegi D, Chang JC, Wiley KJ. The effect of passive smoking and tobacco exposure through breast milk on sudden infant death syndrome. *JAMA* **273**:795–798(1995).

20. Antzelevitch C, Brugada P, Brugada J, Brugada R, Shimizu W, Gussak I, Perez Riera AR. Brugada syndrome: a decade of progress. *Circ Res* **91**:1114–1118(2002).

21. Nademanee K, Veerakul G, Mower M, Likittanasombat K, Krittayapong R, Bhuripanyo K, Sitthisook S, Chaothawee L, Lai MY, Azen SP. Defibrillator versus beta-blockers for unexplained death in Thailand (DEBUT): a randomized clinical trial. *Circulation* **107**:2221–2226(2003).

22. Kostis JB, Rosen RC. Central nervous system effects of beta-adrenergic-blocking drugs: the role of ancillary properties. *Circulation* **75**:204–212(1987).

23. Schmidt G, Malik M, Barthel P, Schneider R, Ulm K, Rolnitzky L, Camm AJ, Bigger JT, Schomig A, Bigger JT Jr. Heart-rate turbulence after ventricular premature beats as a predictor of mortality after acute myocardial infarction. *Lancet* **353**:1390–1396(1999).

ADDITIONAL READING

Josephson ME (Ed). *Clinical Cardiac Electrophysiology Techniques and Interpretations.* Lea & Febiger, Philadelphia, 2002. Textbook provides expert knowledge of cardiac arrhythmias and their therapy.

Verrier RL, Mittleman MA. Sleep-related cardiac risk. In: Kryger MH, Roth T, Dement WC (Eds), *Principles and Practice of Sleep Medicine*, 4th ed. Saunders, Philadelphia, 2005. Review of sleep-state related autonomic nervous system activity and disturbed nighttime breathing as mechanisms of nocturnal cardiac events in sizable patient populations. Presents incidence and triggers of nocturnal ischemia, angina, myocardial infarction, and hypertension. Describes diagnostic opportunities from monitoring nocturnal cardiorespiratory status.

Verrier RL, Nearing BD, Kwaku KF. Noninvasive sudden death risk stratification by ambulatory ECG-based T-wave alternans analysis: evidence and methodological guidelines. *Ann Noninvasive Electrocardiol* **10**:110–120(2005). Guidelines for clinicians for ambulatory electrocardiographic monitoring of T-wave alternans.

PART XIII

SLEEP IN THE OTHER MEDICAL DISORDERS

96

SLEEP AND THE GASTROINTESTINAL TRACT

WILLIAM C. ORR

Lynn Institute for Healthcare Research, Oklahoma University Health Sciences Center, Oklahoma City, Oklahoma

The study of sleep has progressed from a relatively obscure endeavor largely devoted to psychological and psychiatric applications to a more recent burgeoning of work that has elucidated marked alterations in respiratory functioning during sleep, hormonal functioning during sleep, and major health consequences attributable to sleep restriction and/or deprivation. These discoveries have led to a remarkable broadening of the focus and importance of the applications of basic sleep physiology to numerous areas of clinical medicine. Lagging somewhat behind these developments has been the description of gastrointestinal functioning during sleep, and the possible applications of these changes to clinical medicine. Perhaps the most obvious reason for this is the relative inaccessibility of the gastrointestinal (GI) tract to easy study during sleep. Studying GI physiology during sleep in humans requires an unpleasant placement of measuring devices via some external orifice. This alone generally disrupts sleep; however, more recent developments in measurement technology have allowed somewhat more convenient access to the luminal GI tract. As a result of this, there has been a marked increase in studies describing alterations in GI functioning during sleep, and the specific applications of these changes to the practice of gastroenterology.

Recent research has described anatomical pathways that relate the stimulation of various brain nuclei to motor functioning of the GI tract, particularly the colon. These physiologic and anatomic mechanisms have been shown to be altered by stress, and this allows a much more clear-cut understanding and perspective on the relationship between GI phenomena and symptoms. The understanding of these mechanisms, their clinical relevance, and how they may be altered during sleep will form the basis of this chapter.

Among the first issues to bring together the clinical relevance of GI functioning during sleep was the notion of the pathogenesis of duodenal ulcer disease. It was thought that vagal stimulation during sleep was instrumental in producing the hypersecretion of acid, which was thought to be associated with the pathogenesis of duodenal ulcer disease. Some studies have described the hypersecretion of acid during rapid eye movement (REM) sleep, and this prompted a study from our laboratory which involved patients with duodenal ulcer disease and controls. Each of these individuals was studied for several nights with full polysomnographic (PSG) monitoring and the assessment of acid secretion and serum gastrum levels [1]. This study did not document any relationship between acid secretion and the stages of sleep; however, it stimulated numerous other studies that attempted to describe GI physiology and how it may be altered by sleep.

UPPER GI TRACT FUNCTIONING DURING SLEEP

Gastroesophageal Reflux

Clearly the most common and familiar problem related to the upper GI system is gastroesophageal reflux (GER), and its most common symptom, heartburn. It is well established that GER is a common event postprandially and in fact it is a normal physiologic response to gastric distention, which induces a transient relaxation of the lower esophageal sphincter (LES). Heartburn and regurgitation are also

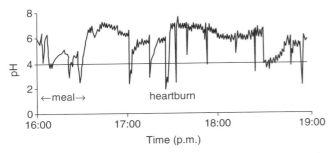

Figure 96.1 The pattern of daytime gastroesophageal reflux (in normal volunteer) is demonstrated in this recording. It reveals several short episodes of reflux that occur postprandially.

well established as the most common symptoms of esophageal mucosal contact. Since the sensation of heartburn is a waking conscious experience, and many reflux events do not necessarily produce a symptom, the actual occurrence of GER during sleep is difficult to estimate by symptoms alone. Gastroesophageal reflux does occur during sleep as has been documented by recent studies, but it is clearly less common than that which occurs in the waking stage [2]. Classic studies that involve the utilization of 24 hour esophageal pH monitoring have established that GER occurs less commonly during the sleeping interval and is generally associated with the prolongation of acid clearance. As is noted in Figure 96.1, waking reflux is generally postprandial and reflux events are rapidly cleared (1–2 minutes). During sleep, however, reflux events are certainly less frequent and generally associated with a longer period of acid contact time (Figure 96.2). Subsequent studies from our laboratory have confirmed these findings, in that we have demonstrated that the complications of reflux that result in discontinuity of the esophageal mucosa are generally associated with an increase in the percent of supine (during the sleeping interval) GER [3]. Clearly, the pattern of GER is different, and the occurrence of acid mucosal contact during sleep appears to be associated with esophageal complications. What are the sleep-related physiologic alterations that facilitate this prolongation of acid mucosal contact?

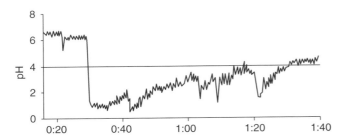

Figure 96.2 This recording demonstrates the prolongation of acid clearance, which is commonly associated with episodes of gastroesophageal reflux that occur during sleep in the supine position.

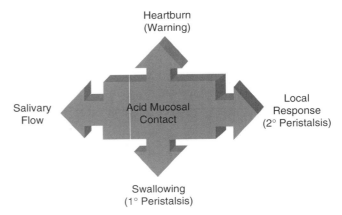

Figure 96.3 Normal defense mechanisms against acid load.

There are secretory, motor, and sensory responses that are associated with acid mucosal contact in the human esophagus (Figure 96.3). Typically, acidification of the distal esophagus will produce a marked increase in the secretion and bicarbonate concentration of saliva. This allows ample buffering potential in order to neutralize the acidic lining of the distal esophagus. Also, in response to an acidic distal esophagus, there is a marked increase in the rate of swallowing, which allows the delivery of the potent buffer of saliva into the distal esophagus. Swallowing and the subsequent primary peristaltic contractions of the esophagus allow the rapid removal of large volumes of refluxate from the distal esophagus. Finally, acid mucosal contact is associated with a sensation of substernal burning, which is perceived as uncomfortable and/or painful. These responses have been determined to be present in a normal waking individual and it is immediately obvious that swallowing and the experience of heartburn are generally assumed to be waking conscious phenomena. The combination of these responses typically results in a rapid clearance of esophageal volume of reflux gastric contents, as well as neutralization of the acidic mucosa. The dependence of this rapid acid clearance response on at least two waking, conscious responses logically raises the question of how these responses may be altered during sleep.

The characteristic responses to acid mucosal contact, which are noted in the waking state, are generally absent during sleep (Figure 96.4). It is clearly these alterations in response to acid mucosal contact during sleep that result in the marked prolongation of acid clearance noted during sleep. A study from our laboratory has demonstrated that the simple infusion of acid into the distal esophagus during polygraphically monitored sleep results in a highly significant prolongation of acid clearance time compared to infusions in the supine waking state [4]. Adding to the risks associated with reflux during sleep is the fact that the swallowing rate is markedly diminished, and salivary flow is essentially absent. Heartburn, of course, being a waking conscious phenomenon, is absent during sleep.

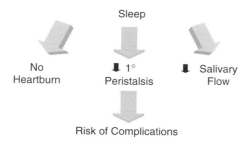

Figure 96.4 Schematic diagram of typical responses to the contact of the esophageal mucosa by refluxed gastric contents.

Thus the combination of these alterations in acid mucosal response associated with sleep establishes a significant risk for the prolongation of acid mucosal contact during sleep.

The prolongation of acid mucosal contact carries with it significant risks. For example, it has been documented that the backdiffusion of hydrogen ions into the esophageal mucosa is directly related to the duration of esophageal acid contact time (Figure 96.5). Thus, extrapolating from this, brief and rapidly cleared episodes of reflux would appear to be relatively benign, while more prolonged episodes of GER would be associated with a greater risk of mucosal damage. An additional risk of prolonged acid mucosal contact relates to the higher risk of the proximal migration and eventual spillover of reflux gastric contents into the tracheobronchial tree. In a recent study from our laboratory, small (1 and 3 mL) volumes of acid were instilled into the esophagus during supine waking and sleep to evaluate the proximal migration of acid infused into the

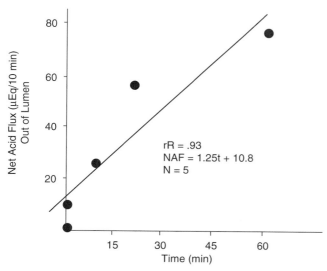

Figure 96.5 This demonstrates the relationship between acid contact time infused into the esophagus and the backdiffusion of hydrogen ions from the lumen of the esophagus into the mucosa. (From Johnson LF, Harmon JW. *J Clin Gastroenterol* **8**:(Suppl 1)26–44(1986).)

distal esophagus during sleep [5]. Esophageal pH sensors were located in both the distal and proximal esophagus and proximal migration was assessed by a drop in the pH in the proximal sensor subsequent to the infusion of acid. It was noted that in the awake, supine position, none of the normal volunteers studied showed evidence of proximal migration of 1 mL. During sleep, however, 40% of these same individuals showed evidence of a proximal migration of acid infused into the distal esophagus during polygraphically determined non-rapid eye movement (NREM) sleep.

Thus it can be concluded from these data that sleep itself induces considerable risk of prolonged acid mucosal contact and facilitates the occurrence of a proximal migration of refluxed gastric contents. Maintaining sleep in response to an episode of GER seems to be a maladaptive response, since an arousal from sleep is required to produce a more rapid clearance from a sleep-related reflux event. The complications of nighttime reflux are significant in that nighttime GER can lead to the development of esophagitis, as well as other complications such as exacerbation of bronchial asthma, chronic cough, and sleep complaints. These complications appear to be primarily related to the presence of significant nighttime reflux and the consequent prolongation of esophageal acid clearance. Of particular interest to the sleep clinician are studies that have been done in patients with obstructive sleep apnea. The frequent occurrence of obesity and the appreciable negative intrathoracic pressures associated with upper airway obstruction are certainly risk factors for the occurrence of nighttime GER. Although studies have not indicated that there is a specific relationship between obstructive events and reflux events, these patients tend to have an overall increase in esophageal acid contact time. Other studies have shown that a reduction in obstructive events, via appropriate CPAP treatment, results in an associated significant reduction in heartburn complaints [6, 7].

Clinical Manifestations of Nighttime Gastroesophageal Reflux

Patients with nighttime heartburn have a number of complaints related to disturbed sleep [8]. In addition to documenting the presence of a number of sleep complaints, these studies documented the fact that in patients with both daytime and nighttime heartburn, the nighttime heartburn was significantly more bothersome. More than 50% of these individuals complained that nighttime heartburn awakened them with GER symptoms and about 30% were awakened by coughing and choking due to regurgitation. About 40% of the patients with nighttime symptoms noted that their heartburn affected their ability to function the next day and about 60% indicated it affected their mood. The use of sleeping pills was also substantially increased in patients with nighttime GER symptoms.

Nighttime GER has been associated with a number of respiratory symptoms, such as wheezing, chronic cough, and hoarseness. Not uncommonly, patients with GER-related asthma or chronic cough will not have heartburn as a symptom. Thus the presence of nighttime reflux cannot be ruled out on the basis of a negative history of nighttime heartburn. Nighttime wheezing is quite common in asthmatics and approximately 41% of asthmatic patients have been shown to have reflux-associated respiratory symptoms [9]. Additional support for this association has been noted in an excellent study recently published by a group of European epidemiologists [10]. In this study they showed that individuals who reported nighttime heartburn at least twice a week had an odds ratio of 2.0 for associated respiratory symptoms such as coughing and wheezing. In addition, this group has noted from their studies that nighttime heartburn is also an independent risk factor for sleep complaints and daytime sleepiness.

Gastric Function During Sleep

The motor function of the stomach serves to deliver gastric contents into the antrum and ultimately into the duodenum at an appropriate rate and pH. Reports of alterations in gastric motility during sleep have been contradictory and there are no particularly conclusive results with regard to either the actual motor functioning of the stomach or its ultimate consequence, gastric emptying. However, the noninvasive recording of the gastric electric pacemaker provides easy access to the measurement of gastric myoelectric activity, which is a fundamental property of the stomach.

By utilizing surface electrodes placed in periumbilical locations, a three cycle per minute electrical rhythm can be detected from the surface of the stomach. By subjecting this measure to spectral analysis, various properties of the gastric electrical rhythm can be determined, such as its amplitude and frequency modulation. The electrical rhythm is a product of the endogenous functioning of the enteric nervous system, and it serves to modulate the contractility of the gastric smooth muscle.

Sleep studies from our laboratory have challenged the traditional belief that the pacemaker activity is without central nervous system (CNS) influence. It has been shown that a significant decline in the power of the three cycle per minute activity is apparent during NREM sleep, and there is a significant recovery during REM sleep to levels comparable to the waking state [11]. These results suggest that NREM sleep is associated with a marked destabilization of the basic gastric electrical rhythm, and that the cerebral activation of REM sleep appears to result in stabilization of this basic gastric pacemaker.

It might be concluded from these results that higher cortical input, or a degree of CNS arousal, must be present in order to stabilize and promote normal gastric functioning,

and consequently normal gastric emptying. Much additional work needs to be done to understand gastric functioning during sleep and its possible clinical consequences.

INTESTINAL MOTILITY AND IRRITABLE BOWEL SYNDROME

Technological and practical difficulties in monitoring intestinal activity in humans have somewhat retarded the study of intestinal activity during sleep. Some data have gradually been appearing in the medical literature due to technological advances in both invasive and noninvasive monitoring of the lower GI tract. Several studies have now demonstrated a decrease in colonic motor activity during sleep and a clear inhibition of the colonic motility index in the transverse, descending, and sigmoid colon. A marked increase in activity was noted upon awakening, or with brief arousals from sleep. This offers a logical explanation of the common urge to defecate upon awakening in the morning [12]. Studies done to date clearly suggest an inhibition of colonic contractile and myoelectric activity during sleep, as well as diminished colonic tone [13]. Rectoanal pressures have been measured continuously during sleep, and results have indicated a decrease in the minute-to-minute variation in the amplitude of spontaneous anal canal activity during sleep [14]. Another study by Rao and colleagues has shed light on intrinsic anorectal functioning [15]. They have indicated that the intrinsic oscillation in rectal motor activity increases by 44% during sleep as compared to waking activity. Of particular importance is that the majority of these contractions were propagated in a retrograde direction. Again, this activity would clearly facilitate rectal continence during periods of depressed consciousness since other studies have documented a marked decrease in anal canal pressure during sleep. Other studies have documented the fact that although anal canal pressure is decreased during sleep, it is maintained at a level somewhat higher than the intrarectal pressure, thus facilitating rectal continence during sleep.

These studies collectively shed important light on the mechanisms of rectal continence during sleep. There appear to be at least two mechanisms that prevent the passive escape of rectal contents during sleep. First, rectal motor activity increases substantially during sleep, but the propagation is retrograde rather than anterograde. Furthermore, these physiologic studies have shown that, even under the circumstances of periodic rectal contractions, the anal canal pressure is consistently above that of the rectum. Both of these mechanisms would tend to protect against rectal leakage during sleep, and alterations in these mechanisms would explain loss of rectal continence during sleep in individuals with diabetes or who have undergone ileoanal anastomosis.

Ambulatory monitoring of the small intestine has been accomplished in patients with irritable bowel syndrome (IBS). Studies have documented that nighttime motor patterns of the small bowel did not differentiate patient groups from controls. In fact, it was noted that there was a notable lack of activity in the small bowel during sleep, which led the investigators to suggest that the changes in motor functioning noted were primarily the result of reactions to "stressful events" during the waking state [16]. Several studies have shown a relationship between poor sleep and pain in IBS patients. Studies have estimated the prevalence of reported sleep complaints to be as high as 25–30% in the population of patients with functional bowel disorders (i.e., irritable bowel syndrome and functional dyspepsia). The high prevalence of sleep complaints in patients with functional bowel disorders has been specifically noted in a study by Fass and colleagues [17] in which they prospectively studied a group of patients and healthy controls utilizing both bowel symptom and sleep questionnaires. In this study patients were divided into groups with functional dyspepsia, IBS, or a combination of dyspepsia and IBS symptoms. This study showed that patients with IBS symptoms alone did not differ from normal controls with regard to the reported incidence of sleep complaints; however, if dyspeptic symptoms were part of the symptom complex, sleep complaints were significantly greater.

A consensus appears to be emerging with regard to subjective and objective parameters of sleep in patients with functional bowel disorders. Compared to normal controls, sleep architecture appears to be quite similar to normals in patients with functional bowel disorders. In this sense, the pattern of behavior in these patients appears to resemble that of many insomniac patients, in which normal sleep patterns are quite exaggerated in terms of actual complaints of prolonged sleep latency and interrupted, nonrefreshing sleep. In a recent report from our laboratory, we confirmed normal sleep architecture in patients with these disorders [18]. In addition, this study did not reveal any abnormalities in cortisol secretion or reports of stress or anxiety. It was clearly demonstrated that although IBS patients did have more sleep complaints, their polysomnographic (PSG) patterns were completely indistinguishable from those of age- and sex-matched controls. In perhaps the largest PSG study to date in IBS patients, we have demonstrated that there is a substantial correlation of subjective sleep complaints and depression. Once again, the PSG parameters were, with the single exception of the REM onset latency, indistinguishable from normals [19].

Recent studies have indicated that the autonomic nervous system appears to be a mediator of visceral pain in patients with functional bowel disorders. Studies have shown some disruption in autonomic functioning during the waking state in patients with IBS, but these studies have had conflicting results. A recent study has suggested the presence of vagal withdrawal during sleep in patients with IBS. This was followed by work from our laboratory, which showed increasing sympathetic tone during REM sleep in patients with IBS. In a series of studies we have evaluated autonomic regulation by conducting spectral analysis of heart rate variability, which allows a determination of the sympathetic and parasympathetic regulation of cardiac activity. Subsequent studies have noted that IBS patients who have dyspeptic symptoms did not appear to have this autonomic dysfunction, but rather it was most notable in patients with IBS, but without dyspeptic symptoms [20].

Collectively, these studies from various sleep investigations in patients with functional bowel disorders suggest not only that there are sleep disturbances noted in this patient population, but also that the sleep disturbances may contribute to altered gastrointestinal functioning. Certainly, these studies confirm the notion that there are central nervous system alterations in patients with functional bowel disorders, and that these alterations are perhaps uniquely identified during sleep. Future studies in sleep and patients with functional bowel disorders will undoubtedly provide additional understanding of the pathophysiology of the brain–gut axis and its alterations during sleep.

CONCLUSIONS

In this chapter, an integration of gastrointestinal functioning has been attempted with regard to its relationship to sleep and how this interaction may lead to sleep complaints of sleep disorders, as well as the pathogenesis of some GI disorders. Considerable data have been presented here to support the notion that sleep-related GER is an important factor not only in the development of esophagitis, but also respiratory complications of GER. Although sensory functioning is markedly altered during sleep with regard to most standard sensory functions (i.e., auditory), there appears to be an enhancement of some visceral sensation during sleep, which would appear to protect the tracheobronchial tree from aspiration of gastric contents refluxed during sleep.

Patients with functional bowel disorders clearly reveal an increase in sleep complaints compared to normal volunteers. The actual mechanisms of these disturbances remain somewhat obscure and studies have not demonstrated any consistent abnormalities in actual sleep patterns of these patients. Some studies have shown that autonomic functioning during sleep, particularly REM sleep, can distinguish patients with IBS and dyspeptic symptoms. Thus the continued study of sleep and GI functioning promises to create a new dimension in the understanding of the pathophysiology of a variety of GI disorders.

REFERENCES

1. Johnson LF, DeMeester TR. Twenty-four hour pH monitoring of the distal esophagus. *Am J Gastroenterol* **62**:325–332(1974).

2. Armstrong RH, Burnap D, Jacobson A, et al. Dreams and acid secretions in duodenal ulcer patients. *New Physician* **33**:241–243(1965).

3. Orr WC, Allen ML, Robinson M. The pattern of nocturnal and diurnal esophageal acid exposure in the pathogenesis of erosive mucosal damage. *Am J Gastroenterol* **89**:509–512(1994).

4. Orr WC, Robinson MG, Johnson LF. Acid clearing during sleep in the pathogenesis of reflux esophagitis. *Dig Dis Sci* **26**:423(1980).

5. Orr WC, Elsenbruch S, Harnish MJ, Johnson LF. Proximal migration of esophageal acid perfusions during waking and sleep. *Am J Gastroenterol* **95**(1):37–42(2000).

6. Ing AJ, Ngu MC, Breslin AB. Obstructive sleep apnea and gastroesophageal reflux. *Am J Med* **108**(Suppl 4a):120S–125S(2000).

7. Green BT, Broughton WA, O'Connor JB. Marked improvement in nocturnal gastroesophageal reflux in a large cohort of patients with obstructive sleep apnea treated with continuous positive airway pressure. *Arch Intern Med* **163**:341–345(2003).

8. Shaker R, Castell DO, Schoenfeld PS, Spechler SJ. Nighttime heartburn is an under-appreciated clinical problem that impacts sleep and daytime function: the results of a Gallup survey conducted on behalf of the American Gastroenterological Association. *Am J Gastroenterol* **98**:1487–1493(2003).

9. Sontag SJ. Gastroesophageal reflux disease and asthma. *J Clin Gastroenterol* **39**(Suppl):S9–S30(2000).

10. Gisalson T, Janson C, Vermeire P, Plaschke P, Bjornsson E, Gislason D, Boman G. Respiratory symptoms and nocturnal gastroesophageal reflux. *Chest* **121**:158–163(2002).

11. Elsenbruch S, Harnish MJ, Orr WC, et al. Disruption of normal gastric myoelectric functioning by sleep. *Sleep* **22**:453A–458A(1999).

12. Narducci F, Bassotti G, Gaburri M, et al. Twenty-four hour manometric recording of colonic motor activity in healthy men. *Gut* **28**:17–25(1987).

13. Steadman CJ, Phillips SF, Camilleri M, et al. Variations of muscle tone in the human colon. *Gastroenterology* **101**:24(1991).

14. Orkin BA, Hanson RB, Kelly KA, et al. Human anal motility while fasting, after feeding, and during sleep. *Gastroenterology* **100**:1016–1023(1991).

15. Rao SS, Welcher K. Periodic rectal motor activity: the intrinsic colonic gatekeeper? *Am J Gastroenterol* **91**:890–897(1996).

16. Kellow JE, Gill RG, Wingate DL. Prolonged ambulant recordings of small bowel motility demonstrate abnormalities in the irritable bowel syndrome. *Gastroenterology* **98**:1208–1218(1990).

17. Fass R, Fullerton S, Tung S, et al. Sleep disturbances in clinic patients with functional bowel disorders. *Am J Gastroenterol* **95**:1195–2000(2000).

18. Elsenbruch S, Harnish MJ, Orr WC. Subjective and objective sleep quality in irritable bowel syndrome. *Am J Gastroenterol* **94**:2447–2452(1999).

19. Robert JJ, Orr WC, Elsenbruch S. Modulation of sleep quality and autonomic functioning by symptoms of depression in women with irritable bowel syndrome. *Dig Dis Sci* **49**(7-8):1250–1258(2004).

20. Thompson JJ, Elsenbruch S, Harnish MJ, et al. Autonomic functioning during REM sleep differentiates IBS symptom subgroups. *Am J Gastroenterol* **97**:3147–3153(2002).

97

RENAL DISEASE

Kathy P. Parker

Nell Hodgson Woodruff School of Nursing, Atlanta, Georgia

INTRODUCTION

End-stage renal disease (ESRD) is a significant health problem in the United States. As of 2003, there were approximately 400,000 patients with renal failure requiring treatment with dialysis or transplantation and this number is expected to approach 700,00 by the year 2010 [1]. Although considerable progress has been made in these treatments, the quality of life of ESRD patients remains poor, an observation likely related to an overall decline in functional status and well-being. In addition, these individuals experience numerous untoward symptoms, including significant sleep disturbance [2]. Primary sleep disorders such as sleep apnea (SA), restless legs syndrome (RLS), and periodic limb movement disorder (PLMD) are also very common. This chapter reviews the major features, etiologies, and treatments of sleep problems in individuals with ESRD and the negative impact that these problems appear to have on important clinical outcomes.

SLEEP COMPLAINTS AND CORRELATES

Sleep complaints, characterized by difficulty initiating and maintaining sleep, problems with restless, jerking legs, and/or daytime sleepiness, are reported by up to 80% of ESRD patients surveyed [3, 4]. These problems are also among the most disturbing symptoms experienced and are consistently cited as major sources of stress and factors negatively affecting life quality [5]. The high prevalence of sleep complaints is likely related to a variety of factors as numerous demographic, clinical, and laboratory correlates have been identified. For example, sleep complaints are more common in elderly ESRD patients. Males are more likely to have sleep complaints than females and whites may have a higher prevalence of restless sleep than blacks. Sleep complaints have also been associated with caffeine intake, pruritus, bone pain, cigarette use, premature discontinuation of dialysis, anxiety, depression, and stress [3, 4, 6]. Although no consistent relationships between subjective sleep complaints and blood urea nitrogen (BUN), creatinine, or Kt/V (a measure of dialysis adequacy) have been detected, anemia has been associated with complaints of poor sleep with improvement noted after treatment with recombinant erythropoietin. Mild hypercalcemia has also been associated with the frequency of subjective insomnia episodes.

GENERAL POLYSOMNOGRAPHIC FEATURES OF SLEEP

The earliest polysomnographic examinations of ESRD patients, done in the late 1960s and early 1970s, found that their sleep was characterized by decreased total sleep time, irregular sleep cycles, and long periods of interspersed waking. More recent studies have confirmed that the sleep of these patients is indeed short and fragmented with total sleep times ranging between 260 and 360 minutes, sleep efficiencies between 66% and 85%, large amounts of wake time (77–135 minutes), and numerous arousals (25–30/hour of sleep). Increased amounts of stage 1 and stage 2 sleep and decreased slow-wave sleep (SWS) and rapid

Sleep: A Comprehensive Handbook, Edited by T. Lee-Chiong.
Copyright © 2006 John Wiley & Sons, Inc.

eye movement (REM) sleep have been reported. Sleep latencies ranged between 10 and 30 minutes and REM latencies were between 92 and 164 minutes. Some of the variability in these reports may be related to differences in the age of subjects and the degree to which variables known to affect sleep quality, such as comorbidity, medications, and the presence or absence of sleep disorders, were controlled. Nonetheless, the data provide convincing evidence that ESRD patients have overall reduced quantity and quality of sleep.

SLEEP APNEA SYNDROME

Sleep apnea is a syndrome characterized by intermittent episodes of breathing cessation during sleep, either from airway collapse (obstructive sleep apnea), cessation of respiratory effort (central sleep apnea), or a combination of the two (mixed type). Prevalence rates in ESRD patients are estimated to be approximately 50% [7], a rate significantly greater than that reported for the general population. The apneas are most commonly of the obstructive type, but central and mixed events have also been observed. Apneas occur in both REM and NREM sleep with oxygen desaturations typically ranging between 80% and 85%. The type and severity of the apneas do not appear to vary substantially with type of dialysis (see Table 97.1) [8, 9] or between nights "on" and nights "off" treatment. No consistent relationships between apnea indices and biochemical measures such as BUN, creatinine, hematocrit, waking arterial blood gases, or dialysis adequacy have been described [8, 10].

The mechanisms underlying the increased prevalence of SA in ESRD patients remain to be well characterized. Hypocapnia from metabolic acidosis and acidemia may change the apnea-P_{CO_2} or hydrogen ion threshold and predispose the patients to an unstable breathing pattern [11, 12]. In addition, accumulation of uremic toxins may affect

TABLE 97.1   Types of Dialysis

Continuous Ambulatory Peritoneal Dialysis (*CAPD*). An ongoing, continuous dialysis process that involves infusion of fluid into the peritoneum, a prolonged dwell period for dialysis, and drainage. The procedure typically involves four exchanges of fluid daily.

Hemodialysis (*HD*). A process of removal of fluid and solutes from the blood through a semipermeable membrane into dialysate by passing the blood though an artificial kidney. Hemodialysis is most commonly delivered to patients three times a week for 3–4 hours, but may also be given more slowly across the day or night by reducing blood and dialysate flows.

Peritoneal Dialysis (*PD*). The process of removal of fluid and wastes from the body using the semipermeable membrane of the peritoneum for diffusion and osmosis.

the central nervous system and result in a reduction of airway muscle tone during sleep, a discoordination of diaphragm and upper airway muscle activity, or an instability of respiratory control [13]. Many patients also have peripheral neuropathy and edema from fluid overload, making them more likely to have upper airway collapse. Anemia, hormone abnormalities, cytokine production during dialysis, and the mechanical effects of peritoneal fluid on diaphragmatic action may all contribute to ventilatory control instability.

Treatment for sleep apnea has been successfully accomplished in ESRD patients with continuous positive airway pressure (CPAP) resulting in stabilization of nocturnal oxygenation, decreases in stage 1 sleep, and reports of increased daytime alertness. However, the long-term impact has not been assessed and improvements in cognition and other aspects of functional status remain to be described. Cure after transplantation has also been reported, providing additional evidence that factors related to renal disease and/or its treatment precipitate SA in this population. A decrease in apnea has also been reported with the use of bicarbonate versus acetate-based dialysate and both slow nocturnal and intensive daily hemodialysis.

RESTLESS LEGS SYNDROME AND PERIODIC LIMB MOVEMENT DISORDER

Restless legs syndrome (RLS) is a disorder characterized by disagreeable sensations that usually occur prior to sleep onset and that cause an almost irresistible urge to move the legs. Periodic limb movement disorder (PLMD) is characterized by periodic episodes of repetitive limb movements that occur during sleep, often disrupting its quality. RLS is an extremely distressing problem experienced by ESRD patients and up to 80% experience symptoms. The prevalence of PLMD in ESRD patients, both in association with RLS and as an independent condition, is also high and may approach 70% [14].

Information regarding the clinical and laboratory correlates of RLS and PLMD in ESRD patients is limited. Higher predialysis urea and creatinine levels have been associated with increased RLS complaints in some, but not all studies. Higher intact parathyroid hormone (PTH) levels have been observed in association with periodic limb movements while lower PTH levels were noted in those with RLS. Although normalization of hematocrit with recombinant erythropoietin has resulted in a significant reduction in periodic limb movements in a sample of dialysis patients [15], no specific relationship between RLS symptoms and anemia has been detected. Additional information suggests that RLS severity is also unrelated to age, gender, body weight, number of years on dialysis, or median, ulnar, and sural nerve amplitudes [16].

Several factors intrinsic to the uremic state may predispose ESRD patients to these disorders. Anemia secondary to decreased production of endogenous erythropoietin and reduced iron stores from dietary restrictions and blood loss during dialysis may be important risk factors. In addition, vitamin deficiencies (particularly water-soluble vitamins such as folate and B_{12}) may occur secondary to poor intake, interference with absorption by drugs or uremia, altered metabolism, and losses to the dialysate. Peripheral neuropathy associated with uremia and/or diabetes and skeletomuscular abnormalities related to secondary hyperparathyroidism may predispose dialysis patients to RLS/PLMD. Alterations of dopamine (DA) and opioid synthesis/metabolism may also play a role. The beneficial effects of treatment with L-dopa and DA_2 agonists in ESRD patients with RLS/PLMD indeed suggest that DA pathways are involved. Abnormalities in the endogenous opioid system have also been reported but clinical trials evaluating both the safety and efficacy of opioids in the treatment of RLS in this population have not been undertaken.

Data regarding the effective treatment of RLS and PLMD in ESRD patients are limited. Reports from three prospective clinical trials of L-dopa cite reduction of RLS symptoms and improved nocturnal sleep, decreased numbers of nocturnal limb movements, and improved subjective quality of life. Although rebound (return) and augmentation (worsening) of symptoms have been reported with L-dopa in other clinical populations, none of these studies mentioned these phenomena. In a double-blind, placebo-controlled, crossover study, treatment with pergolide (a DA_2 agonist) resulted in decreased symptoms but had no effect on objective indicators such as numbers of nocturnal limb movements and sleep architecture. Favorable responses to treatment with clonazepam and clonidine in ESRD patients have also been described. Alleviation of RLS symptoms was reported to occur in one patient after renal transplantation.

EXCESSIVE DAYTIME SLEEPINESS

Excessive daytime sleepiness (EDS), or difficulty maintaining the alert, awake state, has been anecdotally described in ESRD patients for decades. However, although numerous sleep-related investigations conducted with these patients specifically mention EDS as an important complaint, the phenomenon remains to be well studied. To date, only three studies collecting objective measurements of EDS using the Multiple Sleep Latency Test (MSLT) have been conducted, but all demonstrated that the problem is highly prevalent and probably underrecognized (17–19). The high prevalence of SA, PLMD, and RLS in the dialysis populations provides the most parsimonious explanation for EDS.

Arousals from sleep and nocturnal sleep disruption related to these conditions have certainly been implicated as important causes of EDS.

Other factors, however, possibly those related to renal disease and/or its treatment, may also contribute to EDS [2, 7]. The subclinical uremic encephalopathy commonly present in dialysis patients may play a role in making patients more susceptible to sleepiness. For example, mild elevations of BUN and creatinine in renal failure patients have been associated with increased slow-wave activity in the waking EEG and abnormalities in cognitive function. Elevations of parathyroid hormone, the neurotoxic effects of which have been well described, have also been associated with increased waking EEG slow-wave activity in uremic animals and in stable ESRD patients. Low plasma levels of tyrosine (precursor of norepinephrine and DA, neurotransmitters important in neurologic arousal) have also been reported in uremia and may contribute to abnormalities in neurotransmitter synthesis.

Treatment with dialysis may predispose patients to sleepiness. Abnormal production of interleukin-1 and TNF-α, substances with somnogenic properties, has been demonstrated in response to dialysis. Rapid changes in serum electrolyte, acid-base balance, and serum osmolarity occur and may decrease arousal and alertness. Removal of sleep-promoting substances during treatment has also been suggested as a possible mechanism for fragmenting nocturnal sleep and causing daytime fatigue and sleepiness. Treatment may disrupt the circadian pattern sleepiness via inappropriately timed elevations of serum melatonin in response to the hemoconcentration or from changes in the rhythm of body temperature [2, 7]. Other factors, including medications commonly prescribed such as antihypertensives and antidepressants, may also contribute to the EDS experienced by these patients.

IMPACT OF SLEEP PROBLEMS

Sleep abnormalities are associated with reduced quality of life and functional status in ESRD patients. In addition, these problems have been related to several other important clinical outcomes such as the ability to learn and perform home dialysis, quality of family interactions, anxiety, depression, and days of disability. Thus it seems reasonable to suggest that unless sleep problems can be recognized and treated effectively, the rehabilitative potential of therapies for ESRD may be seriously compromised.

The long-term consequences of untreated sleep problems in ESRD patients remain to be described. However, a link between SA and cardiovascular disease has already been established in the general population. Because cardiovascular disease is the leading cause of death in the ESRD population, the detection and appropriate management of

SA may have important effects on survival. The nocturnal hypoxemia associated with apnea may also dampen autonomic reflexes and be linked to the autonomic dysfunction often noted in these patients. In addition, both RLS and PLMD have been associated with increased mortality in ESRD [20].

CONCLUSIONS

Sleep complaints and primary sleep disorders are very prevalent in ESRD patients and appear to have important adverse effects on their overall health and well-being. Therefore the effective assessment and management of these problems has the potential to significantly enhance patient outcomes. Research designed to identify the mechanisms underlying these sleep problems and evaluate the effects of treatment is greatly needed.

REFERENCES

1. USRDS. *United States Renal Data System 2003 Annual Data Report: Atlas of End-Stage Renal Disease in the United States.* National Institutes of Health, National Institute of Diabetes and Digestive and Kidney Diseases, Bethesda, MD, 2003.

2. Parker KP. Sleep disturbances in dialysis patients. *Sleep Med Rev* **7**(2):131–143(2003).

3. Holley JL, Nespor S, Rault R. A comparison of reported sleep disorders in patients on chronic hemodialysis and continuous peritoneal dialysis. *Am J Kidney Dis* **19**(2):156–161(1992).

4. Walker S, Fine A, Kryger MH. Sleep complaints are common in a dialysis unit. *Am J Kidney Dis* **26**(5):751–756(1995).

5. Mucsi I, Molnar MZ, Rethelyi J, Vamos E, Csepanyi G, Tompa G, et al. Sleep disorders and illness intrusiveness in patients on chronic dialysis *Nephrol Dial Transplant* **19**(7):1815–1822(2004).

6. Kutner NG, Bliwise DL, Brogan D, Zhang R. Race and restless sleep complaints in older chronic dialysis patients and nondialysis community controls. *J Gerontol Ser B Psychol Sci Soc Sci* **56**(3):P170–P175(2001).

7. Hanly P. Sleep apnea and daytime sleepiness in end-stage renal disease. *Semin Dial* **17**(2):109–114(2004).

8. Hallett MD, Burden S, Stewart D, Mahony J, Farrell PC. Sleep apnea in ESRD patients on HD and CAPD. *Perit Dial In* **16**(Suppl 1):S429-S433(1996).

9. Wadhwa NK, Mendelson WB. A comparison of sleep-disordered respiration in ESRD patients receiving hemodialysis and peritoneal dialysis. *Adv Perit Dial* **8**:195–198(1992).

10. Wadhwa NK, Seliger M, Greenberg HE, Bergofsky E, Mendelson WB. Sleep related respiratory disorders in end-stage renal disease patients on peritoneal dialysis. *Perit Dial Int* **12**(1):51–56(1992).

11. Kimmel PL, Miller G, Mendelson WB. Sleep apnea syndrome in chronic renal disease. *Am J Med* **86**(3):308–314(1989).

12. Mendelson WB, Wadhwa NK, Greenberg HE, Gujavarty K, Bergofsky E. Effects of hemodialysis on sleep apnea syndrome in end-stage renal disease. *Clin Nephrol* **33**(5):247–251(1990).

13. Fein AM, Niederman MS, Imbriano L, Rosen H. Reversal of sleep apnea in uremia by dialysis. *Arch Intern Med* **147**(7):1355–1356(1987).

14. Pressman MR, Benz RL, Peterson DD. High incidence of sleep disorders in end-stage renal disease patients (abstract). *Sleep Res* **24**:417A(1995).

15. Benz RL, Pressman MR, Hovick ET, Peterson DD. A preliminary study of the effects of correction of anemia with recombinant human erythropoietin therapy on sleep, sleep disorders, and daytime sleepiness in hemodialysis patients (The SLEEPO study). *Am J Kidney Dis* **34**(6):1089–1095(1999).

16. Winkelman JW, Chertow GM, Lazarus JM. Restless legs syndrome in end-stage renal disease. *Am J Kidney Dis* **28**(3):372–378(1996).

17. Stepanski E, Faber M, Zorick F, Basner R, Roth T. Sleep disorders in patients on continuous ambulatory peritoneal dialysis. *J Am Soc Nephrol* **6**(2):192–197(1995).

18. Parker KP, Bliwise DL, Bailey JL, Rye DB. Daytime sleepiness in stable hemodialysis patients. *Am J Kidney Dis* **41**(2):394–402(2003).

19. Hanly PJ, Gabor JY, Chan C, Pierratos A. Daytime sleepiness in patients with CRF: impact of nocturnal hemodialysis. *Am J Kidney Dis* **41**(2):403–410(2003).

20. Unruh ML, Levey AS, D'Ambrosio C, Fink NE, Powe NR, Meyer KB. Restless legs symptoms among incident dialysis patients. association with lower quality of life and shorter survival. *Am J Kidney Dis* **43**(5):900–909(2004).

98

ENDOCRINE AND METABOLIC DISORDERS AND SLEEP

ALEXANDROS N. VGONTZAS AND SLOBODANKA PEJOVIC
Penn State College of Medicine, Hershey, Pennsylvania

GEORGE P. CHROUSOS
National Institutes of Health, Bethesda, Maryland

OBESITY

Obesity is a very common metabolic disorder caused by a combination of environmental, constitutional, and genetic factors. It is also frequently associated with sleep disorders and disturbances and a large proportion of patients evaluated in sleep disorders clinics are obese. Thus it is important for the sleep specialist to be familiar with this disorder and associated sleep problems.

Obesity and Sleep Apnea

Obesity is highly prevalent in western industrialized countries. The prevalence of overweight and obesity has increased significantly in the last two decades in the United States. About 65.1% of American adults are overweight (body mass index (BMI) $\geq$ 25), while 30.4% are obese with a BMI $\geq$ 30 (with 33.2% women and 27.6% men affected). Among children aged 6–19 years, 16% are overweight (BMI $\geq$ 95th percentile of the sex-specific BMI-for-age growth chart) [1]. The association of obesity with sleep apnea was observed and reported since the mid-1960s [2–4]. It was noticed that the onset of sleep apnea frequently follows a marked increase in body weight. In turn, the development of sleep apnea seems to lead to further weight

Correspondence to: Alexandros N. Vgontzas, Penn State College of Medicine, Department of Psychiatry H073, 500 University Drive, Hershey, PA 17033.

gain, which, in a vicious cycle, leads to further deterioration of sleep apnea.

The prevalence of sleep apnea in obese clinical populations appears to be as high as 40% for morbidly obese men (BMI $>$ 39) and 3% for premenopausal morbidly obese women [5]. Another 5–8% demonstrate sleep apneic activity that warrants further follow-up in the sleep disorders clinic and sleep laboratory. The best clinical predictors of sleep apnea in obese populations seem to be severity of snoring, subjectively reported nocturnal breath cessation, and sleep attacks during the day. The high prevalence of sleep apnea in obese men indicates that severely or morbidly obese middle-aged men are at an extremely high risk for sleep apnea and should be routinely referred for a sleep laboratory evaluation to rule out this condition. Physicians assessing severely or morbidly obese women should include a detailed sleep history.

A strong association between obesity and sleep apnea was shown in large studies on the prevalence of sleep apnea in general, randomized samples by Young et al. [6] and Bixler et al. [7]. Interestingly, in samples of the general population, only 65% of those with apnea met the criteria of obesity [7], in contrast to clinical populations where 85% of sleep apneics were obese. It is not clear why obese people are overrepresented in the clinical population; however, one possibility is that nonobese patients with sleep apnea exhibit a lesser degree of daytime sleepiness, which is one of the most frequent symptoms for which patients seek evaluation and treatment.

Among obese patients with sleep apnea, it appears that central obesity is the predominant pattern in body fat distribution. More recently, studies using imaging techniques suggest that, among obese individuals, it is intra-abdominal (visceral) fat and not generalized obesity that predisposes to the development of sleep apnea [8, 9]. One controlled study showed that sleep apneics had a significantly higher amount of visceral fat compared to obese control subjects, although there were no differences between the two groups in terms of BMI or amount of subcutaneous or total body fat [9]. Increased visceral adiposity has been associated with several metabolic abnormalities, including insulin resistance, displipidemia, diabetes mellitus type II, hypertension, and cardiovascular problems [10]. Visceral adiposity also appears to play a strong role in the pathogenesis of sleep apnea and manifestations that frequently accompany this disorder, namely, hypertension and its sequelae [11].

Weight reduction in obese patients with sleep apnea is associated with an improvement in sleep disordered breathing [12], although the effects on daytime sleepiness are inconsistent [13, 14]. Furthermore, most obese people are unable to achieve significant weight loss and, even then, the results are inconsistent. No predictive factors have as yet been detected to guide clinicians on the success or failure of weight loss on sleep apnea. It is possible that the critical factor in determining a successful outcome of weight management is reduction of visceral fat rather than total weight loss. Reduction of visceral fat has been associated with exercise or use of medication (e.g., insulin sensitizing agent, glucosidase inhibitor) [15–17]. These approaches might be of clinical significance given the high rate of failure of treatments such as surgery and a low rate of compliance such as CPAP.

Obesity, Sleep Disturbance, and Daytime Sleepiness

Obesity not associated with sleep apnea is frequently a cause of sleep disruption and daytime sleepiness and fatigue. It has been demonstrated that obese patients, both men and women, without any sleep disordered breathing demonstrate a significant degree of nocturnal sleep disturbance compared with nonobese controls [5]. Specifically, wake time after sleep onset, number of awakenings, and percentage of stage 1 sleep are significantly higher in obese patients than in controls, while rapid eye movement (REM) sleep is significantly lower. More recent studies have shown that obesity is associated with bimodal sleep patterns; that is, some obese patients report short sleep duration whereas others long sleep duration [18, 19]. It is not clear what is different between the two groups. However, it has been suggested that those who sleep longer are morbidly obese [19]. It should be noted that these findings are based on subjective reports and not on objective measures of sleep within a 24 hour sleep–wake cycle.

Other studies either experimental or epidemiologic have shown that short sleep duration is associated with increased appetite and a disturbance of satiety hormones, that is, decreased leptin levels and increased ghrelin levels [18, 20]. Also, in a prospective study, sleep has been associated with weight gain in young adults [21]. Furthermore, in children several cross-sectional studies have shown that short sleep duration is a risk factor for obesity [22–24]. The authors have speculated that sleep loss is one of the paths leading to the current epidemic of obesity in western societies.

It appears that obesity per se is associated with excessive daytime sleepiness and might be a significant factor underlying the current epidemic of fatigue and sleepiness in modern societies. One study using objective measures for assessment of daytime sleepiness demonstrated that obese patients without sleep apnea compared with controls were sleepier during the day as indicated by shorter sleep latencies, decreased wake time after sleep onset, and total wake time, whereas the percentage of sleep time was significantly higher [25]. Two other studies showed similar findings in regard to independent contribution of obesity in sleepiness [26, 27]. These findings are consistent with a recent study on the prevalence of excessive daytime sleepiness in a large general, randomized sample that suggests that obesity is a significant risk factor for excessive daytime sleepiness independent of sleep disordered breathing [28]. Interestingly, in one study postsurgical weight loss was associated with improvement of sleep disordered breathing but not daytime sleepiness. It appears that obesity per se should be considered in the differential diagnosis of patients who present with a primary complaint of daytime sleepiness. Given the impact of daytime sleepiness on the patients' lives, as well as on public safety, the nighttime and daytime sleep patterns of obese patients should be thoroughly evaluated by every physician.

Prader–Willi syndrome, a rare genetic disorder characterized by mental retardation, neonatal hypotonia, extreme hyperphagia, and obesity, is frequently associated with sleep disordered breathing, daytime sleepiness, and early onset of REM sleep [29, 30]. In this case, also, daytime sleepiness and early REM sleep appear to be independent of sleep disordered breathing and even obesity [31].

Obesity, Sleep Apnea, Sleepiness, and Cytokines

New research suggests that the proinflammatory cytokines TNF-α and IL-6 may play a role in mediating sleepiness in patients with sleep apnea and obesity [9, 32]. In controlled studies, it has been shown that sleep apneic men had higher plasma concentrations of the inflammatory, fatigue-causing cytokines TNF-α and IL-6 than nonapneic obese men, who had intermediate values, or lean men, who had the lowest values.

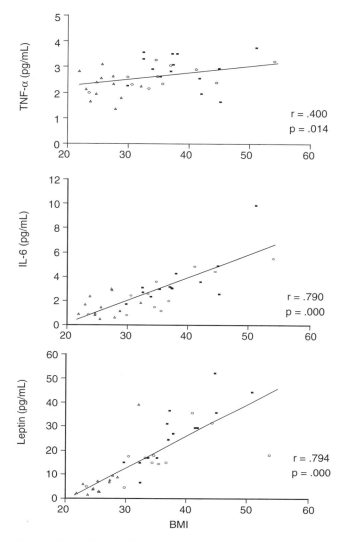

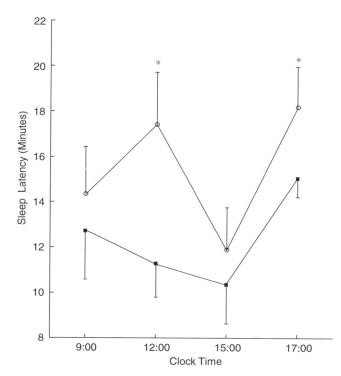

Figure 98.1 Plasma TNF-α, IL-6, or leptin levels (evening values) and BMI are positively correlated. ■ Sleep apneics; ○ obese controls; ▲ lean controls. (Reprinted from [9] by permission of The Endocrine Society.)

Figure 98.2 Sleep latencies during daytime testing with MSLT in the placebo (■) and etanercept (○) conditions. Each data point represents the mean ± SE (*$P < 0.05$ adjusted change between placebo and etanercept). (Reprinted from [33] by permission of The Endocrine Society.)

DIABETES AND INSULIN RESISTANCE

It has been suggested that sleep has a modulating effect on metabolic regulation [34] and that sleep debt has a harmful impact on carbohydrate metabolism [35]. Yet our knowledge on the association between diabetes mellitus and sleep disturbances is limited.

Diabetes, Sleep, and Disorders of Excessive Sleepiness

Few epidemiologic studies have assessed the prevalence of sleep complaints associated with diabetes. In a large study of a random sample of 3201 Swedish men, diabetes was associated with more frequent complaints of difficulty initiating sleep (21.1%), difficulty maintaining sleep (21.9%), and excessive daytime sleepiness (12.2%) [36]. Increased prevalence of sleep disturbances, primarily difficulty initiating sleep, was reported also in a clinical group of 184 diabetics compared to controls [37]. Also, data from a recent prospective study showed that both short and long sleep durations are significant predictors of developing type II diabetes [38].

In sleep laboratory studies, diabetes appears to be associated with sleep disruption, in both children and adults.

Also, BMI positively correlated to the plasma levels of TNF-α and IL-6 [9] (see Figure 98.1). In a recent study, and in order to test the hypothesis that these proinflammatory cytokines are mediators of sleepiness, etanercept, a medication that neutralizes TNF-α, was administered in obese apneic men. There was a significant reduction of sleepiness as measured by the Multiple Sleep Latency Test (MSLT) and a slight but significant reduction of apnea–hypopnea index (AHI) [33]. The latter finding suggests that the reduction of sleepiness was not related to the improvement of apnea but rather to the neutralizing effect of TNF-α and IL-6 (see Figure 98.2).

Specifically, polysomnographically recorded sleep of diabetic patients was associated with more wakefulness, high number of awakenings, and fragmented sleep [39, 40]. The exact cause of this sleep disruption is not well understood. In one study, it was shown that sleep disruption in a type II diabetic population was mediated by physical complications of the disease, such as pain and nocturia, and not by either obesity or emotional adjustment [39]. In children, the sleep disruption was not associated with nocturnal hypoglycemia [40].

Several studies have shown an increased prevalence of sleep apnea and sleep disordered breathing in patients with diabetes mellitus type II. In 1985, Mondini and Guilleminault [41] reported increased frequency of abnormal breathing during sleep in lean and obese diabetics. In 1994, Brooks et al. [42] demonstrated that 70% of obese diabetics had moderate or severe obstructive sleep apnea. In a Chinese population with obstructive sleep apnea, diabetes mellitus was the second most common medical condition (about 10%), next to hypertension, associated with sleep apnea [43]. It appears that obstructive sleep apnea is more prevalent in diabetic patients with autonomic neuropathy than in those without [44]. However, the results of these studies are limited because of a sample size that was either too small or biased, or the absence of an appropriate control group. Two large prospective studies, the one from Sweden and the other from the United States (Nurses' Health Study Cohort) showed that regular snoring is associated with a two- to sevenfold risk for type II diabetes over a 10 year period [45, 46]. Furthermore, Bixlar et al. in a recent analysis of a large random sample of men and women (1741) showed that diabetes, independent of sleep apnea, BMI, and age, was a risk factor for the complaint of excessive daytime sleepiness [28].

In a single study by Honda et al. [47], it was reported that non-insulin dependent diabetes mellitus was more frequent among narcoleptic patients. Also, an earlier report by Roberts [48] indicated "diabetogenic hyperinsulinism" and reduced glucose tolerance in "many narcoleptics." These results are interesting given that these narcoleptic patients were nonobese, as well as in light of the recent findings that (1) proinflammatory cytokines are elevated in disorders of excessive daytime sleepiness, including narcolepsy [9, 32, 49] and (2) hypercytokinemia is associated with metabolic dysregulation [9].

Role of Insulin Resistance in the Pathogenesis of Sleep Apnea

Insulin resistance is defined as the impaired ability of insulin (either endogenous or exogenous) to lower blood glucose [50]. Insulin resistance is very prevalent in the general population (about 25%) and is strongly associated with central obesity [51]. Sleep apnea based on clinical and sleep laboratory criteria is present in 4% of men and 2% of women, whereas milder forms are present in 10–15% of the general population [52]. Based on the above associations, as well as that medical conditions commonly associated with central obesity (i.e., hypertension) are frequent in sleep apnea, it is plausible that there is an association between insulin resistance and sleep apnea.

Earlier studies reported inconsistent results in terms of an association between sleep apnea and insulin resistance. One study showed a modest correlation between severity of sleep apnea and indices of insulin resistance [53]. In contrast, two controlled studies suggested the relation between sleep apnea and plasma insulin levels or insulin resistance reflected the known effects of obesity [54, 55]. This controversy was resolved in a well-controlled study that included three groups—obese sleep apneics, obese without sleep apnea, and normal weight controls [9]. The first two groups were matched for weight, whereas all three groups were matched for age and sex. None of the sleep apneics or the obese controls had developed overt diabetes. Those with sleep apnea had significantly higher levels of fasting plasma insulin and glucose levels compared to their obese controls (see Figure 98.3). Also, the group of sleep apneics demonstrated a higher degree of visceral but not subcutaneous fat compared to their obese controls. Based on these findings, it was suggested that visceral obesity/insulin resistance determined by both genetic and constitutional/environmental factors may be the principal culprits of obstructive sleep apnea, progressively leading to worsening metabolic syndrome manifestations and sleep apnea [56] (see Figure 98.4). Further support of the association of metabolic syndrome and sleep apnea is provided by a recent study that showed that the age distribution of this syndrome is similar to the age distribution of symptomatic

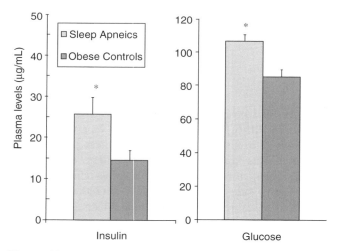

Figure 98.3 Plasma insulin and glucose are higher in sleep apneics than obese controls. $^{*}P < 0.05$. (Reprinted from [9] by permission of The Endocrine Society.)

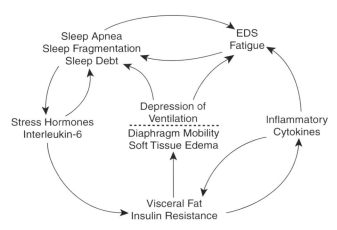

Figure 98.4 A heuristic model of the association between visceral fat/insulin resistance, cytokines, stress hormones, EDS, and sleep apnea. (Reprinted from [56] by permission of Blackwell Publishing.)

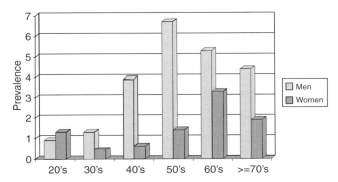

Figure 98.5 The age distribution of sleep apnea by decade (apnea–hypopnea index ≥ 10 and presence of daytime symptoms) peaks in the sixth decade for men (□) and the seventh for women (■). (Reprinted from [57] by permission of Elsevier.)

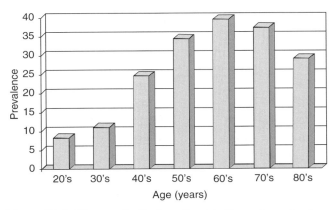

Figure 98.6 The age distribution of metabolic syndrome by decade in men in the U.S. population based on data from the Third National Health and Nutrition Examination Survey, 1988–1994. (Reprinted from [58] by permission of the American Medical Association.)

sleep apnea (i.e., apnea, sleepiness, hypertension) (see Figures 98.5 and 98.6).

Similarly, two subsequent studies that employed larger samples reported on association between sleep apnea and insulin resistance independently of obesity. Importantly, in the first study authors observed that the association between sleep apnea and insulin resistance was present even in nonobese subjects, while in the other study it was reported that insulin resistance was present even in the mild forms of sleep apnea [59, 60]. In a recent large study based on community-dwelling subjects, The Sleep Heart Health Study, sleep disordered breathing was independently associated with glucose intolerance and insulin resistance [61].

Women with polycystic ovary syndrome (PCOS) in whom insulin resistance is a primary pathophysiologic abnormality had a much higher prevalence of sleep apnea and daytime sleepiness [62]. Importantly, insulin resistance was a better predictor than BMI or testosterone of sleep apnea. These findings support the conclusion that sleep apnea in obese patients is a manifestation of a metabolic abnormality, and attempts to understand and correct this metabolic abnormality may lead to more successful, cost-effective, and better tolerated treatments for this disorder.

REPRODUCTION

Male Sex Hormones, Sleep, and Sleep Apnea

Early studies demonstrated that men had sleep that was more disturbed in terms of greater number of awakenings compared to women [63, 64]. Several studies have reported that men compared with age-matched women tend to lose slow-wave sleep (SWS) faster [65, 66] and that they have lower power density on sleep EEG in delta and theta frequency [67]. Also, a recent analysis of men and women from large general population samples showed that men had a poorer quantity (total sleep time) and quality (SWS and REM sleep) of sleep [68, 69].

The early observation that male gender is a risk factor associated with sleep apnea led to the belief that sex hormones play a significant role in the pathogenesis of sleep apnea. In the middle 1980s, investigators demonstrated that in men exogenous administration of testosterone induced sleep apnea [70, 71]. Also, circulating testosterone levels were shown to be associated with upper airway collapsibility in patients with sleep apnea [72]. Furthermore, in women sleep apnea was induced by exogenous administration of testosterone [73] and resolved after removal of testosterone-producing tumors [74]. In eleven hypogonadal men, the administration of testosterone was associated with a significant increase in apnea in the group as a whole, but clinically significant apnea occurred in only three patients

[71]. Also, short-term administration of high-dose testosterone worsens sleep apnea in older men [75]. Men taking testosterone for sexual dysfunction or for "andropause" should be monitored during the course of therapy for the development of symptoms of sleep apnea. Interestingly, agents lowering androgen levels did not improve sleep disordered breathing or awake ventilatory drive [76].

Testosterone levels appear to be low in men with sleep apnea [77]. Conversely, testosterone levels appear to increase following treatment of sleep apnea using nasal CPAP [77]. Interestingly, despite the fall in plasma total and free testosterone levels, there was no increase in basal plasma gonadotropin [luteinizing hormone (LH) or follicle stimulating hormone (FSH)] levels. The mechanism of the drop of testosterone levels in sleep apnea is not well understood; however, it appears that hypoxia and sleep fragmentation may be the principal factors. It is possible that the frequently reported sexual dysfunction in men with sleep apnea may be mediated by the sex hormone changes. In any event, middle-aged men with a complaint of sexual dysfunction and low testosterone levels should be assessed thoroughly for the presence of sleep apnea before the institution of hormone "replacement" therapy.

Female Sex Hormones, Sleep Apnea, and Insomnia

In contrast to the role of male sex hormones, female sex hormones appear to be protective of obstructive sleep apnea. For example, women experience an increase in ventilatory drive during the luteal phase of the menstrual cycle when progesterone levels are the highest [78]. Also, oral progesterone has been associated with slight but definite improvement in ventilatory indices during sleep in both male and female patients with sleep apnea [79].

From clinical and epidemiologic studies, sleep apnea appears to be rather infrequent before menopause, whereas its prevalence reaches almost the prevalence in men matched for age and BMI postmenopausally [52] (see Figure 98.6). Specifically, the prevalence of sleep apnea in premenopausal women is about 0.6%. Interestingly, premenopausal women with sleep apnea are exclusively obese. The prevalence of sleep apnea in postmenopausal women not receiving hormone replacement therapy is significantly higher than the prevalence in postmenopausal women undergoing hormone replacement therapy. In contrast, the prevalence of sleep apnea in postmenopausal women undergoing hormone replacement therapy remains as low as premenopausally (0.5%). The adverse effects of menopause and the protective role of gonadal hormones in sleep apnea were confirmed in two recent studies—the Sleep Heart Health Study as well as in a Wisconsin Cohort [80, 81]. Furthermore, a recent study from the Women's Initiative Hormone Trial reported that estrogen plus progestin

decreased diabetes and insulin resistance in postmenopausal women [82], which might be a mechanism through which hormone therapy protects women from sleep apnea. These findings indicate that, during menopause, there is a significant risk for sleep apnea and that hormone replacement appears to be associated with reduced risk. Although these data suggest that hormone replacement therapy appears to protect women from the development of sleep apnea and its potential cardiovascular sequelae, studies assessing the effects of female hormones after the development of sleep apnea are inconsistent. It is plausible that although female hormones are protective of sleep apnea, they are ineffective once apnea has developed.

Polycystic ovary syndrome (PCOS) is the most common endocrine disorder of premenopausal women and is characterized by chronic hyperandrogenic oligoanovulation and oligoamenorrhea [83]. A recent study demonstrated that premenopausal women with a diagnosis of PCOS are at a much higher risk for development of sleep apnea and daytime sleepiness [62]. Specifically, in a clinical sample, PCOS patients were 30 times more likely to suffer from sleep disordered breathing than the controls (see Figure 98.7). About 17% of PCOS women were recommended treatment for sleep disordered breathing in contrast to only 0.6% of the control group. Obstructive sleep apnea was present even in nonobese PCOS women in contrast to premenopausal controls. In addition, PCOS women more frequently report daytime sleepiness than the controls (80% versus 25%). Furthermore, in this study, the use of oral contraceptives seems to protect PCOS women from development of sleep disordered breathing, which is consistent with the protective role of hormone replacement therapy in postmenopausal women. Findings on the association of PCOS and sleep apnea were confirmed in two other studies [84, 85] that reported a 44% and 21% prevalence of sleep disordered breathing in PCOS women, respectively. Furthermore, it has been shown that sleep apnea in PCOS

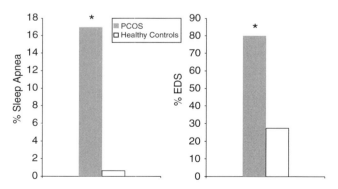

Figure 98.7 Prevalence of obstructive sleep apnea and excessive daytime sleepiness (EDS) in women with PCOS is markedly higher compared to healthy controls. (Reprinted from [56] by permission of Blackwell Publishing.)

women is independent of obesity [85]. Given that PCOS is associated with insulin resistance and central obesity [83], these data suggest that visceral obesity and insulin resistance, determined both by genetic and developmental/environmental factors, are at the basis of the pathogenetic mechanisms leading to sleep apnea.

Insomnia is more prevalent in women compared to men, which is associated with the fact that depression, a major risk factor for insomnia, is twice as prevalent in women compared to men [68]. The increased prevalence of insomnia in women aged 45–54 has been attributed to menopause. However, studies using sleep laboratory measures have not confirmed a deterioration of sleep during postmenopausal years [68, 86]. Further studies are needed to clarify the complex interaction between depression, menopause, hormones, and age in the etiology of insomnia in women. Interestingly, recently, it has been reported that young premenopausal women and postmenopausal women on hormone replacement therapy are more resilient to external sleep disturbances compared to young men and postmenopausal women without hormone replacement therapy, respectively [87, 88].

HYPOTHALAMIC–PITUITARY–ADRENAL AXIS

The hypothalamic–pituitary–adrenal (HPA) axis is a major component of the stress system whose function is to preserve homeostasis. Corticotropin-releasing hormone (CRH), which is produced and released from parvocellular neurons of the paraventricular nucleus in the hypothalamus, is the key regulator of the HPA axis. The release of CRH is followed by the enhanced secretion of adrenocorticotropin hormone (ACTH) from the anterior pituitary and cortisol from the adrenal cortex.

Hypothalamic–Pituitary–Adrenal Axis and Sleep Disturbance

Both CRH and cortisol lead to arousal and sleeplessness in humans and animals [89, 90]. Conversely, sleep, particularly deep sleep, has an inhibitory effect on the stress system of which the HPA axis is a major component [91–93]. More recently, it was demonstrated that chronic, persistent insomnia was associated with 24 hour hypercortisolemia, which correlated with polysomnographic indices of the severity of the disorder [94]. These findings have been confirmed by two studies that found that cortisol was elevated in primary insomnia with objectively documented sleep disturbance during the evening and nighttime [95, 96]. These results indicate that chronic activation of the HPA axis in insomnia places these patients at risk not only for mental disorders (i.e., chronic anxiety and depression) but

also for significant medical morbidity associated with such activation (i.e., hypertension, visceral obesity with its associated metabolic syndrome, osteoporosis, and other). More recently, in a large epidemologic study, an association was demonstrated between insomnia and hypertension [97].

In another study, it was shown that the bolus intravenous administration of CRH during the first sleep cycle in middle-aged healthy men compared to young healthy men was associated with significant sleep disturbance consisting of increased wakefulness and decreased slow-wave sleep [98]. These findings suggest that changes of sleep physiology associated with middle-age play a significant role in the marked increase of prevalence of insomnia in middle-age.

The effects of the exogenously administered corticosteroids and/or their excessive endogenous release on sleep have been recognized for many years. Sleeplessness and insomnia is one of the most frequent side effects of the administration of glucocorticoids [89]. Also, sleep disturbances are frequent in conditions associated with endogenous hypercortisolemia (e.g., melancholic depression).

Adrenal Function, Fatigue, Sleepiness, and Sleep Apnea

Patients with Cushing's syndrome frequently complain of daytime fatigue and sleepiness. In one controlled study, it was demonstrated that Cushing's syndrome is associated with increased frequency of sleep apnea [99]. In that study, about 32% of patients with Cushing's syndrome were diagnosed with at least mild sleep apnea, and about 18% had significant sleep apnea (AHI > 17.5 events/hour). Interestingly, those patients with sleep apnea were not more obese or different in any craniofacial features compared to those patients without sleep apnea. These findings are interesting in light of the new findings that visceral obesity, which is prominent in Cushing's syndrome, is a predisposing factor to sleep apnea. Also, Cushing's syndrome in the absence of sleep apnea was associated with increased sleep fragmentation, increased stage 1 and wake, and decreased delta sleep [100].

Adrenal insufficiency or Addison's disease is associated with chronic fatigue. A recent study indicated that untreated patients with adrenal insufficiency demonstrate increased sleep fragmentation, increased REM latency, and decreased amount of time in REM sleep, findings that may explain the patients' fatigue [101]. These sleep abnormalities were reversed following treatment with a replacement dose of hydrocortisone. These results suggest that cortisol secretion may be needed to facilitate both initiation and maintenance of REM sleep. It should be noted that in normal individuals, exogenous glucocorticoids have been found to reduce REM sleep [102]. The authors interpreted their findings that the inhibitory role of glucocorticoids on REM sleep in normals, along with the permissive role in Addison's patients, demonstrate that some cortisol is needed for REM sleep, with excess cortisol inhibiting REM sleep.

Also, fatigue and excessive daytime sleepiness are prominent in patients with secondary adrenal insufficiency (e.g., steroid withdrawal) and African trypanosomiasis. The latter conditions were associated with elevated TNF-α and IL-6 levels [103, 104], which may be the mediators of the excessive sleepiness and fatigue associated with primary or secondary adrenal insufficiency.

GROWTH HORMONE DISORDERS

Acromegaly, Sleep Apnea, and Sleepiness

Acromegaly is a condition of growth hormone excess in adults characterized by the insidious development of coarsening of facial features, bony growth, and soft tissue swelling. It is usually secondary to growth hormone-producing pituitary adenoma.

Sleep apnea is frequent in acromegaly. Several studies have indicated a prevalence of sleep apnea that ranges from 40% to 91% [105, 106]. Sleep apnea is more common in acromegalic patients over 50 years old.

The etiology of sleep disordered breathing in acromegaly is reported to be associated with macroglossia and craniofacial changes [107]. However, these studies have not been consistent [108].

Interestingly, central sleep apnea has been observed frequently in patients with acromegaly [105]. This finding speaks in favor of the role of central effects of the hormone abnormalities of acromegaly on the respiratory control rather than a mere blocking effect of the peripheral manifestations of the disorder.

Hypertension is quite common in acromegalic patients with sleep apnea. In fact, in one study, it was reported that over 50% of patients with both acromegaly and sleep apnea had hypertension [105]. In contrast, none of the patients who did not have sleep apnea were hypertensive.

Mild growth hormone deficiency has been described in patients having sleep apnea. This deficiency appears to be secondary to sleep fragmentation or obesity associated with sleep apnea. The treatment of sleep apnea appears to improve the growth hormone and IGF-1 levels [36].

Pituitary surgery is the treatment of choice for acromegalic patients. Interestingly, sleep apnea persists despite treatment of acromegaly by pituitary surgery. Nonsurgical methods, such as the use of octreotide, a somatostatin analogue, are potentially useful in reducing growth hormone and improving apnea [109].

Daytime sleepiness has been recognized as a manifestation of acromegaly independently of the presence of sleep apnea. The mechanisms underneath this phenomenon are not well understood. Growth hormone excess, cerebral irradiation, or acromegalic myopathy may explain the sleepiness complaint in the absence of sleep apnea.

THYROID FUNCTION

Hypothyroidism, Sleep, and Sleep Apnea

The first case of sleep apnea in a patient with myxedema was reported in the early 1960s. This report was later confirmed by others, and it has been suggested that myxedema coma was probably due to severe sleepiness secondary to severe sleep apnea.

Because hypothyroidism and sleep apnea share several common clinical features (i.e., fatigue and sleepiness), several studies have attempted to delineate the association between sleep apnea and hypothyroidism. Studies on the prevalence of sleep apnea on diagnosed hypothyroid patients reported as high as a 50% prevalence of obstructive sleep apnea (mild to severe) [110]. However, these studies did not control well for the presence of obesity and male gender. More recent studies, which assessed the prevalence of undiagnosed hypothyroidism in patients referred for evaluation of sleep apnea, reported a rather low prevalence of clinical hypothyroidism (1.2–3.1%) [111, 112]. Based on these findings, the authors concluded that thyroid screening does not appear to be appropriate for patients with suspected or confirmed obstructive sleep apnea in the absence of signs or symptoms consistent with hypothyroidism. However, in most of these studies, the prevalence of hypothyroidism in patients referred for obstructive sleep apnea was higher, although not significant, compared to controls. A recent study reported that subclinical hypothyroidism is more frequent (about 11%) in obese patients referred to a sleep clinic for sleep disordered breathing than previously reported. However, this study did not include a control group [113]. Although several studies have examined the association between thyroid function and sleep disordered breathing, no studies have reported on the prevalence of hypothyroidism in patients with a chief complaint of excessive daytime sleepiness and fatigue. It has been our clinical impression that a number of patients with fatigue and sleepiness not related to nocturnal respiratory disturbances have previously undetectable subclinical hypothyroidism. Until this issue is explored in well-controlled studies, it is clinically prudent to screen patients with a chief complaint of fatigue and sleepiness for thyroid function.

Another, as yet unexplored, possibility is whether sleep apnea is associated with "euthyroid sick syndrome." This is a syndrome associated with mild suppression of thyroid function and is the result of chronic illness [114, 115]. This syndrome resolves with appropriate treatment of the primary condition.

Preliminary data from our center indicate a mild suppression of thyroid function (slightly suppressed TSH, mildly suppressed T_3, T_4) in patients with sleep apnea consistent with the "euthyroid sick syndrome." If these results are confirmed in larger studies, this would suggest that patients with a complaint of fatigue and borderline abnormal thyroid function tests should be thoroughly evaluated for sleep apnea before thyroid replacement therapy is instituted.

The results of thyroid hormone replacement on sleep apnea and hypothyroidism are inconsistent. While several successes have been reported of improving sleep apnea after the institution of thyroid hormone replacement, it appears that, at least in the initial phase of the treatment, both thyroid hormone replacement as well as treatment for sleep apnea, for example, CPAP, should be instituted. These patients should be followed up clinically and in the sleep laboratory to assess whether the adequate thyroid hormone replacement, in terms of dose and duration, abolished also sleep apnea.

The rapid restoration of the euthyroid state in hypothyroid patients may be associated with significant cardiovascular morbidity and mortality, particularly in the elderly or those with cardiovascular disease [116]. These risks appear to be increased in those with undiagnosed sleep apnea.

Patients with hypothyroidism not associated with sleep disordered breathing experience daytime sleepiness and fatigue and a significant reduction of slow-wave sleep [117], which is reversible with treatment [118]. In children with congenital and acquired hypothyroidism, reduced REM sleep and REM density have been reported [117].

Hyperthyroidism and Sleep

Clinical observations suggest an association between hyperthyroidism and sleep disturbance. Hyperthyroid patients complain of insomnia and impairment in mood [119]. Association between primary disorders associated with sleep disturbance (e.g., insomnia and hyperthyroidism) has not been reported. Thus thyroid screening is not indicated in these patients.

In sleep laboratory studies, hyperthyroidism is associated with a significant increase of slow-wave sleep, which after treatment tends to return to normal levels [120].

REFERENCES

1. Hedley AA, Ogden CL, Johnson CL, et al. Prevalence of overweight and obesity among US children, adolescents, and adults, 1999–2002. *JAMA* **291**:2847–2850(2004).

2. Gastaut H, Tassinari CA, Duron B. Polygraphic study of the episodic diurnal and nocturnal (hypnic and respiratory) manifestations of the Pickwick syndrome. *Brain Res* **2**:167–186(1966).

3. Guilleminault C, van den Hoed J, Mitler MM. Clinical overview of the sleep apnea syndromes. In: Guilleminault C, Dement WC (Eds), *Sleep Apnea Syndromes*. Alan R Liss, New York, 1978, pp 1–12.

4. Lugaresi E, Coccagna G, Montavani M. Hypersomnia with periodic apneas. In: Weitzman E (Ed), *Advances in Sleep Research*. Spectrum Publications, New York, 1978, p 4.

5. Vgontzas AN, Tan TL, Bixler EO, et al. Sleep apnea and sleep disruption in obese patients. *Arch Intern Med* **154**:1705–1711(1994).

6. Young T, Palta M, Dempsey J, et al. The occurrence of sleep-disordered breathing among middle aged adults. *N Engl J Med* **328**:1203–1205(1993).

7. Bixler EO, Vgontzas AN, Ten Have T, et al. Effects of age on sleep apnea in men: I. Prevalence and severity. *Am J Respir Crit Care Med* **157**:144–148(1998).

8. Shinohara E, Kihara S, Yamashita S, et al. Visceral fat accumulation as an important risk factor for obstructive sleep apnea syndrome in obese subjects. *J Intern Med* **241**:11–18(1997).

9. Vgontzas AN, Papanicolaou DA, Bixler EO, et al. Sleep apnea and daytime sleepiness and fatigue: relation to visceral obesity, insulin resistance, and hypercytokinemia. *J Clin Endocrinol Metab* **85**:1151–1158(2000).

10. Björntorp P. Metabolic implications of body fat distribution. *Diabetes Care* **14**:1132–1143(1991).

11. Bixler EO, Vgontzas AN, Lin H-M, et al. Association of hypertension and sleep-disordered breathing. *Arch Intern Med* **160**: 2289–2295(2000).

12. Vgontzas AN, Kales A. Sleep and its disorders. *Annu Rev Med* **50**:387–400(1999).

13. Rasheid S, Banasiak M, Gallagher SF, et al. Gastric bypass is an effective treatment for obstructive sleep apnea in patients with clinically significant obesity. *Obes Surg* **13**:58–61(2003).

14. Valencia-Flores M, Orea A, Herrera M, et al. Effect of bariatric surgery on obstructive sleep apnea and hypopnea syndrome, electrocardiogram, and pulmonary arterial pressure. *Obes Surg* **14**:755–762(2004).

15. Fujioka S, Matsuzawa Y, Tokunaga K, et al. Treatment of visceral fat obesity. *Int J Obes* **15**(Suppl 2):59–65(1991).

16. Pasquali R, Gambineri A, Biscotti D, et al. Effect of long-term treatment with metformin added to hypocaloric diet on body composition, fat distribution, and androgen and insulin levels in abdominally obese women with and without the polycystic ovary syndrome. *J Clin Endocrinol Metab* **85**:2767–2774(2000).

17. Shimomura I, Tokunaga K, Kotani K, et al. Marked reduction of acyl-CoA synthetase activity and mRNA in intra-abdominal visceral fat by physical exercise. *Am J Physiol* **265**(1 Pt 1): E44–E50(1993).

18. Taheri S, Lin L, Austin D, et al. Short sleep duration is associated with reduced leptin, elevated ghrelin, and increased body mass index. *PLoS Med* **1**:e62(2004).

19. Vorona RD, Winn MP, Babineau TW, et al. Overweight and obese patients in a primary care population report less sleep than patients with a normal body mass index. *Arch Intern Med* **165**:25–30(2005).

20. Spiegel K, Tasali E, Penev P, et al. Brief communication: sleep curtailment in healthy young men is associated with decreased leptin levels, elevated ghrelin levels, and increased hunger and appetite. *Ann Intern Med* **141**:846–856(2004).

21. Hasler G, Buysse DJ, Klaghofer R, et al. The association between short sleep duration and obesity in young adults: a 13-year prospective study. *Sleep* **27**:661–666(2004).

22. Gupta NK, Mueller WH, Can W, et al. Is obesity associated with poor sleep quality in adolescents? *Am J Hum Biol* **14**:762–768(2002).

23. Sekine M, Yamagami T, Handa K, et al. A dose–response relationship between short sleeping hours and childhood obesity: results of the Toyama Birth Cohort Study. *Child Care Health Dev* **28**:163–170(2002).

24. von Kries R, Toschke AM, Wurmser H, et al. Reduced risk for overweight and obesity in 5- and 6-y-old children by duration of sleep—a cross-sectional study. *Int J Obes Relat Metab Disord* **26**:710–716(2002).

25. Vgontzas AN, Bixler EO, Tan T-L, et al. Obesity without sleep apnea is associated with daytime sleepiness. *Arch Intern Med* **158**:1333–1337(1998).

26. Resta O, Foschino-Barbaro MP, Legari G, et al. Sleep-related breathing disorders, loud snoring and excessive daytime sleepiness in obese subjects. *Int J Obes Relat Metab Disord* **25**:669–675(2001).

27. Resta O, Roschino Barbaro MP, Bonfitto P, et al. Low sleep quality and daytime sleepiness in obese patients without obstructive sleep apnoea syndrome. *J Intern Med* **253**:536–543(2003).

28. Bixler EO, Vgontzas AN, Lin H-M, et al. Excessive daytime sleepiness in a general population sample: the role of sleep apnea, age, obesity, diabetes and depression. *J. Clin Endocrinol Metab.* Aug;**90**:4510–4515(2005).

29. Vela-Bueno A, Kales A, Soldatos CR. Sleep in the Prader–Willi syndrome. *Arch Neurol* **41**:294–296(1984).

30. Vgontzas AN, Kales A, Seip J, et al. Relationship of sleep abnormalities to patient genotypes in Prader–Willi syndrome. *Am J Med Genet* **67**:478–482(1996).

31. Vgontzas AN, Bixler EO, Kales A, Vela-Bueno A. Prader–Willi syndrome: effects of weight loss on sleep-disordered breathing, daytime sleepiness and REM sleep disturbance. *Acta Paediatr* **84**:813–814(1995).

32. Vgontzas AN, Papanicolaou DA, Bixler EO, et al. Elevation of plasma cytokines in disorders of excessive daytime sleepiness: role of sleep disturbance and obesity. *J Clin Endocrinol Metab* **82**:1313–1316(1997).

33. Vgontzas AN, Zoumakis E, Lin HM, et al. Marked decrease in sleepiness in patients with sleep apnea by etanercept, a tumor necrosis factor-alpha antagonist. *J Clin Endocrinol Metab* **89**:4409–4413(2004).

34. Van Cauter E, Spiegel K. Hormones and metabolism during sleep. In: Schwartz WJ (Ed), *Sleep Science: Integrating Basic Research and Clinical Practice (Monographs in Clinical Neuroscience)*, Vol 15. S. Karger Publishers, Farmington, CT, 1997, pp 144–174.

35. Spiegel K, Leproult R, Van Cauter E. Impact of sleep debt on metabolic and endocrine function. *Lancet* **354**:1435–1439(1999).

36. Gislason T, Almqvist M. Somatic diseases and sleep complaints. *Acta Med Scand* **221**:475–481(1987).

37. Sridhar GR, Madhu K. Prevalence of sleep disturbances in diabetes mellitus. *Diabetes Res Clin Pract* **23**:183–186(1994).

38. Ayas NT, White DP, Al-Delaimy WK, et al. A prospective study of self-reported sleep duration and incident diabetes in women. *Diabetes Care* **26**:380–384(2003).

39. Lamond N, Tiggemann M, Dawson D. Factors predicting sleep disruption in type II diabetes. *Sleep* **23**(3):415–416(2000).

40. Matyka KA, Crawford C, Wiggs L, et al. Alterations in sleep physiology in young children with insulin-dependent diabetes mellitus: relationship to nocturnal hypoglycemia. *J Pediatr* **137**(20):233–238(2000).

41. Mondini S, Guilleminault C. Abnormal breathing patterns during sleep in diabetes. *Ann Neurol* **17**(4):391–395(1985).

42. Brooks B, Cistulli PA, Borkman M, et al. Obstructive sleep apnea in obese noninsulin-dependent diabetic patients: effect of continuous positive airway pressure treatment on insulin responsiveness. *J Clin Endocrinol Metab* **79**(6):1681–1685(1994).

43. Ip SM, Tsang WT, Lam WK, et al. Obstructive sleep apnea syndrome: an experience in Chinese adults in Hong Kong. *Chin Med J (Engl)* **111**(3):257–260(1998).

44. Ficker JH, Dertinger SH, Siegfried W. Obstructive sleep apnoea and diabetes mellitus: the role of cardiovascular autonomic neuropathy. *Eur Respir J* **11**(1):14–19(1998).

45. Elmasry A, Janson C, Lindberg E, et al. The role of habitual snoring and obesity in the development of diabetes: a 10-year follow-up study in a male population. *J Intern Med* **248**:13–20(2000).

46. Al-Delaimy WK, Manson JE, Willett WC, et al. Snoring as a risk factor for type II diabetes mellitus: a prospective study. *Am J Epidemiol* **155**:387–393(2002).

47. Honda Y, Doi Y, Ninomiya R, Ninomiya C. Increased frequency of non-insulin-dependent diabetes mellitus among narcoleptic patients. *Sleep* **9**(1 Pt 2):254–259(1986).

48. Roberts HJ. Obesity due to the syndrome of narcolepsy and diabetogenic hyperinsulinism: clinical and therapeutic observations in 252 patients. *J Am Geriat Soc* **15**:721–743(1967).

49. Okun ML, Giese S, Lin L, Einen M, Mignot E, Coussons-Read ME. Exploring the cytokine and endocrine involvement in narcolepsy. *Brain Behav Immun.* Jul;**18**(4):326–332(2004).

50. Garvey WT, Hermayer KL. Clinical implications of the insulin resistance syndrome. *Clin Cornerstone* **1**(3):13–28(1998).

51. Reaven GM. Insulin resistance and human disease: a short story. *J Basic Clin Physiol Pharmacol* **9**:387–406(1998).

52. Bixler EO, Vgontzas AN, Lin H-M, et al. Prevalence of sleep-disordered breathing in women. *Am J Respir Crit Care Med* **162**:1–6(2001).

53. Strohl KP, Novak RD, Singer W, et al. Insulin levels, blood pressure and sleep apnea. *Sleep* **17**:614–618(1994).

54. Davies RJO, Turner R, Crosby J, Stradling JR. Plasma insulin and lipid levels in untreated obstructive sleep apnoea and snoring: their comparison with matched controls and response to treatment. *J Sleep Res* **3**:180–185(1994).

55. Stoohs RA, Facchini F, Guilleminault C. Insulin resistance and sleep-disordered breathing in healthy humans. *Am J Respir Crit Care Med* **154**:170–174(1996).

56. Vgontzas AN, et al. *Metabolic Disturbances in Obesity Versus Sleep Apnoea: The Importance of Visceral Obesity and Insulin Resistance of Internal Medicine* **254**:32–44(2003). Blackwell Publishing, London, 2003.

57. Vgontzas AN, et al. *Sleep Apnea Is a Manifestation of the Metabolic Syndrome.* Elsevier, New York. *Sleep Med Rev* **9**:211–224(2005).

58. Park YW, Zhu S, Palaniappan L, et al. The metabolic syndrome: prevalence and associated risk factor findings in the US population from the Third National Health and Nutrition Examination Survey, 1988–1994. *Arch Intern Med* **163**:427–436(2003).

59. Ip MS, Lam B, Ng MM, et al. Obstructive sleep apnea is independently associated with insulin resistance. *Am J Respir Crit Care Med* **165**:670–676(2002).

60. Punjabi NM, Sorkin JD, Katzel LI, et al. Sleep-disordered breathing and insulin resistance in middle-aged and overweight men. *Am J Respir Crit Care Med* **165**:677–682(2002).

61. Punjabi NM, Shahar E, Redline S, et al. Sleep-disordered breathing, glucose intolerance, and insulin resistance: the Sleep Heart Health Study. *Am J Epidemiol* **160**:521–530(2004).

62. Vgontzas AN, Legro RS, Bixler EO, et al. Polycystic ovary syndrome is associated with obstructive sleep apnea and daytime sleepiness: role of insulin resistance. *J Clin Endocrinol Metab* **86**:517–520(2001).

63. Bixler EO, Kales A, Jacoby JA, et al. Nocturnal sleep and wakefulness: effects of age and sex in normal sleepers. *Int J Neurosci* **23**:33–42(1984).

64. Williams RL, Karacan I, Hursch C. *EEG of Human Sleep: Clinical Applications.* John Wiley & Sons, Hoboken, NJ, 1974.

65. Ehlers CL, Kupfer DJ. Slow-wave sleep: do young adult men and women age differently? *J Sleep Res* **6**:211–215(1997).

66. Hume KI, Van F, Watson A. A field study of age and gender differences in habitual adult sleep. *J Sleep Res* **7**:85–94(1998).

67. Carrier J, Land S, Buysse DJ, et al. The effects of age and gender on sleep EEG power spectral density in the middle years of life (ages 20–60 years old). *Psychophysiology* **38**:232–242(2001).

68. Bixler EO, Vgontzas AN, Lin H-M, et al. Normal sleep and sleep stage patterns: effects of age, BMI and gender. *Sleep* **26**(Suppl):A61–A62(2003).

69. Redline S, Bonekat W, Gottlieb D, et al. Sleep stage distributions in the sleep heart health study (SHHS) cohort. *Sleep* **21**(Suppl):210(1998).

70. Sandblom RE, Matsumoto AM, Schoene RB, et al. Obstructive sleep apnea induced by testosterone administration. *N Engl J Med* **308**:506–510(1983).

71. Schneider BK, Pickett CK, Zwillich CW. Influence of testosterone on breathing during sleep. *J Appl Physiol* **61**:618–623(1986).

72. Cistulli PA, Grunstein RR, Sullivan CE. Effect of testosterone administration on upper airway collapsibility during sleep. *Am J Respir Crit Care Med* **149**:530–532(1994).

73. Johnson MW, Anch AM, Remmers JE. Induction of the obstructive sleep apnea syndrome in a woman by exogenous androgen administration. *Am Rev Respir Dis* **129**:1023–1025(1984).

74. Dexter DD, Dovre EJ. Obstructive sleep apnea due to endogenous testosterone production in a woman. *Mayo Clin Proc* **73**:246–248(1988).

75. Liu PY, Yee B, Wishart SM, et al. The short-term effects of high-dose testosterone on sleep, breathing, and function in older man. *J Clin Endocrinol Metab* **88**:3605–3613(2003).

76. Stewart DA, Grunstein RR, Berthon-Jones M, et al. Androgen blockade does not affect sleep-disordered breathing or chemosensitivity in men with obstructive sleep apnea. *Am Rev Respir Dis* **146**:1389–1393(1992).

77. Grunstein RR, Handelsman DJ, Lawrence S, et al. Neuroendocrine dysfunction in sleep apnea: reversal by nasal continuous positive airway pressure. *J Clin Endocrinol Metab* **68**:352–358(1989).

78. White D, Douglas N, Pickett C, et al. Hypoxic ventilatory response during sleep in normal premenopausal women. *Am Rev Respir Dis* **126**:530–553(1982).

79. Block AJ, Wynne JW, Boysen PG, et al. Menopause, medroxyprogesterone and breathing during sleep. *Am J Med* **70**:506–510(1981).

80. Shahar E, Redline S, Young T, et al. Hormone replacement therapy and sleep-disordered breathing. *Am J Respir Crit Care Med* **167**:1186–1192(2003).

81. Young T, Finn L, Austin D, et al. Menopausal status and sleep-disordered breathing in the Wisconsin Sleep Cohort Study. *Am J Respir Crit Care Med* **167**:1181–1185(2003).

82. Margolis KL, Bonds DE, Rodabough RJ, et al. Effect of oestrogen plus progestin on the incidence of diabetes in postmenopausal women: results from the Women's Health Initiative Hormone Trial. *Diabetologia* **47**:1175–1187(2004).

83. Dunaif A. Insulin resistance and the polycystic ovary syndrome: mechanism and implications for pathogenesis. *Endocr Rev* **18**:774–800(1997).

84. Fogel RBMA, Pillar G, Pittman SD, et al. Increased prevalence of obstructive sleep apnea syndrome in obese women with polycystic ovary syndrome. *J Clin Endocrinol Metab* **86**:1175–1178(2001).

85. Gopal M, Duntley S, Uhles M, et al. The role of obesity in the increased prevalence of obstructive sleep apnea syndrome in patients with polycystic ovarian syndrome. *Sleep Med* **3**:401–404(2004).

86. Young T, Rabago D, Zgierska A, et al. Objective and subjective sleep quality in premenopausal, perimenopausal, and postmenopausal women in the Wisconsin Sleep Cohort Study. *Sleep* **26**:667–672(2003).

87. Moe KE, Larsen LH, Vitiello MV, et al. Estrogen replacement therapy moderates the sleep disruption associated with nocturnal blood sampling. *Sleep* **24**:886–894(2001).

88. Pejovic S, Bixler EO, Lin H-M, et al. Young women compared to men are more resilient to the sleep disturbing effects of external stressors. Abstract presented at the APSS 19th Annual Meeting, Denver, CO, 2005.

89. Chrousos GP, Kattah JC, Beck RW, Cleary PA and the Optic Neuritis Study Group. Side effects of glucocorticoid treatment. *JAMA* **269**:2110–2112(1993).

90. Opp M, Obál F Jr, Krueger JM. Corticotropin-releasing factor attenuates interleukin 1-induced sleep and fever in rabbits. *Am J Physiol* **257**:R528–R535(1989).

91. Bierwolf C, Struve K, Marshall L, et al. Slow wave sleep drives inhibition of pituitary–adrenal secretion in humans. *Neuroendocrinology* **9**:479–484(1997).

92. Vgontzas AN, Mastorakos G, Bixler EO, et al. Sleep deprivation effects on the activity of the hypothalamic–pituitary–adrenal and growth axes: potential clinical implications. *Clin Endocrinol (Oxf)* **51**:205–215(1999).

93. Weitzman ED, Zimmerman JC, Czeisler CA, Ronda J. Cortisol secretion is inhibited during sleep in normal men. *J Clin Endocrinol Metab* **56**:352–358(1983).

94. Vgontzas AN, Bixler EO, Lin H-M, et al. Chronic insomnia is associated with nyctohemeral activation of the hypothalamic–pituitary–adrenal axis: clinical implications. *J Clin Endocrinol Metab* **86**:3787–3794(2001).

95. Rodenbeck A, Huether G, Ruther E, et al. Interactions between evening and nocturnal cortisol secretion and sleep parameters in patients with severe chronic primary insomnia. *Neurosci Lett* **324**:159–163(2002).

96. Rodenbeck A, Cohrs S, Jordan W, et al. The sleep-improving effects of doxepin are paralleled by a normalized plasma cortisol secretion in primary insomnia. A placebo-controlled, double-blind, randomized, cross-over study followed by an open treatment over 3 weeks. *Psychopharmacology* **170**:423–428(2003).

97. Bixler EO, Vgontzas AN, Lin H-M, et al. Insomnia in central Pennsylvania. *J Psychosom Res* **53**:589–592(2002).

98. Vgontzas AN, Bixler EO, Wittman AM, et al. Middle-aged men show higher sensitivity of their sleep to the arousing effects of corticotropin-releasing hormone than young men: clinical implications. *J Clin Endocrinol Metab* **86**:1489–1495(2001).

99. Shipley JE, Schteingart DE, Tandon R, et al. Sleep architecture and sleep apnea in patients with Cushing's disease. *Sleep* **15**:514–518(1992).

100. Friedman TC, García-Borreguero D, Hardwick D, et al. Decreased delta-sleep and plasma delta-sleep-inducing peptide in patients with Cushing syndrome. *Neuroendocrinology* **60**:626–634(1994).

101. García-Borreguero D, Wehr TA, Larrosa O, et al. Glucocorticoid replacement is permissive for rapid eye movement sleep and sleep consolidation in patients with adrenal insufficiency. *J Clin Endocrinol Metab* **85**:4201–4206(2000).

102. Gillin JC, Jacobs LS, Fram DH, Snyder F. Acute effect of a glucocorticoid on normal human sleep. *Nature* **237**:398–399(1972).

103. Papanicolaou DA, Tsigos C, Oldfield EH, Chrousos GP. Acute glucocorticoid deficiency is associated with plasma elevations of interleukin-6: Does the latter participate in the symptomatology of the steroid withdrawal syndrome and adrenal insufficiency? *J Clin Endocrinol Metab* **81**:2303–2306(1996).

104. Reincke M, Heppner C, Petzke F, et al. Impairment of adrenocortical function associated with increased plasma tumor necrosis factor-alpha and interleukin-6 concentrations in African trypanosomiasis. *Neuroimmunomodulation* **1**:14–22(1994).

105. Grunstein RR, Ho KY, Sullivan CE. Acromegaly and sleep apnea. *Ann Intern Med* **115**:527–532(1991).

106. Rosenow F, Reuter S, Deuss U, et al. Sleep apnoea in treated acromegaly: relative frequency and predisposing factors. *Clin Endocrinol (Oxf)* **45**:563–569(1996).

107. Hochban W, Ehlenz K, Conradt R, et al. Obstructive sleep apnea in acromegaly: the role of craniofacial changes. *Eur Respir J* **14**:196–202(1999).

108. Cadieux RJ, Kales A, Santen RJ, et al. Endoscopic findings in sleep apnea associated with acromegaly. *J Clin Endocrinol Metab* **55**:18–22(1982).

109. Grunstein RR. Metabolic aspects of sleep apnea. *Sleep* **19**:S218–S220(1996).

110. Pelttari L, Rauhala E, Polo O, et al. Upper airway obstruction in hypothyroidism. *J Intern Med* **236**:177–181(1994).

111. Kapur VK, Koepsell TD, deMaube J, et al. Association of hypothyroidism and obstructive sleep apnea. *Am J Respir Crit Care Med* **158**:1379–1383(1998).

112. Winkelman JW, Goldman H, Piscatelli N, et al. Are thyroid function tests necessary in patients with suspected sleep apnea? *Sleep* **19**:790–793(1996).

113. Resta O, Pannacciulli N, Di Gioia G, et al. High prevalence of previously unknown subclinical hypothyroidism in obese patients referred to a sleep clinic for sleep disordered breathing. *Nutr Metab Cardiovasc Dis* **14**:248–253(2004).

114. Papanicolaou DA, Wilder RL, Manolagas SC, Chrousos GP. The pathophysiologic roles of interleukin-6 in human disease. *Ann Intern Med* **128**:127–137(1998).

115. Wartofsky L, Burman KD. Alterations in thyroid function in patients with systemic illness: the "euthyroid sick syndrome." *Endocr Rev* **3**:164–217(1982).

116. Grunstein RR, Sullivan CE. Hypothyroidism and sleep apnea: mechanisms and management. *Am J Med* **85**:775–779(1988).

117. Hayashi M, Araki S, Kohyama J, et al. Sleep development in children with congenital and acquired hypothyroidism. *Brain Dev* **19**:43–49(1997).

118. Ruiz-Primo E, Jurado JL, Solis H, et al. Polysomnographic effects of thyroid hormones in primary myxedema. *Electroencephalogr Clin Neurophysiol* **53**:559–564(1982).

119. Huang YR, Wang GH. Study on quality of sleep and mental health in patients with hyperthyroidism. *Chung Hua Hu Li Tsa Chih* **32**:435–439(1997).

120. Dunleavy DLF, Oswald I, Brown P, Strong JA. Hyperthyroidism, sleep and growth hormone. *Electroencephalogr Clin Neurophysiol* **36**:259–263(1974).

99

SLEEP IN FIBROMYALGIA AND CHRONIC PAIN

SUSAN M. HARDING
University of Alabama at Birmingham Sleep–Wake Disorders Center, Birmingham, Alabama

TEOFILO LEE-CHIONG
National Jewish Medical and Research Center, Denver, Colorado

FIBROMYALGIA

Clinical Features

Fibromyalgia (FM) is a syndrome and not a disease. According to the American College of Rheumatology, the case definition requires the presence of 11 of 18 tender points, which must be bilateral, occur above and below the waist, include axial sites, have widespread distribution, and occur for ≥3 months [1]. Although current diagnostic criteria do not include sleep-related abnormalities, sleep disturbance is a major complaint of FM patients.

Fibromyalgia symptom onset can be acute or insidious. Factors associated with symptom onset include an infectious illness, physical or emotional trauma, and stress. Twenty-three percent of FM patients experience seasonal variations in their symptoms, noting fewer symptoms in July and more symptoms in December and January [2]. In one study, poor sleep, stress, exercise, and a cold environment aggravated pain; while rest, warm baths, heat, and relaxation improved pain [3].

Fibromyalgia affects approximately 3% of the population (~6–10 million Americans) and is much more common in women than men. Approximately 75% of FM patients are women. Fibromyalgia is most prevalent in people aged 30–50 years. Fibromyalgia can be classified as primary (when there are no underlying or concomitant medical conditions) or secondary (when other rheumatologic disorders, such as rheumatoid arthritis, are present). Secondary FM accounts for 30% of patients [4].

Approximately 20–50% of patients have depression, and 53–65% of patients report a history of sexual abuse [5].

Pathophysiology

Complex mechanisms are involved in the etiology and maintenance of FM. There is a generalized heightened pain sensitivity arising from pathologic nociceptive processing within the central nervous system (CNS). It primarily involves CNS sensitization of nociceptive neurons in the dorsal horn of the spinal cord and activation of N-methyl-D-aspartate (NMDA) receptors [6]. Neuroendocrine and neuropeptide abnormalities, regional cerebral blood flow alterations, and alterations in the neuromatrix (thalamus, cortex, and limbic systems) are present as well.

There are low serum and cerebrospinal fluid (CSF) levels of serotonin and 5-hydroxyindole-3-acetic acid (5HIAA) and increased CSF levels of dynorphin A, substance P, calcitonin gene-related peptide, and nerve growth factor [7]. These neuropeptide alterations may activate NMDA receptor mechanisms. Sallanon et al. [8] noted that 5-hydroxytryptamine (5-HT), a serotonin precursor, is released from axonal nerve endings in the basal hypothalamus during wakefulness. It may further act as a neurohormone and induce hypnogenic factors that promote slow-wave non-rapid eye movement (NREM) and rapid eye movement (REM) sleep.

Levels of insulin growth hormone-I (somatomedin C) are related to the production of growth hormone (GH),

---

which is primarily secreted during stage 4 of slow-wave NREM sleep. Somatomedin C levels are low in approximately one-third of FM patients [9]. Bennett et al. [10] showed in a placebo-controlled trial that the administration of GH improved sleep in FM patients compared to those receiving placebo.

Cerebral blood flow is also different in FM patients compared to control subjects. FM patients have lower regional cerebral blood flow to the thalamus and caudate nuclei [11]. All of these findings indicate that the pathophysiology of FM includes central sensitization resulting in generalized pain sensitivity.

Sleep in Patients with Fibromyalgia

Patients with FM often report sleep fragmentation, early morning awakenings, unrefreshing sleep, fatigue, and insomnia [12]. Molony et al. [13], for example, noted that FM patients had three times more microarousals per hour (brief sleep interruptions lasting 5–19 seconds) than did control subjects. Nonrestorative sleep is reported in 76% of FM patients [14]. Some FM patients report such restless sleep that they are conscious of dreaming during their sleep.

Primary sleep disorders, especially sleep apnea and restless legs syndrome (RLS) with periodic limb movements of sleep (PLMS), may also be found in FM patients. May et al. [15] screened all new FM patients for sleep apnea and if there was a suspicion of sleep apnea, polysomnography was performed and approximately 2% of women and 44% of men had sleep apnea. Other studies, however, have not supported this increased sleep apnea frequency in men with FM. Donald et al. [16] reported that FM was uncommon (approximately 3%) in patients with disturbed sleep at a respiratory sleep disorders clinic, and Molony et al. [13] found that FM patients had the same frequency of sleep apnea as normal controls. Thus, although sleep apnea is found in selected FM patients, apnea does not appear to be responsible for FM symptoms.

Yunus and Aldag [17] examined 135 consecutive female patients with primary FM, 54 women with rheumatoid arthritis, and 88 pain-free control subjects. FM patients had a higher frequency of RLS (31%) compared to control subjects (2%) and patients with rheumatoid arthritis (15%) [17]. RLS may be more frequent in FM patients.

There may be a relationship between poor sleep quality, pain intensity, and perception of pain in FM patients. Poor sleep quality may contribute to the hypervigilance and abnormal somatic perceptions present in FM patients [14]. Affleck et al. [14] asked 50 women with FM to recall their ratings of subjective sleep quality, pain intensity, and attention to pain for 30 days using a palm-top computer. Poor sleepers reported significantly more pain than those with a relatively high sleep quality. Moreover, ratings indicated that a night of poor sleep tended to be followed by a significantly more painful day and that a painful day tended to be followed by a night of poor sleep [14]. Sleep disturbances are also related to disturbances in mood and cognition [18].

Polysomnographic Features

Polysomnographic features in FM patients include a decrease in total sleep time, long periods of awakenings, a high number of arousals, reductions in delta NREM sleep and REM sleep, and increase in alpha rhythm. Controlled polysomnographic studies also note prolonged sleep onset latencies, an increase in stage 1 NREM sleep, and an increase in wake time after sleep onset in FM patients compared to control subjects [18].

Sleep spindles are important for induction and maintenance of stage 2 NREM sleep and originate in the thalamus. Potentially, FM patients could have disruption of sleep-regulatory mechanisms involving sleep spindle generation in the corticothalamic axis that could impact sleep and pain. Landis et al. [19] examined 37 women with FM and 30 control women without sleep complaints or pain with EEG power spectral analyses and spindle and time domain analyses. The FM patients had fewer sleep spindles per minute of stage 2 NREM sleep and had reduced spindle frequency even when controlling for age, depression, and psychiatric diagnosis. Furthermore, after controlling for age and depression, pain pressure threshold predicted spindle number and spindle frequency activity in stage 2 NREM sleep. These findings support the hypothesis that reduced sleep spindles may impact sensory processing in the thalamus allowing for FM patients to be more vulnerable to sleep–disturbing stimuli [19].

The microstructure of sleep has been examined using power spectral and frequency domain analysis. With this approach, different frequencies are described. Sigma frequency (12–13 Hz) reflects a sleep-protective function and is associated with the perception of greater sleep depth. Beta frequency (14–38 Hz) reflects arousal, is associated with lightened sleep perception, and is seen in depression. Occipital alpha frequency (8–13 Hz) is arousal-related. Some investigators note that depriving normal control subjects of stage 4 NREM sleep with noise gives rise to the perception of unrefreshing sleep, pain, and fatigue [20]. Conversely, deep pain stimulation reduced delta and sigma power and increased beta power on spectral analysis of sleep microstructure. Branco et al. [20] prospectively studied alpha and delta activity and the alpha–delta ratio across sleep cycles in 14 normal controls and in 10 FM patients. Nine of 10 patients exhibited the alpha–delta sleep anomaly, which appeared to increase exponentially through the night; this anomaly was not observed in any of the controls.

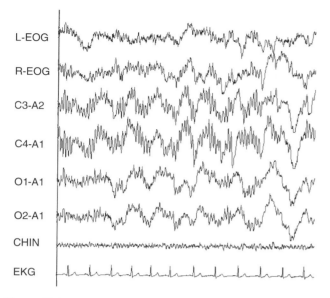

Figure 99.1 Alpha NREM sleep anomaly. This 10 second epoch shows alpha frequency superimposed on delta frequency. Note that, in this example, the alpha frequency is more prominent in the central EEG leads (C3–A2, C4–A1) compared to the occipital EEG leads (O1–A2, O2–A1).

Alpha NREM Sleep Anomaly

Alpha–delta sleep is an abnormal sleep EEG rhythm characterized by alpha activity that is superimposed on delta waves of stages 3 and 4 of NREM sleep. Alpha sleep has been broadened to include alpha intrusion into all stages of NREM sleep. This sleep anomaly is accompanied by indications of vigilance during sleep and the subjective experience of unrefreshing sleep [21]. A 10 second epoch is displayed in Figure 99.1.

Moldofsky and Lue [22] first identified the alpha NREM anomaly in FM patients, referring to it as "alpha intrusion or alpha–delta." Drewes et al. [23] examined spectral EEG patterns in FM patients and observed that they displayed a decreased delta frequency power and an increased power in alpha frequency bands.

Roizenblatt et al. [24] examined the microstructure of sleep using power spectral and frequency domain analyses in 40 FM patients taking no medications for more than 4 weeks and in 43 control subjects. Worsening of morning symptoms was present in 72% of FM patients compared to 12% of control subjects. An increase in the number of tender points was seen in 93% of FM patients after the sleep period. Seventy percent of FM patients reported poor sleep quality, compared to none of the control subjects. The alpha NREM anomaly was identified in 70% of FM patients and in 16% of control subjects. They found three distinct patterns of alpha activity while analyzing the microstructure of sleep: (1) a *phasic* alpha pattern that

was episodic and occurred simultaneously with delta (found in 50% of FM patients and in 70% of control subjects); (2) a *tonic* alpha pattern in which the alpha frequency was continuously present throughout NREM independent of delta activity (20% of FM patients and 9% of control subjects); and (3) a *low* alpha pattern with minimal alpha activity (30% of FM patients and 84% of control subjects). They observed that phasic alpha was associated with decreased sleep efficiency, the feeling of superficial sleep, longer pain duration, and increased morning stiffness and pain. Indeed, nonrestorative sleep was present in 100% of FM patients with the phasic alpha pattern, 25% with the tonic alpha pattern, and 58% with the low alpha pattern. Pain duration (in years) was also longer in FM patients with the phasic alpha pattern compared to the other two alpha patterns. There was no difference in the frequency of the different alpha patterns as to whether the patients were depressed or not.

It has been hypothesized that a decrease in stage 4 NREM sleep is associated with musculoskeletal and mood symptoms. Moldofsky et al. [25] examined six normal controls during two nights of undisturbed sleep, followed by three nights of stage 4 NREM sleep deprivation produced by auditory stimuli and two subsequent nights of undisturbed sleep. During stage 4 NREM sleep deprivation, three subjects experienced an overnight increase in dolorimeter scores (i.e., increased tenderness), musculoskeletal symptoms, and mood disturbances comparable to those of "fibrositis" patients. In contrast, symptom remission and a decrease in dolorimeter scores accompanied the return of undisturbed sleep [25]. Perlis et al. [26] examined the relationships between alpha sleep and the perception of sleep, information processing, muscle tenderness, and arousability in 20 FM patients. They found that alpha sleep correlated with perceived shallow sleep and an increased tendency to display arousal in response to external stimuli [26].

However, not all investigators agree that the alpha NREM sleep anomaly is abnormal. The alpha NREM sleep anomaly is not specific for FM, and not all FM patients have it. In normal subjects, the alpha NREM anomaly may be elicited by auditory stimulation during stage 4 NREM sleep deprivation. Mahowald and Mahowald [27] stated that alpha NREM sleep is neither specific for nor a marker for FM, and that alpha–delta sleep is an atypical and nonspecific sleep pattern with an unclear pathophysiology. Data supports that this phenomenon is not associated with enhanced long-term or short-term memory during sleep and is not necessarily associated with myalgic symptoms. Mahowald and Mahowald [27] found the alpha NREM anomaly in 15% of 240 normal subjects (20% of persons with undisturbed sleep versus 8% of persons with disturbed sleep) and hypothesized that the alpha frequency may actually be a sleep-maintaining process.

Rains and Penzien [28] reported that patients with chronic pain had similar amounts of the alpha NREM sleep anomaly as pain-free patients. Reviewing 1076 consecutive patients at a university-based sleep disorders center using questionnaires, overnight PSG, multiple sleep latency tests, and the Minnesota Multiphasic Personality Inventory, they identified 54 patients (5%) with the alpha NREM sleep anomaly. Less than 40% of subjects with the anomaly had pain syndromes, and the majority of subjects exhibiting this anomaly had a nonpainful medical or psychiatric condition.

Therapy of Fibromyalgia

No consensus regarding optimal management of patients with FM exists. Therapy is often difficult, costly, and of limited success with less than 50% of patients having adequate symptom relief. In a study where 70 FM patients were followed longitudinally for 40 months, 47% of patients reported marked or moderate improvement in the 7-point tenderness scale at 3 year follow-up, while 53% noted no improvement or reported deterioration in their clinical status [29]. Predictors of a favorable outcome included younger age and less severe sleep disturbance.

Therapy of FM patients should be multimodal and flexible. Treatment strategies should be individualized. Note that no therapy is FDA approved for use in FM patients. Potential pharmacologic therapies include antidepressants, analgesics, and muscle relaxants. Management often includes amitriptyline or a selective serotonin reuptake inhibitor if tricyclic antidepressants fail, or if the patient has depression. Occasionally, a combination of a tricyclic antidepressant and a selective serotonin reuptake inhibitor is used. Cyclobenzaprine is used for muscle relaxation. Intermittent use of zolpidem or zalepon may be helpful for patients presenting with acute insomnia. Potential new therapies include sibutramine and tizanidine. Eight weeks of sibutramine, a serotonin norepinephrine–dopamine reuptake inhibitor, improved pain and sleep, although this study was not controlled [30]. Tizanidine, an α_2-adrenergic antagonist, may improve pain but causes drowsiness [31]. Human recombinant growth hormone was investigated in a randomized controlled trial involving 50 FM patients with low IGF-1 levels. Growth hormone resulted in improvements in FM outcome questionnaires and tender-point scores. There was a 6 month lag time for a meaningful clinical response. Relapse occurred with discontinuation of growth hormone therapy [32]. Administration of tropisetron, a 5-HT3-receptor antagonist, resulted in a 39% response rate and a 35% decrease in pain scores [33]. The 5 mg dose decreased symptom and tender points and improved sleep and dizziness.

Nonpharmacologic therapy includes exercise, cognitive–behavioral therapy, biofeedback, hypnotherapy, and even acupuncture. Exercise has a role in FM therapy. In a Cochrane review, Busch et al. [34] examined the outcomes of 16 randomized controlled trials examining more than 700 FM patients and noted that exercise improved aerobic performance and tender-point pressure threshold [34]. Richards and Scott [35] performed a parallel group, randomized trial that included more than 100 FM patients who were randomly assigned to receive either a graded aerobic exercise program or relaxation and flexibility therapy. Exercise reduced tender point count and improved the Fibromyalgia Impact Questionnaire (FIQ) score at 1 year follow-up. However, compliance with exercise was low with only 53% of subjects attending at least one-third of the exercise sessions.

Cognitive–behavioral therapy has been evaluated as a treatment modality for FM. Williams et al. [36] examined 145 FM patients receiving either standard therapy or six cognitive–behavioral therapy sessions in group format. Cognitive–behavioral therapy included progressive muscle relaxation, pacing skills, cued rest, pleasant activities scheduling, relaxation skills, communication skills, assertiveness training, stress management, and problem solving. An improvement in physical function was noted in the cognitive–behavioral therapy group [36].

Several therapies have not prove to be effective in FM patients. Sedatives and benzodiazepines do not provide any specific benefit; however, their intermittent use may alleviate acute sleep disturbances [34]. Although non benzodiazepine agents, including zaleplon and zolpidem, improve subjective sleep and excessive daytime sleepiness, they do not alter the alpha NREM sleep anomaly or pain scores and, like benzodiazepines, should not be used on a chronic basis [37]. Melatonin failed to improve any specific outcome measures in FM patients [38].

Most FM therapeutic trials have significant limitations in study design—small sample sizes, low statistical power, and different subject types. Few studies examined similar or comparable interventions or outcome measures. Large systematic randomized controlled trials of sufficient duration and using standardized outcome measures are lacking.

Effect of Fibromyalgia Therapy on Sleep

Several of the pharmacologic agents evaluated in FM patients have been tested specifically for their potential for improving sleep. However, no treatment has proved to have long-lasting benefits.

Moldofsky and Lue [22] showed that chlorpromazine, which decreases alpha–delta frequencies, reduced pain and trigger-point tenderness. Amitriptyline improved sleep symptom scores but did not have a long-lasting effect. Amitriptyline therapy for 1 month improved symptoms in approximately one-third of FM patients; however, the response declined by 6 months of therapy [39]. Fluoxetine helped sleep scores but not pain scores or the number of tender points [40]. Goldenberg and colleagues [41] noted that treatment with both amitriptyline and fluoxetine resulted in significant improvement in sleep scores among FM patients and appeared to be superior to either drug alone.

Branco et al. [42] examined the effects of trazodone, an antidepressant with an inhibitory effect on 5-HT reuptake, on sleep architecture in 13 patients in a double-blind crossover placebo-controlled trial. Polysomnography was performed with spectral analysis before and after 2 months of placebo or trazodone 150 mg/day. Although trazodone almost doubled the amount of stage 3 and stage 4 NREM sleep over basal and placebo levels, there was no difference in sleep latency, sleep efficiency, or percentage of wake time. Alpha activity was reduced [42]. In another study, trazodone increased delta NREM sleep and decreased the alpha NREM sleep anomaly without altering wake time after sleep onset or sleep latency.

Sodium oxybate was noted to reduce pain and fatigue symptoms and improve sleep quality in FM patients. In a placebo-controlled trial, Scharf et al. [43] noted that the tender-point index decreased from baseline by 8.5 points in the treated group versus 0.4 point in the placebo group. Six of seven pain and fatigue scores (overall pain, pain at rest, pain during movement, end-of-day fatigue, overall fatigue, and morning fatigue) were relieved by 29–33% with sodium oxybate, compared to 6–10% with placebo [43]. Therapy also significantly decreased the amount of alpha NREM sleep, sleep latency, and REM sleep and increased slow-wave NREM sleep percentages.

Management of Sleep in Patients with Fibromyalgia

Potential management strategies in FM patients are displayed in Table 99.1. Important sleep management issues include maintaining good sleep hygiene habits, including

TABLE 99.1 Sleep Management in Fibromyalgia Patients

- Screen, diagnose, and treat primary sleep disorders including:
  Obstructive sleep apnea
  Upper airway resistance syndrome
  Restless legs syndrome
  Circadian rhythm disorders
- Employ sleep hygiene techniques
  Regular sleep–wake and light–dark schedule
  Conducive sleep environment
  Adequate sleep time
  Avoidance of medications and substances that disrupt sleep
- Screen for overlapping conditions
- Screen for depression
- Employ a multidisciplinary team
- Consider prescribing a graded exercise program
- Consider cognitive–behavioral therapy

a regular sleep–wake and light–dark schedule, an adequate sleep time, a sleep-conducive bedroom environment, and avoidance of alcohol, caffeine, nicotine, and other substances that might interfere with sleep. Maladaptive sleep behaviors must be identified and corrected if possible. Medication intake, including over-the-counter agents, should be assessed for their potential to disrupt sleep. Relaxation techniques and a graded exercise program may also be helpful. Actigraphy may aid in identifying issues related to sleep–wake schedules and sleep maintenance issues. Landis et al. [44] reported that actigraphy was useful in identifying total sleep time, sleep efficiency, sleep latency, wake time after sleep onset, and the sleep fragmentation index. They found the sleep quality of FM patients was directly related to total sleep time and inversely related to sleep fragmentation. Fatigue in FM patients was directly related to wake time after sleep onset and inversely related to sleep efficiency [44].

Screening for primary sleep disorders, including obstructive sleep apnea (OSA), restless legs syndrome (RLS) with sleep disturbance, and delayed sleep phase syndrome, is important.

Obstructive sleep apnea can be diagnosed and treated with nasal continuous positive airway pressure (CPAP) or oral appliances that splint the collapsible upper airway in fibromyalgia patients. Even mild inspiratory flow limitation is associated with sleep fragmentation, arousals, fatigue, and excessive daytime sleepiness. This form of sleep disordered breathing is termed upper airway resistance syndrome (UARS). Gold et al. [45] performed a descriptive study of 28 FM patients noting that 27 of them had inspiratory flow limitation associated with arousals. Also, the mean pharyngeal critical pressure of the entire group of FM patients was similar to a UARS patient cohort (-6.5 ± 3.5 cm H_2O compared to -5.8 ± 3.8 cm H_2O; $p = 0.62$). Furthermore, 3 weeks of nasal CPAP therapy improved fatigue, pain, sleep problems, disability, and Rheumatology Distress Index scores. This preliminary study shows that FM patients have findings consistent with UARS and that CPAP therapy has the potential to improve functional symptoms and outcomes. Note that these FM patients did not have typical signs and symptoms of sleep apnea [45].

Needless to say, a multidisciplinary team approach consisting of a sleep specialist, rheumatologist, psychiatrist, clinical psychologist, and other health professionals as necessary is ideal. Social and family support systems should be strengthened. Overlapping conditions often seen in FM patients must be noted and treated. Overlapping conditions include irritable bowel syndrome, chronic headaches, temporomandibular joint dysfunction, autonomic dysfunction, chronic fatigue syndrome, posttraumatic stress syndrome, irritable bladder, and chemical sensitivities.

CHRONIC PAIN SYNDROME

There is a complex relationship between chronic pain (CP) and sleep. Pain can disrupt sleep and poor sleep can increase pain intensity. Therefore pain and disturbed sleep have a bidirectional relationship with multifaceted dynamics.

A 1996 Gallup Poll sponsored by the National Sleep Foundation found that 56 million Americans have nighttime pain that interferes with sleep or is associated with early morning awakenings [46]. Fifty to 70% of patients with chronic pain report a sleep impairment. In addition, depression is present in over 20% of CP patients, and there appears to be a synergistic relationship between pain and depression. Sleep disturbances correlate with higher pain intensity, depression, anxiety, and decreased activity levels.

Hyyppa and Kronholm [47] compared reports of sleep disturbances among 24 patients with FM, 60 patients with a variety of other musculoskeletal disorders (e.g., sciatica, low back pain without sciatica, neck and shoulder syndrome, spondylarthrosis, rheumatoid arthritis) and 91 healthy controls. All patients reported poor sleep and insomnia [47]. In another study, Harman et al. [48] reported that there was no difference in sleep onset, total sleep time, or sleep stage parameters between patients with chronic low back pain and age-matched control subjects with no sleep disorders. However, low sigma power was present in patients with chronic lower back pain, which could provide an explanation for poor sleep quality and less effective sensorimotor gating/during sleep. Depressed patients with chronic lower back pain had more occipital delta, more occipital and central alpha, and more high beta activity than nondepressed patients with chronic lower back pain. Control subjects had more sigma and more low occipital frontal beta activity than chronic lower back pain patients.

Chronic pain may also impact insomnia prevalence. Wilson et al. [49] reported that 64% of 150 consecutive clinic CP patients had significant insomnia. Insomnia prevalence was 84% among patients with major depression and 55% among those without major depression. Patients with major depression and insomnia had the highest level of pain-related impairment [49].

Pain management during sleep includes the use of medications, psychological therapy, and behavioral therapy. Chronic pain might be managed with oral long-acting opioids and implantable delivery systems allowing for epidural and intrathecal drug delivery. One possible consequence of this therapy is the emergence of central sleep apnea [50]. Therefore, patients on chronic opioid therapy should be screened for symptoms of central sleep apnea.

REFERENCES

1. Wolfe F, Smythe HA, Yunus MB, Bennett RM, Bombardier C, Goldenberg DL, Tugwell P, Campbell SM, Abeles M, Clark P, et al. The American College of Rheumatology 1990 Criteria for the Classification of Fibromyalgia. Report of the Multicenter Criteria Committee. *Arthritis Rheum* **33**(2):160–172(1990).
2. Hawley DJ, Wolfe F, Lue FA, Moldofsky H. Seasonal symptom severity in patients with rheumatic diseases: a study of 1,424 patients. *J Rheumatol* **28**(8):1900–1909(2001).
3. Okifuji A, Turk DC. Stress and psychophysiological dysregulation in patients with fibromyalgia syndrome. *Appl Psychophysiol Biofeedback* **27**(2):129–141(2002).
4. Campbell SM, Clark S, Tindall EA, Forehand ME, Bennett RM. Clinical characteristics of fibrositis. I. A "blinded," controlled study of symptoms and tender points. *Arthritis Rheum* **26**(7):817–824(1983).
5. Bradley LA, Alarcon GS. Sex-related influences in fibromyalgia. *Sex Gender Pain* **17**:281–307(2000).
6. Neeck G. Pathogenic mechanisms of fibromyalgia. *Ageing Res Rev* **1**(2):243–255(2002).
7. Bradley LA, McKendree-Smith NL, Alarcon GS, Cianfrini LR. Is fibromyalgia a neurologic disease? *Curr Pain Headache Rep* **6**(2):106–114(2002).
8. Sallanon M, Buda C, Janin M, Jouvet M. Implication of serotonin in sleep mechanisms: induction, facilitation? In: Wauquier A, Monte JM, Gaillard JM, Radulovacki M (Eds), *Sleep: Neurotransmitters and Neuromodulators*. Raven Press, New York, 1985, p 136.
9. Bennett RM, Clark SR, Campbell SM, Burchkhardt CS. Low levels of somatomedin C in patients with the fibromyalgia syndrome: a possible link between sleep and muscle pain. *Arthritis Rheum* **35**:1113–1116(1992).
10. Bennett RM, Clark S, Campbell SM, Walczyk J. A randomized double-blind placebo-controlled study of growth hormone in the treatment of fibromyalgia. *Am J Med* **104**:227–231(1998).
11. Mountz JM, Bradley LA, Modell JG, et al. Fibromyalgia in women. Abnormalities of regional cerebral blood flow in the thalamus and caudate nucleus are associated with low pain threshold levels. *Arthritis Rheum* **38**:926–938(1995).
12. Schaefer KM. Sleep disturbances and fatigue in women with fibromyalgia and chronic fatigue syndrome. *J Obstet Gynecol Neonatal Nurs* **24**:229–233(1995).
13. Molony RR, MacPeek DM, Schiffman PL, Frank M, Neubauer JA, Schwartzberg M, et al. Sleep, sleep apnea, and the fibromyalgia syndrome. *J Rheumatol* **13**:797–800(1986).
14. Affleck G, Urrows S, Tennen H, Higgins P, Abeles M. Sequential daily relations of sleep, pain intensity, and attention to pain among women with fibromyalgia. *Pain* **68**(2-3):363–368(1996).
15. May KP, West SG, Baker MR, Everett DW. Sleep apnea in male patients with the fibromyalgia syndrome. *Am J Med* **94**:505–508(1993).
16. Donald F, Esdaile JM, Kimoff JR, Fitzcharles M-A. Musculoskeletal complaints and fibromyalgia in patients attending a

respiratory sleep disorders clinic. *J Rheumatol* **23**:1612–1616(1996).

17. Yunus MB, Aldag JC. Restless legs syndrome and leg cramps in fibromyalgia syndrome: a controlled study. *BMJ* **312**(7042):1339(1996).

18. Harding SM. Sleep in fibromyalgia patients: subjective and objective findings. *Am J Med Sci* **315**(6):367–376(1998).

19. Landis CA, Lentz MJ, Rothermel J, Buchwald D, Shaver JLF. Decreased sleep spindles and spindle activity in midlife women with fibromyalgia and pain. *Sleep* **27**:741–750(2004).

20. Branco J, Atalaia A, Paiva T. Sleep cycles and alpha–delta sleep in fibromyalgia syndrome. *J Rheumatol* **21**(6):1113–1117(1994).

21. Anch AM, Lue FA, MacLean AW, Moldofsky H. Sleep physiology and psychological aspects of fibrositis (fibromyalgia) syndrome. *Can J Exp Psychol* **45**:179–184(1991).

22. Moldofsky H, Lue FA. The relationship of alpha and delta EEG frequencies to pain and mood in "fibrositis" patients treated with chlorpromazine and L-tryptophan. *Electroencephalogr Clin Neurophysiol* **50**(1-2):71–80(1980).

23. Drewes AM, Gade J, Nielsen KD, Bjerregard K, Taagholt SJ, Svendsen L. Clustering of sleep electroencephalopathic patterns in patients with the fibromyalgia syndrome. *Br J Rheumatol* **34**:1151–1156(1995).

24. Roizenblatt S, Moldofsky H, Benedito-Silva AA, Tufik S. Alpha sleep characteristics in fibromyalgia. *Arthritis Rheum* **44**(1):222–230(2001).

25. Moldofsky H, Scarisbrick P, England R, Smythe H. Musculoskeletal symptoms and non-REM sleep disturbance in patients with "fibrositis syndrome" and healthy subjects. *Psychosom Med* **37**:341–351(1975).

26. Perlis ML, Giles DE, Bootzin RR, Dikman ZV, Fleming GM, Drummond SPA, et al. Alpha sleep and information processing, perception of sleep, pain, and arousability in fibromyalgia. *Int J Neurosci* **89**:265–280(1997).

27. Mahowald ML, Mahowald MW. Nighttime sleep and daytime functioning (sleepiness and fatigue) in less well-defined chronic rheumatic diseases with particular reference to the "alpha–delta NREM sleep anomaly." *Sleep Med* **1**(3):195–207(2000).

28. Rains JC, Penzien DB. Sleep and chronic pain: challenges to the alpha-EEG sleep pattern as a pain specific sleep anomaly. *J Psychosom Res* **54**(1):77–83(2003).

29. Fitzcharles MA, Costa DD, Poyhia R. A study of standard care in fibromyalgia syndrome: a favorable outcome. *J Rheumatol* **30**(1):154–159(2003).

30. Palangio M, Flores JA, Joyal SV. Treatment of fibromyalgia with sibutramine hydrochloride monohydrate: comment on the article by Goldenberg et al. *Arthritis Rheum* **46**(9):2545–2546(2002); author reply p 2546.

31. Henriksson KG, Sorensen J. The promise of N-methyl-D-aspartate receptor antagonists in fibromyalgia. *Rheum Dis Clin North Am* **28**(2):343–351(2002).

32. Bennett RM, Clark SC, Walczyk J. A randomized, double-blind, placebo-controlled study of growth hormone in the treatment of fibromyalgia. *Am J Med* **104**(3):227–231(1998).

33. Farber L, Stratz TH, Bruckle W, Spath M, Pongratz D, Lautenschlager J, Kotter I, Zoller B, Peter HH, Neeck G, Welzel D, Muller W; German Fibromyalgia Study Group. Short-term treatment of primary fibromyalgia with the 5-HT3-receptor antagonist tropisetron. Results of a randomized, double-blind, placebo-controlled multicenter trial in 418 patients. *Int J Clin Pharmacol Res* **21**(1):1–13(2001).

34. Busch A, Schachter CL, Peloso PM, Bombardier C. Exercise for treating fibromyalgia syndrome. *Cochrane Database Syst Rev* **3**:CD003786(2002).

35. Richards SC, Scott DL. Prescribed exercise in people with fibromyalgia: parallel group randomized controlled trial. *BMJ*:**325**(7357):185(2002).

36. Williams DA, Cary MA, Groner KH, Chaplin W, Glazer LJ, Rodriguez AM, Clauw DJ. Improving physical functional status in patients with fibromyalgia: a brief cognitive-behavioral intervention. *J Rheumatol* **29**(6):1280–1286(2002).

37. Moldofsky H, Lue FA, Mously C, Roth-Schechter B, Reynolds WJ. The effect of zolpidem in patients with fibromyalgia: a dose ranging, double blind, placebo controlled, modified cross-over study. *J Rheumatol* **23**(3):529–533(1996).

38. Moldofsky H. Management of sleep disorders in fibromyalgia. *Rheum Dis Clin North Am* **28**(2):353–365(2002).

39. Heymann RE, Helfenstein M, Feldman D. A double-blind, randomized, controlled study of amitriptyline, nortriptyline and placebo in patients with fibromyalgia. An analysis of outcome measures. *Clin Exp Rheumatol* **19**(6):697–702(2001).

40. O'Malley PG, Balden E, Tomkins G, Santoro J, Kroenke K, Jackson JL. Treatment of fibromyalgia with antidepressants: a meta-analysis. *J Gen Intern Med* **15**(9):659–666(2000).

41. Goldenberg D, Mayskiy M, Mossey C, Ruthazer R, Schmid C. A randomized, double-blind crossover trial of fluoxetine and amitriptyline in the treatment of fibromyalgia. *Arthritis Rheum* **39**:1852–1859(1996).

42. Branco JC, Martini A, Palva T. Treatment of sleep abnormalities and clinical complaints in fibromyalgia with trazodone(abstract). *Arthritis Rheum* **39**:591(1996).

43. Scharf MB, Baumann M, Berkowitz DV. The effects of sodium oxybate on clinical symptoms and sleep patterns in patients with fibromyalgia. *J Rheumatol* **30**(5):1070–1074(2003).

44. Landis CA, Frey CA, Lentz MJ, Rothermel J, Buchwald D, Shaver JL. Self-reported sleep quality and fatigue correlates with actigraphy in midlife women with fibromyalgia. *Nurs Res* **52**(3):140–147(2003).

45. Gold AR, Dipalo F, Gold MS, Broderick J. Inspiratory airflow dynamics during sleep in women with fibromyalgia. *Sleep* **27**:459–466(2004).

46. National Sleep Foundation. Sleep and aging. Available at http://sleepfoundation/publications/sleepage.html (accessed October 10, 2003).

47. Hyyppa MT, Kronholm E. Nocturnal motor activity in fibromyalgia patients with poor sleep quality. *J Psychosom Res* **39**:85–91(1995).

48. Harman K, Pivik RT, D'Eon JL, Wilson KG, Swenson JR, Matsunaga L. Sleep in depressed and nondepressed participants with chronic low back pain: electroencephalographic and behavior findings. *Sleep* **25**(7):775–783(2002).

49. Wilson KG, Watson ST, Currie SR. Daily diary and ambulatory activity monitoring of sleep in patients with insomnia associated with chronic musculoskeletal pain. *Pain* **75**(1):75–84(1998).

50. Farney RJ, Walker JM, Cloward TV, Rhondeau S. Sleep-disordered breathing associated with long-term opioid therapy. *Chest* **123**(2):632–639(2003).

100

SLEEP AND THE IMMUNE RESPONSE

JAMES M. KRUEGER AND JEANNINE A. MAJDE
Washington State University, Pullman, Washington

INTRODUCTION

Sleepiness is frequently experienced during acute infections and other inflammatory diseases and encourages the patient to seek a warm bed. Since Hippocrates, physicians (and concerned parents) have recommended getting more sleep during acute illnesses, though its benefits have not been established scientifically. Today the fatigue, sleepiness, and social withdrawal associated with illness onset are considered part of the acute phase response to infectious challenge, along with fever. Over the last two decades, advances in immunology and sleep biochemistry have demonstrated the molecular basis for the association of excess sleep with inflammation. In this chapter we describe changes in sleep that occur during infections. Then, because infection is associated with the activation of the immune response, we outline some changes in the immune response associated with sleepiness and sleep loss. We go on to describe a molecular network that connects the immune response to the brain; this network is responsible for physiological sleep regulation as well as inflammation-induced changes in sleep. Finally, we end with the question: Does sleep really help in recuperation from or prevention of disease?

ENCEPHALITIS LETHARGICA AND SLEEP

An important early paper for sleep research by Von Economo [1] described how encephalitic lesions of the hypothalamus result in permanent changes in sleep for the affected individual. If the lesion was in the anterior hypothalamus, the patient slept less, while if the lesion

was in the posterior hypothalamus, the patient slept more. This work resulted in the concept that sleep was actively regulated. (Von Economo's encephalitis lethargica, first reported in 1916 and still seen clinically, is currently thought to be an autoimmune disease induced by streptococcal infections [2] rather than a virus, as previously thought.)

MICROBIAL CHALLENGE—WHAT DOES IT DO TO SLEEP?

The first systematic study of sleep over the course of an infectious disease dealt with a gram-positive bacterial septicemia in rabbits [3]. Within a few hours of the intravenous injection of live *Staphylococcus aureus*, the animals begin to exhibit more non-rapid eye movement (NREM) sleep and simultaneously less rapid eye movement (REM) sleep, a sleep profile characteristic of acute infections [4]. After a period of about 20 hours, the rabbits go into a period of prolonged reduction of NREM sleep and REM sleep. During this biphasic sleep response, other facets of the acute phase response are evident, including fever, fibrinogenemia, and neutrophilia. In subsequent studies using other bacteria and other routes of administration, this biphasic sleep response of initial NREM sleep enhancement followed by sleep disruption was also evident. However, the specific time courses of the sleep responses are dependent on the capacity of the infectious agent to invade the host, the dose, the time of day, and its route of administration. For instance, after intravenous administration of the nonpathogenic gram-negative bacterium *Escherichia coli*,

Sleep: A Comprehensive Handbook, Edited by T. Lee-Chiong.

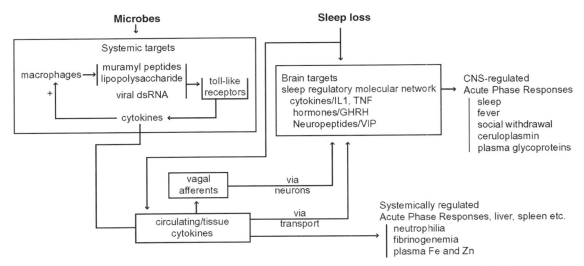

Figure 100.1   Microbes and sleep loss promote sleepiness via the brain's sleep-regulating molecular network. This network is involved in physiological sleep regulation and it includes cytokines. Cytokines are also key regulatory molecules for the host's defense. (IL-1, interleukin-1; TNF, tumor necrosis factor; VIP, vasoactive intestinal polypeptide; GHRH, growth hormone releasing hormone.)

there is a very rapid increase in NREM sleep, but this enhancement only lasts for 4–6 hours. In contrast, intranasal injection of another gram-negative bacterium that is a natural pathogen of rabbits, *Pasteurella multocida,* induced more prolonged changes in sleep (reviewed in [4, 5]).

The mechanisms by which bacteria induce changes in sleep involve macrophage processing of bacteria. Thus macrophages ingest and digest bacteria and in the process release chemically distinctive molecules derived from the bacterial cell walls such as muramyl peptides from peptidoglycan and lipopolysaccharide [4] (see Figure 100.1). These molecules can initiate sleep profiles characteristic of infections in the absence of living bacteria, though their action is shorter. Microbial components are detected by a family of pathogen recognition receptors (primarily toll-like receptors (TLRs), see Figure 100.1) [6] that initiate the production of the intercellular signaling molecules called cytokines. Cytokines are produced by virtually all cells, particularly immunocytes such as macrophages and central nervous system cells such as neurons and glia. Over a hundred cytokines have been defined, both proinflammatory (binding Class I receptors) and anti-inflammatory (binding Class II receptors) [7], and operate in extremely complex networks to initiate inflammation and acquired immunity [6]. Some Class I cytokine receptors share homologies with receptors for classical endocrine hormones such as prolactin and growth hormone [7]. Cytokines within the brain are involved in sleep regulation (see later discussion) while cytokines within the peripheral immune system can induce cytokines in the brain through action at circumventricular organs, vagal afferents, and endothelial transporters [8], thus causing an acute phase response to extraneural inflammatory events such as infections (reviewed in [4]).

Viral infections also induce cytokines and affect sleep. Some viruses may cause brain lesions and thus changes in sleep during these diseases, resulting from direct damage to the brain as well as the virus-induced cytokine response. Enhanced NREM sleep during the early stages of human immunodeficiency virus (HIV) infections, before acquired immunodeficiency syndrome (AIDS) onset, has been described [4]. After AIDS onset, sleep is disrupted. Similarly, another CNS viral disease, rabies, is associated with disrupted sleep (reviewed in [5]). Viruses are also implicated in a wide range of other disorders that involve sleep disruption, such as sudden infant death syndrome, chronic fatigue syndrome, and infectious mononucleosis (reviewed in [4]). However, direct involvement of the viruses in the sleep disruptions seen in these conditions is yet to be demonstrated.

As indicated earlier, in viral diseases that involve CNS damage, it is difficult to distinguish whether the effects of the virus on sleep result from virus-induced lesions or other mechanisms. Several investigators have thus turned to an influenza virus infection model in which the virus localizes to the lungs during the early stages of the disease. In humans, low doses of influenza induce excess behavioral sleep without certain other facets of the acute phase response such as fever. More detailed studies of influenza virus effects on sleep involve animal models, largely rabbits and mice. Large doses of live influenza virus (but not killed virus) given intravenously to rabbits induce a short-term fever and sleep response similar to that induced by *E. coli* [4]. Influenza virus does not completely replicate in rabbits, accounting for its short-term effects. However, there are indications that partial viral replication occurs, with the production of the replication intermediate double-stranded (ds) RNA. There are many common features of the acute phase response to abortive viral infections and to synthetic or viral dsRNA [9], and dsRNA (recognized by TLR3) [6] is probably a primary inducer of the viral acute phase response through its induction of cytokines [10] (Figure 100.1).

Influenza virus replicates completely in the mouse and can cause lethal pneumonitis. This disease is associated with excess NREM sleep and reduced REM sleep that become more marked as the disease progresses [11]. The mice also become severely hypothermic and lose up to 20% of their body weight. Infected mice deficient in the gene for the growth hormone releasing hormone receptor have suppressed NREM sleep compared to controls, while REM sleep is suppressed in the same manner as in controls [12]. On the other hand, mice deficient in the gene for neuronal nitric oxide synthetase show more suppression of REM sleep in response to influenza [13], while mice deficient in the gene for inducible nitric oxide synthetase show increased REM sleep and reduced NREM sleep compared to controls [13]. Mice deficient in the receptor for type I interferons (IFNs), cytokines known for their antiviral activity, show a marked suppression of spontaneous REM sleep [14]. These mice also show altered expression of certain neuropeptides in their hypothalami that may mediate the suppressed REM sleep (cf. Figure 100.2). When infected with low-dose influenza or challenged with dsRNA, these mice show earlier and more intense NREM sleep [15]. REM sleep did not change in response to dsRNA challenge but was suppressed below baseline in IFN receptor-deficient mice infected with influenza [15]. Based on our studies to date in the influenza model, we can say that a product of growth hormone releasing hormone (possibly not growth hormone itself), nitric oxide made by inducible nitric oxide synthetase, and type I IFNs are involved in NREM sleep regulation during infection, while nitric oxide made by neuronal nitric oxide synthetase appears to be involved in influenza-induced REM sleep suppression. In addition, we have shown that type I IFNs are important in spontaneous REM sleep regulation as well as influenza-induced REM sleep regulation. Our studies continue in inflammatory factor-deficient mice to better define the mediators of sleep changes induced by acute infection.

SLEEP LOSS AND EFFECTS ON IMMUNE SYSTEM PARAMETERS

It is widely observed that the intense pressure in developed nations to be productive has resulted in generalized sleep deprivation. Shift workers are particularly subject to chronic sleep deprivation. It is important to determine the consequences of sleep deprivation on public health beyond

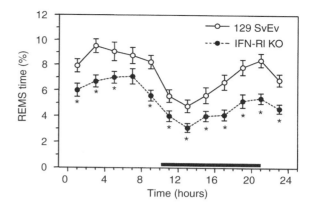

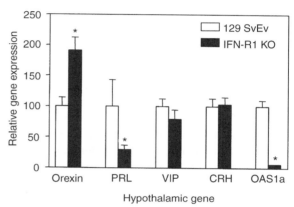

Figure 100.2 Rapid eye movement sleep (REMS) is reduced in mice lacking the interferon type I receptor (IFN-RI KO). The reduction in REMS is associated with enhanced orexin (also called hypocretin), a wake-promoting substance, and reduced prolactin (PRL), a REMS regulatory substance (see Table 100.1). These mice also have reduced 2,5-oligoadenylate synthetase (OAS1a) (an antiviral enzyme) mRNA levels. These molecules are related to each other, for example, IFN-α inhibits orexin expression. Orexin inhibits pituitary release of prolactin. OAS1a inhibits prolactin-mediated activation of STAT-1 and interferon regulatory factor 1 (see [14]).

the obvious association of increased accidents. A few studies have examined the consequences of sleep deprivation on the immune response to vaccinations in healthy individuals. Sleep deprivation for only one night substantially impairs the antibody response to hepatitis A vaccine [16]. Chronic but less profound sleep deprivation substantially slows the response to influenza vaccine [17], though the subjects do eventually achieve antibody levels similar to controls. Studies of the immune effects of acute sleep deprivation in animals have given inconsistent results [4].

More profound chronic sleep loss in rats (total deprivation for 2–3 weeks) results in sepsis and death of the animals, probably as a consequence of bacterial translocation from the intestine [4]. Yolked controls that obtain about 80% of normal sleep survive. These findings suggest an

TABLE 100.1   Sleep Regulatory Substances[a]

NREM Sleep	REM Sleep	Wakefulness
Growth hormone releasing hormone (GHRH)	Vasoactive intestinal polypeptide (VIP)	Corticotropin-releasing hormone (CRH)
Tumor necrosis factor-α (TNF-α)	Prolactin	Hypocretin
Interleukin-1β (IL-1β)	Nitric oxide	Noradrenalin
Adenosine	Acetylcholine	Serotonin
Prostaglandin D_2		Acetylcholine

[a]The substances listed have fulfilled the criteria to implicate them in sleep regulation (see text). Many other substances are likely components of the molecular network regulating sleep (see [19]).

impairment of the innate immune system that normally prevents the escape of normal intestinal flora into the draining lymph nodes. While such chronic studies cannot be performed ethically in human subjects, studies in military trainees subjected to prolonged sleep deprivation (in addition to the other stresses of training) experience profound endocrine and immune changes [18] and, in general, more frequent and severe infections. How much of this effect is sleep related versus stress related cannot be determined.

Examination of specific immune parameters following sleep deprivation indicates that antigen uptake, lymphocyte mitogenesis, phagocytosis, circulating immune complexes, circulating immunoglobulin, secondary antibody responses, natural killer cells, and T lymphocyte populations are altered (reviewed in [4]). Studies of cytokine production in cultured lymphocytes from sleep-deprived subjects show increased IFN, tumor necrosis factor (TNF), and interleukin-1β (IL-1β) production. Circulating cytokines display circadian variation and associations for different sleep stages; IL-1 levels peak at the onset of NREM sleep and TNF levels vary with EEG slow waves [4]. When examined in sleep-deprived subjects, a trend toward increased circulating IL-1 is seen [4]. In sleep apnea, which is associated with sleep deprivation as well as hypoxia, increased TNF levels are seen [4].

MOLECULAR NETWORKS: SLEEP AND IMMUNE RESPONSE MODIFIERS

Research concerning the biochemical regulation of sleep had its origins in the early 1900s in the observation that the transfer of cerebrospinal fluid from sleep-deprived animals to controls enhances sleep in the recipients (reviewed in [19]). Since that time, several sleep regulatory substances (SRSs) have been identified. In order for a substance to be classified as a SRS, the molecule should meet the criteria that have been developed by several investigators. These criteria include: (1) the candidate SRS should enhance sleep, (2) its inhibition should inhibit spontaneous sleep, (3) its levels should vary in brain with sleep propensity, (4) it should act on sleep regulatory circuits to affect sleep, and (5) its levels should vary with pathologies that

affect sleep. Table 100.1 shows the molecules that have thus far met these criteria. Included on this list are two cytokines, IL-1 and TNF, and we focus on this evidence here because these cytokines, as mentioned earlier, are also immune response mediators. Many other cytokines have the capacity to either enhance (e.g., fibroblast growth factor, nerve growth factor, IL-2, IL-6, IL-8, IL-15, IL-18) or inhibit (e.g., IL-4, IL-10, IL-13, insulin-like growth factor) sleep. However, although these molecules may indeed be part of a physiological network of molecules involved in sleep regulation, insufficient information is available to classify them as SRSs. SRSs are also part of the regulatory mechanism by which the brain keeps track of past sleep/wake activity for prolonged periods of time and thereby provide a mechanism for sleep homeostasis [20].

Administration of either TNF or IL-1, whether they are given centrally or systemically, alters NREM sleep. After low doses, NREM sleep is enhanced. Slightly higher doses result in more NREM sleep, accompanied by a reduction of REM sleep. Even higher doses inhibit both NREM and REM sleep [19]. These effects are also time-of-day dependent in that some doses inhibit sleep at one time of the day while they enhance sleep at another time of the day. In contrast, inhibition of either IL-1 or TNF using antibodies, soluble receptors, or inhibitory cytokines (such as the endogenous IL-1 receptor antagonist) inhibit spontaneous sleep. These inhibitors also inhibit the sleep rebound that follows sleep deprivation. Somnogenic doses of either TNF or IL-1 also enhance the amplitudes of EEG slow waves during NREM sleep. Enhanced EEG slow-wave activity is thought to be indicative of deeper sleep since it is also observed after sleep loss when the stimulus threshold needed to awake an individual is higher. Other aspects of physiological sleep also persist in TNF-treated or IL-1-treated animals (e.g., sleep remains episodic and easily reversible).

Brain levels of TNF and IL-1 vary with the time of day. For instance, in rats TNF mRNA and TNF protein levels are higher in the cortex and hypothalamus during the daytime when sleep propensity is highest. In contrast, if animals are sleep deprived, levels of these cytokines in brain increase as does sleep propensity. As already mentioned, circulating levels of TNF vary with pathologies such as sleep apnea, preeclampsia, and insomnia, all of which are associated with changes in sleepiness (reviewed in [19]). Furthermore, infectious challenge is also associated with an upregulation of cytokines and sleepiness as described earlier. Finally, bacterial cell wall products such as muramyl peptides and viral dsRNA [10] also enhance cytokine production including IL-1 and TNF (Figure 100.1). Collectively, such results strongly implicate these cytokines in physiological sleep as well as the sleep responses associated with pathology.

Both TNF and IL-1 act on sleep regulatory circuits to affect sleep. Microinjection of TNF, for example, into the preoptic area of the anterior hypothalamus enhances NREM sleep. In contrast, the injection of the TNF soluble receptor into this area reduces spontaneous sleep. In this same area, IL-1 inhibits wake-active neurons while it enhances sleep-active neurons (reviewed in [19]). Such data suggest that these cytokines are acting on sleep regulatory circuits to induce sleep. However, other evidence suggests they may act elsewhere as well to enhance sleep. Microinjection of either TNF or IL-1 directly onto the cortex unilaterally enhances EEG slow-wave power during NREM sleep on the side injected but not on the opposite side of the brain. Furthermore, similar localized injections of either the TNF soluble receptor or the IL-1 soluble receptor inhibit sleep deprivation-induced increases in EEG slow-wave power during NREM sleep on the side receiving the soluble receptor but not on the other side of the brain [21]. Such state-dependent changes in EEG power suggest that these cytokines can act locally within the cortex to promote functional state changes in small regions of the brain. Such changes may also provide a mechanism by which sleep is targeted to specific areas of brain depending on their prior wakefulness activity (reviewed in [20]).

DOES SLEEP HELP IN COMBATING INFECTIOUS DISEASE?

This question is difficult to address experimentally because it is impossible to isolate sleep *per se* as an independent variable. If one deprives an animal or person of sleep, many physiological systems change, including body temperature, food intake, many hormones, and many immune response parameters. Thus any change in the host's response to infectious challenge occurring during sleep loss may be secondary to these other changes. Nevertheless, the limited evidence to date suggests that there is an association between sleep and morbidity and mortality. For instance, Toth et al. [22] showed that the animals that sleep more during the first few hours after infectious challenge have a higher probability of survival than those that did not.

ACKNOWLEDGMENTS

This work was supported in part by the National Institutes of Health, grant numbers NS25378, NS27250, NS31453, and HD36520.

REFERENCES

1. Von Economo C. Sleep as a problem of localization. *J Nerv Ment Dis* **71**:249–259(1930).

2. Dale RC, Church AJ, Surtees RAH, Lees AJ, Adcock JE, Harding B, Neville BGR, Giovannoni G. Encephalitis lethargica syndrome: 20 new cases and evidence of basal ganglia autoimmunity. *Brain* **127**:21–33(2003).

3. Toth LA, Krueger JM. Alteration of sleep in rabbits by *Staphylococcus aureus* infection. *Infect Immun* **56**:1785–1791(1988).

4. Majde JA, Krueger JM. Neuroimmunology of sleep. In: *Biological Psychiatry*. John Wiley & Sons, London, 2002, pp 1247–1257.

5. Toth LA. Microbial modulation of sleep. In: *Handbook of Behavioral State Control: Cellular and Molecular Mechanisms*. CRC Press, Boca Raton, FL, 1999, pp 641–657.

6. Akira S, Hemmi H. Recognition of pathogen-associated molecular patterns by TLR family. *Immunol Lett* **85**:85–95(2003).

7. Gadina M, Hilton D, Johnston JA, Morinobu A, Lighvani A, Zhou Y-J, Visconti R, O'Shea JJ. Signaling by type I and II cytokine receptors: ten years after. *Curr Opin Immunol* **13**:363–373(2001).

8. Larson SJ, Dunn AJ. Behavioral effects of cytokines. *Brain Behav Immun* **15**:371–387(2001).

9. Traynor TR, Majde JA, Bohnet SG, Krueger JM. Intratracheal double-stranded RNA plus interferon-gamma: a model for analysis of the acute phase response to respiratory viral infections. *Life Sci* **74**:2563–2576(2004).

10. Majde JA. Viral double-stranded RNA, cytokines and the flu. *J Interferon Cytok Res* **20**:259–272(2000).

11. Fang J, Sanborn CK, Renegar KB, Majde JA, Krueger JM. Influenza viral infections enhance sleep in mice. *Proc Soc Exp Biol Med* **210**:242–252(1995).

12. Alt JA, Obal F Jr, Traynor TR, Gardi J, Majde JA, Krueger JM. Alterations in EEG activity and sleep after influenza viral infection in GHRH receptor-deficient mice. *J Appl Physiol* **95**:460–468(2003).

13. Chen L, Duricka D, Nelson S, Mukherjee S, Bohnet SG, Taishi P, Majde JA, Krueger JM. Influenza virus-induced sleep responses in mice with targeted disruptions in neuronal or inducible nitric oxide synthases. *J Appl Physiol* **97**:17–28(2004).

14. Bohnet SG, Traynor TR, Majde JA, Kacsoh B, Krueger JM. Mice deficient in the interferon type I receptor have reduced REM sleep and altered hypothalamic hypocretin, prolactin and 2,5-oligoadenylate synthase expression. *Brain Res* **1027**:117–125(2004).

15. Traynor TR, Majde JA, Bohnet SG, Krueger JM. The role of type I interferons in sleep and body temperature responses to influenza virus infection. Sleep **28**:A41 (2005).

16. Lange T, Perras B, Fehm HL, Born J. Sleep enhances the human antibody response to hepatitis A vaccination. *Psychosom Med* **65**:831–835(2003).

17. Spiegel K, Sheridan JF, Van Cauter E. Effect of sleep deprivation on responses to immunization. *JAMA* **288**:1471–1472(2002).

18. Bernton E, Hoover D, Galloway R, Popp K. Adaptation to chronic stress in military trainees. Adrenal androgens, testosterone, glucocorticoids, IGF-1, and immune function. *Ann NY Acad Sci* **774**:217–231(1995).

19. Obal F Jr, Krueger JM. Biochemical regulation of sleep. *Front Biosci* **8**:511–519(2003).

20. Krueger JM, Obal F Jr. Sleep function. *Front Biosci* **8**:520–550(2003).

21. Yoshida H, Peterfi Z, Garcia-Garcia F, Kirkpatrick R, Yasuda T, Krueger JM. Asymmetries in slow wave sleep EEG induced by local application of TNFα. *Brain Res* **1009**:129–136(2004).

22. Toth LA, Tolley EA. Krueger JM. Sleep as a prognostic indication during infectious disease in rabbits. *Proc Soc Exp Biol Med* **203**:179–192(1993).

PART XIV

SLEEP IN THE NEUROLOGIC DISORDERS

101

ALZHEIMER'S DEMENTIA

Michael V. Vitiello

University of Washington, Seattle, Washington

INTRODUCTION

Alzheimer's disease (AD) is a progressive neurodegenerative disorder that accounts for approximately two-thirds of all dementias worldwide. In the United States it has recently been estimated that between 2 and 4 million older adults have AD. Further estimates suggest that this number is likely to quadruple over the next 50 years. Both normal aging and AD are associated with disturbances in the daily sleep–wake cycle, although the disturbances in AD are typically much more severe. Within the AD population itself, significant sleep disturbances are quite common, affecting as much as half of community- and clinic-based samples.

Certainly the same neurodegenerative mechanism that results in the evolving cognitive deficits seen in AD are a potential primary cause of the sleep disruption seen in this population. However, it is important to recognize that sleep can be disrupted for other additional reasons as well. The major likely causes of sleep disruption in aging and dementia include (1) physiological changes that arise as part of normal, "nonpathological" aging; (2) sleep problems due to comorbid physical or mental health conditions and possibly their treatments; (3) primary sleep disorders; (4) poor "sleep hygiene" or sleep-related habits, which can include patterns of daytime sleeping established over long years of night-shift work that become maladaptive in the retirement years; or commonly, (5) some combination of these factors.

For AD patients, sleep disturbance adds an additional burden to the compromised function and quality of life directly attributable to dementia. For AD caregivers, disturbances in patients' sleep and nighttime behavior, particularly the reduced nighttime sleep time, increased nighttime wakefulness, and wandering that commonly require considerable caregiver attention, with subsequent chronic sleep loss for the caregiver, are a significant source of physical and psychological burden and are often cited as one of the principal reasons for a family's decision to institutionalize a demented person. Consequently, more effective management of sleep disturbances in AD should be a priority for AD treatment research.

BIOLOGICAL BASES OF SLEEP DISTURBANCES IN AD

The sleep disturbances that accompany early stage or mild dementia are remarkable in that they appear to be exacerbations of the sleep changes found with "normal" aging, rather than unique disease-related phenomena [1]. The sleep of AD patients is marked by an increased duration and frequency of awakenings, decreased slow-wave sleep and REM sleep, and more daytime napping. Damage to neuronal pathways that initiate and maintain sleep is the most likely cause of the acceleration of these age-related sleep changes in AD patients. These compromised neural structures include the suprachiasmatic nucleus (SCN) of the hypothalamus, and neuronal pathways originating in

Sleep: A Comprehensive Handbook, Edited by T. Lee-Chiong.
Copyright © 2006 John Wiley & Sons, Inc.

subcortical regions also regulate arousal and sleep–wake cycles, including the cholinergic basal forebrain nuclei, serotoninergic raphe nuclei, dopaminergic nigrostriatal and pallidostriatal pathways, and noradrenergic locus caeruleus and the cortical regions that generate EEG slow-wave activity during sleep [2–5]. Very recent work has suggested that at least some of the sleep disruption seen in AD many be an inherent trait [6], likely linked to APOE status [7], such that AD patients negative for the APOE epsilon4 allele show greater sleep disruption over time.

There is considerable evidence that sleep disturbance grows more severe with increasing severity of AD [1, 8]. However, the moderate, or intermediate, stages of the disease are typically when most other behavioral disturbances occur with peak frequency. Shifts in the basic circadian sleep–wake rhythm of dementia patients can be severe and, in extreme cases, may lead to complete day/night sleep pattern reversals. In end-stage AD, patients may appear to sleep throughout most of the day and night, awakening only for brief periods. However, currently there have been no prospective longitudinal studies of sleep in AD that would inform an accurate understanding of both course and individual differences in sleep disturbances, and this remains an important gap in both the biology and therapeutics of sleep disturbances in demented patients.

Treatment of dementia with oral acetylcholinesterase inhibitors, currently the standard of care for treating cognitive disorders in AD and increasingly used by specialists in treating a wide variety of dementing diseases, may improve sleep patterns in some patients. These agents are believed to act by enhancing cholinergic transmission in the brain. The involvement of both forebrain and brainstem cholinergic nuclei in regulating sleep–wake cycles and arousal forms the rationale for expecting both positive and negative sleep effects with use of these agents.

DIAGNOSIS OF SLEEP DISTURBANCE IN AD

Effective diagnosis and treatment of disturbances of the sleep–wake cycle in AD patients have the potential to reduce the distress caused by these disturbances and to postpone institutionalization that commonly results from these sleep disturbances. While many AD patients develop significant sleep–wake cycle disturbances, treatment research for these problems has been severely hampered. Other than the recent ground-breaking work of Yesavage and colleagues [7] described earlier, there is almost no research to determine which AD patients are at greatest risk for sleep disruption or whether sleep disruptions

make AD symptoms more severe. As described in more detail later, there has also been very little research into possible treatments for sleep disorders in this population.

This dearth of information can be attributed, in part, to deficiencies of the current diagnostic system. In fact, the Food and Drug Administration Psychopharmacological Drugs Advisory Committee has recently emphasized the need for a comprehensive diagnostic system. A key point made by this committee was that behavioral problems associated with dementia (including sleep and chronobiological disturbances) are scientifically and clinically valid targets of pharmacological treatment. However, the diagnostic criteria currently available to define such targets preclude development of FDA-acceptable studies of pharmacological interventions because they do not include the required specific indications for such treatments. Furthermore, the current diagnostic criteria in themselves may hinder researchers in their efforts to achieve greater understanding of the epidemiology and pathophysiology of these disorders.

Recently, at the request of the National Institute of Mental Health and the National Institute on Aging, a working group was formed to address these problems by developing a provisional set of diagnostic criteria for defining sleep disturbance in AD. This working group has attempted to develop better-defined provisional criteria for sleep disturbances in AD [9]. These provisional criteria are based on the best current understanding of sleep–wake cycle disturbances in the AD patient and are designed to correct the limitations of prior criteria. The working group's hope is that these new criteria will help promote state-of-the-art epidemiological, physiological, and especially pharmacological and nonpharmacological treatment research on sleep–wake disturbances associated with AD. How widespread acceptance and use of these criteria will be and whether they will have a beneficial impact on research in the area will await the test of time.

INTERACTIONS BETWEEN BIOLOGICAL CHANGES AND ENVIRONMENTAL SITUATIONS

Studies of patients living in institutional settings provide much of the available information about sleep disorders in cognitively impaired patients [10]. Environmental factors that promote circadian dysregulation in dementia patients living in such settings include light, noise, activity schedules, and the needs of staff (see [3–5]). Ambient light levels are typically too low in many congregate care facilities to support natural light-dependent internal rhythms, and noisy conditions, especially during the night, are both common and inimical to sleep. The staffing schedules and timing of specific activities in many facilities caring for demented persons may be driven less by the needs of the patients and more by compliance with federal and state requirements governing nursing home operations. Regulatory requirements in general fail to incorporate many of the positive evidence-based practices found to be beneficial for demented residents, focusing more attention on feeding and bathing schedules, injury prevention, and detection of medical problems than on sleep and other issues related to quality of life.

OTHER CAUSES OF SLEEP DISTURBANCE IN AD

Population-based studies examining the causes, incidence, and persistence of sleep disturbances in AD patients are lacking; consequently, little is known about the risk factors for their development. Therefore one must look to the literature concerning sleep disturbance in the nondemented elderly and factor this with clinical assessment of individual patients to make informed inferences about the AD population in order to arrive at effective treatment interventions for individual patients (see [11]).

Physical Illness and Related Treatments

Many elderly persons, whether demented or cognitively intact, have medical conditions that disrupt sleep [12]. Untreated insomnia and daytime sleepiness have been associated with nursing home placement and mortality. Medically ill older adults admitted to acute care hospitals are particularly vulnerable to sleep disruptions, which appear to be created as much by the various treatments and procedures, unfamiliar routines, and environmental conditions, as by the pain, anxiety, and discomfort associated with their underlying medical condition. Medical conditions especially likely to disrupt sleep are congestive heart failure, chronic obstructive pulmonary disease, Parkinson's disease, gastroesophageal reflux disease, arthritis, and nocturia.

Many prescription drugs, over-the-counter medications, and social drugs (e.g., caffeine, nicotine, alcohol) can disrupt sleep. However there are no population-based studies relating insomnia or nighttime waking to specific drug classes. Prescribing information for many psychotropic and other drugs often highlights sleep disturbance as a potential side effect in individual patients. In AD, clinical experience dictates that such potential drug effects should always be considered when sleep is disturbed.

Identifying medical disorders that cause poor sleep, followed by changes in management to optimize results, is the best initial approach, but may not be sufficient to reverse an associated sleep problem. Simple measures may, however, be highly effective in selected cases. For example, pre-bedtime use of analgesics may greatly improve the sleep of patients awakened at night by pain. Other possible pharmacological causes of sleep disturbance in the medically ill should also be considered, including high-potency diuretics or drugs with CNS stimulant activity (e.g., caffeine, amphetamines, methylphenidate, and newer stimulants) used too late in the day, and multiple medications with the potential for pharmacokinetic or pharmacodynamic interactions that can affect brain function and sleep rhythms.

Concurrent or Complicating Neuropsychiatric Disorders

Sleep in the elderly may also be affected by psychiatric morbidity. Psychiatric disorders, particularly major depressions, are not only associated with disturbed sleep but can also greatly impact both self-report and objective ratings of sleep quantity and quality. Depressive symptoms are common in older adults, especially among persons who are medically ill, bereaved, or cognitively impaired, but by no means always associated with disrupted sleep. In AD patients seen in clinical psychiatric settings, rates of major depression as high as 86% have been reported, but the majority of studies report rates of 17–29%.

Depression should always be evaluated as a possible contributor to the sleep disturbances encountered in demented individuals. Pharmacological treatment of mood or behavioral disorders associated with sleep disturbances in AD frequently improves sleep patterns, although controlled clinical trials focusing specifically on this dimension are lacking. In psychotic or severely agitated or aggressive patients, antipsychotics are frequently the drugs of choice. Atypical antipsychotic agents with low potential for causing extrapyramidal signs and symptoms are preferred.

Primary Sleep Disorders

In addition to the sleep disturbances that result from normal aging or brain disease, sleep quality may be impaired by primary sleep disorders, some of which themselves may be seen with increasing prevalence with age. Sleep disordered breathing (sleep apnea), restless legs syndrome (RLS), and REM sleep behavior disorder (RBD) are three such primary sleep disorders.

Sleep apnea syndrome is characterized by the repeated cessation or significant diminution of breathing for 10 seconds or longer, resulting in multiple episodes of hypoxemia, multiple brief awakenings, complaints of excessive daytime sleepiness, and impaired daytime functioning. Major risk factors include male gender and obesity. Sleep apnea should be considered in the differential diagnosis when older adults report poor sleep and when cognitive impairment is discovered for the first time. It has been observed that 24–62% of community-dwelling older adults have sleep-related breathing disturbances, although the true clinical implications of these observations are unclear.

Treatment of obstructive sleep apnea includes behavior modification to minimize sleeping on the back, weight loss for obese patients, avoidance of respiratory depressant drugs (hypnotics and alcohol), use of oral appliances such as mandibular advancement devices (MADs), and use of nasal continuous positive airway pressure (CPAP). Apnea is also treated with a variety of surgical interventions including tracheostomy, although these approaches are typically not first-line treatment and carry significant morbidity and mortality risk. For most cases of clinically significant obstructive sleep apnea, CPAP remains the treatment of choice. However, adherence to CPAP can be problematic, particularly in demented patients; those with significant cognitive impairment may be unable to understand the value of treatment or learn to use or tolerate it, and nocturnal confusion may lead to automatic removal of the device.

RLS is characterized by a very strong presleep urge to move one's legs and is often described as a "pulling," "searing," or "crawling," which often leads to significant sleep onset insomnia. RLS has been successfully treated with benzodiazepines and opiates; however, dopaminergic agents are the current treatment of choice.

RBD is characterized by a relative absence of the atonia characteristic of REM sleep. This lack of atonia permits the physical acting out of dream mentation, particularly dreams involving confrontation, aggression, and violence. RBD is seen most frequently in older men. RBD occurs in both acute and chronic forms. Acute RBD can occur during withdrawal from alcohol or sedative–hypnotics. RBD has also been induced by the tricyclics, selective serotonin reuptake inhibitors (SSRIs), and venlafaxine. The chronic form of RBD may occur as part of an identifiable underlying neurological disorder but typically is idiopathic. RBD may also be an initial manifestation of parkinsonism. REM sleep behavior disorder (RBD) is very responsive to clonazepam, although this use has not been FDA approved.

Although it might be expected that the incidence of the primary sleep disorders would increase in demented patients relative to age-matched controls because of the CNS dysfunction underlying these disorders, studies comparing the rates of sleep apnea in dementia patients and aged controls have not found consistent differences. Nevertheless, these conditions may interact with the dementia syndrome to further worsen sleep quality as well as cognitive and functional abilities. For example, some studies have shown that sleep apnea is associated with increased morning confusion in AD patients.

SYMPTOMATIC TREATMENT OF INSOMNIA IN AD PATIENTS

When insomnia is not caused by, or fails to respond to treatment for, another medical or psychiatric condition in dementia, pharmacological treatment with sedating agents may be considered as symptomatic therapy [13]. Controversies regarding the use of sedating medications in demented patients revolve around issues of efficacy and issues of potential toxicity, neither of which have been resolved by appropriately comprehensive empirical study. There is evidence, however, that sedative–hypnotics as a class may be inappropriately prescribed or overprescribed for demented patients.

Several recent studies have now shown that use of prescription drugs does not necessarily improve subjective and objective ratings of sleep quality in community-dwelling or institutionalized older patients [3, 14]. However, no controlled clinical trials have evaluated the efficacy or toxicity of benzodiazepines or the newer nonbenzodiazepines, imidazopyridine hypnotics such as zolpidem or zaleplon, in groups of demented patients. The hazards of excessive sedation for patients with dementia, including increased impairment in cognition, gait, and balance, and the consequent risk of falls have been widely publicized but have been surprisingly poorly studied. Presumably, currently available hypnotics of either benzodiazepine or nonbenzodiazepine classes are effective at least in part because of diffuse effects on brain activity mediated through benzodiazepine receptors that are widely distributed in the brain, rather than by specific effects on a putative "sleep center." Because of this, the common side effects of both classes of hypnotics are part and parcel of their impact on sleep.

There is considerable disagreement in sleep medicine as to whether long-term drug treatment of primary insomnia is effective and safe. If reliable data are sparse regarding the older population at large, they are sparser still regarding the treatment of AD patients. Buysse [15] has recently examined the state-of-the-art pharmacological treatment of chronic insomnia and has proposed a point of view that may be applicable to studies of sleep in AD. Buysse distinguishes insomnia as a symptom or complaint, from insomnia as a disorder or disease that causes functional impairment. He highlights evidence that otherwise healthy patients with insomnia have significant abnormalities in physiological function beyond the domains of complaint and sleep reduction. He develops an approach to insomnia that calls for development of neuropsychobiologic models of insomnia, a reliable and valid nosology, and a sequential program of intervention research beginning with nonpharmacological treatments followed by drug therapies for nonresponders. This view provides a useful framework not only for insomnia but also for understanding and treating sleep disorders in patients with AD.

NEED FOR CONTROLLED CLINICAL TRIALS FOR IMPROVING SLEEP QUALITY IN AD

The absence of controlled clinical trials of symptomatic treatments for insomnia in demented patients represents a serious and continuing gap in knowledge. An exception to this absence is the recently completed multicentered trail of melatonin to improve sleep in AD patients [16]. Based on some promising preliminary results pointing to melatonin's potential efficacy in this arena, the Alzheimer's Cooperative Study, an NIA-funded consortium of AD research centers around the country, undertook the first large, multicenter trial of a sleep therapy in AD patients—specifically melatonin. One hundred fifty seven subjects with AD and sleep disturbance were recruited at 36 different sites and randomized to placebo, 2.5 mg, or 10 mg melatonin and monitored continuously for 2 months by wrist actigraphs. Melatonin failed to improve sleep quality in these severely sleep disturbed AD patients [16]. Based on this and other recent reports that found little beneficial effect of melatonin on either sleep or agitation in severe AD patients, it appears that melatonin is not particularly effective across the broad range of sleep disturbance in AD. While the results of this trail [16] were negative, it is nonetheless noteworthy, not only for providing comprehensive data indicating that melatonin is inappropriate for managing sleep disorders in AD patients, but also as an exemplar of exactly the type of trials that are necessary if efficacious evidence-based treatments are to be developed.

NONPHARMACOLOGICAL APPROACHES TO TREATING SLEEP DISTURBANCE IN AD PATIENTS

In situations where a sleep disturbance is not wholly the result of age-related sleep change, a primary sleep disorder, or a specific medical or psychiatric disorder or complication of dementia, sleep may become chronically disrupted through the development of poor sleep habits, conditioned emotional responses, or poor environmental conditions. These problematic habits and responses interfere with normal regulatory sleep mechanisms and may serve as inhibitors to sleep. A number of behavioral and environmental modification strategies, including sleep hygiene, sleep compression, relaxation training, stimulus control, and multicomponent cognitive–behavioral therapy, have proved effective for enhancing sleep in older adults without dementing diseases [11], and some of their components can be helpful in demented patients [3, 17]. There is also an emerging body of literature indicating that light and exercise may also have beneficial impact on sleep quality in both the healthy elderly and demented patients, possibly through enhancement of the patient's underlying circadian rhythms [4, 5].

Overall, there is good evidence that nonpharmacological treatments can improve sleep quality and reduce sleeping medication use in older adults [11]. Furthermore, there is emerging evidence that similar nonpharmacological approaches can work as well with AD patients [18–21]. McCurry and colleagues have recently reported that a behaviorally based intervention incorporating sleep hygiene education, exercise, and light exposure can be successfully implemented in AD patients [19] and has positive treatment effects on the sleep quality of these AD patients at both post-treatment and follow-up [21]. However, considerably more such data will have to be developed before the practicing physicians most often responsible for treating AD patients can begin to recommend such nonpharmacological treatments as a first-order intervention.

CONCLUSIONS AND AN AGENDA FOR THE FUTURE

Effective evaluation and treatment of disturbed sleep in an AD patient require an appreciation of the many ways that sleep can be disturbed in such individuals and the willingness to parse what those casual agents might be and marshal the most effective treatment for each of them [3, 17, 22]. Accurate assessment of sleep disturbances in demented patients can only be done in the context of appreciating the potential contributing associated medical disorders, current drug treatments, psychopathologies, primary sleep disorders, and behavioral and environmental conditions that may exist. Each AD patient's case is likely to be unique and only accurate identification of underlying causes, their effective treatment, attention to behavioral and environmental conditions and, where possible, their correction, coupled with appropriate and judicious phamacotherapy when necessary, will best address most sleep disturbances in AD patients [3].

This chapter has described the current state-of-the-art for understanding and treating sleep disturbance in AD. What is clear is that the literature available to guide such an understanding and to structure effective treatment of sleep disturbance in AD patients is at present woefully inadequate. This lack has been noted in the recently developed 2003 National Sleep Disorders Research Plan [23]. It is in this context that this chapter concludes with a brief summary (see Table 101.1) of the major gaps in the clinical research literature that will need to be filled before a more definitive guide to treating sleep disturbances in AD can be developed. It is hoped that this list will help guide future research in this important arena of sleep medicine.

TABLE 101.1 Research Agenda for Future Studies of Sleep-Related Issues in Alzheimer's Disease

1. Studies that will lead to a better understanding of both the underlying causes for and the exact nature of the longitudinal changes in sleep that occur during the course of dementia and the impact of those changes on patient quality of life, treatment, and care.

2. Randomized controlled trials of both behavioral and pharmacological treatments for behavioral disturbances using objective measures of sleep daytime function, and impact on caregivers in addition to standardized behavioral and psychiatric outcome measures.

3. Randomized controlled trials to assess comparative efficacy of specific pharmacological and nonpharmacological approaches to improve sleep quality and determination of the long-term efficacy and safety of newer hypnotic agents in demented patients with sustained sleep disorders.

4. Studies to provide a better understanding of the impact of improved sleep quality on the daytime cognitive and physical function of demented patients.

5. Empirical validation of diagnostic criteria and algorithms for assessing and managing sleep problems in dementia.

6. Controlled health services trials of effect of changes in institutional policies affecting the patient milieu, such as staffing, lighting, organized patient activities, and medication protocols, on the sleep of institutionalized dementia patients.

Source: Modified from [3] in the context of the 2003 National Sleep Disorders Research Plan [23].

ACKNOWLEDGMENTS

Supported by Public Health Service Grant K02-MH01158 to MVV.

REFERENCES

1. Vitiello MV, Prinz PN, Williams DE, Frommlet MS, Ries RK. Sleep disturbances in patients with mild-stage Alzheimer's disease. *J Gerontol* **45**(4):M131–M138(1990).

2. Bliwise DL. Dementia. In: Kryger MH, Roth T, Dement W (Eds), *Principles and Practice of Sleep Medicine*, 3rd ed. Saunders, Philadelphia, 2000, pp 1058–1071.

3. Vitiello MV, Borson S. Sleep disturbances in patients with Alzheimer's disease: epidemiology, pathophysiology and management. *CNS Drugs* **15**(10):777–796(2001).

4. Swaab DF, Dubelaar EJ, Scherder EJ, van Someren EJ, Verwer RW. Therapeutic strategies for Alzheimer disease: focus on neuronal reactivation of metabolically impaired neurons. *Alzheimer Dis Assoc Disord* **17**(Suppl 4):S114–S122(2003).

5. Van Someren EJ. Circadian and sleep disturbances in the elderly. *Exp Gerontol* **35**(9-10):1229–1237(2000).

6. Yesavage JA, Taylor JL, Kraemer H, Noda A, Friedman L, Tinklenberg JR. Sleep/wake cycle disturbance in Alzheimer's disease: how much is due to an inherent trait? *Int Psychogeriatr* **14**(1):73–81(2002).

7. Yesavage JA, Friedman L, Kraemer H, Tinklenberg JR, Salehi A, Noda A, Taylor JL, O'Hara R, Murphy G. Sleep/wake disruption in Alzheimer's disease: APOE status and longitudinal course. *J Geriatr Psychiatry Neurol* **17**(1):20–24(2004).

8. Harper DG, Stopa EG, McKee AC, Satlin A, Fish D, Volicer L. Dementia severity and Lewy bodies affect circadian rhythms in Alzheimer disease. *Neurobiol Aging* **25**(6):771–781(2004).

9. Yesavage JA, Friedman L, Ancoli-Israel A, Bliwise DL, Singer C, Vitiello MV, Monjan AA, Lebowitz B. Development of diagnostic criteria for defining sleep disturbance in Alzheimer's disease. *J Geriatr Psychiatry Neurol* **16**(3):131–139(2003).

10. Alessi CA, Schnelle JF. Approches to sleep disorders in the nursing home setting. *Sleep Med Rev* **4**(1):45–56(2000).

11. Vitiello MV. Effective treatment of sleep disturbances in older adults. *Clin Cornerstone* **2**(5):16–27(2000).

12. Vitiello MV, Moe KE, Prinz PN. Sleep complaints cosegregate with illness in older adults: clinical research informed by and informing epidemiological studies of sleep. *J Psychosom Res* **53**(1):555–559(2002).

13. Tariot PN, Ryan JM, Porsteinsson AP, Loy R, Schneider LS. Pharmacologic therapy for behavioral symptoms of Alzheimer's disease. *Clin Geriatr Med* **17**(2):359–376(2001).

14. Vitiello MV. Sleep disorders and aging. *Curr Opin Psychiatry Geriatr Psychiatry* **9**(4):284–289(1996).

15. Buysse DJ. Rational pharmacotherapy for insomnia: time for a new paradigm. *Sleep Med Rev* **4**(6):521–527(2000).

16. Singer C, Tractenberg RE, Kaye J, Schafer K, Gamst A, Grundman M, Thomas R, Thal LJ. Alzheimer's Disease Cooperative Study. A multicenter, placebo-controlled trial of melatonin for sleep disturbance in Alzheimer's disease. *Sleep* **26**(7):893–901(2003).

17. McCurry SM, Reynolds CF, Ancoli-Israel S, Teri L, Vitiello MV. Treatment of sleep disturbance in Alzheimer's disease. *Sleep Med Rev* **4**(6):603–628(2000).

18. Luijpen MW, Scherder EJ, Van Someren EJ, Swaab DF, Sergeant JA. Non-pharmacological interventions in cognitively impaired and demented patients—a comparison with cholinesterase inhibitors. *Rev Neurosci* **14**(4):343–368(2003).

19. McCurry SM, Logsdon GD, Vitiello MV, Gibbons LE, Teri L. Training caregivers to change the sleep hygiene practices of paitients with dementia: the NITE-AD Project. *J Am Geriatr Soc* **51**(10):1455–1460(2003).

20. McCurry SM, Gibbons LE, Logsdon RG, Teri L. Anxiety and nighttime behavioral disturbances. Awakenings in patients with Alzheimer's disease. *J Gerontol Nurs* **30**(1):12–20(2004).

21. McCurry SM, Gibbons LE, Logsdon RG, Vitiello MV, Teri L. Nighttime insomnia treatment and education for Alzheimer's disease (NITE-AD): a randomized controlled trial. *J Am Geriatr Soc* **53**(5):793–802(2005).

22. McCurry SM, Ancoli-Israel S. Sleep dysfunction in Alzheimer's disease and other dementias. *Curr Treat Options Neurol* **5**(3):261–272(2003).

23. 2003 National Sleep Disorders Research Plan. Department of Health and Human Services. National Heart, Lung, and Blood Institute. NIH Publication No.03-5209, July 2003.

ADDITIONAL READING

Teri L, Logsdon RG, McCurry SM. Nonpharmacologic treatment of behavioral disturbance in dementia. *Med Clin North Am* **86**(3):641–656(2002).

102

NEURODEGENERATIVE DISORDERS

Dᴀᴠɪᴅ G. Hᴀʀᴘᴇʀ
Harvard Medical School, Belmont, Massachusetts

INTRODUCTION

Sleep disturbance in neurodegenerative illnesses is a complex, multifactorial entity requiring considerable sophistication in the research designed to study it as well as flexibility in the management and treatment of patients with it. The impact of homeostatic versus circadian disruption, environmental versus neurodegenerative etiologies, and the role played by aging across the continuum of age-associated sleep and circadian changes all need to be accounted for to ultimately understand the etiology of sleep disturbance associated with a particular neurodegenerative illness. Furthermore, sleep disturbance in a patient with a neurodegenerative illness can have a devastating impact in ways that are not always obvious.

CAUSES OF SLEEP DISTURBANCE IN PATIENTS WITH NEURODEGENERATIVE DEMENTIA

Normal human behavior is temporally divided into organized patterns of nocturnal sleep and diurnal alertness by the interaction of two discrete physiological processes—the circadian and the homeostatic—that can best be expressed as mathematical functions regulating sleep tendency [1]. Both of these processes have neuroanatomical substrates in regions of the brain affected by Alzheimer's disease and other neurodegenerative illnesses [2–4]. The circadian, oscillatory rhythm is generated by the suprachiasmatic nucleus (SCN) of the hypothalamus and promotes alertness as a function of time of day, thereby consolidating wakefulness during the diurnal period. Opposing this alerting rhythm is a homeostatic process that builds the need to sleep as a function of the duration of prior wakefulness [5, 6]. The areas of the brain involved in homeostatic regulation are located in structures of the basal forebrain and hypothalamus including the ventrolateral preoptic area (VLPO) and nucleus basalis of Meynart [7]. The interaction between these two processes builds an oscillating pattern of sleep and wakefulness with each occurring at the appropriate time of day as a result of environmental feedback.

However, patients with neurodegenerative illness frequently live in environments lacking stimulation and the cues essential for maintaining healthy sleep–wake rhythms [8]. Therefore it is frequently difficult to determine whether an individual sleep or rhythmicity disturbance experienced is a consequence of neurodegenerative illness, environment, or some combination of both factors [9]. In addition, most neurodegenerative illnesses are age-associated, with their incidence increasing dramatically as patients grow older. Sleep changes that occur as a result of the aging process may be mingled with changes resulting from the more pathological forms of aging as represented in a neurodegenerative illness [10, 11]. Thus, sleep change as a result of aging can become confused with the consequences of neurodegeneration, leading to inappropriate treatment of the underlying disturbance.

Taken together, these issues yield an intricate matrix of causal factors, any one of which could be responsible for the significant disturbance in the continuity and quantity of sleep in patients with neurodegenerative dementia [12].

Identifying the correct etiologic factors is imperative in choosing correct treatment, since many of the pharmacological methods used to aid sleep in patients with transient insomnia have substantial cognitive effects that are undesirable in this patient population [13, 14].

Sleep also has an internal architecture that can become disrupted as a consequence of neurodegenerative illness. Normal sleep contains several stages of slow-wave "deep" sleep alternating with "light" sleep associated with faster EEG activity. Inserted into light sleep are periods of desynchronized, rapid eye movement (REM) essentially indistinguishable from wakefulness except for the presence of pervasive muscle atonia preventing behavioral expression during dreaming. REM sleep appears to be initiated and muscle atonia maintained by pontine and brainstem nuclei in association with higher brain structures. Each of these elements of sleep architecture can be disrupted by the presence of a neurodegenerative dementia with consequences to patient and caregiver.

CONSEQUENCES OF SLEEP DISTURBANCE IN NEURODEGENERATIVE DEMENTIA

Most patients, particularly in the early and middle stages of a neurodegenerative illness, have a caregiver and the consequences of a sleep disturbance impact both members of this patient–caregiver dyad. For the caregiver, the ability to maintain a steady sleep schedule may become impossible due to the need to manage the patient's agitation and restlessness at night. These increased demands invariably lead to some degree of caregiver exhaustion and consequent potentially institutionalization of the patient [15]. For the patient, the impact of sleep loss may be severe in cognitive and behavioral domains. Since sleep deprivation decrements working memory [16] and executive [17] functioning in normal adults in a dose-dependent fashion, cognitive and behavioral symptoms of Alzheimer's disease or other dementias are likely to worsen as sleep deficit increases. The consequences of these additional decrements to the overall disease burden are likely to be substantial [18]. There can also be increased physical danger from falls and physical injury resulting from the nocturnal wandering and agitation that frequently accompany sleep disturbance [13], in addition to the known increase in risk of physical injury resulting from chronic sleep deprivation [19].

Cognitive performance can be substantially decremented not only by sleep deprivation or restriction resulting from the impact of the illness on the sleep homeostatic system or the circadian system, but also by circadian disruption alone without apparent sleep loss. Circadian misalignment induced through a forced desynchrony protocol (maintaining a sleep–wake rhythm significantly different from 24 hours such that the sleep–wake rhythm and physiological circadian rhythm become separate) also has been demonstrated to impair performance on simple cognitive tasks, such as addition performance, during the period when the sleep–wake and circadian rhythms are out of alignment. Motivation also becomes quite impaired in normal subjects under these conditions. To a patient suffering from neurodegenerative illness these outcomes can be particular devastating.

CHARACTERIZATION OF SLEEP DISTURBANCES IN THE DIFFERENT NEURODEGENERATIVE ILLNESSES

There are several progressive neurodegenerative dementias and all of them are characterized by a progressive cognitive and functional decline with increasing neuropathology over the course of the illness. However, while many of them share significant pathological features, there are also substantial differences between them. Given this wide variety of neurodegenerative illnesses, their differing etiologies and courses of neural destruction, it would make sense that although then sleep as disturbances associated with them may appear superficially similar, there are likely to be vast differences in their expression.

There are also great differences in our knowledge of sleep disturbance in these different neurodegenerative dementias. Some of the illnesses have a large quantity of information characterizing the polysomnography, circadian dysregulation, or other factors impacting sleep. However, for others, there is virtually no information available characterizing the sleep disturbance although it is known to be present.

Alzheimer's Disease

The pathological features of Alzheimer's disease (AD) include the aggregation of extraneuronal beta-amyloid into plaques as well as the concomitant intraneuronal neurofibrillary tangles composed of hyperphosphorylated tau protein. The first quantification of sleep disturbance in Alzheimer's disease was in the early 1980s. Polysomnographic measurement of sleep in patients with Alzheimer's disease revealed a reduction in total sleep time and a significant loss of slow-wave and REM sleep when patients were compared to age-matched controls [20]. In addition, the reduction, at least in REM sleep, appeared severity dependent with mild AD patients showing little attenuation of REM sleep and moderate to severe AD patients showing increasing REM sleep loss [21]. Alzheimer's disease also has an association with sleep apnea that appears linked to severity of illness [22], although it is unclear whether the apnea's linkage to severity is based on degeneration in brain regions important to ventilation secondary to AD or whether the apnea-related sleep loss is causing patients with more severe apnea to appear more demented due to cognitive decrements resulting from to sleep restriction [23].

Rhythms of locomotor activity, which can be seen as representative of sleep [24], and the rhythms of sleep and wakefulness in AD patients confirmed the findings of increased activity during the nocturnal period [9, 25] and fragmentation over the 24 hour period [26, 27]. In addition, the circadian stability of the activity rhythms appears to be compromised when patients were studied during the middle to late stage of illness.

More direct measurements of the functioning of the circadian timing system are difficult to perform in this population. Melatonin levels are frequently reduced over 24 hours [28], and rhythmicity is generally lost [29, 30], making melatonin a poor marker for the phase position of the circadian rhythm [31]. As a result, core body temperature has been used to estimate the phase position of the circadian rhythm in Alzheimer's disease even though masking influences on the temperature rhythm interfere with these estimates. The circadian change with perhaps the most significant impact in AD is a phase delay of core body temperature compared to a normal comparison group that has been observed in patients with autopsy-confirmed AD but not seen in patients with frontotemporal dementia [9]. Also, one study has been published employing strict unmasking procedures [11] confirming findings of phase delay in advanced dementia. Activity rhythms also appear to share this phase delay, suggesting that sleep–wakefulness rhythms may be phase delayed in these patients.

Dementia with Lewy Bodies

Dementia with Lewy bodies (DLB) has many pathological similarities to Alzheimer's disease including the development of amyloid plaques in the brain. The phenotypic expression of sleep–wake disturbances is also quite similar between the diseases [32], although the severity of the disturbances may be more pronounced in Lewy body dementia according to caregiver reports [33]. To date, no polysomnographic study has compared the sleep architecture seen in DLB versus AD. However, objective quantitative measures, such as locomotor activity monitoring, in pathologically confirmed cases can show differences between DLB and AD, especially in the quantity of diurnal activity seen, with DLB patients showing markedly reduced diurnal activity [32].

The most striking sleep-associated feature of dementia with Lewy bodies is the emergence of REM sleep behavior disorder (RSBD) in a subpopulation of these patients [34]. This finding of RSBD is almost unique to DLB and often precedes the symptoms of dementia by more than 10 years [35]. The only other neuropathological diagnosis associated with RSBD is multiple system atrophy, a synucleinopathy sharing the significant neuropathological finding of intracellular inclusions composed of alpha synuclein. A recent estimate of the prevalence of RSBD in pathologically confirmed DLB was put at 50%, which yields significant diagnostic value to this parasomnia [35].

Frontotemporal Dementia, Pick's Disease, and Corticobasal Degeneration

Frontotemporal dementia (FTD) and Pick's disease are now nosologically considered subtypes of a single diagnostic entity based on neuropathological patterns of damage [36]. They share the hyperphosphorylation of tau protein and consequent development of neurofibrillary tangles with Alzheimer's disease; however, these illnesses do not generate the amyloid plaques seen in Alzheimer's disease. The sleep disturbances in patients diagnosed with Pick's disease or other frontotemporal dementias have several features in common and can be quite devastating to patients. Polysomnographically, patients show significant loss in total sleep, especially slow-wave sleep, and frequent awakenings throughout the night. While REM sleep was observed in all patients studied, it was remarkably fragmented [37], frequently occurring at sleep onset.

Studies of time-series activity recordings of patients with FTD show a remarkable loss of diurnal–nocturnal rhythmicity with very high fragmentation of the rhythm. While the circadian rhythms of core body temperature in patients with FTD are indistinguishable from those in normal elderly subjects, suggesting that the circadian oscillator is unimpaired, the locomotor activity appears arrhythmic and chaotic [9]. These results suggest that an uncoupling has occurred between the central circadian clock and the expression of that rhythm in a coherent pattern of rest and activity occurring over 24 hours. Therefore treatment approaches based on altering circadian rhythmicity are unlikely to be effective in these patients.

Huntington's Disease

Until recently, very little attention has been paid to the clear circadian and sleep disturbances seen in Huntington's disease. Case reports have described deeply disturbed sleep in these patients but no quantitative work has been done until very recently. A recent study [38] shows deeply disrupted expression of circadian timing via the rhythm of locomotor activity in these patients. In addition, the nocturnal activity was significantly increased in Huntington's patients, although it is difficult to tell whether this effect was due to a shift in the circadian rhythm or wakefulness not attributable to a circadian etiology.

CONCLUSION

Tremendous progress has been made over the last decade in our understanding of the sleep disturbances in patients with different forms of neurodegenerative illnesses, so that this knowledge can now employed in developing innovative treatments. However, much work remains to be done including elaborating the underlying mechanisms behind neurodegenerative damage in the brain and how this damage then leads to the phenotypic expression of sleep and/or circadian disturbances. Increasing our knowledge of these factors will ultimately lead to new insights for treatments.

Another important area for further research is the impact of ameliorating sleep and circadian disturbances on the severity and progression of the illness. It is possible that by treating sleep problems successfully, we can lower the cognitive burden of the disease and possibly impact the disease's course.

REFERENCES

1. Borbely AA. A two process model of sleep regulation. *Hum Neurobiol* **1**(3):195–204(1982).
2. Vitiello MV, Bliwise DL, Prinz PN. Sleep in Alzheimer's disease and the sundown syndrome. *Neurology* **42**(7 Suppl 6):83–93(1992); discussion pp 93–94.
3. Swaab DF, Fliers E, Partiman TS. The suprachiasmatic nucleus of the human brain in relation to sex, age and senile dementia. *Brain Res* **342**(1):37–44(1985).
4. Stopa EG, Volicer L, Kuo-Leblanc V, Harper D, Lathi D, Tate B, Satlin A. Pathologic evaluation of the human suprachiasmatic nucleus in severe dementia. *J Neuropathol Exp Neurol* **58**(1):29–39(1999).
5. Franken P, Tobler I, Borbely AA. Sleep and waking have a major effect on the 24-hr rhythm of cortical temperature in the rat. *J Biol Rhythms* **7**(4):341–352(1992).
6. Jewett ME, Kronauer RE. Interactive mathematical models of subjective alertness and cognitive throughput in humans. *J Biol Rhythms* **14**(6):588–597(1999).
7. Saper CB, Chou TC, Scammell TE. The sleep switch: hypothalamic control of sleep and wakefulness. *Trends Neurosci* 726–731(2001).
8. Shochat T, Martin J, Marler M, Ancoli-Israel S. Illumination levels in nursing home patients: effects on sleep and activity rhythms. *J Sleep Res* **9**(4):373–379(2000).
9. Harper DG, Stopa EG, McKee A, Satlin A, Harlan PC, Goldstein RL, Volicer L. Differential circadian rhythm disturbances in men with Alzheimer disease and frontotemporal degeneration. *Arch Gen Psychiatry* **58**(4):353–360(2001).
10. Ancoli-Israel S, Klauber MR, Jones DW, Kripke DF, Martin J, Mason W, Pat-Horenczyk R, Fell R. Variations in circadian rhythms of activity, sleep, and light exposure related to dementia in nursing-home patients. *Sleep* **20**(1):18–23(1997).
11. Harper DG, Volicer L, Stopa EG, McKee AC, Nitta M, Satlin A. Disturbance of endogenous circadian rhythm in aging and Alzheimer disease. *Am J Geriatr Psychiatry* 359–368(2005).
12. Vitiello MV, Prinz PN. Alzheimer's disease. Sleep and sleep/wake patterns. *Clin Geriatr Med* **5**(2):289–299(1989).
13. McCurry SM, Ancoli-Israel S. Sleep dysfunction in Alzheimer's disease and other dementias. *Curr Treat Options Neurol* **5**(3):261–272(2003).
14. Lawlor BA. Behavioral and psychological symptoms in dementia: the role of atypical antipsychotics. *J Clin Psychiatry* **65**(Suppl 11):5–10(2004).
15. Pollak CP, Perlick D. Sleep problems and institutionalization of the elderly. *J Geriatr Psychiatry Neurol* **4**(4):204–210(1991).
16. Smith ME, McEvoy LK, Gevins A. The impact of moderate sleep loss on neurophysiologic signals during working-memory task performance. *Sleep* **25**(7):784–794(2002).
17. Jones K, Harrison Y. Frontal lobe function, sleep loss and fragmented sleep. *Sleep Med Rev* **5**(6):463–475(2001).
18. Tractenberg RE, Singer CM, Cummings JL, Thal LJ. The sleep disorders inventory: an instrument for studies of sleep disturbance in persons with Alzheimer's disease. *J Sleep Res* **12**(4):331–337(2003).
19. Goldberg R, Shah SJ, Halstead J, McNamara RM. Sleep problems in emergency department patients with injuries. *Acad Emerg Med* **6**(11):1134–1140(1999).
20. Prinz PN, Peskind ER, Vitaliano PP, Raskind MA, Eisdorfer C, Zemcuznikov N, Gerber CJ. Changes in the sleep and waking EEGs of nondemented and demented elderly subjects. *J Am Geriatr Soc* **30**(2):86–93(1982).
21. Vitiello MV, Bokan JA, Kukull WA, Muniz RL, Smallwood RG, Prinz PN. Rapid eye movement sleep measures of Alzheimer's-type dementia patients and optimally healthy aged individuals. *Biol Psychiatry* **19**(5):721–734(1984).
22. Ancoli-Israel S, Klauber MR, Butters N, Parker L, Kripke DF. Dementia in institutionalized elderly: relation to sleep apnea. *J Am Geriatr Soc* **39**(3):258–263(1991).
23. Gehrman PR, Martin JL, Shochat T, Nolan S, Corey-Bloom J, Ancoli-Israel S. Sleep-disordered breathing and agitation in institutionalized adults with Alzheimer disease. *Am J Geriatr Psychiatry* **11**(4):426–433(2003).
24. Ancoli-Israel S, Clopton P, Klauber MR, Fell R, Mason W. Use of wrist activity for monitoring sleep/wake in demented nursing-home patients. *Sleep* **20**(1):24–27(1997).
25. Satlin A, Volicer L, Stopa EG, Harper D. Circadian locomotor activity and core-body temperature rhythms in Alzheimer's disease. *Neurobiol Aging* **16**(5):765–771(1995).
26. Van Someren EJ, Kessler A, Mirmiran M, Swaab DF. Indirect bright light improves circadian rest–activity rhythm disturbances in demented patients. *Biol Psychiatry* **41**(9):955–963(1997).
27. Hatfield CF, Herbert J, van Someren EJ, Hodges JR, Hastings MH. Disrupted daily activity/rest cycles in relation to daily cortisol rhythms of home-dwelling patients with early Alzheimer's dementia. *Brain* **127**(Pt 5):1061–1074(2004).

28. Skene DJ, Vivien-Roels B, Sparks DL, Hunsaker JC, Pevet P, Ravid D, Swaab DF. Daily variation in the concentration of melatonin and 5-methoxytryptophol in the human pineal gland: effect of age and Alzheimer's disease. *Brain Res* **528**(1):170–174(1990).

29. Mishima K, Tozawa T, Satoh K, Matsumoto Y, Hishikawa Y, Okawa M. Melatonin secretion rhythm disorders in patients with senile dementia of Alzheimer's type with disturbed sleep–waking. *Biol Psychiatry* **45**(4):417–421(1999).

30. Wu YH, Feenstra MG, Zhou JN, Liu RY, Torano JS, Van Kan HJ, Fischer DF, Ravid R, Swaab DF. Molecular changes underlying reduced pineal melatonin levels in Alzheimer disease: alterations in preclinical and clinical stages. *J Clin Endocrinol Metab* **88**(12):5898–5906(2003).

31. Lewy AJ, Sack RL. The dim light melatonin onset as a marker for circadian phase position. *Chronobiol Int* **6**(1):93–102(1989).

32. Harper DG, Stopa EG, McKee AC, Satlin A, Fish D, Volicer L. Dementia severity and Lewy bodies affect circadian rhythms in Alzheimer disease. *Neurobiol Aging* **25**(6):771–781(2004).

33. Grace JB, Walker MP, McKeith IG. A comparison of sleep profiles in patients with dementia with Lewy bodies and Alzheimer's disease. *Int J Geriatr Psychiatry* **15**(11):1028–1033(2000).

34. Boeve BF, Silber MH, Ferman TJ, Kokmen E, Smith GE, Ivnik RJ, Parisi JE, Olson EJ, Petersen RC. REM sleep behavior disorder and degenerative dementia: an association likely reflecting Lewy body disease. *Neurology* **51**(2):363–370(1998).

35. Boeve BF, Silber MH, Ferman TJ. REM sleep behavior disorder in Parkinson's disease and dementia with Lewy bodies. *J Geriatr Psychiatry Neurol* **17**(3):146–157(2004).

36. Kersaitis C, Halliday GM, Kril JJ. Regional and cellular pathology in frontotemporal dementia: relationship to stage of disease in cases with and without Pick bodies. *Acta Neuropathol (Berl)* **108**(6):515–523(2004).

37. Pawlak C, Blois R, Gaillard JM, Richard J. La Sommeille dans la maladie Picks. *Encephale* **12**(6):327–334(1986).

38. Morton AJ, Wood NI, Hastings MH, Hurelbrink C, Barker RA, Maywood ES. Disintegration of the sleep–wake cycle and circadian timing in Huntington's disease. *J Neurosci* 157–163(2005).

103

PARKINSON'S DISEASE

Michael H. Silber

Mayo Clinic College of Medicine, Rochester, Minnesota

Idiopathic Parkinson's disease (PD) is the commonest cause of parkinsonism, a clinical syndrome characterized by rest tremor, rigidity, bradykinesia, and abnormal postural reflexes. It is associated with Lewy inclusion bodies in multiple nuclei, especially in the substantia nigra. The frequency of sleep disorders in PD is high and can best be conceptualized under the headings of insomnia, excessive daytime sleepiness, excessive motor activity at night, and nocturnal perceptual and behavioral abnormalities. (See Table 103.1)

INSOMNIA

Problems initiating or maintaining sleep in PD may be due to uncontrolled motor symptoms, the effects of medications, restless legs syndrome, depression, or circadian sleep–wake reversal in patients with superimposed dementia. Bradykinesia and rigidity may prevent patients from changing position in bed, a normal phenomenon that occurs two or three times an hour. This results in muscle pain and stiffness, which may wake the patient from sleep. Parkinsonian tremor may occur during light sleep and periods of wakefulness during the night. Nocturia is common in PD patients and painful early morning dystonia may also contribute to insomnia. Dopaminergic drugs, such as levodopa or dopamine receptor agonists, have alerting effects and thus may cause insomnia if taken before sleep. Patients will often describe feeling wide awake or "wired" and can relate the start of insomnia to the initiation of medication.

Restless legs syndrome (RLS) is probably more common in PD than in age-matched controls, although case–control data is not available. However, both conditions are common in older people and frequently coincide. High-dose levodopa therapy for PD may result in daytime augmentation of RLS, thus exacerbating the problem. Depression is a frequent occurrence in PD and may be more common compared to other disorders with similar degrees of disability. It may be difficult to diagnose as the neurovegetative symptoms of depression may resemble the bradykinesia of PD. When dementia is present in advanced PD, disruption of circadian rhythms may result in sleep–wake reversal and apparent nocturnal insomnia.

The mechanisms of insomnia in PD can usually be elucidated by a careful history from the patient and bedpartner or caregiver. Difficulty turning over during the night and persistent tremor suggest undertreatment of the disorder at night and should be managed by addition of controlled release carbidopa/levodopa or a dopamine agonist before bed. If insomnia is believed due to the alerting effects of an evening dose of medication, then the dose should be reduced or a hypnotic such as a benzodiazepine or sedating antidepressant such as trazodone added. The management of RLS in PD can be challenging, especially if it appears that levodopa augmentation has occurred. Strategies include substituting a dopamine agonist such as pramipexole or ropinirole for levodopa or adding a dopamine agonist, gabapentin, an opioid or a benzodiazepine, especially in the evening. If depression is suspected, a trial of an antidepressant should be undertaken.

Sleep: A Comprehensive Handbook, Edited by T. Lee-Chiong.
Copyright © 2006 John Wiley & Sons, Inc.

TABLE 103.1   Mechanisms of Sleep Disturbances in Parkinson's Disease

Insomnia
   Motor symptoms of PD
   Restless legs syndrome
   Effects of medication
   Depression
   Circadian dysrhythmias
Excessive Daytime Sleepiness
   Intrinsic to PD
   Effect of medication
   Sleep disordered breathing
   Depression
   Circadian dysrhythmias
Excessive Movement at Night
   REM sleep behavior disorder
   Periodic limb movements of sleep
   PD tremor
   Drug-induced dyskinesias
Perceptual or Behavioral Disturbances
   Effects of medication
   Confusion with dementia
   REM sleep behavior disorder

EXCESSIVE DAYTIME SLEEPINESS

Clinic-based studies of consecutive PD patients have found an abnormal Epworth Sleepiness Scale (ESS) score of >10 in about 30–40% of patients [1, 2]. In one study, 76% reported feeling sleepy during the day with 21% falling asleep while driving [2]. In a community-based study, 27% complained of excessive daytime sleepiness (EDS) [3]. Multiple factors interact to cause sleepiness in PD patients, with the three most important being somnolence intrinsic to the disorder itself, the effects of medication, and sleep disordered breathing.

It might intuitively be thought that disrupted nocturnal sleep due to undertreated PD might be an important cause of EDS. However, in several studies, nocturnal sleep complaints such as short sleep duration and frequent arousals did not correlate with daytime sleepiness [1, 3]. In objective studies of PD patients, polysomnography (PSG) measures such as total sleep time and sleep efficiency actually correlated negatively with the subsequent mean Multiple Sleep Latency Test (MSLT) latency and longer sleep latencies correlated positively [4, 5]. In contrast, shorter MSLT latencies were associated with longer duration of PD, lower cognitive functioning, hallucinations, and sleep onset rapid eye movement (REM) periods (SOREMPs) during MSLT naps. Hallucinating PD patients were more likely to have SOREMPs than nonhallucinators [6]. Low cerebrospinal fluid (CSF) hypocretin-1 levels compared to controls were found in 15 patients with advanced PD [7]. This data has led to the hypothesis that an intrinsic, narcolepsy-like condition may

occur in PD, especially as the disease progresses [4, 8]. However, proof of this concept has been difficult to achieve, as almost all advanced PD patients are taking medications that can themselves cause drowsiness, as discussed next.

Concern has been raised that the newer parkinsonian drugs such as pramipexole and ropinirole may cause patients to fall asleep while driving. Dopamine release is generally considered to have alerting effects and thus hypersomnia induced by these drugs is paradoxical and little understood. However, it has become clear that all dopaminergic agents, including levodopa, can in certain patients induce sleep, sometimes, but not always, linked in time to ingestion of a dose of the drug [9]. Despite initial reports, it appears that the newer, nonergot-based agonists do not have a greater propensity to induce sleepiness than the older ergot-based agonists such as pergolide or bromocriptine [10]. While sudden "sleep attacks" without preceding warnings of drowsiness may occasionally occur, this phenomenon is uncommon and nonspecific, also occurring in some patients with sleepiness due to narcolepsy, sleep apnea syndrome, and sleep deprivation. Most, but not all, studies using the ESS [1, 2, 9] and MSLT [5] have shown that total daily dose of dopaminergic agents correlates with severity of sleepiness.

Pulmonary function tests indicate upper airway obstruction during wakefulness in PD patients and thus one might predict a high prevalence of obstructive sleep apnea–hypopnea (OSAH) syndrome in the disorder. Snoring and its severity [1] correlate with ESS scores in PD patients. Two small controlled studies have shown contradictory results, one demonstrating a significant increase in obstructive apneas and hypopneas [11] and the other showing no difference between patients and controls [12]. Larger studies are needed to establish a definite association, but both conditions are common in the elderly. At the very least, the coincidental occurrence of OSAH must be an important consideration in the sleepy PD patient.

In addition to these three major factors, depression may mimic sleepiness in PD, and circadian disruption with day–night reversal may occur in PD patients with dementia. In assessing the sleepy PD patient, a careful history of snoring, snort arousals, and observed apneas should be taken from the patients and bedpartner or caregiver. If OSAH is suspected, PSG should be performed followed by appropriate treatment of sleep disordered breathing. If OSAH is not present, then consideration should be given to reduction of the dose of dopaminergic agents if this can be achieved without compromise of control of the movement disorder. If this is ineffective or not medically possible, then stimulant medication such as modafinil or methylphenidate should be considered. Trials of modafinil in sleepy PD patients have shown significant reduction in the ESS scores [13, 14] but no change in maintenance of wakefulness test latencies [13]. Driving should be restricted in PD patients with uncontrolled sleepiness until the problem has been resolved.

EXCESSIVE MOTOR ACTIVITY AT NIGHT

REM sleep behavior disorder (RSBD) is characterized by often violent dream enactment behavior associated with an absence of the normal skeletal muscle atonia of REM sleep. RSBD is closely linked to the alpha synucleinopathies, disorders characterized by neuronal inclusion bodies staining positive to alpha synuclein, and including PD and dementia with Lewy bodies (DLB). RSBD is present in 15–33% of PD patients assessed in PD clinics [6, 15] and the PSG finding of REM sleep without atonia in 58% [15]. Patients with PD comprise up to 27% of RSBD patients diagnosed in sleep disorders centers [16]. RSBD in the presence of dementia, even without parkinsonism, is highly suggestive of dementia with Lewy bodies. Of 18 reported autopsy studies of patients with RSBD, 15 had Lewy body pathology [17].

Symptoms of RSBD often develop before the onset of parkinsonism or dementia. In a retrospective study, RSBD was recalled as the first manifestation by 52% of patients with RSBD and PD, with the median time from onset of RSBD to onset of other symptoms of the neurologic disorder being 3–4 years [16]. In a prospective study of 29 RSBD patients more than 50 years old with normal neurologic examinations, 65% had developed parkinsonism or dementia after a mean of 13 years from onset [18]. This data suggests that Lewy body pathology may underlie many patients with apparently cryptogenic RSBD and that RSBD may be a very early manifestation of PD and DLB.

A modified PSG with additional upper extremity surface EMG derivation and time synchronized video recording is usually needed to diagnose RSBD. However, in clinical practice, the diagnosis can sometimes be inferred from a typical history in patients with PD. Management includes improving the safety of the bed environment by moving furniture away from the bedside and considering placing a mattress or large pillows on the floor next to the bed. The drug of choice is clonazepam, commencing at 0.25–0.5 mg before bed and increasing as needed and tolerated to 2 mg. Up to 80% of patients appear to respond to this drug in open label trials [16]. However, clonazepam is associated with side effects that may be harmful to PD and DLB patients, including gait disturbances and cognitive impairment, especially in the elderly. Other possible alternative treatments, tested in open label studies, include melatonin (3–12 mg) [19] and pramipexole [20]. Quetiapine may sometimes be useful, especially in DLB patients who also manifest nocturnal confusion.

Other causes of excessive motor behavior during sleep in PD patients include persistent parkinsonian tremor, levodopa-induced dyskinesias, and periodic limb movements of sleep. Periodic limb movements of sleep (PLMS) appear more frequent in PD patients compared to controls [12]. While PLMS can sometimes fragment sleep, they are common in asymptomatic older subjects and should not be considered clinically significant unless most are accompanied by arousals.

PERCEPTUAL AND BEHAVIORAL ABNORMALITIES AT NIGHT

Hallucinations and behavioral problems at night in PD usually result from one of three causes: the effects of dopaminergic drugs, RSBD, and confusion from dementia, which may develop in about one-third of patients. Levodopa and the dopamine agonists can cause a spectrum of perceptual disturbances, ranging from vivid dreams through nightmares to complex visual hallucinations. These hallucinations may occur during the day but are especially common when sensory input is diminished at night. Initially, the patient may retain insight into their unreality, but later this insight is lost. They manifest as vivid, multicolor, sometimes oddly distorted images of animals or people and disappear when the light is switched on. Very similar complex nocturnal visual hallucinations may occur in DLB even without the use of dopaminergic drugs.

Confusional wandering behavior may occur in demented patients with PD. This can usually be differentiated from RSBD by careful questioning of an observer. Confusional wandering usually lasts hours while episodes of RSBD last minutes. Confusional wandering occurs out of bed, while patients with RSBD usually kick and flail their arms in bed. An observer usually describes the wandering patient as awake but confused, while the patient with RSBD appears to be asleep but acting out dreams. If a PSG is performed, the behavior usually arises from NREM sleep and takes place during wakefulness, often with slow EEG rhythms, in contrast to RSBD in which the behavior arises from REM sleep without atonia. It should be recalled that 58% of PD patients have abnormal tone in REM sleep [15] and thus it is essential in PSG studies to record the actual behavior and not rely on the neurophysiological findings alone.

Reduction in dopaminergic therapy should be considered in PD patients who develop hallucinations or confusion, but often a careful balance is needed between control of motor symptoms and reduction of side effects. RSBD should always be considered and a sleep study performed if there is clinical doubt. Often neuroleptic medication is required with the off-label use of new generation agents such as quetiapine that do not worsen parkinsonism.

REFERENCES

1. Hogl B, Seppi K, Brandauer E, et al. Increased daytime sleepiness in Parkinson's disease: a questionnaire survey. *Mov Disord* **18**:319–323(2003).

2. Brodsky MA, Godbold J, Roth T, Olanow CW. Sleepiness in Parkinson's disease: a controlled study. *Mov Disord* **18**:668–672(2003).

3. Tandberg E, Larsen JP, Karlsen K. Excessive daytime sleepiness and sleep benefit in Parkinson's disease: a community-based study. *Mov Disord* **14**:922–927(1999).

4. Rye DB, Bliwise DL, Dihenia B, Gurecki P. Daytime sleepiness in Parkinson's disease. *J Sleep Res* **9**:63–69(2000).

5. Razmy A, Lang AE, Shapiro C. Predictors of impaired daytime sleep and wakefulness in patients with Parkinson disease treated with older (ergot) vs newer (nonergot) dopamine agonists. *Arch Neurol* **61**:97–102(2004).

6. Comella CL, Nardine TM, Diederich NJ, Stebbins GT. Sleep-related violence, injury, and REM sleep behavior disorder in Parkinson's disease. *Neurology* **51**:526–529(1998).

7. Drouot X, Moutereau S, Nguyen JP, et al. Low levels of ventricular CSF orexin/hypocretin in advanced PD. *Neurology* **61**:540–543(2003).

8. Arnulf I, Bonnet AM, Damier P, et al. Hallucinations, REM sleep, and Parkinson's disease: a medical hypothesis. *Neurology* **55**:281–288(2000).

9. Hobson DE, Lang AE, Martin WRW, Razmy A, Rivest J, Fleming J. Excessive daytime sleepiness and sudden-onset sleep in Parkinson disease. A survey by the Canadian Movement Disorders Group. *JAMA* **287**:455–463(2002).

10. Paus S, Brecht HM, Koster J, Seeger G, Klockgether T, Wullner U. Sleep attacks, daytime sleepiness, and dopamine agonists in Parkinson's disease. *Mov Disord* **18**:659–667(2003).

11. Maria B, Sophia S, Michalis M, et al. Sleep breathing disorders in patients with idiopathic Parkinson's disease. *Respir Med* **97**:1151–1157(2003).

12. Wetter TC, Collado-Seidel V, Pollmacher T, Yassouridis A, Trenkwalder C. Sleep and periodic leg movement patterns in drug-free patients with Parkinson's disease and multiple system atrophy. *Sleep* **23**:361–367(2000).

13. Hogl B, Saletu M, Brandauer E, et al. Modafinil for the treatment of daytime sleepiness in Parkinson's disease: a double-blind, randomized, crossover, placebo-controlled polygraphic trial. *Sleep* **25**:905–909(2002).

14. Adler CH, Caviness JN, Hentz JG, Lind M, Tiede J. Randomized trial of modafinil for treating subjective daytime sleepiness in patients with Parkinson's disease. *Mov Disord* **18**:287–293(2003).

15. Gagnon J, Bedard M, Fantini M, et al. REM sleep behavior disorder and REM sleep without atonia in Parkinson's disease. *Neurology* **59**:585–589(2002).

16. Olson EJ, Boeve BF, Silber MH. Rapid eye movement sleep behavior disorder: demographic, clinical and laboratory findings in 93 cases. *Brain* **123**:331–339(2000).

17. Boeve B, Silber MH, Parisi JE, et al. Synucleinopathy pathology and REM sleep behavior disorder plus dementia or parkinsonism. *Neurology* **61**:40–45(2003).

18. Schenck CH, Bundlie SR, Mahowald MW. REM behavior disorder (RBD): delayed emergence of parkinsonism and/or dementia in 65% of older men initially diagnosed with idiopathic RBD, and an analysis of the minimum and maximum tonic and/or phasic electromyographic abnormalities found during REM sleep (abstract). *Sleep* **26**:A316(2003).

19. Boeve BF, Silber MH, Ferman TJ. Melatonin for treatment of REM sleep behavior disorder in neurologic disorders: results in 14 patients. *Sleep Med* **4**:281–284(2003).

20. Fantini M, Gagnon J, Filipini D, Montplaisir J. The effects of pramipexole in REM sleep behavior disorder. *Neurology* **61**:1418–1420(2003).

ADDITIONAL READING

Chokroverty S, Hening WA, Walters AS (Eds). *Sleep and Movement Disorders.* Butterworth-Heinemann, Philadelphia, 2003. This book is a comprehensive and authoritative account of the relationship between sleep and movement disorders, including PD.

Garcia-Borreguero D, Larossa O, Bravo M. Parkinson's disease and sleep. *Sleep Med Rev* **7**:115–129(2003). This is a comprehensive review of the interaction of PD and sleep.

Garcia-Borreguero D, Odin P, Serrano C. Restless legs syndrome and PD. A review of the evidence for a possible association. *Neurology* **61**:S49–S55(2003). This review article explores the evidence for a relationship between the movement disorders of PD and RLS.

Tandberg E, Larsen JP, Karlsen K. A community-based study of sleep disorders in patients with Parkinson's disease. *Mov Dis* **13**:895–899(1998). This is an important study as it is one of the only community-based epidemiologic studies of the frequency of sleep disorders in PD.

104

SEIZURES

Margaret N. Shouse

VA School of Medicine, Los Angeles, California
UCLA Chief, Sleep Disturbance Research VAGLAHS, Sepulveda Campus, CA

INTRODUCTION

Sleep states have a potent effect on the expression or suppression of epileptic seizure manifestations. Epileptic seizure manifestations include various generalized or localized epileptic electroencephalographic (EEG) discharges. Epileptiform EEG activity may consist of interictal discharges (IIDs), which occur between seizures, or of ictal discharges, which typically last longer than IIDs and are frequently associated with clinically evident behavioral seizure events. This chapter describes the distribution of ictal and IID events during non-rapid eye movement (NREM) versus rapid eye movement (REM) sleep in different epileptic seizure syndromes as well as the reciprocal effects of seizures and antiepileptic drugs (AEDs) on sleep and of sleep deprivation on seizures.

SEIZURES AND IIDs IN NREM VERSUS REM SLEEP

During NREM sleep, focal and generalized epileptic discharges are common, even if clinical seizures do not occur. During REM sleep, either generalized epileptiform discharges are infrequent or the degree of spread is restricted. Focal IIDs persist during REM sleep and are highly localized. Clinically evident seizures rarely occur during REM sleep (see [1] or [2] for review). The differential effects of NREM versus REM sleep on interictal and ictal epileptiform discharge depend somewhat on the type and severity of the epileptic seizure syndrome, as summarized in Tables 104.1 and 104.2.

PRIMARY GENERALIZED EPILEPSIES OF GENETIC OR IDIOPATHIC ORIGIN: AWAKENING EPILEPSIES

In the primary generalized epilepsies (PGEs), interictal and ictal EEG discharges begin simultaneously virtually everywhere in the brain. Most PGEs are thought to be genetic with age-related penetrance. The prognosis is good in that PGEs respond well to antiepileptic medications and tend to remit spontaneously (Table 104.2). Examples are childhood and juvenile absence epilepsy and juvenile myoclonic epilepsy (JME). These disorders can occur with or without generalized tonic–clonic convulsions (GTCCs). GTCCs can also occur as the only seizure type.

PGEs are considered "awakening" epilepsies because massive myoclonus and GTCCs occur on awakening from sleep in ~90% of the cases [2, 6]. Nocturnal awakening, naps, and evening relaxation may also be associated with increases in these ictal seizure events. IIDs are frequently detected in transitional arousal states to and from NREM sleep and sometimes between NREM and REM sleep [7]. IIDs are most often detected in relation to "arousal" related events during NREM sleep.

Light NREM Sleep (Stages 1 and 2) Versus Deep NREM Sleep (Stages 3 and 4)

Light NREM sleep is the most likely time to detect IIDs, such as typical spike-and-wave (3/second) and multispike-and-wave or polyspike-and-wave complexes. It is also easier to provoke behavioral arousal from light NREM

Sleep: A Comprehensive Handbook, Edited by T. Lee-Chiong.
Copyright © 2006 John Wiley & Sons, Inc.

TABLE 104.1 Generalized Epileptiform Discharges by Sleep States in Three Epilepsy Syndromes with Convulsions at Different Times in the Sleep–Wake Cycle

Epilepsy Syndrome/Circadian Seizure Rhythm	Interictal		Ictal Discharges	
	NREM	REM	NREM	REM
Primary generalized epilepsies of idiopathic or hereditary origin: awakening epilepsies	Common	Rare	Rare[a]	Rare
Localization epilepsies with known or suspected local pathology: sleep epilepsies	Common	Rare[b]	Common	Rare
Symptomatic epilepsies with extensive encephalopathy	Common	Rare[c]	Common	Rare[c]

[a] Except stage 1 (≥ 5 s trains of spike-wave complexes would likely be associated with an absence seizure in waking).
[b] Note maximal focalization.
[c] As long as REM is intact.
Source: Reference [1] by permission of Lippincott-Raven Publishers.

TABLE 104.2 Characteristics of Epilepsy Syndromes as a Function of the Timing of Generalized Tonic–Clonic Convulsions (GTCCs)

Epilepsy Syndrome	Main IID Type	Seizure Type	Age at Onset	Response to Medication	Spontaneous Remission
Primary Generalized Epilepsies (PGEs): Awakening Epilepsies					
Childhood absence with or without GTCCs	3/second spike-and-wave	Absence seizure, decreased responsiveness with behavioral arrest	~4 years	Good; ethosuximide for absence, valproate or benzodiazepines for GTCCs	May convert to primary GTCCs at 12 years with remission ~25–50 years
Juvenile absence and/or myoclonus (Figure 104.1)	Multi or polyspike-and-wave	Absences and/or bilateral myoclonus	Puberty	Good; valproate	Frequent seizures may require continued medication
GTCCs on awakening	4–5/second spike-and-wave with frontal maximum	GTCCs within 2 hours of awakening	10–20 years	Good, phenobarbital, benzodiazepines	Frequent remission at ~25 years
Localization Related Epilepsies: Sleep Epilepsies					
Benign epilepsy with centrotemporal spikes (BECT)	Centrotemporal slow spikes or sharp waves	Vocalization, salivation, jaw and/or limb clonus, >50% in NREM sleep	1–13 years; more male than female	Good, if needed; carbemazepine or valproate	~Puberty
Electrical status epilepticus during sleep (ESES) (Figure 104.2)	Frontocentral spikes and 1.5–3/second spike-and-wave in <25% of waking; >85% generalized spike and wave during NREM sleep	Partial motor, mostly nocturnal, and atypical absence	~8 years (4–14)	Good; prednisone	~12 years with possible residual IQ; other cognitive and behavior disorders
Landau–Kleffner syndrome (acquired epileptic aphasia)	Temporal (50%) or parietal-occipital foci (30%); can otherwise resemble ESES in NREM sleep	Partial motor, GTCCs, or atypical absences. Few clinical seizures even in sleep	~6.5 years (3–9)	Good; prednisone	~Puberty with possible residual aphasia
Secondary generalized temporal and/or frontal lobe epilepsies (Figures 104.3 and 104.4)	Temporal or frontal spikes, multispike-and-wave and/or sharp waves	Partial complex or typical automatisms with or without secondary GTCCs	Any age, frequently 15–30 years	Good to intractable; carbamazepine, phenytoin, and gabapentin	Unlikely, especially with documented pathology

TABLE 104.2   (*Continued*)

Epilepsy Syndrome	Main IID Type	Seizure Type	Age at Onset	Response to Medication	Spontaneous Remission
Symptomatic Generalized Epilepsies: Diffuse (Sleep and Waking) Epilepsies					
West Syndrome (infantile spasms) Fig 5	Hypsarrhythmia; IIDs most frequent and have a burst–suppression pattern in NREM sleep	"Jack-knife" seizures; rare in sleep	6 months to 1 year	Poor; ACTH, benzodiazepines, vigabatrin	No; often converts to Lennox–Gastau syndrome
Lennox–Gastaut	1.5–2.5 slow spike-and-wave; tonic seizure pattern of 10–25 c/second bursts lasting 1–30 seconds	Tonic, atonic, myoclonic, GTCCs, atypical absence. "Tonic" EEG seizure patterns (often subclinical) are very likely to increase in sleep	1–10 years	Poor; valproate, benzodiazepines, felbamate, lamotrigine; best response if seizures are state dependent	No; seizures and mental retardation are less likely with a discernible state-dependent seizure pattern
Progressive (essential hereditary) myoclonus Lafora–Unverricht–Lundborg syndrome	Generalized spike-and-wave, wave, and multispike-and-wave	Myoclonus and GTCCs plus numerous other subclinical epileptic or nonepileptic symptoms	Childhood to adolescence	Poor; autosomal recessive genetic syndrome complicated by numerous cerebellar and mental symptoms	No; death within 2 years of onset or later depending on age at onset; worst prognosis with earliest onset

Sources: References [2–5].

sleep than from deep NREM or REM sleep. The longest trains of spike-and-wave complexes (>5 seconds) occur in stage 1 sleep. Discharges of similar duration in an awake patient are typically accompanied by a clinically evident absence seizure.

The next most prominent time for spike-and-wave and multispike-and-wave complexes is stage 2 sleep. During this stage, IID complexes are frequent but of short duration. They usually coincide with sleep transients, which have an EEG configuration similar to epileptiform potentials and include sleep spindles, K-complexes, vertex waves, and bursts of slow waves. These events are thought to reflect aborted arousals during sleep, in part because they can be evoked by exogenous stimulation (e.g., a noise) during NREM sleep.

Deep NREM sleep stages are usually considered least likely to activate spike-and-wave and multispike-and-wave complexes. However, some evidence suggests that deep NREM sleep is equally if not more conducive than light NREM sleep to these IIDs. The IID trains are briefer and less organized in deep NREM sleep than in light NREM sleep. This probably results from the prolonged hyperpolarization of thalamocortical cells that generates the delta-wave oscillation of deep NREM sleep and disrupts the rhythmic burst–pause discharge pattern as well as the morphology of IIDs seen in light NREM sleep [1].

Some of the most compelling evidence linking PGEs to sleep-related arousal events derives from Terzano's discovery of "cyclic alternating patterns" (CAPs) of arousal, which pervade sleep [7, 8]. Figure 104.1 illustrates this phenomenon and its provocative relationship to IIDs in juvenile myoclonic epilepsy.

REM Sleep

Generalized IIDs can occur during REM sleep but are uncommon. "Breakthrough" generalized IIDs are less diffuse in REM than NREM sleep. Even when epileptic EEG potentials do occur in REM sleep, there is rarely clinical accompaniment [1, 2].

LOCALIZATION-RELATED EPILEPSIES: SLEEP EPILEPSIES

This group of seizure disorders is thought to result from one or more focal regions of cerebral dysfunction and usually reflects a more severe pathophysiology than that seen in PGE. However, there is significant variation in severity of syndrome defined by medical refractoriness, spontaneous remission, and organicity (Table 104.2). Partial epilepsies that secondarily generalize are often most

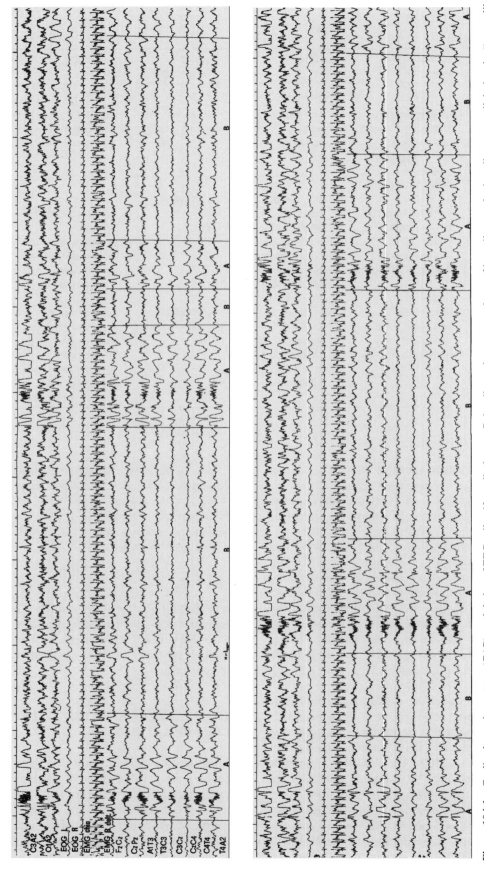

Figure 104.1 Cyclic alternating pattern (CAP) modulation of EEG epileptiform discharges. Polyspike-and-wave complexes of juvenile myoclonic epilepsy prevail during the "arousal" phase (CAP phase A). (Reprinted from [8] by permission of Blackwell Publishing.)

likely to display IIDs and to generate clinically evident seizures, especially GTCCs, during sleep [6] ([1] or [2] for review).

Examples of "benign" localization-related seizure disorders are benign epilepsy with centrotemporal spikes (BECT, also called benign rolandic partial epilepsy), benign occipital localization epilepsies, electrical status epilepticus during sleep (ESES), and Landau–Kleffner syndrome (acquired epileptic aphasia). Like seizure disorders of deep temporal or frontal lobe origin, these epilepsies are associated with diffuse IIDs during NREM sleep (Figure 104.2) and maximal localization during REM sleep. Unlike seizure disorders of temporal or frontal lobe origin, ESES and Landau–Kleffner syndrome are childhood seizure disorders and are considered benign because the seizures are likely to remit spontaneously. The patient

may be referred for a variety of reasons other than seizures (e.g., slow progress in school in ESES, sudden deterioration of language skills related to auditory aphasia in Landau–Kleffner syndrome), and the seizure disorder may only be noted in a polysomnogram. In both ESES and Landau–Kleffner syndrome, cognitive, social, and linguistic anomalies tend to mitigate with seizure remission, but residual "nonepileptic" effects often persist (see [2] for a recent review).

Temporal and frontal lobe epilepsies comprise the largest group of partial seizure disorders and the largest group of "sleep epilepsies." Onset can occur at any age, and the response to treatment is variable. Most clinicians agree that NREM sleep activates seizure events. A range of 45–78% of patients are reported to exhibit IIDs, and up to 58% present partial complex seizures and GTCCs

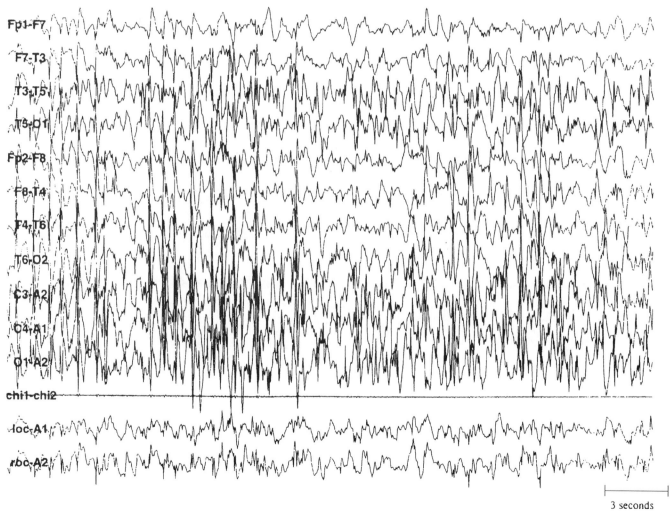

3 seconds

Figure 104.2 This 30-second epoch depicts NREM sleep in an epileptic patient with electrical status epilepticus during sleep (ESES), also known as the syndrome of continuous spike waves during slow-wave sleep (CSWS). The interictal epileptiform discharges (IEDs) are so frequent that NREM sleep cannot be differentiated accurately into stages. (Reprinted from [9] by permission of Elsevier BV.)

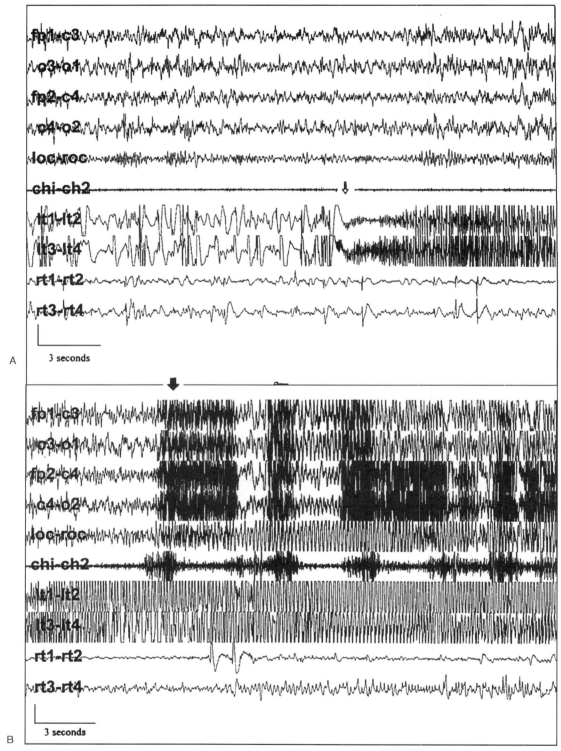

Figure 104.3 Combined scalp electrode (first six channels) and intracranial montage (last four channels). lt1–lt2 and lt3–lt4 indicate left temporal depth electrode contacts. rt1–rt2 and rt3–rt4 indicate right temporal depth electrode contacts. Other depth electrode contacts are not shown. Calibration for scalp electrodes is 100 μV. Each figure represents 30 seconds. The open arrow indicates the intracranial electrode seizure onset in the left temporal depth electrode contacts, characterized by the cessation of interictal epileptiform discharges and their replacement by high-frequency sinusoidal activity. The solid arrow indicates the clinical arousal from sleep noted on the videotape, marked by myogenic artifact. (Reprinted from [10] by permission of Associated Professional Sleep Societies LLC.)

only or mostly during sleep [1, 2, 6]. IIDs, such as anterior spikes or sharp waves, have higher amplitudes, longer durations, and rounder peaks in NREM sleep than in other sleep or waking states.

IIDs and clinically evident seizures can occur at any time in NREM sleep, but some evidence suggests that transitions to or from REM sleep and during deep NREM sleep are the most vulnerable periods for both IID and clinical seizures. Seizure activity may be associated with arousal-related sleep EEG transients and/or CAPs, although seizure onset may induce rather than follow behavioral arousal during sleep (Figure 104.3) [10]. With some exceptions (notably neocortical foci) [2], IID and clinically evident seizures rarely occur in REM sleep. In fact, some authors suggest that optimal "focalization" of IIDs occurs during REM sleep, as seen in Figure 104.4 [11]. The possibility of localizing epileptogenic foci during REM sleep may assist in targeting the epileptogenic region for surgical resection.

Unlike the PGEs in which "awakening" seizure patterns rarely switch to a sleep and waking seizure pattern, localization-related seizures that initially occur only during sleep often develop a random seizure pattern in the sleep–wake cycle later in the clinical course. Several contributing factors have been suggested. Localization-related epilepsies without documented lesions are most likely to retain nocturnal clinical seizure patterns. Frequent seizures may exacerbate cell dysfunction at the seizure focus and can ultimately influence cell discharge patterns in extrafocal brain areas [1, 2].

SYMPTOMATIC GENERALIZED EPILEPSIES: SLEEP AND WAKING EPILEPSIES

This group of seizure disorders is associated with severe, diffuse cerebral encephalopathy, is medically refractory, and does not remit spontaneously (Table 104.2). The underlying pathology can disrupt sleep, dissociate its components, and disperse IIDs and ictal discharges throughout the sleep–wake cycle. Examples are West's syndrome, Lennox–Gastaut syndrome, and progressive myoclonic epilepsies such as Lafora–Unverricht–Lundborg syndrome [3].

When NREM and REM sleep can be differentiated from waking, there is sleep-state modulation of IID and ictal seizure manifestations. For example, hypsarrhythmia, the characteristic EEG manifestation associated with infantile spasms or West's syndrome (Figure 104.5), as well as the atypical or slow spike-and-wave discharges and multiple clinical seizures of Lennox–Gastaut syndrome, show a peak in NREM sleep and subside during intact REM sleep. The prognosis for response to medical treatment is improved if IID or clinical seizures exhibit a sleep–waking state dependency, even if modulation is confined to a suppression of seizure activity in REM sleep [1, 2].

SLEEP ABNORMALITIES

Just as sleep affects seizures, seizures influence sleep. Epileptic patients may have disorganized nocturnal sleep

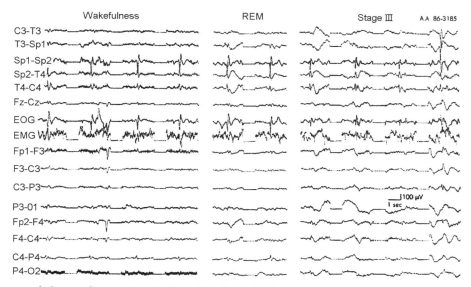

Figure 104.4 Electroencephalogram from scalp and sphenoidal electrodes during wakefulness, sleep stage 3, and REM sleep for a patient with a temporal lobe focus. Sp1 and Sp2 are the left and right sphenoidal electrodes. Note similarity of spiking fields in wakefulness and bilateral spiking in stage 3. Also note maximal localization during REM sleep. (Reprinted from [11] by permission of Lippincott Williams & Wilkins.)

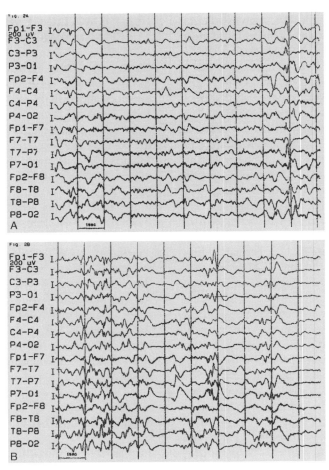

Figure 104.5   Effect of sleep on hypsarrhythmia. EEG from a 15 month old girl with new onset of infantile spasms. (A) EEG during waking shows multifocal high-amplitude sharp and slow-wave discharges, with intermittent periods during which a 5–6 Hz background could be seen. Note that the amplitude of this activity is 100–300 µV. (B) The tracing from the same patient during sleep showed more prominent bursts of spike discharges, with periods of relative attenuation, which was not seen during wakefulness. Note the high-voltage activity is in the range of 200–500 µV. (Reprinted from [2] by permission of Lippincott williams & Wilkins.)

as a result of the effects of recent seizures, antiepileptic medications, and severity of the epileptic disorder. Epileptic patients may, in addition, have a coincident sleep disorder such as nocturnal myoclonus or sleep apnea. These disorders may produce the symptoms of hypersomnolence, insomnolence, or parasomnias. These sleep abnormalities may be the initial symptom resulting in referral [12].

The most frequently reported sleep abnormalities are increased sleep onset latency, increased number and duration of awakenings after sleep onset, reduced sleep efficiency, reduced or abnormal K-complexes and sleep spindles, reduced or fragmented REM sleep, and increased stage shifts. Figure 104.5 illustrates some of these deficits by comparing the statistics of a normal hypnogram (Figure 104.6A) to those of a medically refractory patient during a seizure-free night (Figure 104.6B).

One or more of these sleep abnormalities have been observed in patients with juvenile myoclonic epilepsy, primary generalized tonic–clonic seizure disorders, and localization-related epilepsies of frontal or temporal lobe origin. Reports conflict about the presence of sleep anomalies in patients with pure absence seizures. No sleep abnormalities have been reported in BECT.

Some sleep abnormalities, especially delayed REM sleep onset, can occur as a result of recent convulsions or acute anticonvulsant drugs. Sleep abnormalities are more common in patients with generalized convulsive seizures than in those with simple or partial complex seizures. Sleep disruption is also prominent in patients with frequent or medically refractory seizures. Severe sleep abnormalities occur in symptomatic epilepsies that are associated with significant neurologic deficits. The sleep abnormalities described earlier are exaggerated, and sleep cycles may be poorly organized. In very

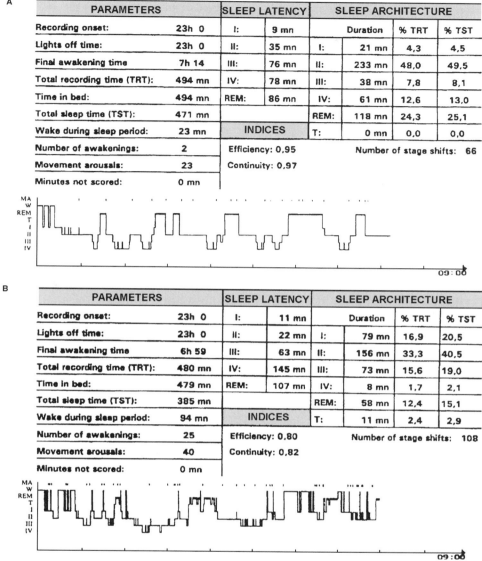

A

PARAMETERS		SLEEP LATENCY		SLEEP ARCHITECTURE			
Recording onset:	23h 0	I:	9 mn		Duration	% TRT	% TST
Lights off time:	23h 0	II:	35 mn	I:	21 mn	4,3	4,5
Final awakening time	7h 14	III:	76 mn	II:	233 mn	48,0	49,5
Total recording time (TRT):	494 mn	IV:	78 mn	III:	38 mn	7,8	8,1
Time in bed:	494 mn	REM:	86 mn	IV:	61 mn	12,6	13,0
Total sleep time (TST):	471 mn			REM:	118 mn	24,3	25,1
Wake during sleep period:	23 mn	INDICES		T:	0 mn	0,0	0,0
Number of awakenings:	2	Efficiency: 0,95		Number of stage shifts: 66			
Movement arousals:	23	Continuity: 0,97					
Minutes not scored:	0 mn						

B

PARAMETERS		SLEEP LATENCY		SLEEP ARCHITECTURE			
Recording onset:	23h 0	I:	11 mn		Duration	% TRT	% TST
Lights off time:	23h 0	II:	22 mn	I:	79 mn	16,9	20,5
Final awakening time	6h 59	III:	63 mn	II:	156 mn	33,3	40,5
Total recording time (TRT):	480 mn	IV:	145 mn	III:	73 mn	15,6	19,0
Time in bed:	479 mn	REM:	107 mn	IV:	8 mn	1,7	2,1
Total sleep time (TST):	385 mn			REM:	58 mn	12,4	15,1
Wake during sleep period:	94 mn	INDICES		T:	11 mn	2,4	2,9
Number of awakenings:	25	Efficiency: 0,80		Number of stage shifts: 108			
Movement arousals:	40	Continuity: 0,82					
Minutes not scored:	0 mn						

Figure 104.6 A comparison of hypnograms and related statistics using Somnis C software (AMISYSE), a semiautomatic method for sleep analysis of polysomnograms. Normally, the stages of sleep vary throughout the night in a somewhat predictable order. Sleep onset typically begins with NREM sleep, which predominates during the first third of the night. Then NREM and REM sleep alternate with a periodicity of about 90 minutes. Finally, REM sleep predominates during the last third of the night and is linked to the circadian rhythm of body temperature. A normal range of percentages for NREM sleep stages 1–4 and for REM sleep in a young adult are: stage 1, 2–5%; stage 2, 45–55%; stage 3, 3–8%; stage 4, 10–15%; REM, 20–25%, the latter occurring in four to six discrete episodes. (A) Normal sleep organization in a typical young adult. Percentages of NREM and REM sleep are normal. The REM sleep latency of 86 minutes and number of REM sleep cycles are within normal limits. Sleep efficiency, continuity, number of movement-related arousals, and awakenings are normal. (B) Disorganized sleep architecture in a young adult with medically intractable temporal lobe epilepsy during a seizure-free recording. Sleep analysis shows a decrease in percentage of stage 4 and REM sleep together with increased stage 1 sleep. The increased number of movement-related arousals, number of awakenings after sleep onset, and stage shifts contribute to the fragmented sleep architecture in a young adult with medically intractable temporal lobe epilepsy during a seizure-free recording. Sleep analysis shows a decrease in percentage of stage 4 and REM sleep together with increased stage 1 sleep. The increased number of movement-related arousals, number of awakenings after sleep onset, and stage shifts contribute to the fragmented sleep architecture. The REM sleep latency (107 minutes) and number of REM sleep cycles (four) are normal, but REM sleep fragmentation is detectable. Sleep efficiency and continuity are decreased. Abbreviations as follows: I, stage 1 sleep; II, stage 2 sleep; III, stage 3 sleep; IV, stage 4 sleep; REM sleep; T, transitional stage; MA, movement arousal; W, wakefulness. Time is indicated by H (hour) or mn (minute). Recording started at 23:00 hours and ended at 07:14 hours. Each line on an x axis divides the hypnogram into intervals of 1 hour. Sleep efficiency = total sleep time/(final awakening time − lights-off time). Sleep continuity = total sleep time/(final awakening time − time of sleep onset). (Reprinted from [1] by permission of Lippincott-Raven Publishers.)

serious epileptic disorders, a sleep cycle may not even be discernible [1, 2].

There are also various nocturnal paroxysmal events that are not clearly defined with regard to etiology, notably partial complex seizures. Differential diagnosis of epileptic versus nonepileptic manifestations depends primarily on the use of extracranial recordings, although intracranial EEG recordings are sometimes necessary. For example, epileptic seizures originating from mesio-orbitofrontal areas often cannot be recorded extracranially, and attacks of uncertain etiology, such as sleepwalking, attacks of screaming, and complex automatisms during sleep may be inaccurately diagnosed as parasomnias. Short-duration paroxysmal dystonia attacks may represent sleep-related frontal lobe seizures, but parasomnias can also coexist with limbic epilepsy without a common etiology [12].

ANTIEPILEPTIC DRUGS

The degree to which seizure patterns are state dependent may predict response to medication. Frequent seizures that are entrained to a specific sleep or arousal state usually respond better to medical treatment than epilepsies in which seizures are randomly distributed across the sleep–wake cycle. However, antiepileptic drugs (AEDs) that reduce seizure frequency may also cause seizures to be randomly dispersed in the sleep–wake cycle. While a correlation between AED effects on sleep and seizure improvement has long been hypothesized, effective AEDs such as those listed in Table 104.2 need not improve short- or long-term sleep quality, as indexed by sleep onset latency, sleep time, sleep efficiency, nocturnal arousals, or percentages of NREM or REM sleep (see [1] or [2] for a review).

SLEEP DEPRIVATION AND SEIZURES

Sleep deprivation can precipitate seizures, but most studies have focused on activation of IIDs for diagnostic purposes. The range of experimentally induced sleep deprivation time is usually briefer in children than adults (3–4 hours versus ~24 hours), although partial sleep deprivation has been studied in adults as well. The results of sleep deprivation have been compared to routine waking EEGs and natural or drug-induced sleep with or without hyperventilation and photic stimulation.

The benefits of sleep deprivation, particularly when compared to natural or drug-induced sleep, have been debated. The majority of findings indicate that sleep deprivation increases activation of IIDs by $50\% \pm 20\%$ in post-deprivation sleep versus ~15% in postdeprivation waking records. Intermediate levels of activation have been reported during natural or drug-induced sleep without sleep deprivation. Sleep deprivation increases activation of IIDs in all seizure types, regardless of the state-dependent or random distribution of seizures in the sleep–wake cycle. Increased activation has been detected at all ages, although some report that children under 10 years old are more responsive than 11–30 year olds, who are in turn more responsive than adults >30 years of age. Unexpected gains in activation have been seen in response to photic stimulation [1, 2].

SUMMARY

1. NREM sleep, awakening from sleep, and other transitional arousal states are conducive to electrographic and clinically evident seizures, whereas REM sleep is not. Epileptic discharges are most focal during REM sleep.

2. Timing is to some extent dependent on the type of seizure disorder. Increased severity of the seizure disorder, indexed by degree of encephalopathy, is associated with increased dispersion of seizure manifestations across the sleep–wake cycle, increased refractoriness to medication, and reduced likelihood of spontaneous remission.

3. Clinical neurophysiological techniques can assess disturbances in nocturnal sleep or excessive daytime sleepiness. Sleep disturbances often parallel the severity of seizure disorders.

4. Antiepileptic drugs can ameliorate seizure-related sleep disturbances, but improvement in sleep architecture is not a critical factor in seizure control.

5. Sleep deprivation may be superior to routine waking EEGs and to natural or drug-induced sleep for activation of epileptic discharges.

ACKNOWLEDGMENT

Supported by the Department of Veterans Affairs.

REFERENCES

1. Shouse MN, da Silva AM, Sammaritano M. Sleep. In: Engel J Jr, Pedley TA (Eds), *Epilepsy: A Comprehensive Textbook*. Lippincott-Raven Publishers, Philadelphia, 1997, pp 1929–1942.

2. Mendez M, Radtke RA. Interactions between sleep and epilepsy. *J Clin Neurophysiol* **18**:106–127(2001).

3. Niedermeyer E. Epileptic seizure disorders. In: Niedermeyer E, Lopez da Silva F (Eds), *Electroencephalography: Basic*

Principles, Clinical Applications and Related Fields, 2nd ed. Urban and Schwarzenberg, Baltimore, 1982, pp 339–428.

4. Dreifuss FE, Porter RJ. Choice of antiepileptic drugs. In: Engel J Jr, Pedley TA (Eds), *Epilepsy: A Comprehensive Textbook.* Lippincott-Raven Publishers, Philadelphia, 1997, pp 1233–1236.

5. Engel J Jr, Pedley T (Eds). *Epilepsy: A Comprehensive Textbook.* Lippincott-Raven Publishers, Philadelphia, 1997, Vols 1–3, pp 1–2976.

6. Janz D. The grand mal epilepsies and sleeping–waking cycle. *Epilepsia* **3**:169–109(1962).

7. Halasz P, Terzano MG, Parrino L. Spike-wave discharge and the microstructure of sleep–wake continuum in idiopathic generalized epilepsy. *Neurophysiol Clin* **32**:38–53(2002).

8. Gigli GL, Calia E, Marciani MG, Mazza S, Mennuni G, Diomedi M, Terzano MG, Janz D. Sleep microstructure and EEG epileptiform activity in patients with juvenile myoclonic epilepsy. *Epilepsia* **33**:799–804(1992).

9. Marzec ML, Malow BA. Approaches to staging sleep in polysomnographic studies with epileptic activity. *Sleep Med* **4**:409–419(2003).

10. Malow BA, Bowes RJ, Ross D. Relationship of temporal lobe seizures to sleep and arousal: a combined scalp–intracranial electrode study. *Sleep* **23**:231–234(2000).

11. Sammaritano MR, Gigli GL, Gotman J. Interictal spiking during wakefulness and sleep and the localization of foci in temporal lobe epilepsy. *Neurology* **41**:290–297(1991).

12. Shouse MN, Mahowald M. Epilepsy, sleep and sleep disorders. In: Kryger MH, Roth T, Dement WC (Eds), *Principles and Practice of Sleep Medicine*, 4th ed. Saunders, Philadelphia, 2004.

105

HEADACHES AND SLEEP

GLEN P. GREENOUGH

Dartmouth–Hitchcock Medical Center, Lebanon, New Hampshire

INTRODUCTION

The relationship between headaches and sleep was first written about in the 19th century. Romberg in 1853 and Living in 1873 each commented on the therapeutic effect of sleep on individual migraine attacks [1]. Since that time multiple layers of complexity have been added to the relationship of sleep and headaches. Sleep may not only alleviate headaches but may play a role in headache genesis. The relationship can be further dissected to reveal that certain headache types may be associated with or evolve from certain sleep stages. Various sleep pathologies or disorders may also lead to the development of headaches arising from sleep.

MIGRAINE HEADACHE

Migraine headaches are typically unilateral throbbing headaches associated with nausea, phonophobia, and photophobia. Migraine headaches have a female preponderance and are believed to be hereditary. While most migraine headaches arise from wakefulness, these headaches may arise from sleep. In patients with nocturnal migraine, the headaches appear to arise from REM sleep. Dexter and Weitzman [2] studied patients with nocturnal migraine headaches and demonstrated that of eight arousals with headache in three patients, six of the headaches arose directly from REM sleep and the remaining two occurred within 9 minutes of REM sleep. To further evaluate the relationship between sleep and migraine headache, one subject slept from 4 p.m. to midnight, yet stage REM-related headaches

continued to occur, suggesting a sleep stage but not circadian linkage for the headaches [3]. From a pathophysiologic standpoint, serotonin may be the key compound linking migraine and REM sleep. The dorsal raphe nucleus, which secretes serotonin, is quiescent during REM sleep. The depletion of serotonin can induce migraine attacks and intravenous serotonin can abort migraine attacks [1].

Many patients with migraine headache report relief or improvement in their headaches with sleep. Inamorato et al. [4] reported pain relief with sleep in the majority of migraine patients surveyed. A prospective trial was carried out by Wilkinson et al. [5] to determine if it was sleep or inactivity that relieved migraine headaches. In this trial, 310 patients were randomized to rest or nap when they had a migraine headache. Resolution of migraine headaches was noted in 50% of those who slept versus 31% in the group that rested but did not sleep.

In contrast, there is also evidence, however, that sleep can precipitate migraine headaches. Inamorato et al. [4] found that 6% of migraine patients reported precipitation of headaches with excess sleep. The increase in slow-wave and REM sleep that characterizes rebound sleep after sleep deprivation may be the mediating factor. In support of this concept, Dexter [6] reported that morning arousals with headache were associated with increased amounts of REM or slow-wave sleep the preceding night.

Multiple studies have documented an increased incidence of parasomnias in patients with migraine. The incidence of somnambulism is increased among migraine patients and ranges from 22% to 71% with most of the studies reporting incidences of 22–30%. An association with night terrors is also suggested in the literature [3].

CLUSTER HEADACHE

In addition to migraine, there is a probable relationship between REM sleep and cluster headaches. Cluster headache disorder is a headache disorder occurring primarily in men that may either be episodic or chronic. Clusters of paroxysmal headaches characterize the episodic type. The clusters may last from 1 week to 1 year, alternating with headache-free periods lasting at least 2 weeks. Individual cluster headaches are characterized by a brief (less than 60 minutes) boring orbital pain often associated with agitation, rhinorrhea, lacrimation, nasal stuffiness, and possibly a partial Horner's syndrome. These headaches often occur at night. Dexter and Weitzman [2] performed polyomnograms and recorded 9 headaches in 3 cluster headache patients and 7 of the 9 arose from REM sleep. Most later studies have replicated this REM association in cluster headache patients.

Pfaffenrath et al. [7] did a similar study in patients with chronic cluster headache but had conflicting results. In chronic cluster headache, patients experience cluster headaches without remissions. In this study, 25 cluster headaches were recorded in 8 patients. Five arose from REM sleep, eleven from stage 2 sleep, four from stage 1 sleep, and two from stage 3 sleep. To reconcile these findings, Sahota and Dexter [3] proposed that this difference might be because episodic cluster headache is REM related and chronic cluster headache is not.

Part of the REM relatedness of cluster headaches may be mediated by sleep disordered breathing. Sleep disordered breathing may be more prominent in REM sleep as opposed to NREM sleep. Obstructive sleep apnea may trigger cluster headaches in predisposed individuals. Kudrow et al. [8] demonstrated that 57% of cluster headaches were REM related in patients with sleep apnea as opposed to 40% in patients without sleep apnea. A study by Chervin et al. [9] noted apnea and hypopnea indexes greater than 5 in 20 of 25 cluster headache patients not selected for sleep complaints. Patients with active cluster headaches in this study had higher end-tidal carbon dioxide levels. Nobre et al. [10] found 5 of 16 patients with episodic cluster headache to have sleep apnea. Two of the patients with sleep apnea experienced headaches in the laboratory and both were associated with desaturations in REM sleep [10]. The reason sleep disordered breathing triggers cluster headaches remains unclear, although hypoxia and hypercapnia have been suggested as possibilities.

The chronobiology of cluster headaches is another interesting feature of this disorder. For a given individual, headaches in this disorder may strike with a certain regularity or periodicity (e.g., the same time every day). The headache clusters also are most likely to occur at the summer and winter solstices, suggesting a link to photoperiod [1]. In further support of this concept is the observation of the activation of the gray matter of the anteroventral hypothalamus, which contains the suprachiasmatic nucleus, during cluster headache attacks [1]. Recent data suggests that the secretion of melatonin, a hormone implicated in circadian rhythm modulation, may be impaired in cluster headache and its replacement may help prevent cluster headaches [11].

CHRONIC PAROXYSMAL HEMICRANIA

Chronic paroxysmal hemicrania is characterized by brief (average 13 minutes each), intense, recurrent (on average 11 times/day) hemicranial pain. Women are more likely to be affected than men. This disorder responds, by definition, to indomethacin. Kayed et al. [12] described one 29 year old patient polysomnographically studied for four nights off medications: 17 of 18 attacks arose from REM sleep. Other abnormalities in this patient include decreased total sleep time, decreased REM sleep time, and REM sleep fragmentation

HYPNIC HEADACHES

Hypnic headaches are a rare disorder that primarily afflicts males over the age of 60 years. Patients present with complaints of diffuse headaches during the night lasting 30–60 minutes associated with nausea. These are differentiated from cluster headache by the absence of autonomic signs, although they respond to lithium carbonate much like cluster headache [13]. Interestingly, they, like paroxysmal hemicrania, may respond to indomethacin [14]. Pinessi et al. [15] report on polysomnography in two patients with hypnic headaches. All four headaches recorded emerged from REM sleep. Dodick [16] reports on polysomnography in three subjects with hypnic headache and the findings varied among the three patients. Interestingly, the only patient to experience a headache during the polysomnogram had his headache in association with a desaturation from obstructive sleep apnea (OSA) during REM sleep. In this patient, the hypnic headaches resolved with nasal continuous positive airway pressure (CPAP) and supplemental oxygen.

IDIOPATHIC INTRACRANIAL HYPERTENSION

Idiopathic intracranial hypertension (IIH), also called pseudotumor cerebri, which occurs primarily in obese women, has as its principal symptoms headaches and papilledema. Evaluation typically reveals an elevated opening pressure with lumbar puncture but normal spinal fluid analysis as well as normal brain imaging. Marcus et al. [17] in a retrospective study demonstrate that the majority of IIH patients have some sleep complaints suggestive of OSA. This is not particularly surprising as obesity is a risk factor common to both disorders. The patient subset that underwent polysomnography was likely to have either OSA or an upper airway resistance syndrome (UARS). Lee et al. [18] retrospectively identified an elevated frequency of sleep apnea among men with IIH and also found that treatment of the sleep apnea may improve the signs and symptoms of IIH in these men. Purvin et al. [19] proposed that the episodic hypoxemia and hypercarbia that characterizes OSA leads to cerebral vasodilation, thereby increasing intracranial pressure.

OBSTRUCTIVE SLEEP APNEA SYNDROME AND HEADACHE

Brief morning headaches are potentially part of the obstructive sleep apnea syndrome. Forty-eight percent of patients with obstructive sleep apnea syndrome complain of headaches upon awakening [20], and these headaches are considered frequent in 18–24% [21, 22]. Pavia et al. [23] polysomnographically studied patients in a headache clinic who reported that 75% or more of their headaches arose from sleep. Seven of 49 patients meeting these criteria were diagnosed with OSA [23]. Pavia et al. [23] and Loh et al. [20] were both able to demonstrate that treatment of the obstructive sleep apnea syndrome leads to improvement or resolution of headaches present upon awakening. As to why obstructive sleep apnea is associated with headaches remains a matter of debate although sleep fragmentation remains a distinct possibility. A possible association of headache complaint with nocturnal desaturation remains controversial with conflicting data available [24, 25].

OTHER SLEEP DISORDERS AND HEADACHE

While the association between obstructive sleep apnea and headache is well recognized, headaches have been shown to be no more common in sleep apnea than in other sleep disorders [21]. Pavia et al. [23], when polysomnographically studying patients presenting to a headache clinic in whom 75% or more of their headaches arose from sleep, found that 8 of 49 met criteria for periodic limb movement disorder. Treatment of the periodic limb movements may lead to some improvement of the headaches but to less of a degree than does treatment of OSA-related hedaches [23]. The restless legs syndrome, which often coexists with periodic limb movement disorder, may be associated with an increased incidence of headaches. Complaints of headache upon awakening and daytime headaches are three to five times more frequent among those afflicted with the restless legs syndrome [26]. Eighty-one percent of narcolepsy patients in one series met International Headache Society criteria for a headache disorder with 54% meeting criteria for migraine [27]. In one series, patients treated with melatonin for delayed sleep phase noted an improvement in headaches of multiple types [28]. Sleep deprivation itself may lead to headaches. Sleep deprivation headaches are characterized by a dull ache, heaviness or pressure in the forehead and/or at the vertex relieved by simple analgesics usually within 1 hour [29].

REFERENCES

1. Dodick DW, Eross EJ, Parish JM. Clincial, anatomical, and physiologic relationship between sleep and headache. *Headache* **43**:282–292(2003).
2. Dexter JD, Weitzman ED. The relationship of nocturnal headaches to sleep stage patterns. *Neurology* **20**:513–518(1970).
3. Sahota PK, Dexter JD. Sleep and headache syndromes: a clinical review. *Headache*. **30**:80–84(1990).
4. Inamorato E, Minatti-Hannuch SN, Zuckerman E. The role of sleep in migraine attacks. *Arq Neuropsiquiatr* **51**(4):429–432(1993).
5. Wilkinson M, Williams K, Leyton M. Observation on the treatment of an acute attack of migraine. *Res Clin Studies Headache* **6**:141–146(1978).
6. Dexter JD. The relationship between stage III and IV and REM sleep and arousals with migraine. *Neurology* **20**:513–518(1979).
7. Pfaffenrath V, Pollmann W, Ruther E. Onset of nocturnal attacks of chronic cluster headache in relation to sleep stages. *Acta Neurol Scand* **73**:403–407(1986).
8. Kudrow L, McGunty DJ, Phillips ER, et al. Sleep apnea in cluster headache. *Cephalalgia* **4**:33–38(1984).
9. Chervin RD, Zallek SN, Lin X, et al. Sleep disordered breathing in patients with cluster headache. *Neurology* **54**(12):2302–2306(2002).
10. Nobre ME, Filho PF, Dominici M. Cluster headache associated with sleep apnoea. *Cephalalgia* **23**:276–279(2003).
11. Weintraub JR. Cluster headaches and sleep disorders. *Curr Pain Headache Rep* **7**:150–156(2003).
12. Kayed K, Godthbsen OB, Sjaastad O. Chronic paroxsysmal hemicrania: "REM sleep locked" nocturnal headache attacks. *Sleep* **1**:91–95(1978).
13. Raskin NH. The hypnic headache syndrome. *Headache* **28**:534–536(1988).
14. Ivanex V, Soler R, Barreiro P. Hypnic headache syndrome: a case with good response to indomethacin. *Cephalalgia* **18**(4):225–226(1998).
15. Pinessi L, Rainero I, Cicolin A, et al. Hypnic headache syndrome: association of the attacks with REM sleep. *Cephalagia* **23**(2):150–154(2003).

16. Dodick DW. Polysomnography in hypnic headache syndrome. *Headache* **40**:748–750(2000).

17. Marcus DM, Lynn J, Miller JJ, et al. Sleep disorders: a risk factor for pseudotumor cerebri? *J Neuroopthalmol* **21**(2):121–123(2001).

18. Lee AG, Golnik K, Kardon R, et al. Sleep apnea and intracranial hypertenison in men. *Opthalmology* **109**(3):482–485(2002).

19. Purvin VA, Kawasaki A, Yee RD. Papilledema and obstructive sleep apnea. *Arch Opthalmol* **118**:1626–1630(2000).

20. Loh NK, Dinner DS, Foldvar N, et al. Do patients with obstructive sleep apnea wake up with headaches? *Arch Intern Med* **159**(15):1765–1768(1999).

21. Aldrich MS, Chauncey JB. Are morning headaches part of obstructive sleep apnea syndrome? *Arch Intern Med* **150**:1265–1267(1990).

22. Poceta JS, Dalessio DJ. Identification and treatment of sleep apnea in patients with chronic headaches. *Headache* **35**:586–589(1995).

23. Pavia T, Farinha A, Martins A, et al. Chronic headaches and sleep disorders. *Arch Intern Med* **157**:1701–1705(1997).

24. Greenough GP, Nowell PN, Sateia MJ. Headache complaints in relation to nocturnal oxygen saturation among patients with sleep apnea syndrome. *Sleep Med* **3**:361–364(2002).

25. Rains J, Penzien D, Mohammad Y. Sleep and headache: morning hedache associated with sleep disordered breathing. *Cephalalgia* **21**:520(2001).

26. Ulfberg J, Nystrom B, Carter N, et al. Prevalence of restless legs syndrome among men aged 18 to 64 years: an association with somatic disease and neuropsychiatric syndromes. *Mov Disord* **16**(6):1159–1163(2001).

27. Dahmen N, Querings K, Grun B. Increased frequency of migraine in narcoleptic patients. *Neurology* **52**(6):1291–1293(1999).

28. Nagtegaal JE, Smits MG, Swart AC, et al. Melatonin-responsive headache in delayed sleep phase syndrome: preliminary observations. *Headache* **38**(4):303–307(1998).

29. Blau JN. Sleep deprivation headache. *Cephalalgia* **10**(4):157–160(1990).

ADDITIONAL READING

Greenough GP. Miscellaneous neurologic disorders and sleep. In: Lee-Chiong TL, Sateia MJ, Carskadon MA (Eds), *Sleep Medicine*. Hanley & Belfus, Philadelphia, 2002, pp 533–539. This chapter touches on the relationship between various neurologic disorders including headache.

Sahota PK, Dexter JD. Sleep and headache syndromes: a clinical review. *Headache* **30**:80–84(1990). This article offers a clinical overview of headaches and sleep disorders.

Weintraub JR. Cluster headaches and sleep disorders. *Curr Pain Headache Rep* **7**:150–156(2003). A current and through review of the ineraction between cluster headache and sleep.

106

CEREBROVASCULAR DISORDERS

BISHOY LABIB AND SARKIS M. NAZARIAN

Central Arkansas Veterans Healthcare System, Little Rock, Arkansas

INTRODUCTION

Obstructive sleep apnea (OSA) and snoring are associated with an increased risk for stroke, and the three conditions share several risk factors. Recent studies, however, strongly suggest that OSA is an independent risk factor for stroke. As population-based studies suggest a prevalence of at least 20% for mild and 6% for moderate OSA worldwide [1], the correct diagnosis and appropriate treatment of OSA are very important components of stroke prevention campaigns.

OSA is a condition characterized by repeated episodes of apnea and hypopnea during sleep, leading to multiple nighttime arousals and daytime somnolence. Apnea is defined as cessation of breathing for 10 seconds, whereas hypopnea is defined as a 30% or greater reduction in airflow accompanied with 4% or greater oxygen desaturation. The apnea–hypopnea index (AHI), also known as the respiratory disturbance index (RDI), is defined as the average number of apneas and hypopneas per hour of sleep. The American Academy of Sleep Task Force defined OSA in 1999 as mild if the AHI was 5–15, moderate if 15–30, and severe if greater than 30 [2].

Pharyngeal patency is maintained by forces that increase the cross-sectional area of the airway, such as the pharyngeal dilator (genioglossus) muscles and increased lung volume, and forces that promote its collapse, such as high inspiratory negative pressures and physical factors (e.g., adipose tissue, which result in a constricted airway) [3]. During sleep, pharyngeal dilator muscle activity and lung volume both decrease, leading to partial or complete airway collapse. The rapid vibration of pharyngeal soft tissues caused by high velocity airflow within a partially collapsed airway results in snoring. When the airway completely collapses and results in obstructive apnea, breathing is established only after arousal from sleep. Depending on the number of apneas, the patient develops progressively worse sleep deprivation, daytime sleepiness, deterioration of intellectual function, tendency to accidents, and decreased quality of life.

Five case–control and three cohort studies since 1985 have demonstrated a relative stroke risk ranging from 1.26 to 3.2 for snorers [4], with the only outlier risk of 10.3 found in the earliest and smallest of these studies [5]. The relationship of OSA documented by polysomnography and stroke has been investigated in three studies with fewer than 50 patients, two with 100–200 patients, and the Sleep Heart Health Study (SHHS), a cross-sectional study of over 6400 patients [6]. The five smaller studies found that OSA with AHI $\geq$ 10 had a prevalence of 32–80% in hospitalized patients with recent stroke. The SHHS found a relative stroke risk of 1.55 in the highest quartile (AHI $>$ 11) compared to the lowest (AHI $<$ 1.4) [6]. The vexing issue of whether OSA is a cause or consequence of stroke is not entirely resolved, but convincing arguments have been made that OSA should improve as motor deficits improve after stroke if it represents a consequence of stroke. In one study, polysomnography repeated 3 months after the initial recording during hospitalization found that the prevalence was 62% compared to 70% in the acute phase [7]. The same study found that the central apnea index was 3.3 per hour in the repeat study compared to 6.2 in the acute phase, and only 6 of the 17 patients with Cheyne–Stokes breathing in the initial study exhibited that pattern in the follow-up study. Although not definitive, these findings suggest that central apnea and Cheyne–Stokes breathing occur as a consequence of stroke and improve at 3 months, whereas OSA shows little improvement and most likely represents a baseline condition.

STROKE TIMING IN RELATION TO THE SLEEP– WAKE CYCLE

Stroke is most common in the early morning hours, especially in the 30 minutes after awakening, and 6 a.m. to noon is the period with the highest incidence of ischemic and hemorrhagic strokes and transient ischemic attacks (TIAs). The usual sleep hours between midnight and 6 a.m. have the lowest stroke risk, as well as having the lowest blood pressure and catecholamine and corticosteroid levels. The stroke risk rises with elevations in these three parameters after awakening. The highest incidence of stroke and sudden cardiac death occurs between 6 a.m. and noon. In the general population, the percentage of vascular deaths between 10 p.m. and 6 a.m. is 24%, compared to 45% between 6 a.m. and 2 p.m., reflecting the overall protective effect of sleep [8]. In patients with OSA, on the other hand, the corresponding figures are 54% and 22%, respectively. The increased vascular risk is shifted to the usual sleep hours in patients with OSA, for whom sleep is risky rather than protective. Although the corresponding stroke data has not been analyzed, it appears to mirror the pattern of vascular deaths in patients with OSA [9].

SLEEP–WAKE AND BREATHING DISORDERS AFTER STROKE

Specific sleep–wake cycle anomalies resulting from stroke injury to specific brain regions have been well documented. Lesions in the midbrain and diencephalon result in a sleep-like awake state and stage I hypersomnia, often accompanied by vertical gaze disturbances. Conversely, lesions in the medial pons cause insomnia with decreased NREM and REM sleep, along with horizontal gaze palsies. Strokes in the medulla frequently result in abnormal breathing and sleep patterns, including central sleep apnea. Reduction of stage II sleep after stroke has been correlated with poor prognosis. Middle cerebral artery infarcts often cause an increase in NREM sleep and a decrease in REM sleep, especially if they are large and involve the right hemisphere. Narcolepsy has been described following vertebrobasilar vascular lesions and after cardiac arrest.

Sleep disordered breathing (SDB), a spectrum of disorders that include snoring, OSA, and central sleep apnea among others, has been documented in the post-stroke

period in several studies. Central sleep apnea usually occurs in the setting of brainstem strokes; unlike OSA, patients with this condition do not attempt to ventilate. Cheyne–Stokes breathing, a cyclic form of breathing with alternating hyperpnea and hypopnea, is usually seen following large hemispheric strokes, as well as noncerebral etiologies such as pulmonary edema and congestive heart failure.

Nocturnal oxygen saturation is lower in stroke victims compared to controls, and 23% of stroke patients in one study were found to spend more than 30 minutes with oxygen saturation less than 90% in one recent study [10]. Another recent study investigating the relationship between SDB and stroke severity found a correlation between high SDB indices (AHI and desaturation index, or DI) and previous stroke, as well as severity of white matter pathology, especially in the frontal lobes and in the basal ganglia [11]. The study does not answer the question of causality, that is, whether the white matter disorder is a result or a cause of SDB.

CEREBRAL AND CARDIAC HEMODYNAMICS IN SLEEP APNEA

The futile attempts by the patient to ventilate during an episode of apnea result in high negative intrathoracic pressures, measured at up to $-80\,\text{cm}\,H_2O$ [12], leading to elevated central venous return and pressure. The increase in central venous pressure (CVP) is directly transmitted to the intracranial compartment. Oxygen desaturation and hypercapnia occur during the period of apnea, and increased cerebral blood flow (CBF) results as a normal physiologic response to hypercapnia. After the apnea is terminated by arousal, the hyperventilation that occurs during snoring restores oxygen and carbon dioxide concentrations to normal, and the CBF falls to baseline levels. The negative intrathoracic pressures result in increased pressure across the aortic and myocardial walls, leading to increased cardiac afterload. The increased CVP, on the other hand, leads to decreased cardiac end-diastolic volume. The overall result is a fall in the cardiac output. Oxygen desaturation and hypercapnia during apnea cause chemoreceptor activation, with resultant compensatory increased sympathetic output, systemic blood pressure, and pulmonary artery pressure. When breathing resumes at the end of the apnea, venous return and cardiac output increase, leading to a further rise in the systemic blood pressure, which can reach up to 240/120 mmHg.

Blood pressure and sympathetic nerve activity both decrease with increasing depth of slow-wave sleep. Rapid eye movement (REM) sleep, on the other hand, is associated with an increase in sympathetic activity, blood pressure, and heart rate to waking levels. The nocturnal fall in sympathetic activity and blood pressure does not occur in patients with OSA, mainly as a result of hypoxia, hypercapnia, and the elimination of sympathetic inhibitory influences from pulmonary afferents during periods of apnea [13]. Recordings from sympathetic afferents to muscle from the peroneal nerve in patients with OSA have shown that sympathetic activity and blood pressure are highest at the end of apneas [13]. Apneas are more common during REM sleep, when muscle tone in the genioglossus is low compared to other sleep epochs. OSA thus further magnifies the physiologic rise in sympathetic activity and blood pressure that occurs during REM sleep. Furthermore, the baseline sympathetic nerve activity in patients with OSA remains elevated during the day, in the absence of apnea.

Vasomotor reactivity, defined as the variation in CBF due to carbon dioxide concentration, is decreased in patients with arterial hypertension and atherosclerosis. Vasomotor reactivity is lower in the morning than in the evening and has been suggested as a cause of the diurnal variation in stroke incidence. It has been hypothesized that patients with OSA have lower vasomotor reactivity than normal patients and thus are more susceptible to stroke, especially in the morning.

HYPERTENSION AND OTHER VASCULAR RISK FACTORS RELATED TO OSA

Sleep apneas cause large increases in heart rate and blood pressure, the latter as high as 40 mmHg in mean arterial pressure (MAP) near the end of the apnea, due to hypoxia-triggered chemoreceptor activation. In addition to these frequent episodes of hypertension during sleep, patients with OSA also develop daytime hypertension [14], even when controlling for age, gender, body mass index (BMI), alcohol consumption, and smoking. The relative risk (RR) for hypertension for AHI > 30 compared to AHI < 1.5 was 1.37 (1.03–1.83) in the Wisconsin Sleep Stroke Study [15]; the same study found a RR of 2.9 for developing hypertension in patients with AHI > 30, adjusting for other vascular risk factors. Several studies have now found that hypertension decreases in OSA patients successfully treated with continuous positive airway pressure (CPAP).

Sleep apnea has been found in several studies to be associated with cardiac arrhythmias, mainly bradyarrhythmias such as second-degree atrioventricular block resulting from hypoxia-induced physiologic diving reflex. The stress imposed on the myocardium by repeated apneic episodes eventually leads to ventricular tachycardias and atrial fibrillation [4]. This latter condition may be explained by the right heart afterload due to pulmonary hypertension and vasoconstriction due to hypoxia, and is four times as common in patients with OSA as in controls. Since atrial fibrillation conveys stroke risk of one order of magnitude greater than in controls, its increased incidence in patients with OSA is consistent with increased stroke risk in this group. Another cardiac condition known to be associated with increased stroke risk is patent foramen ovale (PFO). This condition has been found in about 30% of patients at autopsy, and several studies have found that it is over-represented in patients with stroke, especially in the group without other common stroke risk factors. The increased right-heart pressures during apneas can increase the risk of right-to-left shunting of emboli, and a recent study confirmed that such shunting does indeed occur with apneas that last more than 17 seconds [16].

Elevated fibrinogen, a known risk factor for stroke, is an acute-phase reactant associated with a variety of infectious and inflammatory conditions and is known to peak in the morning hours, as does platelet reactivity. Enhanced thrombogenic tendency due to increased fibrinogen level [17], increased platelet activation and aggregation [18], and faster modified recalcification time, a measure of whole-blood coagulability [19], have been found in patients with OSA, and all improve with CPAP.

A variety of metabolic abnormalities associated with increased risk of vascular disease have been described in patients with OSA. These include elevated leptin levels (higher than in similarly obese individuals without OSA), impaired glucose tolerance, and higher levels of fasting glucose, insulin, and glycosylated hemoglobin.

Finally, chronic cycles of hypoxia and reoxygenation can lead to oxidative stress and reactive oxygen radical generation. Superoxide production can be blunted by CPAP [20]. A variety of vasoactive substances (C-reactive protein, interleukin-6, and tumor necrosis factor alpha) have been shown to be increased in patients with OSA. Along with increased levels of endothelin-1, a potent vasoconstrictor, superoxide and cytokines contribute to vascular endothelial dysfunction and increased susceptibility to vascular disease.

Snoring as a Stroke Risk Factor

Numerous case–control studies suggest that snoring is a stroke risk factor. Early epidemiological studies investigated the relationship between sleep disordered breathing and cerebrovascular disease, using self-reported snoring as the primary exposure variable. Most of these studies showed a strong association of snoring with stroke. They also raised the possibility that snoring is as important a stroke risk factor as hypertension, smoking, and hyperlipidemia. Even when adjusted for possible confounding risk factors such as age, gender, hypertension, and obesity, an independent association between snoring and stroke persisted.

These initial studies were either case–control or cross-sectional, which almost always are dependent on retrospective information and recall bias [5, 21–27]. More-over, these studies could not establish the causal relation-ship between stroke and sleep apnea (i.e., whether OSA is a cause or result of stroke) [28]. Recently, prospective cohort studies such as The Nurses Health Study showed an increased relative risk of 1.33 for stroke in occasional snorers in a more compelling fashion [27]. Another pro-spective cohort study showed that the relative risk for the combined outcome of stroke and ischemic heart disease in habitual snorers was 2.1 times that in nonsnorers [21]. The evidence on snoring as an independent risk factor for stroke remains controversial. Long-term prospective studies in large populations are required in order to determine the actual risk of stroke in snorers.

Obstructive Sleep Apnea as a Stroke Risk Factor

Sleep apnea is strongly associated with some of the leading causes of vascular mortality and morbidity including myo-cardial infarction, cardiac arrhythmia, stroke, and hyperten-sion. The strongest association between sleep apnea and cerebrovascular disease is derived from the results of the Sleep Heart Health Study [6]. This community-based study involved 6424 subjects who underwent overnight polysomnography at home. The study revealed a positive association between the severity of sleep apnea and increas-ing risk for the development of cardiovascular disease (cor-onary artery disease, congestive heart failure, and stroke, independent of known cardiovascular risk factors). A Span-ish prospective study followed 161 patients admitted with newly diagnosed TIA or stroke and previously validated portable respiratory recording was done within 48–72 h after admission and repeated after 3 months [7]. This study showed no significant differences in the severity of sleep apnea according to stroke subtype or the location of stroke. Moreover, the frequency of sleep apnea did not change from baseline to 3 months after stroke. Based on these results, the authors concluded that sleep apnea was present before stroke and possibly acted as a risk factor of cerebrovascular disease.

Overall, given the similar severity and frequency of sleep apnea in various stroke subtypes, the high prevalence of obstructive rather than central sleep apnea in patients with stroke, the association between sleep apnea and stroke favors sleep apnea as a risk factor rather than a consequence of stroke.

Sleep Apnea as a Cause of Dementia and White Matter Disease

Multiple studies show an association of impaired attention and concentration in middle-aged patients with sleep apnea [29, 30]. Impaired executive function can persist even when sleep apnea is treated [31, 32]. These deficits may be

mediated through excessive daytime sleepiness associated with sleep apnea, and with repeated brain exposure to hypoxia. Elderly patients with white matter changes are more likely to have sleep apnea than those without white matter changes. On the other hand, younger patients with sleep apnea show no difference in the prevalence of white matter changes compared to normal controls. The differ-ence between older and younger patients may stem from the duration of exposure to sleep apnea and its cerebral hemodynamic consequences. Obstructive sleep apnea is associated with recurrent nocturnal and persistent daytime hypertension, increased sympathetic activation, reduced cerebral blood flow, and increased platelet activation [33], any or all of which may predispose to both white matter dis-ease and stroke. Cerebral white matter is more vulnerable than gray matter to the effects of persistent hypoxia, hypotension, and hypoperfusion [34–36]. Reduced perfu-sion in areas of white matter disease has been demonstrated [37]. Conversely, sustained hypertension is also known to be a cause of white matter disease through microvascular damage [38]. One study emphasized the association between leukoaraiosis and worse sleep disordered breath-ing (SDB) after stroke especially in men [39]. Another study reported an association between SDB and lacunar stroke [40]. This would be consistent with the hypothesis that it is the white matter disease component of chronic cerebrovascular disease that is specifically associated with SDB.

Treatment of Sleep Apnea in Stroke

There are no prospective studies done to evaluate the need to treat sleep apnea to decrease the risk of stroke, either as a primary or secondary prevention measure. CPAP remains the gold standard medical therapy for OSA. It reduces day-time sleepiness, improves quality of life [41], improves left ventricular function in patients with congestive heart failure [42], decreases blood pressure in patients with hypertension [43], and possibly lowers mortality [44].

One study reported that some younger patients during rehabilitation after stroke tolerated CPAP [45]. Other stu-dies, which included older patients, have shown disappoint-ing results [46].

If CPAP is to influence the outcome of stroke by limiting ischemic damage to the vulnerable areas of brain in the penumbra, it seems likely that its optimal timing would be very early during the ischemia.

Introducing CPAP to an elderly population in an acute stroke setting caused confusion in patients and has proved to be impractical. It may, nonetheless, be beneficial in selected individuals. Hui et al. [47] found that the small minority of patients who tolerated CPAP shortly after a stroke had symptoms suggesting preexisting OSA. This conclusion concurs with studies of patients with

OSA in whom compliance with CPAP is better in those with more severe symptoms, particularly daytime sleepiness [48].

In practice, stroke patients with features of preexisting OSA may be the very ones to benefit from CPAP therapy if the adverse prognosis associated with SDB following stroke is due mainly to pre-stroke OSA.

ROLE OF SLEEP DISORDERS IN RECOVERY FROM STROKE

Obstructive sleep apnea can theoretically worsen stroke outcome by causing unrestful sleep, nocturnal hypoxia, and decreased cardiac output and cerebral perfusion in stroke victims. OSA-related post-stroke complications include delirium, depression, impaired cognition and functional capacity, as well as prolonged hospitalization and rehabilitation [45, 48–52]. A number of studies have found a prevalence of 43–95% of OSA in patients admitted to hospital with stroke [53]. Measures of stroke severity and recovery, such as mortality, length of hospitalization, and improvement in functional scales, have been found to be consistently worse in stroke patients with OSA. A study conducted at Leeds Hospital revealed a highly significant 6 month mortality of 45% in stroke patients with AHI > 10, compared to 25% in patients with AHI < 10 [54]. Length of hospital stay was also prolonged at 43 days in the former group, compared to 24 days in the latter.

Polysomnography carried out on 60 patients admitted to the Stroke Rehabilitation Unit at Toronto General Hospital revealed that 72% had sleep apnea. The group with sleep apnea had worse FIM functional score at admission and discharge (95 versus 80 and 113 versus 102), and a longer overall hospital stay (67 versus 52 days) than patients without OSA [53]. Similarly, a study performed with overnight pulse oximetry on 60 stroke patients in Israel revealed that, in those patients with an initial FIM score of 70 or less, only 2 of 6 with AHI > 10 improved by 30 points, compared to 7 of 9 patients with AHI < 10 [52]. Whether sleep apnea is an independent predictor of poor functional outcome or just a marker for severe stroke is not clear.

CONCLUSION

Sleep disordered breathing is associated with increased risk of myocardial infarction, stroke, and death. Although studies proving direct causality are lacking, available evidence strongly implicates OSA as an independent risk factor for stroke. In addition to the direct adverse effects of OSA on cerebral perfusion and myocardial function, secondary effects such as hyperthrombosis, vascular endothelial injury, free oxygen radicals, and cytokine activation seem to contribute to the overall vascular risk. Central sleep apnea after stroke appears to be secondary to the stroke and tends to improve in tandem with motor deficits. OSA, on the other hand, tends to persist after stroke and adversely affects stroke recovery. CPAP therapy not only reduces vascular risk factors in patients with OSA but also improves their recovery from stroke. Early diagnosis of OSA and its treatment with CPAP offer the hope of decreased morbidity and mortality due to vascular disease.

REFERENCES

1. Young T, Peppard B, Gottlieb D. Epidemiology of obstructive sleep apnea: a population health perspective. *Am J Respir Crit Care Med* **165**:1217–1239(2002).

2. American Academy of Sleep Medicine Task Force. Sleep-related breathing disorders in adults. *Sleep* **22**:667–689(1999).

3. Malhotra A, White DP. Obstructive sleep apnea. *Lancet* **360**:237–245(2002).

4. Yaggi H, Mohsenin V. Obstructive sleep apnoea and stroke. *Lancet Neurol* **3**:333–342(2004).

5. Partinen M, Palomaki H. Snoring and cerebral infarction. *Lancet* **2**:1325–1326(1985).

6. Shahar E, Whitney C, Redline S, et al. Sleep-disordered breathing and cardiovascular disease: cross sectional results of the Sleep Heart Health Study. *Am J Respir Crit Care Med* **163**:19–25(2001).

7. Parra O, Arboix A, Bechich S, et al. Time course of sleep-related breathing disorders in first-ever stroke or transient ischemic attack. *Am J Respir Crit Care Med* **161**:375–380(2000).

8. Gami AS, Howard DE, Olson EJ, Somers VK. Day–night pattern of sudden death in obstructive sleep apnea. *N Engl J Med* **352**:1206–1214(2005).

9. Somers V. *ASA International Stroke Conference*, 2005.

10. Roffe C, Sills S, Halim M, Wilde K, Allen MB, Jones PW, Crome P. Unexpected nocturnal hypoxia in patients with acute stroke. *Stroke* **34**:2641–2645(2003).

11. Harbison J, Gibson GJ, Birchall D, Zammit-Maempel I, Ford GA. White matter disease and sleep-disordered breathing after acute stroke. *Neurology* **61**:959–963(2003).

12. Shiomi T, Guilleminault C, Stoohs R, Schnittger I. Leftward shift of the interventricular septum and pulsus paradoxus in obstructive sleep apnea syndrome. *Chest* **100**:894–902(1991).

13. Somers VK, Dyken ME, Clary MP, Abboud FM. Sympathetic neural mechanisms in obstructive sleep apnea. *J Clin Invest* **96**:1897–1904(1995).

14. Neito F, Young T, Lind B, et al. Association of sleep-disordered breathing, sleep apnea, and hypertension in a large community-based study. *JAMA* **283**:1829–1836(2000).

15. Peppard P, Young T, Palta M, Skatrud J. Prospective study of the association between sleep-disordered breathing and hypertension. *N Engl J Med* **342**:1378–1384(2000).

16. Beelke M, Angeli S, Del Sette M, et al. Obstructive sleep apnea can be provocative for right-to-left shunting through a patent foramen ovale. *Sleep* **25**:856–862(2002).

17. Wessendorf TE, Thilmann AF, Wang YM, Schreiber A, Konietzko N, Teshler H. Fibrinogen levels and obstructive sleep apnea in ischemic stroke. *Am J Respir Crit Care Med* **162**:2039–2042(2000).

18. Bokinski G, Miller M, Ault K, et al. Spontaneous platelet activation and aggregation during obstructive sleep apnea and in response to therapy with nasal continuous positive airway pressure. *Chest* **108**:625–630(1995).

19. Guardiola JJ, Matheson PJ, Clavijo LC, Wilson MA, Fletcher EC. Hypercoagulability in patients with obstructive sleep apnea. *Sleep Med* **2**:517–523(2001).

20. Schulz R, Mahmoudi S, Hattar K, et al. Enhanced release of superoxide from polymorphonuclear neutrophils in obstructive sleep apnea: impact of continuous positive airway pressure therapy. *Am J Respir Crit Care Med* **162**:566–570(2000).

21. Koskenvuo M, Kapri J, Telakivi T, Partinen M, Heikkila K, Sarna S. Snoring as a risk factor for ischemic heart disease and stroke in men. *BMJ* **294**:16–19(1987).

22. Spriggs D, French J, Murdy J, Curless R, Bates D, James O. Snoring increases the risk of stroke and adversely affects prognosis. *Q J Med* **83**:555–562(1992).

23. Palomaki H. Snoring and the risk of ischemic brain infarction. *Stroke* **22**:1021–1025(1991).

24. Smirne S, Palazzi S, Zucconi M, Chierchia S, Ferini-Strambi L. Habitual snoring as a risk factor for acute vascular disease. *Eur Respir J* **6**:1357–1361(1993).

25. Jennum P, Schultz-Larsen K, Davidsen M, Christiansen NJ. Snoring and risk of stroke and ischemic heart disease in a 70 year old population: a 6 year follow up study. *Int J Epidemiol* **23**:1159–1164(1994).

26. Neau J, Meurice J, Paquereau J, Chavagnat J, Ingrand P, Gail R. Habitual snoring is a risk factor for brain infarction. *Acta Neurol Scand* **92**:63–68(1995).

27. Hu F, Willet W, Manson J, et al. Snoring and the risk of cardiovascular disease in women. *J Am Coll Cardiol* **35**:308–313(2000).

28. Askenasy J, Goldhammer I. Sleep apnea as a feature of bulbar stroke. *Stroke* **19**:637–639(1988).

29. Kim H, Young T, Mathews T, Webber S, Woodard A, Palta M. Sleep-disordered breathing and neuropsychological deficits. *Am J Respir Crit Care Med* **156**:1813–1819(1997).

30. Redline S, Strauss ME, Adams N, et al. Neuropsychological function in mild sleep-disordered breathing. *Sleep* **20**:160–167(1997).

31. Sforza E, Krieger J. Daytime sleepiness after long term continuous positive airway pressure (CPAP) treatment in obstructive sleep apnea syndrome. *J Neurol Sci* **110**:21–26(1992).

32. Bedard MA, Montplaisir J, Mallo J, Richer F, Rouleau I. Persistent neuropsychological deficits and vigilance impairment in sleep apnea syndrome after treatment of continuous positive airway pressure (CPAP). *J Clin Exp Neuropsychol* **15**:330–341(1993).

33. Harbison JA, Gibson J. Snoring, sleep apnea and stroke: chicken or scrambled eggs? *Q J Med* **93**:647–654(2000).

34. Macey PM, Henderson LA, Macey KA, et al. Brain morphology associated with obstructive sleep apnea. *Am J Respir Crit Care Med* **166**:1382–1387(2002).

35. Suter OC, Sunthorn T, Kraftsik R, et al. Cerebral hypoperfusion generates cortical watershed microinfarcts in Alzheimer disease. *Stroke* **33**:1986–1992(2002).

36. Wisniewska M, Devuyst G, Bougousslavsky J, Ghika J, Van Melle G. What is the significance of leukoaraiosis in patients with acute ischemic stroke? *Arch Neurol* **57**:925–926(2000).

37. Hatazawa J, Shimosegawa E, Satoh T, Toyoshima H, Okudera T. Subcortical hypoperfusion associated with asymptomatic white matter lesions on magnetic resonance imaging. *Stroke* **28**:1944–1947(1997).

38. O'Sullivan M, Lythgoe DJ, Pereria AC, et al. Patterns of cerebral blood flow reduction in patients with ischemic leukoaraiosis. *Neurology* **59**:321–326(2002).

39. Harbison J, Gibson GJ, Birchall D, Zammit-Maempel I, Ford GA. White matter disease in sleep-disordered breathing after acute stroke. *Neurology* **61**:959–963(2003).

40. Lawrence E, Dundas R, Higgen S, et al. The natural history and associations of sleep-disordered breathing in first ever stroke. *Int J Clin Pract* **9**:584–589(2001).

41. Young T, Peppard B, Gottlieb D. Epidemiology of obstructive sleep apnea: a population health perspective. *Am J Respir Crit Care Med* **165**:1217–1239(2002).

42. Kaneko Y, Floras J, Usui K, et al. Cardiovascular effects of continuous positive airway pressure in patients with heart failure and obstructive sleep apnea. *N Engl J Med* **348**:1233–1241(2003).

43. Pepperell J, Ramdassingh-Dow S, Crosthwaite N, et al. Ambulatory blood pressure after therapeutic and subtherapeutic nasal continuous positive airway pressure for obstructive sleep apnea: a randomized parallel trial. *Lancet* **359**:204–210(2002).

44. Marti S, Sampol G, Munoz X, et al. Mortality in severe sleep Apnea/hypopnoea syndrome patients: impact of treatment. *Eur Respir J* **20**:1511–1518(2002).

45. Wessendorf TE, Wang YM, Thilmann AF, et al. Treatment of obstructive sleep apnea with nasal continuous positive airway pressure in stroke. *Eur Respir J* **18**:623–629(2001).

46. Sandberg O, Franklin KA, Bucht G, Eriksson S, Gustafson Y. Nasal continuous positive airway pressure in stroke patients with sleep apnea: a randomized treatment study. *Eur Respir J* **18**:630–634(2001).

47. Hui DSC, Choy DKL, Wong LKS, et al. Prevalence of sleep-disordered breathing and continuous positive airway pressure compliance. Results in Chinese patients with first ever ischemic stroke. *Chest* **122**:852–860(2002).

48. McArdle N, Devereux G, Heidarnejad H, et al. Long term use of CPAP therapy for sleep apnea/hypopnea syndrome. *Am J Respir Care Med* **159**:1108–1114(1999).

49. Good D, Henkle J, Gelber D, Welsh J, Verhulst S. Sleep-disordered breathing and poor functional outcome after stroke. *Stroke* **27**: 252–259(1996).

50. Iranzo A, Santamaria J, Berenguer J, Sanchez M, Chamorro A. Prevalence and clinical importance of sleep apnea in the first night after cerebral infarction. *Neurology* **58**:911–916(2002).

51. Kaneko Y, Hajek VE, Zivanovic V, Raboud J, Bradly TD. Relationship of stroke apnea to functional capacity and length of hospitalization following stroke. *Sleep* **26**:293–297(2003).

52. Cherkassky T, Oksenberg A, Froom P, Ring H. Sleep-related breathing disorders and rehabilitation outcome of stroke patients. *Am J Phys Med Rehabil* **82**:542–555(2003).

53. Kaneko Y, Hajek V, Zivanovic V, Raboud J, Bradley T. Relationship of sleep apnea to functional capacity and length of hospitalization following stroke. *Sleep* **26**:293–297(2003).

54. Turkington PM, Allgar V, Bamford J, Wanklyn P, Elliott MW. Effect of upper airway obstruction in acute stroke on functional outcome at 6 months. *Thorax* **59**:367–371(2004).

107

BRAIN AND SPINAL CORD INJURY

RICHARD J. CASTRIOTTA

University of Texas Health Science Center at Houston, Houston, Texas

TRAUMATIC BRAIN INJURY

There is a complex relationship between disorders of sleep and traumatic brain injury (TBI). Part of the confusion is a result of the high risk of accidents in those with disorders such as obstructive sleep apnea (OSA), narcolepsy, and shift-work sleep disorder. Thus there is often a question about the existence of a pretraumatic sleep disorder, which may have been the underlying cause of the accident in the first place. Sleepiness causes 42–45% of motor vehicle accidents (MVAs) and 36% of highway fatalities [1]. Those with OSA are 6.3–7.3 times more likely to have MVAs than the general population [2]. In addition to the high frequency of MVAs, those with OSA are much more likely to have accidents in the home or workplace [3]. It is thus not surprising to find a high prevalence of sleep disorders in any group selected because of traumatic brain injury. There is, however, clear evidence to support the development of a number of sleep disorders as a result of traumatic brain injury, with no history of symptoms prior to trauma. These include insomnia, circadian rhythm disorders, periodic limb movement disorder (PLMD), obstructive sleep apnea (OSA), narcolepsy, and post-traumatic hypersomnia (PTH). The latter diagnosis is made when sleepiness begins only after TBI, other causes for sleepiness have been excluded by history and polysomnography, and there is a multiple sleep latency test (MSLT) score <10 minutes with <2 sleep-onset REM periods. Subjective complaints of "sleeping problems" are common, having been found in 47% of 639 patients presenting to a minor head injury clinic [4]. Sleep problems are also common in children after TBI [5].

Case reports of "traumatic narcolepsy" were published as early as 1941 [6], but it was impossible at that time to distinguish sleep apnea from narcolepsy and what is today known as post-traumatic hypersomnia. Many of the early observations were made by French and Italian investigators [7–11]. Several more recent studies have resulted in a general assessment of the problem and a conclusion that excessive daytime somnolence and disorders of initiating and maintaining sleep are common after TBI. The objective measure of daytime sleepiness used is the multiple sleep latency test (MSLT), with excessive daytime sleepiness (EDS) defined as a MSLT score ≤10 minutes and pathological sleepiness as MSLT score ≤5 minutes. In a recently concluded prospective multicenter study of 87 TBI patients, we have found EDS in 23.4% and pathological sleepiness in 12.6%, while an earlier study with 71 TBI subjects [12] found 47% with EDS and 18.3% with pathological sleepiness (MSLT score ≤5 minutes). There was no significant correlation between the MSLT and the Epworth Sleepiness Score (ESS) or Pittsburgh Sleep Quality Index (PSQI). It should be noted, however, that the PSQI has been validated for assessing insomnia in TBI patients [13]. There did not appear to be any relationships between sleep pathology and injury characteristics, Glasgow Coma Scale, length of coma, psychopathology, medications, length of time from injury, or urinary excretion of sulfatoxymelatonin. Among TBI patients with MSLT <10 minutes in that study, 12% had OSA, 24% had PLMD, 3% had narcolepsy, and 61% (28% of all subjects) had PTH. In a prospective study of ten TBI patients with subjective sleepiness (ESS > 10), seven were found to have sleep disordered breathing (SDB = OSA or upper airway resistance syndrome (UARS)), two had narcolepsy, and one had PTH, with pretrauma symptoms in three subjects [14], but in a general TBI population (regardless of symptoms) OSA was found in 21.3%, PTH in 13.5%, PLMD in 7.9%, and narcolepsy in 6.7%. Among 28 TBI patients admitted for rehabilitation and prospectively studied with polysomnography, 36% had >5 apneas + hypopneas/hour and 11% had >10 apneas + hypopneas/hour [15]. When TBI patients were referred for outpatient sleep evaluation in a group of 184 symptomatic patients with head and neck injury referred to a sleep clinic, 32% had SDB, including all 16 with whiplash [16]. Of these, 88% had no preinjury symptoms of SDB, and none had overt pretrauma sleepiness. In this study, 49% had PTH, two patients (1%) had narcolepsy without cataplexy, and eight patients (4%) had "pseudohypersomnia." This was described as a syndrome of apathy with subjective sleepiness (ESS > 10) and MSLT score >10 minutes. This is associated with bilateral paramedian thalamic infarcts, subthalamic and mesencephalic lesions, and "attention deficit." An earlier similar study of 20 TBI patients reported SDB in 40% of subjects [17].

After analysis of these studies (Table 107.1), we can conclude that 44% of all patients with TBI have EDS demonstrated by MSLT score <10 minutes and 22% have pathological sleepiness with MSLT score <5 minutes. Of those who are symptomatic, 34.7% had sleep disordered breathing with overt obstructive sleep apnea (>10 apneas + hypopneas/hour) in 27%. It is unclear how many of these are truly post-traumatic in the absence of cranial nerve or neck injury. When it is looked for, PLMD was present in 9% of TBI cases, but was often accompanied by other conditions such as OSA. Narcolepsy was found in 3% of all TBI patients in the above reports (11 of 316 TBI patients). Like OSA, narcolepsy can be an undiagnosed preexisting condition predisposing to accidental head trauma. However, there have been well-documented case reports of

the complete narcolepsy–cataplexy syndrome presenting after TBI [19–21] without pretrauma signs or symptoms. There are, however, cases of TBI resulting from drivers with SDB and narcolepsy who fell asleep while driving and had pretrauma daytime hypersomnolence [14]. Of the 354 TBI subjects studied by investigators with polysomnography and MSLT, 124 (35%) met the criteria for diagnosis of post-traumatic hypersomnia with onset of symptoms after head injury, MSLT score <10 minutes, and no other cause for EDS. Approximately 20% (32 of 160 TBI subjects) prospectively studied regardless of symptoms had PTH, while 49% of sleepy TBI patients (112 of 227) had PTH. We can conclude, then, that EDS is very common after TBI, and that most sleepy TBI patients will be found to have either SDB or PTH. Narcolepsy is less common, but still 60 times more frequent than in the general population. The significance of PLMD in contributing to sleep problems after TBI remains uncertain, but 9% of TBI patients manifest periodic limb movements with or without other sleep disorders. These appear to respond to pramipexole, while modafinil is effective in PTH and narcolepsy after TBI, and continuous positive airway pressure (CPAP) is effective in those TBI patients with SDB.

There have been no prospective studies and no large case series describing circadian rhythm disturbances after TBI, and none of the above investigations had study designs that would allow an estimation of the prevalence of this potential problem. There have been enough published reports to suggest the existence of a post-traumatic delayed sleep phase syndrome (DSPS) after traumatic brain injury [22–25], and there is one report of 16 patients who developed DSPS after severe whiplash without loss of consciousness who responded to melatonin [26]. One case of non-24-hour sleep–wake (hypernyctohemeral) syndrome after TBI was described as a late complication or more severe form of DSPS and responsive to bright light [27, 28] and have been reported in severely brain-damaged children. None of these

TABLE 107.1   Frequency of Sleep Disorders Associated with Traumatic Brain Injury

Study	Sx's Only[a]	N^b	% MSLT <10 min	% MSLT <5 min	%SDB	%PLMD	%NARC	%PTH
Castriotta [18]	No	87	26.4	12.6	22	7	7	15
Castriotta and Lai [14]	Yes	10			70	0	2	1
Guilleminault et al. [16]	Yes	184			32		1	49
Guilleminault et al. [17]	Yes	20			40			
Masel et al. [12]	No	71	47	18.3	12	24	3	28
Webster et al. [15]	No	28			36			
All studies		400	34.4	16.45	26.75	10.7	3	34.4

[a]"No" indicates prospective study of all TBI patients; "Yes" indicates sample taken from symptomatic patients.
[b]N = sample size.

Abbreviations: MSLT, multiple sleep latency test score; SDB, sleep disordered breathing (obstructive sleep apnea and upper airway resistance syndrome); PLMD, periodic limb movement disorder; NARC, narcolepsy; PTH, post-traumatic hypersomnia.

patients had discernible lesions on MRI or CT scans of brain, but small nonvisualized lesions might have been missed. DSPS might be associated with mesencephalothalamic lesions [29], rostral hypothalamic damage, and lesions of the circadian pacemaker in the suprachiasmatic nucleus [30]. DSPS has been estimated to have a prevalence of 0.7% in middle-aged adults [31] and 7% in adolescents [32] and is thought to have an inherited genetic basis. It is therefore not possible at present to determine whether or not post-traumatic DSPS constitutes a distinct condition (such as PTH) or whether this represents a result of environmental sleep–wake disruption after injury in predisposed individuals.

While EDS remains the most important chronic (>2 years) symptom after TBI, hospitalized patients with TBI of recent onset are more likely to complain of difficulty initiating or maintaining sleep [33]. Insomnia complaints may persist in the outpatient setting [34], and even with minor head injury there can be an increase in sleep interruptions and decreased sleep quality compared to before injury [35, 36]. It has been estimated that 50% of TBI patients in the outpatient rehabilitation setting have difficulty sleeping [37], with 64% of those waking up early and 45% having difficulty falling asleep. When compared to those with orthopedic injuries [38] or general, nonneurologic patients [39], TBI patients are more likely to have difficulty initiating and maintaining sleep. However, in the outpatient rehabilitation setting, twice as many patients with spinal cord injury or musculoskeletal disease have insomnia when compared to post-acute TBI patients [40]. It appears that insomnia after TBI is associated with milder brain injuries and the presence of pain and depression [39, 41]. There are sleep–endocrine alterations months to years after TBI with a decreased nocturnal growth hormone peak and increased maximal prolactin secretion without change in cortisol secretion [42]. This pattern of sleep-associated hormone release is similar to patients who have recovered from depression. One confounding element in determining the effect of TBI on post-traumatic sleep is the presence or absence of assault. In one English study [4], 44% of the TBI patients had been assaulted. The additional trauma of being personally attacked (rather than the less personal affront of a MVA) may contribute to insomnia problems quite distinct from the head injury. It appears that cognitive–behavioral therapy is effective for the management of insomnia after TBI [43].

Jactatio nocturna after TBI has been reported in a patient with global encephalopathy and frontal lobe dysfunction, which was successfully treated with imipramine [44]. Bruxism can occur after TBI and has been treated with botulinum toxin [45]. The incidence of dreaming does not seem to change after TBI, although dreams with sexual content may be reduced while threatening dreams tend to be more frequent [46].

SPINAL CORD INJURY

In comparison to the normal population, patients with spinal cord injury (SCI) have greater difficulty falling asleep, have more frequent awakenings, sleep subjectively less well, sleep more hours, take more and longer naps, and snore more [47]. Respiratory patterning and plasticity are altered after cervical spinal cord injury [48]. Patients with SCI can awaken because of hypercapnia [49]. Odontoid or cervical fractures and cervical hernias can induce OSA and cause EDS [50]. The prevalence of OSA in the SCI population is high, particularly among tetraplegics, where the prevalence was 55% in men and 20% in women from a sample of 50 randomly selected tetraplegic patients [51]. The prevalence of OSA in the nonobese outpatient cervical cord injury population is 15% [52], 25% in those SCI patients more than 40 years old [53] and 40% in the hospitalized SCI population [54]. These patients are at risk for significant nocturnal hypoxemia [55]. In the tetraplegic population there does not appear to be a consistent correlation between apnea–hypopnea index and lesion level, ASIA impairment scale or spirometric values, and daytime complaints were present only in those with >40 apneas + hypopneas/hour. Thus OSA is likely to be underdiagnosed in this group. In addition, it is possible for patients with diffuse lower cervical cord injury to develop delayed apnea and death, especially when C4 is involved. The presence of bradycardia and dyspnea should be viewed as warning signs for this scenario. Whiplash may predispose to SDB [16]. OSA in SCI patients has been successfully treated with CPAP [56].

Periodic limb movements can occur after spinal cord injury [57, 58]. In fact, this can be demonstrated experimentally in rats [59]. With SCI, periodic limb movements can be seen in both REM and NREM sleep [60]. These can be improved with physical activity and are correlated with K-complexes [61, 62].

CONCLUSION

Sleep disorders after traumatic brain and spinal cord injury are common. Approximately 34% of all TBI patients will have excessive daytime somnolence by MSLT, and most of these will have either sleep disordered breathing or post-traumatic hypersomnia. TBI patients should also be evaluated for possible delayed sleep phase syndrome and non-24-hour sleep–wake syndrome. Insomnia is seen more frequently with mild brain injury than severe head trauma and may be linked with depression. The prevalence of OSA is also increased with whiplash and after spinal cord injury, but OSA most frequently is found in tetraplegic patients. Periodic limb movements are likely to be seen after spinal cord injury, and their significance after TBI is uncertain. Given the frequency of treatable sleep disorders in these populations, it would seem appropriate to evaluate most TBI and SCI patients with an eye to identifying and correcting these problems.

REFERENCES

1. Leger D. The cost of sleep-related accidents: a report for the National Commission on Sleep Disorders Research. *Sleep* **17**:84–93(1994).

2. Teran-Santos J, Jimenez-Gomez A, Cordero-Guevara J. The association between sleep apnea and the risk of traffic accidents. *N Engl J Med* **340**:847–851(1999).

3. Horstmann S, Hess CW, Bassetti C, Gugger M, Mathis J. Sleepiness-related accidents in sleep apnea patients. *Sleep* **23**:383–389(2000).

4. Haboubi NHJ, Long J, Koshy M, Ward AB. Short-term sequelae of minor head injury (6 years experience of minor head injury clinic). *Disabil Rehabil* **23**:635–638(2001).

5. Hooper SR, Alexander J, Moore D, Sasser HC, Laurent S, King J, Bartel S, Callahan B. Caregiver reports of common symptoms in children following traumatic brain injury. *Neuro Rehabil* **19**:175–189(2004).

6. Gill AW. Idiopathic and traumatic narcolepsy. *Lancet* **1**:474–476(1941).

7. Barbano G, Bossi L. Rheography in the subjective syndrome of patients with cranial trauma. *Minerva Med* **55**:640–644(1964).

8. Meurice E. Signs of diurnal falling asleep in a group of post-concussion patients compared with a group of normal subjects. *Rev Neurol (Paris)* **115**:524–526(1966).

9. Meurice E. Signs of diurnal sleep in the post-concussional patient (examined in a dark room). *Electroencephalogr Clin Neurophysiol* **23**:287(1967).

10. Petitjean F, Jouvet M. Hypersomnia and increase of cerebral 5-hydroxyindoleacetic acid due to isthmic lesions in the cat. *C R Seances Soc Biol Fil* **164**:2288–2293(1970).

11. Amico G, Pasquali F, Pittaluga E. Pickwickian-narcoleptic disorders after brain concussion. *Riv Sper Freniatr Med Leg Alien Ment* **96**:74–85(1972).

12. Masel BE, Scheibel RS, Kimbark T, Kuna ST. Excessive daytime sleepiness in adults with brain injuries. *Arch Phys Med Rehabil* **82**:1526–1532(2001).

13. Fichtenberg NL, Putnam SH, Mann NR, Zafonte RD, Millard AE. Insomnia screening in postacute traumatic brain injury: utility and validity of the Pittsburgh Sleep Quality Index. *Am J Phys Med Rehabil* **80**:339–345(2001).

14. Castriotta RJ, Lai JM. Sleep disorders associated with traumatic brain injury. *Arch Phys Med Rehabil* **82**:1403–1406(2001).

15. Webster JB, Bell KR, Hussey JD, Natale TK, Lakshminarayan S. Sleep apnea in adults with traumatic brain injury: a preliminary investigation. *Arch Phys Med Rehabil* **82**:316–321(2001).

16. Guilleminault C, Yuen KM, Gulevich MG, Karadeniz D, Leger D, Philip P. Hypersomnia after head–neck trauma. A medico-legal dilemma. *Neurology* **54**:653–659(2000).

17. Guilleminault C, Faull KF, Miles L, van den Hoed J. Posttraumatic excessive daytime sleepiness: a review of 20 patients. *Neurology* **33**:1584–1589(1983).

18. Castriotta RJ. Chest 128(4)sup (2005).

19. Good JL, Barry E, Fishman PS. Posttraumatic narcolepsy: the complete syndrome with tissue typing. *J Neurosurg* **71**:765–767(1989).

20. Lankford DA, Wellman JJ, O'Hara C. Posttraumatic narcolepsy in mild to moderate closed head injury. *Sleep* **17**:528–528(1994).

21. Francisco GE, Ivanhoe CB. Successful treatment of post-traumatic narcolepsy with methylphenidate: a case report. *Am J Phys Med Rehabil* **75**:63–65(1996).

22. Lenard HG, Pennigstorff H. Alterations in the sleep patterns of infants and young children following acute head injuries. *Acta Paediatr Scand* **59**:565–571(1970).

23. Patten SB, Lauderdale WM. Delayed sleep phase disorder after traumatic brain injury. *J Am Acad Child Adolesc Psychiatry* **31**:100–102(1992).

24. Nagtegaal JE, Kerkhof GA, Smits MG, Swart AC, van der Meer YG. Traumatic brain injury-associated delayed sleep phase syndrome. *Funct Neurol* **12**:345–348(1997).

25. Quinto C, Gellido C, Chokroverty S, Masdeu J. Posttraumatic delayed sleep phase syndrome. *Neurology* **54**:250–253(2000).

26. Smits MG, Nagtegaal JE. Post-traumatic delayed sleep phase syndrome. *Neurology* **55**:902–903(2000).

27. Boivin DB, James FO, Santo JB, Caliyurt O, Chalk C. Non-24-hour sleep–wake syndrome following a car accident. *Neurology* **60**:1841–1843(2003).

28. Boivin DB, Caliyurt O, James FO, Chalk C. Association between delayed sleep phase and hypernyctohemeral syndromes: a case study. *Sleep* **27**:417–401(2004).

29. Angeleri F, Quattrini A, Chinzari P. Circadian rhythm of wakefulness and sleep in cats with bilateral lemniscal electrolytic lesions at the mesencephalo-thalamic level. *Boll Soc Ital Biol Sper* **45**:985–989(1969).

30. Cohen RA, Albers HE. Disruption of human circadian and cognitive regulation following a discrete hypothalamic lesion: a case study. *Neurology* **41**:726–729(1991).

31. Ando K, Kripke DF, Ancoli-Israel S. Estimated prevalence of delayed and advanced sleep phase syndromes. *Sleep Res* **24**:392(1995).

32. Pelayo RP, Thorpy MJ, Glovinsky P. Prevalence of delayed sleep phase syndrome among adolescents. *Sleep Res* **17**:392(1988).

33. Cohen M, Oksenberg A, Snir D, Stern MJ, Groswasser Z. Temporally related changes of sleep complaints in traumatic brain injured patients. *J Neurol Neurosurg Psychiatry* **55**:313–315(1992).

34. Tobe EH, Schneider JS, Mrozik T, Lidsky TI. Persisting insomnia following traumatic brain injury. *J Neuropsychiatry Clin Neurosci* **11**:504–506(1999).

35. Parsons LC, Ver Beek D. Sleep–wake patterns following cerebral concussion. *Nurs Res* **31**:260–264(1982).

36. Kaufman Y, Tzischinsky O, Epstein R, Etzioni A, Lavie P, Pillar G. Long-term sleep disturbances in adolescents after minor head injury. *Pediatr Neurol* **24**:129–134(2001).

37. Clinchot DM, Bogner J, Mysiw WJ, Fugate L, Corrigan J. Defining sleep disturbance after brain injury. *Am J Phys Med Rehabil* **77**:291–295(1998).

38. Perlis NL, Artiola L, Giles DE. Sleep complaints in chronic post-concussion syndrome. *Percept Mot Skills* **84**:595–599(1997).

39. Beetar JT, Guilmette TJ, Sparadeo FR. Sleep and pain complaints in symptomatic traumatic brain injury and neurologic populations. *Arch Phys Med Rehabil* **77**:1298–1302(1996).

40. Fichtenberg NL, Zafonte RD, Putnam S, Mann NR, Millard AE. Insomnia in a post-acute brain injury sample. *Brain Inj* **16**:197–206(2002).

41. Fichtenberg NL, Millis SR, Mann NR, Zafonte RD, Millard AE. Factors associated with insomnia among post-acute traumatic brain injury survivors. *Brain Inj* **14**:659–667(2000).

42. Friboes RM, Muller U, Murek H, von Cramon DY, Holsboer F, Steiger A. Nocturnal hormone secretion and the sleep EEG in patients several months after traumatic brain injury. *J Neuropsychiatry Clin Neurosci* **11**:354–360(1999).

43. Ouellet MC, Morin CM. Cognitive behavioral therapy for insomnia associated with traumatic brain injury: a single-case study. *Arch Phys Med Rehabil* **85**:1298–1302(2004).

44. Drake ME. Jactatio nocturna after head injury. *Neurology* **36**:867–868(1986).

45. Ivanhoe CB, Lai JM, Francisco GE. Bruxism after brain injury: successful treatment with botulinum toxin-A. *Arch Phys Med Rehabil* **78**:1272–1273(1997).

46. Benyakar M, Tadir M, Groswasser Z, Stern MJ. Dreams in head-injured patients. *Brain Inj* **2**:351–356(1988).

47. Biering-Sorensen F, Biering-Sorensen M. Sleep disturbances in the spinal cord injured: an epidemiological questionnaire investigation, including a normal population. *Spinal Cord* **39**:505–513(2001).

48. Morris KF, Baekey DM, Nuding SC, Dick TE, Shannon R, Lindsey BG. Plasticity in respiratory control. Invited review: neural network plasticity in respiratory control. *J Appl Physiol* **94**:1242–1252(2003).

49. Ayas NT, Brown R, Shea SA. Hypercapnia can induce arousal from sleep in the absence of altered respiratory mechanoreception. *Am J Respir Crit Care Med* **162**:1001–1008(2000).

50. Hall CS, Danoff D. Sleep attacks: apparent relationship to atlantoaxial dislocation. *Arch Neurol* **25**:58–59(1975).

51. Stockhammer E, Tabon A, Michel F, Eser P, Scheuler W, Bauer W, Baumberger M, Muller W, Kakebeeke TH, Knecht H, Zach GA. Characteristics of sleep apnea syndrome in tetraplegic patients. *Spinal Cord* **40**:286–294(2002).

52. Klefbeck B, Sternberg M, Weinberg J, Levi R, Hultling C, Borg J. Obstructive sleep apneas in relation to severity of cervical spinal cord injury. *Spinal Cord* **36**:621–628(1998).

53. Short DJ, Stradling JR, Williams SJ. Prevalence of sleep apnoea in patients over 40 years of age with spinal cord lesions. *J Neurol Neurosurg Psychiatry* **55**:1032–1036(1992).

54. Burns SP, Little JW, Hussey JD, Lyman P, Lakshninarayanan S. Sleep apnea syndrome in chronic spinal cord injury: associated factors and treatment. *Arch Phys Med Rehabil* **81**:1334–1339(2000).

55. Flavell H, Marshall R, Thornton AT, Clements PL, Antic R, McEvoy RD. Hypoxia episodes during sleep in high tetraplegia. *Arch Phys Med Rehabil* **73**:623–627(1992).

56. Biering-Sorensen M, Norup PW, Jacobsen E, Biering-Sorensen F. Treatment of sleep apnea in spinal cord injured patients. *Paraplegia* **33**:271–273(1995).

57. Yokota T, Hirose K, Tanabe H, Tsukagoshi H. Sleep-related periodic leg movements (nocturnal myoclonus) due to spinal cord lesion. *Neurol Sci* **104**:13–18(1991).

58. Lee MS, Choi YC, Lee SH, Lee SB. Sleep-related periodic leg movements associated with spinal cord lesions. *Mov Disord* **11**:719–722(1996).

59. Esteves AM, de Mello MT, Lancellotti CLP, Natal CL, Tufik S. Occurrence of limb movements during sleep in rats with spinal cord injury. *Brain Res* **1017**:32–38(2004).

60. Dickel MJ, Renfrow SD, Moore PT, Berry RB. Rapid eye movement sleep periodic leg movements in patients with spinal cord injury. *Sleep* **17**:733–738(1994).

61. De Mello MT, Lauro FA, Silva AC, Tufik S. Incidence of periodic limb movements and of the restless legs syndrome during sleep following acute physical activity in spinal cord injury subjects. *Spinal Cord* **34**:294–296(1996).

62. De Mello MT, Silva AC, Rueda AD, Poyares D, Tufik S. Correlation between K complex, periodic leg movements (PLM), and myoclonus during sleep in paraplegic adults before and after an acute physical activity. *Spinal Cord* **35**:248–252(1997).

108

THE BLIND PATIENT

DAMIEN LEGER AND ARNAUD METLAINE

Hotel Dieu de Paris Sleep Center and Université Paris V, René Descartes, Paris, France

INTRODUCTION

According to the World Health Organization, there are approximately 180 million severely visually impaired people worldwide, of whom some 45 million are blind. Besides the many consequences of visual loss, a virtually unknown effect is the desynchronization of biological rhythms, which is related to the absence of light perception. The solar light–dark cycle is considered the most powerful environmental time cue for circadian rhythm entrainment in most animal species and humans [1]. The retinohypothalamic tract (RHT) conducts photic information from the retina to the biological clock located in the suprachiasmatic nucleus (SCN) of the hypothalamus. Blind persons may have defective retinal processing or an impaired RHT and therefore may be unable to exhibit a 24 hour pattern. Almost 50 years ago, the first clinical case was published describing circadian rhythm abnormalities in a totally blind person [2]. Several studies have subsequently observed free-running circadian rhythms of core body temperature, cortisol and melatonin secretion, as well as sleep disorders in blind individuals [3, 4]. Epidemiological studies have more recently observed a higher percentage of insomnia and free-running patterns according to the International Classification of Sleep Disorders (ICSD) in blind people compared to a group of sighted adults. The goal of this chapter is to clarify which kind of sleep disorders may be observed in blind subjects and which treatment may be proposed.

EPIDEMIOLOGY OF SLEEP DISORDERS IN THE BLIND

Very few epidemiological studies have inquired about sleep disorders in the blind. In a large study comparing 1003 blind subjects (totally blind and almost blind) to matched sighted subjects, Leger et al. [5, 6] interviewed blind subjects with braille questionnaires. They reported a much higher prevalence of sleep complaints (difficulty falling asleep, awakenings during sleep, inappropriate early morning awakenings, nonrestorative sleep, poor sleep quality) in the blind subjects: 82.7% of blind subjects were complaining of at least one sleep problem compared with 57.3% of controls ($p < 0.001$); two problems were reported by 22.1% and 15.6%, respectively ($p = 0.0003$); three problems by 18.6% and 12.3%, respectively ($p = 0.0002$); four problems by 14.4% and 4.7%, respectively ($p < 0.001$); and five problems by 7.9% and 3%, respectively ($p < 0.001$). The duration of the sleep complaints was much longer for blind subjects than for controls: 20.5 ± 16.8 years, compared with 9.8 ± 11.25 years ($p < 0.001$). The number of subjects who felt that their sleep difficulties were related to depression or anxiety was similar for blind subjects (65%) and controls (63.2%). Of the blind subjects, 36% had already consulted a physician specifically for sleep problems, compared with 20% of controls.

In the same study, the authors used the minimum criteria outlined in the DSM-IV and ICSD-90, to assess the frequencies of sleep disorders in three groups of subjects: totally

TABLE 108.1   Prevalence of Sleep Disorders in Blind Patients [5, 6]

% of Subjects with Sleep Disorders	Totally Blind (901 = 100%)	Almost Blind (174 = 100%)	Sighted Controls (894 = 100%)	P(Totally Blind Compared to Sighted)
Insomnia				
According to DSM-IV criteria	36	28	26	<0.001
DSM-IV with at least two sleep disorders	28	23	15	<0.001
Psychophysiological insomnia (ICSD-90)	47	39	28	<0.001
Other Sleep Problems				
Chronic sleep apnea syndrome	7	3	4	ns
Periodic leg movement disorder	17	16	11	ns
Free-running patterns (incomplete criteria)	18	13	8	<0.001
Narcolepsy	0.0019	0	0.0025	ns
Irregular sleep–wake patterns	1	1	2	ns
Idiopathic hypersomnia	1	0	0.00012	ns

ICSD = International Classification of Sleep Disorders; ns = nonsignificant.

blind, almost blind, and controls. The results are presented in Table 108.1. Differences between the groups concerned mostly insomnia. They observed a progressive increase in the prevalence of sleep difficulties from controls to almost blind subjects and to totally blind subjects, although there was not a significant difference between the two blind groups.

In another study conducted in a group of 388 blind subjects, Tabandeh and colleagues [7] reported sleep disorders in 48.7% of the blind. It was rated severe in 7.2%, moderate in 13.9%, and mild in 27.6%. The prevalence was higher (65.5%) and the severity most important (12.1%) in the group of blind subjects with no light perception.

CHARACTERISTICS OF SLEEP DISORDERS IN THE BLIND

Symptoms

Insomnia was the main disorder differentiating blind subjects and controls in the epidemiological survey of Leger et al. [6] (see Table 108.1). Blind subjects particularly complained of a too short duration of sleep with an estimated sleep length of 6.9 ± 1.6 hours during weekdays versus 7.5 ± 1.9 hours for controls ($p < 0.001$) and a sleep length of 7.5 ± 1.9 hours during the weekends versus 8 ± 1.5 in controls ($p < 0.001$). They also had more difficulties falling asleep (30 minutes versus 20 minutes for controls ($p < 0.001$)). They reported more night awakenings than sighted subjects (2.08 versus 1.56 ($p < 0.001$)). Almost one-third of blind subjects (31.5%) reported waking more than three times during the night, with difficulty getting back to sleep, compared with 22.5% of controls ($p < 0.001$). Finally, 26% of the totally blind subjects frequently used

drugs (anxiolytics or hypnotics) to promote sleep, compared with 20% of the almost blind and 13% of the controls ($p < 0.001$).

Sleepiness was also reported as a common symptom in this survey: 14.3% of blind subjects compared to 5.5% of controls said they had involuntary daily sleep episodes ($p < 0.001$) and 11.6% of blind subjects versus 6.8% of controls reported napping daily ($p = 0.0002$). In a wrist actimetry study completed on 59 blind subjects, Lockley and colleagues [8] demonstrated that daytime napping was the most sensitive indicator of a circadian rhythm disorder in blind subjects.

Sleep and Its Relation to the Circadian Phase

It is easy to understand that sleep disorders in the blind may be related to the abnormal circadian phase and to the desynchronization of the circadian rhythms. However, it has not been demonstrated that all blind people have circadian rhythm disorders. Lewy and Newsome [9] have proposed a classification of four types of circadian rhythms in the blind: (1) normally entrained to 24 hours; (2) abnormally entrained to 24 hours (phase delayed or phase advanced); (3) free-running at a period different from (usually greater than) 24 hours; and (4) unclassified with no discernible rhythm. If a circadian rhythm disorder is present, it is expected that sleep would be disrupted. ICSD has defined the non-24-hour sleep–wake syndrome. Sack and colleagues [13] have reported that in a group of 20 totally blind persons, half of the subjects were free-running. However, in the epidemiological survey of 1003 blind subjects, only 18% were free-running (Table 108.1). According to Lockley and colleagues [8], napping is the best indicator of circadian disorders in the blind. Subjects with normally entrained rhythms of melatonin have fewer naps of shorter duration than abnormally entrained or free-running subjects [8].

Assessment of Circadian Disorders in the Blind

It is actually still difficult to use melatonin laboratory testing to assess free-running rhythms in blind subjects. You have to repeat salivary or urinary testing several times during the 24 hour cycle in a dim-light environment (and it is not easy in blind subjects) and to renew the testing 1 and 2 weeks after to show the phase shift.

We recommend the use of actigraphy, which is the most simple and powerful device to assess the type and severity of circadian disorders. Temperature actigraphy may also be used at night to calculate the phase shift between the 2 week and 3 week measurements.

Polysomnography

In a polysomnography study completed in 24 free-running blind subjects compared to a group of 24 matched sighted persons, Leger and colleagues [10] found that the blind had a significantly lower total sleep time (289.7 ± 80 minutes compared to 424 ± 56 minutes ($p < 0.001$)) and a lower sleep efficiency ($76.2 \pm 17.4\%$ versus $92.3 \pm 3.6\%$ ($p < 0.001$)) than controls. The REM latency was higher (122.3 ± 79.4 minutes compared to 72.4 ± 10.5 minutes ($p < 0.001$)) and the percentage of REM was lower in blind subjects (12 ± 6 versus $23.7 \pm 4.7\%$ ($p < 0.001$)). Sleep latency and the percentage of slow-wave sleep did not differ significantly between the two groups. However, these polysomnography measurements were made independently of the circadian phase and may not reflect the more severe aspects of sleep in the blind.

SLEEP DISORDERS IN CHILDREN WITH BLINDNESS

Several case reports have focused on children presenting a combination of mental retardation and blindness and have emphasized the presence of important circadian dyschronosis [11]. In a survey conducted on 77 blind children with no mental retardation (ranging from 3 to18 years of age) compared to a matched group of 79 sighted children in the same age group, we found a surprisingly low percentage (8%) of clear circadian dyschronosis in the study population [12]. Mental retardation may limit a blind person's ability to use other cues (zeitgebers) than light to entrain the circadian pacemaker. However, the group of blind children had a higher prevalence of insomnia according to DSM-IV criteria (35.1% compared to 20.2% in the sighted group ($p < 0.001$)) and reported more daily episodes of involuntary sleepiness (13.4% versus 1.3% in the control group ($p < 0.001$)). They also reported more sleepwalking (5% compared to 1% ($p = 0.001$)) and periodic leg movement disorders (6% versus 1% ($p < 0.001$)). One major complaint of blind children was an early awakening particularly on school days. Perhaps situational anxiety related to the need to be ready for school on time was involved in this early awakening.

THERAPIES

General Approach and Education

Most blind patients would be relieved to know that their sleep problem may have an understandable physiological basis. It is often possible that free-running rhythms may be the precipitating or the only cause of sleep disorders. However, we have seen that insomnia is very common in blind subjects and that many other causes may interfere. It is therefore important to be careful to consider other classical treatments of insomnia besides those for circadian disorders.

Melatonin

Melatonin appears to be one of the most appropriate pharmacological tools to entrain the biological clock of blind patients abnormally entrained to 24 hours, or free-running. Sack and colleagues [13] gave seven free-running blind subjects 10 mg of melatonin or placebo 1 hour before bedtime for 3–9 weeks. Six of the seven subjects were entrained and were sleeping at a more normal circadian phase after treatment. Their wake after sleep onset was lower with melatonin than with placebo ($p = 0.02$) and the sleep efficiency was greater ($p = 0.06$). Three subjects were entrained a second time with 10 mg of melatonin and then the dose was gradually reduced to 0.5 mg. They maintained entrainment for 3 months, suggesting a long-term benefit of a dose close to the melatonin physiological range. Lockley and colleagues [14] also treated seven free-running blind patients with a 5 mg dose of melatonin given at 9 p.m. Four of seven were entrained, while three in whom the treatment was initiated during the phase delay portion of the melatonin phase response curve continued to free-run. Lockley et al. [14] suggested that the phase at which treatment is given may be a critical factor in response. However, melatonin agonists are still not on the market in Europe and several countries around the world.

Light Therapy

In sighted individuals, exposure to bright light can reset the circadian pacemaker amplitude and phase independent of the timing of the sleep–wake cycle [15]. Klerman et al. [16] have shown that blind people with no conscious light perception and no pupillary light reflexes may still have light input to their circadian pacemaker despite the loss of light input to their conscious visual system. These blind subjects with no conscious light perception but with normal suppression of melatonin in response to ocular light report no sleep or other circadian-related disturbances [15].

These data support the evidence that light therapy may be proposed to blind patients with circadian disorders even if they have no conscious light perception. It is difficult in the clinical practice to perform a melatonin suppression test, which is used in research [15, 16]. However, one first step may be to avoid the use of dark goggles (frequently given to blind people for esthetic purposes in the morning for a case of phase delay and in the evening for a case of phase advance. Another step would be to apply light therapy according to the usual rules applied in circadian disorders.

CONCLUSION: DREAMS

The question about how blind people dream often comes up when you discuss sleep disorders in the blind. Several studies have been devoted to that topic and they mostly agree that blind and sighted subjects do not differ regarding the content and the characteristics of dreams except for vision [17, 18]. Other case studies also reported that patients suffering blindness as a result of accidents continue to have visual dreams similar to their dream experiences prior to the impairment. We agree with Kerr's assessment [19] when he said: "It seems that visual imagery in dreaming is not a reflection of current visual abilities, but more certainly represent a cognitive processing of re-creating at night a visual aspect of the world which cannot be directly perceived in wakefulness."

REFERENCES

1. Czeisler CA, Kronauer RE, Allan JS, Duffy JF, Jewett ME, Brown EN, Ronda JM. Bright light induction of strong (Type 0) resetting of the human pacemaker. *Science* **244**:1328–1333(1989).

2. Migeon CJ, Tyler FH, Mahoney JP. The diurnal variation of plasma levels and urinary excretion of 17-hydroxycorticosteroids in normal subjects, night workers and blind subjects. *J Clin Endocrinol Metab* **16**:622–633(1956).

3. Arendt J, Adhous M, Wright J. Synchronization of a disturbed sleep–wake cycle in a blind man by melatonin treatment. *Lancet* **i**:772–773(1988).

4. Sack RL, Lewy AJ, Blood ML, Keith LD, Nakagawa H. Circadian rhythm abnormalities in totally blind people: incidence and clinical significance. *J Clin Endocrinol Metab* **75**:127–133(1992).

5. Leger D, Guilleminault C, Defrance R, Domont A, Paillard M. High frequency of sleep/wake disorders in blind individuals. *Lancet* **348**:830–831(1996).

6. Leger D, Guilleminault C, Defrance R, Domont A, Paillard M. Prevalence of sleep/wake disorders in persons with blindness. *Clin Sci* **97**:193–199(1999).

7. Tabandeh H, Lockley SW, Buttery R, Skene DJ, Defrance R, Arendt J. Disturbance of sleep in blindness. *Am J Ophthalmol* **126**:707–712(1998).

8. Lockley SW, Skene DJ, Butler LJ, Arendt J. Sleep and activity rhythms are related to circadian phase in the blind. *Sleep* **22**:616–623(1999).

9. Lewy AJ, Newsome N. Different types of melatonin circadian secretion rhythms in some blind subjects. *J Clin Endocrinol Metab* **56**:1103–1107(1983).

10. Leger D, Guilleminault C, Santos C, Paillard M. Sleep/wake cycles in the dark: sleep recorded by polysomnography in 26 totally blind subjects compared to controls. *Clin Neurophysiol* **113**:1607–1614(2002).

11. Okawa M, Nanami T, Wada S, Shimizu T, Hishikawa Y, Sasaki H, Nagamine H, Takahashi K. Four congenitally blind children with circadian sleep–wake disorder. *Sleep* **10**:101–110(1987).

12. Leger D, Prevot E, Philip P, Yence C, Labaye N, Paillard M, Guilleminault C. Sleep disorders in children with blindness. *Ann Neurol* **46**:648–651(1999).

13. Sack RL, Brandes RW, Kendall BS, Lewy AJ. Entrainment of free-running circadian rhythms by melatonin in totally blind people. *N Engl J Med* **343**:1070–1077(2000).

14. Lockley SW, Skene DJ, James K, Thapan K, Wright J, Arendt J. Melatonin administration can entrain the free-running circadian system of blind subjects. *J Endocrinol* **164**:R1–R6(2000).

15. Czeisler CA, Shanahan TL, Klerman EB, Martens H, Brotman DJ, Emens JS, Klein T, RizzoII JF. Suppression of melatonin secretion in some blind patients by exposure to bright light. *N Engl J Med* **332**:6–11(1995).

16. Klerman EB, Shanahan TL, Brotman DJ, Rimmer DW, Emens JS, RizzoII JF, Czeisler CA. Photic resetting of the human circadian pacemaker in the absence of conscious vision. *J Biol Rhythms* **17**:548–555(2002).

17. Kerr NH. Dreaming, imagery and perception. In: Kryger MH, Roth T, Dement WC (Eds), *Principles and Practice of Sleep Medicine*,3rd ed. Saunders, Philadelphia, 2000, pp 482–490.

18. Leger D, Hommey N, Raffray. Le reve des aveugles. *Neuropsy News* **1**:19–22(2002).

19. Kerr NH, Foulkes D, Schmidt M. The structure of laboratory dream reports in blind and sighted subjects. *J Nerv Ment Dis* **170**:247–264(1982).

PART XV

SLEEP IN THE PSYCHIATRIC DISORDERS

109

SCHIZOPHRENIA

Rachel J. Norwood and Teofilo Lee-Chiong
National Jewish Medical and Research Center, Denver, Colorado

INTRODUCTION

The complicated interface between psychiatric diagnoses and sleep disturbance is well illustrated in schizophrenia, where the psychiatric symptoms are amplified and aggravated by sleep disruption while at the same time the pathology of schizophrenia may contribute to worsened sleep. Schizophrenia is a disabling, chronic disorder with a worldwide lifetime prevalence of approximately 1% and a median age of onset between 15 and 25 years of age [1]. Clinical manifestations can include what are termed "positive symptoms" of psychosis—hallucinations and delusions—as well as the "negative symptoms" of disorganized speech and behavior, affective flattening, avolition, and alogia. The syndrome of schizophrenia typically leads to deterioration in self-care, interpersonal relations and work performance. Successful therapy enables approximately 25% of schizophrenics to return to their preschizophrenic level of social and occupational functioning [1].

Sleep disturbance is an integral symptom of schizophrenia as well as of depression. Approximately 25% of schizophrenic patients experience what is referred to as a "postpsychotic depression" or a "secondary depression" often associated with the extrapyramidal symptoms imposed by antipsychotic medications [2]. Just as with primary depression, these affective symptoms are associated with increased sleep disturbance and, to the extent that a given schizophrenic patient also struggles with depressive symptoms, there will be a corresponding increase in the burden of sleep disturbance. In one study, 26.8% of 358 patients with schizophrenia had consistent depressive symptomatology, with sleep disorders being among the

most common of the presenting depressive symptoms [3]. When the relationship between perceived quality of life (QoL) and subjective quality of sleep (Pittsburgh Sleep Quality Index, PSQI) is evaluated, poor sleepers have lower mean scores on all QoL domains, are more depressed and distressed, and have more adverse effects to medications than good sleepers [4]. In addition to their cognitive impairments, these dysfunctions also cause profound subjective suffering. Among middle-aged and elderly outpatients with schizophrenia and related psychotic disorders, patients assign improvements in sleep and mood as areas of high priority [5].

EARLY INVESTIGATIONS IN SLEEP AND PSYCHOSIS

Early psychiatric theorists postulated a shared mechanism for dreaming and hallucinations. The discovery of the association between dreaming and rapid eye movements during sleep provided an objectively identifiable marker of the physiological state of dreaming. Nevertheless, early investigators lacked standardized criteria to stage sleep and define sleep architecture as well as to diagnose schizophrenia. Therefore early sleep researchers sought abnormalities in rapid eye movements during the sleep of schizophrenic patients. The validity and reliability of these studies were limited because electrophysiologic measures were not yet available, leaving researchers with only patient reports to assess the relationships between dreaming and psychosis—an especially difficult prospect in a subject population defined by disorganized thought and frequent paranoia.

The applicability of these studies was also complicated by the frequently heterogeneous samples of schizophrenics who varied with respect to their stage of illness, age, and medication status.

Still, research forged ahead with quantitative studies and, in 1955, Dement [6] reported that the quantity, quality, and distribution of rapid eye movements during sleep did not differ between chronic unmedicated schizophrenics and normal control subjects, and that, just as in normal subjects, schizophrenics were able to recall dream content when awakened from sleep with rapid eye movements. When total sleep time, rapid eye movement activity, and "dream time" were measured in actively ill schizophrenics, schizophrenics in remission, and normal subjects, the actively ill and recently ill schizophrenics had fewer epochs of sleep with rapid eye movements than normals [7]. Even as monitoring of brain activity has become possible, researchers still have not been able to identify the physiologic correlates of rapid eye movements during waking in acute-phase schizophrenics that would confirm the phenomenologic and physiologic similarities between schizophrenia and dreaming [8]. So, while it is tempting to postulate parallels between psychosis and dreaming as waking and sleeping expressions of the same phenomenon, historical research has not been able to support the concept. Polysomnography has, however, been able to shed some light on the nature of sleep in patients with schizophrenia.

POLYSOMNOGRAPHIC FEATURES

Rapid Eye Movement Sleep

There is a significant variation in rapid eye movement (REM) sleep in schizophrenics compared to normal subjects, with shortened REM sleep latencies and a greater percentage of total sleep time spent in REM. In addition, these differences vary over the natural history of the mental illness. There is a marked reduction in total sleep time and a disproportionately larger reduction in REM sleep during the waxing of psychotic symptoms that then normalizes during waning and perseveres at normal levels during the postpsychotic and remission phases [9]. In unmedicated, recently hospitalized schizophrenics, REM latency was markedly shortened—likely the result of a deficit of slow-wave sleep seen in the first non-rapid eye movement (NREM) period [10]. As treatment progresses, patients may demonstrate a modest increase in REM latency about 4 weeks postpresentation and at 1 year out show futher increases in REM latency, REM time, and average automated REM counts [11]. REM sleep latency correlates with the duration of wakefulness in the first cycle in patients suffering from chronic schizophrenia [12]. When the REM latencies of patients with major depressive disorder, schizophrenia, and schizoaffective disorder and of normal subjects are evaluated, all of the patient groups, including the schizophrenics, have similarly short REM latencies that significantly differ from the controls [13]. This finding is consistent in its REM findings with previous studies but also demonstrates that sleep disturbance is quite common across psychiatric diagnoses. A variety of REM studies have elucidated some data specific to the quality of REM sleep in schizophrenia.

REM Sleep Deprivation Studies As noted earlier, the characteristics of REM sleep change as subjects progress through the phases of schizophrenia. Schizophrenic patients who are actively symptomatic have a reduced total REM sleep and percent REM sleep rebound following sleep deprivation [14, 15].

REM Sleep and Cognitive Function In schizophrenic patients, the minutes of the first period REM sleep and REM density show a negative correlation with performance, while REM minutes occurring after the first REM period correlated positively with neuropsychological performance [16]. REM latency studies also offer some discriminating data between schizophrenia and depression. The REM latency of drug naïve and previously medicated drug-free schizophrenics is inversely correlated with negative schizophrenic symptoms but is unrelated to depressive symptoms [17]. There may be a relationship between other REM sleep abnormalities and suicidal behavior in schizophrenia in that actively psychotic patients with suicidal behavior had significantly increased REM activity and time, perhaps related to alterations in serotonergic function [18].

Slow-Wave Sleep

Patients with schizophrenia also demonstrate alterations in delta, or slow-wave, sleep. Overall they suffer a deficit of slow-wave sleep (SWS) measured as a reduction in the total number of delta half-waves during sleep when compared to normals [19]. Interestingly, slow-wave sleep deficits, as measured by automated delta wave counts, correlates inversely with negative symptoms but not with positive symptoms. When deprived of sleep for extended periods of time, schizophrenic patients have in some instances been shown to fail to generate the slow-wave sleep rebound characteristic of normal subjects [20]. Investigators have examined the relationship between the SWS abnormality and brain morphology and outcomes. Ventricular volumes and ventricle-to-brain ratios of patients with chronic psychotic disorders inversely correlated with duration of stage 3 and stage 4 sleep, implicating brain dysmorphology in the etiology of delta sleep abnormalities [21].

SCHIZOPHRENIA AND SPECIFIC SLEEP DISORDERS

Insomnia

Nonorganic insomnia is a frequent sleep disorder that has a high comorbidity with psychiatric illnesses and this holds true for schizophrenic patients as well. In one sleep outpatient clinic, 41% of the patients demonstrated stress-related and somatoform disorders, 31% had affective disorders, and 1.6% were diagnosed with schizophrenia, compared to a community prevalence of 1% [22]. Insomnia is common in schizophrenia and often becomes severe during exacerbations of psychiatric symptoms. Insomnia can involve disturbances with sleep onset and/or sleep maintenance. The atypical antipsychotics olanzapine, risperidone, and clozapine are helpful in this dimension in that they significantly increase total sleep time and stage 2 sleep. Moreover, olanzapine and risperidone enhance slow-wave sleep. In contrast, the typical antipsychotics haloperidol, thiothixene, and flupentixol significantly reduce stage 2 sleep latency but increase sleep efficiency [23].

Narcolepsy

Narcolepsy and schizophrenia have some degree of reciprocity in terms of phenomenology but whether there is a physiologic commonality remains controversial. Patients with narcolepsy accompanied by unusually prominent hypnagogic hallucinations have mistakenly been diagnosed with schizophrenia. At the same time schizophrenic patients have demonstrated narcolepsy-associated antigens 3.89 times more frequently than normal subjects, leading to the postulate that patients with schizophrenia might have higher rates of associated narcolepsy [24].

Other studies have produced conflicting findings. In one study, 4 of 45 (9%) patients with narcolepsy attending a sleep disorders clinic and none of 50 matched normal controls had experienced psychotic symptoms. However, all four patients were taking amphetamines, and the symptoms resolved when the dose was lowered or treatment was changed to modafinil. These investigators concluded that narcolepsy does not appear to be related to schizophrenia [25].

Obstructive Sleep Apnea Syndrome

Patients with schizophrenia on long-term neuroleptic treatment may have high rates of obstructive sleep apnea, mediated via the weight gain produced by such medications. In one study that evaluated the risk factors for obstructive sleep apnea in 364 patients who were referred to a sleep disorders consultation service from an inpatient psychiatric hospital, patients with schizophrenia were significantly heavier and had higher rates of sleep apnea than did other psychiatric patients. Obesity, male gender, and chronic neuroleptic administration are risk factors for obstructive sleep apnea in psychiatric patients [26].

DREAM CONTENT OF PATIENTS WITH SCHIZOPHRENIA

Studies of the manifest dream content of schizophrenic patients have demonstrated that their dreams reflect their struggles with the subjective experience of their illness. Using the Gottschalk–Gleser Analysis Scales measure, the dream content of schizophrenics was found to differ in both the frequency and intensity of anxious and hostile affects. Patients experience themselves in their dreams more frequently as victims of hostility from outside, which corresponds well with a significantly higher intensity of threat anxieties, while value anxieties (guilt and separation) are found less frequently in their dreams [27]. Another instrument, the standardized Formal Dream Content Rating Scale (FDCRS), evaluates dream-related anxiety, cognitive disturbance, implausibility, involvement, primitivity, and recall, as well as two additional scales measuring emotional expression and duration of dream report. Hadjez et al. [28] used this instrument to compare the manifest dream content of 20 schizophrenic adolescent inpatients whose medications were stable for at least 4 weeks, 21 adolescent inpatients with other mental disorders (nonschizophrenic group), and 31 matched community controls. They observed that the presence of psychopathology, rather than the specific psychiatric disturbance, may be associated with impoverishment of dream content, and that negative, rather than positive, schizophrenic symptomatology may be influential in the dream content of schizophrenic youngsters [28].

BIOLOGY OF SLEEP DISRUPTION IN SCHIZOPHRENIA

The fundamental etiology of schizophrenia is unclear but research has demonstrated apparent pathophysiologic roles anatomically for the limbic system and the basal ganglia. The "dopamine hypothesis of schizophrenia" was the starting point for current understanding of chemical contributors to schizophrenia, positing that schizophrenia was the result of excessive dopaminergic activity. While dopamine continues to be a primary focus for both studies of etiology and development of pharmacologic interventions, other neurotransmitters including other monoamines, particularly serotonin, and hormonal influences are also being studied.

Abnormalities in neuroendocrine release and the consequent impact on sleep have been observed in patients with schizophrenia [29–31]. These include (1) an enhancement of the sleep-related increase in the prolactin level; (2) lower hourly measures of plasma growth hormone from 10:00 p.m. to 8:00 a.m. compared to controls with an absence in the 1:00 a.m. peak seen in controls; and (3) normalization of the failed auditory evoked P50 suppression of schizophrenics by slow-wave but not REM sleep [1]. Shamir et al. [32] demonstrated that schizophrenics have low melatonin levels and that melatonin replacement improved rest-derived sleep efficiency, especially in low efficiency sleepers.

In addition, investigators have found that schizophrenic patients are particularly sensitive to external stimuli as well as to their own thoughts [33]. This causes difficulties with filtering of stimuli, makes it difficult to attend to desired targets, and therefore impairs coding into short-term memory [34]. The etiology of these difficulties cannot be attributed to any single cause but is most likely a pathologic interplay between genetic predisposition, environmental stress, and neurobiologic insult. Among the latter are impaired function of inhibitory neurons; a generalized loss of cortical neuropil, the dendrites and axons that connect neurons; and a decrease in the total number of neurons, particularly in the hippocampus [35–37]. Imaging studies have supported these findings by showing enlarged ventricles and generalized reduction in brain volume via magnetic resonance imaging (MRI) and diminished neuronal content by magnetic resonance spectroscopy [38, 39]. Yeomans [29] contributes to localizing these effects when he argues that disinhibition of the pedunculopontine nucleus (Ch5) and laterodorsal tegmental nucleus (Ch6) triggered the psychotogenic effects of antimuscarinics. He goes on to postulate that Ch5 and Ch6 activation can cause the shortened REM latency seen in schizophrenia. Rye [30] proposed that the pedunculopontine (PPN) region plays a role in the REM sleep abnormalities seen in schizophrenia and other disorders (narcolepsy, RBD). Garcia-Rill [31] suggested that when neurological or psychiatric disorders manifest symptoms related to arousal and sleep–wake control, disturbances of elements of the reticular activating system must be considered responsible.

Medication Effects on Sleep in Schizophrenia

While the endogenous psychopathology of schizophrenia contributes to sleep difficulties, these patients also suffer, and enjoy, iatrogenic influences on their sleep as mediated by their psychotropic medications. As discussed briefly before, sleep and levels of daytime alertness are also influenced by the use of psychotropic medications, some beneficially, some less so. Among the goals of antipsychotic medications are controlling daytime agitation and positive symptoms while preserving energy levels and improving negative symptoms. At night the goal shifts to predictable, restorative sleep.

Currently utilized antipsychotics are divided into two groups—the "typical" and "atypical" agents. The typical agents were the first developed and have as their primary mechanism of action blockade of dopaminergic transmission. Drugs in this group include chlorpromazine, which has low potency, and more potent medications such as haloperidol. The atypical antipsychotics have been introduced over the past 16 years and were designed to improve clinical effects and reduce problematic side effects relative to the typical antipsychotics. They were successful on both counts largely due to their addition of impact on other neurotransmitter systems, particularly serotonin. The interaction between the dopaminergic and serotoninergic effects of these drugs has improved both positive and negative symptoms of schizophrenia while at the same time decreasing patients' risk for neuroleptic-induced movement disorders [40].

Antipsychotic medications have generally been found to improve multiple measures of sleep, including sleep continuity and quantity of slow-wave sleep [41]. In comparing typicals and atypicals, the latter were found to generate significantly more stage 2 sleep, more stable NREM sleep (stages 2, 3, and 4), and less stage 1 sleep than seen in patients treated with the typicals, haloperidol or flupentixol [42]. Neuroleptic effects on REM sleep include lengthened, but not normalized, REM sleep latency in schizophrenics [43]. Patients taking typical antipsychotics regularly report sedation, while with the atypicals there is more variability on the sedation–insomnia continuum [44, 45]. The sedation that does emerge with these medications can be minimized by gradual dose escalation [46]. Treatment with olanzepine has been shown to significantly improve parameters of sleep efficiency and increase delta and REM sleep in patients with schizophrenia [47]. Clozapine, another atypical antipsychotic agent, increased mean total sleep time, sleep efficiency, duration of awakening, and amounts of stage 2 sleep [48].

Finally, comparison has been made between oral and parenteral depot preparations of certain antipsychotic medications. A literature review of 6 depot fluphenazine clinical trials suggested that sleep problems occurred as frequently in schizophrenics on oral fluphenazine as on depot preparations [49].

Sleep disturbances are an especially prominent feature during the prodrome preceding psychotic relapse. Schizophrenia patients with sleep disturbances are at a greater risk for worsening of positive symptoms after antipsychotic discontinuation. Chemerinski et al. [50] noted that sleep quality deteriorated progressively following antipsychotic discontinuation. Total insomnia score prior to antipsychotic withdrawal had a significant effect on the severity of psychotic symptoms, and baseline terminal insomnia had a significant effect on symptoms of disorganization [50].

Conversely, neuroleptic withdrawal was shown to induce reduction in REM sleep and shortening of REM latency [40, 51]. The insomnia that develops during discontinuation of antipsychotic medication or switching from low-potency to high-potency agents can be attributed to either medication withdrawal effects or exacerbation of schizophrenia. Abrupt neuroleptic withdrawal will induce a transient reduction in total sleep that will typically stabilize 2–4 weeks after withdrawal to levels between the values observed during chronic treatment and withdrawal periods [51]. A neuroleptic withdrawal study that recorded sleep at baseline and either at relapse or at 6 weeks (whichever came first) demonstrated global deterioration of sleep on withdrawal from neuroleptics. Compared to nonrelapsers, relapsers had a larger decrease in total sleep time, sleep efficiency, total NREM sleep, and stage 2 sleep. The level of psychosis was inversely correlated with sleep efficiency, total sleep time, and stage 4 sleep in the drug-free patients [40]. Study of schizophrenics on a drug-free weekend schedule of neuroleptics revealed that sleep did not change on the first drug-free night but did change on the second night, indicating that the effects of antipsychotic medications on sleep are not limited to sedation [52]. It has been hypothesized that the rebound insomnia arising from neuroleptic discontinuation is due to a cholinergic hypersensitivity [53].

Neuroleptic Side Effects on Sleep

Akathisia shares some similarities with restless legs syndrome (RLS), clinically and pathophysiologiclly, in that both conditions involve alterations in dopamine and iron metabolism. Akathisia is distinguished from RLS primarily by the sensation of inner restlessness rather than leg discomfort, and by the circadian variability in symptoms seen in RLS. Patients with RLS tend to be more restless in the evening and during repose. Generally, neuroleptic-induced akathisia (NIA) causes less sleep disruption compared to idiopathic RLS [54].

Periodic limb movement disorder (PLMD) is a sleep disorder characterized by periodic involuntary movements of the legs during sleep, which can appear quite similar to NIA that may also manifest as periodic limb movements during sleep [55]. However, NIA does not appear to increase nighttime motor activity as assessed by actigraphy or subjective sleep assessment [56]. Either PLMD or RLS or both have been reported following administration of olanzapine and haloperidol [1]. Long-term use of neuroleptics might contribute to insomnia in elderly patients by giving rise to nocturnal myoclonus or PLMD [57].

Use of Hypnotic Medication as Adjuncts in Schizophrenia Therapy

The reduction in slow-wave sleep secondary to benzodiazepine use may be of concern in schizophrenics, who usually already have decreased delta sleep. Newer hypnotic agents that act more specifically on the benzodiazepine receptor have less effect on slow-wave sleep. Changing from traditional benzodiazepine sleeping medications to zoplicone may reduce the proportion of stage 1 sleep, increase slow-wave amplitude, and decrease the negative symptoms score [58, 59]. In some cases, the sedating effect of the antipsychotic agent may be sufficient for sleep induction if taken at bedtime, thereby eliminating the need for a separate hypnotic medication [60].

CONCLUSION

Sleep disruption is a common and very debilitating comorbid symptom of schizophrenia. As sleep deprivation studies have shown, sleep disruption can aggravate psychosis and the increased susceptibility to external stimuli imposed by schizophrenia means that the psychopathology will increase sleep disruption. For the most part, the medications used to treat schizophrenia improve problematic sleep. However, their discontinuation can, if only temporarily, worsen sleep symptoms. Understanding these complicated interactions between schizophrenia and sleep will lead to improved patient care and decreased long-term morbidity.

REFERENCES

1. Richert A, Baran AS. Schizophrenia. In: Lee-Chiong TL, Sateia MJ, Carskadon MA (Eds), *Sleep Medicine*. Hanley & Belfus, Philadelphia, 2002.

2. Kaplan H, Sadock B. *Synopsis of Psychiatry*, 8 ed. Lippincott/ Williams & Wilkins, Philadelphia, 1997, pp 479–480.

3. Serretti A, Mandelli L, Lattuada E, Smeraldi E. Depressive syndrome in major psychoses: a study on 1351 subjects. *Psychiatry Res* **127**(1-2):85–99(2004).

4. Ritsner M, Kurs R, Ponizovsky A, Hadjez J. Perceived quality of life in schizophrenia: relationships to sleep quality. *Qual Life Res* **13**(4):783–791(2004).

5. Auslander LA, Jeste DV. Perceptions of problems and needs for service among middle-aged and elderly outpatients with schizophrenia and related psychotic disorders. *Commun Ment Health J* **38**(5):391–402(2002).

6. Dement W. Dream recall and eye movements during sleep in schizophrenics and normals. *J Nerv Ment Dis* **122**(3):263–269(1955).

7. Feinberg I, Koresko RL, Gottlieb F, Wender PH. Sleep electroencephalographic and eye-movement patterns in schizophrenic patients. *Compr Psychiatry* **34**:44–53(1964).

8. Rechtschaffen A, Schulsinger F, Mednick SA. Schizophrenia and physiological indices of dreaming. *Arch Gen Psychiatry* **10**:89–93(1964).

9. Kupfer DJ, Wyatt RJ, Scott J, Snyder F. Sleep disturbance in acute schizophrenic patients. *Am J Psychiatry* **126**(9):1213–1223(1970).

10. Hiatt JF, Floyd TC, Katz PH, Feinberg I. Further evidence of abnormal non-rapid-eye-movement sleep in schizophrenia. *Arch Gen Psychiatry* **42**(8):797–802(1985).

11. Keshavan MS, Reynolds CF 3rd, Miewald JM, Montrose DM. A longitudinal study of EEG sleep in schizophrenia. *Psychiatry Res* **59**(3):203–211(1996).

12. Rotenberg VS, Shami E, Barak Y, Indursky P, Kayumov L, Mark M. REM sleep latency and wakefulness in the first sleep cycle as markers of major depression: a controlled study vs. schizophrenia and normal controls. *Prog Neuropsychopharmacol Biol Psychiatry* **26**(6):1211–1215(2002).

13. Zarcone VP Jr, Benson KL, Berger PA. Abnormal rapid eye movement latencies in schizophrenia. *Arch Gen Psychiatry* **44**(1):45–48(1987).

14. Zarcone V, Azumi K, Dement W, Gulevich G, Kraemer H, Pivik T. REM phase deprivation and schizophrenia II. *Arch Gen Psychiatry* **32**(11):1431–1436(1975).

15. Gillin JC, Buchsbaum MS, Jacobs LS, Fram DH, Williams RB, Vaughan TB Jr, Mellon E, Snyder F, Wyatt RJ. Partial REM sleep deprivation, schizophrenia and field articulation. *Arch Gen Psychiatry* **30**(5):653–662(1974).

16. Taylor SF, Goldman RS, Tandon R, Shipley JE. Neuropsychological function and REM sleep in schizophrenic patients. *Biol Psychiatry* **32**(6):529–538(1992).

17. Tandon R, Shipley JE, Taylor S, Greden JF, Eiser A, DeQuardo J, Goodson J. Electroencephalographic sleep abnormalities in schizophrenia. Relationship to positive/negative symptoms and prior neuroleptic treatment. *Arch Gen Psychiatry* **49**(3):185–194(1992).

18. Keshavan MS, Reynolds CF, Montrose D, Miewald J, Downs C, Sabo EM. Sleep and suicidality in psychotic patients. *Acta Psychiatr Scand* **89**(2):122–125(1994).

19. Kajimura N, Kato M, Okuma T, Sekimoto M, Watanabe T, Takahashi K. A quantitative sleep-EEG study on the effects of benzodiazepine and zopiclone in schizophrenic patients. *Schizophr Res* **15**(3):303–312(1995).

20. Luby ED, Caldwell DF. Sleep deprivation and EEG slow wave activity in chronic schizophrenia. *Arch Gen Psychiatry* **17**(3):361–364(1967).

21. Benson KL, Sullivan EV, Lim KO, Lauriello J, Zarcone VP Jr, Pfefferbaum A. Slow wave sleep and computed tomographic measures of brain morphology in schizophrenia. *Psychiatry Res* **60**(2-3):125–134(1996).

22. Saletu-Zyhlarz GM, Arnold O, Saletu B, Anderer P. The key–lock principle in the diagnosis and treatment of non-organic insomnia related to psychiatric disorders: sleep laboratory investigations. *Methods Find Exp Clin Pharmacol* **24**(Suppl D):37–49(2002).

23. Monti JM, Monti D. Sleep in schizophrenia patients and the effects of antipsychotic drugs. *Sleep Med Rev* **8**(2):133–148(2004).

24. Douglass AB, Hays P, Pazderka F, Russell JM. Florid refractory schizophrenias that turn out to be treatable variants of HLA-associated narcolepsy. *J Nerv Ment Dis* **179**(1):12–17(1991); discussion p 18.

25. Vourdas A, Shneerson JM, Gregory CA, Smith IE, King MA, Morrish E, McKenna PJ. Narcolepsy and psychopathology: is there an association? *Sleep Med* **3**(4):353–360(2002).

26. 26. Winkelman JW. Schizophrenia, obesity, and obstructive sleep apnea. *J Clin Psychiatry* **62**(1):8–11(2001).

27. Stompe T, Ritter K, Ortwein-Swoboda G, Schmid-Siegel B, Zitterl W, Strobl R, Schanda H. Anxiety and hostility in the manifest dreams of schizophrenic patients. *J Nerv Ment Dis* **191**(12):806–812(2003).

28. Hadjez J, Stein D, Gabbay U, Bruckner J, Meged S, Barak Y, Elizur A, Weizman A, Rotenberg VS. Dream content of schizophrenic, nonschizophrenic mentally ill, and community control adolescents. *Adolescence* **38**(150):331–342(2003).

29. Yeomans JS. Role of tegmental cholinergic neurons in dopaminergic activation, antimuscarinic psychosis and schizophrenia. *Neuropsychopharmacology* **12**(1):3–16(1995).

30. Rye DB. Contributions of the pedunculopontine region to normal and altered REM sleep. *Sleep* **20**(9):757–788(1997).

31. Garcia-Rill E. Disorders of the reticular activating system. *Med Hypotheses* **49**(5):379–387(1997).

32. Shamir E, Laudon M, Barak Y, Anis Y, Rotenberg V, Elizur A, Zisapel N. Melatonin improves sleep quality of patients with chronic schizophrenia. *J Clin Psychiatry* **61**(5):373–377(2000).

33. Venables PH. Input dysfunction in schizophrenia. *Prog Exp Personality Res* **1**:1–47(1964).

34. Park S, Holzman PS, Goldman-Rakic PS. Spatial working memory deficits in the relatives of schizophrenic patients. *Arch Gen Psychiatry* **52**:821–828(1995).

35. Woo TU, Whitehead RE, Melchitzky DS, Lewis DA. A subclass of prefrontal gamma-aminobutyric acid axon terminals are selectively altered in schizophrenia. *Proc Natl Acad Sci USA* **95**:5341–5346(1998).

36. Benes FM, Kwok EW, Vencent SL, Todtenkopf MS. A reduction of nonpyramidal cells in sector CA2 of schizophrenics and manic depressives. *Biol Psychiatry* **44**:88–97(1998).

37. Jeste DV, Lohr JB. Hippocampal pathologic findings in schizophrenia: a morphometric study. *Arch Gen Psychiatry* **46**:1019–1024(1989).

38. Kubicki M, Shenton ME, Salisbury DF, et al. Voxel-based morphometric analysis of gray matter in first episode schizophrenia. *Neuroimage* **17**:1711–1719(2002).

39. Cecil KM, Lenkinski RE, Gur RE, Gur RC. Proton magnetic resonance spectroscopy in the frontal and temporal lobes of neuroleptic naïve patients with schizophrenia. *Neuropsychopharmacology* **20**:131–140(1999).

40. Neylan TC, van Kammen DP, Kelley ME, Peters JL. Sleep in schizophrenic patients on and off haloperidol therapy. Clinically stable vs relapsed patients. *Arch Gen Psychiatry* **49**(8):643–649(1992).

41. Salin-Pascual RJ, Herrera-Estrella M, Galicia-Polo L, Laurrabaquio MR. Olanzapine acute administration in schizophrenic patients increases delta sleep and sleep efficiency. *Biol Psychiatry* **46**(1):141–143(1999).

42. Wetter TC, Lauer CJ, Gillich G, Pollmacher T. The electroencephalographic sleep pattern in schizophrenic patients treated with clozapine or classical antipsychotic drugs. *J Psychiatr Res* **30**(6):411–419(1996).

43. Taylor SF, Tandon R, Shipley JE, Eiser AS. Effect of neuroleptic treatment on polysomnographic measures in schizophrenia. *Biol Psychiatry* **30**(9):904–912(1991).

44. Keks N, McGrath J, Lambert T, Catts S, Vaddadi K, Burrows G, Varghese F, George T, Hustig H, Burnett P, et al. The Australian multicentre double-blind comparative study of remoxipride and thioridazine in schizophrenia. *Acta Psychiatr Scand* **90**(5):358–365(1994).

45. Mendlewicz J, de Bleeker E, Cosyns P, Deleu G, Lotstra F, Masson A, Mertens C, Parent M, Peuskens J, Suy E, et al. A double-blind comparative study of remoxipride and haloperidol in schizophrenic and schizophreniform disorders. *Acta Psychiatr Scand Suppl* **358**:138–141 (1990).

46. McConville BJ, Sorter MT. Treatment challenges and safety considerations for antipsychotic use in children and adolescents with psychoses. *J Clin Psychiatry* **65**(Suppl 6):20–29 (2004).

47. Muller MJ, Rossbach W, Mann K, Roschke J, Muller-Siecheneder F, Blumler M, Wetzel H, Russ H, Dittmann RW, Benkert O. Subchronic effects of olanzapine on sleep EEG in schizophrenic patients with predominantly negative symptoms. *Pharmacopsychiatry* **37**(4):157–162(2004).

48. Lee JH, Woo JI, Meltzer HY. Effects of clozapine on sleep measures and sleep-associated changes in growth hormone and cortisol in patients with schizophrenia. *Psychiatry Res* **103**(2-3):157–166(2001).

49. Adams CE, Eisenbruch M. Depot fluphenazine for schizophrenia. *Cochrane Database Syst Rev* **2**:CD000307(2000).

50. Chemerinski E, Ho BC, Flaum M, Arndt S, Fleming F, Andreasen NC. Insomnia as a predictor for symptom worsening following antipsychotic withdrawal in schizophrenia. *Compr Psychiatry* **43**(5):393–396(2002).

51. Thaker GK, Wagman AM, Kirkpatrick B, Tamminga CA. Alterations in sleep polygraphy after neuroleptic withdrawal: a putative supersensitive dopaminergic mechanism. *Biol Psychiatry* **25**(1):75–86(1989).

52. Jus K, Jus A, Beland C, Bouchard M, Pires P, Fontaine P, Brunelle R. Sleep analysis during drug-free weekends in chronic schizophrenic patients. *Biol Psychiatry* **11**(6):709–718(1976).

53. van Sweden B. Rebound insomnia in neuroleptic drug withdrawal neurophysiologic characteristics. *Pharmacopsychiatry* **20**(3):116–119(1987).

54. Walters AS, Hening W, Rubinstein M, Chokroverty S. A clinical and polysomnographic comparison of neuroleptic-induced akathisia and the idiopathic restless legs syndrome. *Sleep* **14**(4):339–345(1991).

55. Nishimatsu O, Horiguchi J, Inami Y, Sukegawa T, Sasaki A. Periodic limb movement disorder in neuroleptic-induced akathisia. *Kobe J Med Sci* **43**(5):169–177(1997).

56. Poyurovsky M, Nave R, Epstein R, Tzischinsky O, Schneidman M, Barnes TR, Weizman A, Lavie P. Actigraphic monitoring (actigraphy) of circadian locomotor activity in schizophrenic patients with acute neuroleptic-induced akathisia. *Eur Neuropsychopharmacol* **10**(3):171–176(2000).

57. Staedt J, Dewes D, Danos P, Stoppe G. Can chronic neuroleptic treatment promote sleep disturbances in elderly schizophrenic patients? *Int J Geriatr Psychiatry* **15**(2):170–176(2000).

58. Kato M, Kajimura N, Okuma T, Sekimoto M, Watanabe T, Yamadera H, Takahashi K. Association between delta waves during sleep and negative symptoms in schizophrenia. Pharmaco-EEG studies by using structurally different hypnotics. *Neuropsychobiology* **39**(3):165–172(1999).

59. Kajimura N, Kato M, Okuma T, Onuma T. Effects of zopiclone on sleep and symptoms in schizophrenia: comparison with benzodiazepine hypnotics. *Prog Neuropsychopharmacol Biol Psychiatry* **18**(3):477–490(1994).

60. Jus K, Jus A, Villeneuve A, Pires P, Fontaine P. The utilization of hypnotics in chronic schizophrenics: some critical remarks. *Biol Psychiatry* **14**(6):955–960(1979).

110

MOOD DISORDERS

ROSEANNE ARMITAGE
University of Michigan, Ann Arbor, Michigan

INTRODUCTION

Mood disorders, including major depressive disorder (MDD), dysthymia, and bipolar illness, are among the most prevalent psychiatric illnesses, with an estimated lifetime prevalence of 10–20% in the United States [1, 2]. The greatest age of risk is in young to middle adulthood and women are at about twice the risk of men for developing mood disorders, particularly for MDD. The increased risk in females is believed to be primarily during the reproductive years, from adolescence through menopause [3]. Mood disorders have a slightly lower prevalence in children, estimated to occur in 8–15% of the population. However, there is evidence of a secular trend in depression with an increasingly younger age of onset [4].

Sleep abnormalities are key features of depression and are included in most diagnostic and symptom-severity classification instruments. Sleep complaints are pervasive in those diagnosed with depression with subjective sleep complaints in more than 70% of adults and children with MDD [5, 6]. Sleep disturbances that persist beyond clinical remission of depressive symptoms increase the risk of suicide and the risk of relapse and recurrence of depression [7–9]. Moreover, insomnia of at least 2 weeks duration increases the lifetime risk of developing depression [10].

Difficulty falling asleep, intermittent awakenings, and early morning awakenings are the classic insomnia-type sleep problems in MDD, usually accompanied by complaints of feeling unrested upon awakening. Note, however, that it is estimated that more than 30% of patients with MDD experience hypersomnia [11]. Those with bipolar illness report even more extreme insomnia during manic episodes and often repot sleeping 12–14 hours a night during the depressed phase of the illness. These subjective sleep complaints are often mirrored in the findings of laboratory sleep studies.

LABORATORY SLEEP FINDINGS

Early sleep studies indicated that MDD in adults was associated with prolonged sleep latency, and an increase in both stage 1 sleep and wakefulness after sleep onset. Those with MDD also showed less slow-wave sleep (stages 3 and 4), an early onset of REM sleep (<65 minutes compared to the 80–120 minutes in healthy adults), a prolonged first REM period duration, and an overall increase in the total amount of REM sleep. Increased phasic REM activity has also been reported in MDD (see [5, 12] for a review).

Laboratory findings in childhood and adolescent depression have been much more equivocal, with short REM latency in about only half of published studies and largely restricted to inpatients. Prolonged sleep latency tends to be the most stable sleep finding in young depressed patients [13]. See [6] for a complete review.

Research also suggested that the REM sleep abnormalities persisted into clinical remission and that REM latency was also short (<65 minutes) in the family members of depressed patients [14]. However, most of the recent studies have failed to confirm that any one sleep architectural variable reliably differentiates patients from healthy controls, including several from our own group. Moreover, a meta-analysis of studies published before 1990 indicated that single sleep architectural measure differentiated patients with MDD from those with other psychiatric disorders, bringing into question the specificity of REM sleep abnormalities to depression [15]. Research from the Pittsburgh group has suggested that the sleep abnormalities in depression are best captured by a multivariate approach that considers clusters of multiple sleep measures [16].

Studies that have included quantitative sleep EEG measures such as spectral and period amplitude analyses have been more likely to differentiate patients from controls with some exceptions. Collectively, these studies have shown decreased slow-wave activity (delta in NREM sleep) and increased alpha or beta activity in patients with MDD [17–19], although several studies have pointed to a greater prevalence of slow-wave abnormalities in men than in women with MDD [20, 21]. Our own research group has also shown greater hemispheric asymmetries during sleep in MDD and a reduction in the degree of synchronization of interhemispheric EEG activity and in the relationship between fast and slow frequency EEG throughout sleep [22] both in adults and in children and adolescents [23]. Perhaps most compelling, the quantitative sleep EEG abnormalities also appear to prospectively identify those at greatest risk for developing future depression [24]. Nevertheless, disagreement among studies remains.

MECHANISMS UNDERLYING SLEEP DISTURBANCES

The laboratory studies have led to speculation on the mechanisms underlying sleep disturbances in depression. These theories fall into three general classes: REM impairment, slow-wave activity or homeostatic impairment, and biological rhythm dysregulation. Note that these theoretical positions are not necessarily mutually exclusive and do overlap. It has been suggested that the REM sleep abnormalities in MDD are largely due to increased cholinergic activation. This model, developed by McCarley and colleagues [25–28], assumes that there is a reciprocal interaction of the neurotransmitter system involved in REM sleep regulation, influencing the REM/NREM sleep cycle. Cholinergic neurons in the pontine reticular formation of the brainstem are activated during REM sleep, whereas noradrenergic/serotoninergic neurons in the locus caeruleus and dorsal raphe are deactivated. The imbalance in cholinergic/aminergic neurotransmission would produce disinhibition of REM sleep and result in an early onset of REM and perhaps increased overall REM activity. Support for this view has come from a number of studies on enhanced response to a cholinergic challenge in those with depression using arecholine [29], RS-86 [12, 30, 31], and donepezil [32].

Rather than REM sleep disinhibition, Borbély and colleagues have suggested that an earlier onset of REM sleep is due to a reduction in basic homeostatic sleep drive, reflected in the lower accumulation and faster dissipation of slow-wave activity (SWA) within NREM sleep [33–35]. This extension of the two-process model of sleep regulation assumes that in a healthy brain the propensity of REM sleep increases as SWA dissipates across the night. In the early hours after sleep onset, SWA pressure remains high, resulting in minimal amounts of REM sleep. In depression, there is a reduced buildup of process "S" during the daytime, resulting in a faster dissipation of SWA after sleep onset, an earlier occurrence of REM sleep, and increased REM during the night. The lower amounts of SWA reported in MDD do provide support for this position [20, 21, 35–37]. However, these studies have focused on SWA in baseline sleep, not in a challenge of SWA regulation either through sleep deprivation or restriction. It is the latter that would provide convincing support for SWA impairment. Moreover, our own studies have suggested that SWA is reduced in men but not women with MDD and thus the current theoretical position may require some revision.

The third theoretical position on sleep abnormalities in depression centers on biological rhythm disturbances, citing as evidence the therapeutic effects of sleep deprivation, and symptom relapse after recovery sleep and the efficacy of bright light therapy [38]. Although there are only a few long-term studies of circadian rhythms in mood disorders, there is some evidence for free-running rhythms that are no longer synchronized with sleep time in bipolar illness, and in rapid cyclers in particular [39–41]. Anecdotal evidence of jet lag-induced depression has been taken as further support for this view [42].

Other studies have suggested that a phase advance in the circadian clock underlies the earlier occurrence of REM sleep in depression and is characteristic of bipolar illness [43]. One study showed that the shift from mania to depression was heralded by the most dramatic phase advance of temperature and REM sleep [44].

Phase delay of circadian rhythms has only been supported in the case of seasonal affective disorders [38]. However, a recent study failed to demonstrate an abnormal circadian phase position for dim light melatonin onset, temperature rhythms, and sleep termination in nine patients with MDD [45].

Irregularity in circadian phase and overall damped amplitude have also been identified as potential biological rhythm disturbances in depression [46]. This position is supported by studies of diurnal temperature rhythms [47], blunted amplitude of cortisol rhythms coupled with an overall increase in cortisol levels [48], and erratic phase relationships between fast- and slow-frequency EEG activity during sleep in adults [22] and children with depression [23].

Studies examining rest–activity cycles throughout sleep and wakefulness have also shown damped circadian rhythms in adults and children with depression [49–51]. Our own group has reported that damped circadian rest–activity cycles in childhood depression are also associated with reduced light exposure and less time spent in bright light [52]. Since light is a potent zeitgeber for circadian rhythm regulation, it suggests that those with depression have less overall exposure to entrainment cues as well as endogenous rhythm abnormalities. There is additional evidence that those with depression have irregular exposure and/or withdrawal from social zeitgebers, further contributing to irregularities in rest–activity cycles [53].

LIMITATIONS OF STUDIES

At present, none of these theoretical positions can be ruled out completely in part due to the limitations of existing work, as pointed out by several researchers [30, 54]. Of foremost concern is the small sample size in much of this work. With the exception of just a few studies [20, 21, 37], the majority have included fewer than 20 subjects per group. Thus age or gender main effects and interactions cannot be adequately evaluated nor can the potential influence and clinical features of such things as family history, age of onset, length of current episode of depression, or past medication history be addressed.

Moreover, the definitive studies of biological rhythm dysregulation and REM or NREM sleep impairment have yet to be conducted, largely due to the practical difficulties and ethical considerations in conducting circadian rhythm studies in a large-scale population of symptomatic and unmedicated patients [38].

Regardless, the clinical relevance of the link between sleep disturbances and depression is clear from course of illness data and the prevalence of subjective sleep complaints in this population. The effects of antidepressant medications have also been helpful in contrasting theoretical views of sleep disturbances in depression.

EFFECTS OF ANTIDEPRESSANT MEDICATIONS ON SLEEP

Most antidepressant medications suppress REM sleep, and it has been suggested that the improvement in the symptoms of depression is related to REM sleep deprivation [55]. This view is also consistent with the idea of increased cholinergic activation during sleep in depression. However, not all antidepressants reduce REM sleep [56–58]. Moreover, antidepressant effects are not restricted to REM sleep; they also include alterations in sleep consolidation and sleep architecture that may be relevant both to clinical response and to the need for concomitant or augmentive treatment.

With the exception of iprindole and trimipramine, tricyclic antidepressants (TCAs) are potent REM sleep suppressors, prolonging REM latency and decreasing the total amount of REM sleep [55, 59–62]. Clomipramine and desipramine are among most potent REM-suppressing TCAs [63–66]. Clomipramine, desipramine, and amitriptyline also significantly increase stage 1 sleep and decrease sleep efficiency over baseline levels [67]. Interestingly, clinical response to amitriptyline and clomipramine appears to be related to the degree of REM sleep suppression [68–70], which supports the findings of Vogel and others [71]. As mentioned earlier, response to trimipramine is not related to a decrease in the amount of REM sleep [59, 60, 62].

Monoamine oxidase inhibitors (MAOIs) have also been shown to suppress REM sleep in patients with depression, even eliminating REM sleep in some patients [72]. Relative to TCAs, the effects are delayed; they are, however, associated with a substantial REM rebound upon withdrawal [73, 74]. This class of antidepressants also appears to reduce total sleep time and may decrease sleep efficiency [73]. By contrast, a study published in 1989 indicated that reversible MAOIs, such as moclobemide, may actually enhance REM sleep [75].

Selective serotonin reuptake inhibitors (SSRIs) are also effective antidepressants but act on the aminergic rather than cholinergic neurotransmitter system. Most of the published sleep studies with these agents have focused on fluoxetine although similar sleep effects have also been noted with paroxetine [76]. Fluoxetine decreases total REM sleep time, but it is generally only a 3–5% reduction over the entire night. The effects on REM latency, however, are substantial, virtually doubling the latency compared with baseline [77–80]. Moreover, significantly increased stage 1 sleep and wakefulness (decreased sleep efficiency) have been reported in most studies [78, 80–84]. The sleep effects of fluoxetine were similar in both responders and nonresponders, thus there was no evidence to suggest that clinical response was related to the degree of REM sleep suppression, in contrast to some studies of TCAs [68–70]. A preliminary study suggests that the alerting effects of fluoxetine may also be evident in children and adolescents with depression, although REM sleep measures were largely unaffected by treatment [83]. Although there is some adaptation to fluoxetine, the alerting effects on sleep are still present after 30 weeks of treatment [85]. Buysse and others, however, have shown that the effects of fluoxetine on REM and slow-wave sleep persist for at least 4 weeks postdiscontinuation in women with depression [86].

Several studies have also shown that fluoxetine is associated with significant eye movement and motor abnormalities during sleep, particularly during lighter NREM sleep stages. Fluoxetine-induced oculomotor abnormalities occur in 30–80% of adult patients [87–89] and may also be evident in children and adolescents with depression [83]. Although TCAs have also been reported to exacerbate periodic limb movements during sleep [90], the incidence is stronger for the SSRIs. Two additional reports indicate that fluoxetine or paroxetine exacerbate bruxism [91, 92]. Thus it appears that fluoxetine (and perhaps other SSRIs) exacerbates preexisting conditions, including oculomotor abnormalities in many patients. These adverse effects may also persist postdiscontinuation.

Not all serotoninergic antidepressants are alerting to sleep. Trazodone and nefazodone both impact on serotoninergic neurotransmission but are pharmacologically distinct from the SSRIs, and their effects on sleep also differ dramatically from the SSRIs [57, 72]. For example, both trazodone and nefazodone either have minimal effect on REM sleep or show a small enhancement in REM along with some degree of NREM sleep improvement [57, 72, 80, 83, 84, 93–95].

Bupropion is typically viewed as a dopaminergic antidepressant, although recent preclinical evidence suggests that its effects are more selective for norepinephrine [96]. Two studies of patients with depression also indicate that this antidepressant may not suppress REM sleep [97, 98]. Mianserin is a tetracyclic antidepressant that acts as an antagonist on a number of monoamine receptors, including 5-HT$_2$. Its primary effect on sleep in depression is a significant increase in REM latency with no apparent reduction in total REM time. It does no appear to influence slow-wave sleep, but it does improve sleep continuity. Mianserin, like trazodone, may also be a more sedating antidepressant. Although some studies of venlafaxine and citalopram have been conducted in healthy volunteers, it is the author's opinion that the effects may differ substantially from those observed in patients with depression in whom neurotransmitter and sleep abnormalities already exist. Clearly, additional research on the impact of newer antidepressants on sleep for patients with depression is warranted, particularly in longer-term treatment trials that better reflect clinical practice. Interestingly, there has been a relative paucity of published studies on antidepressant effects on sleep since the beginning of this decade.

Although the laboratory studies are not always in agreement, the clinical relevance of sleep disturbances in depression and how antidepressant treatment impacts on sleep is clear. In addition to the increased risk for relapse, recurrence, and suicide, those with persistent or treatment-induced sleep disturbances are more likely to require adjunctive anxiolytics or hypnotics [99].

REFERENCES

1. Weissman MM, Wickramaratne P, Merikanga KR, et al. Onset of major depression in early adulthood. Increased familial loading and specificity. *Arch Gen Psychiatry* **41**(12):1136–1143(1984).

2. Kessler RC, McGonagle KA, Swartz M, et al. Sex and depression in the National Comorbidity Survey. I: Lifetime prevalence, chronicity and recurrence. *J Affect Disord* **29**:85–96(1993).

3. Kessler RC. Gender differences in major depression. Epidemiological findings. In: Frank E (Ed), *Gender and Its Effects on Psychopathology.* American Psychiatric Press, Washington DC, 2000, pp 61–84.

4. Kovaks M, Gatsonis C. Secular trends in age at onset of major depressive disorder in a clinical sample of children. *J Psychiatr Res* **28**:319–329(1994).

5. Reynolds CF III, Kupfer DJ. Sleep research in affective illness: state of the art circa 1987. *Sleep* **10**(3):199–215(1987).

6. Birmaher B, Heydl P. Biological studies in depressed children and adolescents. *Int J Neuropsychopharmacol* **4**:149–157(2001).

7. Wingard DL, Berkman LF. Mortality risk associated with sleeping patterns among adults. *Sleep* **6**(2):102–107(1983).

8. Ford DE, Kamerow DB. Epidemiologic study of sleep disturbances and psychiatric disorders. An opportunity for prevention? *JAMA* **262**(11):1479–1484(1989).

9. Fawcett J, Scheftner WA, Fogg L, et al. Time-related predictors of suicide in major affective disorder. *Am J Psychiatry* **147**(9):1189–1194(1990).

10. Ohayon MM, Caulet M, Lemoine P. Comorbidity of mental and insomnia disorders in the general population. *Compr Psychiatry* **39**(4):185–197(1998).

11. Jindal RD, Thase ME. Treatment of insomnia associated with clinical depression. *Sleep Med Rev* **8**:19–30(2004).

12. Riemann D, Hohagen F, Bahro M, et al. Sleep in depression: the influence of age, gender and diagnostic subtype on baseline sleep and the cholinergic REM induction test with RS 86. *Eur Arch Psychiatry Clin Neurosci* **243**:279–290(1994).

13. Dahl RE. The regulation of sleep and arousal: development and psychopathology. *Dev Psychopathol* **8**:3–27(1996).

14. Giles DE, Perlis ML, Reynolds CF III, et al. EEG sleep in African-American patients with major depression: a historical case control study. *Depression Anxiety* **8**(2):58–64(1998).

15. Benca RM, Obermeyer WH, Thisted RA, et al. Sleep and psychiatric disorders. A meta-analysis. *Arch Gen Psychiatry* **49**:651–668(1992).

16. Nofzinger EA, Keshavan M, Buysee DJ, et al. The neurobiology of sleep in relation to mental illness. In: Charney DS, Nestler EJ, Bunney BS (Eds), *Neurobiology of Mental Illness.* Oxford University Press, New York, 1999, pp 915–929.

17. Armitage R, Roffwarg HP, Rush AJ, et al. Digital period analysis of sleep EEG in depression. *Biol Psychiatry* **31**:52–68(1992).

18. Armitage R, Roffwarg HP, Rush AJ. Digital period analysis of EEG in depression: periodicity, coherence, and interhemispheric relationships during sleep. *Prog Neuropsychopharmacol Biol Psychiatry* **17**:363–372(1993).

19. Armitage R. Microarchitectural findings in sleep EEG in depression: diagnostic implications. *Biol Psychiatry* **37**:72–84(1995).

20. Armitage R, Hoffmann R, Fitch T, et al. Temporal characteristics of delta activity during NREM sleep in depressed outpatients and healthy adults: group and sex effects. *Sleep* **23**(5):607–617(2000).

21. Armitage R, Hoffmann R, Trivedi M, et al. Slow-wave activity in NREM sleep: sex and age effects in depressed outpatients and healthy controls. *Psychiatry Res* **95**:201–213(2000).

22. Armitage R, Hoffmann RF, Rush AJ. Biological rhythm disturbance in depression: temporal coherence of ultradian sleep EEG rhythms. *Psychol Med* **29**:1435–1448(1999).

23. Armitage R, Emslie GJ, Hoffmann RF, et al. Ultradian rhythms and temporal coherence in sleep EEG in depressed children and adolescents. *Biol Psychiatry* **47**:338–350(2000).

24. Morehouse RL, Kusumakar V, Kutcher SP, et al. Temporal coherence in ultradian sleep EEG rhythms in a never-depressed, high-risk cohort of female adolescents. *Biol Psychiatry* **51**:446–456(2002).

25. McCarley RW, Hobson JA. Neuronal excitability modulation over the sleep cycle: a structural and mathematical model. *Science* **189**:58(1975).

26. McCarley RW. REM sleep and depression: common neurobiological control mechanisms. *Am J Psychiatry* **139**(5):565–570(1982).

27. Massaquoi SC, McCarley RW. The limit cycle reciprocal interaction model of REM cycle control: new neurobiological structure. *J Sleep Res* **1**:138–143(1992).

28. McCarley RW, Massaquoi SC. Neurobiological structure of the revised limit cycle reciprocal interaction model of REM cycle control. *J Sleep Res* **1**:132–137(1992).

29. Gillin JC, Sutton L, Ruiz C, et al. The cholinergic rapid eye movement induction test with arecoline in depression. *Arch Gen Psychiatry* **48**:264–270(1991).

30. Riemann D, Berger M. Sleep, age, depression and the cholinergic REM induction test with RS 86. *Prog Neuropsychopharmacol Biol Psychiatry* **16**:311–316(1992).

31. Riemann D, Hohagen F, Fritsch-Montero R, et al. Cholinergic and noradrenergic neurotransmission: impact on REM sleep regulation in healthy subjects and depressed patients. *Acta Psychiatr Belg* **92**(3):151–171(1992).

32. Perlis ML, Smith MT, Orff HJ, et al. The effects of an orally administered cholinergic agonist on REM sleep in major depression. *Biol Psychiatry* **51**:457–462(2002).

33. Borbély A. A two-process model of sleep regulation. *Hum Neurobiol* **1**:195–204(1982).

34. Borbély AA, Wirz-Justice A. Sleep, sleep deprivation and depression. A hypothesis derived from a model of sleep regulation. *Hum Neurobiol* **1**:205–210(1982).

35. Borbély AA, Tobler I, Loepfe M, et al. All-night spectral analysis of the sleep EEG in untreated depressives and normal controls. *Psychiatry Res* **12**:27–33(1984).

36. Kupfer DJ, Ulrich RF, Coble PA, et al. Application of an automated REM and slow wave sleep analysis: II. Testing the assumptions of the two-process model of sleep regulation in normal and depressed subjects. *Psychiatry Res* **13**(4):335–343(1984).

37. Reynolds CF III, Kupfer DJ, Thase ME, et al. Sleep, gender, and depression: an analysis of gender effects on the electroencephalographic sleep of 302 depressed outpatients. *Biol Psychiatry* **28**:673–684(1990).

38. Wirz-Justice A. Biological rhythms in mood disorders. In: Bloom FE, Kupfer DJ (Eds), *Psychopharmacology: The Fourth Generation of Progress*. Raven Press, New York, 1995, pp 999–1017.

39. Kripke DF, Mullaney DJ, Atkinson M, et al. Circadian rhythm disorders in manic-depressives. *Biol Psychiatry* **13**:335–351(1978).

40. Wehr TA, Sack DA, Duncan WC. Sleep and circadian rhythms in affective patients isolated from external time cues. *Psychiatry Res* **15**:327–339(1985).

41. Mizukawa R, Ishiguro S, Takada H, et al. Long-term observation of a manic–depressive patient with rapid cycles. *Biol Psychiatry* **29**:671–678(1991).

42. Jauhar P, Weller M. Psychiatric morbidity and time zone changes. *Br J Psychiatry* **140**:231–235(1982).

43. Wehr TA, Wirz-Justice A. Circadian rhythm mechanisms in affective illness and in antidepressant drug action. *Pharmacopsychiatry* **15**:31–39(1982).

44. Wehr TA, Goodwin FK. Biological rhythms in manic depressive illness. In: Wehr TA, Goodwin FK (Eds), *Circadian Rhythms in Psychiatry*. Boxwood Press, Pacific Grove, CA, 1983, pp 129–184.

45. Gordijn MCM, Beersma DGM, Korte HJ, et al. Testing the hypothesis of a circadian phase disturbance underlying depressive mood in nonseasonal depression. *J Biol Rhythms* **13**(2):132–147(1998).

46. Siever LJ, Davis KL. Overview: toward a dysregulation hypothesis of depression. *Am J Psychiatry* **142**(9):1017–1031(1985).

47. Tsujimoto T, Yamada N, Shimoda K, et al. Circadian rhythms in depression. Part II: Circadian rhythms in inpatients with various mental disorders. *J Affect Disord* **18**:199–210(1990).

48. Souëtre E, Salvati E, Belugou JL, et al. Circadian rhythms in depression and recovery: evidence for blunted amplitude as the main chronobiological abnormality. *Psychiatry Res* **28**:268–278(1989).

49. Teicher MH, Glod CA, Harper D, et al. Locomotor activity in depressed children and adolescents: I. Circadian dysregulation. *J Am Acad Child Adolesc Psychiatry* **32**(4):760–769(1993).

50. Glod CA, Teicher MH, Polari A, et al. Circadian rest–activity disturbances in children with seasonal affective disorder. *J Am Acad Child Adolesc Psychiatry* **36**(2):188–195(1997).

51. Teicher MH, Glod CA, Magnus E, et al. Circadian rest–activity disturbances in seasonal affective disorder. *Arch Gen Psychiatry* **54**(2):124(1997).

52. Armitage R, Hoffmann R, Emslie G, et al. Rest–activity cycles in childhood and adolescent depression. *J Am Acad Child Adolesc Psychiatry* **43**:761–769(2004).

53. Ehlers CL, Frank E, Kupfer DJ. Social zeitgebers and biological rhythms. A unified approach to understanding the etiology of depression. *Arch Gen Psychiatry* **45**:948–952(1988).

54. Armitage R, Hoffmann R. Sleep EEG, depression and gender. *Sleep Med Rev* **5**:237–246(2001).

55. Vogel G, Neill D, Kors D, et al. REM sleep abnormalities in a new animal model of endogenous depression. *Neurosci Biobehav Rev* **14**(1):77–83(1990).

56. Armitage R. The effects of antidepressants on sleep in patients with depression. *Can J Psychiatry* **45**:803–809(2000).

57. Armitage R, Rush AJ, Trivedi M, et al. The effects of nefazodone on sleep architecture in depression. *Neuropsychopharmacology* **10**(2):123–127(1994).

58. Vogel G, Cohen J, Mullis D, et al. Nefazodone and REM sleep: how do antidepressant drugs decrease REM sleep? *Sleep* **21**(1):70–77(1998).

59. Wiegand M, Berger M, Zulley J, et al. The effect of trimipramine on sleep in patients with major depressive disorder. *Pharmacopsychiatry* **19**:198–199(1986).

60. Ware JC, Brown FW, Moorad PJJ, et al. Effects on sleep: a double-blind study comparing trimipramine to imipramine in depressed insomniac patients. *Sleep* **12**(6):537–549(1989).

61. Feuillade P, Pringuey D, Belugou JL, et al. Trimipramine: acute and lasting effects on sleep in healthy and major depressive subjects. *J Affect Disord* **24**:135–145(1992).

62. Sonntag A, Rothe B, Guldner J, et al. Trimipramine and imipramine exert different effects on the sleep EEG and on nocturnal hormone secretion during treatment of major depression. *Depression* **4**:1–13(1996).

63. Dunleavy DL, Brezinova V, Oswald I, et al. Changes during weeks in effects of tricyclic drugs on the human sleeping brain. *Br J Psychiatry* **120**:663–672(1972).

64. Shipley JE, Kupfer DJ, Griffin SJ, et al. Comparison of effects of desipramine and amitriptyline on EEG sleep of depressed patients. *Psychopharmacology* **85**:14–22(1985).

65. Kupfer DJ, Ehlers CL, Pollock BG, et al. Clomipramine and EEG sleep in depression. *Psychiatry Res* **30**:165–180(1989).

66. Kupfer DJ, Perel JM, Pollock BG, et al. Fluvoxamine versus desipramine: comparative polysomnographic effects. *Biol Psychiatry* **29**:23–40(1991).

67. Shipley JE, Kupfer DJ, Dealy RS, et al. Differential effects of amitriptyline and of zimelidine on the sleep electroencephalogram of depressed patients. *Clin Pharmacol Ther* **36**:251–259(1984).

68. Kupfer DJ, Foster FG, Reich L, et al. EEG sleep changes in REM sleep and clinical depression. *Am J Psychiatry* **133**:622–626(1976).

69. Gillin JC, Wyatt RF, Fram D, et al. The relationship between changes in REM sleep and clinical improvement in depressed patients treated with amitriptyline. *Psychopharmacology* **59**:267–272(1978).

70. Hochli D, Riemann D, Zulley J, et al. Initial REM sleep suppression by clomipramine: a prognostic tool for treatment response in patients with a major depressive disorder. *Biol Psychiatry* **21**:1217–1220(1986).

71. Vogel GW, Buffenstein A, Minter K, et al. Drug effects on REM sleep and on endogenous depression. *Neurosci Biobehav Rev* **14**:49–63(1990).

72. Sharpley AL, Cowen PJ. Effect of pharmacologic treatments on the sleep of depressed patients. *Biol Psychiatry* **37**:85–98(1995).

73. Kupfer DJ, Bowers MB. REM sleep and monoamine oxidase inhibition. *Psychopharmacologia* **27**:183–190(1972).

74. Dunleavy DLF, Oswald I. Phenelzine, mood response, and sleep. *Arch Gen Psychiatry* **28**:353–356(1973).

75. Monti JM. Effect of a reversible monoamine oxidase-A inhibitor (moclobemide) on sleep of depressed patients. *Br J Psychiatry* **155**(Suppl 6):61–65(1989).

76. Staner L, Kerkhofs M, Detroux D, et al. Acute, subchronic and withdrawal sleep EEG changes during treatment with paroxetine and amitriptyline: a double-blind randomized trial in major depression. *Sleep* **18**(6):470–477(1995).

77. Keck PE Jr, Hudson JI, Dorsey CM, et al. Effect of fluoxetine on sleep (letter). *Biol Psychiatry* **29**:618–619(1991).

78. Hendrickse WA, Roffwarg HP, Grannemann BD, et al. The effects of fluoxetine on the polysomnogram of depressed outpatients: a pilot study. *Neuropsychopharmacology* **10**(2): 85–91(1994).

79. Armitage R, Yonkers K, Cole D, et al. A multicenter, double-blind comparison of the effects of nefazodone and fluoxetine on sleep architecture and quality of sleep in depressed outpatients. *J Clin Psychopharmacol* **17**(3):161–168(1997).

80. Gillin JC, Rapaport M, Erman MK, et al. A comparison of nefazodone and fluoxetine on mood and on objective, subjective, and clinician-rated measures of sleep in depressed patients: a double-blind, 8-week clinical trial. *J Clin Psychiatry* **58**(5):185–192(1997).

81. Kerkhofs M, Rielaert C, de Maertelaer V, et al. Fluoxetine in major depression: efficacy, safety and effects on sleep polygraphic variables. *Int Clin Psychopharmacol* **5**:253–260(1990).

82. Keck PE Jr, Hudson JI, Cunningham SL, et al. Polysomnographic features of excessive daytime somnolence associated with monoamine oxidase inhibitor treatment. *Ann Clin Psychiatry* **4**:49–53(1992).

83. Armitage R, Emslie G, Rintelmann J. The effect of fluoxetine on sleep EEG in childhood depression: a preliminary report. *Neuropsychopharmacology* **17**(4):241–245(1997).

84. Rush AJ, Armitage R, Gillin JC, et al. Comparative effects of nefazodone and fluoxetine on sleep in outpatients with major depressive disorder. *Biol Psychiatry* **44**:3–14(1998).

85. Trivedi MH, Rush AJ, Armitage R, et al. Effects of fluoxetine on the polysomnogram in outpatients with major depression. *Neuropsychopharmacology* **20**(5):447–459(1999).

86. Buysse DJ, Kupfer DJ, Cherry C, et al. Effects of prior fluoxetine treatment on EEG sleep in women with recurrent depression. *Neuropsychopharmacology* **21**(2):258–267(1999).

87. Schenck CH, Mahowald MW, Kim SW, et al. Prominent eye movements during NREM sleep and REM sleep behavior disorder associated with fluoxetine treatment of depression and obsessive–compulsive disorder. *Sleep* **15**(3):226–235(1992).

88. Armitage R, Trivedi M, Rush AJ. Fluoxetine and oculomotor activity during sleep in depressed patients. *Neuropsychopharmacology* **12**(2):159–165(1995).

89. Dorsey CM, Lukas SE, Cunningham SL. Fluoxetine-induced sleep disturbance in depressed patients. *Neuropsychopharmacology* **14**(6):437–442(1996).

90. Roth T. Diagnosis and treatment of sleep disorders in the depressed elderly (abstract). Industry Supported Symposium 23—Depression in the Elderly: U.S. and European Perspectives, APA 150th Annual Meeting.

91. Ellison JM, Stanziani P. SSRI-associated nocturnal bruxism in four patients. *J Clin Psychiatry* **54**:432–434(1993).

92. Romanelli F, Adler DA, Bungay KM. Possible paroxetine-induced bruxism. *Ann Pharmacother* **30**:1246–1248(1996).

93. Mouret J, Lemoine P, Minuit MP, et al. Effects of trazodone on the sleep of depressed subjects—a polysomnographic study. *Psychopharmacology* **95**:S37–S43(1988).

94. Sharpley AL, Walsh AES, Cowen PJ. Nefazodone—a novel antidepressant—may increase REM sleep. *Biol Psychiatry* **31**:1070–1073(1992).

95. Scharf MB, McDannold M, Zaretsky N, et al. Evaluation of sleep architecture and cyclic alternating pattern rates in depressed insomniac patients treated with nefazodone hydrochloride. *Am J Ther* **6**(2):77–82(1999).

96. Cooper BR, Wang CM, Cox RF, et al. Evidence that the acute behavioral and electrophysiological effects of bupropion (Wellbutrin®) are mediated by a noradrenergic mechanism. *Neuropsychopharmacology* **11**(2):133–141(1994).

97. Nofzinger EA, Reynolds CF III, Thase ME, et al. REM sleep enhancement by bupropion in depressed men. *Am J Psychiatry* **152**(2):274–276(1995).

98. Ott G, Rao U, Lin K, et al. Effect of treatment with bupropion on EEG sleep: relationship to antidepressant response. *Int J Neuropsychopharmacology* **26**:275–281(2004).

99. Lian J. Nefazodone treatment of depression requires less use of concomitant anxiolytic and sedative/hypnotic drugs. Paper presented at the Annual Meeting of the American Psychiatric Association, San Diego, CA, APA New Research Program and Abstracts.

111

ANXIETY DISORDERS AND SLEEP

MICHAEL WEISSBERG

University of Colorado School of Medicine, Denver, Colorado

INTRODUCTION

Anxiety disorders are among the most prevalent psychiatric conditions worldwide [1] and people who experience excessive anxiety utilize health services at a much higher level than those who are not anxious. The anxious patient is aware of altered physiologic sensations while also cognitively experiencing anxiety, fear, or dread [1]. The experience of anxiety is associated with a reduction in the ability to think clearly, to plan, and to learn [1]. Some have postulated that the cognitive changes associated with chronic anxiety (and post-traumatic stress disorder (PTSD) and depression) might actually be due to cell death in the hippocampus, which makes it even more difficult for the stressed patient to deal with his or her difficulties.

Yet anxiety—the response to real or imagined danger—is a normal and necessary part of life. In general, anxiety has two components—cognitive and the physiologic state of hyperarousal. Anxiety, if manageable, gets us to act—to deal with a perceived threat. But, in some of us, anxiety persists, does not recede, and begins to overwhelm.

It is safe to say that excessive anxiety (and worry and fear) is not conducive to sleep. And the cognitive and physiologic changes associated with excessive anxiety alter sleep—and are made worse by the lack of sleep—in multiple ways. The sleeplessness that anxiety can cause turns back on itself and magnifies anxiety, thus setting up a vicious cycle.

Sleepiness, for instance, increases the likelihood of panic. Certainly anxiety disrupts sleep and insomnia may help precipitate anxiety disorders as well as other psychiatric disease, particularly depression, anxiety disorders, or substance abuse. According to some sleep experts, "sleep complaints may be among the most robust prodromal symptoms reflecting partial depressive or anxiety disorders, which eventually declare themselves as full-blown clinical episodes" [2].

At a clinical level, the sleep specialist must consider anxiety as the possible, or complicating, cause for a number of sleep complaints. For instance, anxiety syndromes in children and adults should be looked for in those complaining of sleep onset or maintenance insomnia, and awakenings with choking, gasping, chest pain, fear, and dread. Sometimes, anxiety might be incorrectly thought of as the cause of insomnia. For instance, the rumination and worry sometimes seen (especially in children) in those who struggle to fall asleep may actually be due to delayed sleep phase syndrome. In addition, "worriers," people who are generally hyperaroused, will also frequently develop psychophysiologic insomnia, which will complicate underlying anxiety disorders. And the fear of not sleeping, of course, increases the hyperarousal in those so prone.

Anxiety as a dimensional symptom is present in many categorical psychiatric diagnoses. It is often part of the presentation of, and comorbid with, other psychiatric problems that are associated with sleep pathology such as bipolar illness, other types of depression, and

Sleep: A Comprehensive Handbook, Edited by T. Lee-Chiong.
Copyright © 2006 John Wiley & Sons, Inc.

schizophrenia—all of which have their own associated effects on sleep. For example, a patient may appear anxious and have a relatively short sleep period but may also be suffering from comorbid bipolar illness. It is believed that approximately 20% of patients who initially present with anxiety go on to develop frank depressive disorders.

However, the categorical diagnostic problems likely to most commonly affect sleep are panic disorder, generalized anxiety disorder, post-traumatic stress disorder and acute stress disorder, performance anxiety, which is part of social phobia, and psychophysiologic insomnia and obsessive–compulsive disorder. Post-traumatic stress and acute stress disorders and psychophysiologic insomnia will be discussed elsewhere. It is useful, therefore, for the sleep clinician to identify patients in which anxiety likely plays an important role.

ANXIETY IN CHILDREN

Besides the usual disorders seen in adults (panic disorder with or without agoraphobia, obsessive–compulsive disorder, generalized anxiety disorder, social phobia), children experience anxiety and worry in other areas, particularly associated with bedtime. Delayed sleep phase may be confused with anxiety as the child struggles unsuccessfully to fall asleep. Limit-setting sleep disorder and sleep onset associate disorder are also associated with child (and parental) worry and anxiety.

Cosleeping raises some interesting issues. In one study, healthy school-aged children who still coslept were more likely than noncosleeper controls to suffer from bedtime resistance, sleep anxiety, night awakenings, and parasomnias than solitary sleepers. However, no significant behavioral problems were found in cosleepers. Cosleeping was associated with low socioeconomic status, one parent who is a shift worker, and one-parent families; one parent who coslept as a child, prolonged breastfeeding, and previous and current sleep problems significantly predicted cosleeping [3].

DIFFERENTIAL DIAGNOSIS

Nocturnal events frequently are hard for patients to describe and for the clinician to differentiate. In addition, multiple mechanisms may exist in the same patient (see Table 111.1). For example, patients may suffer from panic disorder and obstructive sleepapnea syndrome (OSAS) and/or gastroesophageal reflux disease (GERD) and awaken with a variety of symptoms, which must be parsed out. Patient completed diagnostic questionnaires are frequently helpful in identifying the presence of

TABLE 111.1   Differential Diagnoses

A.   Differential Diagnosis of Nocturnal Choking, Gasping, or Coughing

1. Obstructive or central sleep apnea (OSA or CSA)
2. Gastroesophageal reflux disease (GERD)
3. Sleep terrors (may feel like breathing is impaired)
4. Panic disorder
5. Seizures
6. Sleep-related asthma
7. Chronic obstructive pulmonary disease (COPD)

B.   Differential Diagnosis of Nocturnal Anxiety and Panic

1. OSAS/CSA/ GERD
2. Sleep-related abnormal swallowing syndrome (gurgling), sleep choking syndrome (no stridor and high frequency), sleep-related laryngospasm (stridor), Munchausen's stridor
3. Sleep terrors
4. Nightmares
5. Panic disorder
6. PTSD
7. REM sleep behavior disorder
8. Seizures
9. Nocturnal cardiac ischemia

psychiatric disorders in sleep patients, for instance, the Patient Health Questionnaire (PHQ) of the PRIME-MD (see Appendix).

THE DISORDERS

The disorders that create the most difficulty for the sleep clinician (and patient) are panic disorder (PD), because of the often confusing differential of nocturnal choking and/or anxiety, and generalized anxiety disorder (GAD), because people with GAD appear to worry about almost everything. Certainly, patients with anxiety are among the most challenging to work with because they need constant reassurance while treatment is initiated.

Generalized Anxiety Disorder

The usual onset of GAD is during young adulthood, ages 20–35, with a slight predominance in women, although patients usually report that they have been anxious for "most" of their lives. GAD also tends to be a chronic condition [4]. It is not uncommon to see significant life stressors at the onset of GAD; sometimes the clinician is tempted to think of these patients as suffering from a post-traumatic "spectrum" disorder. In fact, GAD is frequently comorbid with other psychiatric disorders, particularly depression, panic disorder, and social phobia.

The symptoms of GAD are the typical ones associated with stress and anxiety. Complaints of excessive fear, worry, and apprehension are often associated with somatic worries and, like depression, trouble with concentration. Physical complaints should predominate for long periods of time (usually at least 6 months) and can include muscle tension, fatigue, irritability, restlessness, trouble falling or staying asleep, or nonrestorative sleep. Because of multiple somatic concerns, patients will visit a variety of healthcare providers—not only psychiatrists.

Panic Disorder with and Without Agoraphobia

Panic episodes are fairly common in the general population and only become a problem if they increase in frequency and patients begin to anticipate further attacks. The first attack of panic occurs usually spontaneously and unpredictably. People experience both physical (precipitated by the "flight or fight" sympathetic reaction) and cognitive symptoms (intense dread, fear of dying, and/or fear of going crazy).

The physical symptoms associated with panic are worrisome to patients and their families: a racing heart and possible palpitations, shortness of breath, numbness and tingling around the mouth and fingers, and feelings of choking accompany the cognitive changes. Attacks can last for minutes to over an hour. As panic recurs patients may begin to associate attacks with specific activities such as crossing bridges, heights, crowds, or being outside [1].

Panic can also occur at night in about a third of patients; in some, they occur exclusively at night; patients then also can become phobic about going to sleep for fear of experiencing further episodes.

Nocturnal panic episodes occur mainly in NREM sleep in stage 2 toward the transition to slow-wave sleep. There appear to be differences in patient characteristics between those who panic during the day and those who panic also at night. It might be that night panic patients have more severe levels of anxiety and experience longer attacks [5]. While some newer findings cast doubt on this difference in groups, other studies suggest that nighttime panic patients are sicker and that this variant of panic disorder has its own biological basis [6]. Panic attacks may precipitate emergency room visits and often lead to expensive cardiac workups.

Physicians often look unsuccessfully for myocardial infarctions, pheochromocytomas, drug intoxications, and thyroid disease. While all of these should be considered, PD is not a diagnosis of exclusion. Furthermore, there is a higher incidence of suicide in those suffering from panic than the general population. Many patients go on to develop frank depression, alcohol abuse, or other substance abuse (e.g., short-acting benzodiazepines). Obsessive–compulsive disorder is also seen in a significant number of patients with panic disorder [1].

This is a familial disorder—the familial risk for panic disorder suggests 40% heritability [7]—and onset is during young adulthood and panic is more prevalent in women. The mood instability experienced in the premenstrual period increases the frequency of attacks and postmenopausal women also appear to be particularly vulnerable. Mitral valve prolapse is now rarely thought to be a significant factor [4].

Panic disorder may be associated with increased hippocampus and locus caeruleus activity, increased adrenergic tone, abnormal benzodiazepine receptors, and a tendency to catastrophize when dealing with life events [1, 4].

Other mechanisms may contribute to panic. Roth and colleagues [8] have pointed out that high-altitude exposure may precipitate symptoms analogous to panic such as breathlessness, palpitations, dizziness, headache, and insomnia. Similar mechanisms are found in three models of panic disorder: hyperventilation (hypoxia leads to hypocapnia), suffocation false alarms (hypoxia counteracted to some extent by hypocapnia), and cognitive distortions (symptoms from hypoxia and hypocapnia interpreted as dangerous). High-altitude exposure in normals also produces respiratory disturbances during sleep analogous to changes found in panic disorder at low altitudes. However, as of now, the evidence for the role of altitude in panic or the development of anxiety is slight at this time [8].

Obsessive–Compulsive Disorder

Two to 30% of the general population suffer from obsessive–compulsive disorder (OCD), an illness with a significant genetic component characterized by obsessions—unwelcome, recurring, and upsetting thoughts usually having to do with cleanliness, sexuality, or aggression—and compulsions—behaviors that patients resort to in order to avoid intense anxiety. Sufferers hope that these rituals, for instance, hand washing, will control the obsessive thoughts. Checking stoves, door locks, and so on are other examples of such behaviors. Pathological doubt, fears of contamination, and intrusive sexual and aggressive thoughts are typical findings. To fulfill OCD criteria, adult patients (not children)—at some point in their illness—must realize that their obsessions and compulsions are unreasonable. OCD "spectrum" consists of other syndromes where obsessions and compulsions are apparent such as compulsive exercising, hair pulling (trichotillomania), nail biting, eating disorders, and Tourette's syndrome [1].

OCD also can be precipitated by stressful life events. It is now clear that OCD is a brain disease—in which serotonin dysregulation appears to play a significant role—which results from abnormal brain functioning, particularly in the striatum. Treatments, both pharmacologic as well as behavioral, also appear to affect this area of the brain. As in GAD and PD, depression and bipolar depression are highly comorbid with OCD. This again highlights an important clinical caveat. The sleep disorder seen in patients with OCD—as in other anxiety disorders—might actually be due to bipolar illness. And the treatment of OCD—high doses of selective serotonin reuptake inhibitors (SSRIs)—can either precipitate mania in those so prone or push a vulnerable patient into a rapid-cycling downhill course. Behavior therapy, particularly response prevention and exposure techniques, are widely used [1].

Phobic Disorder

Phobic people develop fears of a specific object or behavior such as embarrassing themselves in public (social phobia), speaking in groups, answering the telephone, snakes, heights, and so forth. Agoraphobia (the fear of public places) is often a later development of panic disorder. Patients with nocturnal panic may become phobic about going to sleep since they are frightened of an ensuing attack. Interestingly, social phobia is thought to involve performance anxiety, which also plays a role in psychophysiologic insomnia. The avoidance of socializing and a shy demeanor may also be subtle forms of social phobia.

THE BIOLOGY OF ANXIETY

Anxiety disorders, just as in depression, result from the interplay of genes and experience. Areas of concern are the role of genetic transmission, locus caeruleus activity, neurotransmission, and cognitive styles. There is evidence that early stress and particularly early loss can sensitize children to the later development of anxiety syndromes such as panic disorder, PTSD, and depression. While early parental loss or other traumatic experiences can predispose people to the later development of anxiety disorders, our genetic make-up also makes it more or less likely that we will develop anxiety disorders later in life [1].

One area of interest to both sleep and anxiety clinicians is the function of neuropeptide Y. Neuropeptide Y (NPY) appears to act as a neuromodulator and acts on circadian, endocrine, and behavioral circadian processes. NPY is a potent antianxiety agent, while corticotropin-releasing factor (CRF) plays a role in hyperarousal seen in anxiety and depression. It is thought that a balance of the two plays a role in state regulation or dysregulation. Evidence suggests that dysregulation of "sleep and arousal states in depression and anxiety may be consistent with an upset of the balance between hypothalamic neuropeptide systems" [9].

ANXIETY'S EFFECT ON SLEEP

There are interesting findings associated with anxiety in addition to nocturnal awakenings due to panic and insomnia due to excessive nighttime worry. The disorders, clinical complaints, and sleep findings are outlined in Table 111.2. For example, stress that occurs in people who do not reach criteria for an anxiety disorder is likely the most frequent source of temporary sleeplessness and has been shown to be linked with a reduction of slow-wave sleep (SWS). The trepidation concerning a "difficult" next working day is linked with decreased amounts of SWS, increased percentage of stage 2 sleep, bedtime state anxiety, and subjectively poor sleep. Some researchers speculate that moderate "everyday" stress may adversely affect physiological restoration by reducing SWS [13].

Chronic morning headaches have been found to be more likely associated with comorbid anxiety and depressive and insomnia disorders and not specifically related to sleep-related breathing disorders [14]. Confusional arousal, or sleep drunkenness, has been found to be strongly associated with bipolar depression and anxiety disorders as well as OSAS [15].

At least one study found that isolated sleep paralysis (ISP) is more likely present in anxious African-Americans than in anxious Caucasians who are diagnosed with panic disorder or other anxiety disorders. In this study, the authors conclude that "recurrent ISP was found to be more common among African American participants (23%), particularly for those with panic disorder (60%) than in whites with panic disorder" [11].

TREATMENT

Pharmacotherapy, especially SSRIs, and behavioral treatments are useful first-line approaches [1, 4]. Benzodiazepines—long and short acting—offer immediate relief for panic disorder and GAD but have also been associated with problems such as dependence, tolerance, and addiction. For this reason, antidepressants, especially selective serotonin reuptake inhibitors and novel antidepressants such as venlafaxine (which could just as easily be called antianxiety agents), and psychotherapeutic treatments, particularly behavioral treatments such as cognitive–behavioral

TABLE 111.2   Disorders, Complaints, and Sleep Effects [10–13]

Disorder	Clinical Complaints	Sleep Findings
Life stress	Daytime worry associated with life stressor and temporary insomnia; nocturnal awakenings	Bedtime anxiety focuses on the "next day"; increased sleep latency, decreased sleep efficiency, decreased slow-wave sleep
Generalized anxiety disorder	May be precipitated by significant life stressors; chronic daytime anxiety and insomnia; may complain of daytime fatigue and/or EDS; depression may be comorbid; morning headaches	Increased sleep latency, decreased efficiency, decreased SWS and increased stages 1 and 2
Panic disorder	Nocturnal attacks in about a third of panic disorder patients; sudden awakening with choking, anxiety with dread, fears of death, etc.; moderate autonomic arousal; no formed dreams; may develop sleep phobia; sleep paralysis in African Americans? Comorbid depression may be common.	Attacks occur typically in transition between stage 2 and SWS
Post-traumatic stress disorder	Extreme autonomic arousal; multiple complaints of anxiety, reexperiencing, emotional numbing following an excessive traumatic event	Nightmares may occur in REM and in stage 2 sleep; sleep may be very active with much movement; increased REM sleep density may be seen
Obsessive–compulsive disorder	Obsessions and compulsions	Reduced REM sleep latency; may have reduced sleep efficiency and increased sleep latency
OSA	Nonrefreshing sleep, awakening with choking, gasping, terror; EDS, snoring, cognitive deficits, HTN, insomnia; morning headaches	Sleep disruption, airway resistance, apneas, arousals
GERD	Heartburn, insomnia, awakenings with choking, coughing, laryngospasm; may be associated with OSA	Arousals; acid reflux during sleep
Sleep terrors	Sudden awakening with intense fear; confusion to purposeful action; mostly amnestic for episode; no real dream recall; high degree of autonomic arousal	Out of slow-wave sleep; delta hypersynchrony may precede attack; direct SWS to awake transitions
Nightmares	Dream recall; variable autonomic arousal; may be induced by stress, medications such as L-dopa and beta blockers	Mostly occur in REM sleep
REM sleep behavior disorder	Purposeful and sometimes violent action during REM dreaming. Usually starts at least 90 minutes after sleep onset and more often in the latter sleep period	Loss of REM atonia
Sleep related abnormal swallowing syndrome	Coughing, choking, brief arousals with sense of trouble breathing; may complain of insomnia and or awakenings with choking and gurgling	Very little slow-wave sleep, multiple arousals
Seizures	Arousals, possible insomnia, EDS, nightmares, respiratory complaints and stereotypical behaviors	Sleep seizures occur mainly in stages 1 and 2, two hours after bedtime and 4-5 AM in the morning; awakening seizures typically; may show inter-ictal abnormalities in NREM sleep
Nocturnal cardiac ischemia	Can awaken with vise-like chest pain said to occur more frequently in patients with OSA	Either out of REM sleep during early morning or in the beginning of the sleep period due to falls in blood pressure and heart rate.
COPD	Nocturnal insomnia, coughing, anxiety	Frequent arousals

treatment for panic disorder and GAD, should be first-line treatments.

REFERENCES

1. Sadock BJ, Sadock VA. Chapter 16. In: Kaplan H, Sadock B (Eds), *Synopsis of Psychiatry*, 9th ed. Lippincott, Williams & Wilkins, Philadelphia, 2003.

2. Gillin JC. Are sleep disturbances risk factors for anxiety, depressive and addictive disorders? *Acta Psychiatr Scand Suppl* **393**:39–43(1998).

3. Cortesi F, Giannotti F, Sebastiani T, Vagnoni C. Co sleeping and sleep behavior in italian school-aged children. *J Dev Behav Pediatr* **25**(1):28–33(2004).

4. Eisendrath S, Lichtmacher J. Psychiatric disorders. In: Tierney L, McPhee S, Papadakis M (Eds), *Current Medical Diagnosis*

& *Treatment*, 43rd ed. McGraw-Hill/Appleton & Lange, New York, 2004.

5. O'Mahony JF, Ward BG. Differences between those who panic by day and those who also panic by night. *J Behav Ther Exp Psychiatry* **34**(3-4): 239–249(2003).

6. Merritt-Davis O, Balon R. Nocturnal panic: biology, psychopathology, and its contribution to the expression of panic disorder. *Depression Anxiety* **18**(4):221–227(2003).

7. Hamilton SP, et al. Evidence for genetic linkage between a polymorphism in the adenosine 2A receptor and panic disorder. *Neuropsychopharmacology* **29**:558–565(2004).

8. Roth WT, Gomolla A, Meuret AE, Alpers GW, Handke EM, Wilhelm FH. High altitudes, anxiety, and panic attacks: is there a relationship? *Depression Anxiety* **16**(2):51–58(2002).

9. Ehlers CL, Somes C, Seifritz E, Rivier JE. CRF/NPY interactions: a potential role in sleep dysregulation in depression and anxiety. *Depression Anxiety* **6**(1):1–9(1997).

10. American Sleep Disorders Association. *The International Classification of Sleep Disorders, Revised* 1997.

11. Paradis CM, Friedman S, Hatch M. Isolated sleep paralysis in African Americans with panic disorder. *Cultural Diversity Ment Health* **3**(1):69–76(1997).

12. Uhde T. Anxiety disorders. In: Kryger MH, Roth T, Dement W (Eds), *Principles and Practice of Sleep Medicine*, 3rd ed. Saunders, Philadelphia, 2000.

13. Kecklund G, Akerstedt T, Kecklund G. Apprehension of the subsequent working day is associated with a low amount of slow wave sleep. *Biol Psychol* **66**(2):169–176(2004).

14. Ohayon MM. Prevalence and risk factors of morning headaches in the general population. *Arch Intern Med* **164**(1):97–102(2004).

15. Ohayon MM, Priest RG, Zulley J, Smirne S. The place of confusional arousals in sleep and mental disorders: findings in a general population sample of 13,057 subjects. *J Nerv Ment Dis* **188**(6):340–348(2000).

APPENDIX

PATIENT HEALTH QUESTIONNAIRE

This questionnaire is an important part of providing you with the best health care possible. Your answers will help in understanding problems that you may have. Please answer every question to the best of your ability unless you are requested to skip over a question.

DATE _____ NAME _____ AGE _____ SEX: Female   Male

**1. During the *last 4 weeks*, how much have you been bothered
by any of the following problems?**

	Not bothered at all	Bothered a little	Bothered a lot
a. Stomach pain	[ ]	[ ]	[ ]
b. Back pain	[ ]	[ ]	[ ]
c. Pain in your arms, legs, or joints (knees, hips, etc.).	[ ]	[ ]	[ ]
d. Menstrual cramps or other problems with your periods	[ ]	[ ]	[ ]
e. Pain or problems during sexual intercourse.	[ ]	[ ]	[ ]
f. Headaches	[ ]	[ ]	[ ]
g. Chest pain	[ ]	[ ]	[ ]
h. Dizziness	[ ]	[ ]	[ ]
i. Fainting spells.	[ ]	[ ]	[ ]
j. Feeling your heart pound or race	[ ]	[ ]	[ ]
k. Shortness of breath	[ ]	[ ]	[ ]
l. Constipation, loose bowels, or diarrhea	[ ]	[ ]	[ ]
m. Nausea, gas, or indigestion	[ ]	[ ]	[ ]

**2. Over the *last 2 weeks*, how often have you been bothered
by any of the following problems?**

	Not at all	Several days	More than half the days	Nearly every day
a. Little interest or pleasure in doing things	[ ]	[ ]	[ ]	[ ]
b. Feeling down, depressed, or hopeless.	[ ]	[ ]	[ ]	[ ]
c. Trouble falling or staying asleep, or sleeping too much.	[ ]	[ ]	[ ]	[ ]
d. Feeling tired or having little energy	[ ]	[ ]	[ ]	[ ]

851

e. Poor appetite or overeating . . . . [ ]   [ ]   [ ]   [ ]

f. Feeling bad about yourself - or that you are a failure or have
let yourself or your family down . . . . [ ]   [ ]   [ ]   [ ]

g. Trouble concentrating on things, such as reading the
newspaper or watching television. . . . . [ ]   [ ]   [ ]   [ ]

h. Moving or speaking so slowly that other people could have
noticed? Or the opposite - being so fidgety or restless that
you have been moving around a lot more than usual . . . . [ ]   [ ]   [ ]   [ ]

i. Thoughts that you would be better off dead or of hurting
yourself in some way . . . . [ ]   [ ]   [ ]   [ ]

FOR OFFICE CODING:
Som Dis if at least 3 of #1a–m are "a lot" and lack an adequate biol explanation. Maj Dep Syn if #2a or b and 5 or more of #2a–i are at least "More than half the days" (count #2i if present at all).
Other Dep Syn if #2a or b and 2, 3, or 4 of #2a–i are at least "More than half the days" (count #2i if present at all).

		NO	YES
3. Questions about anxiety.			
a. In the *last 4 weeks*, have you had an anxiety attack—suddenly feeling fear or panic?.		[ ]	[ ]

If you checked "NO", go to question #5.

		NO	YES
b. Has this ever happened before?		[ ]	[ ]
c. Do some of these attacks come *suddenly out of the blue*—that is, in situations where you don't expect to be nervous or uncomfortable?		[ ]	[ ]
d. Do these attacks bother you a lot or are you worried about having another attack?.		[ ]	[ ]

		NO	YES
4. Think about your last bad anxiety attack.			
a. Were you short of breath?		[ ]	[ ]
b. Did your heart race, pound, or skip?		[ ]	[ ]
c. Did you have chest pain or pressure?.		[ ]	[ ]
d. Did you sweat?.		[ ]	[ ]
e. Did you feel as if you were choking?.		[ ]	[ ]
f. Did you have hot flashes or chills?.		[ ]	[ ]
g. Did you have nausea or an upset stomach, or the feeling that you were going to have diarrhea?. . .		[ ]	[ ]
h. Did you feel dizzy, unsteady, or faint?		[ ]	[ ]
i. Did you have tingling or numbness in parts of your body?		[ ]	[ ]
j. Did you tremble or shake?.		[ ]	[ ]
k. Were you afraid you were dying?.		[ ]	[ ]

	Not at all	Several days	More than half the days
5. Over the *last 4 weeks*, how often have you been bothered by any of the following problems?			
a. Feeling nervous, anxious, on edge, or worrying a lot about different things . .	[ ]	[ ]	[ ]

If you checked "Not at all", go to question #6.

	Not at all	Several days	More than half the days
b. Feeling restless so that it is hard to sit still.	[ ]	[ ]	[ ]
c. Getting tired very easily	[ ]	[ ]	[ ]
d. Muscle tension, aches, or soreness.	[ ]	[ ]	[ ]
e. Trouble falling asleep or staying asleep	[ ]	[ ]	[ ]
f. Trouble concentrating on things, such as reading a book, watching TV	[ ]	[ ]	[ ]
g. Becoming easily annoyed or irritable.	[ ]	[ ]	[ ]

FOR OFFICE CODING:
Pan Syn if #3a–d are all "Yes" and 4 or more of #4a–k are "Yes."
Other Anx Syn if #5a and answers to 3 or more of #5b–g are "more than half the days."

6. Questions about eating.

	NO	YES
a. Do you often feel that you can't control *what* or how *much* you eat?....................	[ ]	[ ]
b. Do you often eat, *within any 2-hour period*, what most people would regard as an unusually *large* amount of food?..............................	[ ]	[ ]

If you checked "NO" to either #6a or #6b, go to question #9.

	NO	YES
c. Has this been as often, on average, as twice a week for the last 3 months?	[ ]	[ ]

7. In the *last 3 months* have you *often* done any of the following in order to avoid gaining weight?

	NO	YES
a. Made yourself vomit?..	[ ]	[ ]
b. Took more than twice the recommended dose of laxatives?.......................	[ ]	[ ]
c. Fasted—not eaten anything at all for at least 24 hours?........................	[ ]	[ ]
d. Exercised for more than an hour specifically to avoid gaining weight after binge eating?	[ ]	[ ]

8. If you checked "YES" to any of these ways of avoiding gaining weight, were any as often, on average, as twice a week? ....

	NO	YES
	[ ]	[ ]

	NO	YES
9. Do you ever drink alcohol (including beer or wine)?.........................	[ ]	[ ]

If you checked "NO," go to question #11.

10. Have any of the following happened to you *more than once in the last 6 months*?

	NO	YES
a. You drank alcohol even though a doctor suggested that you stop drinking because of a problem with your health....................................	[ ]	[ ]
b. You drank alcohol, were high from alcohol, or hung over while you were working, going to school, or taking care of children or other responsibilities........................	[ ]	[ ]
c. You missed or were late for work, school, or other activities because you were drinking or hung over	[ ]	[ ]
d. You had a problem getting along with other people while you were drinking.................	[ ]	[ ]
e. You drove a car after having several drinks or after drinking too much	[ ]	[ ]

11. If you checked off *any* problems on this questionnaire, how *difficult* have these problems made it for you to do your work, take care of things at home, or get along with other people?

Not difficult at all	Somewhat difficult	Very difficult	Extremely difficult
[ ]	[ ]	[ ]	[ ]

FOR OFFICE CODING:
Bul Ner if #6a, b, and c and #8 are all "Yes"; Bin Eat Dis is the same but #8 either "No" or left blank.
Alc Abu if any of #10a–e are "Yes."

12. During the *last 4 weeks*, how much have you been bothered by any of the following problems?

	Not bothered at all	Bothered a little	Bothered a lot
a. Worrying about your health......................................	[ ]	[ ]	[ ]
b. Your weight or how you look....................................	[ ]	[ ]	[ ]
c. Little or no sexual desire or pleasure during sex........................	[ ]	[ ]	[ ]
d. Difficulties with husband/wife, partner/lover, or boyfriend/girlfriend	[ ]	[ ]	[ ]
e. The stress of taking care of children, parents, or other family members.......	[ ]	[ ]	[ ]
f. Stress at work or outside the home or at school	[ ]	[ ]	[ ]

g. Financial problems or worries . . . . [ ]   [ ]   [ ]

h. Having no one to turn to when you have a problem . . . . [ ]   [ ]   [ ]

i. Something bad that happened *recently* . . . . [ ]   [ ]   [ ]

j. Thinking or dreaming about something terrible that happened to you
in the past—like your house being destroyed, a severe accident, being hit or
assaulted, or being forced to commit a sexual act . . . . [ ]   [ ]   [ ]

	NO	YES
13. In the last year, have you been hit, slapped, kicked, or otherwise physically hurt by someone, or has anyone forced you to have an unwanted sexual act?	[ ]	[ ]

14. What is the most stressful thing in your life right now?

_____

_____

	NO	YES
15. Are you taking any medicine for anxiety, depression or stress?	[ ]	[ ]

16. FOR WOMEN ONLY: Questions about menstruation, pregnancy and childbirth.

a. Which best describes your menstrual periods?

____ Periods are unchanged

____ No periods because pregnant or recently gave birth

____ Periods have become irregular or changed in frequency, duration or amount

____ No periods for at least a year

____ Having periods because taking hormone replacement (estrogen) therapy or oral contraceptive

	NO	YES
b. During the week before your period starts, do you have a *serious* problem with your mood - like depression, anxiety, irritabilityanger or mood swings?	[ ]	[ ]
IF YES: Do these problems go away by the end of your period?	[ ]	[ ]
c. Have you given birth within the last 6 months?	[ ]	[ ]
d. Have you had a miscarriage within the last 6 months?	[ ]	[ ]
e. Are you having difficulty getting pregnant?	[ ]	[ ]

Validation:
Spitzer RL, et al. Validation and utility of a self-report version of PRIME-MD: the PHQ Primary Care Study. *JAMA* **282**(18):1737(1999).

(DO NOT DISTRIBUTE THIS PAGE TO THE PATIENT)

Quick Guide to the Patient Health Questionnaire

Purpose. The Patient Health Questionnaire (PHQ) is designed to facilitate the recognition and diagnosis of the most common mental disorders in primary care patients. For patients with a depressive disorder, a PHQ Depression Severity Index score can be calculated and repeated over time to monitor change.

Who Should Take the PHQ. Ideally, the PHQ should be used with all new patients, all patients who have not completed the questionnaire in the last year, and all patients suspected of having a mental disorder.

Making a Diagnosis. Since the questionnaire relies on patient self-report, definitive diagnoses must be verified by the clinician, taking into account how well the patient understood the questions in the questionnaire, as well as other relevant information from the patient, his or her family, or other sources.

Interpreting the PHQ. To facilitate interpretation of patient responses, all clinically significant responses are found in the column farthest to the right. (The only exception is for suicidal ideation when diagnosing a depressive syndrome.) In addition, the diagnoses of Major Depressive *Disorder* (rather than *Syndrome*) and Other Depressive *Disorder* require ruling out normal bereavement (mild symptoms, duration less than 2 months), a history of a manic episode (Bipolar Disorder) and a physical disorder, medication or other drug as the biological cause of the depressive symptoms. Similarly, the diagnoses of Panic *Disorder* and Other Anxiety *Disorder* require ruling out a physical disorder, medication or other drug as the biological cause of the anxiety symptoms. At the bottom of each page, beginning with "FOR OFFICE CODING," in small type, are criteria for diagnostic judgments for summarizing the responses on that page. The names of the categories are abbreviated, for example, Major Depressive Syndrome is Maj Dep Syn.

Page 1

Somatoform Disorder if at least 3 of #1a–m bother the patient "a lot" and lack an adequate biological explanation.

Major Depressive Syndrome if #2a or b and 5 or more of #2a–i are at least "more than half the days" (count #2i if present at all).

Other Depressive Syndrome if #2a or b and 2, 3, or 4 of #2a–i are at least "more than half the days" (count #2i if present at all).

Note: The diagnoses of Major Depressive *Disorder* and Other Depressive *Disorder* require ruling out normal **bereavement (mild symptoms, duration less than 2 months)**, a history of a **manic** episode (Bipolar Disorder) and a **physical disorder, medication or other drug** as the biological cause of the depressive symptoms.

Page 2

Panic Syndrome if #3a–d are all "Yes" and 4 or more of #4a–k are "Yes."

Other Anxiety Syndrome if #5a and answers to 3 or more of #5b–g are "more than half the days."

Note: The diagnoses of Panic *Disorder* and Other Anxiety *Disorder* require ruling out a **physical disorder, medication or other drug** as the biological cause of the anxiety symptoms.

Page 3

Bulimia Nervosa if #6a,b, and c and #8 are "Yes"; Binge Eating Disorder the same but #8 is either "NO" or left blank.
Alcohol abuse if any of #10a–e are "Yes."

Additional Clinical Considerations. After making a provisional diagnosis with the PHQ, there are additional clinical considerations that may affect decisions about management and treatment:

*Have current symptoms been triggered by psychosocial **stressor(s)**?*

*What is the **duration** of the current disturbance and has the patient received any **treatment** for it?*

*To what extent are the patient's symptoms **impairing** his or her usual work and activities?*

*Is there a **history** of similar episodes, and were they **treated**?*

*Is there a **family history** of similar conditions?*

112

TRAUMA AND POST-TRAUMATIC STRESS DISORDER

GIORA PILLAR
Technion–Israel Institute of Technology, Haifa, Israel

LOUISE HARDER
Spokane, Washington

ATUL MALHOTRA
Brigham and Women's Hospital, Boston, Massachusetts

INTRODUCTION AND DEFINITION

Post-traumatic stress disorder (PTSD), first defined in the DSM-III in 1980, is becoming an increasingly recognized diagnosis, particularly since the terrorist attacks of September 11, 2001. It occurs in individuals exposed to a traumatic event involving actual or threatened death or serious physical injury to the individual or others. Typically the event involves a subjective response of fear, helplessness, or horror. The trauma is one that is beyond the scope of normal life events for the individual, such as rape, burglary, war, torture, accidents, or other catastrophic events. The prevalence of the disorder varies between studies [1] but is still felt to be underdiagnosed by many. In the United States, a lifetime prevalence of PTSD in the general population has been found to be between 1% and 9%, with the condition being twice as common in women as in men. Symptoms that do not meet the full criteria for PTSD appear to be quite common in the general population. In unselected victims of major trauma populations, 20–45% will develop PTSD. Among combat soldiers, a PTSD prevalence of 15–20% has been reported. After exposure to lesser traumas and among well-trained corps, 5–10% may develop PTSD. Over long periods, the prevalence of PTSD in a given traumatized population tends to diminish. Predictive factors related to PTSD are very complex. Research into risk factors is ongoing, and a consensus is emerging that the three factors of genetics, the nature of the trauma, and the recovery environment all contribute to an individual's level of vulnerability or risk for PTSD.

Two aspects of subjective sleep-related symptoms are characteristic of PTSD and thus included in the DSM-IV diagnostic criteria. These are the reexperiencing cluster (criteria B) and the hyperarousal cluster (criteria D). Both are based on subjective reports by the patient and do not require any objective verification to make the diagnosis. Therefore, in the current chapter, we first review the subjective/questionnaire studies of traumatized patients and subsequently review the findings in objective studies of these patients.

QUESTIONNAIRE AND SUBJECTIVE COMPLAINTS

There is abundant evidence based on self-reported symptoms, questionnaires, or structured medical interviews reporting on the association between PTSD and complaints of sleep disturbances. The most frequent self-related complaints of these patients are difficulties initiating and maintaining sleep, shorter sleep duration, restless sleep, daytime fatigue, and especially nightmares. These were reported both in adults [2–8] and in children [9, 10]. Clearly all persons exposed to trauma do not go on to develop

PTSD, but which factors increase the risk of developing this disorder are unknown. Premorbid personality or early childhood events have been proposed as possible predictors [9, 11–13]. There is an emerging consensus that individual genetics, the nature of the trauma, and the post-trauma recovery environment all contribute to an individual's relative risk for PTSD. Psychiatric comorbidities are common in patients with PTSD. According to the National Comorbidity Survey [14], 88.3% of men and 79% of women with PSTD have at least one additional psychiatric disorder. In addition, a significant percentage of both men and women with PTSD have an alcohol or illicit substance abuse problem.

Insomnia, both sleep onset and sleep maintenance, plus nightmares are the most consistent findings in questionnaire-based studies investigating PTSD sleep complaints. Neylan et al. [4] reanalyzed the National Vietnam Veterans Readjustment Study and found that 44% of veterans with PTSD reported difficulty falling asleep, versus 6% of veterans without PTSD. Ninety-one percent with PTSD stated they had difficulty maintaining sleep, compared to 63% of veterans without PTSD. Nightmares were reported by 52% of the veterans with PTSD, and 6% of veterans without this diagnosis. Interviews with 41 survivors of imprisonment by the Japanese during World War II revealed that almost all these individuals, even 40 years after the traumatic event, suffered from sleep disturbances marked by recurrent nightmares [15]. Kuch and Cox [16] reported that 119 out of 124 Holocaust survivors complained of insomniac sleep disturbances (96%), and 103 of these survivors reported recurrent nightmares (83%). Similarly, 42 other survivors of the Holocaust showed decreased sleep quality compared to nontraumatized controls in all dimensions of the Pittsburgh Sleep Quality questionnaire [5]. Sleep disturbances in the period immediately following trauma may also have prognostic significance for the development of PTSD. Klein et al. [17], using the Mini Sleep Questionnaire, assessed 102 traffic accident survivors and 19 matched controls admitted to an orthopedic unit for elective surgery. At 1 week there were no significant differences between the two groups, but using logistic regression, the authors found that those patients with sleep disturbances 1 month post-trauma were significantly more likely to develop PTSD by 12 months post trauma.

While subjective questionnaire-type studies have been the most commonly used research tool in PTSD, they are limited in several ways. First, many studies did not optimally quantify the subjective complaints on a numeric scale and thus the results could be classified as descriptive rather than quantitative. Second, most studies involve small patient samples and frequently lack a control group. Thus considerable bias may be present in these studies. Third, many of the studies examined patients with a known diagnosis of PTSD, rather than looking at an entire group

post-trauma. Therefore distinguishing PTSD-related complaints from preexisting ones becomes difficult if not impossible. Finally, acute stress after trauma, with its attendant sleep complaints, is common and probably a normal response. By DSM-IV criteria, symptoms must persist for 30 days for a diagnosis of PTSD to be made. Despite this, many studies have not differentiated between the acute and chronic reactions to trauma. Nevertheless, insomnia and nightmares are two of the most common complaints of chronic PTSD patients. Patients frequently describe extremely long sleep latencies (frequently more than 2 hours) and estimate their sleep efficiencies as less than 50% (e.g., being awake for more than half of the time in bed). Because of the inherent flaws of subjective questionnaires, several investigators have attempted to quantify the sleep disturbances of PTSD with more objective data, using polysomnographic studies and ambulatory monitoring, as described next.

POLYSOMNOGRAPHIC STUDIES: IMPACT ON SLEEP ARCHITECTURE

Polysomnographic (PSG) studies of PTSD patients' sleep architecture have produced conflicting results. It seems that PTSD patients tend to sleep only slightly worse compared to controls. In many studies, differences in PSG measures between PTSD patients and controls have been negligible or were not statistically significant. Table 112.1 summarizes the findings of studies done by groups who have extensively studied PTSD patients in sleep laboratories. These include Lavie's group in Israel (comprising traffic accidents, war-related and Holocaust survivor patients); Mellman's group in Miami (mainly veterans and survivors of Hurricane Andrew); Ross's group in Philadelphia (veterans); Woodward and colleagues from Palo Alto (veterans); Hurwitz and colleagues from Minneapolis (veterans); and Dow and colleagues from San Diego (veterans).

Summarizing data from 7 PSG studies of 88 PTSD patients in Israel leads to the conclusion that PSG sleep architecture in PTSD patients in their chronic stage does not substantially differ from control subjects. Sleep study data frequently do not objectively support the patients' complaints. When comparing PTSD patients to controls, their sleep latency was 9 minutes longer, and sleep efficiency was 5% lower. The awake time during the night was higher among PTSD patients than controls (12% versus 6%). Thus the differences are subtle though consistently trending toward worse sleep in PTSD patients. There was, however, a very high variability in terms of the sleep stage distribution results. Individuals with PTSD showed that sleep architectures varied between extremely diminished slow-wave sleep to extremely high values. On the basis of

TABLE 112.1   Effect of PTSD on General Sleep Measures: Summary of Studies

Group	N	TST (min)	Sleep Latency (min)	Stage 3/4 (%)	Awakenings n	Awakenings % Sleep	Sleep Efficiency (%)
Lavie, Israel	88	335	24	22		12	84
Controls	60	345	15	19		6	89
Mellman, Miami	34	339	26	15	15	14	85
Controls	18	348	26	15	7	7	88
Ross, Philadelphia	11	378	21	7		6	89
Controls	8	410	10	11		6	93
Woodward, Palo Alto	87	~330	~8	~9			~90
Controls	14	~320	~8	~14			~91
Hurwitz, Minneapolis	18	397	13	13	14	18	82
Controls	10	403	10	13	15	15	85
Gillin, San Diego	14	370	13	6			90
Depressed controls	15	318	38	8			82
Normal controls	12	309	16	10			85

all these studies, slow-wave sleep percentage was slightly higher among PTSD patients than controls (22% versus 19%, Table 112.1). Despite selecting patients without other psychiatric diagnoses or medication usage, the patients showed significantly higher depression and anxiety scores based on several clinical questionnaires. Thus it seems that excluding these factors and examining the net effect of PTSD on sleep is very difficult.

Conceivably, some of the sleep architecture abnormalities might be related to comorbidities rather than to a direct effect of PTSD on sleep. Also, it should be mentioned that the reproducibility of sleep architecture and percentage of slow-wave sleep is variable even among normal populations.

Mellman et al. [3, 18–20] published several papers with PSG data on PTSD patients. Three of these studies were done on veterans with PTSD, and apparently there is some overlap of data within these papers. The fourth one compared 10 patients affected by hurricane Andrew (6 with complete PTSD and 4 with "partial" PTSD) to 9 healthy controls (3 of whom were examined prior to the study). Taking all these studies together, PTSD patients showed more awakenings from sleep and increased nighttime awake, but this only slightly affected the sleep efficiency. Average sleep efficiency was 85% for the study group and 88% for controls. In one of these studies, the results were also compared to a major depression (MD) group. Five out of 24 PTSD patients met criteria for MD as well [20]. Thus, again, controlling for associated psychiatric comorbidity was not ideal. The PTSD group, however, showed reduced total sleep time, reduced sleep efficiency, longer sleep latency, and increases in awakenings compared to both major depression and control groups. The percentage of slow-wave sleep did not differ significantly between the groups.

Hurwitz and colleagues studied 18 PTSD outpatients (Vietnam combat veterans) and 10 controls for two nights

in the lab, followed by a MSLT [21]. They found no significant PSG measure differences between subjects and controls, except for lower arousals/hour from stage 3/4 on night 2, and lower subjectively estimated total sleep time ($p < 0.005$) in the PTSD subjects. No daytime hypersomnolence was detected. They concluded that polysomnographically recorded sleep was notably better than expected in the presence of clinically significant PTSD including typical histories of disrupted sleep. As their subjects consistently underestimated their total sleep time and overestimated their sleep latency, the authors suggested that PTSD patients demonstrated a sleep state misperception.

Ross et al. [22] found no significant sleep architecture differences between 11 PTSD patients, all of whom suffered from concurrent psychiatric diseases, and 8 healthy controls. This finding is surprising, as both depression and anxiety are known to cause sleep disorders. Depression has been consistently reported to be associated with reduced sleep efficiency, reduced total sleep time, reduced slow-wave sleep, increased sleep latency, and some changes in REM sleep (see later discussion). Anxiety can result in similar sleep disturbances.

Woodward and colleagues studied a large group of PTSD patients for several nights. In one study, they found that overall sleep architecture did not differ significantly between 87 PTSD patients and 14 controls [23]. An additional study, in which the authors tested the effect of depression with coexistent PTSD, showed that 17 PTSD patients with depression had a significantly lower percentage of slow-wave sleep than 10 PTSD patients without depression [24]. Sleep latency and sleep efficiency were within normal limits in both groups. Thus PTSD alone did not significantly adversely affect sleep in this study. When coexisting with depression, however, reduced deep sleep was noted.

In another effort to dissociate the effects of depression from that of PTSD on sleep, Dow et al. [25] compared sleep and dreams of three groups of subjects: (1) Vietnam veterans with PTSD and major depression, (2) veterans with depression alone, and (3) veterans with neither PTSD nor depression (i.e., normal controls). Sleep recordings indicated only one significant difference between the PTSD/depressed and solely depressed groups: sleep latency was prolonged in the solely depressed patients compared with the other two groups. The two patient groups differed from controls in the manner already reported for depressed patients (decreased REM latency, increased REM density, reduced total sleep time, reduced sleep efficiency).

Thus, considering all of these studies together, PTSD itself does not dramatically adversely affect objective sleep architecture. On average, PSG recordings show that sleep is somewhat more disturbed in PTSD patients compared to controls, but in many of the comparisons, the differences were not statistically significant. In studies in which such differences were found, they might be related either to the acute phase of the PTSD syndrome or to comorbid psychiatric illness, as will be further discussed later.

AMBULATORY MONITORING

Most evaluations of sleep disturbances in PTSD patients have been done in the lab. To our knowledge, ambulatory PSG techniques have never been used systematically on PTSD patients. Actigraphic monitoring is one objective method of monitoring sleep. This method is advantageous because it allows in-home study, for several consecutive nights, with only minor disruption to the patient's life. Actigraphy is, however, unable to detect stage-specific abnormalities such as awakenings from REM or altered REM latency. Unfortunately, the published data using this method are rather sparse. Kaufman et al. [26] actigraphically studied a group of adolescents complaining of substantial sleep disorders 3 years following minor head injury and compared them with an age- and gender-matched control group. The study group objectively had mild prolonged sleep latency and reduced sleep efficiency [26]. Glod et al. [10] used actigraphic measures to study 19 abused children and compared them to 15 nonabused normal controls and 10 depressed children. These results suggested that abused children had prolonged sleep latency and decreased sleep efficiency. Although the data are limited to these few small studies, sleep latency is likely somewhat longer in post-traumatic children and adolescents, and sleep efficiency is mildly reduced. In the above-mentioned actigraphic study examining adults with PTSD, there were no statistically significant sleep architecture differences compared to a control group, which is consistent with most laboratory investigations using PSG.

EFFECTS ON REM PARAMETERS AND DREAMING

Dream Content and Dream Recall

Most dreams are thought to occur during REM sleep; therefore several investigators have looked at the effects of PTSD on dreams and REM sleep. In a recent large-scale study by Breslau et al. [27], the single objective characteristic of PTSD was increased arousals from REM sleep. The reexperiencing of a traumatic event in the form of repetitive nightmares or anxiety dreams is considered to be one of the cardinal manifestations of PTSD. Dysfunctional REM sleep mechanisms may be involved in the pathogenesis of the post-traumatic anxiety dream and may in fact be relatively specific to this disorder [3, 22, 28].

Subjective reports indicate that trauma-related anxiety dreams are the most consistent complaints among PTSD patients [2–8, 29, 30]. It is difficult to accurately evaluate anxiety dreams and nightmares in the lab. Occasionally, a patient might experience such a dream during a sleep study, making interpretation straightforward [22, 31]. But this is not usually the case. Researchers have attempted to use alternate measures, such as "REM density," or awakening patients out of REM sleep and questioning them about the contents of their dreams. This technique allows assessment of both dream content and dream recall rate. Using this technique, some authors have reported that "less-adjusted" Holocaust survivors had significantly more anxiety dreams—some explicitly related to the Holocaust—than "well-adjusted" survivors or controls [32]. As well, dream hostility was significantly correlated with patients' anxiety and general SCL-90 scores [33]. In addition, patients with war-related PTSD had more aggressive dreams than normal controls. Also, the level of dream aggression was significantly correlated with the clinical condition of the patient [34]. The finding of threatening, anxiety-provoking dreams in PTSD is consistent [6, 35, 36]. Surprisingly, perhaps, although these patients describe vivid and stressful dreams, several reports have shown that they actually have lower dream recall than do controls. Normal patients awakened from REM sleep will recall between 60% and 90%, but PTSD patients remembered only 20–40% of the dreams [31, 37, 38]. Freud theorized that emotional conflicts might be dealt with during dreaming, and that the dream functions to protect sleep. This model was further supported by the hypothesis that the dream develops in the neonate during the arousal of the REM state to protect sleep from burgeoning object-directed impulses [39]. If dreams function to protect sleep, or to solve problems for the patient, then the dream that does the opposite, repeating and disturbing sleep, seems to contradict this theory. It was previously postulated that the low dream recall found in PTSD might result from active attempts to suppress anxiety dreams. When "well-adjusted" and "less-adjusted" Holocaust survivors and controls were compared, dream recall rate was 33.7% in the "well-adjusted" group, 50.5% in the "less-adjusted" group, and 80% in controls. It is possible that the "well-adjusted" Holocaust survivors had a significantly lower dream recall than "less-adjusted" ones because they were better able to actively suppress the anxiety dreams [32]. Furthermore, PTSD patients had elevated awakening thresholds from REM sleep, conceivably due to an additional defense mechanism to avoid recall of these dreams [34]. Other studies have shown no differences in the rate of dream recall between patients with PTSD and controls, but in general the dreams of PTSD patients are more emotional and threatening than those of patients without PTSD.

REM Density

REM density refers to the frequency of rapid eye movements during sleep. It has been previously noted that the hostile and threatening dreams reported by PTSD patients have higher REM density [18]. Therefore measuring REM density might be useful as a surrogate measure of traumatic dreams, without the need to actively awaken subjects and interfere with their natural sleep. Indeed, several relevant studies in PTSD patients have reported increases in REM density [18–20, 24, 31]. However, increased REM density is not specific to PTSD, but also occurs in other psychiatric disorders, particularly in major depression (MD). In fact, Dow and colleagues found increased REM density in both PTSD with MD and MD alone (without PTSD) and speculated that this finding in PTSD reflects actual depression [25]. In contrast, Woodward et al. [24] compared PTSD + MD patients to PTSD patients without MD and found that REM density was similar in the two groups, and they concluded that increased REM density is associated with PTSD specifically, rather than to the MD.

Phasic Events

Ross and colleagues [6] reported that tonic and phasic REM sleep measures in PTSD subjects were significantly elevated. This led them to suggest that dysregulation of the REM sleep control system (particularly phasic event generation) is at the basis of PTSD. These were both eye movements (REM density) and REM sleep phasic leg activity [22, 40]. Increases in REM sleep movements have been reported also by other groups [2, 41]. While normally REM sleep is associated with spinal postsynaptic inhibition of motor neurons (e.g., REM atonia), PTSD patients exhibit bursts of motor unit action potentials. In addition, the finding that these patients show exaggerated abnormal startle response even while awake, might further link PTSD to REM sleep dysfunction [42–45]. PTSD occurring in the aftermath of an overwhelming psychological stressor

could reflect plasticity in brainstem systems controlling REM sleep phasic activity [40]. Lesions of these brain regions in animals (pedunculopontine tegmental nucleus) have been shown to result in dysinhibition of acoustic startle, as well as activation of REM sleep mediated by cholinergic neurons [46]. Thus the CNS processes generating REM sleep may participate in the control of the classical startle response, which may be akin to the startle behavior commonly described in PTSD patients [6]. Therefore it seems possible that "anxiety dreams," increased REM phasic activity, and increased startle response are results of an as yet undefined abnormally activated mechanism in PTSD patients. If this is the case, other changes in REM sleep regulation in these patients would be expected as well. Unfortunately, the available data on REM latencies and amounts in PTSD patients are variable and inconsistent.

REM Latency and Percentage

Speculating that PTSD may be fundamentally a disorder of REM sleep mechanisms, several studies have examined REM parameters in these patients. Studies of Vietnam War veterans [47] and Yom Kippur War veterans [48] showed prolonged REM latencies. Some other studies have shown prolonged REM latency in PTSD [31, 41, 49], but others have instead found rather shortened REM latencies [37, 50, 51]. Interestingly, Mellman et al. [20] in 1997 found that individuals with PTSD had both the highest and the lowest values for REM latency compared to those with MD. In addition, amount of REM sleep as a percentage of total sleep time has varied between different studies. Some have reported shorter REM time during sleep [2, 31, 41, 49], but others have found either normal or prolonged REM duration [3, 22, 40]. The major problem with most of these studies is the lack of control for comorbidities. Since the association between depression and reduced REM latency as well as increased REM percentage is well established, the reports of changes in these variables in PTSD should be carefully taken unless the study was well controlled for depression. Nevertheless, these data suggest that REM sleep latency and percentage may change in PTSD.

EVENTS DURING SLEEP

There are three separate types of events during sleep that we consider when discussing PTSD: those occurring during REM sleep, those involving arousals and body and leg movements, and those occurring during deep sleep.

Movements During Sleep

Several studies of PTSD patients have shown an increased incidence of limb movement during sleep. This activity occurs during both REM and NREM sleep. Using questionnaires, Mellman et al. [3] studied 37 veterans with PTSD and 21 veterans without PTSD with subjective questionnaires and did not find statistically significant differences in sleepwalking or sleep talking frequencies between the two groups. However, those with PTSD reported more body movements and night terrors (defined as screaming or shaking while sleeping) [3]. Many other authors have confirmed an increased EMG, body movements, and periodic limb movements (PLMs) among individuals with PTSD [2, 40, 41, 52, 53].

Limb movements and sleep disruption generally occur in stages 1 and 2 of NREM sleep, and not usually in REM sleep. In addition to these disturbances of light sleep, PTSD patients typically show increases in confusional arousals, sleepwalking, and night terrors, which occur more frequently in deep sleep [10, 54, 55]. PTSD patients commonly suffer from movement disorders in all stages of sleep, which may result from more than one mechanism—a state of hyperarousal arising from associated anxiety or depression, and perhaps an active mechanism to deepen their sleep. The latter might also explain the high frequency of apneas reported in these patients.

Apneas

There are several anecdotal reports on higher frequency of apneas in PTSD patients [34, 56–58]. In addition, Roy et al. [59] found that out of 3719 consecutive Persian Gulf War veterans who were diagnosed with an ICD-9 code (*International Classification of Disease*, 9th ed.) taken from a total of 21,579 Persian Gulf War veterans, 276 had obstructive sleep apnea (7.4%). Krakow et al. [53] reported sleep disordered breathing in 81 of 156 patients (52%). These findings might be only coincidental but, on the other hand, there might be a specific PTSD-related abnormality that results in apneas.

It has been previously suggested that a change in the sleep depth, or in the control of sleep due to an active mechanism in an attempt to suppress anxiety dreams, might result in an increase in the frequency of sleep disordered breathing events [34]. Normally during deep sleep the RDI decreases due to two potential mechanisms. (1) Hypercapnia seen at this stage may protect sleep from apneas (potentially by increasing ventilatory drive and upper airway muscle activation). (2) Since there are no apneas, the patient can fall into slow-wave sleep. In PTSD patients, they might have some underlying hyperventilation due to their anxiety—and thus relative hypocapnia and consequently less protective effect and apneas. This mechanism, however, is currently an untested hypothesis. However, if PTSD does result in increased apnea, then this can, in turn, worsen these patients' sleep. Thus the issue of sleep disordered breathing and respiratory control in PTSD has

not been examined specifically yet, but there is evidence that in some cases PTSD might result in higher vulnerability to apnea, possibly due to a change in mechanisms controlling sleep or ventilation changes.

Awakening Threshold

Based on the insomnia complaints of PTSD patients, some have assumed that their sleep must be a "light" sleep, vulnerable to environmental disturbances. This is congruent with the theory of the "hyperarousal" state they have during their sleep. However, there is evidence that when PTSD patients fall into either slow-wave sleep or REM sleep, they actually have deeper sleep than matched controls. Three independent studies investigating the objectively determined awakening threshold of these patients found that they actually have elevated awakening thresholds [34, 56, 60], both in REM and NREM sleep. This might lead to the conclusion that in spite of their general complaints concerning disturbed sleep, chronic PTSD patients in fact are more difficult to awaken from sleep than normals. A recent report demonstrating fewer arousals from REM and slow-wave sleep in PTSD patients compared to controls is in agreement with this conclusion [21]. This might be a unique specific finding in PTSD and the pathogenesis of this elevated awakening threshold has been postulated to be related to active blocking mechanisms invoked to suppress trauma-related anxiety-provoking materials during sleep [34, 56].

Thus, by undefined active mechanisms, PTSD may deepen their sleep and this in turn can result in the appearance of disorders of arousal. This might also explain the increased percentage of slow-wave sleep reported in several studies in these patients [2, 31, 32], although this finding has not been consistent [3, 18–20].

POSSIBLE PATHOPHYSIOLOGICAL MECHANISMS

Hyperarousal State Versus Sleep Deepening

There are several suggested mechanisms to explain the findings seen in the sleep of PTSD patients. The most common one, which is supported both by laboratory data and by clinical correlates, is the hyperarousal theory. Some laboratory evidence in support of the hyperaroused state in PTSD is given by the findings of the nondiminished noradrenergic production in these patients during the night [19, 61]. Additional support for the role of abnormal noradrenergic CNS function (along with endogenous opioid release) in chronic PTSD was suggested by van der Kolk and colleagues [62]. As inescapable shock in animals leads to both transient catecholamine depletion and subsequent stress-induced analgesia, they postulated that the numbing and catatenoid reactions following trauma in humans are analogous. Although reexposure to trauma may produce a paradoxical sense of calm and control due to endogenous opioid release, a cessation of traumatic stimulation will be followed by symptoms of opioid withdrawal and physiological hyperreactivity mediated by CNS noradrenergic hypersensitivity. This hyperreactivity can, in turn, be temporarily modified by reexposure to trauma. This factor could account for voluntary reexposure to trauma in many traumatized individuals and would provide a complementary formulation to the conventional psychodynamic concept of attempted mastery of the psychosocial meaning of the trauma [62]. Moreover, as increased sensitivity and sensitization of the noradrenergic system may leave the individual in a hyperaroused, sleep-deprived state that worsens over time, it was hypothesized that to calm these symptoms of hyperarousal, PTSD patients often withdraw and use central nervous system depressants [63]. Hence some researchers believe that chronic PTSD patients have simultaneous coexistence of a hyperarousal state (abnormality of the CNS noradrenergic system), with pressure to REM and reexperience the trauma, and the balance between these determines the REM parameters. This hyperarousal state can explain the clinical correlate of these patients in terms of fragmented sleep and tendency toward insomnia.

Although "deepening" of sleep seems in contrast to the "hyperarousal" theory, this is not necessarily the case. As both the hyperarousal theory and the sleep-deepening mechanism are supported by clinical and laboratory findings, it may be possible that they exist simultaneously. Conceivably, the increased tone of the brain's catecholamines sets a "sleep threshold" higher, and thus PTSD patients tend to be insomniac, but once they fall asleep, their sleep is deeper, with an elevated awakening threshold.

If this theory were true, and these patients suffer from a chronic hyperarousal state and, consequently, sleep and REM deprivation, it can further explain the REM sleep abnormalities found in these patients. As 5 out of 11 PTSD patients showed REM onset sleep during MSLTs, it was interpreted to suggest an abnormally increased pressure to REM during the day, due to partial REM deprivation during nights [64]. Moreover, as individuals with PTSD had both the highest and the lowest values for REM latency, it was speculated that this extreme variability might arise from the coexistence of pressure for REM to occur, with heightened arousal at night inhibiting the onset of REM [64]. This, however, cannot explain the findings of increased REM density, REM phasic events, or effects on dream recall. An alternative or complementary theory came from Ross's lab, describing a REM dysfunction theory in PTSD.

REM Dysfunction Theory

Since the frequency of nightmares is greatly increased in PTSD over other psychiatric disorders or in the general population, Ross and colleagues postulated that anxiety dreams, especially those repetitive replicative nightmares, are relatively specific for PTSD. Consequently, it led them to invoke a role for an abnormal REM mechanism in the disorder [6, 22, 65]. As PTSD patients suffer from higher rates of bad dreams despite a lower dream recall rate, it seems reasonable that it reflects a higher rate of the occurrence of these dreams rather than their recollection. Hence dreams that involve reliving of the trauma, experiencing strong emotions, that would have been appropriate short-term reactions to the traumatic event, persist into the chronic stage rather than diminish and disappear as expected from non-PTSD individuals who are exposed to a similar trauma. Moreover, if reexperiencing of the trauma is really a REM sleep disorder, Ross et al. [6] suggested that the flashbacks reported in these patients during the daytime might parallel the curve of REM sleep propensity, as measured by multiple naps. In addition, they found that tonic and phasic REM sleep measures in PTSD subjects were significantly elevated, which is further evidence for dysregulation of the REM sleep control system (particularly phasic event generation) in PTSD (see earlier discussion). These were both eye movements (REM density) and REM sleep phasic leg activity [22, 40]. As previously noted, increases in REM sleep movements have been reported by a number of groups [2, 41]. Finally, an additional finding that might link PTSD to REM sleep dysfunction is the exaggerated abnormal startle response seen in these patients even while awake [42–45]. PTSD occurring in the aftermath of an overwhelming psychological stressor could reflect plasticity in brainstem systems controlling REM sleep phasic activity [40]. Likewise, Kolb [66] postulates that the "constant" symptoms of PTSD are due to the changes in the agonistic neuronal system (as a result of a high-intensity stimulation), which impair cortical control of hindbrain structures concerned with aggressive expression and the sleep–dream cycle.

Therefore, in theory, anxiety-related dreams, increased REM phasic activity, and increased startle response may all be a result of an as yet unknown mechanism in PTSD patients. As some of these may be explained by increased catecholaminic concentration in the brain, it might be the same mechanism responsible for the REM dysfunction and the hyperarousal state.

Comorbidity

In the few studies that have tried to discriminate between the direct effect of the PTSD and that of comorbidity, it has been found that coexisting depression is more likely to cause the sleep disturbances than the PTSD itself [24, 25]. Therefore it seems reasonable that the extreme variability in sleep architecture findings in these patients might arise from different comorbid conditions. For instance, patients with predominant depression might exhibit predominantly short REM latency, increased REM density, and increased REM percentage, while those with predominant anxiety might suffer predominantly from difficulties initiating and maintaining sleep with decreased REM sleep. Additional potential causes of the variability in sleep disorders reported in PTSD might be related to the source of the trauma, medications, or alcohol consumption, or coexisting different sleep disorders (e.g., sleep apnea or leg movements).

The reason why PTSD patients suffer from sleep-state misperception (subjective complaints of disturbed sleep beyond objective findings) is unknown. However, it could also be related to comorbidity as well. For example, in one study the severity of sleep disturbances paralleled the severity of the clinical psychopathological picture of patients with PTSD [34]. In addition, Woodward et al. [24] showed that the variability of REM sleep was dramatically higher in the PTSD with depression group than PTSD alone. That might support the implication that PTSD patients show variable results in different studies due to comorbidity. This is further supported by three additional studies. In the most recent one, decreased REM latency and increased REM density were found in PTSD/depressed patients, as well as in depressed patients without PTSD, compared to the control group [25]. Another one found that decreased REM latency in eight patients with PTSD occurred at frequencies similar to those that have been found in patients with major affective disorder [50]. In a third study, it had been found that PTSD patients had shortened REM latency and altered TRH stimulation test. This was interpreted to suggest that PTSD and depression may share some pathophysiological abnormalities [51]. Thus, although it is possible that the initial trauma itself can change REM parameters, it cannot be refuted that the psychiatric comorbidity is the predominant effect on REM parameters, and that this is the reason for the high variability in findings across studies. Medication effects might also be considered, although the majority of the above-mentioned studies excluded patients who were on chronic medications. One final consideration is whether in fact comorbidities should be excluded, since depression and anxiety may lie upon the causal pathway of PTSD and its complications. Thus comorbidities may not be epiphenomena, but rather mechanistically important in PTSD complications.

CONCLUSIONS

In summary, anxiety dreams, increased REM phasic activity, increased arousals from REM sleep, increased startle response, low dream recall, and possibly elevated awakening thresholds from sleep may characterize PTSD. Other changes sometimes observed in PTSD, namely, changes in REM latency, REM time, and REM density, seem related to frequent comorbidities rather than to a direct effect of the trauma. It is possible, however, that a trauma-related simultaneous coexistence of pressure to REM and REM inhibition forces are responsible for these findings. Further extensive research is needed to examine the effect of the source of the trauma, as well as that of medications, prior to drawing final conclusions.

REFERENCES

1. Breslau N, Kessler R, Chilcoat H, Schultz L, Davis G, Andreski P. Trauma and posttraumatic stress disorder in the community: the 1996 Detroit Area Survey of Trauma. *Arch Gen Psychiatry* **55**:626–632(1988).

2. Lavie P, Hertz G. Increased sleep motility and respiration rates in combat neurotic patients. *Biol Psychiatry* **14**:983–987(1979).

3. Mellman TA, Kulick-Bell R, Ashlock LE, Nolan B. Sleep events among veterans with combat-related posttraumatic stress disorder. *Am J Psychiatry* **152**:110–115(1995).

4. Neylan TC, Marmar CR, Metzler TJ, et al. Sleep disturbances in the Vietnam generation: findings from a nationally representative sample of male Vietnam veterans. *Am J Psychiatry* **155**:929–933(1998).

5. Rosen J, Reynolds CFD, Yeager AL, Houck PR, Hurwitz LF. Sleep disturbances in survivors of the Nazi Holocaust. *Am J Psychiatry* **148**:62–66(1991).

6. Ross RJ, Ball WA, Sullivan KA, Caroff SN. Sleep disturbance as the hallmark of posttraumatic stress disorder. *Am J Psychiatry* **146**:697–707(1989).

7. Sadavoy J. Survivors. A review of the late-life effects of prior psychological trauma. *Am J Geriatr Psychiatry* **5**:287–301(1997).

8. Silva C, McFarlane J, Soeken K, Parker B, Reel S. Symptoms of post-traumatic stress disorder in abused women in a primary care setting. *J Womens Health* **6**:543–552(1997).

9. Famularo R, Fenton T. Early developmental history and pediatric posttraumatic stress disorder. *Arch Pediatr Adolesc Med* **148**:1032–1038(1994).

10. Glod CA, Teicher MH, Hartman CR, Harakal T. Increased nocturnal activity and impaired sleep maintenance in abused children. *J Am Acad Child Adolesc Psychiatry* **36**:1236–1243(1997).

11. Emery PE. The inner world in the outer world: the phenomenology of posttraumatic stress disorder from a psychoanalytic perspective. *J Am Acad Psychoanal* **24**:273–291(1996).

12. Helzer JE, Robins LN, McEvoy L. Post-traumatic stress disorder in the general population. Findings of the epidemiologic catchment area survey. *N Engl J Med* **317**:1630–1634(1987).

13. McCarren M, Goldberg J, Ramakrishnan V, Fabsitz R. Insomnia in Vietnam era veteran twins: influence of genes and combat experience. *Sleep* **17**:456–461(1994).

14. Kessler RC, Sonnega A, Bromet E, Hughes M, Nelson CB. Post-traumatic stress disorder in the National Comorbidity Survey. *Arch Gen Psychiatry* **52**(12):1048–1060(1995).

15. Goldstein G, van Kammen W, Shelly C, Miller DJ, van Kammen DP. Survivors of imprisonment in the Pacific theater during World War II. *Am J Psychiatry* **144**:1210–1213(1987).

16. Kuch K, Cox BJ. Symptoms of PTSD in 124 survivors of the Holocaust. *Am J Psychiatry* **149**:337–340(1992).

17. Klein E, Koren D, Arnon I, Lavie P. No evidence of sleep disturbance in post-traumatic stress disorder: a polysomnographic study in injured victims of traffic accidents. *Isr J Psychiatry Relat Sci* **39**(1):3–10(2002).

18. Mellman TA, David D, Kulick-Bell R, Hebding J, Nolan B. Sleep disturbance and its relationship to psychiatric morbidity after Hurricane Andrew. *Am J Psychiatry* **152**:1659–1663 (1995).

19. Mellman TA, Kumar A, Kulick-Bell R, Kumar M, Nolan B. Nocturnal/daytime urine noradrenergic measures and sleep in combat-related PTSD. *Biol Psychiatry* **38**:174–179(1995).

20. Mellman TA, Nolan B, Hebding J, Kulick-Bell R, Dominguez R. A polysomnographic comparison of veterans with combat-related PTSD, depressed men, and non-ill controls. *Sleep* **20**:46–51(1997).

21. Hurwitz T, Mahowald M, Kuskowski M, Engdahl B. Polysomnographic sleep is not clinically impaired in Vietnam combat veterans with chronic posttraumatic stress disorder. *Biol Psychiatr* **44**:1066–1073(1998).

22. Ross RJ, Ball WA, Dinges DF, et al. Rapid eye movement sleep disturbance in posttraumatic stress disorder. *Biol Psychiatry* **35**:195–202(1994).

23. Woodward SH, Bliwise DL, Friedman MJ, Gusman FD. First night effects in post-traumatic stress disorder inpatients. *Sleep* **19**:312–317(1996).

24. Woodward SH, Friedman MJ, Bliwise DL. Sleep and depression in combat-related PTSD inpatients. *Biol Psychiatry* **39**:182–192(1996).

25. Dow BM, Kelsoe JR Jr, Gillin JC. Sleep and dreams in Vietnam PTSD and depression. *Biol Psychiatry* **39**:42–50(1996).

26. Kaufman Y, Tzichinski O, Epstein R, Etzioni A, Lavie P, Pillar G. Sleep disorders in children in the long term after minimal head injury. *Pediatr Neurol* **24**:129–134(2001).

27. Breslau N, Roth T, Burduvali E, Kapke A, Schultz L, Roehrs T. Sleep in lifetime posttraumatic stress disorder: a community-based polysomnographic study. *Arch Gen Psychiatry* **61**:508–516(2004).

28. van der Kolk B, Blitz R, Burr W, Sherry S, Hartmann E. Nightmares and trauma: a comparison of nightmares after combat with lifelong nightmares in veterans. *Am J Psychiatry* **141**:187–190(1984).

29. Burstein A. Posttraumatic flashbacks, dream disturbances, and mental imagery. *J Clin Psychiatry* **46**:374–378(1985).

30. DeFazio VJ, Rustin S, Diamond A. Symptom development in Vietnam era veterans. *Am J Orthopsychiatry* **45**:158–163(1975).

31. Hefez A, Metz L, Lavie P. Long-term effects of extreme situational stress on sleep and dreaming. *Am J Psychiatry* **144**:344–347(1987).

32. Kaminer H, Lavie P. Sleep and dreaming in Holocaust survivors: dramatic decrease in dream recall in well-adjusted survivors. *J Nerv Ment Dis* **179**:664–669(1991).

33. Lavie P, Kaminer H. Sleep, dreaming and coping style in Holocaust survivors. In: Barrett D (Ed), *Trauma and Dreams*. Harvard University Press, Cambridge, MA, 1996, pp 100–103.

34. Lavie P, Katz N, Pillar G, Zinger Y. Elevated awaking thresholds during sleep: characteristics of chronic war-related posttraumatic stress disorder patients. *Biol Psychiatry* **44**:1060–1065(1998).

35. David D, Mellman TA. Dreams following Hurricane Andrew. American Psychiatric Association 148th Annual Meeting, 1995.

36. Benyakar M, Tadir M, Groswasser Z, Stern MJ. Dreams in head-injured patients. *Brain Inj* **2**:351–356(1988).

37. Greenberg R, Pearlman CA, Gampel D. War neuroses and the adaptive function of REM sleep. *Br J Med Psychol* **45**:27–33(1972).

38. Kramer M, Schoen LS, Kinney L. Psychological and behavioral features of disturbed dreamers. *Psychiatr J Univ Ottawa* **9**:102–106(1983).

39. Stein HH. The dream is the guardian of sleep. *Psychoanal Q* **64**:533–550(1995).

40. Ross RJ, Ball WA, Dinges DF, et al. Motor dysfunction during sleep in posttraumatic stress disorder. *Sleep* **17**:723–732(1994).

41. Schlosberg A, Benjamin M. Sleep patterns in three acute combat fatigue cases. *J Clin Psychiatry* **39**:546–549(1978).

42. Butler RW, Braff DL, Rausch JL, Jenkins MA, Sprock J, Geyer MA. Physiological evidence of exaggerated startle response in a subgroup of Vietnam veterans with combat-related PTSD. *Am J Psychiatry* **147**:1308–1312(1990).

43. Ornitz EM, Pynoos RS. Startle modulation in children with posttraumatic stress disorder. *Am J Psychiatry* **146**:866–870 (1989).

44. Shalev AY, Orr SP, Peri T, Schreiber S, Pitman RK. Physiologic responses to loud tones in Israeli patients with posttraumatic stress disorder. *Arch Gen Psychiatry* **49**:870–875(1992).

45. Shalev AY, Peri T, Orr SP, Bonne O, Pitman RK. Auditory startle responses in help-seeking trauma survivors. *Psychiatry Res* **69**:1–7(1997).

46. Swerdlow NR, Geyer MA. Prepulse inhibition of acoustic startle in rats after lesions of the pedunculopontine tegmental nucleus. *Behav Neurosci* **107**:104–117(1993).

47. Kramer M, Kinney L. Sleep patterns in trauma victims with disturbed dreaming. *Psychiatr J Univ Ottawa* **13**:12–16(1988).

48. Lavie P, Hefez A, Halperin G, Enoch D. Long-term effects of traumatic war-related events on sleep. *Am J Psychiatry* **136**:175–178(1979).

49. Glaubman HM, Miculincer M, Porat A, Wasserman O, Birger M. Sleep of chronic posttraumatic patients. *J Traumatic Stress* **3**:255–263(1990).

50. Kauffman CD, Reist C, Djenderedjian A, Nelson JN, Haier RJ. Biological markers of affective disorders and posttraumatic stress disorder: a pilot study with desipramine. *J Clin Psychiatry* **48**:366–367(1987).

51. Reist C, Kauffmann CD, Chicz-Demet A, Chen CC, Demet EM. REM latency, dexamethasone suppression test, and thyroid releasing hormone stimulation test in posttraumatic stress disorder. *Prog Neuropsychopharmacol Biol Psychiatry* **19**:433–443(1995).

52. Brown TM, Boudewyns PA. Periodic limb movements of sleep in combat veterans with posttraumatic stress disorder. *J Trauma Stress* **9**:129–136(1996).

53. Krakow B, Germain A, Tandberg D, Koss M, Schrader R, Hollifield M, Cheng D, Edmond T. Sleep breathing and sleep movement disorders masquerading as insomnia in sexual-assault survivors. *Compr Psychiatry* **41**(1):49–56(2000).

54. Ghaemi SN, Irizarry MC. Parasomnias as neuropsychiatric complications of electrical injury. *Psychosomatics* **36**:416 (1995).

55. Kravitz M, McCoy BJ, Tompkins DM, et al. Sleep disorders in children after burn injury (see comments). *J Burn Care Rehabil* **14**:83–90(1993).

56. Dagan Y, Lavie P, Bleich A. Elevated awakening thresholds in sleep stage 3-4 in war-related post-traumatic stress disorder. *Biol Psychiatry* **30**:618–622(1991).

57. de Groen JH, Op den Velde W, Hovens JE, Falger PR, Schouten EG, van Duijn H. Snoring and anxiety dreams. *Sleep* **16**:35–36(1993).

58. Youakim JM, Doghramji K, Schutte SL. Posttraumatic stress disorder and obstructive sleep apnea syndrome. *Psychosomatics* **39**:168–171(1998).

59. Roy MJ, Koslowe PA, Kroenke K, Magruder C. Signs, symptoms, and ill-defined conditions in Persian Gulf War veterans: findings from the Comprehensive Clinical Evaluation Program. *Psychosom Med* **60**:663–668(1998).

60. Schoen LKM, Kinney L. Auditory thresholds in the dream disturbed. *Sleep Res* **13**:102(1984).

61. Kosten TR, Mason JW, Giller EL, Ostroff RB, Harkness L. Sustained urinary norepinephrine and epinephrine elevation in post-traumatic stress disorder. *Psychoneuroendocrinology* **12**:13–20(1987).

62. van der Kolk B, Greenberg M, Boyd H, Krystal J. Inescapable shock, neurotransmitters, and addiction to trauma: toward a psychobiology of post-traumatic stress. *Biol Psychiatry* **20**:314–325(1985).

63. Southwick SM, Bremner D, Krystal JH, Charney DS. Psychobiologic research in post-traumatic stress disorder. *Psychiatr Clin North Am* **17**:251–264(1994).

64. Mellman TA. Psychobiology of sleep disturbances in posttraumatic stress disorder. *Ann N Y Acad Sci* **821**:142–149(1997).

65. Ross RJ, Ball WA, Morrison AR. Revising the differential diagnosis of the parasomnias in DSM-IIIR. *Sleep* **12**:287–289(1989).

66. Kolb LC. A neuropsychological hypothesis explaining post-traumatic stress disorders. *Am J Psychiatry* **144**:989–995 (1987).

113

ALCOHOL, ALCOHOLISM, AND SLEEP

MAREN HYDE, TIMOTHY ROEHRS, AND THOMAS ROTH
Henry Ford Hospital, Detroit, Michigan

The effects of ethanol on sleep have been studied extensively since the early experiments by Nathaniel Kleitman. In his 1939 book *Sleep and Wakefulness*, he describes the effects that ethanol administered 60 minutes before bedtime had on body temperature and motility during the sleep of healthy normals. He reported reduced motility during the first half of sleep and increased motility in the second half. In the 1970s alcohol research focused on the sleep of individuals with alcoholism, while recently, with emergence of the field of sleep disorders medicine, research attention has focused on the effects of ethanol on primary sleep disorders such as insomnia and apnea.

A small, water-soluble molecule that quickly distributes throughout the body, ethanol's effects are widespread. It has a negative effect on many organ systems and disrupts almost all neurobiological mechanisms in some way. Therefore it has been difficult to determine the specific mechanisms by which ethanol affects the various aspects of sleep and its disorders. Despite this, it has been recognized that ethanol's sleep effects may play a role in the initiation and maintenance of alcoholism. A better understanding of the ways in which ethanol affects sleep and the adaptive processes that follow will allow for more specific forms of prevention and treatment.

EFFECTS OF ETHANOL ON THE SLEEP OF HEALTHY NORMALS

Ethanol Measurement and Dosing

The measurement and dosing of ethanol is complex and varies from study to study, making comparisons among studies quite difficult. A dose of ethanol that is administered on a gram/kilogram basis can produce various breath ethanol concentrations (BrECs). There are a number of factors that contribute to this, including type of beverage administered, the concentration of ethanol in the beverage, how quickly the beverage is consumed, the contents of the gastrointestinal tract, and the body's total amount of water, which can vary with age, height, and gender [1]. In order to compare and interpret various studies, Table 113.1 provides an estimated dose to BrEC conversion at peak concentrations. The table values represent the administration of ethanol to men who had been fasting, were given a 1:4 ethanol ratio mixed in a tonic water beverage, consumed the ethanol over 30 minutes, and were tested 30 minutes later to allow for ethanol absorption. In order to place the dose and BrEC information from Table 113.1 further into context, the United States considers legal intoxication in most states to be between 0.08% and 0.10% BrEC, and in a minority of states 0.05% BrEC is considered to be impairing. All European countries consider 0.05% or higher BrECs to be the level of legal intoxication. Breath-analyzers within the United States have been calibrated so that a BrEC of 0.10% is equal to a blood ethanol concentration (BEC) of 100 mg ethanol per 100 mL blood [1]. Among European countries and in the United States, breath-analyzer calibrations vary according to industry standards.

Ethanol's Effects on Sleep Physiology

In order to observe the effects of ethanol on sleep, researchers conducting a typical sleep study administer ethanol at night to their subjects about 30–60 minutes before bedtime.

TABLE 113.1   Ethanol Dose and Peak Breath Ethanol Concentration (BrEC)

Dose	Peak BrEC (%)	Number of 12 oz U.S. Beers
0.2 g/kg [2]	0.02	1–2
0.4 g/kg [3]	0.03 (0.011)	2–3
0.6 g/kg [3]	0.05 (0.008)	3–4
0.8 g/kg [3]	0.07 (0.015)	4–5
1.0 g/kg [4]	0.09 (0.005)	5–6

This schedule results in peak concentrations of ethanol in the breath or blood at the time of "lights-out." Ethanol effects in healthy normals have been studied extensively at doses ranging from 0.16 to 1.0 grams of ethanol per kilogram of body weight (g/kg) [5]. These doses correspond to approximately one to six standard drinks (a standard drink is defined as one 12 ounce bottle of beer or wine cooler, one 5 ounce glass of wine, or 1.5 ounces of 80 proof distilled spirits), which can produce BrECs as high as 0.105% [6]. Studies using ethanol doses in this range typically have reported reduced sleep latency compared with no ethanol consumption [5, 7–9]. One study noted an increased sleep time at a low ethanol dose of 0.16 g/kg, but found no effects at higher ethanol doses of 0.32 and 0.64 g/kg [10].

Researchers have analyzed the differential effects of ethanol on the first and second half of the nighttime sleep period. These studies found increased wake or light stage 1 sleep occurring during the second half of the sleep period, especially at higher ethanol doses [9, 11]. This second-half sleep disruption is generally interpreted as a "rebound effect" once the ethanol has been completely metabolized. The term "rebound effect" indicates that certain sleep variables, such as the amount of REM or slow-wave sleep, change in the opposite direction of the ethanol-inducing effects and even exceed normal levels once the ethanol is eliminated [6]. This effect is a result of the body's attempt to maintain a normal sleep pattern despite the presence of ethanol during the first half of the sleep period. Rebound effects are seen with a variety of drugs and with experimental manipulations of sleep. In most experiments, the typical peak BrECs measured at "lights-out" are 0.06–0.08%, and the known ethanol metabolism rate leads to a decrease in BrEC of 0.01–0.02% per hour. At this rate, ethanol is completely metabolized within 4–5 hours of sleep onset; therefore the second-half sleep disruption would correspond with the clearance of ethanol from the body [1].

In addition to these effects on sleep induction and sleep maintenance, research has discovered that ethanol consistently affects the proportions of the various sleep stages. Most studies have reported a dose-dependent suppression of REM sleep at least in the first half of the sleep period and some studies have found increased stage 3/4 sleep in the first half of the sleep period [5]. The studies that have found REM suppression during the first half of the sleep period typically also find a REM rebound during the second half of the night [5]. Thus the overall REM percentage for the complete night did not differ from those subjects who received placebo. Just as there are increased periods of wakefulness or light sleep during the second half of the night, the REM rebound is also related to the ethanol's metabolism and elimination from the body.

Several studies have evaluated the effects of ethanol over repeated nights of administration and have clearly shown that tolerance to ethanol's sedative and sleep stage effects develops within three nights [5]. After that time, the percentages of stage 3/4 sleep and REM sleep return to basal levels. In addition, the discontinuation of nightly ethanol administration resulted in a REM sleep "rebound," which is an increase in REM sleep beyond basal levels [5]. Not all studies have found a REM "rebound" on discontinuation [5, 8] and this variability may be related to several factors, including the degree of REM suppression, the dose and duration of ethanol administration, the basal level of REM sleep, and the extent of prior tolerance to REM suppression.

EFFECTS OF ETHANOL ON THE SLEEP OF INSOMNIACS

Constant difficulty in falling asleep, maintaining sleep, or suffering from nonrestorative sleep is reported by about 10–15% of the U.S. general population [12, 13]. Approximately 30% of individuals with persistent insomnia from the general population have reported using ethanol in the past year to help them sleep, and 67% of those people have reported that ethanol was effective in inducing sleep [12]. Studies of healthy normals sleeping at their usual bedtimes do not adequately represent the hypnotic potential of ethanol in people with insomnia. Healthy, normal people already have an optimal sleep latency and sleep efficiency, and therefore further improvement is difficult to demonstrate. As a result, the reports of ethanol's effects on measures of sleep induction and maintenance in healthy normals are minimal and inconsistent [6]. In addition, the doses administered in these studies are generally much larger (i.e., resulting in BrECs greater than 0.05%) than insomniacs typically report using (i.e., 1–3 drinks).

A recent study compared the effects of an ethanol dose (0.5 g/kg) typically used by insomniacs on their sleep compared to that of age-matched healthy normals [14]. The sleep of the individuals with insomnia was improved with ethanol compared with a placebo, and the sleep disruption of the second half of the night previously reported in normals at higher ethanol doses was not observed. Specifically, the ethanol consumption in those individuals with insomnia increased their slow-wave sleep to the levels of the

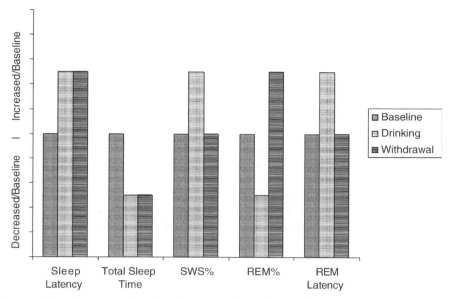

Figure 113.1 Sleep characteristics of patients with alcoholism at baseline, after drinking, and during acute ethanol withdrawal. (Adapted from [15].)

age-matched controls. This ethanol-related improvement in the sleep of insomniacs is supported by self-administration studies. When the opportunity to choose between a previously experienced color-coded ethanol or placebo beverage was presented before sleep, the individuals with insomnia chose ethanol, while the normals chose placebo [14]. The average self-administered nightly dose by those people with insomnia was 0.045 g/kg (up to a possible 0.06 g/kg), which corresponds to the dose that improved the sleep of insomniac patients in the Roehrs et al. [14] study.

These findings from both laboratory and population-based studies, showing a preference for ethanol at bedtime by individuals with insomnia in comparison with noninsomniacs, raise several questions. Does this preference indicate the use of ethanol as self-medication for a sleeping problem, as a way to improve mood, or as a sleep medication that in addition, or subsequently, becomes a "mood-altering" drug [6]? If the motivation for using ethanol is "mood-altering," then what are the "mood-altering" effects that encourage ethanol consumption by the individual with insomnia? Furthermore, does tolerance to ethanol's sedative effects develop in people with insomnia as it does in others? Do people with insomnia successively increase their ethanol doses across nights? Does hypnotic use of ethanol at night generalize to daytime use? And lastly, what are the risks associated with the use of ethanol as a hypnotic? All these issues are yet to be fully addressed. However, there is no question that the ethanol use and sleep relation is dynamic. In other words, disturbed nocturnal sleep increases the likelihood of ethanol use, while ethanol use also has the potential to disturb sleep [6].

EFFECTS OF ETHANOL ON THE SLEEP OF ALCOHOLICS

Several research studies conducted over the past three decades have looked at nocturnal sleep changes in patients with alcoholism after acute ethanol administration and withdrawal. These polysomnographic studies measured sleep characteristics at baseline, after drinking, and during acute ethanol withdrawal (see Figure 113.1). Although these studies vary in methodology and results, some general conclusions can be made regarding the various sleep measures in individuals with alcoholism. First, measures of sleep continuity (i.e., total sleep time and sleep latency) are disturbed on both drinking and withdrawal nights in patients with alcoholism [15]. These findings of increased sleep latency contrast with decreases in sleep latency found in healthy normal men after drinking ethanol. This suggests that tolerance to the sleep-inducing effects of ethanol has developed among individuals with alcoholism, although they are still sensitive to its stimulating effects. Second, the slow-wave sleep percentage (SWS%) increases during drinking and then returns to baseline levels during withdrawal [15]. And lastly, REM sleep is generally suppressed during drinking and returns to baseline levels (with respect to REM latency) or exceeds baseline by "rebounding" (with respect to REM%) during withdrawal [15].

Acute Ethanol Abstinence

The acute ethanol withdrawal phase lasts about 1–2 weeks. Some of the symptoms of withdrawal, such as mood instability, insomnia, and craving, however, can remain even beyond this period of time, a phenomenon that has been called "subacute withdrawal," "protracted withdrawal," and "protracted abstinence" [16, 17]. Acute ethanol abstinence has been characterized by a decrease in the slow-wave sleep percentage [18]. REM sleep is also altered, showing more frequent REM episodes and NREM–REM cycles that are shorter, but without increasing the durations of REM periods [19]. It has been hypothesized that the hallucinations related to alcohol withdrawal are actually the intrusions of REM sleep into wake as a result of REM "rebound" and REM sleep fragmentation [20]. This "rebound" was not found in all patients studied, however, despite showing severe ethanol withdrawal syndrome, including hallucinations [21]. Regardless, it has been hypothesized that REM suppression and a subsequent REM "rebound" is indicative of physical dependence on ethanol.

Sustained Ethanol Abstinence

Polysomnographic studies have found that as long as 1–3 years after the cessation of ethanol consumption, some sleep abnormalities (e.g., brief arousals and REM sleep disruptions) can still persist [22]. Most of the other sleep disturbances that were observed during recent abstinence (e.g., decreased total sleep time, SWS%, and increased sleep latency) appear to return to normal levels with sustained abstinence [15]. The return to basal levels of REM sleep varies considerably across several studies. REM sleep latency may remain at abnormal levels for 9–27 months, and the return of REM% to basal levels can range from 3 to 27 months [15]. However, it is important to note that there are several limitations to the interpretation of these various sustained-abstinence studies. Many of these studies are based on small sample sizes, include only men, do not include control groups, and do not specify their methodology for determining abstinence. Most of the studies do not take into consideration comorbid mental disorders with alcoholism, therefore allowing for the possibility of non-ethanol-related sleep abnormalities to confound the data. Finally, evidence suggests that patients with alcoholism who also have good prognoses tend to sleep better than patients at a high risk for relapse. As a result, good sleepers may be selected for prolonged-abstinence studies, which may underestimate sleep problems among individuals with alcoholism [15].

It has been observed that among individuals with alcoholism sleep problems persist despite sustained abstinence. This suggests three possible explanations: some sleep abnormalities could precede the development of alcoholism and therefore continue into abstinence, the chronic use of ethanol may result in either slowly reversible or irreversible damage to the systems that regulate sleep, and the chronic use of ethanol may be related to persisting medical and psychiatric disorders that interrupt sleep during periods of abstinence [15]. There have been several recent reviews that describe a potential reciprocal relationship between ethanol consumption and insomnia [23–25]. Specifically, the reviews state that insomnia may lead up to initial and persistent drinking, and that heavy consumption of ethanol may interrupt sleep and result in insomnia.

CONCLUSION

Ethanol has been found to have far-reaching effects on sleep and sleep disorders. The sleep of healthy normals appears to be disturbed with acute high ethanol doses, whereas in individuals with insomnia, the lower doses appear to be beneficial. However, the data on healthy normals demonstrate that tolerance to the beneficial effects of ethanol develops quickly. The major concern regarding this is that tolerance development may result in excessive hypnotic ethanol use and may possibly result in excessive daytime use as well.

Individuals with alcoholism commonly have sleep problems, which may occur during active drinking, acute ethanol discontinuation, and prolonged abstinence. Although most sleep abnormalities do improve with time, there are some problems that take months to years to return to baseline levels. Sleep disturbances have been found to develop both before and after the start of alcoholism. It is unknown whether sleep disturbances predispose individuals to develop patterns of drinking that are excessive. An increased likelihood of ethanol use in people with insomnia has been suggested by evidence that ethanol is more reinforcing in nonalcoholic people with insomnia compared to controls. In addition, ethanol is more likely to be used as a sleep aid by people with insomnia rather than people without insomnia.

ACKNOWLEDGMENTS

Supported by NIMH grant No R01 MH59338 awarded to Dr. T. Roth, and NIAAA grant No R01 AA013253 awarded to Dr. T. Roehrs.

REFERENCES

1. Ancoli-Israel S, Roth T. Characteristics of insomnia in the United States: Results of the 1991 Nation Sleep Foundation Survey. I. *Sleep* **22**:S347–S353(2000).
2. Blumenthal SJ, Fine T. Sleep abnormalities associated with mental and addictive disorders: implications for research and

clinical practice. *J Pract Psychiatr Behav Health* **2**:67–79(1996).

3. Brower KJ. Alcohol's effects on sleep in alcoholics. *Alcohol Res Health* **25**:110–125(2001).

4. Dougherty DM, Marsh DM, Moeller G, Chokshi RV, Rosen VC. Effects of moderate and high doses of alcohol on attention, impulsivity, discriminability, and response bias in immediate and delayed memory task performance. *Alcohol Clin Exp Res* **24**:1702–1711(2000).

5. Gross MM, Goodenough DR, Hastey J, Lewis E. Experimental study of sleep in chronic alcoholics before and after four days of heavy drinking with a non-drinking companion. *Ann NY Acad Sci* **215**:254–275(1973).

6. Gross MM, Hastey JM. Sleep disturbances in alcoholism. In: Tarter RE, Sugarman AA (Eds), *Alcoholism: Interdisciplinary Approaches to an Enduring Problem*. Publ, City, 1976, pp 257–307.

7. Johnson LC, Burdick A, Smith J. Sleep during alcohol intake and withdrawal in the chronic alcoholic. *Arch Gen Psychiatry* **22**:406–418(1970).

8. MacLean A, Calms J. Dose–response effects of ethanol on the sleep of young men. *J Stud Alcohol* **43**:434–444(1982).

9. Moskowitz H, Burns M, Williams AF. Skills performance at low blood alcohol levels. *J Stud Alcohol* **46**:482–485(1985).

10. National Institute on Alcohol Abuse and Alcoholism. *Alcohol Alert No. 41: Alcohol and Sleep*. National Institute on Alcohol Abuse and Alcoholism, Bethesda, MD, 1998.

11. Roehrs T, Papineau K, Rosenthal L, Roth T. Ethanol as a hypnotic in insomniacs: self administration and effects of sleep and mood. *Neuropsychopharmacology* **20**:279–286(1999).

12. Roehrs T, Roth T. Sleep, sleepiness, and alcohol use. *Alcohol Res Health* **25**:101–109(2001).

13. Roehrs T, Roth T. Sleep, sleepiness, sleep disorders and alcohol use and abuse. *Sleep Med Rev* **5**:287–297(2001).

14. Roehrs T, Yoon J, Roth T. Nocturnal and next-day effects of ethanol and basal level of sleepiness. *Hum Psychopharm* **6**:307–311(1991).

15. Roehrs T, Zorick F, Roth T. Transient and short-term insomnias. In: Kryger MH, Roth T, Dement WC (Eds), *Principles and Practice of Sleep Medicine*, 3rd ed. Saunders, Philadelphia, 2000, pp 624–632.

16. Rundell JB, Lester BK, Griffiths WJ, Williams HL. Alcohol and sleep in young adults. *Psychopharmacology* **26**:201–218(1972).

17. Satel SL, Kosten TR, Schuckit MA, Fischman MW. Should protracted withdrawal from drugs be included in DSM-IV? *Am J Psychiatry* **150**:695–704(1993).

18. Stone BM. Sleep and low doses of alcohol. *Electroencephalogr Clin Neurophysiol* **48**:706–709(1980).

19. Vitiello MV. Sleep, alcohol, and alcohol abuse. *Addiction Biol* **2**:151–158(1997).

20. Vogel GW. REM deprivation. III. Dreaming and psychosis. *Arch Gen Psychiatry* **18**:312–329(1968).

21. Williams D, MacLean A, Cairns J. Dose–response effects of ethanol on the sleep of young women. *J Stud Alcohol* **44**:515–523(1983).

22. Williams H, Salamy A. Alcohol and sleep. In: Kissin B, Beghleiter H (Eds), *The Biology of Alcoholism*. Plenum Press, New York, 1972, pp 435–483.

23. Williams HL, Rundell OH. Altered sleep physiology in chronic alcoholics: reversal with abstinence. *Alcohol Clin Exp Res* **2**:318–325(1981).

24. Wolin SJ, Mello NK. The effects of alcohol on dreams and hallucinations in alcohol addicts. *Ann NY Acad Sci* **215**:266–302(1973).

25. Zwyghuizen-Doorenbos A, Roehrs T, Lamphere J, Zorick F, Roth T. Increased daytime sleepiness enhances ethanol's sedative effects. *Neuropsychopharmacology* **1**:279–286(1988).

114

DRUGS OF ABUSE AND SLEEP

MAREN HYDE, TIMOTHY ROEHRS, AND THOMAS ROTH
Henry Ford Hospital, Detroit, Michigan

Psychoactive drugs from a variety of drug classes have been used widely in many cultures. Many of these substances, including stimulants, analgesics, and sedative–hypnotics, are used consistently within medical or social norms. However, these psychoactive substances can be used impulsively and in excess for nonmedical purposes, leading to personal and societal harm, which defines drug abuse [1]. Surveys have found that, in the United States, 18% of the general population experiences a substance abuse disorder within their lifetime [2].

Given the diversity of the abused substances, scientists have hypothesized that there may be common biological mechanisms that underlie the reinforcing properties of these drugs, leading to their excessive use. Drug abuse can be analyzed from the perspective of the behavioral mechanisms of positive and negative reinforcement. Positive reinforcement occurs when the presentation of the drug increases the likelihood of a response to obtain the drug again [3]. Another way to describe positive reinforcement is that the pleasurable effects of the drug, in the absence of a deficit state, increase the likelihood of taking the drug again. On the other hand, negative reinforcement occurs when the drug reduces an existing aversive state or lessens a drug-induced aversive state (e.g., withdrawal) [4].

Nearly all drugs of abuse have considerable effects on sleep and wakefulness and on particular stages of sleep. The sleep stage most typically affected by these drugs is rapid eye movement (REM) sleep. It has been hypothesized that sleep and wake changes, although not the primary reinforcing mechanisms, function as contributing factors in maintaining the compulsive and excessive drug use, as well as factors that increase the risk for relapse [1].

THE DRUGS OF ABUSE

Stimulants

The first category of drugs with an extremely high abuse liability is the stimulants, such as amphetamine and its derivatives, cocaine, caffeine, and nicotine. These drugs produce arousal and excitation and have important therapeutic effects for attention deficit disorder, obesity, depression, narcolepsy, and other sleep disorders characterized by excessive sleepiness [5]. Stimulants enhance the activity of the neurotransmitters dopamine and norepinephrine by reuptake blockade and increase release of these neurotransmitters [1, 6]. When taken for short periods of time in moderation, these stimulants are able to counteract excessive daytime sleepiness and sleep loss and improve performance, mood, and endurance [2]. Direct physiological measures, as well as measures of mood and performance, have documented the daytime alerting effects of amphetamines [7]. When administered before sleep, amphetamine and its derivatives delay sleep onset, increase wakefulness during the sleep period, and suppress REM sleep [8]. An increase in slow-wave sleep on the first recovery night and increased amounts of REM sleep and a reduced latency to REM sleep on subsequent nights are characteristic of cessation of chronic amphetamine use [1].

One amphetamine derivative that has become increasingly popular as a recreational drug and drug of abuse is (±)3,4-methylenedioxymethamphetamine (MDMA), or "Ecstasy." This drug has hallucinogenic properties and acts indirectly by stimulating the release of brain monoamines [9]. Exposure to MDMA decreases total sleep time

Sleep: A Comprehensive Handbook, Edited by T. Lee-Chiong.

with a decrease in non-rapid eye movement (NREM) sleep, but no significant effects on REM sleep. Within the NREM sleep, individuals who use MDMA have less stage 2 sleep, but there are no apparent differences in stage 1 or slow-wave sleep (stages 3 and 4) [10]. Anorectics, drugs used to suppress appetite, also enhance central catecholamine activity and are therefore central nervous system stimulants that are amphetamine derivatives. The appetite suppression and stimulant effects of these drugs are inseparable, so insomnia is a frequent side effect of anorectic drug use [11].

Cocaine, which is used medically as a local anesthetic, also has central nervous system stimulant effects, which are the main characteristics that foster its abuse. Cocaine was found to increase fast-frequency EEG activity, which suggests an alerting effect similar to that seen in the studies of daytime effects of amphetamines as described previously [1]. The use of cocaine during the late afternoon and early evening is linked with reduced nocturnal sleep time and REM sleep, as well as an increased latency to REM sleep [12]. Ending chronic cocaine abuse results in increased sleep time and REM rebound, characterized by reduced REM latency and increased REM sleep [1].

Caffeine is the most widely used stimulant in the world. In North America 80–90% of adults report using caffeine on a regular basis [13]. Among caffeine consumers in the United States, the mean daily intake is approximately 280 mg [13, 14]. In the United States, soft drinks and coffee are the major sources of caffeine, with nearly half of caffeine consumers ingesting caffeine from multiple sources, including tea and chocolate [15].

Individuals have very different responses to caffeine, with as little as 250 mg overstimulating some people [16, 17]. Others may be less affected, especially chronic caffeine users who appear to develop tolerance to its stimulating effects. Caffeine intoxication is characterized by restlessness, nervousness, insomnia, excitement, gastrointestinal (GI) disturbances, and flushed face [2]. Although caffeine has a half-life of about 3–7 hours, the effects may last as long as 8–14 hours, which can have significant effects on nighttime sleep, even if the caffeine is consumed in late afternoon. Insomnia is caused by consuming six or more cups of coffee throughout the day, even if it is not taken shortly before bedtime [2]. This may lead to a vicious cycle of consuming caffeine during the day to remain alert, which will end up disrupting sleep at night, and therefore perpetuates daytime sleepiness the following day. Caffeine has been used to produce experimental models of insomnia [17]. However, many people are able to develop tolerance to its stimulating effects and sleep very well.

Nicotine is an addictive, legal, stimulating drug found in tobacco products. Approximately 20% of the adult population in the United States currently smoke cigarettes, 4% smoke cigars or pipes, and 3% use smokeless tobacco [18]. It has been estimated that approximately 20–50% of smokers meet the diagnostic criteria for nicotine dependence. The plasma half-life of nicotine is about 2 hours, so the average smoker who consumes nicotine throughout the day has increasing plasma concentrations, which fall throughout the night. A simple measure of dependence is how long a smoker can go in the morning before the first cigarette [2].

Nicotine has been shown to have sedating effects at low doses and alerting effects at high doses. Both questionnaire and polysomnograghic (PSG) studies have reported an increased sleep latency as well as increased arousals and difficulty staying asleep at night [19]. Some studies have suggested that smoking is a risk factor for sleep disordered breathing [20]. Acute withdrawal from nicotine tends to worsen sleep in smokers and to increase arousals at night followed by sleepiness the next day on the Multiple Sleep Latency Test (MSLT) [21, 22].

Analgesics

Analgesics are drugs that reduce pain, and opiate analgesics have a fairly high abuse liability. The term *opiate* was derived originally from opium, which was acquired from the poppy plant [2]. Although the word *opioid* initially referred to synthetic drugs, the term now refers to natural, semisynthetic, and synthetic substances with opiate-like activity [23–25]. The opiates influence a number of various neurotransmitter systems in the brain, including those that are involved with sleep and wakefulness. They have been shown to produce sedative-like effects on waking EEGs, and individuals taking these drugs also have impaired performance [11]. Opioids decrease REM sleep, as well as reduce total sleep time. Opioids have been used for the treatment of restless legs syndrome (RLS) with and without periodic leg movements (PLMs) [26]. However, tolerance to most of these effects develops quickly with the chronic administration of opioids. The cessation of opiate use leads to REM rebound, an increase in REM sleep, and a shortened latency to the first REM episode [1].

Morphine is the primary active ingredient of opium and was originally named after the Greek god responsible for dreams, Morpheus. Ironically, research has shown that short-term administration of the drug actually reduces REM sleep rather than increasing it and also reduces total sleep time, slow-wave sleep, and sleep efficiency [27]. Heroin is a semisynthetic opiate that is metabolized within the body to morphine, thereby producing the same sleep disturbances as morphine, but with greater effects [2]. Methadone is a synthetic opiate that has similar effects on sleep and wakefulness as morphine. It is rarely abused, and due to its relatively long half-life, it has been used for the treatment of chronic opioid addicts [2]. Immediately after opiate administration before sleep onset, isolated EEG bursts of delta waves are seen on the waking EEG. These findings have been correlated with the behavior of head nodding, which may refer to the street term *being on the nod* [1].

Hallucinogens

Another category of drugs of abuse is the hallucinogens. Their characteristic pharmacological effect is to modify perceptions. Mescalin, psilocybin, and LSD (*d*-lysergic acid diethylamide) are the three classical hallucinogens [1]. After using hallucinogens, a state that is similar to dreaming is experienced. Since REM sleep has been highly correlated with dreaming, scientists have expected the hallucinogens to hasten the onset of REM sleep [1]. The only hallucinogen that has been studied for its effects on sleep is LSD. One study supported the scientific expectation and showed that LSD increased REM sleep early in the night, although the total amount of REM sleep for the night remained unchanged [28].

Marijuana is another drug of abuse with hallucinogenic properties. Its active ingredient is tetrahydrocannabinol (THC), and the effects of this substance on the waking EEG are quite different from those of the classical hallucinogens [29]. The sedating effects of THC occur at much lower doses than the hallucinatory effects. Administration of marijuana or THC to humans is associated with an increase in slow-wave sleep and a reduction in REM sleep [30]. When it is administered chronically, the effects on REM and slow-wave sleep decrease, which indicates the presence of tolerance. Finally, upon discontinuation, increased wakefulness and increased REM sleep time are seen [31].

Sedative–Hypnotics

Sedative–hypnotic drugs including barbiturates, benzodiazepine receptor ligands, and H$_1$ antihistamines have a mixed picture regarding their abuse liability. The barbiturates are well-known drugs of abuse, while the scientific evidence regarding the abuse liability of the benzodiazepine receptor ligands is less compelling. The anxiolytic and sedative effects of barbiturates and benzodiazepine receptor ligands are associated with the facilitation and enhancement of GABA-A receptor function, which is the major inhibitory neurotransmitter system in the central nervous system (CNS). When the GABA-A receptor is occupied, the chloride channel is opened on the neuronal membranes, which allows the negatively charged chloride ions to cross the cell membrane and hyperpolarize the postsynaptic cell [2].

In clinical use, the benzodiazepines are not usually strongly reinforcing or pleasurable for most individuals. Sedative–hypnotic abuse occurs most often in the context of polysubstance abuse, most notably to reduce the adverse effects of stimulants, to self-medicate withdrawal from alcohol or heroin, or to produce intoxication when no other drugs of abuse are available [2]. Other individuals who are also at high risk for sedative–hypnotic abuse include people with a history of sedative–hypnotic abuse or dependence, alcoholics, those who are current substance abusers, and those who are seeking relief from anxiety and depression through pharmacological therapy [2].

All of the hypnotics effectively induce and maintain sleep. The barbiturates suppress REM sleep, to which tolerance develops rapidly, while the benzodiazepine receptor ligands, at clinical doses, have little impact on REM sleep. On discontinuation of barbiturates, a REM rebound, evident in an increased REM percent and reduced REM latency, occurs.

H$_1$ antihistamines are occasionally found to be drugs of abuse. These drugs, which are used for the treatment of allergy, have been associated with sedative effects because they easily pass into the brain and affect the neural structures that are associated with sleep [11]. Although not as potent in hypnotic effects as the traditional benzodiazepine receptor ligands, the sedative potential of H$_1$ antihistamines has made them the most common ingredient of over-the-counter sleeping pills. Overnight sleep studies on antihistamines do not typically show increased sleep times or decreased wakefulness [11]. Daytime studies that use the Multiple Sleep Latency Test clearly exhibit the sedative potential of these drugs, which have also been shown to impair daytime performance. The mechanism of the daytime performance impairment is most likely modulated by the sedative effects of these drugs [11].

A relatively new, commercially available, sedative–hypnotic drug is GHB, or *gamma*-hydroxybutyrate. GHB acts on the GABA-B receptor, producing CNS depressant effects, in many ways similar to the classical sedative–hypnotics such as barbiturates and benzodiazepines [32]. Despite its 30 year history of use, GHB is having increased use as a growth promoter and mild sedative without medical supervision. It has also been labeled as the new recreational "club drug" due to its intoxicating effects, as well as its use in combination with alcohol to leave women more vulnerable to sexual assault [32].

In accordance with its sedative capabilities, GHB has also been used as a sleep-inducing agent [33, 34]. In some cases, insomniacs and others with sleep disturbances have replaced L-tryptophan with GHB, and there have been reports that it is used as a sleep aid for improvement of the aftereffects of methamphetamine [35]. GHB-induced sleep is easily reversed by external stimuli and is virtually identical to normal sleep. Behavior, subjective evaluations, and EEG patterns all suggest that GHB-induced sleep mimics physiological sleep, increasing slow-wave sleep (stages 3 and 4) and REM sleep [36–39]. This differs from the classical sedative–hypnotics, which suppress slow-wave sleep. Thus GHB does not result in REM rebound with nightmares, as is observed with discontinuation of barbiturates [40, 41]. GHB is currently used for the symptomatic treatment of sleepiness and cataplexy in narcoleptic patients. Although this may seem counterintuitive, the ability for GHB to produce physiological sleep during the night allows for the symptoms of narcolepsy (daytime sleepiness, cataplexy, sleep paralysis, and hallucinations) to be alleviated during the day [42].

CONCLUSION

The drugs of abuse all appear to have common biological mechanisms that explain their reinforcing effects as well as altering sleep and wakefulness. Most drugs with high abuse liability change the amount and timing of REM sleep, whether it be the barbiturates and stimulants that decrease REM sleep or the hallucinogen LSD that hastens the onset of REM as well as the amount of REM during the first half of the sleep period. These psychoactive substances affect the neurotransmitters that control sleep and wakefulness. Importantly, tolerance appears to occur during the chronic use of most of these drugs, leading to dose escalation and dependence.

Upon discontinuation of these drugs of abuse, a withdrawal syndrome occurs, with a characteristic REM rebound being observed. Some studies have indicated that the occurrence and intensity of the REM sleep rebound is predictive of the relapse of drug use. Despite this information, further study must be carried out to determine how the sleep–wake changes and specifically the REM sleep changes associated with drugs of abuse contribute to their excessive use.

ACKNOWLEDGMENTS

Supported by NIMH grant No R01 MH59338 awarded to Dr. T. Roth, and NIDA grant No R01 DA17355 awarded to Dr. T. Roehrs.

REFERENCES

1. Roehrs TA. Drugs of abuse. In: Carskadon MA (Ed), *Encyclopedia of Sleep and Dreaming*. Macmillan Publishing, New York, 1993, pp 195–197.

2. Gillin JC, Drummond SPA. Medication and substance abuse. In: Kryger MH, Roth T, Dement WC (Eds), *Principles and Practice of Sleep Medicine*. Saunders, Philadelphia, 2000, pp 1176–1195.

3. Koob GF, Le Moal M. Drug addiction, dysregulation of reward, and allostasis. *Neuropsychopharmacology* **24**:97–129(2001).

4. Wikler A. Dynamics of drug dependence: implications of a conditioning theory for research and treatment. *Arch Gen Psychiatry* **28**:611–616(1973).

5. Mittler MM. Evaluation of treatment with stimulants in narcolepsy. *N Engl J Med* **17**:103–106(1995).

6. Koob GF, Le Moal M. Drug abuse: hedonic homeostatic dysregulation. *Science* **278**:52–58(1997).

7. Mitler MM, Hajdukovic R. Relative efficacy of drugs for the treatment of narcolepsy. *Sleep* **14**:218–220(1991).

8. Rechtschaffen A, Maron L. The effect of amphetamine on the sleep cycle. *Electroencephalogr Clin Neurophysiol* **16**:438–445(1964).

9. Schmidt CJ, Levin JA, Lovenberg W. In vitro and in vivo neurochemical effects of methylenedioxymethamphetamine on striatal monoamine systems in the rat brain. *Biochem Pharmacol* **36**:747–755(1987).

10. Allen RP, McCann UD, Ricaurte GA. Persistent effects of ($\pm$)3, 4-methylenedioxymethamphetamine (MDMA, "Ecstasy") on human sleep. *Sleep* **16**(6):560–564(1993).

11. Roth T. Drugs for medical disorders. In: Carskadon MA (Ed), *Encyclopedia of Sleep and Dreaming*. Macmillan Publishing, New York, 1993, p 197.

12. Watson R, Bakos L, Compton P, Byck R, Gawin F. Cocaine use and withdrawal: the effect on sleep and mood. *Sleep Res* **18**:83(1989).

13. Gilbert RM. Caffeine consumption. In: Spiller GA (Ed), *The Methylxanthine Beverages and Foods: Chemistry, Consumption, and Health Effects*. Alan R Liss, New York, 1984, pp 185–213.

14. Barone JJ, Roberts HR. Caffeine consumption. *Food Chem Toxicol* **34**:119–129(1996).

15. Hughes JR, Oliveto AH. A systematic survey of caffeine intake in Vermont. *Exp Clin Psychopharmacol* **5**:393–398(1997).

16. James JE. Acute and chronic effects of caffeine on performance, mood, headache, and sleep. *Neuropsychobiology* **38**:32–41(1998).

17. Bonnet MH, Arand DL. Caffeine use as a model of acute and chronic insomnia. *Sleep* **15**:526–536(1992).

18. American Psychiatric Association. *Diagnostic and Statistical Manual of Mental Disorders*, 4th ed. American Psychiatric Association, Washington DC, 1994.

19. Davila DG, Hurt RD, Offord KP, et al. Acute effects of transdermal nicotine on sleep architecture, snoring, and sleep-disordered breathing in nonsmokers. *Am J Respir Crit Care Med* **150**:469–474(1994).

20. Wetter DW, Young TB, Bidwell TR, et al. Smoking as a risk factor for sleep-disordered breathing. *Arch Intern Med* **154**:2219–2224(1994).

21. Prosise GL, Bonnet MH, Berry RB, et al. Effects of abstinence from smoking on sleep and daytime sleepiness. *Chest* **105**:1136–1141(1994).

22. Wolter TD, Hauri PJ, Schroeder WJ, et al. Effects of 24-hour nicotine replacement on sleep and daytime activity during smoking cessation. *Prev Med* **25**:601–610(1996).

23. Staedt J, Wassmuth F, Stoppe G, et al. Effects of chronic treatment with methadone and naltrexone on sleep in addicts. *Eur Arch Psychiatry Clin Neurosci* **246**:305–309(1996).

24. Kleber HD. Opioids: detoxification. In: Galanter M, Kleber HD (Eds), *Textbook of Substance Abuse Treatment*. American Psychiatric Press, Washington DC, 1994, pp 191–208.

25. Jaffe JH, Epstein S, Ciraulo DA. Opiods. In: Ciraulo DA, Shader RI (Eds), *Clinical Manual of Chemical Dependence*. American Psychiatric Press, Washington DC, 1991, pp 95–134.

26. Montplaisir J, Lapierre O, Warnes H, et al. The treatment of the restless legs syndrome with or without periodic leg movements in sleep. *Sleep* **15**:391–395(1992).

27. Kay D, Pickworth W, Neider G. Morphine-like insomnia in nondependent human addicts. *Br J Clin Pharmacol* **11**:159–169(1981).

28. Muzio JN, Roffwarg HP, Kaufman MD. Alterations in the nocturnal sleep cycle resulting from LSD. *Electroencephalogr Clin Neurophysiol* **21**:313–324(1966).

29. Fairchild MD, Jenden DJ, Mickey MR, Yale D. EEG effects of hallucinogens and cannabinoids using sleep–waking behavior as baseline. *Pharmacol Biol Behav* **12**:99–105(1979).

30. Pivik RT, Zarcone V, Dement WC, Hollister LE. Delta-9-tetrahydrocannabinol and synhexl: effects on human sleep patterns. *Clin Pharmacol Ther* **13**:426–435(1972).

31. Feinberg I, Jones R, Walker JM, Cavness C, March J. Effects of high dosage delta-9-tetrahydrocannabinol on sleep patterns in man. *Clin Pharmacol Ther* **17**:458–466(1976).

32. Nicholson KL, Balster RL. GHB: a new and novel drug of abuse. *Drug and Alcohol Dependence* **63**:1–22(2001).

33. Chin MY, Kreutzer RA, Dyer JE. Acute poisoning from γ-hydroxybutyrate in California. *West J Med* **156**:380–384(1992).

34. Mack R. Love potion number 8½. *North Carolina Med J* **54**:232–233(1993).

35. Galloway GP, Frederick SL, Staggers FE Jr, Gonzales M, Stalcup SA, Smith DE. Gamma-hydroxybutyrate: an emerging drug of abuse that causes physical dependency. *Addiction* **92**:89–96(1997).

36. Laborit H. Sodium 4-hydroxybutyrate. *Int J Neuropharmacol* **3**:433–452(1964).

37. Yamada Y, Yamamoto J, Fujiki A, Hishikawa Y, Kaneo Z. Effect of butyrolactone and gamma hydroxybutyrate on the EEG and sleep cycle in man. *Electroencephalogr Clin Neurophysiol* **22**:558–562(1967).

38. Vickers MD. Gammahydroxybutyric acid. *Int Anesthesiol Clin* **7**:75–89(1969).

39. Mamelak M, Escriu JM, Stokan O. The effects of gamma-hydroxybutyrate on sleep. *Biol Psychiatry* **12**:273–288(1977).

40. Bonnet MH, Kramer M, Roth T. A dose–response study of the hypnotic effectiveness of alprazolam and diazepam in normal subjects. *Psychopharmacology* **75**:258–261(1981).

41. Grozinger M, Kogel P, Roschke J. Effects of lorazepam on the automatic online evaluation of sleep EEG data in healthy volunteers. *Pharmacopsychiatry* **31**:55–59(1998).

42. Mamelak M. Gammahydroxybutyrate: an endogenous regulator of energy metabolism. *Neurosci Biobehav Rev* **13**:187–198(1989).

PART XVI

SLEEP IN SPECIAL PATIENT GROUPS

115

SLEEP AND THE CAREGIVER

PATRICIA A. CARTER

The University of Texas at Austin School of Nursing, Austin, Texas

INTRODUCTION

Nearly half of the U.S. population—125 million Americans—lives with some type of chronic condition [1]. On average, people with chronic conditions will be cared for at home by a primary caregiver, usually a spouse or daughter, for more than 4 years [1]. More than one-quarter (26.6%) of the adult population has provided care for a chronically ill, disabled, or aged family member or friend during the past year. Based on current census data, that translates into more than 50 million people [2]. Family caregivers assist with physical, psychological, social, and financial needs. Over the past three decades researchers have explored the impact providing assistance (caregiving) to someone with a chronic illness has on the caregiver's overall health and well-being. Researchers have found that caregivers report moderate to severe levels of depression and anxiety [3, 4], poor quality of life [5], and fatigue [6].

A relatively new area of exploration is the relationship between sleep quality and caregiver physical and emotional outcomes. Caregiving is a complex multifaceted environment where many internal and situational factors may impact a caregiver's ability to obtain quality sleep. In fact, not getting enough sleep is a major cause of illness and stress in caregivers. Exhaustion is one of the main complaints, leading to irritability then inappropriate anger, which then lead to more guilt [7]. This chapter will discuss some of the most common factors hypothesized in the research to influence the caregiver's ability to obtain quality sleep. The information presented here is primarily from studies with caregiver–patient dyads focusing on a particular chronic disease (e.g., Alzheimer's disease, cancer, Parkinson's disease, and stroke), where caregiver sleep was assessed.

SITUATIONAL FACTORS

Caregiving requires that the caregiver be able to respond to the needs of the person with chronic illness. Depending on the nature of the illness, those needs may be physical (e.g., pain control, nutritional support, hygiene), psychological (e.g., wandering, forgetfulness, aggression, repetitive behaviors), and/or organizational in nature (e.g., medical appointments, billing). In addition to the needs of the person with chronic illness, the caregiver often assumes the roles of that person while continuing to carry their own (e.g., household chores, financial support, child care). These situational factors may result in stress and anxiety for the caregiver that can impact the caregiver's ability to obtain quality sleep.

A situational factor reported to greatly impact the caregiver's sleep is patient needs. It is well known that patients with chronic illnesses often require 24 hour assistance with their personal needs (eating, toileting, etc.) [3]. In addition, patients with cancer may require medications for pain and other symptoms that must be given at regular intervals (e.g., every 4 hours) to be most effective [8]. Parkinson's caregivers report the most frequent causes for caregiver sleep disruptions are associated with patient motor symptoms (e.g., nocturnal jerks, restless legs) and patient vivid dreams and night terrors [9]. Thommessen and colleagues [10] found caregivers of stroke patients (42%) reported more frequent sleep disruptions as a result of patient needs than did mild dementia (25%) or Parkinson's patient (17%) caregivers.

Despite wide recognition of sleep disruption among family caregivers, few studies have identified specific areas of sleep problems. Carter and Chang [4] found that 95% of 51 caregivers of persons with advanced cancer reported severe sleep problems in the areas of overall sleep quality, sleep duration, and sleep disturbances. Similarly, the most common sleep disturbances reported by Parkinson's caregivers were daytime dysfunction (80%), subjective sleep quality (65%), and sleep latency (55%) [9]. Wilcox and King [11] explored sleep complaints in older women caregivers and found that caregivers reported significantly worse sleep scores in the areas of overall sleep quality, sleep duration, and daytime dysfunction as compared with noncaregiving women. The Pittsburgh Sleep Quality Index (PSQI) was the most frequently used instrument to assess sleep quality in these caregiver studies. Other studies used single item overall sleep quality questions to assess sleep in caregivers.

Social factors that can impact a caregiver's sleep quality include relationship type, relationship quality, social support, and financial resources. For example, researchers have found that high caregiving demands and poor relationship quality results in higher levels of caregiver burden, anxiety, and depression [12]. Anxiety and burden have been associated with sleep disturbances in noncaregiving samples [13].

Caregivers have substantially increased life demands as they try to "pick up the load" of the person diagnosed with chronic illness while continuing to carry their own [12]. Following a diagnosis of chronic illness, caregivers shift the focus of their time and energy toward improving and maintaining the well-being and quality of life of their family member, frequently at the cost of their own self-care [14]. Sleep is often the first area of self-care to be lost when life demands increase. Ironically, sleep is one of the most important areas of self-care if one is going to have the emotional, physical, and mental energy needed to be a caregiver.

INTERNAL FACTORS

Caregiving has physical components; however, it is the emotional strain that often leads to negative outcomes. Stress and strain have been shown to influence sleep quality in many populations [13]. Studies with caregiver samples have begun to show similar findings [11, 12]. Similar to the situational factors, internal factors take several forms; however, they are often much more difficult to separate into distinct, mutually exclusive entities.

Few studies have specifically addressed caregiver sleep quality; however, some studies that focused on caregiver anxiety, worry, grief, or bereavement also reported sleep disturbances in their samples [15]. Studies with adult noncaregiving samples have shown that anxiety, worry, and burden result in poor sleep quality [13].

Physical factors known to impact sleep in healthy adult noncaregivers include (1) exposure to social drugs (e.g., caffeine, nicotine, alcohol) ingested close to bed time, (2) environmental disturbances (e.g., light, noise, temperature, mattress quality), and (3) gender and age [16, 17]. These same factors can impact caregiver sleep.

TREATMENT OPTIONS

Caregiving is a unique situation where the caregiver feels the need to be alert enough to monitor the patient, even at night. This precludes the use of most pharmacological therapies. Behavioral sleep therapies are, however, very useful in this population. Behavioral therapies such as stimulus control, relaxation, and sleep hygiene have been shown to be very useful in improving caregiver sleep quality (self-report and objective) [18]. Traditional delivery methods, however, are not as useful in this overburdened population. Caregivers are not available to meet in the therapist's office once a week for 6–8 weeks. Nor are they available for group therapy sessions, and self-help books are not the first choice for pleasure reading, even if their schedule permits.

Any therapies that focus on the caregiver must also provide a rationale for how the therapy will assist the caregiver to continue in the caregiver role and/or to be a better caregiver. The focus is entirely on the patient, as far as the caregiver is concerned—his/her life has stopped and any and all energy must be focused on the patient's life and well-being [19].

CONCLUSIONS

Many internal and situational factors can influence the caregiver's ability to obtain quality sleep. Caregivers are "on call" 24 hours per day, 7 days per week. Often caregivers do not ask for, do not accept, or do not have access to resources to help with the demands of providing care to a family member with a chronic or disabling condition.

Much research is still needed to fully understand the relationship between caregiving and sleep. This chapter has presented what we know; however, what we do *not* know is a much larger area. The population as a whole is growing older and the medical profession is getting ever better at prolonging life. These two factors alone will lead to more caregivers. In fact, former first lady Rosalyn

Carter stated that "there are four kinds of people—those who have been caregivers, those who are caregivers, those who will be caregivers, and those who need care" [20]. It behooves us to understand the relationship between caregiving and sleep. If we can understand this relationship we can design interventions to prevent chronic sleep deprivation in this population of which we will all be members—that is, if we live long enough.

REFERENCES

1. Partnership for Solutions. Statistics & research prevalence. Available at http://www.partnershipforsolutions.org/statistics/prevalence.cfm (accessed July 10, 2004).

2. National Family Caregivers Association (NFCA). Family caregiving statistics. Available at http://www.nfcacares.org/ (accessed July 10, 2004).

3. Carter P. Caregiver's descriptions of sleep changes and depressive symptoms. *Oncol Nurs Forum* **29**(9):1277–1283(2002).

4. Carter P, Chang B. Sleep and depression in cancer caregivers. *Cancer Nurs* **23**(6):410–415(2000).

5. Weitzner MA, Jacobsen PB, Wagner H Jr, Friedland J, Cox C. The Caregiver Quality of Life Index–Cancer (CQOLC) Scale: development and validation of an instrument to measure quality of life of the family caregiver of patients with cancer. *Qual Life Res* **8**(1-2):55–63(1999).

6. Teel C, Press A. Fatigue among elders in caregiving and non-caregiving roles. *West J Nurs Res* **21**(4):498–520(1999).

7. Family Caregiver Alliance (FCA). A guide to taking care of yourself. Available at http://www.caregiver.org/caregiver/jsp/print_friendly.jsp?nodeid=784 (accessed July 9, 2004).

8. Paice J. Symptom management. In: Miaskowski C, Buchsel P (Eds), *Oncology Nursing: Assessment and Clinical Care*. Mosby, Philadelphia, 1999, p 286.

9. Pal P, Thennarasu K, Fleming J, Schulzer M, Brown T, Calne S. Nocturnal sleep disturbances and daytime dysfunction in patients with Parkinsons disease and their caregivers. *Parkinsonism Relat Disord* **10**:157–168(2004).

10. Thommessen B, Aarsland D, Braekhus A, Oksengaard A, Engedal K, Laake K. The psychosocial burden on spouses of the elderly with stroke, dementia and Parkinsons disease. *Int J Geriatr Psychiatry* **17**:78–84(2002).

11. Wilcox S, King A. Sleep complaints in older women who are family caregivers. *J Gerontol B Psychol Sci Soc Sci* **54B**:189–198(1999).

12. Yates ME, Tennstedt S, Chang BH. Contributors to and mediators of psychological well-being for informal caregivers. *J Gerontol B Psychol Sci Soc Sci* **54B**:12–22(1999).

13. Leger D, Scheuermaier K, Philip P, Paillard M, Guilleminault C. SF-36: evaluation of quality of life in severe and mild insomniacs compared with good sleepers. *Psychosom Med* **63**:49–55(2001).

14. Gallant M, Connell C. The stress process among dementia spouse caregivers. *Res Aging* **20**(3):267–297(1998).

15. Cuthbertson S, Margetts M, Streat S. Bereavement follow-up after critical illness. *Crit Care Med* **28**:1252–1253(2000).

16. Morin C, Espie C. *Insomnia: A Clinical Guide to Assessment and Treatment*. Kluwer Academic/Plenum Publishers, New York, 2003.

17. Spielman A, Conroy D, Glovinsky P. Adult sleep disorders and behavioral sleep medicine. In: Perlis M, Lichstein K (Eds), *Treating Sleep Disorders*. Wiley, Hoboken, NJ, 2003, pp 190–213.

18. King A, Baumann K, O'Sullivan P, Wilcox S, Castro C. Effects of moderate–intensity exercise on physiological, behavioral, and emotional responses to family caregiving: a randomized controlled trial. *J Gerontol A Biol Sci Med Sci* **57A**:M26–M36(2002).

19. Carter P. A not so silent cry for help: older female cancer caregivers' need for information. *J Holist Nurs* **19**(3):271–284(2001).

20. Jewish United Fund (JUF) News and Public Affairs. Elderly and disability. Available at http://www.juf.org/news_public_affairs/article.asp?key=5138 (accessed July 12, 2004).

ADDITIONAL READING

Dodd M, Janson S, Facione N, Faucett J, Froelicher E, Humpherys J, Lee K, Miaskowski C, Puntillo K, Rankin S, Taylor D. Advancing the science of symptom management. *J Adv Nurs* **33**(5):668–676(2001). This report provides current information on the theory of symptom management. Patient symptoms are the most frequent cause of sleep disruption reported by caregivers.

McCurry S, Logsdon R, Vitiello M, Teri L. Treatments of sleep and nighttime disturbances in Alzheimer's disease: a behavior management approach. *Sleep Med* **5**(4):373–377(2004). This report provides a comprehensive description of a successful intervention focused on patient behaviors that can also favorably impact caregiver outcomes.

Nijboer C, Tempelaar R, Triemstra M, van den Bos GA, Sanderman R. The role of social and psychologic resources in caregiving of cancer patients. *Cancer* **91**(5):1029–1039(2001). This report illustrates the importance of caregiver personality and social interactions on depressive symptoms. Depression is a common problem with caregivers that can influence sleep quality.

116

SLEEP IN PATIENTS WITH HIV DISEASE

SUZAN E. JAFFE
Aventura, Florida

HIV DISEASE AND AIDS

The human immunodeficiency virus (HIV) is now well established as the causal agent of acquired immunodeficiency syndrome (AIDS). No one is immune to HIV infection; it can occur in any age group, from newborns to geriatrics. It is also recognized that the blood test of an infected individual will test positive for antibodies to the HIV virus months or years before any symptoms develop.

This period of "asymptomatic" infection is highly individual and is often referred to as the "clinical latency period." Some people may begin to have symptoms within a few months, while others may be symptom-free for more than 10 years.

Even during the clinical latency period, HIV is actively multiplying, infecting, damaging, and eventually killing cells of the immune system. The most obvious effect of HIV infection is a reduction in the number of $CD4^+$ cells found in the blood. These cells are the immune system's key fighters of infection [1, 2].

By destroying or damaging cells of the body's immune system, HIV gradually weakens the body's capacity to fight infections and selected cancers. People diagnosed with HIV infection may suffer life-threatening diseases called opportunistic infections. Microbes such as bacteria, viruses, or parasites, which under normal circumstances do not make healthy people sick, cause opportunistic infections [1, 3, 4].

Prior to the mid-1990s, medical treatments for HIV met with limited success. The introduction of highly active antiretroviral (ARV) therapy in 1996 was a turning point for hundreds of thousands of people with HIV infection who had access to sophisticated health-care systems. Although there is still no cure for HIV/AIDS, ARVs and their use in combination "cocktails" have dramatically reduced mortality and morbidity and prolonged and improved the lives of sufferers [2, 4, 5].

Between 850,000 and 950,000 Americans are living with HIV infection today and suffering from devastating effects on health, performance, and overall quality of life (QoL) [6, 7]. As new therapies extend their life expectancy, long-term adverse impacts on QoL become increasingly important. As will be discussed in this chapter, chronic sleep disturbances and fatigue are highly prevalent in this population and may persist a lifetime [8–10].

DEFINITIONS AND CLASSIFICATIONS

The Centers for Disease Control (CDC) definition of AIDS [11] includes all HIV-infected people who have fewer than 200 $CD4^+$ T cells per cubic millimeter of blood (healthy adults usually have $CD4^+$ cell counts of 1000 or more). In addition, this definition includes 26 clinical conditions that affect people with advanced HIV disease [1, 11].

Children and adolescents with HIV infection may get the same opportunistic infections as adults with the disease. In addition, they also contract severe forms of the bacterial infections all children get, such as conjunctivitis, ear infections, and tonsillitis [12].

Another way of classifying individuals with HIV disease is by categories (see Table 116.1). This classification puts individuals infected with HIV into one of three categories, A, B, or C [13].

Sleep: A Comprehensive Handbook, Edited by T. Lee-Chiong.
Copyright © 2006 John Wiley & Sons, Inc.

TABLE 116.1   CDC Classification Scheme for HIV Disease

Category A

Category A consists of one or more of the conditions listed below in an adolescent or adult (>13 years) with documented HIV infection. Conditions listed in Categories B and C must not have occurred:

- Asymptomatic HIV infection
- Persistent generalized lymphadenopathy
- Acute (primary) HIV infection with accompanying illness or history of acute HIV infection

Category B

Category B consists of symptomatic conditions in an HIV-infected adolescent or adult that are not included among conditions listed in clinical Category C and that meet at least one of the following criteria: (1) the conditions are attributed to HIV infection or are indicative of a defect in cell-mediated immunity; or (2) the conditions are considered by physicians to have a clinical course or to require management that is complicated by HIV infection. Examples of conditions in clinical Category B include, but are not limited to:

- Bacillary angiomatosis
- Candidiasis, oropharyngeal (thrush)
- Candidiasis, vulvovaginal; persistent, frequent, or poorly responsive to therapy
- Cervical dysplasia (moderate or severe) cervical carcinoma in situ
- Constitutional symptoms, such as fever (38.5 °C) or diarrhea lasting more than 1 month
- Hairy leukoplakia, oral
- Herpes zoster (shingles), involving at least two distinct episodes or more than one dermatome
- Idiopathic thrombocytopenic purpura
- Listeriosis
- Pelvic inflammatory disease, particularly if complicated by tubo-ovarian abscess
- Peripheral neuropathy

Category C

Category C includes the clinical conditions listed in the 1993 AIDS surveillance case definition includes HIV-infected adolescents and adults (aged ≥13 years) who have either (1) less than 200 CD4$^+$ T-lymphocytes/cm; (2) a CD4$^+$ T-lymphocyte percentage of total lymphocytes of less than 14%; or (3) any of the following three clinical conditions: pulmonary tuberculosis, recurrent pneumonia, or invasive cervical cancer. For classification purposes, once a Category C condition has occurred, the person will remain in Category C [13].

SLEEP AND HIV DISEASE

HIV infection is now a worldwide pandemic, with tens of millions *infected* and still many millions more *affected* by it. Health-care practitioners in all fields of specialization are among those affected by HIV disease. Sleep disturbance (primarily insomnia, difficulty initiating and maintaining sleep) and fatigue are very common and often disabling symptoms for this population of individuals [7, 14, 15].

The link between HIV infection and sleep was first documented in 1986. At this time, it was noted that during the clinical latency phase of HIV disease (when there are no clinical manifestations of the disease and the immune system is still robust) there are subtle changes in sleep/wake patterns. Early researchers postulated that the interrelationship between hormones, the immune system, and sleep was of major relevance in discerning why individuals with HIV infection progressed at varying rates toward the immunocompromised state [16, 17].

It has been suggested that electrophysiological testing (e.g., polysomnography (PSG)) may be one of the most sensitive indicators of subclinical neurologic injury in otherwise asymptomatic HIV-infected individuals. The earliest PSG studies performed in the mid-1980s found prominent changes in the sleep architecture of HIV-infected men. Regardless of disease severity, early investigators documented sleep/wake pattern changes in these subjects. Results from such studies indicated that sleep architecture differed between the HIV-infected individuals and control subjects, specifically in that wakefulness, slow-wave sleep (SWS), and rapid eye movement (REM) sleep were more evenly dispersed throughout the night. In particular, SWS was prevalent during the second half of recorded sleep. The observed changes in the NREM/REM cycle could not be explained on the basis of depression, anxiety, or other underlying psychopathology [14, 18].

In a retrospective study (1986–1995) an analysis of over 100 polysomnograms (*n* = 51) of HIV-infected men who had been living with HIV for more than 8 years revealed that there was a significant difference in the following sleep parameters: (1) total SWS was increased in long-term survivors; (2) sleep was more efficient in the long-term survivors with fewer arousals; (3) long-term survivors demonstrated a longer total sleep time; and (4) sleep complaints were significantly less frequent in the survivors, even as far back as baseline testing [19]. Figure 116.1 displays the histograms of an HIV-infected man over a period of five years [20].

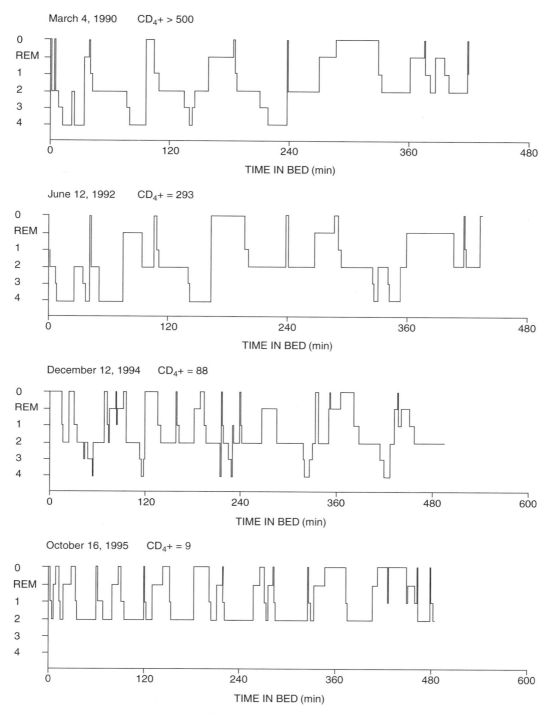

Figure 116.1 Histograms.

It is hypothesized that the observed increase in SWS in those individuals (now the long-term HIV survivors) may be due to an increase in somnogenic cytokines, which have enhanced their ability to fight the HIV infection. The observed decrease in SWS as time passes, with concurrent decrease in $CD4^+$ counts, may correspond to an immune system that is failing.

In summary, a comprehensive review of the published literature has revealed the following major findings in this population:

1. A significant increase in total SWS has been documented repeatedly in studies of HIV-infected men, as early as the first few months following acute infection.

2. Subtle alterations in sleep physiology and mild subjective sleep complaints occur well before clinical manifestation of disease is observed.

3. Deterioration of the immune system is associated with a decline in the quantitative SWS during the night. This decrease in SWS in HIV-infected, symptomatic patients is similar to that found in cancer patients [9].

4. PSG studies indicate that HIV infection disrupts the normal cycle of alternating REM and NREM periods associated with restorative sleep [21].

5. Daytime symptoms, such as fatigue or excessive sleepiness, and difficulties initiating or maintaining sleep are observed in the majority of HIV-infected persons even before a change in their clinical status.

6. As HIV disease progresses the normal cyclic pattern of NREM/REM sleep is destroyed (see Figure 116.1 for representative histograms).

7. The ability to organize the cortical neuronal system to maintain sleep architecture dissipates as HIV disease advances. This is speculated as due to neural microstructure damage.

8. Subjective sleep complaints, including difficulty maintaining sleep and daytime drowsiness, are common during the end stages of HIV disease [19, 22].

A handful of researchers have continued to study this population and have focused on the impact of fatigue and sleep/wake pattern alterations in men and women, children, and adolescents with HIV infection. Some studies have evaluated cognitive function, others psychomotor ability, and others anxiety and depression as cofactors to sleep complaints [23–26].

ANIMAL INVESTIGATIONS AND PATHOPHYSIOLOGY

Seventy years ago, von Economo reported that the hypothalamus was involved in sleep regulation from viral-induced CNS lesions. He concluded that sleep was an active process [27]. Recent literature has suggested close ties between the regulation of sleep, diverse microbial products, and immune system regulators (cytokines) [28]. Such investigations explored the molecular mechanisms responsible for the propensity to sleep. These animal studies primarily challenged rabbits with either viral, fungal, bacterial, or protozoan pathogens. The rabbits showed changes in sleep patterns characterized by increased NREM sleep

(SWS, stages 3 and 4), followed by an inhibition of NREM sleep. Degree of the monitored effects was dependent on the specific microbe, the dose of the infectious substance, and the administration route.

The mobilization of specific and nonspecific immunological pathways has been shown time after time to disrupt the regulation of sleep and wakefulness [29, 30]. Early immunological responses to peripheral HIV infection are known to alter the profile of circulating cytokines (e.g., tumor necrosis factor alpha (TNF-α) and interleukin-1 (IL-1) [31]. TNF-α, a cytokine secreted by macrophages and T cells, mediates inflammatory and immune responses and is associated with wasting in persons with malignancies or AIDS. TNF-α also enhances lymphocytic activity through cytokine activation and interacts with interleukin to produce fever, anorexia, lethargy, and sleep [32].

These acute phase response cytokines are elevated in the blood of HIV-infected individuals and produce fatigue when administered in human and animal models. [33–35]. It is postulated that these substances are involved in the excessive desire to sleep associated with infections, especially those with active infection. Other studies have shown that growth hormone dysregulation may contribute to sleep-related physiology deterioration early in HIV infection [35–37].

HIV surface glycoprotein, gp120, has been postulated to contribute to the sleep abnormalities in HIV-infected individuals [38]. It has been reported that HIV gp120 modifies sleep in freely moving rats and that it also activates the ERK pathway in brain slices. Findings suggest that HIV gp120 increases REM sleep in the rat by specifically affecting the ERK signal transduction pathway [30]. HIV-associated dementia has been detected in 20–30% of patients suffering AIDS. The envelope gp120 derived from HIV seems to play a critical role in the etiology of this dementia [37].

To further determine which components of the virus may be responsible, researchers administered recombinant HIV gp160 or gp41 into rats. Both substances increased NREM sleep, fragmented sleep, altered slow frequency components of the EEG, and induced fever. These data implicate that HIV envelope glycoproteins are capable of altering sleep [30, 33, 37, 38].

It has also been reported that bacterial products such as muramyl peptides and lipopolysaccharides have the capacity to enhance SWS as well as to increase the synthesis and release of IL-1. Darko and colleagues have described sleep pattern changes in feline immunodeficiency virus (FIV)-infected cats. The changes in the sleep of cats with FIV are reported similar to those of humans infected with HIV. For instance, findings include an increase in SWS, a shift of SWS to the later sleep periods, increased arousals, and increased time awake during the sleep period [34].

FATIGUE

In general, fatigue is defined as a vague feeling of tiredness, weariness, or lack of energy. Fatigue must be differentiated from complaints of "drowsiness." A person who complains of drowsiness may be able to take naps or in some cases may fall asleep unexpectedly, such as while driving, sitting in a quiet room, or during a lecture. A person with complaints of fatigue will not have these tendencies but may express indifference to his/her situation or to those close to him/her. If fatigue is not relieved by enough sleep, adequate nutrition, or a decrease in stressful factors, it should be further evaluated.

Fatigue is an almost universal clinical complaint and a troubling symptom in persons of all ages, men and women alike, with HIV/AIDS and can affect all aspects of one's life [39]. Symptoms of fatigue are associated with increased napping, diminished alertness, difficulty falling asleep, driving difficulties, and frequent awakening during sleep. Complaints of increasing daytime drowsiness have contributed to a decrease in driving performance as well [40, 41].

Though it may result in significant disability, fatigue is often overlooked and undertreated in this population [26]. In addition, research has shown that patients seldom discuss fatigue with their physicians [42].

Unfortunately, since fatigue is such a common and often vague complaint, a potentially serious cause, such as undetected HIV infection, may be overlooked. In fact, fatigue alone may be an early indicator of HIV CNS involvement [43, 44].

It seems to be a consensus that the cause of fatigue in HIV-infected individuals is multifactorial. Factors include lack of rest or exercise, improper or inadequate diet, psychological stress (depression and/or anxiety), the use of recreational substances, anemia, thyroid gland disease, hypogonadism, infections, side effects of medications, sleep disturbances, and fever [9, 39, 45].

Some studies have shown that lower CD4 cell counts were related to more daytime sleep, higher evening fatigue, and higher morning fatigue [9, 39]. It has also been documented that HIV-infected women with high fatigue levels have significantly more trouble falling asleep, more nocturnal awakenings, and poorer daytime functioning, as well as a higher rate of depressive symptoms [24]. These symptoms were not correlated with $CD4^+$ counts and corroborated that the complaint of fatigue occurred throughout the severity spectrum of HIV disease.

INSOMNIA

Insomnia is a widespread complaint and underdiagnosed in this population. As stated earlier, insomnia is one of the most common problems associated with HIV disease and, similar to fatigue, is caused by a multitude of precipitating, contributing, and perpetuating factors [15, 46].

Reports of insomnia, specifically difficulty initiating and maintaining sleep occur in a preponderance of HIV-infected individuals prior to an obvious decline in their clinical status. Patients with and without documented objective findings on PSG have complaints of sleep disturbance. Patients with cognitive impairment have also been found to have a higher prevalence of insomnia [15, 23, 26, 47–49].

HIV MEDICATIONS AND SLEEP DISTURBANCES

There have not been many clinical sleep studies exploring how antiretroviral medications affect sleep. However, as early as the introduction of zidovudine (AZT) anecdotal data and small samplings of published data have reported worsening insomnia [50–52]. Newer antiviral medication, such as Efavirenz, has been associated with vivid dreams, difficulty falling asleep, and night awakenings. Patients treated with Efavirenz show low sleep efficiency, increased total time awake and daytime napping. Efavirenz plasma levels were higher in patients with insomnia and/or reduced sleep efficiency. Interestingly all patients had an increase in sleep latency and reduced SWS regardless of whether or not they complained of insomnia [53].

HIV-INFECTED CHILDREN AND SLEEP

Though studies of children with HIV infection are scarce, findings suggest that children infected with HIV experience significant sleep disturbances and that their problems are similar to those recorded in adult populations. One such study described sleep patterns and level of fatigue in children and adolescents (6–18 years old) with HIV infection. This study was conducted within the child's home environment, utilizing a fatigue assessment scale, symptom diary, and wrist actigraphy.

Wrist actigraph provided quantitative movement data to calculate sleep onset, duration, sleep efficiency, and mean activity level. Children in this study were all receiving nucleoside reverse transcriptase inhibitors, either as monotherapy or two to three drug combination therapies. The control group was of similar ethnicity, without chronic illness, and not on any prescribed medications.

Parents reported that infected children slept about 1 hour less per night and had significantly more problems staying asleep. The children also complained of more nightmares and more frequent awakenings. The children napped more as well. The HIV-infected children themselves reported on sleep diary forms the following: increased need for daytime naps and an increased degree of tiredness in the evening hours.

Results demonstrated that HIV-infected subjects had significantly more wake time after sleep onset and more awakenings (stayed awake, once awake) for longer periods, greater daytime fatigue, and a greater level of evening tiredness [12]. Actigraphy data showed that the infected children were in bed longer and had lower sleep efficiency index (SEI). Unfortunately, this study did not obtain PSG/EEG recordings so there is no documentation of sleep staging [12].

MANAGEMENT OF SLEEP DISTURBANCES IN HIV-INFECTED PERSONS

Since daytime fatigue and chronic sleep disturbance are among the most common complaints in this population, clinicians need to take a thorough sleep history. It is up to the health-care provider to initiate this discussion, as research shows that most patients will not report sleep disturbance if not questioned specifically [54, 55].

As patients with HIV/AIDS are living longer with the illness, symptom management is an increasingly important health problem. HIV sleep disturbance may be easy to overlook, underdiagnose, and undertreat due to the many medical and psychosocial problems faced by people with HIV disease. Identifying the cause of disturbances and distinguishing between sleep disturbances and fatigue are critical for effective treatment interventions.

One published study utilized acupuncture to treat a small cohort of HIV-infected persons complaining of insomnia. Sleep activity and subjective sleep quality improved following 5 weeks of acupuncture delivered in this group setting [56]. More research is needed to determine the long-term efficacy of acupuncture in this population and for other alternative therapies.

Stimulus control, sleep restriction, sleep hygiene, behavior modification techniques, relaxation training, and cognitive–behavioral therapy (reversal of negative conditioning) are just some of the standard therapies that should be considered in this population [57, 58]. Clinicians should consider these and other behavioral techniques that have proved beneficial for sleep disturbances in the general population [59] (see Chapters 12, 13, 64, 68–70).

FUTURE CONSIDERATIONS FOR HEALTH-CARE PRACTITIONERS

Could sleep, or more specifically the lack of good sleep, be one of the rate-limiting factors in fighting HIV disease progression? Krueger and colleagues have stated, "sleep and the immune response seem to be so interwoven that sleep might be recognized as a component of the body's fight against infection" [60]. With this notion in mind, one could postulate that falling asleep might allow the body to devote its resources to fighting infection.

The importance of good sleep hygiene in HIV-infected individuals is crucial and may help fortify the immune system. Since sleep is recognized as a restorative process, it is reasonable to suggest that improved sleep quality, or even increased sleep time, could help the immunocompromised patient, though at the moment, this is purely speculation. Health-care practitioners therefore need to recognize the signs and symptoms of sleep disturbances and institute early measures for optimizing their patients' wellbeing [10, 56].

The use of long-term hypnotics or tranquilizers should probably be discouraged, as these drugs are known to alter sleep architecture (decreasing the amount of REM sleep and possibly affecting SWS as well).

Chronic use of sedatives, tranquilizers, antihistamines, or alcohol may also produce nocturnal respiratory depression and worsening sleep apnea (when present), further fragmenting sleep—not to mention an increase in morning lethargy and overall daytime sluggishness. It has also been found that adenotonsillar hypertrophy is an early manifestation of HIV infection and can lead to reductions in upper airway dimensions. Such changes in airway size can increase the risk for sleep disordered breathing in HIV-infected adults compared to the general population [7].

In conclusion, early subjective complaints of sleep disturbance, specifically difficulty maintaining sleep and daytime fatigue, may be early predictors of health outcome in HIV-infected individuals. These subjective complaints combined with the polysomnographic finding of decreased SWS and decreased sleep efficiency appear to accelerate the rate at which an individual progresses along the HIV disease continuum. Review of the literature in this chapter confirms the earliest hypotheses that sleep architecture and subjective sleep complaints are multifactorial in etiology and occur early in the course of HIV disease. Future studies will help decipher the importance of the interrelationship between immune function, pharmacologic and behavioral therapies, and human sleep. Forthcoming data may yield new approaches to manage this progressive and debilitating disease.

REFERENCES

1. Hare BC. Clinical overview of HIV disease. HIV InSite Knowledge Base, UCSF Center for HIV Information, 1998; hivinsite.ucsf.edu/InSite.

2. Rabeneck L, Crane MM, Risser JM, et al. A simple clinical staging system that predicts progression to AIDS using a CD4 count, oral thrush, and night sweats. *J Gen Intern Med* **1**:5–9(1993).

3. Centers for Disease Control and Prevention (CDC). *HIV/AIDS Surveillance Report* **14**:1–40(2002).

4. Fleming PL, HIV prevalence in the United States, 2000, 9th Conference on Retroviruses and Opportunistic Infections, Seattle, WA, Feb 24–28, 2002, Abstract 11.

5. Cruess DG, Antoni MH, Gonzalez J, et al. Sleep disturbance mediates the association between psychological distress and immune status among HIV-positive men and women on combination antiretroviral therapy. *J Psychosom Res* **54**(3):185–189(2003).

6. UNAIDS. Report on the Global AIDS Epidemic, July 2004.

7. RFA-HL-04-010. Inter-relationships of sleep, fatigue and HIV/AIDS. DHHS, NHLBI, NIMH, 2003.

8. Webb A, Norton M. Clinical assessment of symptom-focused health-related quality of life in HIV/AIDS. *J Assoc Nurses AIDS Care* **15**(2):67–78(2004).

9. Groopman JE. Fatigue in cancer and HIV/AIDS. *Oncology* **12**(3):335–344(1998).

10. Hudson A, Kirksey K, Holzemer W. The influence of symptoms on quality of life among HIV-infected women. *West J Nurs Res* **26**(1):9–23(2004).

11. Centers for Disease Control. Revised classification system for HIV infection and expanded surveillance case definition for AIDS among adolescents and adults. *MMWR* **41**:1–19(1992).

12. Franck LS, Johnson LM, Lee KA, et al. Sleep disturbance in children with human immunodeficiency virus infection. *Pediatrics* **104**(5):62–71(1999).

13. Osmond DH. Classification and staging of HIV infection. HIV InSite Knowledge Base, UCSF Center for HIV Information, 1998, L Peiperl and P Volberding (Eds); hivinsite.ucsf.edu/InSite.

14. Norman (Jaffe) SE, Chediak AD, Kiel M, Cohn MA. Sleep disturbances in HIV infected homosexual men. *AIDS* **4**(8):775–781(1990).

15. Robbins JL, Phillips KD, Dugeon WD, Hand GA. Physiological and psychological correlates of sleep in HIV infection. *Clin Nurs Res* **13**(1):33–52(2004).

16. Norman SE, Resnick L, Berger JR, et al. Sleep disturbances in HIV-seropositive patients (letter). *JAMA* **260**(7):922(1988).

17. Kubicki S, Henkes H, Alm D, et al. Polygraphic sleep data in AIDS patients. *Electroencephalogr Electromyogr* **20**:288–294(1989).

18. Brown S, Atkinson, Gutierrez R, et al. Subjective sleep disturbances in HIV illness. University of California at San Diego, La Jolla, CA. *Int Conf AIDS* **6**(2):397 (abstract no. 2173) (1990).

19. Jaffe (Norman) SE, Chediak AD. Long term follow up of sleep & immune function in 11 HIV infected men. *Sleep Res* **25**:433(1996).

20. Jaffe (Norman) SE, Chediak AD. Loss of slow wave sleep and HIV infection: a $5\frac{1}{2}$ year case study. *Sleep Res* **25**:432(1996).

21. Eisen JN, Matlow M, Murphy L, et al. AIDS and sleep. Abstract presented at the 9th European Congress of Sleep Research, Jerusalem, Israel, Sept 4–9, 1988, p 93.

22. Kubicki S, Henkes H, Terstegge x, Ruff B. AIDS related sleep disturbances—a preliminary report. In: S Kubicki, Henkes, Bienzle, Pohle (Eds), *HIV & Nervous System*. Gustav Fischer, Stuttgart, 1998, pp 97–105.

23. Dreher HM. Measuring health status in HIV disease: challenges from a sleep study. *Holist Nurs Pract* **17**(2):81–90(2003).

24. Lee KA, Portillo CJ, Miramontes H. The influence of sleep and activity patterns on fatigue in women with HIV/AIDS. *J Assoc Nurses AIDS Care* **12** (Suppl):19–27(2001).

25. Nokes KM, Kendrew J. Correlates of sleep quality in person with HIV disease. *J Assoc Nurses AIDS Care* **12**(1):17–22(2001).

26. Phillips KD, Sowell RL, Rojas M, et al. Physiological and psychological correlates of fatigue in HIV disease. *Biol Res Nurs* **6**(1):59–74(2004).

27. Von Economo C. Sleep as a problem of localization. *J Nerv Ment Dis* **71**:249–259(1930).

28. Krueger JM. Somnogenic activity of immune response modifiers. *TIPS* **11**:122–126(1990).

29. Pollmacher T, Haack M, Schuld A, Reichenberg A, Yirmiya R. Low levels of circulating inflammatory cytokines—do they affect human brain functions? *Brain Behav Immun* **16**(5):525–532(2002).

30. Diaz-Ruiz O, Navarro L, Mendez-Diaz M, et al. Inhibition of the ERK pathway prevents HIV gp120-induced REM sleep increase. *Brain Res* **913**(1):78–81(2001).

31. Haack M, Pollmacher T, Mullington JM. Diurnal and sleep–wake dependent variations of soluble TNF-α and IL-2 receptors in healthy volunteers. *Brain Behav Immun* **18**(4):361–367(2004).

32. Drexler AM. Tumor necrosis factor: its role in HIV/AIDS. *STEP Perspect* **7**(1):13–15(1995).

33. Gemma C, Opp MR. Human immunodeficiency virus glycoproteins 160 and 41 alter sleep and brain temperature of rats. *J Neuroimmunol* **97**(1-2):94–101(1999).

34. Darko DF, Mitler MM, Propero-Garcia O, Henriksen SJ. Sleep and lentivirus infection: parallel observations obtained from human and animal studies. *SRS Bull* **2**(3):43–51(1996).

35. Hogan D, Hutton LA, Smith EM, Opp MR. Beta (CC)–chemokines as modulators of sleep: implications for HIV-induced alterations in arousal state. *J Neuroimmunol* **119**(2):317–326(2001).

36. Darko DF, Mitler MM, Miller JC. Growth hormone, fatigue, poor sleep, and disability in HIV infection. *Neuroendocrinology* **67**(5):317–324(1998).

37. Sanchez-Alavez M, Criado J, Gomez-Chavarin M, et al. HIV- and FIV-derived gp120 alter spatial memory, LTP, and sleep in rats. *Neurol Dis* **7**(4):384–394(2000).

38. Opp MR, Rady PL Hughes TK, et al. Human immunodeficiency virus envelope glycoprotein 120 alters sleep and induces cytokine mRNA expression in rats. *Am J Physiol* **270**(5 Pt 2):R963–R970(1996).

39. Lee KA, Portillo CJ, Miramontes H. The fatigue experience for women with human immunodeficiency virus. *J Obstet Gynecol Neonatal Nurs* **28**(2):193–200(1999).

40. van Servellen G, Sarna L, Jablonski KJ. Women with HIV: living with symptoms. *West J Nurs Res* **20**(4):448–464(1998).

41. Darko D, McCutchan J, Kripke D, et al. Fatigue, sleep disturbances, disability and indices of progression of HIV infection. *Am J Psychiatry* **149**(4):514–520(1992).

42. Grady C, Anderson R, Chase GA. Fatigue in HIV-infected men receiving investigational interleukin-2. *Nurs Res* **47**(4):227–234(1998).

43. Perkins DO, Leserman J, Stern RA, et al. Somatic symptoms and HIV infection: relationship to depressive symptoms and indicators of HIV disease. *Am J Psychiatry* **152**:1776–1781(1995).

44. Capaldini L. Fatigue and HIV: interview with Lisa Capaldini, M.D. Interview by John S. Jones. *AIDS Treat News* **291**:1–6(1998).

45. Adinolfi A. Assessment and treatment of HIV-related fatigue. *J Assoc Nurses AIDS Care* **12**(Suppl):29–34(2001).

46. Phillips KD. Physiological and pharmacological factors of insomnia in HIV disease. *J Assoc Nurses AIDS Care* **10**(5):93–97(1999).

47. Portillo CJ, Tom L, Lee KA, Miramontes H. Physiological and mental fullness as descriptors that influence sleep in women with HIV. *Holist Nurs Pract* **17**(2):91–98(2003).

48. Norman (Jaffe) SE, Chediak A, Kiel M, et al. HIV infection & sleep: follow-up studies (abstract). *Sleep Res* **19**:339(1990).

49. Norman (Jaffe) SE, Demirozu MC, Chediak AD. Disturbed sleep architecture (NREM/REM sleep cycles) in HIV healthy infected men (abstract). *Sleep Res* **19**:340(1990).

50. Elliott A. Anxiety and HIV infection. *STEP Perspect* **98**(1):11–14(1998).

51. Cooper DA, Gatell JM, Kroon S, et al. Zidovudine in persons with asymptomatic HIV infection and CD4+ cell counts greater than 400 per cubic millimeter. *N Engl J Med* **329**(5):297–303(1993).

52. Louie JK, Hsu LC, Osmond DH, et al. Trends in causes of death among persons with acquired immunodeficiency syndrome in the era of highly active antiretroviral therapy, San Francisco, 1994–1998. *J Infect Dis* **186**(7):1023–1027(2002).

53. Gallego L, Barreiro P, Rio RD, et al. Analyzing sleep abnormalities in HIV-infected patients treated with Efavirenz. *Clin Infect Dis* **38**:430–432(2004).

54. Sciolla A. Sleep disturbance and HIV disease. *Focus* **10**(11):1–4(1995).

55. Newshan G, Sherman DW. Palliative care: pain and symptom management in persons with HIV/AIDS. *Nurs Clin North Am* **34**(1):131–145(1999).

56. Phillips KD, Skelton WD. Effects of individualized acupuncture on sleep quality in HIV disease. *J Assoc Nurses AIDS Care* **12**(1):27–39(2001).

57. Spielman AJ, Saskin P, Thorpy MJ. Treatment of chronic insomnia by restriction of time in bed. *Sleep* **10**(1):45–56(1987).

58. Edinger JD, Sampson WS. A primary care "friendly" cognitive behavioral insomnia therapy. *Sleep* **26**(2):177–182(2003).

59. Edinger JD. Classifying insomnia in a clinically useful way. *Clin Psychiatry* **65**(Suppl 8):36–43(2004).

60. Krueger J, Wlater J, Dinarello C, et al. Sleep-promoting effects of endogenous pyrogen (interleukin-1). *Am J Physiol* **246**:R994–R999(1984).

61. Rubinstein ML, Selwyn PA. High prevalence of insomnia in an outpatient population with HIV infection. *J Acquir Immune Defic Syndr Hum Retrovirol* **19**(3):260–265(1995).

117

THE PATIENT WITH CANCER

Christine A. Engstrom

Department of Veterans Affairs Medical Center, Baltimore, Maryland

INTRODUCTION

Sleep disturbance and sleep deprivation are common problems among cancer patients. Symptom control and symptom management of pain, depression, anxiety, and nausea and vomiting have been researched and described at length in many oncology publications. Cancer-related insomnia has received little attention in the literature compared to other oncology-related symptoms. Insomnia in this patient population may be viewed as a normal and temporary symptom to the cancer diagnosis and treatment of the disease and patients may fail to report their sleep disturbances to their health care provider [1, 2]. The insomnia may be perceived as a secondary symptom resulting from depression or anxiety disorders or associated with the patient's level of fatigue. The health-care providers may have knowledge deficits regarding sleep disorder diagnosis as well as treatment options for this patient population [1, 3].

This chapter will focus on the prevalence of insomnia in cancer patients, selected clinical factors and consequences of sleep disturbances, and practical recommendations for the effective management of insomnia in this patient population.

PREVALENCE OF INSOMNIA IN CANCER PATIENTS

A review of studies conducted on the prevalence of insomnia in cancer patients with newly diagnosed disease or recently treated cancers suggests that between 30% and 50% of the patients report sleep difficulties [3]. Higher prevalence rates (70.7%) as measured by wrist actigraphy was found in a group of patients with various solid tumors with bone metastases receiving radiation treatment for pain [4]. A study of a mixed group of oncology patients, both solid tumors and hematological malignancies at a large urban cancer center, demonstrated that 62% of the cancer patients report moderate to severe sleep disturbance [5]. Breast cancer has been the most researched group of cancer patients demonstrating a high rate of insomnia as compared to other types of cancers with the prevalence ranging from 38% to 61% [6–8].

A large study of 982 patients reported the highest rates of insomnia in lung cancer and breast cancer patients versus gastrointestinal, genitourinary, and nonmelanoma skin cancer patients. The patients reported a combination of excessive fatigue (44% of patients), leg restlessness (41%), insomnia (31%), and excessive sleepiness (28%) [6].

CLINICAL FACTORS AND SLEEP DISTURBANCE

The diagnosis of cancer does not represent a single event, but a complex and multiplicative succession of treatments and cluster of symptoms, each of which can serve as a precipitating factor for sleep disturbances. The triggers for the sleep disturbance may occur at any point in the disease process: at diagnosis or during treatment by surgery, chemotherapy, radiation, or multimodality therapy. It can also occur during palliative care and during the terminal stages of the disease [9] as well as in hospitalization itself due to environmental factors, such as noise and bed

Sleep: A Comprehensive Handbook, Edited by T. Lee-Chiong.
Copyright © 2006 John Wiley & Sons, Inc.

discomfort and psychological factors of anxiety, loneliness, and modification of sleep routine [10].

Surgical treatment modalities may increase the risk to develop sleep disturbances, due to their emotional impact or the direct physical effects of the surgery, such as functional losses with mastectomy or colostomy, or the aesthetic effects of head and neck surgeries. Patients with head and neck surgeries develop obstructive sleep apnea due to enlarged tissues, such as tonsils or the base of the tongue, pharyngeal space narrowing, or decreased muscle tone of the pharyngeal dilator muscles [11].

Postoperative complications were studied on 108 esophageal cancer patients comparing the differences between collar and intrathoracic anastamoses reconstructive procedures. The European Organization for Research and Treatment of Cancer (EORTC) symptom scale was used to measure insomnia, which demonstrated significantly higher rates of insomnia in the intrathoracic collar patients [12].

Systemic chemotherapy and the medications used to minimize the side of effects of the therapy may potentiate or cause sleep disturbances. Many chemotherapeutic regimens contain steroids (glucocorticoids) to reduce inflammation of the disease process or as part of the antinausea regimens. These agents are well known to cause adverse neurological effects such as insomnia, restlessness, and increased motor activity [13].

Chemotherapy regimens that cause premature or aggravated menopausal symptoms because of their potential to cause ovarian failure (tamoxifen) precipitate hot flashes in women with breast cancer [14, 15] as well as in men with prostate cancer treated with androgen ablation (i.e., bilateral orchiectomy, gonadatropin-releasing hormone analogs, antiandrogens) [16, 17]. The hot flashes can interfere with the quality of sleep of these prostate cancer patients [18] and in breast cancer patients receiving treatment [19] as well as in breast cancer survivors [20]. Women who reported alterations in sleep patterns described them as ranging from mild to severe, with the sleep disruption influenced by the number of awakenings, night sweats, and the ability to go back to sleep [19]. It has been reported that the 10 minute periods around hot flashes resulted in more significant wake time, a lower percentage of stage 2 sleep, and a long REM sleep latency compared to nights without hot flashes in a study of breast cancer survivors [15].

It is estimated that 60–80% of advanced cancer patients experience pain [4, 21] with 45% of adult oncology patients experiencing moderate to severe pain on a daily basis [4]. The patients with daily pain who were receiving radiation treatments for bone metastases reported the sleep disturbances were related to the pain on 35% of the awakenings and on 55% of the awakenings for urination [4]. Patients admitted to a palliative care unit for pain control were

TABLE 117.1 Clinical Factors Contributing to Sleep Disturbances

Diagnosis of cancer
Pain
Hospitalization
Chemotherapy
Surgery
Hormonal therapy
  Menopausal symptoms
Medications
  Steroids

surveyed regarding their sleep characteristics; results from this study demonstrated the following: anxiety created more difficulties in falling asleep and produced less restoring sleep and more nightmares; depression was associated with early awakening, nonrestorative sleep, fatigue, and nightmares [22]. These symptoms were supported by a study in which 97 breast cancer patients with advanced disease were studied with results suggesting that these women are at higher risk for having sleeping problems than those who are less educated, in pain, depressed, have bony metastases, or lack social support [23]. Table 117.1 summarizes the clinical factors that contribute to sleep disturbances.

CONSEQUENCES OF SLEEP DISTURBANCE

Fatigue, impaired daytime functioning, and mood disturbances are potential consequences of sleep disturbances in cancer patients (Table 117.2). Fatigue in cancer patients is a multifaceted concept related to the disease process, cancer treatments, medications, activity level, as well as sleep disturbances [3, 4]. Cancer-related fatigue differs in intensity and interference from the fatigue associated with depression. Sleep disturbance is a significant predictor of severe fatigue for cancer patients [5]. Sleep that is inadequate in cancer patients may be important not only to the expression of fatigue, but to the patients' quality of life and their tolerance to treatment, and may influence the development of mood disorders and clinical depression [2].

The impact of sleep disturbances on daytime functioning in cancer patients has not been well researched. However, studies in healthy individuals suggest that sleep loss leads to diminished daytime functioning, including cognitive

TABLE 117.2 Consequences of Sleep Disturbances

Fatigue
Cognitive dysfunction
Alterations in immune function

dysfunction and impaired psychomotor skills. These individuals also report poor concentration and memory as well as difficulty in completing tasks [24]. These findings may translate to cancer patients as well, leading to a decrease in the overall quality of life of these patients.

Savard and Morin [3] have postulated that sleep disturbances in cancer patients may lead to immune downregulation. Results from a sample of women at risk for cervical cancer demonstrated that patients reporting a higher satisfaction with the amount of restful sleep hours had higher levels of helper T cells in the circulating blood [25]. Immune alteration associated with insomnia may have a deleterious effect not only on cancer patients but also on those individuals at risk for cancer.

MANAGEMENT OF SLEEP DISTURBANCES

The number of studies evaluating symptoms and causes of sleep disturbances in cancer patients have increased during the last few years; however, treatment for this symptom has been a neglected problem [1]. Pharmacotherapy with hypnotics has been the mainstay of treatment for this patient population as well as in healthy individuals. Recent literature on cancer patients has begun to describe studies in sleep management training.

A multimodal psychological sleep management program combining relaxation techniques, sleep hygiene, and cognitive techniques was designed to improve sleep latency, sleep duration, sleep efficiency, sleep medication, and daytime dysfunction (Table 117.3). Patients educated on progressive muscle relaxation or autogenic training demonstrate improved sleep quality [26]. Quesnel et al. [27] also demonstrated that insomnia treatment with cognitive–behavioral therapy was associated with significant improvements of mood, general and physical fatigue, and global and cognitive dimensions of quality of life for breast cancer patients.

Exercise with individually tailored programs of tailored walking, cycling, or swimming and modification of activity

TABLE 117.3 Management of Sleep Disorders

Pharmacotherapy
  Hypnotics
Cognitive–behavioral therapy
  Progressive muscle relaxation
  Autogenic training
Sleep hygiene counseling
Sleep restriction
Sleep control
Exercise
Patient education

such as naps during the day may all be effective approaches for the cancer patient. Patients of varied cancer types, stages, and treatment modalities participated in a study to assess the feasibility of a 12 week education and exercise program patterned after a phase II cardiac rehabilitation program to improve selected physiological and psychological measures. The cancer patients demonstrated significant improvements in exercise tolerance and sleep patterns at the completion of the exercise program [28].

The coscientist model as described by Hauri [24] encourages individuals with insomnia to become their own sleep scientists by experimenting with various sleep-promoting behaviors in the following four components: sleep hygiene counseling, relaxation therapy, sleep restriction, and sleep control. Berger et al. [29] used this model in breast cancer patients and found that patients involved in their own care had a higher percentage of adherence to the model than compliance with exercise alone. However, it has also been reported that physical activity and exercise in persons with cancer improve immune system function, cognition, and sleep [30].

It is of interest to assess the psychological morbidity of the surviving partners of terminally ill cancer patients. It is implied that unrelieved psychological symptoms of terminally ill cancer patients will increase the risk of long-term psychological morbidity of their surviving partners. The surviving partners difficulty in falling asleep at night, waking up at night, and intake of sleep pills is related to the cancer patient's level of severe anxiety during the last 3 months of life [31].

CONCLUSIONS

Effective management of sleep disturbances in the cancer patient begins with a specific assessment to include documentation of predisposing factors such as insomnia prior to diagnosis, usual sleep patterns, emotional status, exercise and activity level, as well as other cancer-related symptoms and medications. The interview should focus on the nature of the sleep disturbance as well as the duration and intensity of the sleep problem. It is also important to gain information on the exacerbating and alleviating factors, sleep habits, and response to previous treatments, both pharmacological and nonpharmacological. It is usually not necessary to refer these patients for nocturnal polysomnographic assessment, unless a sleep pathology other than insomnia is suspected [3]. Patients suspected of sleep apnea either related to the surgical treatment of the cancer or the pathology of the cancer may be referred for specific sleep studies.

Available data indicates sleep disturbances are a common complaint in cancer patients. The etiologic factors contributing to the sleep problems are multiplicative and include side effects of treatment, pain, maladaptive sleep behaviors, medications, the diagnosis itself, and the specific treatment of surgery, chemotherapy, radiation, or hormonal therapies. Fatigue, mood disturbance, and a compromised immune system are possible consequences of sleep disturbance.

Sleep disturbances can become an additional chronic problem and it is important to offer alternative treatment options to the patient. Data suggests allowing the patient to be the "driver" of the treatment modality will increase adherence to a sleep intervention. Cognitive–behavior therapy, relaxation therapy, and patient education are among the options found to be effective in treating sleep problems in this patient population.

The potential impact that sleep disturbance has on the health and quality of life of cancer patients should spur health-care providers to routinely assess their patients' sleep patterns and the impact they have on the patients as well as the caregiver and/or family members.

REFERENCES

1. Engstrom CA, et al. Sleep alterations in cancer patients. *Cancer Nurs* **22**(2):143–148(1999).

2. Ancoli-Israel S, Moore PJ, Jones V. The relationship between fatigue and sleep in cancer patients: a review. *Eur J Cancer Care (Engl)* **10**(4):245–255(2001).

3. Savard J, Morin CM. Insomnia in the context of cancer: a review of a neglected problem. *J Clin Oncol* **19**(3):895–908(2001).

4. Miaskowski C, Lee KA. Pain, fatigue, and sleep disturbances in oncology outpatients receiving radiation therapy for bone metastasis: a pilot study. *J Pain Symptom Manage* **17**(5):320–332(1999).

5. Anderson KO, et al. Fatigue and sleep disturbance in patients with cancer, patients with clinical depression, and community-dwelling adults. *J Pain Symptom Manage* **25**(4):307–318(2003).

6. Davidson JR, et al. Sleep disturbance in cancer patients. *Soc Sci Med* **54**(9):1309–1321(2002).

7. Savard J, et al. Prevalence, clinical characteristics, and risk factors for insomnia in the context of breast cancer. *Sleep* **24**(5):583–590(2001).

8. Fortner BV, et al. Sleep and quality of life in breast cancer patients. *J Pain Symptom Manage* **24**(5):471–480(2002).

9. Kyriaki M, et al. The EORTC core quality of life questionnaire (QLQ-C30, version 3.0) in terminally ill cancer patients under palliative care: validity and reliability in a Hellenic sample. *Int J Cancer* **94**(1):135–139(2001).

10. Sheely L. Sleep disturbances in hospitalized patients with cancer. *Oncol Nurs Forum* **23**:109–111(1996).

11. Koliha CA. Obstructive sleep apnea in head and neck cancer patients post treatment . . . something to consider? *ORL Head Neck Nurs* **21**(1):10–14(2003).

12. Schmidt CE, et al. Quality of life associated with surgery for esophageal cancer: differences between collar and intrathoracic anastomoses. *World J Surg* (2004).

13. Lundberg JC, et al. A steroid-induced disorder in a patient with non-Hodgkin's lymphoma. *Cancer Pract* **8**(4):155–159(2000).

14. Knobf MT. Natural menopause and ovarian toxicity associated with breast cancer therapy. *Oncol Nurs Forum* **25**(9):1519–1530(1998); quiz pp 1531–1532.

15. Savard J, et al. The association between nocturnal hot flashes and sleep in breast cancer survivors. *J Pain Symptom Manage* **27**(6):513–522(2004).

16. Stearns V. Management of hot flashes in breast cancer survivors and men with prostate cancer. *Curr Oncol Rep* **6**(4):285–290(2004).

17. Moyad MA. Complementary/alternative therapies for reducing hot flashes in prostate cancer patients: reevaluating the existing indirect data from studies of breast cancer and postmenopausal women. *Urology* **59**(4 Suppl 1):20–33(2002).

18. Penson DF, Litwin MS. The physical burden of prostate cancer. *Urol Clin North Am* **30**(2):305–313(2003).

19. Knobf MT. The menopausal symptom experience in young mid-life women with breast cancer. *Cancer Nurs* **24**(3):201–210(2001); quiz pp 210–211.

20. Harris PF, et al. Prevalence and treatment of menopausal symptoms among breast cancer survivors. *J Pain Symptom Manage* **23**(6):501–509(2002).

21. Caraceni A, et al. Breakthrough pain characteristics and syndromes in patients with cancer pain. An international survey. *Palliat Med* **18**(3):177–183(2004).

22. Mercadante S, Girelli D, Casuccio A. Sleep disorders in advanced cancer patients: prevalence and factors associated. *Support Care Cancer* **12**(5):355–359(2004).

23. Koopman C, et al. Sleep disturbances in women with metastatic breast cancer. *Breast J* **8**(6):362–370(2002).

24. Hauri PJ. Cognitive deficits in insomnia patients. *Acta Neurol Belg* **97**(2):113–117(1997).

25. Savard J, et al. Association between subjective sleep quality and depression on immunocompetence in low-income women at risk for cervical cancer. *Psychosom Med* **61**(4):496–507(1999).

26. Simeit R, Deck R, Conta-Marx B. Sleep management training for cancer patients with insomnia. *Support Care Cancer* **12**(3):176–183(2004).

27. Quesnel C, et al. Efficacy of cognitive–behavioral therapy for insomnia in women treated for nonmetastatic breast cancer. *J Consult Clin Psychol* **71**(1):189–200(2003).

28. Young-McCaughan S, et al. Research and commentary: change in exercise tolerance, activity and sleep patterns, and quality of life in patients with cancer participating in a structured exercise program. *Oncol Nurs Forum* **30**(3):441–454(2003).

29. Berger AM, VonEssen S, Kuhn BR, Piper BF, Agrawal S, Lynch JC, Higginbotham P. Adherence, sleep, and fatigue outcomes after adjuvant breast cancer chemotherapy: results of a feasibility intervention study. *Oncol Nurs Forum* **30**:513–522(2003).

30. McTiernan A. Physical Activity After Cancer: Physiologic Outcomes. *Cancer Invest* **22**:68–81 (2004).

31. Valdimarsdottir U, et al. The unrecognised cost of cancer patients' unrelieved symptoms: a nationwide follow-up of their surviving partners. *Br J Cancer* **86**(10):1540–1545(2002).

118

SLEEP IN THE INTENSIVE CARE UNIT

SAMUEL L. KRACHMAN AND WISSAM CHATILA
Temple University School of Medicine, Philadelphia, Pennsylvania

INTRODUCTION

Since the early 1970s, the most severely ill patients in the hospital are cared for in the intensive care unit (ICU). Included are patients with medical, cardiac, or neurological conditions, as well as postoperative surgical and trauma patients. Yet despite and possibly due to the intensive individualized care that is delivered, patients in the ICU are more susceptible to significant sleep deprivation. The cause appears to be multifactorial and includes the type and severity of the patient's underlying illness [1–9], medications received [10, 11], use of hemodynamic and respiratory monitoring devices, use of mechanical ventilation [12–14], and the ICU environment itself [12–18]. Sleep deprivation in the ICU may increase patient morbidity, in terms of its effect on patient cognition [19, 20] as well as immune function [21–24] and the overall healing process [25–28]. This chapter will review normal sleep physiology and then discuss the above-mentioned factors that may lead to sleep deprivation in the ICU patient. Finally, treatment recommendations will be given to help improve sleep for these critically ill patients.

NORMAL SLEEP

Normal sleep is generally categorized as non-rapid eye movement (NREM) and rapid eye movement (REM) sleep based on differences in a number of physiologic parameters. NREM sleep is further subdivided into stages 1 through 4, representing a continuum of sleep, with stages 3 and 4 (also referred to as delta or slow-wave sleep (SWS)) being a more restful sleep with a higher arousal threshold. Although the brain is very active during REM sleep, this stage is characterized by inhibition of spinal motor neurons leading to muscle atonia (referred to as tonic REM) with intermittent bursts of REM and distal muscle twitches (referred to as phasic REM). REM sleep is also considered to be restful sleep with a variable arousal threshold. Dreaming appears to be associated with REM sleep, which originates in the pons.

Most normal individuals have a mean sleep latency of approximately 10–20 minutes. Sleep progresses through stages 1 and 2, with stages 3 and 4 occurring within 10–25 minutes of sleep onset. Most SWS occurs during the first third of the night, accounting for 13–23% of sleep, and decreasing as a percentage of total sleep time with age [29]. REM sleep is usually first observed 70–90 minutes into sleep, with the first REM period lasting just a few minutes. REM sleep then cycles every 90 minutes, usually alternating with stage 2 sleep (which accounts for 45–55% of total sleep time), with increases in REM sleep length and intensity as the night progresses. REM sleep accounts for approximately 20–25% of total sleep time.

Most young adults sleep approximately 7.5–8 hours per night, yet the variability is quite high from individual to individual and from night to night. Most are able to maintain daytime performance and alertness when their sleep is decreased to not less than 5 hours per night. Besides quantity, continuity of sleep also is important and may affect daytime alertness and task performance [29].

Correspondence to: Samuel L. Krachman, Division of Pulmonary and Critical Care, 767 Parkinson Pavilion, Broad and Tioga Sts. Philadelphia, PA 19140.

IDENTIFYING ABNORMALITIES WITH SLEEP IN THE ICU

Before discussing the observed sleep abnormalities in the ICU patient, it is important to first note the techniques that are used to monitor sleep (Table 118.1). Many of the early studies utilized observational techniques to quantify sleep disruption [19, 30], with only a limited number of studies using both observation and more quantifiable techniques, such as continuous electroencephalogram (EEG)

TABLE 118.1 Studies Monitoring Sleep in the ICU

Study	Patient Illness	Number of Patients	Type of Study	Mechanical Ventilation	Findings
Acute Medical Illness					
Hilton [7]	Respiratory failure	10	EEG	No	↓ TST, SWS, REM, ↑ stage 1
Broughton and Baron [8]	Cardiac	12	EEG	No	↓ TST, REM, ↑ awakenings
Gottschlich et al. [31]	Burn injury	11	EEG	No	↑ Stages 1,2, ↓ SWS, REM
Topf and Thompson [39]	Cardiac	97	Questionnaire	No	Noise stress contributes to disturbed sleep
	Cancer	100	Questionnaire	No	>50% reported moderate to severe sleep deprivation
Aaron et al. [18]	COPD	6	EEG	No	Arousals correlates with noise peaks
Gabor et al. [12]	Respiratory failure/trauma	7	EEG	Yes	Noise and patient care account for <30% of arousals/awakenings
Parthasarathy and Tobin [14]	Respiratory failure	11	EEG	Yes	Mode of mechanical ventilation contributes to sleep deprivation
Fontaine [32]	Trauma	20	EEG/Observational	No	Good correlation of objective and observational assessment of sleep in the ICU
Freedman et al. [13]	Respiratory failure	22	EEG	Yes	Only 12% arousals due to ICU noise
Heller et al. [20]	Medical	100	Observational	No	More delirium patients with less sleep
Helton et al. [19]	Medical	62	Observational	No	Correlation between mental status change and poor sleep
Surgery					
Aurell and Elmqvist [33]	Noncardiac	9	EEG	No	↓ TST, SWS, REM
Knill et al. [34]	Abdominal surgery	6	EEG	No	Fragmented sleep recovered after 2–4 nights
Rosenberg et al. [35]	Abdominal surgery	10	EEG	No	REM rebound 2–3 nights after surgery
Rosenberg-Adamsen et al. [36]	Laproscopic surgery	10	EEG	No	↑ Stage 2, ↓ SWS
Orr and Stahl [37]	Cardiothoracic	9	EEG	No	↓ TST and REM
Medical and Surgical					
Jones et al. [38]	Medical/surgical	100	Questionnaire	No	25% with insufficient sleep
Freedman et al. [15]	Medical/surgical	203	Questionnaire	No	Vital signs and phlebotomy more disruptive than noise

Abbreviations: TST, total sleep time; SWS, slow-wave sleep; REM, rapid eye movement; EEG, electroencephalogram.

recordings [7]. Following the 1970s, more studies utilized continuous EEG recordings to quantitate and characterize the sleep deprivation that was observed, as well as abnormalities in circadian rhythm [8, 14, 31–37]. More recently, studies have incorporated simultaneous continuous recordings of sound levels and EEG to better define the role that environmental noise plays in sleep disturbance [12, 13, 18]. Finally, some studies have used patient perception, based on questionnaires, to access the degree of sleep deprivation that was present during their stay in the ICU [9, 15, 38, 39]. Therefore it is important to take into account the monitoring techniques used in these studies that evaluate sleep deprivation in the ICU patient.

FACTORS AFFECTING SLEEP IN THE ICU

Underlying Chronic Disease

Many patients have an exacerbation of a chronic underlying illness that leads to their admission to the ICU. These include patients with chronic obstructive pulmonary disease (COPD) and congestive heart failure (CHF), both of which have been associated with poor sleep quality. Calverley et al. [1] demonstrated an increased sleep latency and increase in interrupted sleep in 20 patients with severe ($FEV_1 <$ 1.0 L) but stable COPD, as compared to a control group. Others have shown similar findings in regard to disturbed sleep in patients with chronic COPD, with an increase in arousals and sleep stage changes [2, 3]. While some have shown a correlation between the development of hypoxemia and the number of arousals during the night [3], others have not [1, 2], with a variable response to oxygen therapy.

Patients with CHF may also demonstrate disturbed sleep, especially those with Cheyne–Stokes respiration (CSR). Present in approximately 45–56% of patients with a left ventricular ejection fraction <40%, CSR is associated with a crescendo–decrescendo alteration in tidal volume separated by central apneas [4–6]. CSR is associated with a decrease in total sleep time, increase in the number of arousals during the night, and an increase in stage 1 and decrease in stages 3 and 4 of sleep [4–6].

Acute Medical Illness

An acute medical illness, either by itself or in the presence of chronic underlying disease, can lead to severe sleep deprivation in the ICU. Hilton [7] studied 10 patients with acute respiratory failure, which included 48 hours of continuous EEG recordings. Patients demonstrated a decrease in total sleep time, increase in stage 1, and decrease in SWS and REM sleep during the first 24 hours of recordings. Only 50–60% of sleep occurred during the night. There was little improvement in sleep architecture over the subsequent 24 hour period.

Broughton and Baron [8] characterized the sleep quality of 12 patients following an acute myocardial infarction. EEG monitoring was done only nocturnally, both after admission to the ICU, and subsequently up to 13 days later when they were transferred to the ward. The initial night following acute myocardial infarction demonstrated very disturbed sleep, with a decrease in total sleep time, increase in awakenings during the night, and decrease in REM sleep, as compared with normal controls. Sleep quality appeared to improve on subsequent nights, with no further improvement noted when patients were transferred to the wards. These findings suggest that factors other than the environment were responsible for the initial night of disturbed sleep, such as biologic and psychological stresses associated with the acute myocardial infarction.

Significant sleep deprivation has also been reported in patients with severe burn injuries. Gottschlich et al. [31] performed biweekly 24 hour EEG recordings in 11 patients (8 ± 2 years) with severe burn injuries. Although total sleep time appeared adequate (625 ± 32 min), sleep architecture was abnormal with an increase in stages 1 and 2 and a decrease in stages 3, 4, and REM. In 40% of the patients, stages 3 and 4 were absent, as was REM sleep in 19%. Similar findings of sleep deprivation have been noted in trauma patients, with a good correlation between objective and subjective assessment [32]. In addition, Nelson et al. [9] noted that >50% of patients with underlying cancer admitted to the ICU complained of disturbed sleep that they subjectively rated as moderate to severe.

Previously unrecognized sleep disordered breathing may also contribute to the sleep deprivation in some patients admitted to the ICU with an acute medical illness [40, 41]. Buckle et al. [40] described nine patients admitted with respiratory failure who underwent polysomnograms while in the ICU due to a suspicion of sleep disordered breathing. All patients were hypercapnic and hypoxic at the time of admission. Polysomnograms demonstrated central or obstructive sleep apnea, as well as alveolar hypoventilation, often associated with obstructive sleep apnea. Acute sleep deprivation in normal subjects has been shown to decrease central motor output to upper airway muscles in response to certain stimuli such as hypercapnia [42]. In patients with severe sleep disordered breathing this could lead to acute respiratory failure. In addition, the effect of sleep deprivation on upper airway muscle function could affect gas exchange and weanability in other disorders that lead to respiratory failure.

Major Surgery

Patients often require ICU monitoring and care following major surgical procedures, including cardiothoracic and abdominal surgery. Similar to patients with underlying medical illness, the postoperative patient often develops severe sleep deprivation when in the ICU. Aurell and Elmquest [33] studied nine patients who underwent major noncardiac surgery with continuous EEG monitoring, from 2 hours following surgery to the time of discharge from the ICU. Total sleep time was severely decreased with 5 patients not sleeping at all during the first 24 hours after surgery. In those patients who did sleep during the first 24 hours, SWS was abolished and REM sleep significantly reduced.

Studies specifically examining sleep patterns following upper abdominal surgery have also noted significant abnormalities. Knill et al. [34] recorded nocturnal sleep by EEG for 2 nights before and 5–6 nights after major abdominal surgery. The decreases in SWS and REM sleep seen following surgery subsequently recovered, with an eventual REM sleep rebound to levels above the preoperative values. Rosenberg et al. [35] noted a similar postoperative disturbance in sleep architecture with a rebound in REM sleep 2–3 nights following major abdominal surgery. In addition, these increased REM periods were associated with significant oxygen desaturation. Less invasive surgeries appear to be less disruptive to sleep [36]. Rosenberg-Adamsen et al. [36] noted no change in total awake time, number of arousals, or percent REM sleep postoperatively in ten patients who underwent laparoscopic cholecystectomy.

Similar findings have been noted in patients undergoing cardiothoracic surgery. Orr and Stahl [37] studied the sleep patterns of six patients following cardiac surgery and three patients who underwent thoracotomy. EEG was recorded preoperatively, while in the ICU, on the ward, and in three patients up to 5 weeks following surgery. Compared to relatively normal preoperative values, patients were awake most of the night following surgery and in five of the six cardiac surgery patients there was no REM sleep noted until the fourth postoperative night. In three patients REM sleep suppression was still noted 2–4 weeks following surgery. In two of the three thoracotomy patients, sleep architecture returned to normal by the time of discharge from the ICU.

Environmental Factors

Many environmental factors appear to contribute to sleep deprivation in the ICU. Included is the noise level in the ICU, related to monitor and ventilator alarms, as well as conversation by staff members [12, 13, 15–18]. In addition, repeated interruptions by staff members, including nurses, physicians, and lab and x-ray technicians, are an important environmental influence that can lead to sleep deprivation [12, 15]. Loss of the normal light–dark cycle and the type of room the patient is in also contribute to disturbed sleep [12].

Environmental noise in the ICU can be significant and has been objectively measured [16–18]. Bentley et al. [16] found an average level of 53 decibels (dB(A)) in the ICU during the day and 43 dB(A) at night. In comparison, a typical living room during the day is 40–50 dB(A) and a quiet bedroom at night is 20–30 dB(A). In addition, there were repeated peaks in the range of 62–72 dB(A), comparable to a busy city street corner. The U.S. Environmental Protection Agency [43] recommends that noise levels in the hospital not exceed 45 dB(A) during the day and 35 dB(A) at night. Meyers et al. [17] monitored light and sound levels in the ICU, as well as in private rooms, continuously for at least 7 days. Sound levels were significantly less in the private rooms, and there was only a mild decrease in sound intensity at night as compared to the day while patients were in the ICU. Aaron et al. [18] measured the number of sound peaks >80 dB(A) while continuously monitoring EEG tracings for 24–48 hours in six patients in the ICU. There was a significant correlation between the number of arousals and the number of sound peaks >80 dB(A) (Figure 118.1), suggesting environmental noise was directly responsible for the amount of disruptive sleep. In addition, Topf and Thompson [39] noted that perceived noise stress, along with other stress factors, significantly contributed to subjective sleep disruption while in the ICU.

Freedman et al. [15] surveyed 203 patients on the day of discharge from the ICU, to determine the perceived effect of environmental stimuli on sleep disruption. Overall sleep quality was perceived to be much worse in the ICU as compared to home. Activities by staff members, such as taking vital signs and phlebotomy, were considered more disruptive to sleep than the level of noise in the unit (Figure 118.2). The medical ICU was perceived to be more disruptive to sleep than either the surgical or cardiac

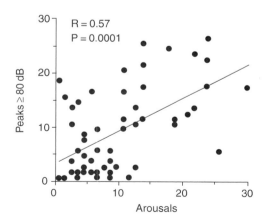

Figure 118.1 Correlation between the number of sound peaks >80 dB(A) and the number of arousals for each 30 minute segment of sleep during the night in six patients in a medical ICU. (From [18], with permission.)

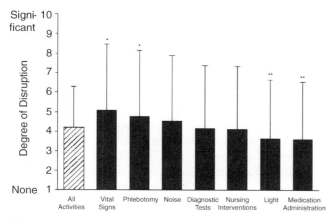

Figure 118.2 Perceived sleep disruption from the ICU environment. There was a significant difference between environmental factors, with recording of vital signs most disruptive to sleep ($p = 0.006$), and light and medication administration less disturbing ($p > 0.05$) than the mean of the other factors. Note that noise in the ICU was not perceived to be disruptive to sleep. (From [15], with permission.)

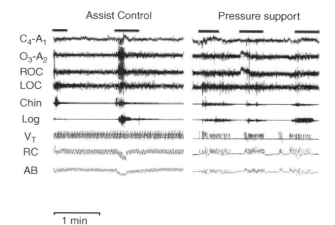

Figure 118.3 Polysomnographic tracing in a representative patient while on assist-control and pressure support ventilation. Note the central apneas while on pressure support ventilation, and the increased number of arousals and awakenings, which are indicated by the horizontal bars. C_4-A_1 and O_3-A_2 represent electroencephalogram; ROC and LOC represent electro-oculogram; Chin and Leg represent electromyograms; integrated tidal volume (V_T), ribcage (RC), and abdominal (AB) excursions recorded using inductive plethysmography. (From [14], with permission.)

intensive care units. In a subsequent study, Freedman et al. [13] continuously monitored noise levels and EEG tracings in 22 patients (20 on mechanical ventilation) in the ICU and found that only 12% of arousals and 17% of awakenings from sleep were secondary to environmental noise. Also of interest was that the majority of total sleep time occurred during the day. Gabor et al. [12] reported similar findings in seven patients in the ICU receiving mechanical ventilation. Together, noise and patient care activities accounted for less than 30% of all arousals and awakenings.

Mechanical Ventilation

Not until recently has it been shown that mechanical ventilation itself may contribute to sleep deprivation in the ICU [12–14]. The fact that approximately 40% of patients in the ICU are on mechanical ventilation [44] demonstrates the importance of these findings. Gabor et al. [12] noted that the majority of patients' arousals and awakenings remained unexplained, and mechanical ventilation may have been partially responsible. Others have noted similar findings in patients receiving mechanical ventilation [13].

Factors other than noise that may be contributing to sleep deprivation in the mechanically ventilated patient include the patient's underlying illness, the presence of an endotracheal tube or mask, suctioning the patient, or the mode of mechanical ventilation that is utilized [14, 45]. Meza et al. [45] demonstrated that healthy subjects, when placed on pressure support ventilation, develop central apneas when their P_{CO_2} decreases below the apneic threshold. Whether these central apneas were associated with arousals and sleep disruption was unknown at the time.

More recently, Parthasarathy and Tobin [14] evaluated whether assist-control ventilation with a backup rate caused less sleep fragmentation than pressure support ventilation in 11 critically ill patients. Six of the 11 patients developed central apneas during pressure support ventilation, but not during assist-control (Figure 118.3). Heart failure was more common among the patients with central apneas (83% versus 20%), and the number of central events decreased with the addition of dead space. In the 11 patients, sleep fragmentation (arousal and awakening frequency) was greater with pressure support ventilation based on the frequency of awakenings. In the six patients who developed central apneas with pressure support, both the arousal and awakening frequency were greater when on pressure support and decreased with the addition of dead space. An increase in tidal volume during sleep, associated with a decrease in P_{CO_2} below the apneic threshold, appeared responsible for the central apneas and associated sleep fragmentation seen with pressure support. Therefore the mode of mechanical ventilation may contribute to the sleep deprivation observed in critically ill patients.

Medications

Most patients in the ICU are on a number of medications, many of which can have an effect on sleep quality. Sedative and narcotic medications are frequently used to control anxiety and pain, respectively; however, the medications themselves may affect normal sleep physiology. Opiates have been shown to decrease REM sleep and increase stage 1 sleep and the number of arousals during the night [10]. Benzodiazepines have an effect on SWS, with an increase during the first 2 hours of sleep seen with a single initial dose, but then a subsequent decrease [11]. Repeated use of benzodiazepines may abolish stage 4 sleep, which is restorative sleep.

Although the effects of medications on normal sleep quality are important, it is also essential to adequately control pain and anxiety in these patients, as these entities significantly contribute to sleep deprivation. Treggiari-Venzi et al. [46] compared the effects of midazolam and propofol on self-reported anxiety, depression, and sleep quality over a 5 day period in 40 nonintubated patients following trauma or surgery. While the level of anxiety and depression remained high in a significant and similar percentage of patients at the end of 5 days, sleep quality tended to improve, with no significant difference between the two groups.

Other medications commonly used in the ICU include intravenous catecholamines for hemodynamic support. Among them are norepinephrine, which as a neurotransmitter normally is associated with the cortical activating system to maintain wakefulness, and dopamine, which has an integral role in behavioral arousal [47]. However, Oldendorf [48] demonstrated that, when administered systemically to rats, these catecholamines do not readily cross the blood–brain barrier. Whether similar limited central effects occur in humans is unknown at this time. Another frequently used medication in the ICU, especially in those on mechanical ventilation, is inhaled beta agonists. In asthmatics, these medications have not been shown to alter total sleep time or sleep efficiency, even when given at bedtime [49].

POTENTIAL IMPORTANCE OF SLEEP IN THE ICU

Healing Process

In both healthy individuals and the critically ill, the exact physiologic importance of sleep remains uncertain. Many believe that sleep is an essential restorative process, in regard to its effects on both protein synthesis and cellular function [25, 26, 29]. Circadian variations are present, with peak activities in both protein synthesis and cellular division noted during sleep [25, 26, 29].

Observed circadian variations in hormone secretion have led to the suggestion that wakefulness enhances catabolism, while sleep shifts metabolism in favor of anabolism. Cortisol and catecholamines are at their highest levels during the day and are known to inhibit protein synthesis. In addition, growth hormone secretion is known to occur during sleep, particularly slow-wave sleep, which stimulates protein and RNA synthesis and amino acid uptake by cells. Garlick et al. [27] studied protein synthesis and catabolism in six obese subjects using $I^{14}C$-leucine over a 24 hour period. Despite

a 33% observed decrease in protein synthesis noted during sleep, the overall contribution of protein oxidation to total energy expenditure decreased from 27% during the day to 13% during the night.

The effects of sleep deprivation on protein metabolism have also been investigated. Scrimshaw et al. [28] examined the effects of 48 hours of sleep deprivation on protein metabolism in a group of normal adult males. An increase in urinary nitrogen excretion suggested the development of a catabolic state. Rechtschaffen et al. [50] found that sleep deprivation in rats led to death, associated with the development of gastric ulcers, internal hemorrhage, and pulmonary edema. Others have disagreed with the restorative hypothesis, noting that circadian peaks occur in the absence of sleep during deprivation studies [51].

Sleep deprivation has also been shown to affect immune system function and thus may impact the ability to resist and fight infection [21–24]. Palmblad et al. [21] noted that 77 hours of sleep deprivation decreased leukocyte phagocytosis in eight normal females, while lymphocyte interferon production (known to be involved with antiviral defense) increased. In a subsequent study, the same group noted that while 48 hours of sleep deprivation decreased lymphocyte DNA synthesis in 12 normal males, neutrophil function was not affected [22]. More recently, Irwin et al. [23] found that in 42 healthy male subjects, early night partial sleep deprivation (from 10 p.m. to 3 a.m.) led to a decrease in natural killer cell and lymphokine-activated killer cell number and function. However, Dinges et al. [24] noted a leukocytosis and increase in natural killer cell activity in 20 healthy volunteers over 64 hours of sleep deprivation. The increase in immune function appeared to parallel changes in neurobehavioral function, suggesting both were associated with a biological pressure to sleep. The noted changes in immune function and neurobehavioral function both normalized during sleep recovery. Although an overall increase in immune function was noted, a transient decrease in natural killer cell number and activity was noted after the first night of sleep deprivation. Despite some conflicting results, these studies suggest that sleep deprivation could potentially affect the healing process in critically ill patients, both in terms of tissue repair and cellular immune function.

ICU Syndrome

Another potential consequence of sleep deprivation in the ICU is the development of mental status changes 3–7 days following admission, in what has been referred to as the "ICU syndrome." Based on observational data, Helton et al. [19] determined that 24% of their ICU patients developed severe sleep deprivation (>50% sleep loss) and 16% moderate sleep deprivation (<50% sleep loss) during the first 5 days following admission. A significant correlation was noted between the severity of sleep deprivation and the development of mental status changes, with 33% of severe and 10% of moderate sleep-deprived patients demonstrating mental status changes. Heller et al. [20] noted delirium in 24% of their patients a mean of 3.8 days following cardiac surgery. Sleep deprivation was based on observational data, with a greater proportion of patients with delirium (27%) having below average sleep compared to the group without delirium (11%). Factors other than poor sleep quality appeared to be important in the development of postoperative delirium, including age, severity of illness, and the amount of time spent on cardiopulmonary bypass.

TREATMENT

It is important that physicians, nurses, and ancillary staff be aware of factors shown both objectively [12–14] and based on patient perception [15] to be important in contributing to sleep deprivation in the ICU. Although noise has been shown to be responsible for <30% of arousals or awakenings in the ICU [12], it still can play a significant role, and modifications to reduce environmental noise should be introduced. These include minimizing talk between staff, especially during the night, and placement of all nonessential monitoring equipment outside the patient's room [52]. In addition, noise-creating procedures should be minimized and planned, when possible, for daytime hours.

Factors shown to be more disruptive than noise in the ICU, such as the taking of vital signs and phlebotomy [15], should be minimized during nighttime hours. Treatment of the underlying acute illness is also important in trying to improve sleep quality. Attempts should be made to simulate the normal 24 hour light–dark cycle, by dimming and turning off lights during the night and encouraging bright light, preferably sunlight, during the day. In healthy subjects, individualized rooms have been shown to quantitatively improve sleep quality in the ICU [12]. Rooms with a large window are recommended to help reduce the development of the ICU syndrome [52]. In addition, adequate pain control and treatment of any associated anxiety should be addressed, as these have been shown to contribute to perceived sleep deprivation [38]. Use of hypnotics that don't affect normal sleep architecture, such as chloral hydrate or zolpidem [53], can be considered on an individual basis [54]. Finally, in patients on mechanical ventilation, choosing the correct mode of ventilation can have a significant effect on sleep quality and continuity [14].

SUMMARY

Sleep deprivation is common in patients admitted to the ICU and appears to be secondary to multiple causes, including the patient's underlying illness, the ICU environment, and, in patients receiving mechanical ventilation, the mode of ventilatory support. Sleep deprivation can affect cognitive behavior, as well as cellular immune function and tissue repair, and thus may have an impact on morbidity and mortality. Therapies should be directed at the underlying causes, with efforts made toward optimizing the sleeping environment and minimizing the number of unnecessary interruptions, particularly during the night. Whether these therapeutic interventions will have an effect on overall outcome has yet to be determined.

REFERENCES

1. Calverley PMA, Brezinov V, Douglas NJ, Catterall JR. The effects of oxygenation on sleep quality in chronic bronchitis and emphysema. *Am Rev Respir Dis* **126**:206–210(1982).

2. Brezinova V, Catterall JR, Douglas NJ Calverley PMA, Flenley DC. Night sleep of patients with chronic ventilatory failure and age matched controls: number and duration of the EEG episodes of intervening wakefulness and drowsiness. *Sleep* **5**:123–130(1982).

3. Fleetham J, West P, Mezon B, Conway W, Roth T, Kryger M. Sleep, arousals, and oxygen desaturation in chronic obstructive pulmonary disease: the effects of therapy. *Am Rev Respir Dis* **126**:429–433(1982).

4. Findley LJ, Zwillich CZ, Ancoli-Israel S, Kripke D, Tisi G, Moser KM. Cheyne–Stokes breathing during sleep in patients with left ventricular heart failure. *S Med J* **78**:11–15(1985).

5. Hanly PJ, Millar TW, Steljes DG, Baert R, Frais MA, Kryger MH. Respiration and abnormal sleep in patients with congestive heart failure. *Chest* **96**:480–488(1989).

6. Krachman SL, D'Alonzo GE, Berger TJ, Eisen HJ. Comparison of oxygen therapy with nasal continuous positive airway pressure on Cheyne–Stokes respiration during sleep in congestive heart failure. *Chest* **116**:1550–1557(1999).

7. Hilton BA. Quantity and quality of patients' sleep and sleep-disturbing factors in a respiratory intensive care unit. *J Adv Nurs* **1**:453–468(1976).

8. Broughton R, Baron R. Sleep patterns in the intensive care unit and on the ward after acute myocardial infarction. *Electroencephalogr Clin Neurophysiol* **45**:348–360(1978).

9. Nelson JE, Meier DE, Oei EJ, Nierman DM, Senzel RS, Manfredi PL, Davis SM, Morrison RS. Self-reported symptom experience of critically ill cancer patients receiving intensive care. *Crit Care Med* **29**:277–282(2001).

10. Bradley CM, Nicholson AN, Viveash JP. Opioids and non-opioids. In: Klepper ID, Saunders LD, Rosen M (Eds), *Ambulatory Anaesthesia and Sedation: Impairment and Recovery*. Blackwell Scientific Publications, Oxford, 1991, pp 218–234.

11. Gallard JM, Blois R. Effect of benzodiazepine antagonist R:15-1788 on flunitrazepam-induced sleep changes. *Br J Clin Pharmacol* **15**:529–536(1983).

12. Gabor JY, Cooper AB, Crombach SA, Lee B, Kadikar N, Bettger HE, Hanly PJ. Contribution of intensive care unit environment to sleep disruption in mechanically ventilated patients and healthy subjects. *Am J Respir Crit Care Med* **167**:708–715(2003).

13. Freedman NS, Gazendam J, Levan L, Pack AI, Schwab RJ. Abnormal sleep/wake cycles and the effect of environmental noise on sleep disruption in the intensive care unit. *Am J Respir Crit Care Med* **163**:451–457(2001).

14. Parthasarathy S, Tobin MJ. Effect of ventilator mode on sleep quality in critically ill patients. *Am J Respir Crit Care Med* **166**:1423–1429(2002).

15. Freedman NS, Kotzer N, Schwab RJ. Patient perception of sleep quality and etiology of sleep disruption in the intensive care unit. *Am J Respir Crit Care Med* **159**:1155–1162(1999).

16. Bentley S, Murphy F, Dudley H. Perceived noise in surgical wards and an intensive care area: an objective analysis. *Br Med J* **2**:1503–1506(1977).

17. Meyers TJ, Eveloff SE, Bauer MS, Schwartz WA, Hill NS, Millman RP. Adverse environmental conditions in the respiratory and medical ICU settings. *Chest* **105**:1211–1216(1994).

18. Aaron JN, Carlisle CC, Carskadon MA, Meyer TJ, Hill NS, Millman RP. Environmental noise as a cause of sleep disruption in an intermediate respiratory care unit. *Sleep* **19**:707–710(2002).

19. Helton MC, Gordon SH, Nunnery SL. The correlation between sleep deprivation and the intensive care unit syndrome. *Heart Lung* **9**(3):464–468(1980).

20. Heller SS, Frank KA, Malm JR, Bowman FO, Harris PD, Charlton MH, Kornfeld DS. Psychiatric complications of open-heart surgery. *N Eng J Med* **283**(19):1015–1020(1970).

21. Palmblad J, Cantell K, Strander H, Froberg J, Karlsson CG, Levi L, Granstrom M, Unger P. Stressor exposure and immunological response in man: interferon-producing capacity and phagocytosis. *Psychosom Res* **20**:193–199(1976).

22. Palmblad J, Petrini B, Wasserman J, Akerstedt T. Lymphocyte and granulocyte reactions during sleep deprivation. *Psychosom Med* **41**(4):273–278(1974).

23. Irwin M, McClintick J, Costlow C, Fortner M, White J, Gillin JC. Partial night sleep deprivation reduces natural killer and cellular immune responses in humans. *FASEB J* **10**:643–653(1996).

24. Dinges DF, Douglas SD, Zaugg L, Campbell DE, McMann JM, Whitehouse WG, Orne EC, Kapoor SC, Icaza E, Orne MT. Leukocytosis and natural killer cell function parallel neurobehavioral fatigue induced by 64 hours of sleep deprivation. *J Clin Invest* **93**:1930–1939(1994).

25. Adam K, Oswald I. Protein synthesis, bodily renewal and the sleep–wake cycle. *Clin Sci* **65**:561–567(1983).

26. Adam K, Oswald I. Sleep helps healing. *Br Med J* **289**:1400–1401(1984).

27. Garlick PJ, Clugston GA, Swick RW, Waterlow JC. Diurnal pattern of protein and energy metabolism in man. *Am J Clin Nutr* **33**:1983–1986(1980).

28. Scrimshaw NS, Habicht JP, Pellet P, Piche ML, Cholakos B. Effects of sleep deprivation and reversal of diurnal activity on protein metabolism of young men. *Am J Clin Nutr* **19**:313–319(1966).

29. Carskadon MA, Dement WC. Normal human sleep: an overview. In: Kryger MH, Roth T, Dement WC (Eds), *Principles and Practice of Sleep Medicine*, 3rd ed. Saunders, Philadelphia, 2000, pp 15–25.

30. Dlin BM, Rosen H, Dickstein K, Lyons JW, Fischer HK. The problems of sleep and rest in the intensive care unit. *Psychosomatics* **12**:155–163(1971).

31. Gottschlich MM, Jenkins ME, Mayes T, Khoury J, Kramer M, Warden GD, Kagan RJ. The 1994 Clinical Research Award. A prospective clinical study of the polysomnographic stages of sleep after burn injury. *J Burn Care Rehabil* **15**:486–492(1994).

32. Fontaine DK. Measurement of nocturnal sleep patterns in trauma patients. *J Acute Crit Care* **18**:402–410(1989).

33. Aurell J, Elmquist D. Sleep in the surgical intensive care unit: continuous polygraphic recording of sleep in nine patients receiving postoperative care. *Br Med J* **290**:1029–1032(1985).

34. Knill RL, Moote CA, Skinner MI, Rose EA. Anesthesia with abdominal surgery leads to intense REM sleep during the first postoperative week. *Anesthesiology* **73**:52–61(1990).

35. Rosenberg J, Wildschiodtz G, Pedersen MH, von Jessen F, Kehlet H. Late postoperative nocturnal episodic hypoxaemia and associated sleep pattern. *Br J Anaesth* **72**:145–150(1994).

36. Rosenberg-Adamsen S, Skarbye M, Wildschiodtz G, Dehlet H, Rosenberg J. Sleep after laparoscopic cholecystectomy. *Br J Anaesth* **77**:772–775(1996).

37. Orr WC, Stahl ML. Sleep disturbances after open heart surgery. *Am J Cardiol* **39**:196–201(1977).

38. Jones J, Hoggart B, Whithey J, Donaghue K, Ellis BW. What the patients say: a study of reactions to an intensive care unit. *Intensive Care Med* **5**:89–92(1979).

39. Topf M, Thompson S. Interactive relationships between hospital patients' noise-induced stress and other stress with sleep. *Heart Lung* **30**:237–243(2001).

40. Buckle P, Pouliot Z, Millar T, Kerr P, Kryger MH. Polysomnography in acutely ill intensive care unit patients. *Chest* **102**:288–291(1992).

41. Hara KS, Shepard JW. Sleep and critical care medicine. In: Martin RJ (Ed), *Cardiorespiratory Disorders During Sleep*, 2nd ed. Futura Publishing, 1990, pp 323–363.

42. Leiter JC, Knuth SL, Barlett D. The effect of sleep deprivation on activity of the genioglossus muscle. *Am Rev Respir Dis* **132**:1242–1245(1985).

43. US Environmental Protection Agency. Information on levels of environmental noise requisite to protect public health and welfare with an adequate margin of safety. US Government Printing Office, Washington DC, 1974.

44. Esteban A, Anzueto A, Alia I, Gordo F, Apezteguia C, Palizas F, Cide D, Goldwaser R, Soto L, Bugedo G, Rodrigo C, Pimentel J, Raimondi G, Tobin MJ. How is mechanical ventilation employed in the intensive care unit? An international

utilization review. *Am J Respir Crit Care Med* **161**:1450–1458(2000).

45. Meza S, Mendez M, Ostrowski M, Younes M. Susceptibility to periodic breathing with assisted ventilation during sleep in normal subjects. *J Appl Physiol* **85**:1929–1940(1998).

46. Treggiari-Venzi M, Borgeat A, Fuchs-Buder T, Gachoud JP, Suter PM. Overnight sedation with midazolam or propofol in the ICU: effects on sleep quality, anxiety and depression. *Intensive Care Med* **22**:1186–1190(1996).

47. Jones BE. Basic mechanisms of sleep–wakes states. In: Kryger MH, Roth T, Dement WC (Eds), *Principles and Practice of Sleep Medicine*, 3rd ed. Saunders, Philadelphia, 2000, pp 134–154.

48. Oldendorf WH. Brain uptake of radiolabeled amino acids, amines, and hexoses after arterial injection. *Am J Physiol* **221**:1629–1639(1971).

49. Neagley SR, White DP, Zwillich CW. Breathing during sleep in stable asthmatic subjects. *Chest* **90**:334–337(1986).

50. Rechtschaffen A, Gilliland MA, Bergmann BM, Winter JB. Physiological correlates of prolonged sleep deprivation in rats. *Science* **221**:182–184(1982).

51. Horne JA. Human sleep and tissue restitution: some qualification and doubts. *Clin Sci* **65**:569–578(1983).

52. Kornfield DS. Psychiatric view of the intensive care unit. *Br Med J* **1**:108–110(1969).

53. Merlotti L, Roehrs T, Koshorek G, Zorick F, Lamphere J, Roth T. The dose effects of zolpidem on the sleep of healthy normals. *J Clin Psychopharmacol* **9**:9–14(1989).

54. Hartmann E, Cravens J. The effects of long term administration of psychotropic drugs on human sleep: V. The effects of chloral hydrate. *Psychopharmacologia* (*Berl*) **33**:219–232(1973).

119

SLEEP AND THE CARDIAC SURGERY PATIENT

NANCY S. REDEKER
University of Medicine and Dentistry of New Jersey, School of Nursing, Newark, New Jersey

CHRISTINE HEDGES
Ann May Center for Nursing Meridian Health, New Jersey

INTRODUCTION

Sleep deprivation, including decreased quantity, increased fragmentation, and decreased quality of sleep, is prevalent in adults who have undergone cardiac surgery and has been shown to be associated with decrements in postoperative physical function and emotional well-being [1, 2]. Disturbed sleep may also have a significant impact on morbidity and quality of life, although these consequences have been understudied. Numerous intrinsic (e.g., age, gender, illness, primary sleep disorders) and extrinsic (e.g., environment, medical treatment) factors may influence sleep patterns in these patients from the preoperative period throughout the recovery period. However, long-term improvements in sleep may occur secondary to the beneficial cardiovascular effects of the cardiac surgical procedure itself (e.g., revascularization or correction of valvular or other structural defects). In this chapter, we discuss the characteristics of sleep disturbance after cardiac surgery and review the literature on the factors that may contribute to sleep. Implications for clinical practice are discussed. Because virtually all of these studies have focused on the sleep of adult patients undergoing coronary artery bypass and valve replacement procedures performed with cardiopulmonary bypass, the focus of this review is on this body of work.

CHARACTERISTICS OF SLEEP DURING THE EARLY POSTOPERATIVE PERIOD

Sleep of hospitalized postoperative cardiac surgical patients is characterized by self-reports of poor nocturnal sleep quality, short duration, high degrees of fragmentation, and large amounts of daytime sleep, especially during the first few postoperative days [3–5]. Cardiac surgical patients have been shown to have as much as half of their total daily sleep time occurring during the day in the very early postoperative period [4–6]. Wrist actigraphic estimates of sleep duration, fragmentation, and temporal patterning corroborated earlier self-reported findings. Studies that employed polysomnography [3, 7–9] documented high degrees of sleep fragmentation and little REM or slow-wave sleep in small groups of patients. Awakenings occurred frequently and were not always associated with external stimuli. Other researchers have also documented high degrees of self-reported sleep disturbance among cardiac surgery patients [10].

CHARACTERISTICS OF SLEEP DURING RECOVERY AFTER CARDIAC SURGERY

Although past studies focused on the early postoperative period during hospitalization and repeatedly demonstrated high degrees of sleep fragmentation during this time period, sleep disturbance appears to be a dynamic phenomenon over the course of recovery after cardiac surgery. Redeker and colleagues [5] reported that although sleep was highly fragmented during the early postoperative period, it became less fragmented and more consolidated during the nocturnal period (reflecting less daytime napping) toward the end of the first postoperative week. Improvements in sleep consolidation and reductions in daytime sleep continued through the sixth month after hospital discharge in a group of female coronary artery bypass patients [5]. These changes were consistent with self-reports of improved sleep on the Sickness Impact Profile sleep–rest subscale [5] and improvements in the sleep scale of the Nottingham Health Profile in another study of men [3]. A more recent study of 72 men and women, of whom the majority had coronary artery bypass, demonstrated that, overall, sleep fragmentation improved from the first postoperative through the eighth postoperative week [11]. Self-reported sleep disturbance returned to preoperative levels at 8 weeks. However, duration of the nocturnal sleep period, efficiency, and nocturnal movement were significantly poorer at 8 weeks than at baseline (the week prior to surgery).

Researchers who compared preoperative and postoperative sleep have found that self-reports of sleep were improved at the sixth [12] and twelfth months [4, 13] compared with preoperative sleep. However, sleep disturbance appears to continue in some patients. As many as 68% of patients report disturbed sleep at 6 months [14]. Hunt and colleagues [1] reported that only 36.7% of coronary artery bypass patients slept "well most of the time" and approximately 40% reported sleeping "poorly" at least most of the time at 12 months. Improvements in sleep beyond the first few postoperative months likely reflect the beneficial effects of correction of the underlying cardiac problem and associated improvements in function and symptoms. Nevertheless, the continued sleep disturbance in some patients suggests the need for continued assessment. Taken together, these findings suggest the need to address patterns of sleep disturbance across the period of recovery.

FACTORS THAT CONTRIBUTE TO POSTOPERATIVE SLEEP DISTURBANCE

Environmental and treatment characteristics and patient attributes appear to contribute to postoperative sleep disturbance, and the contributing factors are likely to be multifactorial. Redeker and colleagues [15] reviewed this literature and presented a conceptual framework to depict these influences. However, they concluded that there has been little examination of the influences of sleep from a multivariate perspective that accounts for the interrelationships among these factors.

Environmental Influences on Sleep

Characteristics of the hospital environment, including lighting, noise, and frequent patient care interactions with health-care providers, appear to contribute to the sleep disturbances of cardiac surgery patients during the early postoperative period. It is likely that the contributions of environmental factors during hospitalization after cardiac surgery are similar to those found in other acute care patients [16]. For example, Simpson and colleagues [17] found positive relationships between the frequency and severity of self-reported sleep-disturbing factors and self-reported sleep disturbance among coronary artery bypass patients, and Redeker and colleagues [18] reported that 48% of a group of 29 postoperative cardiac surgery patients reported that the primary reasons for their sleep disturbance were interventions by health-care providers and environmental noise. However, neither of these studies demonstrated cause and effect, and the findings are likely to be confounded by the coincidence of the environmental influences with the dramatic physiological, emotional, and symptom (e.g., pain) changes associated with the early postoperative period. Reducing external stimuli and improving pain and symptom management may improve sleep during hospitalization and over the course of recovery. However, there have been no systematic studies of the impact of such interventions on sleep during this time period.

Individual Factors that May Influence Sleep After Cardiac Surgery

Characteristics of individual patients, such as primary sleep disorders, comorbid medical or psychiatric illness, aging, and gender, may influence sleep during the postoperative period. However, these potential correlates have been understudied. Evidence for the contributions of these factors is reviewed below.

Primary sleep disorders, such as insomnia and sleep disordered breathing (Cheyne–Stokes breathing and obstructive sleep apnea/hypopnea syndrome), are quite common in middle-aged adults, and there is growing evidence of a higher prevalence of these conditions in patients with hypertension and coronary artery disease. Some of the risk factors for obstructive sleep apnea–hypopnea syndrome (OSAHS) (e.g., obesity and postmenopausal status) are shared by both OSAHS patients and those with coronary disease. Therefore it seems logical that primary sleep disorders might be present preoperatively in cardiac surgery patients and may explain some of the sleep disturbance experienced during the recovery period. This supposition is supported by the findings of a longitudinal study [11] in which preoperative sleep quality, sleep efficiency, and daytime sleep duration predicted sleep patterns during the first, fourth, and eighth postoperative weeks, with more of the variance explained in sleep during the fourth and eighth weeks. Mean sleep efficiency (measured with actigraphy) was 83% preoperatively, and the group mean score on the Pittsburgh Sleep Quality Index was above the established cutoff for disturbed sleep. Therefore it is evident that cardiac surgery patients experience preoperative sleep disturbance and it appears to explain some of the disturbed sleep observed during recovery. To our knowledge, there have been no polysomnographic studies to evaluate the sleep architecture or cardiorespiratory alterations in these patients. Such studies are needed to evaluate the prevalence of sleep disordered breathing, periodic limb movements, or other primary sleep disorders in this population.

Emotional distress, including anxiety and depression, is common among cardiac surgical patients, and such disturbances likely go hand in hand with disturbed sleep among cardiac patients, as in other populations. Disturbed sleep appears to be associated with disturbed mood during the preoperative period [7, 19]. However, studies have not consistently found that emotional distress was associated with postoperative sleep disturbance. Sleep has been shown to be related to anxiety [7] but unrelated to postoperative total mood disturbance [19]. Unstudied is the extent to which pre- or postoperative clinical depression or depressive symptoms are associated with sleep from the preoperative through the postoperative period. Given concerns about the impact of untreated clinical depression on physiological and functional outcomes in cardiac patients and the well-documented relationships among depressive symptoms and insomnia, depression is a potentially important consideration in addressing the sleep of cardiac surgical patients.

The mean age of cardiac surgical patients is over 65 years in many centers. Therefore aging may be an important, albeit understudied, consideration when examining the sleep of postoperative patients. Older, hospitalized cardiac surgery patients (age ≥ 66 years) have reported significantly poorer sleep quality, in addition to less nocturnal sleep time, and longer but not more frequent awakenings, measured with wrist actigraphy, than middle-aged persons (age <66 years) [16]. However, a larger, more recent study [2, 11] did not corroborate these findings.

The patient's gender may be another important consideration relative to sleep after cardiac surgery. However, aging, gender, and comorbidity may be confounded in most correlational analyses, given that women tend to be older and have more severe cardiac disease at the time of the surgical procedure. Few studies have specifically examined sleep and gender in cardiac surgery, and the little available evidence is conflicting. In a recent study, there were no gender differences in pre- or postoperative sleep [11]. However, an earlier study documented that women had poorer scores on the sleep–rest subscale of the Sickness Impact Profile at 1 year after surgery than men [20].

MEDICAL AND SURGICAL TREATMENT

Medical and surgical treatment most certainly has an impact on the sleep of postoperative cardiac surgical patients. The nature of the surgical procedures, anesthesia and pain management practices, use of cardiopulmonary bypass, and postoperative pain and medication management are all likely to influence differences in postoperative sleep. However, there have been no studies that compared the sleep of patients undergoing different surgical or postoperative care procedures. Therefore the potential implications are speculative at best. Dramatic changes in pre- and postoperative management of cardiac surgery patients over the past several years make it difficult to apply the findings of early studies to todays patients. For example, so-called "fast-track" procedures in which faster acting anesthetic agents, more rapid extubation, and shorter intensive care unit stays are used may reduce the negative impact of anesthesia and exposure to the sleep-disturbing environment of the intensive care unit. "Off-pump" procedures performed without cardiopulmonary bypass may reduce the negative impact of cardiopulmonary bypass on the brain and may thereby improve postoperative sleep, compared with traditional "on-pump" procedures. Minimally invasive procedures may improve postoperative sleep by reducing exposure to the hospital environment through reduced length of stay and reducing pain. Improvements in postoperative pain management, earlier ambulation, and earlier hospital discharge are other factors that may have a beneficial impact on postoperative sleep. Understudied are the sleep patterns of patients who have urgent or emergency surgery or those who have complicated recoveries, as the majority of past studies have focused on the sleep of patients with overall positive outcomes, due to the increased feasibility of long-term follow-up of these patients for research purposes.

IMPLICATIONS FOR PATIENT CARE

In the absence of published clinical trials that systematically examined the effects of sleep-promoting interventions provided specifically to patients undergoing cardiac surgery, implications for patient care must be drawn from a summary of the extant descriptive literature. Analysis of existing study findings suggests the following conclusions: (1) many patients experience sleep disturbance, manifested as frequent awakenings, short sleep duration, perceptions of poor sleep quality, and large quantities of daytime napping;

(2) preoperative sleep pattern disturbance appears to predict some of the variance in postoperative sleep problems; (3) sleep patterns appear to be most disturbed during the early postoperative period, a time that also coincides with the excessive noise, lighting, frequent patient care interactions in the hospital environment, as well as pain and other symptoms, and the effects of anesthetic agents and other medications; and (4) sleep patterns appear to improve over the course of recovery and may continue to improve over baseline once the patient has passed the first few postoperative weeks when pain and other sources of distress negatively influence sleep; improvements over baseline may reflect improvements associated with improved cardiovascular function. Less is known, however, about the characteristics of the patient, treatment, or the hospital environment to alterations in these patterns, and these factors are likely to be multifactorial. Lastly, there has been little systematic study of sleep-promoting interventions for patients undergoing cardiac surgery.

IMPLICATIONS FOR PRACTICE

Careful assessment of preoperative patients for primary sleep disorders may assist in identifying those patients who are at highest risk for postoperative sleep disturbance. These patients are likely to benefit from the use of nasal continuous positive airway pressure (NCPAP) for sleep disordered breathing or the judicious use of hypnotics or behavioral strategies to treat insomnia, for example. During the early postoperative period, management of environmental stimuli, including reductions in noise, lighting, and the frequency of intrusive patient care interactions, may facilitate sleep. Adequate medication for pain is also an important consideration. NCPAP should be made available for those patients with sleep disordered breathing during the postoperative period.

Obesity is a shared risk factor for both obstructive sleep apnea–hypopnea syndrome (OSAHS) and coronary artery disease. While weight reduction is an important strategy for reduction of risk of future cardiac morbidity and mortality, it is particularly important for patients who may have OSAHS.

Although sleep disturbance improves over the course of recovery, its persistence in many patients suggests the need for continued assessment and treatment during this time period. Because sleep disorders are often comorbid with psychiatric–mental health disorders, patients who are depressed or anxious may be at particular risk for insomnia. Recent findings on the associations between OSAHS and cardiovascular morbidity and mortality emphasize the importance of evaluation and treatment of this condition.

REFERENCES

1. Hunt JO, Hendrata MV, Myles PS. Quality of life 12 months after coronary artery bypass graft surgery. *Heart Lung J Acute Crit Care* **29**:401–411(2000).

2. Redeker NS, Ruggiero J, Hedges C. Sleep is related to physical function and emotional wellbeing after cardiac surgery. *Nurs Res* **53**:154–162(2004).

3. Edell-Gustafson UM, Hetta JE, Aren CB. Sleep and quality of life assessment in patients undergoing coronary artery bypass grafting. *J Adv Nurs* **29**(5):1213–1220(1999).

4. Hedner J, Caidahl K, Sjoland H, et al. Sleep habits and their association with mortality during 5-year follow-up after coronary artery bypass surgery. *Acta Cardiol* **57**(3):341–348(2002).

5. Redeker NS, Mason DJ, Wykpisz E, et al. Sleep patterns in women after coronary artery bypass surgery. *Appl Nurs Res* **9**:115–122(1996).

6. Knapp-Spooner C, Yarcheski A. Sleep patterns and stress in patients having coronary bypass. *Heart Lung* **21**:342–349(1992).

7. Edell-Gustafson UM, Hetta JE. Anxiety, depression, and sleep in male patients undergoing coronary artery bypass surgery. *Scand J Caring Sci* **13**:137–143(1999).

8. Johns MW, Large AA, Masterson JP, et al. Sleep and delirium after open heart surgery. *Br J Surg* **61**:377–381(1974).

9. Orr WC, Stahl ML. Sleep disturbances after open heart surgery. *Am J Cardiol* **39**:196–201(1977).

10. Simpson T, Lee E, Cameron C. Patients' perceptions of environmental factors that disturb sleep after cardiac surgery. *Am J Crit Care* **5**:173–181(1996).

11. Redeker NS, Ruggiero J, Hedges C. Patterns and predictors of sleep disturbance after cardiac surgery. *Res Nurs Health* **27**:217–224(2004).

12. Lukkarinen H. Quality of life in coronary artery disease. *Nurs Res* **47**(6):337–343(1998).

13. Chocron S, Tatou E, Schjoth B, et al. Perceived health status in patients over 70 before and after open-heart operations. *Age Ageing* **29**(4):329–334(2000).

14. Schaefer KM, Swavely D, Rothenberger C, et al. Sleep disturbances post coronary artery bypass surgery. *Prog Cardiovasc Nurs* **11**:5–14(1996).

15. Redeker NS, Hedges C. Sleep during hospitalization and recovery after cardiac surgery. *J Cardiovasc Nurs* **17**(1):5–68(2002).

16. Redeker NS. Sleep in acute care settings: an integrative review. *J Nurs Scholarship* **32**(1):31–38(2000).

17. Simpson T, Lee ER, Cameron C. Relationships among sleep dimensions and factors that impair sleep after cardiac surgery. *Res Nurs Health* **19**:213–223(1996).

18. Redeker NS, Ruggiero J, Dankanics L, et al. Self-reported sleep of postoperative cardiac surgery patients: Preliminary data. [http://rutgersscholar.rutgers.edu] (2000), cited 2001.

19. Ruggiero J, Redeker N, Cochrane C. Symptom and mood correlates of activity–rest in cardiac surgery patients. *Sleep* **23** (Suppl 2):A352(2000).

20. Kos-Munson BA, Alexander LD, Hinthorn PA, et al. Psychosocial predictors of optimal rehabilitation post-coronary artery bypass surgery. *Scholarly Inquiry Nurs Pract* **2**:171–199(1988).

120

SLEEP DISTURBANCES AFTER NONCARDIAC SURGERY

ISMAIL GÖGENUR AND JACOB ROSENBERG MD, DSc, FRCS, FACS
Copenhagen University Hospital, Gentofte, Denmark

INTRODUCTION

The understanding of postoperative pathophysiological mechanisms in the recent decades has resulted in progress in surgical techniques (minimally invasive surgery) and anesthesiological care, leading to a steady improvement in postoperative outcome. Despite these changes, major surgery is still beset by complications that cannot be attributed solely to inadequate surgical and anesthetic techniques, but rather to the surgical stress response [1]. One result of the complex surgical stress response is the profound sleep disturbances observed in the postoperative period. Patients frequently report subjective sleep problems for up to 4 weeks after noncardiac surgery [2] and polysomnographic measurements have revealed that postoperative sleep is characterized by severe disturbances in sleep architecture with frequent awakenings, lack of slow-wave sleep (SWS) and rapid eye movement (REM) sleep in the first one to two postoperative nights, and subsequent REM rebound in the following nights [3–8]. Sleep disturbances might play a significant role in the development of cardiopulmonary instability, postoperative cognitive disturbances, and fatigue [9]. Recent studies have implied that circadian disturbances may be important in understanding sleep disturbances and the potential clinical implications of these [10]. The pathogenic mechanisms for sleep disturbances after cardiac surgery are probably different from after noncardiac surgery and will not be discussed further in this chapter [1].

We will review in this chapter the present literature on sleep disturbances after noncardiac surgery and evaluate the potential pathogenic mechanisms and clinical implications. Finally, we will discuss potential preventive measures and treatment possibilities of postoperative sleep disturbances.

SLEEP PATTERN AFTER NONCARDIAC SURGERY

In the postoperative period, patients often report sleep problems with frequent awakenings, fragmented sleep with daytime naps, and nightmares [11–13]. Objective sleep measurements by polysomnography have confirmed these findings. Polysomnographic measurement of sleep has been done after major abdominal surgery [3, 5, 7, 8, 14, 15], gynecological surgery [10], orthopedic surgery [3, 16], laparoscopic surgery [17], herniorrhaphy [6], and minor undefined surgery [18] (Table 120.1). There appears to be a direct correlation between magnitude of surgery and the extent of sleep disturbances. After major noncardiac surgery there is almost abolished SWS and REM sleep on the operative night and in most patients also on the second night after surgery (Figure 120.1). On the following two to five nights there is gradual recovery of SWS. There is a pronounced rebound phenomenon with both high intensity and longer duration of REM sleep on the second to fifth postoperative nights (Figure 120.1). The number of REM periods is unaffected, but the duration is markedly increased. Questionnaire studies have shown that nightmares are most frequent on the fourth postoperative night where the REM sleep rebound occurs [12]. Stage 1 and stage 2 sleep dominate, and frequent long awakenings prevent the natural rhythmicity of sleep and distribution of sleep stages. There is considerable interindividual

Sleep: A Comprehensive Handbook, Edited by T. Lee-Chiong.
Copyright © 2006 John Wiley & Sons, Inc.

TABLE 120.1 Clinical Studies with Polysomnography Measurement of Sleep in Patients After Noncardiac Surgery

Study	Type of Surgery	Postoperative Nights	Total Sleep Time	REM Sleep Duration	SWS Duration	Stage 2 Sleep	Number of Patients
Aurell and Elmqvist [3]	Major noncardiac	1–4	↓	↓	↓	↑	9
Ellis and Dudley [5]	Inguinal hernia repair	1,2	↓	↓	↓	↑	8
	Upper abdominal	1,2	↓	↓	↓	↑	4
Knill et al. [7]	Cholecystectomy	1,2,(3)	→	↓	↓	↑	6
		(3),4,5,6	→	↑	→	↓	
	Gastroplasty	1,2,(3)	→	↓	↓	↑	6
		(3),4,5	→	↑	→	→	
Rosenberg et al. [8]	Major abdominal	1,(2)	→	↓	↓	↑	10
		(2),3	→	↑	↓	→	
Rosenberg-Adamsen et al. [17]	Laparoscopic cholecystectomy	1	→	→	↓	↑	10
Kavey and Altshuler [6]	Herniorrhaphy	1,2	↓	↓	↓	↑	10
		3,4	→	→↑	→↓	→	
Lehmkuhl et al. [18]	Minor undefined	1	→	↓	↓	↑	46
Drummond et al. [14]	Major abdominal	1	↓	↓	↓	↑	34
Rahman et al. [15]	Major abdominal	1	↓	↓	↓	0	8
Cronin et al. [10]	Gynecological	1,(2)	→	↓	→	↑→	10
		(2),3	→	↑	→	→↓	
Catley et al. [16]	Upper abdominal	1	0	↓	↓	0	16
	Major orthopedic	1	0	↓	↓	0	16

variability, but overall, total sleep time is reduced the first and second night after surgery.

The surgical stress response is less after laparoscopic surgery and open minor surgery (Figure 120.2), where there are typically only transient SWS disturbances and no REM sleep disturbances [17]. Subjective measures of sleep disturbances were also mainly transient after laparoscopic surgery and there were no differences in nightmares or distressing dreams [19]. In patients undergoing outpatient elective surgery, there was disturbed sleep for 2 days measured by actigraphy, showing increased number of nighttime awakenings and increased overall wake time [20].

The maximum EEG measurement period after major surgery has been the sixth postoperative night and the total duration of sleep disturbances is therefore not known. There have been no systematic studies on sleep during day and evening hours, where the patients might get some of their sleep. Questionnaire studies have shown disturbances 4 weeks after surgery with increased sleep duration and daytime napping [2]. Almost a quarter of the patients undergoing elective major abdominal, vascular, or orthopedic surgery had decreased sleep quality, and a quarter of these had changes for more than 2 weeks after discharge [11]. These factors may contribute to the development of postoperative fatigue, which can persist for

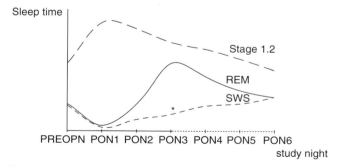

Figure 120.1 Estimated sleep architecture changes in the postoperative period in patients after major noncardiac surgery. PREOPN, preoperative night; PON, postoperative night; REM, rapid eye movement; SWS, slow-wave sleep. *One study showed rebound of SWS on PON3 [8].

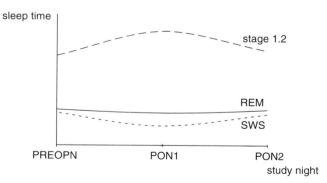

Figure 120.2 Estimated sleep architecture changes in the postoperative period after laparoscopic/minor surgery. PREOPN, preoperative night; PON, postoperative night; REM, rapid eye movement; SWS, slow-wave sleep.

several weeks after operation. Most of the available studies on sleep disturbances have been performed in patients receiving conventional care. Preliminary reports suggest that subjective sleep disturbances may be reduced in patients receiving multimodal analgesia with continuous epidural technique and fast-track perioperative care regimens [21].

FACTORS INFLUENCING SLEEP AFTER SURGERY

Surgical Stress Response

Surgical trauma induces a line of different pathophysiological changes that collectively are called the surgical stress response. The intensity of the stress response is dependent on the magnitude of surgery, and the clinical manifestation ranges from fatal myocardial infarction caused by, for example, increased sympathetic tone and thrombogenesis, to minor wound infections caused by, for example, decreased tissue oxygenation. The surgical stress response was viewed by some as a necessary homeostatic defense mechanism, but this view is changing because of the known adverse effects of the stress response [22]. A profound metabolic–endocrine stress response and an inflammatory stress response are central components of the surgical stress response. The effect of the different stress responses on patients' sleep in the postoperative setting has not been extensively investigated, but from clinical studies it is evident that the magnitude of the surgical trauma, blockade of the neural input by newer anesthetic techniques, and glucocorticoid blockade can inhibit the inflammatory and metabolic–endocrine stress responses [22–25].

Minimally Invasive Surgery

In patients who had their gallbladder removed by laparoscopic surgery, there were no significant changes in the total time awake or number of arousals the first night after surgery compared with the preoperative night [17]. The EEG showed slightly decreased SWS, increased stage 2 sleep, but no REM sleep disturbances. After laparoscopic surgery there is also better pulmonary function with reduced postoperative hypoxemia, less immune suppression, and reduced inflammatory stress response [24, 26]. Reduced inflammatory stress response is not only present in classical major open versus laparoscopic procedures but also in minor procedures as open hernia repair versus laparoscopic repair [27]. Thus it has been shown that the acute inflammatory stress response manifested by CRP, IL-6, and TNF levels were reduced after laparoscopic surgery compared with conventional hernia repair [27].

The effect on sleep of the various mediators of the inflammatory response is difficult to investigate because of the complex inflammatory response to surgery. However, studies in healthy volunteers have shown that administration of IL-6 reduced REM sleep and SWS [28]. In nonsurgical healthy volunteers decreased overall secretion of IL-6 was associated with normal sleep distribution and well-being [29]. Sleep deprivation, however, resulted in increased daytime IL-6, somnolence, and fatigue [30]. The inflammatory response with increased IL-6 after surgery may therefore be responsible for some of the REM disturbances during the first nights, and the sleep deprivation may enforce this effect by increasing daytime IL-6 production. In experimental studies intraventricular TNF injection caused deprivation of REM sleep and increase of light NREM sleep [31]. It is not only the ligand TNF, but also the presence of the TNF receptors (TNFR1) that are believed to regulate sleep. Thus the absence of TNF receptors (TNFR1) in mice also affects normal sleep regulation [32]. In humans, there was a dose–response relationship between injection of endotoxin (simulating an infectious insult and/or surgical trauma) and release of cytokines including TNF-α, and also a dose–response relationship between the effects of cytokines on sleep [33]. Thus, in low concentrations, there was a facilitation of SWS, but in higher concentration, sleep was disrupted with increased awakenings and reduced SWS and REM sleep [33]. Endotoxin injection induced both a cytokine response and an activation of the hypothalamic–pituitary–adrenal (HPA) system. When comparing the sleep modulatory effect of endotoxin injection, it is probably the cytokine response that is most important [33]. Whether the attenuated cytokine response is the reason for the minor sleep disturbances after laparoscopic surgery, compared with more pronounced disturbances after major (open) surgery, is not known.

Neural Blockade

After major abdominal surgery there is a biphasic hypothalamic–pituitary–adrenal response [34]. On the first postoperative day there is an increase in the corticotropin releasing hormone (CRH), adrenocorticotropic hormone (ACTH), and cortisol plasma concentration. Hereafter, on the next 2 days the CRH and ACTH levels decrease below preoperative values and plasma cortisol remains elevated. The CRH/GHRH (growth hormone releasing hormone) balance is thought to be an important key factor in regulation of sleep intensity in humans [35]. Administration of CRH resulted in SWS deprivation, especially when administered in a pulsatile fashion and during the second half of the night [35]. Administration of cortisol, on the other hand, resulted in REM sleep deprivation, probably secondary to stimulation of SWS [36]. A connection between CRH administration and melatonin release has also been described in healthy volunteers, where there was an inhibitory effect of CRH administration on melatonin release [37]. The disturbance in CRH, ACTH, GHRH, and cortisol ratios may therefore play an important role in the observed sleep architecture changes.

Neural blockade by single-dose intraoperative epidural or spinal anesthesia has only a transient stress-reducing effect without prolonged endocrine or metabolic effects [22]. In order to attenuate the endocrine–metabolic stress response, the epidural blockade should be applied for 24–48 hours [22]. The mechanism of action is by reducing the afferent impulses to the central nervous system, thereby reducing the secretion of the catabolically active hormones (CRH, cortisol, glucagon, and catecholamines) and increasing the secretion of anabolic hormones (insulin). The effect on the stress response is also dependent on level of blockade. Thus epidural lumbar blockade for procedures in the lower abdomen are more effective in reducing the stress response than a thoracic blockade for procedures in the upper abdomen.

In conclusion, there is a complex neuroendocrinological stress response in the postoperative period with fluctuations in various hormones resulting in SWS and REM sleep disturbances. The prevention of these deleterious effects by neural blockade can probably have a positive effect on sleep, and preliminary questionnaire studies have supported this [21]. When comparing the immediate neuroendocrine stress response in patients operated on by laparoscopy, minilaparotomy, and conventional laparotomy for benign ovarian cyst surgery, ACTH, cortisol, epinephrine, and norepinephrine secretions were decreased depending on the size of the surgical trauma, suggesting that the size of incision and amount of tissue manipulation are also central in the release mechanisms [38]. The optimal situation for sleep preservation would include reducing the surgical trauma by minimally invasive surgery and prolonged neural blockade. Further studies are needed to clarify these matters.

Glucocorticoid Blockade

The mechanisms behind the stress modulatory effect of glucocorticoids have not yet been fully clarified [39], but several randomized clinical trials have found that glucocorticoids have a positive effect on the surgical stress response and postoperative outcome [23, 25]. Thus single-dose preoperative dexamethasone administration reduced postoperative pain, fatigue, and convalescence after laparoscopic cholecystectomy [23]. The administration of perioperative glucocorticoids was safe and even high-dose methylprednisolone (30–35 mg/kg) has not caused any side effects [40]. The reason why glucocorticoids may reduce fatigue is largely unknown but it is probably by modulation of the inflammatory stress response. Thus preoperative glucocorticoid administration in major abdominal surgery reduced the levels of IL-6 and CRP [41]. As discussed earlier, the reduction in inflammatory cytokines may reduce sleep disturbances, but this has not been studied in patients receiving glucocorticoids. Future studies must clarify if glucocorticoids, in addition to their positive effects on fatigue, also improve subjective sleep quality after surgery. These studies would potentially also reveal if glucocorticoid-induced REM sleep deprivation is subordinate to the potential positive effect of a reduced cytokine response.

Type of Anesthesia

Type of general anesthesia is considered to have only a minor influence on postoperative sleep pattern. The interactions between sleep and anesthesia have not been fully clarified. Preliminary recent experimental studies have suggested that anesthetics (propofolol) and sleep might even have similar regulatory mechanisms, and that anesthetics might facilitate normal sleep in clinical environments where sleep deprivation is common, for instance, the intensive care unit (ICU) [42].

When comparing regional and general anesthesia for inguinal hernia repair [6] and minor orthopedic procedures [12], there was no difference in sleep architecture. Three hours of isoflourane anesthesia in healthy volunteers only caused transient SWS disturbances [43]. When comparing inhalational (halothane) and intravenous (dehydrobenzperidol) anesthesia, SWS and REM sleep were equally inhibited on the first postoperative night [18]. When comparing general and spinal anesthesia, there was no difference in melatonin secretion the first night after orthopedic surgery [44]. Furthermore, it has been shown that nonsurgical acute stress resulted in sleep disturbances comparable with sleep after major noncardiac surgery. Thus sleep after acute myocardial infarction [45], ischemic stroke [46], and congestive heart failure [47] was characterized by reduced SWS and REM sleep, increased stage 1 and stage 2 sleep, and increased awakenings. These changes are very similar to the sleep disturbances that occur after major noncardiac surgery, suggesting that it is the surgery-mediated stress response that triggers the sleep changes rather than effects of anesthesia, analgesia, or environmental factors. In conclusion, anaesthesia per se is not an important pathogenic factor in postoperative sleep disturbances.

Opioids

Opioids suppress REM sleep in healthy nonsurgical volunteers in a dose-dependent fashion. It has been shown that 0.22 mg/kg of intramuscular morphine depressed REM sleep by 50% and that 0.43 mg/kg abolished REM sleep and increased nocturnal awakenings in healthy volunteers

[48]. Laparoscopic surgery, besides reducing the inflammatory stress response, probably also results in less sleep disturbance by reducing postoperative pain and thereby the need for opioids. Reducing postoperative pain with modern multimodal analgesic principles with opioid sparing analgesics (NSAID and acetaminophen) and continuous epidural analgesia would therefore be expected to reduce postoperative sleep disturbances because of the reduced use of opioids. Regional anesthesia for postoperative pain relief reduced the amount of episodic hypoxemic events during the first postoperative night, by reducing the need for systemic opioids, thereby avoiding the opioid-induced sleep-related adverse effects on ventilatory function [16]. There are no such studies on the subsequent postoperative nights. Thus optimizing postoperative analgesia with minimal use of systemic opioids will both have positive effects on sleep architecture as well as sleep-related breathing disturbances like obstructive apnea, paradoxic breathing, and periods of slow ventilatory rate. In recent studies, however, comparing continuous epidural analgesia with a local anesthetic (bupivacaine) or opioid (fentanyl) after gynecological surgery, there were no differences in REM sleep disturbances, indicating that there are additional, unidentified, more powerful REM sleep influencing factors in postoperative patients [4].

Pain

Pain has been reported to be a leading cause of insomnia in medical illness, and in patients with nonspecific pain a high prevalence of subjective sleep disturbances have been reported. There is an increased prevalence of sleep problems in patients with somatic disease, especially in those with pain due to rheumatic complaints [49]. In a community health survey, the variable most strongly connected to sleep problems was pain and most often pain from arthritis [50]. EEG-based studies in patients with musculoskeletal pain, for example, rheumatoid arthritis and fibromyalgia, have found that sleep disturbances were similar to sleep after major surgery [51, 52]. Pain is probably an important cause of nighttime awakenings after surgery [13], and questionnaire studies have shown that improvement in postoperative analgesia, for example, by patient-controlled analgesia, improved subjective sleep quality because of better pain control. However, in a study in gynecological patients undergoing lower abdominal surgery and monitored with polysomnography for three nights, no significant association was found between pain scores, either at rest or when coughing, and any of the monitored sleep variables [4]. These data were obtained in patients who were treated with epidural analgesia (either fentanyl or bupivacaine) for a lower abdominal procedure, and the lack of association could be the result of good pain control with this technique compared with on-demand systemic analgesia. Preliminary

reports have shown that fast-track surgery with multimodal analgesia reduced postoperative pain and fatigue, and future studies should therefore establish if there also is a beneficial effect on sleep architecture [21].

Circadian Disturbances

Melatonin secretion is central for normal sleep regulation, affecting both sleep propensity and sleep architecture. Thus it has been shown that sleep latency and REM sleep density are regulated by the circadian system [53]. There is increasing evidence showing that profound circadian disturbances may occur in the postoperative period. Thus melatonin secretion was impaired for up to 4 days after esophagectomy and this impairment was significantly correlated to the development of postoperative intensive care unit psychosis [54]. Another study examining melatonin secretion after major surgery found a connection between disturbed melatonin secretion and postoperative insomnia in the elderly [55]. In younger women undergoing elective lower abdominal gynecological surgery, there was almost abolished melatonin secretion on the first night, gradually normalizing over the second and third nights [10]. Correspondingly, melatonin secretion has also been found to be disturbed after orthopedic surgery [44]. Reduced nighttime melatonin secretion was also present in nonsurgical patients in the ICU [56]. Thus it was probably the stress response from surgery or an acute severe medical condition (sepsis) that resulted in the observed circadian disturbance in melatonin secretion. Circadian disturbances in the autonomic nervous system have also been shown after surgery, where there is lack of sympathetic withdrawal and lack of increase in parasympathetic tone during nights after major abdominal surgery [57]. When comparing the temperature rhythm on the second night after major abdominal surgery with the preoperative night, there was also a disturbed circadian rhythm [58].

The available results regarding postoperative circadian disturbances are preliminary and include methodological weaknesses. Most studies imply that the disturbances in melatonin secretion after surgery primarily represent an amplitude problem with suppressed melatonin secretion during the night [10, 44, 55, 59], but phase disturbances cannot be excluded [54]. Further studies should clarify the connection between circadian disturbances in other endocrine systems, including release of CRH, cortisol, GH, and GHRH, and the disturbed melatonin secretion. Interestingly, however, preliminary case reports have shown an effect of melatonin substitution on sleep disturbances in the intensive care unit [60] and in the prevention of delirium in the postoperative period [61]. Randomized clinical trials with proper control for confounding factors (light level on the ward, postoperative medications, and nursing interventions) can clarify whether melatonin is superior to placebo in treating sleep disturbances and preventing cognitive disturbances.

Environmental Factors

Environmental stimuli are thought to be an important factor in sleep disruption in the ICU. The environmental stimulus most often cited in the literature is noise. Studies have shown that the noise level in the ICU is very high both during day and night, exceeding the Environmental Protection Agency recommendations [62]. However, patients do not experience noise as being a more sleep-disturbing factor when compared to interruptions from vital signs monitoring and diagnostic testing in the ICU [63]. A study of sleep in mechanically ventilated ICU patients showed that noise was responsible in 12% of overall arousals and 17% of awakenings from sleep [62]. Thus more than 80% of awakenings from sleep were from other causes than noise. In an audio and video surveillance study in the ICU of mechanically ventilated patients, 68% of arousals and awakenings could not be identified, indicating that environmental factors play a limited role in the sleep problems in the ICU [64]. Besides noise and nursing interventions, it has been proved that high room temperature also may have a negative effect on sleep patterns in the postoperative period [65]. Starvation in healthy volunteers impaired sleep and it may have a negative effect also in the postoperative period, because many surgical patients are not eating adequately after operation [66]. Surgical patients often have intravenous catheters but these have not been associated with sleep disturbances in volunteers [67]. Light is known to suppress melatonin secretion and most of the studies showing disturbed melatonin secretion after surgery have controlled for light level during the night, indicating that it may not have a central role. However, light level during the day and evening have not been controlled for and it would be interesting to see if attempts to reset the circadian clock by high-intensity light during day and evening hours, and dim light during the night would have an effect on melatonin secretion and sleep disturbances.

In conclusion, several environmental factors have possible effects on sleep in healthy volunteers, but they probably cannot account for all sleep disturbances after surgery (Figure 120.3).

CLINICAL IMPLICATIONS

Reduced subjective sleep quality extends for several weeks after surgery [2, 12], but the clinical consequences of disturbed sleep in the postoperative period remain unknown [68]. However, the lack of REM sleep the first nights after surgery with subsequent rebound, and lack of SWS for several nights after surgery have various potential

Factors influencing sleep after surgery

| Surgical stress |
| Cortisol, GHRH, CRH |
| Cytokines |
| Sympathetic stimulus | → Postoperative sleep

| Drugs |
| Oploids |
| Benzodiazepines |
| Anesthetics |

| Environmental factors |
| Noise |
| Invasive monitoring |
| Nursing |
| Ambient temperature |
| Light |

| Other factors |
| Pain |
| Circadian disturbances |
| Starvation |
| Psychological factors |

Figure 120.3 Possible pathogenic factors in sleep disturbances after noncardiac surgery.

implications Figure 120.4. In the following, we will discuss the potential implications for cardiopulmonary function, cognitive function, fatigue, and the immune system.

Cardiopulmonary Instability

In normal subjects there are profound hemodynamic changes during different sleep stages [69]. There is a decrease in sympathetic nerve activity with a decline in blood pressure and heart rate from wakefulness to stage 4 NREM sleep. During REM sleep there is a subsequent increase in sympathetic nerve activity and blood pressure, and heart rate increases to levels similar to wakefulness [70]. Sleep onset is associated with decreased catecholamine levels and especially SWS is related to reduced levels of norepinephrine and epinephrine compared with wakeful-

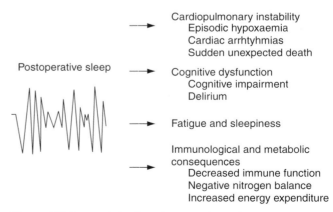

Figure 120.4 Possible clinical implications of sleep disturbances after noncardiac surgery.

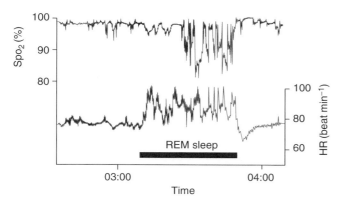

Figure 120.5 Episodic hypoxemia and associated rebound of REM sleep on the second night after operation in a 36 year old woman undergoing gastric surgery. The heart rate (HR) tracing originates from the ECG recording. REM, rapid eye movement; Sp_{O_2}, arterial oxygen saturation. (From [8]. © The Board of Management and Trustees of the British Journal of Anaesthesia. Reproduced by permission of Oxford University Press/British Journal of Anaesthesia.)

ness, stages 1 and 2 sleep and REM sleep [71]. Absent SWS might therefore result in increases in epinephrine and norepinephrine levels. The postoperative period is associated with an adaptive autonomic regulatory mechanism, with persistent increased sympathetic tone and downregulation and desensitization of the lymphocyte betaAR/adenylyl cyclase system for 6 days after major surgery [72]. Circadian disturbance in the autonomic nervous system is also present with lack of sympathetic withdrawal and lack of increase in parasympathetic tone in the night hours on the second postoperative night after major abdominal surgery [57]. REM sleep is associated with intense sympathetic activation on nights with postoperative rebound of REM sleep [7]. The overall increase in sympathetic tone probably results in a greater risk of cardiac arrhythmias and ischemic events [57]. REM sleep rebound would be particularly dangerous because of additional sympathetic activation. During REM sleep the muscular tone in the upper respiratory airway muscles decreases, probably coinciding with impaired central ventilatory regulation, thus leading to episodic desaturations during the middle of the first postoperative week [8] (Figure 120.5). These episodic desaturations may be linked to the occurrence of cardiac arrhythmias and ischemic events [73]. In addition, sleep deprivation can impair ventilatory responses to hypoxia and hypercapnia [74]. Thus the middle of the first postoperative week is a period where patients experience cardiovascular instability with additional ventilatory disturbances resulting in a theoretical increase in the risk of cardiovascular events. This hypothesis is supported by the fact that most postoperative cardiovascular events and unexpected deaths occur in this period [75, 76]. Future studies should clarify the magnitude of this problem, and possible preventive

measures with sympathetic blockade combined with sleep and circadian rhythm restoration must be examined.

Postoperative Cognitive Dysfunction

Impairment in the cognitive function after major surgery occurs frequently [77]. There is a peak incidence in the middle of the first week after surgery but the cognitive impairment may last for several months after operation [78]. After hip fracture repair the incidence may be up to 65% [79]. Many risk factors have been proposed including age >70 years, alcohol abuse, poor preoperative cognitive and functional status, preoperative electrolyte disturbances, noncardiac thoracic surgery, and aortic aneurysm surgery [80]. The clinical impact is substantial with increased morbidity and affected outcome for several months after surgery [81]. The pathogenesis is largely unknown but several factors have been investigated. Postoperative pain [82], hypoxemia, and sleep disturbances may be involved [83, 84]. Recent studies have also suggested that melatonin secretion disturbances may play a role [54]. Whether it is pain per se or the neurohumoral changes caused by pain that result in postoperative delirium is not known. It is known that pain perception is related to sleep architecture [85]. Thresholds for perceiving thermal pain stimuli were lower during stages 1 and 2 sleep compared with SWS and REM sleep [85]. Sleep deprivation also reduced pain tolerance for mechanical and thermal stimuli and when recovering from sleep deprivation, SWS recovery had higher "analgesic" effect than REM sleep recovery [86]. Thus undertreatment with opioids in patients after hip surgery was associated with postoperative delirium [82], suggesting that lack of pain control during the night may result in disturbed sleep and this may increase the risk for postoperative delirium. This hypothesis is supported by a study where delirium was prevented in patients kept asleep during the night with continuous intravenous infusion of an opioid (pethidine) and a hypnotic (flunitrazepam) for the first three nights after major abdominal surgery [87]. These findings may argue for the routine use of hypnotics in the postoperative period as a delirium prevention measure. The effect of most hypnotics on sleep architecture (reduced REM sleep and SWS) is, however, not encouraging. The newer short-acting hypnotics (zaleplon, zolpidem, zopiclone) have less adverse effects on sleep architecture [88] and should probably be preferred in the postoperative setting. Future studies are needed, however, to confirm if there is a clinical effect on subjective sleep quality and/or prevalence of delirium when using hypnotics in the perioperative period.

Experimental hypoxemia to 80% arterial oxygen saturation resulted in decreased cognitive function in healthy volunteers [89]. In a study in patients undergoing noncardiac thoracic surgery, all patients who developed postoperative delirium had low oxygenation and these were treated successfully (delirium disappeared) by supplemental oxygen [83].

Disturbed sleep is probably central for the development of postoperative cognitive dysfunction. Clinical and experimental studies have suggested that there may be an effect of selective SWS and REM sleep deprivation on cognitive dysfunction [90, 91]. Overall sleep deprivation also resulted in regional cerebral perfusion changes in healthy volunteers as well as reduced cognitive function [90, 92]. Since postoperative sleep disturbances may be regarded as a result of profound cerebral functional/metabolic changes caused by the surgical stress, future studies should clarify the isolated role of sleep disturbances versus the overall stress response in cognitive dysfunction after surgery. Preliminary results have shown that multicomponent strategies with nonpharmacological sleeping aids, cognitive stimulation, early mobilization, visual and hearing aids, and rehydration were effective in preventing delirium in nonsurgical patients [93]. However, as mentioned earlier, purely pharmacological intervention with benzodiazepines and opioid analgesia, intended to restore the sleep–wake cycle, may also be effective in preventing delirium in the elderly after abdominal surgery [87].

Reduced sleep during the night and loss of environmental cues to reset the circadian system after surgery may result in phase shifting of the circadian system and hence cognitive disturbances. The negative effect of circadian disturbances on cognitive function are well known in jet lag and circadian rhythm sleep disorders [94]. Melatonin treatment has been shown to have a positive effect on sleep and cognitive function in these situations [94, 95]. As mentioned earlier, several studies have supported that melatonin secretion may be disturbed during the first nights after surgery [10, 44, 54, 55, 59]. Administration of melatonin has in a case report been shown to have an effect on postoperative delirium, implying that the sleep–circadian system interaction may play a role in the development of postoperative cognitive dysfunction [61].

In summary, it is not possible to conclude from the present literature whether it is pain, disturbed sleep architecture, or circadian disturbances that are most important with respect to postoperative cognitive impairment. Considering the minor side effects on sleep pattern and circadian regulation, future studies should include the use of newer hypnotics (zolpidem, zopiclone, zaleplone), chronobiotics (melatonin), and nonopioidanalgetics (epidural analgesia with local anesthetics) when examining postoperative cognitive dysfunction.

Fatigue and Sleepiness

Fatigue frequently occurs after major surgery and may have major implications for recovery. Postoperative fatigue may have a two-stage pathogenesis with different etiological factors in the early and late postoperative periods [96]. Interventions as laparoscopic surgery [97] and growth hormone (GH) supplementation [98], which are both known to result in a modulatory effect on sleep, reduced early and late convalescence, respectively. As discussed earlier, previous studies have shown that various mediators in the inflammatory stress response may produce fatigue as well as sleep disturbances. Fatigue has also been correlated positively to an increase in plasma catecholamines, serum growth hormone, and cardiorespiratory function [99]. Whether postoperative fatigue and sleep disturbances have a common etiological background has not been demonstrated. The relationship between these two phenomena is further complicated by the actual definition of fatigue [100]. In the original fatigue analog scale described in 1982 [101], sleepiness was integrated on a 10 point scale. No attempts to differentiate sleepiness from fatigue were made, and the vast amount of literature concerning postoperative fatigue is based on this scale [102]. However, measures to prevent the stress response will probably reduce both fatigue and sleep disturbances after surgery. This has been implied in preliminary studies demonstrating that fast-track surgery with early mobilization reduced both fatigue and self-reported sleep disturbances [21, 103]. Future studies should clarify the relationship between fatigue and sleepiness after surgery, with special emphasis on the effect of sleep and circadian modulatory therapy on postoperative fatigue.

Immunological and Metabolic Consequences

The clinical consequences of postoperative altered sleep architecture and sleep loss on immune function remain to be established. In animal studies and studies in healthy volunteers, a possible detrimental effect of sleep disturbances on immune function has been shown (Figure 120.5). In experimental studies prolonged sleep deprivation led eventually to death from septicemia [104]. Reduced or fragmented sleep after microbial challenge was associated with poor prognosis, whereas increased sleep was correlated with a favorable outcome [105]. It is believed that it is the stress of infection and trauma that induced an endocrine stress response with HPA axis activation and high levels of corticosterone resulting in changes in cytokine levels (IL-1, IL-1ra, TNF-α). However, recent results have indicated that sleep loss per se may result in reduced immune function independent of HPA axis activation [106]. In human studies, sleep deprivation also resulted in decreased immune function. Sleep increased the antibody response after hepatitis A vaccine when compared with response in sleep-deprived healthy volunteers [107]. Sleep deprivation resulted in up to 50% reduction in natural killer cell activity and 50% decrease in lymphokine killer cell activity in healthy volunteers, and after one night of sleep the natural killer cell activity returned to baseline [108].

Whether the observed immunological changes in healthy volunteers are clinically significant in the postoperative depression of immunological functions remains to be established and should be the focus of future studies. Sleep deprivation also resulted in negative nitrogen balance and increased energy expenditure [109]. Thus some subjects had a 20% increase in nitrogen excretion after 24 hours of sleep deprivation [109]. The endocrine–metabolic stress response described earlier also results in a negative nitrogen balance [22]. The relationship between the negative nitrogen balance resulting from the endocrine–metabolic stress response and the sleep-deprivation-mediated negative nitrogen balance is still unclear.

PREVENTION AND TREATMENT OF POSTOPERATIVE SLEEP DISTURBANCES

The clinical consequence of restoration of sleep disturbances per se compared with reduction of the surgical stress response by minimal invasive surgery, neural blockade, and/or glucocorticoid humoral blockade has not been investigated. Measures to prevent sleep disturbances should in principle be bimodal: on the one hand, by restoring sleep and circadian rhythm by medical and sleep hygiene interventions, and on the other hand, by surgical stress modulation by minimal invasive surgery, neural blockade, and humoral blockade.

Restoration of Sleep by Medical and Sleep Hygiene Measures

It has been reported that about 88% of surgical patients receive sedative drugs during hospitalization [110]. Benzodiazepines, which are most often used, have important influences on sleep architecture with decreased REM sleep and SWS and increased stage 2 sleep [111]. Besides the potential clinical consequences of sleep disturbances as mentioned earlier, the use of benzodiazepines have been associated with increased risk of falls during hospitalization [112]. Restoration of a normal sleep–wake rhythm after major abdominal surgery reduced postoperative delirium [87]. Thus a more rational approach to medical sleep intervention is needed. This could include the newer hypnotics—zopiclone, zaleplon, or zolpidem—which are less likely to alter sleep architecture in nonsurgical patients [88, 113]. Melatonin has a dose-dependent hypnotic effect comparable to certain benzodiazepines (e.g., triazolam) [114] and has less effect on psychomotor performance compared with the newer hypnotics, zopiclone and zaleplon [115]. There have been no studies of hypnotic efficiency of melatonin and the newer hypnotics on postoperative sleep disturbances and sleep architecture. One study has shown a positive effect of melatonin on sleep quality measured

by actigraphy in COPD patients in intensive care units [60]. The possible effect of melatonin and zopiclone/zaleplon/zolpidem on sleep disturbances after surgery is encouraging and should be the focus of future studies.

Besides medical intervention, other measures directed at optimization of the setting in the postoperative intensive care unit and surgical ward is promising. Multicomponent strategies by nonmedical interventions have proved to be efficient in preventing delirium on the medical ward and would probably also improve patients' sleep quality and cognitive function after operations [93]. Reducing unnecessary nursing procedures and noise, increasing light levels during the day and reducing light levels to a minimum during the night, preventing starvation by early enteral feeding and rehydration, and optimizing ambient temperature would probably be feasible methods to increase sleep quality and reduce circadian disturbances.

Reduction of the Postoperative Stress Response

The mediators of the postoperative inflammatory and metabolic–endocrine stress response have several potential influences on sleep after surgery. A modulation of these would therefore probably reduce the severity of sleep disturbances and may have a beneficial effect on postoperative morbidity. Sleep disturbances after minimal invasive abdominal surgery is comparable to sleep disturbances after general anesthesia in healthy volunteers [17]. This is probably because of less pain (and thereby less opioid use) and reduced inflammatory stress response after minimal invasive surgery compared with laparotomies [17]. Therefore it seems rational to reduce the surgical stress whenever possible. Continuous neural blockade with local anesthetics combined with NSAIDs and acetaminophen improves postoperative analgesia and reduces opioid use [116]. Besides optimized analgesic control, continuous neural blockade also provides attenuation of the metabolic–endocrine stress response [116], hence further increasing the beneficial effect on sleep disturbances.

In the last couple of years studies have shown a possible beneficial effect of perioperative glucocorticoid administration on various outcome parameters [23, 25]. Preoperative administration of a single dose of intravenous dexamethasone reduced postoperative fatigue, pain, and duration of convalescence in patients undergoing laparoscopic cholecystectomy [23]. Reduced pain and fatigue after laparoscopic surgery might improve the observed subjective sleep disturbances that can extend for 4 weeks after surgery [2]. The beneficial effects on pain and fatigue might therefore overrule the possible side effects on sleep after glucocorticoid infusion [117]. This, however, needs to be clarified in future studies.

In summary, a multicomponent strategy with minimal invasive surgery, continuous neural blockade, and use of opioid sparing analgesics and perioperative glucocorticoid humoral blockade is promising and may be effective in reducing the profound sleep disturbances observed after surgery.

CONCLUSION

There are profound sleep disturbances in the postoperative period with initial SWS and REM sleep suppression and subsequent REM sleep rebound. It seems that the surgical trauma-induced inflammatory stress response, metabolic–endocrine stress response, circadian disturbances, and postoperative opioid use are the most important factors influencing sleep after surgery. Other factors in the postoperative setting including nursing procedures, noise, room temperature, and light levels are probably also important. The potential clinical implications of postoperative sleep disturbances are numerous. Cardiopulmonary instability with increased risk of hypoxemia and cardiac events may occur because of sleep disturbances. Cognitive dysfunction and fatigue may also be worsened by sleep disturbances. The data so far are limited and further studies should clarify the potential clinical implications.

Most of the sleep studies have been done in a period where modern surgical techniques with minimal invasive surgery, prolonged neural blockade, humoral blockade, accelerated stay regimens, and so on were not employed. The preventive effect of surgical stress reduction by these measures should therefore be examined in the future.

Many surgical patients receive sedatives during their stay in the hospital. The indications for use of hypnotics and the potential adverse effects have not been evaluated fully. Thus indications for use of hypnotics and sedatives and use of newer sleep architecture-preserving hypnotics and chronobiotics should be evaluated in the postoperative setting. These measures should be combined in a multicomponent strategy to prevent sleep disturbances and thereby hopefully improve postoperative mortality, morbidity, and convalescence.

REFERENCES

1. Rosenberg J. Sleep disturbances after non-cardiac surgery. *Sleep Med Rev* **5**:129–137(2001).

2. Gögenur I, Rosenberg-Adamsen S, Kiil C, Kjaersgaard M, Kehlet H, Rosenberg J. Laparoscopic cholecystectomy causes less sleep disturbance than open abdominal surgery. *Surg Endosc* **15**:1452–1455(2001).

3. Aurell J, Elmqvist D. Sleep in the surgical intensive care unit: continuous polygraphic recording of sleep in nine patients receiving postoperative care. *Br Med J* **290**:1029–1032(1985).

4. Cronin AJ, Keifer JC, Davies MF, King TS, Bixler EO. Postoperative sleep disturbance: influences of opioids and pain in humans. *Sleep* **24**:39–44(2001).

5. Ellis BW, Dudley HA. Some aspects of sleep research in surgical stress. *J Psychosom Res* **20**:303–308(1976).

6. Kavey NB, Altshuler KZ. Sleep in herniorrhaphy patients. *Am J Surg* **138**:683–687(1979).

7. Knill RL, Moote CA, Skinner MI, Rose EA. Anesthesia with abdominal surgery leads to intense REM sleep during the first postoperative week. *Anesthesiology* **73**:52–61(1990).

8. Rosenberg J, Wildschiødtz G, Pedersen MH, von Jessen F, Kehlet H. Late postoperative nocturnal episodic hypoxaemia and associated sleep pattern. *Br J Anaesth* **72**:145–150(1994).

9. Kehlet H, Rosenberg J. Surgical stress: pain, sleep and convalescence. In: *Physiology, Stress and Malnutrition: Functional Correlates, Nutritional Intervention*. Lippincot–Raven Publishers, Philadelphia, 1997, pp 95–112.

10. Cronin AJ, Keifer JC, Davies MF, King TS, Bixler EO. Melatonin secretion after surgery. *Lancet* **356**:1244–1245(2000).

11. Beydon L, Rauss A, Lofaso F, Liu N, Cherqui D, Goldenberg F, Bonnet F. Survey of the quality of sleep during the perioperative period. Study of factors predisposing to insomnia. *Ann Fr Anesthesiol Reanim* **13**:669–674(1994).

12. Brimacombe J, Macfie AG. Peri-operative nightmares in surgical patients. *Anaesthesia* **48**:527–529(1993).

13. Closs SJ. Patients' night-time pain, analgesic provision and sleep after surgery. *Int J Nurs Stud* **29**:381–392(1992).

14. Drummond GB, Stedul K, Kingshott R, Rees K, Nimmo AF, Wraith P, Douglas NJ. Automatic CPAP compared with conventional treatment for episodic hypoxemia and sleep disturbance after major abdominal surgery. *Anesthesiology* **96**:817–826(2002).

15. Rahman MQ, Kingshott RN, Wraith P, Adams WH, Drummond GB. Association of airway obstruction, sleep, and phasic abdominal muscle activity after upper abdominal surgery. *Br J Anaesth* **87**:198–203(2001).

16. Catley DM, Thornton C, Jordan C, Lehane JR, Royston D, Jones JG. Pronounced, episodic oxygen desaturation in the postoperative period: its association with ventilatory pattern and analgesic regimen. *Anesthesiology* **63**:20–28(1985).

17. Rosenberg-Adamsen S, Skarbye M, Wildschiødtz G, Kehlet H, Rosenberg J. Sleep after laparoscopic cholecystectomy. *Br J Anaesth* **77**:572–575(1996).

18. Lehmkuhl P, Prass D, Pichlmayr I. General anesthesia and postnarcotic sleep disorders. *Neuropsychobiology* **18**:37–42(1987).

19. Bisgaard T, Klarskov B, Kehlet H, Rosenberg J. Recovery after uncomplicated laparoscopic cholecystectomy. *Surgery* **132**:817–825(2002).

20. Kain ZN, Caldwell-Andrews AA. Sleeping characteristics of adults undergoing outpatient elective surgery: a cohort study. *J Clin Anesth* **15**:505–509(2003).

21. Hjort JD, Sonne E, Basse L, Bisgaard T, Kehlet H. Convalescence after colonic resection with fast-track versus conventional care. *Scand J Surg* **93**:24–28(2004).

22. Kehlet H. Modifications of responses to surgery by neural blockade: clinical implications. In: *Neural Blockade in Clinical Anesthesia*. Lippincott–Raven Publishers, Philadelphia, 1998, pp 129–175.

23. Bisgaard T, Klarskov B, Kehlet H, Rosenberg J. Preoperative dexamethasone improves surgical outcome after laparoscopic cholecystectomy: a randomized double-blind placebo-controlled trial. *Ann Surg* **238**:651–660(2003).

24. Braga M, Vignali A, Zuliani W, Radaelli G, Gianotti L, Martani C, Toussoun G, Di C. Metabolic and functional results after laparoscopic colorectal surgery: a randomized, controlled trial. *Dis Colon Rectum* **45**:1070–1077(2002).

25. Holte K, Kehlet H. Perioperative single-dose glucocorticoid administration: pathophysiologic effects and clinical implications. *J Am Coll Surg* **195**:694–712(2002).

26. Kehlet H. Surgical stress response: does endoscopic surgery confer an advantage? *World J Surg* **23**:801–807(1999).

27. Akhtar K, Kamalky-asl ID, Lamb WR, Laing I, Walton L, Pearson RC, Parrott NR. Metabolic and inflammatory responses after laparoscopic and open inguinal hernia repair. *Ann R Coll Surg Engl* **80**:125–130(1998).

28. Spath-Schwalbe E, Hansen K, Schmidt F, Schrezenmeier H, Marshall L, Burger K, Fehm HL, Born J. Acute effects of recombinant human interleukin-6 on endocrine and central nervous sleep functions in healthy men. *J Clin Endocrinol Metab* **83**:1573–1579(1998).

29. Vgontzas AN, Zoumakis M, Bixler EO, Lin HM, Prolo P, Vela-Bueno A, Kales A, Chrousos GP. Impaired nighttime sleep in healthy old versus young adults is associated with elevated plasma interleukin-6 and cortisol levels: physiologic and therapeutic implications. *J Clin Endocrinol Metab* **88**:2087–2095(2003).

30. Vgontzas AN, Papanicolaou DA, Bixler EO, Lotsikas A, Zachman K, Kales A, Prolo P, Wong ML, Licinio J, Gold PW, Hermida RC, Mastorakos G, Chrousos GP. Circadian interleukin-6 secretion and quantity and depth of sleep. *J Clin Endocrinol Metab* **84**:2603–2607(1999).

31. Kapas L, Krueger JM. Tumor necrosis factor-beta induces sleep, fever, and anorexia. *Am J Physiol* **263**:R703–R707(1992).

32. Deboer T, Fontana A, Tobler I. Tumor necrosis factor (TNF) ligand and TNF receptor deficiency affects sleep and the sleep EEG. *J Neurophysiol* **88**:839–846(2002).

33. Mullington J, Korth C, Hermann DM, Orth A, Galanos C, Holsboer F, Pollmacher T. Dose-dependent effects of endotoxin on human sleep. *Am J Physiol Regul Integr Comp Physiol* **278**:R947–R955(2000).

34. Naito Y, Fukata J, Tamai S, Seo N, Nakai Y, Mori K, Imura H. Biphasic changes in hypothalamo-pituitary-adrenal function during the early recovery period after major abdominal surgery. *J Clin Endocrinol Metab* **73**:111–117(1991).

35. Steiger A, Antonijevic IA, Bohlhalter S, Frieboes RM, Friess E, Murck H. Effects of hormones on sleep. *Horm Res* **49**:125–130(1998).

36. Friess E, Tagaya H, Grethe C, Trachsel L, Holsboer F. Acute cortisol administration promotes sleep intensity in man. *Neuropsychopharmacology* **29**:598–604(2004).

37. Kellner M, Yassouridis A, Manz B, Steiger A, Holsboer F, Wiedemann K. Corticotropin-releasing hormone inhibits melatonin secretion in healthy volunteers—a potential link to low-melatonin syndrome in depression? *Neuroendocrinology* **65**:284–290(1997).

38. Marana E, Scambia G, Maussier ML, Parpaglioni R, Ferrandina G, Meo F, Sciarra M, Marana R. Neuroendocrine stress response in patients undergoing benign ovarian cyst surgery by laparoscopy, minilaparotomy, and laparotomy. *J Am Assoc Gynecol Laparosc* **10**:159–165(2003).

39. Sapolsky RM, Romero LM, Munck AU. How do glucocorticoids influence stress responses? Integrating permissive, suppressive, stimulatory, and preparative actions. *Endocr Rev* **21**:55–89(2000).

40. Sauerland S, Nagelschmidt M, Mallmann P, Neugebauer EA. Risks and benefits of preoperative high dose methylprednisolone in surgical patients: a systematic review. *Drug Saf* **23**:449–461(2000).

41. Schulze S, Andersen J, Overgaard H, Norgard P, Nielsen HJ, Aasen A, Gottrup F, Kehlet H. Effect of prednisolone on the systemic response and wound healing after colonic surgery. *Arch Surg* **132**:129–135(1997).

42. Tung A, Bergmann BM, Herrera S, Cao D, Mendelson WB. Recovery from sleep deprivation occurs during propofol anesthesia. *Anesthesiology* **100**:1419–1426(2004).

43. Moote CA, Knill RL. Isoflurane anesthesia causes a transient alteration in nocturnal sleep. *Anesthesiology* **69**:327–331(1988).

44. Karkela J, Vakkuri O, Kaukinen S, Huang WQ, Pasanen M. The influence of anaesthesia and surgery on the circadian rhythm of melatonin. *Acta Anaesthesiol Scand* **46**:30–36(2002).

45. Broughton R, Baron R. Sleep patterns in the intensive care unit and on the ward after acute myocardial infarction. *Electroencephalogr Clin Neurophysiol* **45**:348–360(1978).

46. Giubilei F, Iannilli M, Vitale A, Pierallini A, Sacchetti ML, Antonini G, Fieschi C. Sleep patterns in acute ischemic stroke. *Acta Neurol Scand* **86**:567–571(1992).

47. Hanly PJ, Millar TW, Steljes DG, Baert R, Frais MA, Kryger MH. Respiration and abnormal sleep in patients with congestive heart failure. *Chest* **96**:480–488(1989).

48. Kay DC, Eisenstein RB, Jasinski DR. Morphine effects on human REM state, waking state and NREM sleep. *Psychopharmacologia* **14**:404–416(1969).

49. Gislason T, Almqvist M. Somatic diseases and sleep complaints. An epidemiological study of 3,201 Swedish men. *Acta Med Scand* **221**:475–481(1987).

50. Moffitt PF, Kalucy EC, Kalucy RS, Baum FE, Cooke RD. Sleep difficulties, pain and other correlates. *J Intern Med* **230**:245–249(1991).

51. Berry RB, Harding SM. Sleep and medical disorders. *Med Clin North Am* **88**:679–703(2004).

52. Moldofsky H. Sleep and pain. *Sleep Med Rev* **5**:385–396(2001).

53. Khalsa SB, Conroy DA, Duffy JF, Czeisler CA, Dijk DJ. Sleep- and circadian-dependent modulation of REM density. *J Sleep Res* **11**:53–59(2002).

54. Miyazaki T, Kuwano H, Kato H, Ando H, Kimura H, Inose T, Ohno T, Suzuki M, Nakajima M, Manda R, Fukuchi M,

Tsukada K. Correlation between serum melatonin circadian rhythm and intensive care unit psychosis after thoracic esophagectomy. *Surgery* **133**:662–668(2003).

55. Leardi S, Tavone E, Cianca G, Barnabei R, Necozione S, Citone G, Simi M. The role of melatonin in the immediate postoperative period in elderly patients. *Minerva Chir* **55**:745–750(2000).

56. Olofsson K, Alling C, Lundberg D, Malmros C. Abolished circadian rhythm of melatonin secretion in sedated and artificially ventilated intensive care patients. *Acta Anaesthesiol Scand* **48**:679–684(2004).

57. Gögenur I, Rosenberg-Adamsen S, Lie C, Rasmussen V, Rosenberg J. Lack of circadian variation in the activity of the autonomic nervous sytem after major abdominal operations. *Eur J Surg* **168**:242–246(2002).

58. Gögenur I, Eversbusch A, Achiam M, Sølving P, Rosenberg J. Disturbed core body temperature rhythm after major surgery. *Sleep and Biological Rhythms* **2**:226–228(2004).

59. Shigeta H, Yasui A, Nimura Y, Machida N, Kageyama M, Miura M, Menjo M, Ikeda K. Postoperative delirium and melatonin levels in elderly patients. *Am J Surg* **182**:449–454(2001).

60. Shilo L, Dagan Y, Smorjik Y, Weinberg U, Dolev S, Komptel B, Shenkman L. Effect of melatonin on sleep quality of COPD intensive care patients: a pilot study. *Chronobiol Int* **17**:71–76(2000).

61. Hanania M, Kitain E. Melatonin for treatment and prevention of postoperative delirium. *Anesth Analg* **94**:338–339(2002).

62. Freedman NS, Gazendam J, Levan L, Pack AI, Schwab RJ. Abnormal sleep/wake cycles and the effect of environmental noise on sleep disruption in the intensive care unit. *Am J Respir Crit Care Med* **163**:451–457(2001).

63. Freedman NS, Kotzer N, Schwab RJ. Patient perception of sleep quality and etiology of sleep disruption in the intensive care unit. *Am J Respir Crit Care Med* **159**:1155–1162(1999).

64. Gabor JY, Cooper AB, Crombach SA, Lee B, Kadikar N, Bettger HE, Hanly PJ. Contribution of the intensive care unit environment to sleep disruption in mechanically ventilated patients and healthy subjects. *Am J Respir Crit Care Med* **167**:708–715(2003).

65. Libert JP, Bach V, Johnson LC, Ehrhart J, Wittersheim G, Keller D. Relative and combined effects of heat and noise exposure on sleep in humans. *Sleep* **14**:24–31(1991).

66. MacFadyen UM, Oswald I, Lewis SA. Starvation and human slow-wave sleep. *J Appl Physiol* **35**:391–394(1973).

67. Kerkhofs M, Linkowski P, Mendlewicz J. Effects of intravenous catheter on sleep in healthy men and in depressed patients. *Sleep* **12**:113–119(1989).

68. Rosenberg J, Rosenberg-Adamsen S, Kehlet H. Postoperative sleep disturbance: causes, factors and effects on outcome. *Eur J Anaesthesiol Suppl* **10**:28–30(1995).

69. Murali NS, Svatikova A, Somers VK. Cardiovascular physiology and sleep. *Front Biosci* **8**:s636–s652(2003).

70. Somers VK, Dyken ME, Mark AL, Abboud FM. Sympathetic-nerve activity during sleep in normal subjects. *N Engl J Med* **328**:303–307(1993).

71. Irwin M, Thompson J, Miller C, Gillin JC, Ziegler M. Effects of sleep and sleep deprivation on catecholamine and interleukin-2 levels in humans: clinical implications. *J Clin Endocrinol Metab* **84**:1979–1985(1999).

72. Amar D, Fleisher M, Pantuck CB, Shamoon H, Zhang H, Roistacher N, Leung DH, Ginsburg I, Smiley RM. Persistent alterations of the autonomic nervous system after noncardiac surgery. *Anesthesiology* **89**:30–42(1998).

73. Rosenberg J, Rasmussen V, von Jessen F, Ullstad T, Kehlet H. Late postoperative episodic and constant hypoxaemia and associated ECG abnormalities. *Br J Anaesth* **65**:684–691(1990).

74. White DP, Douglas NJ, Pickett CK, Zwillich CW, Weil JV. Sleep deprivation and the control of ventilation. *Am Rev Respir Dis* **128**:984–986(1983).

75. Mangano DT. Perioperative cardiac morbidity. *Anesthesiology* **72**:153–184(1990).

76. Rosenberg J, Pedersen MH, Ramsing T, Kehlet H. Circadian variation in unexpected postoperative death. *Br J Surg* **79**:1300–1302(1992).

77. Dyer CB, Ashton CM, Teasdale TA. Postoperative delirium. A review of 80 primary data-collection studies. *Arch Intern Med* **155**:461–465(1995).

78. Selwood A, Orrell M. Long term cognitive dysfunction in older people after non-cardiac surgery. *BMJ* **328**:120–121(2004).

79. Marcantonio ER, Flacker JM, Wright RJ, Resnick NM. Reducing delirium after hip fracture: a randomized trial. *J Am Geriatr Soc* **49**:516–522(2001).

80. Marcantonio ER, Goldman L, Mangione CM, Ludwig LE, Muraca B, Haslauer CM, Donaldson MC, Whittemore AD, Sugarbaker DJ, Poss R. A clinical prediction rule for delirium after elective noncardiac surgery. *JAMA* **271**:134–139(1994).

81. Zakriya K, Sieber FE, Christmas C, Wenz JF Sr, Franckowiak S. Brief postoperative delirium in hip fracture patients affects functional outcome at three months. *Anesth Analg* **98**:1798–1802(2004).

82. Morrison RS, Magaziner J, Gilbert M, Koval KJ, McLaughlin MA, Orosz G, Strauss E, Siu AL. Relationship between pain and opioid analgesics on the development of delirium following hip fracture. *J Gerontol A Biol Sci Med Sci* **58**:76–81(2003).

83. Aakerlund LP, Rosenberg J. Postoperative delirium: treatment with supplementary oxygen. *Br J Anaesth* **72**:286–290(1994).

84. Rosenberg J, Kehlet H. Postoperative mental confusion—association with postoperative hypoxemia. *Surgery* **114**:76–81(1993).

85. Bentley AJ, Newton S, Zio CD. Sensitivity of sleep stages to painful thermal stimuli. *J Sleep Res* **12**:143–147(2003).

86. Onen SH, Alloui A, Gross A, Eschallier A, Dubray C. The effects of total sleep deprivation, selective sleep interruption and sleep recovery on pain tolerance thresholds in healthy subjects. *J Sleep Res* **10**:35–42(2001).

87. Aizawa K, Kanai T, Saikawa Y, Takabayashi T, Kawano Y, Miyazawa N, Yamamoto T. A novel approach to the prevention of postoperative delirium in the elderly after gastrointestinal surgery. *Surg Today* **32**:310–314(2002).

88. Drover DR. Comparative pharmacokinetics and pharmacodynamics of short-acting hypnosedatives: zaleplon, zolpidem and zopiclone. *Clin Pharmacokinet* **43**:227–238(2004).

89. van der Post J, Noordzij LA, de Kam ML, Blauw GJ, Cohen AF, van Gerven JM. Evaluation of tests of central nervous system performance after hypoxemia for a model for cognitive impairment. *J Psychopharmacol* **16**:337–343(2002).

90. Chee MW, Choo WC. Functional imaging of working memory after 24 hr of total sleep deprivation. *J Neurosci* **24**:4560–4567(2004).

91. Ferrara M, De Gennaro L, Bertini M. The effects of slow-wave sleep (SWS) deprivation and time of night on behavioral performance upon awakening. *Physiol Behav* **68**:55–61(1999).

92. Drummond SP, Brown GG. The effects of total sleep deprivation on cerebral responses to cognitive performance. *Neuropsychopharmacology* **25**:S68–S73(2001).

93. Inouye SK, Bogardus ST Jr, Charpentier PA, Leo-Summers L, Acampora D, Holford TR, Cooney LM Jr. A multicomponent intervention to prevent delirium in hospitalized older patients. *N Engl J Med* **340**:669–676(1999).

94. Dagan Y. Circadian rhythm sleep disorders(CRSD). *Sleep Med Rev* **6**:45–54(2002).

95. Herxheimer A, Petrie KJ. Melatonin for the prevention and treatment of jet lag. *Cochrane Database Syst Rev* CD001520.

96. Rubin GJ, Hotopf M. Systematic review and meta-analysis of interventions for postoperative fatigue. *Br J Surg* **89**:971–984(2002).

97. Bardram L, Funch-Jensen P, Jensen P, Crawford ME, Kehlet H. Recovery after laparoscopic colonic surgery with epidural analgesia, and early oral nutrition and mobilisation. *Lancet* **345**:763–764(1995).

98. Kissmeyer-Nielsen P, Jensen MB, Laurberg S. Perioperative growth hormone treatment and functional outcome after major abdominal surgery: a randomized, double-blind, controlled study. *Ann Surg* **229**:298–302(1999).

99. Christensen T, Stage JG, Galbo H, Christensen NJ, Kehlet H. Fatigue and cardiac and endocrine metabolic response to exercise after abdominal surgery. *Surgery* **105**:46–50(1989).

100. Salmon P, Hall GM. A theory of postoperative fatigue: an interaction of biological, psychological, and social processes. *Pharmacol Biochem Behav* **56**:623–628(1997).

101. Christensen T, Bendix T, Kehlet H. Fatigue and cardiorespiratory function following abdominal surgery. *Br J Surg* **69**:417–419(1982).

102. Christensen T. Postoperative fatigue. *Dan Med Bull* **42**:314–322(1995).

103. Brodner G, Van Aken H, Hertle L, Fobker M, Von Eckardstein A, Goeters C, Buerkle H, Harks A, Kehlet H. Multimodal perioperative management—combining thoracic epidural analgesia, forced mobilization, and oral nutrition—reduces hormonal and metabolic stress and improves convalescence after major urologic surgery. *Anesth Analg* **92**:1594–1600(2001).

104. Rechtschaffen A, Gilliland MA, Bergmann BM, Winter JB. Physiological correlates of prolonged sleep deprivation in rats. *Science* **221**:182–184(1983).

105. Toth LA, Tolley EA, Krueger JM. Sleep as a prognostic indicator during infectious disease in rabbits. *Proc Soc Exp Biol Med* **203**:179–192(1993).

106. Hu J, Chen Z, Gorczynski CP, Gorczynski LY, Kai Y, Lee L, Manuel J, Gorczynski RM. Sleep-deprived mice show altered cytokine production manifest by perturbations in serum IL-1ra, TNFa, and IL-6 levels. *Brain Behav Immun* **17**:498–504(2003).

107. Lange T, Perras B, Fehm HL, Born J. Sleep enhances the human antibody response to hepatitis A vaccination. *Psychosom Med* **65**:831–835(2003).

108. Irwin M, McClintick J, Costlow C, Fortner M, White J, Gillin JC. Partial night sleep deprivation reduces natural killer and cellular immune responses in humans. *FASEB J* **10**:643–653(1996).

109. Scrimshaw NS, Habicht JP, Pellet P, Piche ML, Cholakos B. Effects of sleep deprivation and reversal of diurnal activity on protein metabolism of young men. *Am J Clin Nutr* **19**:313–319(1966).

110. Perry SW, Wu A. Rationale for the use of hypnotic agents in a general hospital. *Ann Intern Med* **100**:441–446(1984).

111. Parrino L, Terzano MG. Polysomnographic effects of hypnotic drugs. A review. *Psychopharmacology* **126**:1–16(1996).

112. Mendelson WB. The use of sedative/hypnotic medication and its correlation with falling down in the hospital. *Sleep* **19**:698–701(1996).

113. Monti JM. Effect of zolpidem on sleep in insomniac patients. *Eur J Clin Pharmacol* **36**:461–466(1989).

114. Satomura T, Sakamoto T, Shirakawa S, Tsutsumi Y, Mukai M, Ohyama T, Uchimura N, Maeda H. Hypnotic action of melatonin during daytime administration and its comparison with triazolam. *Psychiatry Clin Neurosci* **55**:303–304(2001).

115. Paul MA, Gray G, Kenny G, Pigeau RA. Impact of melatonin, zaleplon, zopiclone, and temazepam on psychomotor performance. *Aviat Space Environ Med* **74**:1263–1270(2003).

116. Kehlet H. Effect of postoperative pain treatment on outcome-current status and future strategies. *Langenbecks Arch Surg* Aug;**389**(4):244–249(2004).

117. Fehm HL, Benkowitsch R, Kern W, Fehm-Wolfsdorf G, Pauschinger P, Born J. Influences of corticosteroids, dexamethasone and hydrocortisone on sleep in humans. *Neuropsychobiology* **16**:198–204(1986).

121

RELEVANCE OF ANESTHESIOLOGY FOR SLEEP MEDICINE

RALPH LYDIC AND HELEN A. BAGHDOYAN
University of Michigan, Ann Arbor, Michigan

INTRODUCTION

Anesthesia has been known for only about 150 years, but *The New England Journal of Medicine* included anesthesiology in the list of most significant medical developments of the past thousand years [1]. Sleep medicine is more recent than anesthesiology and 2003 marked the 50th anniversary of the discovery of the rapid eye movement (REM) phase of sleep. If one uses 50,000 years ago as the date for the appearance of *Homo sapiens*, then states of anesthesia have been known for 0.3%, and REM sleep for 0.1%, of human existence. To date, there has been little dialogue between anesthesiology and sleep disorders medicine. Clinical and preclinical perspectives are unified by the fact that both sleep and anesthesia are altered arousal states actively generated by the central nervous system [2]. All humans sleep, but the underlying neurobiological mechanisms generating sleep remain incompletely understood. Everyone who has surgery elects to have anesthesia, yet in no case are the mechanisms understood through which anesthetic/analgesic molecules cause unconsciousness. There are no contemporary data that systematically characterize the effect of volatile anesthetics on the sleep of patients without the confounding factors of surgical insult, polypharmacy, trauma, or coexisting disease. This chapter highlights areas of overlap between sleep medicine and anesthesiology.

THE STATE CONCEPT LINKS SLEEP AND ANESTHESIA

The desirable anesthetic state is a constellation of reversible traits that include analgesia, amnesia, unconsciousness, blunted sensory and autonomic reflexes, and skeletal muscle relaxation [3]. In addition to the characteristic of reversibility, another goal of anesthesia is the temporal coordination of the foregoing five traits. Ideally, the onset of these drug-induced traits occurs at approximately the same time. Undesirable anesthetic complications often are characterized by temporal dissociations in the offset of drug-induced traits, such as failure of a seemingly awake, postanesthetic patient to maintain upper airway patency. As with successful anesthesia, normal sleep also requires the temporal coordination of multiple traits. The nosology of many sleep disorders is characterized by the intrusion of sleep traits into the state of wakefulness (e.g., onset of motor atonia during a narcoleptic attack), or the expression of waking traits during sleep (e.g., somnambulism).

Key differences between sleep and anesthesia are summarized in Table 121.1. If the sleep-like state caused by anesthetic molecules is different from naturally occurring sleep, then why include anesthesia in an encyclopedia of sleep medicine? One answer is provided by the mismatch between evidence-based medicine and the persisting lack of scientific rigor concerning states of arousal. Patients are often told "anesthesia will put you to sleep." In

TABLE 121.1 Comparison of Anesthesia and Sleep Onset, Maintenance, and Offset

	Anesthesia	Sleep
Onset	1. Drug-induced 2. Not significantly altered by previous sleep or circadian history 3. Failure to initiate is nonexistent 4. Not significantly altered by environmental factors, such as caffeine intake, environmental temperature, noise, and light	1. Endogenously generated 2. Significantly influenced by circadian phase and duration of prior wakefulness 3. Failure to initiate is a recognized pathology 4. Significantly altered by environmental factors
Maintenance	1. Duration is dependent on agent dose and independent of previous wakefulness 2. Without surgical stimulation, depth of anesthesia can be held constant for long periods of time 3. Failure to maintain is nonexistent 4. Environmental factors (noise, light, temperature) do not alter anesthetic maintenance; sensory input blocked	1. Duration is a function of prior wakefulness and circadian factors 2. Rhythmically oscillates between stages 1, 2, 3, 4, and REM sleep 3. Failure to maintain is a recognized pathology 4. Environmental cues easily disrupt sleep maintenance; sensory input blunted and/or enhanced
Offset	1. Resumption of normal wakefulness is slow (hours to days) 2. Duration of anesthesia and elimination of agent or active metabolites determines timing of wakefulness 3. Reanesthetizing patient easily achieved immediately following offset 4. Offset accompanied by agent side effects (nausea, vomiting, emergence delirium)	1. Resumption of normal wakefulness is rapid (minutes) 2. Timing of waking is modulated by sleep duration and circadian rhythm 3. Immediate initiation of second sleep interval difficult following offset of normal night of sleep 4. Offset normally is associated with reports of feeling rested and refreshed

Source: From [8] with permission.

veterinary medicine, however, "having one's pet put to sleep" is euthanasia. Euphemisms are convenient but denigrate the significance of state-dependent changes in sensory–motor integration, cognition, and autonomic control.

The objective classification of arousal states is based on a constellation of physiological and behavioral traits. Until recently, the core sciences of physics and chemistry have been more advanced than biology regarding state classification based on the time domain. For example, the fundamental steps from oxidation to reduction that occur via mitochondrial enzymes can be considered to transition through different reaction states. The relevance of the state concept was first formalized for sleep neurobiology and medicine as a collection of physiological traits measured at some point in time [4]. This conceptual model identified key criteria for relating cellular mechanisms to arousal states and state-dependent changes in physiological variables. The following paragraphs review criteria that must be satisfied in order to postulate cellular and molecular substrates that cause states of sleep and anesthesia.

The selectivity criterion refers to an organizational scheme for functionally classifying neurons that exhibit increased discharge related to a particular arousal state or autonomic trait. The three major behavioral states (wakefulness, non-rapid eye movement (NREM) sleep, rapid eye movement (REM) sleep) and two neuronal activity profiles (discharging or "on" versus not discharging or "off") permit a classification matrix that can characterize all sleep state-dependent activity patterns so far measured in the brain. Cellular discharge patterns that are independent of arousal state are referred to as state-independent and such cells are not considered as likely regulators of arousal states.

The second and third criteria for evaluating putative neuronal regulators of arousal states refer to the phase relationship between cellular discharge and physiological traits. Physiological traits can occur as phasic bursts during a given state (phasic latency criterion) or can occur tonically throughout an arousal state (tonic latency criterion). During REM sleep, phasic traits include rapid eye movements and muscle twitches, and tonic traits include muscle atonia and cortical EEG activation. Because a cause must precede an effect, in order for a neuronal group to be considered as a regulator of arousal states or physiological traits (e.g., airway obstruction), cell activity (electrical, metabolic, etc.) must change before the tonically or phasically active state-dependent variable.

The third and fourth criteria of periodicity and phasic pattern refer to two different ways that arousal state-dependent neuronal activity can be expressed. Across all arousal states, maximal cell activation may be expressed periodically every time the arousal state occurs, or expressed as a phasically on and off activation during a given arousal state. The behavioral state concept provides a conceptual framework for efforts to synthesize a massive amount of evidence concerning the brain regulation of states of sleep and anesthesia.

PAIN, ANESTHESIOLOGY, AND SLEEP

Pain management is the goal of anesthesiology, and clinical interventions aiming to eliminate acute and chronic pain are directly relevant for sleep medicine. In December 2001, the Joint Commission on Accreditation of Healthcare Organizations (JCAHO) published, in collaboration with the National Pharmaceutical Council (NPC), a monograph entitled *Pain: Current Understanding of Assessment, Management, and Treatment*. JCAHO now requires that all accredited health-care organizations comply with standards for assessing and managing pain. Compliance includes documenting the efficacy of pain treatment plans and educating patients and their families about pain. In March 2003, JCAHO and NPC published a second monograph entitled *Improving the Quality of Pain Management Through Measurement and Action*. Both documents are available in PDF format on the JCAHO web site (http://www.jcaho.org). Although sleep is mentioned frequently in both monographs, gaps in knowledge concerning the interaction between pain and sleep will be readily apparent to sleep disorders physicians. Acutely administered opioids effectively manage pain but inhibit REM sleep [5, 6]. Pain is a complex psychophysiological experience characterized by variability between patients, reciprocal interaction with affective state and coexisting disease, and significant differences between acute and chronic manifestations. The influence of spinal versus supraspinal nociceptive input on sleep also remains poorly understood. The foregoing factors, combined with polypharmacy and pharmacogenetics, preclude simple generalizations about sleep and pain (see Chapter 120 in this volume). The clinical need for effective analgesia and pain management, along with the JCAHO directives, ensure that the topic of pain will become increasingly relevant for sleep disorders medicine.

DEPRESSION OF BREATHING DURING ANESTHESIA, SEDATION, AND SLEEP

As long ago as 1984, pioneering studies showed that anesthetics and sedative–hypnotic drugs exert a more immediate and potent depression of upper airway muscles than the diaphragm (reviewed in [6]). Patients with obstructive sleep apnea (OSA) who undergo anesthesia or sedation are at high risk for developing complications [7]. Childhood OSA is estimated to effect 2% of young children. Brown and colleagues recently reported that some children with OSA have a high incidence of respiratory complications

when administered postoperative opioids and have a reduced opioid requirement for analgesia (reviewed in [6]). The mechanisms underlying an association between frequency of oxygen desaturation and increased sensitivity to opioids in children remain unclear.

Conscious sedation or sedation analgesia refers to a drug-induced dissociated state comprised of some traits characteristic of wakefulness (ability to follow verbal commands) and some traits characteristic of natural sleep (diminished sensory processing, memory impairment, and autonomic depression). Sedation analgesia, rather than general anesthesia, commonly is used in association with discomforting diagnostic procedures, for pediatric patients, in the intensive care unit, and for "alternate site" (non-operating room) care. The current lack of a uniformly accepted technique for monitoring level of sedation illustrates the need for research on levels of consciousness during sedation. Large-scale safety data are not presently available on morbidity and mortality associated with conscious sedation. Office-based liposuction has been associated with death rates that are 50 times higher than current anesthesia-related deaths anticipated by the American Society of Anesthesiologists (reviewed in [8]). Sleep tendency is increased for up to 8 hours following drugs used for ambulatory surgery [9]. The importance of understanding the mechanisms through which anesthetic drugs depress breathing is illustrated by the fact that the majority of all surgeries in the United States currently are performed in an ambulatory environment.

ANESTHETICS, CYTOKINES, AND AROUSAL STATE CONTROL

Cytokines are proteins secreted by virtually all cells; cytokines are involved in regulating the inflammatory response to injury, immune function, and sleep (see Chapter 100 in this volume). Cytokines bind to specific receptors and can alter the release of classical neurotransmitters and hormones. Infection, trauma, surgery, and anesthesia all increase the release of cytokines and alter levels of arousal. The majority of sleep is comprised of the NREM phase and there is good support for the view that NREM sleep is increased by interleukins (IL-1β and IL-1α) and tumor necrosis factors (TNF-α and TNF-β) [10]. Secretion of proinflammatory cytokines is altered by many volatile anesthetic agents and by procedures commonly used in the operating room environment. For example, mechanical ventilation, blood transfusion, intravenous opioids, cardiopulmonary bypass, and even supplemental intraoperative oxygen have been shown to alter proinflammatory cytokines. We are aware of no studies that systematically relate sleep to increases in cytokines caused by the foregoing procedures. Surgery and anesthesia are commonly associated with trauma, and stress disrupts sleep via altered endocrine function of the hypothalamus–pituitary axis. The observation of immune depression associated with sleep deprivation raises important health concerns for anesthesiologists. (see Toth in [2]).

RELEVANCE OF SLEEP FOR PATIENT AND PRACTITIONER

Anesthesiology has assumed a leading role in efforts to understand how sleep and fatigue impact clinical performance [11] and sustainability of clinical careers [12]. Sleep restriction and sleep deprivation are an unavoidable component of the 24 hour clinical coverage provided by anesthesiologists. Sleep restriction and deprivation can impair vigilance and can cause decrements in clinical performance [13]. Sleep deprivation also has a negative effect on medical education [14]. A partnership between sleep medicine and anesthesiology [15] could provide important data on how best to minimize the negative impact of sleep disruption while providing appropriate clinical staffing.

The relevance of fatigue for clinical performance was emphasized by the U.S. Institute of Medicine (IOM) report that medical errors contribute to approximately 98,000 deaths per year [16]. The IOM report has been criticized on a number of issues and subsequent study groups have emphasized flaws in the entire health system as causal contributors. The IOM's Committee on Quality Health Care in America noted that hospitals frequently rely on backup double shifts for nursing staff and very long hours for resident physicians. This is an approach that ignores a large body of work on the effects of fatigue on human performance. Contrary to the common subjective report, humans do not adapt well to fatigue induced by chronic sleep restriction (see Chapter 19 in this volume).

Multiple studies have shown that levels of sleep loss similar to that experienced by medical residents and on-call staff cause performance decrements equivalent to intoxicating blood ethanol levels. Additional data show that driving home on the morning that follows a night shift can be an occupational risk for sleep-deprived medical residents. Post night-call residents and staff have an increased probability of falling asleep while driving. The drowsy driving website (http://www.drowsydriving.org) provides a bibliography for some of these studies. Sleep deprivation causes intervals of sleep that invade ongoing wakefulness and these microsleeps can occur without awareness of the sleep-deprived person. The drowsy driving website contains the complete citation for the 1982 discovery by Bonnet and Moore indicating that one must be asleep for at least 2 min in order to know that one has slept.

Physicians focusing on sleep disorders medicine are certain to encounter other medical specialties asserting that restrictions on clinical hours of service should be abandoned. Operating a motor vehicle involves a lower-level skill set than most clinical procedures yet the negative impact of sleep deprivation is so strong that the U.S. Congress directed the Federal Highway Administration to characterize fatigue among commercial drivers. These studies found that long-haul truck drivers had less sleep than needed for alertness, averaging 5.18 hours in bed and 4.78 hours of sleep per day. The National Highway Traffic Safety Administration estimates that 100,000 police-reported crashes each year are the direct result of driver fatigue. These crashes cause 1550 deaths and 71,000 injuries as well as $12.5 billion in diminished productivity and property loss. The loss of even 1 hour of sleep associated with the shift to daylight savings time has been shown to increase the number of traffic accidents. Effective 4 January 2004, the U.S. Department of Transportation initiated new hours-of-service regulations for truck drivers. These are the first new regulations since 1938 and, contrary to the assertion by some medical specialties, this reflects the growing awareness of the negative impact that sleep deprivation has on safety. The Federal Motor Carrier Safety Administration (http://www.fmcsa.dot.gov) describes the new rules as combining "the best scientific research with real-world analysis to prevent driver fatigue." This includes evidence that time-of-day (i.e., circadian factors) contributes significantly to increased risk for motor vehicle accidents.

On 1 July 2003 the Accreditation Council for Graduate Medical Education (ACGME) issued recommendations and regulations regarding resident work hours. The American Medical Student Association website contains an excellent summary by Rita Kwan and Robert Levy entitled "A Primer on Resident Work Hours" available in PDF format (http://www.amsa.org/hp/rwhprimer2.pdf). Kwan and Levy point out that the ACGME's regulations permit a 10% exception to the 80 hour maximum workweek. Residents must have at least one 24 hour period off per week and work no more than every third night on-call. Between intervals of work, residents are also to be given 10 hours off duty. Currently, the ACGME guidelines contain an averaging procedure that permits a 4 week averaging of weekly hours and on-call time. As long as the monthly average is in line with the ACGME guide, this averaging plan could result in work hours that exceed the 88 hour per week maximum. Residency programs that violate the ACGME rules do so at the risk of losing accreditation and federal funding for resident education.

Societal attitudes about public health and safety and a growing body of knowledge about the impact of fatigue on performance raise important questions about the liability of administrators who are responsible for scheduling work hours. The 2001 Selby road/rail accident is a cautionary tale from the United Kingdom involving one individual who fell asleep while driving. The out-of-control vehicle left the M62 motorway, ending up on the main east coast railway line. A passenger train traveling at greater than 100 mph struck the vehicle. The passenger train was deflected onto a parallel railway track and into the path of an oncoming train laden with 1600 tons of coal. Ten people died and more than 70 people required hospital treatment. The driver survived and was sentenced, by a 10-2 jury majority, to five years in prison. The sentence was upheld by a 2003 appeal ruling that the Selby crash victims were "unlawfully killed." In July 2003, the New Jersey State Senate passed into law a bill that includes "fatigued driving" as part of the vehicular homicide statute (N.J.S.2C:11-5). Fatigue is operationally defined as being without sleep for a period in excess of 24 consecutive hours. A synopsis of the law can be viewed at www.njleg. state.nj.us/2002/Bills/A1500/1347_R2.HTM. This legislation, referred to as Maggie's Law, is named for 20 year old Maggie McDonnell who was killed by a sleep-deprived driver.

SLEEP NEUROBIOLOGY AND MECHANISMS OF ANESTHETIC ACTION

Data have so far failed to support the view that a single cell surface protein or a single portion of the nervous system can cause all the traits comprising an anesthetic state [17]. We postulated more than 10 years ago that neuronal networks that evolved to regulate sleep play a key role in generating states of anesthesia [18, 19]. A logical corollary of our working hypothesis is that sleep neurobiology has the potential to help elucidate the mechanisms by which anesthetic molecules eliminate wakefulness. Likewise, data concerning the mechanisms of anesthetic action are uniquely poised to advance sleep neurobiology. For example, the vast majority of drugs prescribed to produce sleep or used in the operating room as sedative–hypnotics and general anesthetics cause neuronal inhibition by enhancing transmission at the gamma aminobutyric acid$_A$ (GABA$_A$) receptor. GABAergic neurons in the anterior hypothalamus contribute to the generation of NREM sleep by suppressing the activity of a wakefulness-promoting neuronal network. This wakefulness-promoting network includes histaminergic neurons of the posterior hypothalamus, cholinergic neurons of the basal forebrain, and noradrenergic, serotoninergic, and cholinergic neurons in the brainstem that comprise the ascending reticular activating system (reviewed in [20]). The functional contribution of these neuronal networks to arousal state control has been derived by neurobiological studies of sleep. Understanding the effects of anesthetics on these neuronal networks is highly likely to advance both anesthesiology and sleep medicine (reviewed in [6]).

ACKNOWLEDGMENTS

Supported by the NIH Grants HL40881, HL57120, HL65272, and MH45361 and the Department of Anesthesiology.

REFERENCES

1. Editors. Looking back on the millennium in medicine. *N Engl J Med* **342**:42–49(2000).

2. Lydic R, Baghdoyan HA. *Handbook of Behavioral State Control: Cellular and Molecular Mechanisms.* CRC Press, Boca Raton, FL, 1999, pp 1–700. This edited volume summarizes the perspectives of 96 authors at the end of the millennium on the key factors relevant for arousal state control.

3. Trevor AJ, Miller RD. General anesthetics. In: Katsung BG (Ed), *Basic and Clinical Pharmacology*, 8th ed. Appleton & Lange, Norwalk, CT, 2001, pp 419–435. An excellent and accessible handbook.

4. Hobson JA. Toward a cellular neurophysiology of the reticular formation: conceptual and methodological milestones. In: edited by Hobson JA, Brazier MAB (Eds), *The Reticular Formation Revisited*. Raven Press, New York, 1980, pp 1–29. This is an historically significant chapter that identifies criteria required to establish the cellular mechanisms that cause alterations in sleep and wakefulness.

5. Lydic R, Baghdoyan HA. Cholinergic contributions to the control of consciousness. In: Yaksh T, Lynch C, Zapol WM, Maze M, Biebuyck JF, Saidman LJ (Eds), *Anesthesia: Biologic Foundations*. Lippincott–Raven, New York, 1998, pp 433–450. This chapter reviews the role of acetylcholine as a modulator of arousal states.

6. Lydic R, Baghdoyan HA. Sleep, anesthesiology, and the neurobiology of arousal state control. *Anesthesiology* (in press). This paper reviews pontine cholinergic mechanisms that contribute to the regulation of sleep and anesthesia.

7. den Herder C, Schmeck J, Appleboom D, J. K. de Vries N. Risk of general anaesthesia in people with obstructive sleep apnoea. *Br Med J* **329**:955–959(2004). This paper highlights special care needed for patients with OSA who undergo general anesthesia.

8. Lydic R. Pain: a bridge linking anesthesiology and sleep research. *Sleep* **24**:10–12(2001).

9. Lichtor JL, Alessi R, Lane BS. Sleep tendency as a measure of recovery after drugs used for ambulatory surgery. *Anesthesiology* **96**:878–883(2002). This paper is a model for studies that aim to understand the effects of anesthesia-related drugs on sleep.

10. Opp MR. Cytokines and sleep. *Sleep Med Rev* (in press). This paper reviews recent evidence that cytokines contribute to sleep cycle control by altering specific neurotransmitter systems.

11. Weinger MB, Ancoli-Israel S. Sleep deprivation and clinical performance. *JAMA* **287**:955–957(2002). This paper represents

a collaboration between an anesthesiologist working to improve patient safety and a leader in human sleep research. The paper provides data concerning the negative effects of sleep deprivation.

12. Howard SK, Rosekind MR, Katz JD, Berry AJ. Fatigue in anesthesia: implications and strategies for patient and provider safety. *Anesthesiology* **97**:1281–1294(2002). This excellent review demonstrates the power of evidence-based medicine over rule-bound tradition for efforts to understand the impact of fatigue on anesthesia patients and clinical caregivers.

13. Gaba DM, Howard SK. Patient safety: fatigue among clinicians and the safety of patients. *N Engl J Med* **347**:1249–1255(2002). This paper provides an evidence-based view of the impact of fatigue on clinical performance.

14. Veasey SC, Rosen R, Barzansky B, Rosen I, Owens JA. Sleep loss and fatigue in residency training. *JAMA* **288**:1116–1124(2002). A compelling contribution to the literature documenting the need for reform in residency training.

15. Tung A, Mendelson WB. Anesthesia and sleep. *Sleep Med Rev* **8**:213–225(2004). A review resulting from the collaboration between a young and productive anesthesiologist and a psychiatrist with a distinguished record in sleep research.

16. Kohn LT, Corrigan J, Donaldson MS, Richardson WC. *To Err is Human: Building a Safer Health System*. National Academies Press, Washington, DC, 2000. A landmark book with far-reaching implications for all interested in health and society. This book was published during the same year that the World Health Organization ranked the United States 37th in their study of 191 different national health-care systems.

17. Campagna JA, Miller KW, Forman SA. Mechanisms of inhaled anesthetics. *N Engl J Med* **348**:2110–2124(2003). This review provides an update on what is known and what remains to be discovered about the mechanisms of anesthetic action.

18. Lydic R. Reticular modulation of breathing during sleep and anesthesia. *Curr Opin Pulm Med* **2**:474–481(1996). This paper reviews data supporting the probability that neuronal networks that evolved to regulate sleep play a major role in causing states of anesthesia.

19. Lydic R, Biebuyck JF. Sleep neurobiology: relevance for mechanistic studies of anesthesia. *Br J Anaesth* **72**:506–508(1994). The first postulate that neuronal networks that evolved to regulate sleep play a major role in causing states of anesthesia.

20. Baghdoyan HA, Lydic R. Neurotransmitters and neuromodulators regulating sleep. In: Bazil C, Malow B, Sammaritano M (Eds), *Sleep and Epilepsy: The Clinical Spectrum*. Elsevier Science, New York, 2002, pp 17–44. This chapter reviews the neurochemical control of sleep and wakefulness.

122

SLEEP AT HIGH ALTITUDES

WISSAM CHATILA AND SAMUEL KRACHMAN
Temple University School of Medicine, Philadelphia, Pennsylvania

Numerous physiologic adjustments and derangements occur at high altitude resulting in a myriad of symptoms that can be classified into different, yet overlapping, syndromes that are associated with disturbed sleep (Table 122.1). Poor sleep is a prominent manifestation of rapid ascent to high altitude and is in part related to acute changes in ventilatory stimuli, that is, hypoxia and subsequent hypocapnia. The most common syndrome with rapid ascent is acute mountain sickness (AMS) manifesting as insomnia, fatigue, headache, dizziness, dyspnea, anorexia, ataxia, nausea, and fatigue. Unlike high altitude pulmonary edema (HAPE), which manifests 2–4 days after arrival, onset of symptoms of AMS typically begins within 12 hours after the ascent [1]. Although AMS might occur at an altitude as low as 2500 meters, it becomes more prevalent with increasing altitude. Acclimatization to high altitude eventually develops, but prevention and treatment remain important issues particularly for intermittent exposure to high altitude and/or short recreational ascents.

SLEEP QUALITY

Restless sleep and insomnia accompanied by excessive daytime sleepiness are commonly experienced with ascent to high altitude. Objectively, the most consistent finding during sleep at high altitude is an increase in sleep fragmentation and shifts in sleep stage distribution, with the majority

Correspondence to: Wissam Chatila, Division of Pulmonary and Critical Care Medicine, Temple University School of Medicine, 3401 N Broad Street, Philadelphia, PA 19140.

of studies demonstrating increased stage 1 sleep and reductions in stages 2, 3, and 4. The discrepancy between the studies that reported more sleep disruptions [2, 3] versus those that did not [4, 5] are explained by differences in altitude, conditions of testing (hypobaric chambers versus ambient conditions), sample size, and the statistical power necessary to detect changes. In addition, differences in sleep quality responses to atmospheric variation appear to exist between populations [6] and individuals [7], as some experience sleep architecture changes at minimal altitude (1400 m), while others demonstrate similar disturbances only at moderate to extreme elevations (>3000 m). Nonetheless, while duration of sleep does not vary significantly during an ascent, the observed changes in sleep physiology and time awake become more prominent at higher elevations. With more extreme hypoxia, sleep time is dramatically reduced and arousals increase, without a change in the ratio of sleep stages [8]. In contrast, there is no agreement among studies regarding the trend or extent of change in REM sleep at high altitude.

Causes for the various changes in sleep architecture, sleep disruption, and frequent arousals after ascent to a higher altitude have not been completely delineated. Sleep fragmentation is thought to be related to periodic breathing (discussed later) but cannot be explained solely by the respiratory oscillations and respiratory events. First, studies that have supplemented oxygen at high altitude demonstrated resolution of periodic breathing without significantly reducing arousal frequency [2, 9]. Second, although arousals are often associated with periodic breathing (during hyperventilatory phase of the periodic cycle), they also occur during regular breathing at high altitude,

TABLE 122.1   Acute Disorders Reported at High Altitude

Headache
Sleep periodic breathing
Acute mountain sickness (AMS)
High altitude pulmonary edema (HAPE)
High altitude cerebral edema (HACE)
High altitude retinal hemorrhage
Exacerbation of preexisting comorbidities

albeit at lower frequency [10]. Finally, despite the recognized association between acute mountain sickness and high altitude pulmonary edema with poor quality sleep, no correlation has been found between these syndromes and periodic breathing. Rather, nonperiodic and irregular breathing supplant periodic breathing during sleep with these two disorders [11]. Only one study found a trend toward greater periodic breathing and lower nocturnal oxygen saturation in HAPE compared with subjects with and without AMS. However, arterial blood gas analysis before and after sleep recordings indicated that the significantly lower oxygenation in HAPE was due to pulmonary gas exchange disturbances rather than changes in ventilation [12].

ACCLIMATIZATION AND SLEEP QUALITY

Sleep improves with acclimatization and seems to approach that of native highlanders, although no studies have been done to compare sea level residents and highlanders after a prolonged sojourn at high altitude. Salvaggio and colleagues [10] examined sleep architecture in healthy low altitude residents at an elevation of 5050 m and found an increase in stages 3 and 4 sleep after remaining 4 weeks at that elevation when compared to values achieved after the first week of sojourn. However, the percentage of deep sleep had not returned to levels initially found at sea level. On the other hand, despite maintaining similar total sleep times at all three periods (sea level, first week, and fourth week at altitude), the arousal index increased significantly at high altitude and never decreased even after 4 weeks of stay. These findings may be applicable to native highlanders as well, but the degree of their responses is likely to vary according to the level of adaptation and perhaps with ethnic differences. Comparing Han and Tibetans—two different groups of highlanders, with the latter group having a longer history of residence at high altitude—Plywaczewski et al. [6] demonstrated better sleep structure of Tibetans than with Han during acute exposure to a simulated altitude of 5000 m, suggesting a difference in long-term adaptation between these two ethnicities. Interestingly, the acclimatization process takes place during chronic

intermittent exposure but seems to be incomplete, that is, not reaching the level of permanent highlanders [13].

EFFECTS OF HYPOXIA AT HIGH ALTITUDE

At high altitude, as barometric pressure falls the arterial partial pressure oxygen declines, reaching the critical level of 60 mmHg at approximately 2400 m. Without oxygen replacement, hypoxia develops causing sleepiness, impaired judgment, blunted reaction time, tremors, and decreased visual acuity. The resulting hypoxia at high altitude leads to stimulation of the peripheral arterial chemoreceptors and an increase in alveolar ventilation, hence causing hypocapnia and cerebral alkalosis, which in turn further depresses the central hypoxic chemoresponsiveness. With this increase in minute ventilation there is an increase in the work of breathing of an unacclimatized person. In addition, there is an increase in the respiratory muscles workload due to lung stiffness and increased airways resistance from vascular engorgement and hypoxic pulmonary vasoconstriction along with dynamic airway closure.

Although daytime breathing is characterized by hyperventilation, the breathing pattern during sleep is periodic, as initially described in the 19th century [14]. Both hypoxia and hypocapnia contribute to this well-characterized crescendo–decrescendo breathing pattern, also observed in heart failure patients, and known as Cheyne–Stokes respiration (Figure 122.1) [15], but the cycle length of the periodicity at high altitude is shorter when compared to that of patients with congestive heart failure (range between 12 and 34 s versus 40–90 s) [16, 17]. There is an inverse relationship between cycle length and altitude; as

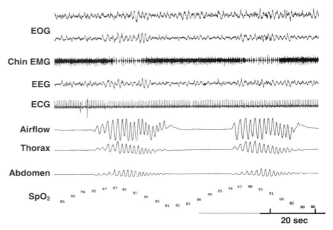

Figure 122.1   Polysomnograph depicting the classic respiratory oscillations of Cheyne–Stokes respiration that are associated with significant oxygen desaturations and intermittently with arousals.

hyperventilation progressively increases during ascent, the cycle length becomes shorter [16]. In addition, there is large interindividual variability of periodic breathing intensity in response to high altitude [5], which is thought to be related to the magnitude of the intrinsic hypoxic drive [11] and affected by age [18].

Long-term acclimatization begins several hours after arriving at a target altitude and continues for days to weeks. The depressed hypoxic drive, which is due to the initial respiratory alkalosis, slowly abates allowing for further increases in ventilation that perpetuate the respiratory alkalosis. In addition, a metabolic acidosis and erythrocytosis compensate for the chronic hypocapnia and lower oxygen carrying capacity of hemoglobin. After approximately 1 month at high altitude, systemic blood pressure, heart rate, and cardiac output decrease to normal levels, but heart rate and arterial blood pressure swings persist during sleep respiratory oscillations in some acclimatized lowlanders [19]. Also, pulmonary hypertension does not resolve because of persistent hypoxia and increased blood viscosity.

Ventilatory acclimatization takes place at high altitude, but it does not restore normal sleep nor does it eliminate periodic breathing [3, 6, 10]. It has even been suggested that periodic breathing might have a role in improving oxygen saturation at high altitude. At an altitude of 5000 m, both Salvaggio et al. [10] in a short-term study and Plywaczewski et al. [6] in a study on Tibetans (possibly the oldest population residing at an altitude between 3700 and 4800 m) demonstrated persistence of periodic breathing and higher mean oxygen saturation compared to the first week at that altitude and compared to acclimatized Han, respectively. Despite the ventilatory variability to altitude between individuals, the respiratory oscillations are invariably observed while transitioning from wake to sleep, occurring during NREM sleep, and often ceasing with REM sleep initiation except at extremely high altitudes [8, 20]. The trigger for periodic breathing seems to be related to the combined effects of hypoxia, which stimulates hyperpnea and alkalosis, and hypocapnia, which initiates apnea. In turn, apnea enhances ventilatory stimulation because of hypoxia and hypercapnia, which consequently perpetuates the vicious circle of Cheyne–Stokes breathing whose axis is determined by the CO_2 apnea threshold. The pivotal role of hypocapnia in generating periodic breathing has been demonstrated in studies that administered CO_2 while maintaining the hypoxia [21] and in those that induced hypocapnia (using passive positive pressure hyperventilation) regardless of arterial oxygenation [22]. In these studies, periodic breathing was abolished with CO_2 administration and was induced by hypocapnia independently of the effect of hypoxia (Figure 122.2). Therefore hypoxia per se is important in initiating periodic breathing, by stimulating ventilation and causing

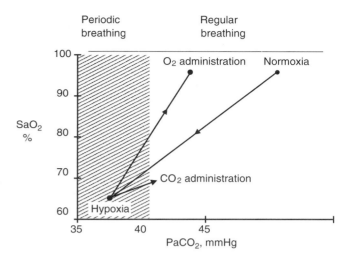

Figure 122.2 Relationship of respiratory pattern to Pa_{CO_2} and Sa_{O_2}, demonstrating that reduction in Pa_{CO_2} is essential in perpetuating the respiratory oscillations in normal subjects during hypoxia. (Reprinted from Weil JV (1985), *Clin Chest Med* **6**:615–621, by permission of Elsevier Inc.)

hypocapnia, but alone is insufficient to perpetuate this pattern of breathing during sleep.

PREVENTION OF HIGH ALTITUDE SLEEP DISTURBANCES AND DISORDERS

Gradual ascent allowing for acclimatization is the best approach for the prevention of high altitude illnesses, thus minimizing certain sleep disturbances. Newcomers to high altitude should start the acclimatization process at approximately 2500 m, and going up no more than 600 m per day, allowing for an extra day for every 600–900 m above 2500 m. For travelers intending to ascent to greater altitude (e.g., 3000–3300 m) in one day, or for those who plan even mild to moderate activity at a slightly lower altitude, returning to lower altitude for sleep and/or prophylaxis is recommended in order to prevent some of the high altitude morbidities (Table 122.1) [23].

For mild to moderate AMS and high altitude sleep disordered breathing, another way to partly acclimatize breathing without descent is to administer oxygen during sleep. It has been estimated that every 1% increase in F_IO_2 above 21% is roughly equivalent to a 300 m drop in altitude [24]. Oxygen reverses periodic breathing and possibly improves sleep quality and symptoms of AMS, although results from various studies have been inconsistent [9, 25].

Slower ascent and prophylaxis, with particular attention to early symptoms, are even more important in persons who had previous HAPE because of a high recurrence rate. Acetazolamide remains the agent of choice for prophylaxis because of its safety profile and effectiveness in preventing AMS. Dexamethasone also is effective for AMS prophylaxis and may have a synergistic effect with acetazolamide in reducing AMS severity, but it is often avoided because of its potential adverse effects and concern for rebound phenomenon when withdrawn at high altitude [23]. Ginkgo biloba has been studied and found to be a safe alternative to acetazolamide for AMS prophylaxis, but no studies are available comparing the two agents [26]. In a blinded randomized placebo trial, slow release theophylline was recently found to be comparable to acetazolamide in almost normalizing breathing during sleep at an altitude of 3500 m in healthy male volunteers [27]. Because of their effects on ventilatory drive and nocturnal oxygenation, alcohol and sedative–hypnotics should be avoided while at high altitude. Regarding HAPE prophylaxis, it is recommended to start nifedipine, preferably the sustained release formulation, 2–3 days before the ascent in susceptible persons, and to continue using nifedipine after reaching the desired elevation. Recently, prophylactic inhalation of long-acting beta-agonist (Salmeterol) has been shown to reduce the incidence of HAPE in a randomized double-blind study, but no comparison was made with the conventional nifedipine prophylaxis [28].

TREATMENT OF HIGH ALTITUDE MORBIDITIES

The management of high altitude related disorders is based on early diagnosis and severity of presentation, realizing that initial presentation does not predict eventual severity. Proceeding with the ascent in the presence of significant symptoms is contraindicated. In general, symptoms start to improve within 24–48 hours of acclimatization or treatment. For persistent sickness, or for severe initial presentation/neurologic changes (e.g., HAPE or HACE), immediate descent is mandatory. Table 122.2 describes treatment guidelines for mild to severe high altitude illnesses.

SUMMARY

Sleep at high altitude is characterized by frequent arousals and poor quality. It is worsened by the development of high altitude illnesses such as AMS and HAPE. Acclimatization and perhaps oxygen administration remain the best and safest strategies to improve sleep quality during ascent, as even low-dose prophylaxis with conventional treatment (acetazolamide, nifedipine) has potentially significant side effects that are poorly tolerated because of concomitant hypoxemia, especially in vulnerable subjects such as those with comorbidities, pregnant women, and children. Familiarizing oneself with high altitude symptoms and different prophylaxis and treatment regimens is crucial to

TABLE 122.2 Treatment Guidelines for High Altitude Related Disorders

High altitude headache and mild acute mountain sickness
   Stop ascent, rest, acclimatize at same altitude
   Acetazolamide, 125–250 mg b.i.d., to speed acclimatization
   Symptomatic treatment as necessary with analgesics
      and antiemetics
   or Descend 500 m or more
Moderate to severe acute mountain sickness
   Low-flow oxygen, if available
   Acetazolamide, 125–250 mg b.i.d., with or without
      dexamethasone, 4 mg p.o., IM, or IV q6h
   Hyperbaric therapy
   or Immediate descent
High altitude cerebral edema
   Immediate descent or evacuation
   Oxygen, 2–4 L/min
   Dexamethasone, 4 mg p.o., IM, or IV q6h
   Hyperbaric therapy
High altitude pulmonary edema
   Minimize exertion and keep warm
   Oxygen, 4–6 L/min until improving, then 2–4 L/min
   Nifedipine, 10 mg p.o. q4h by titration to response, or 10 mg
      p.o. once, followed by 30 mg extended release q12 to 24h
   Hyperbaric therapy
   or Immediate descent
Periodic breathing
   Acetazolamide, 62.5–125 mg at bedtime as needed

Source: From Auerbach PS (2001), *Wilderness Medicine* by permission of Elsevier.

minimize risks of complications that are either related to high altitude illnesses or to their treatments. In the event of the development of a serious life-threatening high altitude illness, pharmacologic and oxygen field treatment are to be initiated immediately, but by no means should substitute for descent to lower altitudes.

REFERENCES

1. Honigman B, Theis MK, Koziol-McLain J, Roach R, Yip R, Houston C, Moore LG, Pearce P. Acute mountain sickness in a general tourist population at moderate altitudes. *Ann Intern Med* **118**:587–592(1993).

2. Reite M, Jackson D, Cahoon RL, Weil JV. Sleep physiology at high altitude. *Electroencephalogr Clin Neurophysiol* **38**:463–471(1975).

3. Selvamurthy W, Raju VRK, Ranganathan S, Hedge KS, Ray US. Sleep patterns at an altitude of 3500 meters. *Int J Biometeor* **30**:123–135(1986).

4. Mizuno K, Asano K, Okudaira N. Sleep and respiration under acute hypobaric hypoxia. *Jpn J Physiol* **43**:161–175(1993).

5. Zielinski J, Koziej M, Makowski M, Sarybaev AS, Tursalieva JS, Sabirov IS, Karamuratov AS, Mirrakhimov MM. The quality

of sleep and periodic breathing in healthy subjects at an altitude of 3200 m. *High Altitude Med Biol* **1**:331–336(2000).

6. Plywaczewski R, Wu TY, Wang XQ, Cheng HW, Sliwiński PS, Zieliński J. Sleep structure and periodic breathing in Tibetans and Han at simulated altitude of 5000 m. *Respir Physiol Neurobiol* **136**:187–197(2003).

7. Kinsman TA, Hahn AG, Gore CJ, Martin DT, Chow CM. Sleep quality responses to atmospheric variation: case studies of two elite female cyclists. *J Sci Med Sport* **6**:436–442(2003).

8. Anholm JD, Powles AC, Downey R III, Houston CS, Sutton JR, Bonnet MH, Cymerman A. Operation Everest II: arterial oxygen saturation and sleep at extreme simulated altitude. *Am Rev Respir Dis* **145**:817–826(1992).

9. Barash IA, Beatty C, Powell FL, Prisk GK, West JB. Nocturnal oxygen enrichment of room air at 3800 meter altitude improves sleep architecture. *High Alt Med Biol* **2**:525–533(2001).

10. Salvaggio A, Insalaco G, Marrone O, Romano S, Braghiroli A, Lanfranchi P, Patruno V, Donner CF, Bonsignore G. Effects of high-altitude periodic breathing on sleep and arterial oxyhaemoglobin saturation. *Eur Respir J* **12**:408–413(1998).

11. Goldenberg F, Richalet JP, Onnen I, Antezana AM. Sleep apneas and high altitude newcomers. *Int J Sports Med* **13**:S34–S36(1992).

12. Eichenberger U, Weiss E, Riemann D, Oelz O, Bartsch P. Nocturnal periodic breathing and the development of acute high altitude illness. *Am J Respir Crit Care Med* **154**:1748–1754(1996).

13. Richalet JP, Vargas Donoso M, Jiménez D, Antezana AM, Hudson C, Cortès G, Osorio J, Leòn A. Chilean miners commuting from sea level to 4500 m: a prospective study. *High Altitude Med Biol* **3**:159–166(2002).

14. Ward M. Periodic respiration: a short historical note. *Ann Coll Surg (Engl)* **52**:330–334(1973).

15. Sutton JR, Houston CS, Mansell AL, et al. Effect of acetazolamide on hypoxemia during sleep at high altitude. *N Engl J Med* **301**:1329–1331(1979).

16. Waggener TB, Brusil PJ, Kronauer RE, Gabel RA, Inbar GF. Strength and cycle time of high-altitude ventilatory patterns in unacclimatized humans. *J Appl Physiol* **56**:576–581(1984).

17. Javaheri S. A mechanism of central sleep apnea in patients with heart failure. *N Engl J Med* **34**:949–954(1999).

18. Arai Y, Tatsumi K, Sherpa NK, Masuyama S, Hasako K, Tanabe N, Takiguchi Y, Kuriyama T. Impaired oxygenation during sleep at high altitude in Sherpa. *Respir Physiol Neurobiol* **133**:131–138(2002).

19. Insalaco G, Romano S, Salvaggio A, Braghiroli A, Lanfranchi P, Patruno V, Marrone O, Bonsignore MR, Donner CF, Bonsignore G. Blood pressure and heart rate during periodic breathing while asleep at high altitude. *J Appl Physiol* **89**:947–955(2000).

20. Normand H, Barragan M, Benoit O, Baillaird O, Raynaud J. Periodic breathing and O_2 saturation in relation to sleep stages at high altitude. *Aviat Space Environ Med* **61**:229–235(1990).

21. Berssenbrugge A, Dempsey J, Iber C, Skatrud J, Wilson P. Mechanisms of hypoxia-induced periodic breathing during sleep in humans. *J Physiol* **343**:507–526(1983).

22. Skatrud JB, Dempsey JA. Interaction of sleep state and chemical stimuli in sustaining rhythmic ventilation. *J Appl Physiol* **55**:813–822(1983).

23. Hackett PH, Roach RC. High altitude medicine. In: Auerbach PS (Ed), *Wilderness Medicine*, 4th ed. Mosby, St Louis, 2001, pp 2–43.

24. West JB. Oxygen enrichment of room air to relieve the hypoxia of high altitude. *Respir Physiol* **99**:225–232(1995).

25. Luks AM, van Melick H, Batarse RR, Powell, Grant I, West JB. Room oxygen enrichment improves sleep and subsequent day-time performance at high altitude. *Respir Physiol* **113**:247–258(1998).

26. Maakestad K, Leadbetter G, Olson S, Hackett PH. Ginkgo biloba reduces incidence and severity of acute mountain sickness. *Wilderness Environ Med* **12**:51–58(2001).

27. Fischer R, Lang SM, Leitl M, Thiere M, Steiner U, Huber RM. Theophylline and acetazolamide reduce sleep-disordered breathing at high altitude. *Eur Respir J* **23**:47–52(2004).

28. Sartori C, Allemann Y, Duplain H, Lepori M, Egli M, Lipp E, Hutter D, Turini P, Hugli O, Cook S, Nicod P, Scherrer U. Salmeterol for the prevention of high-altitude pulmonary edema. *N Engl J Med* **346**:1631–1636(2002).

123

SLEEP AND AVIATION

JOHN A. CALDWELL

Air Force Research Laboratory, Brooks City–Base, Texas

INTRODUCTION

Modern pilots and aircrews must cope with a variety of nonstandard work schedules in order to effectively meet customer/mission demands. Whether the flight missions involve cargo delivery, passenger transport, or the fulfillment of military requirements, aviation personnel are likely to be faced with unpredictable work hours, long duty periods, and circadian disruptions, which all lead to difficulties obtaining adequate sleep [1, 2]. Clearly, such problems must be recognized and mitigated in order to optimize both performance and safety in a wide variety of aviation contexts. This chapter reviews the evidence for sleep-related difficulties in aviation operations and briefly summarizes some of the counterfatigue strategies used by civil and military pilots and crews.

LONG-HAUL COMMERCIAL PILOTS

Several investigations have shown that long-haul pilots and crews (i.e., those staffing international flights) frequently experience shortened sleep, reduced sleep efficiency, and/ or changes in sleep architecture that prevent full recovery during the layovers between flight segments [3–5]. Needless to say, these sleep troubles often lead to serious problems with on-the-job sleepiness due to a combination of homeostatic and circadian factors [6]. Long-haul crews frequently face work-related sleep restrictions associated with late arrivals and early departures while at the same time being confronted with jet-lag-related and shift-lag-related sleep disturbances associated with constant schedule changes. As might be expected based on the circadian factors discussed in earlier chapters, eastward flights are generally more problematic than westward flights [7]. In fact, it has been established that circadian resynchronization rates are 50% slower after eastbound than after westbound travel across multiple time zones, and this in-and-of-itself can interfere with recuperative sleep [8, 9]. Following eastward transitions, sleep patterns are more variable and fragmented [10] primarily because limited layover periods require crew members to attempt to go to sleep at an earlier than normal biological time, which is difficult to accomplish due to the circadian pressure to be awake. In addition, an earlier than normal rise time the following morning creates a significantly shortened sleep period. Following westward transitions, sleep problems are less severe [11], but reduced sleep efficiency and duration have nonetheless been reported despite the fact that the longer biological day makes the layover sleep easier to initiate [12].

SHORT-HAUL COMMERCIAL PILOTS

The pilots and crews responsible for shorter flight trips also can be adversely affected by duty-related sleep difficulties. Early report times, multiple flight legs, and long duty days are just some of the factors that combine to degrade the sleep and alertness of short-haul pilots [13]. In one study, the sleep onset of these pilots was delayed by 12 minutes, the sleep period was shorter than normal, and the wake-up time was almost an hour and a half earlier than normal. Although the average duty day was 10.6 hours, a third of

Sleep: A Comprehensive Handbook, Edited by T. Lee-Chiong.
Copyright © 2006 John Wiley & Sons, Inc.

the duty days were greater than 12 hours. Long duty cycles alone can cause sleep deprivation because, by the time crews complete their postflight paperwork, take a cab to the hotel, grab some dinner, and crawl into bed, there is little chance of squeezing in 8 hours of restful slumber before it's time to wake up, shower, get some coffee, return to the airport, and prepare for the next flight. Similarly, long duty days have been reported by regional airline pilots (with duty periods averaging 11.3 hours) and corporate/executive pilots (who average 9.9 hours per day) [14, 15]. Thus it is not surprising that while long-haul pilots associate their on-the-job fatigue primarily with night flights and jet lag, short-haul pilots attribute their fatigue-related problems more to prolonged duty periods and early wake-ups [16].

MILITARY AVIATORS

Military pilots also suffer sleep disturbances or truncated sleep because of the high demands associated with operational missions. Paul, Pigeau, and Weinberg [17] found that during the days prior to some resupply missions from Canada to the former Yugoslavia, the sleep of Canadian CC-130 pilots steadily decreased from 8.7 hours per day to 6.5 hours per day before departure (largely because the crews were required to report for duty in the early morning hours). These same crews continued to suffer from sleep deprivation immediately after landing at the destination point due to sleep difficulties associated with circadian factors. Such problems are common in the U.S. military as well. Boll et al. [18] indicated that many C-141 pilots slept only 6.7 hours per day during Desert Storm, while pilots slept only 6.4 hours per day during Desert Shield. Neville et al. [1] reported that during the final week of Desert Storm, there were severe sleep deficits in at least one sample of C-141 crews. Despite regulations requiring a minimum crew rest period of 12 hours [19], some pilots obtained no sleep at all during entire 24 hour periods, and several others obtained less than 10 hours of sleep across 48 hour periods. Similarly, Bisson, Lyons, and Hatsel [20] found that C-5 crews during Operation Desert Shield often reported for duty with less than 4 hours of sleep in the previous 12–16 hours because they were simply unable to fall asleep due to a combination of circadian and environmental factors. Insufficient and/or disrupted sleep also were apparently an issue for F-15 pilots during Desert Storm. Cornum [21] found that even though these pilots tended to fly shorter missions than their transport-pilot counterparts, they nevertheless suffered from sleep deprivation and circadian disruptions due to 24 hour taskings, little time off between missions, and long stretches of work days with minimal recuperation time. At one point in the conflict, it was reported that two F-15 pilots flew over 20 hours during a single 24 hour period with little or no sleep.

SLEEP DIFFICULTIES AND SAFETY

Needless to say, sleep disruptions create a potential safety hazard on the flight deck by increasing fatigue levels and making pilots and air crew lethargic, less willing to work cooperatively with other crew members, more distractible, less able to integrate incoming information, and less capable of making the types of higher-level cognitive decisions that are often crucial for flight safety [22, 23]. In addition, it is clear that once pilots reach a certain level of sleepiness, they begin to experience involuntary microsleeps (as measured by electro-oculographic (EOG) and electroencephalographic (EEG) activity) even during critical phases of flight, such as from top-of-descent to landing [24]. In light of such facts, it is clear that efforts must be made to optimize preduty sleep and to implement proven alertness-management strategies to maximize safety on the flight deck [25].

FATIGUE COUNTERMEASURES IN AVIATION

Education

Recent studies have made it clear that sleep restrictions of as little as 1–2 hours will adversely affect vigilance and performance in subsequent duty periods [26, 27]. Therefore, as mentioned earlier, it is critical that aviation personnel be educated about the dangers of fatigue and the importance of sleep and proper sleep hygiene. Ultimately, the pilots themselves and those scheduling the pilots' duty periods must be convinced that quality off-duty sleep is the best possible antifatigue strategy. Thus education programs such as those currently offered by the U.S. Air Force Research Laboratory and the NASA Ames Research Center will continue to warn aircrews that (1) fatigue is a physiological problem that cannot be overcome by motivation, training, or willpower; (2) people cannot reliably judge their own level of fatigue-related impairment; (3) there are wide individual differences in fatigue susceptibility that cannot be reliably predicted; and (4) there is no one-size-fits-all "magic bullet" (other than adequate sleep) that can counter fatigue for every person in every situation. Aircrews will continue to be advised that (1) adequate off-duty sleep must become a priority to ensure on-the-job alertness; (2) an average of 8 hours of sleep per day either in a consolidated block, or in a series of naps, should be the goal even during long trips or on rotating schedules; and (3) there are "good sleep habits" that can optimize sleep quantity and quality (refer to [28]).

On-Board Sleep

For commercial pilots, the Federal Aviation Administration (FAA) requires augmented crews (at least three pilots) and on-board rest facilities for long-haul flights (i.e., those longer than 12 hours) so that pilots can partially attenuate the homeostatic sleep drive between take-offs and landings [29]. Any transoceanic air traveler has no doubt observed one of the pilots leaving the cockpit during the cruise segment of the flight to gain some in-flight sleep (while another pilot has replaced him/her on the flight deck). In some military operations, a similar strategy is sometimes implemented in multicrew aircraft. For instance, in the case of the B-2 bomber, in which the maximum crew complement is only two individuals, one of the pilots may sleep (in a cot located behind the seats) during low-workload flight phases while the other pilot maintains control of the aircraft. Such in-flight sleep is an absolutely essential tool for B-2 missions, which often remain aloft for 44 continuous hours. Clearly, on-board crew sleep is an important aviation fatigue countermeasure in any situation, military or civilian, involving flights longer than 10–12 hours.

Cockpit Naps

A strategy related to on-board crew rest is the cockpit nap in which one pilot actually sleeps in his/her cockpit seat (rather than moving to another part of the aircraft) while the other pilot flies the plane. NASA studies on commercial pilots have shown that cockpit naps of up to 40 minutes in duration are both safe and effective [25], and as a result, many international airlines now authorize cockpit napping on long flights. In some military operations (which are not governed by the FAA), cockpit naps also are used in multicrew aircraft as long as another alert pilot is at the controls. However, cockpit napping is not FAA approved for commercial U.S. pilots despite the proven effectiveness of this countermeasure or the fact that the general public has indicated that they would support the implementation of cockpit napping as an aviation fatigue-mitigation strategy [30]. Efforts are under way to address this discrepancy and to make this proven antifatigue strategy available to U.S. commercial aviators.

Controlled Rest Breaks

Survey data suggest that, when feasible, pilots often use some type of work breaks (chatting, standing up, walking around, etc.) to help sustain alertness during lengthy flights. Although controlled studies of the efficacy of this counter-measure are scarce, Neri et al. [31] reported that Boeing 747 pilots who were offered brief hourly breaks during a 6 hour simulated night flight showed significant postbreak reductions in slow eye movements, theta-band activity, unintended sleep episodes, and subjective sleepiness for up to 15–25 minutes. Benefits were particularly noticeable near the time of the circadian trough. Along similar lines,

Caldwell, Prazinko, and Caldwell [32] found that simply assuming a more upright posture, as opposed to remaining seated, reduced the amount of slow-wave EEG activity and enhanced performance on a 10 minute vigilance task during the later part of a 28 hour sleep-deprivation cycle. Although no information was provided on how long alertness was improved following this manipulation, it is certainly possible that posture-related benefits contributed to the positive effects of rest breaks found earlier by Neri et al. [31]. Thus simply walking off the flight deck for a few minutes each hour during long-duration flights can provide at least a brief respite from operator fatigue.

Proper Crew Work–Rest Scheduling

One method for improving aircrew sleep and alertness is to devise and implement crew schedules that consider the realities of the homeostatic and circadian sleep/alertness mechanisms. Unfortunately, both airline-pilot schedules and military-pilot schedules have traditionally minimized these factors and instead have focused on time-on-task (i.e., boredom effects, etc.). In part, this is because the underlying physiological contributors to human fatigue were poorly understood, but it is also the case that scheduling around circadian and homeostatic factors is both difficult and inconvenient. Although there is little that can be done to solve the "inconvenience issue," the "difficulty issue" is to some extent being mitigated by the development of computerized scheduling tools such as the Fatigue Avoidance Scheduling Tool (FAST) [33] and the System for Aircrew Fatigue Evaluation (SAFE) [34]. These software tools predict alertness decrements (and optimal sleep times) based on such factors as the amount of continuous wakefulness, the circadian phase, and the quality of recovery sleep, and their ease of use facilitates the development of improved work/rest schedules. Although such techniques have not yet been fully validated across a wide range of aviation settings, they hold great promise for the future of aviation alertness management.

Melatonin and Bright Light

In aviation operations involving rapid schedule changes, both melatonin administration and bright light exposure may help to overcome jet lag and shift lag. With regard to melatonin, there is a substantial amount of research to indicate that appropriate administration of this hormone can improve circadian adaptation to new time schedules [35]. There also is evidence that melatonin possesses weak hypnotic or "soporific" properties that may facilitate out-of-phase sleep [36]. Since melatonin is not considered a drug, it is widely available both to commercial pilots and to military aviators, and generally, it can be used by civil aviators with few restrictions. Melatonin's drawback is that it can produce negative effects on alertness and performance if its administration is not properly timed, and most potential users possess insufficient knowledge about circadian rhythms and/or endogenous melatonin secretion to make the best decisions in this regard. As a result, Caldwell [37] has concluded that "currently, melatonin use is unacceptable for aviators" (p. 243). In addition, melatonin is not an approved substance for use in military aviation. As an alternative, some studies have shown that properly timed bright light can facilitate circadian resynchronization after schedule changes [38–40]. However, at present, there remains debate over the optimal levels of light exposure as well as whether personnel are able to readily determine the appropriate times at which bright light exposures should occur. Lewy et al. [41] discusses how to use both melatonin and light to adjust circadian rhythms. However, the difficulties in appropriate timing of both methods suggest that perhaps the safest self-administered strategy for jet-lagged pilots seeking to adjust to new schedules is simply to take advantage of natural environmental lighting, naps, and local zeitgebers [42, 43]. Unfortunately, this option is not as useful for pilots suffering from shift lag, although they can at least be cautioned to *avoid* light exposure (or to minimize it with very dark glasses) before a period of daytime sleep. In addition, whether a person should attempt to adapt to the new environment or a new shift should be considered; if time spent in the new environment or on the new shift is short, adaptation may not be beneficial.

Sleep-Promoting Compounds

In civilian aviation operations, prescription (and over-the-counter) hypnotics and sedatives are discouraged, and several are not typically authorized even to a limited degree. Hypnotics are not authorized for any in-flight use. In the United States, the Federal Air Surgeon has recently approved the use of zolpidem as long as it is not administered more than twice a week or within 24 hours of flight; however, zolpidem cannot be used to aid in overcoming circadian disruptions. In military aviation, the use of hypnotics is slightly more liberal. These compounds are sometimes relied on to promote sleep in extremely stressful or uncomfortable environments, or in situations where the only sleep opportunities occur at times that are problematic from a circadian standpoint. For instance, it is sometimes necessary for military aviators to quickly transition from flying days to flying nights without the luxury of a suitable adjustment period, and they must remain highly vigilant while accomplishing this transition. For such circumstances, it has been shown that nighttime flight performance can be significantly enhanced by utilizing compounds such as temazepam to promote restful daytime sleep [44]. Other military circumstances in which hypnotics may be authorized are those in which the only opportunities for sleep are

brief napping opportunities that occur at the "wrong" circadian times (such as during the midmorning or during the so-called "forbidden sleep zone" from approximately 1800 to 2100 hours). In such cases, it has been shown that the improved sleep quality that can be obtained with the use of zolpidem significantly enhances subsequent performance [45]. However, it must be kept in mind that neither temazepam nor zolpidem should be used in circumstances where the aviator is "on-call" since grogginess after an early awakening could compromise performance. The bottom line is that while hypnotics are not used by military aviators on a frequent or continuous basis, they are sometimes very helpful for maximizing mission success and pilot safety during periods of very high operational tempos. Furthermore, unlike herbal remedies or hormonal remedies, prescription hypnotics are carefully regulated by trained medical personnel.

Alertness-Enhancing Compounds

When, despite everyone's best intentions, sleepiness becomes a problem in the flight environment, caffeine is often used as the "first-line" pharmacological fatigue countermeasure in both civil and military aviation. Numerous studies have shown that caffeine increases vigilance and improves performance in sleep-deprived individuals, especially those who normally do not consume high doses of caffeine [46]. In commercial aviation, caffeine (generally in the form of coffee, tea, or soft drinks) is the only alertness-enhancing substance allowed, although compounds classified as "dietary supplements" are not prohibited. In civil aviation, prescription stimulants are not authorized. Conversely, in military aviation, prescription alertness-enhancing medications are sometimes authorized for highly demanding missions, which simply must be accomplished despite high levels of fatigue. For these situations, dextroamphetamine has long been the drug of choice because of its powerful, reliable, and safe effects on fatigued individuals [47]. Studies have shown that properly administered dextroamphetamine is capable of sustaining pilot performance at near-well-rested levels for over 50–55 hours without sleep. It is for this reason that the U.S. military has authorized the use of Dexedrine® at various times since World War II. More recently, modafinil has been receiving a great deal of attention within the military community as well. In fact, in December of 2003 modafinil was authorized as one counterfatigue strategy that can be used on a limited basis in extended dual-crew bomber missions. Although modafinil has not been as well tested in operational contexts as dextroamphetamine, two studies to date have shown that it is capable of significantly attenuating fatigue-related decrements in pilot performance throughout 30–40 hours of continuous wakefulness [48, 49]. The fact that modafinil has a relatively low abuse potential and the fact that it produces few cardiovascular side effects may make it preferable to dextroamphetamine for some applications. Both modafinil and dextroamphetamine are carefully regulated by trained physicians (flight surgeons), and they are used only when all other counterfatigue strategies have been exhausted.

SUMMARY AND CONCLUSIONS

In summary, it is clear that sleep difficulties have long been a problem in high-tempo aviation contexts, and the pressures of our modern 24/7 society are more likely to exacerbate rather than to alleviate the scheduling challenges that civilian and military pilots face. However, progress toward the widespread development and implementation of scientifically valid sleep/rest- and performance-optimization strategies is already contributing to operational safety throughout the system. Our challenge as sleep and fatigue experts is to bolster our efforts in research and education while maintaining pressure on federal agencies, the commercial aviation industry, and the military leadership to fully recognize the importance of adequate daily sleep and to prioritize the establishment and maintenance of sound crew scheduling practices that will further promote the safety and well-being of aircrews everywhere.

ACKNOWLEDGMENTS

The views and opinions expressed are those of the author and do not necessarily reflect the official positions or policies of the U.S. Air Force or the U.S. Department of Defense.

REFERENCES

1. Neville KJ, Bisson RU, French J, Boll PA, Storm WF. Subjective fatigue of C-141 aircrews during Operation Desert Storm. *Human Factors* **36**(2):339–349(1994).

2. Samel A, Wegmann HM, Vejvoda M. Jet lag and sleepiness in aircrew. *J Sleep Res* **4**(Suppl 2):30–36(1995).

3. Dement WC, Seidel WF, Cohen SA, Bliwise NG, Carskadon MA. Sleep and wakefulness in aircrew before and after transoceanic flights. *Aviat Space Environ Med* **57**(12 Suppl):B14–B28(1986).

4. Nicholson AN, Pascoe PA, Spencer MB, Stone BM, Green RL. Nocturnal sleep and daytime alertness of aircrew after transmeridian flights. *Aviat Space Environ Med* **57**(12 Suppl): B42–B52(1986).

5. Sasaki M, Kurosaki Y, Mori A, Endo S. Patterns of sleep–wakefulness before and after transmeridian flight in commercial pilots. *Aviat Space Environ Med* **57**(12 Suppl):B29–B42(1986).

6. Akerstedt T, Folkard S. Validation of the S and C components of the three-process model of alertness regulation. *Sleep* **18**(1):1–6(1995).

7. Graeber RC, Lauber JK, Connell LJ, Gander PH. International aircrew sleep and wakefulness after multiple time zone flights: a cooperative study. *Aviat Space Environ Med* **57**(12 Suppl): B3–B9(1986).

8. Aschoff J. Problems of re-retrainment of circadian rhythms: asymmetry effect, dissociation, and partition. In: Assenmacher I, Farner DS(Eds), *Environmental Endocrinology*. Springer Verlag, New York, 1975, pp 185–195.

9. Klein KE, Wegmann HM. Significance of circadian rhythms in aerospace operations (AGARDograph No. 247). North Atlantic Treaty Organization Advisory Group for Aerospace Research and Development, Neuilly-sur-Seine, France, 1980.

10. Buck A, Tobler I, Borbely AA. Wrist activity monitoring in air crew members: a method for analyzing sleep quality following transmeridian and north–south flights. *J Biol Rhythms* **4**(1):93–105(1989).

11. Graeber RC, Dement WC, Nicholson AN, Sasaki M, Wegmann HM. International cooperative study of aircrew layover sleep: operational summary. *Aviat Space Environ Med* **57**(12 Suppl):B10–B13(1986).

12. Lowden A, Akerstedt T. Sleep and wake patterns in aircrew on a 2-day layover on westward long distance flights. *Aviat Space Environ Med* **69**(6):596–602(1998).

13. Rosekind MR, Graeber RC, Dinges DF, Connell LJ, Rountree MS, Gillen K. Crew factors in flight operations IX: effects of planned cockpit rest on crew performance and alertness in long-haul operations (NASA/TM-1994-108839). NASA Ames Research Center, Moffett Field, CA, 1994.

14. Co EL, Gregory KB, Johnson JM, Rosekind MR. Crew factors in flight operations XI: a survey of fatigue factors in regional airline operations (NASA/TM-1999-208799). National Aeronautics and Space Administration, 1999.

15. Rosekind MR, Co EL, Gregory KB, Miller DL. Crew factors in flight operations XIII: a survey of fatigue factors in corporate/executive aviation operations (NASA/TM-2000-209610). National Aeronautics and Space Administration, 2000.

16. Bourgeois-Bougrine S, Carbon P, Gounelle C, Mollard R, Coblentz A. Perceived fatigue for short- and long-haul flights: a survey of 739 airline pilots. *Aviat Space Environ Med* **74**(10):1072–1077(2003).

17. Paul MA, Pigeau RA, Weinberg H. CC-130 pilot fatigue during re-supply missions to former Yugoslavia. *Aviat Space Environ Med* **72**(11):965–973(2001).

18. Boll PA, Storm WF, French J, Bisson RU, Armstrong SD, Slater T, Ercoline WE, McDaniel RL. C-141 aircrew sleep and fatigue during the Persian Gulf conflict. In: *Proceedings of the NATO Advisory Group for Aerospace Research and Development (AGARD): Nutrition, Metabolic Disorders and Lifestyle of Aircrew* (Report AGARD-CP533). North Atlantic Treaty Organization Advisory Group for Aerospace Research and Development, Neuilly-sur-Seine, France, pp 29.1–29.11.

19. US Air Force. Flying operations: general flight rules. Air Force Instruction 11-202, 3, Ch 9, Crew rest and flight duty limitations, pp 64–69. US Air Force, 2003. http://www.e-publishing.af.mil/pubfiles/af/11/afi11-202v3/afi11-202v3.pdf.

20. Bisson RU, Lyons TJ, Hatsel C. Aircrew fatigue during Desert Shield C-5 transport operations. *Aviat Space Environ Med* **64**(9 Pt 1):848–853(1993).

21. Cornum KG Extended air combat operations: F-15s over Iraq. In: *Proceedings of the 65th Annual Meeting of the Aerospace Medical Association*, San Antonio, TX. Aerospace Medical Association, Alexandria, VA, 1994, p A49.

22. Petrie KJ. Dawson AG. Symptoms of fatigue and coping strategies in international pilots. *Int J Aviat Psychol* **7**(3):251–258(1997).

23. Ritter RD. "And we were tired": fatigue and aircrew errors. *AES Syst Mag* **Mar**:21–26(1993).

24. Rosekind MR, Gander PH, Miller DL, Gregory KB, Smith RM, Weldon KJ, Co EL, McNally KL, Lebacqz JV. Fatigue in operational settings: examples from the aviation environment. *Human Factors* **36**(2):327–338(1994).

25. Rosekind MR, Smith RM, Miller DL, Co EL, Gregory KB, Webbon LL, Gander PH, Lebacqz JV. Alertness management: strategic naps in operational settings. *J Sleep Res* **4**(Suppl 2): 62–66(1995).

26. Belenky G, Wesensten NJ, Thorne DR, Thomas ML, Sing HC, Redmond DP, Russo MB, Balkin TJ. Patterns of performance degradation and restoration during sleep restriction and subsequent recovery: a sleep dose–response study. *J Sleep Res* **12**(1):1–12(2003).

27. Van Dongen HPA, Maislin G, Mullington JM, Dinges DF. The cumulative cost of additional wakefulness: dose–response effects on neurobehavioral functions and sleep physiology from chronic sleep restriction and total sleep deprivation. *Sleep* **26**(2):117–126(2003).

28. Caldwell JA, Caldwell JL. *Fatigue in Aviation: a Guide to Staying Awake at the Stick*. Ashgate Publishing Limited, Aldershot, Hants, England, 2003.

29. Rosekind MR, Gregory KB, Co EL, Miller DL, Dinges DF. Crew factors in flight operations XII: a survey of sleep quantity and quality in on-board crew rest facilities (NASA/TM-2000-209611). National Aeronautics and Space Administration, 2000.

30. National Sleep Foundation. 2002 "Sleep in America" poll. National Sleep Foundation, Washington DC, 2002.

31. Neri DF, Oyung RL, Colletti LM, Mallis MM, Tam PY, Dinges DF. Controlled breaks as a fatigue countermeasure on the flight deck. *Aviat Space Environ Med* **73**(7):654–664(2002).

32. Caldwell JA, Prazinko BF, Caldwell JL. Body posture affects electroencephalographic activity and psychomotor vigilance task performance in sleep deprived subjects. *Clin Neurophysiol* **114**:23–31(2003).

33. Hursh SR, Redmond DP, Johnson ML, Thorne DR, Belenky TJ, Storm WF, Miller JC, Eddy DR. Fatigue models for applied research in warfighting. *Aviat Space Environ Med* **75**(3)(Sec II, Suppl):A44–A53(2004).

34. Belyavin AJ, Spencer MB. Modeling performance and alertness: the QinetiQ approach. *Aviat Space Environ Med* **75**(3 Sec II, Suppl):A93–A103(2004).

35. Arendt J, Deacon S, English J, Hampton S, Morgan L. Melatonin and adjustment to phase shift. *J Sleep Res* **4**(S2),74–79(1995).

36. Wirz-Justice A, Armstrong SM. Melatonin. Nature's soporific? *J Sleep Res* **5**:137–141(1996).

37. Caldwell JL. The use of melatonin: an information paper. *Aviat Space Environ Med* **71**:238–244(2000).

38. Daan S, Lewy AJ. Scheduled exposure to daylight: a potential strategy to reduce "jet lag" following transmeridian flight. *Psychopharmacol Bull* **20**(3):566–568(1984).

39. Gander PH, Myhre G, Graeber RC, Andersen HT, Lauber JK. Adjustment of sleep and the circadian temperature rhythm after flights across nine time zones. *Aviat Space Environ Med* **60**(8):733–743(1989).

40. Samel A, Wegmann HM. Bright light: a countermeasure for jet lag? *Chronobiol Int* **14**(2):173–183(1997).

41. Lewy AJ, Bauer VK, Ahmed S, Thomas KH, Cutler NL, Singer CM, Moffit MT, Sack RL. The human phase response curve (PRC) to melatonin is about 12 hours out of phase with the PRC to light. *Chronobiol Int* **15**(1):71–83(1998).

42. Stone BM, Turner C. Promoting sleep in shiftworkers and intercontinental travelers. *Chronobiol Int* **14**(2):133–143(1997).

43. Waterhouse J, Reilly T, Atkinson G. Jet lag. *Lancet* **350**:1611–1616(1997).

44. Caldwell JL, Prazinko BF, Rowe T, Norman D, Hall KK, Caldwell JA. Improving daytime sleep with temazepam as a countermeasure for shift lag. *Aviat Space Environ Med* **74**(2):153–163(2003).

45. Caldwell JA, Caldwell JL. Comparison of the effects of zolpidem-induced prophylactic naps to placebo naps and forced-rest periods in prolonged work schedules. *Sleep* **21**(1):79–90(1998).

46. Committee on Military Nutrition Research. *Caffeine for the Sustainment of Mental Task Performance: Formulations for Military Operations*. National Academy Press, Washington DC, 2001.

47. Caldwell JA, Caldwell JL, Darlington KK. The utility of dextroamphetamine for attenuating the impact of sleep deprivation in pilots. *Aviat Space Environ Med* **74**(11):1125–1134(2003).

48. Caldwell JA, Caldwell JL, Smythe NK, Hall KK. A double-blind, placebo-controlled investigation of the efficacy of modafinil for sustaining the alertness and performance of aviators: a helicopter simulator study. *Psychopharmacology* **150**:272–282(2000).

49. Caldwell JA, Caldwell JL, Smith JK, Alvarado LY, Heintz TR, Mylar J, Brown DL. The efficacy of modafinil for sustaining alertness and simulator flight performance in F-117 pilots during 37 hours of continuous wakefulness. U.S. Air Force Research Laboratory Technical Report, No. AFRL-HE-BR-TR-2004-0003. U.S. Air Force Research Laboratory, Brooks City–Base, TX, 2004.

ADDITIONAL READING

Broughton RJ, Ogilvie RD. *Sleep, Arousal, and Performance.* Birkhauser, Boston, 1992.

Folkard S, Monk TH. *Hours of Work*: *Temporal Factors in Work Scheduling.* John Wiley & Sons, Hoboken, NJ, 1985.

Kryger MH, Roth T, Dement WC. *Principles and Practice of Sleep Medicine.* Saunders, Philadelphia, 2000.

Moore-Ede M. *The Twenty-Four Hour Society*: *Understanding Human Limits in a World that Never Stops.* Addison-Wesley Publishing, Reading, MA, 1993.

124

SLEEP, EXERCISE, AND SPORTS

CAROL A. ENDERLIN AND KATHY C. RICHARDS

University of Arkansas for Medical Sciences College of Nursing, Little Rock, Arkansas

INTRODUCTION

Human beings wake and sleep by an internal clock, or circadian rhythm, following a diurnal (daytime) pattern of activity for approximately 8 out of 24 hours. Sleep circadian rhythm disruption is associated with a cluster of psychological and physiological symptoms, including impaired cognition, mood state, neuromuscular coordination, and strength. Since athletes frequently travel over time zones for athletic events, they may experience jet lag, a syndrome resulting from slow circadian rhythm adaptation to rapid time-zone transitions. This syndrome is characterized by sleep impairment and decreased mental and physical performance for five or more days. Severity of symptoms varies with age, chronotype (lark or owl), direction of travel (time gain or loss), and speed of time-zone transitions.

As mental and physical performance can be impaired by sleep circadian rhythm disruption and jet lag, optimal performance may logically be promoted through sleep improvement methods and advantageous timing of events. Pharmacologic sleep improvement methods have included prescription sedatives, hypnotics, and minor tranquilizers. However, potential drug dependency, reduced drug effectiveness, and residual cognitive side effects may outweigh the benefits for some individuals. Nonpharmacologic sleep improvement methods have included hypnotherapy, biofeedback, imagery, relaxation, and breathing techniques, as well as stimulus-control therapy [1]. Unfortunately, these techniques require a period of consistent practice before effects are realized.

Exercise has also been suggested as a sleep improvement method. Numerous studies have shown a therapeutic postexercise effect on sleep, but no consensus as to the timing, frequency, duration, intensity, or type of exercise has been reached.

CIRCADIAN RHYTHMS, JET LAG, AND SPORTS PERFORMANCE

Sports scientists have a possible source of evidence-based guidelines to maximize athletic performance. The theorized impact of time of day on sports performance has been addressed in studies of swimming, cycling, running, football, and basketball. The related effect of jet lag on sports performance has also been investigated in studies of baseball and soccer.

Swimming performance was significantly improved ($p < 0.001$) in several evening events compared to morning start times for various strokes over 100 meters [2], and for freestyle swim times ($p < 0.05$) over 100 and 500 meters [3]. Cycling competitors have also suggested an early-evening advantage for time trials, based on self-report [4]. Additionally, world records in distance running events have been noted as occurring most often in late afternoon and early evening [1].

A review of all major baseball games from 1978 to 1987, assessing for day or night game, failed to support the hypothesis of peak performance in late afternoon [5]. Contradiction was noted as the absence of a consistent night performance advantage [6].

A circadian advantage, equivalent to home game advantage, for teams traveling from the West to East Coast was supported in a review of all National Football League

Sleep: A Comprehensive Handbook, Edited by T. Lee-Chiong.
Copyright © 2006 John Wiley & Sons, Inc.

TABLE 124.1 Summary of Circadian Rhythm and Jet Lag Effects on Sports Performance

Year	Author	Sport	Comparison	Performance
1976	Rodahl et al. [2]	Swimming	Early to late time	PM advantage
1983	Baxter and Reilly [3]	Swimming	Early to late time	PM advantage
1993	Jehue et al. [5]	Baseball	Early to late time	No difference
1994	Atkinson et al. [4]	Cycling	Early to late time	PM advantage
1997	Reilly et al. [1]	Distance running	Early to late time	PM advantage
1993	Jehue et al. [5]	Baseball	Coast to coast travel	No difference
1993	Hill et al. [10]	Soccer	Baseline: post-travel	↑Sleep, ↓strength
1995	Recht et al. [9]	Baseball	Coast to coast travel	WC advantage
1997	Smith et al. [7]	Football	Coast to coast travel	WC advantage
1997	Steenland and Deddens [8]	Basketball	Coast to coast travel	WC advantage

Key: PM = afternoon/evening; WC = West Coast.

Monday Night Football games played from 1970 to 1994. The possibility of these findings being due to confounding variables or to jet lag was considered highly unlikely by the authors [7]. However, the possible superiority of western to eastern teams, engagement in activities that can negatively impact performance by the traveling team, and impaired coaching due to the cognitive effects of jet lag were identified as possible confounding variables [6].

A review of National Basketball results from 1987 to 1995 found a suggestion of circadian rhythm effects upon subanalysis for teams traveling west to east ($p = 0.07$) but no overall significant difference in performance. The same review demonstrated unexpected findings of performance improving significantly with increasing time between games, peaking at 3 days ($p = 0.01$) [8].

Jet Lag and Sports Performance

Jet lag may affect sports performance through temporary desynchrony between an athlete or team's circadian rhythms and the environment. Therefore the effect of performing a sport at atypical times of day may potentially be beneficial or constitute a handicap.

A review of all major baseball games from 1978 to 1987, assessing for home field advantage and day or night game, failed to support the jet lag hypothesis of impaired performance correlating with increased transmeridian travel [5]. Contradictions included patterns of winning percentages between same zone teams, absence of a "dose–response" distance traveled effect, a comparative decreased performance regardless of travel direction, and inconsistent evidence of effective countermeasures [6].

A later analysis of all major baseball games from 1991 to 1993, played within 2 days following transcontinental travel, demonstrated greater home winning percentages ($p = 0.006$) when visiting teams traveled west to east than east to west ($p = 0.10$) [9]. Potential contradiction of these positive findings was the possibility of eastern superiority over western teams [6].

A comparison of pre- and post-transmeridian travel sleep and strength in soccer players reflected significantly decreased vigor on days two and three, increased fatigue and sleep time on day two, earlier retirement on day two ($p < 0.05$), and decreased static grip strength on days one and two ($p < 0.05$). These factors returned to baseline by 3–4 days after arrival [10].

Summary of Circadian Rhythm, Jet Lag, and Sports Performance

Circadian rhythm advantage was suggested for afternoon–evening over morning sports performance, and West to East over East to West Coast travel. However, evidence was not clearly and consistently demonstrated, and findings were attributed to possible multiple noncircadian factors. A jet lag response was similarly unsubstantiated, although there was some evidence of increased performance for west to east over east to west transmeridian travel, and for impaired sleep and strength compared to baseline up to 3 days post-travel. The findings are summarized in Table 124.1.

EXERCISE AND SLEEP

Interrelationships between exercise and sleep have been explored through surveys, polysomnography, acute exercise experiments, retrospective studies, and programs of low, moderate, high, and tailored exercise intensities in adults. The effects of exercise on sleep in specific groups, including female family caregivers, postmenopausal women, and adults with cancer, have also been examined.

Surveys

A survey of elderly adults found a significant correlation between increased sleep problems and decreased exercise ($p < 0.01$), based on self-report of sleep disturbance and frequency of exercise. Multivariate regression analysis also demonstrated a significant independent impact ($p < 0.05$) of exercise on sleeping problems [11].

Another survey of adults showed a significantly reduced risk of disorders in maintaining sleep for adults participating in regular activity at least once weekly (OR 0.67), a regular exercise program (OR 0.71), or a brisk walk exceeding six blocks daily (OR 0.52). Significantly reduced risks of any sleep disorder were associated with regular activity at least once weekly for both sexes (OR 0.62), and a brisk walk exceeding six blocks daily for men (0.45). A significantly reduced risk of nightmares (OR 0.27) was associated with walking at a normal pace over six blocks daily for women [12].

Two studies demonstrated some association between sleep and total daily physical activity. A comparison of sleep diaries to self-reported durations of exercise in college students with normal sleep showed a decreased wake after sleep onset ($p = 0.008$) and increased sleep efficiency ($p = 0.03$) on the most active days. A comparison of sleep diaries to wrist activity and environmental illumination monitors in healthy adults with predominantly normal sleep demonstrated between-subjects correlations of mean total energy expenditure with mean subjective sleep latency ($p = 0.030$), and mean total activity with mean reported insomnia ($p = 0.049$) [13].

Baseline Fitness, Acute Exercise, and Sleep Polysomnography

A comparison of aerobically fit males with sedentary controls demonstrated significantly improved sleep continuity with shorter sleep onset latencies ($p < 0.001$), less wake time after sleep onset ($p < 0.03$), higher sleep efficiency ($p < 0.03$), and fewer discrete sleep episodes during the night ($p < 0.01$) based on pre- and postexercise polysomnography. No significant difference was found between groups for total sleep time. Sleep architectural changes included significantly less stage 1 non-rapid eye movement (NREM) sleep time ($p < 0.005$) and episodes ($p < 0.01$), and an increased number of slow waves ($p < 0.01$–0.02) in aerobically fit as compared to sedentary participants. Findings specific to stage 1 NREM sleep were significant decreases in episodes ($p < 0.003$), shifts ($p < 0.02$), wake time ($p < 0.004$), and epochs (0.008) [14].

Acute and Chronic Exercise

Acute exercise (a single-episode exercise intervention) and chronic exercise (an extended program of exercise as an intervention) effects on sleep have been analyzed and compared. Meta-analysis of acute exercise showed significant increases in stage 3 NREM ($p < 0.05$) and stage 4 NREM sleep ($p < 0.05$), total sleep time, and rapid eye movement

(REM) sleep latency ($p < 0.01$). Significant decreases in sleep onset latency ($p < 0.01$) and REM sleep ($p < 0.05$) were also found [15].

Another meta-analysis of acute exercise and sleep effects indicated a significant mean effect size on slow-wave sleep ($p = 0.005$), reduction in REM ($p < 0.001$), REM latency sleep ($p < 0.001$), and increase in total sleep time ($p < 0.001$). The most consistent moderator variables were time of day and duration of exercise; although implications were unclear for sedentary populations exercising less than 1 hour duration, when reliable sleep effects appear, or for populations exercising at least 4 hours prior to sleep, as impairment of sleep is suggested following late evening exercise. Overall findings were noted as inconsistent with previous narrative reviews and delimited by an exclusive focus on good sleepers [16].

A meta-analytic comparison of acute and chronic exercise effects on sleep showed similar effects on slow-wave sleep, total sleep time, and REM sleep. Meta-analysis of chronic exercise effects on sleep demonstrated increases in slow-wave sleep ($p < 0.05$) and total sleep time ($p < 0.01$), with decreases in sleep onset latency, REM sleep, and awake time ($p < 0.01$) [15].

Intensity of Exercise

Experimental studies of low-, moderate-, and high-intensity chronic exercise have explored sleep-related effects, although the intensities described were not necessarily equivalent by standard exercise parameters. General exercise guidelines specify "low-intensity" exercise as walking at 2–3 miles per hour, cycling on level terrain at 6 miles per hour, performing light stretching exercises, swimming using a float board, and performing light to moderate housework. Balance, including tai chi, and flexibility exercises may also be classified as low intensity [17].

A comparison of tai chi with low-impact (seated exercise, stretching, breathing, relaxation) exercise on sleep in adults with moderate sleep complaints revealed improvements in subjective quality, latency, duration, efficiency, disturbances ($p < 0.001$), and dysfunction ($p < 0.02$). Significant decreases in daytime sleepiness ($p < 0.001$) were also reported in tai chi over low-impact participants. No significant difference was demonstrated between groups for sleep medication use [18].

"Moderate-intensity" exercise is specified as walking at 4 miles per hour, cycling at 8 miles per hour, golf (walking or pulling a cart), light calisthenics, swimming (treading water), and heavy house or yardwork. Moderate-intensity exercise programs are often tailored to a target rate of approximately 70% of the exerciser's maximal heart rate [17].

A comparison of low-impact aerobics for healthy sedentary adults with moderate sleep complaints to nonexercise controls demonstrated significantly improved overall sleep ($p < 0.001$), including self-rated quality ($p = 0.03$), sleep-onset latency ($p = 0.007$), duration ($p = 0.05$), and efficiency ($p = 0.04$) in exercise participants. The findings also suggested a program duration exceeding 8 weeks to achieve desired sleep effects [19]. However, a confounding variable, sunlight, was strongly theorized by dissenting experts as having potentially affected circadian phase-shifting and resultant sleep changes in this study [20, 21].

A stretching and flexibility program comparing pre- and postintervention sleep effects in elderly adults with no sleep problems showed significantly increased sleep efficiency ($p < 0.01$) plus decreased wake time after sleep onset and daytime sleepiness ($p < 0.001$) [22].

A comparison of endurance exercise (primarily brisk walking) in healthy sedentary female family caregivers to nutrition education in similar controls demonstrated significant improvement in rated sleep quality ($p < 0.045$), associated with reductions of perceived stress ($p < 0.04$) in the exercise participants. No improvement in sleep duration or latency was found [23].

A comparison of moderate-intensity exercise in obese postmenopausal women with similar controls receiving stretching and relaxation demonstrated no significant effect on sleep quality overall. Onset of sleep improved for participants performing morning exercise of longer duration ($p \leq 0.05$) and declined with evening exercise of longer duration ($p \leq 0.05$). Decreased use of sleep medication ($p < 0.05$) by the control group was also found, as compared to baseline [24].

"High-intensity" exercise is specified as walking or jogging at 8 miles per hour, cycling at 11–12 miles per hour, swimming one-half mile in 30 minutes, recreational tennis, or hiking. Progressive resistance training, a muscle-strengthening exercise, may be moderate or high in intensity and is usually tailored to the one-repetition maximum, a percentage of the maximal weight an exerciser can lift once [17].

A comparison of high-intensity progressive resistance training (weight-lifting) by adults with depression to similar controls receiving health education demonstrated significant improvements in measures of short- and long-term self-reported subjective sleep quality ($p = 0.07$). No gender effects were found for sleep quality [25].

Exercise and Cancer

Several studies have addressed the exercise and sleep effects in patients with cancer. Comparison of tailored exercise (walking) participants with nonexercise controls in women receiving radiation therapy for breast cancer found a significant decrease in difficulty sleeping ($p = 0.027$) for the exercise group [26].

Another tailored exercise program, for adults with various stages and types of cancer, demonstrated significantly improved self-reported quality of sleep after participation as compared to baseline ($p = 0.03$). However, no objective improvement in sleep was found [27].

Summary of Exercise and Sleep

Based on surveys, improved self-rating of sleep associated with regular exercise participation was suggested, but not consistently demonstrated. Polysomnographic findings suggested baseline aerobic fitness as associated with significantly improved sleep continuity measures, except for total sleep time. Acute exercise was associated with significantly improved sleep continuity and decreased sleep fragmentation, but not with total sleep time. Low-intensity exercise was associated with improved self-reported quality of sleep for most measures of sleep continuity. Moderate-intensity exercise was usually but inconsistently associated with improved sleep; and positive effects were suggested to be mitigated by time of day and duration of exercise, specifically for late evening times and sessions exceeding 1 hour. Moderate-intensity exercise was also associated

TABLE 124.2 Community-Based Studies on Exercise and Sleep

Year	Author	Acuity or Intensity	Exercise Types	Sleep Changes
1996	Kubitz et al. [15]	Acute	Mixed	↑TST, ↓SOL
1996	Kubitz et al. [15]	Chronic	Mixed	↑TST
1997	Youngstedt et al. [16]	Acute	Mixed	↑TST
2004	Li et al. [18]	Low	Tai chi	↑OS, ↑SOL, ↑SE, ↓SD, ↓SDb
1997	King et al. [19]	Moderate	Aerobics, walking	↑OS, ↑QoS, ↑SOL, ↓SD, ↓SDb
2002	Tanaka et al. [22]	Moderate, low	Walking, stretching	↑QoS, ↑SE
2002	King et al. [23]	Moderate	Walking	↑QoS
2003	Tworoger et al. [24]	Moderate	Walking, stationary bike	↔ None
1996	Singh et al. [25]	High	PRT	↑QS
1997	Mock et al. [26]	Tailored	Walking	↓SDb
2003	Young-McCaughan et al. [27]	Tailored	Walk, step, arm	↑QoS

Key: PRT = progressive resistance training; OS = overall sleep; QoS = quality of sleep; SOL = sleep onset latency; TST = total sleep time; SE = sleep efficiency; SD = sleep duration; SDb = sleep disturbance.

with improved sleep where perceived stress was a factor. High-intensity exercise was associated with improved quality of sleep, especially where depression was a factor. Additionally, tailored exercise for cancer patients was associated with decreased difficulty falling asleep and improved self-rated quality of sleep. Characteristics and findings of these studies are summarized in Table 124.2.

CONCLUSIONS

The existence of a circadian advantage or jet lag effect on sports performance remains inconsistently supported. Although, in a global environment of 24-hour a day life styles and rapid transmeridian travel, the implications for athletes and nonathletes justifies pursuing a better understanding of circadian sleep rhythms and their possible influence on performance.

A clear relationship between exercise and sleep remains undetermined. However, some studies suggest the intriguing possibility of utilizing exercise to improve sleep. The benefits of such a sleep-improvement method, in both quality of life and economic terms, certainly warrants further investigation.

REFERENCES

1. Reilly T, Atkinson G, Waterhouse J. *Biological Rhythms and Exercise*. Oxford University Press, New York, 1997.

2. Rodahl A, O'Brien M, Firth RGR. Diurnal variation in performance of competitive swimmers. *J Sports Med* **16**:72–76(1976).

3. Baxter C, Reilly T. Influence of time of day on all-out swimming. *Br J Sports Med* **17**(2):122–127(1983).

4. Atkinson G, Coldwells A, Reilly T, Waterhouse J. The influence of age on diurnal variations in competitive cycling performances. *J Sports Sci* **12**:127(1994).

5. Jehue R, Street D, Huizenga R. Effect of time zone and game time changes on team performance: National Football League. *Med Sci Sports Exerc* **25**:125–131(1993).

6. Youngstedt SD, O'Connor PJ. The influence of air travel on athletic performance. *Sports Med* **28**(3):197–207(1999).

7. Smith RS, Guilleminault C, Efron B. Circadian rhythms and enhanced athletic performance in the National Football League. *Sleep* **20**(5):362–365(1997).

8. Steenland K, Deddens JA. Effect of travel and rest on performance of professional basketball players. *Sleep* **20**(5):366–369(1997).

9. Recht LD, Lew RA, Schwartz WJ. Baseball teams beaten by jet lag. *Nature* **377**:583(1995).

10. Hill DW, Hill CM, Fields KL, Smith JC. Effects of jet lag on factors related to sport performance. *Can J Appl Physiol* **18**(1):91–103(1993).

11. Bazargan M. Self-reported sleep disturbance among African-American elderly: the effects of depression, health status, exercise, and social support. *Int J Aging Hum Dev* **42**(2):143–160(1996).

12. Sherrill DL, Kotchou K, Quan SF. Association of physical activity and human sleep disorders. *Arch Intern Med* **158**:1894–1898(1998).

13. Youngstedt SD, Perlis ML, O'Brien PM, Palmer CR, Smith MT, Orff HJ, Kripke DF. No association of sleep with total daily physical activity in normal sleepers. *Physiol Behav* **78**(3):395–401(2003).

14. Edinger JD, Maorey MC, Sullivan RJ, Higginbotham MB, Marsh GR, Dailey DS, McCall WV. Aerobic fitness, acute exercise and sleep in older men. *Sleep* **16**(4):351–359(1993).

15. Kubitz KA, Landers DM, Petruzzello SJ, Han M. The effects of acute and chronic exercise on sleep. *Sports Med* **21**(4):277–291(1996).

16. Youngstedt SD, O'Conner PJ, Dishman RK. The effects of acute exercise on sleep: a quantitative synthesis. *Sleep* **20**(3):203–214(1997).

17. Beers MH, Berkow MD (Eds). Exercise. In: *The Merck Manual of Geriatrics*. http://www.merck.com/mrkshared/mm_ geriatrics/sec3/ch31.jsp 8/30/04.

18. Li F, Fisher KJ, Harmer P, Irbe D, Tearse RG, Weimer C. Tai chi and self-rated quality of sleep and daytime sleepiness in older adults: a randomized controlled trial. *J Am Geriatr Soc* **52**:892–900(2004).

19. King AC, Oman RF, Brassington GS, Bliwise DL, Haskell WL. Moderate-intensity exercise and self-rated quality of sleep in older adults: a randomized controlled trial. *JAMA* **277**(1):32–37(1997).

20. Buchner DM. Physical activity and quality of life in older adults. *JAMA* **277**(1):64–65(1997).

21. O'Connor PJ, Youngstedt SD. Sleep quality in older adults: effects of exercise training and influence of sunlight exposure (letter to the editor). *JAMA* **277**(13):1034–1035(1997).

22. Tanaka H, Kazuhiko T, Arakawa M, Toguti H, Urasaki C, Yamamoto Y, Uezu E, Hori T, Shirakawa S. Effects of short nap and exercise on elderly people having difficulty in sleeping. *Psychiatry Clin Neurosci* **55**(3):173–174(2001).

23. King AC, Baumann K, O'Sullivan P, Wilcox S, Castro C. Effects of moderate intensity exercise on physiological, behavioral, and emotional responses to family caregiving: a randomized controlled trial. *J Gerontol* **57A**(1):M26–M36(2002).

24. Tworoger SS, Yasui Y, Vitiello MV, Schwartz RS, Ulrich CM, Aiello EJ, Irwin ML, Bowen D, Potter JD, McTiernan A. Effects of a yearlong moderate-intensity exercise and a stretching intervention on sleep quality in postmenopausal women. *Sleep* **26**(7):830–836(2003).

25. Singh NA, Clements KM, Fiatarone MA. Sleep, sleep deprivation, and daytime activities: a randomized controlled trial of the effect of exercise on sleep. *Sleep* **20**(2):95–101(1997).

26. Mock V, Hassey D, Meares CJ, Grimm PM, Dienemann JA, Haisfield-Wolfe ME, Quitasol W, Mitchell S, Chakravarthy A, Gage I. Effects of exercise on fatigue, physical functioning, and emotional distress during radiation therapy for breast cancer. *Oncol Nurs Forum* **24**(6):991–1000(1997).

27. Young-McCaughan S, Mays MZ, Arzola SM, Yoder LH, Dramiga SA, Leclerc KM, Caton JR, Sheffler RL, Nowlin MU. Change in exercise tolerance, activity, and sleep patterns, and quality of life in patients with cancer participating in a structured exercise program. *Oncol Nurs Forum* **30**(3):441–454(2003).

ADDITIONAL READING

Baehr EK, Eastman CI, Revelie W, Losee Olson SH, Wolfe LF, Zee PC. Circadian phase-shifting effects of nocturnal exercise in older compared with young adults. *Am J Physiol Regul Integr Comp Physiol* **10**:R1142–R1150(2003). A comparison of phase-shifting effects of exercise on the circadian system in young versus older adults, using dim-light melatonin onset as a marker.

Ceolim MF, Menna-Barreto L. Sleep/wake cycle and physical activity in healthy elderly people. *Sleep Res Online* **3**(3):87–95(2000). A study of exercise effects on the sleep–wake cycle of older adults, based on sleep logs and actigraphy.

Driver HS, Rogers GG, Mitchel D, Borrow SJ, Allen M, Luus HG, Shapiro CM. Prolonged endurance exercise and sleep disruption. *Med Sci Sports Exerc* **26**(7):903–907(1994). A comparison of pre- and post-ultratriathelon sleep effects in adult male athletes using electrophysiological recordings.

Manifredini R, Manfredini F, Fersini C, Conconi F. Circadian rhythms, athletic performance, and jet lag. *Br J Sports Med* **32**(2):101–106(1998). A review of jet lag, associated symptoms, possible effects on sports performance, and potential alleviating strategies.

O'Connor PJ, Youngstedt SD. Influence of exercise on human sleep. *Exerc Sports Sci Rev* **23**:105–134(1995). A review of the influence of exercise on human sleep, including key concepts, common paradigms, and related research.

125

SLEEP, SLEEP LOSS, AND CIRCADIAN INFLUENCES ON PERFORMANCE AND PROFESSIONALISM OF HEALTH CARE WORKERS

SIGRID CARLEN VEASEY

University of Pennsylvania School of Medicine, Philadelphia, Pennsylvania

INTRODUCTION

Excellence in health care delivery for patients with acute illness, injury, and other medical conditions demands that health care professionals function at optimal cognitive performance. Cognitive performance, however, is highly vulnerable to sleep loss and circadian rhythms, and acute care medicine must be deliverable 24 hours a day, 7 days a week. Acute care nurses, physicians, interns, residents, medical students, and staff are faced with challenges of prolonged wakefulness time, chronic insufficient sleep, and the need to work at times of low circadian wakefulness stimuli and, thus, will likely experience times of impaired cognitive performance. It is essential that the medical community first acknowledges this vulnerability to sleep loss and circadian rhythm changes, then learns how to identify sleep and circadian-based impaired performances, and then understands the involved biological processes, so that we may use our expertise in medicine to begin to test and identify the most effective countermeasures.

SLEEP-RELATED IMPAIRMENTS IN HEALTH CARE WORKERS

Much of the research in this area has focused on residents and physicians; presently, however, these concerns are being addressed in nurses and other health care workers. The medical literature concerning the effects of sleep loss

and call schedules in residents and physicians appears at first glance contradictory: some studies show impairments after sleep loss or call nights, while other studies show no differences [1–33]. Because statistical power was not addressed in most of the studies, it is difficult to interpret negative findings in many studies [1–8]. Moreover, very few studies look at the effects of on-call or sleep loss on health care delivery. There are few validated, simple to use assays of health care delivery and medical error, and thus the majority of study designs have implemented general neurobehavioral assays [9–33]. In general, these tests show that sleep loss in physicians and residents manifests as lengthened reaction times, longer time on task and inaccurate responses to simple repetitive tasks, reduced fine motor control, and impaired spatial memory (see Table 125.1 for a summary of all effects). Time on task and complex reasoning may be most vulnerable to the effects of sleep loss [7, 15, 18, 21, 25]. In contrast, previously learned medical knowledge is less affected by short-term sleep loss [11]. The few studies looking at patient care or simulated patient care have shown negative effects on health care performance of residents and physicians after short-term sleep loss or post-call shifts [12, 13, 19, 27]. Simulated intubations and laparoscopy take longer and show increased error [13, 14]. Another study compared the likelihood of surgical residents' cases being presented at morbidity and mortality rounds, and they found that cases where the surgical resident was on-call the previous night were 45% more likely to be morbidity and mortality

Sleep: A Comprehensive Handbook, Edited by T. Lee-Chiong.
Copyright © 2006 John Wiley & Sons, Inc.

TABLE 125.1 Measures Showing Deterioration with Sleep Loss in Physicians, Physicians-in-Training, and/or Nurses

- Monitoring physiological parameters, particularly for prolonged time periods
- Procedures: increased errors and duration required to complete task; operative technique and postoperative care
- Thoroughness of history and physical exam recording
- Mood, motivation, and professionalism
- Medication error

cases than if the resident was not on-call the night before surgery [12]. Electrocardiogram interpretation and detection of abnormal data in lab reports were poorer in residents following sleep loss [7, 11, 17, 27]. Residents' recording of history and physical exam findings for new admission patients is less thorough when residents have experienced sleep loss [28, 29].

In addition to sleep loss on on-call nights or night shifts, other equally significant sleep disturbances include chronic sleep restriction and long shift or night shift coverage. Average sleep times of 5–6 hours/night can result in similar impairments as seen following 24–48 hours of complete sleep loss [34]. Yet self-recognition of sleepiness and impaired performance is markedly impaired in circumstances of chronic sleep loss. The importance of shift duration has only recently become unveiled for nurses, where the number of medication errors increased threefold in 393 nurses working 12 versus 8 hour shifts [35]. Whether break periods or naps reduce risk for long shifts must now be addressed, as 12 hour shifts are desired by many nurses and hospital employers. Few studies have addressed performance differences across circadian time points, but neurobehavioral assays in young adults show large decrements in cognitive performance at the early morning circadian nadir (2–6 a.m.). Impaired performance parallels circadian reductions in core body temperature.

IDENTIFYING SLEEP LOSS AND FATIGUE IN HEALTH CARE WORKERS

Significant cognitive and mood impairments can be observed in individuals well before sleep ensues. This area has not been extensively studied, but sleepiness in any adult may manifest as a shortened temper, erratic behavior, easy distractibility, poor mood, slowed reactions and speech, and/or lack of interest. Notably, these are all very nonspecific findings. The nonspecificity of symptoms is only one of the challenges in recognizing sleepy workers. The second is that sleepiness and fatigue impair judgment, and thus the sleepy worker will not be able to recognize his/her own impaired behavior [34]. A third challenge to identifying sleepy health care workers is that it is possible

to mask sleepiness with caffeine and other stimulants and thus appear wide awake, while complex cognitive processes are clearly impaired. Thus a most important first step is appreciating the difficulties in detecting sleepiness and fatigue in ourselves and our co-workers. While we await validated assays of fatigue and impaired performance, it is good practice to look back at one's notes on patients charts from previous days to make sure they are thoughtful, coherent, and concise. Unfinished thoughts or sentences that do not flow smoothly may be signs of fatigue and impaired performance. Short of this, health care workers should determine themselves what will satisfy the homeostatic drive to sleep. To accomplish this, workers must allow themselves adequate time on vacation to first recover some sleep debt and then regularly satisfy sleep need for several days to determine the average daily sleep time needed to fulfill sleep need. Alternatively, workers must attempt longer sleep times during the work week.

VARIABLES INVOLVED IN OPTIMAL ALERTNESS

A next step is to understand the biology of alertness, so that we may design more effective schedules and fatigue management programs to correct our performance impairments. Health care worker–patient communication, attention to detail, efficient interpretation and synthesis of patient data, and rapid correct decision-making all require optimal cognitive function (alertness). Alertness is the ability to respond optimally to any situation, including complex reasoning, and this must be distinguished from vigilance (wakefulness stability, ability to rapidly respond to a simple task). The distinction is necessary because health care workers should understand the difference between simply feeling awake and being able to rapidly respond to any complicated problem. Optimal alertness is determined by homeostatic influences, circadian rhythm, and health.

Homeostatic Influences

We now understand that the brain does not function well with prolonged wakefulness, and that sleep serves to restore optimal brain function. Transcriptionally, a stress response in the brain, akin to that seen following ischemia or hypoglycemia, has been shown with prolonged sleep loss in animals [36]. Functional magnetic resonance imaging in humans shows slowed activity in the prefrontal cortex [37]. The brain's response to prolonged sleep loss includes evidence for misfolding of proteins and oxidative injury within neurons [38]. As it turns out, the brain, with minimal energy stores [39] but high metabolic activity, is ill-equipped to remain active for long periods of time, so that with prolonged wakefulness a homeostatic drive to sleep increases exponentially. This homeostatic drive manifests early on as fatigue and later as wakefulness instability. This may first manifest as "zoning out" or loss of concentration for a moment, and with further prolongation of wakefulness, the drive to sleep becomes so pronounced that one may fall asleep anywhere, including at lectures, while writing notes, and even in the operating room assisting in surgery. This homeostatic drive may be countered, in part, with central nervous system stimulants or physical activity (see the management section further on). Stimulants will help reduce the tendency to sleep but, again, may not alter the impaired higher cognitive function, and thus may mask the need for sleep [40]. Walking and increased movement will increase noradrenaline release in the basal forebrain and cortex and very effectively prevent sleep, for a while. Ultimately, the homeostatic power cannot be countered and sleep will ensue [41]. Importantly, impairments from sleep loss are not exclusive to post-call residents. Chronic partial sleep loss (averaging 5–6 hours/night) results in as much impairment and increases homeostatic drive to sleep as much as 1–2 nights of total sleep loss [34]. Indeed, the effects of sleep loss on performance may be just as profound after minimal sleep from one night on-call as when an individual averages 5–6 hours of sleep per night [1]. Thus it is not uncommon that physicians, at all levels of training, caring for acutely ill patients are somewhat sleep deprived.

Circadian Influences

In addition to homeostatic influences, alertness is affected by circadian influences. The circadian influences are more pronounced when the homeostatic drive has not been fully replenished, but even with adequate sleep, there are specific times, relative to a given individual's circadian clock, when alertness is optimal and at other times sleepiness is pronounced. Typically, the nadir for circadian alertness is early morning (2–6 a.m.); ironically, this nadir in circadian alertness is close to the times typically chosen for patient rounds and is the time when important decisions in patient care are made.

Without properly entraining health care workers to night shift work, the workers' normal physiological functions may be significantly disturbed. In addition to behavioral state and mood influences, there are important circadian effects on hormone levels and activity [42, 43]. A regularly timed subjective darkness is essential for the largest pulses of melatonin [44]. Other hormones influenced by the circadian rhythm are cortisol, thyroid stimulating hormone, prolactin, and insulin [42–44].

Cardiopulmonary functions are also influenced by circadian rhythms [45–48]. Bronchial constriction, airways responsiveness, and mucus production are all increased at the circadian dawn and improve midday [45–46]. Blood pressure is lower in the circadian evening to early morning,

and platelet adhesion is reduced during this time [49, 50]. At circadian dawn, blood pressure and heart rate rise, and platelet adhesion increases [51]. The circadian fluctuations in blood pressure, heart rate, and platelet adhesion may explain, in part, the circadian peak for myocardial infarctions and stroke in the morning [52].

Renal function, particularly electrolyte excretion and glomerular filtration rate, varies with the circadian rhythm [53–55]. Immune function, in addition to cell growth and division, is influenced by circadian effector systems [56]. There are numerous circadian influences on gastrointestinal function [57, 58]. One of the challenges with entraining a shift worker to a novel circadian time is that each of the above rhythms is differentially sensitive to a circadian shift, and it is not uncommon to have the above physiological and biological processes unsynchronized in shift workers. This lack of synchronization of effector systems in shift workers contributes to the fatigue.

Sleep Inertia

One of the most profound ways in which sleep impairs cognition is through sleep inertia [59]. Sleep inertia may be defined as a clouded sensorium upon arousal from deep sleep [59]. This may present as confusion, slowed speech, repeating phrases or questions, as if the first response did not register, or having less than relevant questions or comments in a conversation [60]. A particularly vulnerable time for sleep inertia is immediately upon answering a late night or early morning beeper or phone call. If the call is answered from bed, without sitting up and taking a moment to wake up, arousal may not be complete, and the recipient of the phone call may not be able to retain information or synthesize information and may not recall the conversation in the morning. Thus sleep inertia is a serious impairment and a not uncommon scenario. The danger is that a sleepy resident or fellow may have misinterpreted suggestions, or that a sleepy attending makes the wrong suggestion in care and patient care suffers. To some degree, we experience mild sleep inertia many mornings upon arousal, before massively stimulating ourselves with water on the face and/or showers, boluses of caffeine, and the physical activity necessary to get to work. Sleep inertia may be recognized as slowed or slurred speech, repetition of phrases, and speech content that does not flow well. With a clouded sensorium, individuals cannot recognize their own sleep inertia, and thus it is essential that all health care workers, particularly nurses who call residents and physicians at night, understand the importance of sleep inertia and how to recognize the physician with sleep inertia. Sleep inertia is readily reversible with 10 minutes or less of physical activity, caffeine, and other forms of stimulation and should be managed effectively, when possible, prior to making important medical decisions [61].

MANAGEMENT OF SLEEPINESS AND FATIGUE IN HEATH CARE WORKERS

While we wait for details of the specific effects of sleepiness and fatigue on health care workers' performance, we must acknowledge the *potential* of the problem and the *enormity* of that potential problem: medical error resulting in adverse events in our patients. We are obliged, as caring physicians, to determine how best to recognize and manage fatigue. Inadequate sleep should be suspected when personnel exhibit changes in behavior or mood. Falling asleep or "zoning out" unintentionally during activities, including lectures, should be considered a late sign. Brief nap opportunities (the most effective way to combat sleepiness and fatigue) should be available for all health care workers working long shifts or shifts opposite their times of circadian altertness. A brief nap lying down in a quiet dark room will be more restorative than sleeping sitting up in a conference room, and a nap at a circadian nadir may be most restorative. Naps should be less than 2 hours to avoid waking in sleep inertia, or ample time to dissipate sleep inertia must be worked into the schedule [7].

Recently, there has been much publicity concerning the 80 hour workweek for residents [62]. The effectiveness of the 80 hour workweek in reducing physician-in-training fatigue was never studied. It was first employed as a compromise between the 120 hour workweek that residents endured in the mid-1980s and the typical 40 hour workweek. The American Medical Student Association brought the matter to Congress and the Occupational Safety and Health Administration for government regulation of work hours, and the Accreditation Council for Graduate Medical Education was obligated to compromise quickly. Although total work hours are reduced, the increased free time is not spent sleeping, and the vast majority of residents are still averaging 5–6 hours of sleep per night [63]. Thus it is unlikely that changes in work hours alone will have a major positive impact on sleepiness in residents.

Residents, attendings, and nurses are frequently asked to work some night shifts. The circadian rhythm simply cannot adjust rapidly to large (8 hour) shift changes. Thus more thought must go into scheduling of health care workers to provide effective circadian shift changes to optimize job performance. Presently, the major goal is to prevent significant chronic sleep deprivation, and to prevent sleepy workers from falling asleep at the wheel driving home. Individuals covering night shifts intermittently must resort to two sleep periods to obtain adequate sleep times in 24 hours. Because of the challenges with rotating shift work, there may be a role for very short-acting benzodiazepines to promote sleep, for days when the circadian timing for sleep is not present. If prescribed, it is important to begin with the lowest dose, short-acting agents to ensure individuals are fully alert for commutes to work and time on the job. Ideally, workers should be switched to regular nighttime shift work; very close attention to the basics of shift work therapy is essential. These basics include bright lights at work at times of low circadian alertness, avoiding sunlight on the drive home, and ensuring that the house is as dark and as quiet as it would be for late night and early morning hours. These recommendations certainly would impose challenges on families, where nurses and other health care workers are rushing home to take care of children and other dependent family members or to get children off to school.

Indications for stimulant use in fatigued health care workers are not established. Caffeine use is widely prevalent and has been shown in healthy young adults to reduce sleep inertia [61]. What is unclear with any stimulant use is whether vigilance alone improves or whether higher cognitive function also improves. Several reports suggest otherwise [64, 65]. Thus it is possible that a stimulant would keep an individual awake without improving decision-making and attention to details. All of this will require future study.

The real hotbed of controversy now is the use of modafinil. This drug has been marketed as a nonamphetamine stimulant; although research in animal models is controversial, and there are amphetamine-like effects with the drug. Nonetheless, the Food and Drug Administration has just expanded use of modafinil to shift workers, and this would include on-call residents, physicians, and nurses. While there remain a lot of unknowns in this arena, it should be emphasized that this drug is not as effective in alleviating sleepiness as sleep is [65]. A 2 hour nap will have a far greater impact on performance and alertness, and only the 2 hour nap has been shown in chronic sleep loss to improve function [46, 47].

CONCLUSION

Health care workers are beginning to acknowledge that sleep loss may impair performance. The significance of the impaired performance spans lack of sensitivity and reduced compassion to medical error [66]. Acknowledgment of the significance of sleep and circadian impact on health care professionalism is an important step before research can begin to address the many questions raised about this issue. While there are many unknowns, the key messages to all health care workers should be to make sufficient sleep a professional priority and to ensure alertness during work and driving to and from work. Adequate sleep will improve efficiency and performance in health care delivery and safety on the roads. Brief nap times for personnel with long shifts should be scheduled. Nap opportunities in quiet rooms should be provided for all personnel, who have experienced prolonged wakefulness, prior to driving home. Stimulants must be considered second line therapies with some effect on wakefulness and vigilance but unknown effects on health care delivery.

REFERENCES

1. Reznick RK, Folse JR. Effect of sleep deprivation on the performance of surgical residents. *Am J Surg* **154**:520–525(1987).

2. Bartle EJ, Sun JH, Thompson L, Light AI, McCoolC, Heaton S. The effects of acute sleep deprivation during residency training. *Surgery* **104**:311–316(1988).

3. Browne BJ, Van Susteren T, Onsager DR, Simpson D, Salaymeh B, Condon RE. Influence of sleep deprivation on learning among surgical house staff and medical students. *Surgery* **115**:604–610(1994).

4. Christensen EE, Dietz GW, Murry RC, Moore JG. The effect of fatigue on resident performance. *Radiology* **125**:103–105(1977).

5. Bertram DA. Characteristics of shifts and second-year resident performance in an emergency department. *N Y State J Med* **88**:10–15(1988).

6. Klose JK, Wallace-Barnhill GL, Craythorne MWS. Performance test results for anesthesia residents over a five-day week including on-call duty. *Anesthesiology* **63**:A485(1985).

7. Engel W, Seime R, Powell V, D'Alessandri R. Clinical performance of interns after being on call. *South Med J* **80**:761–763(1987).

8. Richardson GS, Wyatt JK, Sullivan JP, Grav EJ, Ward AE, Wolf MA, Czeisler CA. Objective assessment of sleep and alertness in medical house staff and the impact of protected time for sleep. *Sleep* **19**:718–726(1996).

9. Goldman LI, McDonough MT, Rosemond GP. Stresses affecting surgical performance and learning: correlation of heart rate, electrocardiogram and operation simultaneously recorded on videotapes. *J Surg Res* **12**:83–86(1972).

10. Deaconson TF, O'Hair DP, Levy MF, Lee MB, Schueneman AL, Codon RE. Sleep deprivation and resident performance. *JAMA* **260**:1721–1727(1988).

11. Light AI, Sun JH, McCool C, Thompson L, Heaton S, Bartle EJ. The effects of acute sleep deprivation on the level of resident training. *Curr Surg* **46**:29–30(1989).

12. Haynes DF, Schwedler M, Dyslin DC, Rice JC, Kerstein MD. Are postoperative complications related to resident sleep deprivation? *S Med J* **88**:283–289(1995).

13. Taffinder NJ, McManus IC, Gul Y, Russell RC, Darzi A. Effect of sleep deprivation on surgeons' dexterity on laparoscopy simulator. *Lancet* **352**:1191(1998).

14. Grantcharov TP, Bardram L, Funch-Jensen P, Rosenberg J. Laparoscopic performance after one night on call in a surgical department: prospective study. *BMJ* **323**:1222–1223(2001).

15. Godellas CV, Huang R. Factors affecting performance on the American Board of Surgery in-training examination. *Am J Surg* **181**:294–296(2001).

16. Friedman RC, Bigger JT, Kornfeld DS. The intern and sleep loss. *N Engl J Med* **285**:201–203(1971).

17. Poulton EC, Hunt GM, Carpenter A, Edwards RS. The performance of junior hospital doctors following reduced sleep and long hours of work. *Ergonomics* **21**:279–295(1978).

18. Ford CV, Wentz DK. The internship year: a study of sleep, mood states, and psychophysiologic parameters. *South Med J* **77**:1435–1442(1984).

19. Denisco RA, Drummond JN, Gravenstein JS. The effect of fatigue on the performance of a simulated anesthetic monitoring task. *J Clin Monit* **3**:22–24(1987).

20. Orton DI, Gruzelier JH. Adverse changes in mood and cognitive performance of house officers after night duty. *Br Med J* **298**:21–23(1989).

21. Storer JS, Floyd HH, Gill WL, Giusti CW, Ginsberg H. Effects of sleep deprivation on cognitive ability and skills of pediatrics residents. *Acad Med* **64**:29–32(1989).

22. Robbins J, Gottlieb F. Sleep deprivation and cognitive testing in internal medicine house staff. *West J Med* **12**:82–86(1990).

23. Jacques CH, Lynch JC, Samkoff JS. The effects of sleep loss on cognitive performance of resident physicians. *J Fam Pract* **30**:223–229(1990).

24. Rubin R, Orris P, Lau SL, Hryhorczuk DO, Furner S, Letz R. Neurobehavioral effects of the on-call experience in housestaff physicians. *J Occup Med* **33**:13–18(1991).

25. Nelson CS, Dell'Angela K, Jellish WS, Brown IE, Skaredoff M. Residents' performance before and after night call as evaluated by an indicator of creative thought. *J Am Osteopath Assoc* **95**:600–603(1995).

26. Hart RP, Buchsbaum DG, Wade JB, Hamer RM, Kwentus JA. Effect of sleep deprivation on first-year residents' response times, memory, and mood. *J Med Educ* **52**:940–942(1987).

27. Gottlieb DJ, Parenti CM, Peterson CA, Lofgren RP. Effect of a change in housestaff work schedule on resource utilization and patient care. *Arch Intern Med* **151**:2065–2070(1991).

28. Lingenfelser T, Kaschel R, Weber A, Zaiser-Kaschel H, Jakober B, Kuber J. Young hospital doctors after night duty: their task specific cognitive status and emotional condition. *Med Educ* **28**:566–572(1994).

29. Hawkins MR, Vichick DA, Silsby HD, Kruzich DJ, Butler R. Sleep deprivation and performance of house officers. *J Med Educ* **60**:530–535(1985).

30. Leonard C, Fanning N, AttwoodJ, Buckley M. The effect of fatigue, sleep deprivation and onerous working hours on the physical and mental wellbeing of pre-registration house officers. *Ir J Med Sci* **167**:22–25(1998).

31. Smith-Coggins R, Rosekind MR, Hurd S, Buccino KR. Relationship of day versus night sleep to physician performance and mood. *Ann Emerg Med* **24**:928–934(1994).

32. Smith-Coggins R, Rosekind MR, Buccino KR, Dinges DF, Moser RP. Rotating shiftwork schedules: can we enhance physician adaptation to night shifts? *Acad Emerg Med* **4**:951–961(1997).

33. Deary IJ, Tait QR. Effects of sleep disruption on cognitive performance and mood in medical house officers. *Br Med J Clin Res Ed* **295**:1513–1516(1987).

34. Van Dongen HP, Maislin G, Mullington JM, Dinges DF. The cumulative cost of additional wakefulness: dose–response effects on neurobehavioral functions and sleep physiology from chronic sleep restriction and total sleep deprivation. *Sleep* **26**:117–126(2003).

35. Borges FN, Fischer FM. Twelve-hour night shifts of healthcare workers: a risk to the patients? *Chronobiol Int* **20**:351–360(2003).

36. Cirelli C. How sleep deprivation affects gene expression in the brain: a review of recent findings. *J Appl Physiol* **92**:394–400(2002).

37. Drummond SP, Brown GG. The effects of total sleep deprivation on cerebral responses to cognitive performance. *Neuropsychopharmacology* **25**:S68–S73(2001).

38. Silva RH, Abilio VC, Takatsu AL, Kameda SR, Grassl C, Chehin AB, Medrano WA, Calzavara MB, Registro S, Andersen ML, Machado RB, Carvalho RC, Ribeiro Rde A, Tufik S, Frussa-Filho R. Role of hippocampal oxidative stress in memory deficits induced by sleep deprivation in mice. *Neuropharmacology* **46**:895–903(2004).

39. Kong J, Shepel PN, Holden CP, Mackiewicz M, Pack AI, Geiger JD. Brain glycogen decreases with increased periods of wakefulness: implications for homeostatic drive to sleep. *J Neurosci* **22**:5581–5587(2002).

40. Randall DC, Shneerson JM, Plaha KK, File SE. Modafinil affects mood, but not cognitive function, in healthy young volunteers. *Hum Psychopharmacol* **18**:163–173(2003).

41. Borbely AA, Achermann P. Sleep homeostasis and models of sleep regulation. *J Biol Rhythms* **14**:557–568(1999).

42. White HD, Ahmad AM, Guzder R, Wallace AM, Fraser WD, Vora JP. Gender variation in leptin circadian rhythm and pulsatility in adult growth hormone deficiency: effects of growth hormone replacement. *Clin Endocrinol (Oxf)* **58**:482–488(2003).

43. Bacon WL, Vizcarra JA, Morgan JL, Yang J, Liu HK, Long DW, Kirby JD. Changes in plasma concentrations of luteinizing hormone, progesterone, and estradiol-17beta in peripubertal turkey hens under constant or diurnal lighting. *Biol Reprod* **6**:591–598(2002).

44. Tosini G, Fukuhara C. Photic and circadian regulation of retinal melatonin in mammals. *J Neuroendocrinal* **15**:364–369(2003).

45. Spengler CM, Shea SA. Endogenous circadian rhythm of pulmonary function in healthy humans. *Am J Respir Crit Care Med* **162**:1038–1046(2000).

46. Tukek T, Yildiz P, Atilgan D, Tuzcu V, Eren M, Erk O, Demirel S, Akkaya V, Dilmener M, Korkut F. Effect of diurnal variability of heart rate on development of arrhythmia in patients with chronic obstructive pulmonary disease. *Int J Cardiol* **88**:199–206(2003).

47. Calhoun WJ. Nocturnal asthma. *Chest* **123**:399S–405S(2003).

48. Panzer SE, Dodge AM, Kelly EA, Jarjour NN. Circadian variation of sputum inflammatory cells in mild asthma. *J Allergy Clin Immunol* **111**:308–312(2003).

49. Mutoh T, Shibata S, Korf HW, Okamura H. Melatonin modulates the light-induced sympathoexcitation and vagal suppression with participation of the suprachiasmatic nucleus in mice. *J Physiol* **547**:317–332(2003).

50. Mori H, Nakamura N, Tamura N, Sawai M, Tanno T, Narita T, Singh RB, Otsuka K. Circadian variation of basal total vascular tone and chronotherapy in patients with vasospastic angina pectoris. *Biomed Pharmacother* **56**:339s–344s(2002).

51. Walters J, Skene D, Hampton SM, Ferns GA. Biological rhythms, endothelial health and cardiovascular disease. *Med Sci Mon* **9**:RA1–8(2003).

52. Andreotti F, De Luca L, Renda G, Ferro A, Mongiard, Zecchi P, Maseri A. Circadianicity of hemostatic function and coronary vasomotion. *Cardiologia* **44**(S1):245–249(1999).

53. Kokkonen UM, Riskila P, Roihankorpi MT, Soveri T. Circadian variation of plasma atrial natriuretic peptide, cortisol and fluid balance in the goat. *Acta Physiol Scand* **171**:1–8 (2001).

54. Nishimura M, Uzu T, Fujii T, Kimura G. Disturbed circadian rhythm of urinary albumin excretion in non-dipper type of essential hypertension. *Am J Nephrol* **22**:455–462(2002).

55. Koopman MG, Koomen GC, Krediet RT, de Moor EA, Hoek FJ, Arisz L. Circadian rhythm of glomerular filtration rate in normal individuals. *Clin Sci* **77**:105–111(1989).

56. Chacon F, Cano P, Lopez-Varela S, Jimenez V, Marcos A, Esquifino AI. Chronobiological features of the immune system. Effect of calorie restriction. *Eur J Clin Nutr* **56**(Suppl 3):S69–S72(2002).

57. Bubenik GA. Gastrointestinal melatonin: localization, function, and clinical relevance. *Digest Dis Sci* **47**:2336–2348 (2002).

58. Bjarnason GA, Jordan R. Rhythms in human gastrointestinal mucosa and skin. *Chronobiol Int* **19**:129–140(2002).

59. Acherman P, Werth E, Dijk D, Borbely AA. Time course of sleep inertia after nighttime and daytime sleep episodes. *Arch Ital Biol* **134**:109–119(1995).

60. Bruck D, Pisani DL. The effects of sleep inertia on decision-making performance. *J Sleep Res* **8**:95–103(1999).

61. Van Dongen HP, Price NJ, Mullington JM, Szuba MP, Kapoor SC, Dinges DF. Caffeine eliminates psychomotor vigilance deficits from sleep inertia. *Sleep* **24**:813–819(2001).

62. Veasey S, Rosen R, Barzansky B, Rosen I, Owens J. Sleep loss and fatigue in residency training: a reappraisal. *JAMA* **288**:1116–1124(2002).

63. Buysse DJ, Barzansky B, Dinges D, Hogan E, Hunt CE, Owens J, Rosekind M, Rosen R, Simon F, Veasey S, Wiest F. Sleep, fatigue, and medical training: setting an agenda for optimal learning and patient care. *Sleep* **26**:218–225(2003).

64. Wesensten NJ, Belenky G, Kautz MA, Thorne DR, Reichardt RM, Balkin TJ. Maintaining alertness and performance during sleep deprivation: modafinil versus caffeine. *Psychopharmacology (Berl)* **159**:238–247(2002).

65. Batejat DM, Lagarde DP. Naps and modafinil as countermeasures for the effects of sleep deprivation on cognitive performance. *Aviat Space Environ Med* **70**:493–498(1999).

66. Daugherty SR, Baldwin DC Jr, Rowley BD. Learning, satisfaction, and mistreatment during medical internship: a national survey of working conditions. *JAMA* **279**:1194–1199(1998).

126

THE STUDENT WITH SLEEP COMPLAINTS

JAGDEEP BIJWADIA AND DONN DEXTER
Mayo Health System, Eau Claire, Wisconsin

INTRODUCTION

Sleep is a complex process involving multiple systems. In the first two decades of life, when one is most likely to be a student, there are profound changes taking place in physiologic sleep patterns. Both the emergence of well-defined stages and the proportion of sleep that each stage occupies have a distinct ontogenic and developmental pattern. As children mature they assume more adult sleep patterns, longer sleep cycles, and less daytime sleep. In particular, there is a dramatic decrease in the proportion of rapid eye movement (REM) sleep from birth (50% of sleep) through early childhood into adulthood (25–30% of sleep), and an initial preponderance of slow-wave sleep that peaks in early childhood, drops off after puberty, and declines over the life span. Although the developmental significance of the increased percentage of REM sleep in early life remains an area of investigation, accumulated empiric evidence supports the hypothesis that the activation of autonomic and central nervous system processes in REM sleep help to stimulate neuronal development.

In addition to these biologic processes, there are profound psychosocial changes that occur, especially in early student life. Increasing independence from parental supervision, loss of enforcement of sleep/wake times, and peer interactions as well as academic and social pressures unique to particular age groups are examples of changes that result in emergence of patterns that can be unhealthy and lead to significant daytime sleepiness.

Puberty and early adulthood also represent periods when there is growth and maturation of many body systems. The physiologic changes in the neurological and endocrine functions also have profound effects on sleep.

Recent research has focused on the effects of daytime sleepiness in the student population. Poor sleep quality has been linked to increased tension, irritability, depression, more frequent use of alcohol and illicit drugs, accidents, and lowered academic performance. Sleep problems are common and underrecognized in the student group. The influence of sleep on learning and behavior has recently captured the attention of school districts across the United States and school administrators increasingly need to weigh the factual information about the biology of student sleep patterns against the competing demands of teachers' work preference and athletic and after-school activity schedules.

EPIDEMIOLOGY

Excessive sleepiness among children, adolescents, and young adults has become a major international health concern. Reports of poor sleep hygiene and sleep disorders in pediatric groups and adolescents have led to increased concern and speculation about the clinical and functional implications of excessive daytime sleepiness. Defining excessive daytime sleepiness and scientifically evaluating its effects on the behavior, mood, and performance of students are matters of ongoing research and answers are just beginning to emerge.

Previous studies indicate that school-aged children and adolescents may require at least 9.2 hours of sleep per

Sleep: A Comprehensive Handbook, Edited by T. Lee-Chiong.
Copyright © 2006 John Wiley & Sons, Inc.

day for optimal alertness [1]. Results from cross-sectional surveys reflect a corresponding increase in ratings of subjective sleepiness from early school age to early adolescence and from early to later adolescence [2–4]. Multiple studies from diverse geographic locations and cultural backgrounds have shown significant sleep deprivation and sleepiness in junior high and high school students [5–8]. College students are noted for obtaining insufficient sleep during the week and for sleeping long hours during the weekend. Coren noted that in a sample of college students only 36% reported being completely free of any sleep disturbance, while over 30% reported some type of sleep difficulty occurring frequently or always [9]. In a survey of Australian college students it was noted that the most common sleep complaints were difficulty falling asleep (18%), early morning awakening (13%), general sleep difficulties (12%), and difficulty staying asleep (9%). Over 17% of the students in this sample reported symptoms severe enough to meet the criteria of delayed sleep phase syndrome, more than twice the incidence in the general population [10]. Buboltz et al. [11] found similar results in a U.S. college school sample with over 70% of the students reporting some type of sleeping difficulty on a regular basis, the most common being early morning awakening, difficulty falling asleep, and daytime napping.

FACTORS AFFECTING SLEEP IN THE STUDENT POPULATION

Developmental Factors

Middle childhood is a critical period for the development of healthy sleep habits. Between the ages of six and twelve there comes a progressive comprehension of real dangers that may increase nighttime fears resulting in nocturnal awakenings and sleep deprivation. There is also an increasing independence from parental supervision and a shift in responsibility for health habits as children approach adolescence, which may result in decreasing enforcement of sleep times and duration; parents tend to be less aware of sleep problems if they exist. Recent data suggest that individuals may begin to manifest an inherent lifelong circadian sleep phase preference (morningness–eveningness) in childhood [12]. Children who are relatively sleep phase delayed may display sleep onset problems. Around the time of puberty onset adolescents develop an approximately 2 hour physiologically based phase delay relative to sleep cycles in middle childhood. This is a result of hormonal influences on the sleep cycle. Adolescents have a physiologic tendency to develop decreased daytime alertness levels in middle to late puberty. In evaluating the student it is important to recognize and allow for these developmental factors.

Sleep Deprivation

A number of coexisting factors can result in intense pressure and competition for daytime hours. This may result in shortchanging sleep time and thus cause a significant sleep debt. Chronic sleep restriction is a serious problem in the student group and particular groups, including "high achievers" who are engaged in many extracurricular activities, may be at a higher risk. Chronic sleep restriction in adolescents can lead to significant neurobehavioral consequences including negative impact on mood, vigilance, reaction time, attention, memory, behavioral control, and motivation [13, 14]. Involvement in social, athletic, and family activities, as well as parent work schedules, can conflict with time for sleep. Social anxiety and increasing reliance on peer relationships as well as the ready availability of communication modalities like the telephone and computer tend to delay sleep times. Electronic media in general including television, video games, and computers have a similar toll. Academic pressure and the burgeoning volume of homework in many schools also result in a shorter time for sleep. Lastly, there is increasing data that early school start times result in increased daytime sleepiness [15], and altering the times to a later time period has positive effects on academic performance and sleepiness.

Smoking, Alcohol, Caffeine, and Substance Abuse

Caffeine has been called the most widely consumed psychoactive substance on earth. Caffeine is consumed at least weekly by 98% of 5–18 year olds. For adolescents the main dietary source of caffeine is soft drinks to which caffeine has been added. In a recent study by Pollack and Bright, higher caffeine intake in general was associated with shorter nocturnal sleep duration, increased wake time after sleep onset, and increased daytime sleep [16]. The increasing availability of soft drink dispensing machines in schools is apparently welcomed by students and is profitable to school boards. The public health implications of caffeine consumption in adolescents especially in relation to sleep physiology need to be studied further.

Several lines of evidence suggest a relationship between cigarette smoking and sleep disturbance, including the effects of nicotine and nicotine withdrawal on sleep, a tendency for nonsmokers to be more alert in the morning, an association between cigarette smoking and snoring, and a tendency for individuals who engage in one unhealthy behavior to also engage in others. In a sample of 13,831 adolescents, reported by Johnson and Breslau, a total of 25% of adolescents reported smoking at least one cigarette in the past year and 11.6% smoked daily [17]. Studies show smokers are significantly more likely than nonsmokers to report problems going to sleep and staying asleep, daytime sleepiness, minor accidents, depression, and high daily caffeine intake. Shin et al. [18] have reported an association between habitual snoring and cigarette smoking in high school students.

Alcohol consumption among the student population is rising. In a study reported in *Lancet*, sensible levels of consumption were exceeded by 61% of male and 48% of female United Kingdom university students. Sixty percent of male students and 5% of female students also reported having used cannabis once or twice and 20% reported weekly or more frequent use [19]. The nature of associations between sleep problems and use of cigarettes, alcohol, and illicit drugs may be complex. In part, the association between substance abuse and sleep problems may reflect the tendency for psychiatric problems. Whether the independent association of sleep problems with the use of illicit drugs is due to life-style factors associated with substance abuse, such as more erratic sleep–wake schedules due to the social component of drug use, or an effect of the substances themselves is unclear.

Psychiatric Problems

The specific sleep difficulties that are more likely to be reported with psychiatric symptoms among 5–12 year olds include sleep onset delay, enuresis, and subjective daytime sleepiness. Pediatric psychiatric patients frequently report symptoms of bedtime refusal, fear of the dark, fear of dying in sleep nightmares, and night terrors. Adolescents who report sleep difficulties are more likely to report symptoms of depression, anxiety, tension, lethargy, poor self-esteem, daytime stress, negative thoughts, and emotional lability and are more likely to consume nicotine, caffeine, and alcohol. Depression is often overlooked in the student population and sleep disturbance is very prevalent in this group. Up to 2% of children and 8% of adolescents suffer from depression. Two-thirds of depressed children suffer from sleep onset and sleep maintenance insomnia. Approximately 25% complained of hypersomnia [20]. Anxiety disorders, autism, seasonal affective disorders, and post-traumatic stress syndrome can all have profound impact on the sleep of students and should be considered in evaluating this population. Emotional difficulties can lead to sleep problems and alternatively sleep deprivation can disrupt mood and behavior, thereby creating a mutually interactive cycle. Many of the psychoactive and antidepressant medications have independent effects on sleep architecture and should be considered in the evaluation as well.

Specific Sleep Disorders

In addition to factors discussed earlier, the presence of organic sleep disorders is common and needs to be actively searched for in the workup of students with sleep-related symptoms. As in adults, sleep disorders often remain undiagnosed. There has been a detailed discussion of the various specific sleep disorders in other chapters but a few comments about sleep disorders in the student population are important. Whereas the overall behavioral effects of excessive sleepiness in adults is a reduction of physical and mental activity, its effects in the younger population can be increased activity with irritability and other behavioral problems. Dyssomnias caused by extrinsic factors (child-rearing practices, social activities) are particularly common in childhood. While children's sleep patterns may easily be disturbed by environmental influences, they may readily change back toward normal in contrast to adults in whom the factors underlying the sleep problem may become well established and complicated by secondary effects. In younger children, the common sleep disorders include sleepwalking, enuresis, sleep terrors, bruxism, insufficient sleep, and sleep apnea. In the adolescent student chronic sleep deprivation is a serious problem. Delayed sleep phase syndrome, restless legs syndrome, and sleep disorders secondary to psychiatric problems (especially depression and sleep problems associated with substance abuse) need to be strongly considered. In all student populations inadequate sleep hygiene often plays a major role.

CONSEQUENCES OF SLEEP DISORDERS IN STUDENTS

Increased Risk of Unintentional Injuries and Death

Drowsiness or fatigue has been identified as a principal cause in at least 100,000 traffic crashes each year according to the U.S. National Highway Traffic Safety Administration (NHTSA). Sleep-related crashes are most common in the younger population who tend to stay up late, sleep too little, and drive at night. In a North Carolina study, 55% of motor vehicle accidents resulting from falling asleep at the wheel involved people 25 years of age or less. The peak age of occurrence was 20 years of age.

Poor Academic Performance

Although additional studies to determine causal relationships between sleepiness and academic performance are still needed, recent studies have suggested that nocturnal respiratory disturbance is associated with decreased learning in otherwise healthy children. Sleep fragmentation adversely impacts learning and memory, and hypoxemia adversely influences nonverbal skills [21]. Snoring has been associated with poor academic performance. High school students who describe themselves as having academic problems and who are earning C's or below in school report getting less sleep, having later bedtimes, and having more irregular sleep schedules than students reporting higher grades. Recent findings strongly suggest that self-reported shortened total sleep time, erratic sleep–wake schedules, late bed and rise times, and poor sleep quality are negatively associated with academic performance for adolescents from middle school through the college years [22].

Behavioral Problems

Studies suggest sleep loss may be associated with a decreased ability to control, inhibit, or change emotional responses. Impulsivity and inability to stay focused in sleep-deprived students resemble behaviors common in attention deficit hyperactivity disorder. Aggressive behaviors in students transitioning from junior high to senior high school were highly associated with shorter sleep times and later sleep start times.

Increased Likelihood of Stimulant Alcohol and Illicit Drug Use

Use of cigarettes, alcohol, and any illicit drug has been associated with reports of frequent sleep problems by adolescents [17]. Recent studies have found that insomnia predicted subsequent onset of substance abuse. Wolfson and Carskadon identified the sleepy adolescent as a high-risk group with increased likelihood for stimulant use [22].

APPROACH TO MANAGEMENT

Early Diagnosis

Recognizing that basic sleep needs exist and avoiding sleep debt need to be societal priorities in the vulnerable student population. Excessive sleepiness can be an indication of a biological disorder. Recognition of the cause–effect relationship of medical and psychiatric disorders and sleep physiology should be emphasized. Good sleep hygiene measures combined with specific interventions after the early diagnosis of individual sleep disorders are of utmost importance.

Creating Sleep Friendly Learning Institutions

Expanding education regarding sleep disorders among both teachers and health providers is important. Accommodations may be needed for individual students with sleep disorders. Integration in the curriculum of sleep-related education is needed. Driving education should emphasize the dangers of drowsy driving. Some school districts in the United States have taken important first steps like adopting later school start times that have resulted in significant benefits. Edina, Minnesota was voted the healthy sleep capital of the nation in 2000 due to the school districts' efforts to work *with* and not *against* the adolescents'

sleep needs. Much work is still needed and legislation at the national level that is sleep friendly is an urgent need.

New Research and Discovery

Although the last few years have seen great strides for students in terms of sleep disorder research, the National Sleep Foundation has recognized the need for additional studies especially in the fields of the neurobiology, genetics, epidemiology, neurobehavioral, and functional consequences of sleep. More interdisciplinary studies looking at the relationship of sleep to the development, health, and behavior specifically in the adolescent population are needed.

REFERENCES

1. Carskadon MA, Dement WC. Sleepiness in the normal adolescent In: Guilleminault C. *Sleep and Its Disorders in Children.* Raven Press, New York, 1987, pp 53–60.

2. Sadeh A, Raviv A, Gruber R. Sleep patterns and sleep disruptions in school aged children. *Dev Psychol* **36**(3):291–301(2000).

3. Gianotti F, Cortesi F. Sleep patterns and daytime functions in adolescents: an epidemiological survey of Italian high school student population. In: Carskadon MA (Ed), *Adolescent Sleep Patterns: Biological Social and Psychological Influences.* Cambridge University Press, New York, 2002, pp 132–147.

4. Manber R, Pardee RE, Bootzin RR, Kuo T, Rider SP, Bergstrom L. The relationship between sleepiness and sleep patterns in three cohorts of students. *Sleep Res* **24**:105(1995).

5. Takemura T, Funaki K, Kanbayashi T, Kawamoto K, Tsusui K, Saito Y, Aizawa R, Inomata S, Shimizu T. Sleep habits of students attending elementary schools and junior and senior high schools in Akita prefecture. *Psychiatry Clin Neurosci* **56**(3):241–242(2002).

6. Morrison DN, McGee R, Stanton WR. Sleep problems in adolescence. *J Am Acad Child Adolesc Psychiatry* **31**(1):94–99(1992).

7. Anders TF, Carskadon MA, Dement WC, Harvey K. Sleep habits of children and the identification of pathologically sleepy children. *Child Psychiatry Hum Dev* **9**(1):56–389(1978).

8. Kirmil-Gray K, Eagleston JR, Gibson E, Thoresen CE. Sleep disturbance in adolescents: sleep quality, sleep habits, beliefs about sleep and daytime functioning. *J Youth Adolesc* **13**:375–384(1984).

9. Coren S. The prevalence of self reported sleep disturbances in young adults. *Int J Neurosci* **79**(1-2):67–73(1994).

10. Lack LC. Delayed sleep and sleep loss in university students. *J Am Coll Health* **35**(3):105–110(1986).

11. Buboltz WC Jr, Brown F, Soper B. Sleep habits and patterns of college students. A preliminary study. *J Am Coll Health* **50**(3):131–135(2001).

12. Carskadon MA, Acebo C, Richardson GS, Tate BA, Seifer R. An approach to studying circadian rhythms of adolescent humans. *J Biol Rhythms* **12**(3):278–289(1997).

13. Lewin DS, England SJ, Rosen RC. Cognitive and behavioral sequelae of obstructive sleep apnea in children. *Sleep* **22**:s126(1999).

14. Zuckerman B, Stevenson J, Bailey V. Sleep problems in early childhood: continuities, predictive factors and behavioral correlates. *Pediatrics* **80**(5):644–671(1987).

15. Dexter D, Bijwadia J, Schilling D, Applebaugh G. Sleep, sleepiness and school start times: a preliminary study. *WMJ* **102**(1):44–46(2003).

16. Pollack CP, Bright D. Caffeine consumption and weekly sleep patterns in US seventh, eighth and ninth graders. *Pediatrics* **111**(1):42–46(2003).

17. Johnson EO, Breslau N. Sleep problems and substance abuse in adolescence. *Drug Alcohol Depend* **64**(1):1–7(2001).

18. Shin C, Joo S, Kim J, Kim T. Prevalence and correlates of habitual snoring in high school students. *Chest* **124**(5):1709–1715(2003).

19. Webb E, Ashton CH, Kelly P, Kamali F. Alcohol and drug use in UK university students. *Lancet* **348**(9032):922–925(1996).

20. Ivanenko A, McLaughlin C. *Pediatr Clin North Am* **51**:1(2004).

21. Kaemingk KL, Pasvogel AE, Goodwin JL, Mulvaney SA, Martinez F, Enright PL, Rosen GM, Morgan WJ, Fregosi RF, Quan SF. *J Int Neuropsychol Soc* **9**(7):1016–1026(2003).

22. Wolfson RA, Caraskadon MA. Understanding adolescents' sleep patterns and school performance: a critical appraisal. *Sleep Med Rev* **7**(6):491–506(2003).

ADDITIONAL READING

Mindell JA, Owens J. *A Clinical Guide to Pediatric Sleep.* Lippincott williams & Wilkins, Baltimore, 2003.

Caraskadon MA. *Adolescent Sleep Patterns: Biological Social and Psychological Influences.* Cambridge University Press, New York, 2002.

Adolescent Sleep Needs and Patterns: Research Report and Resource Guide. National Sleep Foundation, 2000.

PART XVII

SLEEP ASSESSMENT METHODS

127

THE SLEEP INTERVIEW AND SLEEP QUESTIONNAIRES

CHARLES J. BAE AND JOSEPH A. GOLISH
The Cleveland Clinic Foundation, Cleveland, Ohio

THE SLEEP INTERVIEW

Sleep disorders are common in all races and socioeconomic classes. Disturbed sleep can be the cause of daytime dysfunction of varying degrees. Sleep complaints can be broken down into four types: daytime sleepiness, difficulty falling asleep, difficulty staying asleep, and abnormal behaviors during the night. These complaints can be caused by common sleep disorders such as sleep deprivation, poor sleep hygiene, insomnia, sleep apnea, restless legs syndrome, and parasomnias. As is the case for any medical condition, patients may have multiple sleep problems at the same time.

A comprehensive sleep history is the most important part of the evaluation of patients seen in the sleep clinic. Often the best source of information is the bedpartner if one is available, since patients are often asleep and are not fully aware of what occurs during the night. In addition, because patients will often present with daytime dysfunction, a history of what happens during a typical day is needed to have a complete history of a 24 hour period. The bedpartner may also be able to provide another version of the history regarding daytime function.

A sleep history does not focus only on sleep-related issues, but a review of the general medical history, medications, social history, and family history, needs to be performed. There is no standard physical examination for sleep disorders, and often the examination can be tailored to the specific complaint. This chapter will discuss the sleep interview and various sleep questionnaires that can be used to help assess sleep disorders. The use and indications for sleep studies will be discussed in another chapter.

Many patients go to a sleep clinic because of longstanding problems with sleeping or with daytime sleepiness. However, many patients are not aware that a problem is present and are urged to go (if not dragged) to the sleep clinic by the bedpartner. The chronicity of the sleep problem is a helpful historical detail. Some sleep problems may have been present since childhood, while other problems may have been an issue for only months or years.

Every sleep interview should include an evaluation of sleep habits. This starts with the time that a patient goes to bed and wakes up in the morning, both during the week and weekend. If patients with daytime sleepiness only sleep for 4 hours a night, sleep deprivation is likely to be a contributing factor for their daytime problems. Other sleepy individuals may stay up late during the week and have difficulty waking up in the morning. However, after sleeping in on the weekends, the same patients may feel rested, suggesting that a circadian rhythm sleep disorder such as delayed sleep phase syndrome may be the problem. A helpful question is to ask patients when they would go to bed and start the day if they did not have any commitments to keep.

It is useful to find out how long it takes to fall asleep, and what the patient does during that time if there is difficulty with sleep initiation. This can help start a conversation about other factors that may be contribute to insomnia. Patients may lay in bed for hours tossing and turning, while planning the next day and becoming frustrated due to an inability to fall asleep. Obtaining a description of what

Sleep: A Comprehensive Handbook, Edited by T. Lee-Chiong.
Copyright © 2006 John Wiley & Sons, Inc.

the bedroom looks like and what patients do before getting into bed can help create a mental picture of the environment in which patients are trying to sleep. The number and frequency of nocturnal awakenings should be documented. If a patient snores wakes up multiple times a night, this may be the first clue that sleep disordered breathing is present.

A helpful tool is a sleep log or diary that is completed prior to a visit. This log can be pictorial or word based. A visually based sleep log is easy to fill out and can give the clinician a tremendous amount of information regarding sleep habits. Looking at a sleep log completed over a 1–2 week period can provide information in a pictorial fashion about sleep efficiency (total sleep time divided by the total time in bed), and the number and length of awakenings. Sleep logs can also be used to help patients realize that there are improvements after treatment.

A common complaint is daytime sleepiness. Daytime sleepiness can be caused by a variety of sleep disorders, such as sleep deprivation, obstructive sleep apnea syndrome (OSAS), insomnia, or narcolepsy. Interestingly, even though patients may initially complain of sleepiness, they may be bothered more by fatigue. Daytime fatigue may be more significant than sleepiness, and it may be hard to differentiate the two conditions. The use of two short questionnaires can help patients and clinicians assess the difference between sleepiness and fatigue (see Tables 127.1 and 127.3). It is important to ask whether patients have fallen asleep when driving, and if accidents (or near misses) related to drowsiness have occurred.

Sleep disordered breathing is suspected when snoring is present or if there is a report of witnessed apneas by a bedpartner or friend. Often these patients will wake up multiple times a night, sometimes in association with gasping for air or choking. Other helpful historical facts include amount of recent weight change and whether snoring is changed by body position. A dry mouth upon waking in the morning can be related to nasal congestion, which causes mouth breathing. Daytime sleepiness may or may not be an associated symptom.

When the chief complaint is primarily related to excessive daytime sleepiness, it is important to see if there are other features that may be seen with narcolepsy, such as cataplexy, hypnagogic hallucinations, or sleep paralysis. For a history of cataplexy, ask if the patient has experienced any muscle weakness that is associated with strong emotions such as laughter or anger. Often the weakness may be subtle, to the point that only patients may notice. For example, some patients may experience a brief buckling of their knees or a laxity of their jaw muscles. Patients may have hypnagogic hallucinations, which are things that they may see, hear, or smell that are not actually present as they are falling asleep. Usually upon waking, patients may experience sleep paralysis, which is a brief inability to move even though they know they are fully awake. A good question to ask is how the patient felt with the first few episodes of sleep paralysis, since true episodes are typically associated with a feeling of fear or anxiety.

Patients may complain that they are constantly moving their legs before falling asleep, which may or may not interfere with falling asleep. In addition, the bedpartner may report that the patient is kicking often when asleep.

TABLE 127.1 The Epworth Sleepiness Scale [1]

How likely are you to doze off or fall asleep in the following situations, in contrast to just feeling tired? This refers to your usual way of life in recent times. Even if you have not done some of the things recently, try to work out how they would have affected you. Use the following scale to choose the most appropriate number for each situation:

0 = would **never** doze
1 = **slight** chance of dozing
2 = **moderate** chance of dozing
3 = **high** chance of dozing

Situation	Chance of Dozing			
	never	slight	mod	high
Sitting and reading	☐ 0	☐ 1	☐ 2	☐ 3
Watching TV	☐ 0	☐ 1	☐ 2	☐ 3
Sitting inactive, in a public place (e.g., a theater or meeting)	☐ 0	☐ 1	☐ 2	☐ 3
As a passenger in a car for an hour without a break	☐ 0	☐ 1	☐ 2	☐ 3
Lying down to rest in the afternoon when circumstances permit	☐ 0	☐ 1	☐ 2	☐ 3
Sitting and talking to someone	☐ 0	☐ 1	☐ 2	☐ 3
Sitting quietly after a lunch without alcohol	☐ 0	☐ 1	☐ 2	☐ 3
In a car, while stopped for a few minutes in traffic	☐ 0	☐ 1	☐ 2	☐ 3

These historical findings are suggestive of restless legs syndrome. Ask patients if they have an urge to move their legs because of an uncomfortable sensation in their legs. Moving their legs should partially or completely relieve the sensations temporarily. Additionally, the uncomfortable sensations usually begin or worsen during times of rest or inactivity, which is typically in the evening or at bedtime. It may be helpful to ask if other family members have a similar condition.

Some patients may only have difficulty falling asleep. For these patients it is especially important to inquire about their activities before entering the bedroom and when in the bed. Often they state that they are unable to relax and turn off their thoughts. At the end of the day, these they may not have a bedtime routine and may jump into bed expecting to fall asleep immediately. See how long patients stay in bed before falling asleep. Activities that interfere with falling asleep include watching TV in bed, reading in bed, and checking the clock regularly. It is helpful to know how long the difficulty of falling asleep has been a problem. For those patients who have transient or short-term insomnia, an inciting factor may be elicited. This is in contrast with patients who have chronic insomnia and have difficulty falling asleep due to longstanding behaviors and habits that have been established over time.

There may be a concern about unusual nighttime events which occur during sleep. There can be a variety of behaviors, and the bedpartner may be able to provide valuable information. A history of sleepwalking, sleeptalking or night terrors may be helpful. Patients may act out dreams, and it is important to ask about dream content, since REM sleep behavior disorders are often associated with violent dream content (patients are defending themselves or are being chased).

A complete medical and surgical history can be helpful. Patients with chronic diseases such as hypertension, diabetes mellitus, and obesity may have sleep disturbances related to those conditions and associated comorbidities. Current medications should be reviewed. While the medical history is being reviewed. Many medications can affect the quality of sleep or may cause daytime sleepiness. It may help to find out whether the patient is taking or has tried any medications in the past to help with sleep.

The social history can provide many helpful clues. There may be a great deal of stress at home or at work. Spending time asking about relationships and finding out details about the workplace helps to get a better sense of the daily routine of patients. Inquire about caffeine, tobacco, and alcohol use. An exercise history can be useful as well. The family history should not be overlooked since many sleep disorders can be familial, such as obstructive sleep apnea syndrome, narcolepsy, and restless legs syndrome.

SLEEP QUESTIONNAIRES

There are some commonly used validated questionnaires that can help with the assessment of daytime sleepiness and fatigue: the Epworth Sleepiness Scale, the Stanford Sleepiness Scale, and the Fatigue Severity Scale.

The Epworth Sleepiness Scale (ESS) is a widely used validated subjective self-rating scale of daytime sleepiness. Patients are asked to rate the likelihood of dozing off or falling asleep in eight common situations which involve little stimulation or inactivity [1]. (Table 127.1). The maximum total score is 24, and a score greater than 10 is consistent with significant daytime sleepiness [2]. The ESS has good test–retest reliability and good internal consistency [3]. Even though the ESS is a subjective measure of sleepiness, it can be a useful tool to help assess the effectiveness of a particular treatment over time.

The Stanford Sleepiness Scale (SSS) is another subjective self-rating scale to assess daytime sleepiness which is more situational. Patients are asked to choose 1 of 7 statements that best describes their level of sleepiness or alertness at a specific moment in time [4] (Table 127.2). The primary use of the SSS is before each nap trial of a multiple sleep latency test to see if the subjective assessment of sleepiness or alertness correlates with an objective measure of sleepiness. The SSS has been shown to be a reliable measure of the effects of partial sleep deprivation, though it does not reliably measure performance after sleep deprivation [5].

There are many scales that rate different aspects of fatigue, but the Fatigue Severity Scale (FSS) is a widely used subjective self-rating scale to assess the impact of fatigue on level of functioning in different areas of daily life [6]. Patients are asked to agree or disagree with 9

TABLE 127.2   The Stanford Sleepiness Scale [4]

This is a quick way to assess how alert you are feeling. If it is during the day when you go about your business, ideally you would want a rating of 1. Take into account that most people have two peak times of alertness daily, at about 9 a.m. and 9 p.m. Alertness wanes to its lowest point at around 3 p.m.; after that it begins to build again. Rate your alertness at different times. A rating of 7 is an indication that you have a serious sleep debt and you need more sleep.

1—Feeling active and vital; alert; wide awake.
2—Functioning at a high level, but not at peak; able to concentrate.
3—Relaxed; awake; not at full alertness; responsive.
4—A little foggy; not at peak; let down.
5—Fogginess; beginning to lose interest in remaining awake; slowed down.
6—Sleepiness; prefer to be lying down; fighting sleep; woozy.
7—Almost in reverie; sleep onset soon; lost struggle to remain awake.

TABLE 127.3   The Fatigue Severity Scale [7]

Circle a number from 1 to 7 that indicates the degree of agreement with each statement.
1 indicates strongly disagree, 7 indicates strongly agree.

	Disagree							Agree
1. My motivation is lower when I am fatigued	1	2	3	4	5	6	7	
2. Exercise brings on my fatigue	1	2	3	4	5	6	7	
3. I am easily fatigued	1	2	3	4	5	6	7	
4. Fatigue interferes with my physical conditioning	1	2	3	4	5	6	7	
5. Fatigue causes frequent problems for me	1	2	3	4	5	6	7	
6. My fatigue prevents sustained physical functioning	1	2	3	4	5	6	7	
7. Fatigue interferes with carrying out certain duties and responsibilities	1	2	3	4	5	6	7	
8. Fatigue is among my three most disabling symptoms	1	2	3	4	5	6	7	
9. Fatigue interferes with my work, family, or social life	1	2	3	4	5	6	7	

statements that describe different effects of fatigue on various aspects of their lives [7] (Table 127.3). The sum of the responses are averaged, and the score can range between 1 and 7. Normal controls have a mean of 2.3, compared to a mean of 4.7 in patients with systemic lupus erythematosis and a mean of 4.8 in patients with multiple sclerosis [7]. The FSS has good test–retest reliability and can be used to measure the effect of treatments for fatigue [7].

A questionnaire that is completed the morning after a polysomnogram can provide additional information that may help with the interpretation of the study. An example of a morning questionnaire is seen in Table 127.4. This questionnaire can provide insight into how well patients estimate the time and quality of their sleep. For example, patients may claim that their sleep is quite disturbed, yet the answers on the questionnaire after a polysomnogram may be inconsistent with the results of the sleep study.

A comprehensive sleep interview with the use of appropriate questionnaires will allow clinicians to have a good idea of what sleep disorder or disorders a patient may

TABLE 127.4   Morning Questionnaire

1. How long did it take you to fall asleep last night? _____ h _____ min
2. How does this compare with the length of time it usually takes you to fall asleep at home? (Circle one)
   Much longer   Longer   Same   Shorter   Much shorter
3. How long do you feel you slept last night?_____ h _____ min
4. How long does this compare with the length of time you usually sleep at home? (Circle one)
   Much longer   Longer   Same   Shorter   Much shorter
5. How many times do you remember waking up last night? _____
6. How do you feel right now? (Circle one)
   Very tired and sleepy   Awake but not alert   Rested   Alert and wide awake
7. Do you have any physical complaints this morning? Describe:

_____

8. Rate the quality of your sleep last night by circling a number in each of the categories listed below:

a. Very light	1	2	3	4	5	6	7	Very deep
b. Very short	1	2	3	4	5	6	7	Very long
c. Interrupted	1	2	3	4	5	6	7	Uninterrupted
d. Dreamless	1	2	3	4	5	6	7	Many dreams
e. Restless	1	2	3	4	5	6	7	Restful

9. Do you remember any dreams last night? ☐ Yes   ☐ No
   If yes, please describe:_____

_____

10. What awakened you this morning? (Circle one)
    Noise   Discomfort   Technologist   Spontaneous   Other:_____
11. In general, how long would you say you slept last night compared to your usual sleep at home? (Circle one)
    Much longer   Longer   Same   Shorter   Much shorter

have. It is possible for a patient to have multiple sleep disorders. After a good sleep history and physical exam, sleep testing should be performed if necessary to confirm the diagnoses.

REFERENCES

1. Johns MW. A new method for measuring daytime sleepiness: the Epworth Sleepiness Scale. *Sleep* **14**:540–545(1991).

2. Johns MHB. Daytime sleepiness and sleep habits of Australian workers. *Sleep* **20**:844–849(1997).

3. Johns MW. Reliability and factor analysis of the Epworth Sleepiness Scale. *Sleep* **15**:376–381(1992).

4. Hoddes EDW, Zarcone V. The development and use of the Stanford Sleepiness Scale (abstract). *Psychophysiology* **9**:150(1972).

5. Herscovitch JBR. Sensitivity of the Stanford Sleepiness Scale to the effects of the cumulative partial sleep deprivation and recovery oversleeping. *Sleep* **4**:83–92(1981).

6. Dittner AJW, Brown RG. The assessment of fatigue: a pratical guide for clinicians and researchers. *J Psychosom Res* **56**:157–170(2004).

7. Krupp LBLN, Muir-Nash J, Steinberg AD. The fatigue severity scale: application to patients with multiple sclerosis and systemic lupus erythematosus. *Arch Neurol* **46**:1121–1123(1989).

128

POLYSOMNOGRAPHY

NANCY A. COLLOP

Johns Hopkins University, Baltimore, Maryland

INTRODUCTION

Polysomnography was first described in 1937 [1]. In this original description, brain wave signals were monitored but over time more physiologic variables were added such as an electrocardiogram lead and respiratory signals. In the original polysomnograms, pens with ink recorded the signals on paper, which were reviewed, page by page. After computers became more widely available, conversion to paperless digital recordings ensued. Now the majority of sleep laboratories utilize computerized polysomnography. In this chapter, I will present an overview of polysomnography regarding its technical aspects and indications. Other chapters will provide more details on specific aspects of polysomnography such as sleep staging, respiratory signals, and the unique aspects of monitoring in pediatrics.

CIRCUITRY

Polysomnography is the monitoring of physiologic signals from various organs and transduction of those signals to a recording device. Simplistically, the mechanism in which this is accomplished is shown in Figure 128.1. The electrodes and other monitors such as nasal pressure or microphones are plugged into a jackbox, which can then be plugged into a special connection that routes the signals to a control center. These signals are amplified and filtered as appropriate and sent to the output device. The digitization of the analog signal occurs utilizing sampling of the analog waveform. A sampling rate of a minimum of three times the highest recorded frequency is recommended.

The raw signals from the brain and other organs are very low voltages and therefore require amplification. In most recording environments, there is usually some electrical interference present, therefore the process of common mode rejection is used to "purify" the signal. The two input signals are sent to the amplifier, which will reject similar (common) signals and forward different (unique) signals. Electrodes can either be "referential" in which the input of interest is compared to a relatively inert signal (such as an electrode over the mastoid) or "bipolar" in which two inputs are compared to each other. Most electroencephalographic (EEG) leads are referential and electromyography (EMG) and electrocardiographic (ECG) leads are bipolar.

Another important aspect of the circuitry is filtering of the signals. Filtering allows the signals of interest to be seen most clearly in their appropriate frequency bands. For example, EEG filters are typically set to reduce signals below 0.3 Hz (low frequency filter or LFF) and above 35 Hz (high frequency filter or HFF), whereas EMG signals are set at 10 Hz (LFF) and 100 Hz (HFF). Notch filters may be added to attenuate signals around the bandwidth of power-lines (60 Hz in the USA) if this type of interference is present. Caution is recommended in use of filtering, as overfiltering may eliminate important physiological signals.

METHODOLOGY

Most sleep laboratories utilize similar montages for baseline polysomnography (Table 128.1). It is important that electrodes be properly placed with low impedance and

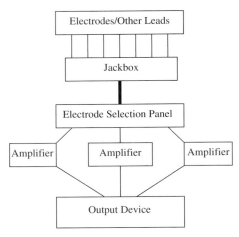

Figure 128.1 Simplistic representation of circuitry for polysomnography. The leads are all plugged into a jackbox, which can easily be disconnected from the remainder of the circuitry to allow the patient to move to the restroom and so on during the study without having to dislodge any leads. The selection panels signal the recording device which electrodes to record and amplification allows for the signals to be readable.

with backup electrodes so the patient doesn't have to be awakened if a signal goes awry. Most polysomnograms will utilize central and occipital EEG leads based on the International 10/20 system (Figure 128.2) [2]. In this system, four anatomic landmarks—the nasion, inion, and right and left preauricular points—are used as references to place electrodes 10% or 20% of the distance apart. Electro-oculographic (EOG) leads are placed slightly above and slightly below the outer canthus of each eye and referenced to the same mastoid lead (left EOG–A1,

TABLE 128.1   Typical Polysomnography Montage

Lead	Name	Type
1	C3–A2 or C4–A1	Electroencephalogram
2	O2–A1 or O1–A2	Electroencephalogram
3	Right eye-A1	Electro-oculogram
4	Left eye-A1	Electro-oculogram
5	Chin muscle	Electromyogram
6	Left leg	Electromyogram
7	Right leg	Electromyogram
8	Lead II	Electrocardiogram
9	Nasal airflow	Thermistor
10	Nasal airflow	Nasal cannula pressure transducer
11	Chest effort	Respiratory inductance plethysmography
12	Abdominal effort	Respiratory inductance plethysmography
13	Sum	Respiratory inductance plethysmography
13	Oxygen saturation	Pulse oximeter

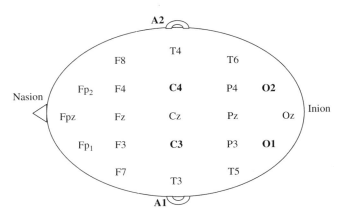

Figure 128.2 The International 10/20 system. Electroencephalographic leads are organized using the four major landmarks and placing the leads 10–20% apart accordingly. The most common leads used in polysomnography are in bold print (C3, C4, A1, A2, O1, O2).

right EOG–A1) or opposite mastoid lead (left EOG–A2, right mastoid lead–A1). Chin (genioglossus) EMG leads are placed either over the front of the chin (below the lower lip), under the chin, or over the masseter muscle and these are referenced to each other. Leg EMG leads are placed over the anterior tibialis muscles. A modified Lead II of a standard electrocardiogram is used to monitor the heart signal. This is done by placing one lead below the right clavicle and the other below the left rib cage.

There are a large variety of choices available to monitor respiration. For airflow, most sleep laboratories use either or both the nasal thermistor and the nasal cannula pressure transducer (NCPT). The thermistor works by sensing changes in temperature during ventilation and is good for detecting apneas but tends to underestimate hypopneas The NCPT measures the pressure change across the nasal inlet using a pressure sensitive transducer and is better for detecting flow limitation and hypopneas [3]. The downside of the NCPT is that it does not work as well in patients with nasal obstruction.

Respiratory effort is usually measured with respiratory inductance plethysmography or with impedance pneumography. The former requires calibration for accuracy and can provide an estimate of tidal volume. The latter provides a less quantitative signal.

Finally, pulse oximetry is used to monitor oxygen saturation continuously throughout the night. The lead is typically placed on the finger or the ear lobe. Oximeters can be set at different sampling intervals and the longer the interval, the lower the accuracy, especially when the patient is having regular fluctuations in the oxygen saturation level as is seen in obstructive sleep apnea [4]. The shorter the sampling interval (e.g., 3 seconds), the better the sensitivity [5].

Most sleep laboratories also utilize video recording to observe the patient and detect position changes or unusual behavior as may occur during a nocturnal seizure. Infrared lighting is used to allow adequate viewing with maintenance of a dark environment.

Once the equipment is set up, machine and biocalibrations should be performed to assure all signals are standardized and working. Impedance for electrodes should be below 5000 ohms and an all channel calibration should be performed by applying a 50 μV negative DC signal and responses checked for each lead. Biocalibrations before the study begins allow the technician to determine if the leads are adequately recording the desired signals. These should include asking the patient to move his/her eyes, legs, and jaw in addition to performing breath holds and nasal only and mouth only breathing.

ARTIFACTS

Artifact recognition is an important skill for both technicians acquiring and scoring the study and for physicians reviewing polysomnograms. Technicians must be able to identify artifacts during the study and correct them and physicians must be able to delineate artifacts from true physiologic signals to avoid misinterpretation. Redundant electrodes may be used and activated by the technician during study acquisition to avoid disturbing the patient if a signal deteriorates. Types of common artifacts include: electrode popping, ECG artifact, respiratory artifact, sweat artifact, or loose/broken electrode.

SCORING AND INTERPRETATION

Following the acquisition of the polysomnogram, the data should be reviewed in detail, first by the technician and then by the interpreting physician. The technician should score each epoch according to the criteria set forth by Rechtschaffen and Kales [6]. Following the sleep staging, all the respiratory signals should be reviewed and breathing events counted. Although most sleep laboratories agree upon a standard definition for an apnea (greater than or equal to 10 seconds of absent airflow), the definition of a hypopnea remains controversial [7]. A hypopnea is often defined as a reduction in airflow with a 4% oxygen desaturation and/or an EEG arousal from sleep. Defining arousals is where most of the controversy arises, because arousals are not consistently scored and many believe because of this inconsistent scoring that they should not be included in the definition [8, 9]. However, the other school of thought is that if the airflow limitation results in sleep disruption, those events should be counted. Currently, the CMS criteria only include hypopneas occurring with a 4% desaturation [10].

Periodic limb movements should also be scored according to standard criteria [11]. The technician should also review the record for arrhythmias, seizure activity, bruxism, or other atypical findings. Once the technician has finalized the scoring, a physician trained in sleep medicine and polysomnography should review the entire polysomnogram and develop an interpretation of the findings.

INDICATIONS

Polysomnography is used to diagnose a number of sleep disorders, the most common of those being obstructive sleep apnea (OSA) [12]. Patients at highest risk for OSA include males and postmenopausal females, the obese, and those with craniofacial abnormalities that compromise the upper airway. In those populations, a history of sleep disordered breathing might include daytime somnolence, habitual snoring, and witnessed apneas during sleep. Cardiovascular disease should heighten the suspicion for OSA as hypertension, coronary heart disease, and stroke have all been associated with OSA. The diagnosis of OSA is typically confirmed with polysomnography when the apnea–hypopnea index is >5–10/hour.

After the confirmation of OSA, many patients require a second polysomnogram to titrate nasal continuous positive airway pressure (CPAP). In many laboratories, one polysomnogram accomplishes both the baseline and CPAP portions of the study (split-night protocol). On rare occasions, some patients may need another night of titration with a bilevel device if CPAP alone is insufficient to treat respiratory events.

Other indications for polysomnography in a patient with OSA include those who are treated with a dental appliance, to assure effectiveness; patients following surgery for OSA; patients who have had substantial weight gain or weight loss since their initial diagnoses; or patients who have a return of symptoms following an initial good response to CPAP or other treatment.

Patients with excessive daytime sleepiness despite adequate sleep amounts often require polysomnography. Aside from OSA, other hypersomnolence diagnoses including narcolepsy and idiopathic hypersomnia often utilize confirmatory polysomnography and a multiple sleep latency test (MSLT). In these disorders, the overnight polysomnogram is used to evaluate for sleep-disrupting events and to document adequate sleep on the night before the MSLT. The MSLT is used to determine the mean sleep latency for a series of naps and to assess for the presence of sleep onset REM periods, suggestive of narcolepsy.

Other sleep disorders that may require polysomnography include parasomnias, particularly those in which the patient or others have described violent or potentially injurious behavior; and nocturnal seizures, in which patients exhibit unusual paroxysmal nocturnal behaviors of a stereotypic or repetitive nature. In these circumstances, video recording and extended montages are often indicated.

SUMMARY

Polysomnography is the standard for the diagnoses of many sleep disorders including, but not limited to, obstructive sleep apnea. Although most sleep laboratories utilize similar montages, equipment and scoring standards may vary. Technicians and physicians skilled in polysomnography are important to assure quality recording and interpretations.

REFERENCES

1. Loomis A, Harvey N, Hobart G. Cerebral states during sleep as studied by human brain potentials. *J Exp Psychol* **21**:127–144(1937).

2. Jasper H. The ten/twenty electrode system of the International Federation. *Electroencephalogr Clin Neurophysiol* **10**:371(1958).

3. Norman R, Ahmad M, Walsleben J, Rapoport D. Detection of respiratory events during NPSG: nasal cannula/pressure sensor versus thermistor. *Sleep* **20**:1175–1184(1997).

4. Farre R, Montserrat JM, Ballester E, Hernandez L, Rotger M, Navajas D. Importance of the pulse oximeter averaging time when measuring oxygen desaturation in sleep apnea. *Sleep* **21**:386–390(1998).

5. Davila D, Richards K, Marshall B, O'Sullivan P, Gregory T, Hernandez V, Rice S. Oximeter performance: the influence of acquisition parameters. *Chest* **122**(5):1654–1660(2002).

6. Rechtschaffen A, Kales A. *A Manual of Standardized Terminology, Techniques and Scoring Systems for Sleep Stages of Human Subjects*. UCLA Brain Information Service/Research Institute, Los Angeles, 1968.

7. Tsai W, Flemons W, Whitelaw W, Remmers J. A comparison of apnea–hypopnea indices derived from different definitions of hypopnea. *Am J Respir Crit Care Med* **159**(1):43–48(1999).

8. Bonnet M, Carley D, Carskadon M, Easton P, Guilleminault C, Harper R, Hayes B, Hirschkowitz M, Ktonas P, Keenan S, Pressman M, Roehrs T, Smith J, Walsh J, Weber S, Westbrook P. ASDA Report: EEG arousals: scoring rules and examples. *Sleep* **15**(2):173–184(1992).

9. Drinnan M, Murray A, Griffiths C, Gibson G. Interobserver variability in recognizing arousal in respiratory sleep disorders. *Am J Respir Crit Care Med* **158**:358–362(1998).

10. Clinical Practice Review Committee—Meoli A, Casey K, Clark R, Coleman J Jr, Fayle R, Troell R, Iber C. Hypopnea in sleep-disordered breathing in adults. *Sleep* **24**:469–470(2001).

11. Bonnet M, Carley D, Carskadon M, Easton P, Guilleminault C, Harper R, Hayes B, Hirschkowitz M, Ktonas P, Keenan S, Pressman M, Roehrs T, Smith J, Walsh J, Weber S, Westbrook P. ASDA Report: Atlas and scoring rules: recording and scoring leg movements. *Sleep* **16**(8):748–759(1993).

12. Kushida C, Littner M, Morgenthaler T, Alessi C, Bailey D, Coleman J, Davila D, Friedman L, Hirshkowitz M, Kapen S, Kramer M, Lee-Chiong T, Loube D, Owens J, Pancer J, Wise M. Practice parameters for the indications for polysomnography and related procedures. Polysomnography Task Force, American Sleep Disorders Association Standards of Practice Committee. *Sleep* **20**(6):406–422(1997).

ADDITIONAL READING

American Academy of Sleep Medicine Task Force. Sleep related breathing disorders in adults: recommendation for syndrome definition and measurement techniques in clinical research. *Sleep* **22**:667–689(1999).

Butkov N. *Atlas of Clinical Polysomnography*. Synapse Media, Medford, OR, 1996.

Carskadon M, Rechtschaffen A. Monitoring and staging human sleep. In: Kryger M, Roth T, Dement W (Eds), *Principles and Practice of Sleep Medicine*, 3rd ed. Saunders, Philadelphia, 2000, pp 1197–1216.

Kennan S. Polysomnographic technique: an overview. In: Chokroverty S (Ed), *Sleep Disorders Medicine: Basic Science, Technical Considerations and Clinical Aspects*, 2nd ed. Butterworth-Heinemann, Boston, 1999, pp 151–169.

Shepard J. *Atlas of Sleep Medicine*. Futura Publishing, Mount Kisko, NY, 1991.

129

PEDIATRIC POLYSOMNOGRAPHY

MAY L. GRIEBEL AND LINDA K. MOYER

Arkansas Children's Hospital, Little Rock, Arkansas

INTRODUCTION

Nowhere in medicine is the maxim that "children are not merely little adults" better demonstrated than in the recording and interpretation of pediatric overnight polysomnograms (OPSGs). The developmental processes underlying all of pediatrics become obstacles, as the polysomnographer faces multiple challenges posed by the evolution of pediatric EEG and sleep staging, changes in respiratory rates and patterns with age, differing presentations and etiologies of disorders such as obstructive sleep disordered breathing in children versus adults, and even the issues of dealing with small children and their sometimes upset families. The complexity of pediatric OPSGs is increased by the reality that there are limited normative data for interpretation of various parameters of the pediatric OPSG. While adult standards are published and generally well accepted, even such basic issues as which events to score on pediatric studies and the clinical significance of their results have not yet been adequately addressed [1].

These difficulties notwithstanding, the importance of accurate performance and interpretation of pediatric polysomnograms to a child's health, behavior, and school performance cannot be underestimated. Numerous studies have demonstrated the lack of predictive value of clinical signs and symptoms in accurately diagnosing obstructive sleep disordered breathing (SDB) in children [2–4]. The importance of polysomnography in infants for the evaluation of apnea and abnormal breathing patterns has been a focus of discussion in the literature [5, 6]. Also, it has been recognized that obstructive SDB is common in certain populations, such as children with Down syndrome, stressing the importance of addressing sleep disordered breathing in children with these diagnoses [7]. At the same time there is growing recognition in the medical community of the neurocognitive and behavioral effects on children of untreated sleep disorders. While long-range physical effects such as growth failure, pulmonary hypertension, and cor pulmonale have been accepted as significant concerns, academic and behavioral effects of poor quality sleep are now coming to the forefront [8–12]. While obstructive sleep disordered breathing has received much of the focus both for long-range growth and health concerns as well as for behavioral and academic issues, the importance of other disorders such as periodic limb movements of sleep (PLMS) is being recognized as an valuable aspect of the pediatric sleep study [13].

Overnight polysomnography is the long recognized "gold standard" for evaluation of sleep disorders including OSDB. The cost, inconvenience, and time required by OSPG have led to the consideration of nap PSG, which has been shown by some to be a reasonable screening tool. However, the difficulty in achieving rapid eye movement (REM) sleep during the day may lead to an underestimation of the severity of SDB and even to missed diagnoses. Thus Marcus et al. [14] recommended over a decade ago that nap PSG be considered as a screening tool with OSPG to be performed if nap PSG is inconclusive. Saeed and colleagues [15] concluded that individual nap parameters were not very sensitive in predicting abnormal OPSGs and suggested that the cost savings and efficiency offered by a nap study would be offset by the need to clarify findings by then performing an OPSG.

Sleep: A Comprehensive Handbook, Edited by T. Lee-Chiong.
Copyright © 2006 John Wiley & Sons, Inc.

PERFORMING OPSGs IN CHILDREN: OVERVIEW

The pediatric sleep center has special features, the most important being technologists who are truly child-oriented and who are also comfortable working with parents. In a tertiary care setting particularly, the children may have a variety of comorbid conditions that complicate the study, such as Prader–Willi or Down syndrome. Children with disorders such as autism present special difficulties. The decor of the environment may be a challenge, as it needs to be acceptable and comfortable for children of all ages, from toddlers to teenagers. Toys, games, and video cassette or DVD players need to be available. In our center, children generally come to the sleep laboratory from the clinic setting to schedule their sleep studies, and at the same time are shown a locally produced video describing an OPSG and then demonstrating the attachment of recording equipment in a young child. The children and their families are encouraged to ask questions, and all efforts are made to allay the child's and parents fears in advance. It is helpful at this point to identify any medications the child is taking and to clarify whether or not those medications need to be stopped prior to the study.

Despite all efforts in advance, however, small children, even including the younger school-aged child, can become upset at the actual reality of the OPSG preparations, and the technologist sometimes must deal with a crying child on one hand and a distraught parent on the other. It is critical that one parent plan to spend the night in the laboratory, as a parent needs to be close at hand for comforting if the child awakens. Thus, beds for the parents must be provided in an area that is at least contiguous and that assures easy access for the parent to the child while providing privacy for the parents. However, it becomes disruptive to the sleep environment to have multiple members from each family in the laboratory area overnight, so it is critical to clarify in advance with families that only one member may stay with the child. It is also important that the families arrive at least 1–2 hours before the desired time of lights out. The process of attaching recording equipment in pediatric patients often progresses much more slowly than with adults, including the not uncommon need to place electrodes more than once, when frightened or lively children have pulled them off. The technologists should involve the parents to try to prevent such delays. Ideally, each child's individual sleep schedule would be followed although practically this is a difficult standard to achieve, even with a 1:1 ratio between patients and technologists. The obstacles to performing OPSGs in children of all ages both successfully and reliably demand a smaller technologist to patient ratio than is standard in adult settings, and most pediatric facilities operate with a technologist to patient ratio of 1:1.

Whether or not to perform at least two nights worth of studies in children has been questioned. Certainly children's sleep anecdotally has been thought to demonstrate what is called "the first-night effect" with increased arousals, decreased total sleep time, and decreased REM percentage. However, in a study of over 80 children at our facility comparing diagnoses and management recommendations made after the first study night in children who then had a second night of study, we found no significant differences in either the diagnoses or management recommendations when the first night was compared to analysis of both nights together (Karlson, unpublished data). Another prospective study of 30 children who had two OSPGs compared the studies on a myriad of parameters and found no significant first-night effect and little clinically significant night-to-night variability, leading to the recommendation that a single OSPG is an adequate assessment for SDB in children [16]. However, a consensus paper recommended that a repeat night of study be considered if the first study is technically inadequate, if at least one REM episode is not recorded during the first night, or if the parents report that the child's sleep was significantly different from that seen in their home setting [17].

PHYSIOLOGICAL PARAMETERS AND THEIR MONITORING

In the child-friendly atmosphere of the pediatric sleep laboratory, children of all ages from infants to toddlers and preschoolers through teenagers can be successfully studied with full multichannel polysomnography. Standardly, a minimum of 16 channels of data are recorded, with DC channels utilized for respiratory effort, ETCO$_2$, and oxygen saturation and AC channels for other parameters such as EEG, EOG, and EKG. Children's movements or voluntary removal of monitoring equipment may necessitate reapplications of the leads during the night.

ELECTROENCEPHALOGRAPHIC AND ELECTRO-OCULOGRAPHIC ACTIVITY

The importance of recording EEG is as significant in the pediatric population as with adults, since staging of sleep is critical to the assessment of a variety of disorders. Many parasomnias are seen primarily in slow-wave sleep, while SDB may be most evident in REM sleep and overlooked in a study with insufficient REM [18]. Arousals and multiple sleep stage changes may suggest upper airway resistance syndrome (UARS) in the snoring child who sleeps restlessly, snorts, and mouth breathes but who lacks frank apneas or hypopneas. While a full seizure montage may not be possible, it is not uncommon for unsuspected EEG abnormalities, which may be clinically significant, to be seen on the OSPG.

The electrodes utilized for recording are commonly the standard 0.6 cm gold cup electrodes often employed in EEG laboratories. We request in advance that our patients arrive with clean hair, having washed out any oils or gels, removed braids or attachments, and dried the hair. After proper skin preparation by the technologist, electrodes are applied using collodion-soaked gauze, which is then dried with pressurized cool air. The electrodes are then filled with an electrolyte gel to facilitate conduction. Another option for electrode placement is the use of an electrolyte paste. However, pasted electrodes are less stable and easier for a child to pull off during the course of the study. Therefore, in an active child, collodion despite its odor is most desirable. If collodion is to be utilized, the room must be well ventilated. Electrodes are placed according to the International 10-20 EEG system using at least one central electrode referenced to the contralateral ear (e.g., C3 to A2) and one occipital electrode referenced also contralaterally (O2 to A1) [19]. The channel with the central electrode is then most useful for sleep activity, which is maximal at the vertex such as K-complexes, while the occipital channel is best for assessment of alpha. Other channels, particularly with a bipolar chain either coronally or longitudinally through the temporal or parasagittal head regions, may be utilized as indicated, particularly for the diagnosis of nocturnal

seizure disorders. Traditionally, because of the length of OPSGs, to save recording paper, paper speed for OPSGs has been set at 10 mm/second as opposed to the standard EEG recording speed of 30 mm/second. One advantage of digital recording is that recorded data can be viewed as if recorded at various speeds, which can be beneficial in analyzing epileptiform activity and also in reviewing subtle respiratory events. Technical aspects of recording, including a discussion of sensitivity settings, use of filters, and artifact recognition, are beyond the scope of this brief review.

Scoring of sleep staging is usually age dependent, with a breakdown of active versus quiet sleep utilized for younger infants and for those children whose EEG abnormalities make more detailed sleep staging impossible. Scoring of standard sleep stages should be possible in children from 6 months or 1 year and older, although the percentages of sleep stages vary from adult standards in infants and small children. Sleep is scored in 30 second epochs, according to standard guidelines of Rechtschaffen and Kales [20]. On the other hand, visualization of subtle respiratory events is often aided by reviewing the digitally obtained recording in longer epochs of 60 or even 90 seconds (see Figures 129.1 and 129.2). Arousals are scored according to guidelines

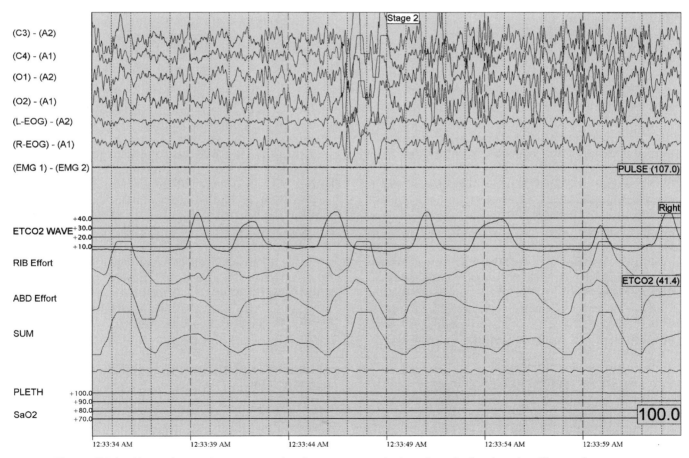

Figure 129.1 Obstructive respiratory events that do not meet standard scoring criteria, viewed as 30 seconds per page.

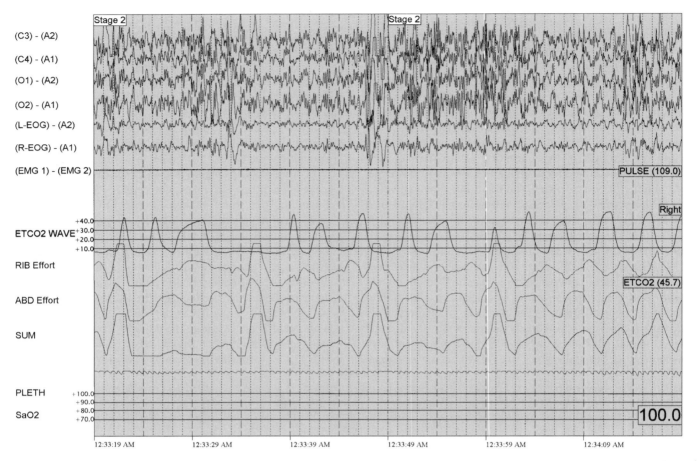

Figure 129.2 Obstructive respiratory events as in Figure 129.1, now recorded with 60 seconds per page, making the repetitive nature of the events more clear.

written by a committee of the American Sleep Disorders Association although they may need to be modified in the future to be more specific for children [21].

Sleep staging also depends on eye movements, and, as with standard adult OPSG, electrodes to record eye movements are placed 1 cm lateral the outer canthus of each eye. The electrodes are usually offset from the horizontal by 1 cm with one placed above and then another below a horizontal line drawn parallel to the floor from the outer canthi. They are then each referred to the ipsilateral ear. Eye movements are essential to the evaluation of sleep onset when slow rolling eye movements are seen, as well as to the very definition of REM sleep. In infants and small children, pre-gelled disposable electrodes can be utilized for the eye leads.

ELECTROMYOGRAPHIC ACTIVITY

While EMG may be recorded from numerous sites including the intercostal muscles, the typical OPSG includes at least chin EMG to assess for muscle activity dropout heralding REM sleep and leads placed commonly over the tibialis anterior muscles to evaluate for periodic limb movements of sleep. EMG is also critical to the evaluation of arousals. Commonly, three chin EMG electrodes are placed, but only two are utilized for recording in a bipolar manner with the third as a backup in the case of electrode malfunction. Four electrodes, two over each anterior tibialis anterior muscle, are commonly placed, referred to the ipsilateral electrode. Upper limb EMG may be added if PLMS are in the differential diagnosis (see Figure 129.3). In addition, continuous videotape and technologists' comments made during recording are extremely beneficial. For example, increases in chin EMG, which are noted by the technologist to follow feedings in infants, can lead to a workup for reflux.

ELECTROCARDIOGRAPHIC ACTIVITY: HEART RATE AND RHYTHM

Only a single channel of EKG is commonly included in the OPSG, typically a modified placement for Lead 2. The determination of normal heart rates varies according to age in pediatrics, and normative data are available to assist

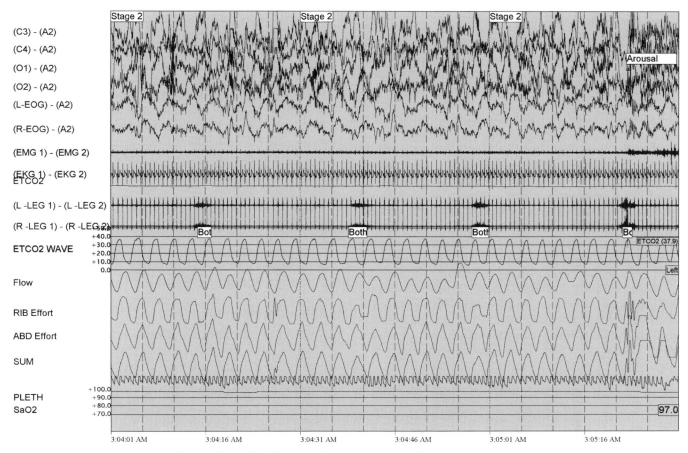

Figure 129.3 PLMS with and without arousals, 90 seconds per page.

interpretation. The EKG allows determination of heart rate and also beat-to-beat variability, which can be important in assessing SDB. Arrhythmias may also be assessed, although their full evaluation requires a standard EKG. It should be noted that sinus arrhythmia is commonly seen in children as a normal pattern and by itself implies no pathology.

RESPIRATORY ACTIVITY, INCLUDING VENTILATION AND OXYGENATION

Features of respiratory activity that are commonly assessed in an OPSG include arterial oxygen saturation, carbon dioxide exchange, nasal and oral airflow, and respiratory effort as reflected in chest and abdominal wall movements.

Oxygen Saturation

Pulse oximetry is the most common method utilized to measure Sao_2 (oxygen saturation) during pediatric OPSGs. Pulse oximetry may, however, be misleading in conditions like sickle cell anemia, since the pulse oximeter readings are based on software assumptions about normal hemoglobin structure. Pulse oximetry probes are now small and lightweight and can be successfully placed on toes or fingers. Many oximeters allow storage of data so that an overnight trend graphed by the hours of the study may be printed out. The computer may also calculate the percentage of time spent at different oxygen saturation levels. Several studies have been performed to assess the range of Sao_2 levels that are normal in infants and children. Overall, these studies suggest that any oxygen desaturations that occur outside the neonatal period are usually less than 10 seconds in duration and also usually represent less than a 4% drop, with Sao_2 levels overall typically more than 94% throughout sleep. Uriel and colleagues [22] found in their study of 70 normal children and adolescents that the mean Sao_2 was 97.2% ± 0.8% with a nadir of 94.6% ± 2.2%. It is recommended that the pulse oximetry signal be separately displayed from the OPSG to help in detection of artifact [17]. The technologists, in addition, should be trained to resite the pulse oximetry probe in the case of abnormal signals and also to document any repositioning of the probe on their worksheets.

Respiratory Effort

Several techniques exist to monitor respiratory effort, but the technique utilized must allow hypopneas to be distinguished as well as apneas and must allow a distinction between central and obstructive events. Respiratory inductive plesthysmography (RIP) can be helpful by allowing identification of out-of-phase movements of the chest and abdomen (i.e., paradoxical breathing), which may be identified in the absence of clear-cut events in an older child with subtle upper airway obstruction. Paradoxical breathing and disorganized breathing can be seen in infants normally, especially in REM sleep, but after 3 years of age they are not a common finding (see Figure 129.4). Although there are no clear diagnostic criteria for upper airway resistance syndrome, in our laboratory we look at paradoxical and disorganized breathing in a child over 3 years of age as being signs of obstructive sleep disordered breathing. It should be noted that both RIP and strain gauges allow semiquantitative assessment of respiratory movement and not of actual effort per se. This failure to assess effort may be misleading, since children may show very little chest expansion during obstructed breathing, resulting in an interpretation of central hypopneas, since airflow and chest/abdominal movement are simultaneously decreased [23]. Intercostal surface EMG may be helpful to estimate effort in this situation but by itself is not adequate. Piezo crystal belts may be utilized if an adequate RIP signal is difficult to maintain, as can be seen with some morbidly obese children. Regardless of the equipment utilized, it is critical that the technologists check belt placement periodically to assure accurate positioning. In a small child especially, one of the belts can easily slip so that both belts will read the same respiratory effort, causing obstructive events to look central. It is generally agreed that measurement of chest and abdominal wall movement only by impedance is inadequate due to poor sensitivity and specificity [19]. Assessment of changes in esophageal pressure as measured by esophageal manometry is the most precise reflection of respiratory effort but generally is not well tolerated or well accepted in most pediatric populations [24].

Airflow Assessment

A variety of methods are available to estimate respiratory airflow. However, the currently utilized technologies

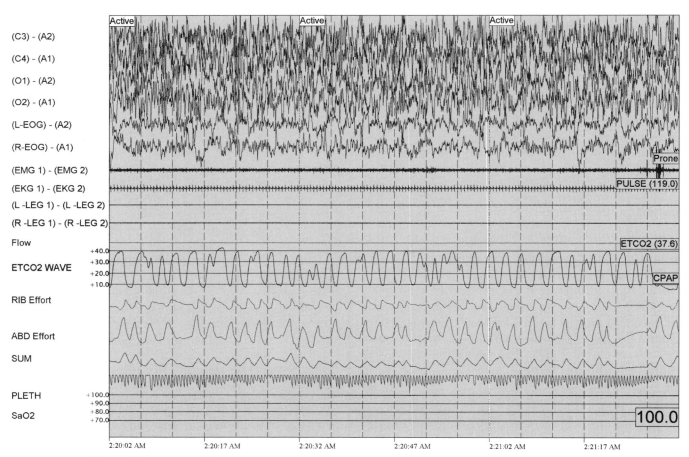

Figure 129.4 Normal disorganized breathing in 11 month old infant, active sleep, 90 seconds per page.

actually do not truly measure airflow, any more than an impedance belt measures effort. The current methods are at best semiquantitative and potentially fraught with error. The technologist must be vigilant in relation to placement of the probe or sensor throughout the night. Since children with obstructed sleep disordered breathing often mouth breath, both oral and nasal airflow must be assessed, and a nasal–oral thermistor or thermocouple may be utilized. Pneumotachnographs are less frequently used, since they commonly require that the patient sleep with a large face mask, which can be frightening to children on their first visit to the sleep laboratory. Recording of nasal end-tidal carbon dioxide ($ETCO_2$) values can increase accuracy of interpretation especially with RIP in infants and children with increased work of breathing, avoiding overinterpretation of events in infants where chest-wall asynchrony is normal and also avoiding underinterpretation of events in children as described in the above paragraph.

Assessment of Ventilation

Measurement of $ETCO_2$ by capnography is very sensitive to changes in CO_2 levels, but it can be technically daunting, especially when patients have an abundance of nasal secretions where patency of the nasal cannula, or sampling tube, is problematic. Because many children at least intermittently mouth breathe, a cannula by the mouth as well as by the nose is helpful. $ETCO_2$ is thought to be a reliable estimate of ventilation, reflecting a good estimate of arterial CO_2 measurements, at least in children without significant pulmonary disease. It has been suggested that transcutaneous CO_2 ($TCCO_2$) should be measured along with $ETCO_2$ to allow more accurate interpretation of hypoventilation, either its presence or its severity [25]. For example, in patients with chronic lung disease, $ETCO_2$ may underestimate $Paco_2$ because of the increased physiologic dead space seen with chronic lung disease. $TCCO_2$ recordings have the benefit of not being clogged by secretions, but may be unreliable in patients with poor perfusion, edema, or skin diseases. They also do not reflect rapid changes. Normative data for $ETCO_2$ values during sleep are limited. One study reported that no normal child had $ETCO_2$ values greater than 50 mmHg for more than 8% of total sleep time [26]. A recent study of 70 children and adolescents suggested that $ETCO_2$ should not be greater than 45 mmHg for more than 10% of sleep time [22].

OTHER CONSIDERATIONS

Many laboratories utilize low-light or infrared cameras and make a video recording of the patient throughout the sleep period. This video, which should also allow audio monitoring, should be available for review along with the polysomnogram data. Some computer programs allow digital audio-visual recording of the patient's sleep; however, unless the program allows the digital recording to be edited, archiving the data may be a problem, as the resulting file with overnight digital audio-visual recording is quite large. Currently, in our lab, the technologist digitally records selected samples of sleep, while at the same time a standard videocassette recording is made, which is then utilized for review as needed. Listening to and watching the child's respiratory pattern, for example, may be critical in some children when obstructive events appear central. Correlating snoring with arousals can be facilitated by time-locked audio-visual recording, or by devoting a channel on the polysomnogram to recording snoring. Correlation of OPSG data including EEG along with audio-visual recording is also critical for evaluating parasomnias and episodes thought to be seizures [27]. In addition to comments made directly on the tracing, we have found it helpful to have the technologist keep an hourly log throughout the study. On this log, the technologist should record behavioral observations and in addition should document Sao_2, $ETCO_2$, heart rate, respiratory rate, body position, snoring level, changes in settings such as the use of different filtering, and interventions undertaken such as institution of supplemental oxygen. In addition to these parameters, the technologist should monitor, assess, and record the child's respiratory patterns throughout the night as listed in Table 129.1. The data recorded by the technologist can serve as a validation of the digitalized data, helps the polysomnographer to evaluate the effect of any changes in the sleep pattern or interventions, and assists in quality control. In addition, both the child, if old enough, and the parents should be asked to rate the child's sleep in the laboratory to help assess whether or not the sleep sample recorded in the laboratory is subjectively reflective of the child's typical sleep in the home setting.

TABLE 129.1 Respiratory Patterns Rated by Polysomnography Technologists

Pattern	Yes/No		Percentage of TST			
Breathing sounds						
Snoring/snorting	y	n	25	50	75	100
Stridor	y	n	25	50	75	100
Gasping	y	n	25	50	75	100
Wheezing	y	n	25	50	75	100
Abnormal use of accessory muscles/retractions						
Suprasternal	y	n	25	50	75	100
Substernal	y	n	25	50	75	100

INTERPRETING AND REPORTING THE OPSG RESULTS

While computer programs may score an entire study quickly, it is imperative that if that capability is utilized, the study be completely reviewed and rescored as necessary by an experienced pediatric polysomnography technologist prior to physician review. The interpreting polysomnographer should then review all of the study. Parameters of sleep architecture which should be part of the analysis are detailed in Table 129.2. The effect of any medications that the child has taken at the time of the study should be taken into consideration as sleep architecture is reviewed. For example, a child who has taken clonidine may have limited REM sleep time. A close analysis of sleep patterns may support a diagnosis of more significant sleep disordered breathing in children than the respiratory disturbance index (RDI) itself may suggest. For example, the arousal index may be elevated, without an obvious scorable cause, while review of the patient's audio-videotape shows those arousals to be clearly associated with the patient's respiratory pattern. Sleep may be fragmented with reduced rapid eye movement sleep and multiple sleep stage shifts. Decreased REM sleep percentages have been reported to occur in infants with obstructive sleep disordered breathing. However, a more recent study found little difference in sleep architecture between 20 children with SDB and 10 controls, although children with SDB had significantly more arousals than did the control population [18]. Printing out a hypnogram, made possible by computerized sleep programs, can be very informative and especially useful in educating a family about their child's sleep disorders.

TABLE 129.2 Important Aspects of Sleep Times and Sleep Architecture

Time at lights off	Sleep stages:	Time in min.	TST%
Time at lights on	Stage 1	Time in min.	TST%
Total recording time	Stage 2	Time in min.	TST%
Total sleep period	Stage 3	Time in min.	TST%
Total sleep time	Stage 4	Time in min.	TST%
Sleep onset latency	REM	Time in min.	TST%
Number of REM periods	NREM	Time in min.	TST%
REM onset latency	Indeterminant	Time in min.	TST%
REM latency minus awake	Technical	Time in min.	TST%
Sleep efficiency	Movement	Time in min.	TST%
Number of awakenings			
Wakefulness after sleep onset			
Number of sleep stage shifts			
Arousal index			

The patient's lowest Sao_2 should be reported, as well as the percentages of sleep time spent with Sao_2 less than 95%, 90%, and 85%. The highest recorded $ETCO_2$ should be given in mmHg, along with the percentage of sleep time spent with $ETCO_2$ more than 50%. The average respiratory and heart rates as well as their ranges should all be included. Any irregularity in the cardiac rhythm should be noted along with any association with respiratory events. Periodic limb movements of sleep as well as associated arousals should be totaled, and indices calculated for both. The total arousal index should be further divided into those related to PLMS, those that are respiratory in nature, and those that appear to be spontaneous.

Further respiratory analysis should focus on the number of apneas and hypopneas and a breakdown of these based on length and type, that is, central, obstructive, or mixed (see Figure 129.5). RDIs should be calculated for the total study, REM sleep alone, and the position in which the child slept, that is, either prone, supine, or side lying. All of these data are best presented in a tabulated fashion. Snoring should be noted as present or absent, with some laboratories quantifying snoring by the estimated percentage of sleep time spent snoring or as mild to moderate or severe. There are, however, no standards for these evaluations. Technologists' comments about the nature of the child's respiratory pattern and observations from the videotapes are also critical, including such patterns as mouth breathing, hyperextension of the neck during sleep, retractions, and the use of accessory muscles. Paradoxical or disorganized breathing should be noted, with the caveat in mind that these patterns can be normal in children less than 3 years of age.

Although reporting raw data and calculated indices is very important, the most important part of the polysomnogram report lies in the clinical interpretation and the recommendations that result from that interpretation. It is here where the pediatric polysomnographer often walks on uncertain turf: there truly is a dearth of large, well-controlled studies covering many aspects of pediatric sleep medicine [1]. For example, what length of respiratory events should be reported? All events greater than 10 seconds, or those greater than 2 times the length of the child's respiratory cycle in a child breathing 60 breaths per minute [23]? How important are arousals, and what arousal index is significant in a 2 month old versus a 6 year old? What are truly normal ranges for RDIs in children versus adults? What levels of apnea or hypopnea indices are *clinically* relevant? Must one score only those respiratory events associated with a drop in oxygen saturation, or can a respiratory event be scored if only associated with an arousal? How does one diagnose upper airway resistance syndrome without esophageal manometry? When are periodic limb movements of sleep significant, and are they independently important or reflective of another disorder? These are only a

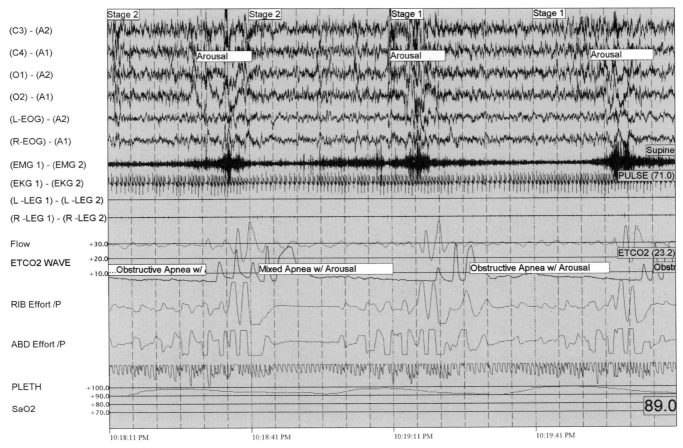

Figure 129.5   Repetitive events with obstruction at 120 seconds per page.

few of the questions about pediatric sleep for which further study is important.

Despite these uncertainties, the pediatric polysomnographer can utilize the available normative data for oxygen saturations and ventilation as reported earlier. Determining which events to score, however, is more complex. The duration of scorable events may need to be shortened from the adult standard of 10 seconds, and also the standard adult norms of scoring events associated with an arousal and a desaturation of greater than 4% may also be inappropriate in some children (see Figure 129.6). Studies have noted that SBD in children is more commonly associated with partial airway obstructions, or hypopneas, than with frank apneas, but it should be noted that the spectrum of sleep disordered breathing in children is wide, including children with obstructive hypoventilation syndrome with no complete obstructions and few scorable events to children with only complete obstructions, something which we rarely see in our laboratory [23]. Scoring of events must be more liberal and should not be limited to those greater than 10 seconds, not just in infants with rapid respiratory rates but even in an older child if those events are associated with oxygen desaturations and arousals [23]. Special interpretive challenges

are posed by children with neuromuscular or neurological diseases, whose neuromotor control of respiratory muscles may be abnormal. Not all researchers have included oxygen desaturations as a necessary feature of SDB in children [28]. In Mauer's study, only 4 of 14 patients with obstructive sleep disordered breathing had a greater than 4% decrease in Sao_2 [28]. Some children may have repetitive events without oxygen desaturations, whereas others may desaturate profoundly with even very short central or obstructive events, particularly children with marginal reserve such as a 2 year old who was an extremely premature newborn who now has chronic lung disease. Recent standards for pediatric polysomnography define an hypopnea as a "50% or greater decrease in the amplitude of the nasal/oral airflow signal, *often* accompanied by hypoxemia or arousal" [17]. Thus, while adult scoring criteria for respiratory events may form a basis for beginning to score and interpret the pediatric polysomnogram, there are often times when those criteria are not applicable and where their use would result in underdiagnosis of significant sleep disordered breathing (see Figure 129.6). There are no widely accepted pediatric equivalents for the standardized adult criteria utilized to score and interpret OPSGs.

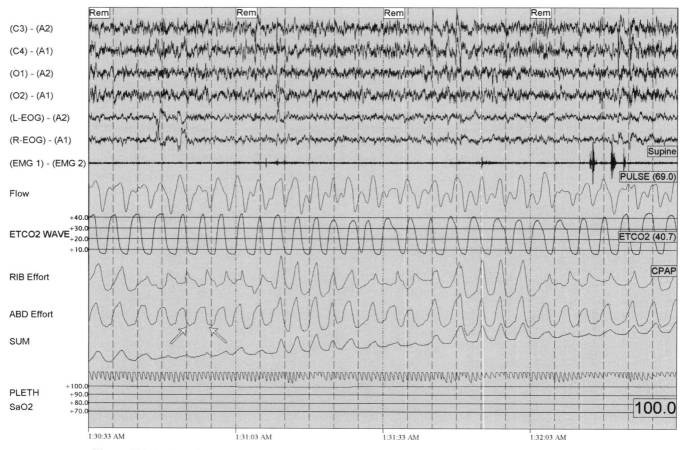

Figure 129.6 Repetitive subtle events with paradox in 14 year old teenager at 120 seconds per page.

Standardization of definitions utilized in pediatric polysomnography would be a tremendous advance.

Whereas periodic limb movements of sleep have been recognized as disrupting sleep in adults and contributing to excessive daytime somnolence, less attention had been directed toward PLMS in children until the association of PLMS with attention deficit hyperactivity disorder (ADHD) became clearer. SDB was shown in a population of 110 children to be associated with PLMS and hyperactivity [9], and PLMS have also been associated with AHDH in children without SDB. In a study by Picchietti and colleagues [13], 9 of 14 children with ADHD had a limb movement index of greater than 5, whereas none of 10 control children displayed PLMS. However, the clear relationship of PLMS, with or without associated arousals, to clinical symptoms in children remains uncertain and is yet another area where further research is warranted.

Scoring the pediatric polysomnogram should be performed by a technologist especially trained in pediatric interpretation and confirmed by a pediatric polysomnographer with careful attention as well to the child's respiratory pattern and behavior as recorded on the audio-videotape. The pediatric polysomnographer should also include in the OPSG report a reflection by the patient's parents as to whether the child's sleep in the laboratory was characteristic of the child's sleep in the home setting. If it were markedly different, a repeat study may be indicated. Labeling a study as normal or abnormal is complicated by the dearth of truly normal data, but a summary of the child's sleep architecture can be presented, followed by a discussion of the child's respiratory pattern both as scored and as seen on their audio-videotape and also as associated with oxygen saturations and ETCO$_2$ data. Our current norms for Sao$_2$ and ETCO$_2$ are statistical norms and are not norms based on how these OPSG measures relate to adverse clinical outcomes. Therefore it is critical to relate the results of the study to the child's clinical picture and symptoms. Other findings such as PLMS, movements compatible with parasomnias, or EEG abnormalities can then be discussed. A summary of the patient's sleep disorder and the recommendations of the polysomnographer for further workup, treatment, and follow-up conclude the report.

The most important aspect of scoring and interpretation of the pediatric polysomnogram is that they both require clinical judgment based on knowledge of pediatric sleep and respiratory patterns. Changes in parameters of sleep as children grow need to be appreciated, and the report should then be based on analysis that is age-specific and not bound by the rules of adult polysomnography.

REFERENCES

1. Marcus CL, England S, Annett RD, Brooks LJ, et al. Cardiorespiratory sleep studies in children: establishment of normative data and polysomnographic predictors of morbidity. *Am J Respir Crit Care Med* **160**:1381–1387(1999).

2. Goldstein NA, Sculerati N, Walsleben JA, et al. Clinical diagnosis of pediatric obstructive sleep apnea validated by polysomnography. *Otolaryngol Head Neck Surg* **111**:611–617(1994).

3. Carroll JL, McColley SA, Marcus CL, Curtis S, Loughlin GM. Inability of clinical history to distinguish primary snoring from obstructive sleep apnea syndrome in children. *Chest* **108**:610–618(1995).

4. Wang RC, Elkins TP, Keech D, Wauquier A, Hubbard D. *Otolaryngol Head Neck Surg* **118**:69–73(1998).

5. Rimell FL, Rosen G, Garcia J. Full polysomnographic evaluation of the infant airway. *Arch Otolaryngol Head Neck Surg* **124**:773–776(1998).

6. Reiterer F, Fox WW. Multichannel polysomnographic recording for evaluation of infant apnea. *Clin Perinatol* **19**:8871–8889(1992).

7. Dyken ME, Lin-Dyken DC, Poulton S, Zimmerman MB, Sedars E. Prospective polysomnographic analysis of obstructive sleep apnea in Down syndrome. *Arch Pediatr Adolesc Med* **157**:655–660(2003).

8. Ali NJ, Pitson D, Stradling JR. Natural history of snoring and related behaviour problems between the ages of 4 and 7 years. *Arch Dis Child* **71**:74–76(1994).

9. Chervin RD, Dillon JE, Bassetti C. Ganoczy DA, Pituch KJ. Symptoms of sleep disorders, inattention, and hyperactivity in children. *Sleep* **20**:1185–1192(1997).

10. Goldstein NA, Post C, Rosenfeld RM, Campbell TF. Impact of tonsillectomy and adenoidectomy on child behavior. *Arch Otolaryngol Head Neck Surg* **126**:494–498(2000).

11. Crabtree VM, Varni JW, Gozal D. Health-related quality of life and depressive symptoms in children with suspected sleep-disordered breathing. *Sleep* **27**:1131–1138(2004).

12. O'Brien LM, Mervis CB, Holbrook CR, Bruner JL, Klaus CJ, Rutherfold J, Raffield TJ, Gozal D. Neurobehavioral implications of habitual snoring in children. *Pediatrics* **114**:44–49(2004).

13. Picchietti DL, Underwood DJ, Farris WA, et al. Further studies on periodic limb movement disorder and restless leg syndrome in children with attention-deficit hyperactivity disorder. *Mov Disord* **6**:1000–1007(1999).

14. Marcus CL, Omlin KJ, Basinski DJ, Bailey SL, et al. Normal polysomnographic values for children and adolescents. *Am Rev Respir Dis* **146**:1235–1239(1992).

15. Saeed MM, Keens TG, Stabile MW, Bolokowicz J, Davidson-Ward SL. Should children with suspected obstructive sleep apnea syndrome and normal nap sleep studies have overnight sleep studies? *Chest* **188**:360–365(2000).

16. Katz ES, Greeene MG, Carson KA, Galster P, Loughlin GM, Carroll JL, Marcus CL. Night-to-night variability of polysomnography in children with suspected obstructive sleep apnea. *J Pediatr* **140**:589–594(2002).

17. Loughlin GM, Brouillette RT, Brook LJ, Carroll JL, et al. Standards and indications for cardiopulmonary sleep studies in children. *Am J Respir Crit Care Med* **153**:866–878(1996).

18. Goh YT, Galster P, Marcus CL. Sleep architecture and respiratory disturbances in children with obstructive sleep apnea. *Am J Respir Crit Care Med* **162**:682–686(2000).

19. Sheldon SH. *Evaluation of Sleep in Infants and Children.* Lippincott-Raven, Philadelphia, 1996, pp 99–111.

20. Rechtschaffen A, Kales A (Eds). *A Manual of Standardized Terminology, Techniques and Scoring System for Sleep Stages of Human Subjects.* UCLA Brain Information Service, NINCS Neurological Information Network, Los Angeles, 1968.

21. Sleep Disorders Task Force. EEG arousals: scoring rules and examples. *Sleep* **12**:174–183(1992).

22. Uriel S, Tauman R, Greenfeld M, Sivan Y. Normal polysomnographic respiratory values in children and adolescents. *Chest* **125**:872–878(2004).

23. Carroll JL, Loughlin GM. Obstructive sleep apnea syndrome in children; diagnosis and management. In: Ferber R, Kryger M (Eds), *Principles and Practice of Sleep Medicine in Children.* Saunders, Philadelphia, 1995, pp 193–216.

24. Guilleminault C, Stoohs R, Clerk A, Cetel M, Maistros P. A cause of excessive daytime sleepiness: the upper airway resistance syndrome. *Chest* **104**:781–787(1993).

25. Morielli A, DesJardins D, Brouillette RT. Transcutaneous and end-tidal carbon dioxide pressures should be measured during pediatric polysomnography. *Am Rev Respir Dis* **148**:1599–1604(1993).

26. Marcus CL, Keens TG, Davidson-Ward SL. Comparison of nap and overnight polysomnography in children. *Pediatr Pulmonol* **13**:16–21(1992).

27. Dyken ME, Yamada T, Lin-Dyken DC. Polysomnographic assessment of spells in sleep: nocturnal seizures versus parasomnias. *Semin Neurol* **21**:377–390(2001).

28. Mauer KW, Staats BA, Olsen KD. Upper airway obstruction and disordered nocturnal breathing in children. *Mayo Clin Proc* **58**:349(1983).

130

INTRODUCTION TO SLEEP ELECTROENCEPHALOGRAPHY

SELIM R. BENBADIS

University of South Florida College of Medicine, Tampa, Florida

INTRODUCTION

Basic Principles

EEG records synaptic potentials from pyramidal cells. It is critical to remember that there is no absolute voltage measurement in clinical EEG. Rather, an EEG trace records potential *differences* between two electrodes. By convention of the amplifiers used in clinical EEG, an upgoing deflection indicates that input 1 (grid 1) is more negative (or less positive) than input 2 (grid 2). This is an arbitrary but critical rule (Figure 130.1). Thus statements like "positivity is up" or "negativity is up" make no sense unless it is stated whether the positivity is at grid 1 or grid 2.

It should be emphasized that the way recordings are displayed can be modified at will with digital acquisition, and this "post-hoc" reformatting is one of the major advantages of modern digital systems.

Types of EEG in Clinical Practice

Routine EEG is typically a brief recording of 20–30 min. The main limitation with EEG is its poor sensitivity for epilepsy. The generally accepted numbers are that the yield of a single routine EEG in epilepsy is ~50% and increases with repeated EEG recordings to reach about 80% by the third recording [1]. For practical purposes, if a diagnosis of epilepsy is strongly suspected clinically, and EEG confirmation or more precise diagnosis is needed, other options should be used. The two options are ambulatory EEG and prolonged EEG-video monitoring.

Ambulatory EEG is to the brain what Holter monitoring is to the heart. Here the patient is hooked up to the EEG and goes home with the intent of recording a seizure or an episode [2]. Ambulatory EEG can occasionally be performed with video, and this option is now emerging.

EEG-video monitoring is the highest level of epilepsy monitoring and the gold standard. This is the basic activity of comprehensive epilepsy centers [3] and certainly the starting point when drugs fail to control seizures [4]. There is no strict "cutoff" for *when* EEG-video monitoring is indicated, but some guidelines state that referral to a specialized epilepsy center is appropriate if seizure control is not achieved within 9 months [5]. As a general rule, prolonged EEG-video monitoring should be obtained on any patient who continues to have seizures frequently (1/week) despite antiepileptic drugs [4]. In the vast majority of situations, this allows one to confirm the diagnosis of epilepsy or to rectify a wrong diagnosis of epilepsy. If epilepsy is confirmed, it is then usually possible to (1) determine whether it is localization-related or generalized; (2) distinguish, among generalized epilepsies, between the "idiopathic" type and the symptomatic (cause known) or cryptogenic (caused unknown); and (3) differentiate, among localization-related epilepsies, between mesiotemporal and extratemporal/neocortical epilepsy. Based on this precise classification of the epilepsy syndrome, treatment options can then be examined. Invasive EEG is

Sleep: A Comprehensive Handbook, Edited by T. Lee-Chiong.
Copyright © 2006 John Wiley & Sons, Inc.

Polarity Convention

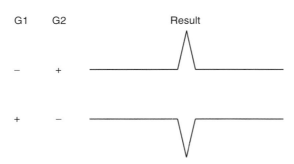

Figure 130.1   Polarity convention. This is an arbitrary but critical rule. By convention of the amplifiers used in clinical EEG, an upgoing deflection indicates that input 1 (grid 1 or G1) is more negative (or less positive) than input 2 (grid 2 or G2). Thus statements like "positivity is up" or "negativity is up" make no sense unless it is stated whether the positivity is at grid 1 or grid 2.

limited to specialized surgical epilepsy centers and is beyond the scope of this chapter (for review, see [6, 7]).

Technical Aspects

In human clinical EEG, electrodes are placed according to a standard system known as the 10-20 system (Figure 130.2). It uses four anatomical landmarks (nasion, inion, and the two preauricular points) from which measurements are made and electrodes are placed at 10% or 20% of the distances.

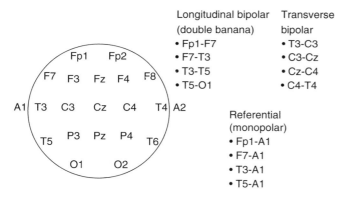

Figure 130.2   The 10–20 system of electrode placement. This uses four anatomical landmarks (nasion, inion, and the two preauricular points) from which measurements are made and electrodes are placed at 10% or 20% of the distances. Letters refer (grossly) to the lobe (i.e., Fp, frontopolar; F, frontal; T, temporal; P, parietal; O, occipital). Odd-numbered electrodes are on the left, even-numbered electrodes on the right, and midline electrodes are designated as "z."

There are two types of montages: bipolar and referential. In a *bipolar* montage, each electrode is linked to the next along a chain (i.e., A–B, B–C, C–D, D–E). The typical longitudinal or anteroposterior bipolar montage is often referred to as a "double banana." Another common bipolar montage is the transverse (from left to right across the head). In a *referential* (or monopolar) montage, each electrode is linked (compared) to a common reference. A helpful analogy is that measuring voltage fields with electrodes is akin to measuring mountain peaks/altitudes with surveyors [8] (Figure 130.3). In terms of localization, maximum voltage (altitude) is indicated by a phase reversal on bipolar montages, or by maximum amplitude on referential montages (Figure 130.3). Contrary to a common misconception, phase reversals are *not at all indicative of an abnormality*, and in fact have nothing to do with the nature of a voltage field (i.e., *what* the discharge is). Instead, phase reversals indicate the maximum (negativity or positivity)

Polarity and phase reversals

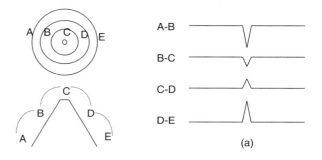

Referential

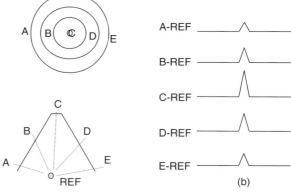

Figure 130.3   Measuring a voltage field is analogous to measuring the altitude of a mountain. (a) On a bipolar montage, a maximum peak (voltage) is indicated by a phase reversal. (b) On a referential montage, a maximum peak (voltage) is indicated by the highest amplitude.

and have everything to do with the location of the discharge (i.e., *where* the discharge is).

The display on EEG typically uses a "paper speed" (obsolete concept with digital recordings) of 3 cm/s (in the United States), that is, 3 times faster than the PSG "paper speed" of 1 cm/s. By convention, in the United States, montages are usually displayed "left over right."

Filters and their principles are similar to those used in polysomnography.

In clinical neurophysiologey (e.g., EEG, PSG), waveforms and discharges are described and characterized by the following parameters:

- Amplitude: how high the voltage is (in μV).
- Duration: how long the discharge is (in ms or seconds).
- Frequency: how frequently a waveform repeats itself (in cycles per seconds or hertz). The frequency is the reciprocal of the duration (e.g., a duration on 200 ms is the same as a frequency 5 Hz.)
- Morphology: the shape and configuration of the discharge (this is qualitative).
- Latency: the delay between an arbitrary event (e.g., stimulus) and another event (in ms).
- Location: where the discharge is.
- Reactivity: what affects the discharge.

ARTIFACTS

Although EEG is designed to record cerebral activity, it also records electrical activities arising from sites other than the brain. The recorded activity that is not of cerebral origin is termed artifact and can be conveniently divided into physiologic and extraphysiologic artifacts. *Physiologic* artifacts are generated by the body but arise from sources other than the brain. *Extraphysiologic* artifacts arise from outside the body (i.e., equipment, environment).

Physiologic Artifacts

Muscle (Electromyogram) Activity Myogenic potentials are probably the most common artifacts and are seen on virtually all EEG performed in clinical practice. Frontalis and temporalis muscles (e.g., clenching of jaw muscles) are particularly common. As a general rule, the potentials generated in the muscles are of shorter duration than those generated in the brain and are identified easily on the basis of duration, morphology, and high frequency of 50–100 Hz (Figures 130.4 and 130.10]. A particular type of muscle artifact is chewing, characterized by rhythmic bursts of muscle maximum in the temporal chains (Figure 130.5).

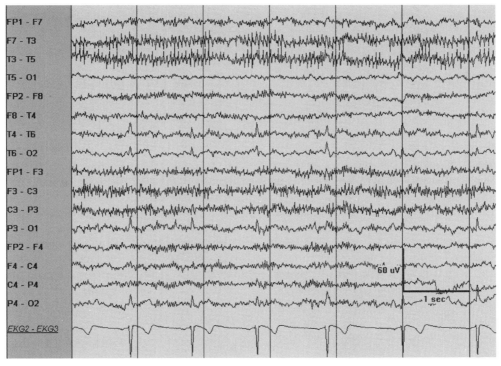

Figure 130.4 Muscle (EMG) and EKG artifact. Muscle artifact is seen as high-amplitude, very fast (>50 Hz), and variable frequency discharges, which are more prominent on the left in this sample. EKG is easily identified in several channels as clearly simultaneous to the EKG channel.

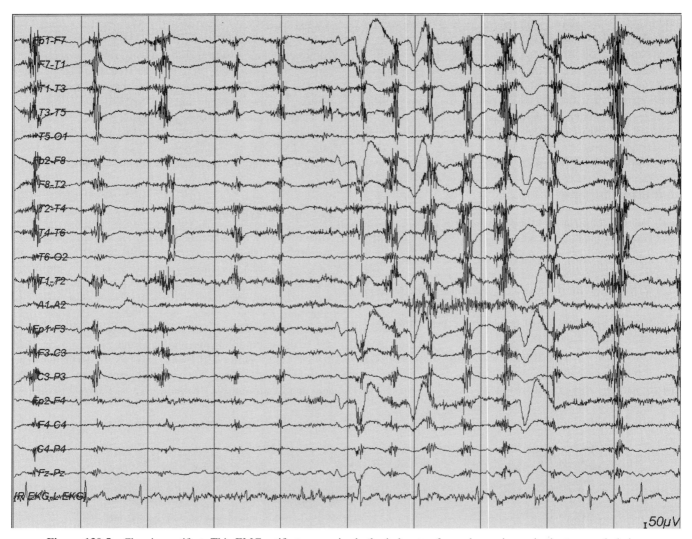

Figure 130.5   Chewing artifact. This EMG artifact occurs in rhythmic bursts of muscle maximum in the temporal chains.

Glossokinetic Artifact   In addition to muscle activity, the tongue (like the eyeball) functions as a dipole, with the tip negative with respect to the base. In this case the tip of the tongue is the most important part because it is more mobile. The artifact produced by the tongue has a broad potential field that drops from frontal to occipital areas, although it is less steep than that produced by eye movement artifacts. The amplitude of the potentials is greater inferiorly than in parasagittal regions; the frequency is variable but usually in the delta range and occurs synchronously when the patient says "Lah-lah-lah-lah," which can be verified by the technologist. Chewing and sucking (pacifier) can produce similar artifacts.

EKG Artifact   Some individual variations in the amount and persistence of EKG artifact are related to the field of the heart potentials over the surface of the scalp. Generally, subjects with short and wide necks have the largest EKG artifacts on their EEG recordings. The voltage and apparent surface of the artifact vary from derivation to derivation and, consequently, from montage to montage. The artifact is observed best in referential montages using ear electrodes (A1 and A2). EKG artifact is recognized easily by its rhythmicity and coincidence with the EKG tracing (Figure 130.4). The situation becomes difficult when abnormal cerebral activity (e.g., sharp or slow waves) appears intermixed with EEG artifact.

Pulse   Pulse artifact occurs when an EEG electrode is placed over a pulsating vessel. The pulsation can cause slow waves that may simulate EEG activity. A direct relationship exists between EKG and the pulse waves. The QRS complex (i.e., electrical component of the heart contraction) happens slightly ahead of the pulse waves 200–300 msec delay after EKG.

Eye Movements   Eye movements are observed on all EEG recordings and are useful in identifying sleep stages. The eyeball acts as a dipole with a positive pole anteriorly (cornea) and a negative pole posteriorly (retina). When the globe rotates about its axis, it generates a large-amplitude alternate current field, which is detectable by any electrodes near the eye. The other source of artifacts comes from EMG potentials from muscles in and around the orbit. Vertical eye movements typically are observed with blinks (i.e., Bell's phenomenon). A blink causes the positive pole (i.e., cornea) to move closer to frontopolar (Fp1–Fp2)

electrodes, producing symmetric downward deflections. During downward eye movement the positive pole (i.e., cornea) of the globe moves away from frontopolar electrodes, producing an upward deflection best recorded in channels 1 and 5 in the bipolar longitudinal montage. Lateral eye movements affect lateral frontal electrodes F7 and F8 (which are just about where "eye electrodes" of the PSG would be). During a left lateral eye movement, the positive pole of the globe moves toward F7 and away from F8. Using a bipolar longitudinal montage, there is a maximum positivity in electrode F7 and maximum negativity simultaneously in electrode F8 (Figure 130.6). A so-called lateral rectus spike (Figure 130.7) may be present in electrode F7. With right lateral eye movement, the opposite occurs.

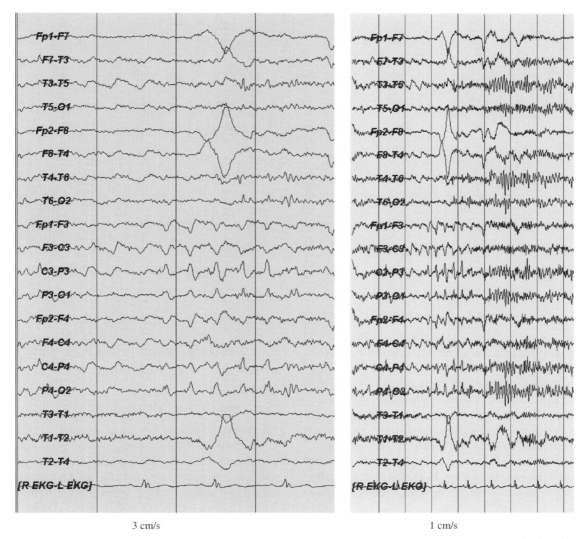

3 cm/s   1 cm/s

Figure 130.6   Eye movements and sawtooth waves in REM sleep. During this rapid lateral eye movement to the left, there is a maximum positivity in electrode F7 and maximum negativity simultaneously in electrode F8. In the vicinity of the rapid eye movement, there are typical sawtooth waves in the central region (C3 and C4).

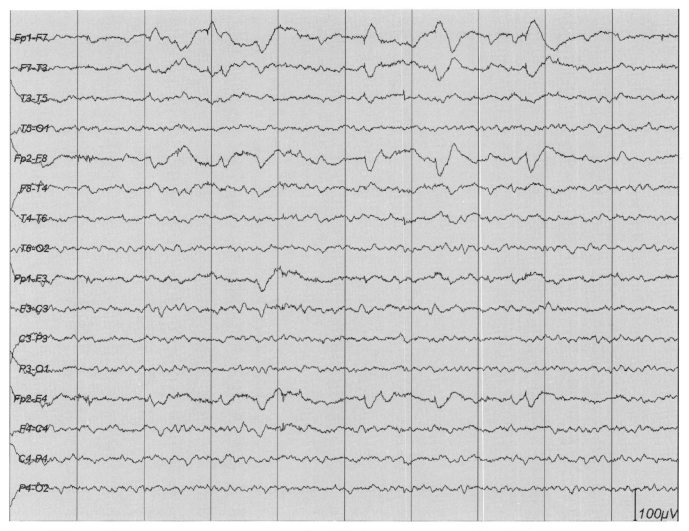

Figure 130.7   Rapid eye movements and lateral rectus spikes. With the eye movements to the left, lateral rectus spikes are seen in electrode F7.

Respiration Artifacts   Respiration can produce two kinds of artifacts. One type is in the form of slow and rhythmic activity, synchronous with the body movements of respiration and mechanically affecting the impedance of (usually) one electrode. The other type can be slow or sharp waves that occur synchronously with inhalation or exhalation and involve those electrodes on which the patient is lying. Several commercially available devices to monitor respiration can be coupled to the EEG machine. As with the EKG, one channel can be dedicated to respiratory movements.

Skin Artifacts   Biological processes may alter impedance and cause artifacts. *Sweat* is a common cause (Figure 130.8). Sodium chloride and lactic acid from sweating

react with metals of the electrodes and produce large and very slow (usually ~ 0.5 Hz) baseline sways.

Extraphysiologic Artifacts

Electrode Artifacts   The most common electrode artifact is the electrode "pop." Morphologically this appears as single or multiple sharp waveforms due to abrupt impedance change. It is identified easily by its characteristic appearance (i.e., abrupt vertical transient that does not modify the background activity) and its usual distribution, which is limited to a single electrode (Figure 130.9). In general, sharp transients that occur at a single electrode (i.e., no field) should be considered artifacts until proved otherwise. At other times, the impedance change is less abrupt,

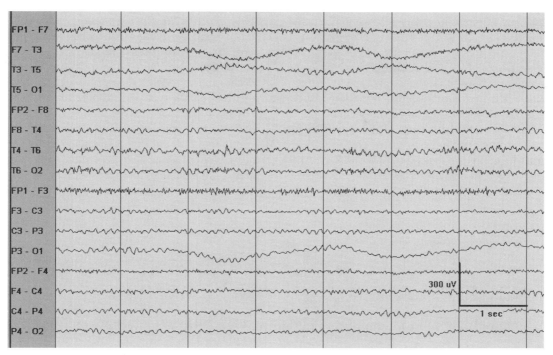

Figure 130.8 Sweat artifact. Note the very slow (0.5 Hz) sways. The slow frequency is similar to slow rolling eye movements, but the distribution is not (in this case it affects electrodes on the left side of the head).

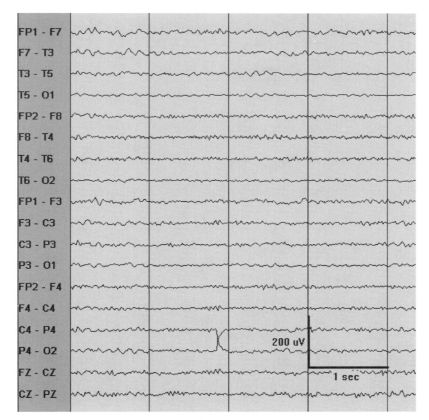

Figure 130.9 Electrode artifact. This typical electrode artifact is the electrode "pop." Note the single sharp waveform with abrupt vertical transient that does not modify the background activity, and its distribution, which is limited to a single electrode (P4).

and the artifact may mimic a low-voltage arrhythmic delta wave.

Alternating Current (60 Hz) Artifact   Adequate grounding on the patient has almost eliminated this type of artifact from power lines. The problem arises when the impedance of one of the active electrodes becomes significantly large between the electrodes and the ground of the amplifier. In this situation, the ground becomes an active electrode that, depending on its location, produces the 60 Hz artifact. The artifact has the exact frequency of 60 Hz and is easily identified by increasing the time base (Figure 130.10).

Movements in the Environment   Movement of other persons around the patient can generate artifacts, usually of capacitive or electrostatic origin. The artifact produced by respirators varies widely in morphology and frequency. Monitoring the ventilator rate in a separate channel helps to identify this type of artifact. Interference from high-frequency radiation from radio, TV, hospital paging systems, and other electronic devices can overload EEG amplifiers. The cutting or coagulating electrode used in the operating room also generates high-voltage high-frequency signals that interfere with the recording system. Touching or hitting electrodes can produce odd waveforms,

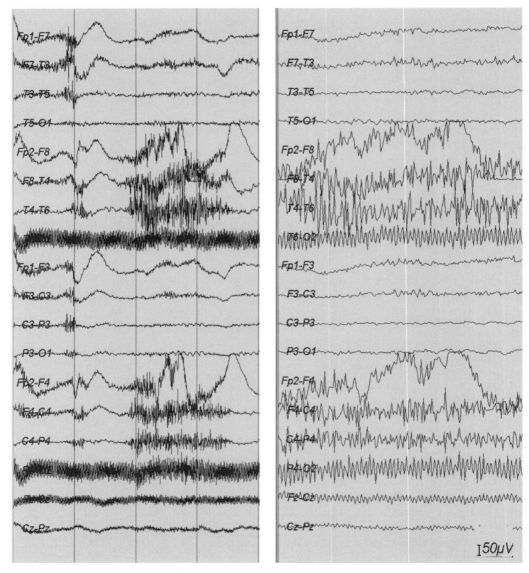

Figure 130.10   The 60 Hz and muscle (EMG) artifact. The 60 Hz artifact is at electrode O2 (thus seen in channels T6–O2 and P4–O2). Note that its amplitude is perfectly regular and exactly and steadily at 60 Hz. EMG (muscle) artifact, by contrast, is of comparable frequency but affects several electrodes, is variable in amplitude, and variable in frequency (not always nor exactly at 60 Hz).

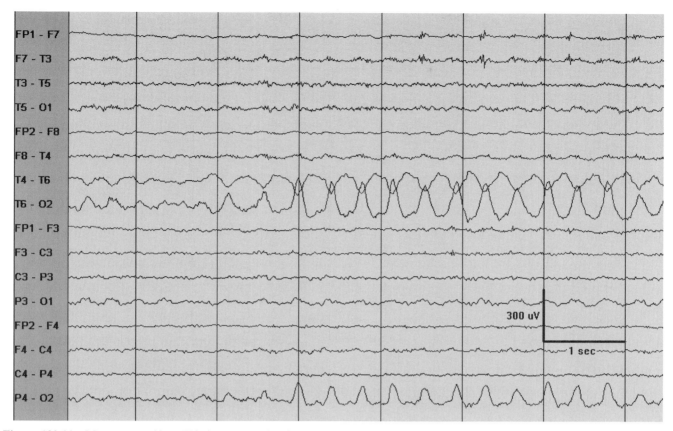

Figure 130.11 Movement artifact. This is an example of rhythmic artifact generated by repetitive movements (shaking the head on the pillow hitting the right posterior electrodes T6 and O2). This should not be mistaken for a focal seizure. Note the "phase reversal" at T6, even though this is an artifact.

and, for example, repetitive head movements can produce rhythmic artifacts (Figure 130.11).

Photic Stimulation

Photic stimulation is performed during routine EEG recordings and can produce some physiologic and some extraphysiologic artifacts, so it is described here separately.

Photic driving is a normal physiologic response, which is actually a visual evoked potential. It is the occipital cortex response to flashing lights and is thus seen in the occipital region (electrodes O1 and O2). It is easily identified because it is "time-locked" with (same frequency as) the strobe light (Figure 130.12).

The *photomyoclonic response* is a special type of EMG artifact that occurs during intermittent photic stimulation. Some subjects contract the frontalis and orbicularis muscles. This superficially resembles normal photic driving but is frontal (Figure 130.13), whereas normal photic driving is occipital. As can be shown by spreading the time base, these contractions occur approximately 50–60 m after each flash.

A *photocell (photoelectric) artifact* can also be seen with photic stimulation. This affects one electrode (Figure 130.14) and is easily identified as it disappears if one blocks the light from the electrode in question.

THE NORMAL EEG

Common Patterns of Wakefulness

Alpha Rhythm   The alpha rhythm (Figure 130.15) is typically what EEG readers identify first. The normal alpha rhythm has the following characteristics:

- Frequency of 8–12 Hz: Lower limit of normal generally accepted in adults and children older than 8 years is 8 Hz.
- Location: Posterior dominant; occasionally, the maximum may be a little more anterior, and it may be more widespread.
- Morphology: Rhythmic, regular, and waxing and waning.

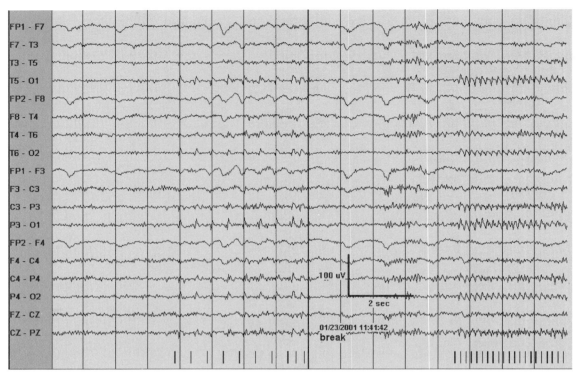

Figure 130.12   Normal photic driving. Bioccipital rhythmic activity time-locked with (same frequency as) the strobe light (the flash frequency is shown as lines at the bottom). Usually, photic driving is seen at several frequencies, such as shown here. Note that only location differentiates this from (frontal) photomyoclonic (or photomyogenic) response (Figure 130.13).

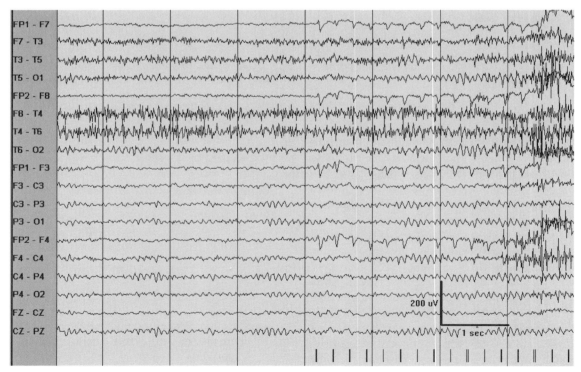

Figure 130.13   Photomyoclonic (or photomyogenic) response. Bifrontal rhythmic activity time-locked with (same frequency as) the strobe light (the flash frequency is shown as lines at the bottom). Note that only location differentiates this from (occipital) photic driving (Figure 130.12).

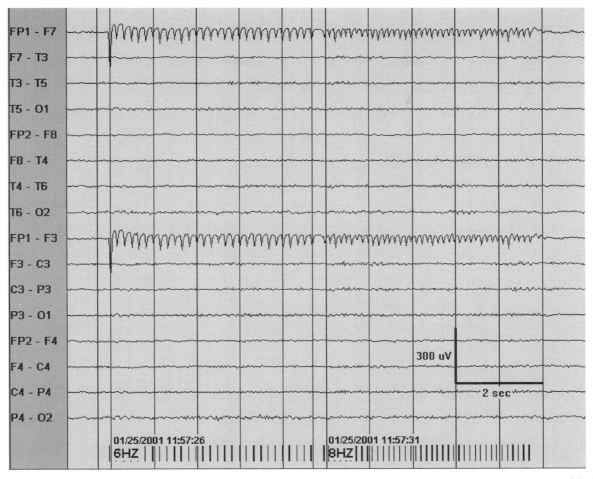

Figure 130.14 Photocell (photoelectric) artifact. This is produced at one single electrode (usually frontal rather than occipital) and is identified by being time-locked with the strobe. It would also disappear if one shielded the electrode in question from the light.

- Amplitude: Generally 20–100 μV.
- Reactivity: Best seen with eyes closed; attenuates with eye opening.

Occasionally the alpha rhythm is of very low amplitude or even not identifiable. This is not in itself abnormal. In addition to amplitude, other characteristics can vary somewhat without being abnormal, including morphology (e.g., spiky), distribution (e.g., widespread), and harmonic frequency (e.g., slow or fast alpha variant).

Beta Activity Normal beta activity (Figure 130.16) has the following characteristics:

- Frequency (by definition) greater than 13 Hz, typically 18–25 Hz.
- Location: Mostly frontocentral but somewhat variable; some describe various types according to location and reactivity—generalized, precentral, and posterior.
- Morphology: Usually rhythmic, waxing and waning, and symmetric.
- Amplitude: Usually in the range of 5–20 μV.
- Reactivity: Most common 18–25 Hz beta activity enhanced during stages 1 and 2 sleep and tends to decrease during deeper sleep stages; central beta activity may be reactive (attenuates) to voluntary movements and proprioceptive stimuli; in infants older than 6 months, onset of sleep marked by increased beta activity in central and postcentral regions.

In healthy individuals, beta activity commonly can be mildly different (<35%) in amplitude between the two hemispheres, which may be caused by differences in skull thickness. The amount and voltage of beta activity is enhanced by commonly used sedative medications (benzodiazepines, barbiturates).

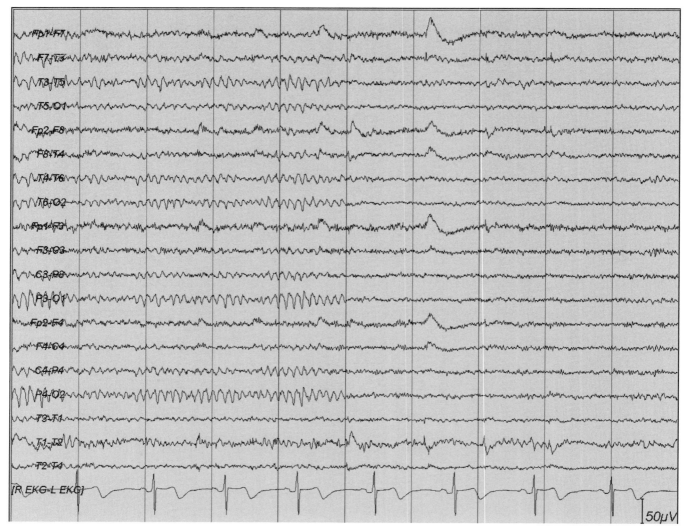

Figure 130.15   Normal alpha rhythm. This sample depicts the typical well-formed sinusoidal rhythmic activity at 9 Hz in the occipital regions.

Mu Rhythm   Characteristics of the mu rhythm (Figure 130.17) are as follows:

- Frequency of 7–11 Hz: Generally in alpha frequency band (8–12 Hz).
- Location: Centroparietal area.
- Morphology: Arch-like shape or like an "m"; most often asymmetric and asynchronous between the two sides and may be unilateral.
- Amplitude: Generally low to medium and comparable to that of the alpha rhythm.
- Reactivity: mu rhythm attenuates with contralateral extremity movement, the thought of a movement, or tactile stimulation.

Asymmetry, unilaterality, or asynchrony of the mu rhythm is not abnormal unless associated with other abnormalities. Very high voltage mu activity may be recorded in the central regions over skull defects and may become sharp in configuration and thus can be mistaken for epileptiform discharges. When mu rhythm is detected in an EEG, it should be verified by testing its reactivity.

Common Patterns of Sleep

Sleep Architecture   Sleep generally is divided in two broad types: non-rapid eye movement (NREM) and rapid eye movement (REM) sleep. On the basis of EEG changes, NREM sleep is divided further into four stages (stage 1, stage 2, stage 3, stage 4). REM sleep is defined not only

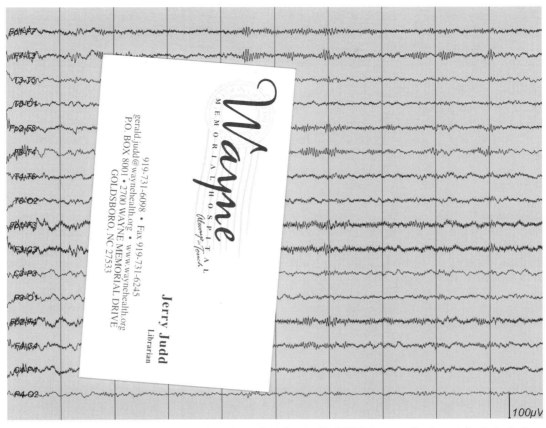

Figure 130.16   Normal beta activity. This sample depicts the typical fast (~18–25 Hz) low-amplitude activity in both frontal regions. As shown here, normal amounts of beta activity are moderate and tend to wax and wane.

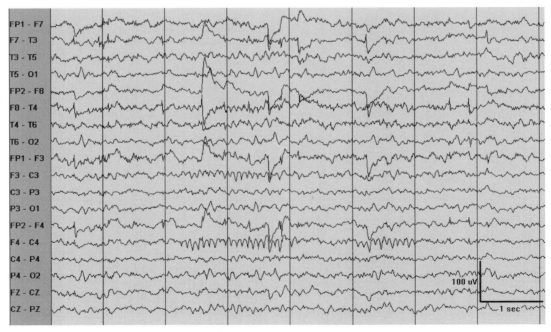

Figure 130.17   Mu rhythm. This sample depicts the typical m-shaped bicentral bursts at a frequency of 8–12 Hz. If tested, this would react (attenuate) to contralateral movement.

by EEG, but also by EMG and eye movements. NREM and REM sleep occur in alternating cycles, each lasting approximately 90–100 minutes, with a total of 4–6 cycles. In general, in the healthy young adult NREM sleep accounts for 75–90% of sleep time (3–5% stage 1, 50–60% stage 2, and 10–20% stages 3 and 4). REM sleep accounts for 10–25% of sleep time.

Total sleep time in the healthy young adult approximates 5–10 hours. In the full-term newborn, sleep cycles last approximately 60 minutes. The newborn sleeps approximately 16–20 hours per day, with a higher proportion (~50%) of REM sleep.

Stage 1 ("Drowsiness") *Slow rolling eye movements* (SREMs) (Figure 130.18) are usually the first evidence of drowsiness seen on the EEG. SREMs of drowsiness most often are horizontal but can be vertical or oblique, and their distribution is similar to eye movements in general. However, they are slow (i.e., typically 0.25–0.5 Hz). Because of their frequency, SREMs superficially resemble sweat artifacts but are easily identified by their nonrandom distribution typical of eye movements (e.g., phase reversals at F7 and F8 if horizontal). SREMs disappear in stage 2 and deeper sleep stages.

Alpha activity dropout (Figure 130.18) typically occurs together with or nearby SREMs. The alpha rhythm gradually becomes slower, less prominent, fragmented, and disappears.

Vertex sharp transients (Figures 130.19 and 130.20), also called vertex waves or V waves, are almost universal. Although they often are grouped together with K-complexes, strictly speaking, vertex sharp transients are distinct from K-complexes because they are briefer in duration, smaller in amplitude, and more focal (i.e., less widespread). Like K-complexes, vertex waves are maximum at the vertex (central midline placement of electrodes (Cz)), so that, depending on the montage, they may be seen on both sides, usually symmetrically, at C3 and C4. Their amplitude is 50–150 μV. They can be contoured sharply and occur in repetitive runs, especially in children. They persist in stage 2 sleep but usually disappear in subsequent stages. Unlike K-complexes, vertex waves do not define stage 2.

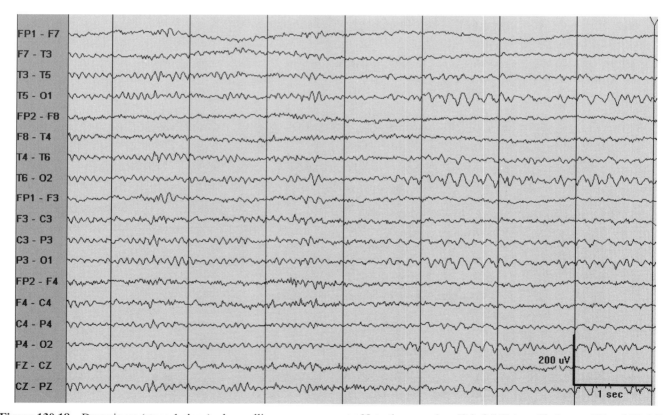

Figure 130.18 Drowsiness (stage 1 sleep): slow rolling eye movements. Note the very slow (0.3–0.5 Hz) oscillations at F7 and F8. Like rapid eye movements, a negativity at F7 occurs at the same time as a positivity at F8 for an eye movement to the right, and vice versa for an eye movement to the left. Also note the attenuation and slowing of the alpha rhythm (alpha dropout) between the first and second half of the sample. This is the other early finding of stage 1 sleep.

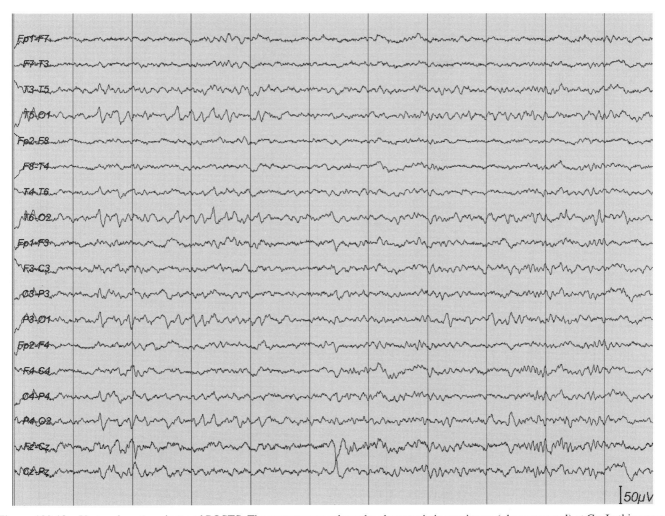

Figure 130.19 Vertex sharp transient and POSTS. The vertex waves show the characteristic maximum (phase reversal) at Cz. In this example, they are very "focal" (restricted to Cz) and are not seen at C3 or C4. POSTS (positive sharp transients of sleep) are seen with their typical bioccipital location and "reversed checkmark" morphology.

Positive occipital sharp transients of sleep (POSTS) (Figure 130.19) are seen very commonly on EEG and have been said to be more common during daytime naps than during nocturnal sleep. Most characteristics of POSTS are contained in their name. They have a positive maximum at the occiput, are contoured sharply, and occur in early sleep (stages 1 and 2). Their morphology is best described as "reversed checkmark," and their amplitude is 50–100 µV. They typically occur in runs of 4–5 Hz and are bisynchronous, although they may be asymmetric. They persist in stage 2 sleep but usually disappear in subsequent stages.

Hypnagogic hypersynchrony (Figure 130.21) is a less common but well recognized normal variant of drowsiness in children aged 3 months to 13 years. This is described as paroxysmal bursts (3–5 Hz) of high-voltage (as high as 350 µV) rhythmic waves, maximally expressed in the prefrontal-central areas.

Stage 2 *Sleep spindles* (Figure 130.22) normally first appear in infants aged 6–8 weeks and are initially asynchronous, becoming synchronous by the age of 2 years. Sleep spindles have a frequency of 12–16 Hz (typically 14 Hz) and are maximal in the central region (vertex), although they occasionally predominate in the frontal regions. They occur in short bursts of waxing and waning fusiform rhythmic activity. Amplitude is usually 20–100 µV. Less typical or "extreme" spindles (described by Gibbs and Gibbs) are unusually high-voltage (100–400 µV) and prolonged (>20 s) spindles located over the frontal regions.

K-complexes (Figure 130.23) are high amplitude (>100 µV), broad (>200 ms), diphasic, transients often associated

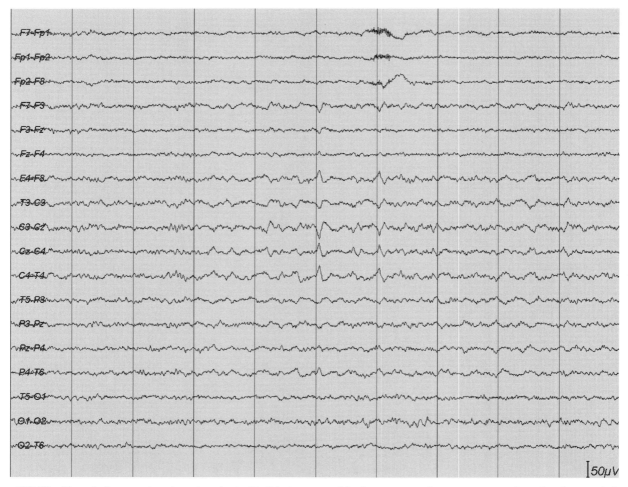

Figure 130.20 Stage 1 sleep: vertex sharp transients. On this transverse bipolar montage, the vertex waves show the characteristic maximum (phase reversal) at Cz. Note that they are also seen frontally with a phase reversal af Fz.

with sleep spindles. Location is frontocentral, with a typical maximum at the midline (central midline electrodes (Cz) or frontal midline electrodes (Fz)). They occur spontaneously or as an arousal response.

Stage 3/4 Slow-wave sleep (SWS), or delta sleep, is characterized, as the name implies, by delta activity. This typically is generalized and polymorphic or semirhythmic (Figure 130.24). By strict sleep staging criteria on polysomnography, SWS is defined by the presence of such delta activity for more than 20% of the time, and an amplitude criterion of at least 75 µV often is applied.

The distinction between stages 3 and 4 is only a quantitative one that has to do with the amount of delta activity. Stage 3 is defined by delta activity that occupies 20–50% of the time, whereas in stage 4 delta activity represents greater than 50% of the time. Sleep spindles and K-complexes may persist in stage 3 and even to some degree in stage 4, but they are not prominent.

As mentioned earlier, SWS usually is not seen during routine EEG, which is too brief a recording. However, it is seen during prolonged EEG monitoring. One important clinical aspect of SWS is that certain parasomnias occur specifically out of this stage and must be differentiated from seizures. These "slow-wave sleep parasomnias" include confusional arousals, night terrors (pavor nocturnus), and sleepwalking (somnambulism).

REM Sleep REM sleep normally is not seen on routine EEG recordings, because the normal latency to REM sleep (100 min) is well beyond the duration of routine EEG recordings (approximately 20–30 min). The appearance of REM sleep during a routine EEG is referred to as sleep-onset REM period (SOREMP) and is abnormal and warrants an MSLT. While not observed on routine EEG, REM sleep commonly is seen during prolonged (>24 h) EEG monitoring.

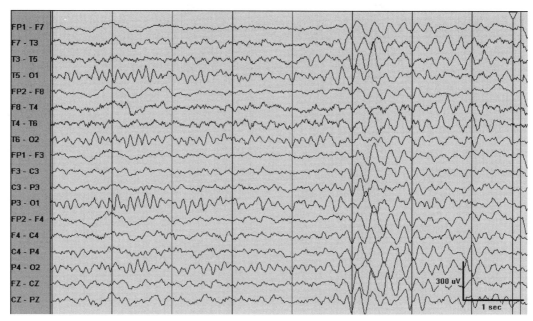

Figure 130.21   Hypnagogic hypersynchrony. This less common but well recognized normal variant of drowsiness is seen in children aged 3 months to 13 years. It consists of paroxysmal bursts (3–5 Hz) of high-voltage (as high as 350 μV) rhythmic waves. Note the normal alpha rhythm in the first two-thirds of the sample, which precedes drowsiness.

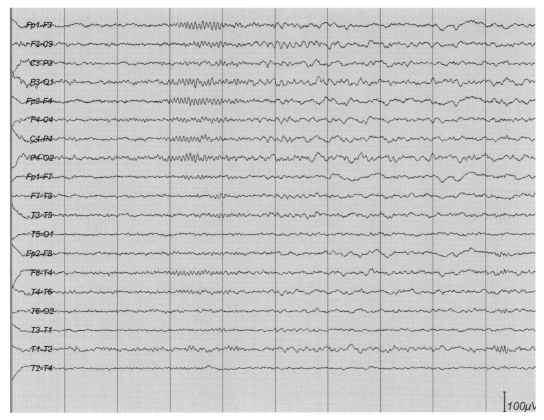

Figure 130.22   Stage 2: sleep spindles. This segment depicts typical short bursts (1–3 seconds) of sinusoidal 12–16 Hz central activity that waxes and wanes.

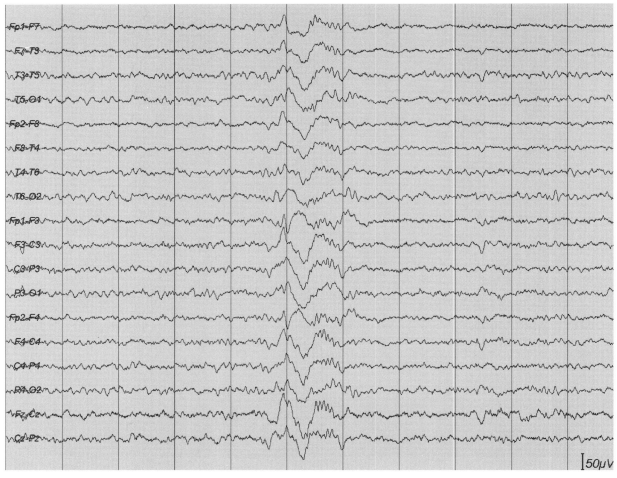

Figure 130.23 Stage 2: K-complexes. This sample shows the typical long-duration (~1 second) diphasic waveform. Note the widespread distribution on this double banana montage, but on a transverse montage one would typically see a maximum (phase reversal) at the midline (Cz, Fz).

By strict sleep staging criteria on polysomnography, REM sleep is defined by (1) rapid eye movements, (2) muscle atonia, and (3) EEG "desynchronization" (compared to slow-wave sleep). Thus two of the three defining characteristics are not cerebral waves and theoretically require monitoring of eye movements (electro-oculogram (EOG)) and muscle tone (electromyogram (EMG)). Fortunately, muscle activity and eye movements can be evaluated on EEG; thus REM sleep is usually not difficult to identify (Figure 130.25). In addition to the three features already named, "sawtooth" waves also are seen in REM sleep.

- *EEG Desynchronization.* The EEG background activity changes from that seen in slow-wave sleep (stage 3 or 4) to faster and lower voltage activity (theta and beta), resembling wakefulness.

- *Rapid Eye Movements.* These are saccadic, predominantly horizontal, and occur in repetitive bursts.

- Despite the lack of a dedicated EMG channel, the muscle atonia that characterizes REM sleep is usually apparent as a general sense of "quiet" muscle activity compared to wakefulness.

- Sawtooth waves are a special type of central theta activity that has a notched morphology, resembling the blade of a saw, and usually occurs close to rapid eye movements (i.e., phasic REM). They are only rarely identifiable on EEG.

The duration of REM sleep increases progressively with each cycle and tends to predominate late in the sleep period into early morning.

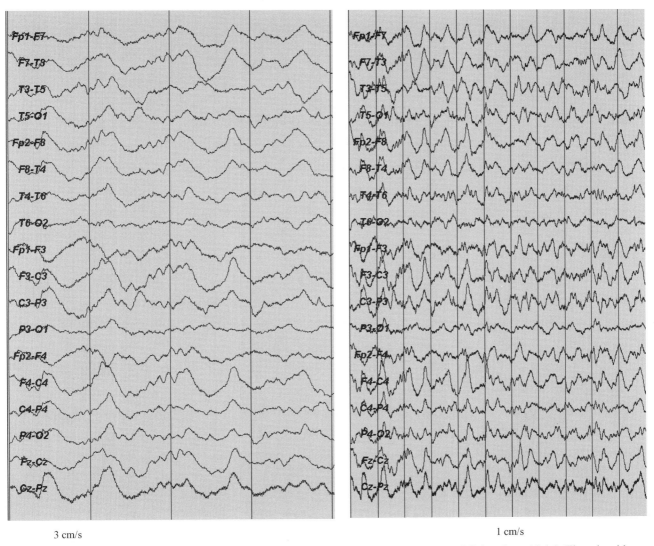

Figure 130.24 Slow-wave sleep. This is the same sample at an EEG time-base (left) and at a PSG time-base (right). There is widespread delta (2 Hz) activity.

Less Common Patterns and Normal Variants

Just like human anatomy, sizes and shapes vary somewhat among individuals, so too do brain waves. There is a wide range of variability, and it is important to read EEG "conservatively" and avoid overreading normal variants [9]. There are many normal variants that have been well described as benign and have no association with epilepsy. These include small sharp spikes (also called benign epileptiform transients of sleep, Figure 130.26), wicket spikes (Figure 130.27), 14 and 6 Hz positive spikes, phantom spike waves, rhythmic midtemporal theta of drowsiness (also called psychomotor variant), and subclinical rhythmic epileptiform discharges of adults. These can be overread as abnormal. Lambda waves (Figure 130.28) are occipital

sharp transients that resemble POSTS but occur in wakefulness when subjects (especially children) scan the environment. Despite these, the most commonly overread patterns are "nameless" fluctuations of background activity [9].

THE ABNORMAL EEG

Like most neurophysiologic tests, EEG is a test of cerebral *function*, and as such is for the most part nonspecific as to etiology. Although earlier investigators have attempted to identify the reliability of EEG in differentiating types of lesions, this has clearly become a senseless and futile exer-

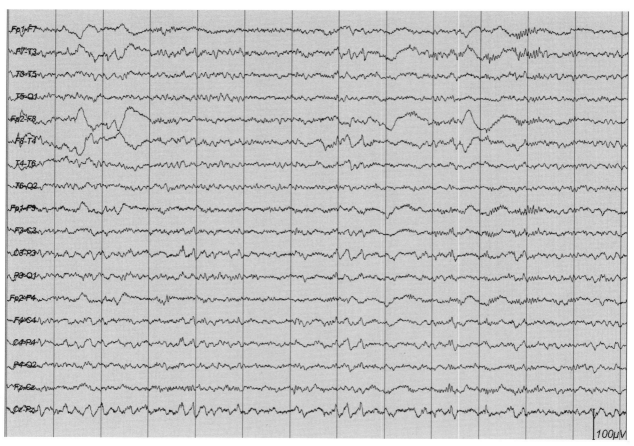

Figure 130.25   REM sleep. This segment shows rapid eye movements (3rd and 10th seconds). Note also the sawtooth waves: notched theta (∼5 Hz) transients in the central regions.

cise in the modern era of neuroimaging [10]. The exercise to describe EEG abnormalities by pathology (stroke, abscess, tumor, etc.), which was common in old EEG texts [11], is clearly obsolete and will not be followed here.

There are many different ways to classify EEG abnormalities. This chapter will use a very practical classification developed at the Cleveland Clinic Foundation [12] and used at many centers. The outline of the classification is shown in Table 130.1.

Epileptiform Abnormalities

Types of Epileptiform Abnormalities   EEG is useful in epilepsy because it is the only test that gives direct evidence for epileptogenicity. Most commonly, clinicians have to rely on *interictal* abnormalities. Interictal epileptiform abnormalities include spikes, sharp waves, and spike–wave complexes. In addition to type and morphology, location is very important and is typically either focal or generalized. Generalized epileptiform discharges (i.e., spikes, sharp waves, polyspikes, spike–wave complexes), as seen

in the primary generalized epilepsies, are usually maximal in the frontal regions, with typical "phase reversals" at the F3 and F4 electrodes or less commonly at F7 and F8.

- *Spikes and sharp waves* are sharp transients that have a very strong association with epilepsy. The two are distinguished only by their duration (spikes <70 ms, sharp waves 70–200 ms), but they have no differences in terms of clinical significance. Several characteristics help distinguish these from benign epileptiform variants but this distinction can at times be difficult. Helpful features that indicate pathologic discharges include high amplitude, which makes them "stand out" from ongoing background activity, and after-going slow waves, which indicate "disruption" of background activity (Figures 130.29 and 130.30). The terms "spikes" and "sharp waves" should be reserved for these abnormalities thought to be pathological and indicative of epileptogenicity. If one wants to use purely descriptive (uncommitted) terms, then "generic" phrases should be used, such as "sharp transients" or "sharply contoured waveforms."

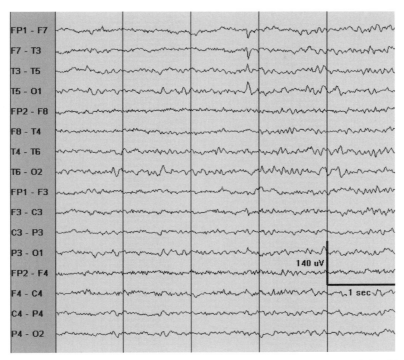

Figure 130.26 Small sharp spikes (SSS) or benign epileptiform transients of sleep (BETS). Typical brief (<50 ms) and low-amplitude (<50 µV) sharp transients in the temporal region, typically in stage 1 or 2 sleep.

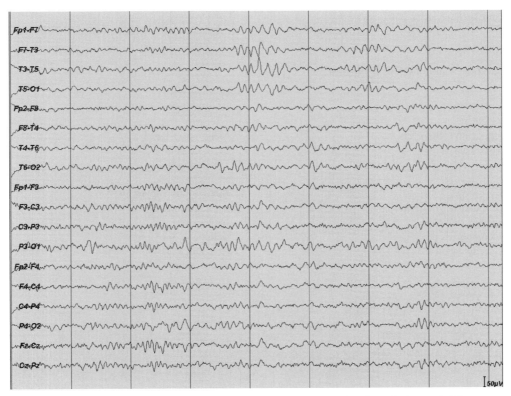

Figure 130.27 Wicket spikes. These are typically sharp transients seen in the temporal regions during wakefulness or stage 1 sleep. They have a symmetric upgoing and downgoing phase, typically arise from an ongoing background activity, and do not disrupt the background (i.e., no after-going slow wave). They are basically fluctuations of background activity.

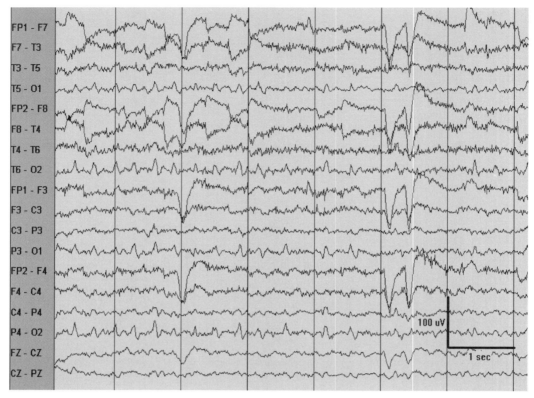

Figure 130.28 Lambda waves. This sample shows the typical occipital sharp transients that resemble POSTS but occur in wakefulness when subjects (especially children) scan the environment.

TABLE 130.1   Cleveland Clinic EEG Classification

Slow activity	*Special patterns*
Background slow (BS)	Asymmetry
Intermittent slow (IS)[a]	Excessive fast
Intermittent rhythmic slow (IRS)[a]	Sleep-onset REM
Continuous slow (CS)[a]	Periodic pattern
Epileptiform patterns[a]	Triphasic waves
Spikes	PLEDs[a]
Sharp waves	Background suppression[a]
Benign epileptiform discharges of childhood (BEDC)	*Special patterns used only in coma*
Spike–wave complexes (SWCs)	Alpha coma
Slow spike–wave complexes (SSWCs)	Spindle coma
3 Hz spike–wave complexes (3 Hz SWCs)	Beta coma
Polyspikes	Theta coma
Hypsarrhythmia	Delta coma
Photoparoxysmal response (PPR)	*ECI*
EEG seizure pattern	
Artifact obscured EEG	

[a] *Abnormalities that require localization*:
- Generalized
- Lateralized
- Multiregional
- Regional
- Multifocal
- Focal

Source: From [12].

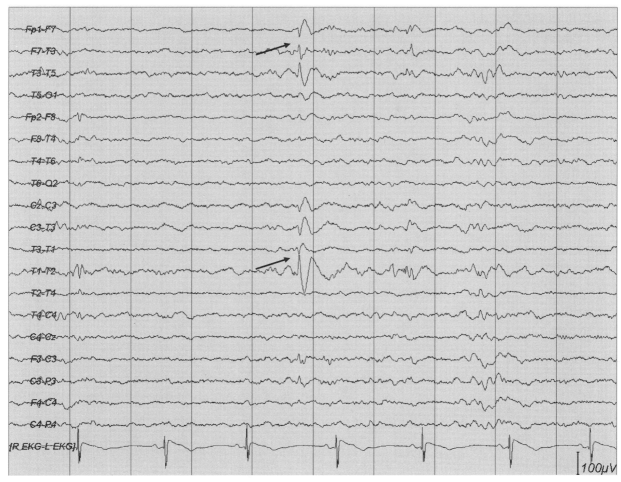

Figure 130.29 Typical left anterior temporal sharp wave. The maximum (negativity) is at F7 and T1 (phase reversal). Note the after-going slow wave. Classification: sharp wave, regional, left frontotemporal.

- *Polyspikes* are usually generalized (rarely focal), although focal spikes can at times have a multiphasic "polyspike-like" morphology. Polyspikes are multiple repetitive spikes occurring at about 20 Hz and are typically seen in the primary generalized epilepsy with myoclonic seizures (Figure 130.31).

- *Spike–wave complexes* (SWCs) are the repetitive occurrence of a spike followed by a slow-wave. Since any significant spike or sharp wave usually is followed by a slow wave (see earlier), a run of 3 seconds is required to classify a record as SWC, as opposed to the categories already mentioned (spike or sharp wave). SWCs can be divided further into two more specific types, as follows:

   3 Hz SWC. This pattern is characterized by a frequency of 2.5–4 Hz and a very monomorphic ("perfectly regular") morphology (Figure 130.32).

   It occurs in very discrete bursts, and between bursts the EEG is normal.

   Slow SWC. This pattern is not only slower (<2.5 Hz) but also more irregular (less monomorphic) than the 3 Hz SWC. Bursts are less discrete than the 3 Hz SWC, and between bursts other abnormalities are seen in symptomatic/cryptogenic epilepsies of the Lennox–Gastaut type.

   Hypsarrhythmia. This is characterized by continuous (during wakefulness) high-amplitude >200 μV (microvolts) generalized and polymorphic slowing with no organized background, and multifocal spikes (Figure 130.33). During NREM sleep, the pattern becomes discontinuous and fragmented, resembling a burst-suppression or pseudoperiodic pattern, while it tends to disappear in REM sleep.

- Electrographic Seizures (Ictal Patterns)

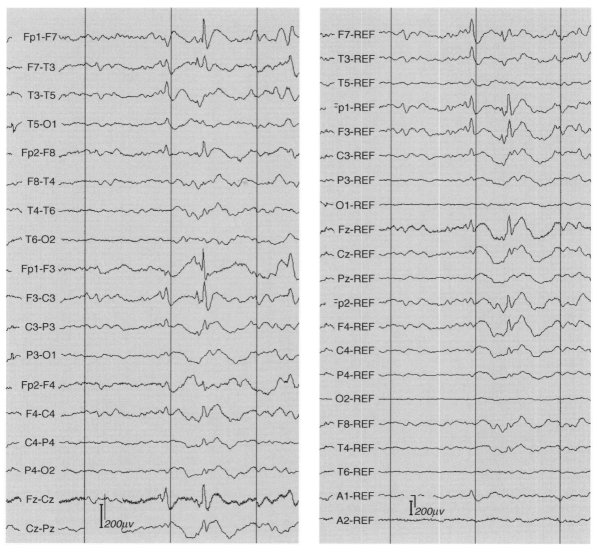

Figure 130.30 This sample shows a left frontal spike in a patient with left frontal lobe epilepsy. The left panel is a double banana, the right is a referential montage showing the same sample. Note that there are two different spikes: the first is maximum at F7, the second at Fp1. Classification: spike, regional left frontal.

Focal seizures are discharges characterized by *rhythmicity* and *evolution* ("buildup") in frequency, amplitude, and distribution. The discharge can consist of rhythmic theta or delta activity, or repetitive spikes or sharp waves, but the most characteristic features of electrographic focal seizures are rhythmicity and evolution (Figure 130.34).

Generalized seizures are also characterized by rhythmic discharges that evolve. They can be variable and there are rhythmic or periodic abnormalities that are not ictal (see later under status epilepticus).

Spike–wave complexes are both an interictal and an ictal pattern. The distinction is arbitrary and based on duration and the presence of detectable clinical symptoms (i.e., alteration of awareness).

Electrodecrement consists of abrupt attenuation ("flattening") of background activity, often preceded by a high-amplitude transient, which can be sharply contoured or broad, or generalized paroxysmal fast activity (GPFA). This typically is associated with infantile spasms and tonic or atonic seizures (Figure 130.35).

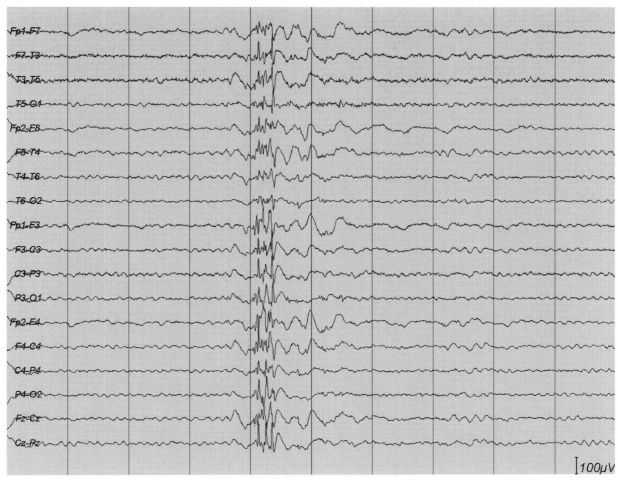

Figure 130.31   Polyspike. This is a typical generalized polyspike in a patient with an idiopathic ("primary") generalized epilepsy. Note that the discharge is followed by about 2 seconds of delta activity. This after-going slow activity indicates substantial disruption of the background and is a significant argument to consider this discharge pathological and not a normal variant. Classification: polyspike, generalized.

Epileptiform discharges have to be differentiated from normal patterns and normal variants that can look like epileptiform, and this can at times be difficult. The overinterpretation of EEG is a common and underreported problem [9, 13, 14] and a major cause for the misdiagnosis of epilepsy. If read conservatively (being careful not to overinterpret normals), the specificity of EEG for epilepsy is very high, that is, greater than 90%. It is generally accepted that <3% of the general population has interictal epileptiform abnormalities, with a slightly higher percentage in children [15] (largely because of benign focal epilepsies in that age group, see later) Because of the relatively frequent misinterpretation (overreading) of benign EEG recordings, it is unfortunately true that "routine interictal EEG is one of the most abused investigations in clinical medicine and is unquestionably responsible for great human suffering" [16]. One of the important remedies to this pitfall is that when in doubt, one should err on the

side of underreading. Clinical experience strongly supports the view that less harm will be done by underinterpreting an abnormality than by overinterpreting a normal pattern.

Electrographic Status Epilepticus   Status epilepticus (SE) is typically an obvious clinical diagnosis, but in some situations an EEG is required to diagnose "nonobvious" SE. Electrographically, SE can take the form of either repetitive discrete seizures or, more commonly, a continued pattern of rhythmic or periodic discharges such as generalized periodic epileptiform discharges (GPEDs) [17, 18]. Unfortunately this pattern (Figure 130.36) is not specific for electrographic SE and can be seen in severe metabolic encephalopathies such as anoxic, uremic, or hepatic disturbances [19, 20, 21]. Thus the final answer as to whether it does or does not represent SE in a given patient often will depend on the clinical circumstances and the response to treatment [21].

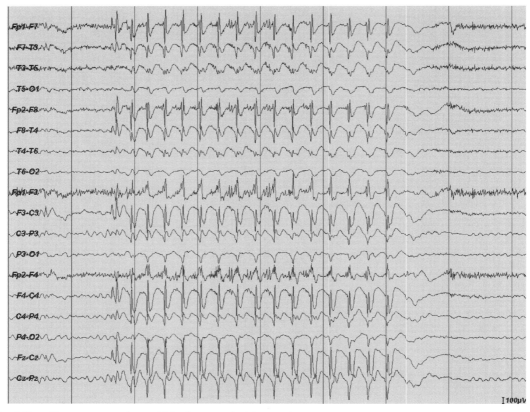

Figure 130.32   3 Hz spike–wave complexes. This shows typical 3 Hz spike–wave complexes in a patient with an idiopathic ("primary") generalized epilepsy. This patient likely has absence seizures. If tested with a clicker, a discharge of this duration (~4.5 seconds) is likely associated with a brief impairment of awareness. Classification: 3 Hz spike–wave complexes, generalized.

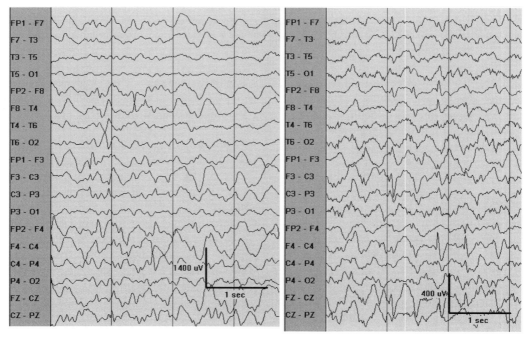

Figure 130.33   Hyparrhythmia. These recordings are from two different patients. Note on both the lack of any reactive background, the very high-amplitude polymorphic delta activity, and the multifocal spikes. Classification: hypsarrhythmia.

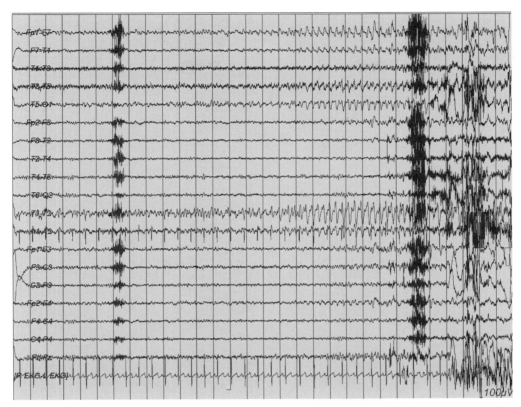

Figure 130.34   Focal seizure. Left temporal seizure (singular), showing the typical rhythmic activity that evolves (build up). Classification: EEG seizure, regional left temporal.

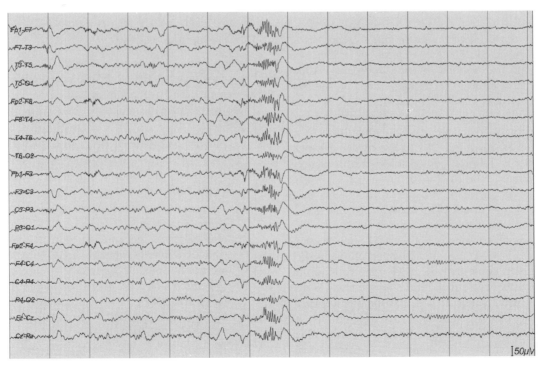

Figure 130.35   Generalized paroxysmal fast activity followed by attenuation (electrodecrement) in a patient with a symptomatic generalized epilepsy of the Lennox–Gastaut type. This could be an interictal (asymptomatic) discharge but could also be an ictal pattern associated with a tonic or atonic seizure. Classification: EEG seizure, generalized.

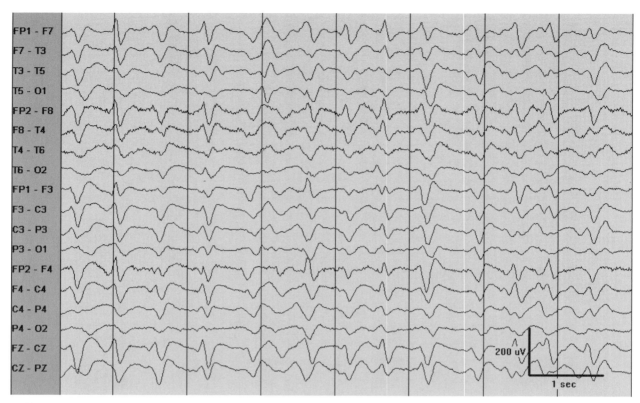

Figure 130.36 Periodic pattern, periodic complexes, or generalized periodic epileptiform discharges (GPEDs). This pattern is equally compatible with some severe metabolic encephalopathies (e.g., anoxic, uremic, or hepatic) or nonconvulsive status epilepticus. Only the clinical situation can distinguish among these possibilities, and when in doubt patients with this pattern should be treated for status epilepticus and monitored for clinical improvement. Classification: periodic pattern, generalized.

EEG-Video Monitoring and the Differential Diagnosis of Epilepsy Prolonged EEG-video monitoring is indicated when seizures do not respond to medications [4]. About 20–30% of patients referred for refractory seizures do not have epilepsy but have psychogenic nonepileptic seizures (PNESs) instead [3, 13]. A small percentage have syncope [14, 22]. Occasionally other paroxysmal conditions can be misdiagnosed as epilepsy, but PNESs are by far the most common condition, followed by syncope. Unfortunately, the current average delay in the diagnosis of PNESs is over 7–9 years and 80% of patients with PNESs have received antiepileptic drugs [23]. This means that EEG-video monitoring is probably underutilized and the "established" diagnosis of seizures is not verified often enough. Once the diagnosis of PNESs is suspected clinically [7], it is usually easily confirmed by EEG-video monitoring, which can even be performed as an outpatient. In the hands of experienced epileptologists, the combined electro-clinical analysis of both the clinical semiology of the "ictus" and the ictal EEG findings allows a definitive diagnosis in nearly all cases. The second reason why medications may fail is that 30% of epilepsies are intractable and require nonpharmacologic treatments [4].

Encephalopathic Patterns

These patterns are associated with *diffuse* (*generalized*) brain dysfunction, that is, diffuse encephalopathies. Again, in general, they are completely nonspecific in terms of etiology. In order of increasing severity, they include:

- *Background Slowing (Classification as "Background slow")*. There is a posterior dominant background that is reactive, but the frequency is too slow for age (<8 Hz after the age of 8) (Figure 130.37).
- *Intermittent Generalized Slowing (Classification as "Intermittent Slow, Generalized")*. There are intermittent bursts of generalized slowing, in the theta or (more commonly) delta range (Figure 130.38), which are not attributable to normal circumstances such as hyperventilation or sleep. A subtype of intermittent generalized slowing is intermittent rhythmic slowing, which is usually frontally predominant and thus often referred as "FIRDA."
- *Continuous Generalized Slowing (Classification as "Continuous Slow, Generalized")*. Here the slowing is continuous or nearly continuous (>80% of the

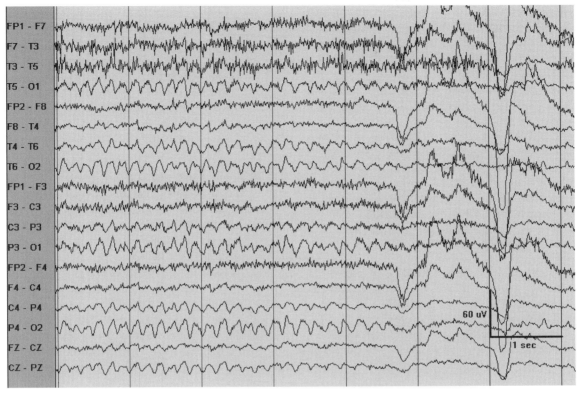

Figure 130.37 Background slowing. Note that there is a posterior dominant background, which is even normally reactive (attenuation with eye opening in the last 2 seconds). However, the frequency is only 6 Hz, which is too slow (for a subject older than 3 years) and thus is evidence for a mild diffuse encephalopathy. Classification: background slow.

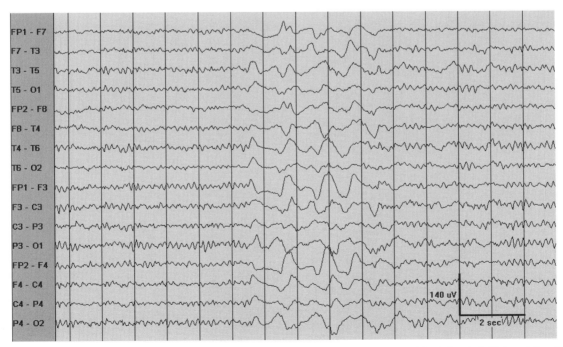

Figure 130.38 Intermittent generalized slowing. Note the brief (~4 sec) burst of polymorphic delta (~2 Hz) activity. When this is intermittent or reactive, it is indicative of a mild to moderate diffuse encephalopathy. Classification: intermittent slow, generalized.

record) and is unreactive. Unreactive means that it does not change with external stimulation, and that there are no state changes such as evidence for drowsiness, sleep, or alerting responses (Figure 130.39).

- Periodic patterns are generalized discharges that occur with periodicity (i.e., at regular intervals) and are often referred to as generalized periodic epileptiform discharges (GPEDs) (Figure 130.35). The discharges are often sharply contoured, and the periodicity is most often 1–3 seconds. The etiology is diverse and includes nonconvulsive status epilepticus [17, 18] as well as severe metabolic encephalopathies such as anoxic, uremic, or hepatic disturbances [19, 20, 21]. Another cause of generalized periodic pattern is Creutzfeldt–Jakob disease.

    Burst-suppression (Figure 130.40) is a subtype of periodic patterns where the activity between complexes is suppressed (i.e., <10 μV). This is indicative of an extremely severe degree of encephalopathy, which in fact immediately precedes eletrocerebral inactivity. It is typically caused by

either drugs (e.g., anesthetic agents), in which case it is reversible, or anoxia.

    Triphasic waves (Figure 130.41) can also be viewed as a subtype of periodic patterns. These are high-amplitude surface-positive sharp transients preceded and followed by lower amplitude negative components (thus "triphasic") and occur semi-periodically at 1–3 seconds. They are typically seen in metabolic encephalopathies, especially hepatic or renal.

- Background suppression is the absence of any cerebral activity greater than 10 μV, including with attempts at activating/stimulating the patient (unreactive). This is often used for severely "flat" EEG recordings that do not meet the criteria for electrocerebral inactivity (see later).

- Electrocerebral inactivity (ECI) (Figure 130.42) is the EEG pattern of brain death. The criteria are precisely defined [24], but it should be remembered that, at least in the United States, brain death is a clinical diagnosis and EEG is a supportive test (i.e., not required) for the diagnosis. The main criteria for ECI include no

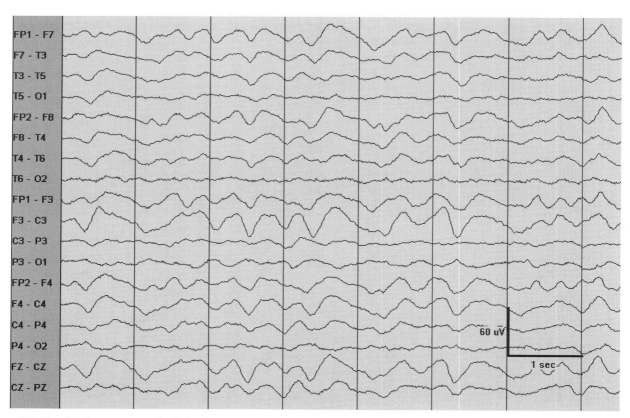

Figure 130.39 Continuous generalized slowing. Note the polymorphic delta (~2 Hz) activity. When this is continuous (>80% of the record) and unreactive, it is indicative of a *severe* diffuse encephalopathy. Classification: continuous slow, generalized.

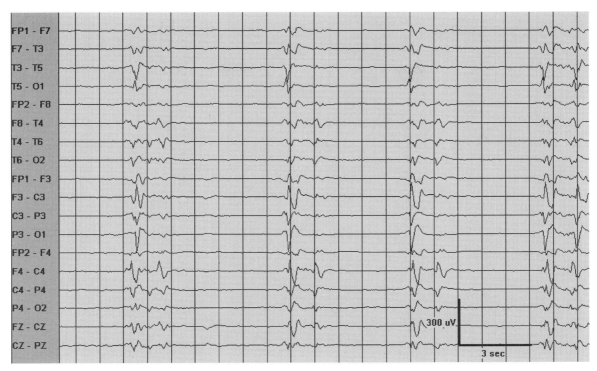

Figure 130.40 Burst-suppression. This sample depicts the typical periods of suppression (activity < 10 μV), which in this example last 3–5 seconds, interrupted by "bursts" of activity. Classification: burst-suppression.

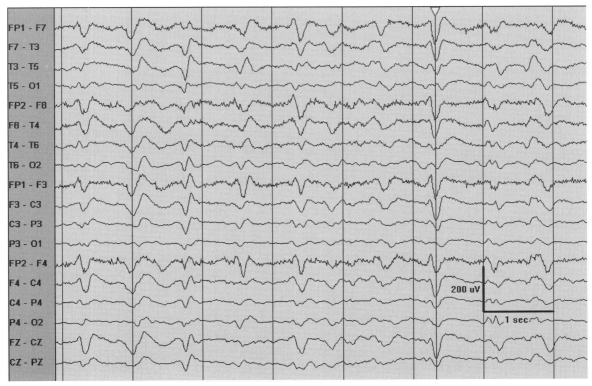

Figure 130.41 Triphasic waves. Note the quasiperiodicity, triphasic morphology with a prominent frontal (anterior) positivity, and the (mild) anteroposterior lag shown by the vertical bold line. Classification: triphasic waves.

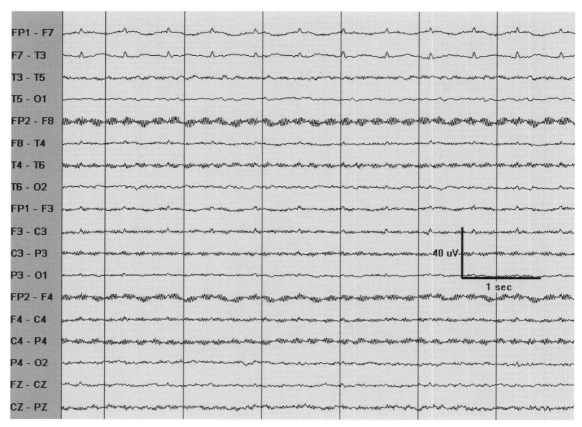

Figure 130.42 Electrocerebral inactivity. There is no cerebral activity greater than 2 μV. Because this is recorded at a sensitivity of 2 μV/mm, there is prominent EKG and 60 Hz artifact. Classification: electrocerebral inactivity.

activity greater than 2 μV with double interelectrode distances, complete lack of reactivity, and exclusion of confounding factors such as sedative drugs and hypothermia.

Patterns of Focal Abnormalities

Slow Activity Abnormal slow activity is by far the most common EEG manifestation of focal brain dysfunction. Like generalized slow activity, focal slow activity can be either intermittent (classification as "Intermittent slow, regional ___") or continuous (classification as "Continuous slow, regional___"). *Continuous* focal slowing is typically in the delta range and polymorphic and is also known as polymorphic delta activity (Figure 130.43). This is a very strong finding, highly associated with the presence of a structural lesion (but completely nonspecific as to the nature of the lesion). Continuous focal slow activity is the only nonepileptiform abnormality that can unequivocally be interpreted as abnormal as an isolated finding. By contrast, intermittent focal slowing (Figure 130.44) is a weak and nonspecific finding, which can even be normal (e.g., "temporal slowing of the elderly").

As outlined above, focal slowing is completely nonspecific as to etiology, and in the era of imaging the EEG has no role in diagnosing the nature of a lesion. Focal slowing is the most common abnormality associated with focal lesions *of any type*, including (but not limited to) neoplastic, vascular, subdural collections, traumatic, and infectious [10].

Amplitude Asymmetry In this discussion, the term asymmetry refers to an asymmetry of *amplitude* and refers to normal rhythms. By contrast, a focal frequency asymmetry would be classified as focal slow.

Destructive lesions can attenuate normal rhythms. However, normal rhythms are never perfectly symmetric in amplitude, so when to consider asymmetries significant is not always clear. A good rule of thumb is that, with very few exceptions, significant focal asymmetries will be associated with slowing. In general, as with other types of focal EEG abnormalities such as slowing, asymmetry is nonspecific as to etiology.

Although asymmetry in amplitude is usually indicative of dysfunction on the side of depressed amplitude, one very common exception to this rule is the so-called breach rhythm [25] (Figure 130.45). This is caused by a skull

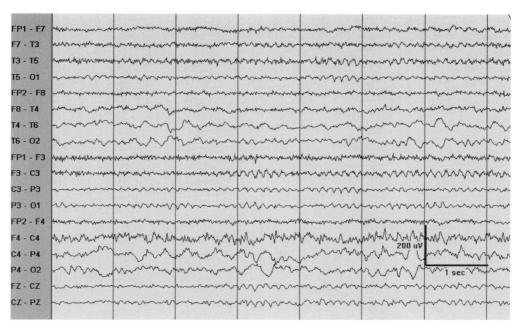

Figure 130.43   Continuous lateralized slowing in the right hemisphere. Note the clear polymorphic delta (~3 Hz) activity over the right hemisphere. Such focal slow activity, when persistent (>80% of the record), is strongly associated with a structural lesion. Classification: continuous slow, lateralized right hemisphere.

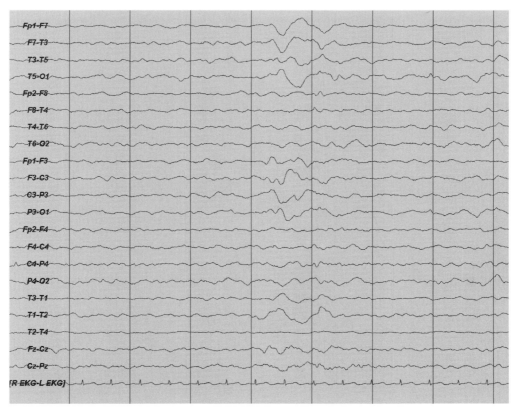

Figure 130.44   Intermittent lateralized slowing in the left hemisphere. Note the brief (~2 second) burst of delta (~2 Hz) activity over the left hemisphere. This type of focal slowing, unlike that showed in Figure 130.42, is indicative of mild dysfunction and is a "weak" (nonspecific) abnormality. Classification: intermittent slow, lateralized left hemisphere.

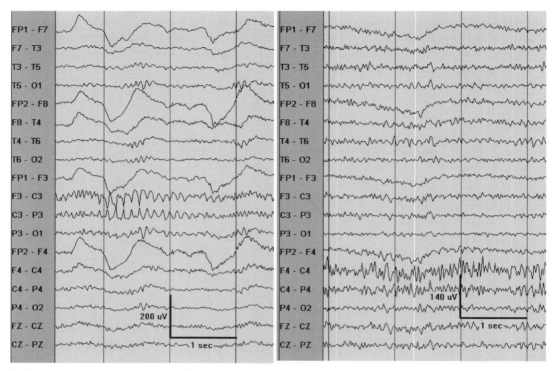

Figure 130.45   Breach rhythm. These are two different examples in two patients who are status post craniotomy. *Left sample*: Note the increase in amplitude very focally at C3, of what is likely a mu rhythm. Classification: asymmetry, increased mu, left central. *Right sample*: Here the increase is in fast (beta) activity at C4. Classification: asymmetry, increased beta, right central. The increased amplitude should not be mistaken for sharp waves.

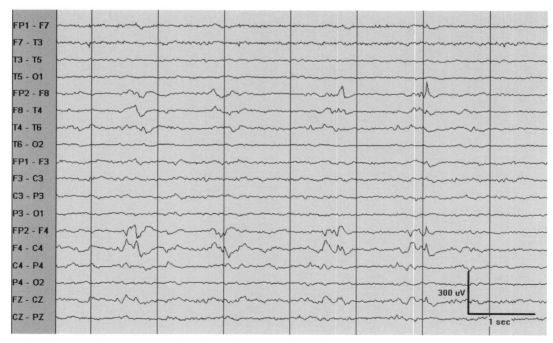

Figure 130.46   PLEDs—periodic lateralized epileptiform discharges. These discharges occur periodically, here with a periodicity of 1–1.5 seconds, and are lateralized (here in the right hemisphere). Note that the discharges look to be of low amplitude but in reality are ~100 μV (see scale). Classification: PLEDs, lateralized right hemisphere.

defect, which attenuates the high-frequency filter function of the intact skull. As a result, faster frequencies (alpha, spindles, beta) are of higher amplitude on the side of the defect. Since morphology is often sharply contoured, determining the epileptogenicity of these discharges can be extremely difficult, and in this situation one should probably err on the conservative side by not interpreting them as epileptiform. Finally, it should be kept in mind that amplitude asymmetries are best interpreted on referential montages, since amplitude is highly dependent on interelectrode distances.

PLEDs Periodic lateralized epileptiform discharges (PLEDs) (Figure 130.46) are a special type of focal abnormality. As implied by their name, they are periodic, lateralized, and often epileptiform in morphology. Periodicity is the most characteristic feature, and the one that sets PLEDs apart from other focal abnormalities. Periodicity refers to a relatively constant interval between two discharges and varies between 0.5 and 3 seconds, most often around 1 second. The epileptiform morphology of the discharges is not invariable, as PLEDs are often closer to slow waves than to sharp waves in morphology.

PLEDs are caused by *acute destructive focal lesions* and are a transitory phenomenon: they tend to disappear in weeks, even if the causal lesion persists. With time, periodicity and duration increase, and the record takes on a less specific focal slow appearance that is more likely to persist. By far the most common etiology is an acute cerebrovascular event, followed by focal encephalitis such as herpes [26–28]. Although the periodic patterns of Creutzfeldt–Jakob disease are usually generalized and bisynchronous, occasionally, especially early in the course, they may be unilateral or markedly asymmetric and thus take on the appearance of PLEDs [29].

As implied by their name, PLEDs have a high association with clinical seizures, and on average about 80% of patients with PLEDs will have clinical seizures. In clinical practice, PLEDs are usually managed as interictal discharges (spikes or sharp waves). They indicate a high risk of focal seizures but are usually not treated with antiepileptic drugs unless there is *clinical* evidence for seizures, although there is some controversy about this [10].

REFERENCES

1. Ajmone-Masan C, Zivin LS. Factors related to the occurrence of typical paroxysmal abnormalities in the EEG records of epileptic patients. *Epilepsia* **11**:361–381(1970).

2. Tatum WO IV, Winters L, Gieron M, Passaro EA, Benbadis S, Ferreira J, Liporace J. Outpatient seizure identification: results of 502 patients using computer-assisted ambulatory EEG. *J Clin Neurophysiol* **18**:14–19(2001).

3. Benbadis SR, O'Neill E, Tatum WO, Heriaud L. Outcome of prolonged EEG-video monitoring at a typical referral epilepsy center. *Epilepsia* **45**:1150–1153(2004).

4. Benbadis SR, Tatum WO IV, Vale FL. When drugs don't work: An algorithmic approach to medically intractable epilepsy. *Neurology* **55**:1780–1784(2000).

5. Anonymous. Patient referral to specialty epilepsy care. *Epilepsia* **31**:S10–S11(1990).

6. Benbadis SR, Wyllie E, Bingaman W. Intracranial EEG and localization studies. In: Wyllie E (Ed), *The Treatment of Epilepsy: Principles and Practice*, 3rd ed. Lippincott, Williams & Wilkins, Philadelphia, 2001, pp 1067–1075.

7. Benbadis SR. Psychogenic non-epileptic seizures. In: Wyllie E (Ed), *The Treatment of Epilepsy: Principles and Practice*. 4th ed. Lippincott, Williams & Wilkins, Philadelphia, in press.

8. Lesser RP, Luders H, Dinner DS, Morris H. An introduction to the basic concepts of polarity and localization. *J Clin Neurophysiol* **2**:45–61(1985).

9. Benbadis SR, Tatum WO. Over-interpretation of EEGs and misdiagnosis of epilepsy. *J Clin Neurophysiol* **20**:42–44(2003).

10. Benbadis SR. Focal disturbances of brain function. In: Levin KH, Lüders HO (Eds), *Comprehensive Clinical Neurophysiology*. Saunders, Philadelphia, 2000, pp 457–467.

11. Niedermeyer E, Lopes da Silva F. *Electroencephalography: Basic Principles, Clinical Applications, and Related Fields*, 3rd ed. Williams & Williams, Baltimore, 1993.

12. Luders H, Noachtar S (Eds.) *Atlas and Classification of Electroencephalography*. Saunders, Philadelphia, 2000.

13. Smith D, Defalla BA, Chadwick DW. The misdiagnosis of epilepsy and the management of refractory epilepsy in a specialist clinic. *QJM* **92**:15–23(1999).

14. Eiris-Punal J, Rodriguez-Nunez A, Fernandez-Martinez N, Fuster M, Castro-Gago M, Martinon JM. Usefulness of the head-upright tilt test for distinguishing syncope and epilepsy in children. *Epilepsia* **42**:709–713(2001).

15. Cavazzuti GB, Capella L, Nalin A. Longitudinal study of epileptiform EEG patterns in normal children. *Epilepsia* **21**:43–55(1980).

16. Chadwick D. Diagnosis of epilepsy. *Lancet* **336**:291–295(1990).

17. Towne AR, Waterhouse EJ, Boggs JG, Garnett LK, Brown AJ, Smith JR Jr, DeLorenzo, RJ. Prevalence of nonconvulsive status epilepticus in comatose patients. *Neurology* **54**:340–345(2000).

18. Treiman DM. Electroclinical features of status epilepticus. *J Clin Neurophysiol* **12**:343–362(1995).

19. Brenner RP, Schaul N. Periodic EEG patterns: classification, clinical correlation, and pathophysiology. *J Clin Neurophysiol* **7**:249–267(1990).

20. Husain AM, Mebust KA, Radtke RA. Generalized periodic epileptiform discharges: etiologies, relationship to status epilepticus, and prognosis. *J Clin Neurophysiol* **16**:51–58(1999).

21. Benbadis SR, Tatum, WO IV. Prevalence of nonconvulsive status epilepticus in comatose patients (letter). *Neurology* **55**:1421–1423(2000).

22. Zaidi A, Clough P, Cooper P, Scheepers B, Fitzpatrick AP. Misdiagnosis of epilepsy: many seizure-like attacks have a cardiovascular cause. *J Am Coll Cardiol* **36**:181–184(2000).

23. Benbadis SR. How many patients with pseudoseizures receive antiepileptic drugs prior to diagnosis? *Eur Neurol* **41**:114–115(1999).

24. American Electroencephalographic Society. Guideline three: minimum technical standards for EEG recording in suspected cerebral death. *J Clin Neurophysiol* **11**:10–13(1994).

25. Cobb WA, Guillof R, Cast J. Breach rhythm: the EEG related to skull defects. *Electroencephalogr Clin Neurophysiol* **47**:251–271(1979).

26. Schaul N, Green L, Peyster R, Gotman J. Structural determinants of elctroencephalographic findings in acute hemispheric lesions. *Ann Neurol* **20**:703–711(1985).

27. Schwartz MS, Prior PF, Scott DF. The occurrence and evolution in the EEG of a lateralized periodic phenomenon. *Brain* **96**:613–622(1973).

28. Snodgrass SM, Tsuburaya K, Ajmone-Marsan C. Clinical significance of periodic lateralized epileptiform discharges: relationship to status epilepticus. *J Clin Neurophysiol* **6**:159–172(1989).

29. Heye N, Cervos-Navarro J. Focal involvement and lateralization in Creutzfeldt–Jakob disease: correlation of clinical, electroencephalographic and neuropathological findings. *Eur Neurol* **32**:289–292(1992).

131

MONITORING RESPIRATION DURING SLEEP

ULYSSES J. MAGALANG, RAMI KHAYAT, AND NAEEM ALI
The Ohio State University Sleep Disorders Center, Columbus, Ohio

INTRODUCTION

The detection of abnormal respiratory events during sleep is essential in the diagnosis of sleep-related breathing disorders (SRBDs). By convention, these respiratory events have to meet certain criteria of frequency and severity. Definitions of these events and the diagnostic criteria for SRBDs are still a matter of debate in the sleep community, and the reader is referred to other chapters of this book for a more detailed discussion. In this chapter, we will discuss methods of monitoring respiration during sleep, an integral part of clinical polysomnography.

Respiratory events are usually classified into obstructive and central events. The hallmark of an obstructive respiratory event is inspiratory flow limitation, which is characterized by an increase in inspiratory effort resulting in a decrease in the intrathoracic pressure without a corresponding increase in airway flow [1]. Central respiratory events are characterized by a reduction or absence of inspiratory effort. Simultaneous recordings of inspiratory effort, airflow, thoracic volume, and noninvasive measurements of blood gases are usually performed during polysomnography to characterize SRBD events.

MONITORING RESPIRATORY EFFORT

During an obstructive respiratory event, pharyngeal collapse occurs with an increasing inspiratory effort. Therefore measurement of the changes in intrathoracic pressure and detection of the activity of inspiratory muscles are important for defining these events.

Esophageal Pressure

The presence of respiratory effort can be determined by measuring pleural pressure during the respiratory cycle. During inspiration, the respiratory muscles generate a decrease in intrathoracic pressure below atmospheric pressure. These changes in pleural pressure during respiration can be detected by esophageal pressure (Pes) monitoring [2]. Pes is usually monitored with either an esophageal balloon [3] or a water-filled catheter [4] positioned in the lower esophagus and connected to a pressure transducer. Both methods require the insertion of a thin catheter via one nostril after topical anesthesia. The use of the water-filled esophageal catheter during polysomnography was reported to result only in minimal effects on sleep architecture [5]. Pes monitoring is considered the reference standard for detecting respiratory effort, as well as the most accurate method in differentiating obstructive from central respiratory events [6]. By definition, Pes monitoring is also required in the detection of respiratory effort-related arousals (RERAs) that characterize the upper airway resistance syndrome (UARS) [7]. A RERA is an event that does not meet criteria for an apnea or hypopnea and is characterized by a pattern of progressively increasing negative Pes, terminated by a sudden change in pressure to a less negative value, and an arousal [6]. The inclusion of UARS in the differential diagnosis is often cited as one of the primary indications for Pes monitoring [5]. Monitoring of Pes has also been suggested to guide continuous positive airway pressure (CPAP), but the validity and practicability of this approach is not established.

Pes monitoring is not widely used in clinical polysomnography. Concerns about patient comfort, safety, and disruptions in sleep architecture, especially with the esophageal balloon [8], are likely factors that limit its use. In addition, nasal pressure monitoring has recently been proposed to be a less invasive alternative to Pes monitoring for the detection of RERAs [9] and appears to have gained a wider acceptance for this purpose (see later discussion).

Diaphragmatic Muscle Activity

Some laboratories utilize surface recordings of the electromyographic activity of the diaphragm (EMG_{di}) as an index of respiratory effort. The EMG_{di} recording is usually obtained using electrodes placed in the seventh and eighth intercostal spaces, 2–3 cm from the costal margin. Electrocardiographic (ECG) artifacts frequently contaminate the EMG_{di} signal. A preliminary study reported that a multistep processing of the EMG_{di} recording resulted in a waveform that correlated well with Pes [10]. However, the study was unblinded and whether the processed EMG_{di} signal will be sufficient to detect RERAs or differentiate between obstructive and central events remains unknown. At present, very little evidence is available to support the use of EMG_{di} monitoring as a sole technique in detecting respiratory effort [6].

Pulse Transit Time

Pulse transit time (PTT) is a new indirect and noninvasive measurement of respiratory effort. The technique usually involves measuring the time that the pulse pressure wave takes to travel from the aortic valve to the periphery as detected by the pulse oximeter probe [11]. The PTT is calculated as the time interval between the R wave on the ECG and the point on the finger pulse waveform that is 50% the height of the maximum value [12]. The PTT is inversely related to blood pressure as increases in blood pressure cause stiffening of the arterial wall and therefore shortening of the PTT. The PTT can then be used to monitor changes in pleural pressure as swings in intrathoracic pressure affect blood pressure. Higher PTT values are associated with increasing respiratory effort as a more negative intrathoracic pressure decreases blood pressure [13]. Continuous monitoring of the PTT signal has been shown to reflect changes in pleural pressure swings, although individual PTT values may not correspond with absolute values of the pleural pressure [14].

Several studies have shown the utility of the PTT signal as a measure of respiratory effort for scoring obstructive apneic and nonapneic respiratory events [12], in differentiating obstructive from central respiratory events [14], and as a marker of arousals [15]. The method is very attractive since no new equipment is required to obtain the necessary signals for calculation of the PTT. Thus some of the most recent polysomnographic recording softwares have incorporated the routine calculation of PTT.

MONITORING AIRFLOW

Detection of changes in flow is essential for the recognition of respiratory events. The only direct method for measuring airflow is pneumotachography. However, indirect measurements of flow are widely used because of better patient tolerance.

Pneumotachograph

The gold standard for measuring flow [6], a pneumotachometer consists of a tube that receives and directs respiratory flow through a low-resistance structure, and a differential pressure transducer that measures the pressure difference across the resistance. The pressure difference is directly proportional to the laminar airflow across the resistance. At higher flow rates, the flow pattern becomes turbulent and the pressure difference loses its proportional relationship to flow. It is necessary to heat and calibrate the pneumotachometer prior to each use as changes in temperature and condensation can alter the relation between pressure and flow. Pneumotachometers are usually attached to a full-face mask in order to detect both oral and nasal inspiratory and expiratory flow. However, the mask–pneumotachograph device is bulky and can impose a resistive load on the patient in some cases. This has excluded the routine use of pneumotachography and promoted the indirect measurements of flow in clinical polysomnography.

Nasal Pressure

A pressure transducer attached to a nasal cannula can detect the pressure changes in the nares during inspiration and expiration and provide a qualitative assessment of flow during polysomnography. Nasal pressure measurement has been shown to have an excellent agreement with a pneumotachograph in identifying respiratory events with a very good inter-/intrarater agreement [16]. Monitoring of nasal pressure significantly improves respiratory event detection compared to thermistors in adults and children [17, 18] and can more reliably detect inspiratory flow limitation [1]. The identification of a plateau on the inspiratory waveform of the nasal pressure signal correlates well with increased upper airway resistance [19]. Nasal pressure can aid in identifying RERAs associated with UARS [9]. Some authors suggest that obtaining the square root of the nasal pressure signal provides a more linear relationship to flow measured by a pneumotachograph but the clinical utility of this transformation remains uncertain [20]. One potential problem with the use of nasal pressure to estimate flow is the possible increase in nasal resistance especially in patients with septal deviation, which may distort the quality of the signal [21]. Mouth breathing significantly compromises the signal. Special cannulas (other than conventional oxygen nasal cannulas) that measure both nasal and oral airflow are commercially available but their validity is not established. In addition, nasal pressure monitoring cannot be used to quantitate flow throughout the night as the signal may change depending on the position of the nasal cannula. Despite these limitations, nasal pressure measurement has now become widely used in clinical polysomnography to identify respiratory events. Some laboratories routinely use nasal pressure measurement via oxygen nasal prongs together with nasobuccal thermistors or thermocouples in the event of mouth breathing.

Thermistors and Thermocouples

Thermistors and thermocouples are sensors that detect the difference in temperature between the relatively warmer exhaled air and the cooler inhaled air. Therefore they provide an indirect qualitative measure of flow. The sensors are usually placed over both the mouth and nostrils. Although they are adequate in detecting obstructive apneas, signals from these sensors have been shown to have a nonlinear relationship to flow measured by a pneumotachometer. Therefore they are generally considered unreliable in detecting hypopneas and RERAs, typically overestimating flow. For example, a 50% reduction in airflow may result in only an 18% reduction in the thermistor signal [22]. The thermistor's signal carries a significant delay from the actual flow owing to the mass of the sensor and the time it takes for temperature change to induce a current in the sensor [23]. Other factors that may compromise the integrity of the signal include skin contact that tends to warm the sensor and excessive room temperature. Most sleep laboratories now primarily use nasal pressure monitoring to detect flow as described earlier and reserve the use of the thermal sensors either as an adjunct or in those patients in whom a nasal cannula does not provide an adequate signal.

MONITORING CHANGES IN VOLUME

Measurement of thoracic volume in sleep studies is mainly performed to infer flow. Changes in the volume of the thorax correspond with the tidal volume of respiration that can be used to characterize respiratory events.

Respiratory Inductance Plethysmography

Respiratory inductance plethysmography (RIP) measures the cross-sectional areas of the rib cage and abdomen by measuring the inductance of sinusoidally arranged insulated wires placed inside two cloth belts that encircle the chest and abdomen. When a current is injected in the wires, voltage changes over time with the movement of the abdomen and chest with respiration. Once the signal is calibrated against a pneumotachograph or a spirometer, the sum of the abdominal and rib cage signals can be used to indirectly measure tidal volume. However, measurement of tidal volume depends on the positions of the thoracic and abdominal bands that frequently change during polysomnography, and therefore detection of respiratory events usually is not based on absolute values of the tidal volume, but rather on relative reductions from baseline. Compared to a pneumotachograph, one study found that the uncalibrated RIP signal was inferior to nasal pressure in identifying respiratory events as well as in inter-rater agreement [16]. The contour of the RIP signals has also been used to identify RERAs [24]. The RIP signal is less prone to artifact than the impedance pneumography signal [25].

Strain Gauges and Piezosensors

Strain gauges are elastic tubes filled with a conducting substance. When an electric current is passed through the conductor, resistance is proportional to the cross-sectional area. Strain gauges are usually placed around the chest and abdomen. Changes in the cross-sectional area of the thorax and abdomen with respirations result in corresponding changes in the resistance to the electric current. The signal provides a qualitative assessment of the changes in the thoracic volume. Use of strain gauges may be associated with multiple technical problems related to displacement, under- or overstretching, and gradual deterioration of the signal through the night. Piezosensors take advantage of the concept of piezoelectricity, wherein mechanical strain in some crystals produces proportional electric polarization. The respiratory belt transducer contains a piezoelectric device that responds linearly to changes in length. It measures changes in thoracic or abdominal circumference during respiration.

Although commonly used in polysomnography because they are relatively inexpensive, the general consensus is that strain gauges and piezosensors are not reliable in detecting hypopneas [6] or in differentiating between central and obstructive sleep apneas [26].

Thoracic Impedance Pneumography

Thoracic impedance pneumography measures changes in electrical impedance of the thoracic cavity to a low-amplitude, high-frequency alternating current applied between two or more electrodes placed on the chest wall. The resistance of the thoracic cavity increases with the increased air volume during inspiration and decreases with reduced air volume during expiration [27]. Therefore the generated signal has an indirect relationship with the actual lung volume and is mainly a qualitative assessment of chest wall movement. The signal is susceptible to interference by patient movements and cardiac pulsations and is not generally reliable in distinguishing central from obstructive events.

Other methods that have been used to estimate volume changes include magnetometers, canopy with neck seal, and a static charge-sensitive bed. However, these methods are not widely used and the reliability of these methods to detect respiratory events is unclear.

MONITORING BLOOD GASES

Direct monitoring of arterial blood gas tension is usually not done during polysomnography because they are invasive, expensive, associated with more complications, and may not adequately track the rapid changes in blood gases that occur during sleep. Therefore noninvasive assessments of blood gases are generally employed during sleep studies.

Pulse Oximetry

Continuous noninvasive monitoring of oxygen saturation (Sao_2) using pulse oximetry is now routinely used in clinical polysomnography. Oxyhemoglobin desaturations are frequently used in the definition of SRBD events [6]. With pulse oximetry, a pulsating vascular bed—typically the ear lobe or fingertip—is placed between a light source and a photodetector. Sao_2 is determined by passing two wavelengths of light (red and infrared) from the light source through body tissue to the photodetector. The calculation of Sao_2 is based on the fact that oxyhemoglobin and reduced hemoglobin have different absorption spectra at these two particular wavelengths; the ratio of absorbencies is then used to generate the Sao_2 value using a calibration curve [28].

Pulse oximeters are widely used because they are easy to use, noninvasive, readily available, and respond rapidly to changes in blood oxygen tension. Because a threshold in the fall of Sao_2 is typically used in the definition of SRBD events [6], knowledge of the factors that affect the pulse oximetry signal is important. First, most pulse oximeters employ an averaging time that can be adjusted on the control panel to minimize artifactual data (typically between 3 and 21 seconds). The longer averaging time is usually employed in applications where artifacts due to movement are common such as in cardiopulmonary exercise tests or in intensive care units. Increasing the averaging time to 12 and 21 seconds during a sleep study resulted in underestimation of oxygen desaturation by up to 60% compared to an averaging time of 3 seconds [29]. The default factory setting of the averaging time of the pulse oximeter needs to be determined and adjusted for polysom-

nography. The shortest averaging time is recommended to keep track of the rapid changes in blood gases. In doing this, however, sleep clinicians and technologists need to be aware that artifactual Sao_2 data will be more frequent. Second, the presence of other hemoglobin species such as carboxyhemoglobin and methemoglobin can cause inaccurate readings since pulse oximetry can only differentiate oxyhemoglobin from reduced hemoglobin. Third, because measurement of Sao_2 depends on a pulsating vascular bed, any condition that restricts blood flow may alter the signal amplitude. Fourth, the accuracy of pulse oximeters deteriorates when Sao_2 falls to 80% or less [30]. Finally, unlike other signals used in polysomnography such as electroencephalography (EEG), there is no standard method that is commercially available currently to determine the frequency response of a particular pulse oximeter.

End-tidal Carbon Dioxide

The end-tidal Pco_2 concentration ($PETCO_2$) is measured as the plateau of the CO_2 level at the end of a complete expiration. $PETCO_2$ is about 1 mmHg less than arterial carbon dioxide tension ($Paco_2$) in healthy subjects [31]. However, $PETCO_2$ measurement during sleep may not reflect changes in $Paco_2$ during spontaneous breathing, or during positive airway pressure therapy [32]. In patients with underlying lung disease, $PETCO_2$ may not accurately measure $Paco_2$ because of ventilation–perfusion abnormalities. For all these reasons, $PETCO_2$ monitoring is usually not used in adult polysomnography. The method is widely used, however, by many pediatric sleep laboratories since it is generally accepted that some measure of CO_2 is required to detect periods of obstructive hypoventilation—prolonged episodes of airflow limitation that is not typical of adult apneic episodes [33].

Transcutaneous Oxygen and Carbon Dioxide Monitoring

Transcutaneous oxygen tension ($Ptco_2$) can be measured using a modified Clark electrode. $Ptco_2$ measurement is influenced by cutaneous perfusion, temperature, and metabolism. More importantly, the method is limited by its slow response time that fails to track rapid changes in blood gases, and the variability of the relation between Pao_2 and $Ptco_2$ values. It requires careful skin preparation, and periodic site changes every 4–6 hours are usually needed to prevent thermal injury [34]. After a site change, a delay in recording is expected during the warm-up period. Transcutaneous carbon dioxide ($Ptcco_2$) can be monitored during sleep using a silver chloride electrode. In adults, $Ptcco_2$ also does not accurately reflect changes in $Paco_2$ during sleep [32]. It also has a slow response time that makes it unsuitable for tracking blood gas changes during

SRBD events. For all these reasons, $Ptco_2$ and $Ptcco_2$ are not usually used during polysomnography.

NONINVASIVE ASSESSMENT OF AIRWAY OBSTRUCTION

During polysomnography, varying degrees of upper airway obstructions occurring during sleep are indirectly inferred from simultaneous recordings of flow, respiratory effort, thoracic volume, and pulse oximetry as described earlier. The quantification of upper airway resistance offers a direct measure of the patency of the upper airway. However, conventional calculation of respiratory resistance requires an esophageal catheter for measurement of pleural pressure and a pneumotachograph to measure flow. As previously mentioned, Pes monitoring is typically not done during polysomnography.

Forced Oscillation Technique (FOT)

The FOT is a noninvasive method for measuring the mechanical impedance of the respiratory system that has recently been applied to the detection of upper airway obstruction during polysomnography [35]. The technique offers a continuous, automatic, real-time, direct measurement of airway impedance without affecting sleep [36]. It involves the application of a nasal mask and the superimposition of low-amplitude pressure oscillations on the nasal pressure and flow signals during spontaneous respirations. The oscillation signal is generated by a CPAP device or by a loudspeaker. Respiratory impedance is computed from the pressure and flow signals recorded at the nasal mask. During upper airway obstructions, FOT measures the impedance of the upper airway segment from the mask to the site of collapse. Because a nasal mask, pneumotachograph, and computational equipment are required to calculate the impedance signal, FOT is not recommended routinely during polysomnography. Proponents of this technique suggest that it could be helpful in differentiating obstructive (high-impedance) from central events (low-impedance), in the diagnosis of UARS, in determining the adequacy of CPAP, and as part of the algorithm of automatic-adjusting CPAP [37]. Air leaks around the mask and mouth could result in falsely low impedance values, while mouth breathing can result in falsely high impedance values. Nonetheless, the FOT appears to be a promising tool in directly characterizing upper airway obstruction during sleep.

CONCLUSION

Monitoring of respiration during sleep allows the assessment of physiological variables that are required to characterize SRBD events. The patency of the upper airway, oxygenation, and the pattern of breathing are usually inferred from simultaneous measurements of different variables. The superiority of nasal pressure to monitor flow over thermal sensors has been accepted. Monitoring of respiratory effort using the PTT and of impedance using the FOT appears to be promising in selected cases. As new techniques of respiratory monitoring emerge, the clinician needs to be familiar with the advantages and shortcomings of each modality.

REFERENCES

1. Clark SA, Wilson CR, Satoh M, Pegelow D, Dempsey JA. Assessment of inspiratory flow limitation invasively and noninvasively during sleep. *Am J Respir Crit Care Med* **158**(3):713–722(1998).

2. Baydur A, Behrakis PK, Zin WA, Jaeger M, Milic-Emili J. A simple method for assessing the validity of the esophageal balloon technique. *Am Rev Respir Dis* **126**(5):788–791(1982).

3. Milic-Emili J, Mead J, Turner JM, Glauser EM. Improved technique for estimating pleural pressure from esophageal balloons. *J Appl Physiol* **19**:207–211(1964).

4. Asher MI, Coates AL, Collinge JM, Milic-Emili J. Measurement of pleural pressure in neonates. *J Appl Physiol* **52**(2):491–494(1982).

5. Chervin RD, Aldrich MS. Effects of esophageal pressure monitoring on sleep architecture. *Am J Respir Crit Care Med* **156**(3 Pt 1):881–885(1997).

6. Sleep-related breathing disorders in adults: recommendations for syndrome definition and measurement techniques in clinical research. The Report of an American Academy of Sleep Medicine Task Force. *Sleep* **22**(5):667–689(1999).

7. Guilleminault C, Stoohs R, Clerk A, Cetel M, Maistros P. A cause of excessive daytime sleepiness. The upper airway resistance syndrome. *Chest* **104**(3):781–787(1993).

8. Chediak AD, Demirozu MC, Nay KN. Alpha EEG sleep produced by balloon catheterization of the esophagus. *Sleep* **13**(4):369–370(1990).

9. Ayappa I, Norman RG, Krieger AC, Rosen A, O'Malley RL, Rapoport DM. Non-invasive detection of respiratory effort-related arousals (RERAs) by a nasal cannula/pressure transducer system. *Sleep* **23**(6):763–771(2000).

10. Stoohs RA, Knaack L, Guilleminault C. Non-invasive estimation of esophageal pressure based on intercostal EMG monitoring. In: *26th Annual International Conference of the IEEE-EMBS*, September 1-5, 2004; San Francisco, CA, pp 3867–3869.

11. Smith RP, Argod J, Pepin JL, Levy PA. Pulse transit time: an appraisal of potential clinical applications. *Thorax* **54**(5):452–457(1999).

12. Argod J, Pepin JL, Smith RP, Levy P. Comparison of esophageal pressure with pulse transit time as a measure of respiratory effort for scoring obstructive nonapneic respiratory events. *Am J Respir Crit Care Med* **162**(1):87–93(2000).

13. Pitson DJ, Sandell A, van den Hout R, Stradling JR. Use of pulse transit time as a measure of inspiratory effort in patients with obstructive sleep apnoea. *Eur Respir J* **8**(10):1669–1674(1995).

14. Argod J, Pepin JL, Levy P. Differentiating obstructive and central sleep respiratory events through pulse transit time. *Am J Respir Crit Care Med* **158**(6):1778–1783(1998).

15. Pitson DJ, Stradling JR. Autonomic markers of arousal during sleep in patients undergoing investigation for obstructive sleep apnoea, their relationship to EEG arousals, respiratory events and subjective sleepiness. *J Sleep Res* **7**(1):53–59(1998).

16. Heitman SJ, Atkar RS, Hajduk EA, Wanner RA, Flemons WW. Validation of nasal pressure for the identification of apneas/hypopneas during sleep. *Am J Respir Crit Care Med* **166**(3):386–391(2002).

17. Norman RG, Ahmed MM, Walsleben JA, Rapoport DM. Detection of respiratory events during NPSG: nasal cannula/pressure sensor versus thermistor. *Sleep* **20**(12):1175–1184(1997).

18. Trang H, Leske V, Gaultier C. Use of nasal cannula for detecting sleep apneas and hypopneas in infants and children. *Am J Respir Crit Care Med* **166**(4):464–468(2002).

19. Johnson PL, Edwards N, Burgess KR, Sullivan CE. Detection of increased upper airway resistance during overnight polysomnography. *Sleep* **28**(1):85–90(2005).

20. Montserrat JM, Farre R, Ballester E, Felez MA, Pasto M, Navajas D. Evaluation of nasal prongs for estimating nasal flow. *Am J Respir Crit Care Med* **155**(1):211–215(1997).

21. Lorino AM, Lorino H, Dahan E, et al. Effects of nasal prongs on nasal airflow resistance. *Chest* **118**(2):366–371(2000).

22. Farre R, Montserrat JM, Rotger M, Ballester E, Navajas D. Accuracy of thermistors and thermocouples as flow-measuring devices for detecting hypopnoeas. *Eur Respir J* **11**(1):179–182(1998).

23. Xiong C, Sjoberg BJ, Sveider P, Ask P, Loyd D, Wranne B. Problems in timing of respiration with the nasal thermistor technique. *J Am Soc Echocardiogr* **6**(2):210–216(1993).

24. Masa JF, Corral J, Martin MJ, et al. Assessment of thoraco-abdominal bands to detect respiratory effort-related arousal. *Eur Respir J* **22**(4):661–667(2003).

25. Cohen KP, Ladd WM, Beams DM, et al. Comparison of impedance and inductance ventilation sensors on adults during breathing, motion, and simulated airway obstruction. *IEEE Trans Biomed Eng* **44**(7):555–566(1997).

26. Boudewyns A, Willemen M, Wagemans M, De Cock W, Van de Heyning P, De Backer W. Assessment of respiratory effort by means of strain gauges and esophageal pressure swings: a comparative study. *Sleep* **20**(2):168–170(1997).

27. Baker LE, Geddes LA, Hoff HE, Chaput CJ. Physiological factors underlying transthoracic impedance variations in respiration. *J Appl Physiol* **21**(5):1491–1499(1966).

28. Wukitsch MW, Petterson MT, Tobler DR, Pologe JA. Pulse oximetry: analysis of theory, technology, and practice. *J Clin Monit* **4**(4):290–301(1988).

29. Farre R, Montserrat JM, Ballester E, Hernandez L, Rotger M, Navajas D. Importance of the pulse oximeter averaging time when measuring oxygen desaturation in sleep apnea. *Sleep* **21**(4):386–390(1998).

30. Jubran A. Advances in respiratory monitoring during mechanical ventilation. *Chest* **116**(5):1416–1425(1999).

31. Jubran A, Tobin MJ. Monitoring during mechanical ventilation. *Clin Chest Med* **17**(3):453–473(1996).

32. Sanders MH, Kern NB, Costantino JP, et al. Accuracy of end-tidal and transcutaneous P_{CO_2} monitoring during sleep. *Chest* **106**(2):472–483(1994).

33. Carroll JL. Obstructive sleep-disordered breathing in children: new controversies, new directions. *Clin Chest Med* **24**(2):261–282(2003).

34. Clark JS, Votteri B, Ariagno RL, et al. Noninvasive assessment of blood gases. *Am Rev Respir Dis* **145**(1):220–232(1992).

35. Farre R, Montserrat JM, Navajas D. Noninvasive monitoring of respiratory mechanics during sleep. *Eur Respir J* **24**(6):1052–1060(2004).

36. Badia JR, Farre R, Rigau J, Uribe ME, Navajas D, Montserrat JM. Forced oscillation measurements do not affect upper airway muscle tone or sleep in clinical studies. *Eur Respir J* **18**(2):335–339(2001).

37. Montserrat JM, Farre R, Navajas D. New technologies to detect static and dynamic upper airway obstruction during sleep. *Sleep Breath* **5**(4):193–206(2001).

132

RECORDING AND MONITORING LIMB MOVEMENTS DURING SLEEP

CRINTZ E. SCOTT

St. Anthony Sleep Disorders Center, Denver, Colorado

Restless legs syndrome (RLS) and periodic limb movement disorder (PLMD) are primary intrinsic sleep disorders that are commonly seen in the general population. Five to 10 of every 100 people experience the symptoms of RLS sometime in their lifetime. Periodic limb movement disorder is typically rare in people under the age of 30 but becomes more common in older populations. Approximately 80% of people with RLS have PLMD, though most people with PLMD do not experience RLS.

RESTLESS LEGS SYNDROME

Patients with RLS complain of an irresistible urge to move their legs while seated or lying down. Some patients complain of a "creepy" or "crawly" sensation in their legs. These sensations may persist for the entire day or may only be present in the evening, making it difficult for the person to sit motionless for any duration. Many patients have reported that the symptoms do subside with walking [1].

PERIODIC LIMB MOVEMENT DISORDER

Periodic limb movement disorder affects patients primarily during sleep. The condition is characterized by periodic episodes of limb movements. Many patients who have the disorder are unaware of the limb movements. The most common complaint from patients with this disorder is excessive daytime sleepiness [1].

RECORDING LIMB MOVEMENTS

Proper electrode placement is essential for recording limb movements [2]. Limb movements can be seen and recorded in the legs as well as the arms. The preferred placement to record leg movements is on the belly of the anterior tibialis muscle group. An effective method to find this site is to have the patient flex and point their big toe while seated. The belly of the muscle is the site where the muscle has the greatest bulge. When monitoring the arms, the extensor digitorium is the best site for recording. To locate it, have the patient extend the forearm with palm downward. Then, feel the upper surface of the forearm as the patient bends the wrist up and down. Amputees require special consideration for such placement. Once the belly is located, two electrode sites are selected. The electrodes should be 2–4 cm apart from one another. If the electrodes are placed too close together, a salt bridge may occur. Limb movements can be seen in one or both limbs; therefore both limbs should be monitored even if one leg channel is available on the polygraph montage.

Surface Electrodes

Surface electrodes traditionally are the most popular choice for limb monitoring in the sleep laboratory. Gold cup, silver–silver chloride, or disposable electrodes may be utilized. They all have proved reliable and provide adequate signals for recording limb movements. When using surface electrodes, it is imperative that the electrode sites are

Sleep: A Comprehensive Handbook, Edited by T. Lee-Chiong.

properly prepped prior to attaching the electrodes to the patient. Mild abrasive skin preparation should be used to rid the skin of excess oils and dirt. Special attention is required for diabetic patients and patients with fragile skin, to avoid abrasions.

Piezoelectric Sensors

There have been many advances in sleep medicine throughout the years, including ancillary equipment that can be used to measure and monitor limb movement. Piezoelectric sensors create a small electric signal when they encounter motion, vibration, or tension. The limb sensor is either worn around the ankle or placed around the anterior tibialis muscle. Piezoelectric sensors are more expensive than surface electrodes but they do not require any skin preparation (Figure 132.1).

Recording and Verification

Proper filter settings are essential for recording limb movements [2]. EMG activity is typically fast with a range of 20–200 cycles/second or hertz (Hz). The filter settings should be set appropriately for the best signal. The low-frequency filter (LFF) should be set at 5 or 10 Hz, while the high-frequency filter (HFF) should be set at >70 Hz. Sensitivity for the limb channels should be adjusted for

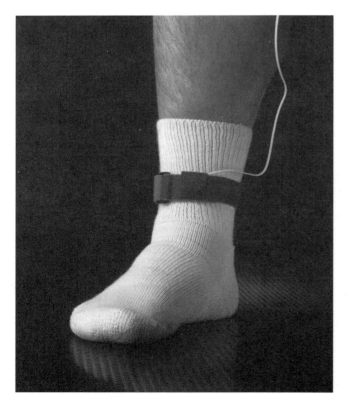

Figure 132.1 Piezoelectric sensor.

optimum recording. The sensitivity should be set at 50 μV/cm for optimal visualization of the limb movement.

Biocalibrations

It is essential to check the integrity of the limb electrode prior to beginning the sleep study. Checking the amplitude of the signal during the physiologic calibrations provides baseline data for the interpretation of the limb movements. The proper technique for performing the physiologic calibrations is to have the patient lie supine in bed without any resistance. The patient is then instructed to dorsiflex and plantarflex the big toe. Each leg should be done individually when using one or two recording channels. The EMG burst should easily be visualized (Figure 132.2).

Event Detection

There are specific rules that are used to score limb movements for PLMD. These include:

1. The limb movement must be 0.5–5.0 seconds in duration to be scored. This is the duration of onset to offset for the individual limb movement.
2. The limb movements recorded during the sleep study should be at least 25% in amplitude of calibrated signal to be counted as a limb movement.
3. The limb movements must be periodic. They need to occur in series of four or more limb movements. The duration between the limb movements must be 5–90 seconds from onset to onset of the limb movements.
4. Limb movements may only be scored during sleep [3] (Figure 132.3).

Event Classification

During the scoring process it is important to analyze the limb movements to assess the potential sleep disturbances since many patients complain of excessive daytime sleepiness. It is helpful to classify limb movements, such as limb movements with an arousal, limb movements without an arousal, or limb movements associated with respiratory events.

Arousal Rules for PLMs

Rules have been established for arousals associated with PLMs. The onset of EEG arousal (standard arousal criteria of 3 seconds or more of clear EEG acceleration and desynchronization) must be concurrent with or after the limb movement. In addition, EEG arousals should follow the limb movement termination by no more than 1–2 seconds.

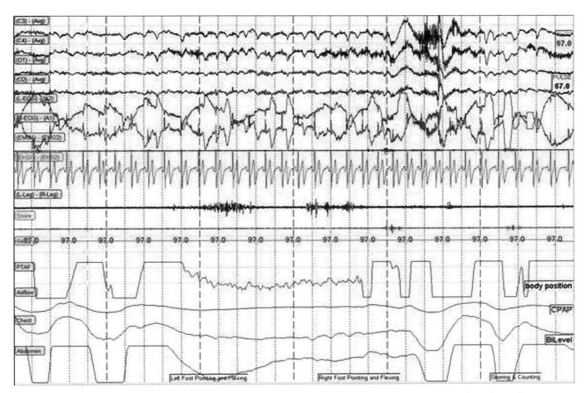

Figure 132.2   Each leg should be done individually when using one or two recording channels.

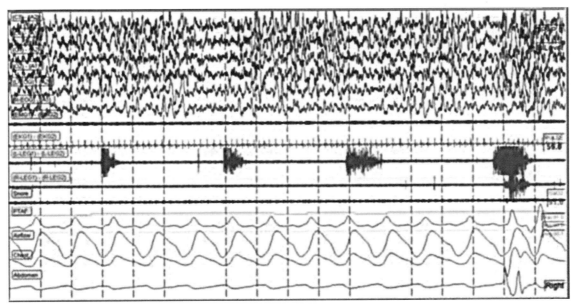

Figure 132.3   Sequencing is essential while scoring limb movements. There must be 5–90 seconds from onset to onset of the limb movements. They need to occur in series of four or more limb movements. The four limb movements recorded in the above 90 second epoch meet these scoring requirements.

Event Analysis and Calculation

Once the limb movements have been scored, the limb movements can be separated into different categories for meaningful interpretation, including (1) number of limb movements, (2) number of limb movements with arousals but not awakenings, (3) number of limb movements with awakenings for one or more epochs, and (4) number of limb movements associated with respiratory events.

The calculation for limb movements is similar to calculating the apnea–hypopnea indices, such as total number/total sleep time, number with arousal/total sleep time, number with awakening/total sleep time, number with respiratory events/total sleep time, and number without arousal/total sleep time.

Artifacts

Artifacts are not uncommon on the limb channels. Due to the wide range of filter settings on the channel and the long lead wires, many artifacts will taint the signal. Some of the most common artifacts are ECG artifact and 60 Hz artifact. ECG artifact may be seen if the interelectrode distance is greater than recommended. Electrical interference or electrode popping can cause 60 Hz artifact.

SUMMARY

It is important to assure that lead placement is correct and biocalibrations are clear from the start of the recording. PLMs are scored according to duration and relationship to arousals, awakenings, and respiratory events.

REFERENCES

1. American Sleep Disorders Association. *International Classification of Sleep Disorders, Revised: Diagnostic and Coding Manual*. American Slep Disorders Association, Rochester, MN 1997, pp 65–71.
2. Marshall B, Davila DG. Monitoring limb movements during sleep. In: Lee-Chiong TL, Sateia MJ, Carskadon MA (Eds), *Sleep Medicine*. Elsevier, Philadelphia, 2002, pp 647–659.
3. Recording and scoring leg movements. The Atlas Task Force. *Sleep* **16**(8):748–759(1993).

133

ACTIGRAPHY

CHRISTINE ACEBO

E.P. Bradley Hospital/Brown Medical School, East Providence, Rhode Island

INTRODUCTION

Actigraphs are small computerized devices that record and store data generated by movement. Modern actigraphs are the size of wristwatches and can record data continuously for days, weeks, or longer. Actigraphy is a general term for any of several systems for recording and analyzing movement. These various systems usually include actigraph devices and procedures and software for collecting data, transferring it to a computer, and analyzing the digital record for levels of activity/inactivity or with scoring algorithms to provide estimates of sleep and wake for aggregated epochs of continuously sampled movement data. Several actigraph systems are commercially available and the components of these systems are often unlike each other in important ways. Beyond cosmetic differences, actigraph devices from individual companies collect movement data in fundamentally distinct ways and have available unique combinations of user-definable settings. These variations have important consequences. For example, if we attach two actigraphs from competing companies to the same moving object, the activity counts recorded by the devices may not be correlated to any degree. Additionally, if we compare two actigraphs from the same company but recording in different "modes," the activity counts recorded by the devices may not be correlated to any degree. For this reason, algorithms derived for estimating sleep and wake from one kind of device will likely be inappropriate for other kinds of devices. The final results for some measures (e.g., minutes of sleep) may be very similar while the results for other measures may be quite different (e.g., number of awakenings or amount/amplitude of

activity). In addition, an algorithm developed for data from one mode of recording will yield different results on data from the same actigraph set for a different mode of recording. For these reasons it is important to understand that generalities about validity of actigraph measures may not apply to all actigraphy systems.

The researcher or clinician interested in using this technology is faced with the task of choosing between a number of overtly attractive systems but finds little consensus in the literature as to which system is best. This chapter will not attempt to answer that question. Rather, the aim is to provide a short review of validity and reliability issues, discussion of especially applicable and especially problematic research and clinical applications, and some recommendations for using actigraphy effectively. For those interested in using this methodology the reviews included in the Additional Reading section can provide an understanding of the range of uses, strengths, and limitations of actigraphy. Thus armed and as an informed consumer, the potential user can query company representatives about the features of their systems and request peer-reviewed publications reporting validity analyses.

VALIDITY OF MEASURES FROM ACTIGRAPH SYSTEMS

Assessment of validity and reliability of measures from the various combinations of device, recording mode, and scoring algorithm has not been completed for all of the current actigraphy systems on the market, and new devices, algorithms, and operating procedures are constantly being

Sleep: A Comprehensive Handbook, Edited by T. Lee-Chiong.
Copyright © 2006 John Wiley & Sons, Inc.

developed. Several reviews have described and summarized the results of laboratory studies assessing the validity of actigraph measures from some of these systems [1–3]. Other systems have no peer-reviewed published validity studies. General conclusions consistent with the reviews are presented here but the reader is encouraged to go to the reviews for additional information.

Overall, assessment studies have shown that epoch-by-epoch agreement for sleep estimated by actigraph algorithms and sleep scored from polysomnographic (PSG) recordings is sufficiently high (more than 0.85 agreement) for normal individuals at a variety of ages and some patient samples. Estimates of wake generally show lower epoch-by-epoch agreement. Correlations between PSG recordings versus actigraphic whole-night measures of sleep and sleep efficiency have generally been sufficiently high (above 0.80) for normal individuals and some patient samples. These results have been relatively consistent regardless of the algorithm/device combination examined.

Accuracy of actigraph sleep and wake measures is generally decreased whenever sleep is disturbed or distorted [4, 5]. Insomnia patients, for example, exhibit more than usual periods of quiet wakefulness likely to be classified as sleep by actigraphy algorithms [6, 7]. Accuracy also decreases for individuals with movement disorders during sleep [7] and those with disturbed sleep due to shift work [8]. The largest discrepancies between actigraph and PSG measures are around transitions—wake to sleep and sleep to wake. Finally, changes in the amount of activity during sleep over the life span may be reflected in differential accuracy of algorithms [9–12]. Some efforts have been made to develop algorithms that are optimized for specific types of individuals, such as very young infants [9], insomnia patients [5], and depressed patients [13]. The increase in accuracy by this means, however, may be offset by limitations in the ability to compare different groups or to use the actigraph measures to screen for the disorder of concern. Validity has not been established for many other clinical samples such as narcoleptics or psychiatric patients on medications.

The results noted above indicate the value of actigraph measures of sleep and sleep efficiency from some systems as reasonable estimates of PSG-determined measures in a variety of samples. They also reveal the limitations of actigraphy for determining the precise timing or amount of waking and the accurate description of sleep and wake for individuals with some sleep disorders or disruptions. Finally, little or no evidence exists for the validity of a large number of other measures routinely provided in the analysis output of actigraph analysis software, including such variables as sleep onset, number of bouts of wake or sleep, longest sleep bout, motionless sleep, movement indices, circadian parameters, and so forth. Additional work is also needed on the development of validated measures of circadian rhythm parameters and measures to describe waking activity.

CAN ACTIGRAPH MEASURES BE USEFUL EVEN IF THEY SHOW LOW VALIDITY?

Scientists in the behavioral sciences routinely use measures that do not show high validity for description and prediction. The error associated with these measures can be mitigated by studying a larger number of individuals, averaging measures over groups of individuals, collecting repeated measures over extended periods of time and/or across multiple treatment conditions, or replicating the study in a new sample. Measures with moderate validity may also provide convergent evidence in studies with multiple methods. Increasing numbers of published reports indicate that actigraph measures can document predicted or predictable differences in sleep–wake patterns between groups of individuals. For example, age- and disease-related changes illustrated by actigraph measures have been consistent with results using other methods. Changes associated with behavioral, drug, light exposure, and other experimental interventions have also been documented successfully with actigraph measures. These and other studies are described and reviewed in the Additional Reading material. Finally, actigraphy is increasingly used to document compliance with fixed sleep schedules imposed during at-home periods before in-lab experimental studies. Knowledge of the sleep history of experimental volunteers is increasingly important for understanding the results of studies investigating the effects of sleep restriction or circadian and homeostatic processes, among others.

ACTIGRAPHY FOR CLINICAL SLEEP DISORDERS

Until recently actigraphy was not considered suitable for diagnosis of sleep disorders [1]. The most current reviews and the latest practice parameters paper [2, 3, 14] have modified that consensus to suggest that although diagnosis is still not appropriate from actigraphy alone for such disorders as sleep disordered breathing or periodic limb movements, actigraphy may be useful for the routine evaluation of insomnia, circadian rhythm disorders, excessive sleepiness, and restless legs/periodic limb movement disorders. Actigraphy may be the only technique possible for monitoring special groups such as children and demented patients. The ability to monitor sleep and sleep scheduling for long periods of time makes actigraphy especially useful as an outcome measure in some treatment protocols.

ARTIFACT CONTROL, PRACTICAL ISSUES, AND SCORING PROCEDURES

The validity studies described earlier were from laboratory studies where control is maximized over situations that lead to artifacts in actigraph records. Recordings made on individuals in their homes and during normal day-to-day activities, on the other hand, often contain numerous artifacts that must be identified and discarded from analysis. Individuals take the actigraph off while bathing, swimming, or during athletic activities, sometimes forgetting to put it on again before going to bed for the night. Individuals at home spend hours quietly watching television or reading, drink alcohol or consume caffeinated drinks before bedtime, use medications that can affect movement or sleep, and sleep with bedpartners. Times when the actigraph is not worn will be scored as sleep by most algorithms, as will quiet waking. Times of excessive movement by the individual or by the bedpartner during sleep periods may or may not be scored correctly as sleep or wake. A common cold during one or two nights of recording may result in an actigraph pattern that is not characteristic of the individual's normal sleep whether or not medication was taken.

A common complaint of new users of this methodology is that they cannot determine where sleep periods start and end simply by looking at the actigraph recording. For example, showers just before bedtime or right after risetime can be remarkably difficult to separate from the sleep period. Some teenagers and young adults spend a remarkable amount of time lying around the house on weekends or evenings, including just before and just after the overnight sleep period; these times of low activity look like (and are scored by algorithms as) multiple or excessively long sleep periods. The lack of procedures to identify such artifacts can make scoring the records a nightmare at best; at worst the data may be entirely unusable. It is heartbreaking to hear from investigators who have collected actigraphy recordings from many individuals and only later realized they couldn't see what was artifact and what was pattern.

Daily logs or diaries provide the most efficient means of documenting events and situations that lead to artifacts in the records and allow the development of scoring procedures that restrict algorithm scoring to documented sleep periods. The log should query information about bedtimes and risetimes, times the actigraph was not worn, alcohol, caffeine, or medication use, illnesses, times sleeping in moving vehicles, and so forth. After actigraph data are retrieved, the record should be printed and carefully compared to the information in the log. As soon as possible, the patient or subject should be questioned about any ambiguous periods on the record or discrepancies between log and record. A feature of some actigraphs that can add information is a button that can be pushed at bedtime and risetime to add an event mark. Some actigraphs can record light level and transpose it on the activity record.

The duration of recordings is an issue that should be considered carefully. The Standards of Practice Committee of the American Academy of Sleep Medicine [14, 15] recommends a minimum of three consecutive 24 hour periods of recording time with the caveat that this may not be long enough for some applications. Longer recording times are requisite for measures that reliably characterize individuals, particularly when sleep patterns are highly variable or when measures with lower validity are of interest [16]. For example, adolescents and young adults may have large weekday/weekend sleep pattern differences; failure to record for an entire week may lead to incomplete or inaccurate description [17]. Shift worker patterns may also be remarkably variable across a week and that variability may itself be an important indicator of dysfunction. Aggregation of measures across 5–7 nights can substantially increase the reliability of measures [16], thus increasing their predictive validity. For practical reasons it is often just as easy to make appointments a week apart for setting up and returning the actigraph as for shorter stints. Finally, if data loss is likely because of low compliance, illness, or other reasons, seven nights of recording may allow the recovery of enough data for analysis even if problems occur.

Actigraphs are mechanical devices and subject to malfunction, leading to data loss. On some devices external buttons may allow a sound check to make sure the actigraph is working. In any case, procedures should be developed for testing actigraphs that yield suspect recordings. Good working relationships with actigraph vendors should be maintained so that broken devices are repaired promptly.

No standard procedures for scoring actigraph records exist. Company manuals for each system provide some guidelines for setting scoring intervals. Scoring rules may also be adapted from descriptions in methods sections of published reports. In any case, scoring should be accomplished by rule-driven procedures with consensus decisions by primary investigators or clinicians for ambiguous nights. Procedures will likely be a result of compromise between software constraints and research or clinical needs. The rapid and ongoing development of new actigraphs and software makes it unlikely that scoring procedures will be standardized in the near future.

CONCLUSIONS

Actigraphy refers to a methodology for recording and analyzing movement from small, computerized devices worn on the body. Assessments of the validity of actigraph measures, though incomplete, generally show that sleep estimated by scoring algorithms is a relatively good proxy for PSG-scored sleep for normal individuals and some patient groups. Accuracy is decreased when sleep is disrupted, however, and though actigraphy may be suitable for documenting and evaluating some sleep disorders, its role in diagnosis is limited. Actigraphy can be very useful in investigations of group differences, sleep pattern variations over time, and the effects of behavioral or treatment interventions. Artifact control is extremely important and a daily diary or log is a necessary adjunct. Duration of recording should be long enough to provide reliable measures and to capture important variations across time.

REFERENCES

1. Sadeh A, Hauri PJ, Kripke DF, Lavie P. The role of actigraphy in the evaluation of sleep disorders. *Sleep* **18**(4):288–302(1995).

2. Sadeh A, Acebo C. The role of actigraphy in sleep medicine. *Sleep Med Rev* **6**:113–124(2002).

3. Ancoli-Israel S, Cole R, Alessi C, Chambers M, Moorcroft W, Pollak CP. The role of actigraphy in the study of sleep and circadian rhythms. *Sleep* **26**:342–392(2003).

4. Kushida C, Chang A, Gadkary C, Guilleminault C, Carrillo O, Dement WC. Comparison of actigraphic, polysomnographic, and subjective assessment of sleep parameters in sleep-disordered patients. *Sleep Med* **2**(5):389–396(2001).

5. Jean-Louis G, von Gizycki H, Zizi F, et al. Determination of sleep and wakefulness with the actigraph data analysis software (ADAS). *Sleep* **19**:739–743(1996).

6. Chambers MJ. Actigraphy and insomnia—a closer look. *Sleep* **17**:405–408(1994).

7. Hauri PJ, Wisbey J. Wrist actigraphy in insomnia. *Sleep* **15**:293–301(1992).

8. Reid K, Dawson D. Correlation between wrist activity monitor and electrophysiological measures of sleep in a simulated shiftwork environment for younger and older subjects. *Sleep* **22**:378–385(1999).

9. Sadeh A, Acebo C, Seifer R, Aytur S, Carskadon MA. Activity-based assessment of sleep–wake patterns during the first year of life. *Infant Behav Dev* **18**:329–337(1995).

10. Jean-Louis G, Kripke DF, Ancoli-Israel S, et al. Sleep duration, illumination, and activity patterns in a population sample: effects of gender and ethnicity. *Biol Psychiatry* **47**(10):921–927(2000).

11. Reyner A, Horne JA. Gender-related and age-related differences in sleep determined by home-recorded sleep logs and actimetry from 400 adults. *Sleep* **18**:127–134(1995).

12. Ancoli-Israel S, Clopton P, Klauber MR, et al. Use of wrist activity for monitoring sleep/wake in demented nursing-home patients. *Sleep* **20**:24–27(1997).

13. Jean-Louis G, Mendlowicz MV, Gillin JC, et al. Sleep estimation from wrist activity in patients with major depression. *Physiol Behav* **70**(1–2):49–53(2000).

14. Littner M, Kushida K, Anderson W, Bailey D, Berry R, Davila D, Hirshkowitz M, Kapen S, Kramer M, Loube D, Wise M, Johnson S. Practice parameters for the role of actigraphy in the study of sleep and circadian rhythms: an update for 2002. *Sleep* **26**:337–341(2003).

15. Standards of Practice Committee A. An American Sleep Disorders Association Report: practice parameters for the use of actigraphy in the clinical assessment of sleep disorders. *Sleep* **18**:285–287(1995).

16. Acebo C, Sadeh A, Seifer R, et al. Estimating sleep patterns with activity monitoring in children and adolescents: how many nights are necessary for reliable measures? *Sleep* **22**:95–103(2003).

17. Carskadon MA, Wolfson AR, Acebo C, et al. Adolescent sleep patterns, circadian timing, and sleepiness at a transition to early school days. *Sleep* **21**:871–881(1998).

ADDITIONAL READING

Ancoli-Israel S, Cole R, Alessi C, Chambers M, Moorcroft W, Pollak CP. The role of actigraphy in the study of sleep and circadian rhythms. *Sleep* **26**:342–392(2003). This article was written under the auspices of the American Academy of Sleep Medicine and updated the information about actigraphy to provide a current state-of-the-art review of actigraphic technologies. The article evaluates newer validity and reliability studies, actigraphy studies of individuals with sleep or circadian rhythm disorders, and studies using actigraphy as treatment outcome measures.

Littner M, Kushida K, Anderson W, Bailey D, Berry R, Davila D, Hirshkowitz M, Kapen S, Kramer M, Loube D, Wise M, Johnson S. Practice parameters for the role of actigraphy in the study of sleep and circadian rhythms: an update for 2002. *Sleep* **26**:337–341(2003). This paper based its recommendations on the 2003 review paper by Ancoli-Israel et al. (2003) (see above) and concluded that actigraphy is not indicated for routine diagnosis, assessment, or management of sleep disorders patients. A number of areas for which actigraphy may be useful include routine evaluation of insomnia, circadian rhythm disorders, excessive sleepiness, restless legs/periodic limb movement disorders, monitoring of special groups (such as children and demented patients), and as an outcome measure in some circumstances.

Sadeh A, Acebo C. The role of actigraphy in sleep medicine. *Sleep Med Rev* **6**:113–124(2002). This review highlights the continued need for additional validity studies and alerts the reader to practical and methodological issues in the use of actigraphy.

Sadeh A, Hauri PJ, Kripke DF, Lavie P. The role of actigraphy in the evaluation of sleep disorders. *Sleep* **18**:288–302(1995). This article was written under the auspices of the American Sleep Disorders Association (now renamed the American Academy of Sleep Medicine) and reviewed technical information, validity and reliability studies, and studies using actigraphy for clinical evaluation and research prior to 1995. The review highlighted problematic methodological issues but concluded that actigraphy showed promise for assessment of some sleep disorders.

134

pH MONITORING AND OTHER ESOPHAGEAL TESTS

GARY R. COTT

National Jewish Medical and Research Center, Denver, Colorado

THE ESOPHAGUS

The esophagus [1] appears to be a very simple organ. It can be characterized as an empty muscular tube with a sphincter at each end that conducts liquids and solids from the pharynx to the stomach and impedes retrograde movement of gastric contents. However, this simplification belies the complex anatomy, physiologic responses, and control mechanisms that are a part of normal esophageal function. Furthermore, esophageal function can be altered by a limited number of disorders that can produce significant symptoms.

Among esophageal disorders, the relationship between gastroesophageal reflux disease (GERD) and sleep has been of particular interest to clinicians and investigators for a variety of reasons. First, in normal subjects almost no esophageal reflux occurs during sleep. However, when nocturnal reflux occurs, physiologic changes associated with sleep may influence the clearance of esophageal acid, upper airway protection, and other defense mechanisms sufficiently to promote an increase in the severity of both esophageal and extraesophageal complications of GERD. Second, nocturnal heartburn has been reported to occur in a large majority of adults with GERD, resulting in sleeping difficulties and impaired daytime function [2]. Third, GERD may result in extraesophageal manifestations (Table 134.1) that can be disruptive of sleep.

A variety of investigative techniques have been employed to study esophageal function. Esophageal pH monitoring has become the most common technique used to investigate the potential role of GERD in producing symptoms. The remaining portions of this chapter are intended to provide a practical review of esophageal pH

monitoring. Several other relevant clinical investigative tools that may aid in the direct assessment of esophageal function are also briefly reviewed.

ESOPHAGEAL pH PROBE MONITORING

Prolonged ambulatory recordings of esophageal pH have been performed in an outpatient setting since the early 1980s. Over the years, increasing knowledge of the clinical utility of these measurements and technologic advancements have resulted in widespread application of ambulatory esophageal pH monitoring in clinical medicine. It has become the best test for quantifying gastroesophageal reflux and determining if symptoms are related to reflux events.

Technical Considerations

The technical components of an ambulatory esophageal pH monitoring system consist in general of one or more electrodes and a battery-powered data-logger or recording device. There are two types of commonly used pH electrodes. Glass electrodes are generally more accurate over a wider pH range (1–12) but are more expensive, fragile, and bulky. Antimony electrodes are less expensive and generally better tolerated by patients but have a shorter life span, are more difficult to sterilize, and may be subject to drift. Comparison studies in patients have yielded similar results with either electrode [3]. Commercially available pH electrodes may be disposable or multiuse with either

Sleep: A Comprehensive Handbook, Edited by T. Lee-Chiong.
Copyright © 2006 John Wiley & Sons, Inc.

TABLE 134.1   Extraesophageal Symptoms and Complications of Gastroesophageal Reflux

Symptoms	Potential Complications
Wheezing	Bronchitis
Chronic cough	Pneumonia
Apnea	Bronchiectasis
Stridor	Asthma
Fatigue	Pulmonary fibrosis
Sore throat	Vocal cord dysfunction
Halitosis	Layrngopharyngeal inflammation
Otalgia	Sinusitis
Cervical/neck pain	
Globus sensation	
Hoarseness	
Nasal drainage	

an internal or external reference. Combination electrodes containing an internal reference can be calibrated without the patient and determine hydrogen ion concentration without an external reference. An external reference pH probe requires the patient to be part of the calibration process with a reference electrode attached to the skin throughout the study. Studies conducted with an external reference are subject to more artifacts due to variations in skin contact by the external reference electrode.

In most cases the pH electrode is part of an approximately 2–3 mm diameter catheter that passes from the esophagus, along the posterior wall of the pharynx, and through the nose. The pH electrode is attached via this catheter to an external data-logger. The lightweight data-logger is worn by the patient and allows the patient to record the occurrence of symptoms, change in body position, and meals—all of which are important in the interpretation of pH data. The data-loggers usually have several channels to allow for simultaneous recording of pH measurements from different positions within the esophagus. A sampling rate of every 6–8 seconds and resolution of 0.1 pH unit are considered sufficient for clinical testing [4]. At the completion of the study, computer software is used to convert the data from the data-logger into appropriate graphic and summary information.

A new radiotelemetry esophageal pH probe (BRAVO™ pH Monitoring System, Medtronics, Minneapolis, MN) requires the attachment of a pH "capsule" to the esophageal mucosa but subsequently allows pH monitoring for up to 48 hours free of the sometimes uncomfortable and potentially embarrassing transnasal catheter. This potentially more comfortable and less obtrusive system may encourage more normal everyday patient activity and thus provide a more realistic representation of associated reflux events. Current contraindications for its specific use include bleeding diathesis, strictures, severe esophagitis, varices, obstructions, pacemakers, or implantable defibrillator devices.

Preparation and Placement

Patient instruction should occur several days prior to the study and include a thorough review of the procedure, the need to limit fluid and food intake for 4–6 hours prior to insertion of the pH catheter, and a discussion of current medication use and what medications should be maintained or discontinued for the study. Depending on the ordering physician's preference, antireflux medications are usually discontinued for diagnostic studies while other medications are maintained even if they have known physiologic effects on the esophagus.

Recommendations vary, but antacids are usually discontinued 6–24 hours, histamine-2 blockers (as well as prokinetic agents and smooth muscle relaxants) 48–72 hours, and proton pump inhibitors at least 7 days prior to the study [4]. If the study is conducted to determine the efficacy of antireflux therapy, these medications should be maintained.

The pH electrode(s) is calibrated prior to insertion using standardized buffer solutions. The pH catheter is then typically inserted through a nostril into the esophagus to the desired depth and taped to the nose to maintain its position. By convention, the distal electrode is placed 5 cm above the superior margin of the lower esophageal sphincter (LES). Theoretically, this distance should avoid electrode displacement into the stomach, especially during swallow-induced esophageal shortening. This is also the standard position for which reference data differentiating normal and abnormal reflux is based. Accurate placement of the pH probe is crucial to correctly interpreting the results of the study. Acid detection drops by 50% at 10 cm above the sphincter and may alter the diagnosis in approximately half of patients compared to studies conducted at 5 cm above the sphincter [5]. Manometric determination of the depth of the LES is the preferred method to establish placement of the electrode. Alternative placement techniques have been used including fluoroscopy or radiograph localization, endoscopic localization, and a pH step-up procedure that utilizes the change in pH between the stomach and body of the esophagus for placement. None of these techniques have proved sufficiently accurate to replace manometric placement [4], although fluoroscopy or radiography localization is often recommended for pediatric or other patients for whom manometry may be contraindicated.

More proximal measurements of pH can be made in an effort to assess the potential for GERD to produce respiratory, laryngeal, pharyngeal, or ear, nose, and throat symptoms. There is no consensus for proximal probe placement and studies have been conducted with pH sensors below the upper esophageal sphincter, above the upper esophageal sphincter, in the hypopharynx, and in the trachea. Although placing the probe in the hypopharynx or trachea may provide better differentiation between healthy individuals and potentially pathologic GERD, considerable drying artifact and variability occur in pH recordings above the upper esophageal sphincter. Thus proximal pH electrodes in most patient studies are placed immediately below the upper esophageal sphincter where hydration is more consistent, resulting in less variance in electrical impedance [6–8]. Dual esophageal pH probes are commercially available with the proximal pH electrode placed at 5, 10, 15, or 20 cm above the distal electrode. Properly positioned dual pH probes with dual channel data-loggers can allow for the simultaneous determination of distal and proximal esophageal reflux and should be considered when assessing for more proximal GERD that may be responsible for extraesophageal manifestations.

Following insertion of the pH probe, the patient should be instructed on use of the data-logger to record symptoms, meals, body position, and other relevant events. This information is often supplemented by having the patient keep a handwritten diary during the test interval to provide more detail regarding the contents of meals, specific nature of symptoms, activities, and so on. To maximize the diagnostic yield of the test the patient should maintain his/her normal routine during the test period. These activities should include meals, taking approved medications, upright activity, and periods of recumbency and sleep. They may also include potential reflux-evoking activities such as exercise or consumption of specific food and beverages known to induce symptoms. The only restricted activities are bathing or similar activities while wearing the data-logger, most gastrointestinal diagnostic tests, tests requiring sedation, pulmonary function testing, and radiologic studies in which the data-logger is not shielded.

Indications and Contraindications

The American Gastroenterological Association has developed guidelines (Table 134.2) to assist the physician in the appropriate use of esophageal pH testing in patient care [9]. Esophageal pH testing was not recommended to detect or verify reflux esophagitis as this is an endoscopic diagnosis; nor was it recommended to evaluate for "alkaline reflux."

Contraindications for this procedure include poor patient cooperation, coumadin or similar anticoagulants, significant coagulopathy, recent gastric surgery, esophageal tumors or ulcers, esophageal varices, obstruction, maxillofacial trauma, cardiac instability, and possibly latex allergy.

Data Interpretation

An accurate interpretation of pH probe data begins with assuring that the various aspects (described earlier) of patient instruction, technical preparation, and probe

TABLE 134.2 Guidelines for the Clinical Use of Esophageal pH Recording

- To document abnormal esophageal acid exposure off antireflux medication in an endoscopy-negative patient being considered for surgical antireflux repair.
- To evaluate patients after antireflux surgery off antireflux medication who are suspected of having ongoing abnormal reflux.
- To evaluate patients with either normal or equivocal endoscopic findings and reflux symptoms that are refractory to proton pump inhibitor therapy.
- To detect refractory reflux in patients with chest pain after cardiac evaluation and a 4 week proton pump inhibitor trial using a symptom reflux association scheme.
- To evaluate a patient with suspected otolaryngologic manifestations (laryngitis, pharyngitis, chronic cough) of gastroesophageal reflux disease after failure to respond to 4 weeks of proton pump inhibitor therapy.
- To document concomitant gastroesophageal reflux disease off antireflux medication in an adult onset, nonallergic asthmatic suspected of having reflux-induced asthma.

placement have been adequately adhered to in order to produce a reliable study. The data tracing should then be screened for potential technical artifacts. Once the reliability of the data for interpretation has been assured, analysis may continue by comparison of patient data to normative data and correlation of reflux episodes to symptoms.

Important considerations for interpreting pH probe data include defining the pH of acid reflux (usually a fall of pH to less than 4.0), identifying when the patient is upright versus recumbent, identifying the time and content of meals, and an adequate study duration. Since the pH probe cannot discriminate between the ingestion of acidic substances and true reflux events, identifying these episodes for potential elimination from the study may improve the reliability of diagnosing GERD [10, 11]. Furthermore, brief periods of normal or "physiologic" reflux can be identified following meals. Study durations of 24 hours generally provide more reliable and reproducible data than 8 or 12 hour studies [12, 13], although 16 hour studies from 1600 to 0800 hours can provide accurate information and improve patient tolerance [14]. In practice, most clinical studies are not performed for a full 24 hours but must be of sufficient duration (18–24 hours) to include major postprandial periods and a full overnight recumbent interval.

Parameters from the analysis of pH probe studies have been established to define normal versus abnormal gastroesophageal reflux into the distal esophagus. The parameters most commonly used include the percentage of time pH < 4.0 during upright, recumbent, and total intervals; the number of reflux episodes; the number of reflux episodes greater than 5 minutes duration; and the duration of the longest episode. Collectively, these parameters are felt to provide information reflecting total esophageal exposure to acid, the frequency of exposure, and the potential for slow esophageal clearance of acid or sustained exposure. Johnson and DeMeester popularized the use of composite scores that take into account all six of these parameters [15, 16]. Jamieson et al. [17] evaluated the sensitivity, specificity, and reproducibility of each of the individual parameters compared to the final composite score and found that the composite score and percent total time pH < 4.0 were the best discriminators of normal versus abnormal gastroesophageal reflux. Similarly, the higher the composite score and/or exposure time, the better the correlation with the appearance of reflux esophagitis [4]. However, the time of acid exposure correlates less well to typical reflux symptoms [18, 19].

The temporal correlation between symptoms and reflux events may be a more important diagnostic tool than a normal versus abnormal incidence of acid reflux in patients suspected of having symptomatic manifestations of GERD. A symptom index (number of reflux-related symptom episodes divided by the total number of symptom episodes) of greater than 50% is defined as "positive" in correlating heartburn or chest pain with reflux [20]. The higher the percentage the better the correlation while an index of less than 25% predicted lack of reflux association [21]. The symptom sensitivity index (symptom-associated reflux events divided by total reflux events) takes into account the total reflux events and is considered positive with values of 10% or higher [22]. However, both of these indexes use arbitrary cutoff points for positive results and do not account for all the potential associations between pain and reflux (positive and negative) in attempting to predict correlations. The symptom–association probability was developed to provide a statistically meaningful cause-and-effect analysis [23]. By statistical convention, symptom–association probability values greater than 95% are positive. Despite the evolution in sophistication of analyzing symptom–reflux associations, the diagnostic accuracy and the predictability for potential therapeutic response of these symptom scoring techniques have not been well established, especially for symptoms believed to be associated with extraesophageal manifestations of GERD.

The interpretation of pH data from the proximal esophagus suffers from poor sensitivity and reproducibility of results [6], in part related to technical aspects as related above. Despite these considerable shortcomings, abnormal levels of proximal acid exposure have in some studies correlated with respiratory symptoms [24] and may predict symptomatic response with GERD therapy [25, 26].

OTHER TESTS

Diagnostic Endoscopy

Upper gastrointestinal (GI) endoscopy [27–29] or esophagoscopy is the diagnostic standard for documenting esophageal mucosal abnormalities. Specific guidelines for endoscopy vary, but usual indications would include the presence of "alarm" symptoms (e.g., dysphagia, odynophagia, weight loss, and GI bleeding) that might signal the presence of infections, ulcers, cancer, strictures, varicies, and so on. Endoscopy is also recommended in patients who have chronic GERD with or without the above symptoms to rule out Barrett's esophagus, a potentially malignant complication of reflux. The presence of reflux esophagitis by endoscopy confirms a subset of patients with GERD showing histopathologic mucosal injury related to reflux. However, the overall sensitivity of endoscopy for diagnosing GERD is relatively low with only 40–60% of patients with abnormal esophageal reflux by pH testing having endoscopic evidence of esophagitis [28].

Esophageal Manometry

Esophageal manometry [30, 31] is used to assess esophageal peristaltic activity as well as upper and lower esophageal sphincter pressure and relaxation characteristics. The low prevalence of clinically significant motility disorders in symptomatic patients and the limited specificity of manometric data have led to the development of practice guidelines (Table 134.3) by the American Gastroenterological Association for the clinical use of this test [32]. Routine manometry including assessment of LES pressures in patients suspected of having GERD does not have diagnostic utility, in part because most acid reflux events in GERD patients occur as a result of a transient inappropriate LES relaxation from a normal baseline sphincter pressure [33].

Esophageal Multichannel Intraluminal Impedance Testing

Multichannel intraluminal impedance (MII) is a new technique recently approved by the United States Food and Drug Administration to detect intraluminal bolus movement [34, 35]. When combined with manometry it can provide information on the functional component of manometrically detected contractions. Together, these two tests can help separate motility abnormalities into those with pressure-only defects and those with pressure and bolus transit defects. When combined with pH testing, MII allows for detection of gastroesophageal reflux as either acid or nonacid reflux. The combination MII–pH test may be particularly useful in characterizing reflux in GERD patients with persistent symptoms despite

TABLE 134.3 Guidelines for the Clinical Use of Esophageal Manometry

- To establish the diagnosis of suspected cases of achalasia or diffuse esophageal spasm. Because of the low prevalence of these diagnoses in patients with esophageal symptoms, more common esophageal disorders should be excluded with barium radiographs or endoscopy before manometric evaluation.

- To detect esophageal motor abnormalities associated with systemic disease (e.g., connective tissue diseases) if their detection would contribute to establishing a multisystem diagnosis or to other aspects of management.

- To place intraluminal devices (e.g., pH probes) when positioning is dependent on the relationship to functional landmarks, such as the lower sphincter.

- For the preoperative assessment of peristaltic function in patients being considered for antireflux surgery; indicated in this setting if uncertainty remains regarding the correct diagnosis.

- Manometry is not indicated for making or confirming a suspected diagnosis of gastroesophageal reflux disease.

- Manometry should not be routinely used as the initial test for chest pain or other esophageal symptoms because of the low specificity of the findings and the low likelihood of detecting a clinically significant motility disorder.

aggressive therapy. Currently, the utility of MII testing is limited by the paucity of centers conducting these studies, the relative complexity of the data obtained, and the lack of studies defining the relevance of test results in regard to prognostic value and predicting therapeutic outcomes. Despite these shortcomings, MII testing shows promise as an important tool in assessing complicated motility and reflux conditions.

REFERENCES

1. Castell DO, Richter JE. *The Esophagus*, 3rd ed. Lippincott williams & Wilkins, Philadelphia, 1999.

2. Shaker R, Castell DO, Schoenfeld PS, Spechler SJ. Nighttime heartburn is an underappreciated clinical problem that impacts sleep and daytime function: the results of a Gallup survey conducted on behalf of the American Gastroenterological Association. *Am J Gastroenterol* **98**(7):1487–1493(2003).

3. Vandenplas Y, Helven R, Goyvaerts H. Comparative study of glass and antimony electrodes for continuous esophageal pH monitoring. *Gut* **32**:708–712(1991).

4. Kahrilas P, Quigley E. Clinical esophageal pH recording: a technical review for practice guideline development. *Gastroenterology* **110**:1982–1996(1996).

5. Anggiansah A, Sumboonnanonda K, Wang J, et al. Significantly reduced acid detection at 10 cm compared with 5 cm above lower esophageal sphincter in patients with acid reflux. *Am J Gastroenterol* **88**:842–846(1993).

6. Vaezi MF, Schroeder PL, Richter JE. Reproducibility of proximal probe pH parameters in 24-hour ambulatory esophageal pH monitoring. *Am J Gastroenterol* **92**:825–829(1997).

7. Dobhan R, Castell DO. Normal and abnormal proximal esophageal acid exposure: results of ambulatory dual probe pH monitoring. *Am J Gastroenterol* **88**:25–29(1993).

8. Jacob P, Kahrilas P, Herzon G. Proximal esophageal pH-metry in patients with "reflux laryngitis." *Gastroenterology* **100**:305–310(1991).

9. American Gastroenterological Association medical position statement: guidelines on the use of esophageal pH recording. *Gastroenterol Clin North Am* **110**(6):1981(1996).

10. Wo JM, Castell DO. Exclusion of meal periods from ambulatory pH monitoring may improve diagnosis of esophageal acid reflux. *Dig Dis Sci* **39**:1601–1607(1994).

11. Ter RB, Johnston BT, Castell DO. Exclusion of the meal period improves the clinical reliability of esophageal pH monitoring. *J Clin Gastroenterol* **25**:314–316(1997).

12. Bianchi PG, Pace F. Comparison of three methods of intra-esophageal pH recording in the diagnosis of gastroesophageal reflux. *Scand J Gastroenterol* **23**:743–750(1988).

13. Johnsson F, Joelsson B. Reproducibility of ambulatory oesophageal pH monitoring in the diagnosis of gastroesophageal pH monitoring. *Gut* **29**:886–889(1988).

14. Dobhan R, Castell DO. Prolonged intraesophageal pH monitoring with 16-hour overnight recording. *Dig Dis Sci* **37**:857–864(1992).

15. Johnson LF, DeMeester TR. Twenty-four-hour pH monitoring of the distal esophagus. *Am J Gastroenterol* **62**:323–332(1974).

16. Johnson LF, DeMeester TR. Development of the 24-hour intraesophageal pH monitoring composite scoring system. *J Clin Gastroenterol* **8**(Suppl 1):52–58(1986).

17. Jamieson JR, Stein HJ, DeMeester TR, et al. Ambulatory 24-hr esophageal pH monitoring: normal values, optimal thresholds, specificity, sensitivity and reproducibility. *Am J Gastroenterol* **87**:1102–1111(1992).

18. Mattioli S, Pilotti V, Spangaro M, et al. Reliability of 24-hour home esophageal pH monitoring in diagnosis of gastroesophageal reflux. *Dig Dis Sci* **34**:71–78(1989).

19. Weusten BLAM, Akkermans LMA, van Berge Henegouwen GP, Smout AJPM. Ambulatory combined oesophageal pressure and pH monitoring: relationships between pathological reflux, oesophageal dysmotility and symptoms of oesophageal dysfunction. *Eur J Gastroenterol Hepatol* **5**:1055–1060(1993).

20. Singh S, Richter JE, Bradley LA, Haile JM. The symptom index. Differential usefulness in suspected acid-related complaints of heartburn and chest pain. *Dig Dis Sci* **38**:1402–1408(1993).

21. Rosen SN, Pope CE. Extended esophageal pH monitoring. An analysis of the literature and assessment of its role in the diagnosis and management of gastroesophageal reflux. *J Clin Gastroenterol* **11**:260–270(1989).

22. Breumelhof R, Smout AJPM. The symptom sensitivity index: a valuable additional parameter in 24-hour esophageal pH recording. *Am J Gastroenterol* **86**:160–164(1991).

23. Weusten BLAM, Roelofs JMM, Akkermans LMA, et al. The symptom association probability: an improved method for symptom analysis of 24-hour esophageal pH data. *Gastroenterology* **107**:1741–1745(1994).

24. Tomonaga T, Awad ZT, Filipi CJ, Hinder RA, Selima M, Tercero F Jr, Marsh RE, Shiino Y, Welch R. Symptom predictability of reflux-induced respiratory disease. *Dig Dis Sci* **47**:19–14(2002).

25. Schnatz PF, Castell JA, Castell DO. Pulmonary symptoms associated with gastroesophageal reflux: use of ambulatory pH monitoring to diagnose and to direct therapy. *Am J Gastroenterol* **91**:1715–1718(1996).

26. Harding S, Richter J, Guzzo M, et al. Asthma and gastroesophageal reflux: acid suppressive therapy improves asthma symptoms. *Am J Med* **100**:395–405(1996).

27. Edmundowicz SA. Endoscopy. In: Castell DO, Richter JE (Eds), *The Esophagus*, 3rd ed. Lippincott williams & Wilkins, Philadelphia, 1999, pp 89–99.

28. Richter JE. Diagnostic tests for gastroesophageal reflux disease. *Am J Med Sci* **326**(5):300–308(2003).

29. DeVault KR, Castell DO. The Practice Parameters Committee of the American College of Gastroenterology: updated guidelines for the diagnosis and treatment of gastroesophageal reflux disease. *Am J Gastroenterol* **94**(6):1434–1442(1999).

30. Adhami T, Shay SS. Esophageal motility in the assessment of esophageal function. *Semin Thorac Cardiovasc Surg* **13**(3):234–240(2001).

31. Murray JA, Clouse RE, Conklin JL. Components of the standard oesophageal manometry. *Neurogastroenterol Motil* **15**:591–606(2003).

32. Kahrilas PJ, Clouse RE, Hogan WJ. Esophageal manometry, American Gastroenterology Association policy and position statement. *Gastroenterology* **107**:1865–1894(1994).

33. Dent J, Holloway RH, Toouli R, et al. Mechanisms of lower esophageal sphincter incompetence in patients with symptomatic gastroesophageal reflux. *Gut* **29**:1020–1028 (1988).

34. Tutuian R, Castell DO. Combined multichannel intraluminal impedance and manometry clarifies esophageal function abnormalities: study in 350 patients. *Am J Gastroenterol* **99**:1011(2004).

35. Shay S, Tutuian R, Sifrim D, et al. Twenty-four hour ambulatory simultaneous impedance and pH monitoring: a multicenter report of normal values from 60 healthy volunteers. *Am J Gastroenterol* **99**:1037(2004).

135

PSYCHOLOGICAL ASSESSMENT OF THE SLEEP PATIENT

AMY B. ROBINSON IKELHEIMER AND BRIAN HOYT
National Jewish Medical and Research Center, Denver, Colorado

Behavioral sleep medicine (BSM) has been described by Stepanski and Perlis [1] as the branch of clinical sleep medicine and health psychology that focuses on the identification of the psychological (e.g., cognitive and/or behavioral) factors that contribute to the development and/or maintenance of sleep disorders. BSM also specializes in developing and providing empirically validated cognitive, behavioral, and/or other nonpharmacologic interventions for the entire spectrum of sleep disorders.

In an effort to evaluate the role of psychological functioning on sleep disorders, mental health providers have been called on to perform psychological assessments of individuals with various sleep problems.

GOALS OF THE PSYCHOLOGICAL ASSESSMENT

The goals of a general psychological assessment of the sleep patient include evaluating psychological factors contributing to sleep problems, diagnosing or ruling out psychiatric disorders, and providing treatment recommendations. A more specialized neuropsychological assessment may be performed when neurocognitive abilities such as attention, concentration, memory, and organization are in question. Figure 135.1 is designed to help providers make decisions regarding referrals for psychological and neuropsychological assessment(s).

Evaluating the psychological components of sleep disorders refers to identifying the behavioral and cognitive factors involved in developing or maintaining the problem. Behavioral factors include *behaviors* such as napping during the day, working in bed, or drinking coffee in the evening (often referred to as "sleep hygiene") [2]. Cognitive factors refer to a person's belief system, that is, how a person interprets, understands, or thinks about life events. For example, a person may worry about having a "bad day" if he/she doesn't sleep for 8 hours on a given night.

Psychiatric disorders are diagnosed using criteria detailed in the *Diagnostic and Statistical Manual of Mental Disorders, 4th Edition (DSM-IV)* [3]. If a person presents with a certain cluster of symptoms (e.g., depressed mood, fatigue, poor appetite, difficulty concentrating, disturbed sleep), a disorder is diagnosed (i.e., major depressive disorder). Accurate diagnosing of the sleep patient can be difficult since disturbed sleep is a symptom of many mood and anxiety disorders. Furthermore, sleep disorders, such as insomnia, are often comorbid with other disorders and can be viewed as primary or secondary disorders [4].

Treatment recommendations are based on empirically validated interventions designed to treat sleep disorders, psychological disorders, and/or maladaptive behaviors and cognitions found during the evaluation. If the sleep disturbance is a symptom of or secondary to another disorder, treatment will initially focus on the primary disorder. Insomnia, however, can be treated independent of other comorbid psychological disorders. In addition, insomnia does not always resolve after treatment of the primary disorder. Zayfert and DeViva [5] found evidence of residual insomnia following successful treatment of post-traumatic stress disorder (PTSD) and suggest that specific insomnia interventions may be required to treat the ongoing sleep problem.

Finally, neuropsychological assessment of the sleep patient typically includes an in-depth evaluation of

Sleep: A Comprehensive Handbook, Edited by T. Lee-Chiong.

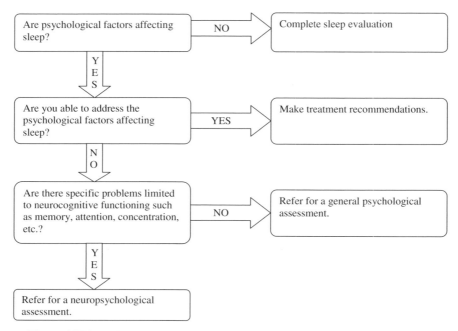

Figure 135.1 When to refer for psychological or neuropsychological assessment.

neurocognitive skills, including intellectual functions, attention, executive functions, learning and memory, language, visuospatial skills, motor skills, and mood/personality. The results of a neuropsychological assessment will provide documentation of current levels of neurocognitive functioning, patterns of cognitive strengths and weaknesses, relationships, between cognitive functioning and the sleep disorder, the impact of cognitive impairments on functioning, and recommendations for compensating for cognitive impairments. Additionally, serial assessments can provide clinicians with important information regarding the functional impact of treatment strategies.

CONDUCTING THE GENERAL PSYCHOLOGICAL ASSESSMENT

During a psychological assessment, the evaluator gathers information most relevant to the patient's sleep and psychological functioning. Rather than relying on any one source, information from the referral source, psychological and sleep-related assessment tools, and clinical interview are used to determine how psychological factors are affecting sleep. A case example is included at the end of this chapter.

Referral Question

To produce the most useful assessment, the evaluator must be familiar with the nature of the problem and clearly understand what the referral source is requesting. Thus by

having a specific question in mind, the referring practitioner can help the evaluator focus on the particular needs of the individual. For example, if the referring practitioner suspects that obesity is contributing to apnea and that anxiety is preventing the patient from using his CPAP machine, the patient may be referred for an assessment of anxiety and weight management issues.

Assessment Tools

Psychological or psychometric instruments, tools, and measures all refer to the wide variety of paper-and-pencil tests that are used to collect information about an individual's psychological functioning. An enormous number of instruments have been developed in order to objectively describe, measure, and quantify different aspects of psychological functioning. Some measure a wide range of psychological characteristics, whereas others focus on specific areas, such as sleep, depression, or anxiety. Test results supplement the more subjective, and sometimes biased, interview data used by clinicians to support hypotheses about a given patient. Quantitative measures also provide objective baselines regarding the severity of symptoms that can serve as a benchmark for measuring progress in treatment.

While giving and scoring a questionnaire may seem routine, assessment tools must only be used by providers well trained in testing, especially test selection and interpretation. A large number of measurement tools exist for any given area, so evaluators need to carefully weigh the strengths and weaknesses of any proposed measure(s). No standard assessment tool or battery of tests has been

TABLE 135.1 General Psychological Assessment Measures

Domain	Instrument
Psychopathology	Minnesota Multiphasic Personality Inventory (MMPI)
	Symptom Checklist-90, Revised (SCL-90)
	Brief Symptom Inventory (BSI)
Functional Status	Profile of Mood States (POMS)
	Medical Outcome Study Short-Form-36 (SF-36)
	Sickness Impact Profile (SIP)
Depression	Beck Depression Inventory (BDI) I/II
	Hamilton Rating Scale
	Zung Self Rating Depression Scale (Zung)
	Multiscore Depression Inventory
	Geriatric Depression Scale (GDS)
Anxiety	Beck Anxiety Inventory (BAI)
	State-Trait Anxiety Inventory (STAI)
	Penn State Worry Questionnaire

identified for the psychological assessment of sleep disorders. Spielman et al. [6] found that, while many sleep centers use some set of structured rating scales and questionnaires in conjunction with interviews for insomnia, the testing varied from center to center. At present, assessment batteries used in sleep settings appear to be chosen according to clinician experience, preference for specific tests, and available resources [7].

The assessment instruments listed in Table 135.1 have been cited in the literature regarding psychological assessment of sleep disorders [6–11]. The specific sleep-related questionnaires covered in Chapter 127 are also used as part of the psychological assessment. If, during the assessment, a specific problem area is suspected, evaluators often follow up with more specific assessment tools. For example, if a patient reports a history of childhood sexual abuse, describes symptoms associated with post-traumatic stress disorder (PTSD), and reveals elevated scores on anxiety measures, a specific measure of PTSD may be employed.

Interview

Structured sleep interviews, inventories, and diaries (see Chapter 127) can guide the practitioner in evaluating the behaviors comprising the sleep problem (e.g., difficulty falling asleep, early morning awakening, difficulty using the CPAP). Information obtained typically includes the onset and duration of the sleep problem, bedtime routines, sleep–wake schedule, weekend/vacation variations, time awake in bed, and number and length of naps. Other factors such as medications, caffeine, nicotine, and alcohol intake, diet, and exercise should be noted as well as behaviors,

thoughts, and emotions occurring before sleep, consequences of disturbed sleep, and how the patient copes with the consequences.

It is useful to have a model of the disorder in mind as pertinent information is collected and synthesized. Spielman and Anderson [12] recommend conceptualizing cases of insomnia into terms of predisposing, precipitating, and perpetuating factors. They suggest that certain individuals, such as those prone to worry, may be *predisposed* to develop insomnia. This person may experience difficulty sleeping when faced with a *precipitating* event, such as starting a new job. Ongoing nighttime worrying about job performance could then *perpetuate* the sleep problem. Harvey [13] proposed a cognitive model of insomnia where anxiety associated with worry about sleep and the consequences of not getting enough sleep (as well as counterproductive behaviors such as drinking alcohol before bed or taking afternoon naps) may, in fact, result in disturbed nighttime sleep and impaired daytime functioning. Both models support the importance of recognizing the behavioral and cognitive factors associated with the sleep problem.

The interview will also include an evaluation of current psychological functioning. The Structured Clinical Interview for DSM-IV Axis I Disorders: Clinician Version (SCID-CV) [14], a semistructured interview, or any modified version, can help guide the clinician through screening the major psychiatric disorders. The clinician will pay special attention to symptoms of concern to the patient or referral source and to any problem areas identified during testing. Past psychiatric history, past and present medical history, family psychiatric and medical history, and current mental status functioning are also included in the psychological assessment.

NEUROPSYCHOLOGICAL ASSESSMENT

Neuropsychology is the study of brain–behavior relationships, with a specific focus on brain or neurocognitive functioning. Neurocognitive impairments in the areas of attention, concentration, vigilance, memory, multitasking, planning, and organization have been associated with sleep disorders. Such difficulties may be related to excessive sleepiness, potential neuropathology, and/or comorbid medical or psychiatric conditions (i.e., depression).

The neuropsychological evaluation typically consists of a battery of neurocognitive tests addressing the specific concerns of the referring physician. Patients may be asked to provide verbal responses (i.e., verbal memory, verbal fluency), to perform motor responses (i.e., visuoconstruction, motor speed), or to perform computerized tasks (i.e., specialized tests of attention). Please see Table 135.2 for a list of measures commonly used in the evaluation of sleep patients [15].

TABLE 135.2 Common Neuropsychological Assessment Measures

Cognitive Domain	Specific Aspects	Instrument
Intellectual functioning	General or estimated level of intelligence	Wechsler Intelligence Scales (WAIS-III, WASI)
Attention	Sustained attention, vigilance	Continuous Performance Test
	Information processing speed	Paced Serial Addition Test
Executive functioning	Cognitive flexibility, set shifting, perseveration	Wisconsin Card Sorting Test
	Sequencing, flexibility	Trailmaking Test
	Response inhibition	Stroop Color and Word Test
Working memory		Auditory Consonant Trigrams
		WAIS-III Digit Span
Memory	List-learning, delayed free recall and recognition	California Verbal Learning Test-II, Rey Auditory Verbal Learning Test
	Visual learning and memory	Rey Complex Figure Test
	Verbal and visual learning and memory	Wechsler Memory Scales (WMS-III)
Language	Verbal fluency	Controlled Oral Word Association Test; Animal Naming
Visuospatial	Visuoconstruction	Rey Complex Figure Test
Motor	Fine motor coordination	Grooved Pegboard
	Simple motor speed	Finger Tapping Test

Impairments in a number of neurocognitive domains have been reported in various sleep disorders. Moderate to severe impairment in attention, memory, and executive functions have been reported in untreated obstructive sleep apnea (OSA) subjects [16]. While many of these cognitive deficits appear to improve following CPAP treatment, deficits in executive functions appear either irreversible or much less responsive to treatment. Reduced vigilance and sustained attention have been reported in narcolepsy patients, while findings regarding attention span and memory have been equivocal [15]. Attention span and vigilance are the two most frequently documented areas of impairment in subjects with insomnia. Findings in other cognitive domains have been equivocal, although a limited number of studies have documented deficits in verbal fluency and executive functions [15]. A number of reported cognitive difficulties may resolve with effective treatment of the underlying sleep disorder; others may need further assessment and treatment.

PUTTING IT TOGETHER: CASE EXAMPLE

Presenting Problem Karen, a 36 year old female, sought treatment for insomnia. Her treating physician was concerned that depression and anxiety were playing a role in her sleep problem.

Assessment Measures An assessment battery including the BSI, SF-36, BDI, and STAI was mailed to the patient before the interview. She had already completed sleep assessment tools for the Sleep Clinic and was asked to bring in the last 2 weeks of her sleep diary. She was also asked to complete the PSWQ following the interview.

History of Presenting Problem Karen reported having difficulty falling and staying asleep 5 nights a week, since losing her job last year. She found a new job immediately, but the sleep problem never improved. Currently, Karen gets into bed anywhere between 10:00 p.m. and midnight and gets out of bed between 6:00 and 9:00 a.m. It takes her a little over an hour to fall asleep and she wakes up 1–2 times a night for 45–90 minutes. In the morning, she wakes up with her alarm but does not feel rested. She naps 3–4 times a week, especially if she has evening plans. Her sleep schedule does not change on weekends or vacations. Karen drinks up to four caffeinated beverages a day and denied the use of tobacco or alcohol. She does not use any type of sleep aid or other medication. She exercises three times a week, usually in the evenings, because she is "too tired" to get up in the morning. She acknowledged being quite distressed about "not sleeping." She becomes anxious before getting into bed and worries about how she will function the next day. Although she generally functions adequately, she has begun to decrease some activities because of fatigue and recently overslept for an important event. Karen noted that her insomnia was worse when she was faced with new or ongoing stressors.

Karen said she felt "anxious, not depressed." She reported, "worrying about everything—the kids, my parents, work, money, my appearance, the yard, the house—everything." She has "always" been a worrier and recognized that her worry was excessive and difficult to control. She acknowledged feeling on edge, irritable, fatigued, and physically tense as well as having disturbed sleep, thus meeting criteria for generalized anxiety disorder (GAD). She believed that anxiety and worry about sleep were interfering with her sleep.

Results of the assessment measures corroborated the diagnosis of insomnia and GAD. Sleep questionnaires given by the Sleep Clinic supported her report of difficulty falling and staying asleep, daytime sleepiness, anxiety about sleep, and maladaptive sleep hygiene practices. In terms of psychological functioning, the BSI showed elevated scores in the areas of anxiety and interpersonal sensitivity (i.e., personal inadequacy, especially in comparison with others). The SF-36 indicated problem areas in role limitations due to physical health problems, social functioning, general mental health, and vitality (energy or fatigue). Measures of depression fell within the normal range according to the BDI and levels of anxiety were elevated according to the STAI. Her score on the PSWQ indicated a moderate to severe degree of worry.

Summary and Recommendations Karen currently meets criteria for insomnia and GAD. She does not meet criteria for depression or any other mood disorder. It is likely that her tendency to worry and ruminate makes her more vulnerable to insomnia. While her sleep problems may have begun during a stressful time, it appears that her more recent worry about sleep as well as poor sleep habits are exacerbating the problem. It is highly recommended that treatment include cognitive–behavioral therapy (CBT) targeting insomnia in conjunction with reduction of rumination and worry associated with GAD.

CONCLUSION

The field of sleep medicine has called on the mental health profession to assess the influence of psychological factors on sleep disorders. Information is obtained from the referral source, paper-and-pencil assessment measures, and the patient. Poor sleep hygiene, anxiety, and perceived stress are examples of behavioral and cognitive factors known to have a negative impact on sleep. Since neurocognitive deficits have been identified in a number of sleep disorders, neuropsychological testing may also prove to be beneficial. The information obtained during a general psychological or a specialized neuropsychological assessment is used to identify and treat the psychological factors maintaining the sleep disorder.

REFERENCES

1. Stepanski EJ, Perlis ML. A historical perspective and commentary on practice issues. In: Perlis ML, Lichstein KL (Eds), *Treating Sleep Disorders: Principles and Practice of Behavioral Sleep Medicine*, John Wiley & Sons, Hoboken, NJ, 2003, pp 3–26.

2. Hauri P. *Current Concepts: The Sleep Disorders*. The Upjohn Company, Kalamazoo, MI, 1977.

3. American Psychiatric Association. *Diagnostic and Statistical Manual of Mental Disorders*, 4th ed. American Psychiatric Association, Washington DC, 1994.

4. Harvey AG. Insomnia: symptom or diagnosis? *Clin Psychol Rev* **21**(7):1037–1059(2001).

5. Zayfert C, DeViva J. Residual insomnia following cognitive behavioral therapy for PTSD. *J Trauma Stress* **17**:69–73(2004).

6. Spielman AJ, Yang CM, Glovinsky PB. Assessment techniques for insomnia. In: Kryger MH, Roth T, Dement WC (Eds), *Sleep Medicine*. Saunders, Philadelphia, 2000, pp 1239–1250.

7. Nowell PD, Buysse DJ, Hall M, et al. Psychometric and psychiatric evaluation. In: Kryger MH, Roth T, Dement WC (Eds), *Sleep Medicine*. Saunders, Philadelphia, 2000, pp 1290–1294.

8. Williamson DA, Veron-Guidry S, Kiper K. Assessment of health-related disorders. In: Bellack A, Hersen M (Eds), *Behavioral Assessment: A Practical Handbook*. Allyn & Bacon, Needham Heights, MA, 1998, pp 256–270.

9. Morin CM. Insomnia: *Psychological Assessment and Management. Treatment Manuals for Practitioners*. The Guilford Press, New York, 1993.

10. Gemmell SB, Souheaver GT. Pschological assessment in sleep disorders. In: Lee-Chiong TL, Sateia MJ, Carskadon MA (Eds), *Sleep Medicine*. Elsevier, Philadelphia, 2002, pp 705–711.

11. Morin CM, Edinger JD. Sleep disorders: evaluation and diagnosis. In: Hensen M, Turner SM (Eds), *Adult Psychopathology and Diagnosis*. John Wiley & Sons, Hoboken, NJ, 2003, pp 583–612.

12. Spielman AJ, Anderson MW. The clinical interview and treatment planning as a guide to understanding the nature of insomnia: the CCNY interviw for insomnia. In: Chokroverty S (Ed), *Sleep Disorders Medicine: Basic Science, Technical Considerations and Clinical Aspects*. Butterworth-Heinemann, Boston, 1999, pp 385–426.

13. Harvey AG. A cognitive model of insomnia. *Behav Res Ther* **40**:869–893(2002).

14. First MB, Spitzer RL, Williams JBW. *Structured Clinical Interview for DSM-IV Axis I Disorders: Clinician Version (SCID-CV) (Users Guide and Interview)*. American Psychiatric Press, Washington DC, 1997.

15. Lezak M, Howieson D, Loring D. *Neuropsychological Assessment*, 4th ed. Oxford University Press, New York, 2004.

16. Decary A, Rouleau I, Montplaisir J. Cognitive deficits associated with sleep apnea syndrome: a proposed neuropsychological test battery. *Sleep* **23**:369–381(2000).

ADDITIONAL READING

Handbook of Psychiatric Measures. American Psychiatric Association, Washington DC, 2000.

Perlis ML, Lichstein KL. *Treating Sleep Disorders: Principles and Practice of Behavioral Sleep Medicine*. John Wiley & Sons, Hoboken, NJ, 2003.

136

OPERATING AND MANAGING A SLEEP DISORDERS CENTER

RICHARD S. ROSENBERG

American Academy of Sleep Medicine, Westchester, Illinois

INTRODUCTION

History of Sleep Disorders Centers

Sleep centers began in academic environments. This is an important consideration when contemplating the current and future status of sleep centers. With the recognition of sleep disorders, researchers were asked to apply their knowledge and techniques to the study of human sleep disorders, often in partnership with senior clinicians. In many cases this occurred in tertiary care hospitals. Early sleep researchers were physiologists, psychologists, and psychiatrists. In collaboration with neurologists, these groups treated patients with narcolepsy and insomnia. With the discovery of sleep disordered breathing, pulmonologists entered the field in large numbers. For the first decade of their existence, research was a primary focus of the sleep disorders centers.

This history developed the perception of sleep medicine as a multidisciplinary field. Physicians and sleep experts bring different perspectives and expertise to the diagnosis and treatment of the patient. The primary tool for the diagnosis of sleep disorders, the polysomnogram, also requires multidisciplinary skills for interpretation. Due to the diverse nature of disorders that lead to a complaint of insomnia or

excessive daytime sleepiness, a variety of treatment options must be available.

Recognition of the huge numbers of patients with sleep disorders forced expansion beyond the academic sleep centers. In order to bring sleep medicine to those who need it, sleep specialists are now trained in accredited fellowship training programs and certified by the American Board of Sleep Medicine. Standards for the professional staff, facilities, equipment, personnel, and operation of the sleep center were developed, and the American Academy of Sleep Medicine (AASM) now offers accreditation to centers that meet these standards and Standards of Practice documents to guide the practice of sleep medicine. The accredited sleep center is envisioned as a resource for the community, providing clinical diagnostic services but also community education, industrial advice, and technical expertise.

Accreditation

The current Standards for Accreditation [1] were developed through review of the literature, surveys of common practices, and consensus. The standards are reviewed and modified annually. The accreditation process includes submission of an application that provides self-assessment; review of the application with requests for supplementary information when necessary; and a site visit to inspect the facility, review documentation, and observe the facility in action.

The standards provide a starting place for the development of a new center. Centers are evaluated during the site visit as failing to meet, meeting, or exceeding

The information in this chapter is believed to be consistent with American Academy of Sleep Medicine (AASM) standards of practice, standards for accreditation and accreditation process; however the information, views, and opinions in this chapter represent only those of the author and do not represent an official position of the Academy or its committees. The author is a full-time employee of the AASM.

standards, and guidelines are available for centers to self-evaluate. There should be no reason to plan a center that fails to meet the AASM Standards for Accreditation.

Clinical Practice Parameters

The current clinical practice parameter publications of the AASM (e.g., the indications for polysomnography [2] or the evaluation of chronic insomnia [3]) are evidence-based, arising from an intensive literature search, development of evidence tables, and review by a committee of experts. The result of this process is a series of documents that provide graded recommendations for the practice of sleep medicine. Standards of care are incorporated into the Standards for Accreditation. Lower levels of evidence provide guidance for the practitioner when the standard treatment fails or in special circumstances.

CONSIDERATIONS FOR CENTER START-UP: PROFESSIONAL ENVIRONMENT

As a subspecialty, sleep disorders medicine requires establishing referral relationships with primary care physicians. Initially, the Standards for Accreditation required sleep specialists to educate colleagues and the public in the recognition of sleep disorders. This is no longer mandated but remains an excellent strategy. A few simple questions can markedly increase the suspicion of sleep disorders. Treatment of apnea, for example, can have beneficial effects on blood pressure, mood, and quality of life—issues of importance to primary care physicians.

In many environments, an otolaryngologist is considered a sleep specialist. This is because snoring is a prevalent complaint leading to sleep center evaluation. The relationship between the sleep center and the local otolaryngologist is often mutually beneficial. The sleep center evaluation determines whether or not the patient has significant sleep apnea. If not, then surgical treatment for snoring may be warranted. If apnea is present, a trial of continuous positive airway pressure (CPAP) is usually the next step. If CPAP is not tolerated, surgical therapy once again becomes a possible alternative.

Sleep specialists hold primary certification in pulmonary medicine, neurology, and psychiatry. A specialist that is certified in one area would need to seek consultative relationships with the other two. In addition, behavioral alternatives for treatment of insomnia are recognized as a necessary part of sleep medicine. A relationship with a clinical psychologist experienced in cognitive–behavioral therapy provides an excellent fit for many centers.

A variety of other factors contribute to the success of a sleep center. Local competition, the age of the population in the catchment area, and the mix of insurance carriers all affect financial success. The location of the center may impact success, as some populations are unwilling to cross real or perceived boundaries. Many centers have expanded to include outreach clinics or a network of centers to improve accessibility for patients.

PERSONNEL

Professional Staff

The director of the sleep disorders center must demonstrate a commitment to the field, to the community, and to the center. The commitment to the field is exemplified by the training, knowledge, and experience required for passing the qualifying examination of the American Board of Sleep Medicine (ABSM). A diplomate of the ABSM is required at each accredited sleep disorders center. The commitment to the community should be reflected in a willingness to engage in educational forums. The commitment to the center should result in a significant time commitment, including development of the center, training of technicians, and quality assurance activities.

The diplomate of the ABSM, if a physician, may also serve as the medical director of the center; this role may be filled by any physician with adequate knowledge of sleep medicine. The medical director is responsible for the care of patients while in the center for testing, the clinical evaluation of patients, and the delivery of appropriate treatment.

Technologists

For most sleep disorders, diagnosis rests on polysomnography. Trained and motivated technologists are the key to quality polysomnographic recordings. Technologists must have a personality and style that puts the patient at ease in what is, for most patients, a highly stressful situation. Center directors must invest time and effort in the recruitment and training of technologists or steal them away from other centers with higher pay and better benefits. For most centers, investing in an experienced chief technologist and developing a training program is the best way to ensure an adequate supply of technologists. This investment must be included in any sleep center business plan.

Support Staff

Sleep disorders medicine is a service industry; failure to answer the telephone, clean the bedrooms, or provide written results in a timely fashion is simply bad business. Word of mouth can make or break a center. Some centers are able to share positions (e.g., the pulmonary function laboratory secretary answers the sleep phone and makes appointments). However, sleep centers frequently experience rapid growth and frequent reassessment of support staffing is recommended.

FACILITY

Bedrooms

Bedrooms should be large and comfortable. One goal of the design of the bedroom is to allow patients to feel at home. Many centers provide bedside tables, reading lamps, televisions, and other amenities. The circumstances of the polysomnogram produce curiosity and inconvenience at best—anxiety and panic at worst. Having a bedroom that approximates home may help to control natural anxiety about the test.

The bedroom should be planned with emergency situations in mind. Access to the bedroom should avoid sharp turns. The doorway should be sufficiently large to allow a cart to be wheeled in and a large patient to be taken out. The door should open out so that a patient who collapses in the doorway does not block the door from opening. Ideally there should be sufficient space on both sides of the bed to allow resuscitation to occur on the floor. Although polysomnographic equipment may need to be visible, often emergency equipment can be hidden behind a panel or in a closet to reduce anxiety.

A private bathroom with shower enhances patient comfort, although many centers require patients to share a bathroom. Clean towels, soap, and shampoo are also appreciated.

The sleep center bedroom must have better soundproofing and temperature control than at home. Traffic noise, especially the arrival of emergency equipment at a hospital, should not disturb the patient. Many patients snore, requiring adequate sound isolation of the bedrooms. Technicians walking down the hallway to care for one of the other patients should not disturb patients. Individual room control of temperature is important to allow patients to keep the room at their own preferred sleeping temperature.

In general, it is not cost-effective to have a single bedroom. Labor costs contribute significantly to the operation of a sleep center. A technician can typically monitor two patients at once [1] (except pediatric patients, demented patients, and other patients requiring a high level of care). Requiring a technician to monitor more than two patients generally results in a lower quality of recording and may place patients at risk when intervention is required.

Control Room

The control room should be as close as possible to the bedrooms to allow technicians rapid access to the patients.

Under no circumstances should the control room be on another floor or on a different wing from the bedrooms.

The control room should be large enough to comfortably hold the recording equipment, communication equipment, and technicians. The control room will be home to the technicians for 8 hours or more. However, they should not feel comfortable enough to nap. Sleepiness countermeasures may include bright lights and exercise equipment.

Waiting Room

A comfortable waiting room is convenient for patients who arrive early for the test, and for patients undergoing a Multiple Sleep Latency Test (MSLT). The MSLT includes four or five naps at 2 hour intervals. Patients should be observed and comfortable in the times between naps.

Equipment

Polygraphic equipment should record and store a minimum of 12 channels per bedroom, although most available equipment provides 21 channels per bedroom. Most systems offer a variety of transducers for recording the electroencephalogram, electro-oculogram, electromyogram, and electrocardiogram. Respiratory monitors include thermistors, which provide a rough estimate of airflow, and the more sensitive pressure transducers. Many centers record both of these signals. Effort belts or respiratory inductance plethysmography is required to distinguish between central and obstructive sleep-related breathing events. A reliable pulse oximeter is also essential. Other transducers that provide useful information include snoring and position monitors.

Continuous auditory monitoring and communication with the patient is essential. An investment in a high-quality intercom system is well worth the price. Continuous visual monitoring is also required. Many polygraph systems include a low-light digital camera and picture-in-picture displays. The Standards for Accreditation require the capability to store video; the addition of a DVD burner to a digital video system meets this requirement.

TREATMENT

The AASM provides clinical practice parameters to guide treatment of sleep disorders. For patients with sleep-related breathing disorders, continuous positive airway pressure (CPAP) is the initial treatment of choice. Many centers have relationships with home durable medical goods providers to ensure that patients are set up with CPAP therapy in a timely manner. One critical aspect of patient tolerance of CPAP is the set-up process. Trained technicians, usually respiratory therapists, must spend time on mask fitting, patient education, and reassurance in order for the therapy to be maximized. Some centers provide CPAP treatment themselves, but there are some legal restrictions on this process.

Pharmacological therapy is typically required for patients with narcolepsy and periodic limb movement disorder. Both disorders require titration of medications and patients may require frequent clinic visits or telephone communication initially. Once a stable dosage is established, annual visits are usually adequate to monitor compliance. An experienced nurse or physician assistant is an invaluable asset for the busy center, especially in the role of patient resource regarding medications. The medical director may set guidelines for therapy and thereafter rely on the nurse's judgment.

Behavioral therapy is required by the Standards for Accreditation, not only for the treatment of most types of insomnia but also for CPAP mask desensitization and in the development of good "sleep hygiene" in patients with undesirable sleep habits. The recent development of the field of behavioral sleep medicine, including training programs and a certification examination, holds promise for developing a cadre of experts as resources for sleep centers.

INTERPRETATION OF POLYSOMNOGRAMS

Who interprets polysomnograms is occasionally an area of contention in sleep centers, as this is the primary source of revenue for sleep medicine specialists. A diplomate of the American Board of Sleep Medicine is clearly qualified to read studies, as are those who are board eligible. The Standards for Accreditation require that the diplomate sign all polysomnogram interpretations. Trainees and other experienced professionals may interpret studies, but the diplomate must be the "gold standard" and efforts to ensure quality must be documented. Typically this involves regular assessment of scoring and continuous review of interpretations.

ADMINISTRATION

Whenever possible, administrative control of the center should reside in the center staff. This maximizes understanding of the special needs of the center, such as overnight work shifts, equipment, and sleep environment. Problems arise when the administrative goals of efficiency and cost-effectiveness do not take into account the usual inefficiencies encountered in the sleep program.

As an example, a busy sleep center may have a waiting list of a month or more. Emergencies or illness will, on occasion, cause a patient to cancel an appointment for a polysomnogram. If adequate secretarial support is not present, it might not be possible to call patients from the waiting list and fill the empty bed. Lost revenue from an empty bed would, in most instances, far exceed the expense of additional secretarial help.

QUALITY CONTROL

Quality control initiatives are an integral part of most hospital environments and are useful in all medical settings. Several measures are particularly relevant to sleep centers:

1. *Scoring Reliability.* In centers with multiple technologists scoring records, it is critical to have as much agreement as possible between scorers to allow for development of criteria for abnormality and for pre- and post-treatment comparisons. Most centers employ an epoch-by-epoch comparison (standard in some computer-based recording systems) and calculate a statistic that can be used to track reliability. The center's diplomate of the American Board of Sleep Medicine should serve as the "gold standard" for comparison. Poor reliability should be countered by training sessions focusing on epochs of disagreement.

2. *Timeliness of Reports.* The time between the sleep study and delivery of care to the patient should be as short as possible. Liability issues regarding patients with established diagnosis but no treatment have arisen in several cases. Excessive delay may indicate insufficient technical, administrative, or professional staff. Tracking this measure with a quality control initiative provides evidence of the need for additional staff.

3. *Patient Satisfaction.* A broader measure that highlights areas of concern and often leads to quality improvement. Given that one goal of the sleep center environment is to put patients at ease, it is important to receive and respond to frequent feedback from patients.

Maintaining high standards of quality and a focus on patient care are the best ways to ensure success.

REFERENCES

1. American Academy of Sleep Medicine. Standards for Accreditation. http://aasmnet.org/standards.asp; 2004.

2. Chesson AL Jr, Ferber RA, Fry JM, et al. Practice parameters for the indications for polysomnography and related procedures. *Sleep* **20**:406–422(1997).

3. Chesson A Jr, Hartse K, Anderson WM, et al. Practice parameters for the evaluation of chronic insomnia. *Sleep* **23**(2):237–241(2000).

137

ACCREDITING A SLEEP PROGRAM

Donna Arand

Kettering Medical Center Sleep Disorders Center, Dayton, Ohio

INTRODUCTION

The decision to pursue accreditation of a sleep disorders center or laboratory is an important one that will impact the development and operation of the program. American Academy of Sleep Medicine (AASM) accreditation is considered the "gold standard" for sleep programs and it provides many advantages. It helps assure quality in delivering sleep medicine services for physicians and patients and it can also be useful in distinguishing a program from others in a competitive marketplace. In addition, many third party payers require a sleep disorders program to be accredited for reimbursement. Pursuing AASM accreditation can seem like a daunting task for the first time applicant; however, the process is actually simple. For the applicant it only involves submission of an application and a site visit. However, there are some general accreditation procedures and issues that are also of interest for the applicant. This chapter will examine various aspects of AASM accreditation including types of accreditation and general issues, basic requirements for accreditation, the application process, and site visit activities.

The author is a former chair of the AASM accreditation committee. Currently the author serves on the AASM board of directors that sets standards for accreditation and also serves as the board liaison to the AASM Accreditation Committee. The information in this chapter is believed to be consistent with American Academy of Sleep Medicine (AASM) standards for accreditation and the accreditation process; however, the information, views, and opinions in this chapter represent only those of the author and do not represent an official statement of the AASM or its committees.

Sleep: A Comprehensive Handbook, Edited by T. Lee-Chiong.
Copyright © 2006 John Wiley & Sons, Inc.

AASM ACCREDITATION

Center Versus Laboratory Accreditation

The AASM accredits both sleep centers and sleep-related breathing laboratories. If the program does not have a Diplomate of the American Board of Sleep Medicine (D.ABSM) on staff or someone accepted to take the sleep boards, they can only apply as a laboratory for sleep-related breathing disorders. If the program has a diplomate on staff at the time of application, they may apply as a center or a laboratory. It is generally more advantageous to apply as a center since the requirements are almost the same for both types of program. The difference is that a laboratory does not need a D.ABSM or someone accepted to take the boards on staff and they can only promote evaluation and treatment of patients with sleep-related breathing disorders. However, since laboratories will find that some of their patients have multiple sleep disorders or that the sleep complaint is not due to sleep disordered breathing, they must be able to recognize all disorders and have a mechanism for treating them. Consequently, after acquiring a D.ABSM or someone accepted to take the exam, many sleep-related breathing laboratories pursue subsequent accreditation as a center since this designation recognizes the ability of the program to evaluate the range of sleep disorders.

Multiple Sites and Accreditation

The sleep program may be housed in one location, where patients are both seen by sleep physicians and where sleep studies are performed, or these may be separate locations. It is not uncommon for patients to be evaluated in the physician's office while the sleep studies are performed at a separate location. Each site where sleep studies are performed must have a separate accreditation, although the sleep physicians may have multiple offsite offices where the sleep patients are seen. Separate recording locations must submit their own accreditation application and go through the site visit process. An exception is made for programs that have recording facilities on different floors in the same building or in adjacent space. Also sleep programs located in hospitals may have separate recording locations if the location is based on the patient population (e.g., psychiatric beds, pediatric beds, inpatient beds). In these latter cases only one accreditation is needed.

AASM STANDARDS FOR ACCREDITATION

The AASM Standards for Accreditation [1] were specifically developed for sleep disorders programs by professionals in the sleep field. In addition, the entire accreditation process is carried out by individuals certified by the American Board of Sleep Medicine. This helps assure that the standards are appropriate to the field and the site visitor has expertise in sleep disorders medicine to effectively evaluate sleep programs.

The standards for accreditation can be found on the AASM website at http://www.aasmnet.org/standards.asp. The standards for center accreditation and sleep–related breathing disorder laboratories are identical with the two exceptions listed earlier. Basically there are requirements for personnel, physical plant, policies and procedures, and daily operations. A few of the requirements are considered extremely important and essential to the operation of a program, so that not meeting them is an automatic fail. However, most of the standards guide daily operations and usually most sleep programs can develop or modify their programs to meet the standard.

Plans for having an accredited sleep program should be part of the initial decision making and development of a program. Programs initially designed to meet the accreditation requirements will make pursuit of accreditation easier. Deciding early that accreditation is a goal for the program will also help avoid potentially costly mistakes that could impede the ability to obtain AASM accreditation for the program.

The personnel requirements for accreditation require that all programs have a medical director. In addition, sleep center applicants must also have a Diplomate of the American Board of Sleep Medicine (D.ABSM) on staff or someone accepted to take the board exam. Often the medical director is also the D.ABSM but another sleep staff professional could fulfill this role. Programs applying as sleep-related breathing disorders laboratories do not need a D.ABSM or someone accepted to take the exam. Specific

duties for the medical director and D.ABSM are included in the standards.

Adequate technical and administrative staff are also required to ensure timeliness of testing and report turnaround. A ratio of one technician to two patients is generally required with occasional exceptions. A registered polysomnographic technician (R.Psg.T.) is not required but is strongly recommended.

There are also requirements for physical space specifically concerning bedroom sizes and distances from the control room. This is an important requirement that needs to be addressed during the initial development of the program. Problems due to small or inappropriate space are not easily rectified and will prevent the facility from being accredited if measures cannot be taken to meet the standards.

There are a number of activities and documentation requirements that are addressed by the standards for accreditation. This includes written policies and procedures for all types of tests performed as well as policies governing all activities of the center. In addition, quality assurance activities and staff education must be ongoing activities that are documented. The majority of the standards deal with the daily activities of a center; however, if a program is not in compliance with the standards, it is usually easy to modify the operation to meet the standard.

THE APPLICATION PROCESS

General Overview

The (AASM) accreditation process is simple. For the applicant it involves submission of an application and a site visit. The application is a simple checklist with appendices that serve as a self-assessment. The site visit consists of a half-day session to review charts, policies, and procedures and interview consultants and administration.

Completing the Application

Once a program is in operation it may begin the accreditation process. There is no minimum amount of time required to be in operation before a program can apply for accreditation. The only requirement is that the program can provide all of the information requested in the application. To begin the accreditation process, either a center or laboratory application must be purchased from the AASM. This can be done on the Web at http://www.aasmnet.org.

The application is primarily a yes or no checklist with some requested appendices. The questions follow each of the standards for accreditation. The yes/no format provides a self-assessment for the sleep program. The applicant should be able to answer yes to every question. If there are any areas where the program does not meet the standard, the program should modify its operation so that it is compliant with the standard.

In addition to a checklist for the standards, the application also includes a checklist of common policies and procedures needed by most sleep programs. This assures that the facility has the essential written policies on file. However, the list is not all inclusive and other policies and procedures may be needed to cover all aspects of a particular program.

The checklists do not take long to complete; however, the information requested for the appendices will take time to compile. The appendices make up the largest part of the application. The appendices primarily include quality assurance summaries, sample patient cases, procedures for sleep tests, and various emergency protocols. Other supporting documentation is also requested but it is a smaller part of the appendices.

The quality assurance summary should include the results of the last two quarterly reviews of initiatives undertaken by the program. A description of the initiative, results, and subsequent action should clearly be stated for each initiative. The report must be signed by the D.ABSM.

The sample cases requested are very important as they reflect the quality of patient care at the program. The cases submitted must be patients who had sleep studies performed at the location applying for accreditation, if studies were done. The sample cases demonstrate the thoroughness of evaluation, testing, and follow-up. They are reviewed with this in mind. The applicant should be careful to remove or completely obscure patient names throughout the case samples to protect patient confidentiality. Consequently, applications should always be reviewed by a sleep staff professional before submission to make sure that the patient cases and information provided are an accurate reflection of the program and that patient confidentiality is protected.

When compiling information for the appendices only the information requested should be submitted. Additional information only slows down the review process and adds to the size of the application. In contrast, missing information may result in the application being returned.

If there are any questions or concerns that arise while completing the application, these can be directed to the national office for assistance and clarification. After the application is completed and all of the information is compiled, the application must be signed by the medical director and others responsible for the operation of the program. When the application is completed, four copies should be sent to the AASM national office along with the site visit fee.

Application Review

Upon receipt of the application at the national office, it is assigned to a reviewer. The application is reviewed by an accreditation committee member for compliance with the standards. The application will be reviewed in detail. If all of the information is complete and consistent with the standards, a site visit will be approved. If important information is missing, the application will be returned. If any information is incomplete or unclear, the reviewer will send a query letter to the facility requesting clarification or more information. These queries are often useful to the facility since they may help identify potential problems in meeting the standards that can be addressed prior to a visit. When all responses are received from the applicant and they indicate compliance with the standards, the program will be approved for a site visit. Once approved for a visit, the center will be contacted by the national office to schedule a site visit. The time from application submission to the actual site visit can be very short, but it depends on the responsiveness of the facility and their availability in accepting the site visit date offered.

THE SITE VISIT

The site visit will consist of a morning session. Typically the visitor arrives at the facility around 8 a.m. A tour of the facility allows the visitor to evaluate the appearance and comfort of the facility from the perspective of the patient.

Review of the policy and procedure manual, patients charts, and interviews with staff will be done in the morning. Policies and procedures are reviewed for completeness, consistency with program operations, and agreement with practice parameter guidelines where they exist. The patient charts are reviewed for completeness of information especially intake and follow-up information, inclusion of outside consultation reports if any, as well as consistency in organization. Review of patient charts should demonstrate that the criteria for the tests were met, there is a written physician order for all tests, and the correct procedures were performed following clinical practice parameters. Evidence that a facility performs all sleep tests at the appropriate sleep/wake time of the patient especially for shift workers should be available.

In the morning the preceding nights study will be reviewed with the D.ABSM or another interpreting physician operating the computer. All interpreting physicians will be asked to operate the computer system in the presence of the site visitor. The program's professional staff will generally be interviewed to determine their role at the center or laboratory, time commitment, and background in sleep medicine. The site visitor will also interview consultants individually to asses the working relationship between the program and the consultant. An individual representing the administration of the program will also be interviewed to determine the degree of support of the parent institution for the sleep program. Any available time during the morning will be used to review random patient charts and any unfinished work from the previous night.

At the completion of the site visit, the visitor discusses the strengths and any weaknesses of the program. The visitor will also answer any questions. However, since the site visitor does not grant accreditation but only makes a recommendation to the AASM Board of Directors, the facility will not know its accreditation status until it is notified by the board.

ACCREDITATION STATUS

The site visitors report and recommendation concerning accreditation are submitted to the chair of the accreditation committee for review. The chair then submits the report and recommendation to the board of directors who grant the accreditation status. A letter is then sent to the facility in about 6–8 weeks announcing the resulting accreditation status. The entire process from submission of the application to receipt of the accreditation notice typically takes a few months.

The program may receive accreditation, accreditation with provisos, or be denied. Accreditation with provisos is given if any deficiencies were noted. The proviso will reference the standard not being met, briefly describe the deficiency, and point out the necessary steps that the facility must take to meet the standards. All provisos must be addressed within 3 months with specific documentation

sent to the national office indicating compliance with the standards.

AASM accreditation is granted for a 5 year period. Any major changes including relocation of the facility or change of a medical director or D.ABSM must be reported to the national office. Such changes must be approved to continue accreditation. Provisos may be issued any time during the 5 year period if there is evidence that a program is not compliant with the standards for accreditation. Since standards may change or be modified annually, it is important that programs review the standards for accreditation periodically to maintain compliance.

REACCREDITATION OF SLEEP PROGRAMS

Programs applying for reaccreditation go through the same application process as new programs. The standards for either centers or sleep-related breathing disorders laboratories are the same for new programs pursuing accreditation or those applying for reaccreditation. The site visit for programs pursuing reaccreditation consists of a half-day visit in the morning or afternoon. All of the same activities are performed. This includes review of random patient charts, interviews, and reviews of policies and procedures as well as changes in the program since the last visit.

REFERENCE

1. AASM Standards for Accreditation (2005). http://www.aasm-net.org/standards.asp.

INDEX

Sleep: A Comprehensive Handbook, Edited by T. Lee-Chiong.
Copyright © 2006 John Wiley & Sons, Inc.